## 3. Solve the related inequality analytically and graphically.

**Illustration:** Solve the inequality $2x^2 - 4x - 6 \le 0$.

**Solution:** As seen in Illustration 2, the solutions of the related equation are $-1$ and 3. We can solve the inequality by performing a sign test on the variable factors, $x + 1$ and $x - 3$.

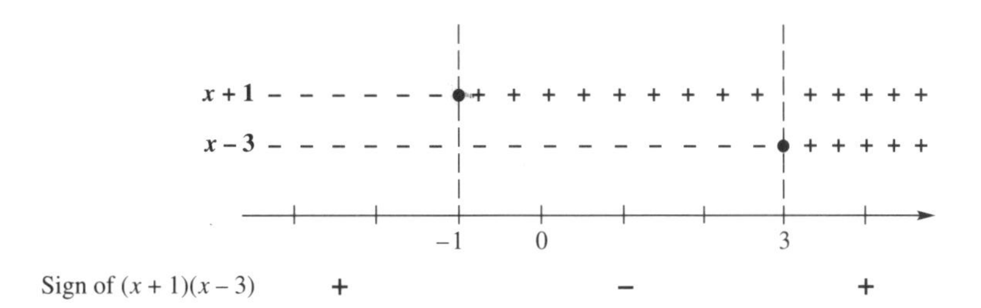


The sign graph shows that the quadratic polynomial is negative or 0 on the interval $[-1, 3]$. The calculator graphs in Illustration 2 support this solution set, since the graph lies *on* or *below* the $x$-axis on this interval.

## 4. Apply analytic and graphical methods to solve an application of that class of function.

**Illustration:** If an object is propelled directly upward from the ground with an initial velocity of 64 feet per second, then (neglecting air resistance) the height of the object $t$ seconds after it is propelled is modeled by $s(t) = -16t^2 + 64t$, where $s(t)$ is in feet. After how many seconds does it reach a height of 28 feet?

### Analytic Solution

We must solve the equation $s(t) = 28$.

| | |
|---|---|
| $-16t^2 + 64t = 28$ | |
| $16t^2 - 64t + 28 = 0$ | Standard form |
| $4t^2 - 16t + 7 = 0$ | Divide by 4. |
| $(2t - 1)(2t - 7) = 0$ | Factor. |
| $t = .5 \quad \text{or} \quad t = 3.5$ | Zero-product property |

The object reaches a height of 28 feet twice, at .5 second (on its way up) and at 3.5 seconds (on its way down).

### Graphing Calculator Solution

Using the intersection-of-graphs method, we see that the graphs of $y = -16x^2 + 64x$ and $y = 28$ shown below intersect at points whose coordinates are $(.5, 28)$ and $(3.5, 28)$, confirming our analytic answer.

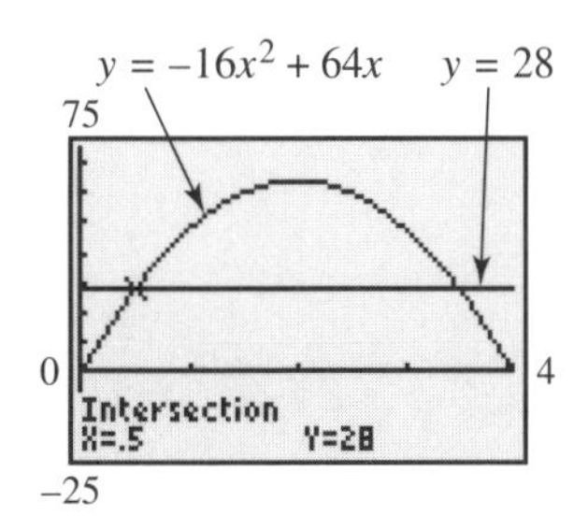


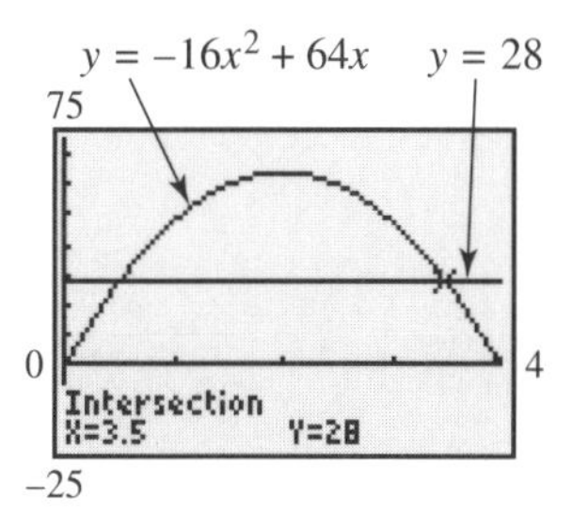

A Graphical Approach to

# PRECALCULUS

Third Edition

John Hornsby

University of New Orleans

Margaret L. Lial

American River College

Gary K. Rockswold

Minnesota State University, Mankato

Boston San Francisco New York
London Toronto Sydney Tokyo Singapore Madrid
Mexico City Munich Paris Cape Town Hong Kong Montreal

*To the memory of Jack Hornsby*
*E.J.H.*

*To the memory of Myrna Rockswold*
*G.K.R.*

*Publisher:* Greg Tobin
*Sponsoring Editor:* Anne Kelly
*Executive Project Manager:* Christine O'Brien
*Assistant Editor:* Jaime Bailey
*Senior Production Supervisor:* Karen Wernholm
*Production Coordination:* Elm Street Publishing Services, Inc.
*Marketing Manager:* Becky Anderson
*Senior Marketing Coordinator:* Julia Coen
*Associate Media Producer:* Jennifer Kerber
*Software Editor/InterAct Math:* John O'Brien
*Senior Software Project Supervisor/TestGen-EQ:* David Malone
*Manufacturing Buyer:* Evelyn Beaton
*Text and Cover Designer:* Barbara T. Atkinson
*Compositor:* The Beacon Group, Inc.
*Illustrations:* Techsetters, Inc.
*Cover Photo:* © Bohdan Hrynewych/Stock Boston/PictureQuest

For permission to use copyrighted material, grateful acknowledgment is made to the copyright holders on page I-20, which is hereby made part of this copyright page.

**Library of Congress Cataloging-in-Publication Data**

Hornsby, E. John
A graphical approach to precalculus.—3rd ed./John Hornsby, Margaret L. Lial, Gary K. Rockswold.
p. cm.
Includes bibliographical references and index.
ISBN 0-201-73511-3
1. Functions. I. Lial, Margaret L. II. Rockswold, Gary K. III. Title.
QA331 .H643 2002
512'.1—dc21 2001056642



1 2 3 4 5 6 7 8 9 10—QWV—05040302

# Contents

## Trigonometric Functions and Applications 566

# Preface

The third edition of this text reflects our ongoing commitment to teaching precalculus mathematics using graphing technology. Our journey began in 1993 when initial drafts of the first edition were developed and class-tested at the University of New Orleans. Two editions and some 10 years later, we remain true to our approach, based on the premise that students use graphing calculators from the first day of class throughout the course. In this third edition, we have polished presentations, developed new features, added an attractive design, and incorporated many helpful suggestions, comments, and ideas from both instructors and students. Gary Rockswold joins us as co-author—we are pleased to add his expertise with applications and the benefits of his classroom teaching experience.

## Our Approach

It is our desire that these books provide instructors with the best possible methods of teaching with a graphing calculator and students with the easiest way to make connections between graphs of functions and their associated equations and inequalities. Using linear functions as the basis of the presentation in Chapter 1, we introduce a four-step process for the study of functions in subsequent chapters. After introducing a class of function:

- We first examine the nature of its graph.
- Next we solve equations analytically and use graphing calculators to find and support solutions using the $x$-intercept method or the intersection-of-graphs method.
- We then solve the associated inequalities analytically and graphically.
- Finally, we apply analytic and graphical methods to modeling and traditional applications.

By using this unifying approach consistently with each class of function, we hope that students will better understand, connect, and ultimately apply concepts and skills.

Because technology is ever-changing, we have not in the past nor do we try in this edition to teach students how to use specific models of graphing calculators. Unlike many texts that use artistic renderings of calculator screens, we use actual graphing calculator screens that can be duplicated using the Texas Instruments TI-83 Plus graphing calculator. In addition, the *Graphing Calculator Manual* that accompanies the text provides students with keystroke operations for many of the more popular models of graphing calculators.

## Content Changes

We have worked hard to fine-tune and polish presentations of topics throughout the text based on user and reviewer feedback. Some of the content changes you may notice include the following:

- Chapter 1 has been streamlined and reorganized for improved continuity.
- The material on graphs of rational functions in Chapter 4 is now covered in two sections.
- Power functions are included in Section 4.4.
- Inverse functions, previously covered in Chapter 4, have been moved to Chapter 5.
- Harmonic motion is covered in new Section 8.8.
- Sum and difference identities are covered in Section 9.2, and Section 9.3 includes double-number and half-number identities as well as new coverage of product-to-sum and sum-to-product identities.
- Parametric equations with trigonometric functions and applications of parametric equations are presented in new Section 10.7.
- Three new appendices on vectors in space, polar form of conic sections, and rotation of axes are included at the back of the text.

## New and Enhanced Features

We believe students and instructors will welcome the following new and enhanced features.

**Function Capsules** This special feature provides a comprehensive, visual introduction to each class of function and also serves as an excellent resource for student reference and review throughout the course. Each function capsule includes traditional and calculator graphs and a calculator table of values, as well as the domain, range, and other specific information about the function. Abbreviated versions of all function capsules are given on the inside back cover of the text.

**New Real-Life Applications** We have provided many new or updated applied problems that focus on real-life applications of mathematics. Since today's students are more visually oriented than ever, new photos and additional figures, diagrams, and tables have been included to enhance applications in examples and exercises. All applications are titled, and an Index of Applications is included at the back of the text.

**Increased Emphasis on Modeling** Many of the new applications feature mathematical models based on real data or real data in table form, thereby providing students with increased opportunities to use, construct, and analyze models. Curve fitting using linear, quadratic, polynomial, exponential, logarithmic, and trigonometric models is also covered.

**Dual Solution Format** To connect analytic methods for solving problems with graphical methods of solution or support, selected examples now provide a graphing calculator solution alongside the traditional analytic solution. The side-by-side format visually unifies the two solution methods for students.

**What Went Wrong?** Using graphing technology to study mathematics opens up a whole new area of error analysis. In anticipation of typical errors, this popular feature from previous editions allows students and instructors to discuss such errors. Answers are now included. (We still do not know the name of the reviewer who suggested this feature several years ago. To this person, we say thank you. Please contact us if you are reading this.)

**Looking Ahead to Calculus** These new margin notes provide glimpses of how the algebraic and trigonometric topics currently being studied are used in calculus.

**Reviewing Basic Concepts** These new sets of exercises appear every few sections and allow students to review and check their understanding of the material in preceding sections. All answers to these problems are included in the answer section at the back of the text.

**Chapter Reviews** Each end-of-chapter Summary now features a section-by-section list of Key Terms and Symbols in addition to Key Concepts. A comprehensive set of Review Exercises and a Chapter Test are also included.

## Continuing Features

We have retained the popular features of previous editions, some of which follow.

**Technology Notes** These marginal notes provide tips to students on how to use graphing calculators more effectively. Many new notes have been included.

**Cautions and Notes** We often give students warnings of common errors and emphasize important ideas in **CAUTION** and **NOTE** comments that appear throughout the exposition.

**For Discussion** Previously titled For Group Discussion, this feature appears within the exposition or in the margins and offers material for instructors and students to discuss or investigate in a classroom setting.

**Ample and Varied Exercise Sets** The text contains a wealth of exercises to provide students with opportunities to practice, apply, and extend concepts and skills. These include writing exercises as well as multiple-choice, matching, true/false, and completion problems. Those problems that focus on conceptual understanding, many of which tie together multiple concepts, are now titled *Concept Check*. More illustrations, diagrams, tables, and graphs now accompany exercises.

**Relating Concepts** As in the previous edition, these exercises appear in selected exercise sets. They tie together topics and highlight the relationships among various concepts and skills.

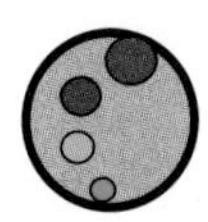

**Chapter Projects** Each chapter concludes with a project that students can complete individually or collaboratively using the material in the chapter. New projects have been provided for Chapters 1, 2, and 5.

**Reference Chapter R** We have retained this helpful student review of basic algebraic concepts. Common formulas from geometry are now grouped in Section R.6 for easy reference.

## Supplements

Our extensive supplements package includes an *Instructor's Edition,* testing materials, solutions manuals, tutorial software, and a state-of-the-art Web site. For more information about any of the following supplements, please contact your Addison-Wesley sales representative.

**FOR THE STUDENT**

**Student's Solutions Manual** ISBN 0-201-79252-4, Norma James, New Mexico State University
Includes detailed solutions to odd-numbered Section and Chapter Review Exercises and to all Relating Concepts, Reviewing Basic Concepts, and Chapter Test problems.

**Graphing Calculator Manual** ISBN 0-201-79253-2
Provides instructions and keystroke operations for TI graphing calculator models. Also contains worked-out examples taken directly from the text.

**InterAct Math® Tutorial Web site**
*www.interactmath.com*
Get practice and tutorial help online! This interactive tutorial Web site provides algorithmically generated practice exercises that correlate directly to the exercises in the text. A detailed worked-out example and guided solution accompany each practice exercise. The Web site recognizes student errors and provides feedback.

**MathXL®**
*www.mathxl.com*
The MathXL Web site provides diagnostic testing and tutorial help, all online, using InterActMath tutorial software and TestGen-EQ testing software. Students can take chapter tests correlated to this textbook, receive individualized study plans based on those test results, work practice problems and receive tutorial instruction for areas in which they need improvement, and take further tests to gauge their progress. Instructors can customize tests and track all student test results, study plans, and practice work. The site is free when an access code is bundled with a new text.

**Addison-Wesley Math Tutor Center**
The Addison-Wesley Math Tutor Center is staffed by qualified mathematics instructors who provide students with tutoring on examples and exercises answered at the back of the textbook. Tutoring is available via toll-free telephone, fax, e-mail, or whiteboard technology—which allows tutors and students to actually see the problems worked while they "talk" in real time over the Internet. This service is available five days a week, seven hours a day. *An access card is required.*

**FOR THE INSTRUCTOR**

**Instructor's Edition** ISBN 0-201-79111-0
This version of the text includes answers to all exercises in addition to the material found in the student text.

**Instructor's Solutions Manual** ISBN 0-201-79273-7, Norma James, New Mexico State University
This manual contains detailed solutions to all exercises in the text and to the Chapter Projects.

**Instructor's Testing Manual** ISBN 0-201-79257-5
This Testing Manual includes four alternative tests and their answers for every chapter.

**TestGen-EQ with QuizMaster-EQ** ISBN 0-201-79274-5
Available on dual-platform Windows/Macintosh CD-ROM, TestGen enables instructors to build, edit, print, and administer tests using a computerized test bank of questions. The program also offers many options for organizing and displaying test banks and tests according to text chapter content. Tests can be printed or saved for online testing via a network or the Web, and the software can generate a variety of grading reports for tests and quizzes.

**MyMathLab**
MyMathLab is a complete, online course for Addison-Wesley mathematics textbooks that integrates interactive, multimedia instruction correlated to the textbook content. MyMathLab is easily customizable to suit the needs of students and instructors and provides a comprehensive and efficient online course-management system that allows for diagnosis, assessment, and tracking of students' progress. For more information, visit www.mymathlab.com or contact your Addison-Wesley sales consultant for a demonstration.

## Acknowledgments

The first and second editions of this text were published after thousands of hours of work, not only by the authors but by class-testers, reviewers, instructors, students, answer-checkers, and editors. To these individuals and all those who have worked in some way on this text over the last 10 years, we are most grateful for your contributions. We could not have done it without you. We especially wish to thank the following individuals who provided valuable input into this edition of the text.

| | |
|---|---|
| Judy Ahrens | Pellissippi State Technical College |
| Randall Allbritton | Daytona Beach Community College |
| Jamie Ashby | Texarkana College |
| Gloria Bass | Mercer University |
| Daniel Biles | Western Kentucky University |
| Linda Buchanan | Howard College |
| Patricia Dueck | Scottsdale Community College |
| Mickle Duggan | East Central University |
| Douglas Dunbar | Okaloosa-Walton Community College |
| William Frederick | Indiana Purdue University, Fort Wayne |
| Kim Gregor | Delaware Technical and Community College |
| W. H. Howland | University of St. Thomas |
| Tuesday J. Johnson | New Mexico State University |
| Cheryl Kane | University of Nebraska |
| Mike Keller | St. John's River Community College |
| Rosemary Kradel | Lehigh Carbon Community College |
| Peggy Miller | University of Nebraska at Kearney |
| Phillip Miller | Indiana University Southeast |
| Stacey McNiel | Lake City Community College |
| Richard Montgomery | The University of Connecticut |
| Michael Nasab | Long Beach City College |
| Jon Odell | Richland Community College |
| Karen Pender | Chaffey College |
| Mary Anne Petruska | Pensacola Junior College |

| | |
|---|---|
| John Putnam | University of Northern Colorado |
| Donna Saye | Georgia Southern University |
| Alicia Schlintz | Meredith College |
| Mike Shirazi | Germanna Community College |
| Jed Soifer | Atlantic Cape Community College |
| Betty Swift | Cerritos College |
| Jennifer Walsh | Daytona Beach Community College |
| Robert Woods | Broome Community College |
| Fred Worth | Henderson State University |

Terry McGinnis has assisted us for many years "behind the scenes" in developing and producing our texts. There is no question that this book is improved because of her attention to detail and consistency, and we are most grateful for her work above and beyond the call of duty.

Kitty Pellissier not only provided excellent help in checking the answers but also reviewed the entire manuscript and provided invaluable suggestions. Donald Ransford and JoAnn Lewin of Edison Community College, Janis Cimperman of Saint Cloud State University, and Steve Ouellette have carefully checked portions of the manuscript for accuracy. Norma James once again prepared excellent solutions manuals. Steven Pusztai of Elm Street Publishing Services expertly guided us through production and did his usual outstanding job. Paul Van Erden prepared yet another accurate, useful index, and Becky Malone compiled the comprehensive Index of Applications. Special thanks also go to Greg Tobin, Anne Kelly, Christine O'Brien, Karen Wernholm, Barbara Atkinson, Becky Anderson, Julia Coen, and Jaime Bailey of Addison-Wesley, who worked long and hard to make this revision a success.

In 1996 in the preface for the first edition, we wrote:

> We hope that this book begins to make a difference in the manner in which algebra is presented and learned as we move into the twenty-first century. We ask that both instructors and students pursue its contents with an open mind, ready to teach and learn in the manner that only now, after so many thousands of years, is possible. We can do so only because we "have stood on the shoulders of giants."

Now that the twenty-first century is upon us, and graphing technology has been embraced in many mathematics classrooms, we are grateful for the small part we have been able to play in its use and acceptance and to all those who have made the journey with us. We hope that you share with us our pride in these books. Please feel free to send your comments via e-mail to *math@aw.com*.

**John Hornsby**
**Marge Lial**
**Gary Rockswold**

# 1 Linear Functions, Equations, and Inequalities

IN AUGUST 2000 Malcolm McKenna published his famous photograph, shown below, of the polar ice pack that was not there. During the past 20 years, average temperatures have risen 7°F over much of the Arctic and ice has melted at an alarming rate. Although a 7° temperature change may not seem like it would have much effect on Earth's climate, it only took an 11° decrease in average global temperatures to cause the last ice age. Calculations show that if all the ice in the Arctic were to melt, sea levels would rise 25 feet. Many cities like Boston, New Orleans, and Santa Barbara would be under water.

This melting phenomenon has also occurred at the Antarctic, where temperatures have risen about 5° in the last 50 years. What will the future bring? Temperatures in the Antarctic have risen, on average, by about $\frac{5}{50}^{\circ}$ or $\frac{1}{10}^{\circ}$ per year. If this trend were to continue for the next hundred years, the average temperature in the Antarctic would rise an additional 10°F, causing major climatic change. Linear functions can help make these types of estimates. In this chapter you will learn how to model many different phenomena with linear functions and equations.

*Source:* Hudson, D., "Clear and Present Danger," *Discover*, January 2001 (Photograph reprinted with permission); Roach, M., "Antarctica's Hot Spot," *Discover*, November 1999.

CHAPTER OUTLINE

# REAL NUMBERS AND COORDINATE SYSTEMS

Sets of Real Numbers • Coordinate Systems • Viewing Windows • Roots • Distance and Midpoint Formulas

## Sets of Real Numbers

The idea of counting goes back to the beginning of civilization. When people first counted they used only the **natural numbers,** written in set notation as

$$\{1, 2, 3, 4, 5, \ldots\}.$$

More recent is the idea of counting *no* object—that is, the idea of the number 0. Including 0 with the set of natural numbers gives the set of **whole numbers,**

$$\{0, 1, 2, 3, 4, 5, \ldots\}.$$

As the need for other kinds of numbers arose, additional sets of numbers were developed. Negative numbers are a relatively recent development in the history of mathematics. The negatives of the natural numbers, included with the set of whole numbers, give the set of **integers,**

$$\{\ldots, -3, -2, -1, 0, 1, 2, 3, \ldots\}.$$

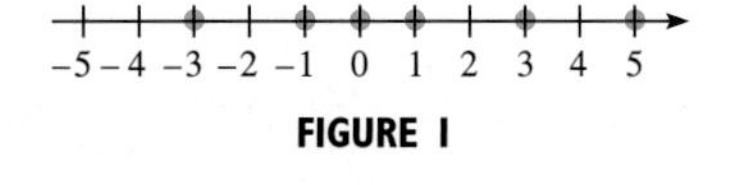


FIGURE 1

Integers can be shown pictorially with a **number line.** The point corresponding to 0 is called the **origin.** The elements of the set $\{-3, -1, 0, 1, 3, 5\}$ are located on the number line in Figure 1.

The result of dividing two integers (with a nonzero divisor) is called a *rational number* or *fraction.* Any **rational number** can be written in the form $\frac{p}{q}$, where $p$ and $q$ are integers with $q \neq 0$. Rational numbers include the natural numbers, whole numbers, and integers. For example, the integer $-3$ is a rational number because it can be written as $\frac{-3}{1}$. Numbers that can be written as repeating or terminating decimals are also rational numbers. For example, $.\overline{6} = .66666\ldots$ represents a rational number that can be expressed as the fraction $\frac{2}{3}$.

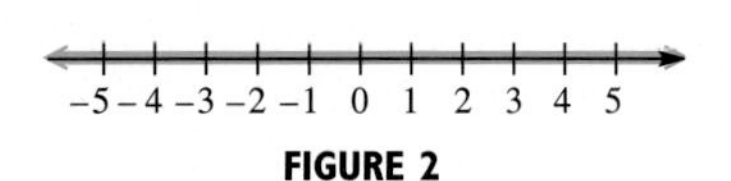


FIGURE 2

The set of all numbers that correspond to points on a number line is called the **real numbers,** as shown in Figure 2. Real numbers can be represented by decimal numbers. Since every fraction has a decimal form—for example, $\frac{1}{4} = .25$—real numbers include rational numbers. However, some real numbers cannot be represented by fractions. These numbers are called **irrational numbers.** The set of irrational numbers includes $\sqrt{3}$ and $\sqrt{5}$ but not $\sqrt{1}$, $\sqrt{4}$, $\sqrt{9}, \ldots$, which equal 1, 2, 3, ..., and hence are rational numbers. Another irrational number is $\pi$, which is approximately equal to 3.14159. The numbers in the set $\left\{-\frac{2}{3}, 0, \sqrt{2}, \sqrt{5}, \pi, 4\right\}$ can be located on a number line, as shown in Figure 3. (Only $\sqrt{2}$, $\sqrt{5}$, and $\pi$ are irrational here. The others are rational.) Note that $\sqrt{2}$ is approximately equal to 1.41 so it is located between 1 and 2, slightly closer to 1.

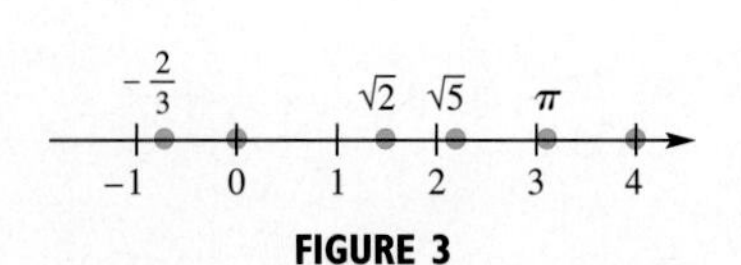


FIGURE 3

## Coordinate Systems

Figure 4 on the next page shows a number line with the points corresponding to several different numbers plotted on the line. A number that corresponds to a particular point on a line is called the **coordinate** of the point. For example, the leftmost plotted point in Figure 4 has coordinate $-4$. Every point on the real number line has a real number coordinate, and every real number corresponds to a point on the real number line. This correspondence is called a **coordinate system.**

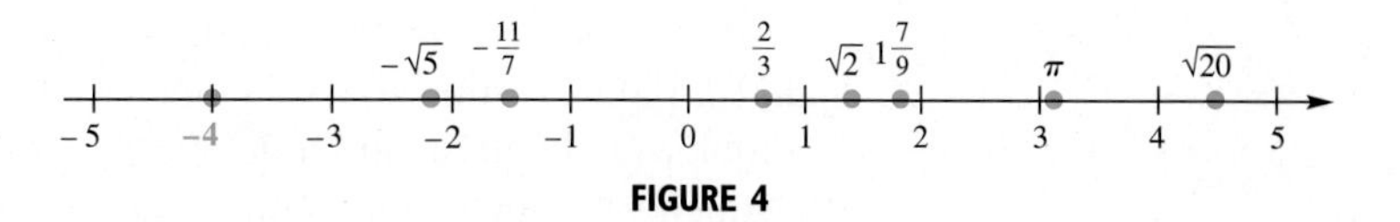


FIGURE 4

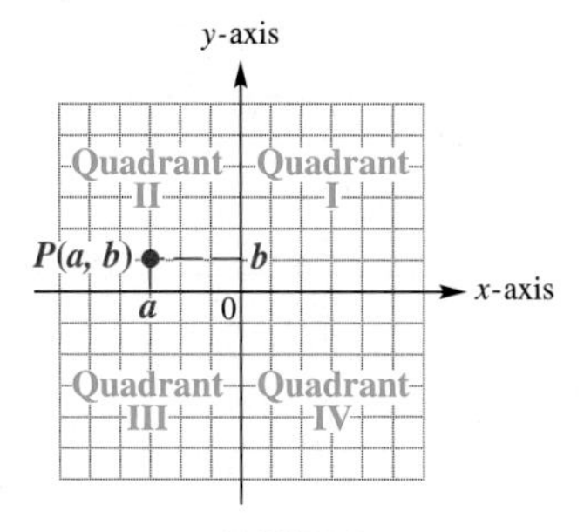


FIGURE 5

If we place two number lines at right angles, intersecting at their origins, we obtain a two-dimensional **rectangular coordinate system.** The number lines intersect at the *origin* of the system, designated 0. The horizontal line is called the ***x*-axis,** and the vertical line is called the ***y*-axis.** On the $x$-axis, positive numbers are located to the right of the origin, and negative numbers are located to the left. On the $y$-axis, positive numbers are located above the origin, negative numbers below.

This rectangular coordinate system is also called the **Cartesian coordinate system,** named after Rene Descartes (1596–1650). The plane into which the coordinate system is introduced is the **coordinate plane,** or ***xy*-plane.** The $x$-axis and $y$-axis divide the plane into four regions, or **quadrants,** as shown in Figure 5. The points on the $x$-axis or $y$-axis belong to no quadrant.

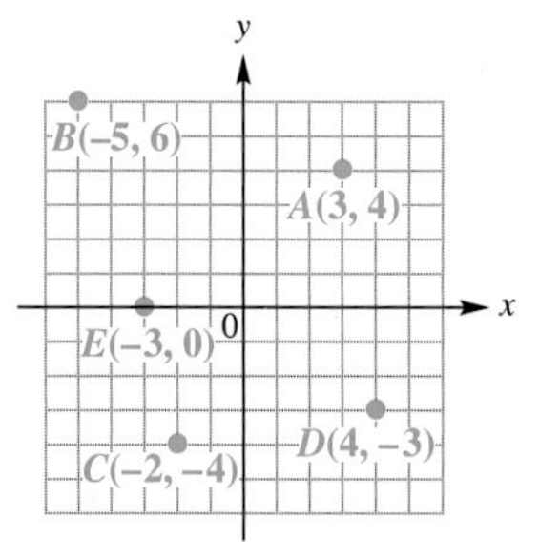


FIGURE 6

Each point $P$ in the $xy$-plane corresponds to a unique ordered pair $(a, b)$ of real numbers. The numbers $a$ and $b$ are the *coordinates* of point $P$. We call $a$ the ***x*-coordinate** and $b$ the ***y*-coordinate.** To locate the point corresponding to the ordered pair $(3, 4)$ on the $xy$-plane, for example, draw a vertical line through 3 on the $x$-axis and a horizontal line through 4 on the $y$-axis. These two lines intersect at point $A$ in Figure 6. Point $A$ corresponds to the ordered pair $(3, 4)$. Also in Figure 6, $B$ corresponds to the ordered pair $(-5, 6)$, $C$ to $(-2, -4)$, $D$ to $(4, -3)$, and $E$ to $(-3, 0)$. The coordinates of the origin are $(0, 0)$. The point $P$ corresponding to the ordered pair $(a, b)$ is often written as $P(a, b)$ as in Figure 5 and referred to as "the point $(a, b)$."

## Viewing Windows

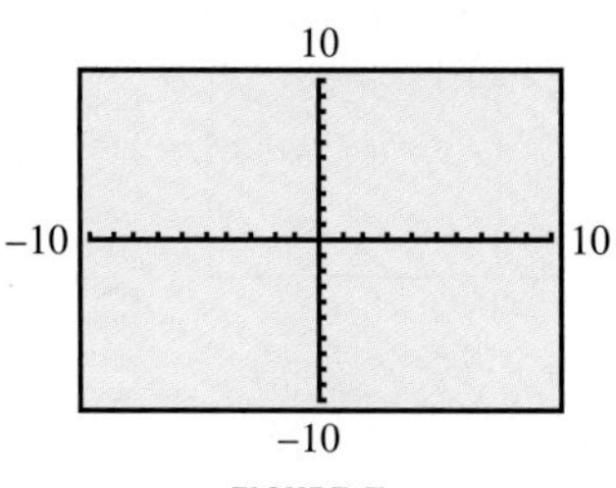


FIGURE 7

The rectangular (Cartesian) coordinate system extends indefinitely in all directions. We can show only a portion of such a system in a text figure. Similar limitations are found in portraying coordinate systems on calculator screens. The most common term used to refer to these limitations is "window." Figure 7 shows a calculator screen that has been set to have a minimum $x$-value of $-10$, a maximum $x$-value of 10, a minimum $y$-value of $-10$, and a maximum $y$-value of 10. Additionally, the tick marks on the axes have been set to be 1 unit apart. Notice, for example, that there are 10 tick marks on the positive $x$-axis. Throughout this book, this window will be called the **standard viewing window.**

To convey important information about a viewing window, we use the following abbreviations:

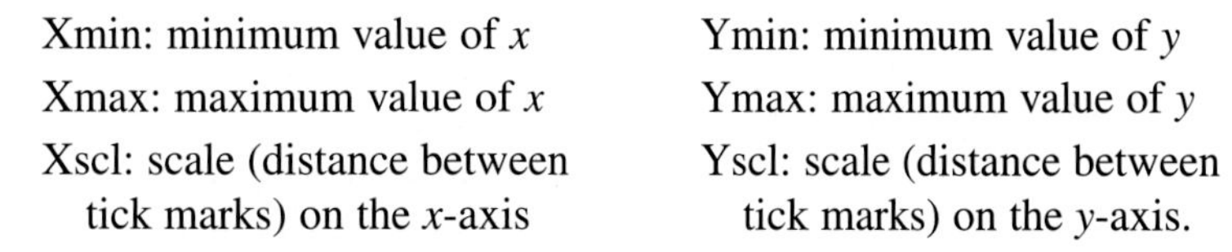

Xmin: minimum value of $x$ | Ymin: minimum value of $y$
Xmax: maximum value of $x$ | Ymax: maximum value of $y$
Xscl: scale (distance between tick marks) on the $x$-axis | Yscl: scale (distance between tick marks) on the $y$-axis.

To further condense this information, we can use the following symbolism.

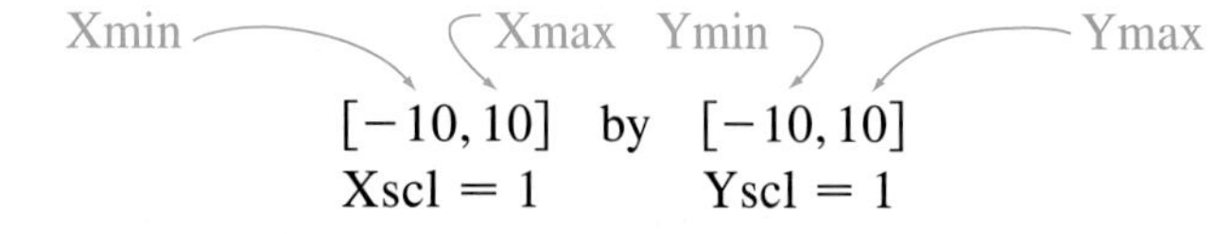


$$[-10, 10] \text{ by } [-10, 10]$$
$$\text{Xscl} = 1 \qquad \text{Yscl} = 1$$

The symbols above indicate the viewing window information for the window in Figure 7.

All calculators have a standard viewing window. Viewing windows may be changed. Figure 8 shows several other viewing windows. Notice that Figures 8(b) and 8(c) look exactly alike, and unless we are told what the settings are, we have no way of distinguishing between them. What are Xscl and Yscl in each figure?

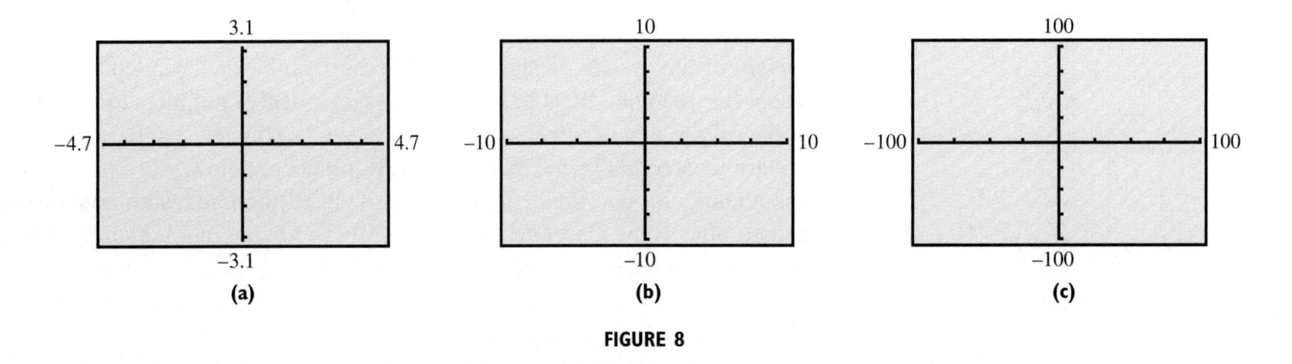


**FIGURE 8**

**TECHNOLOGY NOTE**

You should consult the graphing calculator manual that accompanies this text or your owner's manual to see how to set the viewing window on your screen. Remember that different settings will result in different views of graphs. When you adjust the settings on cameras, telescopes, and binoculars, different views of the subject are obtained; the same goes for graphs generated by calculators.

**What Went WRONG?**

A student learning how to use a graphing calculator could not understand why the axes on the graph were so "thick," as seen in Figure A, while those on a friend's calculator were not, as seen in Figure B.

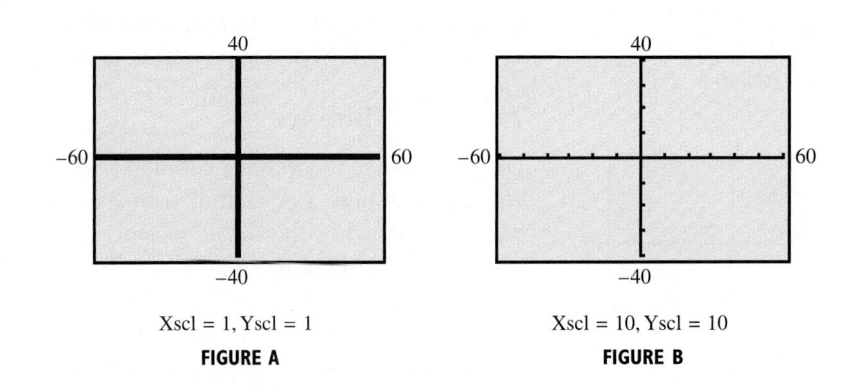


Xscl = 1, Yscl = 1

**FIGURE A**

Xscl = 10, Yscl = 10

**FIGURE B**

What Went Wrong? How can the student correct the problem in Figure A so that the axes look like those in Figure B?

## Roots

Calculators have the capability to express numbers like $\sqrt{2}$, $\sqrt[3]{5}$, and $\pi$ to many decimal places. Rather than asking you to write all these decimal places on your paper, we often ask that answers be rounded. For review, the following table illustrates how to round numbers to the nearest tenth, hundredth, or thousandth.

---

*Answer to What Went Wrong?*

Since Xscl = 1 and Yscl = 1 in Figure A, there are 120 tick marks along the $x$-axis and 80 tick marks along the $y$-axis. The resolution of the graphing calculator screen is not high enough to show all these tick marks, so the axes appear as heavy black lines instead. The values for Xscl and Yscl need to be larger, as in Figure B.

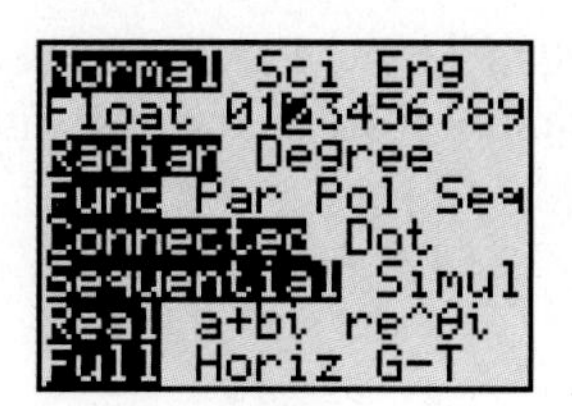


FIGURE 9

FIGURE 10

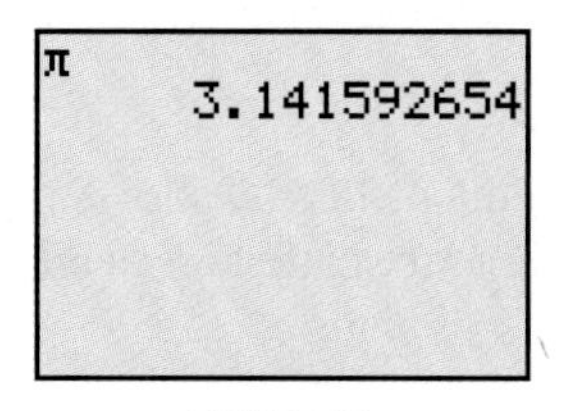


FIGURE 11

**Rounding Numbers**

| Number | Nearest Tenth | Nearest Hundredth | Nearest Thousandth |
|---|---|---|---|
| 1.3782 | 1.4 | 1.38 | 1.378 |
| 201.6666 | 201.7 | 201.67 | 201.667 |
| .0819 | .1 | .08 | .082 |

Many graphing calculators can be set to display a fixed number of decimal places. In Figure 9 a graphing calculator is set to round values to the nearest hundredth (two decimal places), and in Figure 10 the numbers from the above table are rounded to the nearest hundredth. The symbol "≈" is used to indicate that two expressions are *approximately equal.* For example, $\pi \approx 3.14$, but $\pi \neq 3.14$ since $\pi = 3.141592654\ldots$. See Figure 11.

**EXAMPLE 1** *Finding Roots on a Calculator*

Use a calculator to approximate each root to the nearest thousandth. (*Note:* You can use the fact that $\sqrt[n]{a} = a^{1/n}$ to find roots.)

**(a)** $\sqrt{23}$ **(b)** $\sqrt[3]{87}$ **(c)** $\sqrt[4]{12}$

**Solution**

**(a)** The screen in Figure 12(a) shows that an approximation for $\sqrt{23}$, to the nearest thousandth, is 4.796. It is displayed twice, once for $\sqrt{23}$ and once for $23^{1/2}$.

**(b)** To the nearest thousandth $\sqrt[3]{87} \approx 4.431$. See Figure 12(b).

**(c)** Figure 12(c) indicates $\sqrt[4]{12} \approx 1.861$ in three different ways.

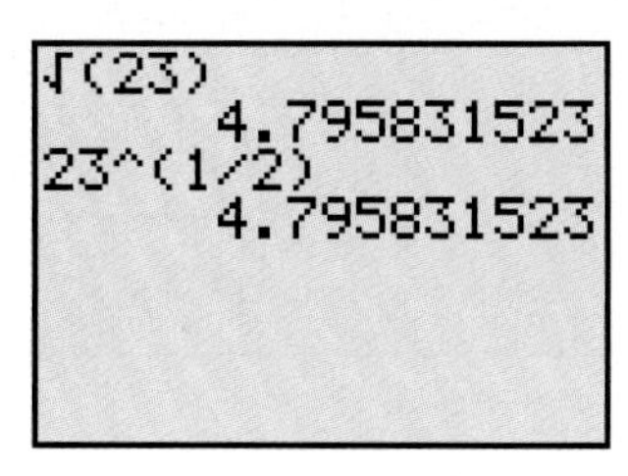


(a)

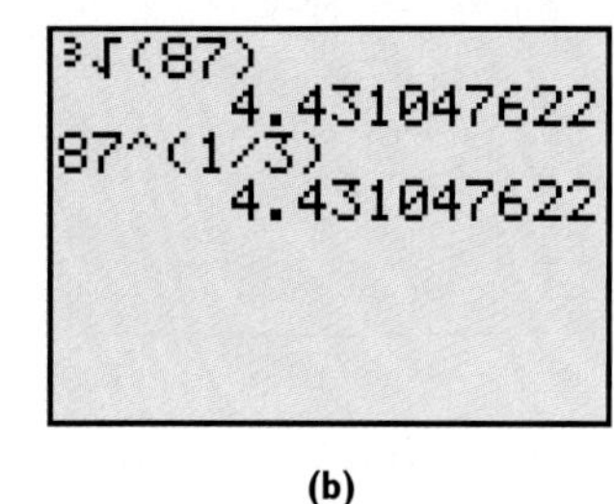


(b)

(c)

FIGURE 12

TECHNOLOGY NOTE

Many graphing calculators have built-in keys for calculating square roots and menus for calculating other types of roots. On some models, roots are located under the MATH menu. Consult the graphing calculator manual that accompanies this text or your owner's manual.

**EXAMPLE 2** *Approximating Expressions with a Calculator*

Approximate each expression to the nearest hundredth.

**(a)** $\dfrac{3.8 - 1.4}{5.4 + 3.5}$ **(b)** $3\pi^4 - 9^2$ **(c)** $\sqrt{(4-1)^2 + (-3-2)^2}$

**Solution**

**(a)** Be sure to insert parentheses around *both* the numerator and the denominator. See Figure 13(a) on the next page. To the nearest hundredth,

$$\frac{3.8 - 1.4}{5.4 + 3.5} \approx .27.$$

**(b)** When entering $\pi$, use the built-in key for $\pi$ rather than 3.14. Many calculators also have a special key to calculate the square of a number. To the nearest hundredth, $3\pi^4 - 9^2 \approx 211.23$. See Figure 13(b).

**(c)** From Figure 13(c), $\sqrt{(4-1)^2 + (-3-2)^2} \approx 5.83$. Notice the careful use of parentheses.

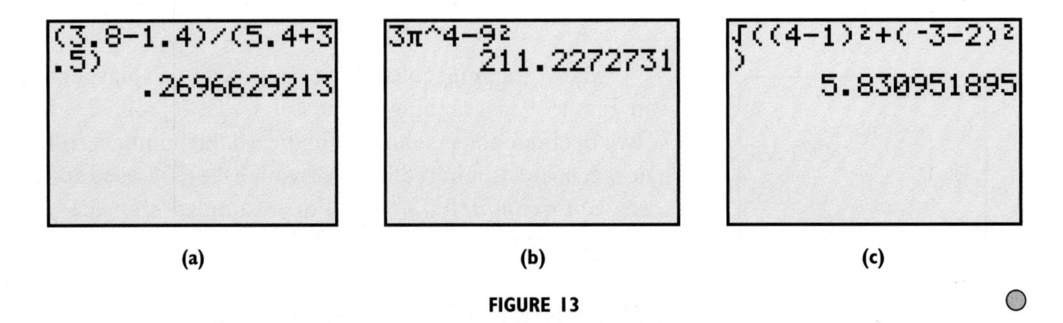


(a) (b) (c)

FIGURE 13

## Distance and Midpoint Formulas

The Pythagorean theorem can be used to calculate the lengths of the sides of a right triangle. In a right triangle, the sides that form the right angle are called *legs* and the side opposite the right angle (the longest side) is called the *hypotenuse*.

**Pythagorean Theorem**

In a right triangle, the sum of the squares of the lengths of the legs is equal to the square of the length of the hypotenuse.

$$a^2 + b^2 = c^2$$

Hypotenuse $c$, Leg $a$, Leg $b$

An outline of a proof of this theorem is found in Exercise 85.

The *converse* of the Pythagorean theorem is also true. That is, if $a$, $b$, and $c$ are lengths of the sides of a triangle and $a^2 + b^2 = c^2$, then the triangle is a right triangle with hypotenuse $c$. For example, if a triangle has sides with lengths 3, 4, and 5, then it is a right triangle with hypotenuse of length 5 because $3^2 + 4^2 = 5^2$.

The Pythagorean theorem can be used to derive a formula to find the distance between two points in the $xy$-plane. Let $P(x_1, y_1)$ and $R(x_2, y_2)$ be any two distinct points in a plane, as shown in Figure 14. Complete a triangle by locating point $Q$ with coordinates $(x_2, y_1)$. Using the Pythagorean theorem, the distance between $P$ and $R$ can be written

$$d(P, R) = \sqrt{(x_2 - x_1)^2 + (y_2 - y_1)^2}.$$

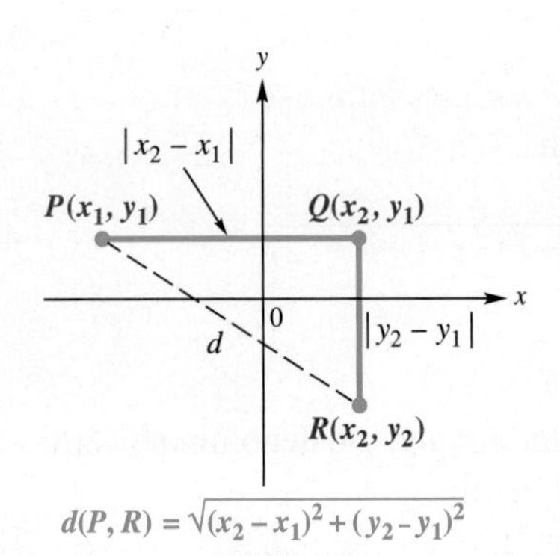


FIGURE 14

**NOTE** Absolute value bars are not necessary in this formula since for all real numbers $a$ and $b$, $|a - b|^2 = (a - b)^2$.

The distance formula can be summarized as follows.

**Distance Formula**

Suppose that $P(x_1, y_1)$ and $R(x_2, y_2)$ are two points in a coordinate plane. Then the distance between $P$ and $R$, written $d(P, R)$, is given by the **distance formula,**

$$d(P, R) = \sqrt{(x_2 - x_1)^2 + (y_2 - y_1)^2}.$$

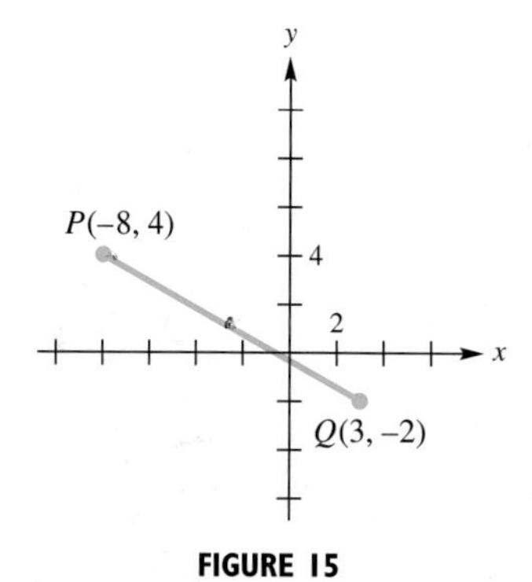

FIGURE 15

**EXAMPLE 3** *Using the Distance Formula*

The line segment shown in Figure 15 joins the points $P(-8, 4)$ and $Q(3, -2)$. Use the distance formula to find $d(P, Q)$.

**Solution** The length of the segment, or $d(P, Q)$, is found as follows.

$$\begin{aligned} d(P, Q) &= \sqrt{[3 - (-8)]^2 + (-2 - 4)^2} \qquad x_1 = -8,\ y_1 = 4,\ x_2 = 3,\ y_2 = -2 \\ &= \sqrt{11^2 + (-6)^2} \\ &= \sqrt{121 + 36} \\ &= \sqrt{157} \end{aligned}$$

**NOTE** As shown in Example 3, it is customary to leave the distance between two points in radical form rather than approximating it with a calculator.

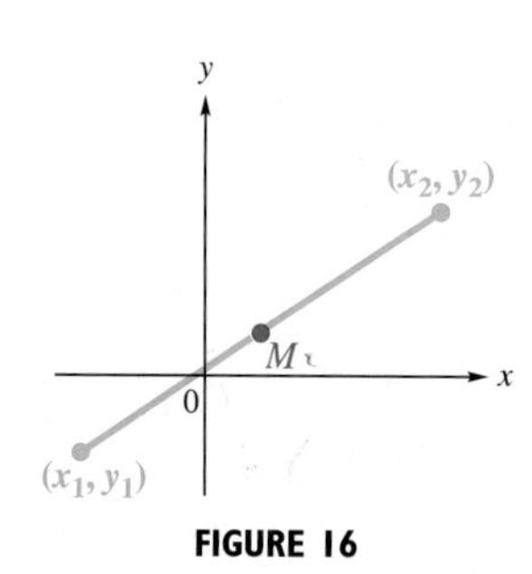

FIGURE 16

The *midpoint* of a line segment is the point on the segment that lies the same distance from both endpoints. Point $M$ in Figure 16 is the midpoint of the segment joining $(x_1, y_1)$ and $(x_2, y_2)$.

**Midpoint Formula**

The **midpoint** of the line segment with endpoints $(x_1, y_1)$ and $(x_2, y_2)$ is

$$\left(\frac{x_1 + x_2}{2}, \frac{y_1 + y_2}{2}\right).$$

The midpoint is found by calculating the *average* of the $x$-coordinates and the *average* of the $y$-coordinates of the endpoints of the segment. In Exercise 86, you are asked to verify that the coordinates above satisfy the definition of midpoint.

**EXAMPLE 4** *Using the Midpoint Formula*

Find the midpoint $M$ of the segment with endpoints $(8, -4)$ and $(-9, 6)$.

**Solution** Let $(x_1, y_1) = (8, -4)$ and $(x_2, y_2) = (-9, 6)$ in the midpoint formula.

$$M = \left(\frac{x_1 + x_2}{2}, \frac{y_1 + y_2}{2}\right) = \left(\frac{8 + (-9)}{2}, \frac{-4 + 6}{2}\right) = \left(-\frac{1}{2}, 1\right)$$

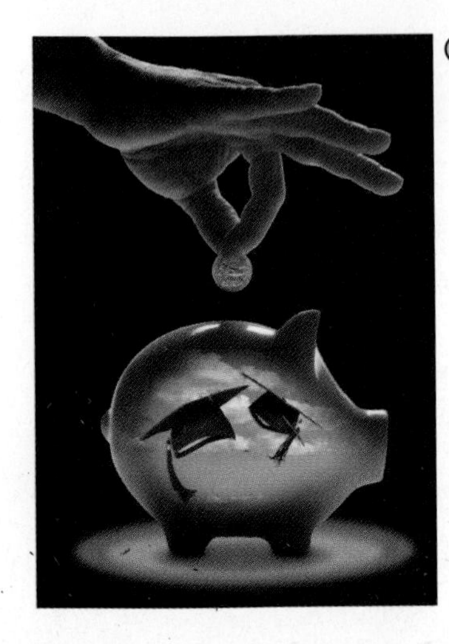

**EXAMPLE 5** *Estimating Tuition and Fees*

In 1995, average tuition and fees at public colleges and universities were \$3077, whereas they were \$3356 in 1999. Use the midpoint formula to estimate tuition and fees in 1997. Compare your estimate to the actual value of \$3197. (*Source:* The College Board.)

**Solution** Notice that the year 1997 lies midway between 1995 and 1999. Therefore, we can use the midpoint formula to find the midpoint of the line segment joining (1995, 3077) and (1999, 3356).

$$\left(\frac{1995 + 1999}{2}, \frac{3077 + 3356}{2}\right) = (1997, 3216.5)$$

The midpoint formula estimates tuition and fees at public colleges and universities to be \$3216.50 in 1997. This is within \$20 of the actual value.

## 1.1 EXERCISES

*For each set, list all elements that belong to the* **(a)** *natural numbers,* **(b)** *whole numbers,* **(c)** *integers,* **(d)** *rational numbers,* **(e)** *irrational numbers, and* **(f)** *real numbers.*

**1.** $\left\{-6, -\frac{12}{4}, -\frac{5}{8}, -\sqrt{3}, 0, .31, .\overline{3}, 2\pi, 10, \sqrt{17}\right\}$

**2.** $\left\{-8, -\frac{14}{7}, -.245, 0, \frac{6}{2}, 8, \sqrt{81}, \sqrt{12}\right\}$

**3.** $\left\{-\sqrt{100}, -\frac{13}{6}, -1, 5.23, 9.\overline{14}, 3.14, \frac{22}{7}\right\}$

**4.** $\{-\sqrt{49}, -.405, -.\overline{3}, .1, 3, 18, 6\pi, 56\}$

*Plot each set of numbers on a number line.*

**5.** $\{-4, -3, -2, -1, 0, 1\}$

**6.** $\{-6, -5, -4, -3, -2\}$

**7.** $\left\{-.5, .75, \frac{5}{3}, 3.5\right\}$

**8.** $\left\{-.6, \frac{9}{8}, 2.5, \frac{13}{4}\right\}$

**9.** Explain the difference between a rational number and an irrational number.

**10.** *Concept Check* Using her calculator, a student found the decimal 1.414213562 when she evaluated $\sqrt{2}$. Is this decimal the exact value of $\sqrt{2}$, or just an approximation of $\sqrt{2}$? Should she write $\sqrt{2} = 1.414213562$ or $\sqrt{2} \approx 1.414213562$?

*Locate each point on a rectangular coordinate system. Identify the quadrant, if any, in which each point lies.*

**11.** $(2, 3)$ **12.** $(-1, 2)$ **13.** $(-3, -2)$ **14.** $(1, -4)$ **15.** $(0, 5)$

**16.** $(-2, -4)$ **17.** $(-2, 4)$ **18.** $(3, 0)$ **19.** $(-2, 0)$ **20.** $(3, -3)$

*Name the possible quadrants in which the point $(x, y)$ can lie if the given condition is true.*

**21.** $xy > 0$ **22.** $xy < 0$ **23.** $\frac{x}{y} < 0$ **24.** $\frac{x}{y} > 0$

**25.** *Concept Check* If the $x$-coordinate of a point is 0, the point must lie on which axis?

**26.** *Concept Check* If the $y$-coordinate of a point is 0, the point must lie on which axis?

*Answer each question about your particular calculator, using a complete sentence or sentences.*

**27.** How do you set the screen in order to obtain the standard viewing window?

**28.** What are the minimum and maximum values of $x$ and $y$ in your standard viewing window?

*Give the values of* Xmin, Xmax, Ymin, *and* Ymax *for each screen, given the values for* Xscl *and* Yscl. *Use the notation described in this section.*

**29.**

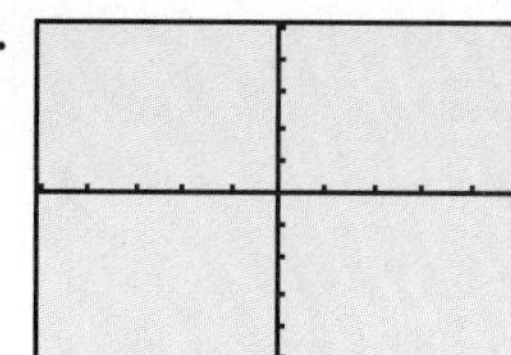

Xscl = 1, Yscl = 5

**30.**

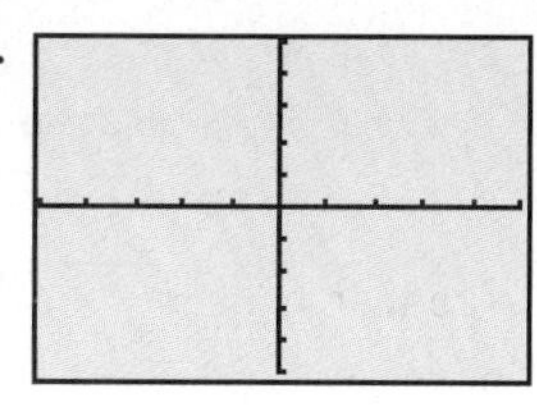

Xscl = 5, Yscl = 1

**31.**

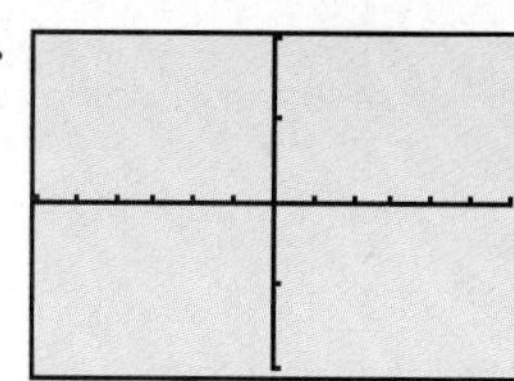

Xscl = 10, Yscl = 50

**32.**

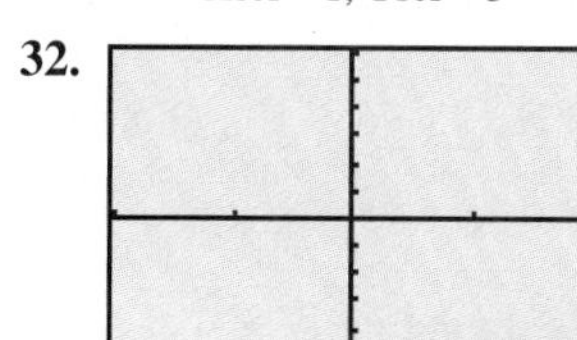

Xscl = 50, Yscl = 10

**33.**

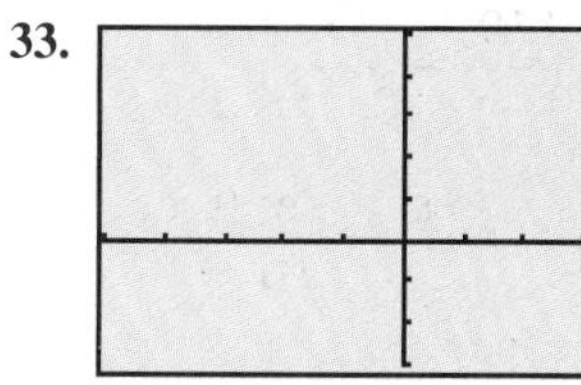

Xscl = 100, Yscl = 100

**34.**

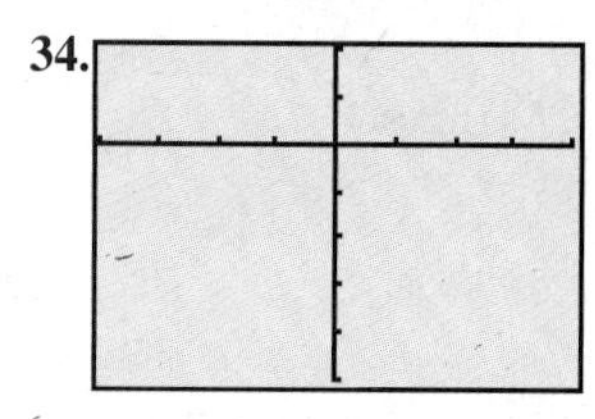

Xscl = 75, Yscl = 75

*Set the viewing window of your calculator to the following specifications. Make a sketch of your window.*

**35.** $[-10, 10]$ by $[-10, 10]$
Xscl = 1 Yscl = 1

**36.** $[-40, 40]$ by $[-30, 30]$
Xscl = 5 Yscl = 5

**37.** $[-5, 10]$ by $[-5, 10]$
Xscl = 3 Yscl = 3

**38.** $[-3.5, 3.5]$ by $[-4, 10]$
Xscl = 1 Yscl = 1

**39.** $[-100, 100]$ by $[-50, 50]$
Xscl = 20 Yscl = 25

**40.** $[-4.7, 4.7]$ by $[-3.1, 3.1]$
Xscl = 1 Yscl = 1

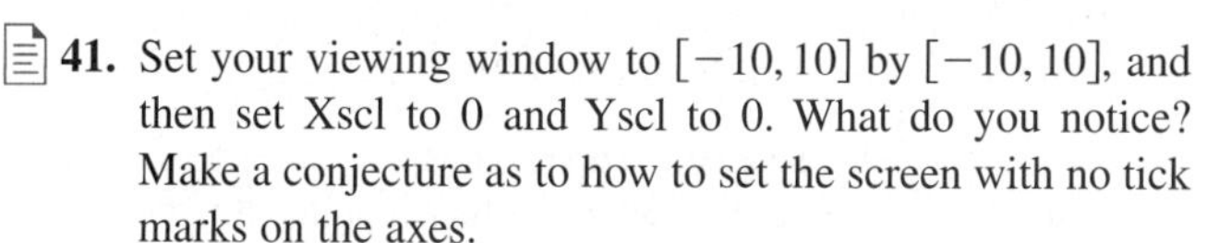

**41.** Set your viewing window to $[-10, 10]$ by $[-10, 10]$, and then set Xscl to 0 and Yscl to 0. What do you notice? Make a conjecture as to how to set the screen with no tick marks on the axes.

**42.** Set your viewing window to $[-50, 50]$ by $[-50, 50]$, and then set Xscl to 1 and Yscl to 1. Observe this screen and describe the appearance of the axes as compared to those seen in the standard window. Why do you think they appear this way? How can you change your scale settings so that this "problem" is alleviated?

*Use a calculator to find a decimal approximation of each root or power. Round answers to the nearest thousandth.*

**43.** $\sqrt{58}$ **44.** $\sqrt{97}$ **45.** $\sqrt[3]{33}$ **46.** $\sqrt[3]{91}$

**47.** $\sqrt[4]{86}$ **48.** $\sqrt[4]{123}$ **49.** $19^{1/2}$ **50.** $29^{1/3}$

*Use a calculator to approximate each expression to the nearest hundredth.*

**51.** $\dfrac{5.6 - 3.1}{8.9 + 1.3}$

**52.** $\dfrac{34 + 25}{23}$

**53.** $\sqrt{\pi^3 + 1}$

**54.** $\sqrt[3]{2.1 - 6^2}$

**55.** $3(5.9)^2 - 2(5.9) + 6$

**56.** $2\pi^3 - 5\pi^2 - 3$

**57.** $\sqrt{(4 - 6)^2 + (7 + 1)^2}$

**58.** $\sqrt{[-1 - (-3)]^2 + (-5 - 3)^2}$

**59.** $\dfrac{\sqrt{\pi - 1}}{\sqrt{1 + \pi}}$

**60.** $\sqrt[3]{4.5 \times 10^5 + 3.7 \times 10^2}$

**61.** $\dfrac{2}{1 - \sqrt[3]{5}}$

**62.** $1 - \dfrac{4.5}{3 - \sqrt{2}}$

*Use the Pythagorean theorem to find the length of the unknown side of the right triangle. In each case, a and b represent the lengt' of the legs and c represents the length of the hypotenuse.*

**63.** $a = 8, b = 15$; find $c$

**64.** $a = 7, b = 24$; find $c$

**65.** $a = 13, c = 85$; find $b$

**66.** $a = 14, c = 50$; find $b$

**67.** $a = 5, b = 8$; find $c$

**68.** $a = 9, b = 10$; find $c$

**69.** $b = \sqrt{13}, c = \sqrt{29}$; find $a$

**70.** $b = \sqrt{7}, c = \sqrt{11}$; find $a$

a c b

Typical Labeling

*Find* **(a)** *the distance between P and Q and* **(b)** *the coordinates of the midpoint of the segment joining P and Q.*

**71.**

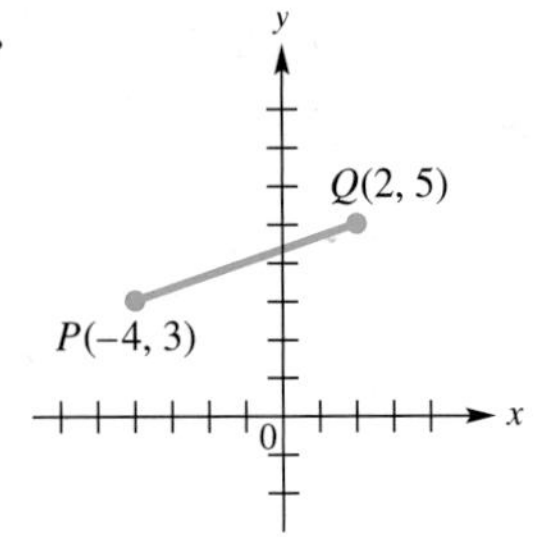


**72.**

y
P(−7, 4)
0
x
Q(6, −2)

**73.** $P(5,7)$, $Q(13,-1)$

**74.** $P(-2,5)$, $Q(4,-3)$

**75.** $P(-8,-2)$, $Q(-3,-5)$

**76.** $P(-6,-10)$, $Q(6,5)$

*Solve each problem.*

**77.** *(Modeling) Tuition and Fees* During the time period 1991–1999, the average annual cost (in dollars) of tuition and fees at private four-year colleges rose in an approximately linear fashion. The graph depicts this growth using a line segment. Use the midpoint formula to approximate the cost during the year 1995.

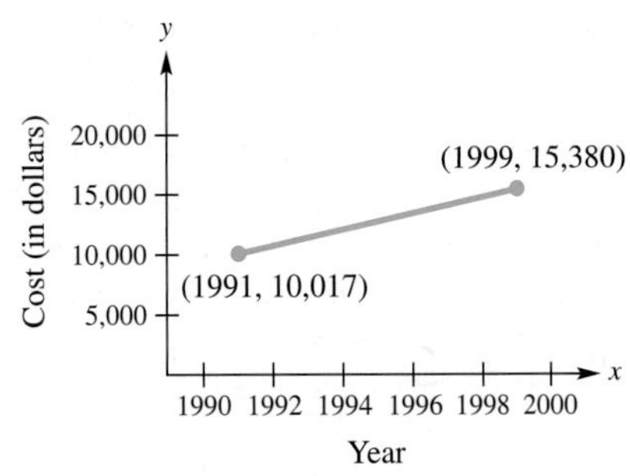


*Source*: The College Board.

**78.** *(Modeling) Two-Year College Enrollment* Enrollments in two-year colleges for 1980, 1990, and 2000 are shown in the table. Use the midpoint formula to estimate the enrollments for 1985 and 1995.

| Year | Enrollment (in millions) |
|---|---|
| 1980 | 4.5 |
| 1990 | 5.2 |
| 2000 | 5.8 |

*Source: Statistical Abstract of the United States.*

**79.** *(Modeling) Poverty Level Income Cutoffs* The table in the next column lists poverty level income cutoffs for a family of four since 1960. Use the midpoint formula to estimate the poverty level cutoffs in 1965 and 1985.

| Year | Income (in dollars) |
|---|---|
| 1960 | 3,022 |
| 1970 | 3,968 |
| 1980 | 8,414 |
| 1990 | 13,359 |
| 1999 | 17,184 |

*Source:* U.S. Bureau of the Census.

**80.** *Concept Check* Suppose that segment $PQ$ has midpoint $M$. The coordinates of $P$ are $(-3,6)$, and the coordinates of $M$ are $(5,8)$. Find the coordinates of $Q$.

**81.** *Geometry* An **isosceles triangle** has at least two sides of equal length. Determine whether the triangle with vertices $(0,0)$, $(3,4)$, and $(7,1)$ is isosceles.

**82.** *Geometry* An **equilateral triangle** has all sides of equal length. Determine whether the triangle with vertices $(-1,-1)$, $(2,3)$, and $(-4,3)$ is equilateral.

**83.** *Distance between Cars* At 9:00 A.M. Car $A$ is traveling north at 50 mph and is located 50 miles south of Car $B$. Car $B$ is traveling west at 20 mph.

**(a)** Let $(0,0)$ be the initial coordinates of Car $B$ in the $xy$-plane, where units are in miles. Plot the locations of each car at 9:00 A.M. and at 11:00 A.M.

**(b)** Find the distance between the cars at 11:00 A.M.

**84.** *Distance between Ships* Two ships leave the same harbor at the same time. The first ship heads north at 20 mph and the second ship heads west at 15 mph. Write an expression that gives the distance between the ships after $t$ hours.

**85.** The figure at the top of the next page is a square made up of four right triangles and a smaller square. Using this figure, the Pythagorean theorem can be proved. Fill in the blanks with the missing information.

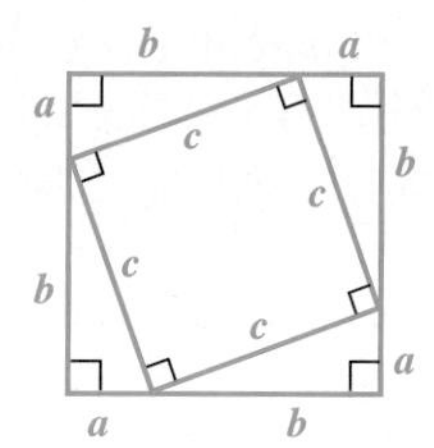


(a) The length of a side of the large square is __________, so its area is (__________)$^2$ or __________.

(b) The area of the large square may also be found by obtaining the sum of the areas of the four right triangles and the smaller square. The area of each right triangle is __________, so the sum of the areas of the four right triangles is __________. The area of the smaller square is __________.

(c) The sum of the areas of the four right triangles and the smaller square is __________.

(d) Since the areas in (a) and (c) represent the area of the same figure, the expressions there must be equal. Set them equal to each other to obtain __________ = __________.

(e) Subtract $2ab$ from each side of the equation in (d) to obtain the desired result, __________ = __________.

**86.** Suppose that the endpoints of a line segment have coordinates $(x_1, y_1)$ and $(x_2, y_2)$.

(a) Show that the distance between the points $(x_1, y_1)$ and $\left(\frac{x_1 + x_2}{2}, \frac{y_1 + y_2}{2}\right)$ is the same as the distance between the points $(x_2, y_2)$ and $\left(\frac{x_1 + x_2}{2}, \frac{y_1 + y_2}{2}\right)$.

(b) Show that the sum of the distances between the points $(x_1, y_1)$ and $\left(\frac{x_1 + x_2}{2}, \frac{y_1 + y_2}{2}\right)$ and between the points $(x_2, y_2)$ and $\left(\frac{x_1 + x_2}{2}, \frac{y_1 + y_2}{2}\right)$ is equal to the distance between the points $(x_1, y_1)$ and $(x_2, y_2)$.

(c) From the results of parts (a) and (b), what conclusion can be made?

# 1.2 INTRODUCTION TO RELATIONS AND FUNCTIONS

Set-Builder Notation and Interval Notation • Relations, Domain, and Range • Functions • Tables • Function Notation

In this section we introduce some important concepts in mathematics: relation, function, domain, and range. To make our work simpler, set notation is useful. We begin by discussing two types: set-builder notation and interval notation.

## Set-Builder Notation and Interval Notation

Suppose we wish to symbolize the set of real numbers greater than $-2$. One way is to write it as $\{x \mid x > -2\}$, read "the set of all $x$ such that $x$ is greater than $-2$." This is called **set-builder notation** since the variable $x$ is used to "build" the set. On a number line, we show the elements of this set by drawing a line from $-2$ to the right. We use a parenthesis at $-2$ to indicate that $-2$ is not an element of the given set. The result, shown in Figure 17, is called the **graph** of the set $\{x \mid x > -2\}$.

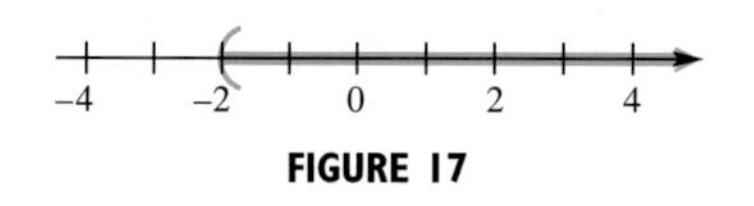


FIGURE 17

The set of numbers greater than $-2$ is an example of an **interval** on the number line. A simplified notation, called **interval notation,** is used for writing intervals. For example, using this notation, the interval of all numbers greater than $-2$ is written $(-2, \infty)$. The **infinity symbol** $\infty$ does not indicate a number; it shows that the interval includes all real numbers greater than $-2$. The left parenthesis indicates that $-2$ is not included. A square bracket, [, would be used if $-2$ was included. A parenthesis is always used next to the infinity symbol in interval notation. The set of all real numbers is written $(-\infty, \infty)$ in interval notation.

The following chart summarizes the names of various types of intervals.

| Type of Interval | Example of Set-Builder Notation | Corresponding Interval Notation | Corresponding Graph |
|---|---|---|---|
| Open interval | $\{x \mid 2 < x < 5\}$ | $(2, 5)$ | 2 5 |
| Closed interval | $\{x \mid 2 \le x \le 5\}$ | $[2, 5]$ | 2 5 |
| Half-open (or half-closed) interval | $\{x \mid 2 < x \le 5\}$ | $(2, 5]$ | 2 5 |
| | $\{x \mid 2 \le x < 5\}$ | $[2, 5)$ | 2 5 |
| Unbounded interval | $\{x \mid x > 2\}$ | $(2, \infty)$ | 2 |
| | $\{x \mid x \ge 2\}$ | $[2, \infty)$ | 2 |
| | $\{x \mid x < 2\}$ | $(-\infty, 2)$ | 2 |
| | $\{x \mid x \le 2\}$ | $(-\infty, 2]$ | 2 |
| All real numbers | $\{x \mid x \text{ is real}\}$ | $(-\infty, \infty)$ | 2 5 |

**CAUTION** Notice how the interval notation for the open interval $(2, 5)$ looks exactly like the notation for the ordered pair $(2, 5)$. We will, when the need arises, distinguish between them by referring to "the interval $(2, 5)$" or "the point $(2, 5)$."

## Relations, Domain, and Range

The bar graph shown in Figure 18 on the next page is typical of the kinds of graphs found in magazines and newspapers. It shows the number of visitors, in millions, to national parks in the United States for selected years between 1990 and 1998. Each year is paired with a number of visitors, and we may depict this information as a set of ordered pairs. The first component represents the year, and the second component represents the number of visitors in millions:

$$\{(1990, 58), (1992, 60), (1994, 63), (1996, 63), (1998, 65)\}.$$

Such a set of ordered pairs is called a *relation*.

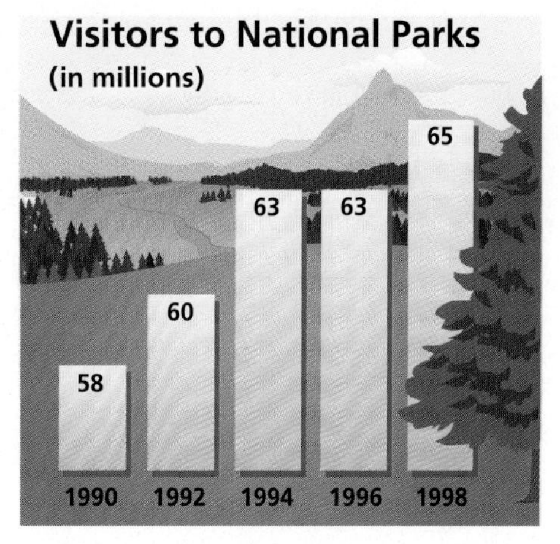


*Source: Statistical Abstract of the United States, 2000.*

**FIGURE 18**

**Relation**

A **relation** is a set of ordered pairs.

If we denote the ordered pairs of a relation by $(x, y)$, then the set of all $x$-values is called the **domain** of the relation and the set of all $y$-values is called the **range** of the relation. For the relation represented by the bar graph in Figure 18,

$$\text{Domain} = \{1990, 1992, 1994, 1996, 1998\}$$

and

$$\text{Range} = \{58, 60, 63, 65\}.$$

Here are three other examples of relations.

$$F = \{(1, 2), (-2, 5), (3, -1)\}$$
$$G = \{(-2, 1), (-1, 0), (0, 1), (1, 2), (2, 2)\}$$
$$H = \{(-4, 1), (-2, 1), (-2, 0)\}$$

For the relations $F$, $G$ and $H$,

| | |
|---|---|
| Domain of $F = \{1, -2, 3\}$ | Range of $F = \{2, 5, -1\}$ |
| Domain of $G = \{-2, -1, 0, 1, 2\}$ | Range of $G = \{1, 0, 2\}$ |
| Domain of $H = \{-4, -2\}$ | Range of $H = \{1, 0\}$. |

Since a relation is a set of ordered pairs, it may be represented graphically in the rectangular coordinate system. The graphs of $F$, $G$, and $H$ are shown in Figure 19 on the next page.

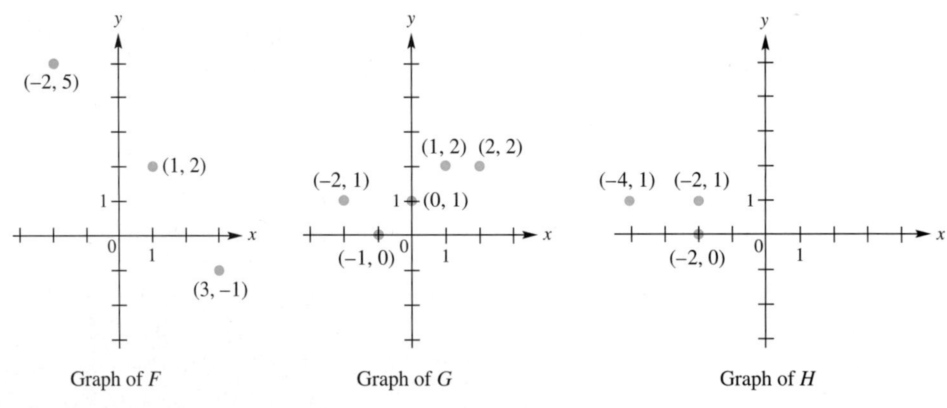


FIGURE 19

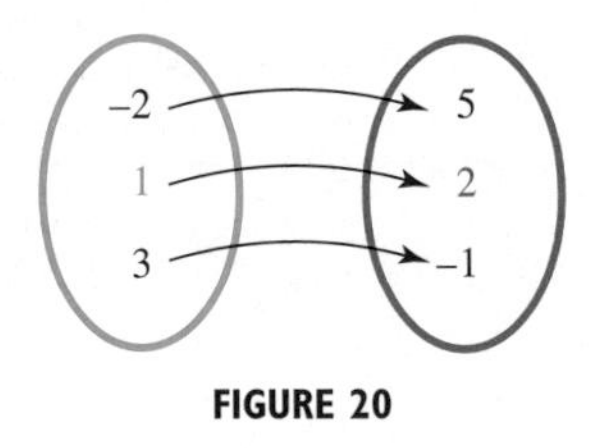


FIGURE 20

A relation can be illustrated with a diagram, as shown in Figure 20 for relation $F$. For example, the arrow from 1 to 2 indicates that the ordered pair $(1, 2)$ belongs to $F$.

Some relations contain infinitely many ordered pairs. For example, let $F$ represent a relation consisting of all ordered pairs having the form $(x, 2x)$, where $x$ is a real number. Since there are infinitely many values for $x$, there are infinitely many ordered pairs in $F$. Examples of these ordered pairs are $(-2, -4)$, $(-1, -2)$, $(0, 0)$, $(1, 2)$, and $(2, 4)$. These five points are plotted in Figure 21(a) and suggest that the graph of $F$ is a line. The graph of $F$ includes all points $(x, y)$, such that $y = 2x$, as shown in Figure 21(b).

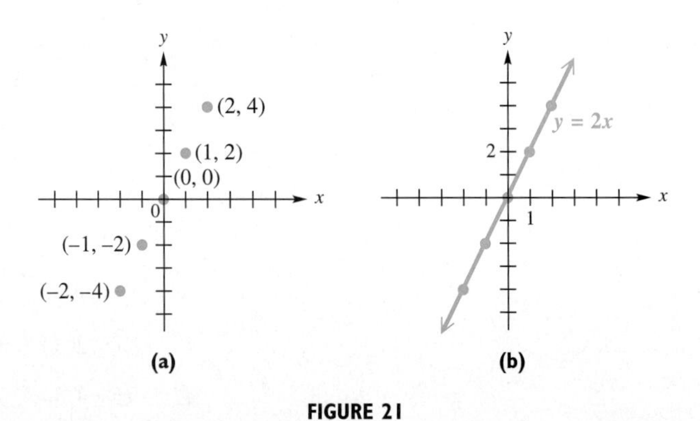


FIGURE 21

A graph of a line or curve in the $xy$-plane represents a relation. The $x$-values of the points in the graph correspond to the domain of the relation, and the $y$-values correspond to the range of the relation.

**EXAMPLE 1** *Determining Domains and Ranges from Graphs*

Give the domain and range of each relation graphed in

**(a)** Figure 22(a). **(b)** Figure 22(b). **(c)** Figure 22(c).

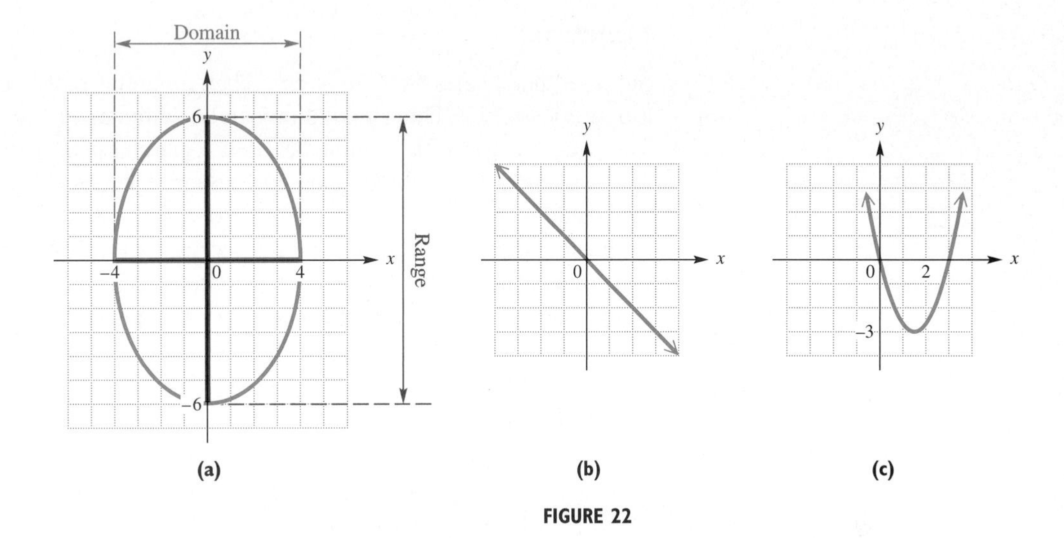


FIGURE 22

**Solution**

**(a)** In Figure 22(a), the $x$-values of the points on the graph include all numbers between $-4$ and 4 inclusive. The $y$-values include all numbers between $-6$ and 6 inclusive. Using interval notation, the domain is $[-4, 4]$, and the range is $[-6, 6]$.

**(b)** In Figure 22(b), the arrowheads indicate that the line extends indefinitely left and right, as well as upward and downward. Therefore, both the domain and range are the set of all real numbers, written $(-\infty, \infty)$.

**(c)** In Figure 22(c), the arrowheads indicate that the graph extends indefinitely left and right, as well as upward. The domain is $(-\infty, \infty)$. Because there is a least $y$-value on the graph, $-3$, the range includes all numbers greater than or equal to $-3$, written $[-3, \infty)$.

**EXAMPLE 2** *Finding Domain and Range from a Calculator Graph*

Figure 23 shows a graph on a screen with viewing window $[-5, 5]$ by $[-5, 5]$, Xscl = 1, Yscl = 1. Give the domain and range of this relation.

**TECHNOLOGY NOTE**

In Figure 23 we see a calculator graph that is formed by a rather jagged curve. These are sometimes called *jaggies* and are typically found on low-resolution graphers, such as graphing calculators. In general, most curves in this book are smooth, and jaggies are just a part of the limitations of technology.

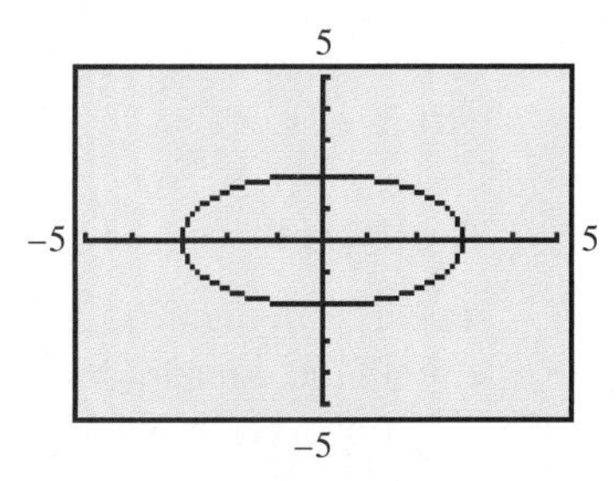


FIGURE 23

**Solution** Since the scales on both axes are 1, we see that the graph *appears* to have minimum $x$-value $-3$, maximum $x$-value 3, minimum $y$-value $-2$, and maximum $y$-value 2. Therefore, observation leads us to conclude that the domain is $[-3, 3]$ and the range is $[-2, 2]$.

## Functions

Suppose that a sales tax rate is 6%. Then a purchase of \$200 results in a sales tax of $.06 \times 200 = \$12$. This calculation can be summarized by the ordered pair (200, 12). The ordered pair (50, 3) indicates that a purchase of \$50 results in a sales tax of \$3. Notice that for each purchase of $x$ dollars, there is exactly one amount of sales tax calculated.

Calculating the sales tax $y$ on a purchase of $x$ dollars results in a set of ordered pairs $(x, y)$, where $y = .06x$. This set of ordered pairs represents a special type of relation called a *function*. To be a function, each $x$-value must correspond to exactly one $y$-value.

**Function**

A **function** is a relation in which each element in the domain corresponds to exactly one element in the range.*

If $x$ represents any element in the domain, $x$ is called the **independent variable.** If $y$ represents any element in the range, $y$ is called the **dependent variable** because the value of $y$ *depends on* the value of $x$. For example, sales tax depends on purchase price.

In most applications of functions, the correspondence between domain and range elements is defined with an equation, like $y = 9x - 5$. The equation is usually solved for $y$, as it is here, because $y$ is the dependent variable. As we choose values from the domain for $x$, we can easily determine the corresponding $y$-values of the ordered pairs of the function. (These equations need not use only $x$ and $y$ as variables; any appropriate letters may be used.)

**EXAMPLE 3** *Deciding Whether Relations are Functions*

Decide whether each relation is a function. Give the domain and range of each relation.

**(a)** $\{(1, 2), (3, 4), (5, 6), (7, 8), (9, 10)\}$

**(b)** $\{(1, 1), (1, 2), (1, 3), (2, 4)\}$

**(c)**

| $x$ | −5 | −4 | −3 | −2 | −1 | 0 | 1 |
|---|---|---|---|---|---|---|---|
| $y$ | 2 | 2 | 2 | 2 | 2 | 2 | 2 |

**(d)** $y = x - 2$

**Solution**

**(a)** The domain is $\{1, 3, 5, 7, 9\}$, and the range is $\{2, 4, 6, 8, 10\}$. Since each element in the domain corresponds to exactly one element in the range, this set is a function. The correspondence is shown below, using $D$ for the domain and $R$ for the range.

$$D = \{1, 3, 5, 7, 9\}$$
$$\downarrow\ \downarrow\ \downarrow\ \downarrow\ \downarrow$$
$$R = \{2, 4, 6, 8, 10\}$$

*An alternative definition of function based on the idea of correspondence is given later in the section.

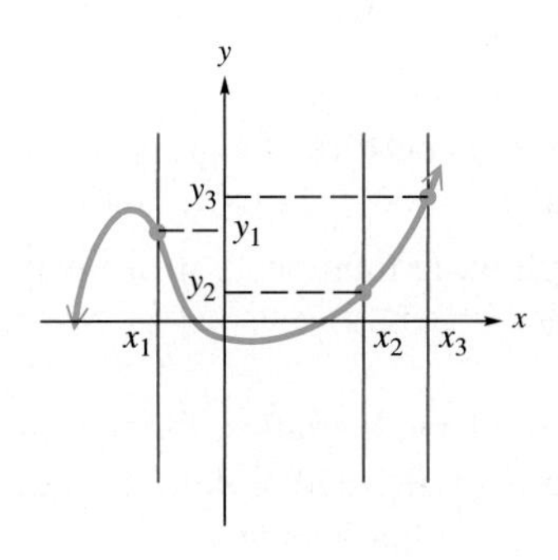


This *is* the graph of a function.

(a)

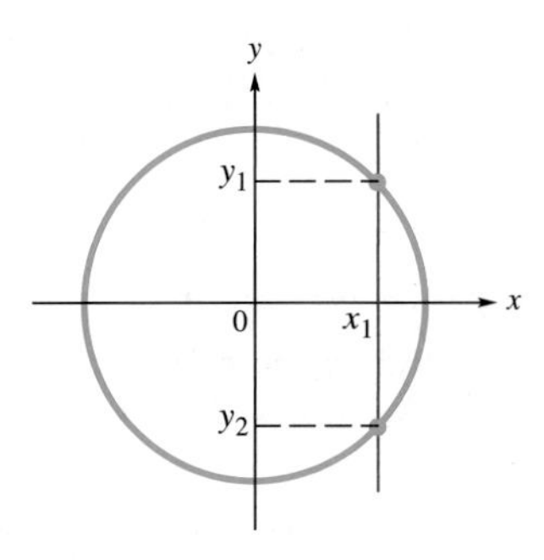


This *is not* the graph of a function.

(b)

**FIGURE 24**

**(b)** The domain here is $\{1, 2\}$, and the range is $\{1, 2, 3, 4\}$. As shown in the correspondence below, one element in the domain, 1, has been assigned three different elements from the range, so this relation is not a function.

$$D = \{1, 2\}$$

$$R = \{1, 2, 3, 4\}$$

**(c)** This is a table of ordered pairs. The domain is $\{-5, -4, -3, -2, -1, 0, 1\}$, and the range is $\{2\}$. Although every element in the domain corresponds to the same range element, this is a function because each element in the domain has exactly one range element assigned to it.

**(d)** Since $y$ is always found by subtracting 2 from $x$, each $x$ corresponds to just one value of $y$, so this relation is a function. Any number can be used for $x$, and each $x$ will give a number 2 smaller for $y$; thus, both the domain and range are the set of real numbers $(-\infty, \infty)$.

There is a quick way to tell whether a given graph is the graph of a function. Figure 24 shows two graphs. In the graph for part (a), each value of $x$ leads to only one value of $y$, so this is the graph of a function. On the other hand, the graph in part (b) is not the graph of a function. For example, if $x = x_1$, the vertical line through $x_1$ intersects the graph at two points, showing that there are two values of $y$ that correspond to this $x$-value. This concept is known as the *vertical line test* for a function.

**Vertical Line Test**

If every vertical line intersects a graph in no more than one point, then the graph is the graph of a function.

**EXAMPLE 4** *Using the Vertical Line Test*

**(a)** Is the graph in Figure 25 the graph of a function?

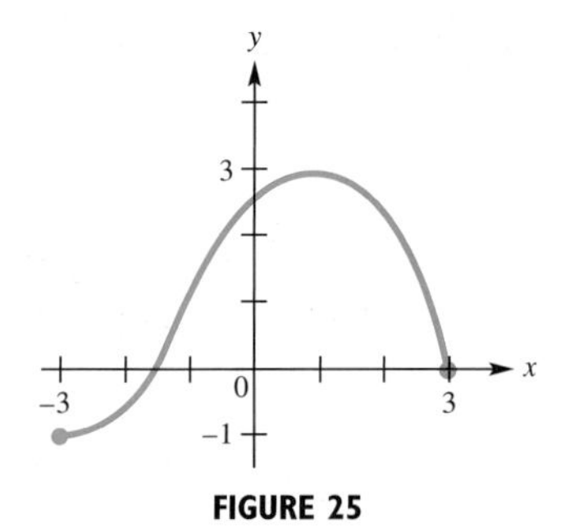


**FIGURE 25**

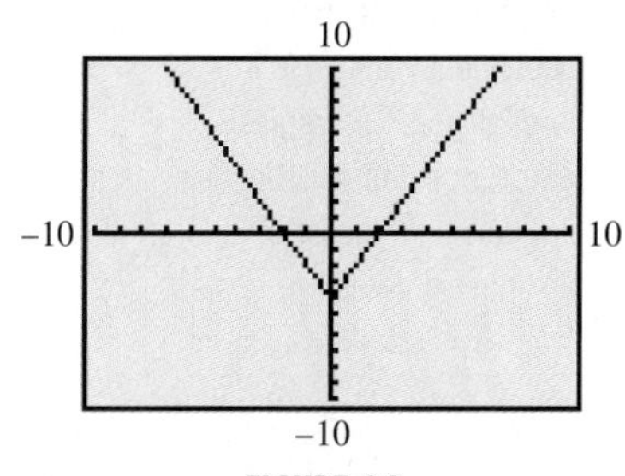


**FIGURE 26**

**(b)** Assuming the graph in Figure 26 extends left and right indefinitely and upward indefinitely, does it appear to be the graph of a function?

**Solution**

**(a)** Any vertical line will intersect the graph at most once. Therefore, the graph satisfies the vertical line test and is the graph of a function.

**(b)** It appears that no vertical line will intersect the graph more than once, so we may conclude that it is the graph of a function.

While the concept of function is crucial to the study of mathematics, the definition of function may vary from text to text. We now give an alternative definition of function that is helpful in understanding the function notation that follows.

**Function (Alternative Definition)**

A function is a correspondence in which each element $x$ from a set called the domain is paired with one and only one element $y$ from a set called the range.

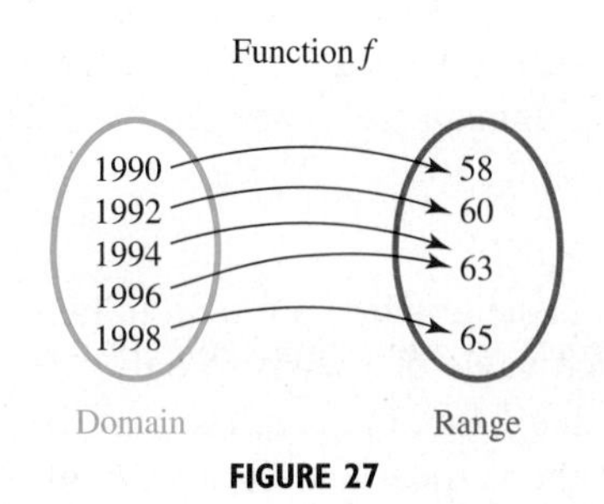


FIGURE 27

This idea of correspondence, or mapping, is illustrated in Figure 27, where the function $f$ consists of the ordered pairs in Figure 18, which gave numbers of visitors to national parks (in millions). Function $f$ pairs 1990 with 58, 1992 with 60, and so on. Each domain value is paired with one and only one range value.

## Tables

A convenient way to display ordered pairs in a function is by using a table. An equation such as $y = 9x - 5$ describes a function. If we choose $x$-values to be $0, 1, 2, \ldots, 6$, then the corresponding $y$-values are

$$y = 9(0) - 5 = -5 \qquad y = 9(1) - 5 = 4$$
$$y = 9(2) - 5 = 13 \qquad y = 9(3) - 5 = 22$$
$$y = 9(4) - 5 = 31 \qquad y = 9(5) - 5 = 40$$
$$y = 9(6) - 5 = 49.$$

These ordered pairs, $(0, -5), (1, 4), \ldots, (6, 49)$, can be organized in a table. A graphing calculator can also generate this table, as shown in Figure 28.

This screen indicates that the table will *start* with 0 and have an *increment* of 1. Both variables appear *automatically.* Tbl-Start represents the initial value of the independent variable (0 in the table in Figure 28) and ΔTbl represents the difference, or increment, between successive values of the independent variable (1 in Figure 28 because 1 − 0 = 1, 2 − 1 = 1, 3 − 2 = 1, and so on).

| $x$ | $y$ |
|---|---|
| 0 | −5 |
| 1 | 4 |
| 2 | 13 |
| 3 | 22 |
| 4 | 31 |
| 5 | 40 |
| 6 | 49 |

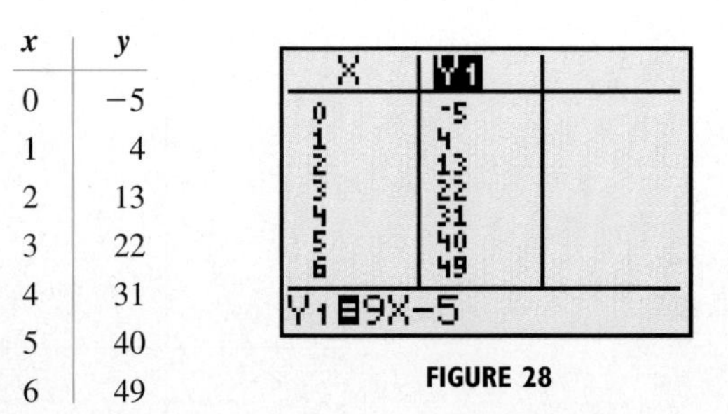


FIGURE 28

## Function Notation

To say that $y$ is a function of $x$ means that for each value of $x$ from the domain of the function, there is exactly one value of $y$. To emphasize that *$y$ is a function of $x$*, or that

$y$ depends on $x$, it is common to write

$$y = f(x),$$

with $f(x)$ read "$f$ of $x$." This notation is called **function notation.** For the function $f$ illustrated in Figure 27,

$$f(1990) = 58 \quad \text{because } (1990, 58) \text{ belongs to the correspondence,}$$
$$f(1992) = 60 \quad \text{because } (1992, 60) \text{ belongs to the correspondence,}$$

and so on. We will often use the variables $f$, $g$, and $h$ to represent functions.

Function notation is used frequently when functions are defined by equations. As an example, for the function defined by $y = 9x - 5$, we may name this function $f$ and write $f(x) = 9x - 5$. Note that $f(x)$ is simply another name for $y$. If $x = 2$, then we find $y$, or $f(2)$, by replacing $x$ with 2.

$$f(2) = 9 \cdot 2 - 5 = 13$$

The statement "if $x = 2$, then $y = 13$" is abbreviated with function notation as

$$f(2) = 13.$$

Also, $f(0) = 9 \cdot 0 - 5 = -5$, and $f(-3) = 9(-3) - 5 = -32$.

These ideas and the symbols used to represent them can be explained as follows.

Name of the function — Defining expression

$$y = f(x) = 9x - 5$$

Value of the function — Name of the independent variable

**CAUTION** The symbol $f(x)$ *does not* indicate "$f$ times $x$," but represents the $y$-value for the indicated $x$-value. As shown above, $f(2)$ is the $y$-value that corresponds to the $x$-value 2.

### EXAMPLE 5 *Using Function Notation*

For each function, find $f(3)$.

**(a)** $f(x) = 3x - 7$

**(b)** The function $f$ depicted in Figure 29

**Looking Ahead to Calculus**

Functions are evaluated frequently in calculus. They can be evaluated in a variety of ways, as illustrated in Example 5.

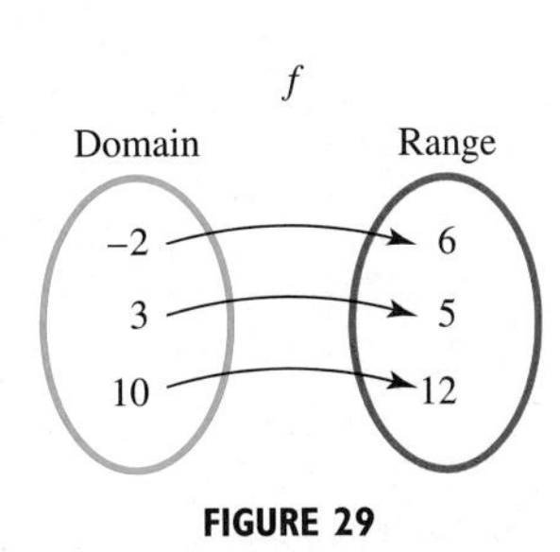


FIGURE 29

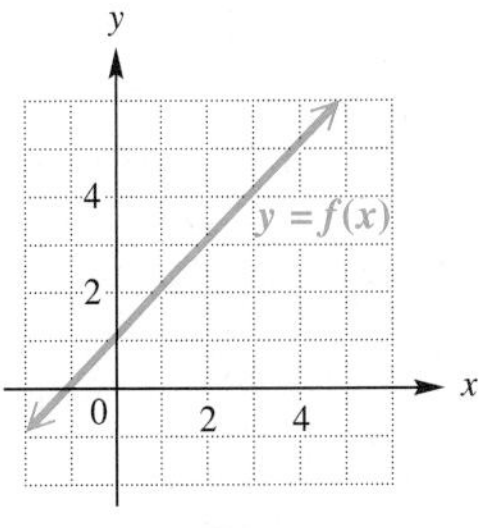


FIGURE 30

**(c)** The function $f$ graphed in Figure 30

**(d)** The function $f$ defined by the table

| $x$ | 1 | 2 | 3 | 4 |
|---|---|---|---|---|
| $f(x)$ | $-15$ | $-12$ | $-9$ | $-6$ |

**Solution**

**(a)** Replace $x$ with 3 to get $f(3) = 3(3) - 7 = 2$.

**(b)** In Figure 29, 3 in the domain is paired with 5 in the range, so $f(3) = 5$.

**(c)** To evaluate $f(3)$, begin by finding 3 on the $x$-axis. See Figure 31. Then move upward until the graph of $f$ is reached. Moving horizontally to the $y$-axis gives 4 for the corresponding $y$-value. Thus, $f(3) = 4$.

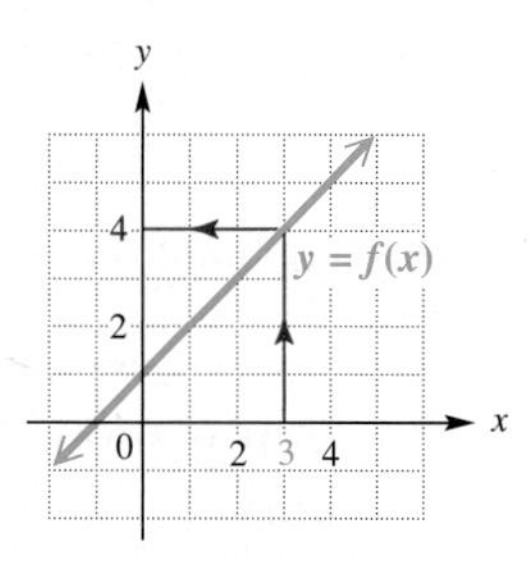


FIGURE 31

**(d)** From the table, $f(3) = -9$.

## 1.2 EXERCISES

*Write each set using interval notation, and then graph it on a number line.*

**1.** $\{x \mid -1 < x < 4\}$

**2.** $\{x \mid x \geq -3\}$

**3.** $\{x \mid x < 0\}$

**4.** $\{x \mid 8 > x > 3\}$

**5.** $\{x \mid 1 \leq x < 2\}$

**6.** $\{x \mid -5 < x \leq -4\}$

*Using the variable x, write each interval using set-builder notation.*

**7.** $(-4, 3)$

**8.** $[2, 7)$

**9.** $(-\infty, -1]$

**10.** $(3, \infty)$

**11.** −2 0 6

**12.** 0 8

**13.** −4 0

**14.** 0 3

**15.** Explain how to determine whether a parenthesis or a square bracket is used when graphing an inequality on a number line.

**16.** *Concept Check* The three-part inequality $a < x < b$ means "$a$ is less than $x$, *and* $x$ is less than $b$." Which one of the following inequalities is not satisfied by some real number $x$?

**A.** $-3 < x < 5$ **B.** $0 < x < 4$

**C.** $-3 < x < -2$ **D.** $-7 < x < -10$

*Determine the domain and range of each relation, and tell whether the relation is a function. Assume that a calculator graph extends indefinitely and a table includes only the points shown.*

**17.** $\{(5, 1), (3, 2), (4, 9), (7, 6)\}$

**18.** $\{(8, 0), (5, 4), (9, 3), (3, 8)\}$

**19.** $\{(4, 1), (3, -5), (-2, 3), (3, 7)\}$

**20.** $\{(0, 5), (1, 3), (0, -4)\}$

**21.**

| $x$ | 1 | 2 | 3 | 4 |
|---|---|---|---|---|
| $y$ | $-5$ | $-5$ | $-5$ | $-5$ |

**22.**

| $x$ | 1 | 1 | 1 | 1 |
|---|---|---|---|---|
| $y$ | 5 | 10 | 15 | 20 |

**23.**

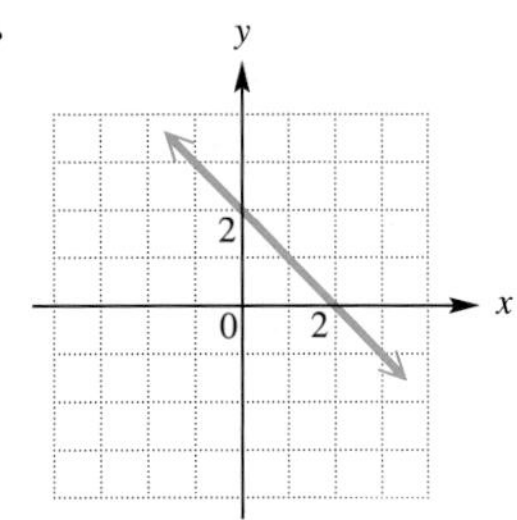


**24.**

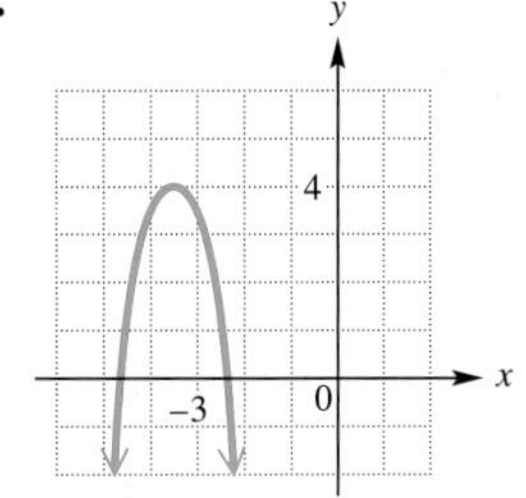


**25.**

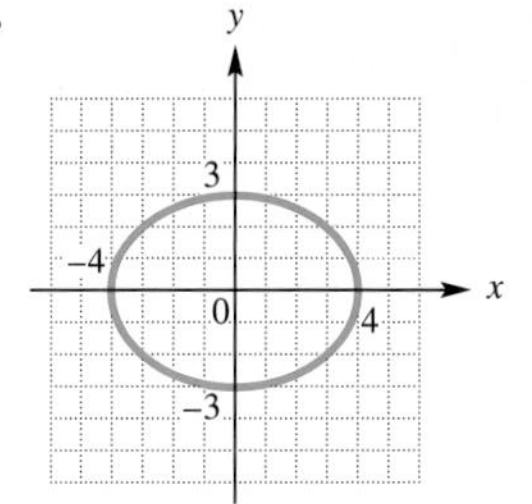


**26.**

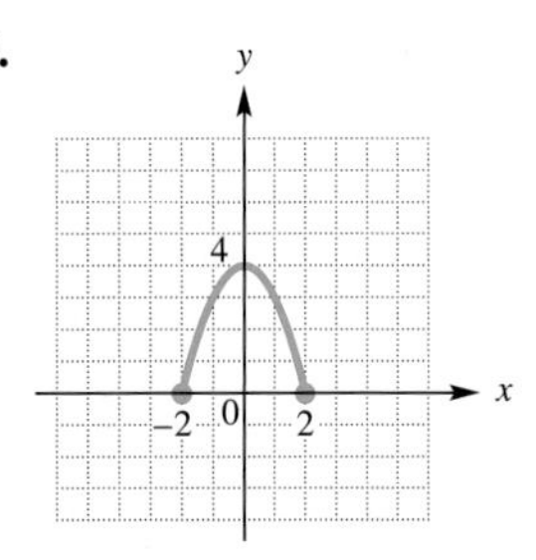


**27.**

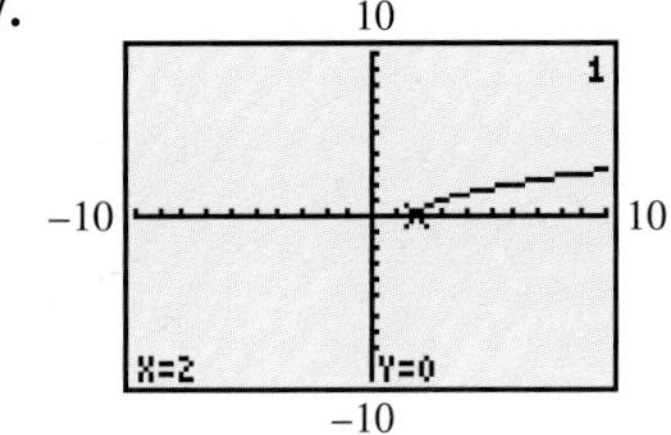


**28.**

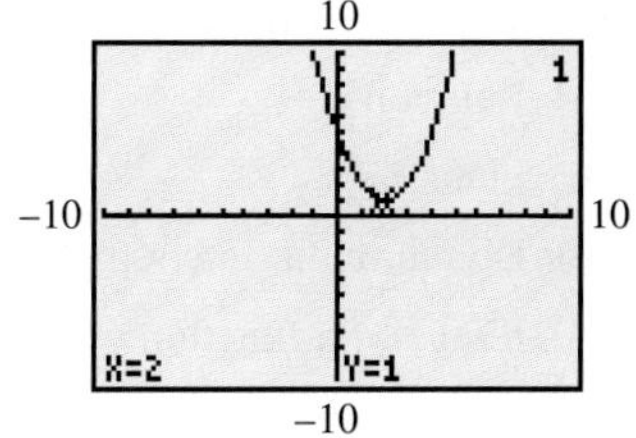


**29.**

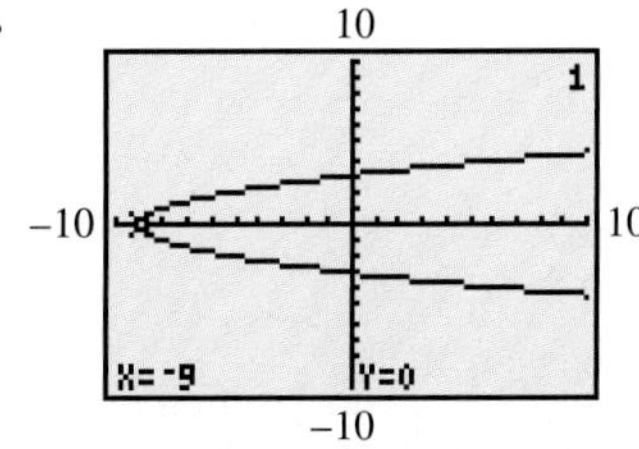


**30.**

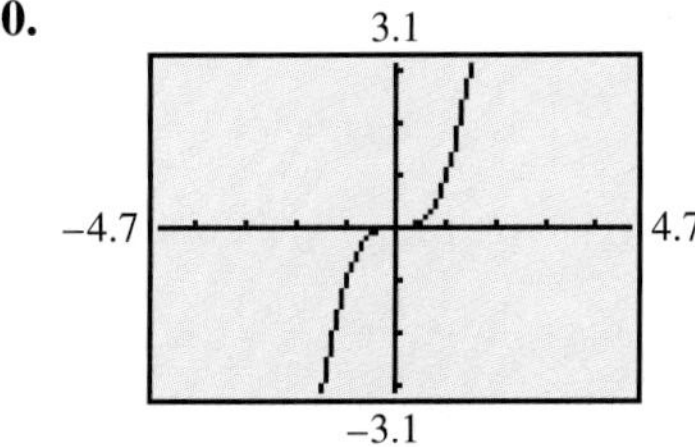


**31.**

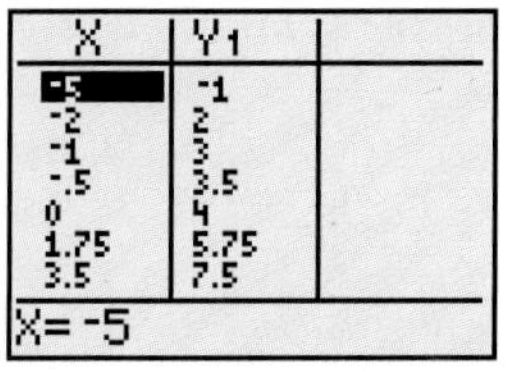

| X | Y1 |
|---|---|
| -5 | -1 |
| -2 | 2 |
| -1 | 3 |
| -.5 | 3.5 |
| 0 | 4 |
| 1.75 | 5.75 |
| 3.5 | 7.5 |

X= -5

**32.**

| X | Y1 |
|---|---|
| -2 | 5 |
| -1 | 0 |
| 0 | -3 |
| 5 | 12 |
| 9 | 60 |
| 10 | 77 |
| 13 | 140 |

X= -2

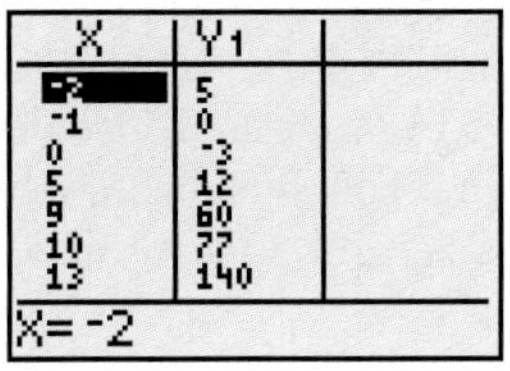

**33.**

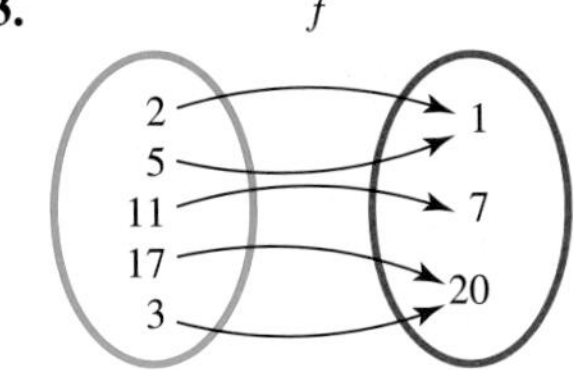


**34.**

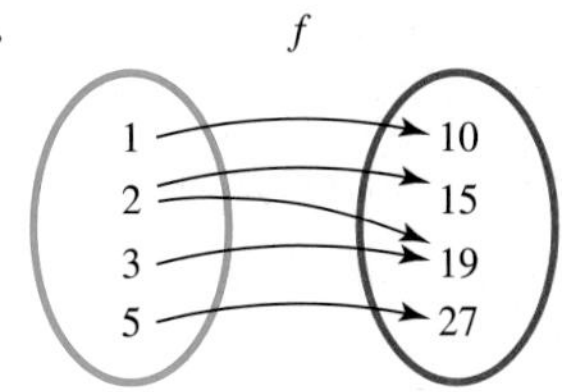


*Find each function value.*

**35.** $f(3)$, if $f(x) = -2x + 9$

**36.** $f(6)$, if $f(x) = -2x + 8$

**37.** $f(11)$, for the function $f$ in Exercise 33

**38.** $f(5)$, for the function $f$ in Exercise 33

*Find $f(x)$ at the indicated value of $x$.*

**39.** $f(x) = 2x, x = -3$

**40.** $f(x) = 3x - 7, x = 9$

**41.** $f(x) = 5, x = 6$

**42.** $f(x) = 1 - \frac{1}{2}x, x = -7$

**43.** $f(x) = x^2, x = -6$

**44.** $f(x) = x^2 + 4x + 6, x = 3$

**45.** $f(x) = \sqrt{x}, x = 100$

**46.** $f(x) = |4 - x|, x = 8$

*Concept Check* *Work each problem related to function notation.*

**47.** If $f(-2) = 3$, identify a point on the graph of $f$.

**48.** If $f(3) = -9.7$, identify a point on the graph of $f$.

**49.** If the point $(7, 8)$ lies on the graph of $f$, then $f(\_\_\_\_) = \_\_\_\_$.

**50.** If the point $(-3, 2)$ lies on the graph of $f$, then $f(\_\_\_\_) = \_\_\_\_$.

*Use the graph of $y = f(x)$ to find each function value:* **(a)** $f(-2)$, **(b)** $f(0)$, **(c)** $f(1)$, *and* **(d)** $f(4)$.

**51.**

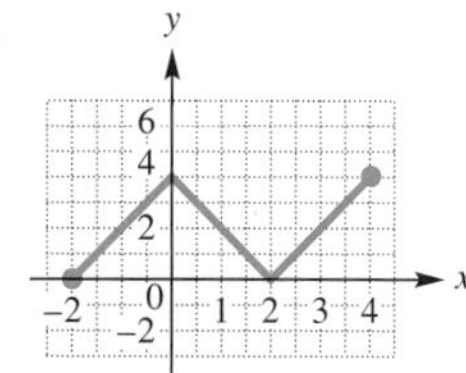

**52.**

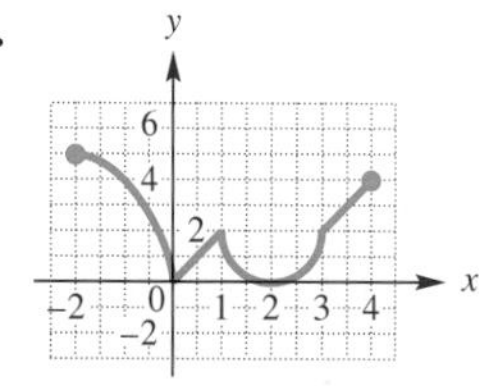

**53.**

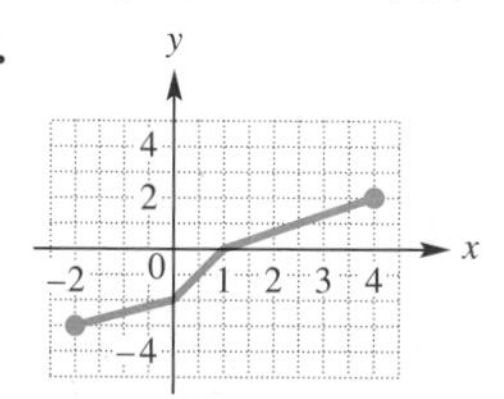

**54.**

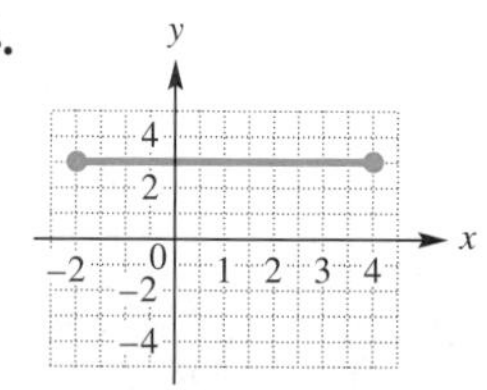

**55.** Explain each term in your own words.

**(a)** Relation

**(b)** Function

**(c)** Domain of a function

**(d)** Range of a function

**56.** *Radio Stations* The function $f$ gives the number $y$ in thousands of radio stations on the air during year $x$. (*Source:* M. Street Corporation.)

$$f = \{(1950, 2.8), (1975, 7.7), (1997, 12.8)\}$$

**(a)** Use a diagram (see Figure 27) to represent $f$.

**(b)** Evaluate $f(1975)$ and explain what it means.

**(c)** Identify the domain and range of $f$.

*Graph $f$ by hand. Start by plotting a few points and then sketch the graph of $f$.*

**57.** $f(x) = x - 3$

**58.** $f(x) = 1 - 2x$

**59.** $f(x) = 3$

**60.** $f(x) = \frac{1}{2}x$

**61.** $f(x) = x^2$

**62.** $f(x) = |x|$

(*Modeling*) *Solve each problem.*

**63.** *Distance to Lightning* When a bolt of lightning strikes in the distance, there is often a delay between seeing the lightning and hearing the thunder. The function defined by $f(x) = \frac{x}{5}$ computes the approximate distance in miles between an observer and a bolt of lightning, when the delay is $x$ seconds. (*Source:* Weidner, R., and R. Sells, *Elementary Classical Physics,* Allyn and Bacon, Inc., 1965.)

**(a)** Find $f(15)$ and interpret the result.

**(b)** Graph $y = f(x)$. Let the domain of $f$ be $0 \le x \le 20$.

**64.** *Air Temperature* When the relative humidity is 100%, air cools 5.8°F for every 1-mile increase in altitude. If the temperature is 80°F on the ground, then $f(x) = 80 - 5.8x$ calculates the air temperature $x$ miles above the ground. Find $f(3)$ and interpret the result. (*Source:* Battan, L., *Weather in Your Life,* W. H. Freeman, 1983.)

**65.** *Sales Tax* If the sales tax rate is 7.5%, write a function $f$ that calculates the sales tax on a purchase of $x$ dollars. What is the sales tax on a purchase of $86?

**66.** *Income and Level of Education* The function $f$ gives the median 1998 individual income (in dollars) by educational attainment for people 15 years old and over. This function is defined by $f(N) = 17{,}462$, $f(H) = 26{,}542$, $f(B) = 45{,}962$, and $f(M) = 55{,}784$, where $N$ denotes no high school diploma, $H$ a high school diploma, $B$ a bachelor's degree, and $M$ a master's degree. (*Source:* U.S. Bureau of the Census.)

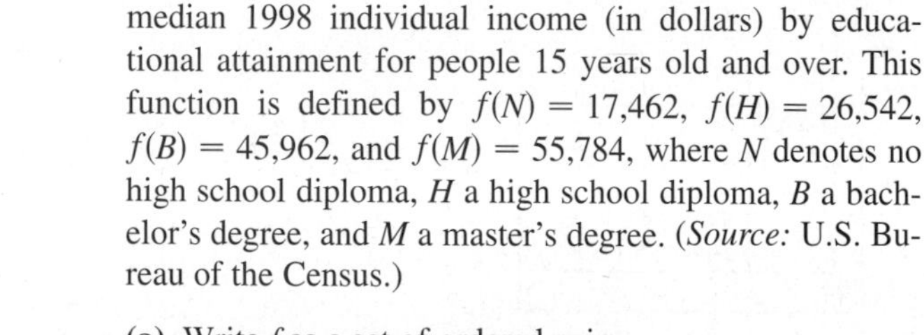

**(a)** Write $f$ as a set of ordered pairs.

**(b)** Give the domain and range of $f$.

**(c)** Discuss the relationship between education and income.

**67.** *Tuition and Fees* If college tuition costs $92 per credit and fees are fixed at $75, write a formula for a function $f$ that calculates the tuition and fees for taking $x$ credits. What is the total cost of taking 11 credits?

**68.** *Converting Units of Measure* Write a formula for a function $f$ that converts $x$ gallons to quarts. How many quarts are there in 19 gallons?

## Reviewing Basic Concepts (Sections 1.1 and 1.2)

1. Plot the points $(-3, 1)$, $(-2, -1)$, $(2, -3)$, $(1, 1)$, and $(0, 2)$. Label each point.
2. Find the length and midpoint of the line segment that connects the points $P(-4, 5)$ and $Q(6, -2)$.
3. Use a calculator to approximate $\frac{\sqrt{5 + \pi}}{\sqrt[3]{3} + 1}$ to the nearest thousandth.
4. The hypotenuse of a right triangle measures 61 inches, and one of its legs measures 11 inches. Find the length of the other leg.
5. Write the sets $\{x \mid -2 < x \leq 5\}$ and $\{x \mid x \geq 4\}$ in interval notation.
6. Determine whether the relation shown in the graph is a function. What is its domain and range?

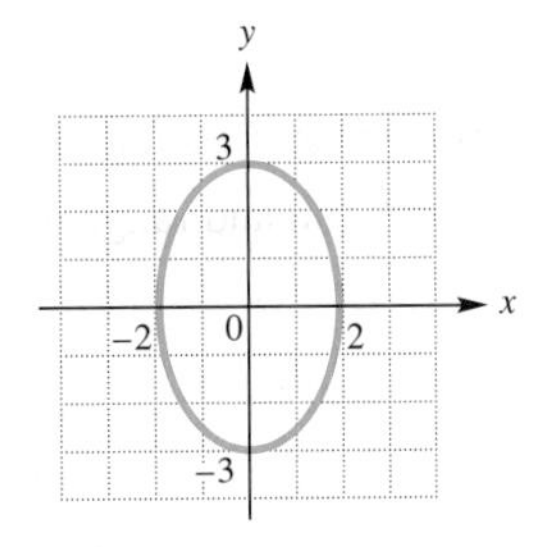


7. Find $f(-5)$ if $f(x) = 3 - 4x$.
8. Use the graph to find $f(2)$ and $f(-1)$.

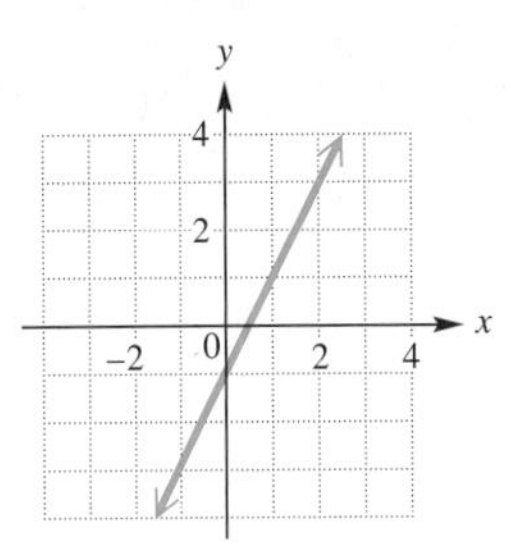


9. Graph $f(x) = \frac{1}{2}x - 1$ by hand.

# 1.3 LINEAR FUNCTIONS

Basic Concepts about Linear Functions • Slope of a Line • Slope-Intercept Form of the Equation of a Line

## Basic Concepts about Linear Functions

By noon, 2 inches of rain had fallen during a storm. Rain continued to fall at the rate of $\frac{1}{2}$ inch per hour in the afternoon. Then, the total rainfall $x$ hours past noon is given by

$$f(x) = \frac{1}{2}x + 2.$$

Rate of rainfall — Hours of rainfall past noon — Amount of rainfall at noon

At 3:00 P.M., the total rainfall equaled

$$f(3) = \frac{1}{2}(3) + 2 = 3.5 \text{ inches.}$$

This function $f$ satisfies the general form $f(x) = ax + b$ (where $a = \frac{1}{2}$ and $b = 2$) and is called a *linear function*.

> **Linear Function**
>
> A function $f$ defined by $\boldsymbol{f(x) = ax + b}$, where $a$ and $b$ are real numbers, is called a **linear function.**

The graph of $f(x) = ax + b$ is a line and can be found by graphing $y = ax + b$. For example, the graph of $f(x) = 3x + 6$ is the line determined by $y = 3x + 6$. An equation such as $y = 3x + 6$ is called a **linear equation in two variables.** A **solution** is an ordered pair $(x, y)$ that makes the equation true. Verify that $(-2, 0)$, $(-1, 3)$, $(0, 6)$, and $(1, 9)$ are all solutions of $y = 3x + 6$.

Graphing linear equations by hand involves plotting points whose coordinates are solutions of the equation, and then connecting them with a straight line. Figure 32(a) shows the ordered pairs just mentioned for the linear equation $y = 3x + 6$. It is accompanied by a *table of values.* Notice that the points appear to lie in a straight line; that is indeed the case. Since we may substitute *any* real number for $x$, we connect these points with a line to obtain the graph of $f(x) = 3x + 6$, as shown in Figure 32(b).

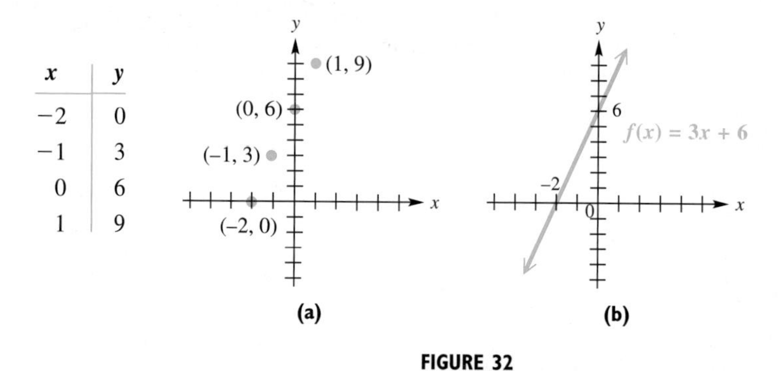


FIGURE 32

To graph the linear function defined by $f(x) = 3x + 6$ on a calculator, we enter $3x + 6$ for one of the $y$-variables. Using the standard viewing window, we get the graph shown in Figure 33(a). A graphing calculator will also give a table of selected points, as shown in Figure 33(b).

**TECHNOLOGY NOTE**

To graph a function $f$ on a graphing calculator, be sure to enter the formula for $f$ and then set an appropriate viewing window.

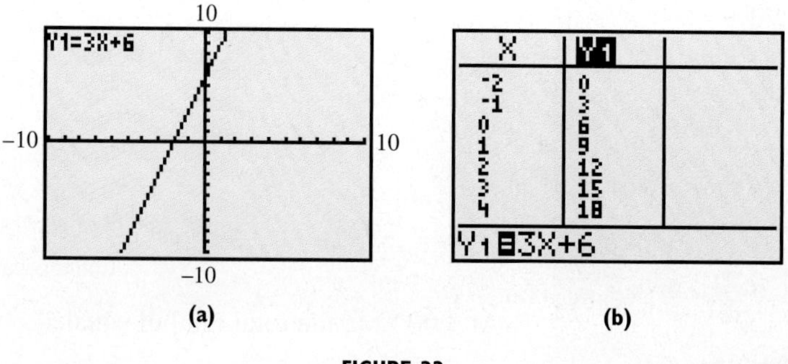


FIGURE 33

From geometry we know that two distinct points determine a line. Therefore, if we know the coordinates of two points, we can graph the line. For the equation $y = 3x + 6$, suppose we let $x = 0$. Then $y = 3(0) + 6 = 6$.

Now, suppose we let $y = 0$ and find $x$.

$$\begin{aligned} y &= 3x + 6 \\ 0 &= 3x + 6 \\ -6 &= 3x \\ x &= -2 \end{aligned}$$

The points $(0, 6)$ and $(-2, 0)$ lie on the graph of $y = 3x + 6$ and are sufficient for obtaining the graph in Figure 32(b). The numbers 6 and $-2$ are called the **$y$-** and **$x$-intercepts** of the line.

**Locating $x$- and $y$-Intercepts**

To find the $x$-intercept of the graph of $y = ax + b$, let $y = 0$ and solve for $x$ (assuming $a \neq 0$). To find the $y$-intercept, let $x = 0$ and solve for $y$.

The $x$-intercept of the graph of a linear function is a value that makes $f(x) = 0$ a true statement; that is, it causes the function value to equal zero. In general, such a number is called a *zero* of the function.

**Zero of a Function**

Let $f$ be a function. Then any number $c$ for which $f(c) = 0$ is true is called a **zero** of the function $f$.

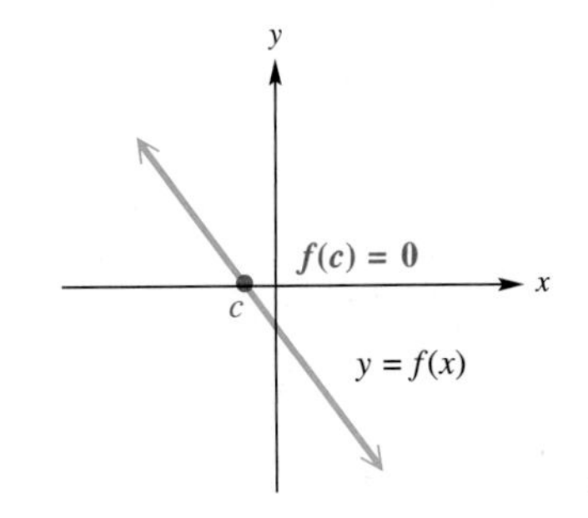


NOTE If $c$ is a zero of $f$, then $f(c) = 0$ and $c$ is an $x$-intercept of the graph of $f$, as illustrated in the figure.

**EXAMPLE 1** *Graphing a Line*

Graph the function defined by $f(x) = -2x + 5$. What is the zero of $f$?

**Analytic Solution**

The graph of $f(x) = -2x + 5$ and its intercepts are shown in Figure 34. The zero of $f$ is 2.5.

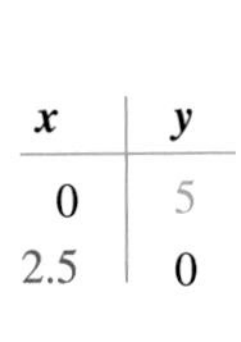

| $x$ | $y$ |
|---|---|
| 0 | 5 |
| 2.5 | 0 |

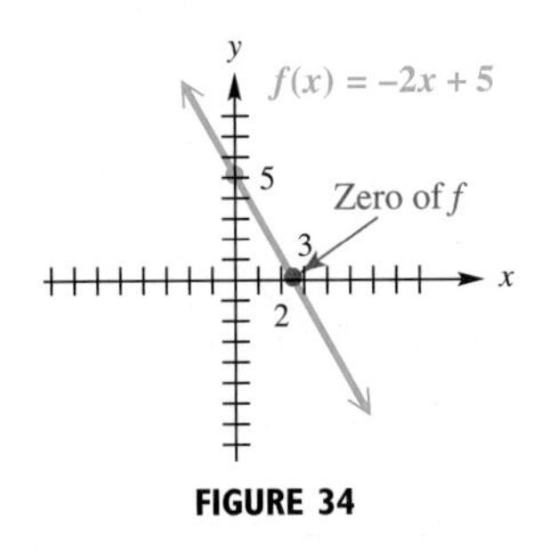


FIGURE 34

**Graphing Calculator Solution**

A calculator graph is shown in Figure 35. The $x$-intercept is 2.5, so 2.5 is the zero of $f$.

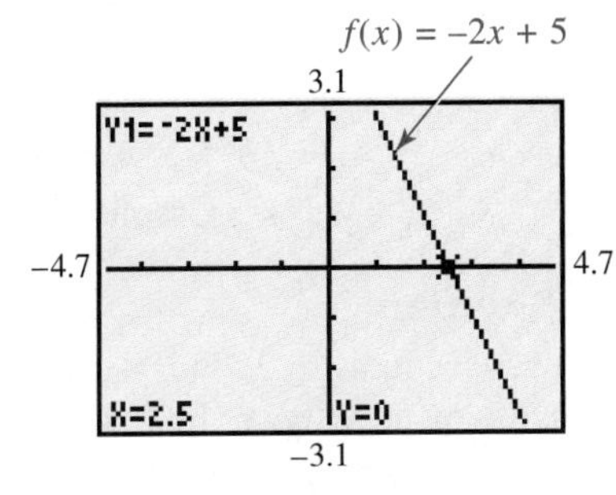


FIGURE 35

**EXAMPLE 2** *Finding a Formula for a Function*

A 100-gallon tank is initially full of water and is being drained at a rate of 5 gallons per minute.

**(a)** Write a formula for a linear function $f$ that models the number of gallons of water in the tank after $x$ minutes.

**(b)** How much water is in the tank after 4 minutes?

**(c)** Use the $x$- and $y$-intercepts to graph $f$. Interpret each intercept.

**Solution**

**(a)** The amount of water in the tank is *decreasing* at a constant rate of 5 gallons per minute, so the constant rate of change is $-5$. The initial amount of water is equal to 100 gallons.

$$f(x) = (\text{constant rate of change})x + (\text{initial amount})$$
$$f(x) = -5x + 100$$

**(b)** After 4 minutes, the tank held $f(4) = -5(4) + 100 = 80$ gallons.

**(c)** The graph of $y = -5x + 100$ has $y$-intercept 100 because $y = 100$ when $x = 0$. To find the $x$-intercept, let $y = 0$ and solve the equation $0 = -5x + 100$ to obtain $x = 20$. The graph of $f$ is shown in Figure 36. The $x$-intercept corresponds to the time in minutes that it takes to empty the tank. The $y$-intercept corresponds to the number of gallons of water in the tank initially.

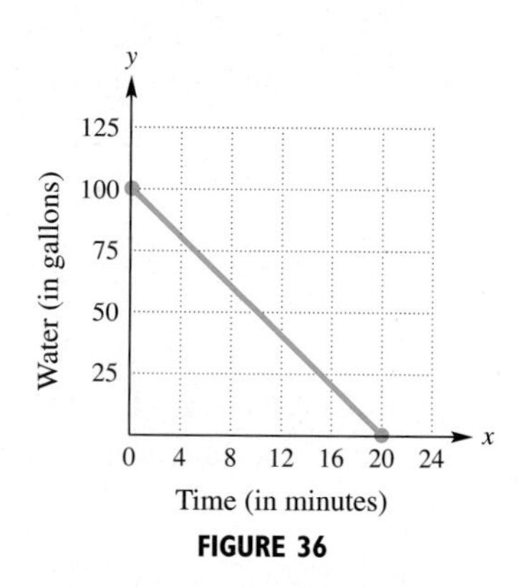


**FIGURE 36**

Suppose that for a linear function defined by $f(x) = ax + b$, we have $a = 0$. Then the function becomes $f(x) = b$, where $b$ is some real number. Its graph is a horizontal line.

**EXAMPLE 3** *Sketching the Graph of $f(x) = b$*

Graph the function defined by $f(x) = -3$.

**Analytic Solution**

Since $y$ always equals $-3$, the $y$-intercept is $-3$. Since the value of $y$ can never be 0, the graph has no $x$-intercept. The only way that a straight line can have no $x$-intercept is for it to be parallel to the $x$-axis, as shown in Figure 37.

**Graphing Calculator Solution**

Using the viewing window $[-5, 5]$ by $[-5, 2]$, we find the same horizontal line. See Figure 38 and compare it to Figure 37.

*(continued)*

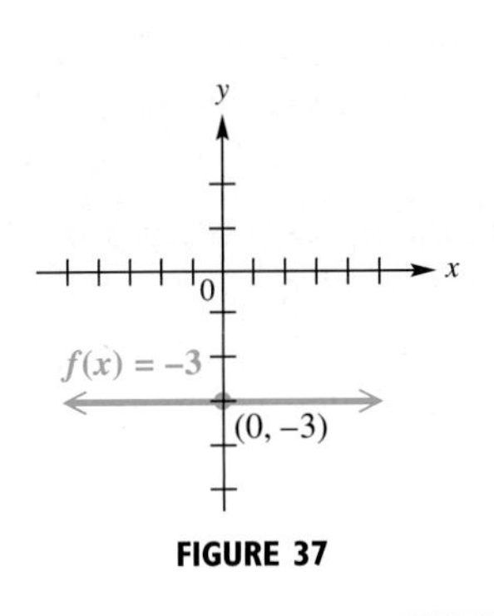


FIGURE 37

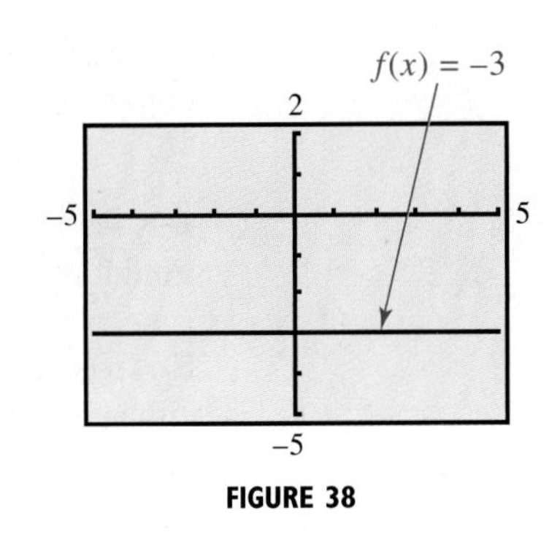


FIGURE 38

The function discussed in Example 3 is a *constant function.*

### Constant Function

A function of the form $f(x) = b$, where $b$ is a real number, is called a **constant function.** Its graph is a horizontal line with $y$-intercept $b$. For $b \neq 0$, it has no $x$-intercept. (Every constant function is also linear.)

Unless otherwise specified, the domain of a linear function will be the set of all real numbers. The range of a nonconstant linear function is also the set of all real numbers. The range of a constant function defined by $f(x) = b$ is $\{b\}$.

Calculators generate graphs by plotting a large number of points and connecting them. The choice of the viewing window may give drastically different views of a graph, so we often discuss choosing appropriate windows. Figure 39 shows three different views of the graph of $f(x) = 3x + 6$.

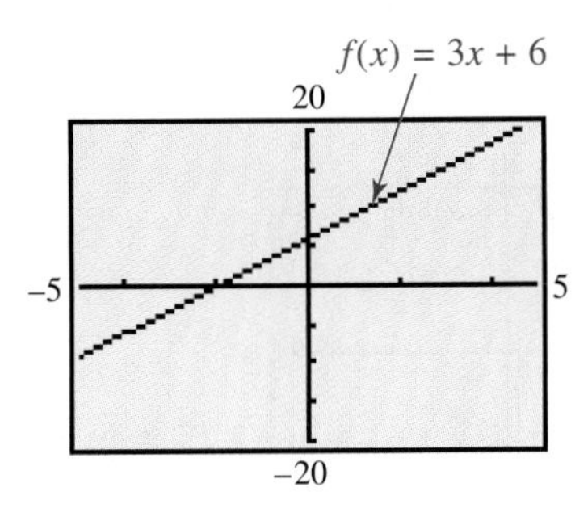


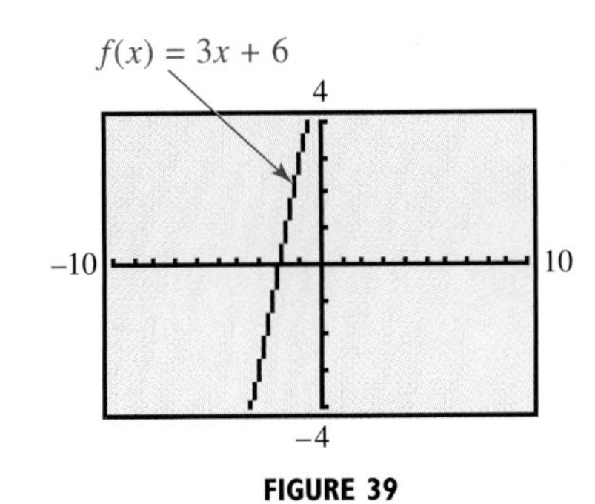


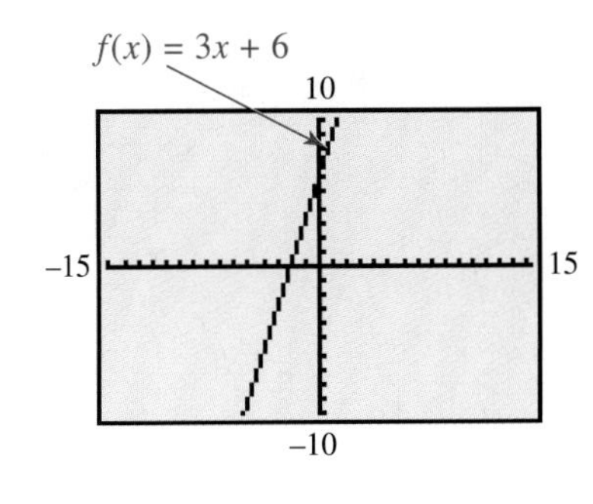


FIGURE 39

We are often interested in choosing a window that shows the most important features of a particular graph. Such a graph is called a **comprehensive graph.*** Keep in mind that the choice of window for a comprehensive graph is not unique—there are many acceptable ones.

*The term *comprehensive graph* was coined by Shoko Aogaichi Brant and Edward A. Zeidman in the text *Intermediate Algebra: A Functional Approach* (HarperCollins College Publishers, 1996), with the assistance of Professor Brant's daughter, Jennifer. The authors thank them for permission to use the terminology in this text.

Each time we introduce a new kind of graph, we will state the requirements for a comprehensive graph. For a line, a comprehensive graph shows all intercepts of the line.

### EXAMPLE 4 *Finding a Comprehensive Graph of a Line*

Find a comprehensive graph of $g(x) = -.75x + 12.5$.

**Solution** The window $[-10, 10]$ by $[-10, 10]$ of Figure 40(a) does not show either intercept, so it will not work. We can increase the window size to $[-10, 20]$ by $[-10, 20]$, for example, to obtain a comprehensive graph. See Figure 40(b).

**TECHNOLOGY NOTE**

One of the most important parts of learning mathematics with the aid of a graphing calculator involves choosing appropriate windows for graphs. Be sure that you know how to change the window on your model.

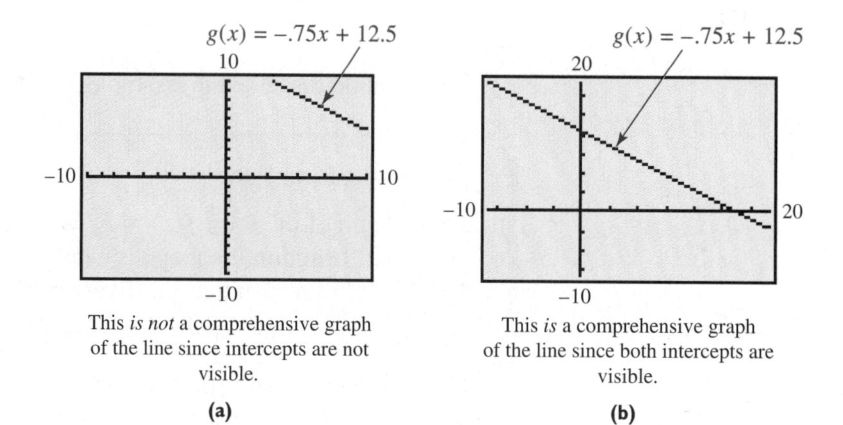


This *is not* a comprehensive graph of the line since intercepts are not visible.

**(a)**

This *is* a comprehensive graph of the line since both intercepts are visible.

**(b)**

**FIGURE 40**

### What Went WRONG?

A student learning to use a graphing calculator attempted to graph $y = \frac{1}{2}x + 15$. However, he obtained the blank screen shown here.

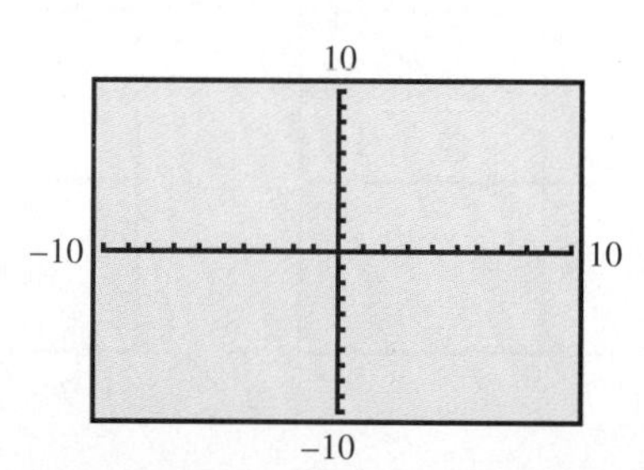


What Went Wrong? How can he obtain a comprehensive graph of this linear function?

---

*Answer to What Went Wrong?*

The $y$-intercept is 15 and the $x$-intercept is $-30$, so the window size must be increased to show the intercepts. For example, a window of $[-40, 40]$ by $[-30, 30]$ would work.

A graphing calculator allows us to locate a point on a graph while displaying its coordinates. We can do this by tracing along the graph or giving the appropriate command by entering the $x$-coordinate of the point. Figure 41 shows some typical screens with designated points identified for the graph of $y = 3x$. Notice that in each case the $y$-value is three times the $x$-value.

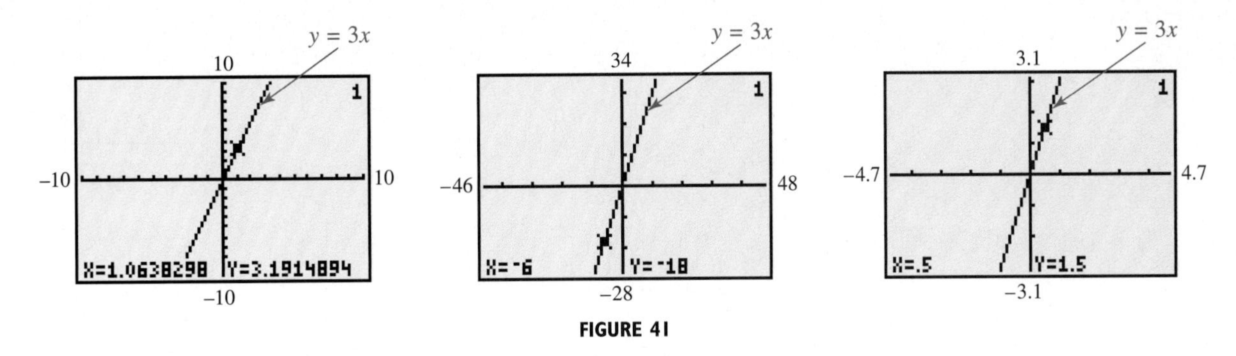


**FIGURE 41**

## Slope of a Line

In 1981 the average annual cost for tuition and fees at private four-year colleges was \$4113. By 1999 this cost had increased to \$15,380. The line graphed in Figure 42 is actually somewhat misleading since it indicates that the increase in cost was the same from year to year. However, we can use the graph to determine the *average* yearly increase in cost. Over the 18-year span, the cost increased \$11,267. Therefore, the average yearly increase was

$$\frac{\$11{,}267}{18} \approx \$626.$$

The number 18 was obtained by subtracting 1999 − 1981, and \$11,267 was found by subtracting \$15,380 − \$4113. This quotient is an illustration of the *slope* of the line joining (1981, 4113) and (1999, 15,380). Slope is important in the study of linear functions.

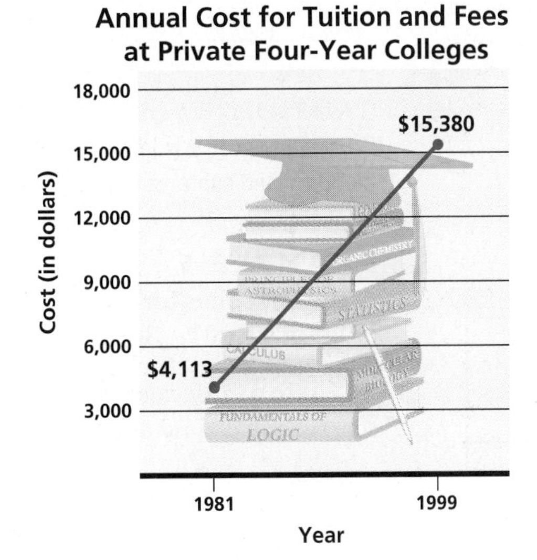


*Source*: The College Board.

**FIGURE 42**

The graph of the line $y = 3x + 1$ is shown in Figure 43. A table of selected points is included.

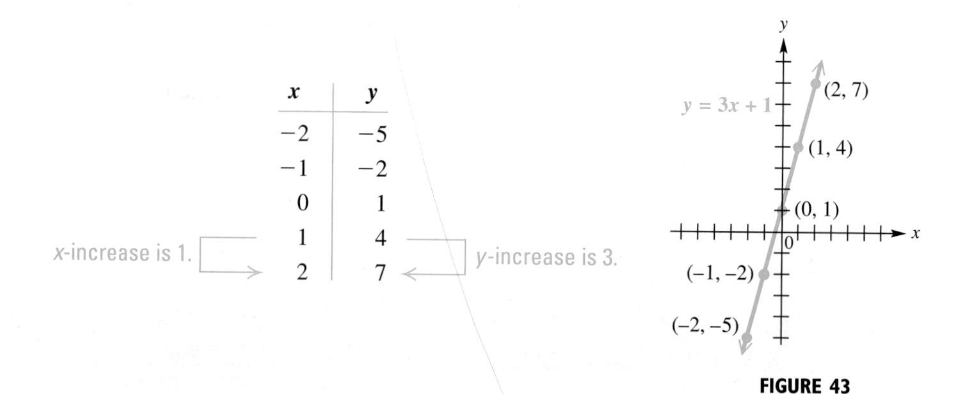


| $x$ | $y$ |
|---|---|
| −2 | −5 |
| −1 | −2 |
| 0 | 1 |
| 1 | 4 |
| 2 | 7 |

**FIGURE 43**

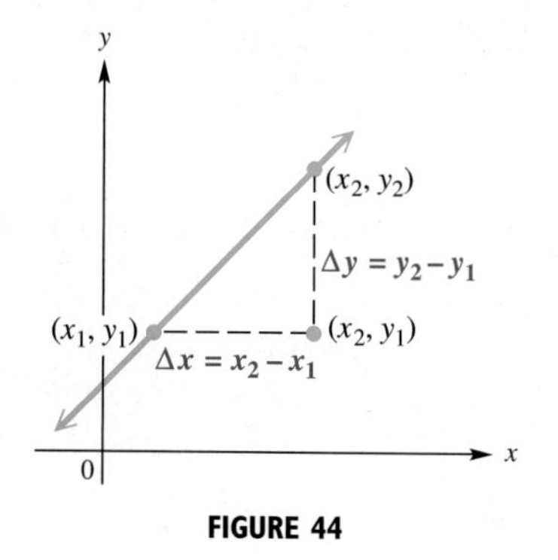


**FIGURE 44**

Notice that for each increase of 1 for the $x$-value, the $y$-value increases by 3. Thus, the slope of the line is 3. Geometrically, the slope is a numerical measure of the steepness of the line. (This may be interpreted as the ratio of *rise* to *run*.) To calculate slope, start with the line through the two distinct points $(x_1, y_1)$ and $(x_2, y_2)$, as shown in Figure 44, where $x_1 \neq x_2$. The difference $x_2 - x_1$ is called the **change in $x$,** denoted $\Delta x$ (read "delta $x$"), where $\Delta$ is the Greek letter *delta*. In the same way, the **change in $y$** is denoted $\Delta y = y_2 - y_1$. The *slope* of a nonvertical line is defined as the quotient of the change in $y$ and the change in $x$, as follows.

**Looking Ahead to Calculus**

The concept of slope of a line is extended in calculus to general curves. The *slope* of a *curve* at a point is understood to mean the slope of the line tangent to the curve at that point.

### Slope

The **slope** $m$ of the line passing through the points $(x_1, y_1)$ and $(x_2, y_2)$ is

$$m = \frac{\Delta y}{\Delta x} = \frac{y_2 - y_1}{x_2 - x_1},$$

where $\Delta x = x_2 - x_1 \neq 0$.

**CAUTION** When using the slope formula, it makes no difference which point is $(x_1, y_1)$ or $(x_2, y_2)$; however, it is important to be consistent. Start with the $x$- and $y$-values of one point (either one) and subtract the corresponding values of the *other* point.

**EXAMPLE 5** *Finding Slope Using the Slope Formula*

A table of points for a linear function $Y_1$ is shown in Figure 45. Determine the slope of the graph of $Y_1$. Sketch the graph by hand.

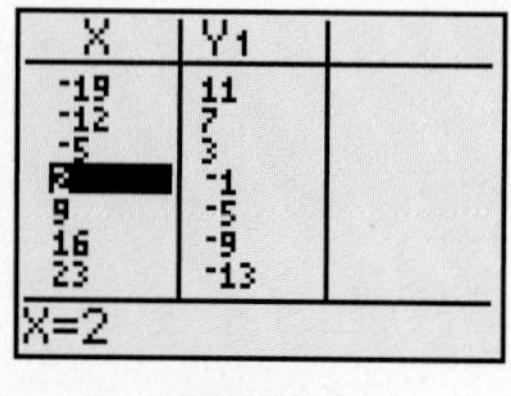

| X | Y1 |
|---|---|
| -19 | 11 |
| -12 | 7 |
| -5 | 3 |
| 2 | -1 |
| 9 | -5 |
| 16 | -9 |
| 23 | -13 |

X=2

**FIGURE 45**

**Solution** Because the slope of a line is the same regardless of the two points chosen, we can choose any two points from Figure 45. If we let

$$(2, -1) = (x_1, y_1) \quad \text{and} \quad (-5, 3) = (x_2, y_2),$$

then

$$m = \frac{y_2 - y_1}{x_2 - x_1} = \frac{3 - (-1)}{-5 - 2} = -\frac{4}{7}.$$

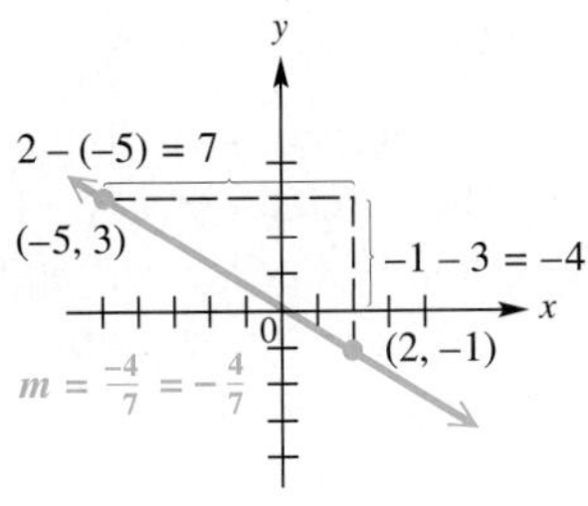

FIGURE 46

See Figure 46. On the other hand, if $(2, -1) = (x_2, y_2)$ and $(-5, 3) = (x_1, y_1)$, then the slope would be

$$m = \frac{-1 - 3}{2 - (-5)} = -\frac{4}{7}.$$

The slope is the same no matter which point is considered first.

We can graph a line by hand if we know the slope and a point that lies on the line.

**EXAMPLE 6** *Using the Slope and a Point to Graph a Line*

Graph the line that passes through $(2, 1)$ and has slope $-\frac{4}{3}$.

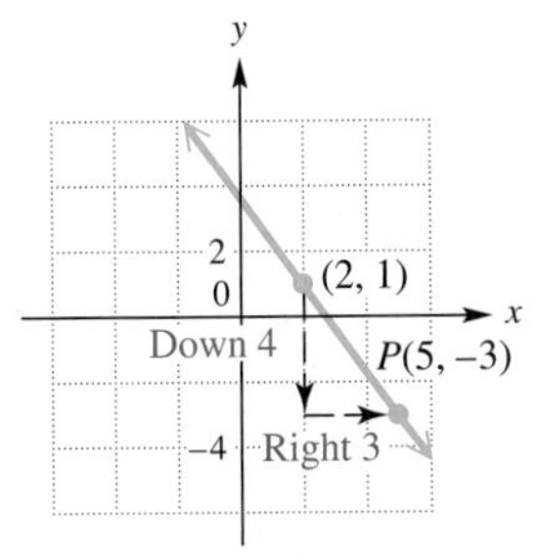

FIGURE 47

**Solution** Start by locating the point $(2, 1)$ on the graph. Find a second point on the line by using the definition of slope.

$$\text{slope} = \frac{\text{change in } y}{\text{change in } x} = \frac{-4}{3}$$

Move *down* 4 units from $(2, 1)$ and then 3 units to the *right* to obtain $P(5, -3)$. Draw a line through this second point $P$ and $(2, 1)$, as shown in Figure 47.

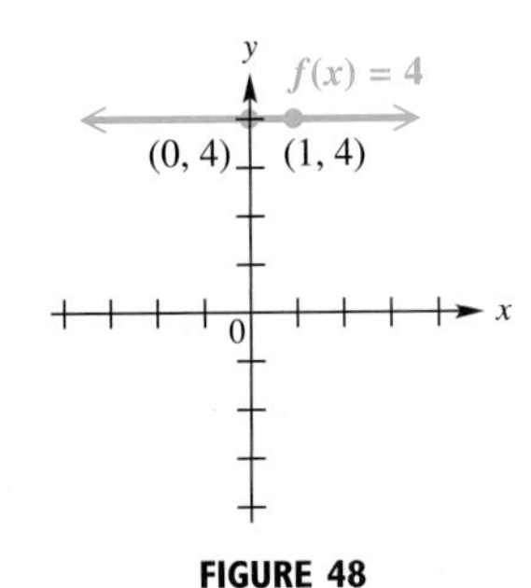

FIGURE 48

The graph of a constant function is a horizontal line. Because there is no change in $y$, the slope of a horizontal line is 0. For example, the graph of $f(x) = 4$ is shown in Figure 48. Using the points $(0, 4)$ and $(1, 4)$, we find its slope to be

$$m = \frac{4 - 4}{1 - 0} = 0.$$

A line with positive slope *rises* from left to right, a line with negative slope *falls* from left to right, and a line with slope 0 is *horizontal*. In general, we have the following.

**Geometric Orientation Based on Slope**

For a line with slope $m$, if $m > 0$, the line rises from left to right. If $m < 0$, it falls from left to right. If $m = 0$, the line is horizontal.

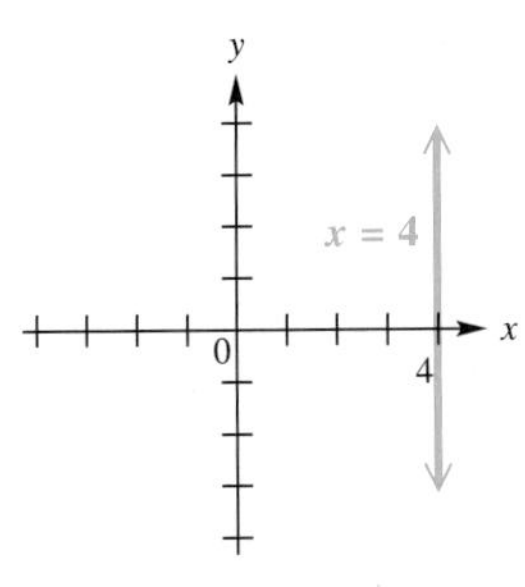

FIGURE 49

In the slope formula we have the condition $\Delta x = x_2 - x_1 \neq 0$. This means that $x_2 \neq x_1$. If we graph a line with two points having the same $x$-values, we get a vertical line. For example, the line with equation $x = 4$ is graphed in Figure 49. Notice that this is *not* the graph of a function, since 4 appears as the first number in more than one ordered pair. If we were to apply the slope formula, the denominator would be 0. As a result, the slope of such a line is *undefined*.

**Vertical Line**

A vertical line with $x$-intercept $a$ has an equation of the form $\boldsymbol{x = a}$. Its slope is undefined.

## Slope-Intercept Form of the Equation of a Line

In Figure 43, the slope $m$ of the line $y = 3x + 1$ is 3 and the $y$-intercept is 1. In general, if $f(x) = ax + b$, then the slope of its graph is $a$ and the $y$-intercept is $b$. To verify this fact, notice that $f(0) = a(0) + b = b$. Thus, the graph of $f$ passes through the point $(0, b)$ and $b$ is the $y$-intercept. Since $f(1) = a(1) + b = a + b$, the graph of $f$ also passes through the point $(1, a + b)$. The slope of the line that passes through the points $(0, b)$ and $(1, a + b)$ is

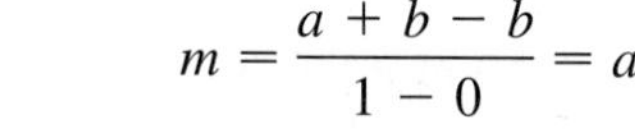

$$m = \frac{a + b - b}{1 - 0} = a.$$

FIGURE 50

See Figure 50.

Because the slope of the graph of $f(x) = ax + b$ is $a$, it is often convenient to use $m$ rather than $a$ in the general form of the equation. Therefore, we will sometimes write

$$f(x) = mx + b \quad \text{or} \quad y = mx + b$$

to indicate a linear function. The slope is $m$, and the $y$-intercept is $b$. This is generally called the *slope-intercept form* of the equation of a line.

**Slope-Intercept Form**

The **slope-intercept form** of the equation of a line is $\mathbf{y = mx + b}$, where $m$ is the slope and $b$ is the $y$-intercept. (Linear functions are often written in the form $f(x) = ax + b$, where $a$ is the slope and $b$ is the $y$-intercept of the graph.)

**EXAMPLE 7** *Matching Graphs with Equations*

Figure 51 shows four calculator graphs of lines. Their equations are

$$y = 2x + 3, \quad y = -2x + 3, \quad y = 2x - 3, \quad \text{and} \quad y = -2x - 3,$$

but not necessarily in this order. Match each equation with its graph.

A. 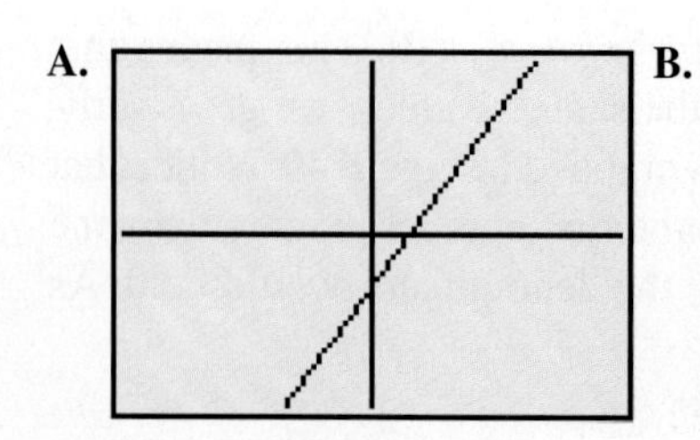

B. 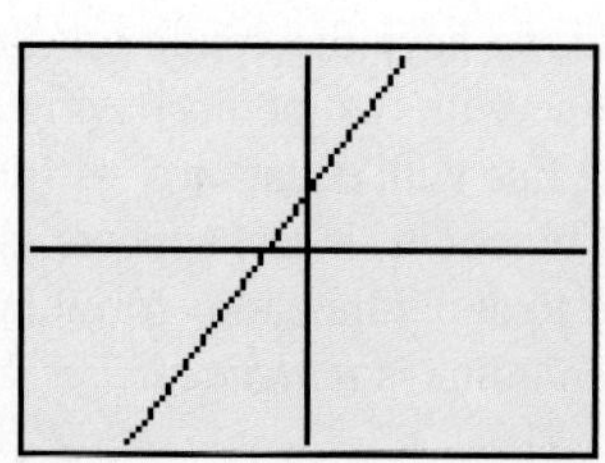

C. 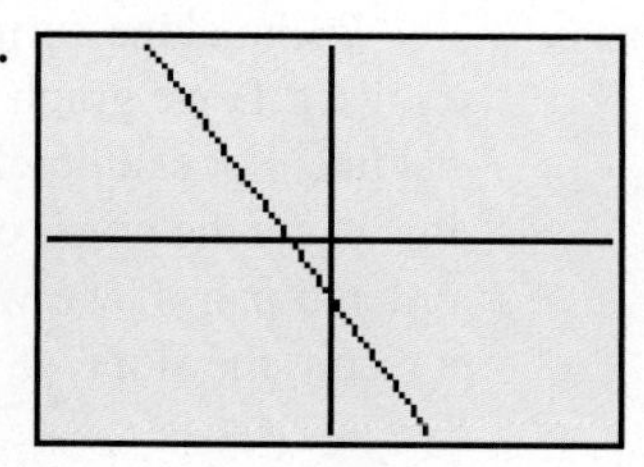

D. 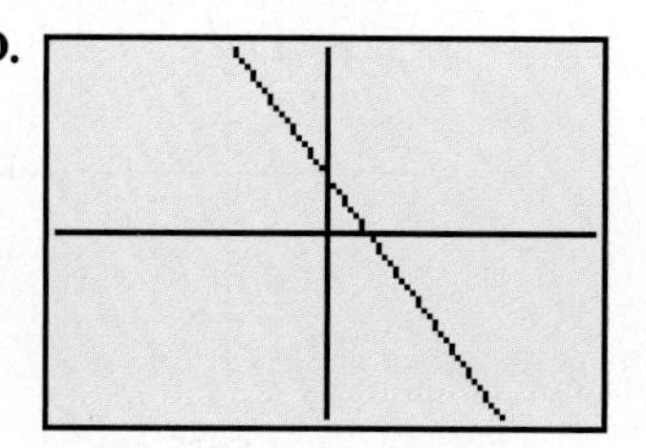

FIGURE 51

**Solution** The sign of $m$ determines whether the graph rises or falls from left to right. Also, if $b > 0$, the $y$-intercept is *above* the $x$-axis, and if $b < 0$, the $y$-intercept

**Looking Ahead to Calculus**

Slope represents a rate of change. In calculus, rates of change for nonlinear functions are studied extensively using the *derivative*.

is *below* the $x$-axis. Therefore,

$y = 2x + 3$ is shown in B, since the graph rises from left to right, and the $y$-intercept is positive;

$y = -2x + 3$ is shown in D, since the graph falls from left to right, and the $y$-intercept is positive.

Similarly, $y = 2x - 3$ is shown in A, and $y = -2x - 3$ is shown in C.

**EXAMPLE 8** *Interpreting Slope*

In 1970 passengers traveled a total of 6.2 billion miles on Amtrak, and in 1997 they traveled 5.2 billion miles. (*Source:* U.S. Department of Transportation.)

**(a)** Find the slope $m$ of the line passing through the points $(1970, 6.2)$ and $(1997, 5.2)$.

**(b)** Interpret the slope.

**Solution**

**(a)** $m = \frac{5.2 - 6.2}{1997 - 1970} = -\frac{1}{27} \approx -.037$

**(b)** Because the slope is negative, the number of passenger-miles traveled on Amtrak *decreased*, on average, by about .037 billion or 37 million miles per year between 1970 and 1997. It is important to realize that the number of passenger-miles did not decrease by exactly 37 million each year. Slope gives an *average* rate of change.

## 1.3 EXERCISES

*Work each problem related to linear functions.*

**(a)** *Evaluate $f(-2)$ and $f(4)$.*

**(b)** *Find the zero of $f$.*

**(c)** *Graph $f$. How can the graph of $f$ be used to determine the zero of $f$?*

**1.** $f(x) = x + 2$
**2.** $f(x) = 2 - \frac{1}{2}x$
**3.** $f(x) = -3x + 2$
**4.** $f(x) = \frac{1}{4}x + \frac{1}{2}$
**5.** $f(x) = \frac{1}{3}x$
**6.** $f(x) = -3x$
**7.** $f(x) = .4x + .15$
**8.** $f(x) = .5 + x$
**9.** $f(x) = \frac{2 - x}{4}$
**10.** $f(x) = \frac{3x + \pi}{2}$

*Graph each linear function by hand. You may wish to support your answer by graphing on a calculator. Also, give the* **(a)** *$x$-intercept,* **(b)** *$y$-intercept,* **(c)** *domain,* **(d)** *range, and* **(e)** *slope of the line.*

**11.** $f(x) = x - 4$
**12.** $f(x) = -x + 4$
**13.** $f(x) = 3x - 6$
**14.** $f(x) = \frac{2}{3}x - 2$
**15.** $f(x) = -\frac{2}{5}x + 2$
**16.** $f(x) = \frac{4}{3}x - 3$
**17.** $f(x) = 3x$
**18.** $f(x) = -.5x$

**19.** *Concept Check* Based on the graphs of the functions in Exercises 17 and 18, what conclusion can you make about one particular point that *must* lie on the graph of the line $y = ax$ (where $b = 0$)?

**20.** *Concept Check* Using the concept of slope and your answer in Exercise 19, give the equation of the line whose graph is shown to the right.

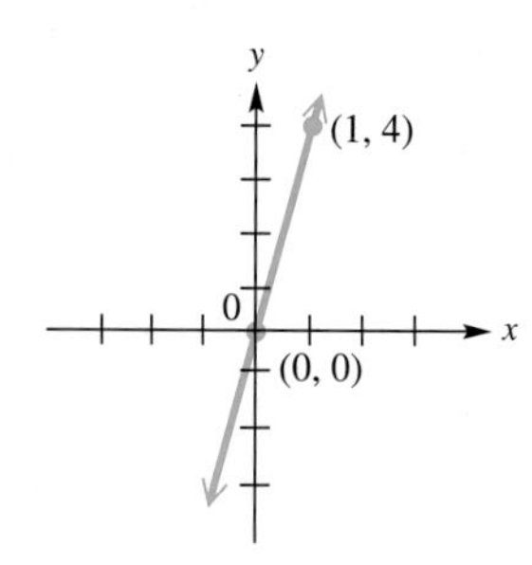


*Graph each line by hand. Also, give the* **(a)** *x-intercept (if any),* **(b)** *y-intercept (if any),* **(c)** *domain,* **(d)** *range, and* **(e)** *slope of the line (if defined).*

**21.** $f(x) = -3$ **22.** $f(x) = 5$ **23.** $x = -1.5$ **24.** $f(x) = \frac{5}{4}$ **25.** $x = 2$ **26.** $x = -3$

**27.** *Concept Check* What special name is given to the functions found in Exercises 21, 22, and 24?

**28.** Give the equation of the line illustrated.

**(a)**

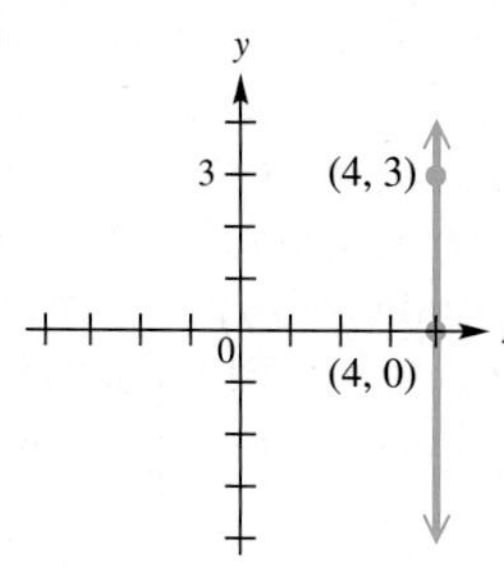


**(b)**

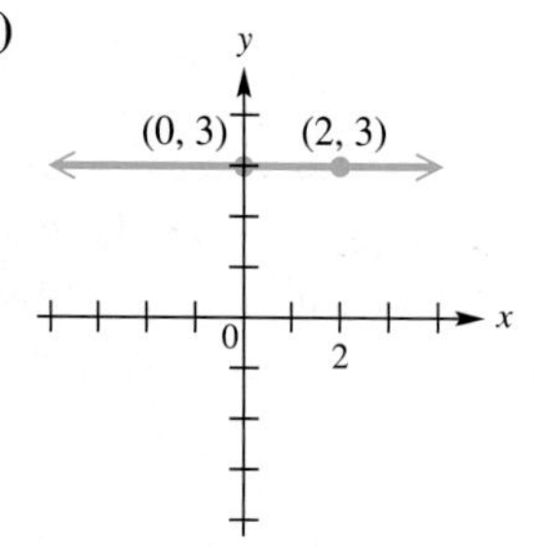


**29.** What is the equation of the $x$-axis?

**30.** What is the equation of the $y$-axis?

*Graph each linear function on a graphing calculator, using the two different windows given. State which window gives the comprehensive graph.*

**31.** $f(x) = 4x + 20$
Window A: $[-10, 10]$ by $[-10, 10]$
Window B: $[-10, 10]$ by $[-5, 25]$

**32.** $f(x) = -5x + 30$
Window A: $[-10, 10]$ by $[-10, 40]$
Window B: $[-5, 5]$ by $[-5, 40]$

**33.** $f(x) = 3x + 10$
Window A: $[-3, 3]$ by $[-5, 5]$
Window B: $[-5, 5]$ by $[-10, 14]$

**34.** $f(x) = -6$
Window A: $[-5, 5]$ by $[-5, 5]$
Window B: $[-10, 10]$ by $[-10, 10]$

*Find the slope of the line that passes through the given points.*

**35.** $(-2, 1)$ and $(3, 2)$ **36.** $(-2, 3)$ and $(-1, 2)$ **37.** $(8, 4)$ and $(-1, -3)$

**38.** $(-4, -3)$ and $(5, 0)$ **39.** $(-6, 5)$ and $(12, 5)$ **40.** $(3, 6)$ and $(3, 1)$

*Concept Check* *Find the slope of the line in each graph.*

**41.**

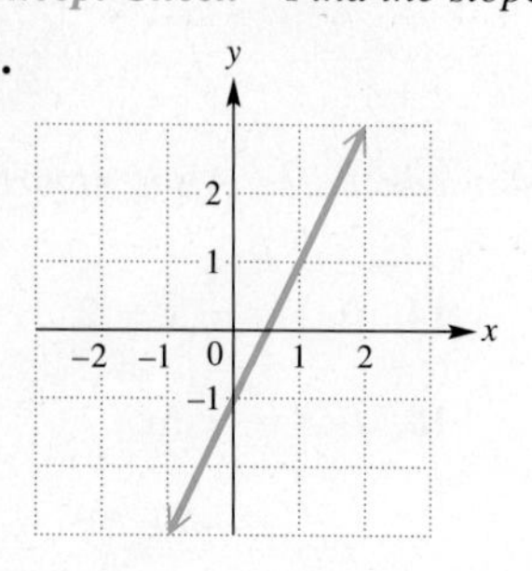


**42.**

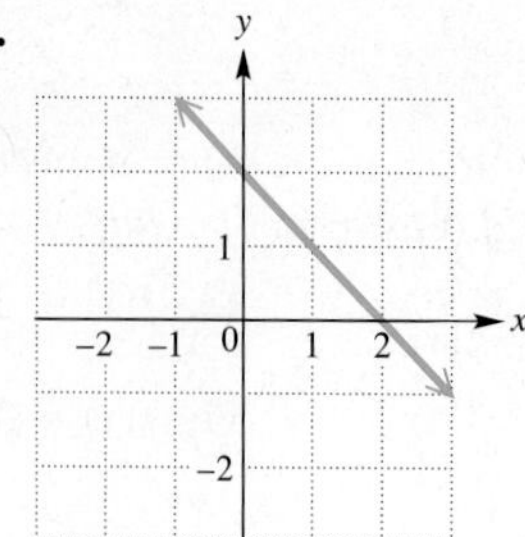


**43.**

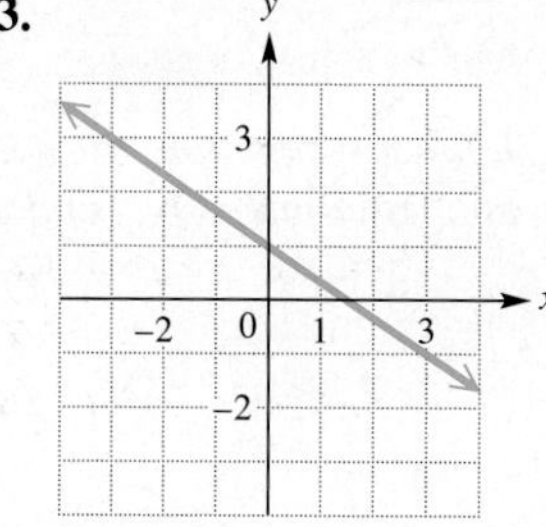

**44.**

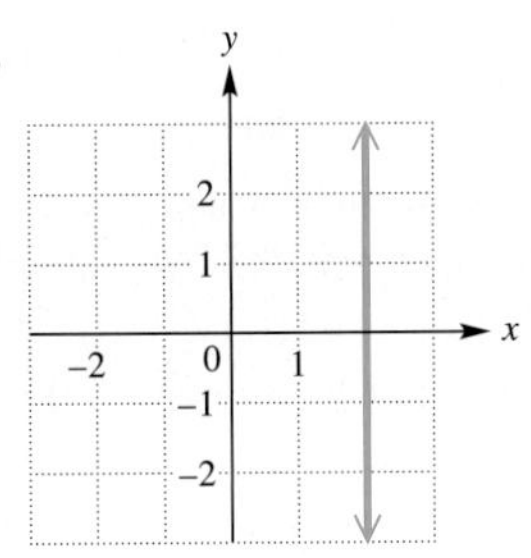


**45.**

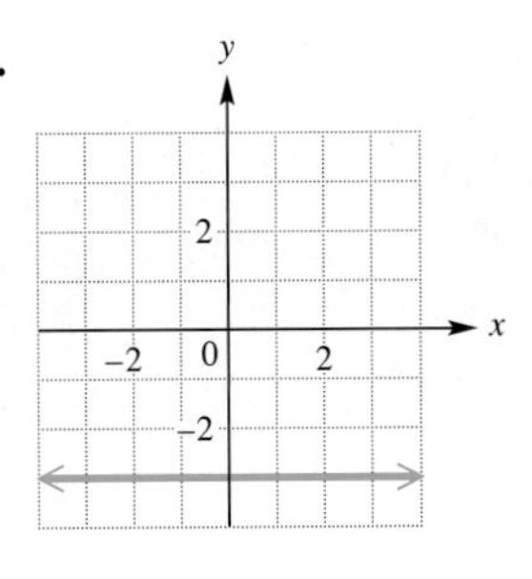


**46.**

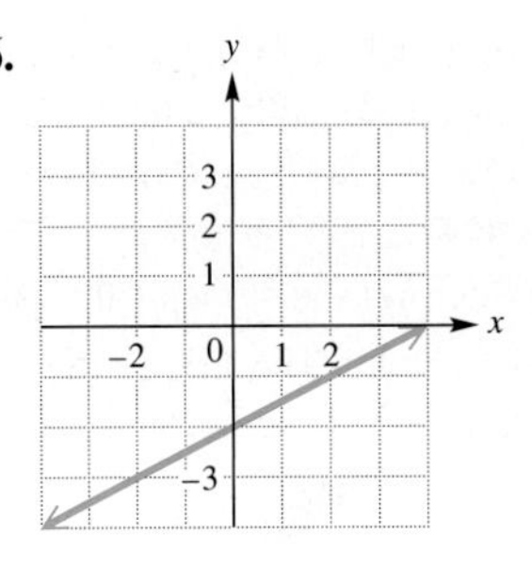


*Concept Check* *A linear function is defined as* $Y_1$ *and a table of points is generated. Find the slope m of the line that is the graph of* $Y_1$. *Then use the table to find the y-intercept b. Finally, give the equation.*

**47.**

| X | Y1 | |
|---|---|---|
| -3 | -10 | |
| -2 | -6 | |
| -1 | -2 | |
| 0 | 2 | |
| 1 | 6 | |
| 2 | 10 | |
| 3 | 14 | |

X=-3

**48.**

| X | Y1 | |
|---|---|---|
| -2 | -11 | |
| -1 | -8 | |
| 0 | -5 | |
| 1 | -2 | |
| 2 | 1 | |
| 3 | 4 | |
| 4 | 7 | |

X=-2

**49.** *Concept Check* Match each equation with the line that would most closely resemble its graph, where $k > 0$.

**(a)** $y = k$ **(b)** $y = -k$ **(c)** $x = k$ **(d)** $x = -k$

**A.**

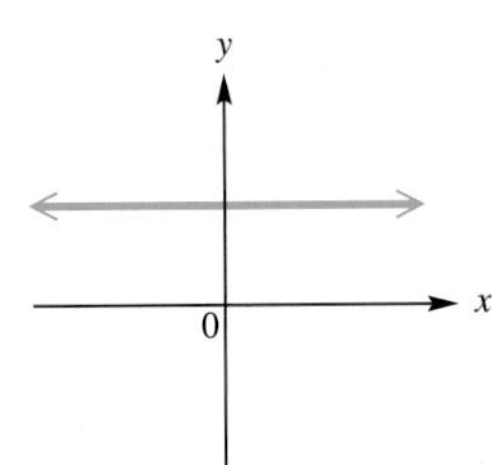


**B.**

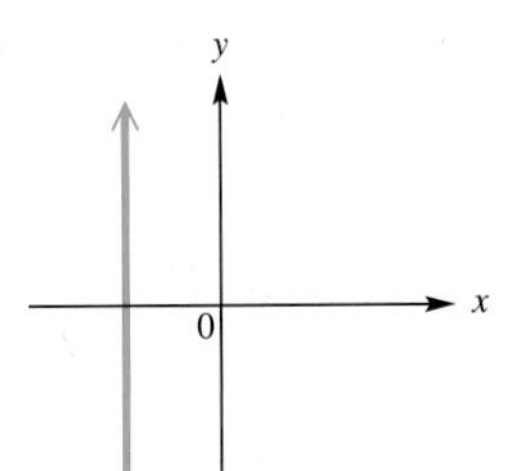


**C.**

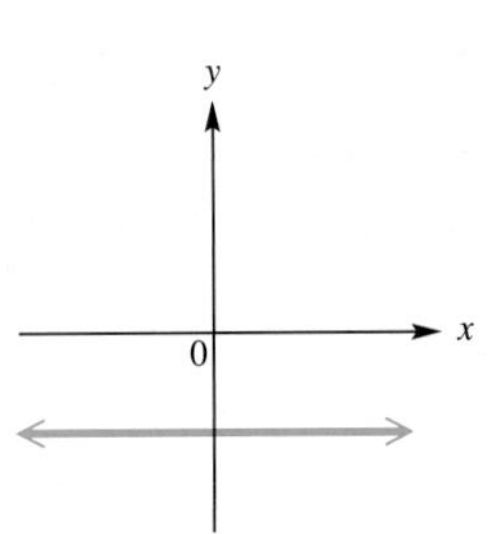


**D.**

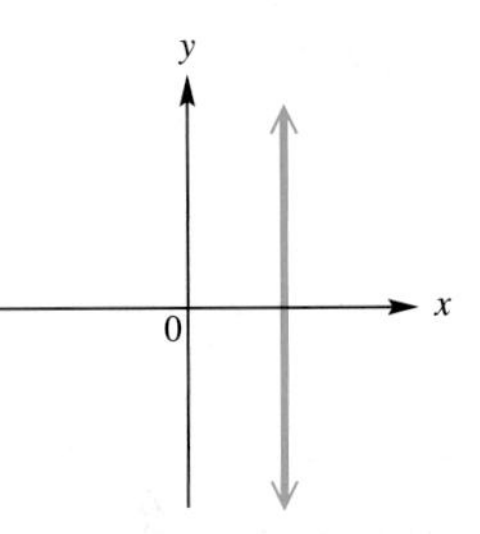


**50.** *Concept Check* Without actually plotting points, sketch by hand a line on a rectangular coordinate system that would resemble the graph of $y = mx + b$ if

**(a)** $m > 0, b > 0$. **(b)** $m > 0, b < 0$. **(c)** $m < 0, b > 0$. **(d)** $m < 0, b < 0$.

*Sketch by hand the graph of the line passing through the given point and having the given slope. Indicate two points on the line.*

**51.** Through $(-1, 3), m = \frac{3}{2}$

**52.** Through $(-2, 8), m = -1$

**53.** Through $(3, -4)$, $m = -\frac{1}{3}$

**54.** Through $(-2, -3), m = -\frac{3}{4}$

**55.** Through $(-1, 4), m = 0$

**56.** Through $\left(\frac{9}{4}, 2\right)$, undefined slope

**57.** Through $(0, -4), m = \frac{3}{4}$

**58.** Through $(0, 5)$, $m = -2.5$

**59.** Give the equation of the line described in Exercise 57.

**60.** Give the equation of the line described in Exercise 58.

*(Modeling)* *Solve each problem.*

**61.** *Rainfall* By noon, 3 inches of rain had fallen during a storm. Rain continued to fall at a rate of $\frac{1}{4}$ inch per hour. Find a formula for a linear function $f$ that models the amount of rainfall $x$ hours past noon. Find the total amount of rainfall by 2:30 P.M.

**62.** *Fuel Consumption* The table shows the distance $y$ traveled in miles by a car burning $x$ gallons of gasoline.

| **x (gallons)** | 5 | 10 | 12 | 16 |
|---|---|---|---|---|
| **y (miles)** | 115 | 230 | 276 | 368 |

**(a)** Find values for $a$ and $b$ so that $f(x) = ax + b$ models the data exactly. That is, find values for $a$ and $b$ so that the graph of $f$ passes through the data points in the table.

**(b)** Interpret the slope of the graph of $f$.

**63.** *Water Flow* The graph gives the number of gallons of water in a small swimming pool after $x$ hours.

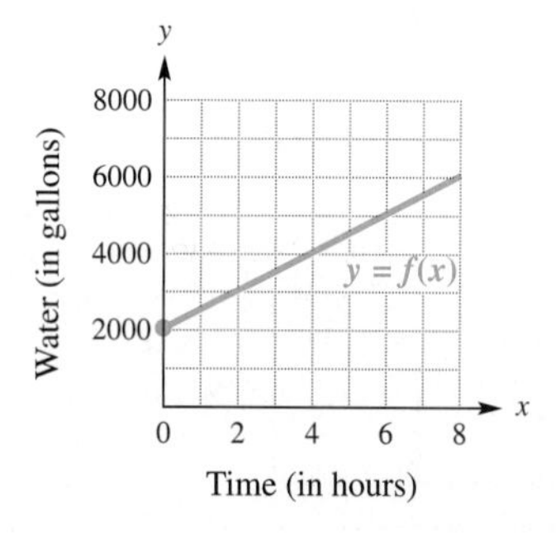


**(a)** Write a formula for $f(x)$.

**(b)** Interpret both the slope and the $y$-intercept.

**(c)** Use the graph to estimate how much water was in the pool after 7 hours. Verify your answer by evaluating $f(x)$.

**64.** *Birth Rate* In 1990 the number of births per 1000 people in the United States was 16.7 and decreasing at .326 birth per 1000 people each year. (*Source:* National Center for Health Statistics.)

**(a)** Write a formula for a linear function $f$ that models the birth rate in year $x$, where $x = 0$ corresponds to 1990, $x = 1$ to 1991, and so on.

**(b)** Estimate the birth rate in 1997 and compare it to the actual value of 14.5.

# 1.4 EQUATIONS OF LINES AND LINEAR MODELS

Point-Slope Form of the Equation of a Line • Other Forms of the Equation of a Line • Parallel and Perpendicular Lines • Linear Models and Regression

## Point-Slope Form of the Equation of a Line

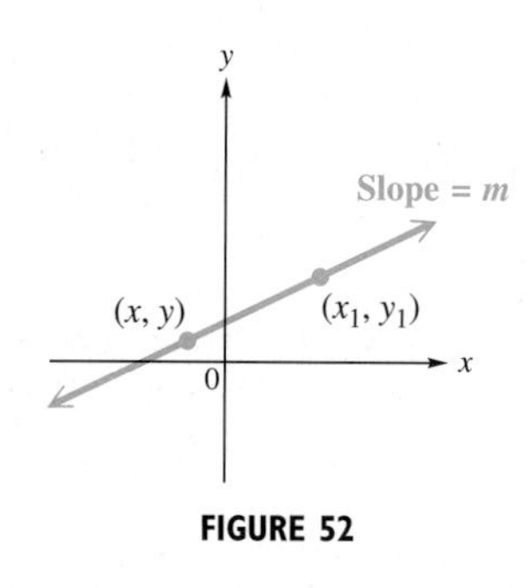


FIGURE 52

The equation of a line can be found if we know a point on the line and the slope of the line. Figure 52 shows a line passing through the fixed point $(x_1, y_1)$ with slope $m$. Let $(x, y)$ be any other point on the line. The slope $m$ of the line is

$$\frac{y - y_1}{x - x_1} = m.$$

Multiplying both sides by $x - x_1$ gives

$$y - y_1 = m(x - x_1).$$

This result is called the **point-slope form** of the equation of a line.

> **Point-Slope Form**
>
> The line with slope $m$ passing through the point $(x_1, y_1)$ has equation
>
> $$\mathbf{y - y_1 = m(x - x_1).}$$

### EXAMPLE 1 *Using Point-Slope Form*

Find the slope-intercept form of the line passing through the two points shown in Figure 53.

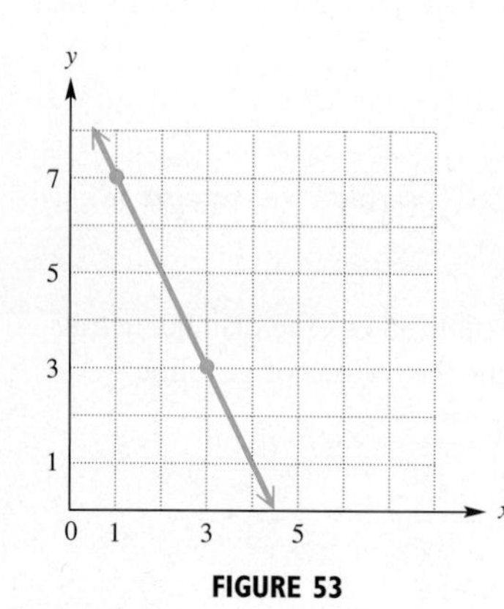


FIGURE 53

**Solution** The points are $(1, 7)$ and $(3, 3)$, so the slope of the line is

$$m = \frac{7 - 3}{1 - 3} = \frac{4}{-2} = -2.$$

**Looking Ahead to Calculus**

In calculus, it is often necessary to find the equation of a line, given its slope and a point on the line. The point-slope form is a valuable tool in these situations.

Now, using either point, say $(1, 7)$, apply the point-slope form.

$$y - y_1 = m(x - x_1) \quad \text{Point-slope form}$$
$$y - 7 = -2(x - 1) \quad y_1 = 7, m = -2, x_1 = 1$$
$$y - 7 = -2x + 2 \quad \text{Distributive property}$$
$$y = -2x + 9 \quad \text{Slope-intercept form}$$

**EXAMPLE 2** *Using Point-Slope Form*

The table in Figure 54 lists points found on the line $y = mx + b$. Find the equation of the line.

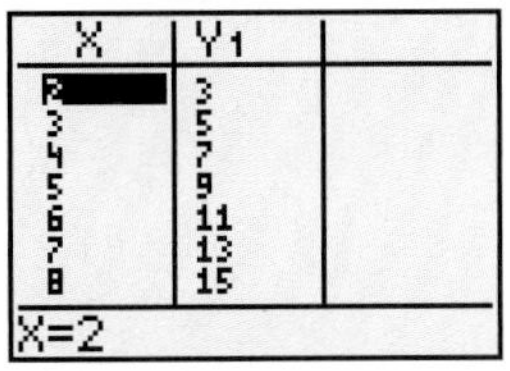

| X | Y1 | |
|---|---|---|
| 2 | 3 | |
| 3 | 5 | |
| 4 | 7 | |
| 5 | 9 | |
| 6 | 11 | |
| 7 | 13 | |
| 8 | 15 | |

X=2

FIGURE 54

**Solution** Find the slope of the line first. Choose any two points, such as $(2, 3)$ and $(6, 11)$.

$$m = \frac{11 - 3}{6 - 2} = \frac{8}{4} = 2$$

Now, use a point on the line, say $(2, 3)$, with $m = 2$ in the point-slope form.

$$y - 3 = 2(x - 2)$$
$$y - 3 = 2x - 4$$
$$y = 2x - 1$$

## Other Forms of the Equation of a Line

The equations found in Examples 1 and 2 were given in $y = mx + b$ form. As mentioned in Section 1.3, this important form is called *slope-intercept form*. In the next example we use the slope-intercept form to model real data.

**EXAMPLE 3** *Interpreting Slope-Intercept Form*

In 1998 there were 47 million people worldwide who had been infected with HIV. At that time the infection rate was 5.8 million per year. (*Source:* World Health Organization.)

**(a)** Find values for $m$ and $b$ so that $y = mx + b$ models the total number of people $y$ in millions that were infected with HIV in year $x$, where $x = 0$ corresponds to 1998, $x = 1$ to 1999, and so on.

**(b)** Estimate the number of people who could be infected in 2002.

**Solution**

**(a)** Since there were 47 million people infected in 1998 and $x = 0$ corresponds to 1998, the line determined by $y = mx + b$ must pass through the point $(0, 47)$. Thus, the $y$-intercept is $b = 47$. The infection rate is 5.8 million per year, so $m = 5.8$. The equation is

$$y = 5.8x + 47.$$

**(b)** The year 2002 corresponds to $x = 4$, so

$$y = 5.8(4) + 47 = 70.2.$$

Thus, according to this model, 70.2 million people could be infected with HIV in 2002.

Another form of the equation of a line is

$$Ax + By = C,$$

which is often referred to as **standard form.** For example, $3x + 2y = 6$ is in standard form. One advantage of standard form is that it allows quick calculation of both intercepts. For example, if we begin with $3x + 2y = 6$, we can find the $x$-intercept by letting $y = 0$ and the $y$-intercept by letting $x = 0$.

| $x$-intercept: $3x + 2(0) = 6$ | $y$-intercept: $3(0) + 2y = 6$ |
|---|---|
| $3x = 6$ | $2y = 6$ |
| $x = 2$ | $y = 3$ |

This information is useful when sketching the graph of the line by hand.

**EXAMPLE 4** *Graphing an Equation in Standard Form*

Graph $3x + 2y = 6$.

**Analytic Solution**

As seen above, the points $(2, 0)$ and $(0, 3)$ correspond to the $x$- and $y$-intercepts. Plot these two points and connect them with a straight line. See Figure 55.

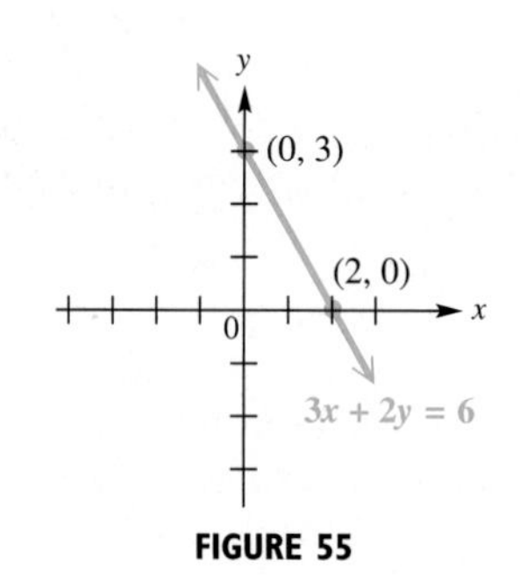


**FIGURE 55**

**Graphing Calculator Solution**

Solve the equation for $y$ so it can be entered in a calculator.

$$\begin{aligned} 3x + 2y &= 6 && \text{Given equation} \\ 2y &= -3x + 6 && \text{Subtract } 3x. \\ y &= -1.5x + 3 && \text{Divide by 2.} \end{aligned}$$

The desired graph is shown in Figure 56.

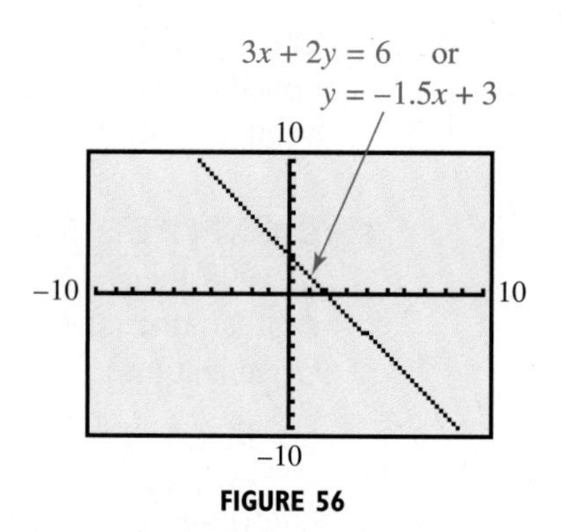


**FIGURE 56**

**NOTE** Because of the usefulness of slope-intercept form when graphing with a graphing calculator, we emphasize slope-intercept form in most of our work.

## Parallel and Perpendicular Lines

**FOR DISCUSSION**

In the standard viewing window of your calculator, graph all four of the following lines.

$$y_1 = 2x - 6 \qquad y_2 = 2x - 2 \qquad y_3 = 2x \qquad y_4 = 2x + 4$$

What is the slope of each line? What geometric term seems to describe the set of lines?

Two lines in a plane are *parallel* if they do not intersect. Although the exercise in the "For Discussion" box does not actually prove the result that follows, it provides visual support.

**Parallel Lines**

Two distinct nonvertical lines are parallel if and only if they have the same slope.

**EXAMPLE 5** *Using the Slope Relationship for Parallel Lines*

Find the equation of the line that passes through the point $(3, 5)$ and is parallel to the line with equation $2x + 5y = 4$. Graph both lines in the standard viewing window.

**Solution** Since the point $(3, 5)$ is on the line, we need only find the slope by writing the given equation in slope-intercept form. (That is, solve for $y$.)

$$2x + 5y = 4$$
$$y = -\frac{2}{5}x + \frac{4}{5}$$

The slope is $-\frac{2}{5}$. Since the lines are parallel, $-\frac{2}{5}$ is also the slope of the line whose equation we must find. Substituting $m = -\frac{2}{5}$, $x_1 = 3$, and $y_1 = 5$ into the point-slope form gives

$$y - y_1 = m(x - x_1) \quad \text{Point-slope form}$$
$$y - 5 = -\frac{2}{5}(x - 3) \quad \text{Substitute.}$$
$$5(y - 5) = -2(x - 3) \quad \text{Multiply by 5.}$$
$$5y - 25 = -2x + 6 \quad \text{Distributive property}$$
$$5y = -2x + 31 \quad \text{Add 25.}$$
$$y = -\frac{2}{5}x + \frac{31}{5}. \quad \text{Slope-intercept form of the desired line}$$

To provide visual support, graph $y_1 = -\frac{2}{5}x + \frac{4}{5}$ and $y_2 = -\frac{2}{5}x + \frac{31}{5}$. Sometimes it is helpful to enter decimal forms of fractions since fractions usually require parentheses to indicate grouping. Therefore, we enter these two equations as follows.

$$y_1 = -.4x + .8 \quad \text{Given line}$$
$$y_2 = -.4x + 6.2 \quad \text{Desired line}$$

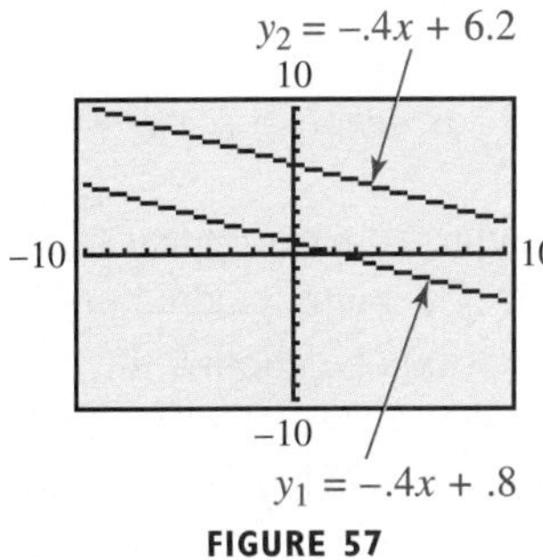

FIGURE 57

As seen in Figure 57, the lines *appear* to be parallel.

**CAUTION** When using a graphing calculator, be aware that visual support (as seen in Example 5) does not necessarily prove the result. For example, Figure 58 shows the graphs of $y_1 = -.5x + 4$ and $y_2 = -.5001x + 2$. Although they *appear* to be parallel by visual inspection, they are *not* parallel because the slope of $y_1$ is $-.5$ and the slope of $y_2$ is $-.5001$.

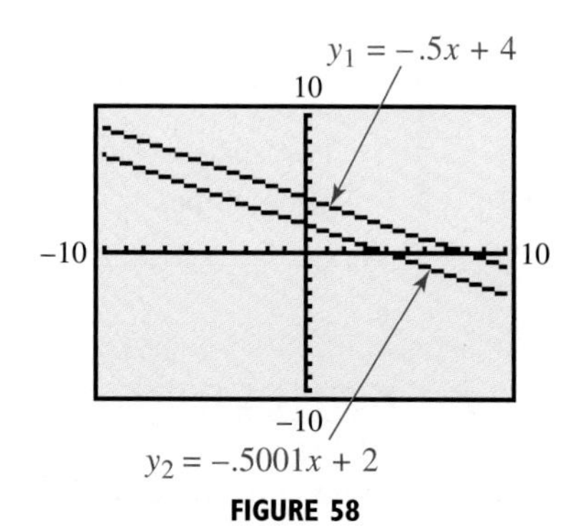


**FIGURE 58**

### FOR DISCUSSION

With your viewing window set for a "square" window, such as $[-9, 9]$ by $[-6, 6]$, graph each pair of lines. Graph each group separately.

| I | II | III | IV |
|---|---|---|---|
| $y_1 = 4x + 1$ | $y_1 = -\frac{2}{3}x + 3$ | $y_1 = 6x - 3$ | $y_1 = \frac{13}{7}x - 3$ |
| $y_2 = -\frac{1}{4}x + 3$ | $y_2 = \frac{3}{2}x - 4$ | $y_2 = -\frac{1}{6}x + 4$ | $y_2 = -\frac{7}{13}x + 4$ |

What geometric term applies to each pair of lines? What is the product of the slopes for each pair of lines?

As in the earlier "For Discussion" box, we have not proved the following result; rather, we have provided visual support for it.

**Perpendicular Lines**

Two lines, neither of which is vertical, are perpendicular if and only if their slopes have product $-1$.

For example, if the slope of a line is $-\frac{3}{4}$, the slope of any line perpendicular to it is $\frac{4}{3}$, since $-\frac{3}{4}\left(\frac{4}{3}\right) = -1$. We often refer to numbers like $-\frac{3}{4}$ and $\frac{4}{3}$ as *negative reciprocals.* A proof of this result is outlined in Exercise 43.

In a **square viewing window,** circles appear to be circular, squares appear to be square, and perpendicular lines appear to be perpendicular. On many calculators, a square viewing window requires that the distance along the $y$-axis be about $\frac{2}{3}$ the distance along the $x$-axis. Examples of square viewing windows on these types of calculators are $[-3, 3]$ by $[-2, 2]$, $[-9, 9]$ by $[-6, 6]$, and $[-15, 15]$ by $[-10, 10]$. Figure 59 illustrates the importance of square viewing windows.

**TECHNOLOGY NOTE**

Many calculators can set a square viewing window automatically. Check the graphing calculator manual that accompanies this text or your owner's manual, or look under the ZOOM menu.

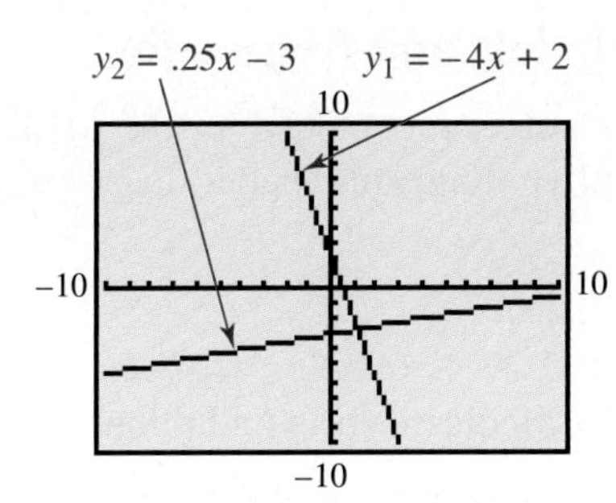


Although the graphs of these lines are perpendicular, they do not appear to be when graphed in a standard viewing window.

**(a)**

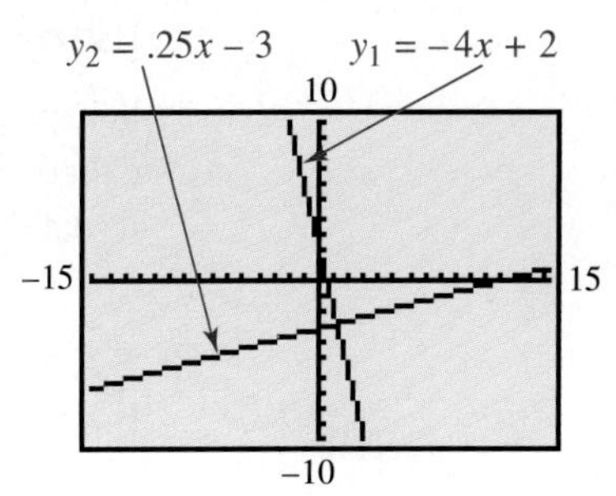


Visual support for perpendicularity is more obvious using a square viewing window.

**(b)**

**FIGURE 59**

**EXAMPLE 6** *Using the Slope Relationship for Perpendicular Lines*

Find the equation of the line that passes through the point $(3, 5)$ and is perpendicular to the line with equation $2x + 5y = 4$. Graph both lines in a square viewing window.

**Solution** In Example 5, we found that the slope of the given line is $-\frac{2}{5}$, so the slope of any line perpendicular to it is $\frac{5}{2}$. Therefore, we use $m = \frac{5}{2}$, $x_1 = 3$, and $y_1 = 5$ in the point-slope form.

$$y - 5 = \frac{5}{2}(x - 3) \quad \text{Point-slope form}$$

$$2(y - 5) = 5(x - 3) \quad \text{Multiply by 2.}$$

$$2y - 10 = 5x - 15 \quad \text{Distributive property}$$

$$2y = 5x - 5 \quad \text{Add 10.}$$

$$y = 2.5x - 2.5 \quad \text{Slope-intercept form of the desired line}$$

Graphing $y_1 = -.4x + .8$ (the slope-intercept form of the given equation) and $y_2 = 2.5x - 2.5$ (the slope-intercept form of the equation just determined) in a *square* viewing window provides visual support for our answer. See Figure 60.

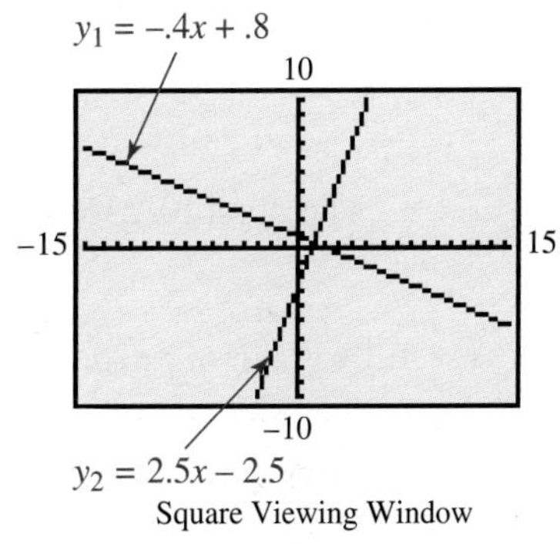


Square Viewing Window

**FIGURE 60**

## Linear Models and Regression

When data points are plotted in the $xy$-plane, the resulting graph is sometimes called a **scatter diagram.** Scatter diagrams are often helpful when analyzing trends in data.

### EXAMPLE 7 *Modeling Medicare Costs with a Linear Function*

Estimates for Medicare costs (in billions of dollars) are shown in the table.

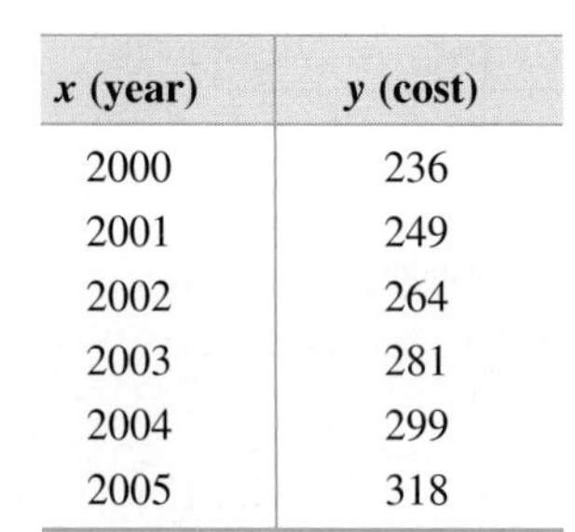

| $x$ (year) | $y$ (cost) |
|---|---|
| 2000 | 236 |
| 2001 | 249 |
| 2002 | 264 |
| 2003 | 281 |
| 2004 | 299 |
| 2005 | 318 |

*Source:* U.S. Health Care Financing Administration.

**(a)** Make a scatter diagram of the data. Let $x = 0$ correspond to 2000, $x = 1$ to 2001, and so on. What type of function might model the data?

**(b)** Find a linear function $f$ that models the data. Graph $f$ and the data in the same viewing window. Interpret the slope $m$.

**(c)** Use $f(x)$ to predict Medicare costs in 2006.

#### Solution

**(a)** Since $x = 0$ corresponds to 2000, $x = 1$ corresponds to 2001, and so on, the data points can be expressed as the ordered pairs

$$(0, 236), \quad (1, 249), \quad (2, 264), \quad (3, 281), \quad (4, 299), \quad \text{and} \quad (5, 318).$$

Scatter diagrams are shown in Figure 61. The data appear to be approximately linear, so a linear function might be appropriate.

**TECHNOLOGY NOTE**

To make a scatter diagram with a graphing calculator, you may need to use the *list* feature by entering the $x$-values in list $L_1$ and the $y$-values in list $L_2$. See Figure 62.

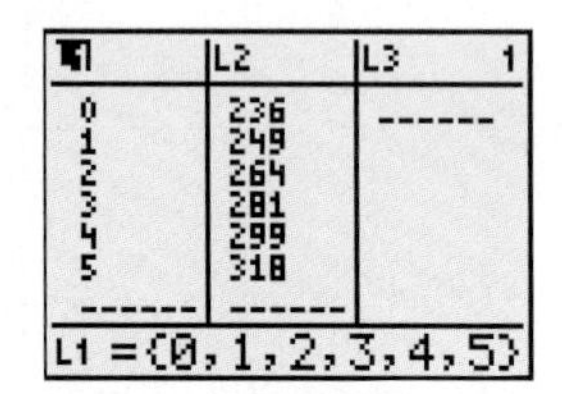


FIGURE 62

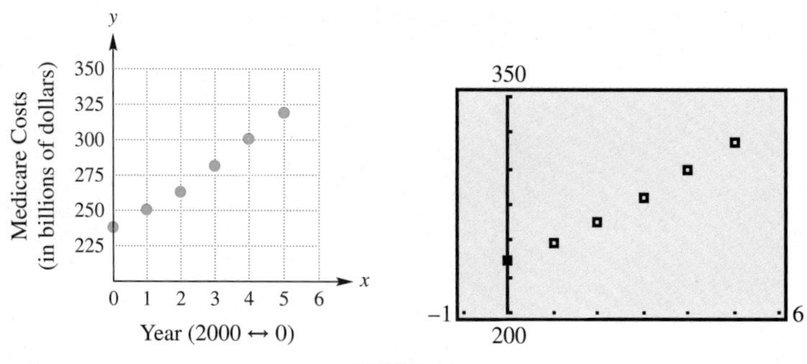


FIGURE 61

**(b)** We start by choosing two data points that the line should pass through. For example, if we use (0, 236) and (4, 299), then the slope of the line is

$$m = \frac{299 - 236}{4 - 0} = 15.75.$$

The point (0, 236) indicates that the $y$-intercept $b$ is 236. Thus,

$$f(x) = 15.75x + 236.$$

A graph of $f$ and the data are shown in Figure 63. The slope $m = 15.75$ indicates that Medicare costs might increase, on average, by \$15.75 billion per year.

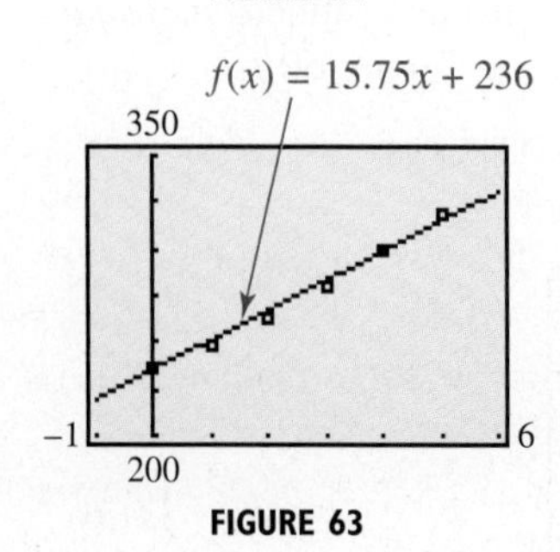


FIGURE 63

**(c)** The value $x = 6$ corresponds to the year 2006. Since

$$f(6) = 15.75(6) + 236 = 330.5,$$

this model predicts that Medicare costs will reach \$330.5 billion in 2006.

**NOTE** It is important to realize that the formula $f(x) = 15.75x + 236$ found in Example 7 is not unique. If two different points are chosen, a different formula for $f(x)$ will result. However, all such formulas for $f(x)$ should be in approximate agreement.

The method for finding the model in Example 7 used algebraic concepts that do not always give a unique line. It is reasonable to expect that a method exists to find the line of "best fit." Graphing calculators are capable of finding this line, called the **least-squares regression line,** by using a technique from statistics known as **least-squares regression.**

**EXAMPLE 8** *Finding the Least-Squares Regression Line*

Use a graphing calculator to find the least-squares regression line that models the Medicare costs presented in Example 7. Graph the data and the line in the same viewing window.

TECHNOLOGY NOTE

To find the equation of a least-squares regression line, refer to the graphing calculator manual that accompanies this text or your owner's manual.

**Solution** Figure 64 shows how a TI-83 Plus graphing calculator finds the regression line for the data in Example 7. In Figure 64(a), the years are entered into list $L_1$, and Medicare costs are entered into list $L_2$. In Figure 64(b), the formula for the regression line is calculated to be

$$y \approx 16.49x + 233.3.$$

Notice that this equation is not exactly the same as the one found for $f(x)$ in Example 7. In Figure 64(c), both the data and regression line are graphed.

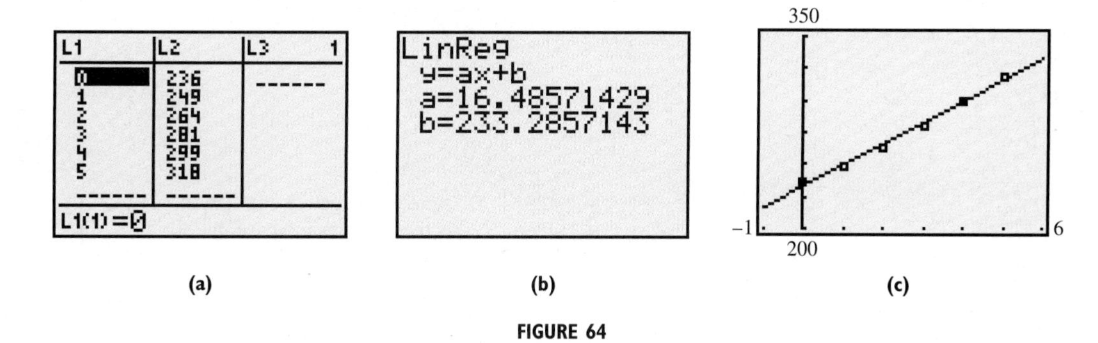


(a) (b) (c)

FIGURE 64

Once an equation for the least-squares regression line has been found, it is reasonable to ask, "Just how good is this line for predictive purposes?" If the line *fits* the observed data points, then future pairs of points might be expected to do so also. One common measure of the strength of the linear relationship in a data set is called the **correlation coefficient,** denoted $r$, where $-1 \leq r \leq 1$. When $r$ is positive and near 1, low $x$-values correspond to low $y$-values and high $x$-values correspond to high $y$-values. For example, there is a positive correlation between years of education $x$ and income $y$. More years of education correlate with higher income. When $r$ is near $-1$, the reverse is true. Low $x$-values correspond to high $y$-values and high $x$-values correspond to low $y$-values. An example of this would be the relation between latitude and average yearly temperature. As latitude increases (moving toward either pole), the average yearly temperature decreases. Therefore, there will be a negative correlation between latitude and average annual temperature. If $r \approx 0$, then there is little or no correlation between the data points. In this case, a least-squares regression line does not provide a suitable model. A summary of these concepts is shown in the table on the next page.

**Correlation Coefficient $r$, $(-1 \leq r \leq 1)$**

| Value of $r$ | Comments | Sample Scatter Diagram |
|---|---|---|
| $r = 1$ | There is an exact linear fit. The line passes through all data points and has a positive slope. | 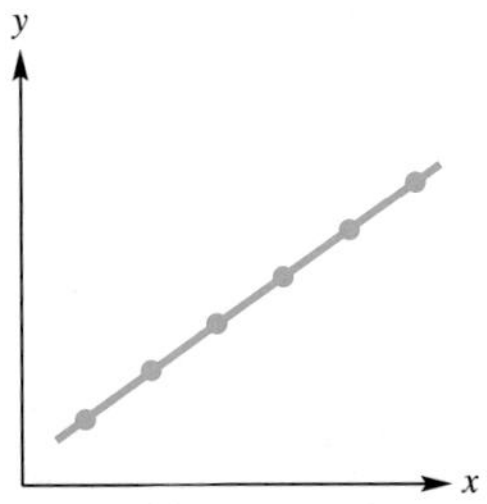 |
| $r = -1$ | There is an exact linear fit. The line passes through all data points and has a negative slope. | 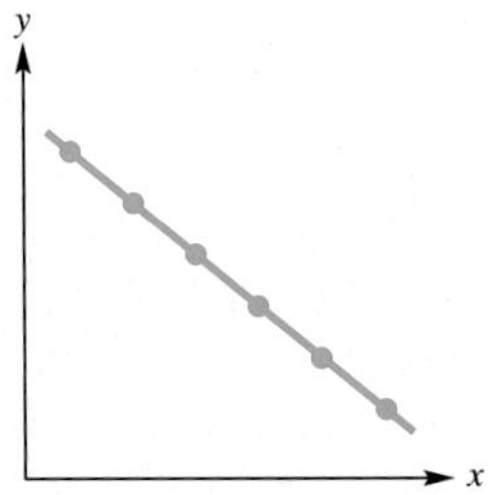 |
| $0 < r < 1$ | There is a positive correlation. As the $x$-values increase, so do the $y$-values. The fit is not exact. | 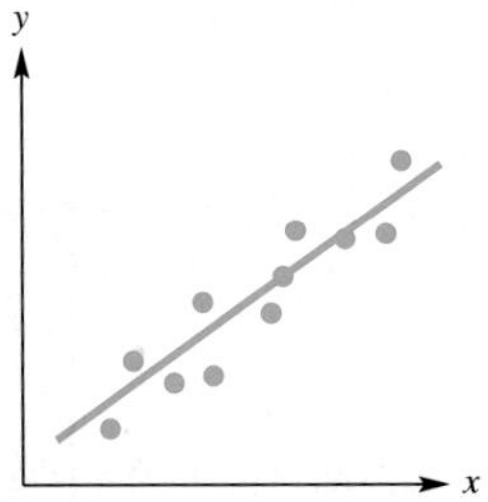 |
| $-1 < r < 0$ | There is a negative correlation. As the $x$-values increase, the $y$-values decrease. The fit is not exact. | 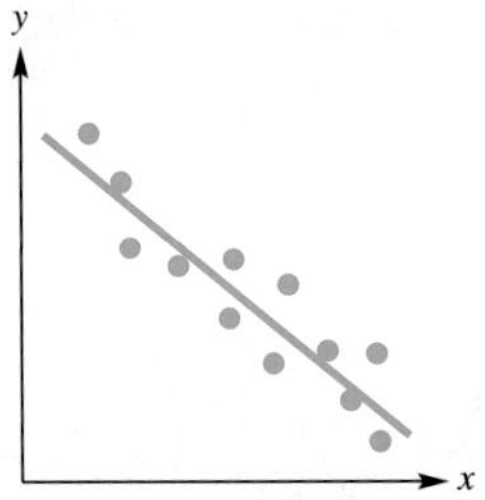 |
| $r = 0$ | There is no correlation. The data has no tendency toward being linear. A regression line predicts poorly. | 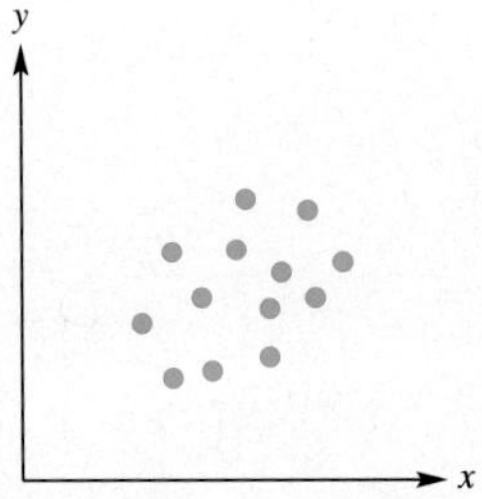 |

There is a difference between correlation and causation. For example, when geese begin to fly north, summer is coming and the weather becomes warmer. Geese flying north correlate with warmer weather. However, geese flying north clearly do not *cause* warmer weather. It is important to remember that correlation does not always indicate the cause.

## EXAMPLE 9 *Predicting Airline Passenger Growth*

The table lists the estimated numbers in millions of airline passengers at some of the fastest growing airports in 1992 and 2005.

**Airline Passengers (in millions)**

| Airport | 1992 | 2005* |
|---|---|---|
| Harrisburg International | .7 | 1.4 |
| Dayton International | 1.1 | 2.4 |
| Austin Robert Mueller | 2.2 | 4.7 |
| Milwaukee General Mitchell | 2.2 | 4.4 |
| Sacramento Metropolitan | 2.6 | 5.0 |
| Fort Lauderdale–Hollywood | 4.1 | 8.1 |
| Washington Dulles | 5.3 | 10.9 |
| Greater Cincinnati | 5.8 | 12.3 |

*Projected
*Source:* Federal Aviation Administration.

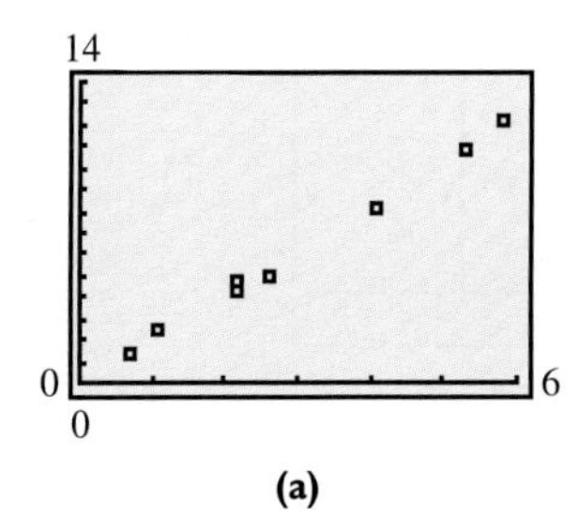


(a)

LinReg
y=ax+b
a=2.081037277
b=-.0931118314
r²=.9961050751
r=.9980506375

(b)

14
0
0
6

(c)

**FIGURE 65**

**(a)** Make a scatter diagram of the data using the 1992 data for $x$-values and the corresponding 2005 data for $y$-values. Predict whether the correlation coefficient will be positive or negative.

**(b)** Use a calculator to find the least-squares regression line.

**(c)** Raleigh-Durham International had 4.9 million passengers in 1992. Assuming Raleigh-Durham International follows the same trend in growth as the other airports, estimate the number of passengers it may have in 2005. Compare this result with the Federal Aviation Administration (FAA) estimate of 10.3 million passengers.

### Solution

**(a)** Plot the data as shown in Figure 65(a). Since increasing $x$-values correspond to increasing $y$-values, the correlation coefficient will be positive.

**(b)** The equation of the regression line is shown in Figure 65(b). The line

$$y = 2.081x - .0931$$

and the data are graphed together in Figure 65(c). Note that $r \approx .998$, which is positive as predicted.

**(c)** When $x = 4.9$, $y = 2.081(4.9) - .0931 \approx 10.1$ million passengers. This is quite close to the FAA prediction of 10.3 million.

## 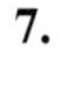 1.4 EXERCISES

*If possible, write the slope-intercept form of the line that passes through the given point with slope m.*

**1.** Through $(1, 3), m = -2$

**2.** Through $(2, 4), m = -1$

**3.** Through $(-5, 4), m = -1.5$

**4.** Through $(-4, 3), m = .75$

**5.** Through $(-8, 1), m = -.5$

**6.** Through $(6, 1), m = 0$

*Concept Check* *Give the slope-intercept form of the line shown in each graph.*

**7.**

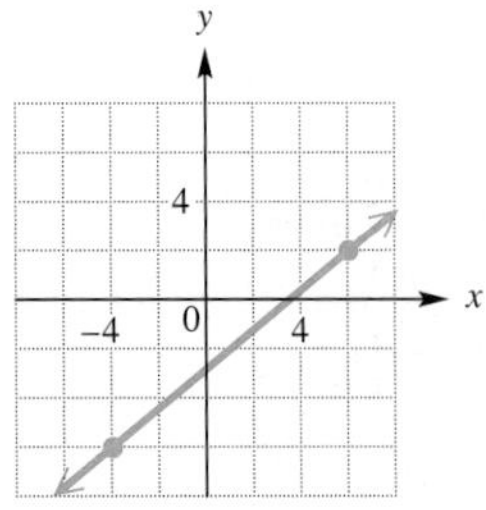


**8.**

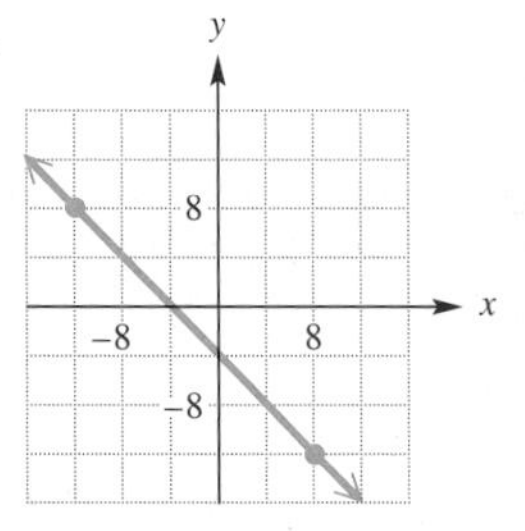


*Find the slope-intercept form of the line satisfying the given conditions.*

**9.** Through $(2, 3.5)$ and $(6, -2.5)$

**10.** Through $(-1, 6.25)$ and $(2, -4.25)$

**11.** Through $(-2, 2)$ and $(-4, -4)$

**12.** Through $(0, -8)$ and $(4, 0)$

**13.** Defined by $Y_1$ and yields the given table of points

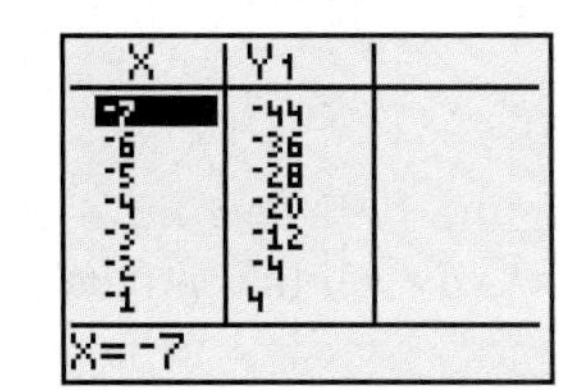

| X | $Y_1$ |
|---|---|
| -7 | -44 |
| -6 | -36 |
| -5 | -28 |
| -4 | -20 |
| -3 | -12 |
| -2 | -4 |
| -1 | 4 |

X=-7

**14.** Defined by $Y_1$ and yields the given table of points

| X | $Y_1$ |
|---|---|
| -2.4 | 5.2 |
| 1.3 | -24.4 |
| 1.75 | -28 |
| 2.98 | -37.84 |
| 13 | -118 |
| -10 | 66 |
| 5 | -54 |

X=-2.4

*Graph each line by hand, finding intercepts to determine two points on the line.*

**15.** $x - y = 4$

**16.** $x + y = 4$

**17.** $3x - y = 6$

**18.** $2x - 3y = 6$

**19.** $2x + 5y = 10$

**20.** $4x - 3y = 9$

*A line having an equation of the form $y = kx$, where $k$ is a real number, $k \neq 0$, will always pass through the origin. To graph such an equation by hand, we must determine a second point and then connect the origin and that second point with a straight line. Use this method to graph each line.*

**21.** $y = 3x$

**22.** $y = -2x$

**23.** $y = -.75x$

**24.** $y = 1.5x$

*Write each equation in the form $y = mx + b$. Then, using your calculator, graph the line in the window indicated.*

**25.** $5x + 3y = 15$
$[-10, 10]$ by $[-10, 10]$

**26.** $6x + 5y = 9$
$[-10, 10]$ by $[-10, 10]$

**27.** $-2x + 7y = 4$
$[-5, 5]$ by $[-5, 5]$

**28.** $-.23x - .46y = .82$
$[-5, 5]$ by $[-5, 5]$

**29.** $1.2x + 1.6y = 5.0$
$[-6, 6]$ by $[-4, 4]$

**30.** $2y - 5x = 0$
$[-10, 10]$ by $[-10, 10]$

*Find the equation of the line satisfying the given conditions, giving it in slope-intercept form if possible.*

**31.** Through $(-1, 4)$, parallel to $x + 3y = 5$

**32.** Through $(3, -2)$, parallel to $2x - y = 5$

**33.** Through $(1, 6)$, perpendicular to $3x + 5y = 1$

**34.** Through $(-2, 0)$, perpendicular to $8x - 3y = 7$

**35.** Through $(-5, 7)$, perpendicular to $y = -2$

**36.** Through $(1, -4)$, perpendicular to $x = 4$

**37.** Through $(-5, 8)$, parallel to $y = -.2x + 6$

**38.** Through $(-4, -7)$, parallel to $x + y = 5$

**39.** Through the origin, perpendicular to $2x + y = 6$

**40.** Through the origin, parallel to $y = -3.5x + 7.4$

**41.** The figure shows the graphs of $y_1 = 2.3x + .57$ and $y_2 = 2.3001x - 4.8$. A student unfamiliar with the concepts presented in this section may conclude that these two lines are parallel. Write a short paragraph explaining why they are or are not parallel.

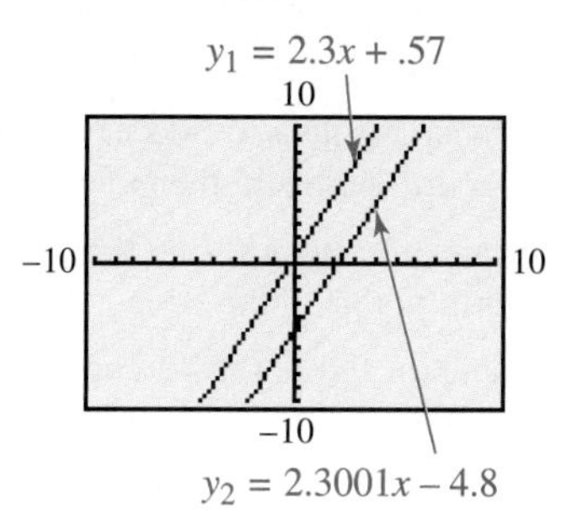


**42.** The figure shows the graphs of $y_1 = -3x + 4$ and $y_2 = \frac{1}{3}x - 4$. A student unfamiliar with the concepts presented in this section may conclude that these two lines are not perpendicular. Write a short paragraph explaining why they are or are not perpendicular. If they *are* perpendicular lines, explain how to set a graphing calculator so that this result may be more easily supported.

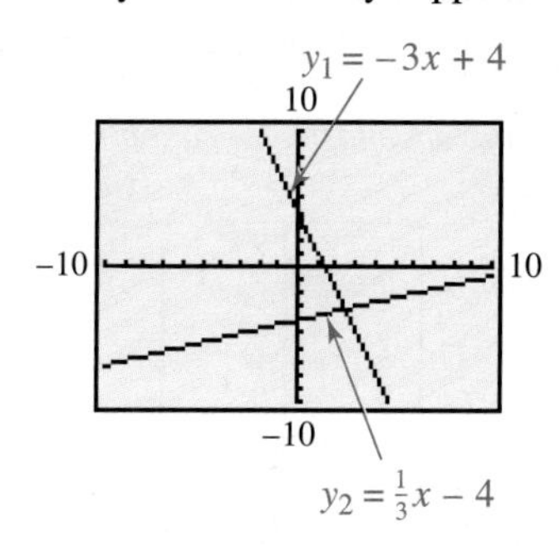


**43.** Refer to the given figure and complete parts (a)–(h) to prove that if two lines are perpendicular, and neither line is parallel to an axis, then the lines have slopes whose product is $-1$.

**(a)** In triangle $OPQ$, angle $POQ$ is a right angle if and only if

$$[d(O, P)]^2 + [d(O, Q)]^2 = [d(P, Q)]^2.$$

What theorem from geometry is this?

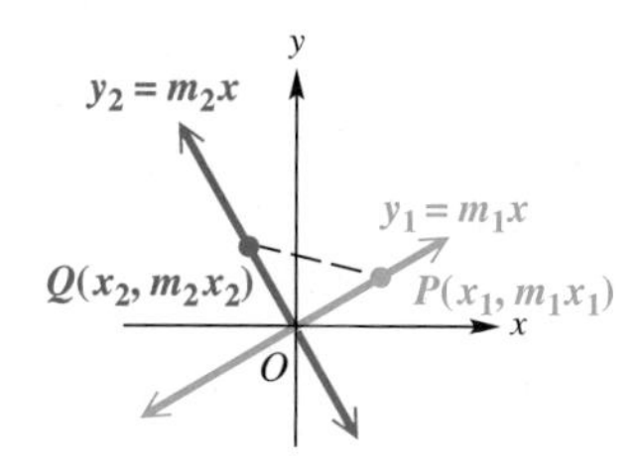


**(b)** Find an expression for the distance $d(O, P)$.

**(c)** Find an expression for the distance $d(O, Q)$.

**(d)** Find an expression for the distance $d(P, Q)$.

**(e)** Use your results from parts (b)–(d) and substitute into the equation in part (a). Simplify to show that this leads to the equation

$$-2m_1m_2x_1x_2 - 2x_1x_2 = 0.$$

**(f)** Factor $-2x_1x_2$ from the final form of the equation in part (e).

**(g)** Use the zero-product property from intermediate algebra to solve the equation in part (f) to show that $m_1m_2 = -1$.

**(h)** State your conclusion based on parts (a)–(g).

**44.** *Concept Check* The following tables show ordered pairs for linear functions defined in $Y_1$ and $Y_2$. Determine whether the lines are parallel, perpendicular, or neither parallel nor perpendicular.

**(a)**

| X | Y1 | Y2 |
|---|---|---|
| 0 | -3 | 4 |
| 1 | 1 | 3.75 |
| 2 | 5 | 3.5 |
| 3 | 9 | 3.25 |
| 4 | 13 | 3 |
| 5 | 17 | 2.75 |
| 6 | 21 | 2.5 |

X=0

**(b)**

| X | Y1 | Y2 |
|---|---|---|
| 0 | -3 | 5 |
| 1 | 2 | 5.2 |
| 2 | 7 | 5.4 |
| 3 | 12 | 5.6 |
| 4 | 17 | 5.8 |
| 5 | 22 | 6 |
| 6 | 27 | 6.2 |

X=0

**(c)**

| X | Y1 | Y2 |
|---|---|---|
| 0 | -3 | 12 |
| 1 | 2 | 17 |
| 2 | 7 | 22 |
| 3 | 12 | 27 |
| 4 | 17 | 32 |
| 5 | 22 | 37 |
| 6 | 27 | 42 |

X=0

*Beginning in this section, we will occasionally include groups of exercises under the heading "Relating Concepts." These exercise groups are designed to be worked in sequential order. Their purpose is to allow you to investigate connections and relationships among various topics that appear in the section and perhaps in earlier sections.*

## RELATING CONCEPTS

For individual or group investigation (Exercises 45–54)

*The table was generated for a linear function. Work Exercises 45–54 in order, to investigate connections between the slope formula, distance formula, midpoint formula, and linear functions.*

| X | Y1 | |
|---|---|---|
| 0 | -6 | |
| 1 | -3 | |
| 2 | 0 | |
| 3 | 3 | |
| 4 | 6 | |
| 5 | 9 | |
| 6 | 12 | |

X=0

**45.** Use the first two points in the table to find the slope of the line.

**46.** Use the second and third points in the table to find the slope of the line.

**47.** Make a conjecture by filling in the blank: If we use any two points on a line to find its slope, we find that the slope is ________ in all cases.

**48.** Use the distance formula to find the distance between the first two points in the table.

**49.** Use the distance formula to find the distance between the second and fourth points in the table.

**50.** Use the distance formula to find the distance between the first and fourth points in the table.

**51.** Add the results in Exercises 48 and 49, and compare the sum to the answer you found in Exercise 50. What do you notice?

**52.** Fill in the blanks, basing your answers on your observations in Exercises 48–51. If points $A$, $B$, and $C$ lie on a line in that order, then the distance between $A$ and $B$ added to the distance between ________ and ________ is equal to the distance between ________ and ________.

**53.** Use the midpoint formula to find the midpoint of the segment joining $(0, -6)$ and $(6, 12)$. Compare your answer to the middle entry in the table. What do you notice?

**54.** If the table were set up to show an $x$-value of 4.5, what would be the corresponding $y$-value?

*(Modeling) Solve each problem.*

**55.** *Distance* A person is riding a bicycle along a straight highway. The graph shows the rider's distance $y$ in miles from an interstate highway after $x$ hours.

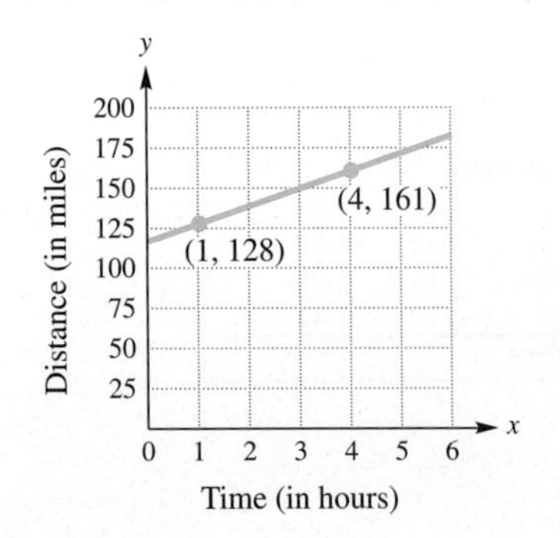


**(a)** Find the slope-intercept form of the line.

**(b)** How fast is the biker traveling?

**(c)** How far was the biker from the interstate highway initially?

**(d)** How far was the biker from the interstate highway after 1 hour and 15 minutes?

**56.** *Water in a Tank* The graph shows the amount of water $y$ in a 100-gallon tank after $x$ minutes have elapsed.

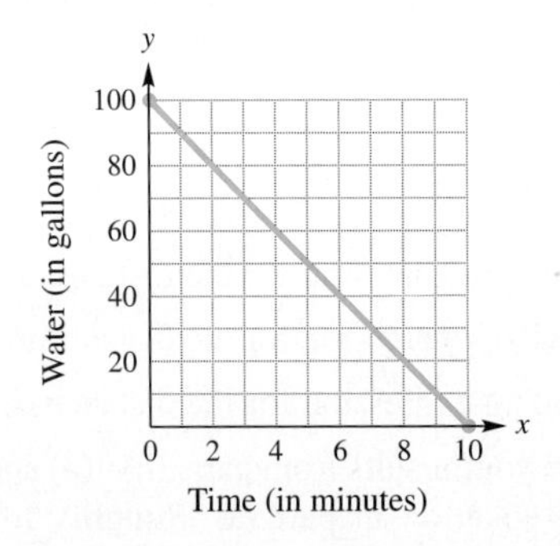


**(a)** Is water entering or leaving the tank? How much water is in the tank after 3 minutes?

**(b)** Find both the $x$- and $y$-intercepts. Interpret their meanings.

**(c)** Find the slope-intercept form of the equation of the line. Interpret the slope.

**(d)** Estimate the $x$-coordinate of the point $(x, 50)$ that lies on the line.

**57.** *Farm Pollution* In 1979 the number of farm pollution incidents reported in England and Wales was 1480. This number increased roughly at a rate of 280 per year after 1979. (*Source:* Mason, C., *Biology of Freshwater Pollution,* John Wiley and Sons, 1991.)

**(a)** Find an equation of a line $y = mx + b$ that models the information, where $y$ represents the number of pollution incidents during the year $x$. Let $x = 0$ correspond to 1979, $x = 1$ to 1980, and so on.

**(b)** Estimate the number of incidents in 1988.

**58.** *Average Wages* The average hourly wage (adjusted to 1982 dollars) was \$8.03 in 1970 and \$7.75 in 1998. (*Source:* U.S. Department of Commerce.)

**(a)** Find an equation of a line that passes through $(1970, 8.03)$ and $(1998, 7.75)$.

**(b)** Interpret the slope.

**(c)** Approximate the hourly wage in 1990. Compare it to the actual value of \$7.52.

**59.** *Temperature Scales* The table shows equivalent temperatures in degrees Celsius and degrees Fahrenheit.

| °F | −40 | 32 | 59 | 95 | 212 |
|---|---|---|---|---|---|
| °C | −40 | 0 | 15 | 35 | 100 |

**(a)** Plot the data by having the $x$-axis correspond to Fahrenheit temperature and the $y$-axis to Celsius temperature. What type of relation exists between the data?

**(b)** Find a function $C$ that uses the Fahrenheit temperature $x$ to calculate the corresponding Celsius temperature. Interpret the slope.

**(c)** If the temperature is 83°F, what is this temperature in degrees Celsius?

**60.** *Population* Asian-American populations (in millions) are shown in the table.

| Year | 1996 | 1998 | 2000 | 2002 |
|---|---|---|---|---|
| Population (in millions) | 9.7 | 10.5 | 11.2 | 12.0 |

*Source:* U.S. Bureau of the Census.

**(a)** Use the points $(1996, 9.7)$ and $(2002, 12.0)$ to find the point-slope form of a line that models the data. Let $(x_1, y_1) = (1996, 9.7)$.

**(b)** Estimate the Asian-American population in 2004 to the nearest tenth of a million.

**61.** *Civilian Labor* The table lists the percent of the female population that worked in the civilian labor force from 1960 to 1999.

| Year | 1960 | 1970 | 1980 | 1990 | 1999 |
|---|---|---|---|---|---|
| Percent (%) | 37.7 | 43.3 | 51.5 | 57.5 | 60.0 |

*Source:* U.S. Bureau of the Census.

**(a)** Find the point-slope form of the line that passes through the points $(1960, 37.7)$ and $(1999, 60.0)$. Let $(x_1, y_1) = (1960, 37.7)$.

**(b)** Use this equation to estimate the percents in 1970, 1980, and 1990 to the nearest tenth of a percent. Compare these estimates to the actual values shown in the table.

**62.** *Civilian Labor* The table lists the percent of the male population that worked in the civilian labor force from 1960 to 1999.

| Year | 1960 | 1970 | 1980 | 1990 | 1999 |
|---|---|---|---|---|---|
| Percent (%) | 83.3 | 79.7 | 77.4 | 76.1 | 74.7 |

*Source:* U.S. Bureau of the Census.

**(a)** Find a point-slope form of the line that passes through $(1960, 83.3)$ and $(1990, 76.1)$.

**(b)** Find a point-slope form of the line that passes through $(1970, 79.7)$ and $(1999, 74.7)$.

**(c)** Discuss how well each line models the data.

**63.** *Tuition and Fees* The table lists the average tuition and fees at private colleges and universities for selected years.

| Year | 1980 | 1985 | 1990 | 1995 | 1999 |
|---|---|---|---|---|---|
| Tuition and Fees (in dollars) | 3617 | 5418 | 9391 | 12,432 | 15,380 |

*Source:* The College Board.

**(a)** Find the equation of the least-squares regression line that models the data.

**(b)** Graph the data and the regression line in the same viewing window.

**(c)** Estimate tuition and fees in 1992, and compare it with the actual value of \$10,498.

**64.** *Tuition and Fees* The table on the next page lists the average tuition and fees at public colleges and universities for selected years.

| Year | 1980 | 1985 | 1990 | 1995 | 1999 |
|---|---|---|---|---|---|
| Tuition and Fees (in dollars) | 804 | 1242 | 1809 | 2860 | 3356 |

*Source:* The College Board.

**(a)** Find the equation of the least-squares regression line that models the data.

**(b)** Graph the data and the regression line in the same viewing window.

**(c)** Estimate tuition and fees in 2003.

**(d)** Discuss the accuracy of using the line to estimate tuition and fees in 1970.

**65.** *Distant Galaxies* In the late 1920s the famous observational astronomer Edwin P. Hubble (1889–1953) determined the distances to several galaxies and the velocities at which they were receding from Earth. Four galaxies with their distances in light-years and velocities in miles per second are listed in the table.

| Galaxy | Distance | Velocity |
|---|---|---|
| Virgo | 50 | 990 |
| Ursa Minor | 650 | 9,300 |
| Corona Borealis | 950 | 15,000 |
| Bootes | 1,700 | 25,000 |

*Source:* Sharov, A., and I. Novikov, *Edwin Hubble, The Discoverer of the Big Bang Universe,* Cambridge University Press, 1993.

**(a)** Let $x$ represent distance and $y$ represent velocity. Find the equation of the least-squares regression line that models the data.

**(b)** If the galaxy Hydra is receding at a speed of 37,000 miles per second, estimate its distance from Earth.

**66.** *Heights and Weights of Men* A sample of 10 adult men gave the following data on their heights and weights.

| Height, $x$ (in inches) | Weight, $y$ (in pounds) |
|---|---|
| 62 | 120 |
| 62 | 140 |
| 63 | 130 |
| 65 | 150 |
| 66 | 142 |
| 67 | 130 |
| 68 | 135 |
| 68 | 175 |
| 70 | 149 |
| 72 | 168 |

**(a)** Find the equation of the least-squares regression line.

**(b)** Using the result of part (a), predict the weight of a man whose height is 60 inches.

**(c)** What would be the predicted weight of a man whose height is 70 inches?

**(d)** Find the correlation coefficient. Interpret your answer.

*(Modeling) The data in Exercises 67 and 68 were adapted from the 1995* Information Please Almanac. *In each case, obtain the least-squares regression line and the correlation coefficient. Make a statement about the correlation.*

**67.** *Gestation Period and Life Span of Animals*

| Animal | Average Gestation or Incubation Period, $x$ (in days) | Record Life Span, $y$ (in years) |
|---|---|---|
| Cat | 63 | 26 |
| Dog | 63 | 24 |
| Duck | 28 | 15 |
| Elephant | 624 | 71 |
| Goat | 151 | 17 |
| Guinea pig | 68 | 6 |
| Hippopotamus | 240 | 49 |
| Horse | 336 | 50 |
| Lion | 108 | 29 |
| Parakeet | 18 | 12 |
| Pig | 115 | 22 |
| Rabbit | 31 | 15 |
| Sheep | 151 | 16 |

**68.** *City Sizes in the World (Population and Area)*

| Rank | City | Population, $x$ (in thousands) | Area, $y$ (in square miles) |
|---|---|---|---|
| 1 | Tokyo-Yokohama, Japan | 28,447 | 1,089 |
| 2 | Mexico City, Mexico | 23,913 | 522 |
| 3 | São Paulo, Brazil | 21,539 | 451 |
| 4 | Seoul, South Korea | 19,065 | 342 |
| 5 | New York, United States | 14,638 | 1,274 |
| 6 | Osaka-Kobe-Kyoto, Japan | 14,060 | 495 |
| 7 | Bombay, India | 13,532 | 95 |
| 8 | Calcutta, India | 12,885 | 209 |
| 9 | Rio de Janeiro, Brazil | 12,788 | 260 |
| 10 | Buenos Aires, Argentina | 12,232 | 535 |

## Reviewing Basic Concepts (Sections 1.3 and 1.4)

**1.** Write the formula for a linear function $f$ with slope 1.4 and $y$-intercept $-3.1$. Find $f(1.3)$.

**2.** Graph $f(x) = -2x + 1$ by hand. State the $x$-intercept, $y$-intercept, domain, range, and slope.

**3.** Find the slope of the line passing through $(-2, 4)$ and $(5, 6)$.

**4.** Give the equations of a vertical line and a horizontal line passing through $(-2, 10)$.

**5.** Let $f(x) = .5x - 1.4$. Use your graphing calculator to graph $f$ in the standard viewing window and to make a table of values for $f$ at $x = -3, -2, -1, \ldots, 3$.

**6.** Give the slope-intercept form of the line shown in the figure.

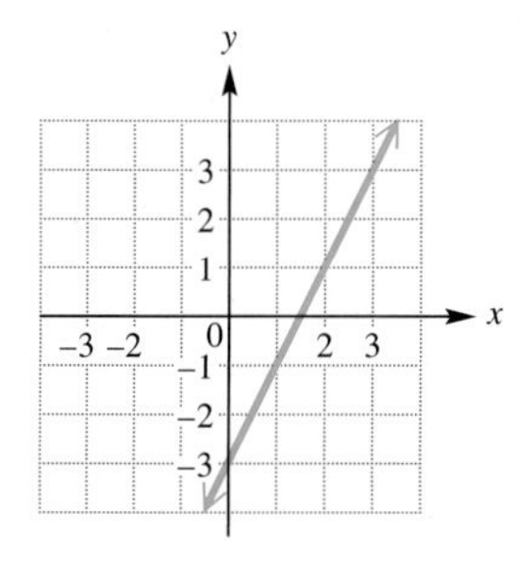


**7.** Find the slope-intercept form of the line passing through $(-2, 4)$ and $(5, 2)$.

**8.** Find the slope-intercept form of the line perpendicular to $3x - 2y = 5$, passing through $(-1, 3)$.

**9.** *Average Household Size* The table lists the average number $y$ of people per household for various years $x$.

| $x$ | 1940 | 1950 | 1960 | 1970 |
|---|---|---|---|---|
| $y$ | 3.67 | 3.37 | 3.33 | 3.14 |

| $x$ | 1980 | 1990 | 1998 |
|---|---|---|---|
| $y$ | 2.76 | 2.63 | 2.62 |

*Source:* U.S. Bureau of the Census.

**(a)** Make a scatter diagram of the data.

**(b)** Decide whether the correlation coefficient is positive, negative, or zero.

**(c)** Find a least-squares regression line that models these data. What is the correlation coefficient?

**(d)** Estimate the number of people per household in 1975, and compare it to the actual value of 2.94.

# LINEAR EQUATIONS AND INEQUALITIES

Solving Linear Equations • Graphical Approaches to Solving Linear Equations • Identities and Contradictions • Solving Linear Inequalities • Graphical Approaches to Solving Linear Inequalities • Three-Part Inequalities

## Solving Linear Equations

In this text we use two distinct approaches to solving equations. The first approach is the *analytic approach*, where paper and pencil are used to transform complicated equations into simpler ones. The second approach is the *graphical approach*, where we often support our analytic solutions by using graphs or tables. Generally the equations encountered in this text can be solved analytically. Occasionally we may encounter an equation that cannot be solved analytically but can be solved graphically. Part of becoming a good mathematics student is learning when and when not to use each approach.

An **equation** is a statement that two expressions are equal. To *solve* an equation means to find all numbers that make the equation a true statement. Such numbers are called **solutions** or **roots** of the equation. A number that is a solution of an equation is said to *satisfy* the equation, and the solutions of an equation make up its **solution set.**

In this section we concentrate on solving *linear equations in one variable*. A linear equation in one variable has one solution.

**Linear Equation in One Variable**

A **linear equation in one variable** is an equation that can be written in the form

$$\mathbf{ax + b = 0}, \quad a \neq 0.$$

One way to solve an equation is to rewrite it as a series of simpler **equivalent equations,** each of which has the same solution set as the original one. Equivalent equations are obtained by using the addition and multiplication properties of equality.

**Addition and Multiplication Properties of Equality**

For real numbers $a$, $b$, and $c$:

**$a = b$ and $a + c = b + c$ are equivalent.**
(*The same number may be added to each side of an equation without changing the solution set.*)

**If $c \neq 0$, then $a = b$ and $ac = bc$ are equivalent.**
(*Each side of an equation may be multiplied by the same nonzero number without changing the solution set.*)

Extending these two properties allows us to subtract the same number from each side of an equation and to divide each side by the same nonzero number.

**EXAMPLE 1** *Solving a Linear Equation*

Solve $10 + 3(2x - 4) = 17 - (x + 5)$.

**Solution** Use the distributive property and then collect like terms to get the following series of simpler, equivalent equations.

$$10 + 3(2x - 4) = 17 - (x + 5)$$
$$10 + 6x - 12 = 17 - x - 5 \quad \text{Distributive property}$$
$$-2 + 7x = 12 \quad \text{Add } x \text{ to each side; combine terms.}$$
$$7x = 14 \quad \text{Add 2 to each side.}$$
$$x = 2 \quad \text{Divide each side by 7.}$$

To check our answer analytically, we substitute 2 for $x$ in the original equation and see if a true statement results.

$$10 + 3(2x - 4) = 17 - (x + 5) \quad \text{Original equation}$$
$$10 + 3(2 \cdot 2 - 4) \stackrel{?}{=} 17 - (2 + 5) \quad \text{Let } x = 2.$$
$$10 + 3(4 - 4) \stackrel{?}{=} 17 - 7$$
$$10 = 10 \quad \text{True; the answer checks.}$$

The solution set is $\{2\}$.

When fractions or decimals appear in an equation, our work can be made simpler by multiplying each side by the least common denominator (LCD) of all the fractions in the equation. Examples 2 and 3 illustrate these types of equations.

**EXAMPLE 2** *Solving a Linear Equation with Fractional Coefficients*

Solve $\dfrac{x+7}{6} + \dfrac{2x-8}{2} = -4.$

**Solution** To eliminate fractions, multiply each side by the LCD, 6.

$$6\left(\frac{x+7}{6} + \frac{2x-8}{2}\right) = 6(-4)$$

$$6\left(\frac{x+7}{6}\right) + 6\left(\frac{2x-8}{2}\right) = 6(-4) \quad \text{Distributive property}$$

$$x + 7 + 3(2x - 8) = -24 \quad \text{Simplify.}$$

$$x + 7 + 6x - 24 = -24 \quad \text{Distributive property}$$

$$7x - 17 = -24 \quad \text{Combine terms.}$$

$$7x = -7 \quad \text{Add 17.}$$

$$x = -1 \quad \text{Divide by 7.}$$

An analytic check will verify that the solution set is $\{-1\}$.

**EXAMPLE 3** *Solving a Linear Equation with Decimal Coefficients*

Solve $.06x + .09(15 - x) = .07(15)$.

**Solution** Since each decimal number is given in hundredths, multiply each side of the equation by 100. (To multiply a number by 100, move the decimal point two places to the right.)

$$.06x + .09(15 - x) = .07(15)$$

$$6x + 9(15 - x) = 7(15) \quad \text{Multiply by 100.}$$

$$6x + 135 - 9x = 105 \quad \text{Distributive property}$$

$$-3x + 135 = 105 \quad \text{Combine terms.}$$

$$-3x = -30 \quad \text{Subtract 135.}$$

$$x = 10 \quad \text{Divide by } -3.$$

An analytic check will verify that the solution set is $\{10\}$.

The equations solved in Examples 1–3 each have one solution and are called **conditional equations.** Later in this section, we will see that some types of equations have no solutions or infinitely many solutions.

## Graphical Approaches to Solving Linear Equations

Since an equation always contains an equals sign, we can think of a linear equation as being in the form

$$f(x) = g(x),$$

where $f(x)$ and $g(x)$ are formulas for linear functions. The equation

$$10 + 3(2x - 4) = 17 - (x + 5)$$

from Example 1 is in this form, where $f(x) = 10 + 3(2x - 4)$ and $g(x) = 17 - (x + 5)$. To find the solution set graphically, we graph $y_1 = f(x)$ and $y_2 = g(x)$ and locate their point of intersection. In general, if $f$ and $g$ are linear functions, then their graphs are lines that intersect at a single point, no point, or infinitely many points, as illustrated in Figure 66.

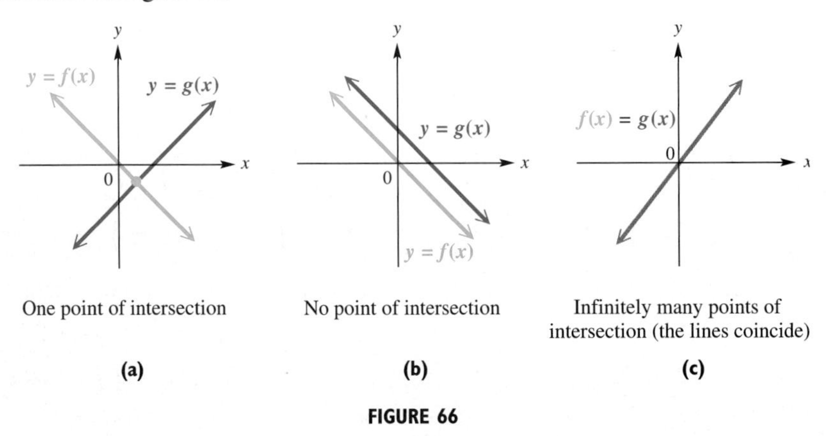


**FIGURE 66**

Figure 67(a) shows graphs of $Y_1 = 10 + 3(2X - 4)$ and $Y_2 = 17 - (X + 5)$ intersecting at the point (2, 10). The $x$-coordinate 2 is the solution of the equation from Example 1, and the $y$-coordinate 10 is the value obtained when either $Y_1$ or $Y_2$ is evaluated at $X = 2$. A table of values can also be used to support our analytic solution from Example 1. In Figure 67(b) a table of values for $Y_1$ and $Y_2$ shows that $Y_1 = Y_2 = 10$ when $X = 2$. Creating a table to find a solution to an equation gives a *numerical solution*. Locating a numerical solution involves searching a table to find an $x$-value where $Y_1 = Y_2$. A table is less practical when the solution is not an integer.

**TECHNOLOGY NOTE**

Graphing calculators often have built-in routines to find the intersection of two graphs. Sometimes this routine is located under the CALC menu. Consult the graphing calculator manual that accompanies this text or your owner's manual.

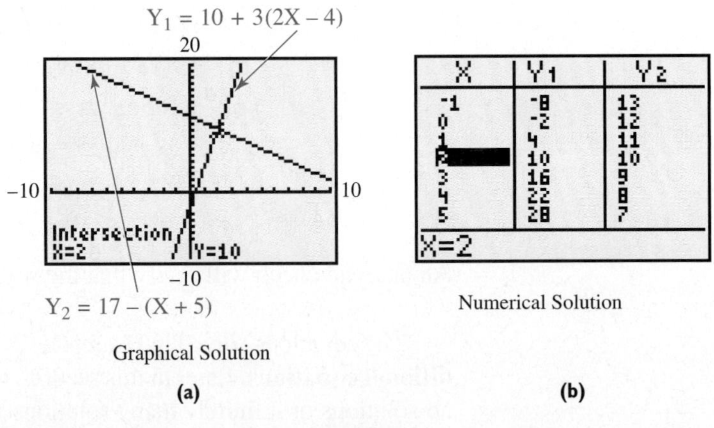


**FIGURE 67**

The graphical method shown in Figure 67(a) is called the *intersection-of-graphs method* and is summarized as follows.

### Intersection-of-Graphs Method of Graphical Solution

To solve the equation $f(x) = g(x)$ graphically, graph

$$y_1 = f(x) \quad \text{and} \quad y_2 = g(x).$$

The $x$-coordinate of any point of intersection of the two graphs is a solution of the equation.

**EXAMPLE 4** *Applying the Intersection-of-Graphs Method*

The percent share of music sales (in dollars) that compact discs (CDs) held from 1987 to 1998 can be modeled by

$$f(x) = 5.91x + 13.7.$$

During the same time period the percent share of music sales that cassette tapes held can be modeled by

$$g(x) = -4.71x + 64.7.$$

In these formulas $x = 0$ corresponds to 1987, $x = 1$ to 1988, and so on. Use the intersection-of-graphs method to estimate the year when sales of CDs equaled sales of cassettes. (*Source:* Recording Industry Association of America.)

**Solution** Solve the linear equation $f(x) = g(x)$ or, equivalently,

$$5.91x + 13.7 = -4.71x + 64.7.$$

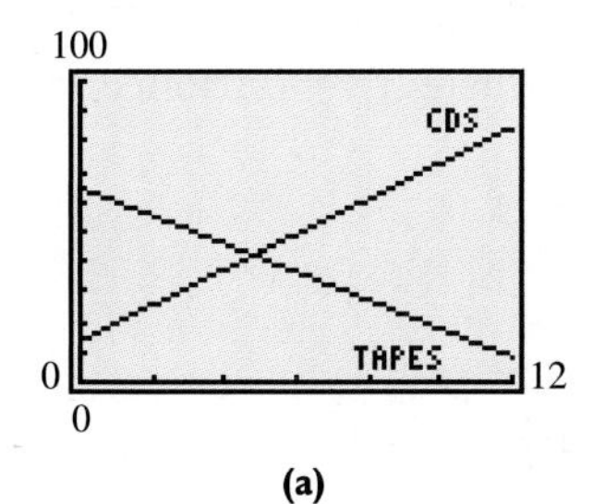

(a)

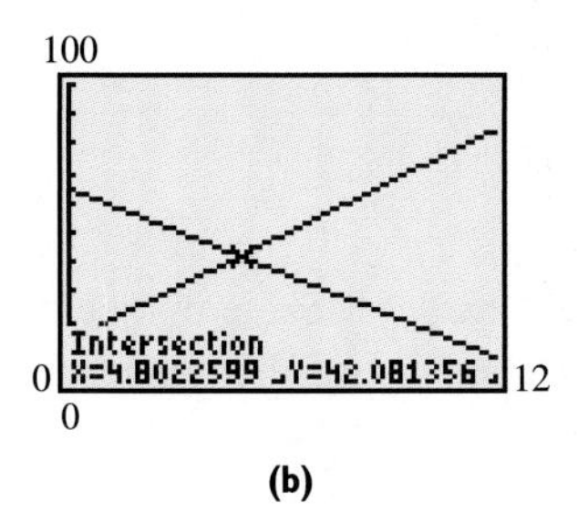

(b)

FIGURE 68

Graph $y_1 = 5.91x + 13.7$ and $y_2 = -4.71x + 64.7$, as shown in Figure 68(a). In Figure 68(b), their graphs intersect near the point $(4.8, 42.1)$. Since $x = 0$ corresponds to 1987 and $1987 + 4.8 \approx 1992$, it follows that in 1992 sales of CDs and cassette tapes were approximately equal. Both shared about 42.1% of the sales in 1992.

There is a second graphical method of solving equations. Suppose that we once again wish to solve

$$f(x) = g(x).$$

By subtracting $g(x)$ from each side, we obtain

$$f(x) - g(x) = 0.$$

Notice that $f(x) - g(x)$ defines a function. Let us call it $F(x)$. Then we only need to solve

$$F(x) = 0$$

to obtain the solution set of the original equation. In Section 1.3 we learned that any number that satisfies this equation is an $x$-intercept of the graph of $y = F(x)$. It is also called a *zero* of $F$. Using this idea, we now state another method of solving an equation graphically.

### $x$-Intercept Method of Graphical Solution

To solve the equation $f(x) = g(x)$ graphically, graph

$$y = f(x) - g(x) = F(x).$$

Any $x$-intercept of the graph of $y = F(x)$ (or zero of the function $F$) is a solution of the equation.

TECHNOLOGY NOTE

Graphing calculators often have built-in programs to locate $x$-intercepts (roots, zeros).

The words *root*, *solution*, and *zero* all refer to the same basic concept. When solving equations remember this: *The real solutions (or roots or zeros) of an equation of the form $f(x) = 0$ correspond to the $x$-intercepts of the graph of $y = f(x)$.*

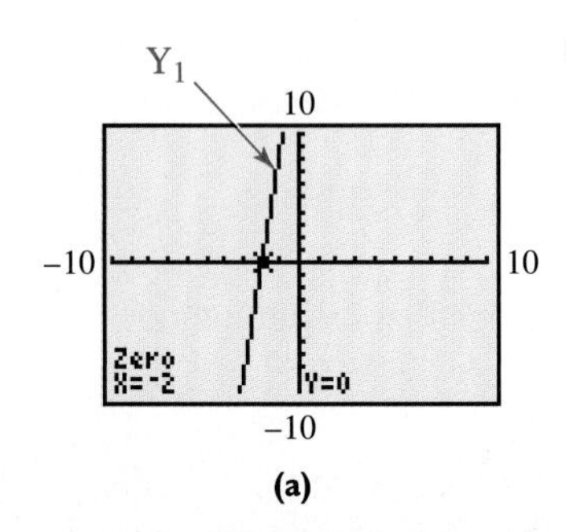


(a)

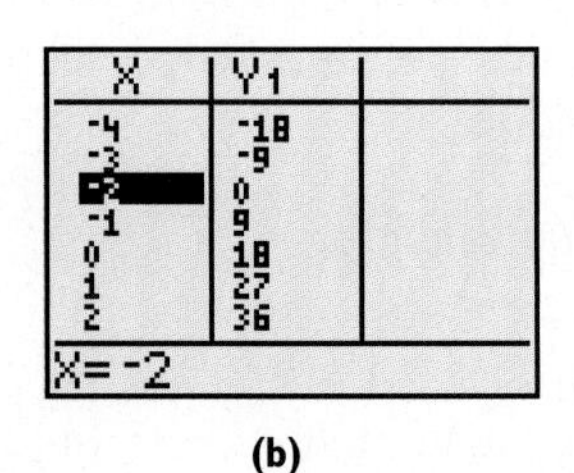

| X | Y1 |
|---|---|
| -4 | -18 |
| -3 | -9 |
| -2 | 0 |
| -1 | 9 |
| 0 | 18 |
| 1 | 27 |
| 2 | 36 |

X=-2

(b)

FIGURE 69

**EXAMPLE 5** *Using the x-Intercept Method*

It can be shown that the solution set of

$$6x - 4(3 - 2x) = 5(x - 4) - 10$$

is $\{-2\}$. Use the $x$-intercept method to solve this equation. Then use a table to obtain a numerical solution.

**Solution** Begin by writing the equation in the form $F(x) = 0$, or

$$6x - 4(3 - 2x) - (5(x - 4) - 10) = 0.$$

Graph $Y_1 = 6X - 4(3 - 2X) - (5(X - 4) - 10)$, and locate any $x$-intercepts or zeros of $Y_1$. Figure 69(a) shows $x$-intercept $-2$, confirming the solution given in the statement of the problem.

Figure 69(b) shows a table of values for the equation $Y_1 = 6X - 4(3 - 2X) - (5(X - 4) - 10)$. To find a solution, look for an $x$-value where $Y_1 = 0$. Notice that $Y_1 = 0$ when $X = -2$.

**FOR DISCUSSION**

In Example 5 we solved a linear equation by letting

$$f(x) = 6x - 4(3 - 2x) \quad \text{and} \quad g(x) = 5(x - 4) - 10$$

and then graphing $F(x) = f(x) - g(x)$. This time, rather than graphing

$$y_1 = f(x) - g(x),$$

graph

$$y_2 = g(x) - f(x).$$

Observe the graph of $y_2$. Does it have the same $x$-intercept as $y_1$? Make a conjecture concerning the order in which $f$ and $g$ are subtracted when using the $x$-intercept method.

**CAUTION** When using the $x$-intercept method, as in Example 5, remember to put parentheses around the expression that is being subtracted. Notice how we used parentheses around $5(x - 4) - 10$ when we determined the expression for $y_1$.

## Identities and Contradictions

Every equation solved thus far in this section has been a conditional equation with one solution. However, since the graphs of two linear functions may not intersect or may coincide (as shown in Figures 66(b) and (c)), there are two other situations that may occur when solving equations.

A **contradiction** is an equation that has no solution. For example, no values for $x$ satisfy the equation $x = x + 3$, because there is no real number equal to itself plus 3. The equation is a contradiction. If we try to solve this equation analytically by subtracting $x$ from each side, we obtain $0 = 3$, a false statement. If we try to solve $x = x + 3$ graphically by letting $y_1 = x$ and $y_2 = x + 3$, we obtain two parallel lines, as shown in Figure 70. Since the lines do not intersect, the solution set is the **empty set** or **null set,** denoted $\emptyset$.

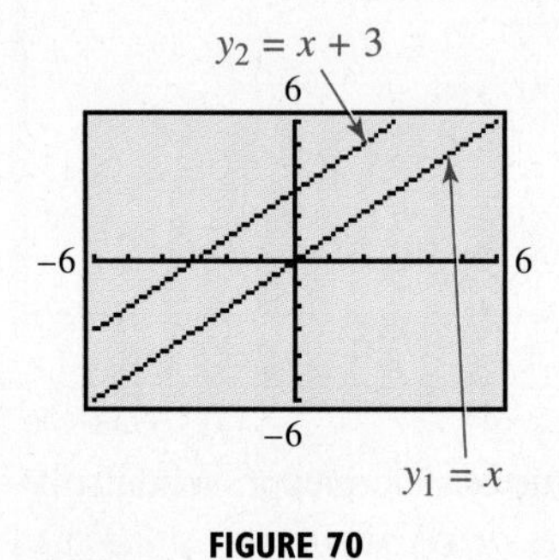


FIGURE 70

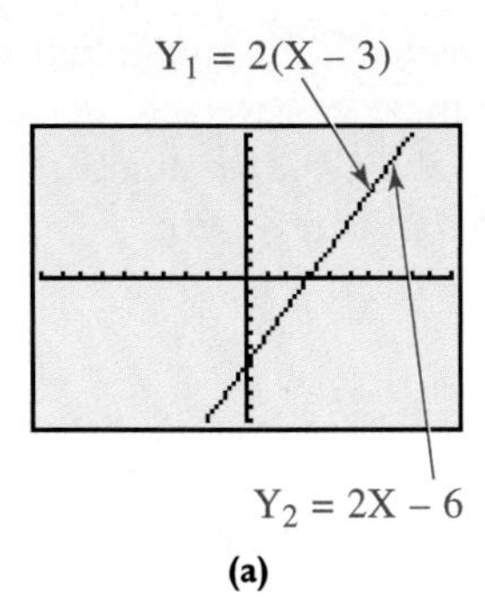


(a)

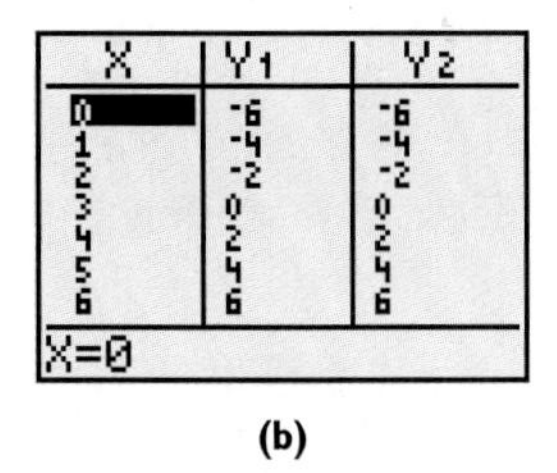

| X | Y1 | Y2 |
|---|---|---|
| 0 | -6 | -6 |
| 1 | -4 | -4 |
| 2 | -2 | -2 |
| 3 | 0 | 0 |
| 4 | 2 | 2 |
| 5 | 4 | 4 |
| 6 | 6 | 6 |

X=0

(b)

**FIGURE 71**

An **identity** is an equation that is true for all values in the domain of its variables. For example, the equation $2(x - 3) = 2x - 6$ is true for all real numbers. If we try to solve this equation analytically, we obtain

$$2(x - 3) = 2x - 6 \quad \text{Given equation}$$
$$2x - 6 = 2x - 6 \quad \text{Distributive property}$$
$$2x = 2x \quad \text{Add 6.}$$
$$0 = 0. \quad \text{Subtract } 2x.$$

The equation $0 = 0$ is always true, indicating that the given equation is an identity. The solution set is all real numbers, or $(-\infty, \infty)$. If we try to solve the equation graphically by letting $Y_1 = 2(X - 3)$ and $Y_2 = 2X - 6$, we obtain two identical lines, shown in Figure 71(a). Since these two lines coincide, rather than intersect at a point, the solution set is all real numbers. The table in Figure 71(b) suggests that $Y_1 = Y_2$ for all $x$-values and that the given equation is an identity.

NOTE Contradictions and identities are *not* linear equations because they cannot be written as $ax + b = 0$ with $a \neq 0$. Contradictions and identities simplify to equations that are either always false or always true. Linear equations always have exactly one solution.

## Solving Linear Inequalities

An equation says that two expressions are equal; an **inequality** says that one expression is greater than, greater than or equal to, less than, or less than or equal to another. As with equations, a value of the variable for which the inequality is true is a solution of the inequality, and the set of all such solutions is the solution set of the inequality. Two inequalities with the same solution set are **equivalent inequalities.**

Inequalities are solved with the following properties of inequality.

**Addition and Multiplication Properties of Inequality**

For real numbers $a$, $b$, and $c$:

(a) **$a < b$ and $a + c < b + c$ are equivalent.**
*(The same number may be added to each side of an inequality without changing the solution set.)*

(b) **If $c > 0$, then $a < b$ and $ac < bc$ are equivalent.**
*(Each side of an inequality may be multiplied by the same positive number without changing the solution set.)*

(c) **If $c < 0$, then $a < b$ and $ac > bc$ are equivalent.**
*(Each side of an inequality may be multiplied by the same negative number without changing the solution set, as long as the direction of the inequality symbol is reversed.)*

Similar properties exist for $>$, $\leq$, and $\geq$.

NOTE Because division is defined in terms of multiplication, the word "multiplied" may be replaced by "divided" in parts (b) and (c) of the properties of inequality. Similarly, in part (a) the words "added to" may be replaced by "subtracted from."

Pay careful attention to part (c); if each side of an inequality is multiplied by a negative number, the direction of the inequality symbol must be reversed. For example, starting with $-3 < 5$ and multiplying each side by the *negative* number $-2$ gives a true result only if the direction of the inequality symbol is reversed:

$$\begin{aligned} -3 &< 5 \\ -3(-2) &> 5(-2) \\ 6 &> -10. \end{aligned}$$

A similar situation exists when dividing each side by a negative number.

A *linear inequality in one variable* is defined as follows.

**Linear Inequality in One Variable**

A **linear inequality in one variable** is an inequality that can be written in one of the following forms, where $a \neq 0$:

$$ax + b > 0 \qquad ax + b < 0 \qquad ax + b \geq 0 \qquad ax + b \leq 0.$$

We solve a linear inequality analytically using essentially the same steps as those used to solve a linear equation. The solution set of a linear inequality is typically an interval of the real number line.

**EXAMPLE 6** *Solving a Linear Inequality*

Solve $3x - 2(2x + 4) \leq 2x + 1$.

**Solution**

$$\begin{aligned} 3x - 2(2x + 4) &\leq 2x + 1 && \text{Given inequality} \\ 3x - 4x - 8 &\leq 2x + 1 && \text{Distributive property} \\ -x - 8 &\leq 2x + 1 && \text{Combine terms.} \\ -3x &\leq 9 && \text{Subtract } 2x\text{; add 8.} \\ x &\geq -3 && \text{Divide by } -3\text{; reverse the direction of the inequality symbol.} \end{aligned}$$

In interval notation the solution set is $[-3, \infty)$.

**FOR DISCUSSION**

Solve each equation or inequality.

$$2(x - 3) + x = 9,$$
$$2(x - 3) + x < 9,$$
$$2(x - 3) + x > 9$$

How can the solution of the equation be used to help solve the two inequalities?

If a linear inequality involves fractions or decimals as coefficients, we use the same procedure as described for linear equations: multiply each side by the least common denominator (which, in the case of decimals, will be a power of 10).

**EXAMPLE 7** *Solving Linear Inequalities*

Solve each inequality.

**(a)** $2x - 3 < \dfrac{x + 2}{-3}$ **(b)** $-3(4x - 4) \geq 4 - (x - 1)$

**Solution**

**(a)** Multiply each side by $-3$ to clear the fraction. Remember to reverse the inequality symbol when multiplying by a negative number.

| | |
|---|---|
| $2x - 3 < \dfrac{x+2}{-3}$ | Given inequality |
| $-6x + 9 > x + 2$ | Multiply by $-3$; reverse the inequality symbol. |
| $9 > 7x + 2$ | Add $6x$. |
| $7 > 7x$ | Subtract 2. |
| $1 > x$ | Divide by 7. |
| $x < 1$ | Rewrite. |

The solution set is $(-\infty, 1)$.

| | |
|---|---|
| **(b)** $-3(4x - 4) \geq 4 - (x - 1)$ | Given inequality |
| $-12x + 12 \geq 4 - x + 1$ | Distributive property |
| $-12x + 12 \geq -x + 5$ | Simplify. |
| $-11x \geq -7$ | Add $x$; subtract 12. |
| $x \leq \dfrac{7}{11}$ | Divide by $-11$; reverse the inequality symbol. |

The solution set is $\left(-\infty, \frac{7}{11}\right]$.

## Graphical Approaches to Solving Linear Inequalities

The two methods of graphical solution of linear equations can be extended to solutions of linear inequalities. Figure 72 shows the distances of two cars from Chicago, Illinois, $x$ hours after traveling in the same direction on a freeway. The distance for Car 1 is denoted $y_1$, and the distance for Car 2 is denoted $y_2$.

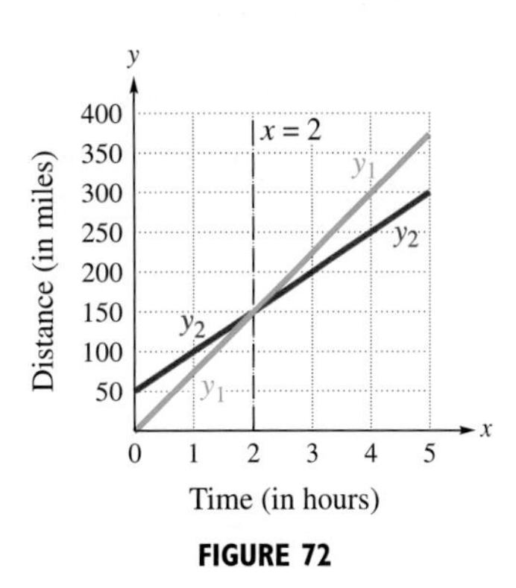


**FIGURE 72**

After $x = 2$ hours, $y_1 = y_2$ and both cars are 150 miles from Chicago. To the left of the dashed vertical line $x = 2$, the graph of $y_1$ is below the graph of $y_2$, so Car 1 is closer to Chicago than Car 2. Thus,

$$y_1 < y_2 \quad \text{when} \quad x < 2.$$

To the right of the dashed vertical line $x = 2$, the graph of $y_1$ is above the graph of $y_2$, so Car 1 is farther from Chicago than Car 2. Thus,

$$y_1 > y_2 \quad \text{when} \quad x > 2.$$

This discussion leads to an extension of the intersection-of-graphs method to linear inequalities.

**Intersection-of-Graphs Method of Solution of a Linear Inequality**

Suppose that $f$ and $g$ are linear functions. The solution set of $f(x) > g(x)$ is the set of all real numbers $x$ such that the graph of $f$ is **above** the graph of $g$. The solution set of $f(x) < g(x)$ is the set of all real numbers $x$ such that the graph of $f$ is **below** the graph of $g$.

If an inequality involves one of the symbols $\geq$ or $\leq$, the same method applies, with the solution of the corresponding equation included in the solution set.

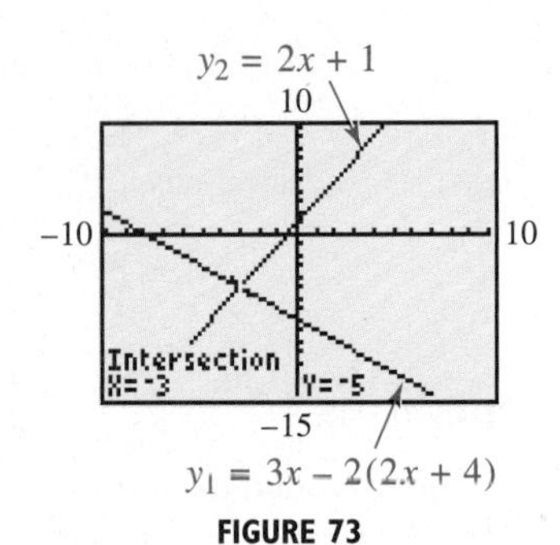

FIGURE 73

### EXAMPLE 8 *Using the Intersection-of-Graphs Method*

The solution set of the inequality

$$3x - 2(2x + 4) \leq 2x + 1,$$

solved in Example 6, is $[-3, \infty)$. Solve this inequality graphically.

**Solution** Graph

$$y_1 = 3x - 2(2x + 4) \quad \text{and} \quad y_2 = 2x + 1.$$

Figure 73 indicates that the point of intersection of the two lines is $(-3, -5)$. Because the graph of $y_1$ is *below* the graph of $y_2$ when $x$ is *greater than* $-3$, the solution set to $y_1 \leq y_2$ is $[-3, \infty)$, agreeing with our analytic solution.

**TECHNOLOGY NOTE**

When solving linear inequalities graphically, the calculator will not determine whether the endpoint of the interval is included or excluded. This must be done by looking at the inequality symbol in the given inequality.

### EXAMPLE 9 *Applying the Intersection-of-Graphs Method*

When the air temperature reaches the dew point, fog may form. This phenomenon also causes clouds to form at higher altitudes. Both the air temperature and the dew point decrease at a constant rate as the altitude above ground level increases. If the ground-level temperature and dew point are $T_0$ and $D_0$, the air temperature can be approximated by

$$T(x) = T_0 - 29x$$

and the dew point by

$$D(x) = D_0 - 5.8x$$

at an altitude of $x$ miles. If $T_0 = 75°\text{F}$ and $D_0 = 55°\text{F}$, determine the altitudes where clouds will not form.

**Solution** Since $T_0 = 75$ and $D_0 = 55$, let $T(x) = 75 - 29x$ and $D(x) = 55 - 5.8x$. Graph $y_1 = 75 - 29x$ and $y_2 = 55 - 5.8x$, as shown in Figure 74. The graphs intersect near $(.86, 50)$. This means that the air temperature and dew point are both 50°F at about .86 mile above ground level. Clouds will not form below this altitude, that is, or when the graph of $y_1$ is above the graph of $y_2$. The solution set is $[0, .86)$, where the endpoint .86 is approximate.

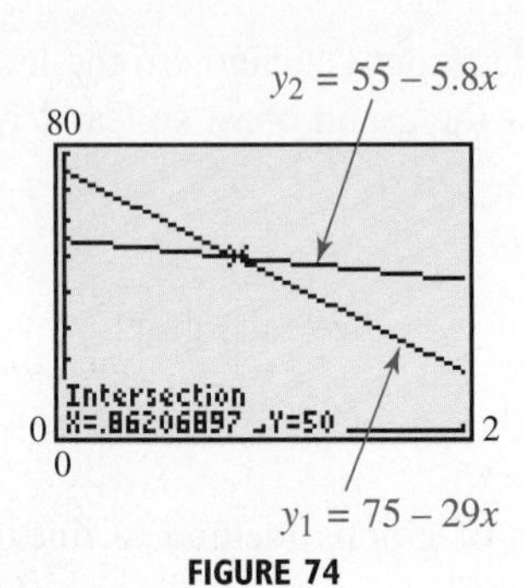

FIGURE 74

In Example 9, the endpoint, .86 mile, is an approximation to the nearest hundredth. In solving inequalities graphically, the appropriate symbol for the endpoint, a parenthesis or a bracket, may not actually be valid because of rounding. As a result, we state the following agreement to be used throughout this text.

**Agreement on Inclusion or Exclusion of Endpoints for Approximations**

When an approximation is used for an endpoint in specifying an interval, we continue to use parentheses in specifying inequalities involving $<$ or $>$, and square brackets in specifying inequalities involving $\leq$ or $\geq$.

The $x$-intercept method can also be used to solve inequalities. For example, to solve $f(x) > g(x)$, we can rewrite the inequality as $f(x) - g(x) > 0$ or $F(x) > 0$, where $F(x) = f(x) - g(x)$. All solutions to the given inequality correspond to the $x$-values where the graph of $y = F(x)$ is above the $x$-axis. This technique is summarized as follows.

TECHNOLOGY NOTE

If two functions defined by $Y_1$ and $Y_2$ are already entered into your calculator, you can enter $Y_3$ as $Y_2 - Y_1$. Then, if you direct the calculator to graph $Y_3$ only, you can solve the equation $Y_1 = Y_2$ by finding the x-intercept of $Y_3$. Consult the graphing calculator manual that accompanies this text or your owner's manual to see how this is accomplished.

***x*-Intercept Method of Solution of a Linear Inequality**

The solution set of $\boldsymbol{F(x) > 0}$ is the set of all real numbers $x$ such that the graph of $F$ is **above** the $x$-axis. The solution set of $\boldsymbol{F(x) < 0}$ is the set of all real numbers $x$ such that the graph of $F$ is **below** the $x$-axis.

Figure 75 illustrates this discussion and summarizes the solution sets for the appropriate inequalities.

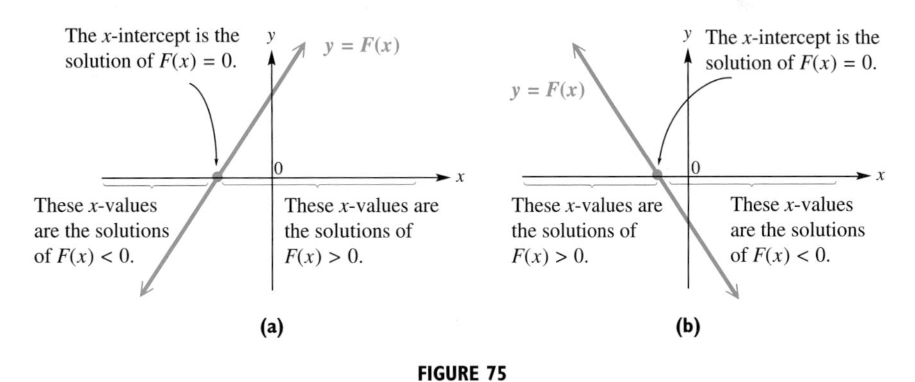


FIGURE 75

**EXAMPLE 10** *Using the x-Intercept Method*

Solve each inequality using the $x$-intercept method.

**(a)** $-2(3x + 1) < 4(x + 2)$ **(b)** $-2(3x + 1) \leq 4(x + 2)$

**(c)** $-2(3x + 1) > 4(x + 2)$ **(d)** $-2(3x + 1) \geq 4(x + 2)$

**Solution**

**(a)** The inequality can be written as $-2(3x + 1) - 4(x + 2) < 0$. Graph the left side of the inequality, as shown in Figure 76. The graph is below the $x$-axis when $x > -1$, so the solution set is $(-1, \infty)$.

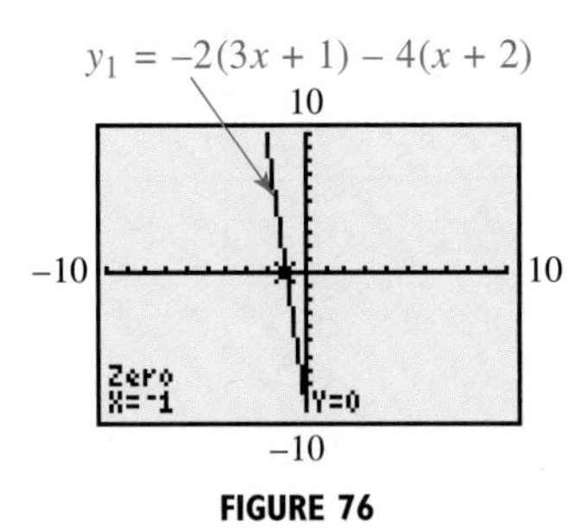


FIGURE 76

**(b)** The solution set is $[-1, \infty)$ because the endpoint at $x = -1$ is included as part of equality.

**(c)** The graph in Figure 76 on the previous page is above the $x$-axis when $x < -1$, so the solution set is $(-\infty, -1)$.

**(d)** The solution set is $(-\infty, -1]$ because the endpoint at $x = -1$ is included as part of equality.

Notice that the same graph was used to solve all four inequalities.

## Three-Part Inequalities

Sometimes a mathematical expression is limited to an interval of values. For example, suppose that $2x + 1$ must satisfy both

$$2x + 1 \geq -3 \quad \text{and} \quad 2x + 1 \leq 2.$$

These two inequalities can be combined into the **three-part inequality**

$$-3 \leq 2x + 1 \leq 2.$$

Three-part inequalities occur in manufacturing, where error tolerances are often maintained. Suppose that aluminum cans are to be constructed with radius 1.4 inches. However, because the manufacturing process is not exact, the radius $r$ actually varies from 1.38 inches to 1.42 inches. We can express possible values for $r$ by using the three-part inequality

$$1.38 \leq r \leq 1.42.$$

$C = 2\pi r$

See the figure. Then the circumference $(C = 2\pi r)$ of the can varies between $2\pi(1.38) \approx 8.67$ inches and $2\pi(1.42) \approx 8.92$ inches. This result is given (approximately) by

$$8.67 \leq C \leq 8.92.$$

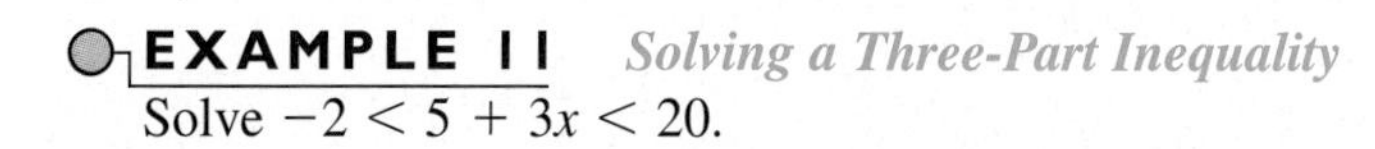

**EXAMPLE 11** *Solving a Three-Part Inequality*

Solve $-2 < 5 + 3x < 20$.

### Analytic Solution

Work with all three parts at the same time.

$$-2 < 5 + 3x < 20$$

$$-7 < 3x < 15 \qquad \text{Subtract 5.}$$

$$-\frac{7}{3} < x < 5 \qquad \text{Divide by 3.}$$

The open interval $\left(-\frac{7}{3}, 5\right)$ is the solution set. Notice that an open interval is used because equality is not included in the three-part inequality.

### Graphing Calculator Solution

Graph

$$y_1 = -2, \quad y_2 = 5 + 3x, \quad \text{and} \quad y_3 = 20,$$

as shown in Figure 77. The $x$-values of the points of intersection are $-\frac{7}{3} = -2.\overline{3}$, and 5, confirming that our analytic work is correct. Notice how the slanted line, $y_2$, lies *between* the graphs of $y_1 = -2$ and $y_3 = 20$ for $x$-values between $-\frac{7}{3}$ and 5.

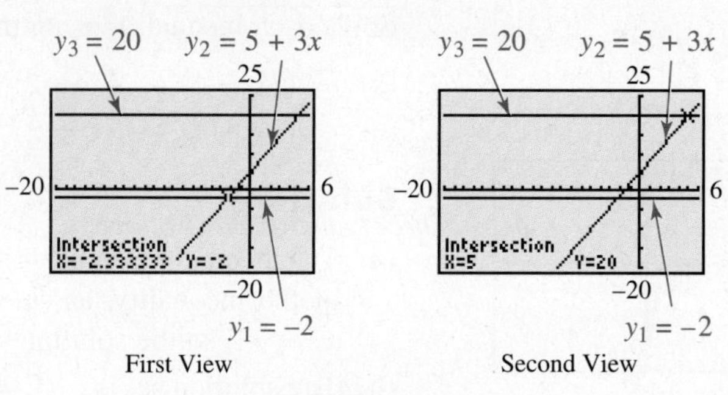


**FIGURE 77**

## 1.5 EXERCISES

*Solve the equation $f(x) = 0$ for the given $f(x)$. That is, find the zero of the function f.*

**1.** $f(x) = -3x - 12$
**2.** $f(x) = 5x - 30$
**3.** $f(x) = 5x$
**4.** $f(x) = -2x$
**5.** $f(x) = 2(3x - 5) + 8(4x + 7)$
**6.** $f(x) = -4(2x - 3) + 8(2x + 1)$
**7.** $f(x) = 3x + 6(x - 4)$
**8.** $f(x) = -8x + .5(2x + 8)$
**9.** $f(x) = 1.5x + 2(x - 3) + 5.5(x + 9)$

**10.** *Concept Check* If $c$ is a zero of the linear function defined by $f(x) = mx + b$, $m \neq 0$, then the point at which the graph intersects the $x$-axis has coordinates (______, ______).

*In Exercises 11 and 12, two linear functions, $y_1$ and $y_2$, are graphed with their point of intersection displayed. Give the solution set of $y_1 = y_2$.*

**11.**

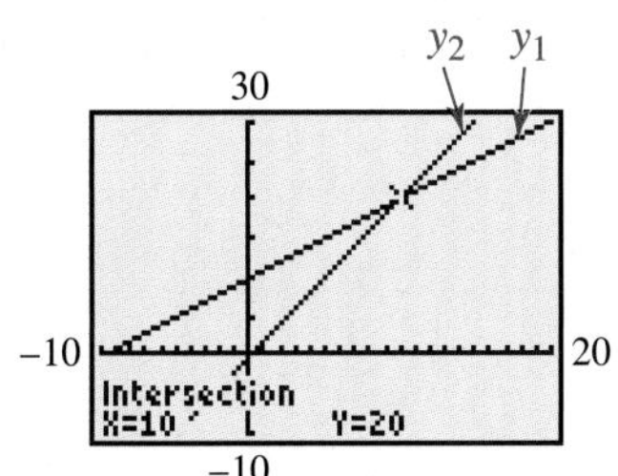


**12.**

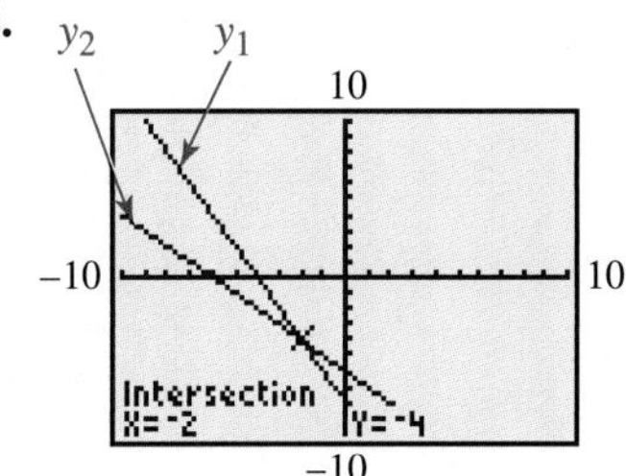


*In Exercises 13 and 14, use the graph of $y = y_1 - y_2$ to solve the equation $y_1 = y_2$, where $y_1$ and $y_2$ represent linear functions.*

**13.**

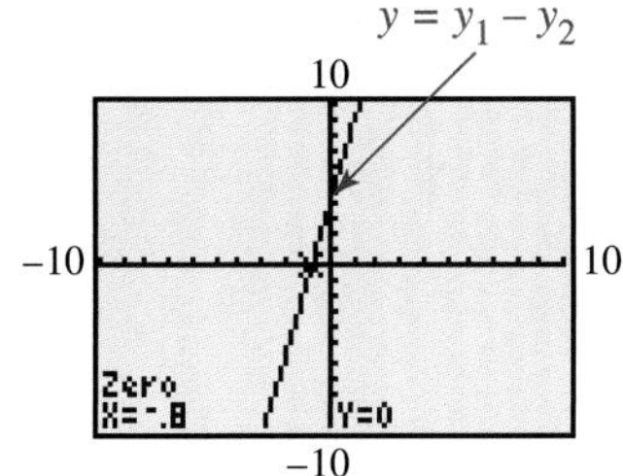


**14.**

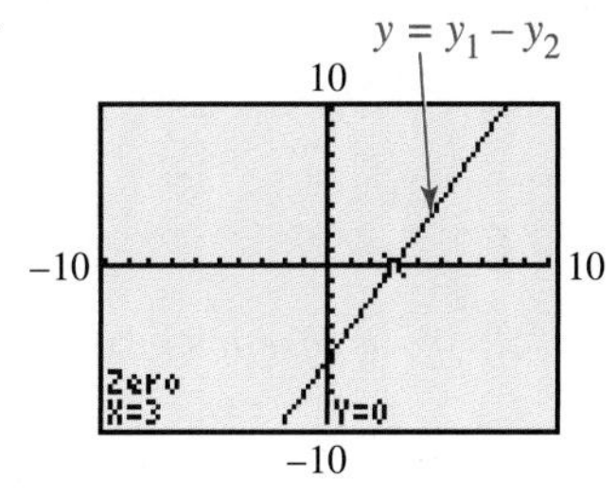


**15.** Interpret the $y$-value shown at the bottom of the screen in Exercise 11.

**16.** If you were asked to solve

$$2x + 3 = 4x - 12$$

by the $x$-intercept method, why would you *not* get the correct answer by graphing $y_1 = 2x + 3 - 4x - 12$?

**17.** *Concept Check* If the $x$-intercept method leads to the graph of a horizontal line above or below the $x$-axis, what is the solution set of the equation? What special name is given to this kind of equation?

**18.** *Concept Check* If the $x$-intercept method leads to a horizontal line that coincides with the $x$-axis, what is the solution set of the equation? What special name is given to this kind of equation?

*Solve each equation analytically. Check it analytically by direct substitution, and then support your solution graphically.*

**19.** $2x - 5 = x + 7$
**20.** $9x - 17 = 2x + 4$
**21.** $.01x + 3.1 = 2.03x - 2.96$
**22.** $.04x + 2.1 = .02x + 1.92$
**23.** $-(x + 5) - (2 + 5x) + 8x = 3x - 5$
**24.** $-(8 + 3x) + 5 = 2x + 3$
**25.** $\dfrac{2x + 1}{3} + \dfrac{x - 1}{4} = \dfrac{13}{2}$
**26.** $\dfrac{x - 2}{4} + \dfrac{x + 1}{2} = 1$

**27.** $.40x + .60(100 - x) = .45(100)$

**28.** $1.30x + .90(.50 - x) = 1.00(50)$

**29.** $2[x - (4 + 2x) + 3] = 2x + 2$

**30.** $6[x - (2 - 3x) + 1] = 4x - 6$

**31.** $\frac{5}{6}x - 2x + \frac{1}{3} = \frac{1}{3}$

**32.** $\frac{3}{4} + \frac{1}{5}x - \frac{1}{2} = \frac{4}{5}x$

**33.** $5x - (8 - x) = 2[-4 - (3 + 5x - 13)]$

**34.** $-[x - (4x + 2)] = 2 + (2x + 7)$

*Each table shows selected ordered pairs for two linear functions* $Y_1$ *and* $Y_2$. *Use the table to solve the given equation.*

**35.** $Y_1 = Y_2$

| X | Y1 | Y2 |
|---|---|---|
| 0 | 4 | 16 |
| 1 | 5 | 14 |
| 2 | 6 | 12 |
| 3 | 7 | 10 |
| 4 | 8 | 8 |
| 5 | 9 | 6 |
| 6 | 10 | 4 |

X=0

**36.** $Y_1 = Y_2$

| X | Y1 | Y2 |
|---|---|---|
| -9 | -35 | -23 |
| -8 | -32 | -22 |
| -7 | -29 | -21 |
| -6 | -26 | -20 |
| -5 | -23 | -19 |
| -4 | -20 | -18 |
| -3 | -17 | -17 |

X= -9

**37.** $Y_1 - Y_2 = 0$

| X | Y1 | Y2 |
|---|---|---|
| 0 | 0 | 3 |
| .5 | 1.5 | 3.5 |
| 1 | 3 | 4 |
| 1.5 | 4.5 | 4.5 |
| 2 | 6 | 5 |
| 2.5 | 7.5 | 5.5 |
| 3 | 9 | 6 |

X=0

**38.** $Y_1 - Y_2 = 0$

| X | Y1 | Y2 |
|---|---|---|
| 5.5 | 8 | 8 |
| 6 | 9 | 8.5 |
| 6.5 | 10 | 9 |
| 7 | 11 | 9.5 |
| 7.5 | 12 | 10 |
| 8 | 13 | 10.5 |
| 8.5 | 14 | 11 |

X=5.5

*Use the intersection-of-graphs method to approximate each solution to the nearest hundredth.*

**39.** $4(.23x + \sqrt{5}) = \sqrt{2}x + 1$

**40.** $9(-.84x + \sqrt{17}) = \sqrt{6}x - 4$

**41.** $2\pi x + \sqrt[3]{4} = .5\pi x - \sqrt{28}$

**42.** $3\pi x - \sqrt[4]{3} = .75\pi x + \sqrt{19}$

**43.** $.23(\sqrt{3} + 4x) - .82(\pi x + 2.3) = 5$

**44.** $-.15(6 + \sqrt{2}x) + 1.4(2\pi x - 6.1) = 10$

*Determine analytically whether each equation is an identity or a contradiction. Use a graph or table to support your answer.*

**45.** $5x + 5 = 5(x + 3) - 10$

**46.** $5 - 4x = 5x - (9 + 9x)$

**47.** $6(2x + 1) = 4x + 8\left(x + \frac{3}{4}\right)$

**48.** $3(x + 2) - 5(x + 2) = -2x - 4$

**49.** $-4[6 - (-2 + 3x)] = 21 + 12x$

**50.** $-3[-5 - (-9 + 2x)] = 2(3x - 1)$

*Use the graph to the right to solve each equation or inequality.*

**51.** $f(x) = g(x)$

**52.** $f(x) > g(x)$

**53.** $f(x) < g(x)$

**54.** $g(x) - f(x) \geq 0$

**55.** $y_1 - y_2 \geq 0$

**56.** $y_2 > y_1$

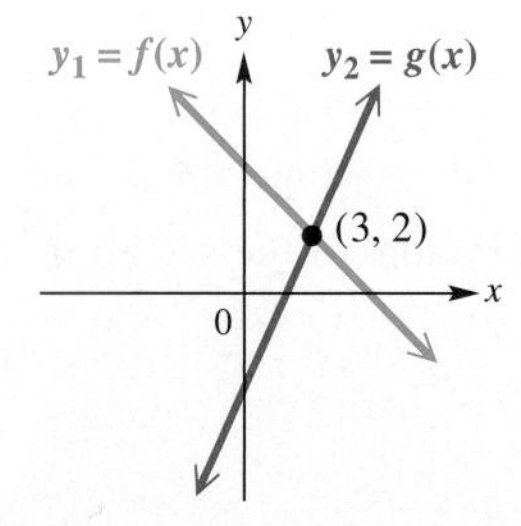


*Refer to the graph of the linear function defined by* $y = f(x)$ *to solve each inequality. Express solution sets in interval notation.*

**57.** **(a)** $f(x) > 0$
**(b)** $f(x) < 0$
**(c)** $f(x) \geq 0$
**(d)** $f(x) \leq 0$

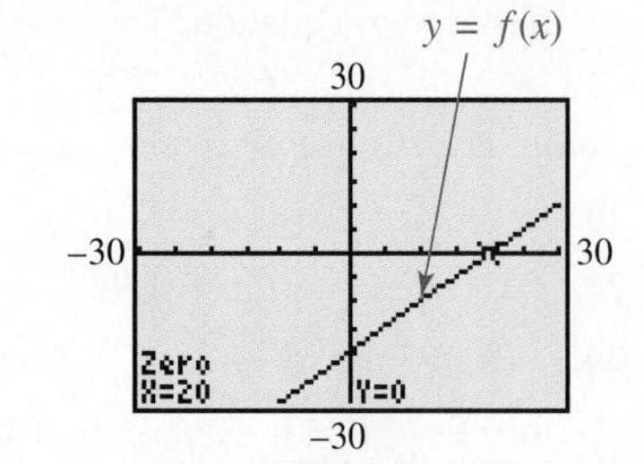


**58.** **(a)** $f(x) < 0$
**(b)** $f(x) \leq 0$
**(c)** $f(x) \geq 0$
**(d)** $f(x) > 0$

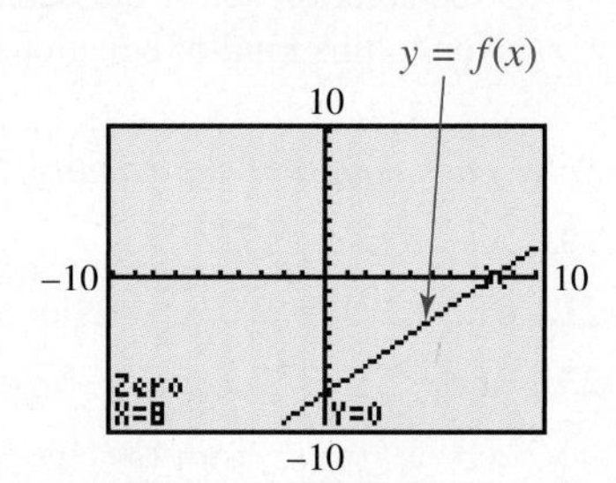

*Concept Check* *In Exercises 59 and 60, f and g are linear functions.*

**59.** If the solution set of $f(x) \geq g(x)$ is $[4, \infty)$, what is the solution set of each equation or inequality?

**(a)** $f(x) = g(x)$

**(b)** $f(x) > g(x)$

**(c)** $f(x) < g(x)$

**60.** If the solution set of $f(x) < g(x)$ is $(-\infty, 3)$, what is the solution set of each equation or inequality?

**(a)** $f(x) = g(x)$

**(b)** $f(x) \geq g(x)$

**(c)** $f(x) \leq g(x)$

*Solve each inequality analytically, writing the solution set in interval notation. Support your answer graphically. (Hint: Once part (a) is done, the answer to part (b) follows from the answer to part (a).)*

**61. (a)** $9 - (x + 1) < 0$
**(b)** $9 - (x + 1) \geq 0$

**62. (a)** $5 + 3(1 - x) \geq 0$
**(b)** $5 + 3(1 - x) < 0$

**63. (a)** $2x - 3 > x + 2$
**(b)** $2x - 3 \leq x + 2$

**64. (a)** $5 - 3x \leq 7 + x$
**(b)** $5 - 3x > 7 + x$

**65. (a)** $10x + 5 - 7x \geq 8(x + 2) + 4$
**(b)** $10x + 5 - 7x < 8(x + 2) + 4$

**66. (a)** $6x + 2 + 10x > -2(2x + 4) + 10$
**(b)** $6x + 2 + 10x \leq -2(2x + 4) + 10$

**67. (a)** $x + 2(-x + 4) - 3(x + 5) < -4$
**(b)** $x + 2(-x + 4) - 3(x + 5) \geq -4$

**68. (a)** $-11x - (6x - 4) + 5 - 3x \leq 1$
**(b)** $-11x - (6x - 4) + 5 - 3x > 1$

*Solve each inequality. Write the solution set in interval notation.*

**69.** $\frac{1}{3}x - \frac{1}{5}x \leq 2$

**70.** $\frac{3x}{2} + \frac{4x}{7} \geq -5$

**71.** $\frac{x - 2}{2} - \frac{x + 6}{3} > -4$

**72.** $\frac{2x + 3}{5} - \frac{3x - 1}{2} < \frac{4x + 7}{2}$

**73.** $.6x - 2(.5x + .2) \leq .4 - .3x$

**74.** $-.9x - (.5 + .1x) > -.3x - .5$

**75.** $-\frac{1}{2}x + .7x - 5 > 0$

**76.** $\frac{3}{4}x - .2x - 6 \leq 0$

**77.** *Distance* The linear function $f$ computes the distance $y$ in miles between a car and the city of Omaha after $x$ hours, where $0 \leq x \leq 6$. The graphs of $f$ and the horizontal lines $y = 100$ and $y = 200$ are shown in the figure. Use the graphs to answer the following.

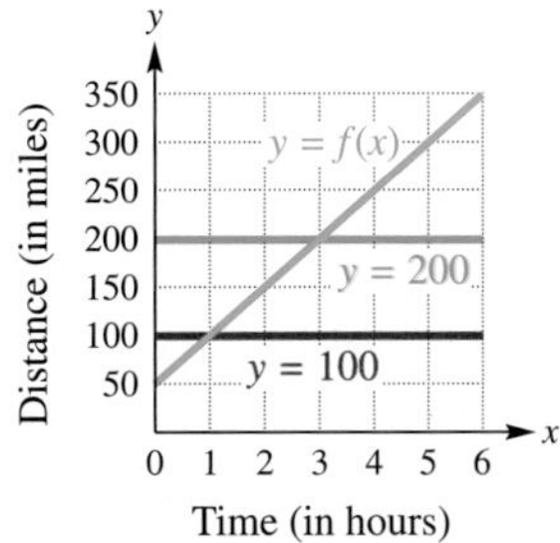


**(a)** Is the car moving toward or away from Omaha? Explain.

**(b)** Determine the times when the car is 100 miles or 200 miles from Omaha.

**(c)** When is the car from 100 to 200 miles (inclusive) from Omaha?

**(d)** When is the car's distance from Omaha greater than 100 miles?

**78.** Use the figure to solve each equation or inequality.

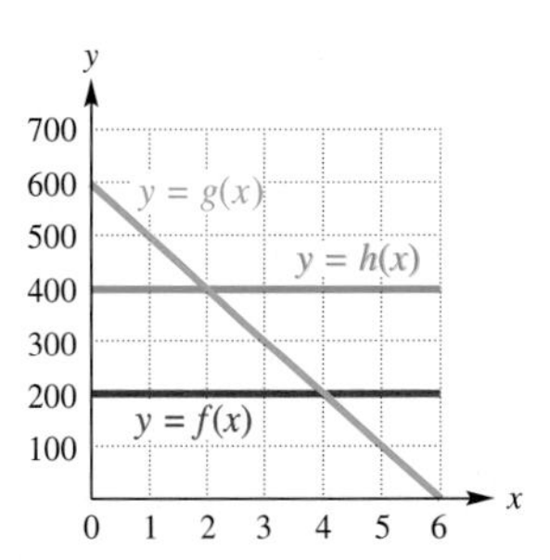


**(a)** $f(x) = g(x)$

**(b)** $g(x) = h(x)$

**(c)** $f(x) < g(x) < h(x)$

**(d)** $g(x) > h(x)$

*Solve each three-part inequality analytically. Support your answer graphically.*

**79.** $4 \le 2x + 2 \le 10$

**80.** $-4 \le 2x - 1 \le 5$

**81.** $-10 > 3x + 2 > -16$

**82.** $4 > 6x + 5 > -1$

**83.** $-3 \le \frac{x - 4}{-5} < 4$

**84.** $1 < \frac{4x - 5}{-2} < 9$

**85.** $\sqrt{2} \le \frac{2x + 1}{3} \le \sqrt{5}$

**86.** $\pi \le 5 - 4x < 7\pi$

## RELATING CONCEPTS

**For individual or group investigation (Exercises 87–90)**

*The solution set of a linear equation is closely related to the solution set of a linear inequality. Work Exercises 87–90 in order, to investigate this connection.*

**87.** Find the solution set of $3.7x - 11.1 = 0$ using the $x$-intercept method. How many solutions are there? How many solutions are there to any linear equation in one variable?

**88.** The solution from Exercise 87 is sometimes called a *boundary number.* Find the solution sets of $3.7x - 11.1 < 0$ and $3.7x - 11.1 > 0$. Write your answers in interval notation. Explain why the *boundary number* is appropriate for the solution you found in Exercise 87.

**89.** Find the solution set of $-4x + 6 = 0$ using the $x$-intercept method. Then find the solution sets of $-4x + 6 < 0$ and $-4x + 6 > 0$.

**90.** Generalize your results from Exercises 87–89 by answering the following.

**(a)** What is the solution set of $ax + b = 0$ if $a \neq 0$?

**(b)** Suppose $a > 0$. What are the solution sets of $ax + b < 0$ and $ax + b > 0$? Write your answers in interval notation.

**(c)** Suppose $a < 0$. What are the solution sets of $ax + b < 0$ and $ax + b > 0$? Write your answers in interval notation.

*(Modeling) Solve each problem.*

**91.** *Clouds and Temperature* Refer to Example 9. Suppose the ground-level temperature is 65°F and the dew point is 50°F.

**(a)** Use the intersection-of-graphs method to estimate the altitudes where clouds will not form.

**(b)** Solve part (a) analytically.

**92.** *Prices of Homes* The median price of a single-family home from 1980 to 1990 can be approximated by

$$P(x) = 3421x + 61{,}000,$$

where $x = 0$ corresponds to 1980 and $x = 10$ to 1990. (*Source:* U.S. Department of Commerce.)

**(a)** Interpret the slope of the graph of $P$.

**(b)** Estimate the years when the median price range was from \$71,000 to \$88,000.

**93.** *Error Tolerances* Suppose that an aluminum can is manufactured so that its radius $r$ can vary from 1.99 inches to 2.01 inches. What range of values is possible for the circumference $C$ of the can? Express your answer using a three-part inequality.

**94.** *Error Tolerances* Suppose that a square picture frame has sides that vary between 9.9 inches and 10.1 inches. What range of values is possible for the perimeter $P$ of the picture frame? Express your answer using a three-part inequality.

**95.** *Data Analysis* The following data are exactly linear.

| $x$ | 0 | 2 | 4 | 6 |
|---|---|---|---|---|
| $y$ | −1.5 | 4.5 | 10.5 | 16.5 |

**(a)** Find a linear function $f$ that models the data.

**(b)** Solve the inequality $f(x) > 2.25$.

**96.** *Data Analysis* The following data are exactly linear.

| $x$ | 1 | 2 | 3 | 4 | 5 |
|---|---|---|---|---|---|
| $y$ | .4 | 3.5 | 6.6 | 9.7 | 12.8 |

**(a)** Find a linear function $f$ that models the data.

**(b)** Solve the inequality $2 \le f(x) \le 8$. Round endpoints to the nearest hundredth.

# 1.6 APPLICATIONS OF LINEAR FUNCTIONS

Problem-Solving Strategies • Applications of Linear Equations • Break-Even Analysis • Direct Variation • Formulas

## Problem-Solving Strategies

Probably the most famous study of problem-solving techniques was developed by George Polya (1888–1985), among whose many publications was the modern classic *How to Solve It*. In this book, Polya proposed a four-step process for problem solving.

**Polya's Four-Step Process for Problem Solving**

1. **Understand the problem.** You cannot solve a problem if you do not understand what you are asked to find. The problem must be read and analyzed carefully. You will probably need to read it several times. After you have done so, ask yourself, "What must I find?"
2. **Devise a plan.** There are many ways to attack a problem and decide what plan is appropriate for the particular problem you are solving. (In this text, the plan will usually be to solve an equation or an inequality.)
3. **Carry out the plan.** Once you know how to approach the problem, carry out your plan. You may run into "dead ends" and unforeseen roadblocks, but be persistent. If you are able to solve a problem without a struggle, it isn't much of a problem, is it?
4. **Look back and check.** Check your answer to see that it is reasonable. Does it satisfy the conditions of the problem? Have you answered all the questions the problem asks? Can you solve the problem in a different way and come up with the same answer?

## Applications of Linear Equations

**EXAMPLE 1** *Determining the Dimensions of a Television Screen*

A new generation of televisions has a 16:9 *aspect ratio*. This means that the length of its rectangular screen is $\frac{16}{9}$ times its width. If the perimeter of the screen is 136 inches, find the length and the width of the screen.

**Analytic Solution**

If we let $x$ represent the width of the screen, then $\frac{16}{9}x$ represents the length. See Figure 78. The formula for the perimeter of a rectangle is $P = 2L + 2W$. We let $P = 136$, $L = \frac{16}{9}x$, and $W = x$, and solve the equation.

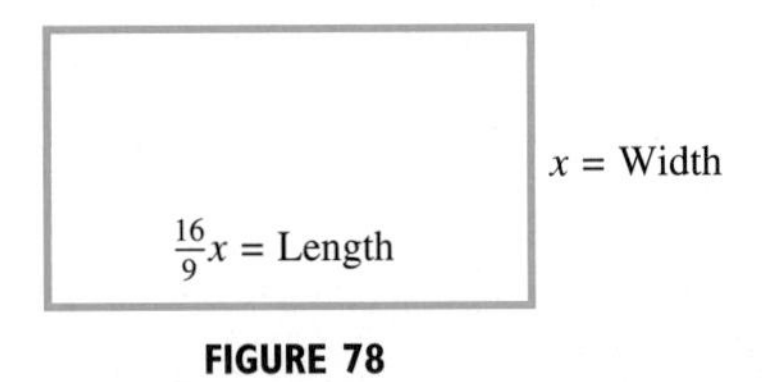


FIGURE 78

**Graphing Calculator Solution**

We graph

$$y_1 = 2\left(\tfrac{16}{9}\right)x + 2x \quad \text{and} \quad y_2 = 136.$$

As seen in Figure 79 on the next page, the point of intersection of the graphs is (24.48, 136). The $x$-coordinate supports our answer of 24.48 inches for the width.

*(continued)*

$$136 = 2\left(\frac{16}{9}x\right) + 2x \qquad P = 2L + 2W$$

$$136 = \frac{32}{9}x + 2x \qquad \text{Multiply.}$$

$$136 = \frac{50}{9}x \qquad \text{Combine terms.}$$

$$x = 24.48 \qquad \text{Multiply by } \tfrac{9}{50}.$$

The width of the screen is 24.48 inches, and the length is $\frac{16}{9}(24.48) = 43.52$ inches. To check this answer, verify that the sum of the lengths of the sides is 136.

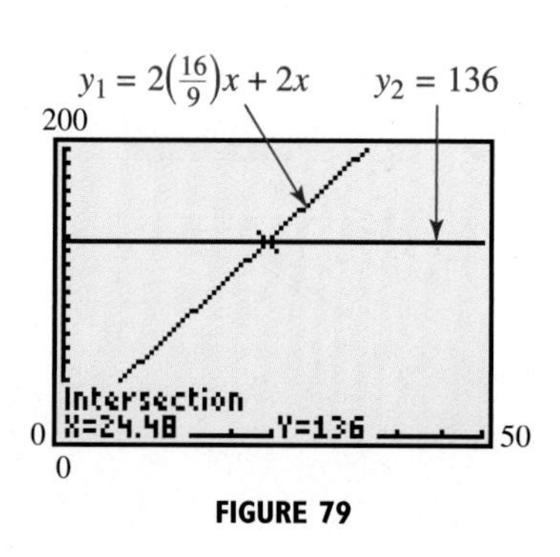


FIGURE 79

**NOTE** If the problem is not set up correctly, a check is worthless.

**FOR DISCUSSION**

Observe Figure 79, and disregard the graph of $y_2 = 136$. What remains is the graph of a linear function that gives the perimeter $y$ of all rectangles satisfying the 16 by 9 aspect ratio as a *function* of its width $x$. Trace along the graph, back and forth, and describe what the display at the bottom represents. Why would nonpositive values of $x$ be meaningless here? If such a television has perimeter 117 inches, what would be its approximate width?

**EXAMPLE 2** *Solving a Mixture-of-Concentrations Problem*

How much pure alcohol should be added to 20 liters of 40% alcohol to increase the concentration to 50% alcohol?

**Solution** We let $x$ represent the number of liters of pure alcohol that must be added. The information can be summarized in a "box diagram," as shown in Figure 80.

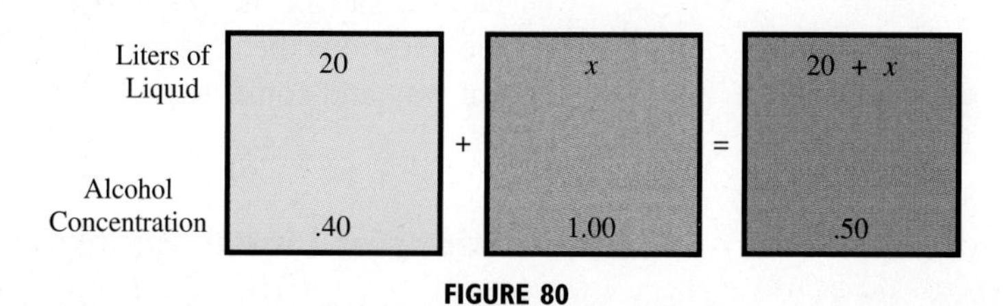


FIGURE 80

In each box, we have the number of liters and the alcohol concentration. We multiply the two items in each box to get the amount of pure alcohol in each case. The amount of pure alcohol on the left must equal the amount of pure alcohol on the right, so the equation to solve is

$$\underbrace{.40(20)}_{\text{Liters of pure alcohol in starting mixture}} + \underbrace{1.00x}_{\text{Liters of pure alcohol added}} = \underbrace{.50(20 + x)}_{\text{Liters of pure alcohol in final mixture}}.$$

$$40(20) + 100x = 50(20 + x) \qquad \text{Multiply by 100.}$$

$$800 + 100x = 1000 + 50x \qquad \text{Distributive property}$$

$$50x = 200 \qquad \text{Subtract } 50x\text{; subtract 800.}$$

$$x = 4 \qquad \text{Divide by 50.}$$

Therefore, 4 liters of pure alcohol must be added.

**EXAMPLE 3** *Examining the Effect of Formaldehyde on the Eyes*

Formaldehyde is a volatile organic compound that has come to be recognized as a highly toxic indoor air pollutant. It is found in building materials such as fiberboard, plywood, foam insulation, and carpeting. When concentrations of formaldehyde in the air exceed 33 $\mu g/ft^3$ (1 $\mu g$ = 1 microgram = $10^{-6}$ gram), a strong odor and irritation to the eyes often occur. One square foot of hardwood plywood paneling can emit 3365 $\mu g$ of formaldehyde per day. (*Source:* Hines, A., T. Ghosh, S. Layalka, and R. Warder, *Indoor Air Quality & Control*, Prentice-Hall, 1993.)

A 4- by 8-foot sheet of this paneling is attached to an 8-foot wall in a room having dimensions 10 by 10 feet.

**(a)** How many cubic feet of air are there in the room?

**(b)** Find the total number of micrograms of formaldehyde that are released into the air by the paneling each day.

**(c)** If there is no ventilation in the room, write a linear function that models the amount of formaldehyde $F$ in the room after $x$ days.

**(d)** How long will it take before a person's eyes become irritated in the room?

**Solution**

**(a)** The volume of the room is $8 \times 10 \times 10 = 800$ cubic feet.

**(b)** The paneling releases 3365$\mu g$ for each square foot of area. The area of the sheet is $4 \times 8 = 32$ square feet, so it will release $32 \times 3365 = 107{,}680$ $\mu g$ of formaldehyde into the air each day.

**(c)** The paneling emits formaldehyde at a constant rate of 107,680 $\mu g$ per day. Thus, $F(x) = 107{,}680x$.

**(d)** We must determine when the concentration exceeds 33 $\mu g/ft^3$. Since the room has 800 cubic feet, this will occur when the total amount reaches $33 \times 800 = 26{,}400$ $\mu g$.

$$107{,}680x = 26{,}400 \qquad \text{Solve } F(x) = 26{,}400.$$

$$x = \frac{26{,}400}{107{,}680} \qquad \text{Divide by 107,680.}$$

$$x \approx .25 \qquad \text{Approximate.}$$

It will take approximately $\frac{1}{4}$ day, or 6 hours.

## Break-Even Analysis

By expressing a company's cost of producing a product and the revenue from selling the product as functions, the company can determine at what point it will break even. In other words, we try to answer the question, "For what number of items sold will the revenue collected equal the cost of producing those items?"

**EXAMPLE 4** *Determining the Break-Even Point*

Peripheral Visions, Inc., produces studio-quality audiotapes of live concerts. The company places an ad in a trade newsletter. The cost of the ad is \$100. Each tape costs \$20 to produce, and the company charges \$24 per tape.

**(a)** Express the cost $C$ as a function of $x$, the number of tapes produced.

**(b)** Express the revenue $R$ as a function of $x$, the number of tapes sold.

**(c)** For what value of $x$ does revenue equal cost?

**(d)** Graph $y_1 = C(x)$ and $y_2 = R(x)$ in an appropriate window to support the answer in part (c).

**(e)** Use a table to support the answer in part (c).

**Solution**

**(a)** The *fixed cost* is \$100, and for each tape produced, the *variable cost* is \$20. Therefore, the total cost $C$ can be expressed as a function of $x$, the number of tapes produced.

$$C(x) = 20x + 100 \quad (C \text{ in dollars})$$

**(b)** Since each tape sells for \$24, the revenue $R$ is given by $R(x) = 24x$ ($R$ in dollars).

**(c)** The company will break even (no profit and no loss) when revenue equals cost, or $R(x) = C(x)$.

$$\begin{aligned} R(x) &= C(x) \\ 24x &= 20x + 100 && \text{Substitute for } R(x) \text{ and } C(x). \\ 4x &= 100 && \text{Subtract } 20x. \\ x &= 25 && \text{Divide by 4.} \end{aligned}$$

When 25 tapes are sold, the company will break even.

**(d)** The graphs shown in Figure 81 confirm our solution of X = 25. The $y$-value of 600 indicates that when 25 tapes are sold, both the cost and the revenue are \$600.

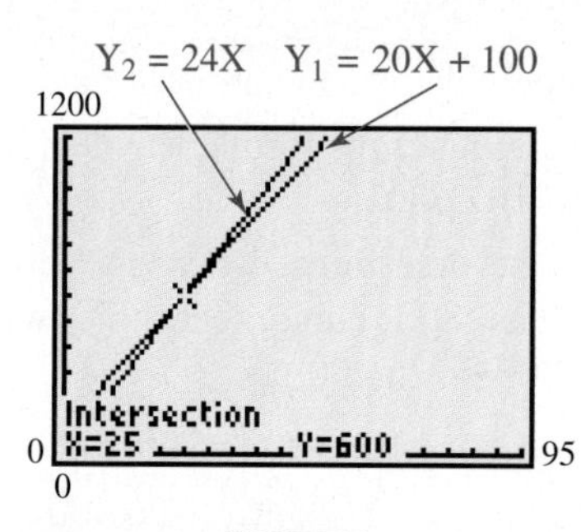


**FIGURE 81**

| X | Y1 | Y2 |
|---|---|---|
| 22 | 540 | 528 |
| 23 | 560 | 552 |
| 24 | 580 | 576 |
| 25 | 600 | 600 |
| 26 | 620 | 624 |
| 27 | 640 | 648 |
| 28 | 660 | 672 |

X=25

**FIGURE 82**

**(e)** The table in Figure 82 shows that when the number of tapes is 25, both function values are 600, numerically supporting our answer in part (c).

## Direct Variation

A common application involving linear functions deals with quantities that *vary directly* (or are in direct proportion).

> **Direct Variation**
>
> A number $y$ **varies directly** with $x$ if there exists a nonzero number $k$ such that
>
> $$y = kx.$$
>
> The number $k$ is called the **constant of variation.**

Notice that the graph of $y = kx$ is simply a straight line with slope $k$, passing through the origin. See Figure 83.

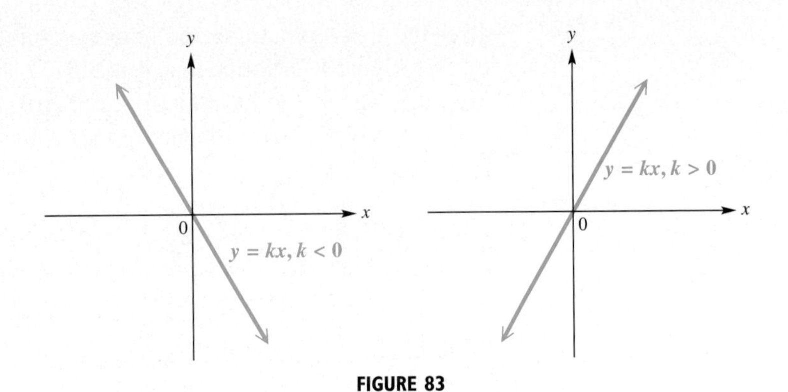


**FIGURE 83**

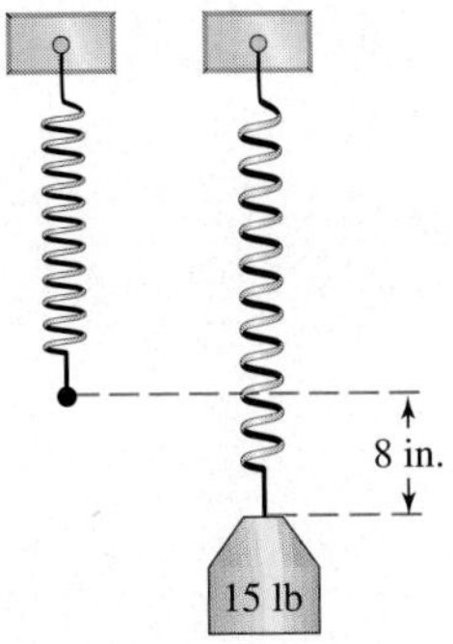

If we divide both sides of $y = kx$ by $x$, we get $\frac{y}{x} = k$, indicating that in a direct variation, the quotient (or proportion) of the two quantities is constant.

8 in.

15 lb

**FIGURE 84**

**EXAMPLE 5** *Solving a Direct Variation Problem (Hooke's Law)*

Hooke's law states that the distance $y$ a spring stretches varies directly with the force $x$ applied. If a force of 15 pounds stretches a spring 8 inches, how much will a force of 35 pounds stretch the spring? See Figure 84.

### Analytic Solution

Using the fact that when $x = 15$, $y = 8$, we can find $k$.

$$y = kx$$

$$8 = k \cdot 15 \qquad \text{Let } x = 15, y = 8.$$

$$k = \frac{8}{15} \qquad \text{Divide by 15.}$$

Therefore, the linear equation $y = \frac{8}{15}x$ describes the relationship between the force $x$ and the distance stretched $y$. To answer the question of the problem, we let $x = 35$. The spring will stretch

$$y = \frac{8}{15}(35) = 18.\overline{6},$$

or approximately 19 inches.

### Graphing Calculator Solution

Locating the point where X = 35, we see in Figure 85 that $Y_1 = 18.\overline{6}$. The table of values in Figure 86 gives numerical support for our analytic solution.

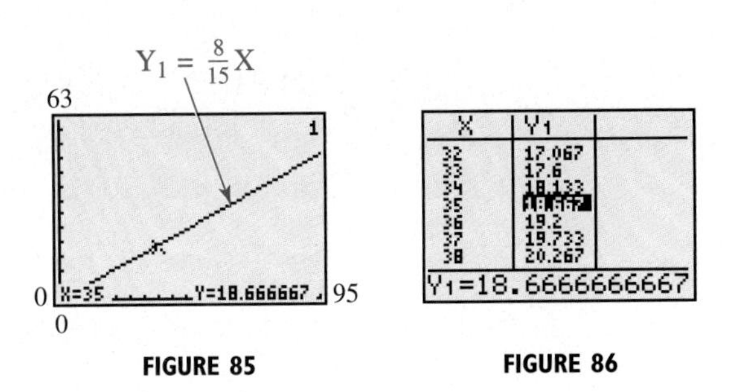


**FIGURE 85** **FIGURE 86**

(Use Figure 86 to determine how far a 33-pound force would stretch the spring.)

## Formulas

In many applications, we use formulas that give a general relationship among several quantities in a problem situation. For example, the formula

$$A = \frac{1}{2}h(b_1 + b_2)$$

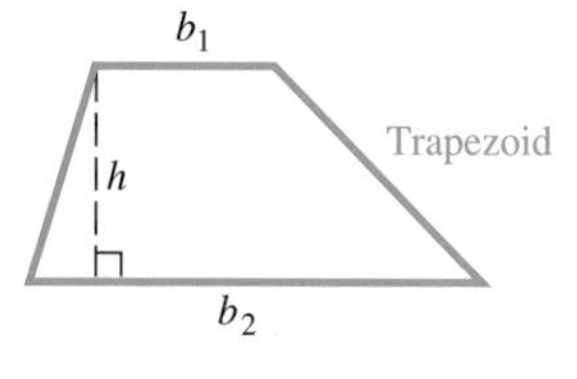


$A = \frac{1}{2}h(b_1 + b_2)$

**FIGURE 87**

gives the area $A$ of a trapezoid in terms of its height $h$ and its two parallel bases $b_1$ and $b_2$. See Figure 87. Notice that $A$ is alone on one side of the formula. Suppose, however, that we want to have the formula arranged in such a way that it is solved for $b_1$. The methods of solving linear equations analytically can be adapted so that this goal is accomplished.

**EXAMPLE 6** *Solving a Formula for a Specified Variable*

A trapezoid has area 169 square inches, height 13 inches, and base 19 inches. Find the length of the other base by solving the formula $A = \frac{1}{2}h(b_1 + b_2)$ for $b_1$ and substituting.

**Solution** We treat the formula as if $b_1$ were the only variable and all other variables are constants.

$$A = \frac{1}{2}h(b_1 + b_2) \quad \text{Given formula}$$

$$2A = h(b_1 + b_2) \quad \text{Multiply by 2.}$$

$$\frac{2A}{h} = b_1 + b_2 \quad \text{Divide by } h.$$

$$\frac{2A}{h} - b_2 = b_1 \quad \text{Subtract } b_2.$$

$$b_1 = \frac{2A}{h} - b_2 \quad \text{Rewrite.}$$

To determine $b_1$, we substitute $A = 169$, $h = 13$, and $b_2 = 19$.

$$b_1 = \frac{2(169)}{13} - 19 = 7 \text{ inches}$$

**EXAMPLE 7** *Solving Formulas for a Specified Variable*

Solve each formula for the specified variable.

**(a)** $P = 2W + 2L$ for $L$ (Perimeter of a rectangle)

**(b)** $C = \frac{5}{9}(F - 32)$ for $F$ (Fahrenheit to Celsius)

**Solution**

**(a)**

$$P = 2W + 2L \quad \text{Given equation}$$

$$P - 2W = 2L \quad \text{Subtract } 2W.$$

$$\frac{P - 2W}{2} = L \quad \text{Divide by 2.}$$

$$L = \frac{P}{2} - W \quad \text{Rewrite.}$$

**(b)**

$$C = \frac{5}{9}(F - 32) \quad \text{Given equation}$$

$$\frac{9}{5}C = F - 32 \quad \text{Multiply by } \tfrac{9}{5}.$$

$$F = \frac{9}{5}C + 32 \quad \text{Add 32; rewrite.}$$

**EXAMPLE 8** *Using Similar Triangles*

A grain bin in the shape of an inverted cone has height 11 feet and radius 3.5 feet, as illustrated in Figure 88. If the grain is 7 feet high in the bin, calculate the volume of the grain.

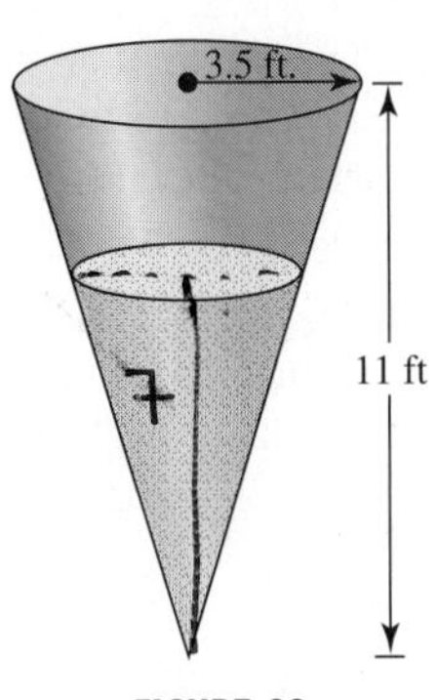


FIGURE 88

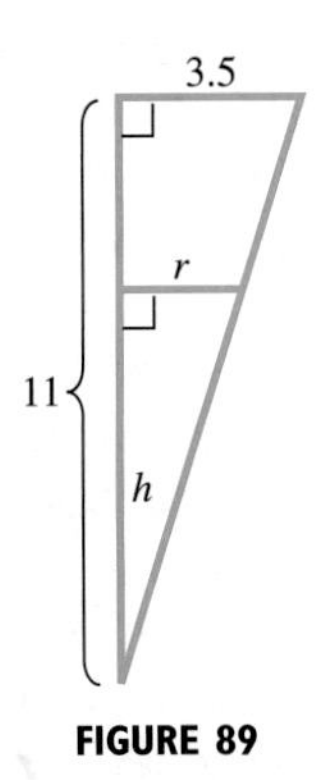


FIGURE 89

**Solution** The volume of a cone is $V = \frac{1}{3}\pi r^2 h$. The height $h$ of the grain in the bin is 7 feet, but we are not given the radius $r$ of the bin at that height. We can find the radius $r$ corresponding to a height of 7 feet by using similar triangles, as illustrated in Figure 89. The sides of similar triangles are proportional.

$$\frac{r}{h} = \frac{3.5}{11}$$

Solving for $r$ and letting $h = 7$ gives

$$r = \frac{3.5}{11}(7) \approx 2.227.$$

The volume of the grain is

$$V = \frac{1}{3}\pi(2.227)^2(7) \approx 36.4 \text{ cubic feet.}$$

## 1.6 EXERCISES

*Solve each formula for the specified variable.*

1. $I = PRT$ for $P$ (Simple interest)
2. $V = LWH$ for $L$ (Volume of a box)

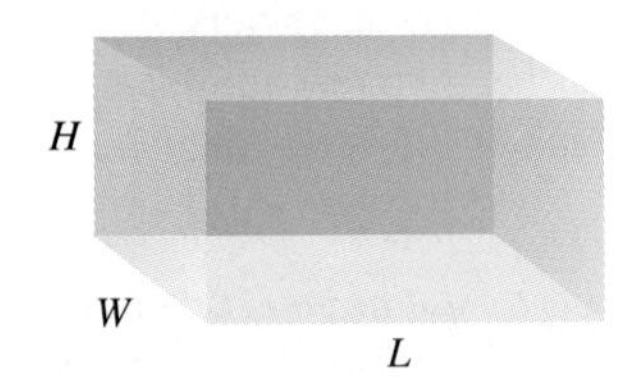


3. $P = 2L + 2W$ for $W$ (Perimeter of a rectangle)
4. $P = a + b + c$ for $c$ (Perimeter of a triangle)
5. $A = \frac{1}{2}h(b_1 + b_2)$ for $h$ (Area of a trapezoid)
6. $S = 2LW + 2WH + 2HL$ for $H$ (Surface area of a rectangular box)
7. $S = 2\pi rh + 2\pi r^2$ for $h$ (Surface area of a cylinder)
8. $V = \frac{1}{3}\pi r^2 h$ for $h$ (Volume of a cone)
9. $F = \frac{9}{5}C + 32$ for $C$ (Celsius to Fahrenheit)

**10.** $s = \frac{1}{2}gt^2$ for $g$ (Distance traveled by a falling object)

*Solve each problem analytically, and support your solution graphically.*

**11.** Dimensions of a Mailing Label The length of a rectangular mailing label is 3 centimeters less than twice the width. The perimeter is 54 centimeters. Find the dimensions of the label.

**12.** Dimensions of a Square If the length of a side of a square is increased by 3 centimeters, the perimeter of the new square is 40 centimeters more than twice the length of the side of the original square. Find the dimensions of the original square.

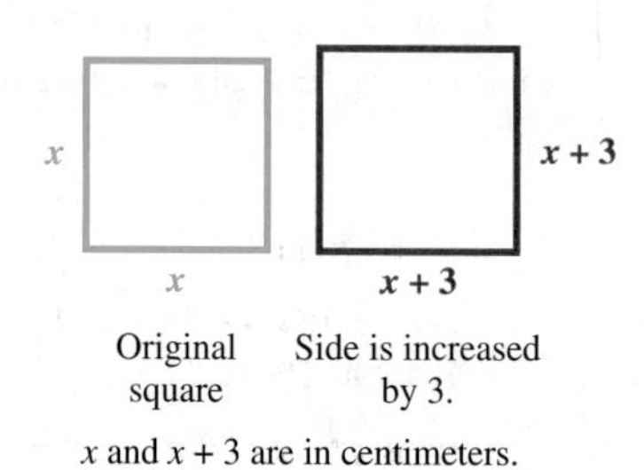


$x$ and $x + 3$ are in centimeters.

**13.** World's Largest Tablecloth The world's largest tablecloth has a perimeter of 28,803.2 inches. It is made of damask and was manufactured by Tonrose Limited of Manchester, England, in June 1988. Its length is 11,757.6 inches more than its width. What is its length in yards? (*Source: Guinness Book of Records,* 1995.)

**14.** Aspect Ratio of a Television Set The aspect ratio of conventional television sets is 4:3. One such model Toshiba television has a rectangular viewing screen with perimeter 98 inches. What are the length and width of the screen? Since televisions are advertised by the diagonal measure of their screens, how would this set be advertised?

*Exercises 15 and 16 involve octane rating of gasoline, a measure of its antiknock qualities. In one measure of octane, a standard fuel is made with only two ingredients: heptane and isooctane. For this fuel, the octane rating is the percent of isooctane. An actual gasoline blend is then compared to a standard fuel. For example, a gasoline with an octane rating of 98 has the same antiknock properties as a standard fuel that is 98% isooctane.*

**15.** Octane Rating of Gasoline How many gallons of 94-octane gasoline should be mixed with 400 gallons of 99-octane gasoline to obtain a mixture that is 97-octane?

**16.** Octane Rating of Gasoline How many gallons of 92-octane and 98-octane gasoline should be mixed together to provide 120 gallons of 96-octane gasoline?

*Solve each mixture problem.*

**17.** Alcohol Mixture A pharmacist wishes to strengthen a mixture from 10% alcohol to 30% alcohol. How much pure alcohol should be added to 7 liters of the 10% mixture?

**18.** Saline Solution Mixture How much water should be added to 8 ml of 6% saline solution to reduce the concentration to 4% acid?

**19.** Acid Mixture How much water should be added to 20 liters of an 18% acid solution to reduce the concentration to 15% acid?

**20.** Antifreeze Mixture An automobile radiator holds 8 liters of fluid. There is currently a mixture in the radiator that is 80% antifreeze and 20% water. How much of this mixture should be drained and replaced by pure antifreeze so that the resulting mixture is 90% antifreeze?

*Solve each problem.*

**21.** Pressure of a Liquid The pressure exerted by a certain liquid at a given point is directly proportional to the depth of the point beneath the surface of the liquid. If the pressure at 30 feet is 13 pounds per square inch, what is the pressure exerted at 70 feet?

**22.** Rate of Nerve Impulses The rate at which impulses are transmitted along a nerve fiber is directly proportional to the diameter of the fiber. If the rate for a certain fiber is 40 meters per second when the diameter is 6 micrometers, what is the rate if the diameter is 8 micrometers?

**23.** Height of a Shadow A certain tree casts a shadow 45 feet long. At the same time, the shadow cast by a vertical stick 2 feet high is 1.75 feet long. How tall is the tree? (*Hint:* Use similar triangles.)

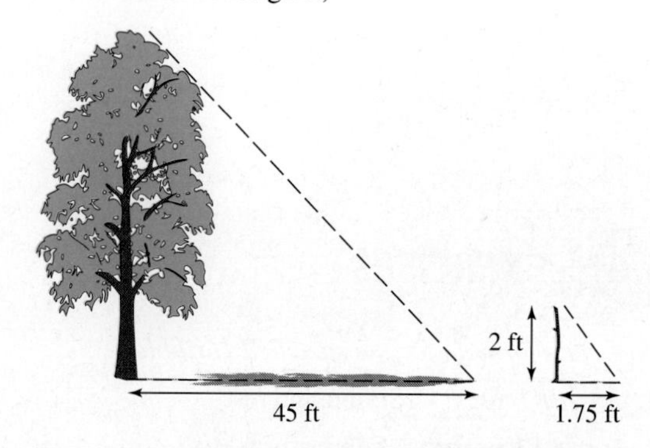


**24.** Volume of Water A water tank in the shape of an inverted cone has height 6 feet and radius 2 feet. If the water level in the tank is 3.5 feet, calculate the volume of the water.

*Biologists use direct variation to estimate the number of individuals of a species in a particular area. They first capture a sample of individuals from the area and mark each specimen with a harmless tag. Then later they return and capture another sample from the same area. They base their estimate on the theory that the proportion of tagged specimens in the new sample is the same as the proportion of tagged individuals in the entire area. Use this idea to work Exercises 25 and 26.*

**25.** *Estimating Fish in a Lake* Biologists tagged 250 fish in City Park Lake on October 12. On a later date they found 7 tagged fish in a sample of 350. Estimate the total number of fish in the lake to the nearest hundred.

**26.** *Estimating Seal Pups in a Breeding Area* According to an actual survey in 1961, to estimate the number of seal pups in a certain breeding area in Alaska, 4963 pups were tagged in early August. In late August, a sample of 900 pups was examined and 218 of these were found to have been tagged earlier. Use this information to estimate, to the nearest hundred, the total number of seal pups in this breeding area. (*Source:* "Estimating the Size of Wildlife Populations," Chatterjee, S., in *Statistics by Example,* 1973, obtained from data in *Transactions of the American Fisheries Society,* July 1968.)

*In Exercises 27–30, find the constant of variation k and the undetermined value in the table if y is directly proportional to x.*

**27.**

| $x$ | 3 | 5 | 6 | 8 |
|---|---|---|---|---|
| $y$ | 7.5 | 12.5 | 15 | ? |

**28.**

| $x$ | 1.2 | 4.3 | 5.7 | ? |
|---|---|---|---|---|
| $y$ | 3.96 | 14.19 | 18.81 | 23.43 |

**29.** Sales tax $y$ on a purchase of $x$ dollars

| $x$ | \$25 | \$55 | ? |
|---|---|---|---|
| $y$ | \$1.50 | \$3.30 | \$5.10 |

**30.** Cost $y$ of buying $x$ compact discs having the same price

| $x$ | 3 | 4 | 5 |
|---|---|---|---|
| $y$ | \$41.97 | \$55.96 | ? |

**31.** *Cost of Tuition* The cost of tuition is directly proportional to the number of credits taken. If 11 credits cost \$720.50, find the cost of taking 16 credits. What is the constant of variation?

**32.** *Strength of a Beam* The maximum load that a horizontal beam can carry is directly proportional to its width. If a beam 1.5 inches wide can support a load of 250 pounds, find the load that a beam of the same type can support if its width is 3.5 inches. What is the constant of variation?

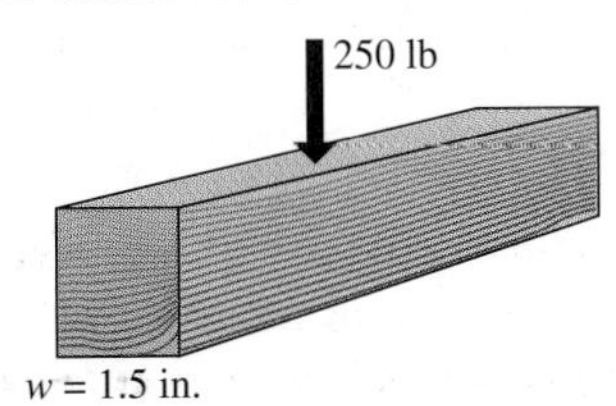


**33.** *Hooke's Law* If a 3-pound weight stretches a spring 2.5 inches, how far will a 17-pound weight stretch the spring?

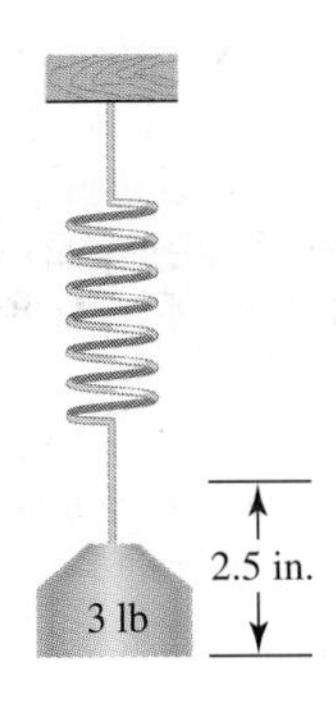


**34.** *Hooke's Law* If a 9.8-pound weight stretches a spring .75 inch, how much weight would be needed to stretch the spring 3.1 inches?

*(Modeling) In Exercises 35–38,*

**(a)** *Express the cost C as a function of x, where x represents the number of items as described.*

**(b)** *Express the revenue R as a function of x.*

**(c)** *Determine analytically the value of x for which revenue equals cost.*

**(d)** *Graph $y_1 = C(x)$ and $y_2 = R(x)$ on the same xy-axes and interpet the graphs.*

**35.** *Stuffing Envelopes* Rebecca Isaac-Fahey stuffs envelopes for extra income during her spare time. Her initial cost to obtain the necessary information for the job was \$200.00. Each envelope costs \$.02, and she gets paid \$.04 per envelope stuffed. Let $x$ represent the number of envelopes stuffed.

**36.** *Copier Service* Gerald Skidmore runs a copying service in his home. He paid \$3500 for the copier and a lifetime service contract. Each sheet of paper he uses costs \$.01, and he gets paid \$.05 per copy. Let $x$ represent the number of copies he makes.

**37.** *Delivery Service* Tom Accardo operates a delivery service in a southern city. His start-up costs amounted to \$2300. He estimates that it costs him (in terms of gasoline, wear and tear on his car, etc.) \$3.00 per delivery. He charges \$5.50 per delivery. Let $x$ represent the number of deliveries he makes.

**38.** *Baking and Selling Cakes* Pat Stone bakes cakes and sells them at county fairs. Her initial cost for the Tomball fair was \$40.00. She figures that each cake costs \$2.50 to make, and she charges \$6.50 per cake. Let $x$ represent the number of cakes sold. (Assume that there were no cakes left over.)

*Work each problem involving grade averaging.*

**39.** *Grade Averaging* Hien has grades of 84, 88, and 92 on his first three algebra tests. What grade on his fourth test will give him an average of 90?

**40.** *Grade Averaging* Jamal scored 78, 94, and 60 on his three trigonometry tests. If his final exam score is to be counted as two test grades in determining his course average, what grade must he make on his final exam to give him on average of 80?

*(Modeling) In Massachusetts, speeding fines are sometimes determined by the linear function defined by*

$$y = 10(x - 65) + 50, \quad x \geq 65,$$

*where y is the cost in dollars of the fine if a person is caught driving x miles per hour. Use this information to work the problems in Exercises 41–44.*

**41.** *Speeding Fines* Radar clocked a driver at 76 mph. How much was the fine?

**42.** *Speeding Fines* While balancing his checkbook, Johnny ran across a canceled check that his wife Gwen had written to the Department of Motor Vehicles for a speeding fine. The check was written for $100. How fast was Gwen driving?

**43.** *Speeding Fines* At what speed do the troopers start giving tickets?

**44.** *Speeding Fines* For what range of speeds is the fine greater than $200? Solve this problem analytically and support graphically.

*(Modeling) In Exercises 45–48, assume that a linear relationship exists between the two quantities.*

**45.** *Solar Heater Production* A company finds that it can produce 10 solar heaters for $7500, while producing 20 heaters costs $13,900.

**(a)** Express the cost, $y$, as a linear function of the number of heaters, $x$.

**(b)** Determine analytically the cost to produce 25 heaters.

**(c)** Support the result of part (b) graphically.

**46.** *Cricket Chirping* At 68°F, a certain species of cricket chirps 24 times per minute. At 40°F, the same cricket chirps 86 times per minute. (*Source:* Bushaw, D., et al., *A Sourcebook of Applications of School Mathematics,* MAA, 1980. Reprinted with permission.)

**(a)** Express the number of chirps, $y$, as a linear function of the Fahrenheit temperature.

**(b)** If the temperature is 60°F, how many times will the cricket chirp per minute?

**(c)** If you count the number of cricket chirps in one-half minute and hear 40 chirps, what is the temperature?

**47.** *Appraised Value of a Home* In 1990 a house was purchased for $120,000. In 2000 it was appraised for $146,000.

**(a)** If $x = 0$ represents 1990 and $x = 10$ represents 2000, express the appraised value of the house, $y$, as a linear function of the number of years, $x$, after 1990.

**(b)** What will the house be worth in the year 2004?

**(c)** What does the slope of the line represent (in your own words)?

**48.** *Depreciation of a Photocopier* A photocopier sold for $3000 in 1994 when it was purchased. Its value in 2002 had depreciated to $600.

**(a)** If $x = 0$ represents 1994 and $x = 8$ represents 2002, express the value of the machine, $y$, as a linear function of the number of years, $x$, after 1994.

**(b)** Graph the function from part (a) in a window $[0, 10]$ by $[0, 4000]$. How would you interpret the $y$-intercept in terms of this particular problem situation?

**(c)** Use your calculator to determine the value of the machine in 1998, and verify this analytically.

*(Modeling) Solve each problem.*

**49.** *Ventilation in a Classroom* Ventilation is an effective method for removing indoor air pollutants. According to the American Society of Heating, Refrigerating and Air-Conditioning Engineers, Inc. (ASHRAE), a non-smoking classroom should have a ventilation rate of 15 cubic feet per minute for each person in the classroom. (*Source:* ASHRAE, 1989.)

**(a)** Write a linear function that describes the total ventilation $V$ (in cubic feet per hour) necessary for a classroom with $x$ people.

**(b)** A common unit of ventilation is an *air change per hour* (ach). One ach is equivalent to exchanging all of the air in a room every hour. If $x$ people are in a classroom having a volume of 15,000 cubic feet, determine how many air exchanges per hour are necessary to keep the room properly ventilated.

**(c)** Find the necessary number of ach $A$ if the classroom has 40 people in it.

**(d)** In areas like bars and lounges that allow smoking, the ventilation rate should be increased to 50 cubic feet per minute per person. Compared to classrooms, ventilation should be increased by what factor in heavy-smoking areas?

**50.** *Snowmaking and Water Consumption* Ski resorts require large amounts of water in order to make snow. Snowmass Ski Area in Colorado plans to pump between 1120 and 1900 gallons of water per minute at least 12 hours per day from Snowmass Creek between mid-October and late December. (*Source:* York Snow Incorporated.)

**(a)** Determine a linear function that calculates the *minimum* amount of water $A$ (in gallons) pumped after $x$ days from mid-October to late December.

**(b)** Find the minimum amount of water pumped in 30 days.

**(c)** Suppose the water being pumped from Snowmass Creek was used to fill swimming pools. The average backyard swimming pool holds 20,000 gallons of water. Determine a linear function that will give the minimum number of pools $P$ that could be filled after $x$ days. How many pools could be filled each day?

**(d)** In how many days could a minimum of 1000 pools be filled?

**51.** *Cancer Risk from Pollutants* The excess lifetime cancer risk $R$ is a measure of the likelihood that an individual will develop cancer from a particular pollutant. For example, if $R = .01$, then a person has a 1% increased chance of developing cancer during a lifetime. This would translate into 1 case of cancer for every 100 people during an average lifetime. For non-smokers exposed to environmental tobacco smoke (passive smokers), $R = 1.5 \times 10^{-3}$. (*Source:* Hines, A., T. Ghosh, S. Layalka, and R. Warder, *Indoor Air Quality & Control,* Prentice-Hall, 1993.)

**(a)** If the average life expectancy is 72 years, what is the excess lifetime cancer risk per year for passive smokers?

**(b)** Write a linear function that gives the expected number of cancer cases $C$ per year if there are $x$ passive smokers.

**(c)** Estimate the number of cancer cases per 100,000 passive smokers.

**(d)** The excess lifetime risk of death from smoking is $R = .44$. Currently, 26% of the U.S. population smokes. If the U.S. population is 281 million, approximate the excess number of deaths caused by smoking each year.

**52.** *Air Pollution* Rework Example 3 if the room has a 10-foot ceiling with a 12- by 14-foot plywood floor. Each square foot of flooring emits 2540 $\mu$g of formaldehyde per day.

**53.** *Expansion and Contraction of Gases* In 1787 Jacques Charles noticed that gases expand when heated and contract when cooled. Suppose that a particular gas follows the model

$$y = \frac{5}{3}x + 455,$$

where $x$ is the temperature in Celsius and $y$ is the volume in cubic centimeters. (*Source:* Bushaw, D., et al., *A Sourcebook of Applications of School Mathematics,* MAA, 1980. Reprinted with permission.)

**(a)** What is the volume when the temperature is 27°C?

**(b)** What is the temperature when the volume is 605 cubic centimeters?

**(c)** Determine what temperature gives a volume of 0 cubic centimeters (that is, absolute zero, or the coldest possible temperature).

**54.** *Tail Length of a Snake* It has been reported that the total length $x$ and the tail length $y$ of females of the snake species *Lampropeltis polyzona* are nearly linearly related by the model

$$y = .134x - 1.18,$$

where $x$ is the length of the snake in millimeters. If a snake of this species measures 1000 millimeters, what is its tail length to the nearest millimeter? (*Source:* Bushaw, D., et al., *A Sourcebook of Applications of School Mathematics,* MAA, 1980. Reprinted with permission.)

## Reviewing Basic Concepts (Sections 1.5 and 1.6)

**1.** Solve $3(x - 5) + 2 = 1 - (4 + 2x)$ analytically. Support your result graphically or numerically.

**2.** Solve $\pi(1 - x) = .6(3x - 1)$ using the intersection-of-graphs method. Round your answer to the nearest thousandth.

**3.** Find the zero of $f(x) = \frac{1}{3}(4x - 2) + 1$ analytically. Use the $x$-intercept method to support your analytic result graphically.

**4.** Determine whether each equation is an identity, a contradiction, or a conditional equation.

**(a)** $4x - 5 = -2(3 - 2x) + 3$

**(b)** $5x - 9 = 5(-2 + x) + 1$

**(c)** $5x - 4 = 3(6 - x)$

**5.** Solve $2x + 3(x + 2) < 1 - 2x$ analytically. Express the solution set in interval notation. Support your result graphically.

**6.** Solve $-5 \leq 1 - 2x < 6$ analytically.

**7.** Use the figure to solve $f(x) = g(x)$ and $f(x) \leq g(x)$.

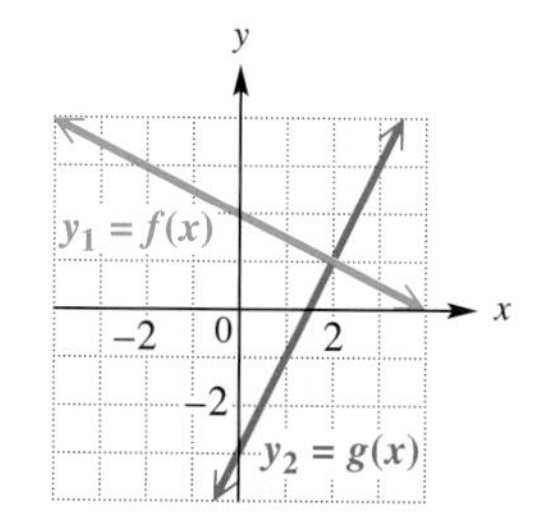


**8.** *Height of a Tree* A tree casts a 27-foot shadow, and a 6-foot person casts a 4-foot shadow. How tall is the tree?

**9.** *(Modeling) Production of Compact Discs* A student is planning to produce and sell compact discs (CDs) with recorded music for \$5.50 each. A computer with a CD burner costs \$2500, and each blank disc costs \$1.50.

**(a)** Find a formula for a function $R$ that calculates the revenue from selling $x$ discs.

**(b)** Find a formula for a function $C$ that calculates the cost of recording $x$ discs. Be sure to include the fixed cost of the computer.

**(c)** Find the break-even point. That is, determine how many discs must be sold for revenue to equal cost.

**10.** Solve $V = \pi r^2 h$ for $h$.

## CHAPTER 1 SUMMARY

**KEY TERMS & SYMBOLS** | **KEY CONCEPTS**

### 1.1 Real Numbers and Coordinate Systems

natural numbers
whole numbers
integers
number line
origin
rational numbers
real numbers
irrational numbers
coordinate
coordinate system
rectangular (Cartesian) coordinate system
$x$-axis
$y$-axis
coordinate ($xy$-) plane
quadrants
$x$-coordinate
$y$-coordinate

Natural numbers $\{1, 2, 3, 4, 5, \ldots\}$
Whole numbers $\{0, 1, 2, 3, 4, 5, \ldots\}$
Integers $\{\ldots, -3, -2, -1, 0, 1, 2, 3, \ldots\}$
Rational numbers $\left\{\frac{p}{q} \,\middle|\, p, q \text{ are integers}, q \neq 0\right\}$
Real numbers $\{x \mid x \text{ corresponds to a point on a number line}\}$
(The real numbers include both the rational and the irrational numbers.)

**CARTESIAN COORDINATE SYSTEM**

Points are located by using ordered pairs in the $xy$-plane.

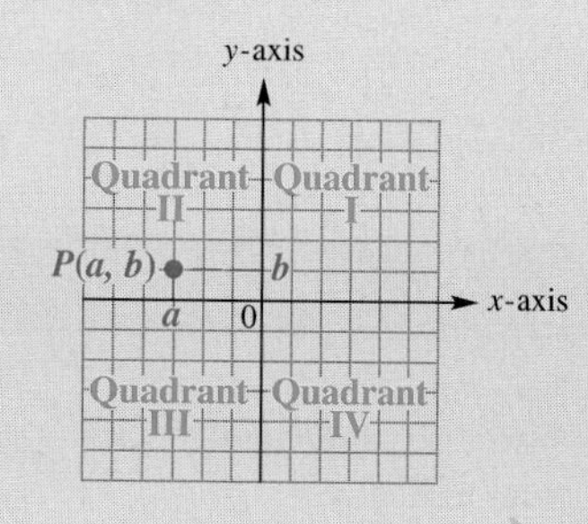

| KEY TERMS & SYMBOLS | KEY CONCEPTS |
|---|---|
| standard viewing window<br>Xmin<br>Xmax<br>Ymin<br>Ymax<br>Xscl<br>Yscl | **ROOTS**<br>The $n$th root of a real number $a$ can be expressed as $\sqrt[n]{a}$ or $a^{1/n}$. Calculators are often used to approximate roots of real numbers.<br><br>**DISTANCE FORMULA**<br>Suppose that $P(x_1, y_1)$ and $R(x_2, y_2)$ are two points in a coordinate plane. Then the distance between $P$ and $R$, written $d(P, R)$, is given by the distance formula,<br>$$d(P,R) = \sqrt{(x_2 - x_1)^2 + (y_2 - y_1)^2}.$$<br>**MIDPOINT FORMULA**<br>The midpoint of the line segment with endpoints $(x_1, y_1)$ and $(x_2, y_2)$ is<br>$$\left(\frac{x_1 + x_2}{2}, \frac{y_1 + y_2}{2}\right).$$ |
| **1.2** Introduction to Relations and Functions<br>set-builder notation<br>graph<br>interval<br>interval notation<br>infinity symbol, $\infty$<br>relation<br>domain<br>range<br>function<br>independent variable<br>dependent variable<br>function notation, $f(x)$<br>vertical line test | **RELATION**<br>A relation is a set of ordered pairs.<br><br>**DOMAIN AND RANGE**<br>If we denote the ordered pairs of a relation by $(x, y)$, the set of all $x$-values is called the domain of the relation, and the set of all $y$-values is called the range of the relation.<br><br>**FUNCTION**<br>A function is a relation in which each element in the domain corresponds to exactly one element in the range.<br><br>**VERTICAL LINE TEST**<br>If every vertical line intersects a graph in no more than one point, then the graph is the graph of a function.<br><br>**INTERVAL NOTATION**<br>Interval notation can be used to denote sets of real numbers.<br>$\{x \mid x \le a\}$ is equivalent to $(-\infty, a]$.<br>$\{x \mid a \le x < b\}$ is equivalent to $[a, b)$.<br>$\{x \mid x > b\}$ is equivalent to $(b, \infty)$.<br><br>**FUNCTION NOTATION**<br>To denote that $y$ is a function of $x$, we write $y = f(x)$. The independent variable is $x$ and the dependent variable is $y$. |

| KEY TERMS & SYMBOLS | KEY CONCEPTS |
|---|---|

## 1.3 Linear Functions

**Key terms & symbols:** linear function; linear equation in two variables; solution; $x$-intercept; $y$-intercept; zero of a function; constant function; comprehensive graph; change in $x$, $\Delta x$; change in $y$, $\Delta y$; slope, $m$; slope-intercept form

**LINEAR FUNCTION**

A function $f$ defined by

$$f(x) = ax + b,$$

where $a$ and $b$ are real numbers, is called a linear function. The graph of a linear function is a line.

**INTERCEPTS**

If the graph of a function $f$ intersects the $x$-axis at $(a, 0)$, then $a$ is an $x$-intercept.
If the graph of a function $f$ intersects the $y$-axis at $(0, b)$, then $b$ is the $y$-intercept.

**ZERO OF A FUNCTION $f$**

Any number $c$ for which $f(c) = 0$ is called a zero of $f$. The real zeros of a function $f$ correspond to the $x$-intercepts on its graph.

**CONSTANT FUNCTION**

A function defined by $f(x) = b$, where $b$ is a real number, is called a constant function. A constant function is a special type of linear function whose graph is a horizontal line.

**SLOPE**

The slope $m$ of the line passing through the points $(x_1, y_1)$ and $(x_2, y_2)$ is

$$m = \frac{\Delta y}{\Delta x} = \frac{y_2 - y_1}{x_2 - x_1}, \quad \Delta x \neq 0.$$

**GEOMETRIC ORIENTATION BASED ON SLOPE**

For a line with slope $m$, if $m > 0$, the line rises from left to right. If $m < 0$, it falls from left to right. If $m = 0$, the line is horizontal.

**VERTICAL LINE**

A vertical line with $x$-intercept $a$ has an equation of the form

$$x = a.$$

Its slope is undefined.

**SLOPE-INTERCEPT FORM**

The slope-intercept form of the equation of a line with slope $m$ and $y$-intercept $b$ is

$$y = mx + b.$$

## 1.4 Equations of Lines and Linear Models

**Key terms & symbols:** point-slope form; standard form; square viewing window; scatter diagram; least-squares regression line

**POINT-SLOPE FORM**

The line with slope $m$ passing through the point $(x_1, y_1)$ has equation

$$y - y_1 = m(x - x_1).$$

**PARALLEL AND PERPENDICULAR LINES**

Two distinct nonvertical lines are parallel if and only if they have the same slope.
Two lines, neither of which is vertical, are perpendicular if and only if their slopes have product $-1$.

| KEY TERMS & SYMBOLS | KEY CONCEPTS |
|---|---|
| least-squares regression<br>correlation coefficient, $r$ | **LINEAR MODELS AND REGRESSION**<br>If a collection of data points approximates a straight line, then we can find the equation of such a line. Choosing two data points, we apply the methods of Sections 1.3 and 1.4 to find this equation. The equation will vary, depending on which two points are chosen.<br><br>The line of best fit can be found using the linear regression feature of a graphing calculator. |
| **1.5** Linear Equations and Inequalities<br>equation<br>solution (root)<br>solution set<br>linear equation in one variable<br>equivalent equations<br>conditional equation<br>contradiction<br>empty (null) set $\emptyset$<br>identity<br>inequality<br>equivalent inequalities<br>linear inequality in one variable<br>three-part inequality | **LINEAR EQUATIONS AND INEQUALITIES**<br>A linear equation can be written in the form $ax + b = 0$, where $a \neq 0$.<br>A linear inequality can be written in one of the following forms:<br>$ax + b < 0$, $ax + b > 0$, $ax + b \leq 0$, or $ax + b \geq 0$.<br><br>**ANALYTIC SOLUTIONS**<br>To solve linear equations, use the addition and multiplication properties of equality. To solve linear inequalities, use the properties of inequality. When multiplying or dividing each side of an inequality by a negative number, be sure to reverse the direction of the inequality.<br><br>**INTERSECTION-OF-GRAPHS METHOD OF GRAPHICAL SOLUTION**<br>To solve the equation $f(x) = g(x)$, graph<br>$y_1 = f(x)$ and $y_2 = g(x)$.<br>The $x$-coordinate of any point of intersection of the two graphs is a solution of the equation. The solution set of $f(x) > g(x)$ is the set of all real numbers $x$ such that the graph of $f$ is **above** the graph of $g$. The solution set of $f(x) < g(x)$ is the set of all real numbers $x$ such that the graph of $f$ is **below** the graph of $g$.<br><br>**$x$-INTERCEPT METHOD OF GRAPHICAL SOLUTION**<br>To solve the equation $f(x) = g(x)$, graph<br>$y = f(x) - g(x) = F(x)$.<br>Any $x$-intercept of the graph of $y = F(x)$ is a solution of the equation. The solution set of $F(x) > 0$ is the set of all real numbers $x$ such that the graph of $F$ is **above** the $x$-axis. The solution set of $F(x) < 0$ is the set of all real numbers $x$ such that the graph of $F$ is **below** the $x$-axis. |
| **1.6** Applications of Linear Functions<br>varies directly<br>constant of variation | **POLYA'S FOUR-STEP PROCESS FOR PROBLEM SOLVING**<br>**1.** Understand the problem. **2.** Devise a plan.<br>**3.** Carry out the plan. **4.** Look back and check.<br><br>**DIRECT VARIATION**<br>A number $y$ varies directly with $x$ if there exists a nonzero number $k$ such that $y = kx$. The number $k$ is called the constant of variation. |

## CHAPTER 1 REVIEW EXERCISES

*Let A represent the point with coordinates* $(-1, 16)$, *and let B represent the point with coordinates* $(5, -8)$.

**1.** Find the exact distance between points $A$ and $B$.

**2.** Find the coordinates of the midpoint of the line segment connecting points $A$ and $B$.

**3.** Find the slope of the line segment $AB$.

**4.** Find the equation of the line passing through points $A$ and $B$. Write it in $y = mx + b$ form.

*Consider the graph of* $3x + 4y = 144$ *in Exercises 5–8.*

**5.** What is the slope of this line?

**6.** What is the $x$-intercept of this line?

**7.** What is the $y$-intercept of this line?

**8.** Give a viewing window that will show a comprehensive graph. (There are many possible such windows.)

**9.** Suppose that $f$ is a linear function such that $f(3) = 6$ and $f(-2) = 1$. Find $f(8)$. (*Hint:* Find a formula for $f(x)$.)

**10.** Find the equation of the line perpendicular to the graph of $y = -4x + 3$, passing through the point $(-2, 4)$. Give it in $y = mx + b$ form.

*For each line shown in Exercises 11 and 12, do the following.*

**(a)** *Find the slope.*

**(b)** *Find the slope-intercept form of the equation.*

**(c)** *Find the midpoint of the segment connecting the two points identified on the graph.*

**(d)** *Find the distance between the two points identified on the graph.*

**11.**

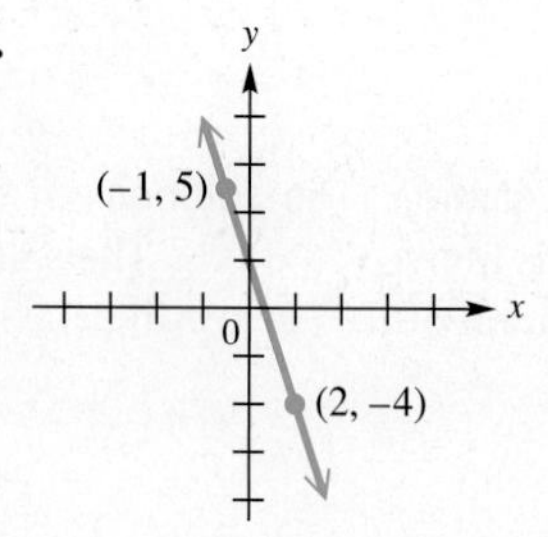


**12.**

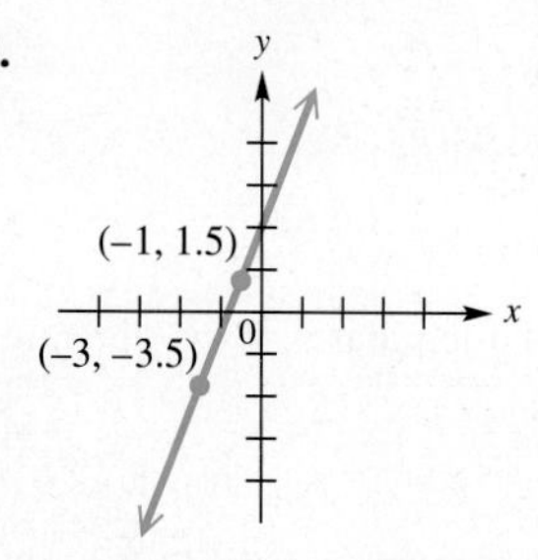


*Concept Check* Choose the letter of the graph that would most closely resemble the graph of $f(x) = mx + b$, given the conditions on $m$ and $b$.

**13.** $m < 0, b < 0$

**14.** $m > 0, b < 0$

**15.** $m < 0, b > 0$

**16.** $m > 0, b > 0$

**17.** $m = 0$

**18.** $b = 0$

**A.**

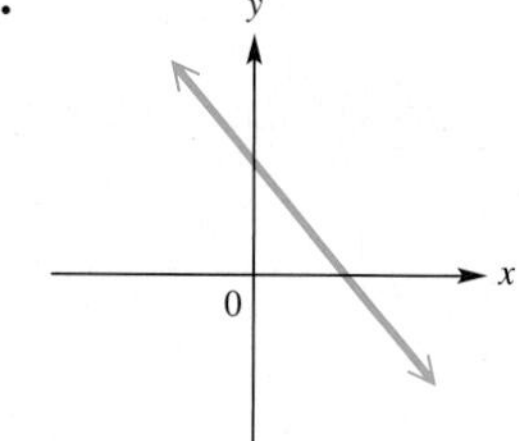


**B.**

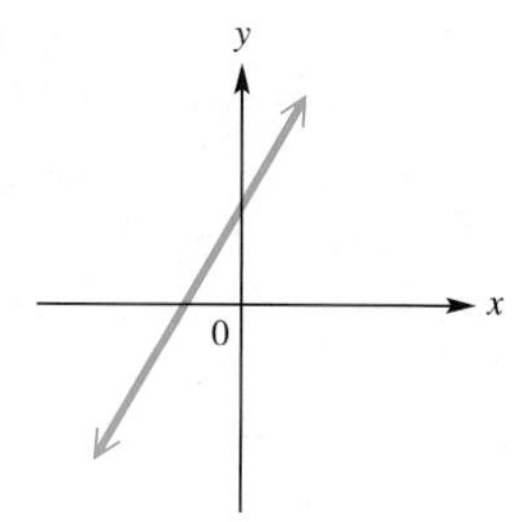


**C.**

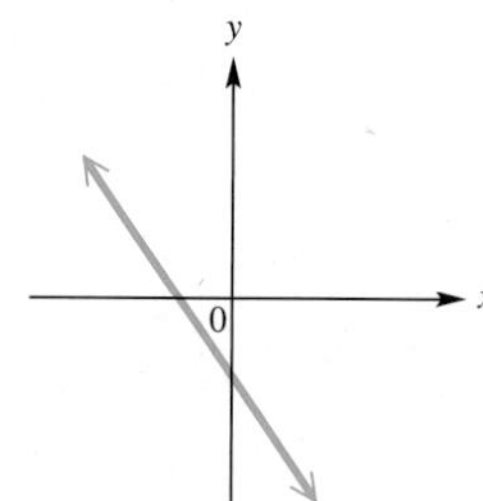


**D.**

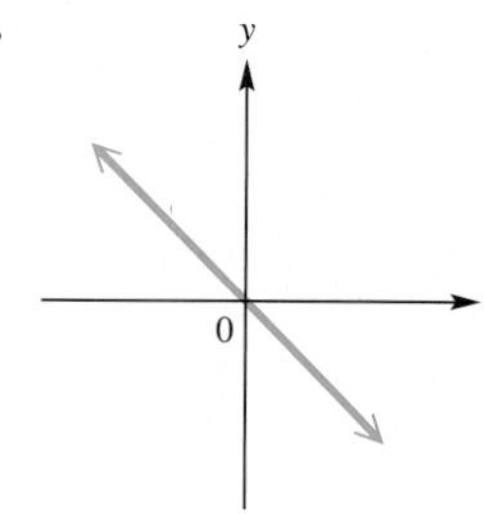


**E.**

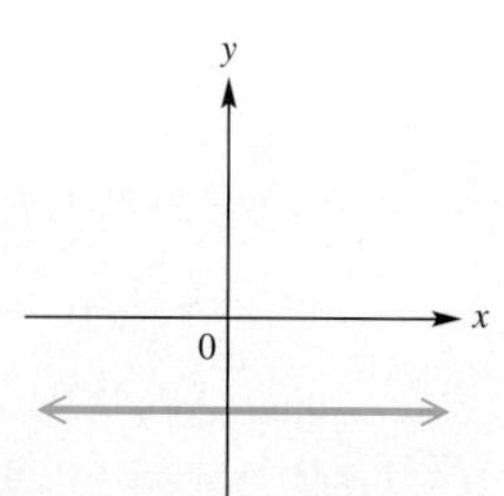


**F.**

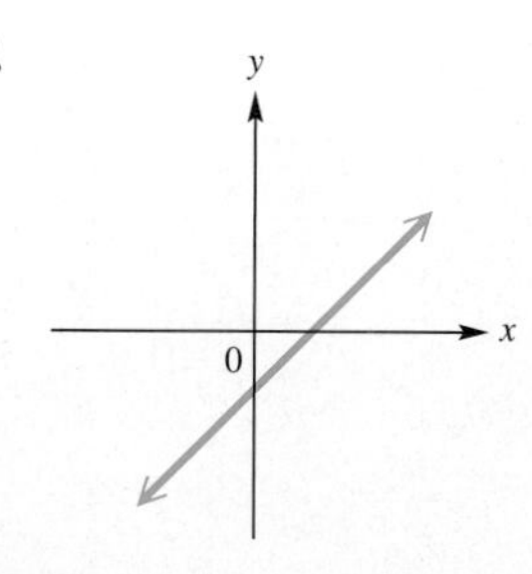


**19.** What are the domain and range of the relation graphed on the screen? (*Hint:* Pay attention to scale.)

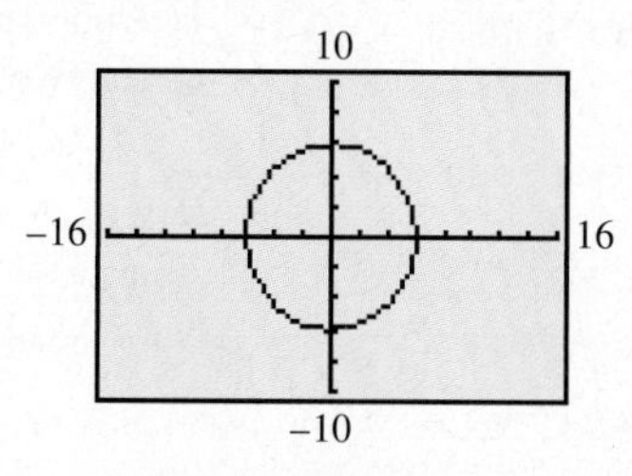

**20.** *Concept Check* True or false? The graphs of the equations $y_1 = 5.001x - 3$ and $y_2 = 5x + 6$ are shown in the screen. From this view, we may correctly conclude that these lines are parallel.

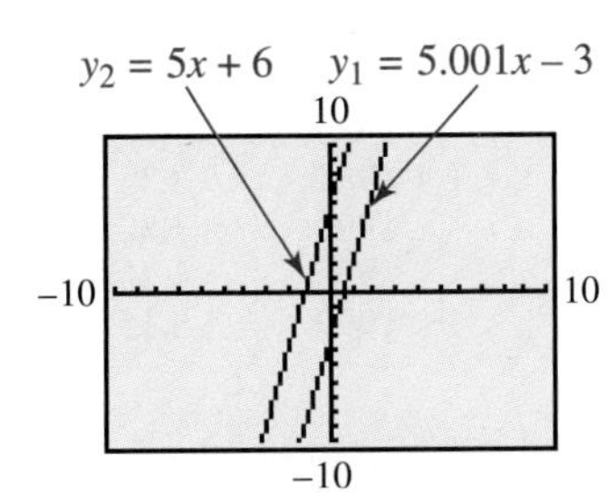


*Refer to the graphs of the linear functions defined by $y_1 = f(x)$ and $y_2 = g(x)$ in the figure to match the solution set in Column II with the equation or inequality in Column I. Choices may be used once, more than once, or not at all.*

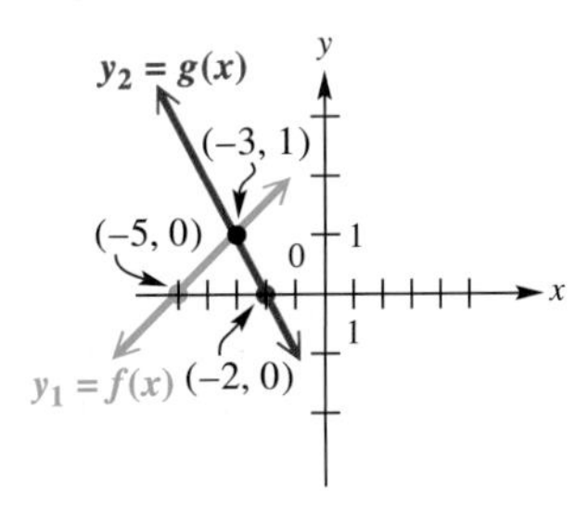


| I | II | |
|---|---|---|
| **21.** $f(x) = g(x)$ | **A.** $(-\infty, -3]$ | **I.** $\{-3\}$ |
| **22.** $f(x) > g(x)$ | **B.** $(-\infty, -3)$ | **J.** $[-3, \infty)$ |
| **23.** $f(x) < g(x)$ | **C.** $\{3\}$ | **K.** $(-3, \infty)$ |
| **24.** $g(x) \geq f(x)$ | **D.** $\{2\}$ | **L.** $\{1\}$ |
| **25.** $y_2 - y_1 = 0$ | **E.** $\{(3, 2)\}$ | **M.** $(-\infty, -5)$ |
| **26.** $f(x) < 0$ | **F.** $\{-5\}$ | **N.** $(-5, \infty)$ |
| **27.** $g(x) > 0$ | **G.** $\{-2\}$ | **O.** $(-\infty, -2)$ |
| **28.** $y_2 - y_1 < 0$ | **H.** $\{0\}$ | **P.** $(-2, \infty)$ |

*Solve each equation or inequality using analytic methods, if possible.*

**29.** $5[3 + 2(x - 6)] = 3x + 1$

**30.** $\frac{x}{4} - \frac{x + 4}{3} = -2$

**31.** $-3x - (4x + 2) = 3$

**32.** $-2x + 9 + 4x = 2(x - 5) - 3$

**33.** $.5x + .7(4 - 3x) = .4x$

**34.** $\frac{x}{4} - \frac{5x - 3}{6} = 2$

**35.** $x - 8 < 1 - 2x$

**36.** $\frac{4x - 1}{3} \geq \frac{x}{5} - 1$

**37.** Solve the inequality $-6 \leq \frac{4 - 3x}{7} < 2$ analytically.

**38.** Consider the linear function defined by

$$f(x) = 5\pi x + (\sqrt{3})x - 6.24(x - 8.1) + (\sqrt[3]{9})x.$$

**(a)** Solve the equation $f(x) = 0$ graphically. Give the solution to the nearest hundredth. Then, give an explanation of how you went about solving this equation graphically.

**(b)** Refer to your graph, and give the solution set of $f(x) < 0$.

**(c)** Refer to your graph, and give the solution set of $f(x) \geq 0$.

*(Modeling) Production of Audiocassettes* *A company produces studio-quality audiocassettes of live concerts. The company places an ad in a trade newsletter. The cost of the ad is \$150. Each tape costs \$30 to produce, and the company charges \$37.50 per tape. Use this information to work Exercises 39–42.*

**39.** Express the company's cost $C$ as a function of $x$, where $x$ is the number of tapes produced and sold.

**40.** Assuming that the company sells $x$ tapes, express the revenue as a function of $x$.

**41.** Determine analytically the value of $x$ for which revenue equals cost.

**42.** The graph shows $y = C(x)$ and $y = R(x)$. Discuss how the graph illustrates when the company is losing money, when it is breaking even, and when it is making a profit.

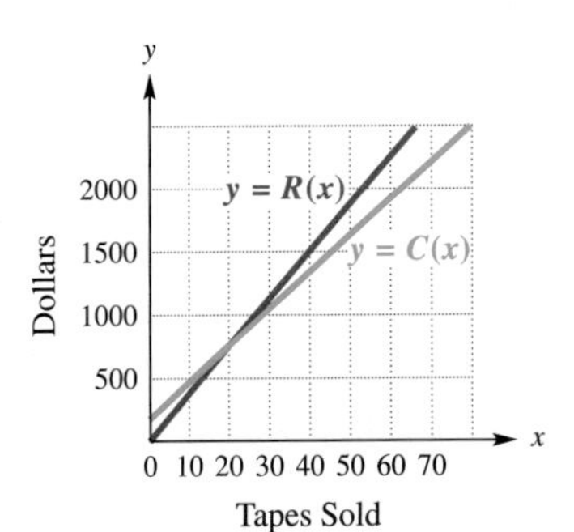

*Solve the formula for the variable indicated.*

**43.** $A = \dfrac{24f}{B(p+1)}$ for $f$ (Approximate annual interest rate)

**44.** $A = \dfrac{24f}{B(p+1)}$ for $B$ (Approximate annual interest rate)

*(Modeling) Solve each problem.*

**45.** *Temperature Levels above Earth's Surface* The linear function defined by

$$f(x) = -3.52x + 58.6$$

gives an approximation for the temperature in degrees Fahrenheit above the surface of Earth, where $x$ is in thousands of feet and $y$ is the temperature. (*Source:* Schwartz, Richard H., *Mathematics and Global Survival,* Ginn Press, 1991.)

**(a)** When the height is 5000 feet, what is the temperature? Solve analytically.

**(b)** When the temperature is $-15°$F, what is the height? Solve analytically.

**(c)** Explain how the answers in parts (a) and (b) can be supported graphically.

**46.** *Prevention of Indoor Pollutants* Kitchen gas ranges are a source of indoor pollutants such as carbon monoxide and nitrogen dioxide. One of the most effective ways of removing contaminants from the air while cooking is to use a *vented* range hood. If a range hood removes $F$ liters of air per second, then the percent $P$ of contaminants that are also removed from the surrounding air can be expressed by the linear function defined by

$$P = 1.06F + 7.18,$$

where $10 \le F \le 75$. What flow rate must a range hood have to remove 50% of the contaminants from the air? (*Source:* Rezvan, R. L., "Effectiveness of Local Ventilation in Removing Simulated Pollutants from Point Sources," in *Proceedings of the Third International Conference on Indoor Air Quality and Climate,* 1984.)

**47.** *Minimum Hourly Wage* The table shows the changes in the minimum hourly wage in the United States for selected years from 1938 to 1999.

| Year | Wage |
|---|---|
| 1938 | $ .25 |
| 1950 | .75 |
| 1956 | 1.00 |
| 1963 | 1.25 |
| 1974 | 2.00 |
| 1980 | 3.10 |
| 1981 | 3.35 |
| 1990 | 3.80 |
| 1991 | 4.25 |
| 1999 | 5.15 |

*Source:* U.S. Labor Department.

**(a)** Find a linear model for the changes in the minimum hourly wage.

**(b)** Use a graphing calculator to plot the data and your linear model. Comment on how well the data fit a linear model.

**(c)** The 1997 minimum wage was $5.15. Use the linear model from part (a) to see how close an approximation it gives for 1997.

**48.** *Indianapolis 500 Pole Speeds* The pole speeds for selected drivers in the Indianapolis 500 from 1980 to 1992 are given in the table. Use the linear regression capability of a graphing calculator to predict the year in which the pole speed should reach 250 mph.*

| Year | Driver | Pole Speed |
|---|---|---|
| 1980 | Johnny Rutherford | 192.256 mph |
| 1981 | Bobby Unser | 200.546 mph |
| 1982 | Rick Mears | 207.004 mph |
| 1983 | Teo Fabi | 207.395 mph |
| 1984 | Tom Sneva | 210.029 mph |
| 1985 | Pancho Carter | 212.583 mph |
| 1986 | Rick Mears | 216.828 mph |
| 1987 | Mario Andretti | 215.390 mph |
| 1988 | Rick Mears | 219.198 mph |
| 1989 | Rick Mears | 223.885 mph |
| 1990 | Emerson Fittipaldi | 225.301 mph |
| 1991 | Rick Mears | 224.113 mph |
| 1992 | Roberto Guerrero | 232.482 mph |

*The authors wish to thank Randall Leigh of the University of Southern Indiana for his input into this exercise.

**49.** *Speed of a Batted Baseball* Suppose a baseball is thrown at 85 mph. The ball will travel 320 feet when hit by a bat swung at 50 mph and 440 feet when hit by a bat swung at 80 mph. Let $y$ be the number of feet traveled by the ball when hit by a bat swung at $x$ mph. Find the equation of the line given by $y = mx + b$ that models the data. (*Note:* This function is valid for $50 \leq x \leq 90$, where the bat is 35 inches long, weighs 32 ounces, and is swung slightly upward to drive the ball at an angle of 35°.) How much farther will a ball travel for each 1-mph increase in the speed of the bat? (*Source:* Adair, Robert K., *The Physics of Baseball,* HarperCollins Publishers, 1990.)

*Solve each application of linear equations.*

**50.** *Dimensions of a Recycling Bin* A recycling bin is in the shape of a rectangular box. Find the height of the box if its length is 18 feet, its width is 8 feet, and its surface area is 496 square feet. (In the figure, let $h$ = height. Assume that the given surface area includes the top lid of the box.)

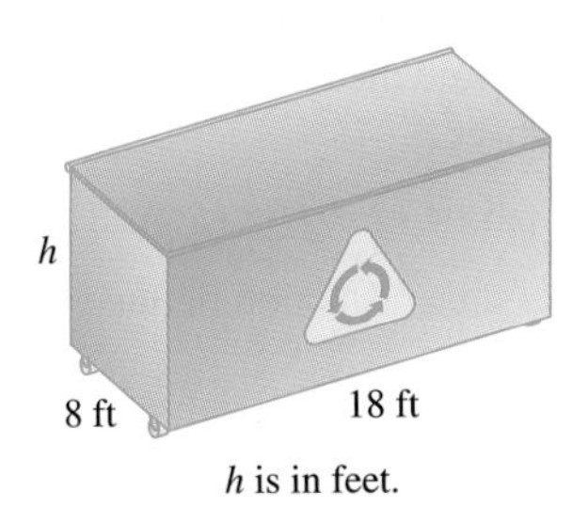

$h$ is in feet.

**51.** *Running Speeds in Track Events* In 1994 Leroy Burrell (USA) set a world record in the 100-meter dash with a time of 9.85 seconds. If this pace could be maintained for an entire 26-mile marathon, how would this time compare to the fastest time for a marathon of 2 hours, 6 minutes, and 50 seconds? (*Hint:* 1 meter ≈ 3.281 feet.) (*Source:* International Amateur Athletic Association.)

**52.** *Temperature of Venus* Venus is our solar system's hottest planet, with a surface temperature of 864°F. What is this temperature in Celsius? $\left(Hint\text{: Use } C = \frac{5}{9}(F - 32).\right)$ (*Source: Guinness Book of Records,* 1995.)

**53.** *Pressure in a Liquid* The pressure on a point in a liquid is directly proportional to the distance from the surface to the point. In a certain liquid, the pressure at a depth of 4 meters is 3000 kilograms per square meter. Find the pressure at a depth of 10 meters.

**54.** *Take-Home Pay* Meredith Many earns take-home pay of $592 a week. If her deductions for taxes, retirement, union dues, and medical plan amount to 26% of her wages, what is her weekly pay before deductions?

**55.** *(Modeling) Math SAT Scores* The table lists average math SAT scores for selected years.

| Year | 1994 | 1995 | 1996 | 1998 | 2000 |
|---|---|---|---|---|---|
| Score | 504 | 506 | 508 | 512 | 514 |

*Source:* The College Board.

**(a)** Find a linear function $f$ that models these data. (Answers may vary.)

**(b)** Use $f$ to approximate the average math SAT score in 1997.

**56.** *(Modeling) Data Analysis* The table lists data that is exactly linear.

| $x$ | −3 | −2 | −1 | 0 | 1 | 2 | 3 |
|---|---|---|---|---|---|---|---|
| $y$ | 6.6 | 5.4 | 4.2 | 3 | 1.8 | .6 | −.6 |

**(a)** Determine the slope-intercept form of the line that passes through these data points.

**(b)** Predict $y$ when $x = -1.5$ and when $x = 3.5$.

**57.** *Alcohol Mixture* A chemist wishes to strengthen a mixture that is 10% alcohol to one that is 30% alcohol. How much pure alcohol should be added to 12 liters of the 10% mixture?

**58.** *Acid Mixture* A student needs 10% hydrochloric acid for a chemistry experiment. How much 5% acid should be mixed with 120 milliliters of 20% acid to get a 10% solution?

**59.** *(Modeling) Videotape Production* A company produces videotapes. The revenue from the sale of $x$ units of tapes is $R(x) = 8x$. The cost to produce $x$ units of tapes is $C(x) = 3x + 1500$. Revenue and cost are in dollars. In what interval will the company at least break even?

**60.** *Intelligence Quotient* A person's intelligence quotient (IQ) is found by multiplying the mental age by 100 and dividing by the chronological age.

**(a)** Jack is 7 years old. His IQ is 130. Find his mental age.

**(b)** If a person is 16 years old with a mental age of 20, what is the person's IQ?

## CHAPTER 1 TEST

1. For each function, determine the

   **(i)** domain. **(ii)** range.

   **(iii)** $x$-intercept(s). **(iv)** $y$-intercept.

   **(a)**

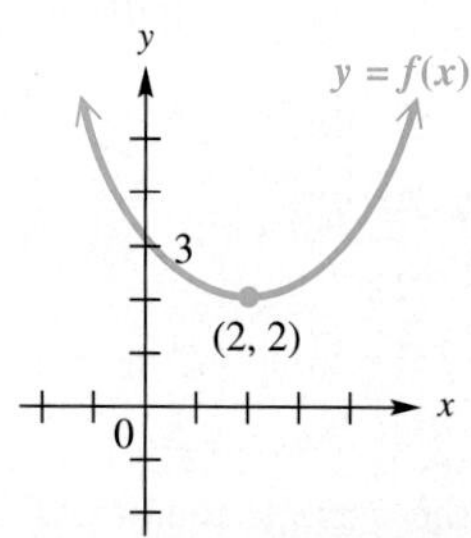


   **(b)**

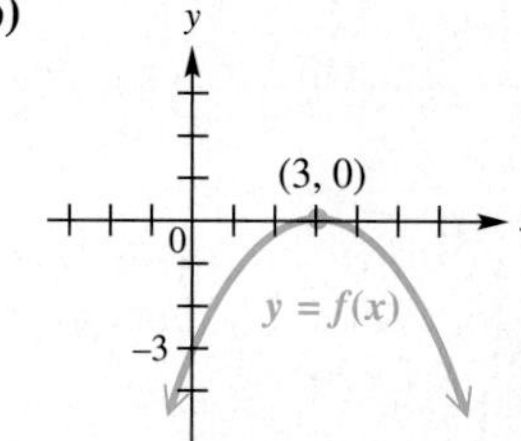


   **(c)**

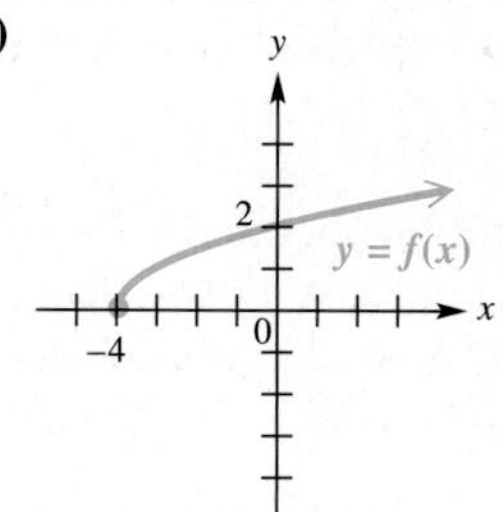


2. Use the figure to solve each equation or inequality.

   **(a)** $f(x) = g(x)$ **(b)** $f(x) < g(x)$

   **(c)** $f(x) \geq g(x)$ **(d)** $y_2 - y_1 = 0$

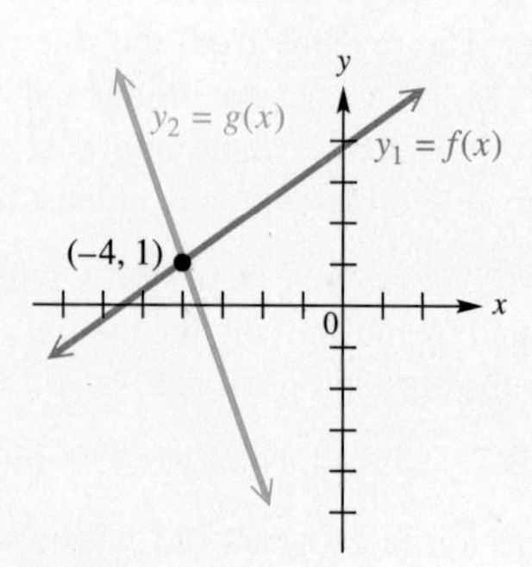


3. Use the screen to solve the equation or inequality. Here, $f$ is a linear function defined over the domain of real numbers.

   **(a)** $y_1 = 0$ **(b)** $y_1 < 0$

   **(c)** $y_1 > 0$ **(d)** $y_1 \leq 0$

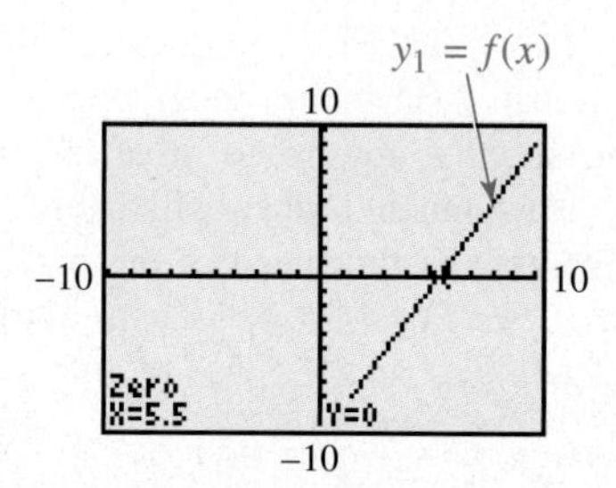


4. Consider the linear functions defined by $f(x) = 3(x - 4) - 2(x - 5)$ and $g(x) = -2(x + 1) - 3$.

   **(a)** Solve $f(x) = g(x)$ analytically, showing all steps. Also show an analytic check.

   **(b)** Graph $y_1 = f(x)$ and $y_2 = g(x)$. Use your result in part (a) to find the solution set of $f(x) > g(x)$. Explain your answer.

   **(c)** Repeat part (b) for $f(x) < g(x)$.

5. Consider the linear function defined by

$$f(x) = -\frac{1}{2}(8x + 4) + 3(x - 2).$$

   **(a)** Solve the equation $f(x) = 0$ analytically.

   **(b)** Solve the inequality $f(x) \leq 0$ analytically.

   **(c)** Graph $y = f(x)$ in an appropriate viewing window, and explain how the graph supports your answers in parts (a) and (b).

6. *(Modeling) Cable Television Rates* The graph depicts how the average monthly rates for cable television increased during the years 1980 through 2000.

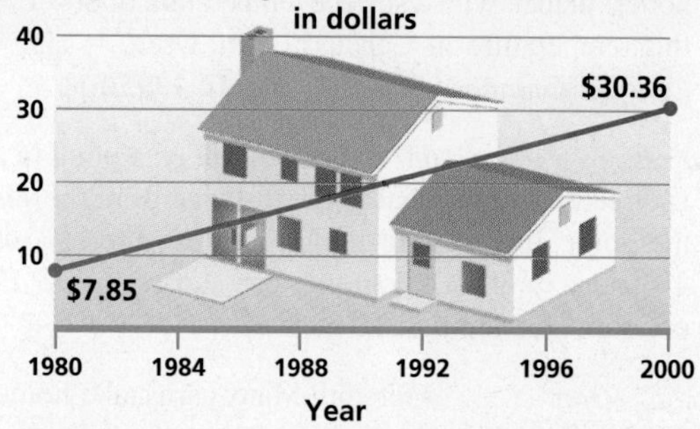


*Sources*: U.S. Bureau of the Census, Paul Kagan Associates.

(a) Use the midpoint formula to approximate the average monthly rate in 1990.

(b) Find the slope of the line, and explain its meaning in the context of this situation.

**7.** Find the equation of the line passing through the point $(-3, 5)$ and

(a) parallel to the line with equation $y = -2x + 4$.

(b) perpendicular to the line with equation $-2x + y = 0$.

**8.** Find the $x$- and $y$-intercepts of the line whose standard form is $3x - 4y = 6$. What is the slope of this line?

**9.** Give the equations of both the horizontal and vertical lines passing through the point $(-3, 7)$.

**10.** *(Modeling) Wind-Chill Factor* The table shows the wind-chill factor for various wind speeds when the Fahrenheit temperature is 40°.

| Wind Speed (mph) | Degrees |
|---|---|
| 10 | 28 |
| 15 | 22 |
| 20 | 18 |
| 25 | 16 |
| 30 | 13 |
| 35 | 11 |

(a) Find the least-squares regression line for the data. Give the correlation coefficient.

(b) Use this line to predict the wind-chill factor when the wind speed is 40 mph.

**11.** *(Modeling) Rainfall* Suppose that during a storm, rain is falling at a rate of 1 inch per hour. The water coming from a circular roof with a radius of 20 feet is running down a downspout that can accommodate 400 gallons of water per hour.

(a) Determine the number of cubic inches of water falling on the roof in one hour. (*Hint:* Each hour a layer of water 1 inch thick falls on the roof.)

(b) One gallon equals about 231 cubic inches. Write a formula for a function $g$ that computes the gallons of water landing on the roof in $x$ hours.

(c) How many gallons of water land on the roof during a 2.5-hour rain storm?

(d) Will one downspout be sufficient to handle this type of rainfall? If not, how many downspouts should there be?

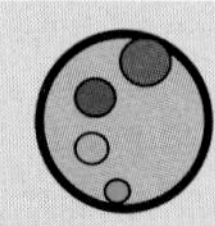

# Chapter 1 Project

## Predicting Heights and Weights of Athletes

In this chapter project we investigate the height–weight relationship among professional athletes. The table lists the heights and weights of 10 female and 10 male professional basketball players.

| Female Basketball Players | | Male Basketball Players | |
|---|---|---|---|
| **Height (in.)** | **Weight (lb)** | **Height (in.)** | **Weight (lb)** |
| 74 | 169 | 74 | 197 |
| 74 | 181 | 75 | 202 |
| 77 | 199 | 71 | 173 |
| 64 | 139 | 77 | 220 |
| 68 | 135 | 83 | 220 |
| 71 | 161 | 83 | 250 |
| 78 | 187 | 81 | 225 |
| 68 | 150 | 76 | 219 |
| 71 | 181 | 79 | 215 |
| 73 | 194 | 85 | 246 |

*Source:* WNBA; NBA.

Notice that taller athletes tend to be heavier, but not always. For example, there is a 64-inch female athlete who weighs more than a 68-inch female athlete.

## Activities

1. Find the least-squares regression line that models the height–weight relationship among the 10 female basketball players. Let $y = mx + b$, where $x$ represents height in inches and $y$ represents weight in pounds. What is the correlation coefficient, and what does it indicate?
2. Use this regression line to predict the weight of a female basketball player who is 75 inches tall.
3. For each 1-inch increase in height, predict the corresponding increase in weight for female basketball players.
4. Now find the least-squares regression line that models the height–weight relationship among the 10 male basketball players. What is the correlation coefficient, and what does it indicate?
5. Predict the weight of a male basketball player who is 80 inches tall.
6. For each 1-inch increase in height, predict the corresponding increase in weight for male basketball players.

# 2 Analysis of Graphs of Functions

A FIGURE HAS *rotational symmetry* about an axis *I* if it coincides with itself by all rotations about *I*. Because of their complete rotational symmetry, the circle in the plane and the sphere in space were considered by the early Greeks to be the most perfect geometric figures. Aristotle assumed a spherical shape for the celestial bodies because any other would detract from their heavenly perfection.

Symmetry has been an important characteristic of art from the earliest times. The art of M.C. Escher (1898–1972) is composed of symmetries and translations. Leonardo da Vinci's sketches indicate a superior understanding of symmetry. Almost all nature exhibits symmetry—from the hexagons of snowflakes to the diatom, a microscopic sea plant. Perhaps the most striking examples of symmetry in nature are crystals. The photos on this page show a diatom and a cross section of a tourmaline.

This chapter introduces, among other things, how symmetry is found in the graphs of certain functions.

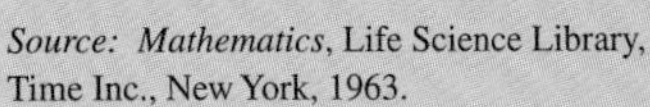

*Source: Mathematics*, Life Science Library, Time Inc., New York, 1963.

CHAPTER OUTLINE

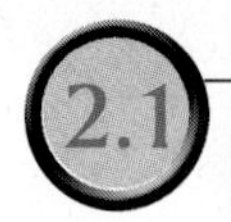

# 2.1 GRAPHS OF BASIC FUNCTIONS AND RELATIONS; SYMMETRY

Continuity • Increasing and Decreasing Functions • The Identity Function • The Squaring Function and Symmetry with Respect to the $y$-Axis • The Cubing Function and Symmetry with Respect to the Origin • The Square Root and Cube Root Functions • The Absolute Value Function • The Relation $x = y^2$ and Symmetry with Respect to the $x$-Axis • Even and Odd Functions

## Continuity

In Chapter 1, our work dealt mainly with linear functions. The graph of a linear function, a straight line, may be drawn by hand over any interval of its domain without picking the pencil up from the paper. In mathematics we say that a function with this property is *continuous* over any interval. The formal definition of continuity requires concepts from calculus, but we can give an informal definition at the precalculus level.

### Continuity

A function is **continuous** over an interval of its domain if its hand-drawn graph over that interval can be sketched without lifting the pencil from the paper.

If a function is not continuous at a point, then it may have a point of *discontinuity* (Figure 1(a)), or it may have a vertical *asymptote* (a vertical line that the graph does not intersect, as in Figure 1(b)). Asymptotes will be discussed further in Chapter 4.

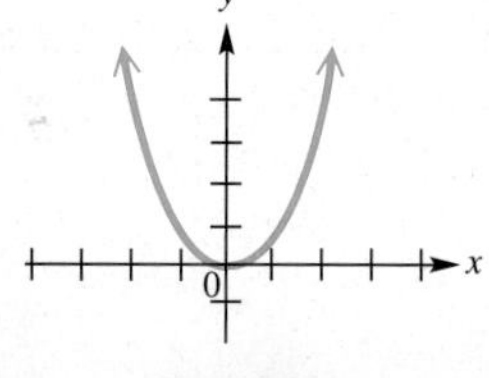


FIGURE 2

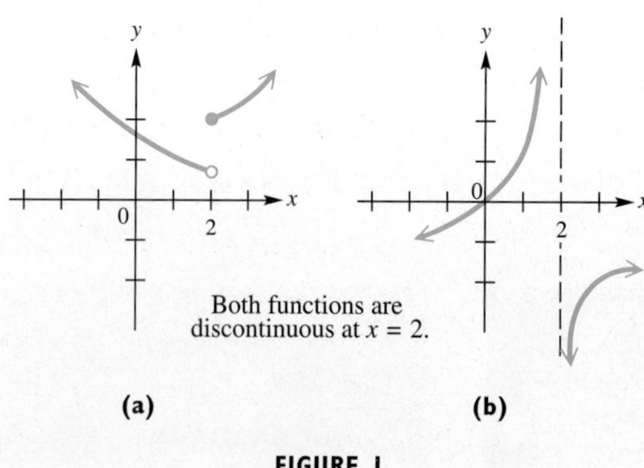


FIGURE 1

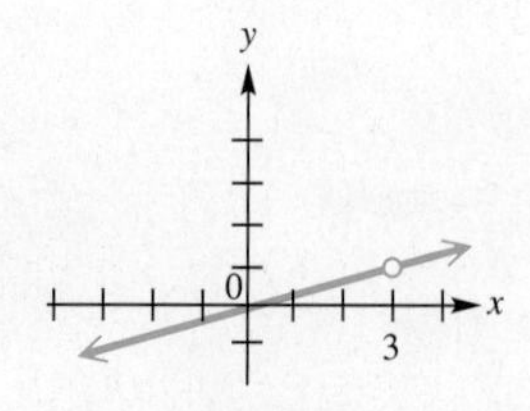


FIGURE 3

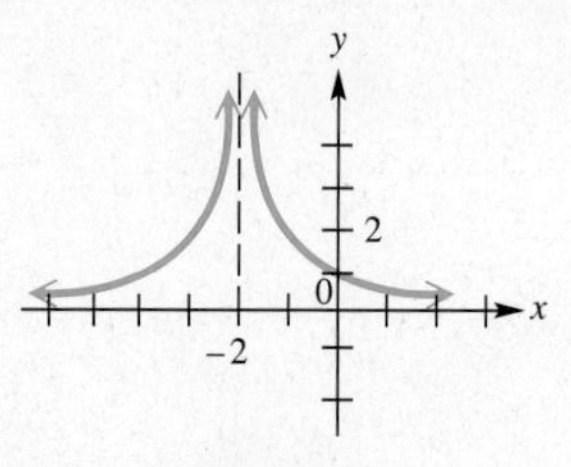


FIGURE 4

**EXAMPLE 1** *Determining Intervals of Continuity*

Figures 2, 3, and 4 show graphs of functions. Indicate the intervals of the domain over which they are continuous.

**Solution** The function in Figure 2 is continuous over the entire domain of real numbers, $(-\infty, \infty)$. The function in Figure 3 has a point of discontinuity at $x = 3$. It is continuous over the interval $(-\infty, 3)$ and the interval $(3, \infty)$. Finally, the function in Figure 4 has a vertical asymptote at $x = -2$, as indicated by the dashed line. It is continuous over the intervals $(-\infty, -2)$ and $(-2, \infty)$.

## Increasing and Decreasing Functions

If a continuous function is not constant over an interval, then its graph will either rise from left to right or fall from left to right or some combination of both. We use the words *increasing* and *decreasing* to describe this behavior. For example, a linear function with positive slope is increasing over its entire domain, while one with negative slope is decreasing. See Figure 5.

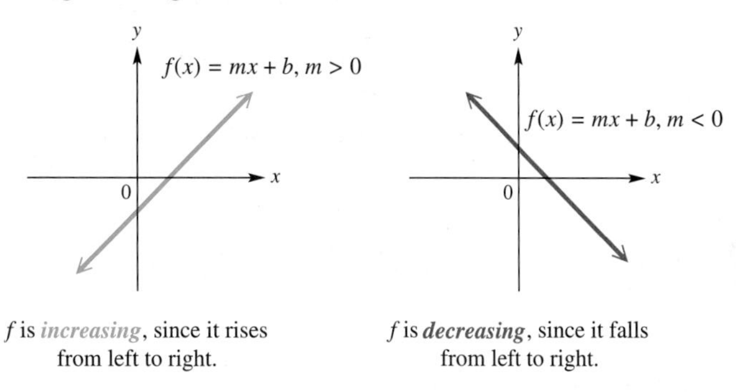


*f* is *increasing*, since it rises from left to right.

*f* is *decreasing*, since it falls from left to right.

**FIGURE 5**

| X | Y1 |
|---|---|
| 0 | -1 |
| 1 | 0 |
| 2 | 1 |
| 3 | 2 |
| 4 | 3 |
| 5 | 4 |
| 6 | 5 |

Y1=X-1

**(a)**

| X | Y2 |
|---|---|
| 0 | 1 |
| 1 | 0 |
| 2 | -1 |
| 3 | -2 |
| 4 | -3 |
| 5 | -4 |
| 6 | -5 |

Y2=-X+1

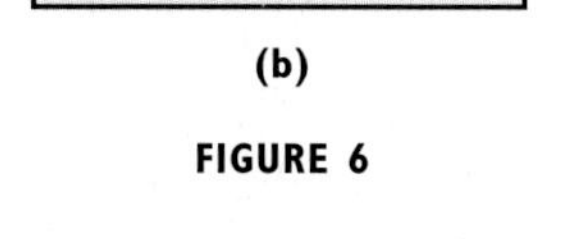

**(b)**

**FIGURE 6**

The table feature of a graphing calculator can be used to illustrate the concepts of increasing and decreasing functions. As seen in Figure 6(a), the graph of $Y_1 = X - 1$ has positive slope ($m = 1$), and as $x$-values increase, range values *increase.* The graph of $Y_2 = -X + 1$ has negative slope ($m = -1$), and as $x$-values increase, range values *decrease*, as shown in Figure 6(b).

Informally speaking, a function *increases* on an interval of its domain if its graph rises from left to right on the interval. A function *decreases* on an interval of its domain if its graph falls from left to right on the interval. A function is *constant* on an interval of its domain if its graph is horizontal on the interval. (While some functions are increasing, decreasing, or constant over their entire domains, many are not. Hence, we speak of increasing, decreasing, or constant over an interval of the domain.)

**Looking Ahead to Calculus**

In calculus, determining where a function increases, decreases, or is constant is important in curve-sketching. The *derivative* of a function provides a formula for determining the slope of a line tangent to the curve. If the slope is positive for a given domain value, the function is increasing at that point; if the slope is negative, the function is decreasing; and if the slope is 0, the function is constant. For example, look at Figure 7(a). Choose a point on the curve between $(x_1, f(x_1))$ and $(x_2, f(x_2))$, and draw a line tangent to the curve at that point. The line rises from left to right, so its slope is positive. This positive slope indicates that the function is increasing at that point. (Extend this idea to Figures 7(b) and 7(c).)

### Criteria for Increasing, Decreasing, and Constant

Suppose that a function *f* is defined over an interval *I*.

**(a)** *f* **increases** on *I* if, whenever $x_1 < x_2$, $f(x_1) < f(x_2)$;

**(b)** *f* **decreases** on *I* if, whenever $x_1 < x_2$, $f(x_1) > f(x_2)$;

**(c)** *f* is **constant** on *I* if, for every $x_1$ and $x_2$, $f(x_1) = f(x_2)$.

Figure 7 illustrates these ideas.

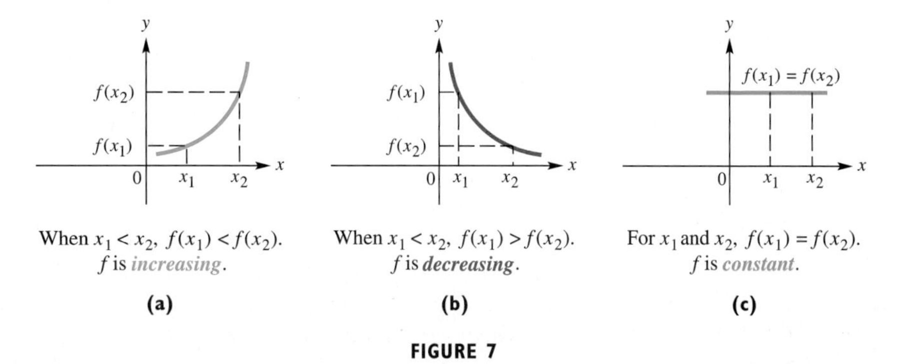


When $x_1 < x_2$, $f(x_1) < f(x_2)$. *f* is *increasing*.

**(a)**

When $x_1 < x_2$, $f(x_1) > f(x_2)$. *f* is *decreasing*.

**(b)**

For $x_1$ and $x_2$, $f(x_1) = f(x_2)$. *f* is *constant*.

**(c)**

**FIGURE 7**

**NOTE** To decide whether a function is increasing, decreasing, or constant on an interval, ask yourself **"What does $y$ do as $x$ goes from left to right?"**

**EXAMPLE 2** *Determining Intervals Over Which a Function Is Increasing, Decreasing, or Constant*

Figure 8 shows the graph of a function. Determine the intervals over which the function is increasing, decreasing, or constant.

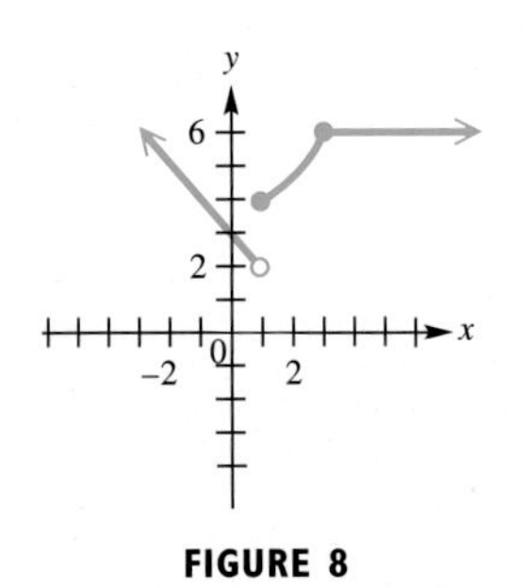


FIGURE 8

**Solution** We should ask, "What is happening to the $y$-values as the $x$-values are getting larger?" Moving from left to right on the graph, we see that on the interval $(-\infty, 1)$, the $y$-values are *decreasing*; on the interval $[1, 3]$, the $y$-values are *increasing*; and on the interval $[3, \infty)$, the $y$-values are *constant* (and equal to 6). Therefore, the function is decreasing on $(-\infty, 1)$, increasing on $[1, 3]$, and constant on $[3, \infty)$.

**CAUTION** A common error involves writing range values ($y$-values) when determining intervals like those in Example 2. Remember that we are determining intervals of the domain, and thus are interested in $x$-values for our interval designations. Range values are *not* written when designating such intervals.

**FOR DISCUSSION**

1. In the standard viewing window of your calculator, enter any linear function defined by $y = mx + b$ with $m > 0$. Now, trace the graph from *left to right*. Watch the $y$-values as the $x$-values get larger. What is happening to $y$? How does this reinforce the concepts presented so far in this section?
2. Repeat this exercise, but with $m < 0$.
3. Repeat this exercise, but with $m = 0$.

The next part of this section introduces several important basic functions and relations.

## The Identity Function

If we let $m = 1$ and $b = 0$ in the general form of the linear function defined by $f(x) = mx + b$, we get the **identity function,** or $f(x) = x$. This function pairs every real number with itself. See Figure 9.

**FUNCTION CAPSULE**

**IDENTITY FUNCTION** $\quad f(x) = x$

Domain: $(-\infty, \infty)$ $\quad$ Range: $(-\infty, \infty)$

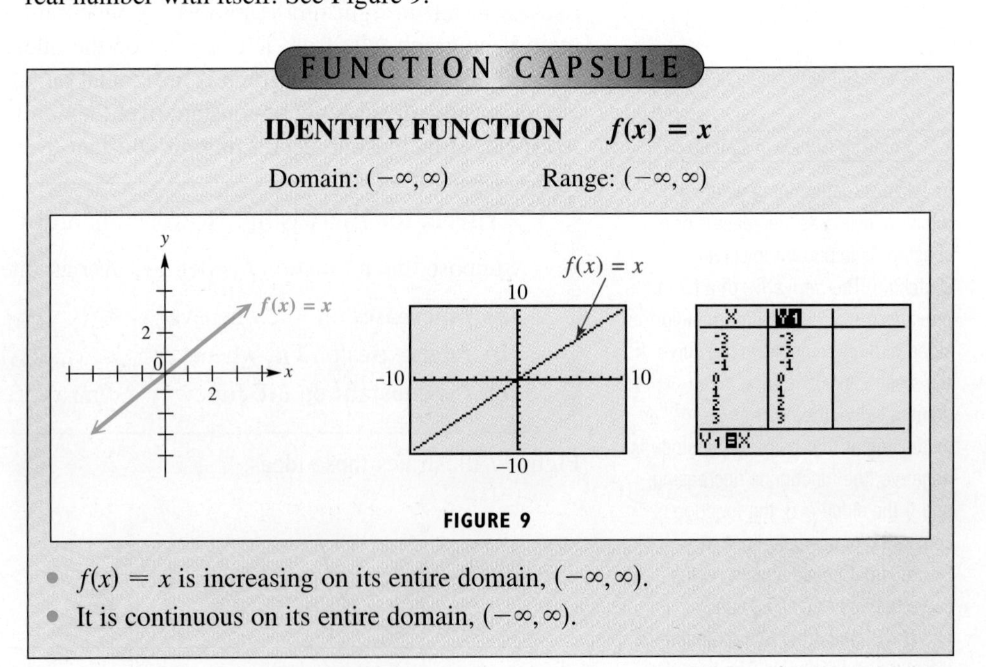


FIGURE 9

- $f(x) = x$ is increasing on its entire domain, $(-\infty, \infty)$.
- It is continuous on its entire domain, $(-\infty, \infty)$.

## The Squaring Function and Symmetry with Respect to the y-Axis

The **degree** of a polynomial function in the variable $x$ is the greatest exponent on $x$ in the polynomial. The identity function is a degree 1 function. The **squaring function,** defined by $f(x) = x^2$, is the simplest degree 2 (or *quadratic*) function. (The word *quadratic* refers to degree 2.) See Figure 10.

The squaring function pairs every real number with its square. Its graph is called a **parabola.** The point at which the graph changes from decreasing to increasing (the point $(0, 0)$) is called the **vertex** of the graph. (For a parabola that opens downward, the vertex is the point at which the graph changes from increasing to decreasing.)

**Looking Ahead to Calculus**

Calculus allows us to find areas of regions in the plane. For example, we can find the area of the region below the graph of $y = x^2$, above the $x$-axis, bounded on the left by the line $x = -2$, and bounded on the right by $x = 2$. Draw a sketch of this region, and notice that due to the symmetry of the graph of $y = x^2$, the desired area is twice that of the area to the right of the $y$-axis. Thus, symmetry can be used to reduce the original problem to an easier one by simply finding the area to the right of the $y$-axis, and then doubling the answer.

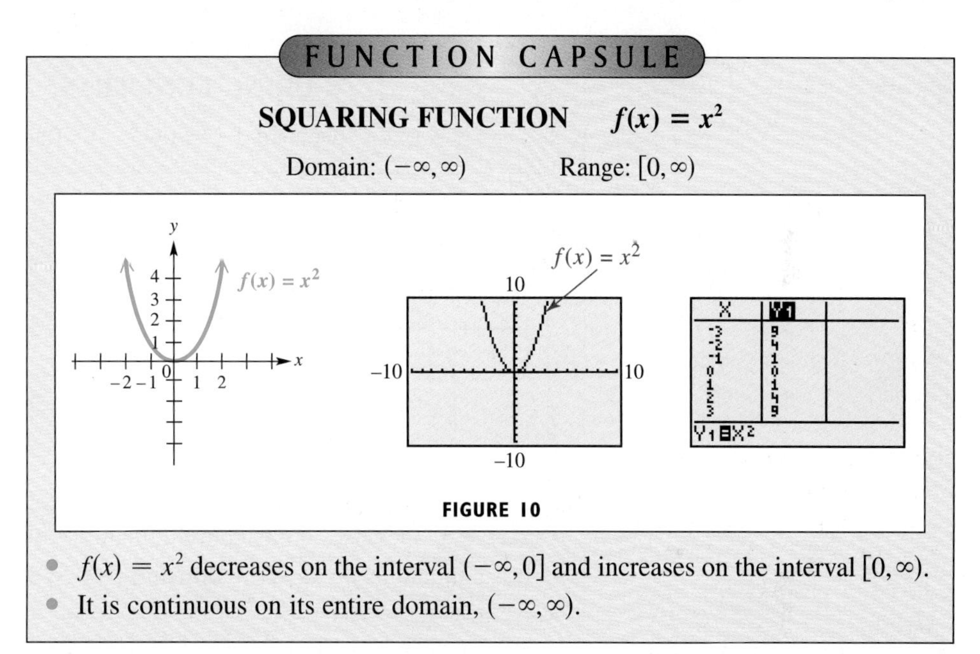


FIGURE 10

- $f(x) = x^2$ decreases on the interval $(-\infty, 0]$ and increases on the interval $[0, \infty)$.
- It is continuous on its entire domain, $(-\infty, \infty)$.

If we were able to "fold" the graph of $f(x) = x^2$ along the $y$-axis, the two halves would coincide exactly. In mathematics we refer to this property as symmetry, and we say that the graph of $f(x) = x^2$ is *symmetric with respect to the y-axis.*

**Symmetry with Respect to the $y$-Axis**

If a function $f$ is defined so that

$$f(-x) = f(x)$$

for all $x$ in its domain, then the graph of $f$ is **symmetric with respect to the $y$-axis.**

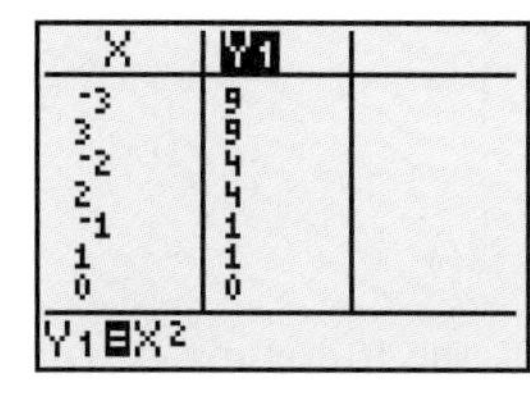

| X | Y1 |
|---|---|
| -3 | 9 |
| 3 | 9 |
| -2 | 4 |
| 2 | 4 |
| -1 | 1 |
| 1 | 1 |
| 0 | 0 |

Y1 = X²

This table supports, but does not *prove*, that the graph of the squaring function is symmetric with respect to the $y$-axis.

To illustrate that the graph of $f(x) = x^2$ is symmetric with respect to the $y$-axis, note that

$$f(-4) = f(4) = 16 \qquad f(-3) = f(3) = 9$$
$$f(-2) = f(2) = 4 \qquad f(-1) = f(1) = 1.$$

This pattern holds for any real number $x$, since

$$f(-x) = (-x)^2 = x^2 = f(x).$$

In general, if a graph is symmetric with respect to the $y$-axis, the following is true: *If $(a, b)$ is on the graph, so is $(-a, b)$.*

## The Cubing Function and Symmetry with Respect to the Origin

The **cubing function,** defined by $f(x) = x^3$, is the simplest degree 3 function. It pairs with each real number the third power, or cube, of the number. See Figure 11. The point at which the graph of the cubing function changes from "opening downward" to "opening upward" (the point $(0, 0)$) is called an **inflection point.**

FUNCTION CAPSULE

**CUBING FUNCTION** $\quad f(x) = x^3$

Domain: $(-\infty, \infty)$ $\qquad$ Range: $(-\infty, \infty)$

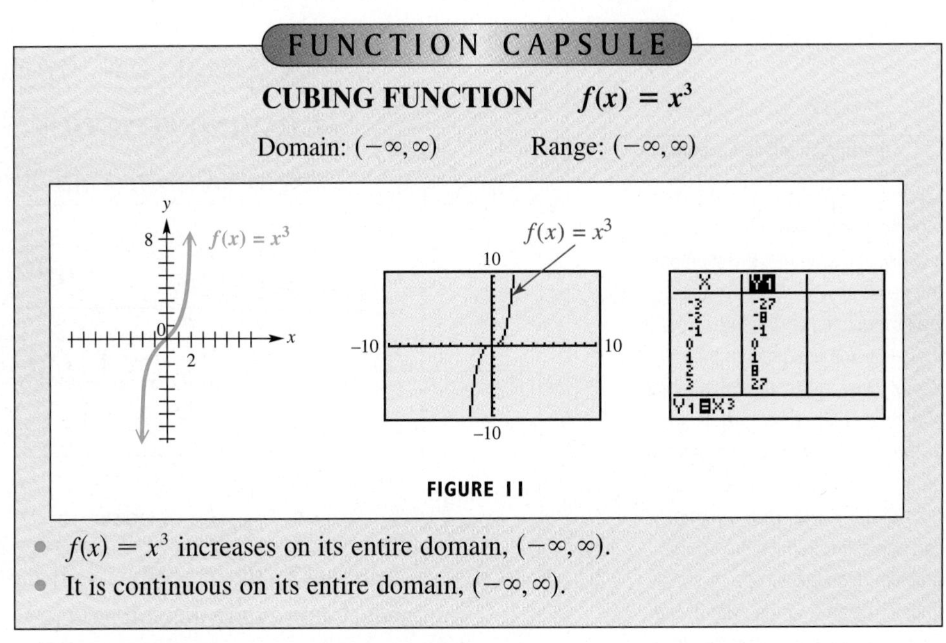


**FIGURE 11**

- $f(x) = x^3$ increases on its entire domain, $(-\infty, \infty)$.
- It is continuous on its entire domain, $(-\infty, \infty)$.

If we were able to "fold" the graph of $f(x) = x^3$ along the $y$-axis and then along the $x$-axis, forming a "corner" at the origin, the two parts of the graph would coincide exactly. We say that the graph of $f(x) = x^3$ is *symmetric with respect to the origin.*

**Symmetry with Respect to the Origin**

If a function $f$ is defined so that

$$f(-x) = -f(x)$$

for all $x$ in its domain, then the graph of $f$ is **symmetric with respect to the origin.**

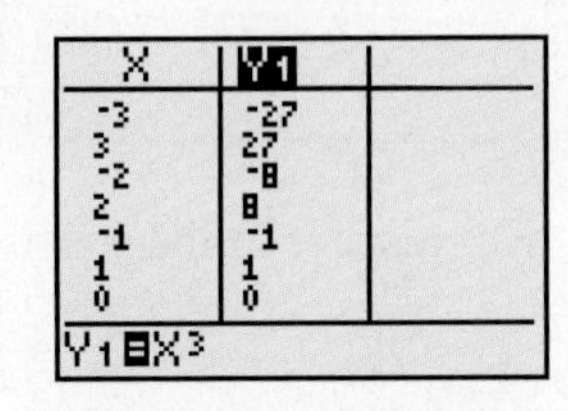

| X | Y1 |
|---|---|
| -3 | -27 |
| 3 | 27 |
| -2 | -8 |
| 2 | 8 |
| -1 | -1 |
| 1 | 1 |
| 0 | 0 |

Y1■X³

This table supports, but does not *prove,* that the graph of the cubing function is symmetric with respect to the origin.

To illustrate that the graph of $f(x) = x^3$ is symmetric with respect to the origin, note that

$$f(-2) = -f(2) = -8 \qquad f(-1) = -f(1) = -1.$$

This pattern holds for any real number $x$, since

$$f(-x) = (-x)^3 = (-1)^3x^3 = -x^3 = -f(x).$$

In general, if a graph is symmetric with respect to the origin, the following is true: *If $(a, b)$ is on the graph, so is $(-a, -b)$.*

**EXAMPLE 3** *Determining Symmetry*

Show that **(a)** $f(x) = x^4 - 3x^2 - 8$ has a graph that is symmetric with respect to the $y$-axis; **(b)** $f(x) = x^3 - 4x$ has a graph that is symmetric with respect to the origin.

### Analytic Solution

**(a)** We must show that $f(-x) = f(x)$ for any $x$.

$$f(-x) = (-x)^4 - 3(-x)^2 - 8 \quad \text{Substitute } -x \text{ for } x.$$
$$= (-1)^4x^4 - 3(-1)^2x^2 - 8$$
$$= x^4 - 3x^2 - 8$$
$$= f(x)$$

This *proves* that the graph is symmetric with respect to the $y$-axis.

**(b)** We must show that $f(-x) = -f(x)$ for any $x$.

$$f(-x) = (-x)^3 - 4(-x) \quad \text{Substitute } -x \text{ for } x.$$
$$= (-1)^3x^3 + 4x$$
$$= -x^3 + 4x \quad *$$
$$= -(x^3 - 4x)$$
$$= -f(x)$$

In the line denoted *, note that the signs of the coefficients are all *opposites* of those in $f(x)$. We completed the solution by factoring out $-1$, showing that the final result is $-f(x)$.

### Graphing Calculator Solution

**(a)** The graph in Figure 12 supports the analytic proof.

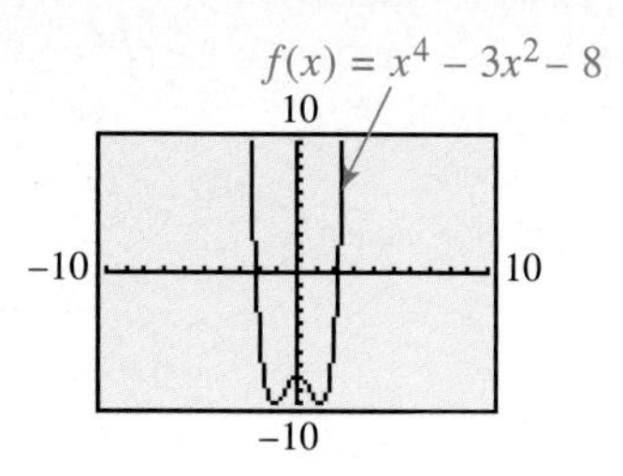


This graph is symmetric with respect to the $y$-axis.

**FIGURE 12**

**(b)** The graph in Figure 13 supports the analytic proof.

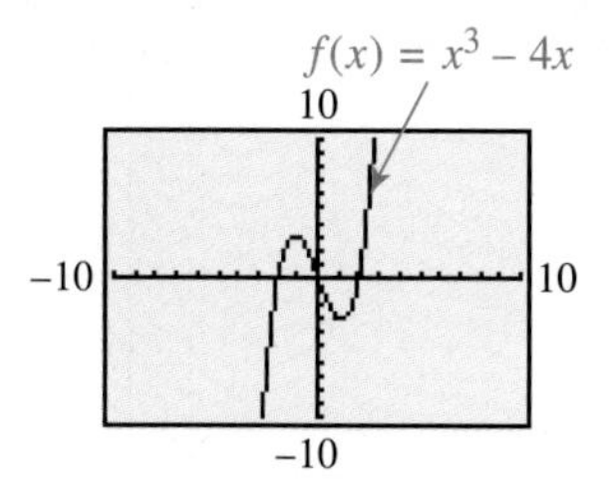


This graph is symmetric with respect to the origin.

**FIGURE 13**

## The Square Root and Cube Root Functions

Some functions involve expressions with radicals. One of these is the **square root function,** defined by $f(x) = \sqrt{x}$. See Figure 14. Notice that for the function value to be a real number, we must have $x \geq 0$. Thus, the domain is restricted to $[0, \infty)$.

**FUNCTION CAPSULE**

**SQUARE ROOT FUNCTION** $\quad f(x) = \sqrt{x}$

Domain: $[0, \infty)$ Range: $[0, \infty)$

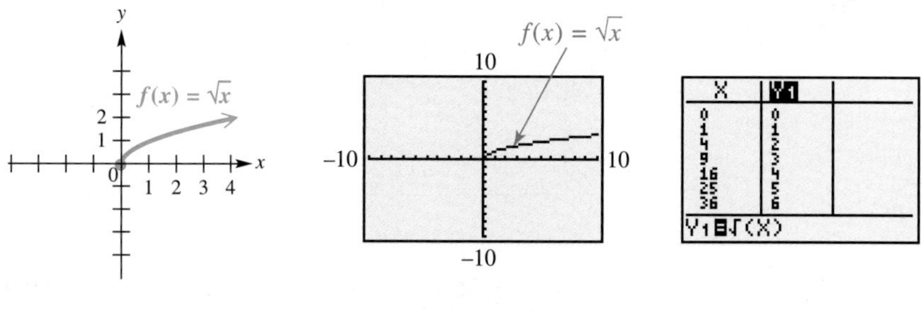

**FIGURE 14**

- $f(x) = \sqrt{x}$ increases on its entire domain, $[0, \infty)$.
- It is continuous on its entire domain, $[0, \infty)$.

TECHNOLOGY NOTE

The definition of a rational exponent allows us to enter $\sqrt{x}$ as $x^{1/2}$ and $\sqrt[3]{x}$ as $x^{1/3}$ when using a calculator. Further discussion of rational exponents can be found in Chapter R.

The **cube root function,** defined by $f(x) = \sqrt[3]{x}$ (Figure 15), differs from the square root function in that *any* real number—positive, 0, or *negative*—has a real cube root, and thus the domain is $(-\infty, \infty)$.

$$\begin{aligned} &\text{When} \quad x > 0, \quad \sqrt[3]{x} > 0, \\ &\text{when} \quad x = 0, \quad \sqrt[3]{x} = 0, \quad \text{and} \\ &\text{when} \quad x < 0, \quad \sqrt[3]{x} < 0. \end{aligned}$$

As a result, the range is $(-\infty, \infty)$.

FUNCTION CAPSULE

**CUBE ROOT FUNCTION** $f(x) = \sqrt[3]{x}$

Domain: $(-\infty, \infty)$ Range: $(-\infty, \infty)$

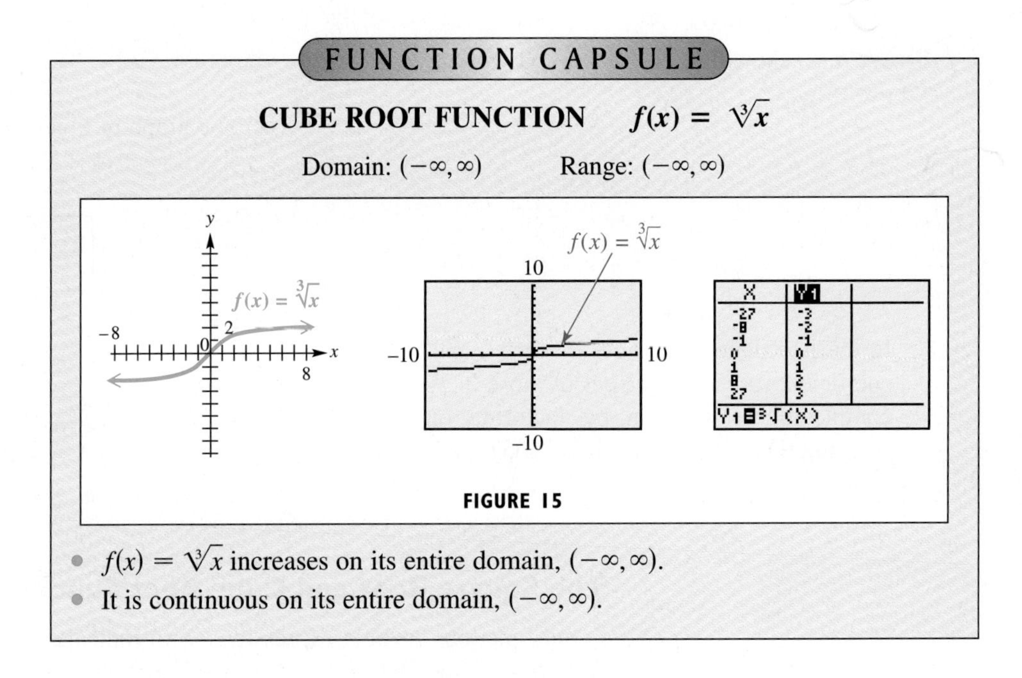


**FIGURE 15**

- $f(x) = \sqrt[3]{x}$ increases on its entire domain, $(-\infty, \infty)$.
- It is continuous on its entire domain, $(-\infty, \infty)$.

TECHNOLOGY NOTE

You should become familiar with the command on your particular calculator that allows you to graph the absolute value function. Consult the graphing calculator manual that accompanies this text or your owner's manual.

## The Absolute Value Function

On a number line, the absolute value of a real number $x$, denoted $|x|$, represents its undirected distance from the origin, 0. The **absolute value function,** which pairs every real number with its absolute value, is defined as follows.

**Absolute Value $|x|$**

$$f(x) = |x| = \begin{cases} x & \text{if } x \geq 0 \\ -x & \text{if } x < 0 \end{cases}$$

Notice that this function is defined in two parts. We use $|x| = x$ if $x$ is positive or 0, and we use $|x| = -x$ if $x$ is negative. Since $x$ can be any real number, the domain of the absolute value function is $(-\infty, \infty)$, but since $|x|$ cannot be negative, the range is $[0, \infty)$. See Figure 16.

**FUNCTION CAPSULE**

**ABSOLUTE VALUE FUNCTION** $f(x) = |x|$

Domain: $(-\infty, \infty)$ Range: $[0, \infty)$

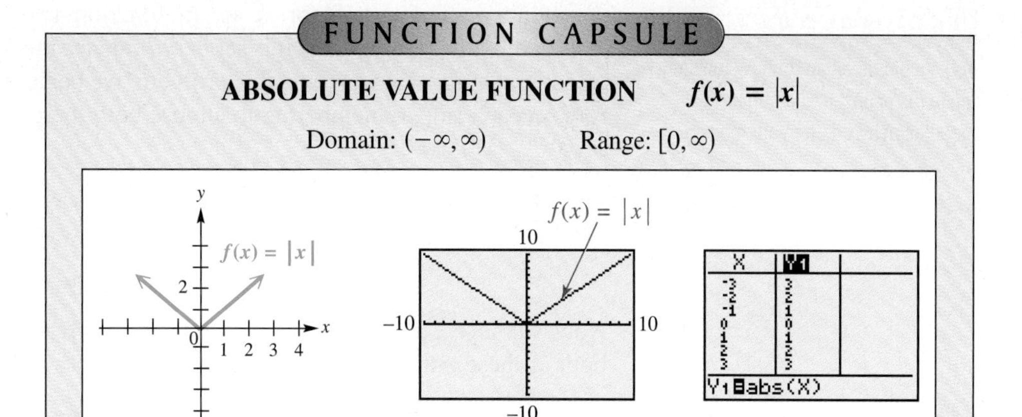


FIGURE 16

- $f(x) = |x|$ decreases on the interval $(-\infty, 0]$ and increases on the interval $[0, \infty)$.
- It is continuous on its entire domain, $(-\infty, \infty)$.

**FOR DISCUSSION**

Based on the functions discussed so far in this section, answer each question.

1. Which functions have graphs that are symmetric with respect to the $y$-axis?
2. Which functions have graphs that are symmetric with respect to the origin?
3. Which functions have graphs with neither of these symmetries?
4. Why is it not possible for the graph of a nonzero function to be symmetric with respect to the $x$-axis?

## The Relation $x = y^2$ and Symmetry with Respect to the $x$-Axis

Recall that a function is a relation where every domain value is paired with one and only one range value. However, there are cases where we are interested in graphing relations that are not functions, and one of the simplest of these is the relation defined by the equation $x = y^2$. Notice that the table of selected ordered pairs below indicates that this relation has two different $y$-values for each positive value of $x$. If we plot these points and join them with a smooth curve, we find that the graph of $x = y^2$ is a parabola opening to the right. See Figure 17.

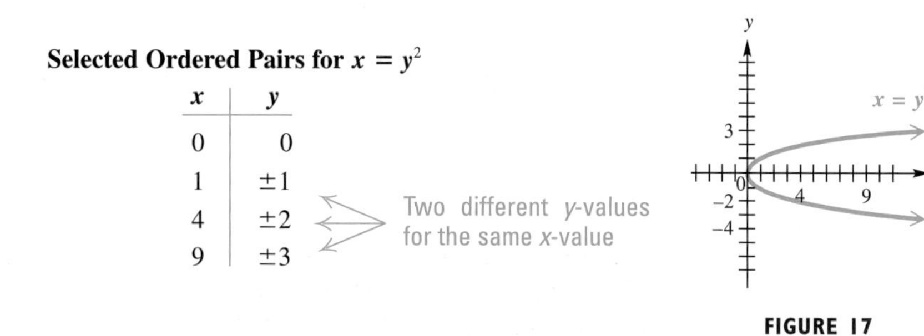

**Selected Ordered Pairs for $x = y^2$**

| $x$ | $y$ |
|---|---|
| 0 | 0 |
| 1 | $\pm 1$ |
| 4 | $\pm 2$ |
| 9 | $\pm 3$ |



FIGURE 17

TECHNOLOGY NOTE

You should be aware of how function mode (as opposed to parametric mode) is activated for your model.

If a graphing calculator is set in function mode, it is not possible to graph $x = y^2$ directly. (However, if it is set in *parametric* mode, such a curve is possible with direct graphing.) To overcome this problem, we begin with $x = y^2$ and take the square root on each side, remembering to choose both the positive and negative square roots of $x$.

$$x = y^2 \qquad \text{Given equation}$$
$$y^2 = x \qquad \text{Transform so that } y \text{ is on the left.}$$
$$y = \pm\sqrt{x} \qquad \text{Take square roots.}$$

Now, we have $x = y^2$ defined by two *functions*, $Y_1 = \sqrt{X}$ and $Y_2 = -\sqrt{X}$. Entering both of these into a calculator gives the graph shown in Figure 18(a).

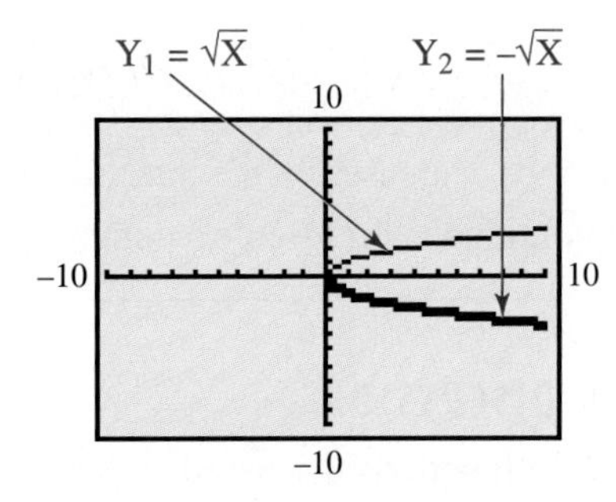


The graph of the relation $x = y^2$ is symmetric with respect to the $x$-axis.

(a)

| X | Y1 | Y2 |
|---|---|---|
| 0 | 0 | 0 |
| 1 | 1 | -1 |
| 4 | 2 | -2 |
| 9 | 3 | -3 |
| 16 | 4 | -4 |
| 25 | 5 | -5 |
| 36 | 6 | -6 |

Y1=√(X)

This screen shows selected ordered pairs for the relation $x = y^2$, where $Y_1 = \sqrt{X}$ and $Y_2 = -\sqrt{X}$.

(b)

**FIGURE 18**

If we were to "fold" the graph of $x = y^2$ along the $x$-axis, the two halves of the parabola would coincide. This graph exhibits *symmetry with respect to the x-axis.* A table supports this symmetry, as seen in Figure 18(b).

### Symmetry with Respect to the $x$-Axis

If replacing $y$ with $-y$ in an equation results in the same equation, then the graph is **symmetric with respect to the $x$-axis.**

To illustrate this, begin with $x = y^2$ and replace $y$ with $-y$.

$$x = (-y)^2$$
$$x = (-1)^2y^2$$
$$x = y^2 \qquad \text{The result is the same equation.}$$

In general, if a graph is symmetric with respect to the $x$-axis, the following is true: *If $(a, b)$ is on the graph, so is $(a, -b)$.*

A summary of the types of symmetry just discussed follows.

| Type of Symmetry | Example | Basic Fact about Points on the Graph |
|---|---|---|
| $y$-axis symmetry | $y$, $x$, 0, $(-a, b)$, $(a, b)$ | If $(a, b)$ is on the graph, so is $(-a, b)$. |
| Origin symmetry | $y$, $x$, 0, $(a, b)$, $(-a, -b)$ | If $(a, b)$ is on the graph, so is $(-a, -b)$. |
| $x$-axis symmetry (not possible for a nonzero function) | $y$, $x$, 0, $(a, b)$, $(a, -b)$ | If $(a, b)$ is on the graph, so is $(a, -b)$. |

**FOR DISCUSSION**

The three functions discussed in Example 4 on the next page are graphed in Figure 19, but not necessarily in the same order as in the example. Without actually using your calculator, identify each function.

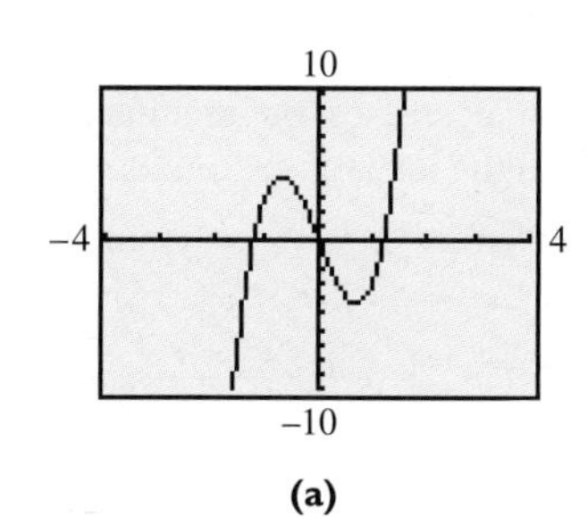


(a)

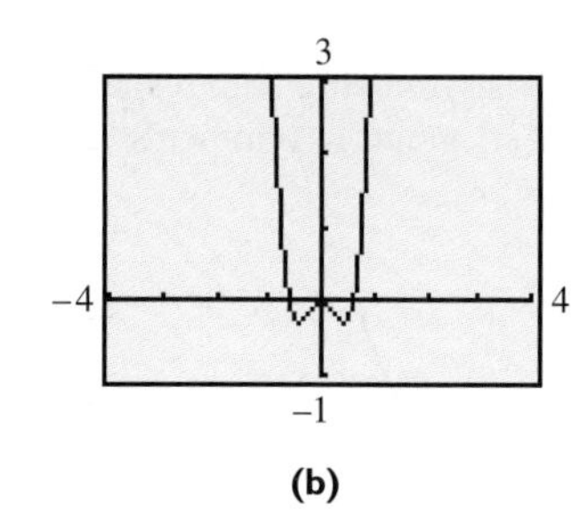


(b)

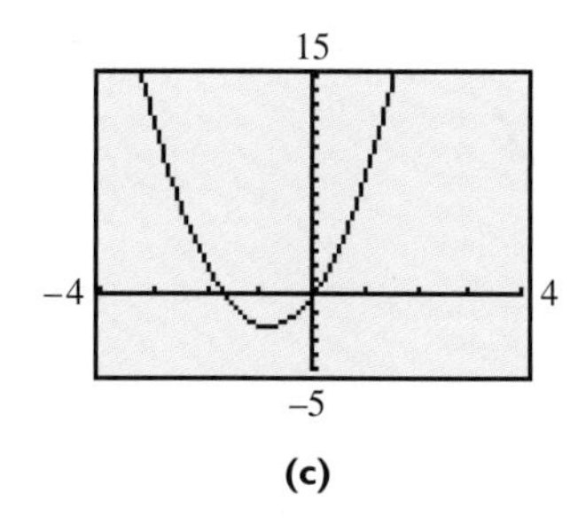


(c)

**FIGURE 19**

## Even and Odd Functions

The concepts of symmetry with respect to the $y$-axis and symmetry with respect to the origin are closely associated with the concepts of *even* and *odd functions.*

**Even and Odd Functions**

A function $f$ is called an **even function** if $f(-x) = f(x)$ for all $x$ in the domain of $f$. (Its graph is symmetric with respect to the $y$-axis.)

A function $f$ is called an **odd function** if $f(-x) = -f(x)$ for all $x$ in the domain of $f$. (Its graph is symmetric with respect to the origin.)

As an illustration, observe the following.

$f(x) = x^2$ is an even function because

$$\begin{aligned} f(-x) &= (-x)^2 \\ &= (-1)^2x^2 \\ &= x^2 \\ &= f(x). \end{aligned}$$

$f(x) = x^3$ is an odd function because

$$\begin{aligned} f(-x) &= (-x)^3 \\ &= (-1)^3x^3 \\ &= -x^3 \\ &= -f(x). \end{aligned}$$

A function may be neither even nor odd; for example, $f(x) = \sqrt{x}$ is neither even nor odd.

**EXAMPLE 4** *Determining Whether Functions Are Even, Odd, or Neither*

Decide whether each function defined is even, odd, or neither.

**(a)** $f(x) = 8x^4 - 3x^2$ **(b)** $f(x) = 6x^3 - 9x$ **(c)** $f(x) = 3x^2 + 5x$

**Solution**

**(a)** Replacing $x$ with $-x$ gives

$$f(-x) = 8(-x)^4 - 3(-x)^2 = 8x^4 - 3x^2 = f(x).$$

Since $f(-x) = f(x)$ for each $x$ in the domain of the function, $f$ is even.

**(b)** Here

$$f(-x) = 6(-x)^3 - 9(-x) = -6x^3 + 9x = -f(x).$$

The function $f$ is odd.

**(c)** $f(-x) = 3(-x)^2 + 5(-x)$

$= 3x^2 - 5x$

Since $f(-x) \neq f(x)$ and $f(-x) \neq -f(x)$, $f$ is neither even nor odd.

## 2.1 EXERCISES

*Concept Check Fill in each blank with the correct response.*

1. The domain and the range of the identity function are both __________.

2. The domain of the squaring function is __________, while its range is __________.

3. The graph of the cubing function changes from "opening downward" to "opening upward" at the point __________.

4. The domain of the square root function is __________, and its range is __________.

5. The cube root function __________ (increases/decreases) on its entire domain.

6. The absolute value function decreases on the interval __________ and increases on the interval __________.

7. The graph of the relation given by $x = y^2$ is symmetric with respect to the __________.

8. The function defined by $f(x) = x^4 + x^2$ is an __________ (even/odd) function.

9. The function defined by $f(x) = x^3 + x$ is an __________ (even/odd) function.

10. If a function is even, its graph is symmetric with respect to the __________; if it is odd, its graph is symmetric with respect to the __________.

*Determine the intervals of the domain over which each function is continuous.*

11.

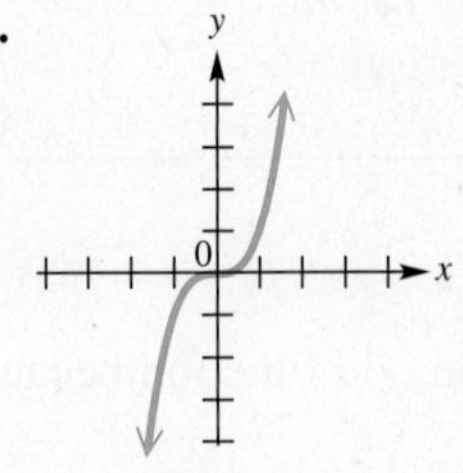


12.

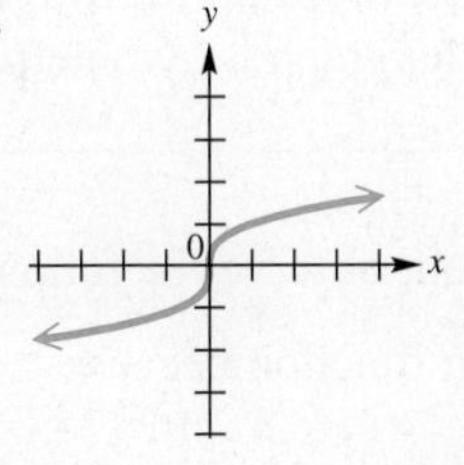


13.

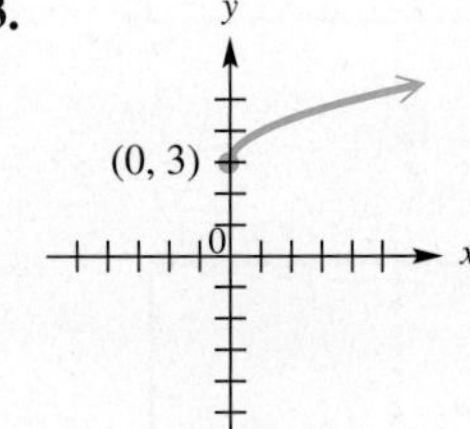


14.

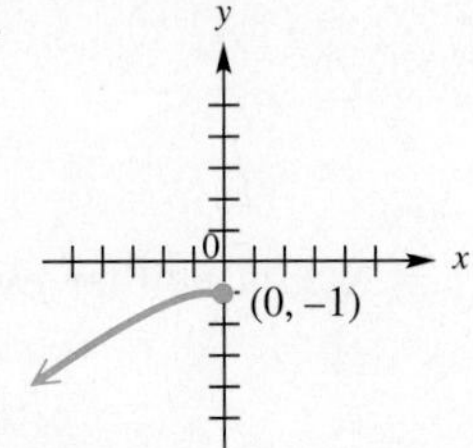


15.

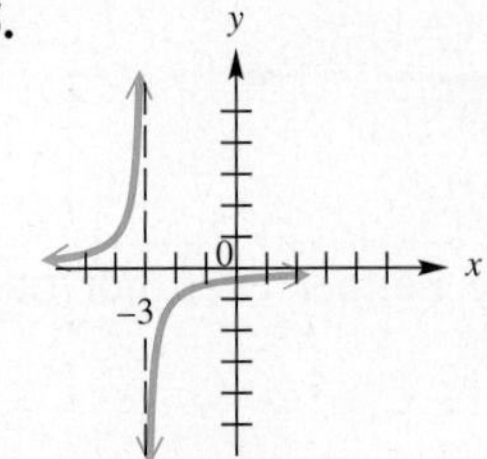


16.

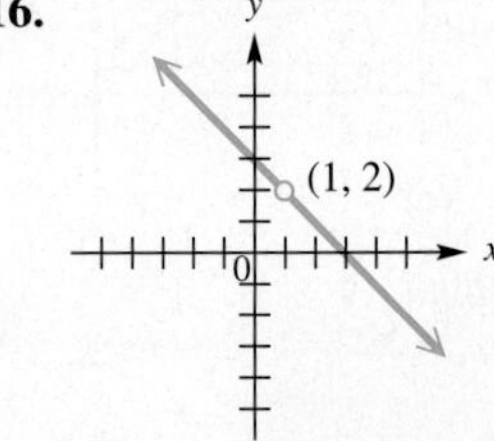

*Determine the intervals of the domain over which each function is* **(a)** *increasing,* **(b)** *decreasing, and* **(c)** *constant. Then give the* **(d)** *domain and* **(e)** *range.*

**17.**

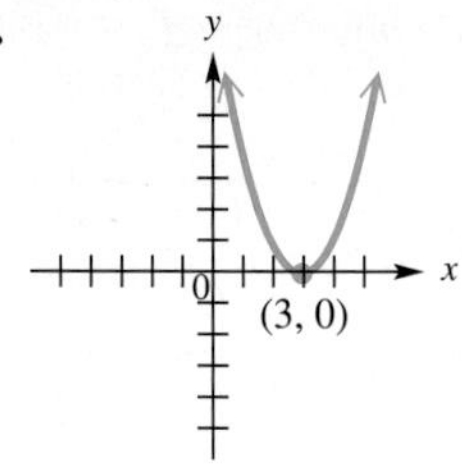


**18.**

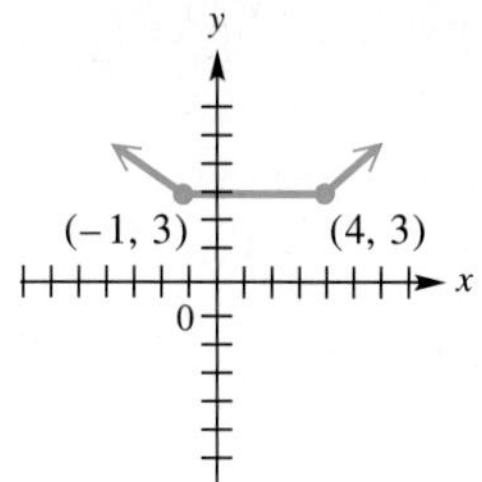


**19.**

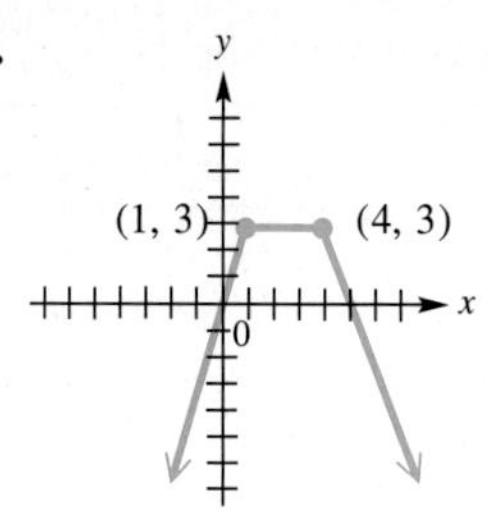


**20.**

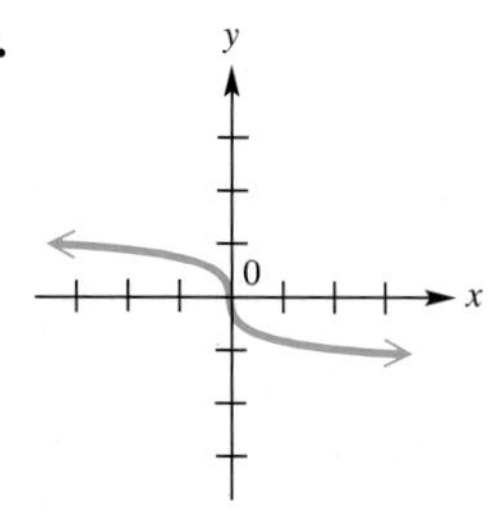


**21.**

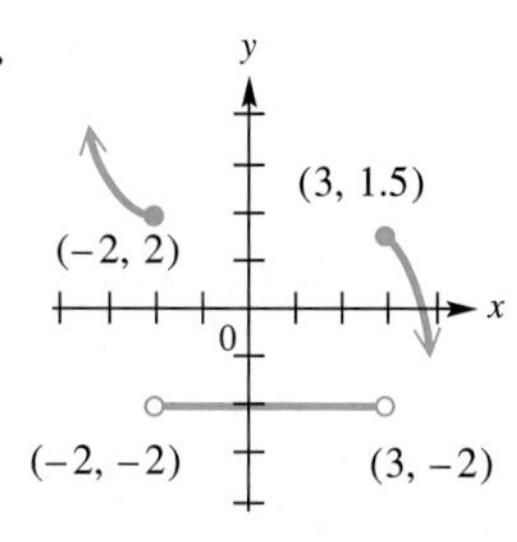


**22.**

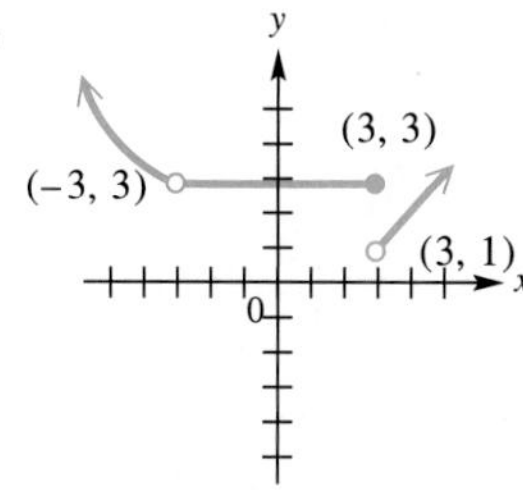


*In Exercises 23–28, graph each function in the standard viewing window of your calculator, and trace from left to right along a representative portion of the specified interval. Then, fill in the blank of this sentence with either* **increasing** *or* **decreasing**:

*OVER THE INTERVAL SPECIFIED, THIS FUNCTION IS __________.*

**23.** $f(x) = x^5; (-\infty, \infty)$

**24.** $f(x) = -x^3; (-\infty, \infty)$

**25.** $f(x) = x^4; (-\infty, 0]$

**26.** $f(x) = x^4; [0, \infty)$

**27.** $f(x) = -|x|; (-\infty, 0]$

**28.** $f(x) = -|x|; [0, \infty)$

*Using visual observation, determine whether each graph is symmetric with respect to the* **(a)** *x-axis,* **(b)** *y-axis,* **(c)** *origin.*

**29.**

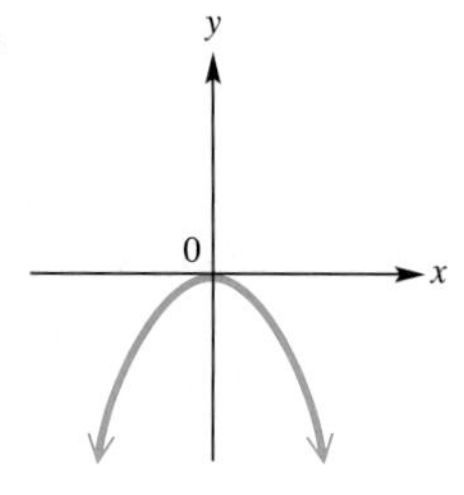


**30.**

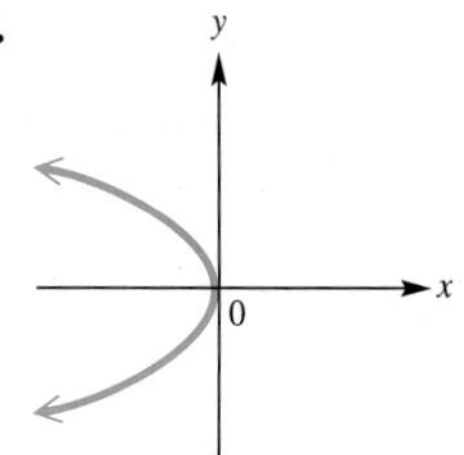


**31.**

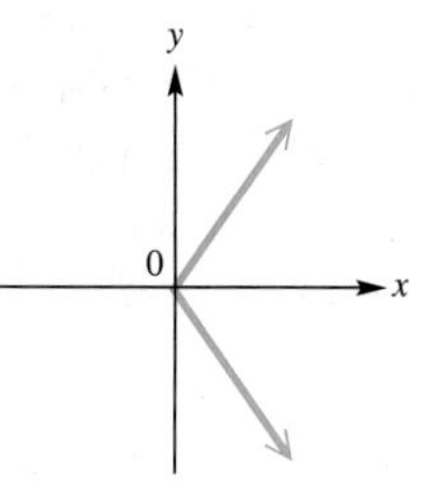


**32.**

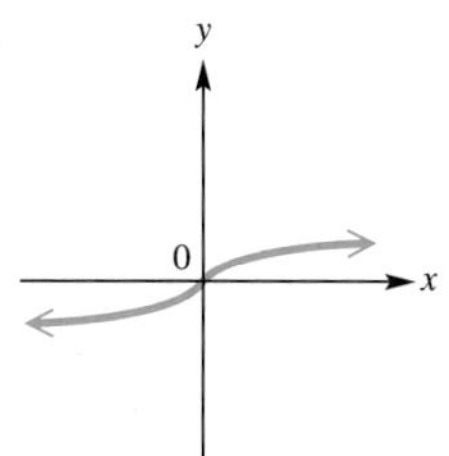


**33.**

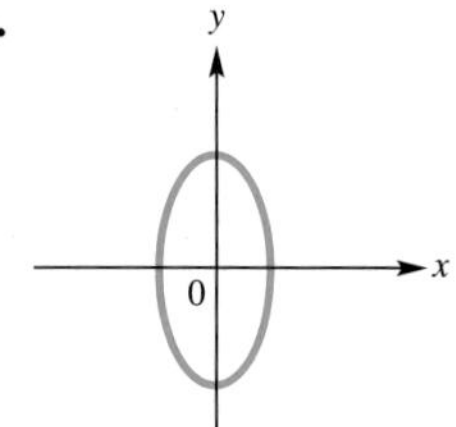


**34.**

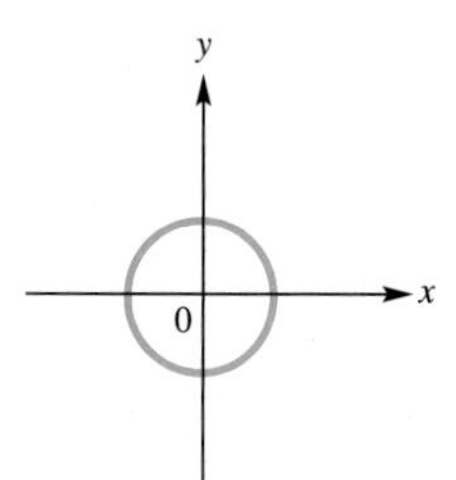


**35.**

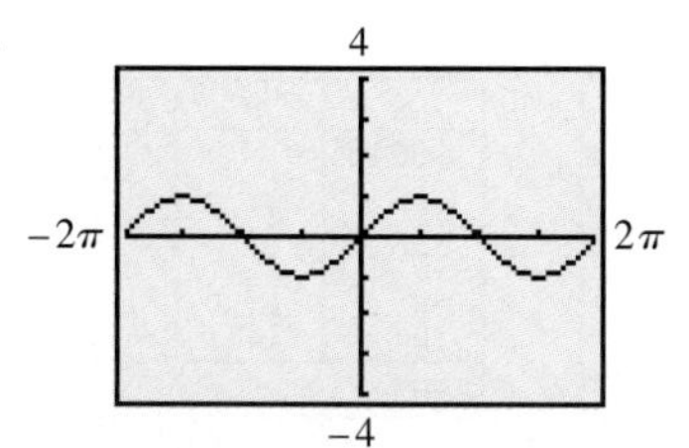


**36.**

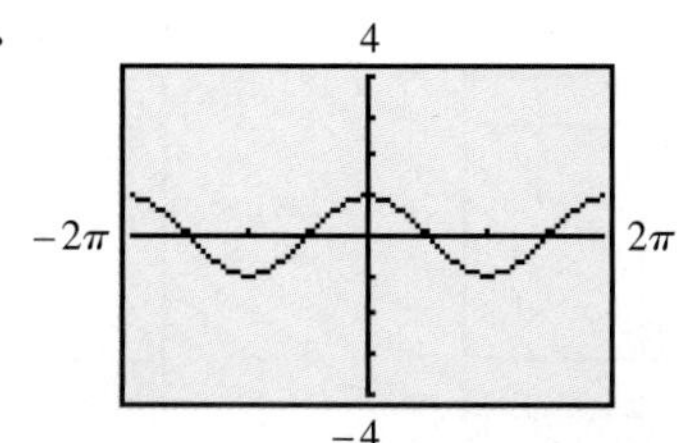

*Concept Check In Exercises 37–42, you are given a calculator graph of a relation that exhibits a type of symmetry. Based on the given information at the bottom of each figure, determine the coordinates of another point that must also lie on the graph.*

**37.** Symmetric with respect to the $y$-axis

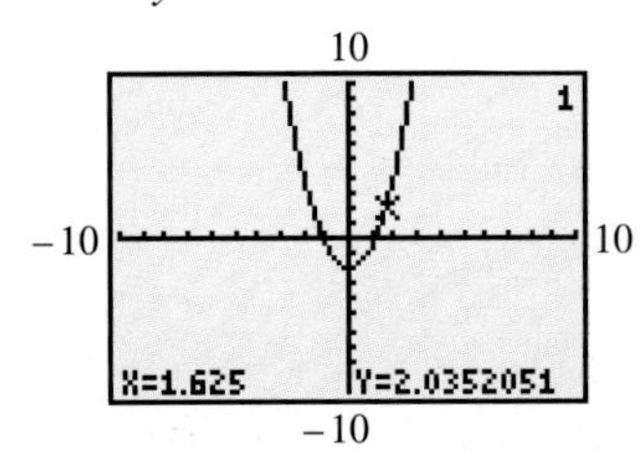


**38.** Symmetric with respect to the $y$-axis

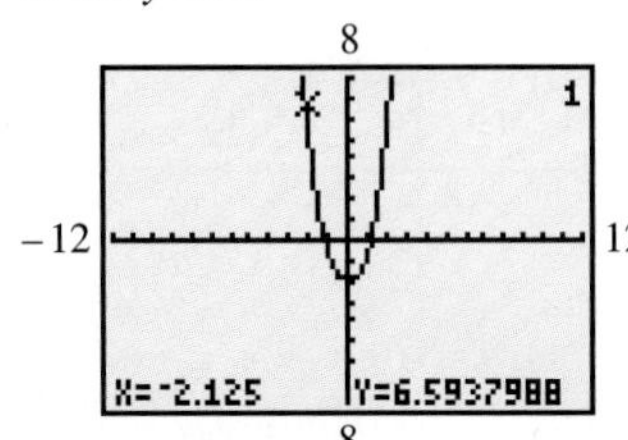


**39.** Symmetric with respect to the $x$-axis

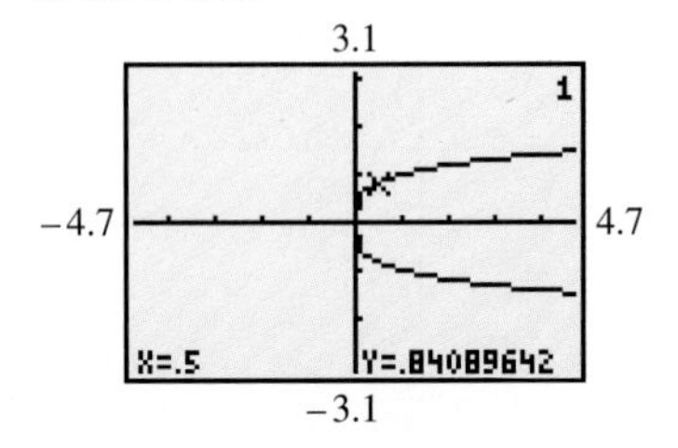


**40.** Symmetric with respect to the $x$-axis

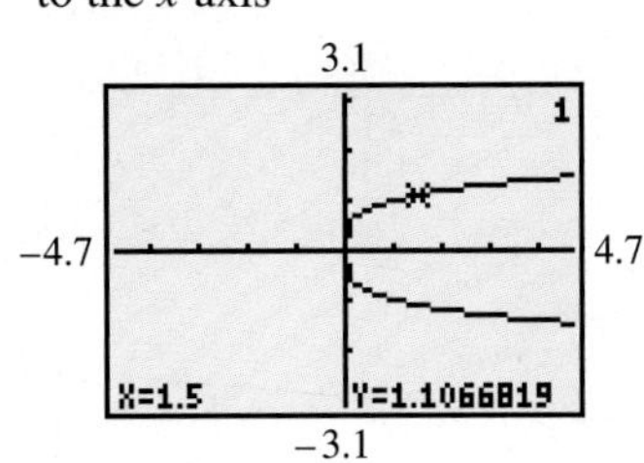


**41.** Symmetric with respect to the origin

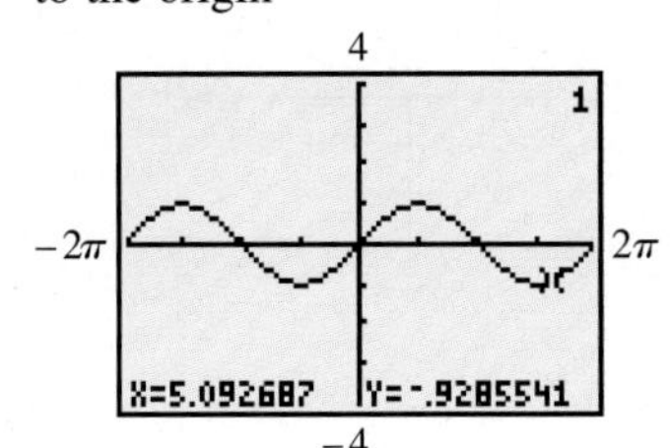


**42.** Symmetric with respect to the origin

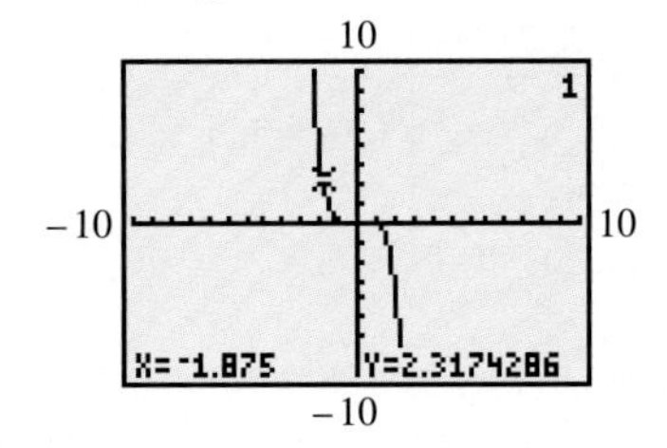


**43.** *Concept Check* Complete the table if $f$ is an even function.

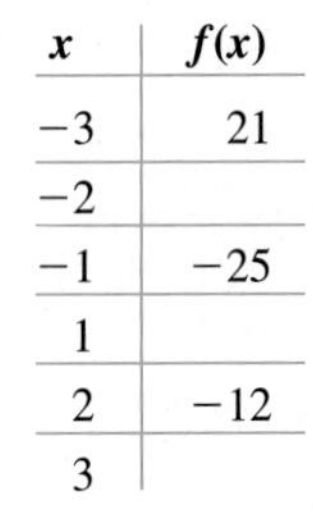

| $x$ | $f(x)$ |
|---|---|
| −3 | 21 |
| −2 | |
| −1 | −25 |
| 1 | |
| 2 | −12 |
| 3 | |

**44.** *Concept Check* Complete the table if $g$ is an odd function.

| $x$ | $g(x)$ |
|---|---|
| −5 | 13 |
| −3 | |
| −2 | −5 |
| 0 | |
| 2 | |
| 3 | −1 |
| 5 | |

**45.** *Concept Check* Complete the left half of the graph of $y = f(x)$ in the figure for each of the following conditions:

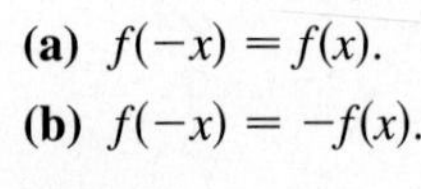

**(a)** $f(-x) = f(x)$.

**(b)** $f(-x) = -f(x)$.

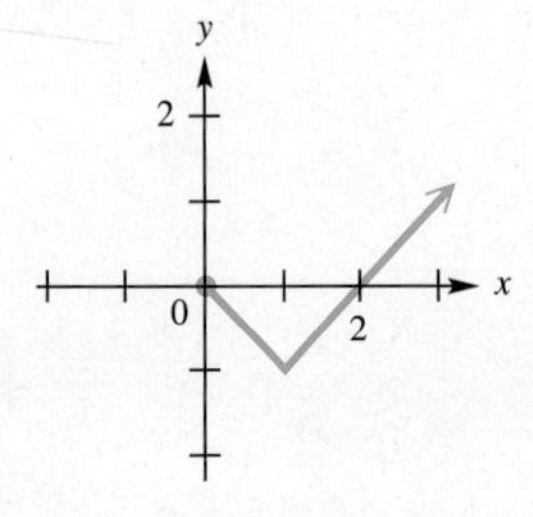


**46.** *Concept Check* Complete the right half of the graph of $y = f(x)$ in the figure for each of the following conditions:

**(a)** $f$ is odd.

**(b)** $f$ is even.

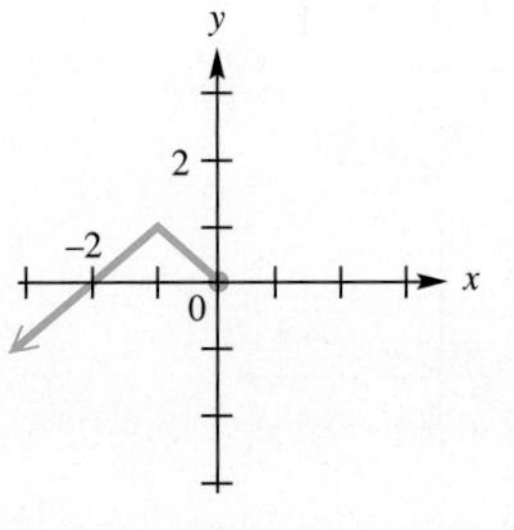


*Based on the ordered pairs seen in each pair of tables, make a conjecture about whether the function defined in* $Y_1$ *is* even, odd, *or* neither even nor odd.

**47.**

| X | Y1 |
|---|---|
| 0 | 0 |
| 1 | 1 |
| 2 | 4 |
| 3 | 9 |
| 4 | 16 |
| 5 | 25 |
| 6 | 36 |

X=0

| X | Y1 |
|---|---|
| 0 | 0 |
| -1 | 1 |
| -2 | 4 |
| -3 | 9 |
| -4 | 16 |
| -5 | 25 |
| -6 | 36 |

X=0

**48.**

| X | Y1 |
|---|---|
| 0 | 0 |
| 1 | 1 |
| 2 | 8 |
| 3 | 27 |
| 4 | 64 |
| 5 | 125 |
| 6 | 216 |

X=0

| X | Y1 |
|---|---|
| 0 | 0 |
| -1 | -1 |
| -2 | -8 |
| -3 | -27 |
| -4 | -64 |
| -5 | -125 |
| -6 | -216 |

X=0

*Prove analytically that each function is even.*

**49.** $f(x) = x^4 - 7x^2 + 6$

**50.** $f(x) = -2x^6 - 8x^2$

*Prove analytically that each function is odd.*

**51.** $f(x) = 3x^3 - x$

**52.** $f(x) = -x^5 + 2x^3 - 3x$

*Use the analytic method of Example 3 to determine whether the graph of the given function is* symmetric with respect to the $y$-axis, symmetric with respect to the origin, *or* neither. *Use your calculator to support your conclusion, using the standard window.*

**53.** $f(x) = -x^3 + 2x$

**54.** $f(x) = x^5 - 2x^3$

**55.** $f(x) = .5x^4 - 2x^2 + 1$

**56.** $f(x) = .75x^2 + |x| + 1$

**57.** $f(x) = x^3 - x + 3$

**58.** $f(x) = x^4 - 5x + 2$

**59.** Refer to Exercises 53–58, and determine which exercises have functions that are:

**(a)** even **(b)** odd **(c)** neither even nor odd.

**60.** Write an explanation of the connections between symmetry with respect to the $y$-axis, symmetry with respect to the origin, even functions, and odd functions.

*In Exercises 61 and 62, you are given a portion of the graph of a function near the origin, along with information regarding symmetry of the graph of the complete function. Sketch by hand a simulated calculator graph, using the window given, to show what each graph would look like.*

**61.** Symmetric with respect to the $y$-axis; window: $[-5, 5]$ by $[0, 5]$

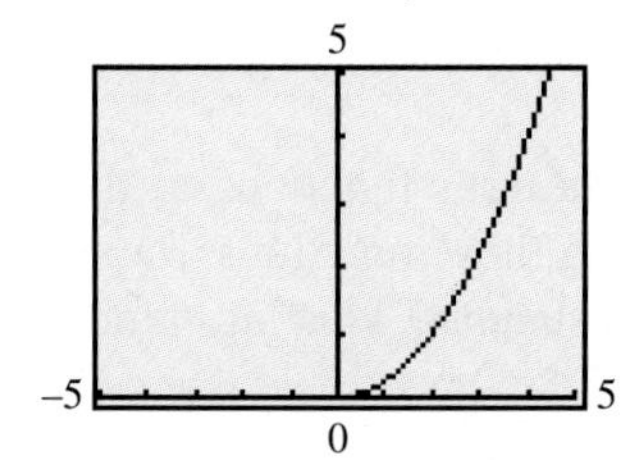


**62.** Symmetric with respect to the origin; window: $[-5, 5]$ by $[-2, 2]$

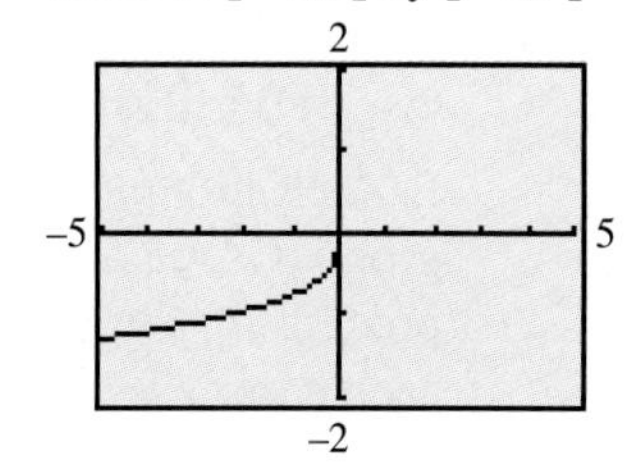


## RELATING CONCEPTS

**For individual or group investigation (Exercises 63–66)**

*The line in the sketch and the line on the screen are said to be* tangent *to the curve at point P.*

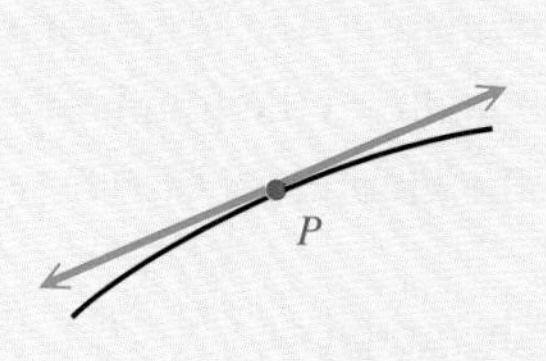


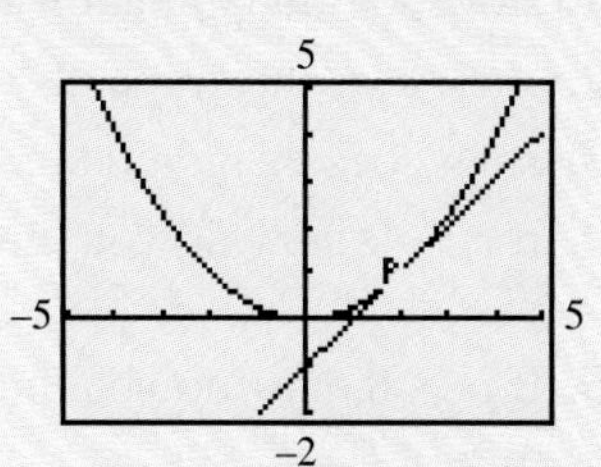


**63.** Sketch by hand the graph of $y = x^2$, and choose a point at which the function is *decreasing*. Draw a tangent line at that point. Is the slope of this line positive, negative, or zero?

**64.** Repeat Exercise 63, but choose a point at which the function is *increasing*.

**65.** Repeat Exercise 63, but choose the point at which the function changes from *decreasing* to *increasing*.

**66.** Based on your answers to Exercises 63–65, what conclusions can be drawn? These conclusions can be verified using methods discussed in calculus.

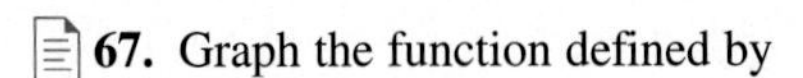

**67.** Graph the function defined by

$$y = \frac{x^2 - 9}{x + 3}$$

in the standard window of your calculator. Try to locate the point for which $x = -3$. What happens? Why do you think this happens? (Functions of this kind, called *rational functions,* will be studied in detail in Chapter 4.)

**68.** Graph the function defined by

$$y = \frac{x^2 - 9}{x + 3}$$

in the standard window of your calculator. Do you think that determining continuity strictly by observation of a calculator-generated graph is foolproof?

## 2.2 VERTICAL AND HORIZONTAL SHIFTS OF GRAPHS

Vertical Shifts • Horizontal Shifts • Combinations of Vertical and Horizontal Shifts • Effects of Shifts on Domain and Range • Horizontal Shifts Applied to Equations for Modeling

In this section we examine how the graphs of the basic functions introduced in the previous section can be shifted vertically and horizontally in the plane.

### Vertical Shifts

**FOR DISCUSSION**

In each group, we give four related functions. Graph the four functions in the first group (Group A), and then answer the questions regarding those functions. Repeat the process for Group B, Group C, and Group D. Use the standard viewing window in each case.

TECHNOLOGY NOTE

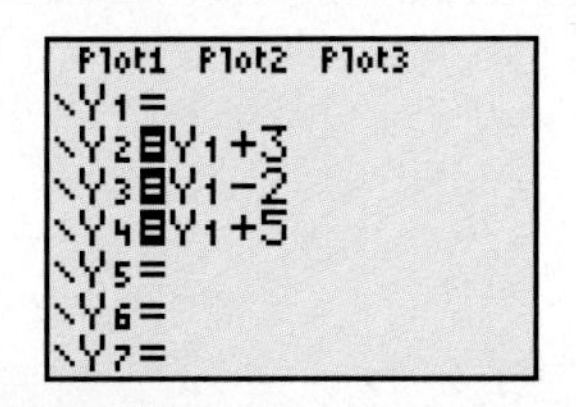


This screen shows how to minimize keystrokes in the activity in the "For Discussion" box. The entry for $Y_1$ can be altered as needed.

| A | B | C | D |
|---|---|---|---|
| $y_1 = x^2$ | $y_1 = x^3$ | $y_1 = \sqrt{x}$ | $y_1 = \sqrt[3]{x}$ |
| $y_2 = x^2 + 3$ | $y_2 = x^3 + 3$ | $y_2 = \sqrt{x} + 3$ | $y_2 = \sqrt[3]{x} + 3$ |
| $y_3 = x^2 - 2$ | $y_3 = x^3 - 2$ | $y_3 = \sqrt{x} - 2$ | $y_3 = \sqrt[3]{x} - 2$ |
| $y_4 = x^2 + 5$ | $y_4 = x^3 + 5$ | $y_4 = \sqrt{x} + 5$ | $y_4 = \sqrt[3]{x} + 5$ |

1. How does the graph of $y_2$ compare to the graph of $y_1$?
2. How does the graph of $y_3$ compare to the graph of $y_1$?
3. How does the graph of $y_4$ compare to the graph of $y_1$?
4. If $c > 0$, how do you think the graph of $y = f(x) + c$ would compare to the graph of $y = f(x)$?
5. If $c > 0$, how do you think the graph of $y = f(x) - c$ would compare to the graph of $y = f(x)$?

Choosing your own value of $c$, support your answers to Items 4 and 5 graphically. (Be sure that your choice is appropriate for the standard window.)

The objective of the preceding discussion activity was to make conjectures about how adding or subtracting a positive constant $c$ affects the graph of a function. In each case, we obtained a vertical *shift*, or **translation,** of the graph of the basic function with which we started. Although our observations were based on the graphs of four basic functions, they can be generalized to any function.

> **Vertical Shifting of the Graph of a Function**
>
> If $c > 0$, the graph of $y = f(x) + c$ is obtained by shifting the graph of $y = f(x)$ *upward* a distance of $c$ units. The graph of $y = f(x) - c$ is obtained by shifting the graph of $y = f(x)$ *downward* a distance of $c$ units.

In Figure 20, we graphically interpret the preceding statement.

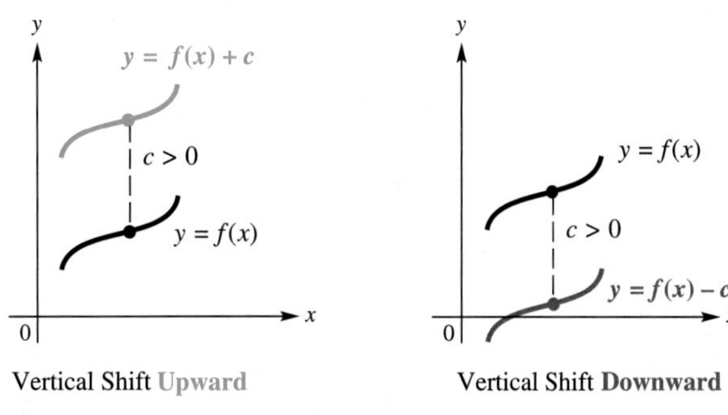


**FIGURE 20**

**EXAMPLE 1** *Recognizing Vertical Shifts*

Give the equation of each function graphed in Figure 21. The displays at the bottom of the screens indicate points for which $x = 0$.

**TECHNOLOGY NOTE**

When using a graphing calculator to investigate shifts of graphs, it is important to use an appropriate window; otherwise, the graph may not appear. For example, the graph of $y = |x| + 12$ does not appear in the viewing window $[-10, 10]$ by $[-10, 10]$. Why not?

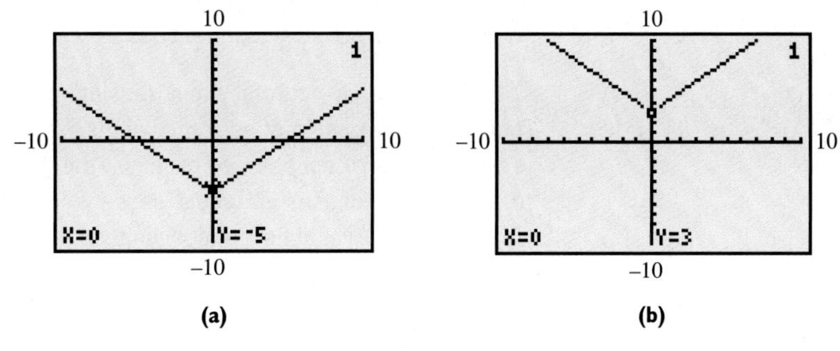


**FIGURE 21**

**Solution** Both graphs are vertical translations of the graph of $y = |x|$. In Figure 21(a), the graph has been shifted 5 units downward, so the equation is $y = |x| - 5$. In Figure 21(b), the graph has been shifted 3 units upward, so the equation is $y = |x| + 3$.

## Horizontal Shifts

**FOR DISCUSSION**

This discussion parallels the one earlier in this section. Follow the same general directions.

| A | B | C | D |
|---|---|---|---|
| $y_1 = x^2$ | $y_1 = x^3$ | $y_1 = \sqrt{x}$ | $y_1 = \sqrt[3]{x}$ |
| $y_2 = (x - 3)^2$ | $y_2 = (x - 3)^3$ | $y_2 = \sqrt{x - 3}$ | $y_2 = \sqrt[3]{x - 3}$ |
| $y_3 = (x - 5)^2$ | $y_3 = (x - 5)^3$ | $y_3 = \sqrt{x - 5}$ | $y_3 = \sqrt[3]{x - 5}$ |
| $y_4 = (x + 4)^2$ | $y_4 = (x + 4)^3$ | $y_4 = \sqrt{x + 4}$ | $y_4 = \sqrt[3]{x + 4}$ |

**1.** How does the graph of $y_2$ compare to the graph of $y_1$?

**2.** How does the graph of $y_3$ compare to the graph of $y_1$?

**3.** How does the graph of $y_4$ compare to the graph of $y_1$?

**4.** If $c > 0$, how do you think the graph of $y = f(x - c)$ would compare to the graph of $y = f(x)$?

**5.** If $c > 0$, how do you think the graph of $y = f(x + c)$ would compare to the graph of $y = f(x)$?

Choosing your own value of $c$, support your answers to Items 4 and 5 graphically. (Again, be sure that your choice is appropriate for the standard window.)

The results of the preceding discussion should remind you of the results found earlier when we shifted graphs of functions vertically. Now, we see how such graphs can be shifted *horizontally*. The observations can be generalized as follows.

### Horizontal Shifting of the Graph of a Function

If $c > 0$, the graph of $y = f(x - c)$ is obtained by shifting the graph of $y = f(x)$ to the *right* a distance of $c$ units. The graph of $y = f(x + c)$ is obtained by shifting the graph of $y = f(x)$ to the *left* a distance of $c$ units.

**CAUTION** Be careful when translating graphs horizontally. In order to determine the direction and magnitude of horizontal shifts, find the value of $x$ that would cause the expression in parentheses to equal 0. For example, the graph of $f(x) = (x - 5)^2$ would be shifted 5 units to the *right*, because $+5$ would cause $x - 5$ to equal 0. On the other hand, the graph of $f(x) = (x + 5)^2$ would be shifted 5 units to the *left*, because $-5$ would cause $x + 5$ to equal 0.

Figure 22 illustrates the effect of a horizontal shift of the graph of a function given by $y = f(x)$.

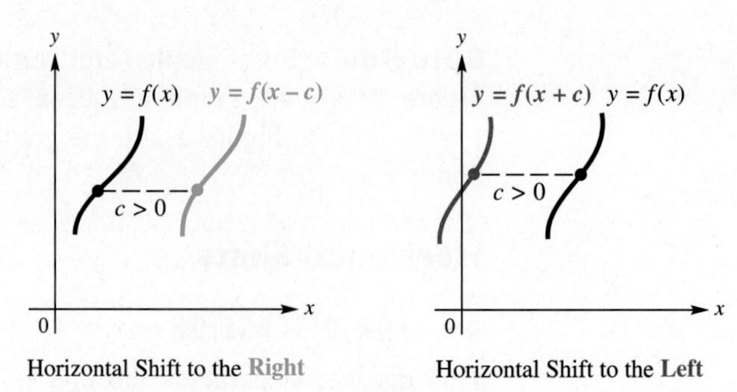


**FIGURE 22**

**EXAMPLE 2** *Recognizing Horizontal Shifts*

Give the equation of each function graphed in Figure 23. The displays at the bottom of the screens indicate points for which $y = 0$.

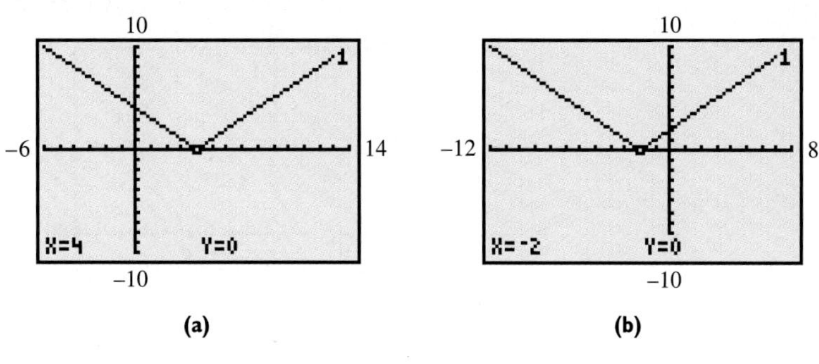


FIGURE 23

**Solution** Both graphs are horizontal translations of the graph of $y = |x|$. In Figure 23(a), the graph has been shifted 4 units to the right, so the equation is $y = |x - 4|$. In Figure 23(b), the graph has been shifted 2 units to the left, so the equation is $y = |x + 2|$. (Be sure that you understand why the signs are as they appear here.)

## Combinations of Vertical and Horizontal Shifts

Now that we have seen how graphs of functions can be shifted vertically or horizontally, we extend these ideas to graphs that are obtained by applying *both* types of translations.

**EXAMPLE 3** *Applying Both Vertical and Horizontal Shifts*

Describe how the graph of $y_2 = |x + 12| - 16$ would be obtained by translating the graph of $y_1 = |x|$. Determine an appropriate viewing window, and support the results by plotting both functions with a graphing calculator.

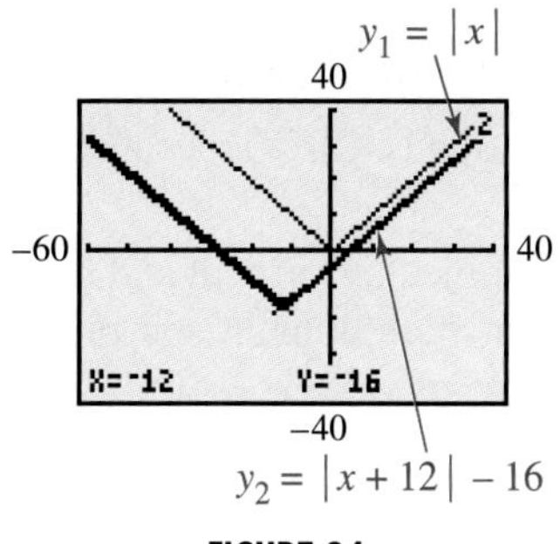


FIGURE 24

**Solution** The function defined by $y_2 = |x + 12| - 16$ is translated 12 units to the *left* (because of the $|x + 12|$) and 16 units *downward* (because of the $-16$) as compared to the graph of $y_1 = |x|$. Because the point at which the graph changes from decreasing to increasing is now $(-12, -16)$, the standard viewing window is not appropriate. We must choose a window that contains the point $(-12, -16)$ in order to obtain a comprehensive graph. While many windows are possible, one such window is shown in Figure 24. The display at the bottom of the screen indicates that the point $(-12, -16)$ lies on the graph.

## Effects of Shifts on Domain and Range

The domains and ranges of functions may or may not be affected by vertical and horizontal shifts. For example, if the domain of a function is $(-\infty, \infty)$, a horizontal shift will not affect the domain. Similarly, if the range is $(-\infty, \infty)$, a vertical shift will not affect the range. However, if the domain is not $(-\infty, \infty)$, a horizontal shift usually affects the domain, and if the range is not $(-\infty, \infty)$, a vertical shift usually affects the range. The next example illustrates this.

**EXAMPLE 4** *Determining Domains and Ranges of Shifted Graphs*

Four functions are graphed in Figure 25. Give the domain and range of each function.

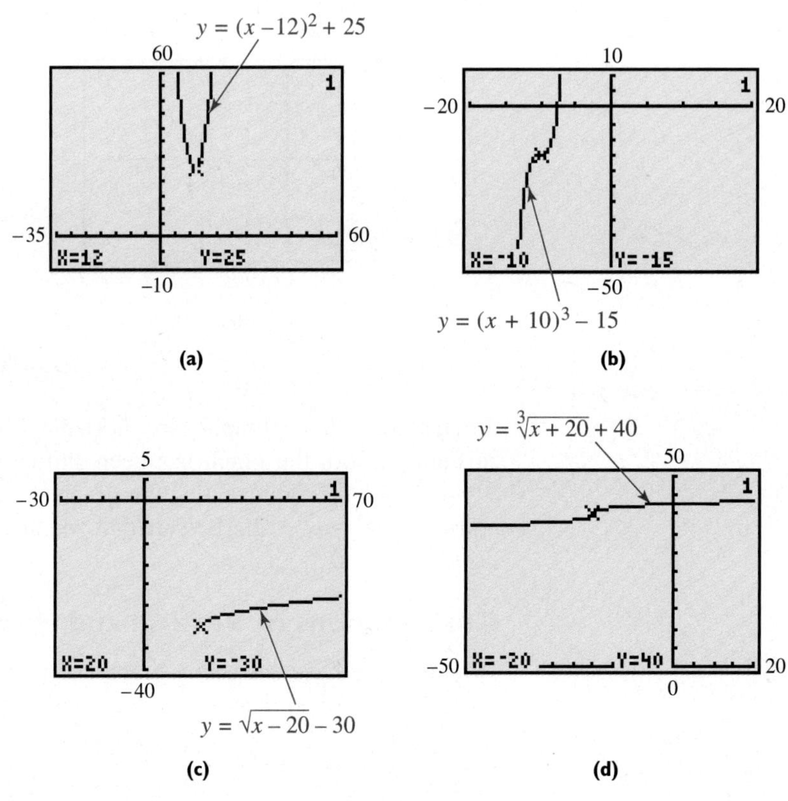


FIGURE 25

### Solution

**(a)** To obtain the graph of $y = (x - 12)^2 + 25$ in Figure 25(a), the graph of $y = x^2$ was translated 12 units to the right and 25 units upward. The original domain $(-\infty, \infty)$ is not affected. However, the range of this function is $[25, \infty)$, because of the vertical translation.

**(b)** The graph of $y = (x + 10)^3 - 15$ in Figure 25(b) is obtained by vertical and horizontal shifts of the graph of $y = x^3$, a function that has both domain and range $(-\infty, \infty)$. Neither is affected here, so the domain and range of $y = (x + 10)^3 - 15$ are also both $(-\infty, \infty)$.

**(c)** The function defined by $y = \sqrt{x}$ has domain $[0, \infty)$. The graph of $y = \sqrt{x - 20} - 30$ in Figure 25(c) is obtained by shifting the basic graph 20 units to the right, so the new domain is $[20, \infty)$. The original range, $[0, \infty)$, has also been affected by the shift of the graph 30 units downward, so the new range is $[-30, \infty)$.

**(d)** The situation in Figure 25(d) is similar to that of Figure 25(b). The graph of $y = \sqrt[3]{x}$, with domain and range both $(-\infty, \infty)$, is shifted 20 units to the left and 40 units upward. Regardless of the direction and size of these shifts, the domain and range are both unaffected and remain $(-\infty, \infty)$.

## Horizontal Shifts Applied to Equations for Modeling

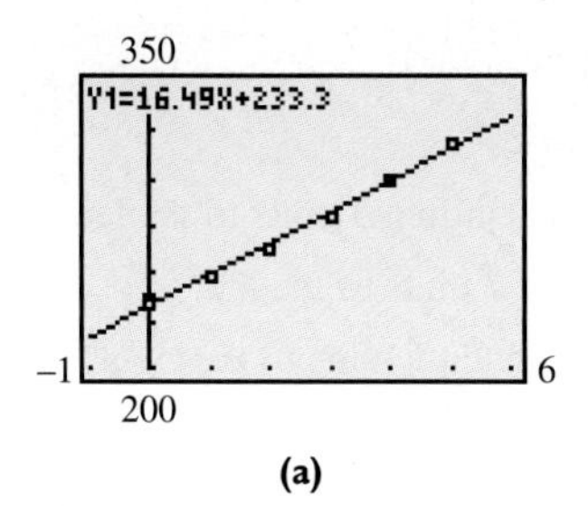


(a)

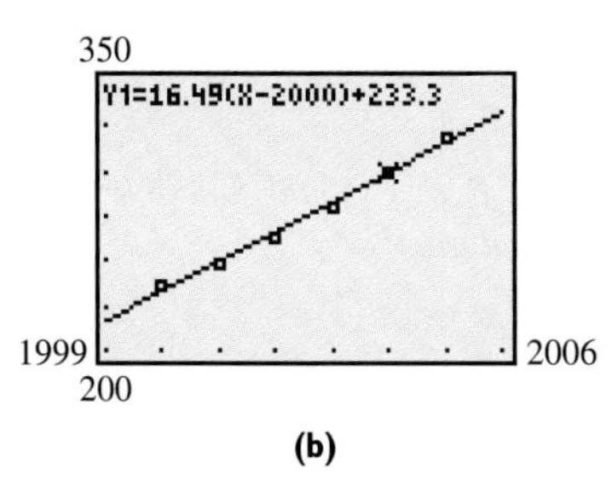


(b)

**FIGURE 26**

In Section 1.4 we examined data that showed estimates for Medicare costs (in billions of dollars) between the years 2000 and 2005. (See Examples 7 and 8 in Section 1.4.) To simplify our work, $x = 0$ corresponded to the year 2000, $x = 1$ to 2001, and so on. We determined that the least-squares regression line has equation

$$y = 16.49x + 233.3.$$

The graph of this line is shown in Figure 26(a). To use the values of the years 2000 through 2005 directly in the regression equation, we must shift the graph of the line 2000 units to the right. The equation of this new line is

$$y = 16.49(x - 2000) + 233.3.$$

Using a graphing calculator to plot the set of points

$$\{(2000, 236),\ (2001, 249),\ (2002, 264),\ (2003, 281),\ (2004, 299),\ (2005, 318)\}$$

and graphing $y = 16.49(x - 2000) + 233.3$ in the window $[1999, 2006]$ by $[200, 350]$ gives the graph shown in Figure 26(b).

**EXAMPLE 5** *Applying a Horizontal Shift to an Equation Model*

The table lists the numbers of juveniles charged in court from 1990 to 1995 in Blue Earth County, Minnesota.

| Year | Juveniles |
|---|---|
| 1990 | 346 |
| 1991 | 379 |
| 1992 | 453 |
| 1993 | 566 |
| 1994 | 681 |
| 1995 | 713 |

*Source:* Blue Earth County Attorney's Office.

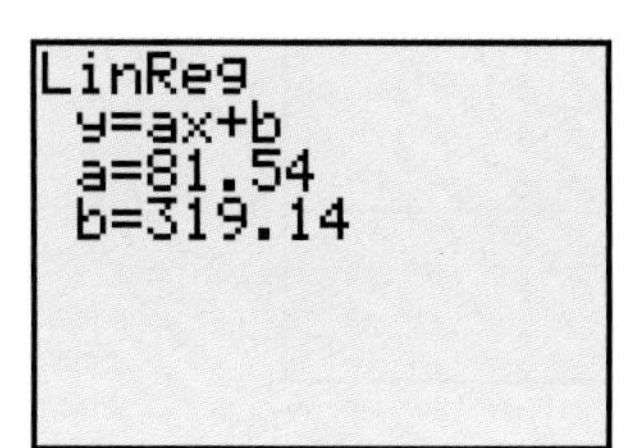


**FIGURE 27**

If we let $x = 0$ correspond to 1990, $x = 1$ correspond to 1991, and so on, then the least-squares regression line is determined to be

$$y = 81.54x + 319.14.$$

(See Figure 27.) Find the corresponding equation that allows direct input of the year.

**Solution** To find this equation, since 1990 corresponds to 0, the graph would have to be shifted 1990 units to the right. Thus, the equation is

$$y = 81.54(x - 1990) + 319.14.$$

## 2.2 EXERCISES

*Write the equation that results in the desired translation.*

1. The squaring function, shifted 3 units upward
2. The cubing function, shifted 2 units downward
3. The square root function, shifted 4 units downward
4. The cube root function, shifted 6 units upward
5. The absolute value function, shifted 4 units to the right
6. The absolute value function, shifted 3 units to the left
7. The cubing function, shifted 7 units to the left
8. The square root function, shifted 9 units to the right
9. Explain how the graph of $g(x) = f(x) + 4$ is obtained from the graph of $y = f(x)$.
10. Explain how the graph of $g(x) = f(x + 4)$ is obtained from the graph of $y = f(x)$.

*Exercises 11–19 are grouped in "threes." For each group of three functions, match the correct graph, A, B, or C, to the function without using your calculator. Then, after you have determined your answers, confirm them with your calculator.*

**11.** $y = x^2 - 3$ **12.** $y = (x - 3)^2$ **13.** $y = (x + 3)^2$

**A.**

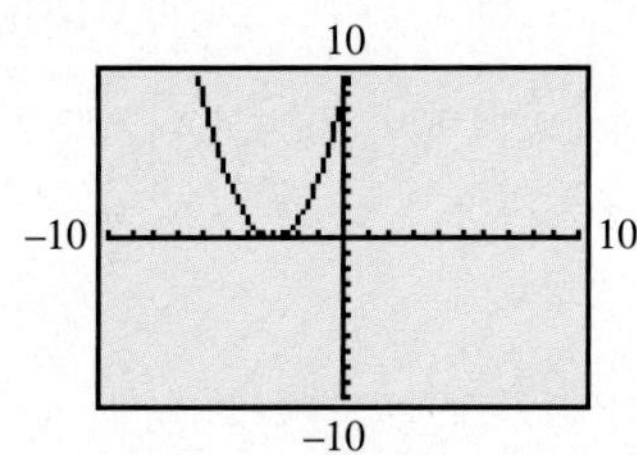


**B.**

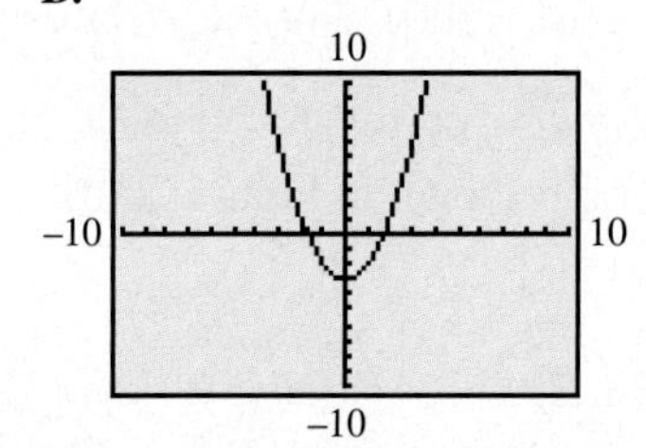


**C.**

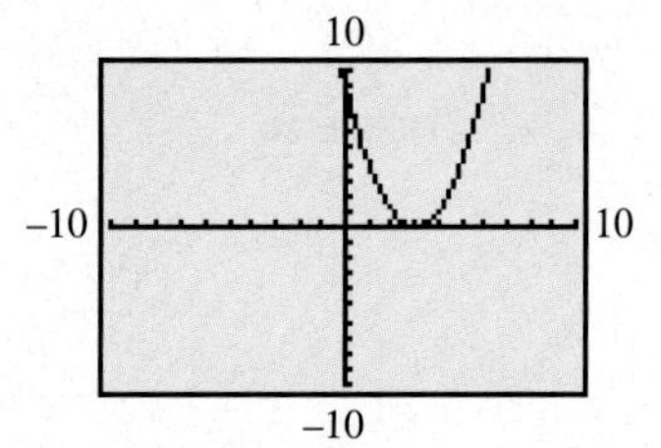


**14.** $y = |x| + 4$ **15.** $y = |x + 4| - 3$ **16.** $y = |x - 4| - 3$

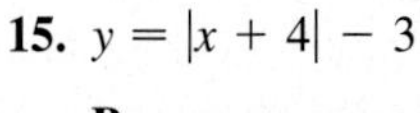

**A.**

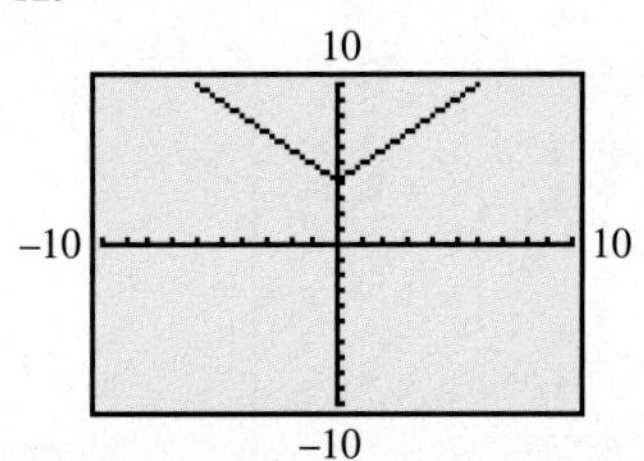


**B.**

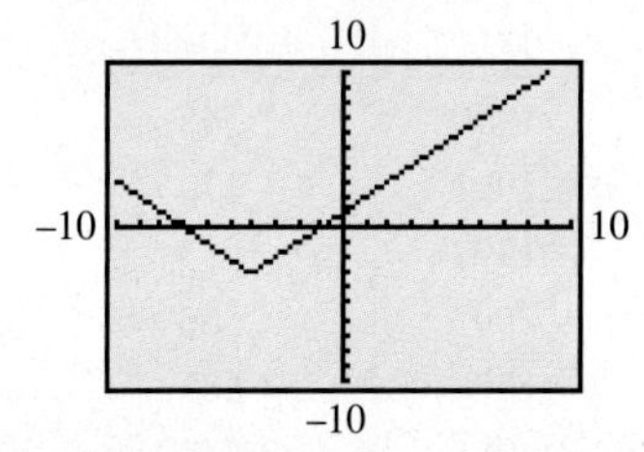


**C.**

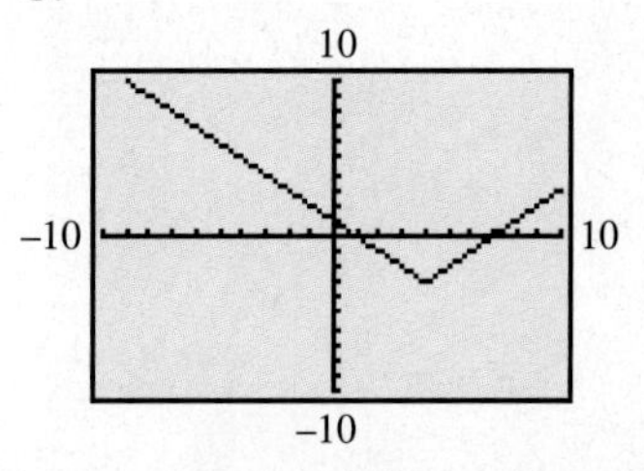


**17.** $y = (x - 3)^3$ **18.** $y = (x - 2)^3 - 4$ **19.** $y = (x + 2)^3 - 4$

**A.**

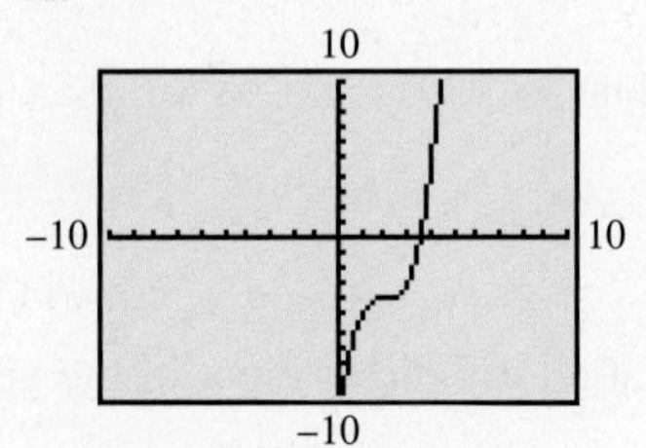


**B.**

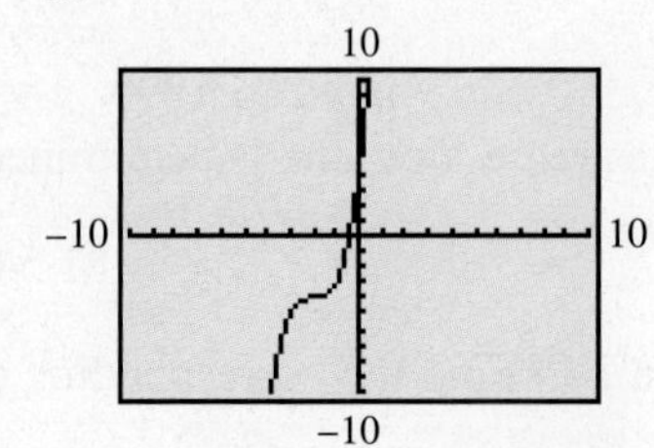


**C.**

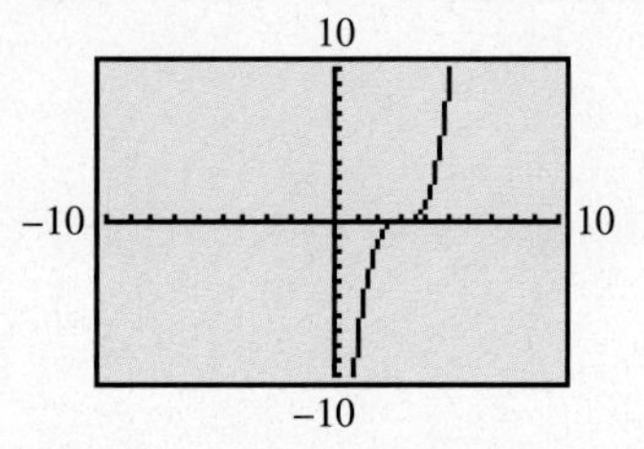


**20.** *Concept Check* In which quadrant does the lowest point on the graph of $y = |x - h|^2 + k$ lie, if $h < 0$ and $k < 0$?

*Use the results of the specified exercises to determine* **(a)** *the domain and* **(b)** *the range of each function.*

**21.** $y = x^2 - 3$ (Exercise 11)
**22.** $y = (x - 3)^2$ (Exercise 12)
**23.** $y = |x + 4| - 3$ (Exercise 15)
**24.** $y = |x - 4| - 3$ (Exercise 16)

*Concept Check* *The function* $Y_2$ *is defined as* $Y_1 + k$ *for some real number k. Based on the table shown, what is the value of k?*

**25.**

| X | Y1 | Y2 |
|---|---|---|
| 0 | 15 | 19 |
| 1 | 10 | 14 |
| 2 | 7 | 11 |
| 3 | 6 | 10 |
| 4 | 7 | 11 |
| 5 | 10 | 14 |
| 6 | 15 | 19 |

X=0

**26.**

| X | Y1 | Y2 |
|---|---|---|
| 0 | -3 | -5 |
| 1 | 2 | 0 |
| 2 | 5 | 3 |
| 3 | 6 | 4 |
| 4 | 5 | 3 |
| 5 | 2 | 0 |
| 6 | -3 | -5 |

X=0

*Concept Check* *The function* $Y_2$ *is defined as* $Y_1 + k$ *for some real number k. Based on the two views of the graphs of* $Y_1$ *and* $Y_2$ *and the displays at the bottoms of the screens, what is the value of k?*

**27.**

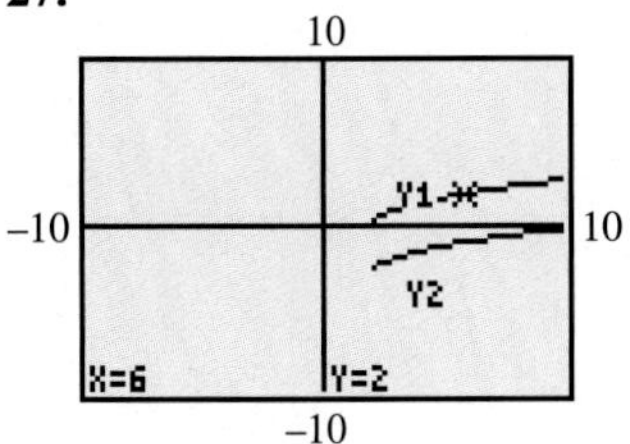


(6, 2) lies on the graph of $Y_1$.
First View

10, −10, 10, −10; Y1, Y2; X=6, Y=−1

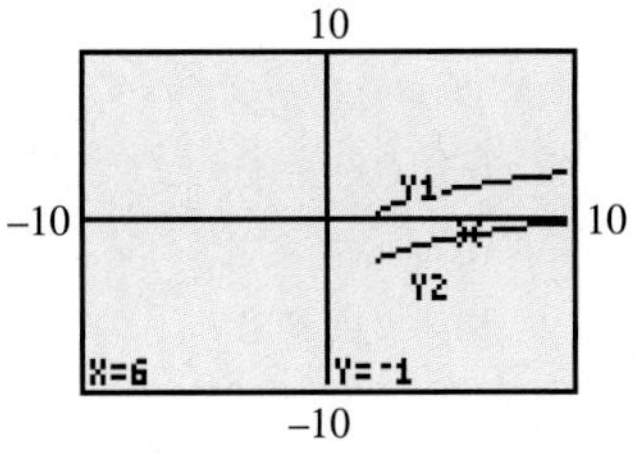

(6, −1) lies on the graph of $Y_2$.
Second View

**28.**

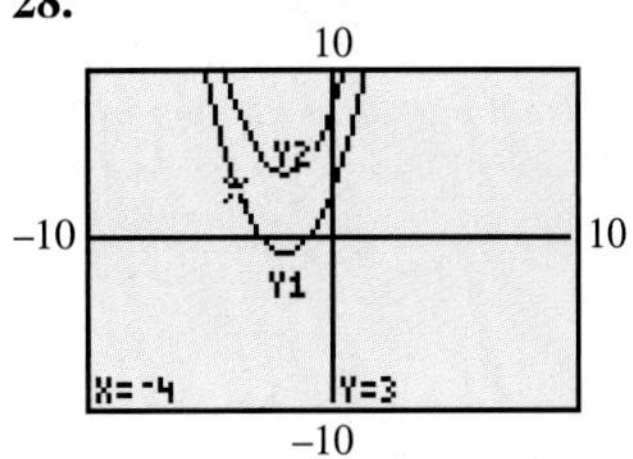


(−4, 3) lies on the graph of $Y_1$.
First View

10, −10, 10, −10; Y2, Y1; X=−4, Y=8

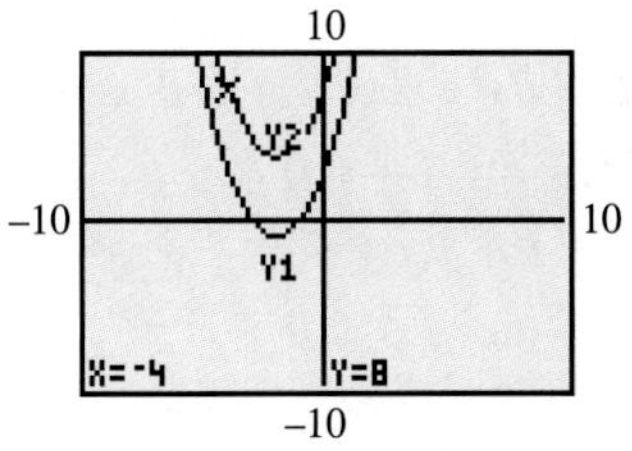

(−4, 8) lies on the graph of $Y_2$.
Second View

*Concept Check* *Suppose that h and k are both positive numbers. Match each equation with the correct graph.*

**29.** $y = (x - h)^2 - k$

**30.** $y = (x + h)^2 - k$

**31.** $y = (x + h)^2 + k$

**32.** $y = (x - h)^2 + k$

**A.**

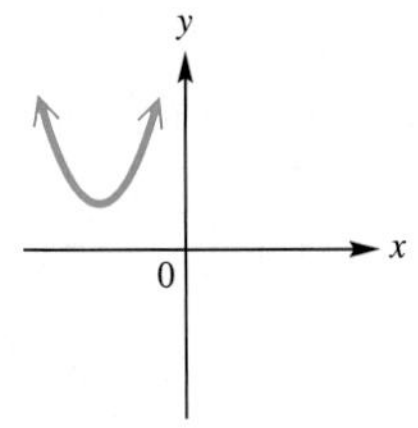

y, x, 0

**B.**

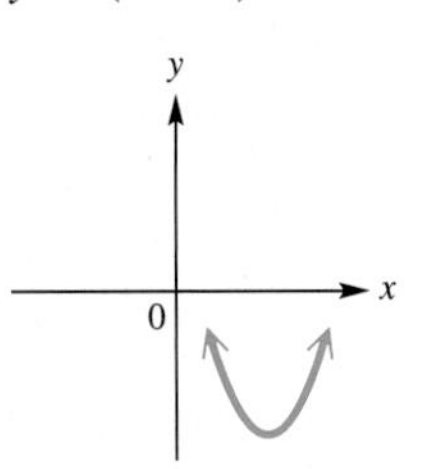


**C.**

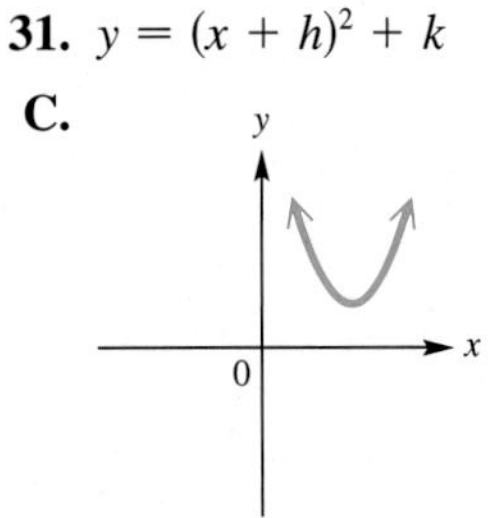


**D.**

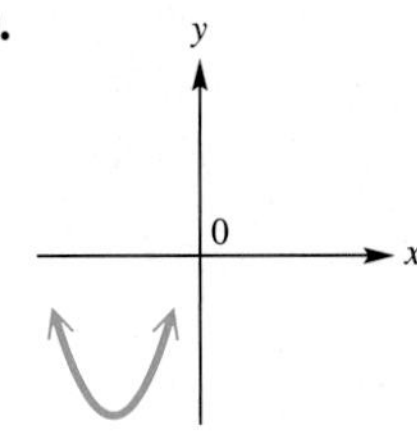


*Concept Check* *Given the graph shown, sketch by hand the graph of each function described, indicating how the three points labeled on the original graph have been translated.*

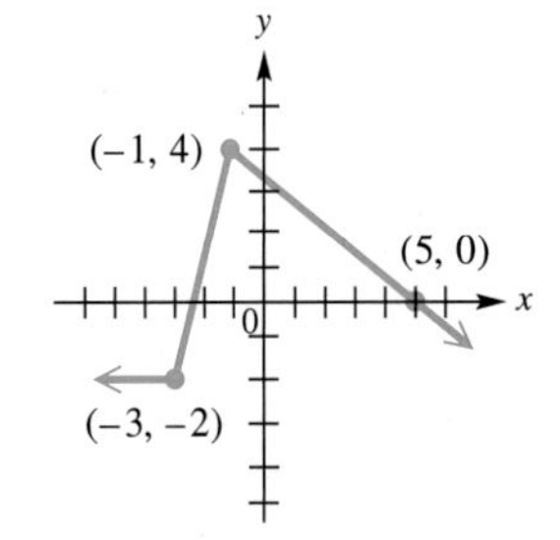


**33.** $y = f(x) + 2$

**34.** $y = f(x) - 2$

**35.** $y = f(x + 2)$

**36.** $y = f(x - 2)$

*Each graph is a translation of the graph of one of the basic functions defined by* $y = x^2$, $y = x^3$, $y = \sqrt{x}$, *or* $y = |x|$. *Find the equation that defines each function. Then, using the concepts of increasing and decreasing functions discussed in Section 2.1, determine the interval of the domain over which the function is* **(a)** *increasing and* **(b)** *decreasing.*

**37.**

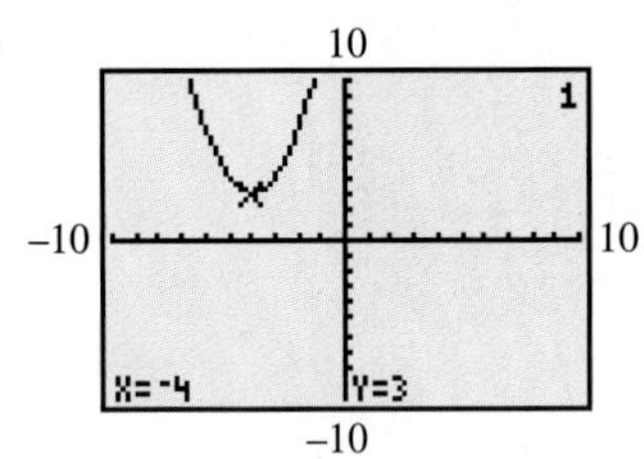


**38.**

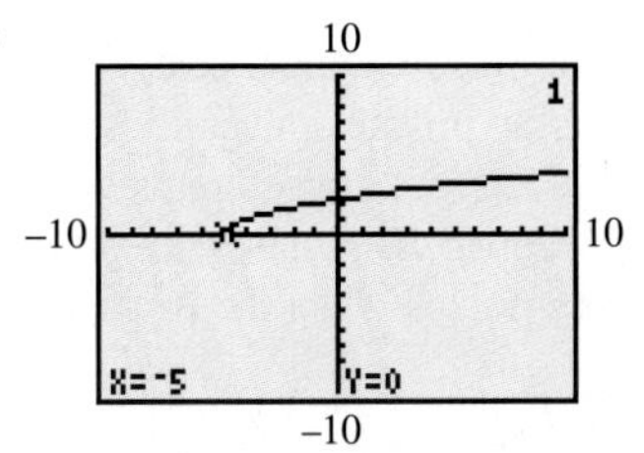

**39.**

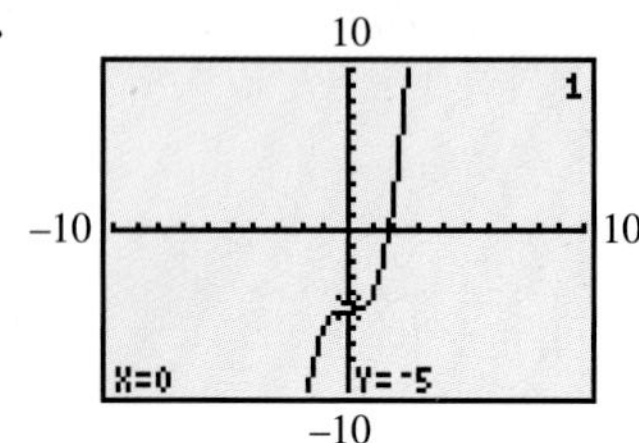


**40.**

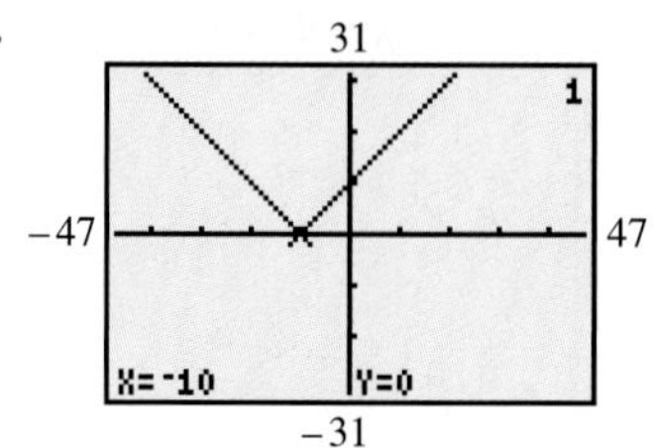


## RELATING CONCEPTS

For individual or group investigation (Exercises 41–44)

*Use the x-intercept method to solve each equation or inequality, using the given graph.*

**41.** **(a)** $f(x) = 0$
**(b)** $f(x) > 0$
**(c)** $f(x) < 0$

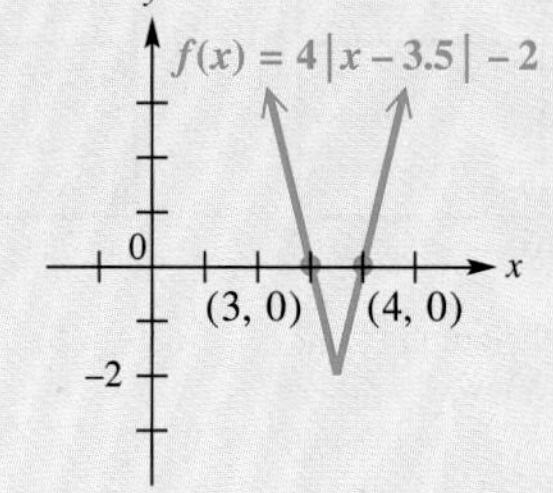


**42.** **(a)** $f(x) = 0$
**(b)** $f(x) > 0$
**(c)** $f(x) < 0$

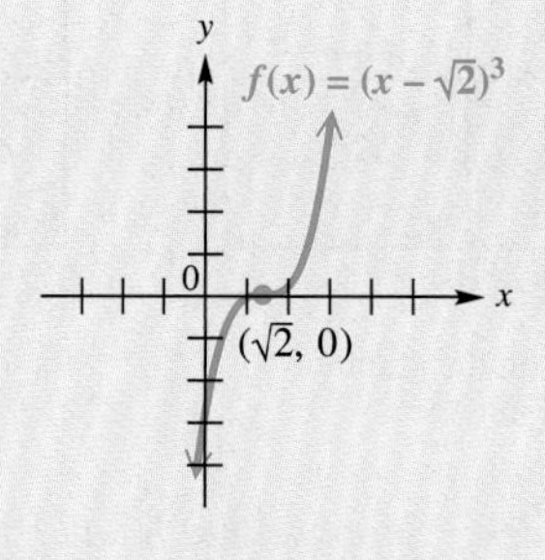


**43.** **(a)** $f(x) = 0$
**(b)** $f(x) \geq 0$
**(c)** $f(x) \leq 0$

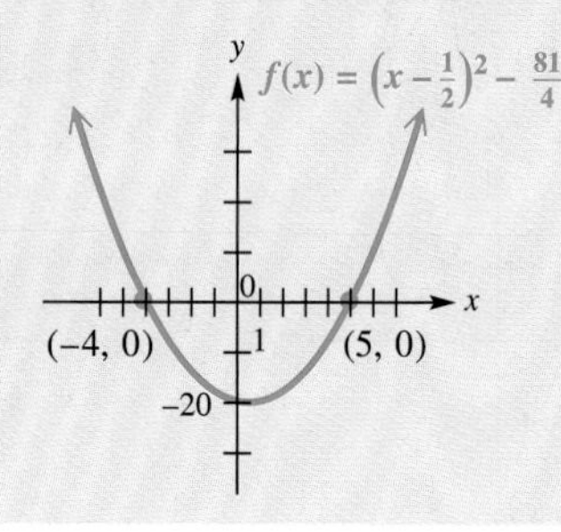


**44.** **(a)** $f(x) = 0$
**(b)** $f(x) \geq 0$
**(c)** $f(x) \leq 0$

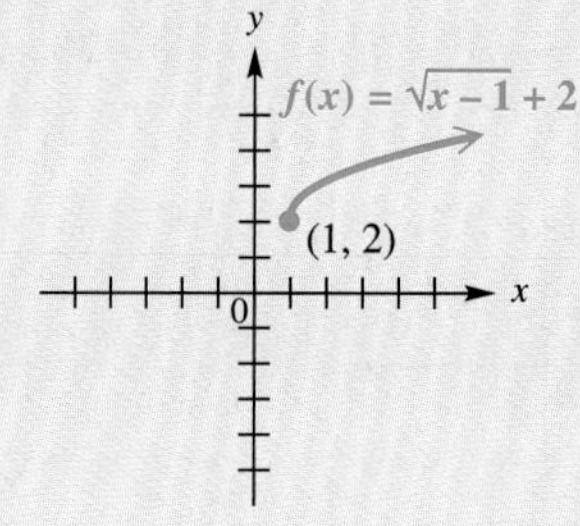


**45.** *Concept Check* The graph shown at the right is a translation of $y = |x|$. What are the values of $h$ and $k$ if the equation is of the form $y = |x - h| + k$?

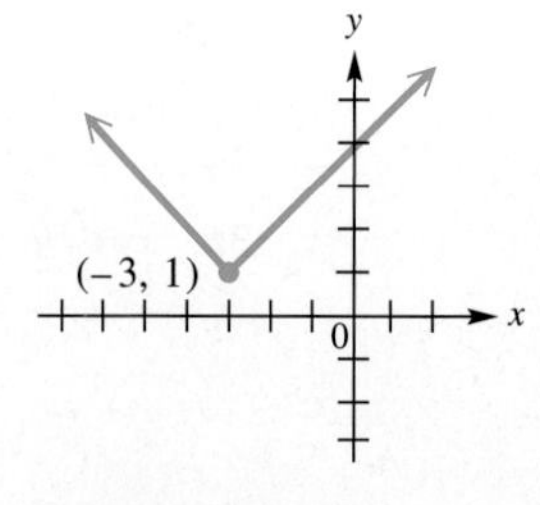


**46.** *Concept Check* Suppose that the graph of $y = x^2$ is translated in such a way that its domain is $(-\infty, \infty)$ and its range is $[38, \infty)$. What values of $h$ and $k$ can be used if the new function is of the form $y = (x - h)^2 + k$?

*(Modeling) Refer to Example 5 to answer Exercises 47–50.*

**47.** *Cost of Private College Education* The linear equation

$$y = 661.4x + 5459.6$$

provides an approximation for the annual cost $y$ (in dollars) for tuition and fees at private universities, where $x = 1$ corresponds to 1995, $x = 2$ corresponds to 1996, and so on. (*Source*: The College Board.) Write an equation that yields the same $y$-values when the exact year is entered.

**48.** *Federal Debt* The linear equation

$$y = 210.5x + 4745.2$$

provides an approximation for the total federal debt $y$ (in billions of dollars) where $x = 0$ corresponds to 1994, $x = 1$ corresponds to 1995, and so on. (*Source:* U.S. Bureau of the Public Debt.) Write an equation that yields the same $y$-values when the exact year is entered.

**49.** *Cost of Public College Education* The table lists the average annual costs (in dollars) of tuition and fees at public four-year colleges for selected years.

| Year | Tuition and Fees (in dollars) |
|---|---|
| 1985 | 1318 |
| 1987 | 1537 |
| 1989 | 1781 |
| 1991 | 2137 |
| 1993 | 2527 |
| 1995 | 2860 |
| 1997 | 3111 |
| 1999 | 3356 |

*Source:* The College Board.

**(a)** Use a calculator to find the least-squares regression line for these data, where $x = 1$ corresponds to 1985, $x = 3$ corresponds to 1987, and so on.

**(b)** Based on your result from part (a), write an equation that yields the same $y$-values when the exact year is entered.

**50.** *Women in the Workforce* The table shows how the percent of population of women in the civilian workforce has changed from 1960 to 1999.

| Year | Percent of Women in the Workforce |
|---|---|
| 1960 | 37.7 |
| 1965 | 39.3 |
| 1970 | 43.3 |
| 1975 | 46.3 |
| 1980 | 51.5 |
| 1985 | 54.5 |
| 1990 | 57.5 |
| 1995 | 58.9 |
| 1999 | 60.0 |

*Source:* U.S. Bureau of Labor Statistics.

**(a)** Use a calculator to find the least-squares regression line for these data, where $x = 0$ corresponds to 1960, $x = 5$ corresponds to 1965, and so on.

**(b)** Based on your result from part (a), write an equation that yields the same $y$-values when the exact year is entered.

## RELATING CONCEPTS

**For individual or group investigation (Exercises 51–58)**

*Recall from Chapter 1 that a unique line is determined by two different points on the line, and that the values of m and b can then be determined for the general form of the linear function defined by $f(x) = mx + b$. Work these exercises in order.*

**51.** Sketch by hand the line that passes through the points $(1, -2)$ and $(3, 2)$.

**52.** Use the slope formula to find the slope of this line.

**53.** Find the equation of this line, and write it in the form $y_1 = mx + b$.

**54.** Keeping the same two $x$-values as indicated in Exercise 51, add 6 to each $y$-value. What are the coordinates of the two new points?

**55.** Find the slope of the line passing through the points determined in Exercise 54.

**56.** Find the equation of this new line, and write it in the form $y_2 = mx + b$.

**57.** Graph both $y_1$ and $y_2$ in the standard viewing window of your calculator, and describe how the graph of $y_2$ can be obtained by vertically translating the graph of $y_1$. What is the value of the constant by which this vertical translation occurs? Where do you think this comes from?

**58.** Fill in the blanks with the correct responses, based on your work in Exercises 51–57.

If the points $(x_1, y_1)$ and $(x_2, y_2)$ lie on a line, then when we add the positive constant $c$ to each $y$-value, we obtain the points $(x_1, y_1 +$ ______$)$ and $(x_2, y_2 +$ ______$)$. The slope of the new line is ____________ (the same as/different from) the slope of the original line. The graph of the new line can be obtained by shifting the graph of the original line ________ units in the ________ direction.

## 2.3 STRETCHING, SHRINKING, AND REFLECTING GRAPHS

Vertical Stretching • Vertical Shrinking • Reflecting Across an Axis • Combining Transformations of Graphs

In the previous section we saw how adding or subtracting a constant can cause a vertical or horizontal shift. Now we will see how multiplying by a constant alters the graph of a function.

### Vertical Stretching

**FOR DISCUSSION**

In each group, we give four related functions. Graph the four functions in the first group (Group A), and then answer the questions regarding those functions. Repeat the process for Group B and Group C. Use the window specified for each group.

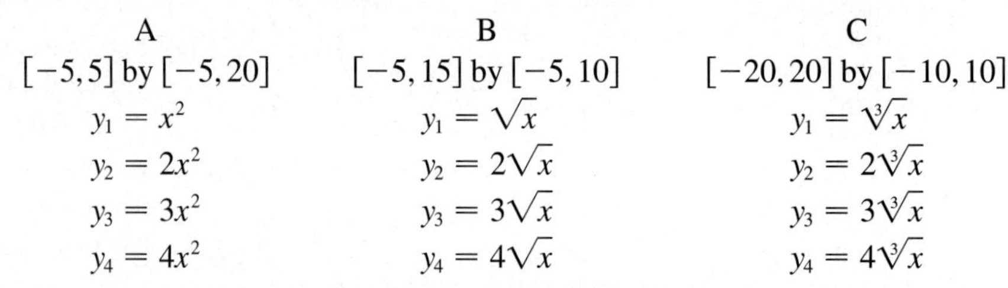

| A | B | C |
|---|---|---|
| $[-5, 5]$ by $[-5, 20]$ | $[-5, 15]$ by $[-5, 10]$ | $[-20, 20]$ by $[-10, 10]$ |
| $y_1 = x^2$ | $y_1 = \sqrt{x}$ | $y_1 = \sqrt[3]{x}$ |
| $y_2 = 2x^2$ | $y_2 = 2\sqrt{x}$ | $y_2 = 2\sqrt[3]{x}$ |
| $y_3 = 3x^2$ | $y_3 = 3\sqrt{x}$ | $y_3 = 3\sqrt[3]{x}$ |
| $y_4 = 4x^2$ | $y_4 = 4\sqrt{x}$ | $y_4 = 4\sqrt[3]{x}$ |

**TECHNOLOGY NOTE**

Plot1 Plot2 Plot3
\Y1=
\Y2=2Y1
\Y3=3Y1
\Y4=4Y1
\Y5=
\Y6=
\Y7=

By defining $Y_1$ as directed in parts A, B, and C, and defining $Y_2$, $Y_3$, and $Y_4$ as shown here, you can minimize your keystrokes. (These graphs will *not* appear unless $Y_1$ is defined.)

1. How does the graph of $y_2$ compare to the graph of $y_1$?
2. How does the graph of $y_3$ compare to the graph of $y_1$?
3. How does the graph of $y_4$ compare to the graph of $y_1$?
4. If we choose $c > 4$, how do you think the graph of $y_5 = c \cdot y_1$ would compare to the graph of $y_4$? Provide support by choosing such a value of $c$.

In each group of functions in the preceding activity, we started with a basic function $y_1$ and observed how the graphs of functions of the form $y = c \cdot y_1$ compared with the graph of $y_1$ for positive values of $c$ that began at 2 and became progressively larger. In each case, we obtained a *vertical stretch* of the graph of the basic function with which we started. These observations can be generalized to any function.

**Vertical Stretching of the Graph of a Function**

If $c > 1$, the graph of $y = c \cdot f(x)$ is obtained by vertically stretching the graph of $y = f(x)$ by a factor of $c$. In general, the larger the value of $c$, the greater the stretch.

y
$y = c \cdot f(x), c > 1$
0
x
$y = f(x)$

**FIGURE 28**

In Figure 28, we graphically interpret the statement above.

**EXAMPLE 1** *Recognizing Vertical Stretches*

Figure 29 shows the graphs of four functions. The graph labeled $y_1$ is that of the function defined by $f(x) = |x|$. The other three functions, $y_2$, $y_3$, and $y_4$, are defined as follows, but not necessarily in the given order:

$$2.4|x|, \quad 3.2|x|, \quad \text{and} \quad 4.3|x|.$$

Determine the correct equation for each graph.

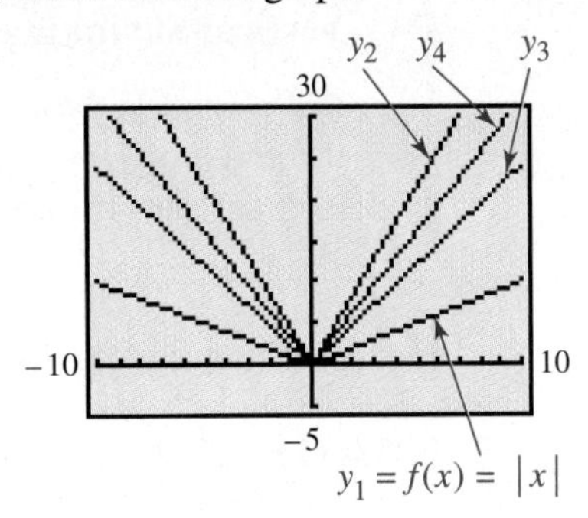


**FIGURE 29**

**Solution** The values of $c$ here are 2.4, 3.2, and 4.3. The vertical heights of the points with the same $x$-coordinates of the three graphs will correspond to the magnitudes of these $c$ values. Thus, the graph just above $y_1 = |x|$ will be that of $y = 2.4|x|$, the "highest" graph will be that of $y = 4.3|x|$, and the graph of $y = 3.2|x|$ will lie "between" the others. Therefore,

$$y_2 = 4.3|x|, \quad y_3 = 2.4|x|, \quad \text{and} \quad y_4 = 3.2|x|.$$

If we were to trace to any point on the graph of $y_1$ and then move to the other graphs one by one, we would see that the $y$-values of the points would be multiplied by the appropriate values of $c$. You may wish to experiment with your calculator in this way.

## Vertical Shrinking

**FOR DISCUSSION**

This discussion parallels the one given earlier in this section. Follow the same general directions. (*Note:* The fractions $\frac{3}{4}$, $\frac{1}{2}$, and $\frac{1}{4}$ may be entered as their decimal equivalents when plotting the graphs.)

| A | B | C |
|---|---|---|
| $[-5, 5]$ by $[-5, 20]$ | $[-5, 15]$ by $[-2, 5]$ | $[-10, 10]$ by $[-2, 10]$ |
| $y_1 = x^2$ | $y_1 = \sqrt{x}$ | $y_1 = \|x\|$ |
| $y_2 = \frac{3}{4}x^2$ | $y_2 = \frac{3}{4}\sqrt{x}$ | $y_2 = \frac{3}{4}\|x\|$ |
| $y_3 = \frac{1}{2}x^2$ | $y_3 = \frac{1}{2}\sqrt{x}$ | $y_3 = \frac{1}{2}\|x\|$ |
| $y_4 = \frac{1}{4}x^2$ | $y_4 = \frac{1}{4}\sqrt{x}$ | $y_4 = \frac{1}{4}\|x\|$ |

1. How does the graph of $y_2$ compare to the graph of $y_1$?
2. How does the graph of $y_3$ compare to the graph of $y_1$?
3. How does the graph of $y_4$ compare to the graph of $y_1$?
4. If we choose $0 < c < \frac{1}{4}$, how do you think the graph of $y_5 = c \cdot y_1$ would compare to the graph of $y_4$? Provide support by choosing such a value of $c$.

TECHNOLOGY NOTE

You can use a screen such as this to minimize your keystrokes in parts A, B, and C. Again, $Y_1$ must be defined in order to obtain the other graphs.

In this "For Discussion" activity, we began with a basic function $y_1$ and observed the graphs of $y = c \cdot y_1$ as we chose progressively smaller values of $c$, with $0 < c < 1$. In each case, the graph of $y_1$ was *vertically shrunk* (or *compressed*). These observations can also be generalized to any function.

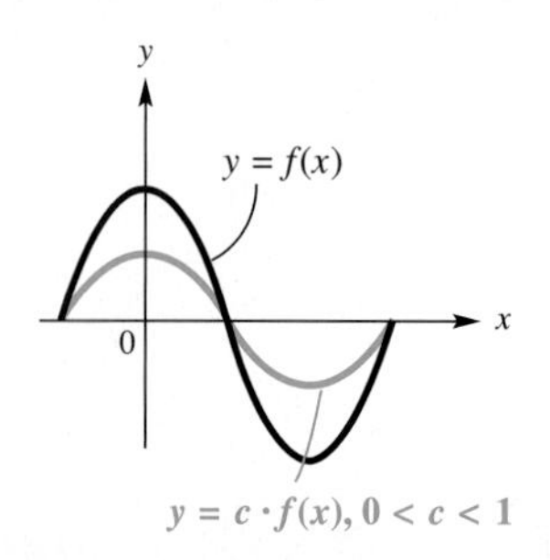


FIGURE 30

**Vertical Shrinking of the Graph of a Function**

If $0 < c < 1$, the graph of $y = c \cdot f(x)$ is obtained by vertically shrinking the graph of $y = f(x)$ by a factor of $c$. In general, the smaller the value of $c$, the greater the shrink.

Figure 30 shows a graphical interpretation of vertical shrinking.

**EXAMPLE 2** *Recognizing Vertical Shrinks*

Figure 31 shows the graphs of four functions. The graph labeled $y_1$ is that of the function defined by $f(x) = x^3$. The other three functions, $y_2$, $y_3$, and $y_4$, are defined as follows, but not necessarily in the given order:

$$.5x^3, \quad .3x^3, \quad \text{and} \quad .1x^3.$$

Determine the correct equation for each graph.

TECHNOLOGY NOTE

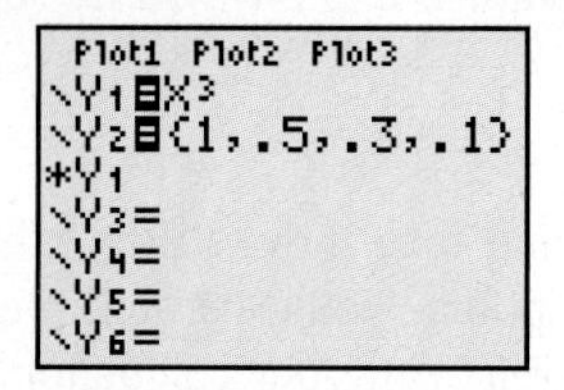


This method of defining $Y_1$ and $Y_2$— using a list of coefficients in $Y_2$— will allow you to duplicate Figure 31.

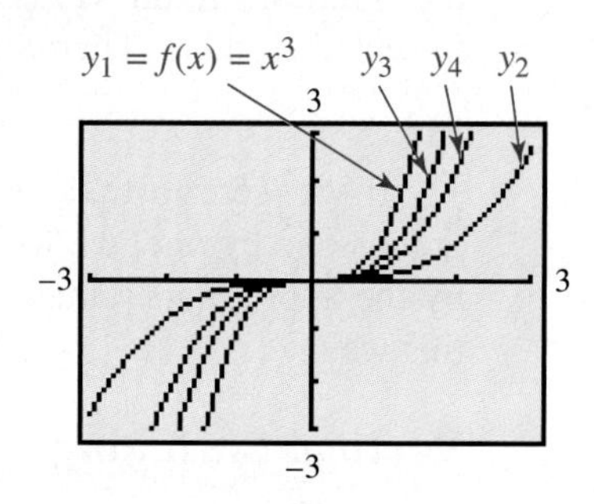


FIGURE 31

**Solution** The smaller the positive value of $c$, where $0 < c < 1$, the more compressed toward the $x$-axis the graph will be. Since we have $c = .5, .3$, and $.1$, the function rules must be as follows:

$$y_2 = .1x^3, \quad y_3 = .5x^3, \quad \text{and} \quad y_4 = .3x^3.$$

## Reflecting Across an Axis

We have seen how graphs can be transformed by shifting, stretching, and shrinking. We now examine how graphs can be reflected across an axis.

TECHNOLOGY NOTE

By defining $Y_1$ as directed in parts A, B, C, and D, and defining $Y_2$ as shown here, you can minimize your keystrokes.

**FOR DISCUSSION**

In each pair, we give two related functions. Graph $y_1 = f(x)$ and $y_2 = -f(x)$ in the standard viewing window, and then answer the questions below for each pair.

| A | B | C | D |
|---|---|---|---|
| $y_1 = x^2$ | $y_1 = \lvert x \rvert$ | $y_1 = \sqrt{x}$ | $y_1 = x^3$ |
| $y_2 = -x^2$ | $y_2 = -\lvert x \rvert$ | $y_2 = -\sqrt{x}$ | $y_2 = -x^3$ |

With respect to the $x$-axis,

1. how does the graph of $y_2$ compare to the graph of $y_1$?
2. how would the graph of $y = -\sqrt[3]{x}$ compare with the graph of $y = \sqrt[3]{x}$, based on your answer to Item 1? Confirm your answer by graphing.

TECHNOLOGY NOTE

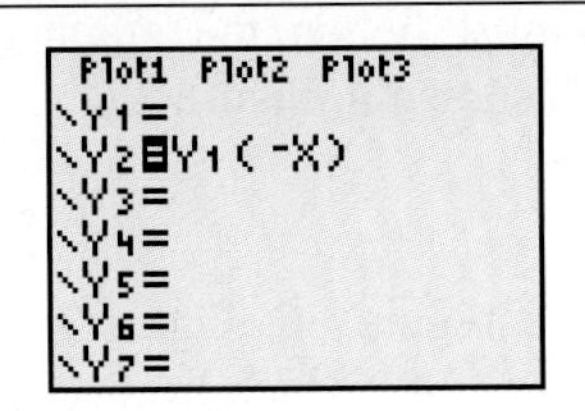


By defining $Y_1$ as directed in parts E, F, and G, and defining $Y_2$ as shown here (using function notation), you can minimize your keystrokes.

Again, in each pair, we give two related functions. Graph $y_1 = f(x)$ and $y_2 = f(-x)$ in the standard viewing window, and then answer the questions below for each pair.

| E | F | G |
|---|---|---|
| $y_1 = \sqrt{x}$ | $y_1 = \sqrt{x - 3}$ | $y_1 = \sqrt[3]{x + 4}$ |
| $y_2 = \sqrt{-x}$ | $y_2 = \sqrt{-x - 3}$ | $y_2 = \sqrt[3]{-x + 4}$ |

With respect to the $y$-axis,

3. how does the graph of $y_2$ compare to the graph of $y_1$?
4. how would the graph of $y = \sqrt[3]{-x}$ compare with the graph of $y = \sqrt[3]{x}$, based on your answer to Item 3? Confirm your answer by graphing.

The results of the preceding discussion can be formally summarized.

**Reflecting the Graph of a Function Across an Axis**

For a function defined by $\boldsymbol{y = f(x)}$,

**(a)** the graph of $\boldsymbol{y = -f(x)}$ is a reflection of the graph of $f$ across the $x$-axis.

**(b)** the graph of $\boldsymbol{y = f(-x)}$ is a reflection of the graph of $f$ across the $y$-axis.

Figure 32 shows how the reflections just described affect the graph of a function in general.

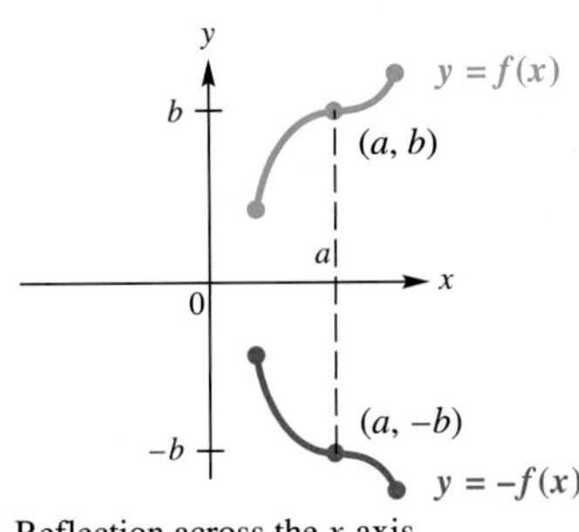


Reflection across the $x$-axis

**(a)**

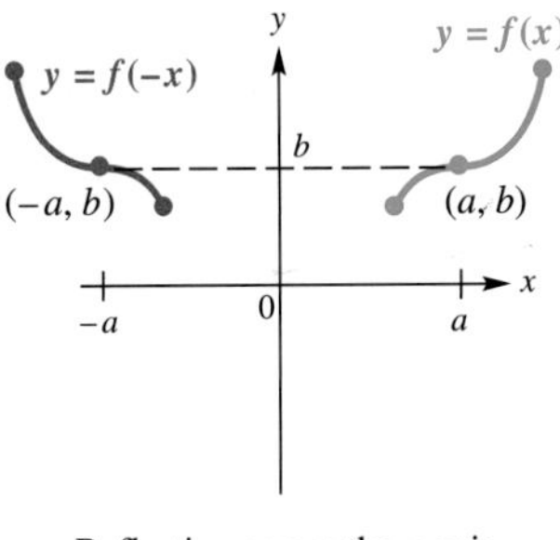


Reflection across the $y$-axis

**(b)**

**FIGURE 32**

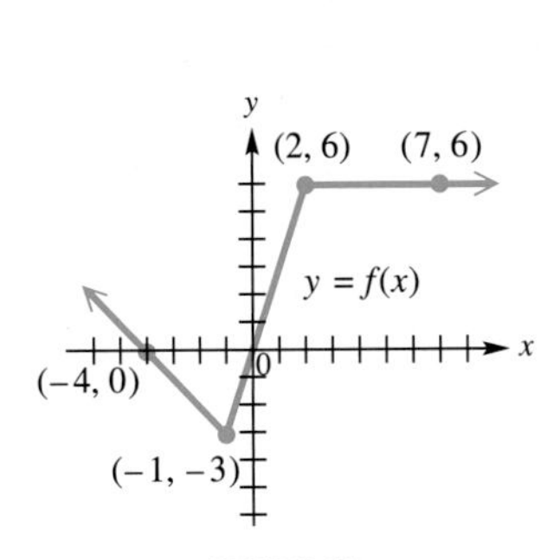


**FIGURE 33**

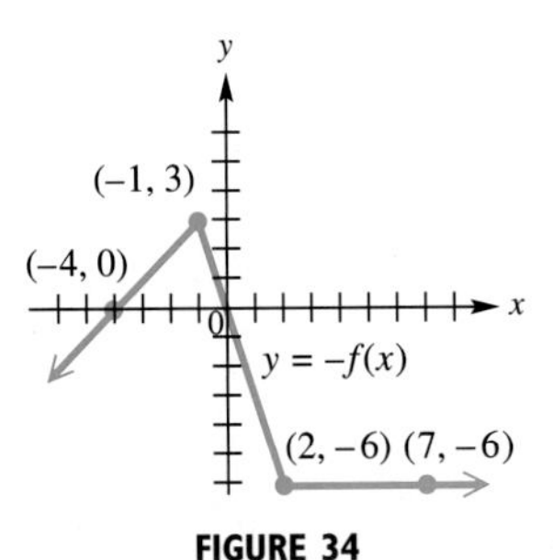


**FIGURE 34**

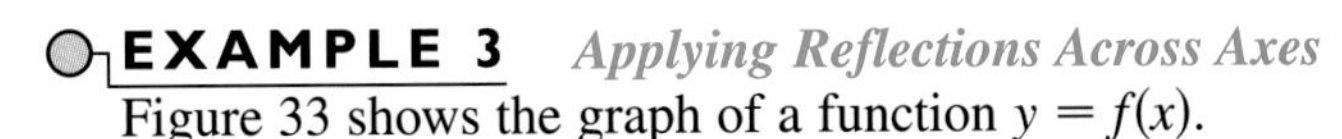

**EXAMPLE 3** *Applying Reflections Across Axes*

Figure 33 shows the graph of a function $y = f(x)$.

**(a)** Sketch the graph of $y = -f(x)$. **(b)** Sketch the graph of $y = f(-x)$.

**Solution**

**(a)** We must reflect the graph across the $x$-axis. This means that if a point $(a, b)$ lies on the graph of $y = f(x)$, then the point $(a, -b)$ must lie on the graph of $y = -f(x)$. Using the labeled points, we find the graph of $y = -f(x)$ in Figure 34.

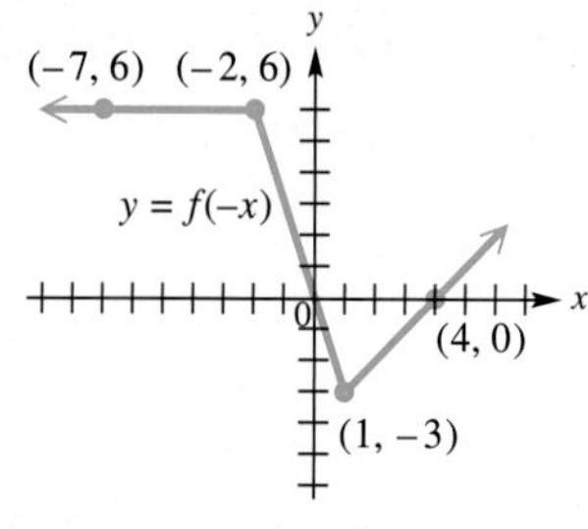


FIGURE 35

**(b)** Here we must reflect the graph across the $y$-axis, meaning that if a point $(a, b)$ lies on the graph of $y = f(x)$, then the point $(-a, b)$ must lie on the graph of $y = f(-x)$. Thus, we obtain the graph of $y = f(-x)$ as shown in Figure 35.

To illustrate reflections on calculator graphs, observe Figure 36. Figure 36(a) shows that $Y_1$ has been defined by $X^2 + 6X + 12$ and $Y_2 = -Y_1$, which means that the graph of $Y_2$ is a reflection of $Y_1$ across the $x$-axis. Figure 36(b) shows the graphs of $Y_1$ and $Y_2$, confirming this fact. Notice that $Y_3 = Y_1(-X)$, indicating that the graph of $Y_3$ is a reflection of $Y_1$ across the $y$-axis. This is confirmed by Figure 36(c).

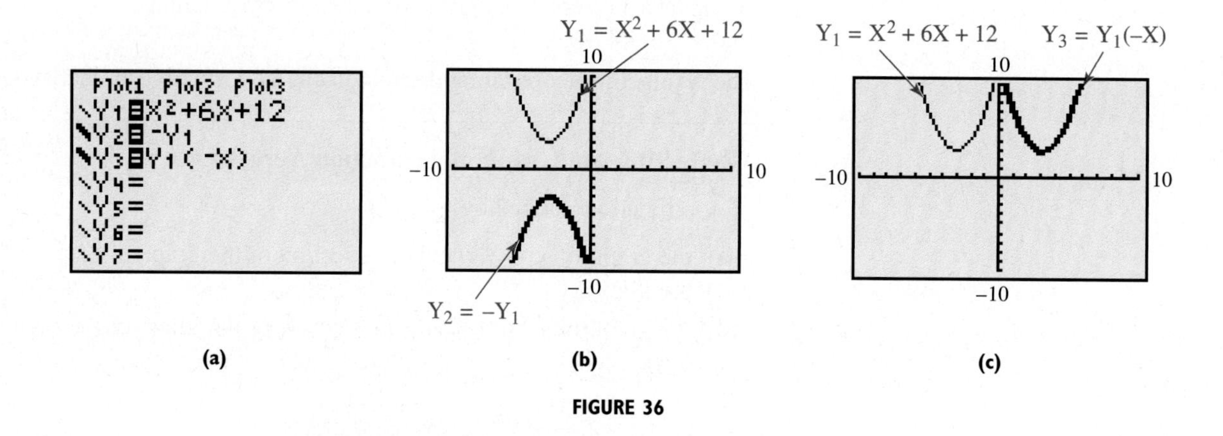


(a) (b) (c)

FIGURE 36

## What Went WRONG?

To see how negative values of $a$ affect the graph of $y = ax^2$, a student entered three functions $Y_1$, $Y_2$, and $Y_3$ as in the accompanying screen. The calculator graphed the first two as shown, but gave a syntax error when it attempted to graph the third.

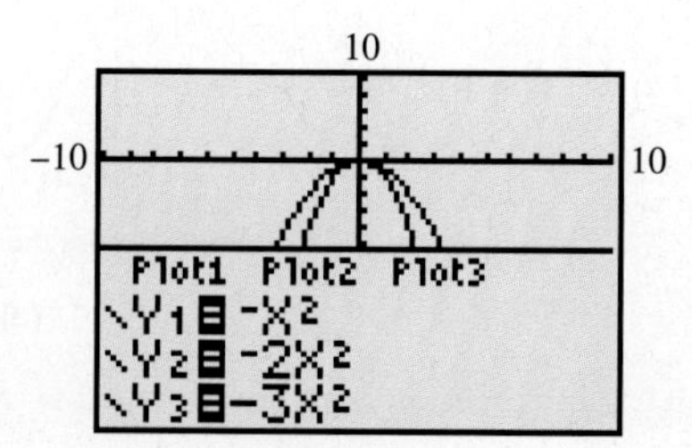


What Went Wrong? What must the student do in order to obtain the desired graph for $y = -3x^2$?

*Answer to What Went Wrong?*

The student used a subtraction sign to define $Y_3$ rather than a negative sign. Notice the difference between the signs in $Y_1$ and $Y_2$ compared to $Y_3$. The student must reenter $Y_3$ using a negative sign.

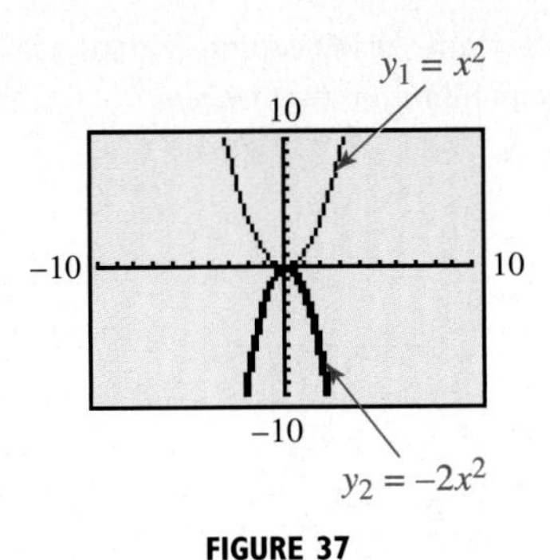


FIGURE 37

## Combining Transformations of Graphs

The graphs of $y_1 = x^2$ and $y_2 = -2x^2$ are shown in the same viewing window in Figure 37. In terms of the types of transformations we have studied, the graph of $y_2$ is obtained by vertically stretching the graph of $y_1$ by a factor of 2 and then reflecting across the $x$-axis. Thus, we have a combination of transformations. As you might expect, we can create an infinite number of functions by vertically stretching or shrinking, shifting upward, downward, left, or right, and reflecting across an axis. The next example investigates examples of this type of function. In determining the order in which the transformations are made, use the order of operations.

**EXAMPLE 4** *Describing Combinations of Transformations of Graphs*

**(a)** Describe how the graph of $y = -3(x - 4)^2 + 5$ can be obtained by transforming the graph of $y = x^2$. Illustrate with a graphing calculator.

**(b)** Give the equation of the function that would be obtained by starting with the graph of $y = |x|$, shifting 3 units to the left, vertically shrinking the graph by a factor of $\frac{2}{3}$, reflecting across the $x$-axis, and shifting the graph 4 units downward, in this order. Illustrate with a graphing calculator.

### Analytic Solution

**(a)** The presence of $(x - 4)^2$ in the definition of the function indicates that the graph of $y = x^2$ must be shifted 4 units to the right. Since the coefficient of $(x - 4)^2$ is $-3$ (a negative number with absolute value greater than 1), the graph is stretched vertically by a factor of 3 and then reflected across the $x$-axis. Finally, the constant $+5$ indicates that the graph is shifted upward 5 units. These ideas are summarized below.

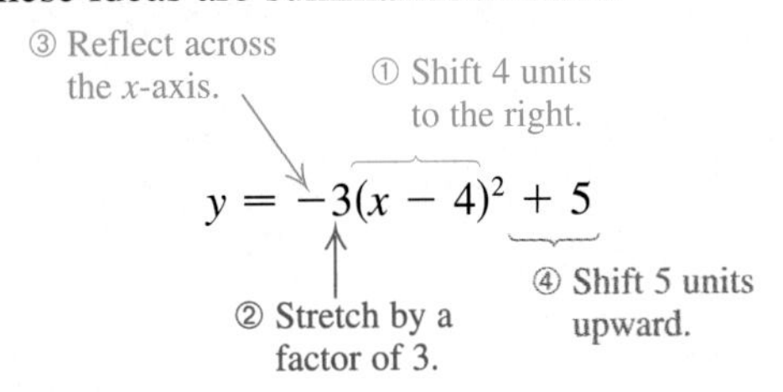


**(b)** Shifting 3 units to the left means that $|x|$ is transformed to $|x + 3|$. Vertically shrinking by a factor of $\frac{2}{3}$ means multiplying $|x + 3|$ by $\frac{2}{3}$, and reflecting across the $x$-axis changes $\frac{2}{3}$ to $-\frac{2}{3}$. Finally, shifting 4 units downward means subtracting 4. Putting this all together leads to the equation

$$y = -\frac{2}{3}|x + 3| - 4.$$

### Graphing Calculator Solution

**(a)** Figure 38 supports the discussion in the analytic solution.

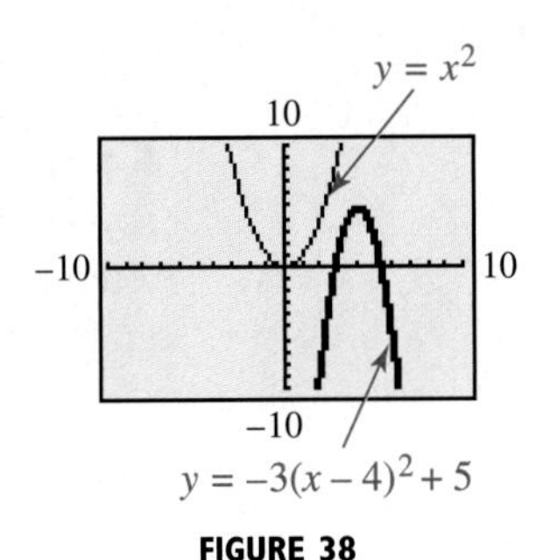


FIGURE 38

**(b)** Figure 39 supports the discussion in the analytic solution.

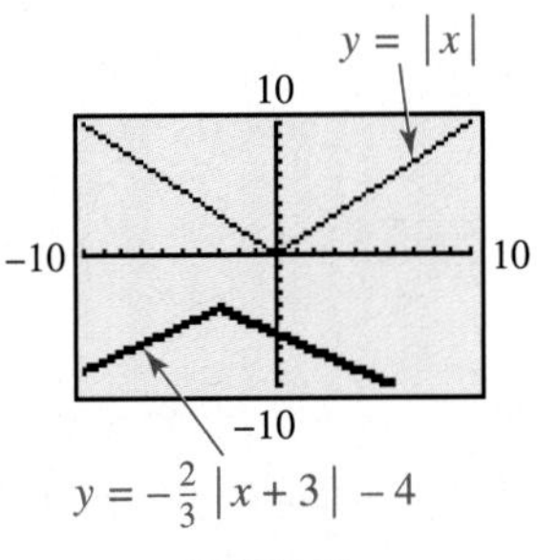


FIGURE 39

**CAUTION** The order in which the transformations are made is important. If they are made in a different order, a different equation can result. See the diagram that follows.

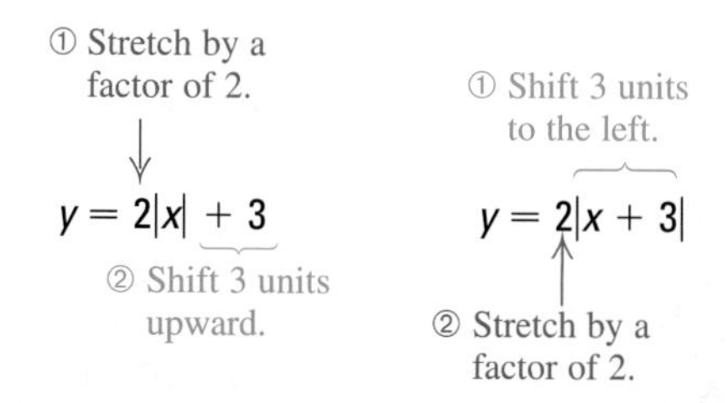


**EXAMPLE 5** *Recognizing a Combination of Transformations*

Figure 40 shows two views of the graph of $y = |x|$ and another graph illustrating a combination of transformations. Find the equation of the transformed graph.

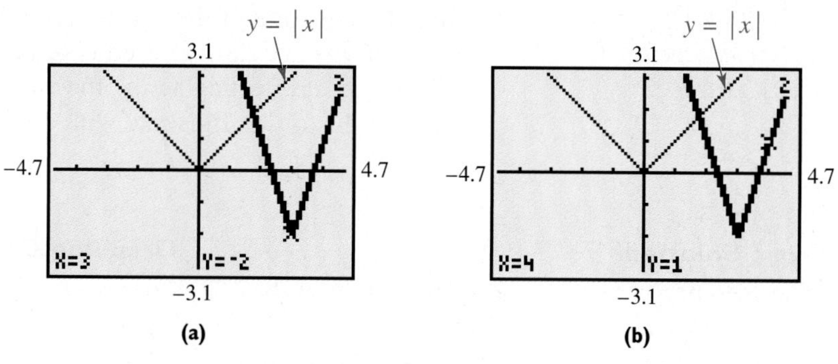


**FIGURE 40**

**Solution** Figure 40(a) shows that the lowest point on the transformed graph has coordinates $(3, -2)$, indicating that the graph has been shifted 3 units to the right and 2 units downward. Figure 40(b) shows that a point on the right side of the transformed graph has coordinates $(4, 1)$, and thus the slope of this ray is

$$m = \frac{-2 - 1}{3 - 4} = \frac{-3}{-1} = 3.$$

Thus, the stretch factor is 3. This information leads to

$$y = 3|x - 3| - 2$$

as the equation of the transformed graph.

## 2.3 EXERCISES

*Write the equation that results in the desired transformation.*

**1.** The squaring function, stretched by a factor of 2

**2.** The cubing function, shrunk by a factor of $\frac{1}{2}$

**3.** The square root function, reflected across the $y$-axis

**4.** The cube root function, reflected across the $x$-axis

**5.** The absolute value function, stretched by a factor of 3 and reflected across the $x$-axis

**6.** The absolute value function, shrunk by a factor of $\frac{1}{3}$ and reflected across the $y$-axis

**7.** The cubing function, shrunk by a factor of .25 and reflected across the $y$-axis

**8.** The square root function, shrunk by a factor of .2 and reflected across the $x$-axis

*Use the concepts of this chapter to draw a rough sketch of the graphs of $y_1$, $y_2$, and $y_3$. Do not plot points. In each case, $y_2$ and $y_3$ can be graphed by one or more of these: a vertical and/or horizontal shift of the graph of $y_1$, a vertical stretch or shrink of the graph of $y_1$, or a reflection of the graph of $y_1$ across an axis. After you have made your sketches, check by graphing them in an appropriate viewing window of your calculator.*

**9.** $y_1 = x$, $y_2 = x + 3$, $y_3 = x - 3$

**10.** $y_1 = x^3$, $y_2 = x^3 + 4$, $y_3 = x^3 - 4$

**11.** $y_1 = |x|$, $y_2 = |x - 3|$, $y_3 = |x + 3|$

**12.** $y_1 = |x|$, $y_2 = |x| - 3$, $y_3 = |x| + 3$

**13.** $y_1 = \sqrt{x}$, $y_2 = \sqrt{x + 6}$, $y_3 = \sqrt{x - 6}$

**14.** $y_1 = |x|$, $y_2 = 2|x|$, $y_3 = 2.5|x|$

**15.** $y_1 = \sqrt[3]{x}$, $y_2 = -\sqrt[3]{x}$, $y_3 = -2\sqrt[3]{x}$

**16.** $y_1 = x^2$, $y_2 = (x - 2)^2 + 1$, $y_3 = -(x + 2)^2$

**17.** $y_1 = |x|$, $y_2 = -2|x - 1| + 1$, $y_3 = -\dfrac{1}{2}|x| - 4$

**18.** *Concept Check* Suppose that the graph of $y = f(x)$ is symmetric with respect to the $y$-axis and it is reflected across the $y$-axis. How will the new graph compare with the original one?

*Fill in each blank with the appropriate response. (Remember that the vertical stretch or shrink factor is positive.)*

**19.** The graph of $y = -4x^2$ can be obtained from the graph of $y = x^2$ by vertically stretching by a factor of ________ and reflecting across the ________-axis.

**20.** The graph of $y = -6\sqrt{x}$ can be obtained from the graph of $y = \sqrt{x}$ by vertically stretching by a factor of ________ and reflecting across the ________-axis.

**21.** The graph of $y = -\frac{1}{4}|x + 2| - 3$ can be obtained from the graph of $y = |x|$ by shifting horizontally ________ units to the ________, vertically shrinking by a factor of ________, reflecting across the ________-axis, and shifting vertically ________ units in the ________ direction.

**22.** The graph of $y = -\frac{2}{5}|-x| + 6$ can be obtained from the graph of $y = |x|$ by reflecting across the ________-axis, vertically shrinking by a factor of ________, reflecting across the ________-axis, and shifting vertically ________ units in the ________ direction.

**23.** The graph of $y = 6\sqrt[3]{x - 3}$ can be obtained from the graph of $y = \sqrt[3]{x}$ by shifting horizontally ________ units to the ________ and stretching vertically by a factor of ________.

**24.** The graph of $y = .5\sqrt[3]{x + 2}$ can be obtained from the graph of $y = \sqrt[3]{x}$ by shifting horizontally ________ units to the ________ and shrinking vertically by a factor of ________.

*Give the equation of each function whose graph is described.*

**25.** The graph of $y = x^2$ is vertically shrunk by a factor of $\frac{1}{2}$, and the resulting graph is shifted 7 units downward.

**26.** The graph of $y = x^3$ is vertically stretched by a factor of 3. This graph is then reflected across the $x$-axis. Finally, the graph is shifted 8 units upward.

**27.** The graph of $y = \sqrt{x}$ is shifted 3 units to the right. This graph is then vertically stretched by a factor of 4.5. Finally, the graph is shifted 6 units downward.

**28.** The graph of $y = \sqrt[3]{x}$ is shifted 2 units to the left. This graph is then vertically stretched by a factor of 1.5. Finally, the graph is shifted 8 units upward.

*Shown on the left is the graph of* $Y_1 = (X - 2)^2 + 1$ *in the standard viewing window of a graphing calculator. Six other functions,* $Y_2$ *through* $Y_7$, *are graphed according to the rules shown in the screen on the right.*

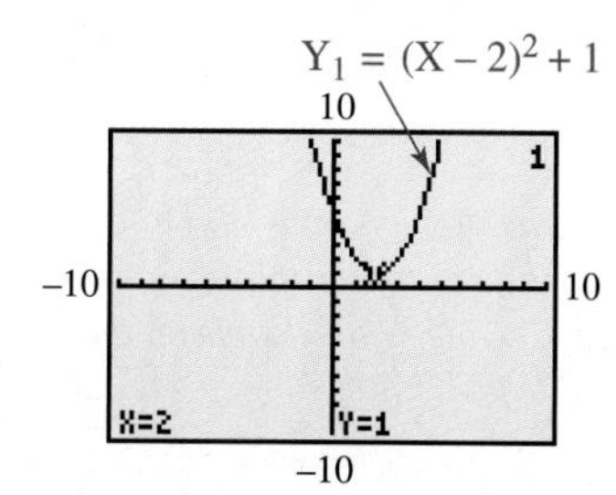


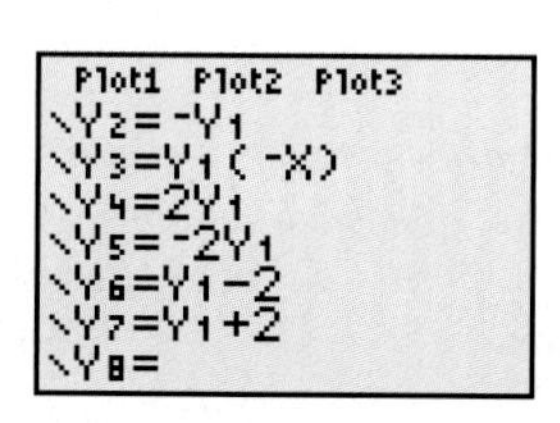


*Match each function with its calculator-generated graph from choices A–F first without using a calculator, by applying the techniques of this chapter. Then, confirm your answer by graphing the function on your calculator.*

**29.** $Y_2$ **30.** $Y_3$ **31.** $Y_4$ **32.** $Y_5$ **33.** $Y_6$ **34.** $Y_7$

**A.**

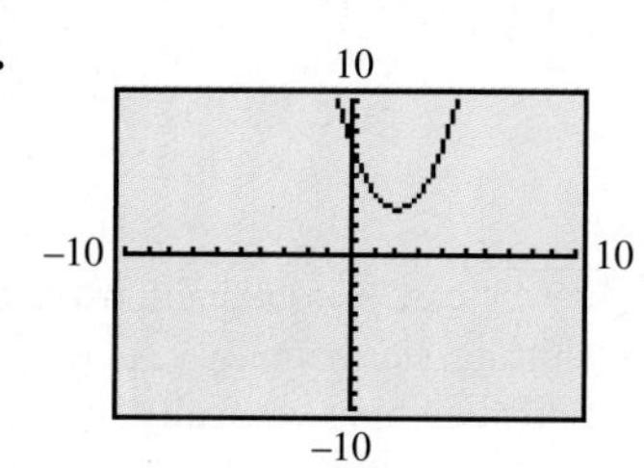


**B.**

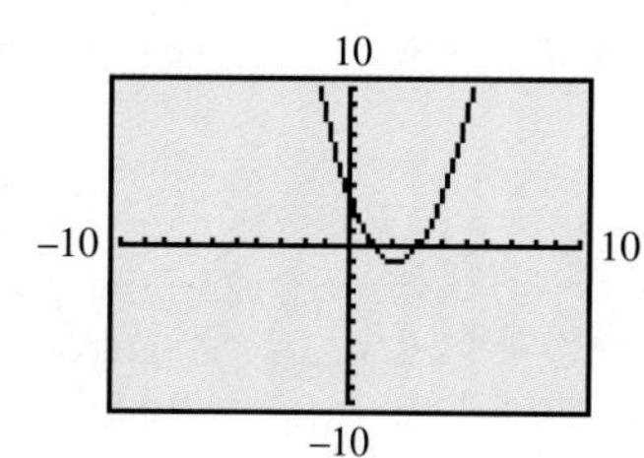


**C.**

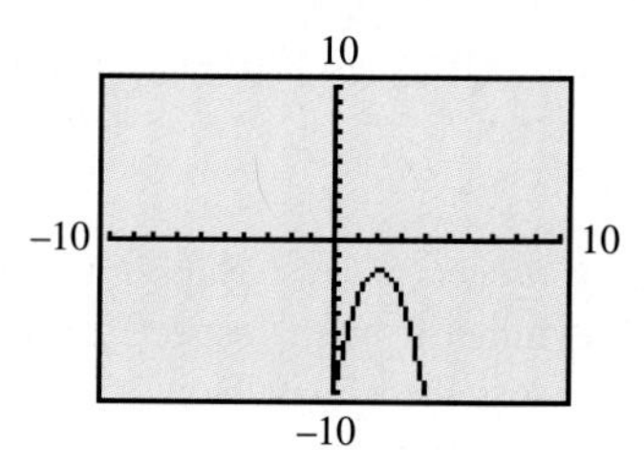


**D.**

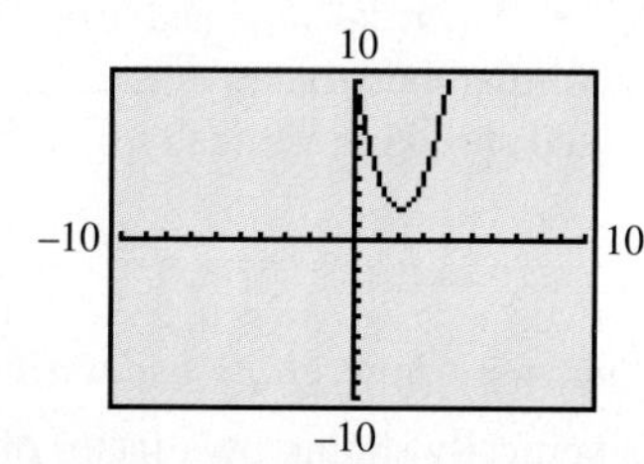


**E.**

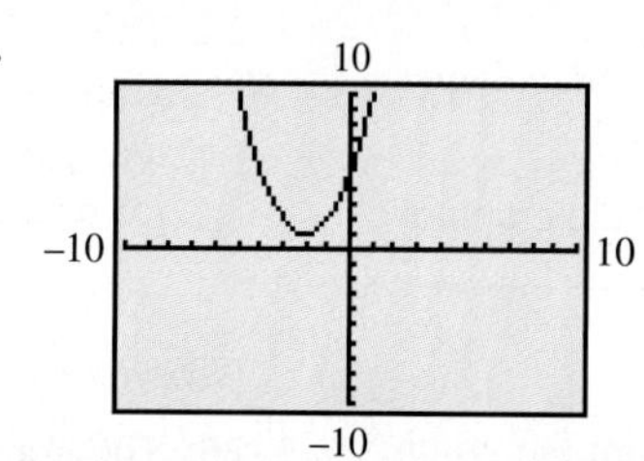


**F.**

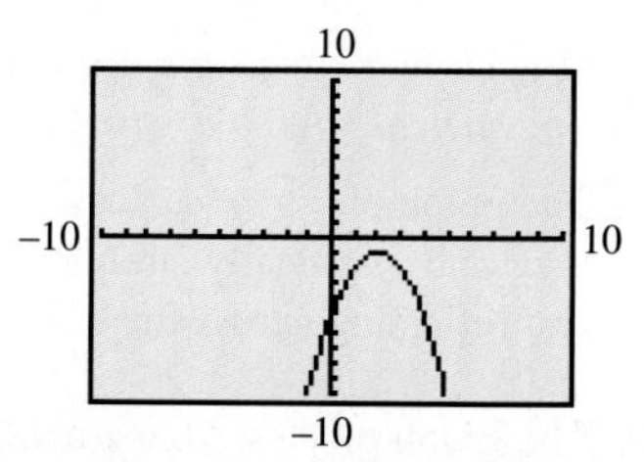


*In Exercises 35 and 36, the graph of* $y = f(x)$ *has been transformed to the graph of* $y = g(x)$. *No shrinking or stretching is involved. Give the equation of* $y = g(x)$.

**35.**

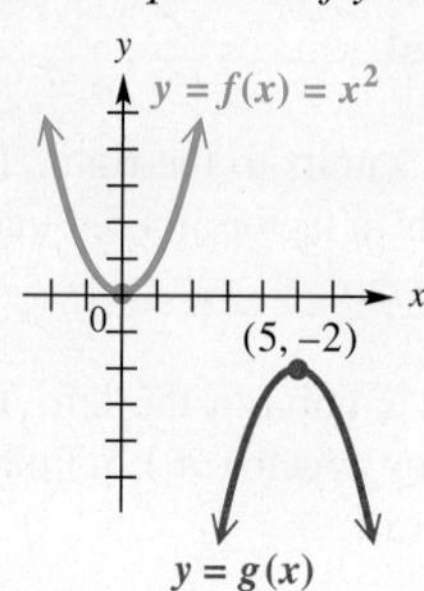


**36.**

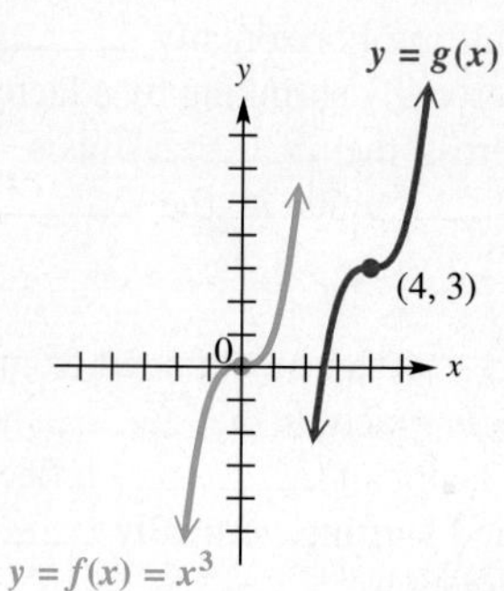

*In Exercises 37–42, each figure shows the graph of a function defined by $y = f(x)$. Sketch by hand the graphs of the functions in parts (a), (b), and (c), and answer the question of part (d).*

**37.** **(a)** $y = -f(x)$ **(b)** $y = f(-x)$ **(c)** $y = 2f(x)$

**(d)** What is $f(0)$?

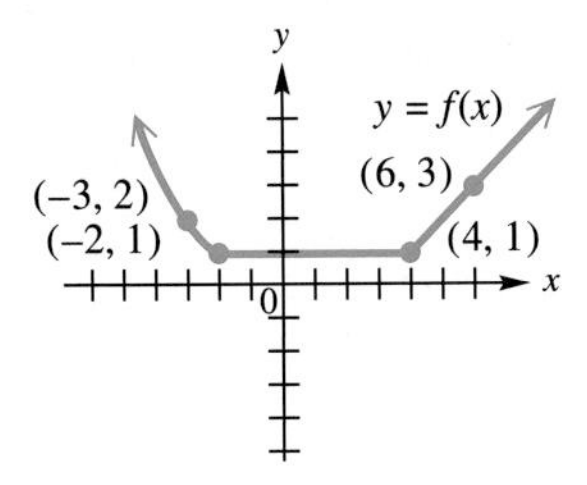


**38.** **(a)** $y = -f(x)$ **(b)** $y = f(-x)$ **(c)** $y = 3f(x)$

**(d)** What is $f(4)$?

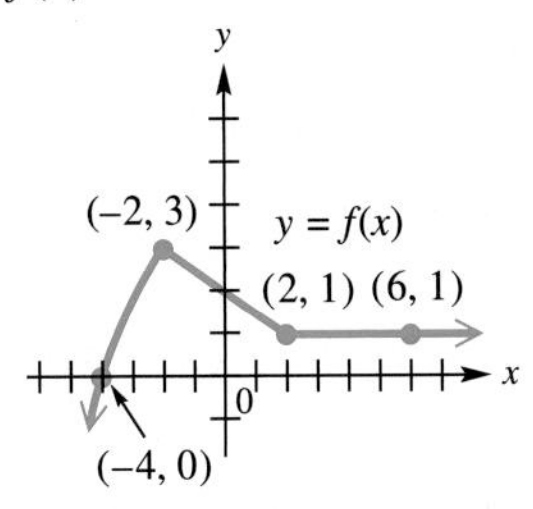


**39.** **(a)** $y = -f(x)$ **(b)** $y = f(-x)$ **(c)** $y = f(x + 1)$

**(d)** What are the $x$-intercepts of $y = f(x - 1)$?

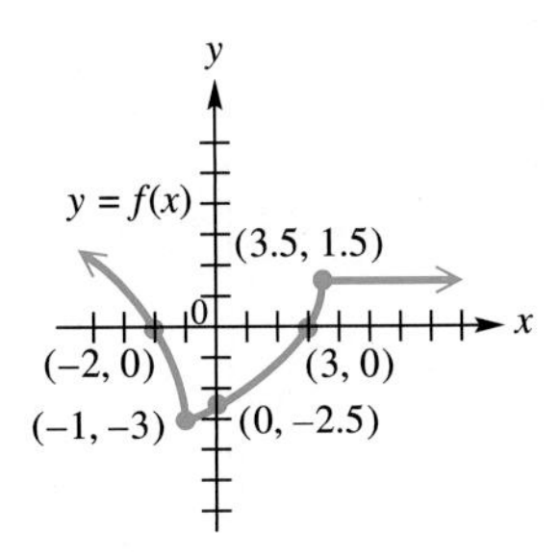


**40.** **(a)** $y = -f(x)$ **(b)** $y = f(-x)$ **(c)** $y = \frac{1}{2}f(x)$

**(d)** On what interval of the domain is $f(x) < 0$?

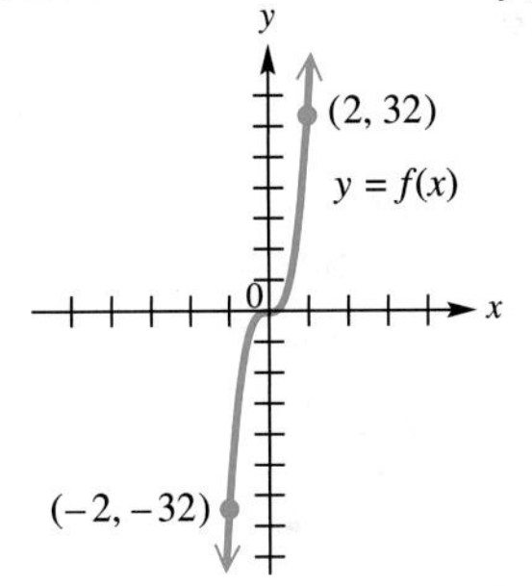


**41.** **(a)** $y = -f(x)$ **(b)** $y = f(-x)$ **(c)** $y = .5f(x)$

**(d)** What symmetry does the graph of $y = f(x)$ exhibit?

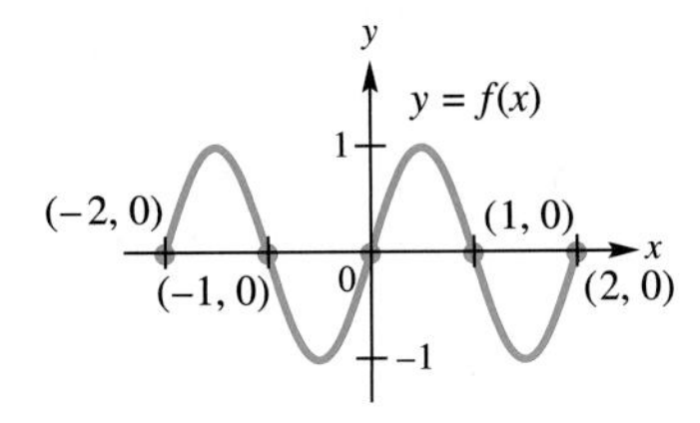


**42.** **(a)** $y = -f(x)$ **(b)** $y = f(-x)$ **(c)** $y = 3f(x)$

**(d)** What symmetry does the graph of $y = f(x)$ exhibit?

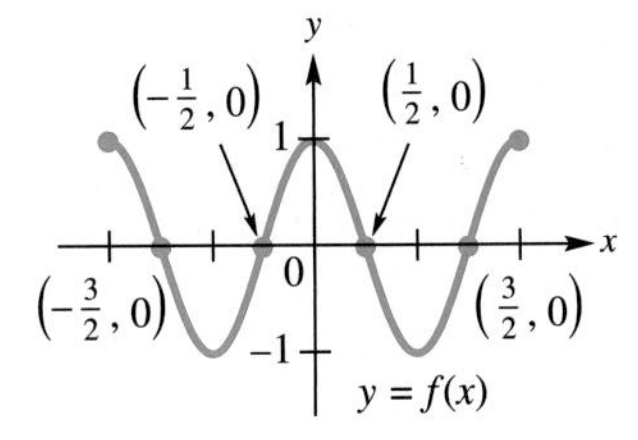


**43.** *Concept Check* If $r$ is an $x$-intercept of the graph of $y = f(x)$, what statement can be made about the $x$-intercept of the graph of each function? (*Hint*: Make a sketch.)

**(a)** $y = -f(x)$ **(b)** $y = f(-x)$

**(c)** $y = -f(-x)$

**44.** *Concept Check* If $b$ is the $y$-intercept of the graph of $y = f(x)$, what statement can be made about the $y$-intercept of the graph of each function? (*Hint:* Make a sketch.)

**(a)** $y = -f(x)$ **(b)** $y = f(-x)$

**(c)** $y = 5f(x)$ **(d)** $y = -3f(x)$

*Concept Check* *The sketch shows an example of a function defined by $y = f(x)$ that increases on the interval $[a, b]$. Use this graph as a visual aid, and apply the concepts of reflection introduced in this section to answer each question. (Make your own sketch if you wish.)*

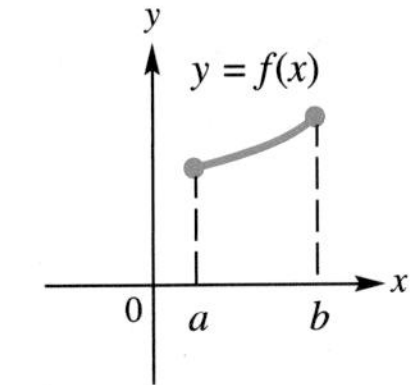


**45.** Does the graph of $y = -f(x)$ increase or decrease on the interval $[a, b]$?

**46.** Does the graph of $y = f(-x)$ increase or decrease on the interval $[-b, -a]$?

**47.** Does the graph of $y = -f(-x)$ increase or decrease on the interval $[-b, -a]$?

**48.** If $c > 0$, does the graph of $y = -c \cdot f(x)$ increase or decrease on the interval $[a, b]$?

*State the intervals over which each function is* **(a)** *increasing,* **(b)** *decreasing, and* **(c)** *constant.*

**49.** The function graphed in Figure 33

**50.** The function graphed in Figure 34

**51.** The function graphed in Figure 35

**52.** $y = -\frac{2}{3}|x + 3| - 4$ (See Figure 39.)

*In Exercises 53–55, each function has a graph with an endpoint (a translation of the point (0, 0)). Enter each into your calculator in an appropriate viewing window, and using your knowledge of the graph of* $y = \sqrt{x}$*, determine the domain and range of the function.* (Hint: *Locate the endpoint.*)

**53.** $y = 10\sqrt{x - 20} + 5$

**54.** $y = -2\sqrt{x + 15} - 18$

**55.** $y = -.5\sqrt{x + 10} + 5$

**56.** *Concept Check* Based on your observations in Exercise 53, what are the domain and range of $f(x) = a\sqrt{x - h} + k$, if $a > 0$, $h > 0$, and $k > 0$?

*Concept Check* *Shown here are the graphs of* $y_1 = \sqrt[3]{x}$ *and* $y_2 = 5\sqrt[3]{x}$*. The point whose coordinates are given at the bottom of the screen lies on the graph of* $y_1$*. Use this graph, not your calculator, to find the value of* $y_2$ *for the same value of x shown.*

**57.**

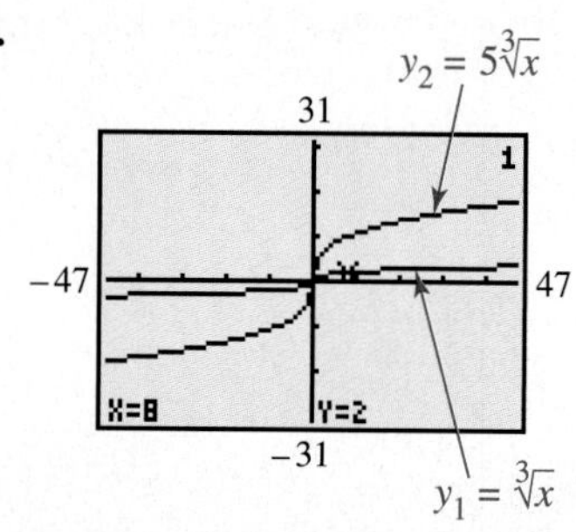


**58.**

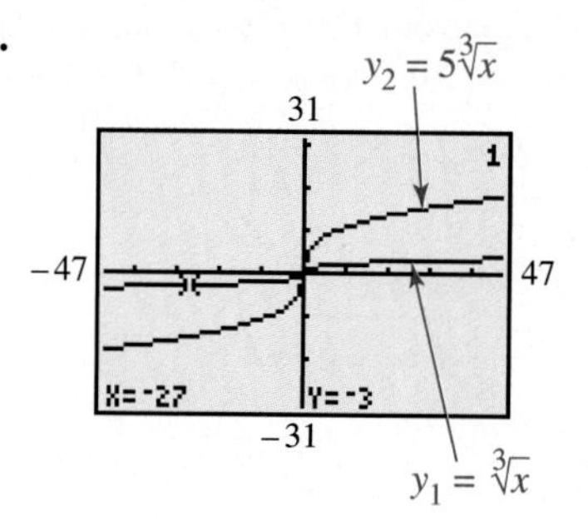


## Reviewing Basic Concepts (Sections 2.1–2.3)

**1.** Suppose that $f$ is defined for all real numbers, and $f(3) = 6$. For the given assumptions, find another function value.

**(a)** The graph of $y = f(x)$ is symmetric with respect to the origin.

**(b)** The graph of $y = f(x)$ is symmetric with respect to the $y$-axis.

**(c)** For all $x$, $f(-x) = -f(x)$.

**(d)** For all $x$, $f(-x) = f(x)$.

**2.** Match each equation in Column I with a description of its graph from Column II as it relates to the graph of $y = x^2$.

| I | II |
|---|---|
| **(a)** $y = (x - 7)^2$ | **A.** A shift of 7 units to the left |
| **(b)** $y = x^2 - 7$ | **B.** A shift of 7 units to the right |
| **(c)** $y = 7x^2$ | **C.** A shift of 7 units upward |
| **(d)** $y = (x + 7)^2$ | **D.** A shift of 7 units downward |
| **(e)** $y = x^2 + 7$ | **E.** A vertical stretch by a factor of 7 |

**3.** Match each equation in parts (a)–(h) with the sketch of its graph in choices A–H on the next page. The basic graph, $y = x^2$, is shown here.

**(a)** $y = x^2 + 2$

**(b)** $y = x^2 - 2$

**(c)** $y = (x + 2)^2$

**(d)** $y = (x - 2)^2$

**(e)** $y = 2x^2$

**(f)** $y = -x^2$

**(g)** $y = (x - 2)^2 + 1$

**(h)** $y = (x + 2)^2 + 1$

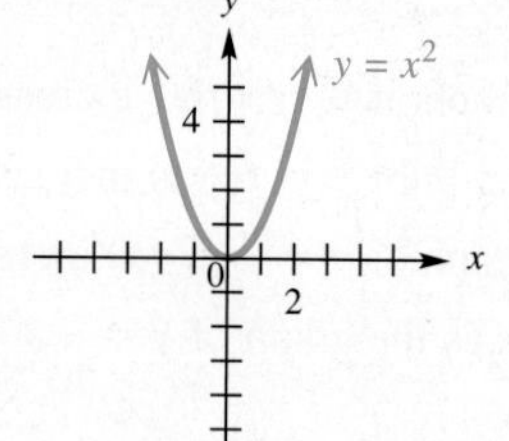

**A.**
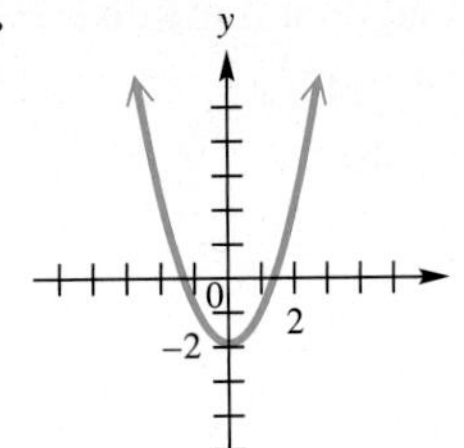


**B.**
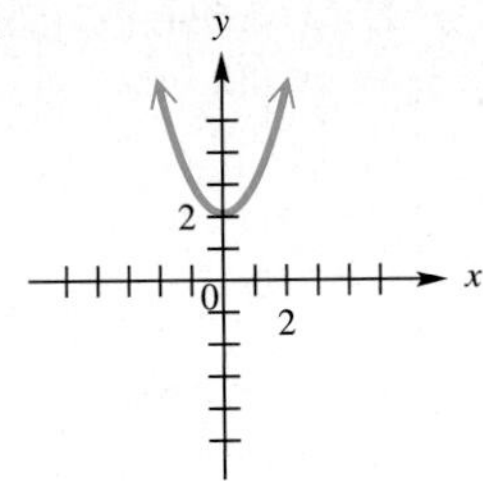


**C.**
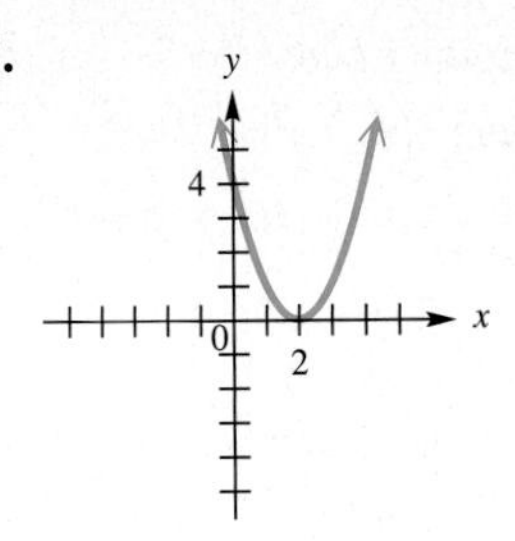


**D.**
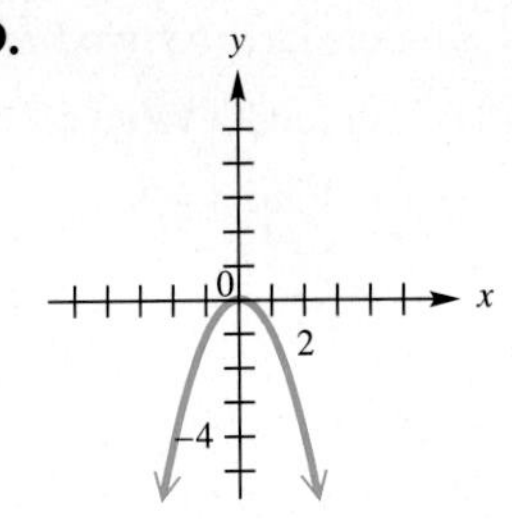


**E.**
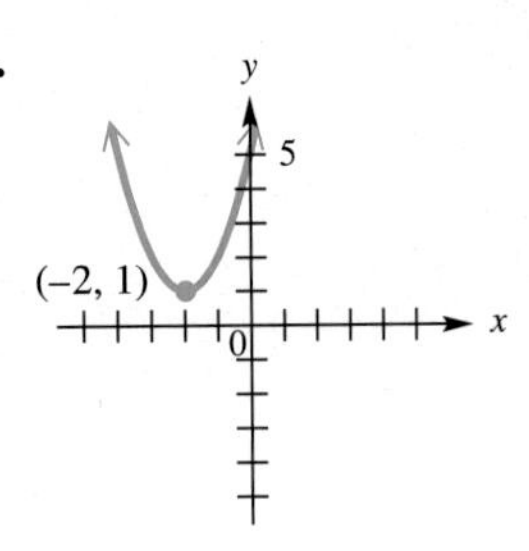


**F.**
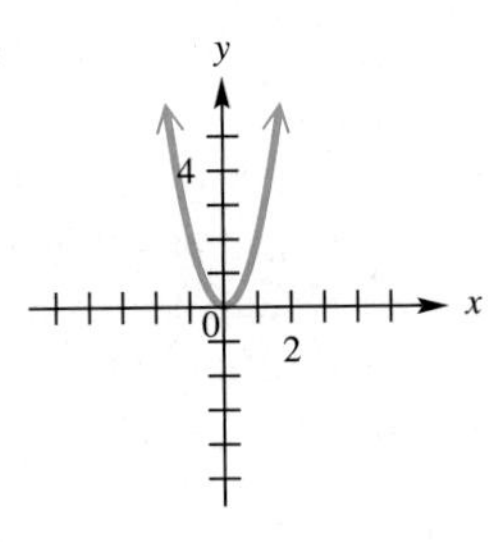


**G.**
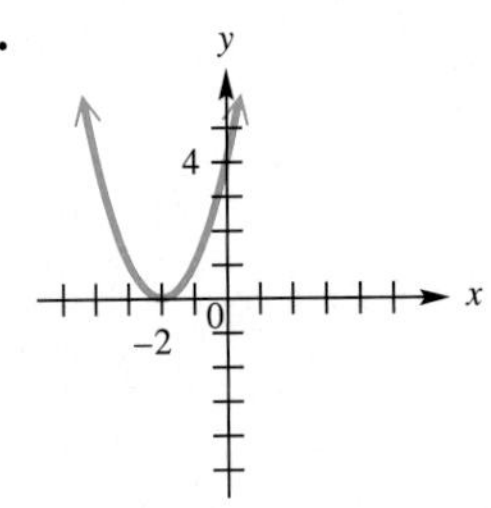


**H.**
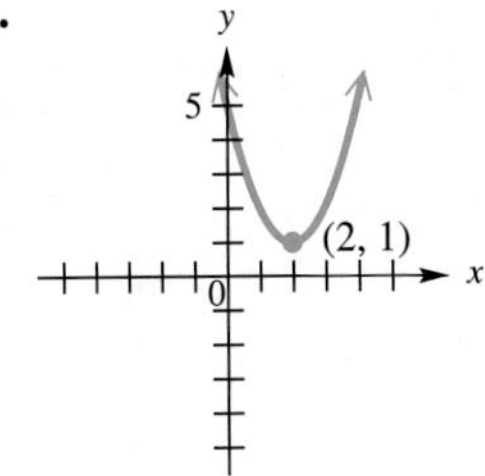


**4.** Match each equation with its calculator graph.

**(a)** $y = \sqrt{x} + 6$ **(b)** $y = \sqrt{x + 6}$ **(c)** $y = \sqrt{x - 6}$

**(d)** $y = \sqrt{x + 2} - 4$ **(e)** $y = \sqrt{x - 2} - 4$

**A.**
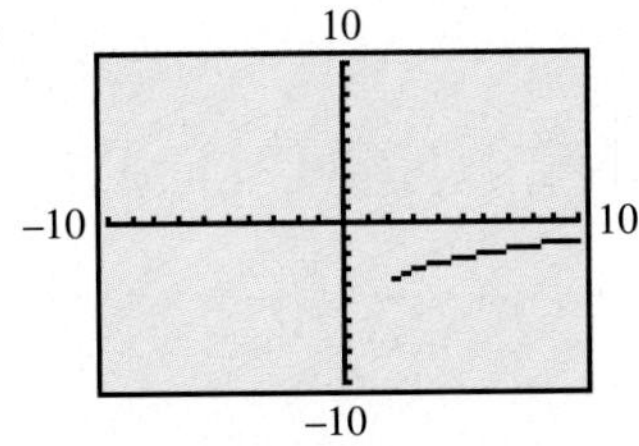


**B.**
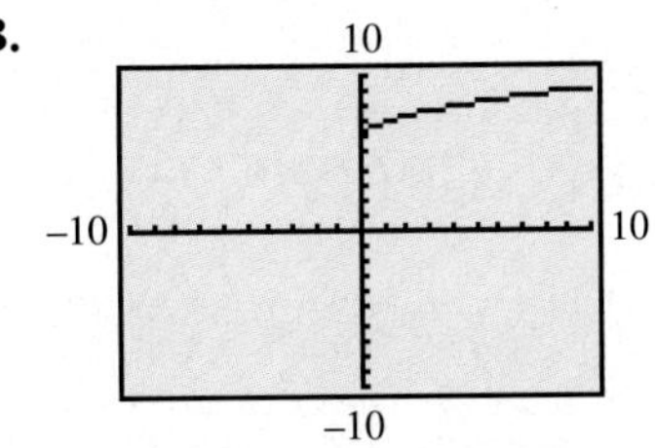


**C.**
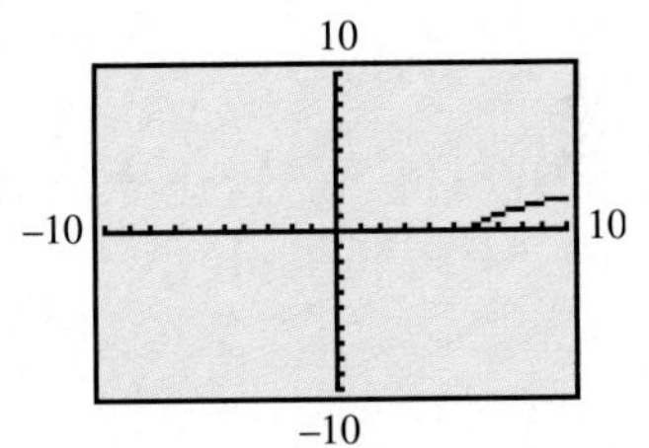


**D.**
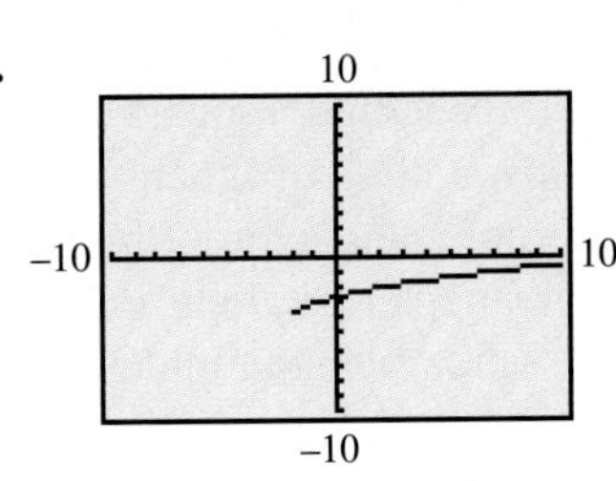


**E.**
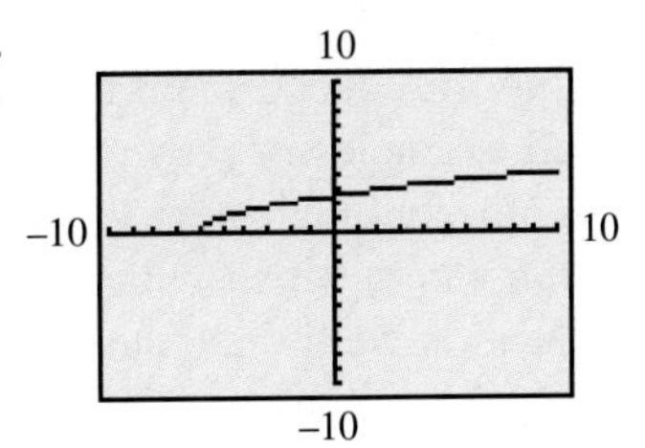


**5.** Each graph is obtained from the graph of $f(x) = |x|$ or $g(x) = \sqrt{x}$ by applying the transformations discussed in Sections 2.2 and 2.3. Describe the transformations, and then give the equation for the graph.

**(a)**
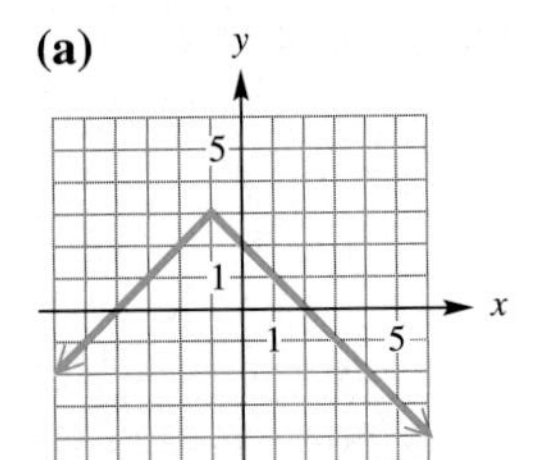


**(b)**
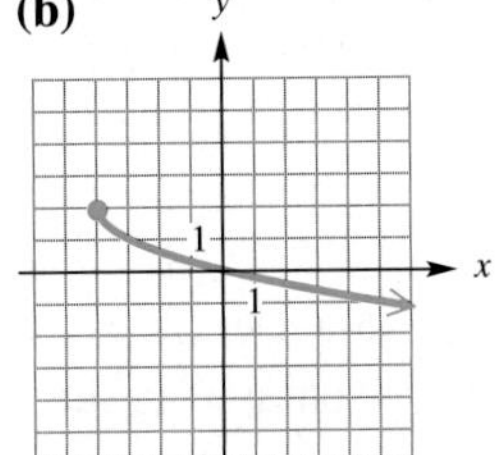


**(c)**
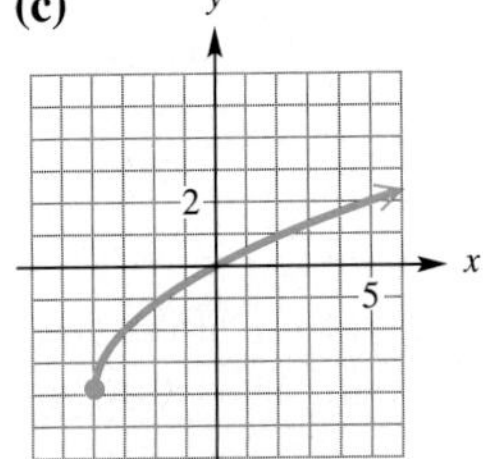


**(d)**
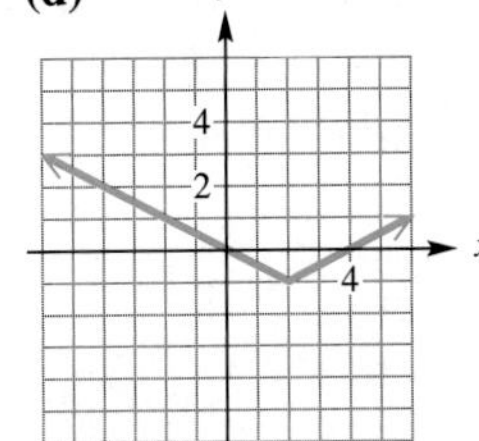

**6.** Consider the two functions in the figure.

**(a)** Find a value of $c$ for which $g(x) = f(x) + c$.

**(b)** Find a value of $c$ for which $g(x) = f(x + c)$.

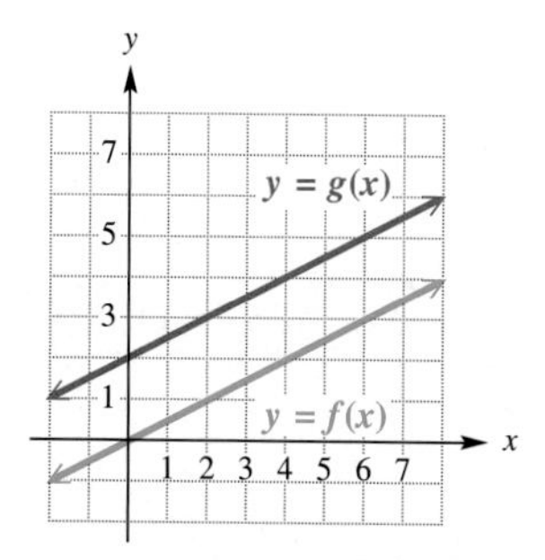


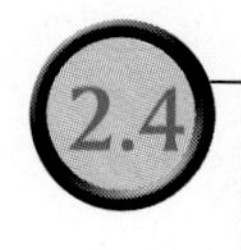

**7.** Suppose the equation $y = F(x)$ is changed to $y = F(x + h)$. How are the graphs of these equations related? Is the graph of $y = F(x) + h$ the same as the graph of $y = F(x + h)$? If not, how do they differ?

**8.** Suppose the equation $y = F(x)$ is changed to $y = c \cdot F(x)$, for some constant $c$. What is the effect on the graph of $y = F(x)$? Discuss the effect depending on whether $c > 0$ or $c < 0$, and $|c| > 1$ or $|c| < 1$.

**9.** Complete the table if **(a)** $f$ is an even function and **(b)** $f$ is an odd function.

| $x$ | $f(x)$ |
|---|---|
| −3 | 4 |
| −2 | −6 |
| −1 | 5 |
| 1 | |
| 2 | |
| 3 | |

**10.** *(Modeling) Carbon Monoxide Levels* The 8-hour maximum carbon monoxide levels (in parts per million) for the United States from 1982 to 1992 can be modeled by the function defined by

$$f(x) = -.012053(x + 1.93342)^2 + 9.07994,$$

where $x = 0$ corresponds to 1982. (*Source:* U.S. Environmental Protection Agency, 1992.) Find a function represented by $g(x)$ that models the same carbon monoxide levels except that $x$ is the actual year between 1982 and 1992. For example, $g(1985) = f(3)$ and $g(1990) = f(8)$. (*Hint:* Use a horizontal translation.)

# 2.4 ABSOLUTE VALUE FUNCTIONS: GRAPHS, EQUATIONS, INEQUALITIES, AND APPLICATIONS

The Graph of $y = |f(x)|$ • Properties of Absolute Value • Equations and Inequalities Involving Absolute Value • An Application Involving Absolute Value

TECHNOLOGY NOTE

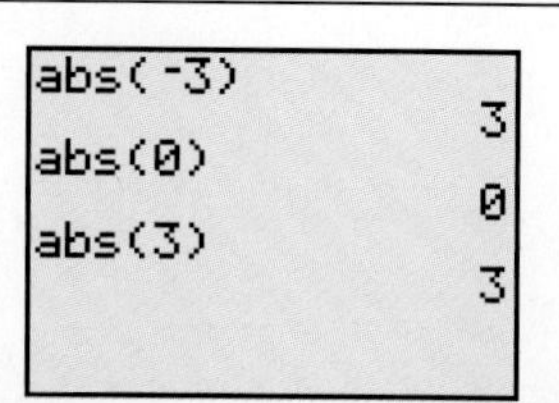


The command abs($x$) is used by some graphing calculators to find absolute value. Consult the graphing calculator manual that accompanies this text or your owner's manual.

Of the basic functions introduced so far in this chapter, the identity, squaring, and cubing functions are all examples of *polynomial functions*, a class of functions that will be examined more closely in Chapter 3. The square root and cube root functions are examples of *root functions*, which will be studied in Chapter 4 (along with *rational functions*). In this section we investigate the absolute value function in detail.

## The Graph of $y = |f(x)|$

The formal definition of the absolute value of a real number was given in Section 2.1. Geometrically, the absolute value of a real number is its undirected distance from 0 on the number line. As a result, the absolute value of a real number is never negative; it is always greater than or equal to 0. Thus, the function defined by

$$f(x) = |x| = \begin{cases} x & \text{if } x \geq 0 \\ -x & \text{if } x < 0, \end{cases}$$

given in Section 2.1, is just an extension of the definition of absolute value. The expression within the absolute value bars, $x$, is the defining expression for the identity function $y = x$.

Now, we extend the concept and consider the definition of a function defined by the *absolute value of a function f*:

$$|f(x)| = \begin{cases} f(x) & \text{if } f(x) \geq 0 \\ -f(x) & \text{if } f(x) < 0. \end{cases}$$

To graph a function of the form $y = |f(x)|$, the first part of the definition indicates that the graph is the same as that of $y = f(x)$ for values of $f(x)$ (that is, range values) that are nonnegative. The second part of the definition indicates that for range values that are negative, the graph of $y = f(x)$ is reflected across the $x$-axis. The domain of $y = |f(x)|$ is the same as the domain of $f$, while the range of $y = |f(x)|$ will be a subset of $[0, \infty)$.

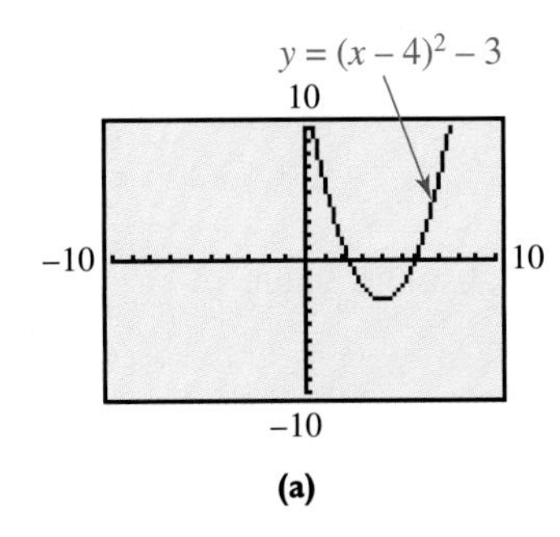

**(a)**

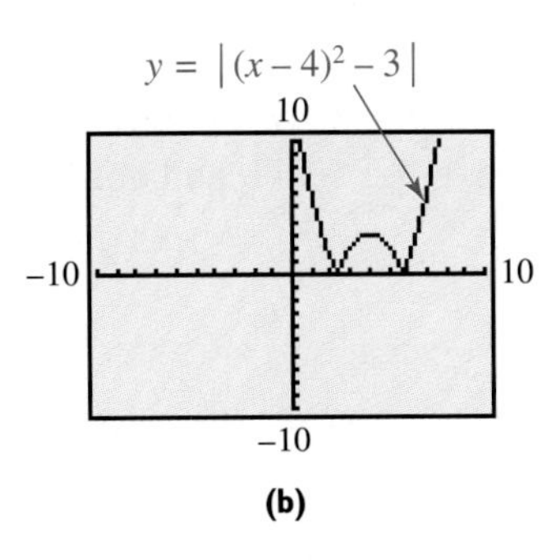

**(b)**

**FIGURE 41**

**FOR DISCUSSION**

Graph $y = (x + 2)^2 + 1$ in the standard window of a graphing calculator. Then fill in the blank with one of the words *above* or *below*: This graph lies entirely _______ the $x$-axis. Discuss what should happen when you graph $y = |(x + 2)^2 + 1|$. Make a conjecture from this exercise. Now, make another conjecture: Suppose a graph lies completely *below* the $x$-axis. What will happen when you graph its absolute value?

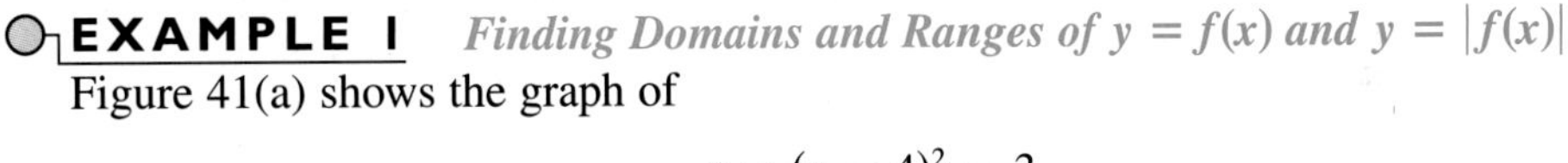

**EXAMPLE 1** *Finding Domains and Ranges of $y = f(x)$ and $y = |f(x)|$*

Figure 41(a) shows the graph of

$$y = (x - 4)^2 - 3,$$

which is the graph of $y = x^2$ shifted 4 units to the right and 3 units downward. Figure 41(b) shows the graph of

$$y = |(x - 4)^2 - 3|.$$

Give the domain and range of each function.

**Solution** Notice that all points with negative $y$-values in the first graph have been reflected across the $x$-axis, while all points with nonnegative $y$-values are the same for both graphs. The domain of each function is $(-\infty, \infty)$. The range of $y = (x - 4)^2 - 3$ is $[-3, \infty)$, while the range of $y = |(x - 4)^2 - 3|$ is $[0, \infty)$.

**EXAMPLE 2** *Sketching the Graph of $y = |f(x)|$ Given the Graph of $y = f(x)$*

Figure 42 shows the graph of a function $y = f(x)$. Use the figure to sketch the graph of $y = |f(x)|$. Give the domain and range of each function.

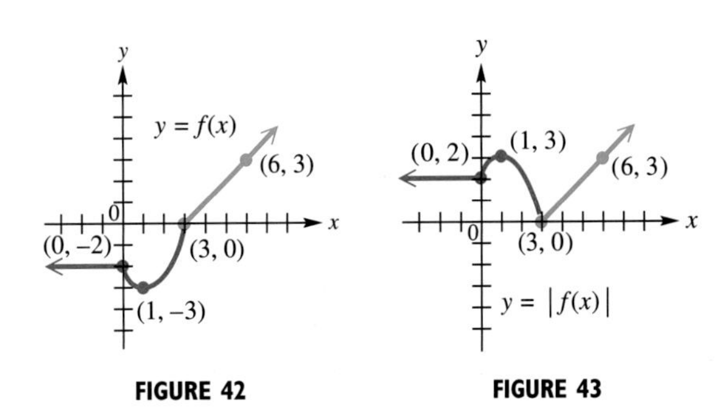

**FIGURE 42** **FIGURE 43**

**Solution** The graph will remain the same for points whose $y$-values are nonnegative, while it will be reflected across the $x$-axis for all other points. Figure 43 shows the graph of $y = |f(x)|$. The domain of both functions is $(-\infty, \infty)$. The range of $y = f(x)$ is $[-3, \infty)$, while the range of $y = |f(x)|$ is $[0, \infty)$.

## Properties of Absolute Value

There are several properties of absolute value that will be useful in our work.

> **Properties of Absolute Value**
>
> For all real numbers $a$ and $b$:
>
> **1.** $|ab| = |a| \cdot |b|$
> The absolute value of a product is equal to the product of the absolute values.
>
> **2.** $\left|\frac{a}{b}\right| = \frac{|a|}{|b|}$ **$(b \neq 0)$**
> The absolute value of a quotient is equal to the quotient of the absolute values.
>
> **3.** $|a| = |-a|$
> The absolute value of a number is equal to the absolute value of its additive inverse.
>
> **4.** $|a| + |b| \geq |a + b|$ **(triangle inequality)**
> The sum of the absolute values of two numbers is greater than or equal to the absolute value of their sum.

These properties can be used to explain the behavior of graphs of functions involving absolute value. For example, consider the function $y = |2x + 11|$ and observe the following sequence of transformations.

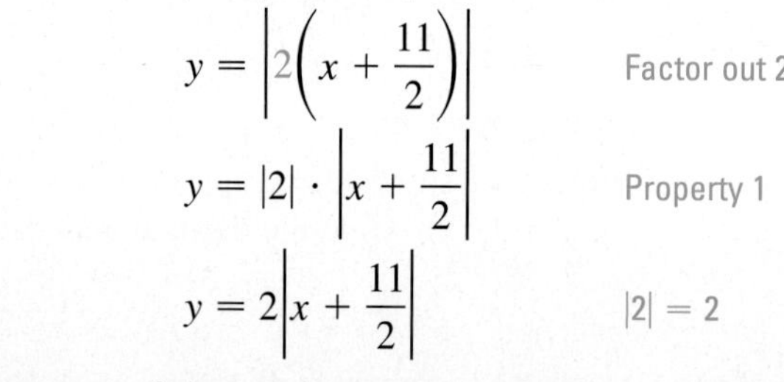

$$y = |2x + 11|$$

$$y = \left|2\left(x + \frac{11}{2}\right)\right| \qquad \text{Factor out 2.}$$

$$y = |2| \cdot \left|x + \frac{11}{2}\right| \qquad \text{Property 1}$$

$$y = 2\left|x + \frac{11}{2}\right| \qquad |2| = 2$$

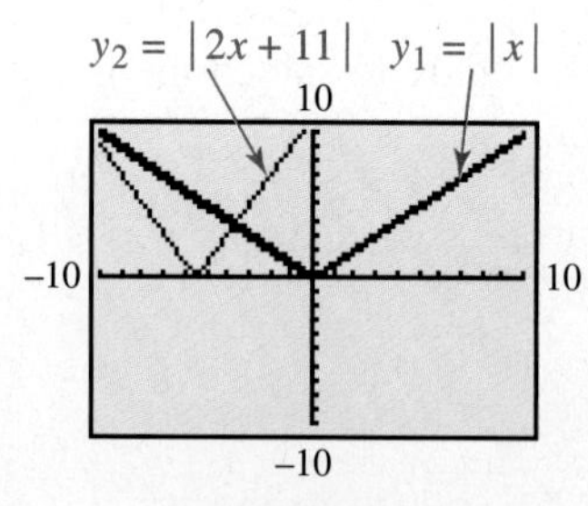


**FIGURE 44**

Using the concepts of the previous two sections, we conclude that the graph of this function can be found by starting with the graph of $y = |x|$, shifting $\frac{11}{2}$ units to the left, and then vertically stretching by a factor of 2. The graphs of $y_1 = |x|$ and $y_2 = |2x + 11|$ in Figure 44 support this statement.

We are often interested in absolute value functions of the form

$$f(x) = |ax + b|,$$

where the expression inside the absolute value bars is linear. For purposes of discussion, a comprehensive graph of $f(x) = |ax + b|$ will include all intercepts and the lowest point on the "V-shaped" graph.

## Equations and Inequalities Involving Absolute Value

We first investigate equations (and related inequalities) that involve absolute values of linear functions and constants.

**EXAMPLE 3** *Solving an Absolute Value Equation*

Solve $|2x + 1| = 7$.

### Analytic Solution

For $|2x + 1|$ to equal 7, $2x + 1$ must be 7 units from 0 on the number line. This can happen only when $2x + 1 = 7$ or $2x + 1 = -7$. Solve this compound equation as follows.

$$2x + 1 = 7 \quad \text{or} \quad 2x + 1 = -7$$
$$2x = 6 \quad \text{or} \quad 2x = -8$$
$$x = 3 \quad \text{or} \quad x = -4$$

Check by substitution in the original equation that the solution set is $\{-4, 3\}$.

### Graphing Calculator Solution

Figure 45 shows that the graphs of $y_1 = |2x + 1|$ and $y_2 = 7$ intersect when $x = -4$ or $x = 3$, confirming the analytic solution.

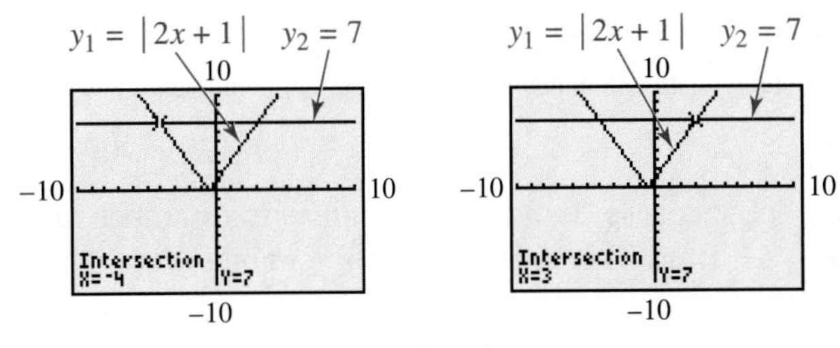


**FIGURE 45**

**CAUTION** A common error when solving $|2x + 1| = 7$ analytically is to consider only one part of the "or" statement, $2x + 1 = 7$. Remember that $2x + 1 = -7$ must also be considered.

In Example 3, we used the property that if $|ax + b| = k$, where $k > 0$, then $ax + b = k$ or $ax + b = -k$. The following summary indicates the general methods for solving absolute value equations and inequalities analytically.

**Looking Ahead to Calculus**

The formal definition of *limit*, one of the most important concepts in calculus, is given below. It uses two absolute value inequalities.

Suppose that a function $f$ is defined at every number in an open interval $I$ containing $a$, except perhaps at $a$ itself. Then the limit of $f(x)$ as $x$ approaches $a$ is $L$, written

$$\lim_{x \to a} f(x) = L$$

if for every $\epsilon > 0$ there exists a $\delta > 0$ such that

$$|f(x) - L| < \epsilon$$

whenever

$$0 < |x - a| < \delta.$$

### Solving Absolute Value Equations and Inequalities

Let $k$ be a positive number.

1. To solve $|ax + b| = k$, solve the compound equation
$$\mathbf{ax + b = k} \quad \text{or} \quad \mathbf{ax + b = -k.}$$
2. To solve $|ax + b| > k$, solve the compound inequality
$$\mathbf{ax + b > k} \quad \text{or} \quad \mathbf{ax + b < -k.}$$
3. To solve $|ax + b| < k$, solve the three-part inequality
$$\mathbf{-k < ax + b < k.}$$

Inequalities involving $\le$ or $\ge$ are solved similarly, using the equality part of the symbol as well.

**EXAMPLE 4** *Solving Absolute Value Inequalities*

Solve each inequality.

**(a)** $|2x + 1| < 7$ **(b)** $|2x + 1| > 7$

### Analytic Solution

**(a)** The expression $2x + 1$ must represent a number that is less than 7 units from 0 on the number line. This means that $2x + 1$ must be *between* $-7$ and 7, which is written as the three-part inequality

$$-7 < 2x + 1 < 7.$$

$$-8 < 2x < 6 \quad \text{Subtract 1 from each part.}$$

$$-4 < x < 3 \quad \text{Divide each part by 2.}$$

The solution set is the open interval $(-4, 3)$.

**(b)** This absolute value inequality must be rewritten as

$$2x + 1 > 7 \quad \text{or} \quad 2x + 1 < -7,$$

because $2x + 1$ must represent a number that is *more* than 7 units from 0 on either side of the number line. Now, solve the compound inequality

$$2x + 1 > 7 \quad \text{or} \quad 2x + 1 < -7.$$

$$2x > 6 \quad \text{or} \quad 2x < -8$$

$$x > 3 \quad \text{or} \quad x < -4$$

The solution set is $(-\infty, -4) \cup (3, \infty)$.

### Graphing Calculator Solution

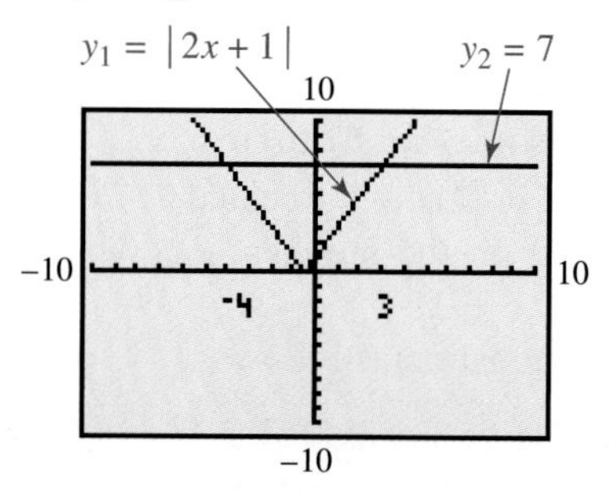


FIGURE 46

**(a)** As seen in Figure 46, the graph of $y_1 = |2x + 1|$ lies *below* the graph of $y_2 = 7$ for $x$-values between $-4$ and 3, so the solution set is $(-4, 3)$.

**(b)** The graph of $y_1 = |2x + 1|$ lies *above* the graph of $y_2 = 7$ for $x$-values less than $-4$ or greater than 3, confirming the analytic result.

An absolute value equation of the form $|ax + b| = k$, where $k < 0$, has solution set $\emptyset$, since the absolute value of a real number cannot be negative. The related inequalities, $|ax + b| > k$ and $|ax + b| < k$, have solution sets $(-\infty, \infty)$ and $\emptyset$, respectively. (What is the solution set of $|ax + b| = 0$?)

## EXAMPLE 5 *Solving Special Cases of Absolute Value Equations and Inequalities*

Solve each equation or inequality.

**(a)** $|3x + 5| = -5$ **(b)** $|3x + 5| < -5$ **(c)** $|3x + 5| > -5$

### Analytic Solution

**(a)** Because the absolute value of an expression will never be $-5$, this equation has no solution. The solution set is $\emptyset$.

**(b)** Using reasoning similar to that in part (a), the absolute value of an expression will never be less than $-5$. Once again, the solution set is $\emptyset$.

### Graphing Calculator Solution

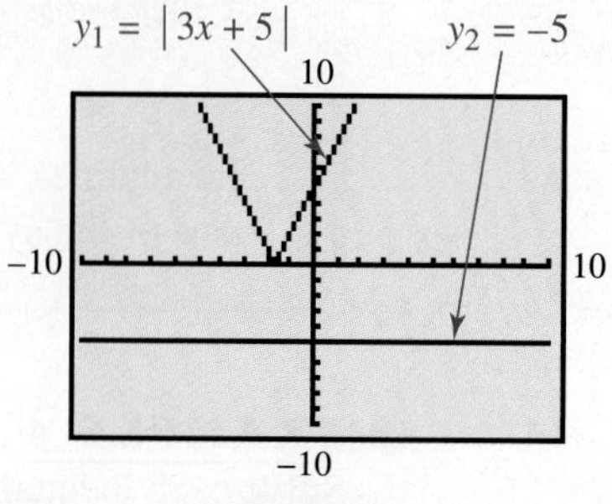


FIGURE 47

*(continued)*

**(c)** Because the absolute value of an expression will always be greater than or equal to 0, the absolute value of an expression will always be greater than $-5$. The solution set is $(-\infty, \infty)$.

**(a)** In Figure 47, we see that the graph of $y_1 = |3x + 5|$ does not intersect the graph of $y_2 = -5$. The solution set is $\emptyset$.

**(b)** The graph of $y_1$ is never below the graph of $y_2$, and thus the solution set of $y_1 < y_2$ is $\emptyset$.

**(c)** The graph of $y_1$ is always above the graph of $y_2$, and thus the solution set of $y_1 > y_2$ is $(-\infty, \infty)$.

If two quantities have the same absolute value, they must either be equal to each other or be negatives of each other. This fact allows us to solve absolute value equations (and related inequalities) of the form $|ax + b| = |cx + d|$.

**Solving $|ax + b| = |cx + d|$**

To solve the equation $|ax + b| = |cx + d|$ analytically, solve the compound equation

$$ax + b = cx + d \quad \text{or} \quad ax + b = -(cx + d).$$

**EXAMPLE 6** *Solving an Equation Involving Two Absolute Value Expressions*

Solve $|x + 6| = |2x - 3|$.

**Analytic Solution**

This equation is satisfied if

$$x + 6 = 2x - 3 \quad \text{or} \quad x + 6 = -(2x - 3).$$

Solve each equation.

$$\begin{aligned} x + 6 &= 2x - 3 & \text{or} \quad x + 6 &= -(2x - 3) \\ 9 &= x & x + 6 &= -2x + 3 \\ & & 3x &= -3 \\ & & x &= -1 \end{aligned}$$

Check each solution by substitution in the original equation. The solution set is $\{-1, 9\}$.

**Graphing Calculator Solution**

Let $y_1 = |x + 6|$ and $y_2 = |2x - 3|$. The equation $y_1 = y_2$ is equivalent to $y_1 - y_2 = 0$, so we graph

$$y_3 = |x + 6| - |2x - 3|$$

and find the $x$-intercepts.

In Figure 48, we see that they are $-1$ and 9.

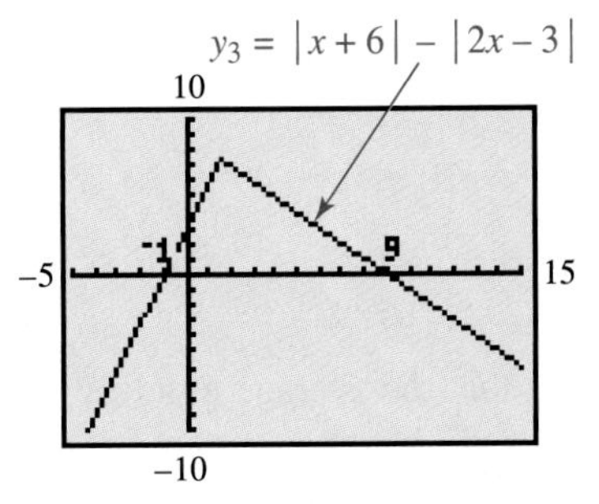


FIGURE 48

**EXAMPLE 7** *Solving Inequalities Involving Two Absolute Value Expressions*

Solve each inequality graphically by referring to Example 6 and Figure 48.

**(a)** $|x + 6| < |2x - 3|$ **(b)** $|x + 6| \geq |2x - 3|$

**Solution**

**(a)** The inequality $y_1 < y_2$ is equivalent to $y_1 - y_2 < 0$, or $y_3 < 0$. In Figure 48 on the previous page, note that the graph of $y_3$ is below the $x$-axis in the interval $(-\infty, -1) \cup (9, \infty)$.

**(b)** The inequality $y_3 \geq 0$ is satisfied by the closed interval $[-1, 9]$.

**EXAMPLE 8** *Solving an Equation Involving a Sum of Absolute Values*

Solve $|x + 5| + |x - 3| = 16$ graphically by the intersection-of-graphs method. Check the solutions by substitution.

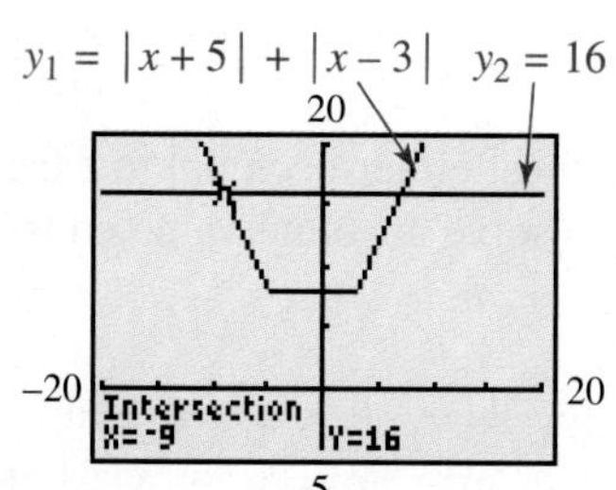


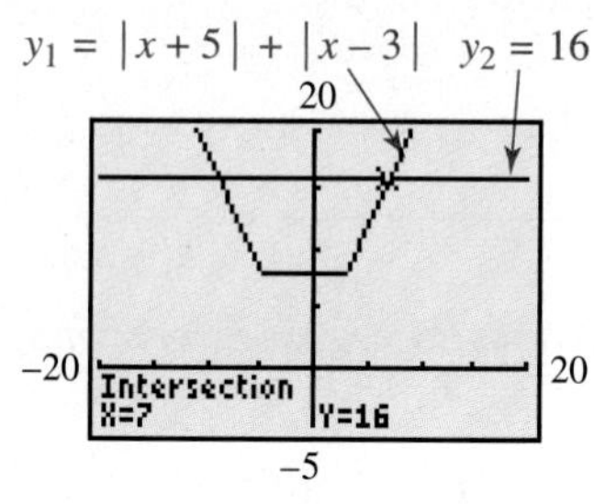


FIGURE 49

**Solution** Let $y_1 = |x + 5| + |x - 3|$ and $y_2 = 16$. Their graphs are shown in Figure 49. Locate the points of intersection of the graphs to find that the $x$-coordinates of the points are $-9$ and 7.

To verify these solutions, substitute them into the original equation.

Let $x = -9$: $\quad |(-9) + 5| + |(-9) - 3| = |-4| + |-12|$

$$= 4 + 12$$
$$= 16. \checkmark$$

Let $x = 7$: $\quad |7 + 5| + |7 - 3| = |12| + |4|$

$$= 12 + 4$$
$$= 16. \checkmark$$

Therefore, the solution set is $\{-9, 7\}$.

## An Application Involving Absolute Value

**EXAMPLE 9** *Analyzing the Temperature Range in Santa Fe*

The inequality $|x - 49| \leq 20$ describes the range of average monthly temperatures $x$ in degrees Fahrenheit for Santa Fe, New Mexico. Solve this inequality, and interpret the result. (*Source:* Miller, A. and J. Thompson, *Elements of Meteorology*, Second Edition, Charles E. Merrill, 1975.)

**Analytic Solution**

To solve $|x - 49| \leq 20$, we must solve the compound inequality

$$-20 \leq x - 49 \leq 20.$$
$$-20 + 49 \leq x - 49 + 49 \leq 20 + 49$$
$$29 \leq x \leq 69$$

This means that the average monthly temperatures in Santa Fe, New Mexico, range from 29°F through 69°F. The average monthly temperatures are always within 20° of 49°F.

**Graphing Calculator Solution**

Figure 50 shows the graphs of $y_1 = |x - 49|$ and $y_2 = 20$. The graph of $y_1$ lies below the graph of $y_2$ when $x$-values are between 29 and 69. This confirms the analytic solution. The endpoints are included because the symbol in the original inequality is $\leq$.

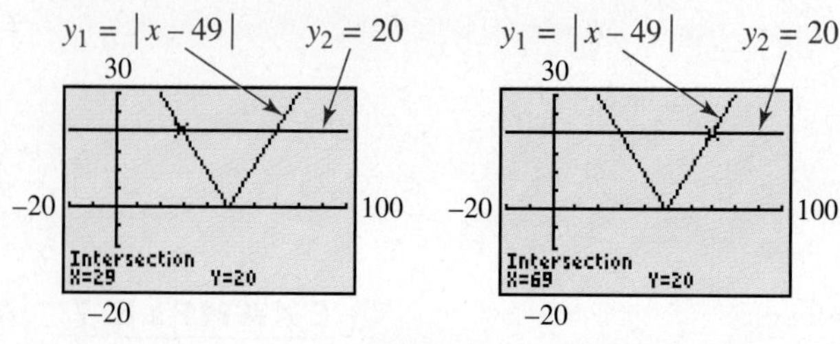


FIGURE 50

## 2.4 EXERCISES

*Concept Check* *Give a short answer to each question.*

**1.** If $f(a) = -5$, what is the value of $|f(a)|$?

**2.** How does the graph of $f(x) = x^2$ compare with the graph of $f(x) = |x^2|$?

**3.** What is the range of $y = |f(x)|$, if $f(x) = -x^2$?

**4.** If the range of $y = f(x)$ is $[-2, \infty)$, what is the range of $y = |f(x)|$?

**5.** If the range of $y = f(x)$ is $(-\infty, -2]$, what is the range of $y = |f(x)|$?

**6.** Why can't the range of $y = |f(x)|$ include $-1$, for any function $f$?

*In Exercises 7–15, you are given graphs of functions $y = f(x)$. Sketch by hand the graph of $y = |f(x)|$.*

**7.**

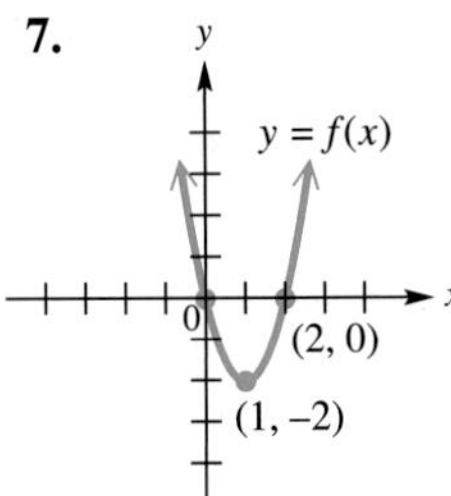


**8.**

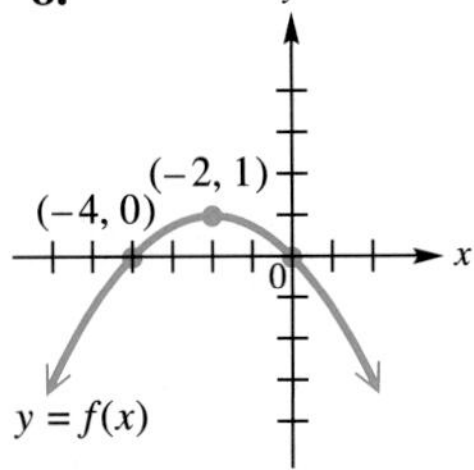


**9.**

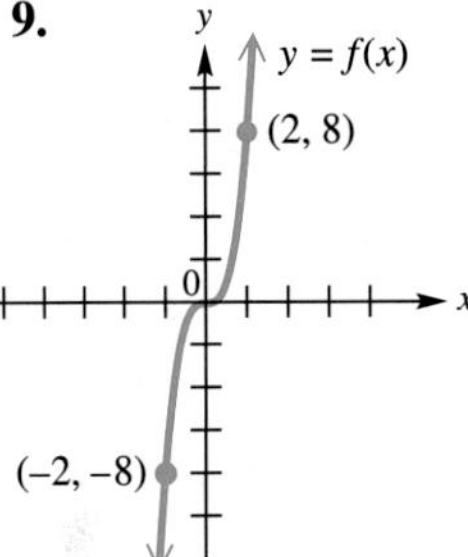


**10.**

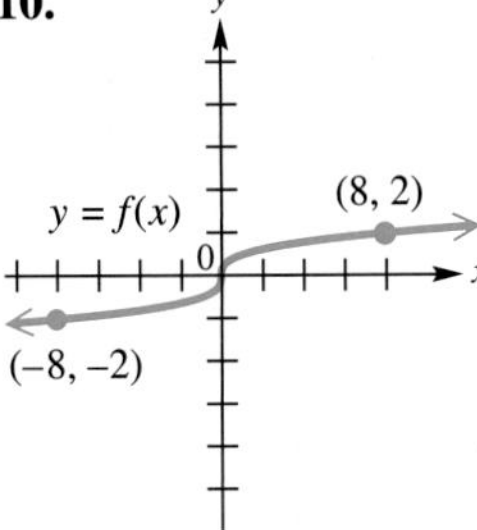


**11.**

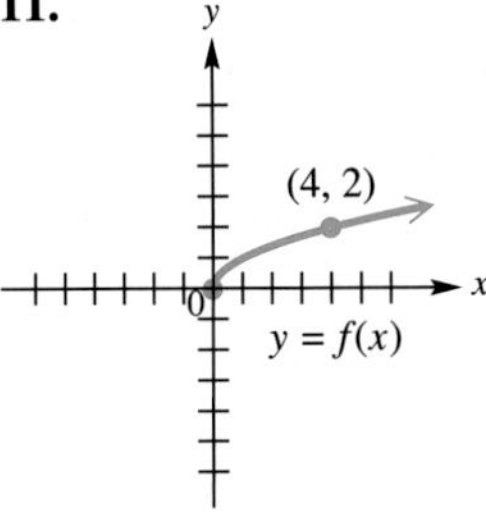


**12.**

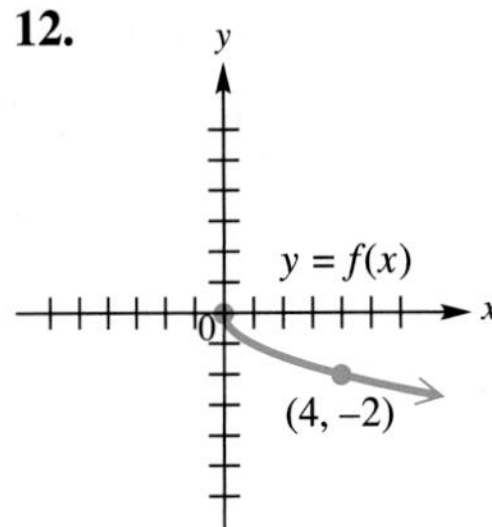


**13.**

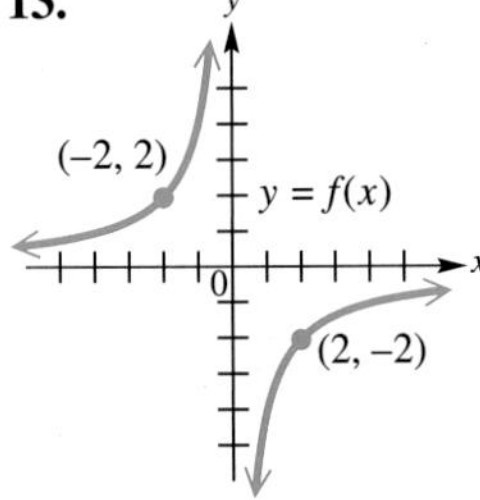


**14.**

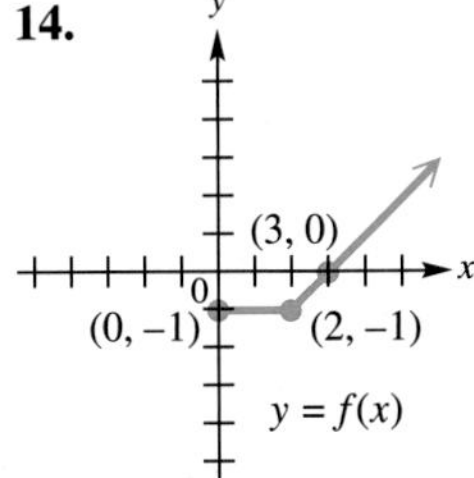


**15.**

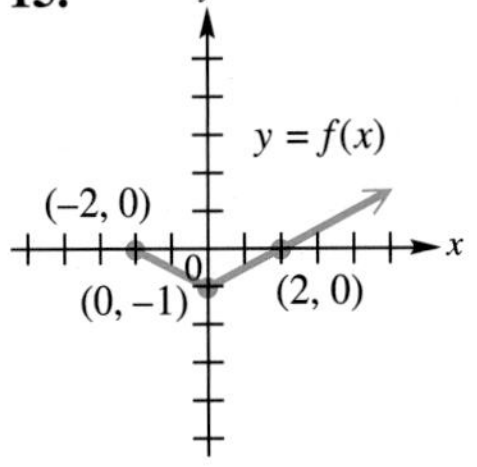


**16.** Explain in your own words the procedure you used to find the graphs of $y = |f(x)|$ in Exercises 7–15.

*In Exercises 17–20, the graph of a function $y = f(x)$ is shown. Match each graph of $y = f(x)$ with the graph of $y = |f(x)|$ from choices A–D.*

**17.**

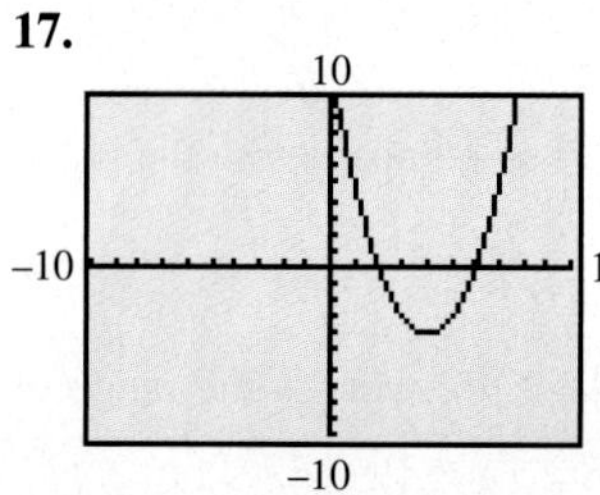


**18.**

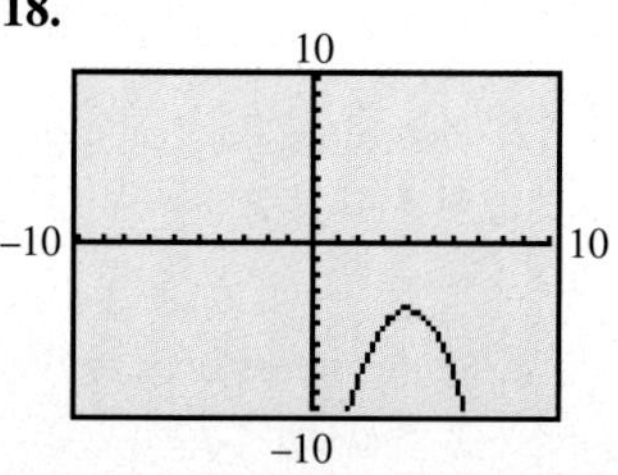


**19.**

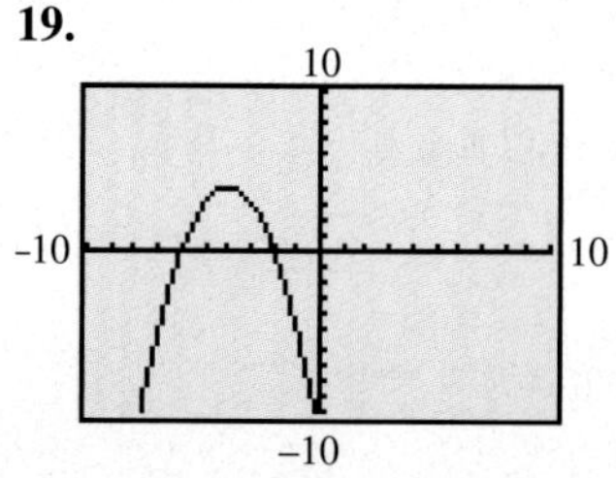


**20.**

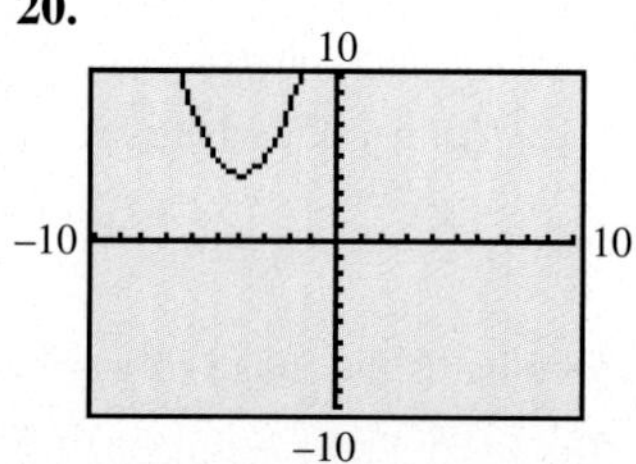


**A.**

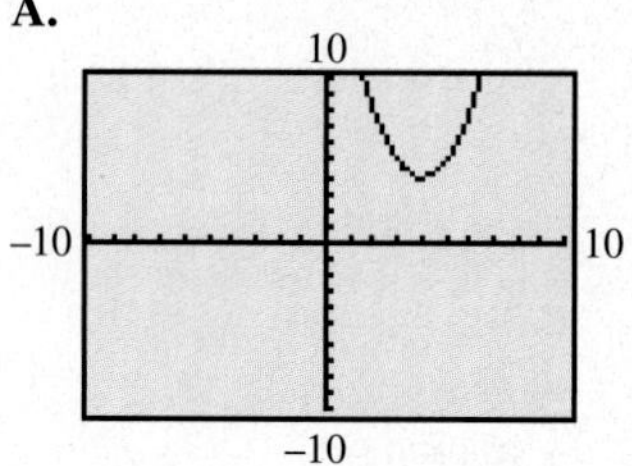


**B.**

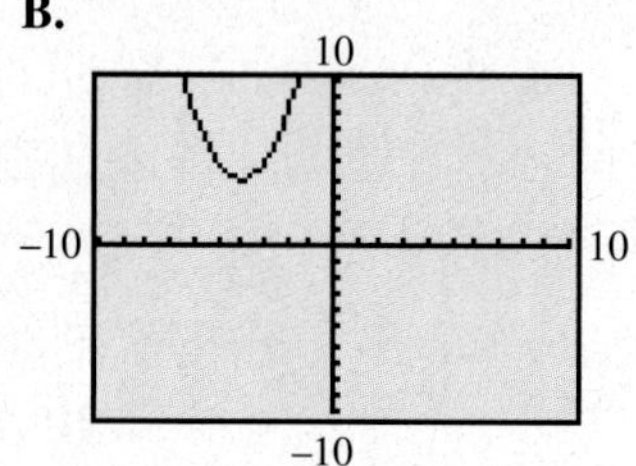


**C.**

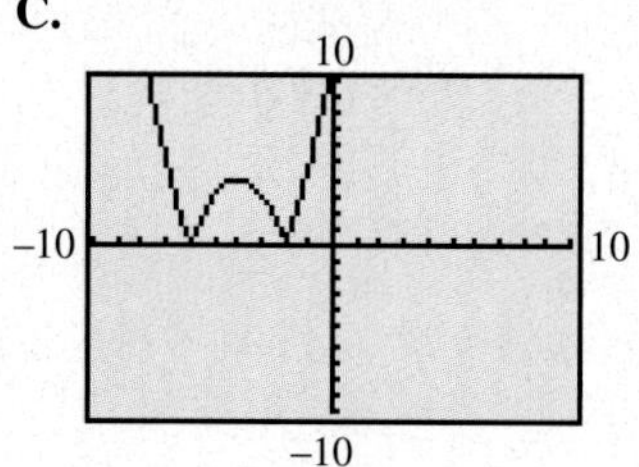


**D.**

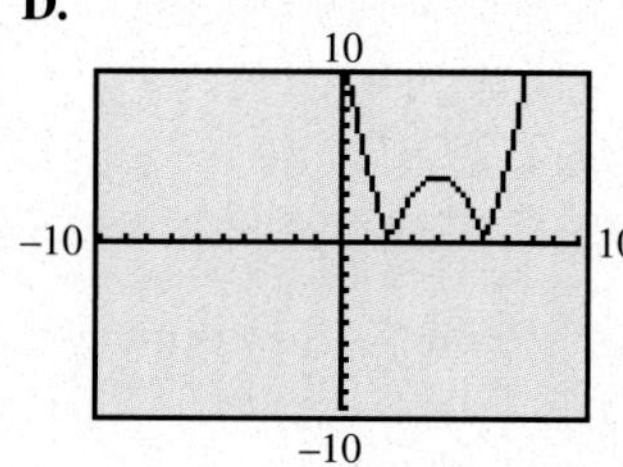


**21.** The graph of a function $y = f(x)$ is shown. Sketch by hand, in order, the graph of each of the following. Use the concept of reflecting introduced in Section 2.3 and the concept of graphing $y = |f(x)|$ introduced in this section.

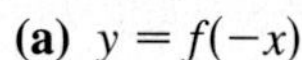

**(a)** $y = f(-x)$

**(b)** $y = -f(-x)$

**(c)** $y = |-f(-x)|$

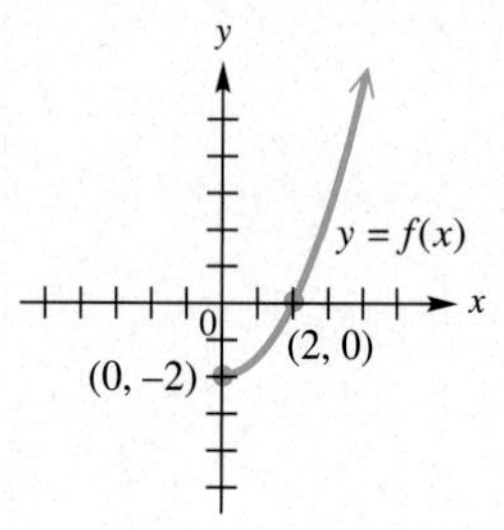


**22.** Repeat Exercise 21 for the graph of $y = f(x)$ shown here.

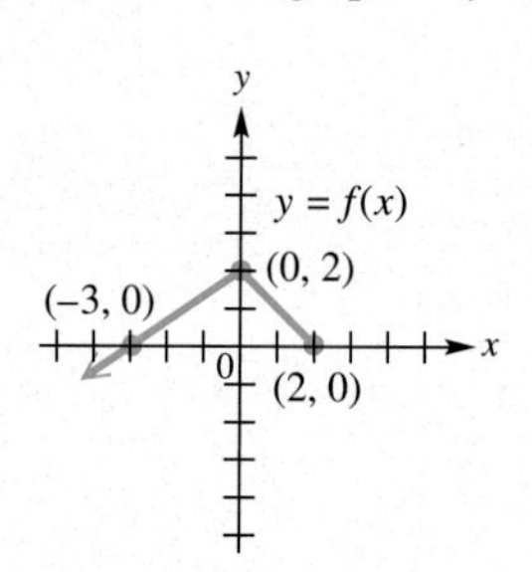


*In Exercises 23 and 24, one graph is that of $y = f(x)$ and the other is that of $y = |f(x)|$. State which is which.*

**23. A.**

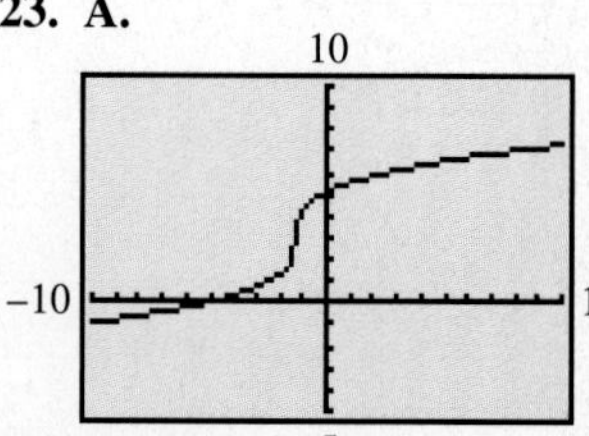


**B.**

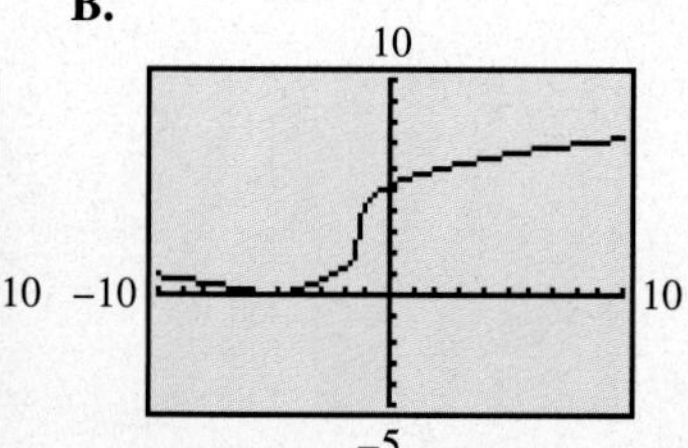


**24. A.**

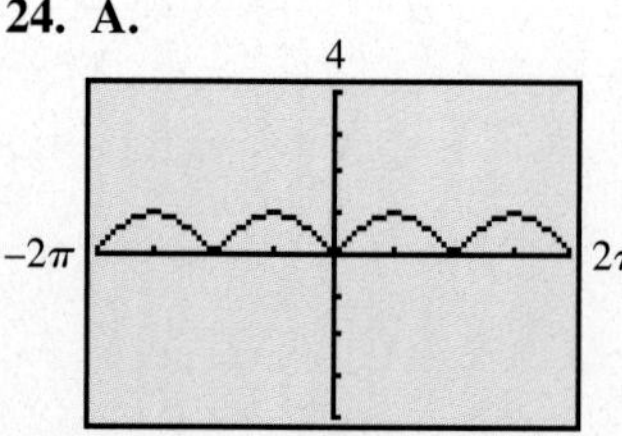


**B.**

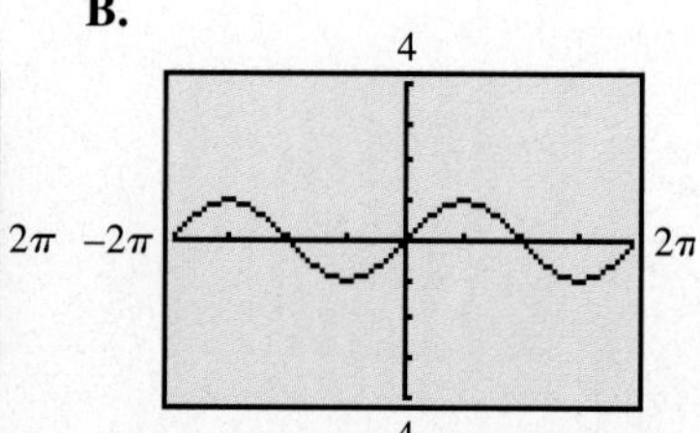


*In Exercises 25–28, use the graph, along with the indicated points, to give the solution set of*

**(a)** $y_1 = y_2$. **(b)** $y_1 < y_2$. **(c)** $y_1 > y_2$.

**25.**

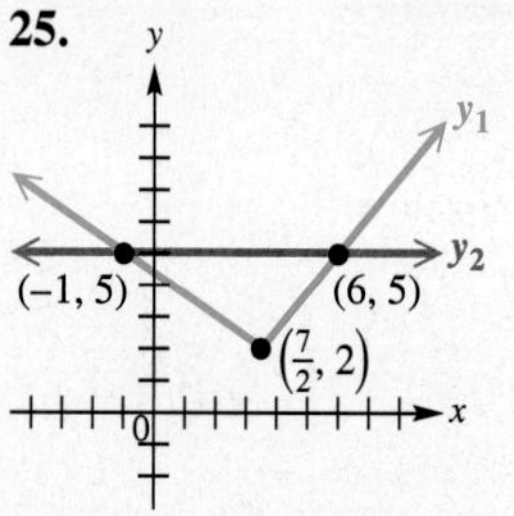


**26.**

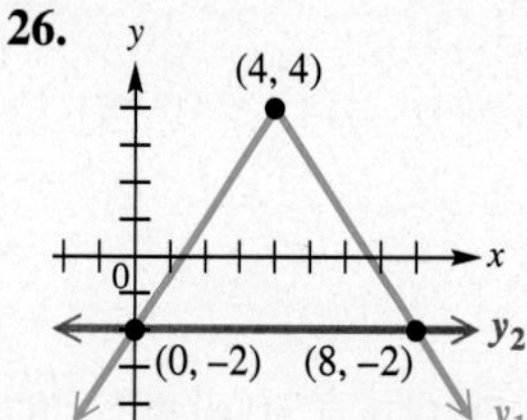


**27.**

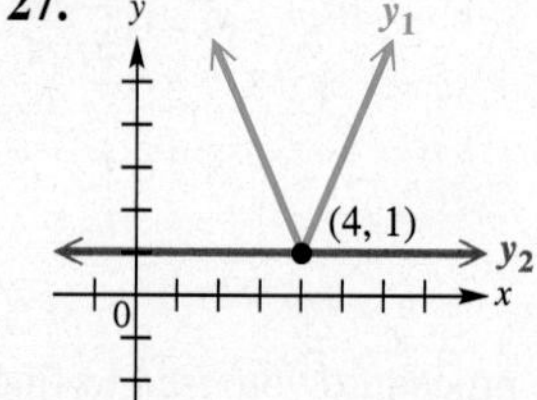


**28.**

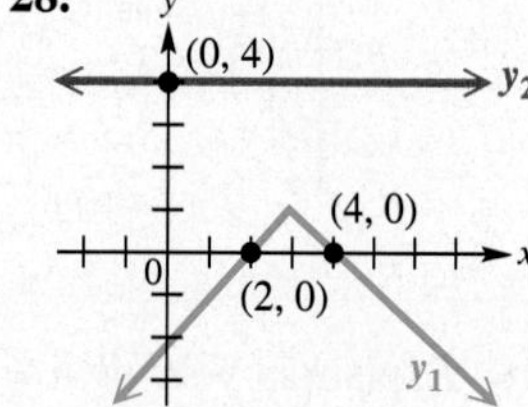

## RELATING CONCEPTS

For individual or group investigation (Exercises 29–34)

*The figure shows the graphs of two functions,* $f(x) = |.5x + 6|$ *and* $g(x) = 3x - 14$.

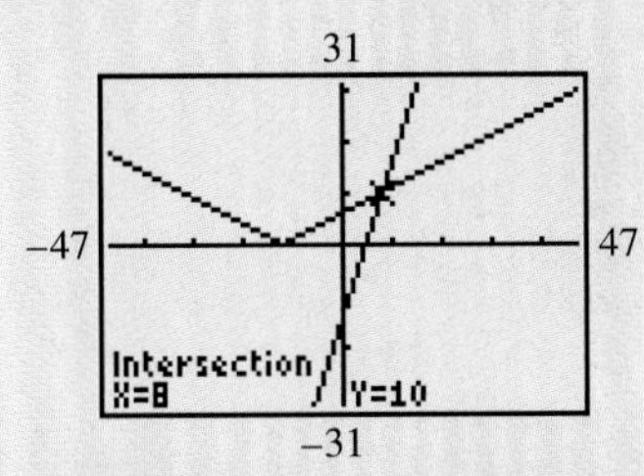


*Work Exercises 29–34 in order, without actually using your graphing calculator.*

**29.** Which graph is that of $y = f(x)$? How do you know?

**30.** Which graph is that of $g(x) = 3x - 14$? How do you know?

**31.** Solve $f(x) = g(x)$ based on the display at the bottom of the screen.

**32.** Solve $f(x) > g(x)$ based on the graphs and the display.

**33.** Solve $f(x) < g(x)$ based on the graphs and the display.

**34.** What is the solution set of

$$|.5x + 6| - (3x - 14) = 0?$$

*Solve each group of equations and inequalities analytically. Support your solutions graphically.*

**35.** **(a)** $|x + 4| = 9$ **(b)** $|x + 4| > 9$ **(c)** $|x + 4| < 9$

**36.** **(a)** $|x - 3| = 5$ **(b)** $|x - 3| > 5$ **(c)** $|x - 3| < 5$

**37.** **(a)** $|-2x + 7| = 3$ **(b)** $|-2x + 7| \geq 3$ **(c)** $|-2x + 7| \leq 3$

**38.** **(a)** $|-3x - 9| = 6$ **(b)** $|-3x - 9| \geq 6$ **(c)** $|-3x - 9| \leq 6$

**39.** **(a)** $|2x + 1| + 3 = 5$ **(b)** $|2x + 1| + 3 \leq 5$ **(c)** $|2x + 1| + 3 \geq 5$

**40.** **(a)** $|4x + 7| = 0$ **(b)** $|4x + 7| > 0$ **(c)** $|4x + 7| < 0$

**41.** **(a)** $|7x - 5| = 0$ **(b)** $|7x - 5| \geq 0$ **(c)** $|7x - 5| \leq 0$

**42.** **(a)** $|\pi x + 8| = -4$ **(b)** $|\pi x + 8| < -4$ **(c)** $|\pi x + 8| > -4$

**43.** **(a)** $|\sqrt{2}x - 3.6| = -1$ **(b)** $|\sqrt{2}x - 3.6| \leq -1$ **(c)** $|\sqrt{2}x - 3.6| \geq -1$

**44.** Explain how to solve an equation of the form $|ax + b| = |cx + d|$ analytically.

*An equation of the form* $|f(x)| = |g(x)|$ *is given.*

**(a)** *Solve the equation analytically and support the solution graphically.* **(b)** *Solve* $|f(x)| > |g(x)|$ *graphically.*

**(c)** *Solve* $|f(x)| < |g(x)|$ *graphically.*

**45.** $|3x + 1| = |2x - 7|$

**46.** $|x - 4| = |7x + 12|$

**47.** $|-2x + 5| = |x + 3|$

**48.** $|-5x + 1| = |3x - 4|$

**49.** $|x - \frac{1}{2}| = |\frac{1}{2}x - 2|$

**50.** $|x + 3| = |\frac{1}{3}x + 8|$

**51.** $|4x + 1| = |4x + 6|$

**52.** $|6x + 9| = |6x - 3|$

**53.** $|.25x + 1| = |.75x - 3|$

**54.** $|.40x + 2| = |.60x - 5|$

*Solve each equation graphically.*

**55.** $|x + 1| + |x - 6| = 11$

**56.** $|2x + 2| + |x + 1| = 9$

**57.** $|x| + |x - 4| = 8$

**58.** $|.5x + 2| + |.25x + 4| = 9$

*(Modeling) Average Temperature* *Each inequality describes the range of average monthly temperatures T in degrees Fahrenheit at a certain location.*

**(a)** *Solve the inequality.*

**(b)** *Interpret the result.*

**59.** $|T - 43| \leq 24$, Marquette, Michigan

**60.** $|T - 62| \leq 19$, Memphis, Tennessee

**61.** $|T - 50| \leq 22$, Boston, Massachusetts

**62.** $|T - 10| \leq 36$, Chesterfield, Canada

**63.** $|T - 61.5| \leq 12.5$, Buenos Aires, Argentina

**64.** $|T - 43.5| \leq 8.5$, Punta Arenas, Chile

*(Modeling) Solve each problem.*

**65.** *Blood Pressure* Systolic blood pressure is the maximum pressure produced by each heartbeat. Both low blood pressure and high blood pressure are cause for medical concern. Therefore, health care professionals are interested in a patient's "pressure difference from normal," or $P_d$. If 120 is considered a normal systolic pressure, $P_d = |P - 120|$, where $P$ is the patient's recorded systolic pressure. For example, a patient with a systolic pressure, $P$, of 113 would have a pressure difference from normal of $P_d = |P - 120| = |113 - 120| = |-7| = 7$.

**(a)** Calculate the $P_d$ value for a woman whose actual systolic pressure is 116 and whose normal value should be 125.

**(b)** If a patient's $P_d$ value is 17 and the normal pressure for his sex and age should be 130, what are the two possible values for his systolic blood pressure?

**66.** *Conversion of Methanol to Gasoline* The industrial process that is used to convert methanol to gasoline is carried out at a temperature range of 680°F to 780°F. Using $F$ as the variable, write an absolute value inequality that corresponds to this range.

**67.** *Weights of Babies* Dr. Cazayoux has found that, over the years, 95% of the babies he has delivered weighed $x$ pounds, where $|x - 8.0| \leq 1.5$. What range of weights corresponds to this inequality?

**68.** *Kite Flying* When a model kite was flown in crosswinds in tests to determine its limits of power extraction, it attained speeds of 98 to 148 feet per second in winds of 16 to 26 feet per second. Using $x$ as the variable in each case, write absolute value inequalities that correspond to these ranges.

*Windchill Factor* *The windchill factor is a measure of the cooling effect that the wind has on a person's skin. It calculates the equivalent cooling temperature if there were no wind. For many years, the table shown here was used to determine the windchill factor for various wind speeds and temperatures.*

| Wind/°F | 40° | 30° | 20° | 10° | 0° | −10° | −20° | −30° | −40° | −50° |
|---|---|---|---|---|---|---|---|---|---|---|
| 5 mph | 37 | 27 | 16 | 6 | −5 | −15 | −26 | −36 | −47 | −57 |
| 10 mph | 28 | 16 | 4 | −9 | −21 | −33 | −46 | −58 | −70 | −83 |
| 15 mph | 22 | 9 | −5 | −18 | −36 | −45 | −58 | −72 | −85 | −99 |
| 20 mph | 18 | 4 | −10 | −25 | −39 | −53 | −67 | −82 | −96 | −110 |
| 25 mph | 16 | 0 | −15 | −29 | −44 | −59 | −74 | −88 | −104 | −118 |
| 30 mph | 13 | −2 | −18 | −33 | −48 | −63 | −79 | −94 | −109 | −125 |
| 35 mph | 11 | −4 | −20 | −35 | −49 | −67 | −82 | −98 | −113 | −129 |
| 40 mph | 10 | −6 | −21 | −37 | −53 | −69 | −85 | −100 | −116 | −132 |

*Source:* Miller, A. and J. Thompson, *Elements of Meteorology*, Second Edition, Charles E. Merrill, 1975.

*Suppose that we wish to determine the difference between two of these entries and are interested only in the magnitude, or absolute value, of this difference. Then, we subtract the two entries and find the absolute value. For example, the difference in windchill factors for wind at 20 mph with a 20°F temperature and wind at 30 mph with a 40°F temperature is $|-10° - 13°| = 23°$F, or equivalently, $|13° - (-10°)| = 23°$F.*

*Find the absolute value of the difference of the two indicated windchill factors.*

**69.** Wind at 15 mph with a 30°F temperature and wind at 10 mph with a −10°F temperature

**70.** Wind at 20 mph with a −20°F temperature and wind at 5 mph with a 30°F temperature

**71.** Wind at 30 mph with a −30°F temperature and wind at 15 mph with a −20°F temperature

**72.** Wind at 40 mph with a 40°F temperature and wind at 25 mph with a −30°F temperature

*Tolerances* *In quality control and other applications, as well as in more advanced mathematics, we often wish to keep the difference between two quantities within some predetermined amount, or tolerance. For example, suppose* $y = 2x + 1$, *and we want* $y$ *to be within* .01 *unit of* 4. *This can be written using absolute value as* $|y - 4| < .01$. *To find the values of* $x$ *that will satisfy this condition on* $y$, *we use properties of absolute value as follows.*

$$|y - 4| < .01$$
$$|2x + 1 - 4| < .01 \quad \text{Substitute } 2x + 1 \text{ for } y.$$
$$|2x - 3| < .01$$
$$-.01 < 2x - 3 < .01$$
$$2.99 < 2x < 3.01 \quad \text{Add 3 to each part.}$$
$$1.495 < x < 1.505 \quad \text{Divide each part by 2.}$$

*By reversing these steps, we can show that keeping* $x$ *between* 1.495 *and* 1.505 *will ensure that the difference between* $y$ *and* 4 *is less than* .01.

*Find the open interval in which* $x$ *must lie in order for the given condition to hold.*

**73.** $y = 2x + 1$ and the difference between $y$ and 1 is less than .1.

**74.** $y = 3x - 6$ and the difference between $y$ and 2 is less than .01.

**75.** $y = 4x - 8$ and the difference between $y$ and 3 is less than .001.

**76.** $y = 5x + 12$ and the difference between $y$ and 4 is less than .0001.

*Solve each equation or inequality graphically. Express solutions or endpoints of intervals rounded to the nearest hundredth if necessary.*

**77.** $|2x + 7| = 6x - 1$

**78.** $-|3x - 12| \geq -x - 1$

**79.** $|x - 4| > .5x - 6$

**80.** $2x + 8 > -|3x + 4|$

**81.** $|3x + 4| < -3x - 14$

**82.** $|x - \sqrt{13}| + \sqrt{6} \leq -x - \sqrt{10}$

# PIECEWISE-DEFINED FUNCTIONS

Graphing Piecewise-Defined Functions • The Greatest Integer Function • Applications of Piecewise-Defined Functions

## Graphing Piecewise-Defined Functions

The absolute value function, defined in Section 2.1, is a simple example of a function defined by different rules over different subsets of its domain. Such a function is called a **piecewise-defined function.** Recall that the domain of $f(x) = |x|$ is $(-\infty, \infty)$. For the interval $[0, \infty)$ of the domain, the rule that we use is $f(x) = x$. On the other hand, for the interval $(-\infty, 0)$, we use the rule $f(x) = -x$. Thus, the graph of $f(x) = |x|$ is composed of two "pieces." One piece comes from the graph of $y = x$ and the other from $y = -x$.

**EXAMPLE 1** *Finding Function Values for a Piecewise-Defined Function*

Consider the piecewise-defined function $f$:

$$f(x) = \begin{cases} x + 2 & \text{if } x \leq 0 \\ \frac{1}{2}x^2 & \text{if } x > 0. \end{cases}$$

Find

**(a)** $f(-3)$. **(b)** $f(0)$. **(c)** $f(3)$. **(d)** Sketch the graph of $f$.

**Solution**

**(a)** Since $-3 \le 0$ (specifically, $-3 < 0$), we use the rule $f(x) = x + 2$. Thus,

$$f(-3) = -3 + 2 = -1.$$

This means the graph of $f$ will contain the point $(-3, -1)$.

**(b)** Since $0 \le 0$ (because $0 = 0$), we again use the rule $f(x) = x + 2$. So we have

$$f(0) = 0 + 2 = 2.$$

The point $(0, 2)$ will lie on the graph of $f$, meaning that the $y$-intercept of the graph will be 2.

**(c)** The number 3 is in the interval $(0, \infty)$, and the second part of the rule $f$ indicates that we must use the rule $f(x) = \frac{1}{2}x^2$. Therefore,

$$f(3) = \frac{1}{2}(3)^2 = \frac{1}{2}(9) = 4.5.$$

**(d)** We graph the ray defined by $y = x + 2$, choosing $x$ so that $x \le 0$, with a solid endpoint at $(0, 2)$. The ray has slope 1 and $y$-intercept 2. Then, graph $y = \frac{1}{2}x^2$ for $x > 0$. This graph will be half of a parabola with an open endpoint at $(0, 0)$. See Figure 51.

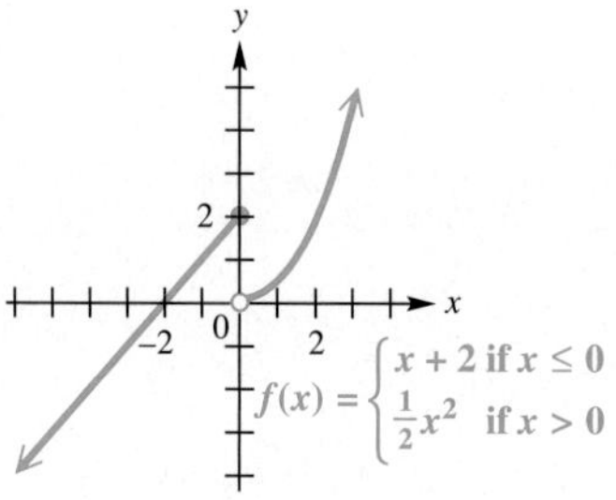

$$f(x) = \begin{cases} x + 2 & \text{if } x \le 0 \\ \frac{1}{2}x^2 & \text{if } x > 0 \end{cases}$$

**FIGURE 51**

To graph a piecewise-defined function with a graphing calculator, we use the *test* feature, which returns a 1 for a true statement or a 0 for a false statement. Figure 52 shows how the function in Example 1 can be entered as $Y_1$.* Notice that the two defining expressions are multiplied by an expression composed of an inequality involving X (which will be 1 or 0, depending on the value of X as the graphing occurs). When $X \le 0$, the expression will be

**FIGURE 52**

$$(X + 2)(1) + \frac{1}{2}X^2(0) = X + 2;$$

when $X > 0$, it will be

$$(X + 2)(0) + \frac{1}{2}X^2(1) = \frac{1}{2}X^2.$$

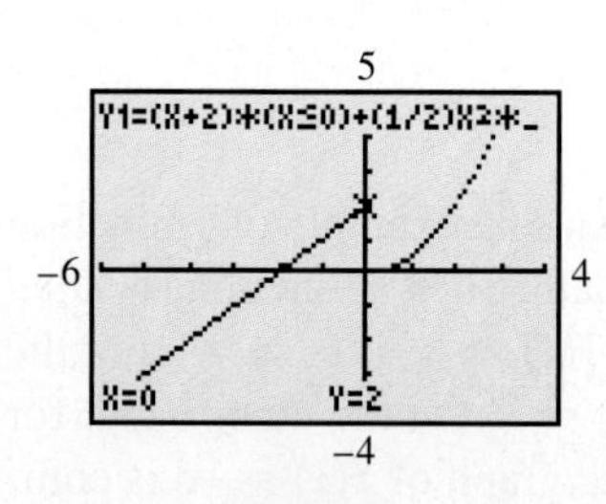


**FIGURE 53**

The graph of $Y_1$ in dot mode is shown in Figure 53.

In general, it is best to graph piecewise-defined functions with the calculator in dot mode, especially when the graph exhibits discontinuities. Otherwise, the calculator may attempt to connect portions of the graph that are actually separate from one another.

**EXAMPLE 2** *Graphing a Piecewise-Defined Function*

Graph the function defined by

$$f(x) = \begin{cases} x + 1 & \text{if } x \le 2 \\ -2x + 7 & \text{if } x > 2. \end{cases}$$

*The user has a choice whether or not to use the multiplication symbol, $*$, when defining the function as $Y_1$.

### Analytic Solution

For $x$-values less than or equal to 2, the graph will be the portion of the line $y = x + 1$ to the left of, and including, the point $(2, 3)$. For $x$-values greater than 2, the graph will be the portion of the line $y = -2x + 7$ to the right of $(2, 3)$. This is a continuous function. See Figure 54.

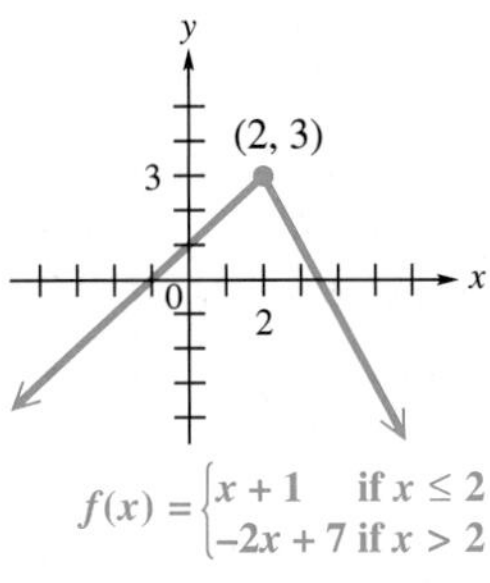


FIGURE 54

### Graphing Calculator Solution

Figure 55 shows how to define $Y_1$ to give the desired graph in Figure 56.

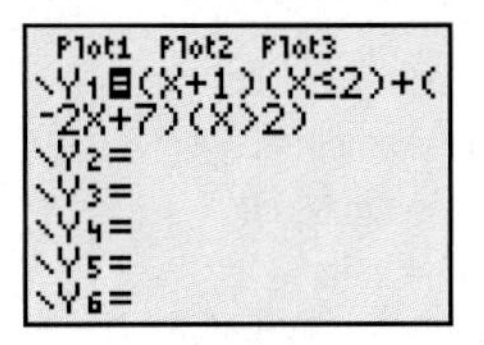


FIGURE 55

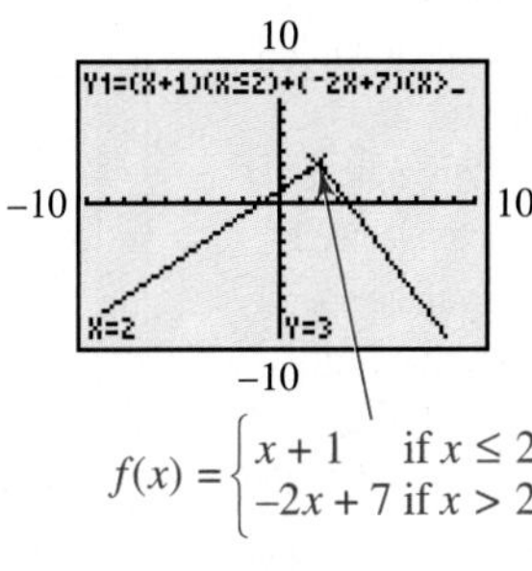


$$f(x) = \begin{cases} x + 1 & \text{if } x \leq 2 \\ -2x + 7 & \text{if } x > 2 \end{cases}$$

FIGURE 56

## The Greatest Integer Function

An important example of a piecewise-defined function is the **greatest integer function.** The notation $[\![x]\!]$ represents the greatest integer less than or equal to $x$. The definition of $[\![x]\!]$ follows.

$$f(x) = [\![x]\!] = \begin{cases} x & \text{if } x \text{ is an integer} \\ \text{the greatest integer less than } x & \text{if } x \text{ is not an integer} \end{cases}$$

**EXAMPLE 3** *Evaluating $[\![x]\!]$*

Evaluate $[\![x]\!]$ for each value of $x$.

**(a)** 4 **(b)** $-5$ **(c)** 2.46 **(d)** $\pi$ **(e)** $-6\frac{1}{2}$

### Analytic Solution

**(a)** $[\![4]\!] = 4$ since 4 is an integer.

**(b)** $[\![-5]\!] = -5$ since $-5$ is an integer.

**(c)** $[\![2.46]\!] = 2$ since 2.46 is not an integer, and 2 is the greatest integer less than 2.46.

**(d)** $[\![\pi]\!] = 3$ since $\pi \approx 3.14$.

**(e)** $\left[\!\left[-6\frac{1}{2}\right]\!\right] = -7$ since $-7$ is the greatest integer less than $-6\frac{1}{2}$.

### Graphing Calculator Solution

**(a)–(e)** The command "int" is used by many graphing calculators for the greatest integer function. See the list in Figure 57.

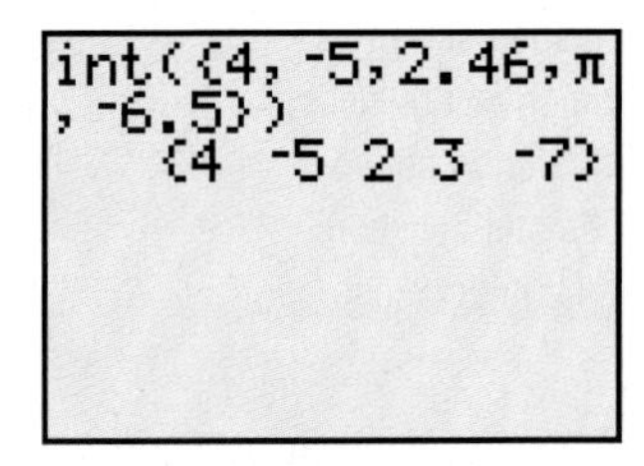


FIGURE 57

The greatest integer function (shown in Figure 58) is called a **step function.** (Why?)

**Looking Ahead to Calculus**

The greatest integer function is used in calculus as a classic example of how the limit of a function may not exist at a particular value in its domain. For a limit to exist, both the left- and right-hand limits must be equal. We can see from the graph of the greatest integer function that for an integer value such as 3, as $x$ approaches 3 from the left, function values are all 2. As $x$ approaches 3 from the right, function values are all 3. Using the precise definitions of calculus, the left- and right-hand limits are not equal, and therefore the limit as $x$ approaches 3 does not exist.

FUNCTION CAPSULE

**GREATEST INTEGER FUNCTION** $f(x) = [\![x]\!]$

Domain: $(-\infty, \infty)$ Range: $\{x \mid x \text{ is an integer}\} = \{\ldots, -3, -2, -1, 0, 1, 2, 3, \ldots\}$

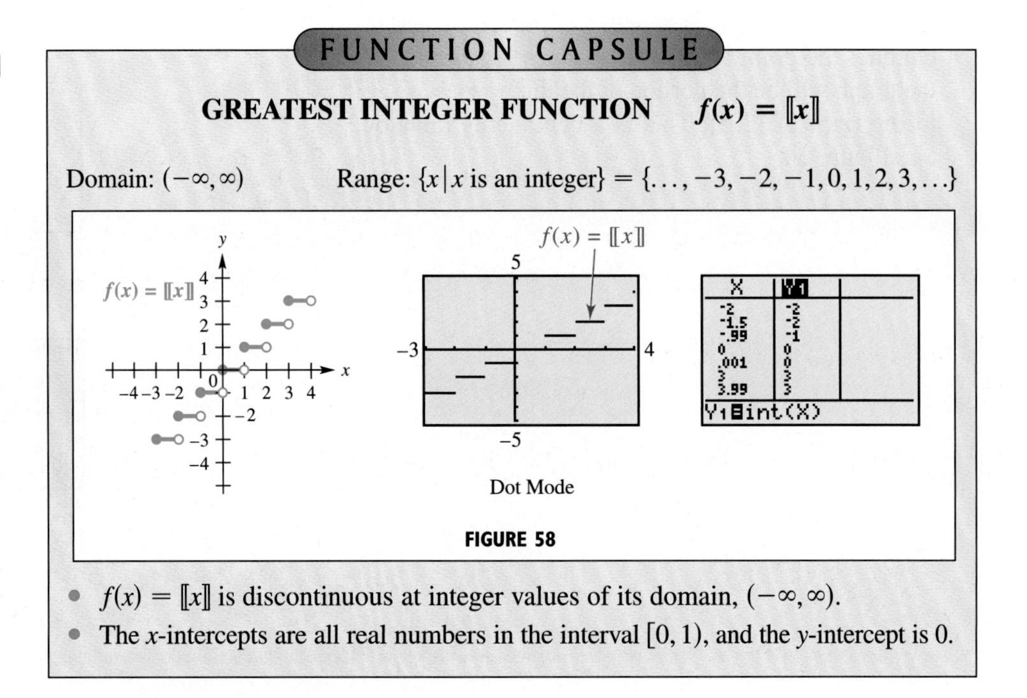


**FIGURE 58**

- $f(x) = [\![x]\!]$ is discontinuous at integer values of its domain, $(-\infty, \infty)$.
- The $x$-intercepts are all real numbers in the interval $[0, 1)$, and the $y$-intercept is 0.

As mentioned earlier, graphing functions with discontinuities often poses problems for graphing calculators. If we graph the greatest integer function with a calculator in *connected mode,* the calculator attempts to connect the portions of the graph at integer values as in Figure 59(a). This is why calculators can also be directed to graph in *dot mode.* In Figure 59(b), we show an accurate graph of $f(x) = [\![x]\!]$ in the standard window, with the calculator in dot mode.

TECHNOLOGY NOTE

You should learn how to activate the greatest integer function command on your particular model. Consult the graphing calculator manual that accompanies this text or your owner's manual. Keep in mind that to obtain an accurate graph of a step function, your calculator must be in dot mode.

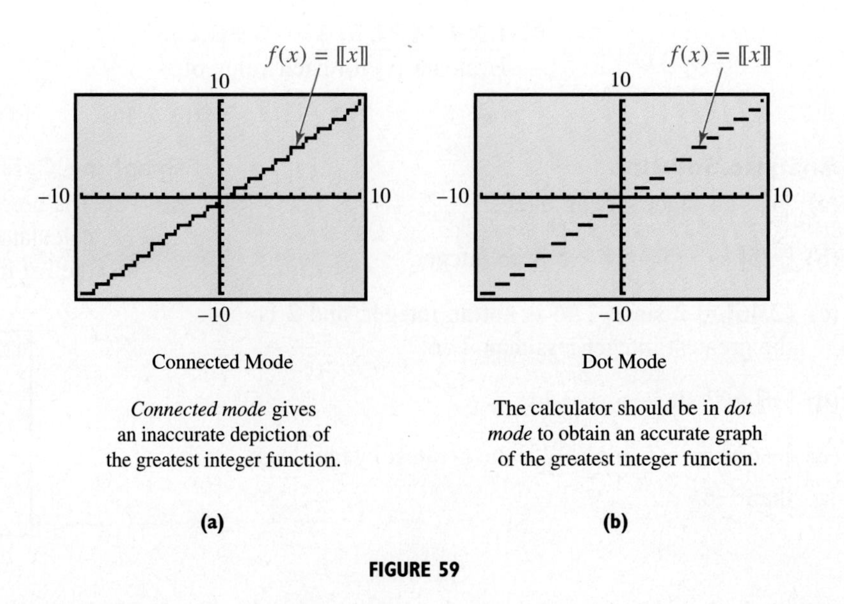


*Connected mode* gives an inaccurate depiction of the greatest integer function.

The calculator should be in *dot mode* to obtain an accurate graph of the greatest integer function.

**(a)** **(b)**

**FIGURE 59**

Notice from the graph in Figure 59(b) that the inclusion or exclusion of the endpoint for each segment is not readily determined from the calculator graph. However, the traditional graph in Figure 58 does show whether endpoints are included or excluded. Once again, we state the following important conclusion.

**A Caution Against Relying Too Heavily on Technology**

Analysis of a calculator graph is often not sufficient to draw correct conclusions. While this technology is incredibly powerful and pedagogically useful, we cannot rely on it alone in our study of the graphs of functions. We must understand the basic concepts of functions as well.

**EXAMPLE 4** *Graphing a Step Function*

Graph the function defined by $y = [\![\frac{1}{2}x + 1]\!]$. Give the domain and range.

### Analytic Solution

Try some values of $x$ in the equation to see how the values of $y$ behave. Some sample ordered pairs are given here.

| $x$ | 0 | $\frac{1}{2}$ | 1 | $\frac{3}{2}$ | 2 | 3 | 4 | $-1$ | $-2$ | $-3$ |
|---|---|---|---|---|---|---|---|---|---|---|
| $y$ | 1 | 1 | 1 | 1 | 2 | 2 | 3 | 0 | 0 | $-1$ |

These ordered pairs suggest that if $x$ is in the interval $[0, 2)$, then $y = 1$. For $x$ in $[2, 4)$, $y = 2$, and so on. The graph is shown in Figure 60. The domain is $(-\infty, \infty)$, and the range is $\{\ldots, -2, -1, 0, 1, 2, \ldots\}$.

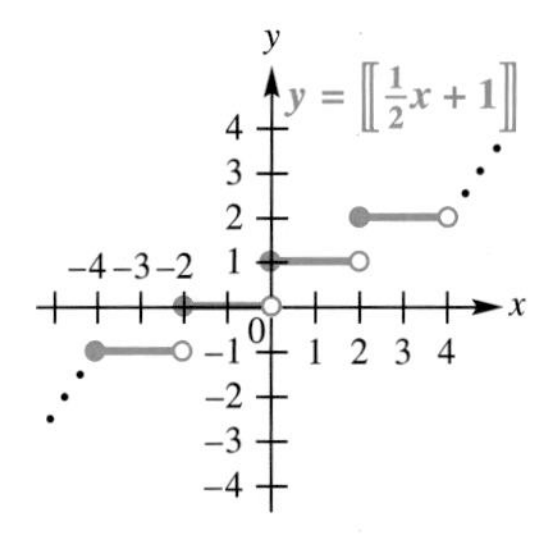


The dots indicate that the graph continues indefinitely in the same pattern.

**FIGURE 60**

### Graphing Calculator Solution

Figure 61(a) shows the graph of $Y_1 = [\![\frac{1}{2}X + 1]\!]$ in the window $[-4, 4]$ by $[-4, 4]$. Scroll up and down, using the table feature to confirm that the range is $\{\ldots, -2, -1, 0, 1, 2, \ldots\}$. See Figure 61(b).

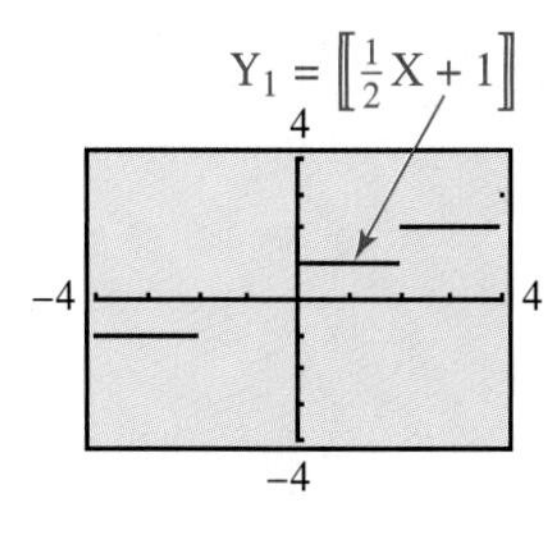


Dot Mode

**(a)**

| X | Y1 |
|---|---|
| 0 | 1 |
| .5 | 1 |
| 1 | 1 |
| 1.5 | 1 |
| 2 | 2 |
| 3 | 2 |
| -1 | 0 |
| -2 | 0 |

Y1■int((1/2)X+1)

**(b)**

**FIGURE 61**

## Applications of Piecewise-Defined Functions

**EXAMPLE 5** *Applying the Greatest Integer Function to Parking Rates*
Downtown Parking charges a \$5 base fee for parking through 1 hour, and \$1 for each additional hour or fraction thereof. The maximum fee for 24 hours is \$15. Sketch a graph of the function that describes this pricing scheme.

**Solution** For any amount of time during and up to the first hour, the rate is \$5. Thus, some sample ordered pairs for the function in the interval $(0, 1]$ would be

$$(.25, 5), (.5, 5), (.75, 5), \text{ and } (1, 5).$$

After the first hour, the price immediately jumps (or steps up) to \$6, and remains \$6 until the time equals 2 hours. It then jumps to \$7 during the third hour, and so on. During the 11th hour, it will have jumped to \$15, and will remain at \$15 for the rest of the 24-hour period. Figure 62 shows the graph of this function for the interval $(0, 24]$. (A closed endpoint is found at the *right* extreme value rather than the left, because the increase does not take effect until time elapses after the hour of use has started.) The range of the function is $\{5, 6, 7, \ldots, 15\}$.

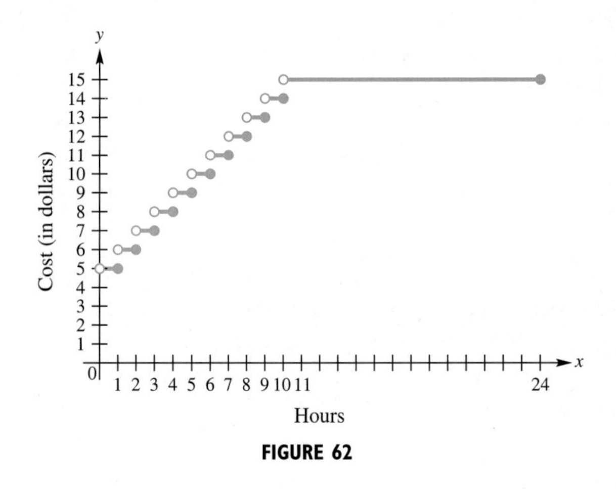


**FIGURE 62**

**EXAMPLE 6** *Using a Piecewise-Defined Function to Analyze Pollution and Brown Trout Populations*
Due to acid rain, the percentage of lakes in Scandinavia that had lost their population of brown trout increased dramatically between 1940 and 1975. Based on a sample of 2850 lakes, this percentage can be approximated by the piecewise-defined function $f$. (*Source:* C. Mason, *Biology of Freshwater Pollution*, John Wiley and Sons, 1991.)

$$f(x) = \begin{cases} \frac{11}{20}(x - 1940) + 7 & \text{if } 1940 \le x < 1960 \\ \frac{32}{15}(x - 1960) + 18 & \text{if } 1960 \le x \le 1975 \end{cases}$$

**(a)** Graph $f$.

**(b)** Determine the percent of lakes that had lost brown trout by 1972.

### Analytic Solution

**(a)** The graph of $f$ consists of two line segments. The first segment is determined by

$$y = \frac{11}{20}(x - 1940) + 7.$$

If $x = 1940$ then $y = 7$, and if $x = 1960$ then $y = 18$. Place a dot at $(1940, 7)$ and an open circle at $(1960, 18)$, since the year 1960 is not in the interval $1940 \leq x < 1960$. These points are endpoints of the line segment.

The second line segment is determined by

$$y = \frac{32}{15}(x - 1960) + 18.$$

If $x = 1960$ then $y = 18$, and if $x = 1975$ then $y = 50$. This segment includes $(1960, 18)$ and $(1975, 50)$, so place a dot at each endpoint. Notice that this fills in the open circle at $(1960, 18)$ from the first line segment. The graph of $f$ is shown in Figure 63.

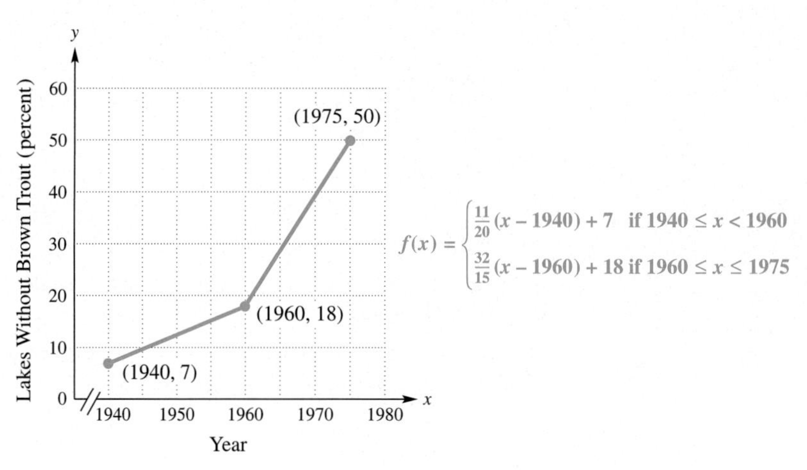


**FIGURE 63**

**(b)** Since $1960 \leq 1972 \leq 1975$, use the second formula to calculate $f(1972)$:

$$f(1972) = \frac{32}{15}(1972 - 1960) + 18 = 43.6.$$

By 1972, about 44% of the lakes had lost their population of brown trout.

### Graphing Calculator Solution

**(a)** Figure 64(a) shows how to enter the piecewise-defined function as $Y_1$. (Three part inequalities cannot be used.) Figure 64(b) shows the resulting graph.

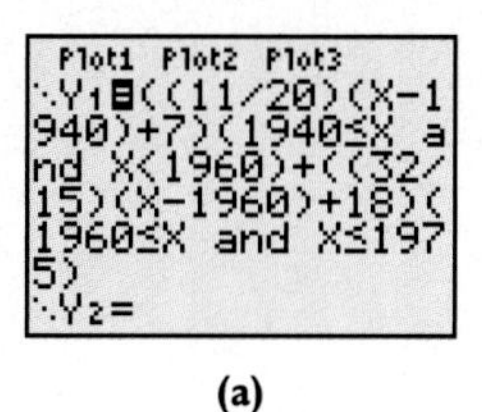


**(a)**

$$Y_1 = \begin{cases} \frac{11}{20}(X - 1940) + 7 \\ \quad \text{if } 1940 \leq X < 1960 \\ \frac{32}{15}(X - 1960) + 18 \\ \quad \text{if } 1960 \leq X \leq 1975 \end{cases}$$

60

Y1=((11/20)(X-1940)+7)(1_

1930 1980

X=1972 Y=43.6

−20

Dot Mode

**(b)**

**FIGURE 64**

**(b)** The display at the bottom of Figure 64(b) confirms the analytic solution.

A table can also be used to confirm the analytic solution. Use the "Ask" option and enter $X = 1972$ to find $Y_1 = 43.6$.

**FOR DISCUSSION**

Refer to Example 6 and determine the percent of lakes that had lost their brown trout population by 1947. Use both analytic and graphical methods.

## 2.5 EXERCISES

*(Modeling) Solve each problem.*

**1.** *Speed Limits* The graph of $y = f(x)$ gives the speed limit $y$ along a rural highway after traveling $x$ miles.

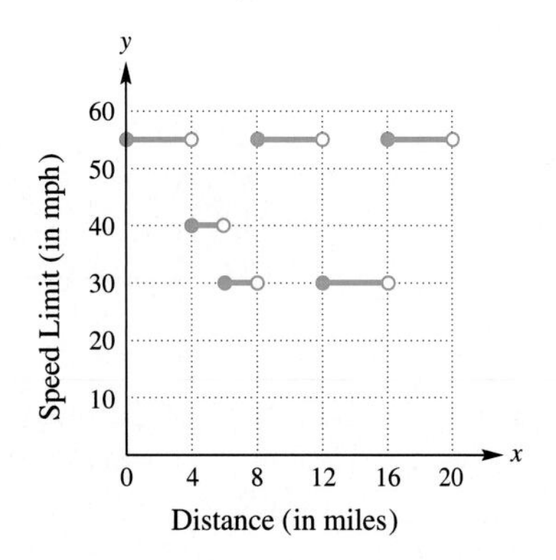


**(a)** What are the highest and lowest speed limits along this stretch of highway?

**(b)** Estimate the miles of highway with a speed limit of 55 mph.

**(c)** Evaluate $f(4)$, $f(12)$, and $f(18)$.

**(d)** At what $x$-values is the graph discontinuous? Interpret each discontinuity.

**2.** *ATM* The graph of $y = f(x)$ depicts the amount of money $y$ in dollars in an automatic teller machine (ATM) after $x$ minutes.

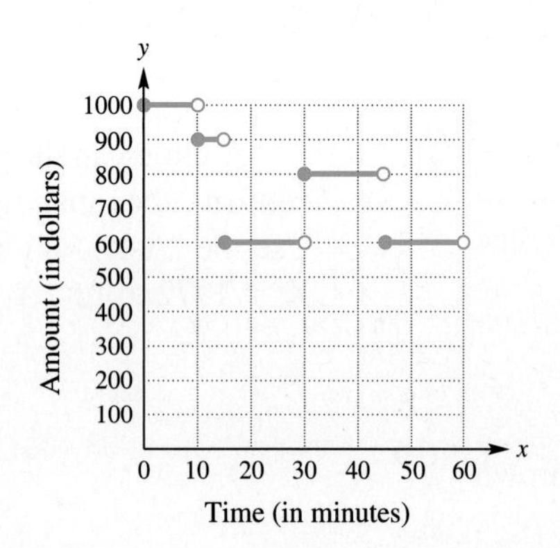


**(a)** Determine the initial and final amounts of money in the ATM.

**(b)** Evaluate $f(10)$ and $f(50)$. Is $f$ continuous?

**(c)** How many withdrawals occurred during this time period?

**(d)** When did the largest withdrawal occur? How much was it?

**(e)** How much was deposited into the machine?

**3.** *Swimming Pool Levels* The graph of $y = f(x)$ represents the amount of water in thousands of gallons remaining in a swimming pool after $x$ days.

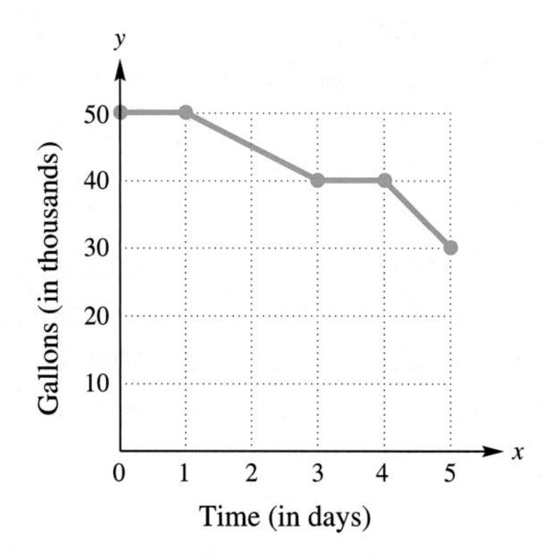


**(a)** Estimate the initial and final amounts of water contained in the pool.

**(b)** When did the amount of water in the pool remain constant?

**(c)** Approximate $f(2)$ and $f(4)$.

**(d)** At what rate was water being drained from the pool when $1 \leq x \leq 3$?

**4.** *Gasoline Usage* The graph shows the gallons of gasoline $y$ in the gas tank of a car after $x$ hours.

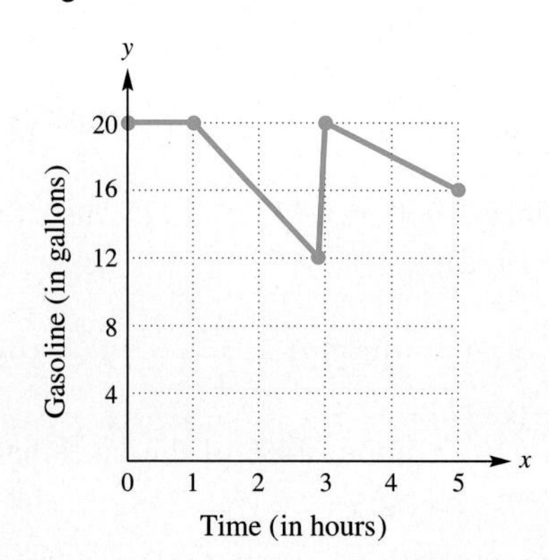


**(a)** Estimate how much gasoline was in the gas tank when $x = 3$.

**(b)** Interpret the graph.

**(c)** During what time interval did the car burn gasoline at the fastest rate?

*For each piecewise-defined function, find* **(a)** $f(-5)$, **(b)** $f(-1)$, **(c)** $f(0)$, *and* **(d)** $f(3)$.

**5.** $f(x) = \begin{cases} 2x & \text{if } x \leq -1 \\ x - 1 & \text{if } x > -1 \end{cases}$

**6.** $f(x) = \begin{cases} x - 2 & \text{if } x < 3 \\ 5 - x & \text{if } x \geq 3 \end{cases}$

**7.** $f(x) = \begin{cases} 2 + x & \text{if } x < -4 \\ -x & \text{if } -4 \leq x \leq 2 \\ 3x & \text{if } x > 2 \end{cases}$

**8.** $f(x) = \begin{cases} -2x & \text{if } x < -3 \\ 3x - 1 & \text{if } -3 \leq x \leq 2 \\ -4x & \text{if } x > 2 \end{cases}$

*Graph each piecewise-defined function.*

**9.** $f(x) = \begin{cases} x - 1 & \text{if } x \leq 3 \\ 2 & \text{if } x > 3 \end{cases}$

**10.** $f(x) = \begin{cases} 6 - x & \text{if } x \leq 3 \\ 3x - 6 & \text{if } x > 3 \end{cases}$

**11.** $f(x) = \begin{cases} 4 - x & \text{if } x < 2 \\ 1 + 2x & \text{if } x \geq 2 \end{cases}$

**12.** $f(x) = \begin{cases} 2x + 1 & \text{if } x \geq 0 \\ x & \text{if } x < 0 \end{cases}$

**13.** $f(x) = \begin{cases} 2 + x & \text{if } x < -4 \\ -x & \text{if } -4 \leq x \leq 5 \\ 3x & \text{if } x > 5 \end{cases}$

**14.** $f(x) = \begin{cases} -2x & \text{if } x < -3 \\ 3x - 1 & \text{if } -3 \leq x \leq 2 \\ -4x & \text{if } x > 2 \end{cases}$

**15.** $f(x) = \begin{cases} -\dfrac{1}{2}x^2 + 2 & \text{if } x \leq 2 \\ \dfrac{1}{2}x & \text{if } x > 2 \end{cases}$

**16.** $f(x) = \begin{cases} x^3 + 5 & \text{if } x \leq 0 \\ -x^2 & \text{if } x > 0 \end{cases}$

*Match each piecewise-defined function with its calculator graph.*

**17.** $f(x) = \begin{cases} x^2 - 4 & \text{if } x \geq 0 \\ -x + 5 & \text{if } x < 0 \end{cases}$

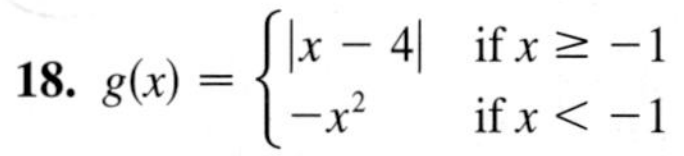

**18.** $g(x) = \begin{cases} |x - 4| & \text{if } x \geq -1 \\ -x^2 & \text{if } x < -1 \end{cases}$

**19.** $h(x) = \begin{cases} 6 & \text{if } x \geq 0 \\ -6 & \text{if } x < 0 \end{cases}$

**20.** $k(x) = \begin{cases} \sqrt{x} & \text{if } x \geq 0 \\ -x^2 & \text{if } x < 0 \end{cases}$

**A.**

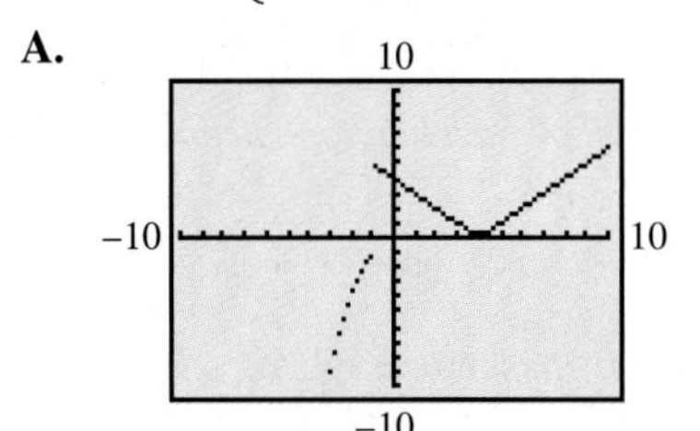


Dot Mode

**B.**

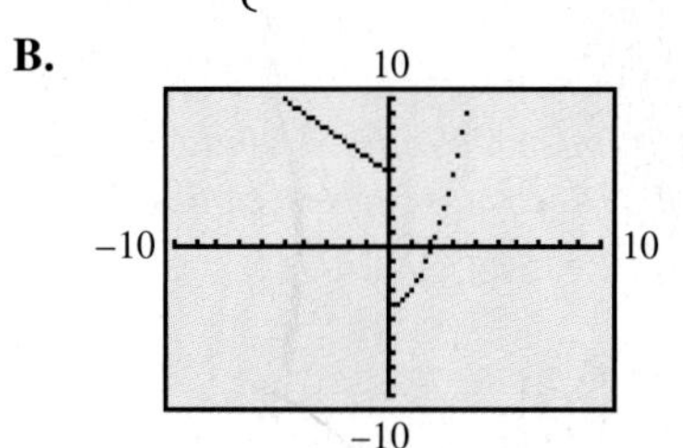


Dot Mode

**C.**

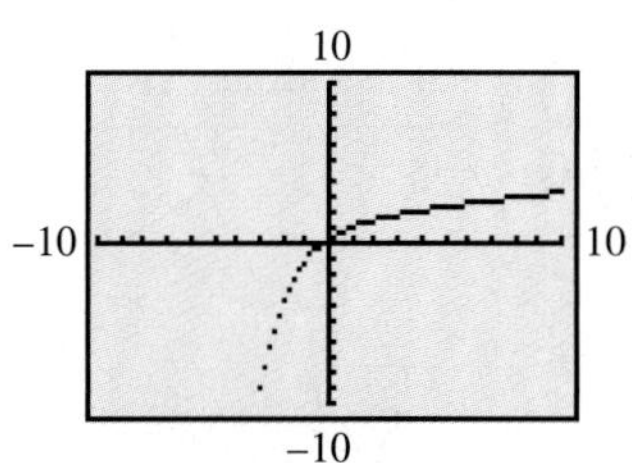


Dot Mode

**D.**

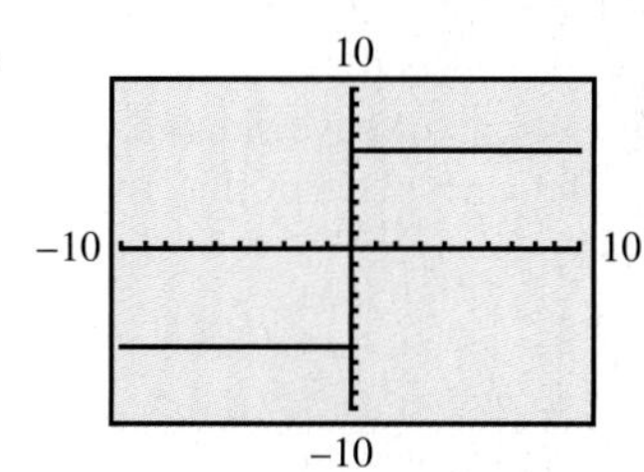


Dot Mode

*Use a graphing calculator to graph each piecewise-defined function in the specified exercise, using the window indicated.*

**21.** Exercise 9, window $[-4, 6]$ by $[-2, 4]$

**22.** Exercise 10, window $[-2, 8]$ by $[-2, 10]$

**23.** Exercise 11, window $[-4, 6]$ by $[-2, 8]$

**24.** Exercise 12, window $[-5, 4]$ by $[-3, 8]$

**25.** Exercise 13, window $[-12, 12]$ by $[-6, 20]$

**26.** Exercise 14, window $[-6, 6]$ by $[-10, 8]$

**27.** Exercise 15, window $[-5, 6]$ by $[-2, 4]$

**28.** Exercise 16, window $[-3, 4]$ by $[-3, 6]$

*Give a formula for a piecewise-defined function f for each graph shown. Give the domain and range.*

**29.**

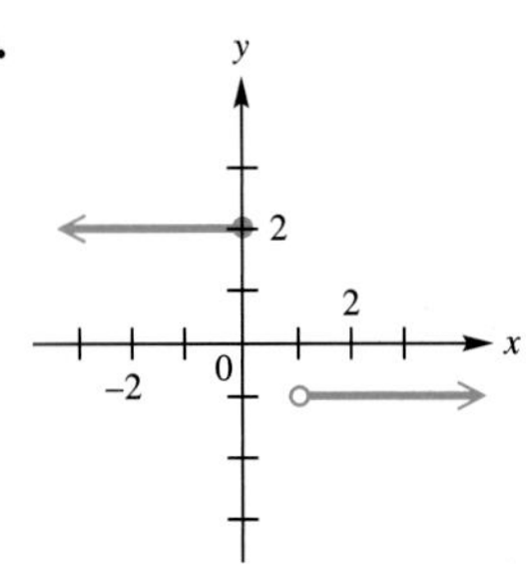

**30.**

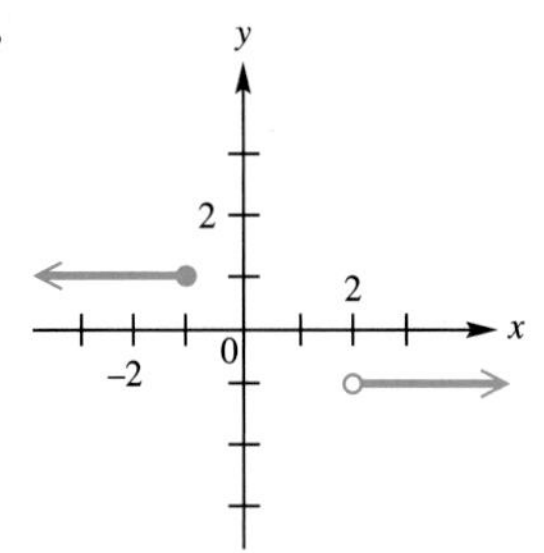

**31.**

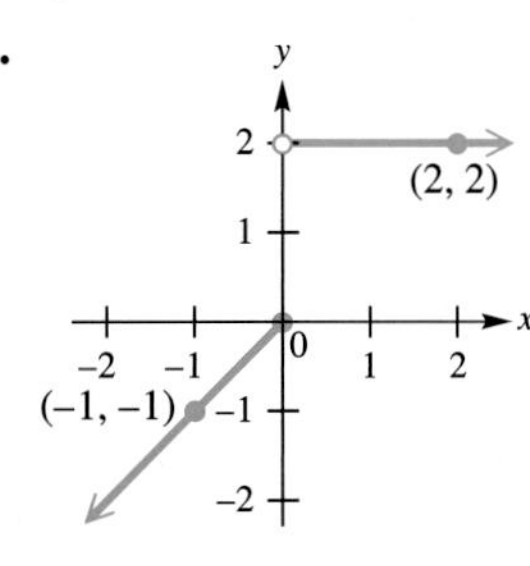


**32.**

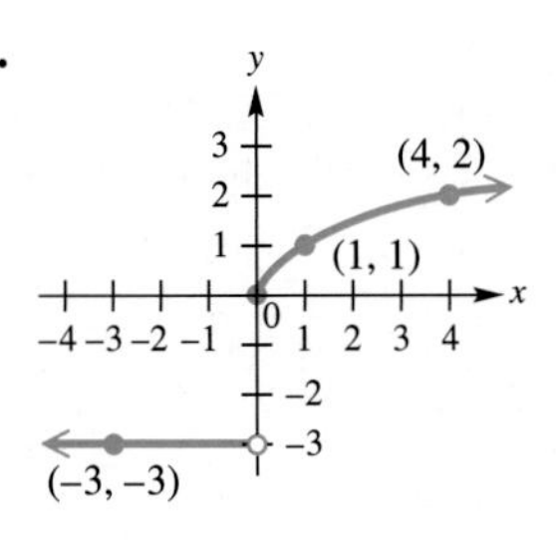


**33.**

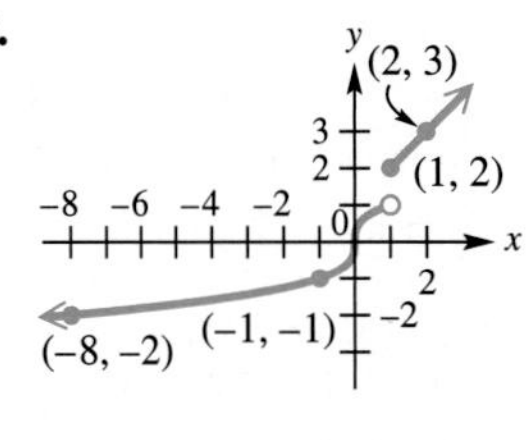


**34.**

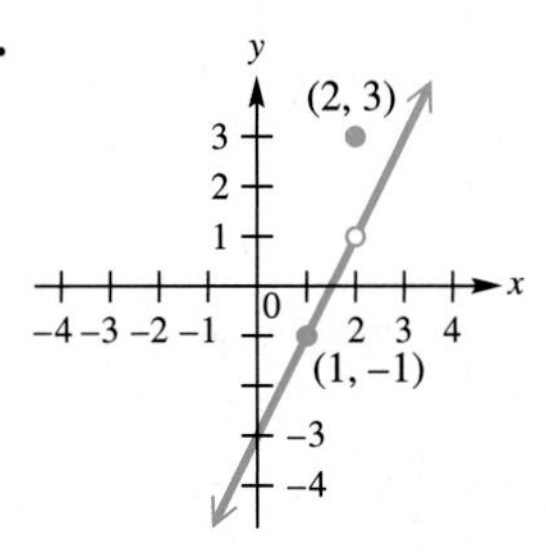


**35.** Why is the following not a piecewise-defined function?

$$f(x) = \begin{cases} x + 7 & \text{if } x \leq 4 \\ x^2 & \text{if } x \geq 4 \end{cases}$$

**36.** Consider the "function" defined in Exercise 35. Suppose you were asked to find $f(4)$. How would you respond?

*Concept Check* *Use the terminology of Sections 2.2 and 2.3 to describe how the graph of the given function can be obtained from the graph of* $y = [\![x]\!]$.

**37.** $y = [\![x]\!] - 1.5$ **38.** $y = [\![-x]\!]$ **39.** $y = -[\![x]\!]$ **40.** $y = [\![x + 2]\!]$

*Use a graphing calculator in dot mode with the window* $[-5, 5]$ *by* $[-3, 3]$ *to create an accurate graph of each function, defined using the greatest integer function. (Refer to your descriptions in Exercises 37–40.)*

**41.** $y = [\![x]\!] - 1.5$ **42.** $y = [\![-x]\!]$ **43.** $y = -[\![x]\!]$ **44.** $y = [\![x + 2]\!]$

*(Modeling) Solve each problem.*

**45.** *Flow Rates* A water tank has an inlet pipe with a flow rate of 5 gallons per minute and an outlet pipe with a flow rate of 3 gallons per minute. A pipe can be either closed or completely open. The graph shows the number of gallons of water in the tank after $x$ minutes have elapsed. Use the concept of slope to interpret each piece of this graph.

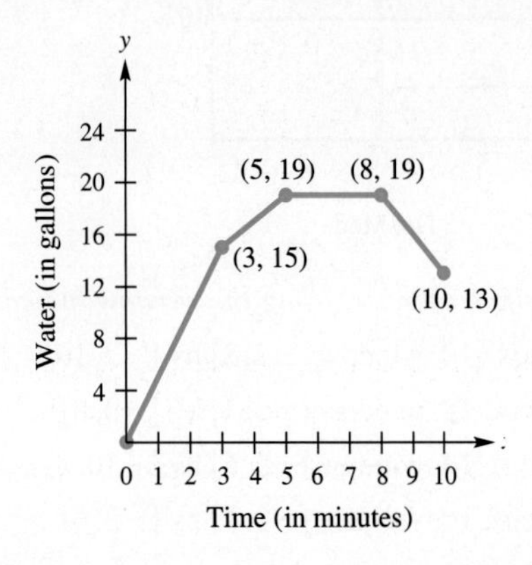


**46.** *Shoe Size* Professional basketball player Shaquille O'Neal is 7 feet 1 inch tall and weighs 325 pounds. The table lists his shoe sizes at certain ages.

| Age | 20 | 21 | 22 | 23 |
|---|---|---|---|---|
| Shoe Size | 19 | 20 | 21 | 22 |

*Source: USA Today.*

**(a)** Write a formula that gives his shoe size at age $x = 20$, 21, 22, and 23.

**(b)** Suppose that after age 23 his shoe size did not change. Sketch a graph of a continuous, piecewise-defined function $f$ that models his shoe size from ages 20 to 26.

**47.** *Insulin Level* When a diabetic takes long-acting insulin, the insulin reaches its peak effect on the blood sugar level in about 3 hours. This effect remains fairly constant for

5 hours, then declines, and is very low until the next injection. In a typical patient, the level of insulin might be given by the following function.

$$f(t) = \begin{cases} 40t + 100 & \text{if } 0 \le t \le 3 \\ 220 & \text{if } 3 < t \le 8 \\ -80t + 860 & \text{if } 8 < t \le 10 \\ 60 & \text{if } 10 < t \le 24 \end{cases}$$

Here $f(t)$ is the blood sugar level, in appropriate units, at time $t$ measured in hours from the time of the injection. Chuck takes his insulin at 6 A.M. Find the blood sugar level at each of the following times.

**(a)** 7 A.M. **(b)** 9 A.M. **(c)** 10 A.M.

**(d)** Noon **(e)** 2 P.M. **(f)** 5 P.M.

**(g)** Midnight

**(h)** Graph $y = f(t)$ for $0 \le t \le 18$.

**48.** *Snow Depth* The snow depth in Michigan's Isle Royale National Park varies throughout the winter. In a typical winter, the snow depth in inches is approximated by the following function.

$$f(x) = \begin{cases} 6.5x & \text{if } 0 \le x \le 4 \\ -5.5x + 48 & \text{if } 4 < x \le 6 \\ -30x + 195 & \text{if } 6 < x \le 6.5 \end{cases}$$

Here $x$ represents the time in months, with $x = 0$ representing the beginning of October, $x = 1$ representing the beginning of November, and so on.

**(a)** Graph the function for $0 \le x \le 6.5$.

**(b)** In what month is the snow the deepest? What is the greatest snow depth?

**(c)** In what months does the snow begin and end?

**49.** *Rabies Cases* The table lists the approximate number of animal rabies cases in the United States from 1988–1992.

| Year | 1988 | 1989 | 1990 | 1991 | 1992 |
|---|---|---|---|---|---|
| Cases | 4800 | 4900 | 5000 | 6700 | 8400 |

*Source:* U.S. Centers for Disease Control and Prevention.

**(a)** Describe the change in the data from one year to the next.

**(b)** Determine a continuous piecewise-defined function $f$ that approximates the data. Let $x = 0$ correspond to the year 1988.

**50.** *Minimum Wage* The table lists the federal minimum wage rates in current dollars for the years 1978–2001. Sketch a graph of the data as a piecewise-defined function. (Assume that wages take effect on Jan. 1 of the year in the first year of the interval.)

| Year(s) | Wage | Year(s) | Wage |
|---|---|---|---|
| 1978 | $2.65 | 1990 | $3.80 |
| 1979 | $2.90 | 1991–95 | $4.25 |
| 1980 | $3.10 | 1996 | $4.75 |
| 1981–89 | $3.35 | 1997–2001 | $5.15 |

*Source:* U.S. Bureau of Labor Statistics.

**51.** *Violent Crimes* The table lists the numbers of victims of violent crime per 1000 people in 1997 by age group using interval notation.

| Age | Crime Rate |
|---|---|
| [12, 16) | 88 |
| [16, 19) | 96 |
| [19, 24) | 68 |
| [24, 35) | 47 |
| [35, 50) | 32 |
| [50, 65) | 15 |
| [65, 90) | 4 |

*Source:* U.S. Department of Justice.

**(a)** Sketch the graph of a piecewise-defined function that models the data, where $x$ represents age.

**(b)** Discuss the impact that age has on the likelihood of being a victim of a violent crime.

**52.** *New Homes* The table lists the numbers in millions of houses built for various time intervals from 1940 to 1990 using interval notation.

| Year | Houses |
|---|---|
| [1940, 1950) | 8.5 |
| [1950, 1960) | 13.6 |
| [1960, 1970) | 16.1 |
| [1970, 1980) | 11.6 |
| [1980, 1990) | 8.1 |

*Source:* U.S. Bureau of the Census, *American Housing Survey.*

**(a)** Sketch the graph of a piecewise-defined function that models the data, where $x$ represents the year.

**(b)** Discuss the trends in housing starts between 1940 and 1990.

**53.** *Cellular Phone Bills* Suppose that the charges for a cellular phone call are \$.50 for the first minute and \$.25 for each additional minute. Assume that a fraction of a minute is rounded up.

**(a)** Determine the cost of a phone call lasting 3.5 minutes.

**(b)** Find a formula for a function $f$ that computes the cost of a telephone call $x$ minutes long, where $0 < x \leq 5$. (*Hint:* Express $f$ as a piecewise-defined function.)

**54.** *Lumber Costs* Lumber that is used to frame walls of houses is frequently sold in multiples of 2 feet. If the length of a board is not exactly a multiple of 2 feet, there is often no charge for the additional length. For example, if a board measures at least 8 feet but less than 10 feet, then the consumer is charged for only 8 feet.

**(a)** Suppose that the cost of lumber is \$.80 every 2 feet. Find a formula for a function $f$ that computes the cost of a board $x$ feet long for $6 \leq x \leq 18$.

**(b)** Graph $f$ using a graphing calculator.

**(c)** Determine the costs of boards with lengths of 8.5 feet and 15.2 feet.

**55.** *Water in a Tank* Sketch a graph that depicts the amount of water in a 100-gallon tank. The tank is initially empty and then filled at a rate of 5 gallons per minute. Immediately after it is full, a pump is used to empty the tank at 2 gallons per minute.

**56.** *Distance from Home* Sketch a graph showing the distance that a person is from home after $x$ hours if that individual drives on a straight road at 40 mph to a park 20 miles away, remains at the park for 2 hours, and then returns home at a speed of 20 mph.

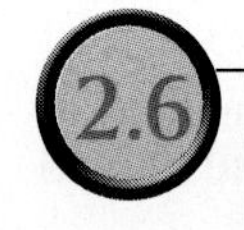


# OPERATIONS AND COMPOSITION

Operations on Functions • The Difference Quotient • Composition of Functions • Applications of Operations and Composition

## Operations on Functions

Just as we add, subtract, multiply, and divide real numbers, we can also perform these operations on functions. Given two functions $f$ and $g$, their sum, written $f + g$, is defined as

$$(f + g)(x) = f(x) + g(x),$$

for all $x$ such that both $f(x)$ and $g(x)$ exist. Similar definitions can be given for the difference $f - g$, product $fg$, and quotient $\frac{f}{g}$ of functions; however, the quotient,

$$\left(\frac{f}{g}\right)(x) = \frac{f(x)}{g(x)},$$

is defined only for those values of $x$ where both $f(x)$ and $g(x)$ exist, with the additional condition $g(x) \neq 0$. The various operations on functions are defined as follows.

**Operations on Functions**

Given two functions $f$ and $g$, then for all values of $x$ for which both $f(x)$ and $g(x)$ are defined, the functions $f + g$, $f - g$, $fg$, and $\frac{f}{g}$ are defined as follows.

| | |
|---|---|
| **Sum** | $(f + g)(x) = f(x) + g(x)$ |
| **Difference** | $(f - g)(x) = f(x) - g(x)$ |
| **Product** | $(fg)(x) = f(x) \cdot g(x)$ |
| **Quotient** | $\left(\frac{f}{g}\right)(x) = \frac{f(x)}{g(x)}, \quad g(x) \neq 0$ |

The domains of $f + g$, $f - g$, and $fg$ include all real numbers in the intersection of the domains of $f$ and $g$, while the domain of $\frac{f}{g}$ includes those real numbers in the intersection of the domains of $f$ and $g$ for which $g(x) \neq 0$.

**EXAMPLE 1** *Using the Operations on Functions*

Let $f(x) = x^2 + 1$ and $g(x) = 3x + 5$. Perform the operations.

**(a)** $(f + g)(1)$ **(b)** $(f - g)(-3)$ **(c)** $(fg)(5)$ **(d)** $\left(\frac{f}{g}\right)(0)$

**Analytic Solution**

**(a)** Since $f(1) = 2$ and $g(1) = 8$, use the above definition to get

$$\begin{aligned}(f + g)(1) &= f(1) + g(1)\\ &= 2 + 8\\ &= 10.\end{aligned}$$

**(b)** $$\begin{aligned}(f - g)(-3) &= f(-3) - g(-3)\\ &= 10 - (-4)\\ &= 14\end{aligned}$$

**(c)** $(fg)(5) = f(5) \cdot g(5) = 26 \cdot 20 = 520$

**(d)** $\left(\frac{f}{g}\right)(0) = \frac{f(0)}{g(0)} = \frac{1}{5}$

**Graphing Calculator Solution**

**(a)–(d)** Figure 65(a) shows $f$ entered as $Y_1$ and $g$ as $Y_2$.

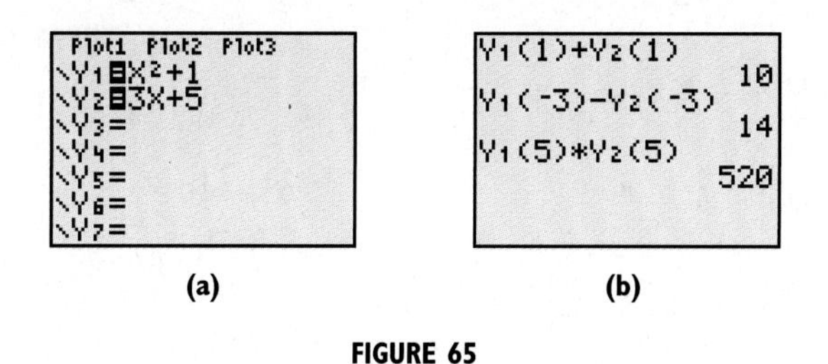


(a) (b)

**FIGURE 65**

Because the calculator is capable of function notation, we can calculate parts (a)–(c) as shown in Figure 65(b). Part (d) can be found similarly.

**EXAMPLE 2** *Using the Operations on Functions*

Let $f(x) = 8x - 9$ and $g(x) = \sqrt{2x - 1}$. Perform the operations.

**(a)** $(f + g)(x)$ **(b)** $(f - g)(x)$ **(c)** $(fg)(x)$ **(d)** $\left(\frac{f}{g}\right)(x)$

**(e)** Give the domains of $f$, $g$, $f + g$, $f - g$, $fg$, and $\frac{f}{g}$.

**Solution**

**(a)** $(f+g)(x) = f(x) + g(x) = 8x - 9 + \sqrt{2x-1}$

**(b)** $(f-g)(x) = f(x) - g(x) = 8x - 9 - \sqrt{2x-1}$

**(c)** $(fg)(x) = f(x) \cdot g(x) = (8x-9)\sqrt{2x-1}$

**(d)** $\left(\dfrac{f}{g}\right)(x) = \dfrac{f(x)}{g(x)} = \dfrac{8x-9}{\sqrt{2x-1}}$

**(e)** The domain of $f$ is $(-\infty, \infty)$, while the domain of $g(x) = \sqrt{2x-1}$ includes just those real numbers that make $2x - 1 \geq 0$; the domain of $g$ is the interval $\left[\frac{1}{2}, \infty\right)$. The domain of $f+g$, $f-g$, and $fg$ is thus $\left[\frac{1}{2}, \infty\right)$. With $\frac{f}{g}$, the denominator cannot be 0, so the value $\frac{1}{2}$ is excluded from the domain. The domain of $\frac{f}{g}$ is $\left(\frac{1}{2}, \infty\right)$.

**EXAMPLE 3** *Evaluating Combinations of Functions*

If possible, use the given representations of functions $f$ and $g$ to evaluate

$$(f+g)(4),\ (f-g)(-2),\ (fg)(1), \text{ and } \left(\frac{f}{g}\right)(0).$$

**(a)**

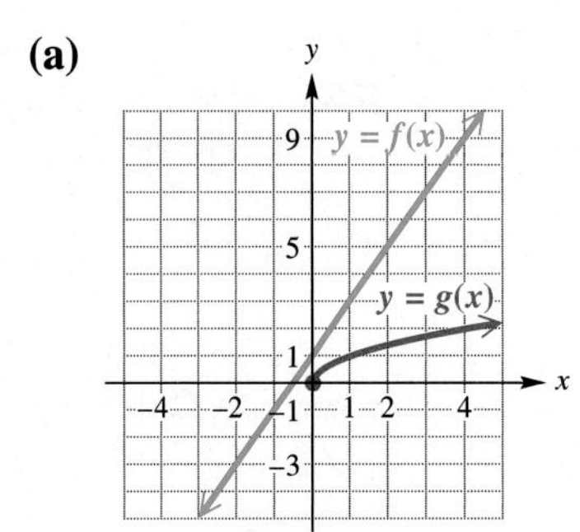


**FIGURE 66**

**(b)**

| $x$ | $f(x)$ | $g(x)$ |
|---|---|---|
| −2 | −3 | undefined |
| 0 | 1 | 0 |
| 1 | 3 | 1 |
| 4 | 9 | 2 |

**(c)** $f(x) = 2x + 1$, $g(x) = \sqrt{x}$

**Solution**

**(a)** In Figure 66, $f(4) = 9$ and $g(4) = 2$. Thus

$$(f+g)(4) = f(4) + g(4) = 9 + 2 = 11.$$

Although $f(-2) = -3$, $g(-2)$ is undefined because $-2$ is not in the domain of $g$. Thus $(f-g)(-2)$ is undefined. The domains of $f$ and $g$ include 1, so

$$(fg)(1) = f(1) \cdot g(1) = 3(1) = 3.$$

The graph of $g$ intersects the origin, so $g(0) = 0$. Thus $\left(\frac{f}{g}\right)(0)$ is undefined.

**(b)** From the table, $f(4) = 9$ and $g(4) = 2$. As in part (a),

$$(f+g)(4) = f(4) + g(4) = 9 + 2 = 11.$$

In the table $g(-2)$ is undefined, so $(f-g)(-2)$ is also undefined. Similarly,

$$(fg)(1) = f(1) \cdot g(1) = 3(1) = 3,$$

and $$\left(\frac{f}{g}\right)(0) = \frac{f(0)}{g(0)} \text{ is undefined since } g(0) = 0.$$

**(c)** Use the formulas $f(x) = 2x + 1$ and $g(x) = \sqrt{x}$.

$$(f + g)(4) = f(4) + g(4) = (2 \cdot 4 + 1) + \sqrt{4} = 9 + 2 = 11$$
$$(f - g)(-2) = f(-2) - g(-2) = [2(-2) + 1] - \sqrt{-2} \text{ is undefined.}$$
$$(fg)(1) = f(1) \cdot g(1) = (2 \cdot 1 + 1)\sqrt{1} = 3(1) = 3$$
$$\left(\frac{f}{g}\right)(0) = \frac{f(0)}{g(0)} \text{ is undefined since } g(0) = 0.$$

## The Difference Quotient

Suppose that the point $P$ lies on the graph of $y = f(x)$, and suppose that $h$ is a positive number. If we let $(x, f(x))$ denote the coordinates of $P$ and $(x + h, f(x + h))$ denote the coordinates of $Q$, then the line joining $P$ and $Q$ has slope

$$m = \frac{f(x + h) - f(x)}{(x + h) - x} = \frac{f(x + h) - f(x)}{h}.$$

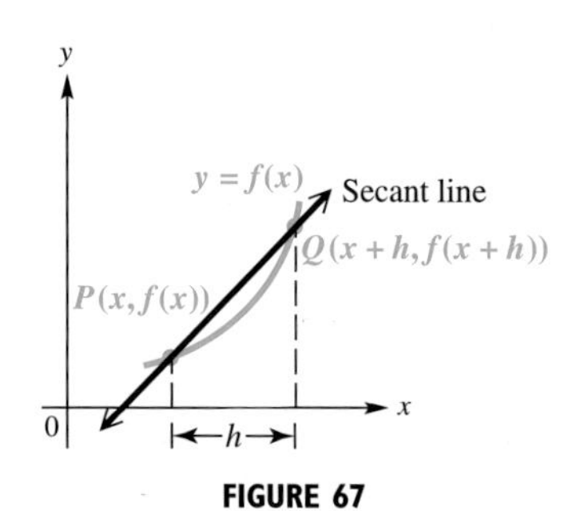


**FIGURE 67**

This expression, called the **difference quotient,** is important in the study of calculus.

Figure 67 shows the graph of the line $PQ$ (called a *secant line*). As $h$ approaches 0, the slope of this secant line approaches the slope of the line tangent to the curve at $P$. Important applications of this idea are developed in calculus, where the concepts of *limit* and *derivative* are investigated.

**EXAMPLE 4** *Finding the Difference Quotient*

Let $f(x) = 2x^2 - 3x$. Find the difference quotient and simplify the expression.

**Solution** To find $f(x + h)$, replace $x$ in $f(x)$ with $x + h$ to get

$$f(x + h) = 2(x + h)^2 - 3(x + h).$$

Then,

$$\frac{f(x + h) - f(x)}{h} = \frac{2(x + h)^2 - 3(x + h) - (2x^2 - 3x)}{h}$$
$$= \frac{2(x^2 + 2xh + h^2) - 3x - 3h - 2x^2 + 3x}{h} \quad \text{Square } x + h\text{; distributive property.}$$
$$= \frac{2x^2 + 4xh + 2h^2 - 3x - 3h - 2x^2 + 3x}{h}$$
$$= \frac{4xh + 2h^2 - 3h}{h} \quad \text{Combine terms.}$$
$$= \frac{h(4x + 2h - 3)}{h} \quad \text{Factor out } h.$$
$$= 4x + 2h - 3. \quad \text{Divide.}$$

**Looking Ahead to Calculus**

The difference quotient is essential in the definition of the *derivative of a function* in calculus. The derivative provides a formula, in function form, for finding the slope of the tangent line to the graph of the function at a given point. Observe Figure 67 and notice that as $h$ approaches 0, the line $PQ$ approaches tangency at point $P$. Thus, the slope of $PQ$ approaches the slope of the tangent line at $P$.

To illustrate, the derivative of $f(x) = x^2 + 3$ is given by $f'(x) = 2x$. Now, $f'(0) = 2(0) = 0$, meaning that the slope of the tangent to $f(x) = x^2 + 3$ at $x = 0$ is 0, implying that the tangent line is horizontal. If you draw this tangent line, you will see that it is the line $y = 3$, which is indeed a horizontal line.

**CAUTION** Notice that $f(x + h)$ is not the same as $f(x) + f(h)$. For $f(x) = 2x^2 - 3x$, as shown in Example 4,

$$f(x + h) = 2(x + h)^2 - 3(x + h) = 2x^2 + 4xh + 2h^2 - 3x - 3h$$

but

$$f(x) + f(h) = (2x^2 - 3x) + (2h^2 - 3h) = 2x^2 - 3x + 2h^2 - 3h.$$

These expressions differ by $4xh$.

## Composition of Functions

The diagram in Figure 68 shows a function $f$ that assigns to each element $x$ of set $X$ some element $y$ of set $Y$. Suppose also that a function $g$ takes each element of set $Y$ and assigns an element $z$ of set $Z$. Using both $f$ and $g$, then, an element $x$ in $X$ is assigned to an element $z$ in $Z$. The result of this process is a new function, $h$, that takes an element $x$ in $X$ and assigns an element $z$ in $Z$. This function $h$ is called the *composition* of functions $g$ and $f$, written $g \circ f$, and is defined as follows.

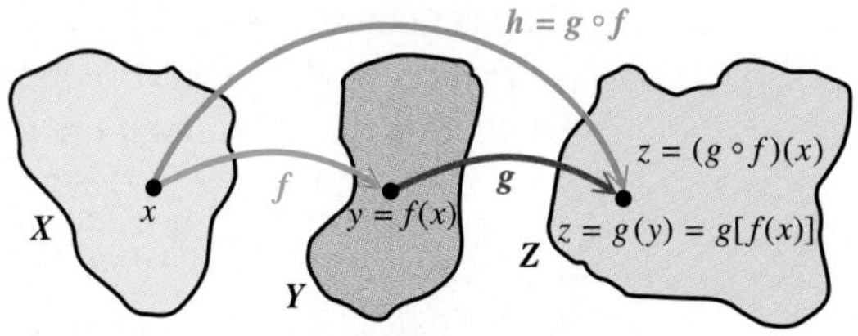

**FIGURE 68**

**Looking Ahead to Calculus**

Finding the derivative of a function in calculus is called differentiation. To differentiate a composite function such as $h(x) = (3x + 2)^4$, we can interpret $h(x)$ as $(f \circ g)(x)$, where $g(x) = 3x + 2$ and $f(x) = x^4$. The chain rule allows the student to differentiate composite functions. Notice the use of the composition symbol and function notation in the following, which comes from the chain rule.

If $h(x) = (f \circ g)(x)$, then $h'(x) = f'[g(x)] \cdot g'(x)$.

**Composition of Functions**

If $f$ and $g$ are functions, then the **composite function,** or **composition,** of $g$ and $f$ is

$$(g \circ f)(x) = g[f(x)]$$

for all $x$ in the domain of $f$ such that $f(x)$ is in the domain of $g$.

As a real-life example of function composition, suppose an oil well off the California coast is leaking, with the leak spreading oil in a circular layer over the water's surface. (See Figure 69.) At any time $t$, in minutes, after the beginning of the leak, the radius of the circular oil slick is $r(t) = 5t$ feet. Since $A(r) = \pi r^2$ gives the area of a circle of radius $r$, the area can be expressed as a function of time by substituting $5t$ for $r$ in $A(r) = \pi r^2$ to get

**FIGURE 69**

$$A(r) = \pi r^2$$
$$A[r(t)] = \pi(5t)^2 = 25\pi t^2.$$

The function $A[r(t)]$ is a composite function of the functions $A$ and $r$.

**EXAMPLE 5** *Evaluating Composite Functions*

Let $f(x) = 2x - 1$ and $g(x) = \dfrac{4}{x - 1}$. Perform the compositions.

**(a)** $(f \circ g)(2)$ **(b)** $(g \circ f)(-3)$

**Analytic Solution**

**(a)** First find $g(2)$. Since $g(x) = \dfrac{4}{x-1}$,

$$g(2) = \frac{4}{2-1} = \frac{4}{1} = 4.$$

Now find $(f \circ g)(2) = f[g(2)] = f(4)$:

$$f[g(2)] = f(4) = 2(4) - 1 = 7.$$

**(b)** Since $f(-3) = 2(-3) - 1 = -7$,

$$\begin{aligned}(g \circ f)(-3) &= g[f(-3)] = g(-7) \\ &= \frac{4}{-7-1} = \frac{4}{-8} \\ &= -\frac{1}{2}.\end{aligned}$$

**Graphing Calculator Solution**

Figure 70 shows how a graphing calculator evaluates the expressions in parts (a) and (b), with $Y_1 = f(x)$ and $Y_2 = g(x)$.

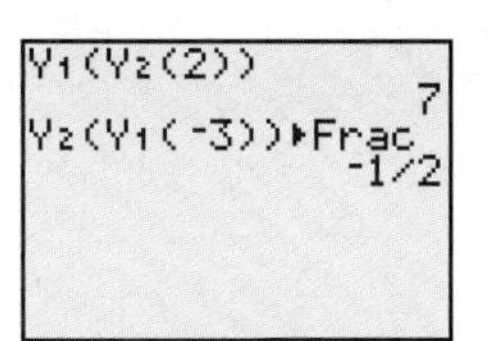


**FIGURE 70**

**EXAMPLE 6** *Finding Composite Functions*

Let $f(x) = 4x + 1$ and $g(x) = 2x^2 + 5x$. Perform the compositions.

**(a)** $(g \circ f)(x)$ **(b)** $(f \circ g)(x)$

**Solution**

**(a)** By definition, $(g \circ f)(x) = g[f(x)]$. Using the given functions,

$$\begin{aligned}(g \circ f)(x) = g[f(x)] &= g(4x+1) && f(x) = 4x+1 \\ &= 2(4x+1)^2 + 5(4x+1) && g(x) = 2x^2 + 5x \\ &= 2(16x^2 + 8x + 1) + 20x + 5 && \text{Square } 4x+1\text{; distributive property.} \\ &= 32x^2 + 16x + 2 + 20x + 5 && \text{Distributive property} \\ &= 32x^2 + 36x + 7. && \text{Combine terms.}\end{aligned}$$

**(b)** If we use the definition above with $f$ and $g$ interchanged, $(f \circ g)(x)$ becomes $f[g(x)]$.

$$\begin{aligned}(f \circ g)(x) &= f[g(x)] \\ &= f(2x^2 + 5x) && g(x) = 2x^2 + 5x \\ &= 4(2x^2 + 5x) + 1 && f(x) = 4x + 1 \\ &= 8x^2 + 20x + 1 && \text{Distributive property}\end{aligned}$$

As Example 6 shows, it is not always true that $f \circ g = g \circ f$. In fact, the composite functions $f \circ g$ and $g \circ f$ are equal only for a special class of functions, discussed in Section 5.1. In Example 6, the domain of both composite functions is $(-\infty, \infty)$.

In calculus it is sometimes necessary to treat a function as the composition of two functions.

**EXAMPLE 7** *Finding Functions That Form a Given Composite Function*

Suppose that $h(x) = (x^2 - 5)^3 - 4(x^2 - 5) + 3$. Find functions $f$ and $g$ so that $(f \circ g)(x) = h(x)$.

**Solution** Note the repeated binomial quantity $x^2 - 5$. If $g(x) = x^2 - 5$ and $f(x) = x^3 - 4x + 3$, then

$$\begin{aligned}(f \circ g)(x) &= f[g(x)] \\ &= f(x^2 - 5) \\ &= (x^2 - 5)^3 - 4(x^2 - 5) + 3 \\ &= h(x),\end{aligned}$$

as required. There are other pairs of functions $f$ and $g$ that also work. For instance, we can use

$$f(x) = (x - 5)^3 - 4(x - 5) + 3 \quad \text{and} \quad g(x) = x^2.$$

## Applications of Operations and Composition

There are many applications in business, economics, physics, and other fields that can be viewed from the perspective of combining functions. For example, in manufacturing, the cost of producing a product usually consists of two parts. One part is a *fixed cost* for designing the product, setting up a factory, training workers, and so on. Usually, the fixed cost is constant for a particular product and does not change as more items are made. The other part of the cost is a *variable cost* per item for labor, materials, packaging, shipping, and so on. The variable cost is often the same per item, so the total amount of variable cost increases as more items are produced.

A *linear cost function* has the form

$$C(x) = mx + b,$$

where $m$ represents the variable cost per item and $b$ represents the fixed cost. The revenue from selling a product depends on the price per item and the number of items sold, as given by the *revenue function*,

$$R(x) = px,$$

where $p$ is the price per item and $R(x)$ is the revenue from the sale of $x$ items. The profit is described by the *profit function*, given by

$$P(x) = R(x) - C(x).$$

**EXAMPLE 8** *Finding and Analyzing Cost, Revenue, and Profit*

Suppose that a businessman invests \$1500 as his fixed cost in a new venture that produces and sells a device that makes programming a VCR easier. Each such device costs \$100 to manufacture.

**(a)** Write a cost function for the product, if $x$ represents the number of devices produced. Assume that the function is linear.

**(b)** Find the revenue function if each device in part (a) sells for \$125.

**(c)** Give the profit function for the item in parts (a) and (b).

**(d)** How many items must be produced and sold before the company makes a profit?

### Analytic Solution

**(a)** Since the cost function is linear, it will have the form $C(x) = mx + b$, with $m = 100$ and $b = 1500$. That is,

$$C(x) = 100x + 1500.$$

**(b)** The revenue function is

$$R(x) = px = 125x. \qquad \text{Let } p = 125.$$

**(c)** Using the results of parts (a) and (b), the profit function is given by

$$\begin{aligned} P(x) &= R(x) - C(x) \\ &= 125x - (100x + 1500) \\ &= 125x - 100x - 1500 \\ &= 25x - 1500. \end{aligned}$$

**(d)** To make a profit, $P(x)$ must be positive. Set $P(x) = 25x - 1500 > 0$ and solve for $x$.

$$\begin{aligned} 25x - 1500 &> 0 \\ 25x &> 1500 \qquad \text{Add 1500.} \\ x &> 60 \qquad \text{Divide by 25.} \end{aligned}$$

At least 61 devices must be sold for the company to make a profit.

### Graphing Calculator Solution

**(d)** Figure 71(a) shows the cost function $C(x)$ entered as $Y_1 = 100X + 1500$ and the revenue function $R(x)$ entered as $Y_2 = 125X$. Then the profit function $P(x)$ can be entered as $Y_3 = Y_2 - Y_1$. The graph of $Y_3$ shows that the $x$-intercept of $Y_3$ is 60, meaning that the company must sell at least 61 devices to earn a profit. The table in Figure 71(b) further confirms this result.

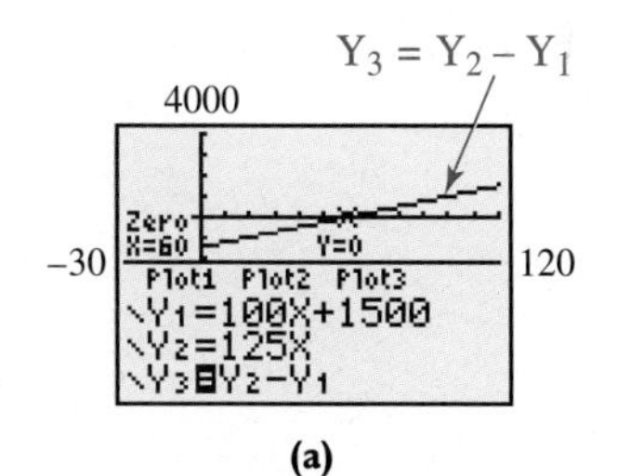


(a)

| X | Y3 |
|---|---|
| 40 | -500 |
| 50 | -250 |
| 59 | -25 |
| 60 | 0 |
| 61 | 25 |
| 70 | 250 |

Y3=Y2−Y1

(b)

**FIGURE 71**

**EXAMPLE 9** *Applying a Difference of Functions to Model the Surface Area of a Ball*

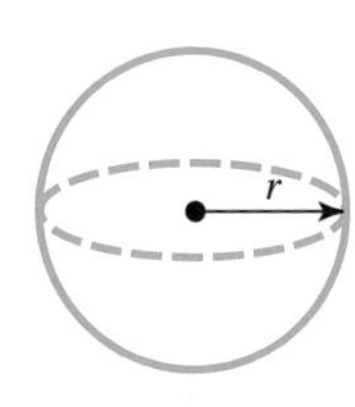


The formula for the surface area $S$ of a sphere is $S = 4\pi r^2$, where $r$ is the radius of the sphere.

**(a)** Construct a model $S(r)$ that describes the amount of surface area gained if the radius $r$ inches of a ball is increased by 2 inches.

**(b)** Determine the amount of extra material needed to manufacture a ball of radius 22 inches as compared to a ball of radius 20 inches.

### Solution

**(a)** In words, we have

surface area gained = larger surface area − smaller surface area.

This translates as a difference of functions:

$$S(r) = 4\pi(r + 2)^2 - 4\pi r^2.$$

**(b)** We let $r = 20$ in the model found in part (a).

$$\begin{aligned} S(r) &= 4\pi(r+2)^2 - 4\pi r^2 \\ S(20) &= 4\pi(20+2)^2 - 4\pi(20)^2 \\ &= 1936\pi - 1600\pi \\ &= 336\pi \\ &\approx 1056 \end{aligned}$$

Thus, it would take about 1056 square inches of extra material. The graph of

$$Y_1 = 4\pi(X+2)^2 - 4\pi X^2$$

is shown in Figure 72. The display at the bottom agrees with the analytic solution.

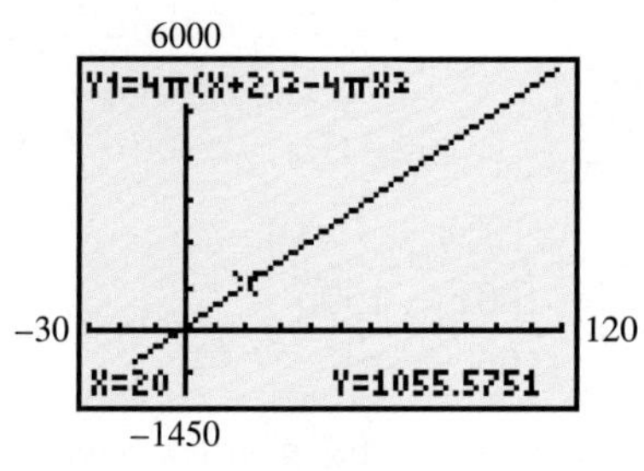


**FIGURE 72**

## 2.6 EXERCISES

*Concept Check* *Let $f(x) = x^2$ and $g(x) = 2x - 5$. Match each function in Group I with the correct expression in Group II.*

| I | | II | |
|---|---|---|---|
| **1.** $(f+g)(x)$ | **2.** $(f-g)(x)$ | **A.** $4x^2 - 20x + 25$ | **B.** $x^2 - 2x + 5$ |
| **3.** $(fg)(x)$ | **4.** $\left(\frac{f}{g}\right)(x)$ | **C.** $2x^2 - 5$ | **D.** $\frac{x^2}{2x - 5}$ |
| **5.** $(f \circ g)(x)$ | **6.** $(g \circ f)(x)$ | **E.** $x^2 + 2x - 5$ | **F.** $2x^3 - 5x^2$ |

*Let $f(x) = 4x^2 - 2x$ and $g(x) = 8x + 1$. Perform the composition or operation indicated.*

**7.** $(f \circ g)(3)$ **8.** $(g \circ f)(-2)$ **9.** $(f \circ g)(x)$ **10.** $(g \circ f)(x)$ **11.** $(f+g)(3)$

**12.** $(f+g)(-5)$ **13.** $(fg)(4)$ **14.** $(fg)(-3)$ **15.** $\left(\frac{f}{g}\right)(-1)$ **16.** $\left(\frac{f}{g}\right)(4)$

**17.** $(f \circ g)(2)$ **18.** $(f \circ g)(-5)$ **19.** $(g \circ f)(2)$ **20.** $(g \circ f)(-5)$

*For each pair of functions,* **(a)** *find $(f+g)(x)$, $(f-g)(x)$ and $(fg)(x)$;* **(b)** *give the domains of the functions in part (a);* **(c)** *find $\frac{f}{g}$ and give its domain;* **(d)** *find $f \circ g$ and give its domain; and* **(e)** *find $g \circ f$ and give its domain.*

**21.** $f(x) = 4x - 1, g(x) = 6x + 3$

**22.** $f(x) = 9 - 2x, g(x) = -5x + 2$

**23.** $f(x) = |x + 3|, g(x) = 2x$

**24.** $f(x) = |2x - 4|, g(x) = x + 1$

**25.** $f(x) = \sqrt[3]{x + 4}, g(x) = x^3 + 5$

**26.** $f(x) = \sqrt[3]{6 - 3x}, g(x) = 2x^3 + 1$

**27.** $f(x) = \sqrt{x^2 + 3}, g(x) = x + 1$

**28.** $f(x) = \sqrt{2 + 4x^2}, g(x) = x$

*Use the graph to evaluate each expression.*

**29.** **(a)** $(f + g)(2)$
**(b)** $(f - g)(1)$
**(c)** $(fg)(0)$
**(d)** $\left(\frac{f}{g}\right)(1)$

**30.** **(a)** $(f + g)(0)$
**(b)** $(f - g)(-1)$
**(c)** $(fg)(1)$
**(d)** $\left(\frac{f}{g}\right)(2)$

**31.** **(a)** $(f + g)(-1)$
**(b)** $(f - g)(-2)$
**(c)** $(fg)(0)$
**(d)** $\left(\frac{f}{g}\right)(2)$

**32.** **(a)** $(f + g)(1)$
**(b)** $(f - g)(0)$
**(c)** $(fg)(-1)$
**(d)** $\left(\frac{f}{g}\right)(1)$

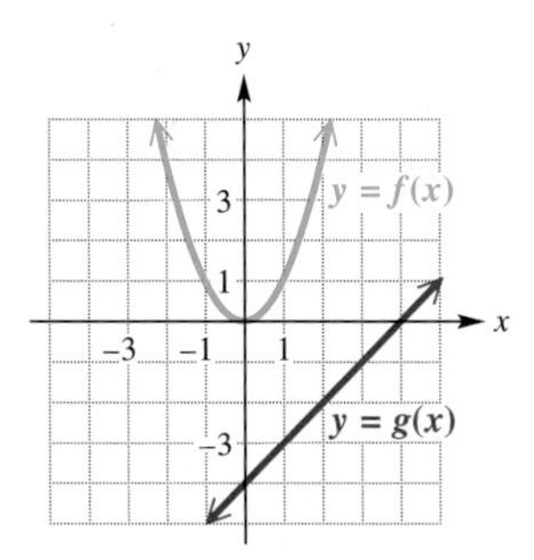


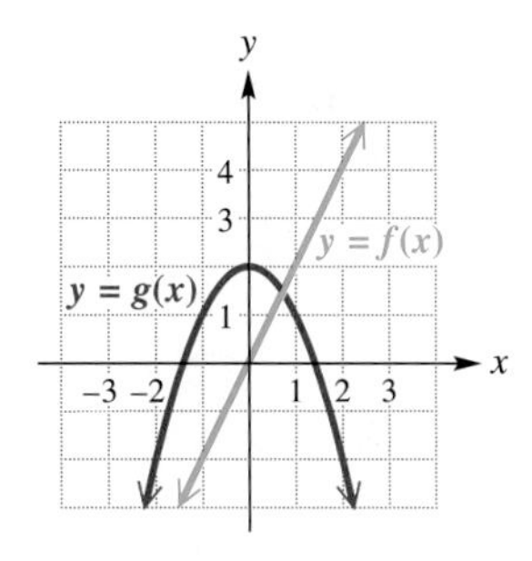


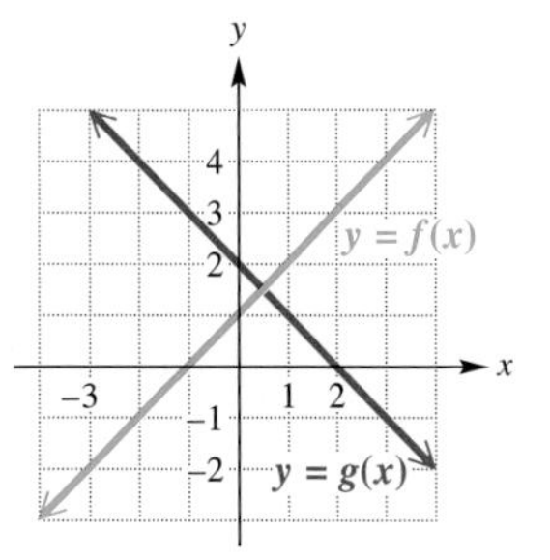


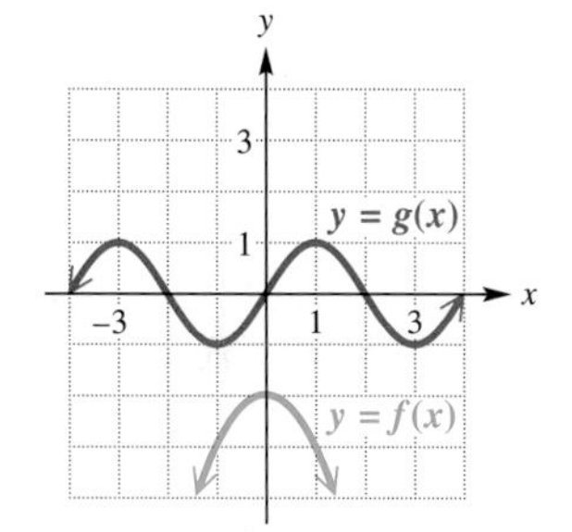


*Use the table to evaluate each expression, if possible.*

**(a)** $(f + g)(2)$ **(b)** $(f - g)(4)$ **(c)** $(fg)(-2)$ **(d)** $\left(\frac{f}{g}\right)(0)$

**33.**

| $x$ | $f(x)$ | $g(x)$ |
|---|---|---|
| −2 | 0 | 6 |
| 0 | 5 | 0 |
| 2 | 7 | −2 |
| 4 | 10 | 5 |

**34.**

| $x$ | $f(x)$ | $g(x)$ |
|---|---|---|
| −2 | −4 | 2 |
| 0 | 8 | −1 |
| 2 | 5 | 4 |
| 4 | 0 | 0 |

**35.** Use the table in Exercise 33 to complete the following table.

| $x$ | $(f + g)(x)$ | $(f - g)(x)$ | $(fg)(x)$ | $\left(\frac{f}{g}\right)(x)$ |
|---|---|---|---|---|
| −2 | | | | |
| 0 | | | | |
| 2 | | | | |
| 4 | | | | |

**36.** Use the table in Exercise 34 to complete the following table.

| $x$ | $(f + g)(x)$ | $(f - g)(x)$ | $(fg)(x)$ | $\left(\frac{f}{g}\right)(x)$ |
|---|---|---|---|---|
| −2 | | | | |
| 0 | | | | |
| 2 | | | | |
| 4 | | | | |

*Use the graph to evaluate each expression.* (Hint: *Extend the ideas of Example 3.*)

**37.** **(a)** $(f \circ g)(4)$
**(b)** $(g \circ f)(3)$
**(c)** $(f \circ f)(2)$

**38.** **(a)** $(f \circ g)(2)$
**(b)** $(g \circ g)(0)$
**(c)** $(g \circ f)(4)$

**39.** **(a)** $(f \circ g)(1)$
**(b)** $(g \circ f)(-2)$
**(c)** $(g \circ g)(-2)$

**40.** **(a)** $(f \circ g)(-2)$
**(b)** $(g \circ f)(1)$
**(c)** $(f \circ f)(0)$

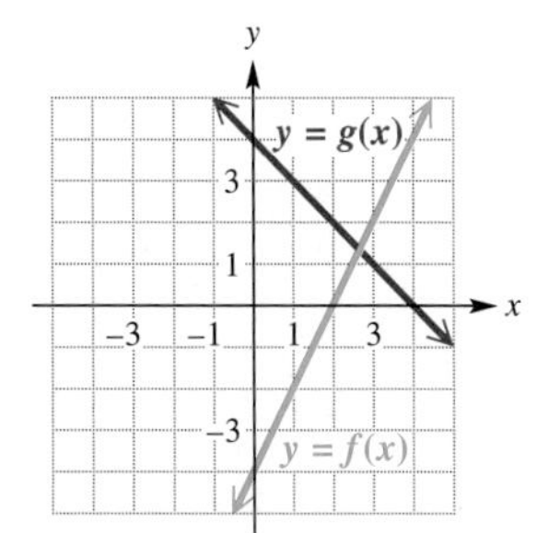


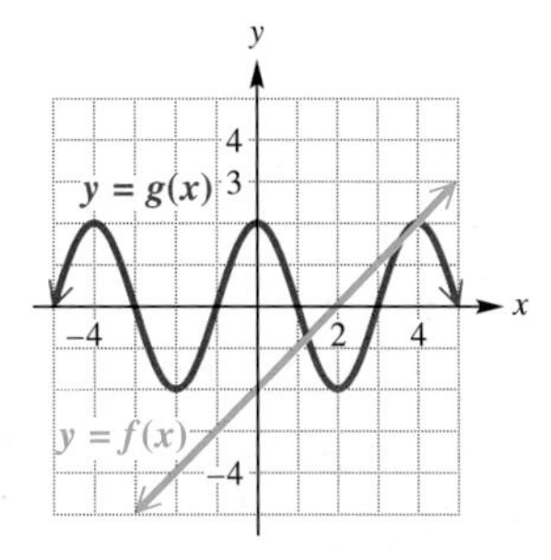


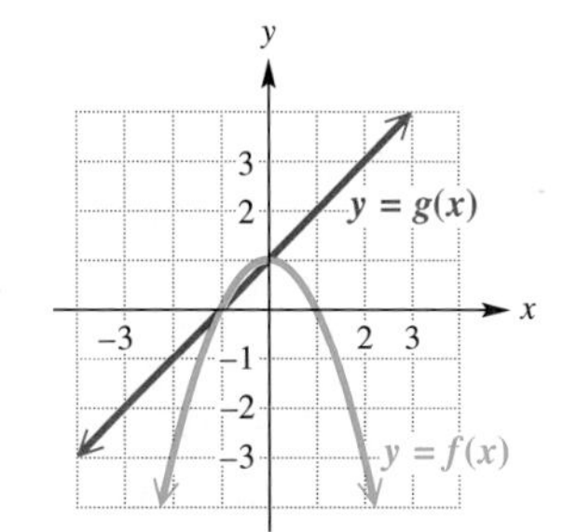


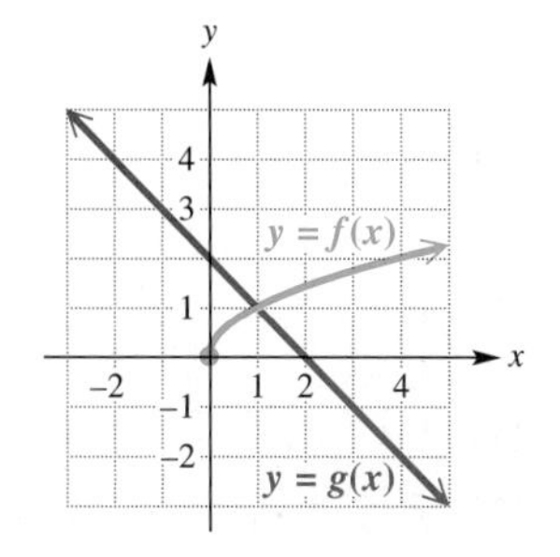

*Tables for the functions f and g are given. Evaluate each expression, if possible.*

**(a)** $(g \circ f)(1)$ **(b)** $(f \circ g)(4)$ **(c)** $(f \circ f)(3)$

**41.**

| $x$ | $f(x)$ | $x$ | $g(x)$ |
|---|---|---|---|
| 1 | 4 | 1 | 2 |
| 2 | 3 | 2 | 3 |
| 3 | 1 | 3 | 4 |
| 4 | 2 | 4 | 5 |

**42.**

| $x$ | $f(x)$ | $x$ | $g(x)$ |
|---|---|---|---|
| 1 | 2 | 2 | 4 |
| 3 | 6 | 3 | 2 |
| 4 | 5 | 5 | 6 |
| 6 | 7 | 7 | 0 |

**43.** Use the tables for $f$ and $g$ in Exercise 41 to complete the composition shown in the diagram.

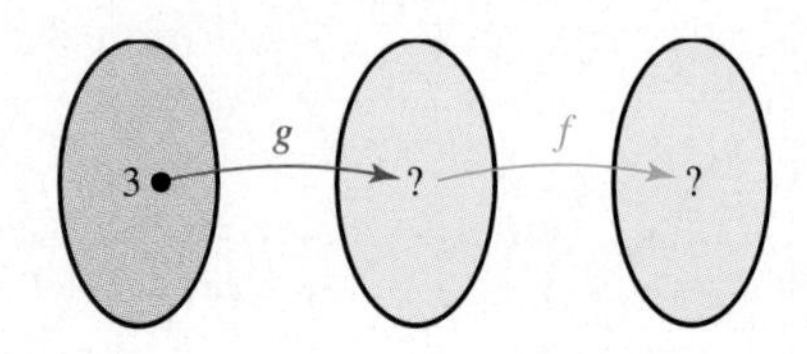


**44.** Use the tables for $f$ and $g$ in Exercise 42 to complete the composition shown in the diagram.

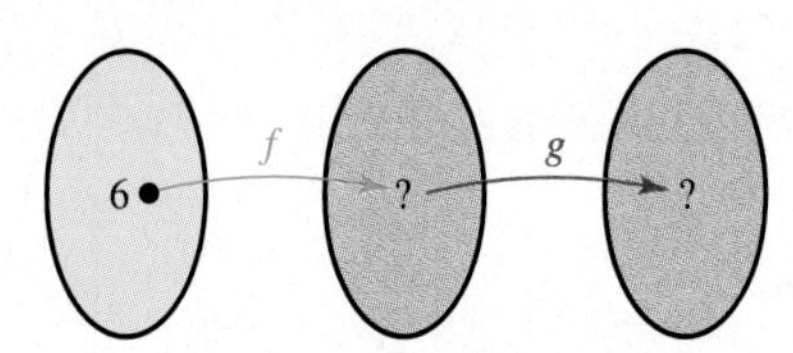


*The graphing calculator screen on the left shows three functions, $Y_1$, $Y_2$, and $Y_3$. The last of these, $Y_3$, is defined as $Y_1 \circ Y_2$, indicated by the notation $Y_3 = Y_1(Y_2)$. The table on the right shows selected values of X, along with the calculated values of $Y_3$. Predict the display for $Y_3$ for the given value of X.*

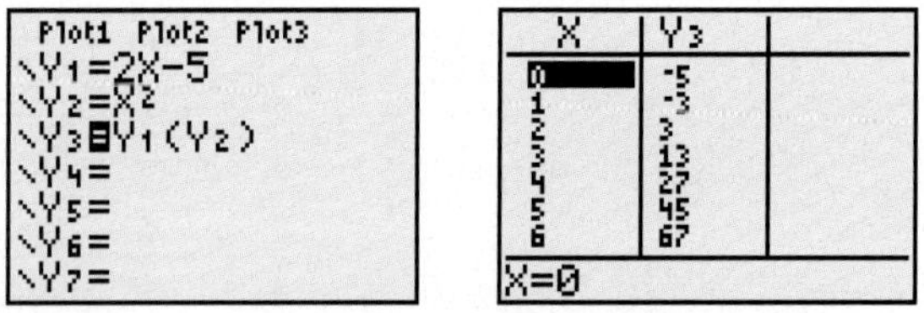

```
Plot1 Plot2 Plot3
\Y1=2X-5
\Y2=X²
\Y3=Y1(Y2)
\Y4=
\Y5=
\Y6=
\Y7=
```

| X | Y3 |
|---|---|
| 0 | -5 |
| 1 | -3 |
| 2 | 3 |
| 3 | 13 |
| 4 | 27 |
| 5 | 45 |
| 6 | 67 |

X=0

**45.** $X = -1$

**46.** $X = -2$

**47.** $X = 7$

**48.** $X = 8$

*For certain pairs of functions f and g, $(f \circ g)(x) = x$ and $(g \circ f)(x) = x$. Show that this is true for the pairs in Exercises 49–52.*

**49.** $f(x) = 4x + 2$, $g(x) = \frac{1}{4}(x - 2)$

**50.** $f(x) = -3x$, $g(x) = -\frac{1}{3}x$

**51.** $f(x) = \sqrt[3]{5x + 4}$, $g(x) = \frac{1}{5}x^3 - \frac{4}{5}$

**52.** $f(x) = \sqrt[3]{x + 1}$, $g(x) = x^3 - 1$

*Functions such as the pairs in Exercises 49–52 are called* inverse functions, *because the result of composition in both directions is the identity function. (Inverse functions will be discussed in detail in Section 5.1.)*

**53.** In a square viewing window, graph $y_1 = \sqrt[3]{x - 6}$ and $y_2 = x^3 + 6$, an example of a pair of inverse functions. Now, graph $y_3 = x$. Describe how the graph of $y_2$ can be obtained from the graph of $y_1$, using the graph $y_3 = x$ as a basis for your description. (*Hint:* Review the terminology of Section 2.3.)

**54.** Repeat Exercise 53 for $y_1 = 5x - 3$ and $y_2 = \frac{1}{5}(x + 3)$.

*Determine the difference quotient*

$$\frac{f(x + h) - f(x)}{h} \quad (h \neq 0)$$

*for each function f. Simplify completely.*

**55.** $f(x) = 4x + 3$

**56.** $f(x) = 5x - 6$

**57.** $f(x) = -6x^2 - x + 4$

**58.** $f(x) = \frac{1}{2}x^2 + 4x$

**59.** $f(x) = x^3$

**60.** $f(x) = -2x^3$

*Consider the function h as defined. Find functions f and g such that $(f \circ g)(x) = h(x)$. (There are many possible ways to do this.)*

**61.** $h(x) = (6x - 2)^2$

**62.** $h(x) = (11x^2 + 12x)^2$

**63.** $h(x) = \sqrt{x^2 - 1}$

**64.** $h(x) = (2x - 3)^3$

**65.** $h(x) = \sqrt{6x} + 12$

**66.** $h(x) = \sqrt[3]{2x + 3} - 4$

*(Modeling) Cost/Revenue/Profit Analysis For each situation, if x represents the number of items produced,* **(a)** *write a cost function,* **(b)** *find a revenue function if each item sells for the price given,* **(c)** *give the profit function,* **(d)** *determine analytically how many items must be produced before a profit is shown (assume whole numbers of items), and* **(e)** *support the result of part (d) graphically.*

**67.** The fixed cost is \$500, the cost to produce an item is \$10, and the selling price of the item is \$35.

**68.** The fixed cost is \$180, the cost to produce an item is \$11, and the selling price of the item is \$20.

**69.** The fixed cost is \$2700, the cost to produce an item is \$100, and the selling price of the item is \$280.

**70.** The fixed cost is \$1000, the cost to produce an item is \$200, and the selling price of the item is \$240.

*(Modeling) Solve each application of operations and composition of functions.*

**71.** *Volume of a Sphere* The formula for the volume of a sphere is $V = \frac{4}{3}\pi r^3$, where $r$ represents the radius of the sphere.

**(a)** Construct a model representing the amount of volume gained when a sphere of radius $r$ inches is increased by 3 inches.

**(b)** Graph the model found in part (a), using $y$ for $V$ and $x$ for $r$, in the window $[0, 10]$ by $[0, 1500]$. What type of function does this appear to be?

**(c)** Use your calculator to graphically find the amount of volume gained when a sphere of 4-inch radius is increased to a 7-inch radius.

**(d)** Verify your conjecture in part (c) analytically.

**72.** *Surface Area of a Sphere* Rework Example 9(a), but consider what happens when the radius is doubled (rather than increased by 2 inches).

**73.** *Dimensions of a Rectangle* Suppose that the length of a rectangle is twice its width. Let $x$ represent the width of the rectangle.

**(a)** Write a formula for the perimeter $P$ of the rectangle in terms of $x$ alone. Then, use $P(x)$ notation to describe it as a function. What type of function is this?

**(b)** Graph the function $P$ found in part (a) in the window $[0, 10]$ by $[0, 100]$. Locate the point for which $x = 4$, and explain what $x$ represents and what $y$ represents.

**(c)** Locate on the graph of $P$ the point with $x$-value 4. Then, sketch a rectangle satisfying the conditions described earlier, and evaluate its perimeter if its width is this $x$-value. Use the standard perimeter formula. How does the result compare with the $y$-value shown on your screen?

**(d)** Locate on the graph of $P$ a point with an integer $y$-value. Describe in words the meaning of the $x$- and $y$-coordinates here.

**74.** *Perimeter of a Square* The perimeter $x$ of a square with side length $s$ is given by the formula

$$x = 4s.$$

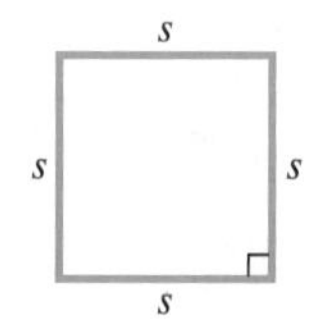


**(a)** Solve for $s$ in terms of $x$.

**(b)** If $y$ represents the area of this square, write $y$ as a function of the perimeter $x$.

**(c)** Use the composite function of part (b) to analytically find the area of a square with perimeter 6.

**(d)** Support the result of part (c) graphically, and explain the result.

**75.** *Area of an Equilateral Triangle* The area of an equilateral triangle with sides of length $x$ is given by the function defined by

$$A(x) = \frac{\sqrt{3}}{4}x^2.$$

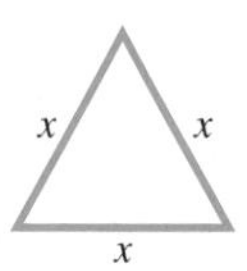


**(a)** Find $A(2x)$, the function representing the area of an equilateral triangle with sides of length twice the original length.

**(b)** Find analytically the area of an equilateral triangle with side length 16. Use the given formula for $A(x)$.

**(c)** Support the result of part (b) graphically.

**76.** *Emission of Pollutants* When a thermal inversion layer is over a city, pollutants cannot rise vertically but are trapped below the layer and must disperse horizontally. Assume that a factory smokestack begins emitting a pollutant at 8 A.M. and that the pollutant disperses horizontally over a circular area. If $t$ represents the time, in hours, since the factory began emitting pollutants ($t = 0$ represents 8 A.M.), assume that the radius of the circle of pollution is $r(t) = 2t$ miles. Let $A(r) = \pi r^2$ represent the area of a circle of radius $r$.

**(a)** Find $(A \circ r)(t)$.

**(b)** Interpret $(A \circ r)(t)$.

**(c)** What is the area of the circular region covered by the layer at noon?

**(d)** Support your result graphically.

**77.** *Hollywood Movies* From 1991 to 1996 the cost to produce and market Hollywood movies increased. In the table, $f$ computes the average cost to produce a movie, and $g$ computes the average cost to market a movie, both in millions of dollars, in the year $x$.

| $x$ | 1991 | 1992 | 1993 |
|---|---|---|---|
| $f(x)$ | 26 | 28 | 30 |
| $g(x)$ | 12 | 14 | 14 |

| $x$ | 1994 | 1995 | 1996 |
|---|---|---|---|
| $f(x)$ | 33 | 37 | 41 |
| $g(x)$ | 17 | 17 | 20 |

*Source:* Motion Picture Association of America.

**(a)** Make a table for a function $h$ that computes the average cost to produce *and* market a movie in the year $x$.

**(b)** Write an equation that relates $f(x)$, $g(x)$, and $h(x)$.

**78.** *Petroleum Spillage* Large amounts of petroleum products enter the oceans each year. In the table, $x$ represents the year and $f$ computes the total amount of petroleum that enters the world's oceans. The function $g$ computes petroleum spillage into the oceans caused by oil tankers. Amounts are given in thousands of tons.

| $x$ | 1973 | 1979 | 1981 | 1983 | 1989 |
|---|---|---|---|---|---|
| $f(x)$ | 6110 | 4670 | 3570 | 3200 | 570 |
| $g(x)$ | 1380 | 900 | 1050 | 1100 | — |

*Source:* Freedman, B., *Environmental Ecology*, Second Edition, Academic Press, 1995.

**(a)** Define a function $h$ by $h(x) = f(x) - g(x)$. Make a table for $h(x)$. What is the domain of $h$?

**(b)** Interpret what $h$ computes.

**79.** *Acid Rain* A common air pollutant responsible for acid rain is sulfur dioxide ($SO_2$). Emissions of $SO_2$ from burning coal during year $x$ are computed by $f(x)$ in the table. Emissions of $SO_2$ from burning oil are computed by $g(x)$. Amounts are given in millions of tons.

| $x$ | 1860 | 1900 | 1940 | 1970 | 2000 |
|---|---|---|---|---|---|
| $f(x)$ | 2.4 | 12.6 | 24.2 | 32.4 | 55.0 |
| $g(x)$ | 0 | .2 | 2.3 | 17.6 | 23.0 |

*Source:* Freedman, B., *Environmental Ecology*, Second Edition, Academic Press, 1995.

**(a)** Evaluate $(f + g)(1970)$.

**(b)** Interpret $(f + g)(x)$.

**(c)** Make a table for $(f + g)(x)$.

**80.** *Methane Emissions* Methane is a greenhouse gas. It lets sunlight into the atmosphere but blocks heat from escaping the earth's atmosphere. Methane is a by-product of burning fossil fuels. In the table, $f$ models the predicted methane emissions in millions of tons produced by developed countries during year $x$. The function $g$ models the same emissions for developing countries.

| $x$ | 1990 | 2000 | 2010 | 2020 | 2030 |
|---|---|---|---|---|---|
| $f(x)$ | 27 | 28 | 29 | 30 | 31 |
| $g(x)$ | 5 | 7.5 | 10 | 12.5 | 15 |

*Source:* Nilsson, A., *Greenhouse Earth*, John Wiley and Sons, 1992.

**(a)** Make a table for a function $h$ that models the total predicted methane emissions for developed *and* developing countries.

**(b)** Write an equation that relates $f(x)$, $g(x)$, and $h(x)$.

**81.** *China's Energy Production* Predicted energy production in China is shown in the figure on the next page. The function $f$ computes total coal production, and the function $g$ computes total coal *and* oil production. Energy units are in million metric tons of oil equivalent (Mtoe). Let the function $h$ compute China's oil production. (*Source:* Pascal, A., *Global Energy: The Changing Outlook*, Organization for Economic Cooperation and Development/International Energy Agency, 1992.)

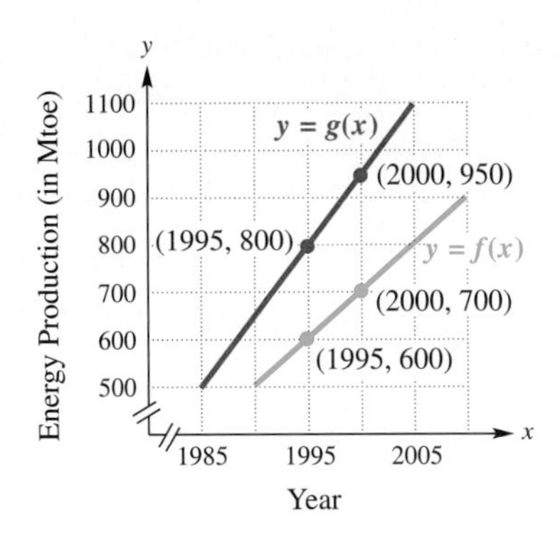


**(a)** Write an equation that relates $f(x)$, $g(x)$, and $h(x)$.

**(b)** Evaluate $h(1995)$ and $h(2000)$.

**(c)** Determine a formula for $h$. (*Hint:* $h$ is linear.)

**82.** *AIDS* During the early years of the AIDS epidemic, cases and cumulative deaths reported for selected years $x$ were modeled quite well by quadratic functions. For 1982–1994, the numbers of AIDS cases are modeled by

$$f(x) = 3200(x - 1982)^2 + 1586$$

and the numbers of deaths are modeled by

$$g(x) = 1900(x - 1982)^2 + 619.$$

| Year | Cases | Deaths |
|---|---|---|
| 1982 | 1,586 | 619 |
| 1984 | 10,927 | 5,605 |
| 1986 | 41,910 | 24,593 |
| 1988 | 106,304 | 61,911 |
| 1990 | 196,576 | 120,811 |
| 1992 | 329,205 | 196,283 |
| 1994 | 441,528 | 270,533 |

*Source:* U.S. Department of Health and Human Services.

**(a)** Graph $h(x) = \frac{g(x)}{f(x)}$ in the window $[1982, 1994]$ by $[0, 1]$. Interpret the graph.

**(b)** Compute the ratio $\frac{\text{deaths}}{\text{cases}}$ for each year. Compare these ratios with the results from part (a).

## Reviewing Basic Concepts (Sections 2.4–2.6)

**1.** Solve the equation in part (a) and the related inequalities in parts (b) and (c).

**(a)** $\left|\frac{1}{2}x + 2\right| = 4$ **(b)** $\left|\frac{1}{2}x + 2\right| > 4$

**(c)** $\left|\frac{1}{2}x + 2\right| \leq 4$

**2.** Given the graph of $y = f(x)$, sketch the graph of $y = |f(x)|$.

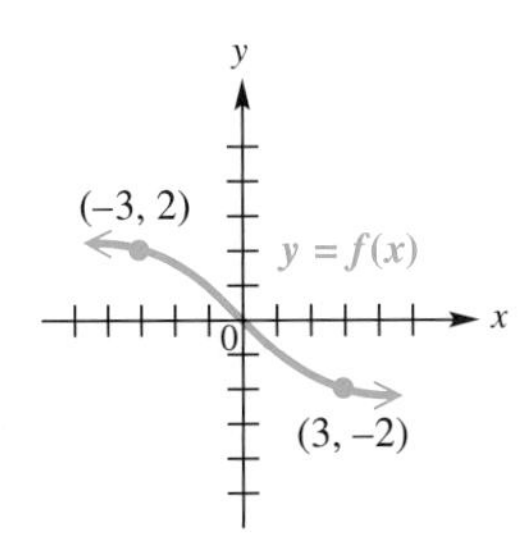


**3.** *(Modeling) Carbon Dioxide Emissions* When humans breathe, carbon dioxide is emitted. In one study, the emission rates of carbon dioxide by college students were measured during both lectures and exams. The average individual rate $R_L$ (in grams per hour) during a lecture class satisfied the inequality $|R_L - 26.75| \leq 1.42$, whereas during an exam the rate $R_E$ satisfied the inequality $|R_E - 38.75| \leq 2.17$. (*Source:* Wang, T. C., *ASHRAE* Trans., 81 (Part 1), 32 (1975).)

**(a)** Find the range of values for $R_L$ and $R_E$.

**(b)** The class had 225 students. If $T_L$ and $T_E$ represent the total amounts of carbon dioxide in grams emitted during a one-hour lecture and exam, respectively, write inequalities that describe the ranges for $T_L$ and $T_E$.

**4.** For $f(x) = \begin{cases} 2x + 3 & \text{if } -3 \leq x < 0 \\ x^2 + 4 & \text{if } x \geq 0 \end{cases}$, find each value.

**(a)** $f(-3)$ **(b)** $f(0)$ **(c)** $f(2)$

**5.** Consider the piecewise-defined function given by

$$f(x) = \begin{cases} -x^2 & \text{if } x \leq 0 \\ x - 4 & \text{if } x > 0 \end{cases}.$$

**(a)** Sketch its graph.

**(b)** Use a graphing calculator to obtain an accurate graph in the window $[-10, 10]$ by $[-10, 10]$.

**6.** Given $f(x) = -3x - 4$ and $g(x) = x^2$. Perform the composition or operation indicated.

**(a)** $(f + g)(1)$ **(b)** $(f - g)(3)$ **(c)** $(fg)(-2)$

**(d)** $\left(\frac{f}{g}\right)(-3)$ **(e)** $(f \circ g)(x)$ **(f)** $(g \circ f)(x)$

**7.** Find functions $f$ and $g$ so that $h(x) = (f \circ g)(x)$, if $h(x) = (x + 2)^4$.

**8.** Find the difference quotient

$$\frac{f(x + h) - f(x)}{h}, \quad h \neq 0,$$

for $f(x) = -2x^2 + 3x - 5$.

**9.** *(Modeling) Textbook Author Royalties* A textbook author invests his royalties in two accounts for 1 year.

**(a)** The first account pays 4% simple interest. If he invests $x$ dollars in this account, write an expression for $y_1$ in terms of $x$, where $y_1$ represents the amount of interest earned.

**(b)** He invests in a second account \$500 more than he invested in the first account. This second account pays 2.5% simple interest. Write an expression for $y_2$, where $y_2$ represents the amount of interest earned.

**(c)** What does $y_1 + y_2$ represent?

**(d)** Graph $y_1 + y_2$ in the window $[0, 1000]$ by $[0, 100]$. Use the graph to find the amount of interest he will receive if he invests \$250 in the first account.

**(e)** Support the result of part (d) analytically.

**10.** *Geometry* The surface area of a cone (excluding the bottom) is given by $S = \pi r\sqrt{r^2 + h^2}$, where $r$ is its radius and $h$ is its height, as shown in the figure. If the height is twice the radius, write a formula for $S$ in terms of $r$.

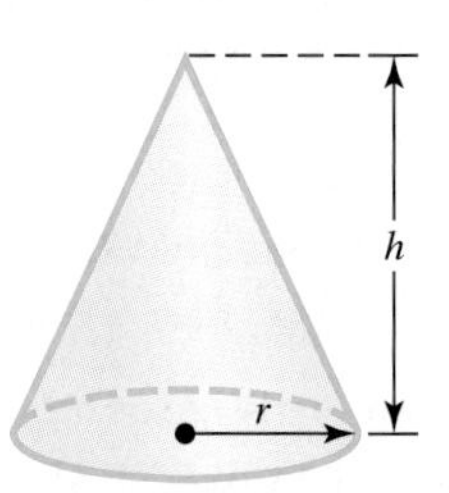


## CHAPTER 2 SUMMARY

**KEY TERMS & SYMBOLS**

### 2.1 Graphs of Basic Functions and Relations; Symmetry

continuity
increasing function
decreasing function
constant function
identity function
degree
squaring function
parabola
vertex
symmetry
cubing function
inflection point
square root function
cube root function
absolute value function
even function
odd function

**KEY CONCEPTS**

**CONTINUITY**

A function is continuous over an interval of its domain if its hand-drawn graph over that interval can be sketched without lifting the pencil from the paper.

**INCREASING, DECREASING, AND CONSTANT FUNCTIONS**

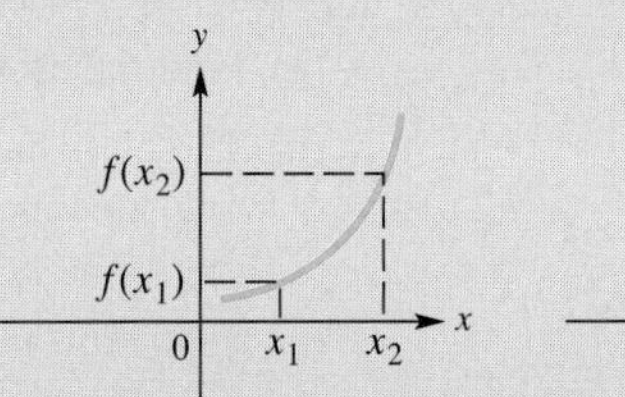


When $x_1 < x_2$, $f(x_1) < f(x_2)$.
$f$ is increasing.

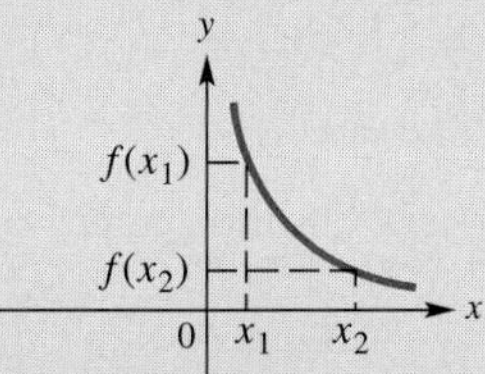


When $x_1 < x_2$, $f(x_1) > f(x_2)$.
$f$ is **decreasing**.

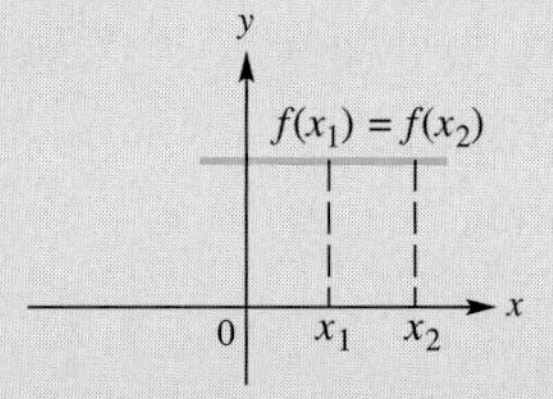


For $x_1$ and $x_2$, $f(x_1) = f(x_2)$.
$f$ is constant.

**TYPES OF SYMMETRY**

| $y$-Axis Symmetry $f(-x) = f(x)$ | Origin Symmetry $f(-x) = -f(x)$ | $x$-Axis Symmetry (not possible for a nonzero function) |
|---|---|---|
| $(-a, b)$, $(a, b)$ | $(a, b)$, $(-a, -b)$ | $(a, b)$, $(a, -b)$ |

| KEY TERMS & SYMBOLS | KEY CONCEPTS |
|---|---|
| | **EVEN AND ODD FUNCTIONS**<br>A function $f$ is called an even function if $f(-x) = f(x)$ for all $x$ in the domain of $f$. (Its graph is symmetric with respect to the $y$-axis.)<br>A function $f$ is called an odd function if $f(-x) = -f(x)$ for all $x$ in the domain of $f$. (Its graph is symmetric with respect to the origin.)<br><br>**BASIC FUNCTIONS AND RELATIONS**<br>• The identity function defined by $f(x) = x$ is increasing and continuous on its entire domain, $(-\infty, \infty)$.<br>• The squaring function defined by $f(x) = x^2$ decreases on the interval $(-\infty, 0]$, increases on the interval $[0, \infty)$, and is continuous on its entire domain, $(-\infty, \infty)$.<br>• The cubing function defined by $f(x) = x^3$ increases and is continuous on its entire domain, $(-\infty, \infty)$.<br>• The square root function defined by $f(x) = \sqrt{x}$ increases and is continuous on its entire domain, $[0, \infty)$.<br>• The cube root function defined by $f(x) = \sqrt[3]{x}$ increases and is continuous on its entire domain, $(-\infty, \infty)$.<br>• The absolute value function defined by $f(x) = \|x\|$ decreases on the interval $(-\infty, 0]$, increases on the interval $[0, \infty)$, and is continuous on its entire domain, $(-\infty, \infty)$. |
| **2.2** Vertical and Horizontal Shifts of Graphs<br>translation | **VERTICAL SHIFTING OF THE GRAPH OF A FUNCTION**<br>If $c > 0$, the graph of $y = f(x) + c$ is obtained by shifting the graph of $y = f(x)$ *upward* a distance of $c$ units. The graph of $y = f(x) - c$ is obtained by shifting the graph of $y = f(x)$ *downward* a distance of $c$ units.<br><br>**HORIZONTAL SHIFTING OF THE GRAPH OF A FUNCTION**<br>If $c > 0$, the graph of $y = f(x - c)$ is obtained by shifting the graph of $y = f(x)$ to the *right* a distance of $c$ units. The graph of $y = f(x + c)$ is obtained by shifting the graph of $y = f(x)$ to the *left* a distance of $c$ units. |
| **2.3** Stretching, Shrinking, and Reflecting Graphs | **VERTICAL STRETCHING OF THE GRAPH OF A FUNCTION**<br>If $c > 1$, the graph of $y = c \cdot f(x)$ is obtained by vertically *stretching* the graph of $y = f(x)$ by a factor of $c$. In general, the larger the value of $c$, the greater the stretch.<br><br>**VERTICAL SHRINKING OF THE GRAPH OF A FUNCTION**<br>If $0 < c < 1$, the graph of $y = c \cdot f(x)$ is obtained by vertically *shrinking* the graph of $y = f(x)$ by a factor of $c$. In general, the smaller the value of $c$, the greater the shrink.<br><br>**REFLECTING THE GRAPH OF A FUNCTION ACROSS AN AXIS**<br>For a function defined by $y = f(x)$,<br>**(a)** the graph of $y = -f(x)$ is a reflection of the graph of $f$ across the $x$-axis.<br>**(b)** the graph of $y = f(-x)$ is a reflection of the graph of $f$ across the $y$-axis. |
| **2.4** Absolute Value Functions: Graphs, Equations, Inequalities, and Applications | **PROPERTIES OF ABSOLUTE VALUE**<br>For all real numbers $a$ and $b$,<br>**1.** $\|ab\| = \|a\| \cdot \|b\|$. **2.** $\left\|\frac{a}{b}\right\| = \frac{\|a\|}{\|b\|}$ $(b \neq 0)$. **3.** $\|a\| = \|-a\|$. **4.** $\|a\| + \|b\| \geq \|a + b\|$ (triangle inequality). |

| KEY TERMS & SYMBOLS | KEY CONCEPTS |
|---|---|
| | **GRAPH OF $y = \lvert f(x)\rvert$**<br>The graph of $y = \lvert f(x)\rvert$ is obtained from the graph of $y = f(x)$ by reflecting the portion of the graph below the $x$-axis across the $x$-axis, and leaving the graph unchanged for the portion on or above the $x$-axis.<br><br>**SOLVING ABSOLUTE VALUE EQUATIONS AND INEQUALITIES**<br>To solve $\lvert ax + b\rvert = k$, $k > 0$, solve the compound equation<br>$ax + b = k$ **or** $ax + b = -k$.<br>To solve $\lvert ax + b\rvert > k$, $k > 0$, solve the compound inequality<br>$ax + b > k$ **or** $ax + b < -k$.<br>To solve $\lvert ax + b\rvert < k$, $k > 0$, solve the three-part inequality<br>$-k < ax + b < k$. |
| **2.5** Piecewise-Defined Functions<br><br>piecewise-defined function<br>greatest integer function<br>step function | **PIECEWISE-DEFINED FUNCTION**<br>A piecewise-defined function is defined by different rules over different subsets of its domain.<br><br>**GREATEST INTEGER FUNCTION**<br>$f(x) = [\![x]\!] = \begin{cases} x & \text{if } x \text{ is an integer} \\ \text{the greatest integer less than } x & \text{if } x \text{ is not an integer} \end{cases}$ |
| **2.6** Operations and Composition<br><br>difference quotient<br>composite function, $g \circ f$ | **OPERATIONS ON FUNCTIONS**<br>Given two functions $f$ and $g$, then for all values for which both $f(x)$ and $g(x)$ are defined, the functions $f + g$, $f - g$, $fg$, and $\frac{f}{g}$ are defined as follows.<br>**Sum** $(f + g)(x) = f(x) + g(x)$<br>**Difference** $(f - g)(x) = f(x) - g(x)$<br>**Product** $(fg)(x) = f(x) \cdot g(x)$<br>**Quotient** $\left(\frac{f}{g}\right)(x) = \frac{f(x)}{g(x)}, \quad g(x) \neq 0$<br><br>The domains of $f + g$, $f - g$, and $fg$ include all real numbers in the intersection of the domains of $f$ and $g$, while the domain of $\frac{f}{g}$ includes those real numbers in the intersection of the domains of $f$ and $g$ for which $g(x) \neq 0$.<br><br>**DIFFERENCE QUOTIENT**<br>$\frac{f(x + h) - f(x)}{h}, \quad h \neq 0$<br><br>**COMPOSITION OF FUNCTIONS**<br>If $f$ and $g$ are functions, then the composite function, or composition, of $g$ and $f$ is<br>$(g \circ f)(x) = g[f(x)]$<br>for all $x$ in the domain of $f$ such that $f(x)$ is in the domain of $g$. |

# CHAPTER 2 REVIEW EXERCISES

*Concept Check Draw sketches of the graphs of the basic functions introduced in Section 2.1, avoiding the temptation to look back at them in the text. Also, do not use your calculator. They are:*

$$f(x) = x, \quad f(x) = x^2, \quad f(x) = x^3,$$
$$f(x) = \sqrt{x}, \quad f(x) = \sqrt[3]{x}, \quad f(x) = |x|.$$

*Use your sketches to determine whether each statement is* true *or* false. *If* false, *tell why.*

1. The range of $f(x) = x^2$ is the same as the range of $f(x) = |x|$.
2. $f(x) = x^2$ and $f(x) = |x|$ increase on the same interval.
3. $f(x) = \sqrt{x}$ and $f(x) = \sqrt[3]{x}$ have the same domain.
4. $f(x) = \sqrt[3]{x}$ decreases on its entire domain.
5. $f(x) = x$ has its domain equal to its range.
6. $f(x) = \sqrt{x}$ is continuous on the interval $(-\infty, 0)$.
7. None of the functions shown decreases on the interval $[0, \infty)$.
8. Both $f(x) = x$ and $f(x) = x^3$ have graphs that are symmetric with respect to the origin.
9. Both $f(x) = x^2$ and $f(x) = |x|$ have graphs that are symmetric with respect to the $y$-axis.
10. None of the graphs shown is symmetric with respect to the $x$-axis.

*In Exercises 11–18, give the interval that describes the following.*

**11.** Domain of $f(x) = \sqrt{x}$ **12.** Range of $f(x) = |x|$

**13.** Range of $f(x) = \sqrt[3]{x}$ **14.** Domain of $f(x) = x^2$

**15.** The largest interval over which $f(x) = \sqrt[3]{x}$ is increasing

**16.** The largest interval over which $f(x) = |x|$ is increasing

**17.** Domain of $x = y^2$ **18.** Range of $x = y^2$

**19.** Consider the function whose graph is shown here.

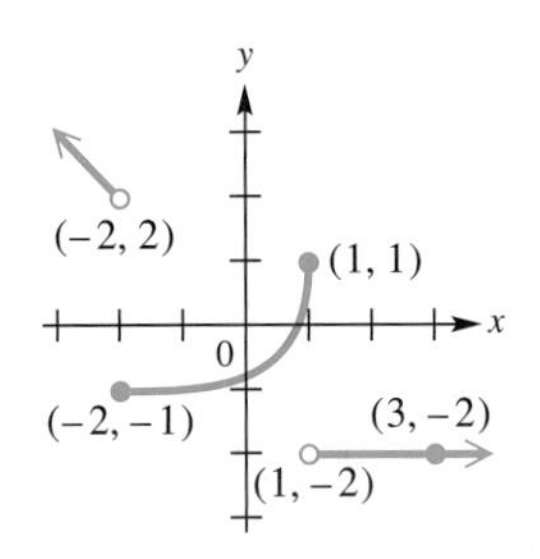


Give the interval(s) over which the function

**(a)** is continuous. **(b)** increases.

**(c)** decreases. **(d)** is constant.

**(e)** What is the domain of the function?

**(f)** What is the range of the function?

**20.** The screen shows the graph of $x = y^2 - 4$. Give the two functions that must be used to graph this relation if the calculator is in function mode.

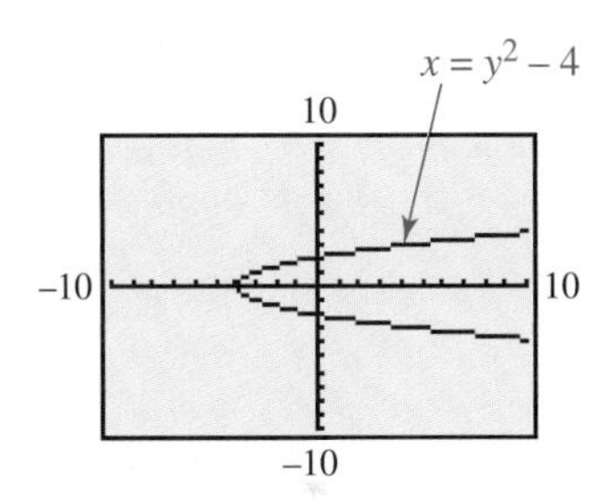


*In Exercises 21–26, determine whether the given relation has* $x$-axis symmetry, $y$-axis symmetry, origin symmetry, *or* none of these *symmetries. (More than one choice is possible.) Also, if the relation is a function, determine whether it is an* even function, *an* odd function, *or* neither.

**21.**

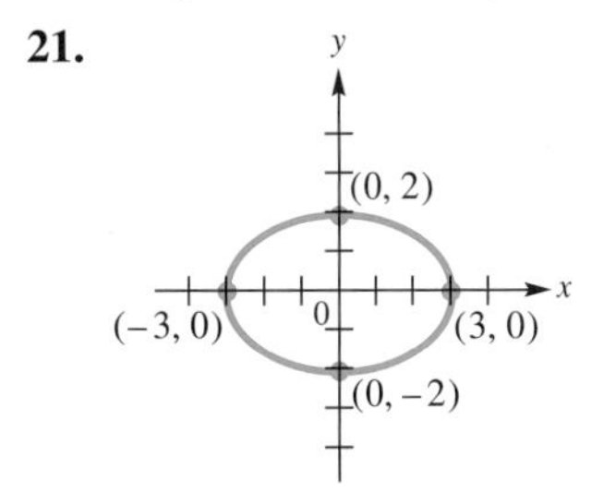


**22.** $F(x) = x^3 - 6$ **23.** $y = |x| + 4$

**24.** $f(x) = \sqrt{x - 5}$ **25.** $y^2 = x - 5$

**26.** $f(x) = 3x^4 + 2x^2 + 1$

*Concept Check Decide whether each statement is* true *or* false. *If* false, *tell why.*

**27.** The graph of a function (except for the constant function defined by $f(x) = 0$) cannot be symmetric with respect to the $x$-axis.

**28.** The graph of an even function is symmetric with respect to the $y$-axis.

**29.** The graph of an odd function is symmetric with respect to the origin.

**30.** If $(a, b)$ is on the graph of an even function, so is $(a, -b)$.

**31.** If $(a, b)$ is on the graph of an odd function, so is $(-a, b)$.

**32.** The constant function defined by $f(x) = 0$ is both even and odd.

**33.** Use the terminology of Sections 2.2 and 2.3 to describe how the graph of $y = -3(x + 4)^2 - 8$ can be obtained from the graph $y = x^2$.

**34.** Give the equation of the function whose graph is obtained by reflecting the graph of $y = \sqrt{x}$ across the $y$-axis, then reflecting across the $x$-axis, shrinking vertically by a factor of $\frac{2}{3}$, and, finally, translating 4 units upward.

*The graph of a function f is shown in the figure. Sketch the graph of each function defined as follows.*

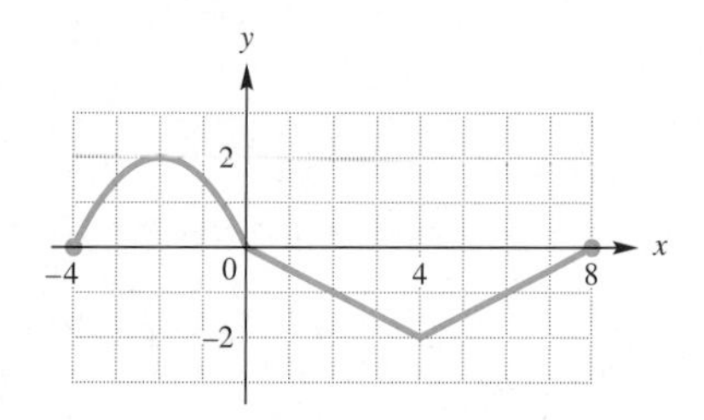


**35.** $y = f(x) + 3$

**36.** $y = f(x - 2)$

**37.** $y = f(x + 3) - 2$

**38.** $y = |f(x)|$

*The graph of a function defined by* $y = f(x)$ *is given. Sketch the graph of* $y = |f(x)|$.

**39.**

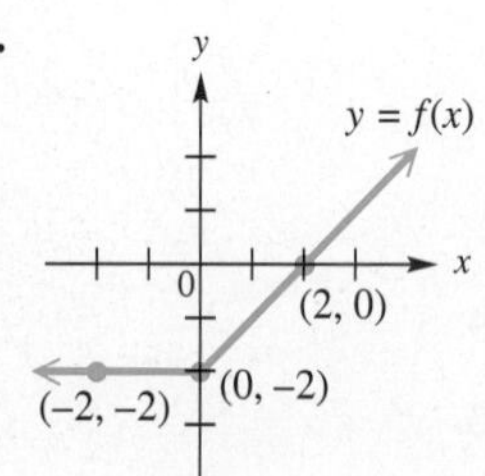


**40.**

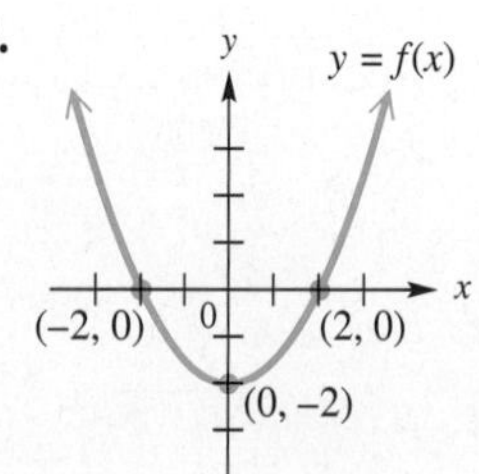


**41.**

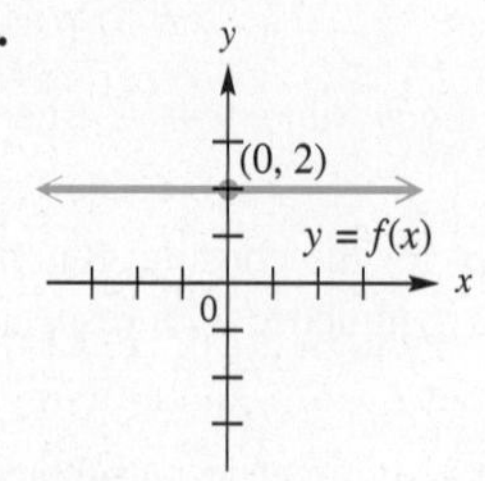


**42.**

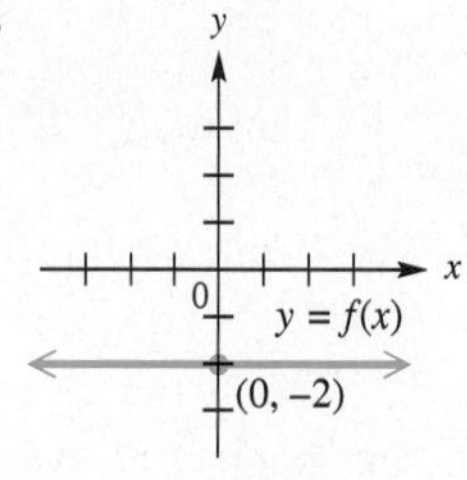


*Solve each equation or inequality analytically.*

**43.** $|4x + 3| = 12$

**44.** $|-2x - 6| + 4 = 1$

**45.** $|5x + 3| = |x + 11|$

**46.** $|2x + 5| = 7$

**47.** $|2x + 5| \leq 7$

**48.** $|2x + 5| \geq 7$

**49.** The graphs of $y_1 = |2x + 5|$ and $y_2 = 7$ are shown, along with the two points of intersection of the graphs. Explain how these screens support the answers in Exercises 46–48.

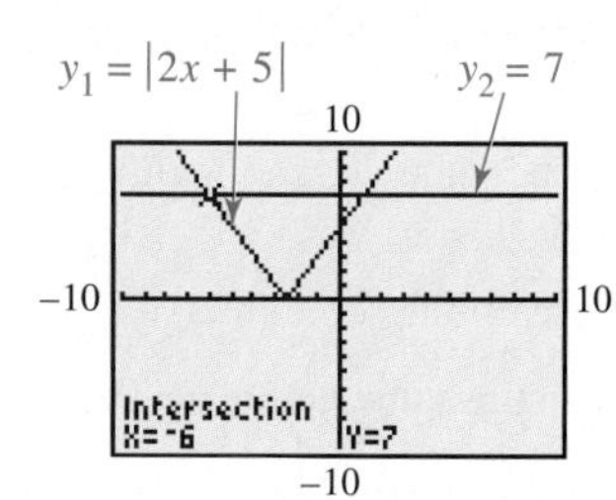


First View

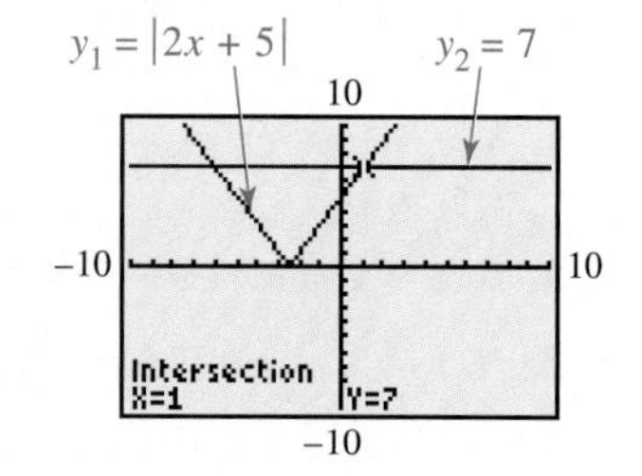


Second View

**50.** Solve the equation $|x + 1| + |x - 3| = 8$ graphically. Then, give an analytic check by substituting the values in the solution set directly into the left-hand side of the equation.

**51.** *(Modeling) Distance from Home* The graph depicts the distance $y$ that a person driving a car on a straight road is from home after $x$ hours. Interpret the graph. What speeds did the car travel?

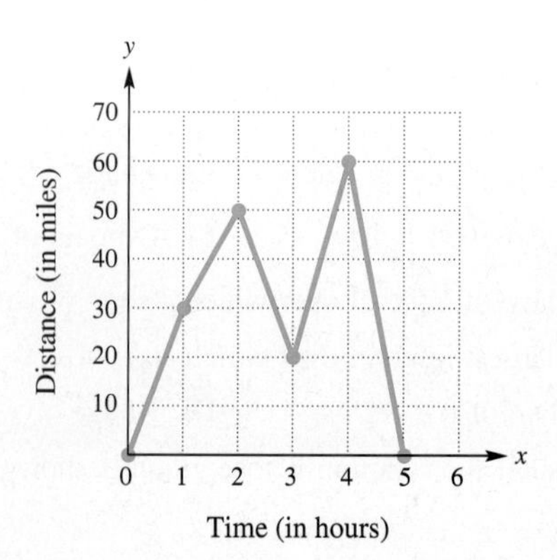


**52.** *(Modeling) Water in a Tank* A 500-gallon water tank is initially full, then emptied at a constant rate of 50 gallons per minute. Then the tank is filled by a pump that outputs 25 gallons of water per minute. Sketch a graph that depicts the amount of water in the tank after $x$ minutes.

*Sketch the graph of each function by hand.*

**53.** $f(x) = \begin{cases} 3x + 1 & \text{if } x < 2 \\ -x + 4 & \text{if } x \geq 2 \end{cases}$

**54.** $f(x) = \begin{cases} |x| & \text{if } x < 3 \\ 6 - x & \text{if } x \geq 3 \end{cases}$

**55.** Graph the function in Exercise 53, using a graphing calculator with the window $[-10, 10]$ by $[-10, 10]$.

**56.** Use a graphing calculator to graph $f(x) = [\![x - 3]\!]$ in the window $[-5, 5]$ by $[-5, 5]$.

*The graphs of functions f and g are shown. Use these graphs to find each value.*

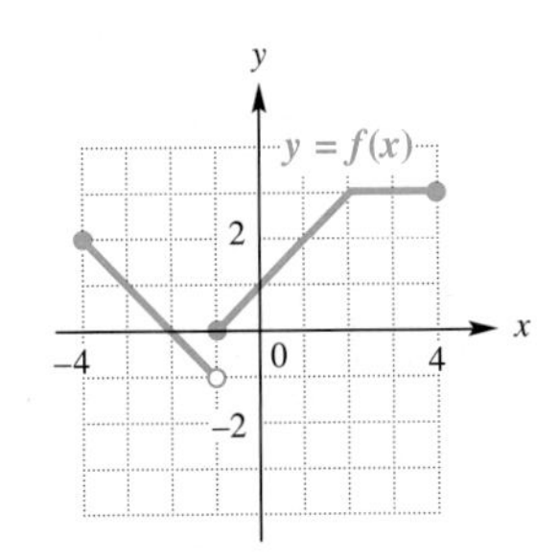


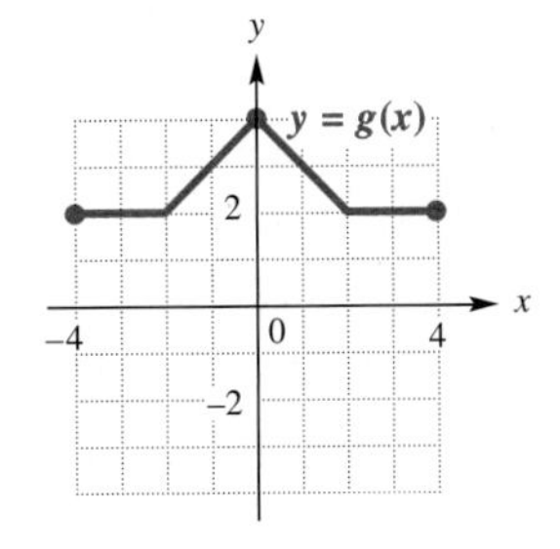


**57.** $(f + g)(1)$

**58.** $(f - g)(0)$

**59.** $(fg)(-1)$

**60.** $\left(\dfrac{f}{g}\right)(2)$

**61.** $(f \circ g)(2)$

**62.** $(g \circ f)(2)$

**63.** $(g \circ f)(-4)$

**64.** $(f \circ g)(-2)$

*Use the table to evaluate each expression, if possible.*

| $x$ | $f(x)$ | $g(x)$ |
|---|---|---|
| $-1$ | 3 | $-2$ |
| 0 | 5 | 0 |
| 1 | 7 | 1 |
| 3 | 9 | 9 |

**65.** $(f + g)(1)$

**66.** $(f - g)(3)$

**67.** $(fg)(-1)$

**68.** $\left(\dfrac{f}{g}\right)(0)$

*Tables for f and g are given. Use them to evaluate each expression in Exercises 69 and 70.*

| $x$ | $f(x)$ |
|---|---|
| $-2$ | 1 |
| 0 | 4 |
| 2 | 3 |
| 4 | 2 |

| $x$ | $g(x)$ |
|---|---|
| 1 | 2 |
| 2 | 4 |
| 3 | $-2$ |
| 4 | 0 |

**69.** $(g \circ f)(-2)$

**70.** $(f \circ g)(3)$

*For the given function, find and simplify* $\dfrac{f(x + h) - f(x)}{h}$.

**71.** $f(x) = 2x + 9$

**72.** $f(x) = x^2 - 5x + 3$

*Find functions f and g such that* $(f \circ g)(x) = h(x)$.

**73.** $h(x) = (x^3 - 3x)^2$

**74.** $h(x) = \dfrac{1}{x - 5}$

*(Modeling) Solve each problem.*

**75.** *Volume of a Sphere* The formula for the volume of a sphere is $V(r) = \frac{4}{3}\pi r^3$, where $r$ represents the radius of the sphere. Construct a model function $D$ representing the amount of volume gained when a sphere of radius $r$ inches is increased by 4 inches.

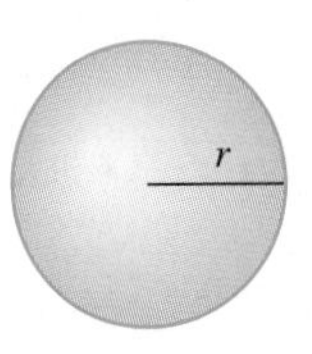


**76.** *Dimensions of a Cylinder* A cylindrical can with a top and bottom makes the most efficient use of materials when its height is the same as the diameter of its top.

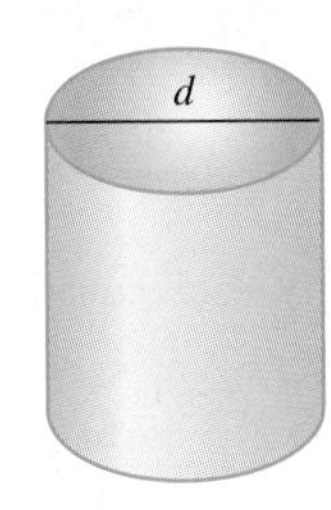


**(a)** Express the volume $V$ of such a can as a function of the diameter $d$ of its top.

**(b)** Express the surface area $S$ of such a can as a function of the diameter $d$ of its top. (*Hint:* The curved side is made from a rectangle whose length is the circumference of the top of the can.)

**77.** *Relationship of Measurement Units* There are 36 inches in 1 yard, and there are 1760 yards in 1 mile. Express the number of inches in $x$ miles by forming two functions and then considering their composition.

**78.** *Perimeter of a Rectangle* Suppose the length of a rectangle is twice its width. Let $x$ represent the width of the rectangle. Write a formula for the perimeter $P$ of the rectangle in terms of $x$ alone. Then use $P(x)$ notation to describe it as a function. What type of function is this?

## CHAPTER 2 TEST

**1.** Match the set described in Column I with the correct interval notation from Column II. Choices in Column II may be used once, more than once, or not at all.

| I | II |
|---|---|
| **(a)** Domain of $f(x) = \sqrt{x} + 3$ | **A.** $[-3, \infty)$ |
| **(b)** Range of $f(x) = \sqrt{x - 3}$ | **B.** $[3, \infty)$ |
| **(c)** Domain of $f(x) = x^2 - 3$ | **C.** $(-\infty, \infty)$ |
| **(d)** Range of $f(x) = x^2 + 3$ | **D.** $[0, \infty)$ |
| **(e)** Domain of $f(x) = \sqrt[3]{x - 3}$ | **E.** $(-\infty, 3)$ |
| **(f)** Range of $f(x) = \sqrt[3]{x} + 3$ | **F.** $(-\infty, 3]$ |
| **(g)** Domain of $f(x) = \lvert x\rvert - 3$ | **G.** $(3, \infty)$ |
| **(h)** Range of $f(x) = \lvert x + 3\rvert$ | **H.** $(-\infty, 0]$ |
| **(i)** Domain of $x = y^2$ | |
| **(j)** Range of $x = y^2$ | |

**2.** The graph of $y = f(x)$ is shown here.

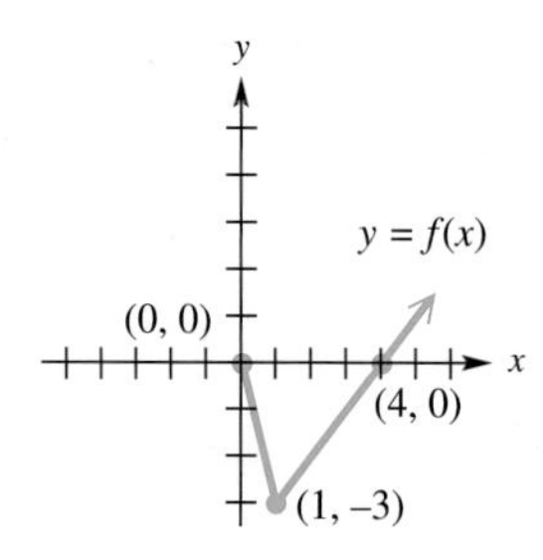


Sketch the graph of each equation. Use ordered pairs to indicate three points on the graph.

**(a)** $y = f(x) + 2$ **(b)** $y = f(x + 2)$

**(c)** $y = -f(x)$ **(d)** $y = f(-x)$

**(e)** $y = 2f(x)$ **(f)** $y = \lvert f(x)\rvert$

**3.** Observe the coordinates displayed at the bottom of the screen showing only the right half of the graph of $y = f(x)$. Answer the following based on your observation.

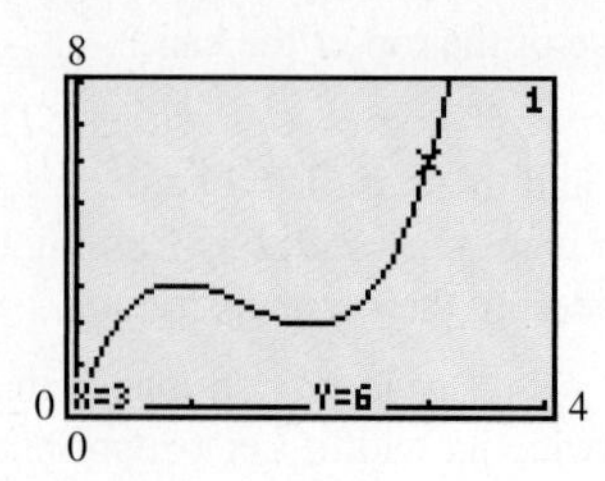


**(a)** If the graph is symmetric with respect to the $y$-axis, what are the coordinates of another point on the graph?

**(b)** If the graph is symmetric with respect to the origin, what are the coordinates of another point on the graph?

**(c)** Suppose the graph is symmetric with respect to the $y$-axis. Sketch a typical viewing window with dimensions $[-4, 4]$ by $[0, 8]$. Then, draw the graph you would expect to see in this window.

**4. (a)** Write a description of how the graph of $y = 4\sqrt[3]{x + 2} - 5$ can be obtained by translation of the graph of $y = \sqrt[3]{x}$.

**(b)** Sketch by hand the graph of $y = -\frac{1}{2}\lvert x - 3\rvert + 2$. Give the domain and range.

**5.** Consider the graph of the function shown here.

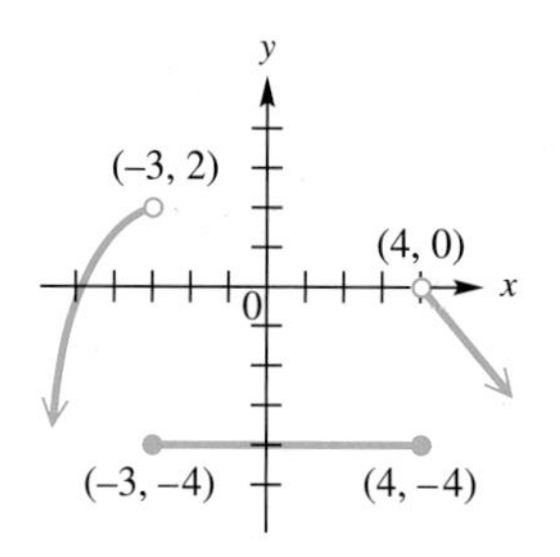


**(a)** Give the interval over which the function is increasing.

**(b)** Give the interval over which the function is decreasing.

**(c)** Give the interval over which the function is constant.

**(d)** Give the intervals over which the function is continuous.

**(e)** What is the domain of this function?

**(f)** What is the range of this function?

**6.** Solve each equation or inequality analytically. Then graph $y_1 = \lvert 4x + 8\rvert$ and $y_2 = 4$ in the standard viewing window of a graphing calculator, and state how the graphs support your solution in each case.

**(a)** $\lvert 4x + 8\rvert = 4$ **(b)** $\lvert 4x + 8\rvert < 4$ **(c)** $\lvert 4x + 8\rvert > 4$

**7.** Given $f(x) = 2x^2 - 3x + 2$ and $g(x) = -2x + 1$. Find the following.

**(a)** $(f - g)(x)$ **(b)** $\left(\frac{f}{g}\right)(x)$

**(c)** The domain of $\frac{f}{g}$ **(d)** $(f \circ g)(x)$

**(e)** $\frac{f(x + h) - f(x)}{h}$ $(h \neq 0)$

**8.** Consider the piecewise-defined function given by

$$f(x) = \begin{cases} -x^2 + 3 & \text{if } x \le 1 \\ \sqrt[3]{x} + 2 & \text{if } x > 1. \end{cases}$$

**(a)** Graph $f$ by hand.

**(b)** Use a graphing calculator to obtain an accurate graph in the window $[-4.7, 4.7]$ by $[-5.1, 5.1]$.

**9.** *(Modeling) Long-Distance Call Charges* A certain long-distance carrier provides service between Podunk and Nowheresville. If $x$ represents the number of minutes for the call, where $x > 0$, then the function $f$ defined by

$$f(x) = .40[\![x]\!] + .75$$

gives the total cost of the call in dollars.

**(a)** Using dot mode and window $[0, 10]$ by $[0, 6]$, graph this function on a graphing calculator.

**(b)** Use the graph to find the cost of a call that is 5.5 minutes long.

**10.** *(Modeling) Cost, Revenue, and Profit Analysis* Tyler McGinnis starts up a small business manufacturing bobble-head figures of famous baseball players. His initial cost is \$3300. Each figure costs \$4.50 to manufacture.

**(a)** Write a cost function $C$, where $x$ represents the number of figures manufactured.

**(b)** Find the revenue function $R$, if each figure in part (a) sells for \$10.50.

**(c)** Give the profit function $P$.

**(d)** How many figures must be produced and sold before Tyler earns a profit?

**(e)** Support the result of part (d) graphically.

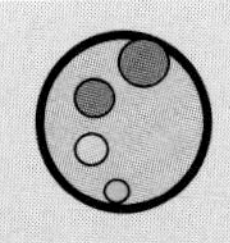

# Chapter 2 Project

## Modeling the Movement of a Cold Front

A weather map of the United States on April 22, 1996, is shown in Figure A. A cold front was traveling in a southeast direction, roughly in the shape of a circular arc passing north of Dallas and west of Detroit. The center of the arc was located near Pierre, South Dakota, with a radius of about 750 miles. Rectangular coordinate axes have been superimposed on the map, with Pierre at the origin. Thus, Pierre has coordinates $(0, 0)$, and the equation of the front can be modeled by

$$f(x) = -\sqrt{750^2 - x^2},$$

where $0 \le x \le 750$. (*Source:* AccuWeather, Inc.)

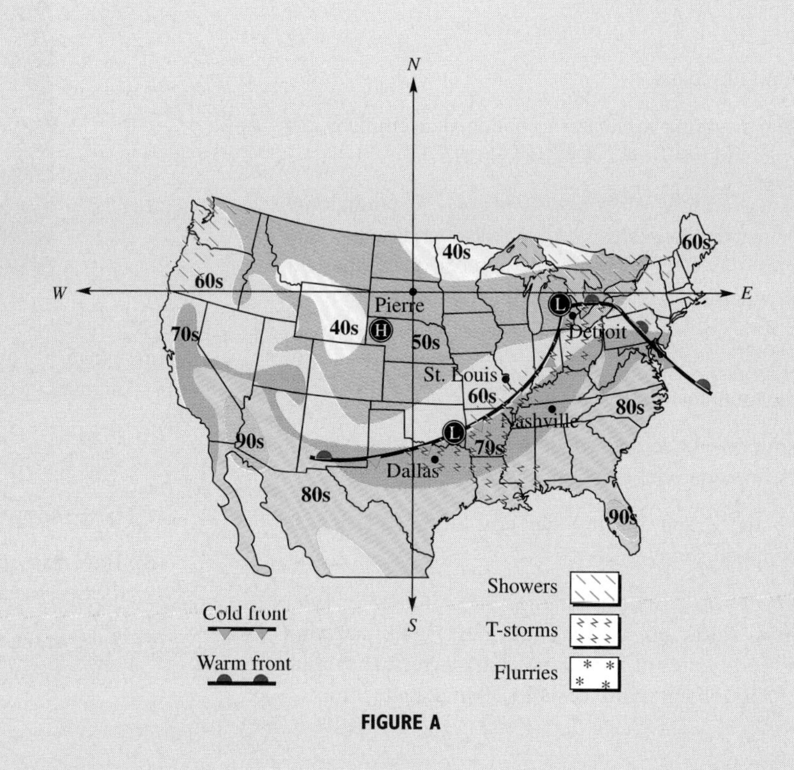


**FIGURE A**

St. Louis is located at $(535, -400)$, and Nashville is at $(730, -570)$. Figure B shows the graph of $f$ along with these two cities plotted as data points. Notice that the cold front had passed through St. Louis but had not yet reached Nashville.

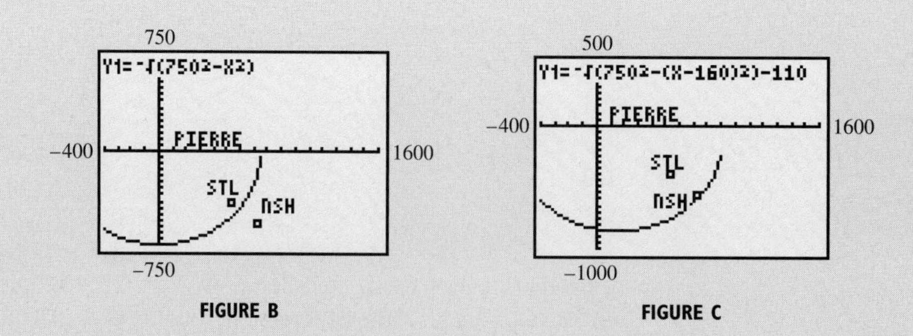


**FIGURE B** **FIGURE C**

During the next 12 hours, the center of the front moved approximately 110 miles south and 160 miles east. If we assume that the front did not change shape, its new position can be modeled by

$$g(x) = -\sqrt{750^2 - (x - 160)^2} - 110.$$

We can see from Figure C that the cold front did indeed reach Nashville after 12 hours.

## Activity

Suppose that a cold front is passing through the United States at noon with a shape described by the graph of

$$f(x) = \frac{1}{20}x^2.$$

Each unit represents 100 miles. Des Moines, Iowa, is located at $(0, 0)$, and the positive $y$-axis points north. See Figure D.

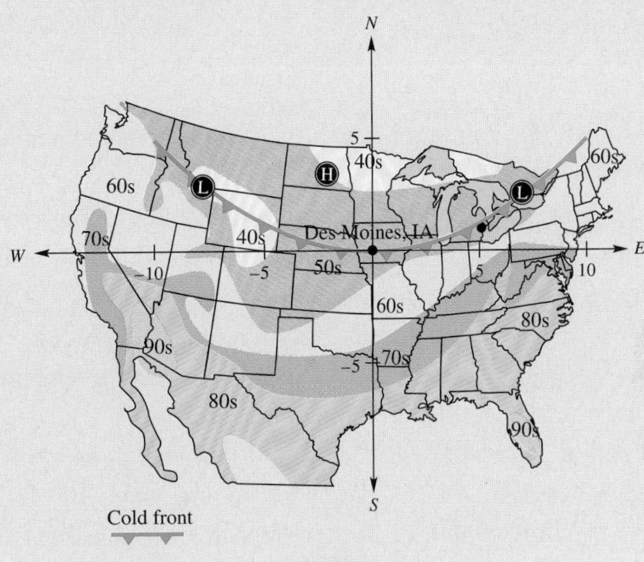


**FIGURE D**

1. If the cold front is moving south at 40 mph for 4 hours and retains its present shape, what would be the equation of its graph at that time?
2. Suppose that by midnight the vertex of the front has moved 250 miles south and 210 miles east of Des Moines, maintaining the same shape. With the class divided into three groups (a), (b), and (c), each group should determine whether the front has reached the designated city.
   **(a)** Columbus, Ohio, which is 550 miles east and 80 miles south of Des Moines
   **(b)** Memphis, Tennessee, which is 190 miles east and 430 miles south of Des Moines
   **(c)** Louisville, Kentucky, which is 420 miles east and 230 miles south of Des Moines

# 3 Polynomial Functions

THE LAST basketball player in the NBA to shoot foul shots underhand was Rick Barry, who retired in 1980. On average, he was able to make about 9 out of 10 shots. Since then, every NBA player has used the overhand style of shooting foul shots—even though this style has often resulted in lower free-throw percentages.

According to Peter Brancazio, a physics professor emeritus from Brooklyn College and author of *Sports Science*, there are good reasons for shooting underhand. An underhand shot obtains a higher arc, and as the ball approaches the hoop, it has a better chance of going through the hoop than does a ball with a flatter arc. If a basketball is tossed at an angle of 32 degrees or less, it will likely hit the back of the rim and bounce out. The optimal angle is greater than 45 degrees and depends on the height at which the ball is released. Lower release points require steeper arcs and increase the chances of the ball passing through the hoop.

Mathematics plays an important role in analyzing applied problems such as shooting foul shots. Whether NBA players choose to agree with Professor Brancazio is another question, but mathematics tells us that steeper arcs are necessary for accurate foul shooting.

*Source:* Rist, C., "The Physics of Foul Shots," *Discover*, October 2000. (Photograph reprinted with permission.)

CHAPTER OUTLINE

# 3.1 COMPLEX NUMBERS

The Number $i$ • Operations with Complex Numbers

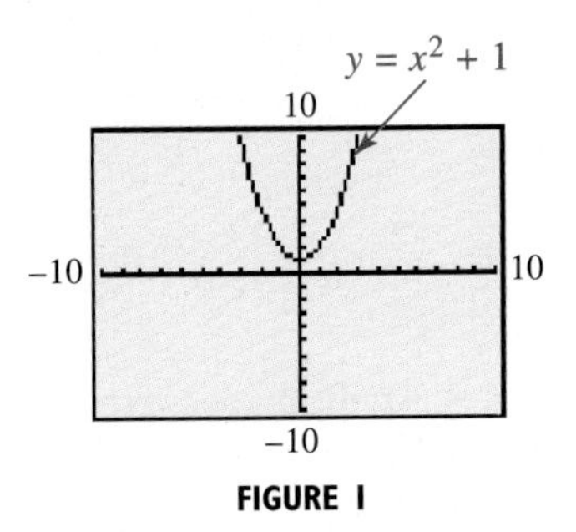


FIGURE 1

## The Number $i$

Our work in the previous chapters involved only real numbers. The set of real numbers, however, does not include all the numbers we need in this chapter. The graph of $y = x^2 + 1$ in Figure 1 does not intersect the $x$-axis, and, therefore, there are no real solutions to the equation $x^2 + 1 = 0$. This equation is equivalent to $x^2 = -1$, and we know from experience that no real number has a square of $-1$. The **complex number system** is an extended number system that includes the set of real numbers as a subset. A basic unit of this new system is $i$, which is defined as the principal square root of $-1$. Thus, $i^2 = -1$.

TECHNOLOGY NOTE

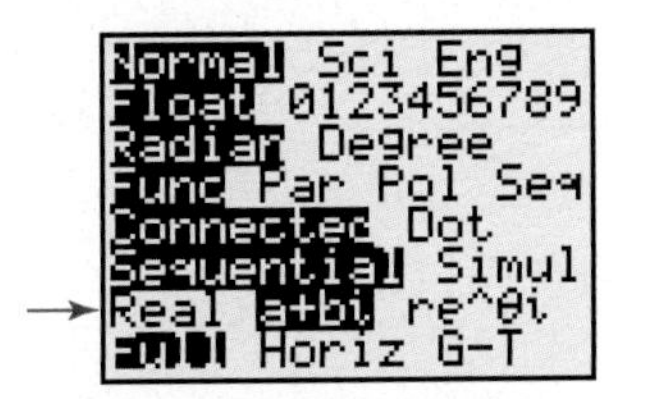


Some graphing calculators such as the TI-83 and TI-83 Plus are capable of complex number operations, as indicated by $a + bi$ here. Consult the graphing calculator manual that accompanies this text or your owner's manual for details.

**Imaginary Number $i$**

$$i = \sqrt{-1} \quad \text{or} \quad i^2 = -1$$

Numbers of the form $a + bi$, where $a$ and $b$ are real numbers, are called **complex numbers.** Each real number is a complex number since a real number $a$ may be thought of as the complex number $a + 0i$. A complex number of the form $a + bi$, where $b$ is nonzero, is called an **imaginary number.** Both the set of real numbers and the set of imaginary numbers are subsets of the set of complex numbers. (See Figure 2.) A complex number that is written in the form $a + bi$ or $a + ib$ is in **standard form.** (The form $a + ib$ is sometimes used to write expressions such as $3 + \sqrt{5}i$, which could be easily mistaken for $3 + \sqrt{5i}$.) The real number $a$ is called the **real part** of $a + bi$, and the real number $b$ is called the **imaginary part.*** For example, in $-7 + 2i$, $-7$ is the real part and 2 is the imaginary part.

**Complex Numbers**

| Imaginary numbers | Real numbers | |
|---|---|---|
| | Rational numbers | Irrational numbers |
| $4i, -11i, 3 + 2i$ | $\frac{4}{9}, -\frac{5}{8}, \frac{11}{7}$ | $-\sqrt{8}$ |
| | Integers $-11, -6, -4$ | $-\sqrt{15}$ |
| | Whole numbers 0 | $\sqrt{23}$ |
| | Natural numbers 1, 2, 3, 4, 5, 37, 40 | $\pi$, $\frac{\pi}{4}$, $e$ |

FIGURE 2

*In some texts, the term $bi$ is defined to be the imaginary part.

The table shows several complex numbers, each with its type and standard form.

| Complex Number | Real or Imaginary | Standard Form |
|---|---|---|
| $-8$ | Real | $-8 + 0i$ |
| $0$ | Real | $0 + 0i$ |
| $3i$ | Imaginary | $0 + 3i$ |
| $-i + 2$ | Imaginary | $2 - i$ |
| $8 + i\sqrt{3}$ | Imaginary | $8 + i\sqrt{3}$ |

Many of the solutions to quadratic equations in Section 3.3 will involve expressions such as $\sqrt{-a}$, for a positive real number $a$, defined as follows.

**$\sqrt{-a}$**

If $a > 0$, then

$$\sqrt{-a} = i\sqrt{a}.$$

**EXAMPLE 1** *Writing $\sqrt{-a}$ as $i\sqrt{a}$*

Write each expression as the product of $i$ and a real number.

**(a)** $\sqrt{-16}$ **(b)** $\sqrt{-80}$

**Analytic Solution**

**(a)** $\sqrt{-16} = \sqrt{-1 \cdot 16}$
$= i\sqrt{16}$
$= 4i$

**(b)** $\sqrt{-80} = i\sqrt{80}$
$= i\sqrt{16 \cdot 5}$
$= 4i\sqrt{5}$

**Graphing Calculator Solution**

Figure 3 confirms the results we found in the analytic solution. The calculator is in complex number mode.

```
√(-16)
                4i
√(-80)
       8.94427191i
4√(5)
        8.94427191
```

FIGURE 3

A product or quotient with a negative radicand is simplified by first rewriting $\sqrt{-a}$ as $i\sqrt{a}$ for a positive number $a$. Then the properties of real numbers are applied, together with the fact that $i^2 = -1$.

**CAUTION** When working with negative radicands, *use the definition $\sqrt{-a} = i\sqrt{a}$ before using any of the other rules for radicals.* In particular, the rule $\sqrt{c} \cdot \sqrt{d} = \sqrt{cd}$ is valid only when $c$ and $d$ are *not* both negative. For example,

$$\sqrt{(-4)(-9)} = \sqrt{36} = 6,$$

while $$\sqrt{-4} \cdot \sqrt{-9} = 2i(3i) = 6i^2 = -6,$$

so $$\sqrt{-4} \cdot \sqrt{-9} \neq \sqrt{(-4)(-9)}.$$

**EXAMPLE 2** *Finding Products and Quotients Involving Negative Radicands*

Multiply or divide as indicated.

**(a)** $\sqrt{-7} \cdot \sqrt{-7}$ **(b)** $\sqrt{-6} \cdot \sqrt{-10}$ **(c)** $\frac{\sqrt{-20}}{\sqrt{-2}}$ **(d)** $\frac{\sqrt{-48}}{\sqrt{24}}$

**Analytic Solution**

**(a)**
$$\begin{aligned}\sqrt{-7} \cdot \sqrt{-7} &= i\sqrt{7} \cdot i\sqrt{7} \\ &= i^2 \cdot (\sqrt{7})^2 \\ &= (-1) \cdot 7 \qquad i^2 = -1 \\ &= -7\end{aligned}$$

**(b)**
$$\begin{aligned}\sqrt{-6} \cdot \sqrt{-10} &= i\sqrt{6} \cdot i\sqrt{10} \\ &= i^2 \cdot \sqrt{60} \\ &= -1 \cdot 2\sqrt{15} \\ &= -2\sqrt{15}\end{aligned}$$

**(c)** $$\frac{\sqrt{-20}}{\sqrt{-2}} = \frac{i\sqrt{20}}{i\sqrt{2}} = \sqrt{\frac{20}{2}} = \sqrt{10}$$

**(d)** $$\frac{\sqrt{-48}}{\sqrt{24}} = \frac{i\sqrt{48}}{\sqrt{24}} = i\sqrt{2}$$

**Graphing Calculator Solution**

The screens in Figure 4 show calculator verification for parts (b) and (d).

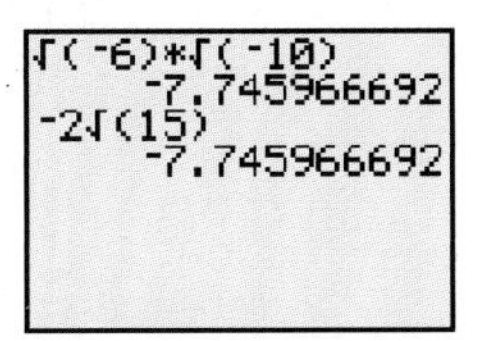


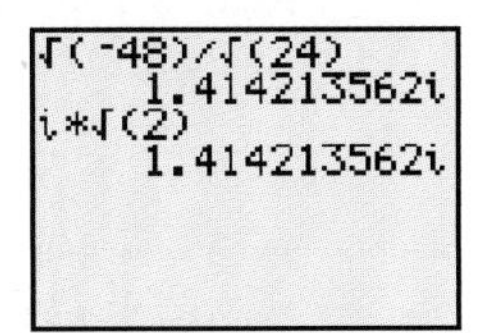


FIGURE 4

## Operations with Complex Numbers

With the definitions $i^2 = -1$ and $\sqrt{-a} = i\sqrt{a}$, for $a > 0$, all properties of real numbers are extended to complex numbers. As a result, complex numbers are added, subtracted, multiplied, and divided using real number properties. We find the sum, difference, and product of two complex numbers just as we would for two binomials.

**EXAMPLE 3** *Adding and Subtracting Complex Numbers*

Find each sum or difference.

**(a)** $(3 - 4i) + (-2 + 6i)$ **(b)** $(-9 + 7i) + (3 - 15i)$

**(c)** $(-4 + 3i) - (6 - 7i)$ **(d)** $(12 - 5i) - (8 - 3i)$

**Analytic Solution**

**(a)**
$$\begin{aligned}(3 - 4i) + (-2 + 6i) &= \underbrace{[3 + (-2)]}_{\text{Add real parts.}} + \underbrace{(-4 + 6)}_{\text{Add imaginary parts.}}i \\ &= 1 + 2i\end{aligned}$$

**(b)** $(-9 + 7i) + (3 - 15i) = -6 - 8i$

**(c)**
$$\begin{aligned}(-4 + 3i) - (6 - 7i) &= \underbrace{(-4 - 6)}_{\text{Subtract real parts.}} + \underbrace{[3 - (-7)]}_{\text{Subtract imaginary parts.}}i \\ &= -10 + 10i\end{aligned}$$

**(d)** $(12 - 5i) - (8 - 3i) = 4 - 2i$

**Graphing Calculator Solution**

Figure 5 shows the calculator results.

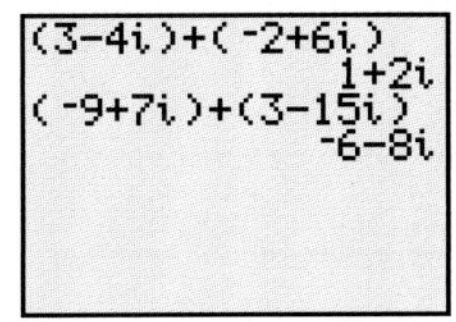


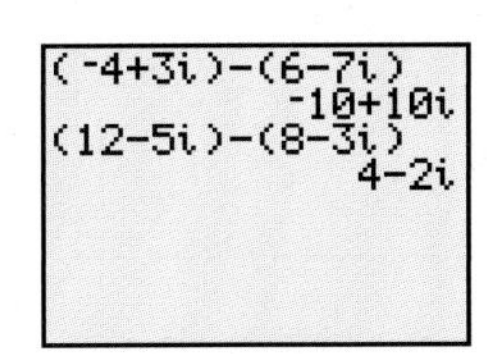


FIGURE 5

The product of two complex numbers is found by multiplying as if the numbers were binomials and using the fact that $i^2 = -1$.

**EXAMPLE 4** *Multiplying Complex Numbers*

Find each product.

**(a)** $(2 - 3i)(3 + 4i)$ **(b)** $(5 - 4i)(7 - 2i)$

**(c)** $(6 + 5i)(6 - 5i)$ **(d)** $(4 + 3i)^2$

**Analytic Solution**

**(a)** 
$$\begin{aligned}(2 - 3i)(3 + 4i) &= 2(3) + 2(4i) - 3i(3) - 3i(4i) && \text{FOIL}\\ &= 6 + 8i - 9i - 12i^2 \\ &= 6 - i - 12(-1) && i^2 = -1\\ &= 18 - i\end{aligned}$$

**(b)** 
$$\begin{aligned}(5 - 4i)(7 - 2i) &= 5(7) + 5(-2i) - 4i(7) - 4i(-2i)\\ &= 35 - 10i - 28i + 8i^2\\ &= 35 - 38i + 8(-1)\\ &= 27 - 38i\end{aligned}$$

**(c)** 
$$\begin{aligned}(6 + 5i)(6 - 5i) &= 6^2 - 25i^2 && \text{Product of the sum and difference of two terms}\\ &= 36 - 25(-1) && i^2 = -1\\ &= 61\end{aligned}$$

**(d)** 
$$\begin{aligned}(4 + 3i)^2 &= 4^2 + 2(4)(3i) + (3i)^2 && \text{Square of a binomial}\\ &= 16 + 24i + (-9)\\ &= 7 + 24i\end{aligned}$$

**Graphing Calculator Solution**

The calculator solutions shown in Figure 6 for parts (a), (c), and (d) give the same results.

```
(2-3i)(3+4i)
           18-i
(6+5i)(6-5i)
             61
(4+3i)²
          7+24i
```

FIGURE 6

A calculator could also be used to evaluate the product in part (b).

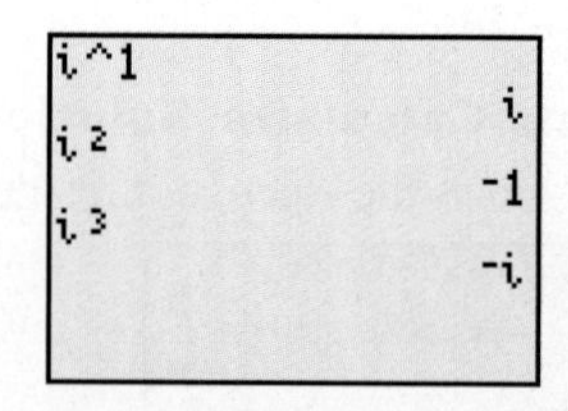


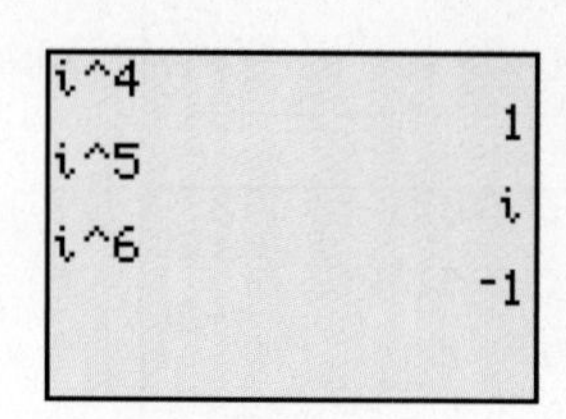


These screens show how powers of $i$ follow a cycle.

By definition, $i^1 = i$ and $i^2 = -1$. Now, observe the following pattern.

$$i^1 = i$$
$$i^2 = -1$$
$$i^3 = i^2 \cdot i = -1 \cdot i = -i$$
$$i^4 = i^3 \cdot i = -i \cdot i = -i^2 = -(-1) = 1$$

Because $i^4 = 1$, any larger power of $i$ may be found by writing the power as a product of two powers of $i$, one exponent being a multiple of 4, and then simplifying.

**EXAMPLE 5** *Simplifying Powers of i*

Simplify each power of $i$.

**(a)** $i^{13}$ **(b)** $i^{56}$ **(c)** $i^{-3}$

**Analytic Solution**

In each case, we use the fact that $i^4 = 1$.

**(a)** $i^{13} = i^{12} \cdot i$
$= (i^4)^3 \cdot i$
$= 1^3 \cdot i$
$= i$

**(b)** $i^{56} = (i^4)^{14} = 1^{14} = 1$

**(c)** $i^{-3} = i^{-4} \cdot i$
$= (i^4)^{-1} \cdot i$
$= (1)^{-1} \cdot i$
$= i$

**Graphing Calculator Solution**

The screen in Figure 7 shows the same results for parts (a) and (b). In the first case, the real part 0 is approximated as $-3 \times 10^{-13}$, and in the second case, the imaginary part 0 is approximated as $-4 \times 10^{-13}$.

Here we see another example of the limitations of technology. The results may be misinterpreted if the mathematical concepts are not understood.

```
i^13
                 -3E-13+i
i^56
                 1-4E-13i
```

FIGURE 7

Example 4(c) illustrates an important property of complex numbers. Notice that the factors $6 + 5i$ and $6 - 5i$ have the same real parts but opposite imaginary parts. Pairs of complex numbers satisfying these conditions are called *complex conjugates,* or simply *conjugates*. The product of a complex number and its conjugate is always a real number.

### Complex Conjugates

The **conjugate** of the complex number $a + bi$ is $a - bi$. Their product is

$$(a + bi)(a - bi) = a^2 + b^2.$$

The table shows several pairs of conjugates and their products.

| Number | Conjugate | Product |
|---|---|---|
| $3 - i$ | $3 + i$ | $(3 - i)(3 + i) = 9 + 1 = 10$ |
| $2 + 7i$ | $2 - 7i$ | $(2 + 7i)(2 - 7i) = 53$ |
| $-6i$ | $6i$ | $(-6i)(6i) = 36$ |

All products are real numbers.

The conjugate of the divisor is used to find the quotient of two complex numbers in standard form. Multiplying both the numerator and the denominator by the conjugate of the denominator gives the quotient.

**EXAMPLE 6** *Dividing Complex Numbers*

Find each quotient.

**(a)** $\dfrac{3 + 2i}{5 - i}$ **(b)** $\dfrac{3}{i}$

**Analytic Solution**

**(a)** Multiply numerator and denominator by the conjugate of $5 - i$.

$$\frac{3+2i}{5-i} = \frac{(3+2i)(5+i)}{(5-i)(5+i)}$$

$$= \frac{15+3i+10i+2i^2}{25-i^2} \quad \text{Multiply.}$$

$$= \frac{13+13i}{26} \quad i^2 = -1$$

$$= \frac{13}{26} + \frac{13i}{26} = \frac{1}{2} + \frac{1}{2}i \quad \frac{a+bi}{c} = \frac{a}{c} + \frac{bi}{c}$$

To check this answer, show that

$$(5-i)\left(\frac{1}{2} + \frac{1}{2}i\right) = 3 + 2i. \quad \text{(Why?)}$$

**(b)** $$\frac{3}{i} = \frac{3(-i)}{i(-i)} \quad -i \text{ is the conjugate of } i.$$

$$= \frac{-3i}{-i^2}$$

$$= \frac{-3i}{1} = -3i \quad -i^2 = -(-1) = 1$$

**Graphing Calculator Solution**

A calculator gives the same quotients. See Figure 8.

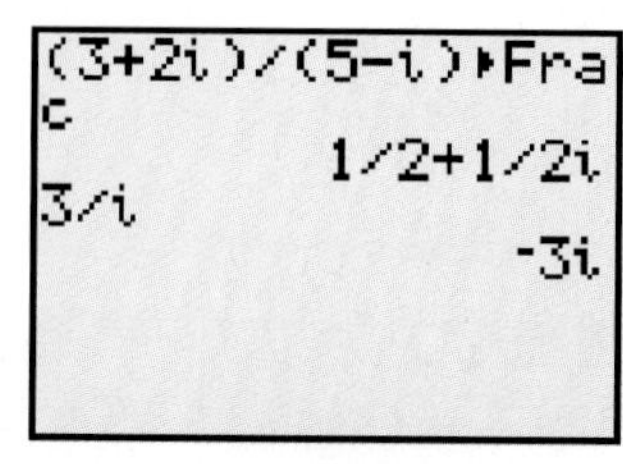


FIGURE 8

Notice that $1/2i$ is actually $\frac{1}{2}i$. Again, we must be careful when interpreting calculator results.

## 3.1 EXERCISES

*Note: For the exercises in this section, use your graphing calculator with complex number capability to confirm your answers when possible.*

Concept Check *For each complex number,* **(a)** *state the real part,* **(b)** *state the imaginary part, and* **(c)** *state whether the number is real or imaginary.*

**1.** $-9i$ **2.** $6$ **3.** $\pi$ **4.** $-\sqrt{7}$

**5.** $i\sqrt{6}$ **6.** $-3i$ **7.** $2 + 5i$ **8.** $-7 - 6i$

*Write each number without a negative radicand.*

**9.** $\sqrt{-100}$ **10.** $\sqrt{-169}$ **11.** $-\sqrt{-400}$ **12.** $-\sqrt{-225}$

**13.** $-\sqrt{-39}$ **14.** $-\sqrt{-95}$ **15.** $5 + \sqrt{-4}$ **16.** $-7 + \sqrt{-100}$

**17.** $9 - \sqrt{-50}$ **18.** $-11 - \sqrt{-24}$ **19.** $\sqrt{-5} \cdot \sqrt{-5}$ **20.** $\sqrt{-20} \cdot \sqrt{-20}$

**21.** $\dfrac{\sqrt{-40}}{\sqrt{-10}}$ **22.** $\dfrac{\sqrt{-190}}{\sqrt{-19}}$

**23.** Explain why a real number must be a complex number, but a complex number need not be a real number.

**24.** If the complex number $a + bi$ is real, then what can be said about the value of $b$?

*Add or subtract as indicated. Write each sum or difference in standard form.*

**25.** $(3 + 2i) + (4 - 3i)$

**26.** $(4 - i) + (2 + 5i)$

**27.** $(-2 + 3i) - (-4 + 3i)$

**28.** $(-3 + 5i) - (-4 + 3i)$

**29.** $(2 - 5i) - (3 + 4i) - (-2 + i)$

**30.** $(-4 - i) - (2 + 3i) + (-4 + 5i)$

*Multiply as indicated. Write each product in standard form.*

**31.** $(2 + 4i)(-1 + 3i)$

**32.** $(1 + 3i)(2 - 5i)$

**33.** $(-3 + 2i)^2$

**34.** $(2 + i)^2$

**35.** $(2 + 3i)(2 - 3i)$

**36.** $(6 - 4i)(6 + 4i)$

**37.** $(\sqrt{6} + i)(\sqrt{6} - i)$

**38.** $(\sqrt{2} - 4i)(\sqrt{2} + 4i)$

**39.** $i(3 - 4i)(3 + 4i)$

**40.** $i(2 + 7i)(2 - 7i)$

**41.** $(6 - i)(1 + i)^2$

**42.** $(3 - 4i)^2(1 + 3i)$

*Simplify each power of i to i, 1, −i, or −1.*

**43.** $i^5$

**44.** $i^8$

**45.** $i^{15}$

**46.** $i^{19}$

**47.** $i^{64}$

**48.** $i^{102}$

**49.** $i^{-6}$

**50.** $i^{-15}$

**51.** $\frac{1}{i^9}$

**52.** $\frac{1}{i^{12}}$

**53.** *Concept Check* Show that $\frac{\sqrt{2}}{2} + \frac{\sqrt{2}}{2}i$ is a square root of $i$.

**54.** *Concept Check* Show that $\frac{\sqrt{3}}{2} + \frac{1}{2}i$ is a cube root of $i$.

*Find the conjugate of each number.*

**55.** $5 - 3i$

**56.** $-3 + i$

**57.** $-18i$

**58.** $\sqrt{7}$

*Divide as indicated. Write each quotient in standard form.*

**59.** $\frac{-19 - 9i}{4 + i}$

**60.** $\frac{-12 - 5i}{3 - 2i}$

**61.** $\frac{1 - 3i}{1 + i}$

**62.** $\frac{-3 + 4i}{2 - i}$

**63.** $\frac{-6 + 8i}{4 + 3i}$

**64.** $\frac{2 - i}{2 + i}$

**65.** $\frac{4 - 3i}{4 + 3i}$

**66.** $\frac{3}{-i}$

**67.** $\frac{-7}{3i}$

**68.** $\frac{-10}{i}$

**69.** Explain why the method of dividing complex numbers (that is, multiplying both the numerator and the denominator by the conjugate of the denominator) works. What property justifies this process?

**70.** Suppose that your friend, Wei Jen Chong, tells you that she has discovered a method of simplifying a positive power of $i$. "Just divide the exponent by 4," she says, "and then look at the remainder. Then, refer to the short table of powers of $i$ in this section. The large power of $i$ is equal to $i$ to the power indicated by the remainder. And if the remainder is 0, the result is $i^0 = 1$." Explain why Wei Jen's method works.

## RELATING CONCEPTS

**For individual or group investigation (Exercises 71–74)**

*Recall that a solution, or root, of an equation is a number that, when substituted for the variable, gives a true statement. In earlier chapters, we considered only real number solutions of equations. In this chapter, we will see that equations may also have solutions that are not real numbers. For the equation*

$$x^3 - x^2 - 7x + 15 = 0,$$

*show that each number is a solution by substituting it for x.*

**71.** The real number $-3$

**72.** The complex number $2 - i$

**73.** The complex number $2 + i$

**74.** What relationship do the solutions in Exercises 72 and 73 have?

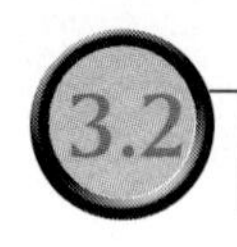

# QUADRATIC FUNCTIONS AND GRAPHS

Completing the Square • Graphs of Quadratic Functions • Vertex Formula • Extreme Values • Applications and Modeling

In Chapter 1 we introduced linear functions, defined by $f(x) = ax + b$. Linear and quadratic functions are examples of *polynomial functions*, which we investigate in general later in this chapter. We often use $P$ to name a polynomial function, and we do so with quadratic functions throughout this chapter.

In Chapter 2 we saw that the graph of $y = x^2$ is a parabola. See Figure 9. The function defined by $P(x) = x^2$ is the simplest example of a quadratic function.

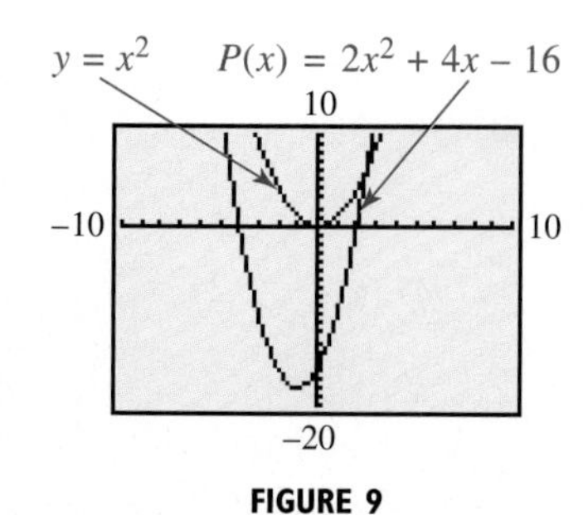


**FIGURE 9**

> **Quadratic Function**
>
> A function defined by
>
> $$\boldsymbol{P(x) = ax^2 + bx + c,}$$
>
> with $a \neq 0$, is called a **quadratic function.**

## Completing the Square

The comprehensive graph of a linear function was defined as a graph that showed all intercepts. For a quadratic function, whose graph is a parabola, a comprehensive graph shows all intercepts and the vertex of the parabola.

Recall from Chapter 2 that the graph of

$$g(x) = a(x - h)^2 + k$$

has the same general shape as the graph of $f(x) = x^2$, but may be stretched, shrunk, reflected, or shifted. For example, consider $P(x) = 2x^2 + 4x - 16$. A comprehensive graph is shown in Figure 9. When compared to the graph of $y = x^2$, also shown in the figure, it appears that the graph of $P$ can be obtained by a vertical stretch with a factor greater than 1, a shift to the left, and a shift downward. It would be possible to determine the sizes of the transformations if the function were written in the form $P(x) = a(x - h)^2 + k$. To transform an equation of the form $P(x) = ax^2 + bx + c$ into this form, we *complete the square*. The steps to do this are summarized below.

> **Completing the Square**
>
> To transform the equation $P(x) = ax^2 + bx + c$ into the form $P(x) = a(x - h)^2 + k$:
>
> **1.** Divide each side of the equation by $a$ so that the coefficient of $x^2$ is 1.
> **2.** Add $-\frac{c}{a}$ to each side.
> **3.** Add to each side the square of half the coefficient of $x$; that is, $\left(\frac{b}{2a}\right)^2$.
> **4.** Factor the right side as the square of a binomial and combine terms on the left.
> **5.** Isolate the term involving $P(x)$ on the left.
> **6.** Multiply each side by $a$.

We now apply this procedure to $P(x) = 2x^2 + 4x - 16$.

$$P(x) = 2x^2 + 4x - 16 \qquad \text{Given equation}$$

$$\frac{P(x)}{2} = x^2 + 2x - 8 \qquad \text{Divide by 2 to make the coefficient of } x^2 \text{ equal to 1.}$$

$$\frac{P(x)}{2} + 8 = x^2 + 2x \qquad \text{Add 8 to each side.}$$

$$\frac{P(x)}{2} + 8 + 1 = x^2 + 2x + 1 \qquad \text{Add } \left[\tfrac{1}{2}(\mathbf{2})\right]^2 = 1 \text{ to each side to complete the square on the right.}$$

$$\frac{P(x)}{2} + 9 = (x + 1)^2 \qquad \text{Combine terms on the left; factor on the right.}$$

$$\frac{P(x)}{2} = (x + 1)^2 - 9 \qquad \text{Subtract 9 from each side.}$$

$$P(x) = 2(x + 1)^2 - 18 \qquad \text{Multiply each side by 2.}$$

From this form of the equation, we can now determine $a$, $h$, and $k$ for the graph of $P(x) = 2x^2 + 4x - 16$ in Figure 9. We see that $a = 2$, $h = -1$, and $k = -18$. Thus, the graph of $P(x)$ is the graph of $y = x^2$ stretched by a factor of 2 and shifted 1 unit left and 18 units downward. The vertex of the parabola has coordinates $(-1, -18)$, the domain is $(-\infty, \infty)$, and the range is $[-18, \infty)$. The function decreases on the interval $(-\infty, -1]$ and increases on the interval $[-1, \infty)$.

**EXAMPLE 1** *Transforming $P(x) = ax^2 + bx + c$ to $P(x) = a(x - h)^2 + k$*

Complete the square on $P(x) = -x^2 - 6x - 8$.

**Looking Ahead to Calculus**

An important concept in calculus is the *definite integral*. The symbol

$$\int_c^d P(x)\,dx$$

represents the area of the region above the $x$-axis and below the graph of $P$ from $x = c$ to $x = d$. For example, in Figure 10 on the next page with

$$P(x) = -x^2 - 6x - 8,$$

$c = -4$, and $d = -2$, calculus provides the tools for determining that the area enclosed by the parabola and the $x$-axis is $\frac{4}{3}$ (square units).

**Solution**

$$P(x) = -x^2 - 6x - 8$$

$$-P(x) = x^2 + 6x + 8 \qquad \text{Divide by } -1.$$

$$-P(x) - 8 = x^2 + 6x \qquad \text{Subtract 8.}$$

$$-P(x) - 8 + 9 = x^2 + 6x + 9 \qquad \text{Add } \left[\tfrac{1}{2}(\mathbf{6})\right]^2 = 9.$$

$$-P(x) + 1 = (x + 3)^2 \qquad \text{Combine terms on the left; factor on the right.}$$

$$-P(x) = (x + 3)^2 - 1 \qquad \text{Subtract 1.}$$

$$P(x) = -(x + 3)^2 + 1 \qquad \text{Multiply by } -1.$$

## Graphs of Quadratic Functions

The function defined by $P(x) = -x^2 - 6x - 8 = -(x + 3)^2 + 1$ from Example 1 can now be graphed easily because it is in the form $P(x) = a(x - h)^2 + k$. Also, recall from Chapter 1 that the $y$-intercept of the graph of an equation is the $y$-value that corresponds to $x = 0$. For a parabola given in the form $P(x) = ax^2 + bx + c$, the $y$-intercept is $P(0) = c$.

**EXAMPLE 2** *Graphing a Quadratic Function*

Graph the function defined by $P(x) = -(x + 3)^2 + 1$. Give the domain and range and the intervals over which the function is increasing or decreasing.

**Analytic Solution**

The vertex is $(-3, 1)$, so the graph is that of $f(x) = x^2$ shifted 3 units to the left and 1 unit upward. From the form $P(x) = -x^2 - 6x - 8$, the $y$-intercept is $P(0) = -8$. Because the coefficient of $(x + 3)^2$ is $-1$, the graph opens downward and has the same shape as that of $f(x) = x^2$. Plot the vertex, the $y$-intercept $-8$, and a few ordered pairs near the vertex, using symmetry about the line $x = -3$ to get additional ordered pairs. For example, use symmetry to get the ordered pair $(-6, -8)$. The line $x = -3$ is called the **axis of symmetry** of the parabola, because if the graph were folded along this line, the two halves would coincide. See the table of values and the graph in Figure 10.

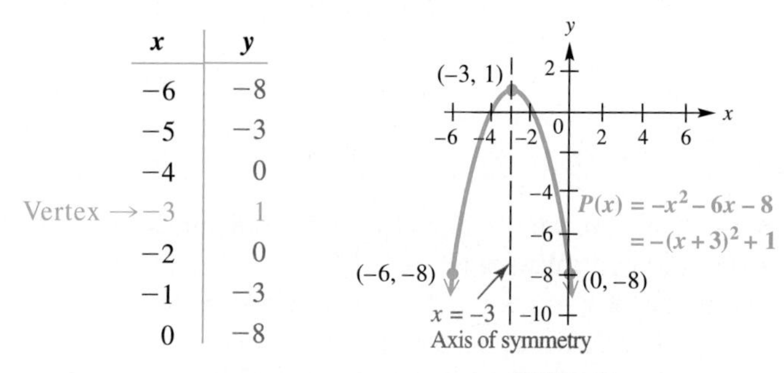

| | $x$ | $y$ |
|---|---|---|
| | $-6$ | $-8$ |
| | $-5$ | $-3$ |
| | $-4$ | $0$ |
| Vertex → | $-3$ | $1$ |
| | $-2$ | $0$ |
| | $-1$ | $-3$ |
| | $0$ | $-8$ |



FIGURE 10

Notice the symmetry in the table. The $y$-values repeat "above" and "below" the entry for the vertex $(-3, 1)$.

**Graphing Calculator Solution**

A calculator graph is shown in Figure 11. Since the vertex is $(-3, 1)$ and the $y$-intercept is $-8$, choose the viewing window $[-8, 2]$ by $[-10, 2]$, which gives a comprehensive graph of the function.

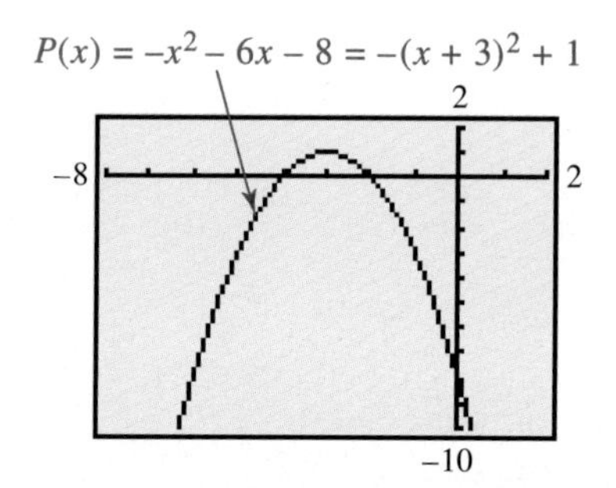


FIGURE 11

The domain is $(-\infty, \infty)$. Since the graph opens downward, the vertex $(-3, 1)$ is the highest point on the graph. Thus, the range is $(-\infty, 1]$. The function increases on the interval $(-\infty, -3]$ and decreases on $[-3, \infty)$.

**NOTE** In the next section we develop analytic methods for finding the $x$-intercepts of the graph of any quadratic function.

**Looking Ahead to Calculus**

The derivative of a function provides a formula for finding the slope of a line tangent to the graph of the function. Using the methods of calculus, the function of Examples 1 and 2,

$$P(x) = -x^2 - 6x - 8,$$

has derivative

$$P'(x) = -2x - 6.$$

If we solve $P'(x) = 0$, we find the $x$-value for which the graph of $P$ has a horizontal tangent. Solve this equation and show that its solution gives the $x$-value of the vertex. Notice that if you draw a tangent line at the vertex, it is a horizontal line with slope 0.

Our discussion up to this point leads to the following generalizations about the graph of a quadratic function in the form $P(x) = a(x - h)^2 + k$.

**Graph of $P(x) = a(x - h)^2 + k$**

The graph of $P(x) = a(x - h)^2 + k$ $(a \neq 0)$

**(a)** is a parabola with vertex $(h, k)$ and the vertical line $x = h$ as axis of symmetry;

**(b)** opens upward if $a > 0$ and downward if $a < 0$;

**(c)** is wider than the graph of $y = x^2$ if $0 < |a| < 1$ and narrower than the graph of $y = x^2$ if $|a| > 1$.

We use the information summarized in the box to graph a quadratic function by hand. First we determine the vertex, plot it, find two or three ordered pairs including the $y$-intercept, and then plot them, using symmetry to locate additional points. We use the sign and value of $a$ to verify that the graph opens in the correct direction and is stretched or shrunk appropriately.

## Vertex Formula

We can determine the coordinates of the vertex of the graph of a quadratic function by completing the square, as shown earlier. Rather than go through that procedure for each individual function, we generalize the result for the standard form of the quadratic function defined by $P(x) = ax^2 + bx + c$, where $a \neq 0$.

$$P(x) = ax^2 + bx + c \qquad \text{Standard form}$$

$$y = ax^2 + bx + c \qquad \text{Replace } P(x) \text{ with } y \text{ to simplify notation.}$$

$$\frac{y}{a} = x^2 + \frac{b}{a}x + \frac{c}{a} \qquad \text{Divide by } a.$$

$$\frac{y}{a} - \frac{c}{a} = x^2 + \frac{b}{a}x \qquad \text{Add } -\frac{c}{a}.$$

$$\frac{y}{a} - \frac{c}{a} + \frac{b^2}{4a^2} = x^2 + \frac{b}{a}x + \frac{b^2}{4a^2} \qquad \text{Add } \left[\frac{1}{2}\left(\frac{b}{a}\right)\right]^2 = \frac{b^2}{4a^2}.$$

$$\frac{y}{a} + \frac{b^2 - 4ac}{4a^2} = \left(x + \frac{b}{2a}\right)^2 \qquad \text{Combine terms on the left; factor on the right.}$$

$$\frac{y}{a} = \left(x + \frac{b}{2a}\right)^2 - \frac{b^2 - 4ac}{4a^2} \qquad \text{Isolate } y\text{-term on the left.}$$

$$y = a\left(x + \frac{b}{2a}\right)^2 + \frac{4ac - b^2}{4a} \qquad \text{Multiply by } a.$$

$$P(x) = a\Bigg[x - \underbrace{\left(-\frac{b}{2a}\right)}_{h}\Bigg]^2 + \underbrace{\frac{4ac - b^2}{4a}}_{k} \qquad \text{Write in the form } P(x) = a(x - h)^2 + k.$$

The final equation shows that the vertex $(h, k)$ can be expressed in terms of $a$, $b$, and $c$. It is not necessary to memorize the expression for $k$, since it is equal to $P(h) = P\left(-\frac{b}{2a}\right)$.

TECHNOLOGY NOTE

Most current models of graphing calculators are capable of determining the coordinates of the "highest point" or "lowest point" in a designated interval of a graph. Consult the graphing calculator manual that accompanies this text or your owner's manual to see if your calculator is capable of this. The highest or lowest point is usually designated with commands like "maximum" or "minimum." With this capability, you can find the coordinates of the vertex of a parabola graphically.

**Vertex Formula**

The vertex of the graph of $P(x) = ax^2 + bx + c$ $(a \neq 0)$ is the point

$$\left(-\frac{b}{2a}, P\left(-\frac{b}{2a}\right)\right).$$

**EXAMPLE 3** *Using the Vertex Formula*

Use the vertex formula to find the coordinates of the vertex of the graph of $P(x) = -.65x^2 + \sqrt{2}x + 4$.

**(a)** Give the exact values of the coordinates.

**(b)** Give the approximate values of the coordinates to the nearest hundredth.

**Analytic Solution**

**(a)** For this function, $a = -.65$ and $b = \sqrt{2}$. Applying the vertex formula,

$$x = -\frac{b}{2a}$$
$$= -\frac{\sqrt{2}}{2(-.65)} = \frac{\sqrt{2}}{2(.65)}$$

and
$$y = P\left(-\frac{b}{2a}\right)$$
$$= -.65\left(\frac{\sqrt{2}}{2(.65)}\right)^2 + \sqrt{2}\left(\frac{\sqrt{2}}{2(.65)}\right) + 4.$$

These are the *exact* values of $x$ and $y$.

**(b)** Using a calculator to approximate these values, we find that to the nearest hundredth,

$$x \approx 1.09 \quad \text{and} \quad y \approx 4.77.$$

**Graphing Calculator Solution**

**(a)** We cannot find exact values of the coordinates of the vertex with a calculator.

**(b)** Graphing the function and using the calculator to locate the highest point on the graph, we see a display of X = 1.0878556 and Y = 4.7692308 in Figure 12. These values agree with our analytic solution.

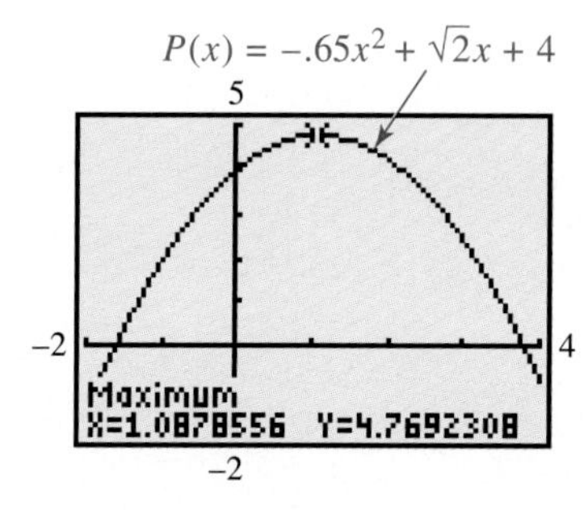


**FIGURE 12**

## Extreme Values

TECHNOLOGY NOTE

Graphing calculators have the capabilities of locating extrema to great accuracy.

The vertex of the graph of $P(x) = ax^2 + bx + c$ is the lowest point on the graph of the function if $a > 0$ and the highest point if $a < 0$. Such points are called **extreme points** (also **extrema**, singular: **extremum**). As we extend our study of polynomial functions, we will examine extrema on a more general basis. For the quadratic function defined by $P(x) = ax^2 + bx + c$,

**(a)** if $a > 0$, the vertex $(h, k)$ is called the *minimum point* of the graph. The *minimum value* of the function is $P(h) = k$.

**(b)** if $a < 0$, the vertex $(h, k)$ is called the *maximum point* of the graph. The *maximum value* of the function is $P(h) = k$.

Figure 13 illustrates these ideas.

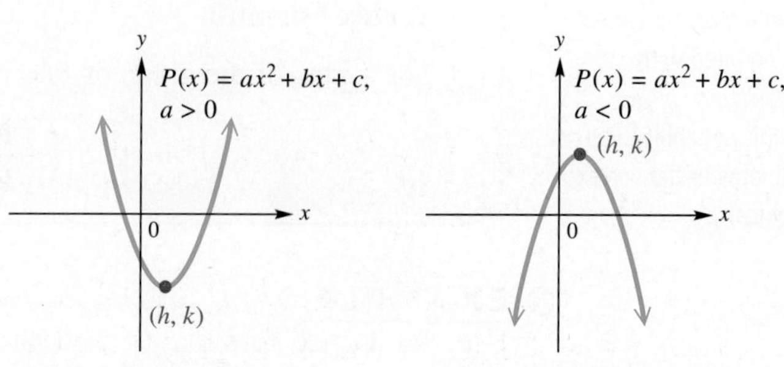


$(h, k)$ is the *minimum point.* $P(h) = k$ is the *minimum value of P.*

$(h, k)$ is the *maximum point.* $P(h) = k$ is the *maximum value of P.*

**FIGURE 13**

**EXAMPLE 4** *Identifying Extreme Points and Extreme Values*

Give the coordinates of the extreme point of the graph of each function and the corresponding maximum or minimum value of the function.

**(a)** $P(x) = 2x^2 + 4x - 16$
**(b)** $P(x) = -x^2 - 6x - 8$

### Analytic Solution

**(a)** In the discussion preceding Example 1, we found that the vertex of the graph of

$$P(x) = 2x^2 + 4x - 16$$

is $(-1, -18)$. The graph opens upward since $a > 0$ (as seen in Figure 9), so the vertex $(-1, -18)$ is the minimum point and $-18$ is the minimum value of the function.

**(b)** We found the vertex $(-3, 1)$ of this parabola in Example 1 and graphed the function in Example 2. Since $a < 0$ in

$$P(x) = -x^2 - 6x - 8,$$

the graph of the parabola opens downward (as seen in Figure 10), so $(-3, 1)$ is the maximum point and 1 is the maximum function value.

### Graphing Calculator Solution

**(a)** The screens in Figure 14 show one way to find the minimum point on the graph of $P(x) = 2x^2 + 4x - 16$. We use the fMin calculator function, which gives the $x$-value where the minimum value of a function occurs. The $y$-value, found by substitution, is the minimum value of the function.

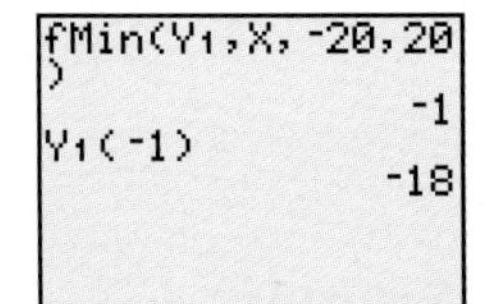


FIGURE 14

**(b)** The screen in Figure 15 shows an alternative way to find the coordinates of an extreme point using a graphing calculator. The results agree with our analytic results.

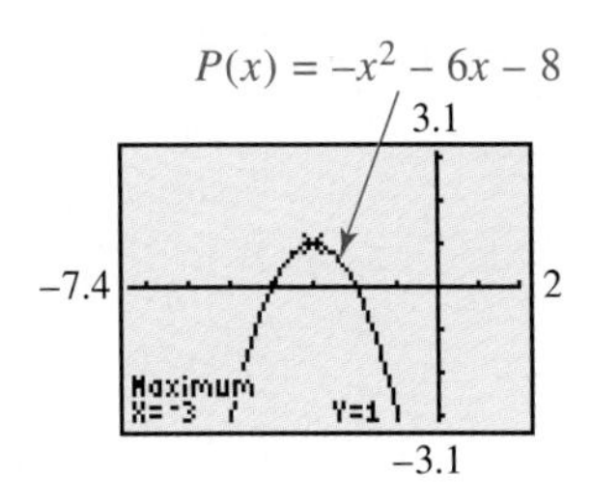


FIGURE 15

## Applications and Modeling

The fact that a quadratic function has either a maximum or a minimum function value makes quadratic functions good models for many applications.

**EXAMPLE 5** *Modeling Drug Prices with a Quadratic Function*

The table on the next page gives data for the percent increase $y$ in price of wholesale drugs for several years from 1990 through 1999, where $x$ is the number of years since 1990. These data are plotted in the scatter diagram in Figure 16.

| Year $x$ | Percent Increase $y$ |
|---|---|
| 1 | 7.2 |
| 3 | 3 |
| 5 | 1.9 |
| 7 | 2.5 |
| 9 | 4.2 |

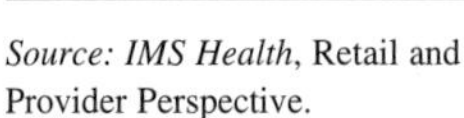

*Source: IMS Health*, Retail and Provider Perspective.

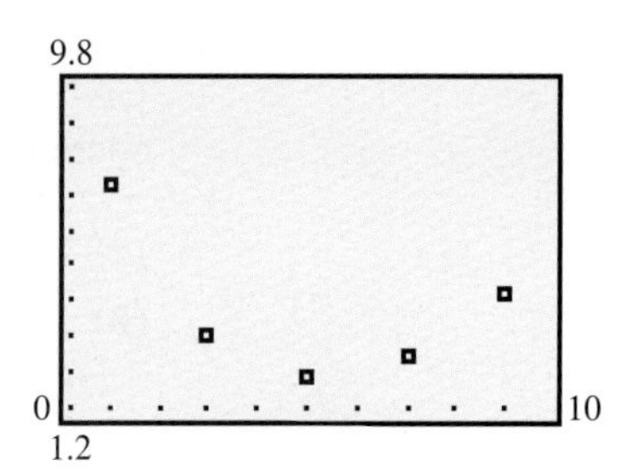


FIGURE 16

The scatter diagram suggests that a quadratic function with a positive value of $a$ (so that the graph opens upward) would be a good model for these data. The function defined by

$$f(x) = .188x^2 - 2.05x + 7.46$$

gives a good approximation of the data.

**(a)** Use $f(x)$ to approximate the year when the percent increase was a minimum.

**(b)** Find the minimum percent increase.

### Analytic Solution

Use the vertex formula.

**(a)** The $x$-value of the minimum point is

$$-\frac{b}{2a} = -\frac{-2.05}{2(.188)} \approx 5.45.$$

Since $x = 5$ corresponds to 1995, the percent increase was at a minimum in 1995.

**(b)** The minimum value is

$$f\left(-\frac{b}{2a}\right) = .188(5.45)^2 - 2.05(5.45) + 7.46$$
$$\approx 1.87.$$

Thus, the minimum percent increase was about 1.87, which closely approximates the data in the table.

### Graphing Calculator Solution

Use a calculator to graph the function, and then use the capabilities of the calculator to get the coordinates of the minimum point. See Figure 17.

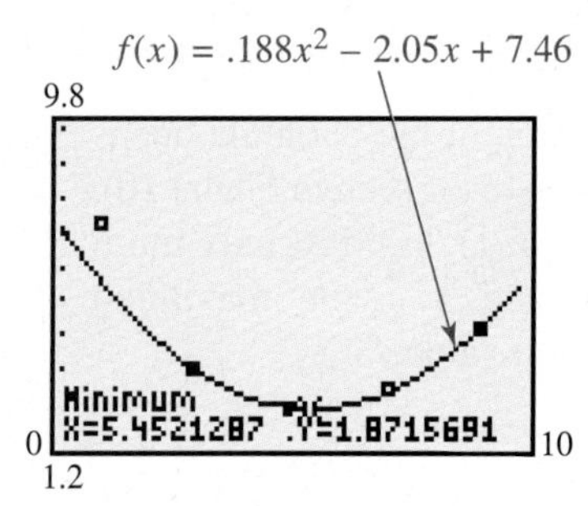


FIGURE 17

An important application of quadratic functions deals with the height of a propelled object as a function of the time elapsed after it is propelled.

### Height of a Propelled Object

If air resistance is neglected, the height $s$ (in feet) of an object propelled directly upward from an initial height $s_0$ feet with initial velocity $v_0$ feet per second is

$$s(t) = -16t^2 + v_0 t + s_0,$$

where $t$ is the number of seconds after the object is propelled.

The coefficient of $t^2$ (that is, $-16$) is a constant based on the gravitational force of Earth. This constant varies on other surfaces, such as the moon or the other planets. Here we have an example of a quadratic function in which height is a function of time. We use $t$ to represent the independent variable; however, when graphing this type of function on a calculator, we will use $x$, since graphing calculators use $x$ for the independent variable. There is no difference between the formulas $s(t) = -16t^2 + v_0t + s_0$ and $s(x) = -16x^2 + v_0x + s_0$.

**EXAMPLE 6** *Solving a Problem Involving Projectile Motion*

A ball is thrown directly upward from an initial height of 100 feet with an initial velocity of 80 feet per second.

**(a)** Give the function that describes the height of the ball in terms of time $t$.

**(b)** Graph this function so that the $y$-intercept, the positive $x$-intercept, and the vertex are visible.

**(c)** The cursor in Figure 18 shows that the point (4.8, 115.36) lies on the graph of the function. What does this mean for this particular situation?

**(d)** After how many seconds does the projectile reach its maximum height? What is this maximum height? Solve analytically and graphically.

**(e)** For what interval of time is the height of the ball greater than 160 feet? Determine the answer graphically.

**(f)** After how many seconds will the ball fall to the ground? Determine the answer graphically.

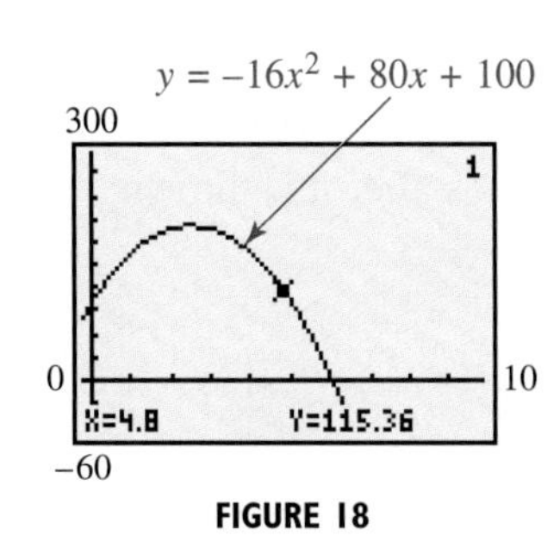


FIGURE 18

**Solution**

**(a)** Use the projectile height function with $v_0 = 80$ and $s_0 = 100$:

$$s(t) = -16t^2 + 80t + 100.$$

**(b)** There are many suitable choices for such a window. One choice is $[0, 10]$ by $[-60, 300]$, as shown in Figure 18. It shows the graph of $y = -16x^2 + 80x + 100$. (Here, $x = t$.)

**(c)** When $x = 4.8$, $y = 115.36$. Therefore, when 4.8 seconds have elapsed, the projectile is at a height of 115.36 feet.

**(d)** To answer this question, find the coordinates of the vertex of the parabola. Using the vertex formula,

$$x = -\frac{b}{2a} = -\frac{80}{2(-16)} = 2.5$$

and $$y = -16(2.5)^2 + 80(2.5) + 100 = 200.$$

Therefore, after 2.5 seconds the ball reaches its maximum height of 200 feet.

Using the capabilities of the calculator, we see that the vertex coordinates are indeed (2.5, 200). See Figure 19.

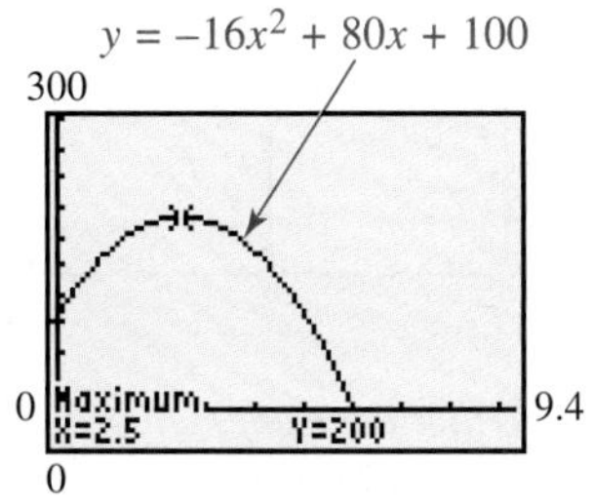

y = −16x² + 80x + 100
300
0 Maximum X=2.5 Y=200 9.4
0

FIGURE 19

**CAUTION** It is easy to misinterpret the graph in Figure 19. This graph does not define the *path* followed by the ball; it defines height as a function of time.

**(e)** With $y_1 = -16x^2 + 80x + 100$ and $y_2 = 160$ graphed, as shown in Figure 20, locate the two points of intersection. The $x$-coordinates for these points are approximately .92 and 4.08. Therefore, between .92 and 4.08 seconds, the ball is more than 160 feet above the ground, that is, $y_1 > y_2$.

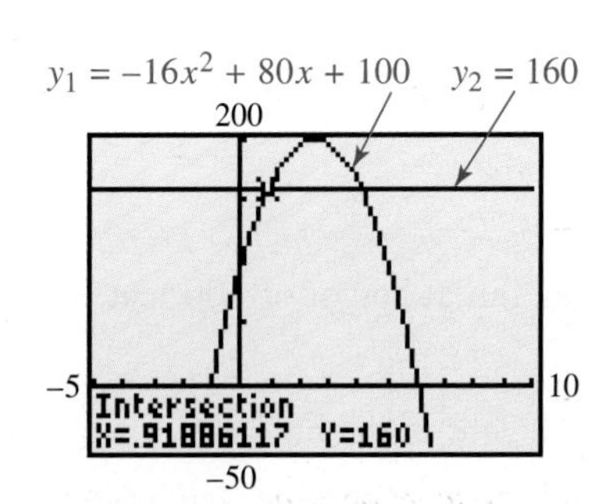


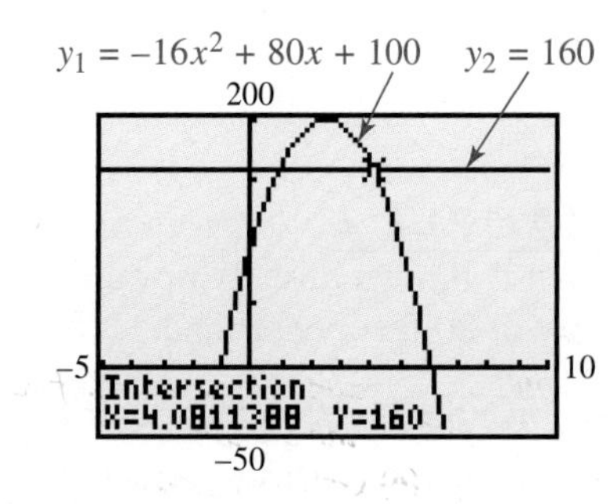


FIGURE 20

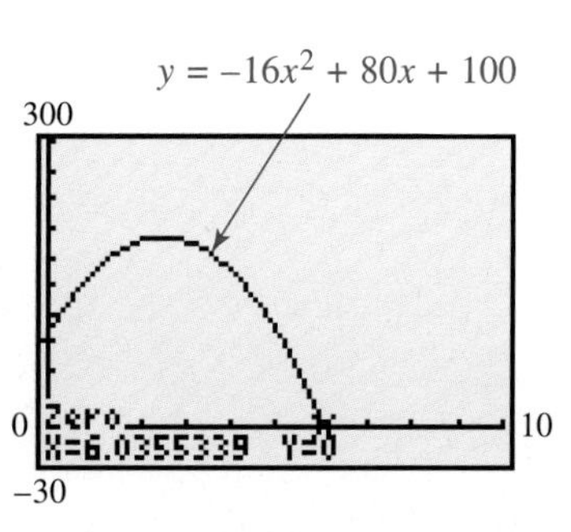


FIGURE 21

**(f)** When the ball falls to the ground, its height will be 0 feet, so find the positive $x$-intercept of the graph. As shown in Figure 21, this intercept is about 6.04, which means that the ball reaches the ground about 6.04 seconds after it is thrown.

## 3.2 EXERCISES

*Concept Check* *The graphs of the functions in Exercises 1–4 are shown in Figures A–D. Match each function with its graph, using the concepts of this section without actually entering the function into your calculator. Then, after you have completed the exercises, check your answers with your calculator. Use the standard viewing window.*

**1.** $y = (x - 4)^2 - 3$ **2.** $y = -(x - 4)^2 + 3$ **3.** $y = (x + 4)^2 - 3$ **4.** $y = -(x + 4)^2 + 3$

**A.** **B.** **C.** **D.**

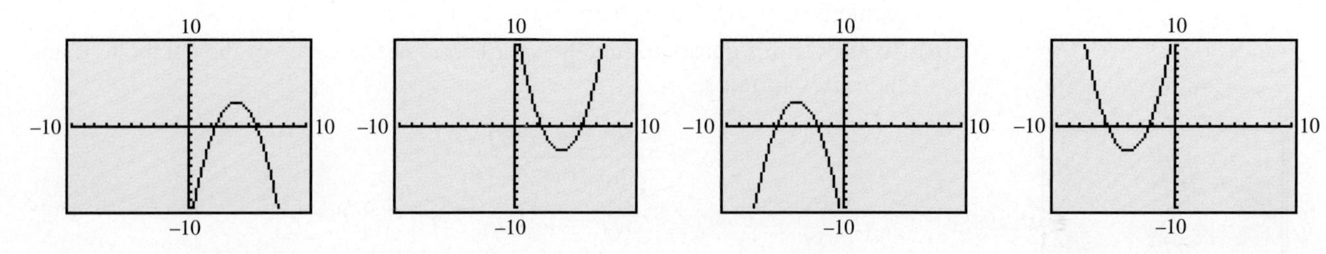


*For each quadratic function defined in Exercises 5–12,* **(a)** *complete the square to write the equation so that the right side is in the form $a(x - h)^2 + k$;* **(b)** *give the vertex of the parabola; and* **(c)** *support your answers using a calculator graph in a viewing window that shows a comprehensive graph of the function.*

**5.** $P(x) = 2x^2 - 2x - 24$ **6.** $P(x) = 3x^2 + 3x - 6$ **7.** $y = x^2 - 2x - 15$

**8.** $y = -x^2 - 3x + 10$ **9.** $f(x) = -2x^2 + 6x$ **10.** $f(x) = 4x^2 - 4x$

**11.** $P(x) = 4x^2 - 22x - 12$ **12.** $P(x) = 6x^2 - 16x - 6$

**13.** *Concept Check* Match each equation with the description of the parabola that is its graph.

**(a)** $y = (x - 4)^2 - 2$

**(b)** $y = (x - 2)^2 - 4$

**(c)** $y = -(x - 4)^2 - 2$

**(d)** $y = -(x - 2)^2 - 4$

**A.** Vertex $(2, -4)$, opens downward

**B.** Vertex $(2, -4)$, opens upward

**C.** Vertex $(4, -2)$, opens downward

**D.** Vertex $(4, -2)$, opens upward

**14.** *Concept Check* In Chapter 2 we saw how certain changes to an equation cause the graph of the equation to be stretched, shrunken, reflected across an axis, or translated vertically or horizontally. It is important to notice that the order in which these changes are done affects the final graph. For example, stretching and then shifting vertically produces a graph that differs from the one produced by shifting vertically, then stretching. To see this, use a graphing calculator to graph $y = 3x^2 - 2$ and $y = 3(x^2 - 2)$, and then compare the results. Are the two equations equivalent algebraically?

*In Exercises 15–20, you are given the equation and graph of a quadratic function. Without using your calculator, do each of the following.* **(a)** *Give the coordinates of the vertex.* **(b)** *Give the domain and range.* **(c)** *Give the equation of the axis of symmetry.* **(d)** *Give the interval over which the function is increasing.* **(e)** *Give the interval over which the function is decreasing.* **(f)** *State whether the vertex is a maximum or minimum point, and give the corresponding maximum or minimum value of the function.*

**15.** $P(x) = (x - 2)^2$

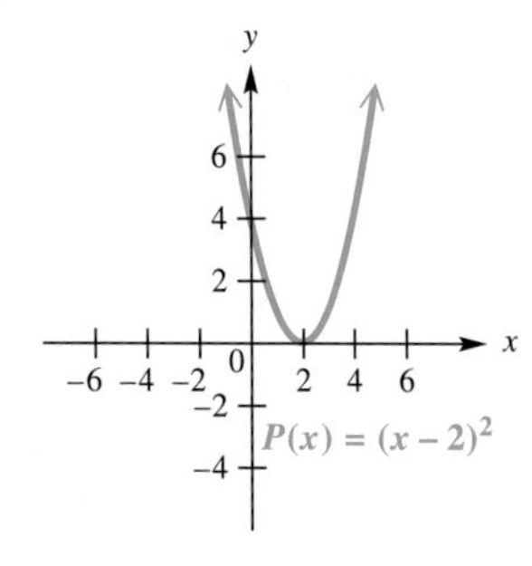


**16.** $P(x) = (x + 4)^2$

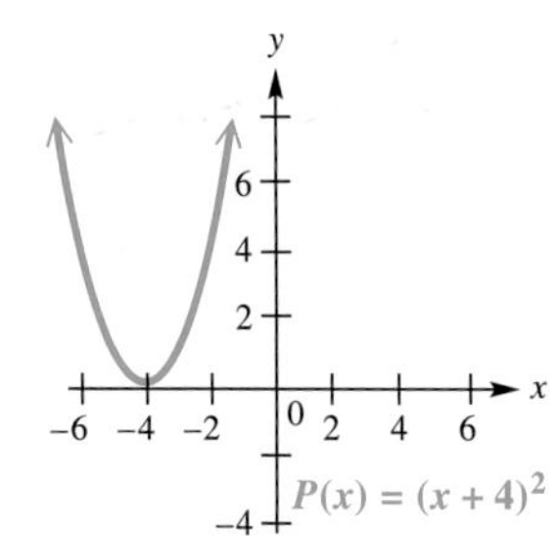


**17.** $y = (x + 3)^2 - 4$

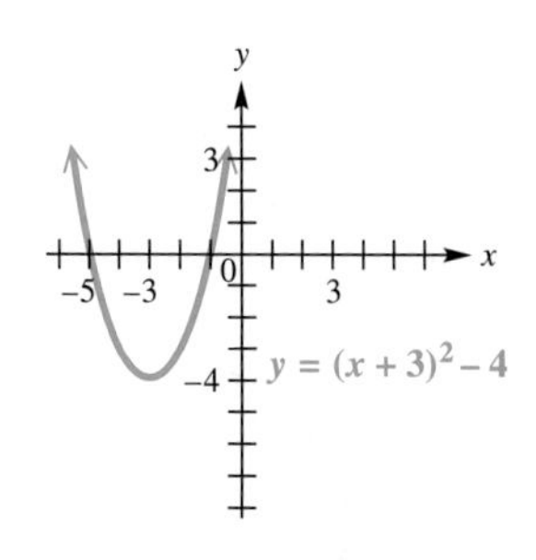


**18.** $P(x) = -.5(x + 1)^2 - 3$

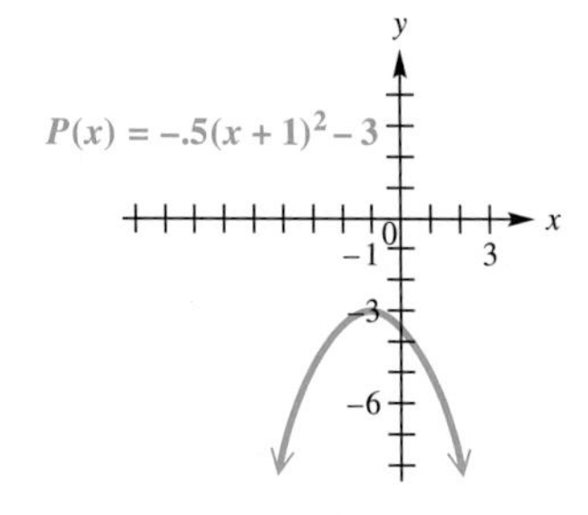


**19.** $f(x) = -2(x + 3)^2 + 2$

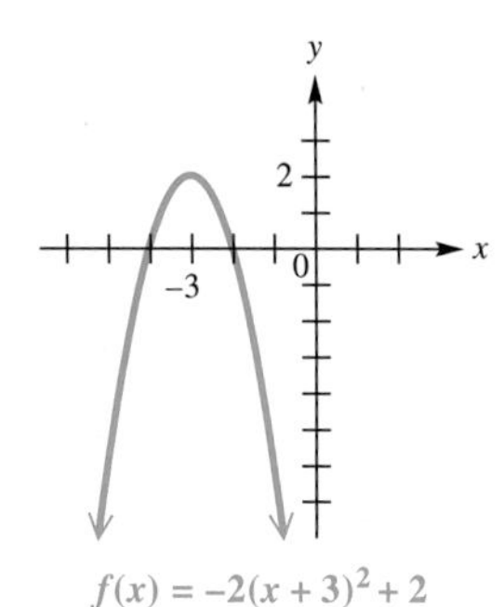


**20.** $f(x) = -3(x - 2)^2 + 1$

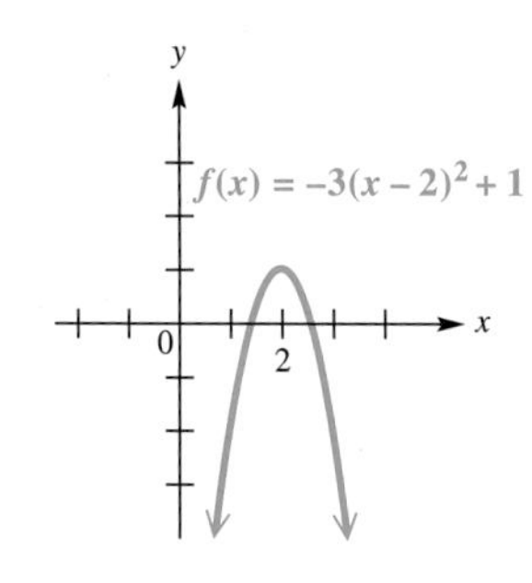


*For each quadratic function defined in Exercises 21–26,* **(a)** *use the vertex formula to find the coordinates of the vertex and* **(b)** *graph the function by hand.*

**21.** $P(x) = x^2 - 10x + 21$

**22.** $P(x) = x^2 - 2x + 3$

**23.** $y = -x^2 + 4x - 2$

**24.** $y = -x^2 + 2x + 1$

**25.** $f(x) = 2x^2 - 4x + 5$

**26.** $f(x) = -3x^2 + 24x - 46$

*Graph each function in a viewing window that will allow you to use your calculator to approximate* **(a)** *the coordinates of the vertex and* **(b)** *the x-intercepts. Give values to the nearest hundredth.*

**27.** $P(x) = -.32x^2 + \sqrt{3}x + 2.86$

**28.** $P(x) = -\sqrt{2}x^2 + .45x + 1.39$

**29.** $y = 1.34x^2 - 3x + \sqrt{5}$

*Concept Check* *The figure shows the graph of a quadratic function $y = f(x)$. Use it to work Exercises 30–32.*

**30.** What is the minimum value of $f(x)$?

**31.** For what value of $x$ is $f(x)$ as small as possible?

**32.** How many real solutions are there to the equation $f(x) = 1$?

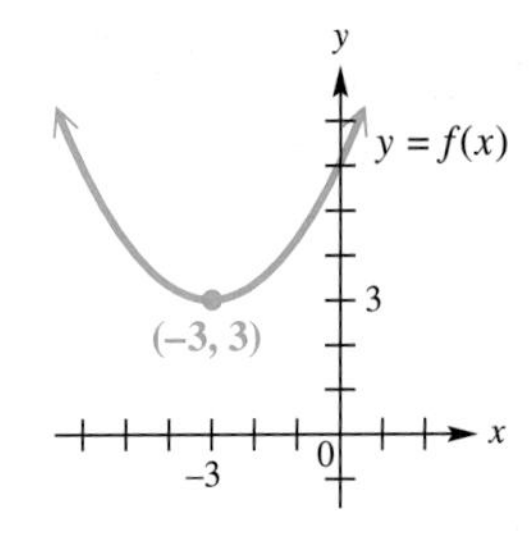


*In each table, $Y_1$ is defined as a quadratic function. Use the table to do the following.* **(a)** *Determine the coordinates of the vertex of the graph.* **(b)** *Determine whether the vertex is a minimum point or a maximum point.* **(c)** *Find the minimum or maximum value of the function.* **(d)** *Determine the range of the function.*

**33.**

| X | Y1 |
|---|---|
| 1 | 15 |
| 2 | 0 |
| 3 | -9 |
| 4 | -12 |
| 5 | -9 |
| 6 | 0 |
| 7 | 15 |

X=1

**34.**

| X | Y1 |
|---|---|
| -6 | 27 |
| -5 | 17 |
| -4 | 11 |
| -3 | 9 |
| -2 | 11 |
| -1 | 17 |
| 0 | 27 |

X= -6

**35.**

| X | Y1 |
|---|---|
| 0 | -7 |
| .5 | -2 |
| 1 | 1 |
| 1.5 | 2 |
| 2 | 1 |
| 2.5 | -2 |
| 3 | -7 |

X=0

**36.**

| X | Y1 |
|---|---|
| -4 | -9.75 |
| -3.5 | -6 |
| -3 | -3.75 |
| -2.5 | -3 |
| -2 | -3.75 |
| -1.5 | -6 |
| -1 | -9.75 |

X= -4

*Curve Fitting* *Exercises 37–42 show scatter diagrams of sets of data. In each case, tell whether a linear or quadratic model is appropriate for the data. For each linear model, decide whether the slope should be positive or negative. For each quadratic model, decide whether the coefficient a should be positive or negative.*

**37.** Social Security assets as a function of time

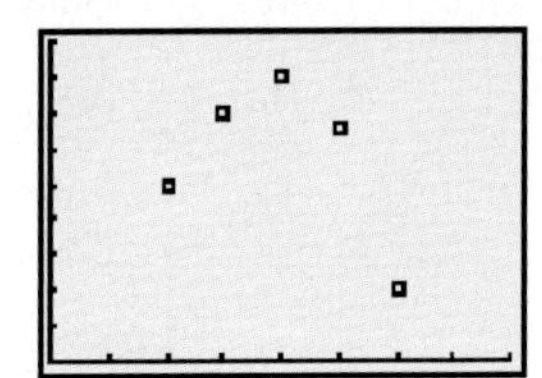

**38.** Growth in science centers/museums as a function of time

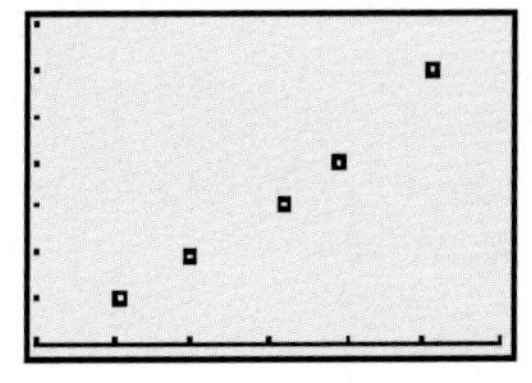

**39.** Value of U.S. salmon catch as a function of time

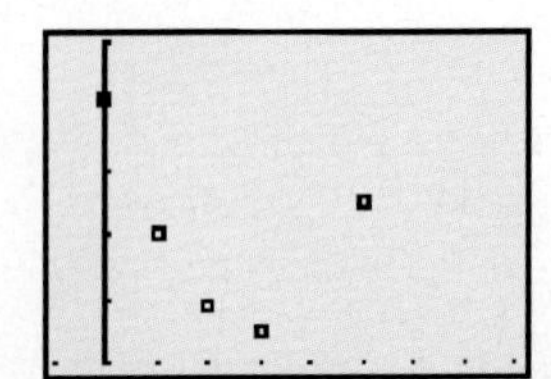

**40.** Height of an object thrown upward from a building as a function of time

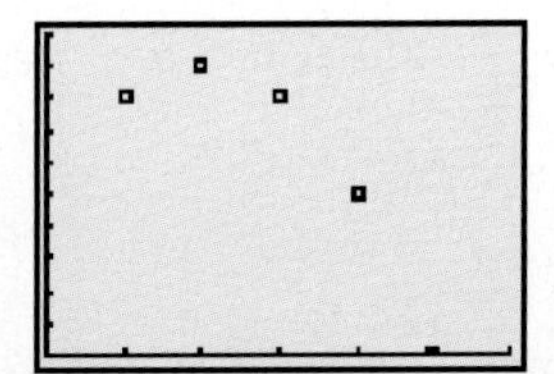

**41.** Number of shopping centers as a function of time

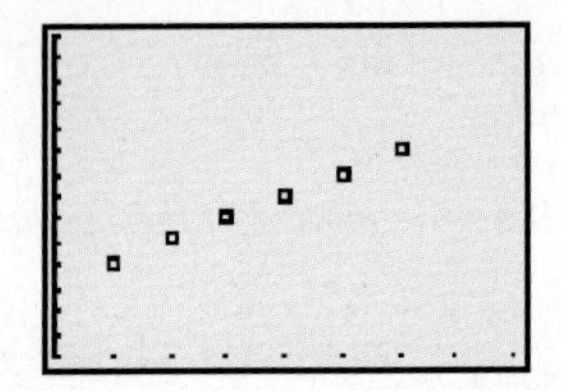

**42.** Newborns with AIDS as a function of time

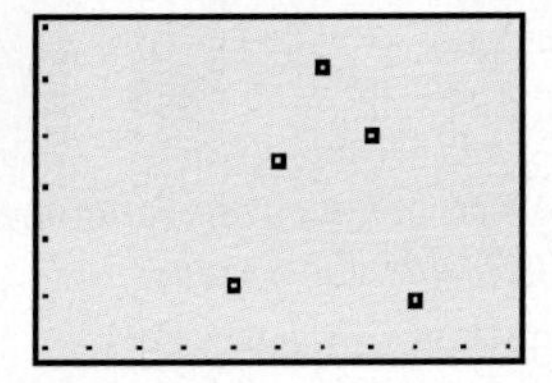

*(Modeling) Solve each problem.*

**43.** *Company Bankruptcy Filings* The number of company bankruptcy filings for selected years between 1990 and 1999 are shown in the table. In the year column, 0 represents 1990, 2 represents 1992, and so on.

| Year | Number of Bankruptcies |
|---|---|
| 0 | 115 |
| 2 | 91 |
| 4 | 70 |
| 6 | 84 |
| 8 | 120 |
| 9 | 145 |

*Source:* www.BankruptcyData.com

**(a)** Plot the data. Use your scatter diagram to decide whether a linear or quadratic function would better model the data. If quadratic, should the coefficient $a$ of $x^2$ be positive or negative?

**(b)** The data are approximated by the quadratic function defined by

$$f(x) = 2.797x^2 - 22.18x + 117.7,$$

where $x = 0$ represents 1990, $x = 2$ represents 1992, and so on. Analytically determine the coordinates of the vertex of the graph to the nearest whole number.

**(c)** Interpret the answer to part (b) as it relates to this application. How does your answer compare to the data given in the table?

**44.** *Social Security Assets* The graph shows how Social Security assets are expected to change as the number of retirees receiving benefits increases.

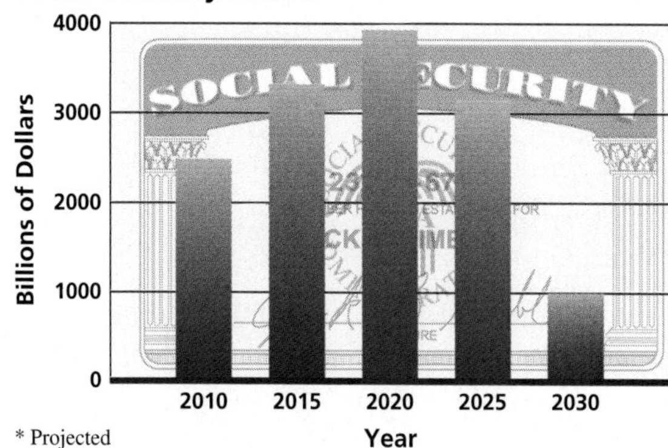


* Projected

*Source:* Social Security Administration.

The graph suggests that a quadratic function would be a good fit to the data. The data are approximated by the function defined by

$$f(x) = -20.57x^2 + 758.9x - 3140.$$

In the model, $x = 10$ represents 2010, $x = 15$ represents 2015, and so on, and $f(x)$ is in billions of dollars.

**(a)** Explain why the coefficient of $x^2$ in the model is negative, based on the graph.

**(b)** Analytically determine the vertex of the graph.

**(c)** Interpret the answer to part (b) as it relates to this application.

**45.** *Domestic Oyster Catch* The U.S. domestic oyster catch (in millions) for the years 1990–1998 can be approximated by the quadratic function defined by

$$f(x) = -.566x^2 + 5.08x + 29.2,$$

where $x = 0$ represents 1990, $x = 1$ represents 1991, and so on. (*Source:* National Marine Fisheries Service.)

**(a)** Since the coefficient of $x^2$ in the model is negative, the graph of this quadratic function is a parabola that opens downward. Will the $y$-value of the vertex of this graph be a maximum or minimum?

**(b)** In what year was the maximum domestic oyster catch? (Round down to the nearest year.) Use the actual $x$-value of the vertex, to the nearest tenth, to find this catch.

**46.** *Tuition Increase* The percent increase for in-state tuition at Iowa public universities during the years 1992–2002 can be modeled by the quadratic function defined by

$$f(x) = .156x^2 - 2.05x + 10.2,$$

where $x = 2$ represents 1992, $x = 3$ represents 1993, and so on. (*Source:* Iowa Board of Regents.)

**(a)** Based on this model, by what percent (to the nearest tenth) did tuition increase in 2001?

**(b)** In what year was the minimum tuition increase? (Round down to the nearest year.) To the nearest tenth, by what percent did tuition increase that year?

*(Modeling) Use the formula given in this section for the height of a propelled object to solve each problem.*

**47.** *Height of a Propelled Rock* A rock is projected upward from ground level with an initial velocity of 90 feet per second. Let $t$ represent the amount of time elapsed after it is thrown.

**(a)** Explain why $t$ cannot be a negative number in this situation.

**(b)** Explain why $s_0 = 0$ in this problem.

**(c)** Give the function $s$ that describes the height of the rock as a function of $t$.

**(d)** How high will the rock be 1.5 seconds after it is thrown?

**(e)** What is the maximum height attained by the rock? After how many seconds will this happen? Determine the answer analytically and graphically.

**(f)** After how many seconds will the rock hit the ground? Determine the answer graphically.

**48.** *Height of a Toy Rocket* A toy rocket is launched from the top of a building 50 feet tall at an initial velocity of 200 feet per second. Let $t$ represent the amount of time elapsed after the launch.

**(a)** Express the height $s$ as a function of the time $t$.

**(b)** Determine both analytically and graphically the time at which it reaches its highest point. How high will it be at that time?

**(c)** For what time interval will the rocket be more than 300 feet above ground level? Determine the answer graphically, and give times to the nearest tenth of a second.

**(d)** After how many seconds will it hit the ground? Determine the answer graphically.

**49.** *Height of a Propelled Ball* A ball is hit upward from ground level with an initial velocity of 150 feet per second.

**(a)** Determine graphically whether the ball will reach a height of 355 feet. If it will, determine the time(s) when this happens. If it will not, explain why, using a graphical interpretation.

**(b)** Repeat part (a) for a ball hit from a height of 30 feet with an initial velocity of 250 feet per second.

**50.** *Height of a Propelled Ball on the Moon* An astronaut on the moon throws a baseball upward. The astronaut is 6 feet 6 inches tall, and the initial velocity of the ball is 30 feet per second. The height of the ball is approximated by the function defined by

$$s(t) = -2.7t^2 + 30t + 6.5,$$

where $t$ is the number of seconds after the ball was thrown.

**(a)** After how many seconds is the ball 12 feet above the moon's surface?

**(b)** How many seconds will it take for the ball to return to the surface?

**(c)** The ball will never reach a height of 100 feet. How can this be determined analytically?

*Find the equation of the quadratic function satisfying the given conditions.* (Hint: *Find $a$, $h$, and $k$ that satisfy $P(x) = a(x - h)^2 + k$.*) *Use your calculator to support your results. Express your answer in the form $P(x) = ax^2 + bx + c$.*

**51.** Vertex $(-1, -4)$; through $(5, 104)$

**52.** Vertex $(-2, -3)$; through $(0, -19)$

**53.** Vertex $(8, 3)$; through $(10, 5)$

**54.** Vertex $(-6, -12)$; through $(6, 24)$

**55.** Vertex $(-4, -2)$; through $(2, -26)$

**56.** Vertex $(5, 6)$; through $(1, -6)$

*Concept Check Sketch a graph that satisfies each set of conditions.*

**57.** Vertex $(-2, -3)$; through $(1, 4)$

**58.** Vertex $(5, 6)$; through $(1, -6)$

**59.** Maximum value of 1 at $x = 3$; $y$-intercept is $-4$

**60.** Minimum value of $-4$ at $x = -3$; $y$-intercept is 3

# 3.3 QUADRATIC EQUATIONS AND INEQUALITIES

Zero-Product Property • Solving $x^2 = k$ • Quadratic Formula and the Discriminant • Solving Quadratic Equations • Solving Quadratic Inequalities • Formulas Involving Quadratics • Another Quadratic Model

In this section we develop general methods of solving quadratic equations and extend them to solving quadratic inequalities.

**Quadratic Equation in One Variable**

An equation that can be written in the form

$$ax^2 + bx + c = 0,$$

where $a$, $b$, and $c$ are real numbers with $a \neq 0$, is a **quadratic equation in standard form.**

## Zero-Product Property

Consider the following problems:

1. Find the zeros of the quadratic function defined by $P(x) = 2x^2 + 4x - 16$.
2. Find the $x$-intercepts of the graph of $P(x) = 2x^2 + 4x - 16$.
3. Find the solution set of the equation $2x^2 + 4x - 16 = 0$.

Each of these problems is solved by finding the numbers that make $2x^2 + 4x - 16$ equal 0. The simplest way to solve a quadratic equation $P(x) = 0$ is by using the zero-product property. However, this property works only for equations where $P(x)$ is factorable.

**Zero-Product Property**

If $a$ and $b$ are complex numbers and $ab = 0$, then $a = 0$ or $b = 0$ or both.

**EXAMPLE 1** *Using the Zero-Product Property (Two Solutions)*

Solve $2x^2 + 4x - 16 = 0$.

**Analytic Solution**

To make our work easier, we first divide each side by 2.

$$2x^2 + 4x - 16 = 0$$

$$x^2 + 2x - 8 = 0 \quad \text{Divide by 2.}$$

$$(x + 4)(x - 2) = 0 \quad \text{Factor.}$$

$$x + 4 = 0 \quad \text{or} \quad x - 2 = 0 \quad \text{Zero-product property}$$

$$x = -4 \quad \text{or} \quad x = 2 \quad \text{Solve each equation.}$$

**Graphing Calculator Solution**

The screens in Figure 22 on the next page show the graph of

$$P(x) = 2x^2 + 4x - 16$$

with the zeros $-4$ and 2. That is, $P(-4) = 0$ and $P(2) = 0$. Since the zeros satisfy $P(x) = 0$, they are also the solutions of the equation and the $x$-intercepts of the graph.

*(continued)*

The solution set of the equation is $\{-4, 2\}$. We can check each solution by substituting it into the given equation. These results also tell us that the zeros of $P(x)$ and the $x$-intercepts of the graph of $P(x)$ are $-4$ and 2.

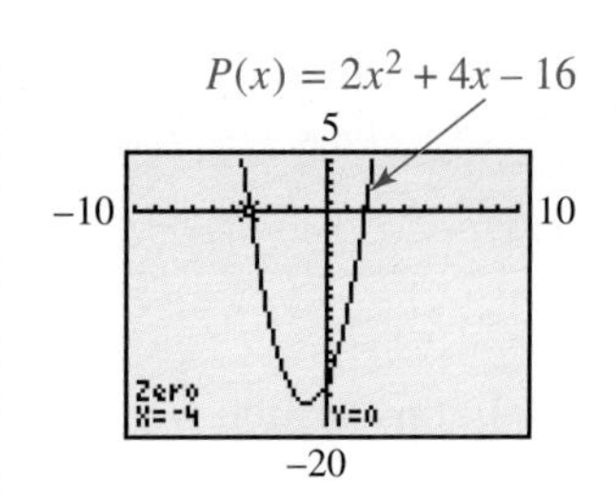


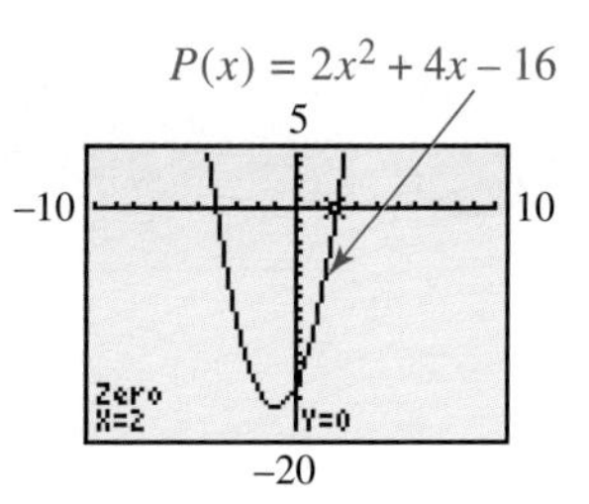


FIGURE 22

**EXAMPLE 2** *Using the Zero-Product Property (One Solution)*

Find all zeros of the quadratic function defined by $P(x) = x^2 - 6x + 9$.

**Analytic Solution**

We must solve $x^2 - 6x + 9 = 0$.

$$x^2 - 6x + 9 = 0$$

$$(x - 3)^2 = 0 \quad \text{Factor.}$$

$$x - 3 = 0 \quad \text{or} \quad x - 3 = 0 \quad \text{Zero-product property}$$

$$x = 3 \quad \text{or} \quad x = 3$$

There is only one *distinct* zero, 3. It is sometimes called a double zero, or double solution (root) of the equation.

**Graphing Calculator Solution**

The graph of the function in Figure 23 shows why there is only one zero: The graph is tangent to the $x$-axis at the vertex, (3, 0). The table also indicates that a zero occurs at $x = 3$, which must be a minimum value. Because of the symmetry shown in the table, 3 is the only zero, which numerically supports our analytic result.

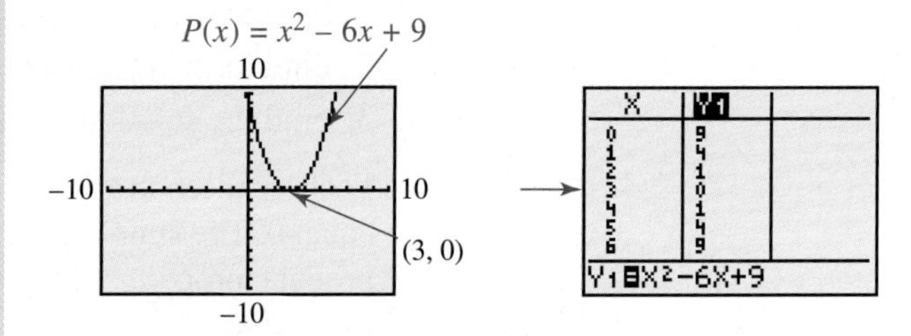


FIGURE 23

**CAUTION** A purely graphical approach may not prove to be as useful with a graph like the one in Figure 23, since the vertex may be slightly above or slightly below the $x$-axis. (See the discussion of "hidden behavior" in Section 3.5.) **This is why we need to understand the algebraic concepts presented in this section—only when we know the mathematics can we use the technology to its utmost.**

**FOR DISCUSSION**

Figure 24 shows possible numbers of $x$-intercepts of the graph of a quadratic function that opens upward.

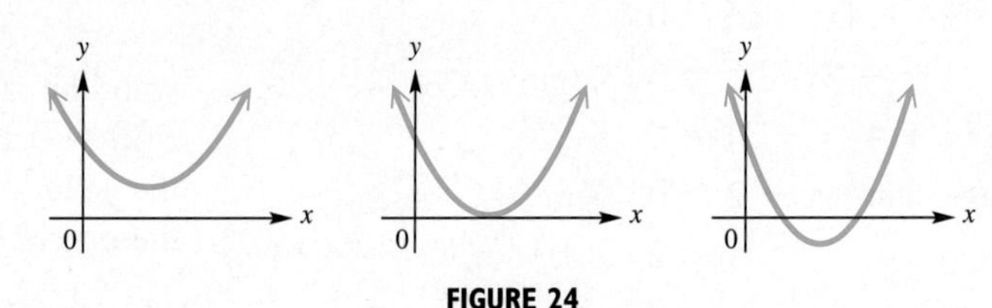


FIGURE 24

Figure 25 shows possible numbers of $x$-intercepts of the graph of a quadratic function that opens downward.

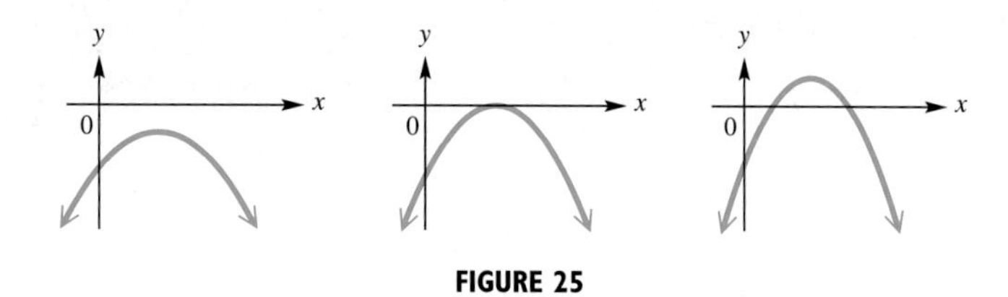

**FIGURE 25**

Use these figures to answer each question.

**1.** What is the maximum number of real solutions of a quadratic equation?

**2.** What is the minimum number of real solutions of a quadratic equation?

**3.** If a quadratic function has only one real zero, what do we know about the vertex of its graph?

The zero-product property is quite limited in its practical application. It is only useful if the quadratic expression can be factored easily. We now develop more general methods for solving quadratic equations.

## Solving $x^2 = k$

We can solve quadratic equations of the form $x^2 = k$, where $k$ is a real number, by factoring, using the following sequence of equivalent equations.

$$x^2 = k$$

$$x^2 - k = 0 \qquad \text{Subtract } k.$$

$$(x - \sqrt{k})(x + \sqrt{k}) = 0 \qquad \text{Factor.}$$

$$x - \sqrt{k} = 0 \quad \text{or} \quad x + \sqrt{k} = 0 \qquad \text{Zero-product property}$$

$$x = \sqrt{k} \quad \text{or} \quad x = -\sqrt{k}$$

We call this result the *square root property* for solving quadratic equations.

**Square Root Property**

The solution set of $x^2 = k$ is

**(a)** $\{\pm\sqrt{k}\}$ **if** $k > 0$. **(b)** $\{0\}$ **if** $k = 0$. **(c)** $\{\pm i\sqrt{|k|}\}$ **if** $k < 0$.

As shown in Figure 26 on the next page, the graph of $y_1 = x^2$ intersects the graph of $y_2 = k$ twice if $k > 0$, once if $k = 0$, and not at all if $k < 0$.

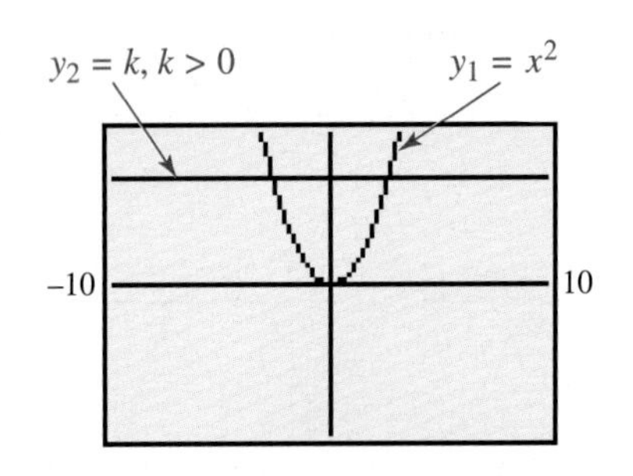


There are two points of intersection if $k > 0$.

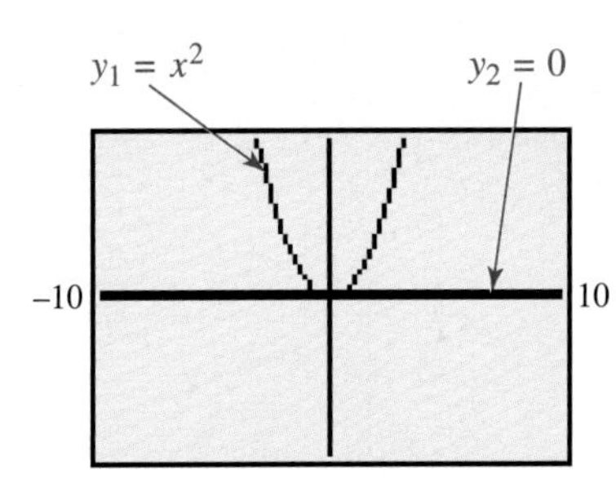


There is only one point of intersection (the origin) if $k = 0$.

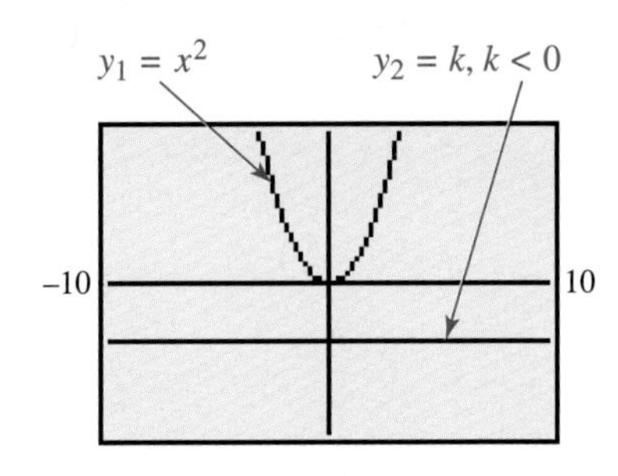


There are no points of intersection if $k < 0$.

**FIGURE 26**

### EXAMPLE 3 *Using the Square Root Property*

Solve each equation.

**(a)** $2x^2 = -10$ **(b)** $2(x - 1)^2 = 14$

**Analytic Solution**

**(a)** Divide each side by 2.

$$2x^2 = -10$$
$$x^2 = -5$$

There is no real number whose square is $-5$. However, this equation has two complex imaginary solutions. Use the square root property to find them.

$$x = \pm\sqrt{-5}$$
$$= \pm i\sqrt{5}$$

The solution set is $\{\pm i\sqrt{5}\}$. These solutions can be checked by substituting $i\sqrt{5}$ and $-i\sqrt{5}$ into the given equation.

**(b)** First solve for the squared quantity. Then use the square root property.

$$2(x - 1)^2 = 14$$
$$(x - 1)^2 = 7 \quad \text{Divide by 2.}$$
$$x - 1 = \pm\sqrt{7} \quad \text{Square root property}$$
$$x = 1 \pm \sqrt{7} \quad \text{Add 1.}$$

The solution set is $\{1 \pm \sqrt{7}\}$.

**Graphing Calculator Solution**

**(a)** Notice that the graphs of $y_1 = 2x^2$ and $y_2 = -10$ in Figure 27 do not intersect. This indicates that there are no *real* solutions.

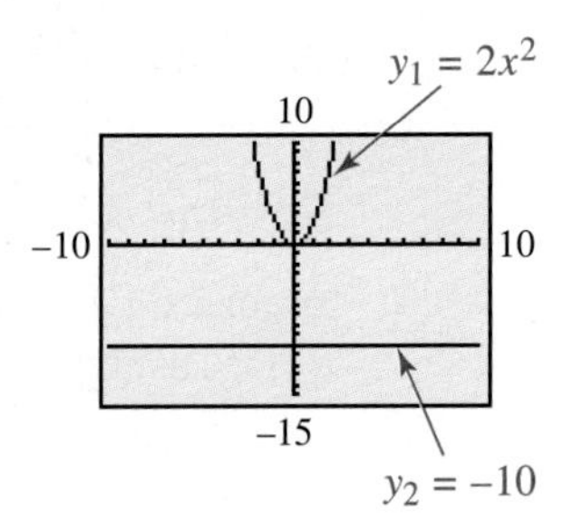


**FIGURE 27**

**(b)** Graph $y_1 = 2(x - 1)^2$ and $y_2 = 14$ and locate the points of intersection. The screens in Figure 28 show the $x$-coordinates of these points, which are approximations of $1 \pm \sqrt{7}$, leading to the same solution set, $\{1 \pm \sqrt{7}\}$.

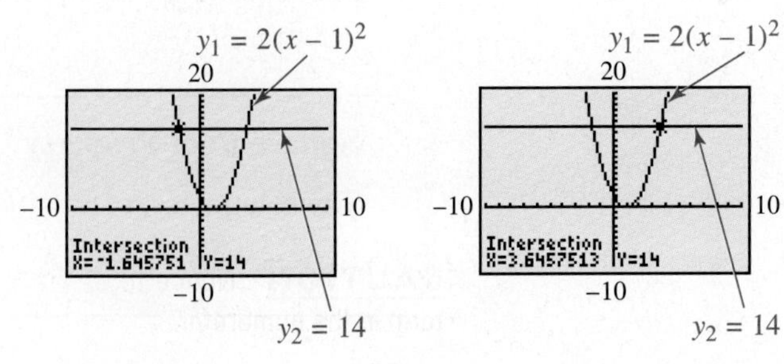


**FIGURE 28**

TECHNOLOGY NOTE

Programs for the various makes and models of graphing calculators are available from the manufacturers, users' groups, and the Web site for this text.

## Quadratic Formula and the Discriminant

If a quadratic equation in standard form, $ax^2 + bx + c = 0$, can be written in the form $a(x - h)^2 + k = 0$, we can solve it as we did the equation in Example 3(b). We saw in Section 3.2 that by the method of completing the square we can write $P(x) = ax^2 + bx + c$ in the form $P(x) = a(x - h)^2 + k$. Following the steps of that method, we can solve any quadratic equation. We start with the quadratic equation

$$ax^2 + bx + c = 0 \quad (a \neq 0),$$

and solve for $x$ in terms of the constants $a$, $b$, and $c$ by completing the square. For now, we assume $a > 0$ and divide each side by $a$ to obtain

$$x^2 + \frac{b}{a}x + \frac{c}{a} = 0.$$

$$x^2 + \frac{b}{a}x = -\frac{c}{a} \qquad \text{Add } -\frac{c}{a}.$$

$$x^2 + \frac{b}{a}x + \frac{b^2}{4a^2} = \frac{b^2}{4a^2} - \frac{c}{a} \qquad \text{Add } \left[\frac{1}{2}\left(\frac{b}{a}\right)\right]^2 = \frac{b^2}{4a^2}.$$

$$\left(x + \frac{b}{2a}\right)^2 = \frac{b^2 - 4ac}{4a^2} \qquad \text{Factor on the left; simplify on the right.}$$

By the square root property, this last statement leads to

$$x + \frac{b}{2a} = \sqrt{\frac{b^2 - 4ac}{4a^2}} \quad \text{or} \quad x + \frac{b}{2a} = -\sqrt{\frac{b^2 - 4ac}{4a^2}}.$$

$$x + \frac{b}{2a} = \frac{\sqrt{b^2 - 4ac}}{2a} \quad \text{or} \quad x + \frac{b}{2a} = \frac{-\sqrt{b^2 - 4ac}}{2a} \qquad \sqrt{4a^2} = 2a \ (a > 0)$$

Adding $-\frac{b}{2a}$ to each side of each result gives

$$x = \frac{-b + \sqrt{b^2 - 4ac}}{2a} \quad \text{or} \quad x = \frac{-b - \sqrt{b^2 - 4ac}}{2a}.$$

It can be shown that these two results are also valid if $a < 0$. A compact form of these two equations, called the *quadratic formula*, follows.

**Quadratic Formula**

The solutions of the quadratic equation $\boldsymbol{ax^2 + bx + c = 0}$, where $a \neq 0$, are

$$\boldsymbol{x = \frac{-b \pm \sqrt{b^2 - 4ac}}{2a}}.$$

**CAUTION** Notice that the fraction bar in the quadratic formula extends under the $-b$ term in the numerator.

The expression under the radical in the quadratic formula, $b^2 - 4ac$, is called the **discriminant.** The value of the discriminant determines whether the quadratic equation has two real solutions, one real solution, or no real solutions. In the latter case, there will be two imaginary solutions. The following chart summarizes how the discriminant affects the number and type of solutions.

**Effect of the Discriminant**

If $a$, $b$, and $c$ are real numbers, $a \neq 0$, then the complex solutions of $ax^2 + bx + c = 0$ are described as follows, based on the value of the discriminant, $b^2 - 4ac$.

| Value of $b^2 - 4ac$ | Number of Solutions | Type of Solutions |
|---|---|---|
| Positive | Two | Real |
| Zero | One (a double solution) | Real |
| Negative | Two | Imaginary |

Furthermore, if $a$, $b$, and $c$ are *integers*, $a \neq 0$, the real solutions are *rational* if $b^2 - 4ac$ is the square of an integer.

**NOTE** The final sentence in the preceding box suggests that the quadratic equation can be solved by factoring if $b^2 - 4ac$ is a perfect square.

## Solving Quadratic Equations

**EXAMPLE 4** *Using the Quadratic Formula*

Solve $x(x - 2) = 2x - 2$ using the quadratic formula.

**Analytic Solution**

Before we can apply the quadratic formula, we must rewrite the equation in the form $ax^2 + bx + c = 0$.

$$x(x - 2) = 2x - 2 \quad \text{Given equation}$$
$$x^2 - 2x = 2x - 2 \quad \text{Distributive property}$$
$$x^2 - 4x + 2 = 0 \quad \text{Subtract } 2x\text{; add 2.}$$

Here $a = 1$, $b = -4$, and $c = 2$. Substitute these values into the quadratic formula.

$$x = \frac{-b \pm \sqrt{b^2 - 4ac}}{2a}$$
$$= \frac{-(-4) \pm \sqrt{(-4)^2 - 4(1)2}}{2(1)} \quad a = 1, b = -4, c = 2$$

**Graphing Calculator Solution**

Using the intersection-of-graphs method, consider the graphs of

$$y_1 = x(x - 2)$$

and

$$y_2 = 2x - 2.$$

(Note that the original form of the equation is $y_1 = y_2$.) Verify that the $x$-coordinates of the points of intersection in Figure 29 are approximations for $2 - \sqrt{2}$ and $2 + \sqrt{2}$.

*(continued)*

$$= \frac{4 \pm \sqrt{16 - 8}}{2}$$

$$= \frac{4 \pm \sqrt{8}}{2}$$

$$= \frac{4 \pm 2\sqrt{2}}{2} \qquad \sqrt{8} = 2\sqrt{2}$$

$$= \frac{2(2 \pm \sqrt{2})}{2} \qquad \text{Factor out 2.}$$

$$= 2 \pm \sqrt{2} \qquad \text{Lowest terms}$$

The solution set is $\{2 + \sqrt{2}, 2 - \sqrt{2}\}$, abbreviated $\{2 \pm \sqrt{2}\}$.

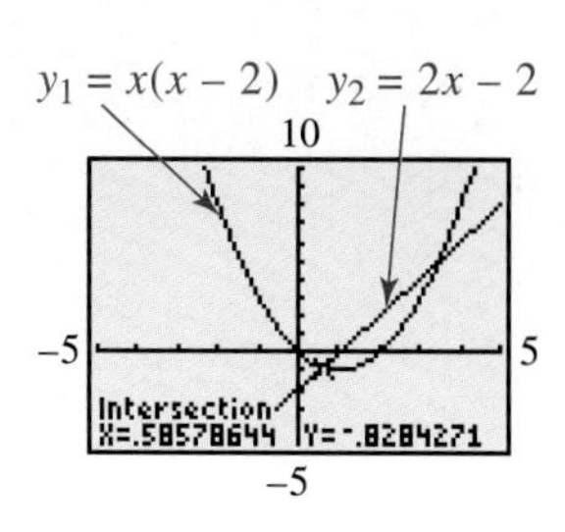


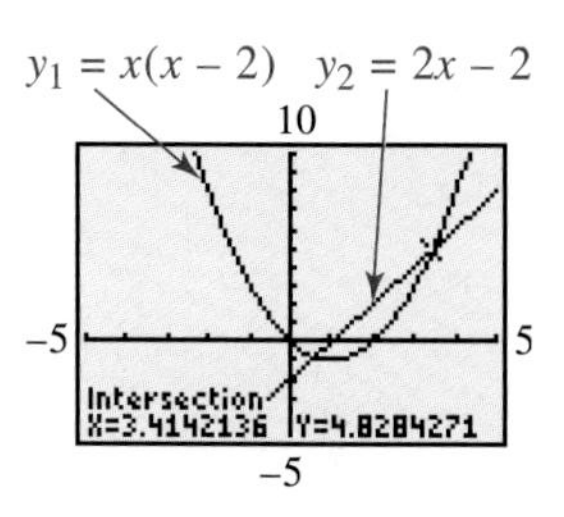


**FIGURE 29**

**EXAMPLE 5** *Using the Quadratic Formula*

Solve $2x^2 - x + 4 = 0$ by the quadratic formula.

**Analytic Solution**

Here we have $a = 2$, $b = -1$, and $c = 4$. By the quadratic formula,

$$x = \frac{-(-1) \pm \sqrt{(-1)^2 - 4(2)(4)}}{2(2)}$$

$$= \frac{1 \pm \sqrt{1 - 32}}{4} = \frac{1 \pm \sqrt{-31}}{4}.$$

Because the discriminant is negative, we know that there are no real solutions. Writing the solutions in $a + bi$ form, we get

$$x = \frac{1 \pm i\sqrt{31}}{4} = \frac{1}{4} \pm i\frac{\sqrt{31}}{4}.$$

The solution set is $\left\{\frac{1}{4} \pm i\frac{\sqrt{31}}{4}\right\}$.

**Graphing Calculator Solution**

If we graph $y = 2x^2 - x + 4$, we see that there are no $x$-intercepts, supporting the analytic result that the solutions are imaginary. See Figure 30.

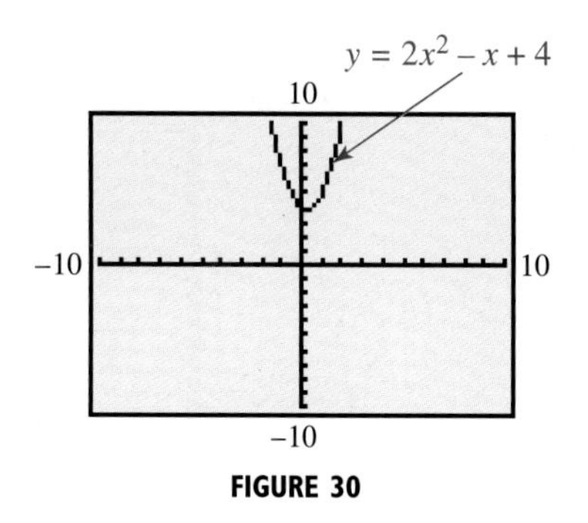


**FIGURE 30**

**FOR DISCUSSION**

Solve $x^2 = 4x - 4$ analytically. Then, graph $y_1 = x^2$ and $y_2 = 4x - 4$, and use the intersection-of-graphs method to support your result.

1. Do you encounter a problem if you use the standard viewing window?
2. Why do you think that analytic methods of solution are essential for understanding graphical methods?

## Solving Quadratic Inequalities

If $P$ is a quadratic function, the solution sets of $P(x) = 0$, $P(x) < 0$, and $P(x) > 0$ can be found graphically based on the following summary.

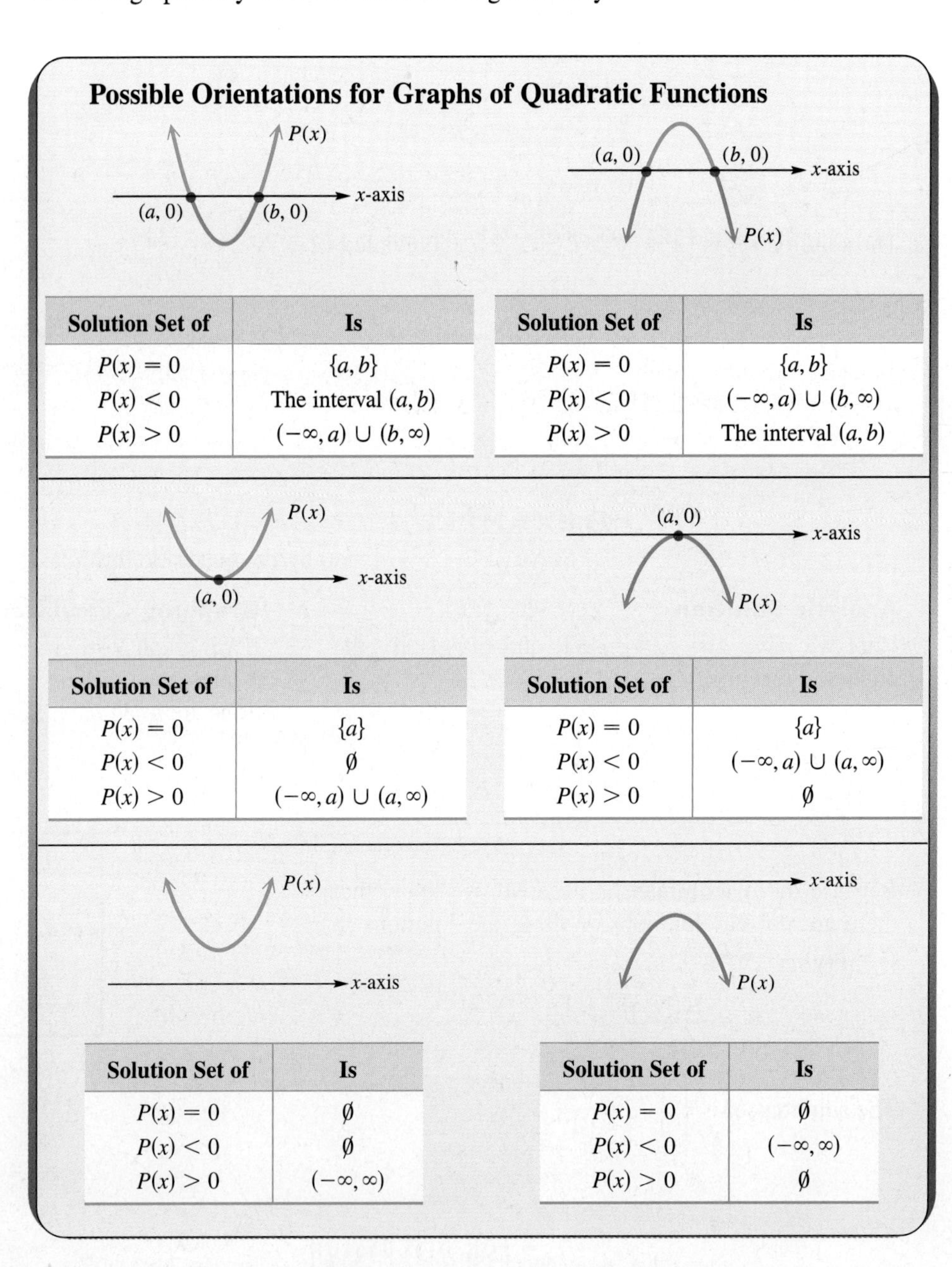


| Solution Set of | Is |
|---|---|
| $P(x) = 0$ | $\{a, b\}$ |
| $P(x) < 0$ | The interval $(a, b)$ |
| $P(x) > 0$ | $(-\infty, a) \cup (b, \infty)$ |

| Solution Set of | Is |
|---|---|
| $P(x) = 0$ | $\{a, b\}$ |
| $P(x) < 0$ | $(-\infty, a) \cup (b, \infty)$ |
| $P(x) > 0$ | The interval $(a, b)$ |

| Solution Set of | Is |
|---|---|
| $P(x) = 0$ | $\{a\}$ |
| $P(x) < 0$ | $\emptyset$ |
| $P(x) > 0$ | $(-\infty, a) \cup (a, \infty)$ |

| Solution Set of | Is |
|---|---|
| $P(x) = 0$ | $\{a\}$ |
| $P(x) < 0$ | $(-\infty, a) \cup (a, \infty)$ |
| $P(x) > 0$ | $\emptyset$ |

| Solution Set of | Is |
|---|---|
| $P(x) = 0$ | $\emptyset$ |
| $P(x) < 0$ | $\emptyset$ |
| $P(x) > 0$ | $(-\infty, \infty)$ |

| Solution Set of | Is |
|---|---|
| $P(x) = 0$ | $\emptyset$ |
| $P(x) < 0$ | $(-\infty, \infty)$ |
| $P(x) > 0$ | $\emptyset$ |

Suppose that the graph of a quadratic polynomial intersects the $x$-axis in two points. Then, the two solutions of the related polynomial *equation* divide the real number line ($x$-axis) into three intervals. Within each interval, the polynomial is either always positive or always negative. This idea is used in explaining how a quadratic inequality may be solved using analytic methods, employing a *sign graph*.

## EXAMPLE 6 *Solving a Quadratic Inequality*

Solve $x^2 - x - 12 < 0$.

### Analytic Solution

We can solve the quadratic equation $x^2 - x - 12 = 0$ by factoring.

$$x^2 - x - 12 = 0$$

$$(x + 3)(x - 4) = 0 \quad \text{Factor.}$$

$$x = -3 \quad \text{or} \quad x = 4 \quad \text{Zero-product property}$$

These two numbers, $-3$ and 4, divide a number line into the three intervals shown in Figure 31. If a point in interval *B*, for example, makes the polynomial $x^2 - x - 12$ negative, then all points in interval *B* will make the polynomial negative.

To find the intervals that make $x^2 - x - 12$ negative $(<0)$, we draw a number line that shows where the factors are positive or negative, as in Figure 31. First we decide on the sign of the factor $x + 3$ in each of the three intervals; then we do the same thing for the factor $x - 4$. The results are shown in Figure 31.

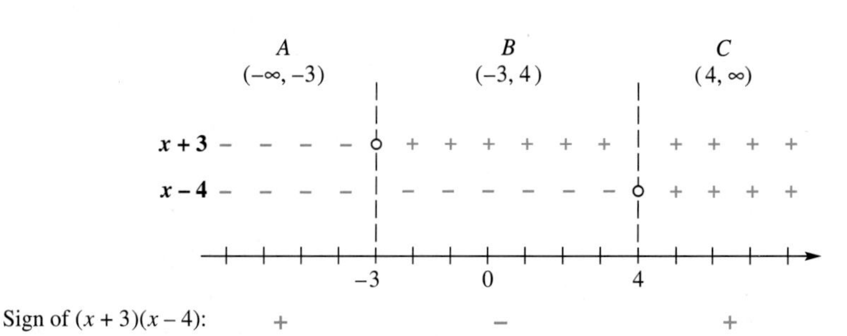


**FIGURE 31**

Now, consider the sign of the product of the two factors in each interval. As Figure 31 shows, both factors are negative in the interval $(-\infty, -3)$; therefore, their product is positive in that interval. For the interval $(-3, 4)$, one factor is positive and the other is negative, giving a negative product. In the last interval, $(4, \infty)$, both factors are positive, so their product is positive. The polynomial $x^2 - x - 12$ is negative (what the original inequality calls for) when the product of its factors is negative, that is, in the interval $(-3, 4)$. Therefore, the solution set is the open interval $(-3, 4)$.

### Graphing Calculator Solution

The graph of $Y_1 = X^2 - X - 12$ is shown in Figure 32. When the graph lies *below* the *x*-axis, the function values are negative $(<0)$. This happens in the interval $(-3, 4)$, which is the same as the analytic result.

The graph of the function is not sufficient to determine whether the endpoints should be included or excluded from the solution set. But the symbol $<$ in the original inequality indicates that the endpoints are excluded.

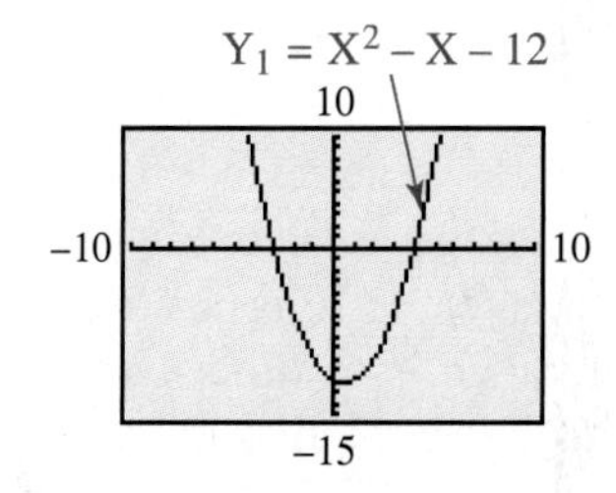


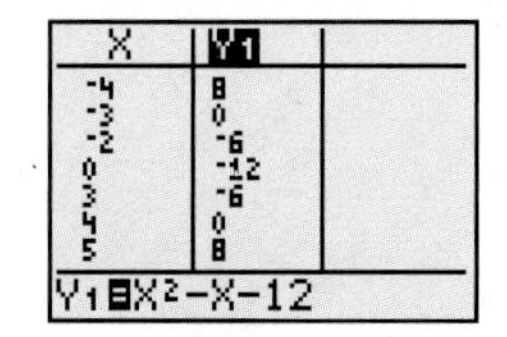

| X | Y1 |
|---|---|
| -4 | 8 |
| -3 | 0 |
| -2 | -6 |
| 0 | -12 |
| 3 | -6 |
| 4 | 0 |
| 5 | 8 |

Y1■X²-X-12

**FIGURE 32**

The table in Figure 32 numerically supports the analytic result. Values of X in the interval $(-3, 4)$ make $Y_1 < 0$, while values outside that interval make $Y_1 > 0$.

**NOTE** Graphical methods for solving inequalities are not sufficient for determining whether endpoints should be included or excluded from solution sets. Therefore, we must make our decision based on the symbol in a given inequality. Either the symbol $<$ or $>$ indicates that the endpoints are excluded, while $\leq$ or $\geq$ indicates that the endpoints are included.

The steps used to solve a quadratic inequality analytically are summarized on the next page.

**Solving a Quadratic Inequality**

1. Solve the corresponding quadratic equation.
2. Identify the intervals determined by the solutions of the equation.
3. Use a sign graph to determine which intervals are in the solution set.
4. Decide whether the endpoints are included or excluded.

**EXAMPLE 7** *Solving a Quadratic Inequality*

Solve $2x^2 \geq -5x + 12$.

### Analytic Solution

First write the quadratic inequality in standard form: $2x^2 + 5x - 12 \geq 0$. The corresponding quadratic equation can be solved by factoring.

$$2x^2 + 5x - 12 = 0$$

$$(2x - 3)(x + 4) = 0$$

$$x = \frac{3}{2} = 1.5 \quad \text{or} \quad x = -4$$

These two numbers divide the number line into the three intervals shown in the sign graph in Figure 33. As the sign graph indicates, the polynomial $2x^2 + 5x - 12$ is positive or 0 in the interval $(-\infty, -4]$ and also in the interval $[1.5, \infty)$. Since both of these intervals belong to the solution set, the result is written as the *union** of the two intervals,

$$(-\infty, -4] \cup [1.5, \infty).$$

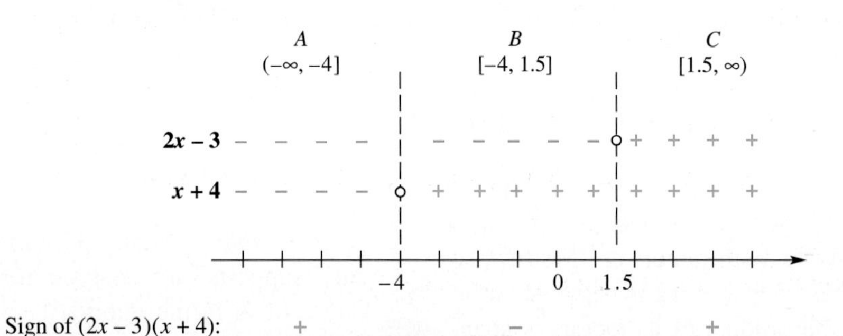


FIGURE 33

### Graphing Calculator Solution

Use the $x$-intercept method here to see the intervals where $2x^2 + 5x - 12 \geq 0$. Write the original inequality in the form $Y_1 - Y_2 = 2X^2 + 5X - 12 \geq 0$, and then find the domain values where the graph of $Y = Y_1 - Y_2$ *lies above or on* the $x$-axis. From the graph, the intervals $(-\infty, -4]$ and $[1.5, \infty)$ give these values, supporting our analytic result. See Figure 34.

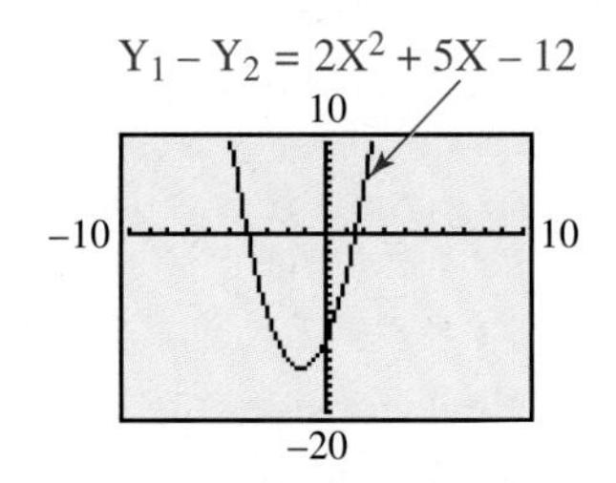


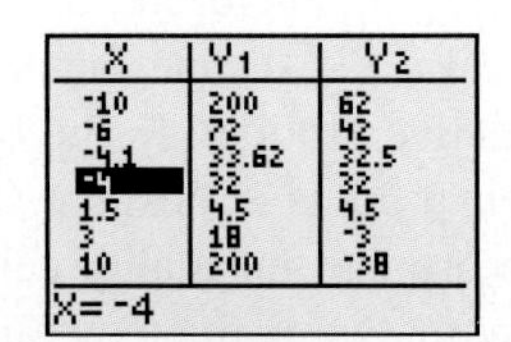


FIGURE 34

The table numerically supports the analytic result because choosing values of X less than or equal to $-4$ *or* greater than or equal to 1.5 gives $Y_1 \geq Y_2$.

*The **union** of sets $A$ and $B$, written $A \cup B$, is defined as

$$A \cup B = \{x \mid x \text{ is an element of } A \text{ or } x \text{ is an element of } B\}.$$

**NOTE** While we solved the quadratic equations in Examples 6 and 7 by factoring, we could also have used the quadratic formula, which will work for *any* quadratic equation, whether or not factoring is applicable.

**FOR DISCUSSION**

1. Graph $P(x) = x^2 - x - 20$ in a window that shows both $x$-intercepts, and, without any analytic work, solve each equation or inequality.
   **(a)** $x^2 - x - 20 = 0$ **(b)** $x^2 - x - 20 < 0$ **(c)** $x^2 - x - 20 > 0$
2. Graph $P(x) = x^2 + x + 20$ in a window that shows the vertex and the $y$-intercept. (Why won't the standard window work?) Then, without any analytic work, solve each equation or inequality for real solutions.
   **(a)** $x^2 + x + 20 = 0$ **(b)** $x^2 + x + 20 < 0$ **(c)** $x^2 + x + 20 > 0$

From the "For Discussion" box, we see that solving a quadratic equation leads easily to the solution sets of the corresponding inequalities.

**EXAMPLE 8** *Solving Quadratic Inequalities*

Use the graph in Figure 35 and the result of Example 4 to solve each inequality.

**(a)** $x^2 - 4x + 2 \leq 0$ **(b)** $x^2 - 4x + 2 \geq 0$

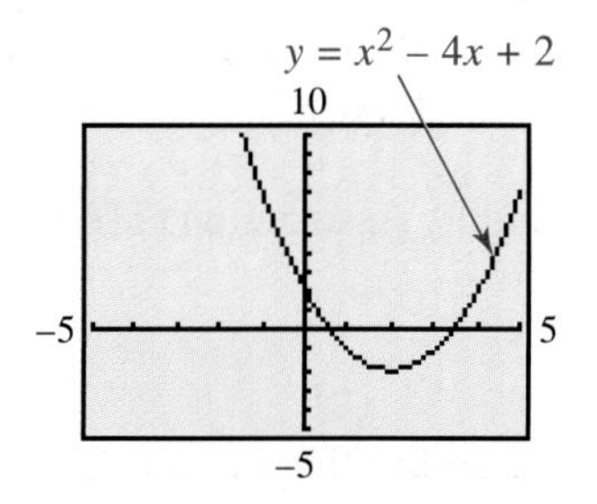


The intercepts are $2 - \sqrt{2} \approx .58578644$ and $2 + \sqrt{2} \approx 3.4142136$.

**FIGURE 35**

**Solution**

**(a)** In Example 4 we found that the exact solutions for $x^2 - 4x + 2 = 0$ are $2 - \sqrt{2}$ and $2 + \sqrt{2}$. Since the graph of $y = x^2 - 4x + 2$ lies below or intersects the $x$-axis (that is, $y \leq 0$) between or at these values, as shown in Figure 35, the solution set for $x^2 - 4x + 2 \leq 0$ is the closed interval $[2 - \sqrt{2}, 2 + \sqrt{2}]$.

**(b)** Figure 35 shows that the graph of $y = x^2 - 4x + 2$ is above or intersects the $x$-axis (that is, $y \geq 0$) on $(-\infty, 2 - \sqrt{2}] \cup [2 + \sqrt{2}, \infty)$, which is the solution set of $x^2 - 4x + 2 \geq 0$.

## Formulas Involving Quadratics

Sometimes we need to solve a formula for a variable that is squared.

**EXAMPLE 9** *Solving for a Squared Variable*

Solve each equation for the indicated variable.

**(a)** $A = \dfrac{\pi d^2}{4}$ for $d$ **(b)** $rt^2 - st = k \quad (r \neq 0)$ for $t$

**Solution**

**(a)** Start by multiplying each side by 4.

$$4A = \pi d^2$$

$$d^2 = \frac{4A}{\pi} \qquad \text{Divide by } \pi.$$

$$d = \pm\sqrt{\frac{4A}{\pi}} \qquad \text{Square root property}$$

$$d = \frac{\pm 2\sqrt{A\pi}}{\pi} \qquad \text{Rationalize the denominator.}$$

**(b)** Because this equation has a term with $t$ as well as one with $t^2$, use the quadratic formula. Subtract $k$ from each side to get

$$rt^2 - st - k = 0.$$

Now, use the quadratic formula to find $t$, with $a = r$, $b = -s$, and $c = -k$.

$$t = \frac{-b \pm \sqrt{b^2 - 4ac}}{2a}$$

$$t = \frac{-(-s) \pm \sqrt{(-s)^2 - 4(r)(-k)}}{2(r)}$$

$$t = \frac{s \pm \sqrt{s^2 + 4rk}}{2r}$$

## Another Quadratic Model

The intersection-of-graphs method is often a good way to find specific function values using a quadratic model.

**EXAMPLE 10** *Modeling Length of Life*

According to recent data from the Teachers Insurance and Annuity Association (TIAA), the survival rate after age 65 can be approximated by

$$S(x) = 1 - .058x - .076x^2,$$

where $x$ is measured in decades. This function gives the probability that an individual who reaches age 65 will live at least $x$ decades ($10x$ years) longer. (*Source:* Lial, M., R. Greenwell, and N. Ritchey, *Calculus with Applications,* Seventh Edition, Addison-Wesley, 2002.) For example,

$$S(1) = 1 - .058(1) - .076(1)^2 = .866.$$

This means that a 65-year old person has a .866 probability of living at least 1 more decade, to age 75.

Find the age for which the survival rate is .5 for people who reach 65. Interpret the result.

**Solution** Let $y_1 = 1 - .058x - .076x^2$ and $y_2 = .5$. Graph both functions and find the positive $x$-value of an intersection point. See Figure 36.

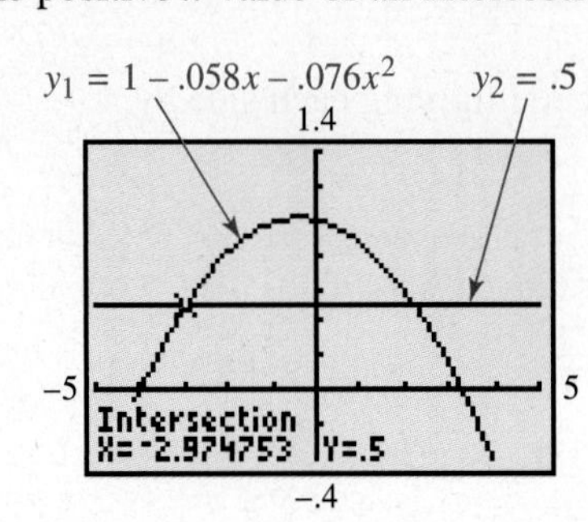


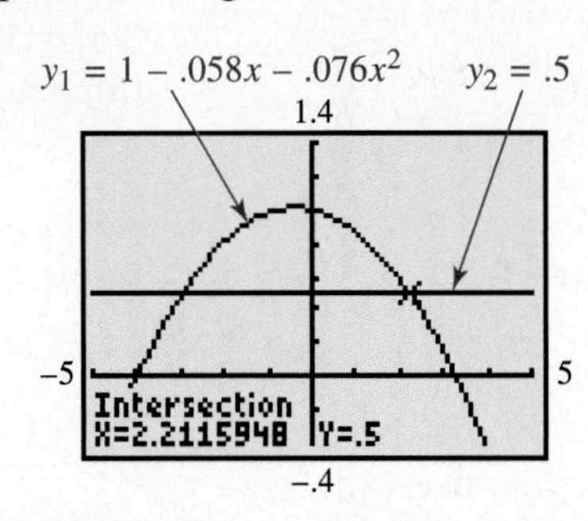


**FIGURE 36**

Figure 36 shows that the graphs intersect at $x \approx -3.0$ and $x \approx 2.2$. This means that half the people who reach age 65 will live about 2.2 decades or 22 years longer, to age 87. (The negative value must be discarded since it does not make sense in the problem.)

## 3.3 EXERCISES

*Concept Check* *Use the concepts of this section to work each problem.*

**1.** Which equation is set up for direct use of the zero-product property? Solve it.

**A.** $3x^2 - 17x - 6 = 0$ **B.** $(2x + 5)^2 = 7$ **C.** $x^2 + x = 12$ **D.** $(3x + 1)(x - 7) = 0$

**2.** Which equation is set up for direct use of the square root property? Solve it.

**A.** $3x^2 - 17x - 6 = 0$ **B.** $(2x + 5)^2 = 7$ **C.** $x^2 + x = 12$ **D.** $(3x + 1)(x - 7) = 0$

**3.** Only one equation does not require step 1 of the method of completing the square described in the previous section. Which one is it? Solve it.

**A.** $3x^2 - 17x - 6 = 0$ **B.** $(2x + 5)^2 = 7$ **C.** $x^2 + x = 12$ **D.** $(3x + 1)(x - 7) = 0$

**4.** Only one equation is set up so that the values of $a$, $b$, and $c$ can be determined immediately. Which one is it? Solve it.

**A.** $3x^2 - 17x - 6 = 0$ **B.** $(2x + 5)^2 = 7$ **C.** $x^2 + x = 12$ **D.** $(3x + 1)(x - 7) = 0$

*Find all solutions, real or imaginary, of each quadratic equation. Use the square root property. For the equations with real solutions, support your answers graphically.*

**5.** $x^2 = 16$ **6.** $x^2 = 144$ **7.** $3x^2 = 27$ **8.** $2x^2 = 48$

**9.** $x^2 = -16$ **10.** $x^2 = -100$ **11.** $x^2 = -18$ **12.** $x^2 = -32$

**13.** $(3x - 1)^2 = 12$ **14.** $(4x + 1)^2 = 20$ **15.** $(5x - 3)^2 = -3$ **16.** $(-2x + 5)^2 = -8$

*Solve each equation by the quadratic formula. Find all solutions. For the equations with real solutions, support your answers graphically by using the x-intercept method.*

**17.** $x^2 - 2x - 4 = 0$ **18.** $x^2 + 8x + 13 = 0$ **19.** $2x^2 + 2x = -1$

**20.** $9x^2 - 12x = -8$ **21.** $x(x - 1) = 1$ **22.** $x(x - 3) = 2$

**23.** $x^2 - 5x = x - 7$ **24.** $11x^2 - 3x + 2 = 4x + 1$ **25.** $4x^2 - 12x = -11$

**26.** $x^2 = 2x - 5$ **27.** $\frac{1}{3}x^2 + \frac{1}{4}x - 3 = 0$ **28.** $\frac{2}{3}x^2 + \frac{1}{4}x = 3$

*If necessary, write each equation so that 0 is on the right side, and then evaluate the discriminant. Use the discriminant to determine the number of real solutions the equation has. If the equation has real solutions, tell whether they are* rational *or* irrational. *Do not actually solve the equation.*

**29.** $x^2 + 8x + 16 = 0$ **30.** $8x^2 = 14x - 3$ **31.** $4x^2 = 6x + 3$

**32.** $2x^2 - 4x + 1 = 0$ **33.** $9x^2 + 11x + 4 = 0$ **34.** $3x^2 = 4x - 5$

*Concept Check* *For each pair of numbers, find the values of a, b, and c for which the quadratic equation $ax^2 + bx + c = 0$ has the given numbers as solutions. Answers may vary.* (Hint: *Use the zero-product property in reverse.*)

**35.** 4, 5 **36.** $-3, 2$ **37.** $1 + \sqrt{2}, 1 - \sqrt{2}$ **38.** $i, -i$

*Concept Check* *Sketch a graph of $f(x) = ax^2 + bx + c$ that satisfies each set of conditions.*

**39.** $a < 0, b^2 - 4ac = 0$ **40.** $a > 0, b^2 - 4ac < 0$ **41.** $a < 0, b^2 - 4ac < 0$

**42.** $a < 0, b^2 - 4ac > 0$ **43.** $a > 0, b^2 - 4ac > 0$ **44.** $a > 0, b^2 - 4ac = 0$

*Concept Check* *Exercises 45–58 refer to the graphs of the quadratic functions f, g, and h shown here.*

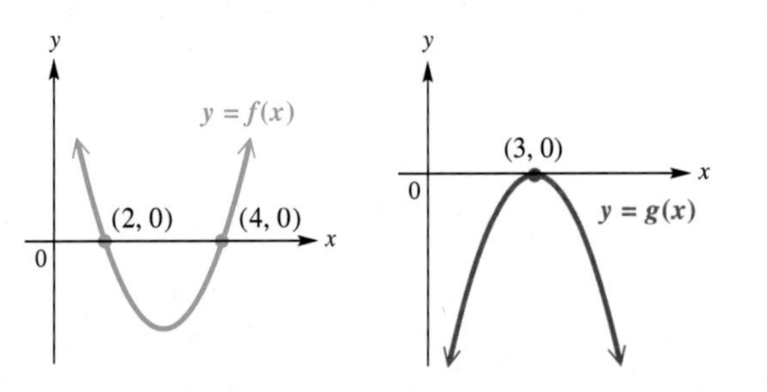


**45.** What is the solution set of $f(x) = 0$?

**46.** What is the solution set of $f(x) < 0$?

**47.** What is the solution set of $f(x) > 0$?

**48.** What is the solution set of $g(x) = 0$?

**49.** What is the solution set of $g(x) < 0$?

**50.** What is the solution set of $g(x) > 0$?

**51.** Solve $h(x) > 0$.

**52.** Solve $h(x) < 0$.

**53.** How many real solutions does $h(x) = 0$ have? How many complex solutions does it have?

**54.** What is the value of the discriminant of $g(x)$?

**55.** What is the $x$-coordinate of the vertex of the graph of $y = f(x)$?

**56.** What is the equation of the axis of symmetry of the graph of $y = g(x)$?

**57.** Does the graph of $y = g(x)$ have a $y$-intercept? If so, is it positive or negative?

**58.** Is the minimum value of $h$ positive or negative?

*Solve each inequality analytically using a sign graph. Support your answer graphically. Give exact values for endpoints.*

**59. (a)** $x^2 + 4x + 3 \geq 0$
**(b)** $x^2 + 4x + 3 < 0$

**60. (a)** $x^2 + 6x + 8 < 0$
**(b)** $x^2 + 6x + 8 \geq 0$

**61. (a)** $2x^2 - 9x > -4$
**(b)** $2x^2 - 9x \leq -4$

**62. (a)** $3x^2 + 13x + 10 \leq 0$
**(b)** $3x^2 + 13x + 10 > 0$

**63. (a)** $2x + 1 \geq x^2$
**(b)** $2x + 1 < x^2$

**64. (a)** $x^2 + 5x < 2$
**(b)** $x^2 + 5x \geq 2$

## RELATING CONCEPTS

**For individual or group investigation (Exercises 65–70)**

*We saw in Chapter 1 that to solve an inequality analytically, we must reverse the direction of the inequality sign if we multiply each side by a negative number. Work Exercises 65–70 in order, to apply this concept to quadratic inequalities.*

**65.** Graph $y_1 = x^2 + 2x - 8$ in the standard viewing window. This function has two integer-valued $x$-intercepts. What are they?

**66.** Based on the graph, what is the solution set of $x^2 + 2x - 8 < 0$?

**67.** Now, graph $y_2 = -y_1 = -x^2 - 2x + 8$ on the same screen. Using the terminology of Chapter 2, how is the graph of $y_2$ obtained by transforming the graph of $y_1$?

**68.** Based on the graph of $y_2$, what is the solution set of $-x^2 - 2x + 8 > 0$?

**69.** How do the two solution sets of the inequalities in Exercises 66 and 68 compare?

**70.** Write a short paragraph explaining how Exercises 65–69 illustrate the property involving multiplying each side of an inequality by a negative number.

*Solve each formula for the indicated variable. Assume that no denominators are* 0.

**71.** $s = \frac{1}{2}gt^2$ for $t$

**72.** $A = \pi r^2$ for $r$

**73.** $F = \frac{kMv^4}{r}$ for $v$

**74.** $s = s_0 + gt^2 + k$ for $t$

**75.** $P = \frac{E^2R}{(r + R)^2}$ for $R$

**76.** $S = 2\pi rh + 2\pi r^2$ for $r$

*Solve each equation for x and then for y.*

**77.** $4x^2 - 2xy + 3y^2 = 2$

**78.** $3y^2 + 4xy - 9x^2 = -1$

*Solve each problem.*

**79.** *(Modeling) Length of Life* Refer to Example 10. Solve $S(x) = 0$ and interpret the result.

**80.** *(Modeling) Medicine* Between 1992 and 1998, the percent of college freshmen who planned to get a professional degree in a medical field can be modeled by

$$f(x) = -.2369x^2 + 1.425x + 6.905,$$

where $x = 0$ represents 1992. (*Source: The American Freshman: National Norms for Fall 1992–1998,* Higher Education Research Institute, UCLA.) Based on this model, in what year did the percent of freshmen planning to get a medical degree reach its maximum? What was the maximum percent?

## Reviewing Basic Concepts (Sections 3.1–3.3)

*Perform each operation. Write answers in standard form.*

**1.** $(5 + 6i) - (2 - 4i) - 3i$

**2.** $i(5 + i)(5 - i)$

**3.** $\dfrac{-10 - 10i}{2 + 3i}$

*Give the following for the function defined by $P(x) = 2x^2 + 8x + 5$.*

**4.** The graph of $P$

**5.** The vertex of the graph and whether it is a maximum or minimum point

**6.** The equation of the axis of symmetry of the graph

**7.** The domain and range of $P$

*Solve each equation or inequality analytically, and support your result graphically.*

**8.** $9x^2 = 25$

**9.** $3x^2 - 5x = 2$

**10.** $-x^2 + x + 3 = 0$

**11.** $3x^2 - 5x - 2 \leq 0$

**12.** $x^2 - x - 3 > 0$

# 3.4 FURTHER APPLICATIONS OF QUADRATIC FUNCTIONS AND MODELS

Applications of Quadratic Functions • Quadratic Models

In Chapter 1 we used linear functions and equations to solve applied problems. We can now solve additional applications using quadratic functions and equations.

## Applications of Quadratic Functions

**EXAMPLE 1** *Finding the Area of a Rectangular Region*

A farmer wishes to enclose a rectangular region. He has 120 feet of fencing and plans to use his barn as one side of the enclosure. See Figure 37. Let $x$ represent the length of one of the parallel sides of the fencing.

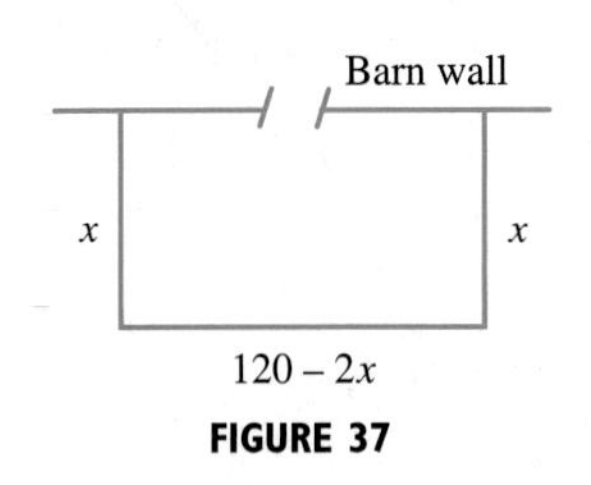

**FIGURE 37**

**(a)** Determine a function $A$ that represents the area of the region in terms of $x$.

**(b)** What are the restrictions on $x$?

**(c)** Graph the function in a viewing window that shows both $x$-intercepts and the vertex of the graph.

**(d)** Figure 38 on the next page shows the cursor at (18, 1512). Interpret this information.

**(e)** What is the maximum area the farmer can enclose?

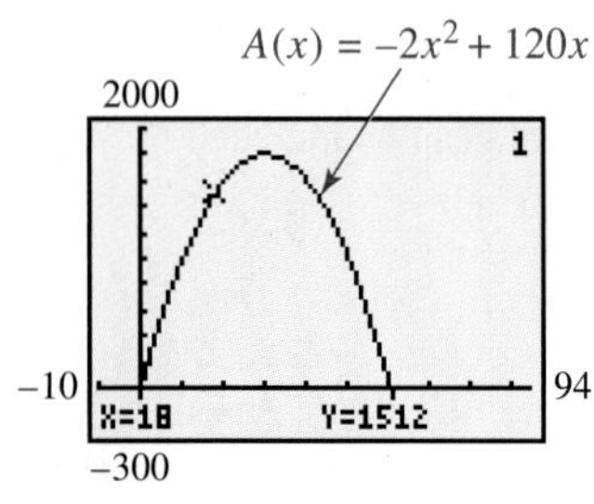


FIGURE 38

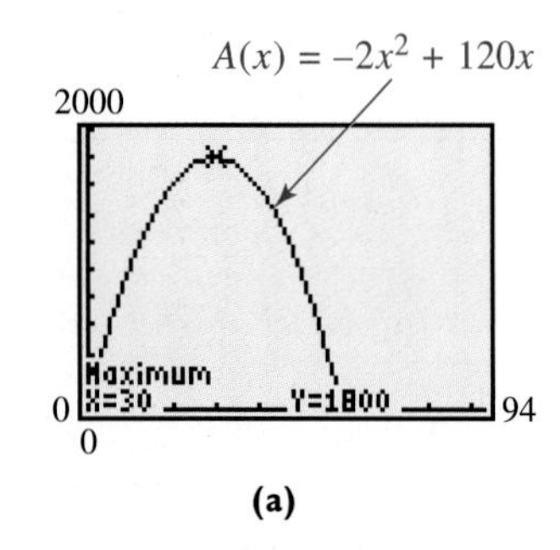


(a)

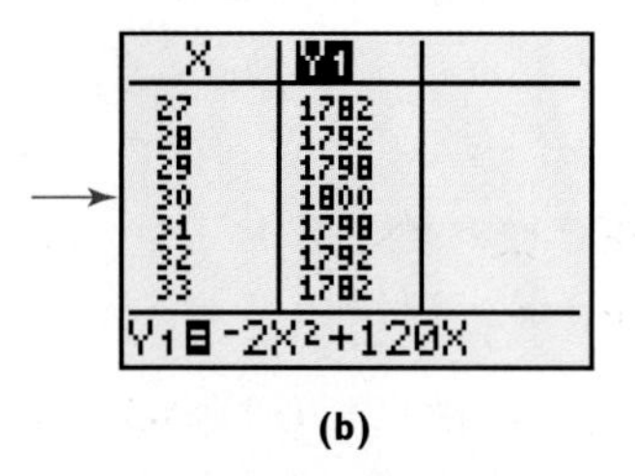


(b)

FIGURE 39

**TECHNOLOGY NOTE**

In the calculator graph in Figure 39(a), the display at the bottom obscures the view of the intercepts. This problem is simple to overcome: just lower the minimum values of $x$ and $y$ enough so that when the display appears, the $x$- and $y$-axes are visible, as in Figure 38.

**Solution**

**(a)** The lengths of the sides of the region bordered by the fencing are $x$, $x$, and $120 - 2x$, as shown in Figure 37. Area = width × length, so the function is

$$A(x) = x(120 - 2x) \quad \text{or} \quad A(x) = -2x^2 + 120x.$$

**(b)** Since $x$ represents a length, $x > 0$. Furthermore, the side of length $120 - 2x > 0$, or $x < 60$. Putting these two restrictions together gives $0 < x < 60$ or $(0, 60)$ as the *theoretical* domain.

**(c)** See Figure 38.

**(d)** If each parallel side of fencing measures 18 feet, the area of the enclosure is 1512 square feet. This can be written $A(18) = 1512$ and checked as follows: If the width is 18 feet, the length is $120 - 2(18) = 84$ feet, and $18 \times 84 = 1512$.

**(e)** The maximum value of the function occurs at the vertex. For $A(x) = -2x^2 + 120x$, use $a = -2$ and $b = 120$ in the vertex formula from Section 3.2. The $x$-value of the vertex is

$$x = -\frac{b}{2a} = -\frac{120}{2(-2)} = 30,$$

and the $y$-value is

$$A(30) = -2(30)^2 + 120(30) = 1800.$$

Therefore, the farmer can enclose a maximum of 1800 square feet if the parallel sides of fencing measure 30 feet.

Alternatively, use a calculator to locate the vertex of the parabola, and observe the $x$- and $y$-values there. The graph in Figure 39(a) confirms the analytic results. The table in Figure 39(b) provides numerical support.

**CAUTION** Be careful when interpreting the meanings of the coordinates of the vertex. The first coordinate, $x$, gives the *domain value* for which the *function value* is a maximum or minimum. Read the problem carefully to determine whether you are asked to find the value of the independent variable $x$, the function value $y$, or both.

### EXAMPLE 2 *Finding the Volume of a Box*

A machine produces rectangular sheets of metal satisfying the condition that the length is three times the width. Furthermore, equal size squares measuring 5 inches on a side can be cut from the corners so that the resulting piece of metal can be shaped into an open box by folding up the flaps. See Figure 40(a) on the next page.

**(a)** Determine a function $V$ that expresses the volume of the box in terms of the width $x$ of the original sheet of metal.

**(b)** What restrictions must be placed on $x$?

**(c)** If specifications call for the volume of such a box to be 1435 cubic inches, what should the dimensions of the original piece of metal be?

**(d)** What dimensions of the original piece of metal will assure a volume greater than 2000 but less than 3000 cubic inches? Solve graphically.

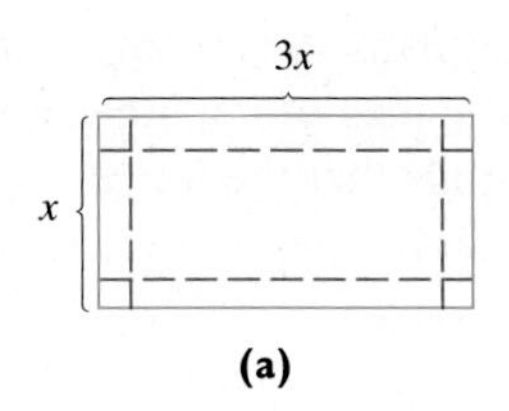


(a)

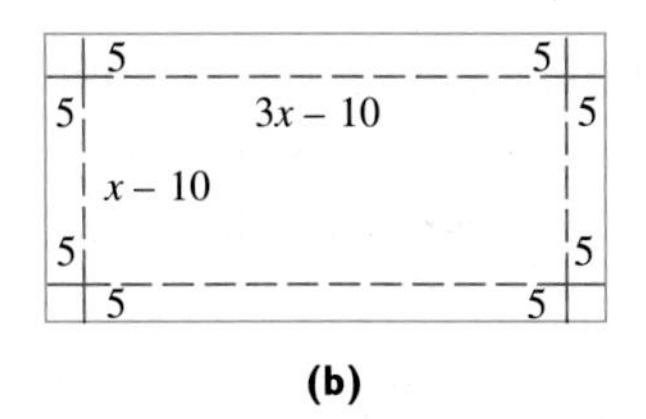


(b)

FIGURE 40

**Solution**

**(a)** If $x$ represents the width, then $3x$ represents the length. Figure 40(b) indicates that the width of the bottom of the box is $x - 10$, the length is $3x - 10$, and the height is 5 inches (the length of the side of each cut-out square). Since volume = length × width × height,

$$V(x) = (3x - 10)(x - 10)(5) = 15x^2 - 200x + 500.$$

**(b)** Since the dimensions of the box represent positive numbers, $3x - 10 > 0$ and $x - 10 > 0$, or

$$x > \frac{10}{3} \quad \text{and} \quad x > 10.$$

Both conditions are satisfied when $x > 10$, so the theoretical domain of $x$ is $(10, \infty)$.

**(c)**

$$15x^2 - 200x + 500 = 1435 \qquad \text{Set } V(x) = 1435.$$

$$15x^2 - 200x - 935 = 0 \qquad \text{Subtract 1435.}$$

$$(15x + 55)(x - 17) = 0 \qquad \text{Factor.}$$

$$15x + 55 = 0 \quad \text{or} \quad x - 17 = 0 \qquad \text{Zero-product property}$$

$$x = -\frac{11}{3} \quad \text{or} \quad x = 17$$

Only 17 satisfies the condition that $x > 10$. The dimensions should be 17 inches by $3(17) = 51$ inches. Since $(51 - 10) \cdot (17 - 10) \cdot 5 = 1435$, this answer is correct.

We can also graph $y_1 = 15x^2 - 200x + 500$ and $y_2 = 1435$ in the same window. As shown in Figure 41, the graphs intersect at the point with $x$-value 17, supporting the analytic result.

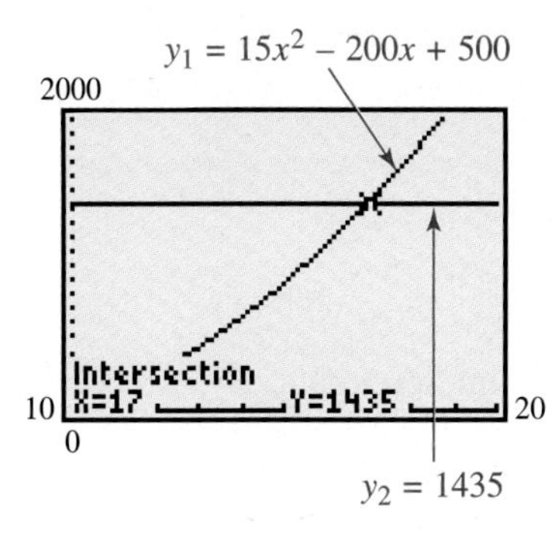


FIGURE 41

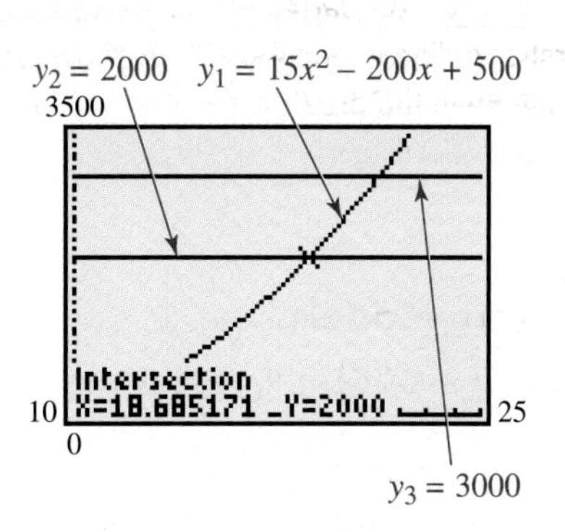


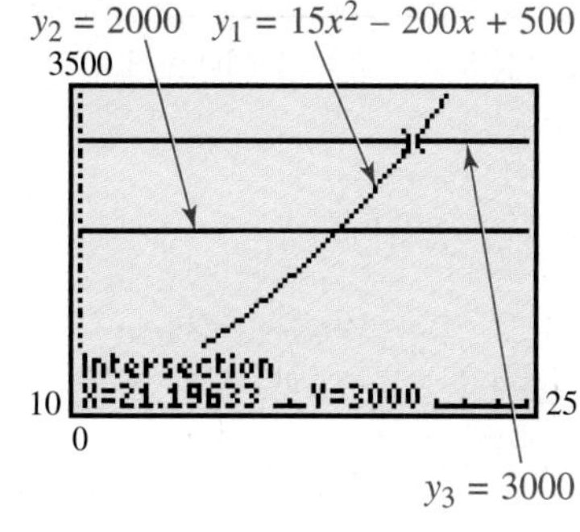


FIGURE 42

**(d)** Using the graphs of the functions $y_1 = 15x^2 - 200x + 500$, $y_2 = 2000$, and $y_3 = 3000$, as shown in Figure 42, we find that the points of intersection of the graphs are *approximately* (18.7, 2000) and (21.2, 3000) for $x > 10$. Therefore, the width of the rectangle should be between 18.7 and 21.2 inches, with the corresponding length three times these values (that is, between $3(18.7) \approx 56.1$ and $3(21.2) \approx 63.6$ inches).

**EXAMPLE 3** *Solving a Problem Requiring the Pythagorean Theorem*

Juan Perez, who is looking for a lot for an office building he plans to construct, finds a piece of property in the shape of a right triangle. To get some idea of its dimensions, he paces the three sides, starting with the shortest side. He finds that the longer leg is approximately 20 meters longer than twice the length of the shorter leg. The length of the hypotenuse is approximately 10 meters longer than the length of the longer leg. Use this information to estimate the lengths of the sides of the triangular lot.

### Analytic Solution

Let $s$ = the length of the shorter leg in meters. Then $2s + 20$ meters represents the length of the longer leg, and $(2s + 20) + 10 = 2s + 30$ meters represents the length of the hypotenuse. See Figure 43.

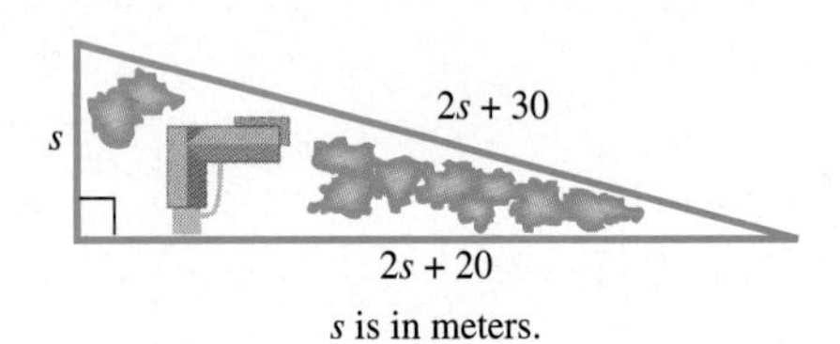


FIGURE 43

Verify that the domain of $s$ is $(0, \infty)$. Use the Pythagorean theorem to write an equation, then solve for $s$.

$$s^2 + (2s + 20)^2 = (2s + 30)^2$$
$$s^2 + 4s^2 + 80s + 400 = 4s^2 + 120s + 900$$
$$s^2 - 40s - 500 = 0$$
$$(s - 50)(s + 10) = 0$$
$$s = 50 \quad \text{or} \quad s = -10$$

The solution $-10$ is not in the domain. The approximate lengths of the sides of the triangular lot are 50 meters, $2(50) + 20 = 120$ meters, and $2(50) + 30 = 130$ meters.

### Graphing Calculator Solution

Replace $s$ with $x$ and, using

$$y_1 = x^2 + (2x + 20)^2$$

and $\quad y_2 = (2x + 30)^2,$

graph $y_1 - y_2$. The $x$-intercept is 50, as shown in Figure 44. We use the $x$-intercept method with $y_1 - y_2$ here because the $y$-values for $y_1$ and $y_2$ are large.

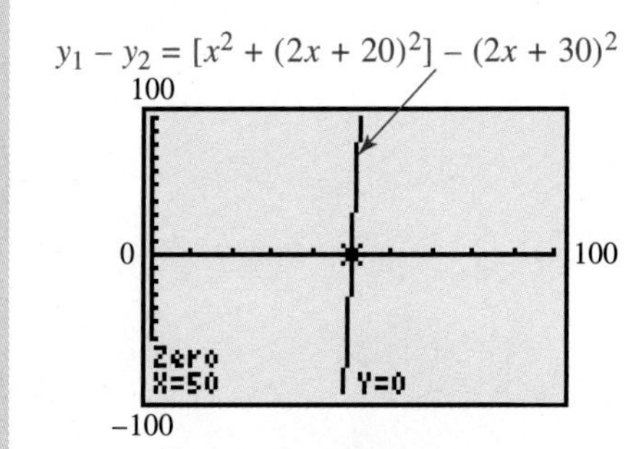


FIGURE 44

## Quadratic Models

In Section 1.4 we used linear models to predict and interpret data. By extending the statistical concept of regression to polynomials, we can sometimes fit data to a quadratic function to provide a quadratic model.

| Year, $x$ | Percent 65 and older, $y$ |
|---|---|
| 1900 | 4.1 |
| 1920 | 4.7 |
| 1940 | 6.8 |
| 1960 | 9.3 |
| 1980 | 11.3 |
| 2000 | 12.4 |
| 2020* | 16.5 |
| 2040* | 20.6 |

*Projected
*Source:* U.S. Bureau of the Census.

**EXAMPLE 4** *Modeling Population Data*

The percent of Americans 65 and older for selected years is shown in the table.

**(a)** Plot the data, letting $x = 0$ correspond to 1900.

**(b)** Find a quadratic function of the form $f(x) = a(x - h)^2 + k$ that models the data by using $(0, 4.1)$ as the vertex and a second point, such as $(140, 20.6)$ to determine $a$.

**(c)** Graph $f$ in the same viewing window as the data. How well does $f$ model the percent of Americans 65 and older?

**(d)** Use the quadratic regression feature of a graphing calculator to determine the quadratic function $g$ that provides the best fit for the data. Graph $g$ with the data.

## Solution

| L1 | L2 | L3 1 |
|---|---|---|
| 0 | 4.1 | ------ |
| 20 | 4.7 | |
| 40 | 6.8 | |
| 60 | 9.3 | |
| 80 | 11.3 | |
| 100 | 12.4 | |
| 120 | 16.5 | |

L1 = {0, 20, 40, 60, ...

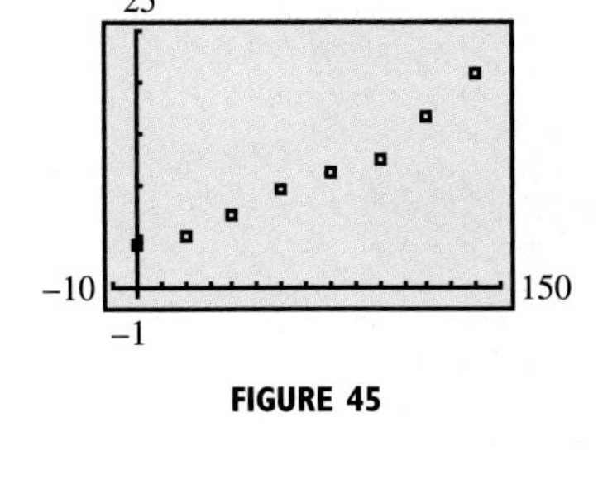


**FIGURE 45**

**(a)** Using the statistical feature of a graphing calculator and letting the $x$-list $L_1$ be $\{0, 20, 40, \ldots, 140\}$ and the $y$-list $L_2$ be $\{4.1, 4.7, 6.8, \ldots, 20.6\}$ gives the scatter diagram shown in Figure 45.

**(b)** Since $(0, 4.1)$ is the lowest point in the scatter diagram of the data, let it correspond to the vertex on the graph of $f$. Substituting 0 for $h$ and 4.1 for $k$ in the given form of the quadratic function $f$, we get

$$f(x) = a(x - 0)^2 + 4.1.$$

To determine $a$, we let $x = 140$ and $f(140) = 20.6$.

$$20.6 = a(140)^2 + 4.1$$
$$20.6 - 4.1 = 19{,}600a$$
$$a = .0008418 \quad \text{Rounded}$$

Using this value for $a$, the quadratic model is

$$f(x) = .0008418(x - 0)^2 + 4.1$$
$$= .0008418x^2 + 4.1.$$

(Note that choosing other second points would produce different models.)

**(c)** Figure 46(a) shows the graph of $f$ plotted with the data points. There is a fairly good fit, especially for the later years.

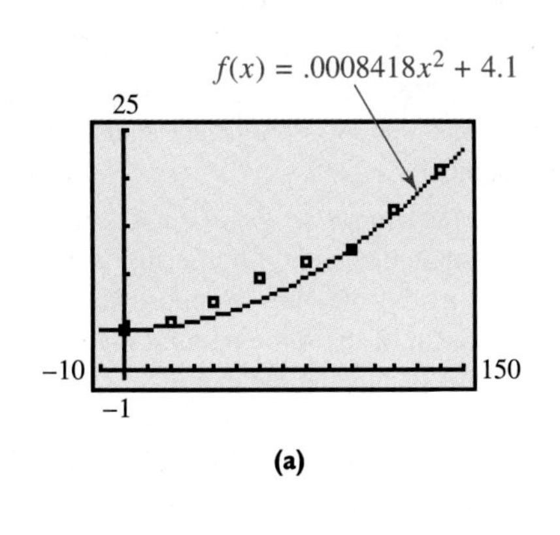


**(a)**

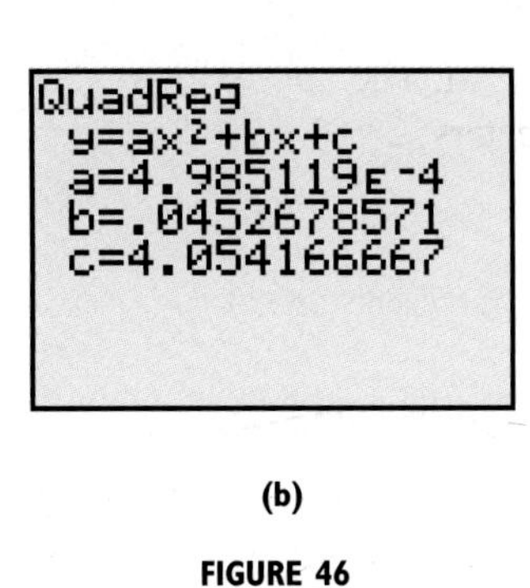


**(b)**

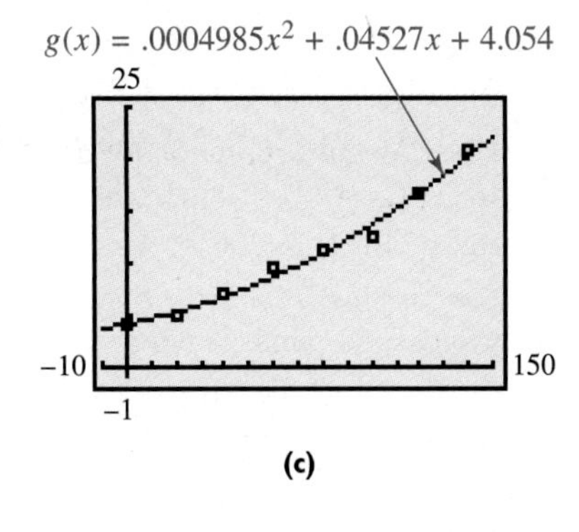


**(c)**

**FIGURE 46**

**(d)** Figure 46(b) shows the calculator quadratic regression model for the data. This model,

$$g(x) = .0004985x^2 + .04527x + 4.054,$$

is graphed in the same window as the data in Figure 46(c). Comparing this graph with the one in Figure 46(a), we see that it is a closer fit to the data. This is because all data are used in determining function $g$, while only the first and last data points were used to determine function $f$.

## 3.4 EXERCISES

*(Modeling) Solve each problem.*

1. *Area of a Parking Lot* For the rectangular parking area of the shopping center shown, which equation says that the area is 40,000 square yards?

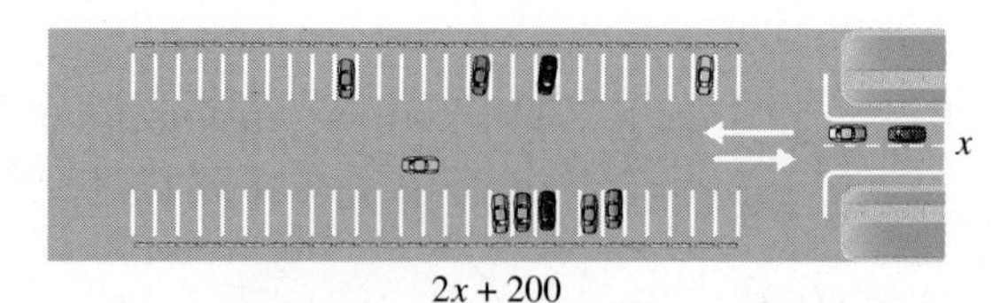


**A.** $x(2x + 200) = 40{,}000$

**B.** $2x + 2(2x + 200) = 40{,}000$

**C.** $x + (2x + 200) = 40{,}000$

**D.** None of the above

2. *Area of a Picture* The mat around the picture shown measures $x$ inches across. Which equation says that the area of the picture itself is 600 square inches?

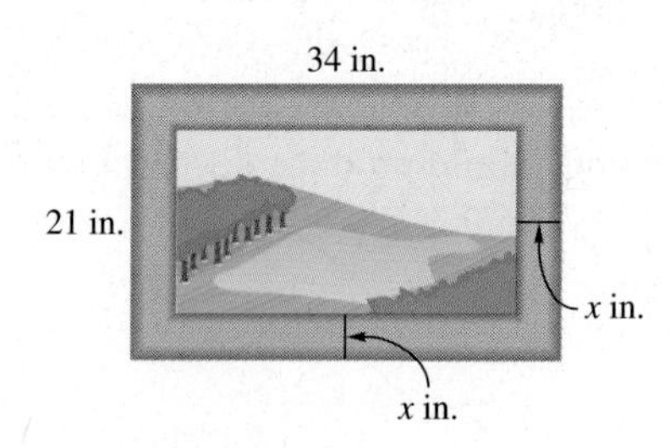


**A.** $2(34 - 2x) + 2(21 - 2x) = 600$

**B.** $(34 - 2x)(21 - 2x) = 600$

**C.** $(34 - x)(21 - x) = 600$

**D.** $x(34)(21) = 600$

3. *Sum of Two Numbers* Suppose that $x$ represents one of two *positive* numbers whose sum is 30.

**(a)** Represent the other of the two numbers in terms of $x$.

**(b)** What are the restrictions on $x$?

**(c)** Determine a function $P$ that represents the product of these two numbers.

**(d)** Determine analytically and support graphically the two such numbers whose product is a maximum. What is this maximum product?

4. *Sum of Two Numbers* Suppose that $x$ represents one of two *positive* numbers whose sum is 45.

**(a)** Represent the other of the two numbers in terms of $x$.

**(b)** What are the restrictions on $x$?

**(c)** Determine a function $P$ that represents the product of these two numbers.

**(d)** For what two such numbers is the product equal to 504? Determine analytically.

**(e)** Determine analytically and support graphically the two such numbers whose product is a maximum.

5. *Area of a Parking Lot* American River College has plans to construct a rectangular parking lot on land bordered on one side by a highway. There are 640 feet of fencing available to fence the other three sides. Let $x$ represent the length of each of the two parallel sides of fencing.

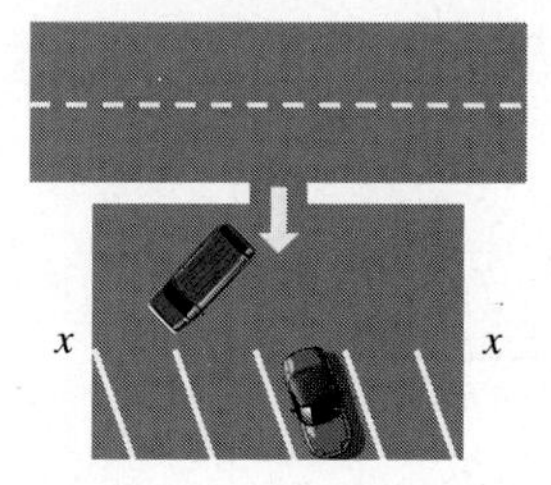


**(a)** Express the length of the remaining side to be fenced in terms of $x$.

**(b)** What are the restrictions on $x$?

**(c)** Determine a function $A$ that represents the area of the parking lot in terms of $x$.

**(d)** Graph the function $A$ from part (c) in a viewing window of $[0, 320]$ by $[0, 55{,}000]$. Determine graphically the values of $x$ that will give an area between 30,000 and 40,000 square feet.

**(e)** What dimensions will give a maximum area, and what will this area be? Determine analytically and support graphically.

6. *Area of a Rectangular Region* A farmer wishes to enclose a rectangular region bordering a river with fencing, as shown in the diagram. Suppose that $x$ represents the length of each of the three parallel pieces of fencing. She has 600 feet of fencing available.

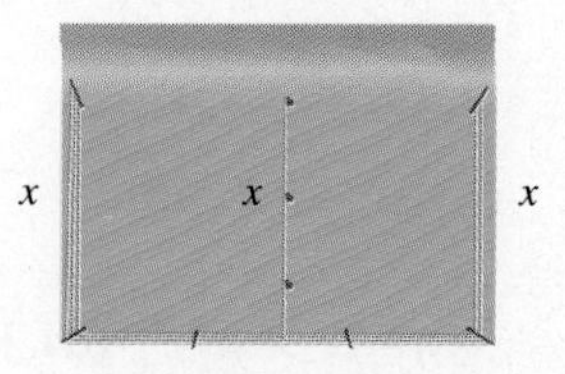


**(a)** What is the length of the remaining piece of fencing in terms of $x$?

**(b)** Determine a function $A$ that represents the total area of the enclosed region. Give any restrictions on $x$.

**(c)** What dimensions for the total enclosed region would give an area of 22,500 square feet? Determine the answer analytically.

(d) Use a graph to find the maximum area that can be enclosed.

**7.** *Volume of a Box* A piece of cardboard is twice as long as it is wide. It is to be made into a box with an open top by cutting 2-inch squares from each corner and folding up the sides. Let $x$ represent the width of the original piece of cardboard.

(a) Represent the length of the original piece of cardboard in terms of $x$.

(b) What will be the dimensions of the bottom rectangular base of the box? Give the restrictions on $x$.

(c) Determine a function $V$ that represents the volume of the box in terms of $x$.

(d) For what dimensions of the bottom of the box will the volume be 320 cubic inches? Determine analytically and support graphically.

(e) Determine graphically (to the nearest tenth of an inch) the values of $x$ if such a box is to have a volume between 400 and 500 cubic inches.

**8.** *Volume of a Box* A piece of sheet metal is 2.5 times as long as it is wide. It is to be made into a box with an open top by cutting 3-inch squares from each corner and folding up the sides. Let $x$ represent the width of the original piece of sheet metal.

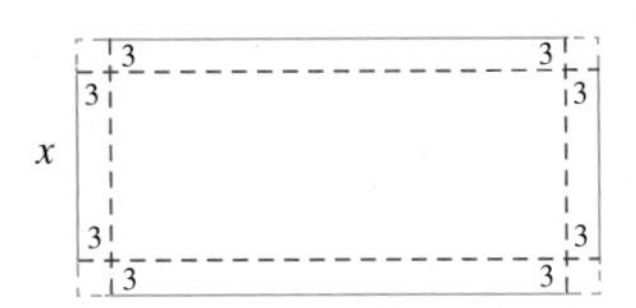


(a) Represent the length of the original piece of sheet metal in terms of $x$.

(b) What are the restrictions on $x$?

(c) Determine a function $V$ that represents the volume of the box in terms of $x$.

(d) For what values of $x$ (that is, original widths) will the volume of the box be between 600 and 800 cubic inches? Determine the answer graphically, and give values to the nearest tenth of an inch.

**9.** *Radius of a Can* A can of S & W garbanzo beans has surface area 600 square inches. Its height is 4.25 inches. What is the radius of the circular top? (*Hint:* The surface area consists of the circular top and bottom and a rectangle that represents the side cut open vertically and unrolled.)

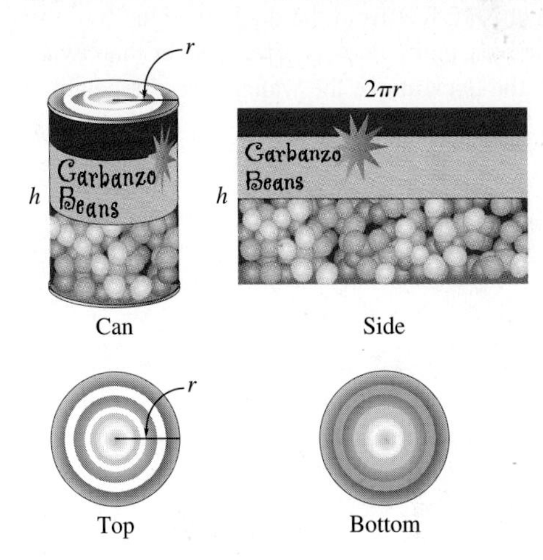


**10.** *Dimensions of a Cereal Box* The volume of a 10-ounce box of Cheerios cereal is 182.742 cubic inches. The width of the box is 3.1875 inches less than the length, and its depth is 2.3125 inches. Find the length and width of the box to the nearest thousandth.

**11.** *Radius Covered by a Circular Lawn Sprinkler* A square lawn has area 800 square feet. A sprinkler placed at the center of the lawn sprays water in a circular pattern that just covers the lawn. What is the radius of the circle?

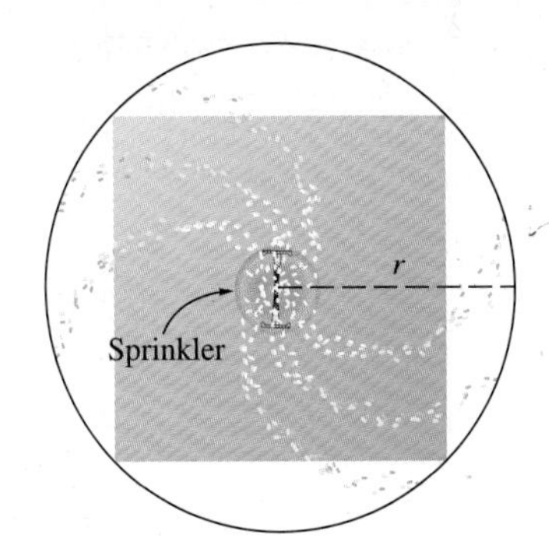


**12.** *Height of a Kite* A kite is flying on 50 feet of string. How high is it above the ground if its height is 10 feet more than the horizontal distance from the person flying it? Assume the string is being released at ground level.

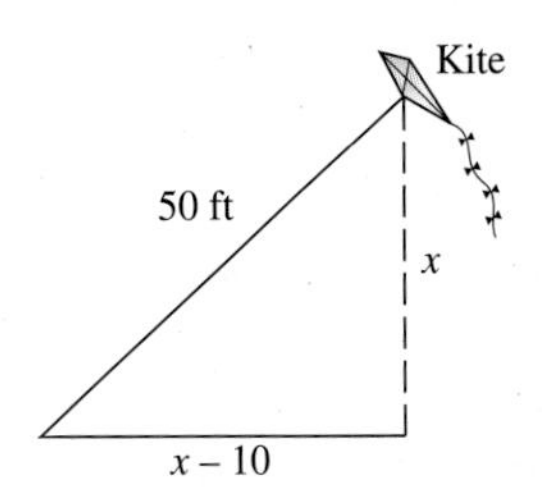

**13.** *Height of a Dock* A boat is being pulled into a dock with a rope attached to the boat at water level. When the boat is 12 feet from the dock, the length of the rope from the boat to the dock is 3 feet longer than twice the height of the dock above the water. Find the height of the dock.

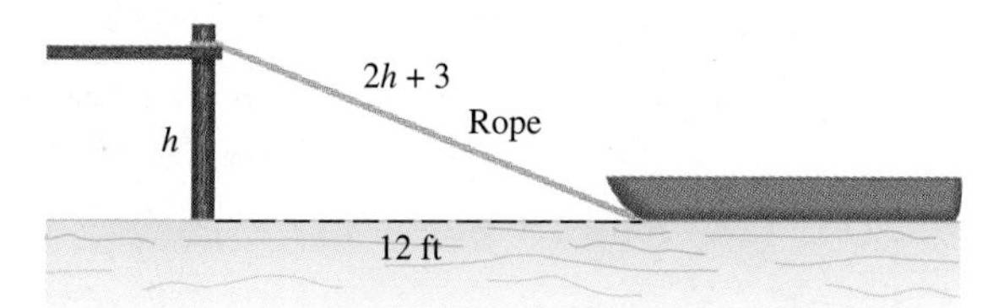


**14.** *Length of a Walkway* The disappearance of wetlands throughout the United States has concerned environmentalists for a number of years. A nature conservancy group has purchased some land that includes a wetland. To make it available for the public to see and to learn about, they decide to construct a raised wooden walkway through the wetland area. The walkway will be a loop that begins and ends at a nature center. To enclose the most interesting part of the wetlands, the walkway will have the shape of a right triangle with one leg 700 yards longer than the other and the hypotenuse 100 yards longer than the longer leg. Find the total length of the walkway.

**15.** *Length of a Ladder* A building is 2 feet from a 9-foot fence that surrounds the property. A worker wants to wash a window in the building 13 feet from the ground. He plans to place a ladder over the fence so it rests against the building. He decides he should place the ladder at least 8 feet from the fence for stability. To the nearest foot, how long a ladder will he need?

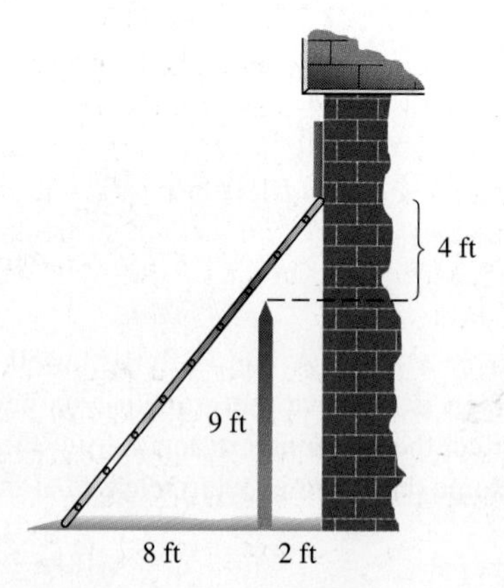


**16.** *Dimensions of a Solar Panel Frame* Christine has a solar panel with a width of 26 inches, as shown in the figure. To get the proper inclination for her climate, she needs a right triangular support frame that has one leg twice as long as the other. To the nearest tenth of an inch, what dimensions should the frame have?

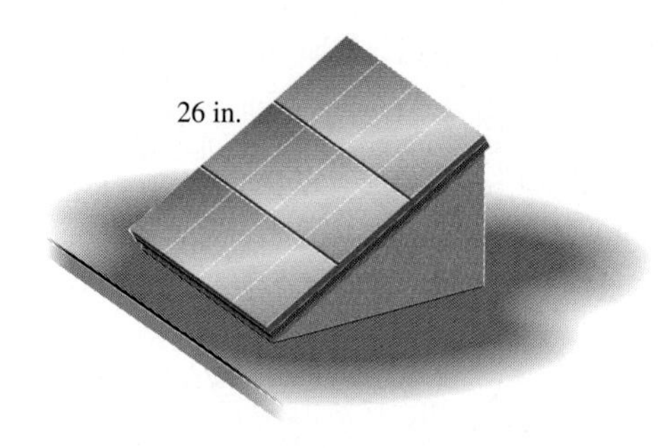


**17.** *Apartment Rental* The manager of an 80-unit apartment complex knows from experience that at a rent of \$400 per month, all units will be rented. However, for each increase of \$20 in rent, he can expect one unit to be vacated. Let $x$ represent the number of \$20 increases over \$400.

**(a)** Express, in terms of $x$, the number of apartments that will be rented if $x$ increases of \$20 are made. (For example, if 3 such increases are made, the number of apartments rented will be $80 - 3 = 77$.)

**(b)** Express the rent per apartment if $x$ increases of \$20 are made. (For example, if he increases rent $\$60 = 3 \times \$20$, the rent per apartment is $400 + 3(20) = \$460$.)

**(c)** Determine a revenue function $R$ in terms of $x$ that will give the revenue generated as a function of the number of \$20 increases.

**(d)** For what number of increases will the revenue be \$37,500?

**(e)** What rent should he change in order to achieve the maximum revenue?

**18.** *Seminar Fee* When *Money Means Power* charges \$600 for a seminar on management techniques, it attracts 1000 people. For each decrease of \$20 in the charge, an additional 100 people will attend the seminar. Let $x$ represent the number of \$20 decreases in the charge.

**(a)** Determine a revenue function $R$ that will give revenue generated as a function of the number of \$20 decreases.

**(b)** Find the value of $x$ that maximizes the revenue. What should the company charge to maximize the revenue?

**(c)** What is the maximum revenue the company can generate?

**19.** *Shooting a Foul Shot* Refer to the chapter introduction. To make a foul shot in basketball, the ball must follow a parabolic arc. This arc depends on both the angle and velocity with which the basketball is released. If a person shoots the basketball overhand from a position 8 feet above the floor, then the path can sometimes be modeled by the parabola

$$y = \frac{-16x^2}{.434v^2} + 1.15x + 8,$$

where $v$ is the initial velocity of the ball in feet per second, as illustrated in the figure. (*Source:* Rist, C., "The Physics of Foul Shots," *Discover,* October 2000.)

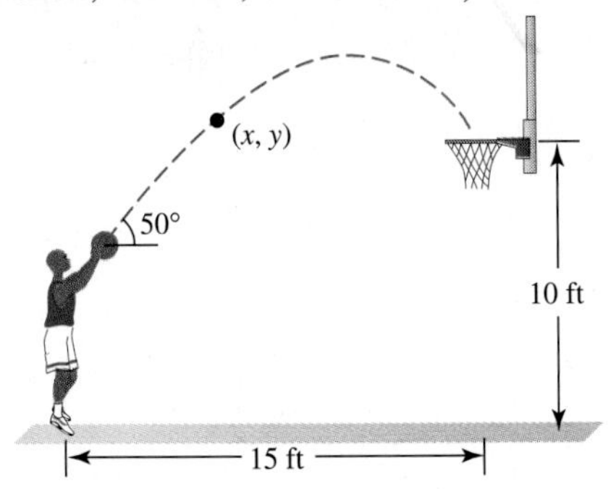


**(a)** If the basketball hoop is 10 feet high and located 15 feet away, what initial velocity $v$ should the basketball have?

**(b)** Check your answer from part (a) graphically. Plot the point $(0, 8)$ where the ball is released and the point $(15, 10)$ where the basketball hoop is. Does your graph pass through both points?

**(c)** What is the maximum height of the basketball?

**20.** *Shooting a Foul Shot* Refer to Exercise 19. If a person releases a basketball underhand from a position 3 feet above the floor, it often has a steeper arc than if it is released overhand and its path can sometimes may be modeled by

$$y = \frac{-16x^2}{.117v^2} + 2.75x + 3.$$

Complete parts (a), (b), and (c) for Exercise 19. Then compare the paths for the overhand shot and the underhand shot.

**21.** *Path of a Frog's Leap* A frog leaps from a stump 3 feet high and lands 4 feet from the base of the stump. We can consider the initial position of the frog to be at $(0, 3)$ and its landing position to be at $(4, 0)$.

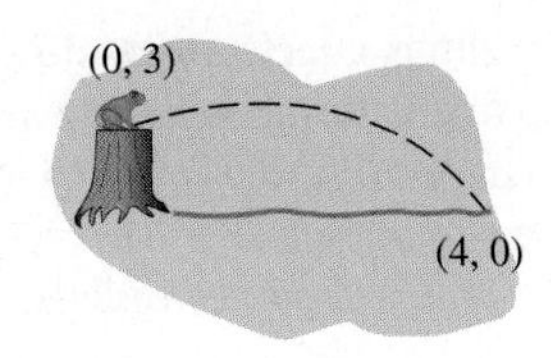


It is determined that the height of the frog as a function of its distance $x$ from the base of the stump is given by

$$h(x) = -.5x^2 + 1.25x + 3,$$

where $h$ is in feet.

**(a)** How high was the frog when its horizontal distance $x$ from the base of the stump was 2 feet?

**(b)** What was the horizontal distance from the base of the stump when the frog was 3.25 feet above the ground?

**(c)** At what horizontal distance from the base of the stump did the frog reach its highest point?

**(d)** What was the maximum height reached by the frog?

**22.** *Path of a Frog's Leap* Refer to Exercise 21. Suppose that the initial position of the frog is $(0, 4)$ and its landing position is $(6, 0)$. The height of the frog in feet is given by

$$h(x) = -\frac{1}{3}x^2 + \frac{4}{3}x + 4.$$

**(a)** What was the horizontal distance $x$ from the base of the stump when the frog reached maximum height?

**(b)** What was the maximum height?

**23.** *Airplane Landing Speed* To determine the appropriate landing speed of a small airplane, the formula

$$D = .1s^2 - 3s + 22$$

is used, where $s$ is the initial landing speed in feet per second and $D$ is the length of the runway in feet. If the landing speed is too fast, the pilot may run out of runway; if the speed is too slow, the plane may stall. If the runway is 800 feet long, what is the appropriate landing speed? What is the landing speed in mph? (*Hint:* There are 5280 feet in one mile.)

**24.** *Accident Rate* According to data from the National Highway Traffic Safety Administration, the accident rate as a function of the age of the driver in years $x$ can be approximated by the function defined by

$$f(x) = 60.0 - 2.28x + .0232x^2$$

for $16 \le x \le 85$. Find both the age at which the accident rate is a minimum and the minimum rate.

**25.** *Carbon Monoxide Exposure* Carbon monoxide (CO) is a dangerous combustion product. It combines with the hemoglobin of the blood to form carboxyhemoglobin (COHb), which reduces transport of oxygen to tissues.

A person's health is affected by both the concentration of carbon monoxide in the air and the exposure time. Smokers routinely have a 4% to 6% COHb level in their blood, which can cause symptoms such as blood flow alterations, visual impairment, and poorer vigilance ability. The quadratic function defined by

$$T(x) = .00787x^2 - 1.528x + 75.89$$

approximates the exposure time in hours necessary to reach this 4% to 6% level, where $50 \le x \le 100$ is the amount of carbon monoxide present in the air in parts per million (ppm). (*Source: Indoor Air Quality Environmental Information Handbook: Combustion Sources,* Report No. DOE/EV/10450-1, U.S. Department of Energy, 1985.)

**(a)** A kerosene heater or a room full of smokers is capable of producing 50 ppm of carbon monoxide. How long would it take for a nonsmoking person to start feeling the above symptoms?

**(b)** Find the carbon monoxide concentration necessary for a person to reach the 4% to 6% COHb level in 3 hours.

**26.** *Carbon Monoxide Exposure* Refer to Exercise 25. High concentrations of carbon monoxide (CO) can cause coma and possible death. The time required for a person to reach a COHb level capable of causing a coma can be approximated by the quadratic function defined by

$$T(x) = .0002x^2 - .316x + 127.9,$$

where $T$ is the exposure time in hours necessary to reach this level and $500 \le x \le 800$ is the amount of carbon monoxide in parts per million (ppm). (*Source: Indoor Air Quality Environmental Information Handbook: Combustion Sources,* Report No. DOE/EV/10450-1, U.S. Department of Energy, 1985.)

**(a)** What is the exposure time when $x = 600$ ppm?

**(b)** Estimate the concentration of CO necessary to produce a coma in 4 hours.

**27.** *Americans Over 100 Years of Age* The table lists the number of Americans (in thousands) who are expected to be over 100 years old for selected years.

| Year | Number (in thousands) |
|---|---|
| 1994 | 50 |
| 1996 | 56 |
| 1998 | 65 |
| 2000 | 75 |
| 2002 | 94 |
| 2004 | 110 |

*Source:* U.S. Bureau of the Census.

**(a)** Plot the data. Let $x = 4$ correspond to the year 1994, $x = 6$ correspond to 1996, and so on.

**(b)** Find a quadratic function defined by

$$f(x) = a(x - h)^2 + k$$

that models the data. Use $(4, 50)$ as the vertex and $(14, 110)$ as the other point to determine $a$.

**(c)** Plot the data together with $f$ in the same window. How well does $f$ model the number of Americans (in thousands) who are expected to be over 100 years old?

**(d)** Use the quadratic regression feature of a graphing calculator to determine the quadratic function $g$ that provides the best fit for the data.

**(e)** Use the functions $f$ and $g$ to predict the number of Americans, in thousands, who will be over 100 years old in the year 2006.

**28.** *Wholesale Drug Prices* The data from Section 3.2, Example 5 is repeated here.

| Year $x$ | Percent Increase $y$ |
|---|---|
| 1 | 7.2 |
| 3 | 3.0 |
| 5 | 1.9 |
| 7 | 2.5 |
| 9 | 4.2 |

*Source: IMS Health,* Retail and Provider Perspective.

**(a)** Plot the data. Then graph the function defined by

$$f(x) = .188x^2 - 2.05x + 7.46$$

given in Example 5 in the same window. Does the function appear to fit the data well?

**(b)** Use the quadratic regression feature of a graphing calculator to find the quadratic function $g$ that best fits the data. Graph this function in the same window as the data. Is it a good fit?

**(c)** Repeat part (b) using all data points but $(1, 7.2)$. Does the function $h$ found with the four remaining data points fit the data as well as function $g$?

**(d)** Find $g(1)$ and $h(1)$. Use your answers to explain which of these functions is a better model for the data.

**29.** *Automobile Stopping Distance* Selected values of the stopping distance $y$ in feet of a car traveling $x$ mph are given in the table.

| Speed (in mph) | Stopping Distance (in feet) |
|---|---|
| 20 | 46 |
| 30 | 87 |
| 40 | 140 |
| 50 | 240 |
| 60 | 282 |
| 70 | 371 |

*Source: National Safety Institute Student Workbook*, 1993, p. 7.

**(a)** Plot the data.

**(b)** The quadratic function defined by

$$f(x) = .056057x^2 + 1.06657x$$

is one model of the data. Find and interpret $f(45)$.

**(c)** Graph the function in the same window as the data. How well does $f$ model the stopping distance?

**30.** *Coast-Down Time* The coast-down time $y$ for a typical car as it drops 10 mph from an initial speed $x$ depends on several factors, such as average drag, tire pressure, and whether the transmission is in neutral. The table gives coast-down time in seconds for a car under standard conditions for selected speeds in miles per hour.

| Initial Speed (in mph) | Coast-Down Time (in seconds) |
|---|---|
| 30 | 30 |
| 35 | 27 |
| 40 | 23 |
| 45 | 21 |
| 50 | 18 |
| 55 | 16 |
| 60 | 15 |
| 65 | 13 |

*Source: Scientific American*, December 1994.

**(a)** Plot the data.

**(b)** Use the quadratic regression feature of a graphing calculator to find the quadratic function $g$ that best fits the data. Graph this function in the same window as the data. Is $g$ a good model for the data?

**(c)** Use $g$ to predict the coast-down time at an initial speed of 70 mph.

**(d)** Use the graph to find the speed that corresponds to a coast-down time of 24 seconds.

# 3.5 HIGHER-DEGREE POLYNOMIAL FUNCTIONS AND GRAPHS

Cubic Functions • Quartic Functions • Extrema • End Behavior • $x$-Intercepts (Real Zeros) • Comprehensive Graphs • Curve Fitting and Polynomial Models

Linear and quadratic functions are examples of *polynomial functions*. A general definition of a polynomial function follows.

**Looking Ahead to Calculus**

In calculus, polynomial functions are used to approximate more complicated functions, such as trigonometric, exponential, or logarithmic functions. For example, the trigonometric function $\sin x$ is approximated by the polynomial

$$x - \frac{x^3}{3!} + \frac{x^5}{5!} - \frac{x^7}{7!}$$

for values of $x$ near 0.

**Polynomial Function**

**A polynomial function of degree $n$ in the variable $x$** is a function defined by

$$P(x) = a_nx^n + a_{n-1}x^{n-1} + \cdots + a_1x + a_0,$$

where each $a_i$ is a real number, $a_n \neq 0$, and $n$ is a whole number.*

*While our definition requires real coefficients, the definition of a polynomial function can be extended to include imaginary numbers as coefficients.

The behavior of the graph of a polynomial function is due largely to the value of the coefficient $a_n$ and the *parity* (that is, "evenness" or "oddness") of the exponent $n$ on the term of highest degree. For this reason, we will refer to $a_n$ as the *leading coefficient*, and $a_nx^n$ as the *dominating term*. The term $a_0$ is the constant term of the polynomial function, and since $P(0) = a_0$, it is the $y$-intercept of the graph.

As we study the graphs of polynomial functions, we use the following general properties.

1. A polynomial function (unless otherwise specified) has domain $(-\infty, \infty)$.
2. The graph of a polynomial function is a smooth, continuous curve with no sharp corners.

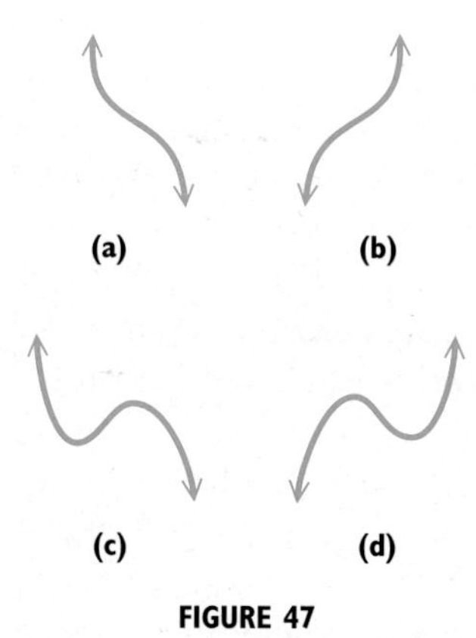

FIGURE 47

## Cubic Functions

A polynomial function of the form $P(x) = ax^3 + bx^2 + cx + d$, $a \neq 0$, is a third-degree, or **cubic function.** We studied the simplest cubic function, defined by $f(x) = x^3$, in Chapter 2. The graph of a cubic function will resemble, in general, one of the shapes shown in Figure 47.

### FOR DISCUSSION

1. Without graphing, which of the shapes in Figure 47 depicts the general form of the graph of $f(x) = x^3$?
2. Using the concepts of reflection of graphs from Chapter 2, which shape in Figure 47 most closely resembles the graph of $g(x) = -x^3$?
3. Graph each function in the standard window, and determine which shape in Figure 47 the graph most closely resembles.
   (a) $P(x) = x^3 + 5x^2 + 5x - 2$
   (b) $f(x) = -x^3 + 5x - 1$
   (c) $g(x) = x^3 + 3x^2 - 3x + 1$

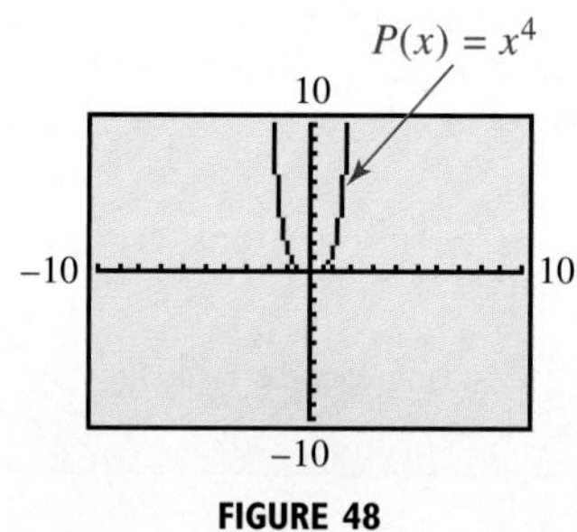

FIGURE 48

## Quartic Functions

A polynomial function of the form $P(x) = ax^4 + bx^3 + cx^2 + dx + e$, $a \neq 0$, is a fourth-degree, or **quartic function**. The simplest quartic function, defined by $P(x) = x^4$, is graphed in Figure 48. Notice that it resembles the graph of the squaring function; however, it is not actually a parabola.

If we graph a quartic function in an appropriate window, the graph will resemble, in general, one of the shapes shown in Figure 49. The dashed portions in (c) and (d) indicate that there may be irregular, but smooth, behavior in those intervals.

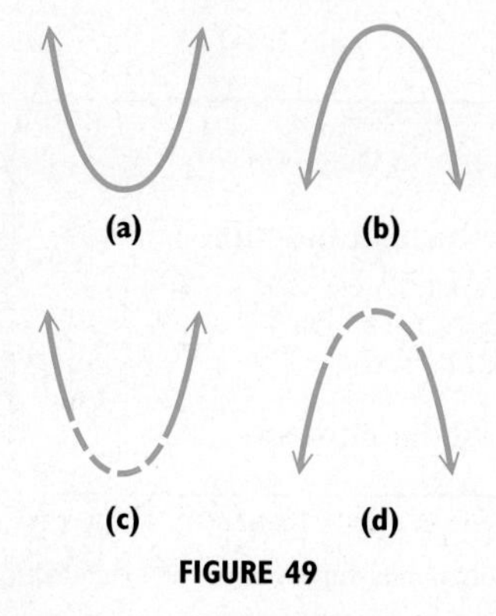

FIGURE 49

### FOR DISCUSSION

1. Which of the shapes in Figure 49 depicts the general form of the graphs of $f(x) = x^2$ and $g(x) = x^4$?
2. Using the concepts of reflection of graphs from Chapter 2, which shape in Figure 49 most closely resembles the graph of $h(x) = -x^4$?

**3.** Graph each function in the specified window, and determine which shape in Figure 49 the graph most closely resembles.

**(a)** $y = 3x^4 + x - 2$; $[-10, 10]$ by $[-10, 10]$

**(b)** $y = -2x^4 - x^3 + x - 3$; $[-10, 10]$ by $[-10, 10]$

**(c)** $y = -x^4 + 12x^3$; $[-15, 15]$ by $[-5000, 2500]$

## Extrema

In Figures 47–49 several graphs have **turning points** where the function changes from increasing to decreasing or vice versa. We first saw this behavior with quadratic functions, where the vertex was a maximum or minimum point on the graph. In general, the highest point at a "peak" is known as a **local maximum point,** and the lowest point at a "valley" is known as a **local minimum point.** Function values at such points are called **local maxima** (plural of maximum) and **local minima** (plural of minimum). Collectively, these values are called *extrema* (plural of *extremum*) as mentioned in Section 3.2. Figure 50 and the accompanying chart illustrate these ideas for typical graphs.

TECHNOLOGY NOTE

The feature described in the Technology Note in Section 3.2 that refers to maxima and minima also applies to polynomial functions of higher degree, provided an appropriate interval is designated.

Refer again to Figure 50(a). Notice that the point $P_2$ is the absolute highest point on the graph, and the range of the function is $(-\infty, y_2]$. We call $P_2$ the **absolute maximum point** on the graph, and $y_2$ the **absolute maximum value** of the function. Because the $y$-values approach $-\infty$, this function has no absolute minimum value. On the other hand, because the graph in Figure 50(b) is that of a function with range $(-\infty, \infty)$, it has neither an absolute maximum nor an absolute minimum.

| Extreme Point | Extreme Point |
|---|---|
| $P_1$ is a local maximum point. The function has a local maximum value of $y_1$ at $x = x_1$. | $P_1$ is a local maximum point. The function has a local maximum value of $y_1$ at $x = x_1$. |
| $P_2$ is a local maximum point. The function has a local maximum value of $y_2$ at $x = x_2$. | $P_2$ is a local minimum point. The function has a local minimum value of $y_2$ at $x = x_2$. |
| $P_3$ is a local minimum point. The function has a local minimum value of $y_3$ at $x = x_3$. | |

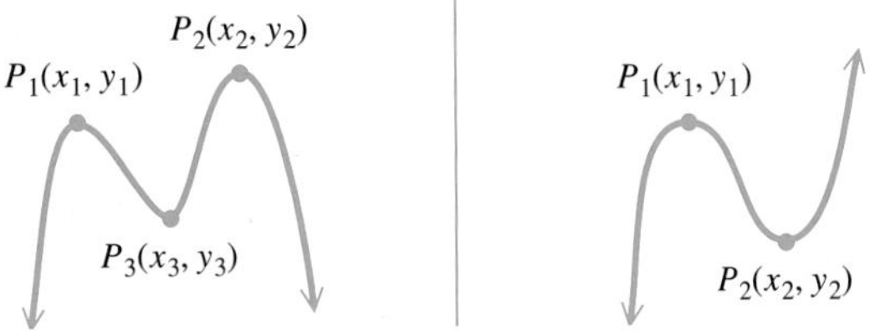

$P_1(x_1, y_1)$ $P_2(x_2, y_2)$ $P_3(x_3, y_3)$

(a)

$P_1(x_1, y_1)$ $P_2(x_2, y_2)$

(b)

**FIGURE 50**

In the next example, we identify extreme points on two graphs. Notice that an extreme point occurs at each turning point.

**Looking Ahead to Calculus**

Suppose we need to find the $x$-coordinates of the two turning points of the graph of

$$f(x) = 2x^3 - 8x^2 + 9.$$

We could use the "maximum" and "minimum" capabilities of a graphing calculator and determine that, to the nearest thousandth, they are 0 and 2.667. In calculus, their exact values can be found by determining the zeros of the derivative function of $f(x)$,

$$f'(x) = 6x^2 - 16x.$$

Factoring would show that the two zeros are 0 and $\frac{8}{3}$, which agree with the approximations found with a graphing calculator.

**EXAMPLE 1** *Identifying Local and Absolute Extreme Points*

Consider the graphs in Figure 51.

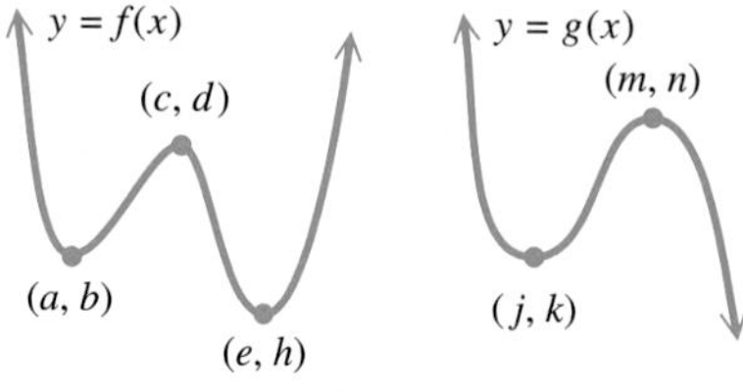


FIGURE 51

**(a)** Identify and classify the local extreme points of $f$.

**(b)** Identify and classify the local extreme points of $g$.

**(c)** Describe the absolute extreme points for $f$ and $g$.

**Solution**

**(a)** The points $(a, b)$ and $(e, h)$ are local minimum points. The point $(c, d)$ is a local maximum point.

**(b)** The point $(j, k)$ is a local minimum point and the point $(m, n)$ is a local maximum point.

**(c)** The absolute minimum point of function $f$ is $(e, h)$, and the absolute minimum value is $h$ since the range of $f$ is $[h, \infty)$. Function $f$ has no absolute maximum value. Function $g$ has no absolute extrema since its range is $(-\infty, \infty)$.

It is possible for the graph of a polynomial function to have a maximum or a minimum point that is not apparent in a particular window. This is an example of *hidden behavior.*

**EXAMPLE 2** *Examining Hidden Behavior*

**(a)** Figure 52(a) shows the graph of $P(x) = x^3 - 2x^2 + x - 2$ in the standard viewing window. Make a conjecture concerning possible hidden behavior, and verify it.

**Solution** In Figure 52(a), because the graph levels off in the domain interval $[0, 2]$, there may be behavior that is not apparent in the given window. By changing the window to $[-2.5, 2.5]$ by $[-4.5, .5]$, we see that there are two extrema there. The local maximum point, as seen in Figure 52(b), has approximate coordinates $(.33, -1.85)$. There is also a local minimum point at $(1, -2)$.

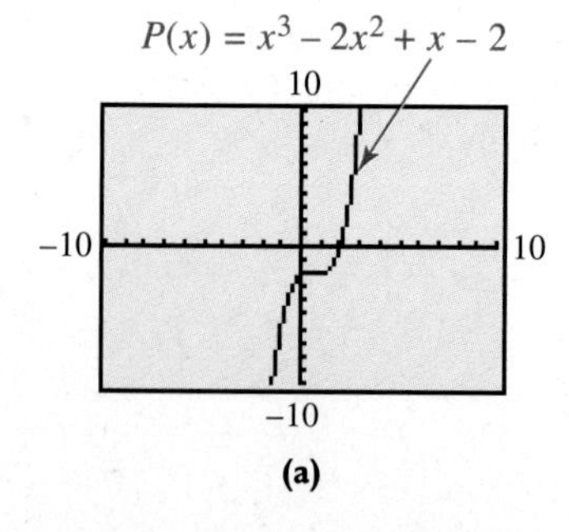


(a)

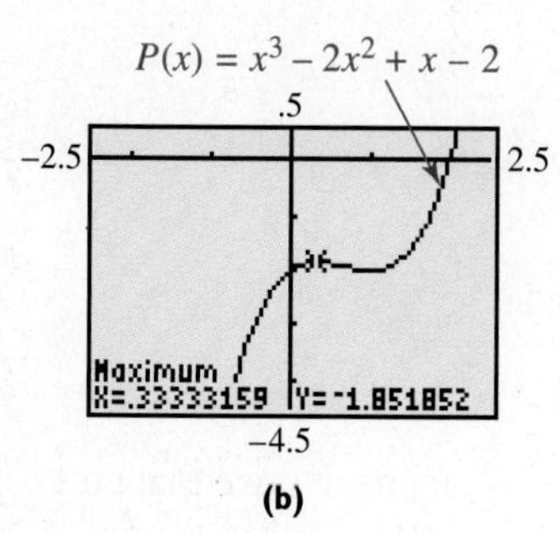


(b)

FIGURE 52

A linear function has degree 1, and since its graph is a straight line, it has no local extrema. A quadratic function has degree 2 and has only one extreme point (its vertex). Extending this idea, we would expect a third-degree polynomial to have at most two local extrema, a fourth-degree polynomial to have at most three local extrema, and so on. This is actually the case.

**Number of Local Extrema**

The number of local extrema of the graph of a polynomial function of degree $n$ is at most $n - 1$.

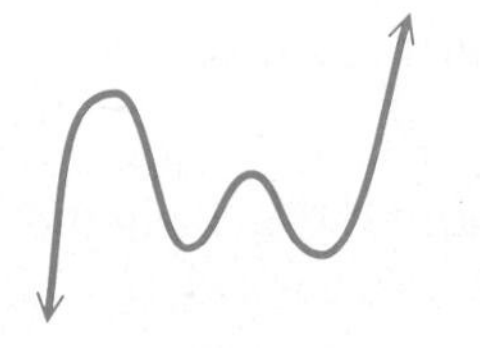

FIGURE 53

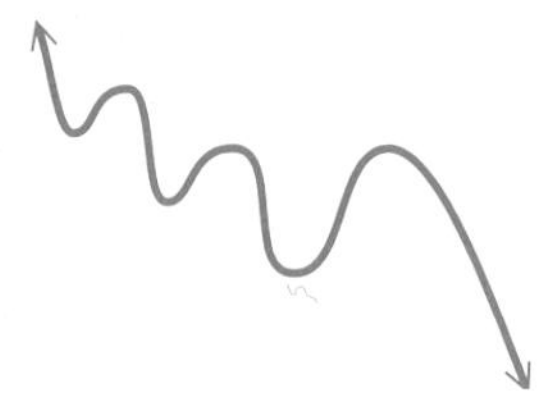

FIGURE 54

Notice that this property states that the number of local extrema is *at most* $n - 1$ for a polynomial function of degree $n$. The graph may have fewer than $n - 1$ local extrema.

**FOR DISCUSSION**

1. How many local extrema does $f(x) = x^3$ have? How does your answer support the property stated above?
2. **(a)** What is the least possible degree of the function in Figure 53?
   **(b)** Explain why this function cannot be of degree 4.
3. Repeat Item 2 for the polynomial function graph in Figure 54.

## End Behavior

If the value of $a$ is positive for the quadratic function defined by $P(x) = ax^2 + bx + c$, the graph opens upward, and if $a$ is negative, the graph opens downward. The sign of $a$ determines the *end behavior* of the graph. If $a > 0$, then as $x$ approaches $-\infty$ or $\infty$ (written $x \rightarrow -\infty$ or $x \rightarrow \infty$), the value of $P(x)$ approaches $\infty$ (written $P(x) \rightarrow \infty$). If $a < 0$, then as $x \rightarrow -\infty$, $P(x) \rightarrow -\infty$, and as $x \rightarrow \infty$, $P(x) \rightarrow -\infty$. In general, the end behavior of the graph of a polynomial function is determined by the sign of the leading coefficient and the parity of the degree.

A cubic function is an example of an odd-degree polynomial function and illustrates the following observations about the end behavior of an odd-degree polynomial function. Similarly, a quartic function is an even-degree polynomial function with the same end behavior as a quadratic function.

**End Behavior of Polynomial Functions**

Suppose that $ax^n$ is the dominating term of a polynomial function $P$ of *odd degree*. (The dominating term is the term of highest degree.)

1. If $a > 0$, then as $x \rightarrow \infty$, $P(x) \rightarrow \infty$, and as $x \rightarrow -\infty$, $P(x) \rightarrow -\infty$. Therefore, the end behavior of the graph is of the type shown in Figure 55(a). We symbolize it as ↙↗.

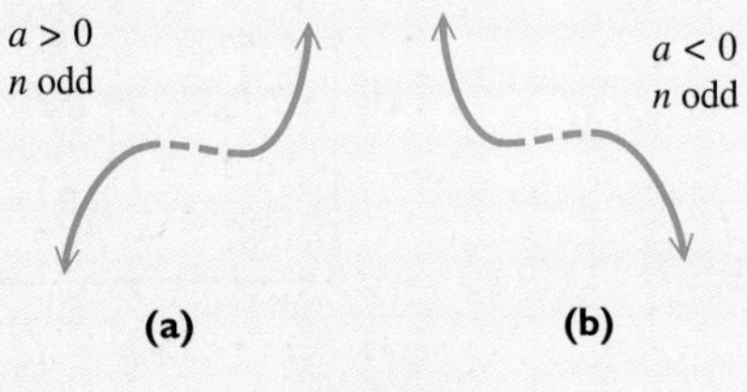


FIGURE 55

2. If $a < 0$, then as $x \rightarrow \infty$, $P(x) \rightarrow -\infty$, and as $x \rightarrow -\infty$, $P(x) \rightarrow \infty$. Therefore, the end behavior of the graph is of the type shown in Figure 55(b). We symbolize it as ↖↘.

*(continued)*

Suppose that $ax^n$ is the dominating term of a polynomial function $P$ of *even degree.*

**1.** If $a > 0$, then as $|x| \to \infty$, $P(x) \to \infty$. Therefore, the end behavior of the graph is of the type shown in Figure 56(a). We symbolize it as ↖↗.

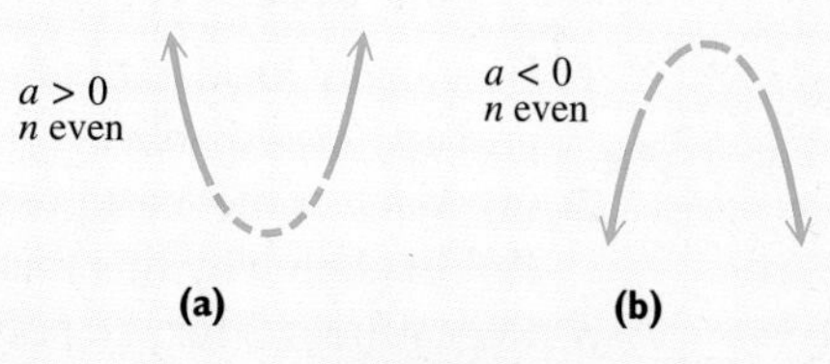


**FIGURE 56**

**2.** If $a < 0$, then as $|x| \to \infty$, $P(x) \to -\infty$. Therefore, the end behavior of the graph is of the type shown in Figure 56(b). We symbolize it as ↙↘.

**EXAMPLE 3** *Determining End Behavior Given the Defining Polynomial*

The graphs of the functions defined as follows are shown in Figure 57.

$$f(x) = x^4 - x^2 + 5x - 4, \qquad g(x) = -x^6 + x^2 - 3x - 4,$$
$$h(x) = 3x^3 - x^2 + 2x - 4, \quad \text{and} \quad k(x) = -x^7 + x - 4.$$

Based on the discussion in the preceding box, match each function with its graph.

**A.**

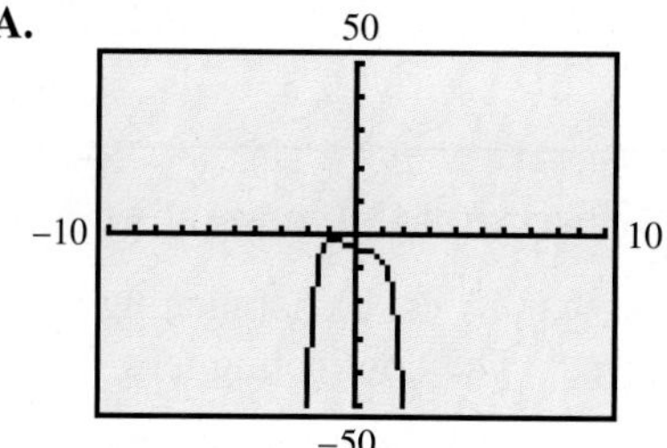


**B.**

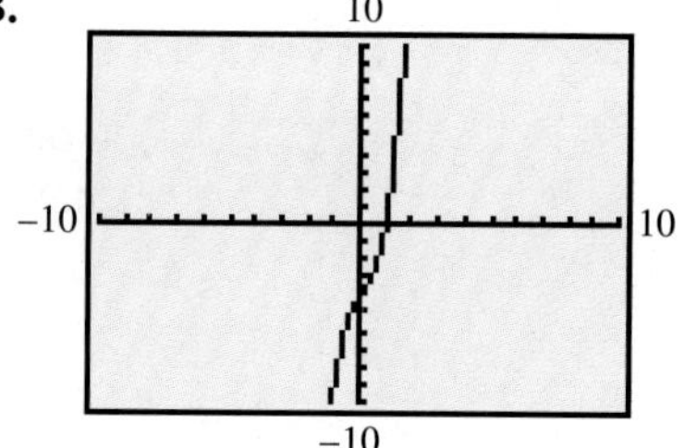


**C.**

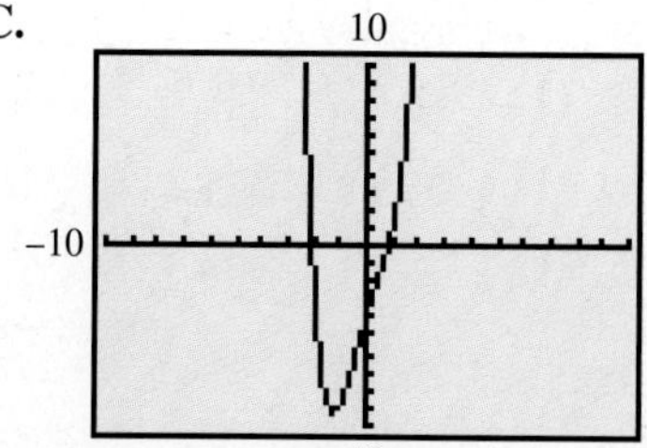


**D.**

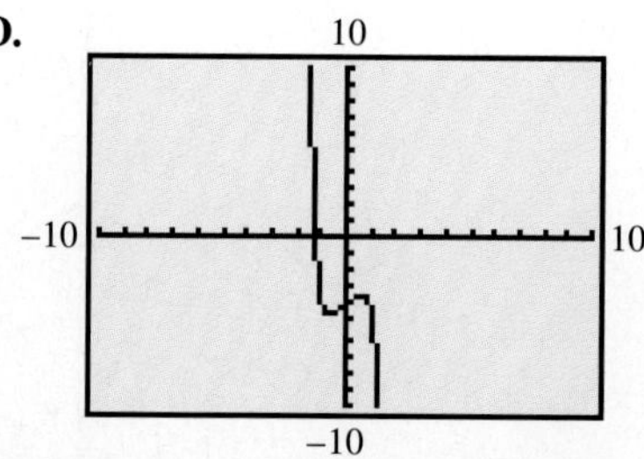


**FIGURE 57**

**Solution** Because $f$ is of even degree with positive leading coefficient, its graph is in C. Because $g$ is of even degree with negative leading coefficient, its graph is in A. Function $h$ has odd degree and the dominating term is positive, so its graph is in B. Because function $k$ has odd degree and a negative dominating term, its graph is in D.

## x-Intercepts (Real Zeros)

A linear function can have no more than one $x$-intercept, and a quadratic function can have no more than two $x$-intercepts. Figure 58 shows how a cubic polynomial may have one, two, or three $x$-intercepts. These observations suggest the following important property of polynomial functions.

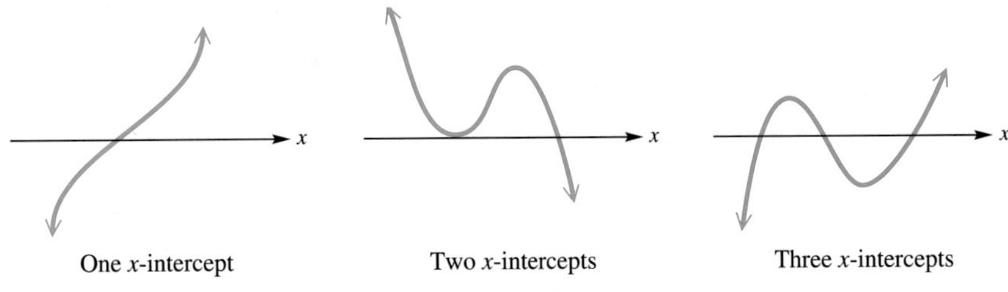


**FIGURE 58**

> **Number of $x$-Intercepts (Real Zeros) of a Polynomial Function**
>
> A polynomial function of degree $n$ will have at most $n$ $x$-intercepts (real zeros).

**EXAMPLE 4** *Determining x-Intercepts Graphically*

Graphically find the $x$-intercepts (that is, the real zeros) of the polynomial function defined by

$$P(x) = x^3 + 5x^2 + 5x - 2.$$

**Solution** If we use the standard window, we get the graph shown in Figure 59. Notice that it has three $x$-intercepts. By using graphing calculator capabilities, we find that the $x$-intercepts (real zeros) are $-2$, approximately $-3.30$, and approximately $.30$.

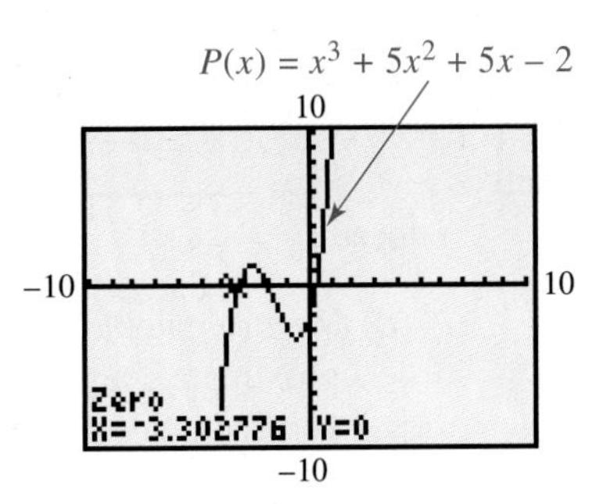


**FIGURE 59**

**EXAMPLE 5** *Analyzing a Polynomial Function*

For the fifth-degree polynomial function defined by

$$P(x) = x^5 + 2x^4 - x^3 + x^2 - x - 4,$$

answer each of the following.

**(a)** Determine its domain. **(b)** Determine its range.

**(c)** Use its graph to find approximations of its local extreme points.

**(d)** Use its graph to find its approximate and/or exact $x$-intercepts.

**Looking Ahead to Calculus**

In calculus, the derivative of a function $f$ is a function $f'$ that gives the slope of $f$ at any value in the domain. At turning points the slope, and thus the derivative, is 0. By solving the equation $f'(x) = 0$, the number and location of the extrema of $f$ can be identified. In Example 5(c), this method would verify that there are only two extrema.

**Solution**

**(a)** Because it is a polynomial function, its domain is $(-\infty, \infty)$.

**(b)** Because it is of odd degree, its range is $(-\infty, \infty)$.

**(c)** From the graph of $P$ in Figure 60, it appears that there are only two extreme points. Using the capabilities of a calculator, we find that the local maximum has approximate coordinates $(-2.02, 10.01)$, and the local minimum has approximate coordinates $(.41, -4.24)$.

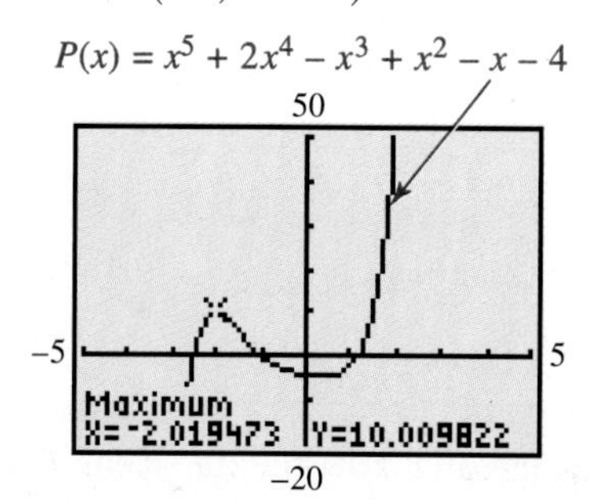


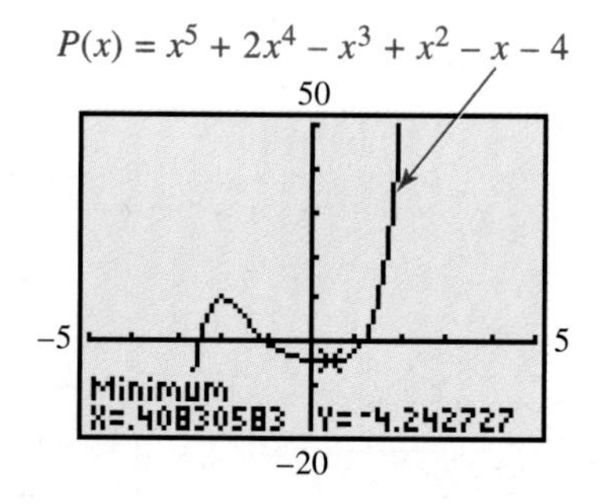


**FIGURE 60**

The preceding results found graphically can also be found using a built-in utility, as shown in the two screens in Figure 61. Here $Y_1$ is defined as

$$X^5 + 2X^4 - X^3 + X^2 - X - 4,$$

so the local maximum is in the interval $[-3, -1]$, and the local minimum is in the interval $[0, 1]$.

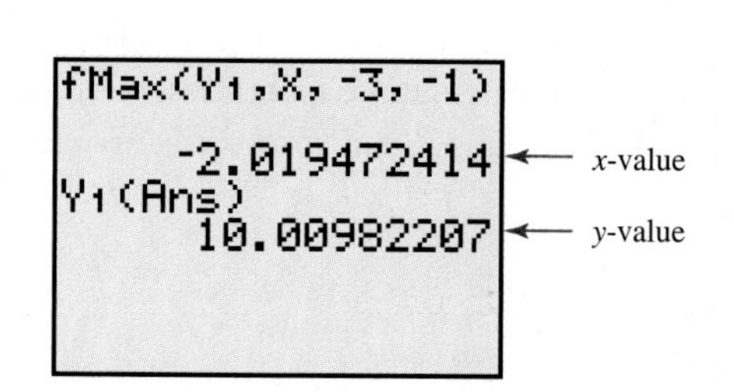


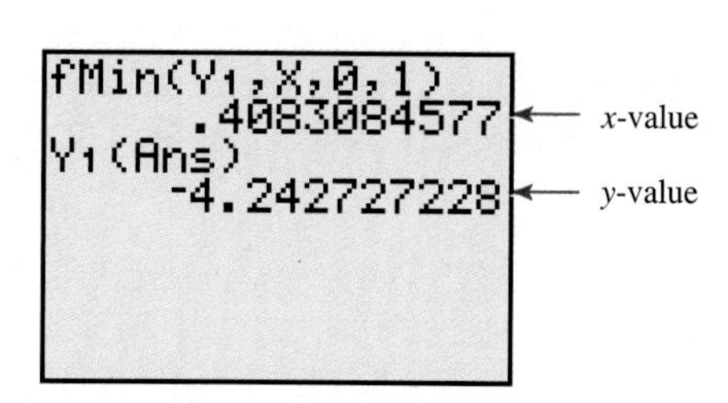


**FIGURE 61**

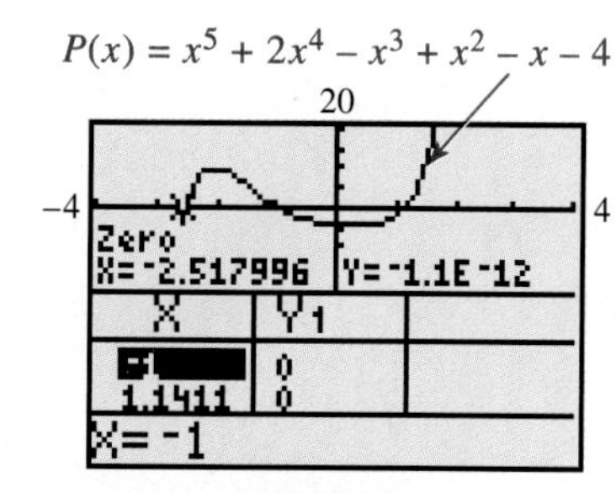


**FIGURE 62**

**(d)** We use calculator methods to find that the $x$-intercepts are $-1$ (exact), 1.14 (approximate), and $-2.52$ (approximate). See Figure 62. The first of these can be verified analytically by evaluating $P(-1)$.

$$P(-1) = (-1)^5 + 2(-1)^4 - (-1)^3 + (-1)^2 - (-1) - 4 = 0$$

This result shows again that an $x$-intercept of the graph of a function is a real zero of the function. This function has only three $x$-intercepts and thus three real zeros, which supports the fact that a polynomial function of degree $n$ will have *at most* $n$ $x$-intercepts. It may have fewer, as in this case.

**EXAMPLE 6** *Analyzing a Polynomial Function*

For the fourth-degree polynomial function defined by

$$P(x) = x^4 + 2x^3 - 15x^2 - 12x + 36,$$

answer each of the following.

**(a)** State the domain.

**(b)** Use its graph to find approximations of its local extreme points. Does it have an absolute minimum? What is the range of the function?

**(c)** Use its graph to find its $x$-intercepts.

**Solution**

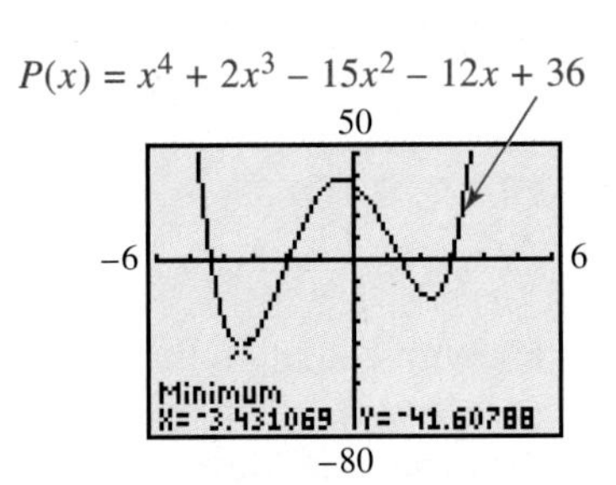


The other two extreme points are (–.38, 38.31) and (2.31, –18.63).

**FIGURE 63**

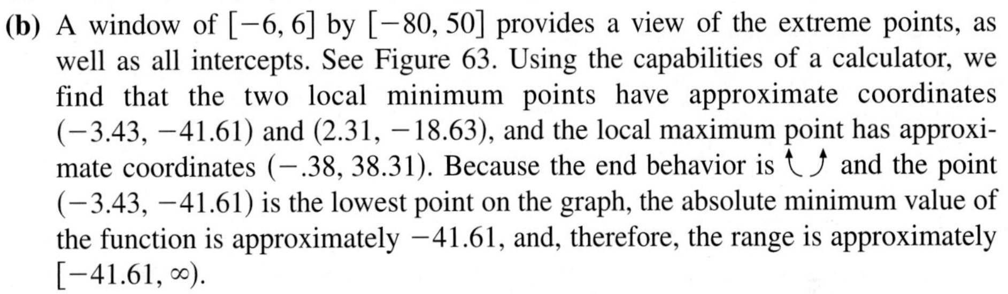

**(a)** Because it is a polynomial function, its domain is $(-\infty, \infty)$.

**(b)** A window of $[-6, 6]$ by $[-80, 50]$ provides a view of the extreme points, as well as all intercepts. See Figure 63. Using the capabilities of a calculator, we find that the two local minimum points have approximate coordinates $(-3.43, -41.61)$ and $(2.31, -18.63)$, and the local maximum point has approximate coordinates $(-.38, 38.31)$. Because the end behavior is ↖ ↗ and the point $(-3.43, -41.61)$ is the lowest point on the graph, the absolute minimum value of the function is approximately $-41.61$, and, therefore, the range is approximately $[-41.61, \infty)$.

**(c)** This function has the maximum number of $x$-intercepts possible, four. Using the calculator capabilities, we find that two exact values for the $x$-intercepts are $-2$ and 3, while, to the nearest hundredth, the other two are $-4.37$ and 1.37.

**What Went WRONG?**

A student graphed $y = .045x^4 - 2x^2 + 2$ in the decimal window of a popular model of graphing calculator. Because the polynomial has even degree and positive leading coefficient, she expected to find end behavior ↖ ↗. However, this graph indicates ↙ ↘ as end behavior.

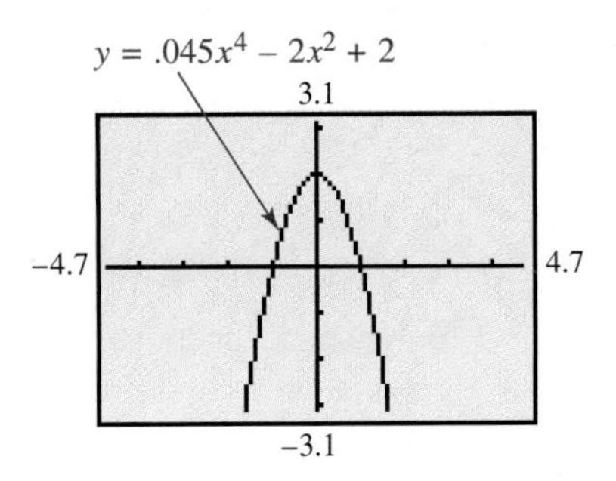


What Went Wrong? Is there a way to graph this function so that the correct end behavior is apparent?

## Comprehensive Graphs

The most important features of the graph of a polynomial function are its intercepts, its extrema, and its end behavior. For this reason, a comprehensive graph of a polynomial function will exhibit the following features:

---

*Answer to What Went Wrong?*

The window must be enlarged. Scrolling through a table should indicate a suitable window. One example is $[-10, 10]$ by $[-25, 10]$.

1. all $x$-intercepts (if any),
2. the $y$-intercept,
3. all extreme points (if any),
4. enough of the graph to exhibit the correct end behavior.

**EXAMPLE 7** *Determining an Appropriate Window for a Comprehensive Graph*

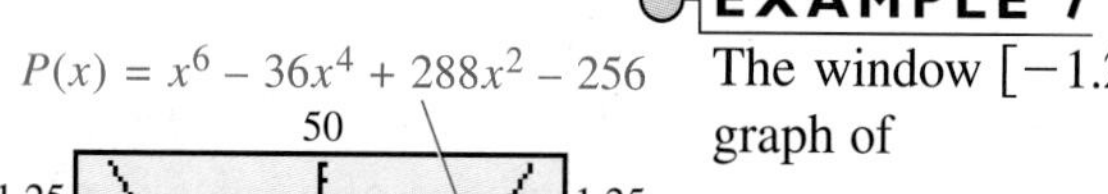


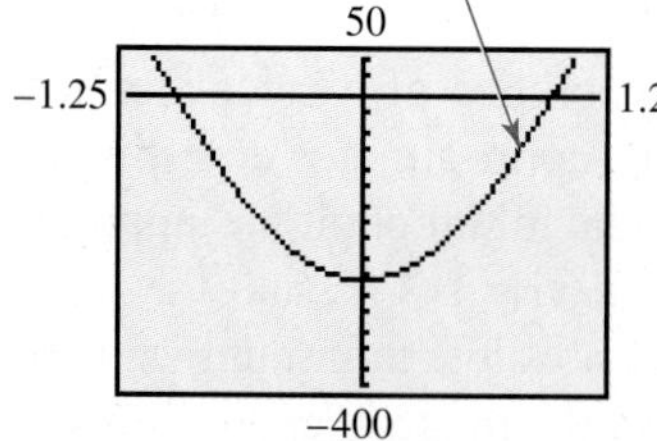


FIGURE 64

The window $[-1.25, 1.25]$ by $[-400, 50]$ is used in Figure 64 to give a view of the graph of

$$P(x) = x^6 - 36x^4 + 288x^2 - 256.$$

Is this a comprehensive graph?

**Solution** Since the function is of even degree and the dominating term has positive coefficient, the end behavior seems to be correct. The $y$-intercept, $-256$, is shown, and two $x$-intercepts are shown. One local minimum is shown. This function may, however, have up to six $x$-intercepts, since it is of degree 6. By experimenting with other viewing windows, we see that a window of $[-8, 8]$ by $[-1000, 600]$ shows a total of five local extrema, and four $x$-intercepts that were not apparent in the earlier figure. See Figure 65. Since there can be no more than five local extrema, this second view (*not* the first view in Figure 64) gives a comprehensive graph.

**FOR DISCUSSION**

How does Example 7 illustrate the warning "Don't always believe what you see"? How does it illustrate the shortcomings of learning technology without learning mathematical concepts?

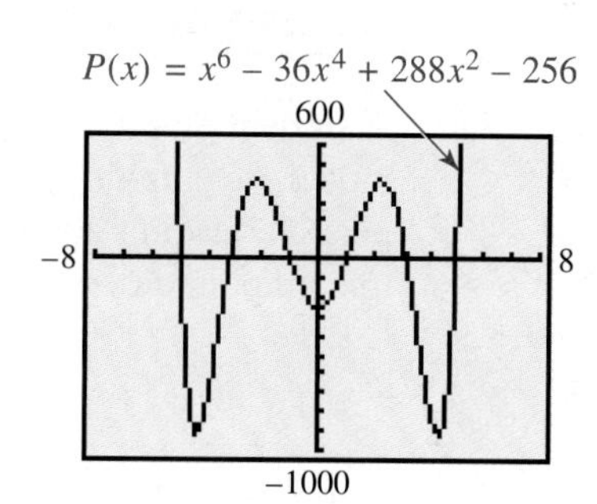


FIGURE 65

## Curve Fitting and Polynomial Models

Polynomial functions are often good choices to model real data. We have used graphing calculators to find linear and quadratic regression equations to model data. Some data can be modeled best using cubic (degree 3) or quartic (degree 4) regression equations.

For example, the table gives the percent of adult female smokers for several years between 1979–1998. The data is graphed in Figure 66(a). The cubic polynomial function defined by

$$f(x) = .0020x^3 - .0686x^2 + .1800x + 30.27,$$

where $x = 0$ corresponds to 1975, $x = 5$ corresponds to 1980, and so on, models the data reasonably well, as shown in Figure 66(b). Figure 66(c) shows the graph of the quartic polynomial function defined by

$$f(x) = -.0007554x^4 + .0440x^3 - .8708x^2 + 6.198x + 16.44,$$

where $x = 0$ corresponds to 1975, and so on. This function does not provide a reasonable model for several reasons. For example, this model indicates that in 1975, when $x = 0$, only 16.44 percent of adult females smoked. Also, it shows the percent of adult female smokers dropping rapidly after 1998, which does not agree with the data.

**Adult Female Smokers**

| Year | Percent |
|---|---|
| 1979 | 29.9 |
| 1985 | 27.9 |
| 1990 | 22.6 |
| 1991 | 23.5 |
| 1992 | 24.6 |
| 1993 | 22.5 |
| 1994 | 23.1 |
| 1995 | 22.6 |
| 1998 | 22.1 |

*Source: Statistical Abstract of the United States*, 2000.

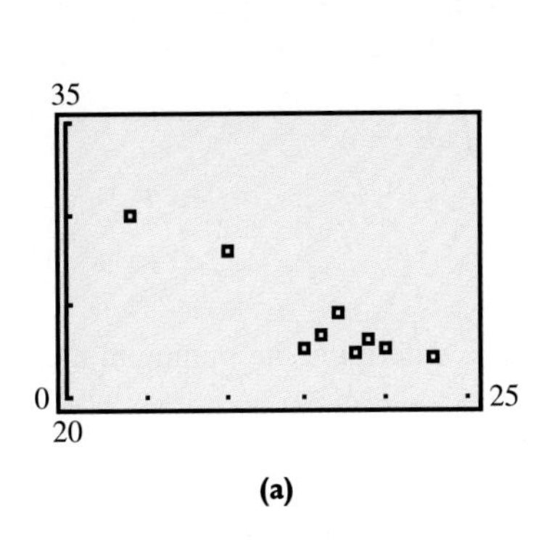


(a)

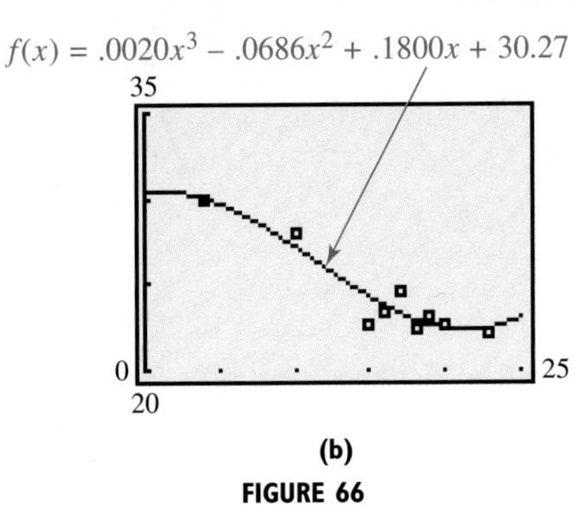


(b)

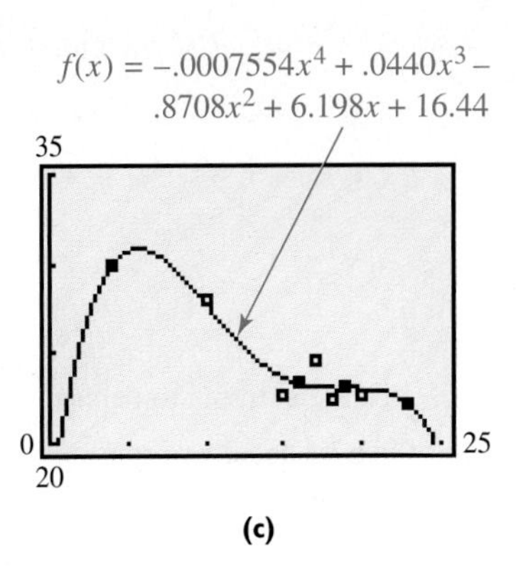


(c)

**FIGURE 66**

## 3.5 EXERCISES

**1.** Using a window of $[-1, 1]$ by $[-1, 1]$, graph the odd-degree polynomial functions

$$y = x, \quad y = x^3, \quad \text{and} \quad y = x^5.$$

Describe the behavior of these functions relative to each other. Predict the behavior of the graph of $y = x^7$ in the same window, and then graph it to support your prediction.

**2.** Repeat Exercise 1 for the even-degree polynomial functions

$$y = x^2, \quad y = x^4, \quad \text{and} \quad y = x^6.$$

Predict the behavior of the graph of $y = x^8$ in the same window, and then graph it to support your prediction.

**3.** The graphs of $f(x) = x^n$ for $n = 3, 5, 7, \ldots$ all resemble each other. Describe how they are the same. As $n$ gets larger, what happens to the graph?

**4.** Repeat Exercise 3 for $f(x) = x^n$, where $n = 2, 4, 6, \ldots$.

## RELATING CONCEPTS

**For individual or group investigation (Exercises 5–8)**

*The concepts of stretching, shrinking, translating, and reflecting graphs presented in Sections 2.2 and 2.3 can be applied to polynomial functions of the form $P(x) = x^n$. For example, the graph of $y = -2(x + 4)^4 - 6$ can be obtained from the graph of $y = x^4$ by shifting 4 units to the left, stretching vertically by a factor of 2, reflecting across the x-axis, and shifting downward 6 units, so the graph should resemble the graph at the right.*

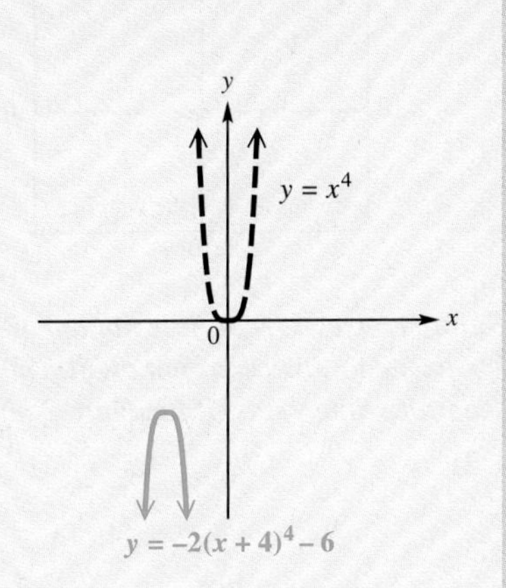


*If we expand the expression $-2(x + 4)^4 - 6$ algebraically, we get*

$$-2x^4 - 32x^3 - 192x^2 - 512x - 518.$$

*Thus, the graph of $y = -2(x + 4)^4 - 6$ is the same as that of*

$$y = -2x^4 - 32x^3 - 192x^2 - 512x - 518.$$

*In Exercises 5–8, two forms of the same polynomial function are given. Sketch by hand the general shape of the graph of the function, using the concepts of Chapter 2, and describe the transformations. Then, support your answer by graphing it on your calculator in a suitable window.*

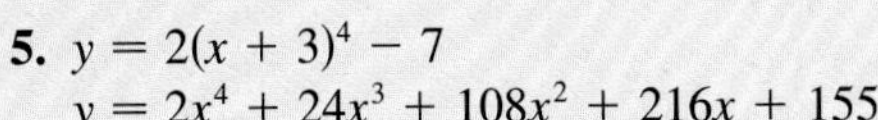

**5.** $y = 2(x + 3)^4 - 7$
$y = 2x^4 + 24x^3 + 108x^2 + 216x + 155$

**6.** $y = -3(x + 1)^4 + 12$
$y = -3x^4 - 12x^3 - 18x^2 - 12x + 9$

**7.** $y = -3(x - 1)^3 + 12$
$y = -3x^3 + 9x^2 - 9x + 15$

**8.** $y = .5(x - 1)^5 + 13$
$y = .5x^5 - 2.5x^4 + 5x^3 - 5x^2 + 2.5x + 12.5$

*Concept Check* *Use the graphs shown here, which include all extrema, for Exercises 9–16.*

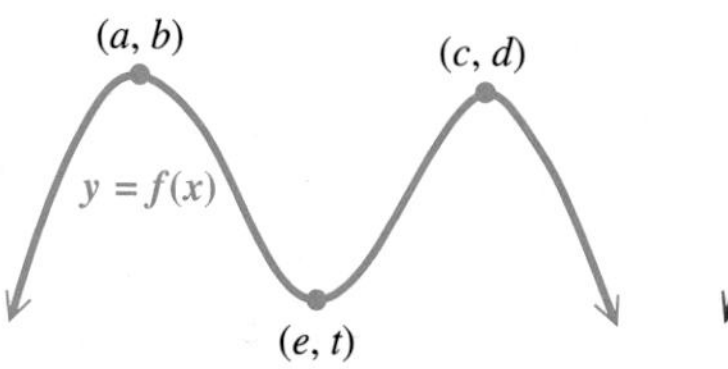


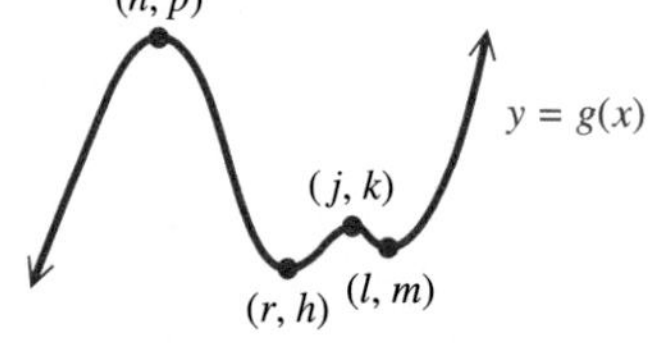


**9.** Use the extrema to determine the minimum degree of $f$.

**10.** Use the extrema to determine the minimum degree of $g$.

**11.** Give all local extreme points of $f$. Tell whether each is a maximum or minimum.

**12.** Give all local extreme points of $g$. Tell whether each is a maximum or minimum.

**13.** Describe all absolute extreme points of $f$.

**14.** Describe all absolute extreme points of $g$.

**15.** Give the local and absolute extreme values of $f$.

**16.** Give the local and absolute extreme values of $g$.

*Each function is graphed in a window that results in hidden behavior. Experiment with various windows to locate the extrema of the function.*

**17.** $y = \frac{1}{3}x^3 - \frac{5}{2}x^2 + 6x - 1$

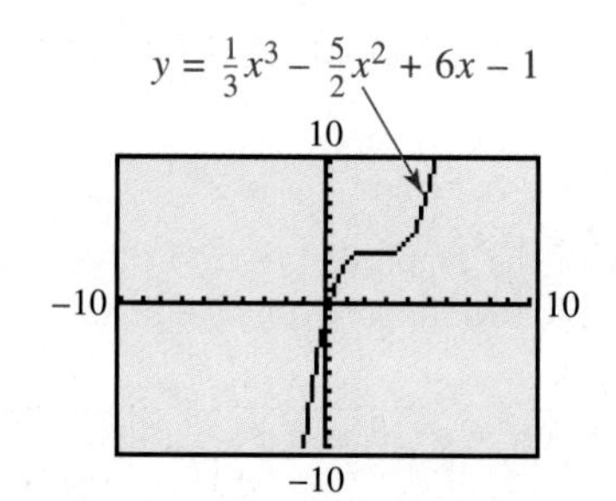


**18.** $y = -\frac{1}{3}x^3 - \frac{9}{2}x^2 - 20x - \frac{59}{3}$



**19.** $y = -x^3 - 11x^2 - 40x - 50$

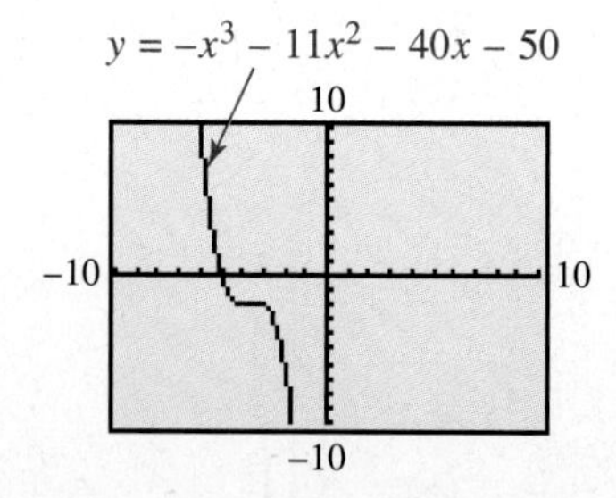


**20.** $y = 2x^3 - 3.3x^2 + 1.8x + 3$

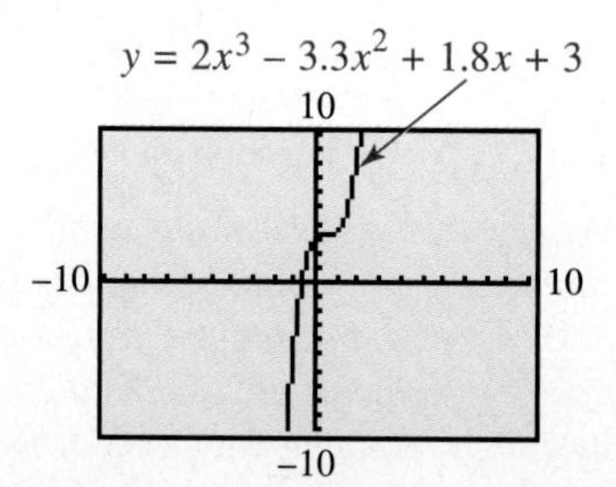


*In Exercises 21–24, it is not apparent from the standard viewing window whether the graph of the quadratic function intersects the x-axis once, twice, or not at all. This is another example of hidden behavior. Experiment with various windows to determine the number of x-intercepts. If there are x-intercepts, give their values to the nearest hundredth.*

**21.** $y = x^2 - 4.25x + 4.515$

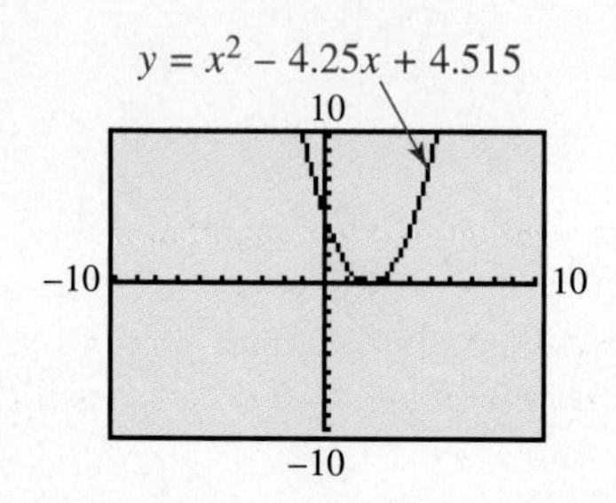


**22.** $y = x^2 + 6.95x + 12.07$

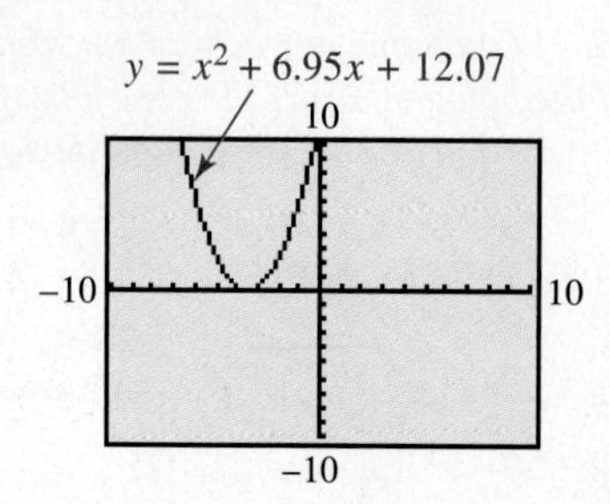

**23.** $y = -x^2 + 6.5x - 10.60$

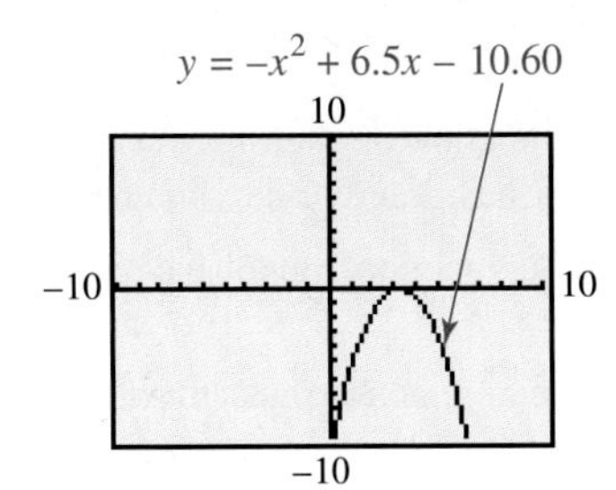


**24.** $y = -2x^2 + .2x - .15$

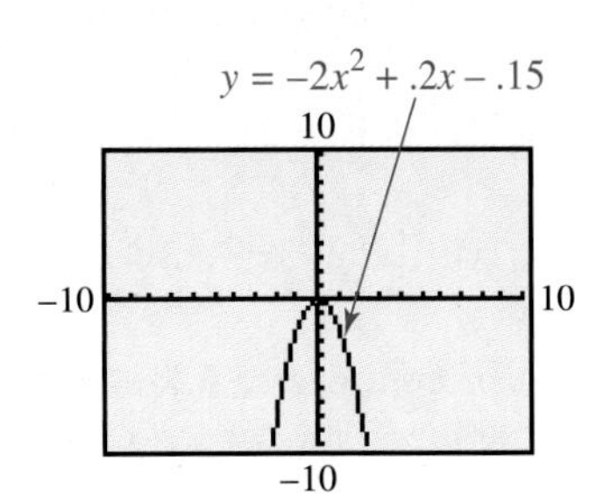


*Use an end behavior diagram (↖↗, ↙↘, ↖↘, or ↙↗) to describe the end behavior of each function. Then, verify your answer by graphing the function on your calculator.*

**25.** $P(x) = \sqrt{5}x^3 + 2x^2 - 3x + 4$

**26.** $P(x) = -\sqrt{7}x^3 - 4x^2 + 2x - 1$

**27.** $P(x) = -\pi x^5 + 3x^2 - 1$

**28.** $P(x) = \pi x^7 - x^5 + x - 1$

**29.** $P(x) = 2.74x^4 - 3x^2 + x - 2$

**30.** $P(x) = \sqrt{6}x^6 - x^5 + 2x - 2$

**31.** $P(x) = -x^4 + x^5 - \pi x^6 - x + 3$

**32.** $P(x) = -x - 3.2x^3 + x^2 - 2.84x^4 + 3$

*The functions in Exercises 33–36 are graphed in Figures A–D. Use end behavior and dominating-term analysis to match each equation with the correct graph.*

**33.** $f(x) = 2x^3 + x^2 - x + 3$

**34.** $g(x) = -2x^3 - x + 3$

**35.** $h(x) = -2x^4 + x^3 - 2x^2 + x + 3$

**36.** $k(x) = 2x^4 - x^3 - 2x^2 + 3x + 3$

**A.**

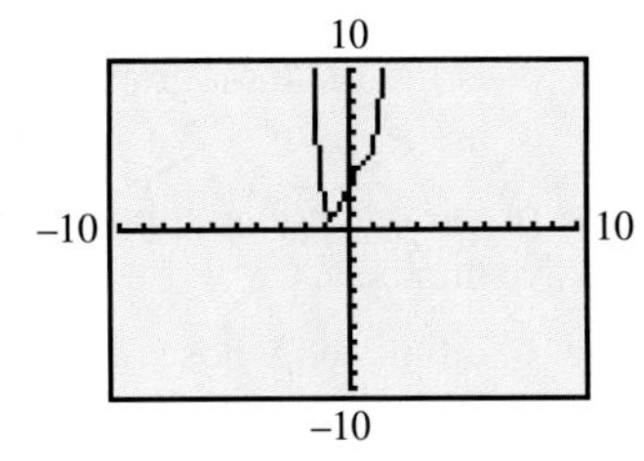


**B.**

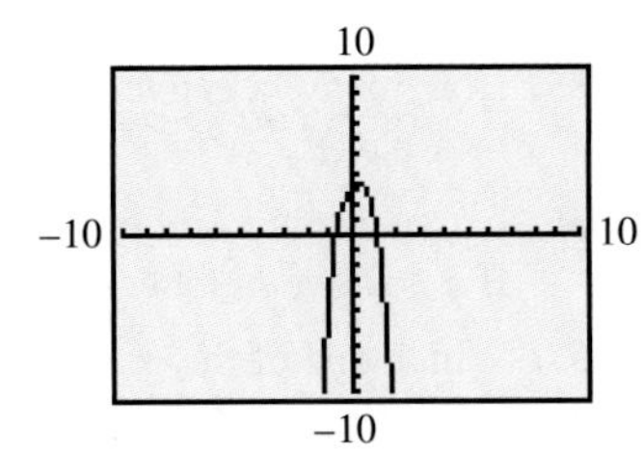


**C.**

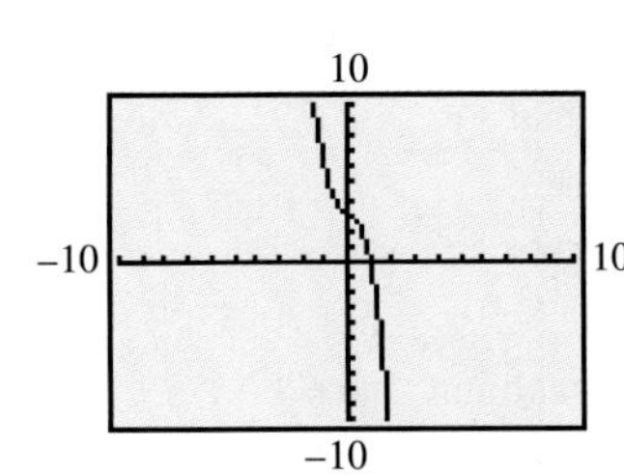


**D.**

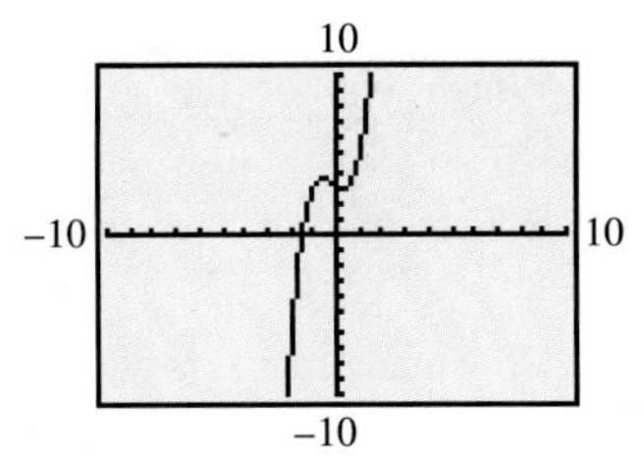


## RELATING CONCEPTS

**For individual or group investigation (Exercises 37–40)**

*Informally, we say that a graph is* concave up *on an interval if the graph would "hold water" there. A graph is* concave down *on an interval if it would "spill water" on that interval. See the figure. (A formal discussion of concavity requires concepts from calculus.) A polynomial graph may change concavity several times throughout its domain.*

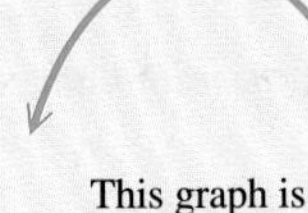

This graph is concave down.

This graph is concave up.

**37.** For a function of the form $P(x) = x^n$, where $n$ is even, discuss the concavity of the graph.

**38.** For a function of the form $P(x) = x^n$, where $n$ is odd, discuss the concavity of the graph.

**39.** Graph $y = x^3 + x$ in the window $[-4, 4]$ by $[-20, 20]$, and discuss the concavity of the graph.

**40.** Explain why an even-degree polynomial function with negative leading coefficient must be concave down for some interval of its domain.

*Concept Check* *Without graphing, answer* true *or* false *to each statement. Then, support your answer, if you wish, by graphing.*

**41.** The function defined by $f(x) = x^3 + 2x^2 - 4x + 3$ has four real zeros.

**42.** The function defined by $f(x) = x^3 + 3x^2 + 3x + 1$ must have at least one real zero.

**43.** If a polynomial function of even degree has a negative leading coefficient and a positive $y$-intercept, it must have at least two real zeros.

**44.** The function defined by $f(x) = 3x^4 + 5$ has no real zeros.

**45.** The function defined by $f(x) = -3x^4 + 5$ has two real zeros.

**46.** The graph of the cubic function defined by $f(x) = x^3 - 3x^2 + 3x - 1 = (x - 1)^3$ has exactly one $x$-intercept.

**47.** A fifth-degree polynomial function cannot have a single real zero.

**48.** An even-degree polynomial function must have at least one real zero.

*Concept Check* *The graphs below show*

$$y = x^3 - 3x^2 - 6x + 8, \qquad y = x^4 + 7x^3 - 5x^2 - 75x,$$
$$y = -x^3 + 9x^2 - 27x + 17, \quad \textit{and} \quad y = -x^5 + 36x^3 - 22x^2 - 147x - 90,$$

*but not necessarily in that order. Assuming that each is a comprehensive graph, answer each question in Exercises 49–58.*

**A.**

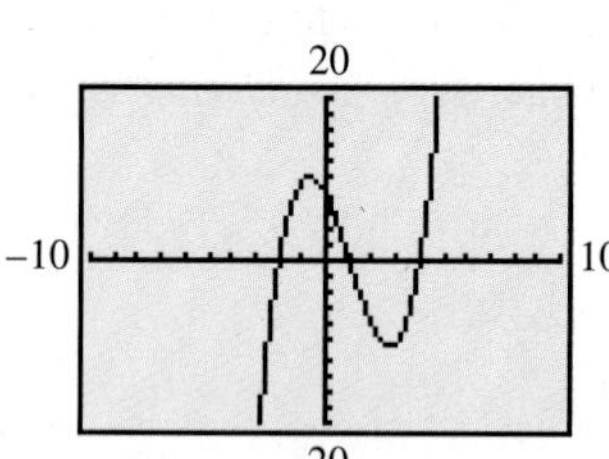


**B.**

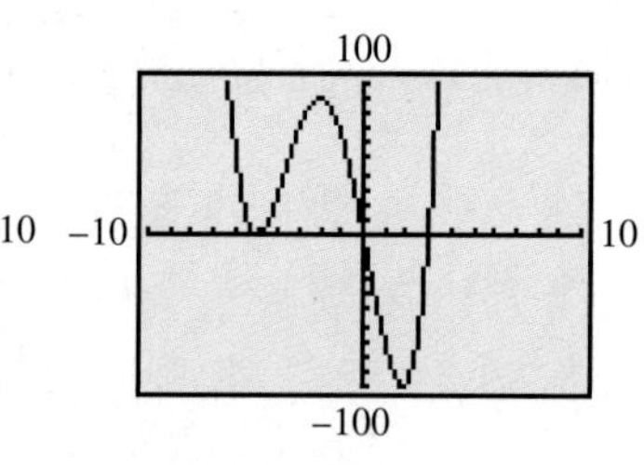


**C.**

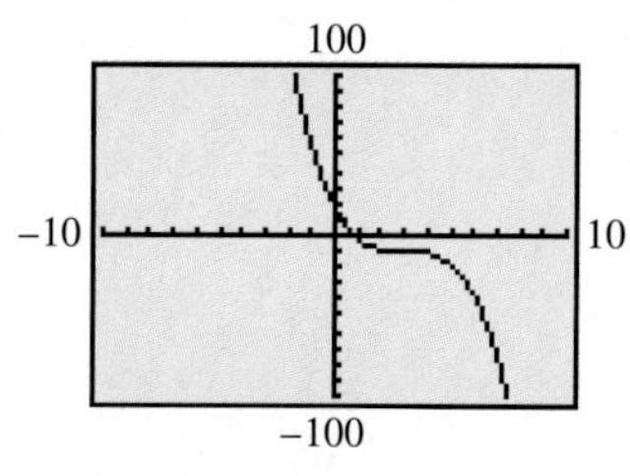


**D.**

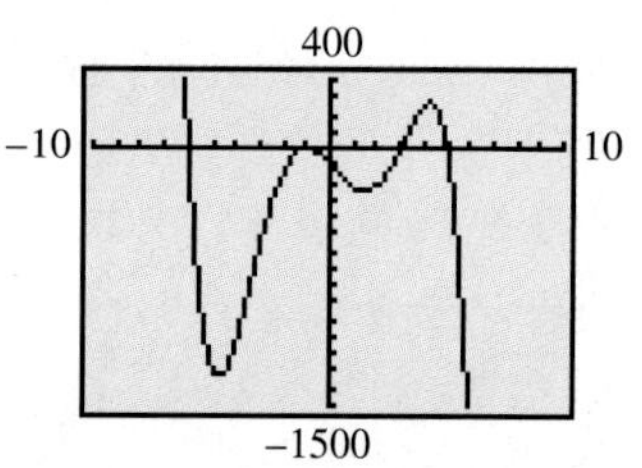


**49.** Which one of the graphs is that of $y = x^3 - 3x^2 - 6x + 8$?

**50.** Which one of the graphs is that of $y = x^4 + 7x^3 - 5x^2 - 75x$?

**51.** How many real zeros does the graph in C have?

**52.** Which one of C and D is the graph of $y = -x^3 + 9x^2 - 27x + 17$?

**53.** Which of the graphs cannot be that of a cubic polynomial function?

**54.** How many positive real zeros does the function graphed in D have?

**55.** How many negative real zeros does the function graphed in A have?

**56.** Is the absolute minimum value of the function graphed in B a positive number or a negative number?

**57.** Which one of the graphs is that of a function whose range is *not* $(-\infty, \infty)$?

**58.** One of the following is an approximation for the local maximum point of the graph in A. Which one is it?

**A.** $(.73, 10.39)$ **B.** $(-.73, 10.39)$

**C.** $(-.73, -10.39)$ **D.** $(.73, -10.39)$

*For the functions defined in Exercises 59–66, use your graphing calculator to find a comprehensive graph and answer each of the following.*

**(a)** *Determine the domain.*

**(b)** *Determine all local minimum points, and tell if any is an absolute minimum point. (Approximate coordinates to the nearest hundredth.)*

**(c)** *Determine all local maximum points, and tell if any is an absolute maximum point. (Approximate coordinates to the nearest hundredth.)*

**(d)** *Determine the range. (If an approximation is necessary, give it to the nearest hundredth.)*

**(e)** *Determine all intercepts. For each function, there is at least one x-intercept that is an integer. For those that are not integers, give an approximation to the nearest hundredth. Determine the y-intercept analytically.*

**(f)** *Give the interval(s) over which the function is increasing.*

**(g)** *Give the interval(s) over which the function is decreasing.*

**59.** $y = -2x^3 - 14x^2 + 2x + 84$

**60.** $y = -3x^3 + 6x^2 + 39x - 60$

**61.** $y = x^5 + 4x^4 - 3x^3 - 17x^2 + 6x + 9$

**62.** $y = -2x^5 + 7x^4 + x^3 - 20x^2 + 4x + 16$

**63.** $y = 2x^4 + 3x^3 - 17x^2 - 6x - 72$

**64.** $y = 3x^4 - 33x^2 + 54$

**65.** $y = -x^6 + 24x^4 - 144x^2 + 256$

**66.** $y = -3x^6 + 2x^5 + 9x^4 - 8x^3 + 11x^2 + 4$

*Determine a window that will provide a comprehensive graph of each polynomial function. (In each case, there are many possible such windows.)*

**67.** $y = 4x^5 - x^3 + x^2 + 3x - 16$

**68.** $y = 3x^5 - x^4 + 12x^2 - 25$

**69.** $y = 2.9x^3 - 37x^2 + 28x - 143$

**70.** $y = -5.9x^3 + 16x^2 - 120$

**71.** $y = \pi x^4 - 13x^2 + 84$

**72.** $y = 2\pi x^4 - 12x^2 + 100$

*(Modeling) Solve each problem.*

**73.** *Government Spending on Research* The table lists the annual amount (in billions of dollars) spent by the federal government on research programs at universities and related institutions.

| Year | Amount | Year | Amount |
|---|---|---|---|
| 1990 | 12.6 | 1995 | 15.7 |
| 1991 | 13.8 | 1996 | 16.3 |
| 1992 | 14.2 | 1997 | 16.7 |
| 1993 | 15.0 | 1998 | 17.1 |
| 1994 | 15.3 | | |

*Source:* National Center for Educational Statistics.

**(a)** Graph the data with the following three function definitions, where $x$ represents the year.

**(i)** $f(x) = .2(x - 1990)^2 + 12.6$

**(ii)** $g(x) = .55(x - 1990) + 12.6$

**(iii)** $h(x) = .0075(x - 1990)^3 + 14.12$

**(b)** Which function definition best models the data?

**74.** *Life Expectancy of Americans* One result of improved health care is that people are living longer. The table lists the number of Americans (in thousands) who are expected to be over 100 years old for selected years.

| Year | Number | Year | Number |
|---|---|---|---|
| 1994 | 50 | 2000 | 75 |
| 1996 | 56 | 2002 | 94 |
| 1998 | 65 | 2004 | 110 |

*Source:* U.S. Bureau of the Census.

**(a)** Use graphing to determine which polynomial best models the number of Americans over 100 years old, where $x = 0$ corresponds to 1994.

**(i)** $f(x) = 6.057x + 44.714$

**(ii)** $g(x) = .4018x^2 + 2.039x + 50.071$

**(iii)** $h(x) = -.06x^3 + .506x^2 + 1.659x + 50.238$

**(b)** Use your choice from part (a) to predict the number of Americans who will be over 100 years old in 2008.

## Reviewing Basic Concepts (Sections 3.4 and 3.5)

**1.** *Dimensions of a Garden* An ecology center wants to set up an experimental garden using 300 meters of fencing to enclose a rectangular area of 5000 square meters.

**(a)** Let $x$ meters represent the width of the garden. Why must the length be $150 - x$ meters?

**(b)** Define an expression $A(x)$ that represents the area of the garden.

**(c)** What are the restrictions on $x$?

**(d)** Use the given values for area and length of fencing to find the dimensions of the garden.

**2.** *Research Funding* National Science Foundation funding for research increased in the decades of the 1980s and 1990s. The table on the next page gives funding at the beginning of each decade since 1951. In the table, years are given as the number of years since 1900. Thus, $x = 51$ corresponds to 1951, $x = 61$ corresponds to 1961, and so on.

| Year $x$ | 51 | 61 | 71 | 81 | 91 | 101 |
|---|---|---|---|---|---|---|
| Dollars $y$ (in billions) | .1 | .2 | .6 | 1.1 | 2.3 | 4.7 |

*Source:* U.S. National Science Foundation.

**(a)** Plot the data.

**(b)** Find a function defined by $f(x) = a(x - h)^2 + k$ that models the data by using $(51, .1)$ as the vertex and $(101, 4.7)$ as the second point.

**(c)** Use the quadratic regression feature of a graphing calculator to determine the quadratic function $g$ that best fits the data.

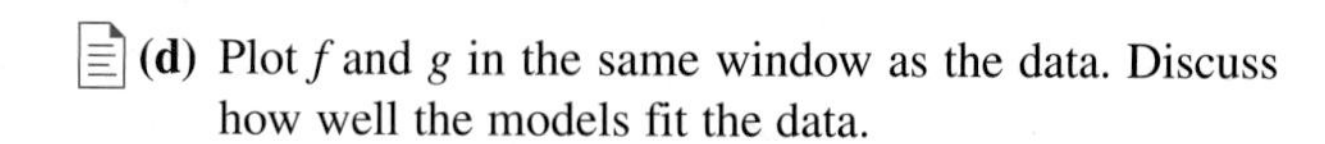

**(d)** Plot $f$ and $g$ in the same window as the data. Discuss how well the models fit the data.

*For* $P(x) = 2x^3 - 9x^2 + 4x + 15$, *answer the following.*

**3.** Predict the end behavior.

**4.** What is the maximum number of extrema the graph of $P$ can have? What is the maximum number of zeros?

*For* $P(x) = x^4 + 4x^3 - 20$, *answer the following.*

**5.** Predict the end behavior.

**6.** Give a comprehensive graph of function $P$.

**7.** Find all extreme points. Tell whether each one is a maximum or minimum and a local or absolute extremum.

**8.** Find all intercepts to the nearest hundredth.

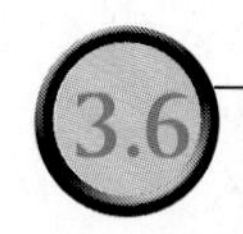

# 3.6 TOPICS IN THE THEORY OF POLYNOMIAL FUNCTIONS (I)

Intermediate Value Theorem • Division of Polynomials and Synthetic Division • Remainder and Factor Theorems

The topics in this section and the next complement the graphical work done in Section 3.5 and help prepare for the work in Section 3.8 (equations, inequalities, and applications of polynomial functions).

## Intermediate Value Theorem

The theorem presented here applies to the zeros of every polynomial function with *real coefficients*. It uses the fact that graphs of polynomial functions are continuous curves, with no gaps or sudden jumps. The proof requires advanced methods.

> **Intermediate Value Theorem**
>
> If $P(x)$ is a polynomial with only real coefficients, and if for real numbers $a$ and $b$, the values $P(a)$ and $P(b)$ are opposite in sign, then there exists at least one real zero between $a$ and $b$.

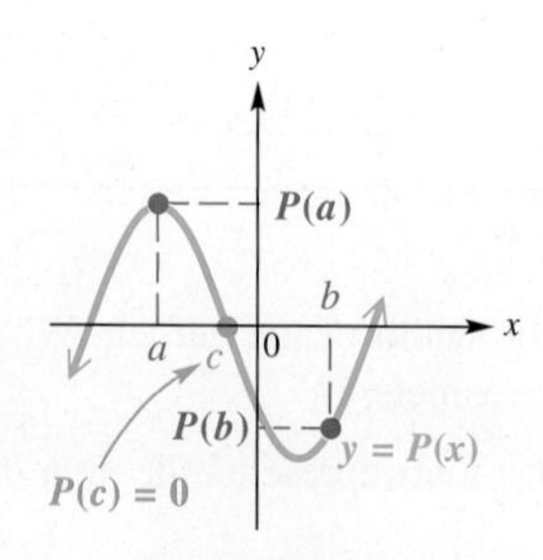


**FIGURE 67**

To see how the intermediate value theorem is applied, note that in Figure 67, $P(a)$ and $P(b)$ are opposite in sign, so 0 is between $P(a)$ and $P(b)$. Then, by this theorem, there must be a number $c$ in $[a, b]$ such that $P(c) = 0$.

**EXAMPLE 1** *Applying the Intermediate Value Theorem*

Show that the polynomial function defined by

$$P(x) = x^3 - 2x^2 - x + 1$$

has a real zero between 2 and 3.

**Analytic Solution**

Using

$$P(x) = x^3 - 2x^2 - x + 1,$$

we begin by evaluating $P(2)$ and $P(3)$.

$$P(2) = 2^3 - 2(2)^2 - 2 + 1 = -1$$
$$P(3) = 3^3 - 2(3)^2 - 3 + 1 = 7$$

Since $P(2) = -1$ and $P(3) = 7$ differ in sign, the intermediate value theorem assures us that there is a real zero between 2 and 3.

**Graphing Calculator Solution**

The graph of $P(x) = x^3 - 2x^2 - x + 1$ in Figure 68 shows that there is an $x$-intercept between 2 and 3, confirming our analytic result. Using the table, we see that the zero lies between 2.246 and 2.247 since there is a sign change in the function values there.

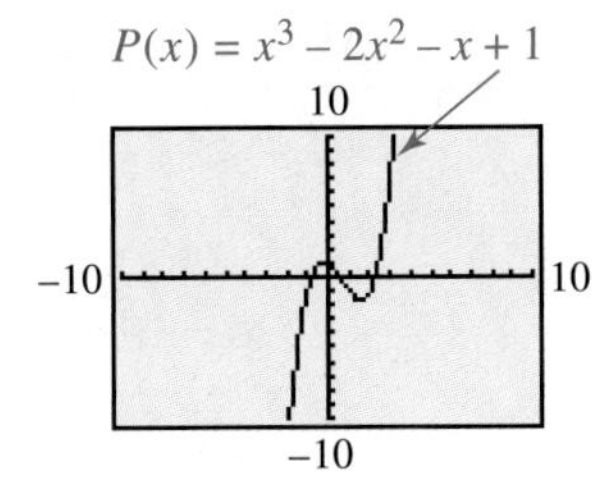


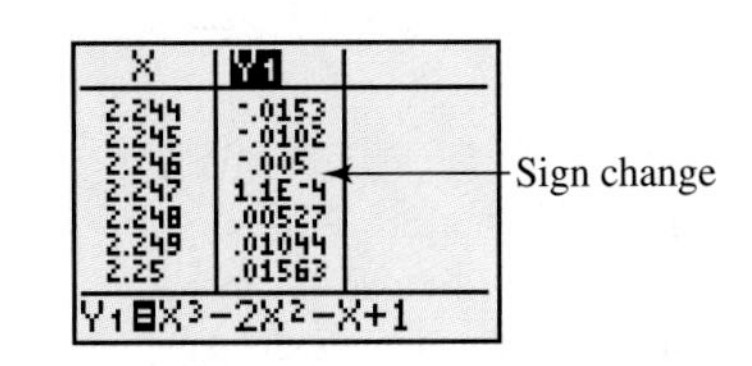


FIGURE 68

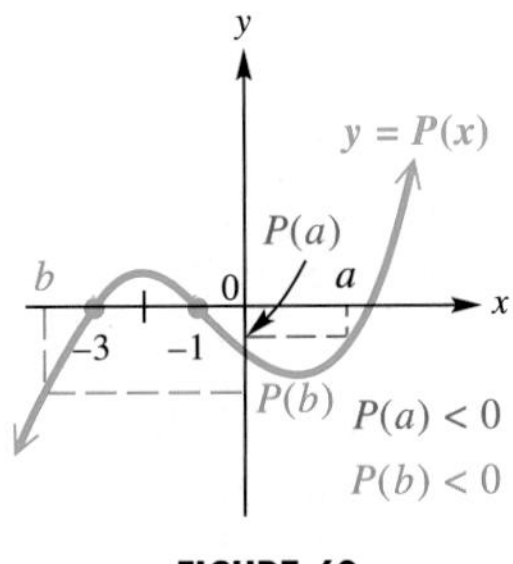


FIGURE 69

**CAUTION** Be careful how you interpret the intermediate value theorem. If $P(a)$ and $P(b)$ are *not* opposite in sign, it does not necessarily mean that there is no zero between $a$ and $b$. For example, in Figure 69, $P(a)$ and $P(b)$ are both negative, but $-3$ and $-1$, which are between $a$ and $b$, are zeros of $P$.

## Division of Polynomials and Synthetic Division

Just as we can use long division to determine whether one whole number is a factor of another, we can also use it to determine whether one polynomial is a factor of another.

**EXAMPLE 2** *Dividing a Polynomial by a Binomial*

Divide the polynomial $3x^3 - 2x + 5$ by $x - 3$. Determine the quotient and the remainder.

**Solution** First make sure that the powers of the variable in the dividend $(3x^3 - 2x + 5)$ are descending, which they are. Insert the term $0x^2$ to act as a placeholder. Then follow the same procedure as for long division of whole numbers.

$$x - 3\overline{)3x^3 + 0x^2 - 2x + 5}$$ ← Missing term ($0x^2$)

Start with $\frac{3x^3}{x} = 3x^2$.

$$\begin{array}{r} 3x^2 \\ x - 3\overline{)3x^3 + 0x^2 - 2x + 5} \\ \underline{3x^3 - 9x^2} \end{array}$$

← $\frac{3x^3}{x} = 3x^2$

← $3x^2(x - 3)$

Subtract by changing the signs on $3x^3 - 9x^2$ and adding.

$$\begin{array}{r} 3x^2 \\ x - 3\overline{)3x^3 + 0x^2 - 2x + 5} \\ \underline{3x^3 - 9x^2} \\ 9x^2 \end{array}$$

← Subtract.

Bring down the next term.

$$
\begin{array}{r@{}l}
 & \quad 3x^2 \\
x - 3 \overline{)} & 3x^3 + 0x^2 - 2x + 5 \\
 & \underline{3x^3 - 9x^2} \\
 & \qquad 9x^2 - 2x \qquad \leftarrow \text{Bring down } -2x.
\end{array}
$$

In the next step, $\dfrac{9x^2}{x} = 9x$.

$$
\begin{array}{r@{}l}
 & \quad 3x^2 + 9x \qquad \leftarrow \dfrac{9x^2}{x} = 9x \\
x - 3 \overline{)} & 3x^3 + 0x^2 - 2x + 5 \\
 & \underline{3x^3 - 9x^2} \\
 & \qquad 9x^2 - 2x \\
 & \qquad \underline{9x^2 - 27x} \qquad \leftarrow 9x(x - 3) \\
 & \qquad\qquad 25x + 5 \leftarrow \text{Subtract and bring down 5.}
\end{array}
$$

Finally, $\dfrac{25x}{x} = 25$.

$$
\begin{array}{r@{}l}
 & \quad 3x^2 + 9x + 25 \leftarrow \dfrac{25x}{x} = 25 \\
x - 3 \overline{)} & 3x^3 + 0x^2 - 2x + 5 \\
 & \underline{3x^3 - 9x^2} \\
 & \qquad 9x^2 - 2x \\
 & \qquad \underline{9x^2 - 27x} \\
 & \qquad\qquad 25x + 5 \\
 & \qquad\qquad \underline{25x - 75} \leftarrow 25(x - 3) \\
 & \qquad\qquad\qquad 80 \leftarrow \text{Subtract.}
\end{array}
$$

The quotient is $3x^2 + 9x + 25$ with a remainder of 80.

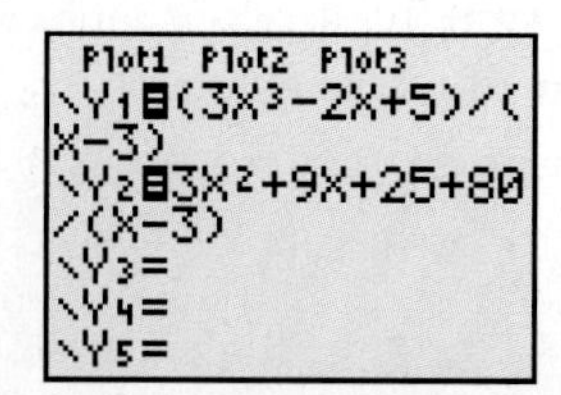


| X | Y1 | Y2 |
|---|---|---|
| 0 | -1.667 | -1.667 |
| 1 | -3 | -3 |
| 2 | -25 | -25 |
| 3 | ERROR | ERROR |
| 4 | 189 | 189 |
| 5 | 185 | 185 |
| 6 | 213.67 | 213.67 |

X=3

A table supports the result of Example 2. Why is there an error message for X = 3?

From the division shown in Example 2, we can make several observations about polynomial division in general. Notice that we divided a cubic polynomial (degree 3) by a linear polynomial (degree 1) and obtained a quadratic polynomial quotient (degree 2). Notice also that $3 - 1 = 2$, so the degree of the quotient polynomial is found by subtracting the degree of the divisor from the degree of the dividend. Also, since the remainder is a nonzero constant 80, we can write it as the numerator of a fraction with denominator $x - 3$ to express the fractional part of the quotient.

$$
\begin{array}{l}\text{Dividend} \rightarrow \\ \text{Divisor} \rightarrow\end{array} \frac{3x^3 - 2x + 5}{x - 3} = \underbrace{3x^2 + 9x + 25}_{\text{Quotient polynomial}} + \underbrace{\frac{80}{x - 3}}_{\text{Fractional part of the quotient}} \begin{array}{l}\leftarrow \text{Remainder} \\ \leftarrow \text{Divisor}\end{array}
$$

In general, the following rules apply to the division of a polynomial by a binomial of the form $x - k$.

**Division of a Polynomial by $x - k$**

1. If the degree $n$ polynomial $P(x)$ is divided by $x - k$, the quotient polynomial, $Q(x)$, has degree $n - 1$.
2. The remainder $R$ is a constant (and may be 0). The complete quotient for $\frac{P(x)}{x-k}$ may be written as

$$\frac{P(x)}{x-k} = Q(x) + \frac{R}{x-k}.$$

Long division of a polynomial by a binomial of the form $x - k$ can be condensed. Using the division performed in Example 2, observe the following.

$$
\begin{array}{rrrrr}
 & & 3x^2 & +\ 9x & +\ 25 \\ \cline{2-5}
x-3\,) & 3x^3 & +\ 0x^2 & -\ 2x & +\ 5 \\
 & 3x^3 & -\ 9x^2 & & \\ \cline{2-3}
 & & 9x^2 & -\ 2x & \\
 & & 9x^2 & -\ 27x & \\ \cline{3-4}
 & & & 25x & +\ 5 \\
 & & & 25x & -\ 75 \\ \cline{4-5}
 & & & & 80
\end{array}
\qquad
\begin{array}{rrrrr}
 & & 3 & 9 & 25 \\ \cline{2-5}
1-3\,) & 3 & 0 & -2 & 5 \\
 & 3 & -9 & & \\ \cline{2-3}
 & & 9 & -2 & \\
 & & 9 & -27 & \\ \cline{3-4}
 & & & 25 & 5 \\
 & & & 25 & -75 \\ \cline{4-5}
 & & & & 80
\end{array}
$$

On the right, exactly the same division is shown written without the variables. All the numbers in color on the right are repetitions of the numbers directly above them, so they can be omitted, as shown on the left below. Since the coefficient of $x$ in the divisor is always 1, it can be omitted too.

$$
\begin{array}{rrrrr}
 & & 3 & 9 & 25 \\ \cline{2-5}
-3\,) & 3 & 0 & -2 & 5 \\
 & & -9 & & \\ \cline{2-3}
 & & 9 & -2 & \\
 & & & -27 & \\ \cline{3-4}
 & & & 25 & 5 \\
 & & & & -75 \\ \cline{4-5}
 & & & & 80
\end{array}
\qquad
\begin{array}{rrrrr}
 & & 3 & 9 & 25 \\ \cline{2-5}
-3\,) & 3 & 0 & -2 & 5 \\
 & & -9 & & \\ \cline{2-3}
 & & 9 & & \\
 & & & -27 & \\ \cline{3-4}
 & & & 25 & \\
 & & & & -75 \\ \cline{4-5}
 & & & & 80
\end{array}
$$

The numbers in color on the left are again repetitions of the numbers directly above them; they may be omitted, as shown on the right.

Now the problem can be condensed. If the 3 in the dividend is brought down to the beginning of the bottom row, the top row can be omitted, since it duplicates the bottom row.

$$
\begin{array}{rrrrr}
-3\,) & 3 & 0 & -2 & 5 \\ \cline{2-5}
 & & -9 & -27 & -75 \\ \cline{2-5}
 & 3 & 9 & 25 & 80
\end{array}
$$

To simplify the arithmetic, we replace subtraction in the second row by addition, and compensate by changing the $-3$ at upper left to its additive inverse, 3. The result of doing all this follows.

Additive inverse ↓

$$
\begin{array}{r|rrrr}
3 & 3 & 0 & -2 & 5 \\
 & & 9 & 27 & 75 \\
\hline
 & 3 & 9 & 25 & 80
\end{array}
\quad \leftarrow \text{Signs changed}
$$

↓ ↓ ↓ ↘

$$\text{Quotient} \rightarrow 3x^2 + 9x + 25 + \frac{80}{x - 3} \quad \leftarrow \text{Remainder}$$

This abbreviated form of long division of polynomials is called **synthetic division.**

## EXAMPLE 3 *Using Synthetic Division*

Use synthetic division to divide $5x^3 - 6x^2 - 28x + 8$ by $x + 2$.

**Solution** To use synthetic division, the divisor must be of the form $x - k$. Writing $x + 2$ as $x - (-2)$ shows that the value of $k$ here is $-2$. Begin by writing

$$-2\overline{)5 \quad -6 \quad -28 \quad 8}.$$

Next, bring down the 5.

$$
\begin{array}{r|rrrr}
-2 & 5 & -6 & -28 & 8 \\
 & & & & \\
\hline
 & 5 & & &
\end{array}
$$

Now, multiply $-2$ by 5 to get $-10$, and add it to $-6$ from the first row. The result is $-16$.

$$
\begin{array}{r|rrrr}
-2 & 5 & -6 & -28 & 8 \\
 & & -10 & & \\
\hline
 & 5 & -16 & &
\end{array}
$$

Next, multiply $-2$ by $-16$ to get 32. Add this to $-28$ from the first row.

$$
\begin{array}{r|rrrr}
-2 & 5 & -6 & -28 & 8 \\
 & & -10 & 32 & \\
\hline
 & 5 & -16 & 4 &
\end{array}
$$

Finally, $-2(4) = -8$. Add this result to 8 to obtain 0.

$$
\begin{array}{r|rrrr}
-2 & 5 & -6 & -28 & 8 \\
 & & -10 & 32 & -8 \\
\hline
 & \underbrace{5 \quad -16 \quad 4}_{\text{Quotient}} & & & 0 \leftarrow \text{Remainder}
\end{array}
$$

The coefficients of the quotient polynomial and the remainder are read directly from the bottom row. Since the degree of the quotient will always be one less than the degree of the polynomial to be divided and here the remainder is 0,

$$\frac{5x^3 - 6x^2 - 28x + 8}{x + 2} = 5x^2 - 16x + 4.$$

Notice that the divisor $x + 2$ is a *factor* of $5x^3 - 6x^2 - 28x + 8$ because the remainder is 0, and so $5x^3 - 6x^2 - 28x + 8 = (x + 2)(5x^2 - 16x + 4)$.

## Remainder and Factor Theorems

In Example 2, we divided $3x^3 - 2x + 5$ by $x - 3$ and obtained a remainder of 80. If we evaluate $P(x) = 3x^3 - 2x + 5$ at $x = 3$, we get

$$P(3) = 3(3)^3 - 2(3) + 5 = 81 - 6 + 5 = 80.$$

Notice that the remainder is equal to $P(3)$. In Example 3 we divided the polynomial $5x^3 - 6x^2 - 28x + 8$ by $x - (-2)$ and obtained a remainder of 0. If we evaluate $P(x) = 5x^3 - 6x^2 - 28x + 8$ at $x = -2$, we get

$$\begin{aligned} P(-2) &= 5(-2)^3 - 6(-2)^2 - 28(-2) + 8 \\ &= -40 - 24 + 56 + 8 \\ &= 0. \end{aligned}$$

Again, the remainder is equal to $P(-2)$. These two examples illustrate an important theorem in the study of polynomial functions, the remainder theorem.

**Remainder Theorem**

If a polynomial $P(x)$ is divided by $x - k$, the remainder is equal to $P(k)$.

**EXAMPLE 4** *Using the Remainder Theorem*

Use the remainder theorem and synthetic division to find $P(-2)$ if

$$P(x) = -x^4 + 3x^2 - 4x - 5.$$

**Analytic Solution**

Use synthetic division to find the remainder when $P(x)$ is divided by $x - (-2)$. Insert 0 for the missing $x^3$-term.

$$\begin{array}{r|rrrrr} -2 & -1 & 0 & 3 & -4 & -5 \\ & & 2 & -4 & 2 & 4 \\ \hline & -1 & 2 & -1 & -2 & -1 \end{array}$$

↑ Remainder

Since the remainder is $-1$, by the remainder theorem $P(-2) = -1$.

**Graphing Calculator Solution**

The graph of $P(x) = -x^4 + 3x^2 - 4x - 5$ in Figure 70 indicates that the point $(-2, -1)$ lies on the graph, so $P(-2) = -1$. Alternatively, the table in Figure 70 shows that $P(-2) = -1$.

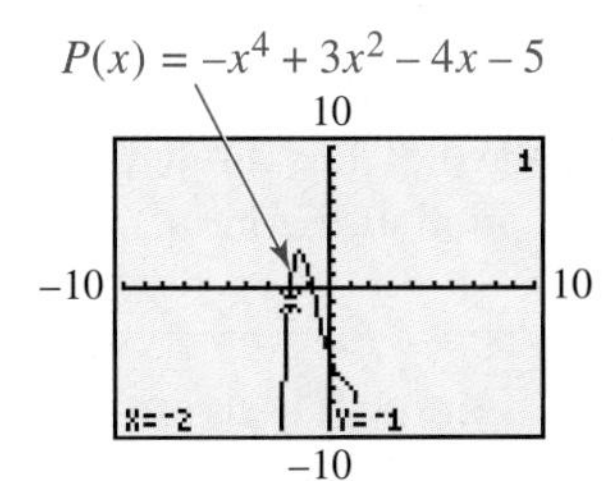


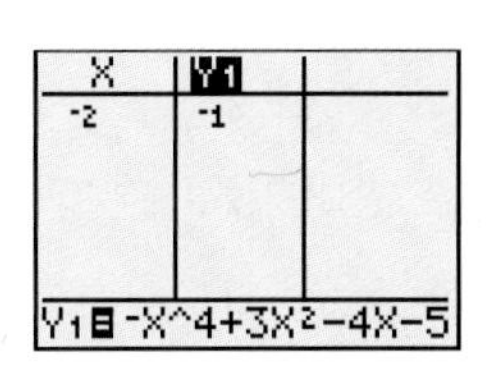


FIGURE 70

**EXAMPLE 5** *Deciding Whether a Number Is a Zero of a Polynomial Function*

Decide whether the given number is a zero of the function defined by the given polynomial.

**(a)** 2; $P(x) = x^3 - 4x^2 + 9x - 10$ **(b)** $-2$; $P(x) = \frac{3}{2}x^3 - x^2 + \frac{3}{2}x$

### Analytic Solution

**(a)** Use synthetic division.

$$\begin{array}{r|rrrr} 2 & 1 & -4 & 9 & -10 \\ & & 2 & -4 & 10 \\ \hline & 1 & -2 & 5 & 0 \end{array}$$

Since the remainder is 0, $P(2) = 0$, and 2 is a zero of the polynomial function defined by $P(x) = x^3 - 4x^2 + 9x - 10$.

**(b)** Remember to use 0 for the missing constant term in the synthetic division.

$$\begin{array}{r|rrrr} -2 & \frac{3}{2} & -1 & \frac{3}{2} & 0 \\ & & -3 & 8 & -19 \\ \hline & \frac{3}{2} & -4 & \frac{19}{2} & -19 \end{array}$$

The remainder is not 0, so $-2$ is not a zero of $P$. In fact, $P(-2) = -19$. From this, we know that the point $(-2, -19)$ lies on the graph of $P$.

### Graphing Calculator Solution

The first screen in Figure 71 shows the polynomials entered as $Y_1$ and $Y_2$. The second screen shows the results of finding $Y_1(k)$ for parts (a) and (b).

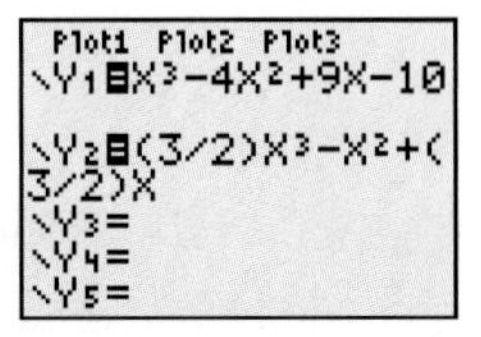


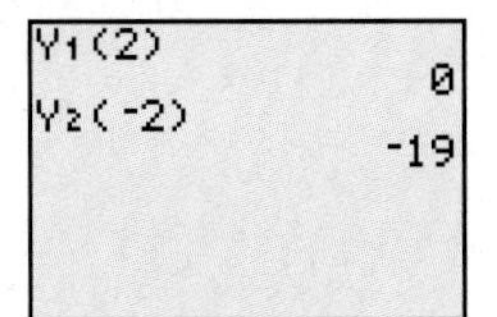


**FIGURE 71**

In Example 5(a), we showed that 2 is a zero of the polynomial function defined by $P(x) = x^3 - 4x^2 + 9x - 10$. The first three numbers in the bottom row of the synthetic division indicate the coefficients of the quotient polynomial, and thus

$$\frac{P(x)}{x - 2} = x^2 - 2x + 5.$$

Multiplying each side of this equation by $x - 2$ gives

$$P(x) = (x - 2)(x^2 - 2x + 5),$$

indicating that $x - 2$ is a *factor* of $P(x)$.

By the remainder theorem, if $P(k) = 0$, then the remainder when $P(x)$ is divided by $x - k$ is 0. This means that $x - k$ is a factor of $P(x)$. Conversely, if $x - k$ is a factor of $P(x)$, then $P(k)$ must equal 0. This is summarized in the following theorem, a corollary of the remainder theorem. (A *corollary* is a theorem that follows directly from another theorem.)

**Factor Theorem**

The polynomial $x - k$ is a factor of the polynomial $P(x)$ if and only if $P(k) = 0$.

**EXAMPLE 6** *Using the Factor Theorem*

Determine whether the second polynomial is a factor of the first.

**(a)** $P(x) = 4x^3 + 24x^2 + 48x + 32$; $x + 2$

**(b)** $P(x) = 2x^4 + 3x^2 - 5x + 7$; $x - 1$

**Solution**

**(a)** We use synthetic division with $k = -2$, since $x + 2 = x - (-2)$.

$$\begin{array}{r|rrrr} -2 & 4 & 24 & 48 & 32 \\ & & -8 & -32 & -32 \\ \hline & 4 & 16 & 16 & 0 \end{array} \leftarrow \text{Remainder is 0.}$$

Since the remainder is 0, $x + 2$ is a factor of $4x^3 + 24x^2 + 48x + 32$. A factored form (but not necessarily *completely* factored form) of the polynomial is $(x + 2)(4x^2 + 16x + 16)$.

**(b)** By the factor theorem, $x - 1$ will be a factor of $P(x)$ if $P(1) = 0$. Use synthetic division and the remainder theorem to decide.

$$\begin{array}{r|rrrrr} 1 & 2 & 0 & 3 & -5 & 7 \\ & & 2 & 2 & 5 & 0 \\ \hline & 2 & 2 & 5 & 0 & 7 \end{array}$$

Since the remainder is 7, $P(1) = 7$, not 0, so $x - 1$ is not a factor of $P(x)$.

**NOTE** An easy way to determine $P(1)$ for a polynomial function $P$ is simply to add the coefficients of $P(x)$. This method works since every power of 1 is equal to 1. For example, using $P(x) = 2x^4 + 3x^2 - 5x + 7$ as shown in Example 6(b), we have $P(1) = 2 + 3 - 5 + 7 = 7$, confirming our result found by synthetic division.

The next example illustrates the close relationship among the ideas of $x$-intercepts of the graph of a polynomial function, real zeros of the function, and solutions of the corresponding polynomial equation.

### EXAMPLE 7 *Examining x-Intercepts, Zeros, and Solutions*

Consider the polynomial function defined by $P(x) = 2x^3 + 5x^2 - x - 6$.

**(a)** Show by synthetic division that $-2$, $-\frac{3}{2}$, and 1 are zeros of $P$, and write $P(x)$ in factored form with all factors linear.

**(b)** Graph $P$ in a suitable viewing window and locate the $x$-intercepts.

**(c)** Solve the polynomial equation $2x^3 + 5x^2 - x - 6 = 0$.

**Solution**

**(a)**

$$\begin{array}{r|rrrr} -2 & 2 & 5 & -1 & -6 \\ & & -4 & -2 & 6 \\ \hline & 2 & 1 & -3 & 0 \end{array} \leftarrow P(-2) = 0$$

Since $P(-2) = 0$, $x + 2$ is a factor, and thus, $P(x) = (x + 2)(2x^2 + x - 3)$. Rather than show that $-\frac{3}{2}$ and 1 are zeros of $P(x)$, we need only show that they are zeros of $2x^2 + x - 3$ by factoring directly, or by synthetic division, as follows.

$$\begin{array}{r|rrr} -\frac{3}{2} & 2 & 1 & -3 \\ & & -3 & 3 \\ \hline & 2 & -2 & 0 \leftarrow P(1) = 0 \end{array} \qquad -\tfrac{3}{2} \text{ is a zero of } 2x^2 + x - 3.$$

$$\downarrow \quad \downarrow$$

$$\begin{array}{r|rr} 1 & 2 & -2 \\ & & 2 \\ \hline & 2 & 0 \end{array} \quad \leftarrow P\left(-\tfrac{3}{2}\right) = 0 \qquad 1 \text{ is a zero of } 2x - 2$$

$\uparrow$

This 2 is the constant factor.

The completely factored form of $P(x)$ is

$$P(x) = 2(x + 2)\left(x + \frac{3}{2}\right)(x - 1),$$

or

$$P(x) = (x + 2)(2)\left(x + \frac{3}{2}\right)(x - 1)$$

$$= (x + 2)(2x + 3)(x - 1).$$

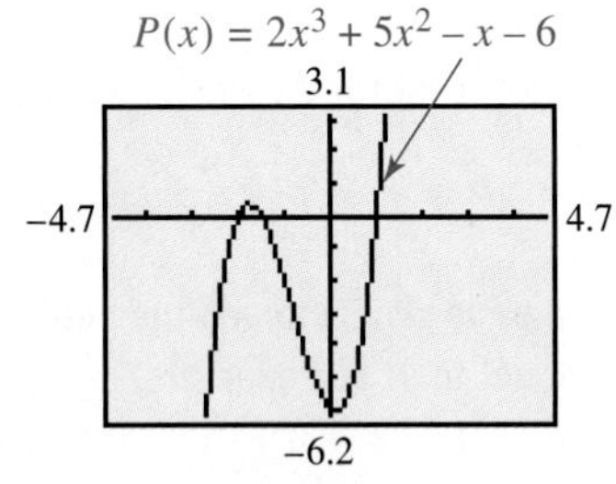


**FIGURE 72**

**(b)** Figure 72 shows the graph of this function. The calculator will determine the $x$-intercepts: $-2$, $-\frac{3}{2}$, and 1.

**(c)** From part (a), the zeros of $P$ are $-2$, $-\frac{3}{2}$, and 1. Because the zeros of $P$ are the solutions of $P(x) = 0$, the solution set is $\{2, -\frac{3}{2}, 1\}$.

## 3.6 EXERCISES

*Use the intermediate value theorem to show that each function has a real zero between the two numbers given. Then, use your calculator to approximate the zero to the nearest hundredth.*

**1.** $P(x) = 3x^2 - 2x - 6$; 1 and 2

**2.** $P(x) = x^3 + x^2 - 5x - 5$; 2 and 3

**3.** $P(x) = 2x^3 - 8x^2 + x + 16$; 2 and 2.5

**4.** $P(x) = 3x^3 + 7x^2 - 4$; $\dfrac{1}{2}$ and 1

**5.** $P(x) = 2x^4 - 4x^2 + 3x - 6$; 1.5 and 2

**6.** $P(x) = x^4 - 4x^3 - x + 1$; .3 and 1

**7.** *Concept Check* Suppose that a polynomial function $P$ is defined in such a way that $P(2) = -4$ and $P(2.5) = 2$. What conclusion does the intermediate value theorem allow you to make?

**8.** Suppose that a polynomial function $P$ is defined in such a way that $P(3) = -4$ and $P(4) = -10$. Can we be certain that there is no zero between 3 and 4? Explain, using a graph.

*Use synthetic division to find each quotient when $P(x)$ is divided by the binomial following it.*

**9.** $P(x) = x^3 + 2x^2 - 17x - 10; \quad x + 5$

**10.** $P(x) = x^4 + 4x^3 + 2x^2 + 9x + 4; \quad x + 4$

**11.** $P(x) = 3x^3 - 11x^2 - 20x + 3; \quad x - 5$

**12.** $P(x) = x^4 - 3x^3 - 5x^2 + 2x - 16; \quad x - 3$

**13.** $P(x) = x^4 - 3x^3 - 4x^2 + 12x; \quad x - 2$

**14.** $P(x) = x^5 - 1; \quad x - 1$

*Use synthetic division to find $P(k)$.*

**15.** $k = 3; \quad P(x) = x^2 - 4x + 5$

**16.** $k = -2; \quad P(x) = x^2 + 5x + 6$

**17.** $k = -2; \quad P(x) = 5x^3 + 2x^2 - x + 5$

**18.** $k = 2; \quad P(x) = 2x^3 - 3x^2 - 5x + 4$

**19.** $k = 2; \quad P(x) = x^2 - 5x + 1$

**20.** $k = 3; \quad P(x) = x^2 - x + 3$

*Use synthetic division to determine whether the given number is a zero of the polynomial.*

**21.** $2; \quad P(x) = x^2 + 2x - 8$

**22.** $-1; \quad P(x) = x^2 + 4x - 5$

**23.** $4; \quad P(x) = 2x^3 - 6x^2 - 9x + 6$

**24.** $-4; \quad P(x) = 9x^3 + 39x^2 + 12x$

**25.** $-.5; \quad P(x) = 4x^3 + 12x^2 + 7x + 1$

**26.** $-.25; \quad P(x) = 8x^3 + 6x^2 - 3x - 1$

## RELATING CONCEPTS

**For individual or group investigation (Exercises 27–32)**

*The close relationships among x-intercepts of a graph of a function, real zeros of the function, and real solutions of the associated equation should, by now, be apparent to you. Using the concepts presented so far in this text, consider the graph of the polynomial function defined by*

$$P(x) = x^3 - 2x^2 - 11x + 12$$

*shown here, and respond to Exercises 27–32.*

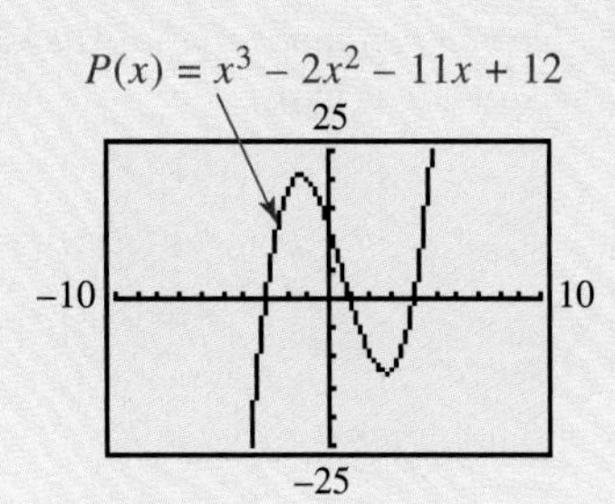

The $x$-intercepts are $-3$, 1, and 4.

**27.** What are the linear factors of $P(x)$?

**28.** What are the solutions of the equation $P(x) = 0$?

**29.** What are the zeros of the function $P$?

**30.** If $P(x)$ is divided by $x - 2$, what is the remainder? What is $P(2)$?

**31.** Give the solution set of $P(x) > 0$, using interval notation.

**32.** Give the solution set of $P(x) < 0$, using interval notation.

*For each polynomial, one zero is given. Find all others analytically.*

**33.** $P(x) = x^3 - 2x + 1; \quad 1$

**34.** $P(x) = 2x^3 + 8x^2 - 11x - 5; \quad -5$

**35.** $P(x) = 3x^3 + 5x^2 - 3x - 2; \quad -2$

**36.** $P(x) = x^3 - 7x^2 + 13x - 3; \quad 3$

*Factor $P(x)$ into linear factors given that $k$ is a zero of $P$.*

**37.** $P(x) = 2x^3 - 3x^2 - 17x + 30; \quad k = 2$

**38.** $P(x) = 2x^3 - 3x^2 - 5x + 6; \quad k = 1$

**39.** $P(x) = 6x^3 + 25x^2 + 3x - 4; \quad k = -4$

**40.** $P(x) = 8x^3 + 50x^2 + 47x - 15; \quad k = -5$

Descartes' rule of signs helps to determine the number of positive and negative real zeros of a polynomial function.

**Descartes' Rule of Signs**

Let $P(x)$ define a polynomial function with real coefficients and a nonzero constant term, with terms in descending powers of $x$.

**(a)** The number of positive real zeros of $P$ either equals the number of variations in sign occurring in the coefficients of $P(x)$, or is less than the number of variations by a positive even integer.

**(b)** The number of negative real zeros of $P$ either equals the number of variations in sign occurring in the coefficients of $P(-x)$, or is less than the number of variations by a positive even integer.

In the theorem, *variation in sign* is a change from positive to negative or negative to positive in successive terms of the polynomial. Missing terms (those with 0 coefficients) are counted as no change in sign and can be ignored. For example, consider the polynomial function defined by $P(x) = x^4 - 6x^3 + 8x^2 + 2x - 1$. $P(x)$ has three variations in sign:

$$\underbrace{+x^4 - 6x^3}_{1} \underbrace{\ + 8x^2}_{2} + \underbrace{2x - 1}_{3}.$$

Thus, by Descartes' rule of signs, $P$ has either 3 or $3 - 2 = 1$ positive real zeros. Since

$$\begin{aligned} P(-x) &= (-x)^4 - 6(-x)^3 + 8(-x)^2 + 2(-x) - 1 \\ &= x^4 + 6x^3 + 8x^2 - 2x - 1 \end{aligned}$$

has only one variation in sign, $P$ has only one negative real zero. If you graph the function in the window $[-5, 5]$ by $[-10, 10]$, you can interpret the theorem in terms of $x$-intercepts. Verify that there are three positive $x$-intercepts and one negative $x$-intercept.

*Use Descartes' rule of signs to determine the possible number of positive and negative real zeros for each function. Then, use a graph to determine the actual numbers of positive and negative real zeros.*

**41.** $P(x) = 2x^3 - 4x^2 + 2x + 7$

**42.** $P(x) = x^3 + 2x^2 + x - 10$

**43.** $P(x) = 5x^4 + 3x^2 + 2x - 9$

**44.** $P(x) = 3x^4 + 2x^3 - 8x^2 - 10x - 1$

**45.** $P(x) = x^5 + 3x^4 - x^3 + 2x + 3$

**46.** $P(x) = 2x^5 - x^4 + x^3 - x^2 + x + 5$

## 3.7 TOPICS IN THE THEORY OF POLYNOMIAL FUNCTIONS (II)

Complex Zeros and the Fundamental Theorem of Algebra • Number of Zeros • Rational Zeros Theorem

### Complex Zeros and the Fundamental Theorem of Algebra

In Example 5 of Section 3.3, we found that the imaginary solutions of the equation $2x^2 - x + 4 = 0$ are $\frac{1}{4} + i\frac{\sqrt{31}}{4}$ and $\frac{1}{4} - i\frac{\sqrt{31}}{4}$. These two solutions are complex conjugates. This is not a coincidence; it can be shown that if $a + bi$ is a zero of a polynominal function with *real* coefficients, then its complex conjugate $a - bi$ is also a zero. This is given as the next theorem. Its proof is left for the exercises.

**Conjugate Zeros Theorem**

If $P(x)$ is a polynomial having only real coefficients, and if $a + bi$ is a zero of $P$, then the conjugate $a - bi$ is also a zero of $P$.

**EXAMPLE 1** *Defining a Polynomial Function Satisfying Given Conditions*

**(a)** Find a cubic polynomial in standard form with real coefficients having zeros 3 and $2 + i$.

**(b)** Find a polynomial function $P$ satisfying the conditions of part (a), with the additional requirement $P(-2) = 4$. Support the result graphically.

**Solution**

**(a)** By the conjugate zeros theorem, $2 - i$ must also be a zero of the function. Since the polynomial will be cubic, it will have three linear factors, and by the factor theorem they must be $x - 3$, $x - (2 + i)$, and $x - (2 - i)$. Therefore, one such cubic polynomial $P(x)$ can be defined as follows:

$$\begin{aligned} P(x) &= (x - 3)[x - (2 + i)][x - (2 - i)] \\ &= (x - 3)(x - 2 - i)(x - 2 + i) \\ &= (x - 3)(x^2 - 4x + 5) \\ &= x^3 - 7x^2 + 17x - 15. \end{aligned}$$

Multiplying this polynomial by any real nonzero constant $a$ will also yield a function satisfying the given conditions, so a more general form is

$$P(x) = a(x^3 - 7x^2 + 17x - 15).$$

**(b)** We must have

$$P(x) = a(x^3 - 7x^2 + 17x - 15)$$

defined in such a way that $P(-2) = 4$. To find $a$, let $x = -2$, and set the result equal to 4. Then solve.

$$\begin{aligned} a[(-2)^3 - 7(-2)^2 + 17(-2) - 15] &= 4 \\ a(-8 - 28 - 34 - 15) &= 4 \\ -85a &= 4 \\ a &= -\frac{4}{85} \end{aligned}$$

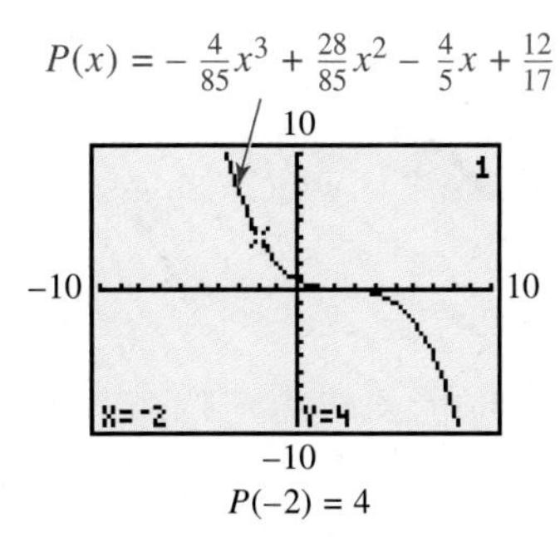


$P(-2) = 4$

**FIGURE 73**

Therefore, the desired function is defined by

$$\begin{aligned} P(x) &= -\frac{4}{85}(x^3 - 7x^2 + 17x - 15) \\ &= -\frac{4}{85}x^3 + \frac{28}{85}x^2 - \frac{4}{5}x + \frac{12}{17}. \end{aligned}$$

We can support this result by graphing $P(x) = -\frac{4}{85}x^3 + \frac{28}{85}x^2 - \frac{4}{5}x + \frac{12}{17}$ and showing that the point $(-2, 4)$ lies on the graph. See Figure 73.

## Number of Zeros

The next theorem says that every polynomial of degree 1 or more has a zero, so every such polynomial can be factored. This theorem was first proved by the mathematician Karl F. Gauss in his doctoral thesis in 1799, when he was 22 years old. Although many proofs of this result have been given, all of them involve mathematics beyond the algebra in this book, so no proof is included here.

**Fundamental Theorem of Algebra**

Every function defined by a polynomial of degree 1 or more has at least one complex (real or imaginary) zero.

From the fundamental theorem, if $P(x)$ is of degree 1 or more, then there is some number $k$ such that $P(k) = 0$. By the factor theorem, then, $P(x) = (x - k) \cdot Q(x)$ for some polynomial $Q(x)$. The fundamental theorem and the factor theorem can be used to factor $Q(x)$ in the same way. Assuming that $P(x)$ has degree $n$, repeating this process $n$ times gives

$$P(x) = a(x - k_1)(x - k_2) \cdots (x - k_n),$$

where $a$ is the leading coefficient of $P(x)$. Each of these factors leads to a zero of $P(x)$, so $P(x)$ has the $n$ zeros $k_1, k_2, k_3, \ldots, k_n$. This result can be used to prove the next theorem. The proof is left for the exercises.

**Number of Zeros Theorem**

A function defined by a polynomial of degree $n$ has at most $n$ distinct complex zeros.

**EXAMPLE 2** *Finding All Zeros of a Polynomial Function*

Find all complex zeros of $P(x) = x^4 - 7x^3 + 18x^2 - 22x + 12$, given that $1 - i$ is a zero.

**Analytic Solution**

This quartic function will have at most four complex zeros. Since $1 - i$ is a zero and the coefficients are real numbers, by the conjugate zeros theorem $1 + i$ is also a zero. The remaining zeros are found by first dividing the original polynomial by $x - (1 - i)$.

$$\begin{array}{r|rrrrr} 1-i & 1 & -7 & 18 & -22 & 12 \\ & & 1-i & -7+5i & 16-6i & -12 \\ \hline & 1 & -6-i & 11+5i & -6-6i & 0 \end{array}$$

**Graphing Calculator Solution**

By the conjugate zeros theorem, since $1 - i$ is a zero, $1 + i$ is also a zero. We can use a graphing calculator to find the real zeros as in Figure 74, which shows the graph of

$$P(x) = x^4 - 7x^3 + 18x^2 - 22x + 12,$$

with the real zeros identified at the bottom.

*(continued)*

Rather than go back to the original polynomial, divide the quotient from the first division by $x - (1 + i)$ as follows.

$$\begin{array}{r|rrrr} 1+i & 1 & -6-i & 11+5i & -6-6i \\ & & 1+i & -5-5i & 6+6i \\ \hline & 1 & -5 & 6 & 0 \end{array}$$

Find the zeros of the function defined by the quadratic polynomial $x^2 - 5x + 6$ by solving the equation $x^2 - 5x + 6 = 0$. By the quadratic formula or by factoring, the other zeros are 2 and 3. Thus, this function has four complex zeros: $1 - i$, $1 + i$, 2, and 3.

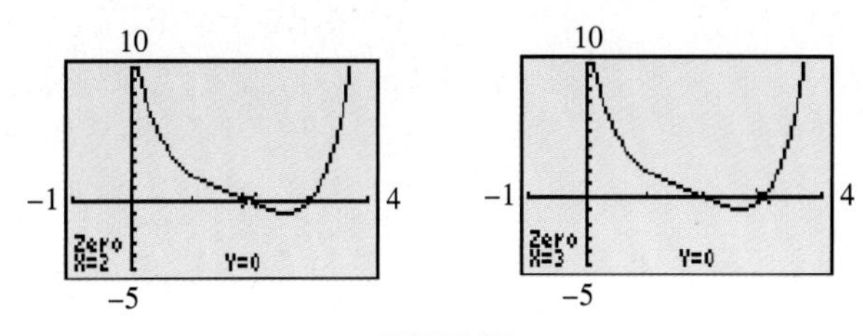

**FIGURE 74**

Thus, the four zeros are $1 - i$, $1 + i$, 2, and 3.

The number of zeros theorem says that a polynomial function of degree $n$ has *at most n* distinct zeros. In the polynomial function defined by

$$\begin{aligned} P(x) &= x^6 + x^5 - 5x^4 - x^3 + 8x^2 - 4x \\ &= x(x + 2)^2(x - 1)^3, \end{aligned}$$

each factor leads to a zero of the function. The factor $x$ leads to a *single* zero, 0, the factor $(x + 2)^2$ leads to a zero of $-2$ appearing *twice*, and the factor $(x - 1)^3$ leads to a zero of 1 appearing *three* times. The number of times a zero appears is referred to as the **multiplicity of the zero**.

**EXAMPLE 3** *Defining a Polynomial Function Satisfying Given Conditions*

Find a polynomial function with real coefficients of lowest possible degree having a zero 2 of multiplicity 3, a zero 0 of multiplicity 2, and a zero $i$ of single multiplicity.

**Solution** By the conjugate zeros theorem, this polynomial function must also have a zero $-i$. This means there are seven zeros, so the lowest possible degree of the polynomial is 7. By the factor theorem, one such polynomial in factored form is

$$P(x) = x^2(x - 2)^3(x - i)(x + i).$$

Multiplying the factors on the right leads to

$$P(x) = x^7 - 6x^6 + 13x^5 - 14x^4 + 12x^3 - 8x^2.$$

This is one of infinitely many such functions. Multiplying $P(x)$ by a nonzero constant will yield another polynomial function satisfying these conditions.

The graph of this function in Figure 75 on the next page shows that it has only two distinct $x$-intercepts, corresponding to the real zeros 0 and 2.

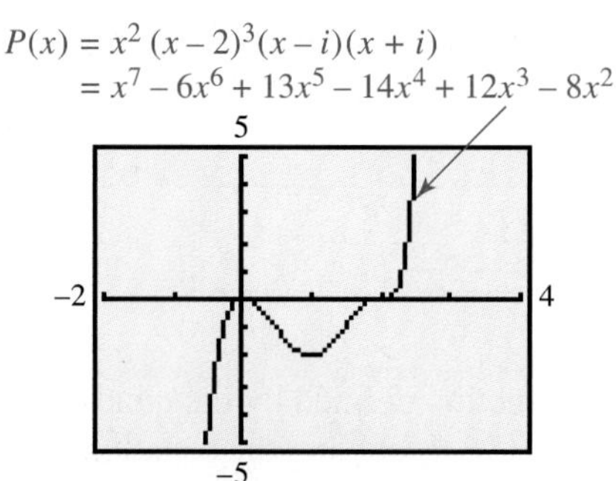


The graph is tangent to the $x$-axis at $x = 0$, and crosses the $x$-axis at $x = 2$.

**FIGURE 75**

## FOR DISCUSSION

Graph each function in the indicated window, and then respond to the items below.

$P(x) = (x + 3)(x - 2)^2$; $[-10, 10]$ by $[-30, 30]$ $\quad$ $P(x) = (x + 3)^2(x - 2)^3$; $[-4, 4]$ by $[-125, 50]$ $\quad$ $P(x) = x^2(x - 1)(x + 2)^2$; $[-4, 4]$ by $[-5, 5]$

1. Describe the behavior of the graph at each $x$-intercept that corresponds to a zero of odd multiplicity.
2. Describe the behavior of the graph at each $x$-intercept that corresponds to a zero of even multiplicity.

The observations in the "For Discussion" box indicate that the behavior of the graph of a polynomial function near an $x$-intercept depends on the parity of multiplicity of the zero that leads to the $x$-intercept. If the zero is of odd multiplicity, the graph will cross the $x$-axis at the corresponding $x$-intercept. If the zero is of even multiplicity, the graph will be tangent to the $x$-axis at the corresponding $x$-intercept (that is, it will touch but not cross the $x$-axis). See Figure 76.

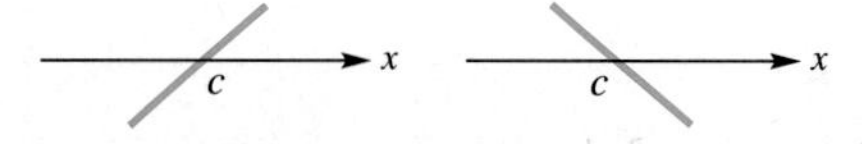


The graph crosses the $x$-axis at $(c, 0)$ if $c$ is a zero of odd multiplicity.

**(a)**

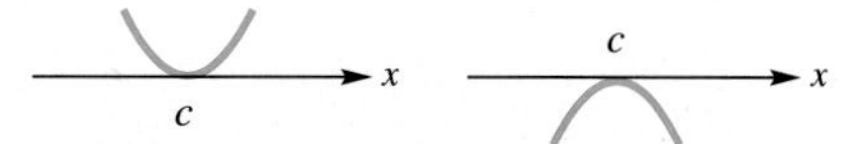


The graph is tangent to the $x$-axis at $(c, 0)$ if $c$ is a zero of even multiplicity.

**(b)**

**FIGURE 76**

By observing the dominating term and noting the parity of multiplicities of zeros of a polynomial function in factored form, we can sketch a rough graph of a polynomial function by hand, as in the next example.

**EXAMPLE 4** *Sketching a Polynomial Function Graph by Hand*

Consider the polynomial function defined by

$$P(x) = -2x^5 - 18x^4 - 38x^3 + 42x^2 + 112x - 96,$$

with factored form

$$P(x) = -2(x + 4)^2(x + 3)(x - 1)^2.$$

Sketch the graph of $P$ by hand. Confirm the result with a calculator.

**Solution** Because the dominating term is $-2x^5$, the end behavior of the graph will be ↖↘. Because $-4$ and 1 are both $x$-intercepts determined by zeros of even multiplicity, the graph will be tangent to the $x$-axis at these $x$-intercepts. Because $-3$ is a zero of multiplicity 1, the graph will cross the $x$-axis at $x = -3$. The $y$-intercept is $-96$. Combining all of this information leads to the rough sketch of the graph in Figure 77(a).

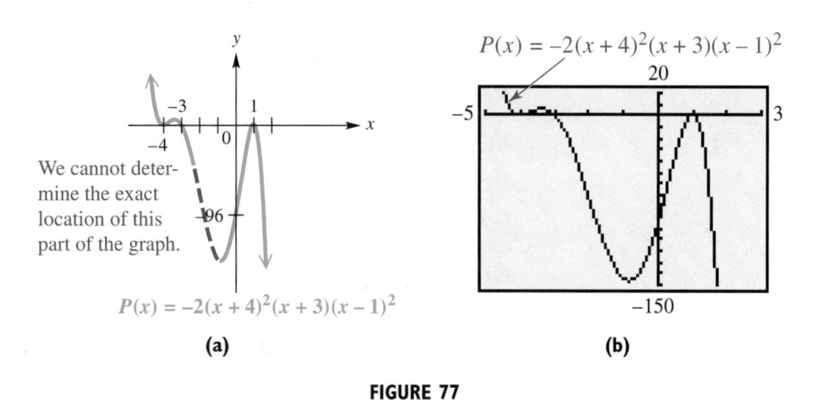


**FIGURE 77**

Notice that the hand-drawn graph does not necessarily give a good indication of local extrema. The calculator graph shown in Figure 77(b) fills in the details.

## Rational Zeros Theorem

The final theorem in this section gives a method to determine all possible candidates for rational zeros of a polynomial function with integer coefficients.

> **Rational Zeros Theorem**
>
> Let $P(x) = a_nx^n + a_{n-1}x^{n-1} + \cdots + a_1x + a_0$, where $a_n \neq 0$, define a polynomial function with integer coefficients. If $\frac{p}{q}$ is a rational number written in lowest terms, and if $\frac{p}{q}$ is a zero of $P$, then $p$ is a factor of the constant term $a_0$, and $q$ is a factor of the leading coefficient $a_n$.

***Proof*** $P(\frac{p}{q}) = 0$ since $\frac{p}{q}$ is a zero of $P(x)$, so

$$a_n\left(\frac{p}{q}\right)^n + a_{n-1}\left(\frac{p}{q}\right)^{n-1} + \cdots + a_1\left(\frac{p}{q}\right) + a_0 = 0.$$

This equation also can be written as

$$a_n\left(\frac{p^n}{q^n}\right) + a_{n-1}\left(\frac{p^{n-1}}{q^{n-1}}\right) + \cdots + a_1\left(\frac{p}{q}\right) + a_0 = 0.$$

Multiply each side of this last result by $q^n$ and add $-a_0q^n$ to each side.

$$a_np^n + a_{n-1}p^{n-1}q + \cdots + a_1pq^{n-1} = -a_0q^n$$

Factoring out $p$ gives

$$p(a_np^{n-1} + a_{n-1}p^{n-2}q + \cdots + a_1q^{n-1}) = -a_0q^n.$$

This result shows that $-a_0q^n$ equals the product of the two factors, $p$ and $(a_np^{n-1} + \cdots + a_1q^{n-1})$. For this reason, $p$ must be a factor of $-a_0q^n$. Since it was assumed that $\frac{p}{q}$ is written in lowest terms, $p$ and $q$ have no common factor other than 1, so $p$ is not a factor of $q^n$. Thus, $p$ must be a factor of $a_0$. In a similar way, it can be shown that $q$ is a factor of $a_n$.

**EXAMPLE 5** *Using the Rational Zeros Theorem*

For the polynomial function defined by $P(x) = 6x^4 + 7x^3 - 12x^2 - 3x + 2$, do the following.

**(a)** List all possible rational zeros.

**(b)** Use a graph to eliminate some of the possible zeros listed in part (a).

**(c)** Find all rational zeros and factor $P(x)$.

**Solution**

**(a)** For a rational number $\frac{p}{q}$ to be a zero, $p$ must be a factor of $a_0 = 2$ and $q$ must be a factor of $a_4 = 6$. Thus, $p$ can be $\pm 1$ or $\pm 2$, and $q$ can be $\pm 1, \pm 2, \pm 3$, or $\pm 6$. The possible rational zeros, $\frac{p}{q}$, are

$$\pm 1, \quad \pm 2, \quad \pm\frac{1}{2}, \quad \pm\frac{1}{3}, \quad \pm\frac{1}{6}, \quad \pm\frac{2}{3}.$$

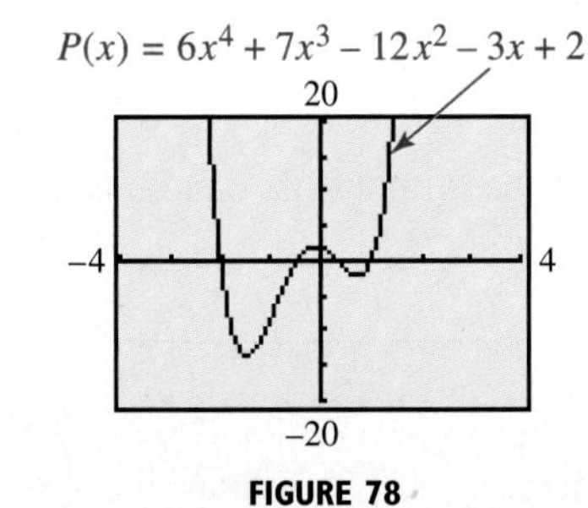

FIGURE 78

**(b)** From Figure 78, we see that the zeros are no less than $-2$ and no greater than 1, so we can eliminate 2. Furthermore, it is obvious that $-1$ is not a zero since the graph does not intersect the $x$-axis at $(-1, 0)$. Note that at this point we have no way of knowing whether the zeros indicated on the graph are rational numbers. They may be irrational.

**(c)** We use the remainder theorem to show that 1 and $-2$ are zeros.

$$\begin{array}{r|rrrrr} 1 & 6 & 7 & -12 & -3 & 2 \\ & & 6 & 13 & 1 & -2 \\ \hline & 6 & 13 & 1 & -2 & 0 \end{array}$$

The 0 remainder shows that 1 is a zero. Now we use the quotient polynomial $6x^3 + 13x^2 + x - 2$ and synthetic division to find that $-2$ is also a zero.

$$\begin{array}{r|rrrr} -2 & 6 & 13 & 1 & -2 \\ & & -12 & -2 & 2 \\ \hline & 6 & 1 & -1 & 0 \end{array}$$

This graph/table verifies that the four zeros of the function of Example 5 are $-2, -\frac{1}{2}, \frac{1}{3}$, and 1.

The new quotient polynomial is $6x^2 + x - 1$. We use the quadratic formula or factor to solve the equation $6x^2 + x - 1 = 0$. The remaining two zeros are the rational numbers $\frac{1}{3}$ and $-\frac{1}{2}$.

We factor the polynomial $P(x)$ in the following way. Since the four zeros of $P(x) = 6x^4 + 7x^3 - 12x^2 - 3x + 2$ are $1, -2, \frac{1}{3}$, and $-\frac{1}{2}$, the factors are $x - 1$,

$x + 2$, $x - \frac{1}{3}$, and $x + \frac{1}{2}$, and

$$P(x) = a(x - 1)(x + 2)\left(x - \frac{1}{3}\right)\left(x + \frac{1}{2}\right).$$

Since the leading coefficient of $P(x)$ is 6, we let $a = 6$. Then,

$$\begin{aligned} P(x) &= 6(x - 1)(x + 2)\left(x - \frac{1}{3}\right)\left(x + \frac{1}{2}\right) \\ &= (x - 1)(x + 2)(3)\left(x - \frac{1}{3}\right)(2)\left(x + \frac{1}{2}\right) \\ &= (x - 1)(x + 2)(3x - 1)(2x + 1). \end{aligned}$$

The polynomial could also be left in the form $P(x) = 6(x - 1)(x + 2)\left(x - \frac{1}{3}\right)\left(x + \frac{1}{2}\right)$.

**CAUTION** The rational zeros theorem has limited usefulness since it gives only possible rational zeros; it does not tell us whether these rational numbers are actual zeros. We must rely on other methods to determine whether or not they are indeed zeros. Furthermore, the function must have integer coefficients. To begin to apply the rational zeros theorem to a polynomial with fractional coefficients, multiply through by the least common denominator of all the fractions. For example, any rational zeros of

$$P(x) = x^4 - \frac{1}{6}x^3 + \frac{2}{3}x^2 - \frac{1}{6}x - \frac{1}{3}$$

will also be rational zeros of

$$Q(x) = 6x^4 - x^3 + 4x^2 - x - 2.$$

The function $Q$ was obtained by multiplying the terms of $P$ by 6.

## 3.7 EXERCISES

*Find a cubic polynomial in standard form with real coefficients, having the given zeros.*

**1.** 4 and $2 + i$

**2.** $-3$ and $6 + 2i$

**3.** 5 and $i$

**4.** $-9$ and $-i$

**5.** 0 and $3 + i$

**6.** 0 and $4 - 3i$

*For Exercises 7–12, find a function P defined by a polynomial of degree 3 with real coefficients that satisfies the given conditions.*

**7.** Zeros of $-3$, $-1$, and 4; $P(2) = 5$

**8.** Zeros of 1, $-1$, and 0; $P(2) = -3$

**9.** Zeros of $-2$, 1, and 0; $P(-1) = -1$

**10.** Zeros of 2, 5, and $-3$; $P(1) = -4$

**11.** Zeros of 4 and $1 + i$; $P(2) = 4$

**12.** Zeros of $-7$ and $2 - i$; $P(1) = 9$

*For each polynomial, one or more zeros are given. Find all remaining zeros.*

**13.** $P(x) = x^3 - x^2 - 4x - 6$; 3 is a zero.

**14.** $P(x) = x^3 - 5x^2 + 17x - 13$; 1 is a zero.

**15.** $P(x) = x^4 + 2x^3 - 10x^2 - 18x + 9$; $-3$ and 3 are zeros.

**16.** $P(x) = 2x^4 - x^3 - 27x^2 + 16x - 80$; $-4$ and 4 are zeros.

**17.** $P(x) = x^4 - x^3 + 10x^2 - 9x + 9$; $3i$ is a zero.

**18.** $P(x) = 2x^4 - 2x^3 + 55x^2 - 50x + 125$; $-5i$ is a zero.

*For Exercises 19–30, find a polynomial function P of lowest possible degree, having real coefficients, with the given zeros.*

**19.** 5 and $-4$

**20.** 6 and $-2$

**21.** $-3$, 2, and $i$

**22.** $1+\sqrt{2}$, $1-\sqrt{2}$, and 3

**23.** $1-\sqrt{3}$, $1+\sqrt{3}$, and 1

**24.** $-2+i$, $-2-i$, 3, and $-3$

**25.** $3+2i$, $-1$, and 2

**26.** 2 and $3i$

**27.** $-1$ and $6-3i$

**28.** $1+2i$ and 2 (multiplicity 2)

**29.** $2+i$ and $-3$ (multiplicity 2)

**30.** 5 (multiplicity 2) and $-2i$

Concept Check *Use the concepts of this section for Exercises 31–36.*

**31.** Show that $-2$ is a zero of multiplicity 2 of $P$, where $P(x) = x^4 + 2x^3 - 7x^2 - 20x - 12$, and find all other complex zeros. Then, write $P(x)$ in factored form.

**32.** Show that $-1$ is a zero of multiplicity 3 of $P$, where $P(x) = x^5 + 9x^4 + 33x^3 + 55x^2 + 42x + 12$, and find all other complex zeros. Then, write $P(x)$ in factored form.

**33.** What are the possible numbers of real zeros (counting multiplicities) for a polynomial function with real coefficients of degree 5?

**34.** Explain why a function defined by a polynomial of degree 4 with real coefficients has either zero, two, or four real zeros (counting multiplicities).

**35.** Explain why it is not possible for a function defined by a polynomial of degree 3 with real coefficients to have zeros of 1, 2, and $1+i$.

**36.** Suppose that $k$, $a$, $b$, and $c$ are real numbers, $a \neq 0$, and a polynomial function defined by $P(x)$ may be expressed in factored form as $(x-k)(ax^2+bx+c)$.

**(a)** What is the degree of $P$?

**(b)** What are the possible numbers of distinct *real* zeros of $P$?

**(c)** What are the possible numbers of *imaginary* zeros of $P$?

**(d)** Use the discriminant to explain how to determine the number and type of zeros $P$ has.

*In Exercises 37–42, a polynomial function is given in both standard form and factored form. Use dominating-term analysis and multiplicity of zeros to draw by hand a rough sketch of the graph of the function. Support your answer using a calculator graph.*

**37.** $P(x) = 2x^3 - 5x^2 - x + 6$
$= (x+1)(2x-3)(x-2)$

**38.** $P(x) = x^3 + x^2 - 8x - 12$
$= (x+2)^2(x-3)$

**39.** $P(x) = x^4 - 18x^2 + 81$
$= (x-3)^2(x+3)^2$

**40.** $P(x) = x^4 - 8x^2 + 16$
$= (x+2)^2(x-2)^2$

**41.** $P(x) = 2x^4 + x^3 - 6x^2 - 7x - 2$
$= (2x+1)(x-2)(x+1)^2$

**42.** $P(x) = 3x^4 - 7x^3 - 6x^2 + 12x + 8$
$= (3x+2)(x+1)(x-2)^2$

*For each polynomial function,* **(a)** *list all possible rational zeros,* **(b)** *use a graph to eliminate some of the possible zeros listed in part (a),* **(c)** *find all rational zeros, and* **(d)** *factor P(x).*

**43.** $P(x) = x^3 - 2x^2 - 13x - 10$

**44.** $P(x) = x^3 + 5x^2 + 2x - 8$

**45.** $P(x) = x^3 + 6x^2 - x - 30$

**46.** $P(x) = x^3 - x^2 - 10x - 8$

**47.** $P(x) = 6x^3 + 17x^2 - 31x - 12$

**48.** $P(x) = 15x^3 + 61x^2 + 2x - 8$

**49.** $P(x) = 12x^3 + 20x^2 - x - 6$

**50.** $P(x) = 12x^3 + 40x^2 + 41x + 12$

*Find all rational zeros of each polynomial function.*

**51.** $P(x) = x^3 + \frac{1}{2}x^2 - \frac{11}{2}x - 5$

**52.** $P(x) = \frac{10}{7}x^4 - x^3 - 7x^2 + 5x - \frac{5}{7}$

**53.** $P(x) = \frac{1}{6}x^4 - \frac{11}{12}x^3 + \frac{7}{6}x^2 - \frac{11}{12}x + 1$

**54.** $P(x) = x^4 - \frac{1}{6}x^3 + \frac{2}{3}x^2 - \frac{1}{6}x - \frac{1}{3}$

**55.** Suppose that $c$ and $d$ represent the complex numbers $c = a + bi$ and $d = m + ni$. Let $\overline{c}$ and $\overline{d}$ represent the complex conjugates of $c$ and $d$, respectively. Prove each statement. (These properties will be used in Exercise 56 to prove the conjugate zeros theorem.)

**(a)** $\overline{c+d} = \overline{c} + \overline{d}$

**(b)** $\overline{cd} = \overline{c} \cdot \overline{d}$

**(c)** $\overline{x} = x$ for any real number $x$

**(d)** $\overline{c^n} = (\overline{c})^n$, where $n$ is a positive integer

**56.** Complete the proof of the conjugate zeros theorem, outlined below. Assume that

$$P(x) = a_n x^n + a_{n-1}x^{n-1} + \cdots + a_1 x + a_0,$$

where all coefficients are real numbers.

**(a)** Suppose the complex number $z$ is a zero of $P$; find $P(z)$.

**(b)** Take the conjugate of each side of the result from part (a).

**(c)** Use generalizations of the properties given in Exercise 55 on the result of part (b) to show that $a_n(\bar{z})^n + a_{n-1}(\bar{z})^{n-1} + \cdots + a_1(\bar{z}) + a_0 = 0$.

**(d)** Why does the result in part (c) mean that $\bar{z}$ is a zero of $P$?

**57.** The function defined by $P(x) = x^2 - x + (i + 1)$ has $i$ as a zero, but does not have its conjugate, $-i$, as a zero. Explain why this does not violate the conjugate zeros theorem.

**58.** For any polynomial $P(x)$ and any complex number $k$, there exists a unique polynomial $Q(x)$ and number $R$ such that

$$P(x) = (x - k) \cdot Q(x) + R.$$

This statement is known as the division algorithm. To prove the remainder theorem (Section 3.6), let $x = k$ in this statement. Write a proof of the remainder theorem.

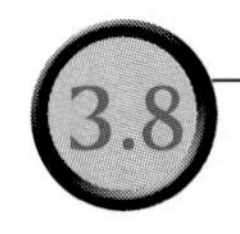

# 3.8 POLYNOMIAL EQUATIONS AND INEQUALITIES; FURTHER APPLICATIONS AND MODELS

Polynomial Equations and Inequalities • Complex *n*th Roots • Applications and Polynomial Models

While methods of solving quadratic equations were known to ancient civilizations, for hundreds of years mathematicians wrestled with finding methods of solving higher-degree equations (analytically, of course). In the 16th century the European mathematicians Scipione del Ferro, Nicolo Fontana (a.k.a. Tartaglia), Girolamo Cardano, and François Viete derived formulas for solving cubic equations. Methods of solving quartics followed. While these methods were quite complicated, they showed that, in theory, third- and fourth-degree polynomial equations could be solved analytically. In 1824 Norwegian mathematician Niels Henrik Abel proved that it is impossible to find a formula that will yield solutions to the general quintic (fifth-degree) equation. A similar result holds for polynomial equations of degree greater than 5.

We can use elementary methods to solve *some* higher-degree polynomial equations analytically, as we show in this section. Graphing calculators support the analytic work and enable us to find accurate approximations of solutions of polynomial equations that cannot be solved easily, or at all, by analytic methods.

## Polynomial Equations and Inequalities

**EXAMPLE 1** *Solving a Polynomial Equation and Associated Inequalities*

**(a)** Solve $x^3 + 3x^2 - 4x - 12 = 0$ using the zero-product property.

**(b)** Graph the equation from part (a).

**(c)** Use the graph from part (b) to solve the inequalities

$$x^3 + 3x^2 - 4x - 12 > 0 \quad \text{and} \quad x^3 + 3x^2 - 4x - 12 \le 0.$$

**Solution**

**(a)** Factor the polynomial.

$$x^3 + 3x^2 - 4x - 12 = 0$$

$$(x^3 + 3x^2) + (-4x - 12) = 0 \quad \text{Group terms with common factors.}$$

$$x^2(x + 3) - 4(x + 3) = 0 \quad \text{Factor out common factors in each group.}$$

$$(x + 3)(x^2 - 4) = 0 \quad \text{Factor out } x + 3.$$

$$(x + 3)(x - 2)(x + 2) = 0 \quad \text{Factor the difference of squares.}$$

$$x = -3 \quad \text{or} \quad x = 2 \quad \text{or} \quad x = -2 \quad \text{Zero-product property}$$

The solution set is $\{-3, -2, 2\}$.

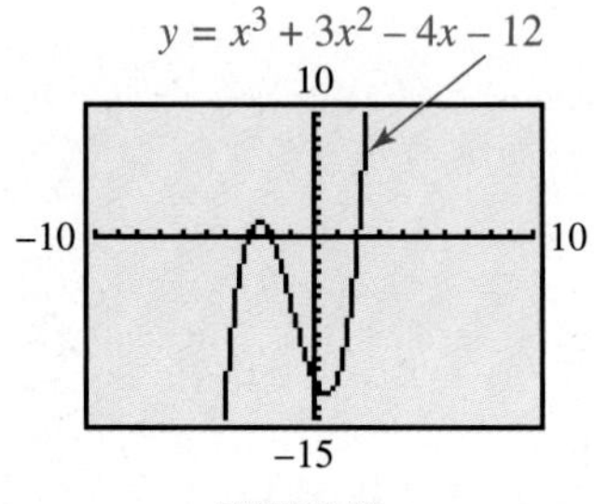

**FIGURE 79**

**(b)** Graph $y = x^3 + 3x^2 - 4x - 12$ and verify that the $x$-intercepts are $-3$, $-2$, and 2, supporting the analytic solution. See Figure 79.

**(c)** Recall that the solution set of $f(x) > 0$ consists of all numbers in the domain of $f$ for which the graph lies *above* the $x$-axis. In Figure 79, this occurs in the intervals $(-3, -2) \cup (2, \infty)$, the solution set of this inequality. To solve $x^3 + 3x^2 - 4x - 12 \le 0$, locate the intervals where the graph lies *below or intersects* the $x$-axis. From the figure, the solution set is $(-\infty, -3] \cup [-2, 2]$. The endpoints are included here due to the $\le$ symbol.

An equation is **quadratic in form** if it can be written as

$$au^2 + bu + c = 0,$$

where $a \ne 0$ and $u$ is an algebraic expression.

**EXAMPLE 2** *Solving an Equation Quadratic in Form and Associated Inequalities*

**(a)** Solve $x^4 - 6x^2 - 40 = 0$ analytically. Find all complex solutions.

**(b)** Graph the equation from part (a), and use the graph to solve the inequalities

$$x^4 - 6x^2 - 40 \ge 0 \quad \text{and} \quad x^4 - 6x^2 - 40 < 0.$$

Give endpoints of intervals in both exact and approximate form.

**Solution**

**(a)** If we let $t = x^2$, then the equation becomes quadratic in the variable $t$. We can solve for $t$ by solving the quadratic equation, and then replacing $t$ with $x^2$ and solving for $x$.

$$x^4 - 6x^2 - 40 = 0$$

$$(x^2)^2 - 6x^2 - 40 = 0$$

$$t^2 - 6t - 40 = 0 \quad \text{Let } t = x^2.$$

$$(t - 10)(t + 4) = 0 \quad \text{Factor.}$$

$$t = 10 \quad \text{or} \quad t = -4 \quad \text{Zero-product property}$$

$$x^2 = 10 \quad \text{or} \quad x^2 = -4 \quad \text{Replace } t \text{ with } x^2.$$

$$x = \pm\sqrt{10} \quad \text{or} \quad x = \pm 2i \quad \text{Square root property}$$

The solution set is $\{-\sqrt{10}, \sqrt{10}, -2i, 2i\}$.

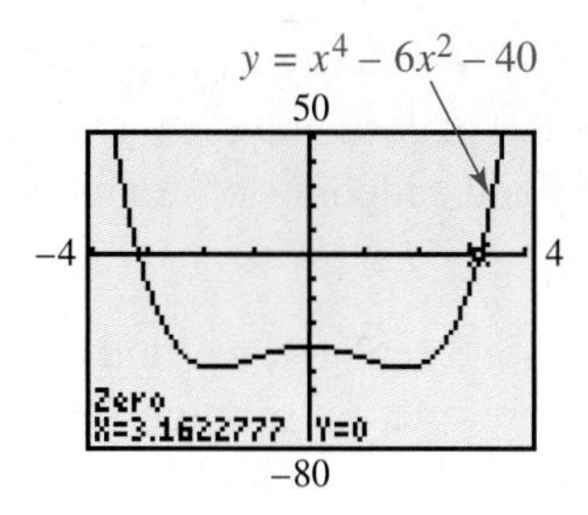


FIGURE 80

**(b)** The graph of $y = x^4 - 6x^2 - 40$ is shown in Figure 80. We find that the $x$-intercepts are approximately $-3.16$ and $3.16$ (approximations of $-\sqrt{10}$ and $\sqrt{10}$). Notice that the graph cannot support the imaginary solutions.

Since the graph lies above or intersects the $x$-axis for real numbers less than or equal to $-\sqrt{10}$ and for real numbers greater than or equal to $\sqrt{10}$, the solution set for $x^4 - 6x^2 - 40 \geq 0$ is

$$\left(-\infty, -\sqrt{10}\right] \cup \left[\sqrt{10}, \infty\right) \leftarrow \text{Exact form}$$

or

$$(-\infty, -3.16] \cup [3.16, \infty). \leftarrow \text{Approximate form}$$

By similar reasoning, the solution set of $x^4 - 6x^2 - 40 < 0$ is

$$\left(-\sqrt{10}, \sqrt{10}\right) \leftarrow \text{Exact form}$$

or

$$(-3.16, 3.16). \leftarrow \text{Approximate form}$$

The imaginary solutions do not affect the solution sets of the inequalities.

**EXAMPLE 3** *Solving a Polynomial Equation*

**(a)** Show that 2 is a real solution of $x^3 + 3x^2 - 11x + 2 = 0$, and then find all solutions of this equation.

**(b)** Support the result of part (a) graphically.

**Solution**

**(a)** Use synthetic division.

$$\begin{array}{r|rrrr} 2 & 1 & 3 & -11 & 2 \\ & & 2 & 10 & -2 \\ \hline & 1 & 5 & -1 & 0 \end{array}$$

$\leftarrow$ $P(2) = 0$ by the remainder theorem.

Coefficients of the quotient polynomial

By the factor theorem, $x - 2$ is a factor of $P(x)$, so

$$P(x) = (x - 2)(x^2 + 5x - 1).$$

To find the other zeros of $P$, solve $x^2 + 5x - 1 = 0$. Use the quadratic formula, with $a = 1$, $b = 5$, and $c = -1$.

$$x = \frac{-5 \pm \sqrt{5^2 - 4(1)(-1)}}{2(1)} = \frac{-5 \pm \sqrt{29}}{2}$$

The solution set is $\left\{\frac{-5 - \sqrt{29}}{2}, \frac{-5 + \sqrt{29}}{2}, 2\right\}$.

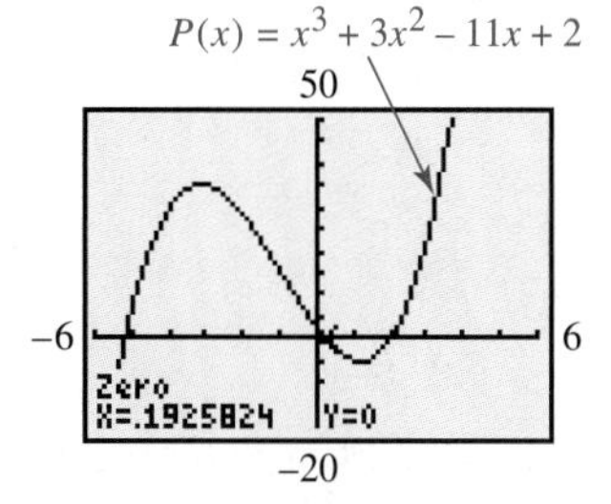


FIGURE 81

**(b)** The graph of $P(x) = x^3 + 3x^2 - 11x + 2$ is shown in Figure 81. The $x$-intercepts are 2, approximately $-5.19$, and approximately $.19$. These latter two approximations correspond to the approximations for $\frac{-5 - \sqrt{29}}{2}$ and $\frac{-5 + \sqrt{29}}{2}$, supporting the analytic results.

The associated inequalities for the equation in Example 3 could be solved using the solutions and graph as in Examples 1 and 2.

**EXAMPLE 4** *Solving an Equation and Associated Inequalities Graphically*

Let $P(x) = 2.45x^3 - 3.14x^2 - 6.99x + 2.58$. Use the graph of $P$ to solve $P(x) = 0$, $P(x) > 0$, and $P(x) < 0$. Express solutions of the equation and endpoints of the intervals for the inequalities to the nearest hundredth.

$P(x) = 2.45x^3 - 3.14x^2 - 6.99x + 2.58$

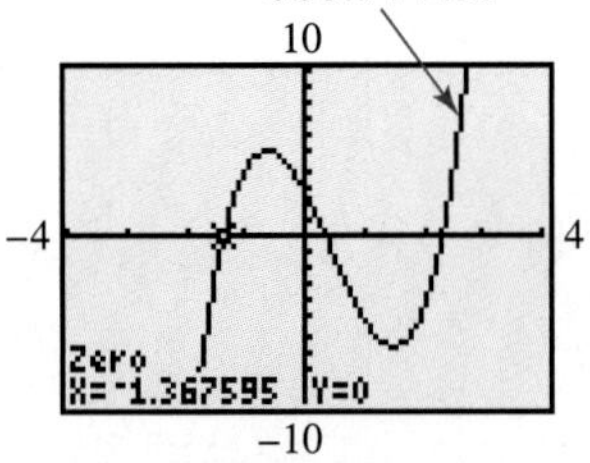

10, −4, 4, −10

Zero X=-1.367595 Y=0

The other two $x$-intercepts are approximately .33 and 2.32.

**FIGURE 82**

**Solution** The graph of $P$ is shown in Figure 82. Using a calculator, we find that the approximate $x$-intercepts are $-1.37$, .33, and 2.32. Therefore, the solution set of the equation $P(x) = 0$ is $\{-1.37, .33, 2.32\}$. Based on the graph, the solution sets of $P(x) > 0$ and $P(x) < 0$ are

$$(-1.37, .33) \cup (2.32, \infty), \quad \text{and} \quad (-\infty, -1.37) \cup (.33, 2.32),$$

respectively.

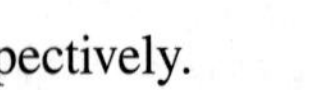

**NOTE** The graphical method of solving $P(x) = 0$ in Example 4 would not have yielded imaginary solutions had there been any. Only real solutions are obtained using this method.

## Complex *n*th Roots

If $n$ is a positive integer and $k$ is a nonzero complex number, then a solution of $x^n = k$ is called an ***n*th root of *k*.** For example, since $-1$ and 1 are solutions of $x^2 = 1$, they are called second or square roots of 1. Similarly, $-2i$ and $2i$ are called square roots of $-4$, since $(\pm 2i)^2 = -4$.

The real numbers $-2$ and 2 are sixth roots of 64, since $(\pm 2)^6 = 64$. However, 64 has four more complex sixth roots. While a complete discussion of the next theorem requires concepts from trigonometry, we state it and use it to solve particular problems involving $n$th roots.

**Complex *n*th Roots Theorem**

If $n$ is a positive integer and $k$ is a nonzero complex number, then the equation $x^n = k$ has *exactly* $n$ complex roots.

**EXAMPLE 5** *Finding nth Roots of a Number*

Find all six complex sixth roots of 64.

**Solution** The roots must be solutions of $x^6 = 64$.

$$x^6 = 64$$

$$x^6 - 64 = 0 \qquad \text{Subtract 64.}$$

$$(x^3 - 8)(x^3 + 8) = 0 \qquad \text{Factor the difference of squares.}$$

$$(x - 2)(x^2 + 2x + 4)(x + 2)(x^2 - 2x + 4) = 0 \qquad \text{Factor the difference of cubes and the sum of cubes.}$$

Now apply the zero-product property to obtain the real roots 2 and $-2$. Setting the quadratic factors equal to 0 and applying the quadratic formula twice gives the remaining four complex roots (all imaginary).

$$x^2 + 2x + 4 = 0 \qquad \text{or} \qquad x^2 - 2x + 4 = 0$$

$$x = \frac{-2 \pm \sqrt{2^2 - 4(1)(4)}}{2(1)} \qquad x = \frac{2 \pm \sqrt{(-2)^2 - 4(1)(4)}}{2(1)}$$

$$= \frac{-2 \pm \sqrt{-12}}{2} \qquad = \frac{2 \pm \sqrt{-12}}{2}$$

$$= \frac{-2 \pm 2i\sqrt{3}}{2} \qquad = \frac{2 \pm 2i\sqrt{3}}{2}$$

$$= -1 \pm i\sqrt{3} \qquad = 1 \pm i\sqrt{3}$$

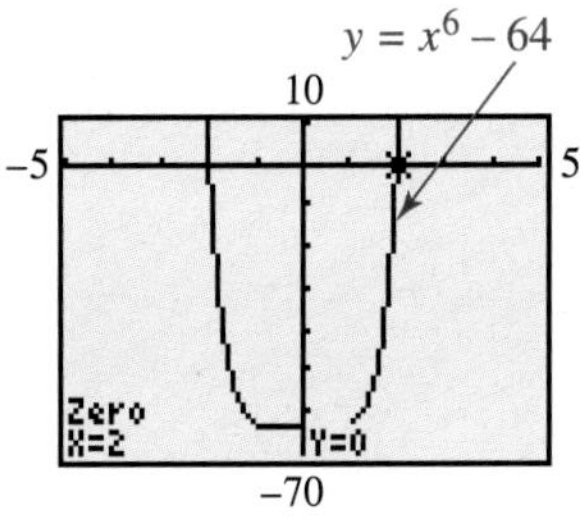

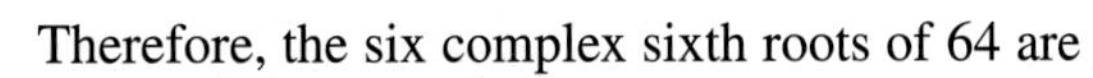

Therefore, the six complex sixth roots of 64 are

$$2, \quad -2, \quad -1 + i\sqrt{3}, \quad -1 - i\sqrt{3}, \quad 1 + i\sqrt{3}, \quad \text{and} \quad 1 - i\sqrt{3}.$$

The graph of $y = x^6 - 64$ confirms the two real sixth roots of 64. See Figure 83. The $x$-intercepts are $-2$ and 2.

$y = x^6 - 64$

FIGURE 83

## Applications and Polynomial Models

In Section 3.4, we used quadratic functions to solve certain types of applied problems. We now use higher-degree polynomial functions similarly.

**EXAMPLE 6** *Using a Polynomial Function to Describe the Volume of a Box*
A box with an open top is to be constructed from a rectangular 12-inch by 20-inch piece of cardboard by cutting equal size squares from each corner and folding up the sides.

**(a)** If $x$ represents the length of the side of each such square, determine a function $V$ that describes the volume of the box in terms of $x$.

**(b)** Graph $V$ in the window $[0, 6]$ by $[0, 300]$, and locate a point on the graph. Interpret the displayed values of $x$ and $y$.

**(c)** Determine the value of $x$ for which the volume of the box is maximized. What is this volume?

**(d)** For what values of $x$ is the volume equal to 200 cubic inches? greater than 200 cubic inches? less than 200 cubic inches?

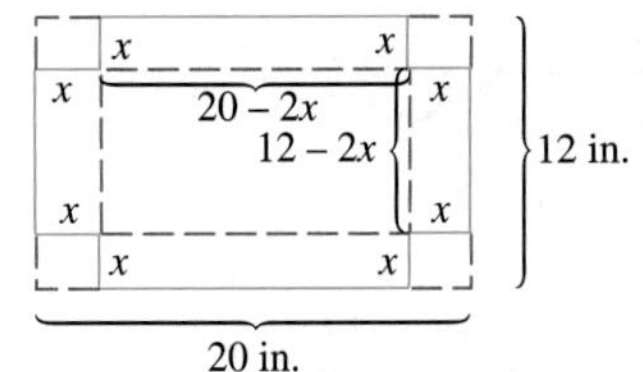


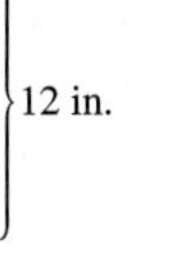

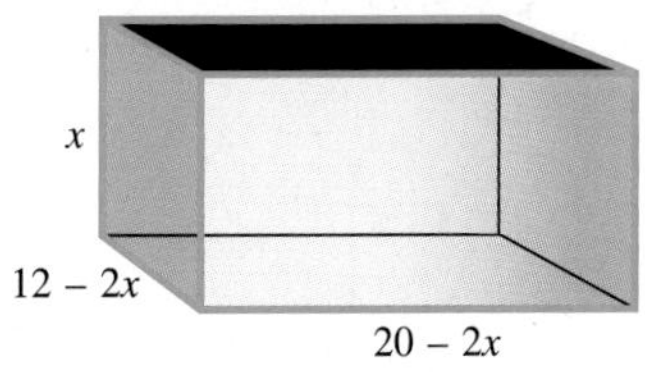

x
12 – 2x
20 – 2x

FIGURE 84

**Solution**

**(a)** As shown in Figure 84, the dimensions (in inches) of the box to be formed will be

$$\text{length} = 20 - 2x, \quad \text{width} = 12 - 2x, \quad \text{height} = x.$$

Furthermore, $x$ must be positive, and both $20 - 2x$ and $12 - 2x$ must be positive, implying that $0 < x < 6$. Since the volume of the box equals length $\times$ width $\times$ height, the desired function is

$$V(x) = (20 - 2x)(12 - 2x)x$$
$$= 4x^3 - 64x^2 + 240x.$$

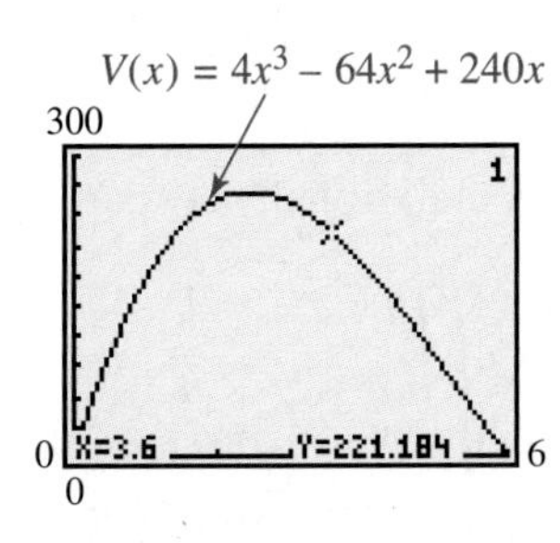


FIGURE 85

**(b)** Figure 85 shows the graph of $V$ with the cursor at the arbitrarily chosen point (3.6, 221.184). This means that when the side of each cut-out square measures 3.6 inches, the volume of the resulting box is 221.184 cubic inches.

**(c)** Use a calculator to find the local maximum point on the graph of $V$. To the nearest hundredth, the coordinates of this point are (2.43, 262.68). See Figure 86. Therefore, when $x \approx 2.43$ is the measure of the side of each cut-out square, the volume of the box is at its maximum, approximately 262.68 cubic inches.

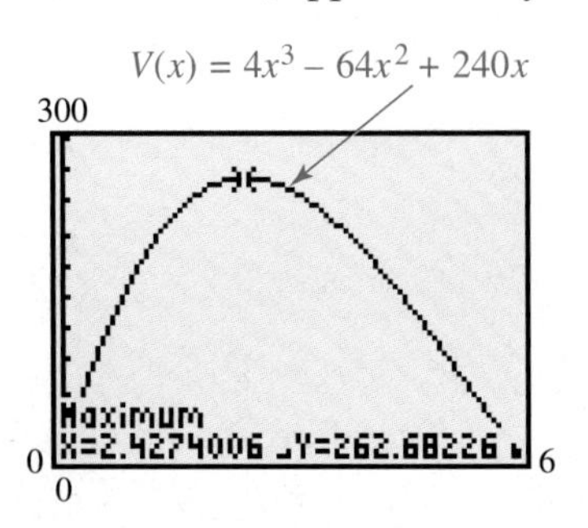


**FIGURE 86**

**(d)** The graphs of $y_1 = V(x)$ and $y_2 = 200$ are shown in Figure 87. The points of intersection of the line and the cubic curve are approximately (1.17, 200) and (3.90, 200), so the volume is equal to 200 cubic inches for $x \approx 1.17$ or 3.90, is greater than 200 cubic inches for $1.17 < x < 3.90$, and is less than 200 cubic inches for $0 < x < 1.17$ or $3.90 < x < 6$.

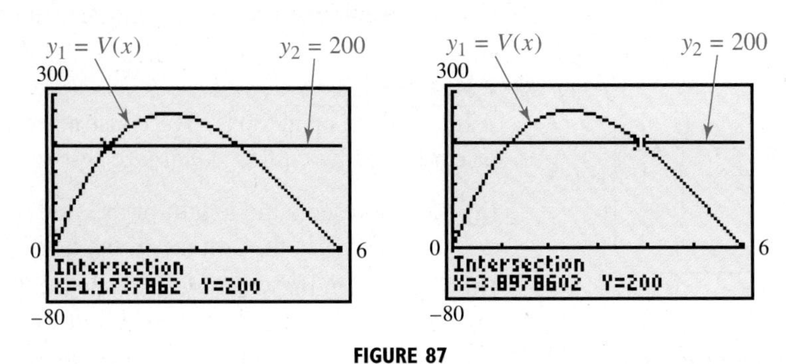


**FIGURE 87**

## FOR DISCUSSION

Refer to Example 6.

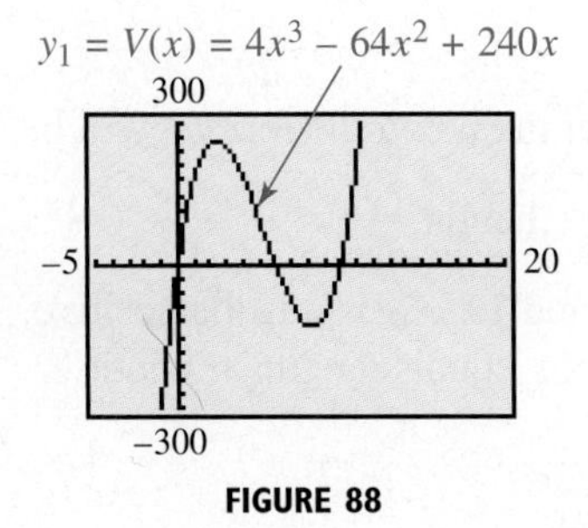


**FIGURE 88**

1. While we can enter $y_1 = (20 - 2x)(12 - 2x)x$ or $y_1 = 4x^3 - 64x^2 + 240x$ to obtain the desired graph, is it necessary to multiply out the factors when graphing to solve the equation? What is a good reason for *not* actually performing the multiplication?
2. Figure 88 shows a comprehensive graph of $V$ in the window $[-5, 20]$ by $[-300, 300]$. Explain why a comprehensive graph was not necessary in solving the problem.
3. Graph $y_3 = y_1 - 200$ in the window $[0, 6]$ by $[-50, 100]$, and explain how part (d) of Example 6 could be solved using the graph of $y_3$.

We introduced polynomial models in Section 3.5, and now we discuss such models in more detail.

**EXAMPLE 7** *Examining Polynomial Models for Debit Card Use*

The table shows the number of transactions, in millions, by users of bank debit cards for selected years.

| Year | 1990 | 1992 | 1995 | 1998 | 2000* |
|---|---|---|---|---|---|
| Transactions (in millions) | 127 | 204 | 829 | 3765 | 6655 |

*Projected
*Source: Statistical Abstract of the United States*, 2000.

**(a)** Using $x = 0$ to represent 1990, $x = 2$ to represent 1992, and so on, use the regression feature of a calculator to determine the quadratic function that best fits the data. Plot the data and the graph.

**(b)** Repeat part (a) for a cubic function.

**(c)** Repeat part (a) for a quartic function.

**(d)** Compare $R^2$ for the three functions found in parts (a)–(c) to decide which function best fits the data.

**Solution**

**(a)** The best-fitting quadratic function is defined by

$$y = 98.08x^2 - 343.7x + 248.6.$$

The regression coordinates screen and the data points with the graph are shown in Figure 89.

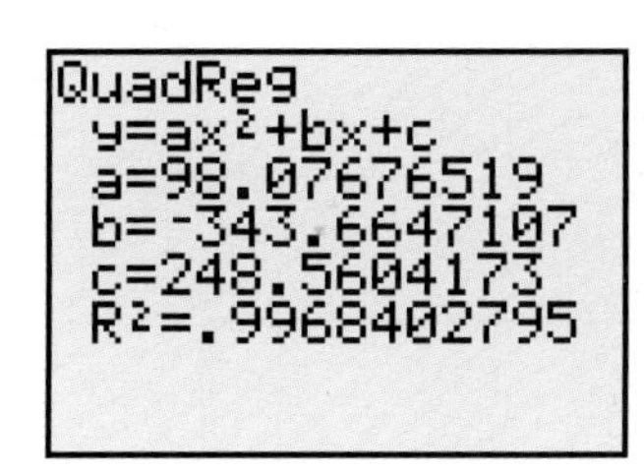


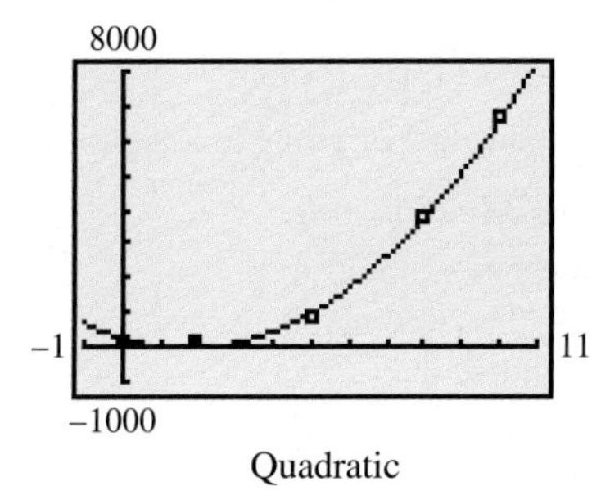


Quadratic

FIGURE 89

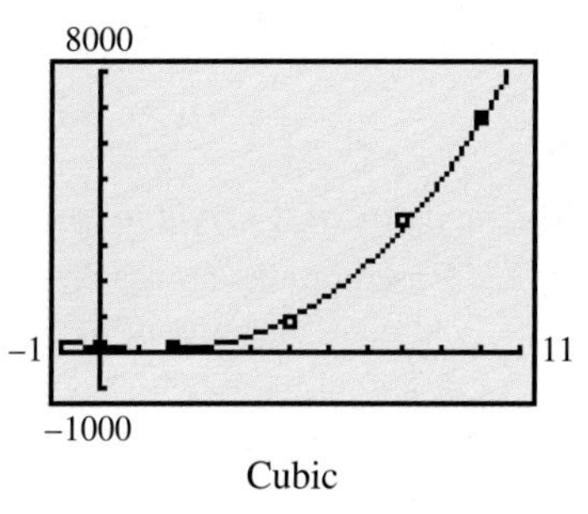


Cubic

FIGURE 90

**(b)** The best-fitting cubic function is defined by

$$y = 3.706x^3 + 42.48x^2 - 142.7x + 170.$$

See Figure 90.

**(c)** The best-fitting quartic function is defined by

$$y = -1.619x^4 + 36.09x^3 - 155.5x^2 + 218.1x + 127.$$

See Figure 91.

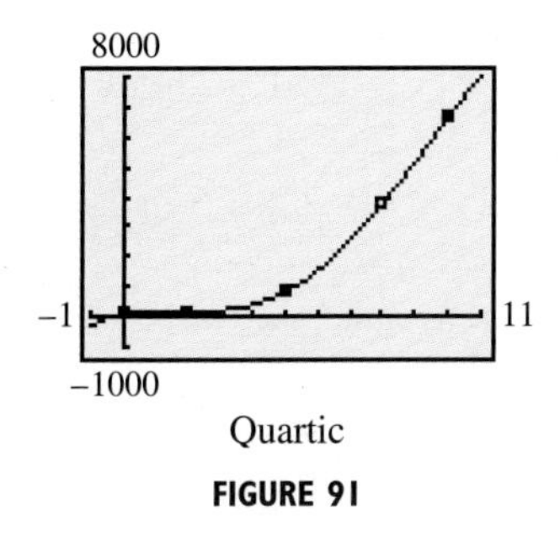


Quartic

FIGURE 91

**(d)** Compare the correlation coefficient values $R^2$ shown with the regression coefficients. The closer $R^2$ is to 1, the better the approximation. Verify that the quartic function provides the best fit.

## 3.8 EXERCISES

*Use factoring to find all complex solutions for each equation.*

**1.** $7x^3 + x = 0$

**2.** $2x^3 - 4x = 0$

**3.** $3x^3 + 2x^2 - 3x - 2 = 0$

**4.** $5x^3 - x^2 + 10x - 2 = 0$

**5.** $x^4 - 11x^2 + 10 = 0$

**6.** $x^4 + x^2 - 6 = 0$

*Solve each equation analytically for all complex solutions, giving* exact *forms in your solution set. Then, graph the left side of the equation as $y_1$ in the suggested viewing window, and support the real solutions using the capabilities of your calculator.*

**7.** $4x^4 - 25x^2 + 36 = 0$; $[-5, 5]$ by $[-5, 100]$

**8.** $4x^4 - 29x^2 + 25 = 0$; $[-5, 5]$ by $[-50, 100]$

**9.** $x^4 - 15x^2 - 16 = 0$; $[-5, 5]$ by $[-100, 100]$

**10.** $9x^4 + 35x^2 - 4 = 0$; $[-3, 3]$ by $[-10, 100]$

**11.** $x^3 - x^2 - 64x + 64 = 0$; $[-10, 10]$ by $[-300, 300]$

**12.** $x^3 + 6x^2 - 100x - 600 = 0$; $[-15, 15]$ by $[-1000, 300]$

**13.** $-2x^3 - x^2 + 3x = 0$; $[-4, 4]$ by $[-10, 10]$

**14.** $-5x^3 + 13x^2 + 6x = 0$; $[-4, 4]$ by $[-2, 30]$

**15.** $x^3 + x^2 - 7x - 7 = 0$; $[-10, 10]$ by $[-20, 20]$

**16.** $x^3 + 3x^2 - 19x - 57 = 0$; $[-10, 10]$ by $[-100, 50]$

**17.** $3x^3 + x^2 - 6x = 0$; $[-4, 4]$ by $[-10, 10]$

**18.** $-4x^3 - x^2 + 4x = 0$; $[-4, 4]$ by $[-10, 10]$

**19.** $3x^3 + 3x^2 + 3x = 0$; $[-5, 5]$ by $[-5, 5]$

**20.** $2x^3 + 2x^2 + 12x = 0$; $[-10, 10]$ by $[-20, 20]$

**21.** $x^4 + 17x^2 + 16 = 0$; $[-4, 4]$ by $[-10, 40]$

**22.** $36x^4 + 85x^2 + 9 = 0$; $[-4, 4]$ by $[-10, 40]$

**23.** $x^6 + 19x^3 - 216 = 0$; $[-4, 4]$ by $[-350, 200]$

**24.** $8x^6 + 7x^3 - 1 = 0$; $[-4, 4]$ by $[-5, 100]$

### RELATING CONCEPTS

**For individual or group investigation (Exercises 25–28)**

*Use the figure at the right to work Exercises 25–28 in order.*

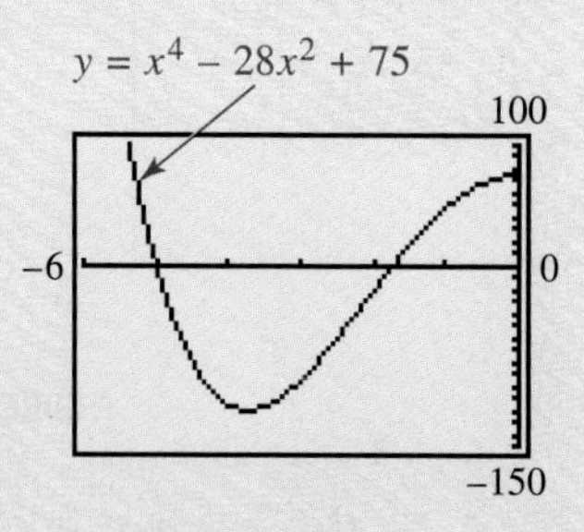


**25.** Based on the discussion of comprehensive graphs of polynomial functions in Section 3.5, is this a comprehensive graph of the function? Why or why not?

**26.** If this function is graphed over the domain of all real numbers, what symmetry is exhibited?

**27.** The $x$-intercepts shown in the window are $-5$ and $-\sqrt{3}$. What is the *complete* solution set of

$$x^4 - 28x^2 + 75 = 0?$$

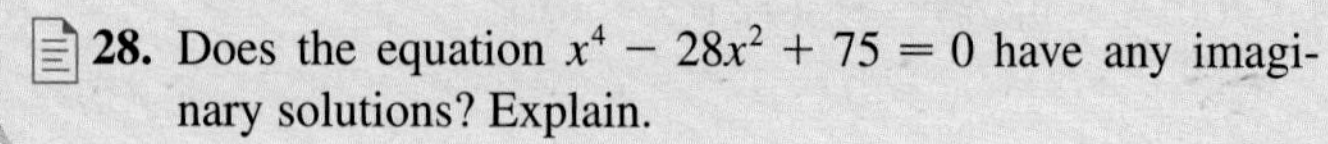

**28.** Does the equation $x^4 - 28x^2 + 75 = 0$ have any imaginary solutions? Explain.

*In Exercises 29–34, a polynomial $P(x)$ is given in expanded and factored forms. Graph the polynomial by hand, as shown in Section 3.7, and solve each equation and inequality.*

**29.** $P(x) = x^3 - 3x^2 - 6x + 8$
$= (x - 4)(x - 1)(x + 2)$

**(a)** $P(x) = 0$ **(b)** $P(x) < 0$ **(c)** $P(x) > 0$

**30.** $P(x) = x^3 + 4x^2 - 11x - 30$
$= (x - 3)(x + 2)(x + 5)$

**(a)** $P(x) = 0$ **(b)** $P(x) < 0$ **(c)** $P(x) > 0$

**31.** $P(x) = 2x^4 - 9x^3 - 5x^2 + 57x - 45$
$= (x - 3)^2(2x + 5)(x - 1)$

**(a)** $P(x) = 0$ **(b)** $P(x) < 0$ **(c)** $P(x) > 0$

**32.** $P(x) = 4x^4 + 27x^3 - 42x^2 - 445x - 300$
$= (x + 5)^2(4x + 3)(x - 4)$

**(a)** $P(x) = 0$ **(b)** $P(x) < 0$ **(c)** $P(x) > 0$

**33.** $P(x) = -x^4 - 4x^3 + 3x^2 + 18x$
$= x(2 - x)(x + 3)^2$

**(a)** $P(x) = 0$ **(b)** $P(x) \geq 0$ **(c)** $P(x) \leq 0$

**34.** $P(x) = -x^4 + 2x^3 + 8x^2$
$= x^2(4 - x)(x + 2)$

**(a)** $P(x) = 0$ **(b)** $P(x) \geq 0$ **(c)** $P(x) \leq 0$

*Use a graphical method to find all* real *solutions of each equation. Express solutions to the nearest hundredth.*

**35.** $.86x^3 - 5.24x^2 + 3.55x + 7.84 = 0$

**36.** $-2.47x^3 - 6.58x^2 - 3.33x + .14 = 0$

**37.** $-\sqrt{7}x^3 + \sqrt{5}x^2 + \sqrt{17} = 0$

**38.** $\sqrt{10}x^3 - \sqrt{11}x - \sqrt{8} = 0$

**39.** $2.45x^4 - 3.22x^3 = -.47x^2 + 6.54x + 3$

**40.** $\sqrt{17}x^4 - \sqrt{22}x^2 = -1$

*Find all n complex solutions of each equation of the form $x^n = k$.*

**41.** $x^2 = -1$ **42.** $x^2 = -4$ **43.** $x^3 = -1$ **44.** $x^3 = -8$

**45.** $x^3 = 27$ **46.** $x^3 = 64$ **47.** $x^4 = 16$ **48.** $x^4 = 81$

*(Modeling) Solve each problem.*

**49.** *Floating Ball* The polynomial

$$f(x) = \frac{\pi}{3}x^3 - 5\pi x^2 + \frac{500\pi d}{3}$$

can be used to find the depth that a ball, 10 centimeters in diameter, sinks in water. The constant $d$ is the density of the ball, where the density of water is 1. The smallest *positive* zero of $f(x)$ equals the depth that the ball sinks. Approximate this depth for each material and interpret the results.

**(a)** A wooden ball with $d = .8$

**(b)** A solid aluminum ball with $d = 2.7$

**(c)** A spherical water balloon with $d = 1$

**50.** *Concentration of Toxin* A survey team measures the concentration (in parts per million) of a particular toxin in a local river. On a normal day, the concentration of the toxin at time $x$ (in hours) after the factory upstream dumps its waste is given by

$$g(x) = -.006x^4 + .14x^3 - .05x^2 + .02x,$$

where $0 \leq x \leq 24$.

**(a)** Graph $y = g(x)$ in the window $[0, 24]$ by $[0, 200]$.

**(b)** Estimate the time at which the concentration is greatest.

**(c)** A concentration greater than 100 parts per million is considered polluted. Using the graph from part (a), estimate the period during which the river is polluted.

**51.** *Floating Ball* Refer to Exercise 49. If a ball has a 20-centimeter diameter, then

$$f(x) = \frac{\pi}{3}x^3 - 10\pi x^2 + \frac{4000\pi d}{3}$$

can be used to determine the depth that it sinks in water. Find the depth that this size ball sinks when $d = .6$.

**52.** *Floating Ball* Refer to Exercise 49. Determine the depth that a pine ball with a 10-centimeter diameter sinks in water, if $d = .55$.

**53.** *Volume of a Box* A rectangular piece of cardboard measuring 12 inches by 18 inches is to be made into a box with an open top by cutting equal size squares from each corner and folding up the sides. Let $x$ represent the length of a side of each such square in inches.

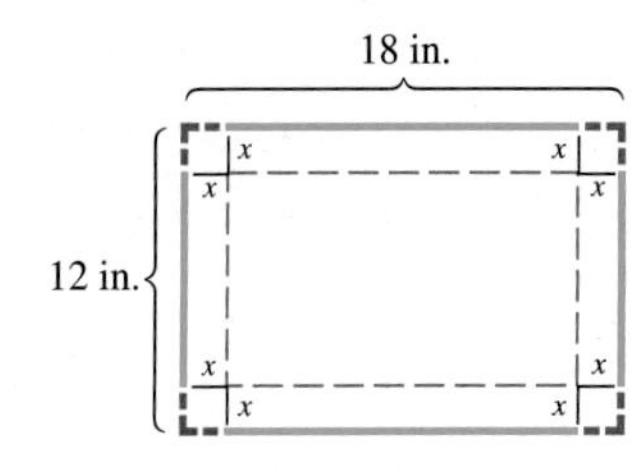


**(a)** Give the restrictions on $x$.

**(b)** Determine a function $V$ that gives the volume of the box as a function of $x$.

**(c)** For what value of $x$ will the volume be a maximum? What is this maximum volume?

**(d)** For what values of $x$ will the volume be greater than 80 cubic inches?

**54.** *Construction of a Rain Gutter* A rectangular piece of sheet metal is 20 inches wide. It is to be made into a rain gutter by turning up the edges to form parallel sides. Let $x$ represent the length of each of the parallel sides in inches. See the figure on the next page.

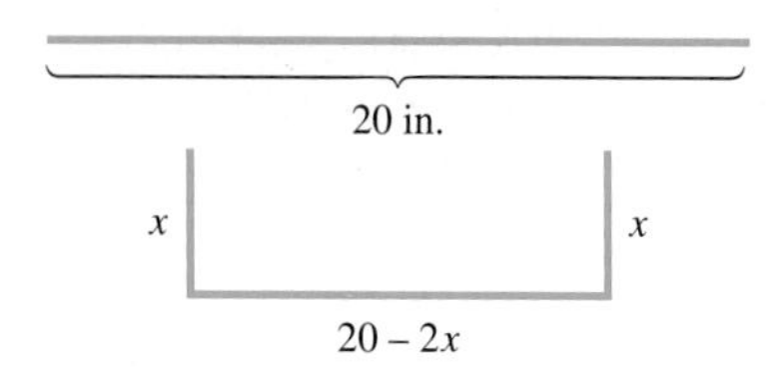


**(a)** Give the restrictions on $x$.

**(b)** Determine a function $A$ that gives the area of a cross section of the gutter.

**(c)** For what value of $x$ will $A$ be a maximum (and thus maximize the amount of water that the gutter will hold)? What is this maximum area?

**(d)** For what values of $x$ will the area of a cross section be less than 40 square inches?

**55.** *Buoyancy of a Spherical Object* It has been determined that a spherical object of radius 4 inches with specific gravity .25 will sink in water to a depth of $x$ inches, where $x$ is the smallest positive root of the equation

$$x^3 - 12x^2 + 64 = 0.$$

To what depth will this object sink given that $x < 10$?

**56.** *Area of a Rectangle* Find the value of $x$ in the figure that will maximize the area of rectangle $ABCD$.

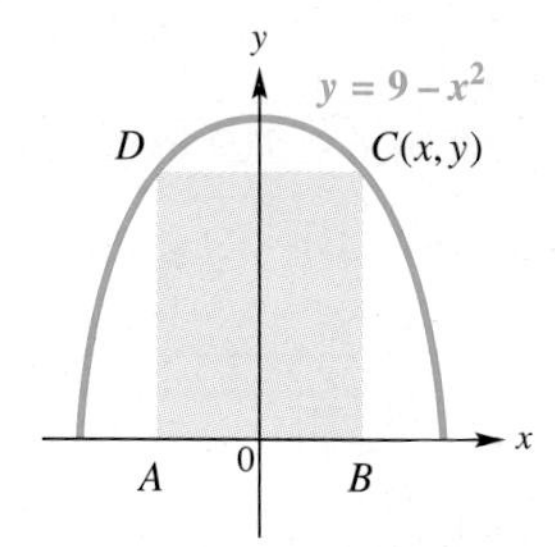


**57.** *Sides of a Right Triangle* A certain right triangle has area 84 square inches. One leg of the triangle measures 1 inch less than the hypotenuse. Let $x$ represent the length of the hypotenuse.

**(a)** Express the length of the leg described above in terms of $x$.

**(b)** Express the length of the other leg in terms of $x$.

**(c)** Write an equation based on the information determined thus far. Square each side, and then write the equation with one side as a polynomial with integer coefficients, in descending powers, and the other side equal to 0.

**(d)** Solve the equation in part (c) graphically. Find the lengths of the three sides of the triangle.

**58.** *Butane Gas Storage* A storage tank for butane gas is to be built in the shape of a right circular cylinder having altitude 12 feet, with a half sphere attached to each end. If $x$ represents the radius of each half sphere, what radius should be used to cause the volume of the tank to be $144\pi$ cubic feet?

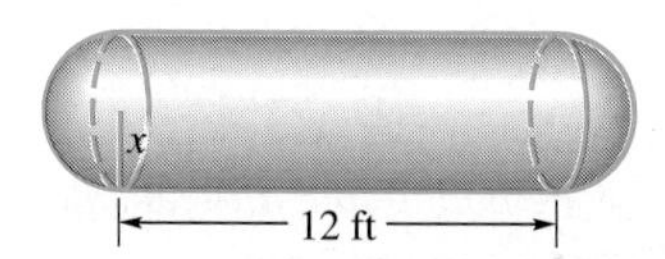


**59.** *Volume of a Box* A standard piece of notebook paper measuring 8.5 inches by 11 inches is to be made into a box with an open top by cutting equal size squares from each corner and folding up the sides. Let $x$ represent the length of a side of each such square in inches.

**(a)** Use the table feature of your graphing calculator to find the maximum volume of the box.

**(b)** Use the table feature to determine when the volume of the box will be greater than 40 cubic inches.

**60.** *Highway Design* To allow enough distance for cars to pass on two-lane highways, engineers calculate minimum sight distances between curves and hills. The table shows the minimum sight distance $y$ in feet for a car traveling at $x$ mph.

| x (in mph) | 20 | 30 | 40 | 50 |
|---|---|---|---|---|
| y (in feet) | 810 | 1090 | 1480 | 1840 |

| x (in mph) | 60 | 65 | 70 |
|---|---|---|---|
| y (in feet) | 2140 | 2310 | 2490 |

*Source:* Haefner, L., *Introduction to Transportation Systems*, Holt, Rinehart and Winston, 1986.

**(a)** Make a scatter diagram of the data.

**(b)** Use the regression feature of a calculator to find the best-fitting linear function for the data. Graph the function with the data.

**(c)** Repeat part (b) for a cubic function.

**(d)** Estimate the minimum sight distance for a car traveling 43 mph using both functions from parts (b) and (c).

**(e)** Which function fits the data better? Why?

**61.** *Water Pollution* In one study, freshwater mussels were used to monitor copper discharge into a river from an electroplating works. Copper in high doses can be lethal to aquatic life. The table lists copper concentrations in mussels after 45 days at various distances downstream from the plant. The concentration $C$ is measured in micrograms of copper per gram of mussel $x$ kilometers downstream.

| $x$ | 5 | 21 | 37 | 53 | 59 |
|---|---|---|---|---|---|
| $C$ | 20 | 13 | 9 | 6 | 5 |

*Sources:* Foster, R., and J. Bates, "Use of mussels to monitor point source industrial discharges," *Eviron. Sci. Technol.*; Mason, C., *Biology of Freshwater Pollution*, John Wiley and Sons, 1991.

**(a)** Make a scatter diagram of the data.

**(b)** Use the regression feature of a calculator to find the best-fitting quadratic function for the data. Graph the function with the data.

**(c)** Repeat part (b) for a cubic function.

**(d)** By comparing graphs of the functions in parts (b) and (c) with the data, decide which function best fits the given data.

**(e)** Concentrations above 10 are lethal to mussels. Find the values of $x$ for which this is the case.

**62.** *Temperature* The cubic polynomial defined by

$$f(x) = -.184x^3 + 1.45x^2 + 10.7x - 27.9$$

models monthly average temperature at Trout Lake, Canada, in degrees Fahrenheit, where $x = 1$ corresponds to January and $x = 12$ represents December. Interpret the zeros of $f$.

**63.** *Cardiac Output* A technique for measuring cardiac output depends on the concentration of a dye in the bloodstream after a known amount is injected into a vein near the heart. For a normal heart, the concentration of dye in the bloodstream at time $x$ (in seconds) is modeled by the function defined by

$$g(x) = -.006x^4 + .140x^3 - .053x^2 + 1.79x.$$

When does the concentration reach 10 units?

**64.** Consider the equation $x^8 = 1$. This equation has eight distinct solutions, each of which is an eighth root of 1. (Roots of 1 are often called *roots of unity.*)

**(a)** Graph $y = x^8 - 1$ to determine the two real eighth roots of unity.

**(b)** Show analytically that $i$ is also an eighth root of unity.

**(c)** Based on the conjugate zeros theorem and the result of part (b), what must be another eighth root of unity? Verify analytically that this is a root.

**(d)** Using concepts from trigonometry, it can be shown that

$$\frac{\sqrt{2}}{2} + \frac{\sqrt{2}}{2}i \quad \text{and} \quad -\frac{\sqrt{2}}{2} + \frac{\sqrt{2}}{2}i$$

are imaginary eighth roots of unity. Based on the conjugate zeros theorem, what two other imaginary numbers must also be eighth roots of unity?

**(e)** List all eight eighth roots of unity.

## Reviewing Basic Concepts (Sections 3.6–3.8)

*For Exercises 1–3, use* $P(x) = 2x^4 - 7x^3 + 29x - 30$.

**1.** Find $P(3)$.

**2.** Is 2 a zero?

**3.** One zero of $P$ is $2 + i$. Factor $P(x)$ into linear factors.

**4.** Find a polynomial $P(x)$ with real coefficients that satisfies these conditions: zeros of $\frac{3}{2}$ and $i$, $P(3) = 15$.

**5.** Find a polynomial $P(x)$ of lowest degree with real coefficients and zeros $-4$ (multiplicity 2) and $1 + 2i$.

**6.** Given $P(x) = 2x^3 + x^2 - 11x - 10$, list all zeros of $P$.

**7.** Find all complex solutions of $3x^4 - 12x^2 + 1 = 0$.

**8.** *(Modeling) Bank Debit Cards* The table shows the number of bank debit cards issued in millions.

| Year | 1990 | 1995 | 1998 | 2000* |
|---|---|---|---|---|
| Cards (in millions) | 9 | 40 | 100 | 126 |

*Projected
*Source: Statistical Abstract of the United States*, 2000.

Let $x = 0$ correspond to 1990, $x = 5$ correspond to 1995, and so on. Determine a cubic model for the data using the regression feature of your calculator. Use the model to estimate $y$ when $x = 9$, and interpret the result.

# CHAPTER 3 SUMMARY

**KEY TERMS & SYMBOLS** | **KEY CONCEPTS**

## 3.1 Complex Numbers

complex number system
complex number
imaginary number
standard form of a complex number
real part
imaginary part
complex conjugates (conjugates)

**IMAGINARY NUMBER $i$**

$$i = \sqrt{-i} \quad \text{or} \quad i^2 = -1$$

**COMPLEX NUMBER**

A number in the form $a + bi$, where $a$ and $b$ are real numbers and $i$ is the imaginary unit, is called a complex number.

$\sqrt{-a}$

If $a > 0$, then $\sqrt{-a} = i\sqrt{a}$.

**COMPLEX CONJUGATE**

The complex conjugate of $a + bi$ is $a - bi$.

**OPERATIONS WITH COMPLEX NUMBERS**

Addition: $(a + bi) + (c + di) = (a + c) + (b + d)i$
Subtraction: $(a + bi) - (c + di) = (a - c) + (b - d)i$
Multiplication: $(a + bi)(c + di) = (ac - bd) + (ad + bc)i$
Division: To divide complex numbers, multiply both the numerator and the denominator by the conjugate of the denominator.

## 3.2 Quadratic Functions and Graphs

quadratic function
axis of symmetry
extreme points (extrema, extremum)

**GRAPH OF $P(x) = a(x - h)^2 + k$**

The graph of $P(x) = a(x - h)^2 + k$ $(a \neq 0)$

**(a)** is a parabola with vertex $(h, k)$ and the vertical line $x = h$ as axis of symmetry;
**(b)** opens upward if $a > 0$ and downward if $a < 0$;
**(c)** is wider than the graph of $y = x^2$ if $0 < |a| < 1$ and narrower than the graph of $y = x^2$ if $|a| > 1$.

**VERTEX FORMULA**

The vertex of the graph of $P(x) = ax^2 + bx + c$ $(a \neq 0)$ is the point $\left(-\frac{b}{2a}, P\left(-\frac{b}{2a}\right)\right)$.

**EXTREME VALUES**

For the quadratic function defined by $P(x) = ax^2 + bx + c$,

**(a)** if $a > 0$, the vertex $(h, k)$ is called the *minimum point* of the graph. The *minimum value* of the function is $P(h) = k$.
**(b)** if $a < 0$, the vertex $(h, k)$ is called the *maximum point* of the graph. The *maximum value* of the function is $P(h) = k$.

## 3.3 Quadratic Equations and Inequalities

quadratic equation in one variable

**ZERO-PRODUCT PROPERTY**

If $a$ and $b$ are complex numbers and $ab = 0$, then $a = 0$ or $b = 0$ or both.

**SQUARE ROOT PROPERTY**

The solution set of $x^2 = k$ is

**(a)** $\{\pm\sqrt{k}\}$ if $k > 0$. **(b)** $\{0\}$ if $k = 0$. **(c)** $\{\pm i\sqrt{|k|}\}$ if $k < 0$.

| KEY TERMS & SYMBOLS | KEY CONCEPTS |
|---|---|

quadratic equation in standard form
discriminant

**QUADRATIC FORMULA**

The solutions of the quadratic equation $ax^2 + bx + c = 0$, where $a \neq 0$, are

$$x = \frac{-b \pm \sqrt{b^2 - 4ac}}{2a}.$$

**EFFECT OF THE DISCRIMINANT**

If $a$, $b$, and $c$ are real numbers, $a \neq 0$, then the complex solutions of $ax^2 + bx + c = 0$ are described as follows, based on the value of the discriminant, $b^2 - 4ac$.

| Value of $b^2 - 4ac$ | Number of Solutions | Type of Solutions |
|---|---|---|
| Positive | Two | Real |
| Zero | One (a double solution) | Real |
| Negative | Two | Imaginary |

**SOLVING A QUADRATIC INEQUALITY**

1. Solve the corresponding quadratic equation.
2. Identify the intervals determined by the solutions of the equation.
3. Use a sign graph to determine which intervals are in the solution set.
4. Decide whether the endpoints are included or excluded.

## 3.4 Further Applications of Quadratic Functions and Models

**QUADRATIC MODEL**

It is possible that a quadratic function can be used to model data. A model can be determined by choosing an appropriate data point $(h, k)$ as vertex, and using another point to find the value of $a$ in $f(x) = a(x - h)^2 + k$.

## 3.5 Higher-Degree Polynomial Functions and Graphs

polynomial function
cubic function
quartic function
turning points
local maximum point
local minimum point
local maximum
local minimum
absolute maximum (minimum) point
absolute maximum (minimum) value
$x \to -\infty$
$x \to \infty$

**EXTREMA OF POLYNOMIAL FUNCTIONS**

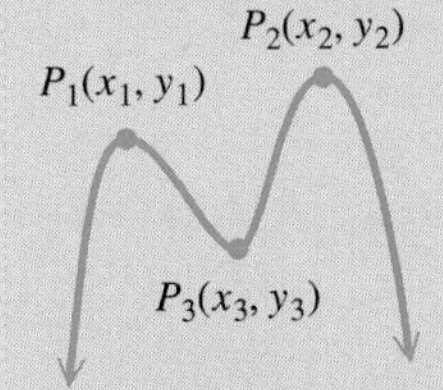


$P_1$ is a local maximum point.
$P_2$ is a local and absolute maximum point.
$P_3$ is a local minimum point.

**NUMBER OF LOCAL EXTREMA**

The number of local extrema of the graph of a polynomial function of degree $n$ is at most $n - 1$.

**END BEHAVIOR OF POLYNOMIAL FUNCTIONS**

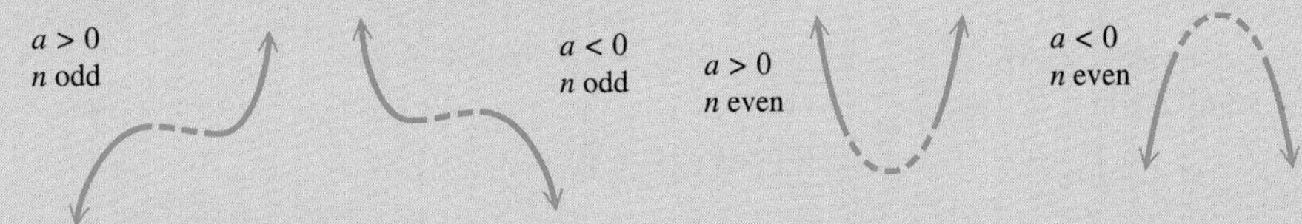


**NUMBER OF $x$-INTERCEPTS (REAL ZEROS) OF A POLYNOMIAL FUNCTION**

A polynomial function of degree $n$ will have at most $n$ $x$-intercepts (real zeros).

| KEY TERMS & SYMBOLS | KEY CONCEPTS |
|---|---|

## 3.6 Topics in the Theory of Polynomial Functions (I)

synthetic division

**INTERMEDIATE VALUE THEOREM**

If $P(x)$ is a polynomial with only real coefficients, and if for real numbers $a$ and $b$, the values $P(a)$ and $P(b)$ are opposite in sign, then there exists at least one real zero between $a$ and $b$.

**DIVISION OF A POLYNOMIAL BY $x - k$**

1. If the degree $n$ polynomial $P(x)$ is divided by $x - k$, the quotient polynomial, $Q(x)$, has degree $n - 1$.
2. The remainder $R$ is a constant (and may be 0). The complete quotient for $\frac{P(x)}{x-k}$ may be written as

$$\frac{P(x)}{x-k} = Q(x) + \frac{R}{x-k}.$$

**REMAINDER THEOREM**

If a polynomial $P(x)$ is divided by $x - k$, the remainder is equal to $P(k)$.

**FACTOR THEOREM**

The polynomial $x - k$ is a factor of the polynomial $P(x)$ if and only if $P(k) = 0$.

## 3.7 Topics in the Theory of Polynomial Functions (II)

multiplicity of a zero

**CONJUGATE ZEROS THEOREM**

If $P(x)$ is a polynomial having only real coefficients, and if $a + bi$ is a zero of $P$, then the conjugate $a - bi$ is also a zero of $P$.

**FUNDAMENTAL THEOREM OF ALGEBRA**

Every function defined by a polynomial of degree 1 or more has at least one complex (real or imaginary) zero.

**NUMBER OF ZEROS THEOREM**

1. A function defined by a polynomial of degree $n$ has at most $n$ distinct complex zeros.
2. A function defined by a polynomial of degree $n$ has exactly $n$ complex zeros if zeros of multiplicity $m$ are counted $m$ times.

**BEHAVIOR OF THE GRAPH OF A POLYNOMIAL FUNCTION NEAR THE $x$-AXIS**

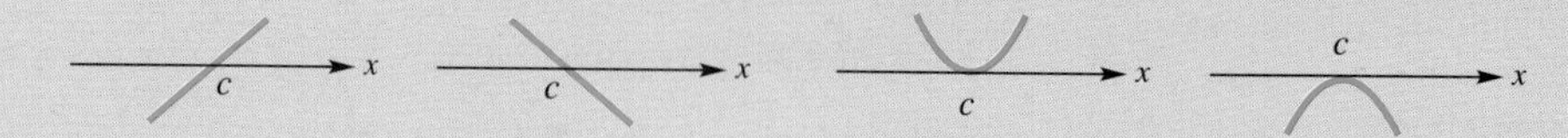


The graph crosses the $x$-axis at $(c, 0)$ if $c$ is a zero of odd multiplicity.

The graph is tangent to the $x$-axis at $(c, 0)$ if $c$ is a zero of even multiplicity.

**RATIONAL ZEROS THEOREM**

Let $P(x) = a_nx^n + a_{n-1}x^{n-1} + \cdots + a_1x + a_0$, where $a_n \neq 0$, define a polynomial function with integer coefficients. If $\frac{p}{q}$ is a rational number written in lowest terms, and if $\frac{p}{q}$ is a zero of $P$, then $p$ is a factor of the constant term $a_0$, and $q$ is a factor of the leading coefficient $a_n$.

| KEY TERMS & SYMBOLS | KEY CONCEPTS |
|---|---|
| **3.8** Polynomial Equations and Inequalities; Further Applications and Models | **COMPLEX *n*TH ROOTS THEOREM** If $n$ is a positive integer and $k$ is a nonzero complex number, then the equation $x^n = k$ has *exactly* $n$ complex roots. |
| quadratic in form | **POLYNOMIAL MODELS** Graphing calculators can be used to find the best-fitting cubic and quartic models for a set of data points. |

## CHAPTER 3 REVIEW EXERCISES

*Let $w = 17 - i$ and let $z = 1 - 3i$. Write each complex number in standard form.*

**1.** $w + z$ **2.** $w - z$ **3.** $wz$

**4.** $w^2$ **5.** $\frac{1}{z}$ **6.** $\frac{w}{z}$

*Consider the function defined by $P(x) = 2x^2 - 6x - 8$ for Exercises 7–16.*

**7.** What is the domain of $P$?

**8.** Determine analytically the coordinates of the vertex of the graph.

**9.** Use an end-behavior diagram to describe the end behavior of the graph of $P$.

**10.** Determine analytically the $x$-intercepts, if any, of the graph of $P$.

**11.** Determine analytically the $y$-intercept of the graph of $P$.

**12.** What is the range of $P$?

**13.** Over what interval is the function increasing? Over what interval is it decreasing?

**14.** Give the solution set of each equation or inequality.

**(a)** $2x^2 - 6x - 8 = 0$ **(b)** $2x^2 - 6x - 8 > 0$
**(c)** $2x^2 - 6x - 8 \leq 0$

**15.** The graph of $P$ is shown here. Explain how the graph supports your solution sets in Exercise 14.

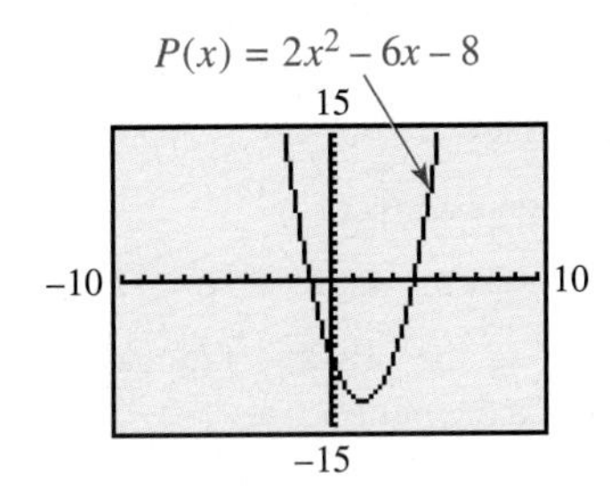


**16.** What is the equation of the axis of symmetry of the graph in Exercise 15?

*Consider the function defined by*

$$P(x) = -2.64x^2 + 5.47x + 3.54$$

*for Exercises 17–20.*

**17.** Use the discriminant to explain how you can determine the number of $x$-intercepts the graph of $P$ will have even before graphing it on your calculator.

**18.** Graph the function in the standard window of your calculator, and use the root-finding capabilities to solve the equation $P(x) = 0$. Express solutions as approximations to the nearest hundredth.

**19.** Use the capabilities of your calculator to find the coordinates of the vertex of the graph. Express coordinates to the nearest hundredth.

**20.** Verify analytically that your answer in Exercise 19 is correct.

**21.** The figure shows the graph of a quadratic function $y = f(x)$.

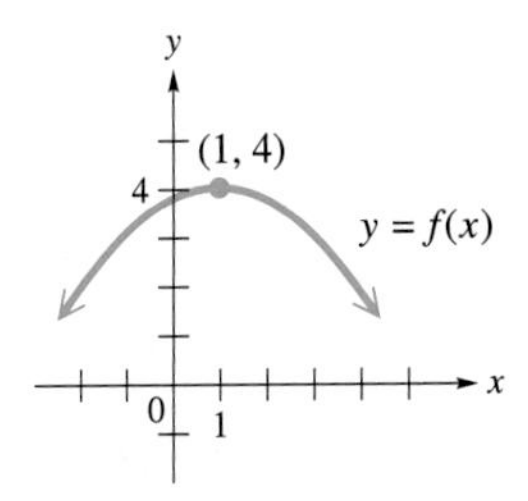


**(a)** What is the maximum value of $f(x)$?

**(b)** For what value of $x$ is $f(x)$ a maximum?

**(c)** How many real solutions are there to the equation $f(x) = 2$?

**(d)** How many real solutions are there to the equation $f(x) = 6$?

*(Modeling) Height of a Projectile* *A projectile is fired vertically upward, and its height $s(t)$ in feet after t seconds is given by the function defined by*

$$s(t) = -16t^2 + 800t + 600.$$

*Graph this function in the window $[0, 60]$ by $[-1000, 11{,}000]$, and use either analytic or graphical methods to answer Exercises 22–26.*

**22.** From what height was the projectile fired?

**23.** After how many seconds will it reach its maximum height?

**24.** What is the maximum height it will reach?

**25.** Between what two times (in seconds, to the nearest tenth) will it be more than 5000 feet above the ground?

**26.** How long will the projectile be in the air? Give your answer to the nearest tenth of a second.

*(Modeling)* *Solve each problem involving modeling with a quadratic function.*

**27.** *Volume of a Box* A piece of cardboard is 3 times as long as it is wide. Equal size squares measuring 4 inches on each side are to be cut from the corners of the piece of cardboard, and the flaps will be folded up to form a box with an open top.

**(a)** Determine a function $V$ that describes the volume of the box as a function of $x$, where $x$ is the original width in inches.

**(b)** What should be the original dimensions of the piece of cardboard if the box is to have a volume of 2496 cubic inches? Solve this problem analytically.

**(c)** Support the answer in part (b) graphically.

**28.** *Concentration of Atmospheric $CO_2$* The International Panel on Climate Change (IPCC) in 1990 published that if current trends of burning fossil fuel and deforestation continue, then future amounts of atmospheric carbon dioxide in parts per million (ppm) will increase, as shown in the table.

| Year | Carbon Dioxide |
|---|---|
| 1990 | 353 |
| 2000 | 375 |
| 2075 | 590 |
| 2175 | 1090 |
| 2275 | 2000 |

*Source:* IPCC.

**(a)** Plot the data. Let $x = 0$ correspond to 1990.

**(b)** Find a function of the form $f(x) = a(x - h)^2 + k$ that models the data. Use $(0, 353)$ as the vertex and $(285, 2000)$ as another point to determine $a$.

**(c)** Use the regression feature of a graphing calculator to find the best-fitting quadratic function $g$ for this data.

*Use synthetic division to find the quotient $Q(x)$ and the remainder R.*

**29.** $\dfrac{x^3 + x^2 - 11x - 10}{x - 3}$

**30.** $\dfrac{3x^3 + 8x^2 + 5x + 10}{x + 2}$

*Use synthetic division to find $P(2)$.*

**31.** $P(x) = -x^3 + 5x^2 - 7x + 1$

**32.** $P(x) = 2x^3 - 3x^2 + 7x - 12$

**33.** $P(x) = 5x^4 - 12x^2 + 2x - 8$

**34.** $P(x) = x^5 + 4x^2 - 2x - 4$

**35.** If $P(x)$ is defined by a polynomial with real coefficients, and $7 + 2i$ is a zero of the function, what other complex number must also be a zero?

*Find a polynomial function with real coefficients and of lowest degree having the given zeros.*

**36.** $-1, 4, 7$

**37.** $8, 2, 3$

**38.** $\sqrt{3}, -\sqrt{3}, 2, 3$

**39.** $-2 + \sqrt{5}, -2 - \sqrt{5}, -2, 1$

**40.** Is $-1$ a zero of $P(x) = 2x^4 + x^3 - 4x^2 + 3x + 1$?

**41.** Is $x + 1$ a factor of $P(x) = x^3 + 2x^2 + 3x + 2$?

**42.** Find a polynomial function $P$ with real coefficients of degree 4 with 3, 1, and $-1 - 3i$ as zeros, and $P(2) = -36$.

**43.** Give an example of a fourth-degree polynomial function having exactly two distinct real zeros, and then graph it, using a calculator.

**44.** Give an example of a cubic polynomial function having exactly one real zero, and then graph it, using a calculator.

**45.** Find all zeros of $P(x) = x^4 - 3x^3 - 8x^2 + 22x - 24$, given that $1 - i$ is a zero.

**46.** Find all zeros of $P(x) = 2x^4 - x^3 + 7x^2 - 4x - 4$, given that 1 and $2i$ are zeros.

**47.** Find all rational zeros of

$$P(x) = 3x^5 - 4x^4 - 26x^3 - 21x^2 - 14x + 8$$

by first listing the possible rational zeros based on the rational zeros theorem.

*Consider the function defined by*

$$P(x) = x^3 - 2x^2 - 4x + 3$$

*for Exercises 48–52.*

**48.** Graph the function in an appropriate window to get a comprehensive graph. Based on the graph, how many real solutions does the equation $x^3 - 2x^2 - 4x + 3 = 0$ have? Use your calculator to find the real root that is an integer.

**49.** Use your answer in Exercise 48 along with synthetic division to factor $x^3 - 2x^2 - 4x + 3$ so that one factor is linear and the other factor is quadratic.

**50.** Find the exact values of any remaining zeros of $P$ analytically.

**51.** Use your calculator to support your answer in Exercise 50.

**52.** Give the solution set of each inequality, using exact values.

**(a)** $x^3 - 2x^2 - 4x + 3 > 0$

**(b)** $x^3 - 2x^2 - 4x + 3 \leq 0$

**53.** Use an analytic method to find all solutions of the equation $x^3 + 2x^2 + 5x = 0$. Then, without graphing, give the exact values of all $x$-intercepts of the graph.

**54.** The graph of $P(x) = x^4 - 5x^3 + x^2 + 21x - 18$ is shown below. Suppose that you know that all zeros of $P$ are integers and each zero has multiplicity 1 or 2. Give the factored form of $P(x)$. (*Hint:* Once you have determined your answer, graph the factored form to see if it matches the graph shown here.)

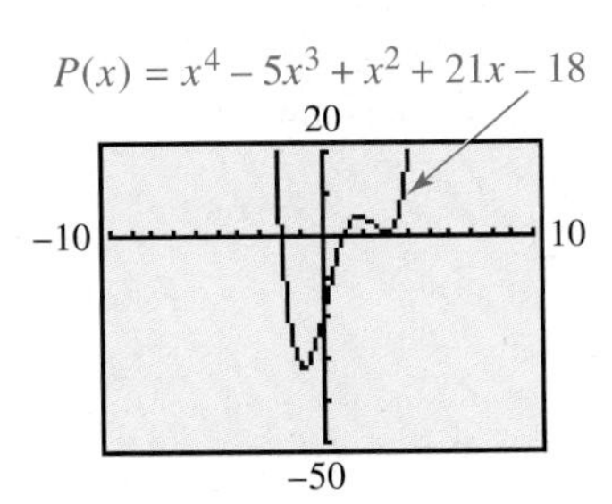


The $x$-intercepts are −2, 1, and 3.

*Concept Check* *Comprehensive graphs of polynomial functions f and g are shown here. They have only real coefficients. Answer Exercises 55–63 based on the graphs.*

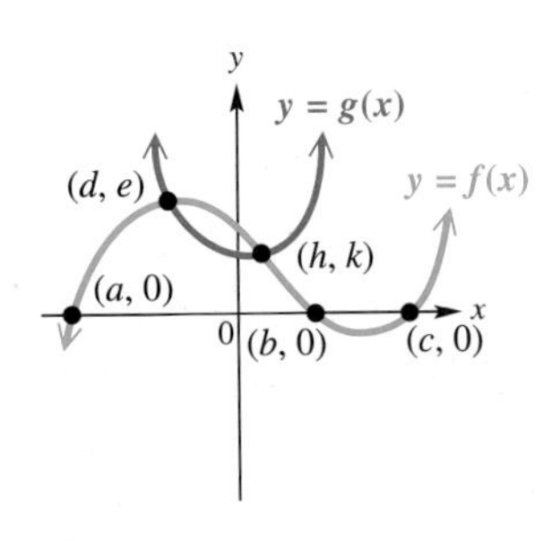


**55.** Is the degree of $g$ even or odd?

**56.** Is the degree of $f$ even or odd?

**57.** Is the leading coefficient of $f$ positive or negative?

**58.** How many real solutions does $g(x) = 0$ have?

**59.** Express the solution set of $f(x) < 0$ in interval form.

**60.** What is the solution set of $f(x) > g(x)$?

**61.** What is the solution set of $f(x) - g(x) = 0$?

**62.** If $r + pi$ is an imaginary solution of $g(x) = 0$, what must be another imaginary solution?

**63.** Suppose that $f$ is of degree 3. Explain why $f$ cannot have imaginary zeros.

**64.** The table shows several ordered pairs that lie on the graph of $Y_1$. Use one of the ordered pairs to find a factor of $Y_1$. Then, find the exact values of all zeros of $Y_1$, and write it as a product of three linear factors.

| X | Y1 |
|---|---|
| -6 | -134 |
| -5 | -51 |
| -4 | 0 |
| -3 | 25 |
| -2 | 30 |
| -1 | 21 |
| 0 | 4 |

Y1=X³−X²−19X+4

**65.** Use the intermediate value theorem to determine which two $x$-values from the table have a zero between them. Explain.

| X | Y1 |
|---|---|
| .6 | .784 |
| .7 | .157 |
| .8 | -.512 |
| .9 | -1.229 |
| 1 | -2 |
| 1.1 | -2.831 |
| 1.2 | -3.728 |

Y1=-X³−5X+4

*Concept Check* *Answer* true *or* false *to each statement in Exercises 66–71.*

**66.** The function defined by $f(x) = 3x^7 - 8x^6 + 9x^5 + 12x^4 - 18x^3 + 26x^2 - x + 500$ has eight $x$-intercepts.

**67.** The function $f$ in Exercise 66 may have up to six local extrema.

**68.** The function $f$ in Exercise 66 has a positive $y$-intercept.

**69.** Based on end behavior of the function $f$ in Exercise 66 and your answer in Exercise 68, the graph must have at least one negative $x$-intercept.

**70.** If a polynomial function of even degree has a positive leading coefficient and a negative $y$-intercept, it must have at least two real zeros.

**71.** Because $-\frac{1}{2} + i\frac{\sqrt{3}}{2}$ is a complex zero of $f(x) = x^2 + x + 1$, another zero must be $\frac{1}{2} + i\frac{\sqrt{3}}{2}$.

*Graph the function defined by*

$$P(x) = -2x^5 + 15x^4 - 21x^3 - 32x^2 + 60x$$

*in the window* $[-8, 8]$ *by* $[-100, 200]$ *to obtain a comprehensive graph. Then, use your calculator and the concepts of this chapter to answer Exercises 72–76.*

**72.** How many local maxima does this function have?

**73.** One local minimum point lies on the $x$-axis and has an integer as its $x$-value. What are its coordinates?

**74.** The greatest $x$-intercept is 5. Therefore, $x - 5$ is a factor of $P(x)$. Use synthetic division to find the quotient polynomial $Q(x)$ obtained when $P(x)$ is divided by $x - 5$.

**75.** What is the range of $P$?

**76.** The graph of $P$ has a local minimum point with a negative $x$-value. Use your calculator to find its coordinates. Express them to the nearest hundredth.

**77.** Solve the equation $3x^3 + 2x^2 - 21x - 14 = 0$ analytically for all complex solutions, giving exact values in your solution set. Then, graph the left side of the equation, and support the real solutions.

**78.** Consider the polynomial function defined by

$$P(x) = -x^4 + 3x^3 + 3x^2 - 7x - 6$$
$$= (-x + 2)(x - 3)(x + 1)^2.$$

Graph the polynomial by hand as explained in Section 3.7, and solve each equation or inequality.

**(a)** $P(x) = 0$ **(b)** $P(x) > 0$ **(c)** $P(x) < 0$

*(Modeling) Solve each application of polynomial functions.*

**79.** *Dimensions of a Cube* After a 2-inch slice is cut off the top of a cube, the resulting solid has volume 32 cubic inches. Find the dimensions of the original cube.

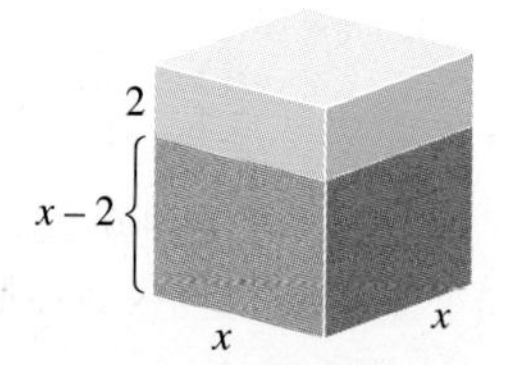


**80.** *Height of a Propelled Object* If air resistance is neglected, an object projected straight upward with an initial velocity of 40 meters per second from a height of 50 meters will be at a height $s$ meters after $t$ seconds, where

$$s(t) = -4.9t^2 + 40t + 50.$$

**(a)** What are the restrictions on $t$?

**(b)** Graph the function $s$ in a suitable window.

**(c)** Use the graph to find the time at which the object reaches its highest point.

**(d)** Use the graph to find the maximum height reached by the object.

**(e)** Write the equation that allows you to find the time at which the object reaches the ground. Then, solve the equation, using the method of your choice. Give your answer to the nearest hundredth of a second.

**81.** *Military Personnel on Active Duty* The number of military personnel on active duty in the United States during the period from 1990 through 1998 is approximated by the cubic model

$$y = .0045x^3 - .072x^2 + .19x + 3.7,$$

where $x = 0$ corresponds to 1990, and $y$ is in millions. Based on this model, how many military personnel were on active duty in 2000? (*Source:* U.S. Department of Defense.)

**82.** *Medicare Beneficiary Spending* Out-of-pocket spending projections for a typical Medicare beneficiary as a share of his or her income are given in the table.

| Year | Percent of Income |
|---|---|
| 1998 | 18.6 |
| 2000 | 19.3 |
| 2005 | 21.7 |
| 2010 | 24.7 |
| 2015 | 27.5 |
| 2020 | 28.3 |
| 2025 | 28.6 |

*Source:* Urban Institute's Analysis of 1998 Medicare Trustees' Report.

Let $x = 0$ represent 1990, so $x = 8$ represents 1998. Use a graphing calculator to do the following.

**(a)** Plot the data.

**(b)** Find a quadratic function to model the data.

**(c)** Find a cubic function to model the data.

**(d)** Graph each function in the same viewing window as the data.

**(e)** Compare the two functions. Which is a better fit for the data?

**(f)** Would a linear model be appropriate for this data? Why or why not?

# CHAPTER 3 TEST

1. Perform each operation with complex numbers. Give answers in standard form.

   (a) $(8 - 7i) - (-12 + 2i)$  (b) $\dfrac{11 + 10i}{2 + 3i}$

   (c) Simplify $i^{65}$.  (d) $2i(3 - i)^2$

2. For the quadratic function defined by $f(x) = -2x^2 - 4x + 6$, do the following.

   (a) Find the vertex analytically.

   (b) Give a comprehensive graph, and use a calculator to support your result in part (a).

   (c) Find the zeros of $f$ and support your result, using a graph for one zero and a table for the other.

   (d) Find the $y$-intercept analytically.

   (e) State the domain and range of $f$.

   (f) Give the interval over which the function is increasing and the interval over which it is decreasing.

3. (a) Solve the quadratic equation $3x^2 + 3x - 2 = 0$ analytically. Give solutions in exact form.

   (b) Graph $f(x) = 3x^2 + 3x - 2$ with a calculator. Use your results in part (a) along with this graph to give the solution set of each inequality. Express endpoints of intervals in exact form.

   (i) $f(x) < 0$  (ii) $f(x) \geq 0$

4. *Dimensions of a Box* The width of a rectangular box is 3 times its height, and its length is 11 inches more than its height. Find the dimensions of the box if its volume is 720 cubic inches.

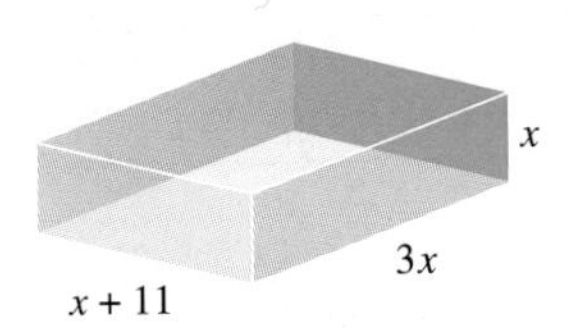


5. *(Modeling) U.S. Air Passengers* The table gives the number of air passengers (in millions) in the United States for selected years.

| Year | 1990 | 1992 | 1994 | 1996 | 1998 | 2000 |
|---|---|---|---|---|---|---|
| Passengers | 470 | 475 | 530 | 580 | 620 | 670 |

*Source:* U.S. Bureau of Transportation Statistics.

   (a) Plot the data. Let $x = 0$ correspond to the year 1990, $x = 2$ correspond to 1992, and so on.

   (b) Find a function of the form $f(x) = a(x - h)^2 + k$ that models the data. Let $(0, 470)$ represent the vertex and use $(10, 670)$ to determine the value of $a$.

   (c) Use the regression feature of your graphing calculator to find the best-fitting quadratic function $g$ for the data. Graph both functions $g$ and $f$ from part (b) in the same window as the data points. Which function is the better fit?

6. *(Modeling) Runner's Distance* A runner is working out on a straight track. The graph shows the runner's distance $y$ in hundreds of feet from the starting line after $x$ minutes.

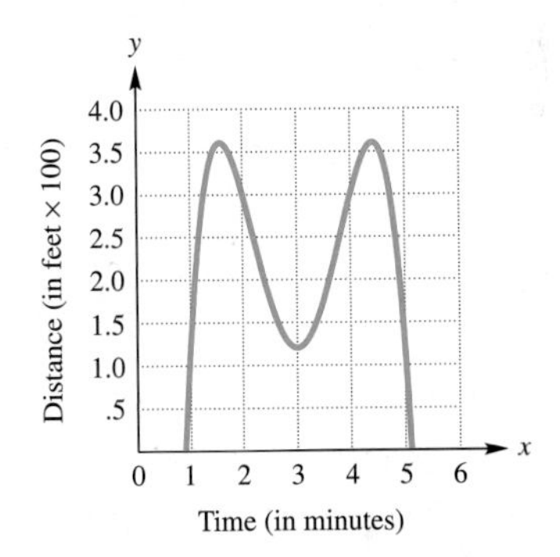


   (a) Estimate the coordinates of the turning points.

   (b) Interpret each turning point.

7. (a) Given that

$$f(x) = x^6 - 5x^5 + 3x^4 + x^3 + 40x^2 - 24x - 72$$

   has 3 as a zero of multiplicity 2, 2 as a single zero, and $-1$ as a single zero, find all other zeros of $f$.

   (b) Give an end-behavior diagram. Use the information from part (a) to sketch the graph of $f$ by hand.

8. For the function defined by

$$f(x) = 4x^4 - 21x^2 - 25,$$

   do the following.

   (a) Find all zeros analytically.

   (b) Find a comprehensive graph of $f$, and support the real zeros found in part (a).

   (c) Discuss the symmetry of the graph of $f$.

   (d) Use the graph and the results of part (a) to find the solution set of each inequality.

   (i) $f(x) \geq 0$  (ii) $f(x) < 0$

**9.** **(a)** Use only a graphical method to find the real solutions of

$$x^5 - 4x^4 + 2x^3 - 4x^2 + 6x - 1 = 0.$$

**(b)** Based on the degree and your answer in part (a), how many imaginary solutions does the equation have?

**10.** *(Modeling) U.S. Air Passengers* Refer to the data in Exercise 5. Again, letting $x = 0$ represent 1990, $x = 2$ represent 1992, and so on, do the following.

**(a)** Use a graphing calculator to find the best-fitting cubic function for the data.

**(b)** Repeat part (a) for a quartic function.

**(c)** Graph the functions of parts (a) and (b) over the data points.

**(d)** Based on the growth through the 1990s, which function gives the best estimate of the number of passengers in 2001?

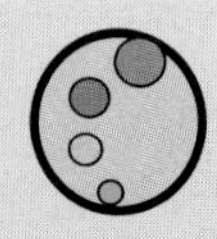

# Chapter 3 Project

## Creating a Social Security Polynomial

Each Social Security number is unique to one person. In this activity you can construct a Social Security polynomial. Let's agree that the polynomial function will be defined as follows, where $a_i$ represents the $i$th digit in the Social Security number that you construct:

$$SSN(x) = (x - a_1)(x + a_2)(x - a_3)(x + a_4)(x - a_5)(x + a_6)(x - a_7)(x + a_8)(x - a_9).$$

For example, if the Social Security number is 539-58-0954, the polynomial is

$$SSN(x) = (x - 5)(x + 3)(x - 9)(x + 5)(x - 8)(x + 0)(x - 9)(x + 5)(x - 4).$$

A comprehensive graph of this function is shown in Figure A. In Figure B, we show a screen obtained by zooming in on the positive zeros, as the comprehensive graph does not show the local behavior well in this region.

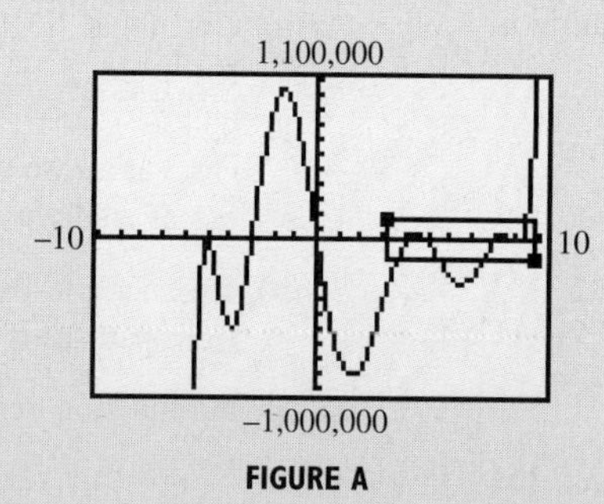


**FIGURE A**

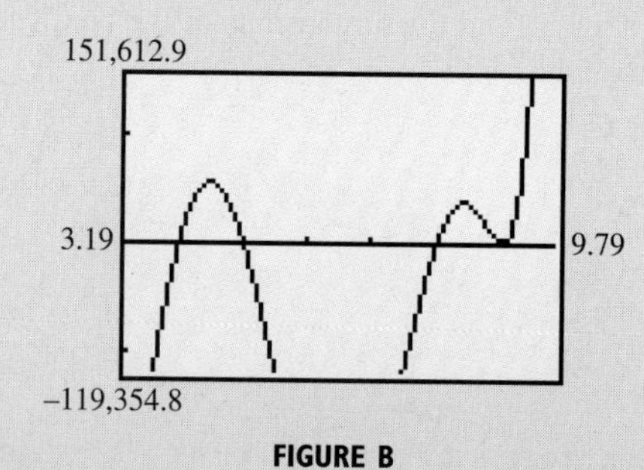


**FIGURE B**

## Activities

*(These activities can be performed by individual students, or by students in groups, where the group chooses digits for a Social Security number to be used.)*

1. Construct your Social Security polynomial $SSN(x)$.
2. What is the degree of this polynomial? What is the dominating term?
3. Construct an end-behavior diagram.
4. List all zeros of multiplicity 1.

5. List all zeros of multiplicity 2.
6. List all zeros of multiplicity 3 or higher.
7. Using your graphing calculator, find a comprehensive graph of $SSN(x)$. It should include all local extreme points, all intercepts, and illustrate end behavior. Sketch the graph on paper, or print it out, using the appropriate computer software. Regardless of which type of graph you submit, give the values of Xmin, Xmax, Xscl, Ymin, Ymax, and Yscl. (*Hint:* Ymin and Ymax will be quite large in comparison to values you might be accustomed to using.) You may need to include "enlargements" of certain parts of the graph, because they may not show enough detail in one window.
8. List the coordinates of each local maximum.
9. List the coordinates of each local minimum.
10. What are the domain and the range of $SSN(x)$?
11. List the intervals of the domain over which $SSN(x)$ is increasing.
12. List the intervals of the domain over which $SSN(x)$ is decreasing.
13. What must be true of the Social Security number that leads to a polynomial whose graph passes through the origin?

# 4 Rational, Power, and Root Functions

**TRAFFIC** congestion is one of the biggest hassles in modern life, and it's getting worse each year. Although the population of the United States has increased by about 20% since 1982, the time spent waiting in traffic has increased by 236%. Today, the average driver spends over 10 hours a year stuck in traffic. Traffic congestion costs Americans $78 billion a year due to wasted fuel and lost time from work. People who are frequently stuck in traffic have a greater risk for high blood pressure, sleep deprivation, and depression.

Traffic is subject to a *nonlinear effect.* According to Joe Sussman, an engineer from Massachusetts Institute of Technology, you can put more cars on the road up to a point. Then, if traffic intensity increases even slightly beyond this point, congestion and waiting time increase dramatically. This nonlinear effect can also occur when you are waiting to enter an amusement park or exit a parking ramp. Mathematics and rational functions play an important role in modeling this type of phenomenon. In this chapter we discuss rational functions and how they apply to real-life situations.

*Source:* Longman, Phillip J., "American Gridlock," *U.S. News & World Report,* May 28, 2001.

CHAPTER OUTLINE

# RATIONAL FUNCTIONS AND GRAPHS

The Reciprocal Function • The Rational Function Defined by $f(x) = \frac{1}{x^2}$ • Mode and Window Choices for Calculator Graphs

A rational expression is a fraction that is the quotient of two polynomials. A function defined by a rational expression is called a *rational function.*

| X | Y1 |
|---|---|
| -1 | -1 |
| -.1 | -10 |
| -.01 | -100 |
| -.001 | -1000 |
| -1E-4 | -10000 |
| -1E-5 | -1E5 |
| -1E-6 | -1E6 |
| $Y_1$=1/X | |

As X approaches 0 from the left, $Y_1 = \frac{1}{X}$ approaches $-\infty$.

| X | Y1 |
|---|---|
| 1 | 1 |
| .1 | 10 |
| .01 | 100 |
| .001 | 1000 |
| 1E-4 | 10000 |
| 1E-5 | 100000 |
| 1E-6 | 1E6 |
| $Y_1$=1/X | |

As X approaches 0 from the right, $Y_1 = \frac{1}{X}$ approaches $\infty$.

**FIGURE 1**

> **Rational Function**
>
> A function $f$ of the form $\frac{p}{q}$ defined by
>
> $$f(x) = \frac{p(x)}{q(x)},$$
>
> where $p(x)$ and $q(x)$ are polynomials, with $q(x) \neq 0$, is called a **rational function.**

Since any values of $x$ such that $q(x) = 0$ are excluded from the domain of a rational function, this type of function often has a *discontinuous* graph, that is, a graph that has one or more breaks in it.

Some examples of rational functions are

$$f(x) = \frac{1}{x}, \quad f(x) = \frac{x+1}{2x^2+5x-3}, \quad \text{and} \quad f(x) = \frac{3x^2-3x-6}{x^2+8x+16}.$$

## The Reciprocal Function

The simplest rational function with a variable denominator is the **reciprocal function,** defined by

$$f(x) = \frac{1}{x}.$$

| X | Y1 |
|---|---|
| 1 | 1 |
| 10 | .1 |
| 100 | .01 |
| 1000 | .001 |
| 10000 | 1E-4 |
| 100000 | 1E-5 |
| 1E6 | 1E-6 |
| $Y_1$=1/X | |

As X approaches $\infty$, $Y_1 = \frac{1}{X}$ approaches 0 through positive values.

| X | Y1 |
|---|---|
| -1 | -1 |
| -10 | -.1 |
| -100 | -.01 |
| -1000 | -.001 |
| -10000 | -1E-4 |
| -1E5 | -1E-5 |
| -1E6 | -1E-6 |
| $Y_1$=1/X | |

As X approaches $-\infty$, $Y_1 = \frac{1}{X}$ approaches 0 through negative values.

**FIGURE 2**

The domain of this function is the set of all real numbers except 0. The number 0 cannot be used as a value of $x$, but it is helpful to find values of $f(x)$ for some values of $x$ close to 0. We can use the table feature of a graphing calculator to do this. The tables in Figure 1 suggest that $|f(x)|$ gets larger and larger as $x$ gets closer and closer to 0, which is written in symbols as

$$|f(x)| \to \infty \quad \text{as} \quad x \to 0.$$

(The symbol $x \to 0$ means that $x$ approaches 0, without necessarily ever being equal to 0.) Since $x$ cannot equal 0, the graph of $f(x) = \frac{1}{x}$ will never intersect the vertical line $x = 0$. This line is called a **vertical asymptote.**

On the other hand, as $|x|$ gets larger and larger, the values of $f(x) = \frac{1}{x}$ get closer and closer to 0, as shown in the tables in Figure 2. Letting $|x|$ get larger and larger without bound (written $|x| \to \infty$) causes the graph of $f(x) = \frac{1}{x}$ to move closer and closer to the horizontal line $y = 0$. This line is called a **horizontal asymptote.**

The graph and important features of $f(x) = \frac{1}{x}$ are summarized here and shown in Figure 3.

> **Looking Ahead to Calculus**
>
> The rational function defined by
>
> $$f(x) = \frac{2}{x+1}$$
>
> in Example 2 has a vertical asymptote at $x = -1$. In calculus, the behavior of the graph of this function for values close to $-1$ is described using *one-sided limits.* As $x$ approaches $-1$ from the *left,* the function values decrease without bound: This is written
>
> $$\lim_{x \to -1^-} f(x) = -\infty.$$
>
> As $x$ approaches $-1$ from the *right,* the function values increase without bound: This is written
>
> $$\lim_{x \to -1^+} f(x) = \infty.$$

**FUNCTION CAPSULE**

**RECIPROCAL FUNCTION** $\quad f(x) = \dfrac{1}{x}$

Domain: $(-\infty, 0) \cup (0, \infty)$ $\qquad$ Range: $(-\infty, 0) \cup (0, \infty)$

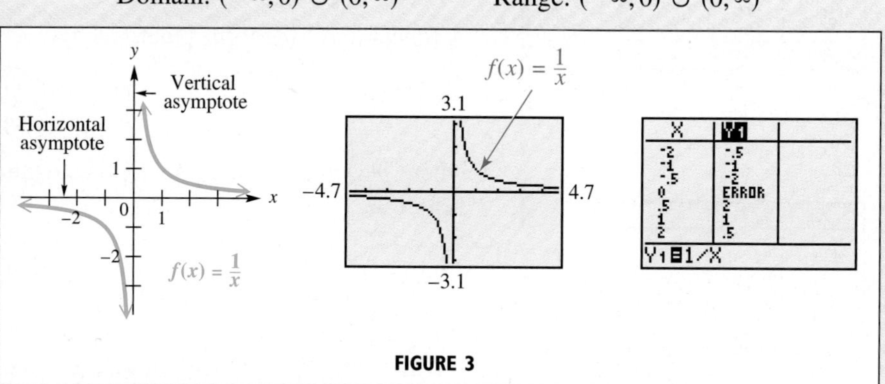


**FIGURE 3**

- $f(x) = \frac{1}{x}$ decreases on the intervals $(-\infty, 0)$ and $(0, \infty)$.
- It is discontinuous at $x = 0$.
- The $y$-axis is a vertical asymptote, and the $x$-axis is a horizontal asymptote.
- It is an odd function, and its graph is symmetric with respect to the origin.

The graph of $y = \frac{1}{x}$ can be translated, and reflected in the same way that we made such transformations on other basic graphs in Chapter 2.

### EXAMPLE 1 *Graphing a Rational Function*

Graph $y = -\dfrac{2}{x}$.

**Analytic Solution**

The expression $-\frac{2}{x}$ can be written as either $2\left(\frac{1}{-x}\right)$ or $-2\left(\frac{1}{x}\right)$, indicating that the graph may be obtained by stretching the graph of $y = \frac{1}{x}$ vertically by a factor of 2 and reflecting it across either the $y$-axis or $x$-axis. The $x$- and $y$-axes remain the horizontal and vertical asymptotes. The domain and range are both still $(-\infty, 0) \cup (0, \infty)$. See Figure 4.

**Graphing Calculator Solution**

Figure 5 shows the graph of $y = -\frac{2}{x}$ in a decimal window.

Later in this section we will discuss why it is important to use a decimal window for this kind of rational function. Using a different window may cause misleading results.

*(continued)*

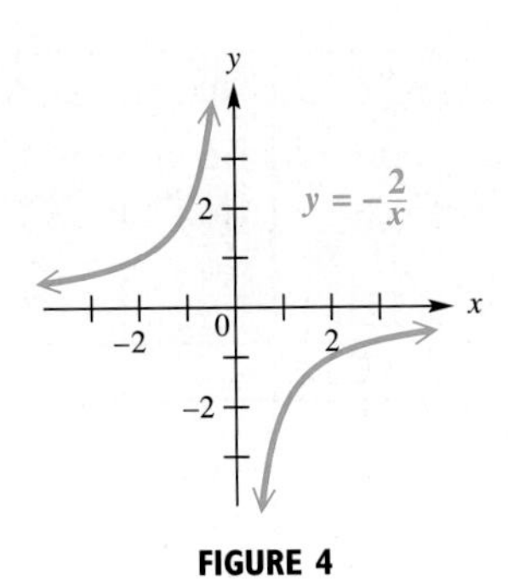


FIGURE 4

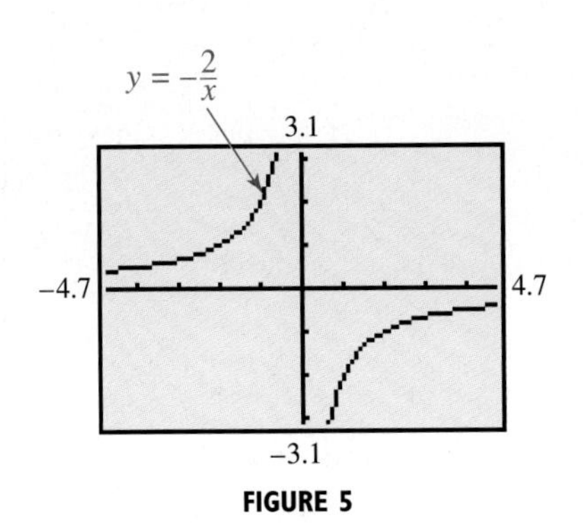


FIGURE 5

**EXAMPLE 2** *Graphing a Rational Function*

Graph $y = \dfrac{2}{x + 1}$.

**Analytic Solution**

The expression $\frac{2}{x + 1}$ can be written as $2\left(\frac{1}{x + 1}\right)$, indicating that the graph may be obtained by shifting the graph of $y = \frac{1}{x}$ to the left 1 unit and stretching it vertically by a factor of 2. The graph is shown in Figure 6. Notice that the horizontal shift affects the domain; the domain of this new function is $(-\infty, -1) \cup (-1, \infty)$. The line $x = -1$ is the vertical asymptote. The range is still $(-\infty, 0) \cup (0, \infty)$.

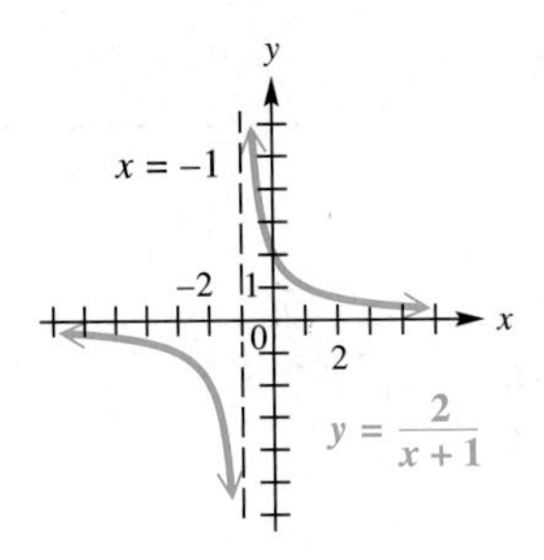


FIGURE 6

**Graphing Calculator Solution**

The graph of $y = \frac{2}{x + 1}$ is shown in a decimal window in Figure 7.

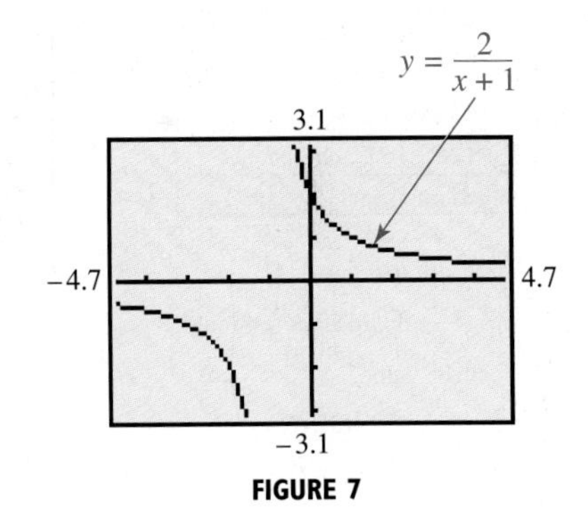


FIGURE 7

## The Rational Function Defined by $f(x) = \dfrac{1}{x^2}$

The rational function defined by

$$f(x) = \frac{1}{x^2}$$

also has domain $(-\infty, 0) \cup (0, \infty)$. We can use the table feature of a graphing calculator to examine values of $f(x)$ for some $x$-values close to 0. See Figure 8 on the next page.

| X | $Y_1$ |
|---|---|
| -1 | 1 |
| -.1 | 100 |
| -.01 | 10000 |
| -.001 | 1E6 |
| -1E-4 | 1E8 |
| -1E-5 | 1E10 |
| -1E-6 | 1E12 |

Y1=1/X²

As X approaches 0 from the left, $Y_1 = \frac{1}{X^2}$ approaches $\infty$.

| X | $Y_1$ |
|---|---|
| 1 | 1 |
| .1 | 100 |
| .01 | 10000 |
| .001 | 1E6 |
| 1E-4 | 1E8 |
| 1E-5 | 1E10 |
| 1E-6 | 1E12 |

Y1=1/X²

As X approaches 0 from the right, $Y_1 = \frac{1}{X^2}$ approaches $\infty$.

**FIGURE 8**

| X | $Y_1$ |
|---|---|
| 1 | 1 |
| 10 | .01 |
| 100 | 1E-4 |
| 1000 | 1E-6 |
| 10000 | 1E-8 |
| 100000 | 1E-10 |
| 1E6 | 1E-12 |

Y1=1/X²

As X approaches $\infty$, $Y_1 = \frac{1}{X^2}$ approaches 0 through positive values.

| X | $Y_1$ |
|---|---|
| -1 | 1 |
| -10 | .01 |
| -100 | 1E-4 |
| -1000 | 1E-6 |
| -10000 | 1E-8 |
| -1E5 | 1E-10 |
| -1E6 | 1E-12 |

Y1=1/X²

As X approaches $-\infty$, $Y_1 = \frac{1}{X^2}$ approaches 0 through positive values.

**FIGURE 9**

The tables suggest that $f(x)$ gets larger and larger as $x$ gets closer and closer to 0. Notice that as $x$ approaches 0 from *either* side, function values are all positive and there is symmetry with respect to the $y$-axis. Thus, we have

$$f(x) \to \infty \quad \text{as} \quad x \to 0.$$

The $y$-axis is the vertical asymptote.

As $|x|$ gets larger and larger, $f(x)$ approaches 0, as suggested by the tables in Figure 9. Again, function values are all positive. The $x$-axis is the horizontal asymptote of the graph.

The graph and important features of $f(x) = \frac{1}{x^2}$ are summarized here and shown in Figure 10.

**FUNCTION CAPSULE**

**RATIONAL FUNCTION** $\quad f(x) = \frac{1}{x^2}$

Domain: $(-\infty, 0) \cup (0, \infty)$ Range: $(0, \infty)$

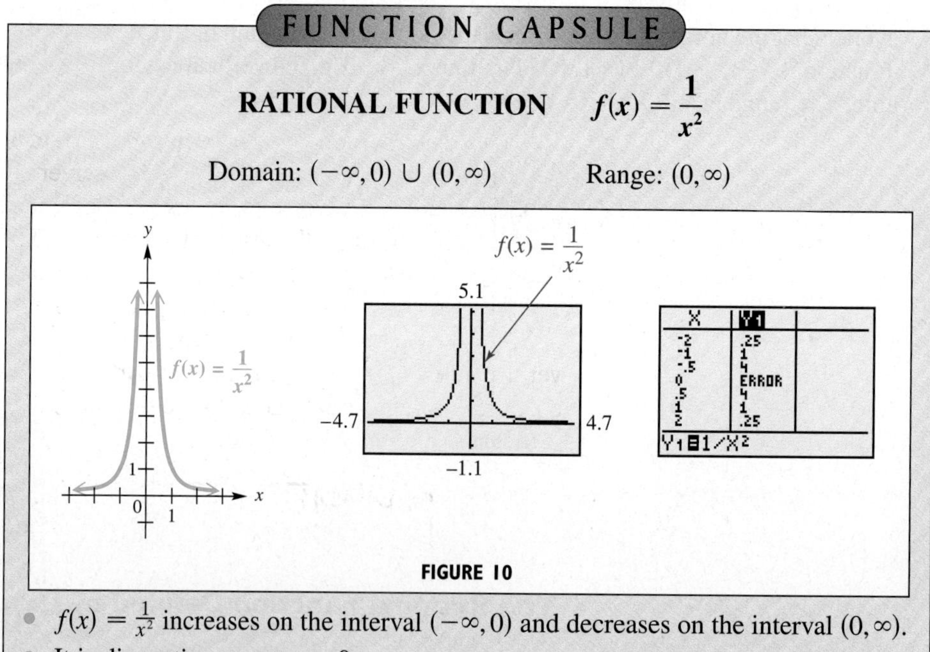


**FIGURE 10**

- $f(x) = \frac{1}{x^2}$ increases on the interval $(-\infty, 0)$ and decreases on the interval $(0, \infty)$.
- It is discontinuous at $x = 0$.
- The $y$-axis is a vertical asymptote, and the $x$-axis is a horizontal asymptote.
- It is an even function, and its graph is symmetric with respect to the $y$-axis.

**EXAMPLE 3** *Graphing a Rational Function*

Graph $y = \frac{1}{(x + 2)^2} - 1$.

**Analytic Solution**

The equation $y = \frac{1}{(x+2)^2} - 1$ is equivalent to

$$y = f(x+2) - 1,$$

where $f(x) = \frac{1}{x^2}$. This indicates that the graph will be shifted 2 units to the left and 1 unit downward. The horizontal shift affects the domain, which is now $(-\infty, -2) \cup (-2, \infty)$, while the vertical shift affects the range, now $(-1, \infty)$. The vertical asymptote has equation $x = -2$, and the horizontal asymptote has equation $y = -1$. The graph is shown in Figure 11.

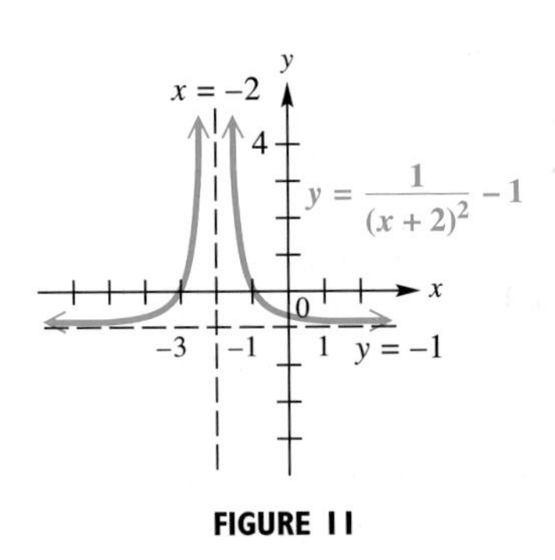


FIGURE 11

**Graphing Calculator Solution**

The graph of

$$y = \frac{1}{(x+2)^2} - 1$$

is shown in a decimal window in Figure 12.

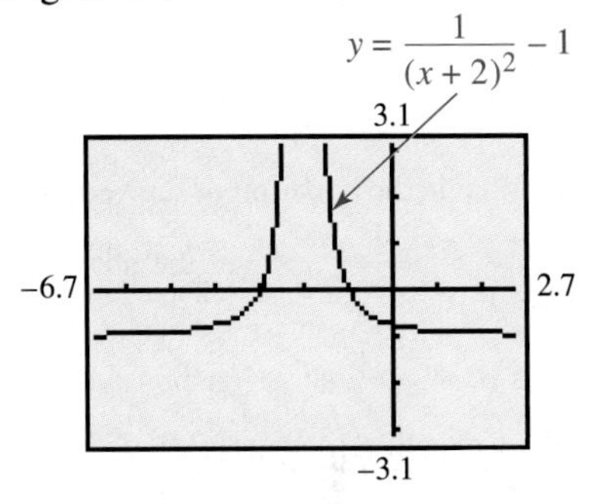


FIGURE 12

A choice of window other than the one here *may* produce an undesired vertical line at $x = -2$. See the explanation below.

## Mode and Window Choices for Calculator Graphs

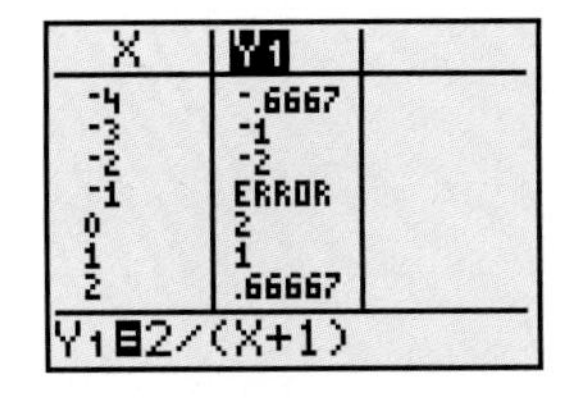

| X | Y1 |
|---|---|
| -4 | -.6667 |
| -3 | -1 |
| -2 | -2 |
| -1 | ERROR |
| 0 | 2 |
| 1 | 1 |
| 2 | .66667 |

Y1=2/(X+1)

$Y_1 = \frac{2}{X+1}$ is undefined at $X = -1$, as indicated by the ERROR message in the table.

In Section 2.5 we studied the graph of the greatest integer function. Because of discontinuities in the graph, we used dot mode (rather than connected mode) to obtain a realistic picture on a graphing calculator screen. Rational functions also pose problems for graphing calculators. In Figure 7 of Example 2, we used a decimal window to accurately illustrate the graph of $y = \frac{2}{x+1}$. The window was set so that the calculator attempted to evaluate $y$ when $x = -1$. Because $y$ is undefined at $x = -1$, the calculator does not attempt to graph a point whose $x$-coordinate is $-1$. Had we used a nondecimal window, such as $[-5, 5]$ by $[-5, 5]$, the calculator would have indicated a vertical line at $x = -1$, which is not part of the graph. Another way to avoid the appearance of the vertical line is to use dot mode. See Figure 13.

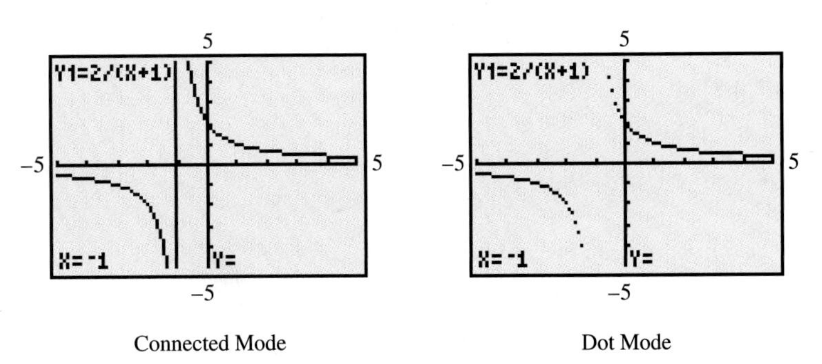


Connected Mode    Dot Mode

The function is not defined for $x = -1$.

FIGURE 13

You may need to experiment with various modes and windows to obtain an accurate depiction of the graph of a rational function.

## 4.1 EXERCISES

*Concept Check* *Provide a short answer to each question.*

1. What is the domain of $f(x) = \frac{1}{x}$? What is its range?

2. What is the domain of $f(x) = \frac{1}{x^2}$? What is its range?

3. What is the interval over which $f(x) = \frac{1}{x}$ increases? decreases? is constant?

4. What is the interval over which $f(x) = \frac{1}{x^2}$ increases? decreases? is constant?

5. What is the equation of the vertical asymptote of the graph of $y = \frac{1}{x-3} + 2$? of the horizontal asymptote?

6. What is the equation of the vertical asymptote of the graph of $y = \frac{1}{(x+2)^2} - 4$? of the horizontal asymptote?

7. Is $f(x) = \frac{1}{x^2}$ an even or odd function? What symmetry does its graph exhibit?

8. Is $f(x) = \frac{1}{x}$ an even or odd function? What symmetry does its graph exhibit?

*Concept Check* *Use the graphs of the rational functions in A–D to answer each question. Give all possible answers, as there may be more than one correct choice.*

9. Which choices have domain $(-\infty, 3) \cup (3, \infty)$?

10. Which choice has range $(-\infty, 3) \cup (3, \infty)$?

11. Which choice has range $(-\infty, 0) \cup (0, \infty)$?

12. Which choices have range $(0, \infty)$?

13. If $f$ represents the function, only one choice has a single solution to the equation $f(x) = 3$. Which one is it?

14. What is the range of the function in A?

**A.**

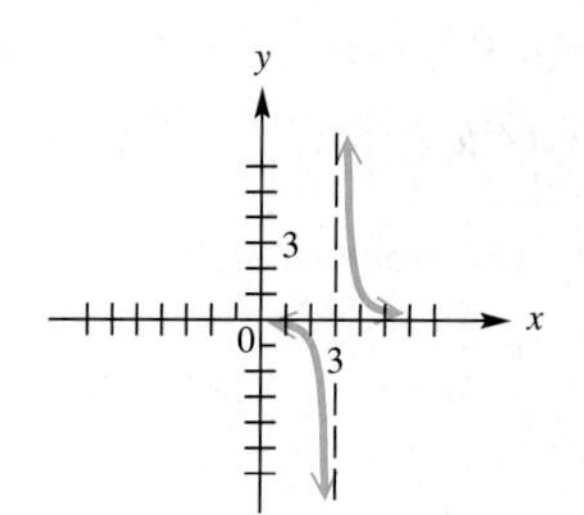

**B.**

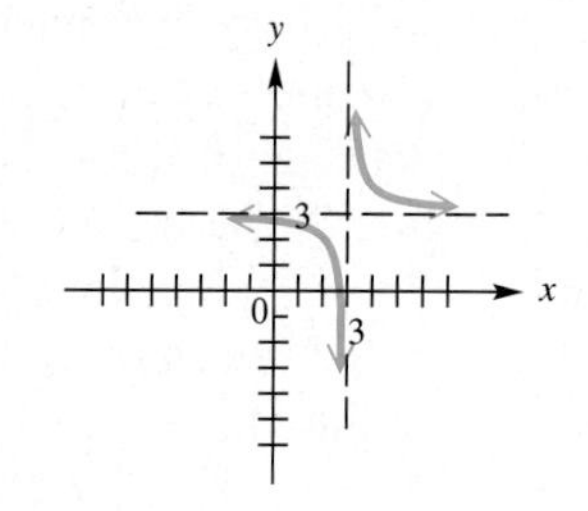

**C.**

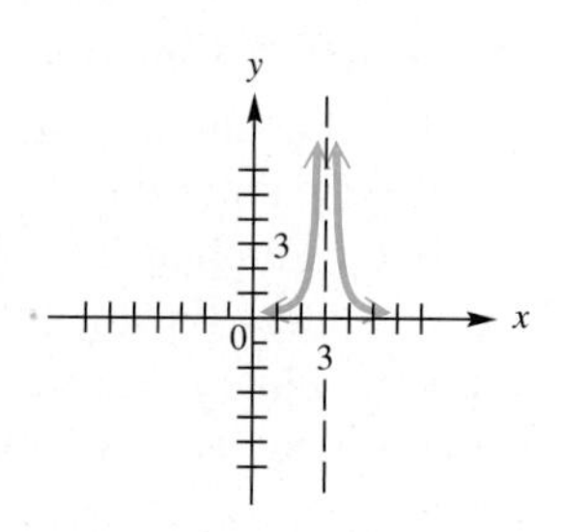

**D.**

*Graph each rational function in the given windows, and determine which window gives the most accurate depiction of the graph.*

15. $f(x) = \frac{1}{x+5}$
    **A.** $[-4.7, 4.7]$ by $[-3.1, 3.1]$
    **B.** $[-14.4, 4.4]$ by $[0, 5]$
    **C.** $[-9.4, 9.4]$ by $[-3.1, 3.1]$

16. $f(x) = \frac{1}{x^2} + 6$
    **A.** $[-4.7, 4.7]$ by $[0, 12.4]$
    **B.** $[-4.7, 4.7]$ by $[-6.2, 6.2]$
    **C.** $[-4.7, 4.7]$ by $[-3.1, 3.1]$

*Explain how the graph of $f$ can be obtained from the graph of $y = \frac{1}{x}$ or $y = \frac{1}{x^2}$. Draw a sketch of the graph of $f$ by hand. Then generate an accurate depiction of the graph using a graphing calculator. Finally, give the domain and range.*

17. $f(x) = \frac{2}{x}$

18. $f(x) = -\frac{3}{x}$

19. $f(x) = \frac{1}{x+2}$

20. $f(x) = \frac{1}{x-3}$

21. $f(x) = \frac{1}{x} + 1$

22. $f(x) = \frac{1}{x} - 2$

23. $f(x) = -\frac{2}{x^2}$

24. $f(x) = \frac{1}{x^2} + 3$

**25.** $f(x) = \dfrac{1}{(x-3)^2}$ **26.** $f(x) = \dfrac{-2}{(x-3)^2}$ **27.** $f(x) = \dfrac{-1}{(x+2)^2} - 3$ **28.** $f(x) = \dfrac{-1}{(x-4)^2} + 2$

*Concept Check The figures in A–D show the four ways that the graph of a rational function can approach the vertical line $x = 2$ as an asymptote. Identify the graph of each rational function in Exercises 29–32.*

**29.** $f(x) = \dfrac{1}{(x-2)^2}$ **30.** $f(x) = \dfrac{1}{x-2}$ **31.** $f(x) = \dfrac{-1}{x-2}$ **32.** $f(x) = \dfrac{-1}{(x-2)^2}$

**A.**

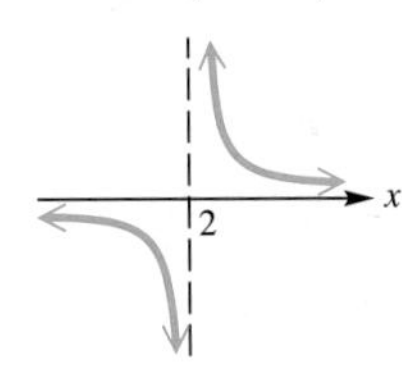


**B.**

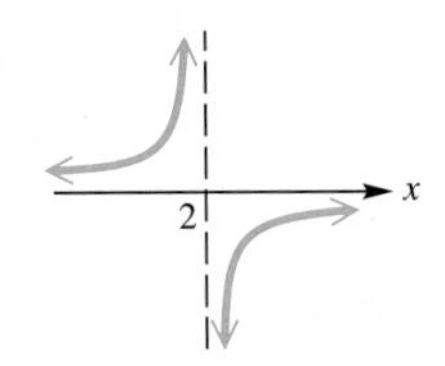


**C.**

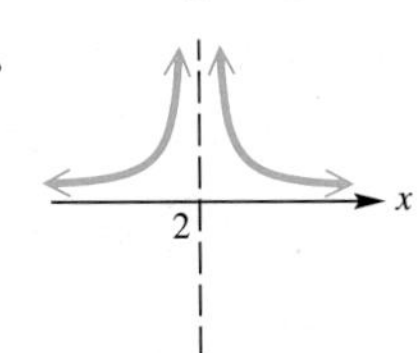


**D.**

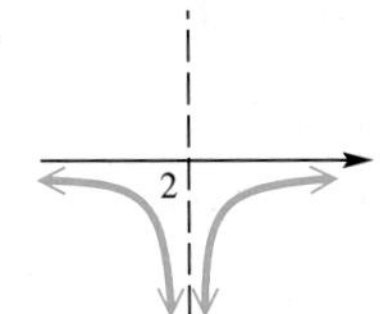


*A rational function of the form*

$$f(x) = \frac{ax+b}{x+c}$$

*can be graphed by using transformations of the graph of $y = \frac{1}{x}$. To accomplish this, first use long division (or synthetic division) to rewrite the rational expression. For example, consider*

$$f(x) = \frac{3x+7}{x+2}: \qquad x+2\overline{\smash{)}3x+7}$$

$$\begin{array}{r} 3 \\ x+2\,)\overline{3x+7} \\ \underline{3x+6} \\ 1 \end{array}$$

*Thus, $\dfrac{3x+7}{x+2} = 3 + \dfrac{1}{x+2}$, so the graph of $f$ is obtained by shifting the graph of $y = \dfrac{1}{x}$ to the left 2 units and upward 3 units.*

*Use this method to rewrite the rational expression, sketch the graph by hand, and generate an accurate depiction of each function using a graphing calculator.*

**33.** $f(x) = \dfrac{x-1}{x-2}$ **34.** $f(x) = \dfrac{x-2}{x-3}$ **35.** $f(x) = \dfrac{-2x-5}{x+3}$

**36.** $f(x) = \dfrac{-2x-1}{x+1}$ **37.** $f(x) = \dfrac{2x-5}{x-3}$ **38.** $f(x) = \dfrac{2x-1}{x-1}$

# MORE ON GRAPHS OF RATIONAL FUNCTIONS

Vertical and Horizontal Asymptotes • Graphing Techniques • Oblique Asymptotes • Graphs with Points of Discontinuity

## Vertical and Horizontal Asymptotes

In the previous section we limited our discussion of rational function graphs to translations of the graphs of $y = \frac{1}{x}$ and $y = \frac{1}{x^2}$. In this section we expand our discussion to include more general rational functions, such as

$$f(x) = \frac{x+1}{2x^2+5x-3}, \quad f(x) = \frac{2x+1}{x-3}, \quad \text{and} \quad f(x) = \frac{x^2+1}{x-2}.$$

Locating asymptotes and finding intercepts is important when graphing such functions. Figure 14 on the next page shows the graphs of these functions. Refer to them

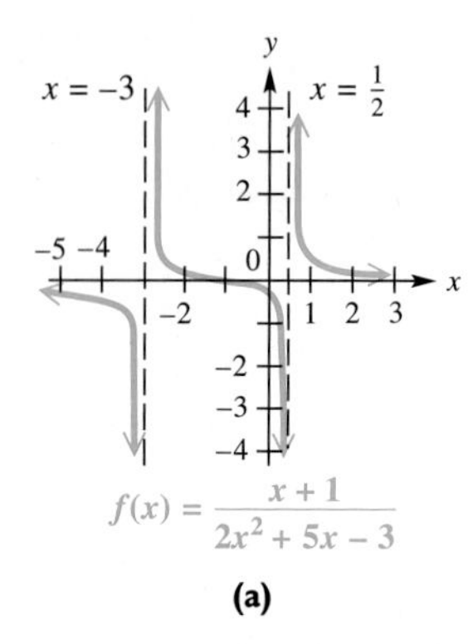

(a)

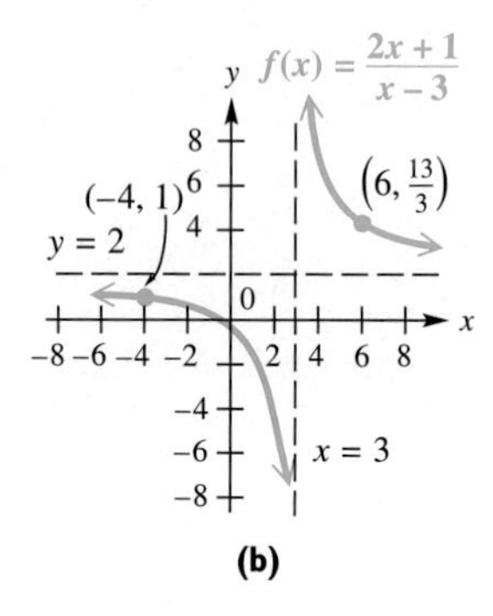

(b)

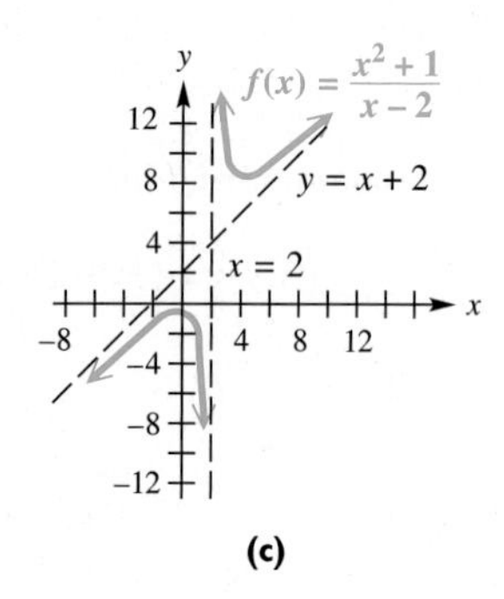

(c)

Typical Rational Function Graphs

**FIGURE 14**

when reading Examples 1–3. In Examples 4, 5, and 7, we discuss the steps for obtaining these graphs.

In our discussion thus far, we have seen how rational functions approach certain values as $|x| \to \infty$ and as $|x| \to a$ if the function is undefined at $x = a$. We now give formal definitions for vertical and horizontal asymptotes.

> **Vertical and Horizontal Asymptotes**
>
> For the rational function $f$ with $f(x) = \frac{p(x)}{q(x)}$, written in lowest terms,
>
> if $|f(x)| \to \infty$ as $x \to a$, then the line $x = a$ is a **vertical asymptote;**
>
> if $f(x) \to a$ as $|x| \to \infty$, then the line $y = a$ is a **horizontal asymptote.**

Vertical asymptotes are found by determining the values of $x$ that make the denominator equal to 0 but do not make the numerator equal to 0. Horizontal asymptotes (and, in some cases, *oblique asymptotes*) are found by considering what happens to $f(x)$ as $|x| \to \infty$.

## EXAMPLE 1 *Finding Asymptotes*

Find the asymptotes of the graph of $f(x) = \dfrac{x + 1}{2x^2 + 5x - 3}$. (See Figure 14(a).)

**Solution** To find the vertical asymptotes, we set the denominator equal to 0 and solve.

$$2x^2 + 5x - 3 = 0$$

$$(2x - 1)(x + 3) = 0 \qquad \text{Factor.}$$

$$2x - 1 = 0 \quad \text{or} \quad x + 3 = 0 \qquad \text{Zero-product property}$$

$$x = \frac{1}{2} \quad \text{or} \quad x = -3$$

The equations of the vertical asymptotes are $x = \frac{1}{2}$ and $x = -3$.

To find the equation of the horizontal asymptote, we divide each term by the variable factor of greatest degree in the expression for $f(x)$. In this case, we divide each term by $x^2$.

$$f(x) = \frac{\frac{x}{x^2} + \frac{1}{x^2}}{\frac{2x^2}{x^2} + \frac{5x}{x^2} - \frac{3}{x^2}} = \frac{\frac{1}{x} + \frac{1}{x^2}}{2 + \frac{5}{x} - \frac{3}{x^2}}$$

As $|x|$ gets larger and larger, the quotients $\frac{1}{x}$, $\frac{1}{x^2}$, $\frac{5}{x}$, and $\frac{3}{x^2}$ all approach 0, and the value of $f(x)$ approaches

$$\frac{0 + 0}{2 + 0 - 0} = \frac{0}{2} = 0.$$

The line $y = 0$ (that is, the $x$-axis) is therefore the horizontal asymptote.

**EXAMPLE 2** *Finding Asymptotes*

Find the asymptotes of the graph of $f(x) = \dfrac{2x + 1}{x - 3}$. (See Figure 14(b).)

**Solution** Set the denominator equal to 0 to find that the vertical asymptote has equation $x = 3$. To find the horizontal asymptote, divide each term in the rational expression by $x$, since the greatest exponent on $x$ in the expression is 1.

$$f(x) = \frac{2x + 1}{x - 3} = \frac{\frac{2x}{x} + \frac{1}{x}}{\frac{x}{x} - \frac{3}{x}} = \frac{2 + \frac{1}{x}}{1 - \frac{3}{x}}$$

As $|x|$ gets larger and larger, both $\frac{1}{x}$ and $\frac{3}{x}$ approach 0, and $f(x)$ approaches

$$\frac{2 + 0}{1 - 0} = \frac{2}{1} = 2,$$

so the line $y = 2$ is the horizontal asymptote.

**EXAMPLE 3** *Finding Asymptotes*

Find the asymptotes of the graph of $f(x) = \dfrac{x^2 + 1}{x - 2}$. (See Figure 14(c).)

**Solution** Setting the denominator equal to 0 shows that the vertical asymptote has equation $x = 2$. If we divide by the variable factor of greatest degree of $x$ as before ($x^2$ in this case), we see that there is no horizontal asymptote because

$$f(x) = \frac{\frac{x^2}{x^2} + \frac{1}{x^2}}{\frac{x}{x^2} - \frac{2}{x^2}} = \frac{1 + \frac{1}{x^2}}{\frac{1}{x} - \frac{2}{x^2}}$$

does not approach any real number as $|x| \to \infty$, since $\frac{1}{0}$ is undefined. This happens whenever the degree of the numerator is greater than the degree of the denominator. In such cases, we divide the denominator into the numerator to write the expression in another form.

$$\begin{array}{r} x + 2 \\ x - 2\overline{)x^2 + 0x + 1} \\ \underline{x^2 - 2x\phantom{ + 1}} \\ 2x + 1 \\ \underline{2x - 4} \\ 5 \end{array}$$

(We could have used synthetic division here.) We can now write the function as

$$f(x) = \frac{x^2 + 1}{x - 2} = x + 2 + \frac{5}{x - 2}.$$

For very large values of $|x|$, $\frac{5}{x - 2}$ is close to 0, and the graph approaches the line $y = x + 2$. This line is an **oblique asymptote** (neither vertical nor horizontal) for the graph of the function.

In general, if the degree of the numerator is exactly one more than the degree of the denominator, a rational function may have an oblique asymptote.* The equation of this asymptote is found by dividing the numerator by the denominator and disregarding the remainder.

The results of Examples 1–3 can be summarized as follows.

**Determining Asymptotes**

To find asymptotes of a rational function defined by a rational expression *in lowest terms,* use the following procedures.

**1. Vertical Asymptotes**
Find any vertical asymptotes by setting the denominator equal to 0 and solving for $x$. If $a$ is a zero of the denominator but not the numerator, then the line $x = a$ is a vertical asymptote.

**2. Other Asymptotes**
Determine any other asymptotes. Consider three possibilities:

**(a)** If the numerator has lower degree than the denominator, there is a horizontal asymptote, $y = 0$ (the $x$-axis).

**(b)** If the numerator and denominator have the same degree and the function is of the form

$$f(x) = \frac{a_n x^n + \cdots + a_0}{b_n x^n + \cdots + b_0}, \quad \text{where } b_n \neq 0,$$

dividing by $x^n$ in the numerator and denominator produces the horizontal asymptote

$$y = \frac{a_n}{b_n}.$$

**(c)** If the numerator is of degree exactly one more than the denominator, there may be an oblique asymptote. To find it, divide the numerator by the denominator and disregard any remainder. Set the rest of the quotient equal to $y$ to get the equation of the asymptote.

**NOTE** The graph of a rational function may have more than one vertical asymptote, or it may have none at all. The graph cannot intersect any vertical asymptote. There can be only one other (nonvertical) asymptote, and the graph *may* intersect that asymptote. This will be seen in Examples 4 and 6.

## Graphing Techniques

As rational functions become more complicated, obtaining accurate calculator graphs becomes more difficult. We suggest sketching graphs by hand before attempting to

*More involved rational functions, such as $f(x) = \dfrac{8x^3 - 1}{x}$, are not covered in this book.

graph them with a calculator, because a calculator can lead to a distorted view if an inappropriate window is used. The procedure for graphing rational functions by hand follows.

**Graphing a Rational Function**

Let $f(x) = \frac{p(x)}{q(x)}$ define a function where the rational expression is written in lowest terms. To sketch its graph, follow these steps.

1. Find any vertical asymptotes.
2. Find any horizontal or oblique asymptotes.
3. Find the $y$-intercept by evaluating $f(0)$.
4. Find the $x$-intercepts, if any, by solving $f(x) = 0$. (These will be the zeros of the numerator, $p$.)
5. Determine whether the graph will intersect its nonvertical asymptote by solving $f(x) = k$, where $k$ is the $y$-value of the horizontal asymptote, or $f(x) = mx + b$, where $y = mx + b$ is the equation of the oblique asymptote.
6. Plot a few selected points, as necessary. Choose an $x$-value in each interval of the domain as determined by the vertical asymptotes and $x$-intercepts.
7. Complete the sketch.

TECHNOLOGY NOTE

Determining intercepts and asymptotes analytically will help to obtain a realistic comprehensive graph of a rational function. Choosing Xmin and Xmax so that the calculator will attempt to compute a value for which the rational function is undefined will eliminate the misleading vertical line that occasionally appears when the calculator is in connected mode.

A comprehensive graph of a rational function exhibits these features:

1. all intercepts, both $x$- and $y$-;
2. location of all asymptotes: vertical, horizontal, and/or oblique;
3. the point at which the graph intersects its nonvertical asymptote (if there is any such point);
4. enough of the graph to exhibit the correct end behavior (i.e., behavior as the graph approaches its nonvertical asymptote).

**EXAMPLE 4** *Graphing a Rational Function with the x-Axis as Horizontal Asymptote*

Graph $f(x) = \dfrac{x + 1}{2x^2 + 5x - 3}$.

**Solution**

**Step 1** As shown in Example 1, the vertical asymptotes have equations $x = \frac{1}{2}$ and $x = -3$.

**Step 2** Again, as shown in Example 1, the horizontal asymptote is the $x$-axis.

**Step 3** The $y$-intercept is $-\frac{1}{3}$, since

$$f(0) = \frac{0 + 1}{2(0)^2 + 5(0) - 3} = -\frac{1}{3}.$$

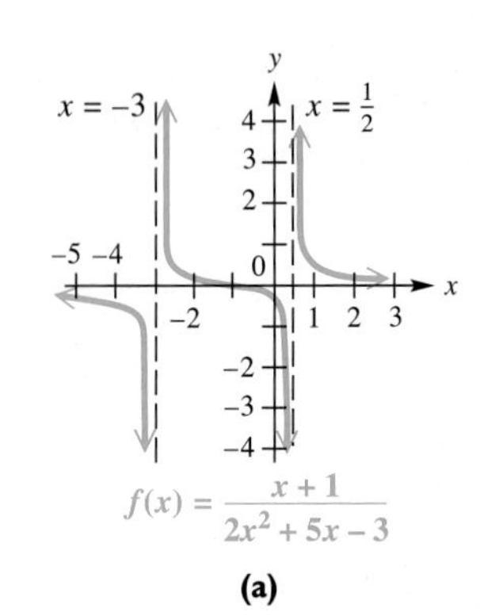


(a)

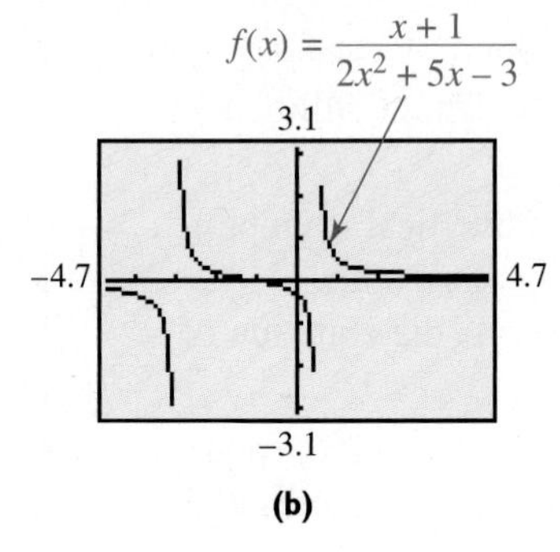


(b)

**FIGURE 15**

**Step 4** The $x$-intercept is found by solving $f(x) = 0$.

$$\frac{x+1}{2x^2+5x-3} = 0$$

$$x + 1 = 0 \qquad \text{If a rational expression is equal to 0, then its numerator must equal 0.}$$

$$x = -1$$

The $x$-intercept is $-1$.

**Step 5** To determine whether the graph intersects its horizontal asymptote, solve

$$f(x) = 0. \leftarrow \text{y-value of horizontal asymptote}$$

Since the horizontal asymptote is the $x$-axis, the solution of this equation was found in Step 4. The graph intersects its horizontal asymptote at $(-1, 0)$.

**Step 6** Plot a point in each of the intervals determined by the $x$-intercepts and vertical asymptotes, $(-\infty, -3)$, $(-3, -1)$, $\left(-1, \frac{1}{2}\right)$ and $\left(\frac{1}{2}, \infty\right)$, to get an idea of how the graph behaves in each region.

**Step 7** Complete the sketch. Keep in mind that the graph approaches its asymptotes as the points on the graph become farther away from the origin. The traditional graph is repeated in Figure 15(a), and a calculator graph is shown in Figure 15(b).

**EXAMPLE 5** *Graphing a Rational Function That Does Not Intersect Its Horizontal Asymptote*

Graph $f(x) = \dfrac{2x+1}{x-3}$.

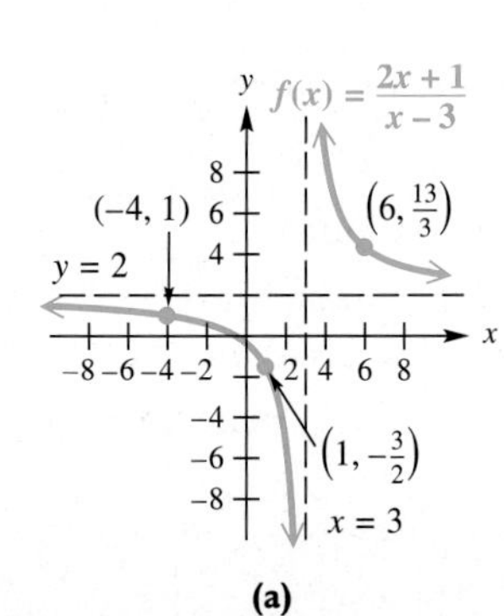


(a)

f(x) = (2x + 1)/(x − 3)

10

−9.4

9.4

−10

(b)

**FIGURE 16**

### Solution

**Steps 1 and 2** As determined in Example 2, the equation of the vertical asymptote is $x = 3$. The horizontal asymptote has equation $y = 2$.

**Step 3** $f(0) = -\frac{1}{3}$, so the $y$-intercept is $-\frac{1}{3}$.

**Step 4** Solve $f(x) = 0$ to find any $x$-intercepts.

$$\frac{2x+1}{x-3} = 0$$

$$2x + 1 = 0$$

$$x = -\frac{1}{2}$$

The $x$-intercept is $-\frac{1}{2}$.

**Step 5** The graph does not intersect its horizontal asymptote since $f(x) = 2$ has no solution. (Verify this.)

**Steps 6 and 7** The points $(-4, 1)$, $\left(1, -\frac{3}{2}\right)$, and $\left(6, \frac{13}{3}\right)$ are on the graph and can be used to complete the sketch, as seen in Figure 16(a). Figure 16(b) shows a calculator graph.

**Looking Ahead to Calculus**

The rational function defined by

$$f(x) = \frac{2x + 1}{x - 3},$$

seen in Example 5, has horizontal asymptote $y = 2$. In calculus, the behavior of the graph of this function as $x$ approaches $-\infty$ and as $x$ approaches $\infty$ is described using *limits at infinity.* As $x$ approaches $-\infty$, $f(x)$ approaches 2 (from below). This is written

$$\lim_{x \to -\infty} f(x) = 2.$$

As $x$ approaches $\infty$, $f(x)$ approaches 2 (from above). This is written

$$\lim_{x \to \infty} f(x) = 2.$$

## EXAMPLE 6 *Graphing a Rational Function That Intersects Its Horizontal Asymptote*

Graph $f(x) = \dfrac{3x^2 - 3x - 6}{x^2 + 8x + 16}$.

### Solution

**Step 1** To find the vertical asymptote(s), solve $x^2 + 8x + 16 = 0$.

$$x^2 + 8x + 16 = 0$$
$$(x + 4)^2 = 0$$
$$x = -4$$

Since the numerator is not 0 when $x = -4$, the only vertical asymptote has equation $x = -4$.

**Step 2** We divide all terms by $x^2$ to get the equation of the horizontal asymptote,

$$y = \frac{3}{1}, \quad \begin{matrix} \leftarrow \text{Leading coefficient of numerator} \\ \leftarrow \text{Leading coefficient of denominator} \end{matrix}$$

or $y = 3$.

**Step 3** The $y$-intercept is $f(0) = -\frac{3}{8}$.

**Step 4** To find the $x$-intercept(s), if any, we solve $f(x) = 0$.

$$\frac{3x^2 - 3x - 6}{x^2 + 8x + 16} = 0$$
$$3x^2 - 3x - 6 = 0 \quad \text{Set the numerator equal to 0.}$$
$$x^2 - x - 2 = 0 \quad \text{Divide by 3.}$$
$$(x - 2)(x + 1) = 0 \quad \text{Factor.}$$
$$x = 2 \quad \text{or} \quad x = -1 \quad \text{Zero-product property}$$

The $x$-intercepts are $-1$ and 2.

**Step 5** We set $f(x) = 3$ and solve to locate the point where the graph intersects the horizontal asymptote.

$$f(x) = \frac{3x^2 - 3x - 6}{x^2 + 8x + 16}$$
$$3 = \frac{3x^2 - 3x - 6}{x^2 + 8x + 16}$$
$$3x^2 - 3x - 6 = 3x^2 + 24x + 48 \quad \text{Multiply by } x^2 + 8x + 16.$$
$$-3x - 6 = 24x + 48 \quad \text{Subtract } 3x^2.$$
$$-27x = 54 \quad \text{Subtract } 24x\text{; add 6.}$$
$$x = -2 \quad \text{Divide by } -27.$$

The graph intersects its horizontal asymptote at $(-2, 3)$.

**Steps 6 and 7** Some other points that lie on the graph are $(-10, 9)$, $\left(-8, 13\frac{1}{8}\right)$, and $\left(5, \frac{2}{3}\right)$. These can be used to complete the hand-drawn graph, as shown in Figure 17(a). The calculator graph in Figure 17(b) confirms our result.

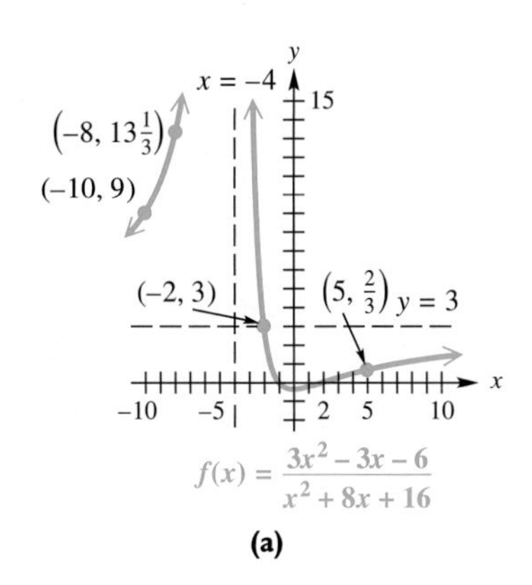

(a)

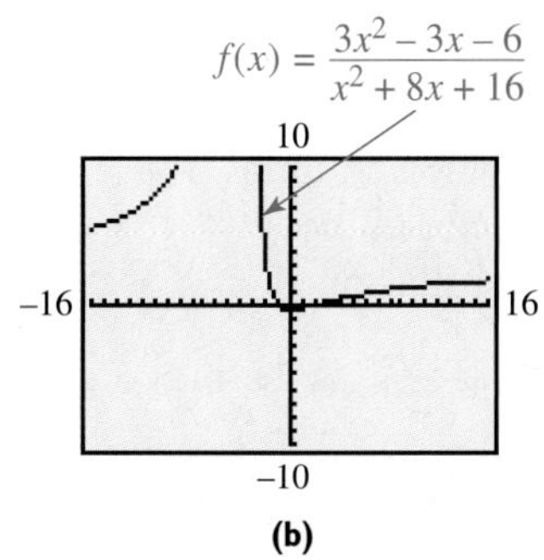

(b)

**FIGURE 17**

Notice the behavior of the graph of the function in Example 6 near the line $x = -4$. As $x \to -4$ from either side, $f(x) \to \infty$. On the other hand, if we examine the behavior of the graph in Figure 16 near the line $x = 3$, $f(x) \to -\infty$ as $x$ approaches 3 from the left, while $f(x) \to \infty$ as $x$ approaches 3 from the right. The behavior of the graph of a rational function near a vertical asymptote $x = a$ will partially depend on the exponent on $x - a$ in the denominator.

### Behavior of Graphs of Rational Functions Near Vertical Asymptotes

Suppose that $f(x)$ is defined by a rational expression in lowest terms. If $n$ is the largest positive integer such that $(x - a)^n$ is a factor of the denominator of $f(x)$, the graph will behave in the manner illustrated.

If $n$ is even:

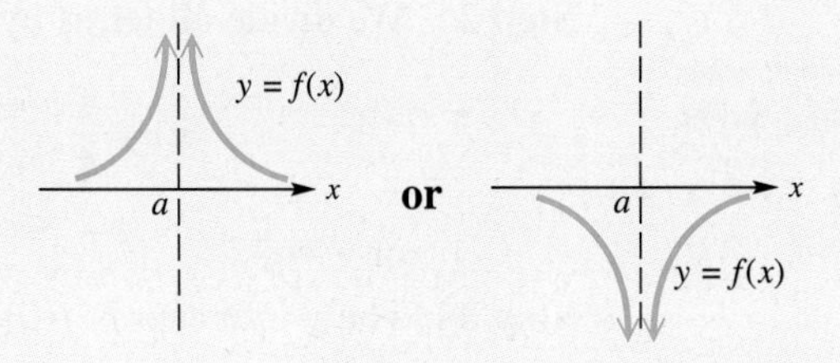


If $n$ is odd:

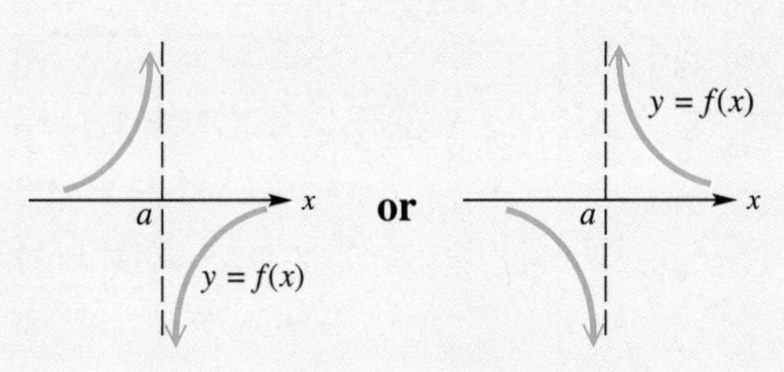


In Section 3.7, we observed that the behavior of the graph of a polynomial function near its zeros is dependent on whether the multiplicity of the zero is even or odd. The same statement can be made for rational functions.

### Behavior of Graphs of Rational Functions Near $x$-Intercepts

Suppose that $f(x)$ is defined by a rational expression in lowest terms. If $n$ is the largest positive integer such that $(x - c)^n$ is a factor of the numerator of $f(x)$, the graph will behave in the manner illustrated.

If $n$ is even:

c x **or** c x

*(continued)*

If $n$ is odd:

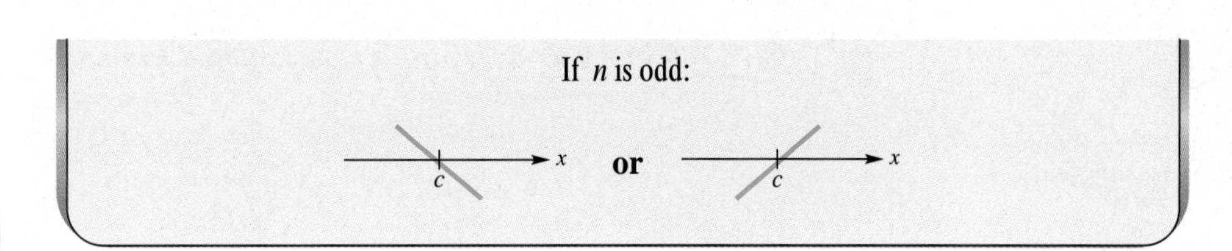


**Looking Ahead to Calculus**

Calculus uses the following notation to describe the asymptotic behavior of the graph of $f(x) = \frac{x^2 + 1}{x - 2}$, seen in Figure 18:

$$\lim_{x \to -\infty} f(x) = -\infty$$

and $$\lim_{x \to \infty} f(x) = \infty.$$

## Oblique Asymptotes

**EXAMPLE 7** *Graphing a Rational Function with an Oblique Asymptote*

Graph $f(x) = \frac{x^2 + 1}{x - 2}$.

**Analytic Solution**

As shown in Example 3, the vertical asymptote has equation $x = 2$, and the graph has an oblique asymptote with equation $y = x + 2$. Refer to the preceding box to determine the behavior near the asymptote $x = 2$. The $y$-intercept is $-\frac{1}{2}$, and the graph has no $x$-intercepts since the numerator, $x^2 + 1$, has no real zeros. The graph does not intersect its oblique asymptote because

$$\frac{x^2 + 1}{x - 2} = x + 2$$

has no solution. Using the $y$-intercept, asymptotes, the points $\left(4, \frac{17}{2}\right)$ and $\left(-1, -\frac{2}{3}\right)$, and the general behavior of the graph near its asymptotes leads to the graph in Figure 18.

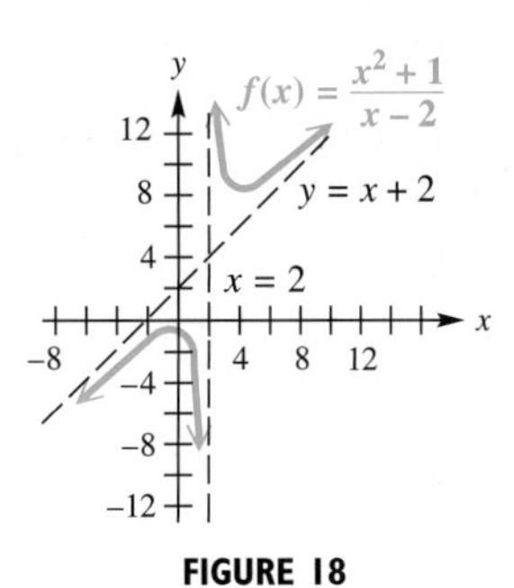


**FIGURE 18**

**Graphing Calculator Solution**

A calculator graph is shown in Figure 19. If we define

$$Y_1 = \frac{X^2 + 1}{X - 2} \quad \text{and} \quad Y_2 = X + 2$$

and observe a table of values as $|X|$ gets larger and larger, we see that for these large values, $Y_1 \approx Y_2$.

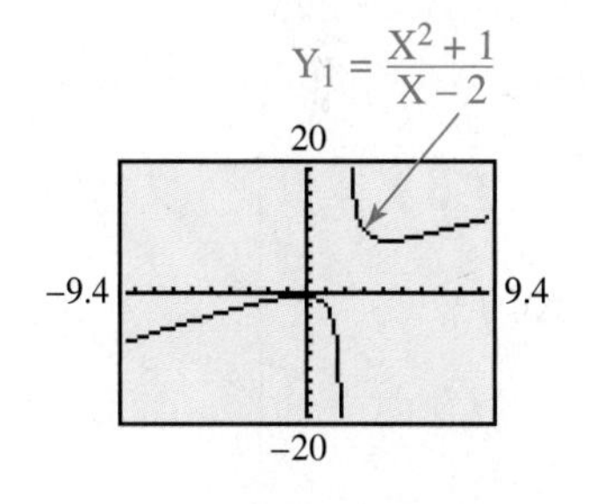


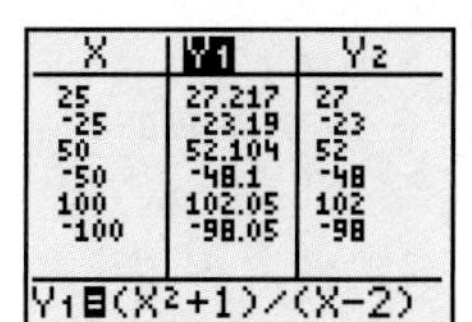

| X | Y1 | Y2 |
|---|---|---|
| 25 | 27.217 | 27 |
| -25 | -23.19 | -23 |
| 50 | 52.104 | 52 |
| -50 | -48.1 | -48 |
| 100 | 102.05 | 102 |
| -100 | -98.05 | -98 |

Y1=(X²+1)/(X-2)

**FIGURE 19**

## Graphs with Points of Discontinuity

As mentioned earlier, a rational function must be defined by an expression in lowest terms before we can use the methods discussed thus far in this section to hand-sketch the graph.

**EXAMPLE 8** *Graphing a Rational Function Defined by an Expression That Is Not in Lowest Terms*

Graph $f(x) = \dfrac{x^2 - 4}{x - 2}$.

### Analytic Solution

Notice that the domain of this function cannot include 2. The rational expression $\frac{x^2 - 4}{x - 2}$ can be written in lowest terms by factoring the numerator, then dividing both the numerator and denominator by $x - 2$.

$$f(x) = \frac{x^2 - 4}{x - 2}$$
$$= \frac{(x + 2)(x - 2)}{x - 2}$$
$$= x + 2 \quad (x \neq 2)$$

Therefore, the graph of this function will be the same as the graph of $y = x + 2$ (a straight line), with the exception of the point with $x$-value 2. A "hole" appears in the graph at $(2, 4)$. See Figure 20.

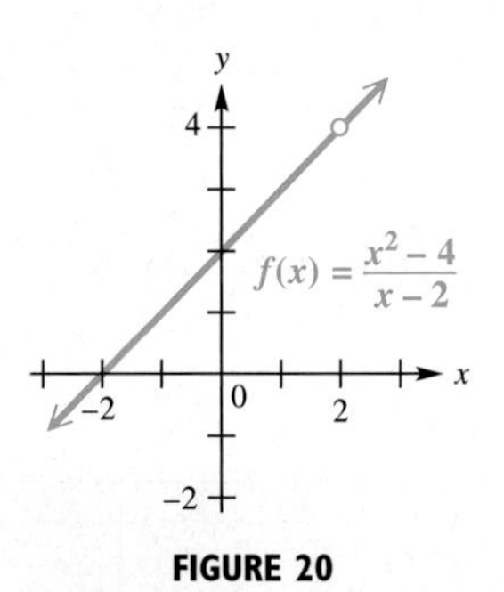


FIGURE 20

### Graphing Calculator Solution

If we set the window of a graphing calculator so that an $x$-value for the location of the tracing cursor is 2, then we can see from the display that the calculator cannot determine a value for Y. We define

$$Y_1 = \frac{X^2 - 4}{X - 2},$$

and graph it in such a window, as in Figure 21. The error message in the table further supports the existence of discontinuity at $X = 2$. (For the table, $Y_2 = X + 2$.)

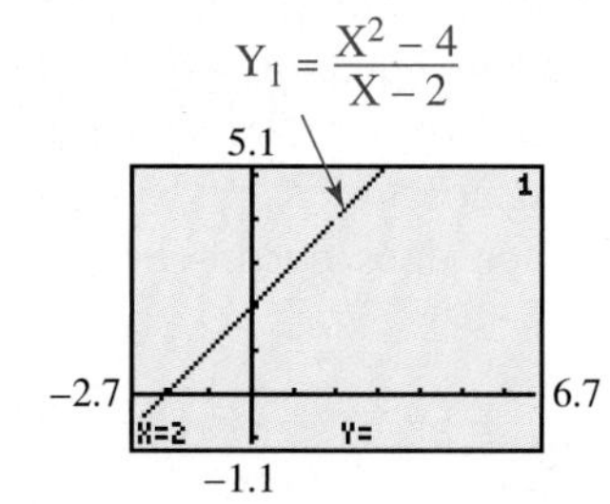


| X | $Y_1$ | $Y_2$ |
|---|---|---|
| -1 | 1 | 1 |
| 0 | 2 | 2 |
| 1 | 3 | 3 |
| 2 | ERROR | 4 |
| 3 | 5 | 5 |
| 4 | 6 | 6 |
| 5 | 7 | 7 |

Y1■(X²-4)/(X-2)

FIGURE 21

Look closely at the graph and notice that there is a visible discontinuity at $X = 2$. The window was chosen in such a way that the "hole" is visible. Other window choices may not show this discontinuity.

**TECHNOLOGY NOTE**

Experiment with your calculator to see if you can obtain a graph like the one in Figure 21, where the point of discontinuity is obvious. Then, experiment with different windows to see if you can find one where the point of discontinuity is not visible.

In summary, rational functions presented in Sections 4.1 and 4.2 fall into these categories.

1. $y = \frac{1}{x}$ and $y = \frac{1}{x^2}$ and their transformations (Section 4.1)
2. Degree of numerator $\leq$ degree of denominator (Examples 1, 2, 4, 5, and 6 of this section)
3. Degree of numerator $>$ degree of denominator (Examples 3 and 7 of this section)
4. Those with common variable factors in numerator and denominator (Example 8 of this section)

**Looking Ahead to Calculus**

Different types of discontinuity are discussed in calculus. The function in Example 8,

$$f(x) = \frac{x^2 - 4}{x - 2},$$

is said to have *removable* discontinuity at $x = 2$ since the discontinuity can be removed by redefining $f$ at 2. The function in Example 7,

$$f(x) = \frac{x^2 + 1}{x - 2},$$

has *infinite* discontinuity at $x = 2$, as indicated by the vertical asymptote there. The greatest integer function, discussed in Section 2.5, has *jump* discontinuities because the function values "jump" from one value to another for integer domain values.

## What Went WRONG?

A student was asked to determine the domain of the rational function defined by

$$f(x) = \frac{x^2 - 40}{x^2 - 4}.$$

Using the techniques of this section, the student determined that $-2$ and 2 are excluded, and gave her answer as

$$(-\infty, -2) \cup (-2, 2) \cup (2, \infty).$$

She then graphed the function in the window $[-9.4, 9.4]$ by $[-6.2, 6.2]$ and obtained the graph shown in this screen.

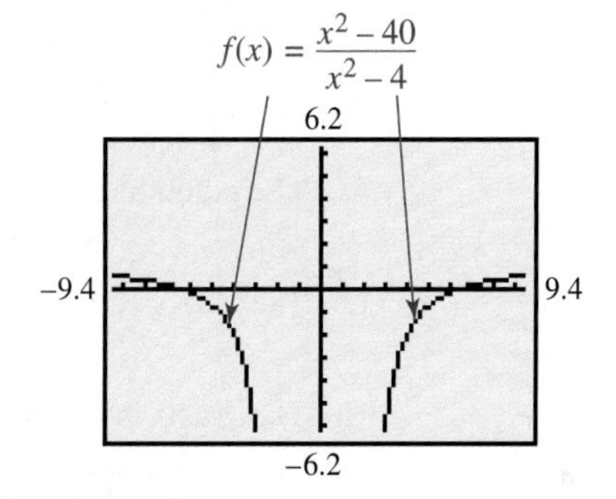


She became puzzled because no portion of the graph appeared between $-2$ and 2. So she changed her answer to

$$(-\infty, -2) \cup (2, \infty).$$

Her instructor counted her second answer incorrect and told her that she should have given her first answer, which was correct.

**What Went Wrong?** Why did her graph not indicate any points between $-2$ and 2?

---

*Answer to What Went Wrong?*

There is a U-shaped portion of the graph between $-2$ and 2, but it appears above the maximum $y$-value shown in the screen (6.2). She should increase the $y$-maximum value. (The window $[-9.4, 9.4]$ by $[-6.2, 18.6]$ works well.)

## 4.2 EXERCISES

*Concept Check* *Match the rational function in Column I with the appropriate description in Column II. Choices in Column II can be used only once.*

| I | II |
|---|---|
| **1.** $f(x) = \frac{x+7}{x+1}$ | **A.** The $x$-intercept is $-3$. |
| **2.** $f(x) = \frac{x+10}{x+2}$ | **B.** The $y$-intercept is 5. |
| **3.** $f(x) = \frac{1}{x+12}$ | **C.** The horizontal asymptote is $y = 4$. |
| **4.** $f(x) = \frac{-3}{x^2}$ | **D.** The vertical asymptote is $x = -1$. |
| **5.** $f(x) = \frac{x^2-16}{x+4}$ | **E.** There is a "hole" in its graph at $x = -4$. |
| **6.** $f(x) = \frac{4x+3}{x-7}$ | **F.** The graph has an oblique asymptote. |
| **7.** $f(x) = \frac{x^2+3x+4}{x-5}$ | **G.** The $x$-axis is its horizontal asymptote. |
| **8.** $f(x) = \frac{x+3}{x-6}$ | **H.** The $y$-axis is its vertical asymptote. |

*Give the equations of any vertical, horizontal, or oblique asymptotes for the graph of each rational function. For those with horizontal asymptotes, graph the function on your calculator and support your answer by tracing and letting $x \to \infty$ and $x \to -\infty$. You should be able to see how the y-value approaches a constant.*

**9.** $f(x) = \frac{3}{x-5}$

**10.** $f(x) = \frac{-6}{x+9}$

**11.** $f(x) = \frac{4-3x}{2x+1}$

**12.** $f(x) = \frac{2x+6}{x-4}$

**13.** $f(x) = \frac{x^2-1}{x+3}$

**14.** $f(x) = \frac{x^2+4}{x-1}$

**15.** $f(x) = \frac{x^2-2x-3}{2x^2-x-10}$

**16.** $f(x) = \frac{3x^2-6x-24}{5x^2-26x+5}$

**17.** *Concept Check* Which function has a graph that does not have a vertical asymptote?

**A.** $f(x) = \frac{1}{x^2+2}$ **B.** $f(x) = \frac{1}{x^2-2}$ **C.** $f(x) = \frac{3}{x^2}$ **D.** $f(x) = \frac{2x+1}{x-8}$

**18.** *Concept Check* Which function has a graph that does not have a horizontal asymptote?

**A.** $f(x) = \frac{2x-7}{x+3}$ **B.** $f(x) = \frac{3x}{x^2-9}$ **C.** $f(x) = \frac{x^2-9}{x+3}$ **D.** $f(x) = \frac{x+5}{(x+2)(x-3)}$

*Graph each rational function using a graphing calculator. First make a rough sketch by hand, using the guidelines of this section, so that it will be easier to find an appropriate window. Your graph should be a comprehensive one.*

**19.** $f(x) = \frac{x+1}{x-4}$

**20.** $f(x) = \frac{x-5}{x+3}$

**21.** $f(x) = \frac{3x}{(x+1)(x-2)}$

**22.** $f(x) = \frac{2x+1}{(x+2)(x+4)}$

**23.** $f(x) = \frac{5x}{x^2-1}$

**24.** $f(x) = \frac{x}{4-x^2}$

**25.** $f(x) = \frac{(x-3)(x+1)}{(x-1)^2}$

**26.** $f(x) = \frac{x(x-2)}{(x+3)^2}$

**27.** $f(x) = \frac{x}{x^2-9}$

**28.** $f(x) = \frac{-5}{2x+4}$

**29.** $f(x) = \frac{1}{x^2+1}$

**30.** $f(x) = \frac{(x-5)(x-2)}{x^2+9}$

**31.** $f(x) = \dfrac{x^2 + 1}{x + 3}$ **32.** $f(x) = \dfrac{2x^2 + 3}{x - 4}$ **33.** $f(x) = \dfrac{x^2 + 2x}{2x - 1}$ **34.** $f(x) = \dfrac{x^2 - x}{x + 2}$

**35.** $f(x) = \dfrac{x^2 - 9}{x + 3}$ **36.** $f(x) = \dfrac{x^2 - 16}{x + 4}$

Concept Check *In each table,* $Y_1$ *is defined by a rational expression of the form* $\dfrac{X - p}{X - q}$. *Use the table to find the values of p and q.*

**37.**

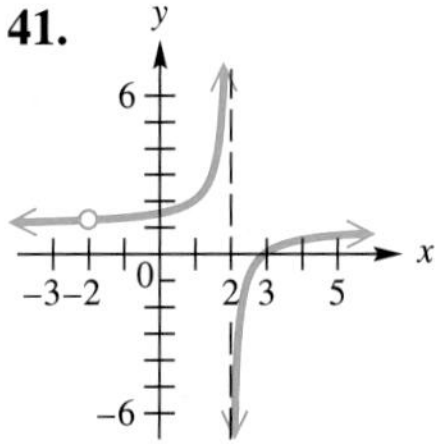

| X | Y1 |
|---|---|
| 0 | 2 |
| 1 | 3 |
| 2 | ERROR |
| 3 | -1 |
| 4 | 0 |
| 5 | .33333 |
| 6 | .5 |

X=0

**38.**

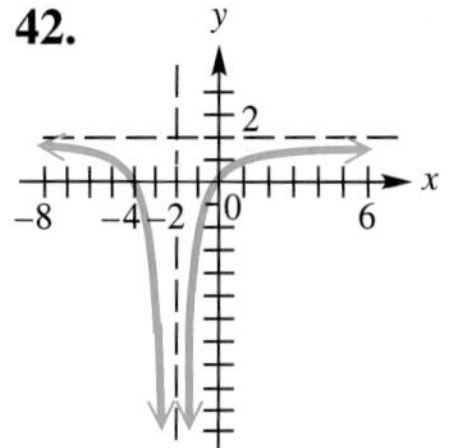

| X | Y1 |
|---|---|
| 0 | 3 |
| 1 | ERROR |
| 2 | -1 |
| 3 | 0 |
| 4 | .33333 |
| 5 | .5 |
| 6 | .6 |

X=0

**39.**

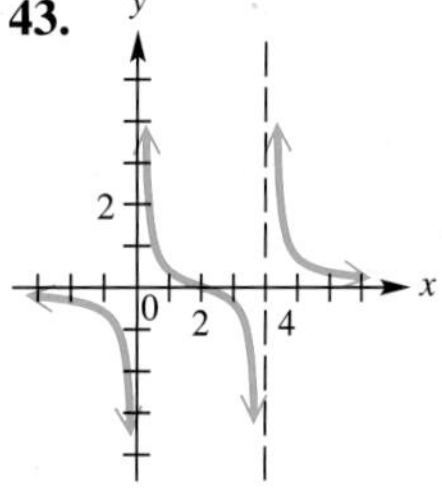

| X | Y1 |
|---|---|
| -3 | .5 |
| -2 | 0 |
| -1 | ERROR |
| 0 | 2 |
| 1 | 1.5 |
| 2 | 1.3333 |
| 3 | 1.25 |

X=-3

**40.**

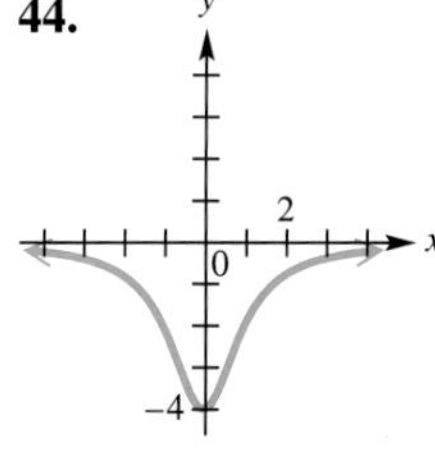

| X | Y1 |
|---|---|
| -7 | .75 |
| -6 | .66667 |
| -5 | .5 |
| -4 | 0 |
| -3 | ERROR |
| -2 | 2 |
| -1 | 1.5 |

X=-7

Concept Check *Find an equation for the rational function graphed.*

**41.** **42.** **43.** **44.**

*Consider*

$$f(x) = \frac{x + 3}{x^2 + x + 4},$$

*which is defined by a rational expression in lowest terms whose denominator is a quadratic polynomial.*

**45.** To find any vertical asymptotes, set the denominator equal to 0 and solve. What are the complex solutions of the equation you solved? What are the real solutions? Does $f$ have any vertical asymptotes?

**46.** With your calculator in connected mode and using a window of $[-10, 10]$ by $[-1, 3]$, graph the function. Why is connected mode acceptable here to get a realistic view of the graph?

## RELATING CONCEPTS

**For individual or group investigation (Exercises 47–50)**

*Recall from Section 2.3 that if we are given the graph of* $y = f(x)$, *we can obtain the graph of* $y = -f(x)$ *by reflecting across the x-axis, and we can obtain the graph of* $y = f(-x)$ *by reflecting across the y-axis. In Exercises 47–50, you are given the graph of a rational function* $y = f(x)$. *Draw a sketch by hand of the graph of* **(a)** $y = -f(x)$ *and* **(b)** $y = f(-x)$.

**47.**

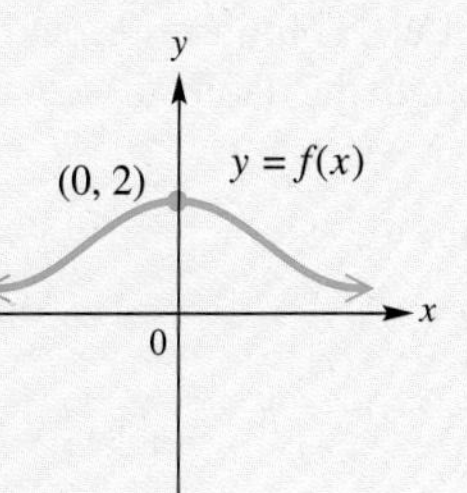


**48.**

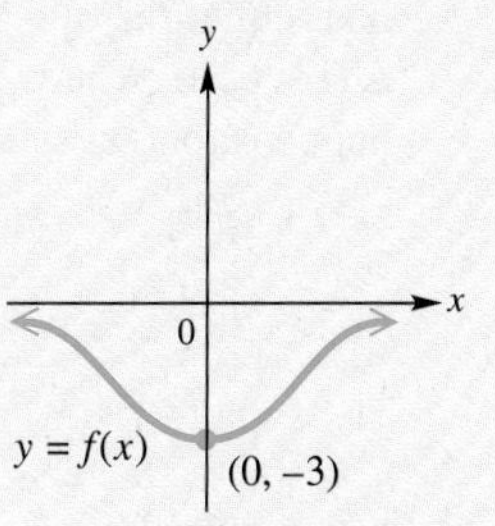


**49.**

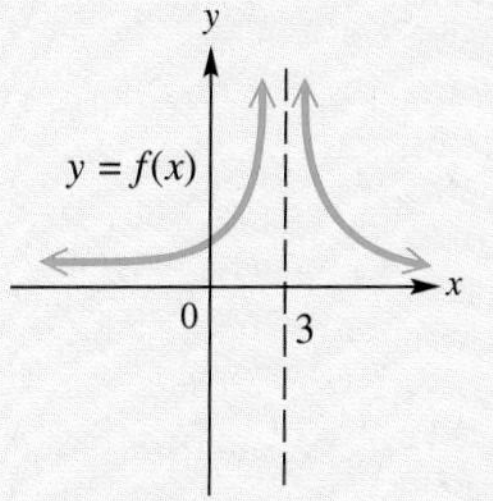


**50.**

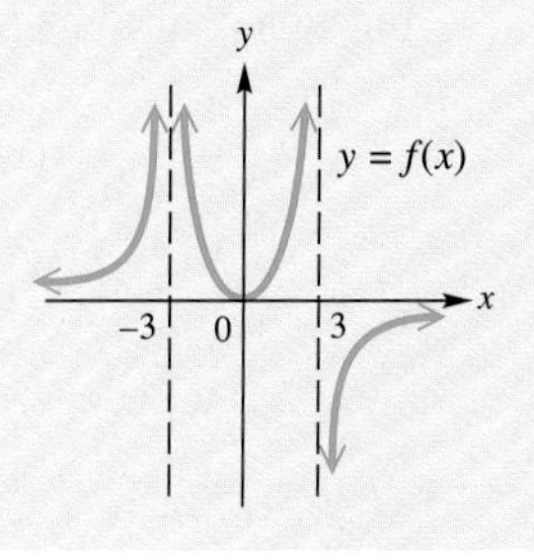

*Each rational function has an oblique asymptote. Determine the equation of this asymptote. Then, use a graphing calculator to graph both the function and the asymptote in the window indicated. (These windows are appropriate for TI-83 calculators.)*

**51.** $f(x) = \dfrac{2x^2 + 3}{4 - x}$; window: $[-18.8, 18.8]$ by $[-50, 25]$

**52.** $f(x) = \dfrac{x^2 + 9}{x + 3}$; window: $[-9.4, 9.4]$ by $[-25, 25]$

**53.** $f(x) = \dfrac{x - x^2}{x + 2}$; window: $[-9.4, 9.4]$ by $[-15, 25]$

**54.** $f(x) = \dfrac{x^2 + 2x}{1 - 2x}$; window: $[-4.7, 4.7]$ by $[-5, 5]$

**55.** *Concept Check* $f(x) = \dfrac{x^5 + x^4 + x^2 + 1}{x^4 + 1}$

becomes $f(x) = x + 1 + \dfrac{x^2 - x}{x^4 + 1}$

after the numerator is divided by the denominator.

**(a)** What is the equation of the oblique asymptote of the graph of the function?

**(b)** Where does the graph of the function intersect its asymptote?

**(c)** As $x \to \infty$, does the graph of the function approach its asymptote from above or below?

**56.** *Concept Check* Consider the rational function defined by

$$f(x) = \frac{x^3 - 4x^2 + x + 6}{x^2 + x - 2}.$$

Divide the numerator by the denominator and use the method of Example 3 to determine the equation of the oblique asymptote. Then, determine the coordinates of the point where the graph of $f$ intersects its oblique asymptote. Use a calculator to support your answer.

**57.** Use long division of polynomials to show that for the function defined by

$$f(x) = \frac{x^4 - 5x^2 + 4}{x^2 + x - 12},$$

if we divide the numerator by the denominator, the quotient polynomial is $x^2 - x + 8$, and the remainder is $-20x + 100$. Then, graph both $f$ and $g$, where $g(x) = x^2 - x + 8$ in the window $[-50, 50]$ by $[0, 1000]$. Comment on the appearance of the two graphs. Explain how the graph of $f$ approaches that of $g$ as $|x|$ gets very large.

**58.** Suppose a friend tells you that the graph of

$$f(x) = \frac{x^2 - 25}{x + 5}$$

has a vertical asymptote with equation $x = -5$. Is this correct? If not, describe the behavior of the graph at $x = -5$.

*Each rational function in Exercises 59 and 60 has a "hole" in its graph.*

**(a)** *Write the fraction in lowest terms, and call this new function $g(x)$.*

**(b)** *Use a table with $Y_1 = f(x)$ and $Y_2 = g(x)$ to support the fact that for all values except those for which $f$ is undefined, $f(x) = g(x)$.*

**(c)** *What are the coordinates of the "hole"? Support your answer with a graph.*

**59.** $f(x) = \dfrac{x^2 - 9}{x + 3}$

**60.** $f(x) = \dfrac{x^2 - 36}{6 - x}$

**61.** Let $f(x) = \dfrac{p(x)}{q(x)}$ define a rational function where the expression is written in lowest terms. Suppose the degree of $p(x)$ is $m$, and the degree of $q(x)$ is $n$. Explain how you would determine the nonvertical asymptote in each situation.

**(a)** $m < n$ **(b)** $m = n$ **(c)** $m = n + 1$

**62.** Use the table feature of a graphing calculator to confirm that the horizontal asymptote of

$$f(x) = \frac{6x^2 + 3}{2x^2}$$

has equation $y = 3$. (*Hint:* Let Tblmin = 0, and let $\Delta$Tbl = 10. Then, scroll through the table to see what happens to $f(x)$ as $x \to \infty$ and $x \to -\infty$.)

# RATIONAL EQUATIONS, INEQUALITIES, APPLICATIONS, AND MODELS

Solving Rational Equations and Inequalities • Applications and Models of Rational Functions • Inverse Variation • Combined and Joint Variation

## Solving Rational Equations and Inequalities

A rational equation (or inequality) is an equation (or inequality) with at least one term having a variable expression in a denominator, or at least one term having a variable expression raised to a negative integer power. Some examples are

$$\frac{x+2}{2x+1}=1, \quad \frac{x+2}{2x+1}\le 1, \quad \frac{x}{x-2}+\frac{1}{x+2}=\frac{8}{x^2-4}, \quad \text{and} \quad 7x^{-4}-8x^{-2}+1=0.$$

When solving rational equations and inequalities, remember that an expression with a variable denominator or with a negative exponent may be undefined for certain values of the variable. Each time we solve such an equation or inequality, we must identify these values (which are actually values at which the associated rational function will have a vertical asymptote or "hole"). The general procedure for solving rational equations is to multiply each side of the equation by the least common denominator (LCD) of all the terms in the equation.

### EXAMPLE 1 *Solving a Rational Equation*

Solve $\dfrac{x+2}{2x+1}=1$.

**Analytic Solution**

Begin by determining that the rational expression is undefined for $x=-\frac{1}{2}$. To clear fractions, multiply each side of the equation by $2x+1$.

$$\left(\frac{x+2}{2x+1}\right)(2x+1)=(1)(2x+1)$$

$$x+2=2x+1$$

$$1=x$$

Check by substituting 1 in the original equation.

$$\frac{1+2}{2(1)+1}=1 \quad ?$$

$$\frac{3}{3}=1 \quad ?$$

$$1=1$$

The solution set is $\{1\}$.

**Graphing Calculator Solution**

Rewrite the equation as

$$\frac{x+2}{2x+1}-1=0,$$

and let

$$y=\frac{x+2}{2x+1}-1.$$

Because graphs of rational functions usually consist of several parts, using the intersection-of-graphs method can be confusing. The $x$-intercept method shows that the zero of the function is 1. See Figure 22.

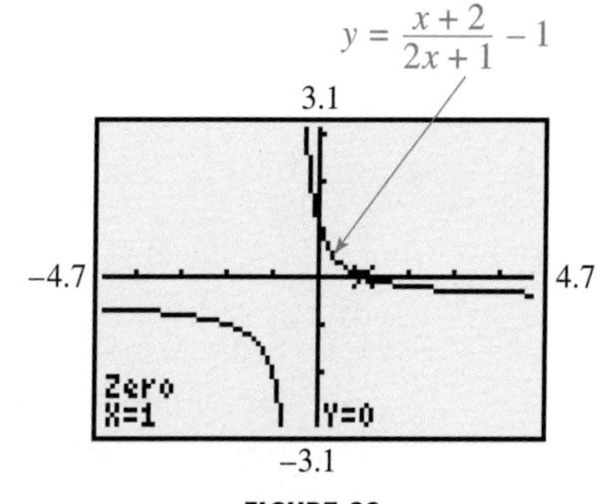


FIGURE 22

To solve the rational inequality

$$\frac{x+2}{2x+1} \leq 1,$$

do not multiply each side by $2x + 1$, since this expression may represent a negative number and multiplying by it would require reversing the inequality symbol. We use sign graphs, first introduced in Section 3.3 to solve quadratic inequalities.

### EXAMPLE 2 *Solving a Rational Inequality*

**(a)** Use a sign graph to solve

$$\frac{x+2}{2x+1} \leq 1.$$

**(b)** Discuss how the graph in Figure 22 supports the solution.

**Solution**

**(a)** We begin by subtracting 1 from each side and combining the terms on the left side to get a single rational expression.

$$\frac{x+2}{2x+1} - 1 \leq 0 \quad \text{Subtract 1.}$$

$$\frac{x+2}{2x+1} - \frac{2x+1}{2x+1} \leq 0 \quad \text{The common denominator is } 2x+1.$$

$$\frac{x+2-(2x+1)}{2x+1} \leq 0 \quad \text{Write as a single rational expression.}$$

$$\frac{x+2-2x-1}{2x+1} \leq 0 \quad \text{Distributive property; be careful with signs.}$$

$$\frac{-x+1}{2x+1} \leq 0$$

To determine the sign graph, we solve the equations

$$-x + 1 = 0 \quad \text{and} \quad 2x + 1 = 0$$

to get $x = 1$ and $x = -\frac{1}{2}$. We use these values to divide the number line into three intervals. Then we complete the sign graph as in Figure 23, and determine the intervals where the quotient is *negative*.

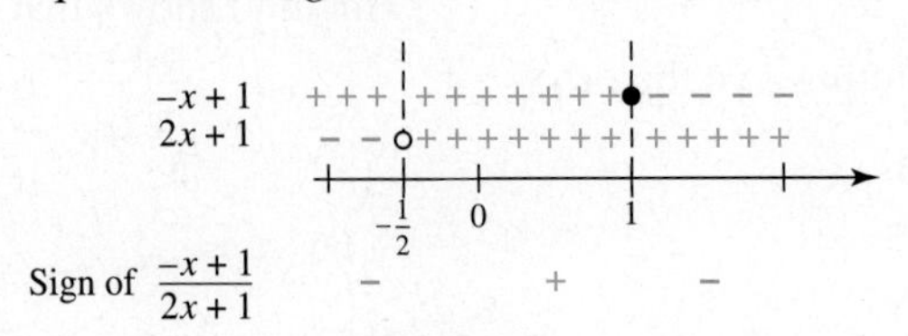


**FIGURE 23**

From the sign graph, we see that values of $x$ in the two intervals $\left(-\infty, -\frac{1}{2}\right)$ and $(1, \infty)$ make the quotient negative, as required. Because of the $\leq$ symbol, the endpoint 1 is included. However, the endpoint $-\frac{1}{2}$ causes the denominator to equal 0, so it is not included in the solution set. Therefore, the solution set is

$$\left(-\infty, -\tfrac{1}{2}\right) \cup [1, \infty).$$

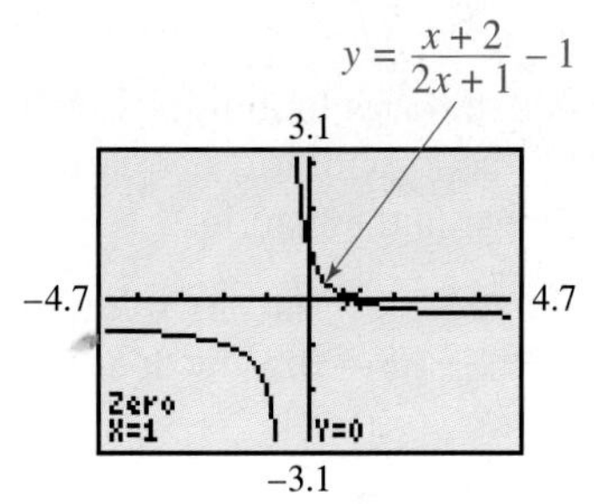


**(b)** The original inequality is equivalent to $\frac{x+2}{2x+1} - 1 \leq 0$. From Figure 22 repeated in the margin, we see that the graph lies *below* the $x$-axis for $x$-values less than the $x$-value of the vertical asymptote $\left(x = -\frac{1}{2}\right)$ as well as those greater than 1, which agrees with the solution set obtained in part (a).

**CAUTION** As suggested by Example 2, be careful with endpoints of intervals when solving rational inequalities.

**EXAMPLE 3** *Solving a Rational Equation*

Solve $\frac{x}{x-2} + \frac{1}{x+2} = \frac{8}{x^2-4}$.

### Analytic Solution

For this equation, $x \neq \pm 2$. (Why?)

$$\frac{x}{x-2} + \frac{1}{x+2} = \frac{8}{x^2-4}$$

$$x(x+2) + 1(x-2) = 8 \quad \text{Multiply by the LCD, } (x-2)(x+2).$$

$$x^2 + 2x + x - 2 = 8 \quad \text{Distributive property}$$

$$x^2 + 3x - 10 = 0 \quad \text{Standard form}$$

$$(x+5)(x-2) = 0 \quad \text{Factor.}$$

$$x + 5 = 0 \quad \text{or} \quad x - 2 = 0 \quad \text{Zero-product property}$$

$$x = -5 \quad \text{or} \quad x = 2$$

The numbers $-5$ and 2 are the *possible* solutions of the equation. Recall that 2 is not in the domain of the original equation and, therefore, must be rejected. (Such a value is called *extraneous.*) The solution set is $\{-5\}$. (The number 2 is a solution of the equation obtained after multiplying by the LCD. Multiplying by $x - 2$ when $x = 2$ is multiplying by 0, which does not lead to an equivalent equation.)

### Graphing Calculator Solution

Rewrite the equation as

$$\frac{x}{x-2} + \frac{1}{x+2} - \frac{8}{x^2-4} = 0,$$

and let $y$ equal the left side of this equation. Figure 24 shows that the zero of the function is $-5$, as expected. The line $x = -2$ is a vertical asymptote, and the graph has a "hole" where $x = 2$.

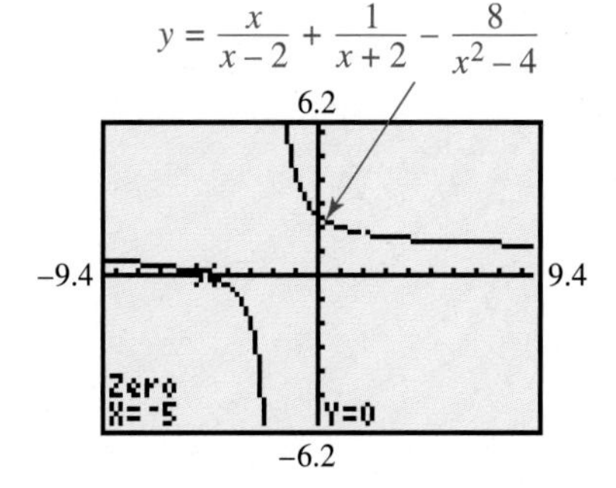


There is a "hole" at (2, 1.75).

**FIGURE 24**

Example 3 illustrates the general procedure for solving rational equations.

**Solving an Equation Involving a Rational Function**

1. Determine all values for which the rational function is undefined.
2. Multiply each side of the equation by the least common denominator of all rational expressions in the equation.
3. Solve the resulting equation.
4. Reject any values found in Step 1.

$y = x^2 + 3x - 10$

10

−10 10

−15

The $x$-intercepts are −5 and 2.

### FOR DISCUSSION

When solving the rational equation in Example 3, our first step was to multiply each side by $(x - 2)(x + 2)$. This eventually led to the quadratic equation $x^2 + 3x - 10 = 0$. The graph of $y = x^2 + 3x - 10$ is shown in the margin.

1. Discuss how the solution of the equation in Example 3 and the extraneous solution found there relate to the solutions of the equation $x^2 + 3x - 10 = 0$. How does the extraneous solution show up on the graph of the parabola?
2. Why should you always pay attention to the domain of a rational equation?

### EXAMPLE 4 *Solving a Rational Equation and Associated Inequalities*

Consider the rational function defined by $f(x) = 7x^{-4} - 8x^{-2} + 1$.

**(a)** Solve the equation $f(x) = 0$.

**(b)** Graph the equation $y = f(x)$. Identify the zeros of $f(x)$.

**(c)** Use the graph to solve the inequalities $f(x) \leq 0$ and $f(x) \geq 0$.

**Solution**

**(a)** The expression for $f(x)$ may be written equivalently as

$$\frac{7}{x^4} - \frac{8}{x^2} + 1.$$

For this expression, $x \neq 0$. Solve the equation $f(x) = 0$.

$$\frac{7}{x^4} - \frac{8}{x^2} + 1 = 0$$

$$7 - 8x^2 + x^4 = 0 \qquad \text{Multiply by } x^4.$$

$$x^4 - 8x^2 + 7 = 0$$

$$(x^2 - 7)(x^2 - 1) = 0$$

$$(x^2 - 7)(x + 1)(x - 1) = 0$$

$$x^2 - 7 = 0 \quad \text{or} \quad x + 1 = 0 \quad \text{or} \quad x - 1 = 0$$

$$x = \pm\sqrt{7} \quad \text{or} \quad x = -1 \quad \text{or} \quad x = 1$$

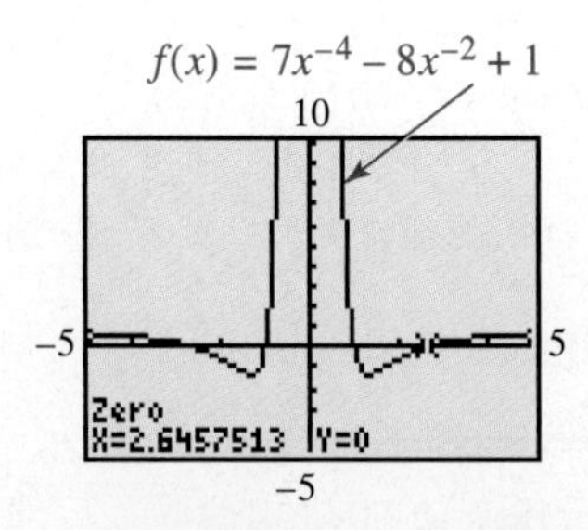


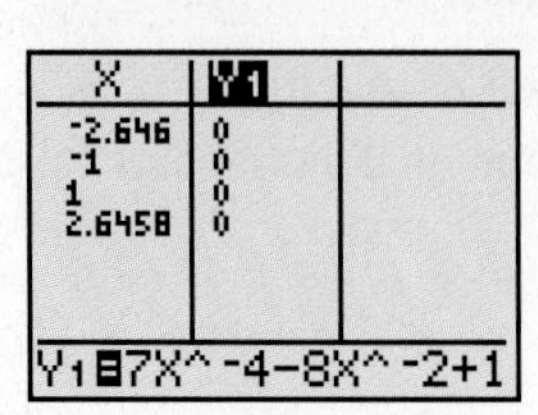


**FIGURE 25**

The solution set is $\{\pm\sqrt{7}, \pm 1\}$.

**(b)** Figure 25 shows that the graph of $f(x) = 7x^{-4} - 8x^{-2} + 1$ has four zeros. The display indicates that one zero is 2.6457513, an approximation for $\sqrt{7}$. The other zeros, $-\sqrt{7}$, $-1$, and 1 can be found similarly. The table provides further support.

**(c)** The solution set of $f(x) \leq 0$ is $[-\sqrt{7}, -1] \cup [1, \sqrt{7}]$ since these are the values for which the graph lies *on* or *below* the $x$-axis. Similarly, the solution set of $f(x) \geq 0$ is

$$(-\infty, -\sqrt{7}] \cup [-1, 0) \cup (0, 1] \cup [\sqrt{7}, \infty).$$

Notice that 0 is not included because it is not in the domain of $f(x)$.

**FOR DISCUSSION**

Tables can be used to numerically support analytic solutions of equations. In Example 4, we show that the equation $7x^{-4} - 8x^{-2} + 1 = 0$ has solutions $-\sqrt{7}$, $-1$, $1$, and $\sqrt{7}$. By creating a table with these four values as in Figure 25, we can show that the function value in each case is 0. The X-values $-2.646$ and $2.6458$ shown in the table are decimal approximations for $-\sqrt{7}$ and $\sqrt{7}$. We must enter these X-values directly as $-\sqrt{7}$ and $\sqrt{7}$; if we use approximations, we will not get 0 for the corresponding $Y_1$ values.

1. Use a table in the manner described to support the solution in Example 1.
2. Use a table to support the single solution in Example 3. What happens when you input the extraneous value 2?

## Applications and Models of Rational Functions

**EXAMPLE 5** *Analyzing Traffic Intensity*

Vehicles arrive randomly at a parking ramp at an average rate of 2.6 vehicles per minute. The parking attendant can admit 3.2 vehicles per minute. However, since arrivals are random, lines form at various times. (*Source:* Mannering, F. and W. Kilareski, *Principles of Highway Engineering and Traffic Control,* Second Edition, John Wiley and Sons, 1998.)

**(a)** The *traffic intensity* $x$ is defined as the ratio of the average arrival rate to the average admittance rate. Determine $x$ for this parking ramp.

**(b)** The average number of vehicles waiting in line to enter the ramp is modeled by

$$f(x) = \frac{x^2}{2(1 - x)},$$

where $0 \le x < 1$ is the traffic intensity. Compute $f(x)$ for this parking ramp.

**(c)** Graph $y = f(x)$. What happens to the number of vehicles waiting as the traffic intensity approaches 1?

### Solution

**(a)** The average arrival rate is 2.6 vehicles and the average admittance rate is 3.2 vehicles, so

$$x = \frac{2.6}{3.2} = .8125.$$

**(b)** In part (a), we found that $x = .8125$. Thus,

$$f(.8125) = \frac{.8125^2}{2(1 - .8125)} \approx 1.76 \text{ vehicles.}$$

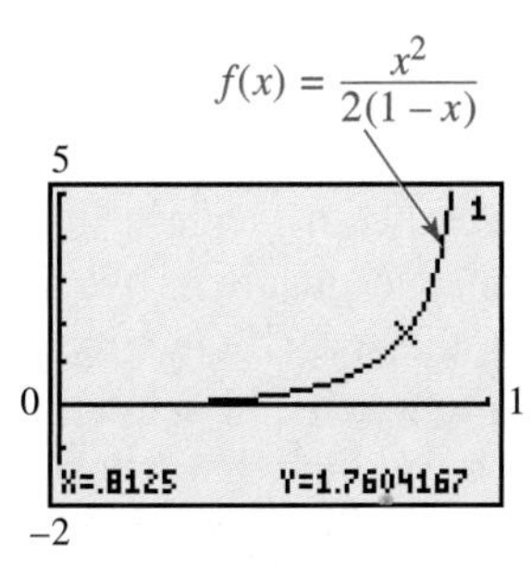


**FIGURE 26**

**(c)** From the graph shown in Figure 26, we see that as $x$ approaches 1, $y = f(x)$ gets very large; that is, the number of waiting vehicles gets very large. This is what we would expect.

**EXAMPLE 6** *Solving a Problem Involving Aluminum Can Manufacture*

A manufacturer wants to construct cylindrical aluminum cans with volume 2000 cubic centimeters (2 liters). What radius and height of the can will minimize the amount of aluminum used? What will this amount be?

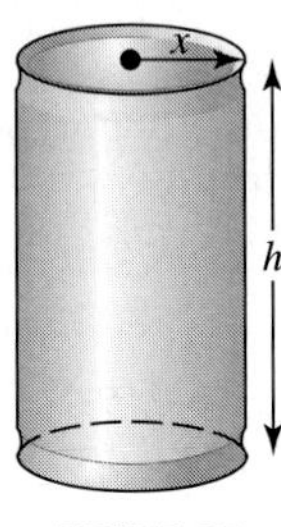


FIGURE 27

**Looking Ahead to Calculus**

The problem in Example 6 here cannot be solved in college algebra without the aid of a computer or a graphing calculator. The traditional calculus method of solving this problem would involve (analytically) finding the appropriate zero of the derivative of $V(x)$:

$$V'(x) = -4000x^{-2} + 4\pi x.$$

**Solution** The two unknowns in this problem are the radius and the height of the can. We label the radius $x$ and the height $h$, as shown in Figure 27. Minimizing the amount of aluminum used requires minimizing the surface area of the can, which we designate $S$. The surface area $S$ is given by the formula

$$S = 2\pi xh + 2\pi x^2. \qquad x \text{ is radius; } h \text{ is height.}$$

The formula involves two variables on the right, so we solve for one in terms of the other to obtain a function of a single variable. Since the volume of the can is to be 2000 cubic centimeters and the formula for volume is $V = \pi x^2 h$ (where $x$ is radius and $h$ is height), we have

$$V = \pi x^2 h$$
$$2000 = \pi x^2 h \qquad \text{Let } V = 2000.$$
$$h = \frac{2000}{\pi x^2}. \qquad \text{Solve for } h.$$

Now we can write the surface area $S$ as a function of $x$ alone.

$$S(x) = 2\pi x\left(\frac{2000}{\pi x^2}\right) + 2\pi x^2 \qquad \text{Substitute for } h.$$
$$= \frac{4000}{x} + 2\pi x^2 \qquad *$$
$$= \frac{4000 + 2\pi x^3}{x} \qquad \text{Combine terms.}*$$

Since $x$ represents the radius, it must be a positive number. We graph the function and find that the local minimum point is approximately (6.83, 878.76). See Figure 28. Therefore, rounded to the nearest hundredth, the radius should be 6.83 centimeters and the height should be

$$\frac{2000}{\pi(6.83)^2}, \quad \text{or} \quad 13.65 \text{ centimeters.}$$

These dimensions lead to a minimum amount of 878.76 cubic centimeters of aluminum used. ◯

$S(x) = \dfrac{4000 + 2\pi x^3}{x}$

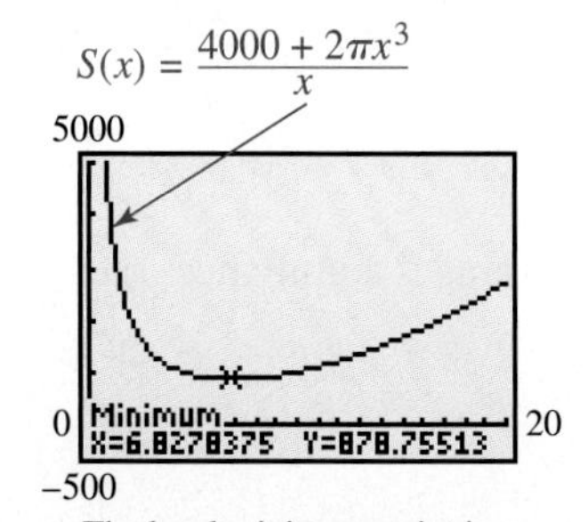


The local minimum point is approximately (6.83, 878.76).

FIGURE 28

## Inverse Variation

Recall from Section 1.6 that when two quantities vary directly, an increase in one quantity results in an increase in the other quantity. When two quantities *vary inversely,* an increase in one quantity results in a decrease in the second quantity. For example, it takes 4 hours to travel 100 miles at 25 miles per hour and 2 hours to travel 100 miles at 50 miles per hour. Greater speed results in less travel time. If $s$ represents the average speed of a car and $t$ is the time to travel 100 miles, then $s \cdot t = 100$ or $t = \frac{100}{s}$. Doubling the speed cuts the time in half, and tripling the speed cuts the time by one-third. The quantities $t$ and $s$ are said to vary inversely. The constant of variation here is 100.

*These steps are not necessary to obtain the appropriate graph.

**Inverse Variation as the *n*th Power**

Let $x$ and $y$ denote two quantities and $n$ be a positive number. Then $y$ is **inversely proportional to the *n*th power** of $x$, or $y$ **varies inversely as the *n*th power** of $x$, if there exists a nonzero number $k$ such that

$$y = \frac{k}{x^n}.$$

If $y = \frac{k}{x}$, then $y$ is **inversely proportional** to $x$, or $y$ **varies inversely** as $x$.

Inverse variation occurs when measuring the intensity of light. If we increase the distance from a light bulb, the intensity of the light decreases.

**EXAMPLE 7** *Modeling the Intensity of Light*

Intensity of light $I$ is inversely proportional to the second power of the distance $d$. (See Figure 29.) The equation

$$I = \frac{k}{d^2}$$

models this phenomenon. At a distance of 3 meters, a 100-watt bulb produces an intensity of .88 watt per square meter. (*Source:* Weidner, R. and R. Sells, *Elementary Classical Physics,* Volume 2, Allyn and Bacon, 1965.)

Find the constant of variation $k$, and then determine the intensity of the light at a distance of 2 meters.

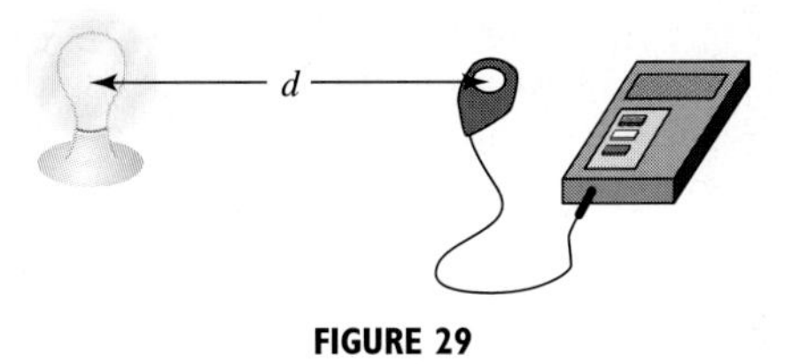


**FIGURE 29**

**Analytic Solution**

Substitute $d = 3$ and $I = .88$ into the variation equation, and solve for $k$.

$$I = \frac{k}{d^2}$$

$$.88 = \frac{k}{3^2}$$

$$k = 7.92$$

**Graphing Calculator Solution**

We must first find $k$ to be 7.92, using the method shown in the analytic solution. The formula

$$I = \frac{7.92}{d^2} \text{ can be written as } f(x) = \frac{7.92}{x^2},$$

a rational function. Enter this function into a calculator, and graph it for positive values of $x$. Calculate $f(2)$ by locating the point having $x = 2$, and determine that $f(2) = 1.98$. See Figure 30 on the next page.

*(continued)*

Since $k = 7.92$, the equation is

$$I = \frac{7.92}{d^2}.$$

Let $d = 2$ and find $I$.

$$I = \frac{7.92}{2^2} = 1.98$$

The intensity at 2 meters is 1.98 watts per square meter.

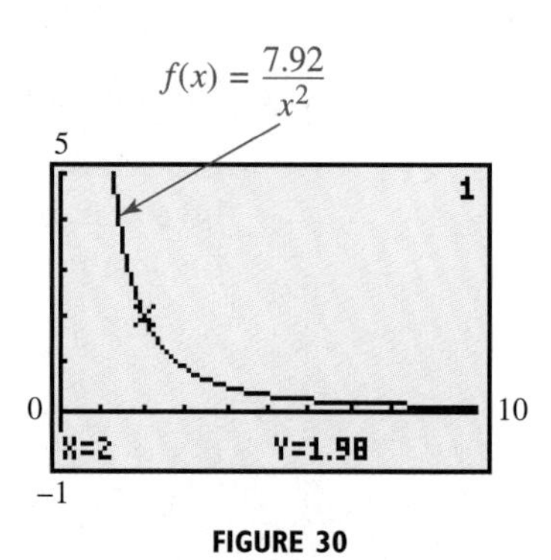

FIGURE 30

## Combined and Joint Variation

One variable may depend on more than one other variable. Such variation is called **combined variation.** More specifically, when a variable depends on the *product* of two or more other variables, we refer to it as *joint variation.* For example, the volume $V$ of a cylinder is given by $V = \pi r^2 h$. We say that $V$ *varies jointly* as $h$ and the square of $r$. The constant of variation here is $\pi$.

**Joint Variation**

Let $m$ and $n$ be real numbers. Then $z$ **varies jointly** as the $n$th power of $x$ and the $m$th power of $y$ if a nonzero real number $k$ exists such that

$$z = kx^n y^m.$$

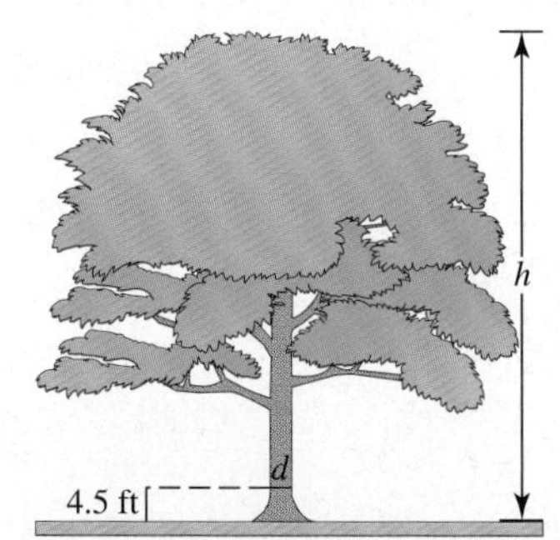

FIGURE 31

**EXAMPLE 8** *Modeling the Amount of Wood in a Tree*

In forestry it is common to estimate the volume of timber in a given area of forest. To do this, formulas are developed to find the amount of wood contained in a tree with height $h$ and diameter $d$. See Figure 31. One study concluded that the volume $V$ of wood in a tree varies jointly as the 1.12 power of $h$ and the 1.98 power of $d$. (The diameter is measured 4.5 feet above the ground.) (*Source:* Ryan, B., B. Joiner, and T. Ryan, *Minitab Handbook,* Duxbury Press, 1985.)

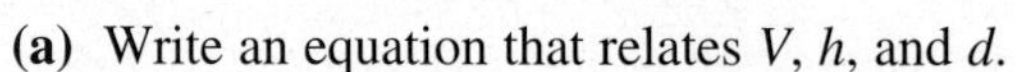

**(a)** Write an equation that relates $V$, $h$, and $d$.

**(b)** A tree with a 13.8-inch diameter and a 64-foot height has a volume of 25.14 cubic feet. Estimate the constant of variation $k$.

**(c)** Estimate the volume of wood in a tree with $d = 11$ inches and $h = 47$ feet.

**Solution**

This screen illustrates how the computations in Examples 8(b) and (c) are made using a graphing calculator.

**(a)** Let $V = kh^{1.12}d^{1.98}$, where $k$ is the constant of variation.

**(b)** Substitute $d = 13.8$, $h = 64$, and $V = 25.14$ into the equation, and solve for $k$.

$$25.14 = k(64^{1.12})(13.8^{1.98})$$

$$k = \frac{25.14}{(64^{1.12})(13.8^{1.98})} \approx .00132$$

Thus $V = .00132h^{1.12}d^{1.98}$.

**(c)** For the given tree, $V = .00132(47^{1.12})(11^{1.98}) \approx 11.4$ cubic feet.

**EXAMPLE 9** *Solving a Combined Variation Problem in Photography*

In the photography formula

$$L = \frac{25F^2}{st},$$

the luminance, $L$ (in foot-candles), varies directly as the square of the F-stop, $F$, and inversely as the product of the film ASA number, $s$, and the shutter speed, $t$. The constant of variation is 25.

Suppose we want to use 200 ASA film and a shutter speed of $\frac{1}{250}$ when 500 foot-candles of light are available. What would be an appropriate F-stop?

**Solution** In the formula, substitute the given values for the variables. Then, solve for $F$.

$$L = \frac{25F^2}{st}$$

$$500 = \frac{25F^2}{200\left(\frac{1}{250}\right)} \qquad L = 500,\ s = 200,\ t = \frac{1}{250}$$

$$400 = 25F^2$$

$$16 = F^2$$

$$4 = F \qquad F > 0$$

An F-stop of 4 would be appropriate.

## 4.3 EXERCISES

*In Exercises 1–9, the graph of a rational function $y = f(x)$ is given. Use the graph to give the solution set of* **(a)** $f(x) = 0$, **(b)** $f(x) < 0$, *and* **(c)** $f(x) > 0$. *Use set notation for part (a) and interval notation for parts (b) and (c).*

**1.**

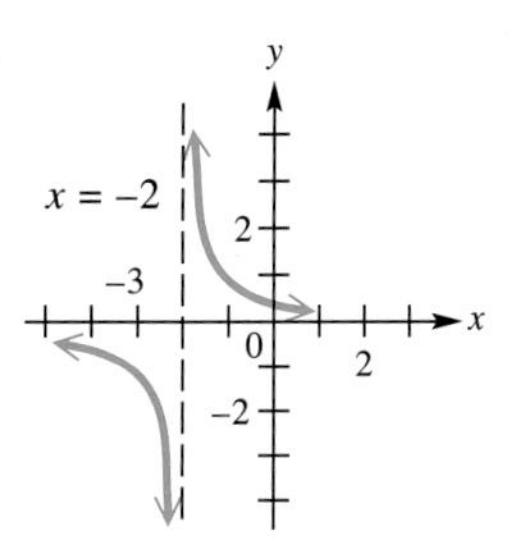


**2.**

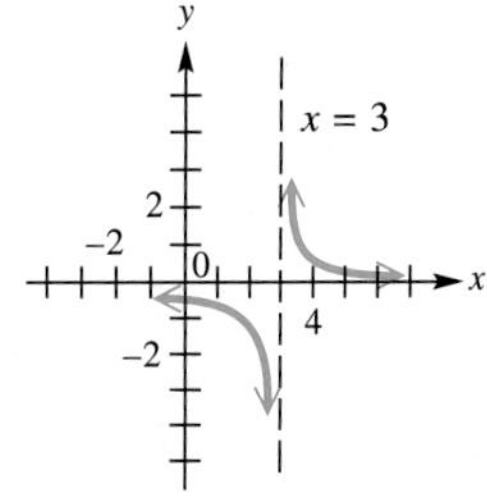


**3.**

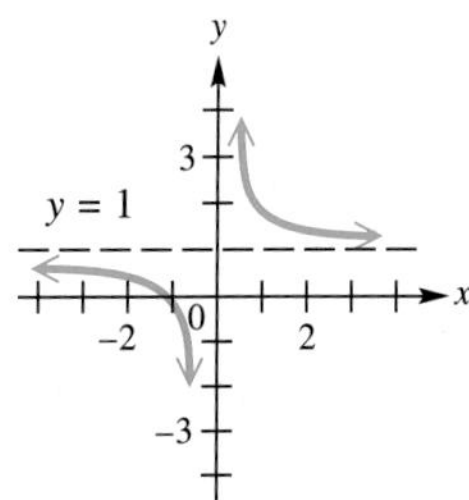


**4.**

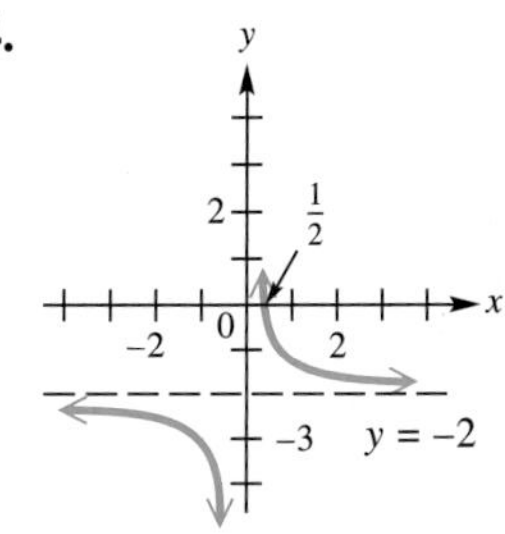


**5.**

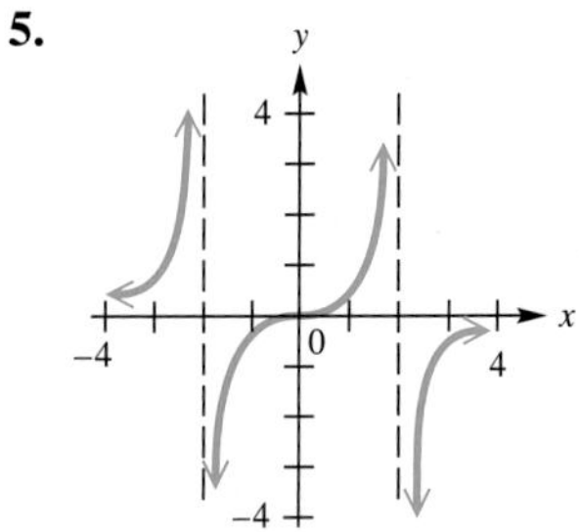


**6.**

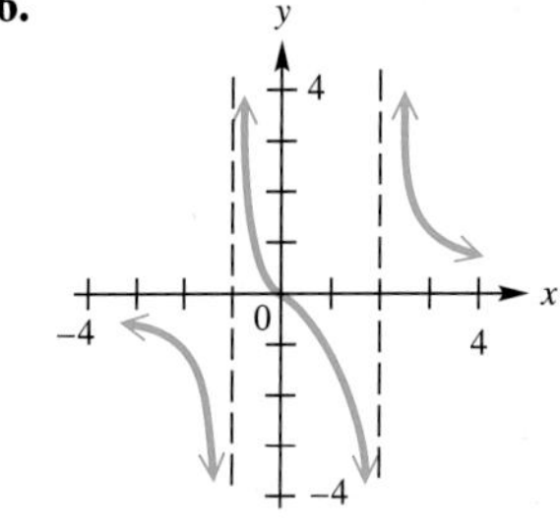

**7.**

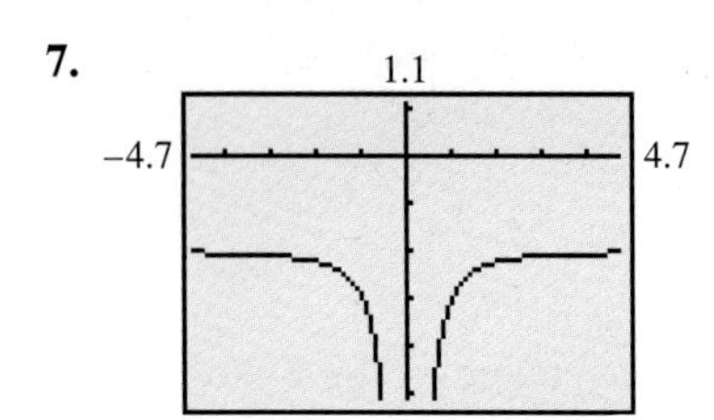


The line $x = 0$ is a vertical asymptote.

**8.**

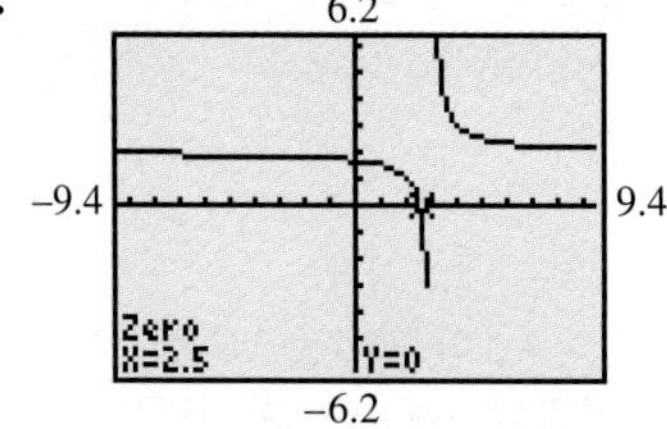


The line $x = 3$ is a vertical asymptote.

**9.**

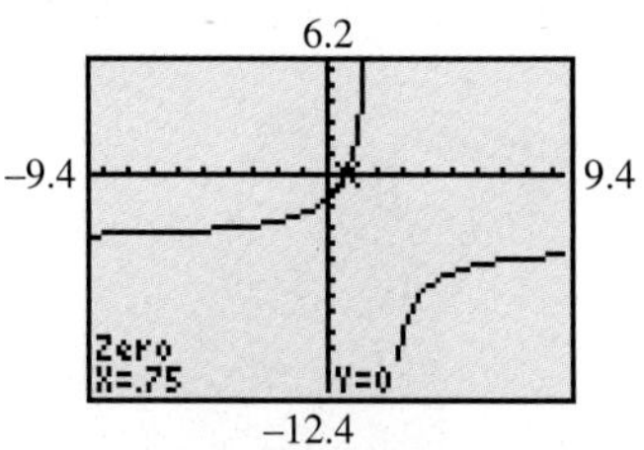


The line $x = 2$ is a vertical asymptote.

**10.** A student attempted to solve the inequality at the right by multiplying each side by $x + 2$. He wrote the solution set as $\left(-\infty, \frac{1}{2}\right]$. Is his solution correct? Explain.

$$\frac{2x-1}{x+2} \le 0$$
$$2x - 1 \le 0$$
$$x \le \frac{1}{2}$$

*Use the methods of Examples 1 and 2 to solve the rational equation and associated inequalities given in Exercises 11–22. Then, support your answer using the x-intercept method with a calculator graph in the suggested window.*

**11.** **(a)** $\frac{x-3}{x+5} = 0$ **(b)** $\frac{x-3}{x+5} \le 0$ **(c)** $\frac{x-3}{x+5} \ge 0$

Window: $[-10, 10]$ by $[-5, 8]$

**12.** **(a)** $\frac{x+1}{x-4} = 0$ **(b)** $\frac{x+1}{x-4} \ge 0$ **(c)** $\frac{x+1}{x-4} \le 0$

Window: $[-10, 10]$ by $[-5, 10]$

**13.** **(a)** $\frac{x-1}{x+2} = 1$ **(b)** $\frac{x-1}{x+2} > 1$ **(c)** $\frac{x-1}{x+2} < 1$

Window: $[-10, 10]$ by $[-5, 10]$

**14.** **(a)** $\frac{x-6}{x+2} = -1$ **(b)** $\frac{x-6}{x+2} < -1$ **(c)** $\frac{x-6}{x+2} > -1$

Window: $[-10, 10]$ by $[-10, 10]$

**15.** **(a)** $\frac{1}{x-1} = \frac{5}{4}$ **(b)** $\frac{1}{x-1} < \frac{5}{4}$ **(c)** $\frac{1}{x-1} > \frac{5}{4}$

Window: $[-5, 5]$ by $[-5, 5]$

**16.** **(a)** $\frac{6}{5-3x} = 2$ **(b)** $\frac{6}{5-3x} \le 2$ **(c)** $\frac{6}{5-3x} \ge 2$

Window: $[-5, 5]$ by $[-5, 5]$

**17.** **(a)** $\frac{4}{x-2} = \frac{3}{x-1}$ **(b)** $\frac{4}{x-2} \le \frac{3}{x-1}$ **(c)** $\frac{4}{x-2} \ge \frac{3}{x-1}$

Window: $[-3, 3]$ by $[-20, 20]$

**18.** **(a)** $\frac{4}{x+1} = \frac{2}{x+3}$ **(b)** $\frac{4}{x+1} < \frac{2}{x+3}$ **(c)** $\frac{4}{x+1} > \frac{2}{x+3}$

Window: $[-8, 5]$ by $[-10, 10]$

**19.** **(a)** $\frac{1}{(x-2)^2} = 0$ **(b)** $\frac{1}{(x-2)^2} < 0$ **(c)** $\frac{1}{(x-2)^2} > 0$

Window: $[-5, 10]$ by $[-5, 10]$

**20.** **(a)** $\frac{-2}{(x+3)^2} = 0$ **(b)** $\frac{-2}{(x+3)^2} > 0$ **(c)** $\frac{-2}{(x+3)^2} < 0$

Window: $[-10, 5]$ by $[-10, 5]$

**21.** **(a)** $\frac{5}{x+1} = \frac{12}{x+1}$ **(b)** $\frac{5}{x+1} > \frac{12}{x+1}$ **(c)** $\frac{5}{x+1} < \frac{12}{x+1}$

Window: $[-10, 10]$ by $[-10, 10]$

**22.** **(a)** $\frac{7}{x+2} = \frac{1}{x+2}$ **(b)** $\frac{7}{x+2} \ge \frac{1}{x+2}$ **(c)** $\frac{7}{x+2} \le \frac{1}{x+2}$

Window: $[-10, 10]$ by $[-10, 10]$

*In some cases, it is possible to solve a rational inequality by simply deciding what sign the numerator and the denominator must have, and then using the rules for quotients of positive and negative numbers to determine the solution set. For example, consider the rational inequality*

$$\frac{1}{x^2 + 1} > 0.$$

*The numerator of the rational expression, 1, is positive, and the denominator, $x^2 + 1$, must always be positive because it is the sum of a nonnegative number, $x^2$, and a positive number, 1. Therefore, the rational expression is the quotient of two positive numbers, which is positive. Because the inequality requires that the rational expression be greater than 0 and this will always be true, the solution set is $(-\infty, \infty)$.*

*Use similar reasoning to solve each inequality.*

**23.** $\dfrac{-1}{x^2 + 2} < 0$

**24.** $\dfrac{5}{x^2 + 2} < 0$

**25.** $\dfrac{-5}{x^2 + 2} > 0$

**26.** $\dfrac{x^4 + 2}{-6} \leq 0$

**27.** $\dfrac{x^4 + 2}{-6} \geq 0$

**28.** $\dfrac{x^4 + x^2 + 3}{x^2 + 2} < 0$

**29.** $\dfrac{x^4 + x^2 + 3}{x^2 + 2} > 0$

**30.** $\dfrac{(x - 1)^2}{x^2 + 4} > 0$

**31.** $\dfrac{(x - 1)^2}{x^2 + 4} \leq 0$

**32.** The graph of $y = f(x)$, where

$$f(x) = \frac{4}{x - 2} - \frac{3}{x - 1} = \frac{x + 2}{x^2 - 3x + 2},$$

is shown in the window $[-4.7, 4.7]$ by $[-3.1, 12.4]$. The solution set of $f(x) \leq 0$ was required in Exercise 17(b) (in a different form, but equivalent to this one). From the graph, it appears that no part of the curve lies below the $x$-axis, which would lead to $\emptyset$ as the solution set. Yet, the solution set of $f(x) \leq 0$ is $(-\infty, -2] \cup (1, 2)$. Use this observation to explain why relying on graphical analysis alone is not sufficient for solving equations and inequalities.

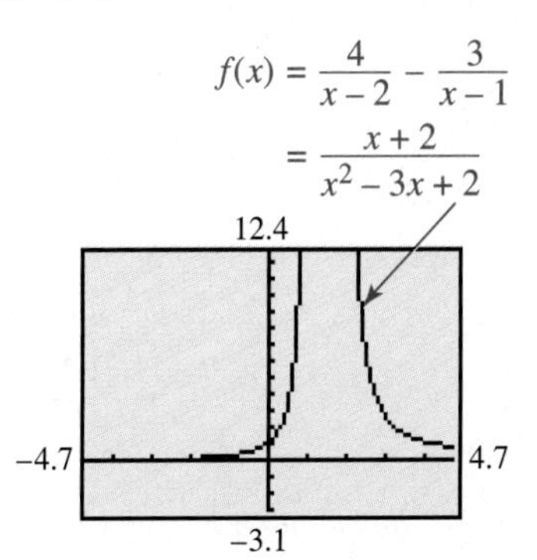


The lines $x = 1$ and $x = 2$ are vertical asymptotes.

*Use the methods explained in Examples 3 and 4 to find all complex solutions for each equation. Support your real solutions with a graph, using an appropriate window.*

**33.** $\dfrac{2x}{x^2 - 1} = \dfrac{2}{x + 1} - \dfrac{1}{x - 1}$

**34.** $\dfrac{8x}{4x^2 - 1} = \dfrac{3}{2x + 1} + \dfrac{3}{2x - 1}$

**35.** $\dfrac{4}{x^2 - 3x} - \dfrac{1}{x^2 - 9} = 0$

**36.** $\dfrac{2}{x^2 - 2x} - \dfrac{3}{x^2 - x} = 0$

**37.** $1 - \dfrac{13}{x} + \dfrac{36}{x^2} = 0$

**38.** $1 - \dfrac{3}{x} - \dfrac{10}{x^2} = 0$

**39.** $1 + \dfrac{3}{x} = \dfrac{5}{x^2}$

**40.** $4 + \dfrac{7}{x} = -\dfrac{1}{x^2}$

**41.** $\dfrac{x}{2 - x} + \dfrac{2}{x} - 5 = 0$

**42.** $\dfrac{2x}{x - 3} + \dfrac{4}{x} - 6 = 0$

**43.** $x^{-4} - 3x^{-2} - 4 = 0$

**44.** $x^{-4} - 5x^{-2} + 4 = 0$

**45.** $\dfrac{1}{x + 2} + \dfrac{3}{x + 7} = \dfrac{5}{x^2 + 9x + 14}$

**46.** $\dfrac{1}{x + 3} + \dfrac{4}{x + 5} = \dfrac{2}{x^2 + 8x + 15}$

**47.** $\dfrac{x}{x - 3} + \dfrac{4}{x + 3} = \dfrac{18}{x^2 - 9}$

**48.** $\dfrac{2x}{x - 3} + \dfrac{4}{x + 3} = \dfrac{24}{9 - x^2}$

**49.** $9x^{-1} + 4(6x - 3)^{-1} = 2(6x - 3)^{-1}$

**50.** $x(x - 2)^{-1} + x(x + 2)^{-1} = 8(x^2 - 4)^{-1}$

## RELATING CONCEPTS

**For individual or group investigation (Exercises 51–56)**

*In Example 2, we used a sign graph to solve a rational inequality. There is another method of solving rational inequalities analytically. For example, suppose that we wish to solve*

$$\frac{x-2}{x+3} \le 2.$$

*Since* $-3$ *is not in the domain, it cannot possibly be part of the solution. We should not multiply each side by* $x + 3$ *because it may be negative, which would require changing the direction of the inequality sign. However,* $(x + 3)^2$ *is positive. (Why?) Therefore, we multiply each side by* $(x + 3)^2$*. This leads to the inequality* $x^2 + 11x + 24 \ge 0$*. (You are asked to provide these steps in Exercise 51.)*

*From the graph, we see that the solution set of* $x^2 + 11x + 24 \ge 0$ *is* $(-\infty, -8] \cup [-3, \infty)$*. Because* $-3$ *must be excluded from the solution set of the original inequality, the solution set of* $\frac{x-2}{x+3} \le 2$ *is* $(-\infty, -8] \cup (-3, \infty)$*.*

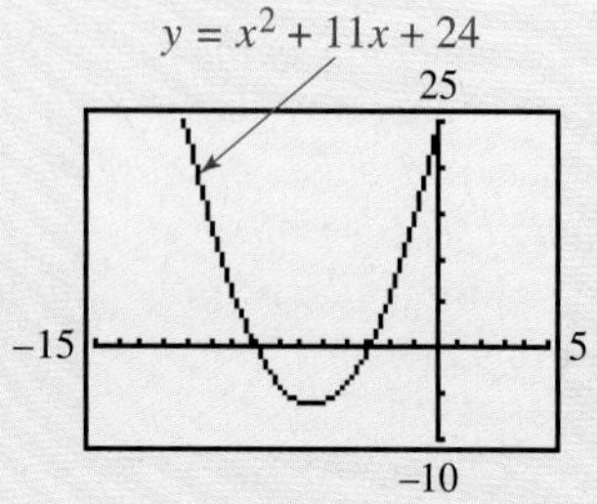

The $x$-intercepts are $-8$ and $-3$.

**51.** Provide the steps needed to show that the original rational inequality

$$\frac{x-2}{x+3} \le 2 \quad (x \ne -3)$$

is equivalent to $x^2 + 11x + 24 \ge 0$ $(x \ne -3)$.

*Use the method described to solve each rational inequality.*

**52.** $\frac{x+5}{x-3} \ge 1$ **53.** $\frac{(x-3)^2}{x+1} \ge x + 1$ **54.** $\frac{(x-6)^2}{x+5} \ge x + 5$ **55.** $\frac{x+3}{(x+1)^2} < \frac{1}{x+1}$ **56.** $\frac{x-4}{(x+2)^2} > \frac{1}{x+2}$

*Solve the equation in part (a) graphically, expressing solutions to the nearest hundredth. Then, use the graph to solve the associated inequalities in parts (b) and (c), expressing endpoints to the nearest hundredth.*

**57.** **(a)** $\frac{\sqrt{2}x + 5}{x^3 - \sqrt{3}} = 0$

**(b)** $\frac{\sqrt{2}x + 5}{x^3 - \sqrt{3}} > 0$

**(c)** $\frac{\sqrt{2}x + 5}{x^3 - \sqrt{3}} < 0$

**58.** **(a)** $\frac{\sqrt[3]{7}x^3 - 1}{x^2 + 2} = 0$

**(b)** $\frac{\sqrt[3]{7}x^3 - 1}{x^2 + 2} > 0$

**(c)** $\frac{\sqrt[3]{7}x^3 - 1}{x^2 + 2} < 0$

*Computers often use rational functions to approximate other types of functions. Use graphing to match each function in Column I with its rational approximation in Column II on the interval* $[1, 15]$*.*

| I | II |
|---|---|
| **59.** $f_1(x) = \sqrt{x}$ | **A.** $r_1(x) = \frac{2 - 2x^2}{3x^2 + 10x + 3}$ |
| **60.** $f_2(x) = \sqrt{4x + 1}$ | **B.** $r_2(x) = \frac{15x^2 + 75x + 33}{x^2 + 23x + 31}$ |
| **61.** $f_3(x) = \sqrt[3]{x}$ | **C.** $r_3(x) = \frac{10x^2 + 80x + 32}{x^2 + 40x + 80}$ |
| **62.** $f_4(x) = \frac{1 - \sqrt{x}}{1 + \sqrt{x}}$ | **D.** $r_4(x) = \frac{7x^3 + 42x^2 + 30x + 2}{2x^3 + 30x^2 + 42x + 7}$ |

*(Modeling) Solve each problem.*

**63.** *Insect Population* Suppose that an insect population in millions is modeled by

$$f(x) = \frac{10x + 1}{x + 1},$$

where $x \geq 0$ is in months.

**(a)** Graph $f$ in the window $[0, 14]$ by $[0, 14]$. Find the equation of the horizontal asymptote.

**(b)** Determine the initial insect population.

**(c)** What happens to the population after several months?

**(d)** Interpret the horizontal asymptote.

**64.** *Fish Population* Suppose that the population of a species of fish in thousands is modeled by

$$f(x) = \frac{x + 10}{.5x^2 + 1},$$

where $x \geq 0$ is in years.

**(a)** Graph $f$ in the window $[0, 12]$ by $[0, 12]$. What is the horizontal asymptote?

**(b)** Determine the initial population.

**(c)** What happens to the population of this fish after many years?

**(d)** Interpret the horizontal asymptote.

**65.** *Time Spent in Line* Suppose the average number of vehicles arriving at the main gate of an amusement park is equal to 10 per minute, while the average number of vehicles being admitted through the gate per minute is equal to $x$. Then the average waiting time in minutes for each vehicle at the gate can be computed by

$$f(x) = \frac{x - 5}{x^2 - 10x},$$

where $x > 10$. (*Source:* Mannering, F. and W. Kilareski, *Principles of Highway Engineering and Traffic Analysis,* Second Edition, John Wiley and Sons, 1998.)

**(a)** Estimate the admittance rate $x$ that results in an average wait of 15 seconds.

**(b)** If one attendant can serve 5 vehicles per minute, how many attendants are needed to keep the average wait to 15 seconds or less?

**66.** *Length of Lines* See Example 5. Determine the traffic intensity $x$ when the average number of vehicles in line equals 3.

**67.** *Construction* Find possible dimensions for a closed box with volume 196 cubic inches, surface area 280 square inches, and length that is twice the width.

**68.** *Volume of a Cylindrical Can* A metal cylindrical can with an *open top* and *closed bottom* is to have a volume of 4 cubic feet. Find the dimensions that require the least amount of material. What would this amount be? (Compare this problem to Example 6.)

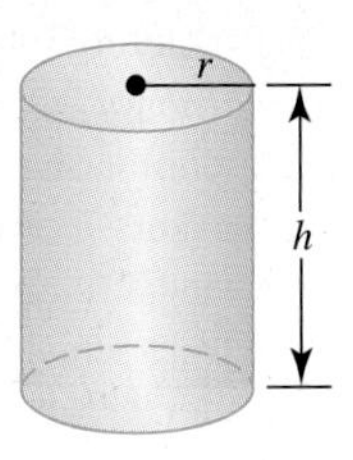


**69.** *Train Curves* When curves are designed for trains, sometimes the outer rail is elevated or *banked* so that a locomotive can safely negotiate the curve at a higher speed. Suppose a circular curve is being designed for a speed of 60 mph. The rational function given by

$$f(x) = \frac{2540}{x}$$

computes the elevation $y$ in inches of the outer track for a curve with a radius of $x$ feet, where $y = f(x)$. (*Source:* Haefner, L., *Introduction to Transportation Systems,* Holt, Rinehart and Winston, 1986.)

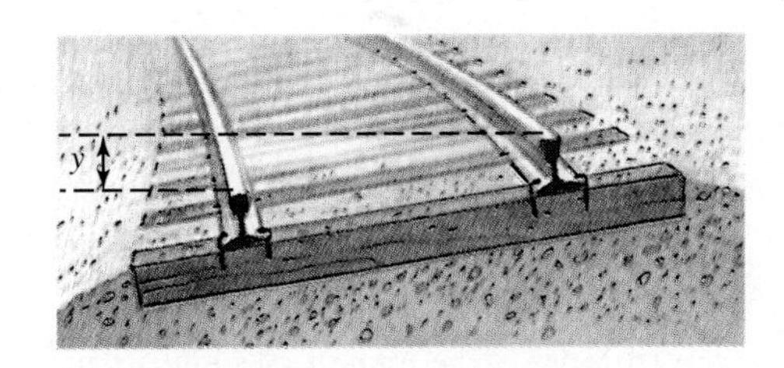


**(a)** Evaluate $f(400)$ and interpret its meaning.

**(b)** Graph $f$ in the window $[0, 600]$ by $[0, 50]$. Discuss how the elevation of the outer rail changes with the radius $x$.

**(c)** Interpret the horizontal asymptote.

**(d)** What radius is associated with an elevation of 12.7 inches?

**70.** *Recycling* A cost-benefit function $C$ computes the cost in millions of dollars of implementing a city recycling project when $x$ percent of the citizens participate, where

$$C(x) = \frac{1.2x}{100 - x}.$$

**(a)** Graph $C$ in the window $[0, 100]$ by $[0, 10]$. Interpret the graph as $x$ approaches 100.

**(b)** If 75% participation is expected, determine the cost for the city.

**(c)** The city plans to spend \$5 million on this recycling project. Estimate graphically the percentage of participation that can be expected.

**(d)** Solve part (c) analytically.

**71.** *Mortality Rates* The table contains incidence ratios by age for deaths from coronary heart disease (CHD) and lung cancer (LC) when comparing smokers (21–39 cigarettes per day) to nonsmokers. For example, an incidence ratio of 10 means that smokers are 10 times more likely than nonsmokers to die from lung cancer between the ages of 55 and 64. If the incidence ratio is $x$, then the percentage $P$ of deaths caused by smoking can be calculated using the rational function defined by

$$P(x) = \frac{x - 1}{x},$$

where the output is in decimal form. (*Source:* Walker, A., *Observation and Inference: An Introduction to the Methods of Epidemiology,* Epidemiology Resources, Inc., 1991.)

| Age | CHD | LC |
|---|---|---|
| 55–64 | 1.9 | 10 |
| 65–74 | 1.7 | 9 |

**(a)** Calculate the percentage of lung cancer deaths in the age group 65–74 that can be attributed to smoking.

**(b)** Determine the percentage of deaths from coronary heart disease in the age group 55–64 that can be attributed to smoking.

**72.** *Braking Distance* The *grade x* of a hill is a measure of its steepness. For example, if a road rises 10 feet for every 100 feet of horizontal distance, then it has an uphill grade of $x = \frac{10}{100}$ or 10%. Grades are typically kept quite small—usually less than 10%. The braking (or stopping) distance $D$ for a car traveling at 50 mph on a wet, uphill grade is given by

$$D(x) = \frac{2500}{30(.3 + x)}.$$

(*Source:* Haefner, L., *Introduction to Transportation Systems,* Holt, Rinehart and Winston, 1986.)

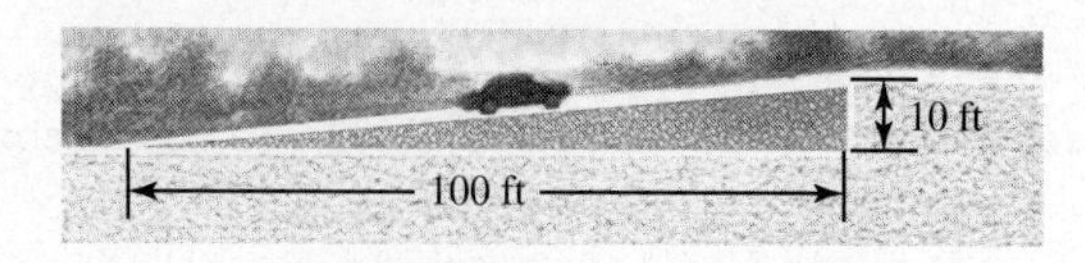


**(a)** Evaluate $D(.05)$ and interpret the result.

**(b)** Describe what happens to braking distance as the hill becomes steeper. Does this agree with your driving experience?

**(c)** Estimate the grade associated with a braking distance of 220 feet.

**73.** *Braking Distance* See Exercise 72. If a car is traveling 50 mph downhill, then its braking distance on a wet pavement is given by

$$D(x) = \frac{2500}{30(.3 + x)},$$

where $x < 0$ for a downhill grade.

**(a)** Evaluate $D(-.1)$ and interpret the result.

**(b)** What happens to braking distance as the downhill grade becomes steeper? Does this agree with your driving experience?

**(c)** The graph of $D$ has a vertical asymptote at $x = -.3$. Give the physical significance of this asymptote.

**(d)** Estimate the grade associated with a braking distance of 350 feet.

**74.** *Slippery Roads* If a car is moving at 50 mph on a level highway, then its braking (or stopping) distance depends on the road conditions. This distance in feet can be computed by

$$D(x) = \frac{2500}{30x},$$

where $x$ is the coefficient of friction between the tires and the road and $0 < x \leq 1$. A smaller value of $x$ indicates that the road is more slippery.

**(a)** Identify and interpret the vertical asymptote.

**(b)** Estimate the coefficient of friction associated with a braking distance of 340 feet.

*Solve each problem involving variation.*

**75.** Suppose $r$ varies directly as the square of $m$ and inversely as $s$. If $r = 12$ when $m = 6$ and $s = 4$, find $r$ when $m = 4$ and $s = 10$.

**76.** Suppose $p$ varies directly as the square of $z$ and inversely as $r$. If $p = \frac{32}{5}$ when $z = 4$ and $r = 10$, find $p$ when $z = 2$ and $r = 16$.

**77.** Let $a$ vary directly as $m$ and $n^2$ and inversely as $y^3$. If $a = 9$ when $m = 4$, $n = 9$, and $y = 3$, find $a$ if $m = 6$, $n = 2$, and $y = 5$.

**78.** If $y$ varies directly as $x$ and inversely as $m^2$ and $r^2$, and $y = \frac{5}{3}$ when $x = 1$, $m = 2$, and $r = 3$, find $y$ if $x = 3$, $m = 1$, and $r = 8$.

**79.** For $k > 0$, if $y$ varies directly as $x$, when $x$ increases, $y$ ________, and when $x$ decreases, $y$ ________.

**80.** For $k > 0$, if $y$ varies inversely as $x$, when $x$ increases, $y$ ________, and when $x$ decreases, $y$ ________.

*In Exercises 81–84, assume that the constant of variation is positive.*

**81.** Let $y$ be inversely proportional to $x$. If $x$ doubles, what happens to $y$?

**82.** Let $y$ vary inversely as the second power of $x$. If $x$ doubles, what happens to $y$?

**83.** Suppose $y$ varies directly as the third power of $x$. If $x$ triples, what happens to $y$?

**84.** Suppose $y$ is directly proportional to the second power of $x$. If $x$ is halved, what happens to $y$?

*Solve each problem involving inverse, joint, or combined variation.*

**85.** *Body Mass Index* In 1998, the federal government developed the *body mass index* (BMI) to determine ideal weights. A person's BMI is directly proportional to his or her weight in pounds and inversely proportional to the square of his or her height in inches. (A BMI of 19 to 25 corresponds to a healthy weight.) A 6-foot-tall person weighing 177 pounds has BMI 24. Find the BMI (to the nearest whole number) of a person whose weight is 130 pounds and whose height is 66 inches. (*Source: Washington Post.*)

**86.** *Volume of a Gas* Natural gas provides 35.8% of U.S. energy. (*Source:* U.S. Energy Department.) The volume of a gas varies inversely as the pressure and directly as the temperature. (Temperature must be measured in *Kelvin* (K), a unit of measurement used in physics.) If a certain gas occupies a volume of 1.3 liters at 300 K and a pressure of 18 newtons per square centimeter, find the volume at 340 K and a pressure of 24 newtons per square centimeter.

**87.** *Electrical Resistance* The electrical resistance $R$ of a wire varies inversely as the square of its diameter $d$. If a 25-foot wire with diameter 2 millimeters has resistance .5 ohm, find the resistance of a wire having the same length and diameter 3 millimeters.

**88.** *Poiseuille's Law* According to Poiseuille's law, the resistance to flow of a blood vessel, $R$, is directly proportional to the length, $l$, and inversely proportional to the fourth power of the radius, $r$. (*Source:* Hademenos, George J., "The Biophysics of Stroke," *American Scientist,* May–June 1997.) If $R = 25$ when $l = 12$ and $r = .2$, find $R$ as $r$ increases to .3, while $l$ is unchanged.

**89.** *Gravity* The weight of an object varies inversely as the square of its distance from the center of Earth. The radius of Earth is approximately 4000 miles. If a person weighs 160 pounds on Earth's surface, what would this individual weigh 8000 miles above the surface of Earth?

**90.** *Hubble Telescope* The brightness or intensity of starlight varies inversely as the square of its distance from Earth. The Hubble Telescope can see stars whose intensities are $\frac{1}{50}$ of the faintest star now seen by ground-based telescopes. Determine how much farther the Hubble Telescope can see into space than ground-based telescopes. (*Source:* National Aeronautics and Space Administration.)

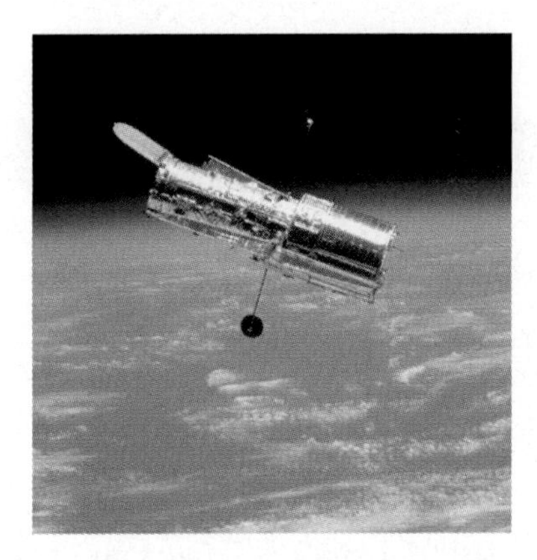

**91.** *Volume of a Cylinder* The volume of a right circular cylinder is jointly proportional to the square of the radius of the circular base and to the height. If the volume is 300 cubic centimeters when the height is 10.62 centimeters and the radius is 3 centimeters, find the volume of a cylinder with radius 4 centimeters and height 15.92 centimeters.

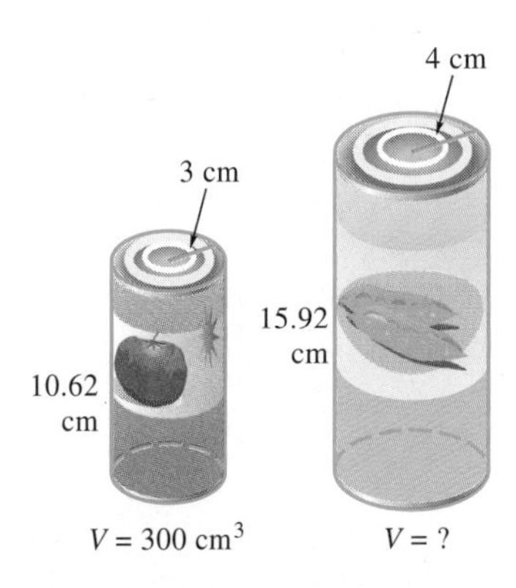


**92.** *Photography* See Example 9. If 125 foot-candles of light are available and an F-stop of 2 is used with 200 ASA film, what shutter speed should be used?

*In each table,* $Y_1 = \dfrac{k}{X}$ *for some value of k. Find the value of k.*

**93.**

| X | $Y_1$ |
|---|---|
| 3 | 1.7 |
| 4 | 1.275 |
| 5 | 1.02 |
| 6 | .85 |
| 7 | .72857 |
| 8 | .6375 |
| 9 | .56667 |

X=3

**94.**

| X | $Y_1$ |
|---|---|
| 5 | .58 |
| 6 | .48333 |
| 7 | .41429 |
| 8 | .3625 |
| 9 | .32222 |
| 10 | .29 |
| 11 | .26364 |

X=5

**95.**

| X | $Y_1$ |
|---|---|
| -3 | .46667 |
| -2 | .7 |
| -1 | 1.4 |
| 0 | ERROR |
| 1 | -1.4 |
| 2 | -.7 |
| 3 | -.4667 |

X= -3

**96.**

| X | $Y_1$ |
|---|---|
| -3 | .86667 |
| -2 | 1.3 |
| -1 | 2.6 |
| 0 | ERROR |
| 1 | -2.6 |
| 2 | -1.3 |
| 3 | -.8667 |

X= -3

## Reviewing Basic Concepts (Sections 4.1–4.3)

**1.** Sketch the graph of $y = \dfrac{1}{x+2} - 3$. Then obtain an accurate representation of the graph using a graphing calculator.

**2.** What is the domain of the rational function defined by $f(x) = \dfrac{3}{x^2 - 1}$?

**3.** What is the equation of the vertical asymptote of the graph of $f(x) = \dfrac{4x + 3}{x - 6}$?

**4.** What is the equation of the horizontal asymptote of the graph of $f(x) = \dfrac{x^2 + 3}{x^2 - 4}$?

**5.** What is the equation of the oblique asymptote of the graph of $f(x) = \dfrac{x^2 + x + 5}{x + 3}$?

**6.** Sketch the graph of $f(x) = \dfrac{3x + 6}{x - 4}$ by hand. Then obtain an accurate representation of the graph using a graphing calculator.

**7.** The graph of a rational function $f$ is shown here.

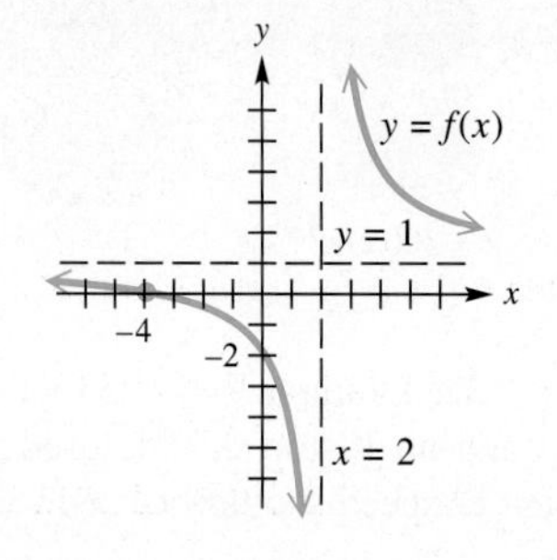


Give the solution set of each equation or inequality.

**(a)** $f(x) = 0$ **(b)** $f(x) > 0$ **(c)** $f(x) < 0$

**8.** Find the solution set of $\dfrac{x + 4}{3x + 1} > 1$.

**9.** Fill in the blanks with the correct responses: $b = \dfrac{24}{h}$ is the formula for the base of a parallelogram with area 24. The base of this parallelogram varies __________ as its __________. The constant of variation is __________.

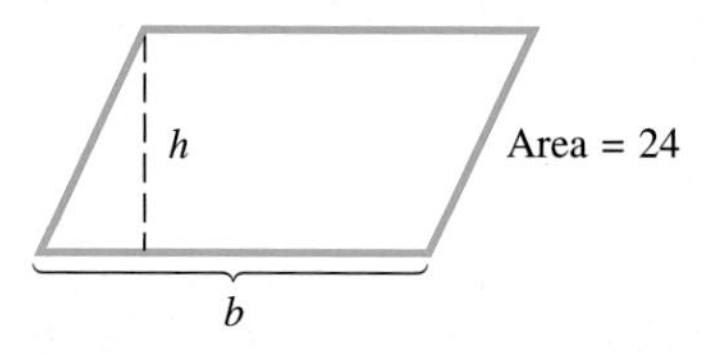


**10.** *Vibrations of a Guitar String* The number of vibrations per second (the pitch) of a steel guitar string varies directly as the square root of the tension and inversely as the length of the string. If the number of vibrations per second is 5 when the tension is 225 kilograms and the length is .60 meter, find the number of vibrations per second when the tension is 196 kilograms and the length is .65 meter.

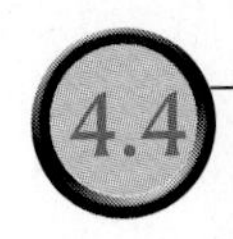

# 4.4 FUNCTIONS DEFINED BY POWERS AND ROOTS

Power and Root Functions • Modeling Using Power Functions • Graphs of $f(x) = \sqrt[n]{ax + b}$ • Graphing Circles and Horizontal Parabolas Using Root Functions

## Power and Root Functions

| Planet | $x$ | $y$ |
|---|---|---|
| Mercury | .387 | .241 |
| Venus | .723 | .615 |
| Earth | 1.00 | 1.00 |
| Mars | 1.52 | 1.88 |
| Jupiter | 5.20 | 11.9 |
| Saturn | 9.54 | 29.5 |

*Source:* Ronan, C., *The Natural History of the Universe*, MacMillan, 1991.

Johannes Kepler (1571–1630) was the first to recognize that the orbits of planets are elliptical, rather than circular. He also found that a *power function* models the relationship between a planet's distance from the sun and its period of revolution. The table lists the average distance $x$ from the sun and the time $y$ in years for several planets to orbit the sun. The distance $x$ has been normalized so that Earth is 1 unit away from the sun. For example, Jupiter is 5.20 times farther from the sun than Earth and requires 11.9 years to orbit the sun.

A scatter diagram of the data in the table is shown in Figure 32. To model the data we might try a polynomial, such as the linear model $f(x) = x$ or the quadratic model $g(x) = x^2$. Figure 33 shows that $f(x) = x$ increases too slowly and $g(x) = x^2$ increases too quickly. To model the data, a new type of function is required. That is, we need a function in the form $h(x) = x^b$, where $1 < b < 2$. In a polynomial, the exponent $b$ must be a *nonnegative integer,* whereas in a power function, $b$ can be *any real number.* See Figure 34.

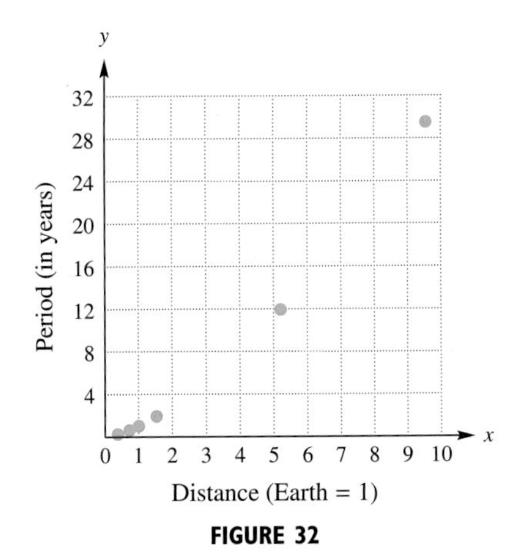


**FIGURE 32**

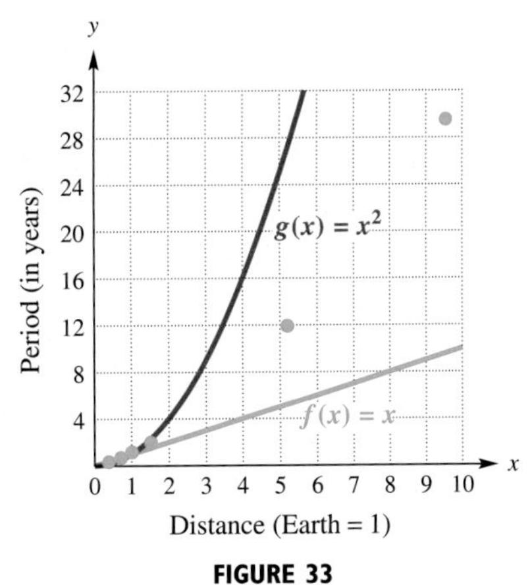


**FIGURE 33**

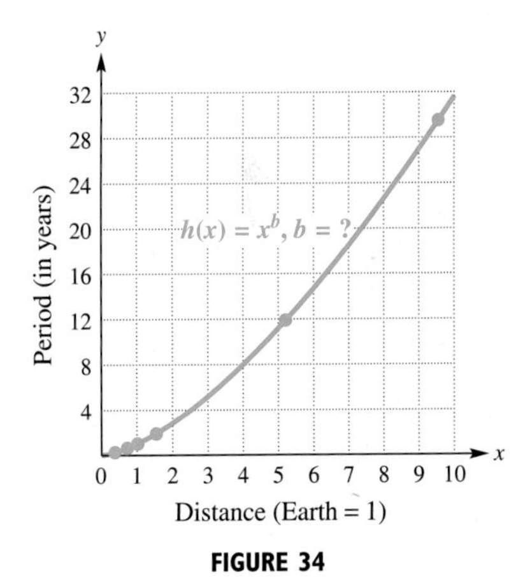


**FIGURE 34**

A *root function* is a special type of power function. These functions are defined as follows.

> **Power and Root Functions**
>
> A function $f$ given by
> $$f(x) = x^b,$$
> where $b$ is a constant, is a **power function.** If $b = \frac{1}{n}$ for some integer $n \geq 2$, then $f$ is a **root function** given by
> $$f(x) = x^{1/n}, \quad \text{or equivalently,} \quad f(x) = \sqrt[n]{x}.$$

Examples of power functions include

$$f(x) = x^2, \quad f(x) = x^{\pi}, \quad f(x) = x^{.4}, \quad \text{and} \quad f(x) = \sqrt[3]{x^2}.$$

Frequently, the domain of a power function $f$ is restricted to nonnegative numbers. Suppose the rational number $\frac{p}{q}$ is written in lowest terms. Then the domain of $f(x) = x^{p/q}$ is all real numbers whenever $q$ is odd and all nonnegative real numbers whenever $q$ is even. If $b$ is an irrational number, the domain of $f(x) = x^b$ is all nonnegative real numbers. For example, the domain of $f(x) = x^{1/3}$ $\left(f(x) = \sqrt[3]{x}\right)$ is all real numbers, whereas the domains of $g(x) = x^{1/2}$ $\left(g(x) = \sqrt{x}\right)$ and $h(x) = x^{\sqrt{2}}$ are all nonnegative numbers.

Graphs of three common power functions are shown in Figures 35–37.

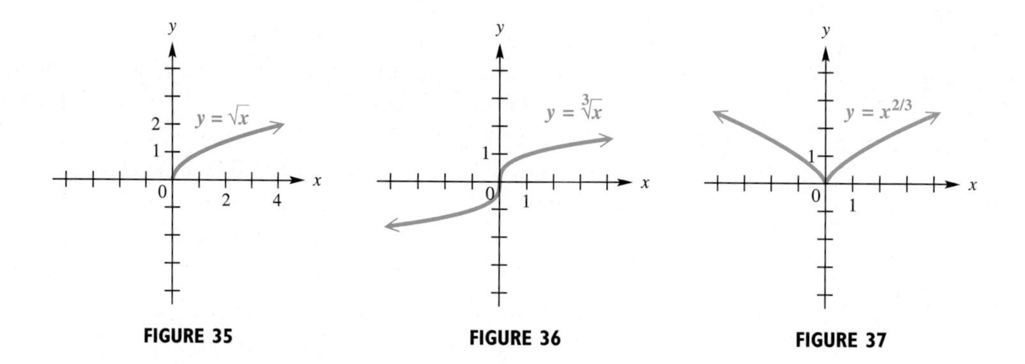


FIGURE 35 FIGURE 36 FIGURE 37

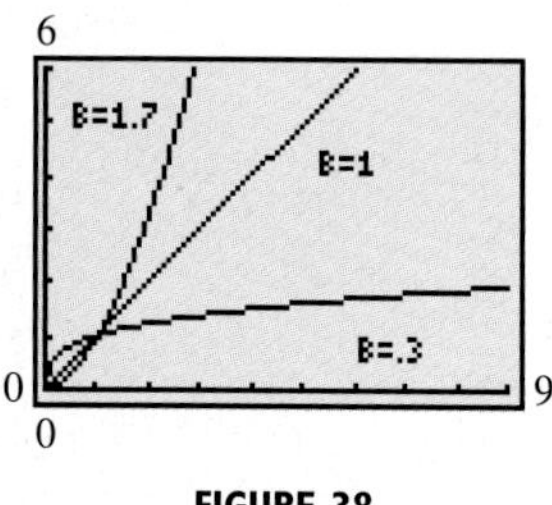


FIGURE 38

**EXAMPLE 1** *Graphing Power Functions*

**(a)** Graph $f(x) = x^b$, where $b = .3$, 1, and 1.7, for $x \geq 0$.

**(b)** Discuss the effect that $b$ has on the graph of $f$.

**Solution**

**(a)** Calculator graphs of $y = x^{.3}$, $y = x^1$, and $y = x^{1.7}$ are shown in Figure 38.

**(b)** Larger values of $b$ cause the graph of $f$ to increase faster.

## Modeling Using Power Functions

**EXAMPLE 2** *Modeling Wing Size of a Bird*

Heavier birds have larger wings with more surface area than do lighter birds. For some species of birds, this relationship can be modeled by

$$S(x) = .2x^{2/3},$$

where $x$ is the weight of the bird in kilograms and $S$ is the surface area of the wings in square meters. (*Source:* Pennycuick, C., *Newton Rules Biology,* Oxford University Press, 1992.) Approximate $S(.5)$ and interpret the result.

**Analytic Solution**

Let $x = .5$ and evaluate $S(.5)$ by direct substitution.

**Graphing Calculator Solution**

The table in Figure 39 shows that when $X = .5$, $Y_1 = .2(.5)^{2/3} \approx .126$.

*(continued)*

$$S(.5) = .2(.5)^{2/3}$$
$$\approx .126.$$

The wings of a bird that weighs .5 kilogram have a surface area of about .126 square meter.

| X | Y1 |
|---|---|
| .5 | .12599 |

Y1=.2*.5^(2/3)

FIGURE 39

| Planet | $x$ | $y$ |
|---|---|---|
| Mercury | .387 | .241 |
| Venus | .723 | .615 |
| Earth | 1.00 | 1.00 |
| Mars | 1.52 | 1.88 |
| Jupiter | 5.20 | 11.9 |
| Saturn | 9.54 | 29.5 |

*Source:* Ronan, C., *The Natural History of the Universe*, MacMillan, 1991.

**EXAMPLE 3** *Modeling the Period of Planetary Orbits*

Use the data in the table to complete the following. (Refer to the opening paragraph of this section.)

**(a)** Make a scatter diagram of the data. Estimate graphically a value for $b$ so that $f(x) = x^b$ models the data.

**(b)** Numerically check the accuracy of $f$.

**(c)** The average distances of Uranus, Neptune, and Pluto from the sun are 19.2, 30.1, and 39.5, respectively. Use $f$ to estimate the periods of revolution for these planets. Compare these answers to the actual values of 84.0, 164.8, and 248.5 years.

**Solution**

**(a)** We make a scatter diagram of the data, and then graph $y = x^b$ for different values of $b$. From the calculator graphs of $y = x^{1.4}$, $y = x^{1.5}$, and $y = x^{1.6}$ in Figures 40–42, we see that $b \approx 1.5$.

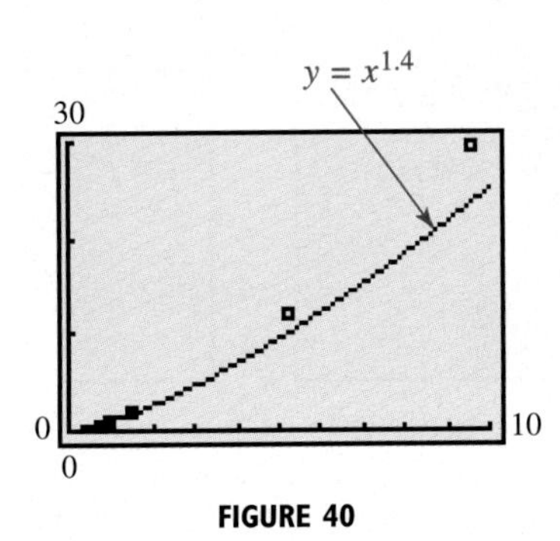


FIGURE 40

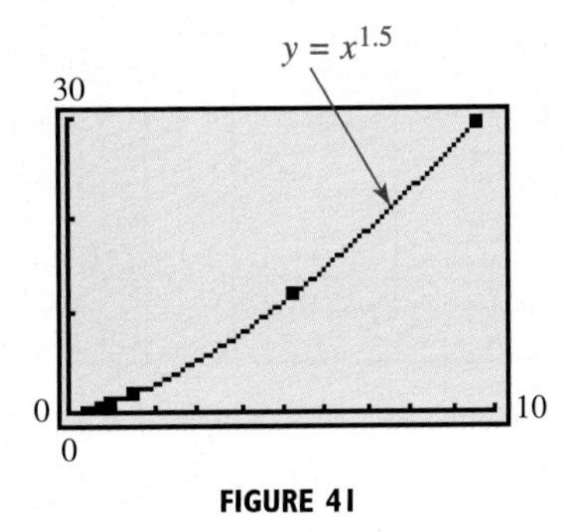


FIGURE 41

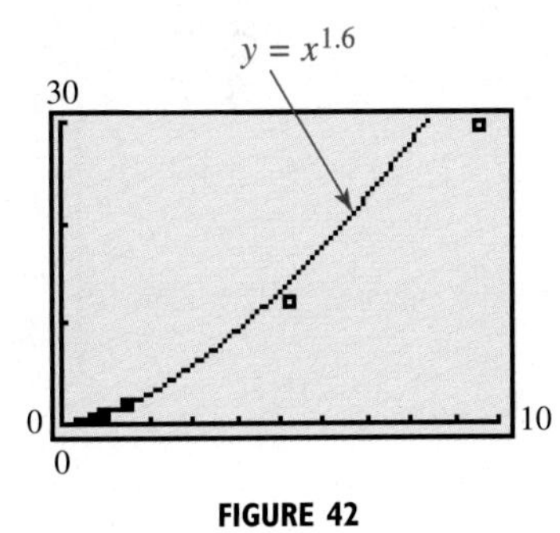


FIGURE 42

**(b)** The values shown in Figure 43 model the data in the table remarkably well.

| X | Y1 |
|---|---|
| .387 | .24075 |
| .723 | .61476 |
| 1 | 1 |
| 1.52 | 1.874 |
| 5.2 | 11.858 |
| 9.54 | 29.466 |

Y1=X^1.5

FIGURE 43

| X | Y1 |
|---|---|
| 19.2 | 84.13 |
| 30.1 | 165.14 |
| 39.5 | 248.25 |

Y1=X^1.5

FIGURE 44

**(c)** To approximate the number of years for Uranus, Neptune, and Pluto to orbit the sun, we evaluate $f(x) = x^{1.5}$ at $x = 19.2$, 30.1, and 39.5, as shown in Figure 44. These values are close to the actual values.

Rather than visually fit a curve to data as we did in Example 3, we can use least-squares regression, first introduced in Section 1.4, to fit the data.

**EXAMPLE 4** *Modeling the Length of a Bird's Wing*

The table lists the weight $W$ and the wingspan $L$ for birds of a particular species.

| $W$ (in kilograms) | .5 | 1.5 | 2.0 | 2.5 | 3.0 |
|---|---|---|---|---|---|
| $L$ (in meters) | .77 | 1.10 | 1.22 | 1.31 | 1.40 |

*Source:* Pennycuick, C., *Newton Rules Biology*, Oxford University Press, 1992.

**(a)** Use power regression to model the data with $L = aW^b$. Graph the data and the equation.

**(b)** Approximate the wingspan for a bird weighing 3.2 kilograms.

**Solution**

**(a)** Let $x$ be the weight $W$ and $y$ be the length $L$. Enter the data in a graphing calculator, and then select power regression (PwrReg), as shown in Figures 45 and 46. The results are shown in Figure 47. Let

$$y = .9674x^{.3326} \quad \text{or} \quad L = .9674W^{.3326}.$$

FIGURE 45

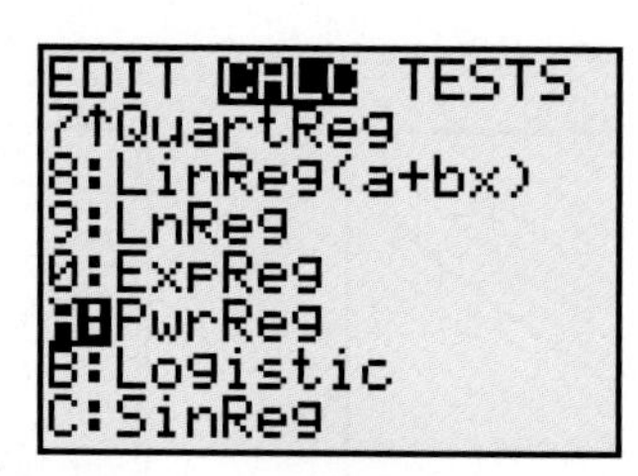


FIGURE 46

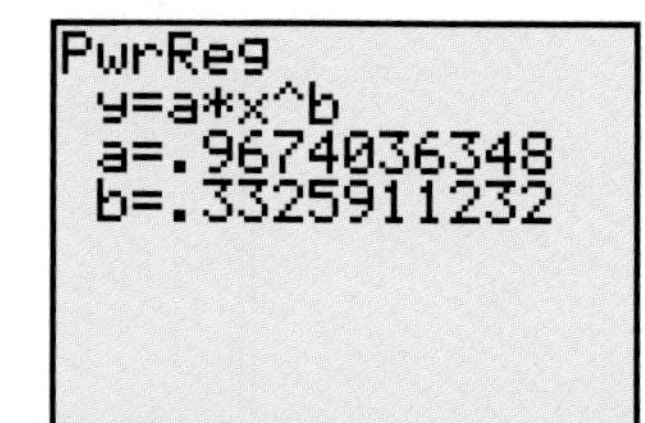


FIGURE 47

The data and equation are graphed in Figure 48.

**(b)** If a bird weighs 3.2 kilograms, this model predicts the wingspan to be

$$L = .9674(3.2)^{.3326} \approx 1.42 \text{ meters}.$$

See the display at the bottom of the screen in Figure 48.

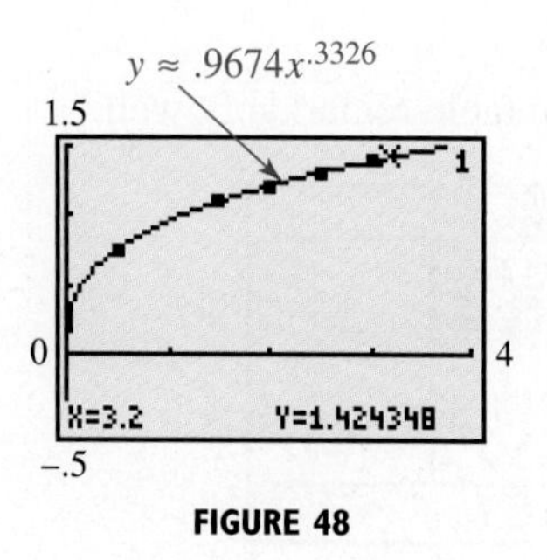


FIGURE 48

## Graphs of $f(x) = \sqrt[n]{ax + b}$

In Section 2.1 we introduced the graph of the *square* root function defined by $f(x) = \sqrt{x}$. When $n$ is even, the graph of the **root function** defined by $f(x) = \sqrt[n]{x}$ resembles the graph of the square root function; as $n$ gets larger, the graph lies closer to the $x$-axis as $x \to \infty$. Figure 49 shows the graphs of $y = \sqrt{x}$, $y = \sqrt[4]{x}$, and $y = \sqrt[6]{x}$ as examples of the graph of $f(x) = \sqrt[n]{x}$, where $n$ is even.

TECHNOLOGY NOTE

The definition of $x^{1/n}$ allows us to use a rational exponent as well as a radical when graphing a root function.

FUNCTION CAPSULE

**ROOT FUNCTION, $n$ EVEN** $\quad f(x) = \sqrt[n]{x}$

Domain: $[0, \infty)$ Range: $[0, \infty)$

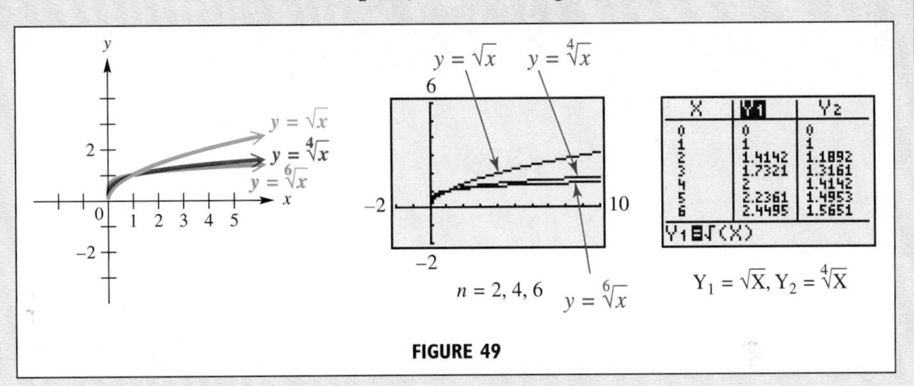


FIGURE 49

- For $n$ even, $f(x) = \sqrt[n]{x}$ increases on its entire domain, $[0, \infty)$.
- It is also continuous on $[0, \infty)$.

A discussion similar to the preceding one pertains to the graph of $f(x) = \sqrt[n]{x}$, where $n$ is odd. We introduced the graph of the *cube* root function defined by $f(x) = \sqrt[3]{x}$ in Section 2.1. When $n$ is odd, the graph of the root function defined by $f(x) = \sqrt[n]{x}$ resembles the graph of the cube root function; for larger $n$, the graph lies closer to the $x$-axis as $|x| \to \infty$. Figure 50 shows the graphs of $y = \sqrt[3]{x}$, $y = \sqrt[5]{x}$, and $y = \sqrt[7]{x}$ as examples of the graph of $f(x) = \sqrt[n]{x}$, where $n$ is odd.

FUNCTION CAPSULE

**ROOT FUNCTION, $n$ ODD** $\quad f(x) = \sqrt[n]{x}$

Domain: $(-\infty, \infty)$ Range: $(-\infty, \infty)$

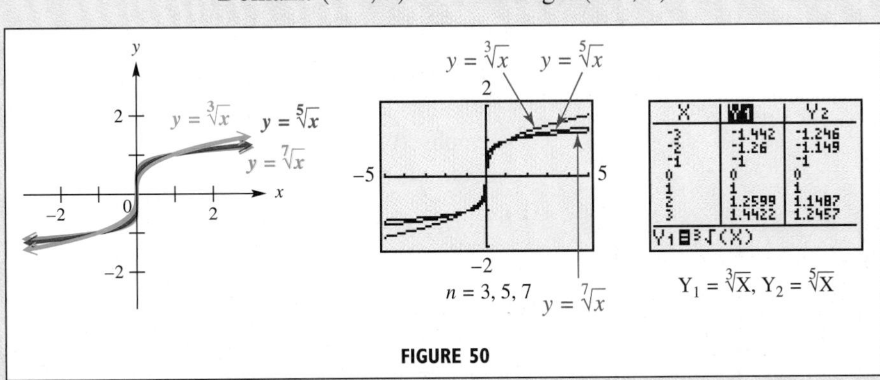


FIGURE 50

- For $n$ odd, $f(x) = \sqrt[n]{x}$ increases on its entire domain, $(-\infty, \infty)$.
- It is also continuous on $(-\infty, \infty)$.

To determine the domain of a function of the form $f(x) = \sqrt[n]{ax + b}$, we must note the parity of $n$. If $n$ is even, then $ax + b$ must be greater than or equal to 0; if $n$ is odd, then $ax + b$ can be any real number.

### EXAMPLE 5 *Finding Domains of Root Functions*

Find the domain of each function.

**(a)** $f(x) = \sqrt{4x + 12}$ **(b)** $g(x) = \sqrt[3]{-8x + 8}$

**Solution**

**(a)** For the function to be defined, $4x + 12$ must be greater than or equal to 0 since this is an even root ($n = 2$).

$$4x + 12 \geq 0$$
$$4x \geq -12$$
$$x \geq -3$$

The domain of $f$ is $[-3, \infty)$.

**(b)** Because $n = 3$ is an odd number, the domain of $g$ is $(-\infty, \infty)$.

### EXAMPLE 6 *Transforming Graphs of Root Functions*

**(a)** Explain how the graph of $y = \sqrt{4x + 12}$ can be obtained from the graph of $y = \sqrt{x}$. Then, graph it in an appropriate window.

**(b)** Repeat part (a) for the graph of $y = \sqrt[3]{-8x + 8}$, as compared to the graph of $y = \sqrt[3]{x}$.

**Solution**

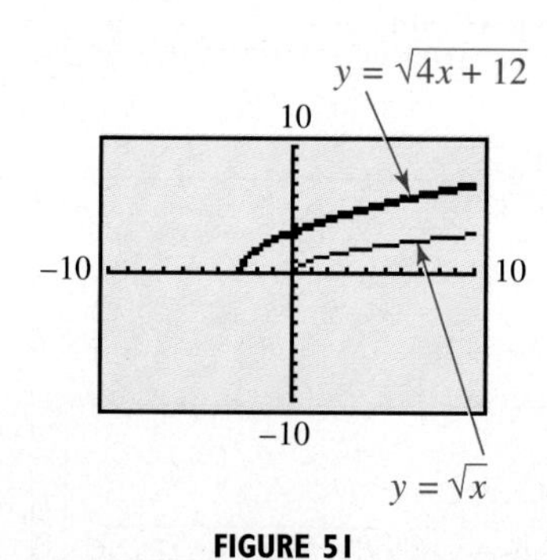


FIGURE 51

**(a)** We begin by writing the expression in an equivalent form.

$$y = \sqrt{4x + 12}$$
$$y = \sqrt{4(x + 3)} \qquad \text{Factor.}$$
$$y = \sqrt{4}\sqrt{x + 3} \qquad \sqrt{ab} = \sqrt{a} \cdot \sqrt{b}$$
$$y = 2\sqrt{x + 3} \qquad \sqrt{4} = 2$$

The graph of this function can be obtained from the graph of $y = \sqrt{x}$ by shifting it 3 units to the left and stretching vertically by a factor of 2. Figure 51 shows both graphs for comparison. Notice that the domain is $[-3, \infty)$, as determined in Example 5(a). The range is $[0, \infty)$.

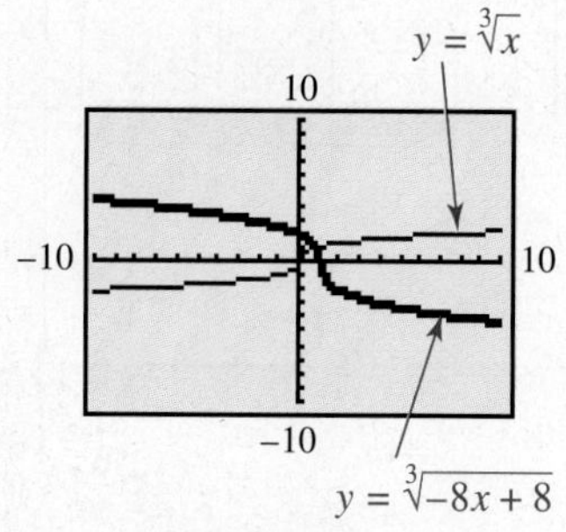

**(b)**

$$y = \sqrt[3]{-8x + 8}$$
$$y = \sqrt[3]{-8(x - 1)} \qquad \text{Factor.}$$
$$y = \sqrt[3]{-8} \cdot \sqrt[3]{x - 1} \qquad \sqrt[3]{ab} = \sqrt[3]{a} \cdot \sqrt[3]{b}$$
$$y = -2\sqrt[3]{x - 1} \qquad \sqrt[3]{-8} = -2$$

The graph can be obtained by shifting the graph of $y = \sqrt[3]{x}$ to the right 1 unit, stretching vertically by a factor of 2, and reflecting across the $x$-axis (because of the negative sign in $-2$). Figure 52 shows both graphs. The domain and range are both $(-\infty, \infty)$.

y = ∛x
10
−10
10
−10
y = ∛−8x + 8

FIGURE 52

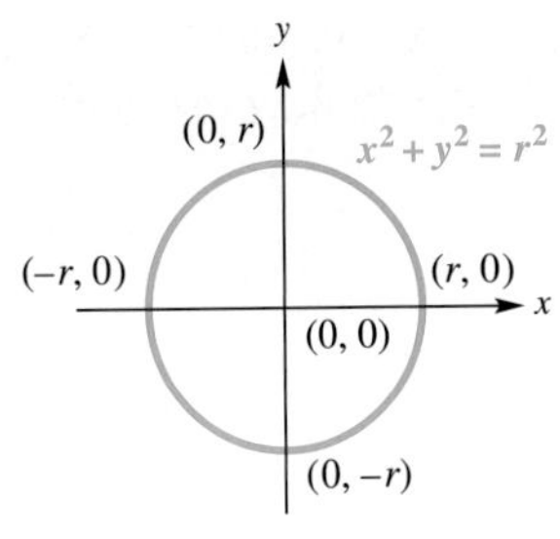


FIGURE 53

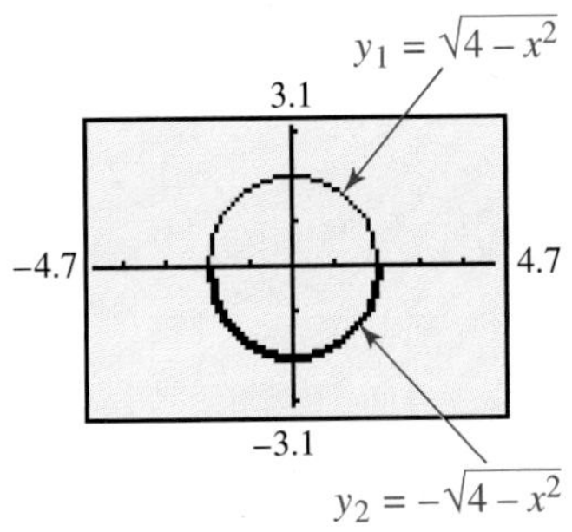


FIGURE 54

TECHNOLOGY NOTE

The graph of $x^2 + y^2 = 36$, formed by the union of the graphs of

$$y_1 = \sqrt{36 - x^2}$$

and $$y_2 = -\sqrt{36 - x^2},$$

is shown here.

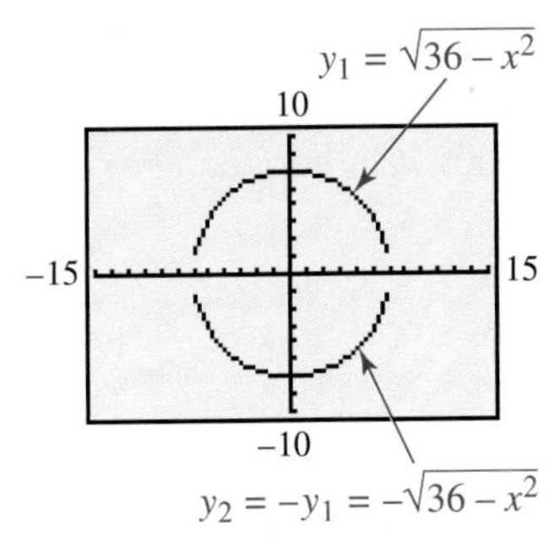


Although a square window is used, it is not a decimal window, and because of the manner in which the calculator plots points in function mode, the two semicircles do not completely "connect." *Mathematically, this is a complete circle,* illustrating how we must understand the concepts to interpret what we see on the screen.

## Graphing Circles and Horizontal Parabolas Using Root Functions

If we consider the graph of all points $(x, y)$ that lie a fixed distance $r$ $(r > 0)$ from the origin, then, by the distance formula,

$$\sqrt{(x - 0)^2 + (y - 0)^2} = r.$$

Simplifying and squaring each side of this equation leads to

$$x^2 + y^2 = r^2.$$

The graph of this set of points is a circle with center at $(0, 0)$ and radius $r$. See Figure 53. While we will study circles in more detail later in the text, this brief introduction allows us to investigate a special type of function defined by a root.

Because a circle is not the graph of a function, a graphing calculator in function mode is not appropriate for graphing a circle. However, if we imagine the circle as being the union of the graphs of two functions—one the top semicircle and the other the bottom semicircle—then we can graph both functions in the same square viewing window to obtain the desired graph. To accomplish this, we solve for $y$ in the equation $x^2 + y^2 = r^2$.

$$x^2 + y^2 = r^2$$
$$y^2 = r^2 - x^2 \qquad \text{Subtract } x^2.$$
$$y = \pm\sqrt{r^2 - x^2} \qquad \text{Square root property}$$

The final result can be interpreted as two equations, both of which define functions:

$$y_1 = \sqrt{r^2 - x^2} \qquad \text{The semicircle above the } x\text{-axis}$$

and $$y_2 = -\sqrt{r^2 - x^2}. \qquad \text{The semicircle below the } x\text{-axis}$$

Since $y_2 = -y_1$, the graph of the "bottom" semicircle is simply a reflection of the graph of the "top" semicircle across the $x$-axis.

**EXAMPLE 7** *Graphing a Circle*

Use a calculator in function mode to graph the circle $x^2 + y^2 = 4$.

**Solution** This graph can be obtained by graphing

$$y_1 = \sqrt{4 - x^2} \quad \text{and} \quad y_2 = -y_1 = -\sqrt{4 - x^2}$$

in the same window. Figure 54 shows the graph of these two root functions, forming a circle. To obtain a graph that is visually in correct proportions, we use a square window. Furthermore, because it is a decimal window, the two semicircles connect at the $x$-axis. (See the Technology Note for a case where this does not happen.)

From the discussion in Sections 3.2 and 3.3, we know that the graph of $y = ax^2 + bx + c$ $(a \neq 0)$ is a parabola with a vertical axis of symmetry. Such an equation defines a function. If we reverse the roles of $x$ and $y$, however, we obtain an equation whose graph is also a parabola, but having a horizontal axis of symmetry. (These, too, will be examined more closely later in the text.) Figure 55 on the next page shows the graph of $x = 2y^2 + 6y + 5$.

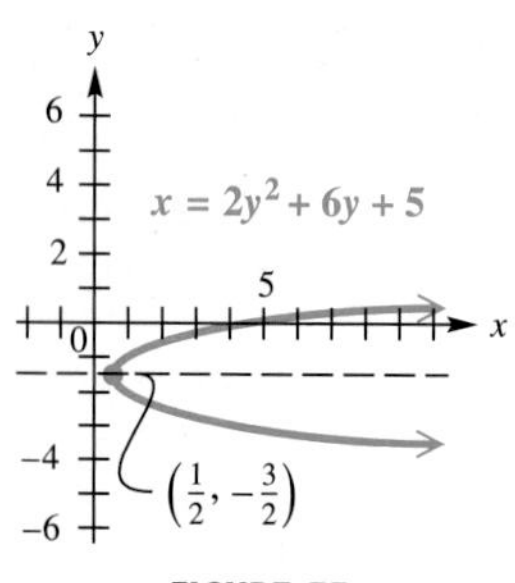

FIGURE 55

Notice that this is not the graph of a function. However, if we consider it to be the union of the graphs of two functions, one the top half-parabola and the other the bottom, then it can be graphed by a calculator in function mode.

### FOR DISCUSSION

The two functions that form the circle in Example 7 both have the same domain.

1. Discuss how a sign graph can be used to solve the inequality $4 - x^2 \geq 0$. How does the solution set of this inequality pertain to the graphs found in Example 7?
2. Graph $y = 4 - x^2$ with a calculator, and use the graph to find the solution set of $4 - x^2 \geq 0$. Discuss how this solution set pertains to the graphs found in Example 7.

### EXAMPLE 8 *Graphing a Horizontal Parabola*

Graph $x = 2y^2 + 6y + 5$ in the window $[-2, 8]$ by $[-4, 2]$.

**Solution** We must solve for $y$ by completing the square.

$$x = 2y^2 + 6y + 5$$

$$\frac{x}{2} = y^2 + 3y + \frac{5}{2} \qquad \text{Divide by 2.}$$

$$\frac{x}{2} - \frac{5}{2} = y^2 + 3y \qquad \text{Subtract } \tfrac{5}{2}.$$

$$\frac{x}{2} - \frac{5}{2} + \frac{9}{4} = y^2 + 3y + \frac{9}{4} \qquad \text{Add } \left[\tfrac{1}{2}(3)\right]^2 = \tfrac{9}{4}.$$

$$\frac{x}{2} - \frac{1}{4} = \left(y + \frac{3}{2}\right)^2 \qquad \text{Combine terms on the left; factor on the right.}$$

$$\left(y + \frac{3}{2}\right)^2 = \frac{x}{2} - \frac{1}{4} \qquad \text{Rewrite.}$$

$$y + \frac{3}{2} = \pm\sqrt{\frac{x}{2} - \frac{1}{4}} \qquad \text{Square root property}$$

$$y = -\frac{3}{2} \pm \sqrt{\frac{x}{2} - \frac{1}{4}} \qquad \text{Subtract } \tfrac{3}{2}.$$

Two functions are now defined. It is easier to use decimal notation when entering the equations into a calculator. Therefore, let

$$y_1 = -1.5 + \sqrt{.5x - .25} \quad \text{and} \quad y_2 = -1.5 - \sqrt{.5x - .25}.$$

The graphs of these two functions together form the parabola with horizontal axis of symmetry $y_3 = -1.5$, shown in Figure 56.

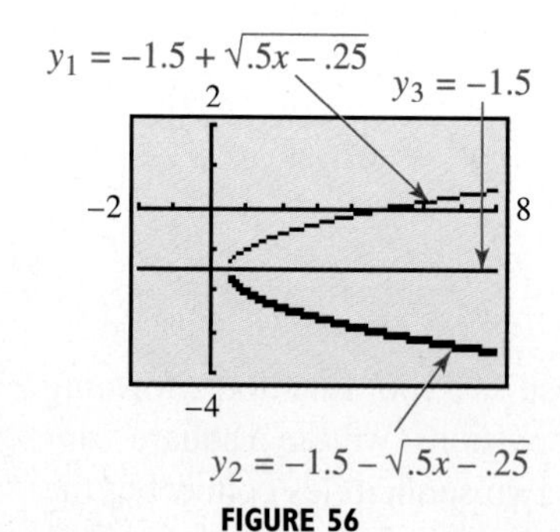

FIGURE 56

### FOR DISCUSSION

The two functions that form the parabola in Example 8 both have the same domain.

1. Solve the inequality $.5x - .25 \geq 0$ analytically. How does the solution set of this inequality pertain to the graphs found in Example 8?
2. Graph $y = .5x - .25$ with a calculator, and use this graph to find the solution set of $.5x - .25 \geq 0$. Discuss how this solution set pertains to the graphs found in Example 8.

## 4.4 EXERCISES

**Note to student: At this point, you may want to review the material on radicals and rational exponents in Chapter R.**

*Evaluate each expression without using a calculator.*

**1.** $\sqrt{169}$ **2.** $-\sqrt[3]{64}$ **3.** $\sqrt[5]{-32}$ **4.** $\sqrt[4]{16}$ **5.** $81^{3/2}$

**6.** $27^{4/3}$ **7.** $125^{-2/3}$ **8.** $(\sqrt[3]{-27})^2$ **9.** $(-1000)^{2/3}$ **10.** $(-125)^{-4/3}$

*The display in the screen on the left illustrates how a calculator computes roots that are rational numbers. The display in the screen on the right illustrates how a calculator gives rational approximations of roots that are irrational numbers.*

```
√(25)
                 5
³√(-27)
                -3
5ˣ√-32
                -2
```

```
³√(-7)
      -1.912931183
5ˣ√36.7
       2.055574459
√(21)
       4.582575695
```

*Use a calculator to find each root. Tell whether the display represents the exact value or a rational approximation of an irrational number. Give as many digits as your display shows.*

**11.** $\sqrt[3]{-4}$ **12.** $\sqrt[5]{-3}$ **13.** $\sqrt[3]{-125}$ **14.** $\sqrt[5]{-243}$

**15.** $\sqrt[6]{9}$ **16.** $\sqrt[4]{12}$ **17.** $\sqrt[3]{18.609625}$ **18.** $\sqrt[5]{286.29151}$

**19.** $\sqrt[3]{-17}$ **20.** $\sqrt[5]{-8}$ **21.** $\sqrt[6]{\pi^2}$ **22.** $\sqrt[6]{\pi^{-1}}$

*The display in the screen on the left illustrates how a calculator computes rational number powers that yield rational number results. The display in the screen on the right illustrates how a calculator gives rational approximations of rational number powers that yield irrational results.*

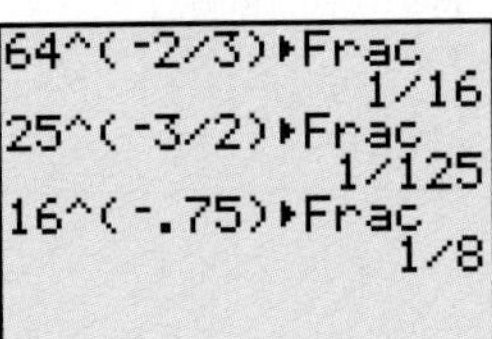


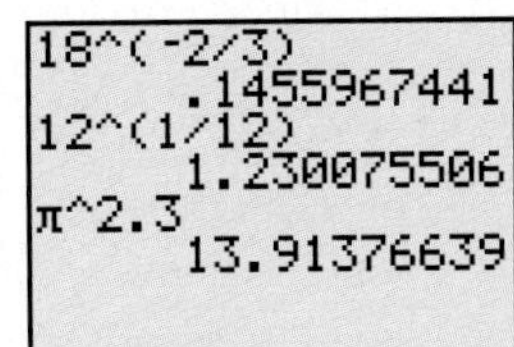


*Use a calculator to find each power. Tell whether the display represents the exact value or a rational approximation of an irrational number. Give as many digits as your display shows.*

**23.** $13^{-1/3}$ **24.** $15^{-1/6}$ **25.** $32^{.2}$ **26.** $81^{.25}$

**27.** $5^{.1}$ **28.** $12^{.37}$ **29.** $\left(\frac{5}{6}\right)^{-1.3}$ **30.** $\left(\frac{4}{7}\right)^{-.6}$

**31.** $\pi^{-3}$ **32.** $(2\pi)^{4/3}$ **33.** $17^{1/17}$ **34.** $17^{-1/17}$

**35.** The screen shows a table of square roots. Use such a table generated by your own calculator to find an approximation of each square root.

(a) $\sqrt{39}$

(b) $\sqrt{143.8}$

(c) $\sqrt{9071}$

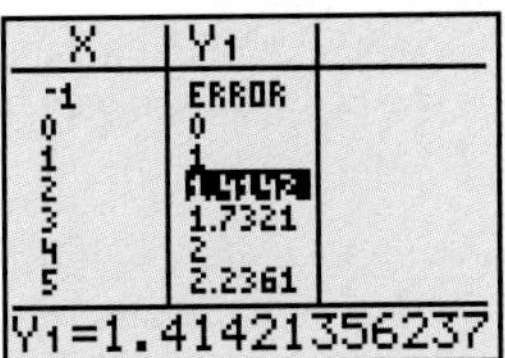


In this table, $Y_1 = \sqrt{X}$. Thus, $\sqrt{2} \approx 1.41421356237$.

**36.** The screen shows a table of cube roots. Use such a table generated by your own calculator to find an approximation of each cube root.

(a) $\sqrt[3]{39}$

(b) $\sqrt[3]{-143.8}$

(c) $\sqrt[3]{9071}$

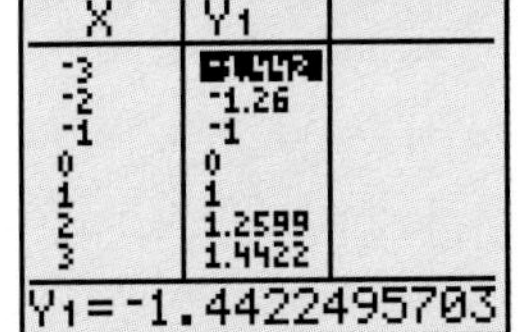


In this table, $Y_1 = \sqrt[3]{X}$. Thus, $\sqrt[3]{-3} \approx -1.4422495703$.

**37.** Consider the expression $16^{-3/4}$.

**(a)** Simplify this expression without using a calculator. Give the answer in both decimal and $\frac{a}{b}$ forms.

**(b)** Write two different radical expressions that are equivalent to it, and use your calculator to evaluate them to show that the result is the same as the decimal form you found in part (a).

**(c)** If your calculator has the capability to convert decimal numbers to fractions, use it to verify your results in part (a).

**38.** Consider the expression $5^{.47}$.

**(a)** Use the exponentiation capability of your calculator to find an approximation. Give as many digits as your calculator displays.

**(b)** Use the fact that $.47 = \frac{47}{100}$ to write the expression as a radical, and then use the root-finding capability of your calculator to find an approximation that agrees with the one found in part (a).

## RELATING CONCEPTS

**For individual or group investigation (Exercises 39–42)**

*Duplicate each screen on your calculator. The screens show multiple ways of finding an approximation for* $\sqrt[6]{9}$. *Work Exercises 39–42 as directed, using your calculator.*

**39.** Use a radical expression to approximate $\sqrt[6]{81}$ to as many decimal places as the calculator will give.

**40.** Use a rational exponent to repeat Exercise 39.

**41.** Use a table to repeat Exercise 39.

**42.** Use the graph of $y = \sqrt[6]{x}$ to repeat Exercise 39.

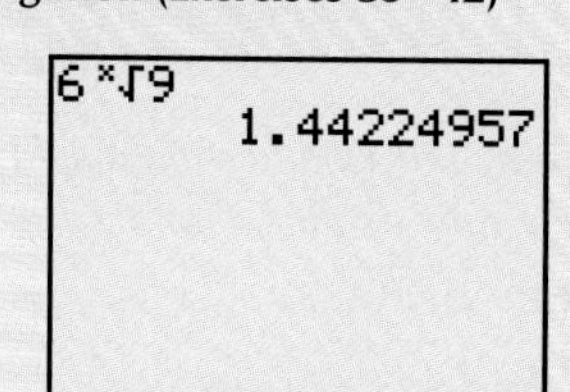


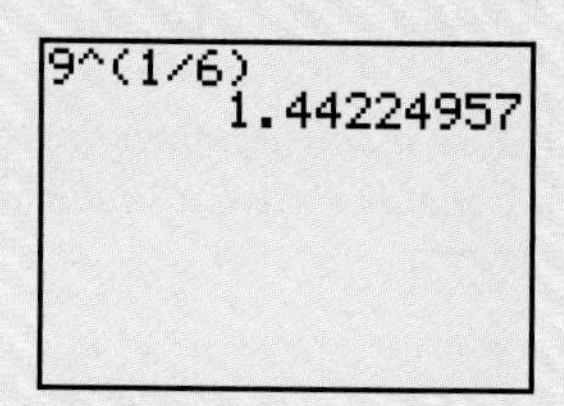


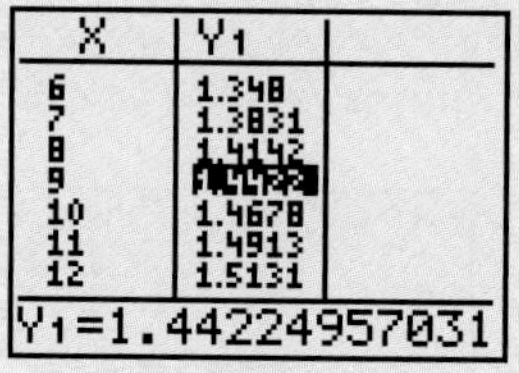


In this table, $Y_1 = \sqrt[6]{X}$.

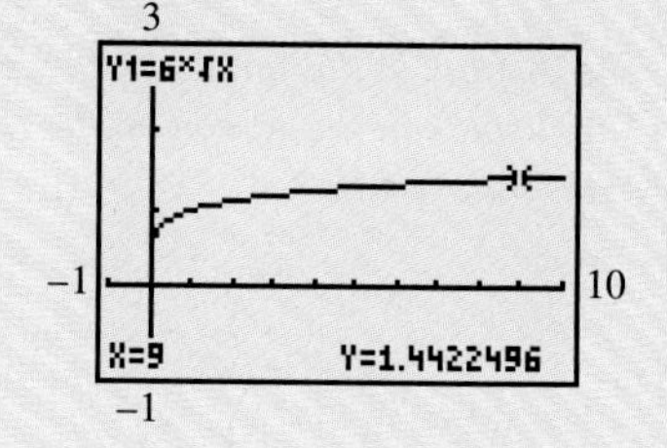


*(Modeling) Solve each problem involving a power function.*

**43.** *Wing Size* Suppose that the surface area $S$ of a bird's wings in square feet can be modeled by

$$S(w) = 1.27w^{2/3},$$

where $w$ is the weight of the bird in pounds. Estimate the surface area of a bird's wings if the bird weighs 4.0 pounds.

**44.** *Wingspan* The wingspan $L$ in feet of a bird weighing $W$ pounds is given by

$$L = 2.43W^{.3326}.$$

Estimate the wingspan of a bird that weighs 5.2 pounds.

**45.** *Planetary Orbits* The formula

$$f(x) = x^{1.5}$$

calculates the number of years it would take for a planet to orbit the sun if its average distance from the sun is $x$ times farther than Earth. If there were a planet located 15 times farther from the sun than Earth, how many years would it take for the planet to orbit the sun?

**46.** *Sight Distance* A formula for calculating the distance one can see from an airplane to the horizon on a clear day is given by

$$f(x) = 1.22x^{.5},$$

where $x$ is the altitude of the plane in feet and $f(x)$ is in miles. If a plane is flying at 30,000 feet, how far can the pilot see?

**47.** *Trout and Pollution* Rainbow trout are sensitive to zinc ions in the water. High concentrations are lethal. The average survival times $x$ in minutes for trout in various concentrations of zinc ions $y$ in milligrams per liter (mg/L) are listed in the table on the next page.

| x (in minutes) | .5 | 1 | 2 | 3 |
|---|---|---|---|---|
| y (in mg/L) | 4500 | 1960 | 850 | 525 |

*Source:* Mason, C., *Biology of Freshwater Pollution*, John Wiley and Sons, 1991.

**(a)** The data can be modeled by $f(x) = ax^b$, where $a$ and $b$ are constants. Determine $a$. (*Hint:* Let $f(1) = 1960$.)

**(b)** Estimate $b$.

**(c)** Evaluate $f(4)$ and interpret the result.

**48.** *Asbestos and Cancer* Insulation workers who were exposed to asbestos and employed before 1960 experienced an increased likelihood of lung cancer. If a group of insulation workers have a cumulative total of 100,000 years of work experience with their first date of employment $x$ years ago, then the number of lung cancer cases occurring within the group can be modeled by

$$N(x) = .00437x^{3.2}.$$

(*Source:* Walker, A., *Observation and Inference: An Introduction to the Methods of Epidemiology,* Epidemiology Resources, Inc., 1991.)

**(a)** Calculate $N(x)$ when $x = 5$, 10, and 20. What happens to the likelihood of cancer as $x$ increases?

**(b)** If $x$ doubles, does the number of cancer cases also double?

**49.** *Fiddler Crab Size* One study of the male fiddler crab showed a connection between the weight of the large claws and its total body weight. For a crab weighing over .75 gram, the weight of its claws can be estimated by

$$f(x) = .445x^{1.25}.$$

The input $x$ is the weight of the crab in grams, and the output $f(x)$ is the weight of the claws in grams. Predict the weight of the claws for a crab that weighs 2 grams. (*Source:* Huxley, J., *Problems of Relative Growth,* Methuen and Co., 1932; Brown, D. and P. Rothery, *Models in Biology: Mathematics, Statistics, and Computing,* John Wiley and Sons, 1993.)

**50.** *Weight and Height of Men* The average weight in pounds for a man can be estimated by

$$f(x) = .117x^{1.7},$$

where $x$ represents the man's height in inches and $f(x)$ is his weight in pounds. What is the average weight of a 68-inch-tall man?

**51.** *Animal Pulse Rate and Weight* According to one model, the rate at which an animal's heart beats varies with its weight. Smaller animals tend to have faster pulses, whereas larger animals tend to have slower pulses. The table lists average pulse rates in beats per minute (bpm) for animals with various weights in pounds (lb). Use regression (or some other method) to find values for $a$ and $b$ so that $f(x) = ax^b$ models these data.

| Weight (in lb) | 40 | 150 | 400 | 1000 | 2000 |
|---|---|---|---|---|---|
| Pulse (in bpm) | 140 | 72 | 44 | 28 | 20 |

*Source:* Pennycuick, C., *Newton Rules Biology*, Oxford University Press, 1992.

**52.** *Animal Pulse Rate and Weight* Use the results of Exercise 51 to calculate the pulse rates for a 60-pound dog and a 2-ton whale.

*Determine the domain of each function defined as follows.*

**53.** $f(x) = \sqrt{5 + 4x}$

**54.** $f(x) = \sqrt{9x + 18}$

**55.** $f(x) = -\sqrt{6 - x}$

**56.** $f(x) = -\sqrt{2 - .5x}$

**57.** $f(x) = \sqrt[3]{8x - 24}$

**58.** $f(x) = \sqrt[5]{x + 32}$

**59.** $f(x) = \sqrt{49 - x^2}$

**60.** $f(x) = \sqrt{81 - x^2}$

**61.** $f(x) = \sqrt{x^3 - x}$

**62.** Explain why determining the domain of a function of the form $f(x) = \sqrt[n]{ax + b}$ requires two different considerations, depending upon the parity of $n$.

*Graph each function in the window specified. The domain of each function was determined in Exercises 53–60. Use the graph to* **(a)** *find the range,* **(b)** *give the interval over which the function is increasing,* **(c)** *give the interval over which the function is decreasing, and* **(d)** *solve graphically the equation $f(x) = 0$.*

**63.** $f(x) = \sqrt{5 + 4x}$;
window: $[-2, 10]$ by $[-10, 10]$

**64.** $f(x) = \sqrt{9x + 18}$;
window: $[-10, 10]$ by $[-10, 10]$

**65.** $f(x) = -\sqrt{6 - x}$;
window: $[-10, 10]$ by $[-10, 10]$

**66.** $f(x) = -\sqrt{2 - .5x}$;
window: $[-10, 10]$ by $[-10, 10]$

**67.** $f(x) = \sqrt[3]{8x - 24}$;
window: $[-10, 10]$ by $[-10, 10]$

**68.** $f(x) = \sqrt[5]{x + 32}$;
window: $[-40, 5]$ by $[-10, 10]$

**69.** $f(x) = \sqrt{49 - x^2}$;
window: $[-14.1, 14.1]$ by $[-9.3, 9.3]$

**70.** $f(x) = \sqrt{81 - x^2}$;
window: $[-14.1, 14.1]$ by $[-9.3, 9.3]$

*Explain how the graph of the given function can be obtained from the graph of the appropriate root function* $\left(y = \sqrt{x}\right.$ *or* $\left.y = \sqrt[3]{x}\right)$.

**71.** $y = \sqrt{9x + 27}$

**72.** $y = \sqrt{16x + 16}$

**73.** $y = \sqrt{4x + 16} + 4$

**74.** $y = \sqrt{32 - 4x} - 3$

**75.** $y = \sqrt[3]{27x + 54} - 5$

**76.** In Example 6(b), we began by factoring $-8$ from the radicand. Repeat the problem, but start by factoring 8 from the radicand. Then, explain how the graph of $y = \sqrt[3]{-8x + 8}$ can alternatively be obtained from the graph of $y = \sqrt[3]{x}$.

*In Exercises 77–84, describe the graph of the equation as either a circle or a parabola with a horizontal axis of symmetry. Then, determine two functions, designated by* $y_1$ *and* $y_2$*, such that their union will give the graph of the given equation. Finally, graph the equation in the given viewing window. (These are appropriate for a TI-83 calculator.)*

**77.** $x^2 + y^2 = 100$;
$[-15, 15]$ by $[-10, 10]$

**78.** $x^2 + y^2 = 81$;
$[-15, 15]$ by $[-10, 10]$

**79.** $(x - 2)^2 + y^2 = 9$;
$[-15, 15]$ by $[-10, 10]$

**80.** $(x + 3)^2 + y^2 = 16$;
$[-9.4, 9.4]$ by $[-6.2, 6.2]$

**81.** $x = y^2 + 6y + 9$;
$[-10, 10]$ by $[-10, 10]$

**82.** $x = y^2 - 8y + 16$;
$[-10, 10]$ by $[-10, 10]$

**83.** $x = 2y^2 + 8y + 1$;
$[-10, 10]$ by $[-10, 10]$

**84.** $x = -3y^2 - 6y + 2$;
$[-9.4, 9.4]$ by $[-6.2, 6.2]$

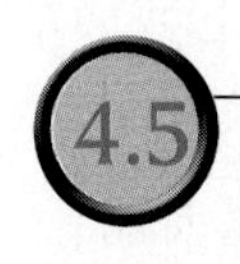

## 4.5 EQUATIONS, INEQUALITIES, AND APPLICATIONS INVOLVING ROOT FUNCTIONS

Equations and Inequalities • Applications

### Equations and Inequalities

We solve equations involving roots, such as

$$\sqrt{11 - x} - x = 1, \quad (5 - 5x)^{1/2} + x = 1, \quad \text{and} \quad \sqrt{2x + 3} - \sqrt{x + 1} = 1,$$

using the following property.

> **Power Property**
>
> If $P$ and $Q$ are algebraic expressions, then every solution of the equation $P = Q$ is also a solution of the equation $P^n = Q^n$, for any positive integer $n$.

**CAUTION** Be careful when using the power property. It does *not* say that the equations $P = Q$ and $P^n = Q^n$ are equivalent; it says only that each solution of the original equation $P = Q$ is also a solution of the new equation $P^n = Q^n$.

When using this property to solve equations, the new equation may have *more* solutions than the original equation. For example, the solution set of the equation $x = -2$ is $\{-2\}$. If we square each side of the equation $x = -2$, we get the new equation $x^2 = 4$, which has solution set $\{-2, 2\}$. Since the solution sets are not equal, the equations are not equivalent. Because of this, when an equation contains radicals or rational exponents, *it is essential to check all proposed solutions in the original equation.*

To solve equations involving roots, follow these steps.

**Solving an Equation Involving a Root Function**

1. Isolate a term involving a root on one side of the equation.
2. Raise each side of the equation to an integer power that will eliminate the radical or rational exponent.
3. Solve the resulting equation. (If a root is still present after Step 2, repeat Steps 1 and 2.)
4. Check each proposed solution in the original equation.

**EXAMPLE 1** *Solving an Equation Involving a Square Root*

Solve $\sqrt{11 - x} - x = 1$.

**Analytic Solution**

$$\sqrt{11 - x} - x = 1$$

$$\sqrt{11 - x} = 1 + x \quad \text{Isolate the radical.}$$

$$\left(\sqrt{11 - x}\right)^2 = (1 + x)^2 \quad \text{Square each side.}$$

$$11 - x = 1 + 2x + x^2$$

$$0 = x^2 + 3x - 10 \quad \text{Standard form}$$

$$0 = (x + 5)(x - 2) \quad \text{Factor.}$$

$$x + 5 = 0 \quad \text{or} \quad x - 2 = 0 \quad \text{Zero-product property}$$

$$x = -5 \quad \text{or} \quad x = 2$$

The proposed solutions are $-5$ and 2. These must be checked in the *original* equation.

$$\sqrt{11 - x} - x = 1 \quad \text{Original equation}$$

Let $x = -5$.

$$\sqrt{11 - (-5)} - (-5) = 1 \quad ?$$

$$\sqrt{16} + 5 = 1 \quad ?$$

$$4 + 5 = 1 \quad ?$$

$$9 = 1 \quad \text{False}$$

Let $x = 2$.

$$\sqrt{11 - 2} - 2 = 1 \quad ?$$

$$\sqrt{9} - 2 = 1 \quad ?$$

$$3 - 2 = 1 \quad ?$$

$$1 = 1 \quad \text{True}$$

Squaring each side of the equation led to the *extraneous* value $-5$, as indicated by the false statement $9 = 1$. Therefore, the only solution of the equation is 2. The solution set is $\{2\}$.

**Graphing Calculator Solution**

The equation in the second line of the analytic solution has the same solution set as the original equation, because each side of the equation has not yet been squared. If we graph

$$y_1 = \sqrt{11 - x} \quad \text{and} \quad y_2 = 1 + x,$$

and observe the $x$-coordinate of the only point of intersection of the graphs, we see that it is 2, confirming the analytic solution. See Figure 57.

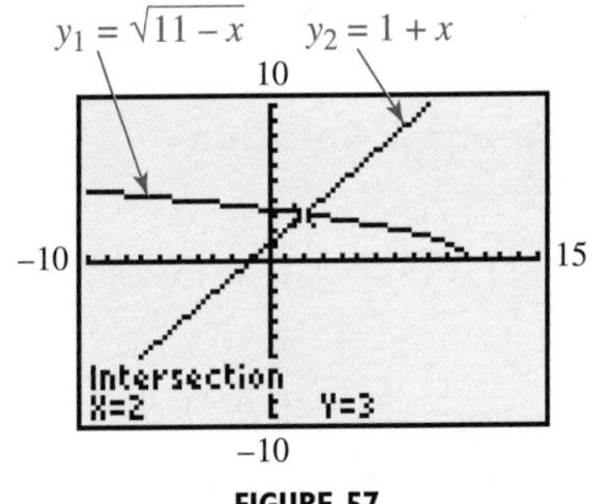


FIGURE 57

**EXAMPLE 2** *Solving an Equation Involving a Rational Exponent*

Solve $(5 - 5x)^{1/2} + x = 1$.

**Analytic Solution**

$$(5 - 5x)^{1/2} + x = 1 \quad \text{Given equation}$$

$$(5 - 5x)^{1/2} = 1 - x \quad \text{Isolate the term with the rational exponent.}$$

$$[(5 - 5x)^{1/2}]^2 = (1 - x)^2 \quad \text{Square each side.}$$

$$5 - 5x = 1 - 2x + x^2$$

$$0 = x^2 + 3x - 4 \quad \text{Standard form}$$

$$0 = (x + 4)(x - 1) \quad \text{Factor.}$$

$$x + 4 = 0 \quad \text{or} \quad x - 1 = 0 \quad \text{Zero-product property}$$

$$x = -4 \quad \text{or} \quad x = 1$$

The proposed solutions are $-4$ and 1. They must be checked in the *given* equation.

$$(5 - 5x)^{1/2} + x = 1$$

Let $x = -4$.

$$[5 - 5(-4)]^{1/2} + (-4) = 1 \quad ?$$

$$25^{1/2} + (-4) = 1 \quad ?$$

$$5 + (-4) = 1 \quad ?$$

$$1 = 1 \quad \text{True}$$

Let $x = 1$.

$$[5 - 5(1)]^{1/2} + 1 = 1 \quad ?$$

$$0^{1/2} + 1 = 1 \quad ?$$

$$0 + 1 = 1 \quad ?$$

$$1 = 1 \quad \text{True}$$

Both proposed solutions are indeed solutions, and the solution set is $\{-4, 1\}$.

**Graphing Calculator Solution**

In Example 1, we used the intersection-of-graphs method. To illustrate the $x$-intercept method, we let

$$y_1 = (5 - 5x)^{1/2} + x$$

and

$$y_2 = 1.$$

We graph $y_3 = y_1 - y_2$ to produce the curve shown in Figure 58. The displays at the bottom of the screens confirm the solutions $-4$ and 1.

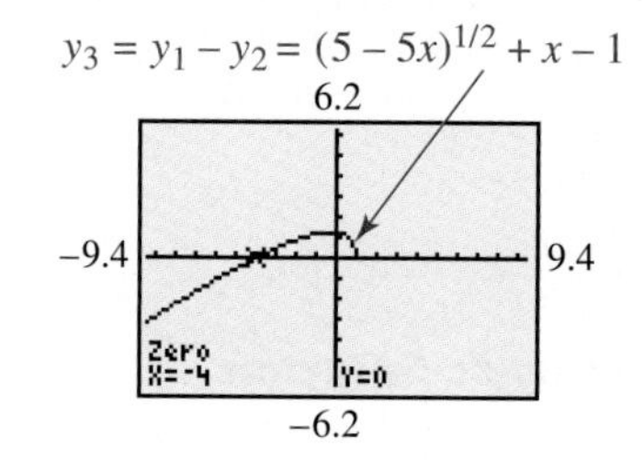


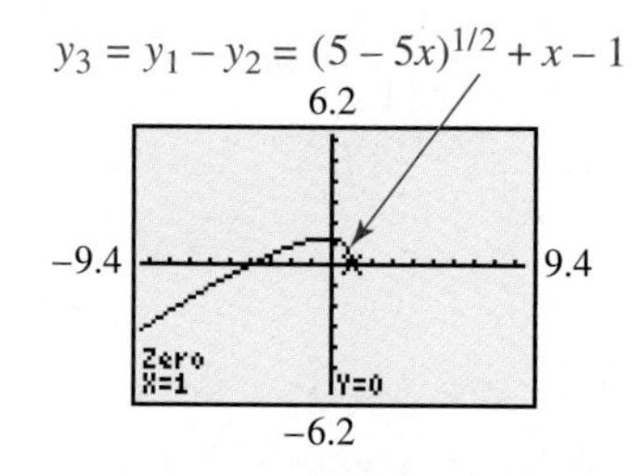


**FIGURE 58**

**EXAMPLE 3** *Solving an Equation Involving Cube Roots*

Solve $\sqrt[3]{x^2 + 3x} = \sqrt[3]{5}$.

**Analytic Solution**

$$\sqrt[3]{x^2 + 3x} = \sqrt[3]{5}$$

$$\left(\sqrt[3]{x^2 + 3x}\right)^3 = \left(\sqrt[3]{5}\right)^3 \quad \text{Cube each side.}$$

$$x^2 + 3x = 5$$

$$x^2 + 3x - 5 = 0 \quad \text{Standard form}$$

Since this equation cannot be solved by factoring, we use the quadratic formula, with $a = 1$, $b = 3$, and

**Graphing Calculator Solution**

The graphs of

$$y_1 = \sqrt[3]{x^2 + 3x} \quad \text{and} \quad y_2 = \sqrt[3]{5}$$

intersect in two points, as seen in Figures 59(a) and 59(b). Figure 59(c) confirms the solutions obtained analytically. The $x$-coordinates of the points of intersection are approximations for $\frac{-3 + \sqrt{29}}{2}$ and $\frac{-3 - \sqrt{29}}{2}$.

*(continued)*

$c = -5$, to find the solutions.

$$x = \frac{-3 \pm \sqrt{3^2 - 4(1)(-5)}}{2(1)}$$
$$= \frac{-3 \pm \sqrt{29}}{2}$$

An analytic check would be quite involved. It can be shown, however, that both of these values are indeed solutions, and the solution set is

$$\left\{\frac{-3 + \sqrt{29}}{2}, \frac{-3 - \sqrt{29}}{2}\right\}.$$

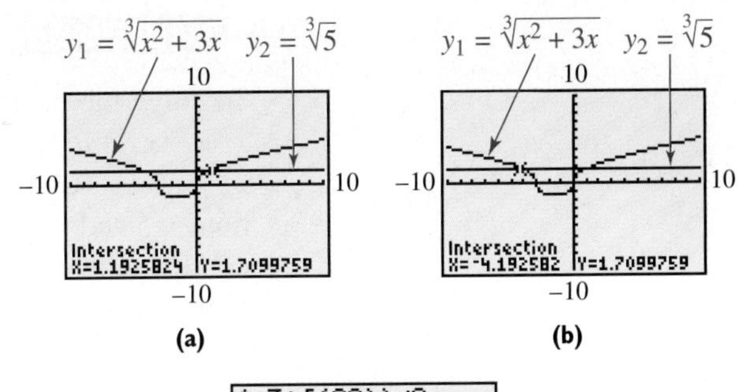


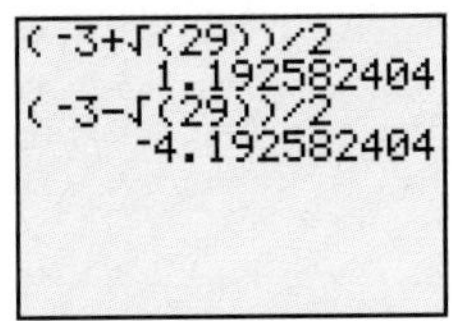


Compare these decimal approximations to the $x$-values in figures (a) and (b).

(c)

**FIGURE 59**

**EXAMPLE 4** *Solving an Equation Involving Roots (Squaring Twice)*

Solve $\sqrt{2x + 3} - \sqrt{x + 1} = 1$.

**Analytic Solution**

Begin by isolating a radical. When a choice must be made as to which radical to isolate, it is usually easier to isolate the more complicated radical first.

$$\sqrt{2x + 3} = 1 + \sqrt{x + 1} \qquad \text{Isolate } \sqrt{2x + 3}.$$
$$\left(\sqrt{2x + 3}\right)^2 = \left(1 + \sqrt{x + 1}\right)^2 \qquad \text{Square each side.}$$
$$2x + 3 = 1 + 2\sqrt{x + 1} + x + 1$$
$$x + 1 = 2\sqrt{x + 1} \qquad \text{Isolate the radical.}$$

One side of the equation still contains a radical; to eliminate it, square each side again.

$$(x + 1)^2 = \left(2\sqrt{x + 1}\right)^2$$
$$x^2 + 2x + 1 = 4(x + 1)$$
$$x^2 + 2x + 1 = 4x + 4$$
$$x^2 - 2x - 3 = 0 \qquad \text{Standard form}$$
$$(x - 3)(x + 1) = 0 \qquad \text{Factor.}$$
$$x - 3 = 0 \quad \text{or} \quad x + 1 = 0 \qquad \text{Zero-product property}$$
$$x = 3 \quad \text{or} \quad x = -1$$

A check shows that both proposed solutions 3 and $-1$ are solutions of the original equation, giving $\{3, -1\}$ as the solution set.

**Graphing Calculator Solution**

Let

$$y_1 = \sqrt{2x + 3} - \sqrt{x + 1}$$

and $y_2 = 1$.

Graph $y_3 = y_1 - y_2$ and find the zeros of the function. As seen in Figure 60, the zeros are $-1$ and 3.

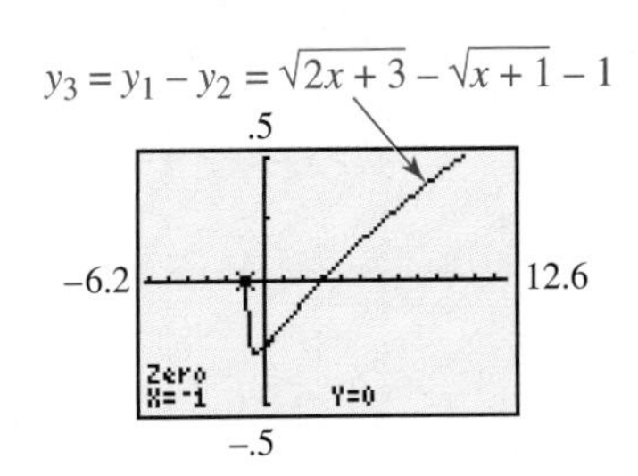


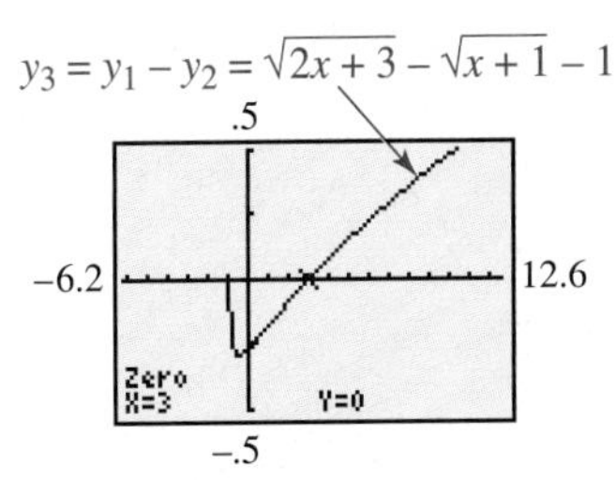


**FIGURE 60**

## What Went WRONG?

To solve the equation $x + 4 = \sqrt{x + 6}$, a student squared each side to get $x^2 + 8x + 16 = x + 6$, and then subtracted $x$ and 6 from each side to obtain $x^2 + 7x + 10 = 0$. Factoring led to $(x + 5)(x + 2) = 0$ and the possible solutions $-5$ and $-2$. He found that, of these, only $-2$ is an actual solution. To support his solution, he entered $Y_1 = X + 4$ and $Y_2 = \sqrt{X + 6}$ into his calculator. Using the intersection-of-graphs method, he obtained the screen on the left.

Feeling great about his success so far, he decided to use the $x$-intercept method to show that $-2$ is a solution. He got the screen shown at the right, and was flabbergasted. Everything had been going great, but now he knew there was a problem.

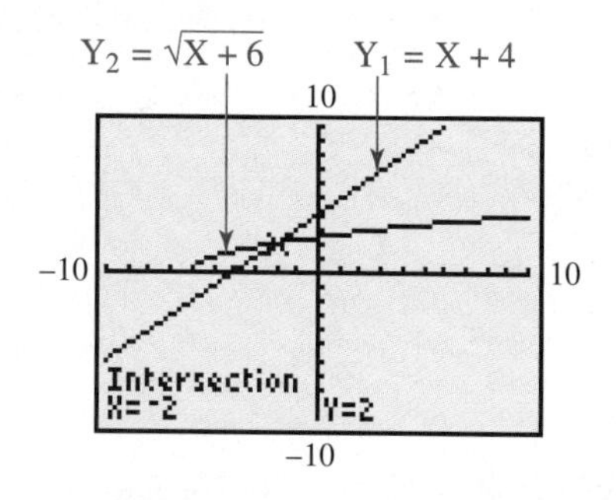


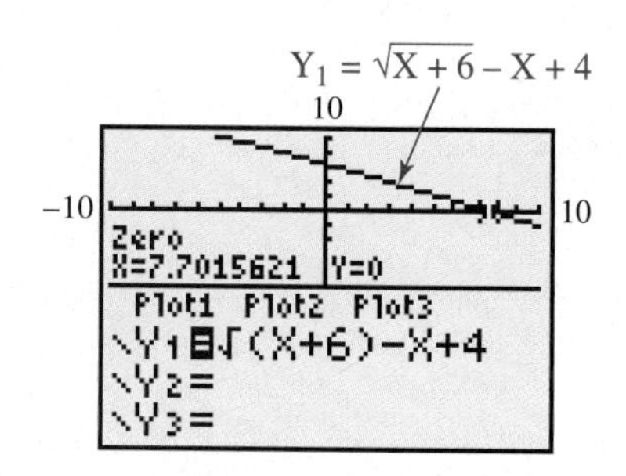


What Went Wrong? How can he correct his work so that the screen on the right shows a zero of $-2$?

**EXAMPLE 5** *Solving Inequalities Involving Roots*

Solve each inequality.

**(a)** $\sqrt[3]{x^2 + 3x} \le \sqrt[3]{5}$ **(b)** $\sqrt{2x + 3} - \sqrt{x + 1} > 1$

**Solution**

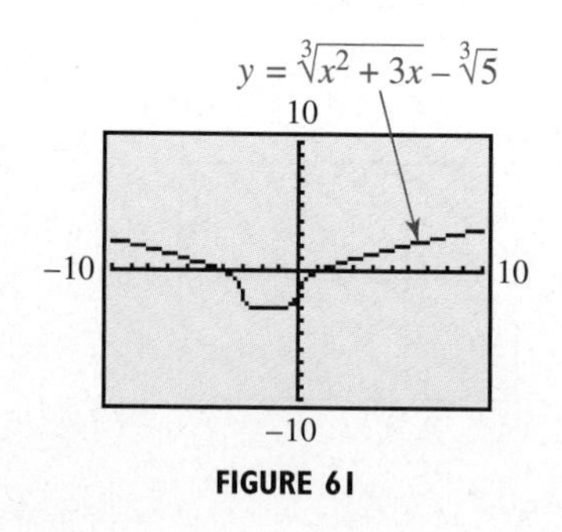


FIGURE 61

**(a)** The associated equation was solved in Example 3, where we found its solutions to be

$$\frac{-3 + \sqrt{29}}{2} \quad \text{and} \quad \frac{-3 - \sqrt{29}}{2}.$$

We use the $x$-intercept method to solve this inequality. As seen in Figure 61, the graph of $y = \sqrt[3]{x^2 + 3x} - \sqrt[3]{5}$ lies on or below the $x$-axis in the interval between the two $x$-intercepts, including the endpoints. Therefore, the solution set of this inequality is

$$\left[\frac{-3 - \sqrt{29}}{2}, \frac{-3 + \sqrt{29}}{2}\right].$$

**(b)** This inequality is equivalent to $\sqrt{2x + 3} - \sqrt{x + 1} - 1 > 0$. The associated equation was solved in Example 4, where we found its solutions to be 3 and $-1$. The graph of $y_3 = \sqrt{2x + 3} - \sqrt{x + 1} - 1$ lies above the $x$-axis for real numbers greater than 3, as we saw in Figure 60. The solution set is $(3, \infty)$.

---

*Answer to What Went Wrong?*

He entered $\sqrt{\ }(X + 6) - X + 4$ rather than $\sqrt{\ }(X + 6) - (X + 4)$. He should insert parentheses around $X + 4$ in $Y_1$.

## Applications

**EXAMPLE 6** *Solving a Cable Installation Problem Involving a Root Function*

A company wishes to run a utility cable from point $A$ on the shore (as shown in Figure 62) to an installation at point $B$ on the island. The island is 6 miles from the shore. It costs \$400 per mile to run the cable on land and \$500 per mile underwater. Assume that the cable starts at $A$ and runs along the shoreline, then angles and runs underwater to the island. Let $x$ represent the distance from $C$ at which the underwater portion of the cable run begins, and let the distance between $A$ and $C$ be 9 miles.

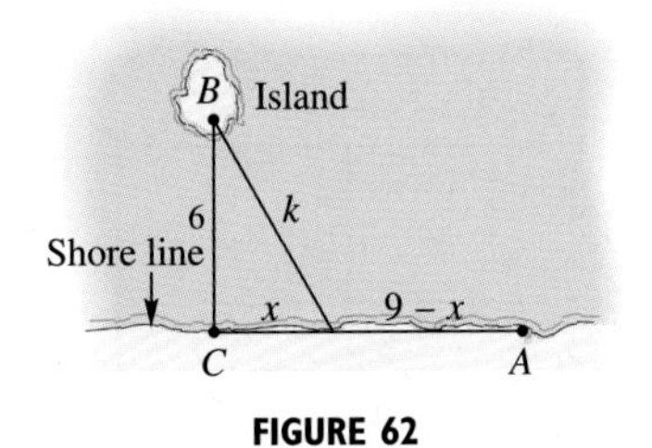


FIGURE 62

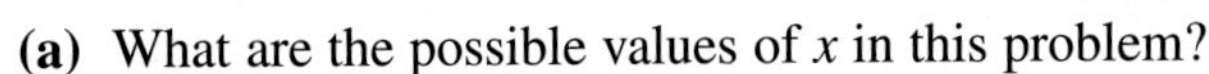

**(a)** What are the possible values of $x$ in this problem?

**(b)** Express the cost of laying the cable as a function of $x$.

**(c)** Find the total cost if 3 miles of cable are on land. Solve analytically and graphically.

**(d)** Find the point at which the line should begin to angle to minimize the total cost. What is this total cost? Solve analytically and graphically.

### Solution

**(a)** The value of $x$ must be a real number greater than or equal to 0 and less than or equal to 9, meaning that $x$ must be in the interval $[0, 9]$.

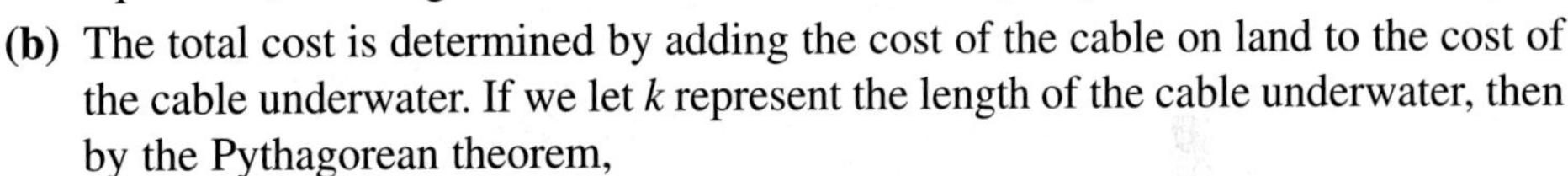

**(b)** The total cost is determined by adding the cost of the cable on land to the cost of the cable underwater. If we let $k$ represent the length of the cable underwater, then by the Pythagorean theorem,

$$k^2 = 6^2 + x^2$$
$$k^2 = 36 + x^2$$
$$k = \sqrt{36 + x^2}. \qquad k > 0$$

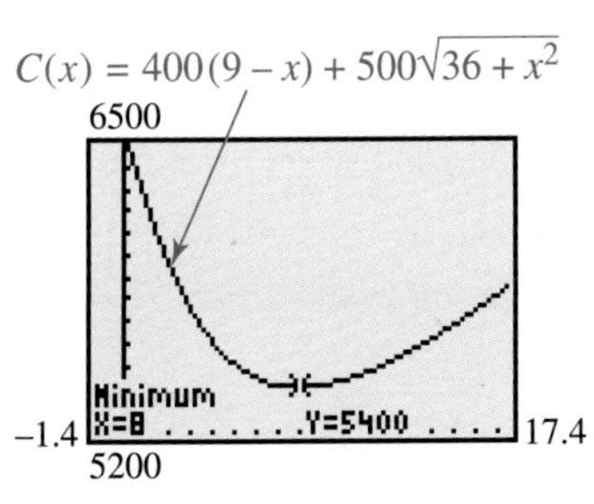


FIGURE 63

The cost of the cable on land is $400(9 - x)$ dollars, while the cost of the cable underwater is $500k$ or $500\sqrt{36 + x^2}$ dollars. Therefore, the total cost is given by

$$C(x) = 400(9 - x) + 500\sqrt{36 + x^2},$$

where $C(x)$ is in dollars.

**(c)** According to Figure 62, if 3 miles of cable are on land, then $3 = 9 - x$, giving $x = 6$. We must evaluate $C(6)$.

$$C(6) = 400(9 - 6) + 500\sqrt{36 + 6^2}$$
$$\approx 5442.64 \text{ dollars}$$

$C(x) = 400(9 - x) + 500\sqrt{36 + x^2}$

FIGURE 64

Figure 63 shows the graph of $C(x) = 400(9 - x) + 500\sqrt{36 + x^2}$. The display at the bottom indicates that $C(6) \approx 5442.64$.

**(d)** The absolute minimum value of the function on the interval $[0, 9]$ is found when $x = 8$, meaning that $9 - 8 = 1$ mile should be along land and $\sqrt{36 + 8^2} = 10$ miles should be underwater. The minimum total cost is given by

$$C(8) = 400(9 - 8) + 500\sqrt{36 + 8^2}$$
$$= 5400.$$

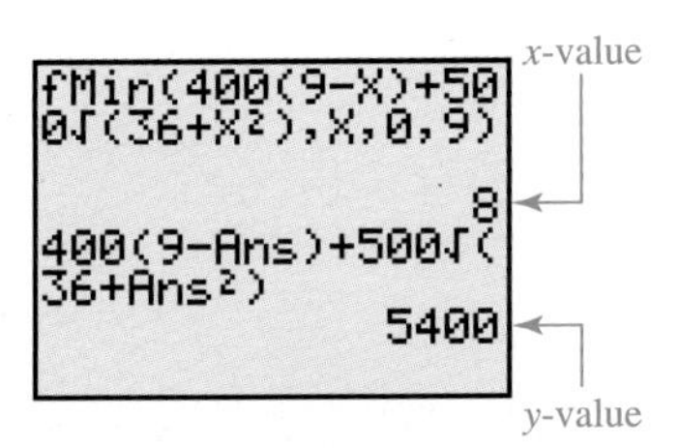


FIGURE 65

The minimum total cost is \$5400.

Figure 64 supports the analytic result, using the graph of $y = C(x)$. Figure 65 shows how the coordinates of the minimum point can be determined without actually graphing the function.

## 4.5 EXERCISES

*Use the calculator graph to find the solution set of the given equation and inequalities.*

**1.** **(a)** $\sqrt{x+5} = x - 1$
**(b)** $\sqrt{x+5} \le x - 1$
**(c)** $\sqrt{x+5} \ge x - 1$

**2.** **(a)** $-\sqrt{x+12} = x$
**(b** $-\sqrt{x+12} \le x$
**(c)** $-\sqrt{x+12} \ge x$

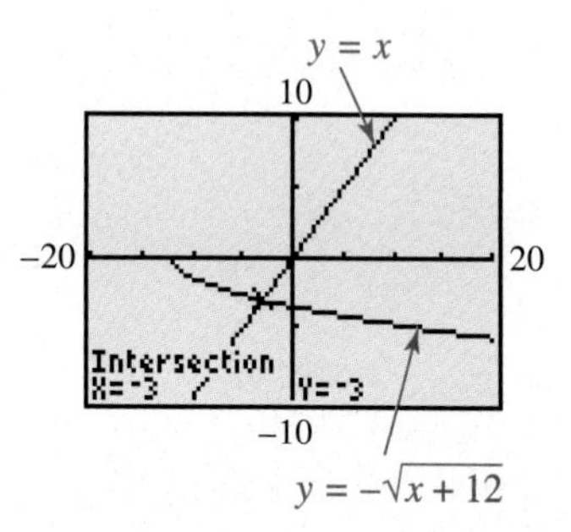


**3.** **(a)** $\sqrt[3]{2-2x} = \sqrt[3]{2x+14}$
**(b)** $\sqrt[3]{2-2x} > \sqrt[3]{2x+14}$
**(c)** $\sqrt[3]{2-2x} < \sqrt[3]{2x+14}$

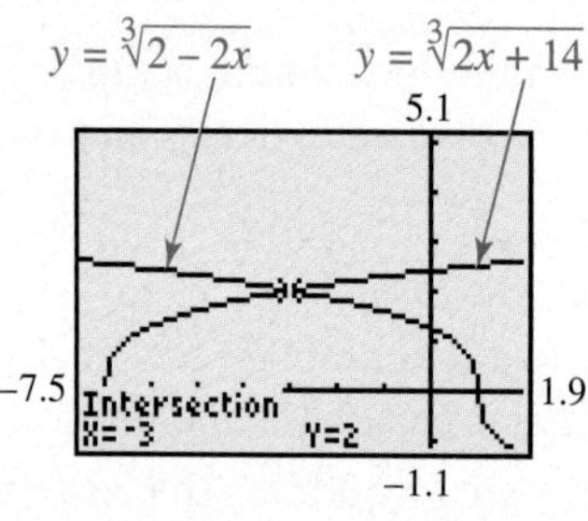


**4.** **(a)** $\sqrt[3]{3x} = \sqrt[3]{7-4x}$
**(b** $\sqrt[3]{3x} > \sqrt[3]{7-4x}$
**(c)** $\sqrt[3]{3x} < \sqrt[3]{7-4x}$

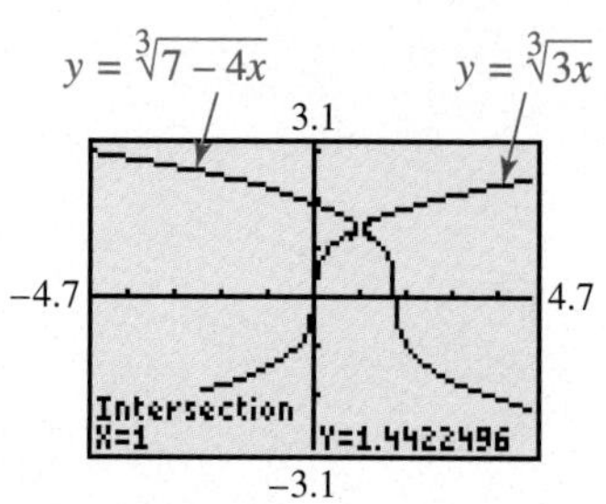


*Consider the rough sketch at the right of the graphs of*

$$y_1 = \sqrt{x} \quad \text{and} \quad y_2 = -x + 6.$$

*There is only one point of intersection of the graphs, and thus only one real solution of the equation $y_1 = y_2$. It can be shown using the methods of Chapter 3 that the solution set of*

$$\sqrt{x} = -x + 6$$

*is {4}, with extraneous value 9.*

*For Exercises 5–9, begin by drawing a rough sketch like the one shown to determine the number of real solutions for the equation $y_1 = y_2$. Then use an analytic method to support your answer. Give the solution set and any extraneous values that may occur.*

**5.** $y_1 = \sqrt{x}$, $y_2 = 2x - 1$

**6.** $y_1 = \sqrt{x}$, $y_2 = x - 6$

**7.** $y_1 = \sqrt{x}$, $y_2 = -x + 3$

**8.** $y_1 = \sqrt{x}$, $y_2 = 3x$

**9.** $y_1 = \sqrt{x}$, $y_2 = x^2$

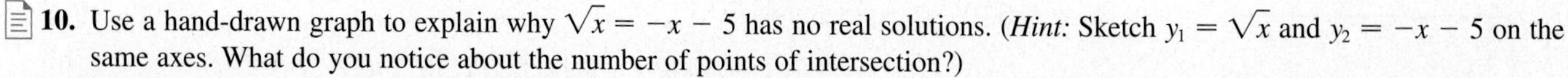

**10.** Use a hand-drawn graph to explain why $\sqrt{x} = -x - 5$ has no real solutions. (*Hint:* Sketch $y_1 = \sqrt{x}$ and $y_2 = -x - 5$ on the same axes. What do you notice about the number of points of intersection?)

*Use an analytic method to solve each equation in part (a) involving square root radicals. Support the solution with a graph. Then use the graph to solve the inequalities in parts (b) and (c).*

**11.** **(a)** $\sqrt{3x+7} = 2$
**(b)** $\sqrt{3x+7} > 2$
**(c)** $\sqrt{3x+7} < 2$

**12.** **(a)** $\sqrt{2x+13} = 3$
**(b)** $\sqrt{2x+13} > 3$
**(c)** $\sqrt{2x+13} < 3$

**13.** **(a)** $\sqrt{4x+13} = 2x - 1$
**(b)** $\sqrt{4x+13} > 2x - 1$
**(c)** $\sqrt{4x+13} < 2x - 1$

**14.** **(a)** $\sqrt{3x+7} = 3x + 5$
**(b)** $\sqrt{3x+7} > 3x + 5$
**(c)** $\sqrt{3x+7} < 3x + 5$

**15.** **(a)** $\sqrt{5x+1} + 2 = 2x$
**(b)** $\sqrt{5x+1} + 2 > 2x$
**(c)** $\sqrt{5x+1} + 2 < 2x$

**16.** **(a)** $\sqrt{3x+4} + x = 8$
**(b)** $\sqrt{3x+4} + x > 8$
**(c)** $\sqrt{3x+4} + x < 8$

**17.** **(a)** $\sqrt{3x-6} + 2 = \sqrt{5x-6}$
**(b)** $\sqrt{3x-6} + 2 > \sqrt{5x-6}$
**(c)** $\sqrt{3x-6} + 2 < \sqrt{5x-6}$

**18.** **(a)** $\sqrt{2x-4} + 2 = \sqrt{3x+4}$
**(b)** $\sqrt{2x-4} + 2 > \sqrt{3x+4}$
**(c)** $\sqrt{2x-4} + 2 < \sqrt{3x+4}$

*Use an analytic method to solve each equation in part (a) involving cube roots and fourth roots. Support the solution with a graph. Then use the graph to solve the inequalities in parts (b) and (c).*

**19.** **(a)** $\sqrt[3]{x^2 - 2x} = \sqrt[3]{x}$
**(b)** $\sqrt[3]{x^2 - 2x} > \sqrt[3]{x}$
**(c)** $\sqrt[3]{x^2 - 2x} < \sqrt[3]{x}$

**20.** **(a)** $\sqrt[3]{4x^2 - 4x + 1} = \sqrt[3]{x}$
**(b)** $\sqrt[3]{4x^2 - 4x + 1} > \sqrt[3]{x}$
**(c)** $\sqrt[3]{4x^2 - 4x + 1} < \sqrt[3]{x}$

**21.** **(a)** $\sqrt[4]{3x + 1} = 1$
**(b)** $\sqrt[4]{3x + 1} > 1$
**(c)** $\sqrt[4]{3x + 1} < 1$

**22.** **(a)** $\sqrt[4]{x - 15} = 2$
**(b)** $\sqrt[4]{x - 15} > 2$
**(c)** $\sqrt[4]{x - 15} < 2$

*Use an analytic method to solve each equation in part (a) involving rational exponents. Support the solution with a graph. Then use the graph to solve the inequalities in parts (b) and (c).*

**23.** **(a)** $(2x - 5)^{1/2} - 2 = (x - 2)^{1/2}$
**(b)** $(2x - 5)^{1/2} - 2 \geq (x - 2)^{1/2}$
**(c)** $(2x - 5)^{1/2} - 2 \leq (x - 2)^{1/2}$

**24.** **(a)** $(x + 5)^{1/2} - 2 = (x - 1)^{1/2}$
**(b)** $(x + 5)^{1/2} - 2 \geq (x - 1)^{1/2}$
**(c)** $(x + 5)^{1/2} - 2 \leq (x - 1)^{1/2}$

**25.** **(a)** $(x^2 + 6x)^{1/4} = 2$
**(b)** $(x^2 + 6x)^{1/4} > 2$
**(c)** $(x^2 + 6x)^{1/4} < 2$

**26.** **(a)** $(x^2 + 2x)^{1/4} = 3^{1/4}$
**(b)** $(x^2 + 2x)^{1/4} > 3^{1/4}$
**(c)** $(x^2 + 2x)^{1/4} < 3^{1/4}$

**27.** **(a)** $(2x - 1)^{2/3} = x^{1/3}$
**(b)** $(2x - 1)^{2/3} > x^{1/3}$
**(c)** $(2x - 1)^{2/3} < x^{1/3}$

**28.** **(a)** $(x - 3)^{2/5} = (4x)^{1/5}$
**(b)** $(x - 3)^{2/5} > (4x)^{1/5}$
**(c)** $(x - 3)^{2/5} < (4x)^{1/5}$

## RELATING CONCEPTS

**For individual or group investigation (Exercises 29–32)**

*Windchill Factor* *The chart describes the windchill factor for various wind speeds and temperatures.*

| Wind/°F | 40° | 30° | 20° | 10° | 0° | −10° | −20° | −30° | −40° | −50° |
|---|---|---|---|---|---|---|---|---|---|---|
| 5 mph | 37 | 27 | 16 | 6 | −5 | −15 | −26 | −36 | −47 | −57 |
| 10 mph | 28 | 16 | 4 | −9 | −21 | −33 | −46 | −58 | −70 | −83 |
| 15 mph | 22 | 9 | −5 | −18 | −36 | −45 | −58 | −72 | −85 | −99 |
| 20 mph | 18 | 4 | −10 | −25 | −39 | −53 | −67 | −82 | −96 | −110 |
| 25 mph | 16 | 0 | −15 | −29 | −44 | −59 | −74 | −88 | −104 | −118 |
| 30 mph | 13 | −2 | −18 | −33 | −48 | −63 | −79 | −94 | −109 | −125 |
| 35 mph | 11 | −4 | −20 | −35 | −49 | −67 | −82 | −98 | −113 | −129 |
| 40 mph | 10 | −6 | −21 | −37 | −53 | −69 | −85 | −100 | −116 | −132 |

*Windchill can be modeled by equations involving radicals, as seen in Exercises 29–32.*

**29.** Consider the equation

$$W = T - \left(\frac{v}{4} + 7\sqrt{v}\right)\left(1 - \frac{T}{90}\right)$$

as a model, where $W$ represents the windchill factor, $T$ represents the temperature, and $v$ represents the wind velocity. Evaluate it for **(a)** $T = -10$ and $v = 30$ and **(b)** $T = -40$ and $v = 5$.

**30.** Consider the equation

$$W = 91.4 - (91.4 - T)(.478 + .301\sqrt{v} - .02v)$$

as a model, where once again $W$ represents the windchill factor, $T$ represents the temperature, and $v$ represents the wind velocity. Repeat parts (a) and (b) of Exercise 29.

**31.** Use the chart to find the windchill factors for the information in parts (a) and (b) of Exercise 29.

**32.** Based on your results in Exercises 29–31, make a conjecture about which formula models the windchill factor better.

*(Modeling)* *Solve each problem involving formulas with radicals.*

**33.** *Velocity of a Meteorite* The velocity $v$ of a meteorite approaching Earth is given by

$$v = \frac{k}{\sqrt{d}},$$

measured in kilometers per second, where $d$ is its distance from the center of Earth and $k$ is a constant. If $k = 350$, what is the velocity of a meteorite that is 6000 kilometers away from the center of Earth? Round to the nearest tenth.

**34.** *Illumination* The illumination $I$ in foot-candles produced by a light source is related to the distance $d$ in feet from the light source by the equation

$$d = \sqrt{\frac{k}{I}},$$

where $k$ is a constant. If $k = 400$, how far from the source will the illumination be 14 foot-candles? Round to the nearest hundredth of a foot.

**35.** *Period of a Pendulum* The period of a pendulum in seconds depends on its length $L$ in feet, and is given by

$$P = 2\pi\sqrt{\frac{L}{32}}.$$

If the length of a pendulum is 5 feet, what is its period? Round to the nearest tenth.

**36.** *Visibility from an Airplane* A formula for calculating the distance $d$ one can see from an airplane to the horizon on a clear day is

$$d = 1.22\sqrt{x},$$

where $x$ is the altitude of the plane in feet and $d$ is given in miles. How far can one see to the horizon in a plane flying at each altitude? Give answers to the nearest mile.

**(a)** 15,000 feet **(b)** 20,000 feet

**37.** *Speed of a Car in an Accident* To estimate the speed $s$ at which a car was traveling at the time of an accident, a police officer drives a car like the one involved in the accident under conditions similar to those during which the accident took place, and then skids to a stop. If the car is driven at 30 miles per hour, the speed at the time of the accident is given by

$$s = 30\sqrt{\frac{a}{p}},$$

where $a$ is the length of the skid marks and $p$ is the length of the marks in the police test. Find $s$ if $a = 900$ feet and $p = 97$ feet.

**38.** *Plant Species and Land Area* A research biologist has shown that the number of different plant species $S$ on a Galápagos Island is related to the area of the island, $A$, by

$$S = 28.6\sqrt[3]{A}.$$

How many plant species would exist on such an island with each area?

**(a)** 100 square miles **(b)** 1500 square miles

*Solve each application of root functions.*

**39.** *(Modeling) Wire between Two Poles* Two vertical poles of lengths 12 feet and 16 feet are situated on level ground, 20 feet apart, as shown in the figure. A piece of wire is to be strung from the top of the 12-foot pole to the top of the 16-foot pole, attached to a stake in the ground at a point $P$ on a line formed by the vertical poles. Let $x$ represent the distance from $P$ to $D$, the base of the 12-foot pole.

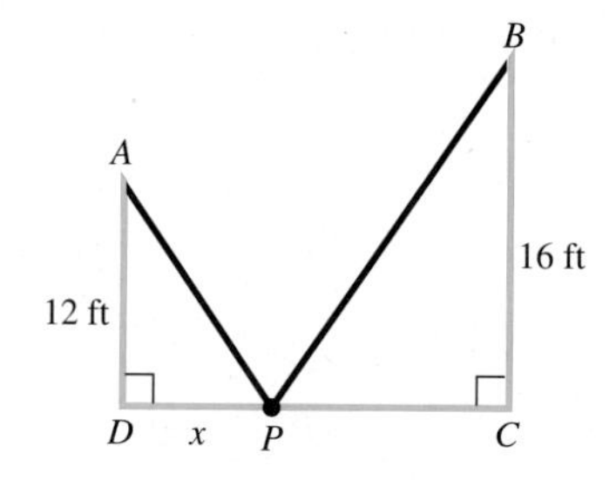


**(a)** Express the distance from $P$ to $C$ in terms of $x$.

**(b)** What are the restrictions on the value of $x$ in this problem?

**(c)** Use the Pythagorean theorem to express the lengths $AP$ and $BP$ in terms of $x$.

**(d)** Form a function $f$ that expresses the total length of the wire used.

**(e)** Graph $f$ in the window $[0, 20]$ by $[0, 50]$. Use your calculator to find $f(4)$, and interpret your result.

**(f)** Find the value of $x$ that will minimize the amount of wire used.

**(g)** Write a short paragraph summarizing what this problem has examined, and the results you have obtained.

**40.** *(Modeling) Wire between Two Poles* Repeat Exercise 39 if the heights of the poles are 9 feet and 12 feet, and the distance between the poles is 16 feet. Let $P$ be $x$ feet from the 9-foot pole. In part (e), use the window $[0, 16]$ by $[0, 50]$.

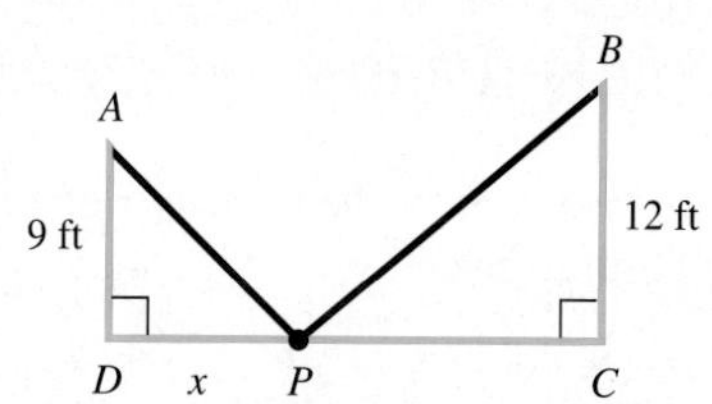

**41.** *Hunter Returns to His Cabin* A hunter is at a point on a river bank. He wants to get to his cabin, located 3 miles north and 8 miles west. He can travel 5 mph on the river but only 2 mph on this very rocky land. How far upriver should he go in order to reach the cabin in a minimum amount of time? (*Hint:* distance = rate × time.)

**42.** *Homing Pigeon Flight* Homing pigeons avoid flying over large bodies of water, preferring to fly around them instead. (One possible explanation is the fact that extra energy is required to fly over water because air pressure drops over water in the daytime.) Assume that a pigeon released from a boat 1 mile from the shore of a lake (point $B$ in the figure) flies first to point $P$ on the shore and then along the straight edge of the lake to reach its home at $L$. If $L$ is 2 miles from point $A$, the point on the shore closest to the boat, and if a pigeon needs $\frac{4}{3}$ as much energy to fly over water as over land, find the location of point $P$ that minimizes the pigeon's energy use.

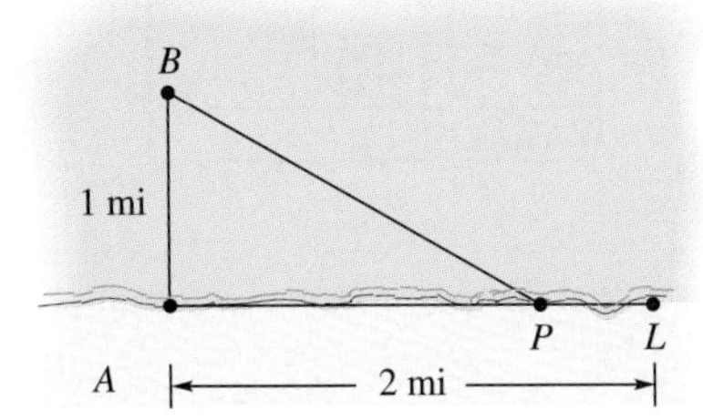


**43.** *Cruise Ship Travel* At noon, the cruise ship *Celebration* is 60 miles due south of the cruise ship *Inspiration* and is sailing north at a rate of 30 mph. If the *Inspiration* is sailing west at a rate of 20 mph, find the time at which the distance $d$ between the ships is a minimum. What is this distance?

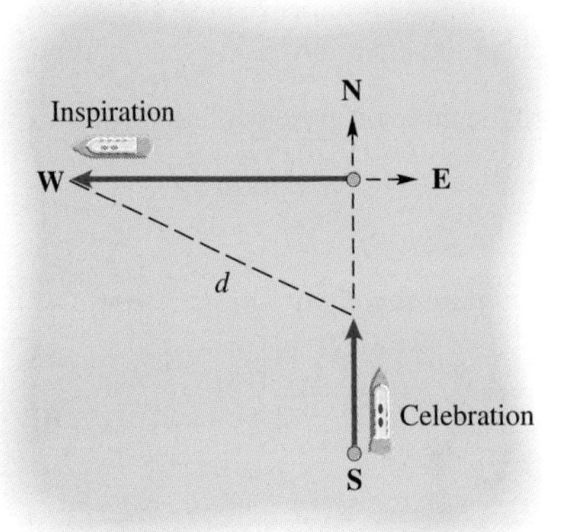


**44.** *Fisherman Returns to Camp* "Mido" Simon is in his bass boat, the *Lesley,* 3 miles from the nearest point on the shore. He wishes to reach his camp at Katie Point, 6 miles farther down the shoreline. If Mido's motor is disabled, and he can row his boat at a rate of 4 mph and walk at a rate of 5 mph, find the least amount of time that he will need to reach the camp.

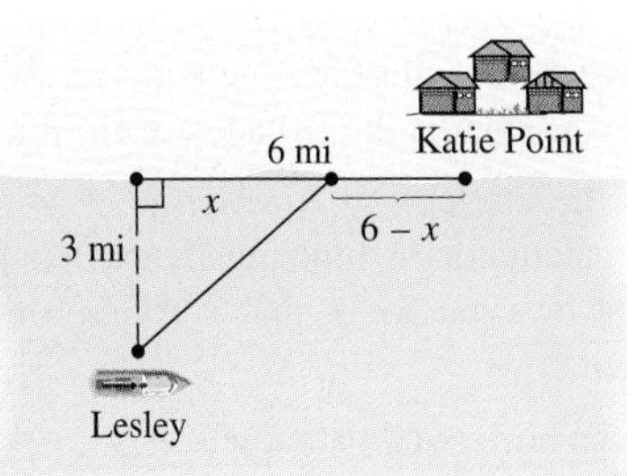


## RELATING CONCEPTS

**For individual or group investigation (Exercises 45–56)**

*Exercises 45–56 incorporate many concepts from Chapter 3 with the method of solving equations involving roots. They should be worked in order. Consider the equation*

$$\sqrt[3]{4x - 4} = \sqrt{x + 1}.$$

**45.** Rewrite the equation, using rational exponents.

**46.** What is the least common denominator of the rational exponents found in Exercise 45?

**47.** Raise each side of the equation in Exercise 45 to the power indicated by your answer in Exercise 46.

**48.** Show that the equation in Exercise 47 is equivalent to $x^3 - 13x^2 + 35x - 15 = 0$.

**49.** Graph the cubic function defined by the polynomial on the left side of the equation in Exercise 48 in the window $[-5, 10]$ by $[-100, 100]$. How many real roots does the equation have?

**50.** Use synthetic division to show that 3 is a zero of $P(x) = x^3 - 13x^2 + 35x - 15$.

**51.** Use the result of Exercise 50 to factor $P(x)$ so that one factor is linear and the other is quadratic.

**52.** Set the quadratic factor of $P(x)$ from Exercise 51 equal to 0, and solve the equation, using the quadratic formula.

**53.** What are the three proposed solutions of the original equation $\sqrt[3]{4x-4} = \sqrt{x+1}$?

**54.** Let $y_1 = \sqrt[3]{4x-4}$ and $y_2 = \sqrt{x+1}$. Graph $y_3 = y_1 - y_2$ in the window $[-2, 20]$ by $[-.5, .5]$ to determine the number of real solutions of the original equation.

**55.** Use both an analytic method and your calculator to find values of the real roots of the original equation.

**56.** Write an explanation of how the solutions of the equation in Exercise 48 relate to the solutions of the original equation. Discuss any extraneous solutions that may be involved.

*Solve each equation involving "nested" radicals for all real solutions analytically. Support your solutions with a graph.*

**57.** $\sqrt{\sqrt{x}} = x$

**58.** $\sqrt[3]{\sqrt[3]{x}} = x$

**59.** $\sqrt{\sqrt{28x+8}} = \sqrt{3x+2}$

**60.** $\sqrt{\sqrt{2x+10}} = \sqrt{2x-2}$

**61.** $\sqrt[3]{\sqrt{32x}} = \sqrt[3]{x+6}$

**62.** $\sqrt[3]{\sqrt{x+63}} = \sqrt[3]{2x+6}$

## Reviewing Basic Concepts (Sections 4.4 and 4.5)

**1.** Use a graphing calculator to graph $y = x^{.7}$, $y = x^{1.2}$, and $y = x^{2.4}$ for $x \geq 0$ on the same screen. Describe what happens as the exponent $n$ increases in value.

**2.** *(Modeling) Wing Size* Suppose that the relationship between the surface area of the wings of a species of bird and the weight of a bird is given by

$$S(w) = .3w^{3/4},$$

where $w$ is the weight of the bird in kilograms, and $S$ is the surface area of the wings in square meters. If a bird weighs .75 kilogram, what is the surface area of the wings?

**3.** To graph the circle with equation $x^2 + y^2 = 16$ using a graphing calculator in function mode, what two expressions must be used for $y_1$ and $y_2$? Graph the circle in a decimal window.

**4.** To graph the horizontal parabola $x = y^2 + 4y + 6$ using a graphing calculator in function mode, what two expressions must be used for $y_1$ and $y_2$? Graph the parabola in a square window.

**5.** Solve the equation $\sqrt{3x+4} = 8 - x$ analytically.

**6.** Use a graph and the solution of the equation in Exercise 5 to solve the inequality $\sqrt{3x+4} > 8 - x$.

**7.** Use a graph and the solution of the equation in Exercise 5 to solve the inequality $\sqrt{3x+4} < 8 - x$.

**8.** Solve the equation $\sqrt{3x+4} + \sqrt{5x+6} = 2$ analytically, and support the solution with a graph.

**9.** Explain why the table shows error messages for all values of $x$ less than 15.

| X | Y1 |
|---|---|
| 12 | ERROR |
| 13 | ERROR |
| 14 | ERROR |
| 15 | 0 |
| 16 | 1 |
| 17 | 1.4142 |
| 18 | 1.7321 |

Y1■√(X−15)

**10.** *(Modeling) Animal Pulse Rates* An eighteenth century study found that the pulse rate of an animal could be approximated by

$$f(x) = \frac{1607}{\sqrt[4]{x^3}},$$

where $x$ is the length of the animal in inches, and $f(x)$ is the approximate number of heartbeats per minute. (*Source:* Lancaster, H., *Quantitative Methods in Biology and Medical Sciences,* Springer-Verlag, 1994.)

**(a)** Use $f$ to estimate the pulse rates of a 2-foot dog and a 5.5-foot person.

**(b)** What length corresponds to a pulse rate of 400 beats per minute?

# CHAPTER 4 SUMMARY

**KEY TERMS & SYMBOLS** | **KEY CONCEPTS**

## 4.1 Rational Functions and Graphs

rational function
discontinuous
reciprocal function
vertical asymptote
horizontal asymptote

**RATIONAL FUNCTION**

A function $f$ of the form $\frac{p}{q}$ defined by

$$f(x) = \frac{p(x)}{q(x)},$$

where $p(x)$ and $q(x)$ are polynomials, is called a rational function.

**RATIONAL FUNCTIONS**

- $f(x) = \frac{1}{x}$ (the reciprocal function) decreases on the intervals $(-\infty, 0)$ and $(0, \infty)$.
- It is discontinuous at $x = 0$.
- The $y$-axis is a vertical asymptote, and the $x$-axis is a horizontal asymptote.
- The graph is symmetric with respect to the origin.

- $f(x) = \frac{1}{x^2}$ increases on the interval $(-\infty, 0)$ and decreases on the interval $(0, \infty)$.
- It is discontinuous at $x = 0$.
- The $y$-axis is a vertical asymptote, and the $x$-axis is a horizontal asymptote.
- The graph is symmetric with respect to the $y$-axis.

## 4.2 More on Graphs of Rational Functions

oblique asymptote

**GRAPHING A RATIONAL FUNCTION**

Let $f(x) = \frac{p(x)}{q(x)}$ define a function where the rational expression is written in lowest terms. To sketch its graph, follow these steps.

1. Find any vertical asymptotes. (See page 282.)
2. Find any horizontal or oblique asymptote. (See page 282.)
3. Find the $y$-intercept by evaluating $f(0)$.
4. Find the $x$-intercepts, if any, by solving $f(x) = 0$. (These will be the zeros of the numerator, $p$.)
5. Determine whether the graph will intersect its nonvertical asymptote by solving $f(x) = k$, where $k$ is the $y$-value of the horizontal asymptote, or $f(x) = mx + b$, where $y = mx + b$ is the equation of the oblique asymptote.
6. Plot a few selected points, as necessary. Choose an $x$-value in each interval of the domain, as determined by the vertical asymptotes and $x$-intercepts.
7. Complete the sketch.

## 4.3 Rational Equations, Inequalities, Applications, and Models

varies inversely
combined variation
varies jointly

**SOLVING AN EQUATION INVOLVING A RATIONAL FUNCTION**

1. Determine all values for which the rational function is undefined.
2. Multiply each side of the equation by the LCD of all rational expressions in the equation.
3. Solve the resulting equation.
4. Reject any values found in Step 1.

| KEY TERMS & SYMBOLS | KEY CONCEPTS |
|---|---|
| | **INVERSE VARIATION AS THE *n*TH POWER**<br>Let $x$ and $y$ denote two quantities and $n$ be a positive number. Then $y$ is inversely proportional to the $n$th power of $x$, or $y$ varies inversely as the $n$th power of $x$, if there exists a nonzero number $k$ such that<br>$$y = \frac{k}{x^n}.$$<br>If $y = \frac{k}{x}$, then $y$ is inversely proportional to $x$, and $y$ varies inversely as $x$.<br>**JOINT VARIATION**<br>Let $m$ and $n$ be real numbers. Then $z$ varies jointly as the $n$th power of $x$ and the $m$th power of $y$ if a nonzero real number $k$ exists such that<br>$$z = kx^n y^m.$$ |
| **4.4** Functions Defined by Powers and Roots<br>power function<br>root function | **POWER AND ROOT FUNCTIONS**<br>A function $f$ given by<br>$$f(x) = x^b,$$<br>where $b$ is a constant, is a power function. If $b = \frac{1}{n}$ for some integer $n \geq 2$, then $f$ is a root function given by<br>$$f(x) = x^{1/n}, \quad \text{or equivalently,} \quad f(x) = \sqrt[n]{x}.$$<br>• For $n$ even, the root function defined by $f(x) = \sqrt[n]{x}$ increases and is continuous on its entire domain, $[0, \infty)$.<br>• For $n$ odd, the root function defined by $f(x) = \sqrt[n]{x}$ increases and is continuous on its entire domain, $(-\infty, \infty)$. |
| **4.5** Equations, Inequalities, and Applications Involving Root Functions | **SOLVING AN EQUATION INVOLVING A ROOT FUNCTION**<br>**1.** Isolate a term involving a root on one side of the equation.<br>**2.** Raise each side of the equation to an integer power that will eliminate the radical or rational exponent.<br>**3.** Solve the resulting equation. (If a root is still present after Step 2, repeat Steps 1 and 2.)<br>**4.** Check each proposed solution in the original equation. |

## CHAPTER 4 REVIEW EXERCISES

*For each rational function, do the following.*

**(a)** *Explain how the graph of the function can be obtained from the graph of* $y = \frac{1}{x}$ *or* $y = \frac{1}{x^2}$.

**(b)** *Sketch the graph by hand.*

**(c)** *Use a graphing calculator to obtain an accurate depiction of the graph.*

**1.** $y = -\dfrac{1}{x} + 6$ **2.** $y = \dfrac{4}{x} - 3$ **3.** $y = -\dfrac{1}{(x-2)^2}$ **4.** $y = \dfrac{2}{x^2} + 1$

**5.** Under what conditions will the graph of a rational function defined by an expression written in lowest terms have an oblique asymptote?

*Use a graphing calculator to graph each function. First make a rough sketch by hand, using the guidelines of Section 4.2, so that it will be easier to find an appropriate window. Your graph should be a comprehensive one. Also, give the equations of any horizontal, vertical, or oblique asymptotes.*

**6.** $f(x) = \dfrac{4x-3}{2x-1}$ **7.** $f(x) = \dfrac{6x}{(x-1)(x+2)}$ **8.** $f(x) = \dfrac{2x}{x^2-1}$ **9.** $f(x) = \dfrac{x^2+4}{x+2}$

**10.** $f(x) = \dfrac{x^2-1}{x}$ **11.** $f(x) = \dfrac{-2}{x^2+1}$ **12.** $f(x) = \dfrac{x^2-1}{x+1}$

*Find an equation for each rational function graphed. (Answers may vary.)*

**13.**

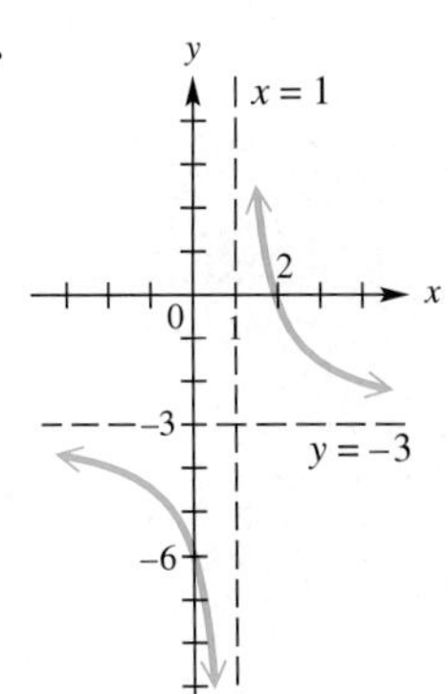


**14.**

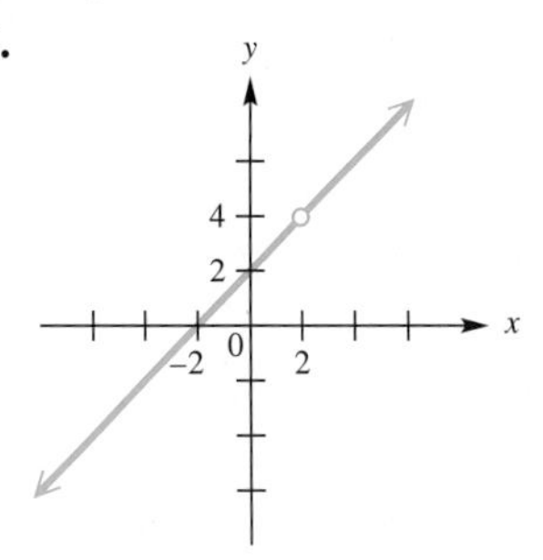


*Solve the rational equation in part (a) analytically. Then, use a graph to determine the solution sets of the associated inequalities in parts (b) and (c).*

**15. (a)** $\dfrac{3x-2}{x+1} = 0$

**(b)** $\dfrac{3x-2}{x+1} < 0$

**(c)** $\dfrac{3x-2}{x+1} > 0$

**16. (a)** $\dfrac{5}{2x+5} = \dfrac{3}{x+2}$

**(b)** $\dfrac{5}{2x+5} < \dfrac{3}{x+2}$

**(c)** $\dfrac{5}{2x+5} > \dfrac{3}{x+2}$

**17. (a)** $\dfrac{3}{x-2} + \dfrac{1}{x+1} = \dfrac{1}{x^2-x-2}$

**(b)** $\dfrac{3}{x-2} + \dfrac{1}{x+1} \leq \dfrac{1}{x^2-x-2}$

**(c)** $\dfrac{3}{x-2} + \dfrac{1}{x+1} \geq \dfrac{1}{x^2-x-2}$

**18. (a)** $1 - \dfrac{5}{x} + \dfrac{6}{x^2} = 0$

**(b)** $1 - \dfrac{5}{x} + \dfrac{6}{x^2} \leq 0$

**(c)** $1 - \dfrac{5}{x} + \dfrac{6}{x^2} \geq 0$

## RELATING CONCEPTS

For individual or group investigation (Exercises 19–22)

*Consider the rational equation below, and work Exercises 19–22 in order.*

$$1 - \frac{2}{x} - \frac{2}{x^2} + \frac{3}{x^3} = 0$$

**19.** Graph

$$f(x) = 1 - \frac{2}{x} - \frac{2}{x^2} + \frac{3}{x^3},$$

and determine its $x$-intercepts graphically. (*Hint:* One is rational and two are irrational.) Give approximations of the irrational solutions correct to the nearest hundredth.

**20.** What is the domain of $f$?

**21.** Use the rational solution determined in Exercise 19 with synthetic division to obtain a quadratic equation that will give the other two solutions in their exact form. (*Hint:* Multiply each side by $x^3$ first.)

**22.** Verify that the approximations found in Exercise 19 are indeed approximations of the two irrational solutions whose exact values were found in Exercise 21.

*A comprehensive graph of a rational function f is shown. Use the graph to find each solution set.*

**23.** $f(x) = 0$

**24.** $f(x) > 0$

**25.** $f(x) < 0$

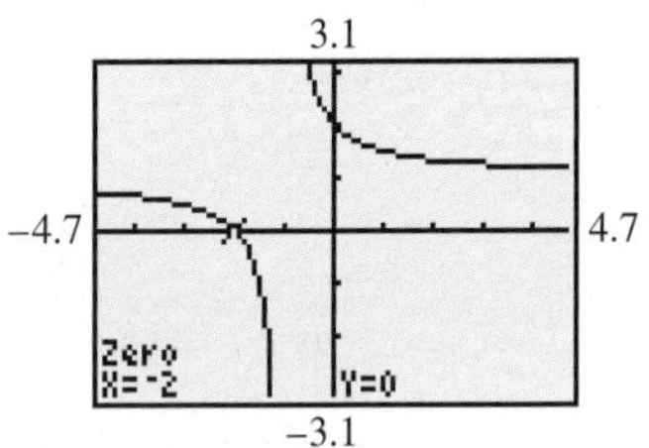


The graph has vertical asymptote $x = -1$.

*(Modeling) Solve each problem involving rational functions.*

**26.** *Environmental Pollution* In situations involving environmental pollution, a cost-benefit model expresses cost as a function of the percentage of pollutant removed from the environment. Suppose a cost-benefit model is expressed as

$$C(x) = \frac{6.7x}{100 - x},$$

where $C(x)$ is the cost, in thousands of dollars, of removing $x$ percent of a certain pollutant.

**(a)** Use a graphing calculator to graph the function in the window $[0, 100]$ by $[0, 150]$.

**(b)** How much would it cost to remove 95% of the pollutant?

**27.** *Antique-Car Competition* Antique-car owners often enter their cars in a *concours d'elegance* in which a maximum of 100 points can be awarded to a particular car. Points are awarded for the general attractiveness of the car. The function defined by

$$C(x) = \frac{10x}{49(101 - x)}$$

expresses the cost, in thousands of dollars, of restoring a car so that it will win $x$ points.

**(a)** Use a graphing calculator to graph the function in the window $[0, 101]$ by $[0, 5]$.

**(b)** How much would an owner expect to pay to restore a car in order to earn 95 points?

**28.** *Concentration of a Drug* The concentration of a drug in a medical patient's bloodstream is given by

$$f(t) = \frac{5}{t^2 + 1},$$

where $t \geq 0$ is in hours and $f(t)$ is in milligrams per liter.

**(a)** Does the concentration of the drug increase or decrease?

**(b)** The patient should not take a second dose until the concentration is below 1.5 milligrams per liter. How long should the patient wait before taking a second dose?

**29.** *Waiting in Line* A parking lot attendant can wait on 40 cars per hour. If cars arrive randomly at a rate of $x$ cars per hour, then the average line length to enter the ramp is given by

$$f(x) = \frac{x^2}{1600 - 40x},$$

where $0 \le x < 40$.

**(a)** Solve the inequality $f(x) \le 8$.

**(b)** Interpret your answer from part (a).

**30.** *Slippery Roads* The coefficient of friction $x$ measures the friction between the tires of a car and the road, where $0 < x \le 1$. A smaller value of $x$ indicates that the road is more slippery. If a car is traveling at 60 mph, then braking distance $D$ in feet is given by

$$D(x) = \frac{120}{x}.$$

**(a)** What happens to the braking distance as the coefficient of friction becomes smaller?

**(b)** Find values for the coefficient of friction $x$ that correspond to a braking distance of 400 feet or more.

*Solve each problem involving variation.*

**31.** Let $y$ be inversely proportional to $x$. When $x = 6$, $y = 5$. Find $y$ when $x = 15$.

**32.** Let $z$ be inversely proportional to the third power of $t$. When $t = 5$, $z = .08$. Find $z$ when $t = 2$.

**33.** Suppose $m$ varies jointly as $n$ and the square of $p$, and inversely as $q$. If $m$ is 20 when $n$ is 5, $p$ is 6, and $q$ is 18, find $m$ when $n$ is 7, $p$ is 11, and $q$ is 2.

**34.** The formula for the height of a right circular cone with volume 100 is

$$h = \frac{\frac{300}{\pi}}{r^2}.$$

The height of this cone varies __________ as the __________ of the __________ of its base. The constant of variation is __________.

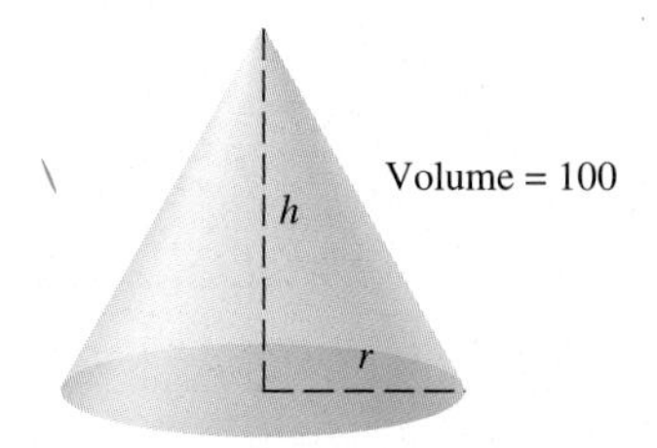


**35.** *Illumination* The illumination produced by a light source varies inversely as the square of the distance from the source. The illumination of a light source at 5 meters is 70 candela. What is the illumination 12 meters from the source?

**36.** *Current Flow* In electric current flow, it is found that the resistance (measured in units called ohms) offered by a fixed length of wire of a given material varies inversely as the square of the diameter of the wire. If a wire .01 inch in diameter has a resistance of .4 ohm, what is the resistance of a wire of the same length and material with a diameter of .03 inch?

**37.** *Simple Interest* Simple interest varies jointly as principal and time. If \$1000 left at interest for 2 years earned \$110, find the amount of interest earned by \$5000 for 5 years.

**38.** *Car Skidding* The force needed to keep a car from skidding on a curve varies inversely as the radius of the curve and jointly as the weight of the car and the square of the speed. It takes 3000 pounds of force to keep a 2000-pound car from skidding on a curve of radius 500 feet at 30 mph. What force is needed to keep the same car from skidding on a curve of radius 800 feet at 60 mph?

**39.** *Sports Arena Construction* The roof of a new sports arena rests on round concrete pillars. The maximum load a cylindrical column of circular cross section can hold varies directly as the fourth power of the diameter and inversely as the square of the height. The arena has 9 meter tall columns that are 1 meter in diameter and will support a load of 8 metric tons. How many metric tons will be supported by a column 12 meters high and $\frac{2}{3}$ meter in diameter?

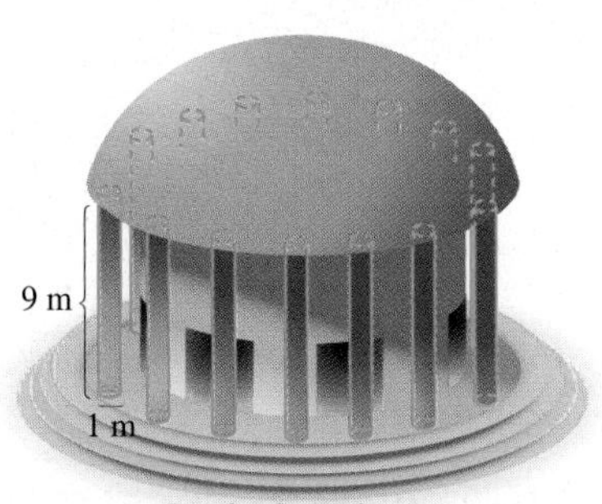

**40.** *Sports Arena Construction* A sports arena requires a beam 16 meters long, 24 centimeters wide, and 8 centimeters high. The maximum load of a horizontal beam that is supported at both ends varies directly as the width and square of the height, and inversely as the length between supports. If a beam of the same material 8 meters long, 12 centimeters wide, and 15 centimeters high can support a maximum of 400 kilograms, what is the maximum load the beam in the arena will support?

**41.** *Weight of an Object* The weight $w$ of an object varies inversely as the square of the distance $d$ between the object and the center of Earth. If a man weighs 90 kilograms on the surface of Earth, how much would he weigh 800 kilometers above the surface? (The radius of Earth is about 6400 kilometers.)

*Suppose that a and b are positive numbers. Sketch the general shape of the graph of each function.*

**42.** $y = -a\sqrt{x}$ **43.** $y = \sqrt[3]{x + a}$ **44.** $y = \sqrt[3]{x - a}$ **45.** $y = -a\sqrt[3]{x} - b$ **46.** $y = \sqrt{x + a} + b$

*Evaluate each expression* without *the use of a calculator. Then, check your results with a calculator.*

**47.** $-(-32)^{1/5}$ **48.** $36^{-3/2}$ **49.** $-1000^{2/3}$ **50.** $(-27)^{-4/3}$ **51.** $16^{3/4}$

*Use your calculator to find each root or power. Tell whether the display represents the exact value or a rational approximation of an irrational number. Give as many digits as your display shows.*

**52.** $\sqrt[5]{84.6}$ **53.** $\sqrt[4]{\frac{1}{16}}$ **54.** $\left(\frac{1}{8}\right)^{4/3}$ **55.** $12^{1/3}$

*Consider the function defined by* $f(x) = -\sqrt{2x - 4}$ *for Exercises 56–59.*

**56.** Find the domain of $f$ analytically.

**57.** Use a graph to determine the range of $f$.

**58.** Give the interval over which the function is increasing (if any).

**59.** Give the interval over which the function is decreasing (if any).

**60.** Determine the expressions for $y_1$ and $y_2$ that are required to graph the circle $x^2 + y^2 = 25$ using a graphing calculator in function mode. Then graph the circle in a square window.

*Solve each equation involving radicals or rational exponents in part (a) analytically. Then, use a graph to determine the solution sets of the associated inequalities in parts (b) and (c).*

**61.** **(a)** $\sqrt{5 + 2x} = x + 1$
**(b)** $\sqrt{5 + 2x} > x + 1$
**(c)** $\sqrt{5 + 2x} < x + 1$

**62.** **(a)** $\sqrt{2x + 1} - \sqrt{x} = 1$
**(b)** $\sqrt{2x + 1} - \sqrt{x} > 1$
**(c)** $\sqrt{2x + 1} - \sqrt{x} < 1$

**63.** **(a)** $\sqrt[3]{6x + 2} = \sqrt[3]{4x}$
**(b)** $\sqrt[3]{6x + 2} \geq \sqrt[3]{4x}$
**(c)** $\sqrt[3]{6x + 2} \leq \sqrt[3]{4x}$

**64.** **(a)** $(x - 2)^{2/3} - x^{1/3} = 0$
**(b)** $(x - 2)^{2/3} - x^{1/3} \geq 0$
**(c)** $(x - 2)^{2/3} - x^{1/3} \leq 0$

## RELATING CONCEPTS

**For individual or group investigation (Exercises 65–68)**

*Exercises 65–68 refer to the functions defined by*

$$y_1 = \sqrt{3x + 12} - 4 \quad \textit{and} \quad y_2 = \sqrt[3]{3x + 12} - 6.$$

*Use a graph to help answer each problem.*

**65.** How many solutions does the equation $y_1 = y_2$ have?

**66.** Based on your answer to Exercise 65, how many $x$-intercepts does the graph of $f(x) = y_1 - y_2$ have?

**67.** Describe the appearance of the graph of $-f(x) = y_2 - y_1$.

**68.** Describe the appearance of the graph of $y = f(-x)$.

*Solve each problem.*

**69.** *Swing of a Pendulum* A simple pendulum will swing back and forth in regular time intervals (periods). Grandfather clocks use pendulums to keep accurate time. The relationship between the length of a pendulum $L$ and the time $T$ for one complete oscillation can be expressed by the equation $L = kT^n$, where $k$ is a constant and $n$ is a positive integer to be determined. The following data were taken for different lengths of pendulums.

| $T$ (in seconds) | $L$ (in feet) |
|---|---|
| 1.11 | 1.0 |
| 1.36 | 1.5 |
| 1.57 | 2.0 |
| 1.76 | 2.5 |
| 1.92 | 3.0 |
| 2.08 | 3.5 |
| 2.22 | 4.0 |

**(a)** As $L$ increases, what happens to $T$?

**(b)** Discuss how $n$ and $k$ can be found.

**(c)** Use the data to approximate $k$, and determine the best value for $n$.

**(d)** Using the values of $k$ and $n$ from part (c), predict $T$ for a pendulum having a length of 5 feet.

**(e)** If the length $L$ of a pendulum doubles, what happens to the period $T$?

**70.** *Volume of a Cylindrical Package* A company plans to package its product in a cylinder that is open at one end. The cylinder is to have a volume of $27\pi$ cubic inches. What radius should the circular bottom of the cylinder have to minimize the cost of the material? (*Hint:* The volume of a circular cylinder is $\pi r^2 h$, where $r$ is the radius of the circular base and $h$ is the height; the surface area of the cylinder open at one end is $2\pi rh + \pi r^2$.)

## CHAPTER 4 TEST

*For each rational function, do the following.*

**(a)** *Sketch its graph.*

**(b)** *Explain how the graph is obtained from the graph of $y = \frac{1}{x}$ or $y = \frac{1}{x^2}$.*

**(c)** *Use a graphing calculator to obtain an accurate depiction of the graph.*

**1.** $f(x) = -\dfrac{1}{x}$

**2.** $f(x) = \dfrac{-1}{x^2} - 3$

**3.** Consider the rational function defined by

$$f(x) = \frac{x^2 + x - 6}{x^2 - 3x - 4} = \frac{(x + 3)(x - 2)}{(x - 4)(x + 1)}.$$

Determine the following analytically.

**(a)** Equations of the vertical asymptotes

**(b)** Equation of the horizontal asymptote

**(c)** $y$-intercept

**(d)** $x$-intercepts, if any

**(e)** Coordinates of the point where the graph of $f$ intersects its horizontal asymptote

**(f)** Then, use a graphing calculator to obtain a comprehensive graph.

**4.** Find the equation of the oblique asymptote of the graph of the rational function defined by

$$f(x) = \frac{2x^2 + x - 3}{x - 2}.$$

Then, graph the function and its asymptote, using a graphing calculator to illustrate an accurate comprehensive graph.

**5.** Consider the rational function defined by

$$f(x) = \frac{x^2 - 16}{x + 4}.$$

**(a)** For what value of $x$ does the graph exhibit a "hole"?

**(b)** Graph the function in a window of a graphing calculator that clearly shows the "hole" in the graph.

**6.** **(a)** Solve the rational equation

$$\frac{3}{x-2}+\frac{21}{x^2-4}=\frac{14}{x+2}$$

analytically.

**(b)** Use the result of part (a) and a graph to find the solution set of

$$\frac{3}{x-2}+\frac{21}{x^2-4}\geq\frac{14}{x+2}.$$

**7.** *(Modeling) Waiting in Line* A small section of highway that is under construction can accommodate at most 40 cars per minute. If cars arrive randomly at an average rate of $x$ vehicles per minute, then average wait $W$ in minutes for a car to pass through this section of highway is approximated by

$$W(x)=\frac{1}{40-x},$$

where $0 \leq x < 40$.

**(a)** Evaluate $W(30)$, $W(39)$, and $W(39.9)$. Interpret the results.

**(b)** Graph $W$ using the window $[0, 40]$ by $[-.5, 1]$. Identify the vertical asymptote. What happens to $W$ as $x$ approaches 40?

**(c)** Find $x$ when the wait is 5 minutes.

**8.** *Measure of Malnutrition* A measure of malnutrition, called the *pelidisi,* varies directly as the cube root of a person's weight in grams, and inversely as the person's sitting height in centimeters. A person with a pelidisi below 100 is considered to be undernourished, while a pelidisi greater than 100 indicates overfeeding. A person who weighs 48,820 grams and has a sitting height of 78.7 centimeters has a pelidisi of 100. Find the pelidisi (to the nearest whole number) of a person whose weight is 54,430 grams and whose sitting height is 88.9 centimeters. Is this individual undernourished or overfed?

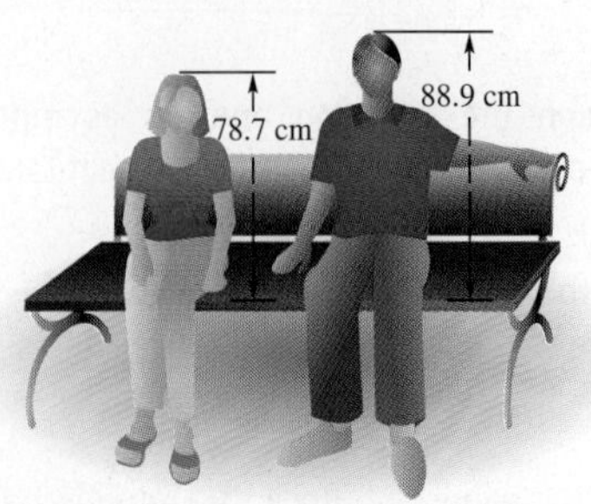


**9.** *Volume of a Cylindrical Can* A manufacturer wants to construct a cylindrical aluminum can with a volume of 4000 cubic centimeters. If $x$ represents the radius of the circular top and bottom of the can, the surface area $S$ as a function of $x$ is given by

$$S(x)=\frac{8000+2\pi x^3}{x}.$$

Use a graph to find the radius that will minimize the amount of aluminum needed, and determine what this amount will be. (*Hint:* Use the window $[-5, 42]$ by $[-1000, 10{,}000]$.)

**10.** Graph the function defined by $f(x) = -\sqrt{5-x}$ in the standard viewing window. Then, do the following.

**(a)** Determine the domain analytically.

**(b)** Use the graph to find the range.

**(c)** This function ______________ (increases/decreases) over its entire domain.

**(d)** Solve the equation $f(x) = 0$ graphically.

**(e)** Solve the inequality $f(x) < 0$ graphically.

**11.** **(a)** Solve the equation $\sqrt{4-x} = x + 2$ analytically, and support the solution(s) with a graph.

**(b)** Use the graph in part (a) to find the solution set of $\sqrt{4-x} > x + 2$.

**(c)** Use the graph in part (a) to find the solution set of $\sqrt{4-x} \leq x + 2$.

**12.** *Laying a Telephone Cable* A telephone company wishes to minimize the cost of laying a cable from point $P$ to point $R$. Points $P$ and $Q$ are directly opposite each other along the banks of a straight river. The river is 300 yards wide. Point $R$ lies on the same side of the river as point $Q$, but 600 yards away. If the cost per yard for the cable is \$125 per yard under the water and \$100 per yard on land, how should the company lay the cable to minimize the cost?

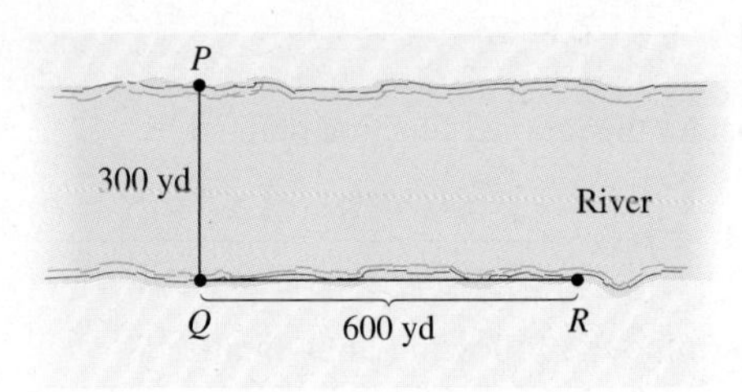

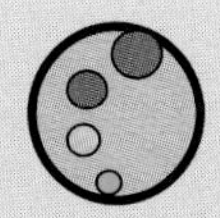

# Chapter 4 Project

## How Rugged Is Your Coastline?*

An interesting feature of coastlines is that their ruggedness is independent of the distance from which they are viewed. From an airplane, we see irregularities as bays, peninsulas, river mouths, and so on. On foot, we see each rock outcropping and creek that makes the coastline appear more rugged. An ant sees every pebble as a mountain to be scaled. An interesting result of this phenomenon is that the total distance you travel along a coastline is dependent on the size of the steps you take. The closer (or smaller) you are, the smaller your steps will be. This means you will have more obstacles in your way, which results in a longer distance to travel. The more rugged the coastline, the longer it will be. In theory, this means that if you could take small enough steps on a rugged enough coastline, the length of the coastline would approach infinity. (If you find this topic interesting, you might want to study *fractal geometry*.)

From a mathematical perspective, we can say that the number of steps needed ($y$) varies inversely with some power of the size of the steps taken ($x$):

$$y = \frac{k}{x^n}.$$

Each coastline will have different values for $k$ and $n$, depending on its ruggedness. To find these values, you must walk the coastline. Gathering such data from a real coastline is, of course, too difficult as a classroom activity, but we can obtain similar data by using maps and "walking" the coastline with compasses (the kind used in geometry for drawing circles).

The data on the next page were gathered by "walking" compasses along the coastline of a map of Lake Tahoe (on the California–Nevada border). For each walk, the compasses were opened to the given step size, and the legs of the compasses were walked around the lake, counting the number of steps until the journey was completed.

*This project is based on an idea presented by Lori Lambertson, of the Nueva School and the Exploratorium in San Francisco.

| Step Size | Number of Steps | Total Length (step size times the number of steps) |
|---|---|---|
| 4 inches | 4.75 | 19 inches |
| 2 inches | 10.75 | 21.5 inches |
| 1 inch | 23.5 | 23.5 inches |
| $\frac{1}{2}$ inch | 51 | 25.5 inches |
| $\frac{1}{4}$ inch | 104 | 26 inches |

Now we can analyze these data. First, enter them into the list editor of a graphing calculator, as shown in Figure A. Next, draw a scatter diagram of the data in an appropriate viewing window, as seen in Figure B. To find an equation to model the data, we solve for $k$ by substituting the coordinates $(1, 23.5)$ for $x$ and $y$ in $y = \frac{k}{x^n}$. (We choose this point because any power of 1 is 1.) This leads to $k = 23.5$.

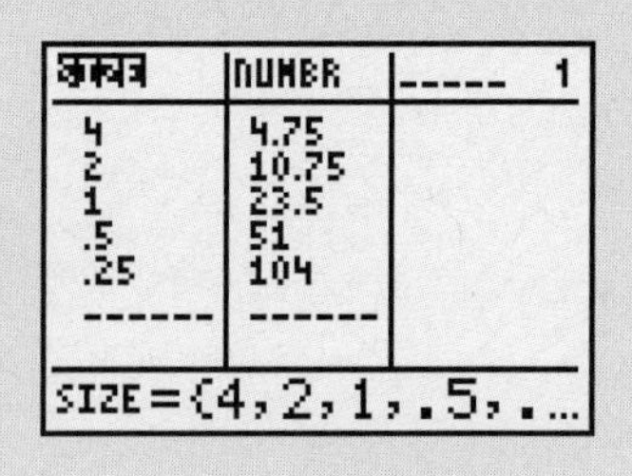

FIGURE A

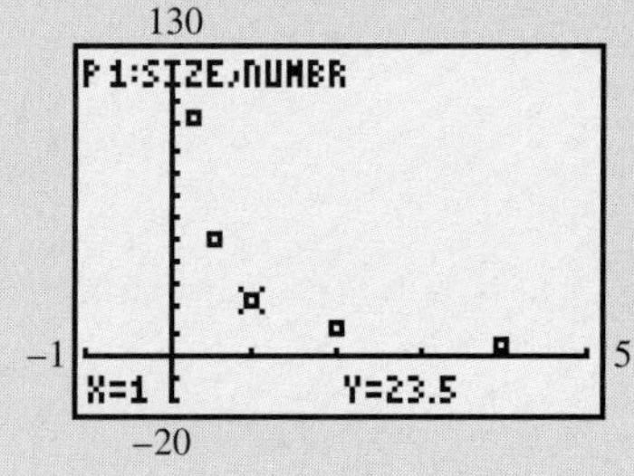

FIGURE B

We now know that the equation will be of the form

$$y = \frac{23.5}{x^n}$$

for some value of $n$. Using the data point $(4, 4.75)$ leads to

$$4.75 = \frac{23.5}{4^n}.$$

Solving this equation for $n$ poses a problem; we have not yet discussed how to solve such an equation analytically. (This is discussed in Chapter 5.) However, the multiline display and editing capabilities of a graphing calculator make it possible to estimate $n$. To do this, enter $23.5/4^x$ onto the home screen, and use guessing to come up with a value of $x$ that leads to approximately 4.75. As seen in the final display of Figure C, a guess of 1.15333 works. Finally, using $n \approx 1.153$, the equation we are looking for is

$$Y_1 = \frac{23.5}{X^{1.153}}.$$

Figure D shows the graph of this equation with the scatter diagram and illustrates a remarkably good fit. The display at the bottom of the screen indicates that when $X = .5$, $Y \approx 52.26$, which is very close to the experimental value of 51 seen in the original table.

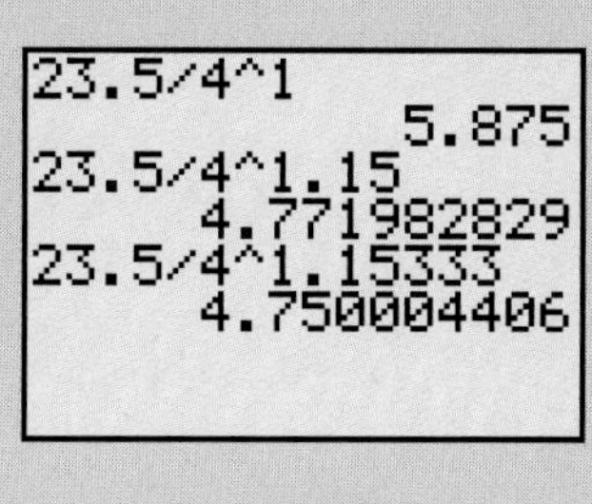


FIGURE C

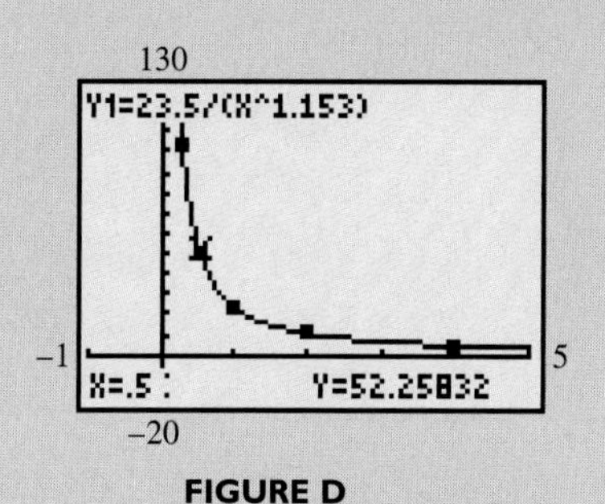


FIGURE D

## Activities

1. Use the equation to determine how long the coastline would be if you "walked" with step sizes of 6 inches, .1 inch, and .01 inch, respectively. Remember that the equation gives you the number of steps, so multiply that value by the step size to get the total length.
2. Gather data from a map of your choice. Using compasses, walk the coastline of your map and count the number of steps. Change the size of the opening of the compasses, and walk the coastline again. After you have done this with four or five different step sizes, find a mathematical model (equation) for your data, using the procedure described above.

# 5 Inverse, Exponential, and Logarithmic Functions

IN 1900, the population of the world was approximately 1.6 billion. At that time the Swedish scientist Svante Arrhenius first predicted a greenhouse effect resulting from emissions of carbon dioxide ($CO_2$) by industrialized countries. In his classic calculation, which made use of logarithms, he predicted that a doubling of carbon dioxide concentrations in the atmosphere would raise the average global temperature by 7°F to 11°F. Today, the world's population is 6 billion and 24 billion tons of carbon dioxide are being added to the atmosphere each year. Carbon dioxide levels are about 364 parts per million (ppm), higher than any time in the previous 160,000 years. Future concentration estimates are shown in the table.

| Year | $CO_2$ (ppm) |
|---|---|
| 2000 | 364 |
| 2050 | 467 |
| 2100 | 600 |
| 2150 | 769 |
| 2200 | 987 |

Finding solutions to environmental problems requires both creativity and mathematics. This chapter presents some of the mathematical tools necessary for modeling this type of phenomenon.

*Source:* Intergovernmental Panel on Climate Change (IPCC).

CHAPTER OUTLINE

# 5.1 INVERSE FUNCTIONS

Inverse Operations • One-to-One Functions • Inverse Functions and Their Graphs • An Application of Inverse Functions

## Inverse Operations

Addition and subtraction are inverse operations: starting with a number $x$, adding 5, and subtracting 5 gives $x$ back as the result. Similarly, some functions are *inverses* of each other. For example, the functions defined by

$$f(x) = 8x \quad \text{and} \quad g(x) = \frac{1}{8}x$$

are inverses of each other with respect to function composition. This means that if a value of $x$ such as $x = 12$ is chosen, then

$$f(12) = 8 \cdot 12 = 96.$$

Calculating $g(96)$ gives

$$g(96) = \frac{1}{8} \cdot 96 = 12.$$

Thus, $$g[f(12)] = 12.$$

Also, $f[g(12)] = 12$. For these functions $f$ and $g$, it can be shown that

$$f[g(x)] = x \quad \text{and} \quad g[f(x)] = x$$

for any value of $x$.

This section shows how to start with a function such as $f(x) = 8x$ and obtain the inverse function, in this case $g(x) = \frac{1}{8}x$. Keep in mind that for a function $f$ to have an inverse function, *$f$ must be one-to-one.*

## One-to-One Functions

For the function $y = 5x - 8$, any two different values of $x$ produce two different values of $y$. On the other hand, for the function $y = x^2$, two different values of $x$ can lead to the *same* value of $y$; for example, both $x = 4$ and $x = -4$ give $y = 4^2 = (-4)^2 = 16$. A function such as $y = 5x - 8$, where different elements from the domain always lead to different elements from the range, is called a *one-to-one function.*

**One-to-One Function**

A function $f$ is a **one-to-one function** if, for elements $a$ and $b$ from the domain of $f$,

$$a \neq b \quad \text{implies} \quad f(a) \neq f(b).$$

**EXAMPLE 1** *Deciding Whether Functions Are One-to-One*

Decide whether each function is one-to-one.

**(a)** $f(x) = -4x + 12$ **(b)** $f(x) = \sqrt{25 - x^2}$

**Solution**

**(a)** For $f(x) = -4x + 12$, two different $x$-values always produce two different $y$-values. To see this, suppose that $a \neq b$. Then, $-4a \neq -4b$, and $-4a + 12 \neq -4b + 12$. Thus, the fact that $a \neq b$ implies that $f(a) \neq f(b)$, so $f$ is one-to-one.

**(b)** If we choose $a = 3$ and $b = -3$, then $3 \neq -3$, but

$$f(3) = \sqrt{25 - 3^2} = \sqrt{25 - 9} = \sqrt{16} = 4$$

and $$f(-3) = \sqrt{25 - (-3)^2} = \sqrt{25 - 9} = 4.$$

Here, even though

$$3 \neq -3, \qquad f(3) = f(-3) = 4.$$

By definition, this is not a one-to-one function.

As shown in Example 1(b), a way to show that a function is *not* one-to-one is to produce a pair of unequal numbers that leads to the same function value. There is also a useful graphical test, called the *horizontal line test*, that tells us whether or not a function is one-to-one.

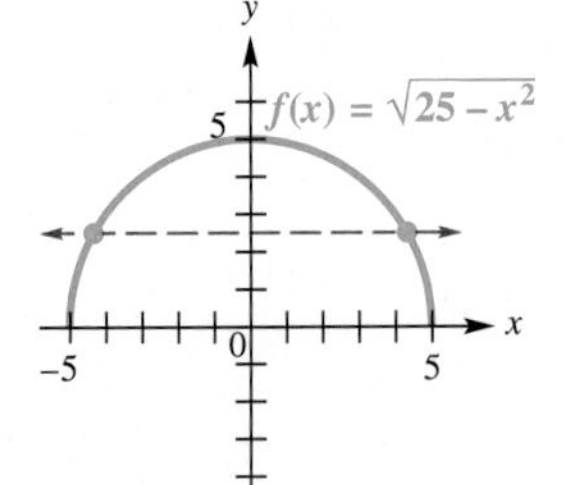

FIGURE 1

**Horizontal Line Test**

If every horizontal line intersects the graph of a function at no more than one point, then the function is one-to-one.

**NOTE** In Example 1(b), the graph of the function is a semicircle, as shown in Figure 1. There are infinitely many horizontal lines that intersect the graph of a semicircle in two points, so the horizontal line test shows that the function is not one-to-one.

**FOR DISCUSSION**

Based on your knowledge of the basic functions studied so far in this text, answer each question. In each case, assume that the function has the largest possible domain.

1. Is a linear function (nonconstant) always one-to-one?
2. Is an odd-degree polynomial function always one-to-one? Why or why not?
3. Is an even-degree polynomial function ever one-to-one? Why or why not?

**EXAMPLE 2** *Using the Horizontal Line Test*

Use the horizontal line test to determine whether the graphs in Figure 2 are graphs of one-to-one functions.

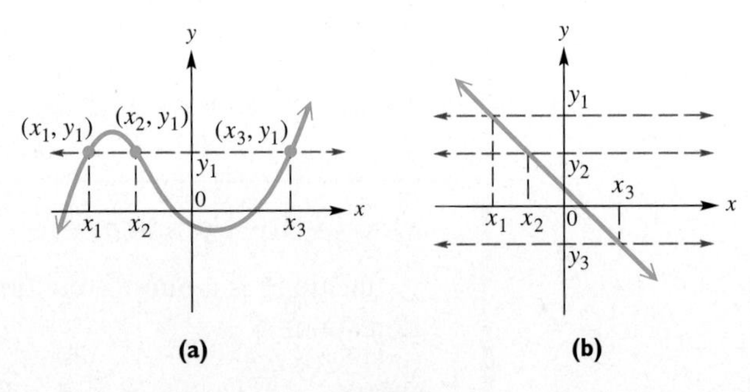

FIGURE 2

**Solution**

**(a)** Each point where the horizontal line intersects the graph in Figure 2(a) has the same value of $y$ but a different value of $x$. Since more than one (here, three) different values of $x$ lead to the same value of $y$, the function is not one-to-one.

**(b)** Every horizontal line will intersect the graph in Figure 2(b) at exactly one point. This function is one-to-one.

**NOTE** A function that is either increasing or decreasing on its entire domain, such as $f(x) = x^3$ or $g(x) = \sqrt{x}$, must be one-to-one.

## Inverse Functions and Their Graphs

As mentioned earlier, certain pairs of one-to-one functions "undo" one another. For example,

$$\text{if} \quad f(x) = 8x + 5 \quad \text{and} \quad g(x) = \frac{x - 5}{8},$$

$$\text{then} \quad f(10) = 8 \cdot 10 + 5 = 85 \quad \text{and} \quad g(85) = \frac{85 - 5}{8} = 10.$$

Starting with 10, we "applied" function $f$ and then "applied" function $g$ to the result, which gave back the number 10. See Figure 3.

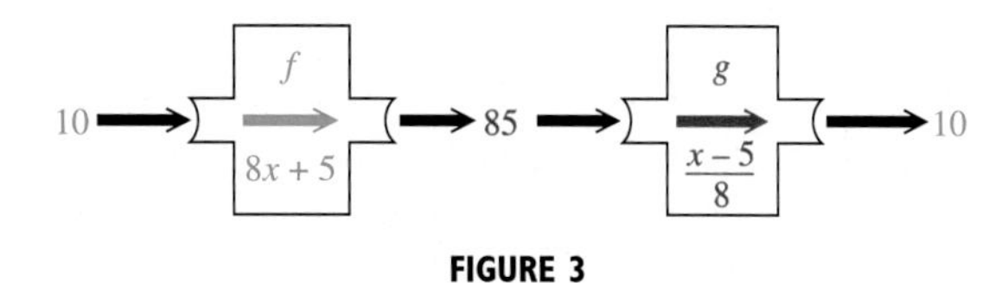


FIGURE 3

Similarly, for these same functions, check that

$$f(3) = 29 \quad \text{and} \quad g(29) = 3,$$
$$f(-5) = -35 \quad \text{and} \quad g(-35) = -5,$$
$$g(2) = -\frac{3}{8} \quad \text{and} \quad f\left(-\frac{3}{8}\right) = 2.$$

In particular, for these functions,

$$f[g(2)] = 2 \quad \text{and} \quad g[f(2)] = 2.$$

In fact for *any* value of $x$,

$$f[g(x)] = x \quad \text{and} \quad g[f(x)] = x,$$

$$\text{or} \quad (f \circ g)(x) = x \quad \text{and} \quad (g \circ f)(x) = x.$$

Because of this property, $g$ is called the *inverse* of $f$.

**Inverse Function**

Let $f$ be a one-to-one function. Then, $g$ is the **inverse function** of $f$ and $f$ is the inverse function of $g$ if

$$(f \circ g)(x) = x \quad \text{for every } x \text{ in the domain of } g,$$

$$\text{and} \quad (g \circ f)(x) = x \quad \text{for every } x \text{ in the domain of } f.$$

### EXAMPLE 3 *Deciding Whether Two Functions Are Inverses*

Let $f$ and $g$ be defined by $f(x) = x^3 - 1$, and $g(x) = \sqrt[3]{x + 1}$. Is $g$ the inverse function of $f$?

**Solution**

As shown in Figure 4, the horizontal line test applied to the graph indicates that $f$ is one-to-one, so it does have an inverse. Now find $(f \circ g)(x)$ and $(g \circ f)(x)$.

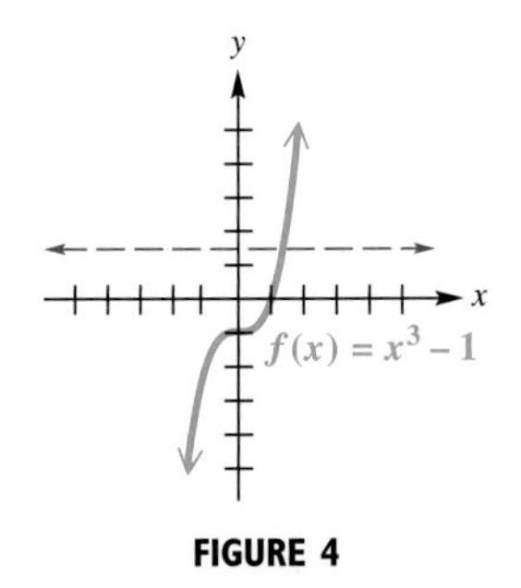


FIGURE 4

$$\begin{aligned}(f \circ g)(x) = f[g(x)] &= \left(\sqrt[3]{x + 1}\right)^3 - 1 \\ &= x + 1 - 1 \\ &= x\end{aligned} \qquad \begin{aligned}(g \circ f)(x) = g[f(x)] &= \sqrt[3]{(x^3 - 1) + 1} \\ &= \sqrt[3]{x^3} \\ &= x\end{aligned}$$

Since $(f \circ g)(x) = x$ and $(g \circ f)(x) = x$, function $g$ is the inverse of function $f$. Also, note that $f$ is the inverse of function $g$.

A special notation is often used for inverse functions: if $g$ is the inverse function of $f$, then $g$ can be written as $f^{-1}$ (read "$f$-inverse"). In Example 3,

$$f^{-1}(x) = \sqrt[3]{x + 1}.$$

**CAUTION** Do not confuse the $-1$ in $f^{-1}$ with a negative exponent. The symbol $f^{-1}(x)$ does not represent $\frac{1}{f(x)}$; rather it represents the inverse function of $f$. Keep in mind that a function $f$ can have an inverse function $f^{-1}$ if and only if $f$ is one-to-one.

The definition of inverse function can be used to show that the domain of $f$ equals the range of $f^{-1}$, and the range of $f$ equals the domain of $f^{-1}$. See Figure 5.

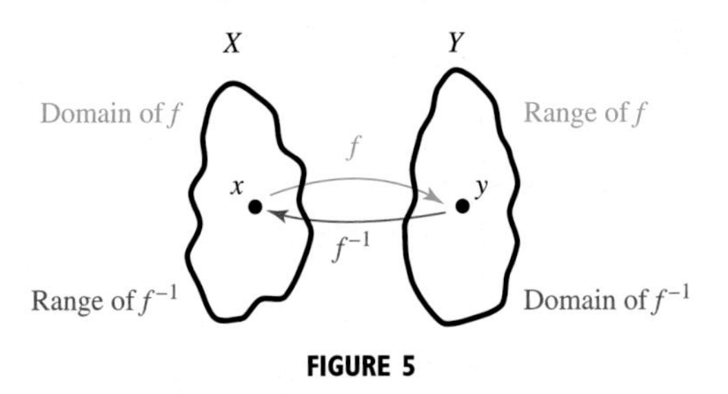


FIGURE 5

For the inverse functions $f$ and $g$ in Figure 3 on the previous page, $f(10) = 85$ and $g(85) = 10$; that is, $(10, 85)$ belongs to $f$ and $(85, 10)$ belongs to $g$. The ordered pairs of the inverse of any one-to-one function $f$ can be found by interchanging the components of the ordered pairs of $f$.

To find the equation that defines the inverse of the one-to-one function $f$, the following procedure produces that interchange of components.

**Finding an Equation for $f^{-1}$, Where $f$ Is One-to-One**

1. Replace $f(x)$ with $y$ in the equation.
2. Interchange $x$ and $y$ in the equation.

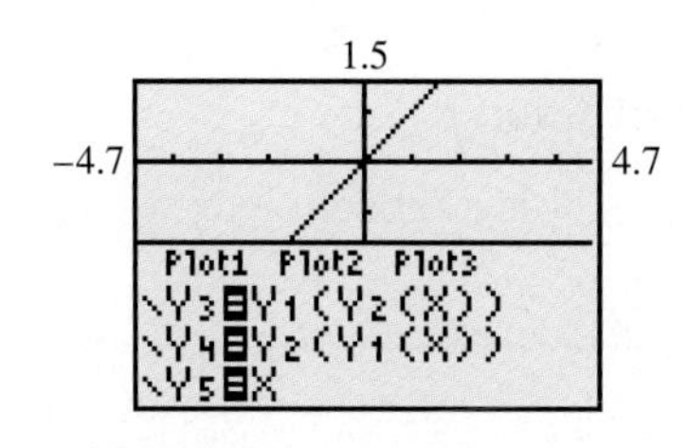


In this screen,
$Y_1 = f(x) = 7X - 2$ and
$Y_2 = f^{-1}(x) = \dfrac{X+2}{7}$.
Graphing both compositions ($Y_3$ and $Y_4$) and the identity function ($Y_5$) leads to coinciding graphs, supporting the discussion in the text.

**3.** Solve for $y$, and replace $y$ with $f^{-1}(x)$.

Any restrictions on $x$ or $y$ should be considered.

For example, we find the equation for the inverse of the one-to-one function defined by $f(x) = 7x - 2$ as follows.

$$y = 7x - 2 \qquad \text{Let } y = f(x). \text{ (Step 1)}$$
$$x = 7y - 2 \qquad \text{Interchange } x \text{ and } y. \text{ (Step 2)}$$
$$x + 2 = 7y \qquad \text{Solve for } y. \text{ (Step 3)}$$
$$\frac{x+2}{7} = y$$
$$f^{-1}(x) = \frac{x+2}{7} \qquad \text{Use inverse notation.}$$

As a check, verify that $(f \circ f^{-1})(x) = (f^{-1} \circ f)(x) = x$.

The inverses discussed in Example 4 are defined as $Y_1$ and $Y_2$ in Figure 6. Notice that the X- and Y-values in the listed ordered pairs are reversed, which supports the results of Example 4. This means that the domain of $f$ is the range of $f^{-1}$, and the range of $f$ is the domain of $f^{-1}$.

## EXAMPLE 4 *Finding the Inverse of a Function*

Find the inverse, if it exists, of $f(x) = \dfrac{4x+6}{5}$.

**Solution** This function is linear and is, therefore, one-to-one. Thus, it has an inverse. Let $f(x) = y$, interchange $x$ and $y$, and then solve for $y$.

$$y = \frac{4x+6}{5}$$
$$x = \frac{4y+6}{5} \qquad \text{Interchange } x \text{ and } y.$$
$$5x = 4y + 6 \qquad \text{Solve for } y.$$
$$5x - 6 = 4y$$
$$y = \frac{5x-6}{4}$$
$$f^{-1}(x) = \frac{5x-6}{4} \qquad \text{Replace } y \text{ with } f^{-1}(x).$$

| X | Y1 |
|---|---|
| 1 | 2 |
| 6 | 6 |
| 11 | 10 |
| -4 | -2 |
| -9 | -6 |
| 51 | 42 |
| 181 | 146 |

Y1=(4X+6)/5

| X | Y2 |
|---|---|
| 2 | 1 |
| 6 | 6 |
| 10 | 11 |
| -2 | -4 |
| -6 | -9 |
| 42 | 51 |
| 146 | 181 |

Y2=(5X−6)/4

**FIGURE 6**

In Example 4, the domain and range of both $f$ and $f^{-1}$ are the set of real numbers. In function $f$, the value of $y$ is found by multiplying $x$ by 4, adding 6 to the product, then dividing that sum by 5. In the equation for the inverse function, $x$ is *multiplied* by 5, then 6 is *subtracted*, and the result is *divided* by 4. This shows how an inverse function is used to "undo" what the function does to the variable $x$.

Suppose $f$ and $f^{-1}$ are inverse functions, and $f(a) = b$ for real numbers $a$ and $b$. Then, by the definition of inverse function, $f^{-1}(b) = a$. This shows that if a point $(a, b)$ is on the graph of $f$, then $(b, a)$ will belong to the graph of $f^{-1}$. As shown in Figure 7, the points $(a, b)$ and $(b, a)$ are reflections of one another across the line $y = x$. Thus, the graph of $f^{-1}$ can be obtained from the graph of $f$ by reflecting the graph of $f$ across the line $y = x$.

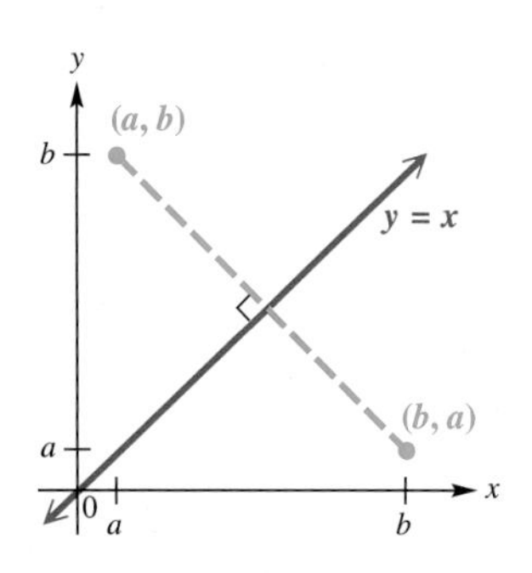


**FIGURE 7**

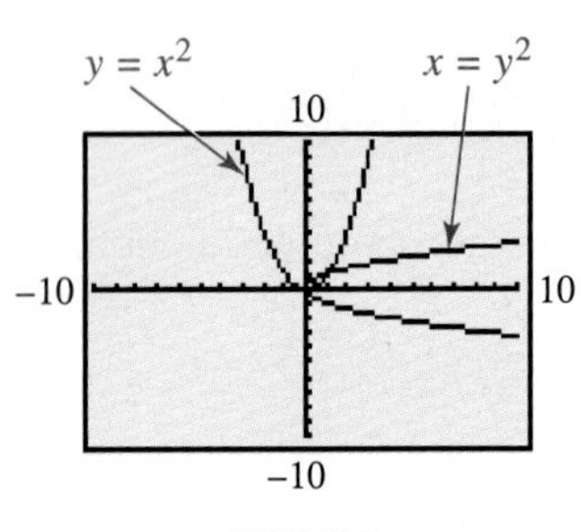

FIGURE 8

**Geometric Relationship between the Graphs of $f$ and $f^{-1}$**

If a function $f$ is one-to-one, the graph of its inverse $f^{-1}$ is a reflection of the graph of $f$ across the line $y = x$.

**CAUTION** Some graphing calculators have the capability of "drawing the inverse" of a function. This feature does not require that the function be one-to-one, however, so the resulting figure may not be the graph of a function. As shown in Figure 8, despite the fact that $y = x^2$ is not one-to-one, the calculator will draw its "inverse," $x = y^2$. Again, it is necessary to understand the mathematics to interpret results correctly.

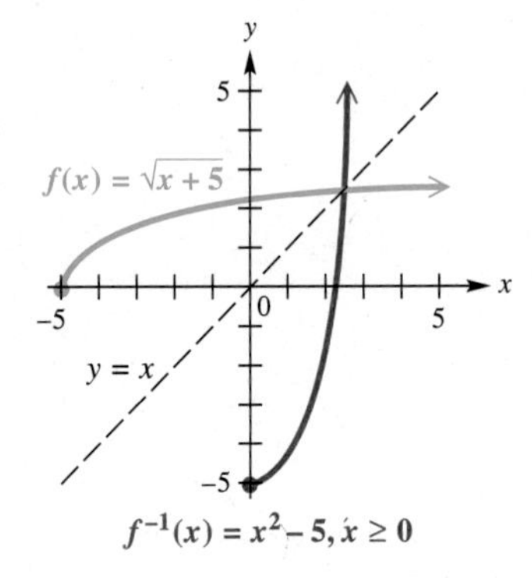


**EXAMPLE 5** *Finding the Inverse of a Function with a Restricted Domain*

Let $f(x) = \sqrt{x + 5}$. Find $f^{-1}(x)$.

**Solution** First, we notice that the domain of $f$ is restricted to the interval $[-5, \infty)$. Since $f$ is an increasing function, as shown in Figure 9, $f$ is one-to-one and thus has an inverse function. We follow the guidelines to find the equation of the inverse.

$$y = \sqrt{x + 5}, \quad x \geq -5 \qquad y = f(x)$$

$$x = \sqrt{y + 5}, \quad y \geq -5 \qquad \text{Interchange } x \text{ and } y.$$

$$x^2 = y + 5 \qquad \text{Square both sides.}$$

$$y = x^2 - 5 \qquad \text{Solve for } y.$$

We cannot simply give $f^{-1}(x)$ as $x^2 - 5$. The domain of $f$ is $[-5, \infty)$, and its range is $[0, \infty)$. The range of $f$ is the domain of $f^{-1}$, so $f^{-1}$ must be defined as

$$f^{-1}(x) = x^2 - 5, \quad x \geq 0.$$

As a check, the range of $f^{-1}$, $[-5, \infty)$, is the domain of $f$. Graphs of $f$ and $f^{-1}$ are shown in Figure 9. The line $y = x$ is included on the graphs to show that the graphs of $f$ and $f^{-1}$ are reflections of one another, that is, mirror images with respect to this line.

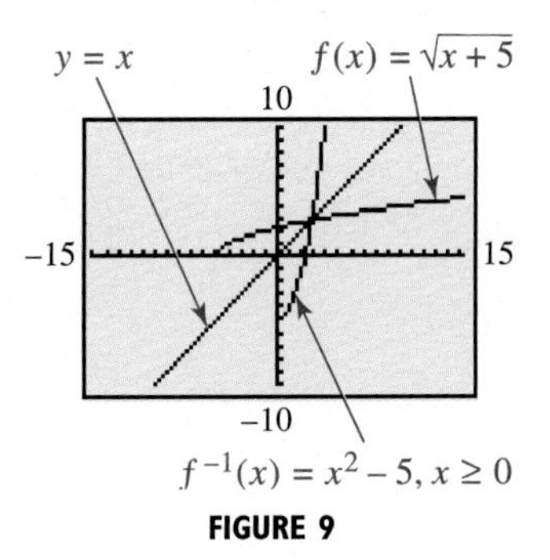

FIGURE 9

**TECHNOLOGY NOTE**

When graphing inverse functions on a calculator as in Figure 9, square windows work best.

**Important Facts about Inverses**

1. If $f$ is one-to-one, then $f^{-1}$ exists.
2. The domain of $f$ is equal to the range of $f^{-1}$, and the range of $f$ is equal to the domain of $f^{-1}$.
3. If the point $(a, b)$ lies on the graph of $f$, then $(b, a)$ lies on the graph of $f^{-1}$, so the graphs of $f$ and $f^{-1}$ are reflections of each other across the line $y = x$.
4. To find the equation for $f^{-1}$, replace $f(x)$ with $y$, interchange $x$ and $y$, and solve for $y$. This gives $f^{-1}(x)$.

## An Application of Inverse Functions

Inverse functions are used to send and receive coded information. The functions are usually very complicated. A simple example might use the function defined by $f(x) = 2x + 5$, which is one-to-one. Suppose that each letter of the alphabet is assigned a numerical value according to its position, as follows.

| | | | | | | | |
|---|---|---|---|---|---|---|---|
| A | 1 | H | 8 | O | 15 | V | 22 |
| B | 2 | I | 9 | P | 16 | W | 23 |
| C | 3 | J | 10 | Q | 17 | X | 24 |
| D | 4 | K | 11 | R | 18 | Y | 25 |
| E | 5 | L | 12 | S | 19 | Z | 26 |
| F | 6 | M | 13 | T | 20 | | |
| G | 7 | N | 14 | U | 21 | | |

**EXAMPLE 6** *Using a Function to Code a Message*

Use the one-to-one function defined by $f(x) = 2x + 5$ and the numerical values above to code the word ALGEBRA.

**Solution** The word ALGEBRA would be encoded as

$$7 \quad 29 \quad 19 \quad 15 \quad 9 \quad 41 \quad 7,$$

because A corresponds to 1, and

$$f(1) = 2(1) + 5 = 7,$$

L corresponds to 12, and

$$f(12) = 2(12) + 5 = 29,$$

and so on. Using the inverse $f^{-1}(x) = \frac{x-5}{2}$ to decode,

$$f^{-1}(7) = \frac{7-5}{2} = 1,$$

which corresponds to A, and

$$f^{-1}(29) = \frac{29-5}{2} = 12,$$

which corresponds to L, and so on. The table feature of a graphing calculator can be very useful for this procedure. Figure 10 shows how decoding can be accomplished for the word ALGEBRA.

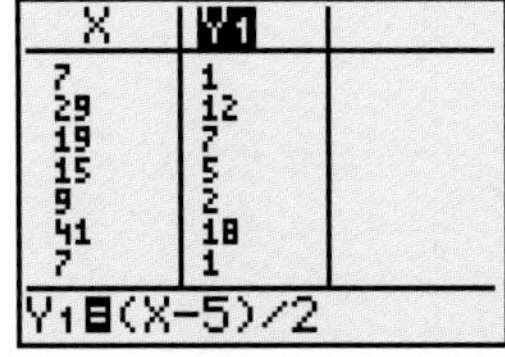

FIGURE 10

## 5.1 EXERCISES

*Concept Check* *Indicate whether the screens suggest that* $Y_1$ *and* $Y_2$ *are inverse functions.*

**1.**

| X | $Y_1$ |
|---|---|
| 1 | 6 |
| 3 | 8 |
| -4 | 1 |
| 9 | 14 |
| 3.2 | 8.2 |
| 0 | 5 |
| -10 | -5 |

X=1

| X | $Y_2$ |
|---|---|
| 6 | 1 |
| 8 | 3 |
| 1 | -4 |
| 14 | 9 |
| 8.2 | 3.2 |
| 5 | 0 |
| -5 | -10 |

X=6

**2.**

| X | $Y_1$ |
|---|---|
| 1 | 7 |
| 3 | 9 |
| -4 | 2 |
| 9 | 15 |
| 3.2 | 9.2 |
| 0 | 6 |
| -10 | -4 |

X=1

| X | $Y_2$ |
|---|---|
| 7 | -1 |
| 9 | -3 |
| 2 | 4 |
| 15 | -9 |
| 9.2 | -3.2 |
| 6 | 0 |
| -4 | 10 |

X=7

**3.**

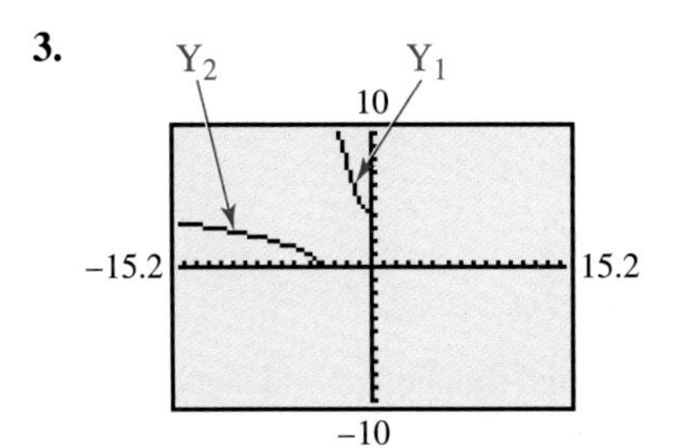


**4.**

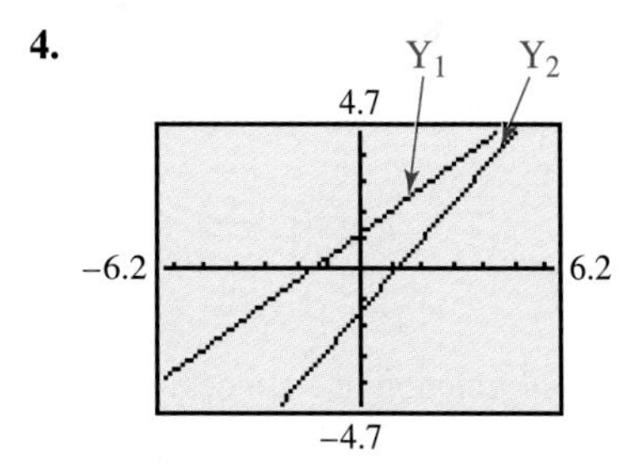

*Concept Check* *Based on the examples and discussion in this section, answer the following.*

**5.** For a function to have an inverse, it must be __________.

**6.** For a function $f$ to be of the type mentioned in Exercise 5, if $a \neq b$, then __________.

**7.** If $f$ and $g$ are inverses, then $(f \circ g)(x) =$ __________, and __________ $= x$.

**8.** The domain of $f$ is equal to the __________ of $f^{-1}$, and the range of $f$ is equal to the __________ of $f^{-1}$.

**9.** If the point $(a, b)$ lies on the graph of $f$, and $f$ has an inverse, then the point __________ lies on the graph $f^{-1}$.

**10.** If $f(x) = x$, then for any function $g$, $(f \circ g)(x) = (g \circ f)(x) =$ __________.

**11.** If a function $f$ has an inverse, then the graph of $f^{-1}$ may be obtained by reflecting the graph of $f$ across the line with equation __________.

**12.** If a function $f$ has an inverse, and $f(-3) = 6$, then $f^{-1}(6) =$ __________.

**13.** If $f(-4) = 16$ and $f(4) = 16$, then $f$ __________ (does/does not) have an inverse because __________.

**14.** If $f$ is a function that has an inverse, and the graph of $f$ lies completely within the second quadrant, then the graph of $f^{-1}$ lies completely within the __________ quadrant.

*Decide whether each function graphed or defined is one-to-one.*

**15.**

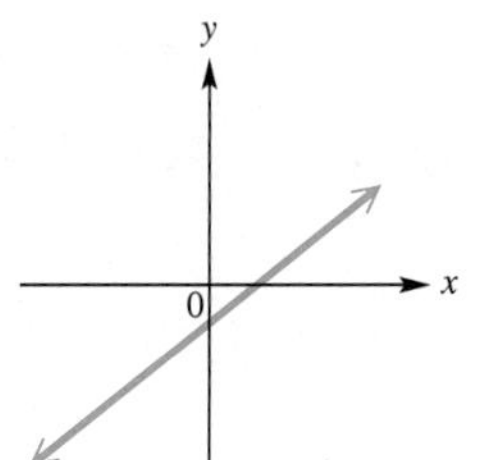


**16.**

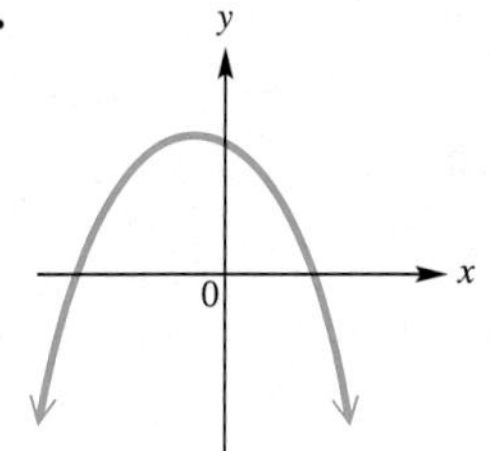


**17.**

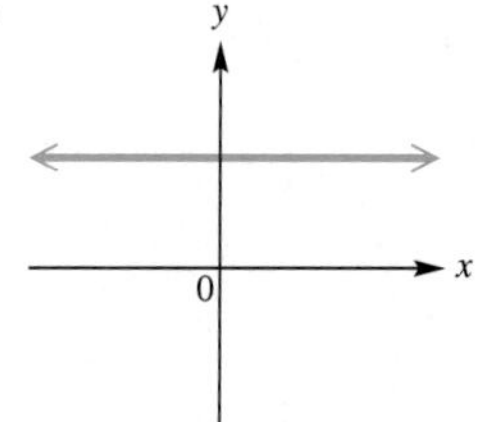


**18.**

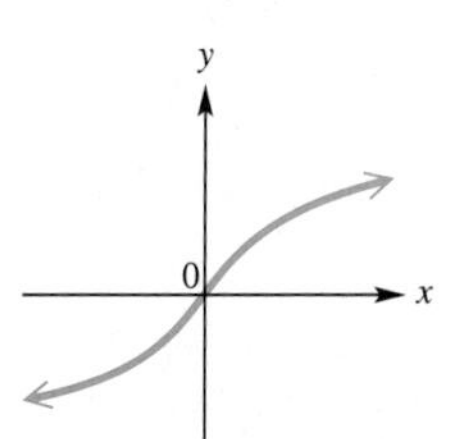


**19.**

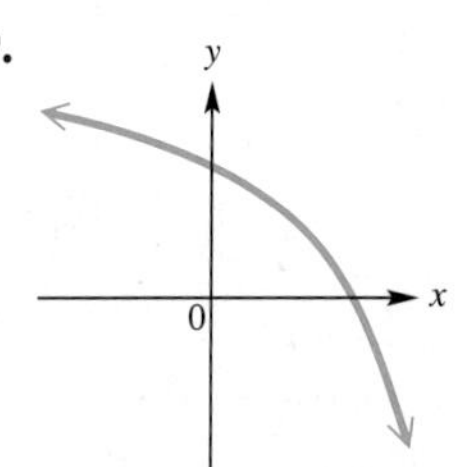


**20.**

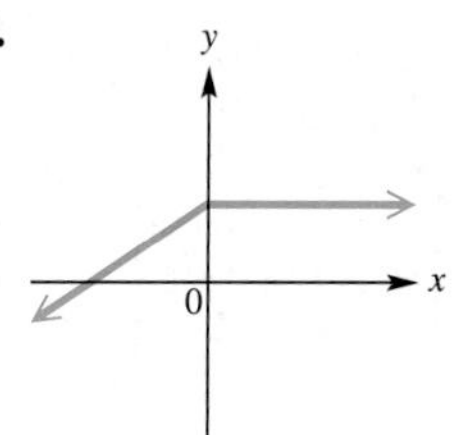


**21.**

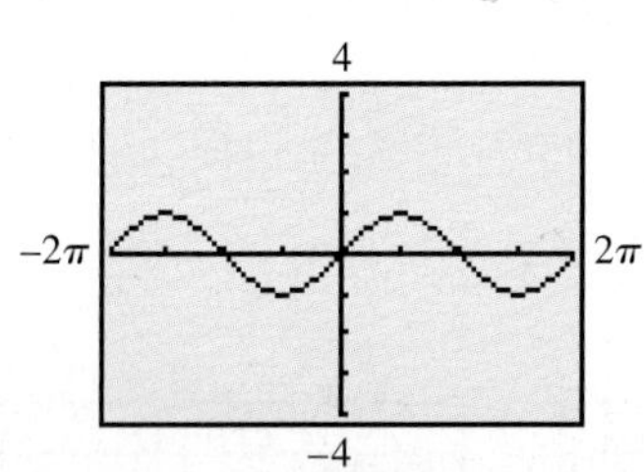


**22.**

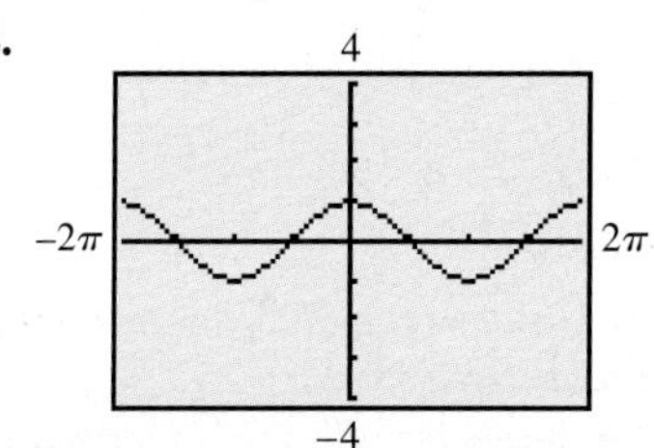


**23.**

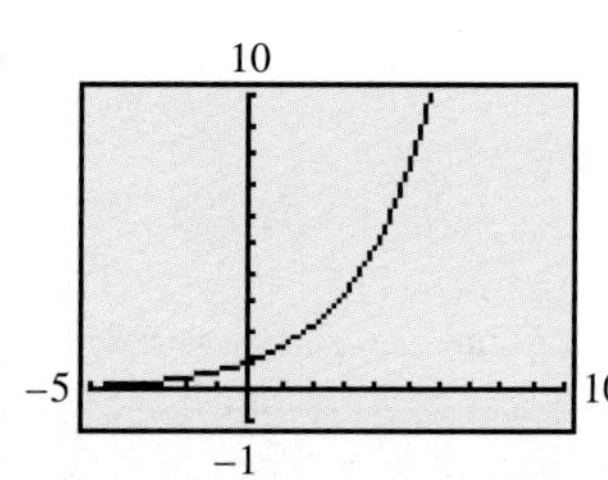


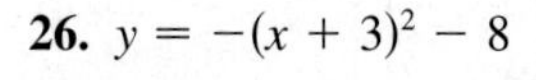

**24.**

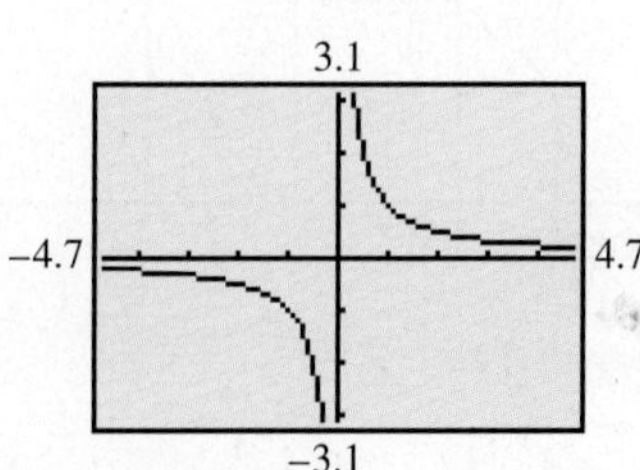


**25.** $y = (x - 2)^2$

**26.** $y = -(x + 3)^2 - 8$

**27.** $y = \sqrt{36 - x^2}$

**28.** $y = -\sqrt{100 - x^2}$

**29.** $y = 2x^3 + 1$

**30.** $y = -\sqrt[3]{x + 5}$

**31.** $y = \dfrac{1}{x + 2}$

**32.** $y = \dfrac{-4}{x - 8}$

**33.** Explain why a polynomial function of even degree cannot be one-to-one.

**34.** Explain why a polynomial function of odd degree *may* not be one-to-one.

*Concept Check* *In Exercises 35–40, an everyday activity is described. Keeping in mind that an inverse operation "undoes" what an operation does, describe the inverse activity.*

**35.** Tying your shoelaces

**36.** Pressing a car's accelerator

**37.** Entering a room

**38.** Climbing the stairs

**39.** Wrapping a package

**40.** Putting on a coat

*Concept Check* *Let $f(x) = x^3$. Evaluate each expression.*

**41.** $f(2)$

**42.** $f(0)$

**43.** $f(-2)$

**44.** $f^{-1}(8)$

**45.** $f^{-1}(0)$

**46.** $f^{-1}(-8)$

*Decide whether the functions in each pair are inverses of each other.*

**47.**

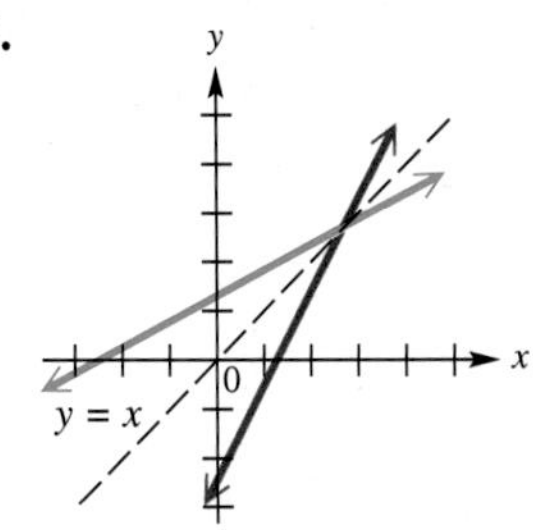


**48.**

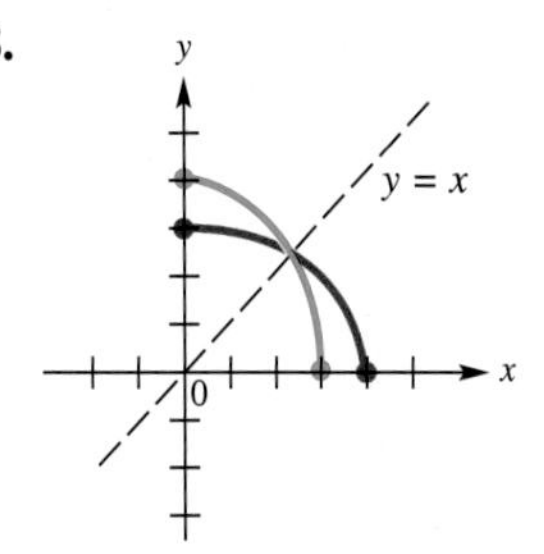


**49.**

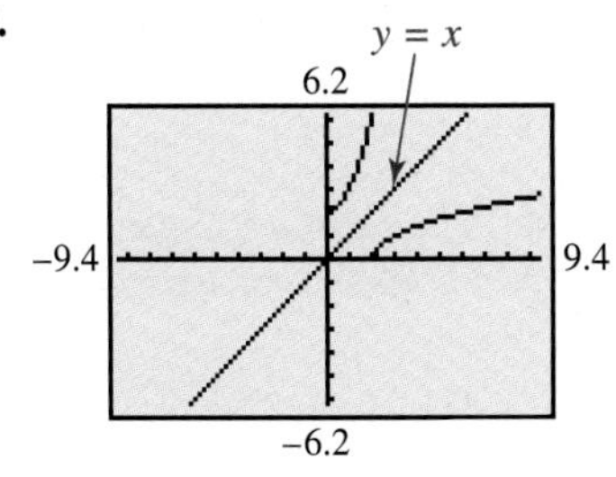


**50.**

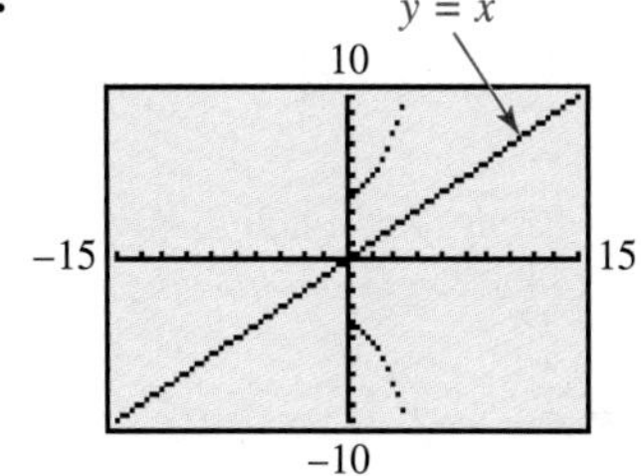


Dot Mode

**51.** $f(x) = -\frac{3}{11}x, \quad g(x) = -\frac{11}{3}x$

**52.** $f(x) = 2x + 4, \quad g(x) = \frac{1}{2}x - 2$

**53.** $f(x) = 5x - 5, \quad g(x) = \frac{1}{5}x + 1$

**54.** $f(x) = 8x - 7, \quad g(x) = \frac{x + 8}{7}$

**55.** $f(x) = \frac{1}{x}, \quad g(x) = \frac{1}{x}$

**56.** $f(x) = \frac{2x + 3}{x - 1}, \quad g(x) = \frac{x + 3}{x - 2}$

**57.** On a square sheet of paper, sketch the graph of any one-to-one function, using the entire sheet. Pattern your axes as shown in the figure. Now, hold the sheet with your left hand at the left end of the $x$-axis and your right hand at the right end of the $x$-axis. Rotate the sheet 180° across the $x$-axis, so that the back of the sheet is now face up. Then, rotate the sheet 90° in a counterclockwise direction. Hold the sheet up to the light, and observe the graph from the back side of the sheet. How does it compare to your original graph?

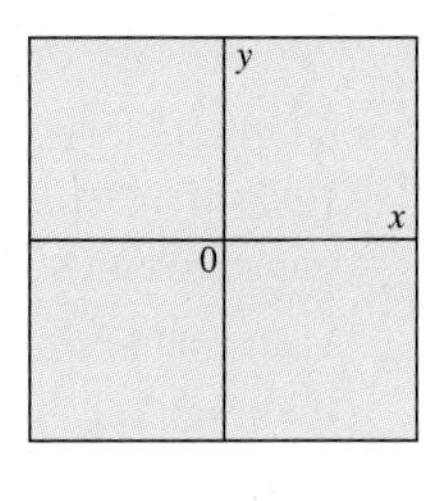


*Graph the inverse of each one-to-one function.*

**58.**

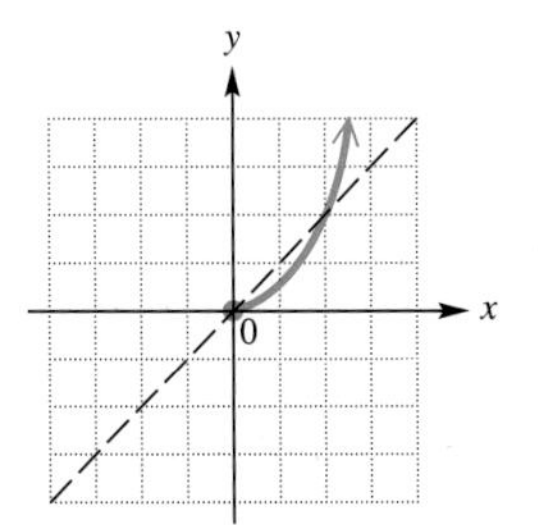


**59.**

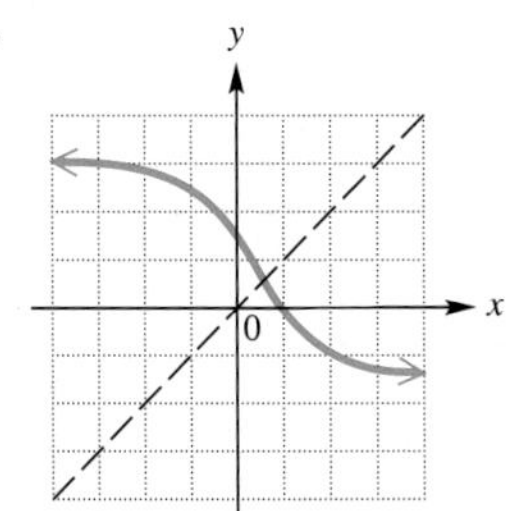


**60.**

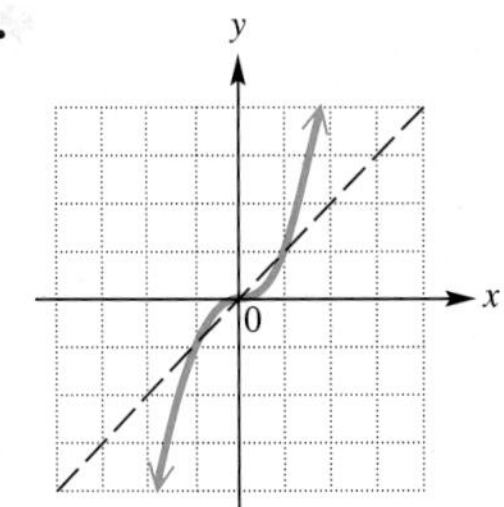

**61.**

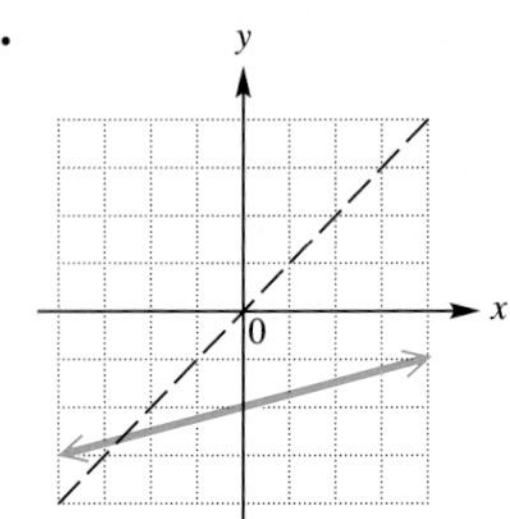


**62.**

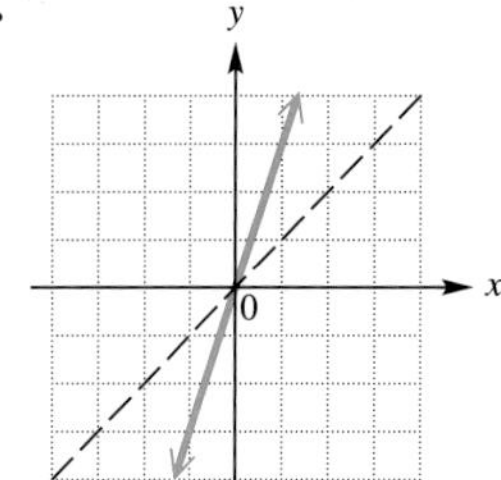


**63.**

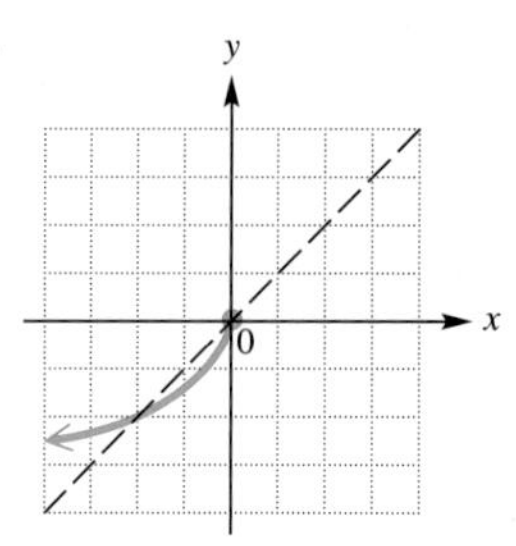


*For each function defined as follows that is one-to-one, write an equation for the inverse function in the form $y = f^{-1}(x)$, and then graph $f$ and $f^{-1}$ on the same axes. Give the domain and range of $f$ and $f^{-1}$. If the function is not one-to-one, say so.*

**64.** $y = 3x - 4$

**65.** $y = 4x - 5$

**66.** $y = x^3 + 1$

**67.** $y = -x^3 - 2$

**68.** $y = x^2$

**69.** $y = -x^2 + 2$

**70.** $y = \dfrac{1}{x}$

**71.** $y = \dfrac{4}{x}$

**72.** $f(x) = \sqrt{6 + x}, x \geq -6$

**73.** $f(x) = -\sqrt{x^2 - 16}, x \geq 4$

Concept Check *Use the concepts of this section to answer the following.*

**74.** Suppose $f(x)$ is the number of cars that can be built for $x$ dollars. What does $f^{-1}(1000)$ represent?

**75.** Suppose $f(r)$ is the volume (in cubic inches) of a sphere of radius $r$ inches. What does $f^{-1}(5)$ represent?

**76.** If a line has nonzero slope $a$, what is the slope of its reflection across the line $y = x$?

**77.** Find $f^{-1}[f(2)]$, where $f(2) = 3$.

*Use a graphing calculator to graph each function defined as follows, using the given viewing window. Use the graph to decide which functions are one-to-one. If a function is one-to-one, give the equation of its inverse function and graph the inverse function on the same coordinate axes.*

**78.** $f(x) = 6x^3 + 11x^2 - x - 6$; $[-3, 2]$ by $[-10, 10]$

**79.** $f(x) = x^4 - 5x^2 + 6$; $[-3, 3]$ by $[-1, 8]$

**80.** $f(x) = \dfrac{x - 5}{x + 3}$; $[-9.4, 9.4]$ by $[-6.2, 6.2]$

**81.** $f(x) = \dfrac{-x}{x - 4}$; $[-4.7, 9.4]$ by $[-6.2, 6.2]$

**82.** If a function is increasing (or decreasing) on its entire domain, then it has an inverse. Explain why this is true.

## RELATING CONCEPTS

**For individual or group investigation (Exercises 83–85)**

*A graphing calculator can be used to support many of the concepts of inverse functions. Work Exercises 83–85 in order.*

**83.** Define $f(x)$ as $y_1 = 2x - 8$ on your calculator. Find $f^{-1}(x)$ analytically, and enter it as $y_2$.

**84.** Enter $y_3$ as $y_1[y_2(x)]$, using function notation on your calculator. Graph only $y_3$. What is the special name given to the function that is graphed?

**85.** Use a table with TblStart = 0 and $\Delta$Tbl = 1 to generate the values of $y_3 = y_1[y_2(x)]$. What is the result?

While a function may not be one-to-one when defined over its "natural" domain, it may be possible to restrict the domain in such a way that it is one-to-one and the range of the function is unchanged. For example, if we restrict the domain of the function $f(x) = x^2$ (which is not one-to-one over $(-\infty, \infty)$) to $[0, \infty)$, we obtain a one-to-one function whose range is still $[0, \infty)$. See the figure. Notice that we could choose to restrict the domain of $f(x) = x^2$ to $(-\infty, 0]$ and also obtain the graph of a one-to-one function, except that it would be the left half of the parabola.

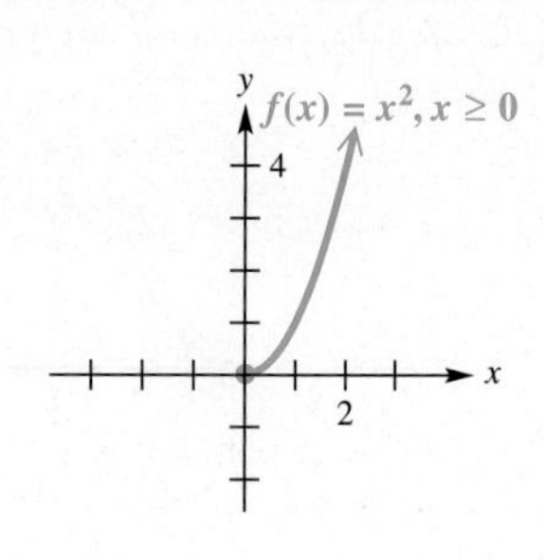

*For each function defined in Exercises 86–91, decide on a suitable restriction on the domain so that the function is one-to-one and the range is not changed. You may wish to use a graph to help you decide.*

**86.** $f(x) = -x^2 + 4$ **87.** $f(x) = (x - 1)^2$ **88.** $f(x) = |x - 6|$

**89.** $f(x) = x^4$ **90.** $f(x) = x^4 + x^2 - 6$ **91.** $f(x) = -\sqrt{x^2 - 16}$

*Using the restrictions given for the functions defined in Exercises 86–89, find a rule for $f^{-1}$.*

**92.** $f(x) = -x^2 + 4, x \geq 0$ **93.** $f(x) = (x - 1)^2, x \geq 1$

**94.** $f(x) = |x - 6|, x \geq 6$ **95.** $f(x) = x^4, x \geq 0$

*Use the alphabet coding assignment given in this section for Exercises 96–99. (See Example 6.)*

**96.** The function defined by $f(x) = 3x - 2$ was used to encode the following message:

37 25 19 61 13 34 22 1 55 1 52 52 25 64 13 10.

Find the inverse function and decode the message.

**97.** The function defined by $f(x) = 2x - 9$ was used to encode the following message:

−5 9 5 5 9 27 15 29 −1 21 19 31 −3 27 41.

Find the inverse function and decode the message.

**98.** Encode the message SEND HELP, using the one-to-one function defined by $f(x) = x^3 - 1$. Give the inverse function that the decoder would need when the message is received.

**99.** Encode the message SAILOR BEWARE, using the one-to-one function defined by $f(x) = (x + 1)^3$. Give the inverse function that the decoder would need when the message is received.

**100.** Why is a one-to-one function essential in this coding/decoding process?

# EXPONENTIAL FUNCTIONS

Real-Number Exponents • Graphs of Exponential Functions • Exponential Equations (Type 1) • The Number $e$ • Compound Interest

## Real-Number Exponents

In Section 4.4, rational numbers were used as exponents in power and root functions. In particular, if $r$ is a rational number and $r = \frac{m}{n}$, then, for appropriate integer values of $m$ and $n$,

$$a^{m/n} = \left(\sqrt[n]{a}\right)^m.$$

For example,

$$16^{3/4} = \left(\sqrt[4]{16}\right)^3 = 2^3 = 8,$$

$$27^{-1/3} = \frac{1}{27^{1/3}} = \frac{1}{\sqrt[3]{27}} = \frac{1}{3},$$

and

$$64^{-1/2} = \frac{1}{64^{1/2}} = \frac{1}{\sqrt{64}} = \frac{1}{8}.$$

```
16^(3/4)
                8
27^(-1/3)▸Frac
              1/3
64^(-1/2)▸Frac
              1/8
```

This screen supports the results in the text.

**NOTE** The expression $a^{m/n}$ also equals $\sqrt[n]{a^m}$, but this form is seldom used because it often requires taking the root of a large number.

In this section, we extend the definition of $a^r$ to include all real (not just rational) values of the exponent $r$. For example, the symbol $2^{\sqrt{3}}$ might be evaluated by approximating the exponent $\sqrt{3}$ by the numbers 1.7, 1.73, 1.732, and so on. Since these decimals approach the value of $\sqrt{3}$ more and more closely, it seems reasonable that $2^{\sqrt{3}}$ should be approximated more and more closely by the numbers $2^{1.7}$, $2^{1.73}$, $2^{1.732}$, and so on. (Recall, for example, that $2^{1.7} = 2^{17/10} = \sqrt[10]{2^{17}}$.) In fact, this is exactly how $2^{\sqrt{3}}$ is defined (in a more advanced course).

With this interpretation of real exponents, all rules and theorems for exponents are valid for real number exponents as well as rational ones. In addition to the usual rules for exponents (see Chapter R), we use several new properties in this chapter. For example, if $y = 2^x$, then each value of $x$ leads to exactly one value of $y$, and, therefore, $y = 2^x$ defines a function. Furthermore,

$$\text{if} \quad 3^x = 3^4, \quad \text{then} \quad x = 4,$$

and for $p > 0$,

$$\text{if} \quad p^2 = 3^2, \quad \text{then} \quad p = 3.$$

Also,

$$4^2 < 4^3, \quad \text{but} \quad \left(\frac{1}{2}\right)^2 > \left(\frac{1}{2}\right)^3.$$

In general, when $a > 1$, increasing the exponent on $a$ leads to a *larger* number, but if $0 < a < 1$, increasing the exponent on $a$ leads to a *smaller* number.

These properties are generalized below. Proofs of the properties are not given here, as they require more advanced mathematics.

**Additional Properties of Exponents**

For any real number $a > 0$, $a \neq 1$, the following statements are true.

**(a)** $a^x$ is a unique real number for all real numbers $x$.
**(b)** $a^b = a^c$ if and only if $b = c$.
**(c)** If $a > 1$ and $m < n$, then $a^m < a^n$.
**(d)** If $0 < a < 1$ and $m < n$, then $a^m > a^n$.

Properties (a) and (b) require $a > 0$ so that $a^x$ is always defined. For example, $(-6)^x$ is not a real number if $x = \frac{1}{2}$. This means that $a^x$ will always be positive, since $a$ is positive. In part (a), $a \neq 1$ because $1^x = 1$ for every real number value of $x$, so each value of $x$ leads to the same real number, 1. For Property (b) to hold, $a$ must not equal 1 since, for example, $1^4 = 1^5$, even though $4 \neq 5$.

## Graphs of Exponential Functions

A graphing calculator can easily find approximations of numbers raised to irrational powers. For example,

$$2^{\sqrt{6}} \approx 5.462228786 \qquad 3^{\sqrt{2}} \approx 4.728804388$$

$$\left(\frac{1}{2}\right)^{\sqrt{3}} \approx .3010237439 \qquad .5^{-\sqrt{2}} \approx 2.665144143.$$

Later we find these approximations using graphs of *exponential functions*.

> **Exponential Function**
>
> If $a > 0$, $a \neq 1$, then
>
> $$f(x) = a^x$$
>
> is the **exponential function** with base $a$.

**NOTE** We do not allow 1 as a base for the exponential function because $1^x = 1$ for all real $x$, and thus it leads to the constant function defined by $f(x) = 1$.

### EXAMPLE 1 *Graphing an Exponential Function*

Graph $f(x) = 2^x$. Determine the domain and range of $f$.

**Solution** Make a table of values, plot the points, and connect them with a smooth curve, as shown in Figure 11(a). As the graph suggests, the domain is $(-\infty, \infty)$ and the range is $(0, \infty)$.

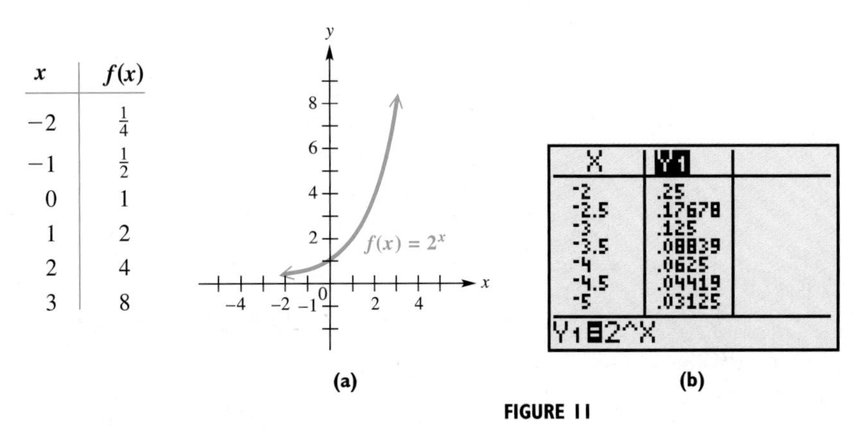

| $x$ | $f(x)$ |
|---|---|
| $-2$ | $\frac{1}{4}$ |
| $-1$ | $\frac{1}{2}$ |
| 0 | 1 |
| 1 | 2 |
| 2 | 4 |
| 3 | 8 |



FIGURE 11

It is difficult to graph values of $a^x$ for $x < -2$. The calculator table in Figure 11(b) shows that $a^x$ continues to decrease as $x$ decreases.

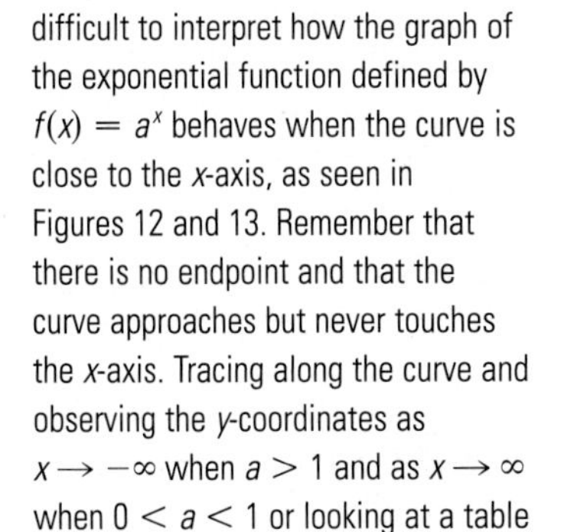

**TECHNOLOGY NOTE**

Because of the limited resolution of the graphing calculator screen, it is difficult to interpret how the graph of the exponential function defined by $f(x) = a^x$ behaves when the curve is close to the $x$-axis, as seen in Figures 12 and 13. Remember that there is no endpoint and that the curve approaches but never touches the $x$-axis. Tracing along the curve and observing the $y$-coordinates as $x \to -\infty$ when $a > 1$ and as $x \to \infty$ when $0 < a < 1$ or looking at a table helps support this fact.

The behavior of the graph of an exponential function depends, in general, on the magnitude of $a$. Figure 12 shows the graphs of $f(x) = a^x$ for $a = 2$, 3, and 4. Based on our earlier discussion, the domain of $f(x) = a^x$ is $(-\infty, \infty)$. From the graphs in Figure 12, we see that the range is $(0, \infty)$, and the function is *increasing* for $a = 2$, 3, and 4. The $x$-axis is the horizontal asymptote as $x \to -\infty$, and the $y$-intercept is 1. As $a$ becomes larger, the graph becomes "steeper."

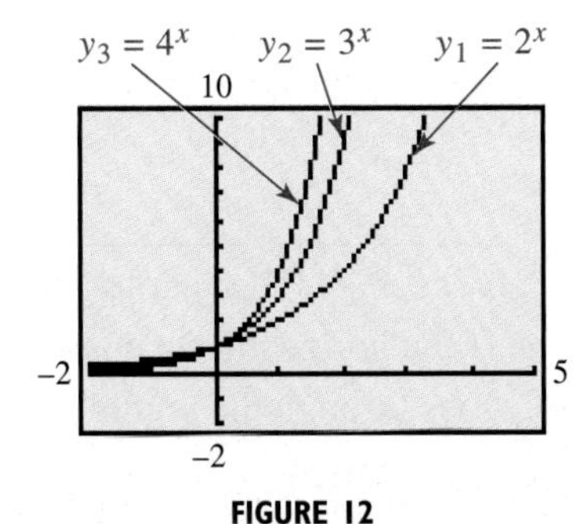


FIGURE 12

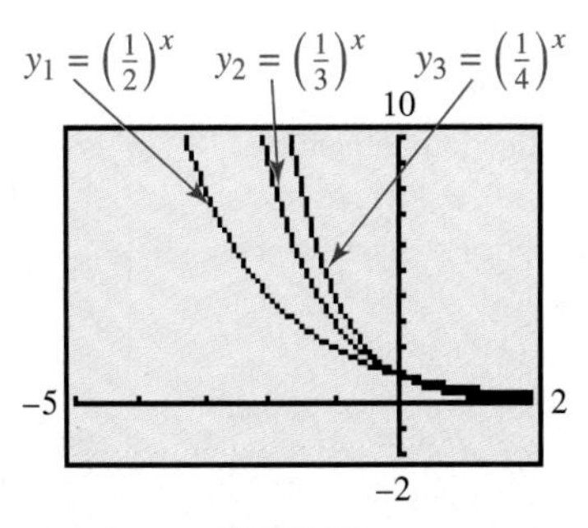


FIGURE 13

Figure 13 on the previous page shows the graphs of $f(x) = a^x$ for $a = \frac{1}{2}, \frac{1}{3}$, and $\frac{1}{4}$. If the base $a$ is between 0 and 1, as $a$ gets closer to 0, the graph becomes steeper to the left of the $y$-axis. Again, the domain is $(-\infty, \infty)$, the range is $(0, \infty)$, and the $y$-intercept is 1. However, the function is *decreasing* for $a = \frac{1}{2}, \frac{1}{3}$, and $\frac{1}{4}$, and the $x$-axis is the horizontal asymptote as $x \rightarrow \infty$. Our observations in Figures 12 and 13 lead to the following generalizations about the graphs of exponential functions.

## FUNCTION CAPSULE

### EXPONENTIAL FUNCTION $f(x) = a^x$, $a > 1$

Domain: $(-\infty, \infty)$ Range: $(0, \infty)$

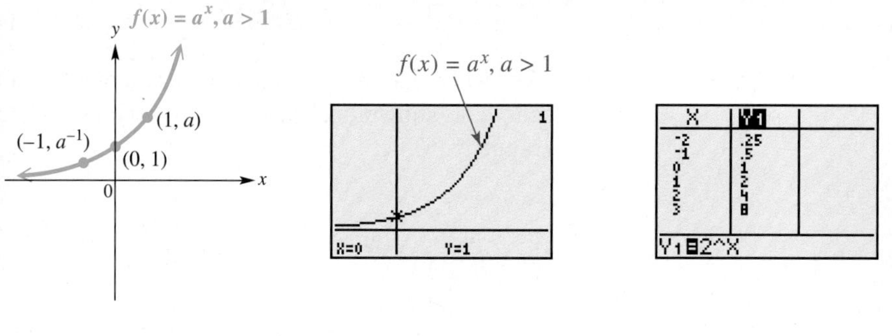

FIGURE 14

- $f(x) = a^x$, $a > 1$, is increasing and continuous on its entire domain, $(-\infty, \infty)$.
- The $x$-axis is the horizontal asymptote as $x \rightarrow -\infty$.
- The graph goes through the points $(-1, a^{-1})$, $(0, 1)$, and $(1, a)$.

### FOR DISCUSSION

**1.** Using a standard viewing window, we cannot observe how the graph of the exponential function defined by $f(x) = 2^x$ behaves for values of $x$ less than about $-3$. Use the window $[-10, 0]$ by $[-.5, .5]$, and then trace to the left to see what happens.

**2.** Repeat Item 1 for the exponential function defined by $f(x) = \left(\frac{1}{2}\right)^x$. Use the window $[0, 10]$ by $[-.5, .5]$, and trace to the right.

**3.** Complete the following statement: For the graph of an exponential function defined by $f(x) = a^x$, $a > 0$, $a \neq 1$, every range value appears exactly once, and thus $f$ is a(n) _____ function. Because of this, $f$ has a(n) _____. (*Hint:* Recall the concepts studied in Section 5.1.)

## FUNCTION CAPSULE

### EXPONENTIAL FUNCTION $f(x) = a^x$, $0 < a < 1$

Domain: $(-\infty, \infty)$ Range: $(0, \infty)$

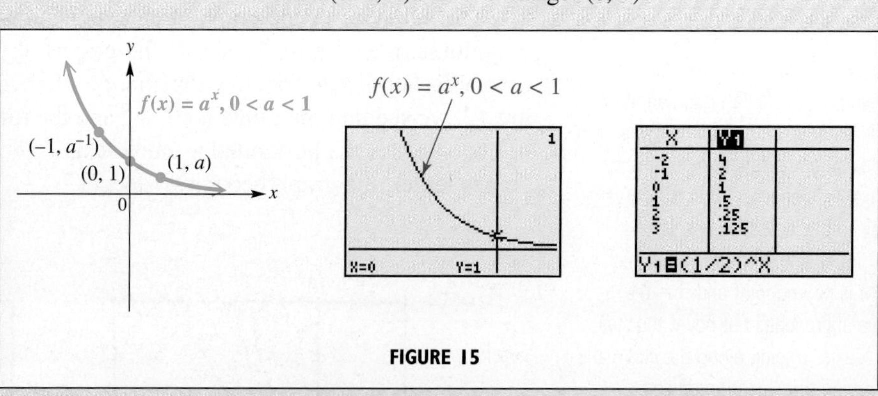


FIGURE 15

- $f(x) = a^x$, $0 < a < 1$, is decreasing and continuous on its entire domain, $(-\infty, \infty)$.
- The $x$-axis is the horizontal asymptote as $x \rightarrow \infty$.
- The graph goes through the points $(-1, a^{-1})$, $(0, 1)$, and $(1, a)$.

**NOTE** Since $a^{-x} = (a^{-1})^x = \left(\frac{1}{a}\right)^x$, the graph of $f(x) = a^{-x}$, $a > 1$, resembles the graph in Figure 15.

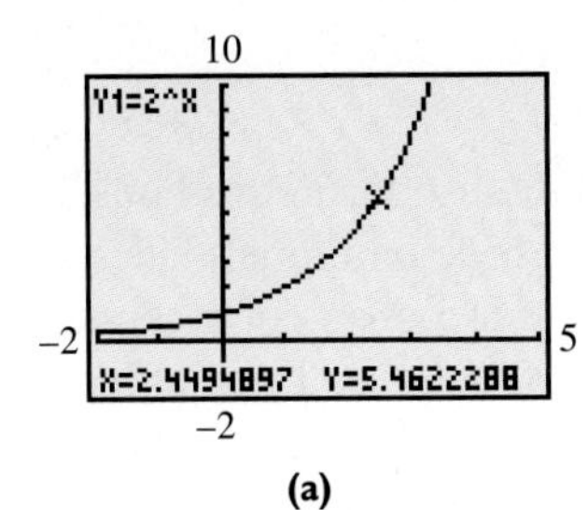


(a)

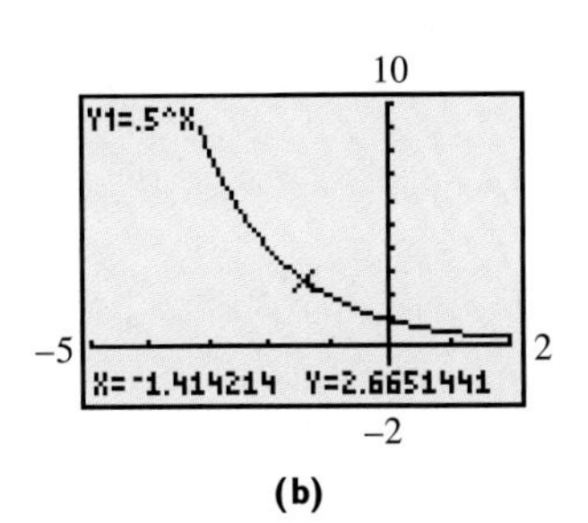


(b)

FIGURE 16

### EXAMPLE 2 *Using Graphs to Evaluate Exponential Expressions*

Use a graph to evaluate each expression.

**(a)** $2^{\sqrt{6}}$ **(b)** $.5^{-\sqrt{2}}$

**Solution**

**(a)** Figure 16(a) at the left shows the graph of $Y_1 = 2^X$, with $Y_1$ evaluated for $X = \sqrt{6} \approx 2.4494897$. From the information at the bottom of the graph, we see that $Y = 2^{\sqrt{6}} \approx 5.4622288$, which confirms our earlier result when we raised numbers to irrational powers.

**(b)** Earlier, we found $.5^{-\sqrt{2}} \approx 2.665144143$. Now, using the graph of $Y_1 = .5^X$ with $X = -\sqrt{2} \approx -1.414214$, we find $Y \approx 2.6651441$, again supporting our earlier result. See Figure 16(b).

### EXAMPLE 3 *Comparing the Graphs of $f(x) = 2^x$ and $g(x) = \left(\frac{1}{2}\right)^x$*

Explain why the graph of $g(x) = \left(\frac{1}{2}\right)^x$ is a reflection across the $y$-axis of the graph of $f(x) = 2^x$.

**Analytic Solution**

We must show that $g(x) = f(-x)$ to prove analytically that the graph of $g$ is a reflection of the graph of $f$ across the $y$-axis.

| | |
|---|---|
| $g(x) = \left(\frac{1}{2}\right)^x$ | Given |
| $= (2^{-1})^x$ | Definition of $a^{-1}$ |
| $= 2^{-x}$ | $(a^m)^n = a^{mn}$ |
| $= f(-x)$ | $f(x) = 2^x$; $f(-x) = 2^{-x}$ |

**Graphing Calculator Solution**

Figure 17 shows a calculator graph of both functions. The graph indicates that $g(x)$ is a reflection across the $y$-axis of $f(x)$.

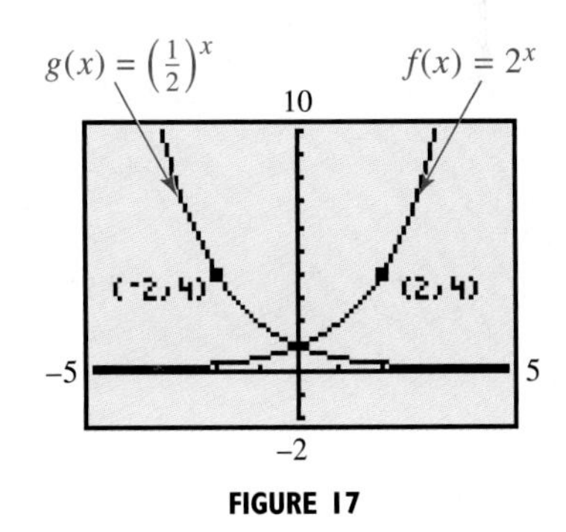


FIGURE 17

### EXAMPLE 4 *Using Translations to Graph an Exponential Function*

Explain how the graph of $y = -2^x + 3$ is obtained from the graph of $y = 2^x$. Give the horizontal asymptote and all intercepts.

**Solution** The graph of $y = -2^x$ is a reflection across the $x$-axis of the graph of $y = 2^x$. (The base is 2, *not* $-2$; negative numbers are not allowed as bases for exponential functions.) The $+3$ indicates that the graph is shifted 3 units upward, so the horizontal asymptote is the line $y = 3$ rather than the $x$-axis. The $y$-intercept is 2. The $x$-intercept can be approximated using a calculator as 1.58, as shown in the calculator graph in Figure 18 on the next page. A traditional graph is also shown.

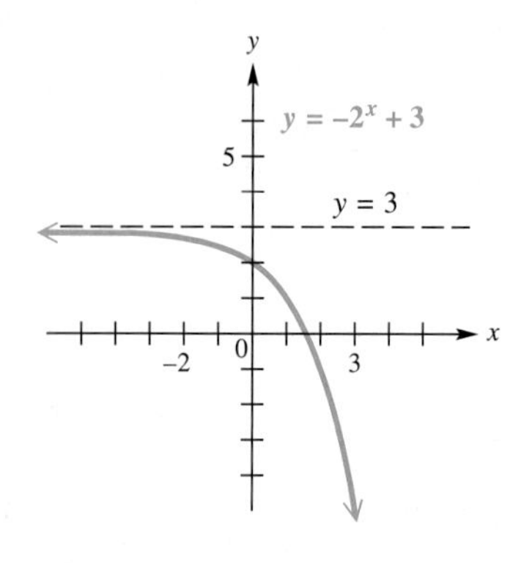


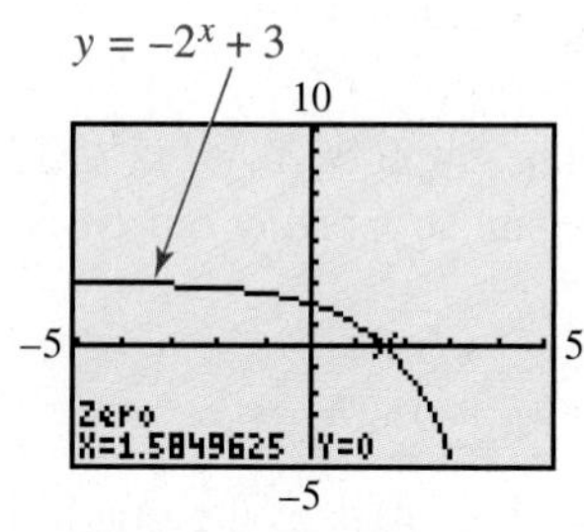


FIGURE 18

**NOTE** Later in this chapter we see how to find the $x$-intercept of the graph in Figure 18 analytically.

## Exponential Equations (Type 1)

The equation $25^x = 125$ is different from any equation studied so far in this book because the variable appears in the exponent. Notice that the base on the left side (25) and the constant on the right side (125) can both be written as powers of a common base, 5. We refer to such equations as Type 1 exponential equations. Thus, a Type 1 exponential equation is one where both sides can be written as powers of the same rational number base. We use Property (b) of the additional properties of exponents listed earlier in this section to solve Type 1 exponential equations.

**EXAMPLE 5** *Solving Type 1 Exponential Equations*

Solve each equation.

**(a)** $25^x = 125$ **(b)** $\left(\frac{1}{3}\right)^x = 81$

**Solution**

**(a)** Because $25 = 5^2$ and $125 = 5^3$, we proceed as follows.

$$25^x = 125$$
$$(5^2)^x = 5^3 \quad \text{Write with the same base.}$$
$$5^{2x} = 5^3 \quad (a^m)^n = a^{mn}$$
$$2x = 3 \quad \text{Set exponents equal.}$$
$$x = \frac{3}{2}$$

We check by substitution.

$$25^{3/2} = 125 \ ? \quad \text{Let } x = \tfrac{3}{2}.$$
$$\left(\sqrt{25}\right)^3 = 125 \ ?$$
$$5^3 = 125 \ ?$$
$$125 = 125 \quad \text{True}$$

The solution set is $\left\{\frac{3}{2}\right\}$.

**(b)**
$$\left(\frac{1}{3}\right)^x = 81$$
$$(3^{-1})^x = 3^4 \quad \text{Write with the same base; } \tfrac{1}{3} = 3^{-1} \text{ and } 81 = 3^4.$$
$$3^{-x} = 3^4 \quad (a^m)^n = a^{mn}$$
$$-x = 4 \quad \text{Set exponents equal.}$$
$$x = -4$$

Check that the solution set is $\{-4\}$.

**EXAMPLE 6** *Solving a Type 1 Exponential Equation*

Solve $1.5^{x+1} = \left(\frac{27}{8}\right)^x$.

**Analytic Solution**

At first glance, this equation may not seem to be of Type 1. However, $1.5 = \frac{3}{2}$ and $\frac{27}{8} = \left(\frac{3}{2}\right)^3$. Thus, each base can be written as a power of $\frac{3}{2}$.

$$1.5^{x+1} = \left(\frac{27}{8}\right)^x$$

$$\left(\frac{3}{2}\right)^{x+1} = \left[\left(\frac{3}{2}\right)^3\right]^x \qquad \text{Write with the same base.}$$

$$\left(\frac{3}{2}\right)^{x+1} = \left(\frac{3}{2}\right)^{3x} \qquad (a^m)^n = a^{mn}$$

$$x + 1 = 3x \qquad \text{Set exponents equal.}$$

$$1 = 2x$$

$$x = \frac{1}{2} \quad \text{or} \quad .5$$

Check that the solution set is $\{.5\}$.

**Graphing Calculator Solution**

Graph

$$y = 1.5^{x+1} - \left(\frac{27}{8}\right)^x$$

and use the $x$-intercept method, as shown in Figure 19. The $x$-intercept is .5, confirming the analytic solution.

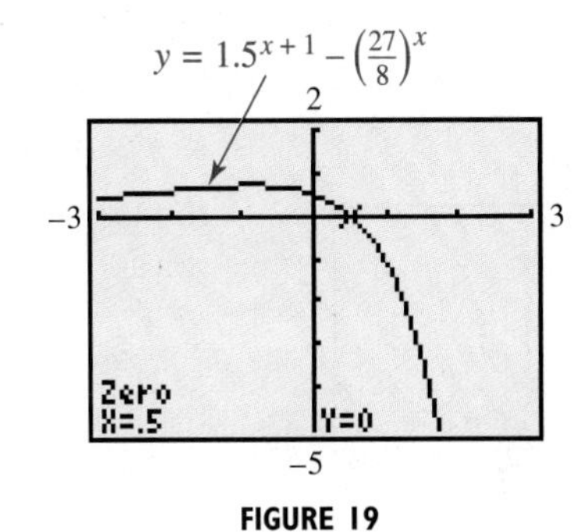


**FIGURE 19**

**EXAMPLE 7** *Using a Graph to Solve Exponential Inequalities*

Use the graph in Figure 19 to solve each inequality.

**(a)** $1.5^{x+1} - \left(\frac{27}{8}\right)^x > 0$ **(b)** $1.5^{x+1} - \left(\frac{27}{8}\right)^x < 0$

**Solution** In Figure 19, the graph of $y = 1.5^{x+1} - \left(\frac{27}{8}\right)^x$ lies *above* the $x$-axis for values of $x$ less than .5 and *below* the $x$-axis for values of $x$ greater than .5. Tracing left and right will support this observation. The solution set for part (a) is $(-\infty, .5)$, and the solution set for part (b) is $(.5, \infty)$.

## The Number e

The most important exponential function has the irrational number $e$ as its base. The number $e$ is named after the Swiss mathematician Leonhard Euler (1707–1783). Many mathematical expressions can be used to approximate $e$, but one of the easiest to illustrate involves the expression $\left(1 + \frac{1}{x}\right)^x$. If $x$ takes on larger and larger values (that is, approaches $\infty$), the expression approaches $e$. The table in Figure 20 illustrates this with $x$ equal to powers of 10.

**Looking Ahead to Calculus**

Let $Y_1 = e$ and $Y_2 = \left(1 + \frac{1}{X}\right)^X$. Define $Y_3$ as $|Y_1 - Y_2|$. As X gets larger, the values of $Y_1$ and $Y_2$ get closer to each other. The table shows how the value of $Y_3$ approaches 0 as X increases. In calculus, $e$ is defined as the *limit* of $\left(1 + \frac{1}{X}\right)^X$ as $X \to \infty$.

| X | Y3 |
|---|---|
| 700 | .00194 |
| 800 | .0017 |
| 900 | .00151 |
| 1000 | .00136 |
| 1100 | .00123 |
| 1200 | .00113 |
| 1300 | .00104 |

Y3=abs(Y1−Y2)

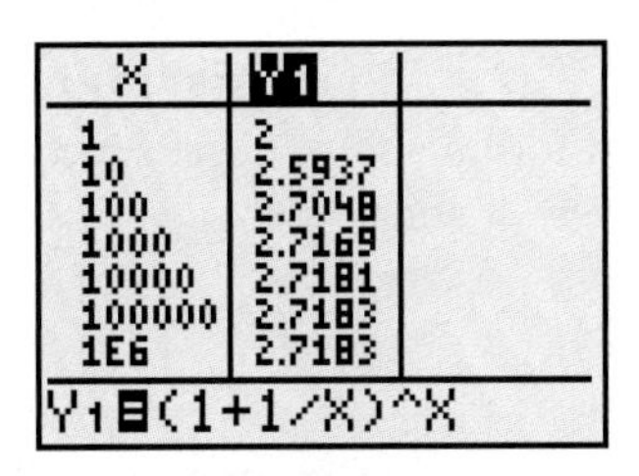

| X | Y1 |
|---|---|
| 1 | 2 |
| 10 | 2.5937 |
| 100 | 2.7048 |
| 1000 | 2.7169 |
| 10000 | 2.7181 |
| 100000 | 2.7183 |
| 1E6 | 2.7183 |

Y1=(1+1/X)^X

**FIGURE 20**

As $x \to \infty$, $\left(1 + \frac{1}{x}\right)^x$ appears to approach some "limiting" number. This irrational number is $e$; to nine decimal places,

$$e \approx \mathbf{2.718281828.}$$

By the definition of a zero exponent, $e^0 = 1$. Both traditional and calculator graphs of $f(x) = e^x$ are shown in Figure 21.

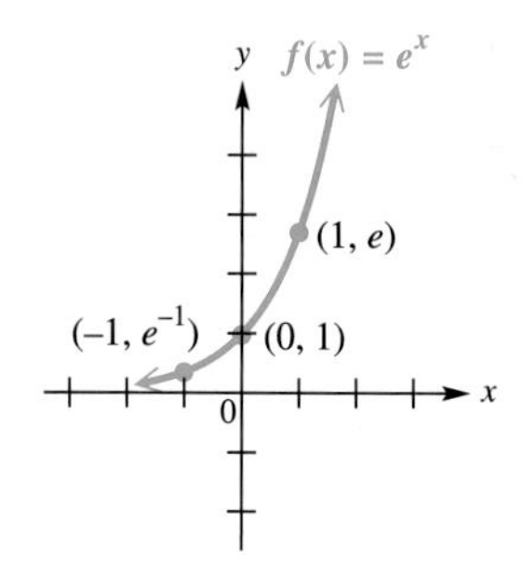


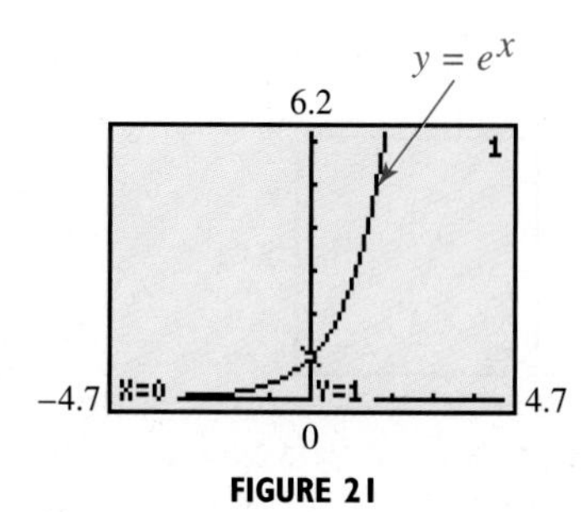


**FIGURE 21**

## Compound Interest

The formula for *compound interest* (interest paid on both principal and interest) is an important application of exponential functions. The formula for simple interest is $I = Prt$, where $P$ is the principal or initial amount of money, $r$ is the rate of interest expressed as a decimal, and $t$ is time in years that the principal earns interest. Suppose $t = 1$ year. Then, at the end of the year, the amount has grown to

$$P + Pr = P(1 + r),$$

the original principal plus the interest. If this amount is left at the same interest rate for another year, the total amount becomes

$$\begin{aligned}[P(1 + r)] + [P(1 + r)]r &= [P(1 + r)](1 + r)\\ &= P(1 + r)^2.\end{aligned}$$

After the third year, this will grow to

$$\begin{aligned}[P(1 + r)^2] + [P(1 + r)^2]r &= [P(1 + r)^2](1 + r)\\ &= P(1 + r)^3.\end{aligned}$$

Continuing in this way produces the following general formula for interest compounded $n$ times per year.

**TECHNOLOGY NOTE**

Because $e$ is such an important base for the exponential function, calculators are programmed to find powers of $e$.

```
e^(1)
         2.718281828
e^(-1)
         .3678794412
e^(2)
         7.389056099
```

> **Compound Interest Formula**
>
> Suppose that a principal of $P$ dollars is invested at an annual interest rate $r$ (in decimal form), compounded $n$ times per year. Then, the amount $A$ accumulated after $t$ years is given by the formula
>
> $$A = P\left(1 + \frac{r}{n}\right)^{nt}.$$

**EXAMPLE 8** *Using the Compound Interest Formula*

Suppose that \$1000 is invested at an annual rate of 8%, compounded quarterly (four times per year). Find the total amount in the account after 10 years if no withdrawals are made.

**Analytic Solution**

Use the compound interest formula, with $P = 1000$, $r = .08$, $n = 4$, and $t = 10$.

$$A = P\left(1 + \frac{r}{n}\right)^{nt}$$

$$A = 1000\left(1 + \frac{.08}{4}\right)^{4 \cdot 10}$$

$$A \approx 2208.039664 \quad \text{Use a calculator.}$$

**Graphing Calculator Solution**

The function

$$Y_1 = 1000\left(1 + \frac{.08}{4}\right)^{4X}$$

is graphed in Figure 22 and evaluated for X = 10. The Y-value at the bottom of the screen is the amount in the account. (Tracing to the right on this graph gives new meaning to the phrase "watching your money grow.")

*(continued)*

To the nearest cent, there will be \$2208.04 in the account after 10 years. This means that

$$\$2208.04 - \$1000 = \$1208.04$$

interest was earned. (Note that 10 years = 40 quarters.)

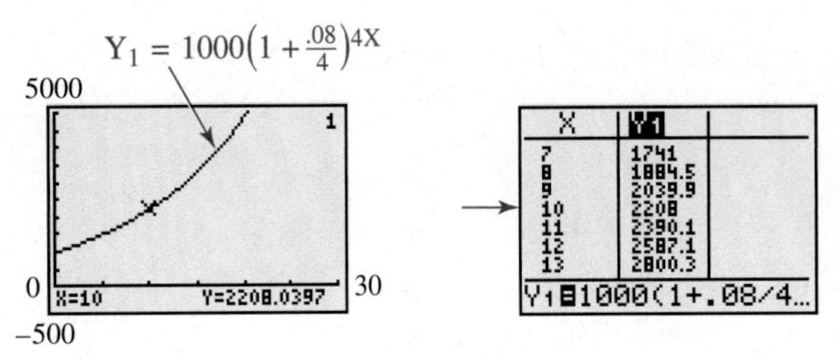


**FIGURE 22**

The table in Figure 22 can be used to numerically determine the amount in the account. Compare the value of $Y_1$ when X = 10 to the value in the analytic solution and the Y-value displayed at the bottom of the calculator graph.

The compound interest formula applies if interest is compounded for a finite number of compounding periods annually. Theoretically, the number of compounding periods per year can get larger and larger (quarterly, monthly, daily, etc.), and if $n$ is allowed to approach infinity, we say that interest is compounded *continuously*.

To derive the formula for continuous compounding, we begin with the compound interest formula,

$$A = P\left(1 + \frac{r}{n}\right)^{nt}.$$

Let $k = \frac{n}{r}$. Then, $n = rk$, and with these substitutions, the formula becomes

$$A = P\left(1 + \frac{1}{k}\right)^{rkt} = P\left[\left(1 + \frac{1}{k}\right)^{k}\right]^{rt}.$$

If $n \to \infty$, $k \to \infty$ as well, and the expression $\left(1 + \frac{1}{k}\right)^k \to e$, as discussed earlier. This leads to the formula $A = Pe^{rt}$.

**Looking Ahead to Calculus**

In calculus, the derivative of a function is a limit that allows us to determine the slope of a tangent line to the graph of the function. For $f(x) = e^x$, the derivative of the function $f$ is $f$ itself: $f'(x) = e^x$. Geometrically this means that the slope of a tangent line to its graph is the same as the $y$-coordinate of the point of tangency. Will this slope ever be negative? Will it ever be 0?

**Continuous Compounding Formula**

If an amount of $P$ dollars is deposited at a rate of interest $r$ compounded continuously for $t$ years, then the final amount $A$ in dollars is

$$A = Pe^{rt}.$$

**EXAMPLE 9** *Solving a Continuous Compounding Problem*

Suppose \$5000 is deposited in an account paying 8% compounded continuously for 5 years. Find the total amount on deposit at the end of 5 years.

**Analytic Solution**

Let $P = 5000$, $r = .08$, and $t = 5$. Then,

$$A = 5000e^{.08(5)} = 5000e^{.4}.$$

**Graphing Calculator Solution**

As Figure 23 on the next page shows, the graph of $Y_1 = 5000e^{.08X}$ indicates that when X = 5, Y ≈ 7459.12.

*(continued)*

Using a calculator, we find, to the nearest cent,

$$A = 5000e^{.4}$$
$$= 7459.12.$$

There will be \$7459.12 in the account after 5 years. (This means that

$$\$7459.12 - \$5000 = \$2459.12$$

interest was earned.)

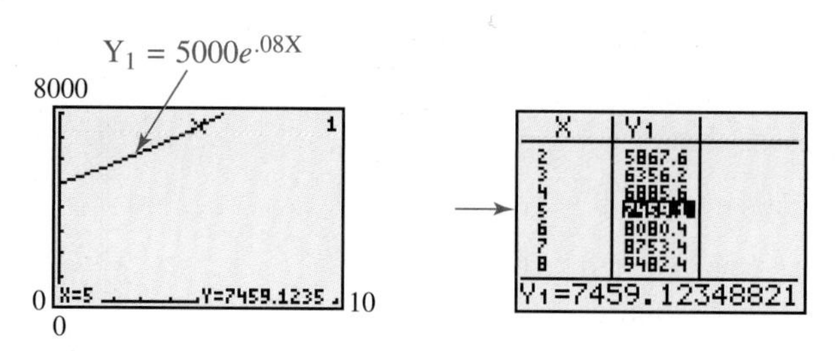


**FIGURE 23**

The analytic result is supported numerically by the table in Figure 23, which shows values of $Y_1 = 5000e^{.08X}$ for several values of X. When $X = 5$, the table gives $Y_1 \approx 7459.12$.

The continuous compounding formula is one example of an exponential growth function with base $e$. Other examples of exponential growth (and decay) are given in Section 5.6.

As the U.S. population ages, there has been increased concern about the incidence of Alzheimer's disease, a disease of aging. Progress toward finding a cure is being reported on many fronts.

**EXAMPLE 10** *Modeling the Risk of Alzheimer's Disease*

The chances of developing Alzheimer's disease increase exponentially after age 65 according to the function defined by

$$f(x) = r(1.149)^x,$$

where $r$ is the risk (in decimal form) at age 65 and $x$ is the number of years greater than 65. (*Source: Parade Magazine,* July 15, 2001.) Compare the risk at age 70 to the risk at age 65.

**Solution** Let $x = 70 - 65 = 5$ and solve for $f(x)$.

$$f(x) = r(1.149)^5 \approx 2.00r \qquad \text{Use a calculator.}$$

Thus, the risk is twice as great at age 70 as at age 65.

## 5.2 EXERCISES

*Use a calculator to find an approximation for each power. Give the maximum number of decimal places that your calculator displays.*

**1.** $2^{\sqrt{10}}$ **2.** $3^{\sqrt{11}}$ **3.** $\left(\frac{1}{2}\right)^{\sqrt{2}}$ **4.** $\left(\frac{1}{3}\right)^{\sqrt{6}}$

**5.** $4.1^{-\sqrt{3}}$ **6.** $6.4^{-\sqrt{3}}$ **7.** $\sqrt{7}^{\sqrt{7}}$ **8.** $\sqrt{13}^{-\sqrt{13}}$

*Use a calculator graph of each exponential function to graphically support the result found in the specified exercise.*

**9.** $y = 2^x$ (Exercise 1) **10.** $y = 3^x$ (Exercise 2)

**11.** $y = \left(\frac{1}{2}\right)^x$ (Exercise 3) **12.** $y = \left(\frac{1}{3}\right)^x$ (Exercise 4)

*Concept Check* *In the figure, the graphs of* $y = a^x$ *for* $a = 1.8, 2.3, 3.2, .4, .75,$ *and* $.31$ *are given. They are identified by letter, but not necessarily in the same order as the values of* a *just given. Use your knowledge of how the exponential function behaves for various powers of* a *to identify each lettered graph.*

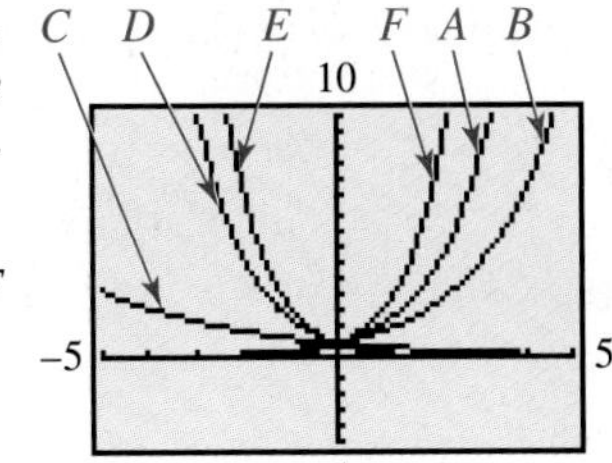


**13.** *A* **14.** *B* **15.** *C* **16.** *D* **17.** *E* **18.** *F*

## RELATING CONCEPTS

**For individual or group investigation (Exercises 19–24)**

*In Exercises 19–24, assume* $f(x) = a^x$*, where* $a > 1$*. Work these exercises in order.*

**19.** Is $f$ a one-to-one function? If so, based on Section 5.1, what kind of related function exists for $f$?

**20.** If $f$ has an inverse function $f^{-1}$, sketch $f$ and $f^{-1}$ on the same axes.

**21.** If $f^{-1}$ exists, find an equation for $y = f^{-1}(x)$ using the method described in Section 5.1. You need not solve for $y$.

**22.** If $a = 10$, what is the equation for $y = f^{-1}(x)$? (You need not solve for $y$.)

**23.** If $a = e$, what is the equation for $y = f^{-1}(x)$? (You need not solve for $y$.)

**24.** If the point $(p, q)$ is on the graph of $f$, then the point __________ is on the graph of $f^{-1}$.

*Graph each function by hand and support your sketch with a calculator graph. Give the domain, range, and equation of the asymptote.*

**25.** $f(x) = 3^x$ **26.** $f(x) = 4^x$ **27.** $f(x) = \left(\frac{1}{3}\right)^x$ **28.** $f(x) = \left(\frac{1}{4}\right)^x$ **29.** $f(x) = \left(\frac{3}{2}\right)^x$

**30.** $f(x) = \left(\frac{2}{3}\right)^x$ **31.** $f(x) = e^x$ **32.** $f(x) = -e^x$ **33.** $f(x) = e^{x+1}$ **34.** $f(x) = e^x - 1$

*Concept Check* *In Exercises 35 and 36, use the graph at the right and follow the directions in parts (a)–(f).*

**35. (a)** Is $a > 1$ or is $0 < a < 1$?

**(b)** Give the domain and range of $f$, and the equation of the asymptote.

**(c)** Sketch the graph of $g(x) = -a^x$.

**(d)** Give the domain and range of $g$, and the equation of the asymptote.

**(e)** Sketch the graph of $h(x) = a^{-x}$.

**(f)** Give the domain and range of $h$, and the equation of the asymptote.

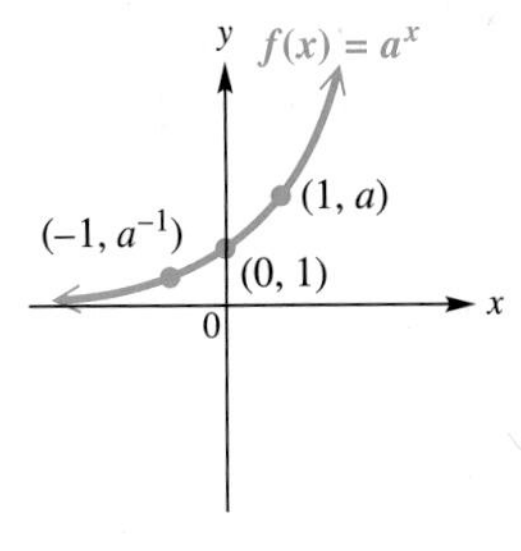


**36. (a)** Is $a > 1$ or is $0 < a < 1$?

**(b)** Give the domain and range of $f$, and the equation of the asymptote.

**(c)** Sketch the graph of $g(x) = a^x + 2$.

**(d)** Give the domain and range of $g$, and the equation of the asymptote.

**(e)** Sketch the graph of $h(x) = a^{x+2}$.

**(f)** Give the domain and range of $h$, and the equation of the asymptote.

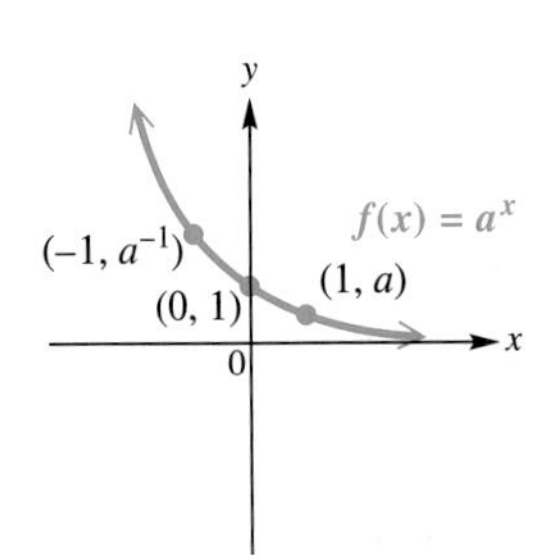

*Explain how the graph of each function can be obtained from the graph of* $y = 2^x$. (Recall that $\left(\frac{1}{2}\right)^x = 2^{-x}$.)

**37.** $y = 2^{x+5} - 3$

**38.** $y = 2^{x-1} + 4$

**39.** $y = \left(\frac{1}{2}\right)^x + 1$

**40.** $y = \left(\frac{1}{2}\right)^x - 6$

**41.** $y = -3 \cdot 2^x$

**42.** $y = -4 \cdot \left(\frac{1}{2}\right)^x$

*Solve each equation.*

**43.** $4^x = 2$

**44.** $125^x = 5$

**45.** $\left(\frac{1}{2}\right)^x = 4$

**46.** $\left(\frac{2}{3}\right)^x = \frac{9}{4}$

**47.** $2^{3-x} = 8$

**48.** $5^{2x+1} = 25$

**49.** $\frac{1}{27} = x^{-3}$

**50.** $\frac{1}{81} = x^{-4}$

*Solve each equation in part (a) analytically. Support your answer with a calculator graph. Then, use the graph to solve the associated inequalities in parts (b) and (c).*

**51.** **(a)** $27^{4x} = 9^{x+1}$
**(b)** $27^{4x} > 9^{x+1}$
**(c)** $27^{4x} < 9^{x+1}$

**52.** **(a)** $32^x = 16^{1-x}$
**(b)** $32^x > 16^{1-x}$
**(c)** $32^x < 16^{1-x}$

**53.** **(a)** $\left(\frac{1}{2}\right)^{-x} = \left(\frac{1}{4}\right)^{x+1}$
**(b)** $\left(\frac{1}{2}\right)^{-x} \geq \left(\frac{1}{4}\right)^{x+1}$
**(c)** $\left(\frac{1}{2}\right)^{-x} \leq \left(\frac{1}{4}\right)^{x+1}$

**54.** **(a)** $\left(\frac{2}{3}\right)^{x-1} = \left(\frac{81}{16}\right)^{x+1}$
**(b)** $\left(\frac{2}{3}\right)^{x-1} \leq \left(\frac{81}{16}\right)^{x+1}$
**(c)** $\left(\frac{2}{3}\right)^{x-1} \geq \left(\frac{81}{16}\right)^{x+1}$

*Use the appropriate compound interest formula to find the amount that will be in each account, given the stated conditions.*

**55.** \$20,000 invested at 3% annual interest for 4 years compounded **(a)** annually; **(b)** semiannually

**56.** \$35,000 invested at 4.2% annual interest for 3 years compounded **(a)** annually; **(b)** quarterly

**57.** \$27,500 invested at 3.95% annual interest for 5 years compounded **(a)** daily ($n = 365$); **(b)** continuously

**58.** \$15,800 invested at 4.6% annual interest for 6.5 years compounded **(a)** quarterly; **(b)** continuously

*In Exercises 59 and 60, decide which of the two plans will provide a better yield. (Interest rates stated are annual rates.)*

**59.** Plan A: \$40,000 invested for 3 years at 4.5%, compounded quarterly

Plan B: \$40,000 invested for 3 years at 4.4%, compounded continuously

**60.** Plan A: \$50,000 invested for 10 years at 4.75%, compounded daily ($n = 365$)

Plan B: \$50,000 invested for 10 years at 4.7%, compounded continuously

*Use the table capabilities of your calculator to work Exercises 61 and 62.*

**61.** *Comparison of Two Accounts* You have the choice of investing \$1000.00 at an annual rate of 5%, compounded either annually or monthly. Let $Y_1$ represent the investment compounded annually, and let $Y_2$ represent the investment compounded monthly. Graph both $Y_1$ and $Y_2$, and observe the slight difference in each curve. Then, use a table to compare the graphs numerically. What is the difference between the returns on each investment after 1 year, 2 years, 5 years, 10 years, 20 years, 30 years, and 40 years?

**62.** *Comparison of Two Accounts* You have the choice of investing \$1000.00 at an annual rate of 7.5%, compounded daily, or at an annual rate of 7.75%, compounded annually. Let $Y_1$ represent the investment at 7.5% compounded daily, and let $Y_2$ represent the investment at 7.75% compounded annually. Graph both $Y_1$ and $Y_2$, and observe the slight difference in each curve. Then, use a table with $Y_3 = Y_1 - Y_2$ to compare the graphs numerically. What is the difference between the returns on each investment after 1 year, 2 years, 5 years, 10 years, 20 years, 30 years, and 40 years? Why does the lower interest rate yield the greater return over time?

*(Modeling) Solve each problem.*

**63.** *Atmospheric Pressure* The function defined by

$$P(x) = 1013e^{-.0001341x}$$

approximates the atmospheric pressure (in millibars) at an altitude of $x$ meters. Use $P$ to predict the pressure at 1500 meters and 11,000 meters and compare to the actual values of 846 millibars and 227 millibars, respectively.

**64.** *World Population Growth* Since 1980, world population in millions closely fits the exponential function defined by

$$y = 4481e^{.0156x},$$

where $x$ is the number of years since 1980.

**(a)** The world population was about 5320 million in 1990. How closely does the function approximate this value?

**(b)** Use this model to approximate the population in 1995.

**(c)** Use this model to predict the population in 2005.

**(d)** Explain why this model may not be accurate for 2005.

**65.** *Traffic Flow* At an intersection cars arrive randomly with an average rate of 30 cars per hour. Highway engineers estimate the likelihood or probability that at least one car will enter the intersection within a period of $x$ minutes using the function defined by

$$f(x) = 1 - e^{-.5x}.$$

(*Source:* Mannering, F. and W. Kilareski, *Principles of Highway Engineering and Traffic Analysis,* Second Edition, John Wiley and Sons, 1998.)

**(a)** Evaluate $f(2)$ and interpret the answer.

**(b)** Graph $f$ for $0 \le x \le 60$. What happens to the likelihood that at least one car enters the intersection during a 60-minute period?

**66.** *Growth of E. coli Bacteria* A type of bacteria that inhabits the intestines of animals is named *E. coli* (*Escherichia coli*). These bacteria are capable of rapid growth and can be dangerous to humans—especially children. In one study, *E. coli* bacteria were capable of doubling in number every 49.5 minutes. Their number $N$ after $x$ minutes can be modeled by

$$N(x) = N_0 e^{.014x}.$$

Suppose $N_0 = 500{,}000$ is the initial number of bacteria per milliliter. (*Source:* Stent, G. S., *Molecular Biology of Bacterial Viruses,* W. H. Freeman, 1963.)

**(a)** Make a conjecture about the number of bacteria per milliliter after 99 minutes. Verify your conjecture.

**(b)** Determine graphically the elapsed time when there were 25 million bacteria per milliliter.

# 5.3 LOGARITHMS AND THEIR PROPERTIES

Definition of Logarithm • Common Logarithms • Natural Logarithms • Properties of Logarithms • Change-of-Base Rule

## Definition of Logarithm

Consider the exponential equation $2^3 = 8$. In this equation, 3 is the exponent to which 2 must be raised to obtain 8. In this context, 3 is called the *logarithm* to the base 2 of 8, abbreviated $3 = \log_2 8$. It is important to remember that *a logarithm is an exponent* and thus has the same properties as exponents.

**Logarithm**

For all positive numbers $a$, where $a \neq 1$,

$$a^y = x \quad \text{is equivalent to} \quad y = \log_a x.$$

A **logarithm** is an exponent, and $\log_a x$ is the exponent to which $a$ must be raised in order to obtain $x$. The number $a$ is called the **base** of the logarithm, and $x$ is called the **argument** of the expression $\log_a x$. The value of $x$ will always be positive.

The table shows several pairs of equivalent statements written in both exponential and logarithmic forms.

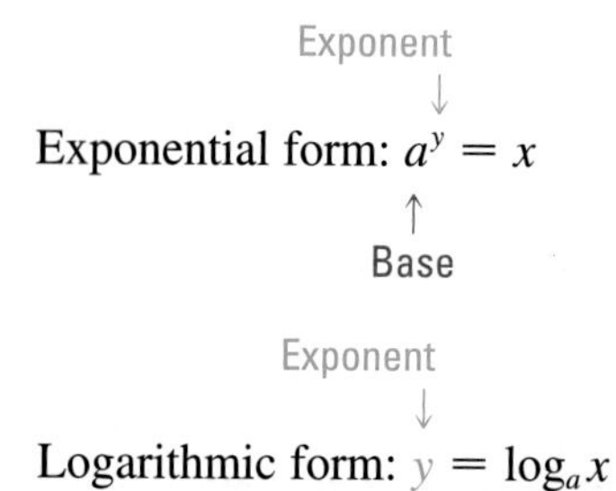


| Exponential Form | Logarithmic Form |
|---|---|
| $2^3 = 8$ | $\log_2 8 = 3$ |
| $\left(\frac{1}{2}\right)^{-4} = 16$ | $\log_{1/2} 16 = -4$ |
| $10^5 = 100{,}000$ | $\log_{10} 100{,}000 = 5$ |
| $3^{-4} = \frac{1}{81}$ | $\log_3\left(\frac{1}{81}\right) = -4$ |
| $5^1 = 5$ | $\log_5 5 = 1$ |
| $\left(\frac{3}{4}\right)^0 = 1$ | $\log_{3/4} 1 = 0$ |

We can solve some logarithmic equations by converting them to Type 1 exponential equations.

**EXAMPLE 1** *Solving Logarithmic Equations*

Solve each equation by first rewriting it in exponential form.

**(a)** $x = \log_8 4$ **(b)** $\log_x 16 = 4$ **(c)** $\log_4 x = \dfrac{3}{2}$

**Solution**

**(a)** Rewrite the equation in exponential form.

$$x = \log_8 4$$
$$8^x = 4 \qquad \text{Exponential form}$$
$$(2^3)^x = 2^2 \qquad \text{Write as a power of the same base, 2.}$$
$$2^{3x} = 2^2 \qquad (a^m)^n = a^{mn}$$
$$3x = 2 \qquad \text{Set exponents equal.}$$
$$x = \frac{2}{3}$$

Since $8^{2/3} = \left(\sqrt[3]{8}\right)^2 = 2^2 = 4$, the solution set is $\left\{\frac{2}{3}\right\}$.

**(b)**
$$\log_x 16 = 4$$
$$x^4 = 16 \qquad \text{Exponential form}$$
$$x = \pm\sqrt[4]{16} \qquad \text{Take fourth roots.}$$
$$x = \pm 2$$

Since the base $x$ must be positive, the solution set is $\{2\}$.

**(c)**
$$\log_4 x = \frac{3}{2}$$
$$x = 4^{3/2} \qquad \text{Exponential form}$$
$$x = 8 \qquad 4^{3/2} = \left(\sqrt{4}\right)^3 = 8$$

Check that the solution set is $\{8\}$.

## Common Logarithms

The two most important bases for logarithms are 10 and $e$. Base 10 logarithms are called **common logarithms.** The common logarithm of $x$ is written $\log x$, where the base is understood to be 10.

> **Common Logarithm**
>
> For all positive numbers $x$, $\quad \mathbf{\log x = \log_{10} x.}$

```
log(6)
         .7781512504
log(100)
                   2
log(.0001)
                  -4
```

(a)

```
10^(log(6))
                   6
10^(log(100))
                 100
10^(log(.0001))
                1E-4
```

(b)

FIGURE 24

Remember that the argument of a common logarithm (and any base logarithm, for that matter) must be a positive number.

Figure 24(a) shows a typical graphing calculator screen with several common logarithms evaluated. The first display indicates that .7781512504 is (approximately) the exponent to which 10 must be raised in order to obtain 6. The second says that 2 is the exponent to which 10 must be raised in order to obtain 100. This is correct, since $100 = 10^2$. The third display indicates that $-4$ is the exponent to which 10 must be raised in order to obtain .0001. Again, this is correct, since $.0001 = 10^{-4}$.

Figure 24(b) shows how the definition of a common logarithm can be applied: Raising 10 to the power $\log x$ gives $x$ as a result. If we try to find the logarithm of a negative number with the calculator in real number mode, it responds with the error message in Figure 25.

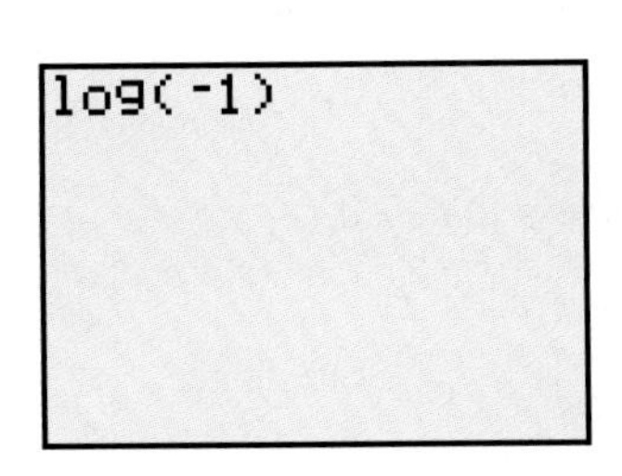


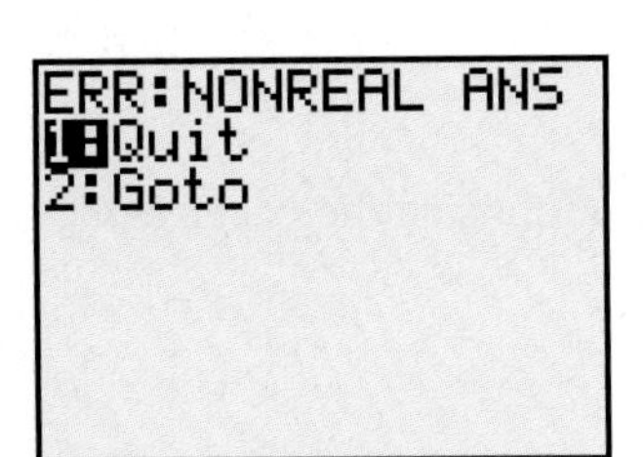


FIGURE 25

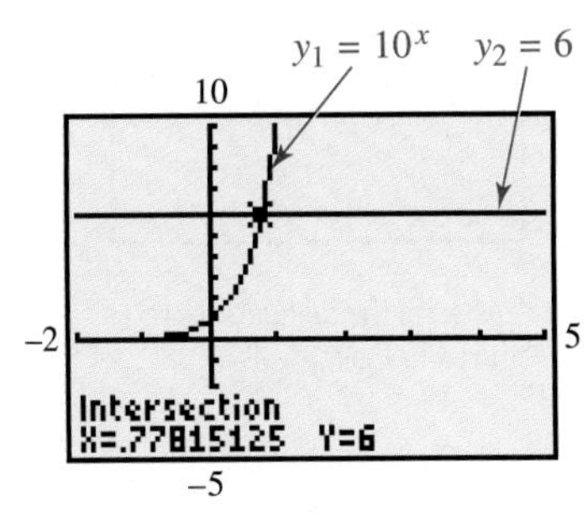


FIGURE 26

In Figure 26, we graphed the functions $y_1 = 10^x$ and $y_2 = 6$. The point of intersection has $x$-value .77815125, which agrees with the value of $\log 6$ in Figure 24(a).

**EXAMPLE 2** *Evaluating Common Logarithms*

Evaluate each expression.

**(a)** $\log 12$ **(b)** $\log .1$ **(c)** $\log \frac{3}{5}$

**Solution** Use a calculator.

**(a)** $\log 12 \approx 1.079181246$ **(b)** $\log .1 = -1$ **(c)** $\log \frac{3}{5} \approx -.2218487496$

Common logarithms were originally used to aid mathematicians and scientists in paper-and-pencil calculations. With the advent of calculators and computers, this use is no longer necessary. However, there are still important applications of common logarithms. For example, in chemistry, the **pH** of a solution is defined as

$$\mathbf{pH = -\log[H_3O^+],}$$

where $[H_3O^+]$ is the hydronium ion concentration in moles per liter.* The pH value is a measure of the acidity or alkalinity of a solution. Pure water has pH 7.0, substances with pH values greater than 7.0 are alkaline, and substances with pH values less than 7.0 are acidic. It is customary to round pH values to the nearest tenth.

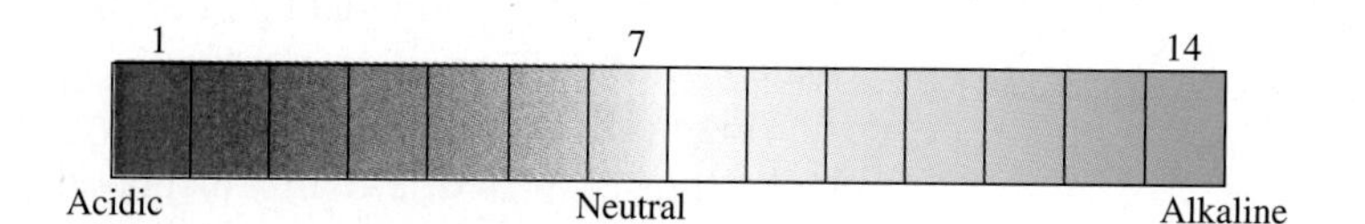


**EXAMPLE 3** *Finding pH and* $[H_3O^+]$

**(a)** Find the pH of a solution with $[H_3O^+] = 2.5 \times 10^{-4}$. Is the solution acidic or alkaline?

**(b)** Find the hydronium ion concentration of a solution with pH $= 7.1$.

```
-log(2.5*10^(-4))
          3.602059991
10^(-7.1)
       7.943282347E-8
```

These are the calculations required in Example 3.

**Solution**

**(a)**

$$\text{pH} = -\log[H_3O^+]$$

$$\text{pH} = -\log(2.5 \times 10^{-4}) \qquad \text{Substitute.}$$

$$\text{pH} \approx 3.6 \qquad \text{Use a calculator.}$$

The solution is acidic because the pH is less than 7.0.

**(b)**

$$\text{pH} = -\log[H_3O^+]$$

$$7.1 = -\log[H_3O^+] \qquad \text{Substitute.}$$

$$-7.1 = \log[H_3O^+] \qquad \text{Multiply by } -1.$$

$$[H_3O^+] = 10^{-7.1} \qquad \text{Exponential form}$$

A graphing calculator has a key marked $10^x$, usually in conjunction with the log $x$ key, that allows you to calculate a power of 10. Use this key to find that

$$[H_3O^+] = 10^{-7.1} \approx 7.9 \times 10^{-8}.$$

## Natural Logarithms

In many practical applications of logarithms, the number $e$ is used as the base. Logarithms to base $e$ are called **natural logarithms,** since they occur in the life sciences and economics in natural situations that involve growth and decay. The natural logarithm of a positive number $x$ is written $\ln x$ (read "el en $x$").

**Looking Ahead to Calculus**

The natural logarithmic function defined by $f(x) = \ln x$ and the reciprocal function defined by $g(x) = \frac{1}{x}$ have an important relationship in calculus. The derivative of the natural logarithmic function is the reciprocal function.

*A *mole* is the amount of substance that contains the same number of molecules as the number of atoms in 12 grams of carbon 12.

**Natural Logarithm**

For all positive numbers $x$, $\ln x = \log_e x$.

Figure 27 shows traditional and calculator graphs of the natural logarithmic function defined by $f(x) = \ln x$.

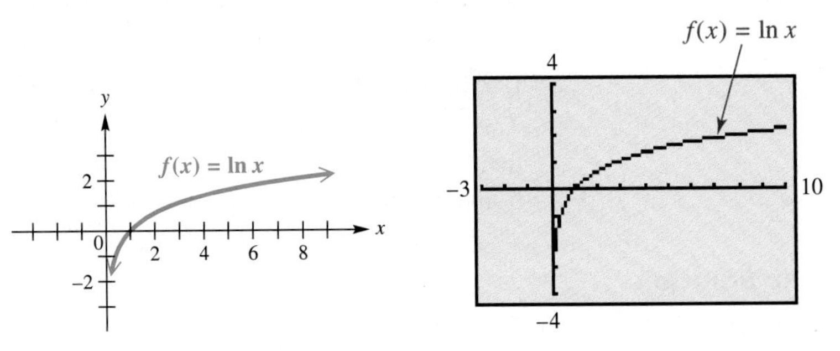


FIGURE 27

Natural logarithms of numbers are found the same way as common logarithms, using a calculator. The natural logarithm key is usually found in conjunction with the $e^x$ key. Figure 28 shows how the natural logarithmic function and the base $e$ exponential function can be applied.

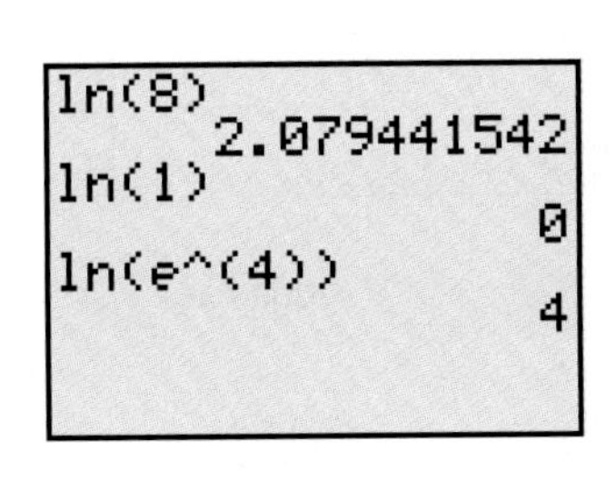


(a)

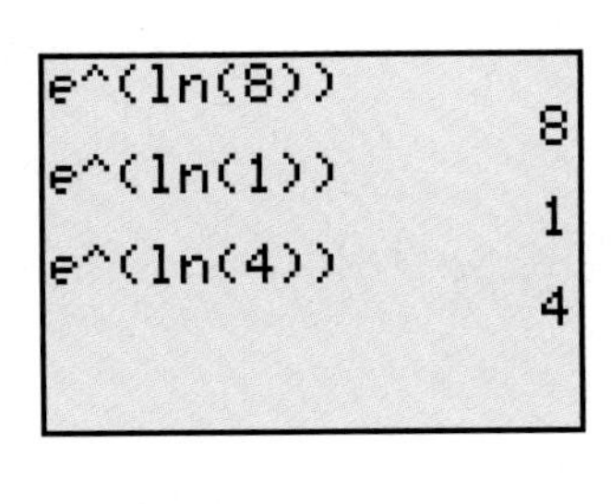


(b)

FIGURE 28

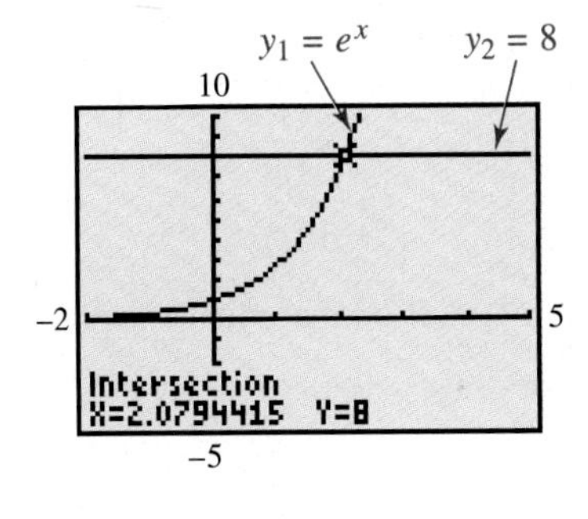


FIGURE 29

The first display in Figure 28(a) indicates that 2.079441542 is (approximately) the exponent to which $e$ must be raised in order to obtain 8. In Figure 29, this is supported by the fact that the graphs of $y_1 = e^x$ and $y_2 = 8$ intersect at a point with $x$-value 2.0794415.

**FOR DISCUSSION**

1. The second display in Figure 28(a) indicates $\ln 1 = 0$. Use your calculator to find $\log 1$. Discuss your results. If 1 is the argument, does it matter what base is used? Why or why not?
2. The third display in Figure 28(a) indicates $\ln(e^4) = 4$. Use the same expression, but replace 4 with the number of letters in your last name. What is the result? Can you make a generalization?
3. Does your generalization apply to common logarithms, too? See Example 2.

### EXAMPLE 4 *Evaluating Natural Logarithms*

Evaluate each expression.

**(a)** $\ln 12$ **(b)** $\ln e^{10}$ **(c)** $\ln \sqrt{e}$

**Solution** Use a calculator. Refer to Item 2 in the "For Discussion" box on the previous page for parts (b) and (c).

**(a)** $\ln 12 \approx 2.48490665$ **(b)** $\ln e^{10} = 10$ **(c)** $\ln \sqrt{e} = .5$

### EXAMPLE 5 *Solving a Continuous Compounding Problem*

The continuous compounding formula $A = Pe^{rt}$ was given in Section 5.2. Suppose that \$1000 is invested at 3% annual interest, compounded continuously. How long will it take for the amount to grow to \$1500?

**Analytic Solution**

Let $P = 1000$, $A = 1500$, and $r = .03$ in the formula.

$$1500 = 1000e^{.03t} \qquad A = Pe^{rt}$$

$$1.5 = e^{.03t} \qquad \text{Divide by 1000.}$$

Because $.03t$ is the exponent to which $e$ must be raised in order to obtain 1.5, it must equal the natural logarithm of 1.5, or $\ln 1.5$.

$$.03t = \ln 1.5 \qquad \text{Rewrite in logarithmic form.}$$

$$t = \frac{\ln 1.5}{.03} \qquad \text{Divide by .03.}$$

$$t \approx 13.5 \qquad \text{Use a calculator.}$$

It will take about 13.5 years to grow to \$1500.

**Graphing Calculator Solution**

Graph $Y_1 = 1000e^{.03X}$ and $Y_2 = 1500$, then find the point of intersection, as shown in Figure 30. The table in Figure 30 gives the same result. When time X is 13.5 years, the amount $Y_1$ is $1499.3 \approx 1500$ dollars.

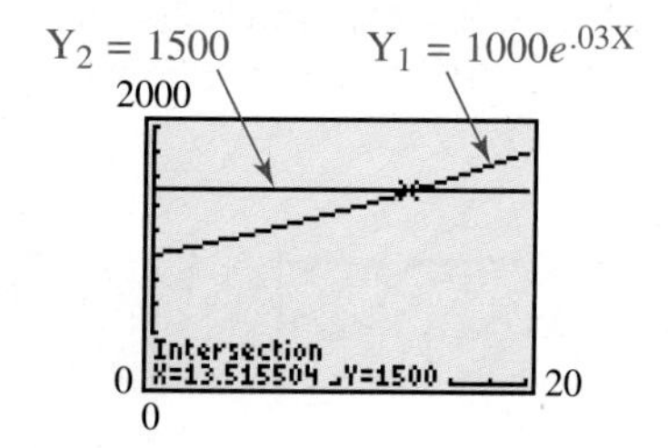


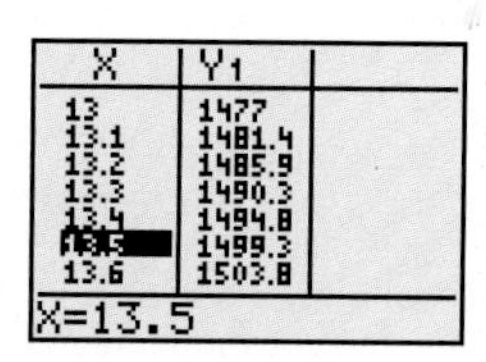


**FIGURE 30**

## Properties of Logarithms

We now formally state several properties of logarithms.

> **Properties of Logarithms**
>
> For $a > 0$, $a \neq 1$, and any real number $k$,
>
> **1.** $\log_a 1 = 0$. **2.** $\log_a a^k = k$. **3.** $a^{\log_a k} = k, \ k > 0$.

Property 1 is true because $a^0 = 1$ for any nonzero value of $a$. Property 2 is verified by writing the equation in exponential form, giving the identity $a^k = a^k$. Property 3 is justified by the fact that $\log_a k$ is the exponent to which $a$ must be raised in order to obtain $k$. Therefore, by the definition, $a^{\log_a k}$ must equal $k$.

Three additional properties that emphasize the close relationship between logarithms and exponents follow.

**Properties of Logarithms (Continued)**

For $x > 0$, $y > 0$, $a > 0$, $a \neq 1$, and any real number $r$,

**Product Rule** **4.** $\log_a xy = \log_a x + \log_a y$.
(The logarithm of the product of two numbers is equal to the sum of the logarithms of the numbers.)

**Quotient Rule** **5.** $\log_a \frac{x}{y} = \log_a x - \log_a y$.
(The logarithm of the quotient of two numbers is equal to the difference between the logarithms of the numbers.)

**Power Rule** **6.** $\log_a x^r = r \log_a x$.
(The logarithm of a number raised to a power is equal to the exponent multiplied by the logarithm of the number.)

The proof of Property 4, the product rule, is given here.

***Proof*** Let

$$m = \log_a x \quad \text{and} \quad n = \log_a y.$$

Then,

$$a^m = x \quad \text{and} \quad a^n = y \qquad \text{Definition of logarithm}$$
$$a^m \cdot a^n = xy \qquad \text{Multiply.}$$
$$a^{m+n} = xy \qquad \text{Add exponents.}$$
$$\log_a xy = m + n. \qquad \text{Definition of logarithm}$$

Since $m = \log_a x$ and $n = \log_a y$,

$$\log_a xy = \log_a x + \log_a y. \qquad \text{Substitute.}$$

Properties 5 and 6, the quotient and power rules, are proved in a similar way. (See Exercises 105 and 106.)

**Looking Ahead to Calculus**

The product rule for logarithms (as well as the quotient and power rules) is proved in calculus using a different method than the one labeled "Proof" here. The typical calculus proof involves the derivative of the base $e$ logarithmic function.

**EXAMPLE 6** *Using the Properties of Logarithms*

Assuming that all variables represent positive real numbers, use the properties of logarithms to rewrite each expression.

**(a)** $\log 8x$ **(b)** $\log_9 \frac{15}{7}$ **(c)** $\log_5 \sqrt{8}$ **(d)** $\log .1$

**(e)** $2^{\log_2 3}$ **(f)** $\log_a \frac{x}{yz}$ **(g)** $\log_a \sqrt[3]{m^2}$ **(h)** $\log_b \sqrt[n]{\frac{x^3y^5}{z^m}}$

**Solution**

**(a)** $\log 8x = \log 8 + \log x$ Product rule

**(b)** $\log_9 \frac{15}{7} = \log_9 15 - \log_9 7$ Quotient rule

**(c)** $\log_5 \sqrt{8} = \log_5 8^{1/2} = \frac{1}{2} \log_5 8$ Power rule

**(d)** $\log .1 = \log \frac{1}{10} = \log 10^{-1} = -1$

**(e)** $2^{\log_2 3} = 3$

**(f)** $\log_a \frac{x}{yz} = \log_a x - (\log_a y + \log_a z) = \log_a x - \log_a y - \log_a z$

**(g)** $\log_a \sqrt[3]{m^2} = \log_a m^{2/3} = \frac{2}{3} \log_a m$

**(h)**

$$
\begin{aligned}
\log_b \sqrt[n]{\frac{x^3y^3}{z^m}} &= \frac{1}{n} \log_b \frac{x^3y^5}{z^m} && \text{Power rule} \\
&= \frac{1}{n}(\log_b x^3 + \log_b y^5 - \log_b z^m) && \text{Use parentheses.} \\
&= \frac{1}{n}(3\log_b x + 5\log_b y - m\log_b z) && \text{Power rule} \\
&= \frac{3}{n}\log_b x + \frac{5}{n}\log_b y - \frac{m}{n}\log_b z && \text{Distributive property}
\end{aligned}
$$

**EXAMPLE 7** *Using the Properties of Logarithms*

Use the properties of logarithms to write each expression as a single logarithm with coefficient 1. Assume that all variables represent positive real numbers.

**(a)** $\log_3(x + 2) + \log_3 x - \log_3 2$

**(b)** $2\log_a m - 3\log_a n$

**(c)** $\frac{1}{2} \log_b m + \frac{3}{2} \log_b 2n - \log_b m^2n$

**Solution**

**(a)** Using the product and quotient rules,

$$\log_3(x + 2) + \log_3 x - \log_3 2 = \log_3 \frac{(x + 2)x}{2}.$$

**(b)** Here we use the power rule and then the quotient rule.

$$2\log_a m - 3\log_a n = \log_a m^2 - \log_a n^3 = \log_a \frac{m^2}{n^3}$$

**(c)**

$$
\begin{aligned}
&\frac{1}{2} \log_b m + \frac{3}{2} \log_b 2n - \log_b m^2n \\
&= \log_b m^{1/2} + \log_b(2n)^{3/2} - \log_b m^2n && \text{Power rule} \\
&= \log_b \frac{m^{1/2}(2n)^{3/2}}{m^2n} && \text{Product and quotient rules} \\
&= \log_b \frac{2^{3/2}n^{1/2}}{m^{3/2}} && \text{Rules for exponents} \\
&= \log_b \left(\frac{2^3n}{m^3}\right)^{1/2} && \text{Rules for exponents} \\
&= \log_b \sqrt{\frac{8n}{m^3}} && \text{Definition of } a^{1/n}
\end{aligned}
$$

**CAUTION** There is no property of logarithms to rewrite a logarithm of a *sum* or *difference*. That is why, in Example 7(a), $\log_3(x + 2)$ was not written as $\log_3 x + \log_3 2$. Remember, $\log_3 x + \log_3 2 = \log_3(x \cdot 2)$. The distributive property does not apply here since $\log_3(x + 2)$ is one term; "log" is *not* a factor.

## Change-of-Base Rule

A natural question to ask at this point is: "Can we use a calculator to find logarithms for bases other than 10 and $e$, such as $\log_5 17$?" The change-of-base rule shows that we can and how to do so.

**Change-of-Base Rule**

For any positive real numbers $x$, $a$, and $b$, where $a \neq 1$ and $b \neq 1$,

$$\log_a x = \frac{\log_b x}{\log_b a}.$$

**FOR DISCUSSION**

1. Without using your calculator, determine exact values of $\log_3 9$ and $\log_3 27$. Then, use the fact that 16 is between 9 and 27 to determine between what two consecutive integers $\log_3 16$ must lie. Finally, use the change-of-base rule to support your answer.
2. Without using your calculator, determine exact values of $\log_5(\frac{1}{5})$ and $\log_5 1$. Then, use the fact that .68 is between $\frac{1}{5}$ and 1 to determine between what two consecutive integers $\log_5 .68$ must lie. Finally, use the change-of-base rule to support your answer.

This rule is proved by using the definition of logarithm to write $y = \log_a x$ in exponential form.

***Proof*** Let

$$y = \log_a x.$$

$$a^y = x \quad \text{Exponential form}$$

$$\log_b a^y = \log_b x \quad \text{Take logarithms on each side.*}$$

$$y \log_b a = \log_b x \quad \text{Power rule}$$

$$y = \frac{\log_b x}{\log_b a} \quad \text{Divide each side by } \log_b a.$$

$$\log_a x = \frac{\log_b x}{\log_b a} \quad \text{Substitute } \log_a x \text{ for } y.$$

Any positive number other than 1 can be used for base $b$ in the change-of-base rule, but usually the only practical bases are $e$ and 10 since calculators typically give logarithms only for these two bases.

**EXAMPLE 8** *Using the Change-of-Base Rule*

Use logarithms and the change-of-base rule to evaluate each expression. Round to four decimal places.

**(a)** $\log_5 17$ **(b)** $\log_2 .1$

**Solution** Set your calculator to round to four decimal places. Figure 31(a) shows how the results can be found with a graphing calculator using common logarithms, while Figure 31(b) uses natural logarithms. The results are the same.

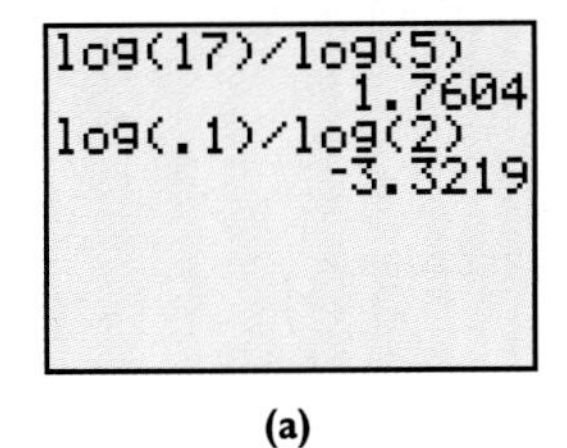


(a)

```
ln(17)/ln(5)
              1.7604
ln(.1)/ln(2)
             -3.3219
```

(b)

**FIGURE 31**

*This follows from the fact that the function defined by $y = \log_a x$ is one-to-one.

**EXAMPLE 9** *Modeling the Diversity of Species*

One measure of the diversity of species in an ecological community is the *index of diversity H*, where

$$H = -(P_1 \log_2 P_1 + P_2 \log_2 P_2 + \cdots + P_n \log_2 P_n),$$

and $P_1, P_2, \ldots, P_n$ are the proportions of a sample belonging to each of $n$ species found in the sample. Find the index of diversity in a community where there are two species, with 90 of one species and 10 of the other.

**Solution** Since there are a total of 100 members in the community, $P_1 = \frac{90}{100} = .9$ and $P_2 = \frac{10}{100} = .1$. First we use the change-of-base rule to find $\log_2 .9$ and $\log_2 .1$. (We use natural logarithms, but common logarithms could be used instead.)

$$\log_2 .9 = \left(\frac{\ln .9}{\ln 2}\right) \approx -.152 \quad \text{and} \quad \log_2 .1 = \left(\frac{\ln .1}{\ln 2}\right)x \approx 3.32$$

Now we substitute to find $H$.

$$\begin{aligned} H &= -(.9 \log_2 .9 + .1 \log_2 .1) \\ &\approx -.9(-.152) - .1(-3.32) \\ &\approx .469 \end{aligned}$$

The interpretation of this index varies depending on the number of species involved. For the case with two species, if they are equally distributed, the measure of diversity is 1. If there is little diversity, $H$ is close to 0. In this example, since $H \approx .5$, there is neither great nor little diversity.

## 5.3 EXERCISES

*For each statement, write an equivalent statement in logarithmic form.*

**1.** $3^4 = 81$

**2.** $2^5 = 32$

**3.** $\left(\frac{1}{2}\right)^{-4} = 16$

**4.** $\left(\frac{2}{3}\right)^{-3} = \frac{27}{8}$

**5.** $10^{-4} = .0001$

**6.** $\left(\frac{1}{100}\right)^{-2} = 10{,}000$

**7.** $e^0 = 1$

**8.** $e^{1/3} = \sqrt[3]{e}$

*For each statement, write an equivalent statement in exponential form.*

**9.** $\log_6 36 = 2$

**10.** $\log_5 5 = 1$

**11.** $\log_{\sqrt{3}} 81 = 8$

**12.** $\log_4 \frac{1}{64} = -3$

**13.** $\log_{10} .001 = -3$

**14.** $\log_3 \sqrt[3]{9} = \frac{2}{3}$

**15.** $\log \sqrt{10} = .5$

**16.** $\ln e^6 = 6$

**17.** Explain in your own words what $\log_a x$ means.

**18.** In the expression $\log_a x$, why can't $x$ be 0? Why can't $x$ be negative?

*Solve each equation by first rewriting it in exponential form.*

**19.** $\log_5 125 = x$

**20.** $\log_3 81 = x$

**21.** $\log_x 3^{12} = 24$

**22.** $\log_x 25 = 6$

**23.** $\log_6 x = -3$

**24.** $\log_4 x = -\frac{1}{6}$

**25.** $\log_x 16 = \frac{4}{3}$

**26.** $\log_{16/25} x = -\frac{3}{2}$

**27.** $\log_x .001 = -3$

**28.** $\log_3(x - 1) = 2$

**29.** $\log_9 \frac{\sqrt[4]{27}}{3} = x$

**30.** $\log_{1/4} \frac{16^2}{2^{-3}} = x$

*Concept Check* *Simplify each expression.*

**31.** **(a)** $3^{\log_3 7}$ **(b)** $4^{\log_4 9}$ **(c)** $12^{\log_{12} 4}$ **(d)** $a^{\log_a k}$ $(k > 0, a > 0, a \neq 1)$

**32.** **(a)** $\log_3 3^{19}$ **(b)** $\log_4 4^{17}$ **(c)** $\log_{12} 12^{1/3}$ **(d)** $\log_a \sqrt{a}$ $(a > 0, a \neq 1)$

**33.** **(a)** $\log_3 1$ **(b)** $\log_4 1$ **(c)** $\log_{12} 1$ **(d)** $\log_a 1$ $(a > 0, a \neq 1)$

*Concept Check* *Evaluate each expression. Do not use a calculator.*

**34.** $\log 10^{1.5}$ **35.** $\log 10^{4.3}$ **36.** $\log 10^{2/5}$ **37.** $\log 10^{\sqrt{3}}$ **38.** $\ln e^{2/3}$

**39.** $\ln e^{.5}$ **40.** $\ln e^{\pi}$ **41.** $\ln e^{\sqrt{6}}$ **42.** $3 \ln e^{1.8}$

*Use a calculator to find a decimal approximation for each common or natural logarithm.*

**43.** $\log 43$ **44.** $\log 1247$ **45.** $\log .783$ **46.** $\log .014$

**47.** $\log 28^3$ **48.** $\log(47 \times 93)$ **49.** $\ln 43$ **50.** $\ln 1247$

**51.** $\ln .783$ **52.** $\ln .014$ **53.** $\ln 28^3$ **54.** $\ln(47 \times 93)$

**55.** **(a)** Use a calculator to find a decimal approximation of each common logarithm: log 2.367, log 23.67, log 236.7, log 2367.

**(b)** Write each number in scientific notation: 2.367, 23.67, 236.7, 2367.

**(c)** Compare the results in part (a) to the expressions in part (b). What similarities and differences do you find?

*Refer to Example 3. For each substance, find the pH from the given hydronium ion* $[H_3O^+]$ *concentration.*

**56.** Grapefruit, $6.3 \times 10^{-4}$ **57.** Limes, $1.6 \times 10^{-2}$

**58.** Crackers, $3.9 \times 10^{-9}$ **59.** Sodium hydroxide (lye), $3.2 \times 10^{-14}$

*Refer to Example 3. Find the hydronium ion* $[H_3O^+]$ *concentration for each substance with the given pH.*

**60.** Soda pop, 2.7 **61.** Wine, 3.4 **62.** Beer, 4.8 **63.** Drinking water, 6.5

*Suppose that \$2500 is invested in an account that pays interest compounded continuously. Find the amount of time that it would take for the account to grow to the given amount at the given rate of interest.*

**64.** \$3000, at 3.75% **65.** \$3500, at 4.25% **66.** \$5000, at 5% **67.** \$5000, at 6%

*Use the product, quotient, and power rules of logarithms to rewrite, if possible, each logarithm. Assume that all variables represent positive real numbers.*

**68.** $\log_3 \frac{2}{5}$ **69.** $\log_4 \frac{6}{7}$ **70.** $\log_2 \frac{6x}{y}$ **71.** $\log_3 \frac{4p}{q}$

**72.** $\log_5 \frac{5\sqrt{7}}{3m}$ **73.** $\log_2 \frac{2\sqrt{3}}{5p}$ **74.** $\log_4(2x + 5y)$ **75.** $\log_6(7m + 3q)$

**76.** $\log_k \frac{pq^2}{m}$ **77.** $\log_z \frac{x^5y^3}{3}$ **78.** $\log_m \sqrt{\frac{r^3}{5z^5}}$ **79.** $\log_p \sqrt[3]{\frac{m^5}{kt^2}}$

*Use the product, quotient, and power rules of logarithms to rewrite each expression as a single logarithm. Assume that all variables represent positive real numbers.*

**80.** $\log_a x + \log_a y - \log_a m$ **81.** $(\log_b k - \log_b m) - \log_b a$

**82.** $2 \log_m a - 3 \log_m b^2$ **83.** $\frac{1}{2} \log_y p^3q^4 - \frac{2}{3} \log_y p^4q^3$

**84.** $2 \log_a(z - 1) + \log_a(3z + 2), z > 1$ **85.** $\log_b(2y + 5) - \frac{1}{2} \log_b(y + 3)$

**86.** $-\frac{2}{3} \log_5 5m^2 + \frac{1}{2} \log_5 25m^2$ **87.** $-\frac{3}{4} \log_3 16p^4 - \frac{2}{3} \log_3 8p^3$

*Use the change-of-base rule to find an approximation for each logarithm.*

**88.** $\log_5 10$

**89.** $\log_9 12$

**90.** $\log_{15} 5$

**91.** $\log_{1/2} 3$

**92.** $\log_{100} 83$

**93.** $\log_{200} 175$

**94.** $\log_{2.9} 7.5$

**95.** $\log_{5.8} 12.7$

*(Modeling) Solve each problem.*

**96.** *Diversity of Species* Suppose a sample of a small community shows two species with 50 individuals each. Find the index of diversity $H$, where

$$H = -(P_1 \log_2 P_1 + P_2 \log_2 P_2 + \cdots + P_n \log_2 P_n).$$

**97.** *Diversity of Species* A virgin forest in northwestern Pennsylvania has four species of large trees with the following proportions of each: hemlock, .521; beech, .324; birch, .081; maple, .074. Find the index of diversity $H$, using the formula in Exercise 96.

**98.** *Diversity of Species* The number of species in a sample is approximated by

$$S(n) = a \ln\left(1 + \frac{n}{a}\right).$$

Here $n$ is the number of individuals in the sample, and $a$ is a constant that indicates the diversity of species in the community. If $a = .36$, find $S(n)$ for each value of $n$. (*Hint:* $S(n)$ must be a whole number.)

**(a)** 100 **(b)** 200 **(c)** 150 **(d)** 10

## RELATING CONCEPTS

**For individual or group investigation (Exercises 99–104)**

*Work Exercises 99–104 in order.*

**99.** Use the terminology of Chapter 2 to explain how the graph of $y = -3^x + 7$ can be obtained from the graph of $y = 3^x$.

**100.** Graph $y_1 = 3^x$ and $y_2 = -3^x + 7$ in the window $[-5, 5]$ by $[-10, 10]$ to support your answer in Exercise 99.

**101.** Use the capabilities of your calculator to find an approximation for the $x$-intercept of the graph of $y_2$ in Exercise 100.

**102.** Solve $0 = -3^x + 7$ for $x$, expressing $x$ in terms of base 3 logarithms.

**103.** Use the change-of-base rule to find an approximation for the solution of the equation in Exercise 102.

**104.** Compare your results in Exercises 101 and 103.

**105.** Prove the quotient rule of logarithms.

**106.** Prove the power rule of logarithms.

## Reviewing Basic Concepts (Sections 5.1–5.3)

**1.** Is the function defined by the table a one-to-one function? Explain your answer.

| $x$ | −2 | 2 | 4 | 6 | 8 |
|---|---|---|---|---|---|
| $y$ | 4 | 4 | 8 | 12 | 16 |

**2.** Give the inverse of each function defined.

**(a)**

| $x$ | 7 | 8 | 9 | 10 |
|---|---|---|---|---|
| $y$ | 12 | 21 | 32 | 45 |

**(b)** $y = \dfrac{x + 5}{4}$

**3.** Graph $f(x) = 2x + 3$ and $f^{-1}(x)$ by hand on the same axes.

**4.** Graph $f(x) = 3^{-x}$ by hand.

**5.** Solve $4^{2x} = 8$ analytically. Use a calculator to support your answer.

**6.** Find the interest earned on \$600 at 4% compounded quarterly for 3 years.

**7.** Evaluate each logarithm without using a calculator.

**(a)** $\log \dfrac{1}{\sqrt{10}}$ **(b)** $2 \ln e^{1.5}$ **(c)** $\log_2 4$

**8.** Use the properties of logarithms to rewrite $\log \dfrac{3x^2}{5y}$.

**9.** Use the properties of logarithms to write

$$\ln 4 + \ln x - 3 \ln 2$$

as a single logarithm.

**10.** How long will it take for \$1500 deposited at 4.5% compounded continuously to grow to \$1650? Give the answer to the nearest tenth of a year.

# 5.4 LOGARITHMIC FUNCTIONS

Graphs of Logarithmic Functions • Applying Earlier Work to Logarithmic Functions • A Logarithmic Model

The exponential function defined by $f(x) = a^x$, $a > 1$, is increasing on its entire domain. If $0 < a < 1$, the function is decreasing on its entire domain. Therefore, for all allowable bases $a$ of $f(x) = a^x$, the graph passes the horizontal line test and is one-to-one, so $f$ has an inverse.

We can find the rule for $f^{-1}$ analytically, using the steps described in Section 5.1.

$$f(x) = a^x \quad \text{Exponential function}$$
$$y = a^x \quad \text{Replace } f(x) \text{ with } y.$$
$$x = a^y \quad \text{Interchange } x \text{ and } y.$$
$$y = \log_a x \quad \text{Write in logarithmic form.}$$
$$f^{-1}(x) = \log_a x \quad y = f^{-1}(x)$$

This final equation indicates that the **logarithmic function** with base $a$ is the inverse of the exponential function with base $a$. To confirm this, use results from Section 5.3 to show that $(f \circ f^{-1})(x) = x$ and $(f^{-1} \circ f)(x) = x$.

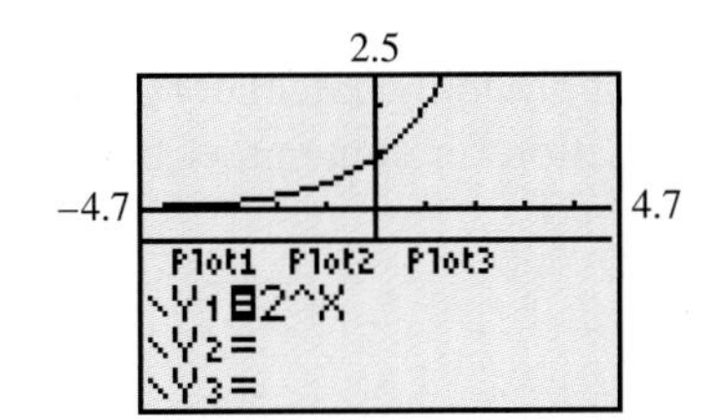


$$(f \circ f^{-1})(x) = f[f^{-1}(x)] = a^{\log_a x} = x$$
$$(f^{-1} \circ f)(x) = f^{-1}[f(x)] = \log_a a^x = x$$

Thus, the functions defined by

$$\boldsymbol{f(x) = a^x \quad \text{and} \quad g(x) = \log_a x}$$

are inverse functions.

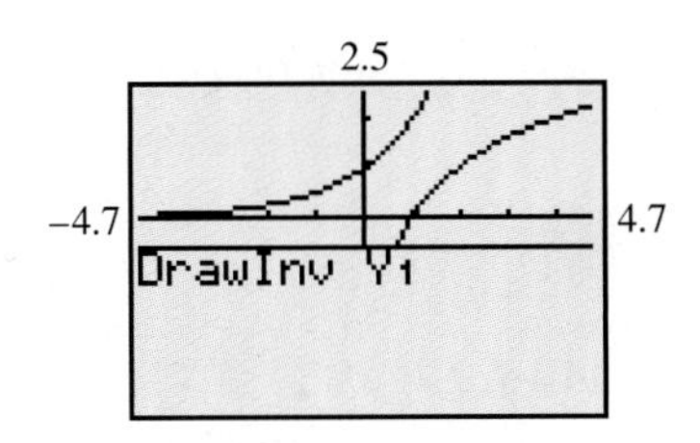


When directed to draw the inverse of $Y_1 = 2^x$, the calculator graphs $y = \log_2 x$.

## Graphs of Logarithmic Functions

Recall from Section 5.1 that the graph of the inverse of a one-to-one function can be obtained by reflecting the graph of the function across the line $y = x$. Figure 32 shows traditional and calculator graphs of the inverse functions $y = 2^x$ and $y = \log_2 x$. These are typical shapes for such graphs where $a > 1$.

| $x$ | $2^x$ |
|---|---|
| −2 | .25 |
| −1 | .5 |
| 0 | 1 |
| 1 | 2 |
| 2 | 4 |

| $x$ | $\log_2 x$ |
|---|---|
| .25 | −2 |
| .5 | −1 |
| 1 | 0 |
| 2 | 1 |
| 4 | 2 |

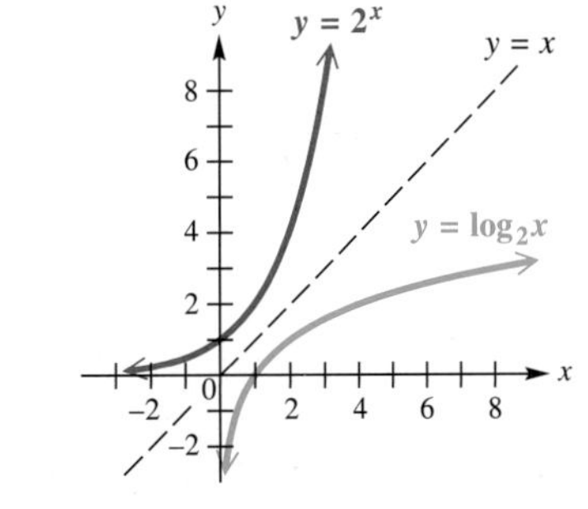


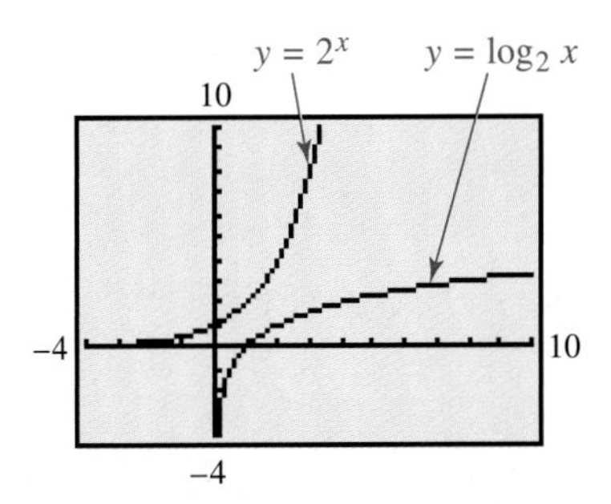


**FIGURE 32**

**TECHNOLOGY NOTE**

In Section 5.2 we saw that it is difficult to interpret from a calculator screen the behavior of the graph of an exponential function as it approaches its horizontal asymptote. A similar problem occurs for graphs of logarithmic functions near their vertical asymptotes. In Figures 34 and 35, be aware that there are no endpoints and that as $x \to 0$ from the right, $y \to -\infty$.

Figure 33 shows traditional and calculator graphs of $y = \left(\frac{1}{2}\right)^x$ and $y = \log_{1/2} x$. These are typical shapes for such graphs where $0 < a < 1$.

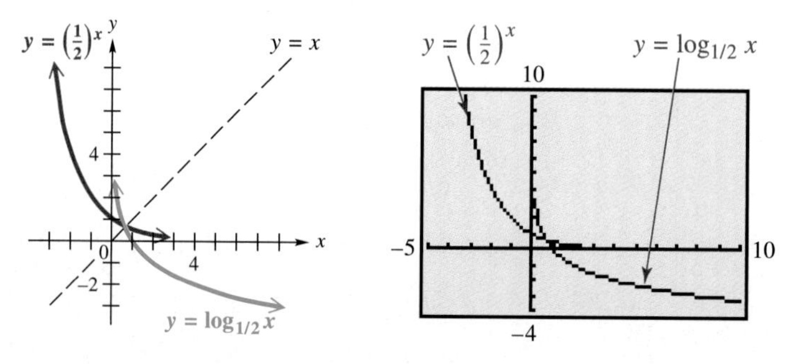


**FIGURE 33**

The most important logarithmic functions are $y = \ln x$ (base $e$) and $y = \log x$ (base 10). These are easily graphed on a graphing calculator, as seen in Figures 34 and 35.

Because $e$ and 10 are the only logarithmic bases on most graphing calculators, we must use the change-of-base rule to graph a logarithmic function for some other base. For example, to graph $y = \log_2 x$, we graph either

$$y = \frac{\log x}{\log 2} \quad \text{or} \quad y = \frac{\ln x}{\ln 2}.$$

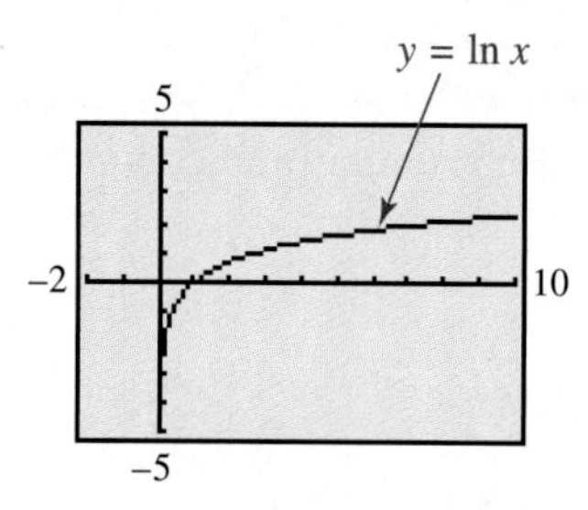


**FIGURE 34**

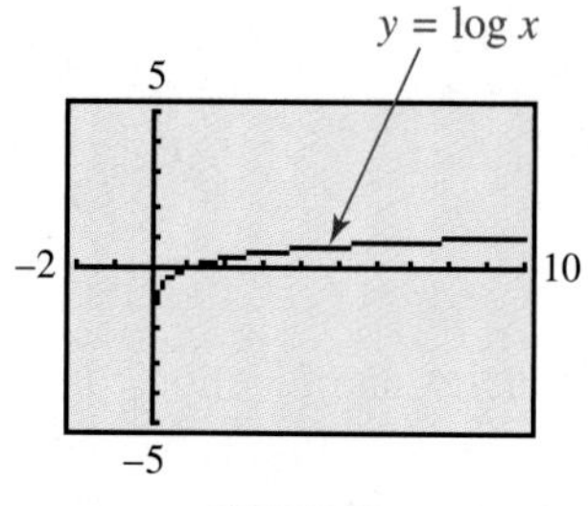


**FIGURE 35**

We now summarize information about the graphs of logarithmic functions.

**FUNCTION CAPSULE**

**LOGARITHMIC FUNCTION** $\quad f(x) = \log_a x, \quad a > 1$

Domain: $(0, \infty)$ $\qquad$ Range: $(-\infty, \infty)$

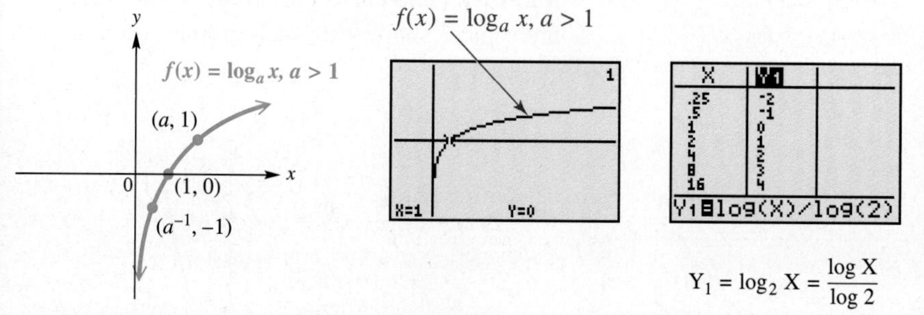

$f(x) = \log_a x, a > 1$

$(a, 1)$, $(1, 0)$, $(a^{-1}, -1)$

$Y_1 = \log_2 X = \frac{\log X}{\log 2}$

**FIGURE 36**

- $f(x) = \log_a x$, $a > 1$, is increasing and continuous on its entire domain, $(0, \infty)$.
- The $y$-axis is a vertical asymptote as $x \to 0$ from the right.
- The graph goes through the points $(a^{-1}, -1)$, $(1, 0)$, and $(a, 1)$.

FUNCTION CAPSULE

**LOGARITHMIC FUNCTION** $f(x) = \log_a x, \quad 0 < a < 1$

Domain: $(0, \infty)$ Range: $(-\infty, \infty)$

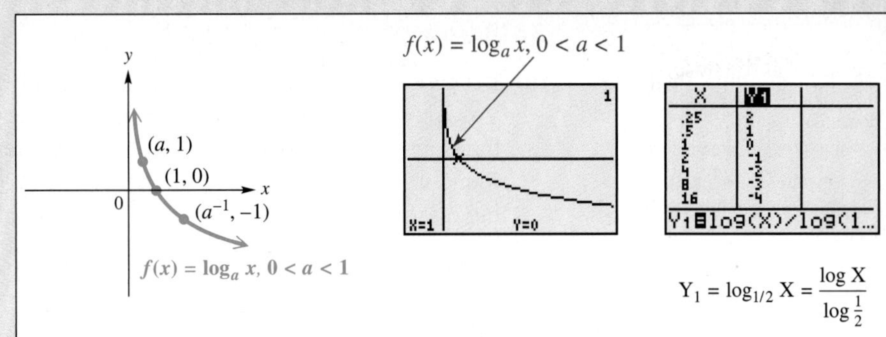


FIGURE 37

- $f(x) = \log_a x$, $0 < a < 1$, is decreasing and continuous on its entire domain, $(0, \infty)$.
- The $y$-axis is a vertical asymptote as $x \to 0$ from the right.
- The graph goes through the points $(a, 1)$, $(1, 0)$, and $(a^{-1}, -1)$.

A function of the form $y = \log_a f(x)$ is defined only for values for which $f(x) > 0$.

**EXAMPLE 1** *Determining Domains of Logarithmic Functions*

Find the domain of each function.

**(a)** $f(x) = \log_2(x - 1)$ **(b)** $f(x) = \log_3 x - 1$

**(c)** $f(x) = \log_3|x|$ **(d)** $f(x) = \ln(x^2 - 4)$

**Solution**

**(a)** Since the argument of a logarithm must be positive, $x - 1 > 0$, or $x > 1$. The domain is $(1, \infty)$.

**(b)** Since we are interested in the logarithm to the base 3 of $x$ (not $x - 1$), we must have $x > 0$. The domain is $(0, \infty)$.

**(c)** $|x| > 0$ is true for all real numbers $x$ except 0. Therefore, the domain is $(-\infty, 0) \cup (0, \infty)$.

**(d)** To solve $x^2 - 4 > 0$ analytically, we use a sign graph, introduced in Section 3.3. We factor $x^2 - 4$ as $(x + 2)(x - 2)$ and then determine the signs of the factors in the intervals $(-\infty, -2)$, $(-2, 2)$, and $(2, \infty)$. See Figure 38. The product $(x + 2)(x - 2)$ is positive in the intervals $(-\infty, -2)$ and $(2, \infty)$. The domain is the union of these two intervals: $(-\infty, -2) \cup (2, \infty)$.

| X | Y1 |
|---|---|
| -2.1 | -.8916 |
| -2.01 | -3.216 |
| -2 | ERROR |
| 0 | ERROR |
| 2 | ERROR |
| 2.01 | -3.216 |
| 2.1 | -.8916 |

Y1=ln(X²-4)

This table numerically supports the result in Example 1(d).

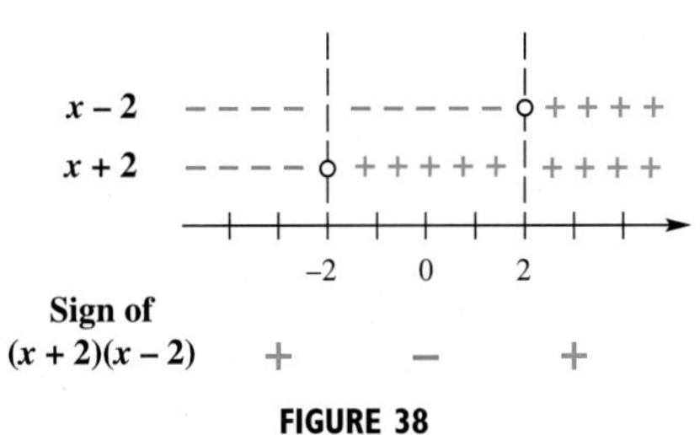


FIGURE 38

| X | $Y_1$ |
|---|---|
| -.5 | ERROR |
| 0 | ERROR |
| .999 | ERROR |
| 1 | ERROR |
| 1.001 | -9.966 |
| 2 | 0 |
| 3 | 1 |

Y1=log(X−1)/log...

This table numerically supports the result in Example 2(a): the domain of $y = \log_2(x - 1)$ (entered here as $Y_1 = \log(X - 1)/\log 2$) consists of real numbers greater than 1.

### EXAMPLE 2 *Graphing Translated Logarithmic Functions*

Graph each function. Give the domain, range, asymptote, and $x$-intercept, and tell whether it is increasing or decreasing on its domain.

**(a)** $y = \log_2(x - 1)$ **(b)** $y = \log_3 x - 1$

**Solution**

**(a)** Because the argument is $x - 1$, the graph of $y = \log_2(x - 1)$ is the graph of $y = \log_2 x$ shifted 1 unit to the right. The vertical asymptote also moves 1 unit to the right, so its equation is $x = 1$. The $x$-intercept is 2. The domain is $(1, \infty)$, as found in Example 1(a), and the range is $(-\infty, \infty)$. The function is always increasing on its domain. Figure 39 shows traditional and calculator graphs. (The calculator graph was obtained using $y = \frac{\log(x - 1)}{\log 2}$.)

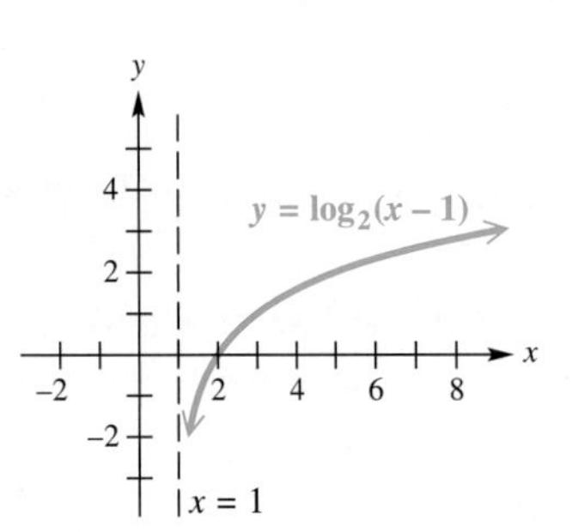


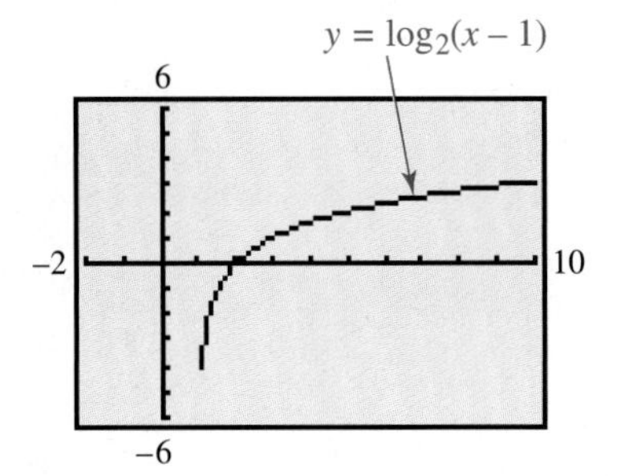


FIGURE 39

**(b)** Here, 1 is subtracted from $\log_3 x$, which causes the graph of $y = \log_3 x$ to be shifted downward 1 unit. The vertical asymptote, the $y$-axis or $x = 0$, is not affected. The $x$-intercept of $y = \log_3 x - 1$ is 3. The domain is $(0, \infty)$, from Example 1(b), and the range is $(-\infty, \infty)$. The function is always increasing. See Figure 40 for the graphs. (The calculator graph was obtained using $y = \frac{\log x}{\log 3} - 1$.)

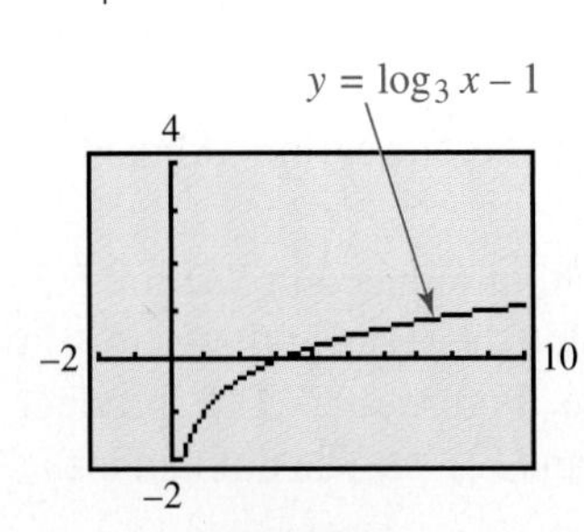


FIGURE 40

### EXAMPLE 3 *Determining Symmetry*

Show analytically that the graph of each function is symmetric with respect to the $y$-axis.

**(a)** $f(x) = \log_3|x|$ **(b)** $f(x) = \ln(x^2 - 4)$

**Solution**

**(a)** For all real numbers $x$, $|-x| = |x|$. Therefore,

$$\begin{aligned} f(-x) &= \log_3|-x| \\ &= \log_3|x| \\ &= f(x). \end{aligned}$$

Because $f(-x) = f(x)$, the graph is symmetric with respect to the $y$-axis. The graph consists of the two parts shown in Figure 41 on the next page, and the $y$-axis serves as both vertical asymptote and axis of symmetry. As we saw in Example 1(c), the domain is $(-\infty, 0) \cup (0, \infty)$.

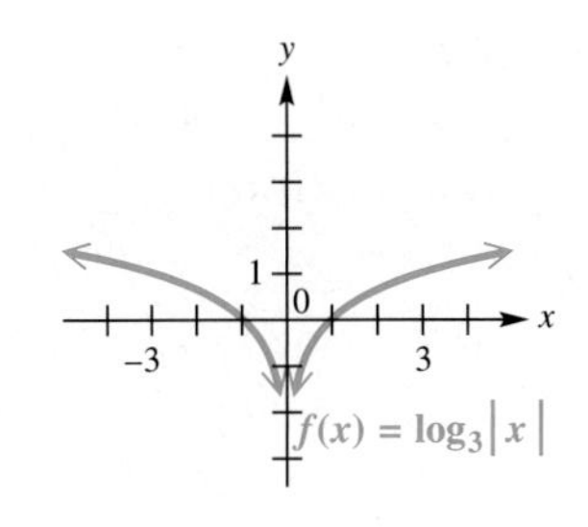


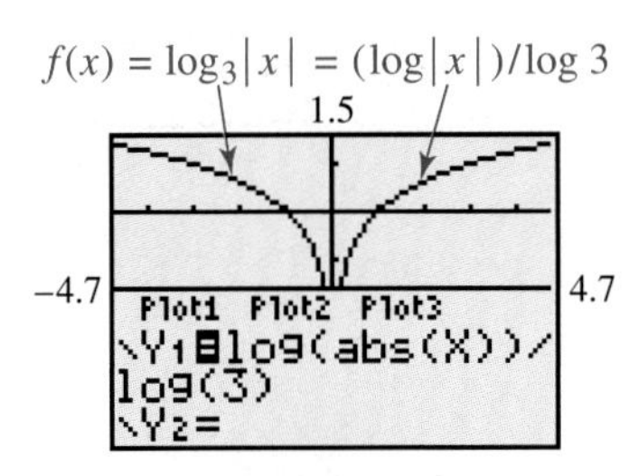


**FIGURE 41**

**(b)** Since $x^2 = (-x)^2$ for all $x$,

$$\ln(x^2 - 4) = \ln[(-x)^2 - 4].$$

Thus, $f(-x) = f(x)$ and the graph is symmetric with respect to the $y$-axis, which is again both vertical asymptote and axis of symmetry. The graph illustrates the domain $(-\infty, -2) \cup (2, \infty)$ from Example 1(d). See Figure 42.

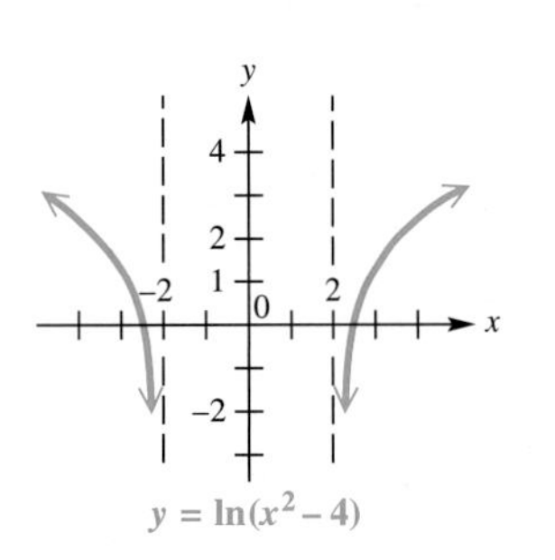


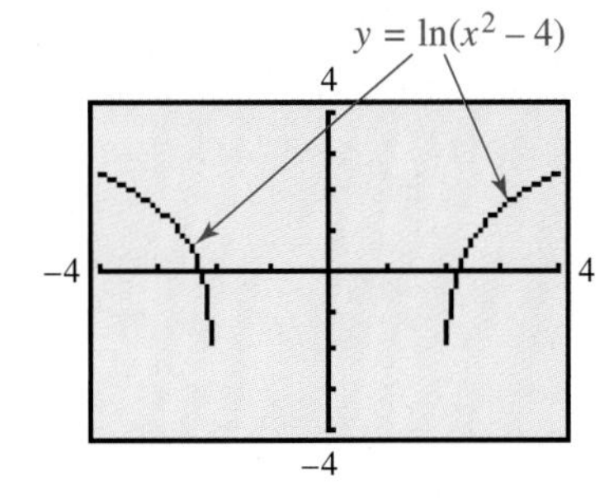


**FIGURE 42**

## FOR DISCUSSION

1. Use a calculator to evaluate $f(2.1), f(2.01), f(2.001)$, and $f(2.0001)$ for $f(x) = \ln(x^2 - 4)$. Describe what happens as $x \to 2$ from the right. How does this relate to the discussion in the Caution?
2. Repeat Item 1 for $f(-2.1)$, $f(-2.01), f(-2.001)$, and $f(-2.0001)$.

**CAUTION** The calculator graphs in Figures 41 and 42 are easy to misinterpret by thinking that there are "endpoints" on the two branches. In Figure 42, the lines $x = -2$ and $x = 2$ are vertical asymptotes, and as $x \to -2$ from the left, $f(x) \to -\infty$, while as $x \to 2$ from the right, $f(x) \to -\infty$. Here, we see another reason for learning the concepts. Calculator graphs are limited in their resolution and may not show important details in certain windows. A calculator table with a small $\Delta x$ often clarifies any misinterpretation.

## What Went WRONG?

A student wanted to support the power rule for logarithms (Property 6 from Section 5.3): $\log_a x^r = r \log_a x$. The student defined $Y_1$ and $Y_2$, as shown in the screen on the left, and expected the two graphs to be the same. However, the graph of $Y_1$, seen in the middle screen, was different from the graph of $Y_2$, seen in the screen on the right.

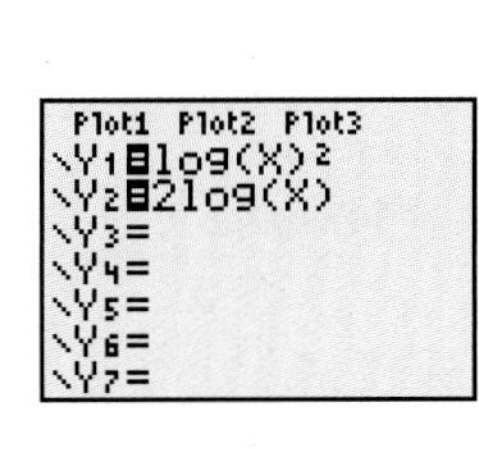


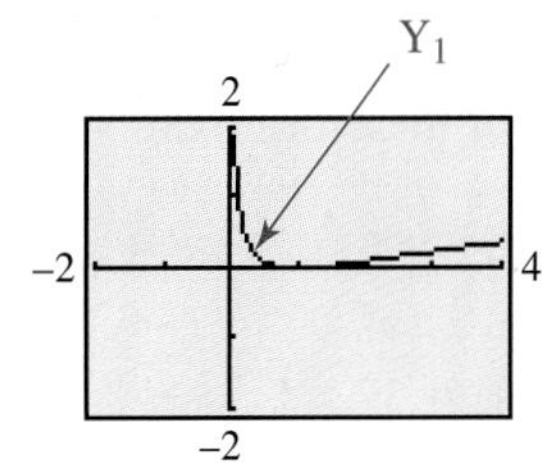


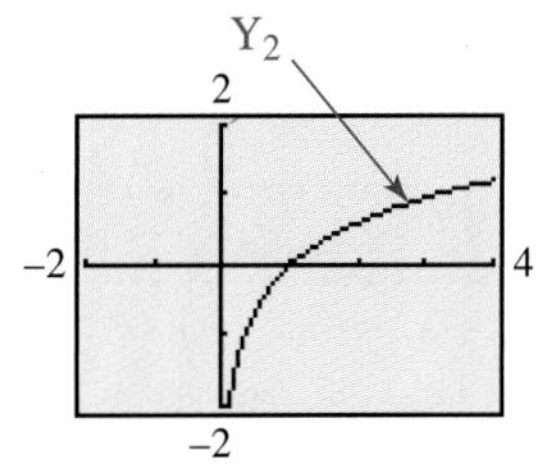


What Went Wrong? How can the student change the input for $Y_1$ to obtain the same graph as that of $Y_2$?

---

*Answer to What Went Wrong?*

With some calculators, the first equation should be entered as $Y_1 = (\log(X^2))(X > 0)$.

## Applying Earlier Work to Logarithmic Functions

**EXAMPLE 4** *Using a Property of Logarithms to Translate a Graph*

Describe how to graph $f(x) = \log 8x$ by translation.

**Solution** By the product rule for logarithms, $\log 8x = \log 8 + \log x$, so we can obtain the graph of $y_2 = \log 8x$ by shifting the graph of $y_1 = \log x$ upward $\log 8 \approx .90309$ unit. Figure 43 shows the graph.

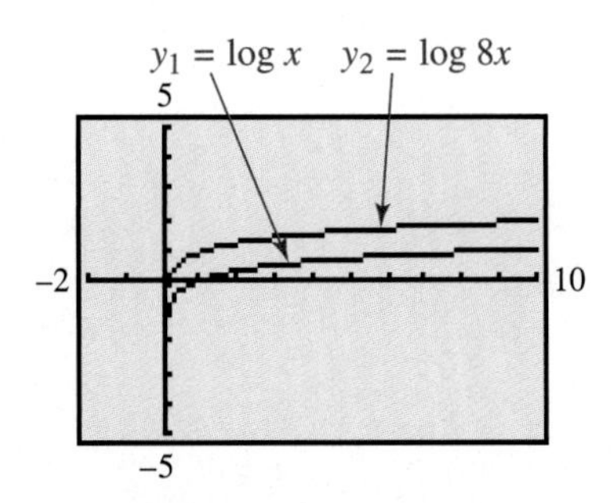


The vertical distance between $y_2$ and $y_1$ is log 8.

**FIGURE 43**

**EXAMPLE 5** *Finding the Inverse of an Exponential Function*

Find the inverse of the function defined by $f(x) = -2^x + 3$ analytically. Discuss the relationship between the two graphs.

**Solution** The function is one-to-one because it is a reflection and translation of the one-to-one function defined by $y = 2^x$. Find the equation of the inverse function.

$$f(x) = -2^x + 3$$

$$y = -2^x + 3 \qquad \text{Replace } f(x) \text{ with } y.$$

$$x = -2^y + 3 \qquad \text{Interchange } x \text{ and } y.$$

$$2^y = -x + 3 \qquad \text{Add } 2^y\text{; subtract } x.$$

$$y = \log_2(-x + 3) \qquad \text{Write in logarithmic form.}$$

$$f^{-1}(x) = \log_2(-x + 3) \qquad \text{Replace } y \text{ with } f^{-1}(x).$$

$f^{-1}(x) = \log_2(-x + 3)$
3.1
−4.7
4.7
−3.1
$f(x) = -2^x + 3$

**FIGURE 44**

Figure 44 shows the graphs of both $f$ and $f^{-1}$. The table compares some of the features of these inverses. Notice in the table how the roles of $x$ and $y$ are reversed in $f$ and $f^{-1}$.

| | Domain | Range | $x$-intercept | $y$-intercept | Asymptote |
|---|---|---|---|---|---|
| $f(x) = -2^x + 3$ | $(-\infty, \infty)$ | $(-\infty, 3)$ | $\log_2 3 \approx 1.58$ | 2 | Horizontal: $y = 3$ |
| $f^{-1}(x) = \log_2(-x + 3)$ | $(-\infty, 3)$ | $(-\infty, \infty)$ | 2 | $\log_2 3 \approx 1.58$ | Vertical: $x = 3$ |

## A Logarithmic Model

**EXAMPLE 6** *Modeling Drug Concentration*

The concentration of a drug injected into the bloodstream decreases with time. The intervals of time in hours when the drug should be administered are given by

$$T = \frac{1}{k} \ln \frac{C_2}{C_1},$$

where $k$ is a constant determined by the drug in use, $C_2$ is the concentration at which the drug is harmful, and $C_1$ is the concentration below which the drug is ineffective. (*Source:* Horelick, B. and S. Koont, "Applications of Calculus to Medicine: Prescribing Safe and Effective Dosage," *UMAP Module 202,* 1977.) Thus, if $T = 4$, the drug should be administered every 4 hours.

For a certain drug, $k = \frac{1}{3}$, $C_2 = 5$, and $C_1 = 2$. How often should the drug be administered?

**Solution** Substitute the given values in the equation of the function.

$$\begin{aligned} T &= \frac{1}{k} \ln \frac{C_2}{C_1} \\ &= \frac{1}{\frac{1}{3}} \ln \frac{5}{2} \\ &\approx 2.75 \end{aligned}$$

The drug should be given about every 2 hours, 45 minutes.

## 5.4 EXERCISES

*The graph of an exponential function $f$ is given, with three points labeled. Sketch the graph of $f^{-1}$ by hand, labeling three points on its graph. For $f^{-1}$, also state the domain, the range, whether it increases or decreases on its domain, and the equation of its vertical asymptote.*

**1.**

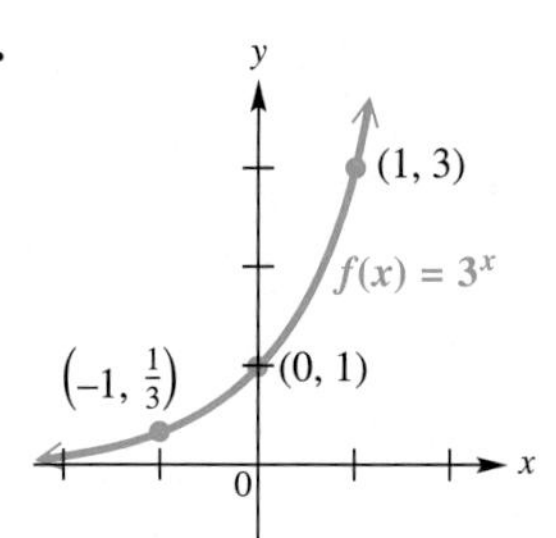


**2.**

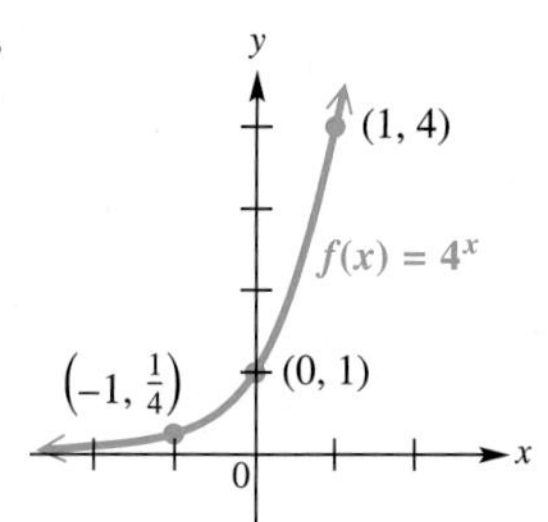


**3.**

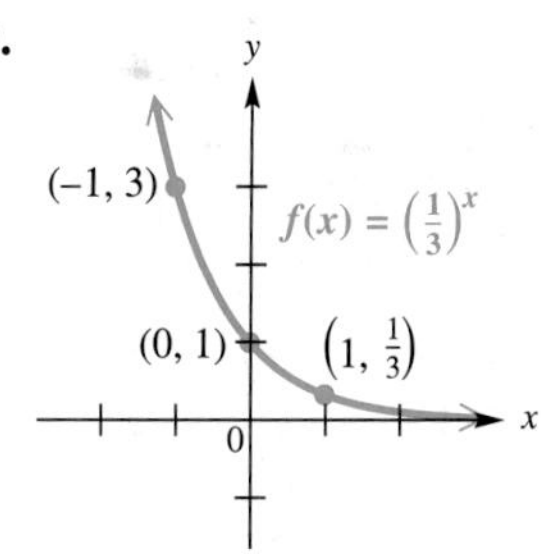


**4.**

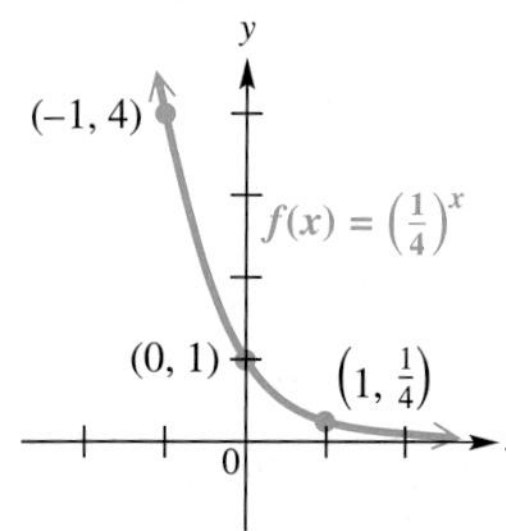


**5.**

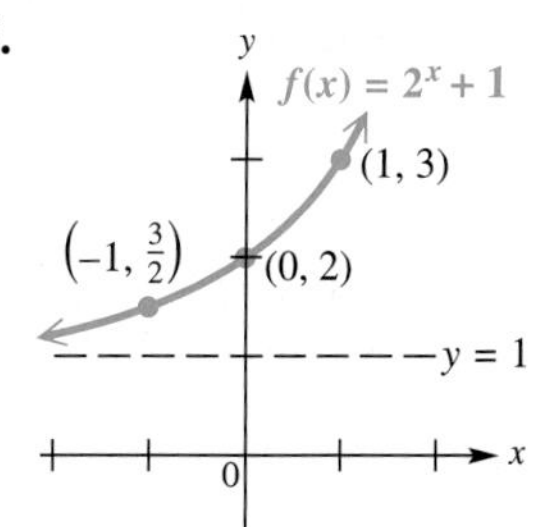


**6.**

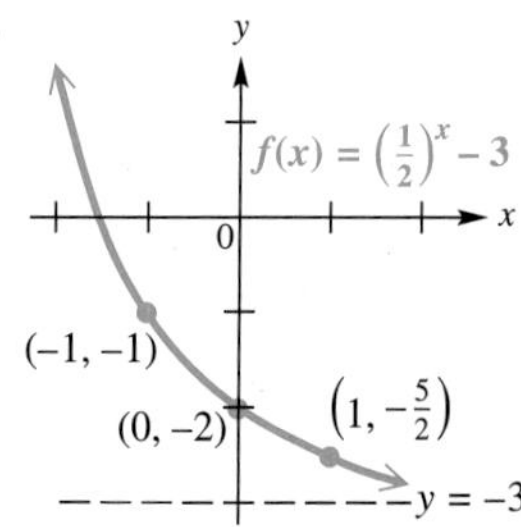


**7.** In Exercises 1–6, each function $f$ is an exponential function. Therefore, each function $f^{-1}$ is a(n) ________ function.

**8.** Compare the characteristics of the graph of $f(x) = \log_a x$ with those of the graph of $f(x) = a^x$ in Section 5.2. Make a list of characteristics that reinforce the idea that these are inverse functions.

*Find the domain of each logarithmic function analytically. You may wish to support your answer graphically.*

**9.** $y = \log 2x$

**10.** $y = \log \dfrac{x}{3}$

**11.** $y = \ln(x^2 + 7)$

**12.** $y = \ln(-x^2 - 4)$

**13.** $y = \log_4(x^2 - 4x - 21)$

**14.** $y = \log_6(2x^2 - 7x - 4)$

**15.** $y = \log(x^3 - x)$

**16.** $y = \log\left(\dfrac{x + 3}{x - 4}\right)$

*Sketch the graph of* $f(x) = \log_2 x$. *Then refer to it and use the techniques of Chapter 2 to graph each function.*

**17.** $f(x) = (\log_2 x) + 3$

**18.** $f(x) = \log_2(x + 3)$

**19.** $f(x) = |\log_2(x + 3)|$

*Sketch the graph of* $f(x) = \log_{1/2} x$. *Then refer to it and use the techniques of Chapter 2 to graph each function.*

**20.** $f(x) = (\log_{1/2} x) - 2$

**21.** $f(x) = \log_{1/2}(x - 2)$

**22.** $f(x) = |\log_{1/2}(x - 2)|$

*Concept Check* *In Exercises 23–30, match the correct graph to each equation, using your knowledge of graphs rather than your calculator.*

**23.** $y = e^x + 3$

**24.** $y = e^x - 3$

**25.** $y = e^{x+3}$

**26.** $y = e^{x-3}$

**27.** $y = \ln x + 3$

**28.** $y = \ln x - 3$

**29.** $y = \ln(x - 3)$

**30.** $y = \ln(x + 3)$

**A.**

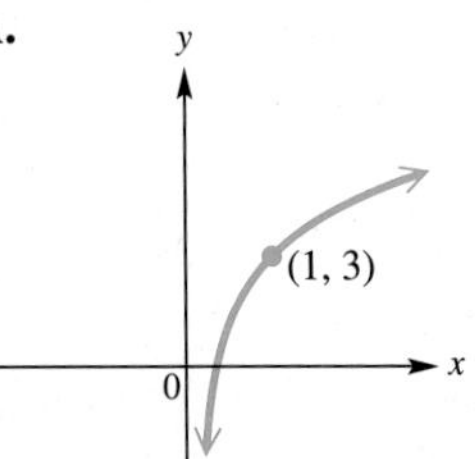


**B.**

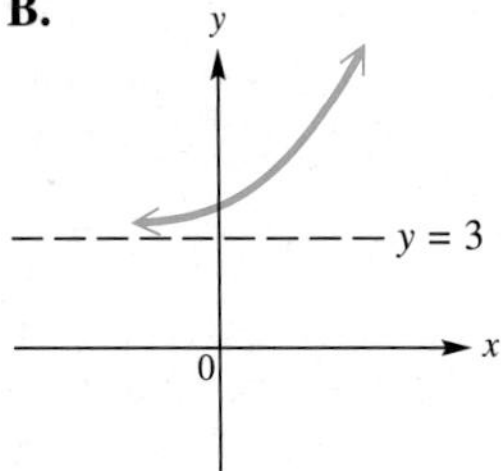


**C.**

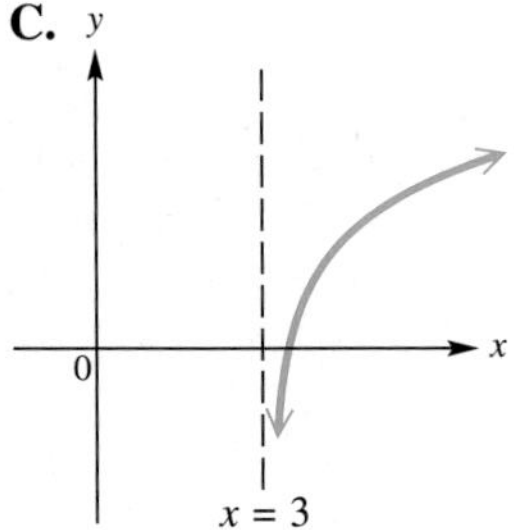


**D.**

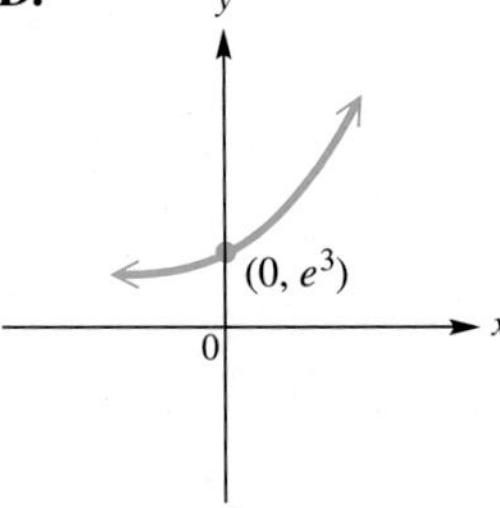


**E.**

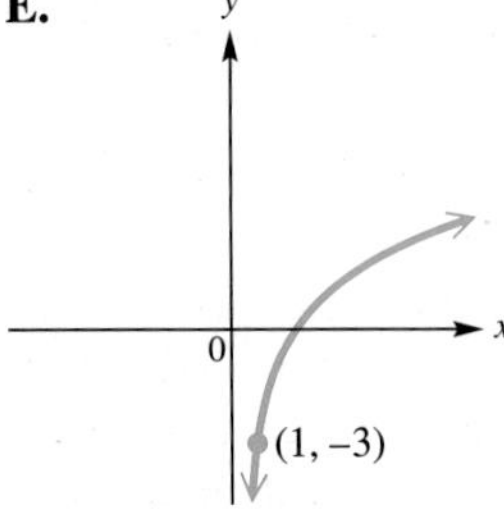


**F.**

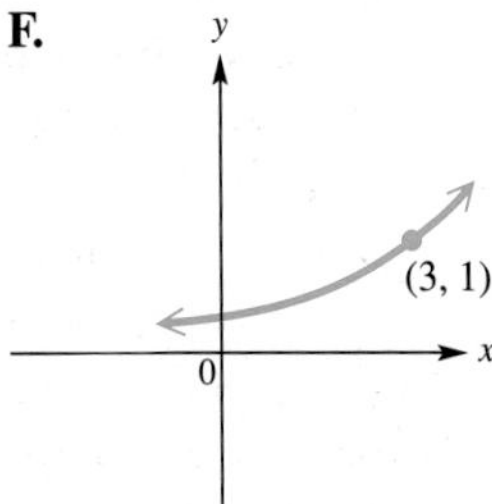


**G.**

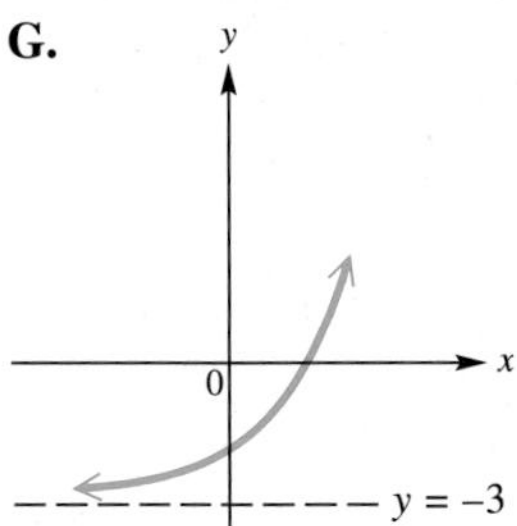


**H.**

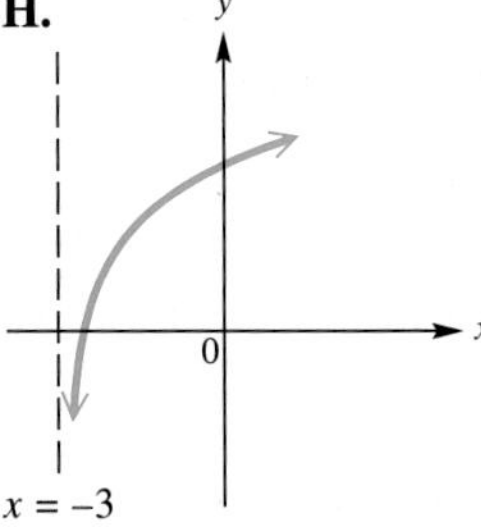


*Explain how the graph of the given function can be obtained from the graph of* $y = \log_2 x$.

**31.** $y = \log_2(x + 4)$

**32.** $y = \log_2(x - 6)$

**33.** $y = 3 \log_2 x + 1$

**34.** $y = -4 \log_2 x - 8$

**35.** $y = \log_2(-x) + 1$

**36.** $y = -\log_2(-x)$

## RELATING CONCEPTS

**For individual or group investigation (Exercises 37–42)**

*Work Exercises 37–42 in order. They apply to the function defined by* $f(x) = \log_4(2x^2 - x)$.

**37.** Determine analytically the domain of $f$.

**38.** Use the change-of-base rule to express, in terms of common logarithms, the equation you would use to graph $f$ on your calculator.

**39.** Give the equation of any vertical asymptotes of the graph of $f$.

**40.** Explain why the graph of $f$ has no $y$-intercept.

**41.** Graph $f$ in the window $[-2.5, 2.5]$ by $[-5, 2.5]$. Based on this graph, give the solution sets of $f(x) = 0$, $f(x) < 0$, and $f(x) > 0$.

**42.** Give the range of $f$.

**43.** Graph $y = \log x^2$ and $y = 2 \log x$ on separate viewing screens. It would seem, at first glance, that by applying the power rule for logarithms, these graphs should be the same. Are they? If not, why not? (*Hint:* Consider the domain in each case.)

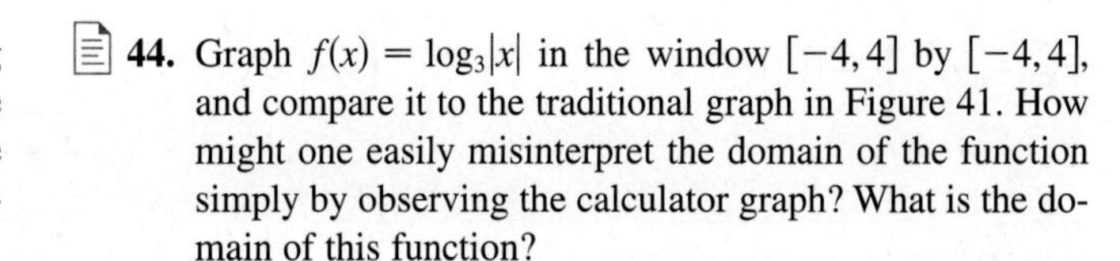

**44.** Graph $f(x) = \log_3|x|$ in the window $[-4, 4]$ by $[-4, 4]$, and compare it to the traditional graph in Figure 41. How might one easily misinterpret the domain of the function simply by observing the calculator graph? What is the domain of this function?

*Evaluate each logarithm in three ways:* **(a)** *Use the definition of logarithm in Section 5.3 to find the exact value analytically.* **(b)** *Support the result of part (a) using the change-of-base rule and common logarithms on your calculator.* **(c)** *Support the result of part (a) by locating the appropriate point on the graph of the function* $y = \log_a x$.

**45.** $\log_9 27$

**46.** $\log_4\left(\frac{1}{8}\right)$

**47.** $\log_{16}\left(\frac{1}{8}\right)$

**48.** $\log_2 \sqrt{8}$

*For each exponential function f, find* $f^{-1}$ *analytically, and graph both f and* $f^{-1}$ *in the same viewing window.*

**49.** $f(x) = 4^x - 3$

**50.** $f(x) = \left(\frac{1}{2}\right)^x - 5$

**51.** $f(x) = -10^x + 4$

**52.** $f(x) = -e^x + 6$

*(Modeling) Solve each problem.*

**53.** *Height of the Eiffel Tower* Paris's Eiffel Tower was constructed in 1889 to commemorate the one hundredth anniversary of the French Revolution. The right side of the Eiffel Tower has a shape that can be approximated by the graph of the function defined by

$$f(x) = -301 \ln \frac{x}{207}.$$

See the figure. (*Source:* Banks, Robert B., *Towing Icebergs, Falling Dominoes, and Other Adventures in Applied Mathematics,* Princeton University Press, 1998.)

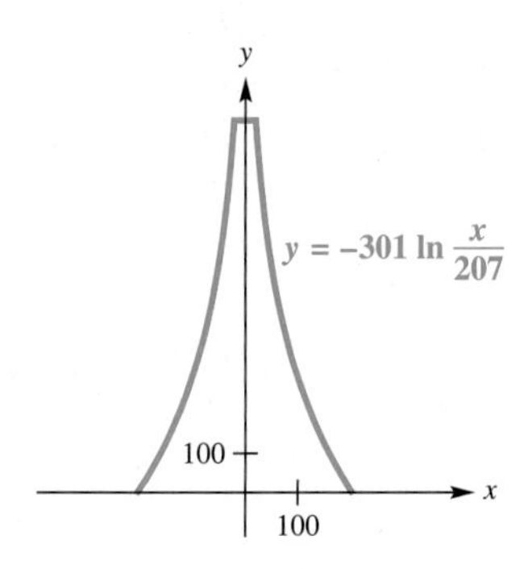


**(a)** Explain why the shape of the left side of the Eiffel Tower has the formula given by $f(-x)$.

**(b)** The short horizontal line at the top of the figure has length 15.7488 feet. Approximately how tall is the Eiffel Tower?

**(c)** Approximately how far from the center of the tower is the point on the right side that is 500 feet above the ground?

**54.** *Age of a Whale* The age in years of a female blue whale is approximated by

$$t = -2.57 \ln\left(\frac{87 - L}{63}\right),$$

where $L$ is its length in feet.

**(a)** How old is a female blue whale that measures 80 feet?

**(b)** The equation that defines $t$ has domain $24 < L < 87$. Explain why.

**55.** *Barometric Pressure* The function defined by

$$f(x) = 27 + 1.105 \log(x + 1)$$

approximates the barometric pressure in inches of mercury at a distance of $x$ miles from the eye of a typical hurricane. (*Source:* Miller, A. and R. Anthes, *Meteorology,* Fifth Edition, Charles E. Merrill, 1985.)

**(a)** Approximate the pressure 9 miles from the eye of the hurricane.

**(b)** The ordered pair $(99, 29.21)$ belongs to this function. What information does it convey?

**56.** *Race Speed* At the World Championship races held at Rome's Olympic Stadium in 1987, American sprinter Carl Lewis ran the 100-meter race in 9.86 seconds. His speed in meters per second after $t$ seconds is closely modeled by the function defined by

$$f(t) = 11.65(1 - e^{-t/1.27}).$$

(*Source:* Banks, Robert B., *Towing Icebergs, Falling Dominoes, and Other Adventures in Applied Mathematics,* Princeton University Press, 1998.)

**(a)** How fast was he running as he crossed the finish line?

**(b)** After how many seconds was he running at the rate of 10 meters per second?

# EXPONENTIAL AND LOGARITHMIC EQUATIONS AND INEQUALITIES

Exponential Equations and Inequalities (Type 2) • Logarithmic Equations and Inequalities • Equations and Inequalities Involving Both Exponentials and Logarithms • Formulas Involving Exponentials and Logarithms

In Section 5.2, we solved Type 1 exponential equations—those with both sides easily written as powers of the same rational number base. We solved simple logarithmic equations in Section 5.3. General methods for solving exponential and logarithmic equations depend on the following properties, which are based on the fact that exponential and logarithmic functions are, in general, one-to-one. Property 1 was used in Section 5.2 to solve Type 1 exponential equations.

**Properties of Logarithmic and Exponential Functions**

For $b > 0$ and $b \neq 1$,

**1.** $b^x = b^y$ **if and only if** $x = y$.

**2.** If $x > 0$ and $y > 0$, then $\log_b x = \log_b y$ **if and only if** $x = y$.

## Exponential Equations and Inequalities (Type 2)

A Type 2 exponential equation or inequality is one in which the exponential expressions cannot easily be written as powers of the same base. Examples of Type 2 equations are $7^x = 12$ and $2^{3x+1} = 3^{4-x}$. The general strategy in solving these equations is to use Property 2 by taking the same-base logarithm of each side (usually either common or natural) and then applying the power rule for logarithms to eliminate the variable exponents. Then, the equation is solved using familiar algebraic techniques.

**EXAMPLE 1** *Solving a Type 2 Exponential Equation*

Solve $7^x = 12$.

**Analytic Solution**

Property 1 cannot be used to solve this equation, so we apply Property 2. While any appropriate base $b$ can be used to apply Property 2, the most practical base to use is base 10 or base $e$. Taking base $e$ (natural) logarithms of each side gives

$$7^x = 12$$

$$\ln 7^x = \ln 12$$

$$x \ln 7 = \ln 12 \quad \text{Power rule}$$

$$x = \frac{\ln 12}{\ln 7}. \quad \text{Divide by } \ln 7.$$

The expression $\frac{\ln 12}{\ln 7}$ is the *exact* solution of $7^x = 12$. Had we used common logarithms, the solution would

**Graphing Calculator Solution**

Using the intersection-of-graphs method, we graph $y_1 = 7^x$ and $y_2 = 12$. The $x$-coordinate of the point of intersection is approximately 1.277, as seen in Figure 45. Figure 46 illustrates the $x$-intercept method of solution. The $x$-intercept of $y = 7^x - 12$ is also approximately 1.277.

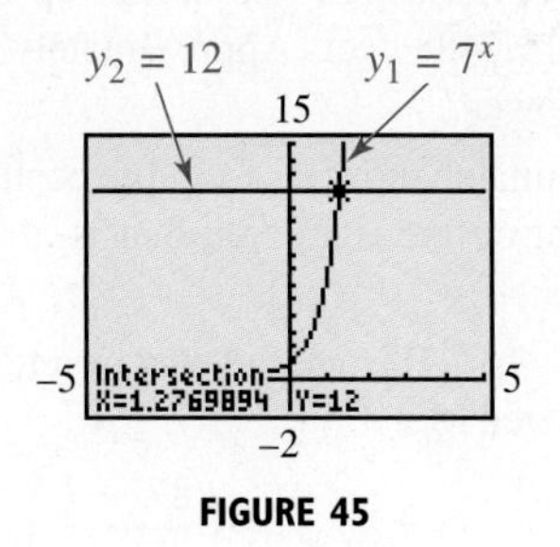


FIGURE 45

*(continued)*

have the form $\frac{\log 12}{\log 7}$. In either case, we use a calculator to find a decimal approximation for the exact solution.

$$\frac{\ln 12}{\ln 7} = \frac{\log 12}{\log 7} \approx 1.277 \qquad \text{Nearest thousandth}$$

The solution set with the exact solution is $\left\{\frac{\ln 12}{\ln 7}\right\}$ or $\left\{\frac{\log 12}{\log 7}\right\}$, while $\{1.277\}$ is an approximate solution set.

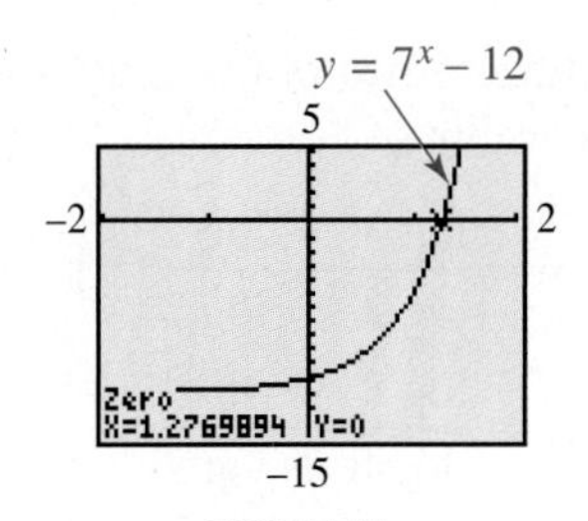


**FIGURE 46**

## EXAMPLE 2 *Solving Type 2 Exponential Inequalities*

```
7^(ln(12)/ln(7))
                12
7^(log(12)/log(7
))
                12
```

To check the analytic solution in Example 1, raise 7 to either of the powers $\frac{\ln 12}{\ln 7}$ or $\frac{\log 12}{\log 7}$, to obtain a result of 12.

**(a)** Use Figure 45 to solve $7^x < 12$. **(b)** Use Figure 46 to solve $7^x > 12$.

### Solution

**(a)** Because the graph of $y_1 = 7^x$ is below the graph of $y_2 = 12$ for all $x$-values less than $\frac{\ln 12}{\ln 7} \approx 1.277$, the solution set is

$$\left(-\infty, \frac{\ln 12}{\ln 7}\right) \quad \text{or} \quad (-\infty, 1.277).$$

**(b)** This inequality is equivalent to $7^x - 12 > 0$. Because the graph of $y = 7^x - 12$ is above the $x$-axis for all values of $x$ greater than $\frac{\ln 12}{\ln 7} \approx 1.277$, the solution set is

$$\left(\frac{\ln 12}{\ln 7}, \infty\right) \quad \text{or} \quad (1.277, \infty).$$

## EXAMPLE 3 *Solving a Type 2 Exponential Equation*

Solve $2^{3x+1} = 3^{4-x}$.

### Analytic Solution

Take logarithms on each side, use the power rule, and solve the resulting linear equation.

| | |
|---|---|
| $2^{3x+1} = 3^{4-x}$ | Given equation |
| $\log 2^{3x+1} = \log 3^{4-x}$ | Take common logarithms on each side. |
| $(3x + 1)\log 2 = (4 - x)\log 3$ | Power rule |
| $3x \log 2 + \log 2 = 4\log 3 - x\log 3$ | Distributive property |
| $3x\log 2 + x\log 3 = 4\log 3 - \log 2$ | Move all $x$-terms to one side. |
| $x(3\log 2 + \log 3) = 4\log 3 - \log 2$ | Factor out $x$. |
| $x = \frac{4\log 3 - \log 2}{3\log 2 + \log 3}$ | Divide by $3\log 2 + \log 3$. |
| $x \approx 1.165$ | Nearest thousandth |

### Graphing Calculator Solution

Locate the point of intersection of the graphs of

$$y_1 = 2^{3x+1}$$

and

$$y_2 = 3^{4-x},$$

and find the $x$-coordinate of the point. As seen in Figure 47 on the next page, the $x$-coordinate is approximately 1.165, supporting the analytic solution.

*(continued)*

A more compact form of the exact solution can be found by using properties of logarithms.

$$x = \frac{\log 3^4 - \log 2}{\log 2^3 + \log 3} \quad \text{Power rule}$$

$$x = \frac{\log 81 - \log 2}{\log 8 + \log 3} \quad 3^4 = 81;\ 2^3 = 8$$

$$x = \frac{\log \frac{81}{2}}{\log 24} \quad \text{Quotient and product rules}$$

The exact solution set is $\left\{\frac{\log(81/2)}{\log 24}\right\}$. A calculator approximation of the solution set is $\{1.165\}$.

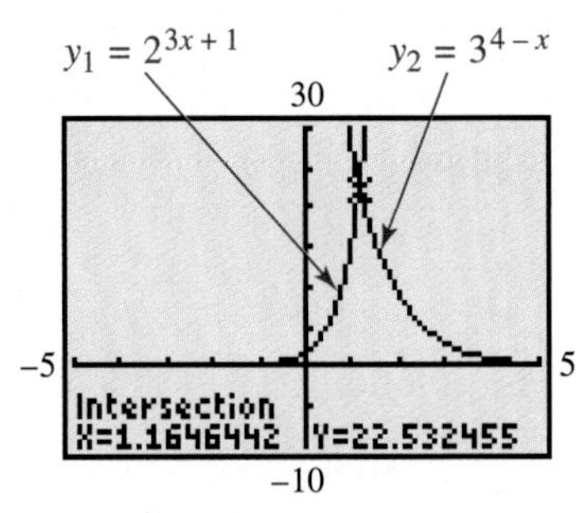


FIGURE 47

**FOR DISCUSSION**

1. Use Figure 47 to determine the solution sets of

$$2^{3x+1} < 3^{4-x} \quad \text{and} \quad 2^{3x+1} > 3^{4-x}.$$

2. Discuss the difference between the exact solution and an approximate solution of an exponential equation. In general, can you find an exact solution from a graph?

## Logarithmic Equations and Inequalities

The properties of logarithms given in Section 5.3 are useful here, as is Property 2 from this section. It is important to note the domain of the variable in the original logarithmic equation, so any extraneous values can be rejected.

**EXAMPLE 4** *Solving a Logarithmic Equation*

Solve $\log_3(x + 6) - \log_3(x + 2) = \log_3 x$.

**Analytic Solution**

The domain must satisfy $x + 6 > 0$, $x + 2 > 0$, and $x > 0$. The intersection of these inequalities gives the domain $(0, \infty)$. Using the quotient property of logarithms, we rewrite the equation as follows and solve.

$$\log_3 \frac{x + 6}{x + 2} = \log_3 x$$

$$\frac{x + 6}{x + 2} = x \quad \text{Property 2}$$

$$x + 6 = x(x + 2) \quad \text{Multiply by } x + 2.$$

$$x + 6 = x^2 + 2x \quad \text{Distributive property}$$

$$x^2 + x - 6 = 0 \quad \text{Standard form}$$

$$(x + 3)(x - 2) = 0 \quad \text{Factor.}$$

$$x = -3 \quad \text{or} \quad x = 2 \quad \text{Zero-product property}$$

**Graphing Calculator Solution**

Figure 48 shows that the $x$-coordinate of the point of intersection of the graphs of

$$y_1 = \log_3(x + 6) - \log_3(x + 2)$$

and

$$y_2 = \log_3 x$$

is 2, which agrees with our analytic solution. Notice that the graphs do not intersect when $x = -3$, supporting the conclusion in the analytic solution that $-3$ is an extraneous value.

*(continued)*

The negative solution ($x = -3$) is not in the domain of $\log_3 x$ in the original equation, so the only valid solution is the positive number 2, giving the solution set $\{2\}$.

To check, let $x = 2$ in the original equation.

$$\log_3(2 + 6) - \log_3(2 + 2) = \log_3 2 \quad ?$$

$$\log_3 \frac{8}{4} = \log_3 2 \quad ?$$

$$\log_3 2 = \log_3 2 \quad \text{True}$$

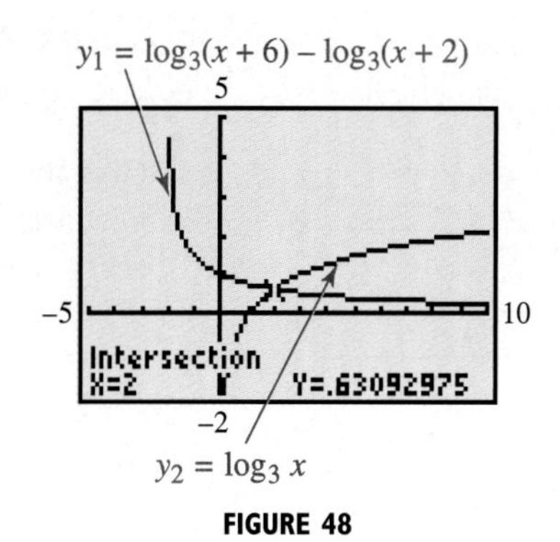


**FIGURE 48**

**NOTE** The equation in Example 4 could also have been solved by transforming so that all logarithmic terms appeared on one side. Then, by applying the properties of logarithms from Section 5.3 and rewriting in exponential form, the same solution set would result. This would also allow for solving graphically by using the $x$-intercept method.

## EXAMPLE 5 *Solving a Logarithmic Equation*

Solve $\log(3x + 2) + \log(x - 1) = 1$.

### Analytic Solution

Find the domain of $x$. For $3x + 2 > 0$, $x > -\frac{2}{3}$ and for $x - 1 > 0$, $x > 1$. Thus, the domain is $x > 1$ or $(1, \infty)$. Recall from Section 5.3 that $\log x$ means $\log_{10} x$.

$$\log(3x + 2) + \log(x - 1) = 1 \quad \text{Given equation}$$

$$\log[(3x + 2)(x - 1)] = 1 \quad \text{Product rule}$$

$$(3x + 2)(x - 1) = 10^1 \quad \text{Exponential form}$$

$$3x^2 - x - 2 = 10 \quad \text{Multiply.}$$

$$3x^2 - x - 12 = 0 \quad \text{Standard form}$$

Since this equation cannot be solved by factoring, use the quadratic formula with $a = 3$, $b = -1$, and $c = -12$ to get

$$x = \frac{1 \pm \sqrt{1 + 144}}{6} = \frac{1 \pm \sqrt{145}}{6}.$$

Since $\frac{1 - \sqrt{145}}{6}$ is less than 1, it is not in the domain and must be discarded, giving the solution set $\left\{\frac{1 + \sqrt{145}}{6}\right\}$.

### Graphing Calculator Solution

We arbitrarily choose to use the $x$-intercept method of solution, and graph

$$y = \log(3x + 2) + \log(x - 1) - 1.$$

As seen in Figure 49, the $x$-intercept is approximately 2.174. Since

$$\frac{1 + \sqrt{145}}{6} \approx 2.174,$$

this confirms the analytic result.

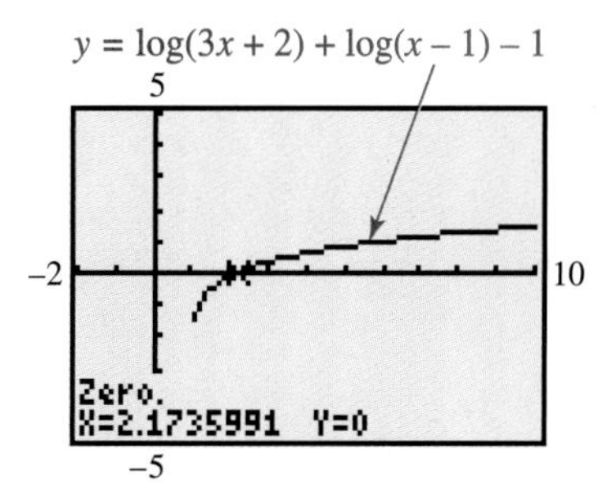


**FIGURE 49**

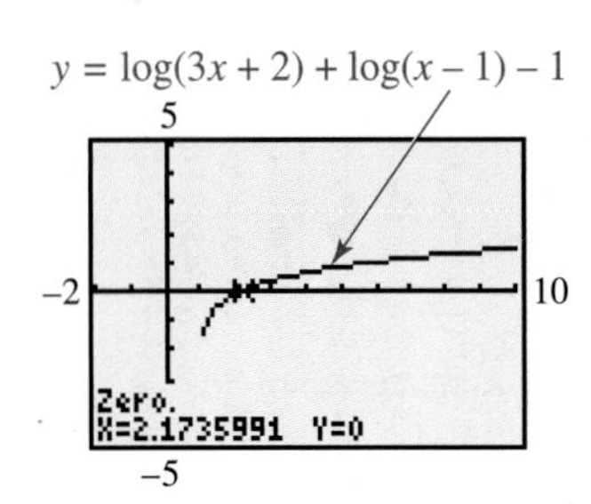


**FOR DISCUSSION**

Figure 49 from Example 5 is repeated in the margin. Use it to answer each item.

1. Give the exact solution set of $\log(3x + 2) + \log(x - 1) - 1 \geq 0$.
2. Give the exact solution set of $\log(3x + 2) + \log(x - 1) - 1 \leq 0$.
   (*Hint:* Pay attention to the domain.)

## Equations and Inequalities Involving Both Exponentials and Logarithms

**EXAMPLE 6** *Solving a More Complicated Exponential Equation*

Solve $e^{-2\ln x} = \frac{1}{16}$.

### Analytic Solution

The domain of $x$ is $(0, \infty)$. Use the power rule for logarithms to rewrite the exponent on the left side of the equation.

$$e^{-2\ln x} = \frac{1}{16}$$

$$e^{\ln x^{-2}} = \frac{1}{16} \qquad \text{Power rule}$$

$$x^{-2} = \frac{1}{16} \qquad a^{\log_a x} = x$$

$$x^{-2} = 4^{-2} \qquad \frac{1}{16} = \frac{1}{4^2} = 4^{-2}; -4 \text{ is not a valid base, since } -4 < 0, \text{ and } x > 0.$$

$$x = 4 \qquad \text{Property 1}$$

An analytic check shows that 4 is indeed a solution. The solution set is $\{4\}$.

### Graphing Calculator Solution

We see that $y = e^{-2\ln x} - \frac{1}{16}$ has $x$-intercept 4 in Figure 50. The solution set is $\{4\}$. (Exercise 31 requires the solutions of the associated inequalities using Figure 50.)

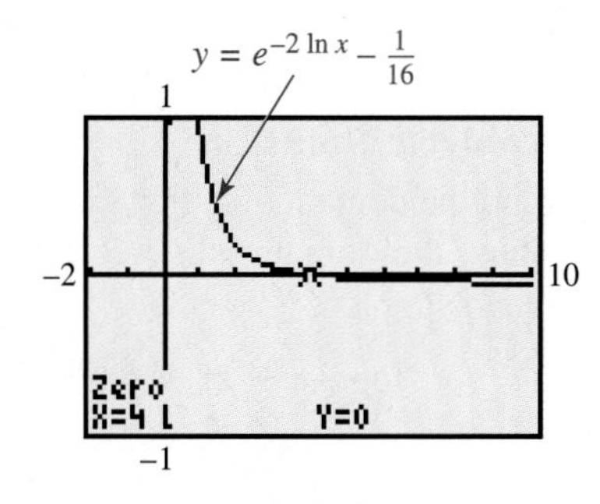


**FIGURE 50**

**EXAMPLE 7** *Solving a More Complicated Logarithmic Equation*

Solve $\ln e^{\ln x} - \ln(x - 3) = \ln 2$.

**Solution** Here, the domain is $(3, \infty)$. On the left, $\ln e^{\ln x} = \ln x$. The equation becomes

$$\ln x - \ln(x - 3) = \ln 2$$

$$\ln \frac{x}{x - 3} = \ln 2 \qquad \text{Quotient rule}$$

$$\frac{x}{x - 3} = 2 \qquad \text{Property 2}$$

$$x = 2x - 6 \qquad \text{Multiply by } x - 3.$$

$$6 = x.$$

To check, let $x = 6$ in the original equation.

$$\ln e^{\ln x} - \ln(x - 3) = \ln 2 \qquad \text{Given equation}$$

$$\ln e^{\ln 6} - \ln(6 - 3) = \ln 2 \quad ? \qquad \text{Let } x = 6.$$

$$\ln 6 - \ln 3 = \ln 2 \quad ?$$

$$\ln \frac{6}{3} = \ln 2 \quad ?$$

$$\ln 2 = \ln 2 \qquad \text{True}$$

The solution set is {6}. In Exercise 32, you are asked to solve the equation graphically, and then use the graph to solve the associated inequalities.

A summary of the methods for solving equations in this section follows.

**Solving Exponential and Logarithmic Equations**

An exponential or logarithmic equation can be solved by changing the equation into one of the following forms, where $a$ and $b$ are real numbers, $a > 0$, and $a \neq 1$.

1. $\mathbf{a^{f(x)} = b}$
   Solve by taking logarithms of each side. (Natural logarithms are the best choice if $a = e$.)
2. $\mathbf{\log_a f(x) = \log_a g(x)}$
   From the given equation, $f(x) = g(x)$, which is solved analytically.
3. $\mathbf{\log_a f(x) = b}$
   Solve by changing to exponential form, $f(x) = a^b$.

## Formulas Involving Exponentials and Logarithms

**EXAMPLE 8** *Solving an Exponential Formula from Psychology*

The strength of a habit is a function of the number of times the habit is repeated. If $N$ is the number of repetitions and $H$ is the strength of the habit, then, according to psychologist C. L. Hull,

$$H = 1000(1 - e^{-kN}),$$

where $k$ is a constant. Solve this formula for $k$.

**Solution** We must first solve the formula for $e^{-kN}$.

$$\frac{H}{1000} = 1 - e^{-kN} \qquad \text{Divide by 1000.}$$

$$\frac{H}{1000} - 1 = -e^{-kN}$$

$$e^{-kN} = 1 - \frac{H}{1000} \qquad \text{Multiply by } -1.$$

Now we can solve for $k$, taking logarithms on each side of the formula and using the fact that $\ln e^x = x$.

$$\ln e^{-kN} = \ln\left(1 - \frac{H}{1000}\right)$$

$$-kN = \ln\left(1 - \frac{H}{1000}\right) \qquad \ln e^x = x$$

$$k = -\frac{1}{N}\ln\left(1 - \frac{H}{1000}\right) \qquad \text{Multiply by } -\tfrac{1}{N}.$$

With the last equation, if one pair of values for $H$ and $N$ is known, $k$ can be found, and the formula can then be used to find either $H$ or $N$ for given values of the other variable.

**EXAMPLE 9** *Solving a Logarithmic Formula from Biology*

The formula

$$S = a\ln\left(1 + \frac{n}{a}\right)$$

gives the number of species in a sample, where $n$ is the number of individuals in the sample, and $a$ is a constant indicating the diversity of species in the community. Solve the formula for $n$.

**Solution** We begin by solving for $\ln\left(1 + \frac{n}{a}\right)$. Then we can change to exponential form and solve the resulting formula for $n$.

$$\frac{S}{a} = \ln\left(1 + \frac{n}{a}\right) \qquad \text{Divide by } a.$$

$$e^{S/a} = 1 + \frac{n}{a} \qquad \text{Exponential form}$$

$$e^{S/a} - 1 = \frac{n}{a} \qquad \text{Subtract 1.}$$

$$n = a(e^{S/a} - 1) \qquad \text{Multiply by } a.$$

Using this formula and given values of $S$ and $a$, we can find the number of individuals in a sample.

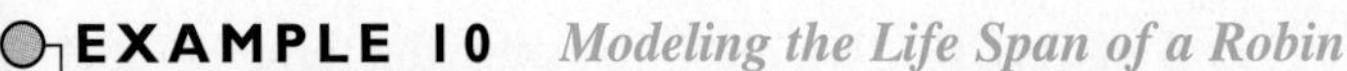

**EXAMPLE 10** *Modeling the Life Span of a Robin*

A study monitored the life spans of 129 robins over a 4-year period. The equation

$$y = \frac{2 - \log(100 - x)}{.42}$$

was developed to calculate the number of years $y$ for $x$ percent of the robin population to die. (*Source:* Lack, D., *The Life of a Robin,* Collins, 1965.) How many robins had died after 6 months?

**Solution** Let $y = \frac{6}{12} = .5$ year, and solve the equation for $x$.

$$
\begin{aligned}
.5 &= \frac{2 - \log(100 - x)}{.42} \\
.21 &= 2 - \log(100 - x) && \text{Multiply by .42.} \\
\log(100 - x) &= 1.79 \\
10^{1.79} &= 100 - x && \text{Exponential form} \\
x &= 100 - 10^{1.79} \approx 38.3
\end{aligned}
$$

About 38% of the robins had died after 6 months.

## 5.5 EXERCISES

*In Exercises 1–4, a logarithmic or exponential function f is graphed. Use the graph to find the solution set of* **(a)** $f(x) = 0$ *and* **(b)** $f(x) > 0$.

**1.**

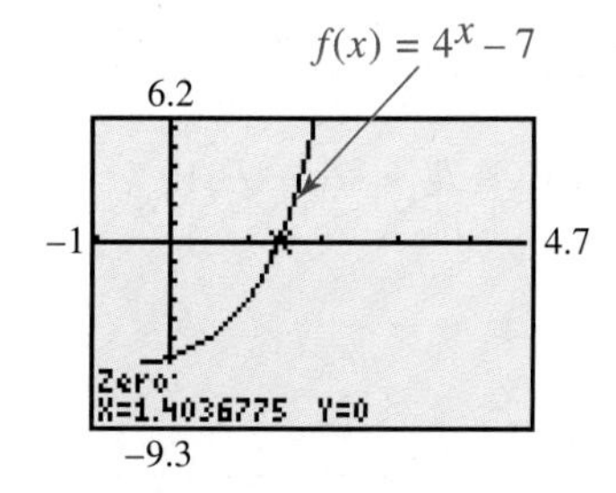


**2.**

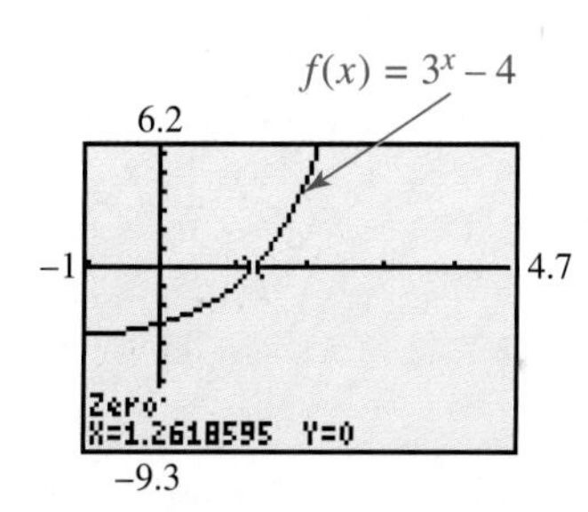


**3.**

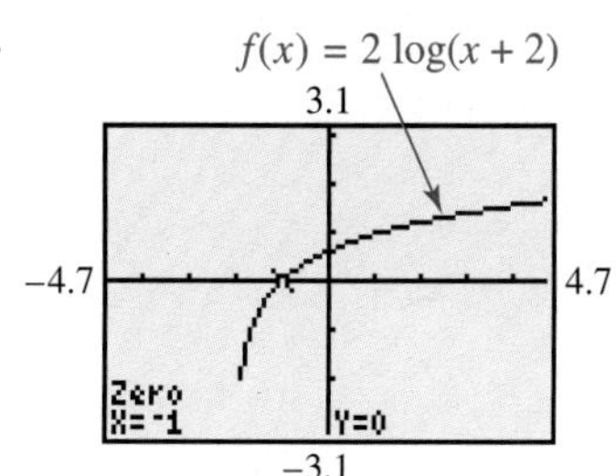


**4.**

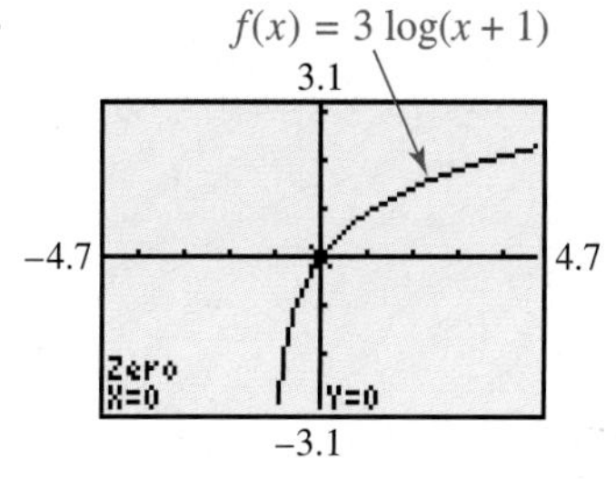


*Solve each equation analytically, and express the solutions in exact form, if possible. Use a graph to support your solutions.*

**5.** $3^x = 10$

**6.** $.5^x = .3$

**7.** $2^{x+3} = 5^x$

**8.** $6^{x+3} = 4^x$

**9.** $e^{x-1} = 4$

**10.** $e^{2-x} = 12$

**11.** $2e^{5x+2} = 8$

**12.** $10e^{3x-7} = 5$

**13.** $2^x = -3$

**14.** $3^x = -6$

**15.** $e^{8x} \cdot e^{2x} = e^{20}$

**16.** $e^{6x} \cdot e^x = e^{21}$

**17.** $\log_3(x + 50) = 2$

**18.** $\log_5(2x - 1) = 3$

**19.** $\log(4 - x) = -1$

**20.** $\log_2(3x + 2) = -2$

**21.** $\log x + \log(x - 21) = 2$

**22.** $\log x + \log(3x - 13) = 1$

**23.** $\ln(4x - 2) = \ln 4 - \ln(x - 2)$

**24.** $\ln(5 + 4x) - \ln(3 + x) = \ln 3$

**25.** $\log_5(x + 2) + \log_5(x - 2) = 1$

**26.** $\log_2 x + \log_2(x - 7) = 3$

**27.** $\ln e^x - \ln e^3 = \ln e^5$

**28.** $\ln e^x - 2 \ln e = \ln e^4$

**29.** $\log_2(\log_2 x) = 1$

**30.** $\log x = \sqrt{\log x}$

**31.** In Example 6, we found that the solution of $e^{-2 \ln x} = \frac{1}{16}$ is 4. Use the solution and Figure 50 to solve each inequality.

**(a)** $e^{-2 \ln x} > \dfrac{1}{16}$

**(b)** $e^{-2 \ln x} < \dfrac{1}{16}$

**32.** In Example 7, we found that the solution of the equation $\ln e^{\ln x} - \ln(x - 3) = \ln 2$ is 6. Complete the following.

**(a)** Solve the equation graphically.

**(b)** Solve the inequality $\ln e^{\ln x} - \ln(x - 3) > \ln 2$, using the graph from part (a).

**(c)** Solve the inequality $\ln e^{\ln x} - \ln(x - 3) < \ln 2$, using the graph from part (a).

**33.** A student told a friend "You must reject any negative solution of an equation involving logarithms." Is this correct? Write an explanation of your answer.

**34.** Use the fact that $\log 10 = 1$ to rewrite the equation $\log(3x + 2) + \log(x - 1) = 1$ with a logarithm on the right side. Then solve the equation using properties of logarithms. (See Example 4.)

**35.** Use a graph to explain why the logarithmic equation

$$\ln x - \ln(x + 1) = \ln 5$$

has no solution.

**36.** Use a graph to explain why

$$\ln x + 1 \geq \ln(x - 4)$$

has solution set $(4, \infty)$.

*Use a graphing calculator to find the solution of each equation. Round your result to the nearest thousandth.*

**37.** $1.5^{\log x} = e^{.5}$

**38.** $1.5^{\ln x} = 10^{.5}$

*Each formula comes from an application of exponential or logarithmic functions. Solve the formula for the indicated variable.*

**39.** $r = p - k \ln t$, for $t$

**40.** $p = a + \dfrac{k}{\ln x}$, for $x$

**41.** $T = T_0 + (T_1 - T_0)10^{-kt}$, for $t$

**42.** $A = \dfrac{Pi}{1 - (1 + i)^{-n}}$, for $n$

**43.** $A = T_0 + Ce^{-kt}$, for $k$

**44.** $y = \dfrac{K}{1 + ae^{-bx}}$, for $b$

**45.** $y = A + B(1 - e^{-Cx})$, for $x$

**46.** $m = 6 - 2.5 \log\left(\dfrac{M}{M_0}\right)$, for $M$

**47.** $\log A = \log B - C \log x$, for $A$

**48.** $d = 10 \log\left(\dfrac{I}{I_0}\right)$, for $I$

## RELATING CONCEPTS

**For individual or group investigation (Exercises 49–54)**

*The methods of solving quadratic equations from Chapter 3 can be applied to equations that are quadratic in form. Consider the equation*

$$e^{2x} - 4e^x + 3 = 0,$$

*and work Exercises 49–54 in order.*

**49.** The expression $e^{2x}$ is equivalent to $(e^x)^2$. Explain why.

**50.** The given equation is equivalent to $(e^x)^2 - 4e^x + 3 = 0$. Factor the left side of this equation.

**51.** Solve the equation in Exercise 50 by the zero-product property. Give exact values.

**52.** Solve the equation in Exercise 50 using a calculator graph of $y = e^{2x} - 4e^x + 3$.

**53.** Use the graph from Exercise 52 to solve the inequality $e^{2x} - 4e^x + 3 > 0$.

**54.** Use the graph from Exercise 52 to solve the inequality $e^{2x} - 4e^x + 3 < 0$.

*In general, it is not possible to find exact solutions analytically for equations that involve exponential or logarithmic functions together with polynomial, radical, and rational functions. However, it is possible to solve them graphically with a calculator, using either the intersection-of-graphs method or the x-intercept method. Solve each equation, using a graphical method, and express solutions to the nearest thousandth if an approximation is appropriate.*

**55.** $x^2 = 2^x$ **56.** $x^2 - 4 = e^{x-4} + 4$ **57.** $\log x = x^2 - 8x + 14$

**58.** $\ln x = -\sqrt[3]{x + 3}$ **59.** $e^x = \dfrac{1}{x + 2}$ **60.** $3^{-x} = \sqrt{x + 5}$

*Use any method (analytic or graphical) to solve each equation. If appropriate, round your solution to the nearest thousandth.*

**61.** $\log_2 \sqrt{2x^2} - 1 = .5$ **62.** $\log x^2 = (\log x)^2$ **63.** $\ln(\ln e^{-x}) = \ln 3$ **64.** $e^{x+\ln 3} = 4e^x$

*(Modeling) Life Span of Robins* *Use the equation*

$$y = \frac{2 - \log(100 - x)}{.42}$$

*from Example 10 for Exercises 65 and 66.*

**65.** Estimate analytically the number of robins that died after 2 years.

**66.** Estimate graphically the number of years for 75% of the robins to die.

*(Modeling) Salinity* *The salinity of the oceans changes with latitude and with depth. In the tropics, the salinity increases on the surface of the ocean due to rapid evaporation. In the higher latitudes, there is less evaporation and rainfall causes the salinity to be less on the surface than at lower depths. The function given by*

$$f(x) = 31.5 + 1.1 \log(x + 1)$$

*models salinity to depths of* 1000 *meters at a latitude of* 57.5°N. *The variable x is the depth in meters, and f(x) is in grams of salt per kilogram of seawater. (Source: Hartman, D.,* Global Physical Climatology, *Academic Press, 1994.)*

**67.** Approximate analytically the depth where the salinity equals 33.

**68.** Estimate graphically the salinity at a depth of 500 meters.

## Reviewing Basic Concepts (Sections 5.4 and 5.5)

**1.** Fill in the blanks: If $f(x) = 3^x$ and $g(x) = \log_3 x$, then functions $f$ and $g$ are ________ functions, and their graphs are ________ with respect to the line with equation ________. The domain of $f$ is the ________ of $g$ and vice versa.

*Let* $f(x) = 2 - \log_2(x - 1)$ *in Exercises 2–5.*

**2.** Graph $f(x)$.

**3.** Give the equation of the asymptote of the graph of $f(x)$ and any intercepts.

**4.** Compare the graph of $f(x)$ to the graph of $g(x) = \log_2 x$.

**5.** Find the inverse of the function defined by $f(x)$.

*Solve each equation.*

**6.** $3^{2x-1} = 4^x$

**7.** $\ln 5x - \ln(x + 2) = \ln 3$

**8.** $10^{5 \log x} = 32$

**9.** $H = 1000(1 - e^{-kN})$, for $N$

**10.** *(Modeling) Caloric Intake* The function defined by

$$f(x) = 280 \ln(x + 1) + 1925$$

models the number of calories consumed daily by a person owning $x$ acres of land in a developing country. (*Source:* Grigg, D., *The World Food Problem,* Blackwell Publishers, 1993.) Estimate the number of acres owned for the average intake to be 2300 calories per day.

# 5.6 FURTHER APPLICATIONS AND MODELING WITH EXPONENTIAL AND LOGARITHMIC FUNCTIONS

Physical Science Applications • Financial Applications • Biological and Medical Applications • Modeling Data with Exponential and Logarithmic Functions

**Looking Ahead to Calculus**

The exponential growth and decay function formulas are studied in calculus in conjunction with the topic known as *differential equations.*

## Physical Science Applications

The formula $A = Pe^{rt}$ for continuous compounding of interest is an example of an exponential growth function. A function of the form

$$A(t) = A_0e^{kt},$$

where $A_0$ represents the initial quantity present, $t$ represents time elapsed, $k > 0$ represents the growth constant associated with the quantity, and $A(t)$ represents the amount present at time $t$, is called an **exponential growth function**. This is an increasing function, because $e > 1$ and $k > 0$. On the other hand, for $k > 0$, a function of the form

$$A(t) = A_0e^{-kt}$$

is an **exponential decay function.** It is a decreasing function because

$$e^{-k} = (e^{-1})^k = \left(\frac{1}{e}\right)^k,$$

and $0 < \frac{1}{e} < 1$. In both cases, we usually restrict $t$ to be nonnegative, giving domain $[0, \infty)$. (Why?) Figure 51 shows graphs of typical growth and decay functions.

If a quantity decays exponentially, the amount of time that it takes to become one-half its original amount is called the **half-life.**

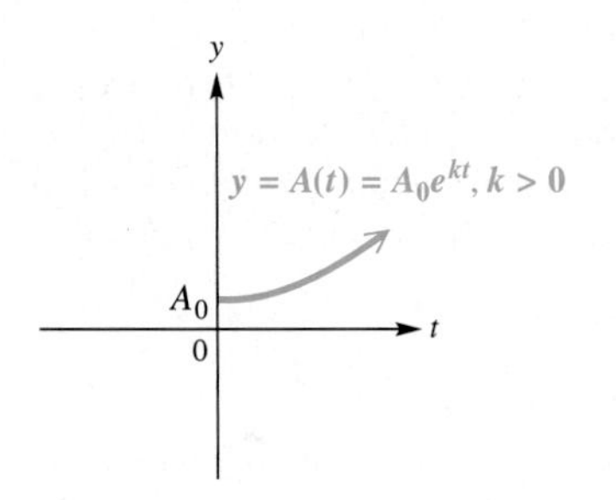


Exponential Growth Function

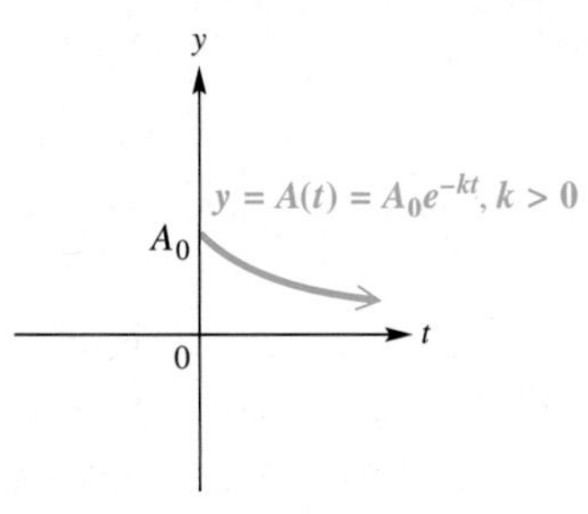


Exponential Decay Function

**FIGURE 51**

### EXAMPLE 1 *Analyzing an Exponential Decay Function Involving Radioactive Isotopes*

Nuclear energy derived from radioactive isotopes can be used to supply power to space vehicles. Suppose that the output of the radioactive power supply for a certain satellite is given by the function defined by

$$y = 40e^{-.004t},$$

where $y$ is measured in watts and $t$ is the time in days.

**(a)** What is the initial output of the power supply?

**(b)** After how many days will the output be reduced to 35 watts?

**(c)** After how many days will the output be half of its initial amount? (That is, what is its half-life?) Support this result with a graph.

#### Solution

**(a)** Let $t = 0$ in the given equation.

$$y = 40e^{-.004t}$$

$$y = 40e^{-.004(0)} \qquad \text{Let } t = 0.$$

$$y = 40e^0$$

$$y = 40 \qquad e^0 = 1$$

The initial output is 40 watts.

**(b)** Let $y = 35$, and solve for $t$.

$$35 = 40e^{-.004t}$$

$$\frac{35}{40} = e^{-.004t} \qquad \text{Divide by 40.}$$

$$\ln\frac{35}{40} = \ln e^{-.004t} \qquad \text{Take logarithms on each side.}$$

$$\ln\frac{35}{40} = -.004t \qquad \ln e^k = k$$

$$t = \frac{\ln\frac{35}{40}}{-.004} \qquad \text{Divide by } -.004.$$

Using a calculator, we find that $t \approx 33.4$, so the output will be reduced to 35 watts in about 33.4 days.

**(c)** Because the initial amount is 40, we must find the value of $t$ for which $y = \frac{1}{2}(40) = 20$.

$$20 = 40e^{-.004t}$$

$$.5 = e^{-.004t} \qquad \text{Divide by 40.}$$

$$\ln .5 = \ln e^{-.004t} \qquad \text{Take logarithms on each side.}$$

$$\ln .5 = -.004t \qquad \ln e^k = k$$

$$t \approx 173 \qquad \text{Divide by } -.004.$$

FIGURE 52

The half-life is approximately 173 days. The graph in Figure 52 also shows that when $x = t = 173$, $y \approx 20 = \frac{1}{2}(40)$.

**EXAMPLE 2** *Using an Exponential Function to Find the Age of a Fossil*

Carbon 14 is a radioactive form of carbon that is found in all living plants and animals. After a plant or animal dies, the radiocarbon disintegrates. Using a technique called *carbon dating*, scientists determine the age of the remains by comparing the amount of carbon 14 present with the amount found in living plants and animals. The amount of carbon 14 present after $t$ years is given by the exponential function defined by $A(t) = A_0e^{-kt}$, with $k = \frac{\ln 2}{5700}$. Find the half-life.

**Solution** Let $A(t) = \frac{1}{2}A_0$ and $k = \frac{\ln 2}{5700}$.

$$\frac{1}{2}A_0 = A_0e^{-\frac{\ln 2}{5700}t}$$

$$\frac{1}{2} = e^{-\frac{\ln 2}{5700}t} \qquad \text{Divide by } A_0.$$

$$\ln\frac{1}{2} = \ln e^{-\frac{\ln 2}{5700}t} \qquad \text{Take logarithms on each side.}$$

$$\ln\frac{1}{2} = -\frac{\ln 2}{5700}t \qquad \ln e^x = x$$

$$-\frac{5700}{\ln 2}\ln\frac{1}{2} = t \qquad \text{Multiply by } -\tfrac{5700}{\ln 2}.$$

$$-\frac{5700}{\ln 2}(\ln 1 - \ln 2) = t \qquad \text{Quotient rule}$$

$$-\frac{5700}{\ln 2}(-\ln 2) = t \qquad \ln 1 = 0$$

$$5700 = t$$

The half-life is 5700 years.

**EXAMPLE 3** *Finding an Exponential Decay Function Given Half-life*

Radium 226, which decays according to the function defined by

$$A(t) = A_0e^{-kt},$$

where $t$ is time in years, has a half-life of about 1612 years. Find $k$. Then find how long it takes a 10-gram sample to decay to 6 grams.

**Solution** The half-life tells us that when $t = 1612$, $A(t) = \frac{1}{2}A_0$. We substitute these values into the given equation.

$$\frac{1}{2}A_0 = A_0e^{-k(1612)}$$

$$\frac{1}{2} = e^{-1612k} \qquad \text{Divide by } A_0.$$

$$\ln\frac{1}{2} = \ln e^{-1612k} \qquad \text{Take logarithms on each side.}$$

$$\ln\frac{1}{2} = -1612k \qquad \ln e^x = x$$

$$k = \frac{\ln\frac{1}{2}}{-1612} \approx .00043$$

Thus, radium 226 decays according to the equation

$$A(t) = A_0e^{-.00043t}.$$

Now we let $A(t) = 6$ and $A_0 = 10$ to find $t$.

$$6 = 10e^{-.00043t} \qquad \text{Substitute.}$$

$$.6 = e^{-.00043t} \qquad \text{Divide by 10.}$$

$$\ln .6 = \ln e^{-.00043t} \qquad \text{Take logarithms on each side.}$$

$$\ln .6 = -.00043t \qquad \ln e^x = x$$

$$t = \frac{\ln .6}{-.00043}$$

$$t \approx 1188$$

The 10-gram sample will decay to 6 grams in about 1188 years.

**EXAMPLE 4** *Measuring Sound Intensity*

The loudness of sounds is measured in a unit called a *decibel*. To measure with this unit, we first assign an intensity of $I_0$ to a very faint sound, called the *threshold sound*.

If a particular sound has intensity $I$, then the decibel rating of this louder sound is

$$d = 10 \log \frac{I}{I_0}.$$

In the movie *Pearl Harbor,* action sequences reached levels of $10^{10.7}I_0$. (*Source:* www.lyhh.org/noise/) Find the decibel rating of these sequences, and compare it to the average rating of 95 decibels for a motorcycle.

**Solution** Substitute $I = 10^{10.7}I_0$ in the equation.

$$d = 10 \log \frac{10^{10.7}I_0}{I_0}$$

$$d = 10 \log 10^{10.7}$$

$$d = 10(10.7) \qquad \log 10^{10.7} = 10.7$$

$$d = 107$$

The action sequences reached 107 decibels. Since 107 decibels is equivalent to $10^{10.7}I_0$ and 95 decibels is equivalent to $10^{9.5}I_0$, their ratio is

$$\frac{10^{10.7}I_0}{10^{9.5}I_0} = 10^{1.2} \approx 15.8.$$

The action sequences are 15.8 times as loud as a motorcycle.

## Financial Applications

The compound interest formulas

$$A = P\left(1 + \frac{r}{n}\right)^{nt} \quad \text{and} \quad A = Pe^{rt}$$

were introduced in Section 5.2. The first applies to interest compounded $n$ times per year; the second applies to interest compounded continuously. We can use logarithms to determine how long it takes for an investment to grow to a desired amount.

**EXAMPLE 5** *Solving Compound Interest Formulas for t*

**(a)** How long will it take \$1000 invested at 6% interest, compounded quarterly, to grow to \$2700?

**(b)** How long will it take for the money in an account that is compounded continuously at 8% interest to double?

**Solution**

**(a)** Here we must find $t$, where $A = 2700$, $P = 1000$, $r = .06$, and $n = 4$.

$$A = P\left(1 + \frac{r}{n}\right)^{nt} \qquad \text{Given formula}$$

$$2700 = 1000\left(1 + \frac{.06}{4}\right)^{4t} \qquad \text{Substitute.}$$

$$2700 = 1000(1.015)^{4t}$$

$$2.7 = 1.015^{4t} \qquad \text{Divide by 1000.}$$

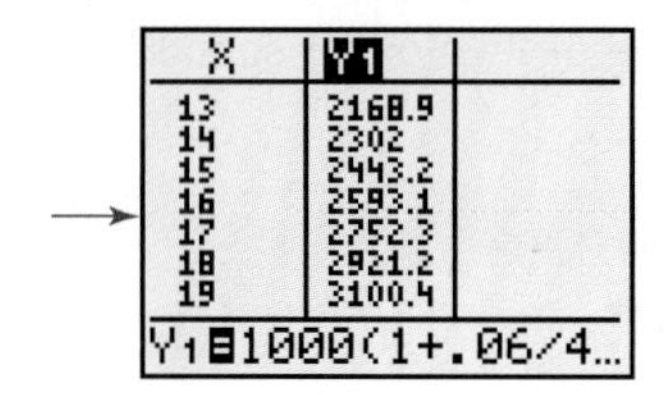


This table numerically supports the result of Example 5(a). Notice that $Y_1 = 2700$ for some X-value between 16 and 17, closer to 17. We found analytically that this value is approximately 16.678.

To solve for $t$, we choose common logarithms.

$$\log 2.7 = \log 1.015^{4t}$$

$$\log 2.7 = 4t \log 1.015 \quad \text{Power rule}$$

$$\frac{\log 2.7}{4 \log 1.015} = t \quad \text{Divide by 4 log 1.015.}$$

$$t \approx 16.678 \quad \text{Use a calculator.}$$

Since interest is compounded quarterly, it will take about $16\frac{3}{4}$ years for the initial amount to grow to \$2700.

**(b)** Use the formula for continuous compounding, $A = Pe^{rt}$, to find the time $t$ that makes $A = 2P$. Substitute $2P$ for $A$ and .08 for $r$; then, solve for $t$.

$$A = Pe^{rt}$$

$$2P = Pe^{.08t} \quad \text{Substitute.}$$

$$2 = e^{.08t} \quad \text{Divide by } P.$$

$$\ln 2 = \ln e^{.08t} \quad \text{Take logarithms on each side.}$$

$$\ln 2 = .08t \quad \ln e^x = x$$

$$\frac{\ln 2}{.08} = t \quad \text{Divide by .08.}$$

$$8.664 \approx t$$

The amount will double in about $8\frac{2}{3}$ years.

If a quantity grows exponentially, the amount of time that it takes to become twice its original amount is called the **doubling time.** This is analogous to half-life for quantities that decay exponentially. In both cases, the actual amount present initially does not affect either the doubling time or half-life.

A loan is **amortized** if both the principal and interest are paid by a sequence of equal periodic payments.

TECHNOLOGY NOTE

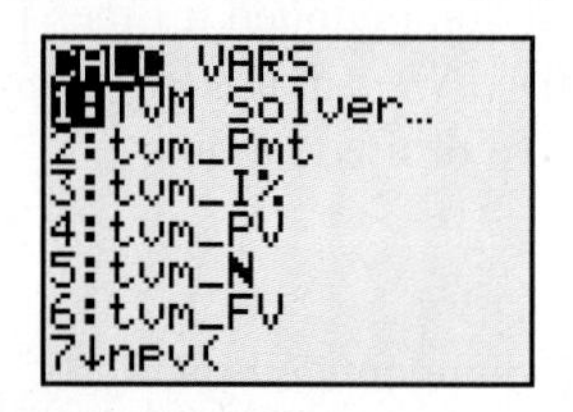


Graphing calculators such as the TI-83 are capable of financial calculations. Consult the graphing calculator manual that accompanies this text or your owner's manual.

**Amortization Payments**

A loan of $P$ dollars at interest rate $i$ per period (as a decimal) may be amortized in $n$ equal periodic payments of $R$ dollars made at the end of each period, where

$$R = \frac{P}{\left[\frac{1-(1+i)^{-n}}{i}\right]}.$$

The total interest $I$ that will be paid during the term of the loan is

$$I = nR - P.$$

**EXAMPLE 6** *Using Amortization to Finance an Automobile*

Following the traditional American process of haggling, you agree to purchase a used SUV for \$24,000. After a down payment of \$4000, the balance will be paid off in 36 equal monthly payments at 8.5% interest per year. Find the amount of each payment. How much interest will you pay over the life of the loan?

**Solution** Use the amortization formula with $P = 24{,}000 - 4000 = 20{,}000$, $i = \frac{.085}{12} \approx .007083$, and $n = 36$.

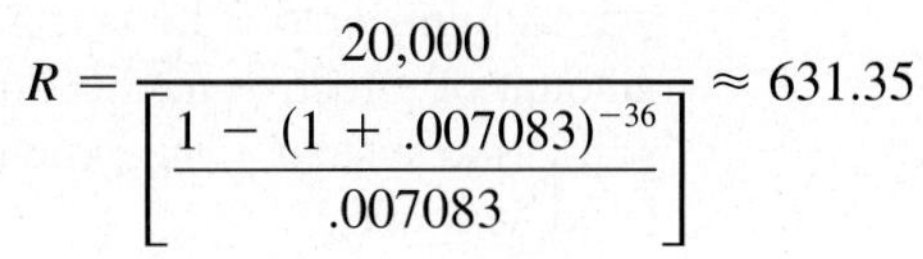

$$R = \frac{20{,}000}{\left[\frac{1 - (1 + .007083)^{-36}}{.007083}\right]} \approx 631.35$$

The monthly payment will be \$631.35. The total interest paid will be 36(\$631.35) − \$20,000 = \$2728.60.

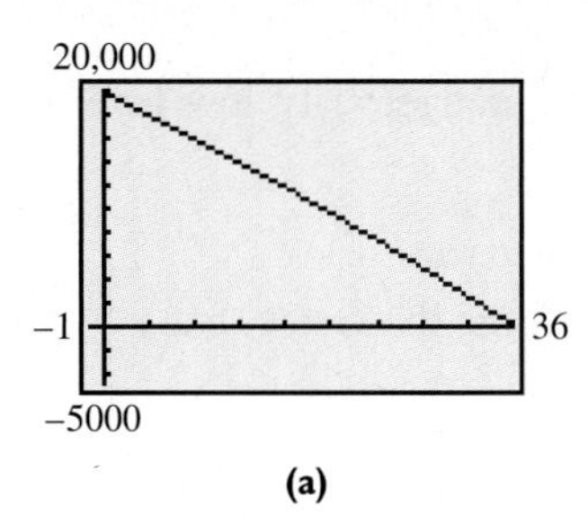


(a)

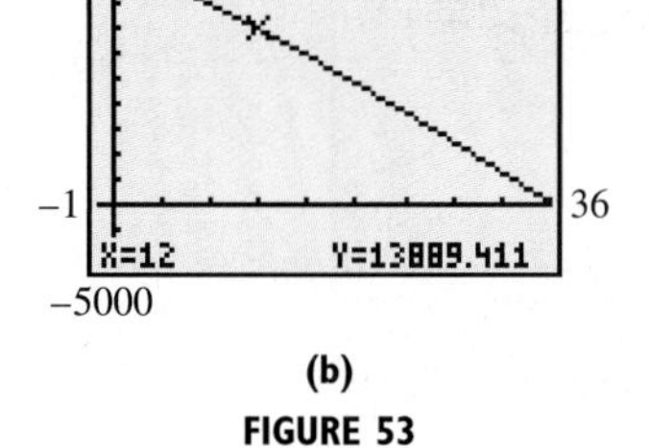


(b)

FIGURE 53

In Example 6, the approximate unpaid balance $y$ after $x$ payments is given by

$$y = R\left[\frac{1 - (1 + i)^{-(n-x)}}{i}\right].$$

For example, the unpaid balance after 1 payment is

$$y = \$631.35\left[\frac{1 - (1 + .007083)^{-35}}{.007083}\right] \approx \$19{,}510.40.$$

A graph of $y$ as a function of $x$ is shown in Figure 53(a). We can find the unpaid balance after any number of payments, $x$, by finding the $y$-value that corresponds to $x$. For example, the remaining balance after 1 year or 12 payments is shown at the bottom of the screen in Figure 53(b). You may be surprised that the remaining balance on a \$20,000 loan after 12 payments is as large as \$13,889.41. This is because most of each early payment on a loan goes toward interest.

## Biological and Medical Applications

**EXAMPLE 7** *Determining Diversity of Insects*

The number of types of insects found in a wooded region is given by

$$N = 200 + 300 \log x,$$

where $x$ is the corresponding number of acres. How much acreage would have 1200 types of insects?

**Solution** Let $N = 1200$, and solve the formula for $x$.

$$1200 = 200 + 300 \log x$$

$$1000 = 300 \log x$$

$$\log x = \frac{1000}{300}$$

$$\log x = \frac{10}{3}$$

$$x = 10^{10/3} \qquad \text{Exponential form}$$

$$x \approx 2154$$

About 2154 acres are needed to support 1200 types of insects.

**EXAMPLE 8** *Determining the Amount of Medication*

When physicians prescribe medication, they must consider how the drug's effectiveness decreases over time. If, each hour, a drug is only 90% as effective as the previous hour, at some point the patient will not be receiving enough medication and must receive another dose. For example, if the initial dose was 200 milligrams and the drug was administered 3 hours ago, the exponential expression $200(.90)^2$ represents the amount of effective medication still available. Thus, $200(.90)^2 = 162$ milligrams are still in the system. (The exponent is the number of hours since the drug was administered, less one.)

How long will it take for this initial dose to reach the dangerously low level of 50 milligrams?

**Analytic Solution**

We must solve the equation $200(.90)^x = 50$.

$$
\begin{aligned}
200(.90)^x &= 50\\
(.90)^x &= .25\\
\log(.90)^x &= \log .25\\
x \log .90 &= \log .25\\
x &= \frac{\log .25}{\log .90} \approx 13.16
\end{aligned}
$$

Since $x$ represents one *less than* the number of hours from when the drug was administered, the drug will reach a level of 50 milligrams in about 14 hours.

**Graphing Calculator Solution**

The graph in Figure 54 confirms the result that $y = 50$ when $x \approx 13.16$.

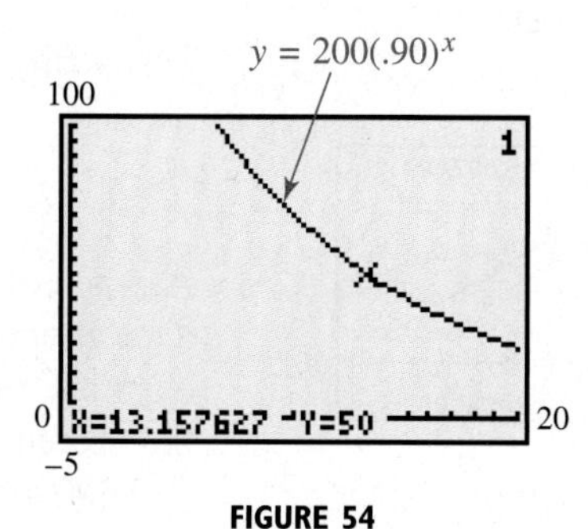


FIGURE 54

## Modeling Data with Exponential and Logarithmic Functions

Some data can be modeled quite well using exponential or logarithmic functions. Graphing calculators can determine such models for appropriate data.

**EXAMPLE 9** *Modeling Atmospheric $CO_2$ Concentrations*

Future concentrations of atmospheric carbon dioxide ($CO_2$) in parts per million (ppm) are shown in the table. (These concentrations assume that current trends continue.) As mentioned in the chapter introduction, $CO_2$ levels in the year 2000 were greater than they have been any time in the previous 160,000 years. The increase in concentrations of $CO_2$ has been accelerated by the burning of fossil fuels and by deforestation.

| Year | 2000 | 2050 | 2100 | 2150 | 2200 |
|---|---|---|---|---|---|
| $CO_2$ (ppm) | 364 | 467 | 600 | 769 | 987 |

*Source:* Turner, R., *Environmental Economics, An Elementary Approach,* Johns Hopkins University Press, 1993.

**(a)** Let $x = 0$ correspond to 2000 and $x = 200$ to 2200. Find values for $C$ and $a$ so that $f(x) = Ca^x$ models these data.

**(b)** Use a graphing calculator with regression capability to find an exponential function $g$ that models all the data, and graph it with the data points. How does function $g$ compare with function $f$ from part (a)?

**(c)** Estimate $CO_2$ concentrations for the year 2025.

### Solution

**(a)** The concentration is 364 when $x = 0$, so $C = 364$. This gives

$$f(x) = 364a^x.$$

Next, we estimate a value for $a$. One possibility for determining $a$ is to require that the graph of $f$ pass through the point (2200, 987), where $x = 200$. It then follows that $f(200) = 987$.

$$364a^x = f(x)$$

$$364 \cdot a^{200} = 987 \qquad f(200) = 987$$

$$a^{200} = \frac{987}{364} \qquad \text{Divide by 364.}$$

$$(a^{200})^{1/200} = \left(\frac{987}{364}\right)^{1/200} \qquad \text{Raise to the } \tfrac{1}{200}\text{th power.}$$

$$a = \left(\frac{987}{364}\right)^{1/200} \qquad \text{Property of exponents}$$

$$a \approx 1.005 \qquad \text{Use a calculator.}$$

Thus $f(x) = 364(1.005)^x$. Answers for $f(x)$ may vary slightly.

**(b)** Figure 55(a) shows the coefficients given by the calculator for exponential regression. Using these values, $g(x) = 364.0(1.005)^x$, which is the same as $f(x)$ from part (a). The data points and the function are graphed in Figure 55(b).

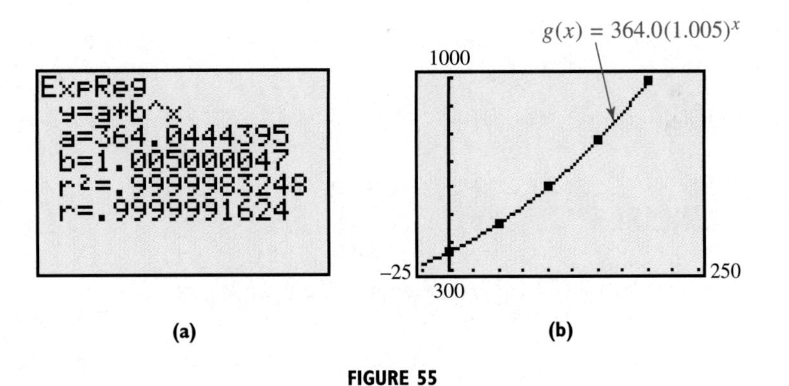


(a) (b)

**FIGURE 55**

**(c)** Since 2025 corresponds to $x = 25$, evaluate $f(25)$.

$$f(25) = 364(1.005)^{25} \approx 412.$$

Concentration of carbon dioxide could reach 412 ppm by 2025.

**EXAMPLE 10** *Modeling Interest Rates*

The table lists the interest rates for certificates of deposit during September 2001.

| Time | 6 months | 1 year | 2.5 years | 5 years |
|---|---|---|---|---|
| Yield (%) | 3.23 | 3.42 | 3.84 | 4.58 |

*Source:* The Associated Press.

**(a)** Make a scatter diagram of the data. What type of function might model these data?

**(b)** Use least-squares regression to obtain a formula, $f(x) = a + b \ln x$, that models the data.

**(c)** Graph $f$ and the data in the same viewing window.

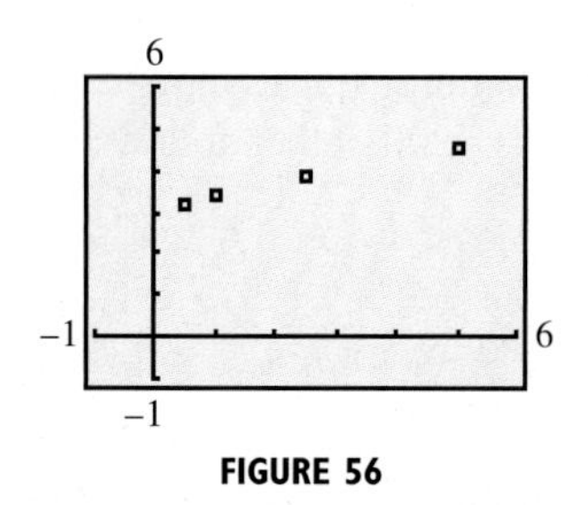


**FIGURE 56**

### Solution

**(a)** Enter the data points (.5, 3.23), (1, 3.42), (2.5, 3.84), and (5, 4.58) into a calculator. A scatter diagram of the data is shown in Figure 56. The data increase but gradually level off. A logarithmic function model may be appropriate.

**(b)** In Figures 57(a) and 57(b), least-squares regression gives a logarithmic function $f$ defined (approximately) by

$$f(x) = 3.51 + .569 \ln x.$$

**(c)** A graph of $f$ and the data are shown in Figure 57(c).

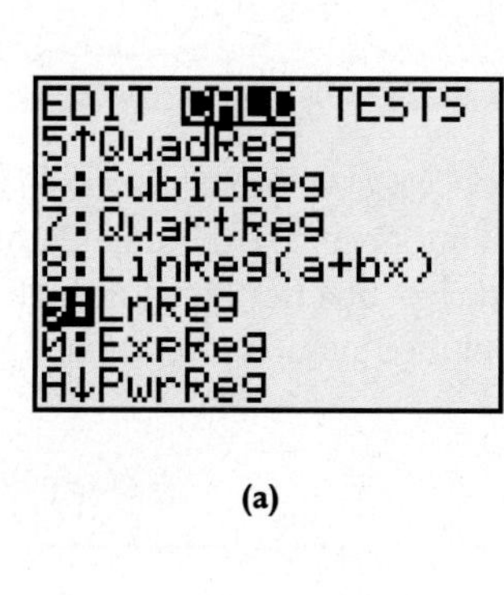


(a)

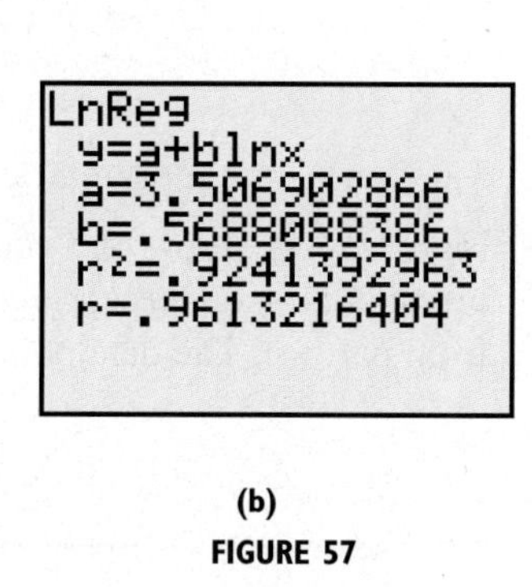


(b)

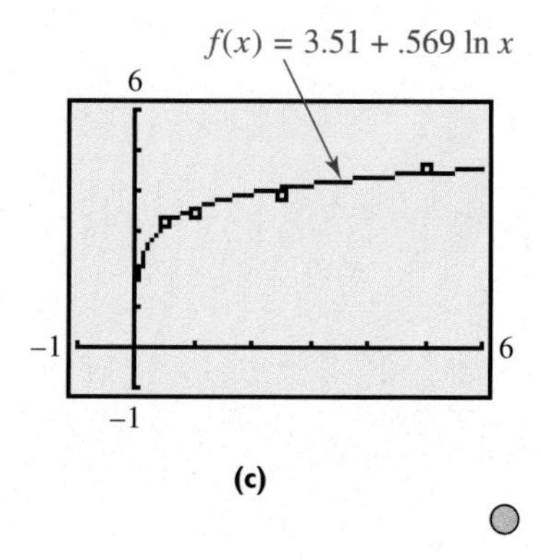


(c)

**FIGURE 57**

## 5.6 EXERCISES

*(Modeling) In Exercises 1–20, solve each problem from the physical sciences.*

*The information in Example 2 allows us to use the function defined by* $A(t) = A_0 e^{-.0001216t}$ *to approximate the amount of carbon 14 remaining in a sample, where t is in years. Use this function in Exercises 1–4.* (Note: $-.0001216 \approx -\frac{\ln 2}{5700}$.)

**1.** *Carbon 14 Dating* Suppose an Egyptian mummy is discovered in which the amount of carbon 14 present is only about one-third the amount found in the atmosphere. About how long ago did the Egyptian die?

**2.** *Carbon 14 Dating* A sample from a refuse deposit near the Strait of Magellan had 60% of the carbon 14 of a contemporary living sample. How old was the sample?

**3.** *Carbon 14 Dating* Paint from the Lascaux caves of France, shown in the photo, contains 15% of the normal amount of carbon 14. Estimate the age of the paint.

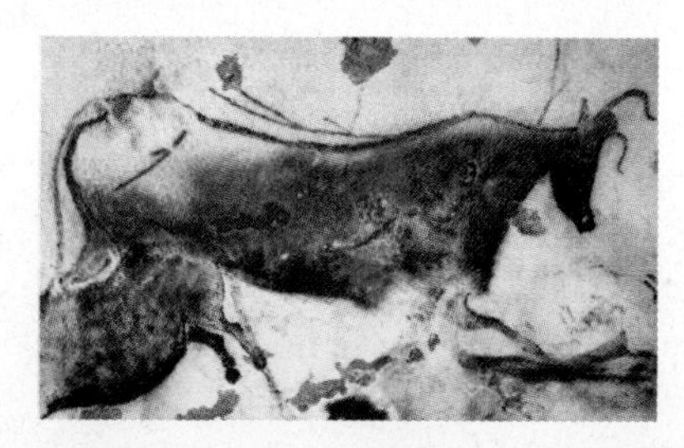

**4.** *Carbon 14 Dating* Estimate the age of a specimen that contains 20% of the carbon 14 of a comparable living specimen.

**5.** *Radioactive Lead 210* The half-life of radioactive lead 210 is 21.7 years.

**(a)** Find an exponential decay model for lead 210.

**(b)** How long will it take a sample of 500 grams to decay to 400 grams?

**(c)** How much of the sample of 500 grams will remain after 10 years?

**6.** *Radioactive Cesium 137* Radioactive cesium 137 was emitted in large amounts in the Chernobyl nuclear power station accident in Russia on April 26, 1986. The amount of cesium 137 remaining after $x$ years in an initial sample of 100 milligrams can be described by

$$A(x) = 100e^{-.02295x}.$$

(*Source:* Mason, C., *Biology of Freshwater Pollution,* John Wiley and Sons, 1991.)

**(a)** How much is remaining after 50 years? Is the half-life of cesium 137 greater or less than 50 years?

**(b)** Estimate graphically the half-life of cesium 137.

**7.** *Radioactive Polonium 210* The table shows the amount $y$ of polonium 210 in grams remaining after $x$ days from an initial sample of 2 milligrams.

| x (days) | 0 | 100 | 200 | 300 |
|---|---|---|---|---|
| y (milligrams) | 2 | 1.22 | .743 | .453 |

**(a)** Use the table to determine mentally if the half-life of polonium 210 is greater or less than 200 days.

**(b)** Find a formula that models the amount $A$ of polonium 210 in the table after $x$ days.

**(c)** Estimate graphically the half-life of polonium 210.

**8.** *Sound Intensity* Use the formula

$$d = 10 \log \frac{I}{I_0}$$

to find the average decibel level for each popular movie with the given intensity $I$. For comparison, a jackhammer or helicopter has a sound level of about 105 decibels.

**(a)** *Armageddon:* $5.012 \times 10^{10} I_0$

**(b)** *Godzilla:* $10^{10} I_0$

**(c)** *Saving Private Ryan:* $6{,}310{,}000{,}000 I_0$

*The intensity of an earthquake, measured on the Richter scale, is given by* $\log \frac{I}{I_0}$, *where* $I_0$ *is the intensity of an earthquake of a certain small size. Use this information in Exercises 9 and 10.*

**9.** *Earthquake Intensity* On July 14, 1991, Peshawar, Pakistan, was shaken by an earthquake that measured 6.6 on the Richter scale.

**(a)** Express this reading in terms of $I_0$.

**(b)** In February of the same year, a quake measuring 6.5 on the Richter scale killed about 900 people in the mountains of Pakistan and Afghanistan. Express the intensity of a 6.5 reading in terms of $I_0$.

**(c)** Compare this with the force of the earthquake with a measure of 6.6.

**10.** *Earthquake Intensity* The San Francisco earthquake of 1906 had a Richter scale rating of 8.3.

**(a)** Express the intensity of this earthquake in terms of $I_0$.

**(b)** In 1989, the San Francisco region experienced an earthquake with a Richter scale rating of 7.1. Express the intensity of this earthquake in terms of $I_0$.

**(c)** Compare the intensities of the two San Francisco earthquakes discussed above.

**11.** *Rock Sample Age* Use the function defined by

$$t = T\frac{\ln\left[1 + 8.33\left(\frac{A}{K}\right)\right]}{\ln 2}$$

to estimate the age $t$ in years of a rock sample, if tests show that $\frac{A}{K}$ is .103 for the sample. Let $T = 1.26 \times 10^9$.

**12.** *Air Pressure* The air pressure in pounds per square inch $h$ feet above sea level is given by

$$P(h) = 14.7e^{-.0000385h}.$$

At approximately what height is the pressure 10% of the pressure at sea level?

**13.** *Magnitude of a Star* The magnitude $M$ of a star is defined by the equation

$$M = 6 - \frac{5}{2}\log\frac{I}{I_0},$$

where $I_0$ is the measure of a just-visible star, and $I$ is the actual intensity of the star being measured. The dimmest stars are of magnitude 6, and the brightest are of magnitude 1. Determine the ratio of light intensities between a star of magnitude 1 and a star of magnitude 3.

*Newton's law of cooling says that the rate at which a body cools is proportional to the difference* $C$ *in temperature between the body and the environment around it. The temperature* $f(t)$ *of the body at time* $t$ *in appropriate units after being introduced into an environment having constant temperature* $T_0$ *is*

$$f(t) = T_0 + Ce^{-kt},$$

*where* $C$ *and* $k$ *are constants. Use this result in Exercises 14–16.*

**14.** *Newton's Law of Cooling* Boiling water, at 100°C, is placed in a freezer at 0°C. The temperature of the water is 50°C after 24 minutes. Find the temperature of the water after 96 minutes.

**15.** *Newton's Law of Cooling* A pot of coffee with a temperature of 100°C is set down in a room with a temperature of 20°C. The coffee cools to 60°C after 1 hour.

**(a)** Write an equation to model the data.

**(b)** Find the temperature after a half hour.

**(c)** How long will it take for the coffee to cool to 50°C? Support your answer graphically.

**16.** *Newton's Law of Cooling* A piece of metal is heated to 300°C and then placed in a cooling liquid at 50°C. After 4 minutes, the metal has cooled to 175°C. Find its temperature after 12 minutes.

**17.** *Greenhouse Gases* Chlorofluorocarbons (CFCs) are gases created by people that increase the greenhouse effect and damage the ozone layer. CFC 12 is one type of chlorofluorocarbon used in refrigeration, air conditioning, and foam insulation. The table lists future concentrations of CFC 12 in parts per billion (ppb), if current trends continue.

| Year | 2000 | 2005 | 2010 | 2015 | 2020 |
|---|---|---|---|---|---|
| CFC 12 (ppb) | .72 | .88 | 1.07 | 1.31 | 1.60 |

*Source:* Turner, R., *Environmental Economics, An Elementary Approach,* Johns Hopkins University Press, 1993.

**(a)** Let $x = 0$ correspond to 2000 and $x = 20$ to 2020. Find values for $C$ and $a$ so that $f(x) = Ca^x$ models these data.

**(b)** Estimate the CFC 12 concentration in 2013.

**18.** *Global Warming* Greenhouse gases such as carbon dioxide trap heat from the sun. Presently, the net incoming solar radiation reaching Earth's surface is approximately 240 watts per square meter (w/m$^2$). Any portion of this amount that is due to greenhouse gases is called *radiative forcing*. The table lists the estimated increase in radiative forcing $R$ over the levels in 1750.

| x (year) | 1800 | 1850 | 1900 | 1950 | 2000 |
|---|---|---|---|---|---|
| $R(x)$ (w/m$^2$) | .2 | .4 | .6 | 1.2 | 2.4 |

*Source:* Nilsson, A., *Greenhouse Earth,* John Wiley and Sons, 1992.

**(a)** Estimate constants $C$ and $k$ so that $R(x) = Ce^{kx}$ models the data. Let $x = 0$ correspond to 1800.

**(b)** Estimate the year when the additional radiative forcing could reach 3 w/m$^2$.

**19.** *Global Warming* See Exercise 18. The relationship between radiative forcing $R$ and the increase in average global temperature $T$ in degrees Fahrenheit can be modeled by

$$T(R) = 1.03R.$$

For example, $T(2) = 1.03(2) = 2.06$ means that if Earth's atmosphere traps an additional 2 watts per square meter, then the average global temperature may increase by 2.06°F, if all other factors remain constant. (*Source:* Clime, W., *The Economics of Global Warming,* Institute for International Economics, 1992.)

**(a)** Use $R(x)$ from Exercise 18 to write an equation for $(T \circ R)(x)$, where $x$ is the year and $x = 0$ corresponds to the year 1800.

**(b)** Evaluate $(T \circ R)(100)$, and interpret its meaning.

**20.** *Reducing Carbon Emissions* When fossil fuels are burned, carbon is released into the atmosphere. Governments could reduce carbon emissions by placing a tax on fossil fuels. For example, a higher tax might lead people to use less gasoline. The cost-benefit equation

$$\ln(1 - P) = -.0034 - .0053x$$

estimates the relationship between a tax of $x$ dollars per ton of carbon and the percent $P$ reduction in emissions of carbon, where $P$ is in decimal form. Determine $P$ when $x = 60$. Interpret the result. (*Source:* Clime, W., *The Economics of Global Warming,* Institute for International Economics, 1992.)

*(Modeling) In Exercises 21–38, solve each financial application.*

**21.** *Interest on an Account* How long will it take for $1000 to grow to $5000 at an interest rate of 3.5% if interest is compounded **(a)** quarterly **(b)** continuously?

**22.** *Interest on an Account* How long will it take for $5000 to grow to $8400 at an interest rate of 6% if interest is compounded **(a)** semiannually **(b)** continuously?

**23.** *Interest on an Account* George Duda wants to buy a $30,000 car. He has saved $27,000. Find the number of years (to the nearest tenth) it will take for his $27,000 to grow to $30,000 at 6% interest, compounded quarterly.

**24.** *Doubling Time* Find the doubling time of an investment earning 2.5% interest if interest is compounded **(a)** quarterly **(b)** continuously?

**25.** *Comparison of Investment* Karen Guardino, who is self-employed, wants to invest $60,000 in a pension plan. One investment offers 7%, compounded quarterly. Another offers 6.75%, compounded continuously. Which investment will earn more interest in 5 years? How much more will the better plan earn?

**26.** *Growth of an Account* See Exercise 25. If Karen chooses the plan with continuous compounding, how long will it take for her $60,000 to grow to $80,000?

*The interest rate stated by a financial institution is sometimes called the* ***nominal rate.*** *If interest is compounded, the actual* *rate is, in general, higher than the nominal rate, and is called the* ***effective rate.*** *If r is the nominal rate and n is the number of times interest is compounded annually, then*

$$R = \left(1 + \frac{r}{n}\right)^n - 1$$

*is the effective rate. Here, R represents the annual rate that the investment would earn if simple interest were paid. Use this formula in Exercises 27 and 28.*

**27.** *Effective Rate* Find the effective rate if the nominal rate is 6% and interest is compounded quarterly.

**28.** *Effective Rate* Find the effective rate if the nominal rate is 4.5% and interest is compounded daily ($n = 365$).

*In the formula* $A = P\left(1 + \frac{r}{n}\right)^{nt}$, *we can interpret P as the* ***present value*** *of A dollars t years from now, earning annual interest r compounded n times per year. In this context, A is called the* ***future value.*** *If we solve the formula for P, we obtain*

$$P = A\left(1 + \frac{r}{n}\right)^{-nt}.$$

*Use this formula in Exercises 29–32.*

**29.** *Present Value* Find the present value of $10,000 5 years from now, if interest is compounded semiannually at 12%.

**30.** *Present Value* Find the present value of $25,000 2.75 years from now, if interest is compounded quarterly at 6%.

**31.** *Future Value* Find the interest rate necessary for a present value of $25,000 to grow to a future value of $31,360, if interest is compounded annually for 2 years.

**32.** *Future Value* Find the interest rate necessary for a present value of $1200 to grow to a future value of $1780, if interest is compounded quarterly for 5 years.

*In Exercises 33–35, use the amortization formulas given in this section to find* **(a)** *the monthly* ($n = 12$) *payment on a loan with the given conditions and* **(b)** *the total interest that will be paid during the term of the loan.*

**33.** *Amortization* $8500 is amortized over 4 years with an interest rate of 7.5%.

**34.** *Amortization* $9600 is amortized over 5 years with an interest rate of 9.2%.

**35.** *Amortization* $125,000 is amortized over 30 years with an interest rate of 7.25%.

**36.** *Amortization Payments* Use the formula

$$y = R\left[\frac{1 - (1 + i)^{-(n-x)}}{i}\right],$$

where $y$ is the unpaid balance after $x$ payments have been made on a loan with $n$ payments of $R$ dollars. Find the balance after 120 payments have been made on the loan in Exercise 35.

**37.** *Growth of an Account* Use the table feature of your graphing calculator to work parts (a) and (b).

**(a)** Find how long it will take $1500 invested at 5.75%, compounded daily, to triple in value. Locate the solution by systematically decreasing ΔTbl. Find the answer to the nearest day. (Find your answer to the nearest day by eventually letting $\Delta\text{Tbl} = \frac{1}{365}$. The decimal part of the solution can be multiplied by 365 to determine the number of days greater than the nearest year. For example, if the solution is determined to be 16.2027 years, then multiply .2027 by 365 to get 73.9855. The solution is then, to the nearest day, 16 years and 74 days.) Confirm your answer analytically.

**(b)** Find how long it will take $2000 invested at 8%, compounded daily, to be worth $5000.

**38.** *Growth of an Account* If interest is compounded continuously and the interest rate is tripled, what effect will this have on the time required for an investment to double?

*(Modeling) In Exercises 39–46, solve each problem from biological science and medicine.*

**39.** *Growth of Bacteria* *Escherichia coli* is a strain of bacteria that occurs naturally in many organisms. Under certain conditions, the number of bacteria present in a colony is approximated by

$$A(t) = A_0 e^{.023t},$$

where $t$ is in minutes. If $A_0 = 2{,}400{,}000$, find the number of bacteria at each time.

**(a)** 5 minutes **(b)** 10 minutes **(c)** 60 minutes

**40.** *Growth of Bacteria* The growth of bacteria in food products makes it necessary to time-date some products (such as milk) so that they will be sold and consumed before the bacteria count becomes too high. Suppose for a product that the number of bacteria present is given by

$$f(t) = 500e^{.1t},$$

under certain storage conditions, where $t$ is time in days after packing of the product and the value of $f(t)$ is in millions.

**(a)** If the product cannot be safely eaten after the bacteria count reaches 3,000,000,000, how long will this take?

**(b)** If $t = 0$ corresponds to January 1, what date should be placed on the product?

*A large cloud of radioactive debris from a nuclear explosion has floated over the Pacific Northwest, contaminating much of the hay supply. Consequently, farmers in the area are concerned that the cows who eat this hay will give contaminated milk. (The tolerance level for radioactive iodine in milk is 0.) The percent of the initial amount of radioactive iodine still present in the hay after t days is approximated by*

$$P(t) = 100e^{-.1t}.$$

*Use this information in Exercises 41 and 42.*

**41.** *Radioactive Iodine Level* Some scientists feel that the hay is safe after the percent of radioactive iodine has declined to 10% of the original amount. Find the number of days before the hay can be used.

**42.** *Radioactive Iodine Level* Other scientists believe that the hay is not safe until the level of radioactive iodine has declined to only 1% of the original level. Find the number of days this would take.

*For Exercises 43 and 44, refer to Example 8.*

**43.** *Drug Level* If 250 milligrams of a drug are administered, and the drug is only 75% as effective each subsequent hour, how much effective medicine will remain in the person's system after 6 hours?

**44.** *Drug Level* A new drug has been introduced which is 80% as effective each hour as the previous hour. A minimum of 20 milligrams must remain in the patient's bloodstream during the course of treatment. If 100 milligrams are administered, how many hours may elapse before another dose is necessary?

**45.** *Epidemics* In 1666 the village of Eyam, located in England, experienced an outbreak of the Great Plague. Out of 261 people in the community, only 83 people survived. The table shows a function $f$ that computes the number of people who had not (yet) been infected after $x$ days.

| $x$ | 0 | 15 | 30 | 45 |
|---|---|---|---|---|
| $f(x)$ | 254 | 240 | 204 | 150 |

| $x$ | 60 | 75 | 90 | 125 |
|---|---|---|---|---|
| $f(x)$ | 125 | 103 | 97 | 83 |

*Source:* Raggett, G., "Modeling the Eyam Plague," *The Institute of Mathematics and Its Applications* 18.

**(a)** Use a table to represent a function $g$ that computes the number of people in Eyam that were infected after $x$ days.

**(b)** Write an equation that shows the relationship between $f(x)$ and $g(x)$.

**(c)** Use graphing to decide which equation represents $g(x)$ better.

$$y_1 = \frac{171}{1 + 18.6e^{-.0747x}} \quad \text{or} \quad y_2 = 18.3(1.024)^x$$

**(d)** Use your results from parts (b) and (c) to find a formula for $f(x)$.

**46.** *Heart Disease Death Rates* The table contains heart disease death rates per 100,000 people in 1996 for selected ages.

| Age | 30 | 40 | 50 | 60 | 70 |
|---|---|---|---|---|---|
| Death Rate | 30.5 | 108.2 | 315 | 776 | 2010 |

*Source:* U.S. Department of Health and Human Services.

**(a)** Make a scatter diagram of the data in the window $[25, 75]$ by $[-100, 2100]$.

**(b)** Find a function $f$ that models the data.

**(c)** Estimate the heart disease death rate for people who are 80 years old.

*(Modeling) In Exercises 47–52, solve each problem from the social sciences.*

**47.** *Social Security* If major reform occurs in the Social Security system, individuals may be able to invest some of their contributions into individual accounts. Many of these accounts would be managed by financial firms that charge fees. The table lists the amount in billions of dollars that may be collected if fees are .93% of the assets each year.

| Year | 2005 | 2010 | 2015 | 2020 |
|---|---|---|---|---|
| Fees (in billions) | 20 | 41 | 80 | 136 |

*Source:* Social Security Advisory Council.

**(a)** Use exponential regression to find $a$ and $b$ so that $f(x) = ab^x$ models the data. Let $x = 5$ correspond to 2005, $x = 10$ to 2010, and so on.

**(b)** Graph $f$ and the data.

**(c)** Estimate the fees in 2013.

**48.** *Telecommuting* Some workers use technology such as fax machines, e-mail, computers, and multiple phone lines to work at home rather than in the office. However, because of the need for teamwork and collaboration in the workplace, fewer employees are telecommuting than expected. The table lists the expected telecommuters in millions from 1997 to 2006.

| Year | 1997 | 1998 | 1999 | 2000 |
|---|---|---|---|---|
| Telecommuters | 9.2 | 9.6 | 10.0 | 10.4 |

| Year | 2001 | 2002 | 2003 | 2004 |
|---|---|---|---|---|
| Telecommuters | 10.6 | 11.0 | 11.1 | 11.2 |

| Year | 2005 | 2006 |
|---|---|---|
| Telecommuters | 11.3 | 11.4 |

*Source: USA Today.*

Find values for $a$ and $b$ so that $f(x) = a + b \ln x$ models the data, where $x = 1$ corresponds to 1997, $x = 2$ to 1998, and so on.

**49.** *Gambling Revenues* Revenues in the United States from all forms of legal gambling increased between 1991 and 1995. The function represented by

$$f(x) = 26.6e^{.131x}$$

models these revenues in billions of dollars. In this model $x$ represents the year, where $x = 0$ corresponds to 1991. (*Source: International Gaming & Wagering Business.*)

**(a)** Estimate gambling revenues in 1995.

**(b)** Determine analytically the year when these revenues reached \$30 billion. (Round to the nearest whole number.)

**(c)** Support your result in part (b) graphically.

**50.** *Population Growth* In 2000 India's population reached 1 billion, and in 2025 it is projected to be 1.4 billion. (*Source:* U.S. Bureau of the Census.)

**(a)** Find values for $C$ and $a$ so that $f(x) = Ca^{x-2000}$ models the population of India in year $x$.

**(b)** Estimate India's population in 2010.

**(c)** Use $f$ to determine the year when India's population might reach 1.5 billion.

**51.** *Midair Near Collisions* The table shows the number of airliner near collisions $y$ in year $x$.

| $x$ | 1989 | 1991 | 1993 | 1995 |
|---|---|---|---|---|
| $y$ | 131 | 78 | 44 | 34 |

*Source:* Federal Aviation Administration.

**(a)** Approximate constants $C$ and $a$ so that $f(x) = Ca^{x-1989}$ models the data.

**(b)** Support your answer graphically, by graphing $f$ and the data.

**52.** *Population Growth* The population of Tennessee in millions is given by

$$P(x) = 4.88e^{.0133x},$$

where $x = 0$ corresponds to 1990, $x = 1$ to 1991, and so on. (*Source:* U.S. Bureau of the Census.)

**(a)** Determine analytically the year when the population of Tennessee was 5.4 million.

**(b)** Solve part (a) graphically.

*(Modeling) In real life, populations of bacteria, insects, and animals do not continue to grow indefinitely. Initially, population growth may be slow. Then, as their numbers increase, so does the rate of growth. After a region has become heavily populated or saturated, the population usually levels off because of limited resources. This type of growth may be modeled by a **logistic function** represented by*

$$f(x) = \frac{c}{1 + ae^{-bx}},$$

*where a, b, and c are positive constants.*

**53.** *Heart Disease* As age increases, so does the likelihood of coronary heart disease (CHD). The fraction of people $x$ years old with some CHD is modeled by

$$f(x) = \frac{.9}{1 + 271e^{-.122x}}.$$

(*Source:* Hosmer, D., and S. Lemeshow, *Applied Logistic Regression,* John Wiley and Sons, 1989.)

**(a)** Evaluate $f(25)$ and $f(65)$. Interpret the results.

**(b)** At what age does this likelihood equal 50%.

**54.** *Tree Growth* The height of a tree in feet after $x$ years is modeled by

$$f(x) = \frac{50}{1 + 47.5e^{-.22x}}.$$

**(a)** Make a table for $f$ starting at $x = 10$, incrementing by 10. What appears to be the maximum height of the tree?

**(b)** Graph $f$ and identify the horizontal asymptote. Explain its significance.

**(c)** After how long was the tree 30 feet tall?

## CHAPTER 5 SUMMARY

**KEY TERMS & SYMBOLS** | **KEY CONCEPTS**

### 5.1 Inverse Functions

one-to-one function
inverse function
$f^{-1}$ ($f$ inverse)

**ONE-TO-ONE FUNCTION**

A function $f$ is a one-to-one function if, for elements $a$ and $b$ from the domain of $f$,

$$a \neq b \quad \text{implies} \quad f(a) \neq f(b).$$

**HORIZONTAL LINE TEST**

If every horizontal line intersects the graph of a function at no more than one point, then the function is one-to-one.

**INVERSE FUNCTION**

Let $f$ be a one-to-one function. Then, $g$ is the inverse function of $f$ and $f$ is the inverse function of $g$ if

$$(f \circ g)(x) = x \quad \text{for every } x \text{ in the domain of } g,$$

and

$$(g \circ f)(x) = x \quad \text{for every } x \text{ in the domain of } f.$$

**FINDING AN EQUATION FOR $f^{-1}$, WHERE $f$ IS ONE-TO-ONE**

1. Replace $f(x)$ with $y$ in the equation.
2. Interchange $x$ and $y$ in the equation.
3. Solve for $y$, and replace $y$ with $f^{-1}(x)$.

Any restrictions on $x$ or $y$ should be considered.

**IMPORTANT FACTS ABOUT INVERSES**

1. If $f$ is one-to-one, then $f^{-1}$ exists.
2. The domain of $f$ is equal to the range of $f^{-1}$, and the range of $f$ is equal to the domain of $f^{-1}$.
3. If the point $(a, b)$ lies on the graph of $f$, then $(b, a)$ lies on the graph of $f^{-1}$, so the graphs of $f$ and $f^{-1}$ are reflections of each other across the line $y = x$.

### 5.2 Exponential Functions

exponential function
$e$

**EXPONENTIAL FUNCTION**

If $a > 0$, $a \neq 1$, then $f(x) = a^x$ is the exponential function with base $a$.

**CHARACTERISTICS OF EXPONENTIAL FUNCTIONS**

$f(x) = a^x, \quad a > 1$

- This function is increasing and continuous on its entire domain, $(-\infty, \infty)$.
- The $x$-axis is the horizontal asymptote as $x \to -\infty$.
- The graph goes through the points $(-1, a^{-1})$, $(0, 1)$, and $(1, a)$.

$f(x) = a^x, \quad 0 < a < 1$

- This function is decreasing and continuous on its entire domain, $(-\infty, \infty)$.
- The $x$-axis is the horizontal asymptote as $x \to \infty$.
- The graph goes through the points $(-1, a^{-1})$, $(0, 1)$, and $(1, a)$.

| KEY TERMS & SYMBOLS | KEY CONCEPTS |
|---|---|

**SOLVING EXPONENTIAL EQUATIONS (TYPE I)**

To solve a Type 1 exponential equation such as $4^x = 8^{2x-3}$, write each base as a power of the same rational number base, apply the power rule for exponents, set exponents equal, and solve the resulting equation.

**COMPOUNDING FORMULAS**

Compounded $n$ times per year: $A = P\left(1 + \frac{r}{n}\right)^{nt}$  Compounded continuously: $A = Pe^{rt}$

## 5.3 Logarithms and Their Properties

logarithm, $\log_a x$
base
argument
common logarithm, $\log x$
pH
natural logarithm, $\ln x$

**LOGARITHM**

For all positive numbers $a$, where $a \neq 1$, $a^y = x$ is equivalent to $y = \log_a x$. A logarithm is an exponent, and $\log_a x$ is the exponent to which $a$ must be raised in order to obtain $x$. The number $a$ is called the base of the logarithm, and $x$ is called the argument of the expression $\log_a x$. The value of $x$ will always be positive.

**COMMON LOGARITHM**

For all positive numbers $x$, $\log x = \log_{10} x$.

**NATURAL LOGARITHM**

For all positive numbers $x$, $\ln x = \log_e x$.

**PROPERTIES OF LOGARITHMS**

For $a > 0$, $a \neq 1$, and any real number $k$,

1. $\log_a 1 = 0$.
2. $\log_a a^k = k$.
3. $a^{\log_a k} = k, \quad k > 0$.

For $x > 0$, $y > 0$, $a > 0$, $a \neq 1$, and any real number $r$,

**Product Rule** 4. $\log_a xy = \log_a x + \log_a y$.
(The logarithm of the product of two numbers is equal to the sum of the logarithms of the numbers.)

**Quotient Rule** 5. $\log_a \frac{x}{y} = \log_a x - \log_a y$.
(The logarithm of the quotient of two numbers is equal to the difference between the logarithms of the numbers.)

**Power Rule** 6. $\log_a x^r = r \log_a x$.
(The logarithm of a number raised to a power is equal to the exponent multiplied by the logarithm of the number.)

**CHANGE-OF-BASE RULE**

For any positive real numbers $x$, $a$, and $b$, where $a \neq 1$ and $b \neq 1$,

$$\log_a x = \frac{\log_b x}{\log_b a}.$$

## 5.4 Logarithmic Functions

logarithmic function

The functions defined by

$$f(x) = a^x \quad \text{and} \quad g(x) = \log_a x$$

are inverses.

| KEY TERMS & SYMBOLS | KEY CONCEPTS |
|---|---|
| | **CHARACTERISTICS OF LOGARITHMIC FUNCTIONS**<br>$f(x) = \log_a x, \quad a > 1$<br>• This function is increasing and continuous on its entire domain, $(0, \infty)$.<br>• The $y$-axis is the vertical asymptote as $x \to 0$ from the right.<br>• The graph goes through the points $(a^{-1}, -1)$, $(1, 0)$, and $(a, 1)$.<br>$f(x) = \log_a x, \quad 0 < a < 1$<br>• This function is decreasing and continuous on its entire domain, $(0, \infty)$.<br>• The $y$-axis is the vertical asymptote as $x \to 0$ from the right.<br>• The graph goes through the points $(a, 1)$, $(1, 0)$, and $(a^{-1}, -1)$. |
| **5.5** Exponential and Logarithmic Equations and Inequalities | **PROPERTIES OF LOGARITHMIC AND EXPONENTIAL FUNCTIONS**<br>For $b > 0$ and $b \neq 1$,<br>**1.** $b^x = b^y$ **if and only if** $x = y$.<br>**2.** If $x > 0$ and $y > 0$, then $\log_b x = \log_b y$ **if and only if** $x = y$.<br>**SOLVING EXPONENTIAL AND LOGARITHMIC EQUATIONS**<br>To solve a Type 2 exponential equation such as $2^x = 3^{x+1}$, take the same base logarithm on each side, apply the power rule for logarithms so that the variables are no longer in the exponents, and solve the resulting equation.<br>An exponential or logarithmic equation can be solved by changing the equation into one of the following forms, where $a$ and $b$ are real numbers, $a > 0$, and $a \neq 1$.<br>**1.** $a^{f(x)} = b$<br>Solve by taking logarithms of each side. (Natural logarithms are the best choice if $a = e$.)<br>**2.** $\log_a f(x) = \log_a g(x)$<br>From the given equation, $f(x) = g(x)$, which is solved analytically.<br>**3.** $\log_a f(x) = b$<br>Solve by changing to exponential form, $f(x) = a^b$. |
| **5.6** Further Applications and Modeling with Exponential and Logarithmic Functions<br>exponential growth function<br>exponential decay function<br>half-life<br>doubling time | **GROWTH** $A(t) = A_0 e^{kt}, \quad k > 0$<br>**DECAY** $A(t) = A_0 e^{-kt}, \quad k > 0$<br>**SOUND INTENSITY** $d = 10 \log \frac{I}{I_0}$<br>**AMORTIZATION PAYMENTS** $R = \dfrac{P}{\left[\dfrac{1 - (1 + i)^{-n}}{i}\right]}$ |

## CHAPTER 5 REVIEW EXERCISES

*Determine whether each function is one-to-one. Assume that graphs shown are comprehensive.*

**1.**

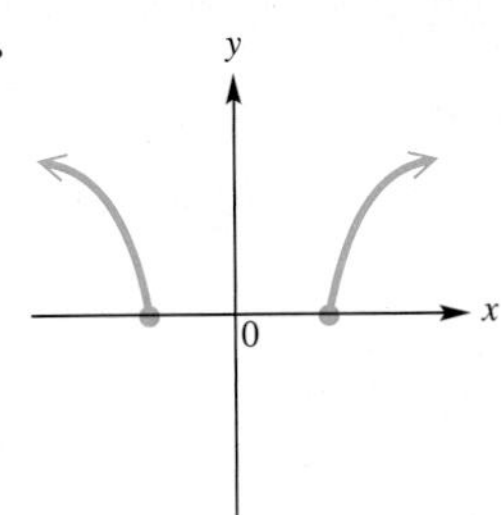


**2.** $f(x) = \sqrt{3x + 2}$

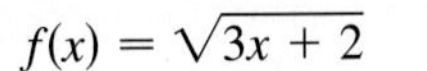

**3.**

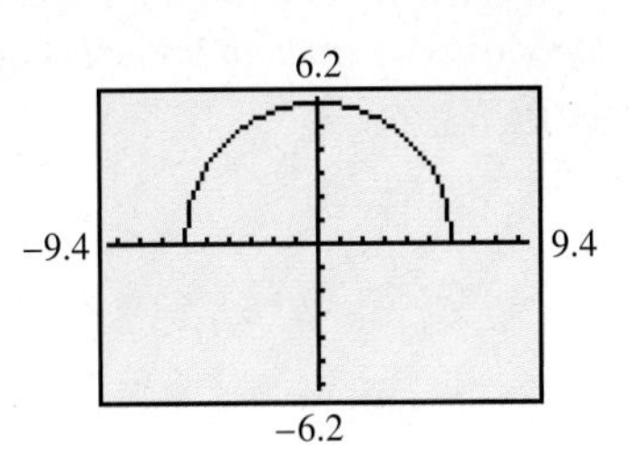


*For Exercises 4–9, consider the function defined by* $f(x) = \sqrt[3]{2x - 7}$.

**4.** What is the domain of $f$?

**5.** What is the range of $f$?

**6.** Explain why $f^{-1}$ exists.

**7.** Find a formula for $f^{-1}(x)$.

**8.** Graph both $f$ and $f^{-1}$ in a square viewing window, along with the line $y = x$. Describe how the graphs of $f$ and $f^{-1}$ are related.

**9.** Verify analytically that

$$(f \circ f^{-1})(x) = x \quad \text{and} \quad (f^{-1} \circ f)(x) = x.$$

*Concept Check* *Match each equation with the graph that most closely resembles its graph. Assume that $a > 1$.*

**10.** $y = a^{x+2}$

**11.** $y = a^x + 2$

**12.** $y = -a^x + 2$

**13.** $y = a^{-x} + 2$

**A.**

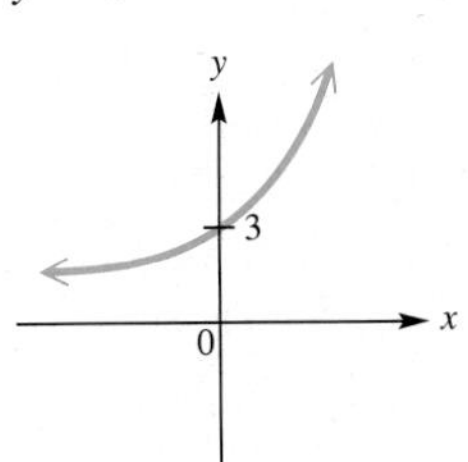


**B.**

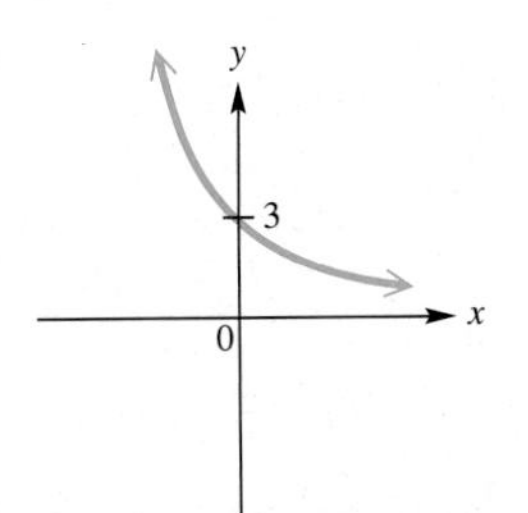


**C.**

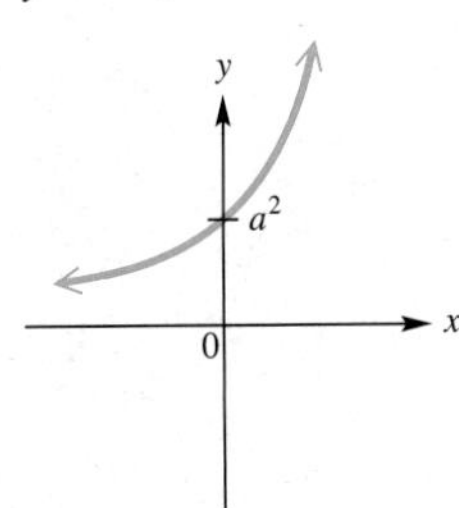


**D.**

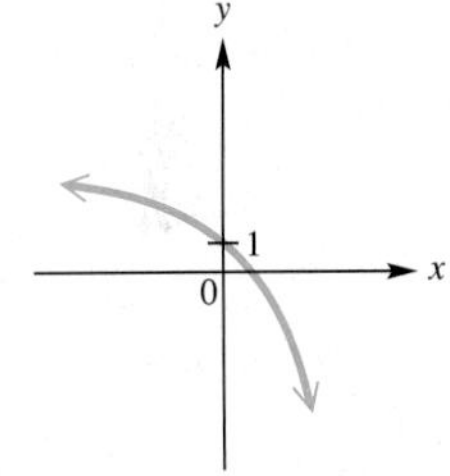


*Concept Check* *Consider the exponential function defined by $y = f(x) = a^x$, graphed at the right. Answer the following based on the graph.*

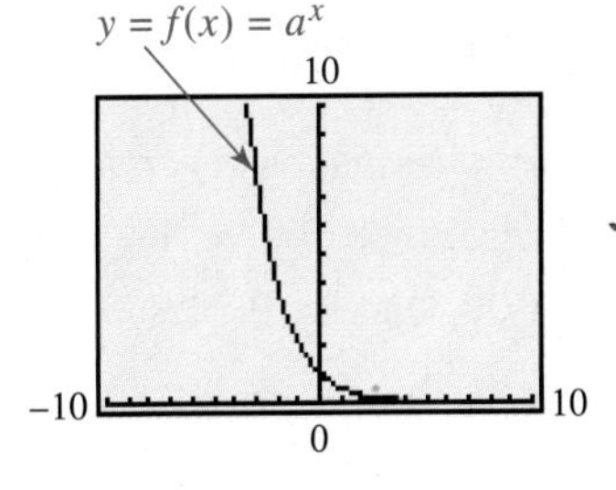


**14.** What is true about the value of $a$ in comparison to 1?

**15.** What is the domain of $f$?

**16.** What is the range of $f$?

**17.** What is the value of $f(0)$?

**18.** Sketch the graph of $y = f^{-1}(x)$ by hand.

**19.** What is the expression that defines $f^{-1}(x)$?

*Solve the equation in part (a) analytically. Then, use a graph to solve the inequality in part (b).*

**20. (a)** $\left(\frac{1}{8}\right)^{-2x} = 2^{x+3}$

**(b)** $\left(\frac{1}{8}\right)^{-2x} \geq 2^{x+3}$

**21. (a)** $3^{-x} = \left(\frac{1}{27}\right)^{1-2x}$

**(b)** $3^{-x} < \left(\frac{1}{27}\right)^{1-2x}$

**22. (a)** $.5^{-x} = .25^{x+1}$

**(b)** $.5^{-x} > .25^{x+1}$

**23. (a)** $.4^x = 2.5^{1-x}$

**(b)** $.4^x < 2.5^{1-x}$

**24.** The graphs of $y = x^2$ and $y = 2^x$ have the points $(2, 4)$ and $(4, 16)$ in common. There is a third point in common to the graphs whose coordinates can be approximated by using a graphing calculator. Find the coordinates, giving as many decimal places as your calculator will display.

*Use a calculator to find an approximation for each logarithm to four decimal places.*

**25.** $\log 58.3$ **26.** $\log .00233$ **27.** $\ln 58.3$ **28.** $\log_2 .00233$

*Evaluate each expression, giving an exact or approximate value as directed. In the case of approximations, give as many decimal places as your calculator displays.*

**29.** $\log_{13} 1$ (exact) **30.** $\ln e^{\sqrt{6}}$ (exact) **31.** $\log_5 5^{12}$ (exact)

**32.** $7^{\log_7 13}$ (exact) **33.** $3^{\log_3 5}$ (exact) **34.** $\log_4 9$ (approximate)

*Concept Check* *In Exercises 35–40, identify the corresponding graph for each function.*

**35.** $f(x) = \log_2 x$ **36.** $f(x) = \log_2(2x)$ **37.** $f(x) = \log_2\left(\frac{1}{x}\right)$

**38.** $f(x) = \log_2\left(\frac{x}{2}\right)$ **39.** $f(x) = \log_2(x - 1)$ **40.** $f(x) = \log_2(-x)$

**A.**

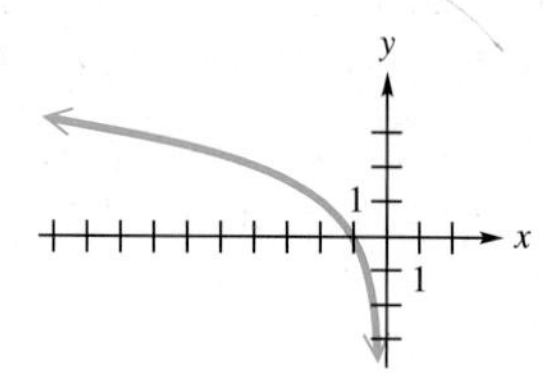


**B.**

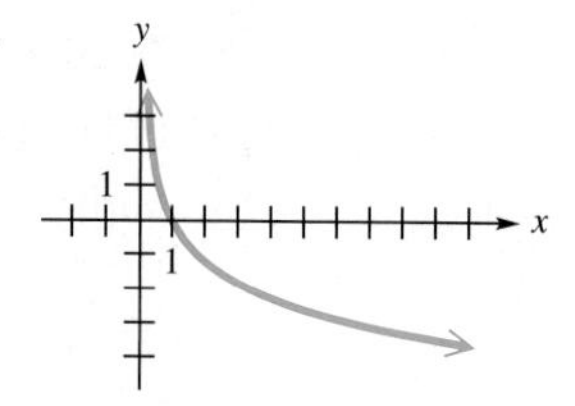


**C.**

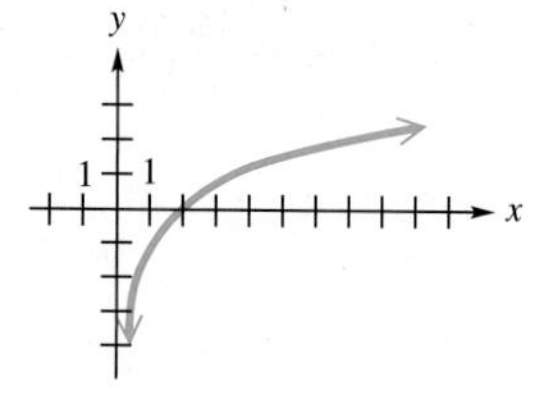


**D.**

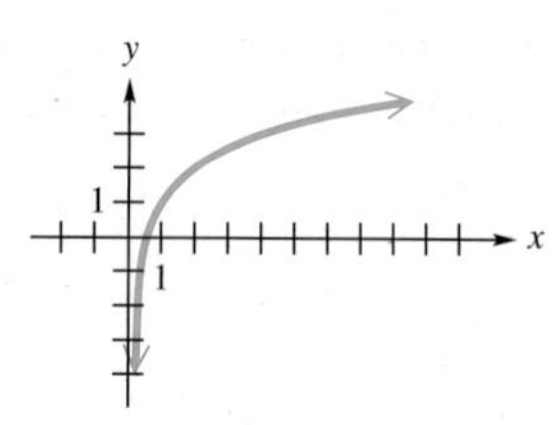


**E.**

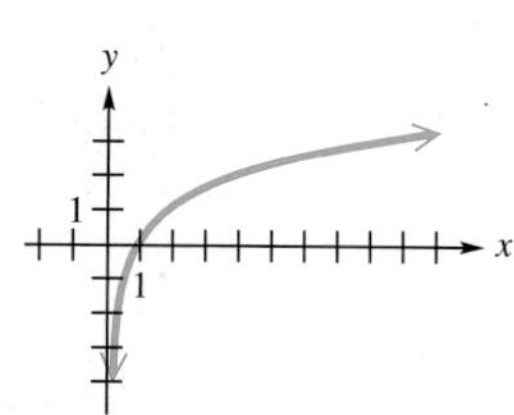


**F.**

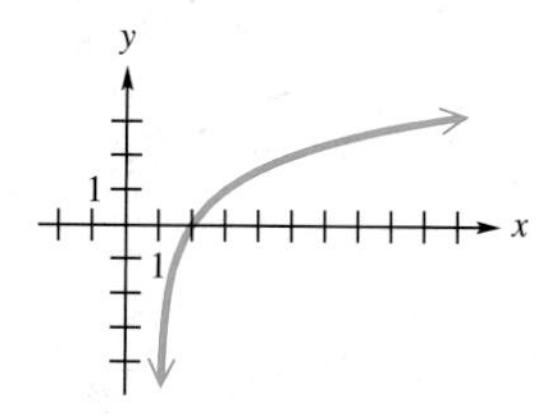


**41.** *Concept Check* What is the base of the logarithmic function whose graph contains the point $(81, 4)$?

**42.** *Concept Check* What is the base of the exponential function whose graph contains the point $\left(-4, \frac{1}{16}\right)$?

*Use properties of logarithms, if possible, to write each logarithm as a sum, difference, or product of logarithms.*

**43.** $\log_3 \frac{mn}{5r}$ **44.** $\log_2 \frac{\sqrt{7}}{15}$ **45.** $\log_5\left(x^2y^4\sqrt[5]{m^3p}\right)$ **46.** $\log_7(7k + 5r^2)$

*Solve the equation in part (a) analytically. Then, solve the inequality in part (b) using a graph.*

**47. (a)** $\log(x + 3) + \log x = 1$
**(b)** $\log(x + 3) + \log x > 1$

**48. (a)** $\ln e^{\ln x} - \ln(x - 4) = \ln 3$
**(b)** $\ln e^{\ln x} - \ln(x - 4) \leq \ln 3$

**49. (a)** $\ln e^{\ln 2} - \ln(x - 1) = \ln 5$
**(b)** $\ln e^{\ln 2} - \ln(x - 1) \geq \ln 5$

*Use one of the methods described in this chapter to solve each equation. Round to the nearest thousandth if necessary.*

**50.** $8^x = 32$ **51.** $\frac{8}{27} = x^{-3}$ **52.** $10^{2x-3} = 17$

**53.** $e^{x+1} = 10$ **54.** $\log_{64} x = \frac{1}{3}$ **55.** $\ln(6x) - \ln(x + 1) = \ln 4$

**56.** $\log_{16}\sqrt{x + 1} = \frac{1}{4}$ **57.** $\ln x + 3\ln 2 = \ln \frac{2}{x}$ **58.** $\ln[\ln(e^{-x})] = \ln 3$

**59.** $N = a + b\ln\frac{c}{d}$, for $c$ **60.** $A = A_0\left(1 + \frac{r}{n}\right)^{nt}$, for $t$

*Use the x-intercept method to estimate the solution(s) of each equation. Round to the nearest thousandth.*

**61.** $\log_{10} x = x - 2$

**62.** $2^{-x} = \log_{10} x$

**63.** $x^2 - 3 = \log x$

*Solve each application.*

**64.** *Interest Rate* To the nearest tenth, what interest rate, compounded annually, will produce \$8780 if \$3500 is left at interest for 10 years?

**65.** *Growth of an Account* Find the number of years (to the nearest tenth) needed for \$48,000 to become \$58,344 at 5% interest, compounded semiannually.

**66.** *Growth of an Account* Lateisha Shaw deposits \$12,000 for 8 years in an account paying 5%, compounded annually. She then leaves the money alone with no further deposits at 6%, compounded annually, for an additional 6 years. Find the total amount on deposit after the entire 14-year period.

**67.** *Growth of an Account* Suppose that \$2000 is invested in an account that pays 3% annually and is left for 5 years.

**(a)** How much will be in the account if interest is compounded quarterly (4 times per year)?

**(b)** How much will be in the account if interest is compounded continuously?

**(c)** To the nearest tenth of a year, how long will it take the \$2000 to triple if interest is compounded continuously?

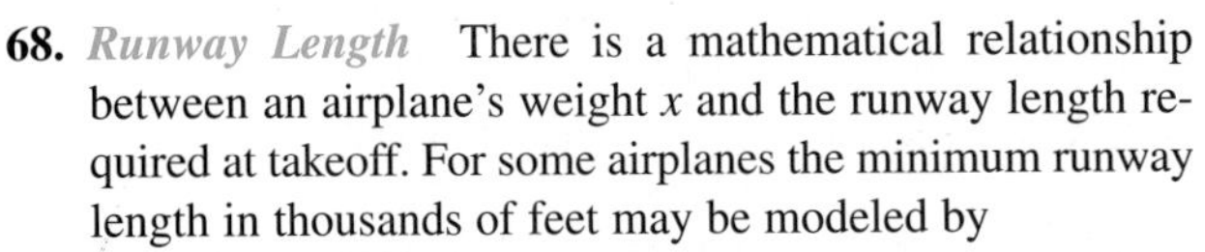

*(Modeling) Solve each problem.*

**68.** *Runway Length* There is a mathematical relationship between an airplane's weight $x$ and the runway length required at takeoff. For some airplanes the minimum runway length in thousands of feet may be modeled by

$$L(x) = 3 \log x,$$

where $x$ is measured in thousands of pounds. (*Source:* Haefner, L., *Introduction to Transportation Systems,* Holt, Rinehart and Winston, 1986.)

**(a)** Graph $L$ in the window $[0, 50]$ by $[0, 6]$. Interpret the graph.

**(b)** If the weight of an airplane increases tenfold from 10,000 to 100,000 pounds, does the length of the required runway also increase by a factor of 10? Explain.

**69.** *Calorie Consumption* The function defined by

$$f(x) = 280 \ln(x + 1) + 1925$$

models the number of calories consumed daily by a person owning $x$ acres of land in a developing country. Find the number of acres owned by someone whose average calorie intake is 2200 calories per day. (*Source:* Grigg, D., *The World Food Problem,* Blackwell Publishers, 1993.)

**70.** *Pollutant Concentration* The concentration of a pollutant, in grams per liter, in the east fork of the Big Weasel River is approximated by

$$P(x) = .04e^{-4x},$$

where $x$ is the number of miles downstream from a paper mill where the measurement is taken. Find the following.

**(a)** $P(.5)$ **(b)** $P(1)$

**(c)** The concentration of the pollutant 2 miles downstream

**(d)** The number of miles downstream where the concentration of the pollutant is .002 gram per liter

**71.** *Repetitive Skills* A person learning certain skills involving repetition tends to learn quickly at first. Then, learning tapers off and approaches some upper limit. Suppose the number of symbols per minute that a textbook typesetter can produce is given by

$$p(x) = 250 - 120(2.8)^{-.5x},$$

where $x$ is the number of months the typesetter has been in training. Find the following.

**(a)** $p(2)$ **(b)** $p(10)$

**(c)** Graph $y = p(x)$ in the window $[0, 10]$ by $[0, 300]$, and support the answer to part (a).

**72.** *Free Fall of a Skydiver* A skydiver in free fall travels at a speed of

$$v(t) = 176(1 - e^{-.18t})$$

feet per second after $t$ seconds. How long will it take for the skydiver to attain a speed of 147 feet per second (100 mph)?

## CHAPTER 5 TEST

1. Match each equation in parts (a)–(d) with its graph.

   **(a)** $y = \log_{1/2} x$ **(b)** $y = e^x$

   **(c)** $y = \ln x$ **(d)** $y = \left(\frac{1}{2}\right)^x$

   **(e)** Which pairs of functions in parts (a)–(d) are inverses?

**A.**

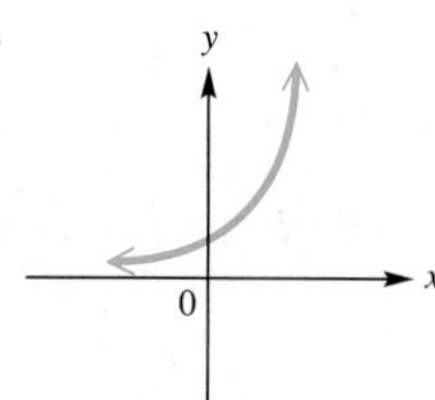


**B.**

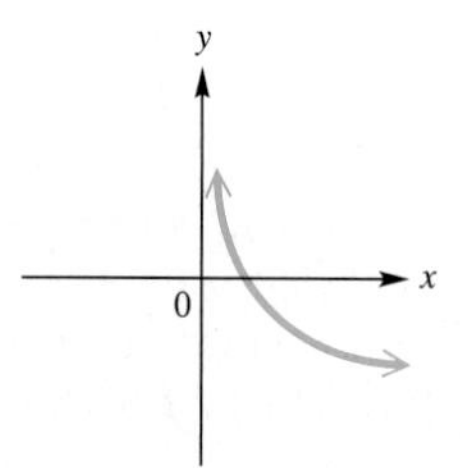


**C.**

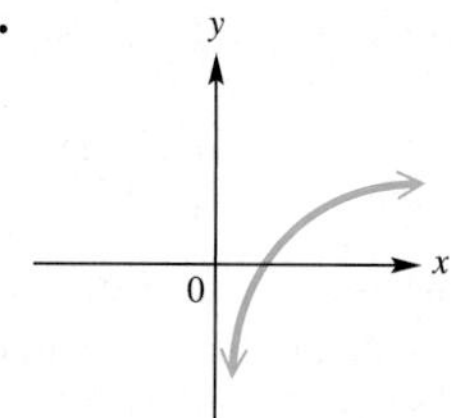


**D.**

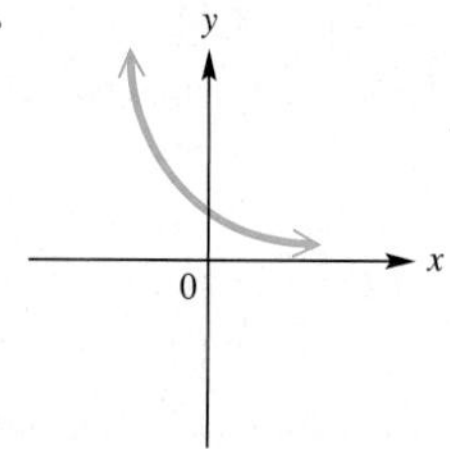


2. Consider the function defined by $f(x) = -2^{x-1} + 8$.

   **(a)** Graph $f$ in the standard viewing window of your calculator.

   **(b)** Give the domain and range of $f$.

   **(c)** Does the graph have an asymptote? If so, is it vertical or horizontal, and what is its equation?

   **(d)** Find the $x$- and $y$-intercepts analytically, and use the graph from part (a) to support your answers graphically.

   **(e)** Find $f^{-1}(x)$.

3. Solve the equation $\left(\frac{1}{8}\right)^{2x-3} = 16^{x+1}$ analytically.

4. *Growth of an Account* Suppose that \$10,000 is invested at 5.5% for 4 years. Find the total amount present at the end of this time period if interest is compounded **(a)** quarterly and **(b)** continuously.

5. One of your friends is taking another mathematics course and tells you, "I have no idea what an expression like $\log_5 27$ really means." Write an explanation of what it means, and tell how you can find an approximation for it with a calculator.

6. Use a calculator to find an approximation of each logarithm to the nearest thousandth.

   **(a)** $\log 45.6$ **(b)** $\ln 470$ **(c)** $\log_3 769$

7. Use the power, quotient, and product rules of logarithms to write $\log \frac{m^3 n}{\sqrt{y}}$ as an equivalent expression.

8. Consider the equation $\log_2 x + \log_2(x + 2) = 3$.

   **(a)** Solve the equation analytically. If there is an extraneous solution, what is it?

   **(b)** To support the solution in part (a), we can graph $y_1 = \log_2 x + \log_2(x + 2) - 3$ and find the $x$-intercept. Write an expression for $y_1$, using the change-of-base rule with base 10, and graph the function.

   **(c)** Use the graph to solve the logarithmic inequality $\log_2 x + \log_2(x + 2) > 3$.

9. Solve $6^{2-x} = 2^{3x+1}$. Give the solution set **(a)** with an exact value, using common logarithms and **(b)** with an approximation to the nearest thousandth.

10. *(Modeling) Population Growth* A population is increasing according to the equation

$$y = 2e^{.02t},$$

where $y$ is in millions and $t$ is in years. Match each question in parts (a)–(d) with one of the solution methods A–D.

   **(a)** How long will it take for the population to triple?

   **(b)** When will the population reach 3 million?

   **(c)** How large will the population be in 3 years?

   **(d)** How large will the population be in 4 months?

   **A.** Evaluate $2e^{.02(1/3)}$. **B.** Solve $2e^{.02t} = 3 \cdot 2$, for $t$.

   **C.** Evaluate $2e^{.02(3)}$. **D.** Solve $2e^{.02t} = 3$, for $t$.

11. *(Modeling) Drug Level in the Bloodstream* After a medical drug is injected into the bloodstream, it is gradually eliminated from the body. Graph each function in parts (a)–(d) on the interval $[0, 10]$. Use $[0, 500]$ for the

range of $A(t)$. Use a graphing calculator to determine the function that best models the amount $A(t)$ (in milligrams) of a drug remaining in the body after $t$ hours if 350 milligrams were initially injected.

**(a)** $A(t) = t^2 - t + 350$ **(b)** $A(t) = 350 \log(t + 1)$

**(c)** $A(t) = 350(.75)^t$ **(d)** $A(t) = 100(.95)^t$

**12.** *(Modeling) Decay of Radium* Radium 226 has a half-life of about 1600 years. An initial sample weighs 2 grams.

**(a)** Find a model for the decay function.

**(b)** Find the amount left after 9600 years.

**(c)** Find the time for the initial amount to decay to .5 gram.

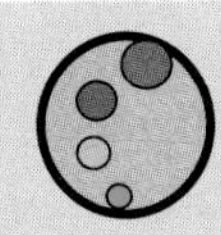

# Chapter 5 Project

## Modeling Motor Vehicle Sales in the United States (with a lesson about careless use of mathematical models)

During the 1990s both the numbers of new motor vehicles sold and the numbers of new trucks sold increased significantly. The table gives approximations for these numbers in 1992, 1996, and 1999. Note that new motor vehicle sales *include* new truck sales.

| Year | Motor Vehicle Sales | Truck Sales |
|---|---|---|
| 1992 | 13,100,000 | 4,900,000 |
| 1996 | 15,500,000 | 6,900,000 |
| 1999 | 17,400,000 | 8,700,000 |

*Source:* U.S. Bureau of Economic Analysis.

Growth in sales can sometimes be modeled by a function having the form $f(x) = a(b)^x$, where $a$ and $b$ are constants. The constant $a$ represents the sales during an initial year, and the variable $x$ corresponds to the number of years that have passed since the initial year. Once a rule for $f(x)$ is found, the constant $b$ can be used to determine how fast sales increased. For example, if $b = 1.03$, then sales increased by 3% per year.

### Activities

1. Find a function defined by $f(x) = a(b)^x$ that models new motor vehicle sales (in millions) in 1992 and in 1999. Let $x = 0$ correspond to 1992 and $x = 7$ to 1999. (*Hint:* To determine values for $a$ and $b$, let $f(0) = 13.1$ and $f(7) = 17.4$.)
2. Now find a function defined by $g(x) = c(d)^x$ that models new truck sales (in millions) in 1992 and in 1999 similarly to how $f(x)$ was found.

3. Use $f(x)$ and $g(x)$ to estimate to the nearest tenth of a percent the annual increases in sales of motor vehicles and in sales of trucks from 1992 to 1999.
4. Estimate sales of new motor vehicles and sales of new trucks in 1996. Compare your results to the actual values shown in the table.
5. Use $f(x)$ and $g(x)$ to predict new motor vehicle sales and new truck sales in 2020. Discuss your results.
6. Graph $f$ and $g$ together in the window $[0, 33]$ by $[0, 80]$. Do the functions $f$ and $g$ give realistic sales figures from 1992 to 2025? Explain. Comment on the following statement: "A model is best when it is used to estimate values between given data points."

# 6 Analytic Geometry

SINCE the beginning of civilization, people have been fascinated by the universe they live in. In 1887, T.H. Huxley wrote:

> The known is finite, the unknown infinite; intellectually we stand on an islet in the midst of an illimitable ocean of inexplicability. Our business in every generation is to reclaim a little more land.

The Greeks, together with early Christian astronomers, believed that Earth was the center of the universe and the sun and planets traveled in circular orbits around Earth. The circle was regarded as the perfect geometric shape. Later, Johannes Kepler determined that the planets move in elliptical, rather than circular orbits, around the sun. Edmund Halley determined that comets also followed elliptical orbits around the sun and accurately predicted the return of Halley's comet. People now know that both celestial objects and satellites move through space in one of three types of paths: elliptical, parabolic, or hyperbolic. These curves are called *conic sections* and were discovered in 200 B.C. by the Greek geometer Apollonius.

The search to understand the universe has been directed toward both the infinite and the infinitesimal. In 1911 Ernest Rutherford determined the basic structure of the atom. Small atomic particles are capable of traveling in trajectories described by conic sections.

In this chapter, we learn about these age-old curves that have had such a profound influence on our understanding of who we are and the cosmos we live in.

*Source:* Boorse, H., L. Motz, and J. Weaver, *The Atomic Scientists,* John Wiley and Sons, 1989; National Council of Teachers of Mathematics, *Historical Topics for the Mathematics Classroom,* Thirty-first Yearbook, Washington, D.C., 1969; Sagan, C., *Cosmos,* Random House, 1980.

CHAPTER OUTLINE

# 6.1 CIRCLES AND PARABOLAS

Conic Sections • Equations and Graphs of Circles • An Application of Circles • Equations and Graphs of Parabolas • An Application of Parabolas

## Conic Sections

Parabolas, circles, ellipses, and hyperbolas form a group of curves known as the **conic sections,** because they are the result of intersecting a cone with a plane. Figure 1 illustrates these curves, which can be defined mathematically using the distance formula,

$$d(P, R) = \sqrt{(x_2 - x_1)^2 + (y_2 - y_1)^2},$$

from Section 1.1.

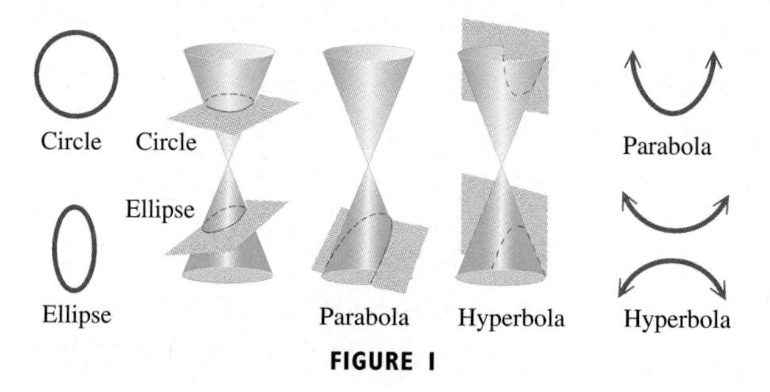


**FIGURE 1**

If the plane intersects the cone at the vertex, the intersections illustrated in Figure 1 are reduced respectively to a point, a line, or two intersecting lines. These are called *degenerate conic sections.*

## Equations and Graphs of Circles

> **Circle**
>
> A **circle** is a set of points in a plane that are equidistant from a fixed point. The distance is called the **radius** of the circle, and the fixed point is called the **center.**

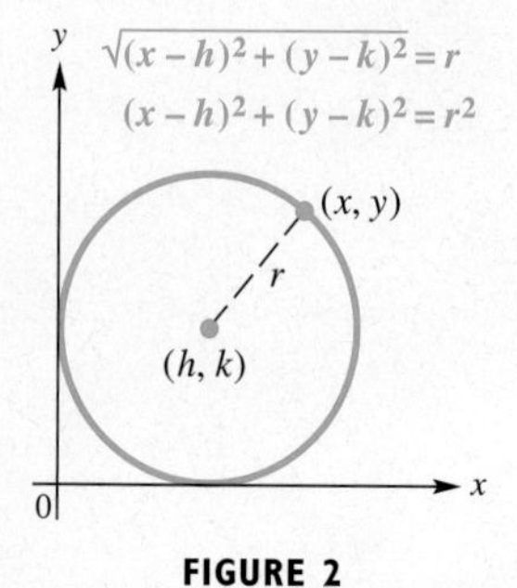


**FIGURE 2**

Suppose a circle has center $(h, k)$ and radius $r > 0$, as shown in Figure 2. Then the distance between the center $(h, k)$ and any point $(x, y)$ on the circle must equal $r$. By the distance formula,

$$\sqrt{(x - h)^2 + (y - k)^2} = r,$$

so

$$(x - h)^2 + (y - k)^2 = r^2 \qquad \text{Square both sides.}$$

is an equation of the circle.

**Center-Radius Form of the Equation of a Circle**

The circle with center $(h, k)$ and radius $r$ has equation

$$(x - h)^2 + (y - k)^2 = r^2,$$

the **center-radius form** of the equation of the circle.

For example, $(x + 4)^2 + (y - 5)^2 = 10$ is the equation of the circle with center $(-4, 5)$ and radius $\sqrt{10}$.

Notice that a circle is the graph of a relation that is not a function, since it does not pass the vertical line test.

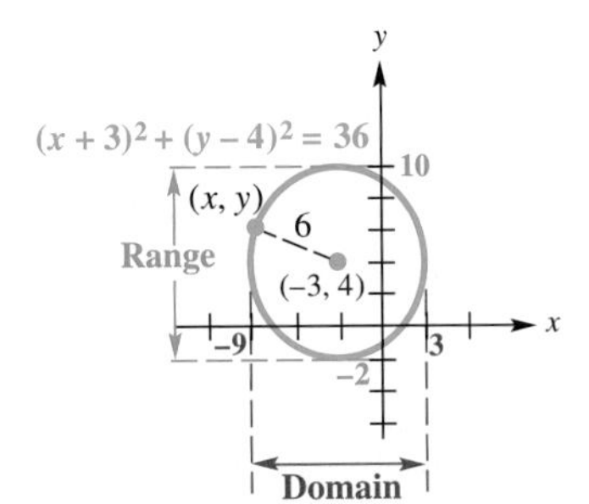

FIGURE 3

**EXAMPLE 1** *Finding the Equation of a Circle*

Find the center-radius form of the equation of a circle with radius 6 and center $(-3, 4)$. Graph the circle by hand, and give the domain and range of the relation.

**Solution** Using the center-radius form with $h = -3$, $k = 4$, and $r = 6$, we find that the equation of the circle is

$$(x - (-3))^2 + (y - 4)^2 = 6^2 \quad \text{or} \quad (x + 3)^2 + (y - 4)^2 = 36.$$

The graph is shown in Figure 3. The domain is $[-9, 3]$, and the range is $[-2, 10]$.

If a circle has center at the origin $(0, 0)$, then its equation is found by using $h = 0$ and $k = 0$ in the center-radius form.

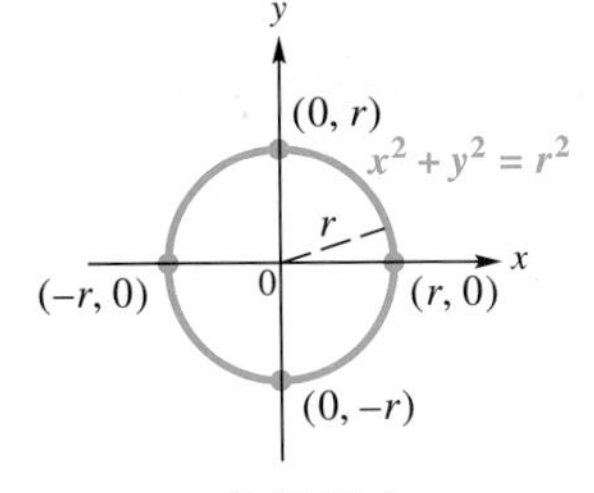

FIGURE 4

**Equation of a Circle with Center at the Origin**

A circle with center $(0, 0)$ and radius $r$ has equation

$$x^2 + y^2 = r^2.$$

See Figure 4.

**EXAMPLE 2** *Finding the Equation of a Circle with Center at the Origin*

Find the equation of a circle with center at the origin and radius 3. Give a traditional graph, and state the domain and range of the relation.

**Solution** Using the form $x^2 + y^2 = r^2$ with $r = 3$, we find that the equation of the circle is $x^2 + y^2 = 9$. The graph is shown in Figure 5. Both the domain and range are $[-3, 3]$.

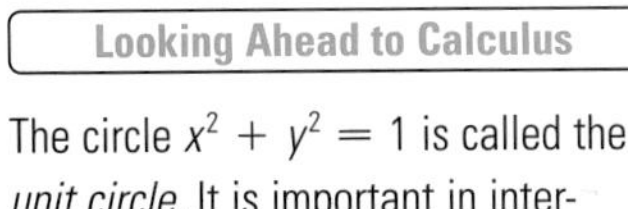
**Looking Ahead to Calculus**

The circle $x^2 + y^2 = 1$ is called the *unit circle*. It is important in interpreting the *trigonometric* or *circular* functions in calculus.

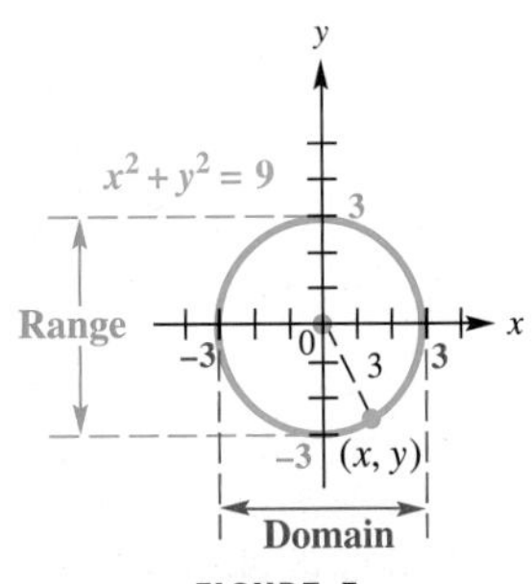

FIGURE 5

TECHNOLOGY NOTE

To obtain an undistorted graph of a circle on a graphing calculator screen, a *square viewing window* must be used. For example, the graph of $x^2 + y^2 = 9$ on a rectangular window looks like an ellipse.

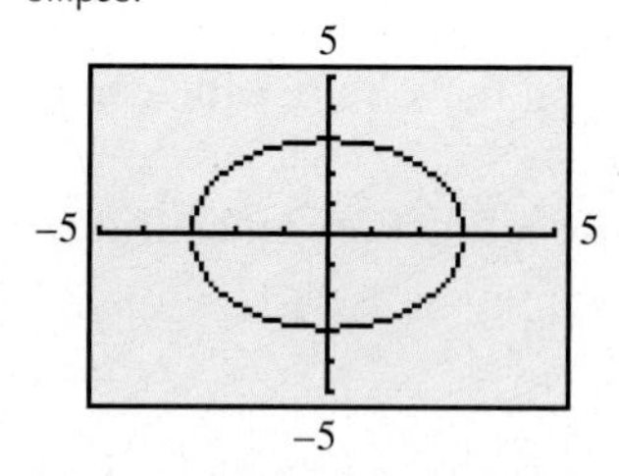


As mentioned in Section 4.4, a graphing calculator in function mode cannot directly graph a circle. We must first solve the equation of the circle for $y$, obtaining two functions $y_1$ and $y_2$. The union of these two graphs will be the graph of the entire circle.

**EXAMPLE 3** *Graphing Circles*

Use a graphing calculator to graph each circle in a square viewing window.

**(a)** $x^2 + y^2 = 9$ **(b)** $(x + 3)^2 + (y - 4)^2 = 36$

**Solution** In both cases, first solve for $y$. Recall that if $k > 0$, $y^2 = k$ has two real solutions, $\sqrt{k}$ and $-\sqrt{k}$.

**(a)** $x^2 + y^2 = 9$

$$y^2 = 9 - x^2 \qquad \text{Subtract } x^2.$$
$$y = \pm\sqrt{9 - x^2} \qquad \text{Take square roots.}$$

Graph $y_1 = \sqrt{9 - x^2}$ and $y_2 = -\sqrt{9 - x^2}$. See Figure 6, and compare to the traditional graph in Figure 5 on the previous page.

TECHNOLOGY NOTE

If you try to graph the circle in Figure 6 without using a decimal window, you may get the figures shown here, where the two graphs do not meet at the points (−3, 0) and (3, 0) as expected. In some cases, a decimal window does not correct this problem. Although the technology is deceiving, mathematically this is a complete circle.

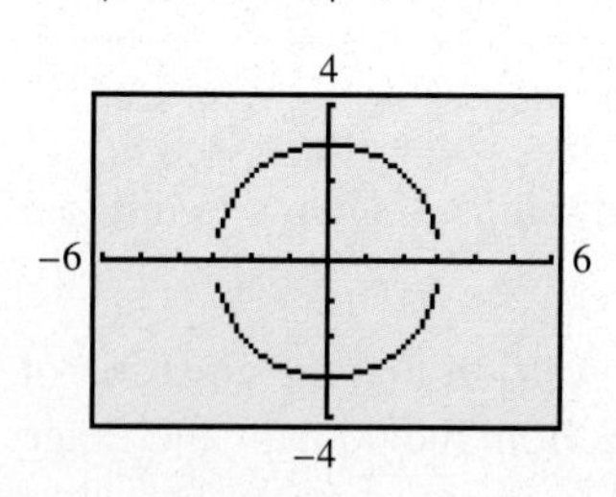


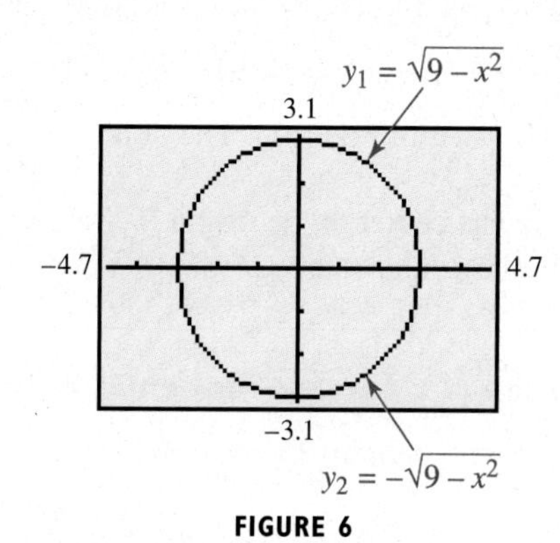


**FIGURE 6**

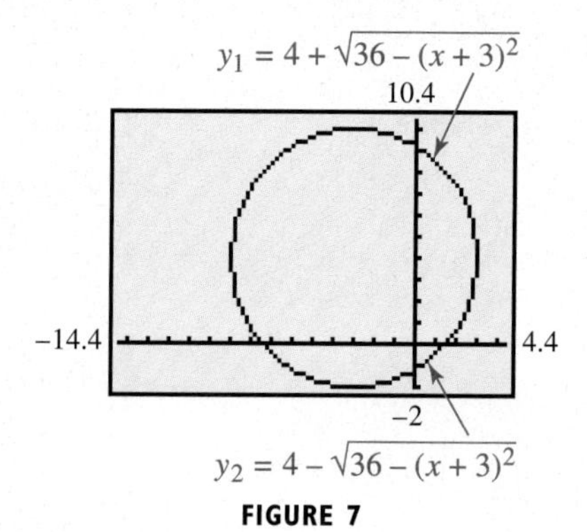


**FIGURE 7**

**(b)** $(x + 3)^2 + (y - 4)^2 = 36$

$$(y - 4)^2 = 36 - (x + 3)^2 \qquad \text{Subtract } (x + 3)^2.$$
$$y - 4 = \pm\sqrt{36 - (x + 3)^2} \qquad \text{Take square roots.}$$
$$y = 4 \pm \sqrt{36 - (x + 3)^2} \qquad \text{Add 4.}$$

The two functions to be graphed are

$$y_1 = 4 + \sqrt{36 - (x + 3)^2} \quad \text{and} \quad y_2 = 4 - \sqrt{36 - (x + 3)^2}.$$

See Figure 7, and compare to Figure 3 on the previous page.

The table feature of a graphing calculator suggests the domain of the two functions used to obtain the graph of the circle $x^2 + y^2 = 9$. Figure 8 shows two views of the table. Notice that the table shows ERROR for $Y_1$ and $Y_2$ when $X < -3$ and when $X > 3$. $Y_1$ and $Y_2$ are evaluated only when $-3 \leq X \leq 3$, which suggests that the domain of both $Y_1$ and $Y_2$ is the closed interval $[-3, 3]$. This result can be supported graphically in a window that evaluates the endpoints of each semicircle. (A decimal window will usually work.)

### FOR DISCUSSION

**1.** In Example 3(b), the final expressions for $y_1$ and $y_2$ have radicand* $36 - (x + 3)^2$. If this were simplified analytically, the result would be $27 - x^2 - 6x$. Use your calculator to replace the given radicands with this simplified result. Do you get the same graphs? Does it matter to the calculator which form you use? What would be a possible drawback of attempting to use the simplified form? (*Hint*: Nobody's perfect.)

**2.** Suppose the functions $y_1$ and $y_2$ from Example 3(a) were graphed in a standard window. What might be a possible misinterpretation by a student who has not studied the mathematical theory along with the technological approach?

| X | Y1 | Y2 |
|---|---|---|
| -6 | ERROR | ERROR |
| -5 | ERROR | ERROR |
| -4 | ERROR | ERROR |
| -3 | 0 | 0 |
| -2 | 2.2361 | -2.236 |
| -1 | 2.8284 | -2.828 |
| 0 | 3 | -3 |
| X= -6 | | |

| X | Y1 | Y2 |
|---|---|---|
| 0 | 3 | -3 |
| 1 | 2.8284 | -2.828 |
| 2 | 2.2361 | -2.236 |
| 3 | 0 | 0 |
| 4 | ERROR | ERROR |
| 5 | ERROR | ERROR |
| 6 | ERROR | ERROR |
| X=6 | | |

**FIGURE 8**

Starting with the center-radius form of the equation of a circle, $(x - h)^2 + (y - k)^2 = r^2$, and squaring $x - h$ and $y - k$ gives the **general form of the equation of a circle,**

$$x^2 + y^2 + cx + dy + e = 0, \qquad (*)$$

where $c$, $d$, and $e$ are real numbers dependent on $h$, $k$, and $r$. Starting with an equation in this form, we can complete the square to get an equation of the form

$$(x - h)^2 + (y - k)^2 = m$$

for some number $m$. If $m > 0$, then $r^2 = m$, and the equation represents a circle with radius $\sqrt{m}$. If $m = 0$, then the equation represents the single point $(h, k)$. If $m < 0$, no points satisfy the equation.

**EXAMPLE 4** *Finding the Center and Radius of a Circle*

Find the center and the radius of the circle with equation

$$x^2 - 6x + y^2 + 4y - 3 = 0.$$

Then, graph the circle, and give the domain and range.

**Analytic Solution**

The goal is to transform the given equation into an equivalent equation of the form $(x - h)^2 + (y - k)^2 = r^2$. First write the given equation with the constant on the right.

$$x^2 - 6x + y^2 + 4y = 3$$

Now complete the square on both $x$ and $y$. To complete the square on $x$, add $\left[\frac{1}{2}(-6)\right]^2 = 9$ to each side, and to complete the square on $y$, add $\left[\frac{1}{2}(4)\right]^2 = 4$ to each side. Insert parentheses as shown.

$$(x^2 - 6x + 9) + (y^2 + 4y + 4) = 3 + 9 + 4$$

Now, factor on the left and add on the right.

$$(x - 3)^2 + (y + 2)^2 = 16 \qquad (\dagger)$$

$$(x - 3)^2 + [y - (-2)]^2 = 4^2 \qquad \text{Write } +2 \text{ as } -(-2) \text{ and } 16 \text{ as } 4^2.$$

The circle has center $(3, -2)$ and radius 4. The graph in Figure 9 on the next page suggests that the domain is $[-1, 7]$, and the range is $[-6, 2]$.

**Graphing Calculator Solution**

To graph the circle on a graphing calculator, use the equation in the line marked (†) in the analytic solution, and solve for $y$ to get

$$y = -2 \pm \sqrt{16 - (x - 3)^2}.$$

Let

$$y_1 = -2 + \sqrt{16 - (x - 3)^2}$$

and

$$y_2 = -2 - \sqrt{16 - (x - 3)^2}$$

to obtain the calculator graph shown in Figure 10 on the next page.

*(continued)*

*The *radicand* is the expression under the radical symbol.

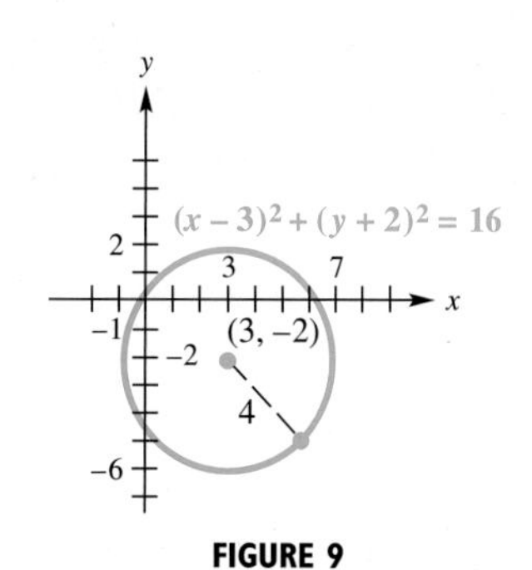


FIGURE 9

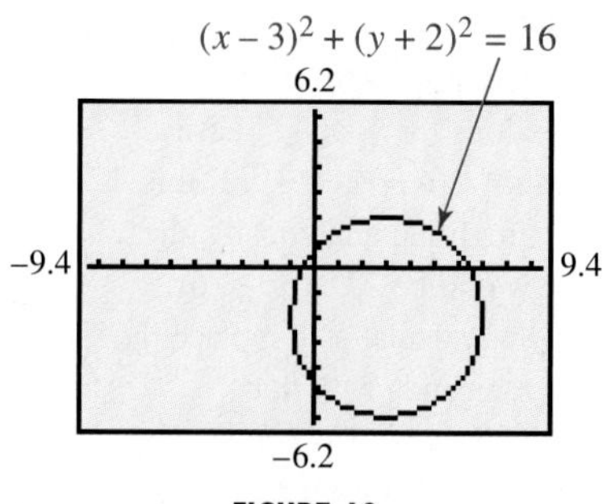


FIGURE 10

**EXAMPLE 5** *Finding the Center and Radius of a Circle*

Find the center and radius of the circle with equation $2x^2 + 2y^2 - 6x + 10y = 1$.

**Solution** Complete the square on both $x$ and $y$.

$$2x^2 + 2y^2 - 6x + 10y = 1$$

$$2(x^2 - 3x \quad) + 2(y^2 + 5y \quad) = 1 \qquad \text{Group terms; factor 2 from each group.}$$

$$2\left(x^2 - 3x + \frac{9}{4}\right) + 2\left(y^2 + 5y + \frac{25}{4}\right) = 1 + 2\left(\frac{9}{4}\right) + 2\left(\frac{25}{4}\right) \qquad \text{Complete the square.}$$

$$2\left(x - \frac{3}{2}\right)^2 + 2\left(y + \frac{5}{2}\right)^2 = 18 \qquad \text{Factor; simplify.}$$

$$\left(x - \frac{3}{2}\right)^2 + \left(y + \frac{5}{2}\right)^2 = 9 \qquad \text{Divide each side by 2.}$$

The graph of the equation is a circle with center $\left(\frac{3}{2}, -\frac{5}{2}\right)$ and radius 3.

## An Application of Circles

Seismologists can locate the epicenter of an earthquake by determining the intersection of three circles. The radii of these circles represent the distances from the epicenter to the three receiving stations. The centers of the circles represent the receiving stations.

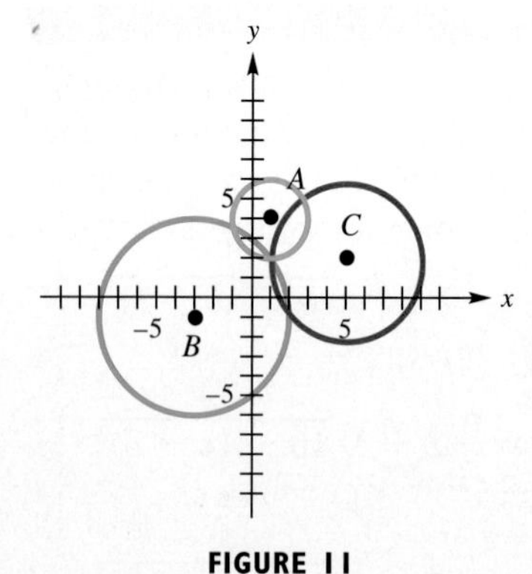


FIGURE 11

**EXAMPLE 6** *Locating the Epicenter of an Earthquake*

Suppose receiving stations $A$, $B$, and $C$ are located on a coordinate plane at the points $(1,4)$, $(-3,-1)$, and $(5,2)$, respectively. Let the distance from the earthquake epicenter to each station be 2 units, 5 units, and 4 units, respectively. See Figure 11. Where on the coordinate plane is the epicenter located?

**Solution** Graphically, the epicenter appears to be located at $(1,2)$. To check this analytically, we find the equation for each circle and substitute $x = 1$ and $y = 2$.

*Station A:*

$$(x-1)^2 + (y-4)^2 = 2^2$$
$$(1-1)^2 + (2-4)^2 \stackrel{?}{=} 4$$
$$0 + 4 \stackrel{?}{=} 4$$
$$4 = 4 \quad \text{True}$$

*Station B:*

$$(x+3)^2 + (y+1)^2 = 5^2$$
$$(1+3)^2 + (2+1)^2 \stackrel{?}{=} 25$$
$$16 + 9 \stackrel{?}{=} 25$$
$$25 = 25 \quad \text{True}$$

*Station C:*

$$(x-5)^2 + (y-2)^2 = 4^2$$
$$(1-5)^2 + (2-2)^2 \stackrel{?}{=} 16$$
$$16 + 0 \stackrel{?}{=} 16$$
$$16 = 16 \quad \text{True}$$

Thus, we can be sure that the epicenter is located at $(1,2)$.

**NOTE** To find the epicenter analytically, we must find the intersections of the circles taken two at a time, and look for the common intersection. This requires a good bit of work. It is easier to use a graphing calculator to graph the three circles and locate the common intersection point.

## Equations and Graphs of Parabolas

The definition of a parabola is also based on distance.

> **Parabola**
>
> A **parabola** is a set of points in a plane equidistant from a fixed point and a fixed line. The fixed point is called the **focus** and the fixed line the **directrix** of the parabola.

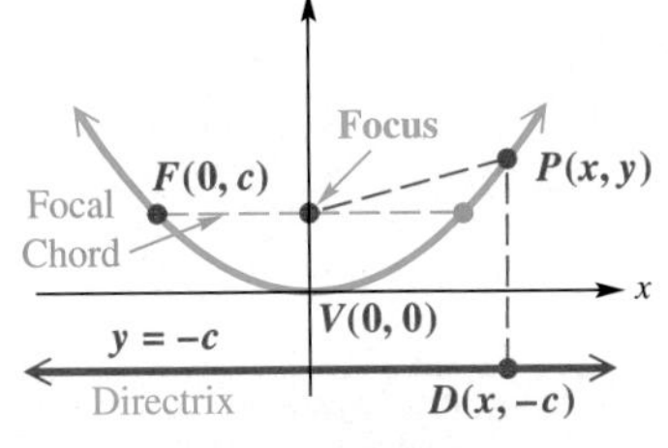


$d(P, F) = d(P, D)$
for all $P$ on the parabola.

**FIGURE 12**

We can find an equation of a parabola from this definition. Let the directrix be the line $y = -c$ and the focus be the point $F$ with coordinates $(0, c)$, as shown in Figure 12.

To find the equation of the set of points that are the same distance from the line $y = -c$ and the point $(0, c)$, choose one such point $P$ and give it coordinates $(x, y)$. Then, since $d(P, F)$ and $d(P, D)$ must be equal, the distance formula gives

$$d(P, F) = d(P, D)$$

$$\sqrt{(x - 0)^2 + (y - c)^2} = \sqrt{(x - x)^2 + (y - (-c))^2} \quad \text{Distance formula}$$

$$\sqrt{x^2 + (y - c)^2} = \sqrt{(y + c)^2}$$

$$x^2 + y^2 - 2yc + c^2 = y^2 + 2yc + c^2 \quad \text{Square both sides; multiply.}$$

$$x^2 = 4cy. \quad \text{Combine terms.}$$

This discussion is summarized as follows.

> **Parabola with a Vertical Axis**
>
> The parabola with focus $(0, c)$ and directrix $y = -c$ has equation
>
> $$x^2 = 4cy.$$
>
> The parabola has vertical axis $x = 0$, opens upward if $c > 0$, and opens downward if $c < 0$.

The *focal chord* through the focus and perpendicular to the axis of symmetry of a parabola has length $|4c|$. To see this, note in Figure 12 that the endpoints of the chord are $(-x, c)$ and $(x, c)$. To find $x$, let $y = c$ in the equation of the parabola and solve for $x$.

$$x^2 = 4cy$$

$$x^2 = 4c^2 \quad \text{Substitute } c \text{ for } y.$$

$$x = |2c| \quad \text{Take the positive square root.}$$

The length of half the focal chord is $|2c|$, so its full length is $|4c|$. This fact is useful when graphing a parabola because it means that the points with coordinates $(-2c, c)$ and $(2c, c)$ lie on the parabola.

If the directrix is the line $x = -c$ and the focus is at $(c, 0)$, then the definition of a parabola and the distance formula leads to the equation of a parabola with a horizontal axis. (See Exercise 96.)

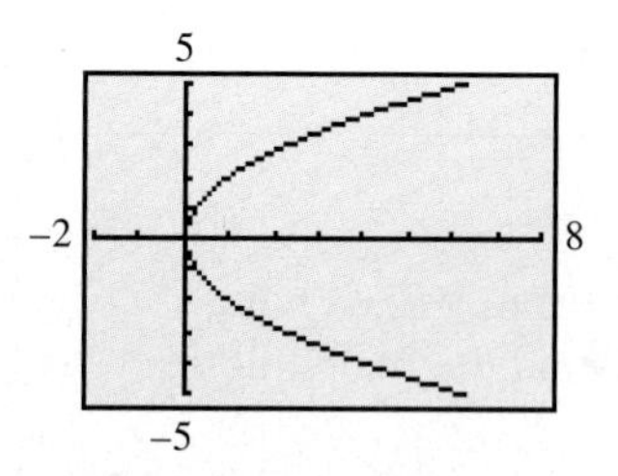


The graphs of the equations $y_1 = 2\sqrt{x}$ and $y_2 = -2\sqrt{x}$ form a horizontal parabola. Here $c = 1 > 0$, so the parabola opens to the right.

**Parabola with a Horizontal Axis**

The parabola with focus $(c, 0)$ and directrix $x = -c$ has equation

$$y^2 = 4cx.$$

The parabola opens to the right if $c > 0$, to the left if $c < 0$, and has horizontal axis $y = 0$.

**NOTE** A parabola with a horizontal axis is not the graph of a function. However, since the the graph of $y^2 = 4cx$ is the union of the graphs of $y_1 = 2\sqrt{cx}$ and $y_2 = -2\sqrt{cx}$, such a parabola can be graphed with a graphing calculator, using the same general procedure discussed for circles in Example 3(a). If $c > 0$, then the parabola opens to the right, and if $c < 0$, then the parabola opens to the left.

**EXAMPLE 7** *Determining Information about Parabolas from Equations*

Find the focus, directrix, vertex, and axis of each parabola. Then use this information to graph each parabola.

**(a)** $x^2 = 8y$ **(b)** $y^2 = -28x$

### Analytic Solution

**(a)** The equation has the form $x^2 = 4cy$, so set $4c = 8$, from which $c = 2$. Since the $x$-term is squared, the parabola is vertical, with focus $(0, c) = (0, 2)$ and directrix $y = -2$. The vertex is $(0, 0)$, and the axis of the parabola is the $y$-axis. See Figure 13, which also shows the endpoints of the focal chord.

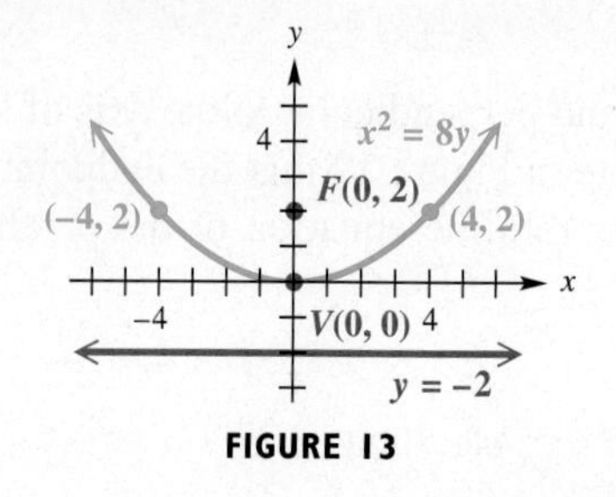


FIGURE 13

### Graphing Calculator Solution

**(a)** Solve the equation for $y$ to obtain $y = \frac{1}{8}x^2$. A calculator graph of this function is shown in Figure 14.

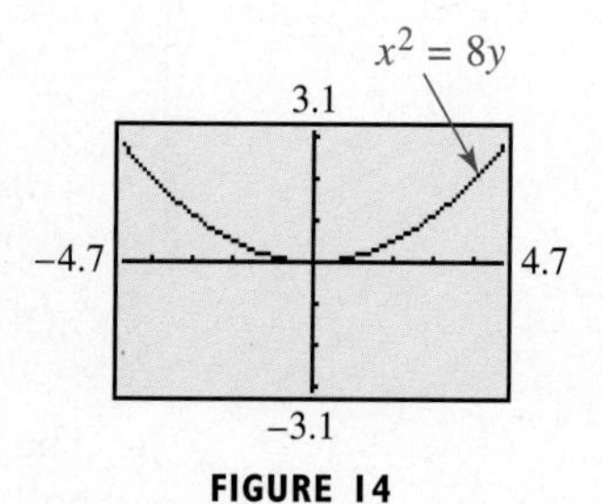


FIGURE 14

*(continued)*

**(b)** This equation has the form $y^2 = 4cx$, with $4c = -28$, so $c = -7$. The parabola is horizontal, with focus $(-7, 0)$, directrix $x = 7$, vertex $(0, 0)$, and $x$-axis as axis of the parabola. The focal chord has endpoints $(-7, -14)$ and $(-7, 14)$. Since $c$ is negative, the graph opens to the left, as shown in Figure 15.

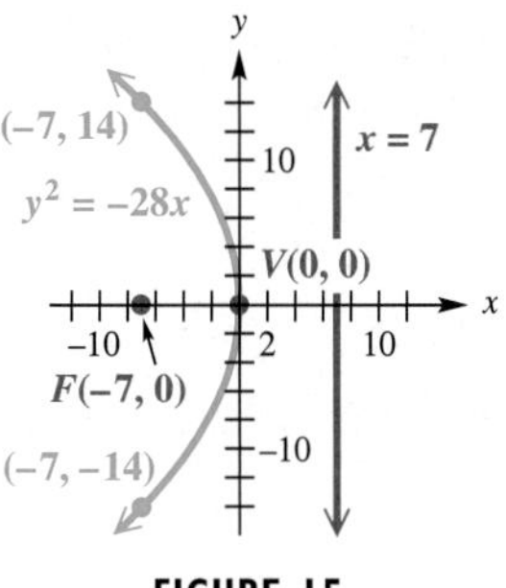


**FIGURE 15**

**(b)** To graph this parabola with a graphing calculator, first transform it into the union of two functions.

$$y^2 = -28x$$
$$y = \pm\sqrt{-28x}$$
$$y_1 = \sqrt{-28x} \quad \text{or} \quad y_2 = -\sqrt{-28x}$$

Note from the equations for $y_1$ and $y_2$ that $x$ must be nonpositive for $y$ to represent a real number. See Figure 16 for the calculator graph.

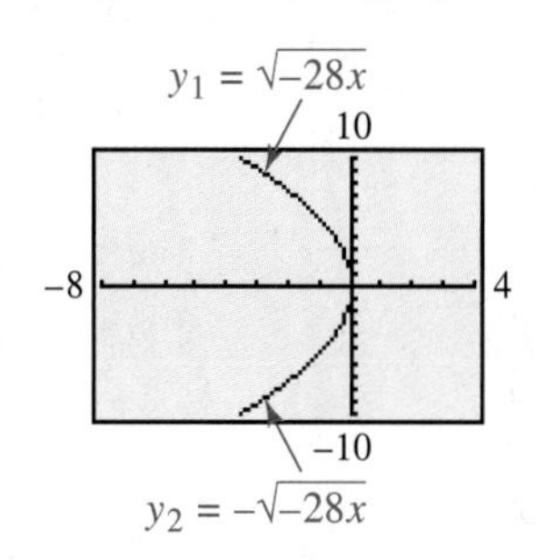


**FIGURE 16**

**EXAMPLE 8** *Writing Equations of Parabolas*

Write an equation for each parabola.

**(a)** Focus $\left(\frac{2}{3}, 0\right)$ and vertex at the origin

**(b)** Vertical axis, vertex at the origin, through the point $(-2, 12)$

**(c)** Vertex $(1, 3)$ and focus $(-1, 3)$

**Solution**

**(a)** Since the focus $\left(\frac{2}{3}, 0\right)$ and the vertex $(0, 0)$ are both on the $x$-axis, the parabola is horizontal. It opens to the right because $c = \frac{2}{3}$ is positive. See Figure 17. The equation, which will have the form $y^2 = 4cx$, is

$$y^2 = 4\left(\frac{2}{3}\right)x \quad \text{or} \quad y^2 = \frac{8}{3}x.$$

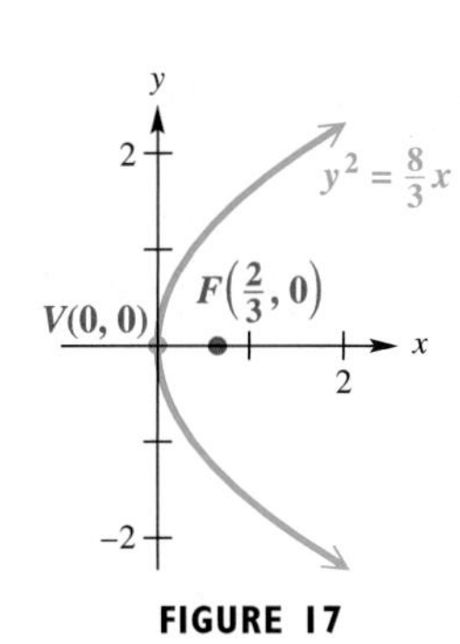


**FIGURE 17**

**(b)** The parabola will have an equation of the form $x^2 = 4cy$ because the axis is vertical and the vertex is $(0, 0)$. Since the point $(-2, 12)$ is on the graph, it must satisfy the equation. Substitute $x = -2$ and $y = 12$ into $x^2 = 4cy$ to get

$$(-2)^2 = 4c(12)$$
$$4 = 48c$$
$$c = \frac{1}{12},$$

which gives
$$x^2 = \frac{1}{3}y \quad \text{or} \quad y = 3x^2$$

as an equation of the parabola.

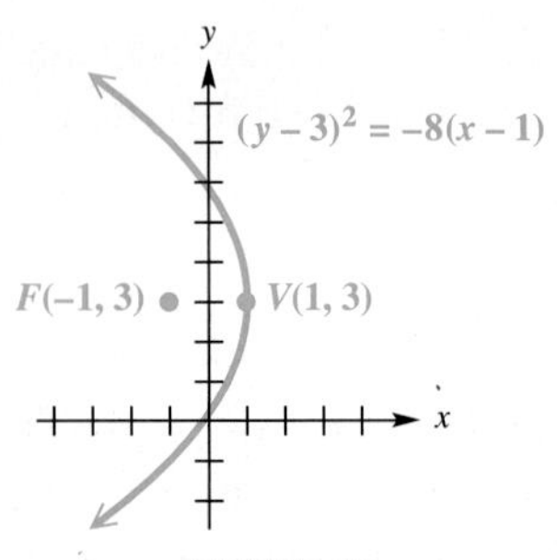

FIGURE 18

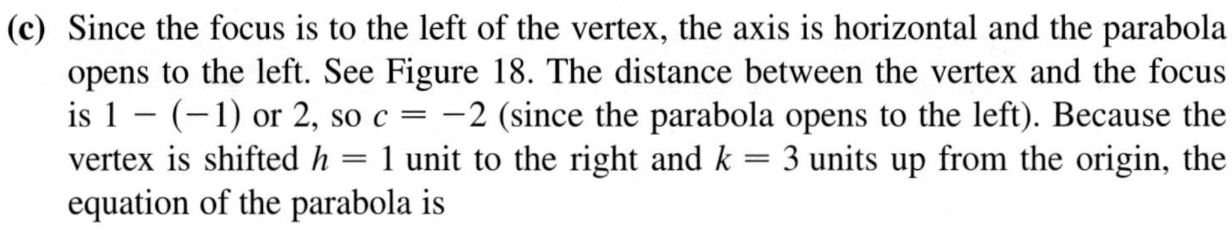

**(c)** Since the focus is to the left of the vertex, the axis is horizontal and the parabola opens to the left. See Figure 18. The distance between the vertex and the focus is $1 - (-1)$ or 2, so $c = -2$ (since the parabola opens to the left). Because the vertex is shifted $h = 1$ unit to the right and $k = 3$ units up from the origin, the equation of the parabola is

$$(y - 3)^2 = 4(-2)(x - 1) \quad \text{Substitute for } c, h, \text{ and } k.$$
$$(y - 3)^2 = -8(x - 1).$$

As indicated in Example 8(c), the equations $y^2 = 4cx$ and $x^2 = 4cy$ can be extended to parabolas having vertices at $(h, k)$ by replacing $x$ by $x - h$ and $y$ by $y - k$ to get the forms

$$(y - k)^2 = 4c(x - h) \quad \text{Horizontal axis}$$

and

$$(x - h)^2 = 4c(y - k), \quad \text{Vertical axis}$$

where the focus is distance $c$ or $-c$ from the vertex. If these equations are solved for $x$ and $y$, respectively, we obtain the equations

$$x = \frac{1}{4c}(y - k)^2 + h$$

and

$$y = \frac{1}{4c}(x - h)^2 + k$$

introduced earlier where $a$ is replaced by $\frac{1}{4c}$.

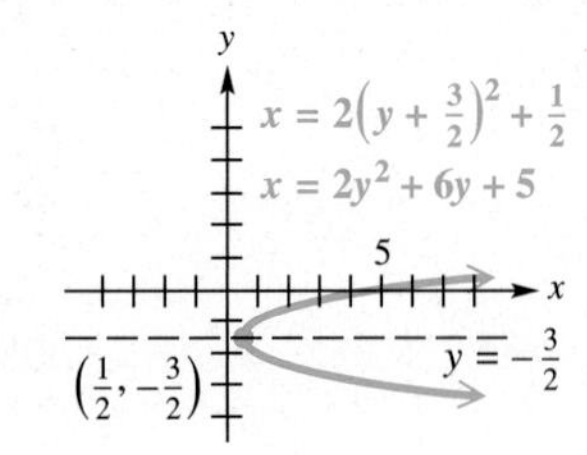

FIGURE 19

Figure 19 shows the graph of $x = 2\left(y + \frac{3}{2}\right)^2 + \frac{1}{2}$. It is a horizontal parabola opening to the right, with vertex $\left(\frac{1}{2}, -\frac{3}{2}\right)$. Refer to Example 8 in Section 4.4 to see a calculator graph of this relation. We started with the equation $x = 2y^2 + 6y + 5$ in that example. Notice that this latter equation is equivalent to the one given above.

## EXAMPLE 9 *Completing the Square to Graph a Horizontal Parabola*

Graph $x = -2y^2 + 4y - 3$. Give the domain and range.

**Solution** Complete the square to express the given equation of the parabola in the form $x = a(y - k)^2 + h$.

$$\frac{x}{-2} = y^2 - 2y + \frac{3}{2} \quad \text{Divide by } -2.$$

$$\frac{x}{-2} - \frac{3}{2} = y^2 - 2y \quad \text{Subtract } \tfrac{3}{2}.$$

$$\frac{x}{-2} - \frac{3}{2} + 1 = y^2 - 2y + 1 \quad \text{Add 1; complete the square.}$$

$$\frac{x}{-2} - \frac{1}{2} = (y - 1)^2 \quad \text{Factor on the right; add on the left.}$$

$$\frac{x}{-2} = (y - 1)^2 + \frac{1}{2} \quad \text{Add } \tfrac{1}{2}.$$

$$x = -2(y - 1)^2 - 1 \quad \text{Multiply by } -2.$$

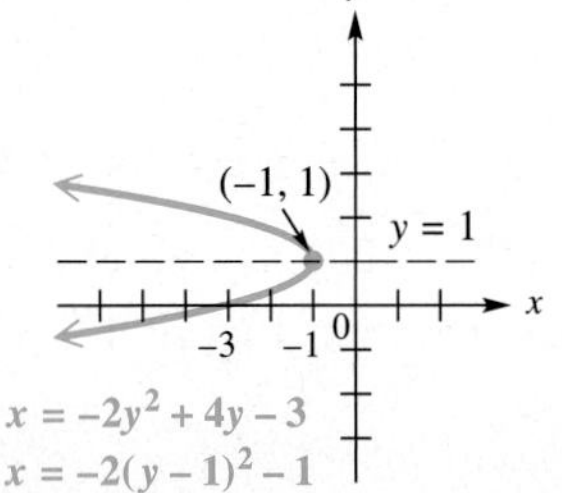

(a)

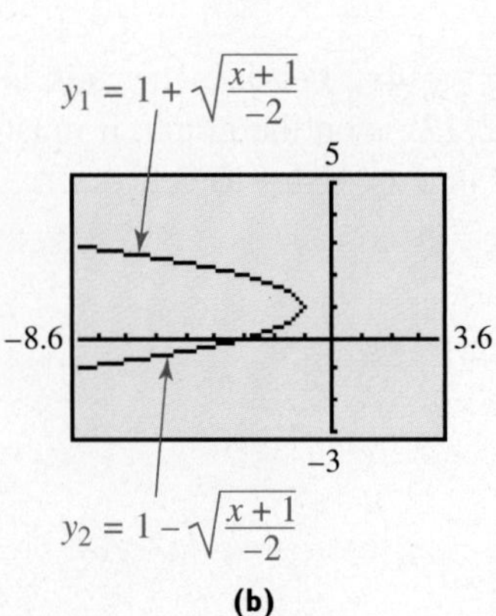

(b)

FIGURE 20

The coefficient $-2$ indicates that the graph opens to the left (the negative $x$-direction) and is narrower than the graph of $x = y^2$. As the traditional graph in Figure 20(a) shows, the vertex is $(-1, 1)$. The domain is $(-\infty, -1]$, and the range is $(-\infty, \infty)$.

To get the calculator graph of the parabola, shown in Figure 20(b), solve

$$x = -2(y - 1)^2 - 1$$

for $y$ to get $\quad y_1 = 1 + \sqrt{\dfrac{x + 1}{-2}} \quad$ or $\quad y_2 = 1 - \sqrt{\dfrac{x + 1}{-2}}.$

## An Application of Parabolas

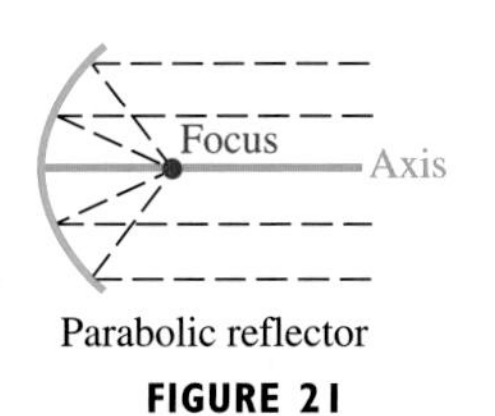

Parabolic reflector

**FIGURE 21**

The geometric properties of parabolas lead to many practical applications. For example, if a light source is placed at the focus of a parabolic reflector, as in Figure 21, light rays reflect parallel to the axis, making a spotlight or flashlight. The process also works in reverse. Light rays from a distant source come in parallel to the axis and are reflected to a point at the focus. This use of parabolic reflection is seen in the satellite dishes used to pick up signals from communications satellites.

**EXAMPLE 10** *Modeling the Reflective Property of Parabolas*

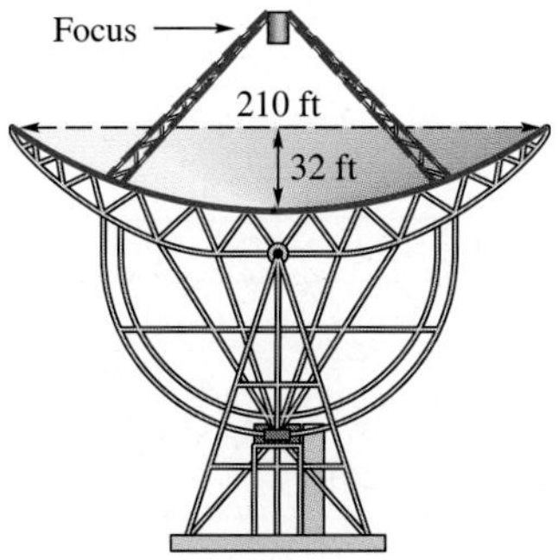

The Parkes radio telescope has a parabolic dish shape with diameter 210 feet and depth 32 feet. Because of this parabolic shape, distant rays hitting the dish are reflected directly toward the focus. A cross section of the dish is shown in Figure 22. (*Source:* Mar, J. and H. Liebowitz, *Structure Technology for Large Radio and Radar Telescope Systems,* MIT Press, 1969.)

Focus 210 ft 32 ft

**FIGURE 22**

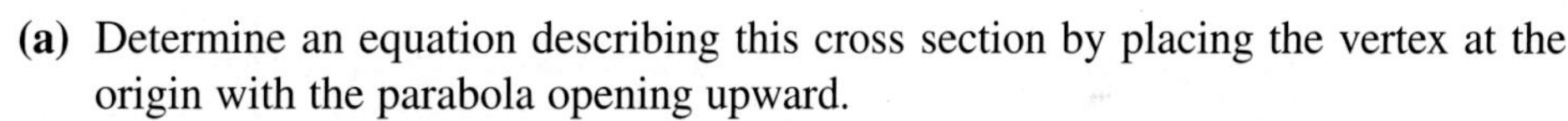

**(a)** Determine an equation describing this cross section by placing the vertex at the origin with the parabola opening upward.

**(b)** The receiver must be placed at the focus of the parabola. How far from the vertex of the parabolic dish should the receiver be located?

### Solution

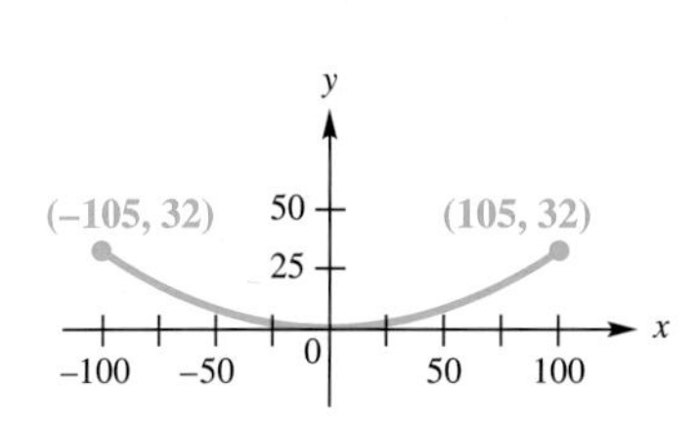


**FIGURE 23**

**(a)** Locate the vertex at the origin, as shown in Figure 23. Then, the form of the parabola will be $y = ax^2$. The parabola must pass through the point $\left(\frac{210}{2}, 32\right) = (105, 32)$. Thus,

$$32 = a(105)^2$$

$$a = \frac{32}{105^2} = \frac{32}{11{,}025},$$

so the cross section can be described by

$$y = \frac{32}{11{,}025}x^2.$$

**(b)** Since $y = \frac{32}{11{,}025}x^2$,

$$4c = \frac{1}{a}$$

$$4c = \frac{11{,}025}{32}$$

$$c = \frac{11{,}025}{128} \approx 86.1.$$

The receiver should be located at $(0, 86.1)$, or 86.1 feet above the vertex.

## 6.1 EXERCISES

**1.** Describe the graph of the equation
$$(x-3)^2+(y-3)^2=0.$$

**2.** Describe the graph of the equation
$$(x-3)^2+(y-3)^2=-1.$$

*Concept Check* *Match each equation in Exercises 3–10 with its calculator graph in A–H, without actually using your calculator. Check your answer by generating a calculator graph of your own. (Every window has* Xscl = Yscl = 1.*)*

**3.** $x=2(y+3)^2-4$ **4.** $y=2(x+3)^2-4$ **5.** $x^2=-3y$ **6.** $y^2=-3x$

**7.** $x^2+y^2=25$ **8.** $(x-3)^2+(y+4)^2=25$ **9.** $(x+3)^2+(y-4)^2=25$ **10.** $x^2+y^2=-4$

**A.**

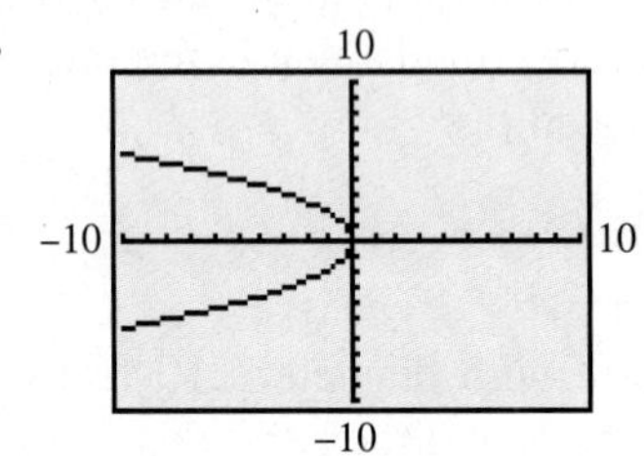


**B.**

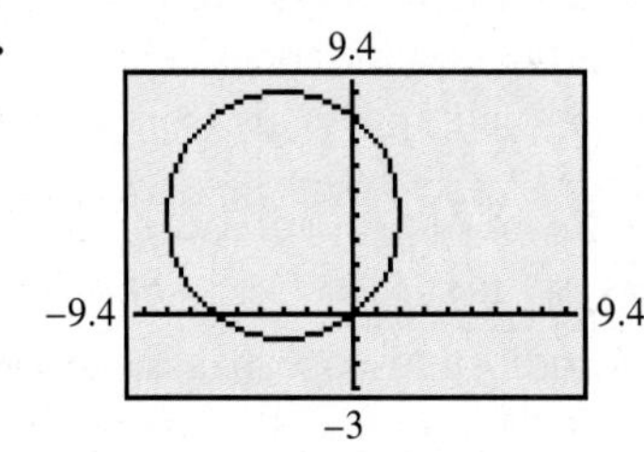


**C.**

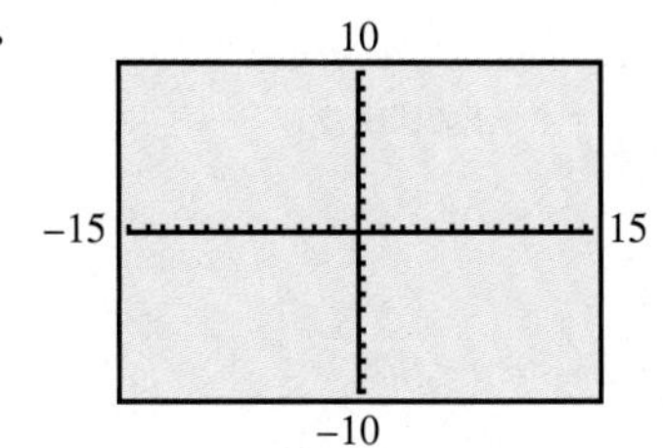


**D.**

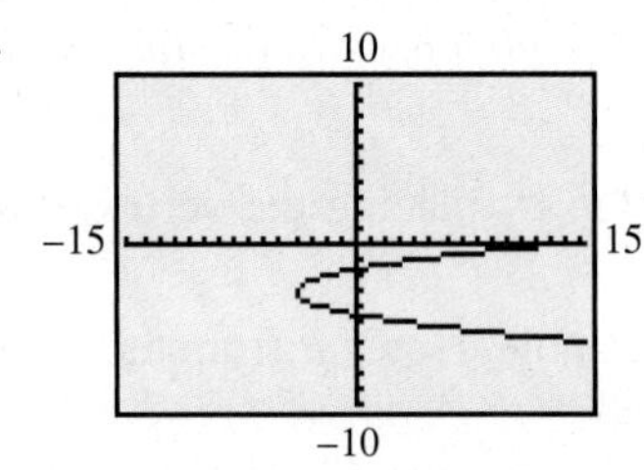


**E.**

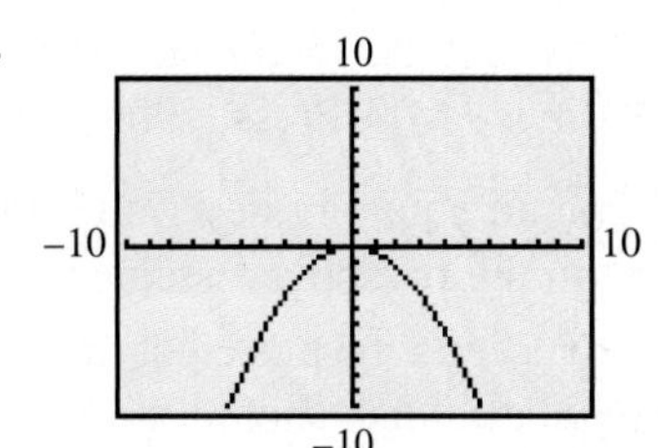


**F.**

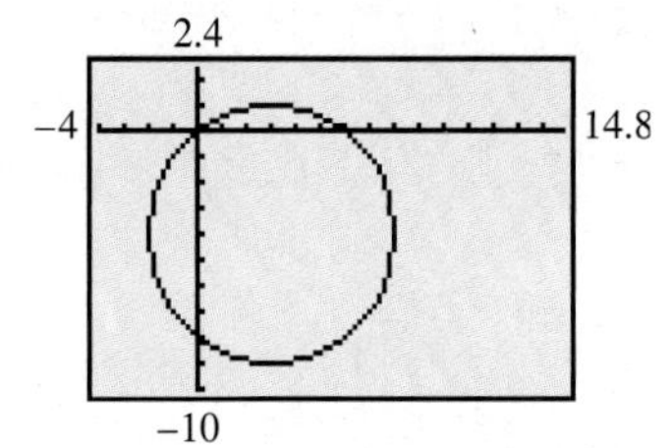


**G.**

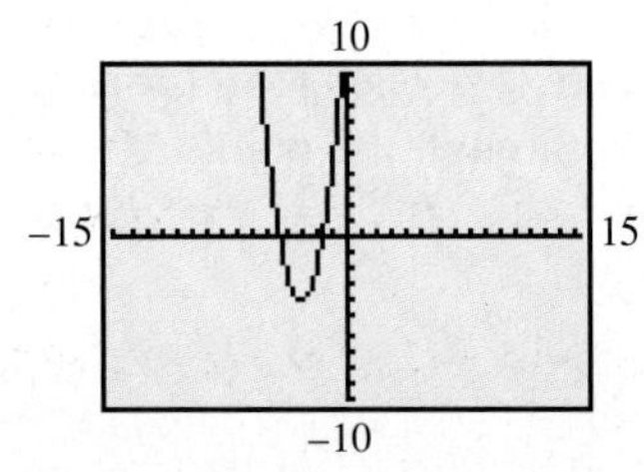


**H.**

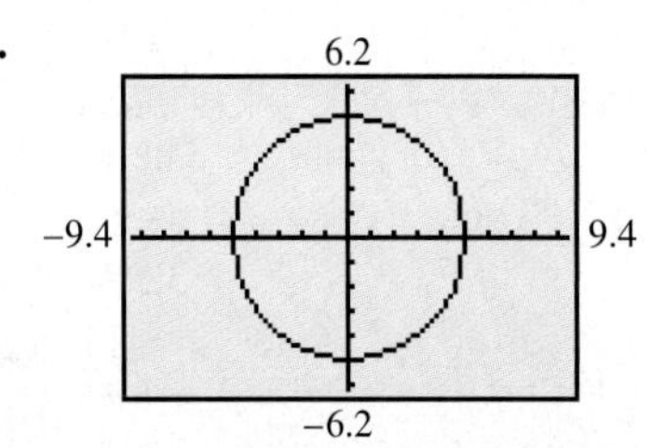


*Find the center-radius form for each circle satisfying the given conditions.*

**11.** Center $(1,4)$; radius 3

**12.** Center $(-2,5)$; radius 4

**13.** Center $(0,0)$; radius 1

**14.** Center $(0,0)$; radius 5

**15.** Center $\left(\frac{2}{3},-\frac{4}{5}\right)$; radius $\frac{3}{7}$

**16.** Center $\left(-\frac{1}{2},-\frac{1}{4}\right)$; radius $\frac{12}{5}$

**17.** Center $(-1,2)$; passing through $(2,6)$

**18.** Center $(2,-7)$; passing through $(-2,-4)$

**19.** Center $(-3,-2)$; tangent to the $x$-axis (*Hint: Tangent to* means touching at one point.)

**20.** Center $(5,-1)$; tangent to the $y$-axis

## RELATING CONCEPTS

**For individual or group investigation (Exercises 21–24)**

*The figure shows a circle and a diameter of the circle. The endpoints of the diameter have coordinates* $(-1, 3)$ *and* $(5, -9)$.

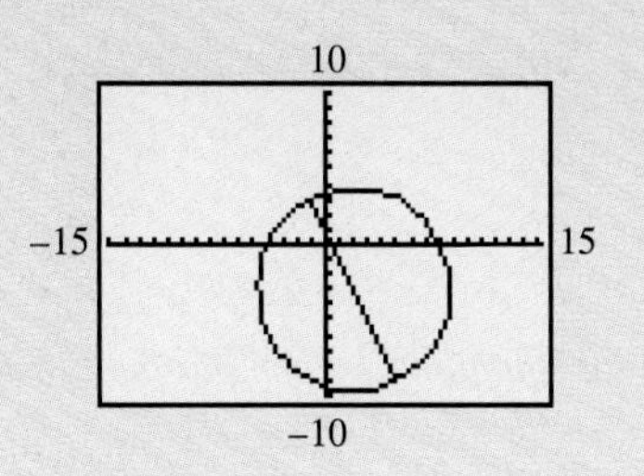


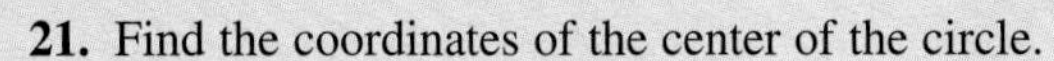

**21.** Find the coordinates of the center of the circle.

**22.** Find the radius of the circle.

**23.** Give the center-radius form of the equation of the circle.

**24.** Give the center-radius form of the equation of the circle with endpoints of a diameter having coordinates $(3, -5)$ and $(-7, 3)$.

*Graph each circle by hand. Give the domain and range.*

**25.** $x^2 + y^2 = 36$

**26.** $(x - 2)^2 + y^2 = 36$

**27.** $(x + 2)^2 + (y - 5)^2 = 16$

**28.** $(x - 5)^2 + (y + 4)^2 = 49$

**29.** $(x - 4)^2 + (y - 3)^2 = 25$

**30.** $(x + 3)^2 + (y + 2)^2 = 36$

*Graph each circle, using a graphing calculator. See Example 3. Use a square viewing window. Give the domain and range.*

**31.** $x^2 + y^2 = 81$

**32.** $x^2 + (y + 3)^2 = 49$

**33.** $(x - 3)^2 + (y - 2)^2 = 25$

**34.** $(x + 2)^2 + (y + 3)^2 = 36$

*Decide whether each equation has a circle as its graph. If it does, give the center and radius.*

**35.** $x^2 + 6x + y^2 + 8y + 9 = 0$

**36.** $x^2 + 8x + y^2 - 6y + 16 = 0$

**37.** $x^2 - 4x + y^2 + 12y = -4$

**38.** $x^2 - 12x + y^2 + 10y = -25$

**39.** $4x^2 + 4x + 4y^2 - 16y - 19 = 0$

**40.** $9x^2 + 12x + 9y^2 - 18y - 23 = 0$

**41.** $x^2 + 2x + y^2 - 6y + 14 = 0$

**42.** $x^2 + 4x + y^2 - 8y + 32 = 0$

**43.** $x^2 - 2x + y^2 + 4y + 9 = 0$

*Refer to Example 6 for Exercises 44 and 45.*

**44.** Show analytically that if three receiving stations at $(1, 4)$, $(-6, 0)$, and $(5, -2)$ record distances to an earthquake epicenter of 4 units, 5 units, and 10 units, respectively, then the epicenter would lie at $(-3, 4)$.

**45.** Three receiving stations record the presence of an earthquake. The locations of the receiving centers and the distances to the epicenter are contained in the equations $(x - 2)^2 + (y - 1)^2 = 25$, $(x + 2)^2 + (y - 2)^2 = 16$, and $(x - 1)^2 + (y + 2)^2 = 9$. Determine the location of the earthquake epicenter.

Concept Check *Each equation in Exercises 46–53 defines a parabola. Without actually graphing, match the equation in Column I with its description in Column II.*

| I | II |
|---|---|
| **46.** $y = (x - 4)^2 - 2$ | **A.** Vertex $(2, -4)$; opens downward |
| **47.** $y = (x - 2)^2 - 4$ | **B.** Vertex $(2, -4)$; opens upward |
| **48.** $y = -(x - 4)^2 - 2$ | **C.** Vertex $(4, -2)$; opens downward |
| **49.** $y = -(x - 2)^2 - 4$ | **D.** Vertex $(4, -2)$; opens upward |
| **50.** $x = (y - 4)^2 - 2$ | **E.** Vertex $(-2, 4)$; opens left |
| **51.** $x = (y - 2)^2 - 4$ | **F.** Vertex $(-2, 4)$; opens right |
| **52.** $x = -(y - 4)^2 - 2$ | **G.** Vertex $(-4, 2)$; opens left |
| **53.** $x = -(y - 2)^2 - 4$ | **H.** Vertex $(-4, 2)$; opens right |

**54.** Concept Check For the graph of $y = a(x - h)^2 + k$, in what quadrant is the vertex if:

**(a)** $h < 0, k < 0$; **(b)** $h < 0, k > 0$; **(c)** $h > 0, k < 0$; **(d)** $h > 0, k > 0$?

**55.** *Concept Check* Repeat parts (a)–(d) of Exercise 54 for the graph of $x = a(y - k)^2 + h$.

*Give the coordinates of the focus, the equation of the directrix, and the axis of each parabola.*

**56.** $x^2 = 16y$

**57.** $x^2 = 4y$

**58.** $x^2 = -\frac{1}{2}y$

**59.** $x^2 = \frac{1}{9}y$

**60.** $y^2 = \frac{1}{16}x$

**61.** $y^2 = -\frac{1}{32}x$

**62.** $y^2 = -16x$

**63.** $y^2 = -4x$

*Write an equation for each parabola with vertex at the origin.*

**64.** Focus $(0, -2)$

**65.** Focus $(5, 0)$

**66.** Focus $\left(-\frac{1}{2}, 0\right)$

**67.** Focus $\left(0, \frac{1}{4}\right)$

**68.** Through $(2, -2\sqrt{2})$; opening to the right

**69.** Through $(\sqrt{3}, 3)$; opening upward

**70.** Through $(\sqrt{10}, -5)$; opening downward

**71.** Through $(-3, 3)$; opening to the left

**72.** Through $(2, -4)$; symmetric with respect to the $y$-axis

**73.** Through $(3, 2)$; symmetric with respect to the $x$-axis

*Graph each parabola by hand, and check using a graphing calculator. Give the coordinates of the vertex, the axis, the domain, and the range.*

**74.** $y = (x + 3)^2 - 4$

**75.** $y = (x - 5)^2 - 4$

**76.** $y = -2(x + 3)^2 + 2$

**77.** $y = \frac{2}{3}(x - 2)^2 - 1$

**78.** $y = x^2 - 2x + 3$

**79.** $y = x^2 + 6x + 5$

**80.** $y = 2x^2 - 4x + 5$

**81.** $y = -3x^2 + 24x - 46$

**82.** $x = y^2 + 2$

**83.** $x = (y + 1)^2$

**84.** $x = (y - 3)^2$

**85.** $x = (y + 2)^2 - 1$

**86.** $x = (y - 4)^2 + 2$

**87.** $x = -2(y + 3)^2$

**88.** $x = \frac{2}{3}y^2 - 4y + 8$

**89.** $x = y^2 + 2y - 8$

**90.** $x = -4y^2 - 4y - 3$

**91.** $x = 2y^2 - 4y + 6$

**92.** $x = -2y^2 + 2y - 3$

**93.** $2x = y^2 - 4y + 6$

**94.** $2x = y^2 - 2y + 9$

**95.** $x = -3y^2 + 6y - 1$

**96.** Prove that the parabola with focus $(c, 0)$ and directrix $x = -c$ has equation $y^2 = 4cx$.

*(Modeling) Solve each problem.*

**97.** *Path of an Object on a Planet* When an object moves under the influence of a constant force (without air resistance), its path is parabolic. This would occur if a ball is thrown near the surface of a planet or other celestial object. Suppose two balls are simultaneously thrown upward at a 45° angle on two different planets. If their initial velocities are both 30 mph, then their $xy$-coordinates in feet at time $x$ in seconds can be expressed by the equation

$$y = x - \frac{g}{1922}x^2,$$

where $g$ is the acceleration due to gravity. The value of $g$ will vary depending on the mass and size of the planet. (*Source:* Zeilik, M., S. Gregory, and E. Smith, *Introductory Astronomy and Astrophysics,* Saunders College Publishers, 1992.)

**(a)** On Earth, $g = 32.2$; on Mars, $g = 12.6$. Find the two equations, and graph on the same screen of a graphing calculator the paths of the two balls thrown on Earth and Mars. Use the window $[0, 180]$ by $[0, 120]$. (*Hint:* If possible, set the mode on your graphing calculator to simultaneous.)

**(b)** Determine the difference in the horizontal distances traveled by the two balls.

**98.** *Path of an Object on a Planet* Refer to Exercise 97. Suppose the two balls are now thrown upward at a 60° angle on Mars and the moon. If their initial velocities are 60 mph, then their $xy$-coordinates in feet at time $x$ in seconds can be expressed by the equation

$$y = \frac{19}{11}x - \frac{g}{3872}x^2.$$

(*Source:* Zeilik, M., S. Gregory, and E. Smith, *Introductory Astronomy and Astrophysics,* Saunders College Publishers, 1992.)

**(a)** On the same screen, graph the paths of the balls if $g = 5.2$ for the moon.

**(b)** Determine the maximum height of each ball to the nearest foot.

**99.** *Design of a Radio Telescope* The U.S. Naval Research Laboratory designed a giant radio telescope weighing 3450 tons. Its parabolic dish had a diameter of 300 feet with focal length (the distance from the focus to the parabolic surface) 128.5 feet. Determine the maximum depth of the 300-foot dish. (*Source:* Mar, J. and H. Liebowitz, *Structure Technology for Large Radio and Radar Telescope Systems,* MIT Press, 1969.)

**100.** *Path of an Alpha Particle* When an alpha particle is moving in a horizontal path along the positive $x$-axis and passes between charged plates, it is deflected in a parabolic path. If the plate is charged with 2000 volts and is .4 meter long, then an alpha particle's path can be described by the equation

$$y = -\frac{k}{2v_0}x^2,$$

where $k = 5 \times 10^{-9}$ is constant and $v_0$ is the initial velocity of the particle. If $v_0 = 10^7$ meters per second, what is the deflection of the alpha particle's path in the $y$-direction when $x = .4$ meter? (*Source:* Semat, H. and J. Albright, *Introduction to Atomic and Nuclear Physics,* Holt, Rinehart and Winston, 1972.)

**101.** *Height of a Bridge's Cable Supports* The cable in the center portion of a bridge is supported as shown in the figure to form a parabola. The center support is 10 feet high, the tallest supports are 210 feet high, and the distance between the two tallest supports is 400 feet. Find the height of the remaining supports, if the supports are evenly spaced. (Ignore the width of the supports.)

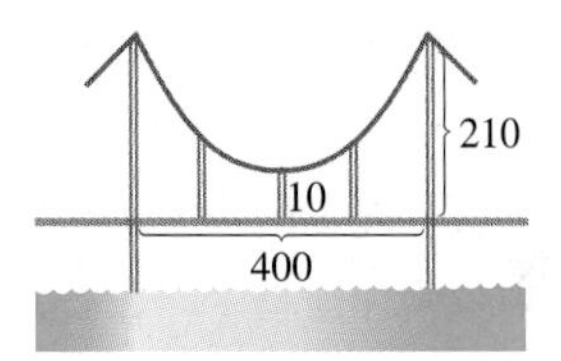


**102.** *Parabolic Arch* An arch in the shape of a parabola has the dimensions shown in the figure. How wide is the arch 9 feet up?

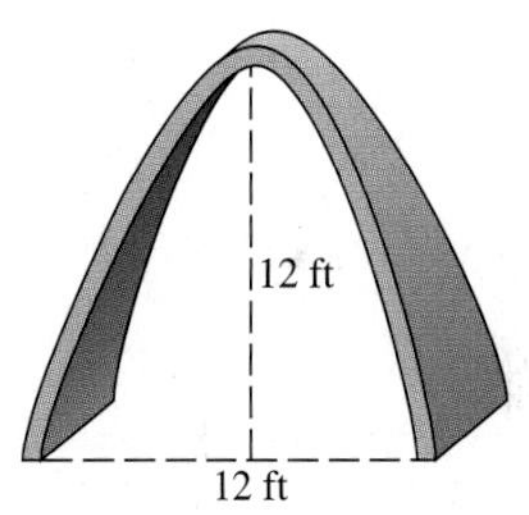


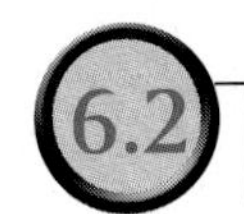

# 6.2 ELLIPSES AND HYPERBOLAS

Equations and Graphs of Ellipses • Applications of Ellipses • Equations and Graphs of Hyperbolas

## Equations and Graphs of Ellipses

The *ellipse* is another second-degree relation whose equation is based on the distance formula.

**Ellipse**

An **ellipse** is the set of all points in a plane the sum of whose distances from two fixed points is constant. Each fixed point is called a **focus** (plural, **foci**) of the ellipse.

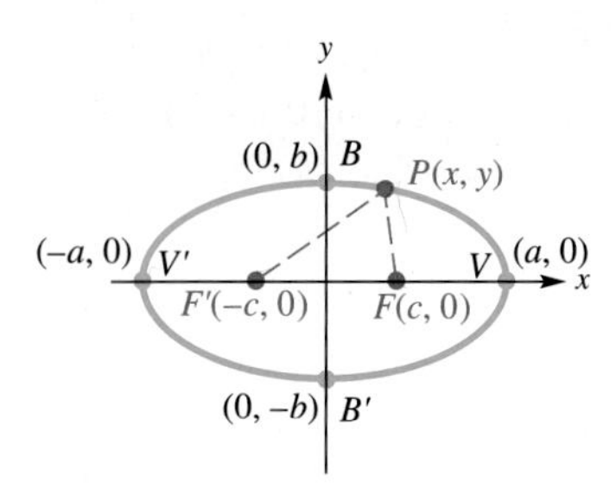


**FIGURE 24**

For example, the ellipse in Figure 24 has foci at points $F$ and $F'$. By the definition, the ellipse consists of all points $P$ such that the sum $d(P, F) + d(P, F')$ is constant. This ellipse has its **center** at the origin. Points $V$ and $V'$ are the **vertices** of the ellipse, and the line segment $VV'$ is the **major axis.** The foci always lie on the major axis. The line segment $BB'$ is the **minor axis.** The major axis has length $2a$, and the minor axis has length $2b$. The graph of an ellipse is not that of a function, since it fails the vertical line test.

To obtain an equation for an ellipse centered at the origin, let the two foci $F$ and $F'$ have coordinates $(c, 0)$ and $(-c, 0)$, respectively, as shown in Figure 24. Let the sum of the distances from any point $P(x, y)$ on the ellipse to the two foci be $2a$. By the distance formula, segment $PF$ has length

$$d(P, F) = \sqrt{(x - c)^2 + y^2},$$

while segment $PF'$ has length

$$d(P, F') = \sqrt{[x - (-c)]^2 + y^2} = \sqrt{(x + c)^2 + y^2}.$$

The sum of the lengths $d(P, F)$ and $d(P, F')$ must be $2a$.

$$\sqrt{(x - c)^2 + y^2} + \sqrt{(x + c)^2 + y^2} = 2a$$

$$\sqrt{(x - c)^2 + y^2} = 2a - \sqrt{(x + c)^2 + y^2} \quad \text{Isolate } \sqrt{(x - c)^2 + y^2}.$$

$$(x - c)^2 + y^2 = 4a^2 - 4a\sqrt{(x + c)^2 + y^2} + (x + c)^2 + y^2 \quad \text{Square both sides.}$$

$$x^2 - 2cx + c^2 + y^2 = 4a^2 - 4a\sqrt{(x + c)^2 + y^2} + x^2 + 2cx + c^2 + y^2$$

$$4a\sqrt{(x + c)^2 + y^2} = 4a^2 + 4cx \quad \text{Isolate } 4a\sqrt{(x + c)^2 + y^2}.$$

$$a\sqrt{(x + c)^2 + y^2} = a^2 + cx \quad \text{Divide by 4.}$$

$$a^2(x^2 + 2cx + c^2 + y^2) = a^4 + 2ca^2x + c^2x^2 \quad \text{Square both sides.}$$

$$a^2x^2 + 2ca^2x + a^2c^2 + a^2y^2 = a^4 + 2ca^2x + c^2x^2 \quad \text{Distributive property}$$

$$a^2x^2 + a^2c^2 + a^2y^2 = a^4 + c^2x^2 \quad \text{Subtract } 2ca^2x.$$

$$a^2x^2 - c^2x^2 + a^2y^2 = a^4 - a^2c^2 \quad \text{Rearrange terms.}$$

$$(a^2 - c^2)x^2 + a^2y^2 = a^2(a^2 - c^2) \quad \text{Factor.}$$

$$\frac{x^2}{a^2} + \frac{y^2}{a^2 - c^2} = 1 \quad (*) \quad \text{Divide by } a^2(a^2 - c^2).$$

Since $B(0, b)$ is on the ellipse in Figure 24, we have

$$d(B, F) + d(B, F') = 2a$$

$$\sqrt{(-c)^2 + b^2} + \sqrt{c^2 + b^2} = 2a$$

$$2\sqrt{c^2 + b^2} = 2a$$

$$\sqrt{c^2 + b^2} = a$$

$$c^2 + b^2 = a^2$$

$$b^2 = a^2 - c^2.$$

Replacing $a^2 - c^2$ with $b^2$ in equation (*) gives

$$\frac{x^2}{a^2} + \frac{y^2}{b^2} = 1,$$

the standard form of the equation of an ellipse centered at the origin with foci on the $x$-axis. If the vertices and foci were on the $y$-axis, an almost identical proof could be used to obtain the standard form

$$\frac{y^2}{a^2} + \frac{x^2}{b^2} = 1.$$

**Standard Forms of Equations for Ellipses**

The ellipse with center at the origin and equation

$$\frac{x^2}{a^2} + \frac{y^2}{b^2} = 1 \quad (a > b)$$

has vertices $(\pm a, 0)$, endpoints of the minor axis $(0, \pm b)$, and foci $(\pm c, 0)$, where $c^2 = a^2 - b^2$. The ellipse with center at the origin and equation

$$\frac{y^2}{a^2} + \frac{x^2}{b^2} = 1 \quad (a > b)$$

has vertices $(0, \pm a)$, endpoints of the minor axis $(\pm b, 0)$, and foci $(0, \pm c)$, where $c^2 = a^2 - b^2$.

**Looking Ahead to Calculus**

The forms for equations of ellipses are developed in calculus, and then methods of calculus are applied to solve problems involving ellipses. For example, differentiation is used to find the slope of the tangent line at a point on the ellipse, so the equation of the tangent line can be found. Similar examples are explored for hyperbolas.

Do not be confused by the two standard forms—in one case $a^2$ is associated with $x^2$; in the other case $a^2$ is associated with $y^2$. However, when graphing an ellipse in a traditional manner, it is necessary only to find the intercepts of the graph—if the positive $x$-intercept is larger than the positive $y$-intercept, the major axis is horizontal, and otherwise it is vertical. When using the relationship $a^2 - b^2 = c^2$, choose $a^2$ and $b^2$ so that $a^2 > b^2$.

**NOTE** An ellipse is symmetric with respect to its major axis, its minor axis, and its center. Also, if $a = b$ in the equation of an ellipse, the ellipse is a circle.

**EXAMPLE 1** *Graphing an Ellipse Centered at the Origin*

Graph $4x^2 + 9y^2 = 36$. Give the domain and range.

**Analytic Solution**

Begin by dividing each side by 36 to get the equation in standard form,

$$\frac{x^2}{9} + \frac{y^2}{4} = 1.$$

This ellipse is centered at the origin, with $x$-intercepts 3 and $-3$ and $y$-intercepts 2 and $-2$. The domain of this relation is $[-3, 3]$, and the range is $[-2, 2]$. See Figure 25 on the next page.

**Graphing Calculator Solution**

Solving the equation for $y$ gives the equations of the two functions

$$y_1 = 2\sqrt{1 - \frac{x^2}{9}} \quad \text{and} \quad y_2 = -2\sqrt{1 - \frac{x^2}{9}}.$$

Graph these in a square window to get the graph shown in Figure 26 on the next page. From the graph the domain is $[-3, 3]$ and the range is $[-2, 2]$.

*(continued)*

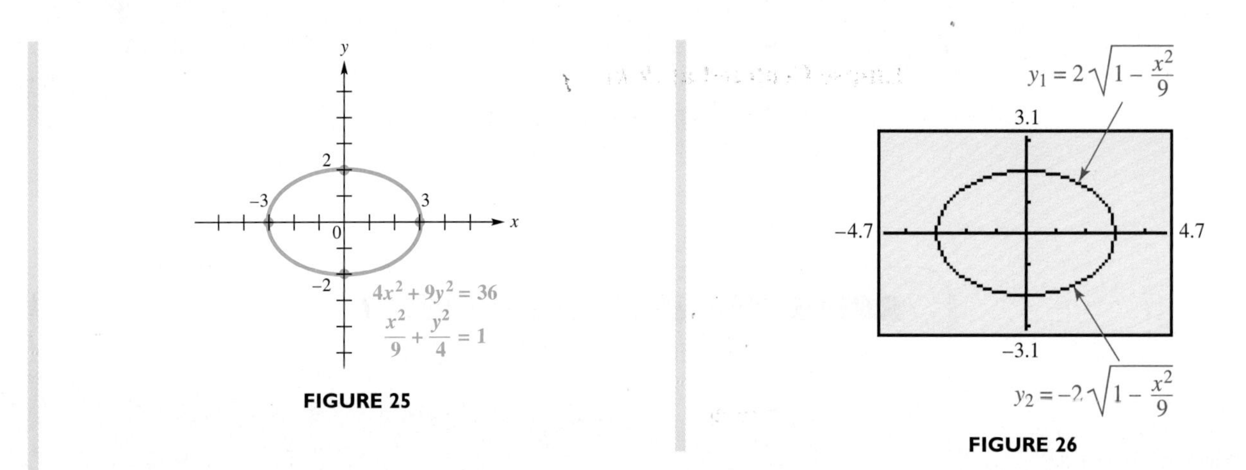


FIGURE 25

FIGURE 26

| X | Y1 | Y2 |
|---|---|---|
| -6 | ERROR | ERROR |
| -5 | ERROR | ERROR |
| -4 | ERROR | ERROR |
| -3 | 0 | 0 |
| -2 | 1.4907 | -1.491 |
| -1 | 1.8856 | -1.886 |
| 0 | 2 | -2 |

Y1=2√(1-X²/9)

| X | Y1 | Y2 |
|---|---|---|
| 0 | 2 | -2 |
| 1 | 1.8856 | -1.886 |
| 2 | 1.4907 | -1.491 |
| 3 | 0 | 0 |
| 4 | ERROR | ERROR |
| 5 | ERROR | ERROR |
| 6 | ERROR | ERROR |

Y2=-2√(1-X²/9)

The tables show that the points of intersection of the two halves of the ellipse are $(-3, 0)$ and $(3, 0)$ in Figure 26. The tables also suggest that the domain of the relation in Example 1 is $[-3, 3]$.

## EXAMPLE 2 *Finding Foci of an Ellipse*

Find the coordinates of the foci of the ellipse in Example 1.

**Solution** Since $9 > 4$, we can find the foci by letting $a^2 = 9$ and $b^2 = 4$ in the equation $c^2 = a^2 - b^2$. Then, we solve for $c$:

$$c^2 = a^2 - b^2 = 9 - 4 = 5$$
$$c = \sqrt{5}.$$

(By definition, $c > 0$. See Figure 24.) The major axis is along the $x$-axis, so the foci have coordinates $(-\sqrt{5}, 0)$ and $(\sqrt{5}, 0)$.

## EXAMPLE 3 *Finding the Equation of an Ellipse*

Find the equation of the ellipse having center at the origin, foci $(0, 3)$ and $(0, -3)$, and major axis of length 8 units. Give the domain and range.

**Solution** Since the major axis is 8 units long, $2a = 8$ or $a = 4$. We use the relationship $a^2 - b^2 = c^2$ to find $b^2$, with $a = 4$ and $c = 3$.

$$a^2 - b^2 = c^2$$
$$4^2 - b^2 = 3^2 \qquad \text{Substitute for } a \text{ and } c.$$
$$16 - b^2 = 9$$
$$b^2 = 7$$

The foci are on the $y$-axis, so we use the larger intercept, $a$, to find the denominator for $y^2$, giving the equation in standard form as

$$\frac{y^2}{16} + \frac{x^2}{7} = 1.$$

A graph of this ellipse is shown in Figure 27. The domain of this relation is $[-\sqrt{7}, \sqrt{7}]$, and the range is $[-4, 4]$.

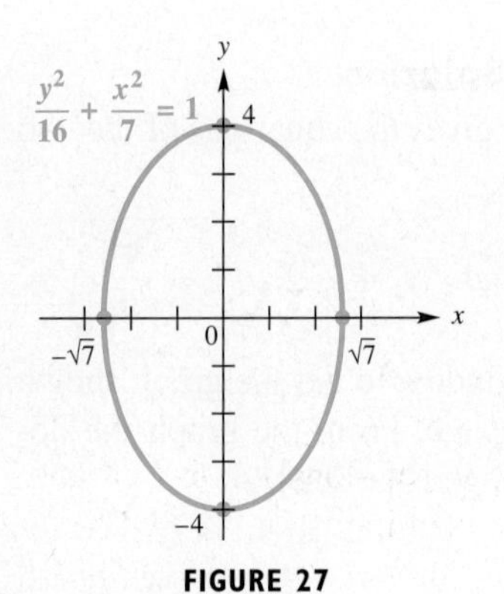


FIGURE 27

Just as a circle need not have its center at the origin, an ellipse also may have its center translated from the origin.

**Ellipse Centered at $(h, k)$**

An ellipse centered at $(h, k)$ with horizontal major axis of length $2a$ and vertical minor axis of length $2b$ has equation

$$\frac{(x-h)^2}{a^2}+\frac{(y-k)^2}{b^2}=1.$$

There is a similar result for an ellipse having a vertical major axis.

The definition of an ellipse can be used to prove the statement above.

**EXAMPLE 4** *Graphing an Ellipse Translated from the Origin*

Graph $\frac{(y+1)^2}{16}+\frac{(x-2)^2}{9}=1$. Give the domain and range.

**Analytic Solution**

The graph of this equation is an ellipse centered at $(2, -1)$. As mentioned earlier, ellipses always have $a > b$. For this ellipse, $a = 4$ and $b = 3$. Since $a = 4$ is associated with $y^2$, the vertices of the ellipse are on the vertical line through $(2, -1)$. Find the vertices by locating two points on the vertical line through $(2, -1)$, one 4 units up from $(2, -1)$ and one 4 units down. The vertices are $(2, 3)$ and $(2, -5)$. Locate two other points on the ellipse by locating points on the horizontal line through $(2, -1)$, one 3 units to the right and one 3 units to the left. The graph is shown in Figure 28. The domain is $[-1, 5]$, and the range is $[-5, 3]$.

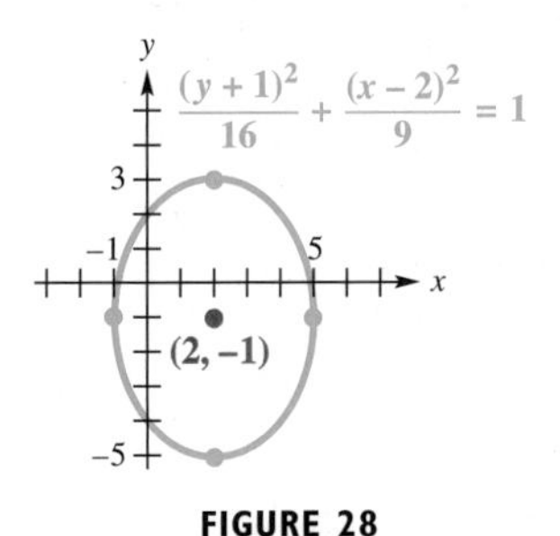


**FIGURE 28**

**Graphing Calculator Solution**

Solve for $y$ in the equation of the ellipse to obtain

$$y=-1\pm 4\sqrt{1-\frac{(x-2)^2}{9}}.$$

The + sign yields the top half of the ellipse, while the − sign yields the bottom half. See Figure 29.

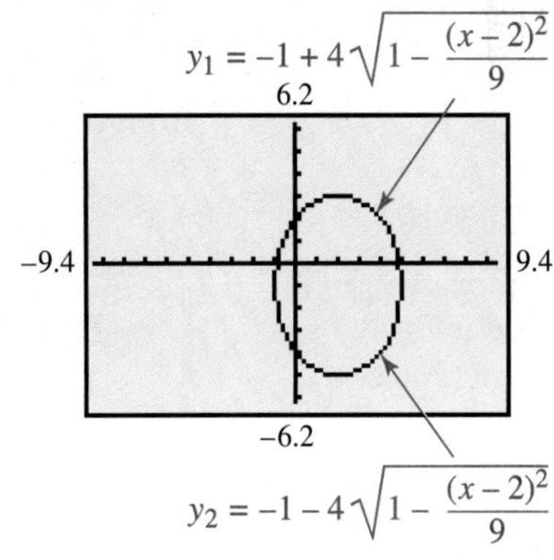


**FIGURE 29**

## Applications of Ellipses

Ellipses have many useful applications. As Earth makes its year-long journey around the sun, it traces an ellipse. Spacecraft travel around Earth in elliptical orbits, and planets make elliptical orbits around the sun.

An interesting recent application is the use of an elliptical tub in the nonsurgical removal of kidney stones, using the reflective property of the ellipse. A lithotripter is a machine that uses shock waves to crush kidney stones. The waves originate at one focus and are reflected to hit the kidney stone, which is positioned at the second focus. See Figure 30.

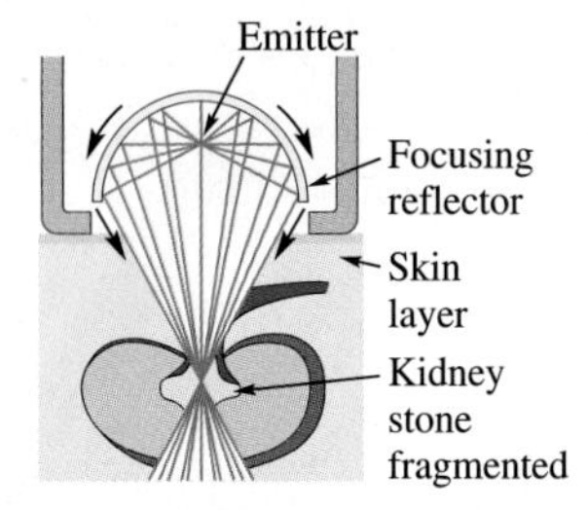


The top of an ellipse is illustrated in this depiction of how a lithotripter crushes kidney stones.

*Source*: Adapted drawing of an ellipse in illustration of a lithotripter. The American Medical Association, *Encyclopedia of Medicine*, 1989.

**(a)**

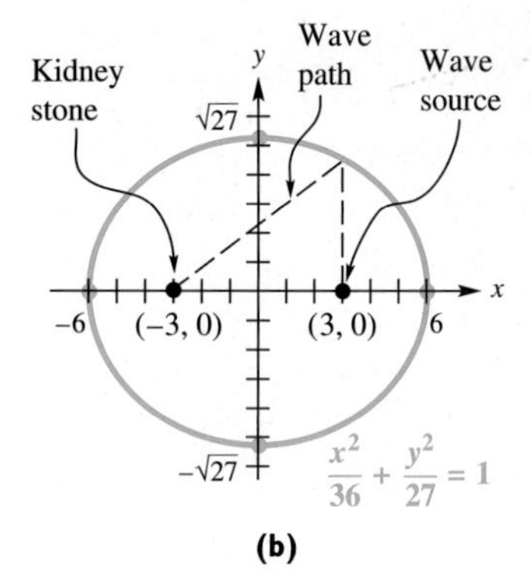


**(b)**

**FIGURE 30**

**EXAMPLE 5** *Modeling the Reflective Property of Ellipses*
If a lithotripter is based on the ellipse

$$\frac{x^2}{36} + \frac{y^2}{27} = 1,$$

determine how many units the kidney stone and the wave source must be placed from the center of the ellipse.

**Solution** The kidney stone and the source of the beam must be placed at the foci, $(c, 0)$ and $(-c, 0)$. Here $a^2 = 36$ and $b^2 = 27$, so

$$c = \sqrt{a^2 - b^2} = \sqrt{36 - 27} = \sqrt{9} = 3.$$

Thus, the foci are $(3, 0)$ and $(-3, 0)$, so the kidney stone and the source both must be placed on a line 3 units from the center. See Figure 30(b).

**FOR DISCUSSION**

What is a "whispering gallery"? How does its special feature correspond to the foci of an ellipse?

## Equations and Graphs of Hyperbolas

An ellipse was defined as the set of all points in a plane the sum of whose distances from two fixed points is constant. A *hyperbola* is defined similarly.

> **Hyperbola**
>
> A **hyperbola** is the set of all points in a plane such that the absolute value of the *difference* of the distances from two fixed points is constant. The two fixed points are called the **foci** of the hyperbola.

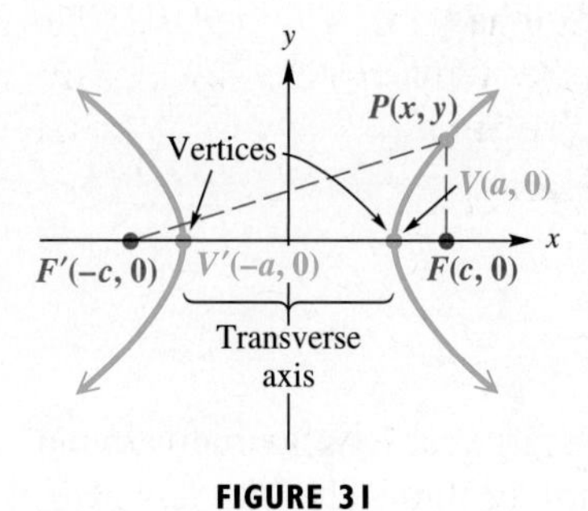


**FIGURE 31**

Suppose a hyperbola has foci at $F'(-c, 0)$ and $F(c, 0)$. See Figure 31. The midpoint of the segment $F'F$ is the **center** of the hyperbola, and the points $V'(-a, 0)$ and $V(a, 0)$ are the **vertices** of the hyperbola. The line segment $V'V$ is the **transverse axis** of the hyperbola.

Using the distance formula in conjunction with the definition of a hyperbola, you can verify the following standard forms of the equations for hyperbolas centered at the origin. (See Exercise 67.)

**Standard Forms of Equations for Hyperbolas**

The hyperbola with center at the origin and equation

$$\frac{x^2}{a^2} - \frac{y^2}{b^2} = 1$$

has vertices $(\pm a, 0)$ and foci $(\pm c, 0)$, where $c^2 = a^2 + b^2$. The hyperbola with center at the origin and equation

$$\frac{y^2}{a^2} - \frac{x^2}{b^2} = 1$$

has vertices $(0, \pm a)$ and foci $(0, \pm c)$, where $c^2 = a^2 + b^2$.

| X | Y1 | Y2 |
|---|---|---|
| 20 | 27.111 | 28 |
| 70 | 97.75 | 98 |
| 120 | 167.85 | 168 |
| 170 | 237.9 | 238 |
| 220 | 307.92 | 308 |
| 270 | 377.94 | 378 |
| 320 | 447.95 | 448 |

X=320

The table in Figure 32 shows numerically how the asymptote with equation $y = \frac{7}{5}x$ approaches the upper-right branch of the hyperbola with equation

$$\frac{x^2}{25} - \frac{y^2}{49} = 1$$

in Example 6.

**FIGURE 32**

Starting with the first equation for a hyperbola and solving for $y$ gives

$$\frac{x^2}{a^2} - \frac{y^2}{b^2} = 1$$

$$\frac{x^2}{a^2} - 1 = \frac{y^2}{b^2}$$

$$\frac{x^2 - a^2}{a^2} = \frac{y^2}{b^2}$$

$$y = \pm\frac{b}{a}\sqrt{x^2 - a^2}. \quad (*)$$

If $x^2$ is very large in comparison to $a^2$, the difference $x^2 - a^2$ would be very close to $x^2$. If this happens, then the points satisfying equation (*) would be very close to one of the lines

$$y = \pm\frac{b}{a}x.$$

Thus, as $|x|$ gets larger and larger, the points of the hyperbola $\frac{x^2}{a^2} - \frac{y^2}{b^2} = 1$ become closer to the lines $y = \pm\frac{b}{a}x$. These lines, called the **asymptotes** of the hyperbola, are very helpful when graphing the hyperbola with traditional graphing methods. The lines are the extended diagonals of the rectangle whose vertices are $(a, b)$, $(-a, b)$, $(a, -b)$, and $(-a, -b)$. This rectangle is called the **fundamental rectangle** of the hyperbola. Similar results hold for a hyperbola of the form $\frac{y^2}{a^2} - \frac{x^2}{b^2} = 1$.

**FOR DISCUSSION**

With your graphing calculator, graph the equation of the asymptote with positive slope for the hyperbola in Example 6, along with $y_1$ as defined in the example. Use various viewing windows to "watch" the hyperbola approach the asymptote.

**EXAMPLE 6** *Using Asymptotes to Graph a Hyperbola*

Sketch the asymptotes and graph the hyperbola $\frac{x^2}{25} - \frac{y^2}{49} = 1$. Give the domain and range.

**Analytic Solution**

For this hyperbola, $a = 5$ and $b = 7$. With these values,

$$y = \pm\frac{b}{a}x \quad \text{becomes} \quad y = \pm\frac{7}{5}x.$$

Choosing $x = 5$ gives $y = \pm 7$. Choosing $x = -5$ also gives $y = \pm 7$. These four points—$(5, 7)$, $(5, -7)$, $(-5, 7)$, and $(-5, -7)$—are the corners of the fundamental rectangle shown in Figure 33.

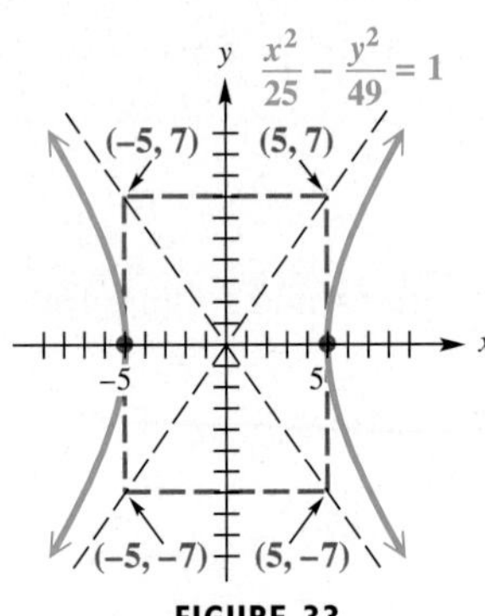


**FIGURE 33**

The extended diagonals of this rectangle are the asymptotes of the hyperbola. The final graph is shown in Figure 33. The hyperbola has $x$-intercepts 5 and $-5$. The domain of this relation is $(-\infty, -5] \cup [5, \infty)$, and the range is $(-\infty, \infty)$.

**Graphing Calculator Solution**

Solve the given equation for $y$ as follows.

$$\frac{x^2}{25} - \frac{y^2}{49} = 1$$

$$-\frac{y^2}{49} = 1 - \frac{x^2}{25} \qquad \text{Subtract } \tfrac{x^2}{25}.$$

$$\frac{y^2}{49} = \frac{x^2}{25} - 1 \qquad \text{Multiply by } -1.$$

$$\frac{y}{7} = \pm\sqrt{\frac{x^2}{25} - 1} \qquad \text{Take square roots.}$$

$$y = \pm 7\sqrt{\frac{x^2}{25} - 1} \qquad \text{Multiply by 7.}$$

The graphs of the hyperbola and its asymptotes are shown in Figure 34. The graph of $y_1$ creates the upper half of the hyperbola, while the graph of $y_2$ creates the lower half.

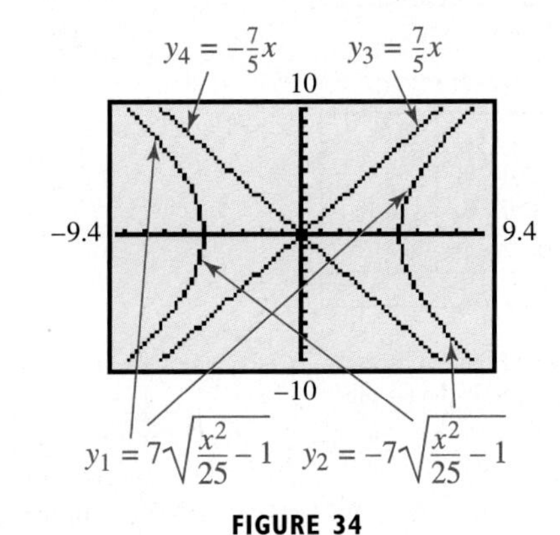


**FIGURE 34**

**EXAMPLE 7** *Graphing a Hyperbola Centered at the Origin*

Graph $25y^2 - 4x^2 = 9$. Give the domain and range.

**Analytic Solution**

Divide each side by 9 to get

$$\frac{25y^2}{9} - \frac{4x^2}{9} = 1.$$

To determine the values of $a$ and $b$, write the equation as

$$\frac{y^2}{\frac{9}{25}} - \frac{x^2}{\frac{9}{4}} = 1.$$

**Graphing Calculator Solution**

To graph this hyperbola using a graphing calculator, solve the given equation for $y$.

$$25y^2 - 4x^2 = 9$$

$$25y^2 = 9 + 4x^2 \qquad \text{Add } 4x^2.$$

$$5y = \pm\sqrt{9 + 4x^2} \qquad \text{Take square roots.}$$

$$y = \pm\frac{1}{5}\sqrt{9 + 4x^2} \qquad \text{Multiply by } \tfrac{1}{5}.$$

*(continued)*

This hyperbola is centered at the origin, has foci on the $y$-axis, and has $y$-intercepts $-\frac{3}{5}$ and $\frac{3}{5}$. Use the points $\left(\frac{3}{2}, \frac{3}{5}\right)$, $\left(-\frac{3}{2}, \frac{3}{5}\right)$, $\left(\frac{3}{2}, -\frac{3}{5}\right)$, and $\left(-\frac{3}{2}, -\frac{3}{5}\right)$ to get the fundamental rectangle shown in Figure 35. Use the diagonals of this rectangle to determine the asymptotes for the graph. The domain is $(-\infty, \infty)$, and the range is $\left(-\infty, -\frac{3}{5}\right] \cup \left[\frac{3}{5}, \infty\right)$.

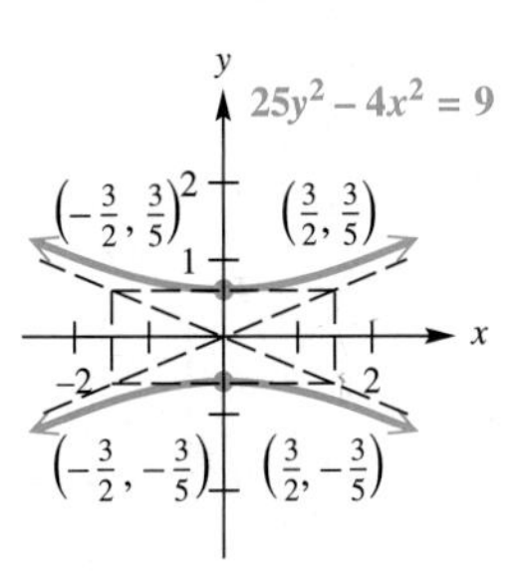


FIGURE 35

The union of the graphs of

$$y_1 = \frac{1}{5}\sqrt{9 + 4x^2} \quad \text{and} \quad y_2 = -\frac{1}{5}\sqrt{9 + 4x^2}$$

is the graph of the hyperbola. See Figure 36.

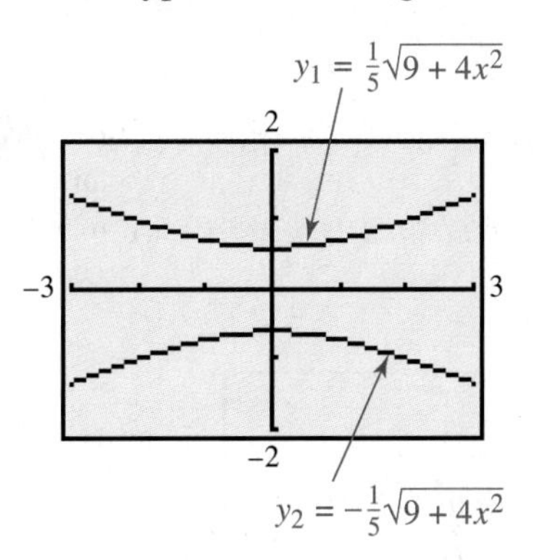


FIGURE 36

Like ellipses, hyperbolas can be translated from the origin.

**EXAMPLE 8** *Graphing a Hyperbola Translated from the Origin*

Graph $\dfrac{(y + 2)^2}{9} - \dfrac{(x + 3)^2}{4} = 1$. Give the domain and range.

**Solution** This hyperbola has the same graph as

$$\frac{y^2}{9} - \frac{x^2}{4} = 1,$$

except that it is centered at $(-3, -2)$. See Figure 37. The domain is $(-\infty, \infty)$, and the range is $(-\infty, -5] \cup [1, \infty)$.

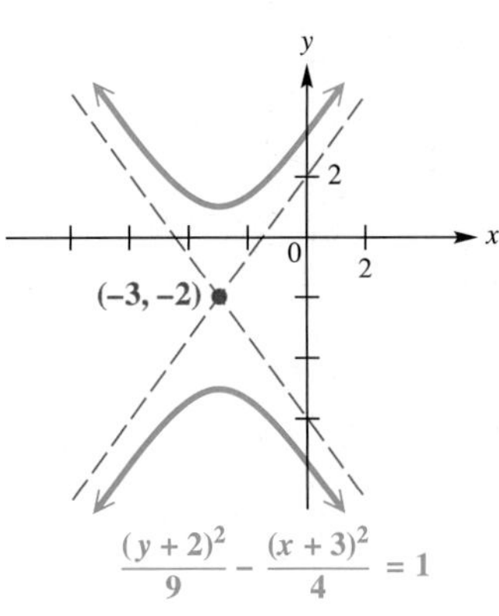


FIGURE 37

**FOR DISCUSSION**

How would you graph the hyperbola in Example 8 on your graphing calculator? Duplicate the calculator graph of that hyperbola as shown here.

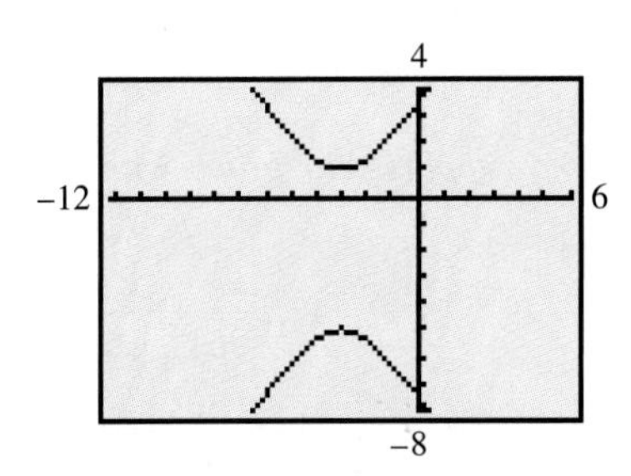


The two branches of a hyperbola are reflections about two different axes, and also about a point. What are the axes and the point for the hyperbola shown in the figure? (This reflecting property of the hyperbola was used on the Hubble space telescope.)

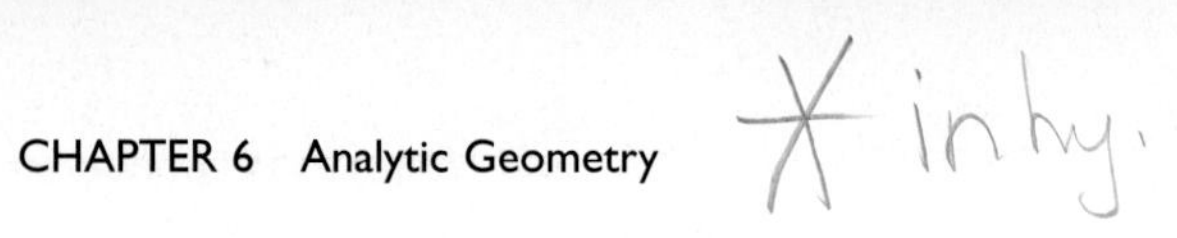

## 6.2 EXERCISES

*Concept Check* *Match each equation in Exercises 1–8 with its calculator graph in A–H. Check your answer by generating a calculator graph of your own. (Every window has* Xscl = Yscl = 1.*)*

1. $\frac{y^2}{16} + \frac{x^2}{4} = 1$

2. $\frac{x^2}{16} + \frac{y^2}{4} = 1$

3. $\frac{x^2}{4} - \frac{y^2}{16} = 1$

4. $\frac{y^2}{4} - \frac{x^2}{16} = 1$

5. $\frac{(y-4)^2}{25} + \frac{(x+2)^2}{9} = 1$

6. $\frac{(y+4)^2}{25} + \frac{(x-2)^2}{9} = 1$

7. $\frac{(x+2)^2}{9} - \frac{(y-4)^2}{25} = 1$

8. $\frac{(x-2)^2}{9} - \frac{(y+4)^2}{25} = 1$

**A.**

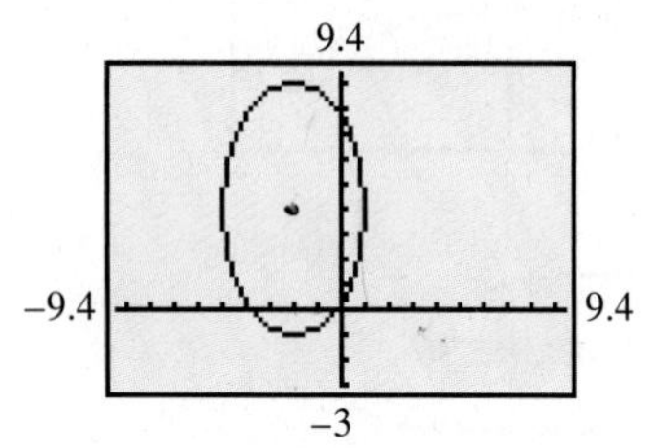


**B.**

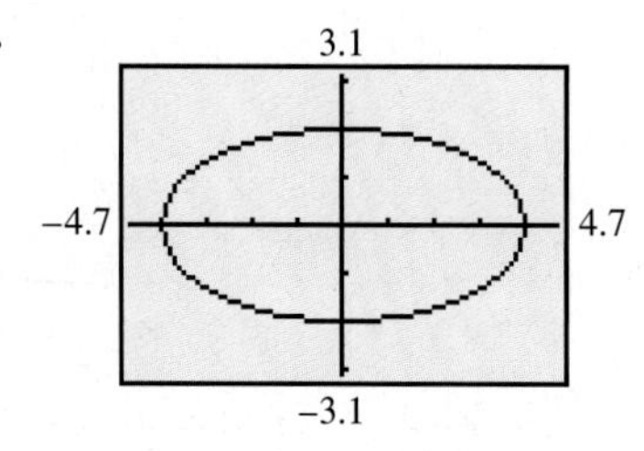


**C.**

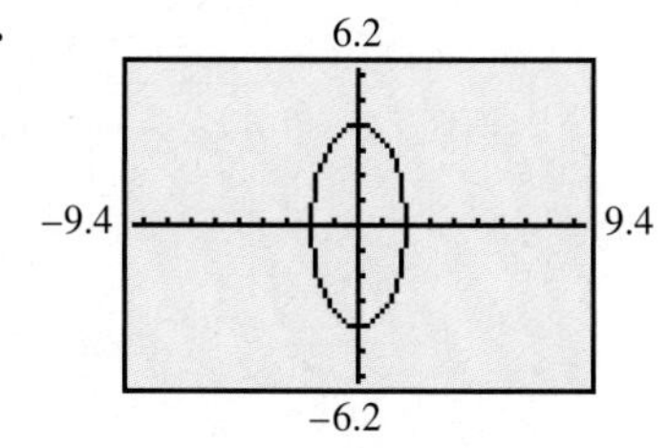


**D.**

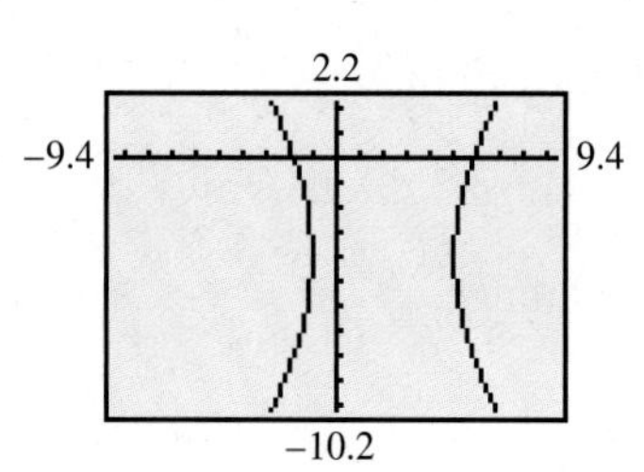


**E.**

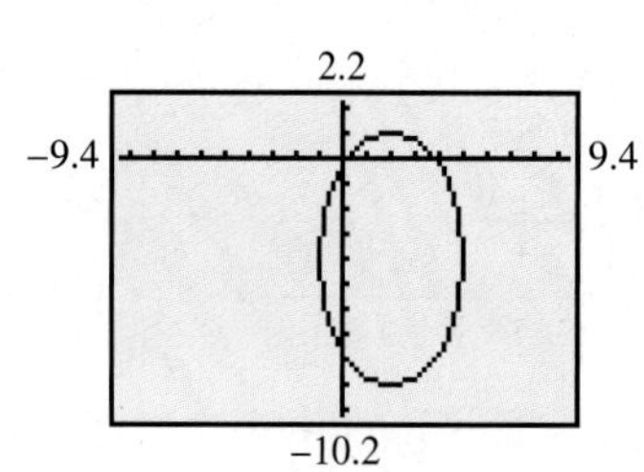


**F.**

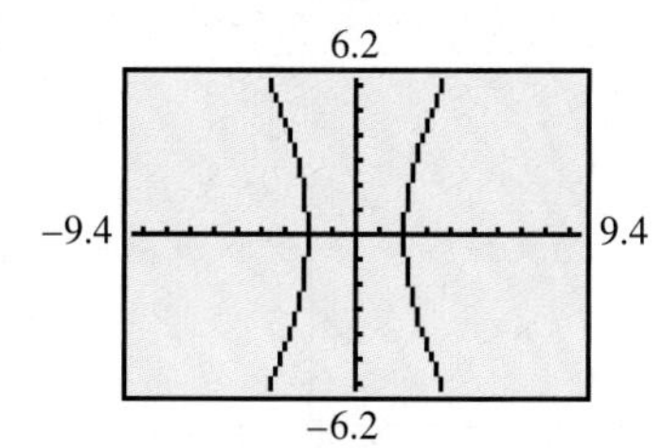


**G.**

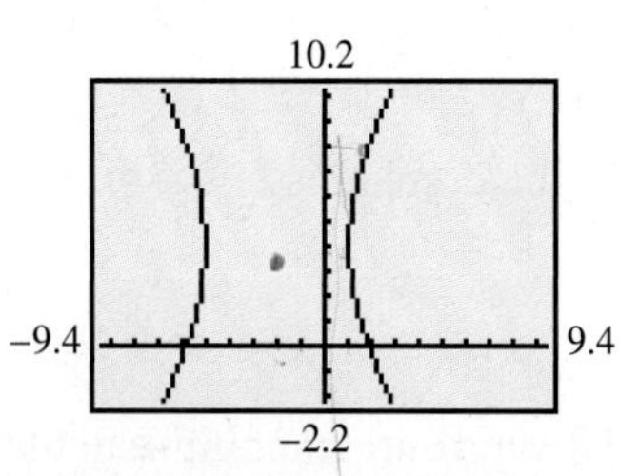


**H.**

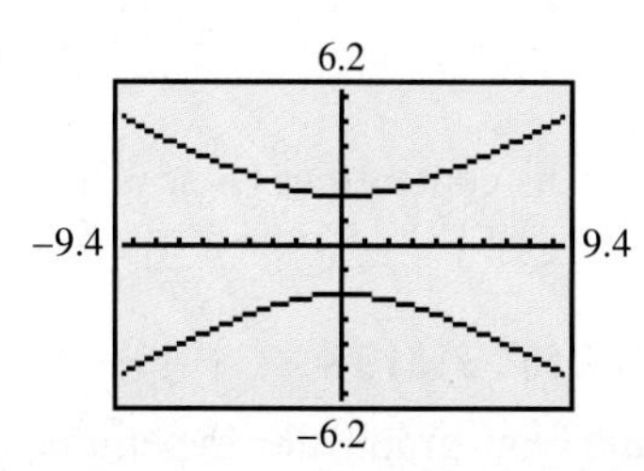


9. Explain how a circle can be interpreted as a special case of an ellipse.

10. *Concept Check* If an ellipse has endpoints and vertices at $(-3, 0)$, $(3, 0)$, $(0, 5)$, and $(0, -5)$, what is its domain? What is its range?

*Graph each ellipse by hand, and support your graph with your calculator. Give the domain and range. Give the foci in Exercises 11–14.*

11. $\frac{x^2}{9} + \frac{y^2}{4} = 1$

12. $\frac{y^2}{36} + \frac{x^2}{16} = 1$

13. $9x^2 + 6y^2 = 54$

14. $12x^2 + 8y^2 = 96$

15. $\frac{25y^2}{36} + \frac{64x^2}{9} = 1$

16. $\frac{16y^2}{9} + \frac{121x^2}{25} = 1$

*Graph each ellipse. Give the domain and range.*

17. $\frac{(y+3)^2}{25} + \frac{(x-1)^2}{9} = 1$

18. $\frac{(y-2)^2}{36} + \frac{(x+3)^2}{16} = 1$

19. $\frac{(x-2)^2}{16} + \frac{(y-1)^2}{9} = 1$

**20.** $\dfrac{(y+2)^2}{36} + \dfrac{(x+3)^2}{25} = 1$

**21.** $\dfrac{(x+1)^2}{64} + \dfrac{(y-2)^2}{49} = 1$

**22.** $\dfrac{(x-4)^2}{9} + \dfrac{(y+2)^2}{4} = 1$

**23.** The ellipse $\dfrac{(y+1)^2}{16} + \dfrac{(x-2)^2}{9} = 1$ is graphed in Figure 28. Give the equations of its horizontal and vertical axes of symmetry.

## RELATING CONCEPTS

**For individual or group investigation (Exercises 24–29)**

*The graph of the ellipse* $\dfrac{(y+1)^2}{16} + \dfrac{(x-2)^2}{9} = 1$ *is the union of the graphs of*

$$y_1 = -1 + \sqrt{16 - \frac{16(x-2)^2}{9}}$$

*and*

$$y_2 = -1 - \sqrt{16 - \frac{16(x-2)^2}{9}}.$$

*The domain of this relation is* $[-1, 5]$. *Use this information to work Exercises 24–29 in order.*

**24.** The relation is defined only when the radicand in $y_1$ and $y_2$ is greater than or equal to 0. Write the inequality that could be solved to find the domain analytically.

**25.** Let $y$ represent the expression in $x$ found in the radicand. What conic section is the graph of this function?

**26.** Graph the function defined by the radicand with a graphing calculator in the window $[-10, 10]$ by $[-10, 20]$.

**27.** Use the graph in Exercise 26 to solve the inequality of Exercise 24.

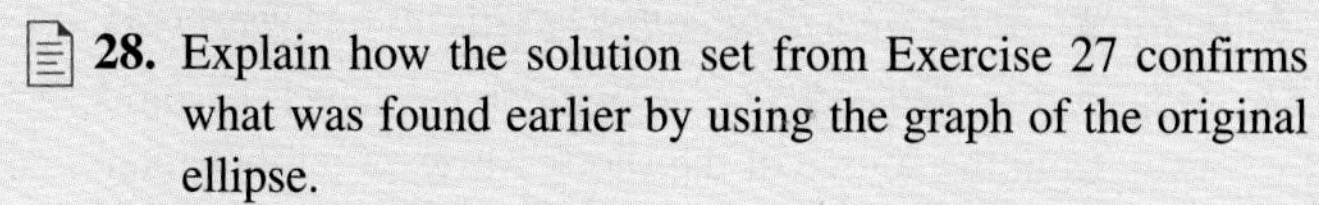

**28.** Explain how the solution set from Exercise 27 confirms what was found earlier by using the graph of the original ellipse.

**29.** Solve the inequality of Exercise 24 analytically, using a sign graph.

**30.** Solve each equation for $y$ and then use a table to find and compare the domains of

$$\frac{x^2}{4} + \frac{y^2}{25} = 1 \quad \text{and} \quad \frac{x^2}{4} - \frac{y^2}{25} = 1.$$

In particular, note when each relation returns an error message. Support your answer graphically.

*Graph each hyberbola. Give the domain and range. Give the center in Exercises 37–40.*

**31.** $\dfrac{x^2}{16} - \dfrac{y^2}{9} = 1$

**32.** $\dfrac{y^2}{9} - \dfrac{x^2}{9} = 1$

**33.** $49y^2 - 36x^2 = 1764$

**34.** $144x^2 - 49y^2 = 7056$

**35.** $\dfrac{4x^2}{9} - \dfrac{25y^2}{16} = 1$

**36.** $x^2 - y^2 = 1$

**37.** $\dfrac{(x-1)^2}{9} - \dfrac{(y+3)^2}{25} = 1$

**38.** $\dfrac{(x+3)^2}{16} - \dfrac{(y-2)^2}{36} = 1$

**39.** $\dfrac{(y-5)^2}{4} - \dfrac{(x+1)^2}{9} = 1$

**40.** $\dfrac{(y+1)^2}{25} - \dfrac{(x-3)^2}{36} = 1$

*Find an equation for each ellipse.*

**41.** $x$-intercepts $\pm 4$; foci $(-2, 0)$ and $(2, 0)$

**42.** $y$-intercepts $\pm 3$; foci $(0, \sqrt{3})$ and $(0, -\sqrt{3})$

**43.** Endpoints of major axis at $(6, 0)$ and $(-6, 0)$; $c = 4$

**44.** Vertices $(0, 5)$ and $(0, -5)$; $b = 2$

**45.** Center $(3, -2)$; $a = 5$; $c = 3$; major axis vertical

**46.** Center $(2, 0)$; minor axis of length 6; major axis horizontal and of length 9

*Find an equation for each hyperbola.*

**47.** $x$-intercepts $\pm 3$; foci $(-4, 0)$ and $(4, 0)$

**48.** $y$-intercepts $\pm 5$; foci $(0, 3\sqrt{3})$ and $(0, -3\sqrt{3})$

**49.** Asymptotes $y = \pm \dfrac{3}{5}x$; $y$-intercepts 3 and $-3$

**50.** $y$-intercept $-2$; center at origin; passing through $(2, 3)$

## RELATING CONCEPTS

**For individual or group investigation (Exercises 51–56)**

*Consider the ellipse and hyperbola defined by*

$$\frac{x^2}{16} + \frac{y^2}{12} = 1 \quad and \quad \frac{x^2}{4} - \frac{y^2}{12} = 1,$$

*respectively.*

**51.** Find the foci of the ellipse. Call them $F_1$ and $F_2$.

**52.** Graph the ellipse with your calculator, and trace to find the coordinates of several points on the ellipse.

**53.** For each of the points $P$, verify that

$$[\text{distance of } P \text{ from } F_1] + [\text{distance of } P \text{ from } F_2] = 8.$$

**54.** Repeat Exercises 51 and 52 for the hyperbola.

**55.** For each of the points $P$ from Exercise 54, verify that

$$|[\text{distance of } P \text{ from } F_1] - [\text{distance of } P \text{ from } F_2]| = 4.$$

**56.** How do Exercises 53 and 55 relate to the definitions of the ellipse and the hyperbola we have given in this section?

*(Modeling) Solve each application of ellipses or hyperbolas.*

**57.** *Shape of a Lithotripter* A patient is placed 12 units away from the source of the shock waves of a lithotripter. The lithotripter is based on an ellipse with a minor axis that measures 16 units. Find an equation of an ellipse that would satisfy this situation.

**58.** *Orbit of Venus* The orbit of Venus is an ellipse, with the sun at one focus. An approximate equation for the orbit is

$$\frac{x^2}{5013} + \frac{y^2}{4970} = 1,$$

where $x$ and $y$ are measured in millions of miles.

**(a)** Find the length of the major axis.

**(b)** Find the length of the minor axis.

**59.** *The Roman Coliseum* The Roman Coliseum is an ellipse with major axis 620 feet and minor axis 513 feet. Find the distance between the foci of this ellipse.

**60.** *The Roman Coliseum* A formula for the approximate circumference of an ellipse is

$$C \approx 2\pi\sqrt{\frac{a^2 + b^2}{2}},$$

where $a$ and $b$ are the lengths shown in the figure in the next column. Use this formula to find the approximate circumference of the Roman Coliseum. (See Exercise 59.)

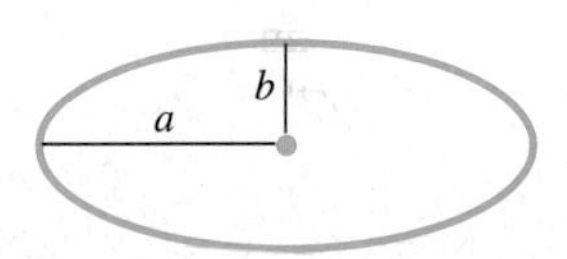


**61.** *Height of an Overpass* A one-way road passes under an overpass in the form of half of an ellipse, 15 feet high at the center and 20 feet wide. Assuming a truck is 12 feet wide, what is the height of the tallest truck that can pass under the overpass?

**62.** *Location-Finding System* Ships and planes often use a location-finding system called LORAN. With this system, a radio transmitter at $M$ on the figure sends out a series of pulses. When each pulse is received at transmitter $S$, it then sends out a pulse. A ship at $P$ receives pulses from both $M$ and $S$. A receiver on the ship measures the difference in the arrival times of the pulses. The navigator then consults a special map, showing certain curves according to the differences in arrival times. In this way, the ship can be located as lying on a portion of what type of curve?

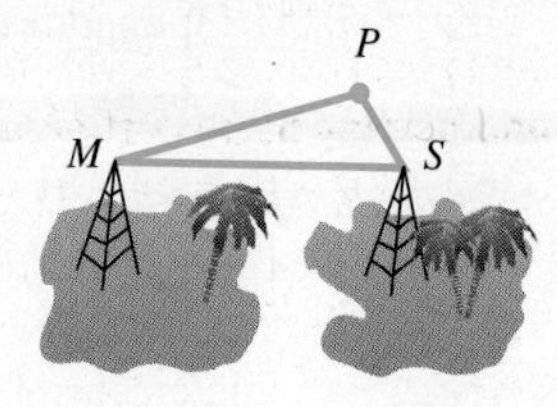

Later

**63.** *Orbit of a Satellite* The coordinates in miles for the orbit of the artificial satellite *Explorer VII* can be described by the equation

$$\frac{x^2}{a^2} + \frac{y^2}{b^2} = 1,$$

where $a = 4465$ and $b = 4462$. Earth's center is located at one focus of its elliptical orbit. (*Source:* Loh, W., *Dynamics and Thermodynamics of Planetary Entry,* Prentice-Hall, 1963; Thomson, W., *Introduction to Space Dynamics,* John Wiley and Sons, 1961.)

**(a)** Graph both the orbit of *Explorer VII* and Earth on the same coordinate axes if the average radius of Earth is 3960 miles. Use the window $[-6750, 6750]$ by $[-4500, 4500]$.

**(b)** Determine the maximum and minimum heights of the satellite above Earth's surface.

**64.** *Design of a Sports Complex* Two buildings in a sports complex are shaped and positioned like a portion of the branches of the hyperbola

$$400x^2 - 625y^2 = 250{,}000,$$

where $x$ and $y$ are in meters.

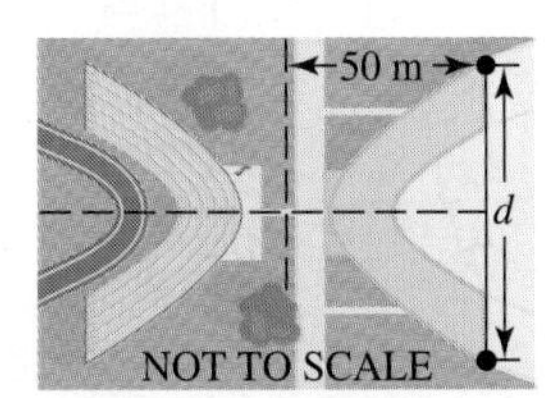


**(a)** How far apart are the buildings at their closest point?

**(b)** Find the distance $d$ in the figure.

**65.** *Structure of an Atom* In 1911, Ernest Rutherford discovered the basic structure of the atom by "shooting" positively charged alpha particles with a speed of $10^7$ meters per second at a piece of gold foil $6 \times 10^{-7}$ meters thick. Only a small percentage of the alpha particles struck a gold nucleus head-on and were deflected directly back toward their source. The rest of the particles often followed a hyperbolic trajectory because they were repelled by positively charged gold nuclei. Thus, Rutherford proposed that the atom was composed of mostly empty space with a small and dense nucleus. The figure in the next column shows an alpha particle $A$ initially approaching a gold nucleus $N$ and being deflected at an angle $\theta = 90°$. $N$ is located at a focus of the hyperbola, and the trajectory of $A$ passes through a vertex of the hyperbola. (*Source:* Semat, H. and J. Albright, *Introduction to Atomic and Nuclear Physics,* Holt, Rinehart and Winston, 1972.)

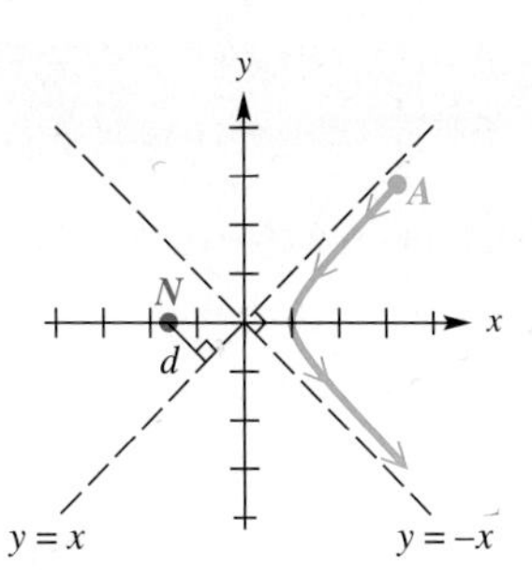


**(a)** Determine the equation of the trajectory of the alpha particle if $d = 5 \times 10^{-14}$ meters.

**(b)** What was the minimum distance between the centers of the alpha particle and the gold nucleus?

**66.** *Sound Detection* Microphones are placed at points $(-c, 0)$ and $(c, 0)$. An explosion occurs at point $P(x, y)$ having positive $x$-coordinate.

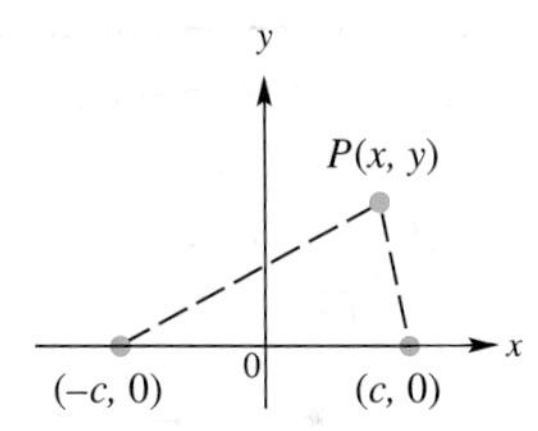


The sound is detected at the closer microphone $t$ seconds before being detected at the farther microphone. Assume that sound travels at a speed of 330 meters per second, and show that $P$ must be on the hyperbola

$$\frac{x^2}{330^2t^2} - \frac{y^2}{4c^2 - 330^2t^2} = \frac{1}{4}.$$

**67.** Suppose a hyperbola has center at the origin, foci at $F'(-c, 0)$ and $F(c, 0)$, and equation $d(P, F') - d(P, F) = 2a$. Let $b^2 = c^2 - a^2$, and show that an equation of the hyperbola is

$$\frac{x^2}{a^2} - \frac{y^2}{b^2} = 1.$$

**68.** Use the definition of an ellipse to find an equation of an ellipse with foci $(3, 0)$ and $(-3, 0)$, where the sum of the distances from any point of the ellipse to the two foci is 10.

**69.** Use the definition of a hyperbola to find an equation of a hyperbola with center at the origin, foci $(-2, 0)$ and $(2, 0)$, and the absolute value of the difference of the distances from any point of the hyperbola to the two foci equal to 2.

## Reviewing Basic Concepts (Sections 6.1 and 6.2)

**1.** Match each definition in A–D with the appropriate conic section.

**(a)** Circle **(b)** Parabola **(c)** Ellipse **(d)** Hyperbola

**A.** A set of points in a plane such that the sum of their distances from two fixed points is constant

**B.** A set of points in a plane that are equidistant from a fixed point

**C.** A set of points in a plane such that the absolute value of the difference of their distances from two fixed points is constant

**D.** A set of points in a plane equidistant from a fixed point and a fixed line

*Graph each conic section by hand and check with your calculator.*

**2.** $12x^2 - 4y^2 = 48$

**3.** $y = 2x^2 + 3x - 1$

**4.** $x^2 + y^2 - 2x + 2y - 2 = 0$

**5.** $4x^2 + 9y^2 = 36$

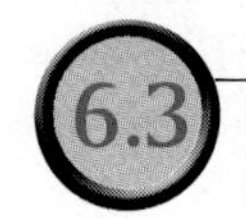

**6.** Given the two vertices and two foci of a conic section, how can you tell whether it is an ellipse or a hyperbola?

*Write an equation for each conic section.*

**7.** Center $(2, -1)$; radius 3

**8.** Foci $(\pm 4, 0)$; major axis length 12

**9.** Vertices $(0, \pm 2)$; foci $(0, \pm 4)$

**10.** Focus $\left(0, \frac{1}{2}\right)$; vertex at the origin

# 6.3 SUMMARY OF THE CONIC SECTIONS

Characteristics • Identifying Conic Sections • Eccentricity

## Characteristics

The conic sections presented in this chapter have equations that can be written in the form

$$Ax^2 + Dx + Cy^2 + Ey + F = 0,$$

where either $A$ or $C$ must be nonzero. The special characteristics of each conic section are summarized below.

| Conic Section | Characteristic | Example |
|---|---|---|
| Parabola | Either $A = 0$ or $C = 0$, but not both | $y = x^2$<br>$x = 3y^2 + 2y - 4$ |
| Circle | $A = C \neq 0$ | $x^2 + y^2 = 16$ |
| Ellipse | $A \neq C$, $AC > 0$ | $\frac{x^2}{16} + \frac{y^2}{25} = 1$ |
| Hyperbola | $AC < 0$ | $x^2 - y^2 = 1$ |

The following chart summarizes our work with conic sections.

| Equation | Graph | Description | Identification |
|---|---|---|---|
| $(x-h)^2+(y-k)^2=r^2$ | 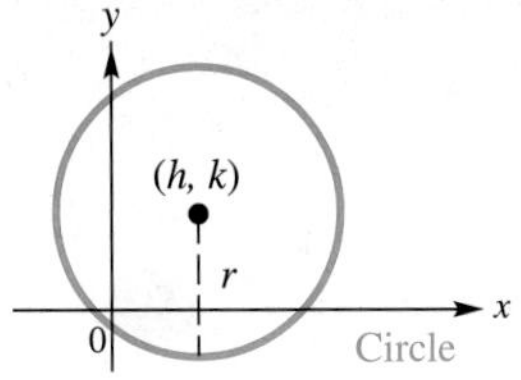 Circle | Center is $(h, k)$, and radius is $r$. | $x^2$- and $y^2$-terms have the same positive coefficient. |
| $y=a(x-h)^2+k$ | 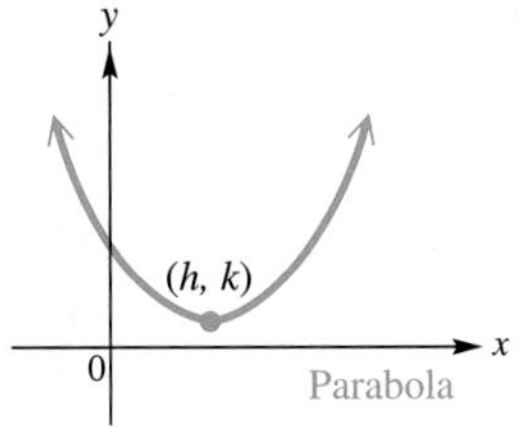 Parabola | Opens upward if $a>0$, downward if $a<0$. Vertex is $(h, k)$. | $x^2$-term; $y$ is not squared. |
| $x=a(y-k)^2+h$ | 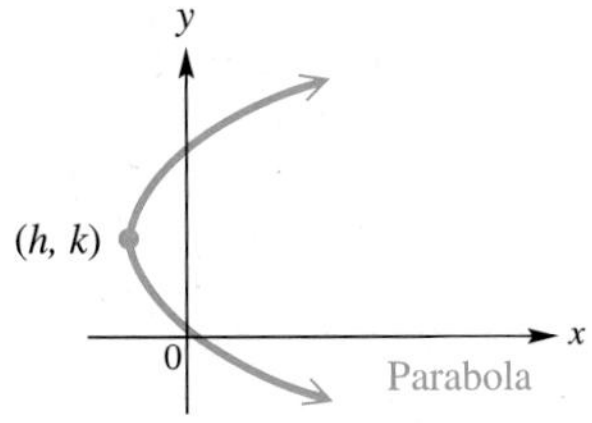 Parabola | Opens to the right if $a>0$, to the left if $a<0$. Vertex is $(h, k)$. | $y^2$-term; $x$ is not squared. |
| $\frac{x^2}{a^2}+\frac{y^2}{b^2}=1$ | 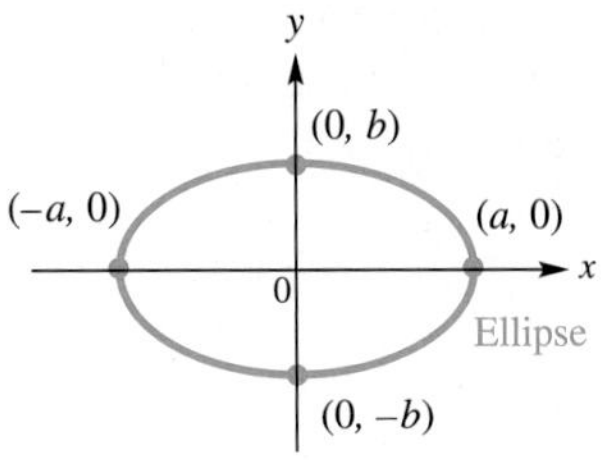 Ellipse | $x$-intercepts are $a$ and $-a$. $y$-intercepts are $b$ and $-b$. | $x^2$- and $y^2$-terms have different positive coefficients. |
| $\frac{x^2}{a^2}-\frac{y^2}{b^2}=1$ | 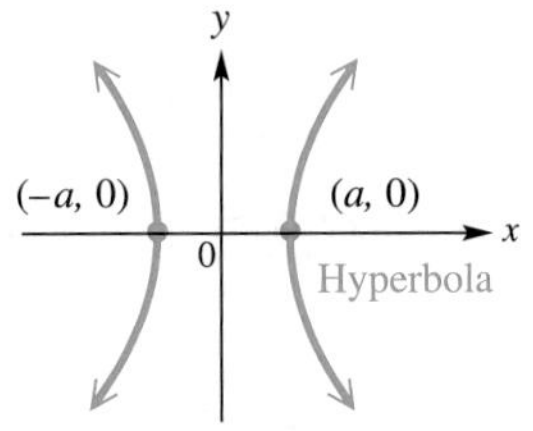 Hyperbola | $x$-intercepts are $a$ and $-a$. Asymptotes found from $(a, b)$, $(a, -b)$, $(-a, -b)$, and $(-a, b)$. | $x^2$ has a positive coefficient. $y^2$ has a negative coefficient. |
| $\frac{y^2}{a^2}-\frac{x^2}{b^2}=1$ | 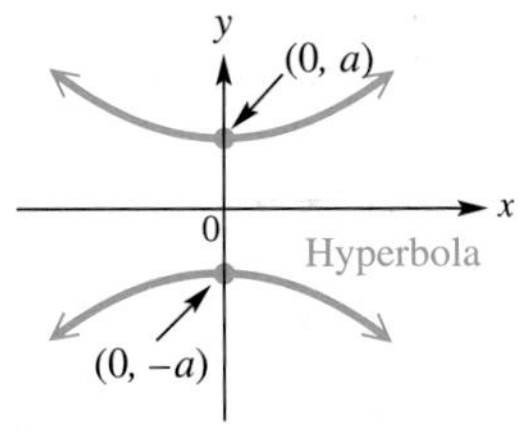 Hyperbola | $y$-intercepts are $a$ and $-a$. Asymptotes found from $(b, a)$, $(b, -a)$, $(-b, -a)$, and $(-b, a)$. | $y^2$ has a positive coefficient. $x^2$ has a negative coefficient. |

## Identifying Conic Sections

To recognize the type of graph that a given conic section has, we sometimes need to transform the equation into a more familiar form.

**EXAMPLE 1** *Determining the Type of Conic Sections from Equations*

Decide on the type of conic section represented by each equation, and give each graph.

**(a)** $x^2 = 25 + 5y^2$ **(b)** $4x^2 - 16x + 9y^2 + 54y = -61$

**(c)** $x^2 - 8x + y^2 + 10y = -41$ **(d)** $x^2 - 6x + 8y - 7 = 0$

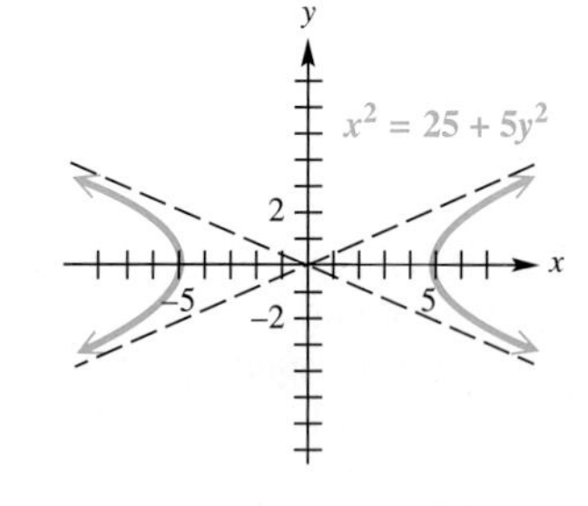


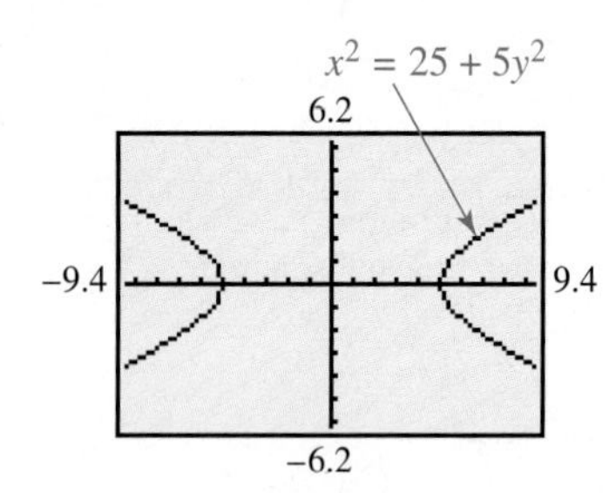


FIGURE 38

**Solution**

**(a)** Rewriting the equation as

$$x^2 - 5y^2 = 25$$

or

$$\frac{x^2}{25} - \frac{y^2}{5} = 1$$

shows that the equation represents a hyperbola centered at the origin, with asymptotes

$$y = \frac{\pm\sqrt{5}}{5}x.$$

The $x$-intercepts are $\pm 5$; both types of graphs are shown in Figure 38.

**(b)** Since the coefficients of the $x^2$- and $y^2$-terms of $4x^2 - 16x + 9y^2 + 54y = -61$ are unequal and both positive, this equation may represent an ellipse but not a circle. (It may also represent a single point or no points at all.) To find out, complete the squares on $x$ and $y$.

$$4(x^2 - 4x \quad\quad) + 9(y^2 + 6y \quad\quad) = -61$$ Factor out 4; factor out 9.

$$4(x^2 - 4x + 4 - 4) + 9(y^2 + 6y + 9 - 9) = -61$$ Add and subtract the same quantity.

$$4(x^2 - 4x + 4) - 16 + 9(y^2 + 6y + 9) - 81 = -61$$ Regroup and distribute.

$$4(x - 2)^2 + 9(y + 3)^2 = 36$$ Add 97; factor.

$$\frac{(x - 2)^2}{9} + \frac{(y + 3)^2}{4} = 1$$ Divide by 36.

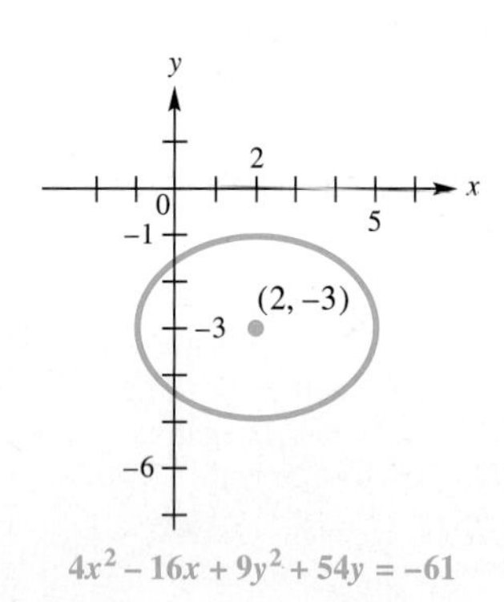


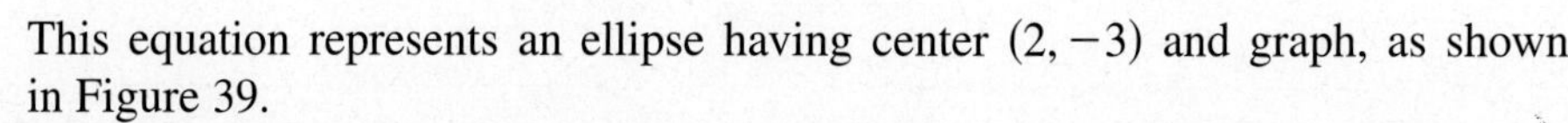
This equation represents an ellipse having center $(2, -3)$ and graph, as shown in Figure 39.

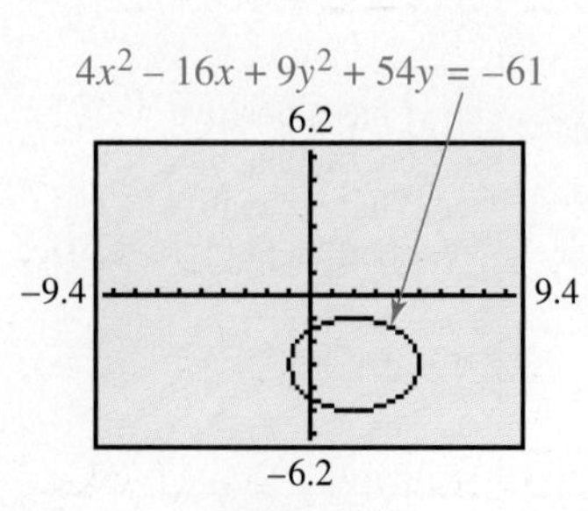


FIGURE 39

**(c)** Complete the squares on both $x$ and $y$.

$$x^2 - 8x + y^2 + 10y = -41$$

$$(x^2 - 8x + 16) + (y^2 + 10y + 25) = -41 + 16 + 25$$

$$(x - 4)^2 + (y + 5)^2 = 0$$

This result shows that the equation is that of a circle with radius 0; that is, the point $(4, -5)$. See Figure 40. Had a negative number been obtained on the right

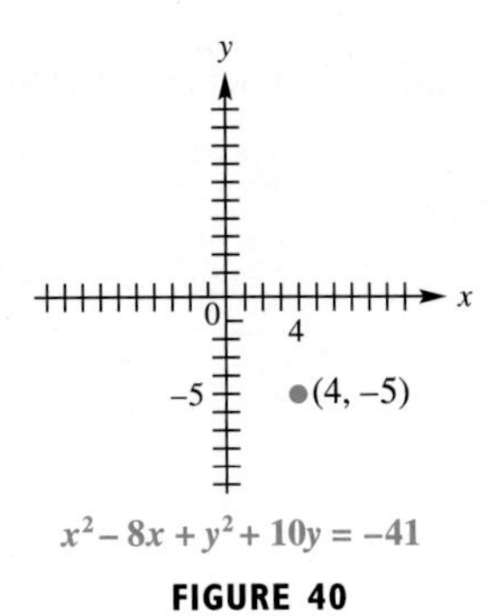


FIGURE 40

(instead of 0), the equation would have represented no points at all, and there would be no graph.

**(d)** Since only one variable of $x^2 - 6x + 8y - 7 = 0$ is squared ($x$, and not $y$), the equation represents a parabola. Rearrange the terms with $y$ (the variable that is not squared) alone on one side. Then, complete the square on the other side of the equation.

$$8y = -x^2 + 6x + 7$$

$$8y = -(x^2 - 6x \qquad) + 7 \qquad \text{Regroup; factor out } -1.$$

$$8y = -(x^2 - 6x + 9) + 7 + 9 \qquad \text{Add 0 in the form } -9 + 9.$$

$$8y = -(x - 3)^2 + 16 \qquad \text{Factor.}$$

$$y = -\frac{1}{8}(x - 3)^2 + 2 \qquad \text{Multiply by } \tfrac{1}{8}.$$

The parabola has vertex $(3, 2)$ and opens downward, as shown in both graphs in Figure 41.

**Looking Ahead to Calculus**

In Figure 41 with

$$f(x) = -\frac{x^2}{8} + \frac{3}{4}x + \frac{7}{8},$$

calculus allows us to use the definite integral from $-1$ to 7 to find that the area below the parabola and above the $x$-axis is $\frac{32}{3}$ (square units).

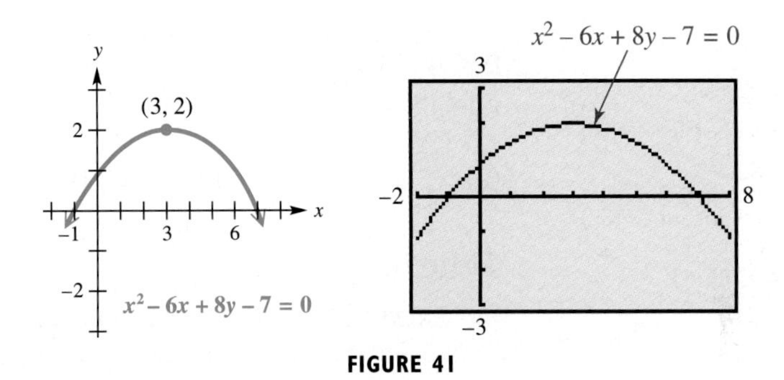


FIGURE 41

## Eccentricity

In Sections 6.1 and 6.2, we introduced definitions of the conic sections. The conic sections (or *conics*) can all be characterized by one general definition.

**Conic**

A **conic** is the set of all points $P(x, y)$ in a plane such that the ratio of the distance from $P$ to a fixed point and the distance from $P$ to a fixed line is constant.

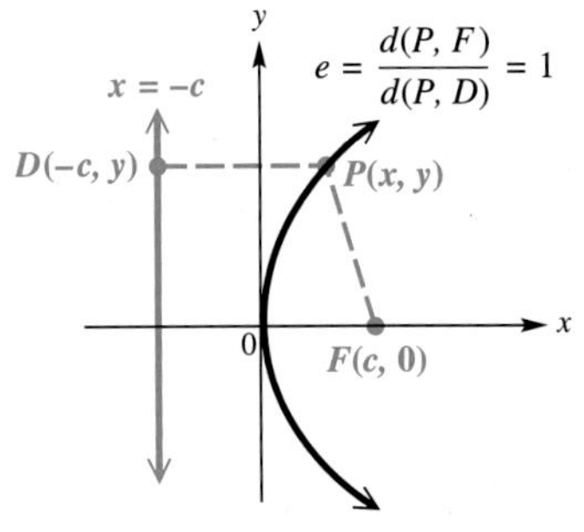


FIGURE 42

As with parabolas, the fixed line is the directrix, and the fixed point is the focus. In Figure 42, the focus is $F(c, 0)$, and the directrix is the line $x = -c$. The constant ratio is called the **eccentricity** of the conic, written $e$. (This is not the same $e$ as the base of natural logarithms.)

By definition, the distances $d(P, F)$ and $d(P, D)$ in Figure 42 are equal if the conic is a parabola. Thus, a parabola always has eccentricity $e = 1$.

For the ellipse and hyperbola, the constant ratio of the definition is $e = \frac{c}{a}$, where $c$ is the distance from the center of the figure to a focus, and $a$ is the distance from the

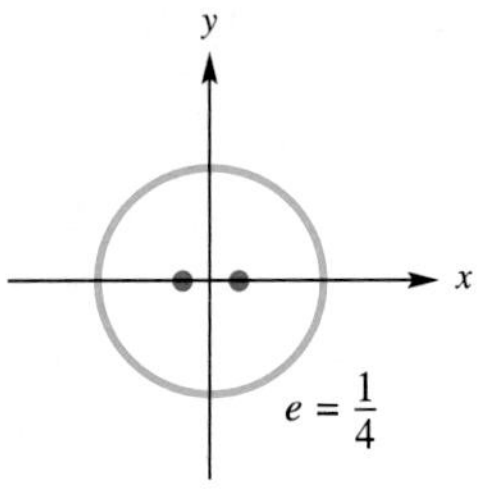


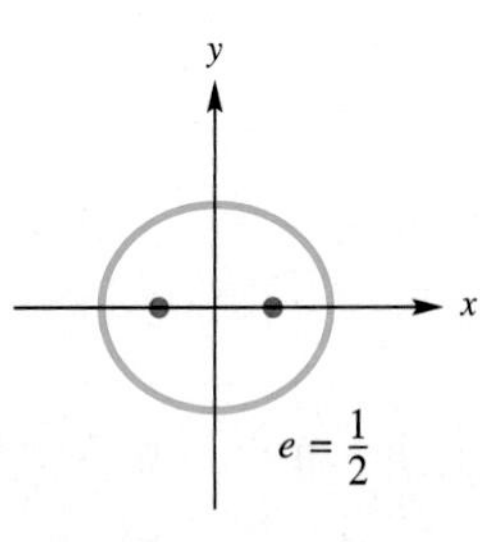


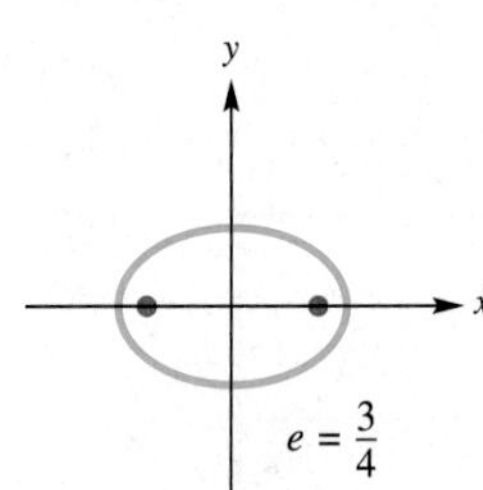


FIGURE 43

center to a vertex. By the definition of an ellipse, $a^2 > b^2$ and $c = \sqrt{a^2 - b^2}$. Thus, for the ellipse,

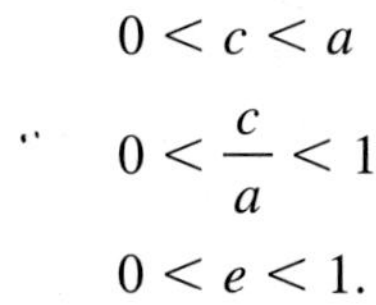

$$0 < c < a$$

$$0 < \frac{c}{a} < 1$$

$$0 < e < 1.$$

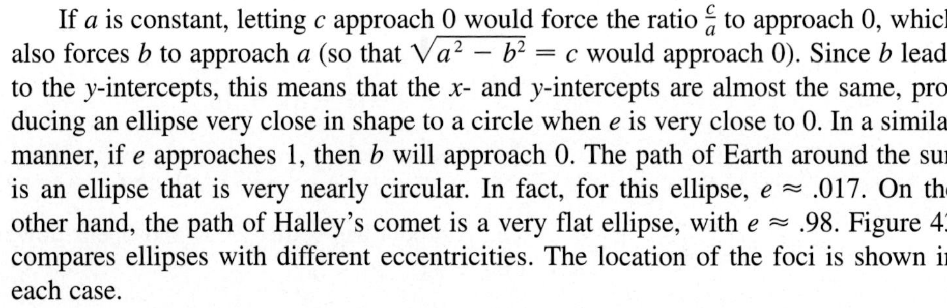

If $a$ is constant, letting $c$ approach 0 would force the ratio $\frac{c}{a}$ to approach 0, which also forces $b$ to approach $a$ (so that $\sqrt{a^2 - b^2} = c$ would approach 0). Since $b$ leads to the $y$-intercepts, this means that the $x$- and $y$-intercepts are almost the same, producing an ellipse very close in shape to a circle when $e$ is very close to 0. In a similar manner, if $e$ approaches 1, then $b$ will approach 0. The path of Earth around the sun is an ellipse that is very nearly circular. In fact, for this ellipse, $e \approx .017$. On the other hand, the path of Halley's comet is a very flat ellipse, with $e \approx .98$. Figure 43 compares ellipses with different eccentricities. The location of the foci is shown in each case.

**EXAMPLE 2** *Finding Eccentricity from Equations of Ellipses*

Find the eccentricity of each ellipse.

**(a)** $\dfrac{x^2}{9} + \dfrac{y^2}{16} = 1$ **(b)** $5x^2 + 10y^2 = 50$

**Solution**

**(a)** Since $16 > 9$, let $a^2 = 16$, which gives $a = 4$. Also,

$$c = \sqrt{a^2 - b^2} = \sqrt{16 - 9} = \sqrt{7}.$$

Finally, $e = \dfrac{c}{a} = \dfrac{\sqrt{7}}{4} \approx .66.$

**(b)** Divide by 50 to obtain $\dfrac{x^2}{10} + \dfrac{y^2}{5} = 1$. Here, $a^2 = 10$, with $a = \sqrt{10}$. Now, find $c$.

$$c = \sqrt{10 - 5} = \sqrt{5} \quad \text{and} \quad e = \frac{\sqrt{5}}{\sqrt{10}} = \frac{1}{\sqrt{2}} \approx .71$$

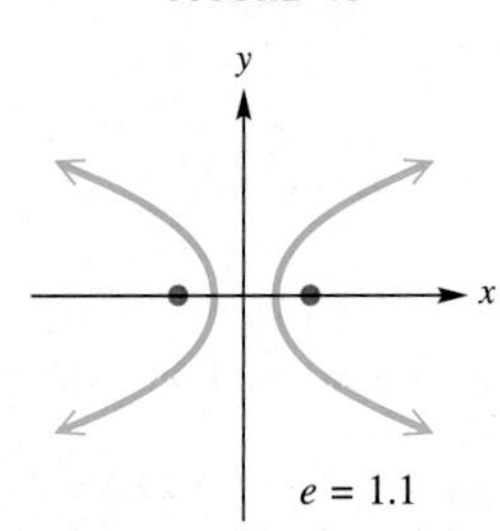


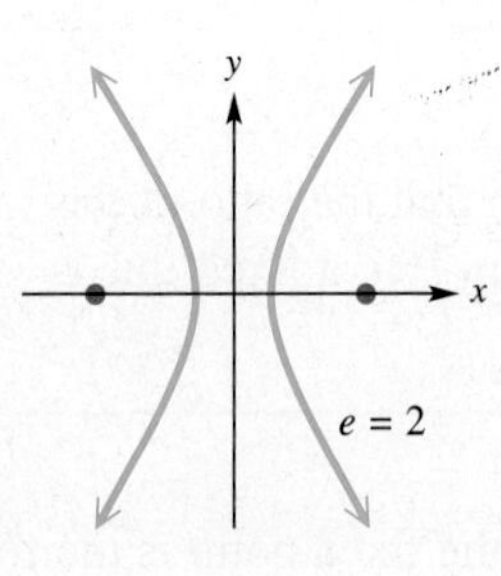


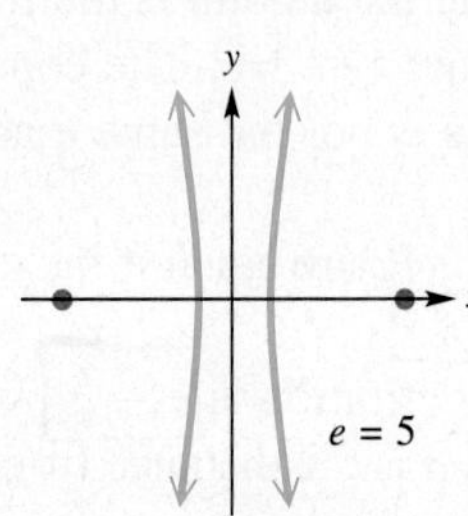


FIGURE 44

As mentioned above, the hyperbola

$$\frac{x^2}{a^2} - \frac{y^2}{b^2} = 1 \quad \text{or} \quad \frac{y^2}{a^2} - \frac{x^2}{b^2} = 1,$$

where $c = \sqrt{a^2 + b^2}$, also has eccentricity $e = \frac{c}{a}$. By definition, $c = \sqrt{a^2 + b^2} > a$, so $\frac{c}{a} > 1$, and for a hyperbola, $e > 1$. Narrow hyperbolas have eccentricity near 1, and wide hyperbolas have large eccentricity. See Figure 44.

**EXAMPLE 3** *Finding Eccentricity from the Equation of a Hyperbola*

Find the eccentricity of the hyperbola

$$\frac{x^2}{9} - \frac{y^2}{4} = 1.$$

**Solution** Here $a^2 = 9$; thus, $a = 3$, $c = \sqrt{9 + 4} = \sqrt{13}$, and

$$e = \frac{c}{a} = \frac{\sqrt{13}}{3} \approx 1.2.$$

The following table summarizes this discussion of eccentricity.

| Conic | Eccentricity |
|---|---|
| Parabola | $e = 1$ |
| Ellipse | $e = \frac{c}{a}$ and $0 < e < 1$ |
| Hyperbola | $e = \frac{c}{a}$ and $e > 1$ |

**FOR DISCUSSION**

Answer the questions in order.

**1.** As the two foci of an ellipse move closer and closer together, what familiar shape does the ellipse begin to resemble?

**2.** For a circle with center at the origin, how do the values of $a$ and $b$ compare?

**3.** What is the value of $c$ for the circle defined above?

**4.** What is the eccentricity of a circle?

### EXAMPLE 4 *Finding Equations of Conics Using Eccentricity*

Find an equation for each conic with center at the origin.

**(a)** Focus $(3, 0)$; eccentricity 2 **(b)** Vertex $(0, -8)$; $e = \frac{1}{2}$

**Solution**

**(a)** Since $e = 2$, which is greater than 1, the conic is a hyperbola with $c = 3$. From $e = \frac{c}{a}$, we find $a$ by substituting $e = 2$ and $c = 3$.

$$e = \frac{c}{a}$$

$$2 = \frac{3}{a}$$

$$a = \frac{3}{2}$$

Now, we find $b$.

$$b^2 = c^2 - a^2 = 9 - \frac{9}{4} = \frac{27}{4}$$

The given focus is on the $x$-axis, so the $x^2$-term is positive, and the equation is

$$\frac{x^2}{\frac{9}{4}} - \frac{y^2}{\frac{27}{4}} = 1 \quad \text{or} \quad \frac{4x^2}{9} - \frac{4y^2}{27} = 1.$$

**(b)** The graph of the conic with vertex $(0, -8)$ and $e = \frac{1}{2}$ is an ellipse because $e = \frac{1}{2} < 1$. From the given vertex, we know that the vertices are on the $y$-axis and $a = 8$. We use $e = \frac{c}{a}$ to find $c$.

$$e = \frac{c}{a}$$

$$\frac{1}{2} = \frac{c}{8}$$

$$c = 4$$

Since $b^2 = a^2 - c^2 = 64 - 16 = 48$, the equation is

$$\frac{y^2}{64} + \frac{x^2}{48} = 1.$$

**EXAMPLE 5** *Applying an Ellipse to the Orbit of a Planet*

The orbit of the planet Mars is an ellipse with the sun at one focus. The eccentricity of the ellipse is .0935, and the closest Mars comes to the sun is 128.5 million miles. (*Source: The World Almanac and Book of Facts,* 2001.) Find the maximum distance of Mars from the sun.

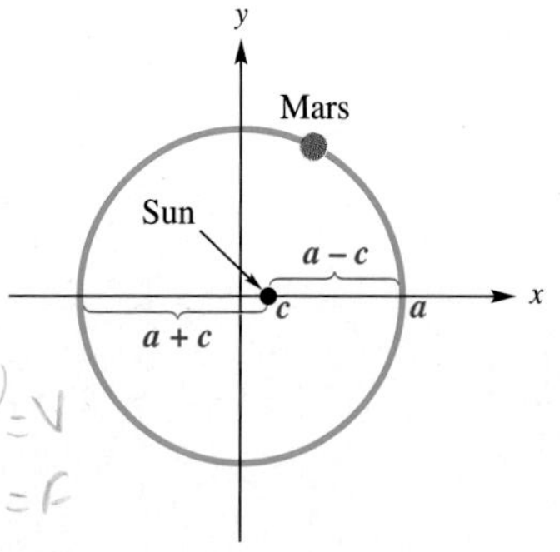

**Solution** Figure 45 shows the orbit of Mars with the origin at the center of the ellipse and the sun at one focus. Mars is closest to the sun when Mars is at the right endpoint of the major axis and farthest from the sun when Mars is at the left endpoint. Therefore, the smallest distance is $a - c$, and the greatest distance is $a + c$. Since $a - c = 128.5$, $c = a - 128.5$. Using this value of $c$ and $e = .0935$, we can find $a$.

Not to Scale

**FIGURE 45**

$$e = \frac{c}{a}$$

$$.0935 = \frac{a - 128.5}{a} \quad \text{Substitute for } e \text{ and } c.$$

$$.0935a = a - 128.5$$

$$.9065a = 128.5$$

$$a \approx 141.8$$

Then $$c = 141.8 - 128.5 = 13.3$$

and $$a + c = 141.8 + 13.3 = 155.1.$$

Thus, the maximum distance of Mars from the sun is about 155.1 million miles.

## 6.3 EXERCISES

*Identify the type of graph that each equation has, without graphing.*

1. $x^2 + y^2 = 144$
2. $(x - 2)^2 + (y + 3)^2 = 25$
3. $y = 2x^2 + 3x - 4$
4. $x = 3y^2 + 5y - 6$
5. $x = -3(y - 4)^2 + 1$
6. $\frac{x^2}{25} + \frac{y^2}{36} = 1$
7. $\frac{x^2}{49} + \frac{y^2}{100} = 1$
8. $x^2 - y^2 = 1$
9. $\frac{x^2}{4} - \frac{y^2}{16} = 1$
10. $\frac{(x + 2)^2}{9} + \frac{(y - 4)^2}{16} = 1$
11. $\frac{x^2}{25} - \frac{y^2}{25} = 1$
12. $y = 4(x + 3)^2 - 7$

*Identify the type of graph for each equation. It may be necessary to transform the equation.*

13. $\frac{x^2}{4} = 1 - \frac{y^2}{9}$
14. $\frac{x^2}{4} = 1 + \frac{y^2}{9}$
15. $\frac{x^2}{4} + \frac{y^2}{4} = -1$
16. $x^2 = 25 + y^2$
17. $x^2 = 25 - y^2$
18. $9x^2 + 36y^2 = 36$
19. $x^2 = 4y - 8$
20. $\frac{(x + 3)^2}{16} + \frac{(y - 2)^2}{16} = 1$
21. $\frac{(x - 4)^2}{8} + \frac{(y + 1)^2}{2} = 0$
22. $y^2 - 4y = x + 4$
23. $(x + 7)^2 + (y - 5)^2 + 4 = 0$
24. $4(x - 3)^2 + 3(y + 4)^2 = 0$

**25.** $3x^2 + 6x + 3y^2 - 12y = 12$

**26.** $2x^2 - 8x + 2y^2 + 20y = 12$

**27.** $x^2 - 6x + y = 0$

**28.** $x - 4y^2 - 8y = 0$

**29.** $4x^2 - 8x - y^2 - 6y = 6$

**30.** $x^2 + 2x = x^2 - 4y - 2$

**31.** $4x^2 - 8x + 9y^2 + 54y = -84$

**32.** $3x^2 + 12x + 3y^2 = -11$

**33.** $6x^2 - 12x + 6y^2 - 18y + 25 = 0$

**34.** $4x^2 - 24x + 5y^2 + 10y + 41 = 0$

**35.** *Concept Check* Suppose that both $A$ and $C$ are 0 in the equation $Ax^2 + Bx + Cy^2 + Dy + E = 0$. What kind of graph does this equation have?

**36.** How can the graph of the equation discussed in Exercise 35 be obtained by intersecting a plane and a cone?

**37.** Identify the type of conic section consisting of the set of all points in the plane for which the sum of the distances from the points $(5, 0)$ and $(-5, 0)$ is 14.

**38.** Identify the type of conic section consisting of the set of all points in the plane for which the absolute value of the difference of the distances from the points $(3, 0)$ and $(-3, 0)$ is 2.

**39.** Identify the type of conic section consisting of the set of all points in the plane for which the distance from the point $(3, 0)$ is one and one-half times the distance from the line $x = \frac{4}{3}$.

**40.** Identify the type of conic section consisting of the set of all points in the plane for which the distance from the point $(2, 0)$ is one-third of the distance from the line $x = 10$.

*Find the eccentricity of each ellipse or hyperbola.*

**41.** $12x^2 + 9y^2 = 36$

**42.** $8x^2 - y^2 = 16$

**43.** $x^2 - y^2 = 4$

**44.** $x^2 + 2y^2 = 8$

**45.** $4x^2 + 7y^2 = 28$

**46.** $9x^2 - y^2 = 1$

**47.** $x^2 - 9y^2 = 18$

**48.** $x^2 + 10y^2 = 10$

*Write an equation for each conic. Each parabola has vertex at the origin, and each ellipse or hyperbola is centered at the origin.*

**49.** Focus $(0, 8)$; $e = 1$

**50.** Focus $(-2, 0)$; $e = 1$

**51.** Focus $(3, 0)$; $e = \dfrac{1}{2}$

**52.** Focus $(0, -2)$; $e = \dfrac{2}{3}$

**53.** Vertex $(-6, 0)$; $e = 2$

**54.** Vertex $(0, 4)$; $e = \dfrac{5}{3}$

**55.** Focus $(0, -1)$; $e = 1$

**56.** Focus $(2, 0)$; $e = \dfrac{6}{5}$

**57.** Vertical major axis of length 6; $e = \dfrac{4}{5}$

**58.** $y$-intercepts $-4$ and 4; $e = \dfrac{7}{3}$

**59.** *Concept Check* Calculator graphs are shown in Figures A–D. Arrange the figures in order so that the first in the list has the smallest eccentricity and the rest have eccentricities in increasing order.

**A.**

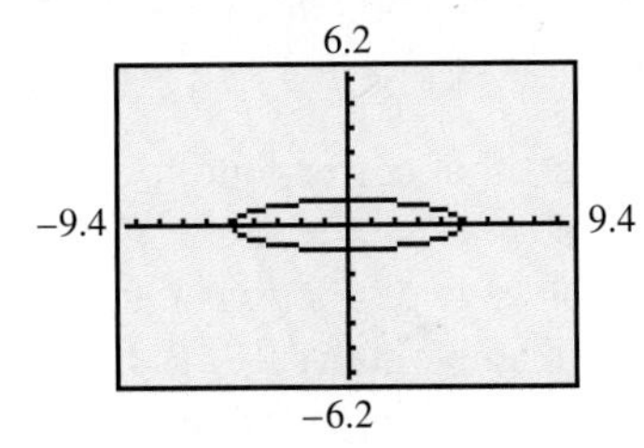


**B.**

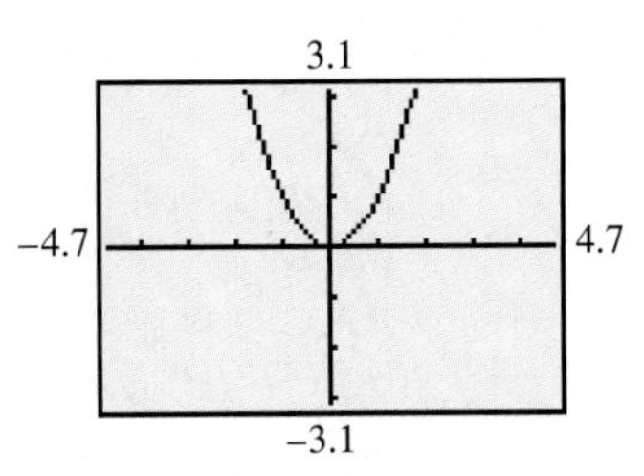


**C.**

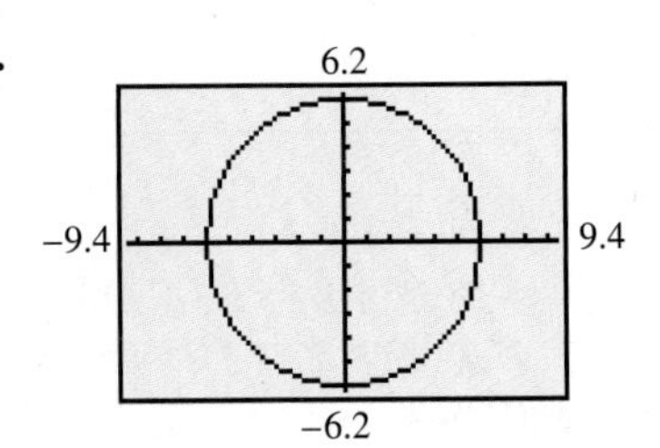


**D.**

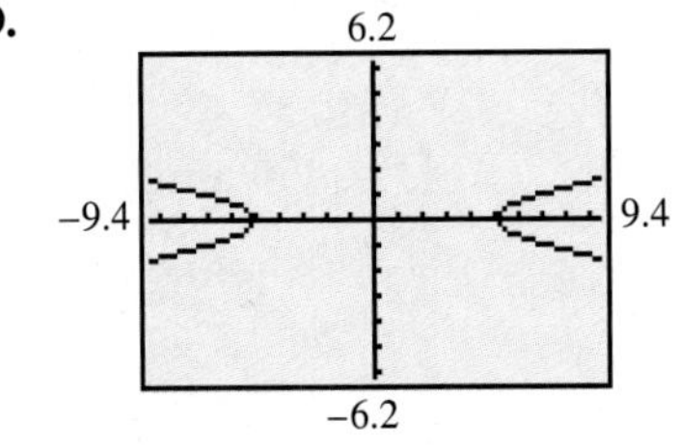

*(Modeling) Solve each application.*

**60.** *Orbit of Mars* The orbit of Mars around the sun is an ellipse with equation

$$\frac{x^2}{5013} + \frac{y^2}{4970} = 1,$$

where $x$ and $y$ are measured in millions of miles. Find the eccentricity of this ellipse.

**61.** *Orbits of Neptune and Pluto* Neptune and Pluto both have elliptical orbits with the sun at one focus. Neptune's orbit has $a = 30.1$ astronomical units (AU) with eccentricity $e = .009$, whereas Pluto's orbit has $a = 39.4$ and $e = .249$. (1 AU is equal to the average distance from Earth to the sun and is approximately 149,600,000 kilometers.) (*Source:* Zeilik, M., S. Gregory, and E. Smith, *Introductory Astronomy and Astrophysics,* Saunders College Publishers, 1992.)

**(a)** Position the sun at the origin, and determine an equation for each orbit.

**(b)** Graph both equations on the same coordinate axes. Use the window $[-60, 60]$ by $[-40, 40]$.

**62.** *Velocity of a Planet in Orbit* The maximum and minimum velocities in kilometers per second of a planet moving in an elliptical orbit can be calculated with the equations

$$v_{max} = \frac{2\pi a}{P}\sqrt{\frac{1+e}{1-e}} \quad \text{and} \quad v_{min} = \frac{2\pi a}{P}\sqrt{\frac{1-e}{1+e}},$$

where $a$ is in kilometers, $P$ is its orbital period in seconds, and $e$ is the eccentricity of the orbit. (*Source:* Zeilik, M., S. Gregory, and E. Smith, *Introductory Astronomy and Astrophysics,* Saunders College Publishers, 1992.)

**(a)** Calculate $v_{max}$ and $v_{min}$ for Earth if $a = 1.496 \times 10^8$ kilometers and $e = .0167$.

**(b)** If an object has a circular orbit, what can be said about its orbital velocity?

**(c)** Kepler showed that the sun is located at a focus of a planet's elliptical orbit. He also showed that a planet's minimum velocity occurs when its distance from the sun is maximum, and a planet's maximum velocity occurs when its distance from the sun is minimum. Where do the maximum and minimum velocities occur in an elliptical orbit?

**63.** *Distance between Halley's Comet and the Sun* The famous Halley's comet last passed Earth in February 1986 and will next return in 2062. Halley's comet has an elliptical orbit of eccentricity .9673 with the sun at one of the foci. The greatest distance of the comet from the sun is 3281 million miles. (*Source: The World Almanac and Book of Facts,* 2001.) Find the least distance between Halley's comet and the sun.

**64.** *Orbit of Earth* The orbit of Earth is an ellipse with the sun at one focus. The distance between Earth and the sun ranges from 91.4 to 94.6 million miles. (*Source: The World Almanac and Book of Facts,* 2001.) Find the eccentricity of Earth's orbit.

# 6.4 PARAMETRIC EQUATIONS

Graphs of Parametric Equations and Their Rectangular Equivalents • Alternative Forms of Parametric Equations • An Application

We have graphed sets of ordered pairs of real numbers that correspond to functions of the form $y = f(x)$ or relations of the form $Ax^2 + Bxy + Cy^2 + Dx + Ey + F = 0$. Another way to determine a set of ordered pairs involves two functions $f$ and $g$ defined by $x = f(t)$ and $y = g(t)$, where $t$ is a real number in some interval $I$. Each value of $t$ leads to a corresponding $x$-value and a corresponding $y$-value, and thus to an ordered pair $(x, y)$.

**TECHNOLOGY NOTE**

In addition to graphing rectangular equations, graphing calculators are capable of graphing plane curves defined by parametric equations. The calculator must be set in parametric mode, and the window requires intervals for the parameter $t$, as well as for $x$ and $y$. Consult the graphing calculator manual that accompanies this text or your owner's manual.

**Parametric Equations of a Plane Curve**

A **plane curve** is a set of points $(x, y)$ such that $x = f(t)$, $y = g(t)$, and $f$ and $g$ are both continuous on an interval $I$. The equations $x = f(t)$ and $y = g(t)$ are **parametric equations** with **parameter $t$.**

## Graphs of Parametric Equations and Their Rectangular Equivalents

**EXAMPLE 1** *Graphing a Plane Curve Defined Parametrically*

Graph the plane curve defined by the parametric equations

$$x = t^2, \quad y = 2t + 3, \quad \text{for } t \text{ in } [-3, 3],$$

and then find an equivalent rectangular equation.

**Analytic Solution**

Make a table of corresponding values of $t$, $x$, and $y$ over the domain of $t$. Then we plot the points as shown in Figure 46. The graph is a portion of a parabola with horizontal axis $y = 3$. The arrowheads indicate the direction the curve traces as $t$ increases.

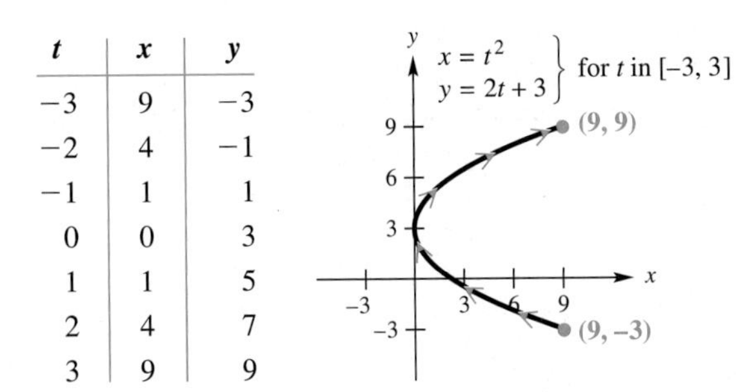

| $t$ | $x$ | $y$ |
|---|---|---|
| −3 | 9 | −3 |
| −2 | 4 | −1 |
| −1 | 1 | 1 |
| 0 | 0 | 3 |
| 1 | 1 | 5 |
| 2 | 4 | 7 |
| 3 | 9 | 9 |



FIGURE 46

To find an equivalent rectangular equation, we eliminate the parameter $t$. For this curve, we begin by solving for $t$ in the second equation $y = 2t + 3$, because it leads to a unique solution.

$$y = 2t + 3$$
$$2t = y - 3$$
$$t = \frac{y-3}{2}$$

Now, we substitute the result in the first equation to get

$$x = t^2 = \left(\frac{y-3}{2}\right)^2 = \frac{(y-3)^2}{4} = \frac{1}{4}(y-3)^2.$$

This is indeed an equation of a horizontal parabola that opens to the right. Because $t$ is in $[-3, 3]$, $x$ is in $[0, 9]$ and $y$ is in $[-3, 9]$. The rectangular equation must be given with its restricted domain as

$$x = \frac{1}{4}(y-3)^2, \quad \text{for } x \text{ in } [0, 9].$$

**Graphing Calculator Solution**

Figure 47 shows a calculator table of values and graph. The formation of the curve as T increases can be seen by using the pause feature and carefully observing the path of the curve.

| T | X1T | Y1T |
|---|---|---|
| -3 | 9 | -3 |
| -2 | 4 | -1 |
| -1 | 1 | 1 |
| 0 | 0 | 3 |
| 1 | 1 | 5 |
| 2 | 4 | 7 |
| 3 | 9 | 9 |

T=-3

$X_{1T} = T^2, Y_{1T} = 2T + 3$
T in [−3, 3]

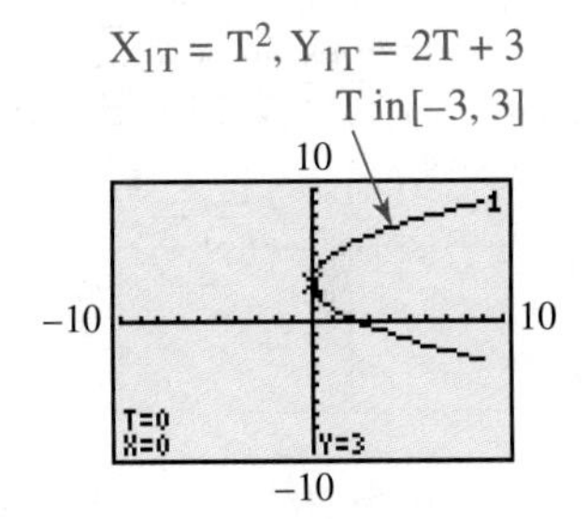



FIGURE 47

The display at the bottom of Figure 47 indicates particular values of T, X, and Y. How do these relate to the parametric and rectangular forms of the equation of this parabola?

**EXAMPLE 2** *Graphing a Plane Curve Defined Parametrically*

Graph the plane curve defined by

$$x = 2t + 5, \quad y = \sqrt{4 - t^2}, \quad \text{for } t \text{ in } [-2, 2],$$

and then find an equivalent rectangular equation.

For the parametric equations $X_{1T} = T^2$ and $Y_{1T} = 2T + 3$ in Example 1, the screens below show the entries that produce the window shown in Figure 47.

```
WINDOW
 Tmin=-3
 Tmax=3
 Tstep=.05
 Xmin=-10
 Xmax=10
 Xscl=1
↓Ymin=-10
```

```
WINDOW
↑Tstep=.05
 Xmin=-10
 Xmax=10
 Xscl=1
 Ymin=-10
 Ymax=10
 Yscl=1
```

**Solution** The graph with $x$ and $y$ in a square window is shown in Figure 48. To get an equivalent rectangular equation, solve the first equation, which does not involve a radical, for $t$, then substitute into the second equation. Since $t$ is in $[-2, 2]$, $x$ is in the interval $[1, 9]$.

$$2t + 5 = x$$
$$2t = x - 5$$
$$t = \frac{x - 5}{2}$$

The rectangular equation is

$$y = \sqrt{4 - t^2}$$
$$= \sqrt{4 - \left(\frac{x - 5}{2}\right)^2}, \quad \text{for } x \text{ in } [1, 9].$$

Squaring both sides and rearranging terms gives

$$y^2 = 4 - \left(\frac{x - 5}{2}\right)^2$$
$$y^2 = 4 - \frac{(x - 5)^2}{4}$$
$$4y^2 = 16 - (x - 5)^2 \qquad \text{Clear the fraction.}$$
$$4y^2 + (x - 5)^2 = 16$$
$$\frac{y^2}{4} + \frac{(x - 5)^2}{16} = 1,$$

which represents the complete ellipse. (By definition, the parametric graph has $y \geq 0$, so it is only the upper half of the ellipse.)

$x = 2t + 5$, $y = \sqrt{4 - t^2}$, $t$ in $[-2, 2]$

6.7

−10.1 10.1

−6.7

**FIGURE 48**

**EXAMPLE 3** *Graphing a Line Defined Parametrically*

Graph the plane curve defined by

$$x = t^2, \quad y = t^2,$$

and then find an equivalent rectangular equation.

**Solution** Figure 49 shows the graph of the curve in a standard window with both $x$ and $y$ in $[-10, 10]$. The graph is a half-line because both $x$ and $y$ must be greater than or equal to 0. Since both $x$ and $y$ equal $t^2$, $y = x$. To be equivalent, however, the rectangular equation must be given as

$$y = x, \quad x \geq 0.$$

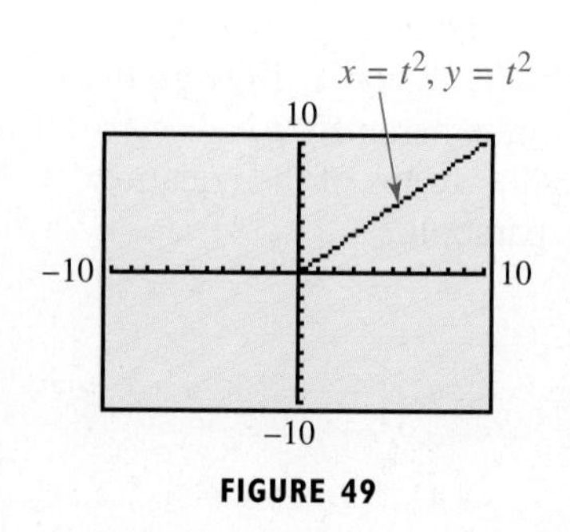


**FIGURE 49**

## Alternative Forms of Parametric Equations

Parametric representations of a curve are not unique. In fact, there are infinitely many parametric representations of a given curve. If the curve can be described by a rectangular equation $y = f(x)$, with domain $X$, then one simple parametric representation is

$$x = t, \quad y = f(t), \quad \text{for } t \text{ in } X.$$

**FOR DISCUSSION**

Consider the rectangular equation

$$y - 1 = (x - 2)^2.$$

Why are the parametric equations

$$x = (t - 2)^{1/2}, \quad y = t + 1,$$

for $t$ in $[2, \infty)$ not equivalent to the rectangular equation?

**EXAMPLE 4** *Finding Alternative Parametric Equation Forms*

Give two parametric representations for the parabola $y = (x - 2)^2 + 1$.

**Solution** The simplest choice is to let

$$x = t, \quad y = (t - 2)^2 + 1, \quad \text{for } t \text{ in } (-\infty, \infty).$$

Another choice that leads to a simpler equation for $y$ is

$$x = t + 2, \quad y = t^2 + 1, \quad \text{for } t \text{ in } (-\infty, \infty).$$

## An Application

An important application of parametric equations is to determine the path of a moving object whose position is given by the function defined by $x = f(t)$, $y = g(t)$, where $t$ represents time. The parametric equations give the position of the object at any time $t$.

**EXAMPLE 5** *Using Parametric Equations to Define the Position of an Object in Motion*

The motion of a projectile moving in a direction at a $\theta = 45°$ angle with the horizontal (neglecting air resistance) is given by

$$x = v_0 \frac{\sqrt{2}}{2} t, \quad y = v_0 \frac{\sqrt{2}}{2} t - 16t^2, \quad \text{for } t \text{ in } [0, k],$$

where $t$ is time in seconds, $v_0$ is the initial speed of the projectile, $x$ and $y$ are in feet, and $k$ is a positive real number. See Figure 50. Find the rectangular form of the equation.

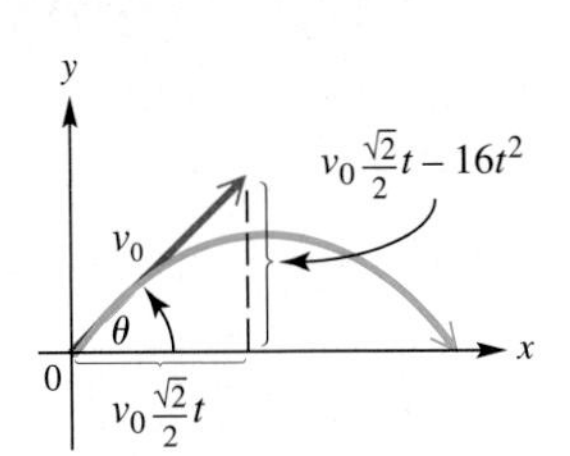


**FIGURE 50**

**Solution** Solving the first equation for $t$ and substituting the result into the second equation gives (after simplification)

$$y = x - \frac{32}{v_0^2}x^2,$$

the equation of a vertical parabola opening downward, as shown in Figure 50.

## 6.4 EXERCISES

*For each plane curve, use a graphing calculator to generate the curve over the interval for the parameter t, using the window specified. Then, find a rectangular equation for the curve.*

1. $x = 2t$, $y = t + 1$, for $t$ in $[-2, 3]$; window: $[-8, 8]$ by $[-8, 8]$
2. $x = t + 2$, $y = t^2$, for $t$ in $[-1, 1]$; window: $[0, 4]$ by $[-2, 2]$
3. $x = \sqrt{t}$, $y = 3t - 4$, for $t$ in $[0, 4]$; window: $[-6, 6]$ by $[-6, 10]$
4. $x = t^2$, $y = \sqrt{t}$, for $t$ in $[0, 4]$; window: $[-2, 20]$ by $[0, 4]$
5. $x = t^3 + 1$, $y = t^3 - 1$, for $t$ in $[-3, 3]$; window: $[-30, 30]$ by $[-30, 30]$
6. $x = 2t - 1$, $y = t^2 + 2$, for $t$ in $[-10, 10]$; window: $[-20, 20]$ by $[0, 120]$

7. $x = 2^t$, $y = \sqrt{3t - 1}$, for $t$ in $\left[\frac{1}{3}, 4\right]$;
window: $[-2, 30]$ by $[-2, 10]$

8. $x = \ln(t - 1)$, $y = 2t - 1$, for $t$ in $(1, 10]$;
window: $[-5, 5]$ by $[-2, 20]$

9. $x = t + 2$, $y = -\frac{1}{2}\sqrt{9 - t^2}$, for $t$ in $[-3, 3]$;
window: $[-6, 6]$ by $[-4, 4]$

10. $x = t$, $y = \sqrt{4 - t^2}$, for $t$ in $[-2, 2]$;
window: $[-6, 6]$ by $[-4, 4]$

11. $x = t$, $y = \frac{1}{t}$, for $t$ in $(-\infty, 0) \cup (0, \infty)$;
window: $[-6, 6]$ by $[-4, 4]$

12. $x = 2t - 1$, $y = \frac{1}{t}$, for $t$ in $(-\infty, 0) \cup (0, \infty)$;
window: $[-6, 6]$ by $[-4, 4]$

*For each plane curve, find a rectangular equation.*

13. $x = 3t^2$, $y = 4t^3$, for $t$ in $(-\infty, \infty)$

14. $x = 2t^3$, $y = -t^2$, for $t$ in $(-\infty, \infty)$

15. $x = t$, $y = \sqrt{t^2 + 2}$, for $t$ in $(-\infty, \infty)$

16. $x = \sqrt{t}$, $y = t^2 - 1$, for $t$ in $[0, \infty)$

17. $x = e^t$, $y = e^{-t}$, for $t$ in $(-\infty, \infty)$

18. $x = e^{2t}$, $y = e^t$, for $t$ in $(-\infty, \infty)$

19. $x = \frac{1}{\sqrt{t + 2}}$, $y = \frac{t}{t + 2}$, for $t$ in $(-2, \infty)$

20. $x = \frac{t}{t - 1}$, $y = \frac{1}{\sqrt{t - 1}}$, for $t$ in $(1, \infty)$

21. $x = t + 2$, $y = \frac{1}{t + 2}$, for $t \neq -2$

22. $x = t - 3$, $y = \frac{2}{t - 3}$, for $t \neq 3$

23. $x = t^2$, $y = 2 \ln t$, for $t$ in $(0, \infty)$

24. $x = \ln t$, $y = 3 \ln t$, for $t$ in $(0, \infty)$

*Give two parametric representations for each plane curve. Use your calculator to verify your results.*

25. $y = 2x + 3$

26. $y = \frac{3}{2}x - 4$

27. $y = \sqrt{3x + 2}$, $x$ in $\left[-\frac{2}{3}, \infty\right)$

28. $y = (x + 1)^2 + 1$

*(Modeling) Solve each application.*

29. *Firing a Projectile* A projectile is fired with an initial velocity of 400 feet per second at an angle of 45° with the horizontal. (See Example 5.)

(a) Find the time when it strikes the ground.

(b) Find the range (horizontal distance covered).

(c) What is the maximum altitude?

30. *Firing a Projectile* If a projectile is fired at an angle of 30° with the horizontal, the parametric equations that describe its motion are

$$x = v_0 \frac{\sqrt{3}}{2} t, \quad y = \frac{v_0}{2} t - 16t^2, \quad \text{for } t \text{ in } [0, \infty).$$

Repeat Exercise 29 if the projectile is fired at 800 feet per second.

31. *Path of a Projectile* A projectile moves so that its position at any time $t$ is given by the equations $x = 60t$ and $y = 80t - 16t^2$. Graph the path of the projectile, and find the equivalent rectangular equation. Use the window $[0, 300]$ by $[0, 200]$.

32. *Path of a Projectile* Repeat Exercise 31 if the path is given by the equations $x = t^2$ and $y = -16t + 64\sqrt{t}$. Use the window $[0, 300]$ by $[0, 200]$.

33. Show that the rectangular equation for the curve defined by

$$x = v_0 \frac{\sqrt{2}}{2} t, \quad y = v_0 \frac{\sqrt{2}}{2} t - 16t^2, \quad \text{for } t \text{ in } [0, k],$$

is $y = x - \frac{32}{v_0^2} x^2$. (See Example 5.)

34. Find the vertex of the parabola given by the rectangular equation of Exercise 33.

**35.** Give two parametric representations of the line through the point $(x_1, y_1)$ with slope $m$.

**36.** Give two parametric representations of the parabola $y = a(x - h)^2 + k$.

## Reviewing Basic Concepts (Sections 6.3 and 6.4)

*Write each equation in standard form, and name the type of conic section defined.*

**1.** $3x^2 + y^2 - 6x + 6y = 0$

**2.** $y^2 - 2x^2 + 8y - 8x - 4 = 0$

**3.** $3y^2 + 12y + 5x = 3$

*Find the eccentricity of each conic section.*

**4.** $x^2 + 25y^2 = 25$

**5.** $8y^2 - 4x^2 = 8$

*Find an equation of each conic.* (Hint: *A sketch may be helpful.*)

**6.** Focus $(-2, 0)$; vertex $(0, 0)$; eccentricity 1

**7.** Foci $(\pm 3, 0)$; major axis length 10

**8.** Foci $(0, \pm 5)$; vertices $(0, \pm 4)$

*Solve each problem.*

**9.** The figure represents an elliptical stone arch, with the dimensions indicated. Find the height $y$ of the arch 6 feet from the center of the base.

**10.** A plane curve is defined by $x = 2t$, $y = \sqrt{t^2 + 1}$, for $t$ in $(-\infty, \infty)$.

**(a)** Graph the curve by hand and support your graph with your calculator.

**(b)** Find an equivalent rectangular equation for the curve.

## CHAPTER 6 SUMMARY

**KEY TERMS & SYMBOLS** | **KEY CONCEPTS**

### 6.1 Circles and Parabolas

conic sections
circle
radius
center
parabola
focus
directrix

**CIRCLE**
A circle is a set of points in a plane that are equidistant from a fixed point. The distance is called the radius of the circle, and the fixed point is called the center.

**CENTER-RADIUS FORM OF THE EQUATION OF A CIRCLE**
The circle with center $(h, k)$ and radius $r$ has equation $(x - h)^2 + (y - k)^2 = r^2$. A circle with center $(0, 0)$ and radius $r$ has equation $x^2 + y^2 = r^2$.

**GENERAL FORM OF THE EQUATION OF A CIRCLE**
For real numbers $c$, $d$, and $e$, the general form of the equation of a circle is

$$x^2 + y^2 + cx + dy + e = 0.$$

**PARABOLA**
A parabola is a set of points in a plane equidistant from a fixed point and a fixed line. The fixed point is called the focus and the fixed line the directrix of the parabola.

| KEY TERMS & SYMBOLS | KEY CONCEPTS |
|---|---|

**PARABOLA WITH A VERTICAL AXIS**

The parabola with focus $(0, c)$ and directrix $y = -c$ has equation $\mathbf{x^2 = 4cy}$. The parabola has vertical axis $x = 0$, opens upward if $c > 0$, and opens downward if $c < 0$.

**PARABOLA WITH A HORIZONTAL AXIS**

The parabola with focus $(c, 0)$ and directrix $x = -c$ has equation $\mathbf{y^2 = 4cx}$. The parabola opens to the right if $c > 0$, to the left if $c < 0$, and has horizontal axis $y = 0$.

**TRANSLATION OF A HORIZONTAL PARABOLA**

The parabola with vertex $(h, k)$ and the horizontal line $y = k$ as axis has an equation of the form $\mathbf{x = a(y - k)^2 + h}$. The parabola opens to the right if $a > 0$ and to the left if $a < 0$.

## 6.2 Ellipses and Hyperbolas

ellipse
focus (foci)
center
vertices
major axis
minor axis
hyperbola
transverse axis
asymptotes
fundamental rectangle

**ELLIPSE**

An ellipse is the set of all points in a plane the sum of whose distances from two fixed points is constant. Each fixed point is called a focus (plural *foci*) of the ellipse.

**STANDARD FORMS OF EQUATIONS FOR ELLIPSES**

The ellipse with center at the origin and equation $\mathbf{\frac{x^2}{a^2} + \frac{y^2}{b^2} = 1}$ has vertices $(\pm a, 0)$, endpoints of the minor axis $(0, \pm b)$, and foci $(\pm c, 0)$, where $c^2 = a^2 - b^2$. The ellipse with center at the origin and equation $\mathbf{\frac{y^2}{a^2} + \frac{x^2}{b^2} = 1}$ has vertices $(0, \pm a)$, endpoints of the minor axis $(\pm b, 0)$, and foci $(0, \pm c)$, where $c^2 = a^2 - b^2$.

**HYPERBOLA**

A hyperbola is the set of all points in a plane such that the absolute value of the *difference* of the distances from two fixed points is constant. The two fixed points are called the foci of the hyperbola.

**STANDARD FORMS OF EQUATIONS FOR HYPERBOLAS**

The hyperbola with center at the origin and equation $\mathbf{\frac{x^2}{a^2} - \frac{y^2}{b^2} = 1}$ has vertices $(\pm a, 0)$ and foci $(\pm c, 0)$, where $c^2 = a^2 + b^2$. The hyperbola with center at the origin and equation $\mathbf{\frac{y^2}{a^2} - \frac{x^2}{b^2} = 1}$ has vertices $(0, \pm a)$ and foci $(0, \pm c)$, where $c^2 = a^2 + b^2$.

## 6.3 Summary of the Conic Sections

conic
eccentricity, $e$

The conic sections in this chapter have equations that can be written in the form

$$\mathbf{Ax^2 + Dx + Cy^2 + Ey + F = 0.}$$

| Conic Section | Characteristic | Example |
|---|---|---|
| Parabola | Either $A = 0$ or $C = 0$, but not both | $y = x^2$<br>$x = 3y^2 + 2y - 4$ |
| Circle | $A = C \neq 0$ | $x^2 + y^2 = 16$ |
| Ellipse | $A \neq C, AC > 0$ | $\frac{x^2}{16} + \frac{y^2}{25} = 1$ |
| Hyperbola | $AC < 0$ | $x^2 - y^2 = 1$ |

**KEY TERMS & SYMBOLS** | **KEY CONCEPTS**

**CONIC**

A conic is the set of all points $P(x, y)$ in a plane such that the ratio of the distance from $P$ to a fixed point and the distance from $P$ to a fixed line is constant. This constant ratio is called the eccentricity of the conic.

| Conic | Eccentricity |
|---|---|
| Parabola | $e = 1$ |
| Ellipse | $e = \frac{c}{a}$ and $0 < e < 1$ |
| Hyperbola | $e = \frac{c}{a}$ and $e > 1$ |

## 6.4 Parametric Equations

plane curve
parametric equations
parameter

**PARAMETRIC EQUATIONS OF A PLANE CURVE**

A plane curve is a set of points $(x, y)$ such that $x = f(t)$, $y = g(t)$, and $f$ and $g$ are both continuous on an interval $I$. The equations $x = f(t)$ and $y = g(t)$ are parametric equations with parameter $t$.

# CHAPTER 6 REVIEW EXERCISES

*Write an equation for the circle satisfying the given conditions. Graph it by hand, and give the domain and range.*

**1.** Center $(-2, 3)$; radius 5

**2.** Center $(\sqrt{5}, -\sqrt{7})$; radius $\sqrt{3}$

**3.** Center $(-8, 1)$; passing through $(0, 16)$

**4.** Center $(3, -6)$; tangent to the $x$-axis

*Find the center and radius of each circle.*

**5.** $x^2 - 4x + y^2 + 6y + 12 = 0$

**6.** $x^2 - 6x + y^2 - 10y + 30 = 0$

**7.** $2x^2 + 14x + 2y^2 + 6y = -2$

**8.** $3x^2 + 3y^2 + 33x - 15y = 0$

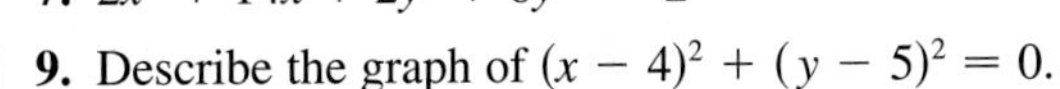

**9.** Describe the graph of $(x - 4)^2 + (y - 5)^2 = 0$.

*Give the focus, directrix, and axis for each parabola, graph it by hand, and give the domain and range.*

**10.** $y^2 = -\dfrac{2}{3}x$

**11.** $y^2 = 2x$

**12.** $3x^2 - y = 0$

**13.** $x^2 + 2y = 0$

*Write an equation for each parabola with vertex at the origin.*

**14.** Focus $(4, 0)$

**15.** Through $(2, 5)$; opening to the right

**16.** Through $(3, -4)$; opening downward

*Write an equation for each parabola.*

**17.** Vertex $(-5, 6)$; focus $(2, 6)$

**18.** Vertex $(4, 3)$; focus $(4, 5)$

*Graph each ellipse or hyperbola by hand, and give the domain, range, and coordinates of the vertices.*

**19.** $\dfrac{y^2}{9} + \dfrac{x^2}{5} = 1$

**20.** $\dfrac{x^2}{16} + \dfrac{y^2}{4} = 1$

**21.** $\dfrac{x^2}{64} - \dfrac{y^2}{36} = 1$

**22.** $\frac{y^2}{25} - \frac{x^2}{9} = 1$

**23.** $\frac{(x-3)^2}{4} + (y+1)^2 = 1$

**24.** $\frac{(x-2)^2}{9} + \frac{(y+3)^2}{4} = 1$

**25.** $\frac{(y+2)^2}{4} - \frac{(x+3)^2}{9} = 1$

**26.** $\frac{(x+1)^2}{16} - \frac{(y-2)^2}{4} = 1$

*Write an equation for each conic section with center at the origin.*

**27.** Ellipse: vertex $(0,4)$; focus $(0,2)$

**28.** Ellipse: $x$-intercept 6; focus $(-2,0)$

**29.** Hyperbola: focus $(0,-5)$; $y$-intercepts $-4$ and 4

**30.** Hyperbola: $y$-intercept $-2$; passing through $(2,3)$

**31.** Focus $(0,-3)$; $e = \frac{2}{3}$

**32.** Focus $(5,0)$; $e = \frac{5}{2}$

**33.** Consider the circle with equation $x^2 + y^2 + 2x + 6y - 15 = 0$.

**(a)** What are the coordinates of the center?

**(b)** What is the radius?

**(c)** What two functions must be graphed to graph this circle with your calculator in function mode?

Concept Check *Match each equation with its graph.*

**34.** $4x^2 + y^2 = 36$

**35.** $x = 2y^2 + 3$

**36.** $(x-1)^2 + (y+2)^2 = 36$

**37.** $\frac{x^2}{36} + \frac{y^2}{9} = 1$

**38.** $(y-1)^2 - (x-2)^2 = 36$

**39.** $y^2 = 36 + 4x^2$

**A.**

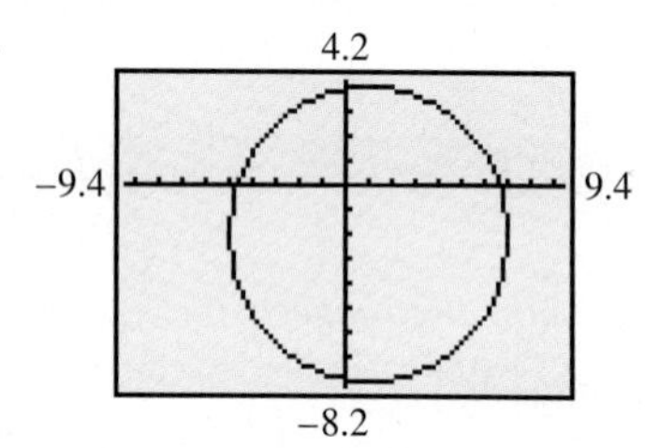


**B.**

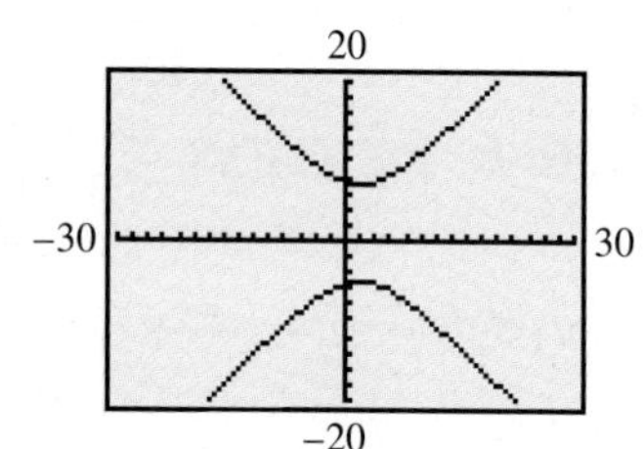


**C.**

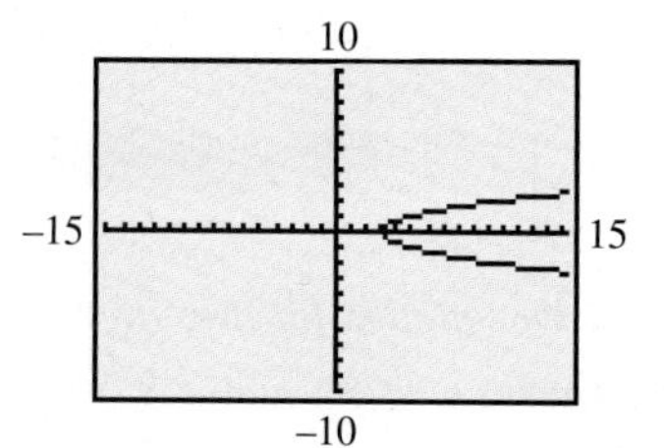


**D.**

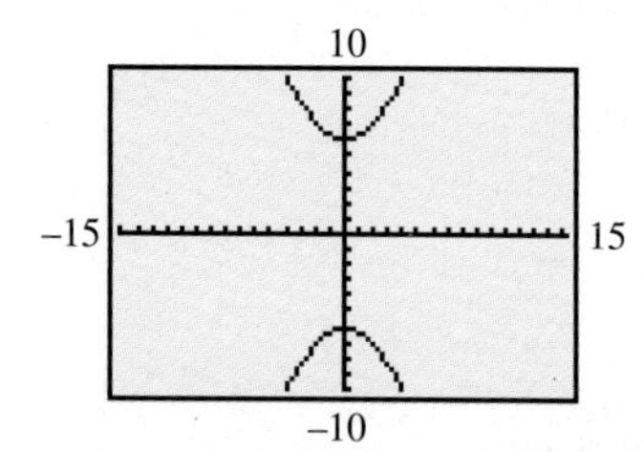


**E.**

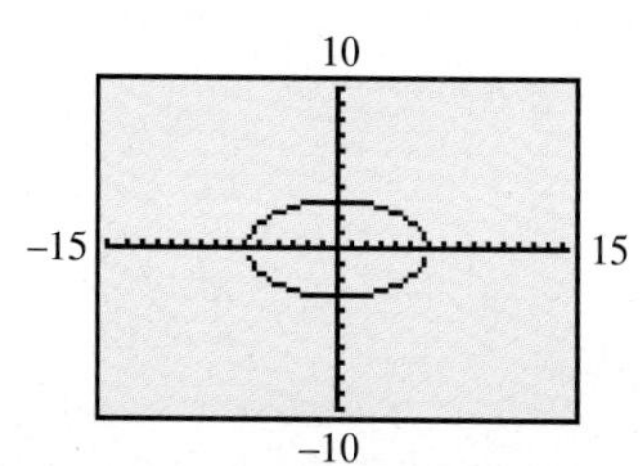


**F.**

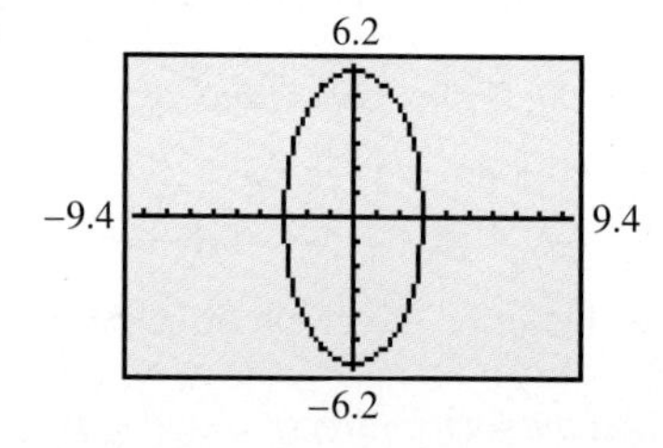


*Find the eccentricity of each ellipse or hyperbola.*

**40.** $9x^2 + 25y^2 = 225$

**41.** $4x^2 + 9y^2 = 36$

**42.** $9x^2 - y^2 = 9$

*Write an equation for each conic section.*

**43.** Parabola with vertex $(-3,2)$ and $y$-intercepts 5 and $-1$

**44.** Hyperbola with foci $(0,12)$ and $(0,-12)$; asymptotes $y = \pm x$

**45.** Ellipse consisting of all points in the plane the sum of whose distances from $(0,0)$ and $(4,0)$ is 8

**46.** Hyperbola consisting of all points in the plane for which the absolute value of the difference of the distances from $(0,0)$ and $(0,4)$ is 2

*Use a graphing calculator to graph each plane curve in the specified window.*

**47.** $x = 4t - 3$, $y = t^2$, for $t$ in $[-3,4]$;
window: $[-20,20]$ by $[-20,20]$

**48.** $x = t^2$, $y = t^3$, for $t$ in $[-2,2]$;
window: $[-15,15]$ by $[-10,10]$

**49.** $x = t + \ln t$, $y = t + e^t$, for $t$ in $(0,2]$;
window: $[-5,5]$ by $[0,10]$

*Find a rectangular equation for each plane curve.*

**50.** $x = 3t + 2,\ y = t - 1$, for $t$ in $[-5, 5]$

**51.** $x = \sqrt{t - 1},\ y = \sqrt{t}$, for $t$ in $[1, \infty)$

**52.** $x = \dfrac{1}{t + 3},\ y = t + 3$, for $t \neq -3$

*(Modeling) Solve each application.*

**53.** *Orbit of Venus* The orbit of Venus is an ellipse with the sun at one of the foci. The eccentricity of the orbit is $e = .006775$, and the major axis has length 134.5 million miles. (*Source: The World Almanac and Book of Facts,* 2001.) Find the smallest and greatest distances of Venus from the sun.

**54.** *Orbit of the Comet Swift-Tuttle* Comet Swift-Tuttle has an elliptical orbit of eccentricity $e = .964$ with the sun at one of the foci. Find the equation of the comet given that the closest it comes to the sun is 89 million miles.

*Trajectory of a Satellite* *When a satellite is near Earth, its orbital trajectory may trace out a hyperbola, parabola, or ellipse. The type of trajectory depends on the satellite's velocity V in meters per second. It will be hyperbolic if $V > k/\sqrt{D}$, parabolic if $V = k/\sqrt{D}$, and elliptic if $V < k/\sqrt{D}$, where $k = 2.82 \times 10^7$ is a constant and D is the distance in meters from the satellite to the center of Earth.* (*Source:* Loh, W., *Dynamics and Thermodynamics of Planetary Entry,* Prentice-Hall, 1963; Thomson, W., *Introduction to Space Dynamics,* John Wiley and Sons, 1961.) Use this information in Exercises 55–57.

**55.** When the artificial satellite *Explorer IV* was at a maximum distance $D$ of $42.5 \times 10^6$ meters from Earth's center, it had a velocity $V$ of 2090 meters per second. Determine the shape of its trajectory.

**56.** If a satellite is scheduled to leave Earth's gravitational influence, its velocity must be increased so that its trajectory changes from elliptic to hyperbolic. Determine the minimum increase in velocity necessary for *Explorer IV* to escape Earth's gravitational influence when $D = 42.5 \times 10^6$ meters.

**57.** Explain why it is easier to change a satellite's trajectory from an ellipse to a hyperbola when $D$ is maximum rather than minimum.

**58.** If $Ax^2 + Cy^2 + Dx + Ey + F = 0$ is the general equation of an ellipse, find its center point by completing the square.

## CHAPTER 6 TEST

**1.** Match each equation with its graph.

**(a)** $4(x + 3)^2 - (y + 2)^2 = 16$

**(b)** $(x - 3)^2 + (y - 2)^2 = 16$

**(c)** $(x + 3)^2 + (y - 2)^2 = 16$

**(d)** $(x + 3)^2 + y = 4$

**(e)** $x + (y - 2)^2 = 4$

**(f)** $4(x + 3)^2 + (y + 2)^2 = 16$

**A.**

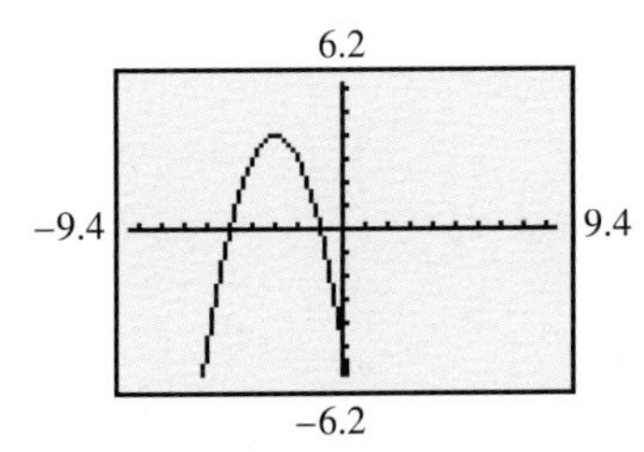


**B.**

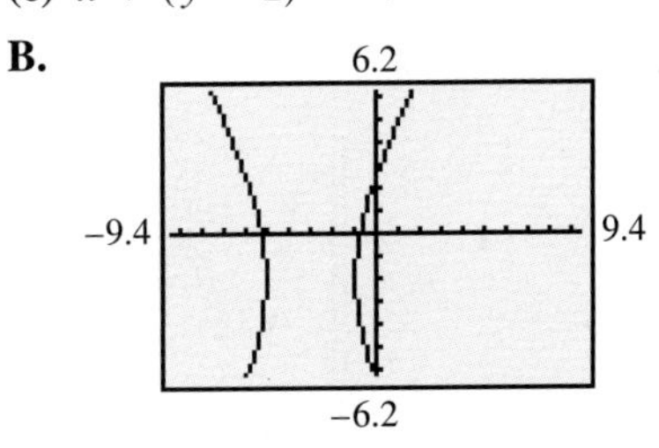


**C.**

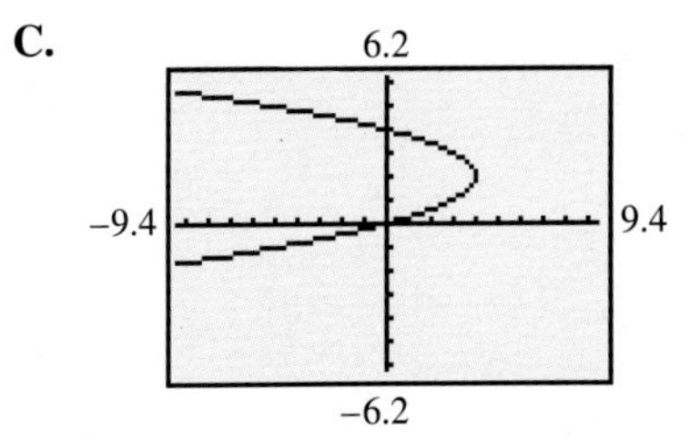


**D.**

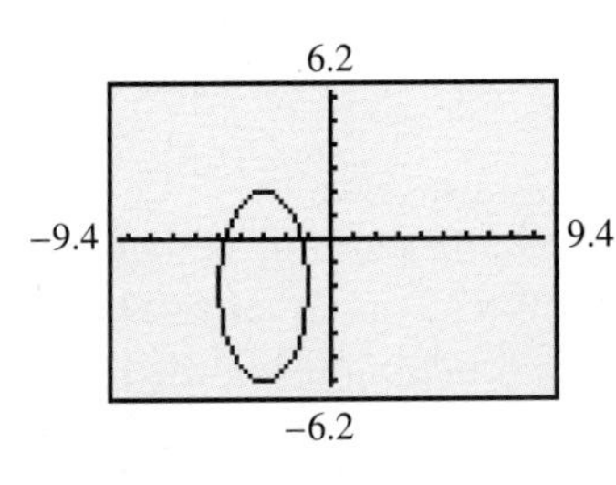


**E.**

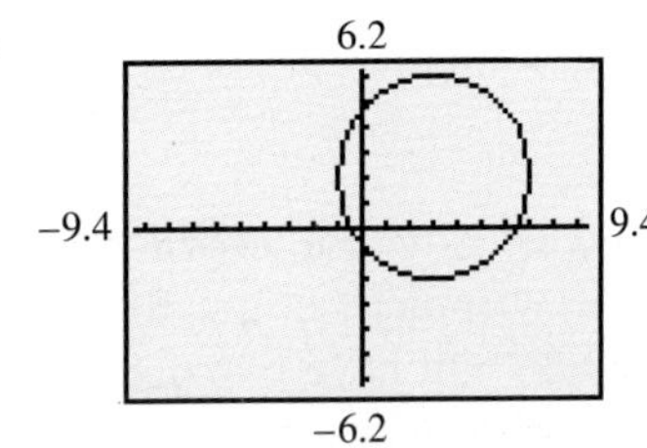


**F.**

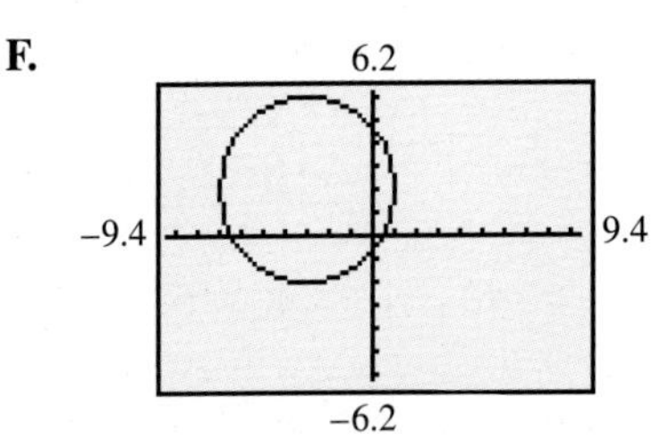

**2.** Give the coordinates of the focus and the equation of the directrix for the parabola with equation $y^2 = \frac{1}{8}x$.

**3.** Graph $y = -\sqrt{1 - \frac{x^2}{36}}$. Is it the graph of a function? Find the domain and range.

**4.** What two functions are used to graph $\frac{x^2}{25} - \frac{y^2}{49} = 1$ on a calculator in function mode?

*Graph each relation by hand. Identify the graph, and give the radius, center, vertices, and foci, as applicable.*

**5.** $\frac{y^2}{4} - \frac{x^2}{9} = 1$

**6.** $x^2 + 4y^2 + 2x - 16y + 17 = 0$

**7.** $y^2 - 8y - 2x + 22 = 0$

**8.** $x^2 + (y - 4)^2 = 9$

**9.** $\frac{(x-3)^2}{49} + \frac{(y+1)^2}{16} = 1$

**10.** $x = 4t^2 - 4$, $y = t - 1$, for $t$ in $[-1.5, 1.5]$

**11.** Write an equation for each conic.

**(a)** Center at the origin; focus $(0, -2)$; $e = 1$

**(b)** Center at the origin; vertical major axis of length 6; $e = \frac{5}{6}$

**12.** *(Modeling) Height of a Bridge's Arch* An arch of a bridge has the shape of the top half of an ellipse. The arch is 40 feet wide and 12 feet high at the center. Find the equation of the complete ellipse. Find the height of the arch 10 feet from the center at the bottom.

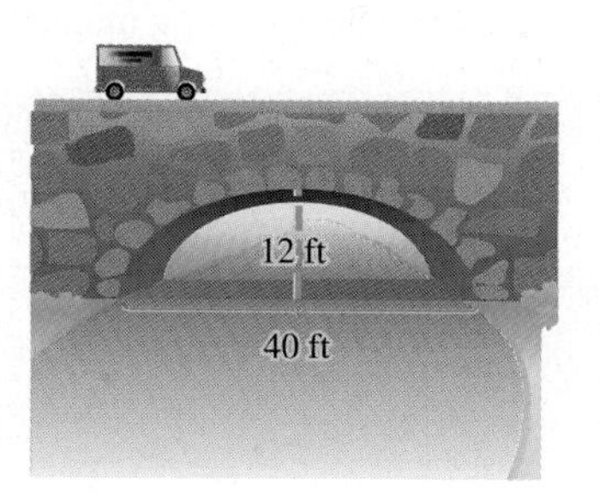


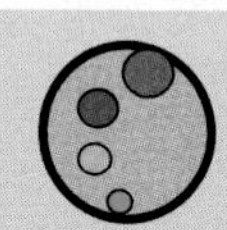

# Chapter 6 Project

## Modeling the Path of a Bouncing Ball

The height of each bounce of a bouncing ball varies quadratically with time. This means that the graph of each bounce will be a parabola. Figure A shows a scatter diagram of a bouncing ball, captured with a CBR (Calculator Based Ranger) connected to a TI-83 graphing calculator. The tick marks on the $x$-axis represent 1-second intervals, and the tick marks on the $y$-axis represent 1-foot intervals.

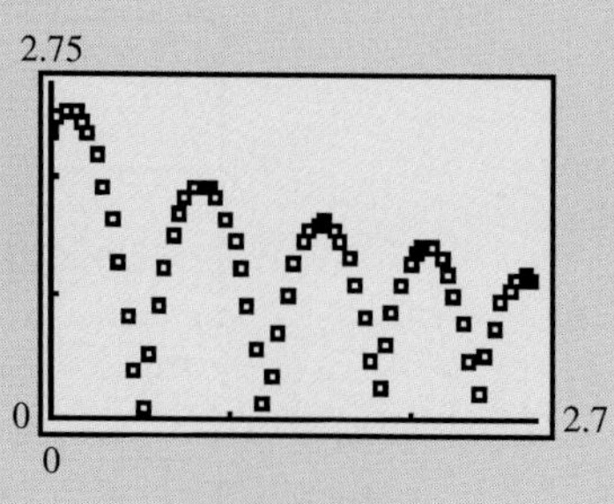


**FIGURE A**

Since each bounce can be modeled with a parabola, the entire path of the ball can be modeled with a piecewise-defined function. The graph consists of all or part of five bounces; therefore, the function will have five pieces. We can find this function, using the data lists that generated the scatter diagram.

| Time | Height | Time | Height | Time | Height |
|---|---|---|---|---|---|
| 0 | 2.37249 | .934912 | 1.80222 | 1.869694 | .60089 |
| .042496 | 2.48284 | .977404 | 1.66124 | 1.912184 | .88197 |
| .084992 | 2.536 | 1.019894 | 1.47971 | 1.954674 | 1.10494 |
| .127488 | 2.53149 | 1.062384 | 1.23962 | 1.997164 | 1.27115 |
| .169984 | 2.46122 | 1.104874 | .92341 | 2.039654 | 1.37565 |
| .21248 | 2.34186 | 1.147364 | .55945 | 2.082144 | 1.41574 |
| .254976 | 2.16168 | 1.189854 | .12928 | 2.124634 | 1.40223 |
| .297472 | 1.92384 | 1.232344 | .37297 | 2.167124 | 1.33602 |
| .339968 | 1.61799 | 1.274834 | .73107 | 2.209614 | 1.20809 |
| .382464 | 1.27476 | 1.317324 | 1.02881 | 2.252104 | 1.02476 |
| .42496 | .85089 | 1.359814 | 1.26845 | 2.294594 | .78512 |
| .467456 | .37882 | 1.402304 | 1.44863 | 2.337084 | .48648 |
| .509952 | .10045 | 1.444794 | 1.54592 | 2.379574 | .20991 |
| .552448 | .53423 | 1.487284 | 1.60808 | 2.422064 | .5108 |
| .594944 | .9153 | 1.529774 | 1.6225 | 2.464554 | .76125 |
| .63744 | 1.23286 | 1.572264 | 1.57295 | 2.507044 | .9635 |
| .679936 | 1.48782 | 1.614754 | 1.47836 | 2.549534 | 1.08467 |
| .722432 | 1.6752 | 1.657244 | 1.32701 | 2.592024 | 1.1689 |
| .764928 | 1.81754 | 1.699734 | 1.10854 | 2.634514 | 1.19683 |
| .807424 | 1.89321 | 1.742224 | .83332 | 2.677004 | 1.168 |
| .84992 | 1.91889 | 1.784714 | .49234 | 2.719494 | 1.0671 |
| .892416 | 1.88826 | 1.827204 | .2509 | 2.761984 | .95134 |

First, we determine the endpoints of each parabola or partial parabola by looking through the data list. For example, to find the endpoints of the second parabola, look in the data list for the first height that is close to 0. We find that the first near-0 height of .10045 occurs at time .509952, and the next near-0 height of .12928 occurs at 1.189854. So, the domain of the second parabola is the interval $(.51, 1.19)$. Its vertex is at time .84992, which corresponds to the largest height, 1.91889, in the list between these endpoints.

The equation of each parabola should be in the form $y = a(x - h)^2 + k$. Since the ball is a projectile in motion, $a$ is half the acceleration due to gravity: $-16$ feet per second per second. The values of $h$ and $k$ are the coordinates of each vertex; that is, $(.84992, 1.91889)$ for the second parabola. The resolution of the calculator screen is such that rounding to hundredths will not result in loss of detail. Thus, the equation and domain of the second parabola are $y = -16(x - .85)^2 + 1.92$, on the interval $(.51, 1.19)$.

## Activities

1. Find the equations and domains of the four remaining parabolas. Write a piecewise-defined function for the five parabolas and graph it. Prepare a report that includes the piecewise-defined function, the graph, and a discussion comparing the graph with the scatter diagram in Figure A.
2. The TI-83 calculator has a graph style called "Path" in which a circular cursor traces the leading edge of the graph and draws a path. If you are using this model, set the graph style to Path and watch the ball bounce. (*Note:* In reality, the ball bounces straight up and down.)
3. Working in groups of 3 or 4, gather your own original data. Write a piecewise-defined function to model the path of the ball. (Different groups should use different balls.)

# 7 Matrices and Systems of Equations and Inequalities

TO MAKE PREDICTIONS and forecasts, professionals in many fields attempt to determine relationships between different factors. These relationships often result in equations containing more than one variable. When quantities are interrelated, *systems of equations* in several variables are used to describe their relationship. As early as 4000 B.C. in Mesopotamia, people were able to solve up to 10 equations having 10 variables. In 1940 John Atanasoff, a physicist from Iowa State University, needed to solve a system of equations containing 29 equations and 29 variables. This need to solve a large *linear system* led Atanasoff and graduate student Clifford Berry to invent the first fully electronic digital computer, dubbed ABC for Atanasoff-Berry Computer. Today's supercomputers are capable of performing billions of calculations in a single second and solving more than 600,000 equations simultaneously.

Professionals throughout the world currently use systems of equations and many of the techniques presented in this chapter to determine traffic flow, forecast gross business sales, analyze data related to the greenhouse effect, and solve other important applications.

*Source:* Tucker, A., A. Bernat, W. Bradley, R. Cupper, and G. Scragg, *Fundamentals of Computing Logic, Problem Solving, Programs, and Computers,* McGraw-Hill, 1995; Lowenstein, Adam, "ISU's Atanasoff Helped Lead World into Computer Age," *The Gazette,* Cedar Rapids, IA, January 1, 1999.

CHAPTER OUTLINE

# 7.1 SYSTEMS OF EQUATIONS

Linear Systems • Substitution Method • Elimination Method • Special Systems • Nonlinear Systems • Applications of Systems

## Linear Systems

The definition of a linear equation given in Chapter 1 can be extended to more variables; any equation of the form

$$a_1x_1 + a_2x_2 + \cdots + a_nx_n = b,$$

for real numbers $a_1, a_2, \ldots, a_n$ (not all of which are 0) and $b$, is a **linear equation** or a **first-degree equation in $n$ unknowns.**

A set of equations is called a **system of equations.** The **solutions** of a system of equations must satisfy every equation in the system. If all the equations in a system are linear, the system is a **system of linear equations,** or a **linear system.**

The solution set of a linear equation in two variables is an infinite set of ordered pairs. Since the graph of such an equation is a straight line, there are three possibilities for the solution set of a system of two linear equations in two variables, as shown in Figure 1. The possible graphs of a linear system in two variables are as follows.

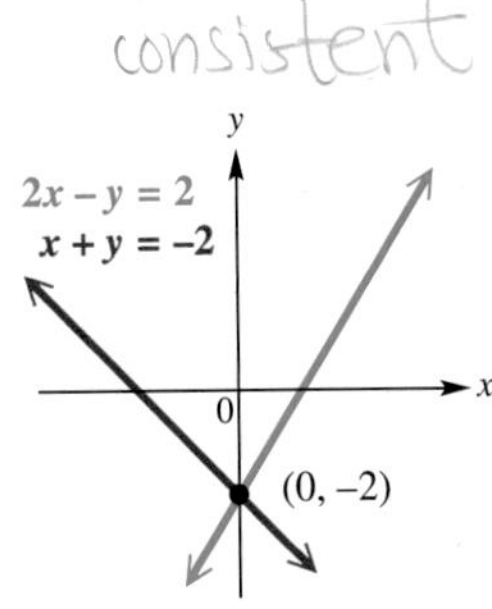


One Solution

(a)

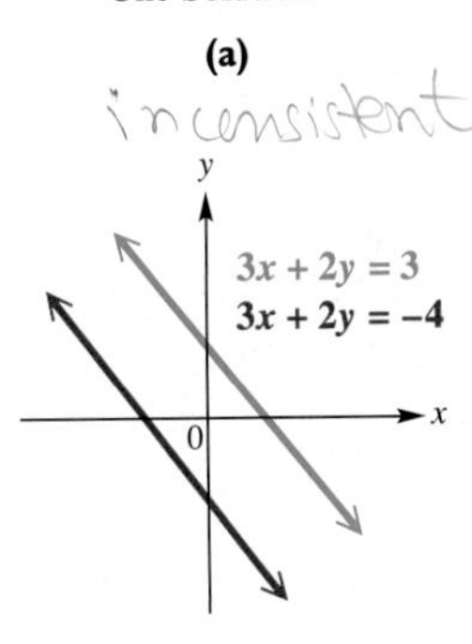


No Solutions

(b)

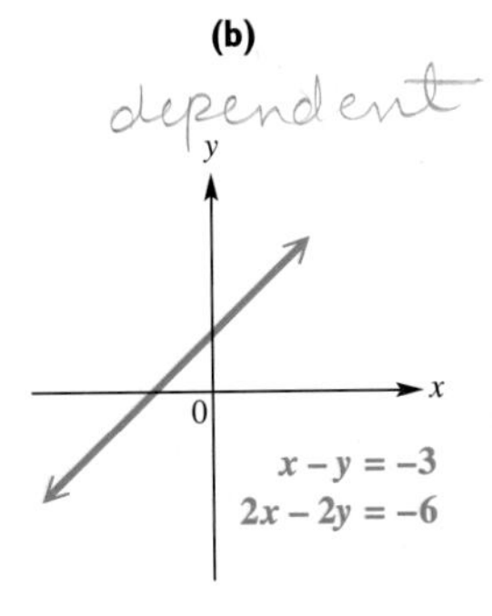


Infinitely Many Solutions

(c)

**FIGURE 1**

1. The graphs of the two equations intersect in a single point. The coordinates of this point give the only solution of the system. In this case, the system is **consistent,** and the equations are **independent.** This is the most common case. See Figure 1(a).
2. The graphs are distinct parallel lines. In this case, the system is said to be **inconsistent.** That is, there is no solution common to both equations. See Figure 1(b).
3. The graphs are the same line. In this case, the equations are said to be **dependent,** and any solution of one equation is also a solution of the other. Thus, there are infinitely many solutions. See Figure 1(c).

## Substitution Method

Although the *number* of solutions of a linear system can usually be seen from the graph of the equations of the system, determining an exact solution from a graph is often difficult. The substitution method is the most general analytic method of solving a system of equations. In a system of two equations with two variables, the **substitution method** involves using one equation to find an expression for one variable in terms of the other, then substituting into the other equation of the system.

**EXAMPLE 1** *Solving a System by Substitution*

Solve the system.

$$3x + 2y = 11 \quad (1)$$

$$-x + y = 3 \quad (2)$$

### Analytic Solution

One way to solve this system is to solve equation (2) for $y$, getting $y = x + 3$. Then substitute $x + 3$ for $y$ in equation (1) to get

$$3x + 2(x + 3) = 11.$$

Now solve this equation for $x$.

$$3x + 2x + 6 = 11$$
$$5x + 6 = 11$$
$$5x = 5$$
$$x = 1$$

Replace $x$ with 1 in $y = x + 3$ to find that $y = 1 + 3 = 4$, so the solution is $(1, 4)$. Check by substituting 1 for $x$ and 4 for $y$ in both equations of the original system. The solution set is $\{(1, 4)\}$.

### Graphing Calculator Solution

To find the solution graphically, graph equations (1) and (2) in an appropriate viewing window and determine the coordinates of the point of intersection. First, solve equation (1) for $y$ to get $Y_1 = -1.5X + 5.5$ and equation (2) for $y$ to get $Y_2 = X + 3$. Graph these in the standard viewing window to find that the point of intersection is $(1, 4)$, as seen in Figure 2. The table in Figure 3 numerically shows that when $X = 1$, both $Y_1$ and $Y_2$ are 4.

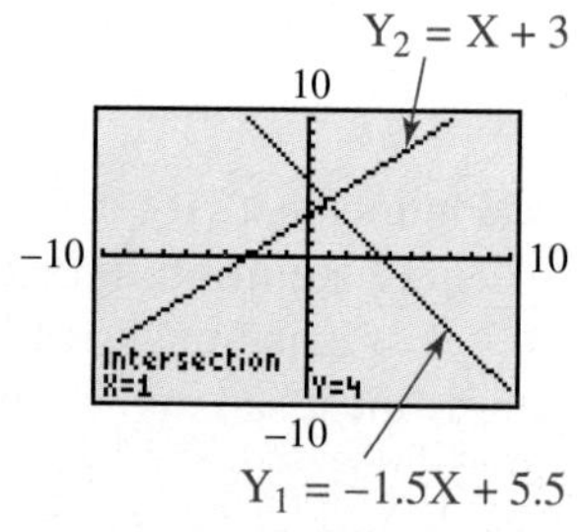

FIGURE 2

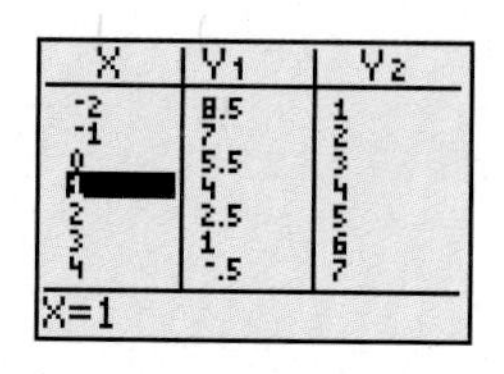

FIGURE 3

## Elimination Method

**Looking Ahead to Calculus**

In calculus, the definite integral (see *Looking Ahead to Calculus*, p. 548),

$$\int_a^b [f(x) - g(x)]\,dx,$$

gives the area between the graphs of $f$ and $g$ from $x = a$ to $x = b$. A system of equations is used to find the $x$-values $a$ and $b$ where the two graphs intersect.

Another method of solving systems of two equations is the **elimination method.** With this method, we first multiply the equations on both sides by suitable numbers, so that when they are added, one variable is eliminated. The result is an equation in one variable that can be solved by methods from Chapter 1. The solution is then substituted into one of the original equations, making it possible to solve for the other variable. In this process, the given system is replaced by new systems that have the same solution set as the original system. Systems that have the same solution set are called **equivalent systems.**

**EXAMPLE 2** *Solving a System by Elimination*

Solve the system.

$$3x - 4y = 1 \quad (1)$$
$$2x + 3y = 12 \quad (2)$$

### Analytic Solution

To eliminate $x$, multiply each side of equation (1) by $-2$ and each side of equation (2) by 3 to get

$$-6x + 8y = -2 \quad (3)$$
$$6x + 9y = 36. \quad (4)$$

Although this new system is not the same as the given system, it will

### Graphing Calculator Solution

Solve equations (1) and (2) for $y$ and graph both in an appropriate viewing window.

$$3x - 4y = 1 \quad (1)$$
$$-4y = -3x + 1$$
$$y = \frac{3}{4}x - \frac{1}{4}$$

$$2x + 3y = 12 \quad (2)$$
$$3y = -2x + 12$$
$$y = -\frac{2}{3}x + 4$$

*(continued)*

have the same solution set. Now, add equations (3) and (4) to eliminate $x$, and solve the result for $y$.

$$\begin{aligned} -6x + 8y &= -2 \\ 6x + 9y &= 36 \\ \hline 17y &= 34 \quad \text{Add.} \\ y &= 2 \end{aligned}$$

Substitute 2 for $y$ in equation (1) or (2).

$$\begin{aligned} 3x - 4(2) &= 1 \quad (1) \\ 3x &= 9 \\ x &= 3 \end{aligned}$$

The solution set is $\{(3, 2)\}$, which can be checked by substituting 3 for $x$ and 2 for $y$ in both equations of the original system.

As seen in Figure 4, the graphs of

$$Y_1 = \tfrac{3}{4}X - \tfrac{1}{4}$$

and

$$Y_2 = -\tfrac{2}{3}X + 4$$

intersect at the point (3, 2). The table in Figure 5 also indicates that the point of intersection is (3, 2).

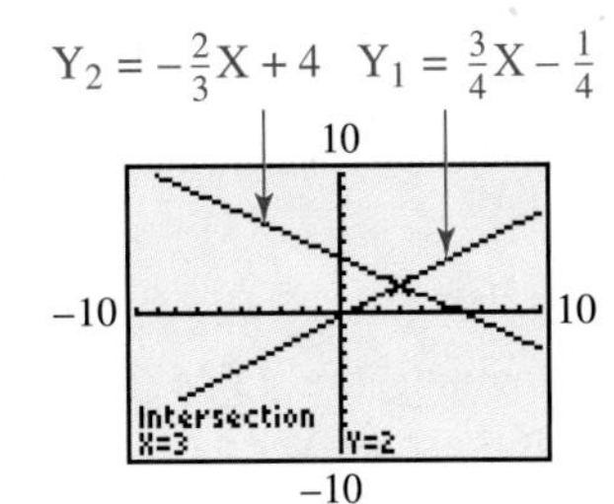


FIGURE 4

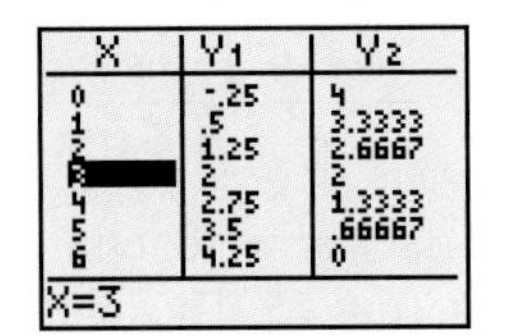

| X | $Y_1$ | $Y_2$ |
|---|---|---|
| 0 | -.25 | 4 |
| 1 | .5 | 3.3333 |
| 2 | 1.25 | 2.6667 |
| 3 | 2 | 2 |
| 4 | 2.75 | 1.3333 |
| 5 | 3.5 | .66667 |
| 6 | 4.25 | 0 |

X=3

FIGURE 5

## What Went WRONG?

To graphically find the solution in Example 2, a student entered the functions shown on the left and got the graph shown on the right.

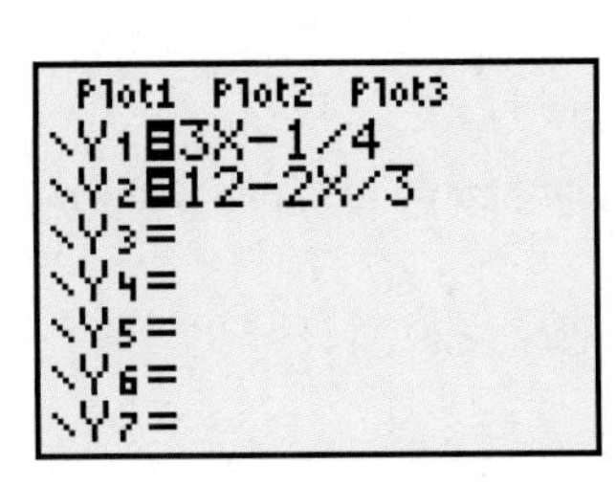


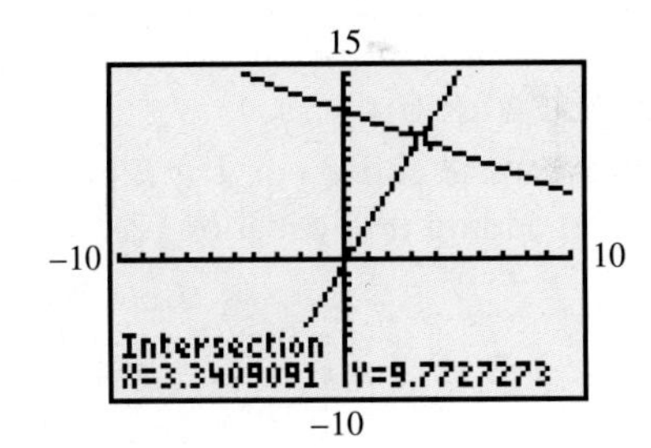


What Went Wrong? What can be done to correct it?

## Special Systems

The examples presented so far in this section have all been systems with a single solution. This is not always the case.

**EXAMPLE 3** *Solving an Inconsistent System*

Solve the system.

$$\begin{aligned} 3x - 2y &= 4 \quad (1) \\ -6x + 4y &= 7 \quad (2) \end{aligned}$$

---

*Answer to What Went Wrong?*

The student neglected to insert parentheses as necessary. $Y_1$ should be entered as $(3X - 1)/4$ and $Y_2$ as $(12 - 2X)/3$.

### Analytic Solution

To solve

$$3x - 2y = 4 \quad (1)$$
$$-6x + 4y = 7, \quad (2)$$

eliminate the variable $x$ by multiplying each side of equation (1) by 2 and then adding the result to equation (2).

$$6x - 4y = 8 \quad \text{Multiply (1) by 2.}$$
$$-6x + 4y = 7 \quad (2)$$
$$0 = 15 \quad \text{False}$$

Both variables were eliminated, leaving the false statement $0 = 15$. This indicates that these two equations have no solutions in common. The system is inconsistent, and the solution set is $\emptyset$.

### Graphing Calculator Solution

In slope-intercept form, equation (1) is written $y = 1.5x - 2$; equation (2) is written $y = 1.5x + 1.75$. Since the lines have the same slope but different $y$-intercepts, they are parallel and have no point of intersection. This supports the analytic conclusion. See Figure 6.

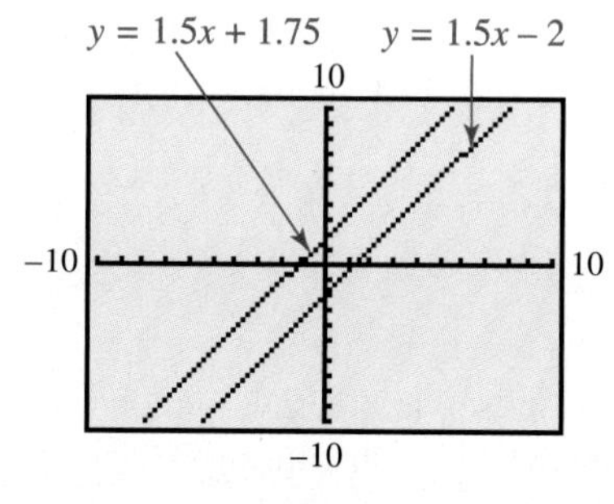


FIGURE 6

**EXAMPLE 4** *Solving a System with Dependent Equations*

Solve the system.

$$-4x + y = 2 \quad (1)$$
$$8x - 2y = -4 \quad (2)$$

### Analytic Solution

We can eliminate either $x$ or $y$ by multiplying each side of equation (1) by 2 and then adding it to equation (2).

$$-8x + 2y = 4 \quad \text{Multiply (1) by 2.}$$
$$8x - 2y = -4 \quad (2)$$
$$0 = 0 \quad \text{True}$$

This true statement, $0 = 0$, indicates that every solution of one equation is also a solution of the other, so the solution set is an infinite set of ordered pairs.

We write the solution of a system of dependent equations as an ordered pair by expressing $x$ in terms of $y$. We can choose either equation, and solve for $x$.

$$-4x + y = 2 \quad (1)$$
$$x = \frac{2 - y}{-4} = \frac{y - 2}{4}$$

We write the solution set as $\left\{\left(\frac{y-2}{4}, y\right)\right\}$. By selecting values for $y$ and calculating the corresponding values for $x$, we can find individual ordered pairs of the solution set. For example, if $y = -2$, then $x = \frac{-2 - 2}{4} = -1$, and the ordered pair $(-1, -2)$ is a solution.

### Graphing Calculator Solution

The graphs of the equations are the same line, since both equations are equivalent to $y = 4x + 2$. See Figure 7. The two equations are dependent. A table would show numerically that $y_1 = y_2$ for *any* selected values of $x$, since the equations are dependent and have the same graph.

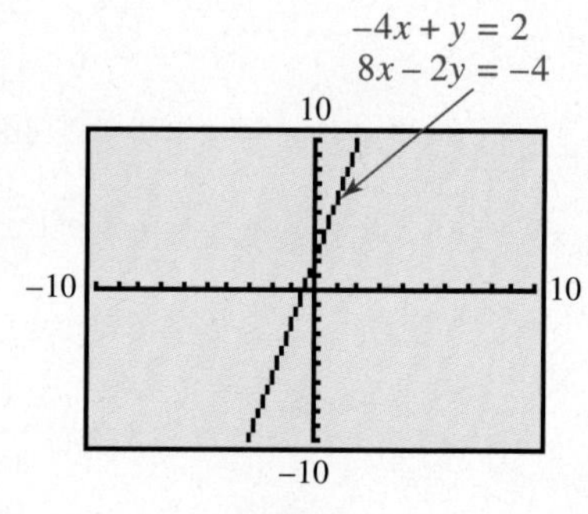


FIGURE 7

**NOTE** In Example 4, we wrote the solution set in a form with the variable $y$ arbitrary. However, we could write the solution set with $x$ arbitrary, that is, $\{(x, 4x + 2)\}$. By selecting values for $x$ and solving for $y$ in this ordered pair, we can find individual solutions. Verify again that $(-1, -2)$ is a solution.

## Nonlinear Systems

A **nonlinear system of equations** is one in which at least one of the equations is not a linear equation.

**EXAMPLE 5** *Solving a Nonlinear System by Substitution*

Solve the system.

$$3x^2 - 2y = 5 \qquad (1)$$
$$x + 3y = -4 \qquad (2)$$

### Analytic Solution

The graph of equation (1) is a parabola, and that of equation (2) is a line. There may be 0, 1, or 2 points of intersection of these graphs.

Although either equation could be solved for either variable, it is simpler here to solve the linear equation (2) for $y$, since $x$ is squared in equation (1).

$$x + 3y = -4 \qquad (2)$$
$$3y = -4 - x$$
$$y = \frac{-4 - x}{3} \qquad (3)$$

Substitute this value of $y$ into equation (1).

$$3x^2 - 2\left(\frac{-4 - x}{3}\right) = 5$$
$$9x^2 - 2(-4 - x) = 15 \qquad \text{Multiply by 3.}$$
$$9x^2 + 8 + 2x = 15 \qquad \text{Distributive property}$$
$$9x^2 + 2x - 7 = 0 \qquad \text{Standard form}$$
$$(9x - 7)(x + 1) = 0 \qquad \text{Factor.}$$
$$9x - 7 = 0 \quad \text{or} \quad x + 1 = 0$$
$$x = \frac{7}{9} \quad \text{or} \quad x = -1$$

Substitute both values of $x$ into equation (3) to find the corresponding $y$-values.

$$y = \frac{-4 - \frac{7}{9}}{3} \quad \text{or} \quad y = \frac{-4 - (-1)}{3}$$
$$y = -\frac{43}{27} \quad \text{or} \quad y = -1$$

Check in the original system that the solution set is $\left\{\left(\frac{7}{9}, -\frac{43}{27}\right), (-1, -1)\right\}$.

### Graphing Calculator Solution

Start by graphing equation (1) as $y_1 = 1.5x^2 - 2.5$ and equation (2) as $y_2 = -\frac{1}{3}x - \frac{4}{3}$. Figure 8 indicates that the points of intersection are $(-1, -1)$ and $(.\overline{7}, 1.\overline{592})$, confirming the analytic solution.

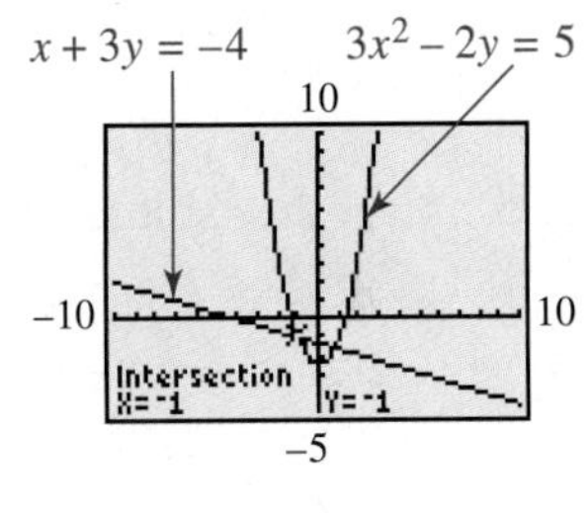


**(a)**

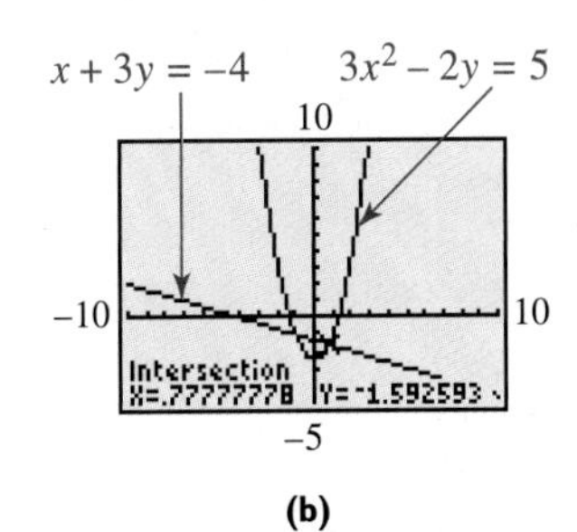


**(b)**

**FIGURE 8**

The decimal values for $x$ and $y$ displayed at the bottom of Figure 8(b) represent the repeating decimal forms for the fractions $\frac{7}{9}$ and $-\frac{43}{27}$, respectively.

A nonlinear system can be solved by the elimination method if we can completely eliminate one variable from the system.

## EXAMPLE 6 *Solving a Nonlinear System by Elimination*

Solve the system.

$$x^2 + y^2 = 4 \qquad (1)$$
$$2x^2 - y^2 = 8 \qquad (2)$$

**Analytic Solution**

The graph of equation (1) is a circle and that of equation (2) is a hyperbola. Visualizing these suggests that there may be 0, 1, 2, 3, or 4 points of intersection.

Add the two equations to eliminate $y^2$.

$$\begin{aligned} x^2 + y^2 &= 4 \\ 2x^2 - y^2 &= 8 \\ \hline 3x^2 \qquad &= 12 \\ x^2 &= 4 \end{aligned}$$

$$x = 2 \quad \text{or} \quad x = -2$$

Substituting into equation (1) gives the corresponding values of $y$.

$$\begin{aligned} 2^2 + y^2 = 4 \quad &\text{or} \quad (-2)^2 + y^2 = 4 \\ y^2 = 0 \quad &\text{or} \quad y^2 = 0 \\ y = 0 \quad &\text{or} \quad y = 0 \end{aligned}$$

Check that the solution set of the system is $\{(2, 0), (-2, 0)\}$.

**Graphing Calculator Solution**

Graph equation (1) as the union of

$$y_1 = \sqrt{4 - x^2} \quad \text{and} \quad y_2 = -\sqrt{4 - x^2},$$

and equation (2) as the union of

$$y_3 = \sqrt{2x^2 - 8} \quad \text{and} \quad y_4 = -\sqrt{2x^2 - 8}.$$

As shown in Figure 9, the two points of intersection are (2, 0) and (−2, 0).

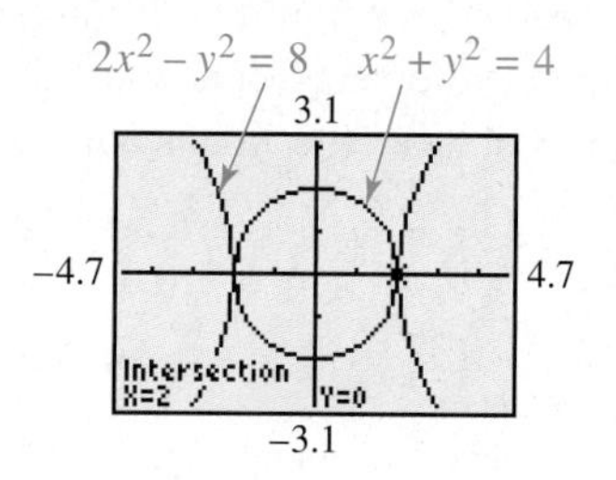


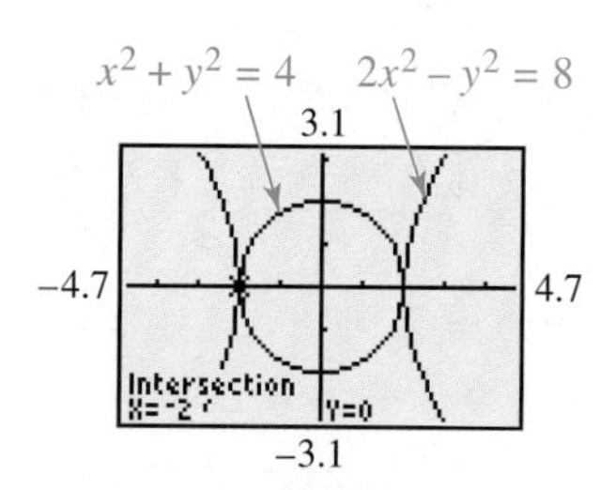


**FIGURE 9**

Many systems are difficult or even impossible to solve using strictly analytic methods. Graphing calculators allow us to solve such systems graphically.

## EXAMPLE 7 *Solving a Nonlinear System Graphically*

Solve the system.

$$y = 2^x \qquad (1)$$
$$|x + 2| - y = 0 \qquad (2)$$

**Solution** Enter equation (1) as $y_1 = 2^x$ and equation (2) as $y_2 = |x + 2|$. As seen in Figure 10, the various windows indicate that there are three points of intersection of the graphs, and thus three solutions. Using the capabilities of the calculator, we find that (2, 4) is an exact solution, and (−2.22, .22) and (−1.69, .31) are approximate solutions. Therefore, the solution set is $\{(2, 4), (-2.22, .22), (-1.69, .31)\}$.

**TECHNOLOGY NOTE**

Example 7 illustrates why proficiency in setting various windows is important when studying algebra with a graphing calculator.

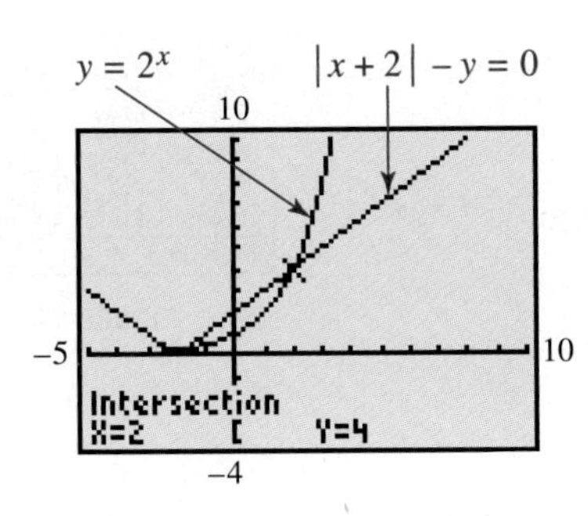


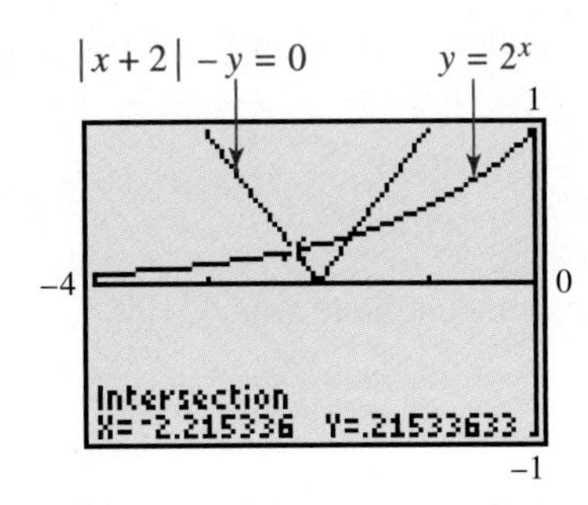


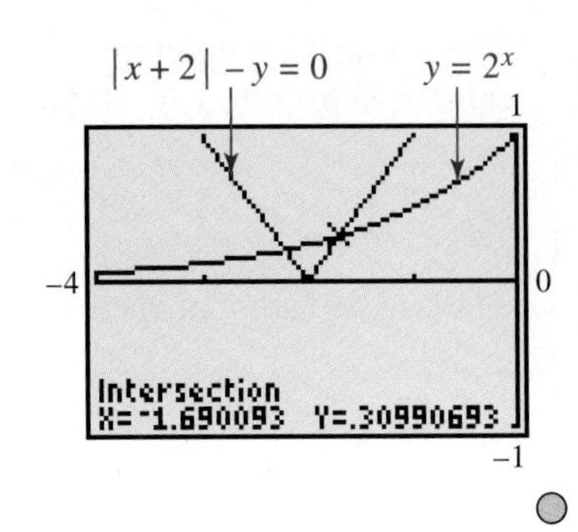


**FIGURE 10**

## Applications of Systems

Many applied problems involve more than one unknown quantity. To solve such a problem using a system, decide on the unknown quantities you are asked to find, and let different variables represent each of these quantities. Then, write a system of equations, and solve it. Be sure that you answer the question(s) posed in the problem, and check to see that your answer is reasonable.

**EXAMPLE 8** *Using a Linear System to Solve an Application*

Landmark Title IX legislation prohibits sex discrimination in sports programs. In 1997 the national average spent on two varsity athletes, one female and one male, was \$6050 for Division I-A schools. However, average expenditures for a male athlete exceeded those for a female athlete by \$3900. Determine how much was spent per varsity athlete for each gender. (*Source: USA Today.*)

### Analytic Solution

Let $x$ represent average expenditures per male athlete and $y$ represent average expenditures per female athlete in 1997. Since the average amount spent on one female and one male athlete was \$6050, one equation is

$$\frac{x+y}{2} = 6050.$$

The expenditures for a male athlete exceeded those for a female athlete by \$3900. Thus, $x - y = 3900$, which gives the system of equations

$$\frac{x+y}{2} = 6050 \qquad (1)$$

$$x - y = 3900. \qquad (2)$$

Multiply equation (1) by 2 and then add the resulting equations.

$$\begin{aligned} x + y &= 12{,}100 && \text{Multiply (1) by 2.} \\ x - y &= 3900 && \text{(2)} \\ \hline 2x &= 16{,}000 && \text{Add.} \\ x &= 8000 \end{aligned}$$

### Graphing Calculator Solution

Solving equations (1) and (2) for $y$ yields

$$Y_1 = 12{,}100 - X$$

and $$Y_2 = X - 3900.$$

Figure 11 shows that their graphs intersect at (8000, 4100), which agrees with the analytic solution. The table in Figure 12 shows that when $X = 8000$, $Y_1 = Y_2 = 4100$.

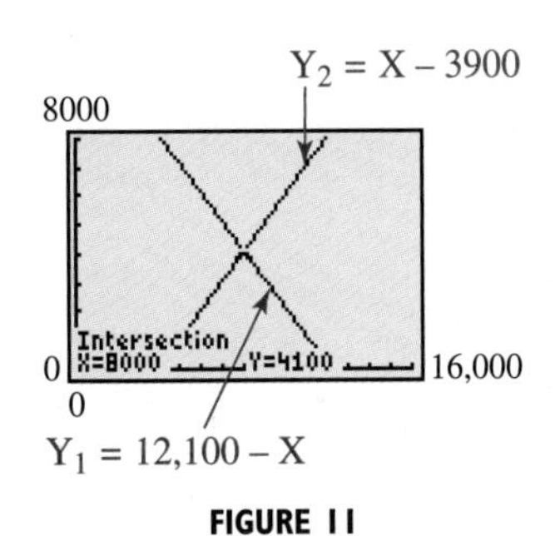


**FIGURE 11**

*(continued)*

The average expenditure per male athlete in 1997 was \$8000. To determine $y$, substitute $x = 8000$ into equation (2).

$$8000 - y = 3900$$
$$y = 4100$$

The average amount spent on each female athlete was \$4100.

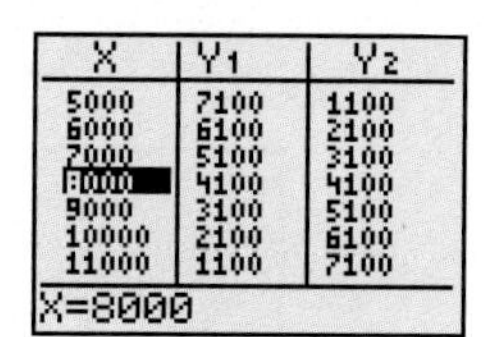

| X | Y1 | Y2 |
|---|---|---|
| 5000 | 7100 | 1100 |
| 6000 | 6100 | 2100 |
| 7000 | 5100 | 3100 |
| 8000 | 4100 | 4100 |
| 9000 | 3100 | 5100 |
| 10000 | 2100 | 6100 |
| 11000 | 1100 | 7100 |

X=8000

**FIGURE 12**

## EXAMPLE 9 *Using a Nonlinear System to Find the Dimensions of a Box*

A box with an open top has a square base and four sides of equal height. The volume of the box is 75 cubic inches, and the surface area is 85 square inches. What are the dimensions of the box?

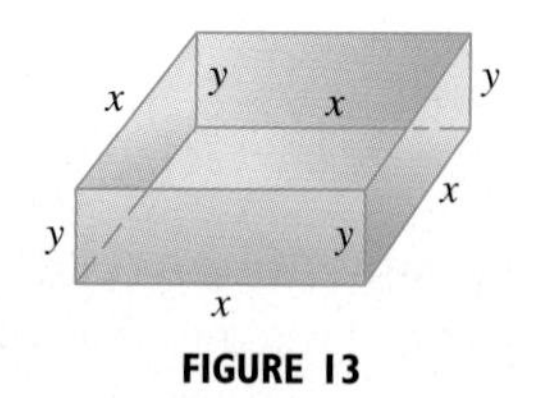


**FIGURE 13**

**Solution** Figure 13 shows a diagram of such a box. If each side of the square base measures $x$ inches and the height measures $y$ inches, then the volume is

$$x^2y = 75 \qquad \text{Volume formula} \qquad (1)$$

and the surface area is

$$x^2 + 4xy = 85. \qquad \text{Sum of areas of base and sides} \qquad (2)$$

We solve equation (1) for $y$ to get $y = \frac{75}{x^2}$, and substitute this into equation (2).

$$x^2 + 4x\left(\frac{75}{x^2}\right) = 85 \qquad (2)$$
$$x^2 + \frac{300}{x} = 85$$
$$x^3 + 300 = 85x \qquad \text{Multiply by } x.$$
$$x^3 - 85x + 300 = 0$$

The solutions of this equation are $x$-intercepts of the graph of $y = x^3 - 85x + 300$. A comprehensive graph of this function indicates that there are three real solutions. However, one of them is negative and must be rejected. As Figure 14 indicates, one positive solution is 5 and the other positive solution is approximately 5.64. By substituting back into equation (1), we find that when $x = 5$, $y = 3$, and when $x \approx 5.64$, $y \approx 2.36$. Therefore, this problem has two solutions: the box may have base 5 inches by 5 inches and height 3 inches, or it may have base 5.64 inches by 5.64 inches and height 2.36 inches.

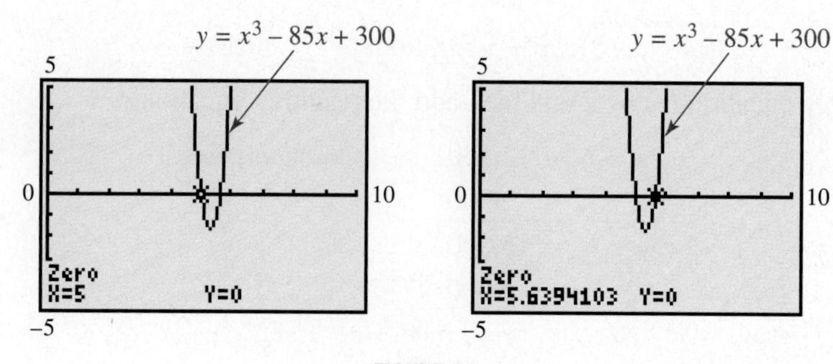


**FIGURE 14**

## 7.1 EXERCISES

*Effects of NAFTA* *In 1994, the North American Free Trade Agreement (NAFTA) made the United States, Mexico, and Canada the largest free-trade zone in the world. "Trade theory predicts that, because of more competition and economies of scale, NAFTA member countries should have faster-growing economies, more jobs, and higher wages, which should reduce migration. However, the U.S. Commission for the Study of International Migration and Cooperative Economic Development warned 'the economic development process itself tends in the short to medium term to stimulate migration'."* (Source: *Martin, P. and E. Midgley, "Immigration to the United States,"* Population Bulletin, *June 1999.) The figure at the right shows the projected levels of migration with and without NAFTA.*

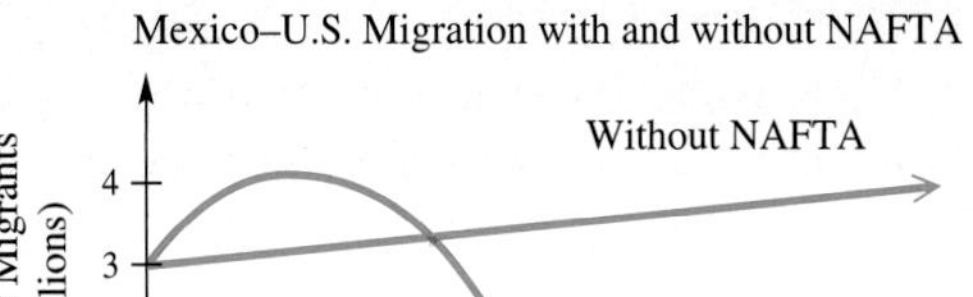
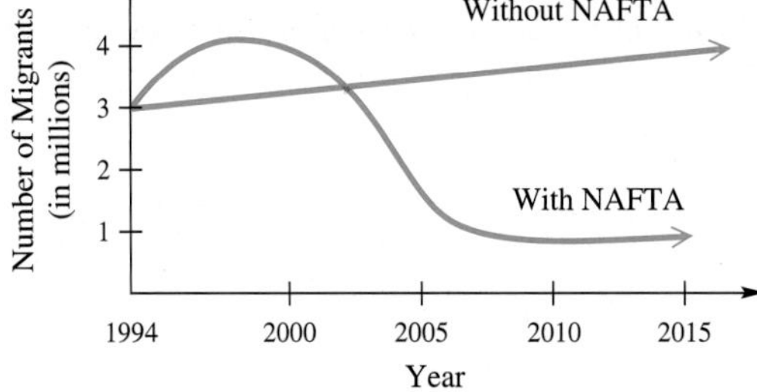


*Source: Population Bulletin*, June 1999.

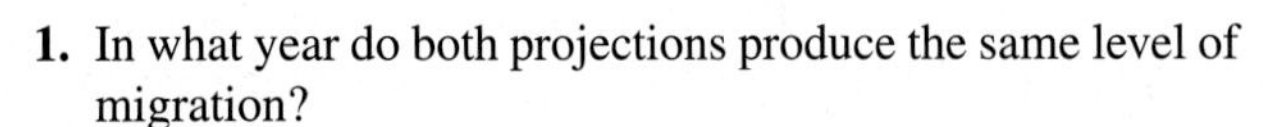

1. In what year do both projections produce the same level of migration?
2. Refer to Exercise 1. In the year that the two projections produced the same level of migration, what was the level?
3. Express as an ordered pair the solution of the system containing the graphs of the two projections.
4. Use the terms *increasing* and *decreasing* to describe the trends for the "With NAFTA" graph.
5. If equations of the form $y = f(t)$ were determined that modeled either of the two graphs, then $t$ would represent ________ and $y$ would represent ________.

6. Explain why each graph is that of a function.

*In Exercises 7–10, a system of equations is given, along with a graph indicating the coordinates of one point of intersection of the graphs of the equations in the system. Show by direct substitution, using the displayed values of x and y, that the indicated point is indeed a solution of the system.*

**7.** $4x - y = 3$
$-2x + 3y = 1$

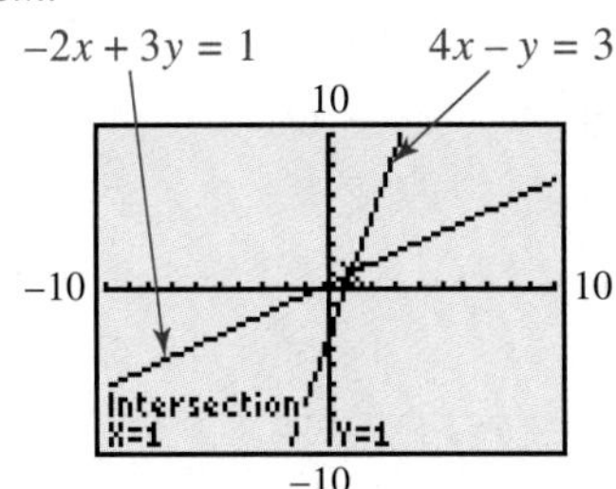


**8.** $5x + 3y = 1$
$-3x - 4y = 6$

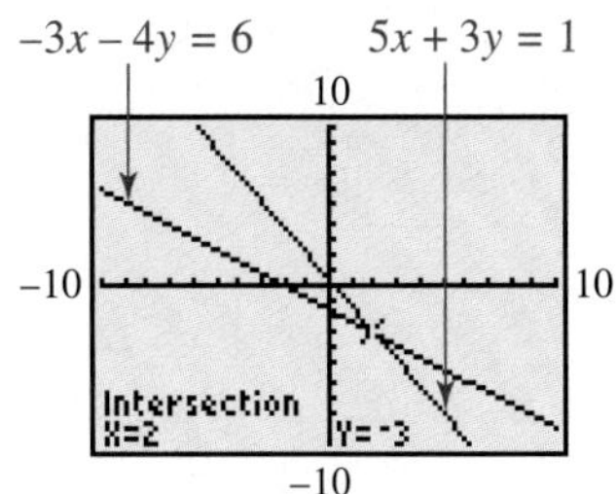


**9.** $y - x = 2$
$x^2 + y^2 = 34$

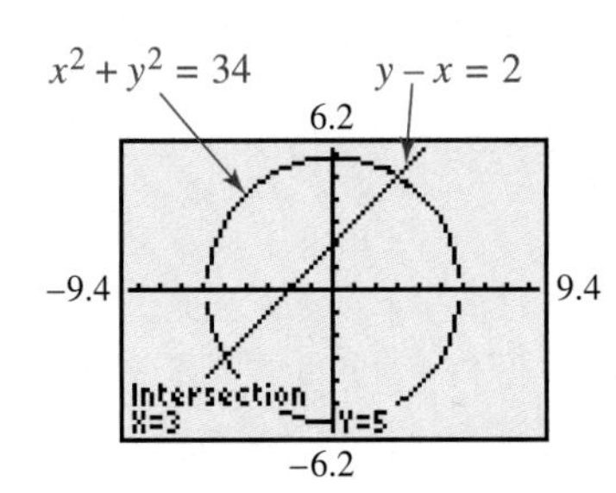


**10.** $x^2 - y^2 = 12$
$x^2 + y^2 = 20$

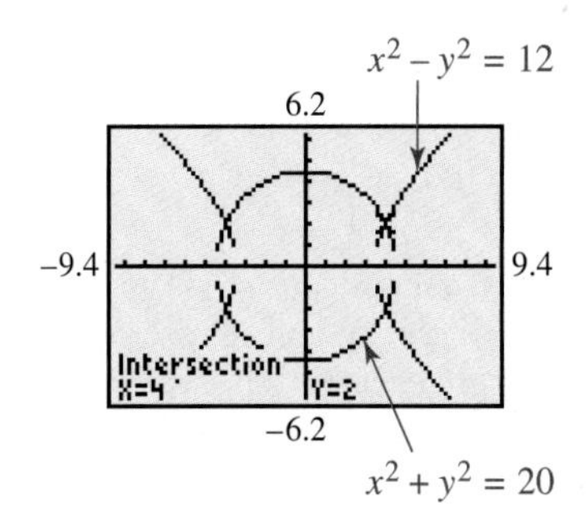


*Solve each system by substitution.*

**11.** $6x - y = 5$
$y = x$

**12.** $5x + y = 2$
$y = -3x$

**13.** $x + 2y = -1$
$2x + y = 4$

**14.** $2x + y = -11$
$x + 3y = -8$

**15.** $y = 2x + 3$
$3x + 4y = 78$

**16.** $y = 4x - 6$
$2x + 5y = -8$

**17.** $3x - 2y = 12$
$5x = 4 - 2y$

**18.** $8x + 3y = 2$
$5x = 17 + 6y$

**19.** $4x - 5y = -11$
$2x + y = 5$

**20.** $7x - y = -10$
$3y - x = 10$

**21.** $4x + 5y = 7$
$9y = 31 + 2x$

**22.** $-2x = 6y + 18$
$-29 = 5y - 3x$

**23.** $3x - 7y = 15$
$3x + 7y = 15$

**24.** $3y = 5x + 6$
$x + y = 2$

**25.** *Concept Check* Only one of the screens gives the correct graphical solution of the system in Exercise 19. Which one is it? (*Hint:* Solve for $y$ in each equation, and use the slope-intercept forms to help you answer the question.)

**A.**

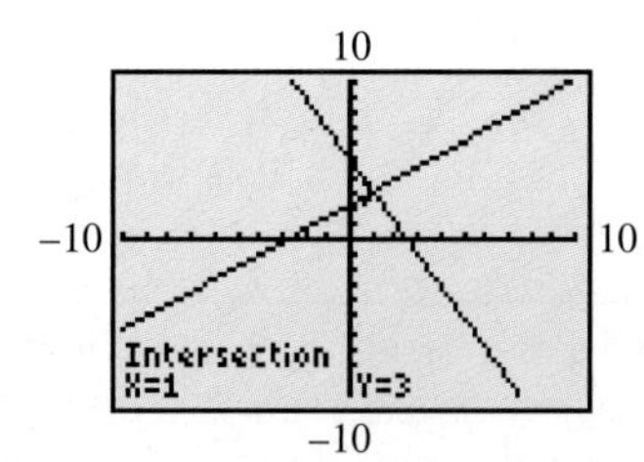


**B.**

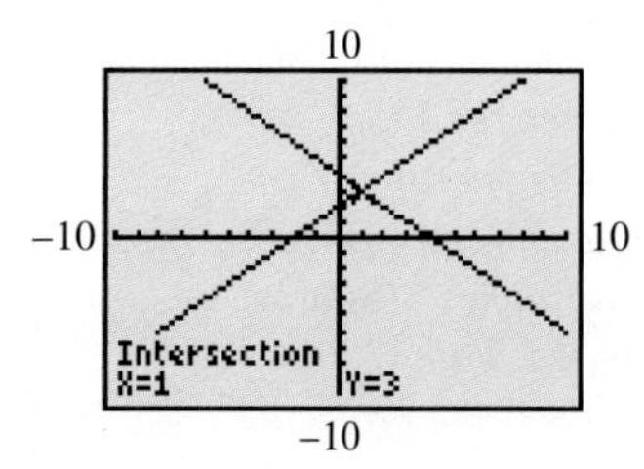


**C.**

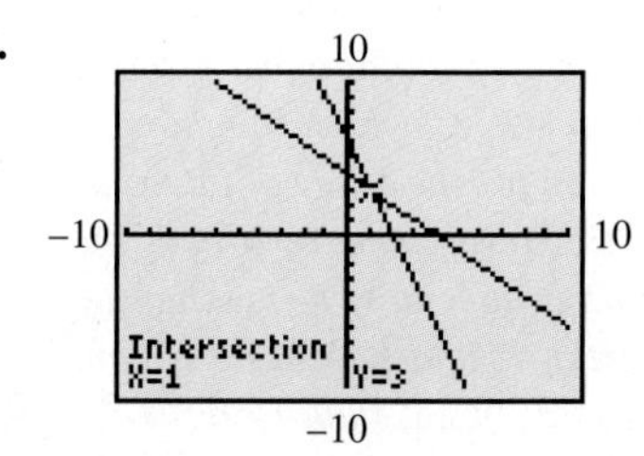


**26.** *Concept Check* Refer to Example 2 in this section. If we began solving the system by eliminating $y$, by what numbers might we have multiplied equations (1) and (2)?

**27.** Explain how one can determine whether a system is inconsistent or has dependent equations when using the substitution or elimination method.

**28.** *Concept Check* For what value(s) of $k$ will the following system of linear equations have no solution? infinitely many solutions?

$$x - 2y = 3$$
$$-2x + 4y = k$$

*Solve each system by elimination.*

**29.** $3x - y = -4$
$x + 3y = 12$

**30.** $2x - 3y = -7$
$5x + 4y = 17$

**31.** $4x + 3y = -1$
$2x + 5y = 3$

**32.** $5x + 7y = 6$
$10x - 3y = 46$

**33.** $12x - 5y = 9$
$3x - 8y = -18$

**34.** $6x + 7y = -2$
$7x - 6y = 26$

**35.** $4x - y = 9$
$-8x + 2y = -18$

**36.** $x + y = 4$
$3x + 3y = 12$

**37.** $9x - 5y = 1$
$-18x + 10y = 1$

**38.** $3x + 2y = 5$
$6x + 4y = 8$

**39.** $3x + y = 6$
$6x + 2y = 1$

**40.** $3x + 5y = -2$
$9x + 15y = -6$

**41.** $\dfrac{x}{2} + \dfrac{y}{3} = 8$
$\dfrac{2x}{3} + \dfrac{3y}{2} = 17$

**42.** $\dfrac{x}{5} + 3y = 31$
$2x - \dfrac{y}{5} = 8$

**43.** $\dfrac{2x - 1}{3} + \dfrac{y + 2}{4} = 4$
$\dfrac{x + 3}{2} - \dfrac{x - y}{3} = 3$

**44.** $\dfrac{x + 6}{5} + \dfrac{2y - x}{10} = 1$
$\dfrac{x + 2}{4} + \dfrac{3y + 2}{5} = -3$

*Use a graphing calculator to solve each system. Express solutions with approximations to the nearest thousandth.*

**45.** $\sqrt{3}x - y = 5$
$100x + y = 9$

**46.** $\dfrac{11}{3}x + y = .5$
$.6x - y = 3$

**47.** $\sqrt{5}x + \sqrt[3]{6}\,y = 9$
$\sqrt{2}x + \sqrt[5]{9}\,y = 12$

**48.** $\pi x + ey = 3$
$ex + \pi y = 4$

**49.** *Concept Check* Determine visually the other solution of the system in Exercise 9. Then support your answer graphically.

**50.** *Concept Check* Use symmetry to determine the other three solutions of the system in Exercise 10.

*Draw a sketch of the two graphs described, with the indicated number of points of intersection. (There may be more than one way to do this in some cases.)*

**51.** A line and a circle; no points

**52.** A line and a circle; one point

**53.** A line and a circle; two points

**54.** A line and a hyperbola; no points

**55.** A line and a hyperbola; one point

**56.** A line and a hyperbola; two points

**57.** A circle and an ellipse; four points

**58.** A parabola and an ellipse; one point

**59.** A parabola and a hyperbola; four points

**60.** A circle and a hyperbola; two points

*Solve each nonlinear system of equations analytically.*

**61.** $y = -x^2 + 2$
$x - y = 0$

**62.** $y = (x - 1)^2$
$x - 3y = -1$

**63.** $3x^2 + 2y^2 = 5$
$x - y = -2$

**64.** $x^2 + y^2 = 5$
$-3x + 4y = 2$

**65.** $x^2 + y^2 = 10$
$2x^2 - y^2 = 17$

**66.** $x^2 + y^2 = 4$
$2x^2 - 3y^2 = -12$

**67.** $x^2 + 2y^2 = 9$
$3x^2 - 4y^2 = 27$

**68.** $2x^2 + 3y^2 = 5$
$3x^2 - 4y^2 = -1$

**69.** $2x^2 + 2y^2 = 20$
$3x^2 + 3y^2 = 30$

**70.** $x^2 + y^2 = 4$
$5x^2 + 5y^2 = 28$

**71.** $y = |x - 1|$
$y = x^2 - 4$

**72.** $2x^2 - y^2 = 4$
$|x| = |y|$ (*Hint:* $|x| = |y|$ is equivalent to $x^2 = y^2$.)

*Solve each system graphically. Give x- and y-coordinates correct to the nearest hundredth.*

**73.** $y = \log(x + 5)$
$y = x^2$

**74.** $y = 5^x$
$xy = 1$

**75.** $y = e^{x+1}$
$2x + y = 3$

**76.** $y = \sqrt[3]{x - 4}$
$x^2 + y^2 = 6$

## RELATING CONCEPTS

**For individual or group investigation (Exercises 77–82)**

*In Example 9, we solved a nonlinear system of equations by solving the related polynomial equation* $x^3 - 85x + 300 = 0$.

**77.** Use synthetic division to show that a real solution of this equation is 5.

**78.** Factor $x^3 - 85x + 300$, using the results of Exercise 77.

**79.** From Exercise 78, show that the other positive solution has an exact value of $\frac{-5 + \sqrt{265}}{2}$.

**80.** Use the graph of $y = x^3 - 85x + 300$ to show that the negative solution of the polynomial equation is approximately $-10.64$.

**81.** What is the exact value of the negative solution?

**82.** Why was the negative solution rejected in the problem solved in Example 9?

*(Modeling) Solve each problem.*

**83.** *Bank Robberies* The total number of bank robberies in 1994 and 1995 was 13,787. From 1994 to 1995 the number of bank robberies declined by 271. (*Source:* Federal Bureau of Investigation.)

**(a)** Write a system of equations whose solution represents the number of robberies committed in each of these years.

**(b)** Solve the system analytically.

**(c)** Interpret the solution.

**84.** *Medical Waste* In 1996 hospital incinerators caused significant toxic pollution. The two states that produced the largest medical waste pollution were New York and California. Together they accounted for 185,000 tons. New York's total exceeded California's by 32,900 tons. (*Source:* Environmental Working Group.)

**(a)** Write a system of linear equations whose solution represents the incinerator pollution from each state.

**(b)** Solve the system analytically.

**(c)** Interpret the solution.

**85.** *Tourists* In 1996 a total of 6.2 million tourists visited Fort Lauderdale, Florida. There were four times as many American tourists as foreign tourists. Determine the number of foreign and American tourists that visited Fort Lauderdale in 1996. Round your answers to the nearest tenth of a million. (*Source:* Greater Fort Lauderdale Convention & Visitor Bureau.)

**86.** *Card Catalogs* Libraries are moving toward computer catalogs. From 1968 to 1996 the number of $3 \times 5$-inch cards sold to libraries by the Library of Congress declined by 78.19 million. The number of cards sold in 1996 was only .72% of the 1968 number. How many cards were sold in 1968 and in 1996? Round your answers to the nearest hundredth of a million. (*Source:* Library of Congress.)

**87.** *Heart Rate* In one study the maximum heart rates of conditioned athletes were examined. A group of athletes was exercised to exhaustion. Let $x$ represent an athlete's heart rate 5 seconds after stopping exercise and $y$ this rate after 10 seconds. It was found that the maximum heart rate $H$ for these athletes satisfied the two equations

$$H = .491x + .468y + 11.2$$
$$H = -.981x + 1.872y + 26.4.$$

If an athlete had maximum heart rate $H = 180$, determine $x$ and $y$ graphically. Interpret your answer. (*Source:* Thomas, V., *Science and Sport,* Faber and Faber, 1970.)

**88.** *Heart Rate* Repeat Exercise 87 for an athlete with maximum heart rate 195.

**89.** *Tuna and Shrimp Consumption* The total amounts (in millions of pounds) of canned tuna and fresh shrimp available for U.S. consumption during the years 1990–1996 are modeled by the linear functions defined by

$$y = -6.393x + 894.9 \quad \text{Tuna}$$
$$y = 19.14x + 746.9. \quad \text{Shrimp}$$

In both equations, $x$ is the number of years since 1990. (*Source:* U.S. National Oceanic and Atmospheric Administration, National Marine Fisheries Service, *Fisheries of the United States,* annual.)

**(a)** Solve this system of equations analytically. Give answers to the nearest tenth.

**(b)** Interpret the solution found in part (a).

**(c)** Use a graphing calculator to support the analytic solution in part (a).

**90.** *Populations of Minorities in the United States* The current and estimated resident populations (in percent) of blacks and Hispanics in the United States for the years 1990–2050 are modeled by the linear functions defined by

$$y = .0515x + 12.3 \quad \text{Blacks}$$
$$y = .255x + 9.01. \quad \text{Hispanics}$$

In each case, $x$ represents the number of years since 1990. (*Source:* U.S. Bureau of the Census, *U.S. Census of Population.*)

**(a)** Solve the system to find the year when these population percents will be equal.

**(b)** What percent of the U.S. resident population will be black or Hispanic in the year found in part (a)?

**(c)** Use a graphing calculator to support the analytic solution in part (a).

**(d)** Which population is increasing more rapidly? (*Hint:* Consider the slopes of the lines.)

**91.** *Geometry* See Example 9. Approximate graphically the radius and height of a cylindrical container with volume 50 cubic inches and a lateral surface area 65 square inches.

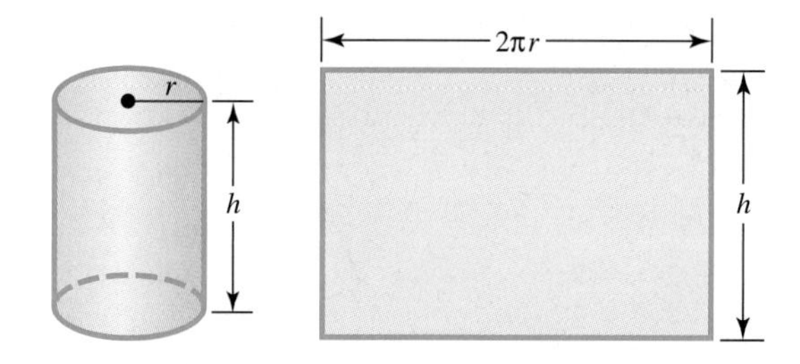


**92.** *Geometry* See Example 9. Determine graphically if it is possible to construct a cylindrical container, *including* the top and bottom, with volume 38 cubic inches and surface area 38 square inches.

**93.** *Roof Truss* The forces or weights $W_1$ and $W_2$ exerted on each rafter for the roof truss shown in the figure are determined by the following system of linear equations. Solve the system.

$$W_1 + \sqrt{2}\, W_2 = 300$$
$$\sqrt{3}\, W_1 - \sqrt{2}\, W_2 = 0$$

150 lb

$W_1$ $W_2$

30° 45°

**94.** *Height and Weight* The relationship between a professional basketball player's height $h$ in inches and weight $w$ in pounds was modeled using two samples of players. The resulting equations that modeled each sample were

$$w = 7.46h - 374$$
$$w = 7.93h - 405.$$

Assume that $65 \le h \le 85$.

**(a)** Use each equation to predict the weight of a professional basketball player who is 6 feet 11 inches.

**(b)** Determine graphically the height where the two models give the same weight.

**(c)** For each model, what change in weight is associated with a 1-inch increase in height?

**95.** *Traffic Control* The figure on the next page shows two intersections labeled A and B that involve one-way streets. The numbers and variables represent the average traffic flow rates measured in vehicles per hour. For example, an average of 500 vehicles per hour enter intersection A from the west, whereas 150 vehicles per hour enter this intersection from the north. A stoplight will control the unknown traffic flow denoted by the variables $x$ and $y$. Use the fact that the number of vehicles entering an intersection must equal the number leaving to determine $x$ and $y$.

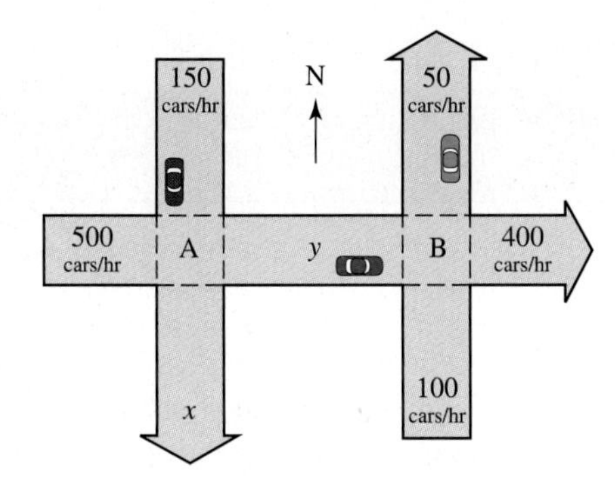


*Break-Even Point* *The break-even point for a company is the point where its costs equal its revenues. If both cost and revenue are expressed as linear equations, the break-even point is the solution of a linear system. In each exercise, C represents the cost in dollars to produce x items, and R represents the revenue in dollars from the sale of x items. Use the substitution method to find the break-even point in each case, that is, the point where C = R. Then, find the value of C and R at that point.*

**96.** $C = 20x + 10{,}000$
$R = 30x - 11{,}000$

**97.** $C = 4x + 125$
$R = 9x - 200$

*Equilibrium Supply and Demand* *In each exercise, p is the price of an item in dollars, while q represents the supply in one equation and the demand in the other. Find the equilibrium price and the equilibrium supply/demand.*

**98.** $p = 80 - \frac{3}{5}q$
$p = \frac{2}{5}q$

**99.** $p = 630 - \frac{3}{4}q$
$p = \frac{3}{4}q$

**100.** Let the supply and demand equations for banana smoothies be

$$\text{supply: } p = \frac{3}{2}q \quad \text{and} \quad \text{demand: } p = 81 - \frac{3}{4}q.$$

**(a)** Graph these equations on the same axes. Let $x = q$, $y = p$, and use the window $[0, 120]$ by $[0, 100]$. Find the point of intersection.

**(b)** Find the equilibrium demand analytically.

**(c)** Find the equilibrium price analytically.

## RELATING CONCEPTS

**For individual or group investigation (Exercises 101–106)**

*The system*

$$\frac{5}{x} + \frac{15}{y} = 16$$
$$\frac{5}{x} + \frac{4}{y} = 5$$

*is not a linear system because the variables appear in the denominators. However, it can be solved in a manner similar to the method for solving a linear system by using a substitution-of-variable technique. Let $t = \frac{1}{x}$ and let $u = \frac{1}{y}$.*

**101.** Write a system of equations in $t$ and $u$ by making the appropriate substitutions.

**102.** Solve the system in Exercise 101 for $t$ and $u$.

**103.** Solve the given system for $x$ and $y$ by using the equations relating $t$ and $x$, and $u$ and $y$.

**104.** Refer to the first equation in the given system, and solve for $y$ in terms of $x$ to obtain a rational function.

**105.** Repeat Exercise 104 for the second equation in the given system.

**106.** Using a viewing window of $[0, 10]$ by $[0, 2]$, show that the point of intersection of the graphs of the functions in Exercises 104 and 105 has the same $x$- and $y$-values as found in Exercise 103.

*Use the substitution-of-variable technique from the preceding Relating Concepts exercises to solve each system analytically.*

**107.** $\frac{2}{x} + \frac{1}{y} = \frac{3}{2}$
$\frac{3}{x} - \frac{1}{y} = 1$

**108.** $\frac{2}{x} + \frac{1}{y} = 11$
$\frac{3}{x} - \frac{5}{y} = 10$

# 7.2 SOLUTION OF LINEAR SYSTEMS BY THE ECHELON METHOD

Geometric Considerations • Analytic Solution of Systems in Three Variables • Applications of Systems • Curve Fitting Using a System

## Geometric Considerations

Our work with systems of equations so far has dealt strictly with systems in two variables. We can extend the ideas of systems of equations to linear equations of the form $Ax + By + Cz = D$. A solution of such an equation is called an **ordered triple** and is denoted $(x, y, z)$. For example, $(1, 2, -4)$ is a solution of $2x + 5y - 3z = 24$. The solution set of such an equation is an infinite set of ordered triples.

In geometry, the graph of a linear equation in three variables is a plane in three-dimensional space. Considering the possible intersections of the planes representing three equations in three unknowns shows that the solution set of such a system may be either a single ordered triple $(x, y, z)$, an infinite set of ordered triples (dependent equations), or the empty set (an inconsistent system). See Figure 15.

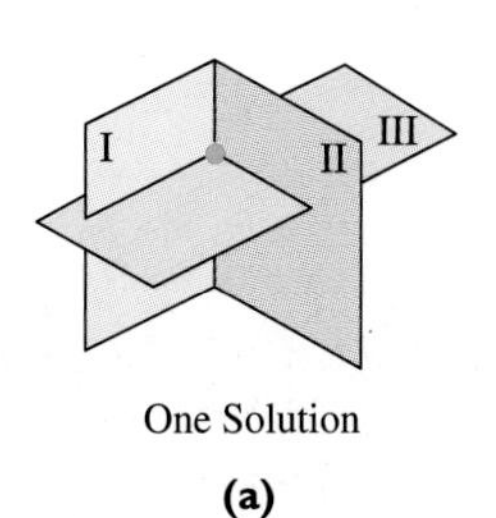

One Solution

(a)

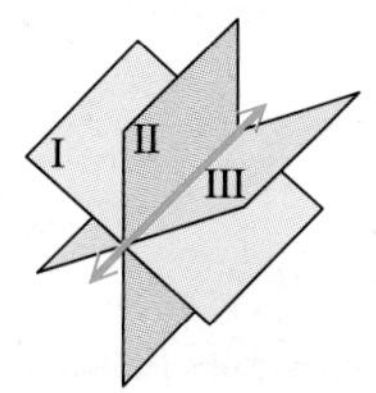

Infinitely Many Solutions

(b)

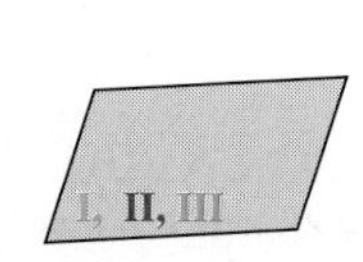

Infinitely Many Solutions

(c)

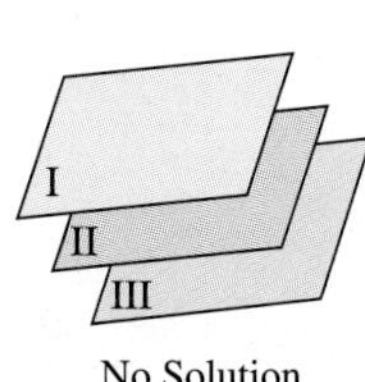

No Solution

(d)

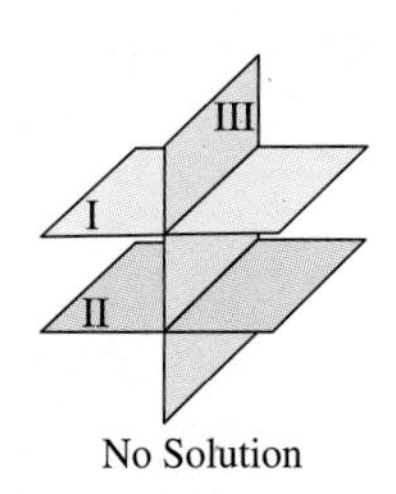

No Solution

(e)

**FIGURE 15**

## Analytic Solution of Systems in Three Variables

A linear system of equations can be solved by repeated use of the elimination method, using the following transformations.

> **Transformations of a System**
>
> To transform a system of linear equations into an equivalent system, apply the following.
>
> **1.** Interchange any two equations.
>
> **2.** Multiply each side of an equation by any nonzero real number.
>
> **3.** Replace any equation by a nonzero multiple of that equation plus a nonzero multiple of any other equation.

A systematic approach for using these transformations to solve a system is called the **echelon method.** The goal of the echelon method is to use the transformations to rewrite the equations of the system until the system has a triangular form. For a system of three equations in three variables, for example, the system should have the form

$$\begin{aligned} x + ay + bz &= c \\ y + dz &= e \\ z &= f, \end{aligned}$$

where $a$, $b$, $c$, $d$, $e$, and $f$ are constants. Then the value of $z$ from the third equation can be substituted back into the second equation to find $y$, and the values of $y$ and $z$ can be substituted into the first equation to find $x$. This is called **back-substitution.**

**EXAMPLE 1** *Solving a System of Three Equations by the Echelon Method*

Solve the system.

$$\begin{aligned} 2x + y - z &= 2 \qquad (1) \\ x + 3y + 2z &= 1 \qquad (2) \\ x + y + z &= 2 \qquad (3) \end{aligned}$$

**Solution** In the first equation, $x$ should have coefficient 1. Use transformation 1 to interchange equations (1) and (2) to get the system

$$\begin{aligned} x + 3y + 2z &= 1 \qquad (2) \\ 2x + y - z &= 2 \qquad (1) \\ x + y + z &= 2. \qquad (3) \end{aligned}$$

Next use transformation 3 to eliminate the $x$-term in equations (1) and (3). Multiply equation (2) by $-2$ and add the results to equation (1) to get an equivalent system. We indicate this process by the notation $-2R_1 + R_2 \rightarrow R_2$. (R stands for row.) Similarly, multiplying equation (2) by $-1$ and adding the results to equation (3) eliminates the $x$-term there.

$$\begin{aligned} x + 3y + 2z &= 1 && \qquad (2) \\ -5y - 5z &= 0 && -2R_1 + R_2 \rightarrow R_2 \qquad (4) \\ -2y - z &= 1 && -1R_1 + R_3 \rightarrow R_3 \qquad (5) \end{aligned}$$

Use transformation 2 to get a coefficient of 1 for $y$ in equation (4); multiply by $-\frac{1}{5}$.

$$\begin{aligned} x + 3y + 2z &= 1 && \qquad (2) \\ y + z &= 0 && -\tfrac{1}{5}R_2 \rightarrow R_2 \qquad (6) \\ -2y - z &= 1 && \qquad (5) \end{aligned}$$

To eliminate the $y$-term in equation (5), multiply equation (6) by 2 and add the results to equation (5).

$$\begin{aligned} x + 3y + 2z &= 1 && \qquad (2) \\ y + z &= 0 && \qquad (6) \\ z &= 1 && 2R_2 + R_3 \rightarrow R_3 \end{aligned}$$

The system is now in triangular form. Substitute 1 for $z$ in equation (6) to get $y = -1$. Finally, substitute 1 for $z$ and $-1$ for $y$ in equation (2) to get $x = 2$. Verify that the ordered triple $(2, -1, 1)$ satisfies the given equations (1)–(3). The solution set of the system is $\{(2, -1, 1)\}$.

Although the echelon method may seem confusing at first, it is systematic and leads to the matrix method we discuss in the next section.

**EXAMPLE 2** *Solving a System of Two Equations with Three Variables*

Solve the system.

$$x + 2y + z = 4 \quad (1)$$
$$3x - y - 4z = -9 \quad (2)$$

**Solution** Geometrically, the solution is the intersection of the two planes given by equations (1) and (2). The intersection of two different nonparallel planes is a line. Thus, if these two planes are not parallel, there will be an infinite number of ordered triples in the solution set, representing the points on the line of intersection.

$$x + 2y + z = 4 \quad (1)$$
$$-7y - 7z = -21 \quad -3R_1 + R_2 \rightarrow R_2$$

$$x + 2y + z = 4 \quad (1)$$
$$y + z = 3 \quad -\tfrac{1}{7}R_2 \rightarrow R_2$$

This is as far as we can go with the echelon method. We solve $y + z = 3$ for $y$ to get $y = 3 - z$. Now, we express $x$ in terms of $z$ by solving equation (1) for $x$ and substituting $3 - z$ for $y$.

$$x + 2y + z = 4 \quad (1)$$
$$x = -2y - z + 4$$
$$x = -2(3 - z) - z + 4 \quad \text{Let } y = 3 - z.$$
$$x = z - 2$$

The solution set is written $\{(z - 2, 3 - z, z)\}$. The system has an infinite number of solutions. For any arbitrary value of $z$, $y = 3 - z$ and $x = z - 2$. For example, if $z = 1$, then $y = 3 - 1 = 2$ and $x = 1 - 2 = -1$, giving the solution $(-1, 2, 1)$. Verify that another solution is $(0, 1, 2)$.

A system like the one in Example 2 occurs when two of the equations in a system of three equations with three variables are dependent. In such a case, there are really only two equations in three variables, and Example 2 illustrates the method of solution. On the other hand, an inconsistent system is indicated by a false statement at some point in the solution, as in Example 3 of Section 7.1.

## Applications of Systems

**EXAMPLE 3** *Solving a System of Equations to Satisfy Feed Requirements*

An animal feed is made from three ingredients: corn, soybeans, and cottonseed. One unit of each ingredient provides units of protein, fat, and fiber, as shown in the table. How many units of each ingredient should be used to make a feed that contains 22 units of protein, 28 units of fat, and 18 units of fiber?

| | Corn | Soybeans | Cottonseed | Total |
|---|---|---|---|---|
| Protein | .25 | .4 | .2 | 22 |
| Fat | .4 | .2 | .3 | 28 |
| Fiber | .3 | .2 | .1 | 18 |

**Solution** Let $x$ represent the number of units of corn, $y$ the number of units of soybeans, and $z$ the number of units of cottonseed that are required. Since the total amount of protein must be 22 units,

$$.25x + .4y + .2z = 22.$$

Also, for the 28 units of fat,

$$.4x + .2y + .3z = 28,$$

and, for the 18 units of fiber,

$$.3x + .2y + .1z = 18.$$

To clear decimals, multiply each side of the first equation by 100, and each side of the second and third equations by 10 to get the system

$$\begin{aligned} 25x + 40y + 20z &= 2200 \\ 4x + 2y + 3z &= 280 \\ 3x + 2y + z &= 180. \end{aligned}$$

Using the methods described in this section, show that $x = 40$, $y = 15$, and $z = 30$. The feed should contain 40 units of corn, 15 units of soybeans, and 30 units of cottonseed to fulfill the given requirements.

**NOTE** The table in Example 3 is useful in setting up the equations of the system, since the coefficients in each equation can be read from left to right. We extend this idea in the next section, where we introduce solution of systems by matrices.

### EXAMPLE 4 *Solving a Feed Requirements Application with Fewer Equations Than Variables*

In Example 3, suppose that only the fat and fiber content of the feed are of interest. How would the solution be changed?

**Solution** We would need to solve the system

$$\begin{aligned} 4x + 2y + 3z &= 280 \\ 3x + 2y + z &= 180, \end{aligned}$$

which does not have a single, unique solution since there are two equations with three variables. Following the procedure outlined in Example 2, we find that the solution set of this system is $\{(100 - 2z, 2.5z - 60, z)\}$. In this applied problem, however, all three variables must be nonnegative, so $z$ must satisfy the conditions

$$100 - 2z \geq 0, \quad 2.5z - 60 \geq 0, \quad z \geq 0.$$

From the first inequality, $z \leq 50$; from the second inequality, $z \geq 24$; thus, $24 \leq z \leq 50$. Only solutions with $z$ in this range are usable.

## Curve Fitting Using a System

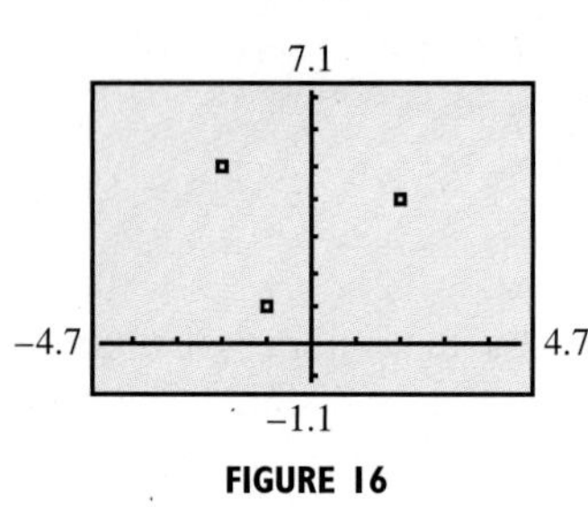

FIGURE 16

Recall from Chapter 3 that the graph of a quadratic function is a parabola with a vertical axis. Given three noncollinear points, we can find the equation of a parabola that passes through them.

**EXAMPLE 5** *Using a System to Fit a Parabola to Three Data Points*

Figure 16 shows three data points: (2, 4), (−1, 1), and (−2, 5). Find the equation of the parabola with vertical axis that passes through these points.

### Analytic Solution

Since the three points lie on the graph of the equation $y = ax^2 + bx + c$, they must satisfy the equation. Substituting each ordered pair into the equation gives three equations with three variables.

$$4 = a(2)^2 + b(2) + c \quad \text{or} \quad 4 = 4a + 2b + c \quad (1)$$
$$1 = a(-1)^2 + b(-1) + c \quad \text{or} \quad 1 = a - b + c \quad (2)$$
$$5 = a(-2)^2 + b(-2) + c \quad \text{or} \quad 5 = 4a - 2b + c \quad (3)$$

This system can be solved by the elimination method. First, eliminate $c$, using equations (1) and (2).

$$4 = 4a + 2b + c \quad (1)$$
$$-1 = -a + b - c \quad \text{Multiply (2) by } -1.$$
$$3 = 3a + 3b \quad (4)$$

Now, use equations (2) and (3) to also eliminate $c$.

$$1 = a - b + c \quad (2)$$
$$-5 = -4a + 2b - c \quad \text{Multiply (3) by } -1.$$
$$-4 = -3a + b \quad (5)$$

Solve the system of equations (4) and (5) in two variables by eliminating $a$.

$$3 = 3a + 3b$$
$$-4 = -3a + b$$
$$-1 = 4b$$
$$-\frac{1}{4} = b$$

Find $a$ by substituting $-\frac{1}{4}$ for $b$ in equation (4), which is equivalent to $1 = a + b$.

$$1 = a + b \quad \text{Multiply (4) by } \tfrac{1}{3}.$$
$$1 = a - \frac{1}{4} \quad \text{Let } b = -\tfrac{1}{4}.$$
$$\frac{5}{4} = a$$

*(continued)*

### Graphing Calculator Solution

Figure 17 shows that when

$$Y_1 = 1.25X^2 - .25X - .5$$

is graphed, the parabola contains the three data points.

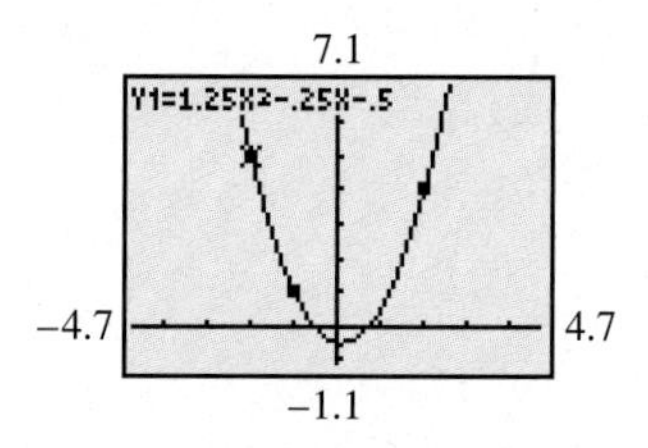

FIGURE 17

The table in Figure 18 further supports the fact that the three points lie on the graph of $Y_1$.

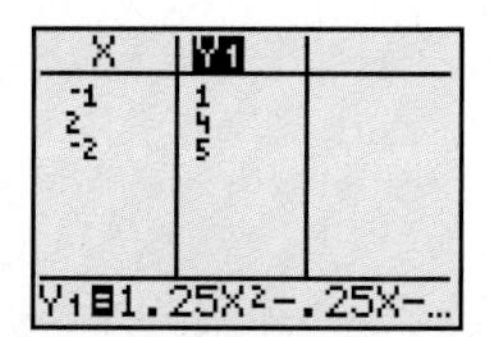

| X | Y1 |
|---|---|
| -1 | 1 |
| 2 | 4 |
| -2 | 5 |

Y1=1.25X²-.25X-...

FIGURE 18

The discussion above confirms that the result found in the analytic solution is correct. A graphing calculator with quadratic regression capability can be used to find the

Finally, find $c$ by substituting $a = \frac{5}{4}$ and $b = -\frac{1}{4}$ in equation (2).

$$1 = a - b + c$$

$$1 = \frac{5}{4} - \left(-\frac{1}{4}\right) + c \qquad a = \frac{5}{4},\ b = -\frac{1}{4}$$

$$1 = \frac{6}{4} + c$$

$$-\frac{1}{2} = c$$

An equation of the parabola is

$$y = \frac{5}{4}x^2 - \frac{1}{4}x - \frac{1}{2}, \quad \text{or equivalently,} \quad y = 1.25x^2 - .25x - .5.$$

quadratic function that exactly fits these three data points. See Section 3.4 for more information on this procedure.

**FOR DISCUSSION**

Suppose you want to find a polynomial function of the form $y = P(x)$ with a graph that goes through the points represented by four given ordered pairs. How would you proceed? What degree polynomial would $P$ be? If you were given five ordered pairs, what degree polynomial would $P$ be? In a practical problem of this type with a set of at least five points, how could you decide which degree polynomial would best fit the data?

## 7.2 EXERCISES

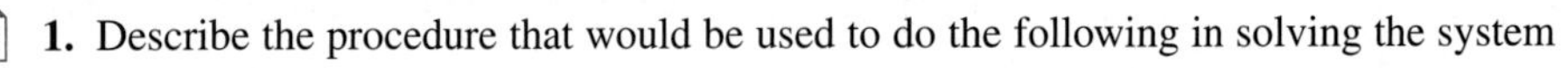

**1.** Describe the procedure that would be used to do the following in solving the system

$$\begin{aligned} x + 2y - 3z &= -10 \qquad (1) \\ 3x - y + z &= 5 \qquad (2) \\ 2x + y - 4z &= -18. \qquad (3) \end{aligned}$$

**(a)** Eliminate $y$ using equations (1) and (2) by multiplying equation (2) by an appropriate constant.

**(b)** Eliminate $x$ using equations (1) and (3) by multiplying equation (1) by an appropriate constant.

**(c)** Eliminate $z$ using equations (2) and (3) by multiplying equation (2) by an appropriate constant.

*Verify that the given ordered triple is a solution of the system by substituting the given values of x, y, and z into each equation.*

**2.** $(-3, 6, 1)$

$$\begin{aligned} 2x + y - z &= -1 \\ x - y + 3z &= -6 \\ -4x + y + z &= 19 \end{aligned}$$

**3.** $\left(\frac{1}{2}, -\frac{3}{4}, \frac{1}{6}\right)$

$$\begin{aligned} 2x + 8y - 6z &= -6 \\ x + y + z &= -\frac{1}{12} \\ x + 3z &= 1 \end{aligned}$$

**4.** $(-.2, .4, .5)$

$$\begin{aligned} 5x - y + 2z &= -.4 \\ x + 4z &= 1.8 \\ -3y + z &= -.7 \end{aligned}$$

*Solve each system analytically.* (Hint: *In Exercises 25–28, let* $t = \frac{1}{x}$, $u = \frac{1}{y}$, *and* $v = \frac{1}{z}$. *Solve for t, u, and v, and then solve for x, y, and z.*)

**5.** $\begin{aligned} x + y + z &= 2 \\ 2x + y - z &= 5 \\ x - y + z &= -2 \end{aligned}$

**6.** $\begin{aligned} 2x + y + z &= 9 \\ -x - y + z &= 1 \\ 3x - y + z &= 9 \end{aligned}$

**7.** $\begin{aligned} x + 3y + 4z &= 14 \\ 2x - 3y + 2z &= 10 \\ 3x - y + z &= 9 \end{aligned}$

**8.** $\begin{aligned} 4x - y + 3z &= -2 \\ 3x + 5y - z &= 15 \\ -2x + y + 4z &= 14 \end{aligned}$

**9.** $\begin{aligned} x + 2y + 3z &= 8 \\ 3x - y + 2z &= 5 \\ -2x - 4y - 6z &= 5 \end{aligned}$

**10.** $\begin{aligned} 3x - 2y - 8z &= 1 \\ 9x - 6y - 24z &= -2 \\ x - y + z &= 1 \end{aligned}$

**11.** $\begin{aligned} x + 4y - z &= 6 \\ 2x - y + z &= 3 \\ 3x + 2y + 3z &= 16 \end{aligned}$

**12.** $\begin{aligned} 4x - 3y + z &= 9 \\ 3x + 2y - 2z &= 4 \\ x - y + 3z &= 5 \end{aligned}$

**13.** $\begin{aligned} 5x + y - 3z &= -6 \\ 2x + 3y + z &= 5 \\ -3x - 2y + 4z &= 3 \end{aligned}$

**14.** $\begin{aligned} 2x - 5y + 4z &= -35 \\ 5x + 3y - z &= 1 \\ x + y + z &= 1 \end{aligned}$

**15.** $\begin{aligned} x - 3y - 2z &= -3 \\ 3x + 2y - z &= 12 \\ -x - y + 4z &= 3 \end{aligned}$

**16.** $\begin{aligned} x + y + z &= 3 \\ 3x - 3y - 4z &= -1 \\ x + y + 3z &= 11 \end{aligned}$

**17.** $\begin{aligned} 2x + 6y - z &= 6 \\ 4x - 3y + 5z &= -5 \\ 6x + 9y - 2z &= 11 \end{aligned}$

**18.** $\begin{aligned} 8x - 3y + 6z &= -2 \\ 4x + 9y + 4z &= 18 \\ 12x - 3y + 8z &= -2 \end{aligned}$

**19.** $\begin{aligned} x + z &= 4 \\ x + y &= 4 \\ y + z &= 4 \end{aligned}$

**20.** $\begin{aligned} x - z &= 2 \\ x + y &= -3 \\ y - z &= 1 \end{aligned}$

**21.** $\begin{aligned} 2x + y - z &= -4 \\ y + 2z &= 12 \\ 2x - z &= -4 \end{aligned}$

**22.** $\begin{aligned} 3x + 2y - z &= -1 \\ 3y + z &= 12 \\ x - 3z &= -3 \end{aligned}$

**23.** $\begin{aligned} 2x + 3y + 4z &= 3 \\ 6x + 3y + 8z &= 6 \\ 6y - 4z &= 1 \end{aligned}$

**24.** $\begin{aligned} 10x + 2y - 3z &= 0 \\ 5x + 4y + 6z &= -1 \\ 6y + 3z &= 2 \end{aligned}$

**25.** $\begin{aligned} \frac{1}{x} + \frac{1}{y} - \frac{1}{z} &= \frac{1}{4} \\ \frac{2}{x} - \frac{1}{y} + \frac{3}{z} &= \frac{9}{4} \\ -\frac{1}{x} - \frac{2}{y} + \frac{4}{z} &= 1 \end{aligned}$

**26.** $\begin{aligned} \frac{3}{x} + \frac{2}{y} - \frac{1}{z} &= \frac{11}{6} \\ \frac{1}{x} - \frac{1}{y} + \frac{3}{z} &= -\frac{11}{12} \\ \frac{2}{x} + \frac{1}{y} + \frac{1}{z} &= \frac{7}{12} \end{aligned}$

**27.** $\begin{aligned} \frac{2}{x} - \frac{2}{y} + \frac{1}{z} &= -1 \\ \frac{4}{x} + \frac{1}{y} - \frac{2}{z} &= -9 \\ \frac{1}{x} + \frac{1}{y} - \frac{3}{z} &= -9 \end{aligned}$

**28.** $\begin{aligned} \frac{5}{x} - \frac{1}{y} - \frac{2}{z} &= -6 \\ -\frac{1}{x} + \frac{3}{y} - \frac{3}{z} &= -12 \\ \frac{2}{x} - \frac{1}{y} - \frac{1}{z} &= 6 \end{aligned}$

**29.** *Concept Check* Consider the linear equation in three variables $x + y + z = 4$. Find a pair of linear equations that, when considered together with the given equation, will form a system having **(a)** exactly one solution, **(b)** no solution, **(c)** infinitely many solutions.

**30.** *Concept Check* Refer to Example 2 in this section. Write the solution set with $x$ arbitrary.

**31.** Give an example, using your immediate surroundings, of three planes that intersect in a single point.

**32.** Give an example, using your immediate surroundings, of three planes that intersect in a line.

*Solve each system in terms of the arbitrary variable z.*

**33.** $\begin{aligned} x - 2y + 3z &= 6 \\ 2x - y + 2z &= 5 \end{aligned}$

**34.** $\begin{aligned} 3x + 4y - z &= 13 \\ x + y + 2z &= 15 \end{aligned}$

**35.** $\begin{aligned} 5x - 4y + z &= 9 \\ x + y &= 15 \end{aligned}$

**36.** $\begin{aligned} 3x - 5y - 4z &= -7 \\ y - z &= -13 \end{aligned}$

**37.** $\begin{aligned} x - y + z &= -6 \\ 4x + y + z &= 7 \end{aligned}$

**38.** $\begin{aligned} 3x - 2y + z &= 15 \\ x + 4y - z &= 11 \end{aligned}$

*Use a system of equations in three variables to solve each problem.*

**39.** *Mixing Waters* A sparkling-water distributor wants to make up 300 gallons of sparkling water to sell for \$6.00 per gallon. She wishes to mix three grades of water selling for \$9.00, \$3.00, and \$4.50 per gallon, respectively. She must use twice as much of the \$4.50 water as the \$3.00 water. How many gallons of each should she use?

**40.** *Coin Collecting* A coin collection contains a total of 29 coins, made up of pennies, nickels, and quarters. The number of quarters is 8 less than the number of pennies. The total face value of the coins is \$1.77. How many of each denomination are there?

**41.** *Pricing Concert Tickets* Frank Capek and his Generation Gap group sells three kinds of concert tickets: "up close," "middle," and "farther back." "Up close" tickets cost \$6 more than "middle" tickets, while "middle" tickets cost \$3 more than "farther back" tickets. Twice the cost of an "up close" ticket is \$3 more than 3 times the cost of a "farther back" seat. Find the price of each kind of ticket.

**42.** *Mixing Glue* A glue company needs to make some glue that it can sell for \$120 per barrel. It wants to use 150 barrels of glue worth \$100 per barrel, along with some glue worth \$150 per barrel, and glue worth \$190 per barrel. It must use the same number of barrels of \$150 and \$190 glue. How much of the \$150 and \$190 glue will be needed? How many barrels of \$120 glue will be produced?

**43.** *Triangle Dimensions* The sum of the measures of the angles of any triangle is 180°. In a certain triangle, the largest angle measures 55° less than twice the medium angle, and the smallest measures 25° less than the medium angle. Find the measures of each of the three angles.

**44.** *Triangle Dimensions* The perimeter of a triangle is 59 inches. The longest side is 11 inches longer than the medium side, and the medium side is 3 inches more than the shortest side. Find the length of each side of the triangle. See the figure.

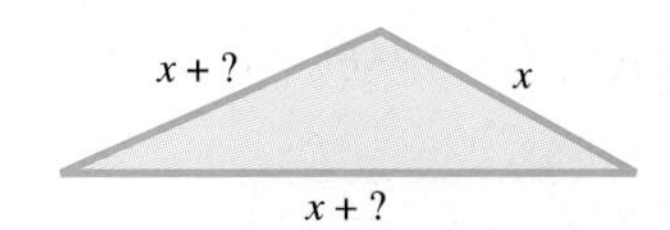


**45.** *Investment Decisions* Norma James invested \$10,000 received in an inheritance in three parts. With one part she bought mutual funds that offered a return of 4% per year. The second part, which amounted to twice the first, was used to buy government bonds paying 4.5% per year. She put the rest into a savings account that paid 2.5% annual interest. During the first year, the total interest was \$415. How much did Norma invest at each rate?

**46.** *Investment Decisions* Tom Accardo won \$100,000 in the Louisiana state lottery. He invested part of the money in real estate with an annual return of 5% and another part in a money market account at 4.5% interest. He invested the rest, which amounted to \$20,000 less than the sum of the other two parts, in certificates of deposit that pay 3.75%. If the total annual interest on the money was \$4450, how much was invested at each rate?

**47.** *Scheduling Deliveries* Doctor Rug sells rug cleaning machines. The EZ model weighs 10 pounds and comes in a 10-cubic-foot box. The compact model weighs 20 pounds and comes in an 8-cubic-foot box. The commercial model weighs 60 pounds and comes in a 28-cubic-foot box. Each of their delivery vans has 248 cubic feet of space and can hold a maximum of 440 pounds. In order for a van to be fully loaded, how many of each model should it carry?

**48.** *Scheduling Production* Ciolino's Furniture makes dining room furniture. A buffet requires 30 hours for construction and 10 hours for finishing. A chair requires 10 hours for construction and 10 hours for finishing. A table requires 10 hours for construction and 30 hours for finishing. The construction department has 350 hours of labor, and the finishing department has 150 hours of labor available each week. How many pieces of each type of furniture should be produced each week if the factory is to run at full capacity?

*Curve Fitting* *Find the equation of the parabola with vertical axis that passes through the data points specified. Then support your answer with a graph.*

**49.** $(2, 9), (-2, 1), (-3, 4)$

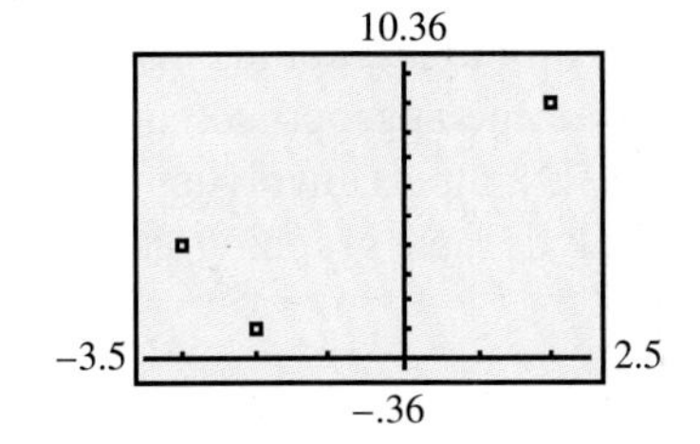


**50.** $(1.5, 6.25), (0, -2), (-1.5, 3.25)$

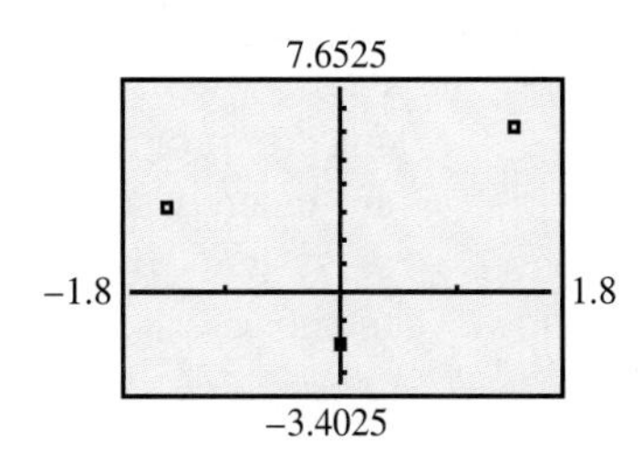

**51.** $(2, 14), (0, 0), (-1, -1)$

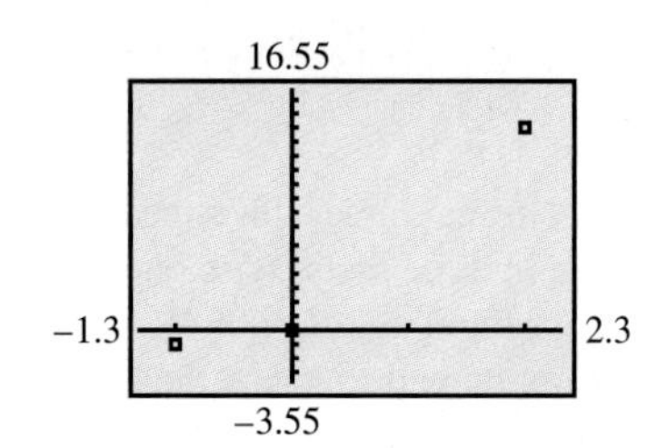


**52.** $(2, 3), (-1, 0), (-2, 2)$

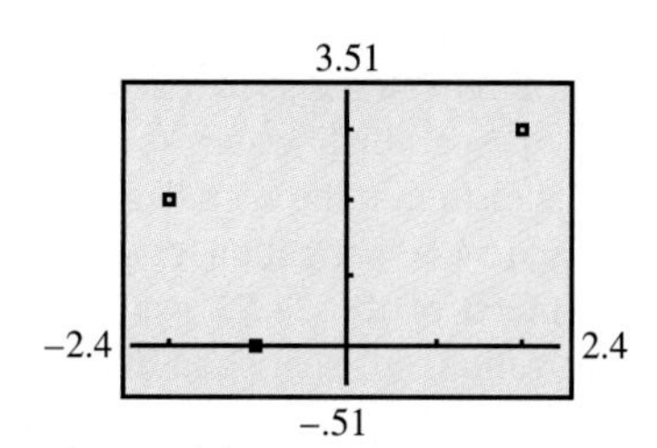


*Curve Fitting* *Given three noncollinear points, there is one and only one circle that passes through them. Knowing that the equation of a circle may be written in the form*

$$x^2 + y^2 + ax + by + c = 0,$$

*find the equation of each circle described or graphed in Exercises 53 and 54.*

**53.** Passing through the points $(2, 1)$, $(-1, 0)$, and $(3, 3)$

**54.**

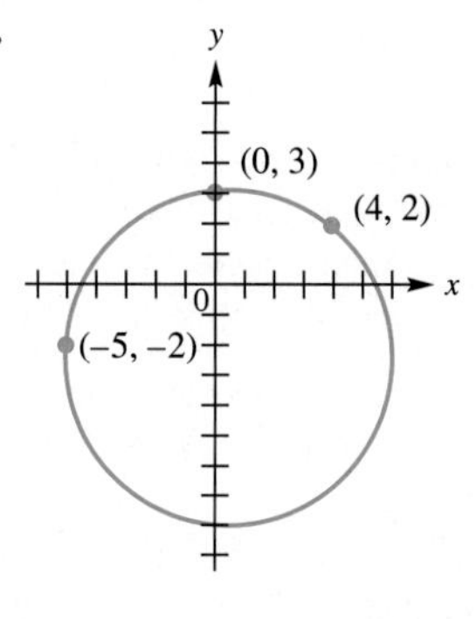


*Position of a Particle* *Suppose that the position of a particle moving along a straight line is given by*

$$s(t) = at^2 + bt + c,$$

*where t is time in seconds and a, b, and c are real numbers.*

**55.** If $s(0) = 5$, $s(1) = 23$, and $s(2) = 37$, find the equation that defines $s(t)$. Then, find $s(8)$.

**56.** If $s(0) = -10$, $s(1) = 6$, and $s(2) = 30$, find the equation that defines $s(t)$. Then, find $s(10)$.

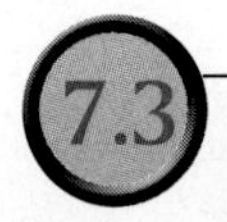

# 7.3 SOLUTION OF LINEAR SYSTEMS BY ROW TRANSFORMATIONS

Matrices and Technology • Matrix Row Transformations • Row Echelon Method • Reduced Row Echelon Method • Special Cases • An Application

**TECHNOLOGY NOTE**

The manner in which matrices are entered and displayed varies greatly among different manufacturers and even among the various models manufactured by the same corporation. As always, refer to the graphing calculator manual that accompanies this text or your owner's manual when necessary.

## Matrices and Technology

In this section and Section 7.6, we show methods for solving linear systems with matrices. Matrix methods are particularly suitable for calculator or computer solutions of large systems of equations that have many unknowns. Although graphing calculators are programmed to do the necessary computations when the appropriate commands are entered, a knowledge of the concepts is essential. Examples involving solution of systems by matrices are presented using both analytic and graphing calculator methods.

## Matrix Row Transformations

The echelon method used to solve linear systems of equations in the previous section can be streamlined using *matrices* (singular: *matrix*). Consider a system of three equations and three unknowns.

$$\begin{aligned} a_1x + b_1y + c_1z &= d_1 \\ a_2x + b_2y + c_2z &= d_2 \\ a_3x + b_3y + c_3z &= d_3 \end{aligned} \quad \text{can be written as} \quad \begin{bmatrix} a_1 & b_1 & c_1 & d_1 \\ a_2 & b_2 & c_2 & d_2 \\ a_3 & b_3 & c_3 & d_3 \end{bmatrix}.$$

Such a rectangular array of numbers enclosed by brackets is called a **matrix.** Each number in the array is an **element** or **entry**. The constants in the last column of the matrix can be set apart from the coefficients of the variables with a vertical line, as shown in the following **augmented matrix**. (Because the matrix of coefficients has an extra column determined by the constants of the system, the coefficient matrix is *augmented.*)

$$\text{Rows} \rightarrow \left[\begin{array}{ccc|c} a_1 & b_1 & c_1 & d_1 \\ a_2 & b_2 & c_2 & d_2 \\ a_3 & b_3 & c_3 & d_3 \end{array}\right]$$

Columns

As an example, the system on the left has the augmented matrix shown on the right.

$$\begin{aligned} x + 3y + 2z &= 1 \\ 2x + y - z &= 2 \\ x + y + z &= 2 \end{aligned} \qquad \left[\begin{array}{ccc|c} 1 & 3 & 2 & 1 \\ 2 & 1 & -1 & 2 \\ 1 & 1 & 1 & 2 \end{array}\right]$$

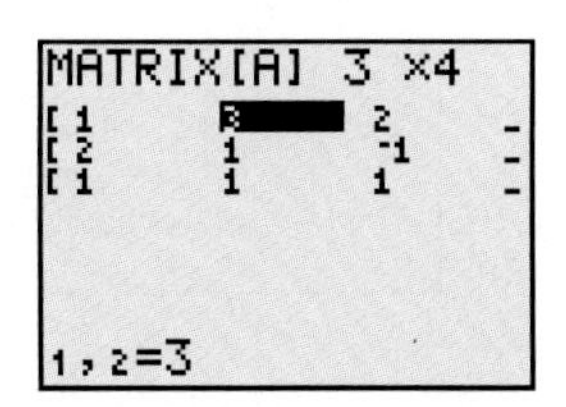


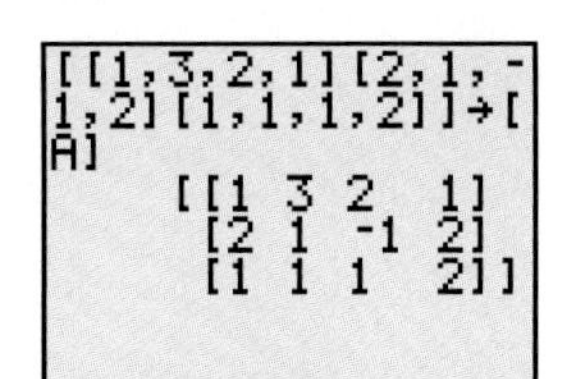


**FIGURE 19**

Figure 19 shows two screens indicating how this matrix can be entered into a graphing calculator. It has 3 rows (horizontal) and 4 columns (vertical), so it is a $3 \times 4$ (read "three by four") matrix. To refer to a number in the matrix, use its row and column numbers. For example, the number 3 is in the first row, second column.

The rows of an augmented matrix can be treated just like the equations of the corresponding system of linear equations. Since the augmented matrix is nothing more than a short form of the system, any transformation of the matrix that results in an equivalent system of equations can be performed.

**TECHNOLOGY NOTE**

Consult the graphing calculator manual that accompanies this text or your owner's manual to see how these transformations are accomplished with your model.

### Matrix Row Transformations

For any augmented matrix of a system of linear equations, the following row transformations will result in the matrix of an equivalent system.

1. Any two rows may be interchanged.
2. The elements of any row may be multiplied by a nonzero real number.
3. Any row may be changed by adding to its elements a multiple of the corresponding elements of another row.

Notice how these transformations relate to the transformations of a system discussed in the previous section.

**EXAMPLE 1** *Using Row Transformations*

Use the matrix row transformations to transform

$$\left[\begin{array}{rrr|r} 1 & 3 & 5 & 2 \\ 0 & 1 & 2 & 3 \\ 1 & -1 & -2 & 4 \end{array}\right]$$

to each matrix.

**(a)** $\left[\begin{array}{rrr|r} 0 & 1 & 2 & 3 \\ 1 & 3 & 5 & 2 \\ 1 & -1 & -2 & 4 \end{array}\right]$ **(b)** $\left[\begin{array}{rrr|r} -2 & -6 & -10 & -4 \\ 0 & 1 & 2 & 3 \\ 1 & -1 & -2 & 4 \end{array}\right]$ **(c)** $\left[\begin{array}{rrr|r} 1 & 3 & 5 & 2 \\ 0 & 1 & 2 & 3 \\ 0 & -4 & -7 & 2 \end{array}\right]$

### Analytic Solution

**(a)** Interchange the first two rows in the given matrix to obtain

$$\left[\begin{array}{rrr|r} 0 & 1 & 2 & 3 \\ 1 & 3 & 5 & 2 \\ 1 & -1 & -2 & 4 \end{array}\right]. \qquad \begin{array}{l} R_2 \rightarrow R_1 \\ R_1 \rightarrow R_2 \end{array}$$

**(b)** Multiply the elements of the first row by $-2$ to obtain

$$\left[\begin{array}{rrr|r} -2 & -6 & -10 & -4 \\ 0 & 1 & 2 & 3 \\ 1 & -1 & -2 & 4 \end{array}\right]. \qquad -2R_1$$

**(c)** Multiply each element of the first row by $-1$, and add it to the corresponding element in the third row. For example, to obtain 0 in row 3, column 1, multiply 1 by $-1$ and add to 1.

$$1(-1) + 1 = 0$$

The final result is

$$\left[\begin{array}{rrr|r} 1 & 3 & 5 & 2 \\ 0 & 1 & 2 & 3 \\ 0 & -4 & -7 & 2 \end{array}\right]. \qquad -1R_1 + R_3$$

(Rows 1 and 2 are unchanged.)

### Graphing Calculator Solution

Figure 20 shows the given augmented matrix, designated [A].

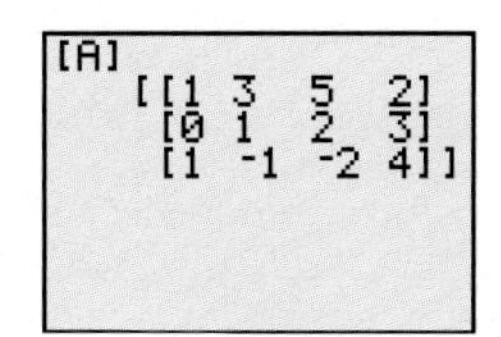


FIGURE 20

The TI-83 calculator uses the commands indicated in the screens below to perform the desired transformations.

**(a)**

```
rowSwap([A],1,2)
  [[0 1  2  3]
   [1 3  5  2]
   [1 -1 -2 4]]
```

**(b)**

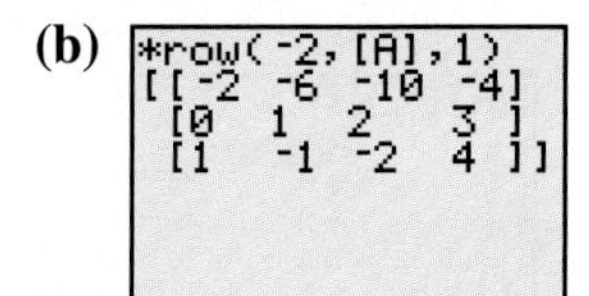


**(c)**

```
*row+(-1,[A],1,3
)
  [[1 3  5  2]
   [0 1  2  3]
   [0 -4 -7 2]]
```

## Row Echelon Method

$$\left[\begin{array}{rrr|r} 1 & 5 & 3 & 7 \\ 0 & 1 & 2 & 9 \\ 0 & 0 & 1 & 4 \end{array}\right]$$

Matrix row transformations are used to transform the augmented matrix of a system into one that is in **echelon** (or **triangular**) **form**. In this form, the matrix will have 1s down the diagonal from upper left to lower right and 0s below each 1, as shown in the matrix in the margin. The 1s lie on the **main diagonal** of the matrix.

Before using matrices to solve a linear system, we must arrange the system in the proper form, with variable terms on the left side of the equation and constant terms on the right. The variable terms must be in the same order in each equation. This method of using matrices to solve a system of linear equations is called the **row echelon method**. It is closely related to the echelon method discussed in the previous section.

**EXAMPLE 2** *Solving a System by the Row Echelon Method*

Solve the system.

$$5x + 2y = 1$$
$$2x - y = 4$$

### Analytic Solution

The augmented matrix for the system is

$$\left[\begin{array}{cc|c} 5 & 2 & 1 \\ 2 & -1 & 4 \end{array}\right].$$

Transform the first row so that the first entry is 1. To do this, multiply row 1 by $\frac{1}{5}$.

$$\left[\begin{array}{cc|c} 1 & \frac{2}{5} & \frac{1}{5} \\ 2 & -1 & 4 \end{array}\right] \quad \frac{1}{5}R_1$$

Now, transform so that the entry below the main diagonal (that is, 2) is 0. Multiply row 1 by $-2$ and add to row 2. Leave row 1 unchanged.

$$\left[\begin{array}{cc|c} 1 & \frac{2}{5} & \frac{1}{5} \\ 0 & -\frac{9}{5} & \frac{18}{5} \end{array}\right] \quad -2R_1 + R_2$$

Finally, multiply row 2 by $-\frac{5}{9}$ to get a 1 in the main diagonal.

$$\left[\begin{array}{cc|c} 1 & \frac{2}{5} & \frac{1}{5} \\ 0 & 1 & -2 \end{array}\right] \quad -\frac{5}{9}R_2$$

This matrix represents the system

$$x + \frac{2}{5}y = \frac{1}{5}$$
$$y = -2.$$

From the second equation, we know that $y = -2$. Use back-substitution to find $x$.

$$x + \frac{2}{5}(-2) = \frac{1}{5}$$
$$x - \frac{4}{5} = \frac{1}{5}$$
$$x = 1$$

The solution set is $\{(1, -2)\}$.

### Graphing Calculator Solution

Figure 21 shows matrix [A], the augmented matrix of this system.

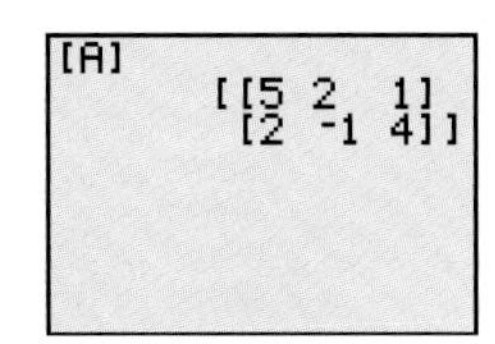


FIGURE 21

Using the ref command on the TI-83, we can find the row echelon form. See Figure 22.

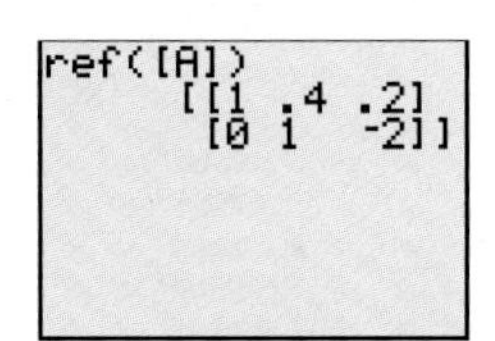


FIGURE 22

The screen in Figure 22 indicates decimal entries, but they can be converted to fractions, as shown in Figure 23.

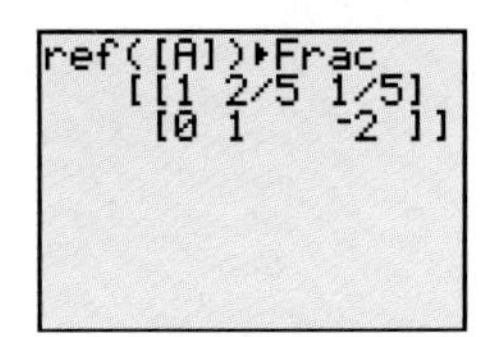


FIGURE 23

The augmented matrix in Figure 23 corresponds to the one in color in the analytic solution. The rest of the solution process is the same.

The row echelon method can be extended to larger systems. The final matrix will always have 0s below the diagonal of 1s to the left of the vertical bar. To transform the matrix, work column by column from upper left to lower right. For each column, first perform the step that gives the 1 on the main diagonal, then obtain 0s below the 1 in that column.

**EXAMPLE 3** *Solving a System by the Row Echelon Method (Analytic)*

Solve the system.

$$\begin{aligned} x - y + 5z &= -6 \\ 3x + 3y - z &= 10 \\ x + 3y + 2z &= 5 \end{aligned}$$

**Solution** First write the augmented matrix of the linear system.

$$\left[\begin{array}{ccc|c} 1 & -1 & 5 & -6 \\ 3 & 3 & -1 & 10 \\ 1 & 3 & 2 & 5 \end{array}\right]$$

There is already a 1 in row 1, column 1. Next, we must obtain 0s in the rest of column 1. First add to row 2 the results of multiplying row 1 by $-3$.

$$\left[\begin{array}{ccc|c} 1 & -1 & 5 & -6 \\ 0 & 6 & -16 & 28 \\ 1 & 3 & 2 & 5 \end{array}\right] \qquad -3R_1 + R_2$$

Now, add to row 3 the results of multiplying row 1 by $-1$.

$$\left[\begin{array}{ccc|c} 1 & -1 & 5 & -6 \\ 0 & 6 & -16 & 28 \\ 0 & 4 & -3 & 11 \end{array}\right] \qquad -1R_1 + R_3$$

To get 1 in row 2, column 2, multiply row 2 by $\frac{1}{6}$.

$$\left[\begin{array}{ccc|c} 1 & -1 & 5 & -6 \\ 0 & 1 & -\frac{8}{3} & \frac{14}{3} \\ 0 & 4 & -3 & 11 \end{array}\right] \qquad \tfrac{1}{6}R_2$$

Next, to obtain 0 in row 3, column 2, add to row 3 the results of multiplying row 2 by $-4$.

$$\left[\begin{array}{ccc|c} 1 & -1 & 5 & -6 \\ 0 & 1 & -\frac{8}{3} & \frac{14}{3} \\ 0 & 0 & \frac{23}{3} & -\frac{23}{3} \end{array}\right] \qquad -4R_2 + R_3$$

Now, multiply the last row by $\frac{3}{23}$ to obtain 1 in row 3, column 3.

$$\left[\begin{array}{ccc|c} 1 & -1 & 5 & -6 \\ 0 & 1 & -\frac{8}{3} & \frac{14}{3} \\ 0 & 0 & 1 & -1 \end{array}\right] \qquad \tfrac{3}{23}R_3$$

While not absolutely necessary, all integer elements can be obtained by multiplying row 2 by 3.

$$\left[\begin{array}{ccc|c} 1 & -1 & 5 & -6 \\ 0 & 3 & -8 & 14 \\ 0 & 0 & 1 & -1 \end{array}\right] \qquad 3R_2$$

The final matrix corresponds to the system of equations

$$\begin{aligned} x - \phantom{3}y + 5z &= -6 && (1)\\ 3y - 8z &= 14 && (2)\\ z &= -1. && (3) \end{aligned}$$

Since $z = -1$ from equation (3), use back-substitution into equation (2) to find that $y = 2$. Then substitute into equation (1) to find that $x = 1$. The solution set of the system is $\{(1, 2, -1)\}$.

(a)

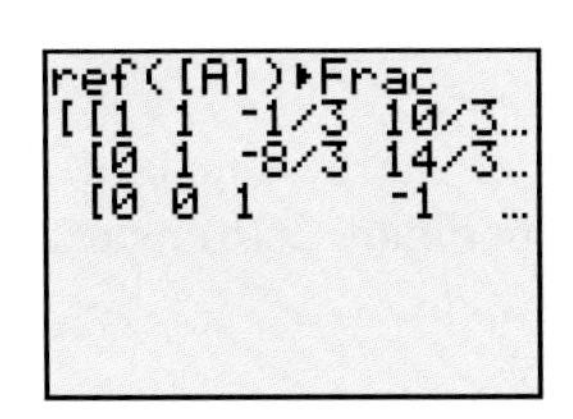


(b)

FIGURE 24

**EXAMPLE 4** *Solving a System by the Row Echelon Method (Graphical)*

Solve the system of Example 3,

$$\begin{aligned} x - \phantom{3}y + 5z &= -6\\ 3x + 3y - \phantom{5}z &= 10\\ x + 3y + 2z &= 5, \end{aligned}$$

using the row reduction capability of a graphing calculator.

**Solution** Enter the augmented matrix of the system as matrix [A]. See Figure 24(a). Then use the ref command to obtain the row echelon form. The matrix in Figure 24(b) indicates that the row echelon form is

$$\begin{aligned} x + y - \frac{1}{3}z &= \frac{10}{3}\\ y - \frac{8}{3}z &= \frac{14}{3}\\ z &= -1. \end{aligned}$$

Using back-substitution as in Example 3, we find that the solution set is $\{(1, 2, -1)\}$.

Figure 25(a) shows the augmented matrix of the system. Figure 25(b) shows the reduced row echelon form of this matrix, which was computed using the rref command. Compare these to the two matrices in color in the discussion.

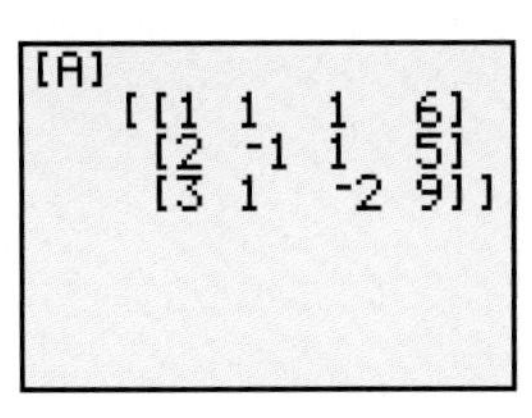


(a)

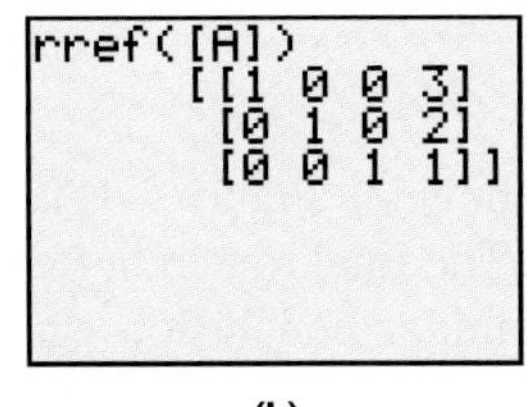


(b)

FIGURE 25

**NOTE** In Example 4, the entries in row 1, columns 2, 3, and 4 are different than the corresponding entries in the row echelon form shown in Example 3. This occurs because the steps were performed in an alternative way. However, the final result is the same.

## Reduced Row Echelon Method

Another matrix method for solving systems is called the **reduced row echelon method.** Earlier, we saw that the row echelon form of a matrix has 1s along the main diagonal and 0s below. The reduced row echelon form has 1s along the main diagonal and 0s both below *and above*. For example, the augmented matrix of the system

$$\begin{aligned} x + y + \phantom{2}z &= 6\\ 2x - y + \phantom{2}z &= 5\\ 3x + y - 2z &= 9 \end{aligned} \quad \text{is} \quad \left[\begin{array}{ccc|c} 1 & 1 & 1 & 6\\ 2 & -1 & 1 & 5\\ 3 & 1 & -2 & 9 \end{array}\right].$$

By using the row transformations described earlier, this augmented matrix can be transformed to

$$\left[\begin{array}{ccc|c} 1 & 0 & 0 & 3\\ 0 & 1 & 0 & 2\\ 0 & 0 & 1 & 1 \end{array}\right],$$

which represents the system

$$x = 3$$
$$y = 2$$
$$z = 1.$$

The solution set is $\{(3, 2, 1)\}$.

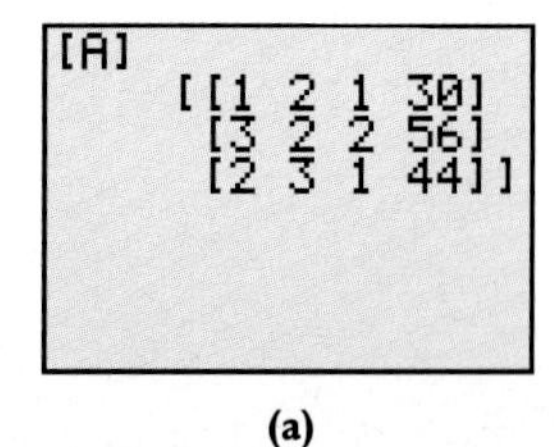


(a)

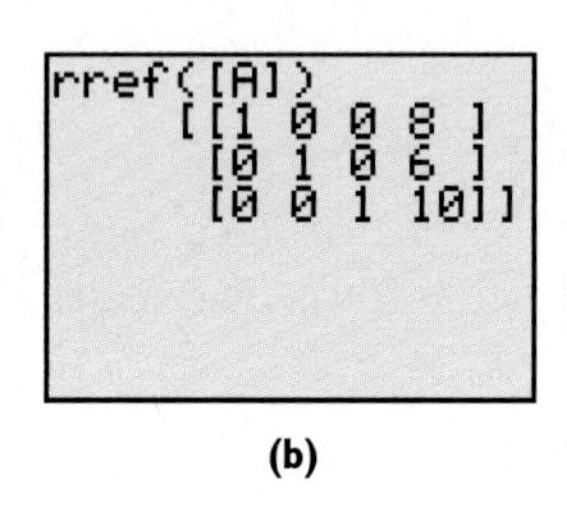


(b)

**FIGURE 26**

**EXAMPLE 5** *Solving a System by the Reduced Row Echelon Method*

Use a graphing calculator with reduced row echelon capability to solve the system.

$$x + 2y + z = 30$$
$$3x + 2y + 2z = 56$$
$$2x + 3y + z = 44$$

**Solution** Figure 26(a) shows the augmented matrix of the system, and Figure 26(b) shows its reduced row echelon form. The matrix in Figure 26(b) represents the system

$$x = 8$$
$$y = 6$$
$$z = 10.$$

Thus, the solution set is $\{(8, 6, 10)\}$.

## Special Cases

Whenever a row of the augmented matrix is of the form

$$0 \quad 0 \quad 0 \quad \ldots \quad | a, \quad \text{where } a \neq 0,$$

the system is inconsistent and there will be no solution since this row corresponds to the equation $0 = a$. A row of the matrix of a linear system in the form

$$0 \quad 0 \quad 0 \quad \ldots \quad | 0$$

indicates that the equations of the system are dependent.

**EXAMPLE 6** *Solving a Special System (No Solution)*

Show that the system is inconsistent.

$$x + 2y + z = 4$$
$$x - 2y + 3z = 1$$
$$2x + 4y + 2z = 9$$

**Analytic Solution**

The augmented matrix of the system is

$$\left[\begin{array}{rrr|r} 1 & 2 & 1 & 4 \\ 1 & -2 & 3 & 1 \\ 2 & 4 & 2 & 9 \end{array}\right].$$

**Graphing Calculator Solution**

The augmented matrix for this system, [A], is shown in Figure 27(a), and the row echelon form is shown in Figure 27(b). The final row indicates an inconsistent system with solution set $\emptyset$. (A similar conclusion can be made if the rref command is used.)

*(continued)*

$$\left[\begin{array}{ccc|c} 1 & 2 & 1 & 4 \\ 1 & -2 & 3 & 1 \\ 0 & 0 & 0 & 1 \end{array}\right] \quad -2R_1 + R_3$$

The final row indicates that the system is inconsistent and has solution set $\emptyset$.

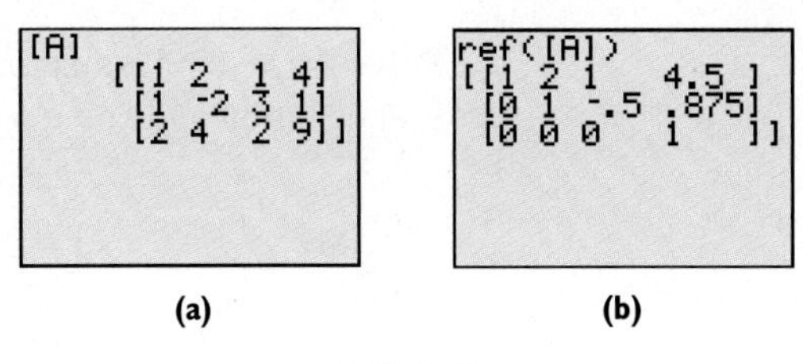


(a) (b)

**FIGURE 27**

## EXAMPLE 7 *Solving a Special System (Dependent Equations)*

Show that the system

$$\begin{aligned} 2x - 5y + 3z &= 1 \\ x + 4y - 2z &= 8 \\ 4x - 10y + 6z &= 2 \end{aligned}$$

has dependent equations. Express the general solution using an arbitrary variable.

**Solution**

$$\left[\begin{array}{ccc|c} 2 & -5 & 3 & 1 \\ 1 & 4 & -2 & 8 \\ 4 & -10 & 6 & 2 \end{array}\right] \quad \text{Augmented matrix}$$

$$\left[\begin{array}{ccc|c} 2 & -5 & 3 & 1 \\ 1 & 4 & -2 & 8 \\ 0 & 0 & 0 & 0 \end{array}\right] \quad -2R_1 + R_3$$

The final row of 0s indicates that the system has dependent equations. The first two rows represent the system

$$\begin{aligned} 2x - 5y + 3z &= 1 \\ x + 4y - 2z &= 8. \end{aligned}$$

Recall from Section 7.2 that a system with two equations and three variables may have an infinite number of solutions. Start with the augmented matrix

$$\left[\begin{array}{ccc|c} 2 & -5 & 3 & 1 \\ 1 & 4 & -2 & 8 \end{array}\right].$$

Interchange rows to obtain a 1 in row 1, column 1.

$$\left[\begin{array}{ccc|c} 1 & 4 & -2 & 8 \\ 2 & -5 & 3 & 1 \end{array}\right]$$

$$\left[\begin{array}{ccc|c} 1 & 4 & -2 & 8 \\ 0 & -13 & 7 & -15 \end{array}\right] \quad -2R_1 + R_2$$

$$\left[\begin{array}{ccc|c} 1 & 4 & -2 & 8 \\ 0 & 1 & -\frac{7}{13} & \frac{15}{13} \end{array}\right] \quad -\tfrac{1}{13}R_2$$

This is as far as you can go with the row echelon method. The equations that correspond to the final matrix are

$$x + 4y - 2z = 8 \quad \text{and} \quad y - \frac{7}{13}z = \frac{15}{13}.$$

Solve the second equation for $y$: $y = \frac{15}{13} + \frac{7}{13}z$. Now, substitute this result for $y$ in the first equation and solve for $x$.

$$x + 4y - 2z = 8$$

$$x + 4\left(\frac{15}{13} + \frac{7}{13}z\right) - 2z = 8$$

$$x + \frac{60}{13} + \frac{28}{13}z - 2z = 8$$

$$x + \frac{60}{13} + \frac{2}{13}z = 8$$

$$x = 8 - \frac{60}{13} - \frac{2}{13}z$$

$$x = \frac{44}{13} - \frac{2}{13}z$$

The solution set written with $z$ arbitrary is

$$\left\{\left(\frac{44 - 2z}{13}, \frac{15 + 7z}{13}, z\right)\right\}.$$

**FOR DISCUSSION**

The three screens shown here depict matrix [A] from Example 7, the row echelon form of [A], and the reduced row echelon form of [A].

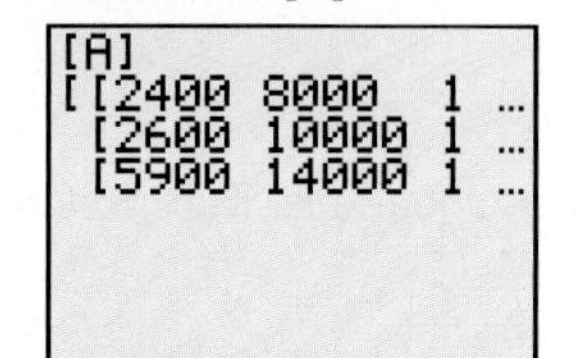


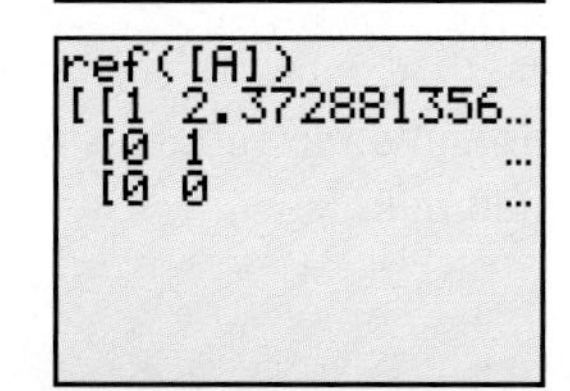


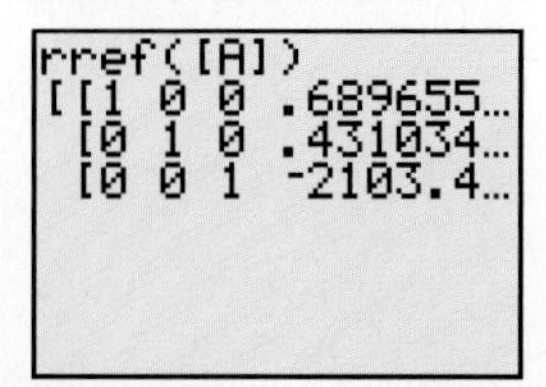


**1.** Use your calculator to re-create the screens shown here. (Note that only a portion of the matrix is visible.)

**2.** Which one of the two bottom screens provides an easier interpretation of the solution of the system? Why?

## An Application

**EXAMPLE 8** *Determining a Model Using Given Data*

Three food shelters are operated by a charitable organization. Three different quantities are computed, which include monthly food costs $F$ in dollars, number of people served per month $N$, and monthly charitable receipts $R$ in dollars. The data are shown in the table.

| Food Costs ($F$) | Number Served ($N$) | Charitable Receipts ($R$) |
|---|---|---|
| 3000 | 2400 | 8000 |
| 4000 | 2600 | 10,000 |
| 8000 | 5900 | 14,000 |

*Source:* Sanders, D., *Statistics: A First Course*, McGraw-Hill, 1995.

**(a)** Model the data using the equation $F = aN + bR + c$, where $a$, $b$, and $c$ are constants.

**(b)** Predict the food costs for a shelter that serves 4000 people and receives charitable receipts of \$12,000. Round your answer to the nearest hundred dollars.

### Solution

**(a)** Since $F = aN + bR + c$, the constants $a$, $b$, and $c$ satisfy the following three equations.

$$3000 = a(2400) + b(8000) + c$$
$$4000 = a(2600) + b(10{,}000) + c$$
$$8000 = a(5900) + b(14{,}000) + c$$

We can rewrite this system as

$$\begin{aligned} 2400a + \quad 8000b + c &= 3000 \\ 2600a + 10{,}000b + c &= 4000 \\ 5900a + 14{,}000b + c &= 8000. \end{aligned}$$

The resulting augmented matrix is

$$A = \left[\begin{array}{ccc|c} 2400 & 8000 & 1 & 3000 \\ 2600 & 10{,}000 & 1 & 4000 \\ 5900 & 14{,}000 & 1 & 8000 \end{array}\right].$$

Using the reduced row echelon method (see the "For Discussion" box on the previous page), we find that $a \approx .6897$, $b \approx .4310$, and $c \approx -2103$; thus,

$$F = .6897N + .4310R - 2103.$$

**(b)** To predict the food costs for a shelter that serves 4000 people and receives charitable receipts of $12,000, let $N = 4000$ and $R = 12{,}000$ and evaluate $F$.

$$F = .6897(4000) + .4310(12{,}000) - 2103 = 5827.8.$$

This model predicts monthly food costs of approximately $5800.

## 7.3 EXERCISES

*Concept Check* *Use the row transformation described to transform each matrix as indicated.*

**1.** $\begin{bmatrix} 2 & 4 \\ 4 & 7 \end{bmatrix}$ $-2R_1$

**2.** $\begin{bmatrix} -1 & 4 \\ 7 & 0 \end{bmatrix}$ $7R_1$

**3.** $\begin{bmatrix} 1 & 5 & 6 \\ -2 & 3 & -1 \\ 4 & 7 & 0 \end{bmatrix}$ $R_1 + R_2$

**4.** $\begin{bmatrix} 2 & 5 & 6 \\ 4 & -1 & 2 \\ 3 & 7 & 1 \end{bmatrix}$ $R_3 + R_1$

**5.** $\begin{bmatrix} -3 & 1 & -4 \\ 2 & 1 & 3 \\ -7 & 5 & 2 \end{bmatrix}$ $-5R_2 + R_3$

**6.** $\begin{bmatrix} 4 & 10 & -8 \\ 7 & 4 & 3 \\ -1 & 1 & 0 \end{bmatrix}$ $-4R_3 + R_2$

*Write the augmented matrix for each system. Do not solve the system.*

**7.** $\begin{aligned} 2x + 3y &= 11 \\ x + 2y &= 8 \end{aligned}$

**8.** $\begin{aligned} 3x + 5y &= -13 \\ 2x + 3y &= -9 \end{aligned}$

**9.** $\begin{aligned} x + 5y &= 6 \\ x &= 3 \end{aligned}$

**10.** $\begin{aligned} 2x + 7y &= 1 \\ 5x &= -15 \end{aligned}$

**11.** $\begin{aligned} 2x + y + z &= 3 \\ 3x - 4y + 2z &= -7 \\ x + y + z &= 2 \end{aligned}$

**12.** $\begin{aligned} 4x - 2y + 3z &= 4 \\ 3x + 5y + z &= 7 \\ 5x - y + 4z &= 7 \end{aligned}$

**13.** $\begin{aligned} x + y &= 2 \\ 2y + z &= -4 \\ z &= 2 \end{aligned}$

**14.** $\begin{aligned} x &= 6 \\ y + 2z &= 2 \\ x - 3z &= 6 \end{aligned}$

*Write the system of equations associated with each augmented matrix.*

**15.** $\left[\begin{array}{cc|c} 2 & 1 & 1 \\ 3 & -2 & -9 \end{array}\right]$

**16.** $\left[\begin{array}{cc|c} 1 & -5 & -18 \\ 6 & 2 & 20 \end{array}\right]$

**17.** $\left[\begin{array}{ccc|c} 1 & 0 & 0 & 2 \\ 0 & 1 & 0 & 3 \\ 0 & 0 & 1 & -2 \end{array}\right]$

**18.** $\left[\begin{array}{ccc|c} 1 & 0 & 1 & 4 \\ 0 & 1 & 0 & 2 \\ 0 & 0 & 1 & 3 \end{array}\right]$

**19.**
```
[A]
 [[3  2  1 1 ]
  [0  2  4 22]
  [-1 -2 3 15]]
```

**20.**
```
[B]
 [[2 1  3  12 ]
  [4 -3 0  10 ]
  [5 0  -4 -11]]
```

*Use one of the methods of this section to solve each system of equations.*

**21.** $\begin{aligned} x + y &= 5 \\ x - y &= -1 \end{aligned}$

**22.** $\begin{aligned} x + 2y &= 5 \\ 2x + y &= -2 \end{aligned}$

**23.** $\begin{aligned} x + y &= -3 \\ 2x - 5y &= -6 \end{aligned}$

**24.** $\begin{aligned} 3x - 2y &= 4 \\ 3x + y &= -2 \end{aligned}$

**25.** $\begin{aligned} 2x - 3y &= 10 \\ 2x + 2y &= 5 \end{aligned}$

**26.** $\begin{aligned} 4x + y &= 5 \\ 2x + y &= 3 \end{aligned}$

**27.** $\begin{aligned} 2x - 3y &= 2 \\ 4x - 6y &= 1 \end{aligned}$

**28.** $\begin{aligned} x + 2y &= 1 \\ 2x + 4y &= 3 \end{aligned}$

**29.** $\begin{aligned} 6x - 3y &= 1 \\ -12x + 6y &= -2 \end{aligned}$

**30.** $\begin{aligned} x - y &= 1 \\ -x + y &= -1 \end{aligned}$

**31.** $\begin{aligned} x + y &= -1 \\ y + z &= 4 \\ x + z &= 1 \end{aligned}$

**32.** $\begin{aligned} x - z &= -3 \\ y + z &= 9 \\ x + z &= 7 \end{aligned}$

**33.** $\begin{aligned} x + y - z &= 6 \\ 2x - y + z &= -9 \\ x - 2y + 3z &= 1 \end{aligned}$

**34.** $\begin{aligned} x + 3y - 6z &= 7 \\ 2x - y + 2z &= 0 \\ x + y + 2z &= -1 \end{aligned}$

**35.** $\begin{aligned} -x + y &= -1 \\ y - z &= 6 \\ x + z &= -1 \end{aligned}$

**36.** $\begin{aligned} x + y &= 1 \\ 2x - z &= 0 \\ y + 2z &= -2 \end{aligned}$

**37.** $\begin{aligned} 2x - y + 3z &= 0 \\ x + 2y - z &= 5 \\ 2y + z &= 1 \end{aligned}$

**38.** $\begin{aligned} 4x + 2y - 3z &= 6 \\ x - 4y + z &= -4 \\ -x + 2z &= 2 \end{aligned}$

**39.** $\begin{aligned} 2.1x + .5y + 1.7z &= 4.9 \\ -2x + 1.5y - 1.7z &= 3.1 \\ 5.8x - 4.6y + .8z &= 9.3 \end{aligned}$

**40.** $\begin{aligned} .1x + .3y + 1.7z &= .6 \\ .6x + .1y - 3.1z &= 6.2 \\ 2.4y + .9z &= 3.5 \end{aligned}$

**41.** $\begin{aligned} 53x + 95y + 12z &= 108 \\ 81x - 57y - 24z &= -92 \\ -9x + 11y - 78z &= 21 \end{aligned}$

**42.** $\begin{aligned} 103x - 886y + 431z &= 1200 \\ -55x + 981y &= 1108 \\ -327x + 421y + 337z &= 99 \end{aligned}$

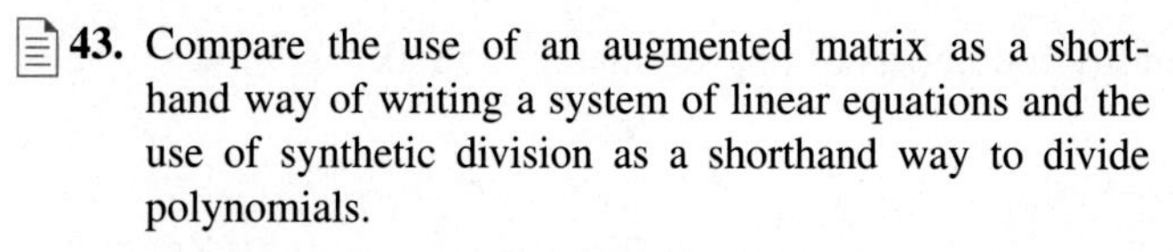

**43.** Compare the use of an augmented matrix as a shorthand way of writing a system of linear equations and the use of synthetic division as a shorthand way to divide polynomials.

**44.** Compare the use of the third row transformation on a matrix and the elimination method of solving a system of linear equations.

*Solve each system using one of the methods of this section. Let $z$ be the arbitrary variable if necessary.*

**45.** $\begin{aligned} x - 3y + 2z &= 10 \\ 2x - y - z &= 8 \end{aligned}$

**46.** $\begin{aligned} 3x + y - z &= 12 \\ x + 2y + z &= 10 \end{aligned}$

**47.** $\begin{aligned} x + 2y - z &= 0 \\ 3x - y + z &= 6 \\ -2x - 4y + 2z &= 0 \end{aligned}$

**48.** $\begin{aligned} 3x + 5y - z &= 0 \\ 4x - y + 2z &= 1 \\ -6x - 10y + 2z &= 0 \end{aligned}$

**49.** $\begin{aligned} x - 2y + z &= 5 \\ -2x + 4y - 2z &= 2 \\ 2x + y - z &= 2 \end{aligned}$

**50.** $\begin{aligned} 3x + 6y - 3z &= 12 \\ -x - 2y + z &= 16 \\ x + y - 2z &= 20 \end{aligned}$

*Use one of the methods of this section to solve each system of four equations in four variables. Express the solutions in the form $(x, y, z, w)$.*

**51.** $\begin{aligned} x + 3y - 2z - w &= 9 \\ 4x + y + z + 2w &= 2 \\ -3x - y + z - w &= -5 \\ x - y - 3z - 2w &= 2 \end{aligned}$

**52.** $\begin{aligned} 3x + 2y - w &= 0 \\ 2x + z + 2w &= 5 \\ x + 2y - z &= -2 \\ 2x - y + z + w &= 2 \end{aligned}$

*Solve each application.*

**53.** *(Modeling) Food Shelters* Three food shelters have monthly food costs $F$ in dollars, number of people served per month $N$, and monthly charitable receipts $R$ in dollars, as shown in the table.

| Food Costs ($F$) | Number Served ($N$) | Charitable Receipts ($R$) |
|---|---|---|
| 1300 | 1800 | 5000 |
| 5300 | 3200 | 12,000 |
| 6500 | 4500 | 13,000 |

**(a)** Model these data using $F = aN + bR + c$, where $a$, $b$, and $c$ are constants.

**(b)** Predict the food costs for a shelter that serves 3500 people and receives charitable receipts of \$12,500. Round your answer to the nearest hundred dollars.

**54.** *(Modeling) Paid Vacation for Employees* More than half of private-sector employees cannot carry vacation days into a new year. The average number $y$ of paid days off for full-time workers at medium to large companies after $x$ years is listed in the table.

| **x (years)** | 1 | 15 | 30 |
|---|---|---|---|
| **y (days)** | 9.4 | 18.8 | 21.9 |

*Source:* Bureau of Labor Statistics.

**(a)** Determine the coefficients for $f(x) = ax^2 + bx + c$ so that $f$ models these data.

**(b)** Graph $f$ with the data in the window $[-4, 32]$ by $[8, 23]$.

**(c)** Estimate the number of paid days off after 3 years of experience. Compare it to the actual value of 11.2 days.

**55.** *Scheduling Production* A company produces two models of bicycles, model $A$ and model $B$. Model $A$ requires 2 hours of assembly time, and model $B$ requires 3 hours of assembly time. The parts for model $A$ cost \$25 per bike; those for model $B$ cost \$30 per bike. If the company has a total of 34 hours of assembly time and \$365 available per day for these two models, what is the maximum number of each model that can be made in a day and use all of the available resources?

**56.** *Scheduling Production* Caltek Computer Company makes two products: computer monitors and printers. Both require time on two machines—monitors: 1 hour on machine $A$ and 2 hours on machine $B$; printers: 3 hours on machine $A$ and 1 hour on machine $B$. Both machines operate 15 hours per day. What is the maximum number of each product that can be produced per day under these conditions?

**57.** *Financing Expansion* To get necessary funds for a planned expansion, a small company took out three loans totaling \$25,000. The company was able to borrow some of the money at 8%. It borrowed \$2000 more than $\frac{1}{2}$ the amount of the 8% loan at 10%, and the rest at 9%. The total annual interest was \$2220. How much did the company borrow at each rate?

**58.** *Investment Decisions* Rick Pal deposits some money in a bank account paying 3% per year. He uses some additional money, amounting to $\frac{1}{3}$ the amount placed in the bank, to buy bonds paying 4% per year. With the balance of his funds, he buys a 4.5% certificate of deposit. The first year, his investments bring a return of \$400. If the total of the investments is \$10,000, how much did he invest at each rate?

*(Modeling) Each set of data can be modeled by $f(x) = ax^2 + bx + c$, where $x$ represents the year.*

**(a)** *Find a linear system whose solution represents values of $a$, $b$, and $c$.*

**(b)** *Find the solution using one of the methods of this section.*

**(c)** *Graph $f$ and the data in the same viewing window.*

**(d)** *Make your own prediction using $f$.*

**59.** *Chronic Health Care* A large percentage of the U.S. population will require chronic health care in the coming decades. The average caregiving age is 50–64, while the typical person needing chronic care is 85 or older. The ratio of potential caregivers to those needing chronic health care will shrink in the coming years, as shown in the table.

| **Year** | 1990 | 2010 | 2030 |
|---|---|---|---|
| **Ratio** | 11 | 10 | 6 |

*Source:* Robert Wood Johnson Foundation, Chronic Care in America: A 21st Century Challenge.

**60.** *Home Health Care* The table shows the cost of Medicare home health care $y$ in billions of dollars during the year $x$. In this table, $x = 0$ corresponds to 1990 and $x = 6$ corresponds to 1996.

| $x$ | 0 | 3 | 6 |
|---|---|---|---|
| $y$ | 3.9 | 10.5 | 18.1 |

*Source:* Health Care Financing Administration.

**61.** *Women in the Military* The table shows the percentage $y$ of the enlisted people in the military who are women. In this table $x = 3$ corresponds to 1973 and $x = 26$ to 1996.

| $x$ | 3 | 18 | 26 |
|---|---|---|---|
| $y$ | 2.2 | 10.4 | 12.8 |

*Source:* Department of Defense.

**62.** *Carbon Dioxide Levels* Carbon dioxide ($CO_2$) is a greenhouse gas. Its concentration in parts per million (ppm) has been measured at Mauna Loa, Hawaii, during past years. The table lists measurements for three selected years.

| **Year** | 1958 | 1973 | 1992 |
|---|---|---|---|
| **$CO_2$ (ppm)** | 315 | 325 | 354 |

*Source:* Nilsson, A., *Greenhouse Earth*, John Wiley and Sons, 1992.

*(Modeling) Traffic Flow* *Each figure shows three one-way streets with intersections A, B, and C. Numbers indicate the average traffic flow in vehicles per minute. The variables x, y, and z denote unknown traffic flows that need to be determined for timing of stoplights.*

**(a)** *If the number of vehicles per minute entering an intersection must equal the number exiting an intersection, verify that the system of linear equations describes the traffic flow.*

**(b)** *Rewrite the system and solve.*

**(c)** *Interpret your solution.*

**63.** A: $x + 5 = y + 7$

B: $z + 6 = x + 3$

C: $y + 3 = z + 4$

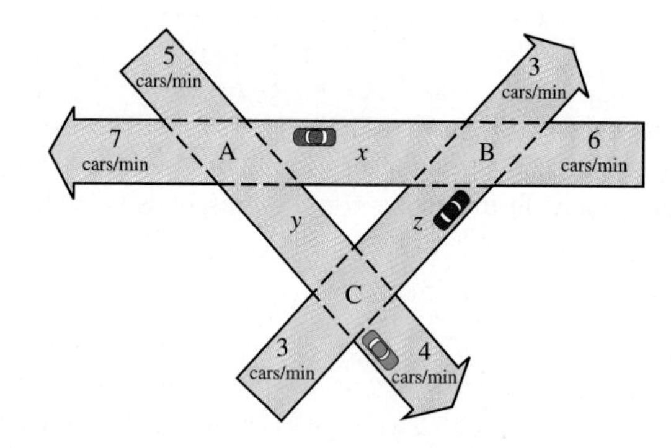


**64.** A: $x + 7 = y + 4$

B: $4 + 5 = x + z$

C: $y + 8 = 9 + 4$

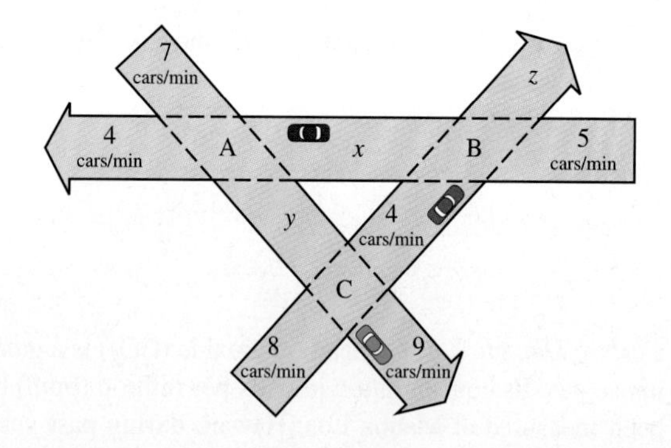


**65.** *(Modeling) Fawn Population* To model the spring fawn count $F$ from the adult antelope population $A$, the precipitation $P$, and the severity of winter $W$, environmentalists have used the equation

$$F = a + bA + cP + dW,$$

where the coefficients $a$, $b$, $c$, and $d$ are constants that must be determined before using the equation. (*Source:* Brase, C. and C. Brase, *Understandable Statistics,* D.C. Heath and Company, 1995; Bureau of Land Management.) The table lists the results of four different years.

| Fawns | Adults | Precip. (in inches) | Winter Severity |
|---|---|---|---|
| 239 | 871 | 11.5 | 3 |
| 234 | 847 | 12.2 | 2 |
| 192 | 685 | 10.6 | 5 |
| 343 | 969 | 14.2 | 1 |

**(a)** Substitute the values for $F$, $A$, $P$, and $W$ from the table for each of the four years into the equation $F = a + bA + cP + dW$, to obtain four linear equations involving $a$, $b$, $c$, and $d$.

**(b)** Write an augmented matrix representing the system, and solve for $a$, $b$, $c$, and $d$.

**(c)** Write the equation for $F$, using the values from part (b) for the coefficients.

**(d)** If a winter has severity 3, adult antelope population 960, and precipitation 12.6 inches, predict the spring fawn count. (Compare this with the actual count of 320.)

**66.** *Computing Time* When computers are programmed to solve large linear systems involved in applications, such as designing aircraft or large electrical circuits, they frequently use elimination with back-substitution. Solving a linear system with $n$ equations and $n$ variables requires a computer to perform a total of

$$T(n) = \frac{2}{3}n^3 + \frac{3}{2}n^2 - \frac{7}{6}n$$

arithmetic operations (additions, subtractions, multiplications, and divisions). (*Source:* Burden, R. and J. Faires, *Numerical Analysis,* PWS-Kent, 1993.)

**(a)** John Atanasoff, the inventor of the modern digital computer, needed to solve a system of 29 linear equations. Evaluate $T(29)$ to find the number of arithmetic operations this would require. Would it be too many to do by hand?

**(b)** Compute $T$ for $n = 10$, 100, 1000, 10,000, and 100,000. List the results in a table.

**(c)** If the number of equations and variables increases by a factor of 10, does the number of arithmetic operations also increase by a factor of 10? Explain.

**(d)** Discuss why supercomputers are needed to solve large systems of linear equations.

## Reviewing Basic Concepts (Sections 7.1–7.3)

*Solve each system using the method indicated.*

**1.** (Elimination)

$$\begin{aligned} 2x - 3y &= 18 \\ 5x + 2y &= 7 \end{aligned}$$

**2.** (Graphical)

$$\begin{aligned} 2x + \phantom{2}y &= -4 \\ -x + 2y &= 2 \end{aligned}$$

**3.** (Substitution)

$$\begin{aligned} 5x + 10y &= 10 \\ x + \phantom{1}2y &= 2 \end{aligned}$$

**4.** (Elimination)

$$\begin{aligned} x - y &= 6 \\ x - y &= 4 \end{aligned}$$

**5.** (Analytic, with graphical support)

$$\begin{aligned} 6x + 2y &= 10 \\ 2x^2 - 3y &= 11 \end{aligned}$$

**6.** (Row echelon method)

$$\begin{aligned} x + y + \phantom{2}z &= 1 \\ -x + y + \phantom{2}z &= 5 \\ y + 2z &= 5 \end{aligned}$$

**7.** (Reduced row echelon method)

$$\begin{aligned} 2x + 4y + 4z &= 4 \\ x + 3y + \phantom{2}z &= 4 \\ -x + 3y + 2z &= -1 \end{aligned}$$

**8.** Solve the system represented by the augmented matrix shown here.

```
[A]
   [[2 1  2 10]
    [1 0  2 5 ]
    [1 -2 2 1 ]]
```

*Solve each problem by writing and solving a system of equations.*

**9.** In 1999, there were about 32 million television sets sold. For every 10 televisions sold with stereo sound, there were about 19 sold without stereo sound. (*Source:* Consumer Electronics.) About how many of each type of television were sold that year?

**10.** A sum of $5000 is invested in three mutual funds that pay 8%, 11%, and 14% interest rates. The amount of money invested in the fund paying 14% equals the total amount of money invested in the other two funds, and the total annual interest from all three funds is $595. Find the amount invested at each rate.

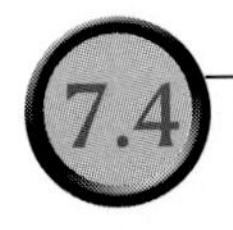

# 7.4 MATRIX PROPERTIES AND OPERATIONS

Terminology of Matrices • Operations on Matrices • Applying Matrix Algebra

## Terminology of Matrices

Matrix notation was used to solve a system of linear equations in the previous section. In this section, we examine algebraic properties of matrices.

Suppose you are the manager of a video store and one day you receive the following products from two distributors: from Wholesale Enterprises, 2 videotapes, 7 DVDs, and 5 video games; from Discount Distributors, 4 videotapes, 6 DVDs, and 9 video games. We can organize the information in a table to make it more understandable.

| | Product | | |
|---|---|---|---|
| **Distributor** | **Videotapes** | **DVDs** | **Video Games** |
| Wholesale Enterprises | 2 | 7 | 5 |
| Discount Distributors | 4 | 6 | 9 |

In fact, as long as we remember what each number represents, we can remove all the labels and write the numbers in the table as a matrix.

$$\begin{bmatrix} 2 & 7 & 5 \\ 4 & 6 & 9 \end{bmatrix}$$

Matrices are classified by their **dimension,** that is, by the number of rows and columns that they contain. For example, the preceding matrix has 2 rows and 3 columns, with dimension $2 \times 3$; a matrix with $m$ rows and $n$ columns has dimension $m \times n$. The number of rows is always given first.

Certain matrices have special names: an $n \times n$ matrix is a **square matrix** of order $n$. Also, a matrix with just one row is a **row matrix,** and a matrix with just one column is a **column matrix.**

TECHNOLOGY NOTE

You should familiarize yourself with the matrix capabilities of your particular calculator. Refer to the graphing calculator manual that accompanies this text or your owner's manual.

Two matrices are equal if they have the same dimension and if corresponding elements, position by position, are equal. Using this definition, the matrices

$$\begin{bmatrix} 2 & 1 \\ 3 & -5 \end{bmatrix} \text{ and } \begin{bmatrix} 1 & 2 \\ -5 & 3 \end{bmatrix}$$

are *not* equal (even though they contain the same elements and are the same dimension), because the corresponding elements differ.

**EXAMPLE 1** *Classifying Matrices by Dimension*

Find the dimension of each matrix, and determine any special characteristics.

**(a)** $\begin{bmatrix} 6 & 5 \\ 3 & 4 \\ 5 & -1 \end{bmatrix}$

**(b)**

```
[A]
     [[1  2   1]
      [1  -2  3]
      [2  4   2]]
```

**(c)** $\begin{bmatrix} 3 \\ -5 \\ 0 \\ 2 \end{bmatrix}$

**(d)** $[1 \quad 6 \quad 5 \quad -2 \quad 5]$

**Solution**

**(a)** This is a $3 \times 2$ matrix, because it has 3 rows and 2 columns.

**(b)** This screen illustrates a $3 \times 3$ matrix. It is also a square matrix.

**(c)** This matrix is a $4 \times 1$ column matrix.

**(d)** $[1 \quad 6 \quad 5 \quad -2 \quad 5]$ is a $1 \times 5$ matrix. It is an example of a row matrix.

**EXAMPLE 2** *Determining Equality of Matrices*

**(a)** If $A = \begin{bmatrix} 2 & 1 \\ p & q \end{bmatrix}$ and $B = \begin{bmatrix} x & y \\ -1 & 0 \end{bmatrix}$, find the values of $x$, $y$, $p$, and $q$ such that $A = B$.

**(b)** Are matrices [A] and [B] equal?

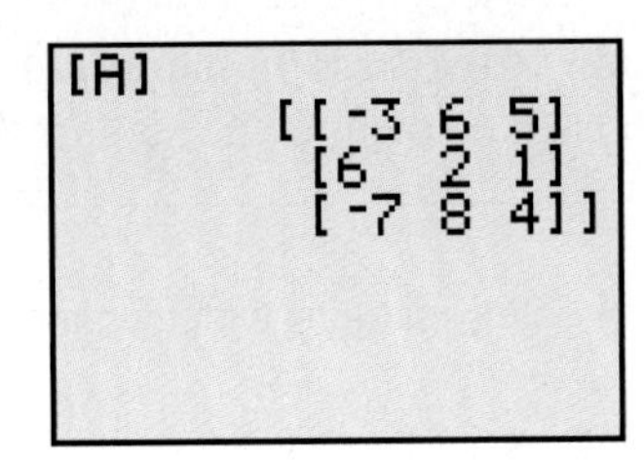


[B]
[[-3 6 5]
[6 2 1]
[-7 4 8]]

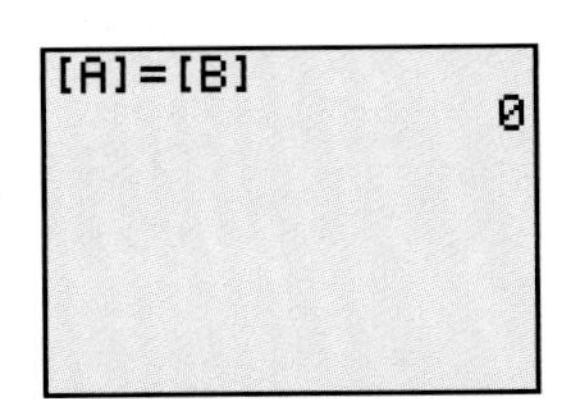


Because the calculator returns 0, the statement is false. (A true statement returns 1.)

**FIGURE 28**

### Solution

**(a)** From the definition of equality, the only way that the statement

$$\begin{bmatrix} 2 & 1 \\ p & q \end{bmatrix} = \begin{bmatrix} x & y \\ -1 & 0 \end{bmatrix}$$

can be true is if $x = 2$, $y = 1$, $p = -1$, and $q = 0$.

**(b)** The matrices are not equal because the entries in the third row do not all match. Figure 28 uses the test feature of a graphing calculator to show that these matrices are not equal.

## Operations on Matrices

**TECHNOLOGY NOTE**

Graphing calculators can perform operations on matrices provided the dimensions of the matrices are compatible for the operation. A dimension error message will occur if the operation cannot be performed.

At the beginning of this section, we gave an example using the matrix

$$\begin{bmatrix} 2 & 7 & 5 \\ 4 & 6 & 9 \end{bmatrix},$$

where the columns represent the numbers of certain products (videotapes, DVDs, and video games, respectively) and the rows represent the two different distributors (Wholesale Enterprises and Discount Distributors, respectively). For example, the element 7 represents 7 DVDs received from Wholesale Enterprises.

Now suppose that on another day, the shipments from these two distributors are described with this matrix:

$$\begin{bmatrix} 3 & 12 & 10 \\ 15 & 11 & 8 \end{bmatrix}.$$

Here, for example, 8 video games were received from Discount Distributors. The total number of products received from the distributors on the two days can be found by adding the corresponding elements of the two matrices.

$$\begin{bmatrix} 2 & 7 & 5 \\ 4 & 6 & 9 \end{bmatrix} + \begin{bmatrix} 3 & 12 & 10 \\ 15 & 11 & 8 \end{bmatrix} = \begin{bmatrix} 2+3 & 7+12 & 5+10 \\ 4+15 & 6+11 & 9+8 \end{bmatrix}$$
$$= \begin{bmatrix} 5 & 19 & 15 \\ 19 & 17 & 17 \end{bmatrix}$$

The 5 in the sum indicates that on the two days, 5 videotapes were received from Wholesale Enterprises. Generalizing from this example leads to the definition of matrix addition.

**Matrix Addition**

The sum of two $m \times n$ matrices $A$ and $B$ is the $m \times n$ matrix $A + B$ in which each element is the sum of the corresponding elements of $A$ and $B$.

**CAUTION** Only matrices with the same dimension can be added.

**EXAMPLE 3** *Adding Matrices*

Find each sum.

**(a)** $\begin{bmatrix} 5 & -6 \\ 8 & 9 \end{bmatrix} + \begin{bmatrix} -4 & 6 \\ 8 & -3 \end{bmatrix}$ **(b)** $\begin{bmatrix} 2 \\ 5 \\ 8 \end{bmatrix} + \begin{bmatrix} -6 \\ 3 \\ 12 \end{bmatrix}$

**(c)** $A + B$, if $A = \begin{bmatrix} 5 & 8 \\ 6 & 2 \end{bmatrix}$ and $B = \begin{bmatrix} 3 & 9 & 1 \\ 4 & 2 & 5 \end{bmatrix}$

**Analytic Solution**

**(a)** $\begin{bmatrix} 5 & -6 \\ 8 & 9 \end{bmatrix} + \begin{bmatrix} -4 & 6 \\ 8 & -3 \end{bmatrix}$

$$= \begin{bmatrix} 5 + (-4) & -6 + 6 \\ 8 + 8 & 9 + (-3) \end{bmatrix}$$

$$= \begin{bmatrix} 1 & 0 \\ 16 & 6 \end{bmatrix}$$

**(b)** $\begin{bmatrix} 2 \\ 5 \\ 8 \end{bmatrix} + \begin{bmatrix} -6 \\ 3 \\ 12 \end{bmatrix} = \begin{bmatrix} -4 \\ 8 \\ 20 \end{bmatrix}$

**(c)** The matrices

$$A = \begin{bmatrix} 5 & 8 \\ 6 & 2 \end{bmatrix}$$

and

$$B = \begin{bmatrix} 3 & 9 & 1 \\ 4 & 2 & 5 \end{bmatrix}$$

have different dimensions. Therefore, the sum $A + B$ does not exist.

**Graphing Calculator Solution**

**(a)** Figure 30 shows the sum of matrices [C] and [D], defined in Figure 29.

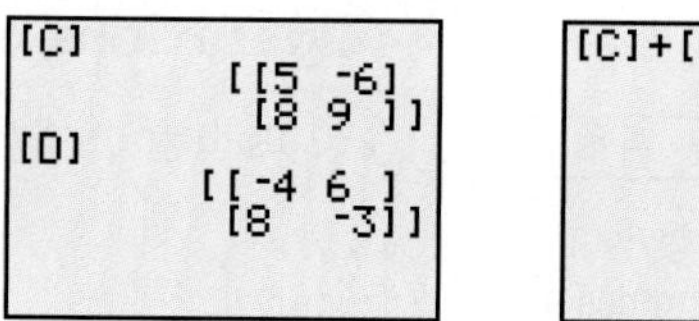


FIGURE 29

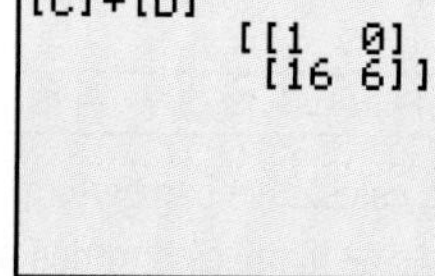


FIGURE 30

**(b)** See Figure 31, which shows the two matrices entered directly on the home screen.

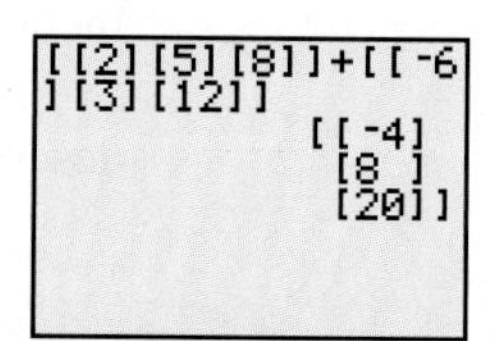


FIGURE 31

**(c)** The calculator returns a dimension error message when directed to add two matrices of different dimensions. See Figure 32.

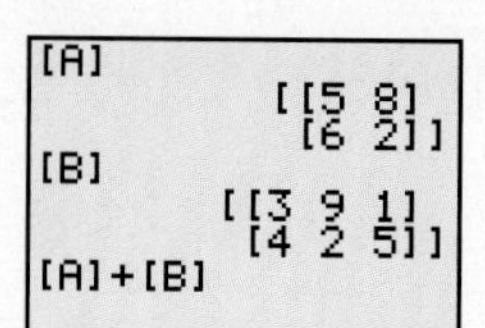


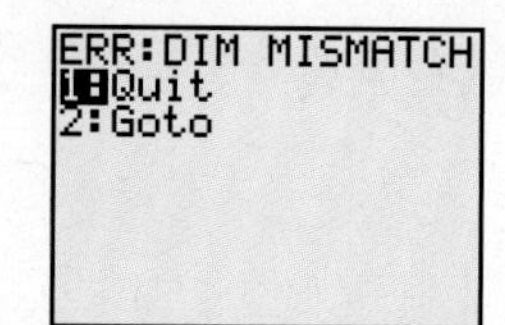


FIGURE 32

Figure 33 illustrates how matrices [A] and –[A] from the discussion are displayed on a graphing calculator screen. Figure 34 shows how a zero matrix appears.

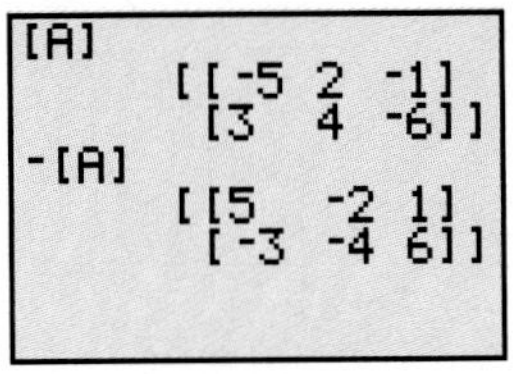

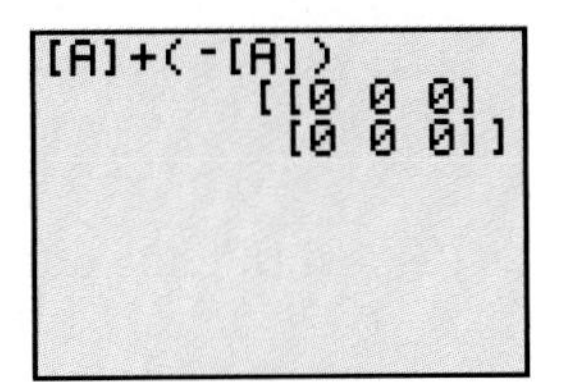


**FIGURE 33**

[A]+(-[A])
[[0 0 0]
[0 0 0]]

**FIGURE 34**

A matrix containing only zero elements is called a **zero matrix.** For example, $[0 \quad 0 \quad 0]$ is the $1 \times 3$ zero matrix, while

$$\begin{bmatrix} 0 & 0 & 0 \\ 0 & 0 & 0 \end{bmatrix}$$

is the $2 \times 3$ zero matrix. A zero matrix can be written with any dimension.

The additive inverse of a real number $a$ is the unique real number $-a$ such that $a + (-a) = 0$ and $-a + a = 0$. Given a matrix $A$, a matrix $-A$ can be found such that $A + (-A) = O$, where $O$ is the appropriate zero matrix, and $-A + A = O$. For example, if

$$A = \begin{bmatrix} -5 & 2 & -1 \\ 3 & 4 & -6 \end{bmatrix}, \quad \text{then} \quad -A = \begin{bmatrix} 5 & -2 & 1 \\ -3 & -4 & 6 \end{bmatrix}.$$

The elements of matrix $-A$ are the additive inverses of the corresponding elements of $A$. (Remember that each element of $A$ is a real number and thus has an additive inverse.) To check, test that $A + (-A)$ equals the zero matrix $O$.

$$A + (-A) = \begin{bmatrix} -5 & 2 & -1 \\ 3 & 4 & -6 \end{bmatrix} + \begin{bmatrix} 5 & -2 & 1 \\ -3 & -4 & 6 \end{bmatrix} = \begin{bmatrix} 0 & 0 & 0 \\ 0 & 0 & 0 \end{bmatrix} = O$$

Then, test that $-A + A$ is also $O$. Matrix $-A$ is the **additive inverse,** or **negative,** of matrix $A$. Every matrix has a unique additive inverse.

## What Went WRONG?

A student entered [A] into his calculator and obtained the first display shown in Figure 35(a). Then he attempted to find the additive inverse of [A], as shown in the second entry in Figure 35(a), and got the error message indicated in Figure 35(b).

**FIGURE 35**

What Went Wrong? How can he correct his error?

Just as subtraction of real numbers is defined in terms of the additive inverse, subtraction of matrices is defined in the same way.

---

*Answer to What Went Wrong?*

The student entered the subtraction symbol, rather than the negative symbol, preceding A. He can obtain $-A$ by using the correct symbol.

**Matrix Subtraction**

If $A$ and $B$ are matrices with the same dimension, then

$$A - B = A + (-B).$$

**EXAMPLE 4** *Subtracting Matrices*

Find each difference.

**(a)** $\begin{bmatrix} -5 & 6 \\ 2 & 4 \end{bmatrix} - \begin{bmatrix} -3 & 2 \\ 5 & -8 \end{bmatrix}$ **(b)** $[8 \quad 6 \quad -4] - [3 \quad 5 \quad -8]$

**(c)** $A - B$, if $A = \begin{bmatrix} -2 & 5 \\ 0 & 1 \end{bmatrix}$ and $B = \begin{bmatrix} 3 \\ 5 \end{bmatrix}$

**Analytic Solution**

**(a)** $\begin{bmatrix} -5 & 6 \\ 2 & 4 \end{bmatrix} - \begin{bmatrix} -3 & 2 \\ 5 & -8 \end{bmatrix}$

$$= \begin{bmatrix} -5 & 6 \\ 2 & 4 \end{bmatrix} + \begin{bmatrix} 3 & -2 \\ -5 & 8 \end{bmatrix}$$

$$= \begin{bmatrix} -2 & 4 \\ -3 & 12 \end{bmatrix}$$

**(b)** $[8 \quad 6 \quad -4] - [3 \quad 5 \quad -8]$
$= [5 \quad 1 \quad 4]$

**(c)** Matrices $A$ and $B$ have different dimensions, so their difference does not exist.

**Graphing Calculator Solution**

**(a)** In Figure 36, the two matrices are defined as [C] and [D]; the difference [C] − [D] is displayed in Figure 37.

```
[C]
        [[-5 6]
         [2  4]]
[D]
        [[-3 2 ]
         [5  -8]]
```

FIGURE 36

```
[C]-[D]
        [[-2 4 ]
         [-3 12]]
```

FIGURE 37

**(b)** See Figure 38.

```
[E]
        [[8 6 -4]]
[F]
        [[3 5 -8]]
[E]-[F]
        [[5 1 4]]
```

FIGURE 38

**(c)** As in Example 3(c), the calculator returns a dimension error message.

If a matrix $A$ is added to itself, each element in the sum is twice as large as the corresponding element of $A$. For example,

$$\begin{bmatrix} 2 & 5 \\ 1 & 3 \\ 4 & 6 \end{bmatrix} + \begin{bmatrix} 2 & 5 \\ 1 & 3 \\ 4 & 6 \end{bmatrix} = \begin{bmatrix} 4 & 10 \\ 2 & 6 \\ 8 & 12 \end{bmatrix} = 2\begin{bmatrix} 2 & 5 \\ 1 & 3 \\ 4 & 6 \end{bmatrix}.$$

In the last expression, the number 2 in front of the matrix is called a **scalar** to distinguish it from a matrix. A scalar is just a special name for a real number. This example suggests the definition of multiplication of a matrix by a scalar.

**Multiplication of a Matrix by a Scalar**

The product of a scalar $k$ and a matrix $A$ is the matrix $kA$, each of whose elements is $k$ times the corresponding element of $A$.

**EXAMPLE 5** *Multiplying Matrices by Scalars*

Perform each multiplication.

**(a)** $5\begin{bmatrix} 2 & -3 \\ 0 & 4 \end{bmatrix}$ **(b)** $\frac{3}{4}\begin{bmatrix} 20 & 36 \\ 12 & -16 \end{bmatrix}$

**Analytic Solution**

**(a)** $$5\begin{bmatrix} 2 & -3 \\ 0 & 4 \end{bmatrix} = \begin{bmatrix} 5(2) & 5(-3) \\ 5(0) & 5(4) \end{bmatrix} = \begin{bmatrix} 10 & -15 \\ 0 & 20 \end{bmatrix}$$

**(b)** $$\frac{3}{4}\begin{bmatrix} 20 & 36 \\ 12 & -16 \end{bmatrix} = \begin{bmatrix} \frac{3}{4}(20) & \frac{3}{4}(36) \\ \frac{3}{4}(12) & \frac{3}{4}(-16) \end{bmatrix} = \begin{bmatrix} 15 & 27 \\ 9 & -12 \end{bmatrix}$$

**Graphing Calculator Solution**

**(a)** Figure 39 shows [A] and 5[A] displayed.

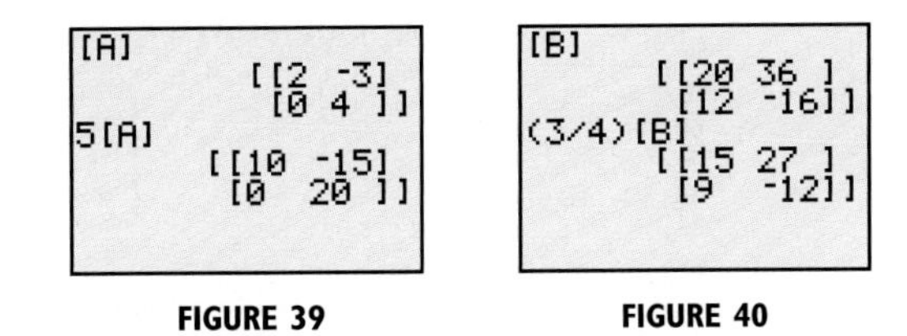


FIGURE 39 FIGURE 40

**(b)** Notice the use of parentheses in Figure 40.

To motivate the definition of multiplying a matrix by another matrix, we return to the video store example. Recall the matrix for the number of each type of product received from the two distributors.

$$\begin{bmatrix} 2 & 7 & 5 \\ 4 & 6 & 9 \end{bmatrix}$$

Now suppose each videotape costs the store \$12, each DVD costs \$18, and each video game costs \$9. To find the total cost of the products from Wholesale Enterprises, we multiply as follows.

| Type of Product | Number of Items | Cost per Item | Total Cost |
|---|---|---|---|
| Videotape | 2 | \$12 | \$ 24 |
| DVD | 7 | \$18 | \$126 |
| Video game | 5 | \$ 9 | \$ 45 |
| | | | \$195 ← Total from Wholesale Enterprises |

The products from Wholesale Enterprises cost a total of \$195. This result is the sum of three products:

$$2(\$12) + 7(\$18) + 5(\$9) = \$195.$$

In the same way, using the second row of the matrix and the three product costs gives the total cost from Discount Distributors.

$$4(\$12) + 6(\$18) + 9(\$9) = \$237$$

The product costs, \$12, \$18, and \$9, can be written as a column matrix.

$$\begin{bmatrix} 12 \\ 18 \\ 9 \end{bmatrix}$$

The total costs from each distributor can also be written as a column matrix.

$$\begin{bmatrix} 195 \\ 237 \end{bmatrix}$$

The product of the matrices

$$\begin{bmatrix} 2 & 7 & 5 \\ 4 & 6 & 9 \end{bmatrix} \quad \text{and} \quad \begin{bmatrix} 12 \\ 18 \\ 9 \end{bmatrix}$$

can be written

$$\begin{bmatrix} 2 & 7 & 5 \\ 4 & 6 & 9 \end{bmatrix}\begin{bmatrix} 12 \\ 18 \\ 9 \end{bmatrix} = \begin{bmatrix} 2 \cdot 12 + 7 \cdot 18 + 5 \cdot 9 \\ 4 \cdot 12 + 6 \cdot 18 + 9 \cdot 9 \end{bmatrix} = \begin{bmatrix} 195 \\ 237 \end{bmatrix}.$$

Each element of the product was found by multiplying the elements of the *rows* of the matrix on the left and the corresponding elements of the *column* of the matrix on the right, and then finding the sum of these products. Notice that the product of a $2 \times 3$ matrix and a $3 \times 1$ matrix is a $2 \times 1$ matrix.

Generalizing from this example gives the definition of matrix multiplication.

**Matrix Multiplication**

The product $AB$ of an $m \times n$ matrix $A$ and an $n \times k$ matrix $B$ is found as follows:

To find the $i$th row, $j$th column element of $AB$, multiply each element in the $i$th row of $A$ by the corresponding element in the $j$th column of $B$. The sum of these products will give the element of row $i$, column $j$ of $AB$. The dimension of $AB$ is $m \times k$.

The product $AB$ can be found only if *the number of columns of A is the same as the number of rows of B*. The final product will have as many rows as $A$ and as many columns as $B$.

**EXAMPLE 6** *Deciding Whether Two Matrices Can Be Multiplied*

Suppose matrix $A$ has dimension $2 \times 2$, while matrix $B$ has dimension $2 \times 4$. Can the product $AB$ be calculated? If so, what is the dimension of the product?

**Solution** The following diagram helps answer these questions.

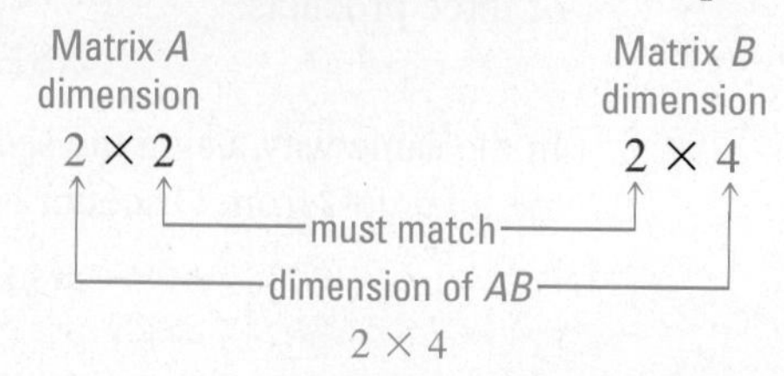

The product $AB$ can be calculated because $A$ has two columns and $B$ has two rows. The dimension of the product is $2 \times 4$. (However, the product $BA$ cannot be found.)

## EXAMPLE 7 *Multiplying Matrices*

Find the product $AB$ of the two matrices

$$A = \begin{bmatrix} -3 & 4 & 2 \\ 5 & 0 & 4 \end{bmatrix} \quad \text{and} \quad B = \begin{bmatrix} -6 & 4 \\ 2 & 3 \\ 3 & -2 \end{bmatrix}.$$

**Solution** Since $A$ has dimension $2 \times 3$ and $B$ has dimension $3 \times 2$, they are compatible for multiplication. The product $AB$ will have dimension $2 \times 2$.

$$\begin{array}{ccc} & A & B \end{array}$$

**Step 1** $\begin{bmatrix} -3 & 4 & 2 \\ 5 & 0 & 4 \end{bmatrix}\begin{bmatrix} -6 & 4 \\ 2 & 3 \\ 3 & -2 \end{bmatrix}$ $(-3)(-6) + 4(2) + 2(3) = 32$

**Step 2** $\begin{bmatrix} -3 & 4 & 2 \\ 5 & 0 & 4 \end{bmatrix}\begin{bmatrix} -6 & 4 \\ 2 & 3 \\ 3 & -2 \end{bmatrix}$ $(-3)(4) + 4(3) + 2(-2) = -4$

**Step 3** $\begin{bmatrix} -3 & 4 & 2 \\ 5 & 0 & 4 \end{bmatrix}\begin{bmatrix} -6 & 4 \\ 2 & 3 \\ 3 & -2 \end{bmatrix}$ $5(-6) + 0(2) + 4(3) = -18$

**Step 4** $\begin{bmatrix} -3 & 4 & 2 \\ 5 & 0 & 4 \end{bmatrix}\begin{bmatrix} -6 & 4 \\ 2 & 3 \\ 3 & -2 \end{bmatrix}$ $5(4) + 0(3) + 4(-2) = 12$

**Step 5** Write the product.

$$\begin{bmatrix} -3 & 4 & 2 \\ 5 & 0 & 4 \end{bmatrix}\begin{bmatrix} -6 & 4 \\ 2 & 3 \\ 3 & -2 \end{bmatrix} = \begin{bmatrix} 32 & -4 \\ -18 & 12 \end{bmatrix}$$

```
[A]
      [[-3 4 2]
       [5  0 4]]
[B]
      [[-6 4 ]
       [2  3 ]
       [3  -2]]
```

FIGURE 41

```
[B]*[A]
  [[38  -24 4 ]
   [9   8   16]
   [-19 12  -2]]
```

FIGURE 42

## EXAMPLE 8 *Multiplying Matrices*

Use a graphing calculator to find the product $BA$ of the two matrices given in Example 7.

**Solution** Define [A] and [B], as shown in Figure 41. Since [B] has dimension $3 \times 2$ and [A] has dimension $2 \times 3$, the dimension of the desired product is $3 \times 3$. Figure 42 shows the actual product.

As shown in Examples 7 and 8,

$$AB \neq BA.$$

This shows that, in general, *matrix multiplication is not commutative*. In fact, in some cases one of the products may be defined while the other is not. For example, see the matrix multiplication shown in Figure 43 on the next page.

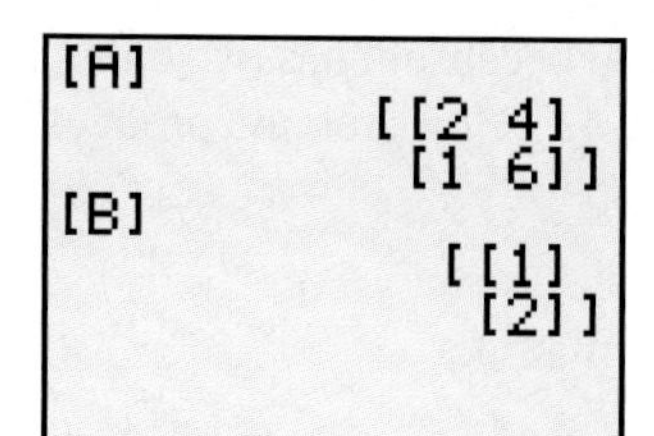


```
[A]*[B]
            [[10]
             [13]]
[B]*[A]
```

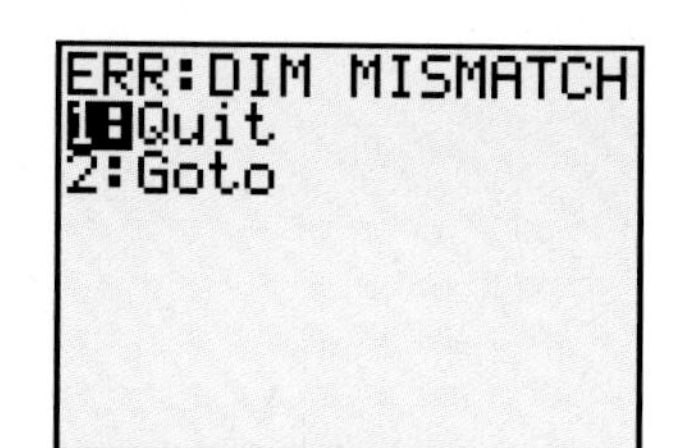


**FIGURE 43**

**EXAMPLE 9** *Multiplying Square Matrices*

Find $AB$ and $BA$, given

$$A = \begin{bmatrix} 1 & 3 \\ -2 & 4 \end{bmatrix} \quad \text{and} \quad B = \begin{bmatrix} 2 & 5 \\ 10 & -3 \end{bmatrix}.$$

**Analytic Solution**

$$\begin{aligned} AB &= \begin{bmatrix} 1 & 3 \\ -2 & 4 \end{bmatrix}\begin{bmatrix} 2 & 5 \\ 10 & -3 \end{bmatrix} \\ &= \begin{bmatrix} 2 + 30 & 5 - 9 \\ -4 + 40 & -10 - 12 \end{bmatrix} \\ &= \begin{bmatrix} 32 & -4 \\ 36 & -22 \end{bmatrix} \end{aligned}$$

$$\begin{aligned} BA &= \begin{bmatrix} 2 & 5 \\ 10 & -3 \end{bmatrix}\begin{bmatrix} 1 & 3 \\ -2 & 4 \end{bmatrix} \\ &= \begin{bmatrix} 2 - 10 & 6 + 20 \\ 10 + 6 & 30 - 12 \end{bmatrix} \\ &= \begin{bmatrix} -8 & 26 \\ 16 & 18 \end{bmatrix} \end{aligned}$$

**Graphing Calculator Solution**

Figure 44 shows matrices [A], [B], and both products.

```
[A]
          [[1  3]
           [-2 4]]
[B]
        [[2  5 ]
         [10 -3]]
```

```
[A]*[B]
        [[32 -4 ]
         [36 -22]]
[B]*[A]
        [[-8 26]
         [16 18]]
```

**FIGURE 44**

## Applying Matrix Algebra

**EXAMPLE 10** *Using Matrix Multiplication to Model Plans for a Subdivision*

A contractor builds three kinds of houses, models X, Y, and Z, with a choice of two styles, colonial or ranch. Matrix $A$ shows the number of each kind of house the contractor is planning to build for a new 100-home subdivision. The amounts for each of the main materials used depend on the style of the house. These amounts are shown in matrix $B$, while matrix $C$ gives the cost in dollars for each kind of material. Concrete is measured here in cubic yards, lumber in 1000 board feet, brick in 1000s, and shingles in 100 square feet.

$$\begin{array}{l} \\ \text{Model X} \\ \text{Model Y} \\ \text{Model Z} \end{array} \begin{array}{c} \text{Colonial} \quad \text{Ranch} \\ \begin{bmatrix} 0 & 30 \\ 10 & 20 \\ 20 & 20 \end{bmatrix} \end{array} = A$$

$$\begin{matrix} & \text{Concrete} & \text{Lumber} & \text{Brick} & \text{Shingles} \\ \text{Colonial} \\ \text{Ranch} \end{matrix} \begin{bmatrix} 10 & 2 & 0 & 2 \\ 50 & 1 & 20 & 2 \end{bmatrix} = B \qquad \begin{matrix} \text{Concrete} \\ \text{Lumber} \\ \text{Brick} \\ \text{Shingles} \end{matrix} \overset{\text{Cost per Unit}}{\begin{bmatrix} 20 \\ 180 \\ 60 \\ 25 \end{bmatrix}} = C$$

**(a)** What is the total cost of materials for all houses of each model?

**(b)** How much of each of the four kinds of material must be ordered?

**(c)** Use a graphing calculator to find the total cost of the materials.

**Solution**

**(a)** To find the cost of materials for each model, first find matrix $AB$, which will give the total amount of each material needed for all houses of each model.

$$AB = \begin{bmatrix} 0 & 30 \\ 10 & 20 \\ 20 & 20 \end{bmatrix} \begin{bmatrix} 10 & 2 & 0 & 2 \\ 50 & 1 & 20 & 2 \end{bmatrix} = \begin{matrix} \text{Concrete} & \text{Lumber} & \text{Brick} & \text{Shingles} \end{matrix} \begin{bmatrix} 1500 & 30 & 600 & 60 \\ 1100 & 40 & 400 & 60 \\ 1200 & 60 & 400 & 80 \end{bmatrix} \begin{matrix} \text{Model X} \\ \text{Model Y} \\ \text{Model Z} \end{matrix}$$

Multiplying $AB$ and the cost matrix $C$ gives the total cost of materials for each model.

$$(AB)C = \begin{bmatrix} 1500 & 30 & 600 & 60 \\ 1100 & 40 & 400 & 60 \\ 1200 & 60 & 400 & 80 \end{bmatrix} \begin{bmatrix} 20 \\ 180 \\ 60 \\ 25 \end{bmatrix} = \overset{\text{Cost}}{\begin{bmatrix} 72{,}900 \\ 54{,}700 \\ 60{,}800 \end{bmatrix}} \begin{matrix} \text{Model X} \\ \text{Model Y} \\ \text{Model Z} \end{matrix}$$

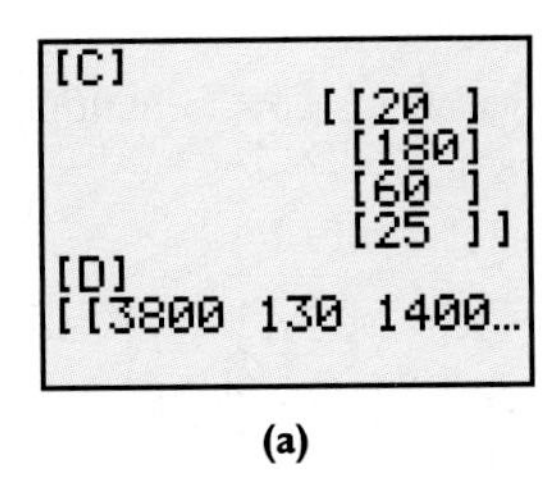


(a)

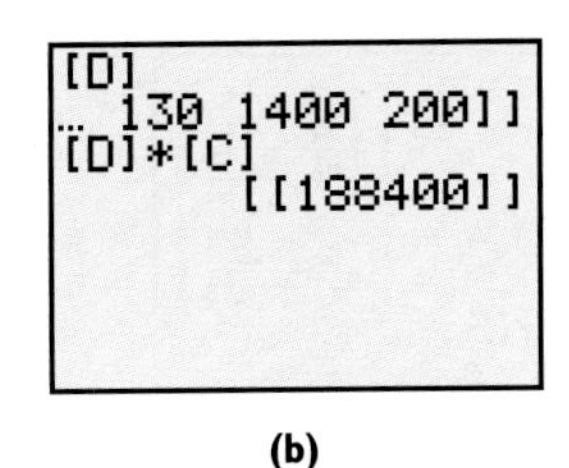


(b)

FIGURE 45

**(b)** The totals of the columns of matrix $AB$ will give a matrix whose elements represent the total amounts of each material needed for the subdivision. Call this matrix $D$, and write it as a row matrix.

$$D = [3800 \quad 130 \quad 1400 \quad 200]$$

**(c)** The total cost of all the materials is given by the product of matrix $C$, the cost matrix, and matrix $D$, the total amounts matrix. To multiply these and get a $1 \times 1$ matrix, representing the total cost, requires multiplying a $1 \times 4$ matrix and a $4 \times 1$ matrix. This is why in part (b) a row matrix was written rather than a column matrix. The total materials cost is given by $DC$. Figure 45 shows how a graphing calculator computes this product. The total cost of the materials is \$188,400. ○

To help keep track of the quantities a matrix represents, let matrix $A$ from Example 10 represent models/styles, matrix $B$ represent styles/materials, and matrix $C$ represent materials/cost. In each case the meaning of the rows is written first and that of the columns second. When the product $AB$ was found in Example 10, the rows of the matrix represented models and the columns represented materials. Therefore, the matrix product $AB$ represents models/materials.

## 7.4 EXERCISES

*Find the dimension of each matrix. Identify any square, column, or row matrices.*

**1.**
```
[A]
        [[-3 6 ]
         [7  -4]]
```

**2.**
```
[A]
        [[2 -8 6 ]
         [1 0  -5]
         [5 -2 3 ]]
```

**3.** $\begin{bmatrix} -6 & 8 & 0 & 0 \\ 4 & 1 & 9 & 2 \\ 3 & -5 & 7 & 1 \end{bmatrix}$

**4.** $\begin{bmatrix} -3 & 4 & 2 & 1 \\ 0 & 8 & 6 & 3 \end{bmatrix}$

**5.** $\begin{bmatrix} 2 \\ 4 \end{bmatrix}$

**6.** $[4 \ 9]$

**7.** $[-9]$

**8.** $\begin{bmatrix} 0 & 0 & 0 & 0 & 0 \\ 0 & 0 & 0 & 0 & 0 \end{bmatrix}$

*Find the value of each variable.*

**9.** $\begin{bmatrix} w & x \\ y & z \end{bmatrix} = \begin{bmatrix} 3 & 2 \\ -1 & 4 \end{bmatrix}$

**10.** $\begin{bmatrix} 2 & 5 & 6 \\ 1 & m & n \end{bmatrix} = \begin{bmatrix} z & y & w \\ 1 & 8 & -2 \end{bmatrix}$

**11.** $\begin{bmatrix} 0 & 5 & x \\ -1 & 3 & y+2 \\ 4 & 1 & z \end{bmatrix} = \begin{bmatrix} 0 & w+3 & 6 \\ -1 & 3 & 0 \\ 4 & 1 & 8 \end{bmatrix}$

**12.** $\begin{bmatrix} 3+x & 4 & t \\ 5 & 8-w & y+1 \\ -4 & 3 & 2r \end{bmatrix} = \begin{bmatrix} 9 & 4 & 6 \\ z+3 & w & 9 \\ p & q & r \end{bmatrix}$

**13.** $\begin{bmatrix} -7+z & 4r & 8s \\ 6p & 2 & 5 \end{bmatrix} + \begin{bmatrix} -9 & 8r & 3 \\ 2 & 5 & 4 \end{bmatrix} = \begin{bmatrix} 2 & 36 & 27 \\ 20 & 7 & 12a \end{bmatrix}$

**14.** $\begin{bmatrix} a+2 & 3z+1 & 5m \\ 8k & 0 & 3 \end{bmatrix} + \begin{bmatrix} 3a & 2z & 5m \\ 2k & 5 & 6 \end{bmatrix} = \begin{bmatrix} 10 & -14 & 80 \\ 10 & 5 & 9 \end{bmatrix}$

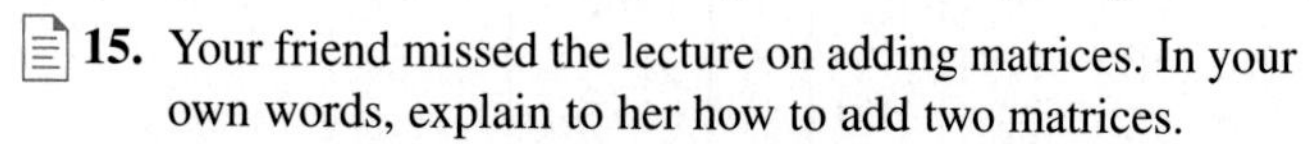

**15.** Your friend missed the lecture on adding matrices. In your own words, explain to her how to add two matrices.

**16.** Explain to a friend in your own words how to multiply a matrix by a scalar.

*Perform each operation, when possible.*

**17.** $\begin{bmatrix} 6 & -9 & 2 \\ 4 & 1 & 3 \end{bmatrix} + \begin{bmatrix} -8 & 2 & 5 \\ 6 & -3 & 4 \end{bmatrix}$

**18.** $\begin{bmatrix} 9 & 4 \\ -8 & 2 \end{bmatrix} + \begin{bmatrix} -3 & 2 \\ -4 & 7 \end{bmatrix}$

**19.** $\begin{bmatrix} -6 & 8 \\ 0 & 0 \end{bmatrix} - \begin{bmatrix} 0 & 0 \\ -4 & -2 \end{bmatrix}$

**20.** $\begin{bmatrix} 1 & -4 \\ 2 & -3 \\ -8 & 4 \end{bmatrix} - \begin{bmatrix} -6 & 9 \\ -2 & 5 \\ -7 & -12 \end{bmatrix}$

**21.**
```
[A]
        [[6 -2]
         [5 4 ]]
[B]
        [[-1 7 ]
         [7  -4]]
[A]+[B]
```

**22.**
```
[A]
        [[12 -5]
         [10 3 ]]
[B]
        [[6  9]
         [-2 0]]
[A]-[B]
```

**23.** $\begin{bmatrix} -8 & 4 & 0 \\ 2 & 5 & 0 \end{bmatrix} + \begin{bmatrix} 6 & 3 \\ 8 & 9 \end{bmatrix}$

**24.** $\begin{bmatrix} 2 \\ 3 \end{bmatrix} - \begin{bmatrix} 8 & 1 \\ 9 & 4 \end{bmatrix}$

**25.** $\begin{bmatrix} 9 & 4 & 1 & -2 \\ 5 & -6 & 3 & 4 \\ 2 & -5 & 1 & 2 \end{bmatrix} - \begin{bmatrix} -2 & 5 & 1 & 3 \\ 0 & 1 & 0 & 2 \\ -8 & 3 & 2 & 1 \end{bmatrix} + \begin{bmatrix} 2 & 4 & 0 & 3 \\ 4 & -5 & 1 & 6 \\ 2 & -3 & 0 & 8 \end{bmatrix}$

**26.** $\begin{bmatrix} 6 & -2 & 4 \\ -2 & 5 & 8 \\ 1 & 0 & 2 \end{bmatrix} + \begin{bmatrix} 3 & 0 & 8 \\ 1 & -2 & 4 \\ 6 & 9 & -2 \end{bmatrix} - \begin{bmatrix} -4 & 2 & 1 \\ 0 & 3 & -2 \\ 4 & 2 & 0 \end{bmatrix}$

**27.** $3\begin{bmatrix} 6 & -1 & 4 \\ 2 & 8 & -3 \\ -4 & 5 & 6 \end{bmatrix} + 5\begin{bmatrix} -2 & -8 & -6 \\ 4 & 1 & 3 \\ 2 & -1 & 5 \end{bmatrix}$

**28.** $4\begin{bmatrix} 1 & -4 \\ 2 & -3 \\ -8 & 4 \end{bmatrix} - 3\begin{bmatrix} -6 & 9 \\ -2 & 5 \\ -7 & -12 \end{bmatrix}$

*Matrices* [A] *and* [B] *are shown on the screen.*

```
[A]
        [[-2 4]
         [0  3]]
[B]
        [[-6 2]
         [4  0]]
```

*Find each matrix.*

**29.** 2[A]

**30.** −3[B]

**31.** 2[A] − [B]

**32.** −2[A] + [B]

**33.**
```
5[A]+.5[B]
```

**34.**
```
-4[A]+1.5[B]
```

**35.** *Concept Check* Based on the screen shown here, what is matrix [A]?

```
[B]
       [[4  6 -5]
        [-6 3 2 ]]
[A]+[B]
     [[6   12 0 ]
      [-10 -4 11]]
```

**36.** *Concept Check* Based on the screen shown here, what is matrix [B]?

```
[A]
          [[3  6 5]
           [-2 1 4]]
[A]-[B]
        [[9  0 -5]
         [-4 6 -3]]
```

*The dimensions of matrices A and B are given. Find the dimensions of the product AB and the product BA if the products are defined. If they are not defined, say so.*

**37.** $A$ is $4 \times 2$; $B$ is $2 \times 4$.

**38.** $A$ is $3 \times 1$; $B$ is $1 \times 3$.

**39.** $A$ is $3 \times 5$; $B$ is $5 \times 2$.

**40.** $A$ is $7 \times 3$; $B$ is $2 \times 7$.

**41.** $A$ is $4 \times 3$; $B$ is $2 \times 5$.

**42.** $A$ is $1 \times 6$; $B$ is $2 \times 4$.

**43.** *Concept Check* The product $MN$ of two matrices can be found only if the number of ________ of $M$ equals the number of ________ of $N$.

**44.** *Concept Check* When finding the product $AB$ of matrices $A$ and $B$, the first row, second column entry is found by multiplying the ________ elements in $A$ and the ________ elements in $B$ and then ________ these products.

*Find each matrix product, whenever possible.*

**45.** $\begin{bmatrix} p & q \\ r & s \end{bmatrix}\begin{bmatrix} a & c \\ b & d \end{bmatrix}$

**46.** $\begin{bmatrix} a & b & c \\ d & e & f \\ g & h & i \end{bmatrix}\begin{bmatrix} x \\ y \\ z \end{bmatrix}$

**47.** $\begin{bmatrix} 3 & -4 & 1 \\ 5 & 0 & 2 \end{bmatrix}\begin{bmatrix} -1 \\ 4 \\ 2 \end{bmatrix}$

**48.** $\begin{bmatrix} -6 & 3 & 5 \\ 2 & 9 & 1 \end{bmatrix}\begin{bmatrix} -2 \\ 0 \\ 3 \end{bmatrix}$

**49.** $\begin{bmatrix} 5 & 2 \\ -1 & 4 \end{bmatrix}\begin{bmatrix} 3 & -2 \\ 1 & 0 \end{bmatrix}$

**50.** $\begin{bmatrix} -4 & 0 \\ 1 & 3 \end{bmatrix}\begin{bmatrix} -2 & 4 \\ 0 & 1 \end{bmatrix}$

**51.** $\begin{bmatrix} 2 & 2 & -1 \\ 3 & 0 & 1 \end{bmatrix}\begin{bmatrix} 0 & 2 \\ -1 & 4 \\ 0 & 2 \end{bmatrix}$

**52.** $\begin{bmatrix} -9 & 2 & 1 \\ 3 & 0 & 0 \end{bmatrix}\begin{bmatrix} 2 \\ -1 \\ 4 \end{bmatrix}$

**53.** $\begin{bmatrix} -2 & -3 & -4 \\ 2 & -1 & 0 \\ 4 & -2 & 3 \end{bmatrix}\begin{bmatrix} 0 & 1 & 4 \\ 1 & 2 & -1 \\ 3 & 2 & -2 \end{bmatrix}$

**54.** $\begin{bmatrix} -1 & 2 & 0 \\ 0 & 3 & 2 \\ 0 & 1 & 4 \end{bmatrix}\begin{bmatrix} 2 & -1 & 2 \\ 0 & 2 & 1 \\ 3 & 0 & -1 \end{bmatrix}$

**55.**
```
[A]
        [[-2 4 1]]
[B]
        [[3 -2 4]
         [2 1  0]
         [0 -1 4]]
[A][B]
```

**56.**
```
[A]
        [[0 3 -4]]
[B]
        [[-2 6 3]
         [0  4 2]
         [-1 1 4]]
[A][B]
```

*Given* $A = \begin{bmatrix} 4 & -2 \\ 3 & 1 \end{bmatrix}$, $B = \begin{bmatrix} 5 & 1 \\ 0 & -2 \\ 3 & 7 \end{bmatrix}$, *and* $C = \begin{bmatrix} -5 & 4 & 1 \\ 0 & 3 & 6 \end{bmatrix}$, *find each product whenever possible.*

**57.** $BA$  **58.** $AC$  **59.** $BC$  **60.** $CB$

**61.** $AB$  **62.** $CA$  **63.** $A^2$  **64.** $A^3$

**65.** *Concept Check* Compare the answers to Exercises 57 and 61, 59 and 60, and 58 and 62. Is matrix multiplication commutative?

**66.** *Concept Check* For any matrices $P$ and $Q$, what must be true for both $PQ$ and $QP$ to exist?

*Solve each problem.*

**67.** *Company Growth* Rite Aid Corporation recently has been buying small, pharmacist-owned stores at a rapid pace, as indicated in the first table. The same data for the Walgreen Company is given in the second table.

**Rite Aid**

| Year | Revenue (millions of dollars) | Net Income (millions of dollars) | Number of Employees |
|---|---|---|---|
| 1998 | 11,375 | 316 | 83,000 |
| 1997 | 6970 | 115 | 73,000 |
| 1996 | 5446 | 159 | 35,700 |
| 1995 | 4534 | 141 | 36,700 |
| 1994 | 4059 | 9 | 27,364 |

*Source:* Hoover's Outline.

**Walgreen Company**

| Year | Revenue (millions of dollars) | Net Income (millions of dollars) | Number of Employees |
|---|---|---|---|
| 1998 | 15,307 | 511 | 90,000 |
| 1997 | 13,363 | 436 | 85,000 |
| 1996 | 11,778 | 372 | 77,000 |
| 1995 | 10,395 | 321 | 68,000 |
| 1994 | 9235 | 282 | 61,900 |

*Source:* Hoover's Outline.

**(a)** Write the numbers in the last three columns of each table as a $5 \times 3$ matrix.

**(b)** Use the matrices from part (a) to write a matrix giving the total amounts for each year in each category for the two companies. What does the element in row 2, column 3 represent?

**(c)** Write a matrix representing the difference between the Walgreen Company matrix data and the Rite Aid matrix data. What does the element in row 1, column 2 represent?

**68.** *Purchasing Costs* The Bread Box, a small neighborhood bakery, sells four main items: sweet rolls, bread, cakes, and pies. The amount of each ingredient (in cups, except for eggs) required for these items is given by matrix $A$.

| | Eggs | Flour | Sugar | Shortening | Milk |
|---|---|---|---|---|---|
| Rolls (doz) | 1 | 4 | $\frac{1}{4}$ | $\frac{1}{4}$ | 1 |
| Bread (loaves) | 0 | 3 | 0 | $\frac{1}{4}$ | 0 |
| Cakes | 4 | 3 | 2 | 1 | 1 |
| Pies (crust) | 0 | 1 | 0 | $\frac{1}{3}$ | 0 |

$= A$

The cost (in cents) for each ingredient when purchased in large lots or small lots is given in matrix $B$.

$$\begin{array}{l} \\ \\ \text{Eggs} \\ \text{Flour} \\ \text{Sugar} \\ \text{Shortening} \\ \text{Milk} \end{array} \begin{array}{c} \text{Cost} \\ \begin{array}{cc} \text{Large lot} & \text{Small lot} \end{array} \\ \begin{bmatrix} 5 & 5 \\ 8 & 10 \\ 10 & 12 \\ 12 & 15 \\ 5 & 6 \end{bmatrix} \end{array} = B$$

**(a)** Use matrix multiplication to find a matrix giving the comparative cost per item for the two purchase options.

**(b)** Suppose a day's orders consist of 20 dozen sweet rolls, 200 loaves of bread, 50 cakes, and 60 pies. Write the orders as a $1 \times 4$ matrix and, using matrix multiplication, write as a matrix the amount of each ingredient needed to fill the day's orders.

**(c)** Use matrix multiplication to find a matrix giving the costs under the two purchase options to fill the day's orders.

**69.** *(Modeling) Predator-Prey Relationship* In certain parts of the Rocky Mountains, deer provide the main food source for mountain lions. When the deer population is large, the mountain lions thrive. However, a large mountain lion population drives down the size of the deer population. Suppose the fluctuations of the two populations from year to year can be modeled with the matrix equation

$$\begin{bmatrix} m_{n+1} \\ d_{n+1} \end{bmatrix} = \begin{bmatrix} .51 & .4 \\ -.05 & 1.05 \end{bmatrix} \begin{bmatrix} m_n \\ d_n \end{bmatrix}.$$

The numbers in the column matrices give the numbers of animals in the two populations after $n$ years and $n + 1$ years, where the number of deer is measured in hundreds.

**(a)** Give the equation for $d_{n+1}$ obtained from the second row of the square matrix. Use this equation to determine the rate the deer population will grow from year to year if there are no mountain lions.

**(b)** Suppose we start with a mountain lion population of 2000 and a deer population of 500,000 (that is, 5000 hundred deer). How large would each population be after 1 year? 2 years?

**(c)** Consider part (b) but change the initial mountain lion population to 4000. Show that the populations would each grow at a steady annual rate of 1.01.

**70.** *(Modeling) Northern Spotted Owl Population* In 1991, the U.S. Fish and Wildlife Service proposed logging restrictions on nearly 12 million acres of Pacific Northwest forest to help save the endangered northern spotted owl. This decision caused considerable controversy between the logging industry and environmentalists. As a result, mathematical ecologists created a mathematical model to analyze population dynamics of the northern spotted owl. (*Source:* Lamberson, R. H., R. McKelvey, B. R. Noon, and C. Voss, "A Dynamic Analysis of Northern Spotted Owl Viability in a Fragmented Forest Landscape," *Conservation Biology*, Vol. 6, No. 4, December, 1992.)

The ecologists divided the female owl population into three categories: juvenile (up to 1 year old), subadult (1 to 2 years old), and adult (over 2 years old). They concluded that the change in the makeup of the northern spotted owl population in successive years could be described by the following matrix equation.

$$\begin{bmatrix} j_{n+1} \\ s_{n+1} \\ a_{n+1} \end{bmatrix} = \begin{bmatrix} 0 & 0 & .33 \\ .18 & 0 & 0 \\ 0 & .71 & .94 \end{bmatrix} \begin{bmatrix} j_n \\ s_n \\ a_n \end{bmatrix}$$

The numbers in the column matrices give the numbers of females in the three age groups after $n$ years and $n + 1$ years. Multiplying the matrices yields

$j_{n+1} = .33a_n$ — Each year 33 juvenile females are born for each 100 adult females.

$s_{n+1} = .18j_n$ — Each year 18% of the juvenile females survive to become adults.

$a_{n+1} = .71s_n + .94a_n.$ — Each year 71% of the subadults survive to become adults and 94% of the adults survive.

**(a)** Suppose there are currently 3000 female northern spotted owls made up of 690 juveniles, 210 subadults, and 2100 adults. Use the matrix equation above to determine the total number of female owls for each of the next 5 years.

**(b)** Using advanced techniques from linear algebra, we can show that in the long run,

$$\begin{bmatrix} j_{n+1} \\ s_{n+1} \\ a_{n+1} \end{bmatrix} \approx .98359 \begin{bmatrix} j_n \\ s_n \\ a_n \end{bmatrix}.$$

What can we conclude about the long-term fate of the northern spotted owl?

**(c)** In this model, the main impediment to the survival of the northern spotted owl is the number .18 in the second row of the $3 \times 3$ matrix. This number is low for two reasons. The first year of life is precarious for most animals living in the wild. In addition, juvenile owls must eventually leave the nest and establish their own territory. If much of the forest near their original home has been cleared, then they are vulnerable to predators while searching for a new home. Suppose that due to better forest management, the number .18 can be increased to .3. Rework part (a) under this new assumption.

*Let*

$$A = \begin{bmatrix} a_{11} & a_{12} \\ a_{21} & a_{22} \end{bmatrix},$$

$$B = \begin{bmatrix} b_{11} & b_{12} \\ b_{21} & b_{22} \end{bmatrix},$$

*and*

$$C = \begin{bmatrix} c_{11} & c_{12} \\ c_{21} & c_{22} \end{bmatrix},$$

*where all the elements are real numbers. Use these matrices to show that each statement is true for* $2 \times 2$ *matrices.*

**71.** $A + B = B + A$ (commutative property)

**72.** $A + (B + C) = (A + B) + C$ (associative property)

**73.** $(AB)C = A(BC)$ (associative property)

**74.** $A(B + C) = AB + AC$ (distributive property)

**75.** $c(A + B) = cA + cB$, for any real number $c$

**76.** $(c + d)A = cA + dA$, for any real numbers $c$ and $d$

**77.** $(cA)d = (cd)A$

**78.** $(cd)A = c(dA)$

# 7.5 DETERMINANTS AND CRAMER'S RULE

Determinants of 2 × 2 Matrices • Determinants of Larger Matrices • Derivation of Cramer's Rule • Using Cramer's Rule to Solve Systems

For convenience, we use subscript notation to name the elements of a matrix as in the following matrix $A$.

$$A = \begin{bmatrix} a_{11} & a_{12} & a_{13} & \cdots & a_{1n} \\ a_{21} & a_{22} & a_{23} & \cdots & a_{2n} \\ a_{31} & a_{32} & a_{33} & \cdots & a_{3n} \\ \vdots & \vdots & \vdots & & \vdots \\ a_{m1} & a_{m2} & a_{m3} & \cdots & a_{mn} \end{bmatrix}$$

With this notation, the row 1, column 1 element is $a_{11}$; the row 2, column 3 element is $a_{23}$; and, in general, the row $i$, column $j$ element is $a_{ij}$. We use subscript notation in this section to define determinants.

## Determinants of 2 × 2 Matrices

Associated with every square matrix $A$ is a real number called the **determinant** of $A$. The symbols $|A|$, $\delta(A)$, and $\det A$ represent the determinant of $A$. In this text, we use **det** $\boldsymbol{A}$.

**Looking Ahead to Calculus**

Determinants are used in calculus to find *vector cross products,* which are used to study the effect of forces in the plane or in space. They are also used to express the results of certain vector operations.

**Determinant of a 2 × 2 Matrix**

The **determinant of a 2 × 2 matrix** $\boldsymbol{A}$**,**

$$A = \begin{bmatrix} a_{11} & a_{12} \\ a_{21} & a_{22} \end{bmatrix},$$

is defined as

$$\det A = a_{11}a_{22} - a_{21}a_{12}.$$

**NOTE** You should be able to distinguish between a matrix and its determinant. A matrix is an array of numbers, while a determinant is a single real number associated with a square matrix.

**EXAMPLE 1** *Evaluating the Determinant of a 2 × 2 Matrix*

Find det $A$ if $A = \begin{bmatrix} -3 & 4 \\ 6 & 8 \end{bmatrix}$.

**Analytic Solution**

Find the determinant by using the definition just given.

$$\det A = \det\begin{bmatrix} -3 & 4 \\ 6 & 8 \end{bmatrix}$$
$$= -3(8) - 6(4)$$
$$= -48$$

**Graphing Calculator Solution**

Graphing calculators with matrix capabilities can compute determinants. Figure 46 shows both [A] and det[A].

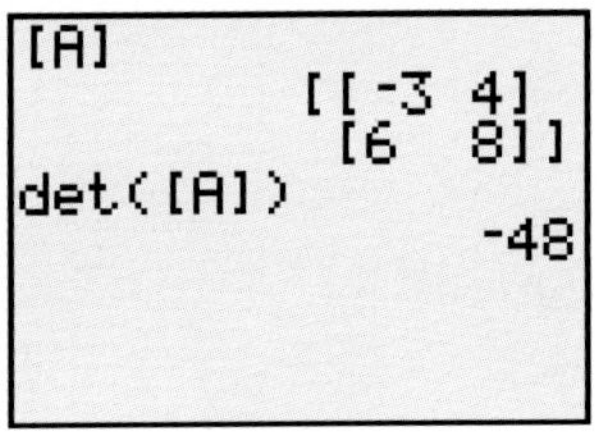

```
[A]
        [[-3 4]
         [6  8]]
det([A])
            -48
```

FIGURE 46

**EXAMPLE 2** *Solving an Equation Involving a Determinant*

If $A = \begin{bmatrix} x & 3 \\ -1 & 5 \end{bmatrix}$ and det $A = 33$, find the value of $x$.

**Solution** Since det $A = 5x - (-1)(3)$, or $5x + 3$, we have

$$5x + 3 = 33$$
$$5x = 30$$
$$x = 6.$$

Check to see that $\det\begin{bmatrix} 6 & 3 \\ -1 & 5 \end{bmatrix} = 33$.

## Determinants of Larger Matrices

**Determinant of a 3 × 3 Matrix**

The **determinant of a 3 × 3 matrix $A$,**

$$A = \begin{bmatrix} a_{11} & a_{12} & a_{13} \\ a_{21} & a_{22} & a_{23} \\ a_{31} & a_{32} & a_{33} \end{bmatrix},$$

is defined as

$$\det A = (a_{11}a_{22}a_{33} + a_{12}a_{23}a_{31} + a_{13}a_{21}a_{32})$$
$$- (a_{31}a_{22}a_{13} + a_{32}a_{23}a_{11} + a_{33}a_{21}a_{12}).$$

A method for calculating $3 \times 3$ determinants is found by rearranging and factoring the terms given in the definition to get

$$\det\begin{bmatrix} a_{11} & a_{12} & a_{13} \\ a_{21} & a_{22} & a_{23} \\ a_{31} & a_{32} & a_{33} \end{bmatrix} = a_{11}(a_{22}a_{33} - a_{32}a_{23}) - a_{21}(a_{12}a_{33} - a_{32}a_{13}) + a_{31}(a_{12}a_{23} - a_{22}a_{13}).$$

Each quantity in parentheses represents the determinant of a $2 \times 2$ matrix that is the part of the $3 \times 3$ matrix remaining when the row and column of the multiplier are eliminated, as shown here.

$$a_{11}(a_{22}a_{33} - a_{32}a_{23}) \quad \begin{bmatrix} a_{11} & a_{12} & a_{13} \\ a_{21} & a_{22} & a_{23} \\ a_{31} & a_{32} & a_{33} \end{bmatrix}$$

$$a_{21}(a_{12}a_{33} - a_{32}a_{13}) \quad \begin{bmatrix} a_{11} & a_{12} & a_{13} \\ a_{21} & a_{22} & a_{23} \\ a_{31} & a_{32} & a_{33} \end{bmatrix}$$

$$a_{31}(a_{12}a_{23} - a_{22}a_{13}) \quad \begin{bmatrix} a_{11} & a_{12} & a_{13} \\ a_{21} & a_{22} & a_{23} \\ a_{31} & a_{32} & a_{33} \end{bmatrix}$$

The determinant of each $2 \times 2$ matrix is called a **minor** of the associated element in the $3 \times 3$ matrix. The symbol $M_{ij}$ represents the minor that results when row $i$ and column $j$ are eliminated. The following list gives some of the minors from the preceding matrix.

| Element | Minor | Element | Minor |
|---|---|---|---|
| $a_{11}$ | $M_{11} = \det\begin{bmatrix} a_{22} & a_{23} \\ a_{32} & a_{33} \end{bmatrix}$ | $a_{22}$ | $M_{22} = \det\begin{bmatrix} a_{11} & a_{13} \\ a_{31} & a_{33} \end{bmatrix}$ |
| $a_{21}$ | $M_{21} = \det\begin{bmatrix} a_{12} & a_{13} \\ a_{32} & a_{33} \end{bmatrix}$ | $a_{23}$ | $M_{23} = \det\begin{bmatrix} a_{11} & a_{12} \\ a_{31} & a_{32} \end{bmatrix}$ |
| $a_{31}$ | $M_{31} = \det\begin{bmatrix} a_{12} & a_{13} \\ a_{22} & a_{23} \end{bmatrix}$ | $a_{33}$ | $M_{33} = \det\begin{bmatrix} a_{11} & a_{12} \\ a_{21} & a_{22} \end{bmatrix}$ |

In a $4 \times 4$ matrix, the minors are determinants of $3 \times 3$ matrices. Similarly, an $n \times n$ matrix has minors that are determinants of $(n - 1) \times (n - 1)$ matrices.

To find the determinant of a $3 \times 3$ or larger square matrix, first choose any row or column. Then, the minor of each element in that row or column must be multiplied by $+1$ or $-1$, depending on whether the sum of the row number and column number is even or odd. The product of a minor and the number $+1$ or $-1$ is called a *cofactor*.

**Cofactor**

Let $M_{ij}$ be the minor for element $a_{ij}$ in an $n \times n$ matrix. The **cofactor** of $a_{ij}$, written $A_{ij}$, is

$$A_{ij} = (-1)^{i+j} \cdot M_{ij}.$$

**EXAMPLE 3** *Finding the Cofactor of an Element*

For the matrix

$$\begin{bmatrix} 6 & 2 & 4 \\ 8 & 9 & 3 \\ 1 & 2 & 0 \end{bmatrix},$$

find the cofactor of each element.

**(a)** 6 **(b)** 3 **(c)** 8

**Solution**

**(a)** Since 6 is in the first row, first column of the matrix, $i = 1$ and $j = 1$. $M_{11} = \det\begin{bmatrix} 9 & 3 \\ 2 & 0 \end{bmatrix} = -6$. The cofactor is

$$(-1)^{1+1}(-6) = 1(-6) = -6.$$

**(b)** Here $i = 2$ and $j = 3$. $M_{23} = \det\begin{bmatrix} 6 & 2 \\ 1 & 2 \end{bmatrix} = 10$. The cofactor is

$$(-1)^{2+3}(10) = -1(10) = -10.$$

**(c)** We have $i = 2$ and $j = 1$. $M_{21} = \det\begin{bmatrix} 2 & 4 \\ 2 & 0 \end{bmatrix} = -8$. The cofactor is

$$(-1)^{2+1}(-8) = -1(-8) = 8.$$

Finally, the determinant of an $n \times n$ matrix is found as follows.

**Finding the Determinant of an $n \times n$ Matrix**

Multiply each element in any row or column of the matrix by its cofactor. The sum of these products gives the value of the determinant.

The process of forming this sum of products is called **expansion by a given row or column.**

**EXAMPLE 4** *Evaluating the Determinant of a 3 × 3 Matrix*

Evaluate $\det\begin{bmatrix} 2 & -3 & -2 \\ -1 & -4 & -3 \\ -1 & 0 & 2 \end{bmatrix}$, expanding by the second column.

In Example 1 we saw that the determinant of a $2 \times 2$ matrix can be found using a graphing calculator. Figure 47 shows how a calculator evaluates the $3 \times 3$ determinant of Example 4.

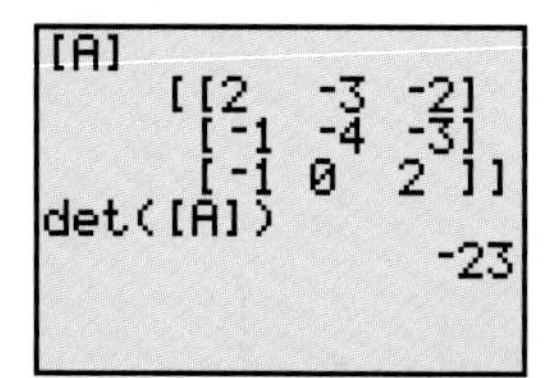


FIGURE 47

**Solution** First find the minor of each element in the second column.

$$M_{12} = \det\begin{bmatrix} -1 & -3 \\ -1 & 2 \end{bmatrix} = -1(2) - (-1)(-3) = -5$$

$$M_{22} = \det\begin{bmatrix} 2 & -2 \\ -1 & 2 \end{bmatrix} = 2(2) - (-1)(-2) = 2$$

$$M_{32} = \det\begin{bmatrix} 2 & -2 \\ -1 & -3 \end{bmatrix} = 2(-3) - (-1)(-2) = -8$$

Now, find the cofactor of each of these minors.

$$A_{12} = (-1)^{1+2} \cdot M_{12} = (-1)^3(-5) = -1(-5) = 5$$

$$A_{22} = (-1)^{2+2} \cdot M_{22} = (-1)^4 \cdot 2 = 1 \cdot 2 = 2$$

$$A_{32} = (-1)^{3+2} \cdot M_{32} = (-1)^5(-8) = -1(-8) = 8$$

Find the determinant by multiplying each cofactor by its corresponding element in the matrix and finding the sum of these products.

$$\det\begin{bmatrix} 2 & -3 & -2 \\ -1 & -4 & -3 \\ -1 & 0 & 2 \end{bmatrix} = a_{12} \cdot A_{12} + a_{22} \cdot A_{22} + a_{32} \cdot A_{32}$$

$$= -3(5) + (-4)2 + 0(8)$$

$$= -15 + (-8) + 0 = -23$$

**CAUTION** Be careful to keep track of all negative signs when evaluating determinants by hand. Write down each step as in the examples. Skipping steps frequently leads to computational errors.

We would have found exactly the same answer using any row or column of the matrix. One reason we used column 2 in Example 4 is that it contains a 0 element, so it was not really necessary to calculate $M_{32}$ and $A_{32}$. One learns quickly that 0s are "friendly" in work with determinants.

Instead of calculating $(-1)^{i+j}$ for a given element, the sign checkerboard in the margin can be used. The signs alternate for each row and column, beginning with + in the first row, first column position. If we expand a $3 \times 3$ matrix about row 3, for example, the first minor would have a + sign associated with it, the second minor a − sign, and the third minor a + sign. This array of signs can be extended for determinants of $4 \times 4$ and larger matrices.

| For 3 × 3 Matrices | | |
|---|---|---|
| + | − | + |
| − | + | − |
| + | − | + |

**EXAMPLE 5** *Evaluating the Determinant of a 4 × 4 Matrix*

Evaluate

$$\det\begin{bmatrix} -1 & -2 & 3 & 2 \\ 0 & 1 & 4 & -2 \\ 3 & -1 & 4 & 0 \\ 2 & 1 & 0 & 3 \end{bmatrix}.$$

**Solution** The determinant of a $4 \times 4$ matrix can be found using cofactors, but the computation is tedious. Using a graphing calculator with matrix capabilities is much easier. Figure 48 indicates that the desired determinant is $-185$.

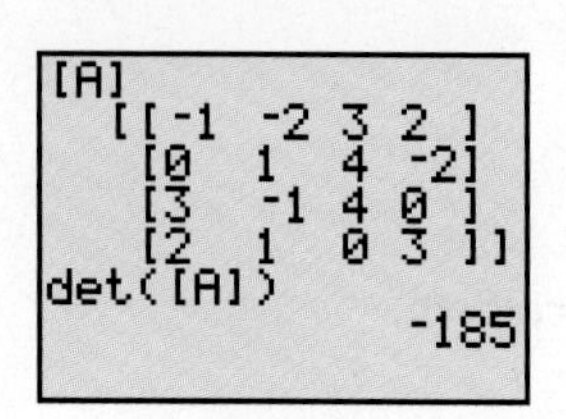


FIGURE 48

## Derivation of Cramer's Rule

Determinants can be used to solve a linear system of equations in the form

$$a_1x + b_1y = c_1 \quad (1)$$
$$a_2x + b_2y = c_2 \quad (2)$$

by elimination as follows.

$$a_1b_2x + b_1b_2y = c_1b_2 \qquad \text{Multiply (1) by } b_2.$$
$$-a_2b_1x - b_1b_2y = -c_2b_1 \qquad \text{Multiply (2) by } -b_1.$$
$$(a_1b_2 - a_2b_1)x = c_1b_2 - c_2b_1 \qquad \text{Add.}$$
$$x = \frac{c_1b_2 - c_2b_1}{a_1b_2 - a_2b_1}, \quad \text{if } a_1b_2 - a_2b_1 \neq 0$$

Similarly,

$$-a_1a_2x - a_2b_1y = -a_2c_1 \qquad \text{Multiply (1) by } -a_2.$$
$$a_1a_2x + a_1b_2y = a_1c_2 \qquad \text{Multiply (2) by } a_1.$$
$$(a_1b_2 - a_2b_1)y = a_1c_2 - a_2c_1 \qquad \text{Add.}$$
$$y = \frac{a_1c_2 - a_2c_1}{a_1b_2 - a_2b_1}, \quad \text{if } a_1b_2 - a_2b_1 \neq 0.$$

Both numerators and the common denominator of these values for $x$ and $y$ can be written as determinants, since

$$c_1b_2 - c_2b_1 = \det\begin{bmatrix} c_1 & b_1 \\ c_2 & b_2 \end{bmatrix}, \quad a_1c_2 - a_2c_1 = \det\begin{bmatrix} a_1 & c_1 \\ a_2 & c_2 \end{bmatrix}, \quad \text{and} \quad a_1b_2 - a_2b_1 = \det\begin{bmatrix} a_1 & b_1 \\ a_2 & b_2 \end{bmatrix}.$$

Using these determinants, the solutions for $x$ and $y$ become

$$x = \frac{\det\begin{bmatrix} c_1 & b_1 \\ c_2 & b_2 \end{bmatrix}}{\det\begin{bmatrix} a_1 & b_1 \\ a_2 & b_2 \end{bmatrix}} \quad \text{and} \quad y = \frac{\det\begin{bmatrix} a_1 & c_1 \\ a_2 & c_2 \end{bmatrix}}{\det\begin{bmatrix} a_1 & b_1 \\ a_2 & b_2 \end{bmatrix}}, \quad \text{if } \det\begin{bmatrix} a_1 & b_1 \\ a_2 & b_2 \end{bmatrix} \neq 0.$$

These results are summarized as **Cramer's rule.**

**Cramer's Rule for 2 × 2 Systems**

For the system

$$a_1x + b_1y = c_1$$
$$a_2x + b_2y = c_2,$$
$$x = \frac{D_x}{D} \quad \text{and} \quad y = \frac{D_y}{D},$$

where

$$D_x = \det\begin{bmatrix} c_1 & b_1 \\ c_2 & b_2 \end{bmatrix}, \quad D_y = \det\begin{bmatrix} a_1 & c_1 \\ a_2 & c_2 \end{bmatrix}, \quad \text{and} \quad D = \det\begin{bmatrix} a_1 & b_1 \\ a_2 & b_2 \end{bmatrix} \neq 0.$$

**CAUTION** Cramer's rule does not apply if $D = 0$. When $D = 0$, the system is inconsistent or has dependent equations. For this reason, evaluate $D$ first.

## Using Cramer's Rule to Solve Systems

**EXAMPLE 6** *Applying Cramer's Rule to a System with Two Equations*

Use Cramer's rule to solve the system

$$5x + 7y = -1$$
$$6x + 8y = 1.$$

### Analytic Solution

By Cramer's rule, $x = \frac{D_x}{D}$ and $y = \frac{D_y}{D}$. Find $D$ first, because if $D = 0$, Cramer's rule does not apply. If $D \neq 0$, then find $D_x$ and $D_y$.

$$D = \det\begin{bmatrix} 5 & 7 \\ 6 & 8 \end{bmatrix} = 5(8) - 6(7) = -2$$

$$D_x = \det\begin{bmatrix} -1 & 7 \\ 1 & 8 \end{bmatrix} = (-1)(8) - 1(7) = -15$$

$$D_y = \det\begin{bmatrix} 5 & -1 \\ 6 & 1 \end{bmatrix} = 5(1) - 6(-1) = 11$$

From Cramer's rule,

$$x = \frac{D_x}{D} = \frac{-15}{-2} = \frac{15}{2}$$

and

$$y = \frac{D_y}{D} = \frac{11}{-2} = -\frac{11}{2}.$$

The solution set is $\left\{\left(\frac{15}{2}, -\frac{11}{2}\right)\right\}$, as can be verified by substituting in the given system.

### Graphing Calculator Solution

Enter $D$, $D_x$, and $D_y$ as [A], [B], and [C], respectively. Then find the desired quotients

$$x = \frac{\det([B])}{\det([A])} \quad \text{and} \quad y = \frac{\det([C])}{\det([A])}.$$

See Figure 49.

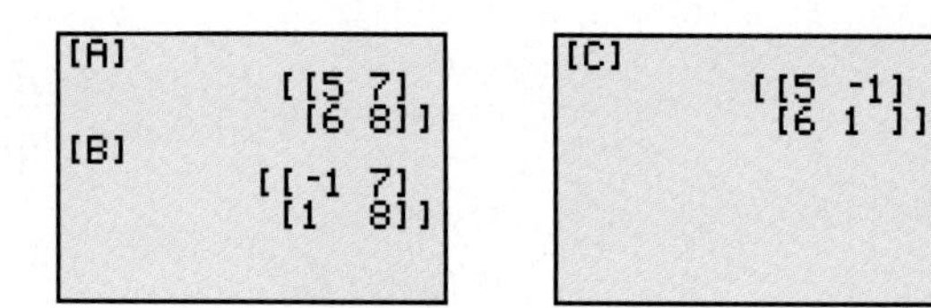


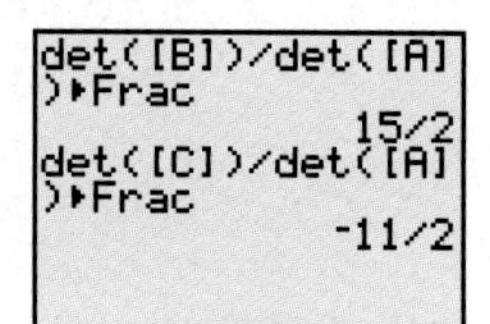


FIGURE 49

Cramer's rule can be generalized to systems of three equations in three variables (or $n$ equations in $n$ variables).

**Cramer's Rule for 3 × 3 Systems**

For the system

$$a_1x + b_1y + c_1z = d_1$$
$$a_2x + b_2y + c_2z = d_2$$
$$a_3x + b_3y + c_3z = d_3,$$

$$x = \frac{D_x}{D}, \quad y = \frac{D_y}{D}, \quad \text{and} \quad z = \frac{D_z}{D},$$

where $D_x = \det\begin{bmatrix} d_1 & b_1 & c_1 \\ d_2 & b_2 & c_2 \\ d_3 & b_3 & c_3 \end{bmatrix}$, $D_y = \det\begin{bmatrix} a_1 & d_1 & c_1 \\ a_2 & d_2 & c_2 \\ a_3 & d_3 & c_3 \end{bmatrix}$,

$D_z = \det\begin{bmatrix} a_1 & b_1 & d_1 \\ a_2 & b_2 & d_2 \\ a_3 & b_3 & d_3 \end{bmatrix}$, and $D = \det\begin{bmatrix} a_1 & b_1 & c_1 \\ a_2 & b_2 & c_2 \\ a_3 & b_3 & c_3 \end{bmatrix} \neq 0.$

**EXAMPLE 7** *Applying Cramer's Rule to a System with Three Equations*

Use Cramer's rule to solve the system.

$$\begin{aligned} x + y - z &= -2 \\ 2x - y + z &= -5 \\ x - 2y + 3z &= 4 \end{aligned}$$

**Solution** Verify that the required determinants are

$$D = \det\begin{bmatrix} 1 & 1 & -1 \\ 2 & -1 & 1 \\ 1 & -2 & 3 \end{bmatrix} = -3, \quad D_x = \det\begin{bmatrix} -2 & 1 & -1 \\ -5 & -1 & 1 \\ 4 & -2 & 3 \end{bmatrix} = 7,$$

$$D_y = \det\begin{bmatrix} 1 & -2 & -1 \\ 2 & -5 & 1 \\ 1 & 4 & 3 \end{bmatrix} = -22, \quad D_z = \det\begin{bmatrix} 1 & 1 & -2 \\ 2 & -1 & -5 \\ 1 & -2 & 4 \end{bmatrix} = -21.$$

Thus,

$$x = \frac{D_x}{D} = \frac{7}{-3} = -\frac{7}{3}, \quad y = \frac{D_y}{D} = \frac{-22}{-3} = \frac{22}{3}, \quad z = \frac{D_z}{D} = \frac{-21}{-3} = 7,$$

so the solution set is $\left\{\left(-\frac{7}{3}, \frac{22}{3}, 7\right)\right\}$.

**NOTE** Example 7 can be solved using a graphing calculator and Cramer's Rule by extending the concepts of the Graphing Calculator Solution in Example 6.

**EXAMPLE 8** *Verifying That Cramer's Rule Does Not Apply*

Show why Cramer's rule does not apply to the system.

$$\begin{aligned} 2x - 3y + 4z &= 10 \\ 6x - 9y + 12z &= 24 \\ x + 2y - 3z &= 5 \end{aligned}$$

**Solution** We must show that $D = 0$. Figure 50 confirms this fact, with $D = \det([A])$. When $D = 0$, the system is either inconsistent or contains dependent equations. We could use the elimination or row echelon methods to tell which is the case. Verify that this system is inconsistent.

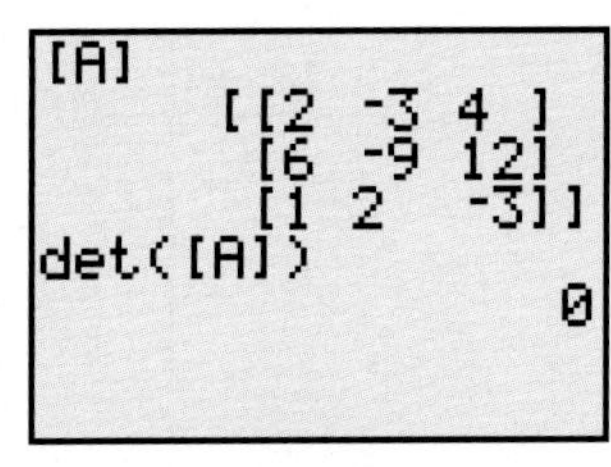


**FIGURE 50**

## 7.5 EXERCISES

*Find each determinant.*

**1.**
```
[A]
       [[-5 9 ]
        [4  -1]]
det([A])
```

**2.**
```
[A]
       [[-1 3]
        [-2 9]]
det([A])
```

**3.** $\det\begin{bmatrix} -1 & -2 \\ 5 & 3 \end{bmatrix}$

**4.** $\det\begin{bmatrix} 6 & -4 \\ 0 & -1 \end{bmatrix}$

**5.** $\det\begin{bmatrix} 9 & 3 \\ -3 & -1 \end{bmatrix}$

**6.** $\det\begin{bmatrix} 0 & 2 \\ 1 & 5 \end{bmatrix}$

**7.** $\det\begin{bmatrix} 3 & 4 \\ 5 & -2 \end{bmatrix}$

**8.** $\det\begin{bmatrix} -9 & 7 \\ 2 & 6 \end{bmatrix}$

*Find the cofactor of each element in the second row for each matrix.*

**9.** $\begin{bmatrix} -2 & 0 & 1 \\ 1 & 2 & 0 \\ 4 & 2 & 1 \end{bmatrix}$

**10.** $\begin{bmatrix} 1 & -1 & 2 \\ 1 & 0 & 2 \\ 0 & -3 & 1 \end{bmatrix}$

**11.** $\begin{bmatrix} 1 & 2 & -1 \\ 2 & 3 & -2 \\ -1 & 4 & 1 \end{bmatrix}$

**12.** $\begin{bmatrix} 2 & -1 & 4 \\ 3 & 0 & 1 \\ -2 & 1 & 4 \end{bmatrix}$

*Find each determinant.*

**13.**
```
[A]
     [[4  -7 8]
      [2   1 3]
      [-6  3 0]]
det([A])
```

**14.**
```
[A]
     [[8 -2 -4]
      [7  0  3 ]
      [5 -1  2 ]]
det([A])
```

**15.** $\det\begin{bmatrix} 1 & 2 & 0 \\ -1 & 2 & -1 \\ 0 & 1 & 4 \end{bmatrix}$

**16.** $\det\begin{bmatrix} 2 & 1 & -1 \\ 4 & 7 & -2 \\ 2 & 4 & 0 \end{bmatrix}$

**17.** $\det\begin{bmatrix} 10 & 2 & 1 \\ -1 & 4 & 3 \\ -3 & 8 & 10 \end{bmatrix}$

**18.** $\det\begin{bmatrix} 7 & -1 & 1 \\ 1 & -7 & 2 \\ -2 & 1 & 1 \end{bmatrix}$

**19.** $\det\begin{bmatrix} 1 & -2 & 3 \\ 0 & 0 & 0 \\ 1 & 10 & -12 \end{bmatrix}$

**20.** $\det\begin{bmatrix} 2 & 3 & 0 \\ 1 & 9 & 0 \\ -1 & -2 & 0 \end{bmatrix}$

**21.** $\det\begin{bmatrix} 3 & 3 & -1 \\ 2 & 6 & 0 \\ -6 & -6 & 2 \end{bmatrix}$

**22.** $\det\begin{bmatrix} 5 & -3 & 2 \\ -5 & 3 & -2 \\ 1 & 0 & 1 \end{bmatrix}$

**23.** $\det\begin{bmatrix} .4 & -.8 & .6 \\ .3 & .9 & .7 \\ 3.1 & 4.1 & -2.8 \end{bmatrix}$

**24.** $\det\begin{bmatrix} -.3 & -.1 & .9 \\ 2.5 & 4.9 & -3.2 \\ -.1 & .4 & .8 \end{bmatrix}$

*Solve each determinant equation for x.*

**25.** $\det\begin{bmatrix} 5 & x \\ -3 & 2 \end{bmatrix} = 6$

**26.** $\det\begin{bmatrix} -.5 & 2 \\ x & x \end{bmatrix} = 0$

**27.** $\det\begin{bmatrix} x & 3 \\ x & x \end{bmatrix} = 4$

**28.** $\det\begin{bmatrix} 2x & x \\ 11 & x \end{bmatrix} = 6$

**29.** $\det\begin{bmatrix} -2 & 0 & 1 \\ -1 & 3 & x \\ 5 & -2 & 0 \end{bmatrix} = 3$

**30.** $\det\begin{bmatrix} 4 & 3 & 0 \\ 2 & 0 & 1 \\ -3 & x & -1 \end{bmatrix} = 5$

**31.** $\det\begin{bmatrix} 5 & 3x & -3 \\ 0 & 2 & -1 \\ 4 & -1 & x \end{bmatrix} = -7$

**32.** $\det\begin{bmatrix} 2x & 1 & -1 \\ 0 & 4 & x \\ 3 & 0 & 2 \end{bmatrix} = x$

*Area of a Triangle* *A triangle with vertices at* $(x_1, y_1)$, $(x_2, y_2)$, *and* $(x_3, y_3)$, *as shown in the figure, has area equal to the absolute value of D, where*

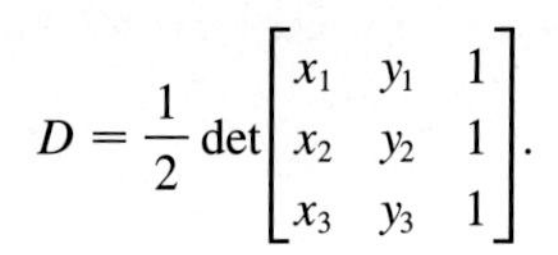

$$D = \frac{1}{2}\det\begin{bmatrix} x_1 & y_1 & 1 \\ x_2 & y_2 & 1 \\ x_3 & y_3 & 1 \end{bmatrix}.$$

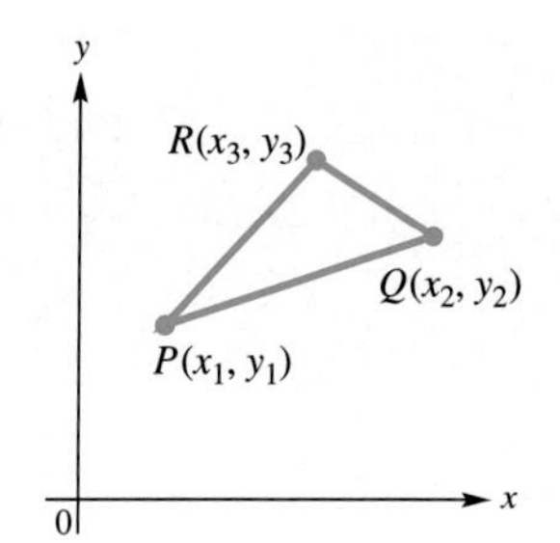


*Use D to find the area of each triangle with coordinates as given.*

**33.** $P(0,0)$, $Q(0,2)$, $R(1,4)$

**34.** $P(0,1)$, $Q(2,0)$, $R(1,5)$

**35.** $P(2,5)$, $Q(-1,3)$, $R(4,0)$

**36.** $P(2,-2)$, $Q(0,0)$, $R(-3,-4)$

**37.** $P(1,2)$, $Q(4,3)$, $R(3,5)$

**38.** Find the area of a triangular lot whose vertices have coordinates in feet of

$$(101.3, 52.7), \quad (117.2, 253.9), \quad \text{and} \quad (313.1, 301.6).$$

(*Source:* Al-Khafaji, A. and J. Tooley, *Numerical Methods in Engineering Practice,* Holt, Rinehart and Winston, 1995.)

*Evaluate each determinant.*

**39.** $\det\begin{bmatrix} 3 & -6 & 5 & -1 \\ 0 & 2 & -1 & 3 \\ -6 & 4 & 2 & 0 \\ -7 & 3 & 1 & 1 \end{bmatrix}$

**40.** $\det\begin{bmatrix} 4 & 5 & -1 & -1 \\ 2 & -3 & 1 & 0 \\ -5 & 1 & 3 & 9 \\ 0 & -2 & 1 & 5 \end{bmatrix}$

**41.** $\det\begin{bmatrix} 4 & 0 & 0 & 2 \\ -1 & 0 & 3 & 0 \\ 2 & 4 & 0 & 1 \\ 0 & 0 & 1 & 2 \end{bmatrix}$

**42.** $\det\begin{bmatrix} -2 & 0 & 4 & 2 \\ 3 & 6 & 0 & 4 \\ 0 & 0 & 0 & 3 \\ 9 & 0 & 2 & -1 \end{bmatrix}$

*Several theorems are useful when calculating determinants. These theorems are true for square matrices of any dimension.*

## Determinant Theorems

1. If every element in a row (or column) of matrix $A$ is 0, then $\det A = 0$.
2. If the rows of matrix $A$ are the corresponding columns of matrix $B$, then $\det B = \det A$.
3. If any two rows (or columns) of matrix $A$ are interchanged to form matrix $B$, then $\det B = -\det A$.
4. Suppose matrix $B$ is formed by multiplying every element of a row (or column) of matrix $A$ by the real number $k$. Then $\det B = k \cdot \det A$.
5. If two rows (or columns) of a matrix $A$ are identical, then $\det A = 0$.
6. Changing a row (or column) of a matrix by adding to it a constant times another row (or column) does not change the determinant of the matrix.

*Use the determinant theorems to find each determinant.*

**43.** $\det\begin{bmatrix} 1 & 0 & 0 \\ 1 & 0 & 1 \\ 3 & 0 & 0 \end{bmatrix}$

**44.** $\det\begin{bmatrix} -1 & 2 & 4 \\ 4 & -8 & -16 \\ 3 & 0 & 5 \end{bmatrix}$

**45.** $\det\begin{bmatrix} 6 & 8 & -12 \\ -1 & 0 & 2 \\ 4 & 0 & -8 \end{bmatrix}$

**46.** $\det\begin{bmatrix} 4 & 8 & 0 \\ -1 & -2 & 1 \\ 2 & 4 & 3 \end{bmatrix}$

**47.** $\det\begin{bmatrix} -4 & 1 & 4 \\ 2 & 0 & 1 \\ 0 & 2 & 4 \end{bmatrix}$

**48.** $\det\begin{bmatrix} 6 & 3 & 2 \\ 1 & 0 & 2 \\ 5 & 7 & 3 \end{bmatrix}$

## RELATING CONCEPTS

**For individual or group investigation (Exercises 49–52)**

*Determinants can be used to find the equation of a line passing through two given points. Exercises 49–52 show how this is done.*

**49.** Expand the determinant in the equation $\det\begin{bmatrix} x & y & 1 \\ 2 & 3 & 1 \\ -1 & 4 & 1 \end{bmatrix} = 0.$

**50.** Find the equation of the line through $(2, 3)$ and $(-1, 4)$. How does your answer compare with the answer to Exercise 49?

**51.** Write the equation of the line through the points $(x_1, y_1)$ and $(x_2, y_2)$, using the point-slope formula.

**52.** Expand the following determinant, and compare the resulting equation with your answer to Exercise 51. What do you find?

$$\det\begin{bmatrix} x & y & 1 \\ x_1 & y_1 & 1 \\ x_2 & y_2 & 1 \end{bmatrix} = 0$$

**53.** *Concept Check* For the following system, match each determinant in (a)–(d) with its equivalent from choices A–D.

$$\begin{aligned} 4x + 3y - 2z &= 1 \\ 7x - 4y + 3z &= 2 \\ -2x + y - 8z &= 0 \end{aligned}$$

**(a)** $D$ **(b)** $D_x$ **(c)** $D_y$ **(d)** $D_z$

**A.** $\det\begin{bmatrix} 1 & 3 & -2 \\ 2 & -4 & 3 \\ 0 & 1 & -8 \end{bmatrix}$ **B.** $\det\begin{bmatrix} 4 & 3 & 1 \\ 7 & -4 & 2 \\ -2 & 1 & 0 \end{bmatrix}$ **C.** $\det\begin{bmatrix} 4 & 1 & -2 \\ 7 & 2 & 3 \\ -2 & 0 & -8 \end{bmatrix}$ **D.** $\det\begin{bmatrix} 4 & 3 & -2 \\ 7 & -4 & 3 \\ -2 & 1 & -8 \end{bmatrix}$

**54.** *Concept Check* For the following system, $D = -43$, $D_x = -43$, $D_y = 0$, and $D_z = 43$. What is the solution set of the system?

$$\begin{aligned} x + 3y - 6z &= 7 \\ 2x - y + z &= 1 \\ x + 2y + 2z &= -1 \end{aligned}$$

*Use Cramer's rule to solve each system of equations. If $D = 0$, use another method to complete the solution.*

**55.** $\begin{aligned} x + y &= 4 \\ 2x - y &= 2 \end{aligned}$

**56.** $\begin{aligned} 3x + 2y &= -4 \\ 2x - y &= -5 \end{aligned}$

**57.** $\begin{aligned} 4x + 3y &= -7 \\ 2x + 3y &= -11 \end{aligned}$

**58.** $\begin{aligned} 4x - y &= 0 \\ 2x + 3y &= 14 \end{aligned}$

**59.** $\begin{aligned} 3x + 2y &= 4 \\ 6x + 4y &= 8 \end{aligned}$

**60.** $\begin{aligned} 1.5x + 3y &= 5 \\ 2x + 4y &= 3 \end{aligned}$

**61.** $\begin{aligned} 2x - 3y &= -5 \\ x + 5y &= 17 \end{aligned}$

**62.** $\begin{aligned} x + 9y &= -15 \\ 3x + 2y &= 5 \end{aligned}$

**63.** $\begin{aligned} 4x - y + 3z &= -3 \\ 3x + y + z &= 0 \\ 2x - y + 4z &= 0 \end{aligned}$

**64.** $\begin{aligned} 5x + 2y + z &= 15 \\ 2x - y + z &= 9 \\ 4x + 3y + 2z &= 13 \end{aligned}$

**65.** $\begin{aligned} 2x - y + 4z &= -2 \\ 3x + 2y - z &= -3 \\ x + 4y + 2z &= 17 \end{aligned}$

**66.** $\begin{aligned} x + y + z &= 4 \\ 2x - y + 3z &= 4 \\ 4x + 2y - z &= -15 \end{aligned}$

**67.** $\begin{aligned} 5x - y &= -4 \\ 3x + 2z &= 4 \\ 4y + 3z &= 22 \end{aligned}$

**68.** $\begin{aligned} 3x + 5y &= -7 \\ 2x + 7z &= 2 \\ 4y + 3z &= -8 \end{aligned}$

**69.** $\begin{aligned} 2x - y + 3z &= 1 \\ -2x + y - 3z &= 2 \\ 5x - y + z &= 2 \end{aligned}$

**70.** $\begin{aligned} -2x - 2y + 3z &= 4 \\ 5x + 7y - z &= 2 \\ 2x + 2y - 3z &= -4 \end{aligned}$

**71.** $\begin{aligned} 3x + 2y - w &= 0 \\ 2x + z + 2w &= 5 \\ x + 2y - z &= -2 \\ 2x - y + z + w &= 2 \end{aligned}$

**72.** $\begin{aligned} x + 2y - z + w &= 8 \\ 2x - y + 2w &= 8 \\ y + 3z &= 5 \\ x - z &= 4 \end{aligned}$

**73.** In your own words, explain what it means when applying Cramer's rule if $D = 0$.

**74.** Describe $D_x$, $D_y$, and $D_z$ in terms of the coefficients and constants in a given system of equations.

# 7.6 SOLUTION OF LINEAR SYSTEMS BY MATRIX INVERSES

Identity Matrices • Multiplicative Inverses of Square Matrices • Solving Linear Systems Using Inverse Matrices • Curve Fitting Using a System

## Identity Matrices

The identity property for real numbers tells us that $a \cdot 1 = a$ and $1 \cdot a = a$ for any real number $a$. If there is to be a multiplicative *identity matrix I*, such that

$$AI = A \quad \text{and} \quad IA = A,$$

for any matrix $A$, then $A$ and $I$ must be square matrices of the same dimension. Otherwise, it would not be possible to find both products. For example, let $A$ be the $2 \times 2$ matrix $A = \begin{bmatrix} a_{11} & a_{12} \\ a_{21} & a_{22} \end{bmatrix}$, and let $I_2 = \begin{bmatrix} x_{11} & x_{12} \\ x_{21} & x_{22} \end{bmatrix}$ represent the $2 \times 2$ identity matrix. (The subscript 2 is used to denote the fact that it is a $2 \times 2$ matrix.) To find $I_2$, use the fact that $I_2A = A$, so

$$\begin{bmatrix} x_{11} & x_{12} \\ x_{21} & x_{22} \end{bmatrix}\begin{bmatrix} a_{11} & a_{12} \\ a_{21} & a_{22} \end{bmatrix} = \begin{bmatrix} a_{11} & a_{12} \\ a_{21} & a_{22} \end{bmatrix}.$$

Figure 51 shows how the identity matrix $I_2$ is displayed on a graphing calculator with matrix capabilities.

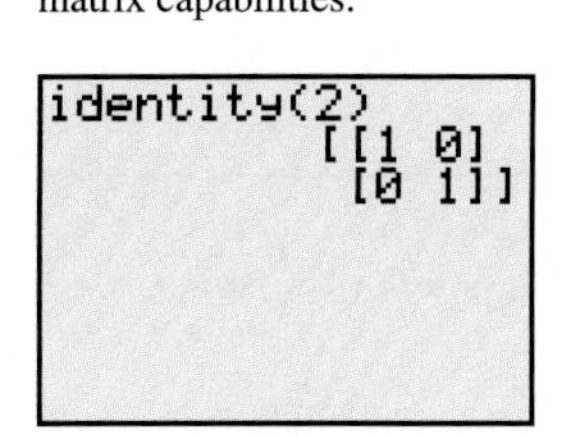


FIGURE 51

Multiplying the two matrices on the left side of this equation and setting the elements of the product matrix equal to the corresponding elements of $A$ gives the following two systems of equations with variables $x_{11}$, $x_{12}$, $x_{21}$, and $x_{22}$.

$$\begin{aligned} x_{11}a_{11} + x_{12}a_{21} &= a_{11} \qquad & x_{21}a_{11} + x_{22}a_{21} &= a_{21} \\ x_{11}a_{12} + x_{12}a_{22} &= a_{12} \qquad & x_{21}a_{12} + x_{22}a_{22} &= a_{22} \end{aligned}$$

Use one of the methods introduced earlier to find that $x_{11} = 1$, $x_{12} = x_{21} = 0$, and $x_{22} = 1$. Therefore, the $2 \times 2$ identity matrix is

$$I_2 = \begin{bmatrix} 1 & 0 \\ 0 & 1 \end{bmatrix}.$$

Check that with this definition of $I_2$, both $AI_2 = A$ and $I_2A = A$.

**EXAMPLE 1** *Using the 2 × 2 Identity Matrix*

Let $A = \begin{bmatrix} -2 & 6 \\ 3 & 5 \end{bmatrix}$. Show that $AI_2 = A$ and $I_2A = A$.

### Analytic Solution

$$\begin{aligned} AI_2 &= \begin{bmatrix} -2 & 6 \\ 3 & 5 \end{bmatrix}\begin{bmatrix} 1 & 0 \\ 0 & 1 \end{bmatrix} \\ &= \begin{bmatrix} -2(1) + 6(0) & -2(0) + 6(1) \\ 3(1) + 5(0) & 3(0) + 5(1) \end{bmatrix} \\ &= \begin{bmatrix} -2 & 6 \\ 3 & 5 \end{bmatrix} = A \end{aligned}$$

### Graphing Calculator Solution

Define [A] as shown in Figure 52(a).

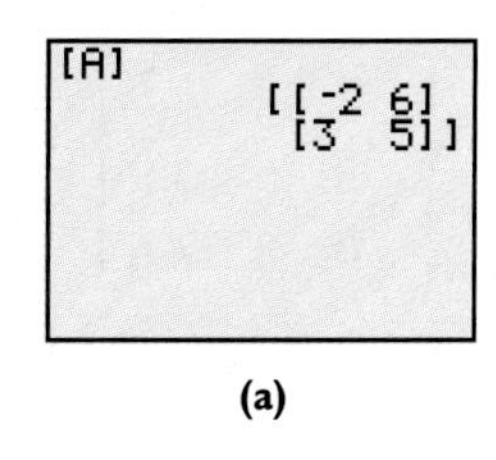


(a)

FIGURE 52

*(continued)*

$$I_2A = \begin{bmatrix} 1 & 0 \\ 0 & 1 \end{bmatrix}\begin{bmatrix} -2 & 6 \\ 3 & 5 \end{bmatrix}$$
$$= \begin{bmatrix} 1(-2) + 0(3) & 1(6) + 0(5) \\ 0(-2) + 1(3) & 0(6) + 1(5) \end{bmatrix}$$
$$= \begin{bmatrix} -2 & 6 \\ 3 & 5 \end{bmatrix} = A$$

Now show that $AI_2 = A$ and $I_2A = A$. See Figure 52(b).

```
[A]*identity(2)
        [[-2 6]
         [3  5]]
identity(2)*[A]
        [[-2 6]
         [3  5]]
```

(b)

FIGURE 52

Figure 53 shows how a graphing calculator displays $I_3$ and $I_4$.

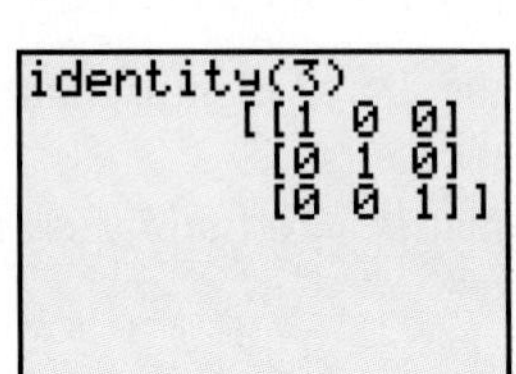


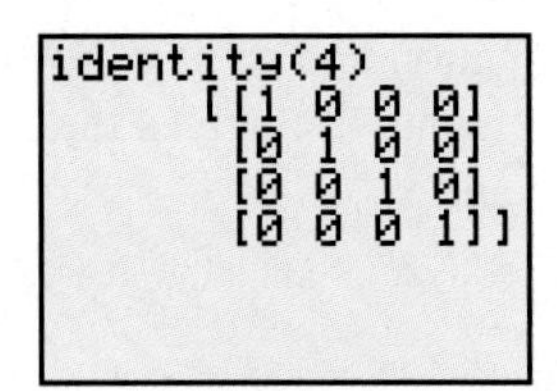


FIGURE 53

The $2 \times 2$ identity matrix found above suggests the following generalization.

**$n \times n$ Identity Matrix**

For any value of $n$ there is an $n \times n$ identity matrix having 1s down the main diagonal and 0s elsewhere. The **$n \times n$ identity matrix** is given by $I_n$, where

$$I_n = \begin{bmatrix} 1 & 0 & \cdots & 0 \\ 0 & 1 & \cdots & 0 \\ \vdots & \vdots & a_{ij} & \vdots \\ 0 & 0 & \cdots & 1 \end{bmatrix}.$$

Here $a_{ij} = 1$ when $i = j$ (the diagonal elements) and $a_{ij} = 0$ otherwise.

**EXAMPLE 2** *Using the 3 × 3 Identity Matrix*

Let $A = \begin{bmatrix} -2 & 4 & 0 \\ 3 & 5 & 9 \\ 0 & 8 & -6 \end{bmatrix}$. Show that $AI_3 = A$.

**Analytic Solution**

Using the $3 \times 3$ identity matrix and the definition of matrix multiplication gives

$$AI_3 = \begin{bmatrix} -2 & 4 & 0 \\ 3 & 5 & 9 \\ 0 & 8 & -6 \end{bmatrix}\begin{bmatrix} 1 & 0 & 0 \\ 0 & 1 & 0 \\ 0 & 0 & 1 \end{bmatrix}$$
$$= \begin{bmatrix} -2 & 4 & 0 \\ 3 & 5 & 9 \\ 0 & 8 & -6 \end{bmatrix} = A.$$

**Graphing Calculator Solution**

See Figure 54.

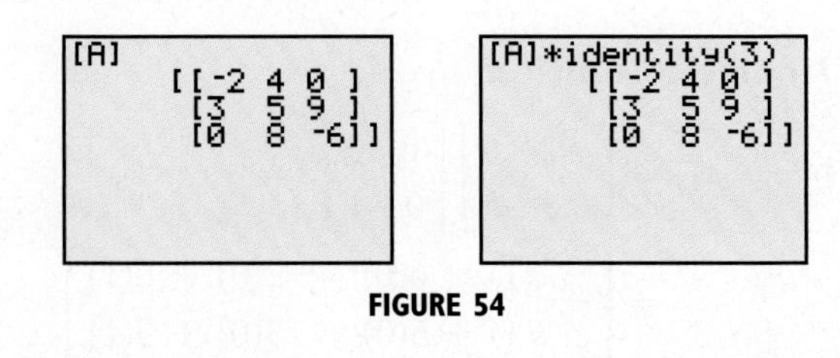


FIGURE 54

**NOTE** Because multiplication by an *identity* matrix is commutative, we can also show that $I_3A = A$ in Example 2.

## Multiplicative Inverses of Square Matrices

Suppose that $A = \begin{bmatrix} a & b \\ c & d \end{bmatrix}$. Is it possible to find a matrix $B$ such that $AB = I_2$ and $BA = I_2$? If so, then $B$ is the inverse of $A$.

Let $B = \begin{bmatrix} x & y \\ z & w \end{bmatrix}$. We must find $x$, $y$, $z$, and $w$ so that $\begin{bmatrix} a & b \\ c & d \end{bmatrix}\begin{bmatrix} x & y \\ z & w \end{bmatrix} = \begin{bmatrix} 1 & 0 \\ 0 & 1 \end{bmatrix}$.

Multiplying the two matrices and setting the elements of the product equal to the corresponding elements in $I_2$ gives the following two systems of equations.

$$\begin{aligned} ax + bz &= 1 \qquad & ay + bw &= 0 \\ cx + dz &= 0 \qquad & cy + dw &= 1 \end{aligned}$$

By solving these two systems, we get the values of $x$, $y$, $z$, and $w$:

$$x = \frac{d}{ad - cb}, \quad y = \frac{-b}{ad - cb}, \quad z = \frac{-c}{ad - cb}, \quad \text{and} \quad w = \frac{a}{ad - cb}.$$

Verify that $AB = I_2$ and that $BA = I_2$. As a result, we can conclude that $B$ is the inverse of $A$, written $A^{-1}$, provided that $\det A = ad - cb \neq 0$.

**Inverse of a 2 × 2 Matrix**

If $A = \begin{bmatrix} a & b \\ c & d \end{bmatrix}$ and $\det A \neq 0$, then

$$A^{-1} = \frac{1}{\det A}\begin{bmatrix} d & -b \\ -c & a \end{bmatrix} \quad \text{or} \quad A^{-1} = \begin{bmatrix} \dfrac{d}{ad - cb} & \dfrac{-b}{ad - cb} \\ \dfrac{-c}{ad - cb} & \dfrac{a}{ad - cb} \end{bmatrix}.$$

If $\det A = 0$, then $A^{-1}$ does not exist, and $A$ is called a **singular matrix**.

**EXAMPLE 3** *Finding the Inverse of a 2 × 2 Matrix*

Find $A^{-1}$, if it exists, for each matrix.

**(a)** $A = \begin{bmatrix} 2 & 3 \\ 1 & -1 \end{bmatrix}$ **(b)** $A = \begin{bmatrix} 3 & -6 \\ 2 & -4 \end{bmatrix}$

**Analytic Solution**

**(a)** First we find $\det A$ and determine whether $A^{-1}$ actually exists.

$$\det A = 2(-1) - 1(3) = -5$$

Since $-5 \neq 0$, $A^{-1}$ exists. By the rule given earlier,

$$A^{-1} = \frac{1}{-5}\begin{bmatrix} -1 & -3 \\ -1 & 2 \end{bmatrix} = \begin{bmatrix} \frac{1}{5} & \frac{3}{5} \\ \frac{1}{5} & -\frac{2}{5} \end{bmatrix}.$$

**Graphing Calculator Solution**

**(a)** See Figure 55, which illustrates both [A] and $[A]^{-1}$.

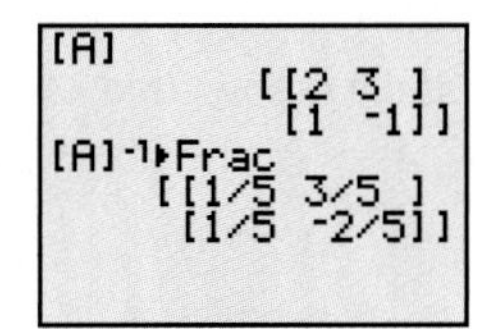


FIGURE 55

*(continued)*

**(b)** Here, $A^{-1}$ does not exist.

$$A = \begin{bmatrix} 3 & -6 \\ 2 & -4 \end{bmatrix}$$

$$\det A = 3(-4) - 2(-6)$$
$$= 0.$$

**(b)** The calculator returns a singular matrix error message when it is directed to find the inverse of a matrix whose determinant is 0. See Figure 56.

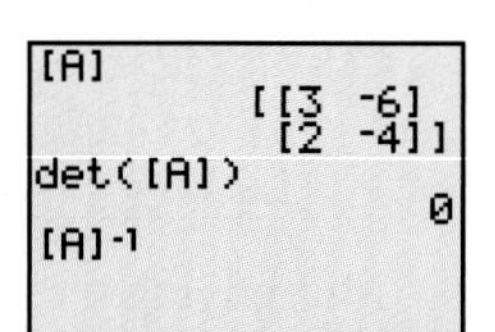


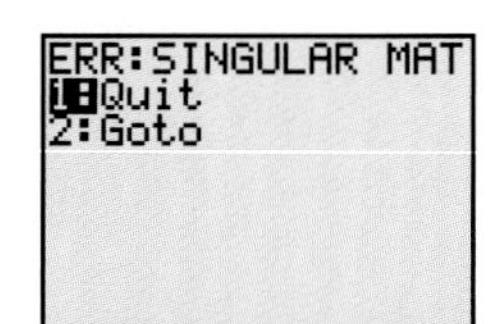


**FIGURE 56**

## EXAMPLE 4 *Verifying Matrix Inverses*

Consider the matrix $A = \begin{bmatrix} 1 & 0 & 1 \\ 2 & -2 & -1 \\ 3 & 0 & 0 \end{bmatrix}$.

**(a)** Show that $\det A \neq 0$. **(b)** Verify that $A^{-1} = \begin{bmatrix} 0 & 0 & \frac{1}{3} \\ -\frac{1}{2} & -\frac{1}{2} & \frac{1}{2} \\ 1 & 0 & -\frac{1}{3} \end{bmatrix}$.

### Analytic Solution

**(a)** Expanding about the third row, we find

$$\det A = 3 \det \begin{bmatrix} 0 & 1 \\ -2 & -1 \end{bmatrix} = 3(0 + 2) = 6.$$

(Because $6 \neq 0$, $A^{-1}$ exists.)

**(b)** First, show that $AA^{-1} = I_3$.

$$AA^{-1} = \begin{bmatrix} 1 & 0 & 1 \\ 2 & -2 & -1 \\ 3 & 0 & 0 \end{bmatrix} \begin{bmatrix} 0 & 0 & \frac{1}{3} \\ -\frac{1}{2} & -\frac{1}{2} & \frac{1}{2} \\ 1 & 0 & -\frac{1}{3} \end{bmatrix}$$

$$= \begin{bmatrix} 0+0+1 & 0+0+0 & \frac{1}{3}+0-\frac{1}{3} \\ 0+1-1 & 0+1+0 & \frac{2}{3}-1+\frac{1}{3} \\ 0+0+0 & 0+0+0 & 1+0+0 \end{bmatrix}$$

$$= \begin{bmatrix} 1 & 0 & 0 \\ 0 & 1 & 0 \\ 0 & 0 & 1 \end{bmatrix}$$

$$= I_3$$

Next, we must show that $A^{-1}A = I_3$. Because $AA^{-1} = A^{-1}A = I_3$, the inverse of $A$ is

$$\begin{bmatrix} 0 & 0 & \frac{1}{3} \\ -\frac{1}{2} & -\frac{1}{2} & \frac{1}{2} \\ 1 & 0 & -\frac{1}{3} \end{bmatrix}.$$

### Graphing Calculator Solution

**(a)** Figure 57 indicates that the determinant is not 0.

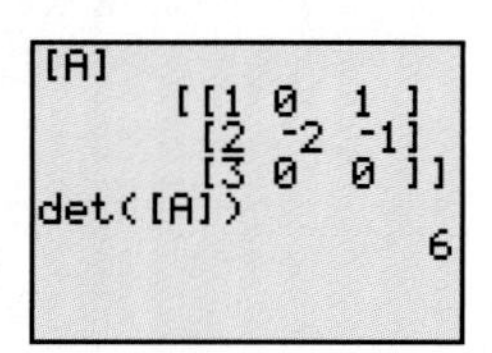


**FIGURE 57**

**(b)** Figure 58 shows that the product of the matrix and its inverse, in both cases, is the identity.

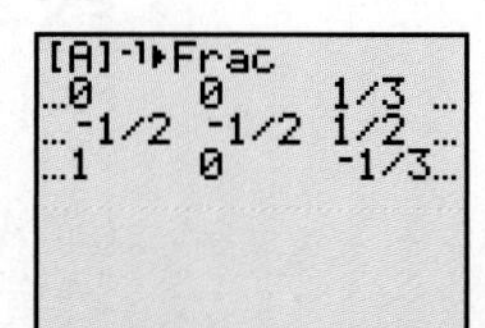


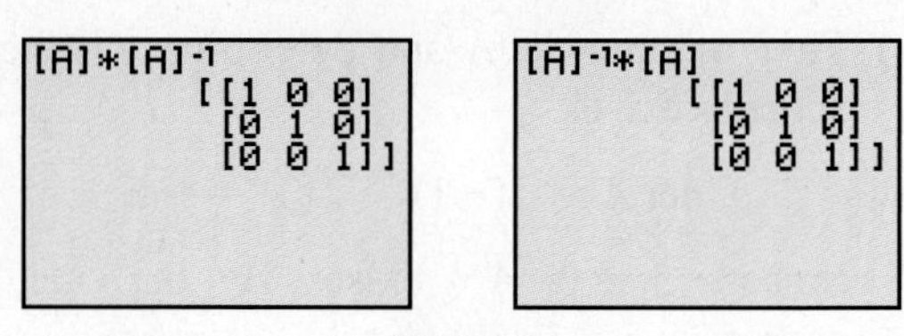


**FIGURE 58**

**NOTE** While there are methods of finding inverses of $3 \times 3$ matrices or greater by hand, we seldom do this because it is so tedious. Calculators and computers perform this function quickly and efficiently.

## Solving Linear Systems Using Inverse Matrices

We used matrices to solve systems of linear equations by the row echelon method in Section 7.3. Another way to use matrices to solve linear systems is to write the system as a matrix equation $AX = B$, where $A$ is the matrix of the coefficients of the variables of the system, $X$ is the matrix of the variables, and $B$ is the matrix of the constants. Matrix $A$ is called the **coefficient matrix.**

To solve the matrix equation $AX = B$, first see if $A^{-1}$ exists. Assuming $A^{-1}$ exists, use the facts that $A^{-1}A = I$ and $IX = X$ to get

$$\begin{aligned} AX &= B & \\ A^{-1}(AX) &= A^{-1}B & &\text{Multiply each side by } A^{-1}. \\ (A^{-1}A)X &= A^{-1}B & &\text{Associative property} \\ IX &= A^{-1}B & &\text{Multiplicative inverse property} \\ X &= A^{-1}B. & &\text{Identity property} \end{aligned}$$

**CAUTION** When multiplying by matrices on each side of a matrix equation, be careful to multiply in the same order on each side since multiplication of matrices is not commutative (unlike multiplication of real numbers).

**EXAMPLE 5** *Solving a System of Equations Using a Matrix Inverse*

Use the inverse of the coefficient matrix to solve the system.

$$\begin{aligned} 2x + y + 3z &= 1 \\ x - 2y + z &= -3 \\ -3x + y - 2z &= -4 \end{aligned}$$

**Analytic Solution**

First we find the determinant of the coefficient matrix to be sure that it is not 0. Using one of the methods described earlier, we find that for

$$A = \begin{bmatrix} 2 & 1 & 3 \\ 1 & -2 & 1 \\ -3 & 1 & -2 \end{bmatrix},$$

$\det A = -10 \neq 0$. Therefore, $A^{-1}$ exists.

With

$$X = \begin{bmatrix} x \\ y \\ z \end{bmatrix} \quad \text{and} \quad B = \begin{bmatrix} 1 \\ -3 \\ -4 \end{bmatrix},$$

**Graphing Calculator Solution**

Entering the coefficient matrix as [A], we find that det([A]) $= -10$ and $[A]^{-1}$ exists, as shown in Figure 59(a) on the next page. Then we enter the column of constants as [B] and find the matrix product $[A]^{-1}[B]$, as shown in Figure 59(b). The display shows the column matrix representing $x = 4$, $y = 2$, and $z = -3$. The solution set is $\{(4, 2, -3)\}$.

*(continued)*

we evaluate $A^{-1}B$:

$$A^{-1}B = \begin{bmatrix} 2 & 1 & 3 \\ 1 & -2 & 1 \\ -3 & 1 & -2 \end{bmatrix}^{-1} \begin{bmatrix} 1 \\ -3 \\ -4 \end{bmatrix}$$

$$= \begin{bmatrix} -\frac{3}{10} & -\frac{1}{2} & -\frac{7}{10} \\ \frac{1}{10} & -\frac{1}{2} & -\frac{1}{10} \\ \frac{1}{2} & \frac{1}{2} & \frac{1}{2} \end{bmatrix} \begin{bmatrix} 1 \\ -3 \\ -4 \end{bmatrix}$$

$$= \begin{bmatrix} 4 \\ 2 \\ -3 \end{bmatrix}.$$

Since $X = A^{-1}B$, we have $x = 4$, $y = 2$, and $z = -3$. The solution set is $\{(4, 2, -3)\}$.

```
[A]
    [[2  1  3 ]
     [1  -2 1 ]
     [-3 1  -2]]
det([A])
                -10
[B]
```

(a)

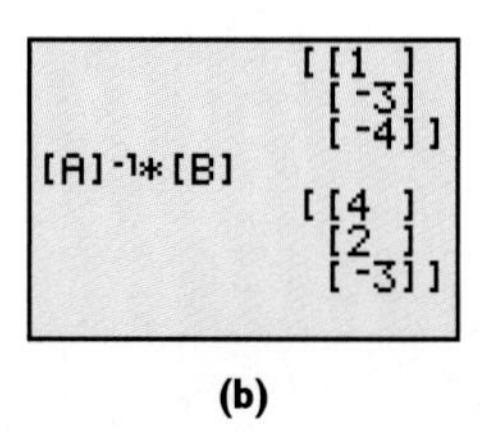


(b)

FIGURE 59

**CAUTION** Always evaluate the determinant of the coefficient matrix *before* using the inverse matrix method. If the determinant is 0, the system is either inconsistent or has dependent equations.

## Curve Fitting Using a System

**EXAMPLE 6** *Using a System to Find the Equation for a Cubic Polynomial*

Figure 60 shows four views of the graph of a polynomial function of the form $P(x) = ax^3 + bx^2 + cx + d$. Use the points indicated to write a system of four equations in the variables $a$, $b$, $c$, and $d$, and then use the inverse matrix method to solve the system. What is the equation that defines this graph?

**FOR DISCUSSION**

1. Using the window $[-5, 5]$ by $[-50, 50]$, reproduce the graph in Figure 60, based on the result of Example 6.
2. If your calculator has *cubic regression,* use it to find the same results as in Example 6.

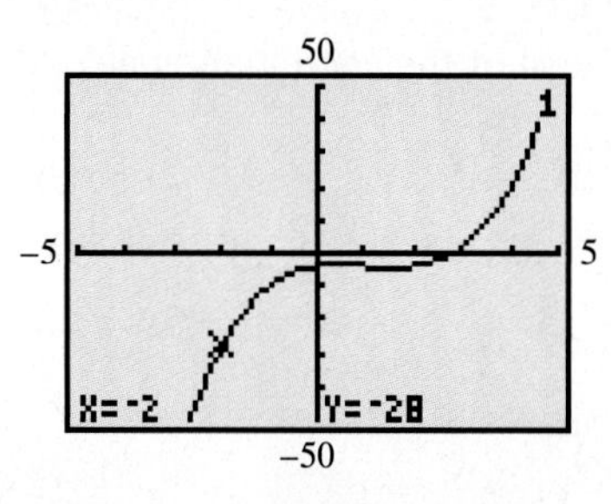


First View

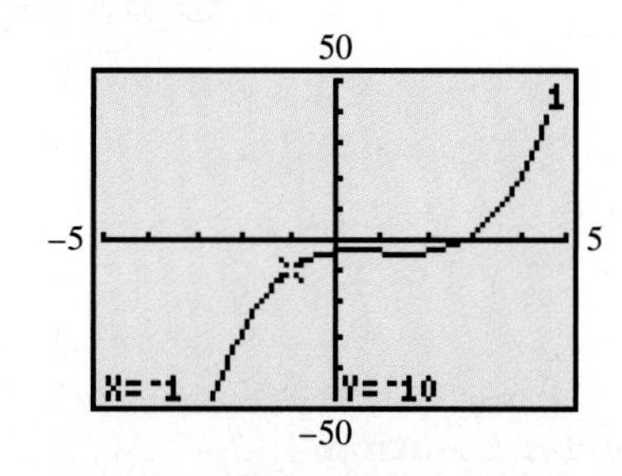


Second View

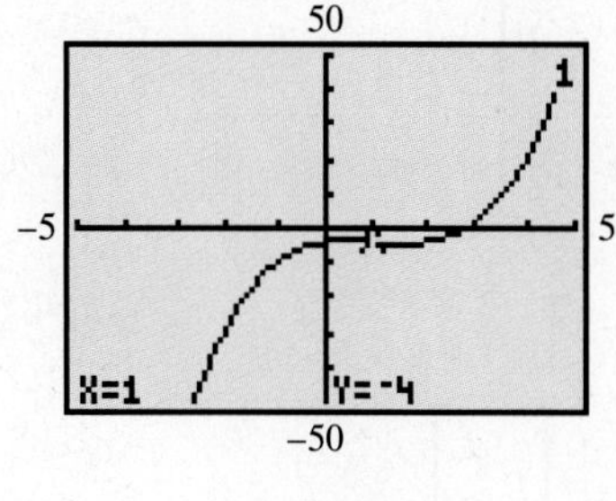


Third View

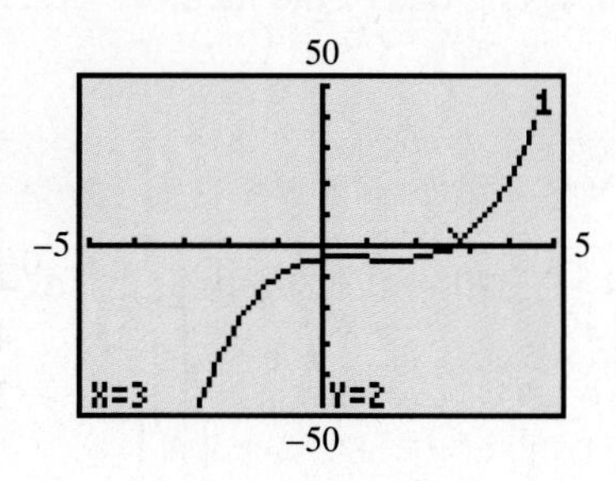


Fourth View

FIGURE 60

**Solution** We see from the graph that $P(-2) = -28$, $P(-1) = -10$, $P(1) = -4$, and $P(3) = 2$. From the first of these, we get

$$P(-2) = a(-2)^3 + b(-2)^2 + c(-2) + d = -28$$

or, equivalently,

$$-8a + 4b - 2c + d = -28.$$

Similarly, from the others, we find the following equations.

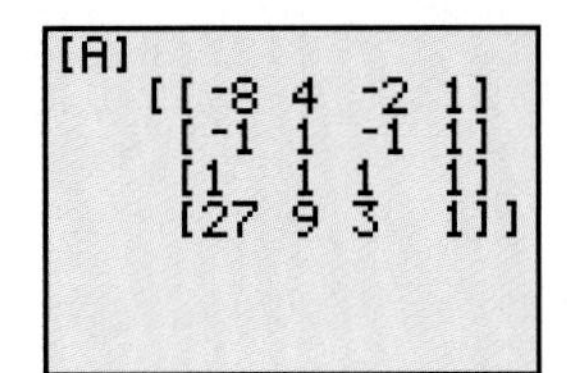


$$\begin{aligned} &\text{From } P(-1) = -10: && -a + b - c + d = -10 \\ &\text{From } P(1) = -4: && a + b + c + d = -4 \\ &\text{From } P(3) = 2: && 27a + 9b + 3c + d = 2 \end{aligned}$$

Now we must solve the system formed by these four equations:

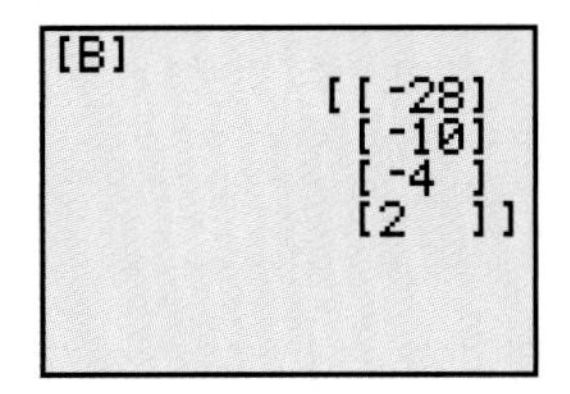


$$\begin{aligned} -8a + 4b - 2c + d &= -28 \\ -a + b - c + d &= -10 \\ a + b + c + d &= -4 \\ 27a + 9b + 3c + d &= 2. \end{aligned}$$

We use the inverse matrix method to solve the system. We let

$$A = \begin{bmatrix} -8 & 4 & -2 & 1 \\ -1 & 1 & -1 & 1 \\ 1 & 1 & 1 & 1 \\ 27 & 9 & 3 & 1 \end{bmatrix}, \quad X = \begin{bmatrix} a \\ b \\ c \\ d \end{bmatrix}, \quad \text{and} \quad B = \begin{bmatrix} -28 \\ -10 \\ -4 \\ 2 \end{bmatrix}.$$

[A]-1*[B]
[[1 ]
[-3]
[2 ]
[-4]]

FIGURE 61

Because $\det A = 240 \neq 0$, a unique solution exists. Based on our discussion earlier, $X = A^{-1}B$. Figure 61 shows how the matrix inverse method yields the values $a = 1$, $b = -3$, $c = 2$, and $d = -4$. The polynomial is defined by $P(x) = x^3 - 3x^2 + 2x - 4$.

## 7.6 EXERCISES

*Determine whether A and B are inverses by calculating AB and BA.*

**1.**
[A]
[[5 7]
[2 3]]
[B]
[[3 -7]
[-2 5 ]]

**2.**
[A]
[[2 3]
[1 1]]
[B]
[[-1 3 ]
[1 -2]]

**3.** $A = \begin{bmatrix} -1 & 2 \\ 3 & -5 \end{bmatrix}; B = \begin{bmatrix} -5 & -2 \\ -3 & -1 \end{bmatrix}$

**4.** $A = \begin{bmatrix} 2 & 1 \\ 3 & 2 \end{bmatrix}; B = \begin{bmatrix} 2 & 1 \\ -3 & 2 \end{bmatrix}$

**5.** $A = \begin{bmatrix} 0 & 1 & 0 \\ 0 & 0 & -2 \\ 1 & -1 & 0 \end{bmatrix}; B = \begin{bmatrix} 1 & 0 & 1 \\ 1 & 0 & 0 \\ 0 & -1 & 0 \end{bmatrix}$

**6.** $A = \begin{bmatrix} 1 & 2 & 0 \\ 0 & 1 & 0 \\ 0 & 1 & 0 \end{bmatrix}; B = \begin{bmatrix} 1 & -2 & 0 \\ 0 & 1 & 0 \\ 0 & -1 & 1 \end{bmatrix}$

**7.** $A = \begin{bmatrix} -1 & -1 & -1 \\ 4 & 5 & 0 \\ 0 & 1 & -3 \end{bmatrix}; B = \begin{bmatrix} 15 & 4 & -5 \\ -12 & -3 & 4 \\ -4 & -1 & 1 \end{bmatrix}$

**8.** $A = \begin{bmatrix} 1 & 3 & 3 \\ 1 & 4 & 3 \\ 1 & 3 & 4 \end{bmatrix}; B = \begin{bmatrix} 7 & -3 & -3 \\ -1 & 1 & 0 \\ -1 & 0 & 1 \end{bmatrix}$

**9.** Under what condition will the inverse of a square matrix not exist?

**10.** Explain why a $2 \times 2$ matrix will not have an inverse if either a column or a row contains all 0s.

*For each matrix, find $A^{-1}$, if it exists.*

**11.** $A = \begin{bmatrix} 3 & 7 \\ 2 & 5 \end{bmatrix}$

**12.** $A = \begin{bmatrix} -5 & 3 \\ -8 & 5 \end{bmatrix}$

**13.** $A = \begin{bmatrix} -1 & -2 \\ 3 & 4 \end{bmatrix}$

**14.** $A = \begin{bmatrix} -1 & 2 \\ -2 & -1 \end{bmatrix}$

**15.** $A = \begin{bmatrix} -6 & 4 \\ -3 & 2 \end{bmatrix}$

**16.** $A = \begin{bmatrix} 5 & 10 \\ -3 & -6 \end{bmatrix}$

**17.** $A = \begin{bmatrix} .6 & .2 \\ .5 & .1 \end{bmatrix}$

**18.** $A = \begin{bmatrix} .8 & -.3 \\ .5 & -.2 \end{bmatrix}$

**19.**

```
[A]
   [[1 0  0]
    [0 -1 0]
    [1 0  1]]
```

**20.**

```
[A]
   [[1 3 3]
    [1 4 3]
    [1 3 4]]
```

**21.** $A = \begin{bmatrix} -1 & -1 & -1 \\ 4 & 5 & 0 \\ 0 & 1 & -3 \end{bmatrix}$

**22.** $A = \begin{bmatrix} 2 & 0 & 4 \\ 3 & 1 & 5 \\ -1 & 1 & -2 \end{bmatrix}$

**23.** $A = \begin{bmatrix} -.4 & .1 & .2 \\ 0 & .6 & .8 \\ .3 & 0 & -.2 \end{bmatrix}$

**24.** $A = \begin{bmatrix} .8 & .2 & .1 \\ -.2 & 0 & .3 \\ 0 & 0 & .5 \end{bmatrix}$

**25.** $A = \begin{bmatrix} 2 & 1 & 2 \\ 5 & 10 & 5 \\ 3 & 6 & 3 \end{bmatrix}$

**26.** $A = \begin{bmatrix} 5 & -3 & 2 \\ -5 & 3 & -2 \\ 1 & 0 & 1 \end{bmatrix}$

**27.** $A = \begin{bmatrix} \sqrt{2} & .5 \\ -17 & \frac{1}{2} \end{bmatrix}$

**28.** $A = \begin{bmatrix} \frac{2}{3} & .7 \\ 22 & \sqrt{3} \end{bmatrix}$

**29.** $A = \begin{bmatrix} 1.4 & .5 & .59 \\ .84 & 1.36 & .62 \\ .56 & .47 & 1.3 \end{bmatrix}$

**30.** $A = \begin{bmatrix} \frac{1}{2} & \frac{1}{4} & \frac{1}{3} \\ 0 & \frac{1}{4} & \frac{1}{3} \\ \frac{1}{2} & \frac{1}{2} & \frac{1}{3} \end{bmatrix}$

## RELATING CONCEPTS

**For individual or group investigation (Exercises 31–40)**

*One method for finding matrix inverses analytically is developed by using a system of linear equations. To see how this is done, work Exercises 31–34 in order. Use the matrices*

$$A = \begin{bmatrix} 2 & 4 \\ 1 & -1 \end{bmatrix} \quad \text{and} \quad A^{-1} = \begin{bmatrix} x & y \\ z & w \end{bmatrix}.$$

**31.** Find the product $AA^{-1}$.

**32.** Set $AA^{-1} = I_2$ and set corresponding elements equal to get four equations.

**33.** Write the equations from Exercise 32 as two systems of two equations in two variables. Solve the systems.

**34.** From Exercise 33, what is matrix $A^{-1}$?

*It was shown that the inverse of matrix* $A = \begin{bmatrix} a & b \\ c & d \end{bmatrix}$ *is*

$$A^{-1} = \begin{bmatrix} \dfrac{d}{ad - bc} & \dfrac{-b}{ad - bc} \\ \dfrac{-c}{ad - bc} & \dfrac{a}{ad - bc} \end{bmatrix}.$$

*Work Exercises 35–40 in order, to relate the concepts in this section and a topic studied earlier in this chapter.*

**35.** With respect to the matrix $A = \begin{bmatrix} a & b \\ c & d \end{bmatrix}$, what do we call $ad - bc$?

**36.** Refer to $A^{-1}$ as given here, and write it using determinant notation.

**37.** Write $A^{-1}$ using scalar multiplication, where the scalar is $\frac{1}{\det A}$.

**38.** Explain in your own words how the inverse of matrix $A$ can be found using a determinant.

**39.** Use the method described here to find the inverse of $A = \begin{bmatrix} 4 & 2 \\ 7 & 3 \end{bmatrix}$.

**40.** Complete the following statement: The inverse of a $2 \times 2$ matrix $A$ does not exist if the determinant of $A$ has value __________. (*Hint:* Look at the denominators in $A^{-1}$ as given here.)

*Solve each system using the matrix inverse method.*

**41.** $\begin{aligned} 2x - y &= -8 \\ 3x + y &= -2 \end{aligned}$

**42.** $\begin{aligned} x + 3y &= -12 \\ 2x - y &= 11 \end{aligned}$

**43.** $\begin{aligned} 2x + 3y &= -10 \\ 3x + 4y &= -12 \end{aligned}$

**44.** $\begin{aligned} 2x - 3y &= 10 \\ 2x + 2y &= 5 \end{aligned}$

**45.** $\begin{aligned} 2x - 5y &= 10 \\ 4x - 5y &= 15 \end{aligned}$

**46.** $\begin{aligned} 2x - 3y &= 2 \\ 4x - 6y &= 1 \end{aligned}$

**47.** $\begin{aligned} 2x + 4z &= 14 \\ 3x + y + 5z &= 19 \\ -x + y - 2z &= -7 \end{aligned}$

**48.** $\begin{aligned} 3x + 6y + 3z &= 12 \\ 6x + 4y - 2z &= -4 \\ y - z &= -3 \end{aligned}$

**49.** $\begin{aligned} x + 3y + z &= 2 \\ x - 2y + 3z &= -3 \\ 2x - 3y - z &= 34 \end{aligned}$

**50.** $\begin{aligned} x + y - z &= 6 \\ 2x - y + z &= -9 \\ x - 2y + 3z &= 1 \end{aligned}$

**51.** $\begin{aligned} x + 3y - 2z - w &= 9 \\ 4x + y + z + 2w &= 2 \\ -3x - y + z - w &= -5 \\ x - y - 3z - 2w &= 2 \end{aligned}$

**52.** $\begin{aligned} 3x + 2y - w &= 0 \\ 2x + z + 2w &= 5 \\ x + 2y - z &= -2 \\ 2x - y + z + w &= 2 \end{aligned}$

**53.** $\begin{aligned} x - \sqrt{2}y &= 2.6 \\ .75x + y &= -7 \end{aligned}$

**54.** $\begin{aligned} 2.1x + y &= \sqrt{5} \\ \sqrt{2}x - 2y &= 5 \end{aligned}$

**55.** $\begin{aligned} \pi x + ey + \sqrt{2}z &= 1 \\ ex + \pi y + \sqrt{2}z &= 2 \\ \sqrt{2}x + ey + \pi z &= 3 \end{aligned}$

**56.** $\begin{aligned} (\log 2)x + (\ln 3)y + (\ln 4)z &= 1 \\ (\ln 3)x + (\log 2)y + (\ln 8)z &= 5 \\ (\log 12)x + (\ln 4)y + (\ln 8)z &= 9 \end{aligned}$

*Curve Fitting* *In Exercises 57 and 58, use the method of Example 6 to find the cubic polynomial $P(x)$ that defines the curve shown in the four figures.*

**57.**

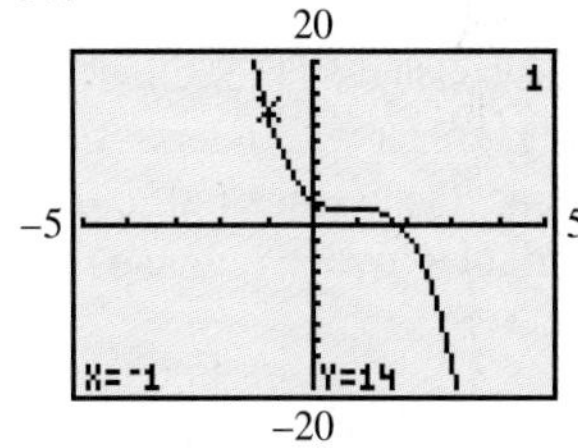


First View

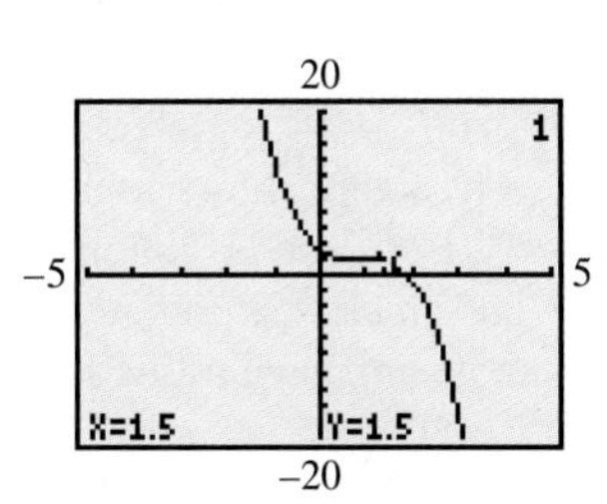


Second View

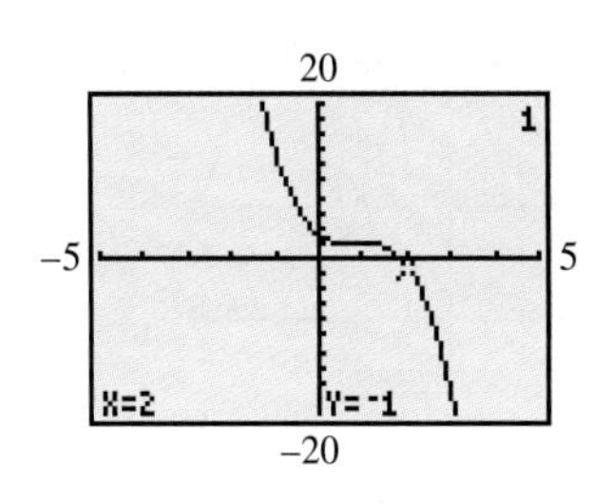


Third View

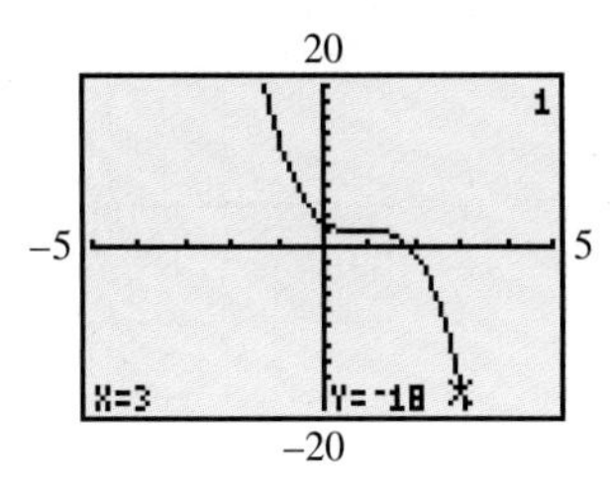


Fourth View

**58.**

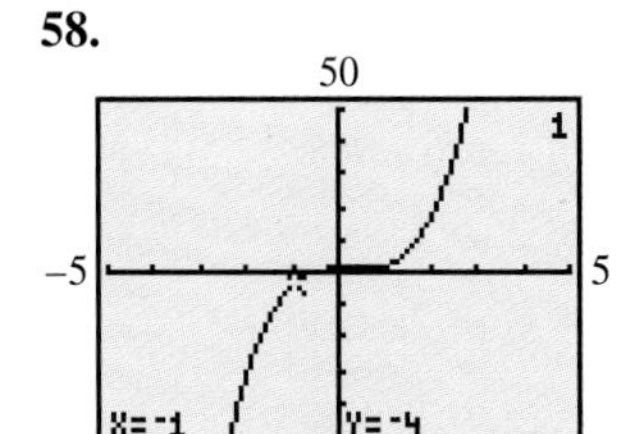


First View

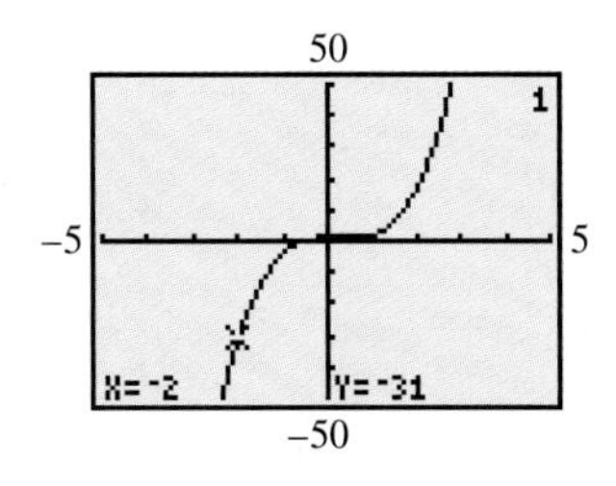


Second View

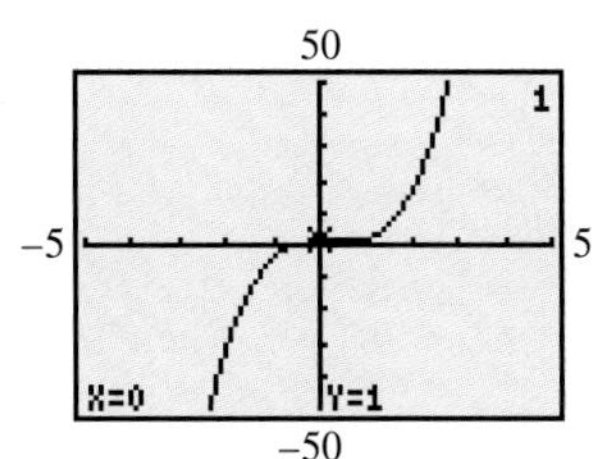


Third View

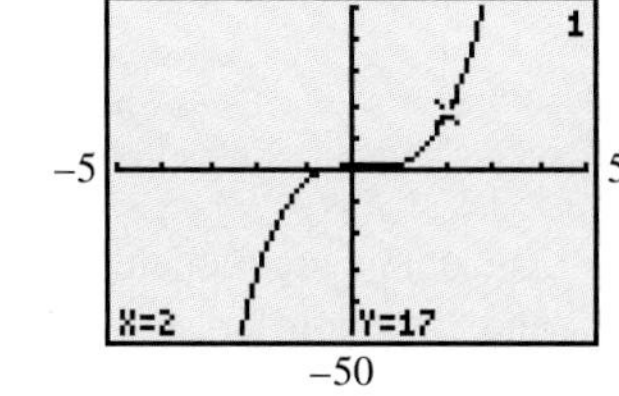


Fourth View

**59.** *Curve Fitting* Find the fourth-degree polynomial $P(x)$ satisfying the following conditions: $P(-2) = 13$, $P(-1) = 2$, $P(0) = -1$, $P(1) = 4$, $P(2) = 41$.

**60.** *Curve Fitting* Find the fifth-degree polynomial $P(x)$ satisfying the following conditions: $P(-2) = -8$, $P(-1) = -1$, $P(0) = -4$, $P(1) = -5$, $P(2) = 8$, $P(3) = 167$.

*(Modeling)* *Solve each problem.*

**61.** *Determining Prices* A group of students bought 3 soft drinks and 2 boxes of popcorn at a movie for \$8.50. A second group bought 4 soft drinks and 3 boxes of popcorn for \$12.

**(a)** Find a matrix equation $AX = B$ whose solution gives the individual prices of a soft drink and a box of popcorn. Solve this matrix equation using $A^{-1}$.

**(b)** Could these prices be determined if both groups had bought 3 soft drinks and 2 boxes of popcorn for \$8.50? Try to calculate $A^{-1}$ and explain your results.

**62.** *Cost of CDs* A music store has compact discs that sell for three prices marked A, B, and C. The last column in the table on the next page shows the total cost of a purchase. Use this information to determine the cost of one CD of each type by setting up a matrix equation and solving it with an inverse.

| A | B | C | Total |
|---|---|---|---|
| 2 | 3 | 4 | \$120.91 |
| 1 | 4 | 0 | \$62.95 |
| 2 | 1 | 3 | \$79.94 |

**63.** *Tire Sales* In one study the relationship among annual tire sales $T$ in thousands, automobile registrations $A$ in millions, and personal disposable income $I$ in millions of dollars was investigated. Representative data for three different years are shown in the table.

| $T$ | $A$ | $I$ |
|---|---|---|
| 10,170 | 113 | 308 |
| 15,305 | 133 | 622 |
| 21,289 | 155 | 1937 |

*Source:* Jarrett, J., *Business Forecasting Methods*, Basil Blackwell, 1991.

The data were modeled by $T = aA + bI + c$, where $a$, $b$, and $c$ are constants.

**(a)** Use the data to write a system of linear equations, whose solution gives $a$, $b$, and $c$.

**(b)** Solve this linear system. Write a formula for $T$.

**(c)** If $A = 118$ and $I = 311$, predict $T$. (The actual value for $T$ was 11,314.)

**64.** *Plate Glass Sales* The amount of plate glass sales $G$ can be affected by the number of new building contracts $B$ issued and automobiles $A$ produced, since plate glass is used in buildings and cars. A plate glass company in California wanted to forecast sales. The table contains sales data for three consecutive years. All units are in millions.

| $G$ | $A$ | $B$ |
|---|---|---|
| 603 | 5.54 | 37.1 |
| 657 | 6.93 | 41.3 |
| 779 | 7.64 | 45.6 |

*Source:* Makridakis, S., and S. Wheelwright, *Forecasting Methods for Management*, John Wiley and Sons, 1989.

The data were modeled by $G = aA + bB + c$, where $a$, $b$, and $c$ are constants.

**(a)** Write a system of linear equations whose solution gives $a$, $b$, and $c$.

**(b)** Solve this linear system. Write a formula for $G$.

**(c)** For the following year, it was estimated that $A = 7.75$ and $B = 47.4$. Predict $G$. (The actual value for $G$ was \$878 million.)

**65.** *Home Prices* Real estate companies sometimes study how the selling price of a home is related to its size and condition. The table contains data for sales of three homes. Price $P$ is measured in thousands of dollars, home size $S$ in square feet, and condition $C$ is rated on a scale from 1 to 10, where 10 represents excellent condition. The variables were found to be related by the linear equation $P = a + bS + cC$.

| $P$ | $S$ | $C$ |
|---|---|---|
| 122 | 1500 | 8 |
| 130 | 2000 | 5 |
| 158 | 2200 | 10 |

**(a)** Use the table to write a system of linear equations, whose solution gives $a$, $b$, and $c$.

**(b)** Estimate the selling price of a home with 1800 square feet and condition 7.

**66.** *Traffic Flow* Refer to Exercises 63 and 64 in Section 7.3. The figure shows four one-way streets with intersections A, B, C, and D. Numbers indicate the average traffic flow in vehicles per minute. The variables $x_1$, $x_2$, $x_3$, and $x_4$ denote unknown traffic flows.

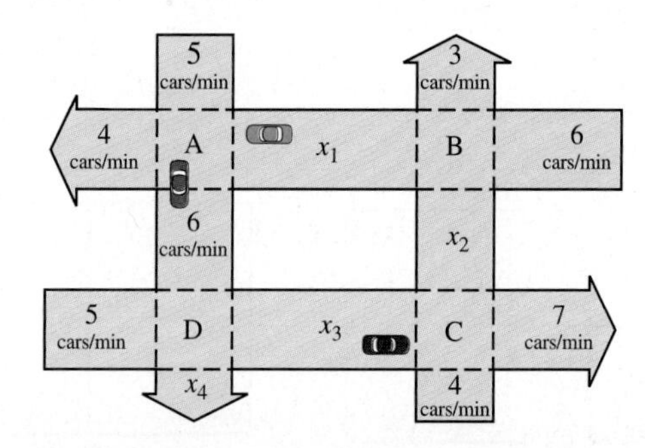


**(a)** The number of vehicles per minute entering an intersection equals the number exiting an intersection. Verify that the given system of linear equations describes the traffic flow.

$$\text{A: } x_1 + 5 = 4 + 6$$
$$\text{B: } x_2 + 6 = x_1 + 3$$
$$\text{C: } x_3 + 4 = x_2 + 7$$
$$\text{D: } 6 + 5 = x_3 + x_4$$

**(b)** Write the system as $AX = B$, and solve using $A^{-1}$.

**(c)** Interpret your results.

*Each graphing calculator screen shows* $[A]^{-1}$ *for some matrix* $[A]$. *Find matrix* $[A]$.

**67.**

```
[A]-1
      [[5  -9]
       [-1 2 ]]
```

**68.**

```
[A]-1▸Frac
    [[3/20  1/4]
     [-1/20 1/4]]
```

**69.**

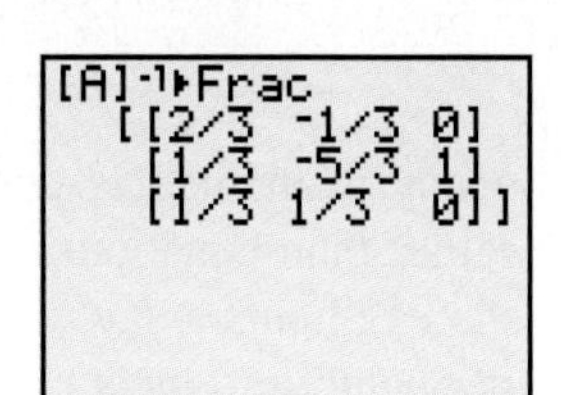

```
[A]-1▸Frac
 [[2/3 -1/3 0]
  [1/3 -5/3 1]
  [1/3 1/3  0]]
```

**70.**

```
[A]-1
       [[0 0 1]
        [0 1 0]
        [1 0 0]]
```

**71.** Let $A = \begin{bmatrix} a & 0 & 0 \\ 0 & b & 0 \\ 0 & 0 & c \end{bmatrix}$, where $a$, $b$, and $c$ are nonzero real numbers. Find $A^{-1}$.

**72.** Let $A = \begin{bmatrix} 1 & 0 & 0 \\ 0 & 0 & -1 \\ 0 & 1 & -1 \end{bmatrix}$. Show that $A^3 = I_3$, and use this result to find the inverse of $A$.

## Reviewing Basic Concepts (Sections 7.4–7.6)

*Given matrices*

$$A = \begin{bmatrix} -5 & 4 \\ 2 & -1 \end{bmatrix}, \quad B = \begin{bmatrix} 0 & -2 \\ 3 & -4 \end{bmatrix}, \quad C = \begin{bmatrix} 2 & -3 & 1 \\ -2 & 1 & 0 \\ 0 & -1 & 4 \end{bmatrix}, \quad D = \begin{bmatrix} 0 & 4 & 1 \\ 0 & 2 & 2 \\ 1 & 1 & 1 \end{bmatrix},$$

*find the following, if possible.*

1. $A - B$
2. $-3B$
3. $A^2$
4. $CD$
5. $\det A$
6. $\det C$
7. $A^{-1}$
8. $C^{-1}$
9. *Roof Trusses* Linear systems occur in the design of roof trusses for new homes and buildings. The simplest type of roof truss is a triangle. The truss shown in the figure is used to frame roofs of small buildings. If a 100-pound force is applied at the peak of the truss, then the forces or weights $W_1$ and $W_2$ exerted parallel to each rafter of the truss are determined by the following linear system of equations.

$$\frac{\sqrt{3}}{2}(W_1 + W_2) = 100$$
$$W_1 - W_2 = 0$$

Solve the system using Cramer's rule to find $W_1$ and $W_2$. (*Source:* Hibbeler, R., *Structural Analysis,* Prentice-Hall, 1995.)

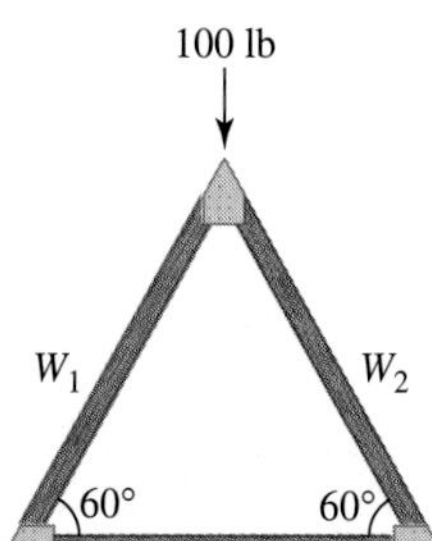


10. Solve the system using the inverse matrix method.

$$\begin{aligned} 2x + y + 2z &= 10 \\ y + 2z &= 4 \\ x - 2y + 2z &= 1 \end{aligned}$$

# SYSTEMS OF INEQUALITIES AND LINEAR PROGRAMMING

Solving Linear Inequalities • Solving Systems of Inequalities • Linear Programming

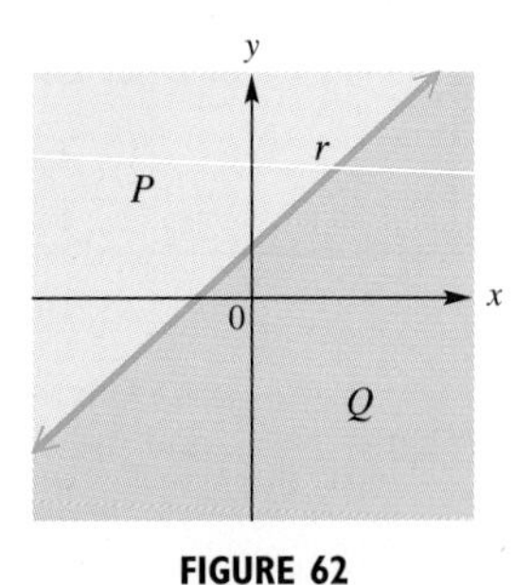


**FIGURE 62**

Many mathematical descriptions of real situations are best expressed as inequalities rather than equations. For example, a firm might be able to use a machine *no more* than 12 hours a day, while production of *at least* 500 cases of a certain product might be required to meet a contract. The simplest way to see the solution of an inequality in two variables is to draw its graph.

A line divides a plane into three sets of points: the points of the line itself and the points belonging to the two regions determined by the line. Each of these two regions is called a **half-plane.** In Figure 62, line $r$ divides the plane into three different sets of points: line $r$, half-plane $P$, and half-plane $Q$. The points on $r$ belong neither to $P$ nor to $Q$. Line $r$ is the **boundary** of each half-plane.

## Solving Linear Inequalities

A **linear inequality in two variables** is an inequality of the form

$$Ax + By \leq C,$$

where $A$, $B$, and $C$ are real numbers, with $A$ and $B$ not both equal to 0. (The symbol $\leq$ could be replaced with $\geq$, $<$, or $>$.) The graph of a linear inequality is a half-plane, perhaps with its boundary. For example, to graph the linear inequality

$$3x - 2y \leq 6,$$

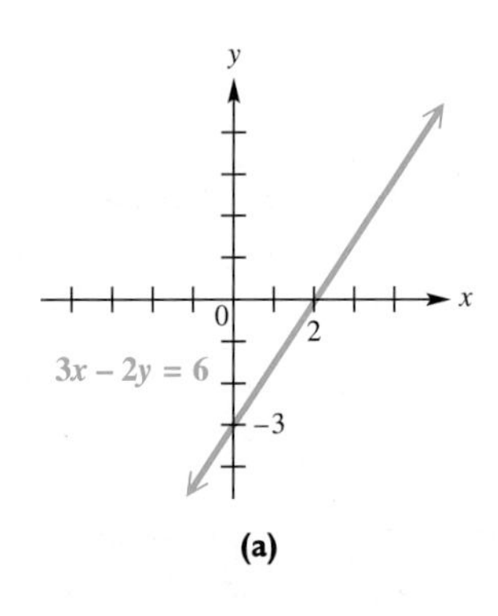


**(a)**

first graph the boundary, $3x - 2y = 6$, as shown in Figure 63(a). Since the points of the line $3x - 2y = 6$ satisfy $3x - 2y \leq 6$, this line is part of the solution. To decide which half-plane (the one above the line $3x - 2y = 6$ or the one below the line) is part of the solution, solve the original inequality for $y$.

$$\begin{aligned} 3x - 2y &\leq 6 \\ -2y &\leq -3x + 6 \\ y &\geq \frac{3}{2}x - 3 \qquad \text{Multiply by } -\tfrac{1}{2}\text{; change } \leq \text{ to } \geq. \end{aligned}$$

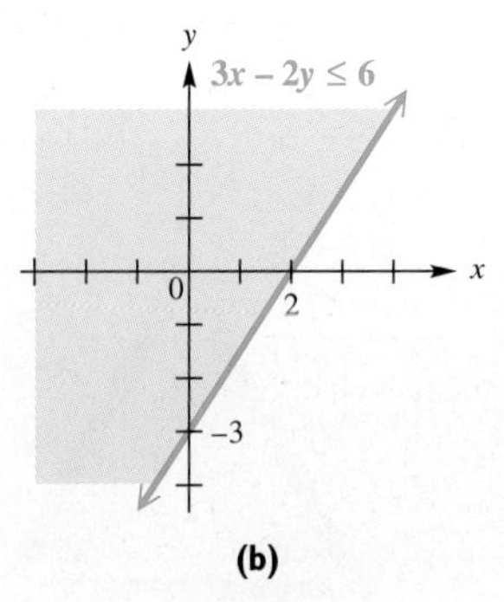


**(b)**

**FIGURE 63**

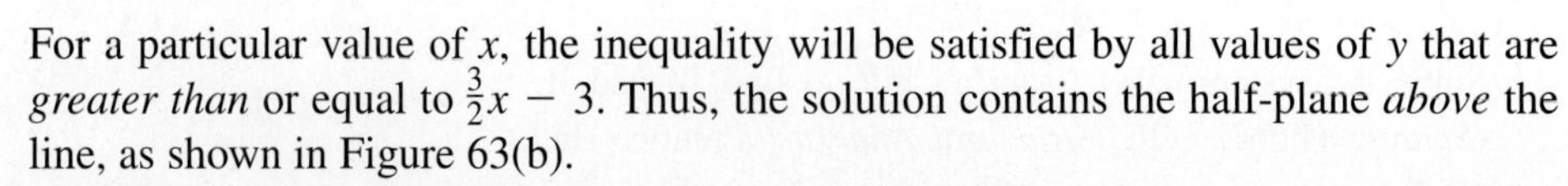

For a particular value of $x$, the inequality will be satisfied by all values of $y$ that are *greater than* or equal to $\frac{3}{2}x - 3$. Thus, the solution contains the half-plane *above* the line, as shown in Figure 63(b).

The screens in Figure 64 on the next page are calculator graphs of $3x - 2y = 6$ and $3x - 2y \leq 6$. We first solved $3x - 2y = 6$ for $y$ to get

$$y = \frac{3}{2}x - 3.$$

Then with $y_1 = \frac{3}{2}x - 3$, we found the graph in Figure 64(a). We directed the calculator to shade *above* the graph to get the region indicated in Figure 64(b).

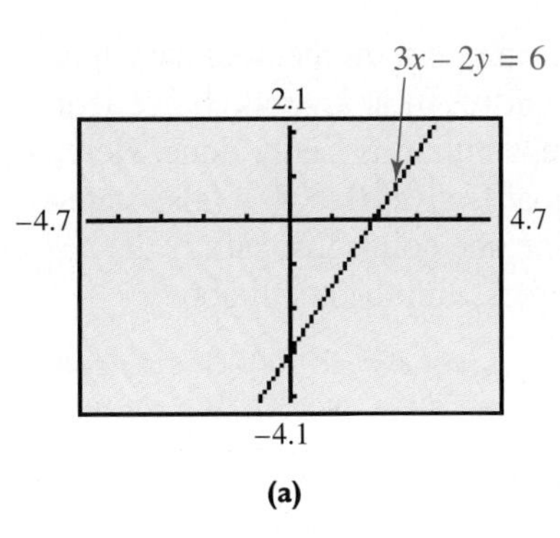

(a)

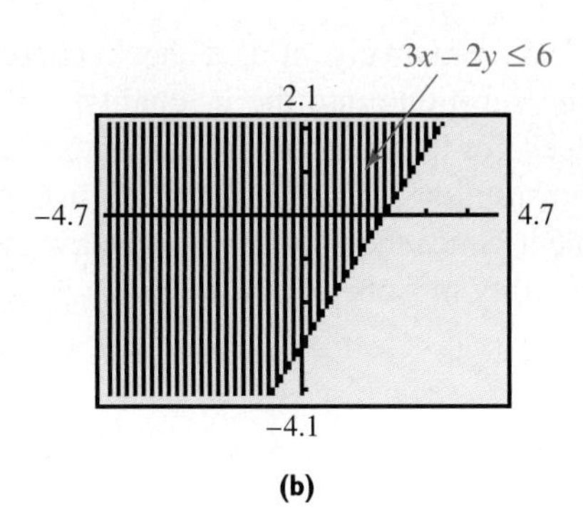

(b)

**FIGURE 64**

**CAUTION** To determine whether to shade above or below a boundary line, solve the equation for $y$. Then,

1. if the symbol is $>$ or $\geq$, shade *above* the boundary.
2. if the symbol is $<$ or $\leq$, shade *below* the boundary.

**EXAMPLE 1** *Graphing a Linear Inequality*

Graph $x + 4y > 4$.

**Analytic Solution**

The boundary here is the line $x + 4y = 4$. Since the points on this line do not satisfy $x + 4y > 4$, make the line dashed, as in Figure 65. To decide which half-plane represents the solution, solve for $y$.

$$x + 4y > 4$$
$$4y > -x + 4$$
$$y > -\frac{1}{4}x + 1$$

Since $y$ is *greater than* $-\frac{1}{4}x + 1$, the graph of the solution is the half-plane above the boundary, as shown in Figure 65.

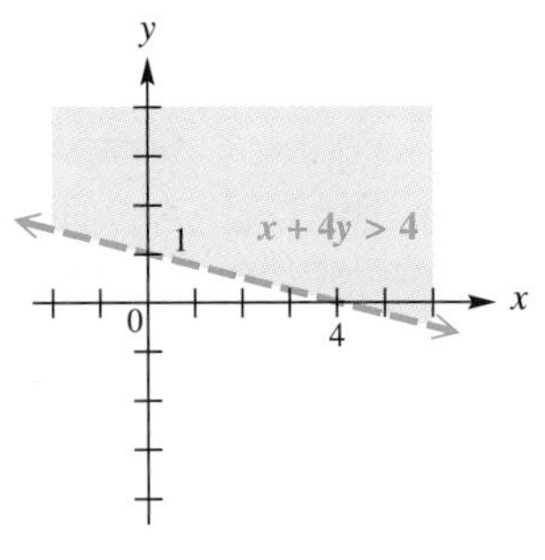

**FIGURE 65**

**Graphing Calculator Solution**

Solve the corresponding equation (the boundary) for $y$ to get

$$y = -\frac{1}{4}x + 1.$$

Graph this boundary, and use the appropriate commands for your calculator to shade the region above the boundary. See Figure 66. (Notice that the calculator does not tell you which region to shade.)

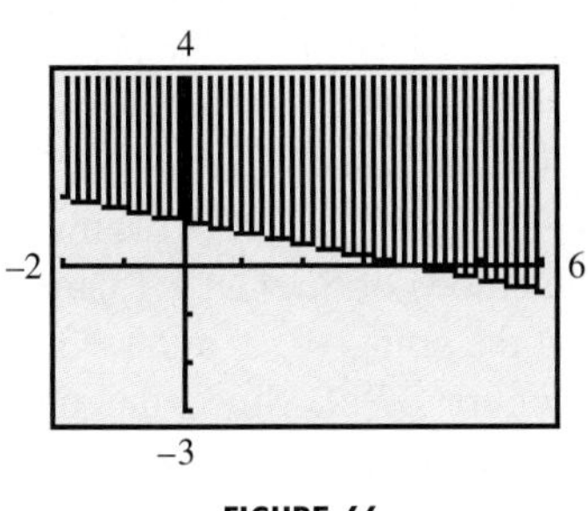

**FIGURE 66**

*(continued)*

Alternatively, or as a check, choose a point not on the boundary line and substitute into the inequality. The point (0, 0) is a good choice if it does not lie on the boundary, since the substitution is easily done. Here, substitution of (0, 0) into the original inequality gives $0 > 4$, a false statement. Since the point (0, 0) is below the line, the points that satisfy the inequality must be above the line, which agrees with the result above.

The calculator graph does not distinguish between solid boundary lines and dashed boundary lines. We must understand the mathematics to correctly interpret a calculator graph of an inequality.

The methods used to graph linear inequalities can be used for other inequalities of the form $y \leq f(x)$, as summarized here. (Similar statements can be made for $<$, $>$, and $\geq$.)

**Graphing an Inequality**

1. For a function $f$, the graph of $y < f(x)$ consists of all the points that are *below* the graph of $y = f(x)$; the graph of $y > f(x)$ consists of all the points that are *above* the graph of $y = f(x)$.
2. If the inequality is not or cannot be solved for $y$, choose a test point not on the boundary. If the test point satisfies the inequality, the graph includes all points on the same side of the boundary as the test point. Otherwise, the graph includes all points on the other side of the boundary.

## Solving Systems of Inequalities

The solution set of a **system of inequalities,** such as

$$x > 6 - 2y$$
$$x^2 < 2y,$$

is the intersection of the solution sets of its members. We find this intersection by graphing the solution sets of both inequalities on the same coordinate axes and identifying, by shading, the region common to both graphs.

**EXAMPLE 2** *Graphing a System of Two Inequalities*

Graph the solution set of the system.

$$x > 6 - 2y$$
$$x^2 < 2y$$

**Analytic Solution**

Figures 67(a) and (b) show the graphs of $x > 6 - 2y$ and $x^2 < 2y$. The methods of Section 7.1 can be used to show that the boundaries intersect at the points (2, 2) and $\left(-3, \frac{9}{2}\right)$. The solution set of the system is shown in Figure 67(c). Since the points on the boundaries of $x > 6 - 2y$ and $x^2 < 2y$ do not belong to the graph of the solution, the boundaries are dashed.

**Graphing Calculator Solution**

As usual, solve each equation for $y$ first. Enter the boundary equations,

$$y_1 = \frac{6 - x}{2}$$

and

$$y_2 = \frac{x^2}{2}.$$

*(continued)*

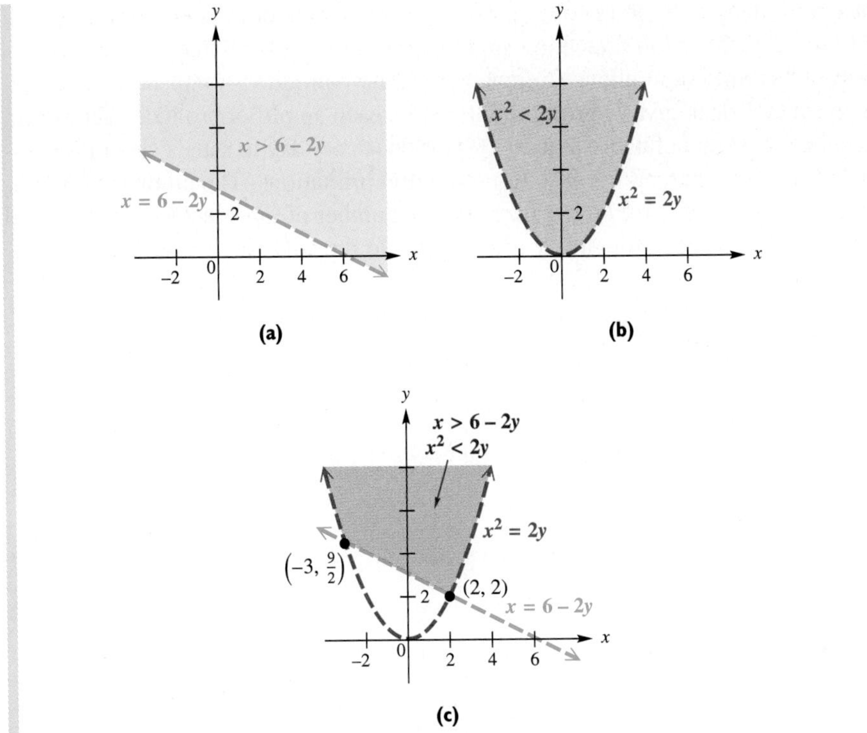

**FIGURE 67**

In both inequalities, $y$ is *greater than* an expression involving $x$. This can be determined by solving for $y$ in each case. Therefore, use the capability of your calculator to shade above each boundary. See Figure 68.

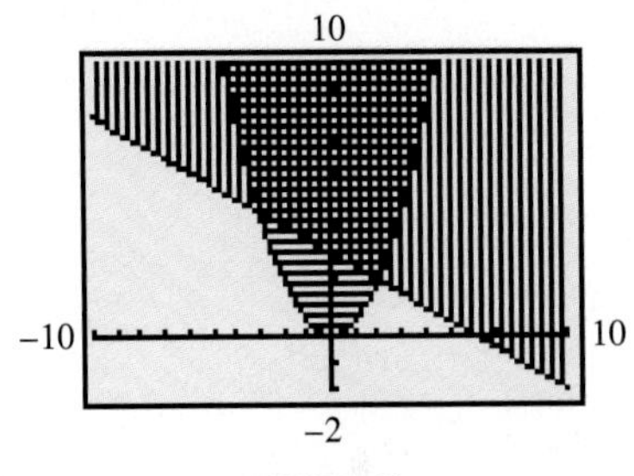


**FIGURE 68**

The cross-hatched region supports the solution shown in Figure 67(c).

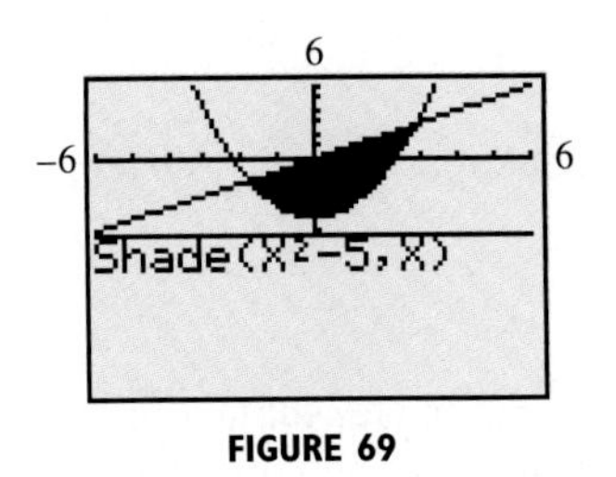


**FIGURE 69**

**EXAMPLE 3** *Graphing a System of Two Inequalities*

Graph the solution set of the system.

$$y > x^2 - 5$$
$$y < x$$

**Solution** This system can be graphed with a graphing calculator by using the appropriate commands to shade the region *above* $Y_1 = X^2 - 5$ and *below* $Y_2 = X$. See Figure 69.

**EXAMPLE 4** *Graphing a System of Three Inequalities*

Graph the solution set of the system.

$$y \geq 2^x$$
$$9x^2 + 4y^2 \leq 36$$
$$2x + y < 1$$

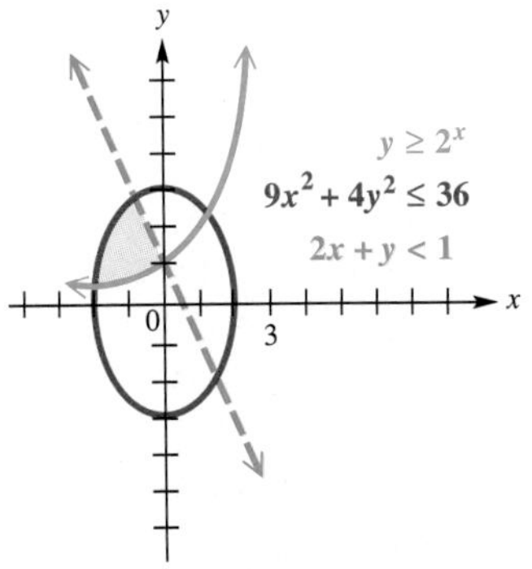

**Solution** Graph the three inequalities on the same axes and shade the region common to all three, as shown in Figure 70. Two boundary lines are solid and one is dashed.

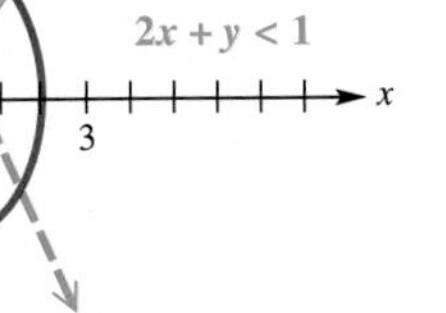


**FIGURE 70**

## Linear Programming

An important application of mathematics to business and social science is called *linear programming*. We use **linear programming** to find an optimum value, for example, minimum cost or maximum profit. It was first developed during World War II to solve problems in allocating supplies for the U.S. Air Force.

Suppose that the Audiophone Company makes two products—tape decks and amplifiers. Each tape deck gives a profit of \$30, while each amplifier produces \$70 profit. The company must manufacture at least 10 tape decks per day to satisfy one of its customers, but no more than 50 because of production limitations. The number of amplifiers produced cannot exceed 60 per day, and the number of tape decks cannot exceed the number of amplifiers. How many of each should the company manufacture to obtain maximum profit?

First, we translate the statement of the problem into symbols.

Let $x$ = number of tape decks to be produced daily,

and $y$ = number of amplifiers to be produced daily.

The company must produce at least 10 tape decks (10 or more), so

$$x \geq 10.$$

Since no more than 50 tape decks may be produced

$$x \leq 50.$$

No more than 60 amplifiers may be made in one day, so

$$y \leq 60.$$

The number of tape decks may not exceed the number of amplifiers, so

$$x \leq y.$$

The numbers of tape decks and of amplifiers cannot be negative, so

$$x \geq 0 \quad \text{and} \quad y \geq 0.$$

These restrictions, or **constraints,** that are placed on production form the system of inequalities

$$x \geq 10, \quad x \leq 50, \quad y \leq 60, \quad x \leq y, \quad x \geq 0, \quad y \geq 0.$$

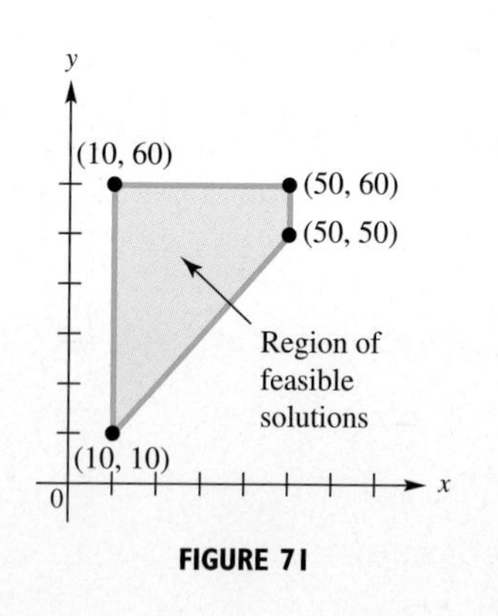


FIGURE 71

To find the maximum possible profit that the company can make, subject to these constraints, we sketch the graph of each constraint. The only feasible values of $x$ and $y$ are those that satisfy all constraints—that is, the values that lie in the intersection of the graphs of the constraints. The intersection is shown in Figure 71. Any point lying inside the shaded region or on the boundary in Figure 71 satisfies the restrictions as to the number of tape decks and amplifiers that may be produced. (For practical purposes, however, only points with integer coordinates are useful.) This region is called the **region of feasible solutions.**

Each tape deck gives a profit of \$30, so the daily profit from production of $x$ tape decks is $30x$ dollars. Also, the profit from production of $y$ amplifiers will be $70y$ dollars per day. Total daily profit is thus

$$\text{profit} = 30x + 70y.$$

This equation defines the function to be maximized, called the **objective function.**

We can now state the problem of the Audiophone Company as follows: Find values of $x$ and $y$ in or on the boundary of the shaded region of Figure 71 that will produce the maximum possible value of $30x + 70y$. To locate the point $(x, y)$ that gives the maximum profit, we add to the graph of Figure 71 lines corresponding to profits of \$0, \$1000, \$3000, and \$7000.

$$30x + 70y = 0$$
$$30x + 70y = 1000$$
$$30x + 70y = 3000$$
$$30x + 70y = 7000$$

For instance, each point on the line $30x + 70y = 3000$ corresponds to production values that yield a profit of \$3000.

Figure 72 shows the region of feasible solutions together with these lines. The lines are parallel, and the higher the line, the higher the profit. The line $30x + 70y = 7000$ has the highest profit but does not contain any points of the region of feasible solutions. To find the feasible solution of greatest profit, we lower the line $30x + 70y = 7000$ until it contains a feasible solution—that is, until it just touches the region of feasible solutions. This occurs at point $A$, a **vertex** (or **corner point**) of the region. See Figure 73. Since the coordinates of this point are (50, 60), the maximum profit is obtained when 50 tape decks and 60 amplifiers are produced each day. This maximum profit will be $30(50) + 70(60) = 5700$ dollars per day.

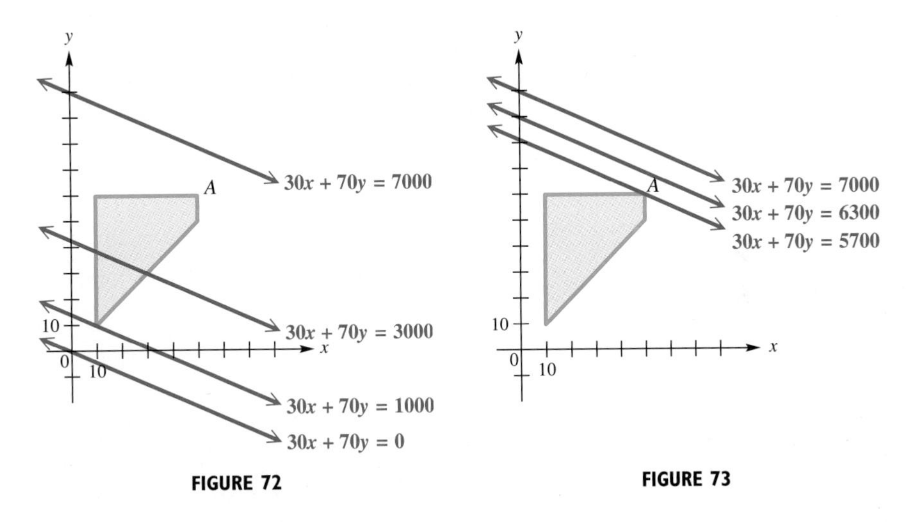


FIGURE 72

FIGURE 73

The result observed in Figure 73 holds for *every* linear programming problem.

**Fundamental Theorem of Linear Programming**

If the optimal value for a linear programming problem exists, it occurs at a vertex of the region of feasible solutions.

To solve a linear programming problem in general, use the following steps.

**EXAMPLE 5** *Finding a Minimum Cost Model*

Robin takes vitamin pills each day. She wants at least 16 units of Vitamin A, at least 5 units of Vitamin $B_1$, and at least 20 units of Vitamin C. She can choose between red pills, costing 10¢ each, that contains 8 units of A, 1 of $B_1$, and 2 of C; and blue pills, costing 20¢ each, that contain 2 units of A, 1 of $B_1$, and 7 of C. How many of each pill should she buy to minimize her cost and yet fulfill her daily requirements?

**Solution** Let $x$ represent the number of red pills to buy, and let $y$ represent the number of blue pills to buy. Then the cost in pennies per day is given by

$$\text{cost} = 10x + 20y.$$

Since Robin buys $x$ of the 10¢ pills and $y$ of the 20¢ pills, she gets 8 units of Vitamin A from each red pill and 2 units of Vitamin A from each blue pill. Altogether, she gets $8x + 2y$ units of A per day. Since she wants at least 16 units,

$$8x + 2y \geq 16.$$

Each red or blue pill supplies 1 unit of Vitamin $B_1$. Robin wants at least 5 units per day, so

$$x + y \geq 5.$$

For Vitamin C, the inequality is

$$2x + 7y \geq 20.$$

Also, $x \geq 0$ and $y \geq 0$, since Robin cannot buy negative numbers of the pills.

Total cost of the pills is minimized by using the solution of the system of inequalities formed by the constraints. (See Figure 74.) The solution to this minimizing problem will also occur at a vertex point. By substituting the coordinates of the vertex points in the cost function, we find the lowest cost.

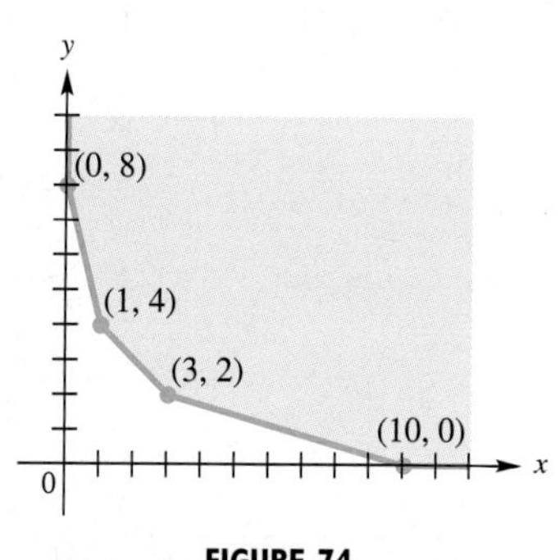

FIGURE 74

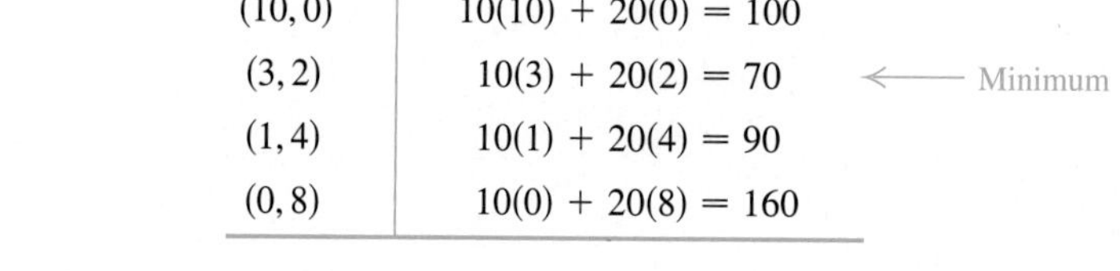

| Point | Cost = $10x + 20y$ | |
|---|---|---|
| (10, 0) | $10(10) + 20(0) = 100$ | |
| (3, 2) | $10(3) + 20(2) = 70$ | ← Minimum |
| (1, 4) | $10(1) + 20(4) = 90$ | |
| (0, 8) | $10(0) + 20(8) = 160$ | |

Robin's best choice is to buy 3 red pills and 2 blue ones, for a total cost of 70¢ per day. She receives the minimum of Vitamins $B_1$ and C, and an excess of Vitamin A.

To solve a linear programming problem in general, use the following steps.

### Solving a Linear Programming Problem

1. Write the objective function and all necessary constraints.
2. Graph the region of feasible solutions.
3. Identify all vertices or corner points.
4. Find the value of the objective function at each vertex.
5. The solution is given by the vertex producing the optimal value of the objective function.

## 7.7 EXERCISES

*Use traditional graphing methods to graph each inequality.*

1. $x \le 3$
2. $y \le -2$
3. $x + 2y \le 6$
4. $x - y \ge 2$
5. $2x + 3y \ge 4$
6. $4y - 3x < 5$
7. $3x - 5y > 6$
8. $x < 3 + 2y$
9. $5x \le 4y - 2$
10. $2x > 3 - 4y$
11. $y < 3x^2 + 2$
12. $y \le x^2 - 4$
13. $y > (x - 1)^2 + 2$
14. $y > 2(x + 3)^2 - 1$
15. $x^2 + (y + 3)^2 \le 16$
16. $(x - 4)^2 + (y + 3)^2 \le 9$
17. In your own words, explain how to determine whether the boundary of the graph of an inequality is solid or dashed.
18. When graphing $y > 3x - 6$, would you shade above or below the line $y = 3x - 6$? Explain your answer.

*Concept Check* *Use the concepts of this section to work Exercises 19–22.*

19. For $Ax + By \ge C$, if $B > 0$, would you shade above or below the boundary line?
20. For $Ax + By \ge C$, if $B < 0$, would you shade above or below the boundary line?
21. Which one of the following is a description of the graph of the inequality
$$(x - 5)^2 + (y - 2)^2 < 4?$$
   **A.** The region inside a circle with center $(-5, -2)$ and radius 2
   **B.** The region inside a circle with center $(5, 2)$ and radius 2
   **C.** The region inside a circle with center $(-5, -2)$ and radius 4
   **D.** The region outside a circle with center $(5, 2)$ and radius 4
22. Which one of the following inequalities satisfies the following description: the region outside an ellipse centered at the origin, with $x$-intercepts 4 and $-4$, and $y$-intercepts 9 and $-9$?
   **A.** $\frac{x^2}{4} + \frac{y^2}{9} > 1$  **B.** $\frac{x^2}{16} - \frac{y^2}{81} > 1$
   **C.** $\frac{x^2}{16} + \frac{y^2}{81} > 0$  **D.** $\frac{x^2}{16} + \frac{y^2}{81} > 1$

*Concept Check* *In Exercises 23–26, match the inequality with the appropriate calculator graph. Do not use your calculator; instead, use your knowledge of the concepts involved in graphing inequalities.*

23. $y \le 3x - 6$
24. $y \ge 3x - 6$
25. $y \le -3x - 6$
26. $y \ge -3x - 6$

**A.**
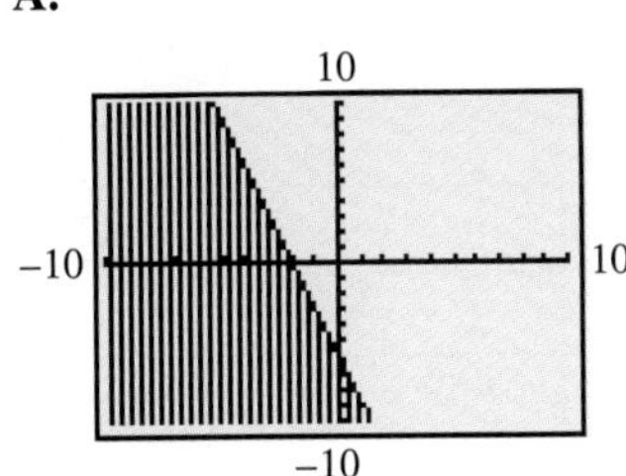


**B.**
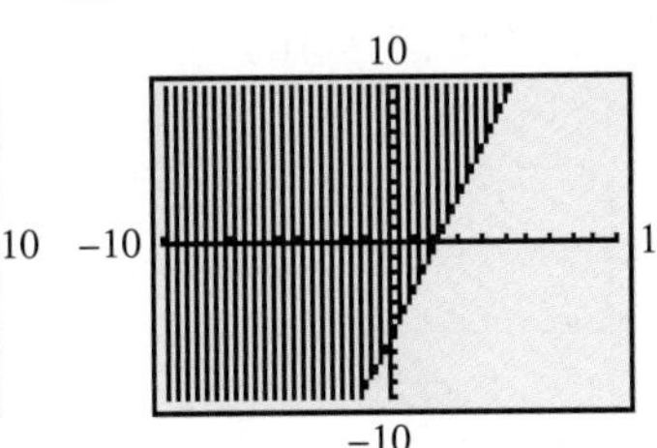


**C.**
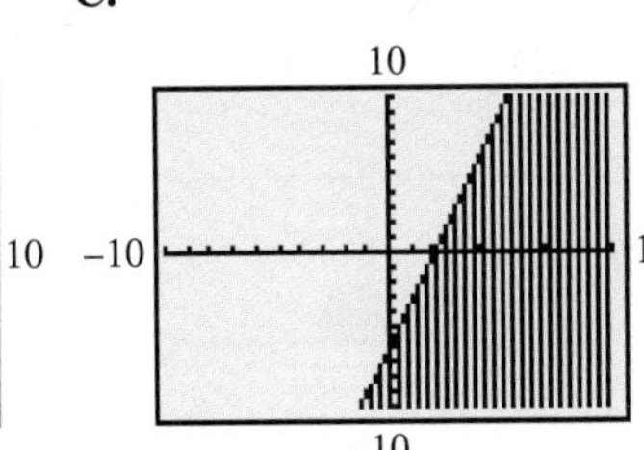


**D.**
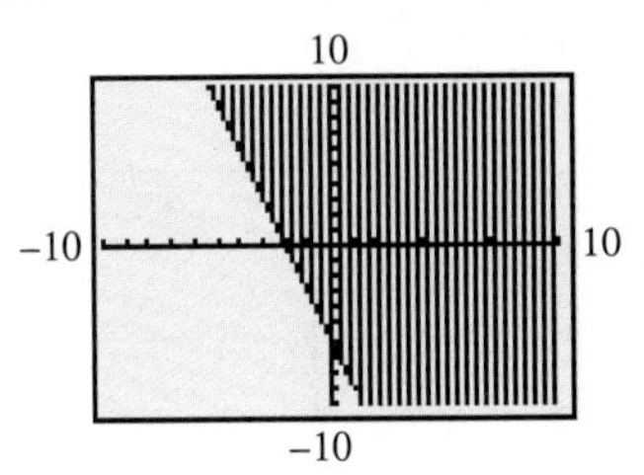


*Use traditional graphing methods to graph each system of inequalities.*

27. $x + y \ge 0$
    $2x - y \ge 3$
28. $x + y \le 4$
    $x - 2y \ge 6$
29. $2x + y > 2$
    $x - 3y < 6$
30. $4x + 3y < 12$
    $y + 4x > -4$
31. $3x + 5y \le 15$
    $x - 3y \ge 9$
32. $y \le x$
    $x^2 + y^2 < 1$
33. $4x - 3y \le 12$
    $y \le x^2$
34. $y \le -x^2$
    $y \ge x^2 - 6$
35. $x + y \le 9$
    $x \le -y^2$
36. $x + 2y \le 4$
    $y \ge x^2 - 1$
37. $y \le (x + 2)^2$
    $y \ge -2x^2$
38. $x - y < 1$
    $-1 < y < 1$

**39.** $x + y \leq 36$
$-4 \leq x \leq 4$

**40.** $y > x^2 + 4x + 4$
$y < -x^2$

**41.** $y \geq (x - 2)^2 + 3$
$y \leq -(x - 1)^2 + 6$

**42.** $x \geq 0$
$x + y \leq 4$
$2x + y \leq 5$

**43.** $3x - 2y \geq 6$
$x + y \leq -5$
$y \leq 4$

**44.** $-2 < x < 3$
$-1 \leq y \leq 5$
$2x + y < 6$

**45.** $-2 < x < 2$
$y > 1$
$x - y > 0$

**46.** $x + y \leq 4$
$x - y \leq 5$
$4x + y \leq -4$

**47.** $x \leq 4$
$x \geq 0$
$y \geq 0$
$x + 2y \geq 2$

**48.** $2y + x \geq -5$
$y \leq 3 + x$
$x \leq 0$
$y \leq 0$

**49.** $2x + 3y \leq 12$
$2x + 3y > -6$
$3x + y < 4$
$x \geq 0$
$y \geq 0$

**50.** $y \geq 3^x$
$y \geq 2$

**51.** $y \leq \left(\frac{1}{2}\right)^x$
$y \geq 4$

**52.** $\ln x - y \geq 1$
$x^2 - 2x - y \leq 1$

**53.** $y \leq \log x$
$y \geq |x - 2|$

**54.** $e^{-x} - y \leq 1$
$x - 2y \geq 4$

**55.** *Concept Check* Which one of the following is a description of the solution set of the following system?

$$x^2 + 4y^2 < 36$$
$$y < x$$

**A.** All points outside the ellipse $x^2 + 4y^2 = 36$ and above the line $y = x$

**B.** All points outside the ellipse $x^2 + 4y^2 = 36$ and below the line $y = x$

**C.** All points inside the ellipse $x^2 + 4y^2 = 36$ and above the line $y = x$

**D.** All points inside the ellipse $x^2 + 4y^2 = 36$ and below the line $y = x$

**56.** *Concept Check* Fill in each blank with the appropriate response. The graph of the system

$$y > x^2 + 2$$
$$x^2 + y^2 < 16$$
$$y < 7$$

consists of all points ________ (above/below) the parabola $y = x^2 + 2$, ________ (inside/outside) the circle $x^2 + y^2 = 16$, and ________ (above/below) the line $y = 7$.

*Concept Check In Exercises 57–60, match each system of inequalities with the appropriate calculator graph. Do not use your calculator; instead, use your knowledge of the concepts involved in graphing systems of inequalities.*

**57.** $y \geq x$
$y \leq 2x - 3$

**58.** $y \geq x^2$
$y < 5$

**59.** $x^2 + y^2 \leq 16$
$y \geq 0$

**60.** $y \leq x$
$y \geq 2x - 3$

**A.**

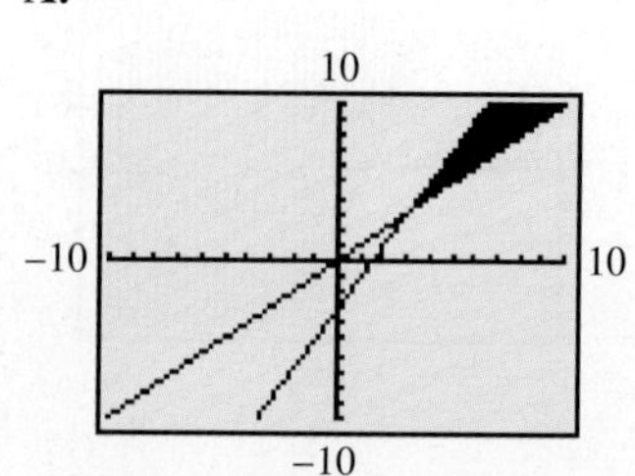


**B.**

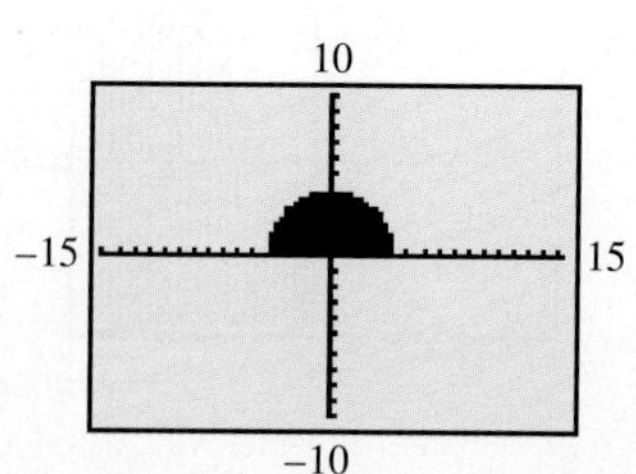


**C.**

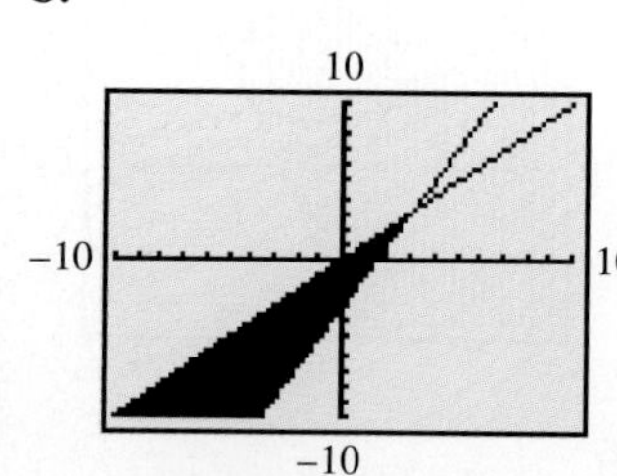


**D.**

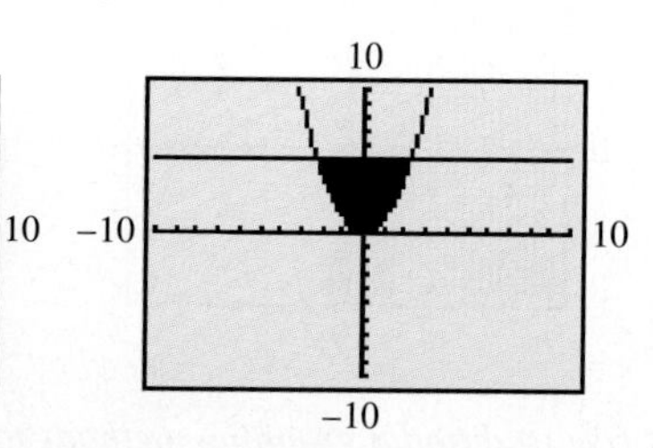


*Use the shading capabilities of your graphing calculator to graph each inequality or system of inequalities.*

**61.** $3x + 2y \geq 6$

**62.** $y \leq x^2 + 5$

**63.** $x + y \geq 2$
$x + y \leq 6$

**64.** $y \geq |x + 2|$
$y \leq 6$

**65.** $y \geq 2^x$
$y \leq 8$

**66.** $y \leq x^3 + x^2 - 4x - 4$

**67.** *Concept Check* Find a system of linear inequalities for which the graph is the region in the first quadrant between and inclusive of the pair of lines $x + 2y - 8 = 0$ and $x + 2y = 12$.

**68.** *Cost of Vitamins* The figure shows the region of feasible solutions for the vitamin problem of Example 5 and the straight line graph of all combinations of red and blue pills for which the cost is 40 cents.

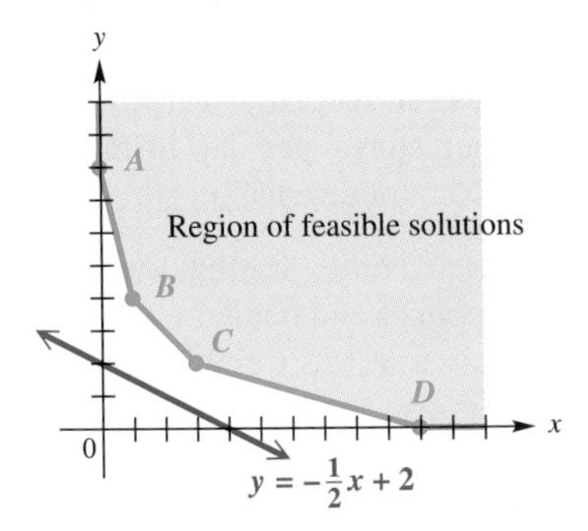


**(a)** The cost function is $C = 10x + 20y$. Give the linear equation (in slope-intercept form) of the line of constant cost $c$.

**(b)** As $c$ increases, does the line of constant cost move up or down?

**(c)** By inspection, find the vertex of the region of feasible solutions that gives the optimal solution.

*The graphs show regions of feasible solutions. Find the maximum and minimum values of the given expressions.*

**69.** $3x + 5y$

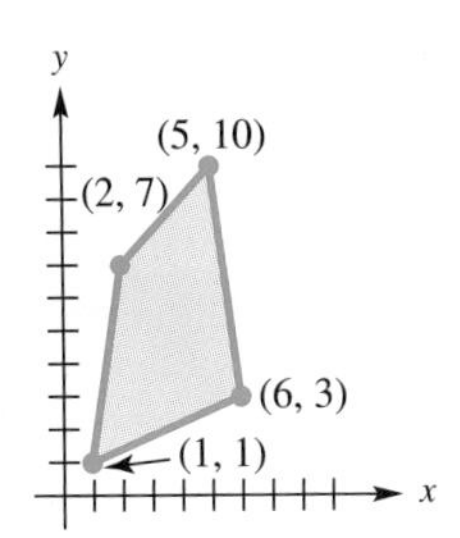


**70.** $6x + y$

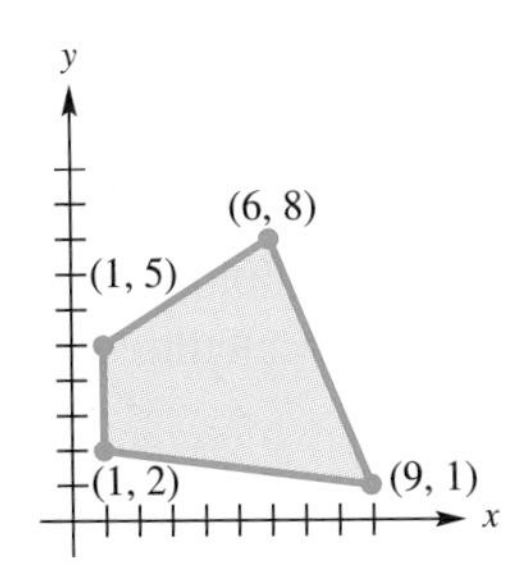


*Find the maximum and minimum values of each objective function over the region of feasible solutions shown.*

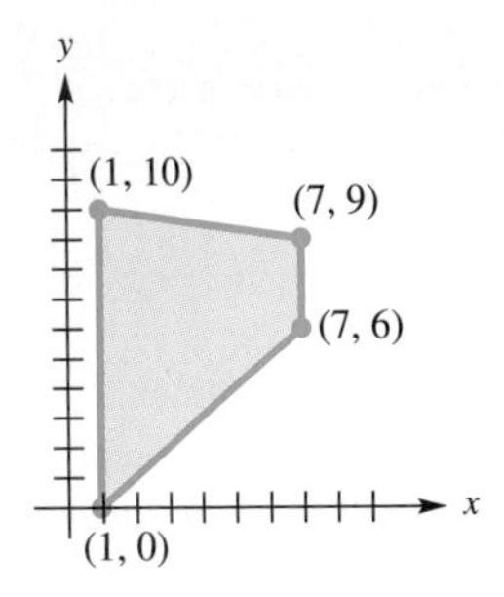


**71.** $3x + 5y$

**72.** $5x + 5y$

**73.** $10y$

**74.** $3x - y$

*Solve each problem.*

**75.** *Inquiries about Displayed Products* A wholesaler of party goods wishes to display her products at a convention of social secretaries in such a way that she gets the maximum number of inquiries about her whistles and hats. Her booth at the convention has 12 square meters of floor space to be used for display purposes. A display unit for hats requires 2 square meters, and for whistles, 4 square meters. Experience tells the wholesaler that she should never have more than a total of 5 units of whistles and hats on display at one time. If she receives three inquiries for each unit of hats and two inquiries for each unit of whistles on display, how many of each should she display in order to get the maximum number of inquiries? What is that maximum number?

**76.** *Profit from Farm Animals* Farmer Jones raises only pigs and geese. She wants to raise no more than 16 animals, with no more than 12 geese. She spends \$50 to raise a pig and \$20 to raise a goose. She has \$500 available for this purpose. Find the maximum profit she can make if she makes a profit of \$80 per goose and \$40 per pig. Indicate how many pigs and geese she should raise to achieve this maximum.

**77.** *Shipment Costs* A manufacturer of refrigerators must ship at least 100 refrigerators to its two West Coast warehouses. Each warehouse holds a maximum of 100 refrigerators. Warehouse A holds 25 refrigerators already, while warehouse B has 20 on hand. It costs \$12 to ship a refrigerator to warehouse A and \$10 to ship one to warehouse B. How many refrigerators should be shipped to each warehouse to minimize cost? What is the minimum cost?

**78.** *Diet Requirements* Theo, who is dieting, requires two food supplements, I and II. He can get these supplements from two different products, $A$ and $B$, as shown in the table.

| Supplement (grams/serving) | I | II |
|---|---|---|
| Product $A$ | 3 | 2 |
| Product $B$ | 2 | 4 |

Theo's physician has recommended that he include at least 15 grams of each supplement in his daily diet. If product $A$ costs 25¢ per serving and product $B$ costs 40¢ per serving, how can he satisfy his requirements most economically?

**79.** *Gasoline Revenues* A manufacturing process requires that oil refineries manufacture at least 2 gallons of gasoline for each gallon of fuel oil. To meet winter demand for fuel oil, at least 3 million gallons a day must be produced. The demand for gasoline is no more than 6.4 million gallons per day. If the price of gasoline is \$1.90 and the price of fuel oil is \$1.50 per gallon, how much of each should be produced to maximize revenue?

**80.** *Manufacturing Revenues* A machine shop manufactures two types of bolts. Each can be made on any of three groups of machines, but the time required on each group differs, as shown in the table.

| | Machine Group | | |
|---|---|---|---|
| **Bolt** | **I** | **II** | **III** |
| Type A | .1 min | .1 min | .1 min |
| Type B | .1 min | .4 min | .5 min |

Production schedules are made up one day at a time. In a day, there are 240, 720, and 160 minutes available, respectively, on these machines. Type A bolts sells for 10¢ and Type B bolts for 12¢. How many of each type of bolt should be manufactured per day to maximize revenue? What is the maximum revenue?

**81.** *Aid to Disaster Victims* Earthquake victims in China need medical supplies and bottled water. Each medical kit measures 1 cubic foot and weighs 10 pounds. Each container of water is also 1 cubic foot but weights 20 pounds. The plane can carry only 80,000 pounds, with total volume 6000 cubic feet. Each medical kit will aid 4 people, while each container of water will serve 10 people. How many of each should be sent in order to maximize the number of people aided? How many people will be aided?

**82.** *Aid to Disaster Victims* If each medical kit could aid 6 people instead of 4, how would the results in Exercise 81 change?

## 7.8 PARTIAL FRACTIONS

Decomposition of Rational Expressions • Distinct Linear Factors • Repeated Linear Factors • Distinct Linear and Quadratic Factors • Repeated Quadratic Factors

**Looking Ahead to Calculus**

To find the integral of a rational function, it is often necessary to express the function as the sum of two or more rational functions, using one of the methods discussed in this section.

### Decomposition of Rational Expressions

The sums of rational expressions are found by combining two or more rational expressions into one rational expression. Here, the reverse process is considered: Given one rational expression, express it as the sum of two or more rational expressions. A special type of sum of rational expressions is called the **partial fraction decomposition;** each term in the sum is a **partial fraction.** The technique of decomposing a rational expression into partial fractions is useful in calculus and other areas of mathematics.

To form a partial fraction decomposition of a rational expression, follow these steps.

**Partial Fraction Decomposition of $\frac{f(x)}{g(x)}$**

**Step 1** If $\frac{f(x)}{g(x)}$ is not a proper fraction (a fraction with the numerator of lower degree than the denominator), divide $f(x)$ by $g(x)$. For example,

$$\frac{x^4 - 3x^3 + x^2 + 5x}{x^2 + 3} = x^2 - 3x - 2 + \frac{14x + 6}{x^2 + 3}.$$

Then, apply the following steps to the remainder, which is a proper fraction.

**Step 2** Factor $g(x)$ completely into factors of the form $(ax + b)^m$ or $(cx^2 + dx + e)^n$, where $cx^2 + dx + e$ is irreducible and $m$ and $n$ are integers.

**Step 3** **(a)** For each distinct linear factor $(ax + b)$, the decomposition must include the term

$$\frac{A}{ax + b}.$$

**(b)** For each repeated linear factor $(ax + b)^m$, the decomposition must include the terms

$$\frac{A_1}{ax + b} + \frac{A_2}{(ax + b)^2} + \cdots + \frac{A_m}{(ax + b)^m}.$$

**Step 4** **(a)** For each distinct quadratic factor $(cx^2 + dx + e)$, the decomposition must include the term

$$\frac{Bx + C}{cx^2 + dx + e}.$$

**(b)** For each repeated quadratic factor $(cx^2 + dx + e)^n$, the decomposition must include the terms

$$\frac{B_1x + C_1}{cx^2 + dx + e} + \frac{B_2x + C_2}{(cx^2 + dx + e)^2} + \cdots + \frac{B_nx + C_n}{(cx^2 + dx + e)^n}.$$

**Step 5** Use algebraic techniques to solve for the constants in the numerators of the decomposition.

To find the constants in Step 5, the goal is to get a system of equations with as many equations as there are unknowns in the numerators. One method for finding these equations is to substitute values for $x$ on each side of the rational equation formed in Steps 3 or 4.

## Distinct Linear Factors

**EXAMPLE 1** *Finding a Partial Fraction Decomposition*

Find the partial fraction decomposition of

$$\frac{2x^4 - 8x^2 + 5x - 2}{x^3 - 4x}.$$

**Solution** The given fraction is not a proper fraction; the numerator has higher degree than the denominator. Perform the division.

$$\begin{array}{r} 2x \phantom{- 8x^2 + 5x - 2} \\ x^3 - 4x \overline{)2x^4 - 8x^2 + 5x - 2} \\ \underline{2x^4 - 8x^2} \phantom{+ 5x - 2} \\ 5x - 2 \end{array}$$

The quotient is $\dfrac{2x^4 - 8x^2 + 5x - 2}{x^3 - 4x} = 2x + \dfrac{5x - 2}{x^3 - 4x}$. Now, work with the remainder fraction. Factor the denominator as $x^3 - 4x = x(x + 2)(x - 2)$. Since the factors are distinct linear factors, use Step 3(a) to write the decomposition as

$$\frac{5x - 2}{x^3 - 4x} = \frac{A}{x} + \frac{B}{x + 2} + \frac{C}{x - 2}, \qquad (1)$$

where $A$, $B$, and $C$ are constants that need to be found. Multiply each side of equation (1) by $x(x + 2)(x - 2)$ to get

$$5x - 2 = A(x + 2)(x - 2) + Bx(x - 2) + Cx(x + 2). \qquad (2)$$

Equation (1) is an identity, since both sides represent the same rational expression. Thus, equation (2) is also an identity. Equation (1) holds for all values of $x$ except 0, $-2$, and 2. However, equation (2) holds for all values of $x$. In particular, substituting 0 for $x$ in equation (2) gives $-2 = -4A$, so $A = \frac{1}{2}$. Similarly, choosing $x = -2$ gives $-12 = 8B$, so $B = -\frac{3}{2}$. Finally, choosing $x = 2$ gives $8 = 8C$, so $C = 1$. The remainder rational expression can be written as the following sum of partial fractions:

$$\frac{5x - 2}{x^3 - 4x} = \frac{1}{2x} + \frac{-3}{2(x + 2)} + \frac{1}{x - 2},$$

and the given rational expression can be written as

$$\frac{2x^4 - 8x^2 + 5x - 2}{x^3 - 4x} = 2x + \frac{1}{2x} + \frac{-3}{2(x + 2)} + \frac{1}{x - 2}.$$

Check the work by combining the terms on the right.

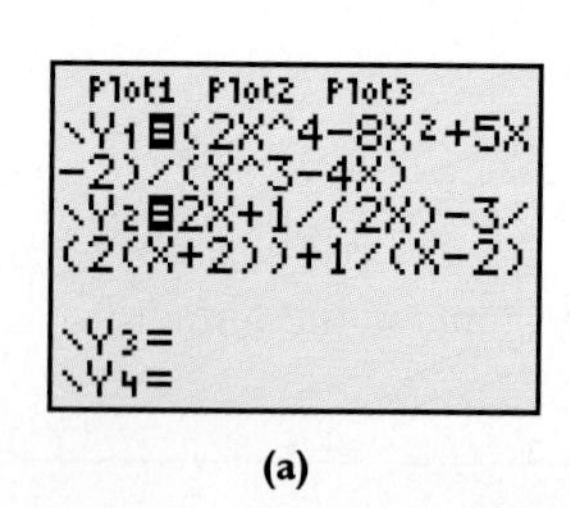


(a)

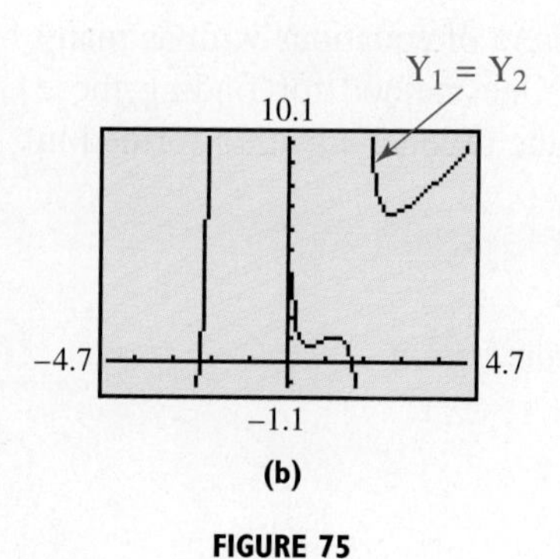


(b)

**FIGURE 75**

A graphing calculator can be used to check the work in Example 1. Define $Y_1$ using the original rational expression, and define $Y_2$ using the partial fraction decomposition:

$$Y_1 = \frac{2X^4 - 8X^2 + 5X - 2}{X^3 - 4X}, \quad Y_2 = 2X + \frac{1}{2X} + \frac{-3}{2(X + 2)} + \frac{1}{X - 2}.$$

See Figure 75(a). Now graph both $Y_1$ and $Y_2$ on the same screen. The graphs should be indistinguishable. See Figure 75(b).

**NOTE** The method of graphical support just described is not foolproof; however, it gives a fairly accurate portrayal of whether the partial fraction decomposition is correct. A table can be used for support as well.

## Repeated Linear Factors

**EXAMPLE 2** *Finding a Partial Fraction Decomposition*

Find the partial fraction decomposition of

$$\frac{2x}{(x-1)^3}.$$

**Solution** This is a proper fraction. The denominator is already factored with repeated linear factors. We write the decomposition as shown, by using Step 3(b).

$$\frac{2x}{(x-1)^3} = \frac{A}{x-1} + \frac{B}{(x-1)^2} + \frac{C}{(x-1)^3}$$

We clear denominators by multiplying each side of this equation by $(x-1)^3$.

$$2x = A(x-1)^2 + B(x-1) + C$$

Substituting 1 for $x$ leads to $C = 2$, so

$$2x = A(x-1)^2 + B(x-1) + 2. \qquad (1)$$

We substituted the only root, and we still need to find values for $A$ and $B$. However, *any* number can be substituted for $x$. For example, when we choose $x = -1$ (because it is easy to substitute), equation (1) becomes

$$\begin{aligned} -2 &= 4A - 2B + 2 \\ -4 &= 4A - 2B \\ -2 &= 2A - B. \qquad (2) \end{aligned}$$

Substituting 0 for $x$ in equation (1) gives

$$\begin{aligned} 0 &= A - B + 2 \\ 2 &= -A + B. \qquad (3) \end{aligned}$$

Now, we solve the system of equations (2) and (3) to get $A = 0$ and $B = 2$. The partial fraction decomposition is

$$\frac{2x}{(x-1)^3} = \frac{2}{(x-1)^2} + \frac{2}{(x-1)^3}.$$

We needed three substitutions because there were three constants to evaluate, $A$, $B$, and $C$. To check this result, we could combine the terms on the right.

## Distinct Linear and Quadratic Factors

**EXAMPLE 3** *Finding a Partial Fraction Decomposition*

Find the partial fraction decomposition of

$$\frac{x^2 + 3x - 1}{(x+1)(x^2+2)}.$$

**Solution** This denominator has distinct linear and quadratic factors, where neither is repeated. Since $x^2 + 2$ cannot be factored, it is irreducible. The partial fraction decomposition is

$$\frac{x^2 + 3x - 1}{(x + 1)(x^2 + 2)} = \frac{A}{x + 1} + \frac{Bx + C}{x^2 + 2}.$$

Multiply each side by $(x + 1)(x^2 + 2)$ to get

$$x^2 + 3x - 1 = A(x^2 + 2) + (Bx + C)(x + 1). \qquad (1)$$

First, substitute $-1$ for $x$ to get

$$(-1)^2 + 3(-1) - 1 = A[(-1)^2 + 2] + 0$$
$$-3 = 3A$$
$$A = -1.$$

Replace $A$ with $-1$ in equation (1) and substitute any value for $x$. For instance, if $x = 0$, then

$$0^2 + 3(0) - 1 = -1(0^2 + 2) + (B \cdot 0 + C)(0 + 1)$$
$$-1 = -2 + C$$
$$C = 1.$$

Now, letting $A = -1$ and $C = 1$, substitute again in equation (1), using another number for $x$. For $x = 1$,

$$3 = -3 + (B + 1)(2)$$
$$6 = 2B + 2$$
$$B = 2.$$

Using $A = -1$, $B = 2$, and $C = 1$, the partial fraction decomposition is

$$\frac{x^2 + 3x - 1}{(x + 1)(x^2 + 2)} = \frac{-1}{x + 1} + \frac{2x + 1}{x^2 + 2}.$$

Again, this work can be checked by combining terms on the right. ○

For fractions with denominators that have quadratic factors, another method is often more convenient. The system of equations is formed by equating coefficients of like terms on each side of the partial fraction decomposition. For instance, in Example 3, after each side was multiplied by the common denominator, the equation was

$$x^2 + 3x - 1 = A(x^2 + 2) + (Bx + C)(x + 1).$$

Multiplying on the right and collecting like terms, we have

$$x^2 + 3x - 1 = Ax^2 + 2A + Bx^2 + Bx + Cx + C$$
$$x^2 + 3x - 1 = (A + B)x^2 + (B + C)x + (C + 2A).$$

Now, equating the coefficients of like powers of $x$ gives the three equations

$$1 = A + B$$
$$3 = B + C$$
$$-1 = C + 2A.$$

Solving this system of equations for $A$, $B$, and $C$ would give the partial fraction decomposition. The next example uses a combination of the two methods.

## Repeated Quadratic Factors

**EXAMPLE 4** *Finding a Partial Fraction Decomposition*

Find the partial fraction decomposition of

$$\frac{2x}{(x^2+1)^2(x-1)}.$$

**Solution** This expression has both a linear factor and a repeated quadratic factor. By Steps 3(a) and 4(b),

$$\frac{2x}{(x^2+1)^2(x-1)} = \frac{Ax+B}{x^2+1} + \frac{Cx+D}{(x^2+1)^2} + \frac{E}{x-1}.$$

Multiplying each side by $(x^2+1)^2(x-1)$ leads to

$$2x = (Ax+B)(x^2+1)(x-1) + (Cx+D)(x-1) + E(x^2+1)^2. \qquad (1)$$

If $x = 1$, equation (1) reduces to $2 = 4E$, or $E = \frac{1}{2}$. Substituting $\frac{1}{2}$ for $E$ in equation (1) and combining terms on the right gives

$$2x = \left(A + \frac{1}{2}\right)x^4 + (-A+B)x^3 + (A-B+C+1)x^2$$
$$+ (-A+B+D-C)x + \left(-B-D+\frac{1}{2}\right). \qquad (2)$$

To get additional equations involving the unknowns, equate the coefficients of like powers of $x$ on each side of equation (2). Setting corresponding coefficients of $x^4$ equal, $0 = A + \frac{1}{2}$ or $A = -\frac{1}{2}$. From the corresponding coefficients of $x^3$, $0 = -A + B$, which means that since $A = -\frac{1}{2}$, $B = -\frac{1}{2}$. Using the coefficients of $x^2$, $0 = A - B + C + 1$. Since $A = -\frac{1}{2}$ and $B = -\frac{1}{2}$, $C = -1$. Finally, from the coefficients of $x$, $2 = -A + B + D - C$. Substituting for $A$, $B$, and $C$ gives $D = 1$. With

$$A = -\frac{1}{2}, \quad B = -\frac{1}{2}, \quad C = -1, \quad D = 1, \quad \text{and} \quad E = \frac{1}{2},$$

the given fraction has the partial fraction decomposition

$$\frac{2x}{(x^2+1)^2(x-1)} = \frac{-\frac{1}{2}x - \frac{1}{2}}{x^2+1} + \frac{-x+1}{(x^2+1)^2} + \frac{\frac{1}{2}}{x-1}$$

or

$$\frac{2x}{(x^2+1)^2(x-1)} = \frac{-(x+1)}{2(x^2+1)} + \frac{-x+1}{(x^2+1)^2} + \frac{1}{2(x-1)}.$$

In summary, to solve for the constants in the numerators of a partial fraction decomposition, use either of the following methods or a combination of the two.

**Techniques for Decomposition into Partial Fractions**

**Method 1 For Linear Factors**

1. Multiply each side of the rational expression by the common denominator.
2. Substitute the zero of each factor in the resulting equation. For repeated linear factors, substitute as many other numbers as necessary to find all the constants in the numerators. The number of substitutions required will equal the number of constants $A, B, \ldots$.

**Method 2 For Quadratic Factors**

1. Multiply each side of the rational expression by the common denominator.
2. Collect terms on the right side of the resulting equation.
3. Equate the coefficients of like terms to get a system of equations.
4. Solve the system to find the constants in the numerators.

## 7.8 EXERCISES

*Find the partial fraction decomposition for each rational expression.*

**1.** $\dfrac{5}{3x(2x+1)}$

**2.** $\dfrac{3x-1}{x(x+1)}$

**3.** $\dfrac{4x+2}{(x+2)(2x-1)}$

**4.** $\dfrac{x+2}{(x+1)(x-1)}$

**5.** $\dfrac{x}{x^2+4x-5}$

**6.** $\dfrac{5x-3}{(x+1)(x-3)}$

**7.** $\dfrac{2x}{(x+1)(x+2)^2}$

**8.** $\dfrac{2}{x^2(x+3)}$

**9.** $\dfrac{4}{x(1-x)}$

**10.** $\dfrac{4x^2-4x^3}{x^2(1-x)}$

**11.** $\dfrac{4x^2-x-15}{x(x+1)(x-1)}$

**12.** $\dfrac{2x+1}{(x+2)^3}$

**13.** $\dfrac{x^2}{x^2+2x+1}$

**14.** $\dfrac{3}{x^2+4x+3}$

**15.** $\dfrac{2x^5+3x^4-3x^3-2x^2+x}{2x^2+5x+2}$

**16.** $\dfrac{6x^5+7x^4-x^2+2x}{3x^2+2x-1}$

**17.** $\dfrac{x^3+4}{9x^3-4x}$

**18.** $\dfrac{x^3+2}{x^3-3x^2+2x}$

**19.** $\dfrac{-3}{x^2(x^2+5)}$

**20.** $\dfrac{2x+1}{(x+1)(x^2+2)}$

**21.** $\dfrac{3x-2}{(x+4)(3x^2+1)}$

**22.** $\dfrac{3}{x(x+1)(x^2+1)}$

**23.** $\dfrac{1}{x(2x+1)(3x^2+4)}$

**24.** $\dfrac{x^4+1}{x(x^2+1)^2}$

**25.** $\dfrac{3x-1}{x(2x^2+1)^2}$

**26.** $\dfrac{3x^4+x^3+5x^2-x+4}{(x-1)(x^2+1)^2}$

**27.** $\dfrac{-x^4-8x^2+3x-10}{(x+2)(x^2+4)^2}$

**28.** $\dfrac{x^2}{x^4-1}$

**29.** $\dfrac{5x^5+10x^4-15x^3+4x^2+13x-9}{x^3+2x^2-3x}$

**30.** $\dfrac{3x^6+3x^4+3x}{x^4+x^2}$

*Determine whether each partial fraction decomposition is correct by graphing the left side and the right side of the equation on the same coordinate axes and observing whether the graphs coincide.*

**31.** $\dfrac{4x^2 - 3x - 4}{x^3 + x^2 - 2x} = \dfrac{2}{x} + \dfrac{-1}{x - 1} + \dfrac{3}{x + 2}$

**32.** $\dfrac{1}{(x - 1)(x + 2)} = \dfrac{1}{x - 1} - \dfrac{1}{x + 2}$

**33.** $\dfrac{x^3 - 2x}{(x^2 + 2x + 2)^2} = \dfrac{x - 2}{x^2 + 2x + 2} + \dfrac{2}{(x^2 + 2x + 2)^2}$

**34.** $\dfrac{2x + 4}{x^2(x - 2)} = \dfrac{-2}{x} + \dfrac{-2}{x^2} + \dfrac{2}{x - 2}$

## Reviewing Basic Concepts (Sections 7.7 and 7.8)

*Graph the solution set of each inequality or system of inequalities.*

**1.** $-2x - 3y \le 6$

**2.** $x - y < 5$
$x + y \ge 3$

**3.** $y \ge x^2 - 2$
$x + 2y \ge 4$

**4.** $x^2 + y^2 \le 25$
$x^2 + y^2 \ge 9$

**5.** Which one of the following systems is represented by the given graph?

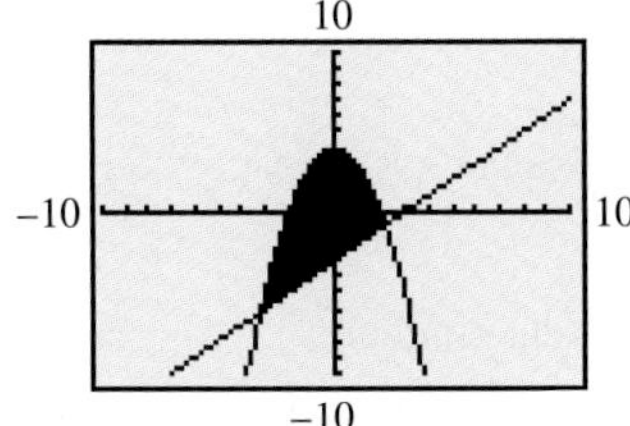


**A.** $y > x - 3$
$y < -x^2 + 4$

**B.** $y < x - 3$
$y < -x^2 + 4$

**C.** $y > x - 3$
$y > -x^2 + 4$

**D.** $y < x - 3$
$y > -x^2 + 4$

**6.** Find the minimum value of $2x + 3y$ subject to the following constraints:

$$x \ge 0$$
$$y \ge 0$$
$$x + y \ge 4$$
$$2x + y \le 8.$$

**7.** The graph shows a region of feasible solutions. Find the maximum and minimum values of $P = 3x + 5y$ over this region.

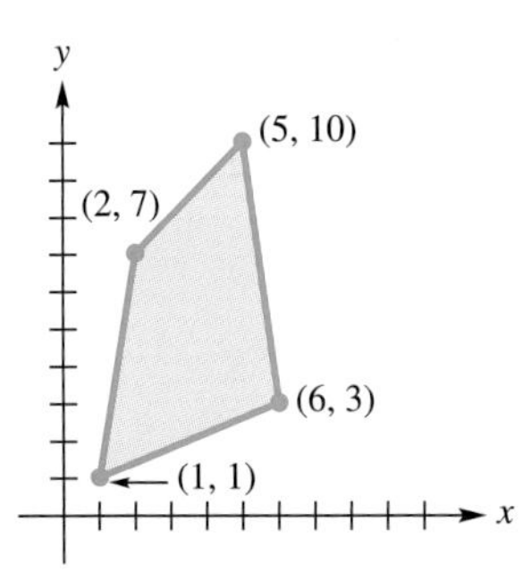


**8.** *Minimizing Cost* Two substances, X and Y, are found in pet food. Each substance contains the ingredients A and B. Substance X is 20% ingredient A and 50% ingredient B. Substance Y is 50% ingredient A and 30% ingredient B. The cost of substance X is \$2 per pound and the cost of substance Y is \$3 per pound. The pet store needs at least 251 pounds of ingredient A and at least 200 pounds of ingredient B. If cost is to be minimal, how many pounds of each substance should be ordered? Find the minimum cost.

*Decompose each rational expression into partial fractions.*

**9.** $\dfrac{10x + 13}{x^2 - x - 20}$

**10.** $\dfrac{2x^2 - 15x - 32}{(x - 1)(x^2 + 6x + 8)}$

## CHAPTER 7 SUMMARY

**KEY TERMS & SYMBOLS** | **KEY CONCEPTS**

### 7.1 Systems of Equations

linear equation (or first-degree equation) in $n$ unknowns
system of equations
solutions
system of linear equations (linear system)
consistent system
independent equations
inconsistent system
dependent equations
substitution method
elimination method
equivalent systems
nonlinear system of equations

Systems of equations in two variables may be solved by the substitution method, the elimination method, or graphically, using the intersection-of-graphs method.

**SUBSTITUTION METHOD** (for a system of two equations in two variables)
Solve one equation for one variable in terms of the other. Substitute for that variable in the other equation, and solve for its value. Then substitute that value into the other equation to find the value of the remaining variable.

**ELIMINATION METHOD** (for a system of two equations in two variables)
Multiply one or both equations by appropriate numbers so that the sum of the coefficients of one of the variables is 0. Add the equations, and solve for the value of the remaining variable. Then substitute that value into either of the given equations to solve for the value of the other variable.

### 7.2 Solution of Linear Systems by the Echelon Method

ordered triple
echelon method
back-substitution

**TRANSFORMATIONS OF A SYSTEM**
To transform a system of linear equations into an equivalent system, apply the following.

1. Interchange any two equations.
2. Multiply each side of an equation by any nonzero real number.
3. Replace any equation by a nonzero multiple of that equation plus a nonzero multiple of any other equation.

### 7.3 Solution of Linear Systems by Row Transformations

matrix (matrices)
element (entry)
augmented matrix
echelon (triangular) form
main diagonal
row echelon method
reduced row echelon method

**MATRIX ROW TRANSFORMATIONS**
For any augmented matrix of a system of linear equations, the following row transformations will result in the matrix of an equivalent system.

1. Any two rows may be interchanged.
2. The elements of any row may be multiplied by a nonzero real number.
3. Any row may be changed by adding to its elements a multiple of the corresponding elements of another row.

(Note the similarities between these transformations and the ones described in Section 7.2.)

### 7.4 Matrix Properties and Operations

dimension
square matrix
row matrix

**MATRIX ADDITION**
The sum of two $m \times n$ matrices $A$ and $B$ is the $m \times n$ matrix $A + B$ in which each element is the sum of the corresponding elements of $A$ and $B$.

**MATRIX SUBTRACTION**
If $A$ and $B$ are matrices with the same dimension, then $\boldsymbol{A - B = A + (-B)}$.

| KEY TERMS & SYMBOLS | KEY CONCEPTS |
|---|---|
| column matrix<br>zero matrix<br>additive inverse (negative)<br>scalar | **SCALAR MULTIPLICATION**<br>If $k$ is a scalar and $A$ is a matrix, then $kA$ is the matrix of the same dimension where each entry of $A$ is multiplied by $k$.<br><br>**MATRIX MULTIPLICATION**<br>The product $AB$ of an $m \times n$ matrix $A$ and an $n \times k$ matrix $B$ is found as follows. To find the $i$th row, $j$th column element of $AB$, multiply each element in the $i$th row of $A$ by the corresponding element in the $j$th column of $B$. The sum of these products will give the element of row $i$, column $j$ of $AB$. The dimension of $AB$ is $m \times k$. |
| **7.5** Determinants and Cramer's Rule<br><br>determinant<br>det $A$<br>minor<br>cofactor<br>expansion by a row or column | **DETERMINANT OF A 2 × 2 MATRIX**<br>The determinant of a $2 \times 2$ matrix $A$, where $A = \begin{bmatrix} a_{11} & a_{12} \\ a_{21} & a_{22} \end{bmatrix}$, is a real number defined as $\mathbf{det}\, \boldsymbol{A} = \boldsymbol{a_{11}a_{22} - a_{21}a_{12}}$.<br><br>**COFACTOR**<br>Let $M_{ij}$ be the minor for element $a_{ij}$ in an $n \times n$ matrix. The cofactor of $a_{ij}$ is $\boldsymbol{A_{ij} = (-1)^{i+j} \cdot M_{ij}}$.<br><br>**FINDING THE DETERMINANT OF AN $n \times n$ MATRIX**<br>Multiply each element in any row or column of the matrix by its cofactor. The sum of these products gives the value of the determinant.<br><br>**CRAMER'S RULE FOR 2 × 2 SYSTEMS**<br>For the system<br>$$a_1x + b_1y = c_1$$ $$a_2x + b_2y = c_2,$$ $$\boldsymbol{x = \frac{D_x}{D}} \quad \text{and} \quad \boldsymbol{y = \frac{D_y}{D}},$$<br>where $D_x = \det\begin{bmatrix} c_1 & b_1 \\ c_2 & b_2 \end{bmatrix}$, $D_y = \det\begin{bmatrix} a_1 & c_1 \\ a_2 & c_2 \end{bmatrix}$, and $D = \det\begin{bmatrix} a_1 & b_1 \\ a_2 & b_2 \end{bmatrix} \neq 0$.<br>Cramer's rule can be extended to $3 \times 3$ and larger systems. |
| **7.6** Solution of Linear Systems by Matrix Inverses<br><br>identity matrix, $I_n$<br>inverse matrix<br>singular matrix<br>coefficient matrix | **INVERSE OF A 2 × 2 MATRIX**<br>If $A = \begin{bmatrix} a & b \\ c & d \end{bmatrix}$ and $\det A \neq 0$, then<br>$$\boldsymbol{A^{-1} = \frac{1}{\det A}\begin{bmatrix} d & -b \\ -c & a \end{bmatrix} = \begin{bmatrix} \frac{d}{ad - cb} & \frac{-b}{ad - cb} \\ \frac{-c}{ad - cb} & \frac{a}{ad - cb} \end{bmatrix}}.$$<br>To find the multiplicative inverse of a $3 \times 3$ (or larger) matrix $A$, we usually use a calculator. |

| KEY TERMS & SYMBOLS | KEY CONCEPTS |
|---|---|
| **7.7** Systems of Inequalities and Linear Programming<br>half-plane<br>boundary<br>linear inequality in two variables<br>system of inequalities<br>linear programming<br>constraints<br>region of feasible solutions<br>objective function<br>vertex (corner point) | **GRAPHING AN INEQUALITY**<br>**1.** For a function $f$, the graph of $y < f(x)$ consists of all the points that are *below* the graph of $y = f(x)$; the graph of $y > f(x)$ consists of all the points that are *above* the graph of $y = f(x)$.<br>**2.** If the inequality is not or cannot be solved for $y$, choose a test point not on the boundary. If the test point satisfies the inequality, the graph includes all points on the same side of the boundary as the test point. Otherwise, the graph includes all points on the other side of the boundary.<br>**SOLVING A LINEAR PROGRAMMING PROBLEM**<br>**1.** Write the objective function and all necessary constraints.<br>**2.** Graph the region of feasible solutions.<br>**3.** Identify all vertices or corner points.<br>**4.** Find the value of the objective function at each vertex.<br>**5.** The solution is given by the vertex producing the optimal value of the objective function. |
| **7.8** Partial Fractions<br>partial fraction decomposition<br>partial fraction | To solve for the constants in the numerators of a partial fraction decomposition, use either of the following methods or a combination of the two.<br>**METHOD 1 FOR LINEAR FACTORS**<br>**1.** Multiply each side of the rational expression by the common denominator.<br>**2.** Substitute the zero of each factor in the resulting equation. For repeated linear factors, substitute as many other numbers as necessary to find all the constants in the numerators. The number of substitutions required will equal the number of constants, $A, B, \ldots$.<br>**METHOD 2 FOR QUADRATIC FACTORS**<br>**1.** Multiply each side of the rational expression by the common denominator.<br>**2.** Collect terms on the right side of the resulting equation.<br>**3.** Equate the coefficients of like terms to get a system of equations.<br>**4.** Solve the system to find the constants in the numerators. |

## CHAPTER 7 REVIEW EXERCISES

*Solve each system. Identify any systems with dependent equations or any inconsistent systems.*

**1.** $4x - 3y = -1$
$3x + 5y = 50$

**2.** $\frac{x}{2} - \frac{y}{5} = \frac{11}{10}$
$2x - \frac{4y}{5} = \frac{22}{5}$

**3.** $4x + 5y = 5$
$3x + 7y = -6$

**4.** $y = x^2 - 1$
$x + y = 1$

**5.** $x^2 + y^2 = 2$
$3x + y = 4$

**6.** $x^2 + 2y^2 = 22$
$2x^2 - y^2 = -1$

**7.** $x^2 - 4y^2 = 19$
$x^2 + y^2 = 29$

**8.** $xy = 4$
$x - 6y = 2$

**9.** $x^2 - y^2 = 32$
$x = y + 4$

**10.** Use your calculator with window $[-18, 18]$ by $[-12, 12]$ to answer the following.

**(a)** Do the circle $x^2 + y^2 = 144$ and the line $x + 2y = 8$ have any points in common?

**(b)** Approximate any intersection points to the nearest tenth.

**(c)** Find the solution set of the system.

**11.** Consider the system in Exercise 5.

**(a)** To graph the first equation, what two functions must you enter into your calculator?

**(b)** To graph the second equation, what function must you enter?

**(c)** What would be an appropriate window in which to graph this system?

**12.** Can a system of two linear equations in two variables have exactly two solutions? Explain.

**13.** Can a system consisting of two equations in three variables have a unique solution? Explain.

*Use the row echelon method to solve each system. Identify any systems with dependent equations or any inconsistent systems.*

**14.** $\begin{aligned} 2x - 3y + z &= -5 \\ x + 4y + 2z &= 13 \\ 5x + 5y + 3z &= 14 \end{aligned}$

**15.** $\begin{aligned} x - 3y &= 12 \\ 2y + 5z &= 1 \\ 4x + z &= 25 \end{aligned}$

**16.** $\begin{aligned} x + y - z &= 5 \\ 2x + y + 3z &= 2 \\ 4x - y + 2z &= -1 \end{aligned}$

**17.** $\begin{aligned} 5x - 3y + 2z &= -5 \\ 2x + y - z &= 4 \\ -4x - 2y + 2z &= -1 \end{aligned}$

*Use the reduced row echelon method to solve each system. Identify any systems with dependent equations or any inconsistent systems.*

**18.** $\begin{aligned} 2x + 3y &= 10 \\ -3x + y &= 18 \end{aligned}$

**19.** $\begin{aligned} 3x + y &= -7 \\ x - y &= -5 \end{aligned}$

**20.** $\begin{aligned} x - z &= -3 \\ y + z &= 6 \\ 2x - 3z &= -9 \end{aligned}$

**21.** $\begin{aligned} 2x - y + 4z &= -1 \\ -3x + 5y - z &= 5 \\ 2x + 3y + 2z &= 3 \end{aligned}$

*Perform each operation if possible. If not possible, say so.*

**22.** $\begin{bmatrix} -5 & 4 & 9 \\ 2 & -1 & -2 \end{bmatrix} + \begin{bmatrix} 1 & -2 & 7 \\ 4 & -5 & -5 \end{bmatrix}$

**23.** $\begin{bmatrix} 3 \\ 2 \\ 5 \end{bmatrix} - \begin{bmatrix} 8 \\ -4 \\ 6 \end{bmatrix} + \begin{bmatrix} 1 \\ 0 \\ 2 \end{bmatrix}$

**24.** $\begin{bmatrix} 2 & 5 & 8 \\ 1 & 9 & 2 \end{bmatrix} - \begin{bmatrix} 3 & 4 \\ 7 & 1 \end{bmatrix}$

**25.** $3\begin{bmatrix} 2 & 4 \\ -1 & 4 \end{bmatrix} - 2\begin{bmatrix} 5 & 8 \\ 2 & -2 \end{bmatrix}$

**26.** $-1\begin{bmatrix} 3 & -5 & 2 \\ 1 & 7 & -4 \end{bmatrix} + 5\begin{bmatrix} 0 & 2 \\ -1 & 3 \end{bmatrix}$

**27.** $10\begin{bmatrix} 2x + 3y & 4x + y \\ x - 5y & 6x + 2y \end{bmatrix} + 2\begin{bmatrix} -3x - y & x + 6y \\ 4x + 2y & 5x - y \end{bmatrix}$

**28.** *Concept Check* Complete the following sentence. The sum of two $m \times n$ matrices $A$ and $B$ is found by ________.

*Perform each operation if possible. If not possible, say so.*

**29.** $\begin{bmatrix} -8 & 6 \\ 5 & 2 \end{bmatrix}\begin{bmatrix} 3 & -1 \\ 7 & 2 \end{bmatrix}$

**30.** $\begin{bmatrix} 3 & 2 & -1 \\ 4 & 0 & 6 \end{bmatrix}\begin{bmatrix} -2 & 0 \\ 0 & 2 \\ 3 & 1 \end{bmatrix}$

**31.** $\begin{bmatrix} 1 & -2 & 4 & 2 \\ 0 & 1 & -1 & 8 \end{bmatrix}\begin{bmatrix} -1 \\ 2 \\ 0 \\ 1 \end{bmatrix}$

**32.** $\begin{bmatrix} 1 & 2 & 5 \\ -3 & 4 & 7 \\ 0 & 2 & -1 \end{bmatrix}\begin{bmatrix} 4 & 2 & 3 \\ 10 & -5 & 6 \end{bmatrix}$

**33.** $\begin{bmatrix} 4 & 2 & 3 \\ 10 & -5 & 6 \end{bmatrix}\begin{bmatrix} 1 & 2 & 5 \\ -3 & 4 & 7 \\ 0 & 2 & -1 \end{bmatrix}$

**34.** $\begin{bmatrix} 3 & -1 & 0 \end{bmatrix}\begin{bmatrix} 1 & 3 & 2 \\ 2 & -4 & 0 \\ 5 & 7 & 3 \end{bmatrix}$

*Find AB and BA to determine whether A and B are inverses.*

**35.** $A = \begin{bmatrix} 3 & 2 \\ 13 & 9 \end{bmatrix}; B = \begin{bmatrix} 9 & -2 \\ -13 & 3 \end{bmatrix}$

**36.** $A = \begin{bmatrix} 1 & 0 \\ 2 & -3 \end{bmatrix}; B = \begin{bmatrix} 1 & 0 \\ \frac{2}{3} & -\frac{1}{3} \end{bmatrix}$

**37.** $A = \begin{bmatrix} 2 & 0 & 6 \\ 0 & 1 & 0 \\ 1 & 0 & 1 \end{bmatrix}; B = \begin{bmatrix} -1 & 0 & \frac{3}{2} \\ 0 & 1 & 0 \\ \frac{1}{4} & 0 & -1 \end{bmatrix}$

**38.** $A = \begin{bmatrix} 1 & 0 & 2 \\ 0 & 2 & 4 \\ 0 & 0 & 1 \end{bmatrix}; B = \begin{bmatrix} 1 & 0 & -2 \\ 0 & \frac{1}{2} & -2 \\ 0 & 0 & 1 \end{bmatrix}$

*For each matrix, find $A^{-1}$, if it exists.*

**39.** $A = \begin{bmatrix} 6 & 3 \\ 10 & 5 \end{bmatrix}$

**40.** $A = \begin{bmatrix} -4 & 2 \\ 0 & 3 \end{bmatrix}$

**41.** $A = \begin{bmatrix} 2 & 0 \\ -1 & 5 \end{bmatrix}$

**42.** $A = \begin{bmatrix} 2 & 0 & 4 \\ 1 & -1 & 0 \\ 0 & 1 & -2 \end{bmatrix}$

**43.** $A = \begin{bmatrix} 2 & -1 & 0 \\ 1 & 0 & 1 \\ 1 & -2 & 0 \end{bmatrix}$

**44.** $A = \begin{bmatrix} 2 & 3 & 5 \\ -2 & -3 & -5 \\ 1 & 4 & 2 \end{bmatrix}$

*Use the inverse matrix method to solve each system if possible. Identify any systems with dependent equations or any inconsistent systems.*

**45.** $x + y = 4$
$2x + 3y = 10$

**46.** $5x - 3y = -2$
$2x + 7y = -9$

**47.** $2x + y = 5$
$3x - 2y = 4$

**48.** $x - 2y = 7$
$3x + y = 7$

**49.** $x + 2y = -1$
$3y - z = -5$
$x + 2y - z = -3$

**50.** $3x - 2y + 4z = 1$
$4x + y - 5z = 2$
$-6x + 4y - 8z = -2$

**51.** $x + y + z = 1$
$2x - y = -2$
$3y + z = 2$

**52.** $x = -3$
$y + z = 6$
$2x - 3z = -9$

*Evaluate each determinant.*

**53.** $\det\begin{bmatrix} -1 & 8 \\ 2 & 9 \end{bmatrix}$

**54.** $\det\begin{bmatrix} -2 & 4 \\ 0 & 3 \end{bmatrix}$

**55.** $\det\begin{bmatrix} -2 & 4 & 1 \\ 3 & 0 & 2 \\ -1 & 0 & 3 \end{bmatrix}$

**56.** $\det\begin{bmatrix} -1 & 2 & 3 \\ 4 & 0 & 3 \\ 5 & -1 & 2 \end{bmatrix}$

*Solve each determinant equation for x.*

**57.** $\det\begin{bmatrix} -3 & 2 \\ 1 & x \end{bmatrix} = 5$

**58.** $\det\begin{bmatrix} 3x & 7 \\ -x & 4 \end{bmatrix} = 8$

**59.** $\det\begin{bmatrix} 2 & 5 & 0 \\ 1 & 3x & -1 \\ 0 & 2 & 0 \end{bmatrix} = 4$

**60.** $\det\begin{bmatrix} 6x & 2 & 0 \\ 1 & 5 & 3 \\ x & 2 & -1 \end{bmatrix} = 2x$

*Exercises 61 and 62 refer to the system*

$$3x - y = 28$$
$$2x + y = 2.$$

**61.** Suppose you are asked to solve this system using Cramer's rule.

**(a)** What is the value of $D$?

**(b)** What is the value of $D_x$?

**(c)** What is the value of $D_y$?

**(d)** Find $x$ and $y$, using Cramer's rule.

**62.** Suppose you are asked to solve this system using the inverse matrix method.

**(a)** What is $A$?

**(b)** What is $B$?

**(c)** Explain how you would go about solving the system using these matrices.

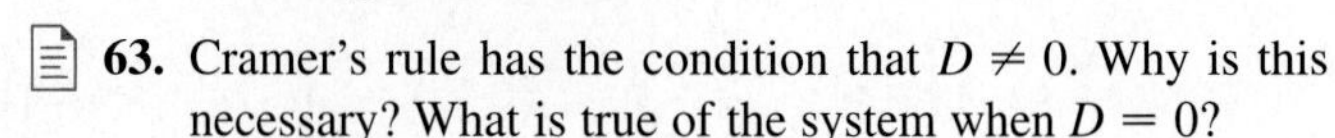

**63.** Cramer's rule has the condition that $D \neq 0$. Why is this necessary? What is true of the system when $D = 0$?

*Solve each system, if possible, using Cramer's rule. Identify any systems with dependent equations or any inconsistent systems.*

**64.** $3x + y = -1$
$5x + 4y = 10$

**65.** $3x + 7y = 2$
$5x - y = -22$

**66.** $2x - 5y = 8$
$3x + 4y = 10$

**67.** $3x + 2y + z = 2$
$4x - y + 3z = -16$
$x + 3y - z = 12$

**68.** $5x - 2y - z = 8$
$-5x + 2y + z = -8$
$x - 4y - 2z = 0$

**69.** $-x + 3y - 4z = 2$
$2x + 4y + z = 3$
$3x - z = 9$

*Use a system of equations to solve each problem.*

**70.** *Determining the Contents of a Meal* A cup of uncooked rice contains 15 grams of protein and 810 calories. A cup of uncooked soybeans contains 22.5 grams of protein and 270 calories. How many cups of each should be used for a meal containing 9.5 grams of protein and 324 calories?

**71.** *Determining Order Quantities* A company sells CDs for 40¢ each and 3.5-inch diskettes for 30¢ each. The company receives $38 for an order of 100 CDs and diskettes. However, the customer neglected to specify how many of each to send. Determine the number of CDs and the number of diskettes that should be sent.

**72.** *Mixing Teas* Three kinds of tea worth $4.60, $5.75, and $6.50 per pound are to be mixed to get 20 pounds of tea worth $5.25 per pound. The amount of $4.60 tea used is to be equal to the total amount of the other two kinds together. How many pounds of each tea should be used?

**73.** *Mixing Solutions of a Drug* A 5% solution of a drug is to be mixed with some 15% solution and some 10% solution to get 20 milliliters of 8% solution. The amount of 5% solution used must be 2 milliliters more than the sum of the other two solutions. How many milliliters of each solution should be used?

**74.** *(Modeling) Blood Pressure* In one study of adult males, it was believed that systolic blood pressure $P$ was affected by both age $A$ in years and weight $W$ in pounds. This was modeled by $P = a + bA + cW$, were $a$, $b$, and $c$ are constants. The table lists three individuals with representative blood pressures for the group. (*Source:* Brase, C.H. and C.P. Brase, *Understandable Statistics,* D.C. Heath and Company, 1995.)

| $P$ | $A$ | $W$ |
|---|---|---|
| 113 | 39 | 142 |
| 138 | 53 | 181 |
| 152 | 65 | 191 |

**(a)** Use the table to approximate values for the constants $a$, $b$, and $c$.

**(b)** Estimate a typical systolic blood pressure for an individual who is 55 years old and weighs 175 pounds.

**75.** *Curve Fitting* Find the equation of the quadratic polynomial that defines the curve shown in the figures.

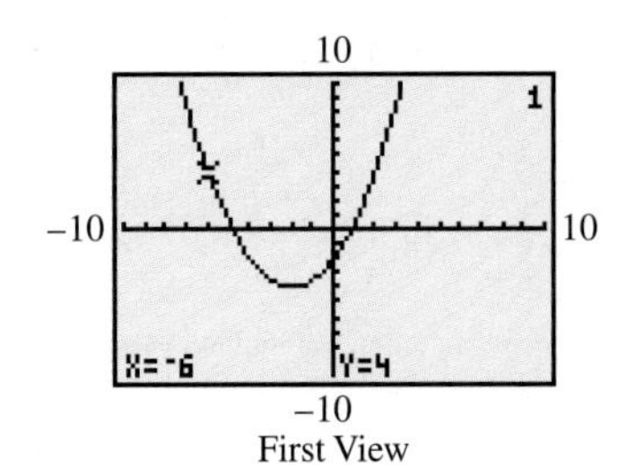

First View

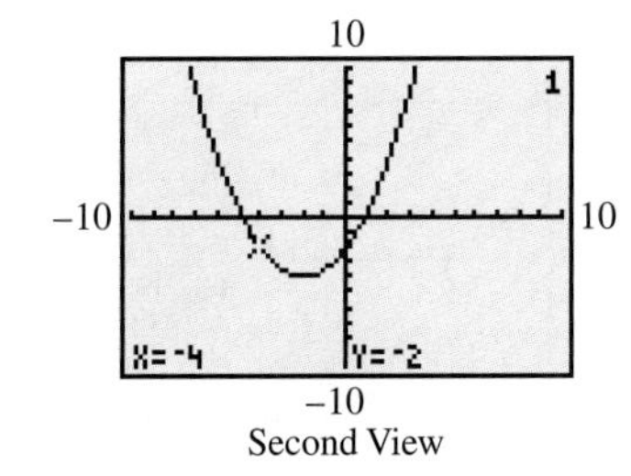

Second View

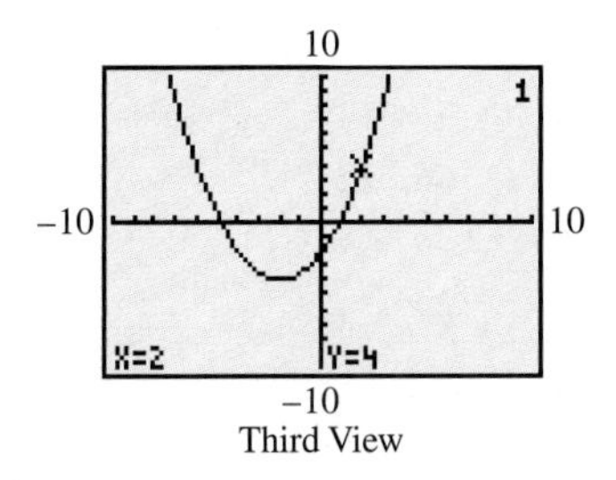

Third View

**76.** *Curve Fitting* Find the equation for the polynomial function of degree 3 whose graph passes through the points $(-2, 1)$, $(-1, 6)$, $(2, 9)$, and $(3, 26)$.

*Graph the solution set of each system.*

**77.** $x + y \leq 6$
$2x - y \geq 3$

**78.** $y \leq \frac{1}{3}x - 2$
$y^2 \leq 16 - x^2$

**79.** Find $x \geq 0$ and $y \geq 0$ such that

$$3x + 2y \leq 12$$
$$5x + y \geq 5$$

and $2x + 4y$ is maximized.

**80.** Find $x \geq 0$ and $y \geq 0$ such that

$$x + y \leq 50$$
$$2x + y \geq 20$$
$$x + 2y \geq 30$$

and $4x + 2y$ is minimized.

**81.** *Company Profit* A small company manufacturers two products—radios and CD players. Each radio results in a profit of $15, and each CD player provides a profit of $35. Due to demand, the company must produce at least 5 and not more than 25 radios per day. The number of radios cannot exceed the number of CD players, and the number of CD players cannot exceed 30. How many of each should the company manufacture to obtain maximum profit? What will that profit be?

*Find the partial fraction decomposition of each rational expression.*

**82.** $\dfrac{5x - 2}{x^2 - 4}$

**83.** $\dfrac{x + 2}{x^3 + 2x^2 + x}$

**84.** $\dfrac{x + 2}{x^3 - x^2 + 4x}$

## CHAPTER 7 TEST

**1.** Consider the system of equations

$$x^2 - 4y^2 = -15$$
$$3x + y = 1.$$

**(a)** What type of graph does each equation have?

**(b)** How many points of intersection of the two graphs are possible?

**(c)** Solve the system.

**(d)** Support the solutions using a graphing calculator.

**2.** Solve the system using one of the row echelon methods.

$$\begin{aligned} 2x + y + z &= 3 \\ x + 2y - z &= 3 \\ 3x - y + z &= 5 \end{aligned}$$

**3.** Perform each matrix operation if possible.

**(a)** $3\begin{bmatrix} 2 & 3 \\ 1 & -4 \\ 5 & 9 \end{bmatrix} - \begin{bmatrix} -2 & 6 \\ 3 & -1 \\ 0 & 8 \end{bmatrix}$

**(b)** $\begin{bmatrix} 1 \\ 2 \end{bmatrix} + \begin{bmatrix} 4 \\ -6 \end{bmatrix} + \begin{bmatrix} 2 & 8 \\ -7 & 5 \end{bmatrix}$

**(c)** $\begin{bmatrix} 2 & 1 & -3 \\ 4 & 0 & 5 \end{bmatrix}\begin{bmatrix} 1 & 3 \\ 2 & 4 \\ 3 & -2 \end{bmatrix}$

**4.** Suppose $A$ and $B$ are both $n \times n$ matrices.

**(a)** Can $AB$ be found?

**(b)** Can $BA$ be found?

**(c)** Does $AB = BA$? Explain why or why not.

**(d)** If $A$ is $n \times n$ and $C$ is $m \times n$, can either $AC$ or $CA$ be found?

**5.** Evaluate each determinant.

**(a)** $\det\begin{bmatrix} 4 & 9 \\ -5 & -11 \end{bmatrix}$

**(b)** $\det\begin{bmatrix} 2 & 0 & 8 \\ -1 & 7 & 9 \\ 12 & 5 & -3 \end{bmatrix}$

**6.** Solve the system using Cramer's rule.

$$\begin{aligned} 2x - 3y &= -33 \\ 4x + 5y &= 11 \end{aligned}$$

**7.** Consider the system of equations

$$\begin{aligned} x + y - z &= -4 \\ 2x - 3y - z &= 5 \\ x + 2y + 2z &= 3. \end{aligned}$$

**(a)** Write the matrix of coefficients $A$, the matrix of variables $X$, and the matrix of constants $B$ for this system.

**(b)** Find $A^{-1}$.

**(c)** Use the matrix inverse method to solve the system.

**(d)** If the first equation in the system above is replaced by $.5x + y + z = 1.5$, the system cannot be solved using the matrix inverse method. Explain why.

**8.** *(Modeling) Voter Turnout* The table shows the percent $y$ of voter turnout in the United States for three presidential elections in year $x$, where $x = 0$ corresponds to 1900. Find a quadratic function defined by $f(x) = ax^2 + bx + c$ that models these data. Graph $f$ together with the data.

| $x$ | 24 | 60 | 96 |
|---|---|---|---|
| $y$ | 48.9 | 62.8 | 48.8 |

*Source:* Committee for the Study of the American Electorate.

**9.** The solution set of a system of inequalities is shown. Which system is it?

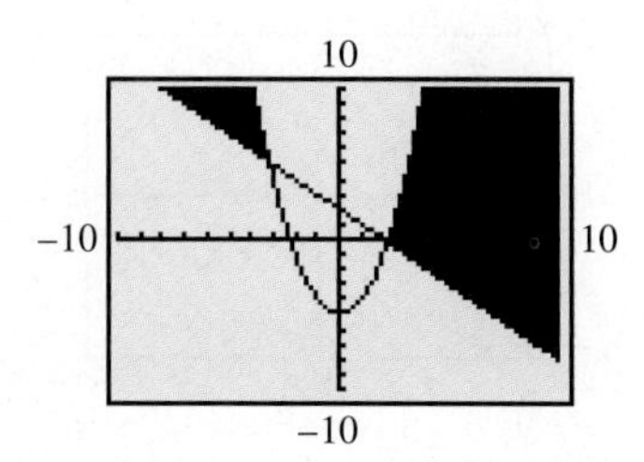


**A.** $y > 2 - x$, $y > x^2 - 5$

**B.** $y > 2 - x$, $y < x^2 - 5$

**C.** $y < 2 - x$, $y < x^2 - 5$

**D.** $y < 2 - x$, $y > x^2 - 5$

**10.** *Storage Capacity* An office manager wants to buy filing cabinets. Cabinet X costs \$100, requires 6 square feet of floor space, and holds 8 cubic feet. Cabinet Y costs \$200, requires 8 square feet of floor space, and holds 12 cubic feet. No more than \$1400 can be spent, and the office has room for no more than 72 square feet of cabinets. The office manager wants the maximum storage capacity within the limits imposed by funds and space. How many of each type of cabinet should be bought?

*Find the partial fraction decomposition for each rational expression.*

**11.** $\dfrac{7x - 1}{x^2 - x - 6}$

**12.** $\dfrac{x^2 - 11x + 6}{(x + 2)(x - 2)^2}$

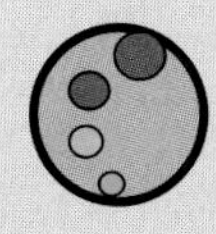

# Chapter 7 Project

## Finding a Polynomial Whose Graph Passes Through Any Number of Given Points

In the Chapter 3 Project, we found a ninth-degree polynomial with zeros that depended on the digits of a Social Security number. In this project, we'll learn to find a polynomial that goes through a given set of points in the plane. Recall that three points define a second-degree polynomial, four points define a third-degree polynomial, five points define a fourth-degree polynomial, and so on. One important restriction on the points is that they must each have a different $x$-coordinate. Since polynomials define functions, no two distinct points can have the same $x$-coordinate.

In the following example, we use the same Social Security number (539-58-0954) that we used in the Chapter 3 Project to find an eighth-degree polynomial that lies on the nine points with $x$-coordinates from 1 to 9 and with $y$-coordinates that are the digits of the Social Security number. The nine ordered pairs are

$$(1,5), (2,3), (3,9), (4,5), (5,8), (6,0), (7,9), (8,5), \text{ and } (9,4).$$

These nine ordered pairs are entered in the lists shown in Figure A. To get the scatter diagram of the data in Figure C, we enter the window settings shown in Figure B.

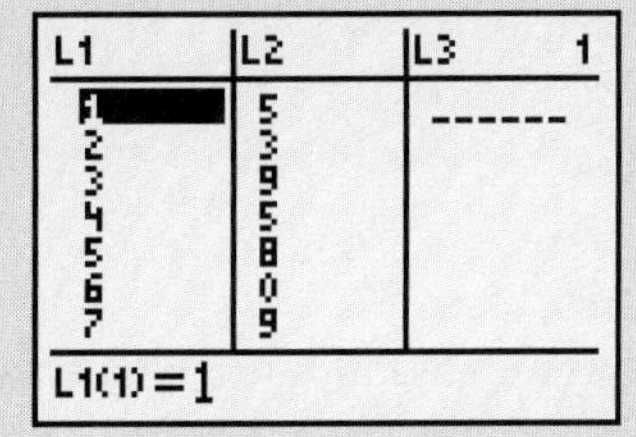


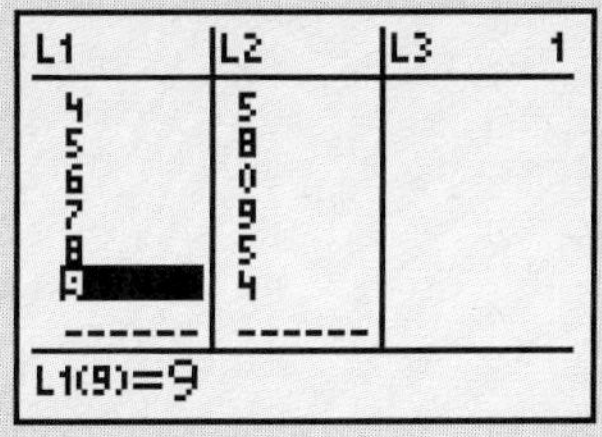


**FIGURE A**

WINDOW
Xmin=-2
Xmax=10
Xscl=1
Ymin=-5
Ymax=15
Yscl=1
Xres=1

**FIGURE B**

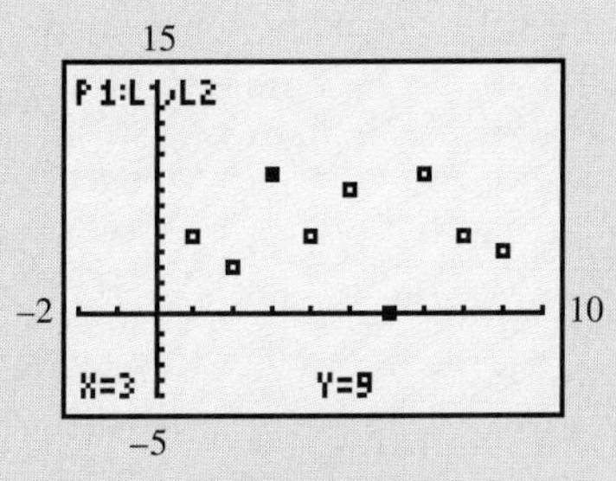


**FIGURE C**

We now use the nine ordered pairs to write a system of nine equations with nine unknowns by substituting each pair of $x$- and $y$-values into the general eighth-degree polynomial equation

$$ax^8 + bx^7 + cx^6 + dx^5 + ex^4 + fx^3 + gx^2 + hx + i = y. \qquad (*)$$

$$\begin{aligned}
a1^8 + b1^7 + c1^6 + d1^5 + e1^4 + f1^3 + g1^2 + h1^1 + i &= 5\\
a2^8 + b2^7 + c2^6 + d2^5 + e2^4 + f2^3 + g2^2 + h2^1 + i &= 3\\
a3^8 + b3^7 + c3^6 + d3^5 + e3^4 + f3^3 + g3^2 + h3^1 + i &= 9\\
a4^8 + b4^7 + c4^6 + d4^5 + e4^4 + f4^3 + g4^2 + h4^1 + i &= 5\\
a5^8 + b5^7 + c5^6 + d5^5 + e5^4 + f5^3 + g5^2 + h5^1 + i &= 8\\
a6^8 + b6^7 + c6^6 + d6^5 + e6^4 + f6^3 + g6^2 + h6^1 + i &= 0\\
a7^8 + b7^7 + c7^6 + d7^5 + e7^4 + f7^3 + g7^2 + h7^1 + i &= 9\\
a8^8 + b8^7 + c8^6 + d8^5 + e8^4 + f8^3 + g8^2 + h8^1 + i &= 5\\
a9^8 + b9^7 + c9^6 + d9^5 + e9^4 + f9^3 + g9^2 + h9^1 + i &= 4
\end{aligned}$$

Next, we write an equivalent matrix equation $AX = B$ as follows.

$$\begin{bmatrix}
1 & 1 & 1 & 1 & 1 & 1 & 1 & 1 & 1\\
2^8 & 2^7 & 2^6 & 2^5 & 2^4 & 2^3 & 2^2 & 2 & 1\\
3^8 & 3^7 & 3^6 & 3^5 & 3^4 & 3^3 & 3^2 & 3 & 1\\
4^8 & 4^7 & 4^6 & 4^5 & 4^4 & 4^3 & 4^2 & 4 & 1\\
5^8 & 5^7 & 5^6 & 5^5 & 5^4 & 5^3 & 5^2 & 5 & 1\\
6^8 & 6^7 & 6^6 & 6^5 & 6^4 & 6^3 & 6^2 & 6 & 1\\
7^8 & 7^7 & 7^6 & 7^5 & 7^4 & 7^3 & 7^2 & 7 & 1\\
8^8 & 8^7 & 8^6 & 8^5 & 8^4 & 8^3 & 8^2 & 8 & 1\\
9^8 & 9^7 & 9^6 & 9^5 & 9^4 & 9^3 & 9^2 & 9 & 1
\end{bmatrix}
\begin{bmatrix} a\\ b\\ c\\ d\\ e\\ f\\ g\\ h\\ i \end{bmatrix}
=
\begin{bmatrix} 5\\ 3\\ 9\\ 5\\ 8\\ 0\\ 9\\ 5\\ 4 \end{bmatrix}$$

To solve the matrix equation for the matrix of variables, enter matrices $A$ and $B$ into the matrix editor of your graphing calculator. (See Figures D and E.) Entries can be typed in exponential form; the calculator will evaluate each entry. Although the calculator displays large entries in scientific notation, it stores the exact value as shown in Figure D, for the sixth row, first column entry of the matrix. (The screen is not large enough to display the entire matrix at one time.)

FIGURE D

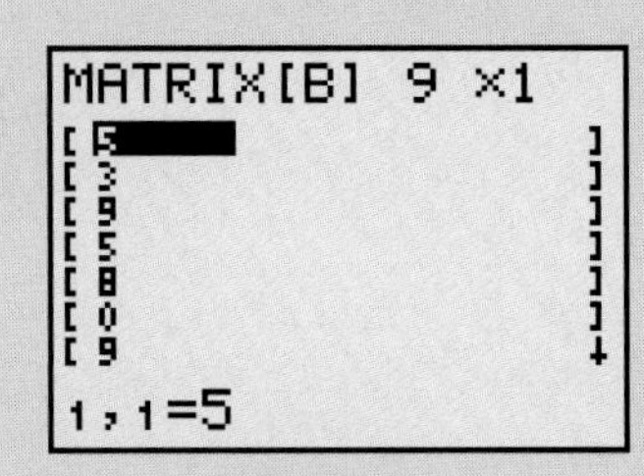


FIGURE E

Solve the system by multiplying the inverse of matrix $A$ times $B$, storing the solution as [C], as shown in Figures F and G. (You will need to scroll down to see the entire matrix [C].)

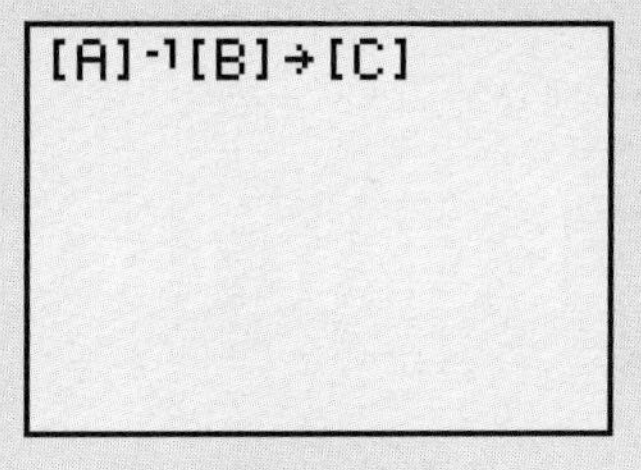


FIGURE F

```
[[.0180803572 ]
 [-.7175595273]
 [11.94826394 ]
 [-108.4666672]
 [583.4623289 ]
 [-1889.097924]
 [3556.071343 ]↓
```

FIGURE G

The entries in [C] are $a$, $b$, $c$, and so on, the coefficients of the polynomial equation (*). We can use the rows of [C] to enter the equation of the polynomial in the Y = menu, using the notation in Figure H. Be sure to use the matrix menu to enter [C] for each coordinate [C] $(m, n)$. Figure I shows the graph of the polynomial in the same window as the scatter diagram in Figure C. In Figure J, the window setting has been changed to show a graph that includes all local minima and maxima.

FIGURE H

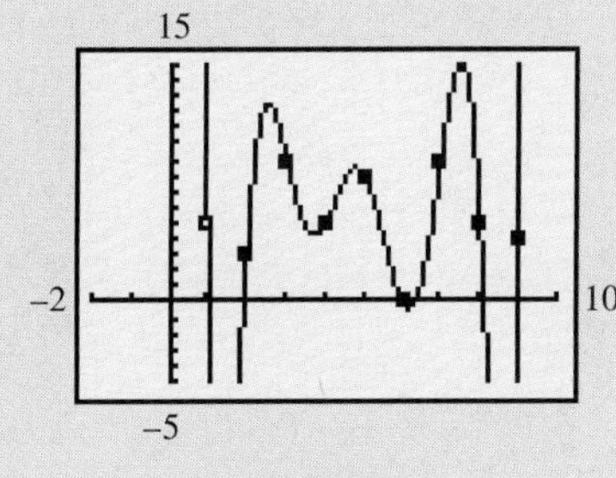


FIGURE I

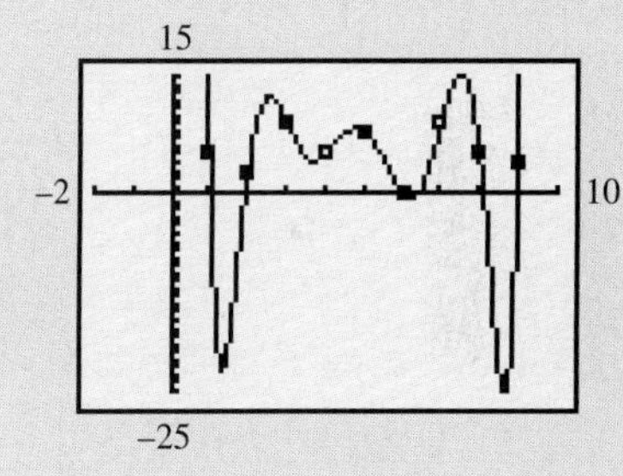


FIGURE J

## Activities

1. Use the process outlined above to construct a second-degree polynomial with three ordered pairs. Use the digits 1, 2, and 3 as the $x$-coordinates and the first three digits of a Social Security number as the $y$-coordinates of the ordered pairs.
2. Construct the eighth-degree polynomial $f(x)$ that lies on the nine points with $x$-coordinates from 1 to 9 and $y$-coordinates that are the digits of a Social Security number.
3. Using your graphing calculator, find a comprehensive graph of $f(x)$, including all local extreme points, all intercepts, and clear end behavior. Sketch the graph on paper or print it using appropriate computer software. Label the extreme points with their coordinates.

# 8 Trigonometric Functions and Applications

TRIGONOMETRY HAS been used for thousands of years to solve problems related to astronomy, surveying, and construction. Trigonometry originated when people tried to correlate shadow lengths with the time of day. Prior to the fifteenth century, astronomy had the greatest influence on the development of trigonometry. It was not until the thirteenth century that astronomy and trigonometry could be regarded as separate entities. The Greek astronomer Hipparchus is usually given credit for first studying the trigonometric properties of angles. The word *trigonometry* is derived from two Greek words that mean "triangle measurement." While the early development of trigonometry was based on angles and triangles, modern-day applications often use trigonometric functions and graphs.

In this chapter we use trigonometric functions not only to solve problems involving triangles but also to solve real-life problems involving periodic motion and rotation. Some topics in which trigonometry is used today are average monthly temperatures, daylight hours, tides, phases of the moon, and household electricity. Trigonometric functions have been essential to the development of our modern society.

*Source:* Freebury, H., *A History of Mathematics,* MacMillan, 1961; *Historical Topics for the Mathematics Classroom, Thirty-first Yearbook,* NCTM.

CHAPTER OUTLINE

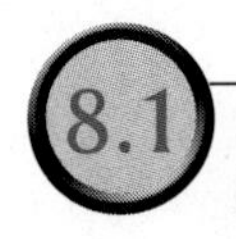


# ANGLES AND THEIR MEASURES

Basic Terminology • Degree Measure • Standard Position and Coterminal Angles • Radian Measure • Arc Lengths and Sectors • Angular and Linear Speed

## Basic Terminology

Line $AB$
$A$ $B$

Segment $AB$
$A$ $B$

Ray $AB$
$A$ $B$

**FIGURE 1**

Two distinct points $A$ and $B$ determine a line called **line *AB*.** The portion of the line between $A$ and $B$, including points $A$ and $B$, is **segment *AB*.** The portion of line $AB$ that starts at $A$ and continues through $B$, and on past $B$, is called **ray *AB*.** Point $A$ is the *endpoint of the ray*. (See Figure 1.)

An **angle** is formed by rotating a ray around its endpoint. The ray in its initial position is called the **initial side** of the angle, and the ray in its location after the rotation is the **terminal side** of the angle. The endpoint of the ray is the **vertex** of the angle. Figure 2 shows the initial and terminal sides of an angle with vertex $A$.

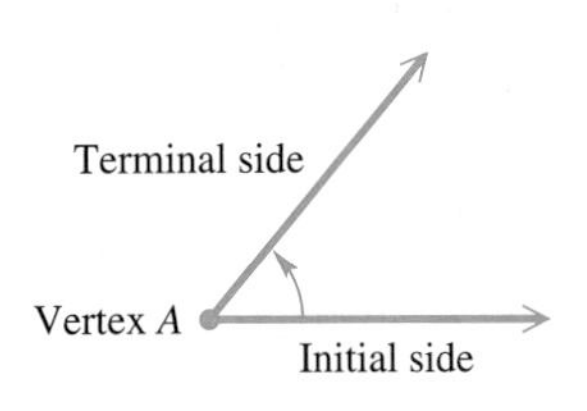


**FIGURE 2**

If the rotation is counterclockwise, the angle is **positive.** If the rotation is clockwise, the angle is **negative.** Figure 3 shows two angles, one positive and one negative.

An angle can be named by using the name of its vertex. For example, either angle in Figure 3 can be called angle $C$. Alternatively, an angle can be named using three letters, with the vertex letter in the middle. Thus, either angle also could be named angle $ACB$ or angle $BCA$.

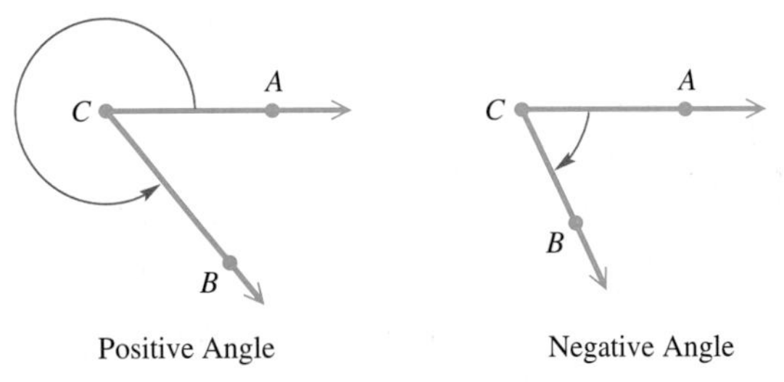


**FIGURE 3**

## Degree Measure

Two systems are commonly used to measure the size of angles. The most common unit of measure is the **degree.** (The other common unit of measure, called the *radian*, is discussed later in this section.) Degree measure was developed by the Babylonians, 4000 years ago.* To use degree measure, we assign 360 degrees to a complete rotation of a ray. In Figure 4, notice that the terminal side of the angle corresponds to its initial side when it makes a complete rotation. One degree, written $1^\circ$, represents $\frac{1}{360}$ of a rotation. Therefore, $90^\circ$ represents $\frac{90}{360} = \frac{1}{4}$ of a complete rotation, and $180^\circ$ represents $\frac{180}{360} = \frac{1}{2}$ of a complete rotation. An angle measuring between $0^\circ$ and $90^\circ$ is an **acute angle.** An angle measuring exactly $90^\circ$ is a **right angle.** An angle measuring more than $90^\circ$ but less than $180^\circ$ is an **obtuse angle,** and an angle of exactly $180^\circ$ is a **straight angle.** See Figure 5 on the next page.

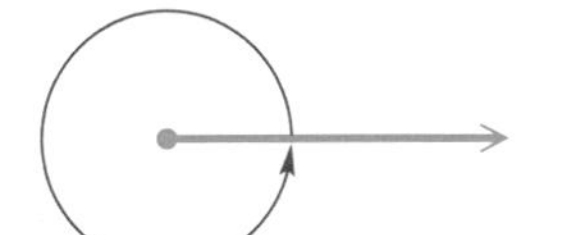

A complete rotation of a ray gives an angle whose measure is $360^\circ$.

**FIGURE 4**

*The Babylonians were the first to subdivide the circumference of a circle into 360 parts. There are various theories as to why the number 360 was chosen. One is that it is approximately the number of days in a year, and it has many divisors, which makes it convenient to work with. Another involves a roundabout theory dealing with the length of a Babylonian mile.

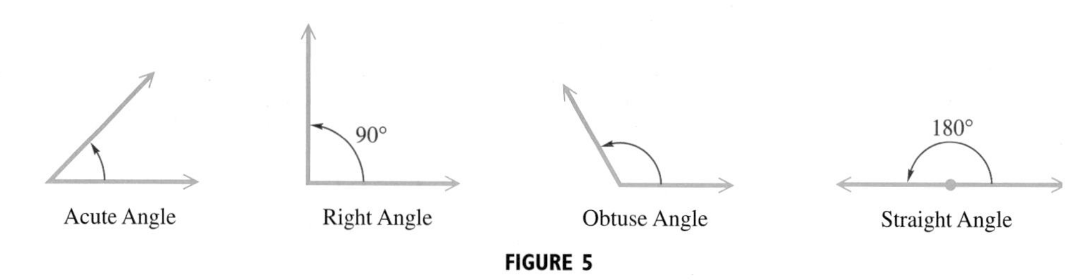


**FIGURE 5**

If the sum of the measures of two positive angles is 90°, the angles are called **complementary.** Two positive angles with measures whose sum is 180° are **supplementary.**

**EXAMPLE 1** *Finding Measures of Complementary and Supplementary Angles*

Find the measure of each angle in Figure 6.

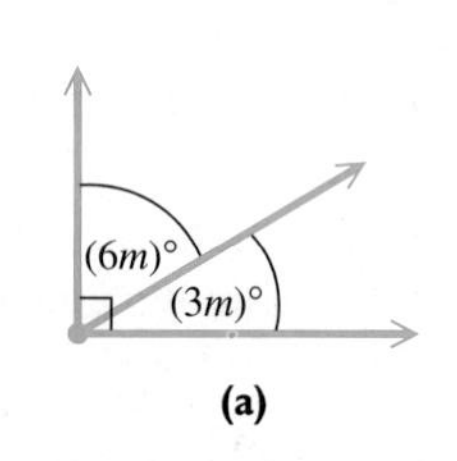


**(a)**

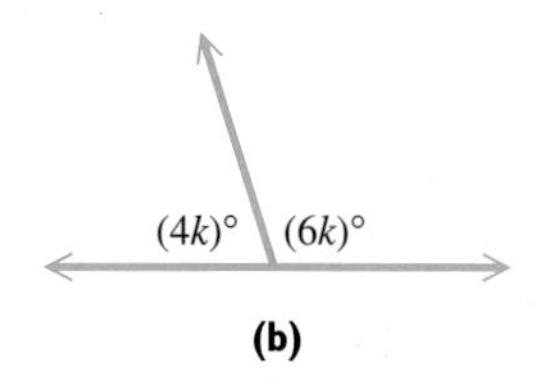


**(b)**

**FIGURE 6**

### Solution

**(a)** In Figure 6(a), since the two angles form a right angle (as indicated by the ⌝ symbol),

$$\begin{aligned} 6m + 3m &= 90 \\ 9m &= 90 \\ m &= 10. \end{aligned}$$

The two angles have measures of $6(10) = 60°$ and $3(10) = 30°$.

**(b)** The angles in Figure 6(b) are supplementary, so

$$\begin{aligned} 4k + 6k &= 180 \\ 10k &= 180 \\ k &= 18. \end{aligned}$$

These angle measures are $4(18) = 72°$ and $6(18) = 108°$.

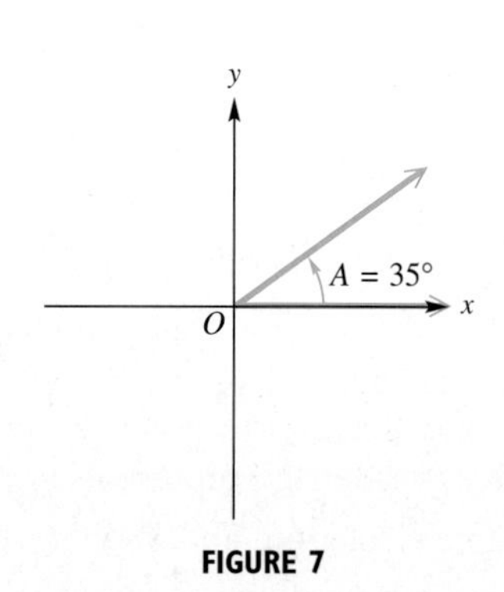


**FIGURE 7**

Do not confuse an angle with its measure. Angle $A$ of Figure 7 is a rotation; the measure of the rotation is 35°. This measure is often expressed by saying that $m(\text{angle } A)$ is 35°, where $m(\text{angle } A)$ is read "the measure of angle $A$." It is convenient, however, to abbreviate $m(\text{angle } A) = 35°$ as simply angle $A = 35°$.

Traditionally, portions of a degree have been measured with minutes and seconds. One **minute,** written $1'$, is $\frac{1}{60}$ of a degree.

$$1' = \frac{1}{60}^{\circ} \quad \text{or} \quad 60' = 1°$$

One **second,** $1''$, is $\frac{1}{60}$ of a minute.

$$1'' = \frac{1}{60}' = \frac{1}{3600}^{\circ} \quad \text{or} \quad 60'' = 1'$$

The measure $12° \, 42' \, 38''$ represents 12 degrees, 42 minutes, 38 seconds.

**EXAMPLE 2** *Calculating with Degrees, Minutes, and Seconds*

Perform each calculation.

**(a)** $51° \, 29' + 32° \, 46'$ **(b)** $90° - 73° \, 12'$

**Analytic Solution**

**(a)** Add the degrees and the minutes separately.

$$\begin{array}{r} 51° \; 29' \\ + \; 32° \; 46' \\ \hline 83° \; 75' \end{array}$$

Since $75' = 60' + 15' = 1° \, 15'$, the sum is written

$$\begin{array}{r} 83° \\ + \quad 1° \; 15' \\ \hline 84° \; 15'. \end{array}$$

**(b)**
$$\begin{array}{r} 89° \; 60' \\ - \; 73° \; 12' \\ \hline 16° \; 48' \end{array} \qquad \text{Write } 90° \text{ as } 89° \, 60'.$$

**Graphing Calculator Solution**

Your calculator can be in either degree or radian mode. Figure 8 shows the calculations.

```
51°29'+32°46'▸DM
S
          84°15'0"
90°0'-73°12'▸DMS
          16°48'0"
```

FIGURE 8

Because calculators are now so prevalent, angles are commonly measured in decimal degrees. For example, 12.4238° represents

$$12.4238° = 12 \frac{4238}{10{,}000}°.$$

**EXAMPLE 3** *Converting between Decimal Degrees and Degrees, Minutes, Seconds*

**(a)** Convert 74° 8′ 14″ to decimal degrees. Round to the nearest thousandth of a degree.

**(b)** Convert 34.817° to degrees, minutes, and seconds.

**Analytic Solution**

**(a)** Since $1' = \frac{1}{60}°$ and $1'' = \frac{1}{3600}°$,

$$\begin{aligned} 74° \, 8' \, 14'' &= 74° + \frac{8}{60}° + \frac{14}{3600}° \\ &\approx 74° + .1333° + .0039° \\ &\approx 74.137° \quad \text{(rounded).} \end{aligned}$$

**(b)**
$$\begin{aligned} 34.817° &= 34° + .817° \\ &= 34° + .817(60') && 1° = 60' \\ &= 34° + 49.02' \\ &= 34° + 49' + .02' \\ &= 34° + 49' + .02(60'') && 1' = 60'' \\ &= 34° + 49' + 1.2'' \\ &= 34° \, 49' \, 1.2'' \end{aligned}$$

**Graphing Calculator Solution**

**(a)** The first two results on the screen in Figure 9 show how 74° 8′ 14″ is converted to decimal degrees. (The second result was obtained by fixing the display to three decimal places.)

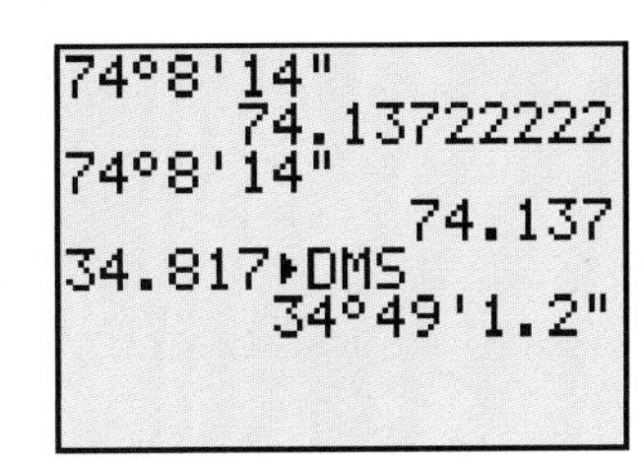


FIGURE 9

**(b)** The final display in Figure 9 shows how to convert decimal degrees to degrees, minutes, and seconds.

## Standard Position and Coterminal Angles

An angle is in **standard position** if its vertex is at the origin and its initial side is along the positive $x$-axis. The angles in Figures 10(a) and 10(b) are in standard position. An angle in standard position is said to lie in the quadrant in which its terminal side lies. For example, an acute angle is in quadrant I (Figure 10(a)) and an obtuse angle is in quadrant II (Figure 10(b)). Figure 10(c) shows ranges of angle measures for each quadrant when $0° < \theta^* < 360°$. Angles in standard position having their terminal sides along the $x$-axis or $y$-axis, such as angles with measures 90°, 180°, 270°, and so on, are called **quadrantal angles.**

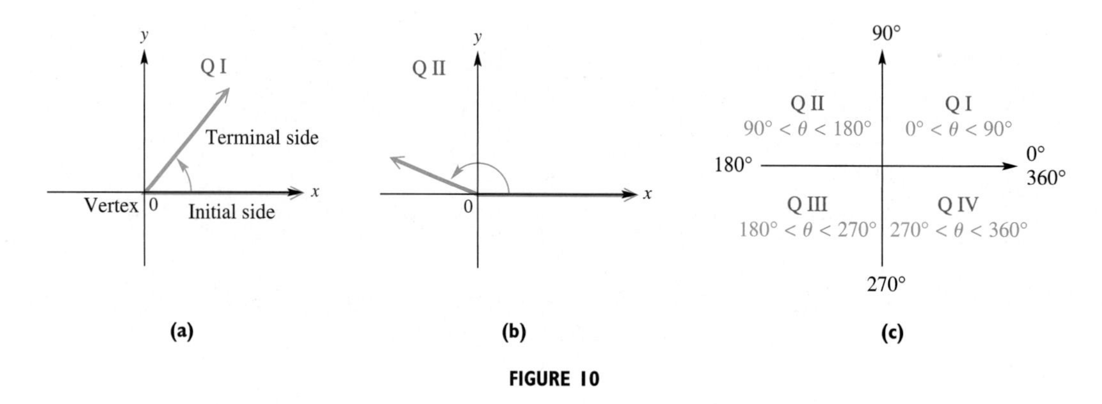


FIGURE 10

A complete rotation of a ray results in an angle measuring 360°. By continuing the rotation, angles of measure larger than 360° can be produced. The angles in Figure 11(a) have measures 60° and 420°. These two angles have the same initial side and the same terminal side, but different amounts of rotation. Such angles are called **coterminal angles.** The measures of coterminal angles differ by a multiple of 360°. As shown in Figure 11(b), angles with measures 110° and 830° are coterminal.

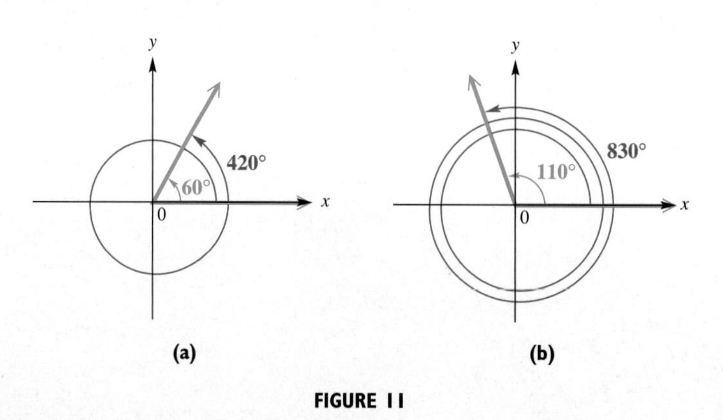


FIGURE 11

**EXAMPLE 4** *Finding Measures of Coterminal Angles*

Find the angle of smallest possible positive measure coterminal with each angle.

**(a)** 908° **(b)** −75°

---

*Greek letters such as $\alpha$ (alpha), $\beta$ (beta), and $\theta$ (theta) are often used to name angles.

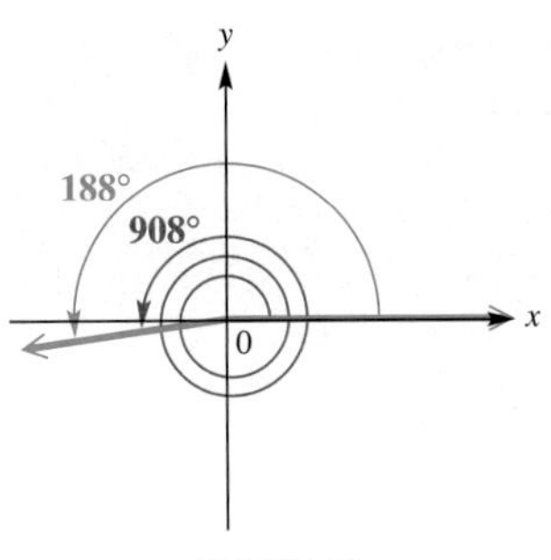


FIGURE 12

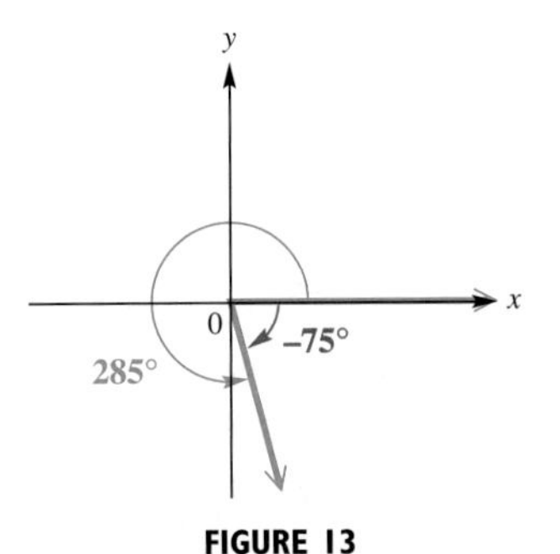


FIGURE 13

**Solution**

**(a)** Add or subtract 360° as many times as needed to obtain an angle with measure greater than 0° but less than 360°. Since

$$908° - 2 \cdot 360° = 908° - 720° = 188°,$$

an angle of 188° is coterminal with an angle of 908°. See Figure 12.

**(b)** Use a rotation of $360° + (-75°) = 285°$. See Figure 13.

Sometimes it is necessary to find an expression that will generate all angles coterminal with a given angle. For example, any angle coterminal with 60° can be obtained by adding an appropriate integer multiple of 360° to 60°. Let $n$ represent any integer; then the expression

$$60° + n \cdot 360°$$

represents all such coterminal angles. The following table shows a few possibilities.

| Value of $n$ | Angle Coterminal with 60° |
|---|---|
| 2 | $60° + 2 \cdot 360° = 780°$ |
| 1 | $60° + 1 \cdot 360° = 420°$ |
| 0 | $60° + 0 \cdot 360° = 60°$ (the angle itself) |
| −1 | $60° + (-1) \cdot 360° = -300°$ |

## Radian Measure

In work involving applications of trigonometry, angles are often measured in degrees. In more advanced work in mathematics, *radian measure* of angles is preferred.

Figure 14 shows an angle $\theta$ in standard position along with a circle of radius $r$. The vertex of $\theta$ is at the center of the circle. Angle $\theta$ intercepts an arc on the circle equal in length to the radius of the circle. So, angle $\theta$ is said to have a measure of 1 radian.

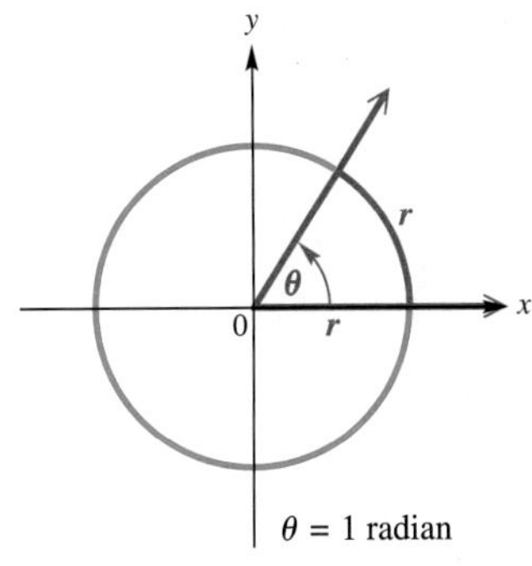


FIGURE 14

**Radian**

An angle with its vertex at the center of a circle that intercepts an arc on the circle equal in length to the radius of the circle has a measure of **1 radian.**

It follows that an angle of measure 2 radians intercepts an arc equal in length to twice the radius of the circle, an angle of measure $\frac{1}{2}$ radian intercepts an arc equal in length to half the radius of the circle, and so on.

The circumference of a circle—the distance around the circle—is given by $C = 2\pi r$, where $r$ is the radius of the circle. The formula $C = 2\pi r$ shows that the radius can be laid off $2\pi$ times around a circle. Therefore, an angle of 360°, which corresponds to a complete circle, intercepts an arc equal in length to $2\pi$ times the radius of the circle. Thus, an angle of 360° has measure $2\pi$ radians:

$$360° = 2\pi \text{ radians}.$$

An angle of 180° is half the degree measure of an angle of 360°, so an angle of 180° has half the radian measure of an angle of 360°.

$$180° = \frac{1}{2}(2\pi) \text{ radians} = \pi \text{ radians} \quad \text{Degree/radian relationship}$$

We can use the relationship $180° = \pi$ radians to develop a method for converting between degrees and radians as follows.

$$180° = \pi \text{ radians}$$

$$1° = \frac{\pi}{180} \text{ radian} \quad \text{Divide by 180.} \qquad \text{or} \qquad 1 \text{ radian} = \frac{180°}{\pi} \quad \text{Divide by } \pi.$$

**Converting between Degrees and Radians**

1. Multiply a radian measure by $\frac{180°}{\pi}$ and simplify to convert to degrees.
2. Multiply a degree measure by $\frac{\pi}{180}$ radian and simplify to convert to radians.

**EXAMPLE 5** *Converting Degrees to Radians*

Convert each degree measure to radians.

**(a)** 45° **(b)** 249.8°

**Analytic Solution**

**(a)** $45° = 45\left(\frac{\pi}{180} \text{ radian}\right)$ Multiply by $\frac{\pi}{180}$ radian.

$= \frac{\pi}{4}$ radian

**(b)** $249.8° = 249.8\left(\frac{\pi}{180} \text{ radian}\right)$

$\approx 4.360$ radians Round to nearest thousandth.

**Graphing Calculator Solution**

For Figure 15, the calculator is in radian mode. Note that when *exact* values involving $\pi$ are required, such as $\frac{\pi}{4}$ in part (a), calculator approximations are not acceptable.

```
45°
         .7853981634
π/4
         .7853981634
249.8°
         4.359832471
```

**FIGURE 15**

**EXAMPLE 6** *Converting Radians to Degrees*

Convert each radian measure to degrees.

**(a)** $\frac{9\pi}{4}$ **(b)** 4.25 (Give the answer in decimal degrees.)

**Analytic Solution**

(a) $\frac{9\pi}{4} = \frac{9\pi}{4}\left(\frac{180°}{\pi}\right)$ Multiply by $\frac{180°}{\pi}$.

$= 405°$

(b) $4.25 = 4.25\left(\frac{180°}{\pi}\right)$

$\approx 243.5°$

**Graphing Calculator Solution**

Figure 16 shows how a calculator in degree mode converts radian measures to decimal degrees.

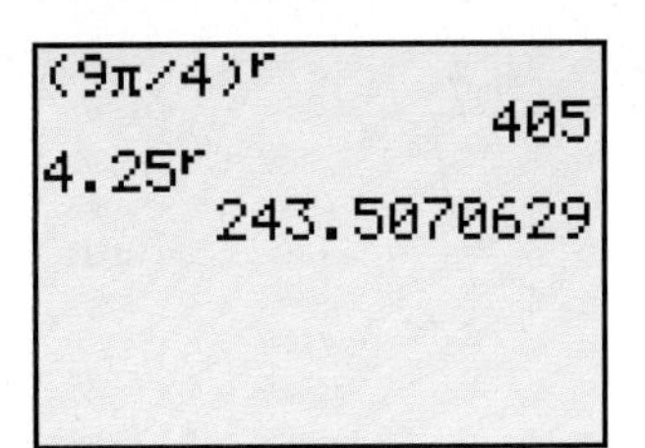


FIGURE 16

If no unit of measure is specified for an angle, radian measure is understood.

**CAUTION** Figure 17 shows angles measuring 30 radians and 30°. These angle measures are not at all close, because 1 radian $\approx$ 57.3°.

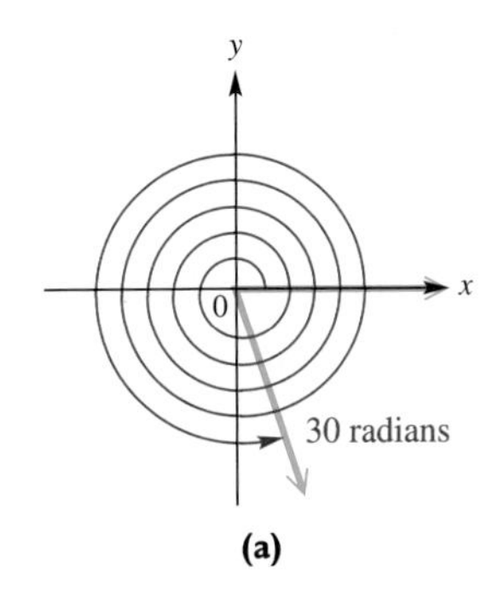


(a)

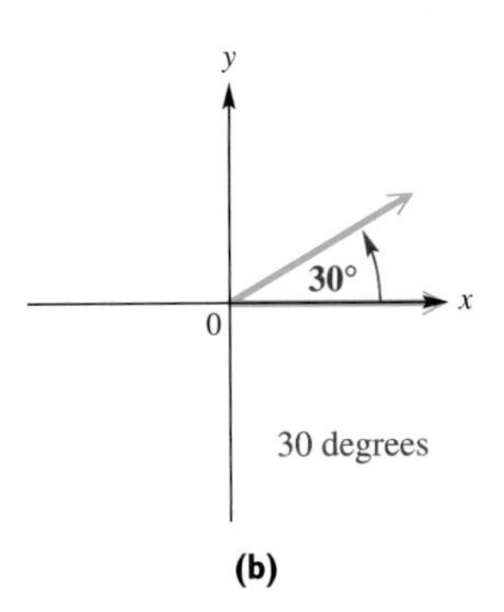


(b)

FIGURE 17

The following table and Figure 18 give some equivalent angles measured in degrees and radians. Keep in mind that $180° = \pi$ radians. Then, you can easily reproduce the rest of the table.

**Equivalent Angle Measures in Degrees and Radians**

| Degrees | Radians | | Degrees | Radians | |
|---|---|---|---|---|---|
| | Exact | Approximate | | Exact | Approximate |
| 0° | 0 | 0 | 90° | $\frac{\pi}{2}$ | 1.5708 |
| 30° | $\frac{\pi}{6}$ | .5236 | 180° | $\pi$ | 3.1416 |
| 45° | $\frac{\pi}{4}$ | .7854 | 270° | $\frac{3\pi}{2}$ | 4.7124 |
| 60° | $\frac{\pi}{3}$ | 1.0472 | 360° | $2\pi$ | 6.2832 |

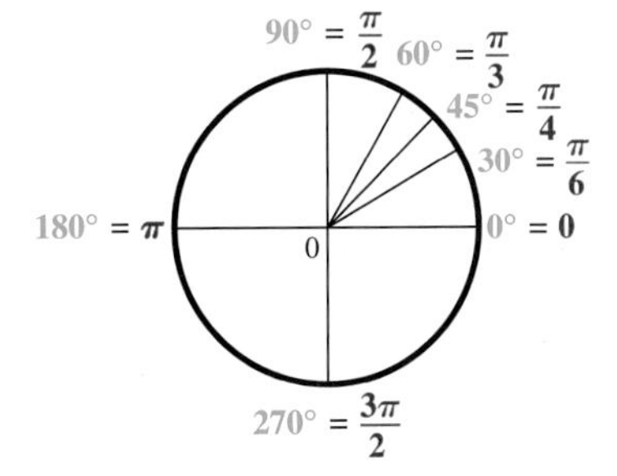


FIGURE 18

We use radian measure to simplify certain formulas, two of which follow. Both would be more complicated if expressed in degrees.

## Arc Lengths and Sectors

We use the first formula to find the length of an arc of a circle. The formula is derived from the fact (proved in geometry) that the length of an arc is proportional to the measure of its central angle.

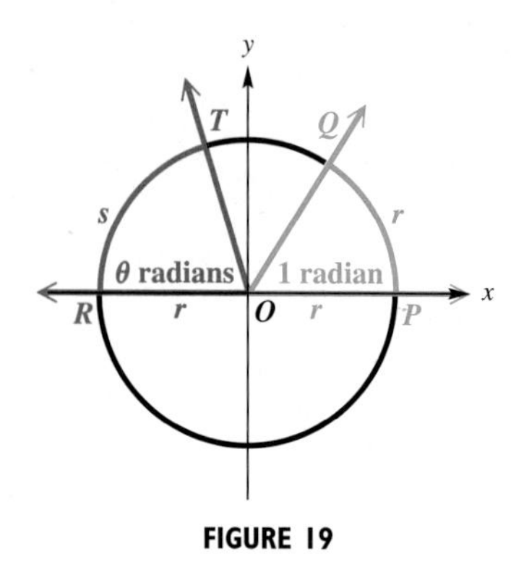

FIGURE 19

In Figure 19, angle $QOP$ has measure 1 radian and intercepts an arc of length $r$ on the circle. Angle $ROT$ has measure $\theta$ radians and intercepts an arc of length $s$ on the circle. Since the lengths of the arcs are proportional to the measures of their central angles,

$$\frac{s}{r} = \frac{\theta}{1}.$$

Multiplying each side by $r$ gives the following result.

### Arc Length

The length $s$ of the arc intercepted on a circle of radius $r$ by a central angle of measure $\theta$ radians is given by the product of the radius and the radian measure of the angle, or

$$s = r\theta, \quad \theta \text{ in radians.}$$

**CAUTION** When applying the formula $s = r\theta$, the value of $\theta$ *must be expressed in radians.*

**EXAMPLE 7** *Finding Arc Length*

A circle has radius 25 inches. Find the length of an arc intercepted by a central angle of 45°.

**Solution** First convert 45° to radian measure.

$$45° = 45\left(\frac{\pi}{180}\right) = \frac{\pi}{4} \text{ radian}$$

The arc length $s$ is given by

$$s = r\theta = 25\left(\frac{\pi}{4}\right) = 6.25\pi \text{ inches.}$$

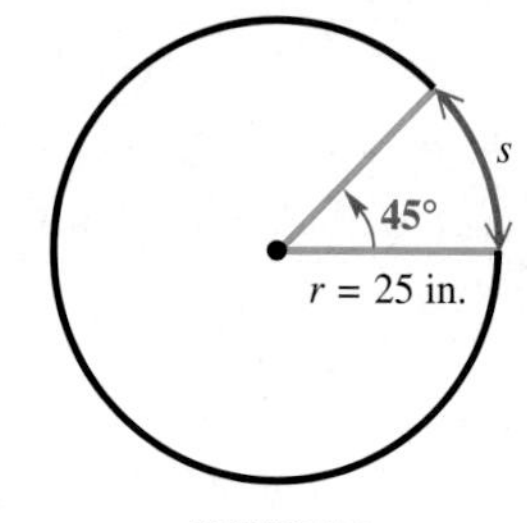

FIGURE 20

The arc length $s$ shown in Figure 20 is $6.25\pi \approx 19.6$ inches.

**EXAMPLE 8** *Using Latitudes to Find the Distance between Two Cities*

Reno, Nevada, is approximately due north of Los Angeles. The latitude of Reno is 40° N, while that of Los Angeles is 34° N. (The N in 34° N means *north* of the equator.) If the radius of Earth is 6400 kilometers, find the north–south distance between the two cities.

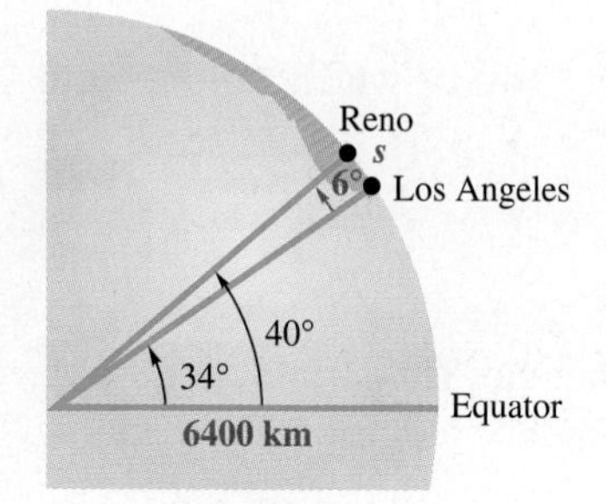

FIGURE 21

**Solution** Latitude gives the measure of a central angle with vertex at Earth's center whose initial side goes through the equator and whose terminal side goes through the given location. As shown in Figure 21, the central angle between Reno and Los Angeles is $40° - 34° = 6°$. The distance between the two cities can be found by the formula $s = r\theta$, after 6° is first converted to radians.

$$6° = 6\left(\frac{\pi}{180}\right) = \frac{\pi}{30} \text{ radian}$$

The distance between the two cities is

$$s = r\theta = 6400\left(\frac{\pi}{30}\right) \approx 670 \text{ km}.$$

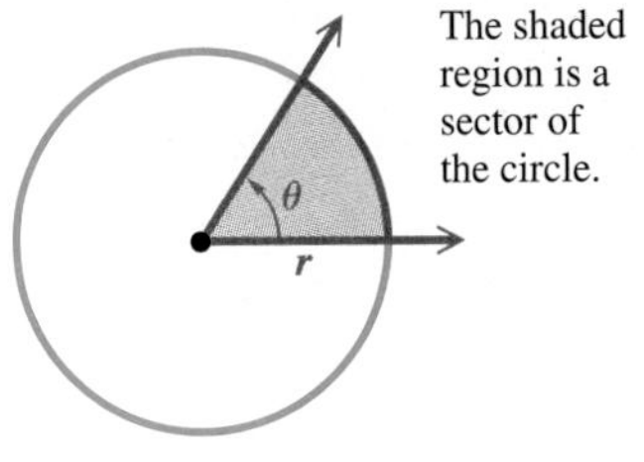


The shaded region is a sector of the circle.

FIGURE 22

A **sector of a circle** is the portion of the interior of a circle intercepted by a central angle. Think of it as a "piece of pie." See Figure 22. The interior of a circle can be thought of as a sector with measure $2\pi$ radians. If a central angle for a sector has measure $\theta$ radians, then the sector makes up the fraction $\frac{\theta}{2\pi}$ of a complete circle. The area inside a circle with radius $r$ is $A = \pi r^2$. Therefore, the area of the sector is given by the product of the fraction $\frac{\theta}{2\pi}$ and the total area, $\pi r^2$.

$$\text{area of sector} = \frac{\theta}{2\pi}(\pi r^2) = \frac{1}{2}r^2\theta, \quad \theta \text{ in radians}$$

This discussion is summarized as follows.

**Area of a Sector**

The area of a sector of a circle of radius $r$ and central angle $\theta$ is given by

$$A = \frac{1}{2}r^2\theta, \quad \theta \text{ in radians.}$$

**CAUTION** As in the formula for arc length, the value of $\theta$ *must be in radians* when using this formula for the area of a sector.

**EXAMPLE 9** *Finding the Area of a Sector-Shaped Field*

Figure 23 shows a field in the shape of a sector of a circle. Find the area of the field.

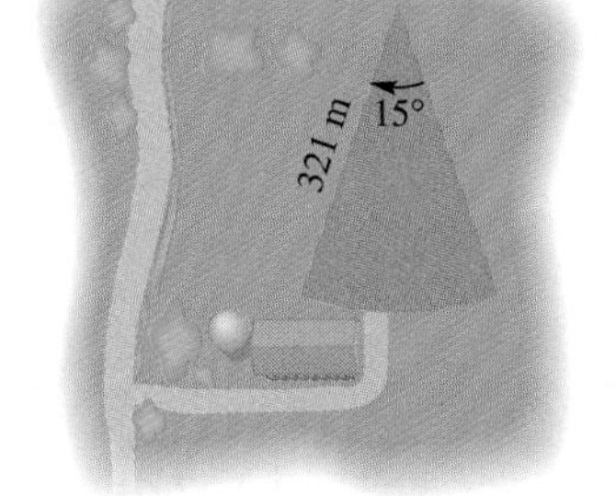


FIGURE 23

**Solution** First, convert 15° to radians.

$$15° = 15\left(\frac{\pi}{180}\right) = \frac{\pi}{12} \text{ radian}$$

Now use the formula for the area of a sector.

$$A = \frac{1}{2}r^2\theta = \frac{1}{2}(321)^2\left(\frac{\pi}{12}\right) \approx 13{,}500 \text{ square meters}$$

## Angular and Linear Speed

The human joint that can be flexed the fastest is the wrist, which can rotate through 90°, or $\frac{\pi}{2}$ radians, in .045 second while holding a tennis racket. **Angular speed** $\omega$ (omega) measures the speed of rotation and is defined by

$$\omega = \frac{\theta}{t},$$

where $\theta$ is the angle of rotation and $t$ is time. The angular speed of a human wrist swinging a tennis racket is

$$\frac{\frac{\pi}{2}}{.045} \approx 35 \text{ radians per second.}$$

The **linear speed** $v$ at which the tip of the racket travels as a result of flexing the wrist is given by

$$v = r\omega,$$

where $r$ is the radius (distance) from the tip of the racket to the wrist joint. If $r = 2$ feet, then the speed at the tip of the racket is

$$v = r\omega \approx 2(35) = 70 \text{ feet per second},$$

or about 48 mph. In a tennis serve the arm rotates at the shoulder, so the final speed of the racket is considerably faster. (*Source:* Cooper, J. and R. Glassow, *Kinesiology*, 2nd ed., C.V. Mosby, 1968.)

**EXAMPLE 10** *Finding Angular Speed of a Pulley and Linear Speed of a Belt*

A belt runs a pulley of radius 6 centimeters at 80 revolutions per minute.

**(a)** Find the angular speed of the pulley in radians per second.

**(b)** Find the linear speed of the belt in centimeters per second.

**Solution**

**(a)** In 1 minute, the pulley makes 80 revolutions. Each revolution is $2\pi$ radians, for a total of

$$80(2\pi) = 160\pi \text{ radians per minute.}$$

Since there are 60 seconds in one minute, we find $\omega$, the angular speed in radians per second, by dividing $160\pi$ by 60.

$$\omega = \frac{160\pi}{60} = \frac{8\pi}{3} \text{ radians per second}$$

**(b)** The linear speed of the belt will be the same as that of a point on the circumference of the pulley. Thus,

$$v = r\omega = 6\left(\frac{8\pi}{3}\right) = 16\pi \approx 50.3 \text{ centimeters per second.}$$

Suppose that an object is moving at a constant speed. If it travels a distance $s$ in time $t$, then its linear speed $v$ is given by

$$v = \frac{s}{t}.$$

If the object is traveling in a circle, then $s = r\theta$, where $r$ is the radius of the circle and $\theta$ is the angle of rotation. Thus, we can write

$$v = \frac{r\theta}{t}.$$

We use these formulas in the next example.

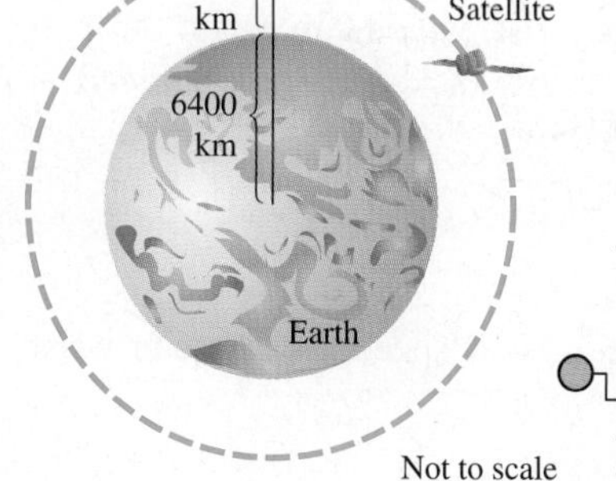

**FIGURE 24**

**EXAMPLE 11** *Finding Linear Speed and Distance Traveled by a Satellite*

A satellite traveling in a circular orbit 1600 kilometers above the surface of Earth takes two hours to make an orbit. Assume that the radius of Earth is 6400 kilometers. See Figure 24.

**(a)** Find the linear speed of the satellite.

**(b)** Find the distance the satellite travels in 4.5 hours.

### Solution

**(a)** The distance of the satellite from the center of Earth is

$$r = 1600 + 6400 = 8000 \text{ kilometers.}$$

For one orbit, $\theta = 2\pi$, and

$$s = r\theta = 8000(2\pi) \text{ kilometers.}$$

Since it takes 2 hours to complete an orbit, the linear speed is

$$v = \frac{s}{t} = \frac{8000(2\pi)}{2} = 8000\pi \approx 25{,}000 \text{ kilometers per hour.}$$

**(b)** $s = vt = 8000\pi(4.5) = 36{,}000\pi \approx 110{,}000$ kilometers

The formulas for angular and linear speed are summarized below.

| Angular Speed | Linear Speed |
|---|---|
| $\omega = \frac{\theta}{t}$ | $v = \frac{s}{t}$ |
| ($\omega$ in radians per unit time, $\theta$ in radians) | $v = \frac{r\theta}{t}$ |
| | $v = r\omega$ |

## 8.1 EXERCISES

*Find* **(a)** *the complement and* **(b)** *the supplement of each angle.*

**1.** 30° **2.** 60° **3.** 45° **4.** $\frac{\pi}{3}$ **5.** $\frac{\pi}{4}$ **6.** $\frac{\pi}{12}$

**7.** *Concept Check* An angle measures $x$ degrees.

**(a)** What is the measure of its complement?

**(b)** What is the measure of its supplement?

**8.** *Concept Check* An angle measures $x$ radians.

**(a)** What is the measure of its complement?

**(b)** What is the measure of its supplement?

*Find the degree measure of the smaller angle formed by the hands of a clock at the following times.*

**9.**

**10.**

*Find the measure of each angle in Exercises 11–14.*

**11.**

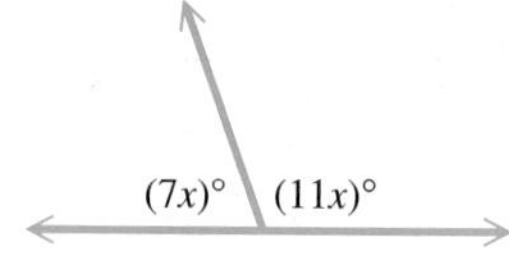


**12.**

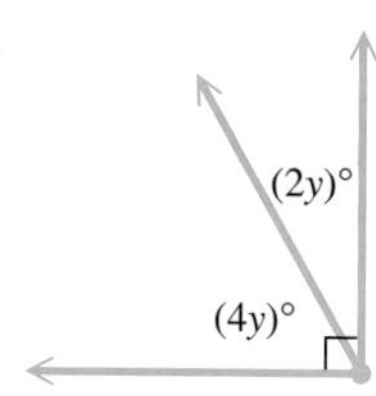


**13.** Supplementary angles with measures $6x - 4$ and $8x - 12$ degrees

**14.** Complementary angles with measures $9z + 6$ and $3z$ degrees

*Perform each calculation.*

**15.** $62° 18' + 21° 41'$ **16.** $75° 15' + 83° 32'$ **17.** $71° 18' - 47° 29'$

**18.** $47° 23' - 73° 48'$ **19.** $90° - 72° 58' 11''$ **20.** $90° - 36° 18' 47''$

*Convert each angle measure to decimal degrees. Use a calculator, and round to the nearest thousandth of a degree if necessary.*

**21.** $20° 54'$ **22.** $38° 42'$ **23.** $91° 35' 54''$ **24.** $34° 51' 35''$

*Convert each angle measure to degrees, minutes, and seconds. Use a calculator as necessary.*

**25.** 31.4296° **26.** 59.0854° **27.** 89.9004° **28.** 102.3771°

Concept Check *Sketch each angle in standard position. Draw an arrow representing the correct amount of rotation. Find the measure of two other angles, one positive and one negative, that are coterminal with the given angle. Give the quadrant of each angle.*

**29.** 75° **30.** 89° **31.** 174° **32.** 234° **33.** $-61°$ **34.** $-159°$

*Find the angle of smallest positive measure that is coterminal with the given angle.*

**35.** $-40°$ **36.** $-98°$ **37.** 450° **38.** 539°

**39.** $-\frac{\pi}{4}$ **40.** $-\frac{\pi}{3}$ **41.** $-\frac{3\pi}{2}$ **42.** $-\pi$

*Give an expression that generates all angles coterminal with each angle. Let n represent any integer.*

**43.** 30° **44.** 45° **45.** $-90°$ **46.** $-135°$

**47.** $\frac{\pi}{4}$ **48.** $\frac{\pi}{6}$ **49.** $-\frac{3\pi}{4}$ **50.** $-\frac{7\pi}{6}$

*Convert each degree measure to radians. Leave answers as multiples of π.*

**51.** 60° **52.** 90° **53.** 150° **54.** 270°

*Convert each radian measure to degrees.*

**55.** $\frac{\pi}{3}$ **56.** $\frac{8\pi}{3}$ **57.** $\frac{7\pi}{4}$ **58.** $\frac{2\pi}{3}$ **59.** $\frac{11\pi}{6}$ **60.** $\frac{15\pi}{4}$

*Convert each degree measure to radians. Round to the nearest hundredth.*

**61.** 39° **62.** 74° **63.** $139° 10'$ **64.** $174° 50'$ **65.** 64.29° **66.** 122.62°

*Convert each radian measure to degrees. Write answers to the nearest minute.*

**67.** 2 **68.** 5 **69.** 1.74 **70.** .3417

**71.** *Concept Check* The figure shows the same angles measured in both degrees and radians. Complete the missing measures.

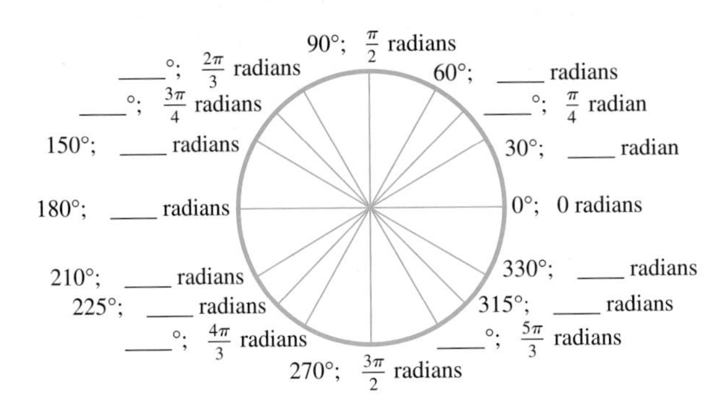


**72.** *Railroad Engineering* The term *grade* has several different meanings in construction work. Some engineers use the term *grade* to represent $\frac{1}{100}$ of a right angle and express grade as a percent. For instance, an angle of .9° would be referred to as a 1% grade. (*Source:* Hay, W., *Railroad Engineering,* John Wiley and Sons, 1982.)

**(a)** By what number should you multiply a grade to convert it to radians?

**(b)** In a rapid-transit rail system, the maximum grade allowed between two stations is 3.5%. Express this angle in degrees and radians.

*Concept Check* *Find the exact length of each arc intercepted by the given central angle.*

**73.**

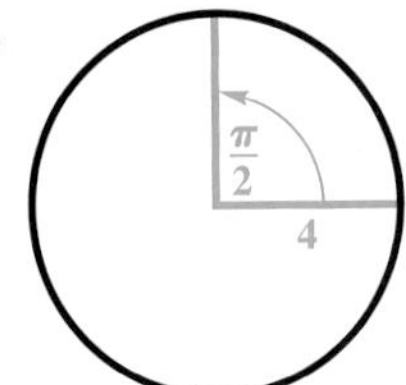


**74.**

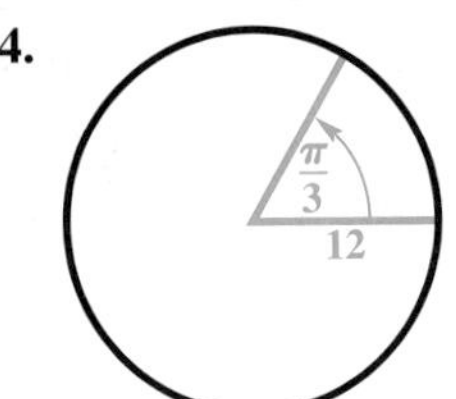


*Concept Check* *Find the radius of each circle.*

**75.**

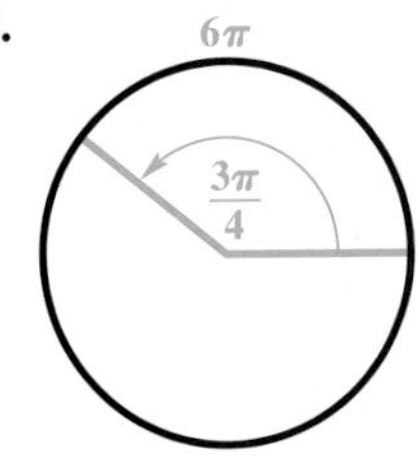


**76.**

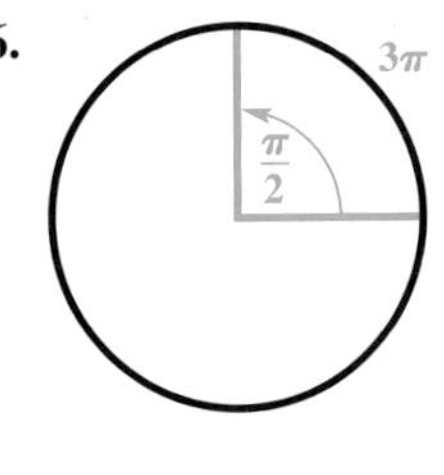


*Concept Check* *Find the measure of each central angle (in radians).*

**77.**

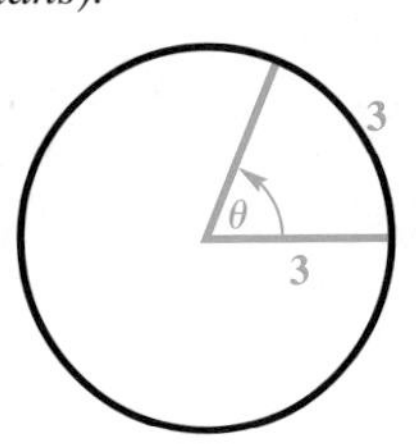


**78.**

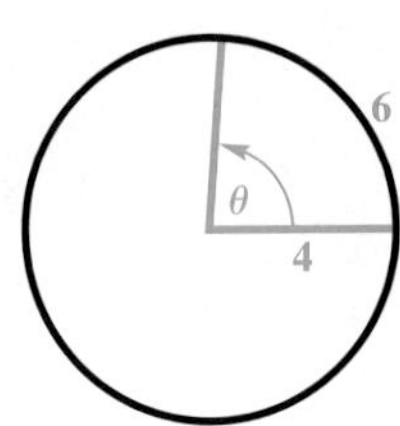


*Find the length of each arc intercepted by a central angle $\theta$ in a circle of radius r.*

**79.** $r = 12.3$ centimeters; $\theta = \dfrac{2\pi}{3}$ radians

**80.** $r = .892$ centimeter; $\theta = \dfrac{11\pi}{10}$ radians

**81.** $r = 4.82$ meters; $\theta = 60°$

**82.** $r = 71.9$ centimeters; $\theta = 135°$

*Distance between Cities* *Find the distance in kilometers between the pair of cities whose latitudes are given. Assume that the cities are on a north–south line, and that the radius of Earth is 6400 kilometers. Round answers to the nearest hundred kilometers.*

**83.** Madison, South Dakota, 44° N, and Dallas, Texas, 33° N

**84.** Charleston, South Carolina, 33° N, and Toronto, Ontario, 43° N

**85.** New York City, New York, 41° N, and Lima, Peru, 12° S

**86.** Halifax, Nova Scotia, 45° N, and Buenos Aires, Argentina, 34° S

*Solve each problem.*

**87.** *Pulley Raising a Weight*

**(a)** How many inches will the weight in the figure rise if the pulley is rotated through an angle of 71° 50′?

**(b)** Through what angle, to the nearest minute, must the pulley be rotated to raise the weight 6 inches?

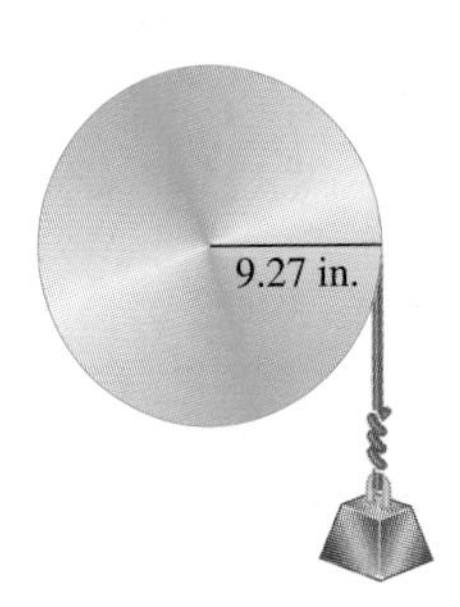

**88.** *Pulley Raising a Weight* Find the radius of the pulley in the figure if a rotation of 51.6° raises the weight 11.4 centimeters.

**89.** *Rotating Wheels* The rotation of the smaller wheel in the figure causes the larger wheel to rotate. Through how many degrees will the larger wheel rotate if the smaller one rotates through 60.0°?

**90.** *Rotating Wheels* Find the radius of the larger wheel in the figure if the smaller wheel rotates 80.0° when the larger wheel rotates 50.0°.

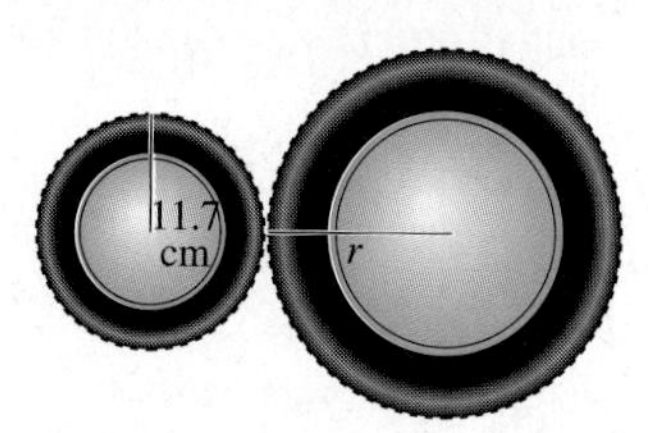


**91.** *Bicycle Chain Drive* The figure shows the chain drive of a bicycle. How far will the bicycle move if the pedals are rotated through 180°? Assume the radius of the bicycle wheel is 13.6 inches.

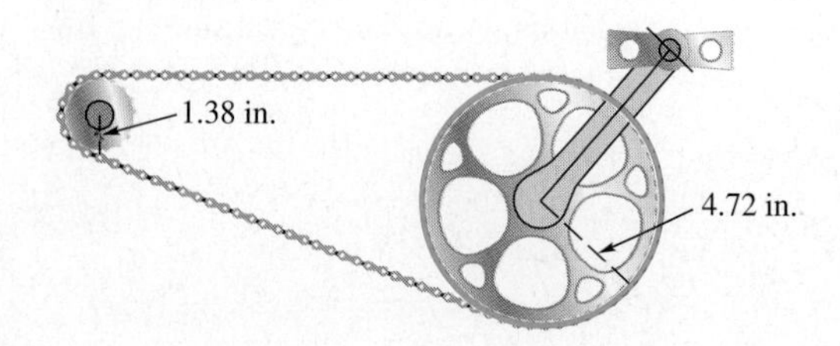


**92.** *Pickup Truck Speedometer* The speedometer of a small pickup truck is designed to be accurate with tires of radius 14 inches.

**(a)** Find the number of rotations of a tire in 1 hour if the truck is driven 55 mph.

**(b)** Suppose that oversize tires of radius 16 inches are placed on the truck. If the truck is now driven for 1 hour with the speedometer reading 55 mph, how far has the truck gone? If the speed limit is 55 mph, does the driver deserve a speeding ticket?

*Find the area of a sector of a circle having radius r and central angle θ.*

**93.** $r = 29.2$ meters; $\theta = \frac{5\pi}{6}$ radians

**94.** $r = 59.8$ kilometers; $\theta = \frac{2\pi}{3}$ radians

**95.** $r = 12.7$ centimeters; $\theta = 81°$

**96.** $r = 18.3$ meters; $\theta = 125°$

*Concept Check* *Find the area of each sector. Express answers in terms of π.*

**97.**

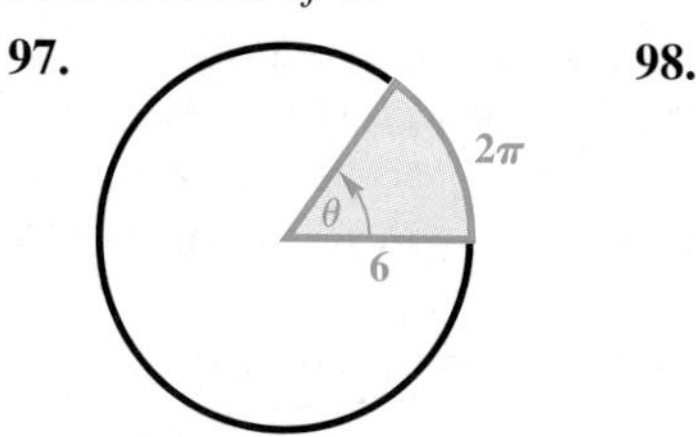


**98.**

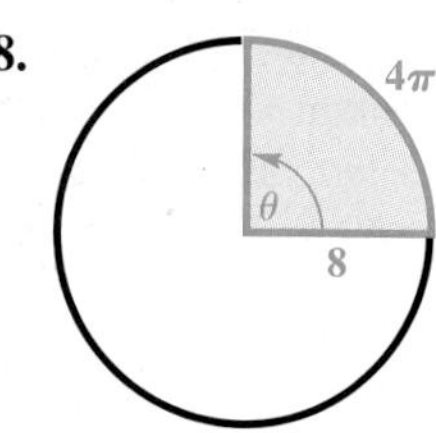


*Solve each problem.*

**99.** *Doppler Radar* Radar is used to identify severe weather. If Doppler radar can detect weather within a 240-mile radius and creates a new image every 48 seconds, find the area scanned by the radar in 1 second.

**100.** *Land Required for a Solar-Power Plant* A 300-megawatt solar-power plant requires approximately 950,000 square meters of land area to collect the required amount of energy from sunlight.

**(a)** If this land area is circular, what is its radius?

**(b)** If this land area is a 35° sector of a circle, what is its radius?

**101.** *Area Cleaned by a Windshield Wiper* The Ford Model A, built from 1928 to 1931, had a single windshield wiper on the driver's side. The total arm and blade was 10 inches long and rotated back and forth through an angle of 95°. The shaded region in the figure is the por-

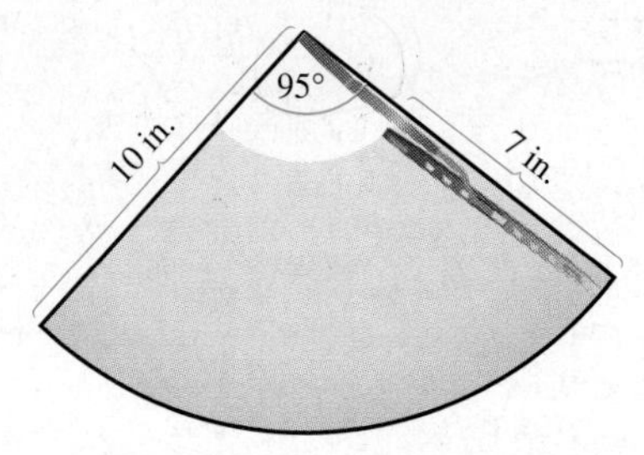

tion of the windshield cleaned by the 7-inch wiper blade. What is the area of the region cleaned?

**102.** *Measures of a Structure* The figure shows Medicine Wheel, a Native American structure in northern Wyoming. This circular structure is perhaps 2500 years old. There are 27 aboriginal spokes in the wheel, all equally spaced.

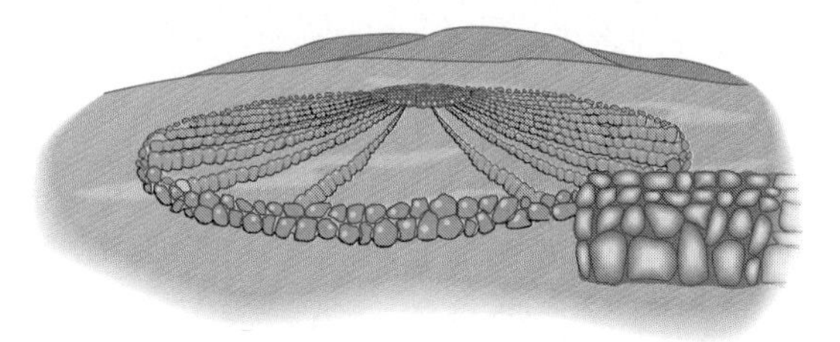

**(a)** Find the measure of each central angle in degrees and in radians.

**(b)** If the radius of the wheel is 76 feet, find the circumference.

**(c)** Find the length of each arc intercepted by consecutive pairs of spokes.

**(d)** Find the area of each sector formed by consecutive spokes.

**103.** *Size of a Corral* The unusual corral in the figure is separated into 26 areas, many of which approximate sectors of a circle. Assume that the corral has a diameter of 50 meters.

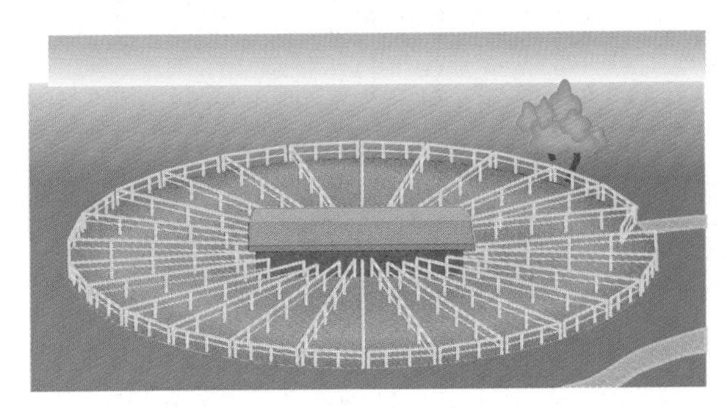

**(a)** Find the central angle for each region, assuming that the 26 regions are all equal sectors, with the fences meeting at the center.

**(b)** What is the area of each sector?

**104.** *Area of a Lot* A frequent problem in surveying city lots and rural lands adjacent to curves of highways and railways is that of finding the area when one or more of the boundary lines is the arc of a circle. Find the area of the lot shown in the figure at the top of the next column. (*Source:* Anderson, J. and E. Michael, *Introduction to Surveying*, McGraw-Hill, 1985.)

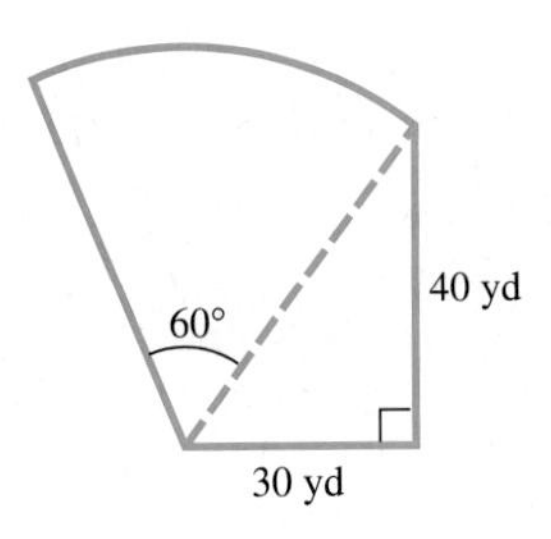


**105.** *Concept Check* Consider the area-of-a-sector formula $A = \frac{1}{2}r^2\theta$. What well-known formula corresponds to the special case $\theta = 2\pi$?

**106.** *Revolution of Earth* Earth revolves on its axis once every 24 hours. Assuming that Earth's radius is 6400 kilometers, find the following.

**(a)** Angular speed of Earth in radians per day

**(b)** Linear speed at the North Pole or South Pole

**(c)** Linear speed at Quito, Ecuador, a city on the equator

**107.** *Bicycle Tire* A bicycle has a tire 26 inches in diameter that is rotating at 15 radians per second. Approximate the speed of the bicycle in feet per second and in miles per hour.

**108.** *Skateboard Wheels* The wheels on a skateboard have diameter 2.25 inches. If a skateboarder is traveling downhill at 15 mph, determine the angular speed of the wheels in radians per second.

**109.** *Pulley* The pulley shown has a radius of 12.96 centimeters. Suppose that it takes 18 seconds for 56 centimeters of belt to go around the pulley. Find the angular speed of the pulley in radians per second.

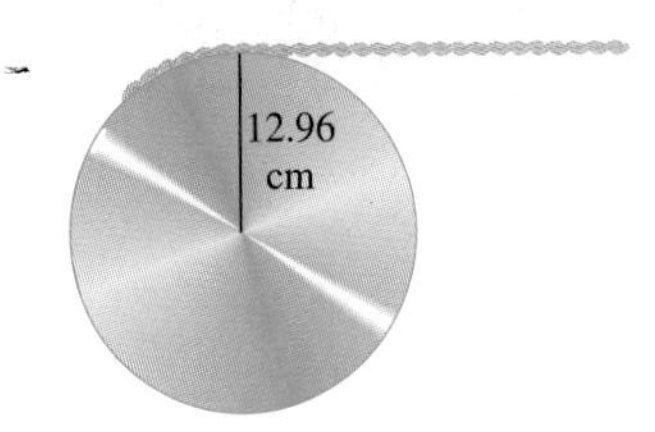


**110.** *Pair of Pulleys* The two pulleys in the figure have radii 15 centimeters and 8 centimeters, respectively. The larger pulley rotates 25 times in 36 seconds. Find the angular speed of each pulley in radians per second.

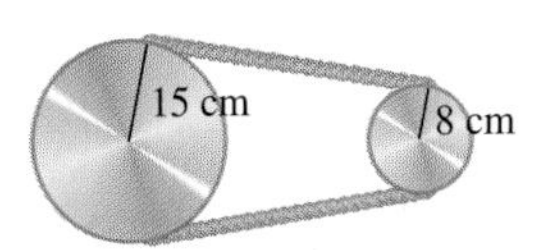


**111.** *Propeller* A 90-horsepower outboard motor at full throttle rotates its propeller 5000 revolutions per minute.

Find the angular speed of the propeller in radians per second. What is the linear speed in inches per second of a point at the tip of the propeller if its diameter is 10 inches?

**112.** *Golf Club* The shoulder joint can rotate at about 25 radians per second. Assuming that a golfer's arm is straight and the distance from the shoulder to the club head is 5 feet, estimate the linear speed of the club head from shoulder rotation. (*Source:* Cooper, J. and R. Glassow, *Kinesiology,* Second Edition, C.V. Mosby, 1968.)

**113.** *Orbit of Earth* Earth travels about the sun in an orbit that is almost circular. Assume that the orbit is a circle, with radius 93,000,000 miles. (See the figure.)

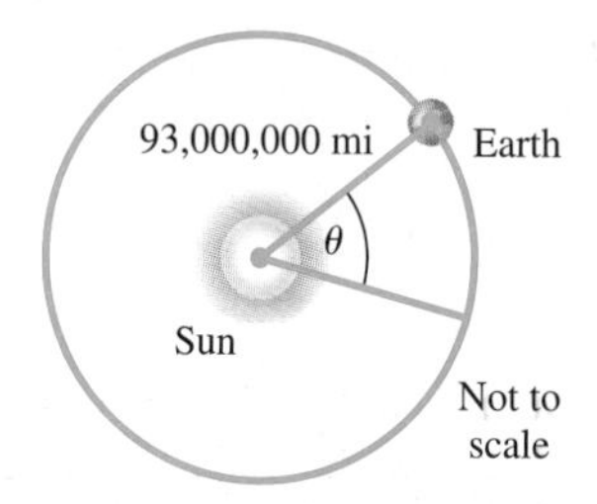


**(a)** Assume that a year is 365 days, and find $\theta$, the angle formed by Earth's movement in 1 day.

**(b)** Give the angular speed in radians per hour.

**(c)** Find the linear speed of Earth in miles per hour.

**114.** *Nautical Miles* *Nautical miles* are used by ships and airplanes. They are different from *statute miles,* which equal 5280 feet. A nautical mile is defined to be the arc length along the equator intercepted by a central angle $AOB$ of 1 minute, as illustrated in the figure in the next column. If the equatorial radius of Earth is 3963 miles, use the arc length formula to approximate the number of statute miles in 1 nautical mile. Round your answer to two decimal places.

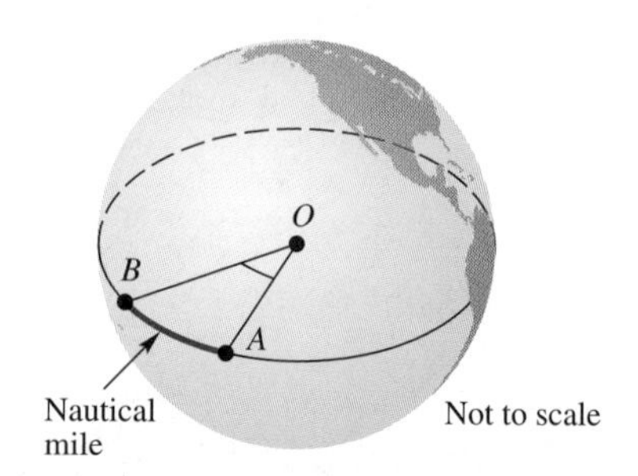


**115.** *Circumference of Earth* The first accurate estimate of the distance around Earth was done by the Greek astronomer Eratosthenes (276–195 B.C.), who noted that the noontime position of the sun at the summer solstice differed by 7° 12′ from the city of Syene to the city of Alexander. (See the figure.) The distance between these two cities is 496 miles. Use the arc length formula to estimate the radius of Earth. Then find the circumference of Earth. (*Source:* Zeilik, M., *Introductory Astronomy and Astrophysics,* Third Edition, Saunders College Publishers, 1992.)

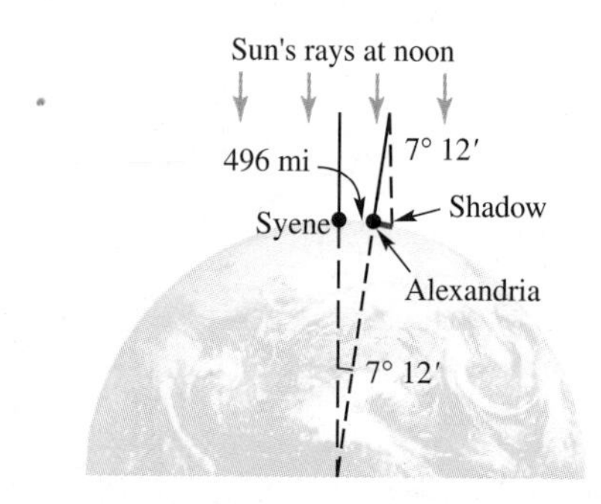


**116.** *Diameter of the Moon* The distance to the moon is approximately 238,900 miles. Use the arc length formula to estimate the diameter $d$ of the moon if angle $\theta$ in the figure is measured to be .517°.

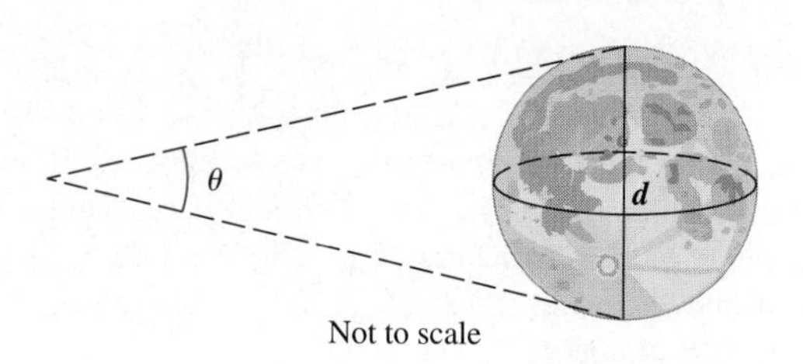

# 8.2 TRIGONOMETRIC FUNCTIONS AND FUNDAMENTAL IDENTITIES

Trigonometric Functions • Quadrantal Angles • Reciprocal Identities • Signs and Ranges of Function Values • Pythagorean Identities • Quotient Identities

## Trigonometric Functions

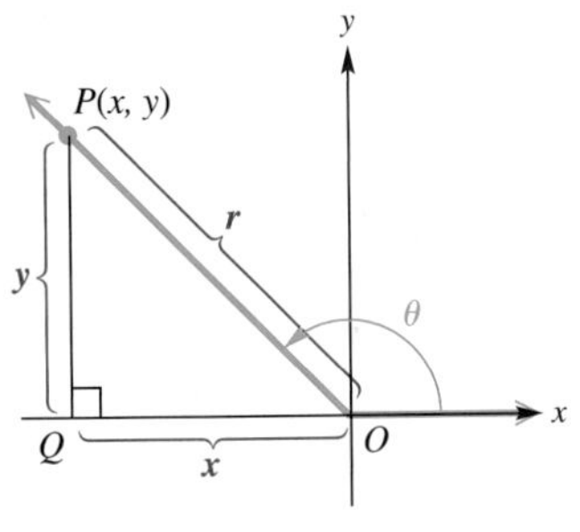


FIGURE 25

To define the six *trigonometric functions*, we start with an angle $\theta$ in standard position. We then choose any point $P$ having coordinates $(x, y)$ on the terminal side of angle $\theta$. (The point $P$ must not be the vertex of the angle.) See Figure 25. A perpendicular from $P$ to the $x$-axis at point $Q$ determines a right triangle having vertices at $O$, $P$, and $Q$. The distance $r$ from $P(x, y)$ to the origin, $(0, 0)$, can be found using the distance formula.

$$r = \sqrt{(x-0)^2 + (y-0)^2} = \sqrt{x^2 + y^2}$$

Notice that $r > 0$ since distance is never negative.

The six trigonometric functions of angle $\theta$ are **sine, cosine, tangent, cotangent, secant,** and **cosecant.** In the following definitions, we use the customary abbreviations for the names of these functions.

### Trigonometric Functions

Let $(x, y)$ be a point other than the origin on the terminal side of an angle $\theta$ in standard position. The distance from the point to the origin is $r = \sqrt{x^2 + y^2}$. The six trigonometric functions of $\theta$ are as follows.

$$\sin\theta = \frac{y}{r} \qquad \cos\theta = \frac{x}{r} \qquad \tan\theta = \frac{y}{x} \quad (x \neq 0)$$

$$\csc\theta = \frac{r}{y} \quad (y \neq 0) \qquad \sec\theta = \frac{r}{x} \quad (x \neq 0) \qquad \cot\theta = \frac{x}{y} \quad (y \neq 0)$$

**NOTE** Although Figure 25 shows a second quadrant angle, these definitions apply to any angle $\theta$. Because of the restrictions on the denominators in the definitions of tangent, cotangent, secant, and cosecant, some angles have undefined function values.

### EXAMPLE 1 *Finding Function Values of an Angle*

The terminal side of an angle $\theta$ in standard position passes through the point $(8, 15)$. Find the values of the six trigonometric functions of angle $\theta$.

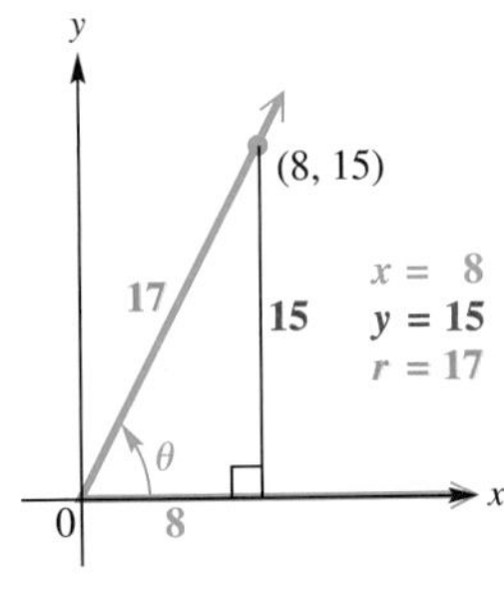


FIGURE 26

**Solution** Figure 26 shows angle $\theta$ and the triangle formed by dropping a perpendicular from the point $(8, 15)$ to the $x$-axis. The point $(8, 15)$ is 8 units to the right of the $y$-axis and 15 units above the $x$-axis, so $x = 8$ and $y = 15$. Since $r = \sqrt{x^2 + y^2}$,

$$r = \sqrt{8^2 + 15^2} = \sqrt{64 + 225} = \sqrt{289} = 17.$$

We can now find the values of the six trigonometric functions of angle $\theta$.

$$\sin\theta = \frac{y}{r} = \frac{15}{17} \qquad \cos\theta = \frac{x}{r} = \frac{8}{17} \qquad \tan\theta = \frac{y}{x} = \frac{15}{8}$$

$$\csc\theta = \frac{r}{y} = \frac{17}{15} \qquad \sec\theta = \frac{r}{x} = \frac{17}{8} \qquad \cot\theta = \frac{x}{y} = \frac{8}{15}$$

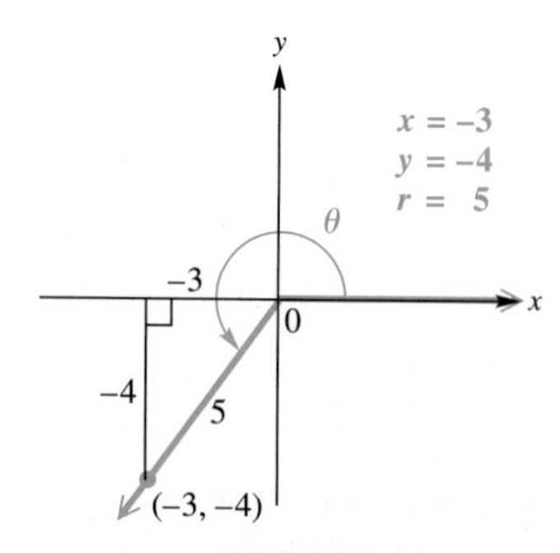

FIGURE 27

**EXAMPLE 2** *Finding Function Values of an Angle*

The terminal side of angle $\theta$ in standard position passes through $(-3, -4)$. Find the values of the six trigonometric functions of $\theta$.

**Solution** As shown in Figure 27, $x = -3$ and $y = -4$. The value of $r$ is

$$r = \sqrt{(-3)^2 + (-4)^2} = \sqrt{25} = 5.$$

(Remember that $r > 0$.) By the definitions of the trigonometric functions,

$$\sin\theta = \frac{-4}{5} = -\frac{4}{5} \qquad \cos\theta = \frac{-3}{5} = -\frac{3}{5} \qquad \tan\theta = \frac{-4}{-3} = \frac{4}{3}$$

$$\csc\theta = \frac{5}{-4} = -\frac{5}{4} \qquad \sec\theta = \frac{5}{-3} = -\frac{5}{3} \qquad \cot\theta = \frac{-3}{-4} = \frac{3}{4}.$$

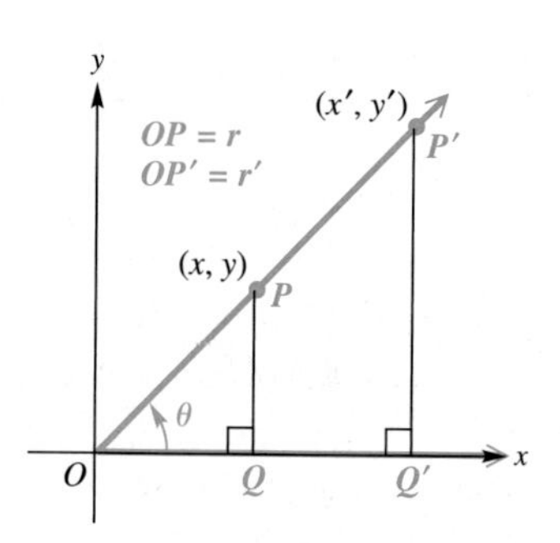

FIGURE 28

We can find the six trigonometric functions from *any* point other than the origin on the terminal side of an angle. To see why any point may be used, refer to Figure 28, which shows an angle $\theta$ and two distinct points on its terminal side. Point $P$ has coordinates $(x, y)$, and point $P'$ (read "$P$-prime") has coordinates $(x', y')$. Let $r$ be the length of the hypotenuse of triangle $OPQ$, and let $r'$ be the length of the hypotenuse of triangle $OP'Q'$. Since corresponding sides of similar triangles are in proportion,

$$\frac{y}{r} = \frac{y'}{r'},$$

so $\sin\theta = \frac{y}{r}$ is the same no matter which point is used to find it. Similar results hold for the other five functions.

We can also find the trigonometric function values of an angle if we know the equation of the line coinciding with the terminal ray. Recall that the graph of the equation

$$Ax + By = 0$$

is a line that passes through the origin. If we restrict $x$ to have only nonpositive or only nonnegative values, we obtain as the graph a ray with endpoint at the origin. For example, the graph of $x + 2y = 0$, $x \geq 0$, shown in Figure 29, is a ray that can serve as the terminal side of an angle in standard position. By choosing any point on the ray, we can find the trigonometric function values of the angle.

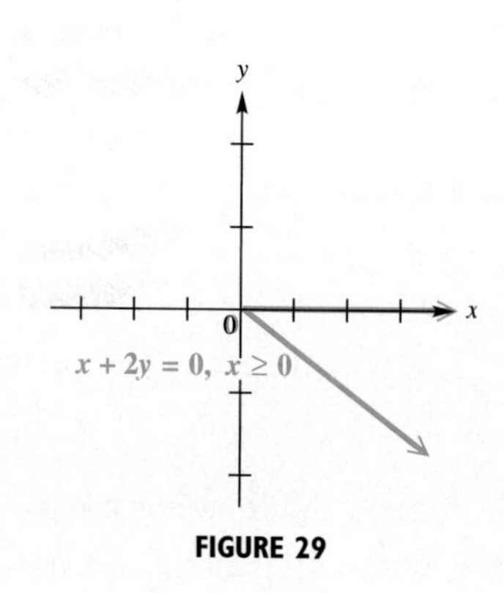

FIGURE 29

**EXAMPLE 3** *Finding Function Values of an Angle*

Find the six trigonometric function values of the angle $\theta$ in standard position, if the terminal side of $\theta$ is defined by $x + 2y = 0$, $x \geq 0$.

### Analytic Solution

The angle is shown in Figure 30. We can use *any* point except $(0, 0)$ on the terminal side of $\theta$ to find the trigonometric function values. We choose $x = 2$ and find the corresponding $y$-value.

$$\begin{aligned} x + 2y &= 0,\ x \geq 0 \\ 2 + 2y &= 0 \\ 2y &= -2 \\ y &= -1 \end{aligned}$$

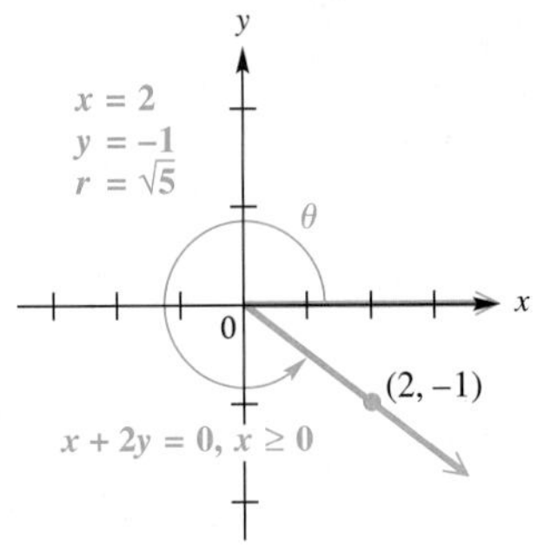

FIGURE 30

The point $(2, -1)$ lies on the terminal side, and the corresponding value of $r$ is $r = \sqrt{2^2 + (-1)^2} = \sqrt{5}$. Now we use the definitions of the trigonometric functions.

$$\sin \theta = \frac{y}{r} = \frac{-1}{\sqrt{5}} = \frac{-1}{\sqrt{5}} \cdot \frac{\sqrt{5}}{\sqrt{5}} = -\frac{\sqrt{5}}{5}$$

$$\cos \theta = \frac{x}{r} = \frac{2}{\sqrt{5}} = \frac{2}{\sqrt{5}} \cdot \frac{\sqrt{5}}{\sqrt{5}} = \frac{2\sqrt{5}}{5}$$

$$\tan \theta = \frac{y}{x} = -\frac{1}{2} \qquad \csc \theta = \frac{r}{y} = -\sqrt{5}$$

$$\sec \theta = \frac{r}{x} = \frac{\sqrt{5}}{2} \qquad \cot \theta = \frac{x}{y} = -2$$

### Graphing Calculator Solution

Figure 31 shows the graph of

$$x + 2y = 0,\ x \geq 0.$$

We can graph the ray by entering the equation as

$$Y_1 = (-1/2)X \quad \text{or} \quad Y_1 = -.5X$$

with the restriction $X \geq 0$. We used the capability of the calculator to find the $y$-value for $X = 2$ which produced the point and values of X and Y shown on the screen.

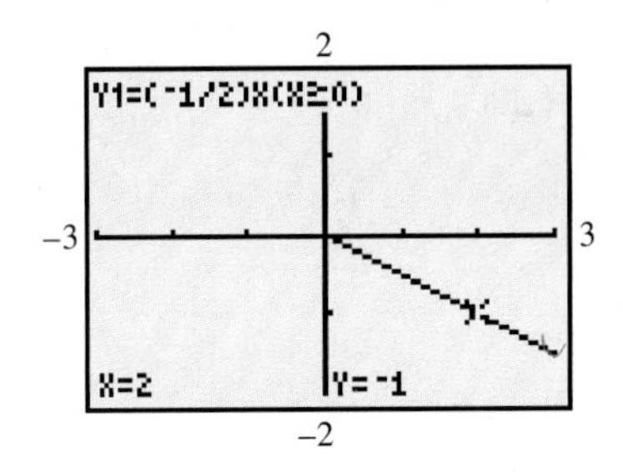


FIGURE 31

We can now use these values of $x$ and $y$ and the definitions of the trigonometric functions to find the six function values of $\theta$, as shown in the analytic solution.

**FOR DISCUSSION**

Either individually or with a group of students, rework Example 3 using a different value for $x$. Find the corresponding $y$-value, and then show that the six trigonometric function values you obtain are the same as the ones shown.

Recall that when the equation of a line is written in the form $y = mx + b$, the coefficient of $x$ is the slope of the line. In Example 3, $x + 2y = 0$ can be written as $y = -\frac{1}{2}x$, so the slope is $-\frac{1}{2}$. Notice that $\tan \theta = -\frac{1}{2}$. In general, it is true that $m = \tan \theta$.

**NOTE** The trigonometric function values we found in Examples 1–3 are *exact.* If we were to use a calculator to approximate these values, the decimal results would not be acceptable if exact values were required.

## Quadrantal Angles

If the terminal side of an angle in standard position lies along the $y$-axis, any point on this terminal side has $x$-coordinate 0. Similarly, an angle with terminal side on the $x$-axis has $y$-coordinate 0 for any point on the terminal side. Since the values of $x$ and $y$ appear in the denominators of some trigonometric functions, some trigonometric function values of quadrantal angles (i.e., those with terminal side on an axis) are undefined.

### EXAMPLE 4 *Finding Function Values of Quadrantal Angles*

Find the values of the six trigonometric functions for each angle.

**(a)** An angle of 90°

**(b)** An angle $\theta$ in standard position with terminal side through $(-3, 0)$

#### Analytic Solution

**(a)** First, we must select any point on the terminal side of a 90° angle. We select the point $(0, 1)$, as shown in Figure 32(a). Here $x = 0$ and $y = 1$, so $r = 1$. Then, by the definitions,

$$\sin 90° = \frac{1}{1} = 1 \qquad \csc 90° = \frac{1}{1} = 1$$

$$\cos 90° = \frac{0}{1} = 0 \qquad \sec 90° = \frac{1}{0} \quad \text{(undefined)}$$

$$\tan 90° = \frac{1}{0} \quad \text{(undefined)} \qquad \cot 90° = \frac{0}{1} = 0.$$

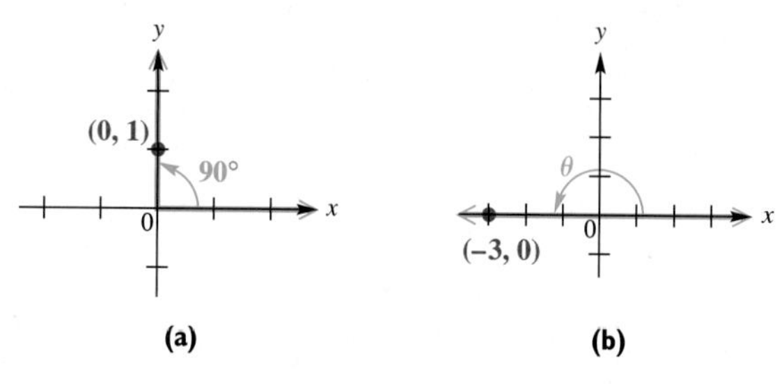


FIGURE 32

**(b)** Figure 32(b) shows the angle. Here, $x = -3$, $y = 0$, and $r = 3$, so the trigonometric functions have the following values.

$$\sin \theta = \frac{0}{3} = 0 \qquad \csc \theta = \frac{3}{0} \quad \text{(undefined)}$$

$$\cos \theta = \frac{-3}{3} = -1 \qquad \sec \theta = \frac{3}{-3} = -1$$

$$\tan \theta = \frac{0}{-3} = 0 \qquad \cot \theta = \frac{-3}{0} \quad \text{(undefined)}$$

#### Graphing Calculator Solution

A calculator set in degree mode (first screen in Figure 33) returns the correct values for sin 90° and cos 90° in part (a). The last screen shows an ERROR message for tan 90°, because 90° is not in the domain of the tangent function. There are no calculator keys for finding the function values of cotangent, secant, or cosecant. Later in this section we review how to find these function values with a calculator.

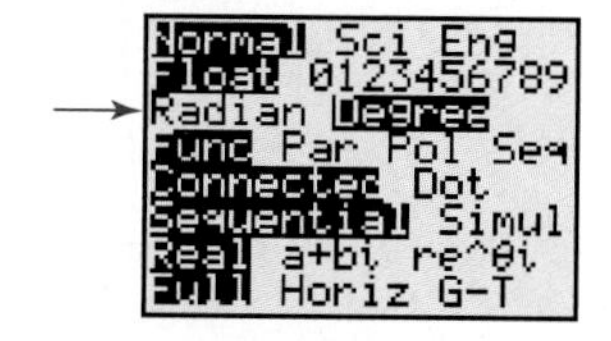


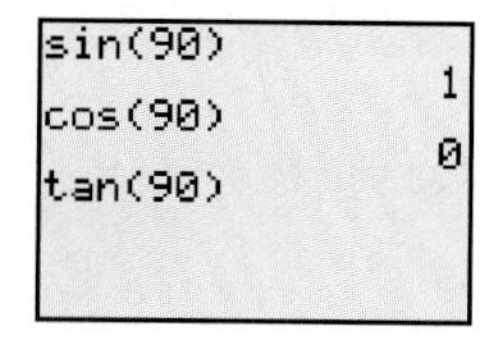

FIGURE 33

**FOR DISCUSSION**

Refer to Example 4 and discuss how you would find the trigonometric function values of an angle of $\frac{\pi}{2}$ radians.

The conditions under which the trigonometric function values of quadrantal angles are undefined are summarized here.

**Undefined Function Values**

If the terminal side of a quadrantal angle lies along the $y$-axis, the tangent and secant functions are undefined. If it lies along the $x$-axis, the cotangent and cosecant functions are undefined.

Since the most commonly used quadrantal angles are 0°, 90°, 180°, 270°, and 360°, the values of the functions of these angles are summarized in the following table.

| $\theta$ | $\sin\theta$ | $\cos\theta$ | $\tan\theta$ | $\cot\theta$ | $\sec\theta$ | $\csc\theta$ |
|---|---|---|---|---|---|---|
| 0° | 0 | 1 | 0 | Undefined | 1 | Undefined |
| 90° | 1 | 0 | Undefined | 0 | Undefined | 1 |
| 180° | 0 | −1 | 0 | Undefined | −1 | Undefined |
| 270° | −1 | 0 | Undefined | 0 | Undefined | −1 |
| 360° | 0 | 1 | 0 | Undefined | 1 | Undefined |

The values given in this table can also be found with a calculator that has trigonometric function keys. First, make sure the calculator is set in *degree mode*.

**CAUTION** One of the most common errors involving calculators in trigonometry occurs when the calculator is set for the wrong mode. Be sure that you know how to set your calculator in either radian or degree mode.

## Reciprocal Identities

Identities are equations that are true for all values of the variables for which all expressions are defined. Both $(x + y)^2 = x^2 + 2xy + y^2$ and $2(x + 3) = 2x + 6$ are examples of identities.

The definitions of the trigonometric functions were written so that functions in the same column are reciprocals of each other. Since $\sin\theta = \frac{y}{r}$ and $\csc\theta = \frac{r}{y}$,

$$\sin\theta = \frac{1}{\csc\theta} \quad \text{and} \quad \csc\theta = \frac{1}{\sin\theta},$$

provided $\sin\theta \neq 0$. Also, $\cos\theta$ and $\sec\theta$ are reciprocals, as are $\tan\theta$ and $\cot\theta$. In summary, we have the **reciprocal identities** that hold for any angle $\theta$ that does not lead to a 0 denominator.

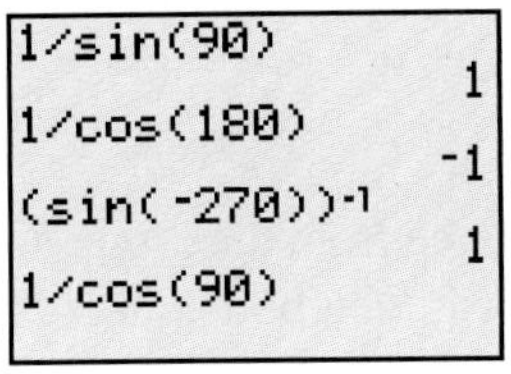

```
1/sin(90)
                1
1/cos(180)
               -1
(sin(-270))-1
                1
1/cos(90)
```

csc 90° = 1, sec 180° = −1, csc (−270°) = 1

Degree Mode

(a)

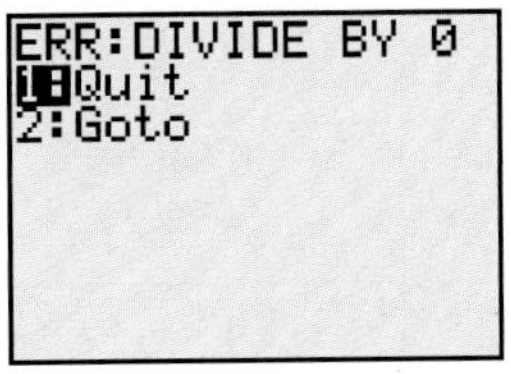

```
ERR:DIVIDE BY 0
1:Quit
2:Goto
```

(b)

**FIGURE 34**

**Reciprocal Identities**

$$\sin\theta = \frac{1}{\csc\theta} \qquad \cos\theta = \frac{1}{\sec\theta} \qquad \tan\theta = \frac{1}{\cot\theta}$$

$$\csc\theta = \frac{1}{\sin\theta} \qquad \sec\theta = \frac{1}{\cos\theta} \qquad \cot\theta = \frac{1}{\tan\theta}$$

Figure 34(a) shows how csc 90°, sec 180°, and csc(−270°) are found, using the reciprocal identities and the reciprocal key of a graphing calculator in degree mode. Be sure *not* to use the *inverse trigonometric function* keys to find the reciprocal function values. Attempting to find sec 90° by entering $\frac{1}{\cos 90°}$ produces an ERROR message, indicating the reciprocal is undefined. See Figure 34(b). Compare these results with those in the table of quadrantal angle function values.

**NOTE** Identities can be written in different forms. For example,

$$\sin\theta = \frac{1}{\csc\theta}$$

can also be written

$$\csc\theta = \frac{1}{\sin\theta} \quad \text{and} \quad (\sin\theta)(\csc\theta) = 1.$$

You should become familiar with all forms of these identities.

**EXAMPLE 5** *Using the Reciprocal Identities*

Find each function value.

**(a)** $\cos\theta$, if $\sec\theta = \frac{5}{3}$ **(b)** $\sin\theta$, if $\csc\theta = -\frac{\sqrt{12}}{2}$

**Solution**

**(a)** Since $\cos\theta$ is the reciprocal of $\sec\theta$,

$$\cos\theta = \frac{1}{\sec\theta} = \frac{1}{\frac{5}{3}} = \frac{3}{5}.$$

**(b)** Since $\sin\theta = \frac{1}{\csc\theta}$,

$$\sin\theta = \frac{1}{-\frac{\sqrt{12}}{2}} = \frac{-2}{\sqrt{12}} = \frac{-2}{2\sqrt{3}}$$
$$= \frac{-1}{\sqrt{3}} = -\frac{\sqrt{3}}{3}.$$

## Signs and Ranges of Function Values

In the definitions of the trigonometric functions, $r$ is the distance from the origin to the point $(x, y)$, so $r > 0$. If we choose a point $(x, y)$ in quadrant I, then both $x$ and $y$ will be positive, so the values of all six functions will be positive in quadrant I.

A point $(x, y)$ in quadrant II has $x < 0$ and $y > 0$. This makes the values of sine and cosecant positive for quadrant II angles, while the other four functions take on negative values. Similar results can be obtained for the other quadrants, as summarized here.

**Signs of Function Values**

| $\theta$ in Quadrant | $\sin\theta$ | $\cos\theta$ | $\tan\theta$ | $\cot\theta$ | $\sec\theta$ | $\csc\theta$ |
|---|---|---|---|---|---|---|
| I | + | + | + | + | + | + |
| II | + | − | − | − | − | + |
| III | − | − | + | + | − | − |
| IV | − | + | − | − | + | − |

$x < 0, y > 0, r > 0$
II
Sine and cosecant positive

$x > 0, y > 0, r > 0$
I
All functions positive

$x < 0, y < 0, r > 0$
III
Tangent and cotangent positive

$x > 0, y < 0, r > 0$
IV
Cosine and secant positive

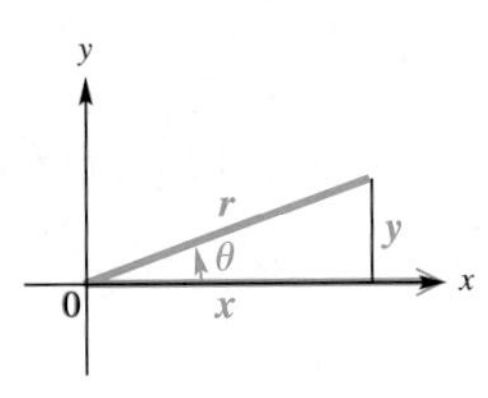


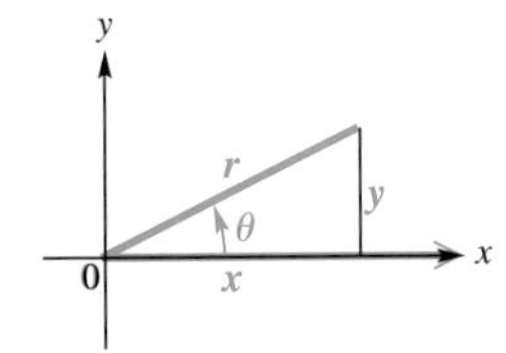


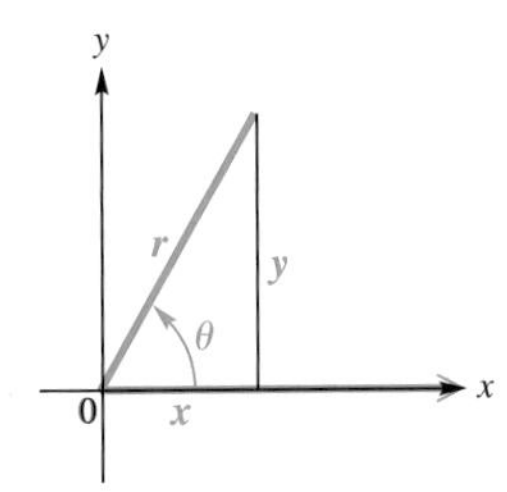


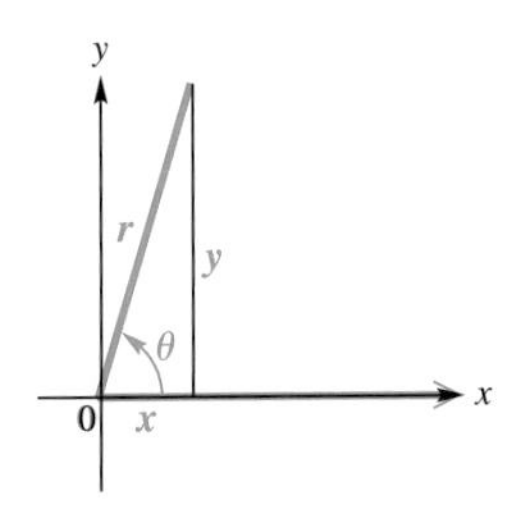


**FIGURE 35**

## EXAMPLE 6 *Identifying the Quadrant of an Angle*

Identify the quadrant (or quadrants) of any angle $\theta$ that satisfies $\sin \theta > 0$, $\tan \theta < 0$.

**Solution** Since $\sin \theta > 0$ in quadrants I and II, while $\tan \theta < 0$ in quadrants II and IV, both conditions are met only in quadrant II.

Figure 35 shows an angle $\theta$ as it increases in measure from near 0° toward 90°. In each case, the value of $r$ is the same. As the measure of the angle increases, $y$ increases but never exceeds $r$, so $y \leq r$. Dividing each side by the positive number $r$ gives $\frac{y}{r} \leq 1$.

In a similar way, angles in quadrant IV suggest that

$$-1 \leq \frac{y}{r},$$

so

$$-1 \leq \frac{y}{r} \leq 1.$$

Since $\frac{y}{r} = \sin \theta$,

$$-1 \leq \sin \theta \leq 1$$

for any angle $\theta$. In the same way,

$$-1 \leq \cos \theta \leq 1.$$

The tangent of an angle is defined as $\frac{y}{x}$. It is possible that $x < y$, $x = y$, or $x > y$. For this reason, $\frac{y}{x}$ can take any value, so $\tan \theta$ can be any real number, as can $\cot \theta$.

The functions $\sec \theta$ and $\csc \theta$ are reciprocals of the functions $\cos \theta$ and $\sin \theta$, respectively, making

$$\sec \theta \leq -1 \text{ or } \sec \theta \geq 1 \qquad \text{and} \qquad \csc \theta \leq -1 \text{ or } \csc \theta \geq 1.$$

In summary, the ranges of the trigonometric functions are as follows.

### Ranges of Trigonometric Functions

For any angle $\theta$ for which the indicated functions exist:

**1.** $-1 \leq \sin \theta \leq 1$ and $-1 \leq \cos \theta \leq 1$;

**2.** $\tan \theta$ and $\cot \theta$ may equal any real number;

**3.** $\sec \theta \leq -1$ or $\sec \theta \geq 1$ and $\csc \theta \leq -1$ or $\csc \theta \geq 1$.

(Notice that $\sec \theta$ and $\csc \theta$ are *never* between $-1$ and 1.)

## EXAMPLE 7 *Deciding Whether a Value Is in the Range of a Trigonometric Function*

Decide whether each statement is *possible* or *impossible*.

**(a)** $\sin \theta = \sqrt{8}$ **(b)** $\tan \theta = 110.47$ **(c)** $\sec \theta = .6$

**Solution**

**(a)** For any value of $\theta$, $-1 \leq \sin \theta \leq 1$. Since $\sqrt{8} > 1$, it is impossible to find a value of $\theta$ with $\sin \theta = \sqrt{8}$.

**(b)** Tangent can equal any value. Thus, $\tan \theta = 110.47$ is possible.

**(c)** Since $\sec \theta \leq -1$ or $\sec \theta \geq 1$, the statement $\sec \theta = .6$ is impossible.

## Pythagorean Identities

We derive three new identities from the relationship $x^2 + y^2 = r^2$. Dividing each side by $r^2$ gives

$$\frac{x^2}{r^2} + \frac{y^2}{r^2} = \frac{r^2}{r^2},$$

or

$$\left(\frac{x}{r}\right)^2 + \left(\frac{y}{r}\right)^2 = 1.$$

Since $\cos\theta = \frac{x}{r}$ and $\sin\theta = \frac{y}{r}$, this result becomes

$$(\cos\theta)^2 + (\sin\theta)^2 = 1,$$

or, as it is usually written,

$$\sin^2\theta + \cos^2\theta = 1.$$

Starting with $x^2 + y^2 = r^2$ and dividing through by $x^2$ gives

$$\frac{x^2}{x^2} + \frac{y^2}{x^2} = \frac{r^2}{x^2}$$

$$1 + \left(\frac{y}{x}\right)^2 = \left(\frac{r}{x}\right)^2$$

$$1 + (\tan\theta)^2 = (\sec\theta)^2$$

or

$$1 + \tan^2\theta = \sec^2\theta.$$

On the other hand, dividing through by $y^2$ leads to

$$1 + \cot^2\theta = \csc^2\theta.$$

These three identities are called the **Pythagorean identities** since the original equation that led to them, $x^2 + y^2 = r^2$, comes from the Pythagorean theorem.

TECHNOLOGY NOTE

Powers of trigonometric functions are not usually entered on calculators as they are written mathematically. For example, we write $\sin^2\theta$ to represent the square of $\sin\theta$, but this would be entered as $(\sin\theta)^2$ on many calculators.

**Pythagorean Identities**

$\boldsymbol{\sin^2\theta + \cos^2\theta = 1}$ $\qquad$ $\boldsymbol{1 + \tan^2\theta = \sec^2\theta}$ $\qquad$ $\boldsymbol{1 + \cot^2\theta = \csc^2\theta}$

As before, we have given only one form of each identity. However, algebraic transformations produce equivalent identities. For example, by subtracting $\sin^2\theta$ from each side of $\sin^2\theta + \cos^2\theta = 1$, we get the equivalent identity

$$\cos^2\theta = 1 - \sin^2\theta.$$

You should be able to transform these identities quickly, and also recognize their equivalent forms.

## Quotient Identities

Recall that $\sin\theta = \frac{y}{r}$ and $\cos\theta = \frac{x}{r}$. Consider the quotient of $\sin\theta$ and $\cos\theta$, where $\cos\theta \neq 0$.

$$\frac{\sin\theta}{\cos\theta} = \frac{\frac{y}{r}}{\frac{x}{r}} = \frac{y}{r} \div \frac{x}{r} = \frac{y}{r} \cdot \frac{r}{x} = \frac{y}{x} = \tan\theta$$

Similarly, $\frac{\cos\theta}{\sin\theta} = \cot\theta$, for $\sin\theta \neq 0$. Thus, we have the **quotient identities.**

**Looking Ahead to Calculus**

The reciprocal, Pythagorean, and quotient identities are used repeatedly in calculus to find limits, derivatives, and integrals of trigonometric functions. These identities are also used to rewrite expressions in a form that permits simplifying a square root. For example, if $a \geq 0$ and $x = a \sin \theta$,

$$\begin{aligned}\sqrt{a^2 - x^2} &= \sqrt{a^2 - a^2 \sin^2 \theta} \\ &= \sqrt{a^2(1 - \sin^2 \theta)} \\ &= \sqrt{a^2 \cos^2 \theta} \\ &= a|\cos \theta|.\end{aligned}$$

**Quotient Identities**

$$\frac{\sin \theta}{\cos \theta} = \tan \theta \qquad \frac{\cos \theta}{\sin \theta} = \cot \theta$$

**EXAMPLE 8** *Finding Other Function Values Given One Value and the Quadrant*

Find $\sin \theta$ and $\cos \theta$, if $\tan \theta = \frac{4}{3}$ and $\theta$ is in quadrant III.

**Solution** Since $\theta$ is in quadrant III, $\sin \theta$ and $\cos \theta$ will both be negative. It is tempting to say that since $\tan \theta = \frac{\sin \theta}{\cos \theta}$ and $\tan \theta = \frac{4}{3}$, then $\sin \theta = -4$ and $\cos \theta = -3$. This is *incorrect*, however, since both $\sin \theta$ and $\cos \theta$ must be in the interval $[-1, 1]$.

We use the Pythagorean identity $\tan^2 \theta + 1 = \sec^2 \theta$ to find $\sec \theta$, and then the reciprocal identity $\cos \theta = \frac{1}{\sec \theta}$ to find $\cos \theta$.

$$1 + \tan^2 \theta = \sec^2 \theta$$

$$1 + \left(\frac{4}{3}\right)^2 = \sec^2 \theta \qquad \tan \theta = \frac{4}{3}$$

$$1 + \frac{16}{9} = \sec^2 \theta$$

$$\frac{25}{9} = \sec^2 \theta$$

$$-\frac{5}{3} = \sec \theta \qquad \text{Choose the negative square root since } \sec \theta \text{ is negative when } \theta \text{ is in quadrant III.}$$

$$-\frac{3}{5} = \cos \theta \qquad \text{Secant and cosine are reciprocals.}$$

Since $\sin^2 \theta = 1 - \cos^2 \theta$,

$$\sin^2 \theta = 1 - \left(-\frac{3}{5}\right)^2 \qquad \cos \theta = -\frac{3}{5}$$

$$\sin^2 \theta = 1 - \frac{9}{25}$$

$$\sin^2 \theta = \frac{16}{25}$$

$$\sin \theta = -\frac{4}{5}. \qquad \text{Choose the negative square root.}$$

Therefore, we have $\sin \theta = -\frac{4}{5}$ and $\cos \theta = -\frac{3}{5}$.

**NOTE** Example 8 can also be worked by drawing $\theta$ in standard position in quadrant III, finding $r$ to be 5, and then using the definitions of $\sin \theta$ and $\cos \theta$ in terms of $x$, $y$, and $r$.

## 8.2 EXERCISES

*Sketch an angle $\theta$ in standard position such that $\theta$ has the smallest possible positive measure, and the given point is on the terminal side of $\theta$.*

**1.** $(5, -12)$

**2.** $(-12, -5)$

*Find the values of the six trigonometric functions for the angles in standard position having the following points on their terminal sides. Rationalize denominators when applicable.*

**3.** $(-3, 4)$ **4.** $(-4, -3)$ **5.** $(0, 2)$ **6.** $(-4, 0)$ **7.** $(1, \sqrt{3})$ **8.** $(-2\sqrt{3}, -2)$

**9.** For any nonquadrantal angle $\theta$, $\sin\theta$ and $\csc\theta$ will have the same sign. Explain why this is so.

**10.** *Concept Check* If the terminal side of an angle $\theta$ is in quadrant III, what is the sign of each of the trigonometric function values of $\theta$?

*Concept Check* *Suppose that the point $(x, y)$ is in the indicated quadrant. Decide whether the given ratio is positive or negative.* (Hint: *It may be helpful to draw a sketch.*)

**11.** II, $\frac{x}{r}$ **12.** III, $\frac{y}{r}$ **13.** IV, $\frac{y}{x}$ **14.** IV, $\frac{x}{y}$

*In Exercises 15–18, an equation with a restriction on x is given. This is an equation of the terminal side of an angle $\theta$ in standard position. Sketch the smallest positive such angle $\theta$, and find the values of the six trigonometric functions of $\theta$.*

**15.** $2x + y = 0, x \geq 0$ **16.** $3x + 5y = 0, x \geq 0$ **17.** $-6x - y = 0, x \leq 0$ **18.** $-5x - 3y = 0, x \leq 0$

**19.** Find the six trigonometric function values of the quadrantal angle $\theta = 540°$.

*Use the trigonometric function values of quadrantal angles given in this section to evaluate each expression. An expression such as* $\cot^2 90°$ *means* $(\cot 90°)^2$, *which is equal to* $0^2 = 0$.

**20.** $3 \sec 180° - 5 \tan 360°$

**21.** $4 \csc 270° + 3 \cos 180°$

**22.** $\tan 360° + 4 \sin 180° + 5 \cos^2 180°$

**23.** $2 \sec 0° + 4 \cot^2 90° + \cos 360°$

**24.** $\sin^2 180° + \cos^2 180°$

**25.** $\sin^2 360° + \cos^2 360°$

*If $n$ is an integer, $n \cdot 180°$ represents an integer multiple of $180°$, $(2n + 1) \cdot 90°$ represents an odd integer multiple of $90°$, and so on. Decide whether each expression is equal to 0, 1, $-1$, or is undefined.*

**26.** $\cos[(2n + 1) \cdot 90°]$ **27.** $\tan(n \cdot 180°)$ **28.** $\cos[(2n + 1) \cdot 180°]$ **29.** $\sin(n \cdot 180°)$

*Provide conjectures in Exercises 30–33.*

**30.** The angles 15° and 75° are complementary. With your calculator determine sin 15° and cos 75°. Make a conjecture about the sines and cosines of complementary angles, and test your hypothesis with other pairs of complementary angles. (*Note:* This relationship will be discussed in detail in the next section.)

**31.** The angles 25° and 65° are complementary. With your calculator determine tan 25° and cot 65°. Make a conjecture about the tangents and cotangents of complementary angles, and test your hypothesis with other pairs of complementary angles. (*Note:* This relationship will be discussed in detail in the next section.)

**32.** With your calculator determine sin 10° and sin(−10°). Make a conjecture about the sines of an angle and its negative, and test your hypothesis with other angles. Also, use a geometry argument with the definition of $\sin\theta$ to justify your hypothesis. (*Note:* This relationship will be discussed in detail in Section 9.1.)

**33.** With your calculator determine cos 20° and cos(−20°). Make a conjecture about the cosines of an angle and its negative, and test your hypothesis with other angles. Also, use a geometry argument with the definition of $\cos\theta$ to justify your hypothesis. (*Note:* This relationship will be discussed in detail in Section 9.1.)

*In Exercises 34–39, set your graphing calculator in parametric and degree modes. Set the window and functions as shown here, and graph. A circle of radius* 1 *will appear on the screen. Trace to move a short distance around the circle. In the screen, the point on the circle corresponds to an angle* T = 25°. *Since* $r = 1$, cos 25° *is* X = .90630779, *and* sin 25° *is* Y = .42261826.

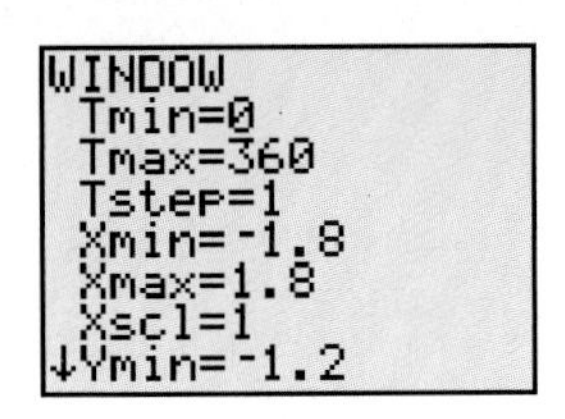


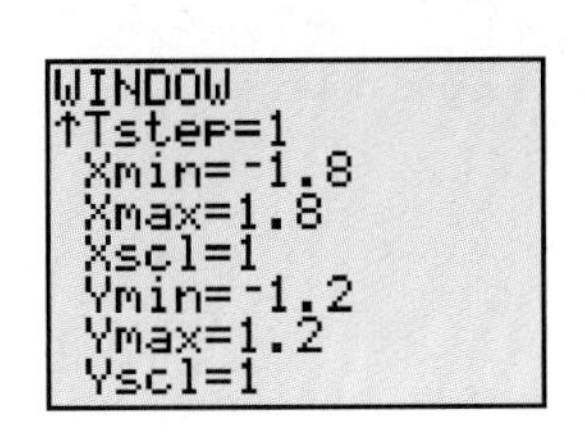


This screen is a continuation of the previous one.

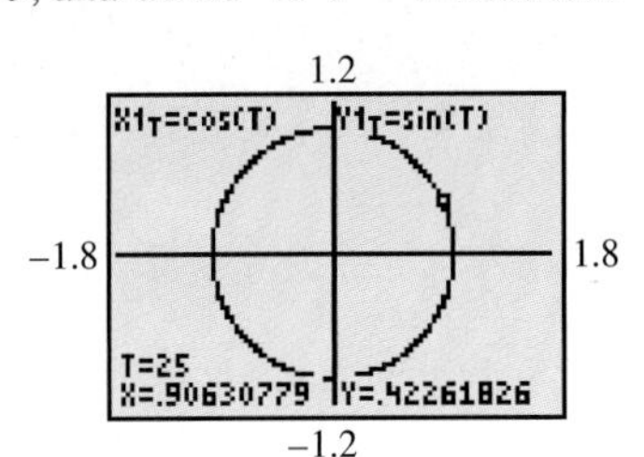


**34.** Use the right- and left-arrow keys to move to the point corresponding to 20°. What are cos 20° and sin 20°?

**35.** For what angle T, $0° \leq T \leq 90°$, is cos T ≈ .766?

**36.** For what angle T, $0° \leq T \leq 90°$, is sin T ≈ .574?

**37.** For what angle T, $0° \leq T \leq 90°$, does cos T = sin T?

**38.** As T increases from 0° to 90°, does the cosine increase or decrease? What about the sine?

**39.** As T increases from 90° to 180°, does the cosine increase or decrease? What about the sine?

**40.** *Concept Check* What positive number $a$ is its own reciprocal? Find a value of $\theta$ for which $\sin \theta = \csc \theta = a$.

**41.** *Concept Check* What negative number $a$ is its own reciprocal? Find a value of $\theta$ for which $\cos \theta = \sec \theta = a$.

*Use the appropriate reciprocal identity to find each function value. Rationalize denominators when applicable. In Exercises 46 and 47, use a calculator.*

**42.** $\cos \theta$, if $\sec \theta = -2.5$

**43.** $\cot \theta$, if $\tan \theta = -\frac{1}{5}$

**44.** $\sin \theta$, if $\csc \theta = \sqrt{15}$

**45.** $\tan \theta$, if $\cot \theta = -\frac{\sqrt{5}}{3}$

**46.** $\sin \theta$, if $\csc \theta = 1.42716321$

**47.** $\cos \theta$, if $\sec \theta = 9.80425133$

**48.** Can a given angle $\theta$ satisfy both $\sin \theta > 0$ and $\csc \theta < 0$? Explain.

**49.** Explain what is wrong with the following item that appears on a trigonometry test:

Find $\sec \theta$, given that $\cos \theta = \frac{3}{2}$.

*Find the tangent of each angle.*

**50.** $\cot \theta = -3$

**51.** $\cot \theta = \frac{\sqrt{3}}{3}$

**52.** $\cot \theta = .4$

*Find a value of each variable that makes the statement true.*

**53.** $\tan(3\theta - 4°) = \dfrac{1}{\cot(5\theta - 8°)}$

**54.** $\sec(2\theta + 6°) \cos(5\theta + 3°) = 1$

*Identify the quadrant or quadrants for the angle satisfying the given conditions.*

**55.** $\sin \theta > 0$; $\cos \theta < 0$

**56.** $\cos \theta > 0$; $\tan \theta > 0$

**57.** $\tan \theta > 0$; $\cot \theta > 0$

**58.** $\tan \theta < 0$; $\cot \theta < 0$

*Give the signs of the sine, cosine, and tangent functions for each angle.*

**59.** 129°

**60.** 183°

**61.** 298°

**62.** 412°

**63.** −82°

**64.** −121°

*Concept Check* *Without using a calculator, decide which is greater.*

**65.** sin 30° or tan 30°

**66.** sin 20° or sin 21°

**67.** sin 33° or sec 33°

*Decide whether each statement is* possible *or* impossible.

**68.** $\sin \theta = 2$

**69.** $\cos \theta = -1.001$

**70.** $\tan \theta = .92$

**71.** $\cot \theta = -12.1$

**72.** $\sec \theta = 1$

**73.** $\tan \theta = 1$

**74.** $\sin \theta = \frac{1}{2}$ and $\csc \theta = 2$

**75.** $\tan \theta = 2$ and $\cot \theta = -2$

*Use identities to find each function value. Use a calculator in Exercises 82 and 83.*

**76.** $\tan \theta$, if $\sec \theta = 3$, with $\theta$ in quadrant IV

**77.** $\sin \theta$, if $\cos \theta = -\frac{1}{4}$, with $\theta$ in quadrant II

**78.** $\csc \theta$, if $\cot \theta = -\frac{1}{2}$, with $\theta$ in quadrant IV

**79.** $\sec \theta$, if $\tan \theta = \frac{\sqrt{7}}{3}$, with $\theta$ in quadrant III

**80.** $\cos \theta$, if $\csc \theta = -4$, with $\theta$ in quadrant III

**81.** $\sin \theta$, if $\sec \theta = 2$, with $\theta$ in quadrant IV

**82.** $\cot \theta$, if $\csc \theta = -3.5891420$, with $\theta$ in quadrant III

**83.** $\tan \theta$, if $\sin \theta = .49268329$, with $\theta$ in quadrant II

*In Exercises 84 and 85, each graphing calculator screen is obtained for a particular stored value of* X. *What will the screen display for the value of the expression in the final line of the display?*

**84.**

```
sin(X)
                .8
(cos(X))²
```

**85.**

```
tan(X)
                 2
(1/cos(X))²
```

**86.** Concept Check Does there exist an angle $\theta$ with $\cos \theta = -.6$ and $\sin \theta = .8$?

*Find all trigonometric function values for each angle. Use a calculator in Exercises 91 and 92.*

**87.** $\tan \theta = -\frac{15}{8}$, with $\theta$ in quadrant II

**88.** $\cos \theta = -\frac{3}{5}$, with $\theta$ in quadrant III

**89.** $\tan \theta = \sqrt{3}$, with $\theta$ in quadrant III

**90.** $\sin \theta = \frac{\sqrt{5}}{7}$, with $\tan \theta > 0$

**91.** $\cot \theta = -1.49586$, with $\theta$ in quadrant IV

**92.** $\sin \theta = .164215$, with $\theta$ in quadrant II

*Work each problem.*

**93.** Derive the identity $1 + \cot^2 \theta = \csc^2 \theta$ by dividing $x^2 + y^2 = r^2$ by $y^2$.

**94.** Using a method similar to the one given in this section showing that $\frac{\sin \theta}{\cos \theta} = \tan \theta$, show that $\frac{\cos \theta}{\sin \theta} = \cot \theta$.

**95.** Concept Check True or false: For all angles $\theta$, $\sin \theta + \cos \theta = 1$. If false, give an example showing why it is false.

**96.** Concept Check True or false: Since $\cot \theta = \frac{\cos \theta}{\sin \theta}$, if $\cot \theta = \frac{1}{2}$ with $\theta$ in quadrant I, then $\cos \theta = 1$ and $\sin \theta = 2$. If false, explain why.

## Reviewing Basic Concepts (Sections 8.1 and 8.2)

**1.** Give the complement and the supplement of each angle.

**(a)** $35°$ **(b)** $\frac{\pi}{4}$

**2.** Convert $32.25°$ to degrees, minutes, and seconds.

**3.** Convert $59° 35' 30''$ to decimal degrees.

4. Find the angle of smallest positive measure coterminal with each angle.

   **(a)** $560°$  **(b)** $-\frac{2\pi}{3}$

5. Convert each angle measure as directed.

   **(a)** $240°$ to radians  **(b)** $\frac{3\pi}{4}$ radians to degrees

6. Suppose a circle has radius 3 centimeters.

   **(a)** What is the exact length of an arc intercepted by a central angle of $120°$?

   **(b)** What is the exact area of the sector formed in part (a)?

7. An angle $\theta$ is in standard position and its terminal side passes through the point $(-2, 5)$. Find the exact value of each of the six trigonometric functions of $\theta$.

8. Find the exact values of the trigonometric functions of a $270°$ angle.

9. Decide whether the following statements are *possible* or *impossible*.

   **(a)** $\cos\theta = \frac{3}{2}$  **(b)** $\tan\theta = 300$  **(c)** $\csc\theta = 5$

10. If $\theta$ is a quadrant III angle with $\sin\theta = -\frac{2}{3}$, use identities to find the other trigonometric function values of $\theta$.

# 8.3 EVALUATING TRIGONOMETRIC FUNCTIONS

Definitions of the Trigonometric Functions • Trigonometric Function Values of Special Angles • Cofunction Identities • Reference Angles • Special Angles as Reference Angles • Finding Function Values with a Calculator • Finding Angle Measures

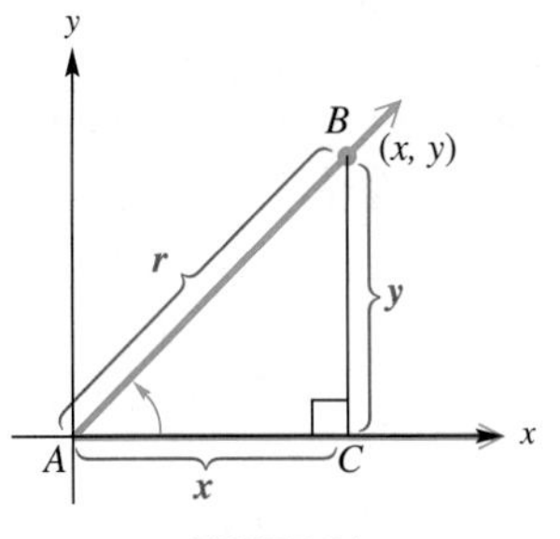


**FIGURE 36**

## Definitions of the Trigonometric Functions

In Section 8.2 we used angles in standard position to define the trigonometric functions. There is another way to approach them: as ratios of the sides of right triangles. Figure 36 shows an acute angle $A$ in standard position. The definitions of the trigonometric function values of angle $A$ require $x$, $y$, and $r$. As drawn in Figure 36, $x$ and $y$ are the lengths of the two legs of the right triangle $ABC$, and $r$ is the length of the hypotenuse.

The side of length $y$ is called the **side opposite** angle $A$, and the side of length $x$ is called the **side adjacent** to angle $A$. We use the lengths of these sides to replace $x$ and $y$ in the definitions of the trigonometric functions, and the length of the hypotenuse to replace $r$, to get the following right-triangle-based definitions.

**Right-Triangle-Based Definitions of Trigonometric Functions**

For any acute angle $A$ in standard position,

$$\sin A = \frac{y}{r} = \frac{\text{side opposite}}{\text{hypotenuse}} \qquad \csc A = \frac{r}{y} = \frac{\text{hypotenuse}}{\text{side opposite}}$$

$$\cos A = \frac{x}{r} = \frac{\text{side adjacent}}{\text{hypotenuse}} \qquad \sec A = \frac{r}{x} = \frac{\text{hypotenuse}}{\text{side adjacent}}$$

$$\tan A = \frac{y}{x} = \frac{\text{side opposite}}{\text{side adjacent}} \qquad \cot A = \frac{x}{y} = \frac{\text{side adjacent}}{\text{side opposite}}.$$

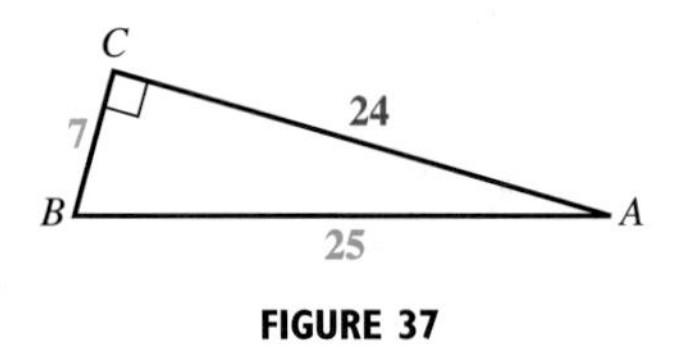

FIGURE 37

**EXAMPLE 1** *Finding Trigonometric Function Values of an Acute Angle*
Find the values of sin $A$, cos $A$, and tan $A$ in the right triangle in Figure 37.

**Solution** The length of the side opposite angle $A$ is 7, the length of the side adjacent to angle $A$ is 24, and the length of the hypotenuse is 25. Use the relationships given in the box.

$$\sin A = \frac{\text{side opposite}}{\text{hypotenuse}} = \frac{7}{25} \qquad \cos A = \frac{\text{side adjacent}}{\text{hypotenuse}} = \frac{24}{25} \qquad \tan A = \frac{\text{side opposite}}{\text{side adjacent}} = \frac{7}{24}$$

**NOTE** Because the cosecant, secant, and cotangent ratios are the reciprocals of the sine, cosine, and tangent values, in Example 1 we can conclude that $\csc A = \frac{25}{7}$, $\sec A = \frac{25}{24}$, and $\cot A = \frac{24}{7}$.

## Trigonometric Function Values of Special Angles

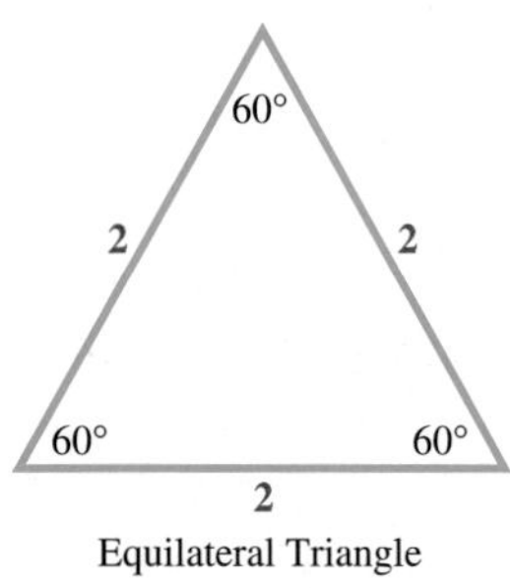

Equilateral Triangle

(a)

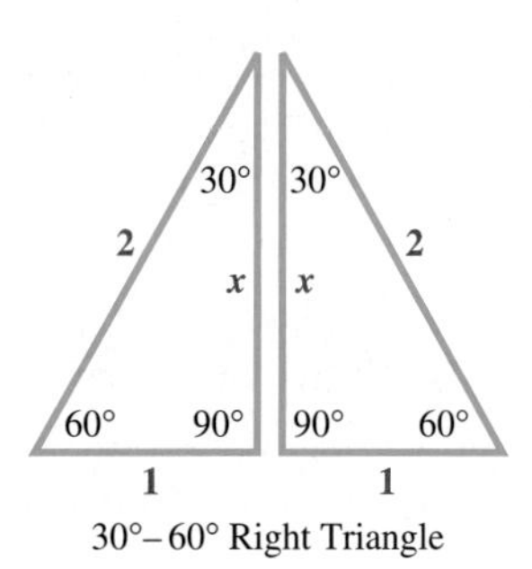

30°–60° Right Triangle

(b)

FIGURE 38

Certain special angles, such as 30°, 45°, and 60°, occur so often in trigonometry and in more advanced mathematics that they deserve special study. We start with an equilateral triangle, a triangle with all sides of equal length. Each angle of such a triangle measures 60°. While the results we will obtain are independent of the length, for convenience we choose the length of each side to be 2 units. See Figure 38(a).

Bisecting one angle of this equilateral triangle leads to two right triangles, each of which has angles of 30°, 60°, and 90°, as shown in Figurc 38(b). Since the hypotenuse of one of these right triangles has length 2, the shortest side will have length 1. (Why?) If $x$ represents the length of the medium side, then, by the Pythagorean theorem,

$$\begin{aligned} 2^2 &= 1^2 + x^2 \\ 4 &= 1 + x^2 \\ 3 &= x^2 \\ \sqrt{3} &= x. \end{aligned}$$

Choose the positive root. (Why?)

Figure 39 summarizes our results using a 30°–60° right triangle. As shown in the figure, the side opposite the 30° angle has length 1; that is, for the 30° angle,

$$\text{hypotenuse} = 2, \quad \text{side opposite} = 1, \quad \text{side adjacent} = \sqrt{3}.$$

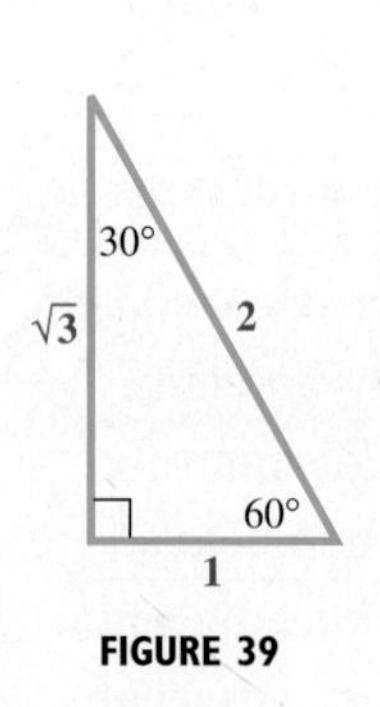

FIGURE 39

Now we use the definitions of the trigonometric functions.

$$\sin 30° = \frac{\text{side opposite}}{\text{hypotenuse}} = \frac{1}{2} \qquad \csc 30° = \frac{2}{1} = 2$$

$$\cos 30° = \frac{\text{side adjacent}}{\text{hypotenuse}} = \frac{\sqrt{3}}{2} \qquad \sec 30° = \frac{2}{\sqrt{3}} = \frac{2\sqrt{3}}{3}$$

$$\tan 30° = \frac{\text{side opposite}}{\text{side adjacent}} = \frac{1}{\sqrt{3}} = \frac{\sqrt{3}}{3} \qquad \cot 30° = \frac{\sqrt{3}}{1} = \sqrt{3}$$

**EXAMPLE 2** *Finding Trigonometric Function Values for 60°*
Find the six trigonometric function values for a 60° angle.

**Solution** Refer to Figure 39 to find the following ratios.

$$\sin 60^\circ = \frac{\sqrt{3}}{2} \qquad \cos 60^\circ = \frac{1}{2} \qquad \tan 60^\circ = \sqrt{3}$$

$$\csc 60^\circ = \frac{2\sqrt{3}}{3} \qquad \sec 60^\circ = 2 \qquad \cot 60^\circ = \frac{\sqrt{3}}{3}.$$

45°–45° Right Triangle

**FIGURE 40**

We find the values of the trigonometric functions for 45° by starting with a 45°–45° right triangle, as shown in Figure 40. This triangle is isosceles, and for convenience, we choose the lengths of the equal sides to be 1 unit. (As before, the results are independent of the length of the equal sides of the right triangle.) Since the shorter sides each have length 1, if $r$ represents the length of the hypotenuse, then

$$1^2 + 1^2 = r^2$$
$$2 = r^2$$
$$\sqrt{2} = r.$$

Now we use the measures indicated on the 45°–45° right triangle in Figure 40.

$$\sin 45^\circ = \frac{1}{\sqrt{2}} = \frac{\sqrt{2}}{2} \qquad \cos 45^\circ = \frac{1}{\sqrt{2}} = \frac{\sqrt{2}}{2} \qquad \tan 45^\circ = \frac{1}{1} = 1$$

$$\csc 45^\circ = \frac{\sqrt{2}}{1} = \sqrt{2} \qquad \sec 45^\circ = \frac{\sqrt{2}}{1} = \sqrt{2} \qquad \cot 45^\circ = \frac{1}{1} = 1$$

The function values for $30^\circ = \frac{\pi}{6}$, $45^\circ = \frac{\pi}{4}$, and $60^\circ = \frac{\pi}{3}$ are summarized in the table that follows.

**TECHNOLOGY NOTE**

Using a graphing calculator to find the decimal value for cos 30° gives a decimal approximation for $\frac{\sqrt{3}}{2}$. The decimal is an approximation, while $\frac{\sqrt{3}}{2}$ is exact. In general, unless a trigonometric function value is a rational number, a graphing calculator will only provide a decimal approximation for the irrational function value. (The TI-92 is an exception.)

**Function Values for Special Angles**

| $\theta$ | $\sin\theta$ | $\cos\theta$ | $\tan\theta$ | $\cot\theta$ | $\sec\theta$ | $\csc\theta$ |
|---|---|---|---|---|---|---|
| $30^\circ = \frac{\pi}{6}$ | $\frac{1}{2}$ | $\frac{\sqrt{3}}{2}$ | $\frac{\sqrt{3}}{3}$ | $\sqrt{3}$ | $\frac{2\sqrt{3}}{3}$ | 2 |
| $45^\circ = \frac{\pi}{4}$ | $\frac{\sqrt{2}}{2}$ | $\frac{\sqrt{2}}{2}$ | 1 | 1 | $\sqrt{2}$ | $\sqrt{2}$ |
| $60^\circ = \frac{\pi}{3}$ | $\frac{\sqrt{3}}{2}$ | $\frac{1}{2}$ | $\sqrt{3}$ | $\frac{\sqrt{3}}{3}$ | 2 | $\frac{2\sqrt{3}}{3}$ |

## Cofunction Identities

In a right triangle $ABC$, with right angle $C$, the acute angles $A$ and $B$ are complementary. See right triangle $ABC$ in Figure 41. The length of the side opposite angle $A$ is $a$, and the length of the side opposite angle $B$ is $b$. The length of the hypotenuse is $c$. In this triangle, $\sin A = \frac{a}{c}$. Since $\cos B$ is also equal to $\frac{a}{c}$,

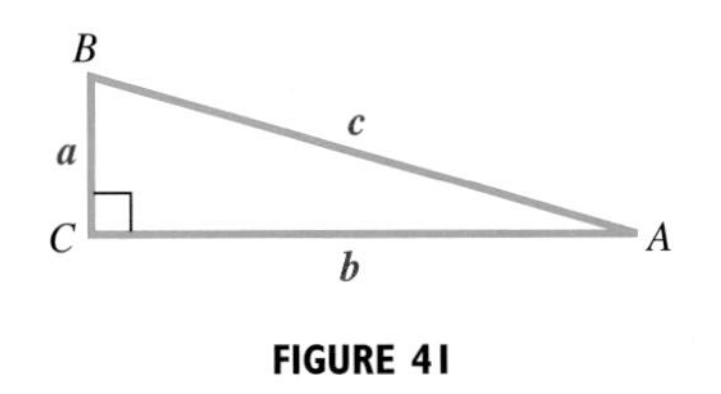


**FIGURE 41**

$$\sin A = \frac{a}{c} = \cos B.$$

In a similar manner,

$$\tan A = \frac{a}{b} = \cot B \quad \text{and} \quad \sec A = \frac{c}{b} = \csc B.$$

Since angles $A$ and $B$ are *complementary* and $\sin A = \cos B$, the functions *sine* and *cosine* are called **cofunctions.** Also, tangent and cotangent are cofunctions, as are secant and cosecant. Since angles $A$ and $B$ are complementary, $A + B = 90°$, or $B = 90° - A$, giving $\sin A = \cos B = \cos(90° - A)$.

Similar results, called the **cofunction identities,** are true for the other trigonometric functions.

**Cofunction Identities**

If $A$ is an acute angle measured in degrees, then

$$\sin A = \cos(90° - A) \quad \csc A = \sec(90° - A) \quad \tan A = \cot(90° - A)$$
$$\cos A = \sin(90° - A) \quad \sec A = \csc(90° - A) \quad \cot A = \tan(90° - A).$$

If $A$ is an acute angle measured in radians, then

$$\sin A = \cos\left(\frac{\pi}{2} - A\right) \qquad \csc A = \sec\left(\frac{\pi}{2} - A\right)$$
$$\cos A = \sin\left(\frac{\pi}{2} - A\right) \qquad \sec A = \csc\left(\frac{\pi}{2} - A\right)$$
$$\tan A = \cot\left(\frac{\pi}{2} - A\right) \qquad \cot A = \tan\left(\frac{\pi}{2} - A\right).$$

**NOTE** These identities actually apply to all angles (not just acute angles). However, for our present discussion we will only need them for acute angles.

**EXAMPLE 3** *Writing Functions in Terms of Cofunctions*

Write each expression in terms of its cofunction.

**(a)** $\cos 52° 16'$ **(b)** $\tan \frac{\pi}{6}$

**Solution**

**(a)** Since $\cos A = \sin(90° - A)$,

$$\cos 52° 16' = \sin(90° - 52° 16') = \sin 37° 44'.$$

**(b)** Since $\tan A = \cot\left(\frac{\pi}{2} - A\right)$,

$$\tan \frac{\pi}{6} = \cot\left(\frac{\pi}{2} - \frac{\pi}{6}\right) = \cot \frac{\pi}{3}.$$

## Reference Angles

Associated with every nonquadrantal angle in standard position is a positive acute angle called its *reference angle*. A **reference angle** for an angle $\theta$, written $\theta'$, is the positive acute angle made by the terminal side of angle $\theta$ and the $x$-axis. Figure 42 shows several angles $\theta$ (each less than one complete counterclockwise revolution) in quadrants II, III, and IV, with the reference angle $\theta'$ also shown. In quadrant I, $\theta$ and

$\theta'$ are the same. If an angle $\theta$ is negative or has measure greater than 360°, its reference angle is found by first finding its coterminal angle that is between 0° and 360°, and then using the diagrams in Figure 42.

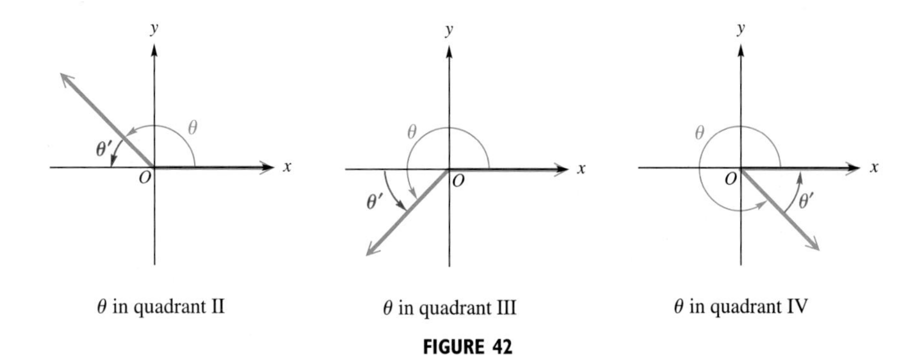


FIGURE 42

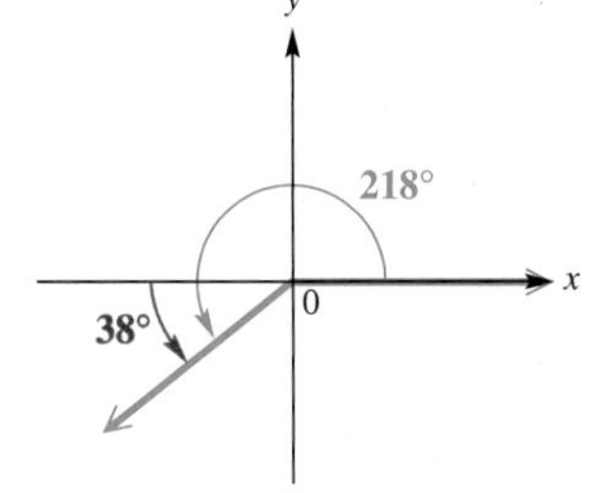


$218° - 180° = 38°$

FIGURE 43

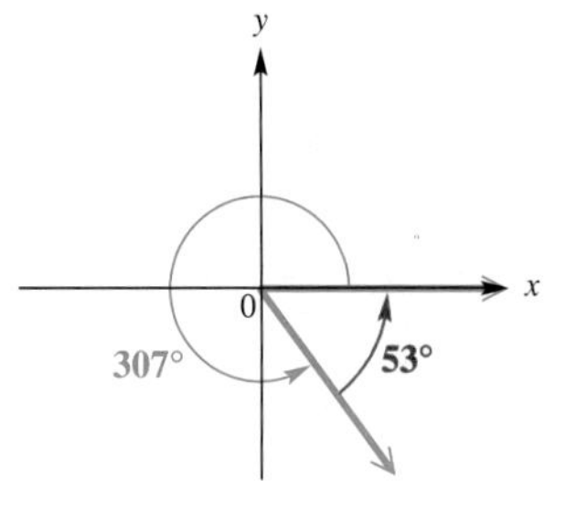


$360° - 307° = 53°$

FIGURE 44

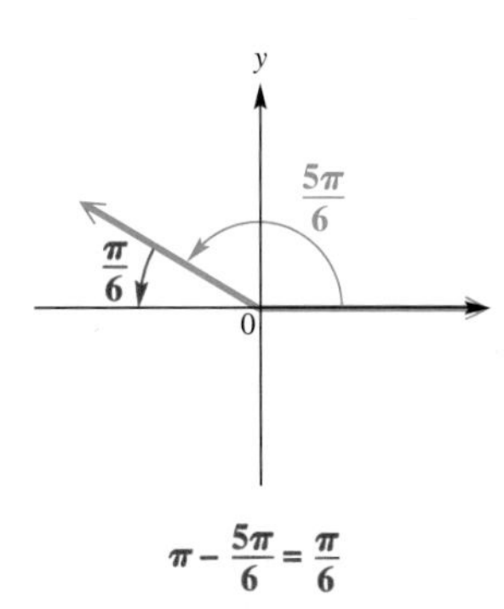


$\pi - \frac{5\pi}{6} = \frac{\pi}{6}$

FIGURE 45

**CAUTION** A common error is to find the reference angle by using the terminal side of $\theta$ and the $y$-axis. *The reference angle is always found with reference to the x-axis.*

**EXAMPLE 4** *Finding Reference Angles*

Find the reference angle for each angle.

**(a)** 218° **(b)** 1387° **(c)** $\frac{5\pi}{6}$

**Solution**

**(a)** As shown in Figure 43, the positive acute angle made by the terminal side of this angle and the $x$-axis is $218° - 180° = 38°$. For $\theta = 218°$, the reference angle is $\theta' = 38°$.

**(b)** First find a coterminal angle to 1387° between 0° and 360°. Divide 1387° by 360° to get a quotient of about 3.9. Begin by subtracting 360° three times (because of the 3 in 3.9): $1387° - 3 \cdot 360° = 307°$. The reference angle for 307° (and thus for 1387°) is $360° - 307° = 53°$. See Figure 44.

**(c)** The reference angle is $\pi - \frac{5\pi}{6} = \frac{\pi}{6}$. See Figure 45.

## Special Angles as Reference Angles

We can now find exact trigonometric function values of angles with reference angles of 30°, 60°, or 45°.

**EXAMPLE 5** *Finding Trigonometric Function Values of a Quadrant III Angle*

Find the values of the trigonometric functions for 210°.

**Solution** An angle of 210° is shown in Figure 46 on the next page. The reference angle is $210° - 180° = 30°$. To find the trigonometric function values of 210°, choose point $P$ on the terminal side of the angle so that the distance from the origin $O$ to $P$ is 2. By the results from 30°–60° right triangles, the coordinates of point $P$ become

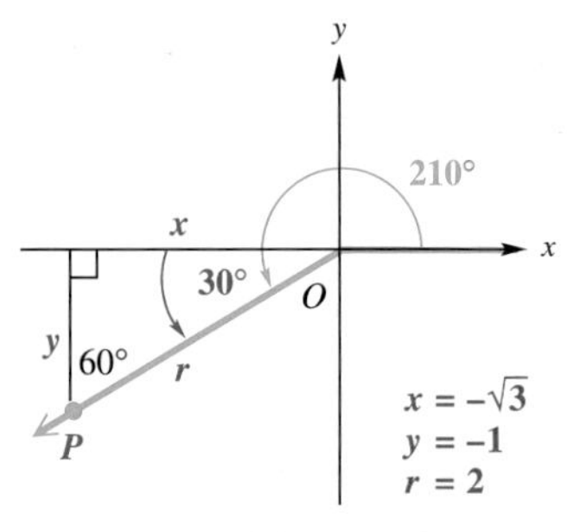


FIGURE 46

$(-\sqrt{3}, -1)$, with $x = -\sqrt{3}$, $y = -1$, and $r = 2$. Then, by the definitions of the trigonometric functions,

$$\sin 210° = -\frac{1}{2} \qquad \cos 210° = -\frac{\sqrt{3}}{2} \qquad \tan 210° = \frac{\sqrt{3}}{3}$$

$$\csc 210° = -2 \qquad \sec 210° = -\frac{2\sqrt{3}}{3} \qquad \cot 210° = \sqrt{3}.$$

Notice in Example 5 that the trigonometric function values of 210° correspond in absolute value to those of its reference angle 30°. The signs are different for the sine, cosine, secant, and cosecant functions because 210° is a quadrant III angle. These results suggest a shortcut for finding the trigonometric function values of a nonacute angle, using the reference angle. In Example 5, the reference angle for 210° is 30°. Using the trigonometric function values of 30°, and choosing the correct signs for a quadrant III angle, we obtain the same results found in Example 5.

Similarly, the values of the trigonometric functions for any nonquadrantal angle $\theta$ can be determined by finding the function values for an angle between 0° and 90°.

**Finding Trigonometric Function Values for Any Nonquadrantal Angle**

1. If $\theta > 360°$, or if $\theta < 0°$, find a coterminal angle by adding or subtracting 360° as many times as needed to get an angle greater than 0° but less than 360°.
2. Find the reference angle $\theta'$.
3. Find the trigonometric function values for reference angle $\theta'$.
4. Determine the correct signs for the values found in Step 3. (Use the table of signs in Section 8.2 if necessary.) This gives the values of the trigonometric functions for angle $\theta$.

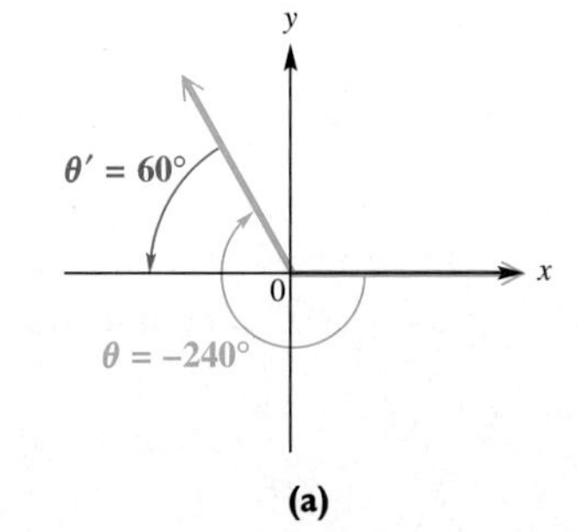


(a)

(b)

FIGURE 47

**EXAMPLE 6** *Finding Trigonometric Function Values Using Reference Angles*

Find the exact value of each expression.

**(a)** $\cos(-240°)$ **(b)** $\tan 675°$

**Solution**

**(a)** Since an angle of $-240°$ is coterminal with an angle of $-240° + 360° = 120°$, the reference angle is $180° - 120° = 60°$, as shown in Figure 47(a). Since the cosine is negative in quadrant II,

$$\cos(-240°) = \cos 120° = -\cos 60° = -\frac{1}{2}.$$

**(b)** Begin by subtracting 360° to get a coterminal angle between 0° and 360°.

$$675° - 360° = 315°$$

As shown in Figure 47(b), the reference angle is $360^\circ - 315^\circ = 45^\circ$. An angle of $315^\circ$ is in quadrant IV, so the tangent will be negative, and

$$\tan 675^\circ = \tan 315^\circ = -\tan 45^\circ = -1.$$

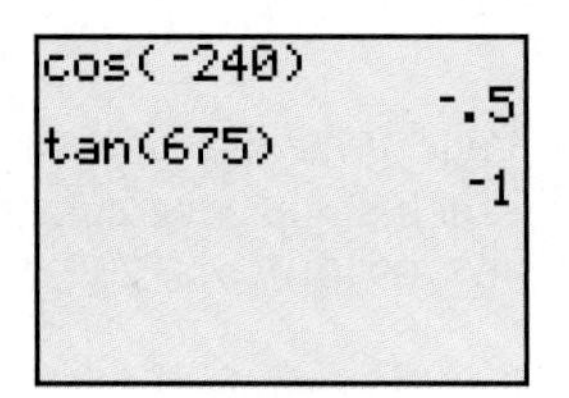


Degree Mode

**FIGURE 48**

## Finding Function Values with a Calculator

Calculators are capable of finding trigonometric function values. For example, the values of $\cos(-240^\circ)$ and $\tan 675^\circ$, found analytically in Example 6, are found with a calculator as shown in Figure 48.

**CAUTION** We know that angles can be measured in either degrees or radians. When evaluating trigonometric functions of angles given in degrees, remember that the calculator must be set in *degree mode*.

When exact values cannot be found, the calculator will provide an approximation.

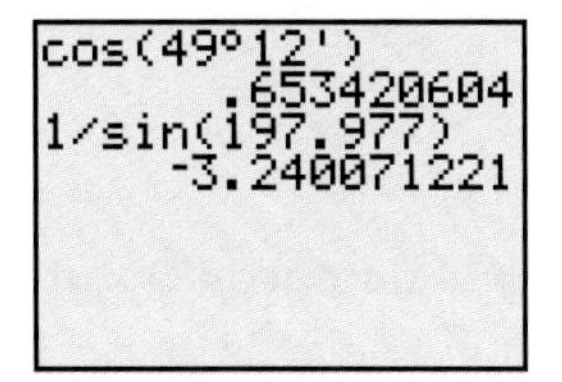


Degree Mode

**(a)**

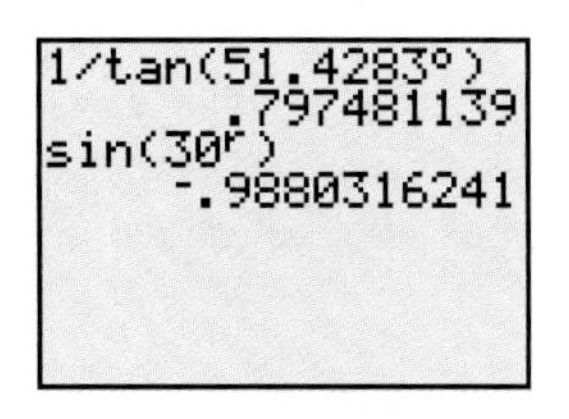


**(b)**

**FIGURE 49**

**EXAMPLE 7** *Finding Trigonometric Function Values with a Calculator*

Approximate the value of each expression.

**(a)** $\cos 49^\circ 12'$ **(b)** $\csc 197.977^\circ$ **(c)** $\cot 51.4283^\circ$ **(d)** $\sin 30$

**Solution** Figure 49(a) shows how to approximate the expressions in parts (a) and (b). The calculator is in degree mode. Notice that to find a cosecant value, we take the reciprocal of the sine value.

Figure 49(b) shows how to approximate the expressions in parts (c) and (d). Here, we use the degree symbol (°) and the radian symbol ($^r$), which overrides the mode of the calculator.

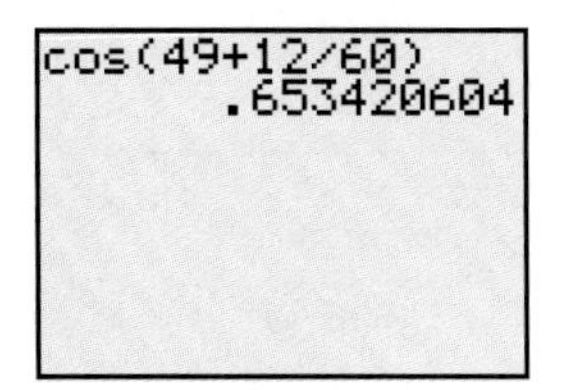


Degree Mode

**FIGURE 50**

**NOTE** An alternative method of finding a value such as $\cos 49^\circ 12'$ is to use the fact that there are 60 minutes in one degree. Thus, as shown in Figure 50,

$$\cos 49^\circ 12' = \cos\left(49 + \frac{12}{60}\right)^\circ \approx .653420604.$$

## Finding Angle Measures

Sometimes it is necessary to find an angle measure when we know a trigonometric function value of that angle and information about the quadrant in which the angle lies. A complete discussion of inverse trigonometric functions is necessary to understand *why* a calculator gives an inverse function value in a particular quadrant. We examine inverse trigonometric functions in detail in the next chapter. For now, we introduce using inverse trigonometric functions with calculators. (You should locate these functions on your particular model. They are usually designated $\sin^{-1}$, $\cos^{-1}$, and $\tan^{-1}$.)

**TECHNOLOGY NOTE**

Students often confuse the symbols for the inverse trigonometric functions with the reciprocal functions. For example, $\sin^{-1} x$ represents an angle whose sine is $x$, not the reciprocal of $\sin x$ (which is $\csc x$). To find reciprocal function values, use the function together with the reciprocal function of the calculator.

**EXAMPLE 8** *Using Inverse Trigonometric Functions to Find Angles*

**(a)** Use a calculator to find an angle $\theta$ in degrees that satisfies $\sin \theta \approx .9677091705$.

**(b)** Use a calculator to find an angle $\theta$ in radians that satisfies $\tan \theta \approx .25$.

```
sin⁻¹(.9677091705
)
                75.4
tan⁻¹(.25)
      .2449786631
```

FIGURE 51

**Solution**

**(a)** With the calculator in degree mode, we find that an angle $\theta$ having sine value .9677091705 is 75.4°. (While there are infinitely many such angles, the calculator only gives this one.) We write this result as $\sin^{-1} .9677091705 \approx 75.4°$. See the first calculation in Figure 51.

**(b)** With the calculator in radian mode, we find $\tan^{-1} .25 \approx .2449786631$. See the second calculation in Figure 51.

## EXAMPLE 9 *Finding Angle Measures*

Find all values of $\theta$, if $\theta$ is in the interval $[0°, 360°)$ and $\cos \theta = -\frac{\sqrt{2}}{2}$.

**Analytic Solution**

Since cosine is negative, $\theta$ must lie in quadrant II or III. Since the absolute value of $\cos \theta$ is $\frac{\sqrt{2}}{2}$, the reference angle $\theta'$ must be $\cos^{-1} \frac{\sqrt{2}}{2} = 45°$. The two possible angles $\theta$ are sketched in Figure 52.

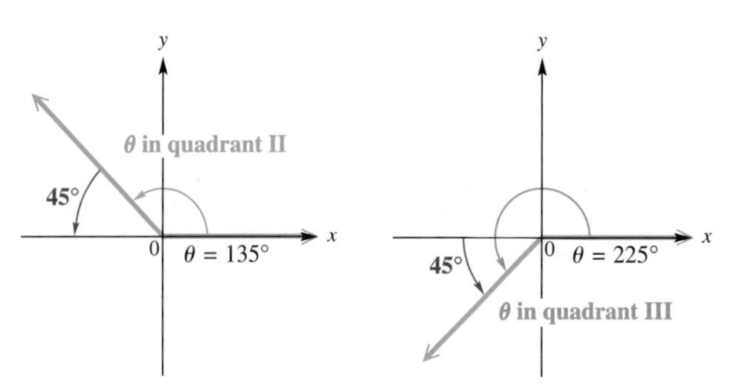


FIGURE 52

The quadrant II angle $\theta$ must equal $180° - 45° = 135°$, and the quadrant III angle $\theta$ must equal $180° + 45° = 225°$.

**Graphing Calculator Solution**

The screen in Figure 53 shows how the inverse cosine function is used to find the two values in $[0°, 360°)$ for which $\cos \theta = -\frac{\sqrt{2}}{2}$. Notice that $\cos^{-1}\left(-\frac{\sqrt{2}}{2}\right)$ yields only one value, 135°; to find the other value, we must use the reference angle concept.

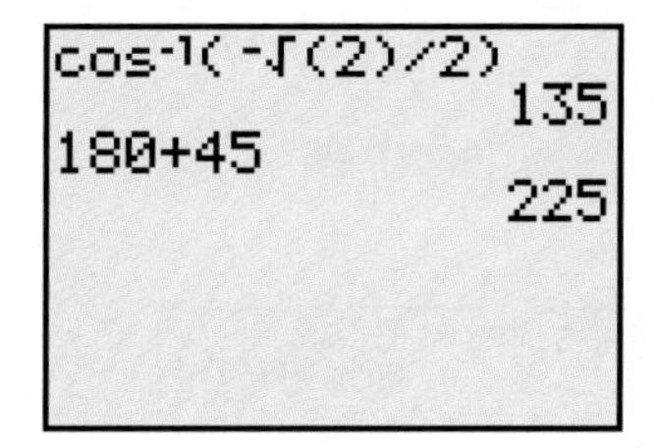


Degree Mode

FIGURE 53

## EXAMPLE 10 *Finding Angle Measures*

Find two angles in the interval $[0, 2\pi)$ that satisfy $\cos \theta \approx .3623577545$.

**Solution** With the calculator in radian mode, we find that one such $\theta$ is

$$\cos^{-1} .3623577545 \approx 1.2.$$

Since $1.2 < \frac{\pi}{2}$, $\theta$ is in quadrant I. We must find another value of $\theta$ that satisfies the given condition. This other value of $\theta$ will have its reference angle $\theta'$ equal to 1.2 and must be in quadrant IV, since the angle given by the calculator is in quadrant I and cosine is also positive in quadrant IV. The other value of $\theta$ is

$$2\pi - 1.2 \approx 5.083185307.$$

See Figure 54, where the last two lines are a check.

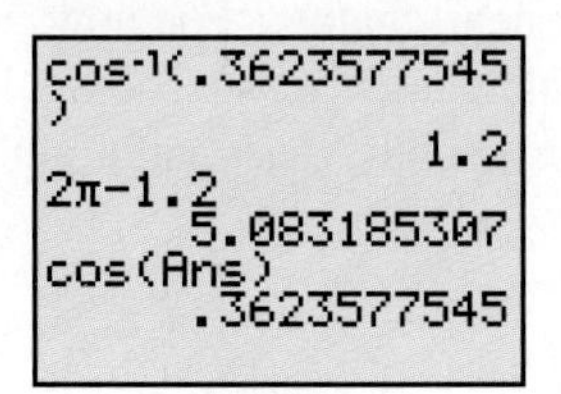


Radian Mode

FIGURE 54

### What Went WRONG?

A student was asked to find cot 20°. He set his calculator in degree mode and, knowing that cotangent is the reciprocal of tangent, produced the screen shown here. His teacher, however, told him that his answer was wrong, because he was confused about notation. The correct answer is cot 20° ≈ 2.747477419.

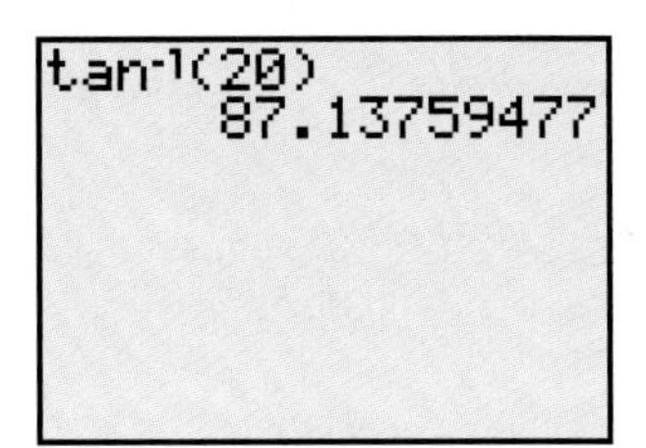


What Went WRONG? How can he obtain the correct answer?

## 8.3 EXERCISES

*Find the exact values or expressions for the six trigonometric functions of angle A.*

**1.**

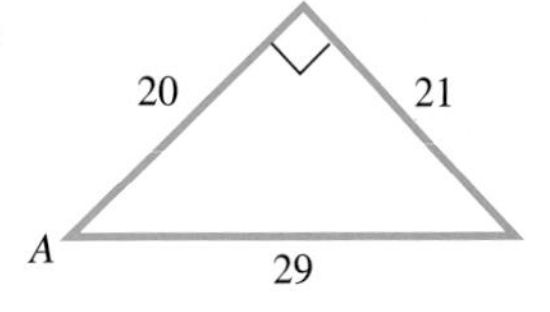


**2.**

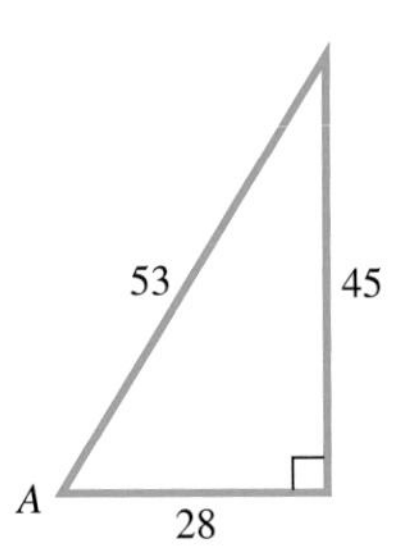


**3.**

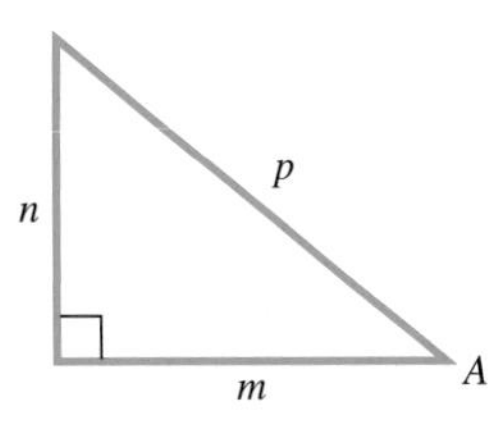


**4.**

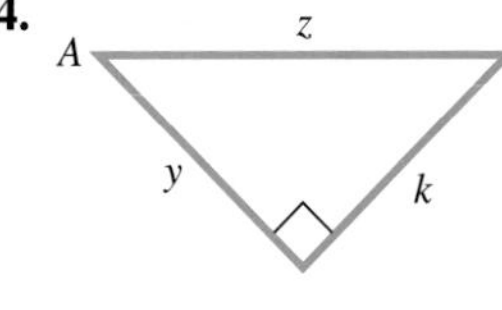


*For each expression,* **(a)** *give the exact value and* **(b)** *if the exact value is irrational, use your calculator to support your answer in part (a) by finding a decimal approximation.*

**5.** tan 30° **6.** cot 30° **7.** sin 30° **8.** cos 30°

**9.** sec 30° **10.** csc 30° **11.** csc 45° **12.** sec 45°

**13.** cos 45° **14.** cot 45° **15.** $\sin \frac{\pi}{3}$ **16.** $\cos \frac{\pi}{3}$

**17.** $\tan \frac{\pi}{3}$ **18.** $\cot \frac{\pi}{3}$ **19.** $\sec \frac{\pi}{3}$ **20.** $\csc \frac{\pi}{3}$

---

*Answer to What Went Wrong?*

The student used the *inverse tangent function* ($\tan^{-1}$) rather than finding the *reciprocal* of the tangent of 20°. If he enters 1/tan(20°), he will obtain the correct answer.

**21.** A student was asked to give the exact value of sin 45°. Using his calculator, he gave the answer .7071067812. The teacher did not give him credit. Why?

**22.** A student was asked to give an approximate value of sin 45. With her calculator in degree mode, she gave the value .7071067812. The teacher did not give her credit. What was her error?

*Write each expression in terms of its cofunction.*

**23.** cot 73° **24.** sec 39° **25.** sin 38° **26.** cos 19° **27.** tan 25° 43′

**28.** sin 38° 29′ **29.** $\cos \frac{\pi}{5}$ **30.** $\sin \frac{\pi}{3}$ **31.** tan .5 **32.** csc .3

*Give the reference angle for each angle measure.*

**33.** 98° **34.** 212° **35.** 230° **36.** 130° **37.** −135° **38.** −60°

**39.** 750° **40.** 480° **41.** $\frac{4\pi}{3}$ **42.** $\frac{7\pi}{6}$ **43.** $-\frac{4\pi}{3}$ **44.** $-\frac{7\pi}{6}$

**45.** *Concept Check* In Example 5, why was 2 a good choice for $r$? Could any other positive number be used?

**46.** Explain how the reference angle is used to find values of the trigonometric functions for an angle in quadrant III.

*Complete the table with* exact *trigonometric function values using the methods of this section. Do* not *use a calculator.*

| | $\theta$ | $\sin\theta$ | $\cos\theta$ | $\tan\theta$ | $\cot\theta$ | $\sec\theta$ | $\csc\theta$ |
|---|---|---|---|---|---|---|---|
| **47.** | 30° | $\frac{1}{2}$ | $\frac{\sqrt{3}}{2}$ | | | $\frac{2\sqrt{3}}{3}$ | 2 |
| **48.** | 45° | | | 1 | 1 | | |
| **49.** | 60° | | $\frac{1}{2}$ | $\sqrt{3}$ | | 2 | |
| **50.** | 120° | $\frac{\sqrt{3}}{2}$ | | $-\sqrt{3}$ | | | $\frac{2\sqrt{3}}{3}$ |
| **51.** | 135° | $\frac{\sqrt{2}}{2}$ | $-\frac{\sqrt{2}}{2}$ | | | $-\sqrt{2}$ | $\sqrt{2}$ |
| **52.** | 150° | | $-\frac{\sqrt{3}}{2}$ | $-\frac{\sqrt{3}}{3}$ | | | 2 |
| **53.** | 210° | $-\frac{1}{2}$ | | $\frac{\sqrt{3}}{3}$ | $\sqrt{3}$ | | −2 |
| **54.** | 240° | $-\frac{\sqrt{3}}{2}$ | $-\frac{1}{2}$ | | | −2 | $-\frac{2\sqrt{3}}{3}$ |

*Find the* exact *values of the six trigonometric functions for each angle.*

**55.** 300° **56.** 315° **57.** 405° **58.** 420°

**59.** $\frac{11\pi}{6}$ **60.** $\frac{5\pi}{3}$ **61.** $-\frac{7\pi}{4}$ **62.** $-\frac{4\pi}{3}$

*Use a calculator to find a decimal approximation for each value. Give as many digits as your calculator displays.*

**63.** tan 29° **64.** sin 38° **65.** cot 41° 24′ **66.** csc 145° 45′

**67.** sec 183° 48′ **68.** cos 421° 30′ **69.** tan(−80° 6′) **70.** sin(−317° 36′)

**71.** sin 2.5 **72.** cos 3.8 **73.** tan 5 **74.** sec 10

*For each expression,* **(a)** *write the function in terms of a function of the reference angle,* **(b)** *give the exact value, and* **(c)** *use a calculator to show that the decimal value or approximation for the given function is the same as the decimal value or approximation for your answer in part (b).*

**75.** $\sin \frac{7\pi}{6}$ **76.** $\cos \frac{5\pi}{3}$ **77.** $\tan \frac{3\pi}{4}$ **78.** $\sin \frac{5\pi}{3}$ **79.** $\cos \frac{7\pi}{6}$ **80.** $\tan \frac{4\pi}{3}$

*Find all values of θ, if θ is in the interval* $[0°, 360°)$ *and has the given function value.*

**81.** $\sin \theta = \frac{1}{2}$ **82.** $\cos \theta = \frac{\sqrt{3}}{2}$ **83.** $\tan \theta = -\sqrt{3}$

**84.** $\sec \theta = -\sqrt{2}$ **85.** $\cot \theta = -\frac{\sqrt{3}}{3}$ **86.** $\cos \theta = \frac{\sqrt{2}}{2}$

*Find all values of θ, if θ is in the interval* $[0°, 360°)$ *and has the given function value. Give approximations to as many decimal places as your calculator displays.*

**87.** $\cos \theta \approx .68716510$ **88.** $\cos \theta \approx .96476120$ **89.** $\sin \theta \approx .41298643$

**90.** $\sin \theta \approx .63898531$ **91.** $\tan \theta \approx .87692035$ **92.** $\tan \theta \approx 1.2841996$

*Find two angles in the interval* $[0, 2\pi)$ *that satisfy the given equation. Give calculator approximations to as many digits as your calculator displays.*

**93.** $\tan \theta \approx .21264138$ **94.** $\cos \theta \approx .78269876$ **95.** $\cot \theta \approx .29949853$ **96.** $\csc \theta \approx 1.0219553$

## RELATING CONCEPTS

**For individual or group investigation (Exercises 97–108)**

*In a square window of your calculator that gives a good picture of the first quadrant, graph the line* $y = \sqrt{3}x$ *with* $x \geq 0$*. Then, trace to any point on the line. See the figure. What we see is a simulated view of an angle in standard position, with terminal side in quadrant I. Store the values of x and y in convenient memory locations. For these exercises, we call them* $x_1$ *and* $y_1$*. Work the exercises in order.*

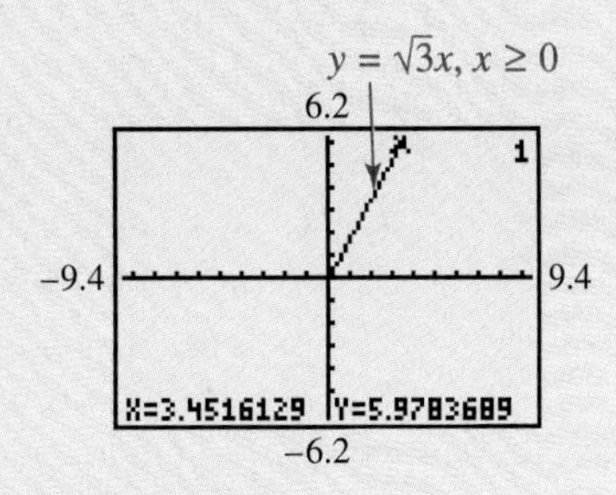


**97.** Find the value of $\sqrt{x_1^2 + y_1^2}$ and store it in a convenient memory location. (Call it $r$.) What does this number mean geometrically?

**98.** With your calculator in degree mode, find $\tan^{-1}\left(\frac{y_1}{x_1}\right)$.

**99.** With your calculator in degree mode, find $\sin^{-1}\left(\frac{y_1}{r}\right)$.

**100.** With your calculator in degree mode, find $\cos^{-1}\left(\frac{x_1}{r}\right)$.

**101.** Your answers in Exercises 98–100 should all be the same. How does this answer relate to the angle formed on your screen?

**102.** Find the value of $\frac{y_1}{x_1}$. Now, square it. What do you get? What is the exact value of $\frac{y_1}{x_1}$?

**103.** Look at the equation of the line you graphed, and make a conjecture: The ________ of a line passing through the origin is equal to the ________ of the angle it forms with the positive $x$-axis.

**104.** Find the value of

$$\left(\frac{x_1}{r}\right)^2 + \left(\frac{y_1}{r}\right)^2.$$

What identity does this illustrate?

**105.** Find csc 60°. Then, find the value of $\frac{r}{y_1}$. Do they agree?

**106.** Graph $y_2 = \sqrt{1 - x^2}$ as a second curve in the same viewing window. This is one-half of a circle centered at the origin with radius 1. Now, use the intersection feature of your calculator to determine the $x$- and $y$-coordinates of the points of intersection of the two graphs. What are they?

**107.** Find a calculator value for cos 60°. How does it compare to the $x$-coordinate of the point you found in Exercise 106? Why is this so?

**108.** Find a calculator approximation for sin 60°. How does it compare to the $y$-coordinate of the point you found in Exercise 106? Why is this so?

*(Modeling) Speed of Light When a light ray travels from one medium, such as air, to another medium, such as water or glass, the speed of the light changes, and the direction in which the ray is traveling changes. (This is why a fish under water is in a different position than it appears to be.) These changes are given by Snell's law*

$$\frac{c_1}{c_2} = \frac{\sin \theta_1}{\sin \theta_2},$$

*where $c_1$ is the speed of light in the first medium, $c_2$ is the speed of light in the second medium, and $\theta_1$ and $\theta_2$ are the angles shown in the figure. (Source:* The Physics Classroom, www.glenbrook.k12.il.us*) In Exercises 109–114, assume that $c_1 = 3 \times 10^8$ meters per second.*

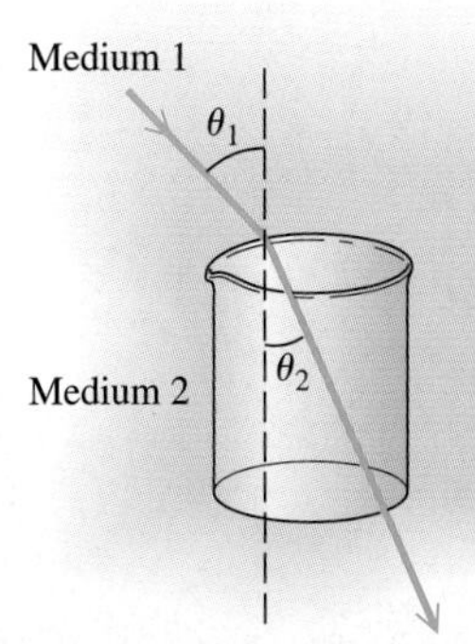


*Find the speed of light in the second medium.*

**109.** $\theta_1 = 46°$; $\theta_2 = 31°$

**110.** $\theta_1 = 39°$; $\theta_2 = 28°$

*Find $\theta_2$ for the following values of $\theta_1$ and $c_2$. Round to the nearest degree.*

**111.** $\theta_1 = 40°$; $c_2 = 1.5 \times 10^8$ meters per second

**112.** $\theta_1 = 62°$; $c_2 = 2.6 \times 10^8$ meters per second

*(Modeling) Fish's View of the World The figure shows a fish's view of the world above the surface of the water. (Source:* Walker, Jearl, "The Amateur Scientist," *Scientific American,* March 1984.*) Suppose that a light ray comes from the horizon, enters the water, and strikes the fish's eye.*

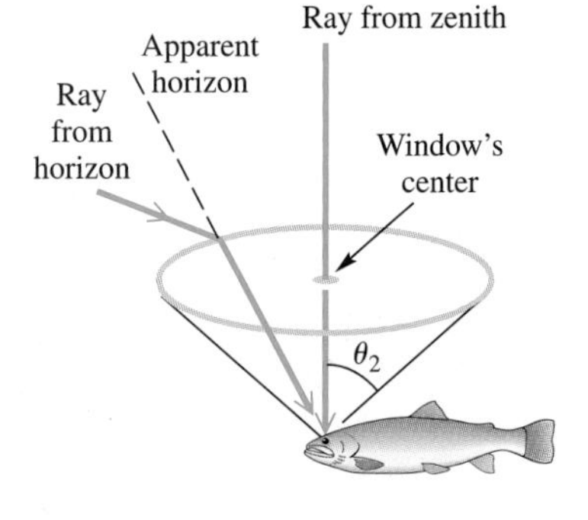


**113.** Assume that this ray gives a value of 90° for angle $\theta_1$ in the formula for Snell's law. (In a practical situation, this angle would probably be a little less than 90°.) The speed of light in water is about $2.254 \times 10^8$ meters per second. Find angle $\theta_2$.

**114.** Suppose an object is located at a true angle of 29.6° above the horizon. Find the apparent angle above the horizon to a fish.

**115.** *(Modeling) Braking Distance* If air resistance is ignored, the braking distance $D$ (in feet) for an automobile to change its velocity from $V_1$ to $V_2$ (feet per second) can be modeled using the equation

$$D = \frac{1.05(V_1^2 - V_2^2)}{64.4(K_1 + K_2 + \sin \theta)}.$$

$K_1$ is a constant determined by the efficiency of the brakes and tires, $K_2$ is a constant determined by the rolling resistance of the automobile, and $\theta$ is the grade of the highway. (*Source:* Mannering, F. and W. Kilareski, *Principles of Highway Engineering and Traffic Analysis,* Second Edition, John Wiley and Sons, 1998.)

**(a)** Compute the number of feet required to slow a car from 55 to 30 mph while traveling uphill with a grade of $\theta = 3.5°$. Let $K_1 = .4$ and $K_2 = .02$. (*Hint:* Change miles per hour to feet per second.)

**(b)** Repeat part (a) with $\theta = -2°$.

**(c)** How is braking distance affected by the grade $\theta$? Does this agree with your driving experience?

**116.** *(Modeling) Car's Speed at Collision* Refer to Exercise 115. An automobile is traveling at 90 mph on a highway with a downhill grade of $\theta = -3.5°$. The driver sees a stalled truck in the road 200 feet away and immediately applies the brakes. Assuming that a collision cannot be avoided, how fast (in miles per hour) is the car traveling when it hits the truck? (Use the same values for $K_1$ and $K_2$ as in Exercise 115.)

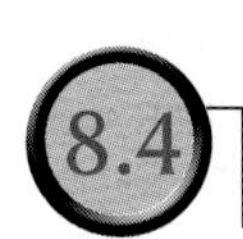

## 8.4 APPLICATIONS OF RIGHT TRIANGLES

Significant Digits • Solving Triangles • Angles of Elevation or Depression • Bearing • Further Applications

### Significant Digits

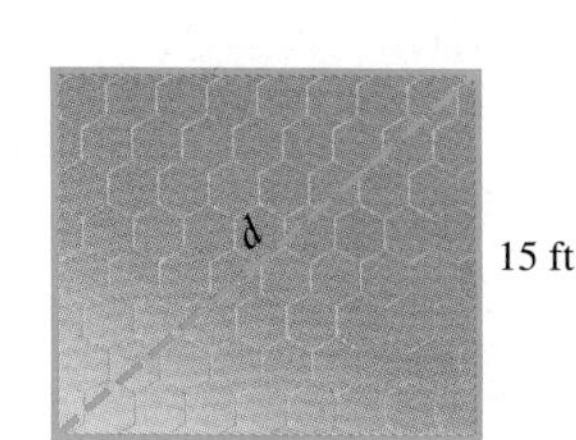


**FIGURE 55**

Suppose we quickly measure a room as 15 feet by 18 feet. See Figure 55. To calculate the length of a diagonal of the room, we use the Pythagorean theorem.

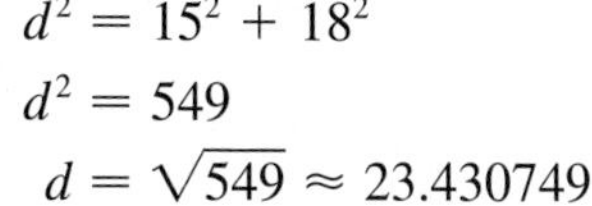

$$d^2 = 15^2 + 18^2$$
$$d^2 = 549$$
$$d = \sqrt{549} \approx 23.430749$$

Should this answer be given as the length of the diagonal of the room? Of course not. The number 23.430749 contains six decimal places, while the original data of 15 feet and 18 feet are only accurate to the nearest foot. Since the results of a problem can be no more accurate than the least accurate number in any calculation, we really should say that the diagonal of the 15- by 18-foot room is about 23 feet.

If a wall measured to the nearest foot is 18 feet long, this actually means that the wall has length between 17.5 feet and 18.5 feet. If the wall is measured more accurately as 18.3 feet long, then its length is really between 18.25 feet and 18.35 feet. A measurement of 18.00 feet would indicate that the length of the wall is between 17.995 feet and 18.005 feet. The measurement 18 feet is said to have two *significant digits* of accuracy; 18.0 has three significant digits, and 18.00 has four.

A **significant digit** is a digit obtained by actual measurement. A number that represents the result of counting, or a number that results from theoretical work and is not the result of a measurement, is an **exact number.** For example, there are 50 states in the United States, so 50 is an exact number.

Most values of trigonometric functions are approximations, and virtually all measurements are approximations. When performing calculations involving approximate numbers, start by determining the number that has the least number of significant digits. Round your final answer to the same number of significant digits as this number. Remember that *your answer is no more accurate than the least accurate number in your calculation.*

Use the following table to determine the significant digits in angle measure.

**Significant Digits for Angles**

| Number of Significant Digits | Angle Measure to Nearest: |
|---|---|
| 2 | Degree |
| 3 | Ten minutes, or nearest tenth of a degree |
| 4 | Minute, or nearest hundredth of a degree |
| 5 | Tenth of a minute, or nearest thousandth of a degree |

For example, an angle measuring $52° \, 30'$ has three significant digits (assuming that $30'$ is measured to the nearest ten minutes).

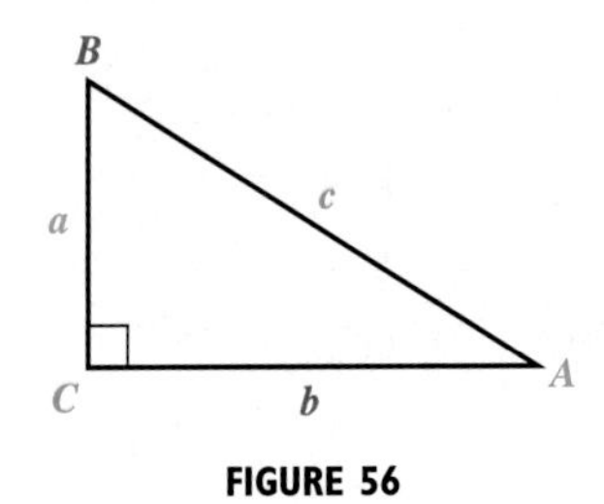

**FIGURE 56**

## Solving Triangles

To *solve a triangle* means to find the measures of all the angles and sides of the triangle. When solving triangles, a labeled sketch is an important aid. As shown in Figure 56, we use $a$ to represent the length of the side opposite angle $A$, $b$ for the length of the side opposite angle $B$, and so on. In a right triangle, the letter $c$ is reserved for the hypotenuse.

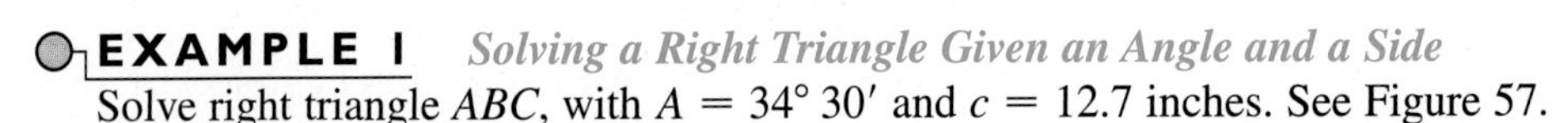

**EXAMPLE 1** *Solving a Right Triangle Given an Angle and a Side*

Solve right triangle $ABC$, with $A = 34° \, 30'$ and $c = 12.7$ inches. See Figure 57.

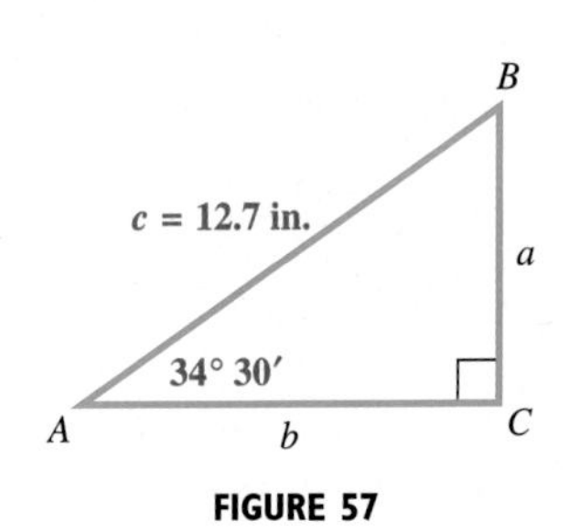

**FIGURE 57**

**Solution** To solve the triangle, find the measures of the remaining sides and angles. To find the value of $a$, use a trigonometric function involving the known values of angle $A$ and side $c$. Since the sine of angle $A$ is given by the quotient of the side opposite $A$ and the hypotenuse, use $\sin A$.

$$\sin A = \frac{a}{c}$$

$$\sin 34° \, 30' = \frac{a}{12.7} \qquad A = 34° \, 30', \, c = 12.7$$

$$a = 12.7 \sin 34° \, 30' \qquad \text{Multiply by 12.7.}$$

$$a \approx 12.7(.56640624) \qquad \text{Use a calculator.}$$

$$a \approx 7.19 \text{ inches} \qquad \text{Three significant digits}$$

To find the value of $b$, we could use the Pythagorean theorem. It is better, however, to use the information given in the problem rather than a result just calculated. If a mistake is made in finding $a$, then $b$ also would be incorrect. Also, rounding more than once may cause the result to be less accurate. Use $\cos A$.

$$\cos A = \frac{b}{c}$$

$$\cos 34° \, 30' = \frac{b}{12.7}$$

$$b = 12.7 \cos 34° \, 30'$$

$$b \approx 10.5 \text{ inches}$$

**Looking Ahead to Calculus**

The derivatives of the parametric equations $x = f(t)$ and $y = g(t)$ often represent the rate of change of physical quantities, such as velocities. In such cases, the derivatives are called *related rates* because a change in one causes a related change in the other. Determining these rates in calculus often requires solving right triangles. Many problems that require the maximum or minimum value of some quantity also involve solving a right triangle.

TECHNOLOGY NOTE

Once you have mastered the material on solving right triangles, you may wish to write a program that will accomplish this goal. You will need to consider the various cases of what is given and what must be found.

Once $b$ is found, the Pythagorean theorem can be used as a check. All that remains to solve triangle $ABC$ is to find the measure of angle $B$. Since $A = 34° 30'$ and $A + B = 90°$,

$$B = 90° - A$$
$$B = 89° 60' - 34° 30'$$
$$B = 55° 30'.$$

**NOTE** In Example 1, we could have found the measure of angle $B$ first, and then used the trigonometric function values of $B$ to find the unknown sides. The process of solving a right triangle (like many problems in mathematics) can usually be done in several ways, each producing the correct answer. To maintain accuracy, always use given information as much as possible, and avoid rounding off in intermediate steps.

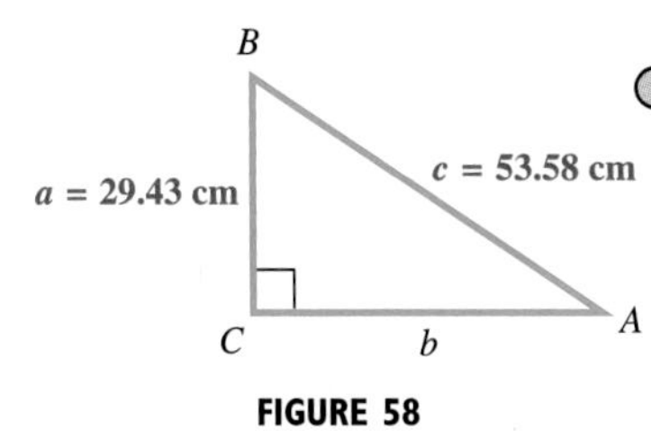

FIGURE 58

**EXAMPLE 2** *Solving a Right Triangle Given Two Sides*

Solve right triangle $ABC$ if $a = 29.43$ centimeters and $c = 53.58$ centimeters.

**Solution** We draw a sketch showing the given information, as in Figure 58. One way to begin is to find angle $A$ by using the sine function.

$$\sin A = \frac{\text{side opposite}}{\text{hypotenuse}} = \frac{29.43}{53.58}$$

Using the inverse sine function on a calculator, we find that $A \approx 33.32°$. The measure of $B$ is approximately $90° - 33.32° = 56.68°$.

We now find $b$ from the Pythagorean theorem, using $a^2 + b^2 = c^2$, or $b^2 = c^2 - a^2$. Since $c = 53.58$ and $a = 29.43$,

$$b^2 = 53.58^2 - 29.43^2$$
$$b \approx 44.77 \text{ centimeters.}$$

## Angles of Elevation or Depression

Many applications of right triangles involve angles of elevation or depression. The **angle of elevation** from point $X$ to point $Y$ (above $X$) is the acute angle formed by ray $XY$ and a horizontal ray with endpoint at $X$. See Figure 59(a). The **angle of depression** from point $X$ to point $Y$ (below $X$) is the acute angle formed by ray $XY$ and a horizontal ray with endpoint $X$. See Figure 59(b).

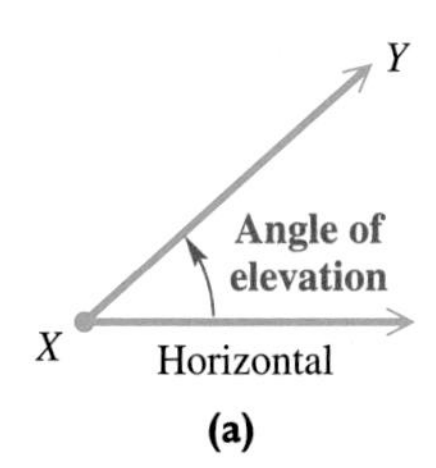

(a)

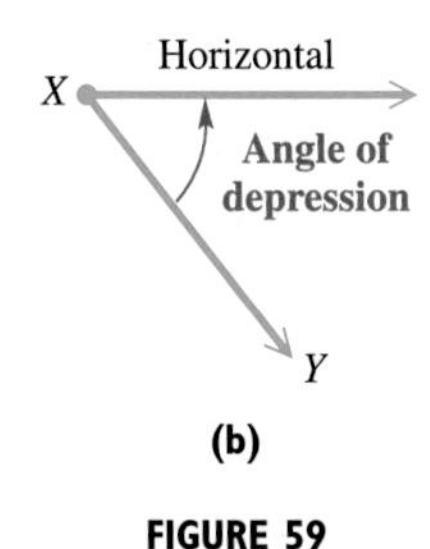

(b)

FIGURE 59

**CAUTION** Be careful when interpreting the angle of depression. Both the angle of elevation *and* the angle of depression are measured between the line of sight and the *horizontal.*

To solve applied trigonometry problems, follow the same procedure as solving a triangle.

**Solving an Applied Trigonometry Problem**

1. Draw a sketch, and label it with the given information. Label the quantity to be found with a variable.
2. Use the sketch to write an equation relating the given quantities to the variable.
3. Solve the equation, and check that your answer makes sense.

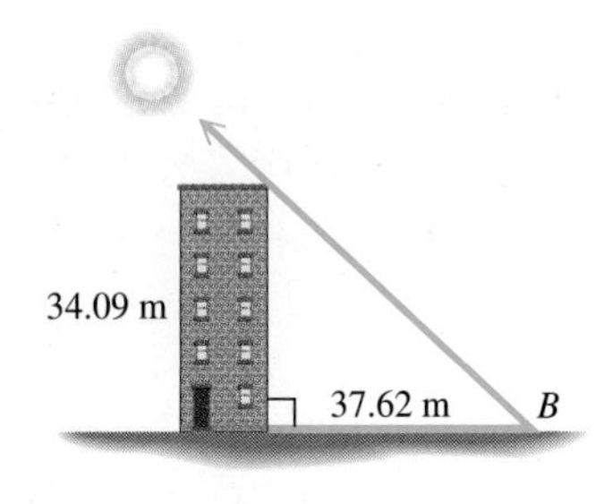


FIGURE 60

**EXAMPLE 3** *Finding the Angle of Elevation When Lengths Are Known*

The length of the shadow of a building 34.09 meters tall is 37.62 meters. Find the angle of elevation of the sun.

**Solution** As shown in Figure 60, the angle of elevation of the sun is angle $B$. Since the side opposite $B$ and the side adjacent to $B$ are known, use the tangent ratio to find $B$.

$$\tan B = \frac{34.09}{37.62}, \quad \text{so } B = \tan^{-1}\frac{34.09}{37.62} \approx 42.18°$$

The angle of elevation of the sun is 42.18°.

## Bearing

Other applications of right triangles involve **bearing,** an important idea in navigation. There are two methods for expressing bearing. When a single angle is given, such as 164°, it is understood that the bearing is measured in a clockwise direction from due north. Several sample bearings using this first method are shown in Figure 61.

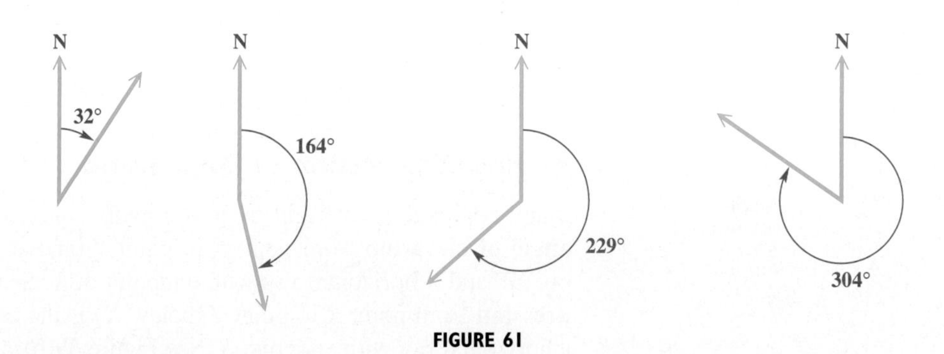


FIGURE 61

FIGURE 62

**EXAMPLE 4** *Solving a Problem Involving Bearing (First Method)*

Radar stations $A$ and $B$ are on an east–west line, with $A$ west of $B$, 3.7 kilometers apart. Station $A$ detects a plane at $C$, on a bearing of 61°. Station $B$ simultaneously detects the same plane, on a bearing of 331°. Find the distance from $A$ to $C$.

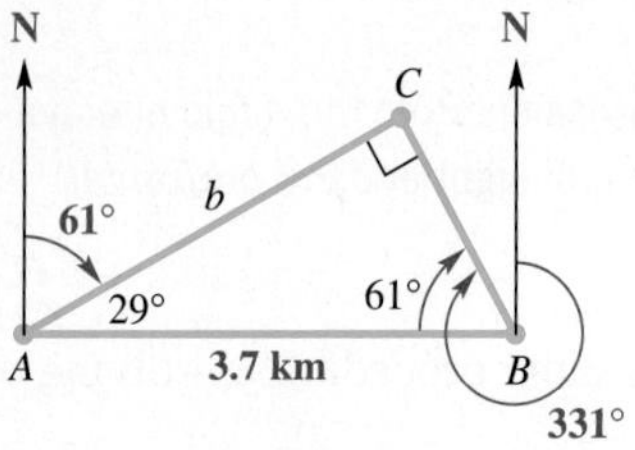

**Solution** Draw a sketch showing the given information, as in Figure 62. Since a line drawn due north is perpendicular to an east–west line, right angles are formed at $A$ and $B$, so angles $CAB$ and $CBA$ can be found. Angle $C$ is a right angle because angles $CAB$ and $CBA$ are complementary. Find distance $b$ by using the cosine function.

$$\cos 29^\circ = \frac{b}{3.7}$$

$$3.7 \cos 29^\circ = b$$

$$b \approx 3.2 \text{ kilometers}$$ Use a calculator; round to the nearest tenth.

**CAUTION** The importance of a correctly labeled sketch when solving applications like that in Example 4 cannot be overemphasized. Some of the necessary information is often not directly stated in the problem and can only be determined from the sketch.

The second method for expressing bearing starts with a north–south line and uses an acute angle to show the direction, either east or west, from this line. Figure 63 shows several sample bearings using this system. Either N or S always comes first, followed by an acute angle, and then E or W.

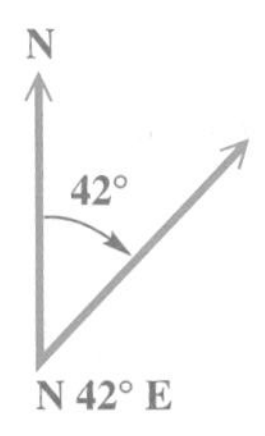

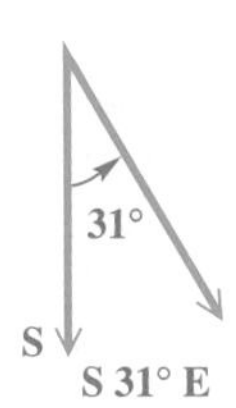

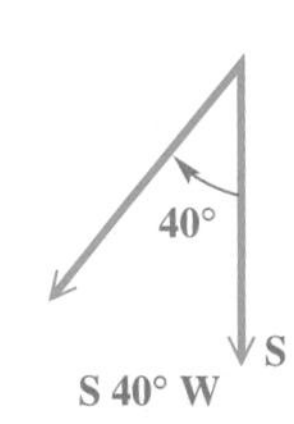

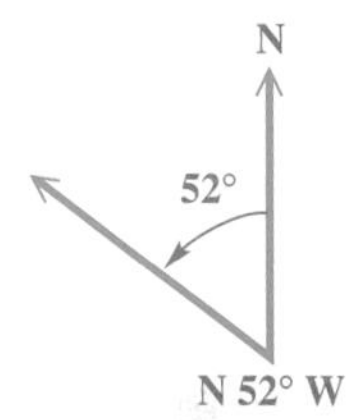


FIGURE 63

## EXAMPLE 5 *Solving a Problem Involving Bearing (Second Method)*

The bearing from $A$ to $C$ is S 52° E. The bearing from $A$ to $B$ is N 84° E. The bearing from $B$ to $C$ is S 38° W. A plane flying at 250 mph takes 2.4 hours to go from $A$ to $B$. Find the distance from $A$ to $C$.

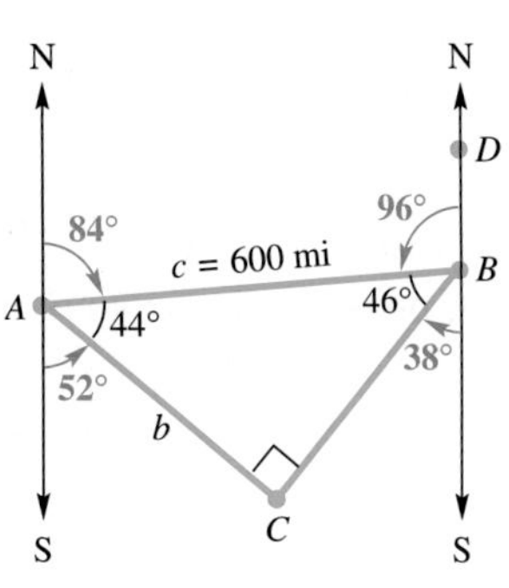


FIGURE 64

**Solution** Make a sketch. First draw the two bearings from point $A$. Choose a point $B$ on the bearing N 84° E from $A$, and draw the bearing to $C$. Point $C$ will be located where the bearing lines from $A$ and $B$ intersect, as shown in Figure 64.

Since the bearing from $A$ to $B$ is N 84° E, angle $ABD$ is $180^\circ - 84^\circ = 96^\circ$. Thus, angle $ABC$ is 46°. Also, angle $BAC$ is $180^\circ - (84^\circ + 52^\circ) = 44^\circ$. Angle $C$ is $180^\circ - (44^\circ + 46^\circ) = 90^\circ$. From the statement of the problem, a plane flying at 250 mph takes 2.4 hours to go from $A$ to $B$. The distance from $A$ to $B$ is the product of rate and time, or

$$c = \text{rate} \times \text{time} = 250(2.4) = 600 \text{ miles}.$$

To find $b$, the distance from $A$ to $C$, use the sine function. (The cosine function could also be used.)

$$\sin 46^\circ = \frac{b}{c}$$

$$\sin 46^\circ = \frac{b}{600}$$

$$600 \sin 46^\circ = b$$

$$b \approx 430 \text{ miles}$$

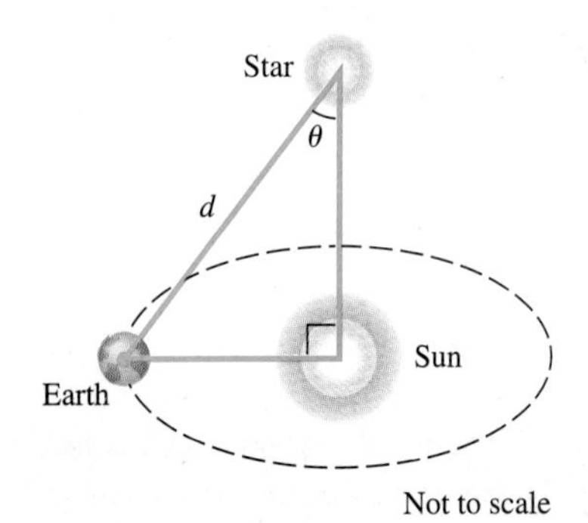


FIGURE 65

## Further Applications

For centuries astronomers wanted to know how far it was to the stars. Not until 1838 did the astronomer Friedrich Bessel determine the distance to a star called 61 Cygni. He used a *parallax** method that relied on the measurement of small angles. See Figure 65. As Earth revolves around the sun, the observed parallax of 61 Cygni is $\theta \approx .0000811°$. Because stars are so distant, parallax angles are very small. (*Source:* Freebury, H., *A History of Mathematics,* MacMillan, 1961; Zeilik, M. et al., *Introductory Astronomy and Astrophysics,* Third Edition, Saunders College Publishers, 1992.)

### EXAMPLE 6 *Calculating the Distance to a Star*

One of the nearest stars is Alpha Centauri, which has parallax $\theta \approx .000212°$.

**(a)** Calculate the distance to Alpha Centauri if the Earth–Sun distance is 93,000,000 miles.

**(b)** A light-year is defined to be the distance that light travels in 1 year and equals about 5.9 trillion miles. Find the distance to Alpha Centauri in light-years.

#### Solution

**(a)** Let $d$ represent the distance between Earth and Alpha Centauri. From Figure 65, it can be seen that

$$\sin \theta = \frac{93{,}000{,}000}{d} \quad \text{or} \quad d = \frac{93{,}000{,}000}{\sin \theta}.$$

Substituting for $\theta$ gives

$$d = \frac{93{,}000{,}000}{\sin .000212°} \approx 2.51 \times 10^{13} \text{ miles.}$$

**(b)** This distance equals $\dfrac{2.51 \times 10^{13}}{5.9 \times 10^{12}} \approx 4.3$ light-years.

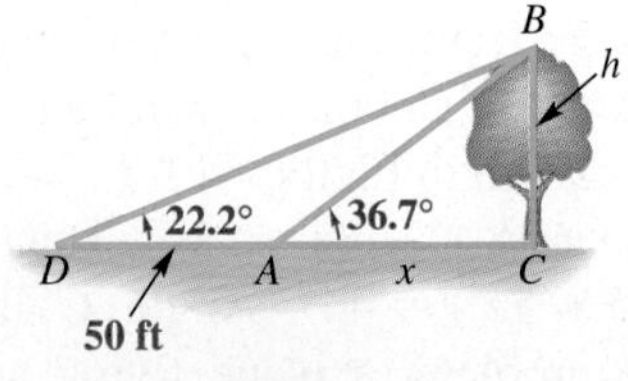


FIGURE 66

### EXAMPLE 7 *Solving a Problem Involving Angle of Elevation*

Francisco needs to know the height of a tree. From a given point on the ground, he finds that the angle of elevation to the top of the tree is 36.7°. He then moves back 50 feet. From the second point, the angle of elevation to the top of the tree is 22.2°. See Figure 66. Find the height of the tree.

#### Analytic Solution

Figure 66 shows two unknowns: $x$, the distance from the center of the trunk of the tree to the point where the first observation was made, and $h$, the height of the tree. Since nothing is given about the length of the hypotenuse of either triangle $ABC$ or triangle $BCD$, we use a ratio that does

#### Graphing Calculator Solution†

In Figure 67(a), we have superimposed Figure 66 on coordinate axes with the origin at $D$. By definition,

*(continued)*

*You observe parallax when you ride in a car and see a nearby object apparently moving backward with respect to more distant objects.

†*Source:* Adapted with permission from "Letter to the Editor," by Robert Ruzich (*Mathematics Teacher,* Volume 88, Number 1). Copyright © 1995 by the National Council of Teachers of Mathematics.

not involve the hypotenuse—tangent. (Refer to Figure 67(a) in the Graphing Calculator Solution.)

In triangle $ABC$, $\tan 36.7° = \frac{h}{x}$ or $h = x \tan 36.7°$.

In triangle $BCD$, $\tan 22.2° = \frac{h}{50 + x}$

$$h = (50 + x) \tan 22.2°.$$

Since each of these expressions equals $h$, the expressions must be equal, so

$$x \tan 36.7° = (50 + x) \tan 22.2°.$$

Now we solve for $x$.

$$x \tan 36.7° = 50 \tan 22.2° + x \tan 22.2°$$ Distributive property

$$x \tan 36.7° - x \tan 22.2° = 50 \tan 22.2°$$ Move $x$ terms to one side.

$$x(\tan 36.7° - \tan 22.2°) = 50 \tan 22.2°$$ Factor out $x$ on the left.

$$x = \frac{50 \tan 22.2°}{\tan 36.7° - \tan 22.2°}$$ Divide by the coefficient of $x$.

We saw above that $h = x \tan 36.7°$. Substituting for $x$,

$$h = \left(\frac{50 \tan 22.2°}{\tan 36.7° - \tan 22.2°}\right) \tan 36.7°.$$

From a calculator,

$$\tan 36.7° \approx .74537703$$
$$\tan 22.2° \approx .40809244$$

so

$$\tan 36.7° - \tan 22.2° \approx .74537703 - .40809244 = .33728459$$

and

$$h \approx \left(\frac{50(.40809244)}{.33728459}\right).74537703$$
$$\approx 45 \text{ (rounded)}.$$

The height of the tree is approximately 45 feet.

the tangent of the angle between the $x$-axis and the graph of a line with equation $y = mx + b$ is the slope of the line, $m$. So for line $DB$, $m = \tan 22.2°$. Since the $y$-intercept $b$ is 0 here, the equation of line $DB$ is $y_1 = (\tan 22.2°)x$. Similarly, the equation of line $AB$ is $y_2 = (\tan 36.7°)x + b$. However, here $b \neq 0$, so we use the point $A(50, 0)$ and the point-slope form to find the equation.

$$y_2 - y_1 = m(x - x_1)$$
$$y_2 - 0 = m(x - 50)$$ Let $x_1 = 50$, $y_1 = 0$.
$$y_2 = [\tan 36.7°](x - 50)$$

Lines $y_1$ and $y_2$ are graphed in Figure 67(b). The $y$-coordinate of the point of intersection of the graphs of these two lines gives the length of $BC$, or $h$. From the information at the bottom of the screen, we see that $h \approx 45$ (rounded), which agrees with our analytic result.

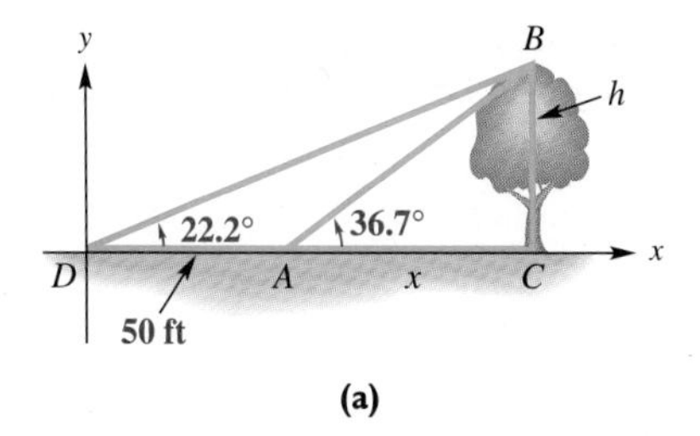


(a)

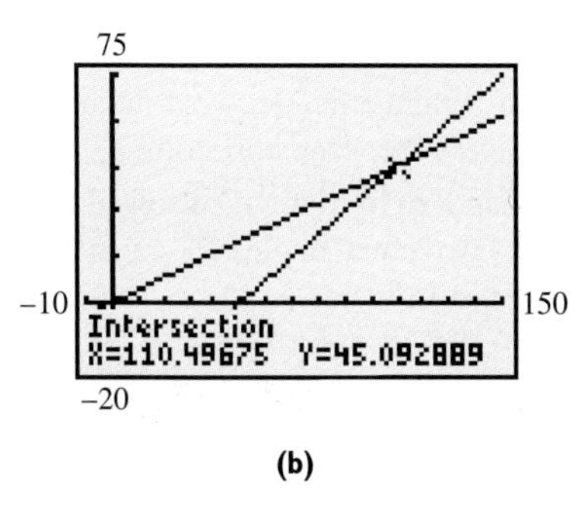


(b)

FIGURE 67

NOTE In practice, we usually do not write down the intermediate calculator approximation steps. However, we have done this in Example 7 so that you may follow the steps more easily.

## 8.4 EXERCISES

*Solve each right triangle.*

**1.**

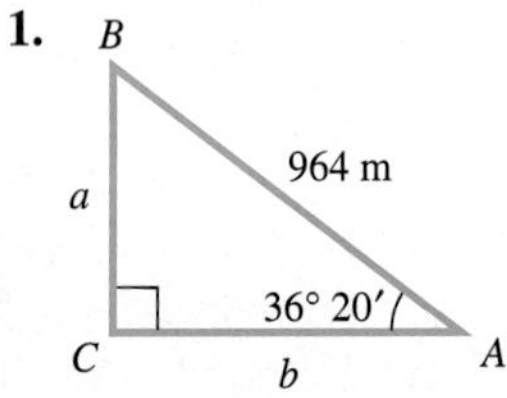


**2.**

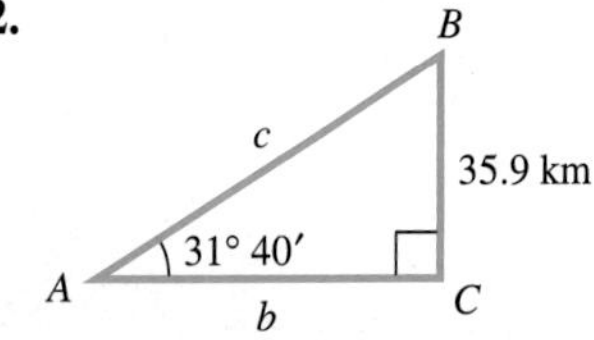


**3.**

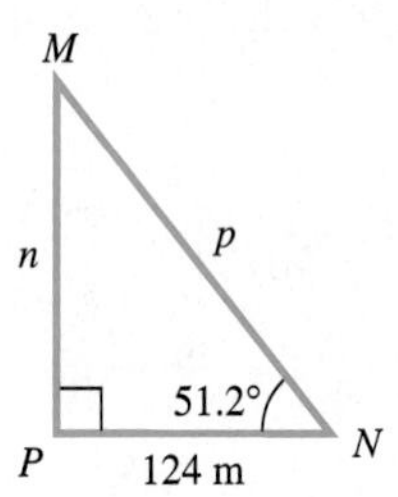


**4.**

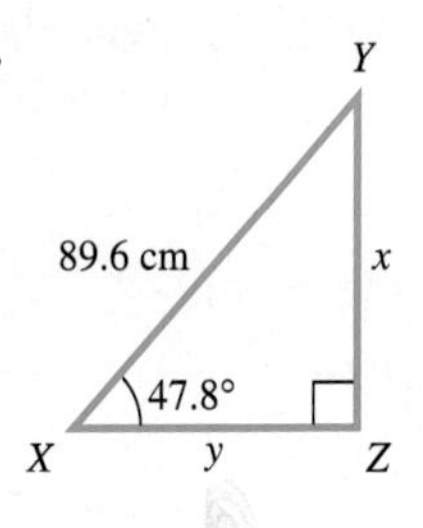


**5.**

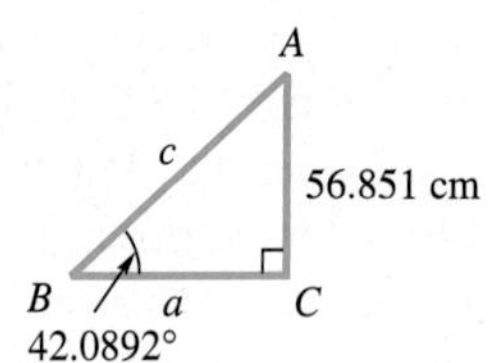


**6.**

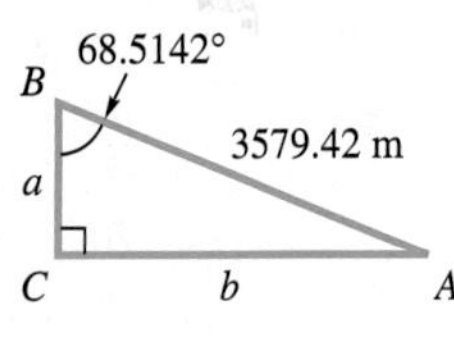


*Solve each right triangle. In each case, $C = 90°$. If angle information is given in degrees and minutes, give answers in the same way. If given in decimal degrees, do likewise in answers. When two sides are given, give angles in degrees and minutes.*

**7.** $A = 28.00°$; $c = 17.4$ feet

**8.** $B = 46.00°$; $c = 29.7$ meters

**9.** $B = 73.00°$; $b = 128$ inches

**10.** $A = 61° \, 00'$; $b = 39.2$ centimeters

**11.** $a = 76.4$ yards; $b = 39.3$ yards

**12.** $a = 958$ meters; $b = 489$ meters

**13.** *Concept Check* If we are given an acute angle and a side in a right triangle, what unknown part of the triangle requires the least work to find?

**14.** Can a right triangle be solved if we are given measures of its two acute angles and no side lengths? Explain.

*Find the measure of the angle formed by the line passing through the origin and the positive part of the x-axis. Use the displayed values at the bottom of the screen.*

**15.**

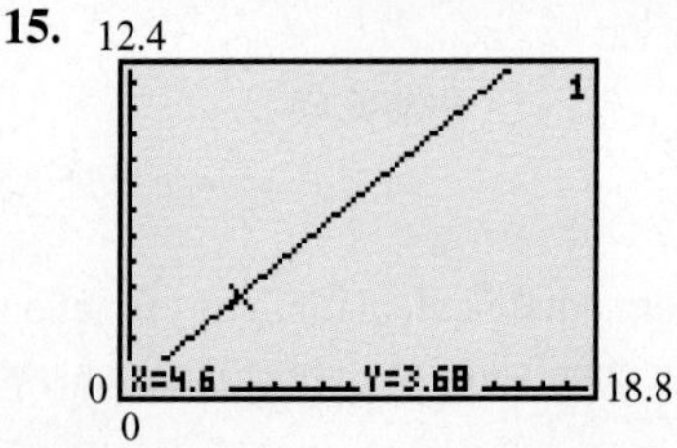


**16.**

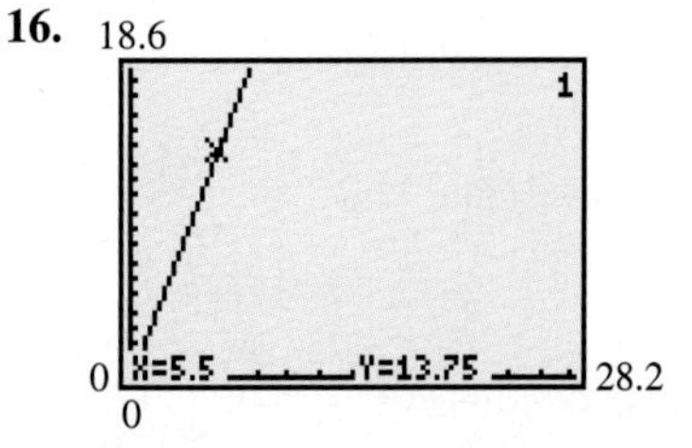


**17.** Explain why the angle of depression $DAB$ has the same measure as the angle of elevation $ABC$ in the figure.

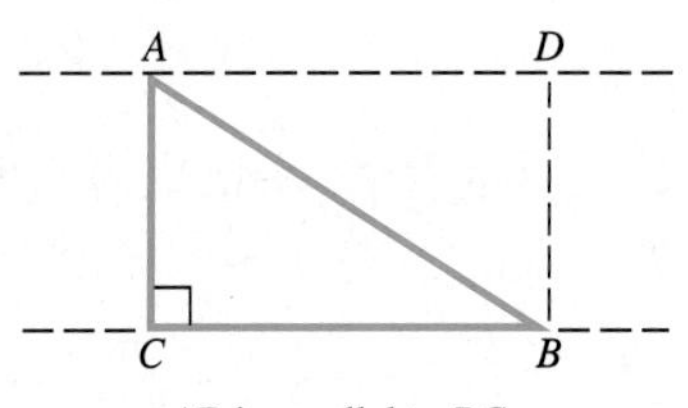


$AD$ is parallel to $BC$.

**18.** Why is angle $CAB$ *not* an angle of depression in the figure for Exercise 17.

**19.** *Concept Check* When bearing is given as a single angle measure, how is the angle represented in a sketch?

**20.** *Concept Check* When bearing is given as N (or S), then the angle measure, then E (or W), how is the angle represented in a sketch?

*Solve each problem.*

**21.** *Height of a Ladder on a Wall* A 13.5-meter fire truck ladder is leaning against a wall. Find the distance $d$ the ladder goes up the wall (above the top of the fire truck) if the ladder makes an angle of $43° \, 50'$ with the horizontal.

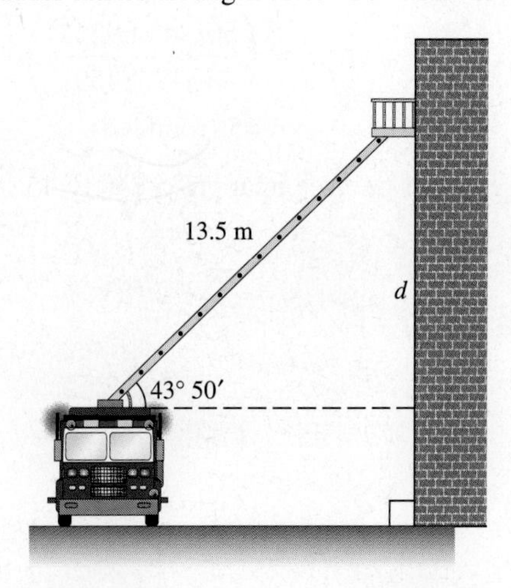


**22.** *Length of a Guy Wire* A weather tower used to measure wind speed has a guy wire attached to it 175 feet above the ground. The angle between the wire and the vertical tower is 57°. Approximate the length of the guy wire.

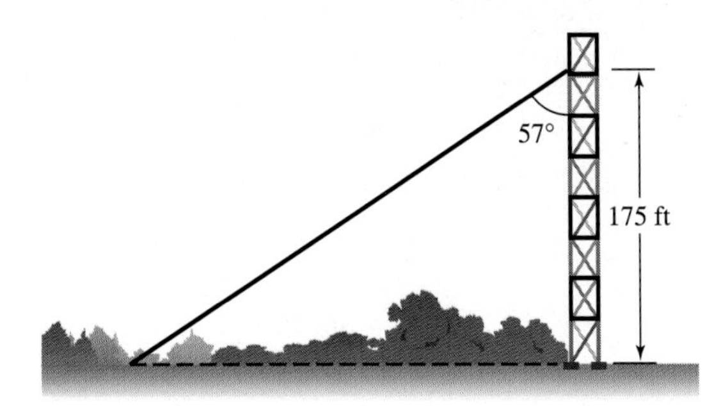


**23.** *Length of a Shadow* Suppose that the angle of elevation of the sun is 23.4°. Find the length of the shadow cast by Diane Carr, who is 5.75 feet tall.

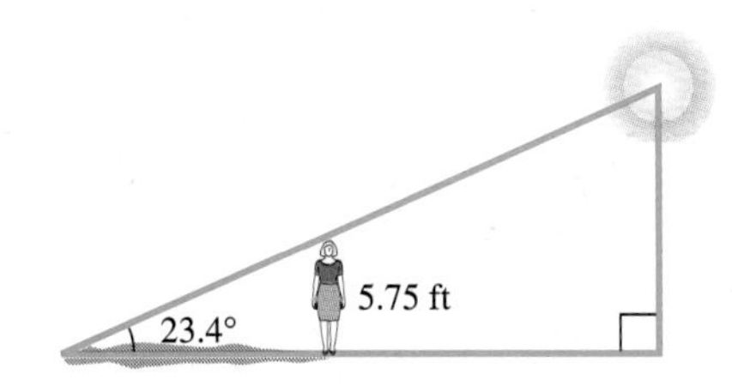


**24.** *Height of a Tower* The shadow of a vertical tower is 40.6 meters long when the angle of elevation of the sun is 34.6°. Find the height of the tower.

**25.** *Height of a Building* From a window 30 feet above the street, the angle of elevation to the top of the building across the street is 50.0° and the angle of depression to the base of this building is 20.0°. Find the height of the building across the street.

**26.** *Height of a Building* The angle of elevation from the top of a small building to the top of a nearby taller building is 46° 40′, while the angle of depression to the bottom is 14° 10′. If the smaller building is 28.0 meters high, find the height of the taller building.

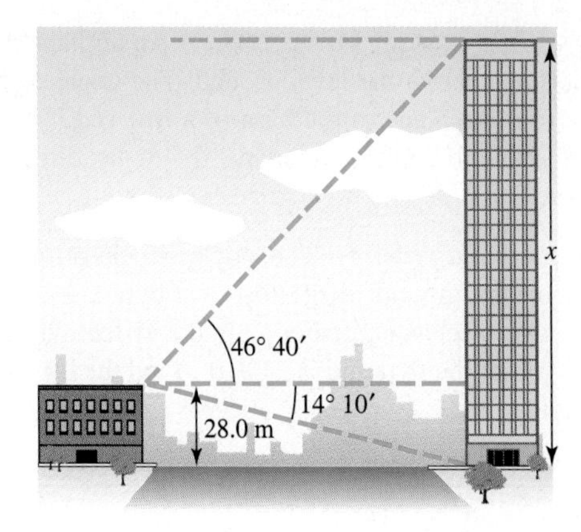


**27.** *Angle of a Television Lens* A television camera is to be mounted on a bank wall so as to have a good view of the head teller. Find the angle of depression of the lens.

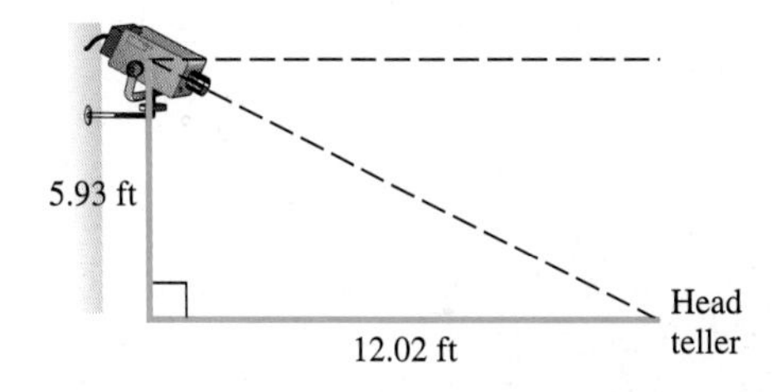


**28.** *Angle of Depression of a Floodlight* A company safety committee has recommended that a floodlight be mounted in a parking lot so as to illuminate the employee exit. Find the angle of depression of the light.

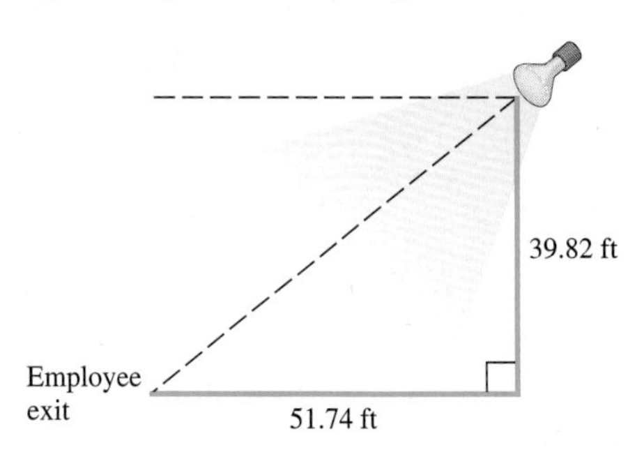


**29.** *Distance through a Tunnel* A tunnel is to be dug from $A$ to $B$. Both $A$ and $B$ are visible from $C$. If $AC$ is 1.4923 miles and $BC$ is 1.0837 miles, and if $C$ is 90°, find the measures of angles $A$ and $B$.

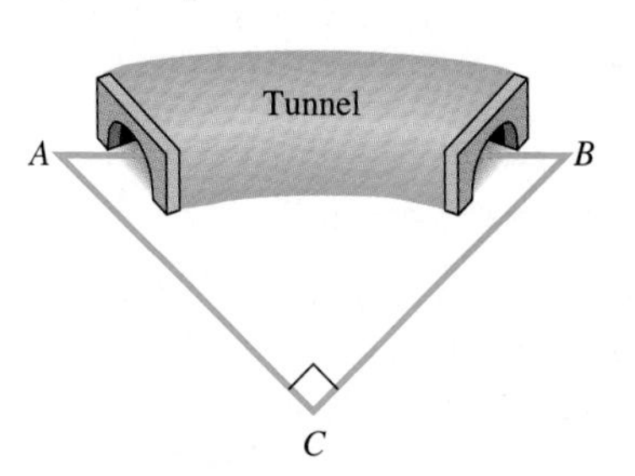

**30.** *Angle of Depression from a Plane* An airplane is flying 10,500 feet above the level ground. The angle of depression from the plane to the base of a tree is 13° 50′. How far horizontally must the plane fly to be directly over the tree?

**31.** *Height of a Pyramid* The angle of elevation from a point on the ground to the top of a pyramid is 35° 30′. The angle of elevation from a point 135 feet farther back to the top of the pyramid is 21° 10′. Find the height of the pyramid.

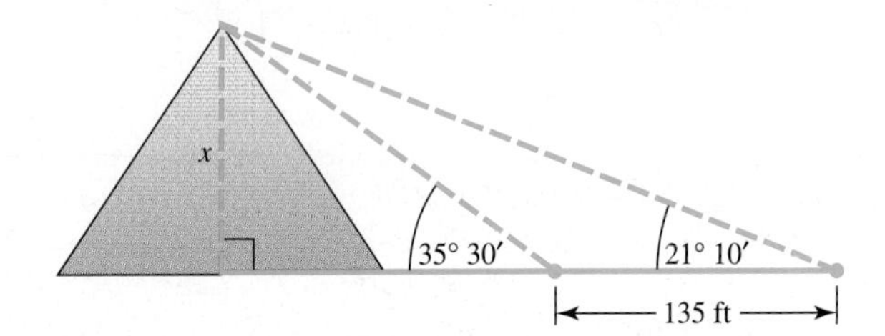


**32.** *Distance between a Whale and a Lighthouse* Debbie Glockner-Ferrari, a whale researcher, is watching a whale approach directly toward her as she observes from the top of a lighthouse. When she first begins watching the whale, the angle of depression of the whale is 15° 50′. Just as the whale turns away from the lighthouse, the angle of depression is 35° 40′. If the height of the lighthouse is 68.7 meters, find the distance traveled by the whale as it approaches the lighthouse.

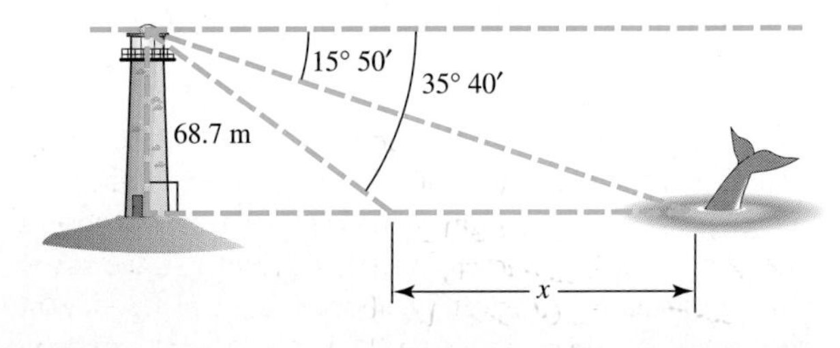


**33.** *Height of an Antenna* An antenna is on top of the center of a house. The angle of elevation from a point on the ground 28.0 meters from the center of the house to the top of the antenna is 27° 10′, and the angle of elevation to the bottom of the antenna is 18° 10′. Find the height of the antenna.

**34.** *Height of Mt. Whitney* The angle of elevation from Lone Pine to the top of Mt. Whitney is 10° 50′. Van Dong Le, traveling 7.00 kilometers from Lone Pine along a straight, level road toward Mt. Whitney, finds the angle of elevation to be 22° 40′. Find the height of the top of Mt. Whitney above the level of the road.

**35.** *Length of a Side of a Lot* A piece of land has the shape shown in the figure at the top of the next column. Find $x$.

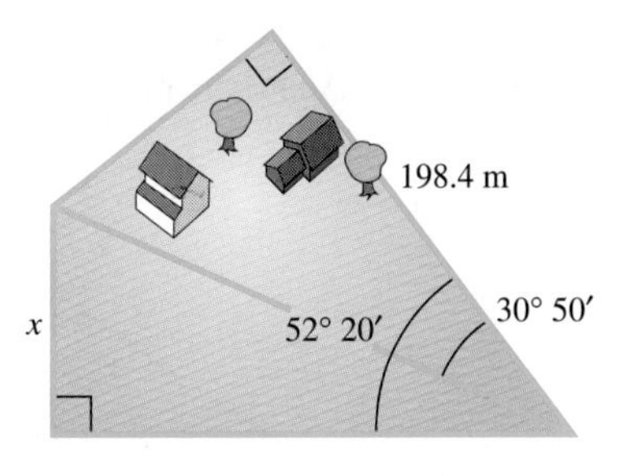


**36.** Find the value of $x$ in the figure.

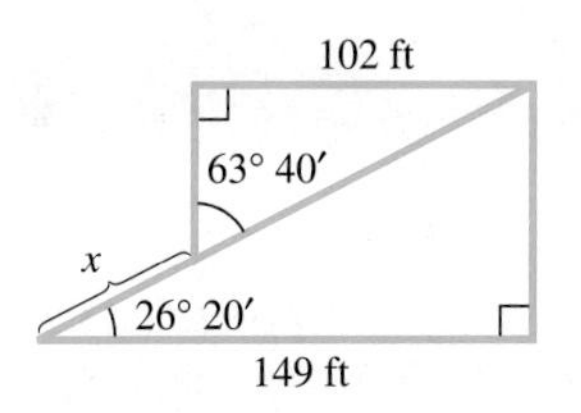


**37.** *Distance between Two Ships* A ship leaves port and sails on a bearing of N 28° 10′ E. Another ship leaves the same port at the same time and sails on a bearing of S 61° 50′ E. If the first ship sails at 24.0 mph and the second sails at 28.0 mph, find the distance between the two ships after 4 hours.

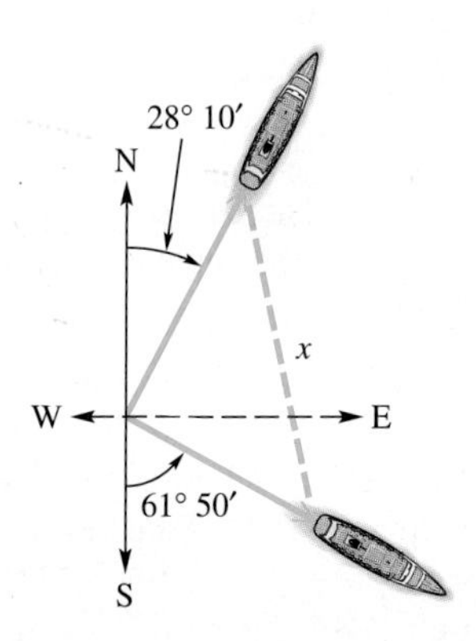


**38.** *Distance between Transmitters* Radio direction finders are set up at two points $A$ and $B$, which are 2.50 miles apart on an east–west line. From $A$, it is found that the bearing of a signal from a radio transmitter is N 36° 20′ E, while from $B$ the bearing of the same signal is N 53° 40′ W. Find the distance of the transmitter from $B$.

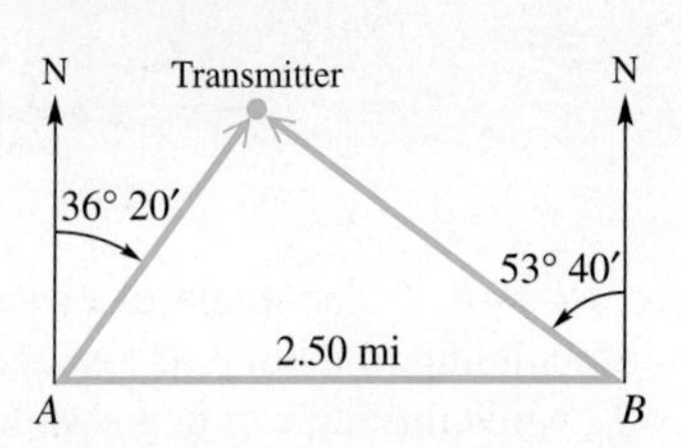

**39.** *Transmitter Distance* In Exercise 38, find the distance of the transmitter from $A$.

**40.** *Distance Traveled by a Plane* A plane flies 1.5 hours at 110 mph on a bearing of 40°. It then turns and flies 1.3 hours at the same speed on a bearing of 130°. How far is the plane from its starting point?

**41.** *Distance Traveled by a Ship* A ship travels 50 kilometers on a bearing of 27°, and then travels on a bearing of 117° for 140 kilometers. Find the distance between the starting point and the ending point.

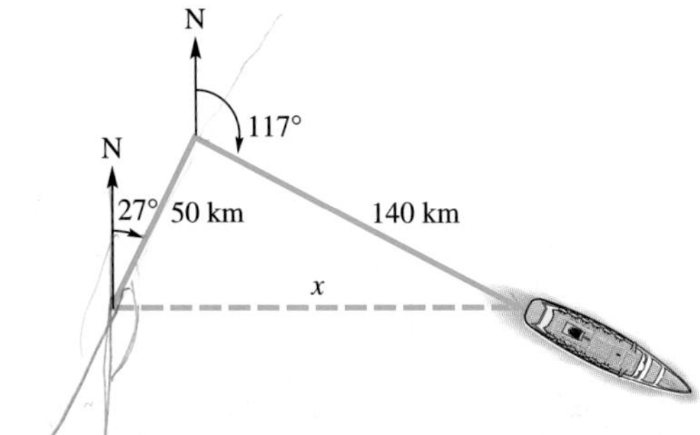


**42.** *Distance between Two Ships* Two ships leave a port at the same time. The first ship sails on a bearing of 40° at 18 knots (nautical miles per hour) and the second at a bearing of 130° at 26 knots. How far apart are they after 1.5 hours?

**43.** *Cloud Ceiling* The U.S. Weather Bureau defines a *cloud ceiling* as the altitude of the lowest clouds that cover more than half the sky. To determine a cloud ceiling, a powerful searchlight projects a circle of light vertically on the bottom of the cloud. An observer sights the circle of light in the crosshairs of a tube called a *clinometer*. A pendant hanging vertically from the tube and resting on a protractor gives the angle of elevation. Find the cloud ceiling if the searchlight is located 1000 feet from the observer and the angle of elevation is 30.0° as measured with a clinometer at eye-height 6 feet. (Assume three significant digits.)

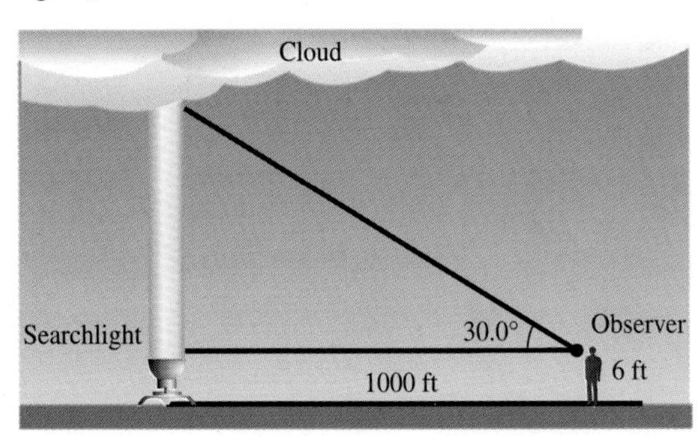


**44.** *Height of Mt. Everest* The highest mountain peak in the world is Mt. Everest, located in the Himalayas. The height of this enormous mountain was determined in 1856 by surveyors using trigonometry long before it was first climbed in 1953. This difficult measurement had to be done from a great distance. At an altitude of 14,545 feet on a different mountain, the straight line distance to the peak of Mt. Everest is 27.0134 miles and its angle of elevation is $\theta = 5.82°$. (*Source:* Dunham, W., *The Mathematical Universe,* John Wiley and Sons, 1994.)

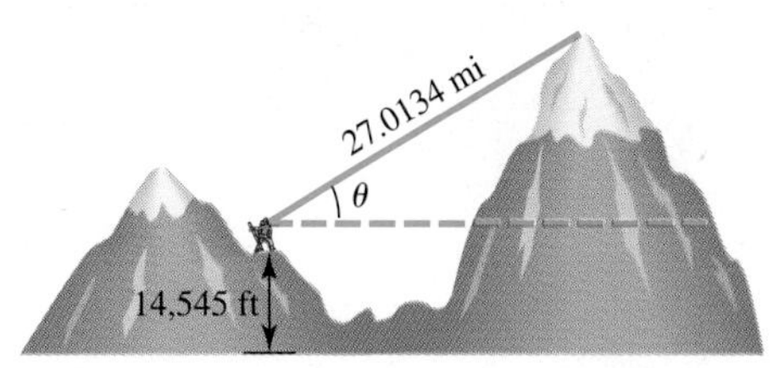


(a) Approximate the height (in feet) of Mt. Everest.

(b) In the actual measurement, Mt. Everest was over 100 miles away and the curvature of Earth had to be taken into account. Would the curvature of Earth make the peak appear taller or shorter than it actually is?

**45.** *Distance Across a Lake* To find the distance $RS$ across a lake, a surveyor lays off $RT = 53.1$ meters, with angle $T = 32° \ 10'$ and angle $S = 57° \ 50'$. Find length $RS$.

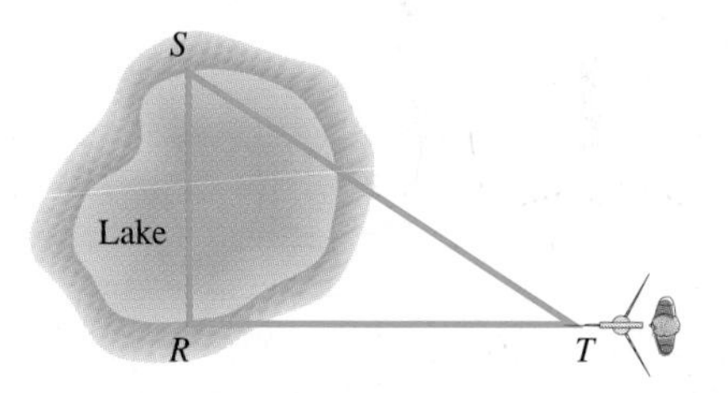


**46.** *Error in Measurement* A degree may seem like a very small unit, but an error of one degree in measuring an angle may be very significant. For example, suppose a laser beam directed toward the visible center of the moon misses its assigned target by 30 seconds. How far is it (in miles) from its assigned target? Take the distance from the surface of Earth to that of the moon to be 234,000 miles. (*Source: A Sourcebook of Applications of School Mathematics* by Donald Bushaw et al. Copyright © 1980 by The Mathematical Association of America.)

**47.** *(Modeling) Stopping Distance on a Curve* When an automobile travels along a circular curve, objects like trees and buildings situated on the inside of the curve can obstruct a driver's vision. These obstructions prevent the driver from seeing sufficiently far down the highway to ensure a safe stopping distance. In the figure on the next page, the *minimum* distance $d$ that should be cleared on the inside of the highway is modeled by the equation

$$d = R\left(1 - \cos\frac{\beta}{2}\right).$$

(*Source:* Mannering, F. and W. Kilareski, *Principles of Highway Engineering and Traffic Analysis,* Second Edition, John Wiley and Sons, 1998.)

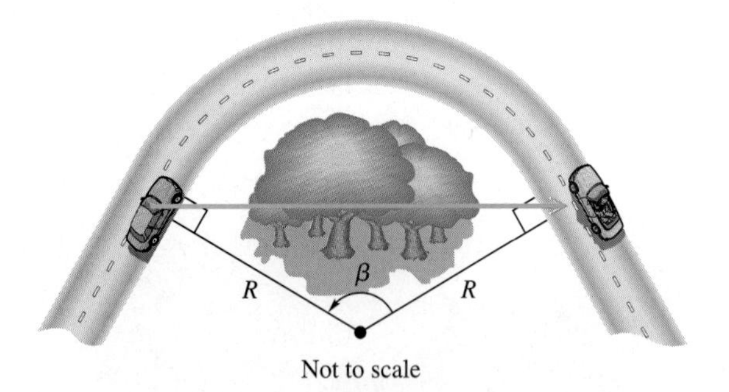


**(a)** It can be shown that if $\beta$ is measured in degrees, then $\beta \approx \frac{57.3S}{R}$, where $S$ is safe stopping distance for the given speed limit. Compute $d$ for a 55 mph speed limit if $S = 336$ feet and $R = 600$ feet.

**(b)** Compute $d$ for a 65 mph speed limit if $S = 485$ feet and $R = 600$ feet.

**(c)** How does the speed limit affect the amount of land that should be cleared on the inside of the curve?

**48.** *(Modeling) Highway Curve Design* Highway curves are sometimes banked so that the outside of the curve is slightly elevated or inclined above the inside of the curve, as shown in the figure. This inclination is called the *superelevation*. It is important that both the curve's radius and superelevation are correct for a given speed limit. The relationship between a car's velocity $v$ in feet per second, the safe radius $r$ of the curve in feet, and the superelevation $\theta$ in degrees is modeled by

$$r = \frac{v^2}{4.5 + 32.2 \tan \theta}.$$

(*Source:* Mannering, F. and W. Kilareski, *Principles of Highway Engineering and Traffic Analysis,* Second Edition, John Wiley and Sons, 1998.)

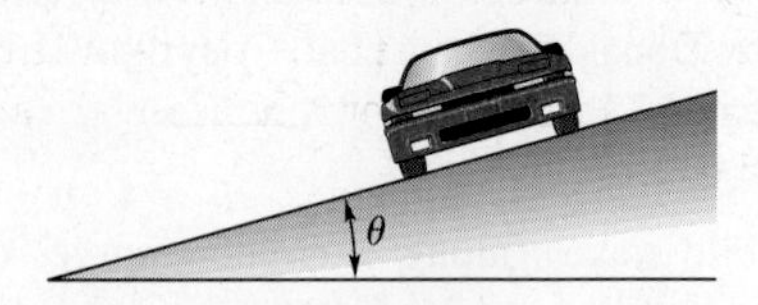


**(a)** A curve has a speed limit of 66 feet per second (45 mph) and a superelevation of $\theta = 3°$. Approximate the safe radius $r$.

**(b)** Find $r$ if $\theta = 5°$ and $v = 66$.

**(c)** Make a conjecture about how increasing $\theta$ with $v = 66$ affects the safe radius $r$.

**(d)** Find $v$ if $r = 1150$ and $\theta = 2.1°$.

**49.** *(Modeling) Distance between the Sun and a Star* Suppose that a star forms an angle $\theta$ with respect to Earth and the sun. Let the coordinates of Earth be $(x, y)$, the star be $(0, 0)$, and the sun be $(x, 0)$. See the figure.

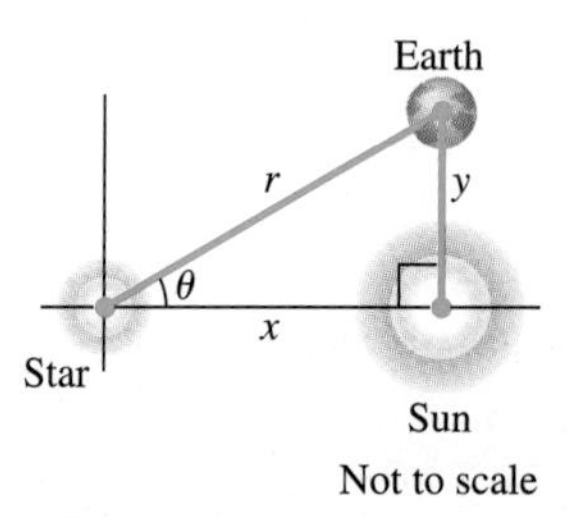


Find an equation for $x$, the distance between the sun and the star, as follows.

**(a)** Write an equation involving a trigonometric function that relates $x$, $y$, and $\theta$.

**(b)** Solve your equation for $x$.

**50.** *Area of a Solar Cell* A solar cell converts the energy of sunlight directly into electrical energy. The amount of energy a cell produces depends on its area. Suppose a solar cell is hexagonal, as shown in the figure. Express its area in terms of $\sin \theta$ and any side $x$. (*Hint:* Consider one of the six equilateral triangles from the hexagon. See the figure.) (*Source:* Kastner, B., *Space Mathematics,* NASA, 1985.)

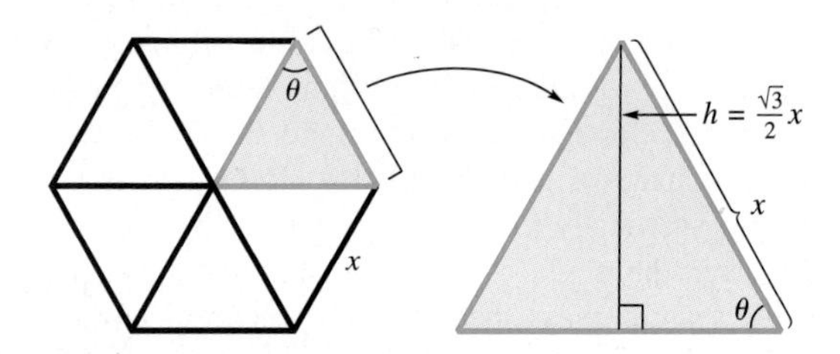


*Find the exact value of each part labeled with a variable in each figure.*

**51.**

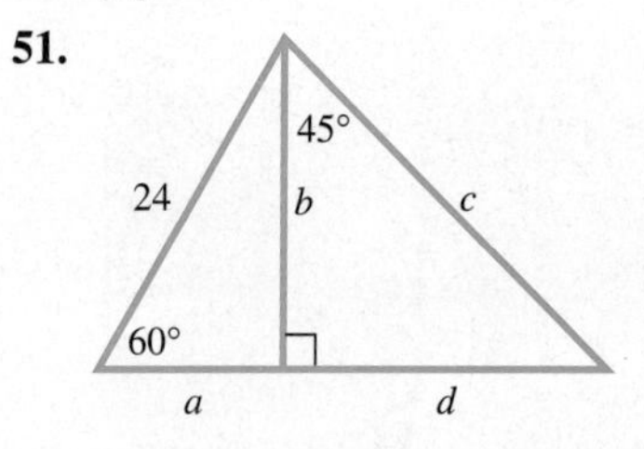


**52.**

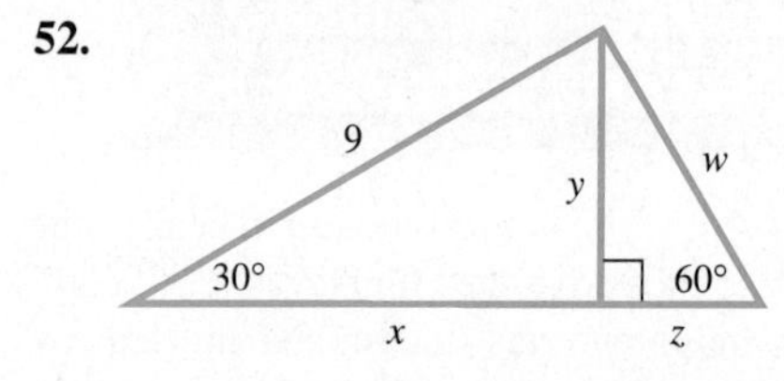

**53.**

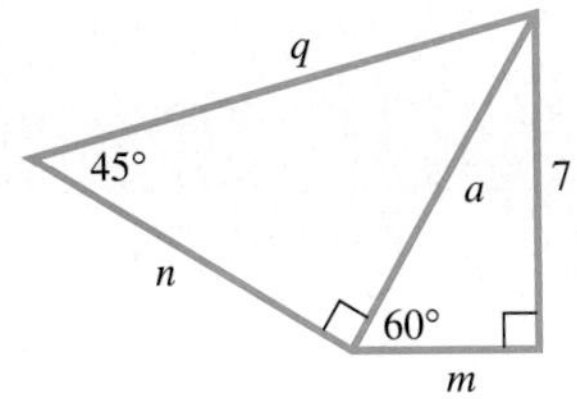


**54.**

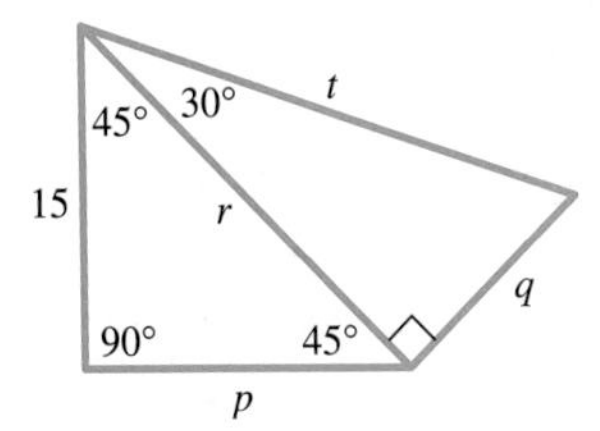


*Find a formula for the area of each figure in terms of s.*

**55.**

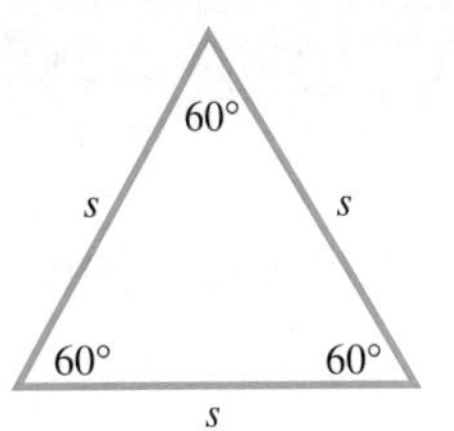


**56.**

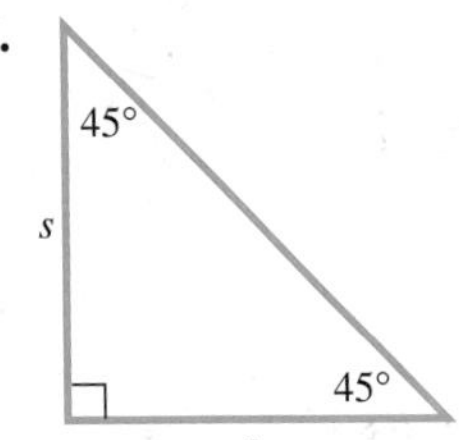


## Reviewing Basic Concepts (Sections 8.3 and 8.4)

**1.** Find the six trigonometric function values of angle $A$ in the triangle shown.

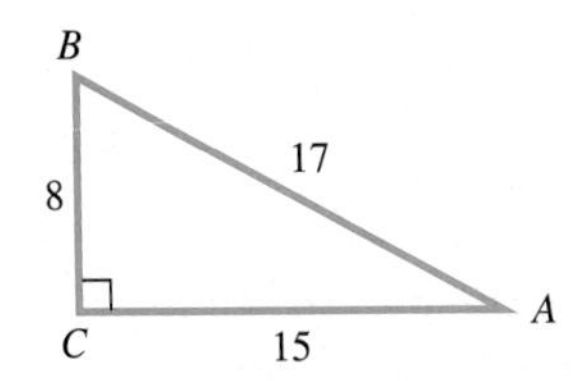


**2.** Complete the following table of exact function values.

| $\theta$ | $\sin\theta$ | $\cos\theta$ | $\tan\theta$ | $\cot\theta$ | $\sec\theta$ | $\csc\theta$ |
|---|---|---|---|---|---|---|
| $30° = \frac{\pi}{6}$ | | | | | | |
| $45° = \frac{\pi}{4}$ | | | | | | |
| $60° = \frac{\pi}{3}$ | | | | | | |

**3.** Write each expression using its cofunction.

**(a)** $\sin 27°$ **(b)** $\tan \dfrac{\pi}{5}$

**4.** Give the reference angle for each angle measure.

**(a)** $100°$ **(b)** $-365°$ **(c)** $\dfrac{8\pi}{3}$

**5.** Find the exact values for the six trigonometric functions of $315°$.

**6.** Use a calculator to find an approximation for each function value.

**(a)** $\sin 46°\ 30'$ **(b)** $\tan(-100°)$ **(c)** $\csc 4$

**7.** Find all values of $\theta$, if $\theta$ is in the interval $[0°, 360°)$ and $\tan\theta = -\frac{\sqrt{3}}{3}$.

**8.** Find two angles in the interval $[0, 2\pi)$ that satisfy $\sin\theta = .68163876$.

**9.** *Aerial Photography* An aerial photograph is taken directly above a building. The length of the building's shadow is 48 feet when the angle of elevation of the sun is 35.3°. Estimate the height of the building.

**10.** *Height of Mt. Kilimanjaro* From a point $A$ the angle of elevation of Mount Kilimanjaro in Africa is 13.7° and from a point $B$ directly behind $A$, the angle of elevation is 10.4°. If the distance between $A$ and $B$ is 5 miles, approximate the height of Mount Kilimanjaro to the nearest hundred feet.

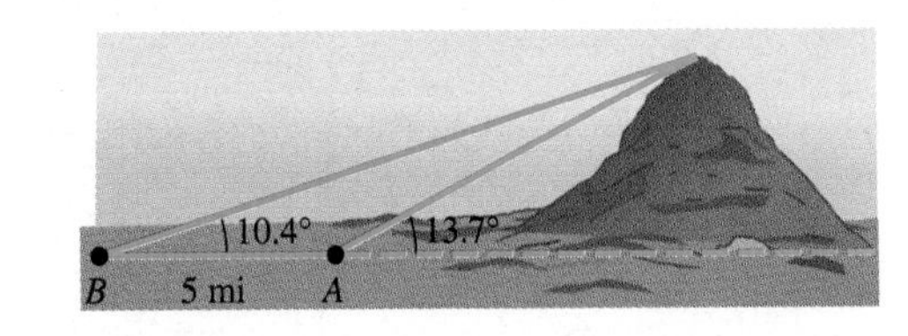

# 8.5 THE CIRCULAR FUNCTIONS

Circular Functions • Applications of Circular Functions

## Circular Functions

In Section 8.2, we defined the six trigonometric functions in such a way that the domain of each function was a set of *angles* in standard position. These angles can be measured in degrees or in radians. In advanced courses, such as calculus, it is necessary to modify the trigonometric functions so that their domains consist of sets of *real numbers* rather than angles. Because the interpretations of these functions are based on the graph of the *unit circle* $x^2 + y^2 = 1$, we refer to them as **circular functions.**

To define the values of the circular functions for any real number $s$, we use the unit circle, shown in Figure 68. The **unit circle** has its center at the origin and radius one unit (hence the name unit circle). We start at the point $(1, 0)$ and measure an arc of length $s$ along the circle, as in Figure 68. If $s > 0$, the arc is measured in a counterclockwise direction, and if $s < 0$, the direction is clockwise. (If $s = 0$, then no arc is measured.) Let the endpoint of this arc be at the point $(x, y)$. Then the six circular functions of the real number $s$ are defined as follows.

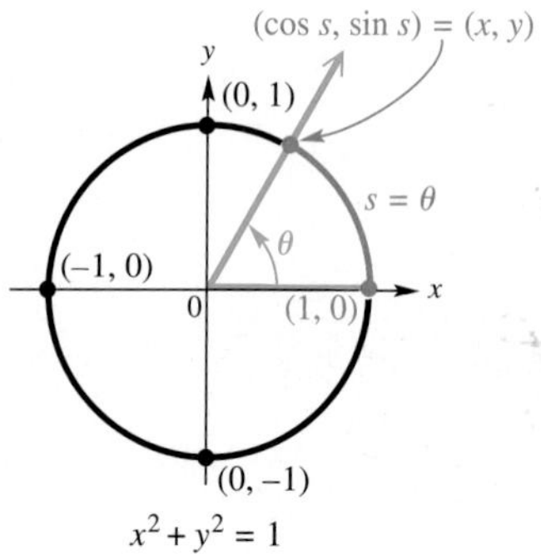

(cos $s$, sin $s$) = $(x, y)$
$y$
(0, 1)
$s = \theta$
$\theta$
(−1, 0)
0
(1, 0)
$x$
(0, −1)
$x^2 + y^2 = 1$

**FIGURE 68**

### Circular Functions

$$\sin s = y \qquad \cos s = x \qquad \tan s = \frac{y}{x} \quad (x \neq 0)$$

$$\csc s = \frac{1}{y} \quad (y \neq 0) \qquad \sec s = \frac{1}{x} \quad (x \neq 0) \qquad \cot s = \frac{x}{y} \quad (y \neq 0)$$

**Looking Ahead to Calculus**

If you plan to go on to calculus, you must become familiar with radian measure. In calculus, the trigonometric or circular functions are always understood to have real number domains.

The circular functions (functions of real numbers) are closely related to the trigonometric functions of angles measured in radians. To see this, let us assume that angle $\theta$ is in standard position, superimposed on the unit circle, as shown in Figure 68. Suppose further that $\theta$ is the *radian* measure of this angle. Using the arc length formula $s = r\theta$ with $r = 1$, we have $s = \theta$. Thus, the length of the intercepted arc is the real number that corresponds to the radian measure of $\theta$. Using the definitions of the trigonometric functions, we have

$$\sin \theta = \frac{y}{r} = \frac{y}{1} = y = \sin s, \quad \text{and} \quad \cos \theta = \frac{x}{r} = \frac{x}{1} = x = \cos s,$$

and so on. As shown here, the trigonometric functions and the circular functions lead to the same function values, provided we think of the angles in radian measure. This leads to the following important result concerning evaluation of circular functions.

### Evaluating a Circular Function

Circular function values of real numbers are obtained in the same manner as trigonometric function values of angles measured in radians. This applies to both methods of finding exact values (such as reference angle analysis) and calculator approximations. Calculators must be in radian mode when finding circular function values.

## EXAMPLE 1 *Evaluating Circular Functions*

Evaluate $\sin \frac{3\pi}{2}$, $\cos \frac{3\pi}{2}$, and $\tan \frac{3\pi}{2}$.

### Analytic Solution

Evaluating a circular function at the real number $\frac{3\pi}{2}$ is equivalent to evaluating it at $\frac{3\pi}{2}$ radians. An angle of $\frac{3\pi}{2}$ radians intersects the unit circle at the point $(0, -1)$, as shown in Figure 69. Since

$$\sin s = y, \quad \cos s = x, \quad \text{and} \quad \tan s = \frac{y}{x},$$

it follows that

$$\sin \frac{3\pi}{2} = -1, \quad \cos \frac{3\pi}{2} = 0, \quad \text{and}$$

$$\tan \frac{3\pi}{2} \text{ is undefined.}$$

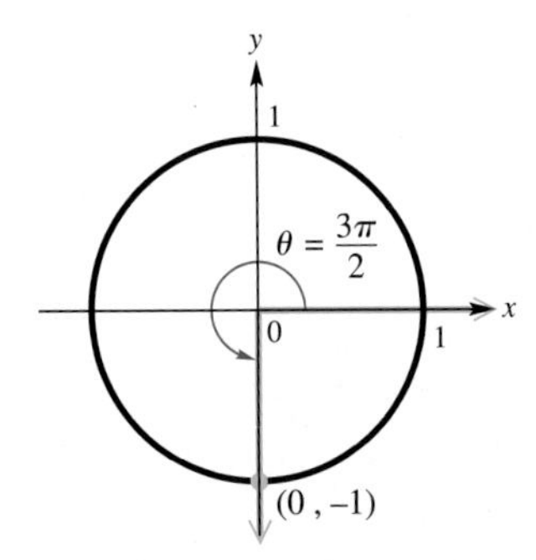


FIGURE 69

### Graphing Calculator Solution

To evaluate a circular function at a real number $s$, be sure that your calculator is set in radian mode. From Figure 70,

$$\sin \frac{3\pi}{2} = -1 \quad \text{and} \quad \cos \frac{3\pi}{2} = 0.$$

However, the evaluation of $\tan \frac{3\pi}{2}$ results in an error.

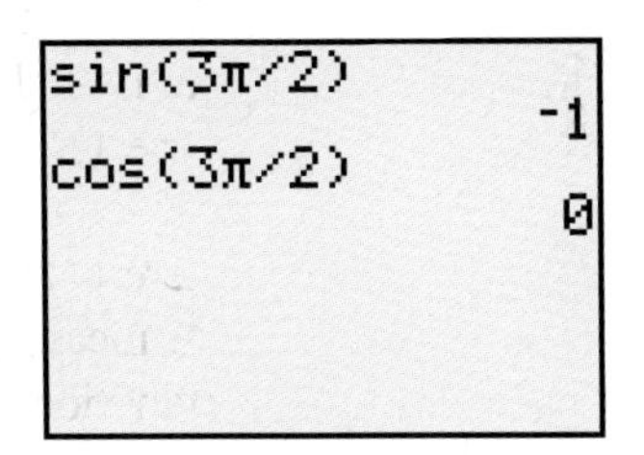


Radian Mode

FIGURE 70

Note that $\tan \frac{3\pi}{2}$ is undefined because

$$\tan \frac{3\pi}{2} = \frac{\sin \frac{3\pi}{2}}{\cos \frac{3\pi}{2}} = \frac{1}{0}.$$

## EXAMPLE 2 *Evaluating Circular Functions*

Evaluate the six circular functions at $s = \frac{5\pi}{6}$.

### Analytic Solution

From the properties of 30°–60° triangles, an angle of $\frac{5\pi}{6}$ radians intersects the unit circle at the point $\left(-\frac{\sqrt{3}}{2}, \frac{1}{2}\right)$. See Figure 71 on the next page. Evaluating the six circular functions at the real number $\frac{5\pi}{6}$ results in

$$\sin \frac{5\pi}{6} = \frac{1}{2} \qquad \csc \frac{5\pi}{6} = 2$$

$$\cos \frac{5\pi}{6} = -\frac{\sqrt{3}}{2} \qquad \sec \frac{5\pi}{6} = -\frac{2}{\sqrt{3}} = -\frac{2\sqrt{3}}{3}$$

$$\tan \frac{5\pi}{6} = -\frac{1}{\sqrt{3}} = -\frac{\sqrt{3}}{3} \qquad \cot \frac{5\pi}{6} = -\sqrt{3}.$$

### Graphing Calculator Solution

Be sure to set your calculator in radian mode. The results in Figure 72 on the next page are decimal approximations for our analytic results. Notice that the reciprocal identities have been used to evaluate the cosecant, secant, and cotangent functions because many calculators do not have keys for these functions.

*(continued)*

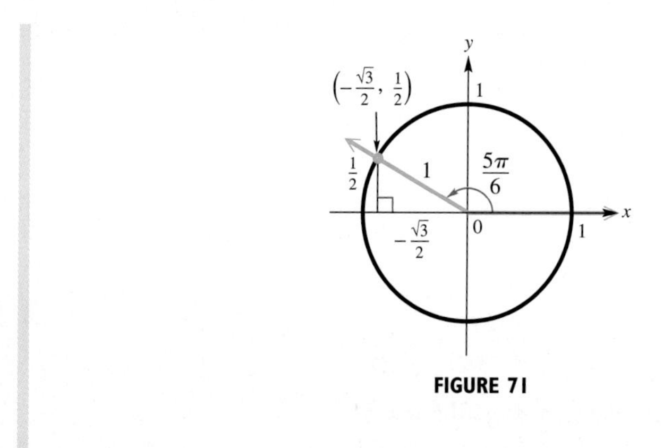


FIGURE 71

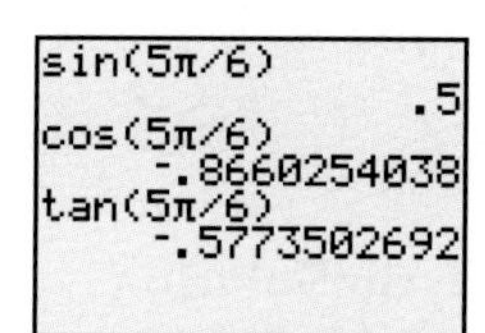


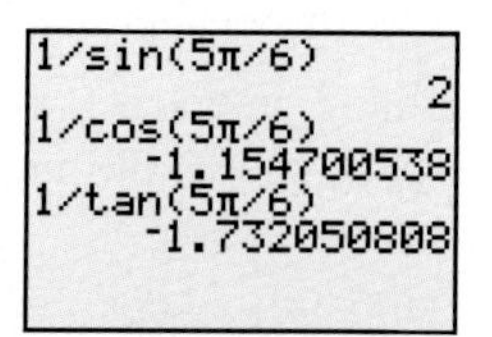


Radian Mode

FIGURE 72

**CAUTION** A common error in trigonometry is using calculators in degree mode when radian mode should be used. Remember, *if you are finding a circular function value of a real number, the calculator must be in radian mode.*

### FOR DISCUSSION

Graph the unit circle $x^2 + y^2 = 1$ by graphing $y_1 = \sqrt{1 - x^2}$ and $y_2 = -\sqrt{1 - x^2}$ in a square viewing window. Now, graph the line $y_3 = \frac{\sqrt{3}}{3}x$. See Figure 73.

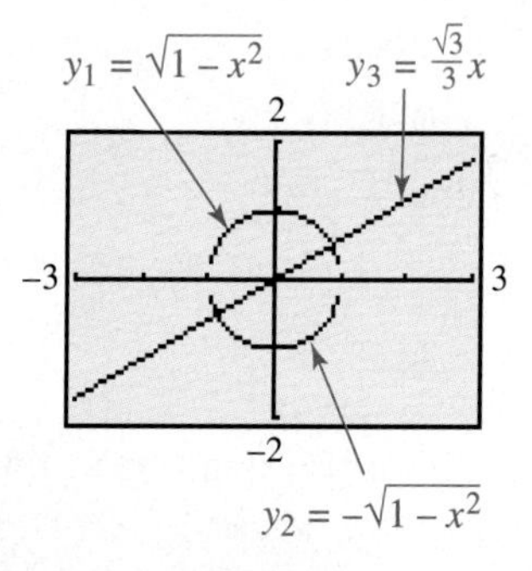


FIGURE 73

1. Explain why the line $y_3 = \frac{\sqrt{3}}{3}x$ makes an angle of $\frac{\pi}{6}$ radian with the positive $x$-axis.
2. Use your calculator to find the point of intersection of the line $y_3 = \frac{\sqrt{3}}{3}x$ and the unit circle in quadrant I.
3. What is the length of the arc from the point $(1, 0)$ to this point of intersection?
4. Find $\cos \frac{\pi}{6}$ using your calculator. How does this compare to the $x$-coordinate of the point of intersection found in Item 2?
5. Find $\sin \frac{\pi}{6}$ using your calculator. How does this compare to the $y$-coordinate of the point of intersection found in Item 2?
6. If $(x, y)$ denotes the point found in Item 2, find approximations for $\frac{y}{x}$ and for $\frac{\sqrt{3}}{3}$. How do they compare? How does the tangent of the angle relate to the slope of the line?

In Figure 74, the graph of the unit circle $x^2 + y^2 = 1$ includes important information. For several special angles, degree and radian measures are given for the first counterclockwise revolution, and the coordinates of the points on the circle are also given. These coordinates can be used to find the trigonometric (or circular) functions of common angles.

For example, if $\theta = 120°$ or $\frac{2\pi}{3}$ radians is in standard position, then its terminal side intersects the unit circle at the point $\left(-\frac{1}{2}, \frac{\sqrt{3}}{2}\right)$. It follows that $\cos \frac{2\pi}{3} = -\frac{1}{2}$ and $\sin \frac{2\pi}{3} = \frac{\sqrt{3}}{2}$. Other trigonometric function values can be found similarly.

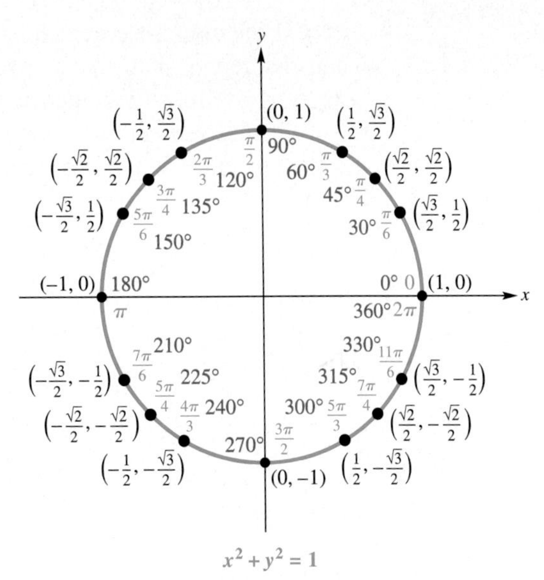


**FIGURE 74**

**EXAMPLE 3** *Finding Circular Function Values*

**(a)** Use Figure 74 to find the exact values of $\cos \frac{7\pi}{4}$ and $\sin \frac{7\pi}{4}$.

**(b)** Use Figure 74 to find the exact value of $\tan\left(-\frac{5\pi}{3}\right)$.

**(c)** Approximate cos 1.85 to four decimal places.

**Solution**

**(a)** In Figure 74, we can see that the terminal side of $\frac{7\pi}{4}$ radians intersects the unit circle at $\left(\frac{\sqrt{2}}{2}, -\frac{\sqrt{2}}{2}\right)$. Thus,

$$\cos \frac{7\pi}{4} = \frac{\sqrt{2}}{2} \quad \text{and} \quad \sin \frac{7\pi}{4} = -\frac{\sqrt{2}}{2}.$$

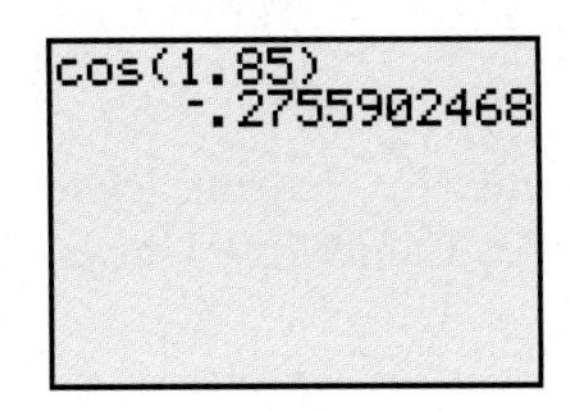


Radian Mode

**FIGURE 75**

**(b)** The angles of $-\frac{5\pi}{3}$ radians and $\frac{\pi}{3}$ radians are coterminal. Their terminal sides intersect the unit circle at $\left(\frac{1}{2}, \frac{\sqrt{3}}{2}\right)$, so

$$\tan\left(-\frac{5\pi}{3}\right) = \tan \frac{\pi}{3} = \frac{\frac{\sqrt{3}}{2}}{\frac{1}{2}} = \sqrt{3}.$$

**(c)** No degree symbol is shown, so we evaluate cos 1.85 in radian mode. To four decimal places, $\cos 1.85 \approx -.2756$. See Figure 75.

**EXAMPLE 4** *Finding a Number Given a Circular Function Value*

Approximate to three decimal places the value of $s$ in the interval $\left[0, \frac{\pi}{2}\right]$ if $\cos s = .457$.

**Solution** With the calculator set in radian mode, Figure 76 shows that $s = \cos^{-1} .457 \approx 1.096$.

cos⁻¹(.457) 1.09617687

Radian Mode

**FIGURE 76**

**FOR DISCUSSION**

A convenient way to see the sine, cosine, and tangent trigonometric ratios geometrically is shown in Figure 77 on the next page for $\theta$ in quadrants I and II. The circle shown is the unit circle, which has radius 1. By remembering this figure and the segments that represent the sine, cosine, and tangent functions, you can quickly recall the properties of the trigonometric functions. Horizontal line segments to the left of the origin and vertical line segments below the $x$-axis represent negative values. Note that the tangent line must be tangent to the circle at $(1, 0)$, no matter which quadrant $\theta$ lies in.

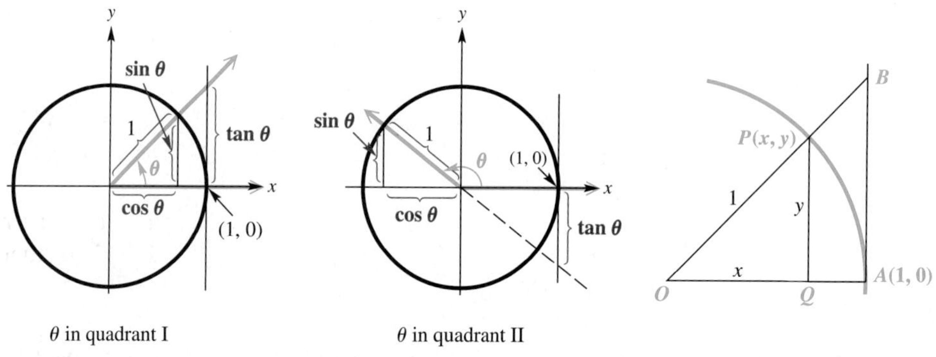


**FIGURE 77**

**FIGURE 78**

1. Label the triangles as shown in Figure 78. Use the definitions of the trigonometric functions and similar triangles to show that $PQ = \sin\theta$, $OQ = \cos\theta$, and $AB = \tan\theta$.
2. Sketch similar figures for $\theta$ in quadrants III and IV.

## Applications of Circular Functions

Because the moon orbits Earth, we observe different phases of the moon during the period of a month. In Figure 79, $t$ is called the *phase angle*. The *phase F* of the moon is computed by

$$F(t) = \frac{1}{2}(1 - \cos t),$$

and gives the fraction of the moon's face that is illuminated by the sun. (*Source:* Duffet-Smith, P., *Practical Astronomy with Your Calculator*, Cambridge University Press, 1988.)

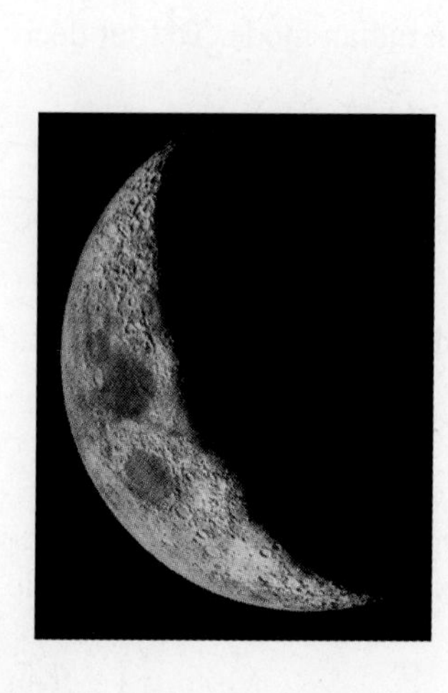

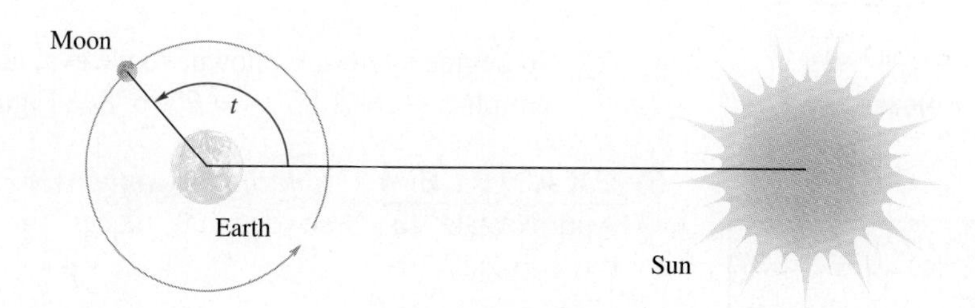


**FIGURE 79**

**EXAMPLE 5** *Modeling the Phases of the Moon*

Let $F(t) = \frac{1}{2}(1 - \cos t)$. Evaluate each expression and interpret the result.

**(a)** $F(0)$ **(b)** $F\left(\frac{\pi}{2}\right)$ **(c)** $F(\pi)$ **(d)** $F\left(\frac{3\pi}{2}\right)$

**Solution**

**(a)** $F(0) = \frac{1}{2}(1 - \cos 0) = \frac{1}{2}(1 - 1) = 0$. When $\theta = 0$, the moon is located between Earth and the sun. Since $F = 0$, the face of the moon is not visible, which corresponds to a *new moon*.

**(b)** $F\left(\frac{\pi}{2}\right) = \frac{1}{2}\left(1 - \cos \frac{\pi}{2}\right) = \frac{1}{2}(1 - 0) = \frac{1}{2}$. When $t = \frac{\pi}{2}$, $F = \frac{1}{2}$. Thus, half the face of the moon is visible. This phase is called the *first quarter*.

**(c)** $F(\pi) = \frac{1}{2}(1 - \cos \pi) = \frac{1}{2}[1 - (-1)] = 1$. When $t = \pi$, Earth is between the moon and the sun. Since $F = 1$, the face of the moon is completely visible, which corresponds to a *full moon*.

**(d)** $F\left(\frac{3\pi}{2}\right) = \frac{1}{2}\left(1 - \cos \frac{3\pi}{2}\right) = \frac{1}{2}(1 - 0) = \frac{1}{2}$. When $t = \frac{3\pi}{2}$, $F = \frac{1}{2}$. Thus, half the face of the moon is visible. This phase is called the *last quarter*.

Because the values of the circular functions repeat every $2\pi$, they are used to model natural phenomena, such as ocean tides, that repeat at regular intervals.

**EXAMPLE 6** *Modeling Tides*

Figure 80 shows a function $f$ that models the tides in feet at Clearwater Beach, Florida, $x$ hours after midnight starting on August 26, 1998. (*Source:* Pentcheff, D., *WWW Tide and Current Predictor.*)

**(a)** Find the time between high tides.

**(b)** What is the difference in water levels between high tide and low tide?

**(c)** The tides can be modeled by

$$f(x) = .6 \cos[.511(x - 2.4)] + 2.$$

Estimate the tides when $x = 10$.

**FIGURE 80**

**Solution**

**(a)** A high tide corresponds to a peak on the graph. The time between peaks is 12.3 hours, since

$$14.7 - 2.4 = 12.3 \quad \text{and} \quad 27 - 14.7 = 12.3.$$

**(b)** High tides were 2.6 feet and low tides were 1.4 feet. Their difference is 1.2 feet.

**(c)** $f(10) = .6 \cos[.511(10 - 2.4)] + 2 \approx 1.56$ feet

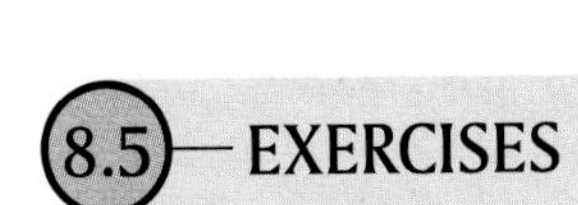

## 8.5 EXERCISES

*Find the exact value for each expression.*

**1.** $\sin \frac{7\pi}{6}$ **2.** $\cos \frac{5\pi}{3}$ **3.** $\tan \frac{3\pi}{4}$ **4.** $\cos \frac{7\pi}{6}$ **5.** $\sec \frac{2\pi}{3}$ **6.** $\csc \frac{11\pi}{6}$

**7.** $\cot \frac{5\pi}{6}$ **8.** $\cos\left(-\frac{4\pi}{3}\right)$ **9.** $\sin\left(-\frac{5\pi}{6}\right)$ **10.** $\tan \frac{17\pi}{3}$ **11.** $\sec \frac{23\pi}{6}$ **12.** $\csc \frac{13\pi}{3}$

*Approximate each expression to four decimal places.*

**13.** $\sin .6109$ **14.** $\sin .8203$ **15.** $\cos(-1.1519)$ **16.** $\cos(-5.2825)$

**17.** tan 4.0203 **18.** tan 6.4752 **19.** csc(−9.4946) **20.** csc 1.3875

**21.** sec 2.8440 **22.** sec(−8.3429) **23.** cot 6.0301 **24.** cot 3.8426

*Concept Check* *Each figure in Exercises 25–28 shows angle $\theta$ in standard position with its terminal side intersecting the unit circle. Evaluate the six trigonometric functions of $\theta$.*

**25.**

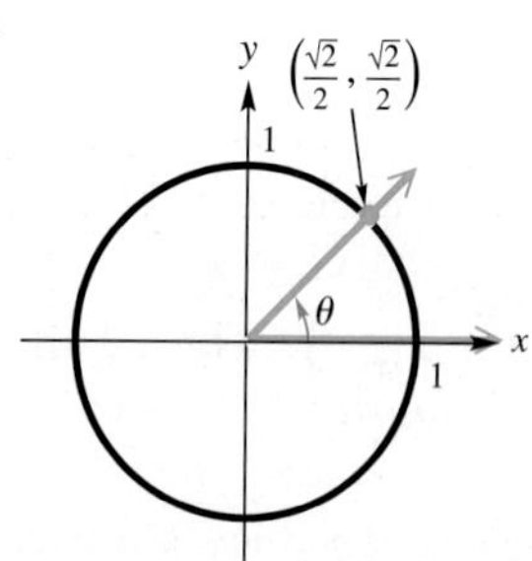


**26.**

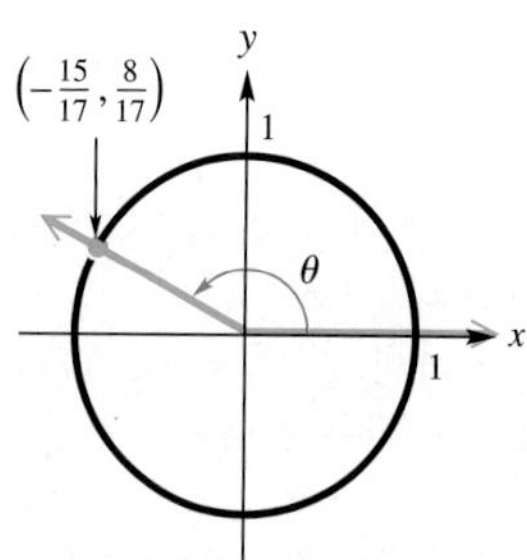


**27.**

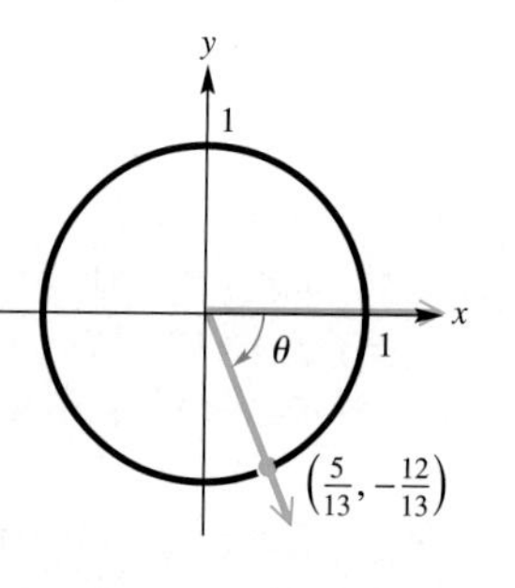


**28.**

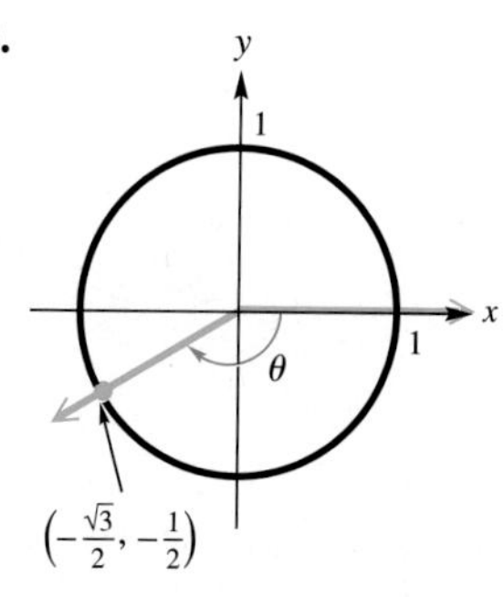


*Concept Check* *The figure displays a unit circle and an angle of* 1 *radian. The tick marks on the circle are spaced at every two-tenths radian. Use the figure to estimate each value.*

**29.** cos .8

**30.** sin 4

**31.** An angle whose cosine is −.65

**32.** An angle whose sine is −.95

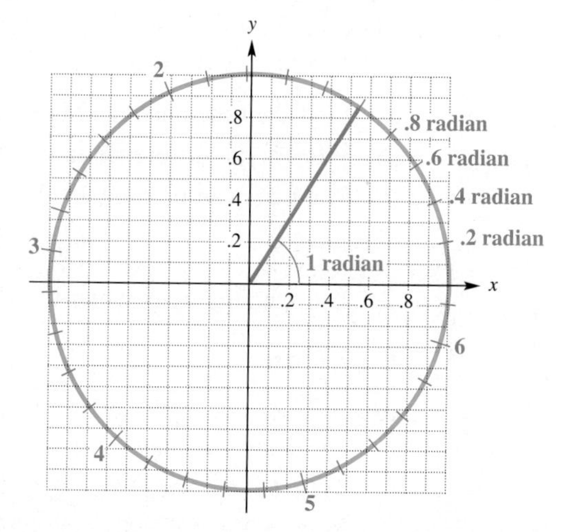


*Approximate to four decimal places the value of s in the interval* $\left[0, \frac{\pi}{2}\right]$ *that makes each statement true.*

**33.** $\tan s = .21264138$ **34.** $\cos s = .78269876$ **35.** $\sin s = .99184065$

**36.** $\cot s = .29949853$ **37.** $\cot s = .09637041$ **38.** $\csc s = 1.0219553$

*Concept Check* *In Exercises 39 and 40, each graphing calculator screen shows a point on the unit circle. What is the length of the shortest arc of the circle from* (1, 0) *to the point?*

**39.**

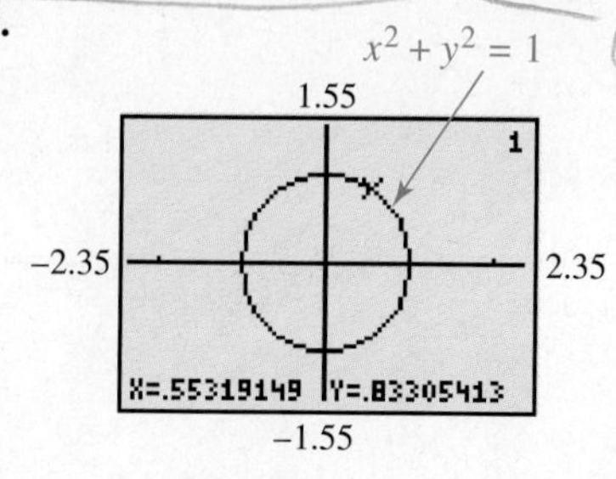


**40.**

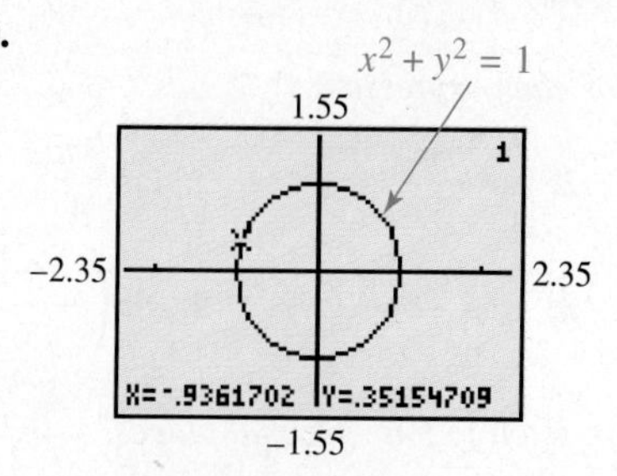

*Suppose an arc of length s lies on the unit circle $x^2 + y^2 = 1$, starting at the point $(1, 0)$ and terminating at the point $(x, y)$. (See Figure 68.) Approximate coordinates for $(x, y)$ to four decimal places.*

**41.** $s = 2.5$ **42.** $s = 3.4$ **43.** $s = -7.4$ **44.** $s = -3.9$

*Concept Check* *For each value of s, decide in which quadrant an angle of s radians lies by evaluating sin s and cos s.*

**45.** $s = 51$ **46.** $s = 49$ **47.** $s = 65$ **48.** $s = 79$

*(Modeling)* *Solve each problem.*

**49.** *Daylight Hours* The average number of daylight hours at San Antonio, Texas, can be modeled by

$$f(x) = 1.95 \cos\left[\frac{\pi}{6}(x - 6.6)\right] + 12.15,$$

where $x = 1$ corresponds to January 1, $x = 2$ to February 1, and so on. Evaluate $f(7)$ and interpret the result.

**50.** *Temperature in Fairbanks* The temperature in Fairbanks, Alaska, is modeled by

$$T(x) = 37 \sin\left[\frac{2\pi}{365}(x - 101)\right] + 25,$$

where $T(x)$ is the temperature in degrees Fahrenheit on day $x$, with $x = 1$ corresponding to January 1 and $x = 365$ corresponding to December 31. Use a calculator to estimate the temperature on each day. (*Source:* Lando, B. and C. Lando, "Is the Graph of Temperature Variation a Sine Curve?" *The Mathematics Teacher,* September 1977.)

**(a)** March 1 (day 60) **(b)** April 1 (day 91)

**(c)** Day 150 **(d)** June 15

**(e)** September 1 **(f)** October 31

**51.** *Maximum Temperatures* The maximum afternoon temperature $t$ (in degrees Fahrenheit) in a given city might be modeled by

$$t = 60 - 30 \cos\frac{x\pi}{6},$$

where $x$ represents the month, with $x = 0$ corresponding to January, $x = 1$ to February, and so on. Find the maximum afternoon temperature for each month.

**(a)** January **(b)** April **(c)** May

**(d)** June **(e)** August **(f)** October

**52.** *Voltage* Electric ranges and ovens often use a higher voltage than that found in normal household outlets. This voltage can be modeled by

$$V(t) = 310 \sin 120\pi t,$$

where $t$ represents time in seconds. Evaluate $V\left(\frac{1}{240}\right)$ and interpret the result.

**53.** *Phases of the Moon* Refer to Example 5. Find all phase angles $t$ that correspond to a full moon and all phase angles that correspond to a new moon. Assume that $t$ can be any angle.

**54.** *Daylight Hours* The ability to calculate the number of daylight hours $H$ at any location is important in estimating the potential solar energy production. $H$ can be calculated using the formula

$$\cos(.1309H) = -\tan D \tan L,$$

where $D$ is the declination of the sun and $L$ is the latitude. Use radian mode to calculate the shortest and longest days in Minneapolis, Minnesota, if its latitude is $L = 44.88°$, the shortest day occurs when $D = -23.44°$, and the longest day occurs when $D = 23.44°$. (*Source:* Winter, C., R. Sizmann, and Vant-Hunt (Editors), *Solar Power Plants,* Springer-Verlag, 1991.)

# 8.6 GRAPHS OF THE SINE AND COSINE FUNCTIONS

Periodic Functions • Graph of the Sine Function • Graph of the Cosine Function • Graphing Techniques, Amplitude, and Period • Translations • Determining a Trigonometric Model Using Curve Fitting

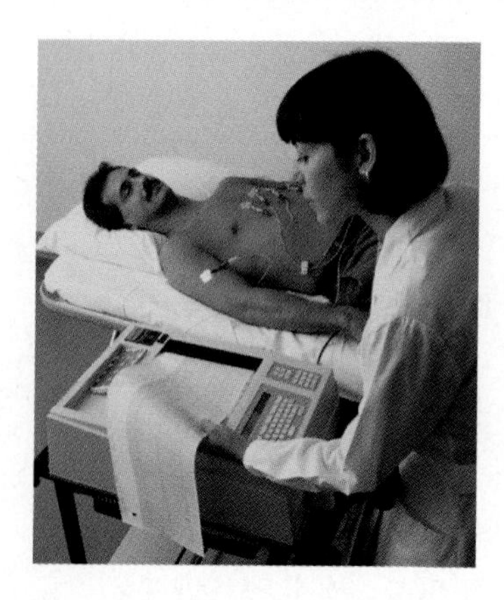

## Periodic Functions

Many things in daily life repeat with a predictable pattern: in warm areas electricity use goes up in summer and down in winter, the price of fresh fruit goes down in summer and up in winter, and attendance at amusement parks increases in spring and declines in autumn. Because the sine and cosine functions repeat their values over and over in a regular pattern, they are examples of *periodic functions*. Figure 81 shows a periodic graph that represents a normal heartbeat.

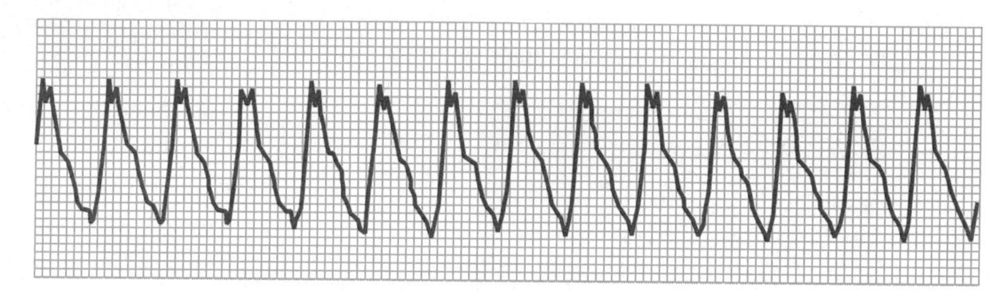

FIGURE 81

**Looking Ahead to Calculus**

Periodic functions are used throughout calculus, so you will need to know their characteristics. One use of these functions is to describe the location of a point in the plane using *polar coordinates*, an alternative to rectangular coordinates. (See Chapter 10.)

**Periodic Function**

A **periodic function** is a function $f$ such that

$$f(x) = f(x + np),$$

for every real number $x$ in the domain of $f$, every integer $n$, and some positive real number $p$. The smallest possible positive value of $p$ is the **period** of the function.

The circumference of the unit circle is $2\pi$, so the smallest value of $p$ for which the sine and cosine functions repeat is $2\pi$. Therefore, the sine and cosine functions are periodic functions with period $2\pi$.

## Graph of the Sine Function

In the previous section we saw that for a real number $s$, the point on the unit circle corresponding to $s$ has coordinates $(\cos s, \sin s)$. See Figure 82, and trace along the circle to verify the results shown in the table.

FIGURE 82

| As $s$ Increases from | $\sin s$ | $\cos s$ |
| --- | --- | --- |
| $0$ to $\frac{\pi}{2}$ | Increases from 0 to 1 | Decreases from 1 to 0 |
| $\frac{\pi}{2}$ to $\pi$ | Decreases from 1 to 0 | Decreases from 0 to $-1$ |
| $\pi$ to $\frac{3\pi}{2}$ | Decreases from 0 to $-1$ | Increases from $-1$ to 0 |
| $\frac{3\pi}{2}$ to $2\pi$ | Increases from $-1$ to 0 | Increases from 0 to 1 |

Any letter can be used instead of $s$, so to avoid confusion when graphing the sine function, we use $x$ rather than $s$; this corresponds to our usual choice of letters in the $xy$-coordinate system. Selecting key values of $x$ and finding the corresponding values of $\sin x$ leads to the table in Figure 83. The increment in the table is $\frac{\pi}{2}$.

To obtain the traditional graph of a portion of the sine function shown in Figure 83, we plot the points from the table of values and join them with a smooth curve. Since $y = \sin x$ is periodic and has $(-\infty, \infty)$ as its domain, the graph continues in the same pattern in both directions. This graph is called a **sine wave** or **sinusoid.** You should learn this shape and be able to sketch it quickly. The sine function is an odd function. For all $x$, $\sin(-x) = -\sin x$.

TECHNOLOGY NOTE

Graphing calculators often have a window designated for graphing circular functions. We refer to the window $[-2\pi, 2\pi]$ by $[-4, 4]$ with Xscl $= \frac{\pi}{2}$ and Yscl $= 1$ as the **trig viewing window.** Your model may use a different "standard" viewing window for the graphs of trigonometric functions.

FUNCTION CAPSULE

**SINE FUNCTION** $\quad f(x) = \sin x$

Domain: $(-\infty, \infty)$ $\quad$ Range: $[-1, 1]$

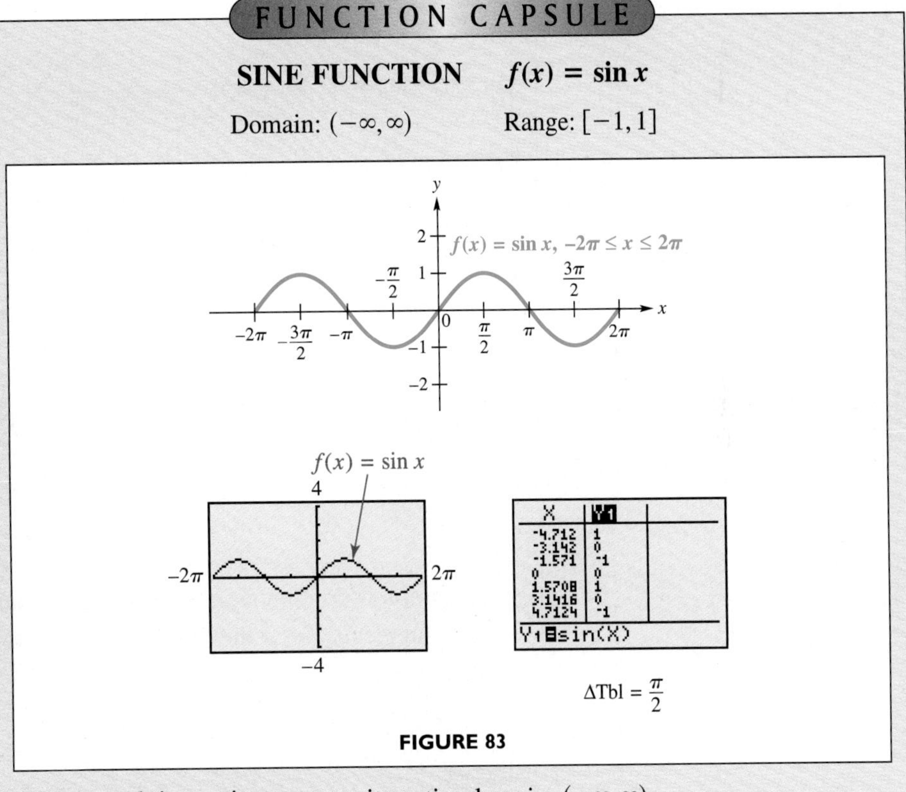


FIGURE 83

- The graph is continuous over its entire domain, $(-\infty, \infty)$.
- Its $x$-intercepts are of the form $n\pi$, where $n$ is an integer.
- Its period is $2\pi$.
- The graph is symmetric with respect to the origin, and it is an odd function.

A comprehensive graph of a sinusoid will consist of at least one period of the graph. It will show the extreme points.

## Graph of the Cosine Function

The graph of $y = \cos x$ can be found in much the same way as the graph of $y = \sin x$. A table of values is shown on the next page in Figure 84 for $y = \cos x$, using the same values for $x$ as before. Figure 84 also shows a traditional graph of $y = \cos x$. Notice

**Looking Ahead to Calculus**

The discussion of the derivative of a function in calculus shows that for the sine function, the slope of the tangent line at any point $x$ is given by $\cos x$. For example, look at the graph of $y = \sin x$ and notice that a tangent line at $x = \pm\frac{\pi}{2}, \pm\frac{3\pi}{2}, \pm\frac{5\pi}{2}, \ldots$ will be horizontal and thus have slope 0. Now look at the graph of $y = \cos x$ and see that for these values, $\cos x = 0$.

that it has the same shape as the graph of $y = \sin x$. It is, in fact, the graph of the sine function shifted, or translated, $\frac{\pi}{2}$ units to the left. The cosine function is an even function. For all $x$, $\cos(-x) = \cos x$.

**FUNCTION CAPSULE**

**COSINE FUNCTION** $\quad f(x) = \cos x$

Domain: $(-\infty, \infty)$ $\quad$ Range: $[-1, 1]$

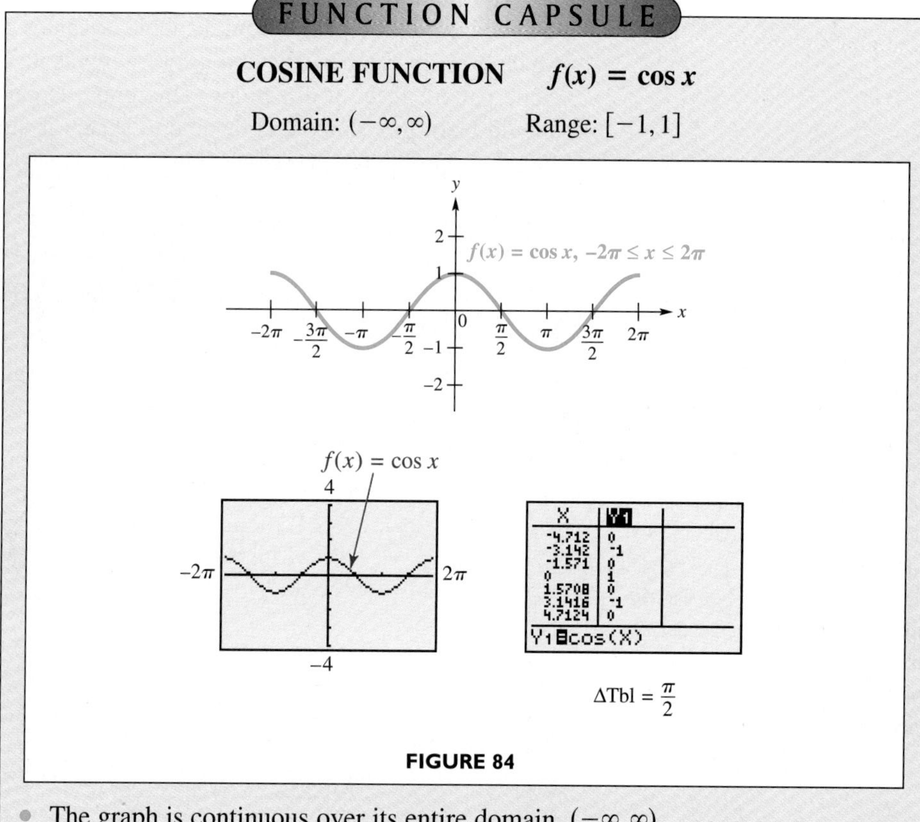


**FIGURE 84**

- The graph is continuous over its entire domain, $(-\infty, \infty)$.
- Its $x$-intercepts are of the form $(2n + 1)\frac{\pi}{2}$, where $n$ is an integer.
- Its period is $2\pi$.
- The graph is symmetric with respect to the $y$-axis, and it is an even function.

## FOR DISCUSSION

*Note: The discussion that follows is specific to TI-83 and TI-83 Plus calculators. It may be modified to apply to other models of graphing calculators.*

If you have a TI-83 or TI-83 Plus calculator, adjust the settings to correspond to the following five screens.

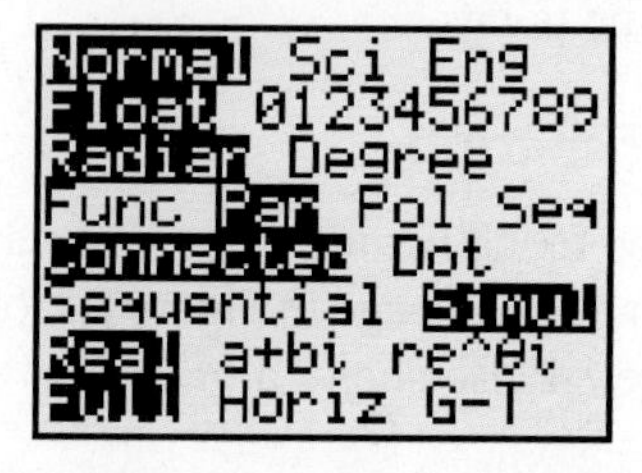

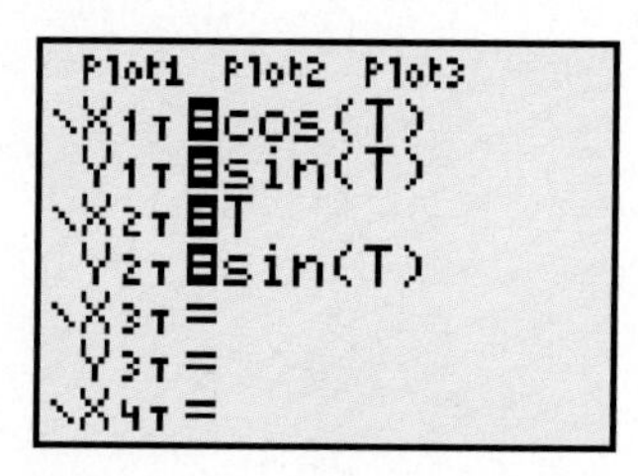

```
WINDOW
 Tmin=0
 Tmax=6.2831853...
 Tstep=.0785398...
 Xmin=-1
 Xmax=6.2831853...
 Xscl=1.5707963...
↓Ymin=-2.5
```

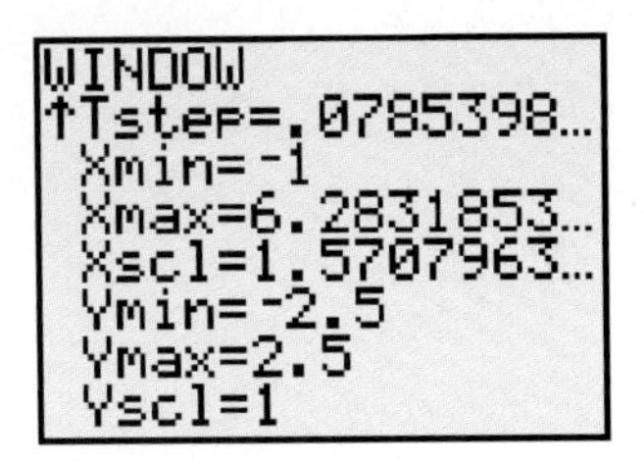


In the two screens above, Tmax is $2\pi$, Tstep is $\frac{\pi}{40}$, Xmax is $2\pi$, and Xscl is $\frac{\pi}{2}$. Now graph these two equations (which are in parametric form), and watch as the unit circle and the sine function are graphed simultaneously. Press the (TRACE) key once to get the screen shown in Figure 85, and then press the up-arrow key to get the screen shown in Figure 86. The screen in Figure 85 gives a unit circle interpretation of $\cos 0 = 1$ and $\sin 0 = 0$. The screen in Figure 86 gives a rectangular coordinate graph interpretation of $\sin 0 = 0$.

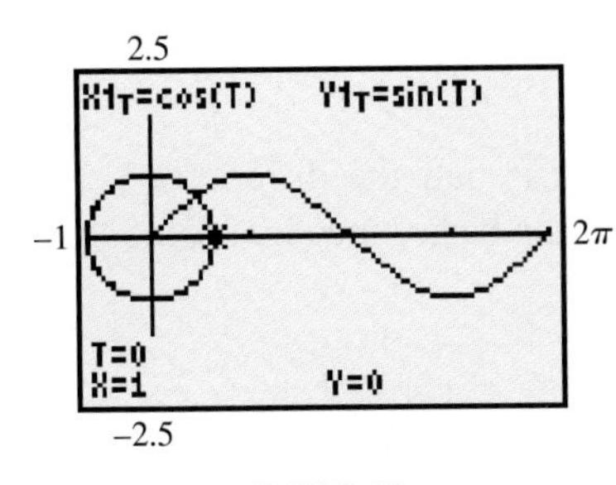


FIGURE 85

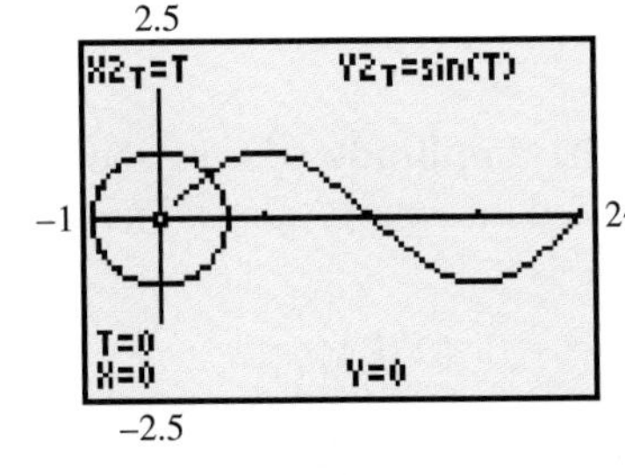


FIGURE 86

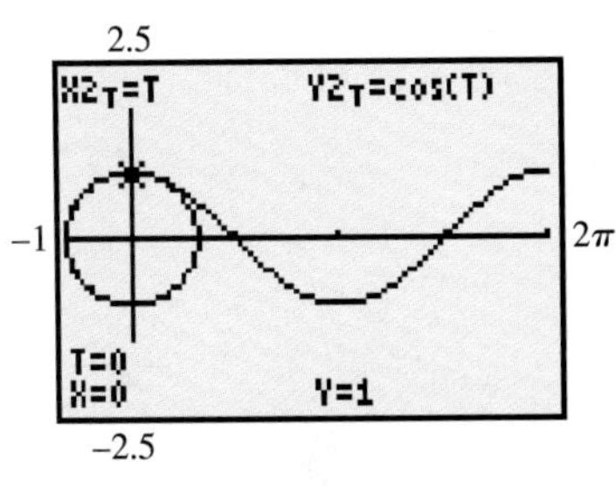


FIGURE 87

Now go back and redefine $Y_{2T}$ as cos(T). Graph both equations again; the second screen will look like Figure 87 after the (TRACE) and up-arrow keys are pressed. This screen indicates that $\cos 0 = 1$.

1. On the unit circle, let $T = 2$. What values of X and Y are displayed? Interpret these values.
2. On the sine graph, trace to the point where $T = 1.9$. What values of X and Y are displayed? Interpret these values with an equation in X and Y.
3. Repeat Item 2, but use the cosine graph.

## Graphing Techniques, Amplitude, and Period

The examples that follow show graphs that are "stretched" or "compressed" either vertically, horizontally, or both when compared with the graphs of $y = \sin x$ or $y = \cos x$.

### EXAMPLE 1 *Graphing $y = a \sin x$*

Graph $y = 2 \sin x$, and compare to the graph of $y = \sin x$.

**Analytic Solution**

For a given value of $x$, the value of $y$ is twice as large as it would be for $y = \sin x$, as shown in the table of values on the next page. The only change in the graph is the range, which becomes $[-2, 2]$. See Figure 88 on the next page, which includes a graph of $y = \sin x$ for comparison.

**Graphing Calculator Solution**

Define

$$Y_1 = 2 \sin X,$$

and direct the calculator to use a thick line to graph it in the trig

*(continued)*

| $x$ | $0$ | $\frac{\pi}{2}$ | $\pi$ | $\frac{3\pi}{2}$ | $2\pi$ |
|---|---|---|---|---|---|
| $\sin x$ | $0$ | $1$ | $0$ | $-1$ | $0$ |
| $2 \sin x$ | $0$ | $2$ | $0$ | $-2$ | $0$ |

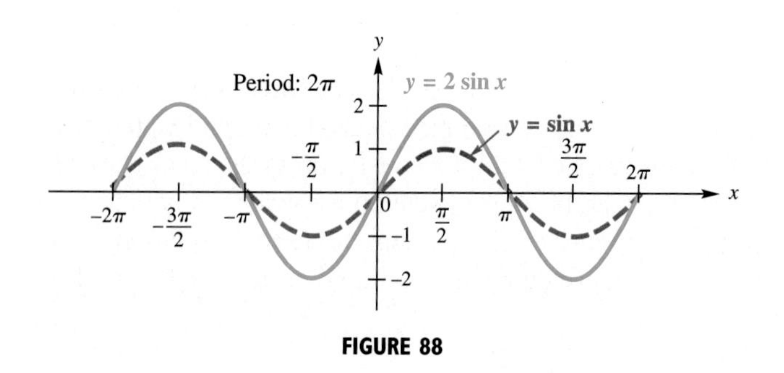


FIGURE 88

viewing window. See Figure 89. For comparison, a thin-line graph is shown for

$$Y_2 = \sin X.$$

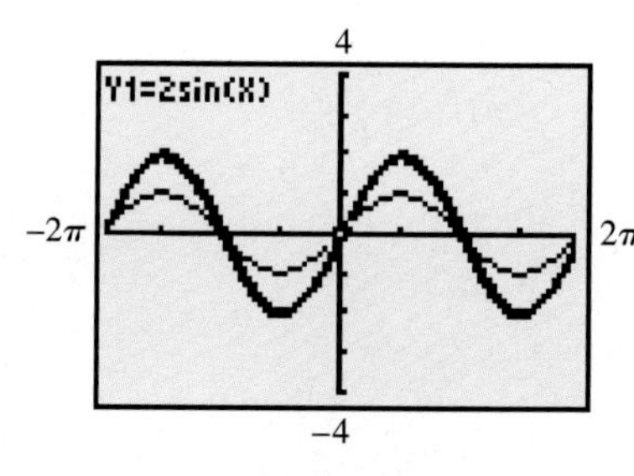


FIGURE 89

The **amplitude** of a periodic function is half the difference between the maximum and minimum values. Thus, for both the basic sine and cosine functions, the amplitude is

$$\frac{1}{2}[1-(-1)]=\frac{1}{2}(2)=1.$$

Generalizing from Example 1 gives the following.

**Amplitude**

The graph of $y = a \sin x$ or $y = a \cos x$, with $a \neq 0$, will have the same shape as the graph of $y = \sin x$ or $y = \cos x$, respectively, except with range $[-|a|, |a|]$. The amplitude is $|a|$.

No matter what the value of the amplitude, the periods of $y = a \sin x$ and $y = a \cos x$ are still $2\pi$. Now consider $y = \sin 2x$. We can complete a table of values for the interval $[0, 2\pi]$.

| $x$ | $0$ | $\frac{\pi}{4}$ | $\frac{\pi}{2}$ | $\frac{3\pi}{4}$ | $\pi$ | $\frac{5\pi}{4}$ | $\frac{3\pi}{2}$ | $\frac{7\pi}{4}$ | $2\pi$ |
|---|---|---|---|---|---|---|---|---|---|
| $\sin 2x$ | $0$ | $1$ | $0$ | $-1$ | $0$ | $1$ | $0$ | $-1$ | $0$ |

Note that one complete cycle occurs in $\pi$ units, not $2\pi$ units. Therefore, the period here is $\pi$, which equals $\frac{2\pi}{2}$. What about $y = \sin 4x$? Look at the next table.

| $x$ | $0$ | $\frac{\pi}{8}$ | $\frac{\pi}{4}$ | $\frac{3\pi}{8}$ | $\frac{\pi}{2}$ | $\frac{5\pi}{8}$ | $\frac{3\pi}{4}$ | $\frac{7\pi}{8}$ | $\pi$ |
|---|---|---|---|---|---|---|---|---|---|
| $\sin 4x$ | $0$ | $1$ | $0$ | $-1$ | $0$ | $1$ | $0$ | $-1$ | $0$ |

These values suggest that a complete cycle is achieved in $\frac{\pi}{2}$ or $\frac{2\pi}{4}$ units, which is reasonable since

$$\sin\left(4 \cdot \frac{\pi}{2}\right) = \sin 2\pi = 0.$$

In general, the graph of a function of the form $y = \sin bx$ or $y = \cos bx$, for $b > 0$, will have a period different from $2\pi$ when $b \neq 1$. To see why this is so, remember that the values of $\sin bx$ or $\cos bx$ will take on all possible values as $bx$ ranges from 0 to $2\pi$. Therefore, to find the period of either of these functions, we must solve the compound inequality

$$0 \leq bx \leq 2\pi$$

$$0 \leq x \leq \frac{2\pi}{b}. \quad \text{Divide by the positive number } b.$$

Thus, the period is $\frac{2\pi}{b}$. By dividing the interval $\left[0, \frac{2\pi}{b}\right]$ into four equal parts, we obtain the values for which $\sin bx$ or $\cos bx$ is $-1$, 0, or 1. These values will give minimum points, $x$-intercepts, and maximum points on the graph. Once these points are determined, we can sketch the graph by joining the points with a smooth sinusoidal curve. (If a function has $b < 0$, then the identities of the next chapter can be used to write the function as one in which $b > 0$.)

**NOTE** To divide an interval into four equal parts, follow these steps.

1. Find the midpoint of the interval by adding the $x$-values of the endpoints and dividing by 2.
2. Find the two midpoints of the intervals found in Step 1, using the same procedure.

### EXAMPLE 2 *Graphing $y = \sin bx$*

Graph $y = \sin 2x$, and compare to the graph of $y = \sin x$.

**Analytic Solution**

For this function, $b = 2$, so the period is $\frac{2\pi}{2} = \pi$. Therefore, the graph will complete one period over the interval $[0, \pi]$.

The endpoints are 0 and $\pi$, and the three points in the middle are

$$\frac{1}{2}\left(0 + \frac{\pi}{2}\right), \quad \frac{1}{2}(0 + \pi), \quad \text{and} \quad \frac{1}{2}\left(\frac{\pi}{2} + \pi\right),$$

which give the following $x$-values.

| $0$ | $\frac{\pi}{4}$ | $\frac{\pi}{2}$ | $\frac{3\pi}{4}$ | $\pi$ |
|---|---|---|---|---|
| ↑ | ↑ | ↑ | ↑ | ↑ |
| Left endpoint | First quarter point | Midpoint | Third quarter point | Right endpoint |

**Graphing Calculator Solution**

Figure 91 on the next page shows the graph of

$$Y_1 = \sin 2X$$

as a thick line and the graph of

$$Y_2 = \sin X$$

as a thin line in the trig viewing window.

*(continued)*

We plot the points from the table of values given earlier, and join them with a smooth sinusoidal curve. More of the graph can be sketched by repeating this cycle, as shown in Figure 90. The amplitude is not changed. The graph of $y = \sin x$ is included for comparison.

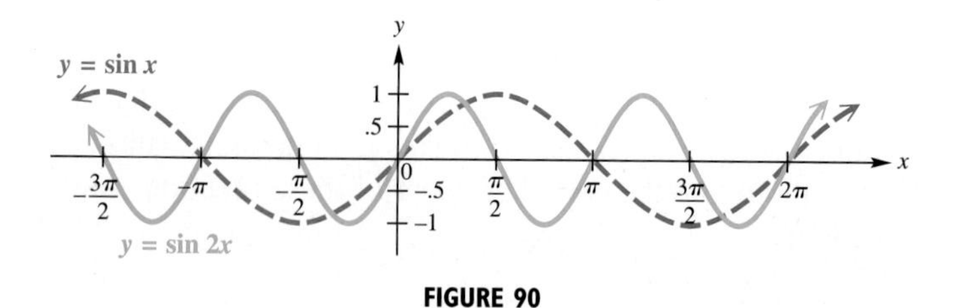


FIGURE 90

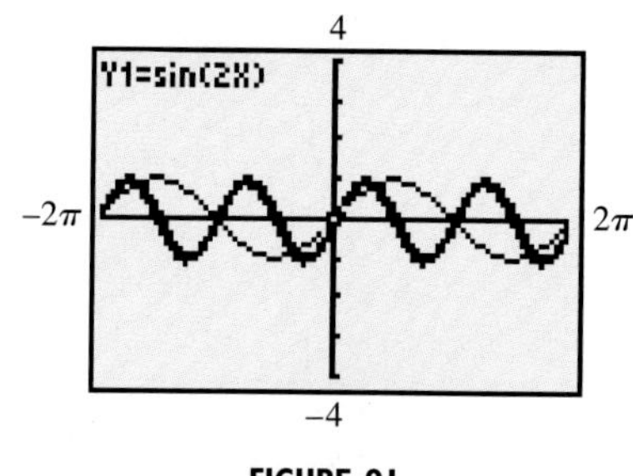


FIGURE 91

The graph of $y = \sin 2x$ oscillates twice as fast as the graph of $y = \sin x$.

Generalizing from Example 2 leads to the following result.

**Period**

For $b > 0$, the graph of $y = \sin bx$ will look like that of $y = \sin x$, but with period $\frac{2\pi}{b}$. Also, the graph of $y = \cos bx$ will look like that of $y = \cos x$, but with period $\frac{2\pi}{b}$.

**EXAMPLE 3** *Graphing $y = \cos bx$*

Graph $y = \cos \frac{2}{3}x$ over one period.

**Analytic Solution**

The period is $\frac{2\pi}{\frac{2}{3}} = 3\pi$. Evaluate the function at the $x$-values

$$0, \quad \frac{3\pi}{4}, \quad \frac{3\pi}{2}, \quad \frac{9\pi}{4}, \quad \text{and} \quad 3\pi$$

to find minimum points, maximum points, and $x$-intercepts. The graph is shown in Figure 92.

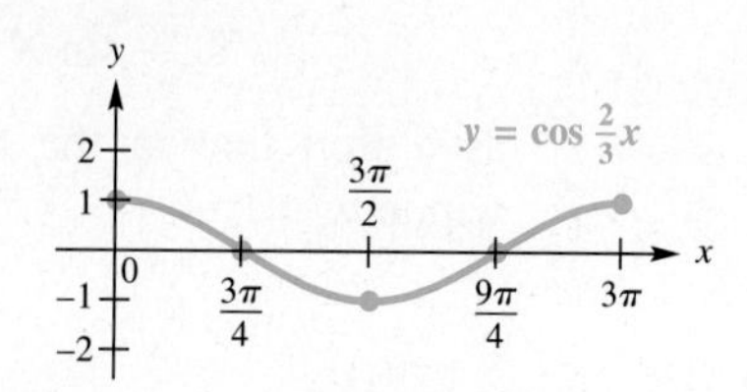


FIGURE 92

**Graphing Calculator Solution**

Use the window $[0, 3\pi]$ by $[-2, 2]$, with Xscl $= \frac{3\pi}{4}$ and Yscl $= 1$ to obtain the graph shown in Figure 93.

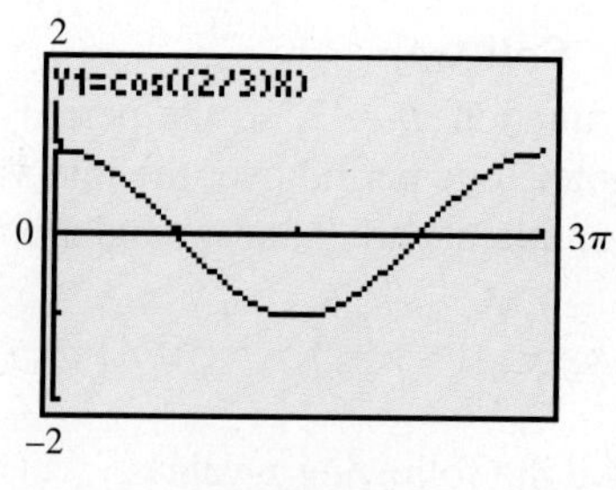


FIGURE 93

We chose Xscl $= \frac{3\pi}{4}$ so that $x$-intercepts, maximums, and minimums coincide with tick marks on the axis.

The method used in the analytic solutions in Examples 1–3 is summarized as follows.

> **Sketching Traditional Graphs of the Sine and Cosine Functions**
>
> To graph $y = a \sin bx$ or $y = a \cos bx$, with $b > 0$,
>
> 1. Find the period, $\frac{2\pi}{b}$. Start at 0 on the $x$-axis, and lay off a distance of $\frac{2\pi}{b}$.
> 2. Divide the interval into four equal parts. (See the Note preceding Example 2.)
> 3. Evaluate the function for each of the five $x$-values resulting from Step 2. The points will be maximum points, minimum points, and $x$-intercepts.
> 4. Plot the points found in Step 3, and join them with a sinusoidal curve with amplitude $|a|$.
> 5. Draw additional cycles of the graph, to the right and to the left, as needed.

The function in Example 4 has both amplitude and period affected by constants.

**EXAMPLE 4** *Graphing $y = a \sin bx$*

Graph $y = -2 \sin 3x$ using the preceding guidelines.

**Solution**

**Step 1** For this function, $b = 3$, so the period is $\frac{2\pi}{3}$. The function will be graphed over the interval $\left[0, \frac{2\pi}{3}\right]$.

**Step 2** Divide the interval $\left[0, \frac{2\pi}{3}\right]$ into four equal parts to get the $x$-values $0, \frac{\pi}{6}, \frac{\pi}{3}, \frac{\pi}{2}$, and $\frac{2\pi}{3}$.

**Step 3** Make a table of values determined by the $x$-values from Step 2.

| $x$ | $0$ | $\frac{\pi}{6}$ | $\frac{\pi}{3}$ | $\frac{\pi}{2}$ | $\frac{2\pi}{3}$ |
|---|---|---|---|---|---|
| $3x$ | $0$ | $\frac{\pi}{2}$ | $\pi$ | $\frac{3\pi}{2}$ | $2\pi$ |
| $\sin 3x$ | $0$ | $1$ | $0$ | $-1$ | $0$ |
| $-2 \sin 3x$ | $0$ | $-2$ | $0$ | $2$ | $0$ |

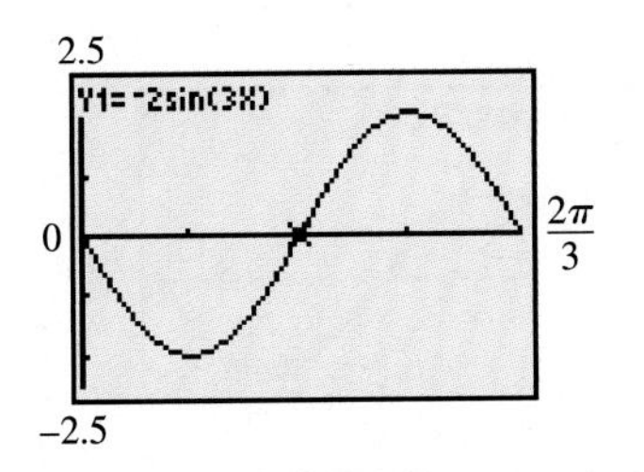


FIGURE 94

**Step 4** Plot the points $(0, 0)$, $\left(\frac{\pi}{6}, -2\right)$, $\left(\frac{\pi}{3}, 0\right)$, $\left(\frac{\pi}{2}, 2\right)$, and $\left(\frac{2\pi}{3}, 0\right)$, and join them with a sinusoidal curve with amplitude 2. See Figure 94.

**Step 5** The graph can be extended by repeating the cycle.

Notice that when $a$ is negative, the graph of $y = a \sin bx$ is a reflection across the $x$-axis of the graph of $y = |a| \sin bx$.

## Translations

In general, the graph of the function defined by $y = f(x - d)$ is translated *horizontally* when compared to the graph of $y = f(x)$. The translation is $d$ units to the right if $d > 0$ and $|d|$ units to the left if $d < 0$. See Figure 95. With trigonometric functions, a horizontal translation is called a **phase shift.** In the function $y = f(x - d)$, the expression $x - d$ is called the **argument.**

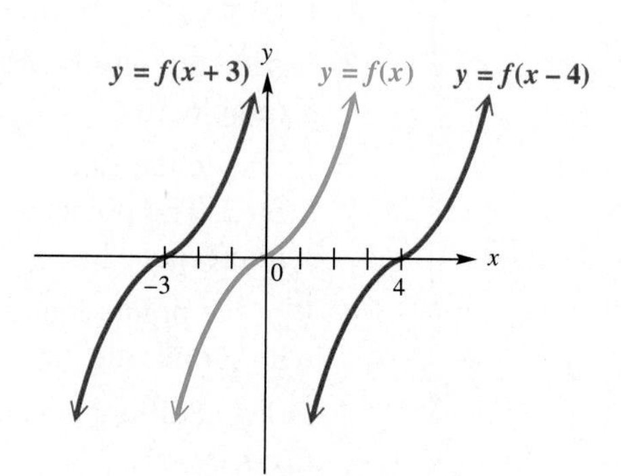


Horizontal Translations of $y = f(x)$

**FIGURE 95**

### EXAMPLE 5 *Graphing $y = \sin(x - d)$*

Graph $y = \sin\left(x - \frac{\pi}{3}\right)$, and compare to the graph of $y = \sin x$.

**Analytic Solution**

The argument $x - \frac{\pi}{3}$ indicates that the graph will be translated $\frac{\pi}{3}$ units to the *right* (the phase shift) as compared to the graph of $y = \sin x$. In Figure 96 the graph of $y = \sin x$ is shown as a dashed curve, and the graph of $y = \sin\left(x - \frac{\pi}{3}\right)$ is shown as a solid curve. First graph the basic circular function $y = \sin x$, and then graph the desired function by using the appropriate translation.

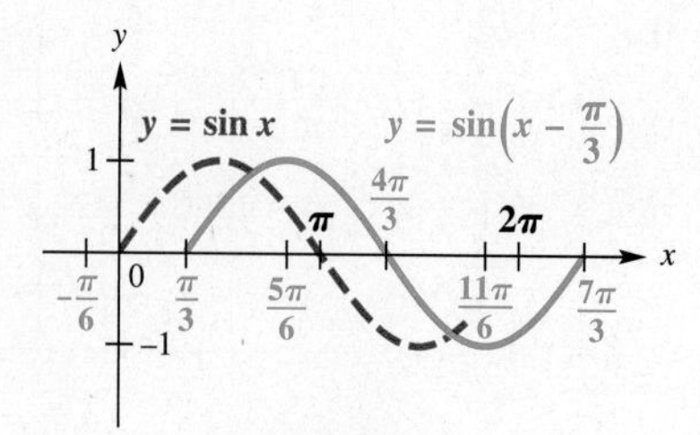


**FIGURE 96**

**Graphing Calculator Solution**

Figure 97 shows the graph of

$$Y_1 = \sin X$$

as a thin line and the graph of

$$Y_2 = \sin\left(X - \tfrac{\pi}{3}\right)$$

as a thick line. Note the horizontal translation.

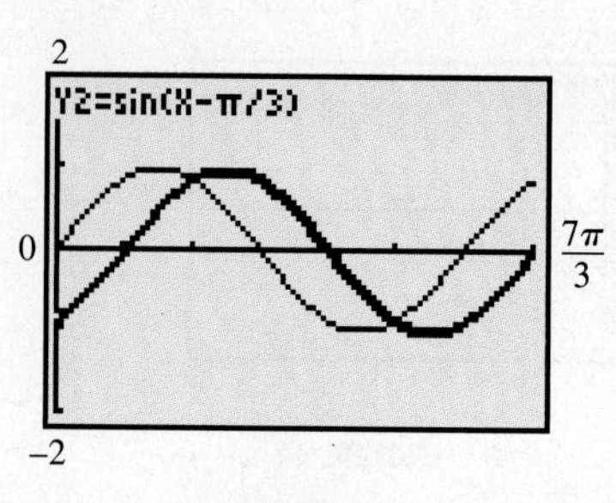


**FIGURE 97**

The graph of a function of the form $y = c + f(x)$ is translated *vertically* as compared with the graph of $y = f(x)$. See Figure 98. The translation is $c$ units upward if $c > 0$ and $|c|$ units downward if $c < 0$.

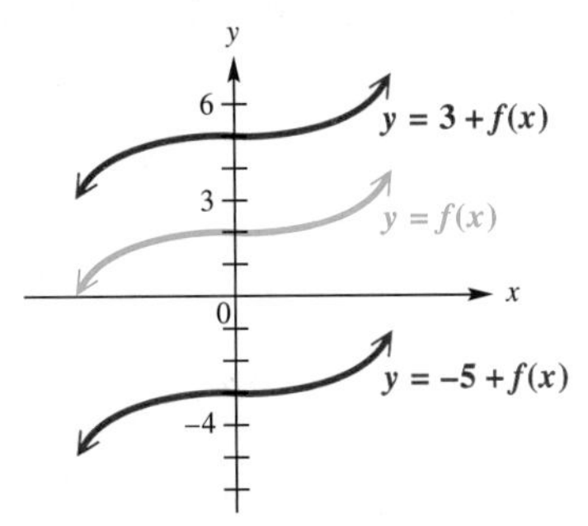


Vertical Translations of $y = f(x)$

**FIGURE 98**

**EXAMPLE 6** *Graphing $y = c + a \cos bx$*

Graph $y = 3 - 2 \cos 3x$.

### Analytic Solution

The values of $y$ will be 3 greater than the corresponding values of $y$ in $y = -2 \cos 3x$. This means that the graph of $y = 3 - 2 \cos 3x$ is the same as the graph of $y = -2 \cos 3x$, vertically translated 3 units upward. Since the period of $y = -2 \cos 3x$ is $\frac{2\pi}{3}$, the key points have $x$-values

$$0, \quad \frac{\pi}{6}, \quad \frac{\pi}{3}, \quad \frac{\pi}{2}, \quad \text{and} \quad \frac{2\pi}{3}.$$

The key points are shown on the graph in Figure 99, along with more of the graph, sketched using the fact that the function is periodic.

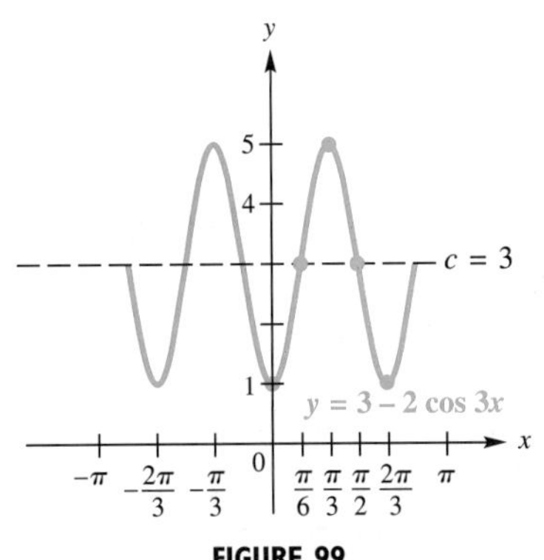


**FIGURE 99**

### Graphing Calculator Solution

Figure 100 shows the graph of

$$Y_1 = 3 - 2 \cos 3X$$

as a thick line. For comparison, the graph of

$$Y_2 = -2 \cos 3X$$

is shown as a thin line. Note the vertical translation. The thick graph is centered vertically on the horizontal axis $y = 3$.

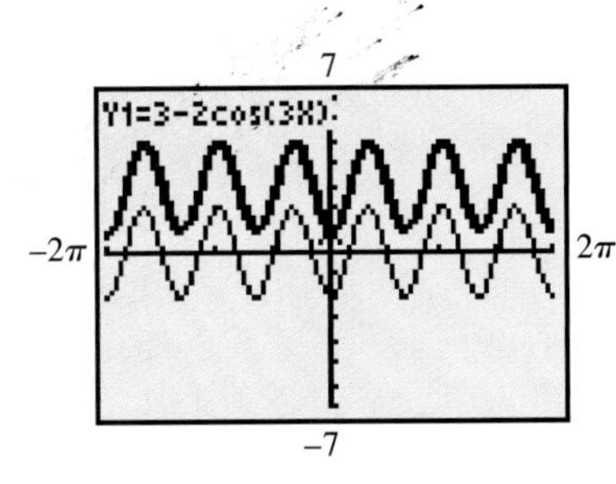


**FIGURE 100**

In the next example, we graph a function that involves all the types of stretching, compressing, and translating presented in this section.

**EXAMPLE 7** *Graphing $y = c + a \sin b(x - d)$*

Graph $y = -1 + 2\sin(4x + \pi)$.

**Analytic Solution**

First write the expression in the form $c + a \sin b(x - d)$ by factoring 4 out of the argument.

$$y = -1 + 2\sin\left[4\left(x + \frac{\pi}{4}\right)\right]$$

The amplitude is 2, the period is $\frac{2\pi}{4} = \frac{\pi}{2}$, and the graph is translated downward 1 unit and $\frac{\pi}{4}$ unit to the left as compared to the graph of $y = 2\sin 4x$. Since the graph is translated $\frac{\pi}{4}$ unit to the left, start at the $x$-value $0 - \frac{\pi}{4} = -\frac{\pi}{4}$. The first period will end at $-\frac{\pi}{4} + \frac{\pi}{2} = \frac{\pi}{4}$. Use the $x$-values

$$-\frac{\pi}{4}, \quad -\frac{\pi}{8}, \quad 0, \quad \frac{\pi}{8}, \quad \text{and} \quad \frac{\pi}{4}$$

to complete one period of the graph. Figure 101 shows two periods, ranging from $-\frac{\pi}{2}$ to $\frac{\pi}{2}$.

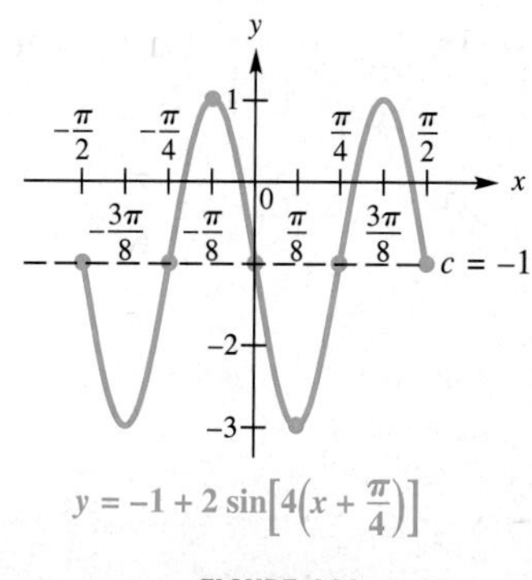


FIGURE 101

**Graphing Calculator Solution**

Enter

$$Y_1 = -1 + 2\sin\left(4\left(X + \frac{\pi}{4}\right)\right)$$

and graph using the window $\left[-\frac{\pi}{2}, \frac{\pi}{2}\right]$ by $[-4, 2]$, with Xscl $= \frac{\pi}{8}$ and Yscl $= 1$. See Figure 102, which also shows the graph of

$$Y_2 = -1,$$

the horizontal axis of the graph of $Y_1$. With Xscl $= \frac{\pi}{8}$, it is easier to identify $x$-intercepts, maximums, and minimums in the graph.

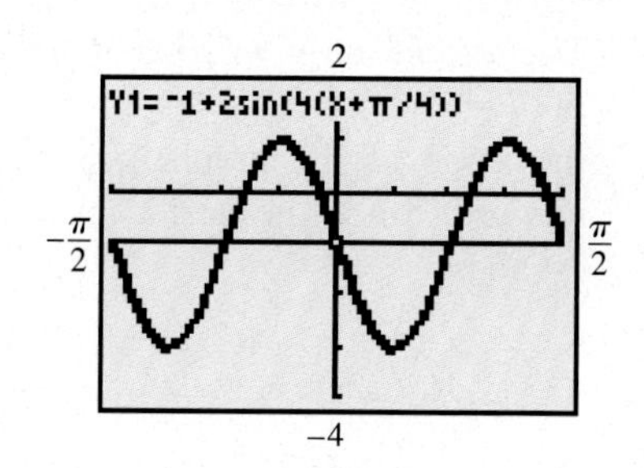


FIGURE 102

## Determining a Trigonometric Model Using Curve Fitting

A sinusoidal function is often a good approximation of a set of data points from a real-life situation.

**EXAMPLE 8** *Modeling Temperature with a Sine Function*

The maximum average monthly temperature in New Orleans is 82°F and the minimum is 54°F. The table shows the average monthly temperatures.

The scatter diagram for a 2-year interval in Figure 103 strongly suggests that the temperatures can be modeled with a sine curve.

| Month | °F | Month | °F |
|---|---|---|---|
| Jan | 54 | July | 82 |
| Feb | 55 | Aug | 81 |
| Mar | 61 | Sept | 77 |
| Apr | 69 | Oct | 71 |
| May | 73 | Nov | 59 |
| June | 79 | Dec | 55 |

*Source:* Miller, A. and J. Thompson, *Elements of Meteorology*, Charles E. Merrill, 1975.

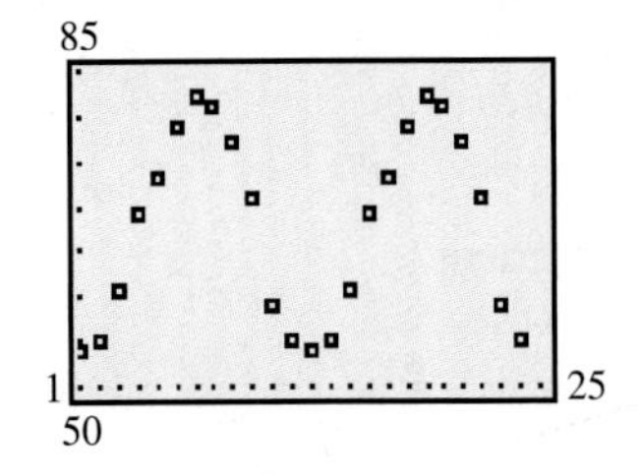


FIGURE 103

**(a)** Using only the maximum and minimum temperatures, determine a function of the form $f(x) = a \sin[b(x - d)] + c$, where $a$, $b$, $c$, and $d$ are constants, that models the average monthly temperature in New Orleans. Let $x$ represent the month, with January corresponding to $x = 1$.

**(b)** On the same coordinate axes, graph $f$ for a 2-year period together with the actual data values found in the table.

**(c)** Use the *sine regression* feature of a graphing calculator to determine a second model for these data.

## Solution

**(a)** We can use the maximum and minimum average monthly temperatures to find the amplitude $a$.

$$a = \frac{82 - 54}{2} = 14$$

The average of the maximum and minimum temperatures is a good choice for $c$. The average is

$$\frac{82 + 54}{2} = 68.$$

Since the coldest month is January, when $x = 1$, and the hottest month is July, when $x = 7$, we should choose $d$ to be about 4. The table shows that temperatures are actually a little warmer after July than before, so we experiment with values just greater than 4 to find $d$. Trial and error using a calculator leads to $d = 4.2$. Since temperatures repeat every 12 months, $b$ is $\frac{2\pi}{12} = \frac{\pi}{6}$. Thus,

$$f(x) = a \sin[b(x - d)] + c = 14 \sin\left[\frac{\pi}{6}(x - 4.2)\right] + 68.$$

**(b)** See Figure 104(a). The figure also shows the graph of $y = 14 \sin \frac{\pi}{6}x + 68$ for comparison. The horizontal translation of the model is fairly obvious here.

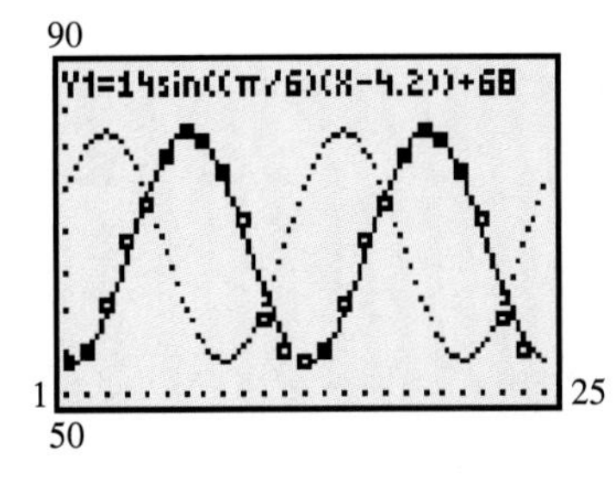


(a)

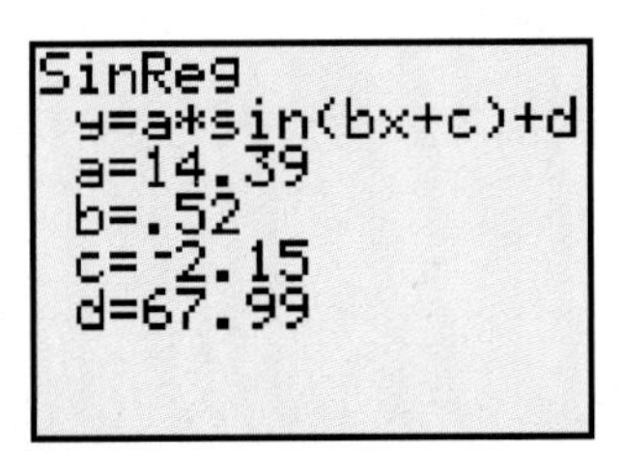


Values are rounded to the nearest hundredth.

(b)

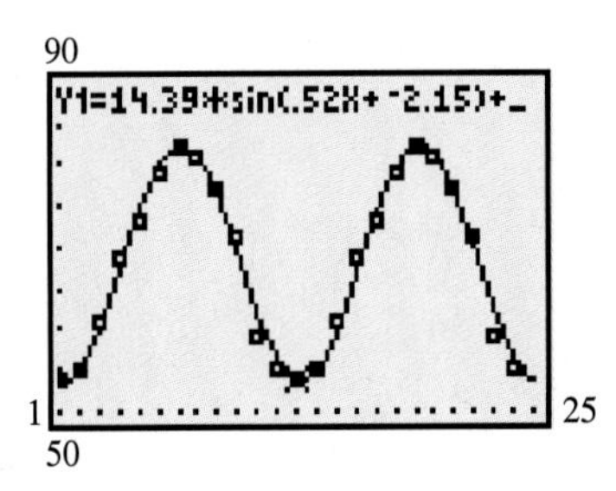


(c)

FIGURE 104

**(c)** Using two years of the given data, the screen in Figure 104(b) shows the equation of the model, while Figure 104(c) shows its graph along with the data points.

## What Went WRONG?

After studying the concepts of this section, a student graphed $y = -1 + 3 \sin \frac{1}{2}x$ in the window $[-2\pi, 2\pi]$ by $[-4, 4]$. He knew the amplitude should be 3 and the period should be $4\pi$. However, his graph looked like this:

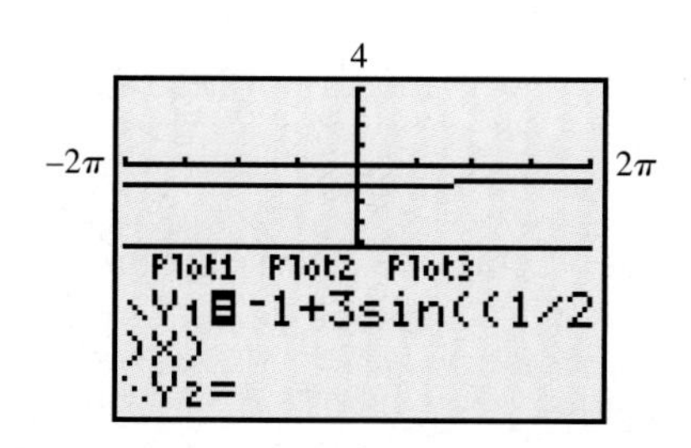


**What Went WRONG?** How can he obtain the correct graph?

## 8.6 EXERCISES

*Concept Check* *Without using a calculator, match each function defined in Exercises 1–8 with its graph in A–H.*

**1.** $y = \sin x$

**2.** $y = \cos x$

**3.** $y = -\sin x$

**4.** $y = -\cos x$

**5.** $y = \sin 2x$

**6.** $y = \cos 2x$

**7.** $y = 2 \sin x$

**8.** $y = 2 \cos x$

**A.**

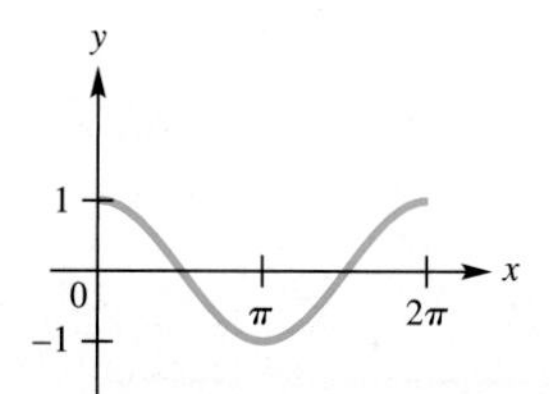


**B.**

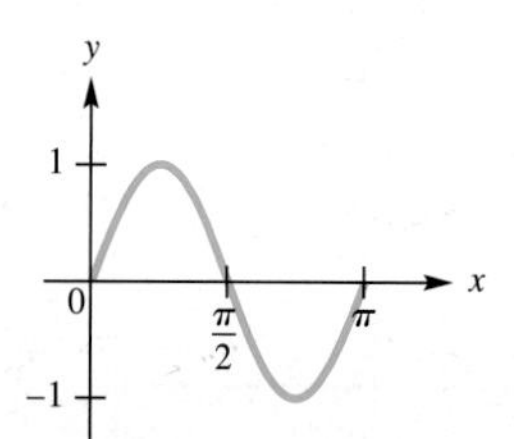


**C.**

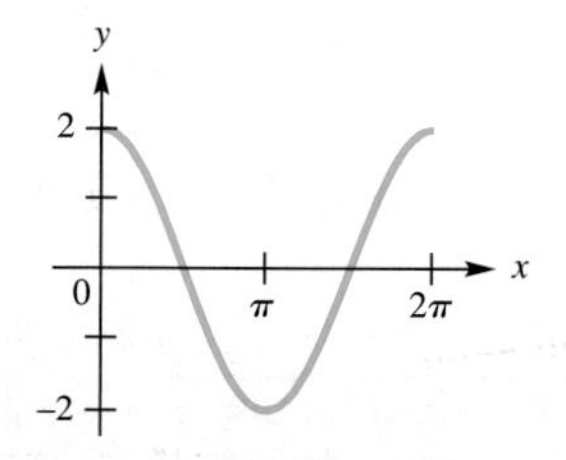


**D.**

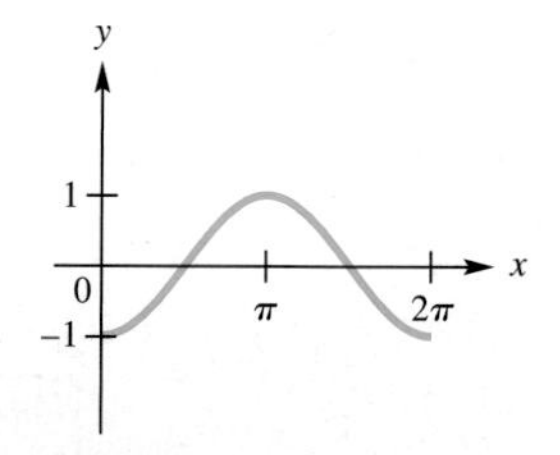


**E.**

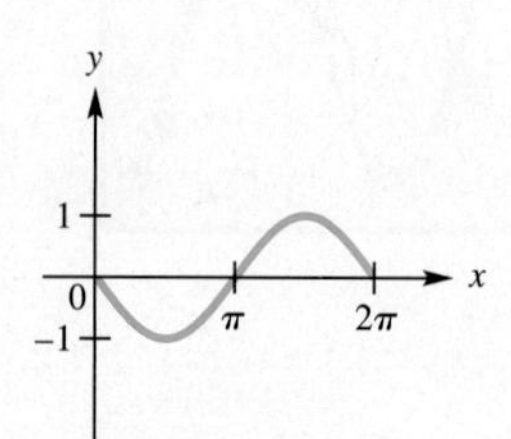


**F.**

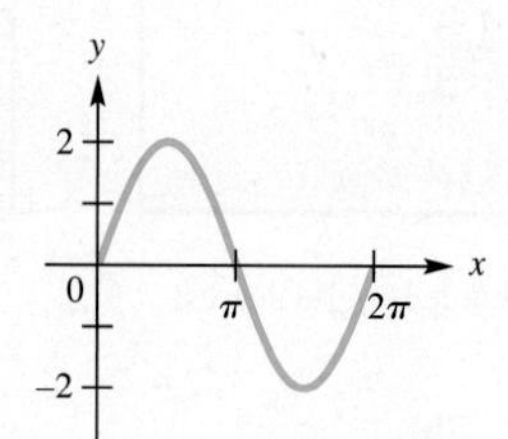


**G.**

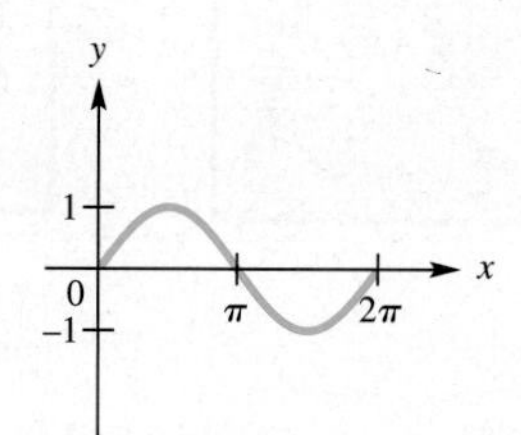


**H.**

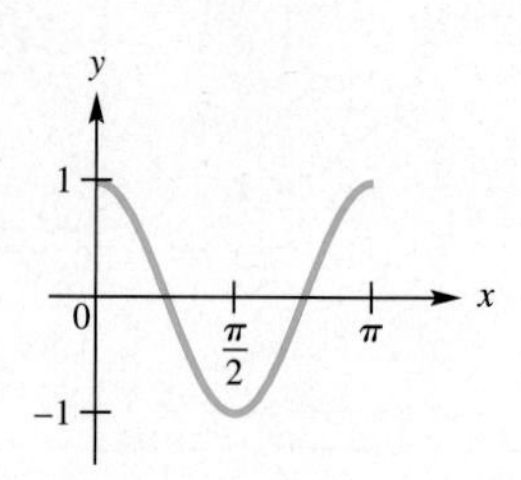


---

*Answer to What Went Wrong?*

He graphed the function while the calculator was in *degree* mode. To obtain the correct graph, he should change the mode to *radian*.

*Concept Check* *Without using a calculator, match each function defined in Exercises 9–16 with its graph in A–H.*

**9.** $y = \sin\left(x - \frac{\pi}{4}\right)$ **10.** $y = \sin\left(x + \frac{\pi}{4}\right)$ **11.** $y = \cos\left(x - \frac{\pi}{4}\right)$ **12.** $y = \cos\left(x + \frac{\pi}{4}\right)$

**13.** $y = 1 + \sin x$ **14.** $y = -1 + \sin x$ **15.** $y = 1 + \cos x$ **16.** $y = -1 + \cos x$

**A.**

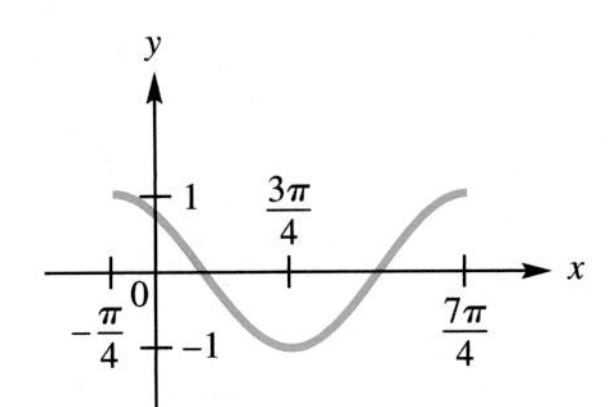


**B.**

**C.**

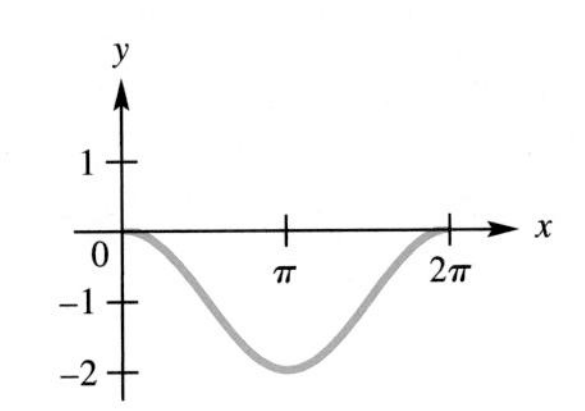


**D.**

**E.**

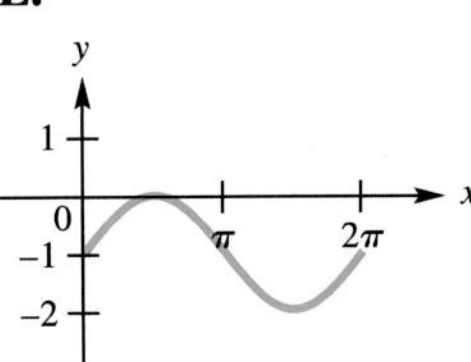


**F.**

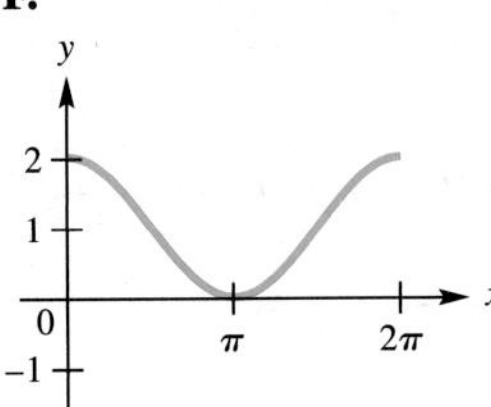


**G.**

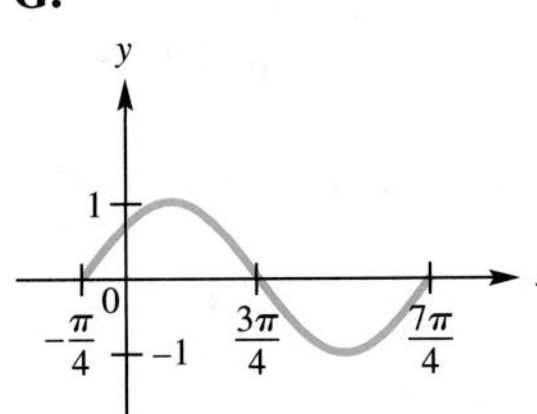


**H.**

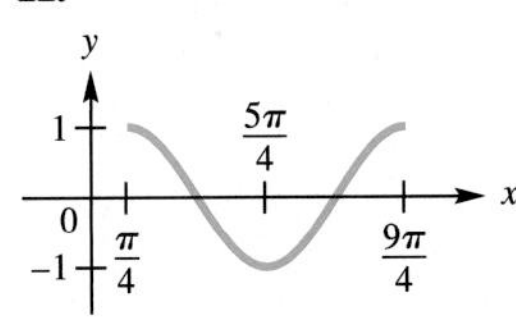


*Concept Check* *Match each function defined in Column I with the appropriate description in Column II.*

| I | II |
|---|---|
| **17.** $y = 3\sin(2x - 4)$ | **A.** Amplitude = 2; period = $\frac{\pi}{2}$; phase shift = $\frac{3}{4}$ |
| **18.** $y = 2\sin(3x - 4)$ | **B.** Amplitude = 3; period = $\pi$; phase shift = 2 |
| **19.** $y = 4\sin(3x - 2)$ | **C.** Amplitude = 4; period = $\frac{2\pi}{3}$; phase shift = $\frac{2}{3}$ |
| **20.** $y = 2\sin(4x - 3)$ | **D.** Amplitude = 2; period = $\frac{2\pi}{3}$; phase shift = $\frac{4}{3}$ |

*Concept Check* *In Exercises 21 and 22, give the equation of a sine function having the given graph.*

**21.**

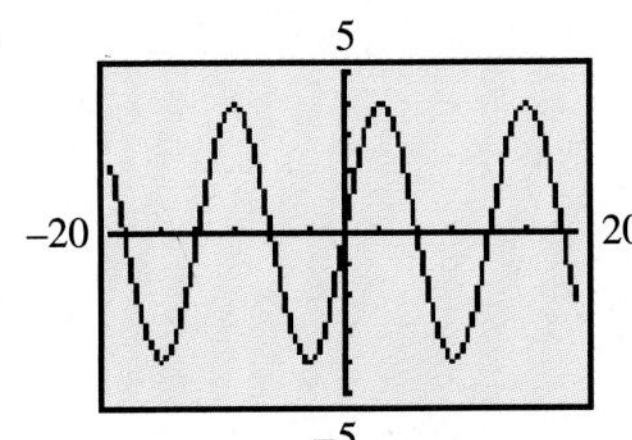


```
WINDOW
 Xmin=-20
 Xmax=20
 Xscl=3.1415926...
 Ymin=-5
 Ymax=5
 Yscl=1
 Xres=1
```

**22.**

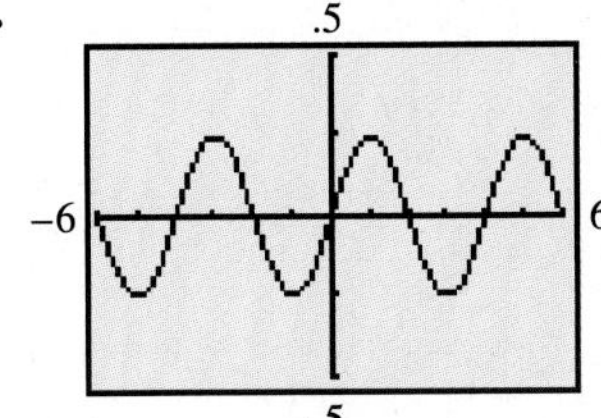


```
WINDOW
 Xmin=-6
 Xmax=6
 Xscl=1
 Ymin=-.5
 Ymax=.5
 Yscl=.25
 Xres=1
```

## RELATING CONCEPTS

**For individual or group investigation (Exercises 23–30)**

*Consider the function defined by* $f(x) = -5 + 3\sin\left[2\left(x - \frac{\pi}{2}\right)\right]$. *Answer Exercises 23–30 in order.*

**23. (a)** Because the maximum value of the sine function is ________, the maximum value of $\sin\left[2\left(x - \frac{\pi}{2}\right)\right]$ is ________, the maximum value of $3\sin\left[2\left(x - \frac{\pi}{2}\right)\right]$ is ________, and thus the maximum value of $-5 + 3\sin\left[2\left(x - \frac{\pi}{2}\right)\right]$ is ________.

**(b)** Because the minimum value of the sine function is ________, the minimum value of $\sin\left[2\left(x - \frac{\pi}{2}\right)\right]$ is ________, the minimum value of $3\sin\left[2\left(x - \frac{\pi}{2}\right)\right]$ is ________, and thus the minimum value of $-5 + 3\sin\left[2\left(x - \frac{\pi}{2}\right)\right]$ is ________.

**24.** Based on the answers in Exercise 23, what is the range of $f$?

**25.** Why will the trig viewing window as defined in the text not provide a comprehensive graph of $f$?

**26.** To obtain a comprehensive graph of $f$, Ymax must be at least ________ and Ymin must be at most ________.

**27.** Explain why using Xmin $= -2\pi$ and Xmax $= 2\pi$ will show exactly four periods of the graph of $f$.

**28.** Use your calculator to graph $f$ in the window $[-2\pi, 2\pi]$ by $[-10, 5]$.

**29.** Look at the calculator graph from Exercise 28. A "border" of empty space appears above and below the graph. If you did not want such a border to appear, what Ymin and Ymax values should you use so that a comprehensive graph would fit?

**30.** Evaluate the approximate values of the function for $x = -2$ and for $x = -2 + \pi$. What is the value in each case? Why is this so?

*Find the* **(a)** *amplitude,* **(b)** *period,* **(c)** *phase shift (if any),* **(d)** *vertical translation (if any), and* **(e)** *range of each function.*

**31.** $y = 2\sin(x - \pi)$

**32.** $y = \frac{2}{3}\cos\left(x + \frac{\pi}{2}\right)$

**33.** $y = 4\cos\left(\frac{1}{2}x + \frac{\pi}{2}\right)$

**34.** $y = -\cos\left[\frac{2}{3}\left(x - \frac{\pi}{3}\right)\right]$

**35.** $y = 2 - \sin\left(3x - \frac{\pi}{5}\right)$

**36.** $y = -1 + \frac{1}{2}\cos(2x - 3\pi)$

*In Exercises 37–52, give a traditional or calculator graph, as directed by your instructor. Graph each function over a two-period interval.*

**37.** $y = \sin\left(x - \frac{\pi}{4}\right)$

**38.** $y = \cos\left(x - \frac{\pi}{3}\right)$

**39.** $y = 2\cos\left(x - \frac{\pi}{3}\right)$

**40.** $y = 3\sin\left(x - \frac{3\pi}{2}\right)$

*Graph each function over a one-period interval.*

**41.** $y = -4\sin(2x - \pi)$

**42.** $y = 3\cos(4x + \pi)$

**43.** $y = \frac{1}{2}\cos\left(\frac{1}{2}x - \frac{\pi}{4}\right)$

**44.** $y = -\frac{1}{4}\sin\left(\frac{3}{4}x + \frac{\pi}{8}\right)$

*Graph each function over a two-period interval.*

**45.** $y = 1 - \frac{2}{3}\sin\frac{3}{4}x$

**46.** $y = -1 - 2\cos 5x$

**47.** $y = 1 - 2\cos\frac{1}{2}x$

**48.** $y = -3 + 3\sin\frac{1}{2}x$

*Graph each function over a one-period interval.*

**49.** $y = -3 + 2\sin\left(x + \frac{\pi}{2}\right)$

**50.** $y = 4 - 3\cos(x - \pi)$

**51.** $y = \frac{1}{2} + \sin\left[2\left(x + \frac{\pi}{4}\right)\right]$

**52.** $y = -\frac{5}{2} + \cos\left[3\left(x - \frac{\pi}{6}\right)\right]$

*Tides for Kahului Harbor* *The graph shows the tides for Kahului Harbor (on the island of Maui, Hawaii). To identify high and low tides and times for other Maui areas, the following adjustments must be made.*

| | |
|---|---|
| Hana: | High, +40 minutes, +.1 foot; Low, +18 minutes, −.2 foot |
| Makena: | High, +1:21, −.5 foot; Low, +1:09, −.2 foot |
| Maalaea: | High, +1:52, −.1 foot; Low, +1:19, −.2 foot |
| Lahaina: | High, +1:18, −.2 foot; Low, +1:01, −.1 foot |

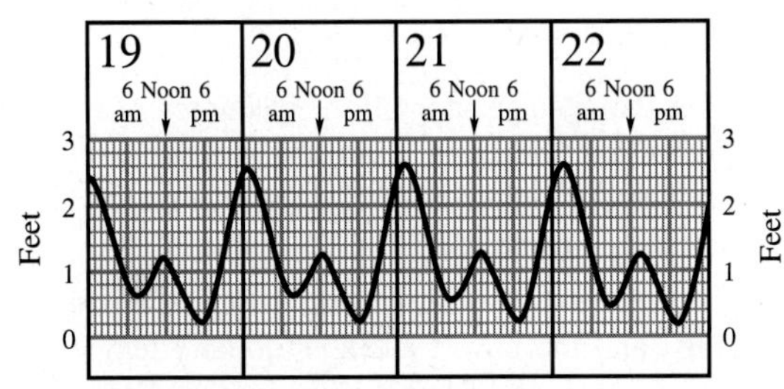


*Source: Maui News.* Original chart prepared by Edward K. Noda and Associates.

*Use the graph to work Exercises 53–58.*

**53.** The graph is an example of a periodic function. What is the period (in hours)?

**54.** What is the amplitude?

**55.** At what time on January 20, 1997, was low tide at Kahului? What was the height?

**56.** Repeat Exercise 55 for Maalaea.

**57.** At what time on January 22, 1997, was high tide at Kahului? What was the height?

**58.** Repeat Exercise 57 for Lahaina.

*(Modeling)* *Solve each problem.*

**59.** *Average Annual Temperature* Scientists believe that the average annual temperature in a given location is periodic. The average temperature at a given place during a given season fluctuates as time goes on, from colder to warmer, and back to colder. The graph shows an idealized description of the temperature (in °F) for the last few thousand years of a location at the same latitude as Anchorage, AK.

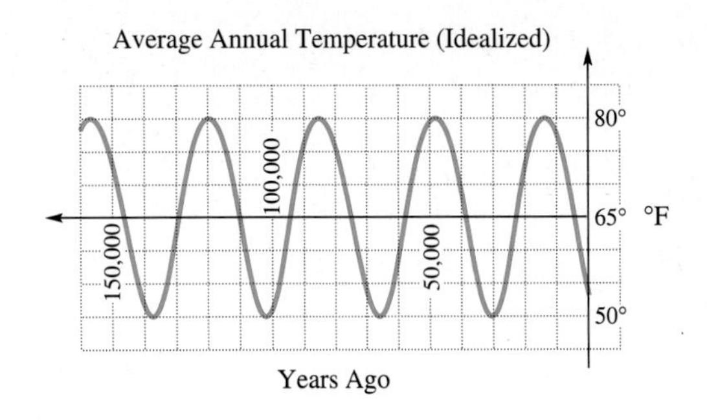


**(a)** Find the highest and lowest average annual temperatures recorded.

**(b)** Use these two numbers to find the amplitude.

**(c)** Find the period of the function.

**(d)** What is the trend of the temperature now?

**60.** *Blood Pressure Variation* The graph gives the variation in blood pressure for a typical person. Systolic and diastolic pressures are the upper and lower limits of the periodic changes in pressure that produce the pulse. The length of time between peaks is called the period of the pulse.

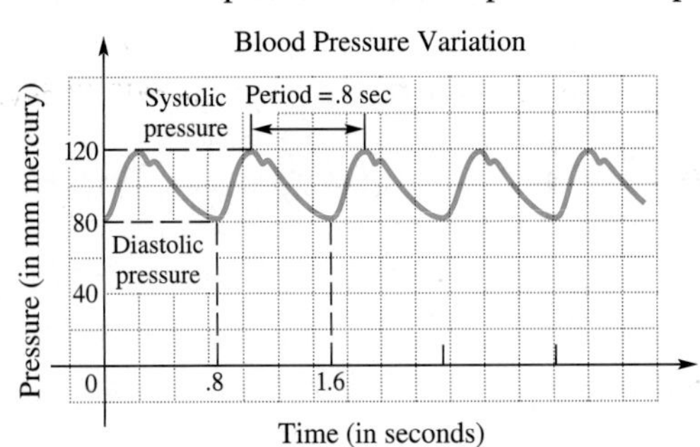


**(a)** Find the amplitude of the graph.

**(b)** Find the pulse rate (the number of pulse beats in one minute) for this person.

**61.** *Activity of a Nocturnal Animal* Many activities of living organisms are periodic. For example, the graph shows the time that a certain nocturnal animal begins its evening activity.

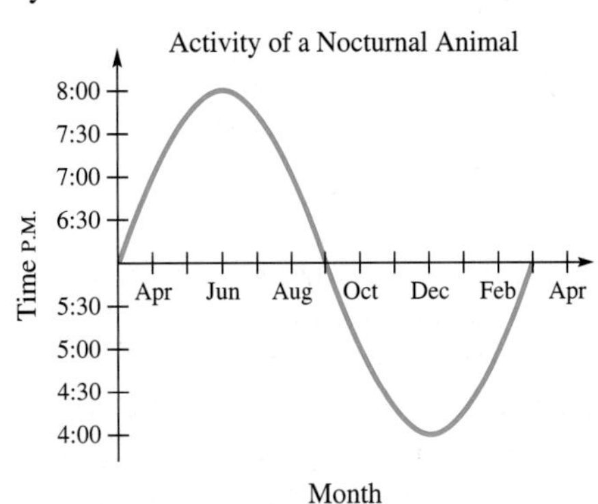

**(a)** Find the amplitude of this graph.

**(b)** Find the period.

**62.** *Voltage* The voltage $E$ in an electrical circuit is modeled by

$$E = 5 \cos 120\pi t,$$

where $t$ is measured in seconds.

**(a)** Find the amplitude and the period.

**(b)** How many cycles are completed in one second? (The number of cycles (periods) completed in one second is the *frequency* of the function.)

**(c)** Find $E$ when $t = 0, .03, .06, .09, .12$.

**(d)** Graph $E$ for $0 \leq t \leq \frac{1}{30}$.

**63.** *Voltage* For another electrical circuit, the voltage $E$ is modeled by

$$E = 3.8 \cos 40\pi t,$$

where $t$ is time measured in seconds.

**(a)** Find the amplitude and the period.

**(b)** Find the frequency. See Exercise 62(b).

**(c)** Find $E$ when $t = .02, .04, .08, .12, .14$.

**(d)** Graph one period of $E$.

**64.** *Atmospheric Carbon Dioxide* At Mauna Loa, Hawaii, atmospheric carbon dioxide levels in parts per million (ppm) have been measured regularly since 1958. The function defined by

$$L(x) = .022x^2 + .55x + 316 + 3.5 \sin(2\pi x)$$

can be used to model these levels, where $x$ is in years and $x = 0$ corresponds to 1960. (*Source:* Nilsson, A., *Greenhouse Earth,* John Wiley and Sons, 1992.)

**(a)** Graph $L$ for $15 \leq x \leq 35$. (*Hint:* For the range use $325 \leq y \leq 365$.)

**(b)** When do the seasonal maximum and minimum carbon dioxide levels occur?

**(c)** $L$ is the sum of a quadratic function and a sine function. What is the significance of each of these functions? Discuss what physical phenomena may be responsible for each function.

**65.** *Atmospheric Carbon Dioxide* Refer to Exercise 64. The carbon dioxide content in the atmosphere at Barrow, Alaska, in parts per million (ppm) can be modeled using the function defined by

$$C(x) = .04x^2 + .6x + 330 + 7.5 \sin(2\pi x),$$

where $x = 0$ corresponds to 1970. (*Source:* Zeilik, M., S. Gregory, and E. Smith, *Introductory Astronomy and Astrophysics,* Fourth Edition, Saunders College Publishers, 1998.)

**(a)** Graph $C$ for $5 \leq x \leq 25$. (*Hint:* Use $320 \leq y \leq 380$.)

**(b)** Discuss possible reasons why the amplitude of the oscillations in the graph of $C$ is larger than the amplitude of the oscillations in the graph of $L$ in Exercise 64, which models Hawaii.

**(c)** Define a new function $C$ that is valid if $x$ represents the actual year where $1970 \leq x \leq 1995$.

**66.** *Average Monthly Temperatures* The average monthly temperature (in °F) in Vancouver, Canada, is shown in the table.

| Month | °F | Month | °F |
|---|---|---|---|
| Jan | 36 | July | 64 |
| Feb | 39 | Aug | 63 |
| Mar | 43 | Sept | 57 |
| Apr | 48 | Oct | 50 |
| May | 55 | Nov | 43 |
| June | 59 | Dec | 39 |

*Source:* Miller, A. and J. Thompson, *Elements of Meteorology*, Charles E. Merrill, 1975.

**(a)** Plot the average monthly temperatures over a 2-year period by letting $x = 1$ correspond to the month of January during the first year. Do the data seem to indicate a translated sine graph?

**(b)** The highest average monthly temperature is 64°F in July, and the lowest average monthly temperature is 36°F in January. Their average is 50°F. Graph the data together with the line $y = 50$. What does this line represent with regard to temperature in Vancouver?

**(c)** Approximate the amplitude, period, and phase shift of the translated sine wave indicated by the data.

**(d)** Determine a sinusoidal function with the form $f(x) = a \sin[b(x - d)] + c$, where $a$, $b$, $c$, and $d$ are constants, that models the data.

**(e)** Graph $f$ together with the data on the same coordinate axes. How well does $f$ model the given data?

**(f)** Use the sine regression capability of a graphing calculator to find the equation of a sine curve that fits these data.

**67.** *Average Monthly Temperatures* The average monthly temperatures (in °F) in Phoenix, Arizona, are shown in the table.

| Month | °F | Month | °F |
|---|---|---|---|
| Jan | 51 | July | 90 |
| Feb | 55 | Aug | 90 |
| Mar | 63 | Sept | 84 |
| Apr | 67 | Oct | 71 |
| May | 77 | Nov | 59 |
| June | 86 | Dec | 52 |

*Source:* Miller, A. and J. Thompson, *Elements of Meteorology*, Charles E. Merrill, 1975.

**(a)** Predict the average yearly temperature and compare it to the actual value of 70°F.

**(b)** Plot the average monthly temperature over a 2-year period by letting $x = 1$ correspond to January of the first year.

**(c)** Determine a sinusoidal function with the form $f(x) = a\cos[b(x - d)] + c$, where $a$, $b$, $c$, and $d$ are constants, that models the data.

**(d)** Graph $f$ together with the data on the same coordinate axes. How well does $f$ model the data?

**(e)** Use the sine regression capability of a graphing calculator to find the equation of a sine curve that fits these data.

*In an article entitled "I Found Sinusoids in My Gas Bill" (Mathematics Teacher, January 2000). Cathy G. Schloemer presents the following graph that accompanied her gas bill.*

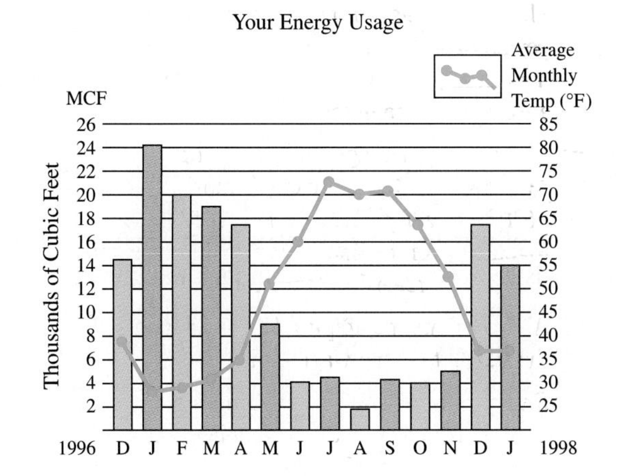


*Notice that two sinusoids are suggested here: one for the behavior of the average monthly temperature and another for gas use in MCF (thousands of cubic feet).*

*Use this graph to answer Exercises 68 and 69.*

**68.** If January 1997 is represented by $x = 1$, the data of estimated ordered pairs (month, temperature) is given in the list shown on the two screens.

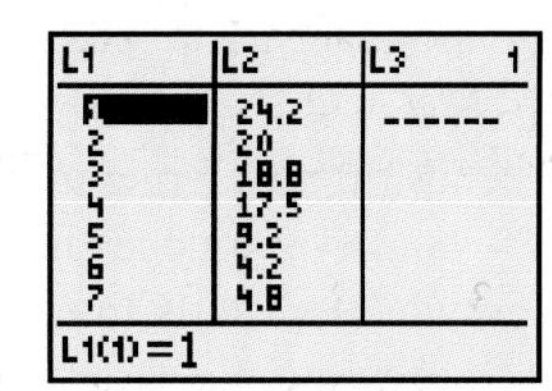

| L1 | L2 | L3 |
|---|---|---|
| 1 | 28 | ------ |
| 2 | 29 | |
| 3 | 31 | |
| 4 | 35 | |
| 5 | 51 | |
| 6 | 60 | |
| 7 | 73 | |

L1(1) = 1

| L1 | L2 | L3 |
|---|---|---|
| 7 | 73 | |
| 8 | 70 | |
| 9 | 71 | |
| 10 | 64 | |
| 11 | 53 | |
| 12 | 37 | |
| | ------ | |

L1(13) =

Use the sine regression feature of a graphing calculator to find a sine function that fits these data points. Then make a scatter diagram, and graph the function.

**69.** Again, if January 1997 is represented by $x = 1$, the data of estimated ordered pairs (month, gas use in MCF) is given in the list shown on the two screens.

| L1 | L2 | L3 |
|---|---|---|
| 1 | 24.2 | ------ |
| 2 | 20 | |
| 3 | 18.8 | |
| 4 | 17.5 | |
| 5 | 9.2 | |
| 6 | 4.2 | |
| 7 | 4.8 | |

L1(1) = 1

| L1 | L2 | L3 |
|---|---|---|
| 7 | 4.8 | |
| 8 | 1.8 | |
| 9 | 4.8 | |
| 10 | 4 | |
| 11 | 5 | |
| 12 | 17.5 | |
| | ------ | |

L1(13) =

Use the sine regression feature of a graphing calculator to find a sine function that fits these data points. Then make a scatter diagram, and graph the function.

## Reviewing Basic Concepts (Sections 8.5 and 8.6)

**1.** Give the exact coordinates of the point on the unit circle that corresponds to the given value of $s$.

**(a)** $-2\pi$ **(b)** $\frac{5\pi}{4}$ **(c)** $\frac{5\pi}{2}$

**2.** Give the values of the trigonometric functions of $-\frac{5\pi}{2}$. State which functions are undefined.

**3.** Give the exact values of the trigonometric functions of each number.

**(a)** $\frac{7\pi}{6}$ **(b)** $-\frac{2\pi}{3}$

**4.** Give calculator approximations for the six trigonometric function values of 2.25.

**5.** Based on the screen shown here, in which quadrant does the point on the unit circle corresponding to the real number 100 lie?

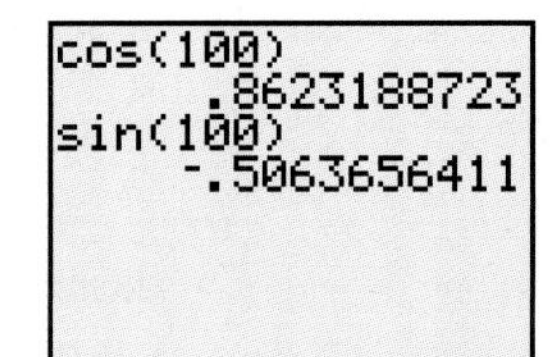

```
cos(100)
          .8623188723
sin(100)
         -.5063656411
```

*In Exercises 6 and 7, give a traditional or calculator graph, as directed by your instructor.*

**6.** Graph one period of $y = -\cos x$. State the period and the amplitude.

**7.** Graph $y = 3 \sin(\pi x + \pi)$ on the interval $-2 \leq x \leq 2$. State the amplitude, period, and phase shift.

**8.** *(Modeling) Daylight Hours* The graph of

$$f(t) = 6.5 \sin\left[\frac{\pi}{6}(x - 3.65)\right] + 12.4$$

models the daylight hours at 60°N latitude, where $x = 1$ corresponds to January 1, $x = 2$ to February 1, and so on.

**(a)** Estimate the maximum and minimum number of daylight hours.

**(b)** Interpret the amplitude and period.

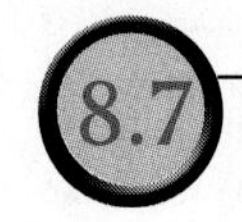

## 8.7 GRAPHS OF THE OTHER CIRCULAR FUNCTIONS

Graphs of the Cosecant and Secant Functions • Graphs of the Tangent and Cotangent Functions • Addition of Ordinates

### Graphs of the Cosecant and Secant Functions

Since cosecant values are reciprocals of the corresponding sine values, the period of the function $y = \csc x$ is $2\pi$, the same as for $y = \sin x$. When $\sin x = 1$, the value of $\csc x$ is also 1, and when $0 < \sin x < 1$, then $\csc x > 1$. Also, if $-1 < \sin x < 0$, then $\csc x < -1$. (Verify these statements.) As $|x|$ approaches 0, $|\sin x|$ approaches 0, and $|\csc x|$ gets larger and larger. The graph of $y = \csc x$ approaches the vertical line $x = 0$ but never touches it, so the line $x = 0$ is a *vertical asymptote.* In fact, the lines $x = n\pi$, where $n$ is any integer, are all vertical asymptotes. Using this information and plotting a few points shows that the graph takes the shape of the solid curve shown in Figure 105. To show how the two graphs are related, the graph of $y = \sin x$ is also shown, as a dashed curve.

**TECHNOLOGY NOTE**

We use dot mode to graph the cosecant and secant functions on a graphing calculator to get an accurate picture. (If connected mode is used, the calculator will attempt to connect points that are actually separated by vertical asymptotes.) We enter $\csc x$ as $\frac{1}{\sin x}$ or $(\sin x)^{-1}$, and $\sec x$ as $\frac{1}{\cos x}$ or $(\cos x)^{-1}$.

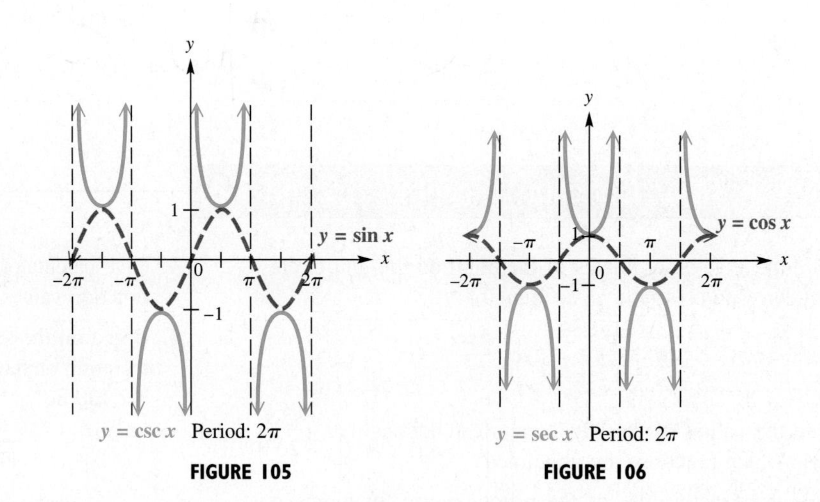


FIGURE 105

FIGURE 106

A similar analysis for the secant leads to the solid curve shown in Figure 106. The dashed curve, $y = \cos x$, is shown so that the relationship between these two reciprocal functions can be seen.

| X | $Y_1$ | $Y_2$ |
| --- | --- | --- |
| 0 | ERROR | ERROR |
| .3927 | 2.6131 | -2.613 |
| .7854 | 1.4142 | -1.414 |
| 1.1781 | 1.0824 | -1.082 |
| 1.5708 | 1 | -1 |
| 1.9635 | 1.0824 | -1.082 |
| 2.3562 | 1.4142 | -1.414 |

X=0

For X in radians, the $Y_1$ and $Y_2$ columns give values of $Y_1 = \csc X$ and $Y_2 = \csc(-X)$ $\left(\Delta\text{Tbl} = \frac{\pi}{8}\right)$. The results suggest that, like the sine function, $\csc(-X) = -\csc X$.

**FUNCTION CAPSULE**

## COSECANT FUNCTION $f(x) = \csc x$

Domain: $\{x \mid x \neq n\pi$, where $n$ is an integer$\}$ Range: $(-\infty, -1] \cup [1, \infty)$

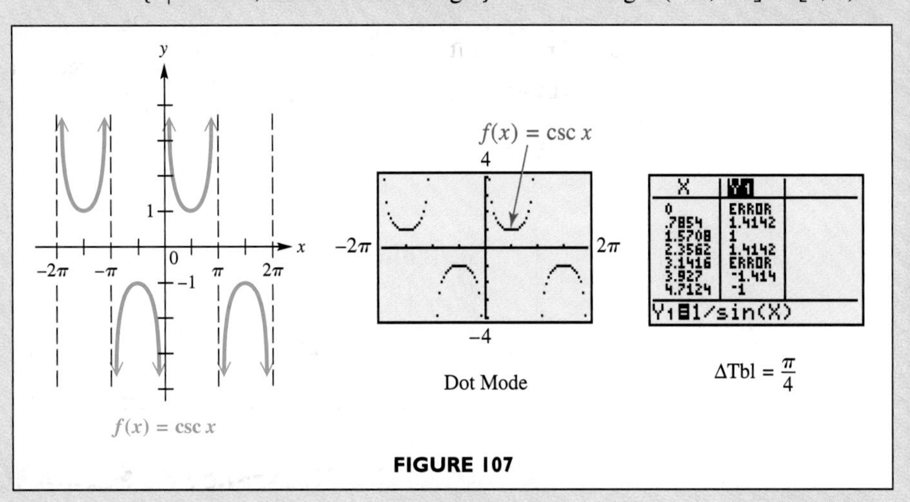


**FIGURE 107**

- The graph is discontinuous at values of $x$ of the form $x = n\pi$ and has vertical asymptotes at these values.
- There are no $x$-intercepts.
- Its period is $2\pi$.
- Its graph has no amplitude, since there are no maximum and minimum values.
- The graph is symmetric with respect to the origin, and it is an odd function.

| X | $Y_1$ | $Y_2$ |
| --- | --- | --- |
| 0 | 1 | 1 |
| .3927 | 1.0824 | 1.0824 |
| .7854 | 1.4142 | 1.4142 |
| 1.1781 | 2.6131 | 2.6131 |
| 1.5708 | ERROR | ERROR |
| 1.9635 | -2.613 | -2.613 |
| 2.3562 | -1.414 | -1.414 |

X=0

The table shows values of $Y_1 = \sec X$ and $Y_2 = \sec(-X)$. These values suggest that, like the cosine function, $\sec(-X) = \sec X$.

**FUNCTION CAPSULE**

## SECANT FUNCTION $f(x) = \sec x$

Domain: $\left\{x \mid x \neq (2n + 1)\frac{\pi}{2}\text{, where } n \text{ is an integer}\right\}$ Range: $(-\infty, -1] \cup [1, \infty)$

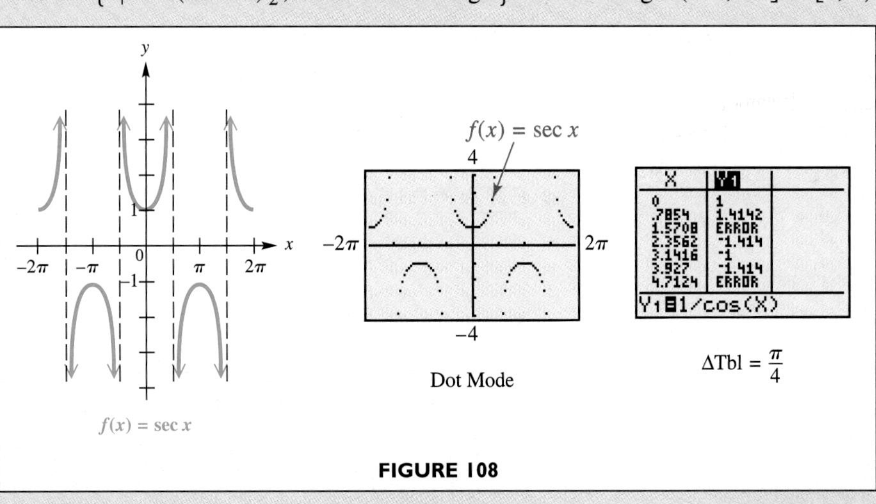


**FIGURE 108**

*(continued)*

*(Secant Function, continued)*

- The graph is discontinuous at values of $x$ of the form $x = (2n + 1)\frac{\pi}{2}$ and has vertical asymptotes at these values.
- There are no $x$-intercepts.
- Its period is $2\pi$.
- Its graph has no amplitude, since there are no maximum and minimum values.
- The graph is symmetric with respect to the $y$-axis, and it is an even function.

**Sketching Traditional Graphs of the Cosecant and Secant Functions**

To graph $y = a \csc bx$ or $y = a \sec bx$, with $b > 0$,

**1.** Graph the corresponding reciprocal function as a guide, using a dashed curve.

| To Graph | Use as a Guide |
|---|---|
| $y = a \csc bx$ | $y = a \sin bx$ |
| $y = a \sec bx$ | $y = a \cos bx$ |

**2.** Sketch the vertical asymptotes. They will have equations of the form $x = k$, where $k$ is an $x$-intercept of the graph of the guide function.

**3.** Sketch the graph of the desired function by drawing the typical U-shaped branches between the adjacent asymptotes. The branches will be above the graph of the guide function when the guide function values are positive and below the graph of the guide function when the guide function values are negative. The graph will resemble those in Figures 107 and 108 on the previous page.

Like the sine and cosine functions, the cosecant and secant function graphs can be translated vertically and horizontally. The period of both functions is $2\pi$.

**FOR DISCUSSION**

If we know the location of the vertical asymptote $x = k$ with smallest positive value of $k$, along with the period of the function, we can determine the locations of all vertical asymptotes. Discuss how you might go about finding this smallest such positive value in Example 1.

### EXAMPLE 1 *Graphing $y = a \sec bx$*

Graph $y = 2 \sec \frac{1}{2}x$ using the preceding guidelines.

**Solution**

**Step 1** This function involves the secant, so the corresponding reciprocal function involves the cosine. The guide function to graph is

$$y = 2 \cos \frac{1}{2}x.$$

Using the guidelines of Section 8.6, we find that one period of the graph lies along the interval that satisfies the inequality

$$0 \leq \frac{1}{2}x \leq 2\pi, \quad \text{or} \quad [0, 4\pi].$$

Dividing this interval into four equal parts gives the key points

$$(0, 2), \qquad (\pi, 0), \qquad (2\pi, -2), \qquad (3\pi, 0), \qquad \text{and} \qquad (4\pi, 2),$$

which are joined with a smooth dashed curve to indicate that this graph is only a guide. An additional period is graphed as seen in Figure 109(a).

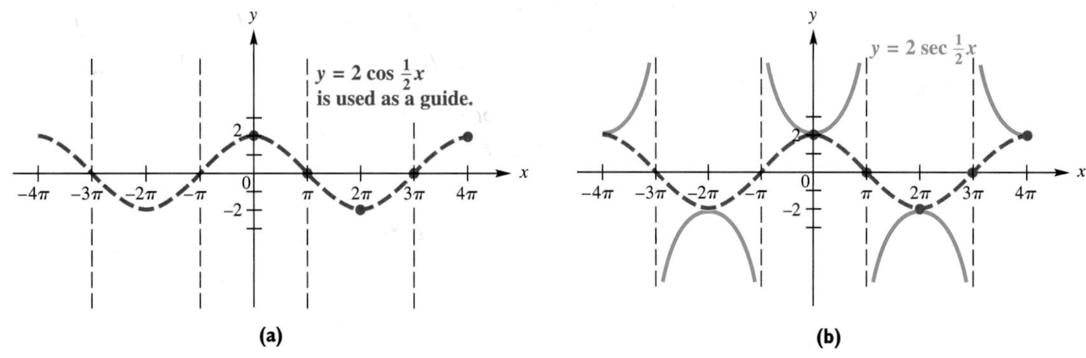


FIGURE 109

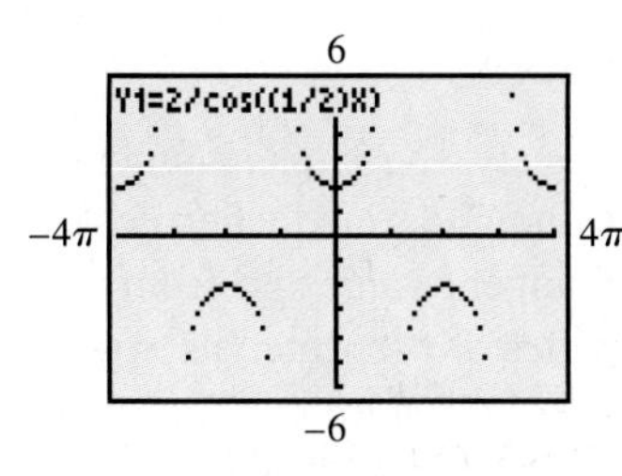


Dot Mode

Compare this graph to the blue graph in Figure 109(b).

**Step 2** We sketch the vertical asymptotes as shown in Figure 109(a). These occur at $x$-values for which the guide function equals 0, such as

$$x = -3\pi, \quad x = -\pi, \quad x = \pi, \quad \text{and} \quad x = 3\pi.$$

**Step 3** We sketch the graph of $y = 2 \sec \frac{1}{2}x$ by drawing the typical U-shaped branches, approaching the asymptotes. See Figure 109(b).

## EXAMPLE 2 *Graphing $y = a \csc(x - d)$*

Graph $y = \dfrac{3}{2} \csc\left(x - \dfrac{\pi}{2}\right)$.

### Analytic Solution

First we graph the corresponding reciprocal function

$$y = \frac{3}{2} \sin\left(x - \frac{\pi}{2}\right).$$

Alternatively, we can analyze the function as follows. Compared with the graph of $y = \csc x$, the graph of $y = \frac{3}{2} \csc\left(x - \frac{\pi}{2}\right)$ has phase shift $\frac{\pi}{2}$ units to the right. Thus, the asymptotes are the lines $x = \frac{\pi}{2}, \frac{3\pi}{2}$, and so on. Also, there are no values of $y$ between $-\frac{3}{2}$ and $\frac{3}{2}$. As shown in Figure 110 on the next page, this is related to the increased amplitude of $y = \frac{3}{2} \sin x$ com-

### Graphing Calculator Solution

Figure 111 on the next page shows graphs of

$$Y_1 = \frac{3}{2}\left[1/\sin\left(X - \frac{\pi}{2}\right)\right].$$

Connected mode draws vertical lines appearing between the portions of the graph, while dot mode does not.

*(continued)*

pared with $y = \sin x$. (Amplitude does not apply to the secant or cosecant functions; it enters only indirectly from the corresponding cosine or sine graphs.) This means that the graph goes through the points $\left(\pi, \frac{3}{2}\right)$, $\left(2\pi, -\frac{3}{2}\right)$, and so on. Two periods are shown in Figure 110.

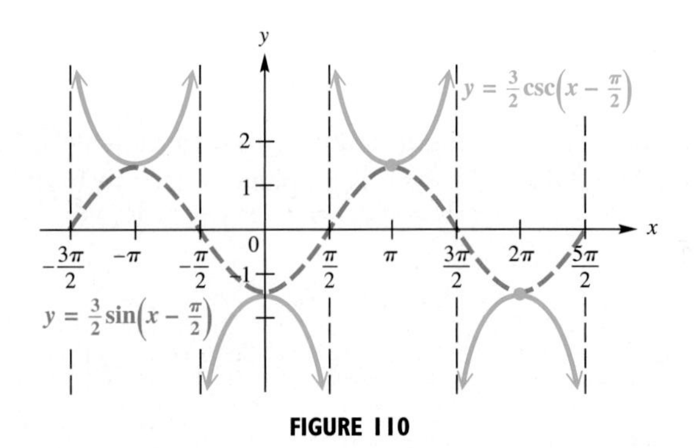


**FIGURE 110**

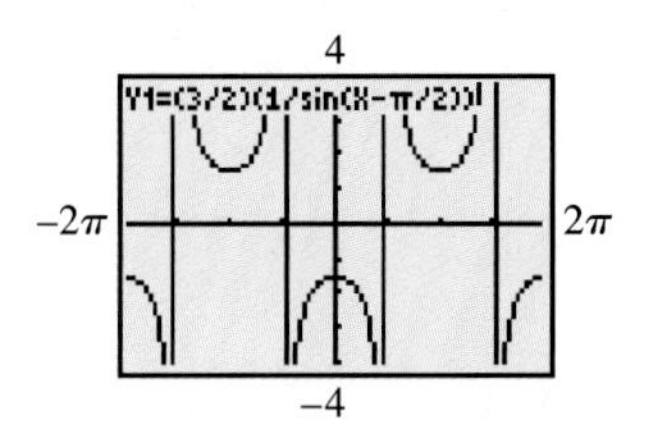


Connected Mode

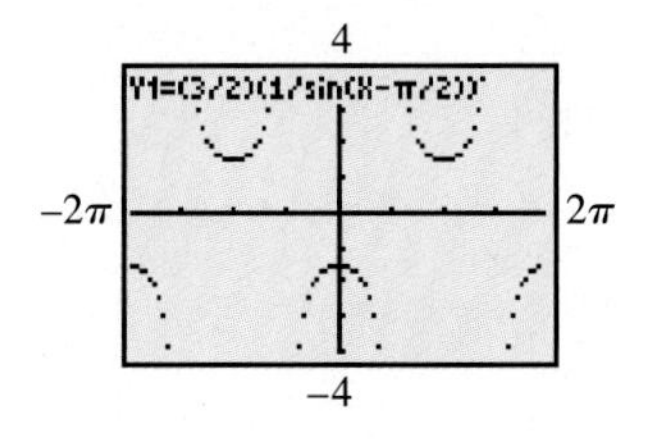


Dot Mode

**FIGURE 111**

## Graphs of the Tangent and Cotangent Functions

Unlike the four functions whose graphs we studied previously, the tangent function has period $\pi$. Tangent values are 0 when sine values are 0, and undefined when cosine values are 0. (Verify this.) As $x$-values go from $-\frac{\pi}{2}$ to $\frac{\pi}{2}$, tangent values go from $-\infty$ to $\infty$ and increase throughout the interval. Those same values are repeated as $x$ goes from $\frac{\pi}{2}$ to $\frac{3\pi}{2}$, $\frac{3\pi}{2}$ to $\frac{5\pi}{2}$, and so on. The graph of $y = \tan x$ from $-\pi$ to $\frac{3\pi}{2}$ is shown in Figure 112.

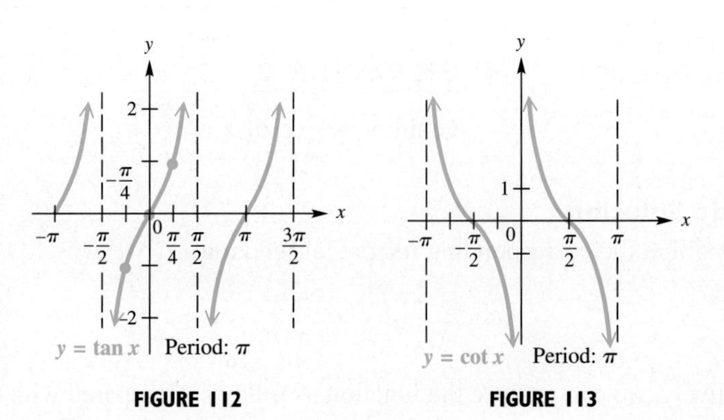


**FIGURE 112** **FIGURE 113**

The cotangent function also has period $\pi$. Cotangent values are 0 when cosine values are 0, and undefined when sine values are 0. (Verify this also.) As $x$-values go from 0 to $\pi$, cotangent values go from $\infty$ to $-\infty$ and decrease throughout the interval. Those same values are repeated as $x$ goes from $\pi$ to $2\pi$, $2\pi$ to $3\pi$, and so on. The graph of $y = \cot x$ from $-\pi$ to $\pi$ is shown in Figure 113.

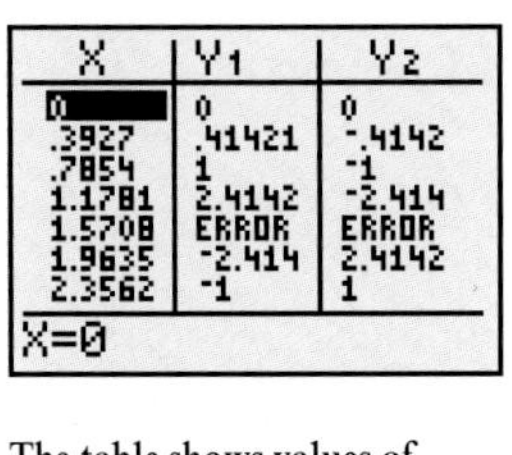

| X | Y1 | Y2 |
|---|---|---|
| 0 | 0 | 0 |
| .3927 | .41421 | -.4142 |
| .7854 | 1 | -1 |
| 1.1781 | 2.4142 | -2.414 |
| 1.5708 | ERROR | ERROR |
| 1.9635 | -2.414 | 2.4142 |
| 2.3562 | -1 | 1 |

X=0

The table shows values of $Y_1 = \tan X$ and $Y_2 = \tan(-X)$. These values suggest that $\tan(-X) = -\tan X$.

## FUNCTION CAPSULE

### TANGENT FUNCTION $f(x) = \tan x$

Domain: $\{x \mid x \neq (2n + 1)\frac{\pi}{2}$, where $n$ is an integer$\}$ Range: $(-\infty, \infty)$

$f(x) = \tan x, -\frac{\pi}{2} < x < \frac{\pi}{2}$

$f(x) = \tan x$

Dot Mode

| X | Y1 |
|---|---|
| 0 | 0 |
| .3927 | .41421 |
| .7854 | 1 |
| 1.1781 | 2.4142 |
| 1.5708 | ERROR |
| 1.9635 | -2.414 |
| 2.3562 | -1 |

Y1=tan(X)

$\Delta\text{Tbl} = \frac{\pi}{8}$

**FIGURE 114**

- The graph is discontinuous at values of $x$ of the form $x = (2n + 1)\frac{\pi}{2}$ and has vertical asymptotes at these values.
- The $x$-intercepts are of the form $x = n\pi$.
- Its period is $\pi$.
- Its graph has no amplitude, since there are no minimum and maximum values.
- The graph is symmetric with respect to the origin, and it is an odd function.

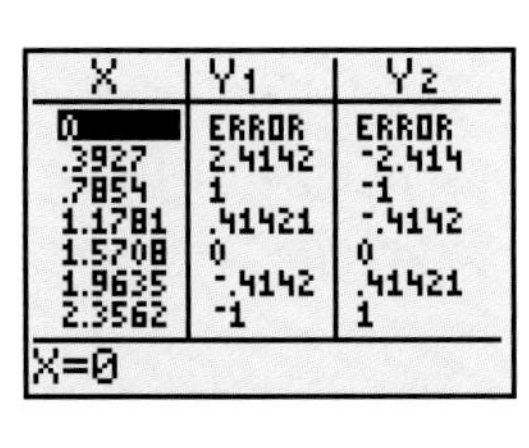

| X | Y1 | Y2 |
|---|---|---|
| 0 | ERROR | ERROR |
| .3927 | 2.4142 | -2.414 |
| .7854 | 1 | -1 |
| 1.1781 | .41421 | -.4142 |
| 1.5708 | 0 | 0 |
| 1.9635 | -.4142 | .41421 |
| 2.3562 | -1 | 1 |

X=0

In this table, $Y_1 = \cot X$ and $Y_2 = \cot(-X)$. As with the tangent function, the table suggests that $\cot(-X) = -\cot X$.

## FUNCTION CAPSULE

### COTANGENT FUNCTION $f(x) = \cot x$

Domain: $\{x \mid x \neq n\pi$, where $n$ is an integer$\}$ Range: $(-\infty, \infty)$

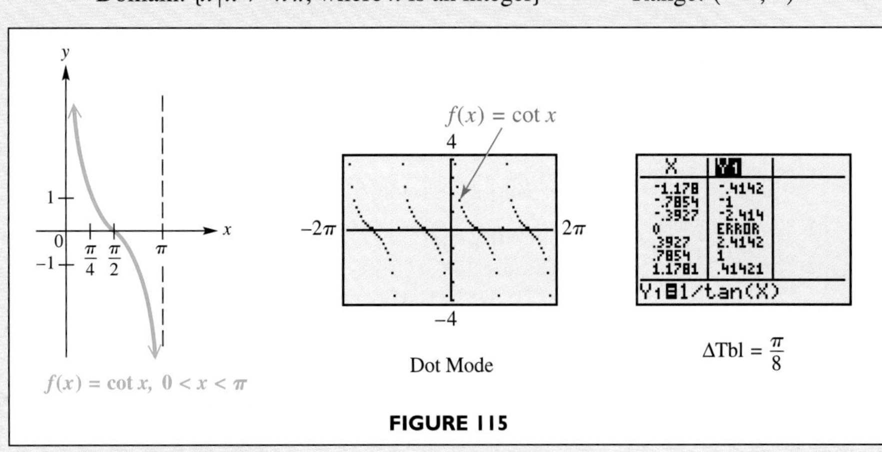


| X | Y1 |
|---|---|
| -1.178 | -.4142 |
| -.7854 | -1 |
| -.3927 | -2.414 |
| 0 | ERROR |
| .3927 | 2.4142 |
| .7854 | 1 |
| 1.1781 | .41421 |

**FIGURE 115**

- The graph is discontinuous at values of $x$ of the form $x = n\pi$ and has vertical asymptotes at these values.
- The $x$-intercepts are of the form $x = (2n + 1)\frac{\pi}{2}$.

*(continued)*

*(Cotangent Function, continued)*

- Its period is $\pi$.
- Its graph has no amplitude, since there are no minimum and maximum values.
- The graph is symmetric with respect to the origin, and it is an odd function.

**Sketching Traditional Graphs of the Tangent and Cotangent Functions**

To graph $y = a \tan bx$ or $y = a \cot bx$, with $b > 0$,

1. The period is $\frac{\pi}{b}$. To locate two adjacent vertical asymptotes, solve the following equations for $x$:

$$\text{For } y = a \tan bx: \quad bx = -\frac{\pi}{2} \quad \text{and} \quad bx = \frac{\pi}{2}.$$

$$\text{For } y = a \cot bx: \quad bx = 0 \quad \text{and} \quad bx = \pi.$$

2. Sketch the two vertical asymptotes found in Step 1.
3. Divide the interval formed by the vertical asymptotes into four equal parts.
4. Evaluate the function for the first-quarter point, midpoint, and third-quarter point, using the $x$-values found in Step 3.
5. Join the points with a smooth curve, approaching the vertical asymptotes. Indicate additional asymptotes and periods of the graph as necessary.

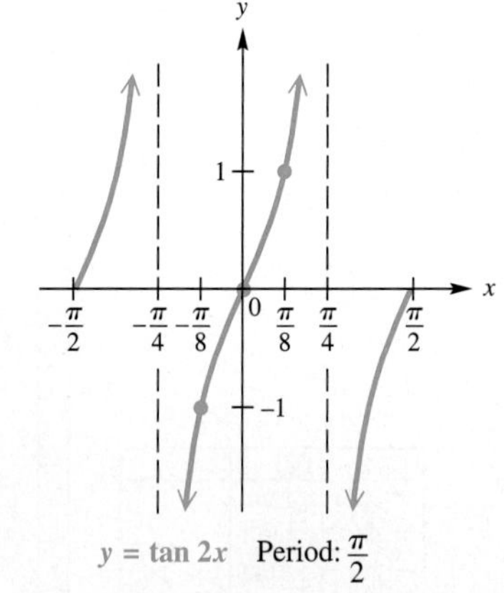


$y = \tan 2x$ Period: $\frac{\pi}{2}$

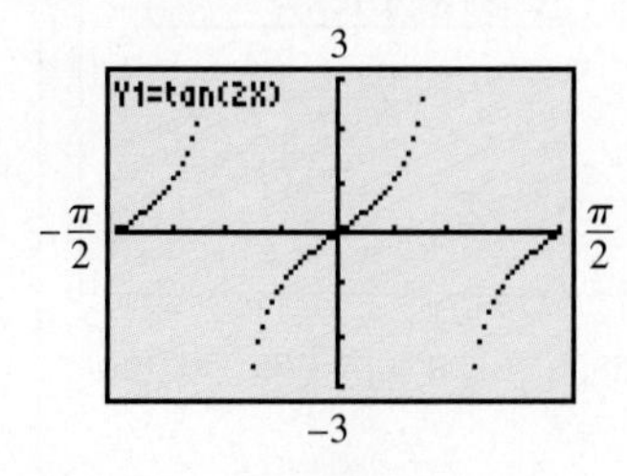


Dot Mode

**FIGURE 116**

**EXAMPLE 3** *Graphing $y = \tan bx$*

Graph $y = \tan 2x$ using the preceding guidelines.

**Solution**

**Step 1** The period of this function is $\frac{\pi}{2}$. To locate two adjacent vertical asymptotes, solve $2x = -\frac{\pi}{2}$ and $2x = \frac{\pi}{2}$ (since this is a tangent function). The two asymptotes have equations $x = -\frac{\pi}{4}$ and $x = \frac{\pi}{4}$.

**Step 2** Sketch the two vertical asymptotes $x = \pm\frac{\pi}{4}$, as shown in Figure 116.

**Step 3** Divide the interval $\left(-\frac{\pi}{4}, \frac{\pi}{4}\right)$ into four equal parts to get the following key $x$-values.

$$\text{first-quarter value: } -\frac{\pi}{8}, \quad \text{middle value: } 0, \quad \text{third-quarter value: } \frac{\pi}{8}$$

**Step 4** Evaluate the function for the $x$-values found in Step 3.

| $x$ | $-\frac{\pi}{8}$ | 0 | $\frac{\pi}{8}$ |
|---|---|---|---|
| $2x$ | $-\frac{\pi}{4}$ | 0 | $\frac{\pi}{4}$ |
| $\tan 2x$ | $-1$ | 0 | 1 |

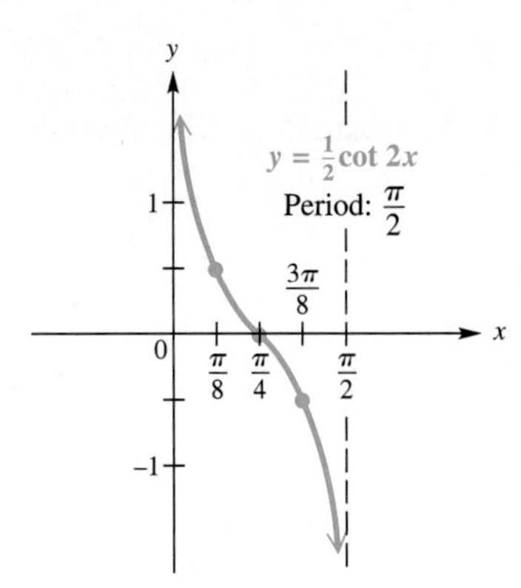


**Step 5** Join these points with a smooth curve, approaching the vertical asymptotes. See Figure 116. Another period has been graphed, one half-period to the left and one half-period to the right.

**EXAMPLE 4** *Graphing $y = a \cot bx$*

Graph $y = \dfrac{1}{2}\cot 2x$.

**Solution** Because this function involves the cotangent, we can locate two adjacent asymptotes by solving the equations $2x = 0$ and $2x = \pi$. The lines $x = 0$ (the $y$-axis) and $x = \frac{\pi}{2}$ are two such asymptotes. We divide the interval $0 < x < \frac{\pi}{2}$ into four equal parts, getting key $x$-values of $\frac{\pi}{8}$, $\frac{\pi}{4}$, and $\frac{3\pi}{8}$. Evaluating the function at these $x$-values gives the following key points.

$$\left(\frac{\pi}{8}, \frac{1}{2}\right), \quad \left(\frac{\pi}{4}, 0\right), \quad \left(\frac{3\pi}{8}, -\frac{1}{2}\right)$$

Joining these points with a smooth curve approaching the asymptotes gives the graph shown in Figure 117.

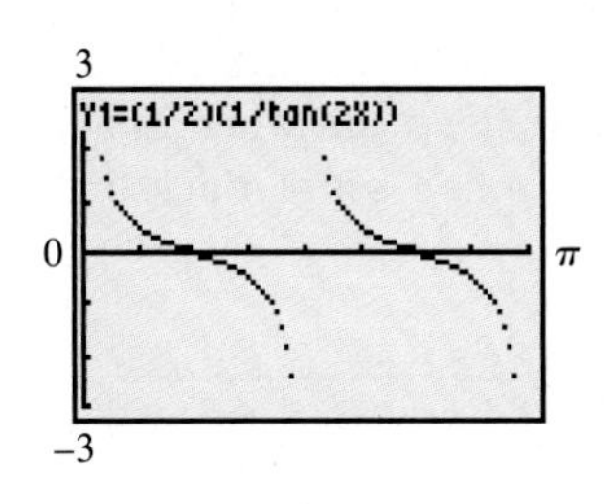


Dot Mode

**FIGURE 117**

Like the other circular functions, the graphs of the tangent and cotangent functions may be translated horizontally as well as vertically.

**EXAMPLE 5** *Graphing a Tangent Function with a Vertical Translation*

Graph $y = 2 + \tan x$.

**Analytic Solution**

Every value of $y$ for this function will be 2 units more than the corresponding value of $y$ in $y = \tan x$, causing the graph of $y = 2 + \tan x$ to be translated 2 units upward compared with the graph of $y = \tan x$. See Figure 118.

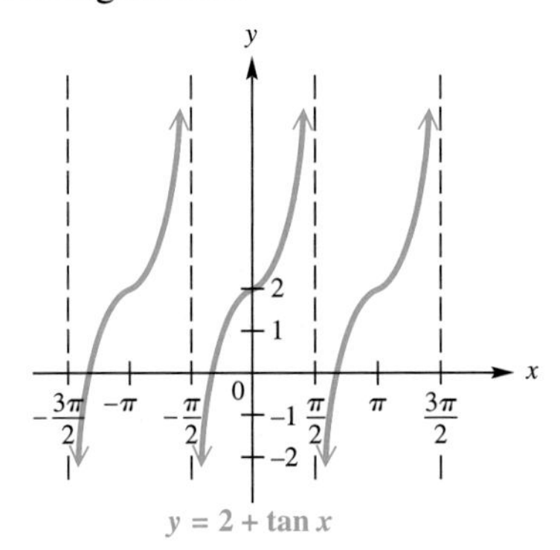


**FIGURE 118**

**Graphing Calculator Solution**

To see the vertical translation, observe the coordinates displayed at the bottoms of the screens in Figures 119 and 120. For $x = \frac{\pi}{4} \approx .78539816$,

$$Y_1 = \tan X = 1,$$

while for the same $x$-value,

$$Y_2 = 2 + \tan X = 2 + 1 = 3.$$

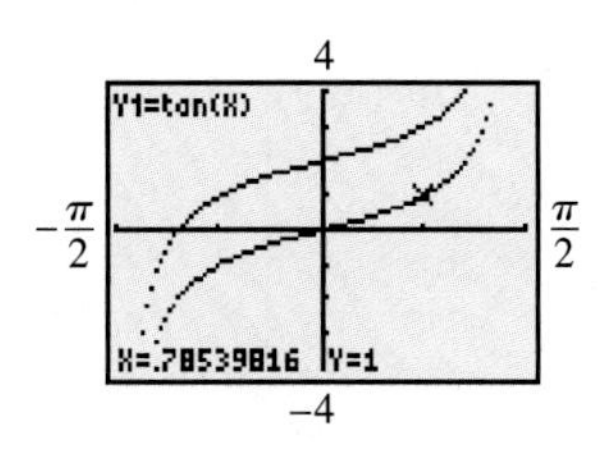


Dot Mode

**FIGURE 119**

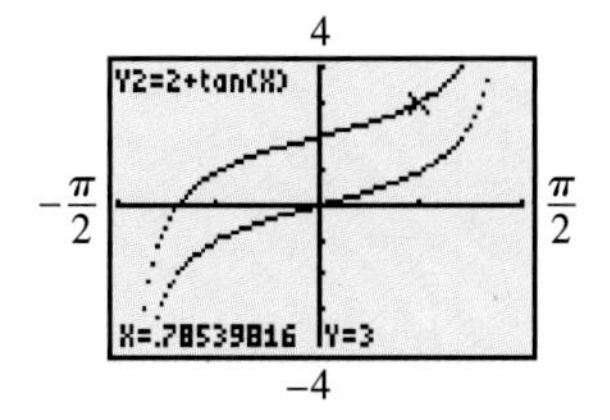


Dot Mode

**FIGURE 120**

## Addition of Ordinates

New functions can be formed by adding or subtracting other functions. A function formed by combining two other functions, such as

$$y = \cos x + \sin x,$$

has historically been graphed using a method known as *addition of ordinates*. (The $x$-value of a point is called its *abscissa*, while its $y$-value is called its *ordinate*.) To apply this method to this function, we would graph the functions $y = \cos x$ and $y = \sin x$. Then, for selected values of $x$, we would add $\cos x$ and $\sin x$, and plot the points $(x, \cos x + \sin x)$. Connecting the selected points with a typical sinusoidal curve would give the graph of the desired function. While this method illustrates some valuable concepts involving the arithmetic of functions, it is time-consuming.

With graphing calculators, this technique is more easily illustrated. Let $Y_1 = \cos X$, $Y_2 = \sin X$, and $Y_3 = Y_1 + Y_2$. Figure 121 shows the result when $Y_1$ and $Y_2$ are graphed with thin lines, and $Y_3 = \cos X + \sin X$ is graphed with a thick line. Notice that for $X = \frac{\pi}{6} \approx .52359878$, $Y_1 + Y_2 = Y_3$.

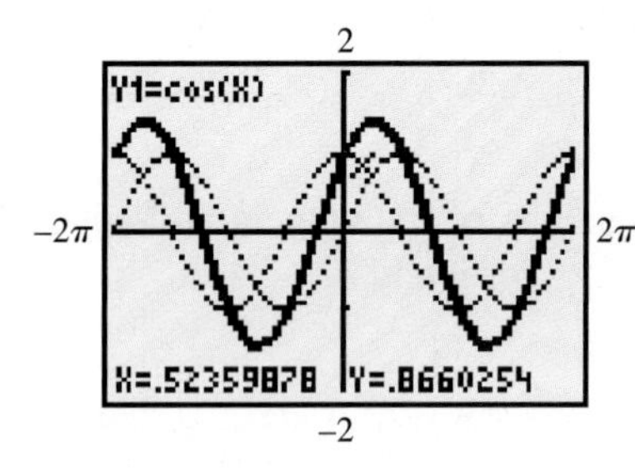

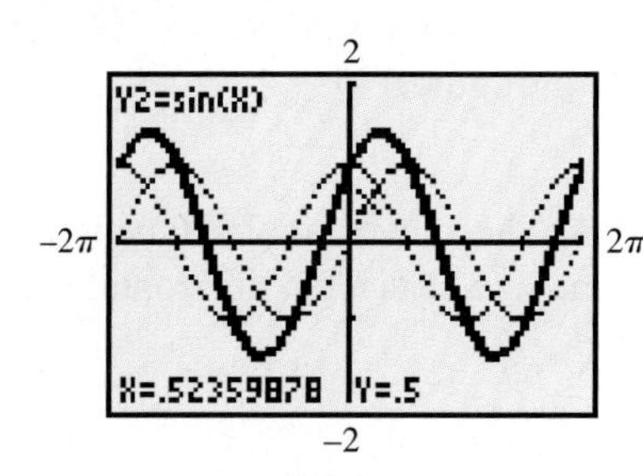

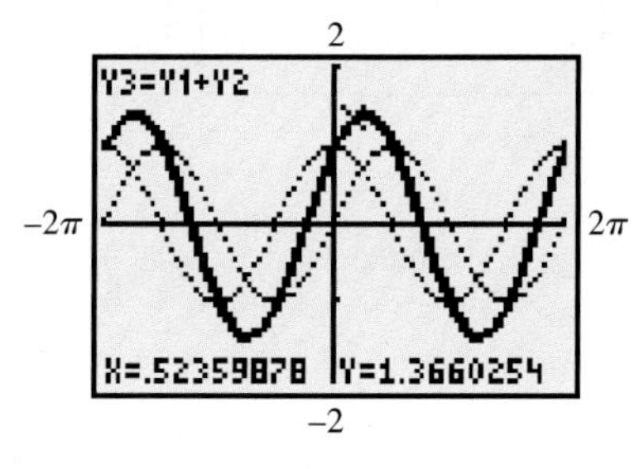


**FIGURE 121**

## 8.7 EXERCISES

*Concept Check* *Without using a calculator, match each function defined in Exercises 1–6 with its graph in A–F.*

**1.** $y = -\csc x$

**2.** $y = -\sec x$

**3.** $y = -\tan x$

**4.** $y = -\cot x$

**5.** $y = \tan\left(x - \frac{\pi}{4}\right)$

**6.** $y = \cot\left(x - \frac{\pi}{4}\right)$

**A.**

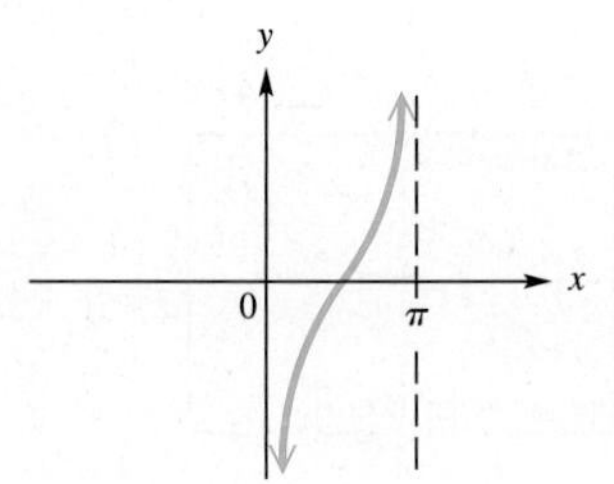


**B.**

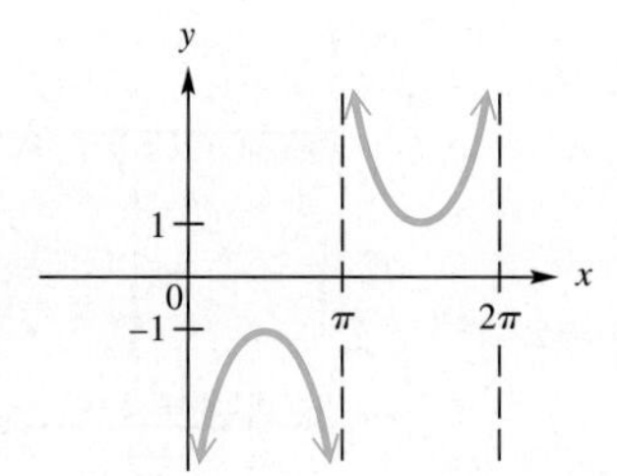


**C.**

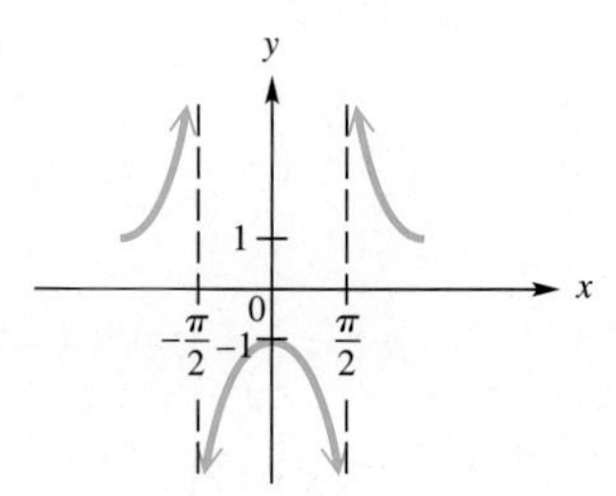


**D.**

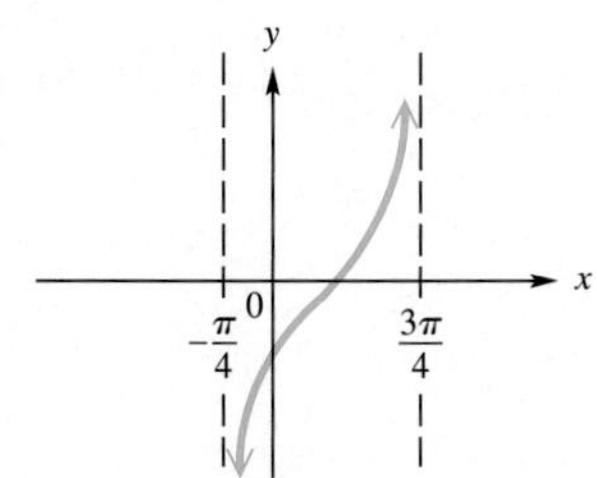


**E.**

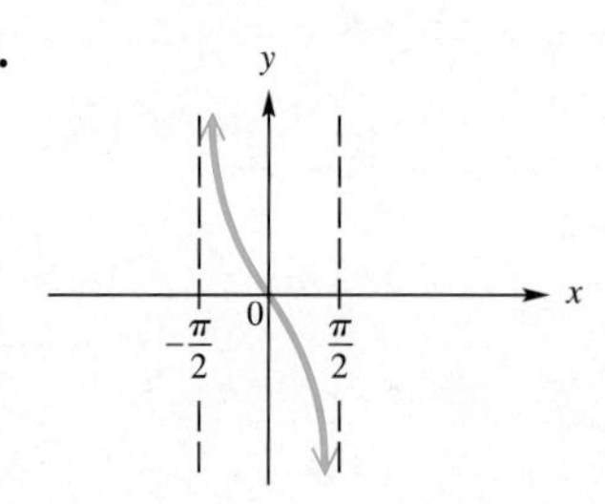


**F.**

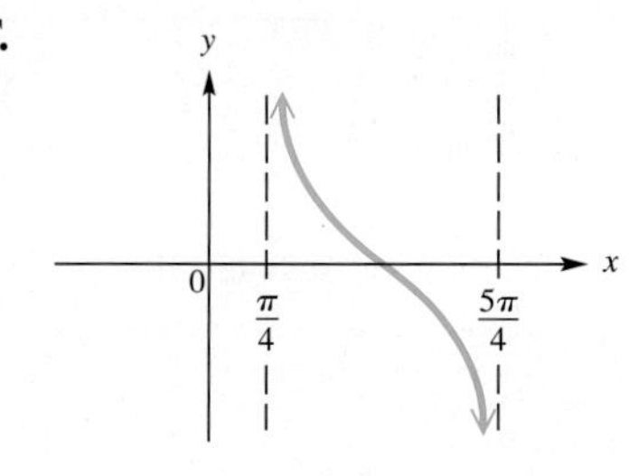

*Concept Check* *Tell whether each statement is* true *or* false. *If* false, *tell why.*

**7.** The smallest positive number $k$ for which $x = k$ is an asymptote for the tangent function is $\frac{\pi}{2}$.

**8.** The smallest positive number $k$ for which $x = k$ is an asymptote for the cotangent function is $\frac{\pi}{2}$.

**9.** The tangent and secant functions are undefined for the same values.

**10.** The secant and cosecant functions are undefined for the same values.

**11.** The graph of $y = \tan x$ in Figure 114 suggests that $\tan(-x) = \tan x$ for all $x$ in the domain of $\tan x$.

**12.** The graph of $y = \sec x$ in Figure 108 suggests that $\sec(-x) = \sec x$ for all $x$ in the domain of $\sec x$.

*Find the* **(a)** *period,* **(b)** *phase shift (if any), and* **(c)** *range of each function.*

**13.** $y = 2\csc \frac{1}{2}x$

**14.** $y = 3\csc 2x$

**15.** $y = -2\sec\left(x + \frac{\pi}{2}\right)$

**16.** $y = -\frac{3}{2}\sec(x - \pi)$

**17.** $y = \frac{5}{2}\cot\left[\frac{1}{3}\left(x - \frac{\pi}{2}\right)\right]$

**18.** $y = -3\tan\left[\frac{1}{2}\left(x + \frac{\pi}{4}\right)\right]$

**19.** $f(x) = \frac{1}{2}\sec(2x + \pi)$

**20.** $f(x) = -\frac{1}{3}\csc\left(\frac{1}{2}x - \frac{\pi}{2}\right)$

**21.** $y = -1 - \tan\left(x + \frac{\pi}{4}\right)$

**22.** $y = 2 + \cot\left(2x - \frac{\pi}{3}\right)$

*In Exercises 23–44, give a traditional or calculator graph, as directed by your instructor. Graph each function over a one-period interval.*

**23.** $y = \csc\left(x - \frac{\pi}{4}\right)$

**24.** $y = \sec\left(x + \frac{3\pi}{4}\right)$

**25.** $y = \sec\left(x + \frac{\pi}{4}\right)$

**26.** $y = \csc\left(x + \frac{\pi}{3}\right)$

**27.** $y = \sec\left(\frac{1}{2}x + \frac{\pi}{3}\right)$

**28.** $y = \csc\left(\frac{1}{2}x - \frac{\pi}{4}\right)$

**29.** $y = 2 + 3\sec(2x - \pi)$

**30.** $y = 1 - 2\csc\left(x + \frac{\pi}{2}\right)$

**31.** $y = 1 - \frac{1}{2}\csc\left(x - \frac{3\pi}{4}\right)$

**32.** $y = 2 + \frac{1}{4}\sec\left(\frac{1}{2}x - \pi\right)$

*Graph each function over a two-period interval.*

**33.** $y = \tan(2x - \pi)$

**34.** $y = \tan\left(\frac{x}{2} + \pi\right)$

**35.** $y = \cot\left(3x + \frac{\pi}{4}\right)$

**36.** $y = \cot\left(2x - \frac{3\pi}{2}\right)$

**37.** $y = 1 + \tan x$

**38.** $y = -2 + \tan x$

**39.** $y = 1 - \cot x$

**40.** $y = -2 - \cot x$

**41.** $y = -1 + 2\tan x$

**42.** $y = 3 + \frac{1}{2}\tan x$

**43.** $y = -1 + \frac{1}{2}\cot(2x - 3\pi)$

**44.** $y = -2 + 3\tan(4x + \pi)$

*Solve each problem.*

**45.** Simultaneously graph $y = \tan x$ and $y = x$ in the window $[-1, 1]$ by $[-1, 1]$. Write a sentence or two describing the relationship of $\tan x$ and $x$ for small $x$-values.

**46.** Between each pair of successive asymptotes, a portion of the graph of $y = \sec x$ or $y = \csc x$ resembles a parabola. Can each of these portions actually be a parabola? Explain.

**47.** *(Modeling) Distance of a Rotating Beacon* A rotating beacon is located at point $A$ next to a long wall. The beacon is 4 meters from the wall. The distance $d$ is given by

$$d = 4\tan 2\pi t,$$

where $t$ is time measured in seconds since the beacon started rotating. (When $t = 0$, the beacon is aimed at point $R$. When the beacon is aimed to the right of $R$, the value of $d$ is positive; $d$ is negative if the beacon is aimed to the left of $R$.)

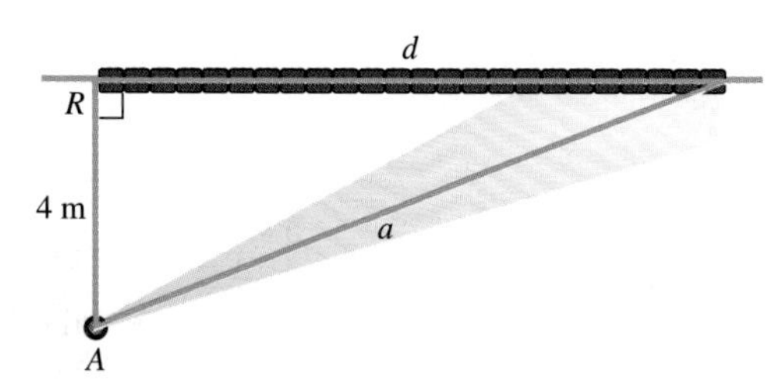

Find $d$ for each time.

**(a)** $t = 0$ **(b)** $t = .4$ **(c)** $t = .8$

**(d)** $t = 1.2$

**(e)** Why is .25 a meaningless value for $t$?

**48.** *(Modeling) Distance of a Rotating Beacon* In the figure for Exercise 47, the distance $a$ is given by

$$a = 4|\sec 2\pi t|.$$

Find a for each time.

**(a)** $t = 0$ **(b)** $t = .86$ **(c)** $t = 1.24$

*Use a graphing calculator to graph* $Y_1$, $Y_2$, *and* $Y_1 + Y_2$ *on the same screen. Evaluate each of the three functions at* $x = \frac{\pi}{6}$, *and verify that* $Y_1(\frac{\pi}{6}) + Y_2(\frac{\pi}{6}) = (Y_1 + Y_2)(\frac{\pi}{6})$.

**49.** $Y_1 = \sin X$; $Y_2 = \sin 2X$

**50.** $Y_1 = \cos X$; $Y_2 = \cos 2X$

**51.** $Y_1 = \sin 2X$; $Y_2 = \cos \frac{1}{2}X$

**52.** $Y_1 = \tan X$; $Y_2 = \tan 2X$

# HARMONIC MOTION

Simple Harmonic Motion • Damped Oscillatory Motion

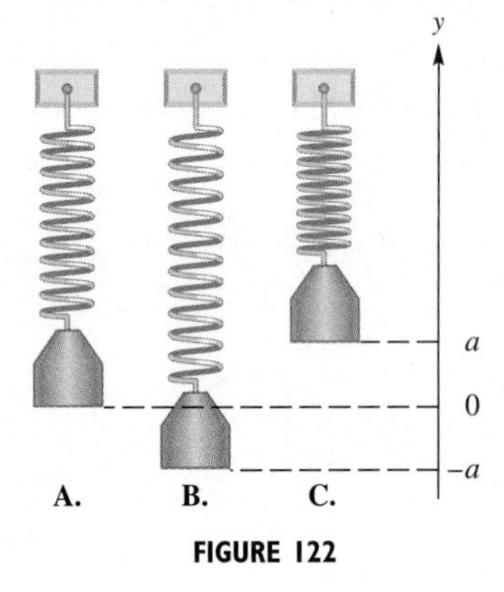


FIGURE 122

## Simple Harmonic Motion

In part A of Figure 122, a spring with a weight attached to its free end is in equilibrium (or rest) position. If the weight is pulled down $a$ units and released (part B of the figure), the spring's elasticity causes the weight to rise $a$ units ($a > 0$) above the equilibrium position, as seen in part C, and then oscillate about the equilibrium position. If friction is neglected, this oscillatory motion is described mathematically by a sinusoid. Other applications of this type of motion include sound, electric current, and electromagnetic waves. We have seen examples of these applications earlier in this chapter.

To develop a general equation for such motion, consider Figure 123. Suppose the point $P(x, y)$ moves around the circle counterclockwise at a uniform angular speed $\omega$. Assume that at time $t = 0$, $P$ is at $(a, 0)$. The angle swept out by ray $OP$ at time $t$ is given by $\theta = \omega t$. The coordinates of point $P$ at time $t$ are

$$x = a \cos \theta = a \cos \omega t \quad \text{and} \quad y = a \sin \theta = a \sin \omega t.$$

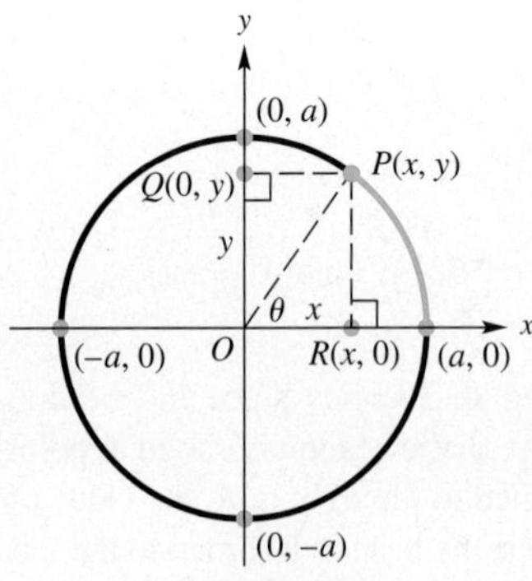


FIGURE 123

As $P$ moves around the circle from the point $(a, 0)$, the point $Q(0, y)$ oscillates back and forth along the $y$-axis between the points $(0, a)$ and $(0, -a)$. Similarly, the point $R(x, 0)$ oscillates back and forth between $(a, 0)$ and $(-a, 0)$. This oscillatory motion is called **simple harmonic motion.**

The amplitude of the motion is $|a|$, and the period $\frac{2\pi}{\omega}$. The moving points $P$ and $Q$ or $P$ and $R$ complete one oscillation or cycle per period. The number of cycles per unit of time, called the **frequency,** is the reciprocal of the period, $\frac{\omega}{2\pi}$, where $\omega > 0$.

**Simple Harmonic Motion**

The position of a point oscillating about an equilibrium position at time $t$ is modeled by either

$$s(t) = a \cos \omega t \quad \text{or} \quad s(t) = a \sin \omega t,$$

where $a$ and $\omega$ are constants, with $\omega > 0$. The amplitude of the motion is $|a|$, the period is $\frac{2\pi}{\omega}$, and the frequency is $\frac{\omega}{2\pi}$.

### EXAMPLE 1 *Modeling the Motion of a Spring*

Suppose that an object is attached to a coiled spring such as the one in Figure 122. It is pulled down a distance of 5 units from its equilibrium position, and then released. The time for one complete oscillation is 4 seconds.

**(a)** Give an equation that models the position of the object at time $t$.

**(b)** Determine the position at $t = 1.5$ seconds.

**(c)** Find the frequency.

**Solution**

**(a)** When the object is released at $t = 0$, the distance of the object from the equilibrium position is 5 inches below equilibrium. If $s(t)$ is to model the motion, then $s(0)$ must equal $-5$. We use

$$s(t) = a \cos \omega t,$$

with $a = -5$. We choose the cosine function because $\cos \omega(0) = \cos 0 = 1$, and $-5 \cdot 1 = -5$. (Had we chosen the sine function, a phase shift would have been required.) The period is 4, so

$$\frac{2\pi}{\omega} = 4, \quad \text{or} \quad \omega = \frac{\pi}{2}.$$

Thus, the motion is modeled by

$$s(t) = -5 \cos \frac{\pi}{2} t.$$

**(b)** After 1.5 seconds, the position is

$$s(1.5) = -5 \cos\left[\frac{\pi}{2}(1.5)\right] \approx 3.54 \text{ inches.}$$

Since $3.54 > 0$, the object is above the equilibrium position.

**(c)** The frequency is the reciprocal of the period, or $\frac{1}{4}$.

### EXAMPLE 2 *Analyzing Harmonic Motion*

Suppose that an object oscillates according to the model

$$s(t) = 8 \sin 3t,$$

where $t$ is in seconds and $s(t)$ is in feet. Analyze the motion.

**Solution** The motion is harmonic because the model is of the form $s(t) = a \sin \omega t$. Because $a = 8$, the object oscillates 8 feet in either direction from its starting point. The period $\frac{2\pi}{3} \approx 2.1$ is the time, in seconds, it takes for one complete oscillation. The frequency is the reciprocal of the period, so the object completes $\frac{3}{2\pi} \approx .48$ oscillation per second.

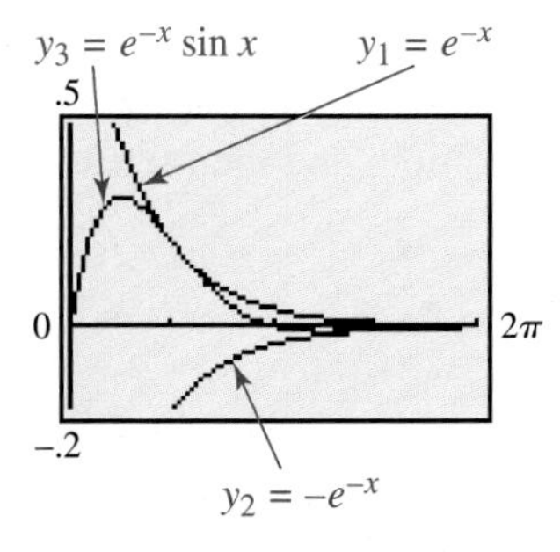


FIGURE 124

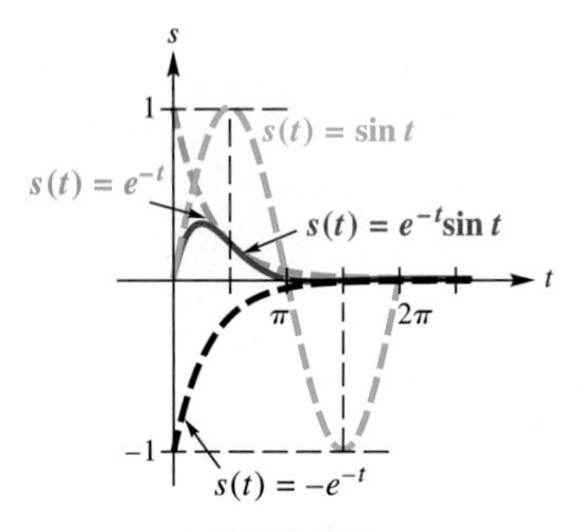


FIGURE 125

## Damped Oscillatory Motion

In the example of the stretched spring, we disregarded the effect of friction. Friction causes the amplitude of the motion to diminish gradually until the weight comes to rest. In this situation, we say that the motion has been *damped* by the force of friction. Most oscillatory motions are damped, and the decrease in amplitude follows the pattern of exponential decay. A typical example of **damped oscillatory motion** is provided by the function defined by

$$s(t) = e^{-t}\sin t.$$

Figure 124 shows how the graph of $y_3 = e^{-x}\sin x$ is bounded above by the graph of $y_1 = e^{-x}$ and below by the graph of $y_2 = -e^{-x}$. The damped motion curve dips below the $x$-axis at $x = \pi$ but stays above the graph of $y_2$. Figure 125 shows a traditional graph, along with the graph of $s = \sin t$.

Shock absorbers are put on an automobile in order to damp oscillatory motion. Instead of oscillating up and down for a long while after hitting a bump or pothole, the oscillations of the car are quickly damped out for a smoother ride.

## 8.8 EXERCISES

*(Modeling) Springs Suppose that a weight on a spring has initial position $s(0)$ and period $P$.*

**(a)** *Find a function $s$ given by $s(t) = a\cos(\omega t)$ that models the displacement of the weight.*

**(b)** *Evaluate $s(1)$. Is the weight moving upward, downward, or neither when $t = 1$? Support your results graphically or numerically.*

**1.** $s(0) = 2$ inches; $P = .5$ second

**2.** $s(0) = 5$ inches; $P = 1.5$ seconds

**3.** $s(0) = -3$ inches; $P = .8$ second

**4.** $s(0) = -4$ inches; $P = 1.2$ seconds

*(Modeling) Music A note on the piano has given frequency $F$. Suppose the maximum displacement at the center of the piano wire is given by $s(0)$. Find constants $a$ and $b$ so that the equation $s(t) = a\cos\omega t$ models this displacement. Graph $f$ in the viewing window $[0, .05]$ by $[-.3, .3]$.*

**5.** $F = 27.5$; $s(0) = .21$

**6.** $F = 110$; $s(0) = .11$

**7.** $F = 55$; $s(0) = .14$

**8.** $F = 220$; $s(0) = .06$

*(Modeling) Solve each problem.*

**9.** *Particle Movement* Write the equation and then determine the amplitude, period, and frequency of the simple harmonic motion of a particle moving uniformly around a circle of radius 2 units, with angular speed **(a)** 2 radians per second, and **(b)** 4 radians per second.

**10.** *Pendulum* What are the period $P$ and frequency $T$ of oscillation of a pendulum of length $\frac{1}{2}$ foot? (*Hint:* $P = 2\pi\sqrt{\frac{L}{32}}$, where $L$ is the length of the pendulum in feet and $P$ is in seconds.)

**11.** *Pendulum* How long should the pendulum be to have period 1 second? (See Exercise 10.)

**12.** *Spring* The formula for the up and down motion of a weight on a spring is given by

$$s(t) = a\sin\sqrt{\frac{k}{m}}\,t.$$

If the spring constant $k$ is 4, what mass $m$ must be used to produce a period of 1 second?

**13.** *Spring* A spring with spring constant $k = 2$ and a 1-unit mass $m$ attached to it is stretched and then allowed to come to rest.

**(a)** If the spring is stretched $\frac{1}{2}$ foot and released, what are the amplitude, period, and frequency of the resulting oscillatory motion?

**(b)** What is the equation of the motion?

**14.** *Spring* The position of a weight attached to a spring is

$$s(t) = -5 \cos 4\pi t$$

inches after $t$ seconds.

**(a)** What is the maximum height that the weight rises above the equilibrium position?

**(b)** What are the frequency and period?

**(c)** When does the weight first reach its maximum height?

**(d)** Calculate and interpret $s(1.3)$.

**15.** *Spring* The position of a weight attached to a spring is

$$s(t) = -4 \cos 10t$$

inches after $t$ seconds.

**(a)** What is the maximum height that the weight rises above the equilibrium position?

**(b)** What are the frequency and period?

**(c)** When does the weight first reach its maximum height?

**(d)** Calculate and interpret $s(1.466)$.

**16.** *Spring* A weight attached to a spring is pulled down 3 inches below the equilibrium position.

**(a)** Assuming that the frequency is $\frac{6}{\pi}$ cycles per second, determine a trigonometric model that gives the position of the weight at time $t$ seconds.

**(b)** What is the period?

**17.** *Spring* A weight attached to a spring is pulled down 2 inches below the equilibrium position.

**(a)** Assuming that the period is $\frac{1}{3}$ second, determine a trigonometric model that gives the position of the weight at time $t$ seconds.

**(b)** What is the frequency?

**18.** Use a graphing calculator to graph $y_1 = e^{-t} \sin t$, $y_2 = e^{-t}$, and $y_3 = -e^{-t}$ in the viewing window $[0, \pi]$ by $[-.5, .5]$.

**(a)** Find the $x$-intercepts of the graph of $y_1$. Explain the relationship of these $x$-intercepts with those of the graph of $y = \sin x$.

**(b)** Find any points of intersection of $y_1$ and $y_2$ or $y_1$ and $y_3$. How are these points related to the graph of $y = \sin x$?

## Reviewing Basic Concepts (Sections 8.7 and 8.8)

*In Exercises 1–4, give a traditional or calculator graph, as directed by your instructor.*

**1.** Graph one period of $y = \csc\left(x - \frac{\pi}{4}\right)$. State the period, phase shift, domain, and range.

**2.** Graph one period of $y = \sec\left(x + \frac{\pi}{4}\right)$. State the period, phase shift, domain, and range.

**3.** Graph one period of $y = 2 \tan x$. State the period, domain, and range.

**4.** Graph one period of $y = 2 \cot x$. State the period, domain, and range.

**5.** *(Modeling) Spring* The height of a weight attached to a spring is

$$s(t) = -4 \cos 8\pi t$$

inches after $t$ seconds.

**(a)** Find the maximum height that the weight rises above the equilibrium position of $y = 0$.

**(b)** When does the weight first reach its maximum height, if $t \geq 0$?

**(c)** What are the frequency and period?

# CHAPTER 8 SUMMARY

**KEY TERMS & SYMBOLS** | **KEY CONCEPTS**

## 8.1 Angles and Their Measures

line
segment
ray
angle
initial side
terminal side
vertex
positive angle
negative angle
degree (°)
acute angle
right angle
obtuse angle
straight angle
complementary angles
supplementary angles
minute (′)
second (″)
angle in standard position
quadrantal angle
coterminal angle
radian
sector of a circle
angular speed, $\omega$
linear speed, $v$

An angle that has its vertex at the center of a circle and that intercepts an arc on the circle equal in length to the radius of the circle has a measure of 1 radian.

**DEGREE/RADIAN RELATIONSHIP** $180° = \pi$ radians

**CONVERTING BETWEEN DEGREES AND RADIANS**

1. Multiply a radian measure by $\frac{180°}{\pi}$ and simplify to convert to degrees.
2. Multiply a degree measure by $\frac{\pi}{180}$ radian and simplify to convert to radians.

**ARC LENGTH**

The length $s$ of the arc intercepted on a circle of radius $r$ by a central angle of measure $\theta$ radians is given by the product of the radius and the radian measure of the angle, or

$$s = r\theta, \quad \theta \text{ in radians.}$$

**AREA OF A SECTOR**

The area of a sector of a circle of radius $r$ and central angle $\theta$ is given by

$$A = \frac{1}{2}r^2\theta, \quad \theta \text{ in radians.}$$

**ANGULAR AND LINEAR SPEED**

| Angular Speed | Linear Speed |
|---|---|
| $\omega = \frac{\theta}{t}$ <br> ($\omega$ in radians per unit time, $\theta$ in radians) | $v = \frac{s}{t}$ <br> $v = \frac{r\theta}{t}$ <br> $v = r\omega$ |

## 8.2 Trigonometric Functions and Fundamental Identities

sine
cosine
tangent
cotangent
secant
cosecant

**TRIGONOMETRIC FUNCTIONS (STANDARD POSITION ANGLE APPROACH)**

Let $(x, y)$ be a point other than the origin on the terminal side of an angle $\theta$ in standard position. Let $r = \sqrt{x^2 + y^2}$, the distance from the origin to $(x, y)$. Then

$$\sin\theta = \frac{y}{r} \qquad \cos\theta = \frac{x}{r} \qquad \tan\theta = \frac{y}{x}\ (x \neq 0)$$

$$\csc\theta = \frac{r}{y}\ (y \neq 0) \qquad \sec\theta = \frac{r}{x}\ (x \neq 0) \qquad \cot\theta = \frac{x}{y}\ (y \neq 0).$$

**RECIPROCAL IDENTITIES**

$$\sin\theta = \frac{1}{\csc\theta} \qquad \cos\theta = \frac{1}{\sec\theta} \qquad \tan\theta = \frac{1}{\cot\theta}$$

$$\csc\theta = \frac{1}{\sin\theta} \qquad \sec\theta = \frac{1}{\cos\theta} \qquad \cot\theta = \frac{1}{\tan\theta}$$

**PYTHAGOREAN IDENTITIES**

$$\sin^2\theta + \cos^2\theta = 1 \qquad 1 + \tan^2\theta = \sec^2\theta \qquad 1 + \cot^2\theta = \csc^2\theta$$

**KEY TERMS & SYMBOLS** | **KEY CONCEPTS**

**QUOTIENT IDENTITIES**

$$\frac{\sin\theta}{\cos\theta} = \tan\theta \qquad \frac{\cos\theta}{\sin\theta} = \cot\theta$$

**SIGNS OF TRIGONOMETRIC FUNCTIONS**

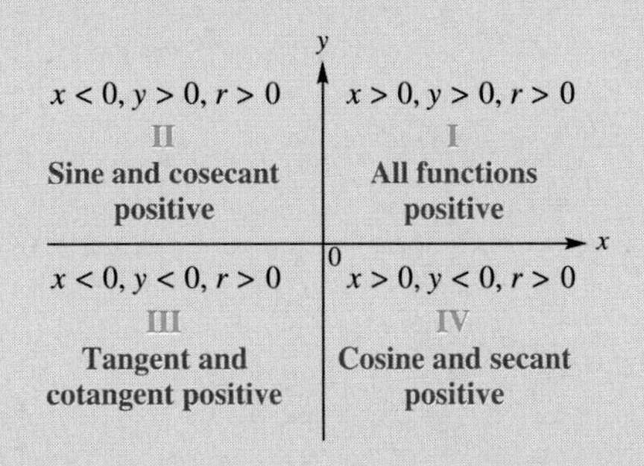


## 8.3 Evaluating Trigonometric Functions

side opposite
side adjacent
reference angle

**TRIGONOMETRIC FUNCTIONS (RIGHT TRIANGLE APPROACH)**

For any acute angle $A$ in standard position,

$$\sin A = \frac{y}{r} = \frac{\text{side opposite}}{\text{hypotenuse}} \qquad \csc A = \frac{r}{y} = \frac{\text{hypotenuse}}{\text{side opposite}}$$

$$\cos A = \frac{x}{r} = \frac{\text{side adjacent}}{\text{hypotenuse}} \qquad \sec A = \frac{r}{x} = \frac{\text{hypotenuse}}{\text{side adjacent}}$$

$$\tan A = \frac{y}{x} = \frac{\text{side opposite}}{\text{side adjacent}} \qquad \cot A = \frac{x}{y} = \frac{\text{side adjacent}}{\text{side opposite}}.$$

**FUNCTION VALUES FOR SPECIAL ANGLES**

| $\theta$ | $\sin\theta$ | $\cos\theta$ | $\tan\theta$ | $\cot\theta$ | $\sec\theta$ | $\csc\theta$ |
|---|---|---|---|---|---|---|
| $30° = \frac{\pi}{6}$ | $\frac{1}{2}$ | $\frac{\sqrt{3}}{2}$ | $\frac{\sqrt{3}}{3}$ | $\sqrt{3}$ | $\frac{2\sqrt{3}}{3}$ | 2 |
| $45° = \frac{\pi}{4}$ | $\frac{\sqrt{2}}{2}$ | $\frac{\sqrt{2}}{2}$ | 1 | 1 | $\sqrt{2}$ | $\sqrt{2}$ |
| $60° = \frac{\pi}{3}$ | $\frac{\sqrt{3}}{2}$ | $\frac{1}{2}$ | $\sqrt{3}$ | $\frac{\sqrt{3}}{3}$ | 2 | $\frac{2\sqrt{3}}{3}$ |

**COFUNCTION IDENTITIES**

If $A$ is an acute angle measured in degrees,

$$\sin A = \cos(90° - A) \qquad \csc A = \sec(90° - A)$$

$$\cos A = \sin(90° - A) \qquad \sec A = \csc(90° - A)$$

$$\tan A = \cot(90° - A) \qquad \cot A = \tan(90° - A).$$

If $A$ is measured in radians replace 90° with $\frac{\pi}{2}$.

| KEY TERMS & SYMBOLS | KEY CONCEPTS |
|---|---|

**REFERENCE ANGLES**

$\theta'$ is the reference angle for angle $\theta$. (If $\theta$ is acute, then $\theta' = \theta$.)

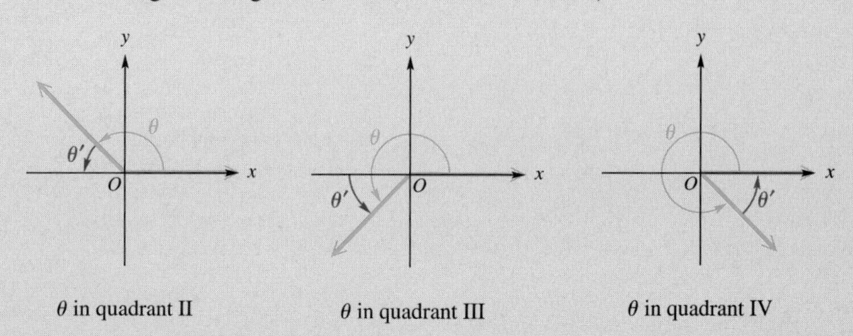


$\theta$ in quadrant II $\theta$ in quadrant III $\theta$ in quadrant IV

**FINDING TRIGONOMETRIC FUNCTION VALUES FOR NONQUADRANTAL ANGLES**

1. Add or subtract 360° as many times as needed to get an angle of at least 0° but less than 360°.
2. Find the reference angle $\theta'$.
3. Find the trigonometric function values for $\theta'$.
4. Determine the correct signs for the values found in Step 3.

## 8.4 Applications of Right Triangles

significant digit
exact number
angle of elevation
angle of depression
bearing

**SOLVING AN APPLIED TRIGONOMETRY PROBLEM**

1. Draw a sketch, and label it with the given information. Label the quantity to be found with a variable.
2. Use the sketch to write an equation relating the given quantities to the variable.
3. Solve the equation, and check that your answer makes sense.

## 8.5 The Circular Functions

unit circle
circular functions

**TRIGONOMETRIC FUNCTIONS (UNIT CIRCLE APPROACH)**

For $s$, $x$, and $y$ as shown in the figure, the circular functions are

$$\sin s = y \qquad \cos s = x \qquad \tan s = \frac{y}{x} \quad (x \neq 0)$$

$$\csc s = \frac{1}{y} \quad (y \neq 0) \qquad \sec s = \frac{1}{x} \quad (x \neq 0) \qquad \cot s = \frac{x}{y} \quad (y \neq 0).$$

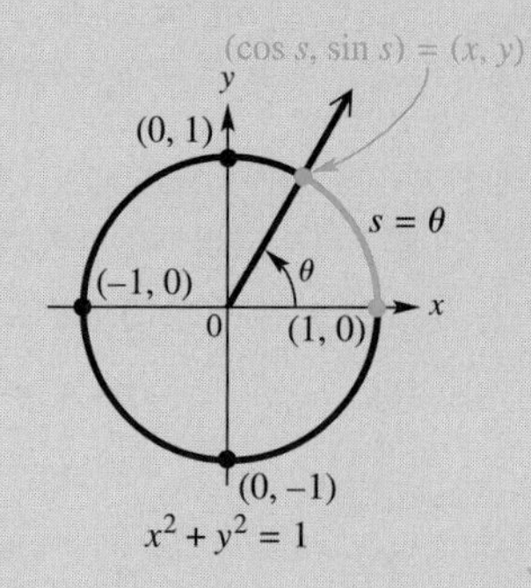

| KEY TERMS & SYMBOLS | KEY CONCEPTS |
|---|---|

## 8.6 Graphs of the Sine and Cosine Functions

periodic function
period
sine wave (sinusoid)
amplitude
phase shift
argument

**SINE AND COSINE FUNCTIONS**

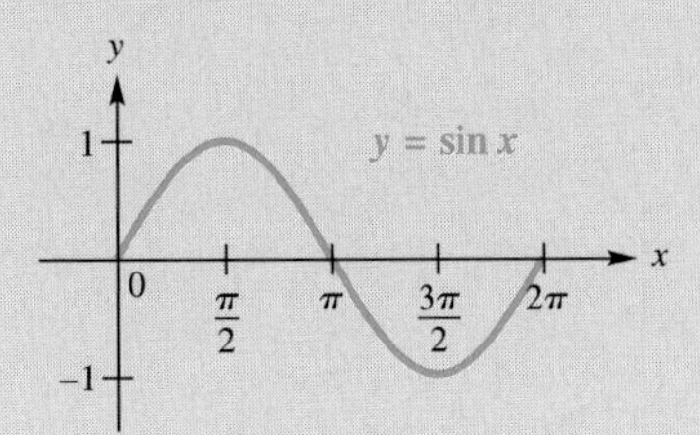


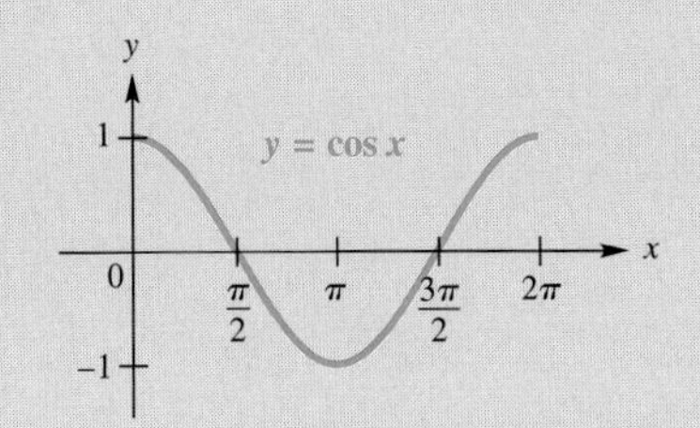


**Domain:** $(-\infty, \infty)$

**Range:** $[-1, 1]$

**Amplitude:** 1

**Period:** $2\pi$

**Domain:** $(-\infty, \infty)$

**Range:** $[-1, 1]$

**Amplitude:** 1

**Period:** $2\pi$

Assume $b > 0$. The graph of

$$y = c + a \sin[b(x - d)] \quad \text{or} \quad y = c + a \cos[b(x - d)]$$

has

1. amplitude $|a|$, 2. period $\frac{2\pi}{b}$,
3. vertical translation $c$ units upward if $c > 0$ or $|c|$ units downward if $c < 0$,
4. phase shift $d$ units to the right if $d > 0$ or $|d|$ units to the left if $d < 0$.

## 8.7 Graphs of the Other Circular Functions

**COSECANT AND SECANT FUNCTIONS**

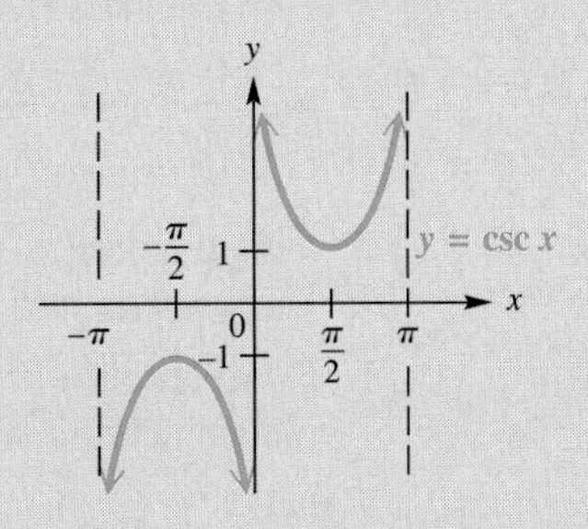


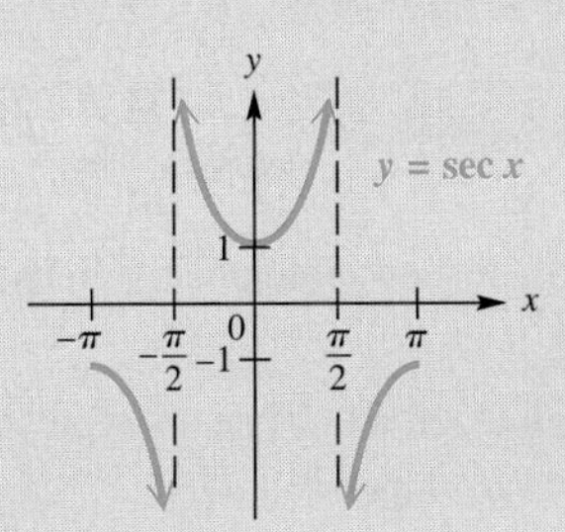


**Domain:** $\{x \mid x \neq n\pi$, where $n$ is an integer$\}$

**Range:** $(-\infty, -1] \cup [1, \infty)$

**Period:** $2\pi$

**Domain:** $\{x \mid x \neq (2n + 1)\frac{\pi}{2}$, where $n$ is an integer$\}$

**Range:** $(-\infty, -1] \cup [1, \infty)$

**Period:** $2\pi$

| KEY TERMS & SYMBOLS | KEY CONCEPTS |
|---|---|
| | **TANGENT AND COTANGENT FUNCTIONS** |

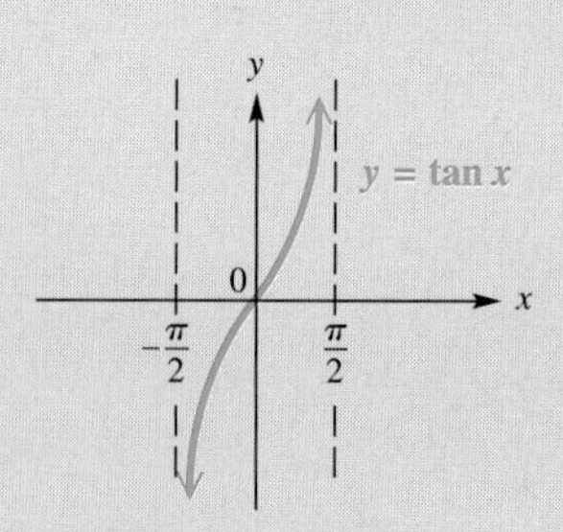


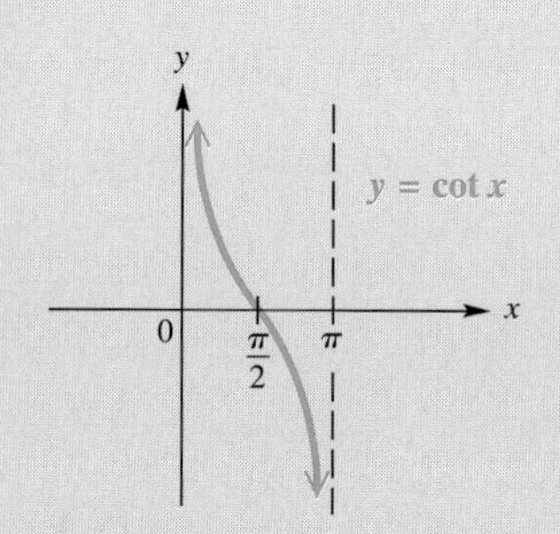


**Domain:** $\left\{x \,|\, x \neq (2n+1)\frac{\pi}{2},\text{ where } n \text{ is an integer}\right\}$

**Range:** $(-\infty, \infty)$

**Period:** $\pi$

**Domain:** $\{x \,|\, x \neq n\pi, \text{ where } n \text{ is an integer}\}$

**Range:** $(-\infty, \infty)$

**Period:** $\pi$

**8.8** Harmonic Motion

simple harmonic motion
frequency
damped oscillatory motion

**SIMPLE HARMONIC MOTION**

The position of a point oscillating about an equilibrium position at time $t$ is modeled by either

$$s(t) = a \cos \omega t \quad \text{or} \quad s(t) = a \sin \omega t,$$

where $a$ and $\omega$ are constants, with $\omega > 0$. The amplitude of the motion is $|a|$, the period is $\frac{2\pi}{\omega}$, and the frequency is $\frac{\omega}{2\pi}$.

## CHAPTER 8 REVIEW EXERCISES

**1.** Find the angle of smallest possible positive measure coterminal with an angle of $-174°$.

**2.** Let $n$ represent any integer, and write an expression for all angles coterminal with an angle of $270°$.

*Solve each problem.*

**3.** *Rotating Pulley* A pulley is rotating 320 times per minute. Through how many degrees does a point on the edge of the pulley move in $\frac{2}{3}$ second?

**4.** *Rotating Propeller* The propeller of a speedboat rotates 650 times per minute. Through how many degrees will a point on the edge of the propeller rotate in 2.4 seconds?

**5.** Which is larger—an angle of $1°$ or an angle of 1 radian? Discuss and justify your answer.

**6.** Consider each angle in standard position having the given radian measure. In what quadrant does the terminal side lie?

**(a)** 3 **(b)** 4 **(c)** $-2$ **(d)** 7

*Convert each degree measure to radians. Leave answers as multiples of $\pi$.*

**7.** $120°$ **8.** $800°$

*Convert each radian measure to degrees.*

**9.** $\frac{5\pi}{4}$ **10.** $-\frac{6\pi}{5}$

*Concept Check Suppose the tip of the minute hand of a clock is two inches from the center of the clock. For each duration, determine the distance traveled by the tip of the minute hand.*

**11.** 20 minutes

**12.** 3 hours

*Solve each problem. Use a calculator as necessary.*

**13.** *Arc Length* The radius of a circle is 15.2 cm. Find the length of an arc of the circle intercepted by a central angle of $\frac{3\pi}{4}$ inches.

**14.** *Area of a Sector* A central angle of $\frac{7\pi}{4}$ radians forms a sector of a circle. Find the area of the sector if the radius of the circle is 28.69 inches.

**15.** *Height of a Tree* A tree 2000 yards away subtends an angle of $1° 10'$. Find the height of the tree to two significant digits.

**16.** *Concept Check* Find the measure of the central angle $\theta$ (in radians) and the area of the sector.

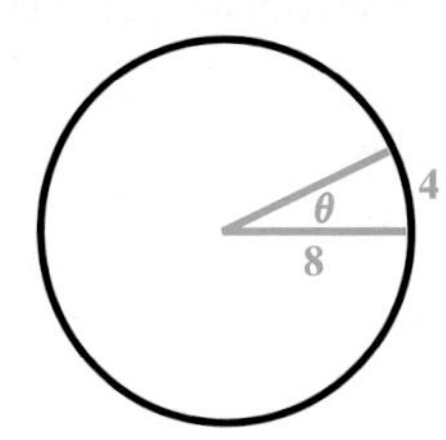


*Find the trigonometric function values of each angle. If a value is undefined, say so.*

**17.**

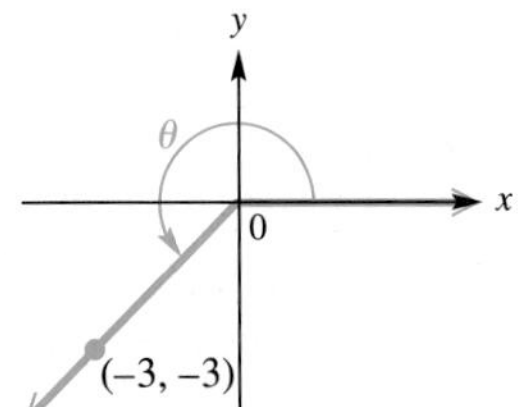


**18.**

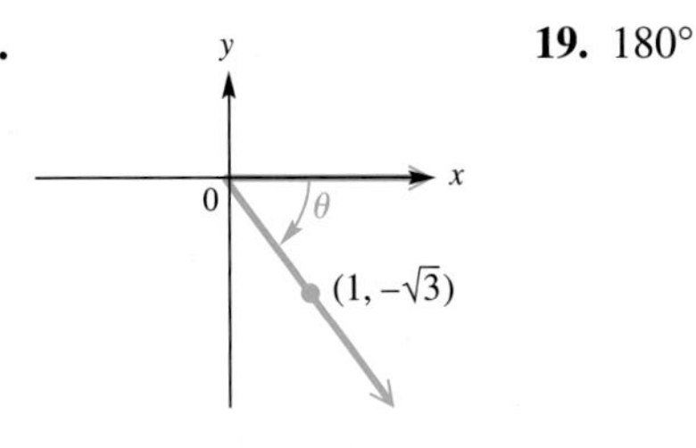


**19.** 180°

*Find the values of all six trigonometric functions for an angle in standard position having each point on its terminal side.*

**20.** $(3, -4)$ **21.** $(9, -2)$ **22.** $\left(-2\sqrt{2}, 2\sqrt{2}\right)$

**23.** *Concept Check* If the terminal side of a quadrantal angle lies along the $y$-axis, which of its trigonometric functions are undefined?

*Decide whether each statement is* possible *or* impossible.

**24.** $\sec\theta = -\frac{2}{3}$ **25.** $\tan\theta = 1.4$

*Find all six trigonometric function values for each angle. Rationalize denominators when applicable.*

**26.** $\sin\theta = \frac{\sqrt{3}}{5}$ and $\cos\theta < 0$

**27.** $\cos\theta = -\frac{5}{8}$, with $\theta$ in quadrant III

**28.** If, for some particular angle $\theta$, $\sin\theta < 0$ and $\cos\theta > 0$, in what quadrant must $\theta$ lie? What is the sign of $\tan\theta$?

**29.** Explain how you would find the cotangent of an angle $\theta$ whose tangent is 1.6778490 using a calculator. Then find $\cot\theta$.

*Find the values of the six trigonometric functions for each angle A.*

**30.**

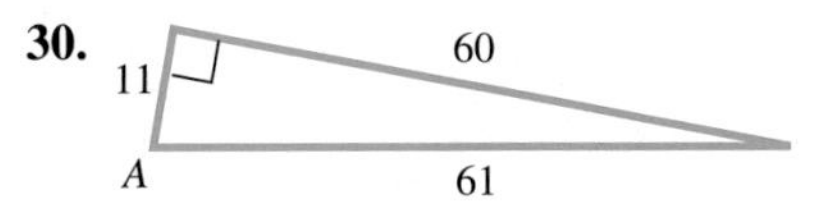


**31.**

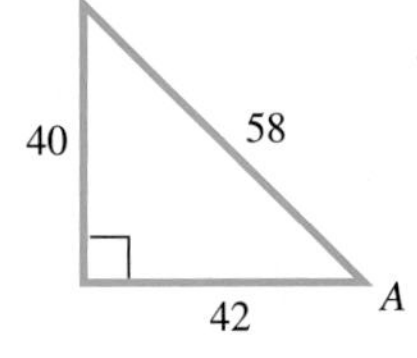


**32.** Explain why, in the figure, the cosine of angle $A$ is equal to the sine of angle $B$.

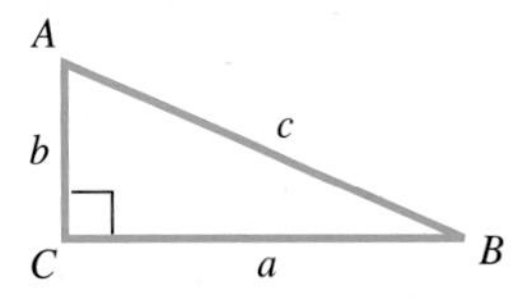


*Find the values of the six trigonometric functions for each angle. Give exact values. Do not use a calculator. Rationalize denominators when applicable.*

**33.** 300° **34.** −225° **35.** −390°

*Use a calculator to find each value.*

**36.** $\sin 72° 30'$

**37.** $\sec 222° 30'$

**38.** $\cot 305.6°$

**39.** $\tan 11.7689°$

**40.** *Concept Check* Which one of the following cannot be *exactly* determined using the methods of this chapter?

**A.** $\cos 135°$ **B.** $\cot(-45°)$

**C.** $\sin 300°$ **D.** $\tan 140°$

*Approximate to the nearest tenth of a degree each value of θ, where θ is in the interval* $[0°, 90°)$.

**41.** $\sin \theta = .82584121$

**42.** $\cot \theta = 1.1249386$

**43.** A student wants to use a calculator to find the value of $\cot 25°$. However, instead of entering $\frac{1}{\tan 25}$, he enters $\tan^{-1} 25$. Assuming the calculator is in degree mode, will this produce the correct answer? Explain.

**44.** For $\theta = 1997°$, use a calculator to find $\cos \theta$ and $\sin \theta$. Use your results to decide what quadrant the angle lies in.

*Solve each right triangle.*

**45.**

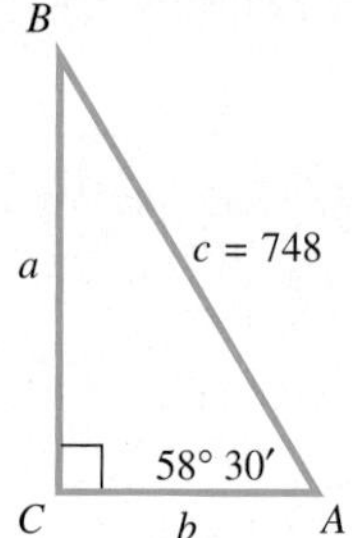


**46.** $A = 39.72°$; $b = 38.97$ m

*Solve each problem.*

**47.** *Height of a Tower* The angle of elevation from a point 93.2 feet from the base of a tower to the top of the tower is 38° 20′. Find the height of the tower.

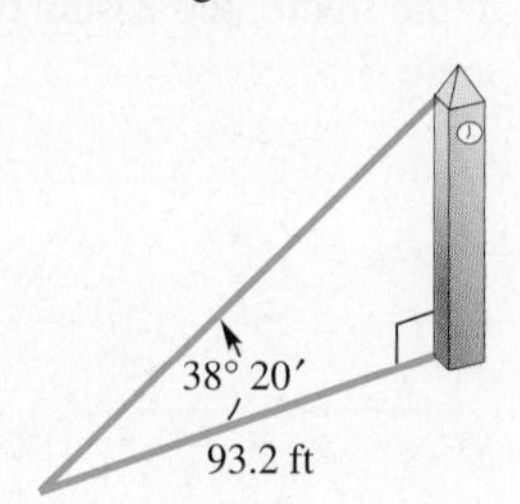


**48.** *Height of a Tower* The angle of depression of a television tower to a point on the ground 36.0 meters from the bottom of the tower is 29.5°. Find the height of the tower. See the figure at the top of the next column.

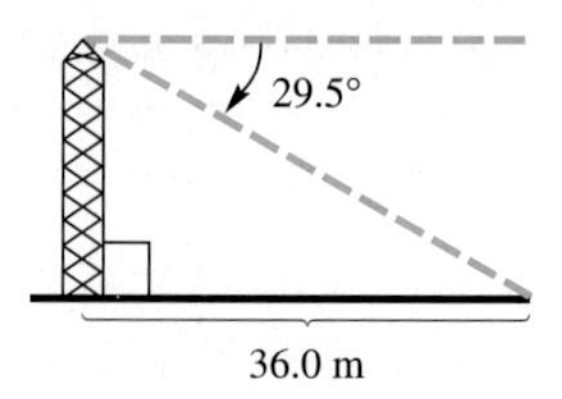


**49.** *Distance between Two Points* The bearing of $B$ from $C$ is 254°. The bearing of $A$ from $C$ is 344°. The bearing of $A$ from $B$ is 32°. The distance from $A$ to $C$ is 780 meters. Find the distance from $A$ to $B$.

**50.** *Distance a Ship Sails* The bearing from point $A$ to point $B$ is S 55° E and from point $B$ to point $C$ is N 35° E. If a ship sails from $A$ to $B$, a distance of 80 kilometers, and then from $B$ to $C$, a distance of 74 kilometers, how far is it from $A$ to $C$?

**51.** *Distance between Two Cars* Two cars leave an intersection at the same time. One heads due south at 55 mph. The other travels due west. After two hours, the bearing of the car headed west from the car headed south is 324°. How far apart are they at that time?

**52.** *Distance Traveled by a Sailboat* From the top of a building that overlooks an ocean, an observer watches a boat sailing directly toward the building. If the observer is 150 feet above sea level, and if the angle of depression of the boat changes from 27° to 39° during the period of observation, approximate the distance that the boat travels.

**53.** *Height of a Triangle* Find the measure of $h$ to the nearest unit.

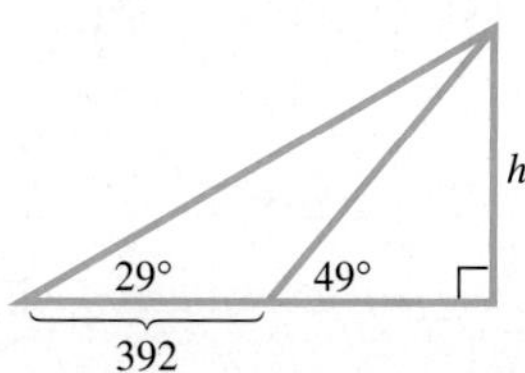


**54.** *(Modeling) Fundamental Surveying Problem* The first fundamental problem of surveying is to determine the coordinates of a point $Q$ given the coordinates of a point $P$, the distance between $P$ and $Q$, and the bearing $\theta$ from $P$ to $Q$. See the figure. (*Source:* Mueller, I. and K. Ramsayer, *Introduction to Surveying,* Frederick Ungar Publishing, 1979.)

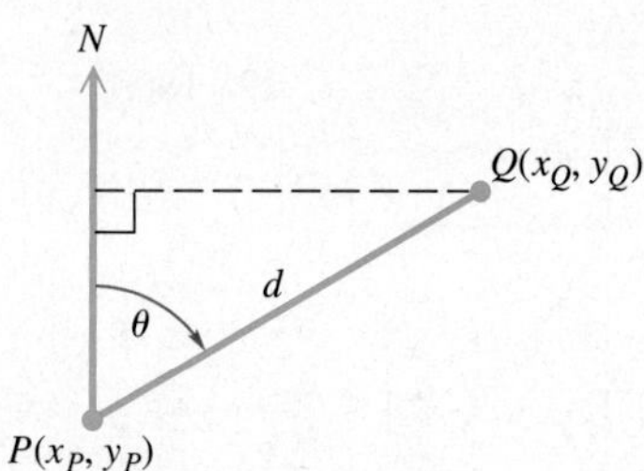

**(a)** Find a formula for the coordinates $(x_Q, y_Q)$ of the point $Q$ given $\theta$, the coordinates $(x_P, y_P)$ of $P$, and the distance $d$ between $P$ and $Q$.

**(b)** Use your formula to determine the coordinates of the point $(x_Q, y_Q)$ if $(x_P, y_P) = (123.62, 337.95)$, $\theta = 17° \, 19' \, 22''$, and $d = 193.86$ feet.

*Find the exact function value. Do not use a calculator.*

**55.** $\cos \dfrac{2\pi}{3}$

**56.** $\tan\left(-\dfrac{7\pi}{3}\right)$

**57.** $\csc\left(-\dfrac{11\pi}{6}\right)$

*Approximate each circular function value to four decimal places.*

**58.** $\cos(-.2443)$

**59.** $\cot 3.0543$

**60.** Use a calculator to approximate $s$ in the interval $\left[0, \frac{\pi}{2}\right]$ if $\sin s = .49244294$.

*Solve each problem.*

**61.** *Atmospheric Effect on Sunlight* The shortest path for the sun's rays through Earth's atmosphere occurs when the sun is directly overhead. Disregarding the curvature of Earth, as the sun moves lower on the horizon, the distance that sunlight passes through the atmosphere increases by a factor of $\csc \theta$, where $\theta$ is the angle of elevation of the sun. This increased distance reduces both the intensity of the sun and the amount of ultraviolet light that reaches Earth's surface. See the figure. (*Source:* Winter, C., R. Sizmann, and Vant-Hunt (Editors), *Solar Power Plants,* Springer-Verlag, 1991.)

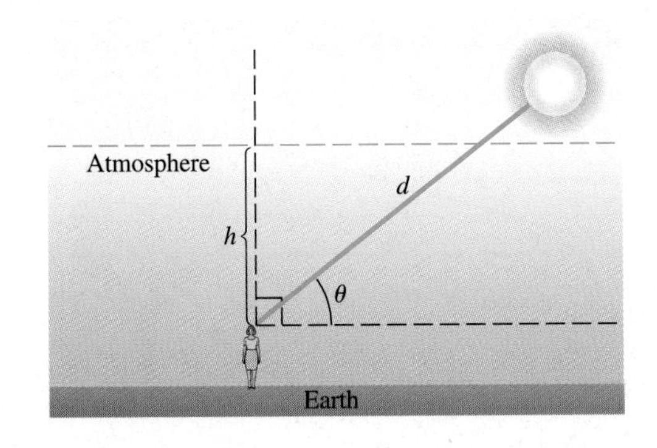


**(a)** Verify that $d = h \csc \theta$.

**(b)** Determine $\theta$ when $d = 2h$.

**(c)** The atmosphere filters out the ultraviolet light that causes skin to burn. Compare the difference between sunbathing when $\theta = \frac{\pi}{2}$ and $\theta = \frac{\pi}{3}$. Which measure gives less ultraviolet light?

**62.** Find the measure (in both degrees and radians) of the angle $\theta$ formed in the screen by the line passing through the origin and the positive part of the $x$-axis. Use the displayed values of $x$ and $y$ at the bottom of the screen.

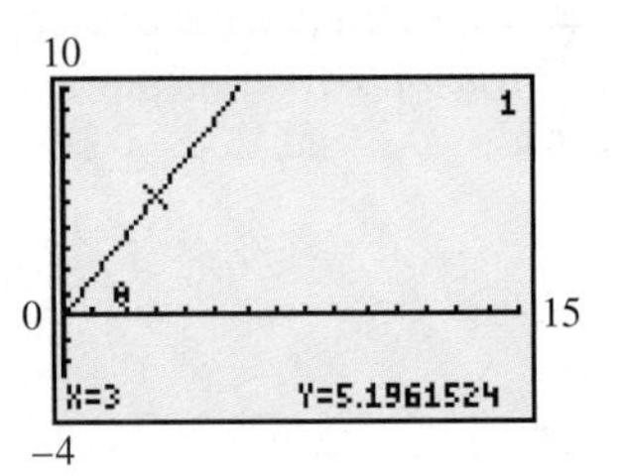


**63.** *Concept Check* The screen shows a point on the unit circle. What is the length of the shortest arc of the circle from $(1, 0)$ to the point?

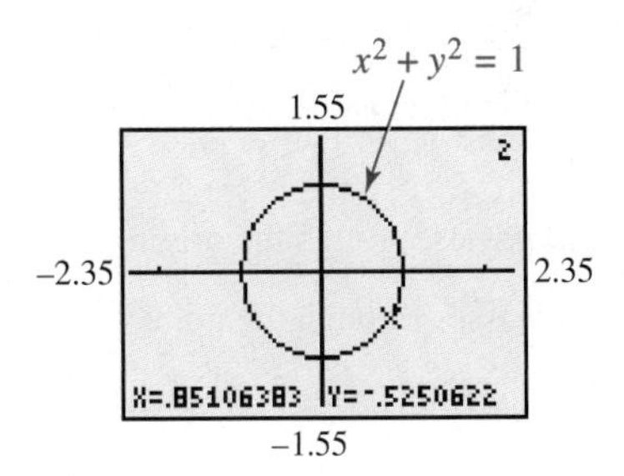


**64.** *Concept Check* Which one of the following is true about the graph of $y = 4 \sin 2x$?

**A.** It has amplitude 2 and period $\dfrac{\pi}{2}$.

**B.** It has amplitude 4 and period $\pi$.

**C.** Its range is $[0, 4]$.

**D.** Its range is $[-4, 0]$.

*For each defined function, give the amplitude, period, vertical translation, and phase shift, as applicable.*

**65.** $y = 2 \sin x$

**66.** $y = \tan 3x$

**67.** $y = -\dfrac{1}{2} \cos 3x$

**68.** $y = 2 \sin 5x$

**69.** $y = 1 + 2 \sin \frac{1}{4}x$

**70.** $y = 3 - \frac{1}{4} \cos \frac{2}{3}x$

**71.** $y = 3 \cos\left(x + \frac{\pi}{2}\right)$

**72.** $y = -\sin\left(x - \frac{3\pi}{4}\right)$

**73.** $y = \frac{1}{2} \csc\left(2x - \frac{\pi}{4}\right)$

**74.** Suppose that $f$ is a sinusoidal function with period $\pi$ and $f\left(\frac{6\pi}{5}\right) = 1$. Explain why $f\left(-\frac{4\pi}{5}\right) = 1$.

*In Exercises 75–80, give a traditional or calculator graph, as directed by your instructor. Graph each defined function over a one-period interval.*

**75.** $y = 3 \cos 2x$

**76.** $y = \frac{1}{2} \cot 3x$

**77.** $y = \cos\left(x - \frac{\pi}{4}\right)$

**78.** $y = \tan\left(x - \frac{\pi}{2}\right)$

**79.** $y = 1 + 2 \cos 3x$

**80.** $y = -1 - 3 \sin 2x$

*Concept Check Identify one circular function that satisfies the description.*

**81.** Period is $\pi$, $x$-intercepts are of the form $n\pi$, where $n$ is an integer

**82.** Period is $2\pi$, passes through the origin

**83.** Period is $2\pi$, passes through the point $\left(\frac{\pi}{2}, 0\right)$

**84.** Period is $2\pi$, domain is $\{x \mid x \neq n\pi$, where $n$ is an integer$\}$

**85.** Period is $\pi$, function is decreasing on the interval $0 < x < \pi$

**86.** Period is $2\pi$, has vertical asymptotes of the form $x = (2n + 1)\frac{\pi}{2}$, where $n$ is an integer

*For each graph, give the equation of a sine function having that graph.*

**87.**

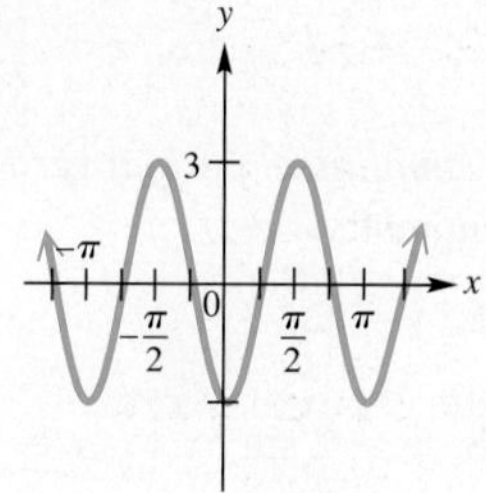

**88.**

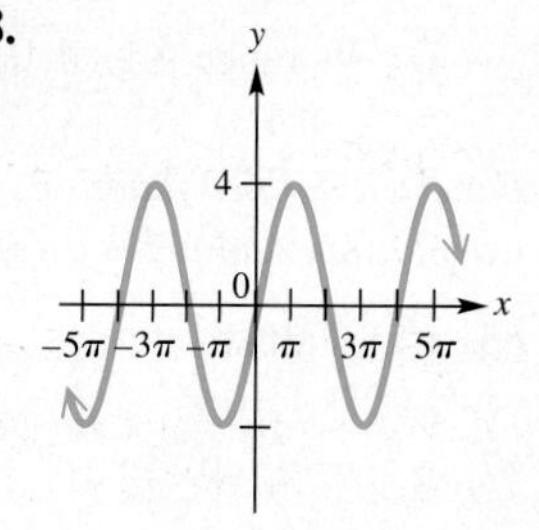

**89.**

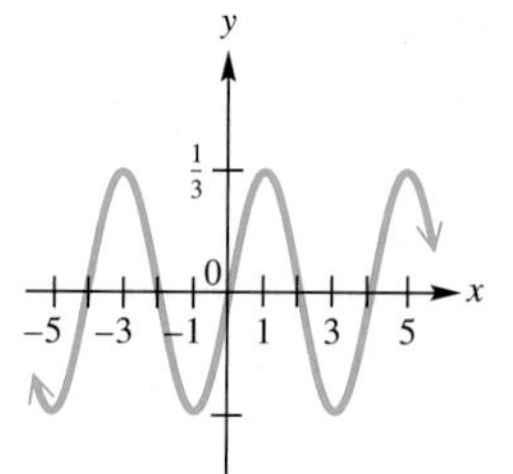

**90.**

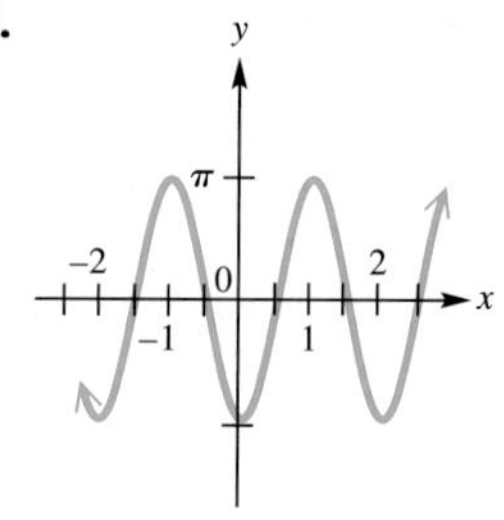

*Solve each problem.*

**91.** *(Modeling) Average Monthly Temperature* The average monthly temperature (in °F) in Chicago, Illinois, is shown in the table.

| Month | °F | Month | °F |
|---|---|---|---|
| Jan | 25 | July | 74 |
| Feb | 28 | Aug | 75 |
| Mar | 36 | Sept | 66 |
| Apr | 48 | Oct | 55 |
| May | 61 | Nov | 39 |
| June | 72 | Dec | 28 |

*Source:* Miller, A. and J. Thompson, *Elements of Meteorology*, Charles E. Merrill, 1975.

**(a)** Plot the average monthly temperature over a 2-year period by letting $x = 1$ correspond to January of the first year.

**(b)** Determine a modeling function of the form

$$f(x) = a \sin[b(x - d)] + c,$$

where $a$, $b$, $c$, and $d$ are constants.

**(c)** Explain the significance of each constant.

**(d)** Graph $f$ together with the data on the same coordinate axes. How well does $f$ model the data?

**(e)** Use the sine regression capability of a graphing calculator to find the equation of a sine curve that fits these data.

**92.** *Viewing Angle to an Object* Let a person $h_1$ feet tall stand $d$ feet from an object $h_2$ feet tall, where $h_2 > h_1$. Let $\theta$ be the angle of elevation to the top of the object. See the figure.

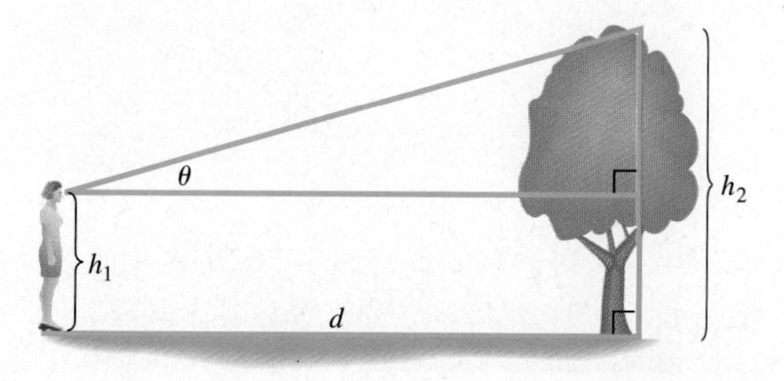

(a) Show that $d = (h_2 - h_1)\cot\theta$.

(b) Let $h_2 = 55$ and $h_1 = 5$. Graph $d$ for the interval $0 < \theta \leq \frac{\pi}{2}$.

**93.** *(Modeling) Pollution Trends* The amount of pollution in the air fluctuates with the seasons. It is lower after heavy spring rains and higher after periods of little rain. Additionally, the long-term trend is upward. An idealized graph of this situation is shown in the figure.

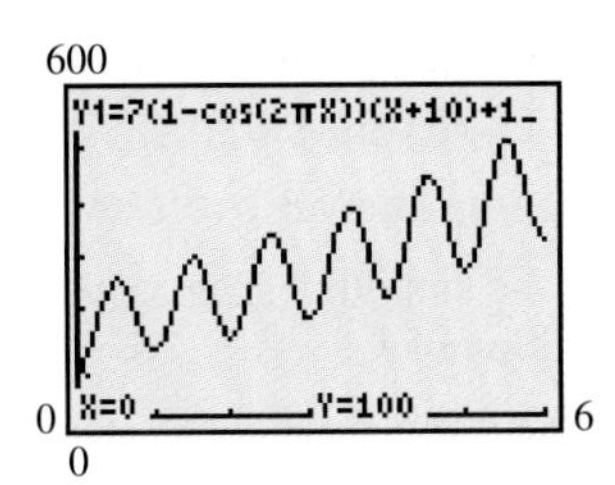


Circular functions can be used to model the fluctuating part of the pollution levels. Exponential functions can be used to model long-term growth. The pollution level in a certain area might be given by

$$y = 7(1 - \cos 2\pi x)(x + 10) + 100e^{.2x},$$

where $x$ is time in years, with $x = 0$ representing January 1 of the base year. July 1 of the same year would be represented by $x = .5$, October 1 of the following year would be represented by $x = 1.75$, and so on. Find the pollution level on each date.

(a) January 1, base year

(b) July 1, base year

(c) January 1, following year

(d) July 1, following year

*(Modeling) Harmonic Motion An object in simple harmonic motion has position function s inches from an initial point, and t is the time in seconds. Find the amplitude, period, and frequency for each function.*

**94.** $s(t) = 3 \cos 2t$

**95.** $s(t) = 4 \sin \pi t$

**96.** In Exercise 94, what does the period represent? What does the amplitude represent?

**97.** In Exercise 95, what does the frequency represent? Find the position of the object from the initial point at 1.5 seconds, 2 seconds, and 3.25 seconds.

## CHAPTER 8 TEST

**1.** Find the angle of smallest positive measure coterminal with $-157°$.

**2.** *Rotating Tire* A tire rotates 450 times per minute. Through how many degrees does a point on the edge of the tire move in 1 second?

**3.** (a) Convert 120° to radians.

(b) Convert $\frac{9\pi}{10}$ radians to degrees.

**4.** A central angle of a circle with radius 150 centimeters cuts off an arc of 200 centimeters. Find each measure.

(a) The radian measure of the angle

(b) The area of a sector with that central angle

**5.** Find the length of the arc intercepted by an angle of 36° in a circle with radius 12 inches.

**6.** Use the figure to find the following.

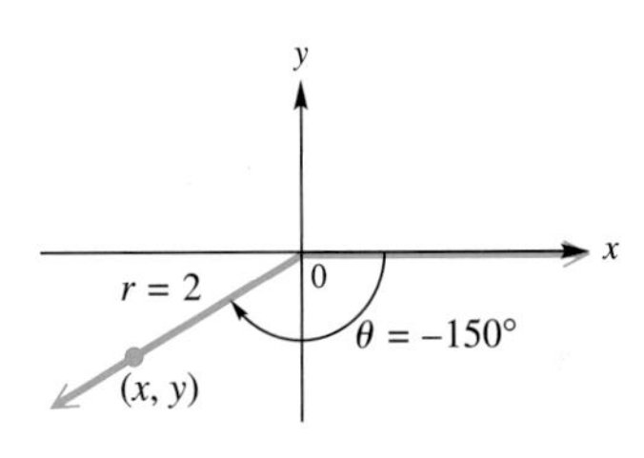


(a) The smallest positive angle coterminal with $\theta$

(b) The values of $x$ and $y$

(c) The values of the six trigonometric functions of $\theta$

(d) The radian measure of $\theta$

**7.** Use a calculator to approximate $s$ in the interval $\left[0, \frac{\pi}{2}\right]$, if $\sin s = .82584121$.

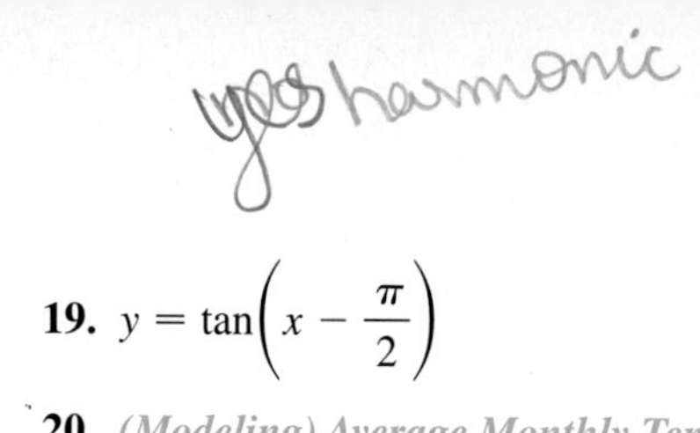

**8.** If $\cos\theta < 0$ and $\cot\theta > 0$, in what quadrant does $\theta$ lie?

**9.** If $(2, -5)$ is on the terminal side of an angle $\theta$ in standard position, find $\sin\theta$, $\cos\theta$, and $\tan\theta$.

**10.** If $\cos\theta = \frac{4}{5}$ and $\theta$ is in quadrant IV, find the values of the other trigonometric functions of $\theta$.

**11.** Find the exact value of each variable in the figure.

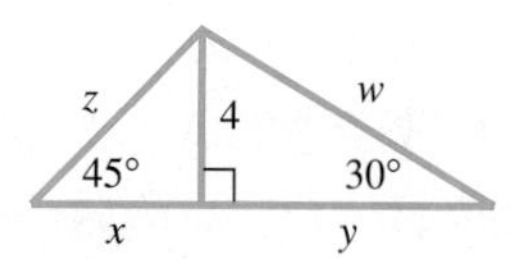


**12.** Find the exact value of $\cot(-750°)$.

**13.** Use a calculator to approximate the following.

**(a)** $\sin 78° 21'$ **(b)** $\tan 11.7689°$

**(c)** $\sec 58.9041°$

**14.** Solve the triangle.

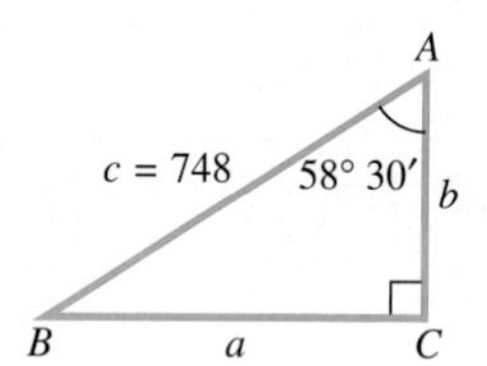


**15.** *Height of a Flagpole* To measure the height of a flagpole, Amado Carillo found that the angle of elevation from a point 24.7 feet from the base to the top is 32° 10′. What is the height of the flagpole?

**16.** *Distance of a Ship from a Pier* A ship leaves a pier on a bearing of S 55° E and travels for 80 kilometers. It then turns and continues on a bearing of N 35° E for 74 kilometers. How far is the ship from the pier?

*Graph each defined function over a two-period interval. Identify asymptotes when applicable.*

**17.** $y = -1 + 2\sin(x + \pi)$ **18.** $y = -\cos 2x$

**19.** $y = \tan\left(x - \frac{\pi}{2}\right)$

**20.** *(Modeling) Average Monthly Temperature* The average monthly temperature (in °F) in Austin, Texas, can be modeled using the trigonometric function defined by

$$f(x) = 17.5\sin\left[\frac{\pi}{6}(x - 4)\right] + 67.5,$$

where $x$ is the month and $x = 1$ corresponds to January. (*Source:* Miller, A. and J. Thompson, *Elements of Meteorology,* Charles E. Merrill, 1975.)

**(a)** Graph $f$ over the interval $1 \le x \le 25$.

**(b)** Determine the amplitude, period, phase shift, and vertical translation of $f$.

**(c)** What is the average monthly temperature for the month of December?

**(d)** Determine the maximum and minimum average monthly temperatures and the months when they occur.

**(e)** What would be an approximation for the average *yearly* temperature in Austin? How is this related to the vertical translation of the sine function in the formula for $f$?

*Spring The height of a weight attached to a spring is*

$$s(t) = -3\cos 2\pi t$$

*inches after t seconds.*

**21.** Find the maximum height that the weight rises above the equilibrium position of $y = 0$.

**22.** When is the first time that the weight reaches its maximum height, if $t \ge 0$?

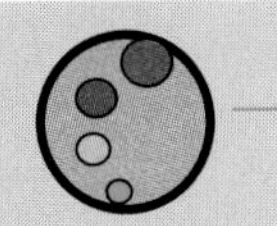

# Chapter 8 Project

## Modeling Sunset Times

Sunset time at a location varies depending on the time of year. For example, see the table for sunset times at a location of 40° N on the longitude of Greenwich, England. (*Note:* Times are for the first day of each month.)

| Month | Sunset | Month | Sunset |
|---|---|---|---|
| Jan | 4:46 P.M. | July | 7:33 P.M. |
| Feb | 5:19 P.M. | Aug | 7:14 P.M. |
| Mar | 5:52 P.M. | Sept | 6:32 P.M. |
| Apr | 6:24 P.M. | Oct | 5:42 P.M. |
| May | 6:55 P.M. | Nov | 4:58 P.M. |
| June | 7:23 P.M. | Dec | 4:35 P.M. |

*Source: World Almanac and Book of Facts,* 2000.

Since the pattern for sunset time repeats itself every year, the data is periodic (with period 1 year or 12 months). This means that there exists a transformation of a sine function of the form

$$y = a \sin[b(x - d)] + c$$

that fits the data closely.

## Activities

To find the function that fits the given data, follow these steps.

1. Use your graphing calculator to make a scatter diagram of the data. Although not required, plot two years of sunsets to display two full periods. For the independent variable, months, use the number of each month, 1 through 24 for two periods. For the dependent variable, sunset, the times must be converted from hours and minutes to decimal hours. For example: 4:46 becomes $4 + \frac{46}{60}$, which the calculator converts to 4.77 hours. Round all times to the nearest hundredth.
2. Find the amplitude of the function. Use the definition: The amplitude is half the difference between the maximum and minimum function values. Which of the constants in the form $y = a \sin[b(x - d)] + c$ gives the amplitude?
3. Find the period. Use the fact that the smallest positive number $p$ for which $f(x) = f(x + np)$ is the period. Use the period to find the value of $b$.
4. Find the phase shift. Consider $\sin 0 = 0$. The corresponding point of the period of the data would be the spring equinox, March 21. Since the given sunset data is for the first day of each month, March 21 corresponds to $3 + \frac{21}{31} \approx 3.68$.
5. Find the vertical shift, the average of the maximum and minimum times.

6. Use the results of Items 2–5 to give the function that models the times for sunset. Verify your answer by graphing the function and the scatter diagram in the same window.
7. Use sine regression to find a function that models the data. The form of this function will be different than the one you found above. Are the two forms nearly equivalent?

# 9 Trigonometric Identities and Equations

IN 1831 Michael Faraday discovered that when a wire passes by a magnet, a small electric current is produced in the wire. This phenomenon became known as Faraday's law. Since then, people have used this property to generate massive amounts of electricity by simultaneously rotating thousands of wires near large electromagnets. The electricity supplied to most homes is produced by electric generators that rotate at 60 cycles per second. Because of this rotation, electric current alternates its direction in electrical wires and can be modeled accurately by either the sine or cosine function.

Explanations of musical phenomena such as tone and frequency require a mathematical understanding of sound. Music is made up of sound waves that cause rapid increases and decreases in air pressure on a person's eardrum. Sound involves periodic motion through the air, and this motion can be modeled using trigonometric equations and graphs.

Understanding electric current and sound waves requires knowledge not only of the trigonometric functions presented in the previous chapter, but also identities that relate the trigonometric functions to each other, as discussed in this chapter.

*Source:* Weidner, R. and R. Sells, *Elementary Classical Physics,* Vol. 2, Allyn & Bacon, 1973; Pierce, J., *The Science of Musical Sound,* Scientific American Books, 1992.

CHAPTER OUTLINE

# 9.1 TRIGONOMETRIC IDENTITIES

Fundamental Identities • Using the Fundamental Identities • Verifying Identities

**Looking Ahead to Calculus**

Much of our work with identities in this chapter is preparation for calculus. Some calculus problems are simplified by making an appropriate trigonometric substitution. For example, if $x = 3 \tan \theta$, then

$$\begin{aligned}\sqrt{9 + x^2} &= \sqrt{9 + (3 \tan \theta)^2}\\ &= \sqrt{9 + 9 \tan^2 \theta}\\ &= \sqrt{9(1 + \tan^2 \theta)}\\ &= 3\sqrt{1 + \tan^2 \theta}\\ &= 3\sqrt{\sec^2 \theta}.\end{aligned}$$

In the interval $\left(0, \frac{\pi}{2}\right)$, the value of $\sec \theta$ is positive, giving

$$\sqrt{9 + x^2} = 3 \sec \theta.$$

## Fundamental Identities

In Chapter 8 we used the definitions of the trigonometric functions to derive the following identities, which hold for the circular functions as well.

**Reciprocal Identities** $\quad \cot \theta = \dfrac{1}{\tan \theta} \qquad \sec \theta = \dfrac{1}{\cos \theta} \qquad \csc \theta = \dfrac{1}{\sin \theta}$

**Quotient Identities** $\quad \tan \theta = \dfrac{\sin \theta}{\cos \theta} \qquad \cot \theta = \dfrac{\cos \theta}{\sin \theta}$

**Pythagorean Identities** $\quad \sin^2 \theta + \cos^2 \theta = 1 \qquad 1 + \tan^2 \theta = \sec^2 \theta$

$$1 + \cot^2 \theta = \csc^2 \theta$$

**NOTE** The forms of the identities given above are the most commonly recognized forms. Throughout this chapter, it will be necessary to also recognize alternative forms of these identities. For example, two other forms of $\sin^2 \theta + \cos^2 \theta = 1$ are

$$\sin^2 \theta = 1 - \cos^2 \theta \quad \text{and} \quad \cos^2 \theta = 1 - \sin^2 \theta.$$

You should be able to transform the fundamental identities, using algebraic transformations.

Another group of identities known as the **negative-number** or **negative-angle identities** will prove useful throughout this chapter. To illustrate these identities, see the unit circle in Figure 1. As suggested in the figure, an angle $\theta$ (in radians) determines an arc of length $\theta$ terminating at a point $(x, y)$. Angle $-\theta$ determines a corresponding arc of length $-\theta$ terminating at the point $(x, -y)$.

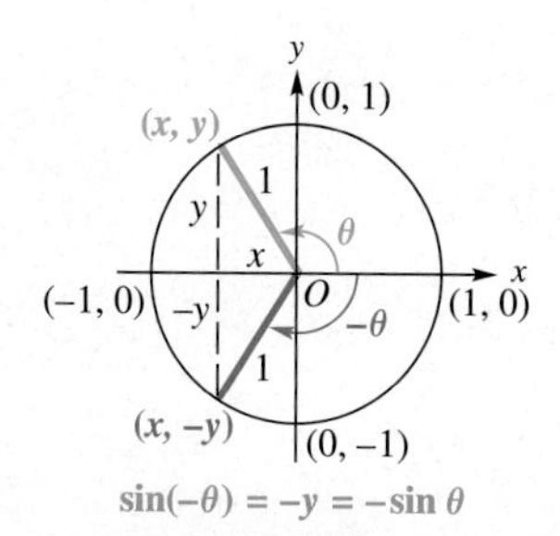


$\sin(-\theta) = -y = -\sin \theta$

**FIGURE 1**

$$\sin(-\theta) = -y \quad \text{and} \quad \sin \theta = y,$$

so $\sin(-\theta)$ and $\sin \theta$ are negatives of each other, or

$$\sin(-\theta) = -\sin \theta.$$

The figure shows an arc $\theta$ in quadrant II, but the same result holds for $\theta$ in any quadrant. Also, by definition,

$$\cos(-\theta) = x \quad \text{and} \quad \cos \theta = x,$$

so

$$\cos(-\theta) = \cos \theta.$$

We use these two identities for $\sin(-\theta)$ and $\cos(-\theta)$ to find $\tan(-\theta)$ in terms of $\tan \theta$:

$$\tan(-\theta) = \frac{\sin(-\theta)}{\cos(-\theta)} = \frac{-\sin \theta}{\cos \theta} = -\frac{\sin \theta}{\cos \theta}$$

or

$$\tan(-\theta) = -\tan \theta.$$

Similar reasoning gives the remaining negative-number identities.

$$\csc(-\theta) = -\csc \theta \qquad \sec(-\theta) = \sec \theta \qquad \cot(-\theta) = -\cot \theta$$

| X | $Y_1$ | $Y_2$ |
|---|---|---|
| -2.356 | -.7071 | .70711 |
| -1.571 | -1 | 1 |
| -.7854 | -.7071 | .70711 |
| 0 | 0 | 0 |
| .7854 | .70711 | -.7071 |
| 1.5708 | 1 | -1 |
| 2.3562 | .70711 | -.7071 |

$Y_2$=sin(-X)

The table shows X (in radians), $Y_1 = \sin X$, and $Y_2 = \sin(-X)$, supporting the negative-number identity for sine. Similar tables would support the identities for the other functions.

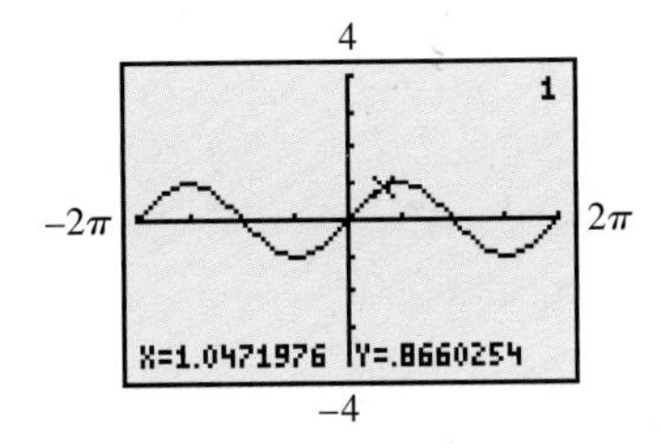


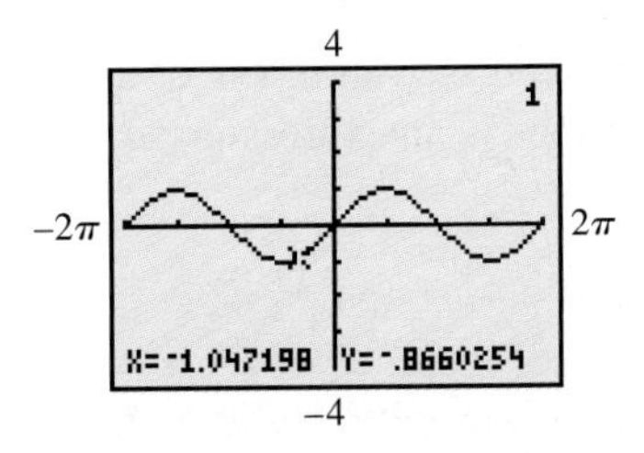


The graph of $Y_1 = \sin X$ also supports the negative-number identity for sine.

As a group, the identities given in this section are called the **fundamental identities.**

**Fundamental Identities**

**Reciprocal Identities**

$$\cot\theta = \frac{1}{\tan\theta} \qquad \sec\theta = \frac{1}{\cos\theta} \qquad \csc\theta = \frac{1}{\sin\theta}$$

**Quotient Identities**

$$\tan\theta = \frac{\sin\theta}{\cos\theta} \qquad \cot\theta = \frac{\cos\theta}{\sin\theta}$$

**Pythagorean Identities**

$$\sin^2\theta + \cos^2\theta = 1 \qquad 1 + \tan^2\theta = \sec^2\theta \qquad 1 + \cot^2\theta = \csc^2\theta$$

**Negative-Number Identities**

$$\sin(-\theta) = -\sin\theta \qquad \cos(-\theta) = \cos\theta \qquad \tan(-\theta) = -\tan\theta$$
$$\csc(-\theta) = -\csc\theta \qquad \sec(-\theta) = \sec\theta \qquad \cot(-\theta) = -\cot\theta$$

## Using the Fundamental Identities

Any trigonometric function of a number or angle can be expressed in terms of any other function.

**EXAMPLE 1** *Expressing One Function in Terms of Another*

Express $\cos x$ in terms of $\tan x$.

**Solution** Since $\sec x$ is related to both $\cos x$ and $\tan x$ by identities, start with $1 + \tan^2 x = \sec^2 x$. Then take reciprocals to get

$$\frac{1}{1 + \tan^2 x} = \frac{1}{\sec^2 x}$$

$$\frac{1}{1 + \tan^2 x} = \cos^2 x$$

$$\pm\sqrt{\frac{1}{1 + \tan^2 x}} = \cos x \qquad \text{Take the square root of each side.}$$

$$\cos x = \frac{\pm 1}{\sqrt{1 + \tan^2 x}}$$

$$\cos x = \frac{\pm\sqrt{1 + \tan^2 x}}{1 + \tan^2 x}. \qquad \text{Rationalize the denominator.}$$

Choose the + sign or the − sign, depending on the quadrant of $x$.

## What Went WRONG?

To verify the result in Example 1, a student entered the functions $y_1 = \cos x$, $y_2 = \dfrac{\sqrt{1 + (\tan x)^2}}{1 + (\tan x)^2}$, and $y_3 = \dfrac{-\sqrt{1 + (\tan x)^2}}{1 + (\tan x)^2}$. She expected the graphs to coincide, but instead got the graph shown here.

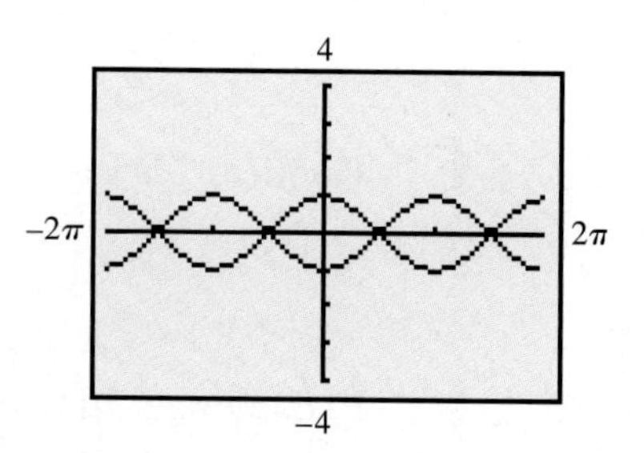


What Went Wrong? Why was she mistaken in her expectation?

Each of $\tan \theta$, $\cot \theta$, $\sec \theta$, and $\csc \theta$ can easily be expressed in terms of $\sin \theta$ and/or $\cos \theta$. For this reason, we often make such substitutions in an expression so that the expression can be simplified.

### EXAMPLE 2 *Rewriting an Expression in Terms of Sine and Cosine*

Use the fundamental identities to write $\tan \theta + \cot \theta$ in terms of $\sin \theta$ and $\cos \theta$, and then simplify the expression.

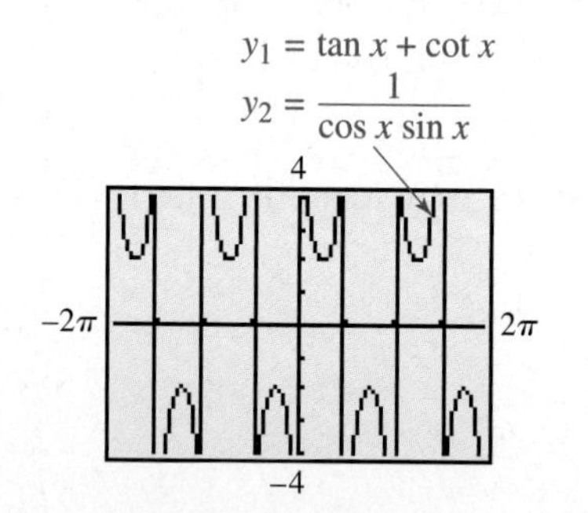


The graph supports the result in Example 2. The graphs of $y_1$ and $y_2$ appear to be identical.

**Solution** From the quotient identities,

$$\tan \theta + \cot \theta = \frac{\sin \theta}{\cos \theta} + \frac{\cos \theta}{\sin \theta}.$$

Simplify this expression by adding the two fractions on the right side, using the common denominator $\cos \theta \sin \theta$.

$$= \frac{\sin^2 \theta}{\cos \theta \sin \theta} + \frac{\cos^2 \theta}{\cos \theta \sin \theta}$$

$$= \frac{\sin^2 \theta + \cos^2 \theta}{\cos \theta \sin \theta}$$

$$\tan \theta + \cot \theta = \frac{1}{\cos \theta \sin \theta} \qquad \sin^2 \theta + \cos^2 \theta = 1$$

---

*Answer to What Went Wrong?*

The graph of $y_2$ always lies *above* the $x$-axis, while that of $y_3$ always lies *below* the $x$-axis. The graph of $y_1 = \cos x$ lies both above and below the $x$-axis. There are values of $x$ for which $y_2 \neq \cos x$ and also values of $x$ for which $y_3 \neq \cos x$.

**CAUTION** When working with trigonometric expressions and identities, be sure to write the argument of the function. For example, we would *not* write $\sin^2 + \cos^2 = 1$; an argument such as $\theta$ is necessary in this identity.

## Verifying Identities

**Looking Ahead to Calculus**

Trigonometric identities are used in calculus to simplify trigonometric expressions, determine derivatives of trigonometric functions, and change the form of some integrals.

One of the skills required for more advanced work in mathematics (and especially in calculus) is the ability to use identities to write expressions in alternative forms. This skill is developed by using the fundamental identities to verify that an equation is an identity (for those values of the variable for which it is defined). Here are some hints to help you get started.

**Verifying Identities**

1. Learn the fundamental identities. Whenever you see either side of a fundamental identity, the other side should come to mind. Also, be aware of equivalent forms of the fundamental identities. For example, $\sin^2\theta = 1 - \cos^2\theta$ is an alternative form of $\sin^2\theta + \cos^2\theta = 1$.
2. Try to rewrite the more complicated side of the equation so that it is identical to the simpler side.
3. It is sometimes helpful to express all functions in the equation in terms of sine and cosine and then simplify the result.
4. Usually, any factoring or indicated algebraic operations should be performed. For example, the expression $\sin^2 x + 2\sin x + 1$ can be factored as $(\sin x + 1)^2$. The sum or difference of two expressions, such as

$$\frac{1}{\sin\theta} + \frac{1}{\cos\theta},$$

can be added or subtracted in the same way as any other rational expressions.

$$\begin{aligned}\frac{1}{\sin\theta} + \frac{1}{\cos\theta} &= \frac{\cos\theta}{\sin\theta\cos\theta} + \frac{\sin\theta}{\sin\theta\cos\theta}\\ &= \frac{\cos\theta + \sin\theta}{\sin\theta\cos\theta}\end{aligned}$$

5. As you select substitutions, keep in mind the side you are not changing, because it represents your goal. For example, to verify the identity

$$\tan^2 x + 1 = \frac{1}{\cos^2 x},$$

try to think of an identity that relates $\tan x$ to $\cos x$. Here, since $\sec x = \frac{1}{\cos x}$ and $\sec^2 x = \tan^2 x + 1$, the secant function is the best link between the two sides.

6. If an expression contains $1 + \sin x$, multiplying both numerator and denominator by $1 - \sin x$ would give $1 - \sin^2 x$, which could be replaced with $\cos^2 x$. Similar results for $1 - \sin x$, $1 + \cos x$, and $1 - \cos x$ may be useful.

**TECHNOLOGY NOTE**

To confirm the "overlap" of two graphs like the ones indicated in Figure 2, trace to any X-value of $Y_1$ and then move the tracing cursor to $Y_2$. There should be no change in the value of Y for that particular value of X.

**CAUTION** Verifying identities is not the same as solving equations. Techniques used in solving equations, such as adding the same term to each side, or multiplying each side by the same term, are not useful when working with identities because you are starting with a statement (to be verified) that may not be true.

### EXAMPLE 3 *Verifying an Identity (Working with One Side)*

Verify that the given equation is an identity.

$$\cot s + 1 = \csc s(\cos s + \sin s)$$

#### Analytic Solution

We use the fundamental identities to rewrite one side of the equation so that it is identical to the other side. Since the right side is more complicated, we work with it. We use the third hint, changing all functions on the right to sine or cosine.

$$\cot s + 1 = \csc s(\cos s + \sin s) \qquad \text{Given equation}$$

| Steps | Reasons |
|---|---|
| Right side of given equation ↓ $\csc s(\cos s + \sin s) = \dfrac{1}{\sin s}(\cos s + \sin s)$ | $\csc s = \dfrac{1}{\sin s}$ |
| $= \dfrac{\cos s}{\sin s} + \dfrac{\sin s}{\sin s}$ | Distributive property |
| $= \cot s + 1$ ↑ Left side of given equation | $\dfrac{\cos s}{\sin s} = \cot s$; $\dfrac{\sin s}{\sin s} = 1$ |

The given equation is an identity because the right side equals the left side.

#### Graphing Calculator Support*

We graph the two functions in the same window, with

$$Y_1 = \cot X + 1 = \frac{1}{\tan X} + 1$$

and

$$Y_2 = \csc X(\cos X + \sin X) = \frac{\cos X + \sin X}{\sin X}.$$

Note that we rewrote each expression in a form that can be entered in the calculator. The graphs coincide, as shown in Figure 2, which supports the analytic result. Numeric support is provided by the table in Figure 2.

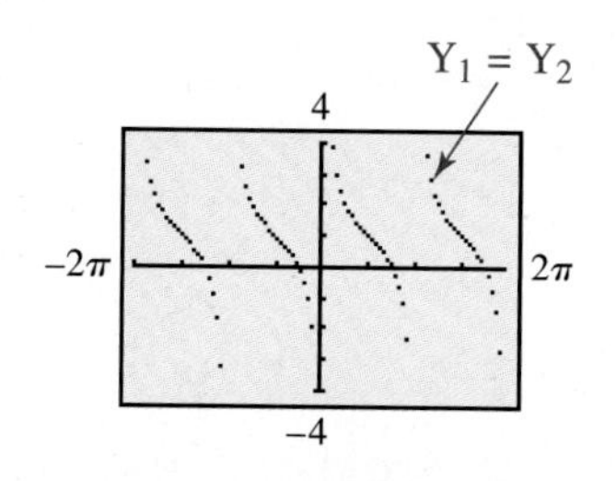


Dot Mode

| X | $Y_1$ | $Y_2$ |
|---|---|---|
| 0 | ERROR | ERROR |
| .3927 | 3.4142 | 3.4142 |
| .7854 | 2 | 2 |
| 1.1781 | 1.4142 | 1.4142 |
| 1.5708 | ERROR | 1 |
| 1.9635 | .58579 | .58579 |
| 2.3562 | 0 | 0 |

X=1.570796326795

**FIGURE 2**

*To verify an identity, we must provide an analytical argument. *A graph can only support, not prove, an identity,* since we can graph only a finite interval of the domain.

### FOR DISCUSSION

We know that $\cot \frac{\pi}{2} + 1 = 0 + 1 = 1$ and $\csc \frac{\pi}{2}\left(\cos \frac{\pi}{2} + \sin \frac{\pi}{2}\right) = 1(0 + 1) = 1$. In the table in Figure 2, why don't the two Y-values agree? How does the answer reinforce the warning that we must understand the mathematics to interpret the calculator results?

### EXAMPLE 4 *Verifying an Identity (Working with One Side)*

Verify that the given equation is an identity.

$$\frac{\tan t - \cot t}{\sin t \cos t} = \sec^2 t - \csc^2 t$$

#### Analytic Solution

Since the left side is more complicated, transform it to equal the right side.

$$\frac{\tan t - \cot t}{\sin t \cos t} = \frac{\tan t}{\sin t \cos t} - \frac{\cot t}{\sin t \cos t} \qquad \frac{a-b}{c} = \frac{a}{c} - \frac{b}{c}$$

$$= \tan t \cdot \frac{1}{\sin t \cos t} - \cot t \cdot \frac{1}{\sin t \cos t} \qquad \frac{a}{b} = a \cdot \frac{1}{b}$$

$$= \frac{\sin t}{\cos t} \cdot \frac{1}{\sin t \cos t} - \frac{\cos t}{\sin t} \cdot \frac{1}{\sin t \cos t} \qquad \tan t = \frac{\sin t}{\cos t};\ \cot t = \frac{\cos t}{\sin t}$$

$$= \frac{1}{\cos^2 t} - \frac{1}{\sin^2 t}$$

$$= \sec^2 t - \csc^2 t \qquad \frac{1}{\cos^2 t} = \sec^2 t;\ \frac{1}{\sin^2 t} = \csc^2 t$$

The hint about writing all trigonometric functions in terms of sine and cosine was used in the third line of the solution. Since the left side does indeed equal the right side, the given equation is an identity.

#### Graphing Calculator Support

Graph the two functions in the same window. The two graphs should coincide, as in Figure 3. The table shows the values of the two functions for selected values of X. For these X-values, the function values agree, further supporting the identity.

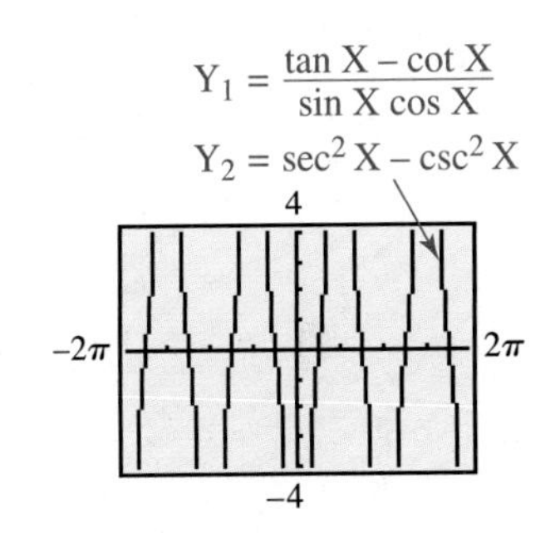


| X | Y1 | Y2 |
|---|---|---|
| [illegible] | 5.6569 | 5.6569 |
| -.9817 | 1.7934 | 1.7934 |
| -.7854 | 0 | 0 |
| -.589 | -1.793 | -1.793 |
| -.3927 | -5.657 | -5.657 |
| -.1963 | -25.23 | -25.23 |
| 0 | ERROR | ERROR |

X=-1.1780972451

**FIGURE 3**

If both sides of an identity appear to be equally complex, the identity can be verified by working independently on the left side and on the right side, until each side is changed into some common third result. *Each step, on each side, must be reversible.* With all steps reversible, the procedure is as follows.

$$\text{left} = \text{right}$$

common third expression

The left side leads to the third expression, which leads back to the right side. This procedure is just a shortcut for the procedure used in Example 4: the left side is changed into the right side, but by going through an intermediate step.

### EXAMPLE 5 *Verifying an Identity (Working with Both Sides)*

Verify that the given equation is an identity.

$$\frac{\sec \alpha + \tan \alpha}{\sec \alpha - \tan \alpha} = \frac{1 + 2 \sin \alpha + \sin^2 \alpha}{\cos^2 \alpha}$$

**Solution** Both sides appear equally complex, so we verify the identity by changing each side into a common third expression. We work first on the left, multiplying numerator and denominator by $\cos \alpha$.

$$\frac{\sec \alpha + \tan \alpha}{\sec \alpha - \tan \alpha} = \frac{(\sec \alpha + \tan \alpha)\cos \alpha}{(\sec \alpha - \tan \alpha)\cos \alpha} \qquad \frac{\cos \alpha}{\cos \alpha} = 1\text{; multiplicative identity}$$

$$= \frac{\sec \alpha \cos \alpha + \tan \alpha \cos \alpha}{\sec \alpha \cos \alpha - \tan \alpha \cos \alpha} \qquad \text{Distributive property}$$

$$= \frac{1 + \tan \alpha \cos \alpha}{1 - \tan \alpha \cos \alpha} \qquad \sec \alpha \cos \alpha = 1$$

$$= \frac{1 + \frac{\sin \alpha}{\cos \alpha} \cdot \cos \alpha}{1 - \frac{\sin \alpha}{\cos \alpha} \cdot \cos \alpha} \qquad \tan \alpha = \frac{\sin \alpha}{\cos \alpha}$$

$$= \frac{1 + \sin \alpha}{1 - \sin \alpha}$$

Now we work on the right side of the given equation. Begin by factoring.

$$\frac{1 + 2 \sin \alpha + \sin^2 \alpha}{\cos^2 \alpha} = \frac{(1 + \sin \alpha)^2}{\cos^2 \alpha} \qquad a^2 + 2ab + b^2 = (a + b)^2$$

$$= \frac{(1 + \sin \alpha)^2}{1 - \sin^2 \alpha} \qquad \cos^2 \alpha = 1 - \sin^2 \alpha$$

$$= \frac{(1 + \sin \alpha)^2}{(1 + \sin \alpha)(1 - \sin \alpha)} \qquad 1 - \sin^2 \alpha = (1 + \sin \alpha)(1 - \sin \alpha)$$

$$= \frac{1 + \sin \alpha}{1 - \sin \alpha} \qquad \text{Lowest terms}$$

We have shown that

$$\frac{\sec \alpha + \tan \alpha}{\sec \alpha - \tan \alpha} = \frac{1 + \sin \alpha}{1 - \sin \alpha} = \frac{1 + 2 \sin \alpha + \sin^2 \alpha}{\cos^2 \alpha},$$

verifying that the original equation is an identity.

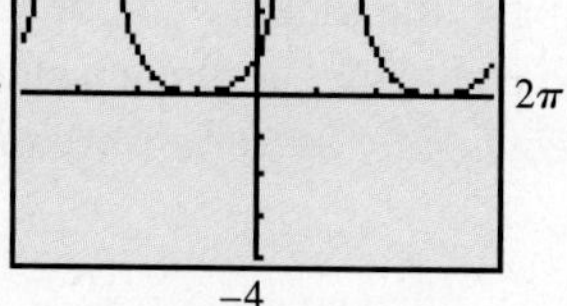

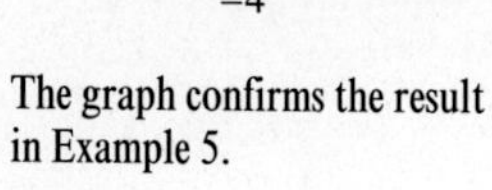


The graph confirms the result in Example 5.

**FOR DISCUSSION**

In Chapter 8, we introduced the cofunction relationships. For example, the functions sine and cosine are cofunctions. The equation

$$\frac{\cot x}{\csc x} = \cos x$$

represents an identity.

**1.** Replace each function in this identity with its cofunction.

**2.** Graph the two sides of the new equation from Item 1 as functions $y_1$ and $y_2$.

**3.** You should have found that $y_1 = y_2$. Verify this analytically.

**4.** Use cofunctions to rewrite the identity in Example 3.

**5.** Verify this new identity analytically and graphically.

There are usually several ways to verify a given identity. For instance, another way to begin verifying the identity in Example 5 is to work on the left as follows.

$$\frac{\sec \alpha + \tan \alpha}{\sec \alpha - \tan \alpha} = \frac{\frac{1}{\cos \alpha} + \frac{\sin \alpha}{\cos \alpha}}{\frac{1}{\cos \alpha} - \frac{\sin \alpha}{\cos \alpha}} \quad \text{Fundamental identities}$$

$$= \frac{\frac{1 + \sin \alpha}{\cos \alpha}}{\frac{1 - \sin \alpha}{\cos \alpha}} \quad \text{Add and subtract fractions.}$$

$$= \frac{1 + \sin \alpha}{1 - \sin \alpha} \quad \text{Divide fractions.}$$

Compare this with the result shown in Example 5 for the right side to see that the two sides indeed agree.

**EXAMPLE 6** *Applying a Pythagorean Identity to Radios*

Tuners in radios select a radio station by adjusting the frequency. A tuner may contain an inductor $L$ and a capacitor $C$, as illustrated in Figure 4. The energy stored in the inductor at time $t$ is given by

$$L(t) = k \sin^2(2\pi F t)$$

and the energy stored in the capacitor is given by

$$C(t) = k \cos^2(2\pi F t),$$

where $F$ is the frequency of the radio station and $k$ is a constant. The total energy $E$ in the circuit is given by $E(t) = L(t) + C(t)$. Show that $E$ is a constant function. (*Source:* Weidner, R. and R. Sells, *Elementary Classical Physics,* Vol. 2, Allyn & Bacon, 1973.)

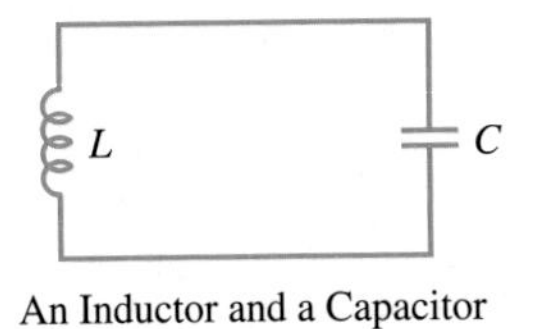


An Inductor and a Capacitor

**FIGURE 4**

**Solution**

$$\begin{aligned} E(t) &= L(t) + C(t) && \text{Given equation} \\ &= k \sin^2(2\pi F t) + k \cos^2(2\pi F t) && \text{Substitute.} \\ &= k[\sin^2(2\pi F t) + \cos^2(2\pi F t)] && \text{Factor.} \\ &= k(1) && \sin^2 \theta + \cos^2 \theta = 1 \ (\theta = 2\pi F t) \\ &= k && k \text{ is constant.} \end{aligned}$$

## 9.1 EXERCISES

*Concept Check* *Use the negative-number identities to decide whether each function is even or odd.*

**1.** $\sin x$ **2.** $\cos x$ **3.** $\tan x$ **4.** $\cot x$ **5.** $\sec x$ **6.** $\csc x$

*Concept Check* *For each expression in Column I, choose the expression from Column II that completes a fundamental identity.*

| I | II |
|---|---|
| **7.** $\dfrac{\cos x}{\sin x} =$ | **A.** $\sin^2 x + \cos^2 x$ |
| **8.** $\tan x =$ | **B.** $\cot x$ |
| **9.** $\cos(-x) =$ | **C.** $\sec^2 x$ |
| **10.** $\tan^2 x + 1 =$ | **D.** $\dfrac{\sin x}{\cos x}$ |
| **11.** $1 =$ | **E.** $\cos x$ |

*Concept Check* *For each expression in Column I, choose the expression from Column II that completes an identity. You may have to rewrite one or both expressions.*

| I | II |
|---|---|
| **12.** $-\tan x \cos x =$ | **A.** $\dfrac{\sin^2 x}{\cos^2 x}$ |
| **13.** $\sec^2 x - 1 =$ | **B.** $\dfrac{1}{\sec^2 x}$ |
| **14.** $\dfrac{\sec x}{\csc x} =$ | **C.** $\sin(-x)$ |
| **15.** $1 + \sin^2 x =$ | **D.** $\csc^2 x - \cot^2 x + \sin^2 x$ |
| **16.** $\cos^2 x =$ | **E.** $\tan x$ |

*Use the negative-number identities to write each expression as a trigonometric function of a positive number. (For example,* $\sin(-3.4) = -\sin 3.4$.*)*

**17.** $\cos(-4.38)$

**18.** $\cos(-5.46)$

**19.** $\sin(-.5)$

**20.** $\sin(-2.5)$

**21.** $\tan\left(-\dfrac{\pi}{7}\right)$

**22.** $\cot\left(-\dfrac{4\pi}{7}\right)$

**23.** A student writes "$1 + \cot^2 = \csc^2$." Comment on this student's work.

**24.** Another student makes the following claim: "Since $\sin^2\theta + \cos^2\theta = 1$, I should be able to also say that $\sin\theta + \cos\theta = 1$ if I take the square root of both sides." Comment on this student's statement.

*Complete this table so that each function in Exercises 25–30 is expressed in terms of the functions given across the top.*

| | $\sin\theta$ | $\cos\theta$ | $\tan\theta$ | $\cot\theta$ | $\sec\theta$ | $\csc\theta$ |
|---|---|---|---|---|---|---|
| **25.** $\sin\theta$ | $\sin\theta$ | $\pm\sqrt{1-\cos^2\theta}$ | $\dfrac{\pm\tan\theta\sqrt{1+\tan^2\theta}}{1+\tan^2\theta}$ | | | $\dfrac{1}{\csc\theta}$ |
| **26.** $\cos\theta$ | | $\cos\theta$ | $\dfrac{\pm\sqrt{\tan^2\theta+1}}{\tan^2\theta+1}$ | | $\dfrac{1}{\sec\theta}$ | |
| **27.** $\tan\theta$ | | | $\tan\theta$ | $\dfrac{1}{\cot\theta}$ | | |
| **28.** $\cot\theta$ | | | $\dfrac{1}{\tan\theta}$ | $\cot\theta$ | $\dfrac{\pm\sqrt{\sec^2\theta-1}}{\sec^2\theta-1}$ | |
| **29.** $\sec\theta$ | | $\dfrac{1}{\cos\theta}$ | | | $\sec\theta$ | |
| **30.** $\csc\theta$ | $\dfrac{1}{\sin\theta}$ | | | | | $\csc\theta$ |

*Each expression simplifies to a constant, a single trigonometric function, or a power of a trigonometric function. Use the fundamental identities to simplify each expression. (You may wish to use a graph to support your result.)*

**31.** $\tan\theta\cos\theta$

**32.** $\cot\alpha\sin\alpha$

**33.** $\dfrac{\sin\beta\tan\beta}{\cos\beta}$

**34.** $\dfrac{\csc\theta\sec\theta}{\cot\theta}$

**35.** $\sec^2 x - 1$

**36.** $\csc^2 t - 1$

**37.** $\dfrac{\sin^2 x}{\cos^2 x} + \sin x\csc x$

**38.** $\dfrac{1}{\tan^2\alpha} + \cot\alpha\tan\alpha$

## RELATING CONCEPTS

**For individual or group investigation (Exercises 39–44)**

*Work Exercises 39–44 in order.*

**39.** Graph $y = (\sec x + \tan x)(1 - \sin x)$.

**40.** The graph should look like the graph of a trigonometric function. Identify the trigonometric function.

**41.** Graph the function you identified in Exercise 40.

**42.** Do your graphs in Exercises 39 and 41 suggest an identity? If so, what is it?

**43.** Verify your answer for Exercise 42 analytically.

**44.** Repeat Exercises 39–43, using the function

$$y = \frac{\cos x + 1}{\sin x + \tan x}.$$

What is the identity?

*In Exercises 45–62, verify that each equation is an identity.*

**45.** $\dfrac{\cot \theta}{\csc \theta} = \cos \theta$

**46.** $\dfrac{\tan \theta}{\sec \theta} = \sin \theta$

**47.** $\cos^2 \theta (\tan^2 \theta + 1) = 1$

**48.** $\dfrac{\cos \theta + 1}{\tan^2 \theta} = \dfrac{\cos \theta}{\sec \theta - 1}$

**49.** $\dfrac{\tan^2 \gamma + 1}{\sec \gamma} = \sec \gamma$

**50.** $\sin^2 \beta(1 + \cot^2 \beta) = 1$

**51.** $\sin^2 \alpha + \tan^2 \alpha + \cos^2 \alpha = \sec^2 \alpha$

**52.** $\cot s + \tan s = \sec s \csc s$

**53.** $\dfrac{\sin^2 \gamma}{\cos \gamma} = \sec \gamma - \cos \gamma$

**54.** $\dfrac{\cos \alpha}{\sec \alpha} + \dfrac{\sin \alpha}{\csc \alpha} = \sec^2 \alpha - \tan^2 \alpha$

**55.** $\dfrac{\cos \theta}{\sin \theta \cot \theta} = 1$

**56.** $\sin^4 \theta - \cos^4 \theta = 2 \sin^2 \theta - 1$

**57.** $\tan^2 \gamma \sin^2 \gamma = \tan^2 \gamma + \cos^2 \gamma - 1$

**58.** $(1 - \cos^2 \alpha)(1 + \cos^2 \alpha) = 2 \sin^2 \alpha - \sin^4 \alpha$

**59.** $\dfrac{(\sec \theta - \tan \theta)^2 + 1}{\sec \theta \csc \theta - \tan \theta \csc \theta} = 2 \tan \theta$

**60.** $\dfrac{1}{1 - \sin \theta} + \dfrac{1}{1 + \sin \theta} = 2 \sec^2 \theta$

**61.** $\dfrac{1}{\tan \alpha - \sec \alpha} + \dfrac{1}{\tan \alpha + \sec \alpha} = -2 \tan \alpha$

**62.** $\dfrac{\csc \theta + \cot \theta}{\tan \theta + \sin \theta} = \cot \theta \csc \theta$

**63.** A student claims that the equation

$$\cos \theta + \sin \theta = 1$$

is an identity, since by letting $\theta = \frac{\pi}{2}$, we get $0 + 1 = 1$, a true statement. Comment on this student's reasoning.

**64.** Explain why the method described in the text for working on both sides of an identity to show that each side is equal to the same expression is a valid method of verifying an identity. When using this method, what must be true about each step taken? (*Hint:* See the discussion preceding Example 5.)

**65.** When does $\sin x = \sqrt{1 - \cos^2 x}$?

**66.** An equation that is an identity has an infinite number of solutions. If an equation has an infinite number of solutions, is it necessarily an identity? Explain your answer.

*(Modeling) Work each problem.*

**67.** *Intensity of a Lamp* According to Lambert's law, the intensity of light from a single source on a flat surface at point $P$ is given by

$$I = k \cos^2 \theta,$$

where $k$ is a constant. (*Source:* Winter, C., *Solar Power Plants,* Springer-Verlag, 1991.)

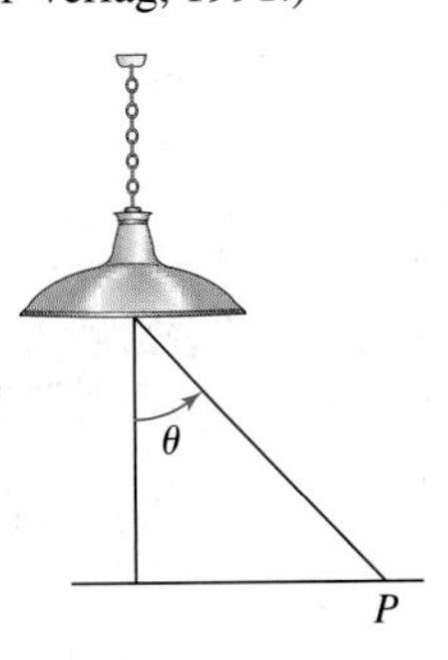


**(a)** Write $I$ in terms of the sine function.

**(b)** Explain why the maximum value of $I$ occurs when $\theta = 0$.

**68.** *Oscillating Spring* The distance or displacement $y$ of a weight attached to an oscillating spring from its natural position is modeled by

$$y = 4\cos(2\pi t),$$

where $t$ is time in seconds. Potential energy is the energy of position and is given by

$$P = ky^2,$$

where $k$ is a constant. The weight has the greatest potential energy when the spring is stretched the most. (*Source:* Weidner, R., and R. Sells, *Elementary Classical Physics,* Vol. 1, Allyn & Bacon, 1973.)

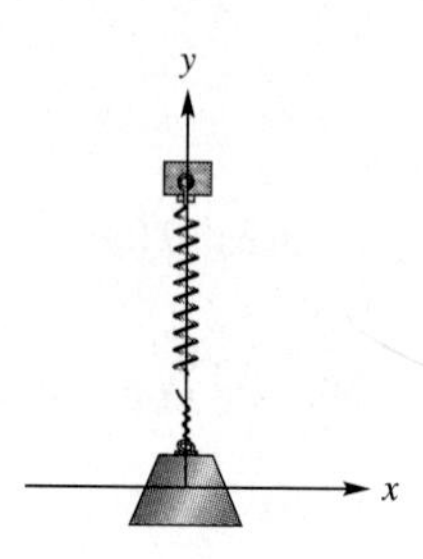


**(a)** Write an expression for $P$ that involves the cosine function.

**(b)** Use a fundamental identity to write $P$ in terms of $\sin(2\pi t)$.

**69.** *Energy in an Oscillating Spring* Refer to Exercise 68. Two types of mechanical energy are kinetic energy and potential energy. Kinetic energy is the energy of motion, and potential energy is the energy of position. A stretched spring has potential energy, which is converted to kinetic energy when it is released. If the potential energy of a weight attached to a spring is

$$P(t) = k\cos^2(4\pi t),$$

where $k$ is a constant and $t$ is time in seconds, then its kinetic energy is given by

$$K(t) = k\sin^2(4\pi t).$$

The total mechanical energy $E$ is given by the equation $E(t) = P(t) + K(t)$.

**(a)** If $k = 2$, graph $P$, $K$, and $E$ in the window $[0, .5]$ by $[-1, 3]$, with Xscl $= .25$ and Yscl $= 1$. Interpret the graph.

**(b)** Make a table of $K$, $P$, and $E$ starting at $t = 0$, incrementing by .05. Interpret the results.

**(c)** Use a fundamental identity to derive a simplified expression for $E(t)$.

**70.** *Radio Tuners* Refer to Example 6. Let the energy stored in the inductor be given by

$$L(t) = 3\cos^2(6{,}000{,}000t)$$

and the energy in the capacitor be given by

$$C(t) = 3\sin^2(6{,}000{,}000t),$$

where $t$ is time in seconds. The total energy $E$ in the circuit is given by $E(t) = L(t) + C(t)$.

**(a)** Graph $L$, $C$, and $E$ in the window $[0, 10^{-6}]$ by $[-1, 4]$, with Xscl $= 10^{-7}$ and Yscl $= 1$. Interpret the graph.

**(b)** Make a table of $L$, $C$, and $E$ starting at $t = 0$, incrementing by $10^{-7}$. Interpret your results.

**(c)** Use a fundamental identity to derive a simplified expression for $E(t)$.

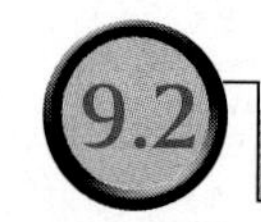

# 9.2 SUM AND DIFFERENCE IDENTITIES

Cosine Sum and Difference Identities • Sine and Tangent Sum and Difference Identities

As mentioned earlier, the identities presented in this chapter hold true whether the arguments represent real numbers or degrees.

## Cosine Sum and Difference Identities

**FOR DISCUSSION**

Use different values for $A$ and $B$ to investigate whether the given statement is true for all values of $A$ and $B$.

**1.** $\cos(A - B) = \cos A - \cos B$ **2.** $\cos(A + B) = \cos A + \cos B$

The results of the preceding "For Discussion" should convince you that the cosine of the difference (or sum) of two numbers is not, in general, equal to the difference (or sum) of the cosines of the numbers. To derive an identity for $\cos(A - B)$ in terms of functions of $A$ and $B$, start by locating angles $A$ and $B$ in standard position on a unit circle, with $B < A$. Let $S$ and $Q$ be the points where the terminal sides of angles $A$ and $B$, respectively, intersect the circle. Locate point $R$ on the unit circle so that angle $POR$ equals the difference $A - B$. See Figure 5.

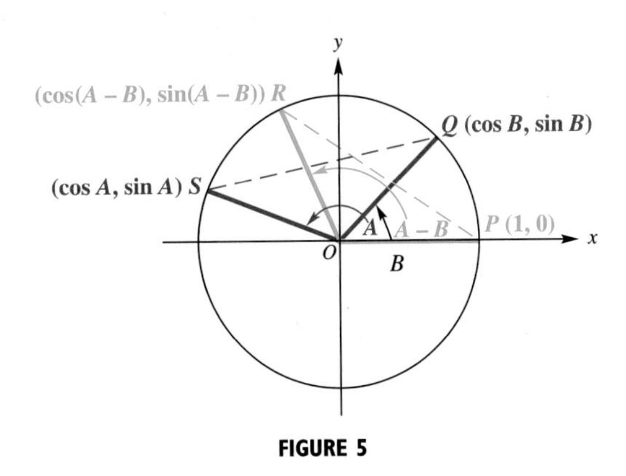


**FIGURE 5**

Point $Q$ is on the unit circle, so by the work with circular functions in Chapter 8, the $x$-coordinate of $Q$ is given by the cosine of angle $B$, while the $y$-coordinate of $Q$ is given by the sine of angle $B$.

$$Q \text{ has coordinates } (\cos B, \sin B).$$

In the same way,

$$S \text{ has coordinates } (\cos A, \sin A),$$

and

$$R \text{ has coordinates } (\cos(A - B), \sin(A - B)).$$

Angle $SOQ$ also equals $A - B$. Since the central angles $SOQ$ and $POR$ are equal, chords $PR$ and $SQ$ are equal. By the distance formula, since $PR = SQ$,

$$\sqrt{[\cos(A - B) - 1]^2 + [\sin(A - B) - 0]^2} = \sqrt{(\cos A - \cos B)^2 + (\sin A - \sin B)^2}.$$

Squaring both sides and clearing parentheses gives

$$\cos^2(A - B) - 2\cos(A - B) + 1 + \sin^2(A - B)$$
$$= \cos^2 A - 2\cos A \cos B + \cos^2 B + \sin^2 A - 2\sin A \sin B + \sin^2 B.$$

Since $\sin^2 x + \cos^2 x = 1$ for any value of $x$, rewrite the equation as

$$2 - 2\cos(A - B) = 2 - 2\cos A \cos B - 2\sin A \sin B$$
$$\cos(A - B) = \cos A \cos B + \sin A \sin B.$$

This is an identity for $\cos(A - B)$. Although Figure 5 shows angles $A$ and $B$ in the second and first quadrants, respectively, this result is the same for any values of these angles.

To find a similar expression for $\cos(A + B)$, rewrite $A + B$ as $A - (-B)$ and use the identity for $\cos(A - B)$.

$$\begin{aligned}\cos(A + B) &= \cos[A - (-B)]\\ &= \cos A \cos(-B) + \sin A \sin(-B)\\ &= \cos A \cos B + \sin A(-\sin B)\\ \cos(A + B) &= \cos A \cos B - \sin A \sin B\end{aligned}$$

The two formulas we have just derived are summarized as follows.

**Cosine of a Sum or Difference**

$$\cos(A - B) = \cos A \cos B + \sin A \sin B$$
$$\cos(A + B) = \cos A \cos B - \sin A \sin B$$

These identities are important in calculus and other areas of mathematics and are useful in some applications. Although a calculator can be used to find an approximation for cos 15°, for example, the method shown in Example 1 can be applied to get an exact value, as well as to practice using the sum and difference identities.

**EXAMPLE 1** *Finding Exact Cosine Function Values*

Find the *exact* value of the following.

**(a)** $\cos 15°$ **(b)** $\cos \frac{5\pi}{12}$ **(c)** $\cos 87° \cos 93° - \sin 87° \sin 93°$

**Analytic Solution**

**(a)** To find cos 15°, we write 15° as the sum or difference of two angles with known function values. Since we know the exact trigonometric function values of both 45° and 30°, we write 15° as 45° − 30°. (We could also use 60° − 45°.) Then we use the identity for the cosine of the difference of two angles.

$$\begin{aligned}\cos 15° &= \cos(45° - 30°)\\ &= \cos 45° \cos 30° + \sin 45° \sin 30° \quad \text{Cosine difference identity}\\ &= \frac{\sqrt{2}}{2} \cdot \frac{\sqrt{3}}{2} + \frac{\sqrt{2}}{2} \cdot \frac{1}{2}\\ &= \frac{\sqrt{6} + \sqrt{2}}{4}\end{aligned}$$

**Graphing Calculator Support**

The calculator screen in Figure 6(a) supports the analytic solution for part (b) by giving the same approximation for both $\cos \frac{5\pi}{12}$ and $\frac{\sqrt{6} - \sqrt{2}}{4}$. Alternatively, in Figure 6(b), with $y = \cos x$, we entered $\frac{5\pi}{12}$ for $x$ (which appears as the approximation 1.3089969), and the corresponding $y$-value is the calculator approximation for $\frac{\sqrt{6} - \sqrt{2}}{4}$ shown in Figure 6(a).

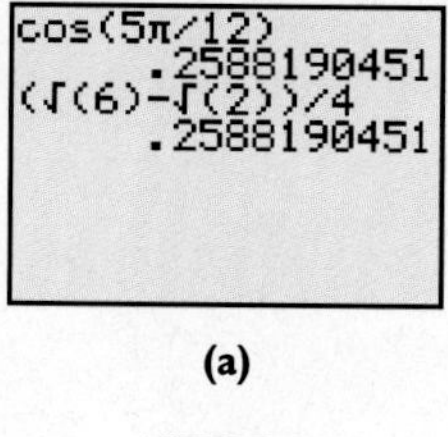


(a)

**FIGURE 6**

*(continued)*

**(b)** $\cos \frac{5\pi}{12} = \cos\left(\frac{\pi}{6} + \frac{\pi}{4}\right)$

$= \cos \frac{\pi}{6} \cos \frac{\pi}{4} - \sin \frac{\pi}{6} \sin \frac{\pi}{4}$ Cosine sum identity

$= \frac{\sqrt{3}}{2} \cdot \frac{\sqrt{2}}{2} - \frac{1}{2} \cdot \frac{\sqrt{2}}{2}$

$= \frac{\sqrt{6} - \sqrt{2}}{4}$

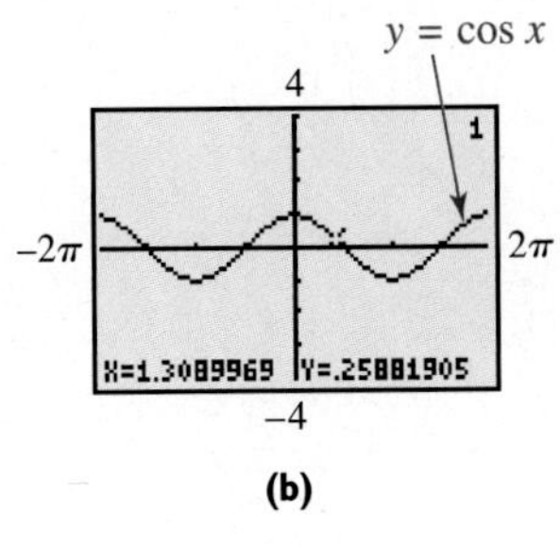


(b)

**FIGURE 6**

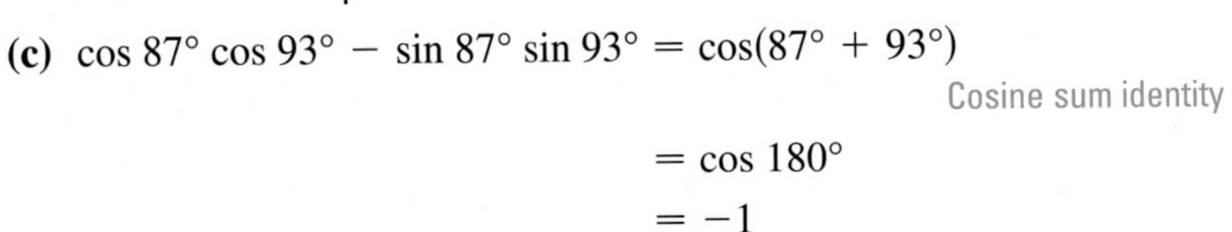

**(c)** $\cos 87° \cos 93° - \sin 87° \sin 93° = \cos(87° + 93°)$ Cosine sum identity

$= \cos 180°$

$= -1$

**NOTE** In Example 1(b), we used the fact that $\frac{5\pi}{12} = \frac{\pi}{6} + \frac{\pi}{4}$. At first glance, this sum may not be obvious. Think of the values $\frac{\pi}{6}$ and $\frac{\pi}{4}$ in terms of fractions with denominator 12: $\frac{\pi}{6} = \frac{2\pi}{12}$ and $\frac{\pi}{4} = \frac{3\pi}{12}$. The following list may help you with problems of this type.

$$\frac{\pi}{3} = \frac{4\pi}{12}, \quad \frac{\pi}{4} = \frac{3\pi}{12}, \quad \frac{\pi}{6} = \frac{2\pi}{12}$$

Using this list, for example, we see that $\frac{\pi}{12} = \frac{\pi}{3} - \frac{\pi}{4}$ (or $\frac{\pi}{4} - \frac{\pi}{6}$).

If one of the angles $A$ or $B$ in the identities for $\cos(A + B)$ and $\cos(A - B)$ is a quadrantal angle, then the identity allows us to write the expression in terms of a single function of $A$ or $B$.

**EXAMPLE 2** *Reducing* $\cos(A - B)$ *to a Function of a Single Variable*

Write $\cos(180° - \theta)$ as a trigonometric function of $\theta$.

**Solution** Replace $A$ with 180° and $B$ with $\theta$ in the difference identity.

$$\begin{aligned} \cos(180° - \theta) &= \cos 180° \cos \theta + \sin 180° \sin \theta \\ &= (-1)\cos \theta + (0)\sin \theta \\ &= -\cos \theta \end{aligned}$$

## Sine and Tangent Sum and Difference Identities

We can use the cosine sum and difference identities to derive similar identities for sine and tangent. By the cofunction relationship, $\sin \theta = \cos\left(\frac{\pi}{2} - \theta\right)$. Now, we replace $\theta$ with $A + B$.

$$\sin(A + B) = \cos\left[\frac{\pi}{2} - (A + B)\right] = \cos\left[\left(\frac{\pi}{2} - A\right) - B\right]$$

Using the formula for $\cos(A - B)$ from the previous discussion gives

$$\sin(A + B) = \cos\left(\frac{\pi}{2} - A\right)\cos B + \sin\left(\frac{\pi}{2} - A\right)\sin B$$

or $\sin(A + B) = \sin A \cos B + \cos A \sin B.$

(Cofunction relationships were used in the last step.)

Now we write $\sin(A - B)$ as $\sin[A + (-B)]$ and use the identity for $\sin(A + B)$ to get

$$\begin{aligned} \sin(A - B) &= \sin[A + (-B)] \\ &= \sin A \cos(-B) + \cos A \sin(-B) && \text{Identity for } \sin(A + B) \\ \sin(A - B) &= \sin A \cos B - \cos A \sin B. && \text{Negative-number identities} \end{aligned}$$

**Sine of a Sum or Difference**

$$\sin(A + B) = \sin A \cos B + \cos A \sin B$$
$$\sin(A - B) = \sin A \cos B - \cos A \sin B$$

To derive the identity for $\tan(A + B)$ using the identities for $\sin(A + B)$, $\cos(A + B)$, and the identity $\tan A = \frac{\sin A}{\cos A}$, we start with

$$\begin{aligned} \tan(A + B) &= \frac{\sin(A + B)}{\cos(A + B)} \\ &= \frac{\sin A \cos B + \cos A \sin B}{\cos A \cos B - \sin A \sin B}. \end{aligned}$$

Now we express this result in terms of the tangent function by multiplying both numerator and denominator by $\frac{1}{\cos A \cos B}$.

$$\begin{aligned} \tan(A + B) &= \frac{\dfrac{\sin A \cos B + \cos A \sin B}{1}}{\dfrac{\cos A \cos B - \sin A \sin B}{1}} \cdot \frac{\dfrac{1}{\cos A \cos B}}{\dfrac{1}{\cos A \cos B}} \\ &= \frac{\dfrac{\sin A \cos B}{\cos A \cos B} + \dfrac{\cos A \sin B}{\cos A \cos B}}{\dfrac{\cos A \cos B}{\cos A \cos B} - \dfrac{\sin A \sin B}{\cos A \cos B}} \\ &= \frac{\dfrac{\sin A}{\cos A} + \dfrac{\sin B}{\cos B}}{1 - \dfrac{\sin A}{\cos A} \cdot \dfrac{\sin B}{\cos B}} \\ \tan(A + B) &= \frac{\tan A + \tan B}{1 - \tan A \tan B} && \tan\theta = \frac{\sin\theta}{\cos\theta} \end{aligned}$$

Replacing $B$ with $-B$ and using the fact that $\tan(-B) = -\tan B$ gives the identity for the tangent of the difference of two numbers.

**Tangent of a Sum or Difference**

$$\tan(A + B) = \frac{\tan A + \tan B}{1 - \tan A \tan B} \qquad \tan(A - B) = \frac{\tan A - \tan B}{1 + \tan A \tan B}$$

**EXAMPLE 3** *Finding Exact Sine and Tangent Function Values*

Find the *exact* value of the following.

**(a)** $\sin 75°$ **(b)** $\tan \frac{7\pi}{12}$ **(c)** $\sin 40° \cos 160° - \cos 40° \sin 160°$

### Analytic Solution

**(a)**
$$\begin{aligned}\sin 75° &= \sin(45° + 30°)\\ &= \sin 45° \cos 30° + \cos 45° \sin 30° \quad \text{Sine sum identity}\\ &= \frac{\sqrt{2}}{2} \cdot \frac{\sqrt{3}}{2} + \frac{\sqrt{2}}{2} \cdot \frac{1}{2}\\ &= \frac{\sqrt{6} + \sqrt{2}}{4}\end{aligned}$$

**(b)**
$$\begin{aligned}\tan \frac{7\pi}{12} &= \tan\left(\frac{\pi}{3} + \frac{\pi}{4}\right)\\ &= \frac{\tan \frac{\pi}{3} + \tan \frac{\pi}{4}}{1 - \tan \frac{\pi}{3} \tan \frac{\pi}{4}} \quad \text{Tangent sum identity}\\ &= \frac{\sqrt{3} + 1}{1 - \sqrt{3} \cdot 1}\\ &= \frac{\sqrt{3} + 1}{1 - \sqrt{3}} \cdot \frac{1 + \sqrt{3}}{1 + \sqrt{3}} \quad \text{Rationalize the denominator.}\\ &= \frac{\sqrt{3} + 3 + 1 + \sqrt{3}}{1 - 3}\\ &= \frac{4 + 2\sqrt{3}}{-2}\\ &= \frac{2(2 + \sqrt{3})}{-2} \quad \text{Factor out 2.}\\ &= -2 - \sqrt{3} \quad \text{Lowest terms}\end{aligned}$$

**(c)**
$$\begin{aligned}\sin 40° \cos 160° - \cos 40° \sin 160° &= \sin(40° - 160°) \quad \text{Sine difference identity}\\ &= \sin(-120°)\\ &= -\sin 120°\\ &= -\frac{\sqrt{3}}{2}\end{aligned}$$

### Graphing Calculator Support

The screen in Figure 7(a) supports the analytic solution for part (b) by giving the same approximation for both $\tan \frac{7\pi}{12}$ and $-2 - \sqrt{3}$. Figure 7(b) indicates that the point $(1.8325957, -3.732051)$, which approximates $\left(\frac{7\pi}{12}, -2 - \sqrt{3}\right)$, lies on the graph of $y = \tan x$, further showing that $\tan \frac{7\pi}{12} = -2 - \sqrt{3}$.

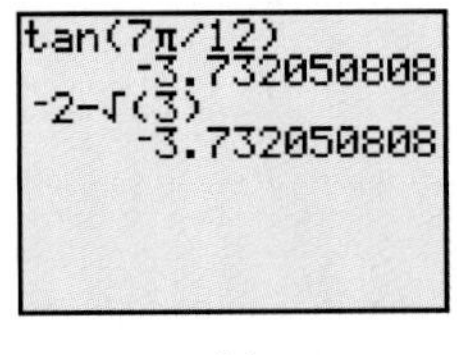


(a)

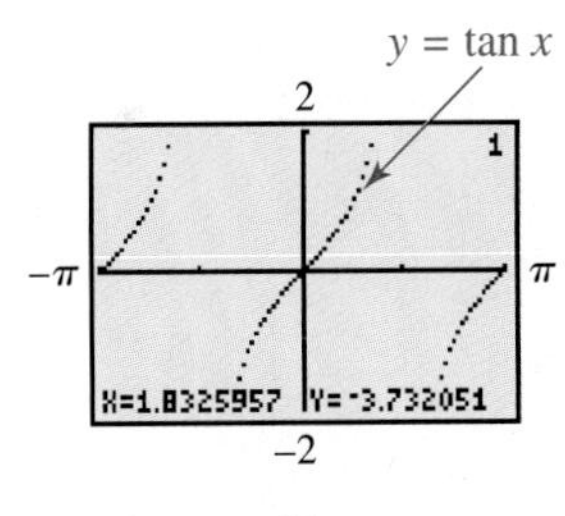


(b)

**FIGURE 7**

To support the result in part (c), we can enter

$\sin 40° \cos 160° - \cos 40° \sin 160°$

and show that its approximation agrees with that of $-\frac{\sqrt{3}}{2}$.

**EXAMPLE 4** *Writing Functions as Expressions Involving Functions of θ*

Write each function as an expression involving functions of $\theta$.

**(a)** $\sin(30° + \theta)$ **(b)** $\tan(45° - \theta)$ **(c)** $\sin(180° + \theta)$

**Solution**

**(a)** Using the identity for $\sin(A + B)$ gives

$$\sin(30° + \theta) = \sin 30° \cos\theta + \cos 30° \sin\theta$$
$$= \frac{1}{2}\cos\theta + \frac{\sqrt{3}}{2}\sin\theta.$$

**(b)** $\tan(45° - \theta) = \dfrac{\tan 45° - \tan\theta}{1 + \tan 45° \tan\theta} = \dfrac{1 - \tan\theta}{1 + \tan\theta}$

**(c)** $\sin(180° + \theta) = \sin 180° \cos\theta + \cos 180° \sin\theta$
$= 0 \cdot \cos\theta + (-1)\sin\theta$
$= -\sin\theta$

**EXAMPLE 5** *Finding Function Values and the Quadrant of A + B*

Suppose that $A$ and $B$ are angles in standard position, with $\sin A = \frac{4}{5}$, $\frac{\pi}{2} < A < \pi$, and $\cos B = -\frac{5}{13}$, $\pi < B < \frac{3\pi}{2}$. Find the following.

**(a)** $\sin(A + B)$ **(b)** $\tan(A + B)$ **(c)** The quadrant of $A + B$

**Solution**

**(a)** The identity for $\sin(A + B)$ requires $\sin A$, $\cos A$, $\sin B$, and $\cos B$. Two of these values are given. The two missing values, $\cos A$ and $\sin B$, can be found, using the identity $\sin^2 x + \cos^2 x = 1$. For $\cos A$,

$$\sin^2 A + \cos^2 A = 1$$

$$\frac{16}{25} + \cos^2 A = 1 \qquad \sin A = \tfrac{4}{5}$$

$$\cos^2 A = \frac{9}{25}$$

$$\cos A = -\frac{3}{5}. \qquad \text{Since } \tfrac{\pi}{2} < A < \pi,\ \cos A < 0.$$

In the same way, $\sin B = -\frac{12}{13}$. Now, use the formula for $\sin(A + B)$.

$$\sin(A + B) = \frac{4}{5}\left(-\frac{5}{13}\right) + \left(-\frac{3}{5}\right)\left(-\frac{12}{13}\right)$$
$$= -\frac{20}{65} + \frac{36}{65} = \frac{16}{65}$$

**(b)** Use the values of sine and cosine from part (a) to get $\tan A = -\frac{4}{3}$ and $\tan B = \frac{12}{5}$. Then,

$$\tan(A + B) = \frac{-\frac{4}{3} + \frac{12}{5}}{1 - \left(-\frac{4}{3}\right)\left(\frac{12}{5}\right)} = \frac{\frac{16}{15}}{1 + \frac{48}{15}} = \frac{\frac{16}{15}}{\frac{63}{15}} = \frac{16}{63}.$$

**(c)** From the results of parts (a) and (b), we find that $\sin(A + B)$ is positive and $\tan(A + B)$ is also positive. Therefore, $A + B$ must be in quadrant I, since it is the only quadrant in which both sine and tangent are positive.

**EXAMPLE 6** *Applying the Cosine Difference Identity to Voltage*

Common household electric current is called *alternating current* because the current alternates direction within the wires. The voltage $V$ in a typical 115-volt outlet can be expressed by the equation $V = 163 \sin \omega t$, where $\omega$ is the angular velocity (in radians per second) of the rotating generator at the electrical plant and $t$ is time measured in seconds. (*Source:* Bell, D., *Fundamentals of Electric Circuits,* Fourth Edition, Prentice-Hall, 1988.)

**(a)** It is essential for electric generators to rotate at precisely 60 cycles per second so household appliances and computers will function properly. Determine $\omega$ for these electric generators.

**(b)** Graph $V$ on the interval $0 \le t \le .05$.

**(c)** For what value of $\phi$ will the graph of $V = 163 \cos(\omega t - \phi)$ be the same as the graph of $V = 163 \sin \omega t$?

**Solution**

**(a)** Each cycle is $2\pi$ radians at 60 cycles per second, so $\omega = 60(2\pi) = 120\pi$ radians per second.

**(b)** $V = 163 \sin \omega t = 163 \sin 120\pi t$. Because the amplitude is 163 here, we choose $-200 \le V \le 200$ for the range, as shown in Figure 8.

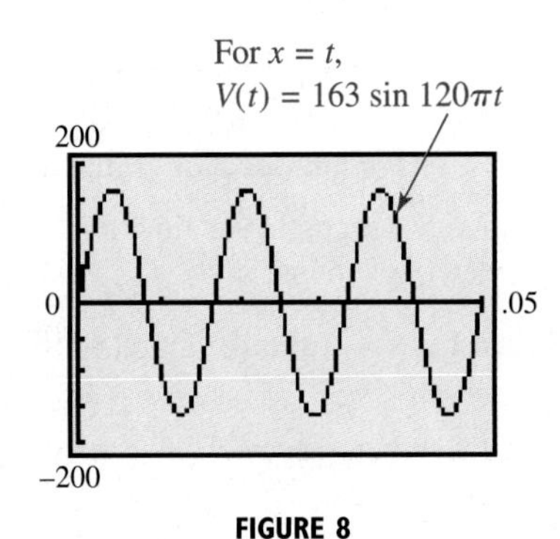


**FIGURE 8**

**(c)** Since $\cos\left(x - \frac{\pi}{2}\right) = \cos\left(\frac{\pi}{2} - x\right) = \sin x$, we choose $\phi = \frac{\pi}{2}$.

## 9.2 EXERCISES

*Concept Check* *Match each expression in Column I with the correct expression in Column II to form an identity.*

| I | II |
|---|---|
| **1.** $\cos(x + y) =$ | **A.** $\cos x \cos y + \sin x \sin y$ |
| **2.** $\cos(x - y) =$ | **B.** $\sin x \sin y - \cos x \cos y$ |
| **3.** $\sin(x + y) =$ | **C.** $\sin x \cos y + \cos x \sin y$ |
| **4.** $\sin(x - y) =$ | **D.** $\sin x \cos y - \cos x \sin y$ |
| | **E.** $\cos x \sin y - \sin x \cos y$ |
| | **F.** $\cos x \cos y - \sin x \sin y$ |

*Use the appropriate sum or difference identity to find the exact value of each expression.*

**5.** $\sin \frac{\pi}{12}$ **6.** $\tan \frac{\pi}{12}$ **7.** $\sin\left(-\frac{5\pi}{12}\right)$ **8.** $\tan\left(-\frac{5\pi}{12}\right)$

**9.** $\sin\left(\frac{13\pi}{12}\right)$ **10.** $\cos\left(\frac{13\pi}{12}\right)$ **11.** $\cos 75°$ **12.** $\sin 105°$

**13.** $\tan 105°$ **14.** $\sin(-15°)$ **15.** $\cos(-15°)$ **16.** $\tan(-75°)$

*Use the appropriate sum or difference identity to find the exact value of each expression.*

**17.** $\cos \frac{\pi}{3} \cos \frac{2\pi}{3} - \sin \frac{\pi}{3} \sin \frac{2\pi}{3}$ **18.** $\cos \frac{7\pi}{8} \cos \frac{\pi}{8} + \sin \frac{7\pi}{8} \sin \frac{\pi}{8}$ **19.** $\sin 76° \cos 31° - \cos 76° \sin 31°$

**20.** $\sin 40° \cos 50° + \cos 40° \sin 50°$ **21.** $\frac{\tan 80° + \tan 55°}{1 - \tan 80° \tan 55°}$ **22.** $\frac{\tan 80° - \tan(-55°)}{1 + \tan 80° \tan(-55°)}$

*Use the appropriate sum or difference identity to write the given expression as a function of x alone.*

**23.** $\cos\left(\frac{\pi}{2} - x\right)$ **24.** $\cos(\pi - x)$ **25.** $\cos\left(\frac{3\pi}{2} + x\right)$ **26.** $\sin(\pi - x)$

**27.** $\sin(\pi + x)$ **28.** $\tan(2\pi - x)$ **29.** $\cos(135° - x)$ **30.** $\tan(45° + x)$

**31.** $\sin(45° + x)$ **32.** $\tan(\pi + x)$ **33.** $\tan(\pi - x)$ **34.** $\sin\left(\frac{3\pi}{2} - x\right)$

**35.** Explain how the identities for $\sec(A + B)$, $\csc(A + B)$, and $\cot(A + B)$ can be found by using the sum identities given in this section.

## RELATING CONCEPTS

**For individual or group investigation (Exercises 36–41)**

**36.** Use a calculator in radian mode to graph the function $y_1 = \cos\left(x + \frac{\pi}{2}\right)$ in the trig viewing window.

**37.** Explain how the graph in Exercise 36 can be obtained by a translation of the graph of $y = \cos x$.

**38.** Find an approximation for $\cos\left(1 + \frac{\pi}{2}\right)$ from the graph in Exercise 36.

**39.** Use the identity for $\cos(A + B)$ to show analytically that $\cos\left(x + \frac{\pi}{2}\right) = -\sin x$.

**40.** Graph $y_2 = -\sin x$ in the trig viewing window. How does this graph support your work in Exercise 39?

**41.** Find an approximation for $-\sin 1$ from the graph in Exercise 40. How does it compare to the result in Exercise 38?

*Suppose that A and B are angles in standard position. Use the given information to find* **(a)** $\sin(A + B)$, **(b)** $\sin(A - B)$, **(c)** $\tan(A + B)$, **(d)** $\tan(A - B)$, **(e)** *the quadrant of* $A + B$, *and* **(f)** *the quadrant of* $A - B$.

**42.** $\cos A = \frac{3}{5}$, $\sin B = \frac{5}{13}$, $0 < A < \frac{\pi}{2}, 0 < B < \frac{\pi}{2}$

**43.** $\sin A = \frac{3}{5}$, $\sin B = -\frac{12}{13}$, $0 < A < \frac{\pi}{2}$, $\pi < B < \frac{3\pi}{2}$

**44.** $\cos A = -\frac{8}{17}$, $\cos B = -\frac{3}{5}$, $\pi < A < \frac{3\pi}{2}, \pi < B < \frac{3\pi}{2}$

**45.** $\cos A = -\frac{15}{17}$, $\sin B = \frac{4}{5}$, $\frac{\pi}{2} < A < \pi$, $0 < B < \frac{\pi}{2}$

*Verify that each equation is an identity.*

**46.** $\sin(x + y) + \sin(x - y) = 2 \sin x \cos y$

**47.** $\tan(x - y) - \tan(y - x) = \frac{2(\tan x - \tan y)}{1 + \tan x \tan y}$

**48.** $\frac{\cos(A - B)}{\cos A \sin B} = \tan A + \cot B$

**49.** $\frac{\sin(A + B)}{\cos A \cos B} = \tan A + \tan B$

**50.** $\frac{\sin(A - B)}{\sin B} + \frac{\cos(A - B)}{\cos B} = \frac{\sin A}{\sin B \cos B}$

**51.** $\frac{\tan(A + B) - \tan B}{1 + \tan(A + B) \tan B} = \tan A$

**52.** $\frac{\sin(x - y)}{\sin(x + y)} = \frac{\tan x - \tan y}{\tan x + \tan y}$

*Exercises 53 and 54 refer to Example 6.*

**53.** How many times does the current oscillate in .05 sec?

**54.** What are the maximum and minimum voltages in this outlet? Is the voltage always equal to 115 volts?

*(Modeling) Solve each problem.*

**55.** *Back Stress* If a person bends at the waist with a straight back making an angle of $\theta$ degrees with the horizontal, then the force $F$ exerted on the back muscles can be modeled by the equation

$$F = \frac{.6W \sin(\theta + 90°)}{\sin 12°},$$

where $W$ is the weight of the person. (*Source:* Metcalf, H., *Topics in Classical Biophysics,* Prentice-Hall, 1980.)

**(a)** Calculate $F$ when $W = 170$ pounds and $\theta = 30°$.

**(b)** Use an identity to show that $F$ is approximately equal to $2.9W \cos \theta$.

**(c)** For what value of $\theta$ is $F$ maximum?

**56.** *Back Stress* Refer to Exercise 55.

**(a)** Suppose a 200-pound person bends at the waist so that $\theta = 45°$. Estimate the force exerted by the person's back muscles.

**(b)** Approximate graphically the value of $\theta$ that results in the back muscles exerting a force of 400 pounds.

**57.** *Sound Waves* Sound is a result of waves applying pressure to a person's eardrum. For a pure sound wave radiating outward in a spherical shape, the trigonometric function

$$P = \frac{a}{r}\cos\left(\frac{2\pi r}{\lambda} - ct\right)$$

can be used to model the sound pressure at a radius of $r$ feet from the source: $t$ is time in seconds, $\lambda$ is length of the sound wave in feet, $c$ is speed of sound in feet per second, and $a$ is maximum sound pressure at the source measured in pounds per square foot. (*Source:* Beranek, L., *Noise and Vibration Control,* Institute of Noise Control Engineering, Washington, D.C., 1988.) Let $\lambda = 4.9$ feet and $c = 1026$ feet per second.

**(a)** Let $a = .4$ pound per square foot. Graph the sound pressure at distance $r = 10$ feet from its source over the interval $0 \le t \le .05$. Describe $P$ at this distance.

**(b)** Now let $a = 3$ and $t = 10$. Graph the sound pressure for $0 \le r \le 20$. What happens to pressure $P$ as radius $r$ increases?

**(c)** Suppose a person stands at a radius $r$ so that $r = n\lambda$, where $n$ is a positive integer. Use the difference identity for cosine to simplify $P$ in this situation.

**58.** *Voltage of a Circuit* When the two voltages

$$V_1 = 30 \sin 120\pi t \quad \text{and} \quad V_2 = 40 \cos 120\pi t$$

are applied to the same circuit, the resulting voltage $V$ will be equal to their sum. (*Source:* Bell, D., *Fundamentals of Electric Circuits,* Second Edition, Reston Publishing Company, 1981.)

**(a)** Graph $V = V_1 + V_2$ over the interval $0 \le t \le .05$.

**(b)** Use the graph to estimate values for $a$ and $\phi$ so that $V = a \sin(120\pi t + \phi)$.

**(c)** Use identities to verify that your expression for $V$ is valid.

## Reviewing Basic Concepts (Sections 9.1 and 9.2)

**1.** Write $\dfrac{\csc x}{\cot x} - \dfrac{\cot x}{\csc x}$ in terms of $\sin x$ and $\cos x$.

**2.** Use a sum or difference identity to find an exact value of $\tan\left(-\frac{\pi}{12}\right)$.

**3.** Find an exact value of

$$\cos 18° \cos 108° + \sin 18° \sin 108°.$$

**4.** Use a sum or difference identity to write $\sin\left(x - \frac{\pi}{4}\right)$ as a function of $x$.

**5.** Given $\sin A = \frac{2}{3}$ and $\cos B = -\frac{1}{2}$, where $A$ is in quadrant II and $B$ is in quadrant III, find $\sin(A + B)$, $\cos(A - B)$, and $\tan(A - B)$.

*Verify that each equation is an identity.*

**6.** $\csc^2 \theta - \cot^2 \theta = 1$

**7.** $\dfrac{\sin t}{1 - \cos t} = \dfrac{1 + \cos t}{\sin t}$

**8.** $\dfrac{\cot A - \tan A}{\csc A \sec A} = \cos^2 A - \sin^2 A$

**9.** $\dfrac{\sin(x - y)}{\sin x \sin y} = \cot y - \cot x$

**10.** *(Modeling) Voltage* A coil of wire rotating in a magnetic field induces a voltage

$$e = 20 \sin\left(\frac{\pi t}{4} - \frac{\pi}{2}\right).$$

Use an identity to express this in terms of $\cos \dfrac{\pi t}{4}$.

## 9.3 FURTHER IDENTITIES

Double-Number Identities • Product-to-Sum and Sum-to-Product Identities • Half-Number Identities

We can now use the sum and difference identities from the previous section to derive several additional identities.

### Double-Number Identities

Some special cases of the identities for the sum of two numbers are used often enough to be expressed separately. These identities, called the **double-number identities,** result from the sum identities when $A = B$ so that $A + B = 2A$. For example, if $B = A$, then the identity $\cos(A + B) = \cos A \cos B - \sin A \sin B$ gives an identity for $\cos 2A$.

$$\begin{aligned} \cos 2A &= \cos(A + A) \\ &= \cos A \cos A - \sin A \sin A \\ \cos 2A &= \cos^2 A - \sin^2 A \end{aligned}$$

Two other useful forms of this identity are obtained by substituting either $\cos^2 A = 1 - \sin^2 A$ or $\sin^2 A = 1 - \cos^2 A$. Replace $\cos^2 A$ with $1 - \sin^2 A$ to obtain

$$\begin{aligned} \cos 2A &= \cos^2 A - \sin^2 A \\ &= (1 - \sin^2 A) - \sin^2 A \\ \cos 2A &= 1 - 2\sin^2 A, \end{aligned}$$

or replace $\sin^2 A$ with $1 - \cos^2 A$ to obtain

$$\begin{aligned} \cos 2A &= \cos^2 A - (1 - \cos^2 A) \\ &= \cos^2 A - 1 + \cos^2 A \\ \cos 2A &= 2\cos^2 A - 1. \end{aligned}$$

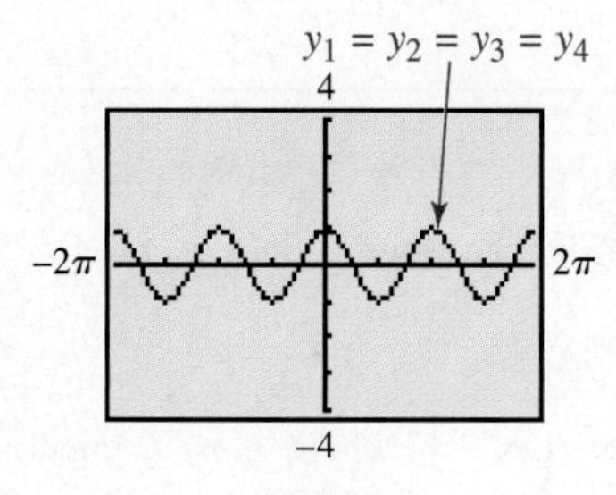

FIGURE 9

These double-number identities for cosine can be supported by graphing the four functions $y_1 = \cos 2x$, $y_2 = \cos^2 x - \sin^2 x$, $y_3 = 1 - 2\sin^2 x$, and $y_4 = 2\cos^2 x - 1$ in the same viewing window. The graphs coincide, supporting our analytic work above. Figure 9 shows this single graph in the trig viewing window.

We find $\sin 2A$ with the identity $\sin(A + B) = \sin A \cos B + \cos A \sin B$, with $B = A$.

$$\begin{aligned}\sin 2A &= \sin(A + A)\\ &= \sin A \cos A + \cos A \sin A\\ \sin 2A &= 2 \sin A \cos A\end{aligned}$$

Similarly, we use the identity for $\tan(A + B)$ to find $\tan 2A$.

$$\begin{aligned}\tan 2A &= \tan(A + A)\\ &= \frac{\tan A + \tan A}{1 - \tan A \tan A}\\ \tan 2A &= \frac{2 \tan A}{1 - \tan^2 A}\end{aligned}$$

The double-number identities are summarized here.

**FOR DISCUSSION**

1. Support the identity for $\sin 2A$ by graphing both

$$y_1 = \sin 2x$$

and $$y_2 = 2 \sin x \cos x$$

in the same viewing window. The graphs should coincide.

2. Repeat Item 1 for $\tan 2A$, by graphing

$$y_1 = \tan 2x$$

and $$y_2 = \frac{2 \tan x}{1 - \tan^2 x}.$$

**Double-Number Identities**

$$\cos 2A = \cos^2 A - \sin^2 A \qquad \cos 2A = 1 - 2 \sin^2 A$$

$$\cos 2A = 2 \cos^2 A - 1 \qquad \sin 2A = 2 \sin A \cos A$$

$$\tan 2A = \frac{2 \tan A}{1 - \tan^2 A}$$

We can find function values of $2\theta$ if we know a trigonometric function of $\theta$ and the quadrant of $\theta$.

**EXAMPLE 1** *Finding Function Values of $2\theta$, Given Information about $\theta$*

Given $\cos \theta = \frac{3}{5}$ and $\sin \theta < 0$, find $\sin 2\theta$, $\cos 2\theta$, and $\tan 2\theta$.

**Solution** To find $\sin 2\theta$, we must first find the value of $\sin \theta$, using the identity $\sin^2 \theta + \cos^2 \theta = 1$.

$$\sin^2 \theta + \left(\frac{3}{5}\right)^2 = 1 \qquad \cos \theta = \tfrac{3}{5}$$

$$\sin^2 \theta = \frac{16}{25}$$

$$\sin \theta = -\frac{4}{5} \qquad \text{Choose the negative square root, since } \sin \theta < 0.$$

Using the double-number identity for sine,

$$\sin 2\theta = 2 \sin \theta \cos \theta = 2\left(-\frac{4}{5}\right)\left(\frac{3}{5}\right) = -\frac{24}{25}.$$

Now, we find $\cos 2\theta$, using the first form of the identity. (Any form may be used.)

$$\cos 2\theta = \cos^2 \theta - \sin^2 \theta = \frac{9}{25} - \frac{16}{25} = -\frac{7}{25}$$

The value of $\tan 2\theta$ can be found in either of two ways. We can use the double-number identity and the fact that $\tan\theta = \frac{\sin\theta}{\cos\theta} = \frac{-\frac{4}{5}}{\frac{3}{5}} = -\frac{4}{3}$.

$$\tan 2\theta = \frac{2\tan\theta}{1-\tan^2\theta} = \frac{2\left(-\frac{4}{3}\right)}{1-\frac{16}{9}} = \frac{-\frac{8}{3}}{-\frac{7}{9}} = \frac{24}{7}$$

Alternatively, we can find $\tan 2\theta$ by finding the quotient of $\sin 2\theta$ and $\cos 2\theta$.

$$\tan 2\theta = \frac{\sin 2\theta}{\cos 2\theta} = \frac{-\frac{24}{25}}{-\frac{7}{25}} = \frac{24}{7}$$

We can now verify additional identities involving double-number identities using the hints given in Section 9.1.

**EXAMPLE 2** *Verifying a Double-Number Identity*

Verify that the given equation is an identity.

$$\cot x \sin 2x = 1 + \cos 2x$$

**Analytic Solution**

We start by working on the left side, using the hint from Section 9.1 about writing all functions in terms of sine and cosine.

$$\cot x \sin 2x = \frac{\cos x}{\sin x} \cdot \sin 2x \qquad \cot x = \frac{\cos x}{\sin x}$$

$$= \frac{\cos x}{\sin x}(2\sin x\cos x) \qquad \sin 2x = 2\sin x\cos x$$

$$= 2\cos^2 x$$

$$= 1 + \cos 2x \qquad 2\cos^2 x - 1 = \cos 2x$$

The final step in this procedure illustrates the importance of being able to recognize alternative forms of identities.

**Graphing Calculator Support**

To support our analytic work graphically, we show that the graphs of

$$y_1 = \cot x \sin 2x$$

and

$$y_2 = 1 + \cos 2x$$

coincide. Notice that the graph of $y_2$ is obtained from that of $y = \cos x$ with period changed to $\frac{2\pi}{2} = \pi$, shifted 1 unit upward. See Figure 10.

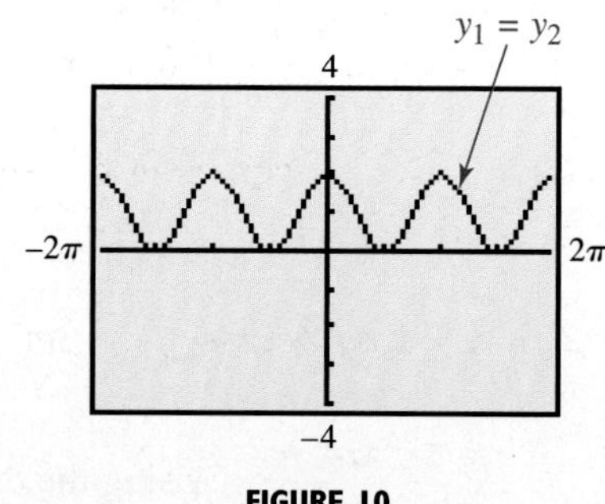


FIGURE 10

**EXAMPLE 3** *Simplifying Expressions Using Double-Number Identities*

Simplify each expression.

**(a)** $\cos^2 7x - \sin^2 7x$ **(b)** $\sin 15° \cos 15°$

## Solution

**(a)** This expression suggests one of the double-number identities for the cosine: $\cos 2A = \cos^2 A - \sin^2 A$. Substituting $7x$ for $A$ gives

$$\cos^2 7x - \sin^2 7x = \cos 2(7x) = \cos 14x.$$

**(b)** If this expression were $2 \sin 15° \cos 15°$, we could apply the identity for $\sin 2A$ directly since $\sin 2A = 2 \sin A \cos A$. We can still apply the identity with $A = 15°$ by writing the multiplicative identity element 1 as $\frac{1}{2}(2)$.

$$\sin 15° \cos 15° = \frac{1}{2}(2) \sin 15° \cos 15° \qquad \text{Multiply by 1 in the form } \tfrac{1}{2}(2).$$

$$= \frac{1}{2}(2 \sin 15° \cos 15°) \qquad \text{Associative property}$$

$$= \frac{1}{2} \sin(2 \cdot 15°) \qquad 2 \sin A \cos A = \sin 2A, \text{ with } A = 15°$$

$$= \frac{1}{2} \sin 30°$$

$$= \frac{1}{2} \cdot \frac{1}{2} \qquad \sin 30° = \tfrac{1}{2}$$

$$= \frac{1}{4}$$

Identities involving larger multiples of the variable can be derived by repeated use of the double-number identities and the Pythagorean identities.

### EXAMPLE 4 *Deriving a Multiple-Number Identity*

Write $\sin 3x$ in terms of $\sin x$.

#### Analytic Solution

$$\sin 3x = \sin(2x + x)$$

$$= \sin 2x \cos x + \cos 2x \sin x \qquad \text{Sine sum identity}$$

$$= (2 \sin x \cos x) \cos x + (\cos^2 x - \sin^2 x) \sin x \qquad \text{Double-number identities}$$

$$= 2 \sin x \cos^2 x + \cos^2 x \sin x - \sin^3 x$$

$$= 2 \sin x(1 - \sin^2 x) + (1 - \sin^2 x) \sin x - \sin^3 x \qquad \cos^2 x = 1 - \sin^2 x$$

$$= 2 \sin x - 2 \sin^3 x + \sin x - \sin^3 x - \sin^3 x \qquad \text{Distributive property}$$

$$= 3 \sin x - 4 \sin^3 x$$

#### Graphing Calculator Support

The table in Figure 11 numerically supports the analytic solution. Here,

$$Y_2 = 3 \sin X - 4 \sin^3 X.$$

Although the table gives strong support to the identity, it does not verify it because it does not list every possible X-value.

| X | Y1 | Y2 |
|---|---|---|
| -1.178 | .38268 | .38268 |
| -.7854 | -.7071 | -.7071 |
| -.3927 | -.9239 | -.9239 |
| 0 | 0 | 0 |
| .3927 | .92388 | .92388 |
| .7854 | .70711 | .70711 |
| 1.1781 | -.3827 | -.3827 |

Y1 = sin(3X)

FIGURE 11

The next example applies a multiple-number identity to answer a question about electric current.

**EXAMPLE 5** *Determining Wattage Consumption*

If a toaster is plugged into a common household outlet, the wattage consumed is not constant. Instead, it varies at a high frequency according to the model

$$W = \frac{V^2}{R},$$

where $V$ is the voltage and $R$ is a constant that measures the resistance of the toaster in ohms. (*Source:* Bell, D., *Fundamentals of Electric Circuits,* Fourth Edition, Prentice-Hall, 1998.) Graph the wattage $W$ consumed by a typical toaster with $R = 15$ and $V = 163 \sin 120\pi t$ over the interval $0 \le t \le .05$. How many oscillations are there?

**Solution** Substituting the given values into the wattage equation gives

$$W = \frac{V^2}{R} = \frac{(163 \sin 120\pi t)^2}{15}.$$

To determine the range for $W$, we note that $\sin 120\pi t$ has maximum value 1, so the expression for $W$ has maximum value $\frac{163^2}{15} \approx 1771$. The minimum value is 0. The graph in Figure 12 shows that there are six oscillations.

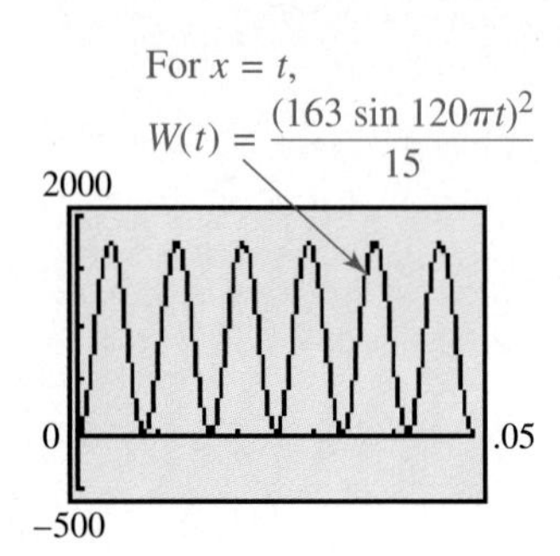


**FIGURE 12**

**What Went WRONG?**

To verify the proposed identity $\cos x = \cos 4x$, a student graphed both equations in the same screen, with the result shown on the left. He then compared the functions, using the table feature of his calculator, getting the result shown on the right. He was confused, because the graph indicates that the equation is not an identity, while the table indicates that it might be one.

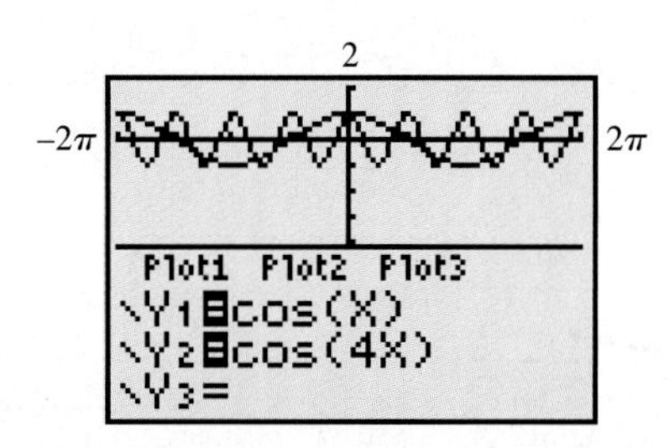


| X | Y1 | Y2 |
|---|---|---|
| 0 | 1 | 1 |
| 2.0944 | -.5 | -.5 |
| 4.1888 | -.5 | -.5 |
| 6.2832 | 1 | 1 |
| 8.3776 | -.5 | -.5 |
| 10.472 | -.5 | -.5 |
| 12.566 | 1 | 1 |

X=0

$\Delta\text{Tbl} = \frac{2\pi}{3} \approx 2.0944$

Tblmin = 0

What Went Wrong? What should the student conclude? Why?

---

*Answer to What Went Wrong?*

He should conclude that $\cos x = \cos 4x$ is *not* an identity, based on the graph. The problem with the table is that a $\Delta$Tbl of $\frac{2\pi}{3}$ produces values where the two functions coincide. Choosing a different interval, such as $\frac{\pi}{2}$, would show different function values at multiples of $\frac{\pi}{2}$.

**Looking Ahead to Calculus**

The product-to-sum identities are used in calculus to find *integrals* of functions that are products of trigonometric functions. One classic calculus text includes the following example:

Evaluate $\int \cos 5x \cos 3x \, dx$.

The first solution line reads: "We may write $\cos 5x \cos 3x = \frac{1}{2}(\cos 8x + \cos 2x)$."

## Product-to-Sum and Sum-to-Product Identities

Because they make it possible to rewrite a product as a sum, the identities for $\cos(A + B)$ and $\cos(A - B)$ are used to derive a group of identities useful in calculus.

Adding the identities for $\cos(A + B)$ and $\cos(A - B)$ gives

$$\begin{aligned} \cos(A + B) &= \cos A \cos B - \sin A \sin B \\ \cos(A - B) &= \cos A \cos B + \sin A \sin B \\ \hline \cos(A + B) + \cos(A - B) &= 2 \cos A \cos B \end{aligned}$$

or

$$\cos A \cos B = \frac{1}{2}[\cos(A + B) + \cos(A - B)].$$

Similarly, subtracting $\cos(A + B)$ from $\cos(A - B)$ gives

$$\sin A \sin B = \frac{1}{2}[\cos(A - B) - \cos(A + B)].$$

Using the identities for $\sin(A + B)$ and $\sin(A - B)$ in the same way, we get two more identities. Those and the previous ones are now summarized.

**Product-to-Sum Identities**

$$\cos A \cos B = \frac{1}{2}[\cos(A + B) + \cos(A - B)]$$

$$\sin A \sin B = \frac{1}{2}[\cos(A - B) - \cos(A + B)]$$

$$\sin A \cos B = \frac{1}{2}[\sin(A + B) + \sin(A - B)]$$

$$\cos A \sin B = \frac{1}{2}[\sin(A + B) - \sin(A - B)]$$

**EXAMPLE 6** *Using a Product-to-Sum Identity*

Rewrite $\cos 2\theta \sin \theta$ as the sum or difference of two functions.

**Solution** By the identity for $\cos A \sin B$, with $2\theta = A$ and $\theta = B$,

$$\begin{aligned} \cos 2\theta \sin \theta &= \frac{1}{2}[\sin(2\theta + \theta) - \sin(2\theta - \theta)] \\ &= \frac{1}{2}\sin 3\theta - \frac{1}{2}\sin \theta. \end{aligned}$$

From these new identities we can derive another group of identities that are used to rewrite sums of trigonometric functions as products. We summarize these identities without proof.

**Sum-to-Product Identities**

$$\sin A + \sin B = 2 \sin\left(\frac{A + B}{2}\right) \cos\left(\frac{A - B}{2}\right)$$

$$\sin A - \sin B = 2 \cos\left(\frac{A + B}{2}\right) \sin\left(\frac{A - B}{2}\right)$$

$$\cos A + \cos B = 2 \cos\left(\frac{A + B}{2}\right) \cos\left(\frac{A - B}{2}\right)$$

$$\cos A - \cos B = -2 \sin\left(\frac{A + B}{2}\right) \sin\left(\frac{A - B}{2}\right)$$

**EXAMPLE 7** *Using a Sum-to-Product Identity*

Write $\sin 2\gamma - \sin 4\gamma$ as a product of two functions.

**Solution** Use the identity for $\sin A - \sin B$, with $2\gamma = A$ and $4\gamma = B$.

$$\begin{aligned}\sin 2\gamma - \sin 4\gamma &= 2 \cos\left(\frac{2\gamma + 4\gamma}{2}\right) \sin\left(\frac{2\gamma - 4\gamma}{2}\right)\\ &= 2 \cos \frac{6\gamma}{2} \sin\left(\frac{-2\gamma}{2}\right)\\ &= 2 \cos 3\gamma \sin(-\gamma)\\ &= -2 \cos 3\gamma \sin \gamma \qquad \sin(-\gamma) = -\sin \gamma\end{aligned}$$

## Half-Number Identities

**Looking Ahead to Calculus**

Half-number identities for sine and cosine are used in calculus when eliminating the $xy$-term from an equation in the form $Ax^2 + Bxy + Cy^2 + Dx + Ey + F = 0$, so the type of conic section it represents can be determined.

From the alternative forms of the identity for $\cos 2A$, we derive three additional identities for $\sin \frac{A}{2}$, $\cos \frac{A}{2}$, and $\tan \frac{A}{2}$. These are known as **half-number identities.**

To derive the identity for $\sin \frac{A}{2}$, start with the following double-number identity for cosine and solve for $\sin x$.

$$\begin{aligned}\cos 2x &= 1 - 2 \sin^2 x\\ 2 \sin^2 x &= 1 - \cos 2x\\ \sin x &= \pm\sqrt{\frac{1 - \cos 2x}{2}}\end{aligned}$$

Now, let $2x = A$ so that $x = \frac{A}{2}$, and substitute into this last expression.

$$\sin \frac{A}{2} = \pm\sqrt{\frac{1 - \cos A}{2}}$$

The $\pm$ sign in this identity indicates that the appropriate sign is chosen depending on the quadrant of $\frac{A}{2}$. For example, if $\frac{A}{2}$ is a third-quadrant number on the unit circle, we choose the negative sign since the sine function is negative there.

The identity for $\cos \frac{A}{2}$ is derived in a similar way, starting with the double-number identity $\cos 2x = 2\cos^2 x - 1$. Solve for $\cos x$.

$$\cos 2x + 1 = 2\cos^2 x$$

$$\cos x = \pm\sqrt{\frac{1 + \cos 2x}{2}}$$

$$\cos \frac{A}{2} = \pm\sqrt{\frac{1 + \cos A}{2}} \qquad \text{Replace } x \text{ with } \tfrac{A}{2}.$$

The $\pm$ sign is chosen as previously described.

Finally, an identity for $\tan \frac{A}{2}$ comes from the half-number identities for sine and cosine.

$$\tan \frac{A}{2} = \frac{\pm\sqrt{\frac{1 - \cos A}{2}}}{\pm\sqrt{\frac{1 + \cos A}{2}}} \qquad \tan \tfrac{A}{2} = \frac{\sin \frac{A}{2}}{\cos \frac{A}{2}}$$

$$\tan \frac{A}{2} = \pm\sqrt{\frac{1 - \cos A}{1 + \cos A}} \qquad \pm \text{ chosen depending on quadrant of } \tfrac{A}{2}$$

We can derive an alternative identity for $\tan \frac{A}{2}$, using the fact that $\tan \frac{A}{2} = \frac{\sin \frac{A}{2}}{\cos \frac{A}{2}}$.

$$\tan \frac{A}{2} = \frac{\sin \frac{A}{2}}{\cos \frac{A}{2}}$$

$$= \frac{2 \sin \frac{A}{2} \cos \frac{A}{2}}{2\cos^2 \frac{A}{2}} \qquad \text{Multiply by } 2\cos \tfrac{A}{2} \text{ in numerator and denominator.}$$

$$= \frac{\sin\left[2\left(\frac{A}{2}\right)\right]}{1 + \cos\left[2\left(\frac{A}{2}\right)\right]} \qquad \text{Double-number identities}$$

$$\tan \frac{A}{2} = \frac{\sin A}{1 + \cos A}$$

From this identity for $\tan \frac{A}{2}$, we can also derive

$$\tan \frac{A}{2} = \frac{1 - \cos A}{\sin A}.$$

See Exercise 35. These last two identities for $\tan \frac{A}{2}$ do not require a sign choice as the first one does.

**Half-Number Identities**

$$\cos \frac{A}{2} = \pm\sqrt{\frac{1 + \cos A}{2}} \qquad \sin \frac{A}{2} = \pm\sqrt{\frac{1 - \cos A}{2}}$$

$$\tan \frac{A}{2} = \pm\sqrt{\frac{1 - \cos A}{1 + \cos A}} \qquad \tan \frac{A}{2} = \frac{\sin A}{1 + \cos A} \qquad \tan \frac{A}{2} = \frac{1 - \cos A}{\sin A}$$

In Example 1 of Section 9.2, we showed that the exact value of $\cos 15°$ or, equivalently, $\cos \frac{\pi}{12}$ is $\frac{\sqrt{6}+\sqrt{2}}{4}$. We did this by using the identity for $\cos(A - B)$. Another form of the exact value of $\cos \frac{\pi}{12}$ can be found by using the identity for $\cos \frac{A}{2}$.

### EXAMPLE 8 *Using a Half-Number Identity to Find an Exact Value*

Find the exact value of $\cos \frac{\pi}{12}$.

**Solution**

$$\cos \frac{\pi}{12} = \cos \frac{\frac{\pi}{6}}{2} \qquad \frac{\pi}{12} = \frac{\frac{\pi}{6}}{2}$$

$$= \sqrt{\frac{1+\cos\frac{\pi}{6}}{2}} \qquad \text{Half-number identity for cosine}$$

$$= \sqrt{\frac{1+\frac{\sqrt{3}}{2}}{2}} \qquad \cos\frac{\pi}{6} = \frac{\sqrt{3}}{2}$$

$$= \sqrt{\frac{\left(1+\frac{\sqrt{3}}{2}\right)\cdot 2}{2\cdot 2}} \qquad \text{Multiply by } \tfrac{2}{2} \text{ under the radical.}$$

$$= \frac{\sqrt{2+\sqrt{3}}}{2} \qquad \sqrt{\frac{a}{b}} = \frac{\sqrt{a}}{\sqrt{b}}$$

Verify that the final expression, $\frac{\sqrt{2+\sqrt{3}}}{2}$, has a calculator approximation of .9659258263, the same as for $\frac{\sqrt{6}+\sqrt{2}}{4}$.

### EXAMPLE 9 *Finding Function Values of $\frac{x}{2}$, Given Information about x*

Given $\cos x = \frac{2}{3}$, with $\frac{3\pi}{2} < x < 2\pi$, find $\cos \frac{x}{2}$, $\sin \frac{x}{2}$, and $\tan \frac{x}{2}$.

**Solution** Since

$$\frac{3\pi}{2} < x < 2\pi,$$

and

$$\frac{3\pi}{4} < \frac{x}{2} < \pi, \qquad \text{Divide by 2.}$$

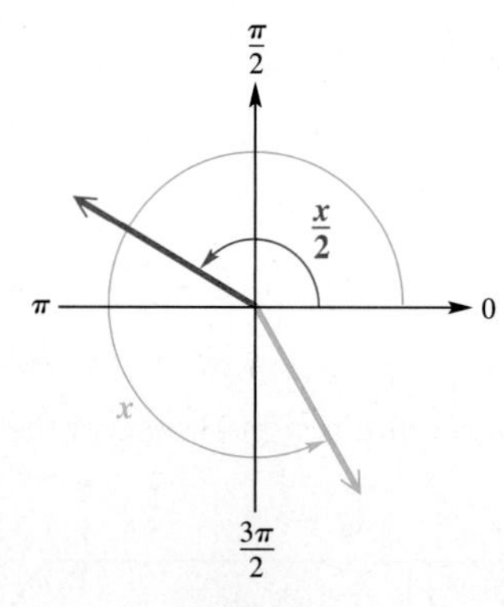

FIGURE 13

$\frac{x}{2}$ terminates in quadrant II. See Figure 13. In this quadrant, the value of $\cos \frac{x}{2}$ is negative and the value of $\sin \frac{x}{2}$ is positive. Now use the appropriate half-number identities.

$$\sin \frac{x}{2} = \sqrt{\frac{1-\frac{2}{3}}{2}} = \sqrt{\frac{1}{6}} = \frac{\sqrt{6}}{6}$$

$$\cos \frac{x}{2} = -\sqrt{\frac{1+\frac{2}{3}}{2}} = -\sqrt{\frac{5}{6}} = -\frac{\sqrt{30}}{6}$$

$$\tan \frac{x}{2} = \frac{\sin\frac{x}{2}}{\cos\frac{x}{2}} = \frac{\frac{\sqrt{6}}{6}}{-\frac{\sqrt{30}}{6}} = -\frac{\sqrt{5}}{5}$$

Notice that it is not necessary to use a half-number identity for $\tan \frac{x}{2}$ once we find $\sin \frac{x}{2}$ and $\cos \frac{x}{2}$. However, using this identity would provide an excellent check.

**EXAMPLE 10** *Simplifying Expressions Using Half-Number Identities*

Simplify each expression.

**(a)** $\pm\sqrt{\dfrac{1+\cos 12x}{2}}$ **(b)** $\dfrac{1-\cos 5\alpha}{\sin 5\alpha}$

**Solution**

**(a)** This matches part of the identity for $\cos \frac{A}{2}$.

$$\cos\frac{A}{2} = \pm\sqrt{\frac{1+\cos A}{2}}$$

Replace $A$ with $12x$ to get

$$\pm\sqrt{\frac{1+\cos 12x}{2}} = \cos\frac{12x}{2} = \cos 6x.$$

**(b)** Use the third identity for $\tan \frac{A}{2}$ to obtain

$$\frac{1-\cos 5\alpha}{\sin 5\alpha} = \tan\frac{5\alpha}{2}.$$

## 9.3 EXERCISES

*Use the identities in this section to find values of the sine and cosine functions for each number.*

**1.** $2\theta$, given $\sin\theta = \frac{2}{5}$ and $\cos\theta < 0$

**2.** $2\beta$, given $\cos\beta = -\frac{12}{13}$ and $\sin\beta > 0$

**3.** $2x$, given $\tan x = 2$ and $\cos x > 0$

**4.** $2x$, given $\tan x = \frac{5}{3}$ and $\sin x < 0$

**5.** $2\alpha$, given $\sin\alpha = -\frac{\sqrt{5}}{7}$ and $\cos\alpha > 0$

**6.** $2\alpha$, given $\cos\alpha = \frac{\sqrt{3}}{5}$ and $\sin\alpha > 0$

*Use an identity to write each expression as a single trigonometric function or as a single number.*

**7.** $\cos^2 15° - \sin^2 15°$

**8.** $\dfrac{2\tan 15°}{1-\tan^2 15°}$

**9.** $1 - 2\sin^2 15°$

**10.** $1 - 2\sin^2 22.5°$

**11.** $2\cos^2 67.5° - 1$

**12.** $\cos^2\frac{\pi}{8} - \frac{1}{2}$

**13.** $\dfrac{\tan 51°}{1-\tan^2 51°}$

**14.** $\dfrac{\tan 34°}{2(1-\tan^2 34°)}$

**15.** $\frac{1}{4} - \frac{1}{2}\sin^2 47.1°$

**16.** $\frac{1}{8}\sin 29.5° \cos 29.5°$

*Graph each expression and use the graph to conjecture an identity. Then verify your conjecture.*

**17.** $\cos^4 x - \sin^4 x$

**18.** $\dfrac{4\tan x\cos^2 x - 2\tan x}{1-\tan^2 x}$

*Use the method of Example 4 to do the following. Then, support your result graphically, using the trig viewing window of your calculator.*

**19.** Express $\cos 3x$ in terms of $\cos x$.

**20.** Express $\tan 3x$ in terms of $\tan x$.

**21.** Express $\tan 4x$ in terms of $\tan x$.

**22.** Express $\cos 4x$ in terms of $\cos x$.

*Use a half-number identity to find an exact value for each trigonometric function.*

**23.** $\sin \frac{\pi}{12}$ **24.** $\cos \frac{\pi}{8}$ **25.** $\tan\left(-\frac{\pi}{8}\right)$ **26.** $\cos 67.5°$ **27.** $\sin 67.5°$ **28.** $\tan 195°$

*Use a half-number identity to find an exact value for each function, given the information about x.*

**29.** $\cos \frac{x}{2}$, given $\cos x = \frac{1}{4}$, with $0 < x < \frac{\pi}{2}$

**30.** $\sin \frac{x}{2}$, given $\cos x = -\frac{5}{8}$, with $\frac{\pi}{2} < x < \pi$

**31.** $\tan \frac{x}{2}$, given $\sin x = \frac{3}{5}$, with $\frac{\pi}{2} < x < \pi$

**32.** $\cos \frac{x}{2}$, given $\sin x = -\frac{4}{5}$, with $\frac{3\pi}{2} < x < 2\pi$

**33.** $\tan \frac{x}{2}$, given $\tan x = \frac{\sqrt{7}}{3}$, with $\pi < x < \frac{3\pi}{2}$

**34.** $\tan \frac{x}{2}$, given $\tan x = -\frac{\sqrt{5}}{2}$, with $\frac{\pi}{2} < x < \pi$

**35.** Use the identity

$$\tan \frac{A}{2} = \frac{\sin A}{1 + \cos A}$$

to derive the equivalent identity

$$\tan \frac{A}{2} = \frac{1 - \cos A}{\sin A}$$

by multiplying both the numerator and denominator by $1 - \cos A$.

**36.** Consider the expression $\tan\left(\frac{\pi}{2} + x\right)$.

**(a)** Why can't we use the identity for $\tan(A + B)$ to express it as a function of $x$ alone?

**(b)** Use the identity $\tan \theta = \frac{\sin \theta}{\cos \theta}$ to rewrite the expression in terms of sine and cosine.

**(c)** Use the result of part (b) to show that

$$\tan\left(\frac{\pi}{2} + x\right) = -\cot x.$$

**37.** The identity

$$\tan \frac{A}{2} = \pm\sqrt{\frac{1 - \cos A}{1 + \cos A}}$$

can be used to show that $\tan 22.5° = \sqrt{3 - 2\sqrt{2}}$, and the identity

$$\tan \frac{A}{2} = \frac{\sin A}{1 + \cos A}$$

can be used to show that $\tan 22.5° = \sqrt{2} - 1$. Verify that these answers are the same, without using a calculator. (*Hint:* If $a > 0$ and $b > 0$ and $a^2 = b^2$, then $a = b$.)

**38.** Explain how you could use an identity of this section to find the exact value of $\sin 7.5°$.

*Use an identity to write each expression as a single trigonometric function.*

**39.** $\sqrt{\frac{1 - \cos 40°}{2}}$ **40.** $\sqrt{\frac{1 + \cos 76°}{2}}$ **41.** $\sqrt{\frac{1 - \cos 147°}{1 + \cos 147°}}$

**42.** $\sqrt{\frac{1 + \cos 165°}{1 - \cos 165°}}$ **43.** $\frac{1 - \cos 59.74°}{\sin 59.74°}$ **44.** $\frac{\sin 158.2°}{1 + \cos 158.2°}$

*Verify that each equation is an identity:*

**45.** $\sin 4\alpha = 4 \sin \alpha \cos \alpha \cos 2\alpha$

**46.** $\frac{1 + \cos 2x}{\sin 2x} = \cot x$

**47.** $\frac{2 \cos 2\alpha}{\sin 2\alpha} = \cot \alpha - \tan \alpha$

**48.** $\sin 4\gamma = 4 \sin \gamma \cos \gamma - 8 \sin^3 \gamma \cos \gamma$

**49.** $\sin 2\alpha \cos 2\alpha = \sin 2\alpha - 4 \sin^3 \alpha \cos \alpha$

**50.** $\cos 2x = \frac{1 - \tan^2 x}{1 + \tan^2 x}$

**51.** $\tan s + \cot s = 2 \csc 2s$

**52.** $\dfrac{\cot \alpha - \tan \alpha}{\cot \alpha + \tan \alpha} = \cos 2\alpha$

**53.** $\sec^2 \dfrac{x}{2} = \dfrac{2}{1 + \cos x}$

**54.** $\cot^2 \dfrac{x}{2} = \dfrac{(1 + \cos x)^2}{\sin^2 x}$

*Write each expression as a sum or difference of trigonometric functions.*

**55.** $2 \sin 58° \cos 102°$

**56.** $5 \cos 3x \cos 2x$

**57.** $2 \cos 85° \sin 140°$

**58.** $\sin 4x \sin 5x$

*Write each expression as a product of trigonometric functions.*

**59.** $\cos 4x - \cos 2x$

**60.** $\cos 5t + \cos 8t$

**61.** $\sin 25° + \sin(-48°)$

**62.** $\sin 102° - \sin 95°$

**63.** $\cos 4x + \cos 8x$

**64.** $\sin 9B - \sin 3B$

*(Modeling) Solve each problem.*

**65.** ***Railroad Curves*** In the United States, circular railroad curves are designated by the *degree of curvature,* the central angle subtended by a chord of 100 feet. (*Source:* Hay, W., *Railroad Engineering,* John Wiley and Sons, 1982.)

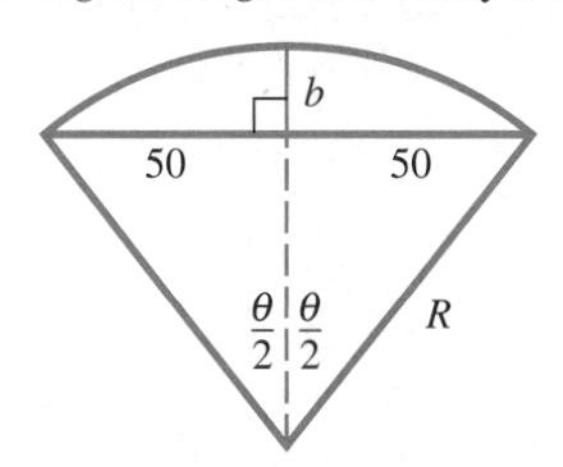


**(a)** Use the figure to write an expression for $\cos \frac{\theta}{2}$.

**(b)** Use the result of part (a) and the third half-number identity for tangent to write an expression for $\tan \frac{\theta}{4}$.

**(c)** If $b = 12$, what is the measure of angle $\theta$ to the nearest degree?

**66.** ***Distance Traveled by a Stone*** The distance $D$ of an object thrown (or propelled) from height $h$ (feet) at angle $\theta$ with initial velocity $v$ (feet per second) is modeled by the formula

$$D = \frac{v^2 \sin \theta \cos \theta + v \cos \theta \sqrt{(v \sin \theta)^2 + 64h}}{32}.$$

(*Source:* Kreighbaum, E. and K. Barthels, *Biomechanics,* Allyn & Bacon, 1996.)

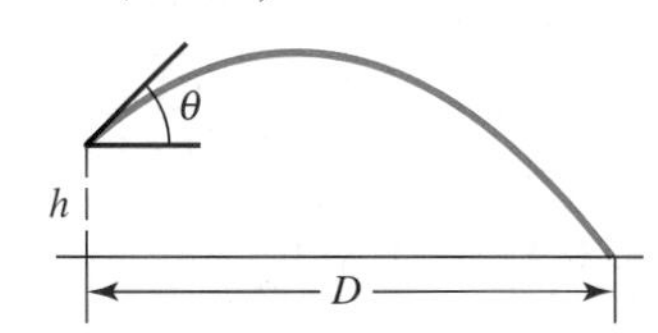


**(a)** Find $D$ when $h = 0$; that is, when the object is propelled from the ground.

**(b)** Suppose a car driving over loose gravel kicks up a small stone at a velocity of 36 feet per second (about 25 mph) and an angle $\theta = 30°$. How far will the stone travel?

**67.** ***Determining Wattage*** Amperage is a measure of the amount of electricity that is moving through a circuit, while voltage is a measure of the force pushing the electricity. The wattage $W$ consumed by an electrical device can be determined by calculating the product of amperage $I$ and voltage $V$. (*Source:* Wilcox, G. and C. Hesselberth, *Electricity for Engineering Technology,* Allyn & Bacon, 1970.)

**(a)** A household circuit has voltage

$$V = 163 \sin 120\pi t$$

when an incandescent light bulb is turned on with amperage

$$I = 1.23 \sin 120\pi t.$$

Graph the wattage $W = VI$ consumed by the light bulb over the interval $0 \le t \le .05$.

**(b)** Determine the maximum and minimum wattages used by the light bulb.

**(c)** Use identities to determine values for $a$, $c$, and $\omega$ so that $W = a \cos(\omega t) + c$.

**(d)** Check your answer in part (c) by graphing both expressions for $W$ on the same coordinate axes.

**(e)** Use the graph from part (a) to estimate the average wattage used by the light. How many watts do you think this incandescent light bulb is rated for?

**68.** ***Relating Voltage and Wattage*** Refer to Exercise 67. Suppose that voltage for an electric heater is given by

$$V = a \sin(2\pi\omega t)$$

and amperage by

$$I = b \sin(2\pi\omega t),$$

where $t$ is time in seconds.

**(a)** Find the period of the graph for the voltage.

**(b)** Show that the graph of the wattage $W = VI$ will have half the period of the voltage. Interpret this result.

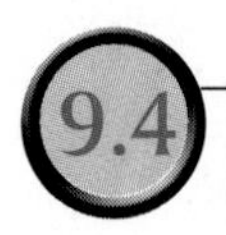

# 9.4 THE INVERSE CIRCULAR FUNCTIONS

Inverse Sine Function • Inverse Cosine Function • Inverse Tangent Function • Remaining Inverse Trigonometric Functions • Inverse Function Values

We discussed inverse functions in Section 5.1. We now apply those ideas to the circular (trigonometric) functions.

**Looking Ahead to Calculus**

The inverse trigonometric functions are used in calculus to express the solutions of trigonometric equations and integrate certain rational functions.

## Inverse Sine Function

From Figure 14 and the horizontal line test, note that the function defined by $y = \sin x$ is not one-to-one. By suitably restricting the domain of the sine function, however, a one-to-one function can be defined. We choose the interval $\left[-\frac{\pi}{2}, \frac{\pi}{2}\right]$ for this restriction. This gives the portion of the graph shown in color in Figure 14.

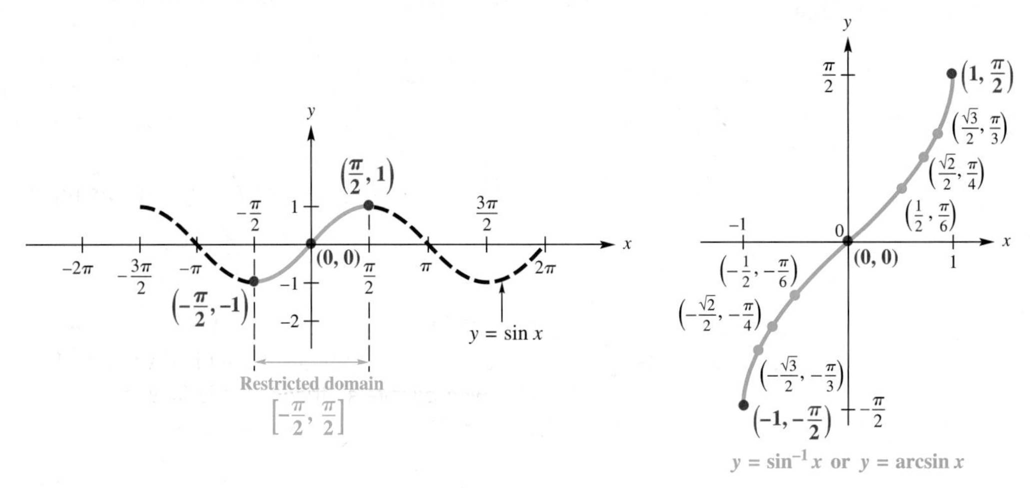


FIGURE 14 FIGURE 15

Reflecting the graph of $y = \sin x$ on the restricted domain across the line $y = x$ gives the graph of the inverse function, shown in Figure 15. Some key points are labeled on the graph. The equation of the inverse of $y = \sin x$ is found by interchanging $x$ and $y$ to get $x = \sin y$. This equation is solved for $y$ by writing $y = \sin^{-1} x$ (read "inverse sine of $x$"). As Figure 15 shows, the domain of $y = \sin^{-1} x$ is $[-1, 1]$, while the restricted domain of $y = \sin x$, $\left[-\frac{\pi}{2}, \frac{\pi}{2}\right]$, is the range of $y = \sin^{-1} x$. An alternative notation for $\sin^{-1} x$ is $\arcsin x$. Both notations will be used in this book.

**Inverse Sine Function**

$\boldsymbol{y = \sin^{-1} x}$ or $\boldsymbol{y = \arcsin x}$ means that $x = \sin y$, for $-\frac{\pi}{2} \le y \le \frac{\pi}{2}$.

We can think of $y = \sin^{-1} x$ or $y = \arcsin x$ as "$y$ is the number in the interval $\left[-\frac{\pi}{2}, \frac{\pi}{2}\right]$ whose sine is $x$."

**EXAMPLE 1** *Finding Inverse Sine Values*

Find $y$ in each equation.

**(a)** $y = \arcsin \frac{1}{2}$ **(b)** $y = \sin^{-1}(-1)$ **(c)** $y = \sin^{-1}(-2)$

**Analytic Solution**

**(a)** The graph of the function defined by $y = \arcsin x$ (Figure 15) shows that the point $\left(\frac{1}{2}, \frac{\pi}{6}\right)$ lies on the graph. Therefore,

$$\arcsin \frac{1}{2} = \frac{\pi}{6}.$$

Alternatively, we can think of $y = \arcsin \frac{1}{2}$ as "$y$ is the number in $\left[-\frac{\pi}{2}, \frac{\pi}{2}\right]$ whose sine is $\frac{1}{2}$." Then we can rewrite the equation as $\sin y = \frac{1}{2}$. Since $\sin \frac{\pi}{6} = \frac{1}{2}$ and $\frac{\pi}{6}$ is in the range of the arcsin function, $y = \frac{\pi}{6}$.

**(b)** Writing the alternative equation, $\sin y = -1$, shows that $y = -\frac{\pi}{2}$. This can be verified by noticing that the point $\left(-1, -\frac{\pi}{2}\right)$ is on the graph of $y = \sin^{-1} x$.

**(c)** Because $-2$ is not in the domain of the inverse sine function, $\sin^{-1}(-2)$ does not exist.

**Graphing Calculator Solution**

To find these values with a graphing calculator, we graph $Y_1 = \sin^{-1} X$ and locate the points with X-values $\frac{1}{2}$ and $-1$. Figure 16(a) shows that when $X = \frac{1}{2}$, $Y = \frac{\pi}{6} \approx .52359878$. Similarly, Figure 16(b) shows that when $X = -1$, $Y = -\frac{\pi}{2} \approx -1.570796$.

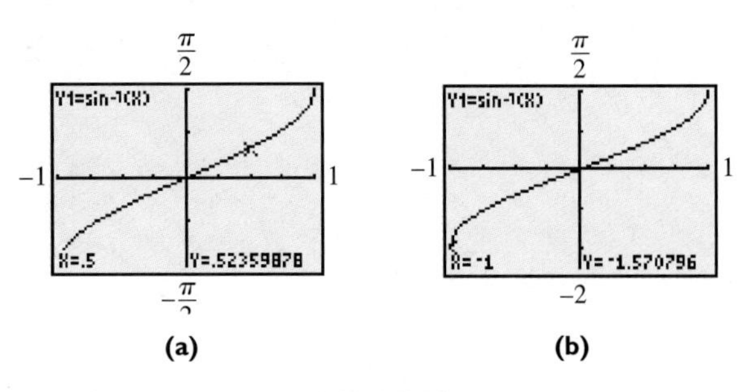


**FIGURE 16**

Since $\sin^{-1}(-2)$ does not exist, a calculator will give an error message for this input.

**CAUTION** In Example 1(b), it is tempting to give the value of $\sin^{-1}(-1)$ as $\frac{3\pi}{2}$, since $\sin \frac{3\pi}{2} = -1$. Notice, however, that $\frac{3\pi}{2}$ is not in the range of the inverse sine function. Be certain (in dealing with *all* inverse trigonometric functions) that the number given for an inverse function value is in the range of the particular inverse function being considered.

**FUNCTION CAPSULE**

**INVERSE SINE FUNCTION** $\quad y = \sin^{-1} x$ or $y = \arcsin x$

Domain: $[-1, 1]$ $\qquad$ Range: $\left[-\frac{\pi}{2}, \frac{\pi}{2}\right]$

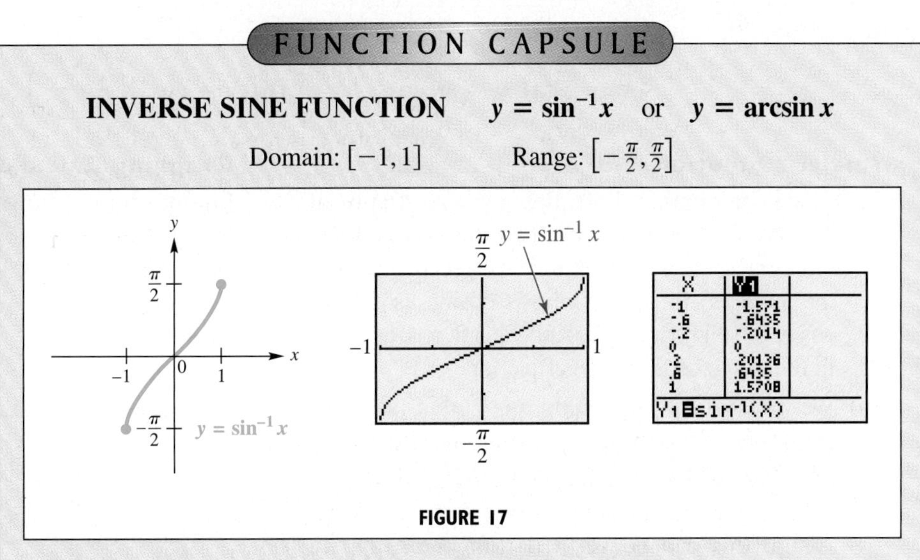


**FIGURE 17**

- The inverse sine function is increasing and continuous on its domain $[-1, 1]$.
- Its $x$-intercept is 0, and its $y$-intercept is 0.
- Its graph is symmetric with respect to the origin, so $\sin^{-1}(-x) = -\sin^{-1} x$.

**FOR DISCUSSION**

Earlier in this section, we stated that $y = \sin^{-1} x$ means "$y$ is the number in the interval $\left[-\frac{\pi}{2}, \frac{\pi}{2}\right]$ whose sine is $x$." Make a similar statement for

$$y = \cos^{-1} x.$$

## Inverse Cosine Function

The function $y = \cos^{-1} x$ (or $y = \arccos x$) is defined by restricting the domain of the function $y = \cos x$ to the interval $[0, \pi]$ as in Figure 18, and then reversing the roles of $x$ and $y$. See Figure 18. The graph of $y = \cos^{-1} x$ is shown in Figure 19. Again, some key points are shown on the graph.

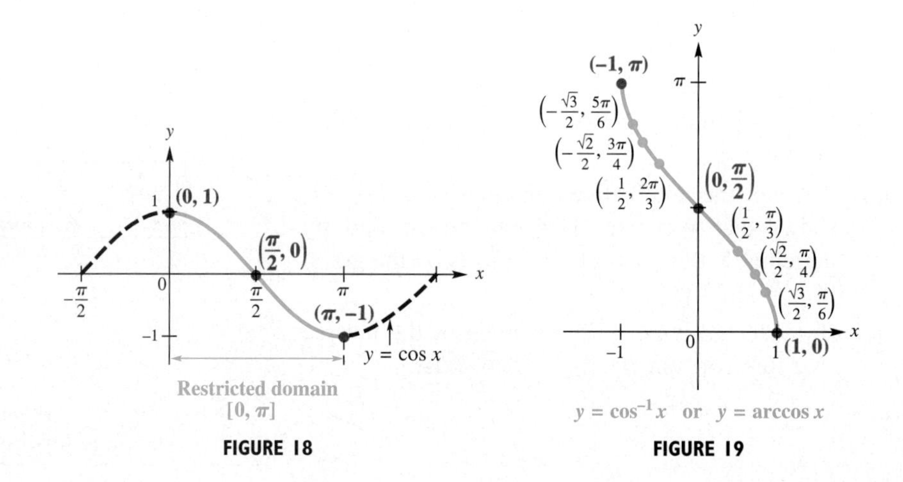


FIGURE 18

FIGURE 19

**Inverse Cosine Function**

$y = \cos^{-1} x$ or $y = \arccos x$ means that $x = \cos y$, for $0 \le y \le \pi$.

### EXAMPLE 2 *Finding Inverse Cosine Values*

Find $y$ in each equation.

**(a)** $y = \arccos 1$ **(b)** $y = \cos^{-1}\left(-\frac{\sqrt{2}}{2}\right)$

**Analytic Solution**

**(a)** Since the point $(1, 0)$ lies on the graph of $y = \arccos x$ in Figure 19, the value of $y$ is 0. Alternatively, we can think of $y = \arccos 1$ as "$y$ is the number in $[0, \pi]$ whose cosine is 1," or $\cos y = 1$. Then $y = 0$, since $\cos 0 = 1$ and 0 is in the range of the arccos function.

**(b)** We must find the value of $y$ that satisfies $\cos y = -\frac{\sqrt{2}}{2}$, where $y$ is in the interval $[0, \pi]$, the range of the function $y = \cos^{-1} x$. The only value for $y$ that satisfies these conditions is $\frac{3\pi}{4}$. Again, this can be verified from the graph in Figure 19.

**Graphing Calculator Solution**

Figure 20(a) shows that when $X = 1$, $Y = 0$ on the graph of $Y_1 = \cos^{-1} X$. Similarly, Figure 20(b) shows that when $X = -\frac{\sqrt{2}}{2} \approx -.7071068$, $Y = \frac{3\pi}{4} \approx 2.3561945$.

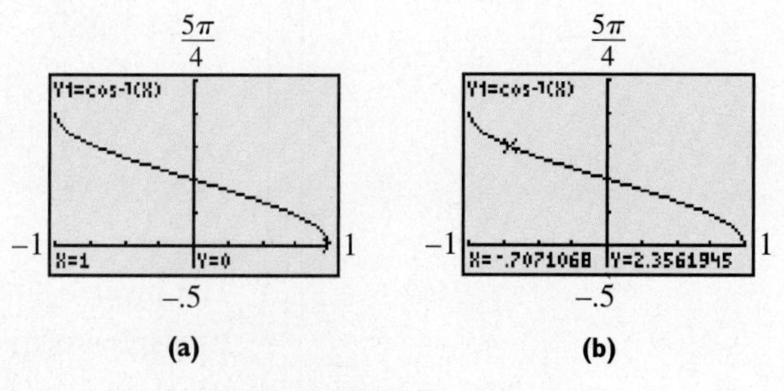


FIGURE 20

**FUNCTION CAPSULE**

**INVERSE COSINE FUNCTION** $\quad y = \cos^{-1} x \quad$ or $\quad y = \arccos x$

Domain: $[-1, 1]$ $\qquad$ Range: $[0, \pi]$

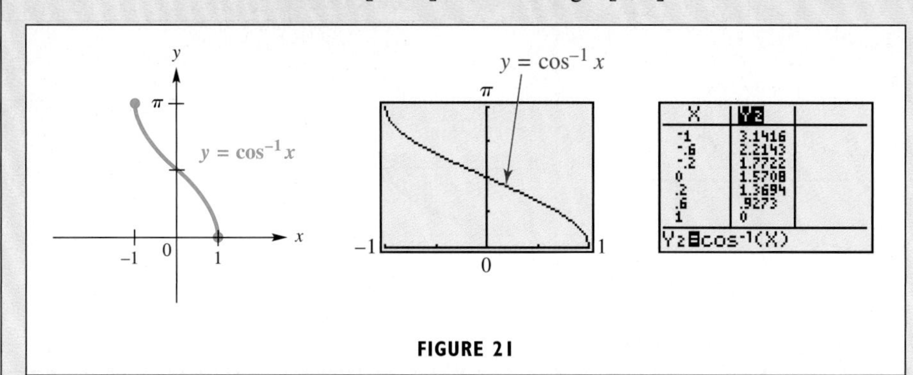


**FIGURE 21**

- The inverse cosine function is decreasing and continuous on its domain $[-1, 1]$.
- Its $x$-intercept is 1, and its $y$-intercept is $\frac{\pi}{2}$.
- Its graph is not symmetric with respect to the $y$-axis or the origin.

## Inverse Tangent Function

Restricting the domain of the function $y = \tan x$ to the open interval $\left(-\frac{\pi}{2}, \frac{\pi}{2}\right)$ yields a one-to-one function. By interchanging the roles of $x$ and $y$, we obtain the inverse tangent function given by $y = \tan^{-1} x$ or $y = \arctan x$. Figure 22 shows the graph of the restricted tangent function. Figure 23 gives the graph of $y = \tan^{-1} x$.

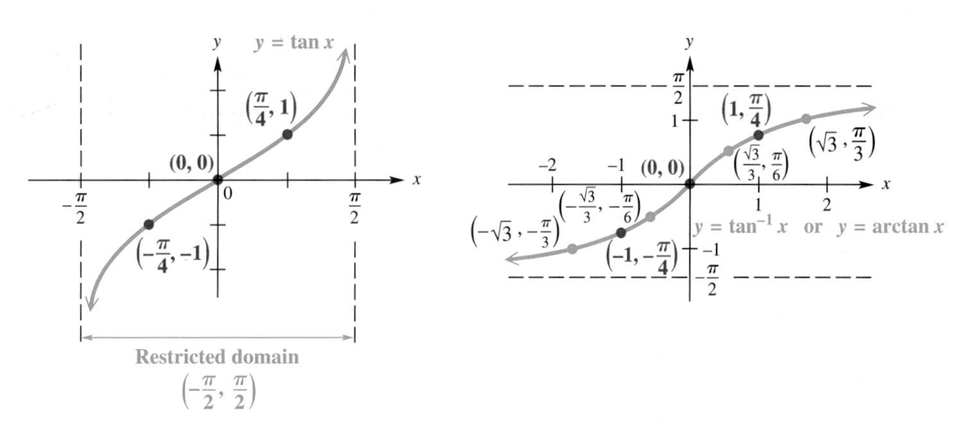


**FIGURE 22** $\qquad$ **FIGURE 23**

**Inverse Tangent Function**

$y = \tan^{-1} x$ or $y = \arctan x$ means $x = \tan y$, for $-\frac{\pi}{2} < y < \frac{\pi}{2}$.

FUNCTION CAPSULE

**INVERSE TANGENT FUNCTION** $y = \tan^{-1} x$ or $y = \arctan x$

Domain: $(-\infty, \infty)$ Range: $\left(-\frac{\pi}{2}, \frac{\pi}{2}\right)$

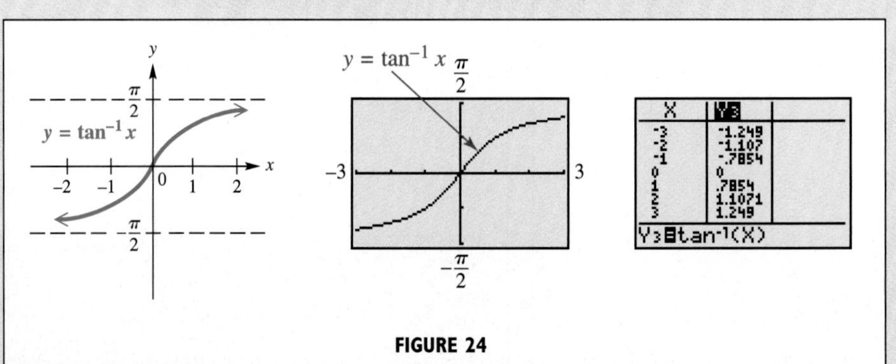


FIGURE 24

- The inverse tangent function is increasing and continuous on its domain $(-\infty, \infty)$.
- Its $x$-intercept is 0, and its $y$-intercept is 0.
- Its graph is symmetric with respect to the origin, so $\tan^{-1}(-x) = -\tan^{-1} x$.
- The lines $y = \frac{\pi}{2}$ and $y = -\frac{\pi}{2}$ are horizontal asymptotes.

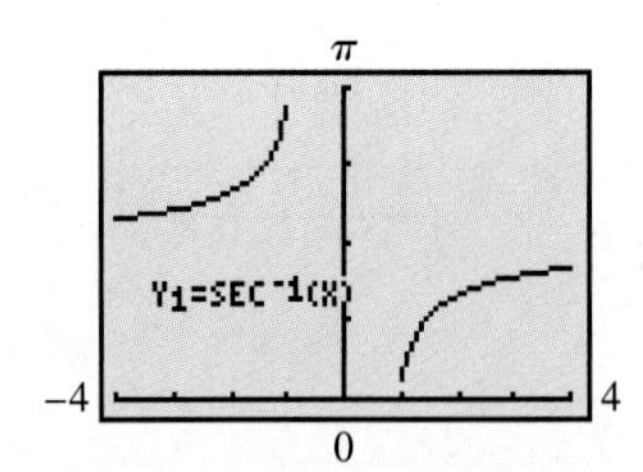


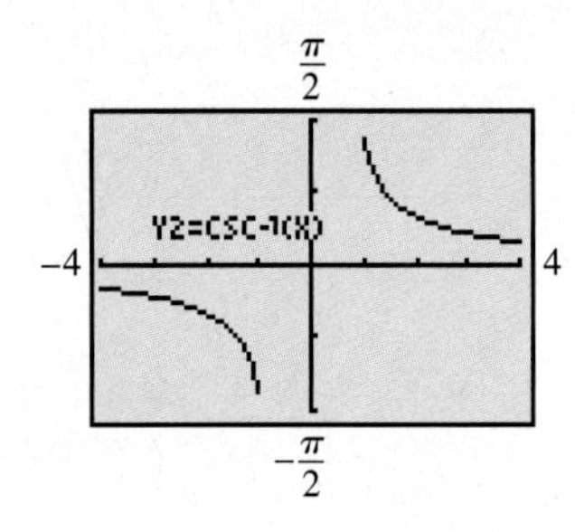


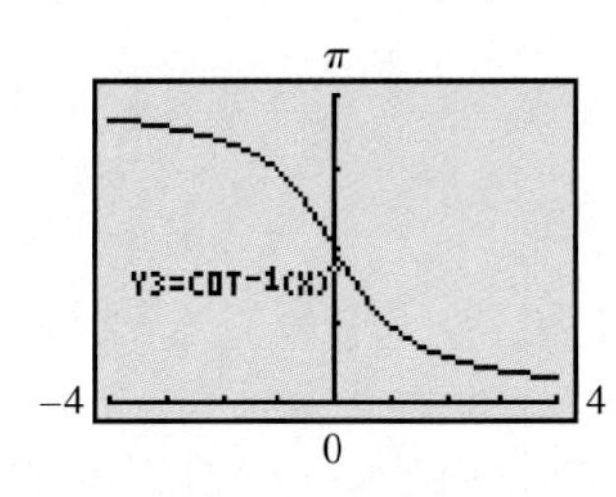


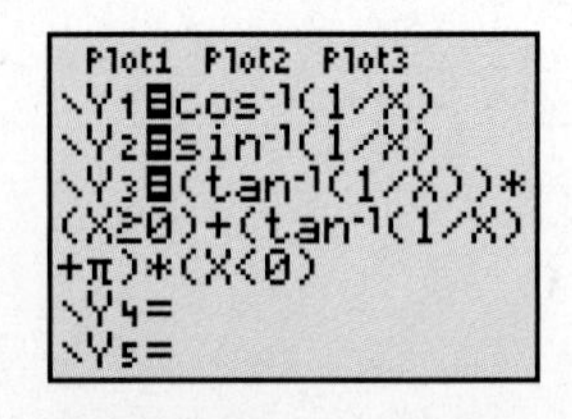


The first three screens show the graphs of the three remaining inverse circular functions. The last screen shows how they are defined.

FIGURE 25

## Remaining Inverse Trigonometric Functions

The remaining three inverse trigonometric functions are defined similarly; their graphs are shown in Figure 25. All six inverse trigonometric functions with their domains and ranges are given in the table.*

| Function | Domain | Range | |
|---|---|---|---|
| | | Interval | Quadrants of the Unit Circle |
| $y = \sin^{-1} x$ | $[-1, 1]$ | $\left[-\frac{\pi}{2}, \frac{\pi}{2}\right]$ | I and IV |
| $y = \cos^{-1} x$ | $[-1, 1]$ | $[0, \pi]$ | I and II |
| $y = \tan^{-1} x$ | $(-\infty, \infty)$ | $\left(-\frac{\pi}{2}, \frac{\pi}{2}\right)$ | I and IV |
| $y = \cot^{-1} x$ | $(-\infty, \infty)$ | $(0, \pi)$ | I and II |
| $y = \sec^{-1} x$ | $(-\infty, -1] \cup [1, \infty)$ | $[0, \pi], y \neq \frac{\pi}{2}$ | I and II |
| $y = \csc^{-1} x$ | $(-\infty, -1] \cup [1, \infty)$ | $\left[-\frac{\pi}{2}, \frac{\pi}{2}\right], y \neq 0$ | I and IV |

## Inverse Function Values

The inverse circular functions are formally defined with real number ranges. However, there are times when it may be convenient to find degree-measured angles equivalent to these real number values. It is also often convenient to think in terms of the unit

*The inverse secant and inverse cosecant functions are sometimes defined with different ranges. We use intervals that match their reciprocal functions (except for one missing point).

circle and choose the inverse function values based on the quadrants given in the preceding table.

**EXAMPLE 3** *Finding Inverse Function Values (Degree-Measured Angles)*
Find the *degree measure* of $\theta$ in the following.

**(a)** $\theta$, if $\theta = \arctan 1$ **(b)** $\theta$, if $\theta = \sec^{-1} 2$

**Solution**

**(a)** Here $\theta$ must be in $(-90°, 90°)$, but since $1 > 0$, $\theta$ must be in quadrant I. The alternative statement, $\tan \theta = 1$, leads to $\theta = 45°$.

**(b)** Write the equation as $\sec \theta = 2$. For $\sec^{-1} x$, $\theta$ is in quadrant I or II. Because 2 is positive, $\theta$ is in quadrant I and $\theta = 60°$, since $\sec 60° = 2$. Note that $60°$ (the degree equivalent of $\frac{\pi}{3}$) is in the range of the inverse secant function.

The inverse trigonometric function keys on a calculator give results in the proper quadrant for the inverse sine, inverse cosine, and inverse tangent functions, according to the definitions of these functions. For example, on a calculator, in degrees,

$$\sin^{-1} .5 = 30°, \qquad \sin^{-1}(-.5) = -30°,$$
$$\tan^{-1}(-1) = -45°, \quad \text{and} \quad \cos^{-1}(-.5) = 120°.$$

Finding $\cot^{-1} x$, $\sec^{-1} x$, and $\csc^{-1} x$ with a calculator is not as straightforward, because these functions must be expressed in terms of $\tan^{-1} x$, $\cos^{-1} x$, and $\sin^{-1} x$, respectively. If $y = \sec^{-1} x$, for example, then $\sec y = x$, which must be written as a cosine function as follows:

$$\text{If } \sec y = x, \text{ then } \frac{1}{\cos y} = x, \text{ or } \cos y = \frac{1}{x}.$$

From this last statement,

$$y = \cos^{-1} \frac{1}{x}.$$

In summary, to find $\sec^{-1} x$, we find $\cos^{-1} \frac{1}{x}$. Similar statements apply to $\csc^{-1} x$ and $\cot^{-1} x$. There is one additional consideration with $\cot^{-1} x$. Since we take the inverse tangent of the reciprocal to find inverse cotangent, the calculator gives values of inverse cotangent with the same range as inverse tangent, $\left(-\frac{\pi}{2}, \frac{\pi}{2}\right)$, which is not the correct range for inverse cotangent. For inverse cotangent, the proper range must be considered and the results adjusted accordingly.

**EXAMPLE 4** *Finding Inverse Function Values with a Calculator*
**(a)** Find $y$ in radians if $y = \csc^{-1}(-3)$.
**(b)** Find $\theta$ in degrees if $\theta = \text{arccot}(-.3541)$.

FIGURE 26

**Solution**

**(a)** With the calculator in radian mode, enter $\csc^{-1}(-3)$ as $\sin^{-1}\left(-\frac{1}{3}\right)$ to get $y \approx -.3398369095$. See Figure 26.

**(b)** Set the calculator in degree mode. A calculator gives the inverse tangent value of a negative number as a quadrant IV angle. The restriction on the range of arccot means that $\theta$ must be in quadrant II, so enter

$$\text{arccot}(-.3541) \quad \text{as} \quad \tan^{-1}\left(\frac{1}{-.3541}\right) + 180°.$$

As shown in Figure 26 on the previous page, $\theta \approx 109.4990544°$.

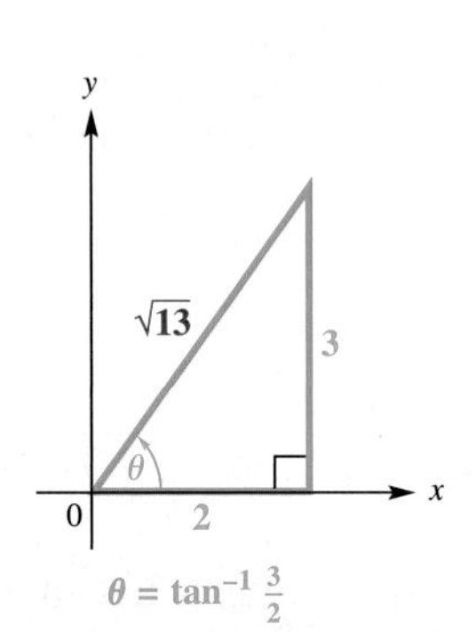


FIGURE 27

## EXAMPLE 5 *Finding Function Values Using Definitions of the Trigonometric Functions*

Evaluate each expression without using a calculator.

**(a)** $\sin\left(\tan^{-1} \frac{3}{2}\right)$ **(b)** $\tan\left(\cos^{-1}\left(-\frac{5}{13}\right)\right)$

**Solution**

**(a)** Let $\theta = \tan^{-1} \frac{3}{2}$ so $\tan \theta = \frac{3}{2}$. The inverse tangent function yields values only in quadrants I and IV, and since $\frac{3}{2}$ is positive, $\theta$ is in quadrant I. Sketch $\theta$ in quadrant I, and label a triangle, as shown in Figure 27. The hypotenuse is $\sqrt{13}$ and the value of sine is the quotient of the side opposite and the hypotenuse, so

$$\sin\left(\tan^{-1} \frac{3}{2}\right) = \sin \theta = \frac{3}{\sqrt{13}} = \frac{3\sqrt{13}}{13}.$$

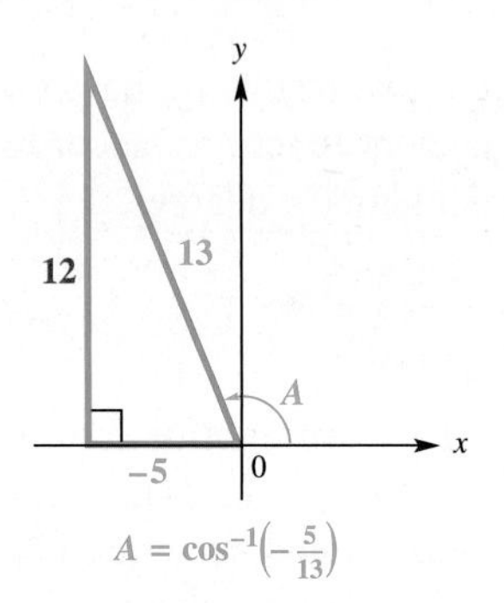


FIGURE 28

**(b)** Let $A = \cos^{-1}\left(-\frac{5}{13}\right)$. Then, $\cos A = -\frac{5}{13}$. Since $\cos^{-1} x$ for a negative value of $x$ is in quadrant II, sketch $A$ in quadrant II, as shown in Figure 28. From the triangle in Figure 28,

$$\tan\left(\cos^{-1}\left(-\frac{5}{13}\right)\right) = \tan A = -\frac{12}{5}.$$

## EXAMPLE 6 *Finding Function Values Using Identities*

Evaluate each expression without using a calculator.

**(a)** $\cos\left(\arctan \sqrt{3} + \arcsin \frac{1}{3}\right)$ **(b)** $\tan\left(2 \arcsin \frac{2}{5}\right)$

**Solution**

**(a)** Let $A = \arctan \sqrt{3}$ and $B = \arcsin \frac{1}{3}$, so $\tan A = \sqrt{3}$ and $\sin B = \frac{1}{3}$. Sketch both $A$ and $B$ in quadrant I, as shown in Figure 29. Now, use the identity for $\cos(A + B)$.

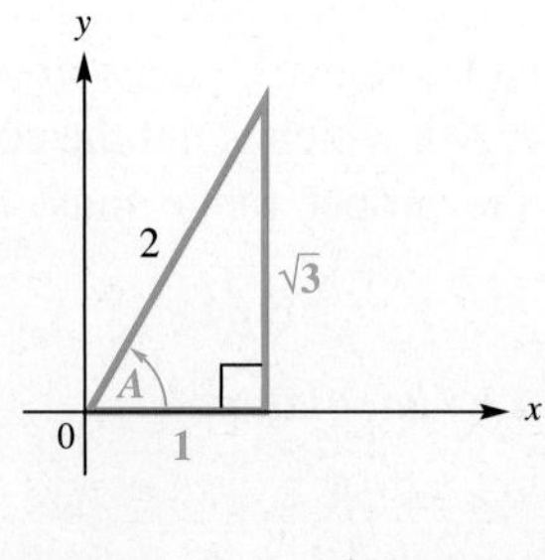


$$\cos(A + B) = \cos A \cos B - \sin A \sin B$$

$$\cos\left(\arctan \sqrt{3} + \arcsin \frac{1}{3}\right) = \cos\left(\arctan \sqrt{3}\right) \cos\left(\arcsin \frac{1}{3}\right) - \sin\left(\arctan \sqrt{3}\right) \sin\left(\arcsin \frac{1}{3}\right) \quad (1)$$

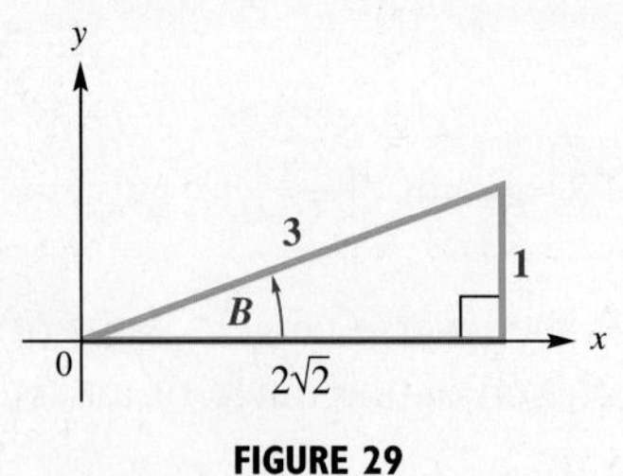


FIGURE 29

From Figure 29,

$$\cos\left(\arctan \sqrt{3}\right) = \cos A = \frac{1}{2}, \quad \cos\left(\arcsin \frac{1}{3}\right) = \cos B = \frac{2\sqrt{2}}{3},$$

$$\sin\left(\arctan \sqrt{3}\right) = \sin A = \frac{\sqrt{3}}{2}, \quad \sin\left(\arcsin \frac{1}{3}\right) = \sin B = \frac{1}{3}.$$

Substitute these values into equation (1) to get

$$\cos\left(\arctan\sqrt{3} + \arcsin\frac{1}{3}\right) = \frac{1}{2}\cdot\frac{2\sqrt{2}}{3} - \frac{\sqrt{3}}{2}\cdot\frac{1}{3} = \frac{2\sqrt{2} - \sqrt{3}}{6}.$$

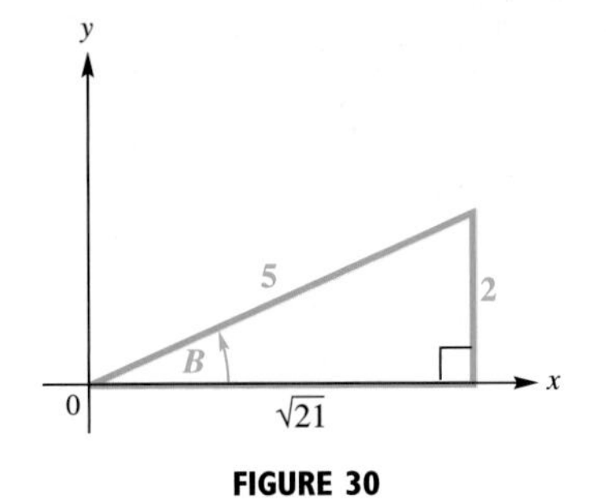

FIGURE 30

**(b)** Let $\arcsin\frac{2}{5} = B$. Then, from the double-number tangent identity,

$$\tan\left(2\arcsin\frac{2}{5}\right) = \tan 2B = \frac{2\tan B}{1 - \tan^2 B}.$$

Since $\arcsin\frac{2}{5} = B$, $\sin B = \frac{2}{5}$. Sketch a triangle in quadrant I, find the length of the third side, and then find $\tan B$. From the triangle in Figure 30, $\tan B = \frac{2}{\sqrt{21}}$, and

$$\tan\left(2\arcsin\frac{2}{5}\right) = \frac{2\left(\frac{2}{\sqrt{21}}\right)}{1 - \left(\frac{2}{\sqrt{21}}\right)^2} = \frac{\frac{4}{\sqrt{21}}}{1 - \frac{4}{21}} = \frac{\frac{4}{\sqrt{21}}}{\frac{17}{21}} = \frac{4\sqrt{21}}{17}.$$

While the work shown in Examples 5 and 6 does not rely on a calculator, we can support our analytic work with one. By entering $\cos\left(\arctan\sqrt{3} + \arcsin\frac{1}{3}\right)$ from Example 6(a) into a calculator, we get the approximation .1827293862, the same approximation as when we enter $\frac{2\sqrt{2} - \sqrt{3}}{6}$ (the exact value obtained analytically). Similarly, we obtain the same approximation when we evaluate $\tan\left(2\arcsin\frac{2}{5}\right)$ and $\frac{4\sqrt{21}}{17}$, supporting our answer in Example 6(b).

## EXAMPLE 7 *Writing Function Values in Terms of u*

Write each expression as an algebraic expression in $u$.

**(a)** $\sin(\tan^{-1} u)$ **(b)** $\cos(2\sin^{-1} u)$

### Solution

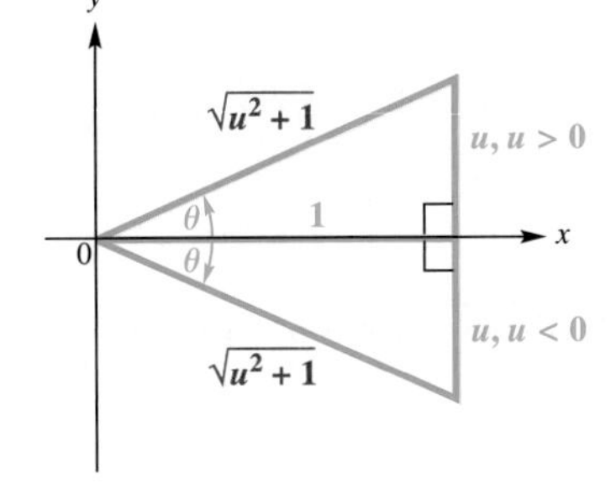

FIGURE 31

**(a)** Let $\theta = \tan^{-1} u$, so $\tan\theta = u$. Here, $u$ may be positive or negative. Since $-\frac{\pi}{2} < \tan^{-1} u < \frac{\pi}{2}$, sketch $\theta$ in quadrants I and IV and label two triangles, as shown in Figure 31. Since sine is given by the quotient of the side opposite and the hypotenuse,

$$\sin(\tan^{-1} u) = \sin\theta = \frac{u}{\sqrt{u^2 + 1}} = \frac{u\sqrt{u^2 + 1}}{u^2 + 1}.$$

The result is positive when $u$ is positive and negative when $u$ is negative.

**(b)** Let $\theta = \sin^{-1} u$, so $\sin\theta = u$. To find $\cos 2\theta$, use the identity $\cos 2\theta = 1 - 2\sin^2\theta$.

$$\cos 2\theta = 1 - 2\sin^2\theta = 1 - 2u^2$$

**What Went WRONG?**

A student found sin 1.74 on her calculator (set for radians). She then found the inverse sine of the answer, but got 1.401592654 instead of 1.74, which she expected.

What Went Wrong (if anything)?

*Answer to What Went Wrong?*

Her thinking was incorrect. 1.74 is not in the range $\left[-\frac{\pi}{2}, \frac{\pi}{2}\right]$ of the inverse sine function. Her answer is actually $\pi - 1.74$, which places the result in the correct interval. Note that $\frac{\pi}{2} \approx 1.57$.

**EXAMPLE 8** *Finding the Optimal Angle of Elevation of a Shot Put*

The optimal angle of elevation $\theta$ for a shot-putter should aim for to throw the greatest distance depends on the velocity $v$ of the throw and the initial height $h$ of the shot. See Figure 32. One model for $\theta$ that achieves this greatest distance is

$$\theta = \arcsin\left(\sqrt{\frac{v^2}{2v^2 + 64h}}\right).$$

(*Source:* Townend, M. Stewart, *Mathematics in Sport,* Chichester, Ellis Horwood Limited, 1984.)

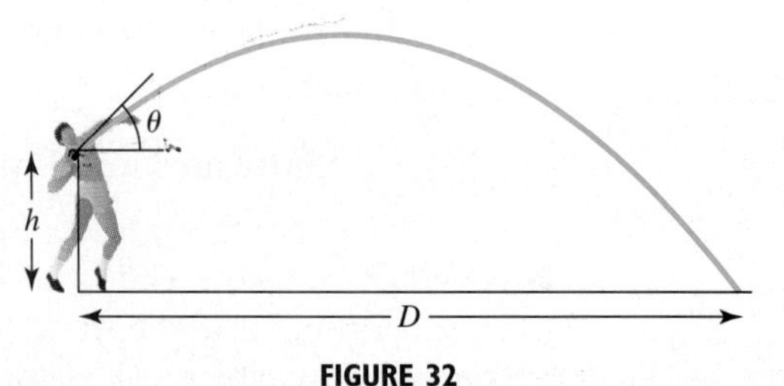


**FIGURE 32**

Suppose a shot-putter can consistently throw the steel ball with $h = 7.6$ feet and $v = 42$ feet per second. At what angle should he throw the ball to maximize distance?

**Solution** To find this angle, substitute and use a calculator in degree mode.

$$\theta = \arcsin\left(\sqrt{\frac{42^2}{2(42^2) + 64(7.6)}}\right) \approx 41.5° \qquad h = 7.6,\ v = 42$$

## 9.4 EXERCISES

*Concept Check* *In Exercises 1–4, write short answers and fill in the blanks.*

**1.** Consider the inverse sine function, defined by $y = \sin^{-1} x$ or $y = \arcsin x$.

**(a)** What is its domain?

**(b)** What is its range?

**(c)** For this function, as $x$ increases, $y$ increases. Therefore, it is a(n) ________________ function. (decreasing/increasing)

**(d)** Why is $\arcsin(-2)$ not defined?

**2.** Consider the inverse cosine function, defined by $y = \cos^{-1} x$ or $y = \arccos x$.

**(a)** What is its domain?

**(b)** What is its range?

**(c)** For this function, as $x$ increases, $y$ decreases. Therefore, it is a(n) ________________ function. (decreasing/increasing)

**(d)** $\text{Arccos}\left(-\frac{1}{2}\right) = \frac{2\pi}{3}$. Why is $\arccos\left(-\frac{1}{2}\right)$ not equal to $-\frac{4\pi}{3}$?

**3.** Consider the inverse tangent function, defined by $y = \tan^{-1} x$ or $y = \arctan x$.

**(a)** What is its domain?

**(b)** What is its range?

**(c)** For this function, as $x$ increases, $y$ increases. Therefore, it is a(n) ________________ function. (decreasing/increasing)

**(d)** Is there any real number $x$ for which $\arctan x$ is not defined? If so, what is it (or where are they)?

**4.** Consider the three other inverse trigonometric functions, as defined in this section.

**(a)** Give the domain and range of the inverse cosecant function.

**(b)** Give the domain and range of the inverse secant function.

**(c)** Give the domain and range of the inverse cotangent function.

5. *Concept Check* Is $\sec^{-1} a$ calculated as $\cos^{-1} \frac{1}{a}$ or as $\frac{1}{\cos^{-1} a}$?

6. *Concept Check* For positive values of $a$, $\cot^{-1} a$ is calculated as $\tan^{-1} \frac{1}{a}$. How is $\cot^{-1} a$ calculated for negative values of $a$?

*Find the exact value of each real number y. Do not use a calculator.*

7. $y = \tan^{-1} 1$
8. $y = \sin^{-1} 0$
9. $y = \cos^{-1}(-1)$
10. $y = \arctan(-1)$
11. $y = \sin^{-1}(-1)$
12. $y = \cos^{-1} \frac{1}{2}$
13. $y = \arctan 0$
14. $y = \arcsin\left(-\frac{\sqrt{3}}{2}\right)$
15. $y = \arccos 0$
16. $y = \tan^{-1}(-1)$
17. $y = \sin^{-1} \frac{\sqrt{2}}{2}$
18. $y = \cos^{-1}\left(-\frac{1}{2}\right)$
19. $y = \arccos\left(-\frac{\sqrt{3}}{2}\right)$
20. $y = \arcsin\left(-\frac{\sqrt{2}}{2}\right)$
21. $y = \cot^{-1}(-1)$
22. $y = \sec^{-1}(-\sqrt{2})$
23. $y = \csc^{-1}(-2)$
24. $y = \text{arccot}(-\sqrt{3})$
25. $y = \text{arcsec} \frac{2\sqrt{3}}{3}$
26. $y = \csc^{-1} \sqrt{2}$

*Give the degree measure of θ. Do not use a calculator.*

27. $\theta = \arctan(-1)$
28. $\theta = \arccos\left(-\frac{1}{2}\right)$
29. $\theta = \arcsin\left(-\frac{\sqrt{3}}{2}\right)$
30. $\theta = \arcsin\left(-\frac{\sqrt{2}}{2}\right)$
31. $\theta = \cot^{-1}\left(-\frac{\sqrt{3}}{3}\right)$
32. $\theta = \sec^{-1}(-2)$
33. $\theta = \csc^{-1}(-2)$
34. $\theta = \csc^{-1}(-1)$

*Use a calculator to give each value in decimal degrees.*

35. $\theta = \sin^{-1}(-.13349122)$
36. $\theta = \cos^{-1}(-.13348816)$
37. $\theta = \arccos(-.39876459)$
38. $\theta = \arcsin .77900016$
39. $\theta = \csc^{-1} 1.9422833$
40. $\theta = \cot^{-1} 1.7670492$

*Use a calculator to give each real number value. (Be sure the calculator is in radian mode.)*

41. $y = \arctan 1.1111111$
42. $y = \arcsin .81926439$
43. $y = \cot^{-1}(-.92170128)$
44. $y = \sec^{-1}(-1.2871684)$
45. $y = \arcsin .92837781$
46. $y = \arccos .44624593$

*Draw by hand the graph of each inverse function as defined in the text.*

47. $y = \cot^{-1} x$
48. $y = \csc^{-1} x$
49. $y = \sec^{-1} x$
50. $y = \text{arccsc } 2x$
51. $y = \text{arcsec } \frac{1}{2}x$

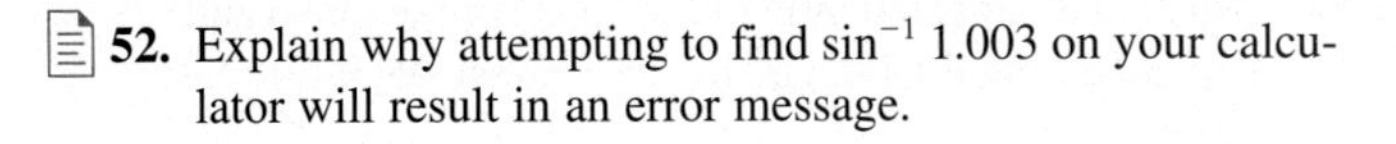

52. Explain why attempting to find $\sin^{-1} 1.003$ on your calculator will result in an error message.

53. Explain why you are able to find $\tan^{-1} 1.003$ on your calculator. Why is this situation different from the one described in Exercise 52?

## RELATING CONCEPTS

**For individual or group investigation (Exercises 54–56)***

54. Consider the function defined by $f(x) = 3x - 2$ and its inverse $f^{-1}(x) = \frac{x+2}{3}$. Simplify $f[f^{-1}(x)]$ and $f^{-1}[f(x)]$. What do you notice in each case? What would the graph look like in each case?

55. Use a graphing calculator to graph $y = \tan(\tan^{-1} x)$ in the standard viewing window, using radian mode. How does this compare to the graph you described in Exercise 54?

56. Use a graphing calculator to graph $y = \tan^{-1}(\tan x)$ in the standard viewing window, using radian and dot modes. Why does this graph not agree with the graph you found in Exercise 55?

*The authors wish to thank Carol Walker of Hinds Community College for making a suggestion on which these exercises are based.

*Give the exact value of each expression without using a calculator.*

**57.** $\tan\left(\arccos \frac{3}{4}\right)$

**58.** $\sin\left(\arccos \frac{1}{4}\right)$

**59.** $\cos(\tan^{-1}(-2))$

**60.** $\sec\left(\sin^{-1}\left(-\frac{1}{5}\right)\right)$

**61.** $\sin\left(2 \tan^{-1} \frac{12}{5}\right)$

**62.** $\cos\left(2 \sin^{-1} \frac{1}{4}\right)$

**63.** $\cos\left(2 \arctan \frac{4}{3}\right)$

**64.** $\tan\left(2 \cos^{-1} \frac{1}{4}\right)$

**65.** $\sin\left(2 \cos^{-1} \frac{1}{5}\right)$

**66.** $\cos(2 \tan^{-1}(-2))$

**67.** $\sec(\sec^{-1} 2)$

**68.** $\csc(\csc^{-1} \sqrt{2})$

**69.** $\cos\left(\tan^{-1} \frac{5}{12} - \tan^{-1} \frac{3}{4}\right)$

**70.** $\cos\left(\sin^{-1} \frac{3}{5} + \cos^{-1} \frac{5}{13}\right)$

**71.** $\sin\left(\sin^{-1} \frac{1}{2} + \tan^{-1}(-3)\right)$

**72.** $\tan\left(\cos^{-1} \frac{\sqrt{3}}{2} - \sin^{-1}\left(-\frac{3}{5}\right)\right)$

*Use a calculator to find each value. Give answers as real numbers.*

**73.** $\cos(\tan^{-1} .5)$

**74.** $\sin(\cos^{-1} .25)$

**75.** $\tan(\arcsin .12251014)$

**76.** $\cot(\arccos .58236841)$

*Write each expression as an algebraic expression in $u$, $u > 0$.*

**77.** $\sin(\arccos u)$

**78.** $\tan(\arccos u)$

**79.** $\cot(\arcsin u)$

**80.** $\cos(\arcsin u)$

**81.** $\sin\left(\sec^{-1} \frac{u}{2}\right)$

**82.** $\cos\left(\tan^{-1} \frac{3}{u}\right)$

**83.** $\tan\left(\sin^{-1} \frac{u}{\sqrt{u^2 + 2}}\right)$

**84.** $\sec\left(\cos^{-1} \frac{u}{\sqrt{u^2 + 5}}\right)$

**85.** $\sec\left(\operatorname{arccot} \frac{\sqrt{4 - u^2}}{u}\right)$

**86.** $\csc\left(\arctan \frac{\sqrt{9 - u^2}}{u}\right)$

*(Modeling) Solve each problem.*

**87.** *Angle of Elevation of a Shot Put* Refer to Example 8.

**(a)** What is the optimal angle when $h = 0$?

**(b)** Fix $h$ at 6 feet and regard $\theta$ as a function of $v$. As $v$ gets larger and larger, the graph approaches a horizontal asymptote. Find the equation of that asymptote.

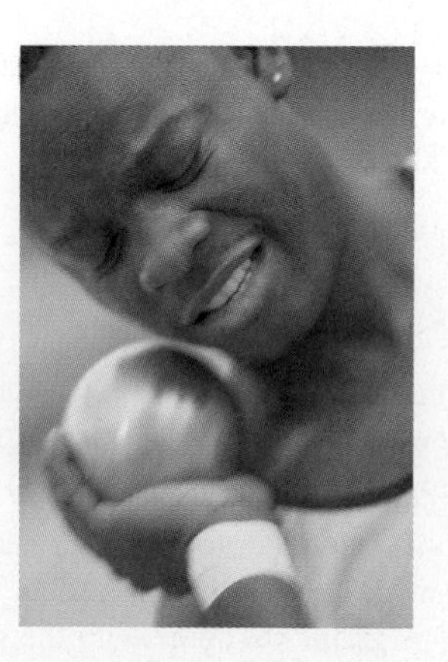

**88.** *Angle of Elevation of a Plane* Suppose an airplane flying faster than sound goes directly over you. Assume that the plane is flying at a constant altitude. At the instant you feel the sonic boom from the plane, the angle of elevation to the plane is given by

$$\alpha = 2 \arcsin \frac{1}{m},$$

where $m$ is the *Mach number* of the plane's speed. (The Mach number is the ratio of the speed of the plane and the speed of sound.) Find $\alpha$ to the nearest degree for each value of $m$.

**(a)** $m = 1.2$

**(b)** $m = 1.5$

**(c)** $m = 2$

**(d)** $m = 2.5$

**89.** *Viewing Angle of an Observer* A painting 3 feet high and 6 feet from the floor will cut off an angle $\theta$ to an observer, where

$$\theta = \tan^{-1}\left(\frac{3x}{x^2 + 4}\right).$$

Assume that the observer is $x$ feet from the wall where the painting is displayed and that the eyes of the observer are 5 feet above the ground. See the figure.

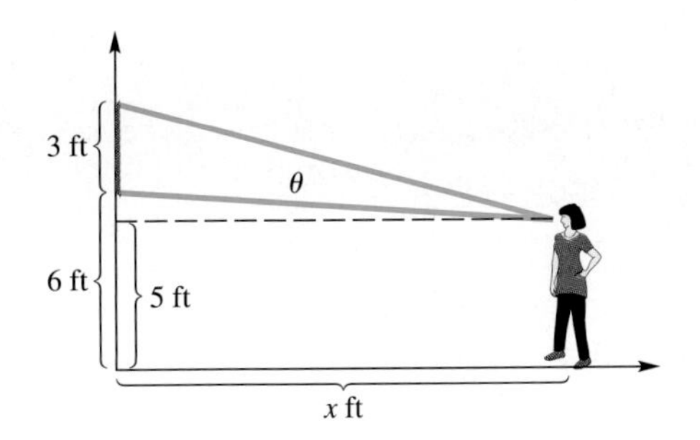


Find the value of $\theta$ for each value of $x$. Round to the nearest degree.

**(a)** $x = 3$ **(b)** $x = 6$ **(c)** $x = 9$

**(d)** Derive the formula given above. (*Hint:* Use right triangles and the identity for $\tan(\theta + \alpha)$.)

**(e)** Graph the function for $\theta$ with a graphing calculator and determine the distance that maximizes the angle.

**90.** *Landscaping Formula* A shrub is planted in a 100-foot wide space between buildings measuring 75 feet and 150 feet tall. The location of the shrub determines how much sun it receives each day. Show that if $\theta$ is the angle in the figure and $x$ is the distance of the shrub from the taller building, then the value of $\theta$ (in radians) is given by

$$\theta = \pi - \arctan\left(\frac{75}{100 - x}\right) - \arctan\left(\frac{150}{x}\right).$$

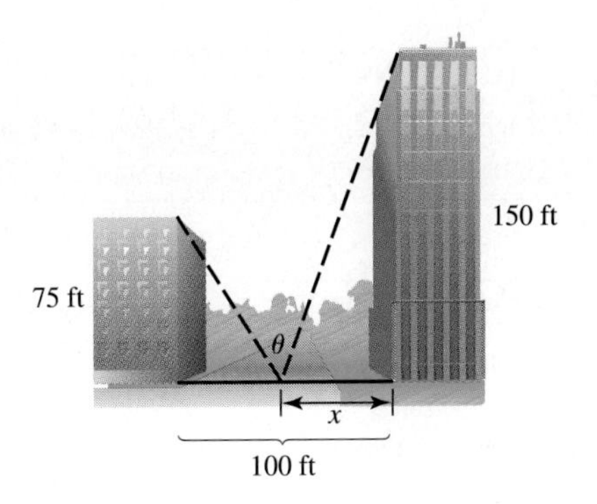


**91.** *Communications Satellite Coverage* The figure shows a stationary communications satellite positioned 20,000 miles above the equator. What percent of the equator can be seen from the satellite? The diameter of Earth is 7927 miles at the equator.

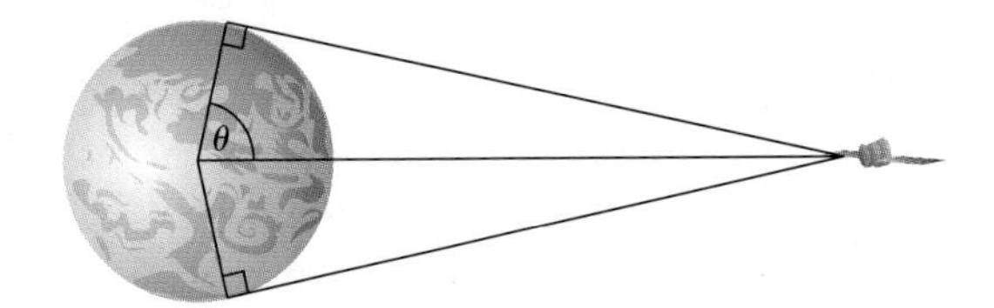


## Reviewing Basic Concepts (Sections 9.3 and 9.4)

*Use identities to work Exercises 1–5.*

**1.** Find $\tan x$, given $\cos 2x = -\frac{5}{12}$ and $\frac{\pi}{2} < x < \pi$.

**2.** Find $\sin 2\theta$, $\cos 2\theta$, and $\tan 2\theta$, if $\sin \theta = -\frac{1}{3}$, and $\theta$ is in quadrant III.

**3.** Find the exact value of $\sin 75°$.

**4.** Write $2 \sin 25° \cos 150°$ as a sum or difference of trigonometric functions.

**5.** Verify each identity.

**(a)** $\sin^2 \dfrac{x}{2} = \dfrac{\tan x - \sin x}{2 \tan x}$

**(b)** $\dfrac{\sin 2x}{2 \sin x} = \cos^2 \dfrac{x}{2} - \sin^2 \dfrac{x}{2}$

**6.** Find the exact value of each real number $y$.

**(a)** $y = \arccos \dfrac{\sqrt{3}}{2}$ **(b)** $y = \sin^{-1}\left(-\dfrac{\sqrt{2}}{2}\right)$

**7.** Find the exact value of each angle $\theta$ measured in degrees.

**(a)** $\theta = \arccos .5$ **(b)** $\theta = \cot^{-1}(-1)$

**8.** Graph $y = 2 \csc^{-1} x$ by hand.

**9.** Find the exact value of each expression.

**(a)** $\cot\left(\arcsin\left(-\dfrac{2}{3}\right)\right)$

**(b)** $\cos\left(\tan^{-1} \dfrac{5}{12} - \sin^{-1} \dfrac{3}{5}\right)$

**10.** Write $\sin(\text{arccot } u)$ as an algebraic expression in $u$, $u > 0$.

# TRIGONOMETRIC EQUATIONS AND INEQUALITIES (I)

Equations Solvable by Linear Methods • Equations Solvable by Factoring • Equations Solvable by the Quadratic Formula • Using Trigonometric Identities to Solve Equations

**Looking Ahead to Calculus**

There are many instances in calculus where it is necessary to solve trigonometric equations. Examples include finding values for which a derivative is 0, solving related rate problems, and solving optimization problems.

Earlier in this chapter, we studied trigonometric equations that were identities. We now consider trigonometric equations that are *conditional*; that is, equations that are satisfied by some values but not others.

## Equations Solvable by Linear Methods

Conditional equations with trigonometric (or circular) functions can usually be solved using analytic methods and trigonometric identities.

**EXAMPLE 1** *Solving a Trigonometric Equation by Linear Methods*

Solve $2 \sin x - 1 = 0$ over the interval $[0, 2\pi)$.

**Analytic Solution**

Since this equation involves the first power of $\sin x$, it is linear in $\sin x$. Thus, we solve it by using the usual method for solving a linear equation.

$$2 \sin x - 1 = 0$$

$$2 \sin x = 1 \qquad \text{Add 1.}$$

$$\sin x = \frac{1}{2} \qquad \text{Divide by 2.}$$

The two values of $x$ in the interval $[0, 2\pi)$ that have sine value $\frac{1}{2}$ are $\frac{\pi}{6}$ and $\frac{5\pi}{6}$. These are obtained by finding

$$\sin^{-1}\frac{1}{2}$$

and

$$\pi - \sin^{-1}\frac{1}{2}.$$

We can also determine this by using the unit circle or the reference angle analysis discussed in Chapter 8. Therefore, the solution set in the specified interval is $\{\frac{\pi}{6}, \frac{5\pi}{6}\}$.

**Graphing Calculator Solution**

We graph $y = 2 \sin x - 1$ over the desired interval and use the capabilities of the calculator to find that the $x$-intercepts have the same decimal approximations as $\frac{\pi}{6}$ and $\frac{5\pi}{6}$. See Figure 33.

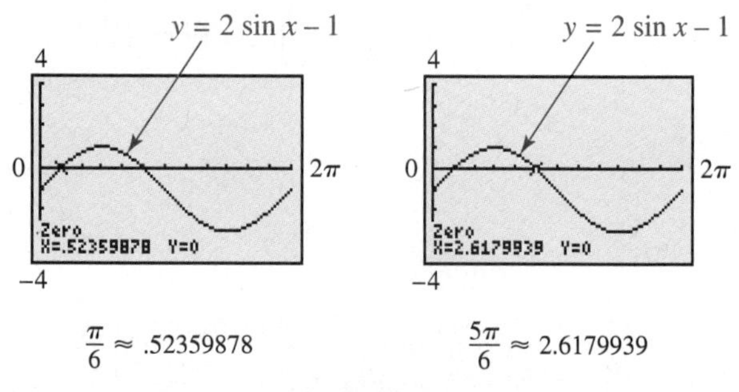


**FIGURE 33**

We used Xscl $= \frac{\pi}{6}$ in Figure 33 so we can see that the graph intersects the $x$-axis at the first and fifth tick marks representing $\frac{\pi}{6}$ and $\frac{5\pi}{6}$. This gives a quick method of predicting the exact values of the solutions from the graph.

**EXAMPLE 2** *Solving Trigonometric Inequalities*

Solve **(a)** $2 \sin x - 1 > 0$ and **(b)** $2 \sin x - 1 < 0$ over the interval $[0, 2\pi)$.

**Solution** To solve the inequality in part (a), we must identify the $x$-values in the interval $[0, 2\pi)$ for which the graph is *above* the $x$-axis. Similarly, to solve the inequality in part (b), we must identify the $x$-values for which the graph is *below* the $x$-axis. The graph in Figure 33 indicates the following solutions in $[0, 2\pi)$.

**(a)** The solution set of $2 \sin x - 1 > 0$ is $\left(\frac{\pi}{6}, \frac{5\pi}{6}\right)$.

**(b)** The solution set of $2 \sin x - 1 < 0$ is $\left[0, \frac{\pi}{6}\right) \cup \left(\frac{5\pi}{6}, 2\pi\right)$.

## Equations Solvable by Factoring

**EXAMPLE 3** *Solving a Trigonometric Equation by Factoring*

Solve $\sin x \tan x = \sin x$ over the interval $[0°, 360°)$.

**Solution**

$$\sin x \tan x = \sin x$$

$$\sin x \tan x - \sin x = 0 \qquad \text{Subtract } \sin x.$$

$$\sin x(\tan x - 1) = 0 \qquad \text{Factor.}$$

$$\sin x = 0 \quad \text{or} \quad \tan x - 1 = 0 \qquad \text{Zero-product property}$$

$$\tan x = 1$$

$$x = 0° \quad \text{or} \quad x = 180° \qquad x = 45° \quad \text{or} \quad x = 225°$$

The solution set is $\{0°, 45°, 180°, 225°\}$.

**CAUTION** There are four solutions in Example 3. Trying to solve the equation by dividing each side by $\sin x$ would give just $\tan x = 1$, which would give $x = 45°$ or $x = 225°$. The other two solutions would not appear. The missing solutions are the ones that make the divisor, $\sin x$, equal 0. For this reason, we must avoid dividing by a variable expression.

**EXAMPLE 4** *Solving a Trigonometric Equation and Associated Inequalities*

Solve **(a)** $\tan^2 x + \tan x - 2 = 0$ and **(b)** $\tan^2 x + \tan x - 2 > 0$ over the interval $[0, 2\pi)$.

**Solution**

**(a)** This equation is quadratic in the term $\tan x$ and can be solved by factoring.

$$\tan^2 x + \tan x - 2 = 0$$

$$(\tan x - 1)(\tan x + 2) = 0 \qquad \text{Factor.}$$

$$\tan x - 1 = 0 \quad \text{or} \quad \tan x + 2 = 0 \qquad \text{Zero-product property}$$

$$\tan x = 1 \quad \text{or} \quad \tan x = -2$$

The solutions for $\tan x = 1$ in the interval $[0, 2\pi)$ are $x = \frac{\pi}{4}$ or $\frac{5\pi}{4}$. To solve $\tan x = -2$ in the interval, we use a calculator set in radian mode. We find that $\tan^{-1}(-2) \approx -1.107148718$. However, due to the way the calculator determines this value, it is not in the desired interval. Because the period of the tangent function is $\pi$, we add $\pi$ and then $2\pi$ to $\tan^{-1}(-2)$ to obtain the solutions in the desired interval.

$$x = \tan^{-1}(-2) + \pi \approx 2.034443936$$

$$x = \tan^{-1}(-2) + 2\pi \approx 5.176036589$$

The solution set is

$$\Bigg\{\underbrace{\frac{\pi}{4}, \quad \frac{5\pi}{4}}_{\text{Exact values}}, \quad \underbrace{2.03, \quad 5.18}_{\text{Approximate values to the nearest hundredth}}\Bigg\}.$$

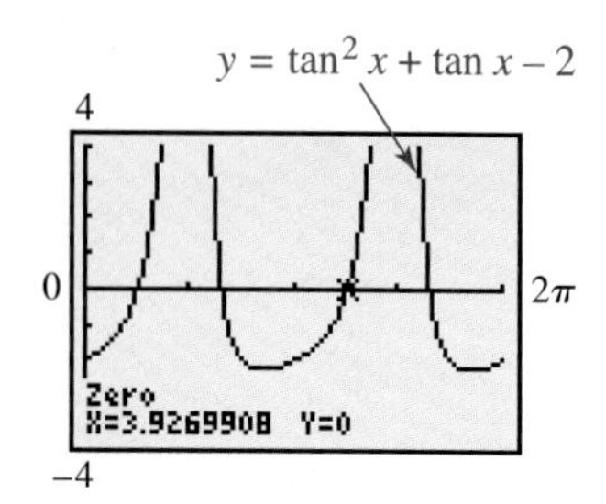


FIGURE 34

**(b)** From the graph of $y = \tan^2 x + \tan x - 2$ as seen in Figure 34, there are four $x$-intercepts in the interval $[0, 2\pi)$. (The figure shows the solution $\frac{5\pi}{4}$, since a decimal approximation of this number is 3.9269908. The other three solutions can be found similarly.) The graph lies *above* the $x$-axis for the following subset of the interval $[0, 2\pi)$:

$$\left(\frac{\pi}{4}, \frac{\pi}{2}\right) \cup \left(\frac{\pi}{2}, 2.03\right) \cup \left(\frac{5\pi}{4}, \frac{3\pi}{2}\right) \cup \left(\frac{3\pi}{2}, 5.18\right).$$

This is the solution set of the inequality. Notice that tangent is not defined when $x = \frac{\pi}{2}$ and when $x = \frac{3\pi}{2}$, so these values must be excluded from the solution set.

**FOR DISCUSSION**

What is the solution set of $\tan^2 x + \tan x - 2 < 0$ over the interval $[0, 2\pi)$? (*Hint:* Refer to Figure 34.)

## Equations Solvable by the Quadratic Formula

**EXAMPLE 5** *Solving a Trigonometric Equation Using the Quadratic Formula*

Solve $\cot x(\cot x + 3) = 1$ over the interval $[0, 2\pi)$.

**Solution** We multiply the factors on the left and subtract 1 to get the equation in standard quadratic form.

$$\cot^2 x + 3 \cot x - 1 = 0$$

Since this equation cannot be solved by factoring, we use the quadratic formula, with $a = 1$, $b = 3$, $c = -1$, and $\cot x$ as the variable.

$$\cot x = \frac{-3 \pm \sqrt{9 + 4}}{2} = \frac{-3 \pm \sqrt{13}}{2}$$

$$\cot x \approx -3.302775638 \quad \text{or} \quad \cot x \approx .3027756377 \qquad \text{Use a calculator.}$$

Since we cannot find inverse cotangent values directly on a calculator, we use the fact that $\cot x = \frac{1}{\tan x}$, and take reciprocals to get

$$\tan x \approx -.3027756377 \quad \text{or} \quad \tan x \approx 3.302775638$$
$$x \approx -.2940013018 \quad \text{or} \quad x \approx 1.276795025.$$

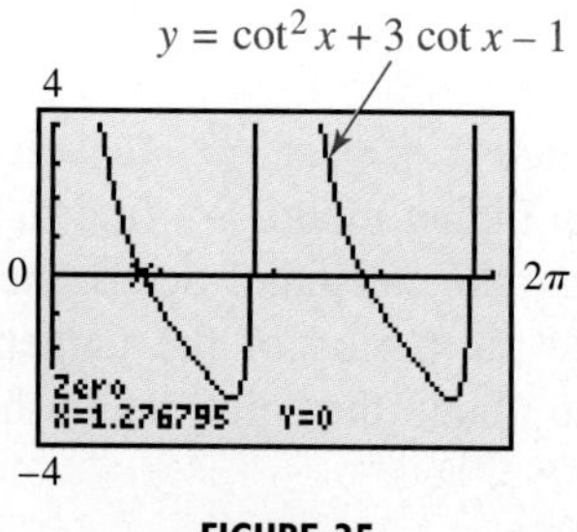


FIGURE 35

The first of these, $-.2940013018$, is not in the desired interval. Since the period of the cotangent function is $\pi$, we add $\pi$ and then $2\pi$ to $-.2940013018$ to get 2.847591352 and 5.989184005.

The second value, 1.276795025, is in the desired interval. We add $\pi$ to it to get another solution in the interval: $1.276795025 + \pi$. Rounding to the nearest hundredth, the four solutions in the interval are 1.28, 2.85, 4.42, and 5.99, and the solution set is $\{1.28, 2.85, 4.42, 5.99\}$.

The graph of $y = \cot^2 x + 3 \cot x - 1$, shown in Figure 35, indicates the solution 1.276795025. The others can be found similarly.

## Using Trigonometric Identities to Solve Equations

Recall that squaring both sides of an equation, such as $\sqrt{x + 4} = x + 2$, will yield all solutions but may also give extraneous values. (In this equation, 0 is a solution, while $-3$ is extraneous. Verify this.) The same situation may occur when trigonometric equations are solved in this manner.

**EXAMPLE 6** *Solving a Trigonometric Equation by Squaring*

Solve $\tan x + \sqrt{3} = \sec x$ over the interval $[0, 2\pi)$.

### Analytic Solution

Since the tangent and secant functions are related by the identity $1 + \tan^2 x = \sec^2 x$, one method of solving this equation is to square both sides and express $\sec^2 x$ in terms of $\tan^2 x$.

$$\tan x + \sqrt{3} = \sec x$$

$$\tan^2 x + 2\sqrt{3}\tan x + 3 = \sec^2 x \quad \text{Square each side.}$$

$$\tan^2 x + 2\sqrt{3}\tan x + 3 = 1 + \tan^2 x \quad \text{Substitute.}$$

$$2\sqrt{3}\tan x = -2 \quad \text{Combine terms.}$$

$$\tan x = -\frac{1}{\sqrt{3}} = -\frac{\sqrt{3}}{3}$$

The possible solutions in the given interval are $\frac{5\pi}{6}$ and $\frac{11\pi}{6}$. Check $\frac{5\pi}{6}$ first.

Left side: $\tan x + \sqrt{3} = \tan\frac{5\pi}{6} + \sqrt{3} = -\frac{\sqrt{3}}{3} + \sqrt{3} = \frac{2\sqrt{3}}{3}$

Right side: $\sec x = \sec\frac{5\pi}{6} = \frac{-2\sqrt{3}}{3}$ ← Not equal

The check shows that $\frac{5\pi}{6}$ is not a solution. Now, check $\frac{11\pi}{6}$.

Left side: $\tan\frac{11\pi}{6} + \sqrt{3} = -\frac{\sqrt{3}}{3} + \sqrt{3} = \frac{2\sqrt{3}}{3}$

Right side: $\sec\frac{11\pi}{6} = \frac{2\sqrt{3}}{3}$ ← Equal

This solution satisfies the equation, so the solution set is $\left\{\frac{11\pi}{6}\right\}$.

### Graphing Calculator Solution

We will use the $x$-intercept method. Graph

$$y = \tan x + \sqrt{3} - \sec x$$

over the desired interval, as in Figure 36. The graph shows that the only $x$-intercept on that interval is 5.7595865, which is an approximation for $\frac{11\pi}{6}$, the solution found analytically.

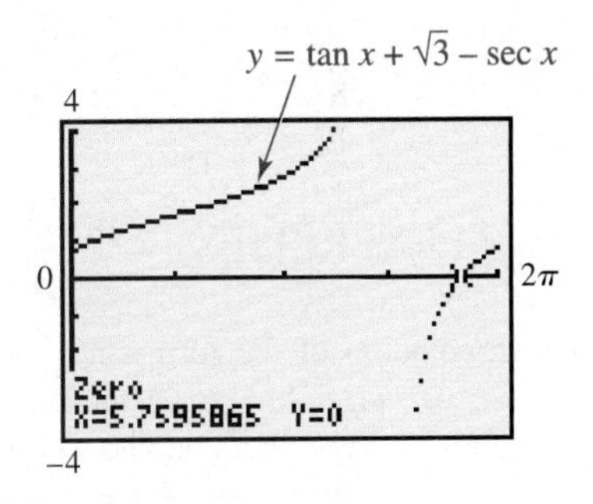


Dot Mode; Radian Mode

**FIGURE 36**

The methods for solving trigonometric equations illustrated in the examples can be summarized as follows.

### Solving a Trigonometric Equation Analytically

1. Decide whether the equation is linear or quadratic, so you can determine the solution method.
2. If only one trigonometric function is present, first solve the equation for that function.
3. If more than one trigonometric function is present, rearrange the equation so that one side equals 0. Then try to factor and set each factor equal to 0 to solve.
4. If the equation is quadratic in form, but not factorable, use the quadratic formula. Check that solutions are in the desired interval.

*(continued)*

5. Try using identities to change the form of the equation. It may be helpful to square both sides of the equation first. If this is done, check for extraneous solutions.

**Solving a Trigonometric Equation Graphically**

1. For an equation of the form $f(x) = g(x)$, use the intersection-of-graphs method.
2. For an equation of the form $f(x) = 0$, use the $x$-intercept method.

## 9.5 EXERCISES

1. An equation of the form $\sin x - b = 0$ has either zero, one, or two solutions in the interval $[0, 2\pi)$. For what values of $b$ will there be two solutions? One solution? No solutions? (*Hint:* Look at the graph of $y = \sin x$.)

2. Suppose an equation of the form $\tan x - b = 0$ has the solution $x = a$ in the interval $\left(-\frac{\pi}{2}, \frac{\pi}{2}\right)$. Give an expression for all solutions to the equation.

*Solve each equation for solutions over the interval* $[0, 2\pi)$ *by first solving for the trigonometric function.*

3. $2 \cot x + 1 = -1$
4. $\sin x + 2 = 3$
5. $2 \sin x + 3 = 4$
6. $2 \sec x + 1 = \sec x + 3$
7. $(\cot x - 1)(\sqrt{3} \cot x + 1) = 0$
8. $(\csc x + 2)(\csc x - \sqrt{2}) = 0$
9. $\cos^2 x + 2 \cos x + 1 = 0$
10. $4(1 + \sin x)(1 - \sin x) = 3$
11. $(\cot x - \sqrt{3})(2 \sin x + \sqrt{3}) = 0$
12. $\tan \theta + 1 = \sqrt{3} + \sqrt{3} \cot \theta$
13. $\tan \theta - \cot \theta = 0$
14. $2 \sin \theta - 1 = \csc \theta$
15. $\cos^2 x = \sin^2 x$
16. $\cos^2 x - \sin^2 x = 1$
17. $\csc^2 x = 2 \cot x$

*In Exercises 18–26, solve* **(a)** $f(x) = 0$, **(b)** $f(x) > 0$, *and* **(c)** $f(x) < 0$ *over the interval* $[0, 2\pi)$.

18. $f(x) = -2 \cos x + 1$
19. $f(x) = 2 \sin x + 1$
20. $f(x) = \tan^2 x - 3$
21. $f(x) = \sec^2 x - 1$
22. $f(x) = 2 \cos^2 x - \sqrt{3} \cos x$
23. $f(x) = 2 \sin^2 x + 3 \sin x + 1$
24. $f(x) = \sin^2 x \cos x - \cos x$
25. $f(x) = \sin^2 x \cos^2 x$
26. $f(x) = 2 \tan^2 x \sin x - \tan^2 x$

### RELATING CONCEPTS

For individual or group investigation (Exercises 27–31)

*Work Exercises 27–31 in order.*

27. Write the equation $\tan^3 x = 3 \tan x$ with 0 on one side.
28. Solve the equation you wrote in Exercise 27 over $[0, 2\pi)$ by factoring.
29. Verify that the solutions you found in Exercise 28 all satisfy the equation.
30. Solve $\tan^3 x = 3 \tan x$ by dividing each side by $\tan x$.
31. Do your answers in Exercises 28 and 30 agree? Explain.

*In Exercises 32–37, give solutions over the interval $[0, 2\pi)$ as approximations to the nearest hundredth when exact values cannot be determined. You will need to use the quadratic formula. Give answers for Exercises 38–40 in degrees.*

**32.** $3 \sin^2 x - \sin x = 2$

**33.** $9 \sin^2 x = 6 \sin x + 1$

**34.** $\tan^2 x + 4 \tan x + 2 = 0$

**35.** $3 \cot^2 x - 3 \cot x = 1$

**36.** $2 \cos^2 x + 2 \cos x = 1$

**37.** $\sin^2 x - 2 \sin x + 3 = 0$

**38.** $\sec^2 \theta = 2 \tan \theta + 4$

**39.** $\cot \theta + 2 \csc \theta = 3$

**40.** $2 \sin \theta = 1 - 2 \cos \theta$

*Solve each equation graphically over the interval $[0, 2\pi)$. Express solutions to the nearest hundredth.* (Hint: *In Exercise 44, the equation has three solutions.)*

**41.** $\cot x + 2 \csc x = 3$

**42.** $2 \sin x = 1 - 2 \cos x$

**43.** $\sin^3 x + \sin x = 1$

**44.** $2 \cos^3 x + \sin x = -1$

**45.** $e^x = \sin x + 3$

**46.** $\ln x = \cos x$

**47.** Explain what is wrong with the following solution for all $x$ over the interval $[0, 2\pi)$ of the following equation.

$$\sin^2 x - \sin x = 0$$

$$\sin x - 1 = 0 \qquad \text{Divide by } \sin x.$$

$$\sin x = 1 \qquad \text{Add 1.}$$

$$x = \frac{\pi}{2}$$

*(Modeling) Solve each problem.*

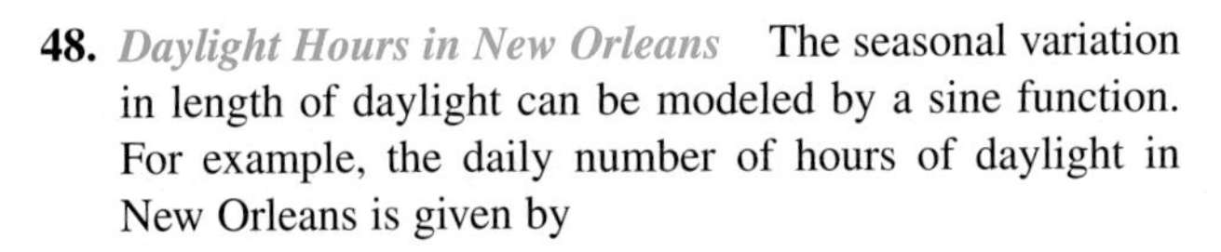

**48.** *Daylight Hours in New Orleans* The seasonal variation in length of daylight can be modeled by a sine function. For example, the daily number of hours of daylight in New Orleans is given by

$$h = \frac{35}{3} + \frac{7}{3} \sin \frac{2\pi x}{365},$$

where $x$ is the number of days after March 21 (disregarding leap year). (*Source:* Bushaw, Donald et al., *A Sourcebook of Applications of School Mathematics.* Copyright © 1980 by The Mathematical Association of America.)

**(a)** On what date will there be about 14 hours of daylight?

**(b)** What date has the least number of hours of daylight?

**(c)** When will there be about 10 hours of daylight?

**49.** *Mach Number for a Plane* An airplane flying faster than sound sends out sound waves that form a cone. The cone intersects the ground to form a hyperbola. As this hyperbola passes over a particular point on the ground, a sonic boom is heard at that point. If $\alpha$ is the angle at the vertex of the cone, then

$$\sin \frac{\alpha}{2} = \frac{1}{m},$$

where $m$ is the Mach number for the speed of the plane. (See Section 9.4, Exercise 88). We assume $m > 1$. Find the measure of $\alpha$, in degrees, if $m = 1.5$.

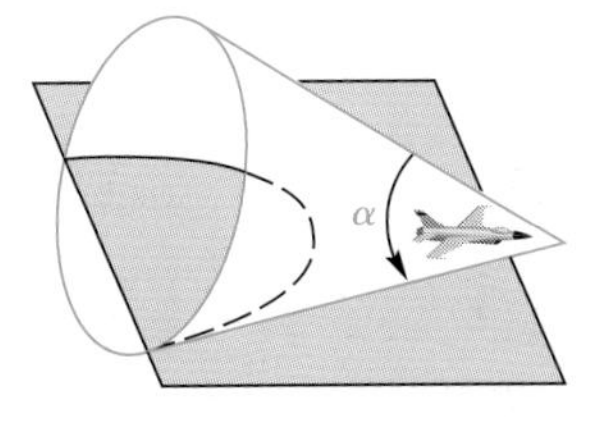


**50.** *Distance of a Particle from a Starting Point* A particle moves along a straight line. The distance of the particle from a starting point at time $t$ is given by

$$s(t) = \sin t + 2 \cos t.$$

Find a value of $t$ that satisfies each equation.

**(a)** $s(t) = \dfrac{2 + \sqrt{3}}{2}$

**(b)** $s(t) = \dfrac{3\sqrt{2}}{2}$

**51.** *Accident Reconstruction* The model

$$.342D \cos \theta + h \cos^2 \theta = \frac{16D^2}{V^2}$$

is used for reconstructing accidents in which a vehicle vaults into the air after hitting an obstruction. $V$ is the velocity in feet per second of the vehicle when it hits the obstruction, $D$ is the distance (in feet) from the obstruction to the vehicle's landing point, and $h$ is the difference in height (in feet) between the landing point and the takeoff point. Angle $\theta$ is the takeoff angle, the angle between the horizontal and the path of the vehicle. Find $\theta$ to the nearest degree if $V = 60$, $D = 80$, and $h = 2$.

**52.** *Maximum Viewing Angle* The bottom of a 10-foot-high movie screen is located 2 feet above the eyes of the viewers, all of whom are sitting at the same level. A viewer seated 5 feet from the screen has the maximum viewing angle, given by $x$ in the equation

$$\frac{\tan x + .4}{1 - .4 \tan x} = 2.4.$$

Find the maximum viewing angle (in degrees).

# 9.6 TRIGONOMETRIC EQUATIONS AND INEQUALITIES (II)

Equations and Inequalities Involving Multiple-Number Identities • Equations and Inequalities Involving Half-Number Identities • An Application

Some trigonometric equations and inequalities involve functions of multiple-numbers or half-numbers, as seen in the following examples.

## Equations and Inequalities Involving Multiple-Number Identities

**EXAMPLE 1** *Solving an Equation Using a Multiple-Number Identity*

Solve $\cos 2x = \cos x$ over the interval $[0, 2\pi)$.

**Analytic Solution**

First change $\cos 2x$ to a trigonometric function of $x$. Use the identity $\cos 2x = 2\cos^2 x - 1$ so that the equation involves only the cosine of $x$. Then factor as in the previous section.

$$\cos 2x = \cos x$$

$$2\cos^2 x - 1 = \cos x \qquad \text{Substitute.}$$

$$2\cos^2 x - \cos x - 1 = 0 \qquad \text{Standard form}$$

$$(2\cos x + 1)(\cos x - 1) = 0 \qquad \text{Factor.}$$

$$2\cos x + 1 = 0 \quad \text{or} \quad \cos x - 1 = 0 \qquad \text{Zero-product property}$$

$$\cos x = -\frac{1}{2} \quad \text{or} \quad \cos x = 1$$

In the required interval,

$$x = \frac{2\pi}{3} \quad \text{or} \quad x = \frac{4\pi}{3} \quad \text{or} \quad x = 0.$$

The solution set is $\left\{0, \frac{2\pi}{3}, \frac{4\pi}{3}\right\}$.

**Graphing Calculator Solution**

With the calculator in radian mode, graph $y = \cos 2x - \cos x$ in an appropriate window, and find the $x$-intercepts. The display in Figure 37 shows that one $x$-intercept is 2.0943951, which is an approximation for $\frac{2\pi}{3}$. The other two $x$-intercepts correspond to 0 and $\frac{4\pi}{3}$.

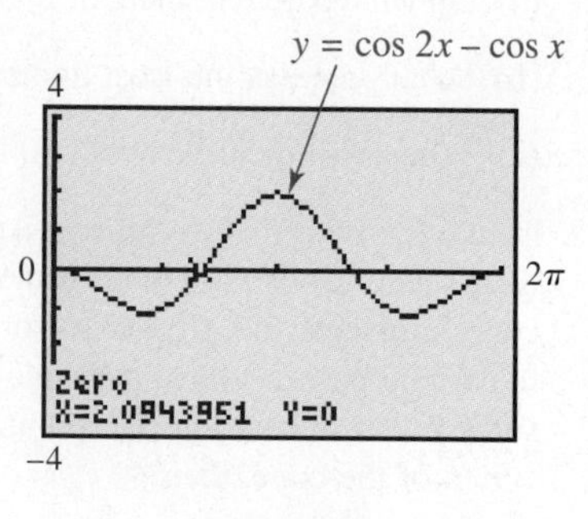


Radian Mode

**FIGURE 37**

**CAUTION** In the analytic solution in Example 1, cos $2x$ cannot be changed to cos $x$ by dividing by 2 since 2 is *not* a factor of cos $2x$.

$$\frac{\cos 2x}{2} \neq \cancel{\cos x}$$

The only way to change cos $2x$ to a trigonometric function of $x$ is by using one of the identities for cos $2x$.

## EXAMPLE 2 *Solving Inequalities Involving a Function of 2x*

Refer to Example 1 and Figure 37 to solve each inequality over the interval $[0, 2\pi)$.

**(a)** $\cos 2x < \cos x$ **(b)** $\cos 2x > \cos x$

### FOR DISCUSSION

Refer to Example 2 and discuss how the solution sets would be affected with the following minor changes.

**1.** Solve $\cos 2x \leq \cos x$.

**2.** Solve $\cos 2x \geq \cos x$ over the interval $[0, 4\pi)$.

### Solution

**(a)** The given inequality is equivalent to $\cos 2x - \cos x < 0$. We determine the solution set of this inequality by finding the $x$-values where the graph of the equation $y = \cos 2x - \cos x$ is *below* the $x$-axis. From Figure 37, we see that this occurs in the interval $[0, 2\pi)$ when $x$ is in the interval $\left(0, \frac{2\pi}{3}\right) \cup \left(\frac{4\pi}{3}, 2\pi\right)$.

**(b)** Here, we want the $x$-values where the graph of $y = \cos 2x - \cos x$ is *above* the $x$-axis, so the solution set is $\left(\frac{2\pi}{3}, \frac{4\pi}{3}\right)$.

Some equations require an additional step to adjust the solution interval.

## EXAMPLE 3 *Solving an Equation Using a Multiple-Number Identity*

Solve $4 \sin x \cos x = \sqrt{3}$ over the interval $[0, 2\pi)$.

### Analytic Solution

The identity $2 \sin x \cos x = \sin 2x$ is useful here.

$$4 \sin x \cos x = \sqrt{3}$$

$$2(2 \sin x \cos x) = \sqrt{3} \qquad 4 = 2 \cdot 2$$

$$2 \sin 2x = \sqrt{3} \qquad 2 \sin x \cos x = \sin 2x$$

$$\sin 2x = \frac{\sqrt{3}}{2} \qquad \text{Divide by 2.}$$

From the given domain, $0 \leq x < 2\pi$, the domain for $2x$ is $0 \leq 2x < 4\pi$. We list all solutions in this interval.

$$2x = \frac{\pi}{3}, \frac{2\pi}{3}, \frac{7\pi}{3}, \frac{8\pi}{3}$$

$$x = \frac{\pi}{6}, \frac{\pi}{3}, \frac{7\pi}{6}, \frac{4\pi}{3}$$

We found the final two solutions for $2x$ by adding $2\pi$ to $\frac{\pi}{3}$ and $\frac{2\pi}{3}$, respectively, giving the solution set for $x$, $\left\{\frac{\pi}{6}, \frac{\pi}{3}, \frac{7\pi}{6}, \frac{4\pi}{3}\right\}$.

### Graphing Calculator Solution

We use the intersection-of-graphs method. Figure 38 indicates that the $x$-coordinate of one point of intersection is .52359878, an approximation for $\frac{\pi}{6}$. The other three intersection points correspond to the remaining solutions, $\frac{\pi}{3}$, $\frac{7\pi}{6}$, and $\frac{4\pi}{3}$.

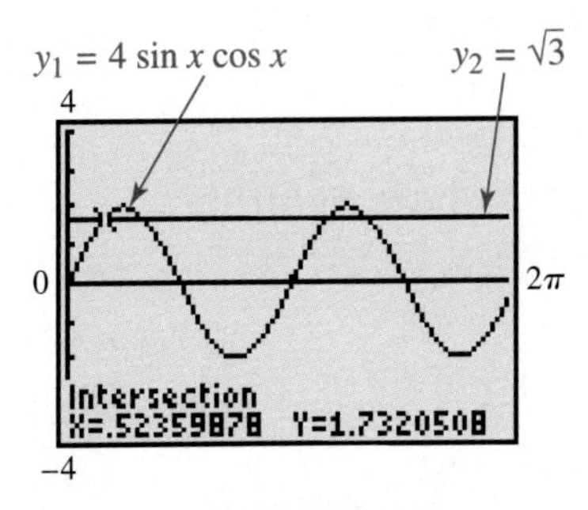


FIGURE 38

**EXAMPLE 4** *Solving an Equation That Involves Squaring Both Sides*

Solve $\tan 3x + \sec 3x = 2$ over the interval $[0, 2\pi)$.

**Solution** Since the tangent and secant functions are related by the identity $1 + \tan^2 \theta = \sec^2 \theta$, we can begin by expressing everything in terms of secant.

$$\tan 3x + \sec 3x = 2$$

$$\tan 3x = 2 - \sec 3x \qquad \text{Subtract } \sec 3x.$$

$$\tan^2 3x = 4 - 4 \sec 3x + \sec^2 3x \qquad \text{Square each side; } (a-b)^2 = a^2 - 2ab + b^2.$$

$$\sec^2 3x - 1 = 4 - 4 \sec 3x + \sec^2 3x \qquad \text{Replace } \tan^2 3x \text{ with } \sec^2 3x - 1.$$

$$0 = 5 - 4 \sec 3x$$

$$4 \sec 3x = 5$$

$$\sec 3x = \frac{5}{4}$$

$$\frac{1}{\cos 3x} = \frac{5}{4} \qquad \sec \theta = \frac{1}{\cos \theta}$$

$$\cos 3x = \frac{4}{5} \qquad \text{Use reciprocals.}$$

We multiply the inequality $0 \le x < 2\pi$ by 3 to find that the interval for $3x$ is $[0, 6\pi)$. Using a calculator and the fact that cosine is positive in quadrants I and IV, we get

$$3x \approx .64350111, 5.6396842, 6.9266864, 11.922870, 13.209872, 18.206055.$$

Dividing by 3 gives

$$x \approx .21450037, 1.8798947, 2.3088955, 3.9742898, 4.4032906, 6.0686849.$$

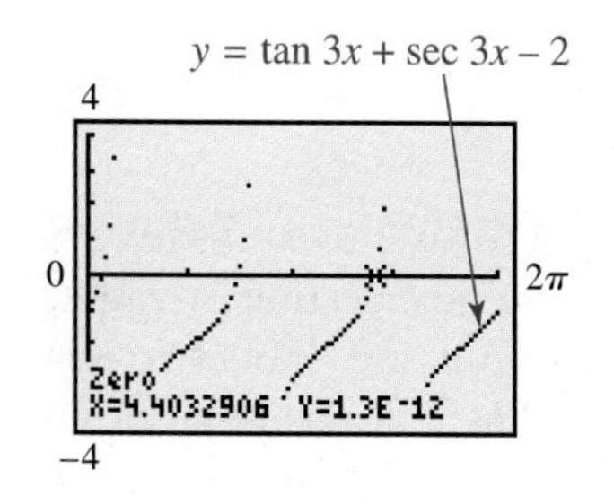


Dot Mode

**FIGURE 39**

Recall from Section 4.5 that when both sides of an equation are squared, there may be extraneous solutions. From the graph of $y = \tan 3x + \sec 3x - 2$ in Figure 39, we see that in the interval $[0, 2\pi)$, there are only three $x$-intercepts. One of these is approximately 4.4032906, which is one of the six possible solutions given above. (See the display in the figure.) The other two can also be verified by a calculator. They are .21450037 and 2.3088955. To the nearest thousandth, the solution set over the interval $[0, 2\pi)$ is $\{.215, 2.309, 4.403\}$.

## Equations and Inequalities Involving Half-Number Identities

**EXAMPLE 5** *Solving an Equation Using a Half-Number Identity*

Solve $2 \sin \frac{x}{2} = 1$ over the interval $[0, 2\pi)$.

### Analytic Solution

Write the interval $[0, 2\pi)$ as the inequality

$$0 \le x < 2\pi.$$

The corresponding interval for $\frac{x}{2}$ is

$$0 \le \frac{x}{2} < \pi. \qquad \text{Divide by 2.}$$

### Graphing Calculator Solution

Graph $y = 2 \sin \frac{x}{2} - 1$ in a window with $x$ in the interval $[0, 2\pi)$. The $x$-intercepts are the solutions found analytically. The display in Figure 40 shows the solution 1.0471976, which approximates $\frac{\pi}{3}$. The other intercept represents the solution $\frac{5\pi}{3}$.

*(continued)*

To find all values of $\frac{x}{2}$ in the interval $[0, \pi)$ that satisfy the given equation, first solve for $\sin \frac{x}{2}$.

$$2 \sin \frac{x}{2} = 1$$

$$\sin \frac{x}{2} = \frac{1}{2} \quad \text{Divide by 2.}$$

The two numbers in the interval $[0, \pi)$ that have sine value $\frac{1}{2}$ are $\frac{\pi}{6}$ and $\frac{5\pi}{6}$, so

$$\frac{x}{2} = \frac{\pi}{6} \quad \text{or} \quad \frac{x}{2} = \frac{5\pi}{6}$$

$$x = \frac{\pi}{3} \quad \text{or} \quad x = \frac{5\pi}{3}. \quad \text{Multiply by 2.}$$

The solution set in the given interval is $\left\{\frac{\pi}{3}, \frac{5\pi}{3}\right\}$.

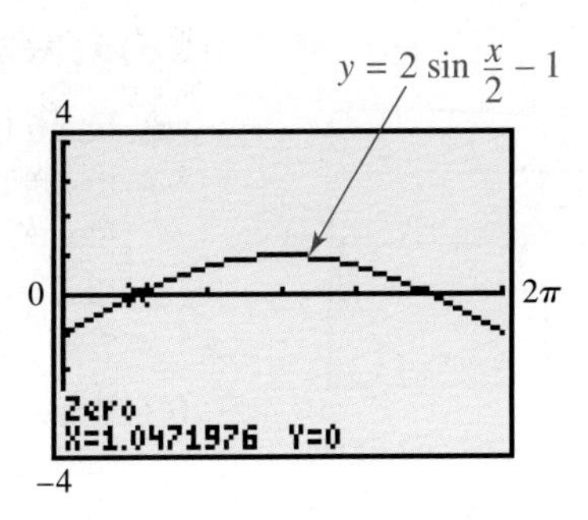


FIGURE 40

Using Xscl $= \frac{\pi}{3}$ in Figure 40 makes it possible to support the exact solutions by counting the tick marks from 0 on the graph.

**EXAMPLE 6** *Solving an Inequality Involving a Function of $\frac{x}{2}$*

Solve $2 \sin \frac{x}{2} > 1$ over the interval $[0, 2\pi)$.

**Solution** Referring to Figure 40, we see that the graph of $y = 2 \sin \frac{x}{2} - 1$ lies *above* the $x$-axis between $\frac{\pi}{3}$ and $\frac{5\pi}{3}$, so the solution set is the open interval $\left(\frac{\pi}{3}, \frac{5\pi}{3}\right)$.

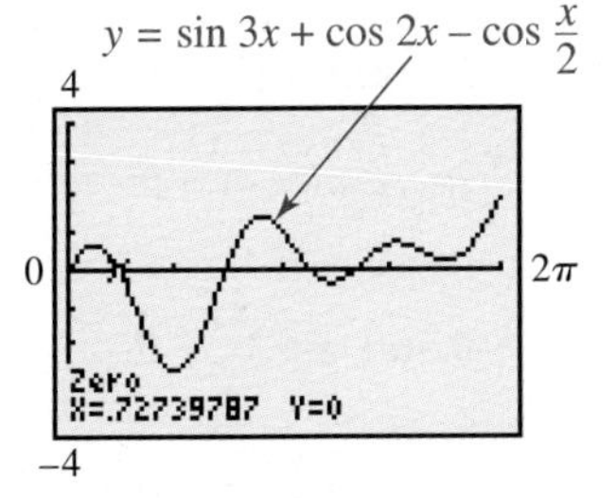


FIGURE 41

**EXAMPLE 7** *Solving an Equation Using Only a Graphical Approach*

Solve $\sin 3x + \cos 2x = \cos \frac{x}{2}$ over the interval $[0, 2\pi)$.

**Solution** We will solve this equation using the $x$-intercept method. The graph of $y = \sin 3x + \cos 2x - \cos \frac{x}{2}$ is shown in Figure 41. The least positive solution, .72739787, is indicated on the screen. The solution 0 can be verified by direct substitution. The solution set over the interval $[0, 2\pi)$ is $\{0, .727, 2.288, 3.524, 4.189\}$.

## An Application

As mentioned in the chapter opener, music is closely related to mathematics.

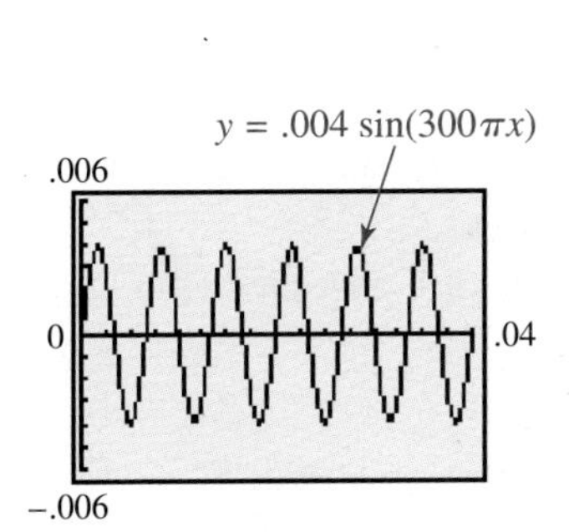


FIGURE 42

**EXAMPLE 8** *Describing a Musical Tone from a Graph*

A basic component of music is a pure tone. The graph in Figure 42 models the sinusoidal pressure $y = P$ in pounds per square foot from a pure tone at time $x = t$ in seconds.

**(a)** The frequency of a pure tone is often measured in hertz. One hertz is equal to one cycle per second and is abbreviated Hz. What is the frequency $f$ in hertz of the pure tone shown in the graph?

**(b)** The time for the tone to produce one complete cycle is called the *period.* Approximate the period $T$ in seconds of the pure tone.

**(c)** An equation for the graph is $y = .004 \sin(300\pi x)$. Use a calculator to estimate all solutions to the equation $y = .004$ on the interval $[0, .02]$.

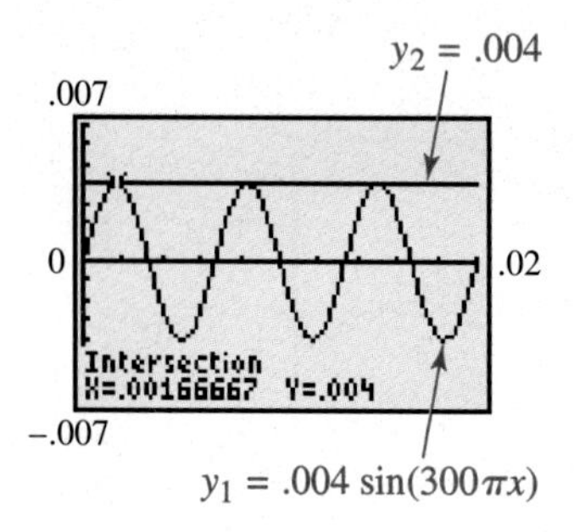


**FIGURE 43**

### Solution

**(a)** From the graph in Figure 42 on the previous page, we see that there are 6 cycles in .04 second. This is equivalent to $6/.04 = 150$ cycles per second. The pure tone has a frequency of $f = 150$ hertz.

**(b)** Six periods cover a time of .04 second. One period would be equal to $T = \frac{.04}{6} = \frac{1}{150}$ or $.00\overline{6}$ second.

**(c)** If we reproduce the graph in Figure 42 on a calculator as $y_1$ and also graph the second function as $y_2 = .004$, we can determine that the approximate values of $x$ at the points of intersection of the graphs on the interval $[0, .02]$ are

$$.0017, \quad .0083, \quad \text{and} \quad .015.$$

See Figure 43. 

## 9.6 EXERCISES

*Concept Check* *Use the concepts of this section to answer each question.*

**1.** Suppose you are solving a trigonometric equation for solutions in $[0, 2\pi)$, and your work leads to

$$2x = \frac{2\pi}{3}, 2\pi, \frac{8\pi}{3}.$$

What are the corresponding values of $x$?

**2.** Suppose you are solving a trigonometric equation for solutions in $[0°, 360°)$, and your work leads to

$$\frac{1}{3}\theta = 45°, 60°, 75°, 90°.$$

What are the corresponding values of $\theta$?

*Solve each equation in part (a) analytically. Then, use a graph to solve each inequality in part (b). In all cases, solve over the interval* $[0, 2\pi)$.

**3. (a)** $\cos 2x = \frac{\sqrt{3}}{2}$ **(b)** $\cos 2x > \frac{\sqrt{3}}{2}$

**4. (a)** $\cos 2x = -\frac{1}{2}$ **(b)** $\cos 2x > -\frac{1}{2}$

**5. (a)** $\sin 3x = -1$ **(b)** $\sin 3x < -1$

**6. (a)** $\sin 3x = 0$ **(b)** $\sin 3x < 0$

**7. (a)** $\sqrt{2} \cos 2x = -1$ **(b)** $\sqrt{2} \cos 2x \leq -1$

**8. (a)** $2\sqrt{3} \sin 2x = \sqrt{3}$ **(b)** $2\sqrt{3} \sin 2x \leq \sqrt{3}$

**9. (a)** $\sin \frac{x}{2} = \sqrt{2} - \sin \frac{x}{2}$ **(b)** $\sin \frac{x}{2} > \sqrt{2} - \sin \frac{x}{2}$

**10. (a)** $\sin x = \sin 2x$ **(b)** $\sin x > \sin 2x$

*Solve each equation over the interval* $[0, 2\pi)$.

**11.** $\sin \frac{x}{2} = \cos \frac{x}{2}$

**12.** $\sec \frac{x}{2} = \cos \frac{x}{2}$

**13.** $\sin^2 \frac{x}{2} - 1 = 0$

**14.** $\sin x \cos x = \frac{1}{4}$

**15.** $\sin 2x = 2 \cos^2 x$

**16.** $\csc^2 \frac{x}{2} = 2 \sec x$

**17.** $\cos x - 1 = \cos 2x$

**18.** $1 - \sin x = \cos 2x$

*Solve each equation over the interval* $[0°, 360°)$.

**19.** $\sqrt{2} \sin 3\theta - 1 = 0$

**20.** $-2 \cos 2\theta = \sqrt{3}$

**21.** $\cos \frac{\theta}{2} = 1$

**22.** $\sin \frac{\theta}{2} = 1$

**23.** $2\sqrt{3} \sin \frac{\theta}{2} = 3$

**24.** $2\sqrt{3} \cos \frac{\theta}{2} = -3$

**25.** $2 \sin \theta = 2 \cos 2\theta$

**26.** $\cos \theta = \sin^2 \dfrac{\theta}{2}$

**27.** $2 - \sin 2\theta = 4 \sin 2\theta$

**28.** $4 \cos 2\theta = 8 \sin \theta \cos \theta$

**29.** $2 \cos^2 2\theta = 1 - \cos 2\theta$

**30.** $\sin \theta - \sin 2\theta = 0$

*Use a graphical method to solve each equation over the interval* $[0, 2\pi)$.

**31.** $\sin x + \sin 3x = \cos x$

**32.** $\sin 3x - \sin x = 0$

**33.** $\cos 2x + \cos x = 0$

**34.** $\sin 4x + \sin 2x = 2 \cos x$

**35.** $\cos \dfrac{x}{2} = 2 \sin 2x$

**36.** $\sin \dfrac{x}{2} + \cos 3x = 0$

**37.** What is wrong with the following solution? Solve $\tan 2\theta = 2$ over the interval $[0, 2\pi)$.

$$\tan 2\theta = 2$$

$$\frac{\tan 2\theta}{2} = \frac{2}{2}$$

$$\tan \theta = 1$$

$$\theta = \frac{\pi}{4} \quad \text{or} \quad \theta = \frac{5\pi}{4}$$

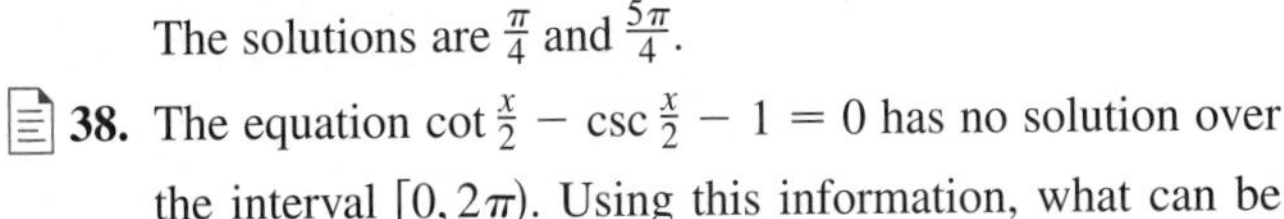

The solutions are $\frac{\pi}{4}$ and $\frac{5\pi}{4}$.

**38.** The equation $\cot \frac{x}{2} - \csc \frac{x}{2} - 1 = 0$ has no solution over the interval $[0, 2\pi)$. Using this information, what can be said about the graph of

$$y = \cot \frac{x}{2} - \csc \frac{x}{2} - 1$$

over this interval? Confirm your answer by actually graphing the function over the interval.

*(Modeling) Solve each problem.*

**39.** *Inducing Voltage* A coil of wire rotating in a magnetic field induces a voltage given by

$$e = 20 \sin\left(\frac{\pi t}{4} - \frac{\pi}{2}\right),$$

where $t$ is time in seconds. Find the smallest positive time to produce each voltage.

**(a)** 0 **(b)** $10\sqrt{3}$

**40.** *Nautical Mile* The British nautical mile is defined as the length of a minute of arc of a meridian. Since Earth is flat at its poles, the nautical mile, in feet, is given by

$$L = 6077 - 31 \cos 2\theta,$$

where $\theta$ is the latitude in degrees. (See the figure.) (*Source:* Bushaw, Donald et al., *A Sourcebook of Applications of School Mathematics.* Copyright © 1980 by The Mathematical Association of America. Reprinted by permission.)

A nautical mile is the length on any of the meridians cut by a central angle of measure 1 minute.

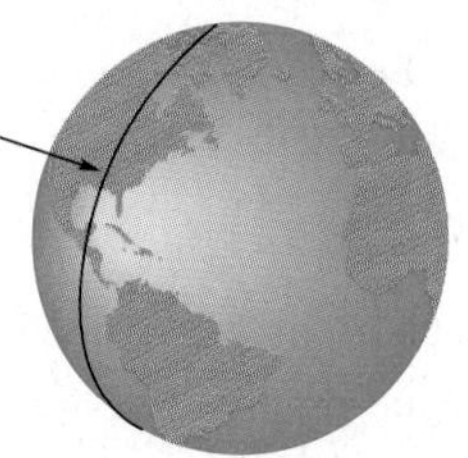

**(a)** Find the latitude between 0° and 90° at which the nautical mile is 6074 feet.

**(b)** At what latitude between 0° and 180° is the nautical mile 6108 feet?

**(c)** In the United States, the nautical mile is defined everywhere as 6080.2 feet. At what latitude between 0° and 90° does this agree with the British nautical mile?

**41.** *Ear Pressure from a Pure Tone* No musical instrument can generate a true pure tone. A pure tone has a unique, constant frequency and amplitude that sounds rather dull and uninteresting. The pressures caused by pure tones on the eardrum are sinusoidal. The change in pressure $P$ in pounds per square feet on a person's eardrum from a pure tone at time $t$ in seconds can be modeled using the equation

$$P = A \sin(2\pi ft + \phi),$$

where $f$ is the frequency in cycles per second and $\phi$ is the phase angle. When $P$ is positive, there is an increase in pressure and the eardrum is pushed inward; when $P$ is negative, there is a decrease in pressure and the eardrum is pushed outward. (*Source:* Roederer, J., *Introduction to the Physics and Psychophysics of Music,* The English Universities Press, 1973.)

**(a)** Middle C has frequency 261.63 cycles per second. Graph this tone with $A = .004$ and $\phi = \frac{\pi}{7}$ in the window $[0, .005]$ by $[-.005, .005]$.

**(b)** Determine analytically the values of $t$ for which $P = 0$ in $[0, .005]$ and support your answers graphically.

**(c)** Determine graphically when $P < 0$ on $[0, .005]$.

**(d)** Would an eardrum hearing this tone be vibrating outward or inward when $P < 0$?

**42.** *Ear Pressure from a Vibrating String* If a string with a fundamental frequency of 110 Hz is plucked in the middle, it will vibrate at the odd harmonics of 110, 330, 550, ... Hz but not at the even harmonics of 220, 440, 660, ... Hz. The resulting pressure $P$ caused by the string can be approximated with the equation

$$P = .003 \sin 220\pi t + \frac{.003}{3} \sin 660\pi t + \frac{.003}{5} \sin 1100\pi t + \frac{.003}{7} \sin 1540\pi t.$$

(*Source:* Benade, A., *Fundamentals of Musical Acoustics,* Oxford University Press, 1976; Roederer, J., *Introduction to the Physics and Psychophysics of Music*, The English Universities Press, 1973.)

**(a)** Graph $P$ in the window $[0, .03]$ by $[-.005, .005]$.

**(b)** Use the graph to describe the shape of the sound wave that is produced.

**(c)** At lower frequencies, the inner ear will hear a tone only when the eardrum is moving outward. (See Exercise 41.) Determine the times on the interval $[0, .03]$ when this will occur.

**43.** *Hearing Beats in Music* Musicians sometimes tune instruments by playing the same tone on two different instruments and listening for a phenomenon known as *beats.* Beats occur when two tones vary in frequency by only a few hertz. When the two instruments are in tune, the beats disappear. The ear hears beats because the pressure slowly rises and falls as a result of this slight variation in the frequency. This phenomenon can be seen using a graphing calculator. (*Source:* Pierce, J., *The Science of Musical Sound,* Scientific American Books, 1992.)

**(a)** Consider two tones with frequencies of 220 and 223 Hz and pressures

$$P_1 = .005 \sin 440\pi t \quad \text{and} \quad P_2 = .005 \sin 446\pi t,$$

respectively. Graph the pressure $P = P_1 + P_2$ felt by an eardrum over the one-second interval $[.15, 1.15]$. How many beats are there in one second?

**(b)** Repeat part (a) with frequencies of 220 and 216.

**(c)** Determine a simple way to find the number of beats per second if the frequency of each tone is given.

**44.** *Hearing Different Tones* Small speakers like those found in older radios and telephones often cannot vibrate slower than 200 Hz—yet 35 keys on a piano have frequencies below 200 Hz. When a musical instrument creates a tone of 110 Hz, it also creates tones at 220, 330, 440, 550, 660, ... Hz. A small speaker cannot reproduce the 110-Hz vibration but it can reproduce the higher frequencies, which are called the upper harmonics. The low tones can still be heard because the speaker produces *difference tones* of the upper harmonics. The difference between consecutive frequencies is 110 Hz, and this difference tone will be heard by a listener. We can model this phenomenon using a graphing calculator. (*Source:* Benade, A., *Fundamentals of Musical Acoustics,* Oxford University Press, 1976.)

**(a)** In the window $[0, .03]$ by $[-2, 2]$, graph the upper harmonics represented by the pressure

$$P = \frac{1}{2}\sin[2\pi(220)t] + \frac{1}{3}\sin[2\pi(330)t] + \frac{1}{4}\sin[2\pi(440)t].$$

**(b)** Estimate all $t$-coordinates where $P$ is maximum.

**(c)** What does a person hear in addition to the frequencies of 220, 330, and 440 Hz?

**(d)** Graph the pressure produced by a speaker that can vibrate at 110 Hz and above.

## Reviewing Basic Concepts (Sections 9.5 and 9.6)

*Solve each equation over the interval* $[0, 2\pi)$.

**1.** $\cos 2x = \frac{\sqrt{3}}{2}$

**2.** $2 \sin x + 1 = 0$

**3.** $(\tan x - 1)(\cos x - 1) = 0$

**4.** $2 \cos^2 x = \sqrt{3} \cos x$

*Solve each equation over the interval* $[0°, 360°)$.

**5.** $3\cot^2\theta - 3\cot\theta = 1$

**6.** $4\cos^2\theta + 4\cos\theta - 1 = 0$

**7.** $2\sin\theta - 1 = \csc\theta$

**8.** $\sec^2\dfrac{\theta}{2} = 2$

*Solve each equation graphically over the interval* $[0, 2\pi)$.

**9.** $x^2 + \sin x - x^3 - \cos x = 0$

**10.** $x^3 - \cos^2 x = \dfrac{1}{2}x - 1$

## CHAPTER 9 SUMMARY

| KEY TERMS & SYMBOLS | KEY CONCEPTS |
|---|---|
| **9.1** Trigonometric Identities | **FUNDAMENTAL IDENTITIES**<br>Reciprocal Identities $\cot\theta = \dfrac{1}{\tan\theta}$ $\sec\theta = \dfrac{1}{\cos\theta}$ $\csc\theta = \dfrac{1}{\sin\theta}$<br>Quotient Identities $\tan\theta = \dfrac{\sin\theta}{\cos\theta}$ $\cot\theta = \dfrac{\cos\theta}{\sin\theta}$<br>Pythagorean Identities $\sin^2\theta + \cos^2\theta = 1$ $1 + \tan^2\theta = \sec^2\theta$ $1 + \cot^2\theta = \csc^2\theta$<br>Negative-Number Identities<br>$\sin(-\theta) = -\sin\theta$ $\cos(-\theta) = \cos\theta$ $\tan(-\theta) = -\tan\theta$<br>$\csc(-\theta) = -\csc\theta$ $\sec(-\theta) = \sec\theta$ $\cot(-\theta) = -\cot\theta$ |
| **9.2** Sum and Difference Identities | **SUM AND DIFFERENCE IDENTITIES**<br>$\cos(A - B) = \cos A\cos B + \sin A\sin B$<br>$\cos(A + B) = \cos A\cos B - \sin A\sin B$<br>$\sin(A + B) = \sin A\cos B + \cos A\sin B$<br>$\sin(A - B) = \sin A\cos B - \cos A\sin B$<br>$\tan(A + B) = \dfrac{\tan A + \tan B}{1 - \tan A\tan B}$ $\tan(A - B) = \dfrac{\tan A - \tan B}{1 + \tan A\tan B}$ |
| **9.3** Further Identities | **DOUBLE-NUMBER IDENTITIES**<br>$\cos 2A = \cos^2 A - \sin^2 A$ $\cos 2A = 1 - 2\sin^2 A$<br>$\cos 2A = 2\cos^2 A - 1$ $\sin 2A = 2\sin A\cos A$<br>$\tan 2A = \dfrac{2\tan A}{1 - \tan^2 A}$ |

| KEY TERMS & SYMBOLS | KEY CONCEPTS |
|---|---|

### PRODUCT-TO-SUM IDENTITIES

$$\cos A \cos B = \frac{1}{2}[\cos(A + B) + \cos(A - B)]$$

$$\sin A \sin B = \frac{1}{2}[\cos(A - B) - \cos(A + B)]$$

$$\sin A \cos B = \frac{1}{2}[\sin(A + B) + \sin(A - B)]$$

$$\cos A \sin B = \frac{1}{2}[\sin(A + B) - \sin(A - B)]$$

### SUM-TO-PRODUCT IDENTITIES

$$\sin A + \sin B = 2 \sin\left(\frac{A + B}{2}\right) \cos\left(\frac{A - B}{2}\right)$$

$$\sin A - \sin B = 2 \cos\left(\frac{A + B}{2}\right) \sin\left(\frac{A - B}{2}\right)$$

$$\cos A + \cos B = 2 \cos\left(\frac{A + B}{2}\right) \cos\left(\frac{A - B}{2}\right)$$

$$\cos A - \cos B = -2 \sin\left(\frac{A + B}{2}\right) \sin\left(\frac{A - B}{2}\right)$$

### HALF-NUMBER IDENTITIES

$$\cos\frac{A}{2} = \pm\sqrt{\frac{1 + \cos A}{2}} \qquad \sin\frac{A}{2} = \pm\sqrt{\frac{1 - \cos A}{2}}$$

$$\tan\frac{A}{2} = \pm\sqrt{\frac{1 - \cos A}{1 + \cos A}} \qquad \tan\frac{A}{2} = \frac{\sin A}{1 + \cos A} \qquad \tan\frac{A}{2} = \frac{1 - \cos A}{\sin A}$$

(The sign is chosen based on the quadrant of $\frac{A}{2}$.)

## 9.4 The Inverse Circular Functions

$\sin^{-1} x$ or $\arcsin x$
$\cos^{-1} x$ or $\arccos x$
$\tan^{-1} x$ or $\arctan x$
$\cot^{-1} x$ or $\operatorname{arccot} x$
$\sec^{-1} x$ or $\operatorname{arcsec} x$
$\csc^{-1} x$ or $\operatorname{arccsc} x$

### INVERSE CIRCULAR FUNCTIONS

| | | Range | |
|---|---|---|---|
| **Function** | **Domain** | **Interval** | **Quadrants of the Unit Circle** |
| $y = \sin^{-1} x$ | $[-1, 1]$ | $\left[-\frac{\pi}{2}, \frac{\pi}{2}\right]$ | I and IV |
| $y = \cos^{-1} x$ | $[-1, 1]$ | $[0, \pi]$ | I and II |
| $y = \tan^{-1} x$ | $(-\infty, \infty)$ | $\left(-\frac{\pi}{2}, \frac{\pi}{2}\right)$ | I and IV |
| $y = \cot^{-1} x$ | $(-\infty, \infty)$ | $(0, \pi)$ | I and II |
| $y = \sec^{-1} x$ | $(-\infty, -1] \cup [1, \infty)$ | $[0, \pi], y \neq \frac{\pi}{2}$ | I and II |
| $y = \csc^{-1} x$ | $(-\infty, -1] \cup [1, \infty)$ | $\left[-\frac{\pi}{2}, \frac{\pi}{2}\right], y \neq 0$ | I and IV |

| KEY TERMS & SYMBOLS | KEY CONCEPTS |
|---|---|
| **9.5** Trigonometric Equations and Inequalities (I) | **SOLVING A TRIGONOMETRIC EQUATION ANALYTICALLY**<br>**1.** Decide whether the equation is linear or quadratic, so you can determine the solution method.<br>**2.** If only one trigonometric function is present, first solve the equation for that function.<br>**3.** If more than one trigonometric function is present, rearrange the equation so that one side equals 0. Then try to factor and set each factor equal to 0 to solve.<br>**4.** If the equation is quadratic in form, but not factorable, use the quadratic formula. Check that solutions are in the desired interval.<br>**5.** Try using identities to change the form of the equation. It may be helpful to square both sides of the equation first. If this is done, check for extraneous solutions.<br>**SOLVING A TRIGONOMETRIC EQUATION GRAPHICALLY**<br>**1.** For an equation of the form $f(x) = g(x)$, use the intersection-of-graphs method.<br>**2.** For an equation of the form $f(x) = 0$, use the $x$-intercept method. |
| **9.6** Trigonometric Equations and Inequalities (II) | Use the multiple-number and half-number identities to rewrite the equations. Then, use the steps given in Section 9.5. |

## CHAPTER 9 REVIEW EXERCISES

**1.** Give all the trigonometric functions $f$ that satisfy the condition $f(-x) = -f(x)$.

**2.** Give all the trigonometric functions $f$ that satisfy the condition $f(-x) = f(x)$.

*Use the negative-number identities to write each expression as a function of a positive number.*

**3.** $\cos(-3)$ **4.** $\sin(-3)$ **5.** $\tan(-3)$ **6.** $\sec(-3)$ **7.** $\csc(-3)$ **8.** $\cot(-3)$

*Concept Check* *For each expression in Group I, give the letter of the expression in Group II that completes an identity.*

| I | | II | |
|---|---|---|---|
| **9.** $\sec x =$ | **10.** $\tan x =$ | **A.** $\dfrac{1}{\sin x}$ | **B.** $\dfrac{1}{\cos x}$ |
| **11.** $\cot x =$ | **12.** $\tan^2 x + 1 =$ | **C.** $\dfrac{\sin x}{\cos x}$ | **D.** $\dfrac{1}{\cot^2 x}$ |
| **13.** $\tan^2 x =$ | | **E.** $\dfrac{1}{\cos^2 x}$ | **F.** $\dfrac{\cos x}{\sin x}$ |

*Use identities to write each expression in terms of* $\sin\theta$ *and* $\cos\theta$*, and simplify.*

**14.** $\sec^2\theta - \tan^2\theta$ **15.** $\dfrac{\cot\theta}{\sec\theta}$ **16.** $\tan^2\theta(1 + \cot^2\theta)$ **17.** $\csc\theta + \cot\theta$

**18.** Use the trigonometric identities to find the remaining five trigonometric functions of $x$, given $\cos x = \frac{3}{5}$ and $x$ is in quadrant IV.

**19.** Given $\tan x = -\frac{5}{4}$, where $\frac{\pi}{2} < x < \pi$, use the trigonometric identities to find the other trigonometric functions of $x$.

*Concept Check For each expression in Group I, choose the expression from Group II that completes an identity. One of the choices from Group II will not be used.*

| I | | II | |
|---|---|---|---|
| **20.** $\cos 210° =$ | **21.** $\sin 35° =$ | **A.** $\sin(-35°)$ | **B.** $\cos 55°$ |
| **22.** $\tan(-35°) =$ | **23.** $-\sin 35° =$ | **C.** $\sqrt{\frac{1 + \cos 150°}{2}}$ | **D.** $2 \sin 150° \cos 150°$ |
| **24.** $\cos 35° =$ | **25.** $\cos 75° =$ | **E.** $\cos 150° \cos 60° - \sin 150° \sin 60°$ | **F.** $\cot(-35°)$ |
| **26.** $\sin 75° =$ | **27.** $\sin 300° =$ | **G.** $\cos^2 150° - \sin^2 150°$ | **H.** $\sin 15° \cos 60° + \cos 15° \sin 60°$ |
| **28.** $\cos 300° =$ | | **I.** $\cos(-35°)$ | **J.** $\cot 125°$ |

*Verify analytically that each equation is an identity.*

**29.** $\sin^2 x - \sin^2 y = \cos^2 y - \cos^2 x$

**30.** $2 \cos^3 x - \cos x = \frac{\cos^2 x - \sin^2 x}{\sec x}$

**31.** $-\cot \frac{x}{2} = \frac{\sin 2x + \sin x}{\cos 2x - \cos x}$

**32.** $\frac{\sin^2 x}{2 - 2 \cos x} = \cos^2 \frac{x}{2}$

**33.** $\frac{\sin 2x}{\sin x} = \frac{2}{\sec x}$

**34.** $2 \cos A - \sec A = \cos A - \frac{\tan A}{\csc A}$

**35.** $\frac{2 \tan B}{\sin 2B} = \sec^2 B$

**36.** $1 + \tan^2 \alpha = 2 \tan \alpha \csc 2\alpha$

**37.** $\frac{\sin t}{1 - \cos t} = \cot \frac{t}{2}$

**38.** $\frac{2 \cot x}{\tan 2x} = \csc^2 x - 2$

**39.** $\tan \theta \sin 2\theta = 2 - 2 \cos^2 \theta$

**40.** $2 \tan x \csc 2x - \tan^2 x = 1$

*Give the exact value of each expression without using a calculator.*

**41.** $\sin^{-1} \frac{\sqrt{2}}{2}$

**42.** $\arccos\left(-\frac{1}{2}\right)$

**43.** $\arctan \frac{\sqrt{3}}{3}$

**44.** $\sec^{-1}(-2)$

**45.** $\operatorname{arccsc} \frac{2\sqrt{3}}{3}$

**46.** $\cot^{-1}(-1)$

*Give the degree measure of $\theta$. Do not use a calculator.*

**47.** $\theta = \arccos \frac{1}{2}$

**48.** $\theta = \arcsin\left(-\frac{\sqrt{3}}{2}\right)$

**49.** $\theta = \tan^{-1} 0$

*Use a calculator to give the degree of measure of $\theta$.*

**50.** $\theta = \arcsin(-.66045320)$

**51.** $\theta = \cot^{-1} 4.5046388$

**52.** Explain why $\sin^{-1}(-3)$ is not defined.

**53.** Explain why the expression $\cos(\arccos x)$ is always equal to $x$, but $\arccos(\cos x)$ may not equal $x$.

*Evaluate each expression without a calculator.*

**54.** $\sin\left(\sin^{-1} \frac{1}{2}\right)$

**55.** $\sin\left(\cos^{-1} \frac{3}{4}\right)$

**56.** $\cos(\arctan 3)$

**57.** $\sec\left(2 \sin^{-1}\left(-\frac{1}{3}\right)\right)$

**58.** $\cos^{-1}\left(\cos \frac{3\pi}{2}\right)$

**59.** $\tan\left(\sin^{-1} \frac{3}{5} + \cos^{-1} \frac{5}{7}\right)$

*Write each expression as an algebraic expression in u, $u > 0$.*

**60.** $\sin(\tan^{-1} u)$

**61.** $\cos\left(\arctan \dfrac{u}{\sqrt{1-u^2}}\right)$

**62.** $\tan\left(\arccos \dfrac{u}{\sqrt{u^2+1}}\right)$

*Solve each equation for solutions in the interval $[0, 2\pi)$.*

**63.** $\sin^2 x = 1$

**64.** $2 \tan x - 1 = 0$

**65.** $3 \sin^2 x - 5 \sin x + 2 = 0$

**66.** $\tan x = \cot x$

**67.** $5 \cot^2 x + 3 \cot x = 2$

**68.** $\sec \dfrac{x}{2} = \cos \dfrac{x}{2}$

**69.** $\sin 2x = \cos 2x + 1$

**70.** $2 \sin 2x = 1$

**71.** $\sin 2x + \sin 4x = 0$

**72.** $\cos x - \cos 2x = 2 \cos x$

**73.** $\tan 2x = \sqrt{3}$

**74.** $\cos^2 \dfrac{x}{2} - 2 \cos \dfrac{x}{2} + 1 = 0$

**75.** Use the results of Exercise 69 and a calculator graph of

$$f(x) = \sin 2x - \cos 2x - 1$$

to find the solution set of each inequality over the interval $[0, 2\pi)$.

**(a)** $\sin 2x > \cos 2x + 1$  **(b)** $\sin 2x < \cos 2x + 1$

*(Modeling) Solve each problem.*

**76.** *Viewing Angle of an Observer* A 10-foot-wide blackboard is situated 5 feet from the left wall of a classroom. See the figure. A student sitting next to the wall $x$ feet from the front of the classroom has a viewing angle of $\theta$ radians.

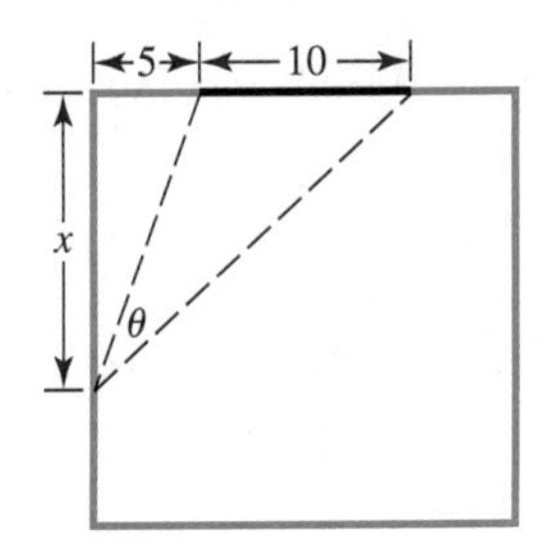


**(a)** Show that the value of $\theta$ is given by the function defined by $\theta = f(x)$, where

$$f(x) = \arctan\left(\frac{15}{x}\right) - \arctan\left(\frac{5}{x}\right).$$

**(b)** Graph $f(x)$ with a graphing calculator to estimate the value of $x$ that maximizes the viewing angle.

**77.** *Motion Formed by a Moving Arm* Suppose that the following equation describes the motion formed by a rhythmically moving arm:

$$y = \frac{1}{3} \sin \frac{4\pi t}{3}.$$

Here, $t$ is time (in seconds) and $y$ represents the measure of the angle formed by the upper arm and a vertical line. (See the figure in the next column.)

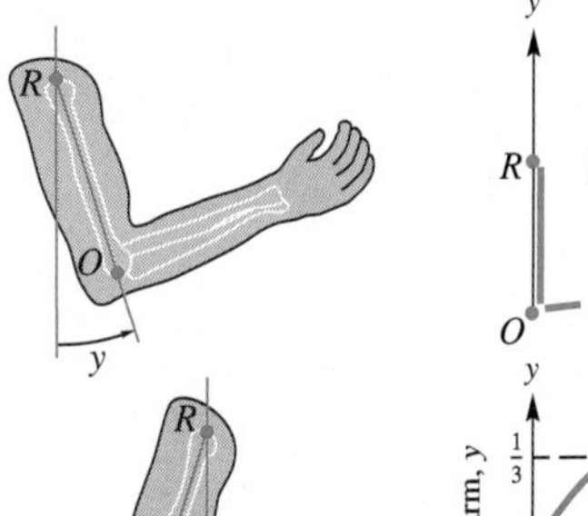


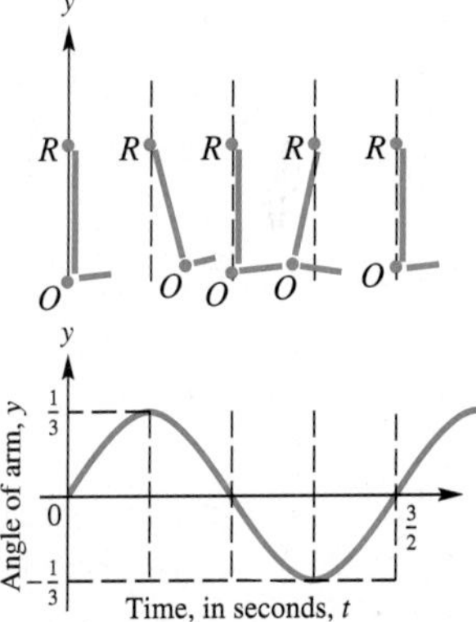


This graph shows the relationship between angle $y$ and time $t$ in seconds.

*Source:* De Sapio, Rodolfo. *Calculus for the Life Sciences.* Copyright © 1978 by W.H. Freeman and Company. Reprinted by permission.

**(a)** Solve the equation for $t$.

**(b)** At what time does the arm form an angle of .3 radian?

*Alternating Electric Current The study of alternating electric current requires the solutions of equations of the form $i = I_{max} \sin 2\pi ft$, for time t in seconds, where i is instantaneous current in amperes, $I_{max}$ is maximum current in amperes, and f is the number of cycles per second. Find the smallest positive value of t, given the following data.*

**78.** $i = 40$, $I_{max} = 100$, $f = 60$

**79.** $i = 50$, $I_{max} = 100$, $f = 120$

**80.** *Speed of Light in Different Mediums* Snell's Law states that

$$\frac{c_1}{c_2} = \frac{\sin \theta_1}{\sin \theta_2},$$

where $c_1$ is the speed of light in one medium, $c_2$ is the speed of light in a second medium, and $\theta_1$ and $\theta_2$ are the angles shown in the figure on the next page. Suppose a light is shining up through water into the air as in the figure.

As $\theta_1$ increases, $\theta_2$ approaches 90°, at which point no light will emerge from the water. Assume the ratio $\frac{c_1}{c_2}$ in this case is .752. For what value of $\theta_1$ does $\theta_2 = 90°$? This value of $\theta_1$ is called the *critical angle* for water.

**81.** *Speed of Light in Different Mediums* Refer to Exercise 80. What happens when $\theta_1$ is greater than the critical angle?

## RELATING CONCEPTS

**For individual or group investigation (Exercises 82–86)**

*The angle between the positive x-axis and a line is called the* angle of inclination *of the line. See the figure on the left. $\theta$ is the angle of inclination of line l. Work Exercises 82–86 in order.*

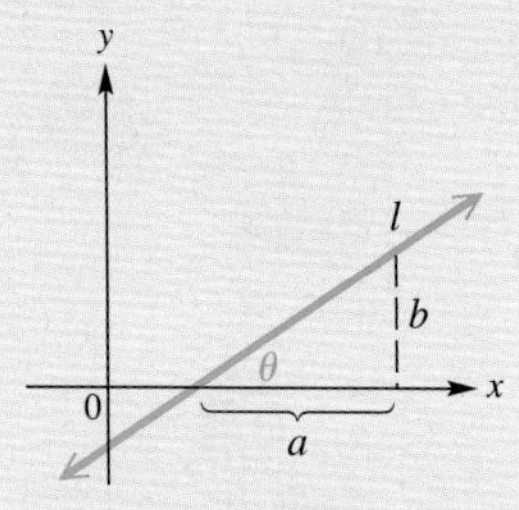


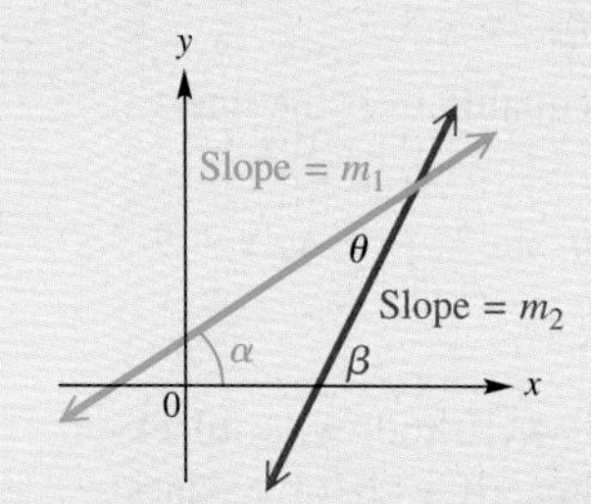


**82.** In the figure on the left, what is $\tan\theta$? What is the slope of line $l$? What conclusion can you draw from your answers?

**83.** In the figure on the right, how are angles $\alpha$, $\beta$, and $\theta$ related?

**84.** Solve the equation from Exercise 83 for $\theta$.

**85.** Take the tangent of each side of the equation from Exercise 84, and use a difference identity to rewrite it.

**86.** Rewrite the equation from Exercise 85, using the slopes $m_1$ and $m_2$.

## CHAPTER 9 TEST

*Remember to set your calculator for radian or degree measure as required.*

**1.** Given $\sin y = -\frac{2}{3}$, $\cos x = -\frac{1}{5}$, and $\frac{\pi}{2} < x < \pi$ and $\pi < y < \frac{3\pi}{2}$, find the exact values for the following.

**(a)** $\sin(x + y)$ **(b)** $\cos(x - y)$ **(c)** $\tan\dfrac{y}{2}$ **(d)** $\cos 2x$

**2.** Express $\tan^2 x - \sec^2 x$ in terms of $\sin x$ and $\cos x$, and simplify.

**3.** Graph $y = \sec x - \sin x \tan x$, and use the graph to conjecture an identity. Verify your conjecture analytically.

**4.** Verify that each equation is an identity.

**(a)** $\sec^2 B = \dfrac{1}{1 - \sin^2 B}$ **(b)** $\cos 2A = \dfrac{\cot A - \tan A}{\csc A \sec A}$

**5.** Use an identity to write each expression as a trigonometric function of $\theta$ alone.

**(a)** $\cos(270° - \theta)$ **(b)** $\sin(\pi + \theta)$

**6.** Graph $f(x) = \sin^{-1}(x - 1)$.

**(a)** Give the domain and range of $f$.

**(b)** What are the intercepts of the graph?

**(c)** Can you find $\sin^{-1} 2$? Explain why or why not.

**7.** Find the exact value of each expression.

**(a)** $\arccos\left(-\dfrac{1}{2}\right)$ **(b)** $\tan^{-1} 0$ **(c)** $\csc^{-1}\dfrac{2\sqrt{3}}{3}$

**(d)** $\cos\left(\arcsin\dfrac{2}{3}\right)$ **(e)** $\sin\left(2\cos^{-1}\dfrac{1}{3}\right)$

**8.** Write $\tan(\arcsin u)$ as an algebraic expression in $u$.

**9.** Solve each equation over the indicated interval.

**(a)** $\sin^2\theta = \cos^2\theta + 1$, $[0, 2\pi)$

**(b)** $\csc^2\theta - 2\cot\theta = 4$, $[0°, 360°)$

**(c)** $\cos x = \cos 2x$, $[0, 2\pi)$

**(d)** $2\sqrt{3}\sin\dfrac{\theta}{2} = 3$, $[0°, 360°)$

**10.** *(Modeling) Migratory Animal Count* The number of migratory animals (in hundreds) for one year counted at a certain checkpoint is given by

$$T(t) = 50 + 50 \cos\left(\frac{\pi}{6}t\right),$$

where $t$ is time in months, with $t = 0$ corresponding to July.

**(a)** Give the domain and range of $T$.

**(b)** Graph $T(t)$ over its domain.

**(c)** Use the graph to determine the maximum and minimum values of $T$ and when they occur.

**(d)** Find $T(3)$ analytically, and support your result graphically. Use symmetry to find $T(9)$.

**(e)** When does the count reach 75 hundred animals?

**(f)** Write the equation

$$T = 50 + 50 \cos\left(\frac{\pi}{6}t\right)$$

as an equation involving arccos by solving for $t$.

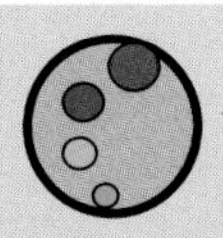

# Chapter 9 Project

## Modeling a Damped Pendulum

A damped pendulum oscillates back and forth. The period (time of each swing) remains constant, but the amplitude diminishes at a constant rate. Therefore, the composition of an exponential function multiplied by a sine function provides a mathematical formula that models the pendulum's behavior. While several forms are possible for such a function, we use

$$y = r^x a \sin[b(x - d)] + c,$$

where $x$ represents time, $a$ represents amplitude, $b$ represents $\frac{2\pi}{\text{period}}$, $c$ represents vertical shift, $d$ represents horizontal phase shift, and $r$ represents rate at which amplitude diminishes.

The data in the table were gathered from an object oscillating for 10 seconds with a CBR data collector.

| Time (seconds) | Distance (from probe in feet) | Time (seconds) | Distance (from probe in feet) | Time (seconds) | Distance (from probe in feet) | Time (seconds) | Distance (from probe in feet) | Time (seconds) | Distance (from probe in feet) |
|---|---|---|---|---|---|---|---|---|---|
| .1 | 2.11 | 2.1 | 2.77 | 4.1 | 3.20 | 6.1 | 3.40 | 8.1 | 3.35 |
| .2 | 2.34 | 2.2 | 2.97 | 4.2 | 3.37 | **6.2** | **3.43** | 8.2 | 3.28 |
| .3 | 2.62 | 2.3 | 3.23 | 4.3 | 3.47 | 6.3 | 3.41 | 8.3 | 3.17 |
| .4 | 2.96 | 2.4 | 3.44 | **4.4** | **3.52** | 6.4 | 3.35 | 8.4 | 3.04 |
| .5 | 3.28 | 2.5 | 3.56 | 4.5 | 3.50 | 6.5 | 3.21 | 8.5 | 2.90 |
| .6 | 3.54 | **2.6** | **3.61** | 4.6 | 3.42 | 6.6 | 3.07 | 8.6 | 2.77 |
| .7 | 3.69 | 2.7 | 3.59 | 4.7 | 3.29 | 6.7 | 2.91 | 8.7 | 2.66 |
| **.8** | **3.81** | 2.8 | 3.51 | 4.8 | 3.11 | 6.8 | 2.75 | 8.8 | 2.59 |
| .9 | 3.78 | 2.9 | 3.37 | 4.9 | 2.92 | 6.9 | 2.62 | **8.9** | **2.56** |
| 1.0 | 3.65 | 3.0 | 3.15 | 5.0 | 2.74 | 7.0 | 2.54 | 9.0 | 2.59 |
| 1.1 | 3.48 | 3.1 | 2.93 | 5.1 | 2.57 | **7.1** | **2.50** | 9.1 | 2.65 |
| 1.2 | 3.24 | 3.2 | 2.69 | 5.2 | 2.46 | 7.2 | 2.54 | 9.2 | 2.76 |
| 1.3 | 2.95 | 3.3 | 2.59 | **5.3** | **2.42** | 7.3 | 2.60 | 9.3 | 2.88 |
| 1.4 | 2.64 | 3.4 | 2.37 | 5.4 | 2.44 | 7.4 | 2.72 | 9.4 | 3.01 |
| 1.5 | 2.41 | **3.5** | **2.33** | 5.5 | 2.55 | 7.5 | 2.87 | 9.5 | 3.13 |
| 1.6 | 2.28 | 3.6 | 2.34 | 5.6 | 2.66 | 7.6 | 3.01 | 9.6 | 3.23 |
| **1.7** | **2.17** | 3.7 | 2.43 | 5.7 | 2.83 | 7.7 | 3.15 | 9.7 | 3.30 |
| 1.8 | 2.20 | 3.8 | 2.58 | 5.8 | 3.00 | 7.8 | 3.27 | **9.8** | **3.32** |
| 1.9 | 2.37 | 3.9 | 2.77 | 5.9 | 3.17 | 7.9 | 3.35 | 9.9 | 3.30 |
| 2.0 | 2.46 | 4.0 | 2.99 | 6.0 | 3.31 | **8.0** | **3.37** | 10.0 | 3.23 |

Figure A is a scatter diagram of the data.

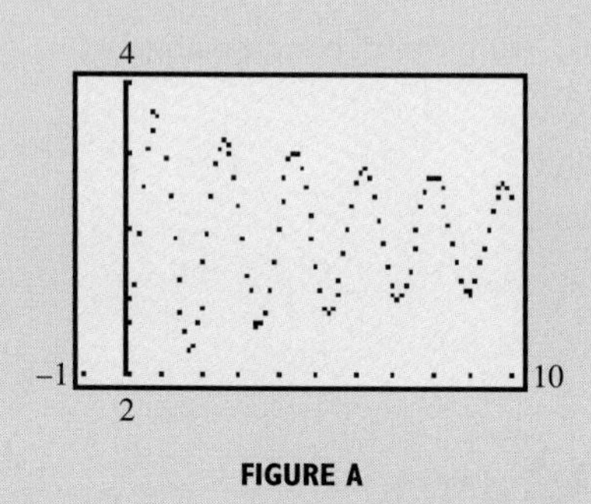


**FIGURE A**

## Activities

1. Reproduce the scatter diagram in Figure A, but to avoid entering so many data points into your calculator by hand, enter only those that correspond to the maximum and minimum distances. (These are printed in bold type.) Use the window $[0, 10]$ by $[-1, 5]$ with Xscl = Yscl = 1.
2. Find a function that models the data by determining the values of $a$, $b$, $c$, $d$, and $r$ in the equation given earlier. (*Hints:* **(a)** Find the amplitude by calculating the distance between the extreme heights and dividing by two. **(b)** Find the period by calculating the time difference between successive maximums (or minimums). **(c)** Find the horizontal phase shift by finding the time of the first data point at the mean height. **(d)** Find the amplitude diminishing rate by calculating the ratios of the first two consecutive maximums.)
3. Graph the function from Activity 2 on top of your scatter diagram. Is it a good fit?
4. If you have access to a CBL or CBR data collector, work in small groups to gather your own data. For the pendulum, use any object that swings or vibrates. The data given earlier was gathered by clamping a meter stick to a secure beam, then pulling and releasing the free end of the meter stick to cause it to vibrate.

# 10 Applications of Trigonometry; Vectors

HIGH-RESOLUTION computer graphics and complex numbers make it possible to produce beautiful shapes called *fractals.* Benoit B. Mandelbrot first used the term *fractal* in 1975. Largely because of his efforts, fractal geometry has become a new field of study. At its basic level, a fractal is a unique, enchanting geometric figure with an endless self-similarity property. A fractal image repeats itself infinitely with ever-decreasing dimensions. If you look at smaller and smaller portions of a fractal image, you will continue to see the whole—much like looking into two parallel mirrors that are facing each other.

Although most current applications of fractals are related to creating fascinating images and pictures, fractals do have tremendous potential in applied science. The example of a fractal shown in the figure is an amazing graphical solution to a difficult problem first presented by Sir Arthur Cayley in 1879. This fractal, called *Newton's basins of attraction for the cube roots of unity,* is discussed in Exercise 53, Section 10.5.

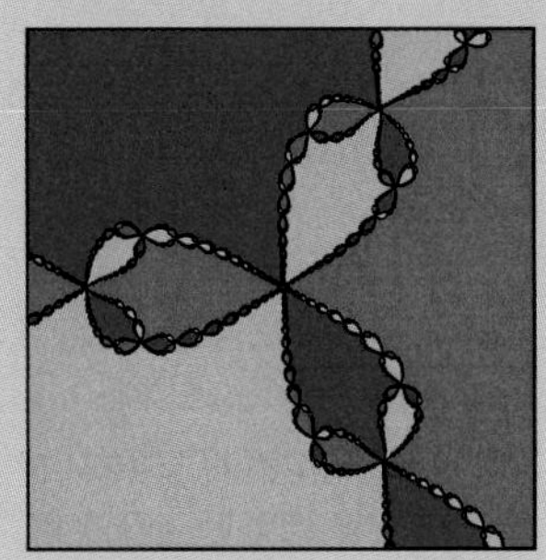

*Source:* Kincaid, D. R. and Cheney, E. W., *Numerical Analysis: Mathematics of Scientific Computing,* 1st edition, © 1991. Reprinted with permission of Brooks/Cole, an imprint of the Wadsworth Group, a division of Thomson Learning.

*Sources:* Crownover, R., *Introduction to Fractals and Chaos,* Jones and Bartlett, 1995.
Kline, M., *Mathematics: The Loss of Certainty,* Oxford University Press, 1980.
Lauwerier, H., *Fractals,* Princeton University Press, 1991. Reprinted with permission.

CHAPTER OUTLINE

# 10.1 THE LAW OF SINES

Congruency and Oblique Triangles • Derivation of the Law of Sines • Applications • Ambiguous Case

Until now, our work with triangles has been limited to right triangles. However, the concepts developed in Chapter 8 can be extended to *all* triangles. Every triangle has three sides and three angles. If any three of the six measures of a triangle are known (provided at least one measure is the length of a side), then the other three measures can be found. This is called *solving the triangle.*

## Congruency and Oblique Triangles

The following axioms from geometry allow us to prove that two triangles are congruent (that is, their corresponding sides and angles are equal).

**Congruence Axioms**

| | |
|---|---|
| **Side-Angle-Side (SAS)** | If two sides and the included angle of one triangle are equal, respectively, to two sides and the included angle of a second triangle, then the triangles are congruent. |
| **Angle-Side-Angle (ASA)** | If two angles and the included side of one triangle are equal, respectively, to two angles and the included side of a second triangle, then the triangles are congruent. |
| **Side-Side-Side (SSS)** | If three sides of one triangle are equal, respectively, to three sides of a second triangle, then the triangles are congruent. |

Throughout this chapter, keep in mind that whenever SAS, ASA, or SSS is given, the triangle is unique. We continue to label triangles as we did earlier with right triangles: side $a$ opposite angle $A$, side $b$ opposite angle $B$, and side $c$ opposite angle $C$.

A triangle that is not a right triangle is called an **oblique triangle.** The measures of the three sides and the three angles of a triangle can be found if at least one side and any other two measures are known. There are four possible cases.

**Data Required for Solving Oblique Triangles**

| | |
|---|---|
| **Case 1** | One side and two angles are known (SAA or ASA). |
| **Case 2** | Two sides and one angle not included between the two sides are known (SSA). This case may lead to more than one triangle. |
| **Case 3** | Two sides and the angle included between the two sides are known (SAS). |
| **Case 4** | Three sides are known (SSS). |

**NOTE** If we know three angles of a triangle, we cannot find unique side lengths since AAA assures us only of similarity, not congruence. For example, there are infinitely many triangles $ABC$ with $A = 35°$, $B = 65°$, and $C = 80°$.

Cases 1 and 2, discussed in this section, require the *law of sines.* Cases 3 and 4, discussed in the next section, require the *law of cosines.*

## Derivation of the Law of Sines

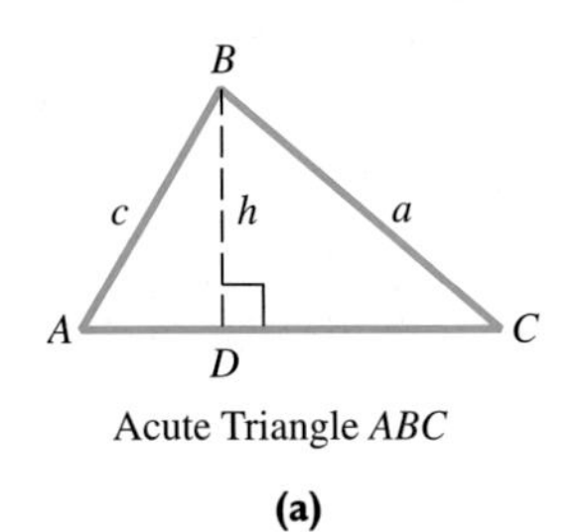

Acute Triangle $ABC$

**(a)**

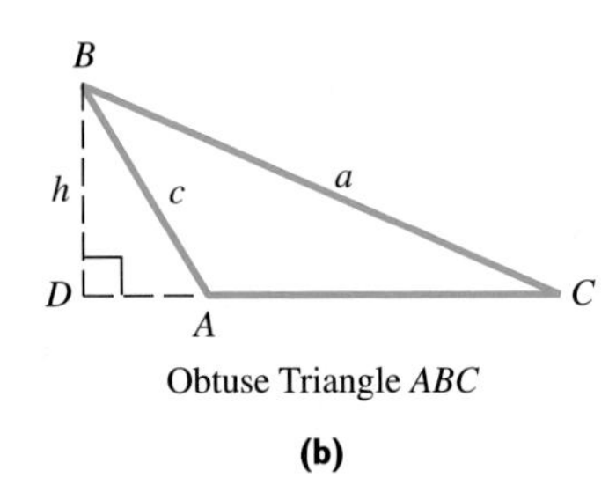

Obtuse Triangle $ABC$

**(b)**

**FIGURE 1**

To derive the law of sines, we start with an oblique triangle, such as the *acute triangle* in Figure 1(a) or the *obtuse triangle* in Figure 1(b). The following discussion applies to both triangles. First, construct the perpendicular from $B$ to side $AC$. Let $h$ be the length of this perpendicular. Then $c$ is the hypotenuse of right triangle $ADB$, and $a$ is the hypotenuse of right triangle $BDC$. By results from Chapter 8,

$$\text{in triangle } ADB, \qquad \sin A = \frac{h}{c} \quad \text{or} \quad h = c \sin A,$$

$$\text{in triangle } BDC, \qquad \sin C = \frac{h}{a} \quad \text{or} \quad h = a \sin C.$$

Since $h = c \sin A$ and $h = a \sin C$,

$$a \sin C = c \sin A,$$

or, upon dividing each side by $\sin A \sin C$,

$$\frac{a}{\sin A} = \frac{c}{\sin C}.$$

In a similar way, by constructing perpendicular lines from the other vertices, it can be shown that

$$\frac{a}{\sin A} = \frac{b}{\sin B} \quad \text{and} \quad \frac{b}{\sin B} = \frac{c}{\sin C}.$$

This discussion proves the following theorem.

**Law of Sines**

In any triangle $ABC$, with sides $a$, $b$, and $c$,

$$\frac{a}{\sin A} = \frac{b}{\sin B}, \qquad \frac{a}{\sin A} = \frac{c}{\sin C}, \qquad \text{and} \qquad \frac{b}{\sin B} = \frac{c}{\sin C}.$$

This can be written in compact form as

$$\frac{a}{\sin A} = \frac{b}{\sin B} = \frac{c}{\sin C}.$$

When solving for an angle, the alternative form of the law of sines,

$$\frac{\sin A}{a} = \frac{\sin B}{b} = \frac{\sin C}{c},$$

is convenient to use.

**NOTE** When using the law of sines, a good strategy is to select an equation so that the unknown variable is in the numerator and all other variables are known.

## Applications

If two angles and a side are known (Case 1, SAA or ASA), then the law of sines can be used to solve the triangle.

**EXAMPLE 1** *Using the Law of Sines to Solve a Triangle (SAA)*

Solve triangle $ABC$ if $A = 32.0°$, $B = 81.8°$, and $a = 42.9$ centimeters.

**Solution** Start by drawing a triangle, roughly to scale, and labeling the given parts as in Figure 2. Since the values of $A$, $B$, and $a$ are known, use the part of the law of sines that involves these variables, and then solve for $b$.

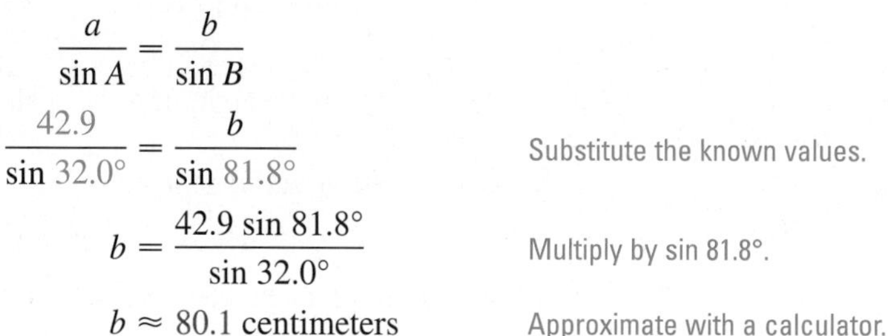

FIGURE 2

$$\frac{a}{\sin A} = \frac{b}{\sin B}$$

$$\frac{42.9}{\sin 32.0°} = \frac{b}{\sin 81.8°} \quad \text{Substitute the known values.}$$

$$b = \frac{42.9 \sin 81.8°}{\sin 32.0°} \quad \text{Multiply by } \sin 81.8°.$$

$$b \approx 80.1 \text{ centimeters} \quad \text{Approximate with a calculator.}$$

To find $C$, use the fact that the sum of the angles of any triangle is 180°.

$$A + B + C = 180°$$

$$C = 180° - A - B$$

$$C = 180° - 32.0° - 81.8° = 66.2°$$

Now, use the law of sines again to find $c$. (Why does the Pythagorean theorem not apply?)

$$\frac{a}{\sin A} = \frac{c}{\sin C}$$

$$\frac{42.9}{\sin 32.0°} = \frac{c}{\sin 66.2°}$$

$$c = \frac{42.9 \sin 66.2°}{\sin 32.0°}$$

$$c \approx 74.1 \text{ centimeters}$$

**CAUTION** When solving oblique triangles such as the one in Example 1, a properly labeled sketch is helpful to set up the correct equation.

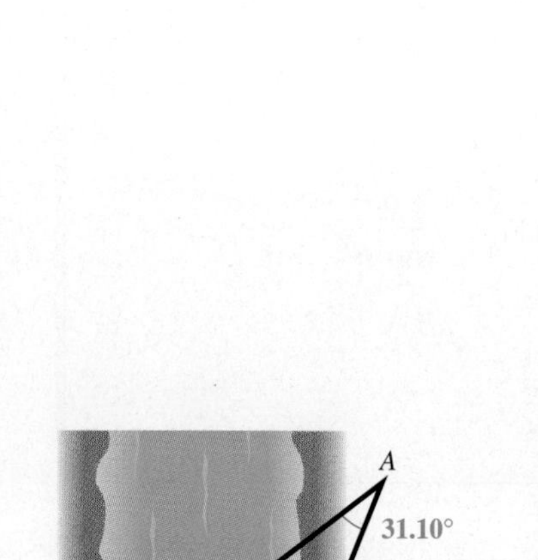

FIGURE 3

**EXAMPLE 2** *Using the Law of Sines in an Application (ASA)*

A surveyor wishes to measure the distance across the Big Muddy River. See Figure 3. The surveyor determines that $C = 112.90°$, $A = 31.10°$, and $b = 347.6$ feet. Find the distance $a$.

**Analytic Solution**

To use the law of sines, one side and the angle opposite it must be known. Since $b$ is the only side whose length is given, angle $B$ must be found before the law of sines can be used.

$$B = 180° - A - C$$
$$B = 180° - 31.10° - 112.90° = 36.00°$$

Now use the form of the law of sines involving $A$, $B$, and $b$ to find $a$.

$$\frac{a}{\sin A} = \frac{b}{\sin B}$$

$$\frac{a}{\sin 31.10°} = \frac{347.6}{\sin 36.00°} \qquad \text{Substitute.}$$

$$a = \frac{347.6 \sin 31.10°}{\sin 36.00°} \qquad \text{Multiply by } \sin 31.10°.$$

$$a \approx 305.5 \text{ feet} \qquad \text{Use a calculator.}$$

**Graphing Calculator Solution**

Triangle $ABC$ in Figure 3 can be solved using a graphing calculator program, as shown in Figure 4. Programs such as this one are available from users' groups or from the Web site for this text.

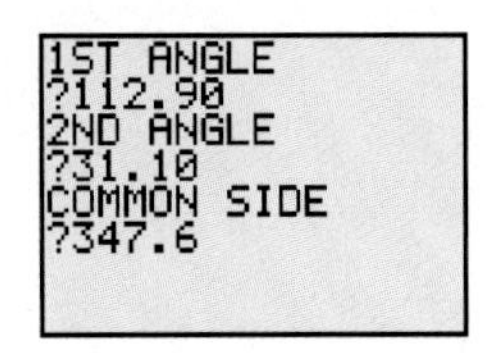


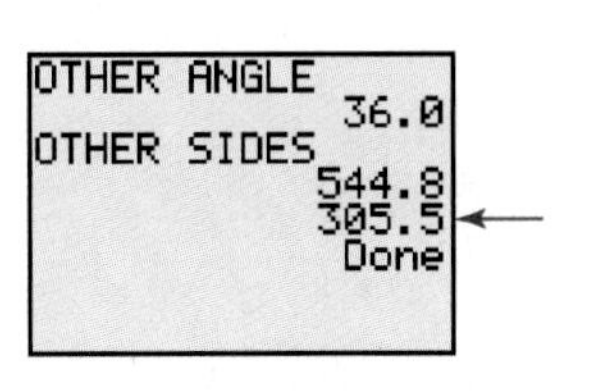


FIGURE 4

The next example involves the concept of bearing, first discussed in Chapter 8.

**EXAMPLE 3** *Using the Law of Sines in an Application (ASA)*

Two ranger stations are on an east–west line 110 miles apart. A forest fire is located on a bearing of N 42° E from the western station at $A$ and a bearing of N 15° E from the eastern station at $B$. How far is the fire from the western station? (See Figure 5.)

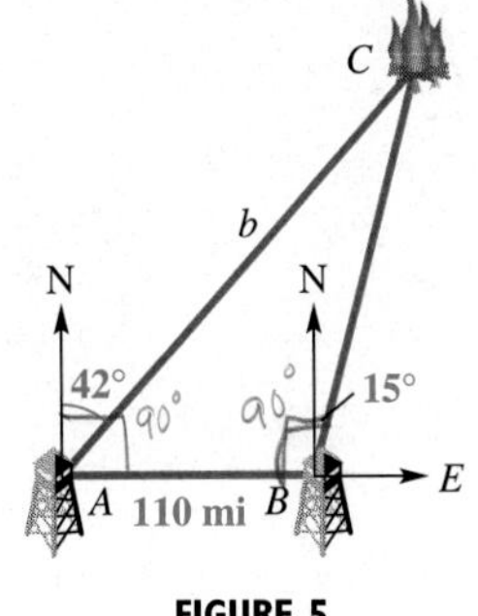


FIGURE 5

**Solution** Figure 5 shows the two stations at points $A$ and $B$ and the fire at point $C$. Angle $BAC = 90° - 42° = 48°$, the obtuse angle at $B$ equals $90° + 15° = 105°$, and the third angle, $C$ equals $180° - 105° - 48° = 27°$. Using the law of sines to find side $b$ gives

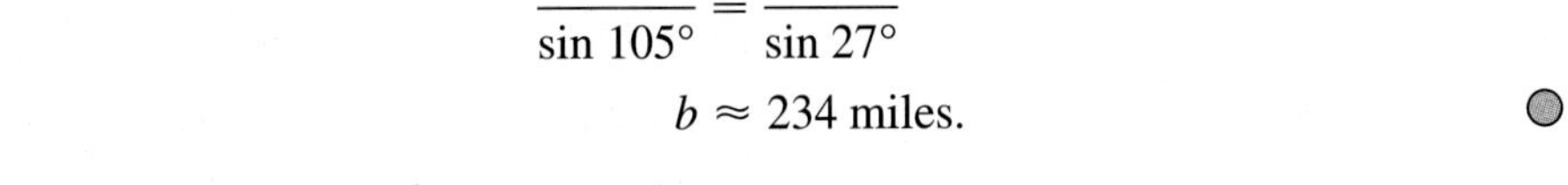

$$\frac{b}{\sin 105°} = \frac{110}{\sin 27°}$$
$$b \approx 234 \text{ miles.}$$

## Ambiguous Case

If we are given the lengths of two sides and the angle opposite one of them (Case 2, SSA), it is possible that 0, 1, or 2 such triangles exist. (Recall that there is no SSA congruence theorem.) To illustrate, suppose that the measure of acute angle $A$ of triangle $ABC$, the length of side $a$, and the length of side $b$ are given. Draw angle $A$ having a terminal side of length $b$. Now draw a side of length $a$ opposite angle $A$. The table on the next page shows possible outcomes. This situation (SSA) is called the **ambiguous case** of the law of sines.

If angle $A$ is acute, there are four possible outcomes.

| Number of Possible Triangles | Sketch | Conditions Necessary for Case to Hold |
|---|---|---|
| 0 | 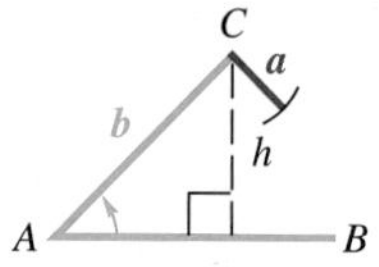 | $\sin B > 1$, $a < h < b$ |
| 1 | 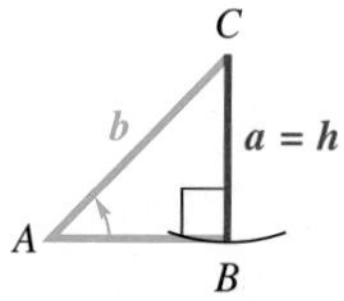 | $\sin B = 1$, $a = h < b$ |
| 1 | 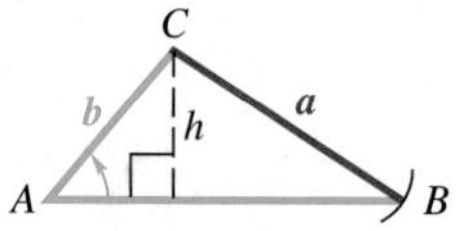 | $0 < \sin B < 1$, $a \geq b$ |
| 2 | 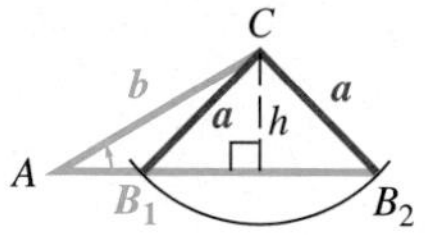 | $0 < \sin B_2 < 1$, $a < b$ |

If angle $A$ is obtuse, there are two possible outcomes.

| Number of Possible Triangles | Sketch | Conditions Necessary for Case to Hold |
|---|---|---|
| 0 | 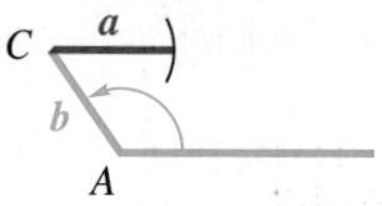 | $\sin B \geq 1$, $a \leq b$ |
| 1 | 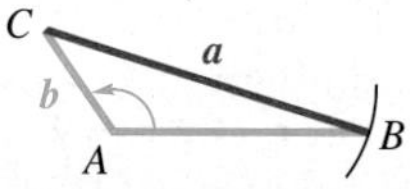 | $0 < \sin B < 1$, $a > b$ |

We can apply the law of sines to the values of $a$, $b$, and $A$ and use some basic properties of geometry and trigonometry to determine which situation applies. The following basic facts should be kept in mind.

1. For any angle $\theta$ of a triangle, $0 < \sin \theta \leq 1$. If $\sin \theta = 1$, then $\theta = 90°$ and the triangle is a right triangle.
2. $\sin \theta = \sin(180° - \theta)$ (That is, supplementary angles have the same sine value.)
3. The smallest angle is opposite the shortest side, the largest angle is opposite the longest side, and the middle-valued angle is opposite the intermediate side (assuming the triangle has sides that are all of different lengths).

**EXAMPLE 4** *Solving the Ambiguous Case (No Such Triangle)*

Solve triangle $ABC$ if $B = 55° \, 40'$, $b = 8.94$ meters, and $a = 25.1$ meters.

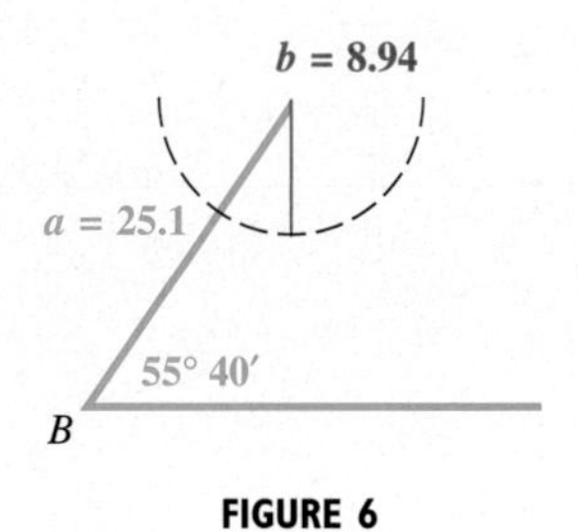


FIGURE 6

**TECHNOLOGY NOTE**

Because there are 60′ in 1°, the expression sin 55° 40′ in Example 4 can be evaluated by entering sin(55 + 40/60) into your calculator. Be sure to use degree mode.

**Solution** Since we are given $B$, $b$, and $a$, we use the law of sines to find $A$.

$$\frac{\sin A}{a} = \frac{\sin B}{b}$$

$$\frac{\sin A}{25.1} = \frac{\sin 55^\circ 40'}{8.94} \qquad \text{Substitute the given values.}$$

$$\sin A = \frac{25.1 \sin 55^\circ 40'}{8.94}$$

$$\sin A \approx 2.3184379$$

Since $\sin A$ cannot be greater than 1, there can be no such angle $A$ and thus, no triangle with the given information. An attempt to sketch such a triangle leads to the situation shown in Figure 6.

**NOTE** In the ambiguous case we are given SSA; that is, two sides and an angle opposite one of the sides. For example, suppose that $b$, $c$, and angle $C$ are given. This situation represents the ambiguous case because angle $C$ is opposite side $c$.

**EXAMPLE 5** *Solving the Ambiguous Case (Two Triangles)*

Solve triangle $ABC$ if $A = 55.3°$, $a = 22.8$ feet, and $b = 24.9$ feet.

**Solution** To begin, use the law of sines to find angle $B$.

$$\frac{\sin A}{a} = \frac{\sin B}{b}$$

$$\frac{\sin 55.3^\circ}{22.8} = \frac{\sin B}{24.9}$$

$$\sin B = \frac{24.9 \sin 55.3^\circ}{22.8}$$

$$\sin B \approx .8978678$$

Since $\sin B \approx .8978678$, to the nearest tenth one value of $B$ is

$$B_1 = 63.9^\circ.$$

Supplementary angles have the same sine value, so another *possible* value of $B$ is

$$B_2 = 180^\circ - 63.9^\circ = 116.1^\circ.$$

To see if $B_2 = 116.1°$ is a valid possibility, simply add 116.1° to the measure of $A$, 55.3°. Since $116.1° + 55.3° = 171.4°$, and this sum is less than 180°, it is a valid angle measure for this triangle.

Now separately solve triangles $AB_1C_1$ and $AB_2C_2$ shown in Figure 7. Begin with $AB_1C_1$. Find $C_1$ first.

$$C_1 = 180^\circ - A - B_1 = 60.8^\circ$$

Now, use the law of sines to find $c_1$.

$$\frac{a}{\sin A} = \frac{c_1}{\sin C_1}$$

$$\frac{22.8}{\sin 55.3^\circ} = \frac{c_1}{\sin 60.8^\circ}$$

$$c_1 = \frac{22.8 \sin 60.8^\circ}{\sin 55.3^\circ}$$

$$c_1 \approx 24.2 \text{ feet}$$

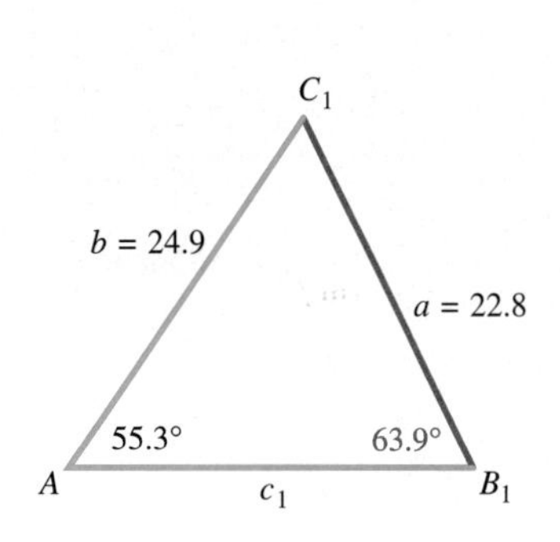


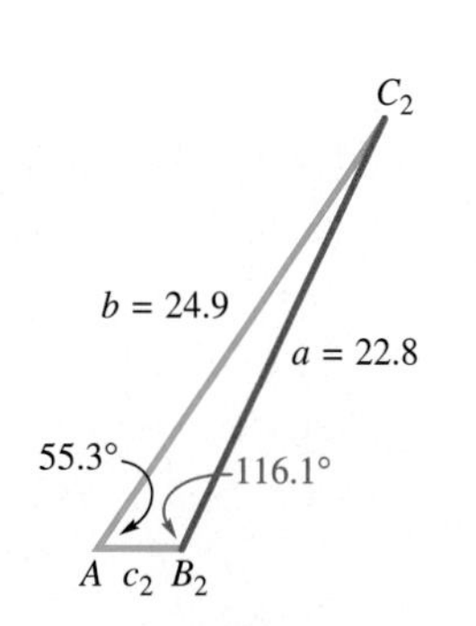


FIGURE 7

To solve triangle $AB_2C_2$, first find $C_2$.

$$C_2 = 180° - A - B_2 = 8.6°$$

By the law of sines,

$$\frac{a}{\sin A} = \frac{c_2}{\sin C_2}$$

$$\frac{22.8}{\sin 55.3°} = \frac{c_2}{\sin 8.6°}$$

$$c_2 = \frac{22.8 \sin 8.6°}{\sin 55.3°}$$

$$c_2 \approx 4.15 \text{ feet.}$$

The ambiguous case results in zero, one, or two triangles. The following guidelines can be used to determine how many triangles there are.

**Number of Triangles Satisfying the Ambiguous Case (SSA)**

Let sides $a$ and $b$ and angle $A$ be given in triangle $ABC$. (The law of sines can be used to calculate the value of $\sin B$.)

1. If $\sin B > 1$, then *no triangle* satisfies the given conditions.
2. If $\sin B = 1$, then *one triangle* satisfies the given conditions and $B = 90°$.
3. If $0 < \sin B < 1$, then either *one or two triangles* satisfy the given conditions.
   (a) If $\sin B = k$, then let $B_1 = \sin^{-1} k$ and use $B_1$ for $B$ in the first triangle.
   (b) Let $B_2 = 180° - B_1$. If $A + B_2 < 180°$, then a second triangle exists. In this case, use $B_2$ for $B$ in the second triangle.

**NOTE** The above guidelines can be applied whenever two sides and an angle opposite one of the sides are given.

**EXAMPLE 6** *Solving the Ambiguous Case (One Triangle)*

Solve triangle $ABC$, given $A = 43.5°$, $a = 10.7$ inches, and $c = 7.2$ inches.

**Solution** To find angle $C$, use the law of sines.

$$\frac{\sin C}{7.2} = \frac{\sin 43.5°}{10.7}$$

$$\sin C = \frac{7.2 \sin 43.5°}{10.7} \approx .46319186$$

$$C \approx 27.6° \qquad \text{Use inverse sine function.}$$

This is the acute angle. The other possible value of $C$ is

$$C = 180° - 27.6° = 152.4°.$$

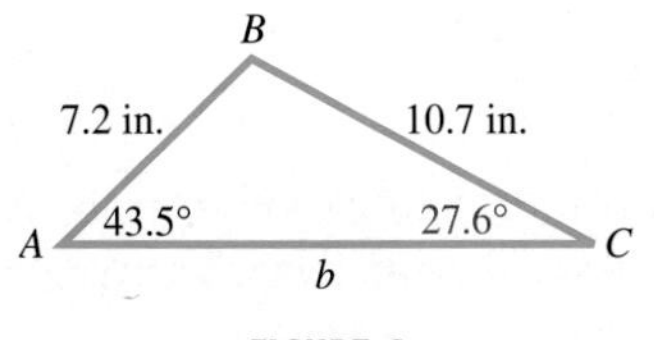


FIGURE 8

However, when we add this possible obtuse angle to the given angle $A = 43.5°$, we obtain

$$152.4° + 43.5° = 195.9°,$$

which is greater than 180°. So there can be only one triangle. See Figure 8. Then

$$B = 180° - 27.6° - 43.5° = 108.9°,$$

and side $b$ can be found with the law of sines.

$$\frac{b}{\sin B} = \frac{a}{\sin A}$$

$$\frac{b}{\sin 108.9°} = \frac{10.7}{\sin 43.5°}$$

$$b = \frac{10.7 \sin 108.9°}{\sin 43.5°}$$

$$b \approx 14.7 \text{ inches}$$

**EXAMPLE 7** *Analyzing Data Involving an Obtuse Angle*

Without using the law of sines, explain why $A = 104°$, $a = 26.8$ meters, and $b = 31.3$ meters cannot be valid for a triangle $ABC$.

**Solution** Since $A$ is an obtuse angle it is the largest angle and so the largest side of the triangle must be $a$. However, we are given $b > a$; thus $B > A$, which is impossible if $A$ is obtuse. Therefore, no such triangle $ABC$ exists.

**What Went WRONG?**

A student used the law of sines to solve for angle $B$ in a triangle $ABC$, where $b = 23$, $c = 14$, and $C = 52°$. He entered the calculation shown on the first screen and got the results in the second screen.

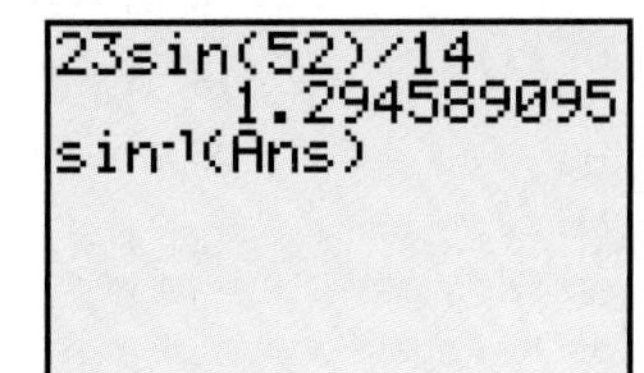

What Went WRONG? What is the solution for angle $B$?

---

*Answer to What Went Wrong?*

Since $\sin B > 1$, there is no solution for angle $B$. Triangle $ABC$ cannot exist.

## 10.1 EXERCISES

1. *Concept Check* Consider this oblique triangle.

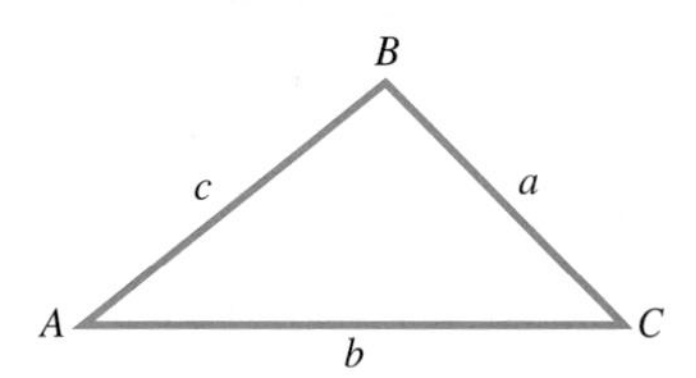


Which one of the following proportions is *not* valid?

A. $\frac{a}{b} = \frac{\sin A}{\sin B}$ B. $\frac{a}{\sin A} = \frac{b}{\sin B}$

C. $\frac{\sin A}{a} = \frac{b}{\sin B}$ D. $\frac{\sin A}{a} = \frac{\sin B}{b}$

2. *Concept Check* Which two of the following situations do not provide sufficient information for solving a triangle by the law of sines?

A. You are given two angles and the side included between them.

B. You are given two angles and a side opposite one of them.

C. You are given two sides and the angle included between them.

D. You are given three sides.

*Concept Check* *Given the following angles and sides, decide if solving triangle ABC results in the ambiguous case.*

3. $A$, $B$, and $a$
4. $A$, $C$, and $c$
5. $a$, $b$, and $c$
6. $A$, $a$, and $b$
7. $B$, $b$, and $c$
8. $A$, $b$, and $c$
9. $C$, $a$, and $c$
10. $A$, $B$, and $b$

***To the Student: Calculator Considerations***
*When making approximations, do not round off during intermediate steps—wait until the final step to give the appropriate approximation.*

*Solve each triangle.*

11. 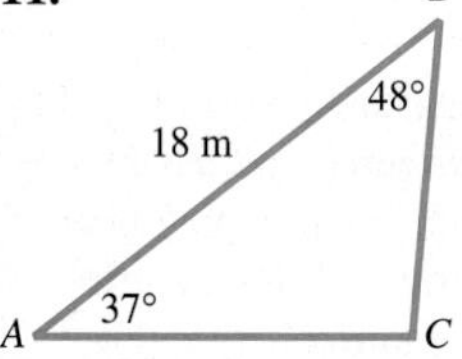


12. 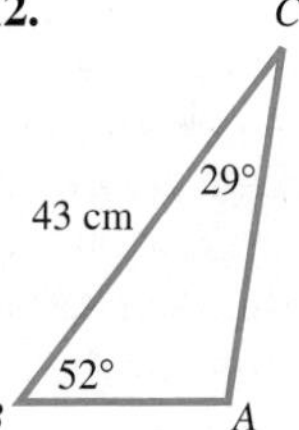


13. 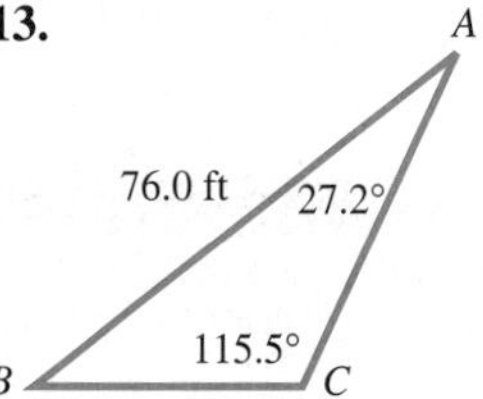


14. 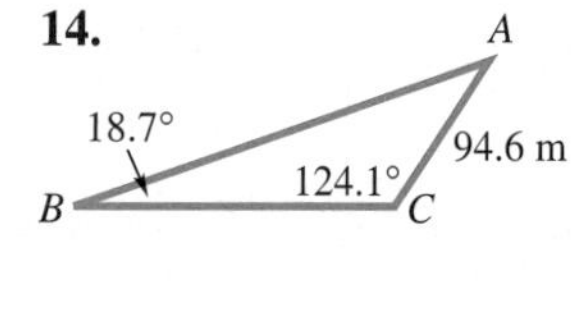


15. $A = 37°$, $C = 95°$, $c = 18$ meters
16. $B = 52°$, $C = 29°$, $a = 43$ centimeters
17. $C = 74.08°$, $B = 69.38°$, $c = 45.38$ meters
18. $A = 87.2°$, $b = 75.9$ yards, $C = 74.3°$
19. $B = 38° \, 40'$, $a = 19.7$ centimeters, $C = 91° \, 40'$
20. $B = 20° \, 50'$, $C = 103° \, 10'$, $b = 132$ feet
21. $A = 35.3°$, $B = 52.8°$, $b = 675$ feet
22. $A = 68.41°$, $B = 54.23°$, $a = 12.75$ feet

*Find the unknown angles in each triangle that exists.*

23. $A = 29.7°$, $b = 41.5$ feet, $a = 27.2$ feet
24. $B = 48.2°$, $a = 890$ centimeters, $b = 697$ centimeters
25. $B = 74.3°$, $a = 859$ meters, $b = 783$ meters
26. $C = 82.2°$, $a = 10.9$ kilometers, $c = 7.62$ kilometers
27. $A = 142.13°$, $b = 5.432$ feet, $a = 7.297$ feet
28. $B = 113.72°$, $a = 189.6$ yards, $b = 243.8$ yards

*Solve each triangle.*

29. $A = 42.5°$, $a = 15.6$ feet, $b = 8.14$ feet
30. $C = 52.3°$, $a = 32.5$ yards, $c = 59.8$ yards
31. $B = 72.2°$, $b = 78.3$ meters, $c = 145$ meters
32. $C = 68.5°$, $c = 258$ centimeters, $b = 386$ centimeters
33. $A = 38° \, 40'$, $a = 9.72$ kilometers, $b = 11.8$ kilometers
34. $C = 29° \, 50'$, $a = 8.61$ meters, $c = 5.21$ meters

**35.** $B = 32° 50'$, $a = 7540$ centimeters, $b = 5180$ centimeters

**36.** $C = 22° 50'$, $b = 159$ millimeters, $c = 132$ millimeters

**37.** In Example 1, we ask the question "Why does the Pythagorean theorem not apply?" Answer this question.

**38.** A trigonometry student makes the statement "If we know *any* two angles and one side of a triangle, then the triangle is uniquely determined." Is this a valid statement? Explain, referring to the congruence axioms given in this section.

**39.** Explain why the law of sines cannot be used to solve a triangle if we are given the lengths of the three sides of a triangle.

**40.** If $a$ is twice as long as $b$, is angle $A$ twice as large as angle $B$? Explain.

*Solve each problem.*

**41.** *Distance across a River* To find the distance $AB$ across a river, a distance $BC = 354$ meters is laid off on one side of the river. It is found that $B = 112° 10'$ and $C = 15° 20'$. (See the figure.) Find $AB$.

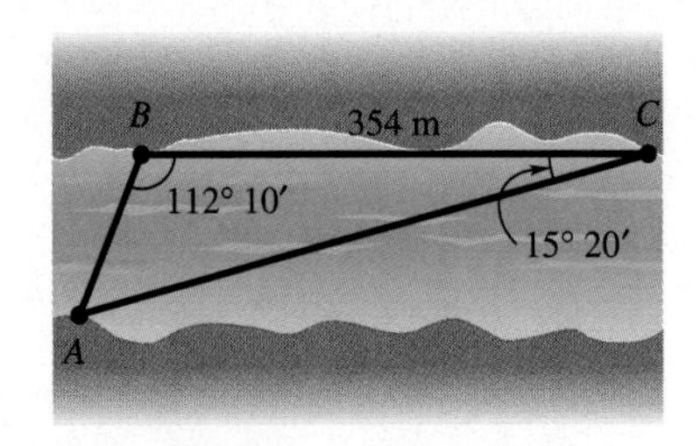


**42.** *Distance across a Canyon* To determine the distance $RS$ across a deep canyon, Joanna lays off a distance $TR = 582$ yards. She then finds that $T = 32° 50'$ and $R = 102° 20'$. (See the figure.) Find $RS$.

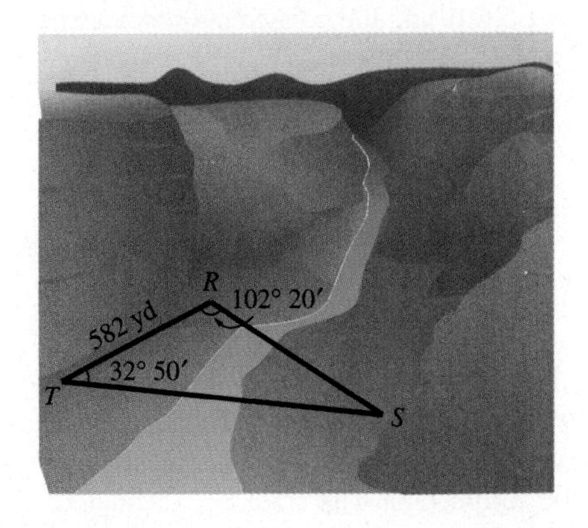


**43.** *Distance between Radio Direction Finders* Radio direction finders are at points $A$ and $B$, which are 3.46 miles apart on an east–west line, with $A$ west of $B$. From $A$ the bearing of a certain radio transmitter is 47.7°, and from $B$ the bearing is 302.5°. Find the distance of the transmitter from $A$.

**44.** *Distance a Ship Travels* A ship is sailing due north. At a certain point, the bearing of a lighthouse 12.5 kilometers away is N 38.8° E. Later on, the captain notices that the bearing of the lighthouse has become S 44.2° E. How far did the ship travel between the two observations of the lighthouse?

**45.** *Measurement of a Folding Chair* A folding chair is to have a seat 12.0 inches deep with angles, as shown in the figure. How far down from the seat should the crossing legs be joined? (*Hint:* Find $x$ in the figure.)

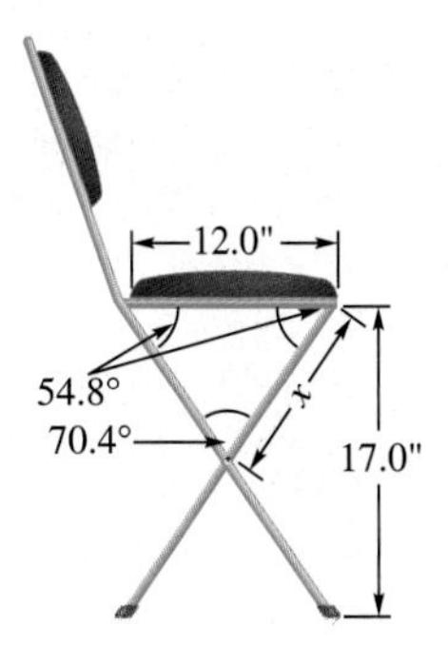


**46.** *Distance across a River* Mark notices that the bearing of a tree on the opposite bank of a river flowing north is 115.45°. Lisa is on the same bank as Mark, but 428.3 meters away. She notices that the bearing of the tree is 45.47°. The two banks are parallel. What is the distance across the river?

**47.** *Angle Formed by Centers of Gears* Three gears are arranged, as shown in the figure. Find angle $\theta$.

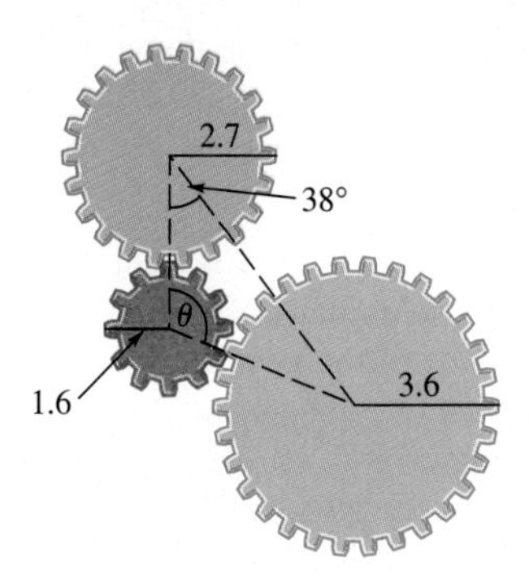


**48.** *Distance between Atoms* Three atoms with atomic radii 2.0, 3.0, and 4.5 are arranged as in the figure on the next page. Find the distance between the centers of atoms $A$ and $C$.

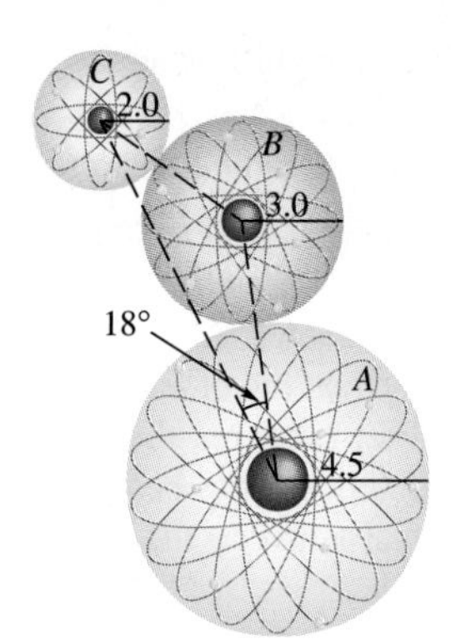


**49.** *Distance between a Ship and a Lighthouse* The bearing of a lighthouse from a ship was found to be N 37° E. After the ship sailed 2.5 miles due south, the new bearing was N 25° E. Find the distance between the ship and the lighthouse at each location.

**50.** *Height of a Balloon* A balloonist is directly above a straight road 1.5 miles long that joins two towns. She finds that the town closer to her is at an angle of depression of 35° and the farther town is at an angle of depression of 31°. How high above the ground is the balloon?

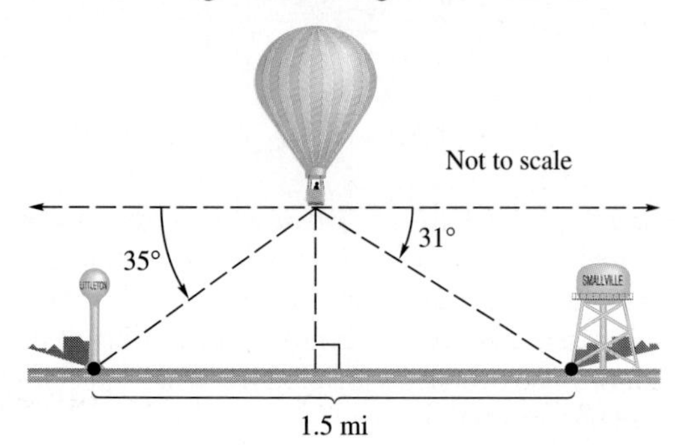


**51.** *Distance from Ship to Shore* From shore station A, a ship C is observed in the direction N 22.4° E. The same ship is observed to be in the direction N 10.6° W from shore station B, located at a distance of 25.5 kilometers exactly southeast of A. Find the distance of the ship from station A.

**52.** *Height of a Helicopter* A helicopter is sighted at the same time by two ground observers who are 3 miles apart on the same side of the helicopter. (See the figure.) They report the angles of elevation as 20.5° and 27.8°. How high is the helicopter?

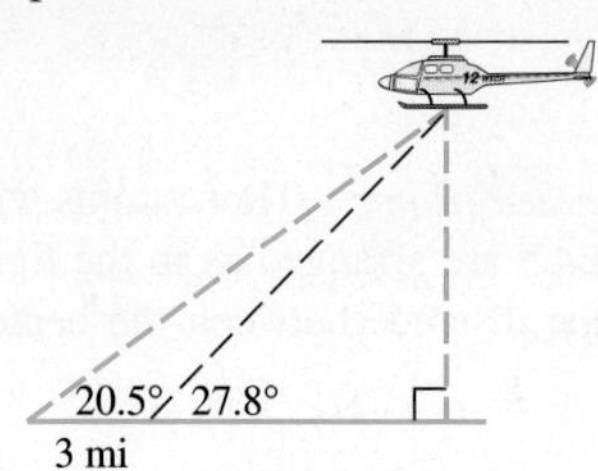


**53.** *Distance from a Rocket to a Radar Station* A rocket tracking facility has two radar stations $T_1$ and $T_2$, placed 1.73 kilometers apart, that lock onto the rocket and continuously transmit the angles of elevation to a computer. Find the distance to the rocket from $T_1$ at the moment when the angles of elevation are 28.1° and 79.5°, as shown in the figure.

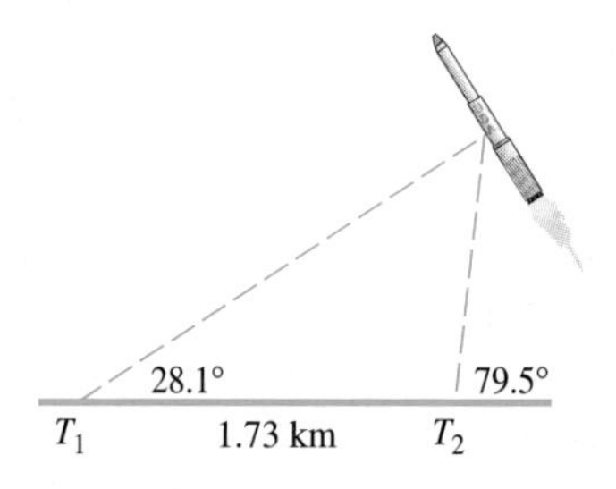


**54.** *Height of a Clock Tower* A surveyor standing 48.0 meters from the base of a building measures the angle to the top of the building and finds it to be 37.4°. The surveyor then measures the angle to the top of a clock tower on the building, finding that it is 45.6°. Find the height of the clock tower.

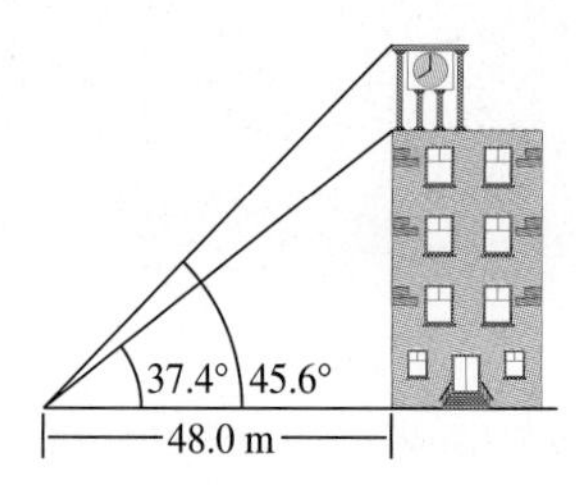


**55.** *Distance between a Pin and a Rod* A slider crank mechanism is shown in the figure. Find the distance between the wrist pin $W$ and the connecting rod center $C$.



**56.** *Path of a Satellite* A satellite is traveling in a circular orbit 1600 kilometers above Earth. It will pass directly over a tracking station at noon. The satellite takes 2 hours to make a complete orbit. Assume that the radius of Earth is 6400 kilometers. The tracking antenna is aimed 30° above the horizon. See the figure on the next page. At what time (before noon) will the satellite pass through the beam of the antenna? (*Source: Space Mathematics* by B. Kastner, Ph.D. Copyright © 1972 by the National Aeronautics and Space Administration. Courtesy of NASA.)

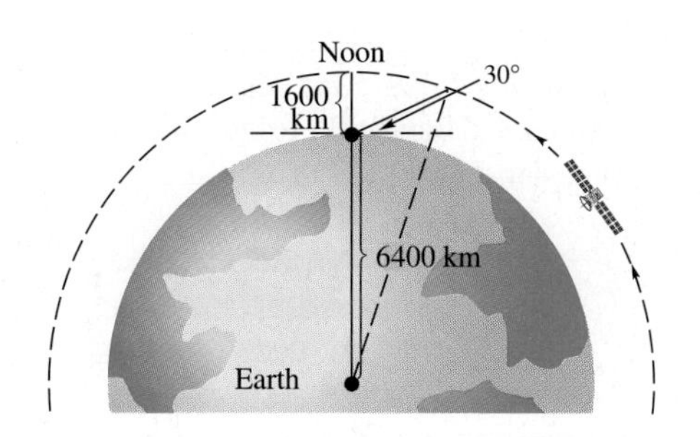


**57.** *Concept Check* In the left figure below, a line of length $a$ is to be drawn from the point $(3, 4)$ to the positive $x$-axis in order to form a triangle. For what value(s) of $a$ can you draw the following?

**(a)** Two triangles **(b)** Exactly one triangle

**(c)** No triangle

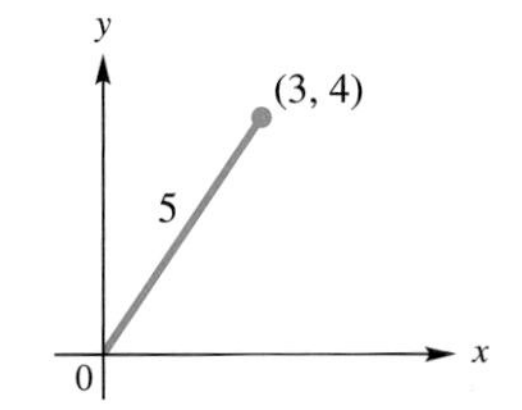


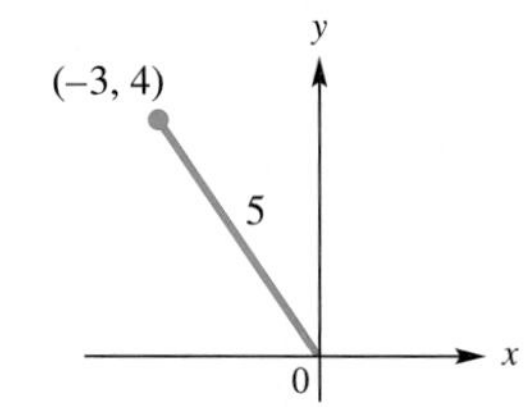


**58.** *Concept Check* In the right figure above, a line of length $a$ is to be drawn from the point $(-3, 4)$ to the positive $x$-axis in order to form a triangle. For what value(s) of $a$ can you draw the following?

**(a)** Two triangles **(b)** Exactly one triangle

**(c)** No triangle

**59.** Apply the law of sines to the following: $a = \sqrt{5}$, $c = 2\sqrt{5}$, $A = 30°$. What is the value of sin $C$? What is the measure of $C$? Based on its angle measures, what kind of triangle is triangle $ABC$?

**60.** Explain the condition that must exist to determine that there is no triangle satisfying the given values of $a$, $b$, and $B$, once the value of sin $A$ is found.

**61.** Without using the law of sines, explain why no triangle $ABC$ exists satisfying $A = 103° 20'$, $a = 14.6$ feet, and $b = 20.4$ feet.

**62.** Apply the law of sines to the data given in Example 7. Describe in your own words what happens when you try to find the measure of angle $B$, using a calculator.

**63.** *Survey of Property* A surveyor reported the following data about a piece of property: "The property is triangular in shape, with dimensions shown in the figure." Use the law of sines to see whether such a piece of property could exist.

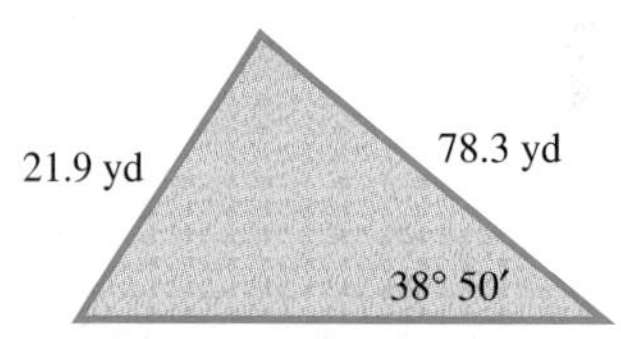


Can such a triangle exist?

**64.** *Survey of Property* The surveyor tries again: "A second triangular piece of property has dimensions shown in the figure." This time it turns out that the surveyor did not consider every possible case. Use the law of sines to show why.

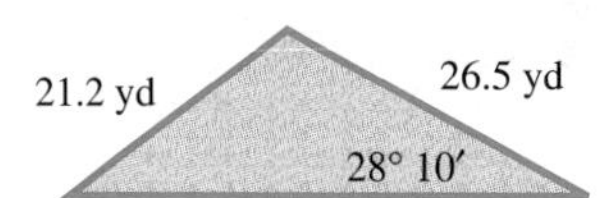


## RELATING CONCEPTS

**For individual or group investigation (Exercises 65–69)**

*In any triangle, the longest side is opposite the largest angle. This result from geometry can be proven using trigonometry. To prove it for acute triangles, work Exercises 65–69 in order. (The case for obtuse triangles will be considered in the Section 10.2 Exercises.)*

**65.** Is the graph of the function $y = \sin x$ increasing or decreasing over the interval $\left[0, \frac{\pi}{2}\right]$?

**66.** Suppose angle $A$ is the largest angle of an acute triangle, and let $B$ be an angle smaller than $A$. Explain why $\dfrac{\sin B}{\sin A} < 1$.

**67.** Solve for $b$ in the first form of the law of sines.

**68.** Use the results in Exercise 66 and 67 to show that $b < a$.

**69.** Use the result proved in Exercises 65–68 to explain why no triangle $ABC$ satisfies $A = 83°$, $a = 14$, and $b = 20$.

**70.** *Distance to the Moon* Since the moon is a relatively close celestial object, its distance can be measured directly by taking two photographs of the moon at precisely the same time from two different locations. The moon will have a different angle of elevation at each location. On April 29, 1976, at 11:35 A.M., the lunar angles of elevation during a partial solar eclipse at Bochum in upper Germany and at Donaueschingen in lower Germany were measured as 52.6997° and 52.7430°, respectively. (See the figure.) The two cities are 398 kilometers apart. Calculate the distance to the moon from Bochum on this day and compare it with the actual value of 406,000 kilometers. Disregard the curvature of Earth in this calculation. (*Source:* Scholosser, W., T. Schmidt-Kaler, and E. Milone, *Challenges of Astronomy,* Springer-Verlag, 1991.)

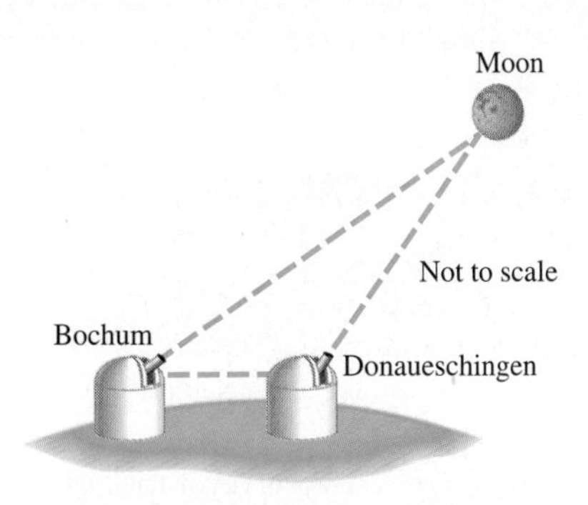


**71.** *Ground Distances Measured by Aerial Photography* The distance shown in an aerial photograph is determined by both the focal length of the lens and the tilt of the camera from the perpendicular to the ground. A camera lens with a 12-inch focal length has an angular coverage of 60°. If an aerial photograph is taken with this camera tilted 35° at an altitude of 5000 feet, calculate the distance $d$ in miles that will be shown in this photograph. (See the figure.) (*Sources:* Brooks, R., and D. Johannes, *Photoarchaeology,* Dioscorides Press, 1990; Moffitt, F., *Photogrammetry,* International Textbook Company, 1967.)

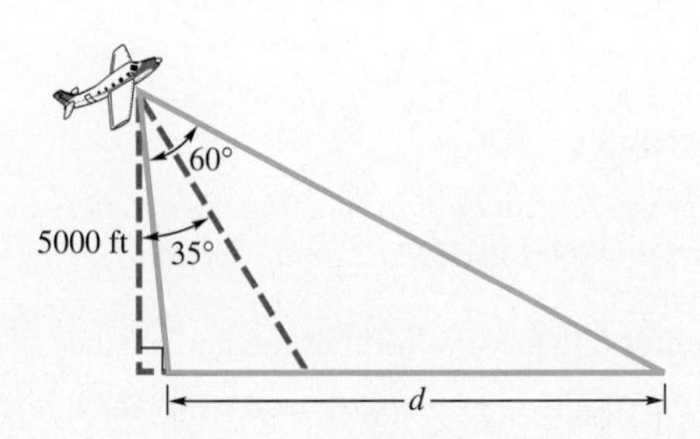


**72.** *U.S. Flag* The flag of the United States includes the colors red, white, and blue. Which color is predominant? Clearly the answer is either red or white. (It can be shown that only 18.73% of the total area is blue.) (*Sources: Slicing Pizzas, Racing Turtles, and Further Adventures in Applied Mathematics,* Banks, R., Princeton University Press, 1999; Schneider, D.)

**(a)** Let $R$ denote the radius of the circumscribing circle of a five-pointed star appearing on the American flag. The star can be decomposed into ten congruent triangles. In the figure, $r$ is the radius of the circumscribing circle of the pentagon in the interior of the star. Show that the area of a star is

$$A = \left[5 \frac{\sin A \sin B}{\sin(A + B)}\right] R^2.$$

(*Hint:* $\sin C = \sin[180° - (A + B)] = \sin(A + B)$.)

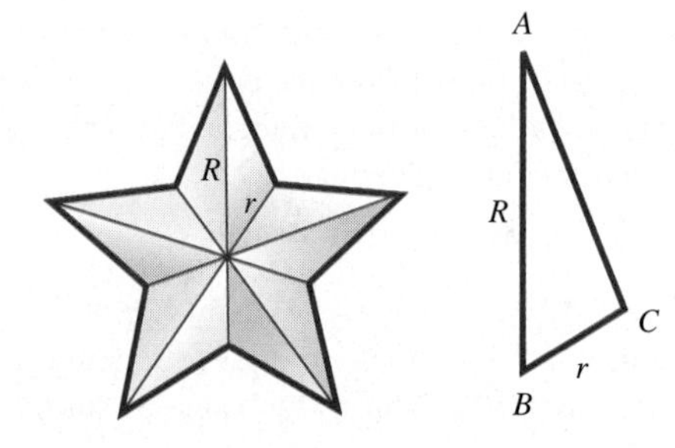


**(b)** Angles $A$ and $B$ have values 18° and 36°, respectively. Express the area of a star in terms of its radius, $R$.

**(c)** To determine whether red or white is predominant, we must know the measurements of the flag. Consider a flag of width 10 inches, length 19 inches, length of each upper stripe 11.4 inches, and radius $R$ of the circumscribing circle of each star .308 inch. The thirteen stripes consist of six matching pairs of red and white stripes and one additional red, upper stripe. Therefore, we must compare the area of a red, upper stripe with the total area of the 50 white stars.

**(i)** Compute the area of the red, upper stripe.

**(ii)** Compute the total area of the 50 white stars.

**(iii)** Which color occupies the greatest area on the flag?

# 10.2 THE LAW OF COSINES AND AREA FORMULAS

Derivation of the Law of Cosines • Applications • Area Formulas

As mentioned in Section 10.1, if we are given two sides and the included angle (Case 3) or three sides (Case 4) of a triangle, then a unique triangle is formed. These are the SAS and SSS cases, respectively. Both cases require using the law of cosines, introduced in this section.

Remember the following property of triangles when applying the law of cosines.

> **Triangle Side Length Restriction**
>
> In any triangle, the sum of the lengths of any two sides must be greater than the length of the remaining side.

For example, it would be impossible to construct a triangle with sides of lengths 3, 4, and 10. See Figure 9.

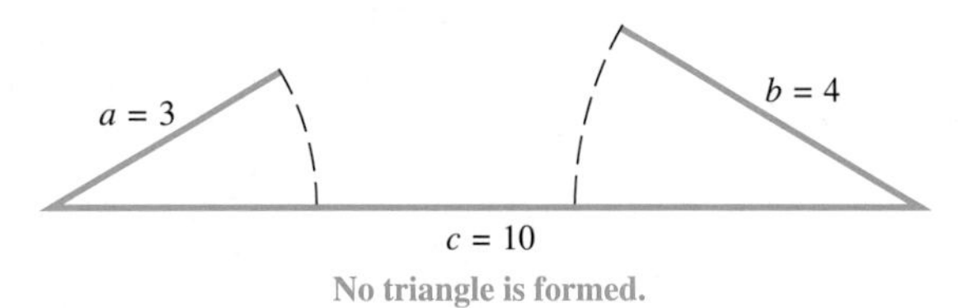


**FIGURE 9**

## Derivation of the Law of Cosines

To derive the law of cosines, let $ABC$ be any oblique triangle. Choose a coordinate system so that vertex $B$ is at the origin and side $BC$ is along the positive $x$-axis. See Figure 10.

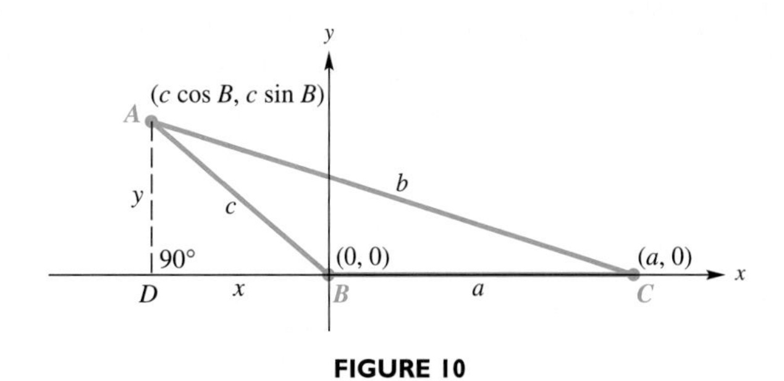


**FIGURE 10**

Let $(x, y)$ be the coordinates of vertex $A$ of the triangle. For angle $B$, whether obtuse or acute,

$$\sin B = \frac{y}{c} \quad \text{and} \quad \cos B = \frac{x}{c}.$$

(Here we assume that $x$ is negative if $B$ is obtuse.) From these results,

$$y = c \sin B \quad \text{and} \quad x = c \cos B,$$

so the coordinates of point $A$ become $(c \cos B, c \sin B)$.

Point $C$ has coordinates $(a, 0)$, and $AC$ has length $b$. By the distance formula,

$$b = \sqrt{(c \cos B - a)^2 + (c \sin B - 0)^2}$$

$$b^2 = (c \cos B - a)^2 + (c \sin B)^2 \qquad \text{Square both sides.}$$

$$= c^2 \cos^2 B - 2ac \cos B + a^2 + c^2 \sin^2 B$$

$$= a^2 + c^2(\cos^2 B + \sin^2 B) - 2ac \cos B$$

$$= a^2 + c^2(1) - 2ac \cos B$$

or $\quad b^2 = a^2 + c^2 - 2ac \cos B.$

This result is one form of the law of cosines. In our work, we could just as easily have placed $A$ or $C$ at the origin. This would have given the same result, but with the variables rearranged.

**Law of Cosines**

In any triangle $ABC$, with sides $a$, $b$, and $c$,

$$\mathbf{a^2 = b^2 + c^2 - 2bc \cos A,}$$

$$\mathbf{b^2 = a^2 + c^2 - 2ac \cos B,}$$

$$\mathbf{c^2 = a^2 + b^2 - 2ab \cos C.}$$

**NOTE** If we let $C = 90°$ in the third form of the law of cosines, we have $\cos C = \cos 90° = 0$, and the formula becomes $c^2 = a^2 + b^2$, the familiar equation of the Pythagorean theorem. Thus, the Pythagorean theorem is a special case of the law of cosines.

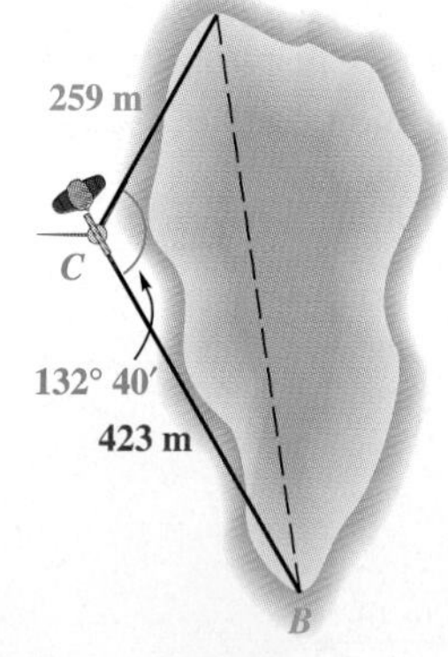

FIGURE 11

## Applications

**EXAMPLE 1** *Using the Law of Cosines in an Application (SAS)*

A surveyor wishes to find the distance between two points $A$ and $B$ on opposite sides of a lake. While standing at point $C$, she finds that $AC = 259$ meters, $BC = 423$ meters, and angle $ACB$ measures $132° 40'$. Find the distance $AB$. (See Figure 11.)

**Solution** The law of cosines can be used here, since we know the lengths of two sides of the triangle and the measure of the included angle.

$$AB^2 = 259^2 + 423^2 - 2(259)(423) \cos 132° 40'$$

$$AB^2 \approx 394{,}510.6 \qquad \text{Use a calculator.}$$

$$AB \approx 628 \qquad \text{Take the square root of each side.}$$

The distance between the points is approximately 628 meters.

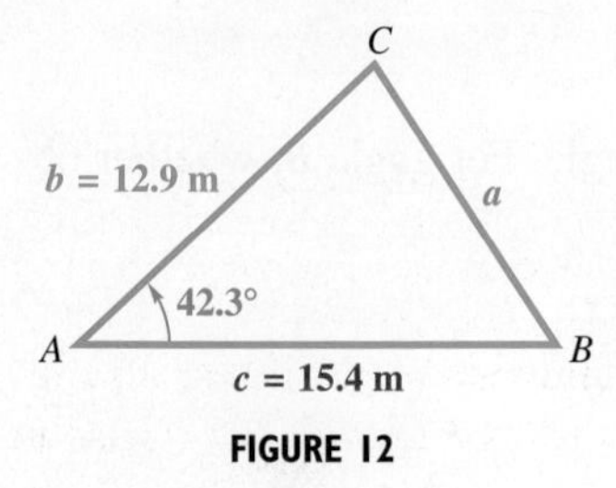

FIGURE 12

**EXAMPLE 2** *Using the Law of Cosines to Solve a Triangle (SAS)*

Solve triangle $ABC$ if $A = 42.3°$, $b = 12.9$ meters, and $c = 15.4$ meters. (See Figure 12.)

**Solution** We start by finding $a$ with the law of cosines.

$$a^2 = b^2 + c^2 - 2bc \cos A$$
$$a^2 = 12.9^2 + 15.4^2 - 2(12.9)(15.4) \cos 42.3°$$
$$a^2 \approx 109.7$$
$$a \approx 10.47 \text{ meters}$$

We now must find the measures of angles $B$ and $C$. Of the two remaining angles, $B$ must be the smaller since it is opposite the shorter of the two sides $b$ and $c$. Therefore, it cannot be obtuse.

$$\frac{\sin 42.3°}{10.47} = \frac{\sin B}{12.9} \quad \text{Law of sines}$$
$$\sin B = \frac{12.9 \sin 42.3°}{10.47}$$
$$B \approx 56.0° \quad \text{Use inverse sine function.}$$

The easiest way to find $C$ is to subtract the measures of $A$ and $B$ from 180°.

$$C = 180° - A - B = 81.7°$$

**TECHNOLOGY NOTE**

Graphing calculators can be programmed to solve triangles. See Example 3.

**CAUTION** Had we chosen to use the law of sines to find $C$ rather than $B$ in Example 2, we would not have known whether $C$ equals 81.7° or its supplement, 98.3°.

**EXAMPLE 3** *Using the Law of Cosines to Solve a Triangle (SSS)*

Solve triangle $ABC$ if $a = 9.47$ feet, $b = 15.9$ feet, and $c = 21.1$ feet.

**Analytic Solution**

Given the lengths of three sides of the triangle, we can use the law of cosines to solve for any angle of the triangle. We solve for $C$, the largest angle, using the law of cosines. We will know that $C$ is obtuse if $\cos C < 0$.

$$c^2 = a^2 + b^2 - 2ab \cos C \quad \text{Law of cosines}$$
$$\cos C = \frac{a^2 + b^2 - c^2}{2ab} \quad \text{Solve for } \cos C.$$
$$\cos C = \frac{9.47^2 + 15.9^2 - 21.1^2}{2(9.47)(15.9)} \quad \text{Substitute.}$$
$$\cos C \approx -.34109402 \quad \text{Use a calculator.}$$

Using the inverse cosine function, we find obtuse angle $C$.

$$C \approx 109.9°$$

We can use either the law of sines or the law of cosines to find $B \approx 45.1°$. (Verify this.) Since $A = 180° - B - C$,

$$A \approx 25.0°.$$

**Graphing Calculator Solution**

This triangle can also be solved using a program on a graphing calculator. In this case, the inputs are the three sides. See Figure 13(a). Figure 13(b) shows the output.

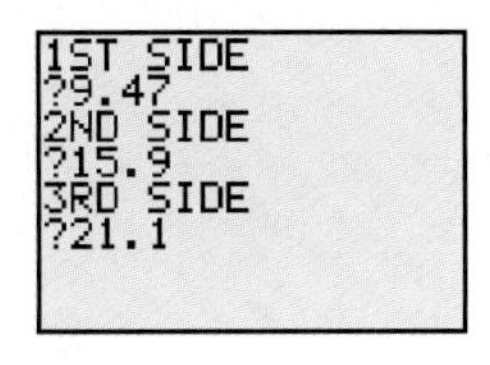


(a)

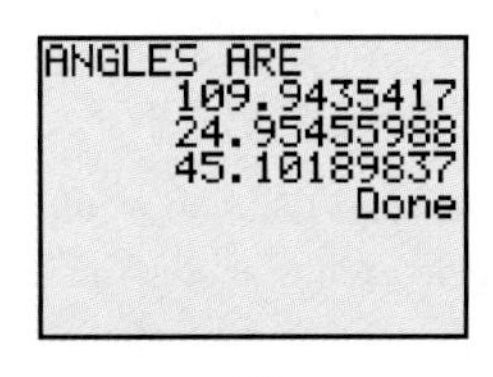


(b)

**FIGURE 13**

Trusses are frequently used to support roofs on buildings, as illustrated in Figure 14(a) on the next page. The simplest type of roof truss is a triangle, as shown in

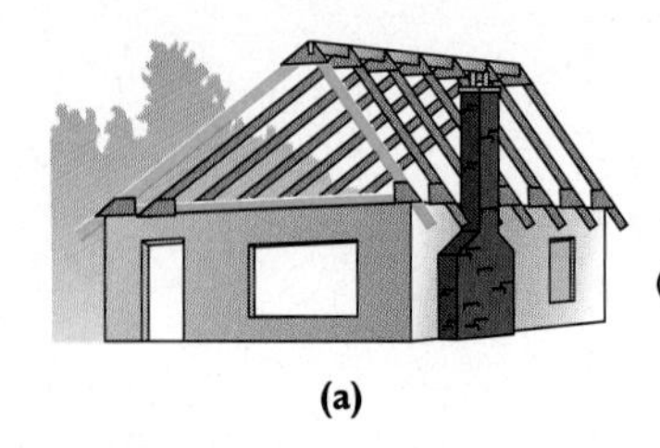

(a)

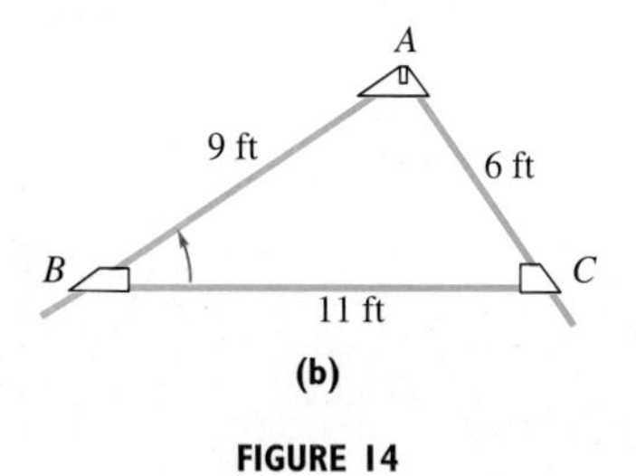


(b)

FIGURE 14

Figure 14(b). One basic task when constructing a roof truss is to cut the ends of the rafters so that the roof has the correct slope. (*Source:* Riley, W., L. Sturges, and D. Morris, *Statics and Mechanics of Materials,* John Wiley and Sons, 1995.)

**EXAMPLE 4** *Designing a Roof Truss (SSS)*

Find angle $B$ for the truss shown in Figure 14(b).

**Solution** Let $a = 11$, $b = 6$, and $c = 9$ in the law of cosines.

$$b^2 = a^2 + c^2 - 2ac \cos B \quad \text{Law of cosines}$$

$$2ac \cos B = a^2 + c^2 - b^2 \quad \text{Transpose terms.}$$

$$\cos B = \frac{a^2 + c^2 - b^2}{2ac} \quad \text{Divide by } 2ac.$$

$$\cos B = \frac{11^2 + 9^2 - 6^2}{2(11)(9)} \quad \text{Substitute.}$$

$$\cos B \approx .8384 \quad \text{Use a calculator.}$$

Thus, $B \approx \cos^{-1} .8384 \approx 33°$.

Four possible cases can occur when solving an oblique triangle. These cases are summarized in the following table, along with a suggested procedure for solving in each case. There are other procedures that will work, but we give the one that is usually most efficient. In all four cases, it is assumed that the given information actually produces a triangle.

| Oblique Triangle | Suggested Procedure for Solving |
|---|---|
| **Case 1:** One side and two angles are known. **(SAA or ASA)** | **1.** Find the remaining angle using the angle sum formula ($A + B + C = 180°$). **2.** Find the remaining sides using the law of sines. |
| **Case 2:** Two sides and one angle (not included between the two sides) are known. **(SSA)** | *This is the ambiguous case; there may be no triangle, one triangle, or two triangles.* **1.** Find an angle using the law of sines. **2.** Find the remaining angle using the angle sum formula. **3.** Find the remaining side using the law of sines. *If two triangles exist, repeat Steps 2 and 3.* |
| **Case 3:** Two sides and the included angle are known. **(SAS)** | **1.** Find the third side using the law of cosines. **2.** Find the smaller of the two remaining angles using the law of sines. **3.** Find the remaining angle using the angle sum formula. |
| **Case 4:** Three sides are known. **(SSS)** | **1.** Find the largest angle using the law of cosines. **2.** Find either remaining angle using the law of sines. **3.** Find the remaining angle using the angle sum formula. |

**FOR DISCUSSION**

Discuss the steps that you would take to solve each triangle. You do not actually need to solve the triangle.

1. $A = 32°, B = 67°, a = 20$
2. $b = 22, c = 15, A = 39°$
3. $a = 5, c = 7, C = 22°$
4. $b = 13, c = 15, B = 22°$
5. $a = 5, b = 7, c = 9$

## Area Formulas

The law of cosines can be used to derive a formula for the area of a triangle given the lengths of the three sides. This formula is known as **Heron's formula,** named after the Greek mathematician Heron of Alexandria, who lived around A.D. 75. It is found in his work *Metrica.* Heron's formula can be used for the case SSS.

**Heron's Area Formula (SSS)**

If a triangle has sides of lengths $a$, $b$, and $c$, and if the **semiperimeter** is

$$s = \frac{1}{2}(a + b + c),$$

then the area of the triangle is

$$\mathcal{A} = \sqrt{s(s - a)(s - b)(s - c)}.$$

**EXAMPLE 5** *Using Heron's Formula to Find an Area*

The distance "as the crow flies" from Los Angeles to New York is 2451 miles, from New York to Montreal is 331 miles, and from Montreal to Los Angeles is 2427 miles. What is the area of the triangular region having these three cities as vertices? (Ignore the curvature of Earth.)

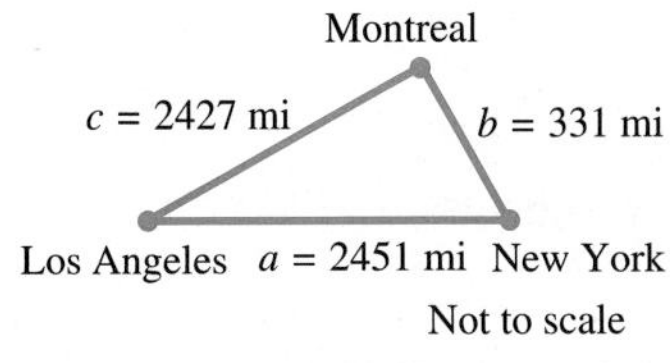


FIGURE 15

**Solution** In Figure 15 we let $a = 2451$, $b = 331$, and $c = 2427$. Then, the semiperimeter $s$ is

$$s = \frac{1}{2}(2451 + 331 + 2427) = 2604.5.$$

Using Heron's formula, the area $\mathcal{A}$ is

$$\mathcal{A} = \sqrt{s(s - a)(s - b)(s - c)}$$
$$\mathcal{A} = \sqrt{2604.5(2604.5 - 2451)(2604.5 - 331)(2604.5 - 2427)}$$
$$\mathcal{A} \approx 401{,}700 \text{ square miles.}$$

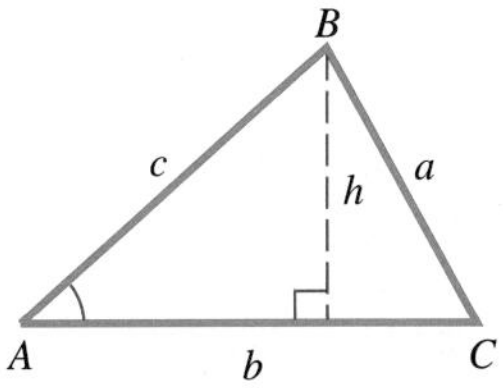

FIGURE 16

If we know the measures of two sides of a triangle and the angle included between them, we can find the area $\mathcal{A}$ of the triangle. The area $\mathcal{A}$ of any triangle is given by $\mathcal{A} = \frac{1}{2}bh$, where $b$ is its base and $h$ is its height. Using trigonometry, we can find a formula for the area of the triangle shown in Figure 16.

$$\sin A = \frac{h}{c} \quad \text{or} \quad h = c \sin A$$

Thus, the area equals

$$\mathcal{A} = \frac{1}{2}bh = \frac{1}{2}bc \sin A.$$

Since the labels for the vertices in triangle $ABC$ could be rearranged, three area formulas can be written. Notice that these formulas can be applied when given SAS.

**Area of a Triangle (SAS)**

In any triangle $ABC$, the area $\mathscr{A}$ is given by any of the following formulas:

$$\mathscr{A} = \frac{1}{2}bc \sin A, \quad \mathscr{A} = \frac{1}{2}ab \sin C, \quad \text{and} \quad \mathscr{A} = \frac{1}{2}ac \sin B.$$

That is, the area is half the product of the lengths of two sides and the sine of the angle included between them.

**EXAMPLE 6** *Finding the Area of a Triangle (SAS)*

Find the area of triangle $ABC$ in Figure 17.

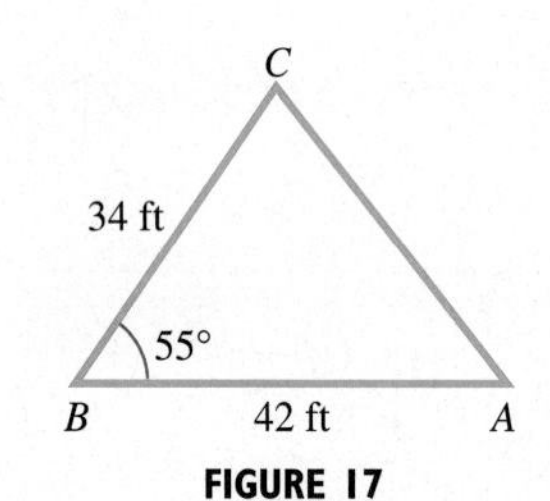

FIGURE 17

**Solution** We are given $B = 55°$, $a = 34$ feet, and $c = 42$ feet. Thus, the area $\mathscr{A}$ is

$$\mathscr{A} = \frac{1}{2}ac \sin B \qquad \text{Area formula}$$

$$= \frac{1}{2}(34)(42)\sin 55° \qquad \text{Substitute.}$$

$$\approx 585 \text{ square feet.} \qquad \text{Approximate.}$$

**EXAMPLE 7** *Finding the Area of a Triangle (ASA)*

Find the area of triangle $ABC$ if $A = 24° 40'$, $b = 27.3$ centimeters, and $C = 52° 40'$.

**Solution** Before the area formula can be used, we must find either $a$ or $c$. Since the sum of the measures of the angles of any triangle is 180°,

$$B = 180° - 24° 40' - 52° 40'$$
$$= 102° 40'.$$

Now, the law of sines can be used to find $a$.

$$\frac{a}{\sin A} = \frac{b}{\sin B}$$

$$\frac{a}{\sin 24° 40'} = \frac{27.3}{\sin 102° 40'}$$

Verify that $a \approx 11.7$ centimeters. Now, we find the area.

$$\mathscr{A} = \frac{1}{2}ab \sin C$$

$$= \frac{1}{2}(11.7)(27.3)\sin 52° 40'$$

$$\approx 127 \text{ square centimeters}$$

## What Went WRONG?

A student evaluates the expression in Example 7 to calculate the area of the triangle. The calculator screen shows the result, which does not equal 127 square centimeters.

What Went WRONG? What should the student do to correct the situation?

## 10.2 EXERCISES

*Concept Check* *Assume triangle ABC has standard labeling and complete the following.*

**(a)** *Determine whether* SAA, ASA, SSA, SAS, *or* SSS *is given.*

**(b)** *Decide whether the law of sines or the law of cosines should be used to begin solving the triangle.*

**1.** $a$, $b$, and $C$
**2.** $A$, $C$, and $c$
**3.** $a$, $b$, and $A$
**4.** $a$, $b$, and $c$
**5.** $A$, $B$, and $c$
**6.** $a$, $c$, and $A$
**7.** $a$, $B$, and $C$
**8.** $b$, $c$, and $A$

*Solve each triangle. Approximate values to the nearest tenth.*

**9.**

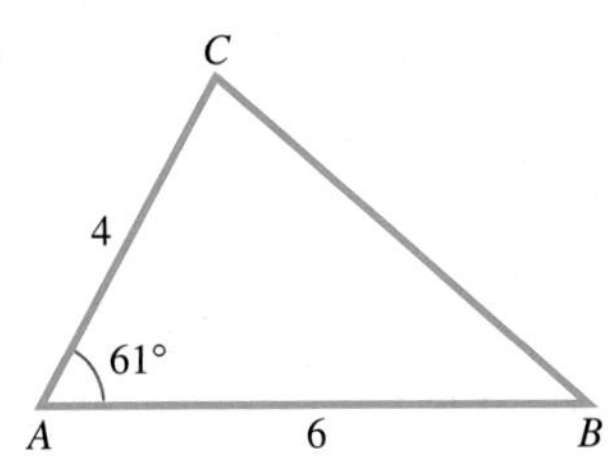


**10.**

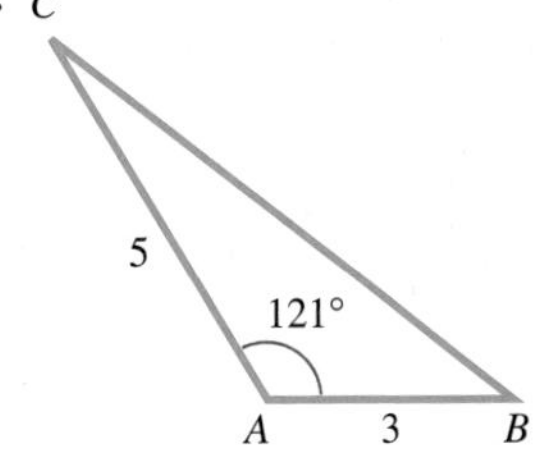


**11.**

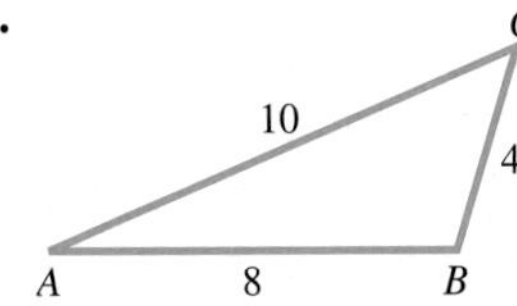


**12.**

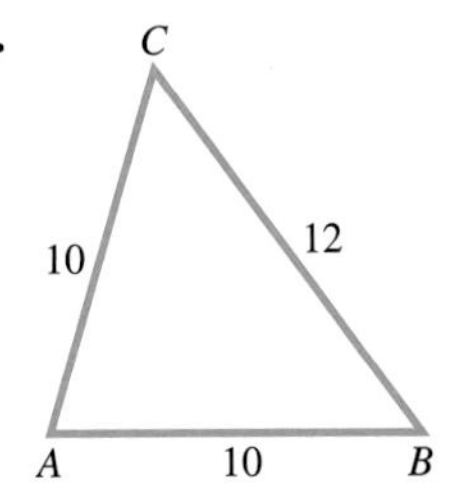


**13.**

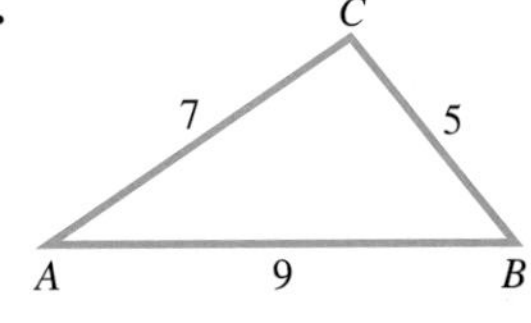


**14.**

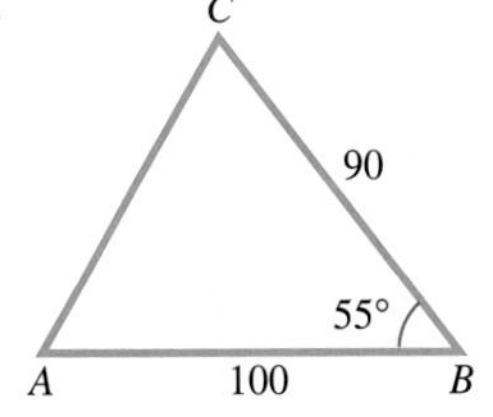


---

*Answer to What Went Wrong?*

The calculator should be set to degree mode rather than radian mode.

*Solve each triangle.*

**15.** $C = 28.3°$, $b = 5.71$ inches, $a = 4.21$ inches

**16.** $A = 41.4°$, $b = 2.78$ yards, $c = 3.92$ yards

**17.** $C = 45.6°$, $b = 8.94$ meters, $a = 7.23$ meters

**18.** $A = 67.3°$, $b = 37.9$ kilometers, $c = 40.8$ kilometers

**19.** $a = 9.3$ centimeters, $b = 5.7$ centimeters, $c = 8.2$ centimeters

**20.** $a = 28$ feet, $b = 47$ feet, $c = 58$ feet

**21.** $a = 42.9$ meters, $b = 37.6$ meters, $c = 62.7$ meters

**22.** $a = 189$ yards, $b = 214$ yards, $c = 325$ yards

**23.** $AB = 1240$ feet, $AC = 876$ feet, $BC = 965$ feet

**24.** $AB = 298$ meters, $AC = 421$ meters, $BC = 324$ meters

**25.** $A = 80° 40'$, $b = 143$ centimeters, $c = 89.6$ centimeters

**26.** $C = 72° 40'$, $a = 327$ feet, $b = 251$ feet

**27.** $B = 74.80°$, $a = 8.919$ inches, $c = 6.427$ inches

**28.** $C = 59.70°$, $a = 3.725$ miles, $b = 4.698$ miles

**29.** $A = 112.8°$, $b = 6.28$ meters, $c = 12.2$ meters

**30.** $B = 168.2°$, $a = 15.1$ centimeters, $c = 19.2$ centimeters

**31.** $a = 3.0$ feet, $b = 5.0$ feet, $c = 6.0$ feet

**32.** $a = 4.0$ feet, $b = 5.0$ feet, $c = 8.0$ feet

**33.** Refer to Figure 9. If you attempt to find any angle of a triangle with the values $a = 3$, $b = 4$, and $c = 10$ with the law of cosines, what happens?

**34.** "The shortest distance between two points is a straight line." Explain how this relates to the geometric property that states that the sum of the lengths of any two sides of a triangle must be greater than the remaining side.

*Solve each problem.*

**35.** *Distance across a Lake* Points $A$ and $B$ are on opposite sides of Lake Yankee. From a third point, $C$, the angle between the lines of sight to $A$ and $B$ is $46.3°$. If $AC$ is 350 meters long and $BC$ is 286 meters long, find $AB$.

**36.** *Diagonals of a Parallelogram* The sides of a parallelogram are 4.0 centimeters and 6.0 centimeters. One angle is $58°$ while another is $122°$. Find the lengths of the diagonals of the parallelogram.

**37.** *Flight Distance* Airports $A$ and $B$ are 450 kilometers apart, on an east–west line. Tom flies in a northeast direction from $A$ to airport $C$. From $C$ he flies 359 kilometers on a heading of $128° 40'$ to $B$. How far is $C$ from $A$?

**38.** *Distance between Two Ships* Two ships leave a harbor together, traveling on courses that have an angle of $135° 40'$ between them. If they each travel 402 miles, how far apart are they?

**39.** *Distance between a Ship and a Rock* A ship is sailing east. At one point, the bearing of a submerged rock is $45° 20'$. After the ship has sailed 15.2 miles, the bearing of the rock has become $308° 40'$. Find the distance of the ship from the rock at the latter point.

**40.** *Distance between a Ship and a Submarine* From an airplane flying over the ocean, the angle of depression to a submarine lying just under the surface is $24° 10'$. At the same moment, the angle of depression from the airplane to a battleship is $17° 30'$. (See the figure.) The distance from the airplane to the battleship is 5120 feet. Find the distance between the battleship and the submarine. (Assume the airplane, submarine, and battleship are in a vertical plane.)

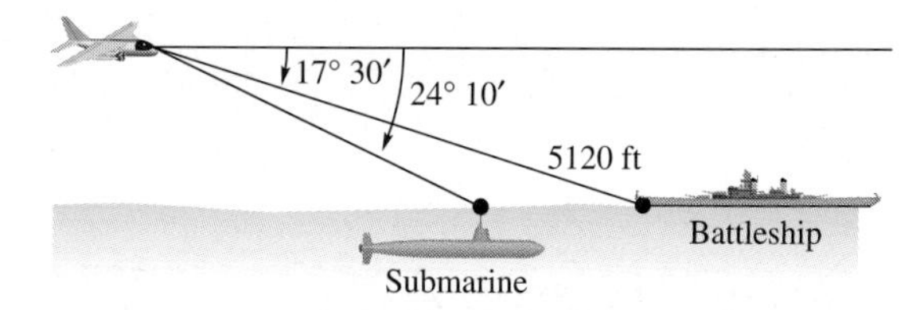


**41.** *Distance between Two Boats* Two boats leave a dock together. Each travels in a straight line. The angle between their courses measures $54° 10'$. One boat travels 36.2 kilometers per hour, and the other travels 45.6 kilometers per hour. How far apart will they be after 3 hours?

**42.** *Truss Construction* A triangular truss is shown in the figure. Find angle $\theta$.

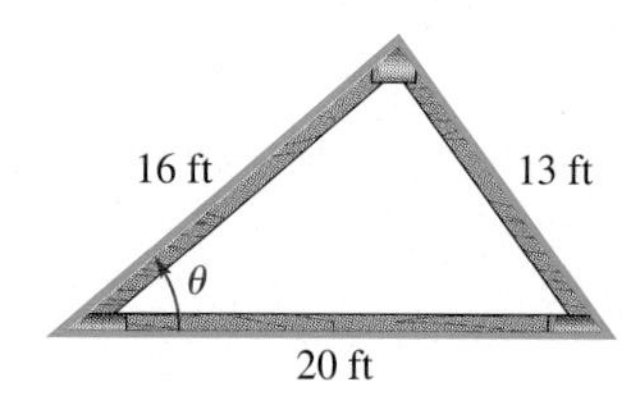


**43.** *Distance between Points on a Crane* A crane with a counterweight is shown in the figure. Find the horizontal distance between points $A$ and $B$.

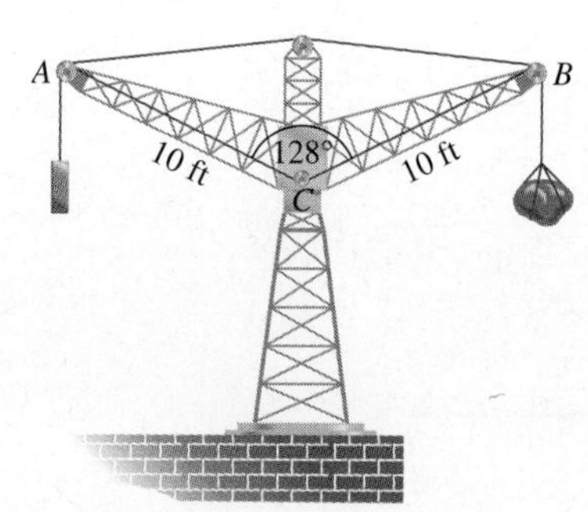


**44.** *Distance between a Beam and Cables* A weight is supported by cables attached to both ends of a balance beam, as shown in the figure on the next page. What angles are formed between the beam and the cables?

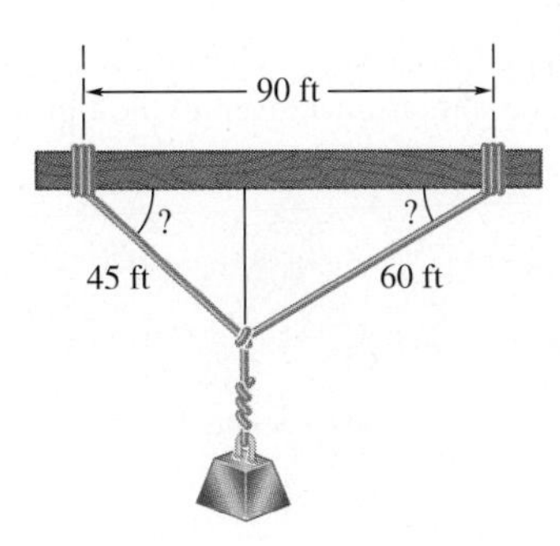


**45.** *Length of a Tunnel* To measure the distance through a mountain for a proposed tunnel, a point $C$ is chosen that can be reached from each end of the tunnel. (See the figure.) If $AC = 3800$ meters, $BC = 2900$ meters, and angle $C = 110°$, find the length of the tunnel.

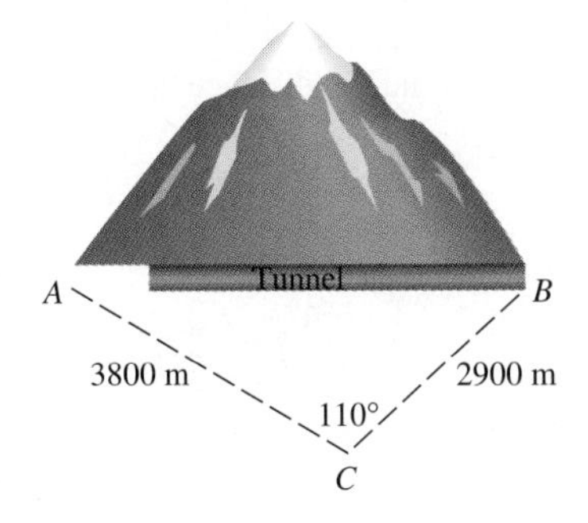


**46.** *Distance on a Baseball Diamond* A baseball diamond is a square, 90 feet on a side, with home plate and the three bases as vertices. The pitcher's rubber is located 60.5 feet from home plate. Find the distance from the pitcher's rubber to each of the bases.

**47.** *Distance between Ends of the Vietnam Memorial* The Vietnam Veterans' Memorial in Washington, D.C., is V-shaped with equal sides of length 246.75 feet, and the angle between these sides measuring 125° 12′. Find the distance between the ends of the two sides. (*Source:* Pamphlet obtained at Vietnam Veterans' Memorial.)

**48.** *Distance between a Ship and a Point* Starting at point $A$, a ship sails 18.5 kilometers on a bearing of 189°, then turns and sails 47.8 kilometers on a bearing of 317°. Find the distance of the ship from point $A$.

**49.** *Bearing of One Town to Another* Two towns 21 miles apart are separated by a dense forest. (See the figure.) To travel from town $A$ to town $B$, a person must go 17 miles on a bearing of 325°, then turn and continue for 9 miles to reach town $B$. Find the bearing of $B$ from $A$.

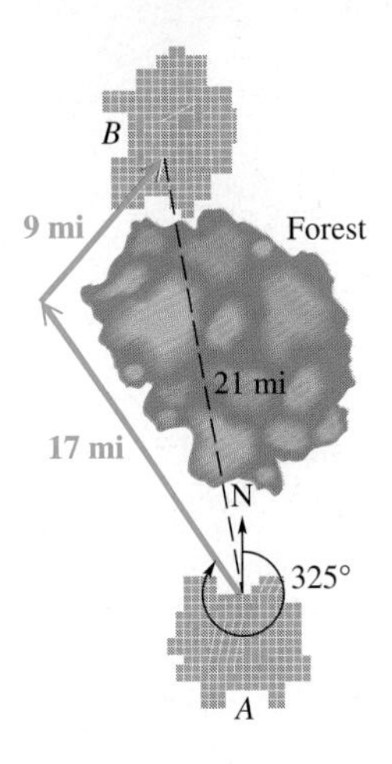


**50.** *Distance between Two Factories* Two factories blow their whistles at exactly 5:00. A man hears the two blasts at 3 seconds and 6 seconds after 5:00, respectively. The angle between his lines of sight to the two factories is 42.2°. If sound travels 344 meters per second, how far apart are the factories?

**51.** *Distance between a Satellite and a Tracking Station* A satellite traveling in a circular orbit 1600 kilometers above Earth is due to pass directly over a tracking station at noon. (See the figure.) Assume that the satellite takes 2 hours to make an orbit and that the radius of Earth is 6400 kilometers. Find the distance between the satellite and the tracking station at 12:03 P.M. (*Source: Space Mathematics* by B. Kastner, Ph.D. Copyright © 1972 by the National Aeronautics and Space Administration. Courtesy of NASA.)

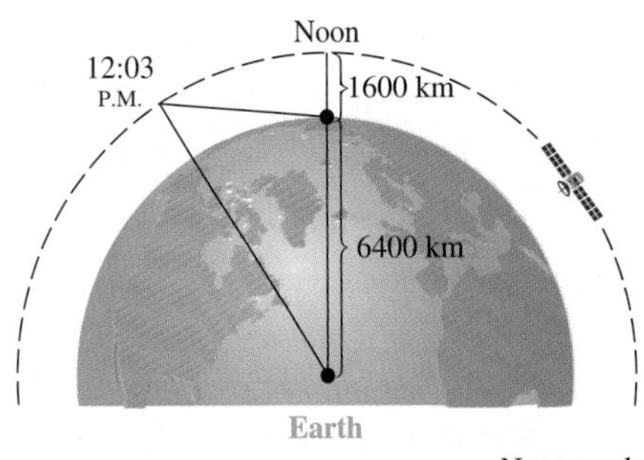


**52.** *Path of a Ship* A ship sailing due east in the North Atlantic has been warned to change course to avoid icebergs. The captain turns and sails on a bearing of 62°, then changes course again to a bearing of 115° until the ship reaches its original course. (See the figure.) How much farther did the ship have to travel to avoid the icebergs?

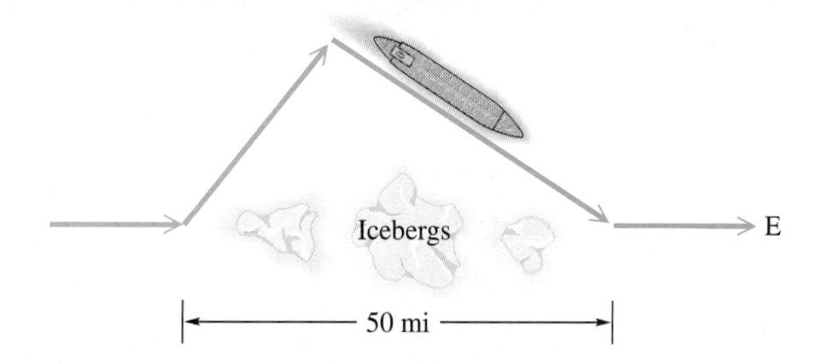


**53.** *Angle in a Parallelogram* A parallelogram has sides of length 25.9 centimeters and 32.5 centimeters. The longer diagonal has length 57.8 centimeters. Find the angle opposite the longer diagonal.

**54.** *Distance between an Airplane and a Mountain* A person in a plane flying straight north observes a mountain at a bearing of 24.1°. At that time, the plane is 7.92 kilometers from the mountain. A short time later, the bearing to the mountain becomes 32.7°. How far is the airplane from the mountain when the second bearing is taken?

**55.** *Layout for a Playhouse* The layout for a child's playhouse has the dimensions given in the figure. Find $x$.

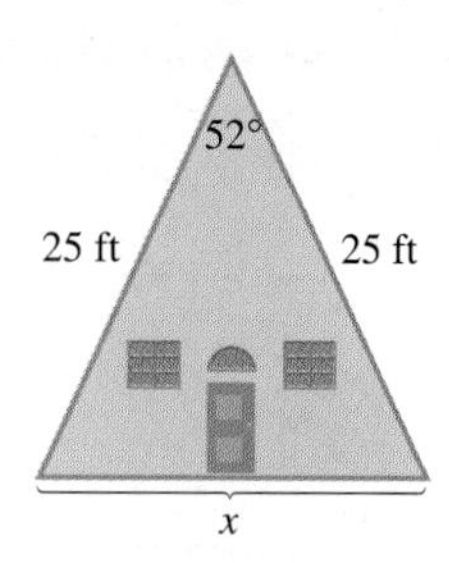


**56.** *Distance between Two Towns* To find the distance between two small towns, an electronic distance measuring (EDM) instrument is placed on a hill from which both towns are visible. The distance to each town from the EDM and the angle between the two lines of sight are measured. (See the figure.) Find the distance between the towns.

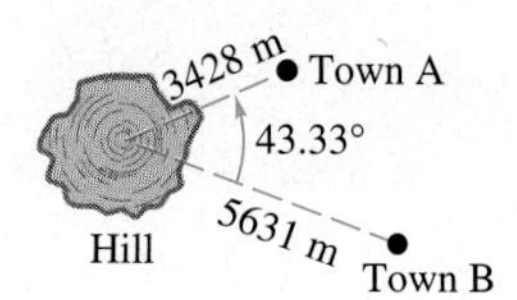


## RELATING CONCEPTS

For individual or group investigation (Exercises 57–60)

*In any triangle, the longest side is opposite the largest angle. This result from geometry was proven for acute triangles in Exercises 65–69 in Section 10.1. To prove it for obtuse triangles, work Exercises 57–60 in order.*

**57.** Suppose angle $A$ is the largest angle of an obtuse triangle. Why is $\cos A$ negative?

**58.** Consider the law of cosines expression for $a^2$, and show that $a^2 > b^2 + c^2$.

**59.** Use the result of Exercise 58 to show that $a > b$ and $a > c$.

**60.** Use the result of Exercise 59 to explain why no triangle $ABC$ satisfies $A = 103°$, $a = 25$, and $c = 30$.

*Find the area of each triangle by using an area formula.*

**61.** $A = 42.5°$, $b = 13.6$ meters, $c = 10.1$ meters

**62.** $B = 124.5°$, $a = 30.4$ centimeters, $c = 28.4$ centimeters

**63.** $a = 12$ meters, $b = 16$ meters, $c = 25$ meters

**64.** $a = 154$ centimeters, $b = 179$ centimeters, $c = 183$ centimeters

**65.** $a = 76.3$ feet, $b = 109$ feet, $c = 98.8$ feet

**66.** $a = 22$ inches, $b = 45$ inches, $c = 31$ inches

**67.** $a = 25.4$ yards, $b = 38.2$ yards, $c = 19.8$ yards

**68.** $a = 15.89$ inches, $b = 21.74$ inches, $c = 10.92$ inches

*Solve each problem.*

**69.** *Area of a Metal Plate* A painter is going to apply a special coating to a triangular metal plate on a new building. Two sides measure 16.1 meters and 15.2 meters. She knows that the angle between these sides is 125°. What is the area of the surface she plans to cover with the coating?

**70.** *Area of a Triangular Lot* A real estate agent wants to find the area of a triangular lot. A surveyor takes measurements and finds that two sides are 52.1 meters and 21.3 meters, and the angle between them is 42.2°. What is the area of the lot?

71. *Cans of Paint Needed for a Job* A painter needs to cover a triangular region 75 meters by 68 meters by 85 meters. A can of paint covers 75 square meters of area. How many cans (to the next higher number of cans) will be needed?

72. *Area of the Bermuda Triangle* Find the area of the Bermuda Triangle, if the sides of the triangle have approximate lengths 850 miles, 925 miles, and 1300 miles.

73. *Perfect Triangles* A *perfect triangle* is a triangle whose sides have whole number lengths and whose area is numerically equal to its perimeter. Show that the triangle with sides of lengths 9, 10, and 17 is perfect.

74. *Heron Triangles* A *Heron triangle* is a triangle having integer sides and area. Show that each of the following is a Heron triangle.

(a) $a = 11, b = 13, c = 20$

(b) $a = 13, b = 14, c = 15$

(c) $a = 7, b = 15, c = 20$

## RELATING CONCEPTS

**For individual or group investigation (Exercises 75–82)**

*In Exercises 33–38 of Section 7.5, we showed how determinants can be used to find the area $\mathcal{A}$ of a triangle in a coordinate plane, using the coordinates of the vertices of the triangle. Consider the triangle from Exercise 35 in that section: triangle PQR has vertices at $P(2, 5)$, $Q(-1, 3)$, and $R(4, 0)$.*

75. Sketch the triangle by hand in a coordinate plane.

76. Use the distance formula to find the lengths $PQ$, $QR$, and $PR$.

77. What is the semiperimeter of triangle $PQR$?

78. Use Heron's formula to find the area $\mathcal{A}$ of triangle $PQR$.

79. According to the original problem in Section 7.5, the area $\mathcal{A}$ of the triangle is given by the formula

$$\mathcal{A} = \frac{1}{2}\det\begin{bmatrix} 2 & 5 & 1 \\ -1 & 3 & 1 \\ 4 & 0 & 1 \end{bmatrix}.$$

Evaluate the area, using this formula. Does it agree with your answer in Exercise 78?

80. Find the measure of angle $PQR$ by using the law of cosines and the lengths found in Exercise 76.

81. Use the second area formula given in this section, along with lengths $PQ$, $QR$, and the measure of angle $PQR$ from Exercise 80, to compute the area of triangle $PQR$. Does it agree with your answers in Exercises 78 and 79?

82. Write a short paragraph explaining what you have learned by working Exercises 75–81.

# VECTORS AND THEIR APPLICATIONS

Basic Terminology • Algebraic Interpretation of Vectors • Operations with Vectors • Dot Product and the Angle between Vectors • Applications of Vectors

## Basic Terminology

Many quantities involve magnitudes, such as 45 pounds or 60 mph. These quantities are called **scalars** and can be represented by real numbers. Other quantities, called **vector quantities,** involve both magnitude and direction. Typical vector quantities are velocity, acceleration, and force. For example, traveling 50 mph *east* represents a vector quantity.

A vector quantity is often represented with a directed line segment, called a **vector.** The length of the vector represents the **magnitude** of the vector quantity. The direction of the vector, indicated with an arrowhead, represents the direction of the quantity. For example, the vector in Figure 18 represents a force of 10 pounds applied at an angle of 30° from the horizontal.

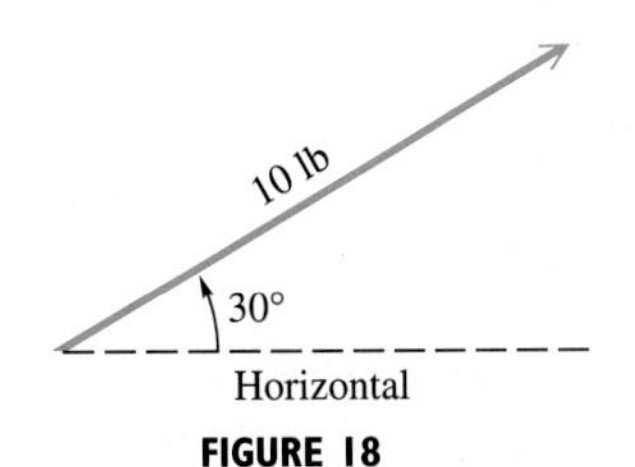


**FIGURE 18**

The symbol for a vector is often printed in boldface type. To write vectors by hand, it is customary to use an arrow over the letter or letters. Thus, **OP** and $\overrightarrow{OP}$ both represent vector **OP**. Vectors may be named with either one lowercase or uppercase letter, or two uppercase letters. When two letters are used, the first indicates the **initial point** and the second indicates the **terminal point** of the vector. Knowing these points gives the direction of the vector. For example, vectors **OP** and **PO** in Figure 19 are not the same vectors. They have the same magnitude, but *opposite directions.* The magnitude of vector **OP** is written $|\mathbf{OP}|$.

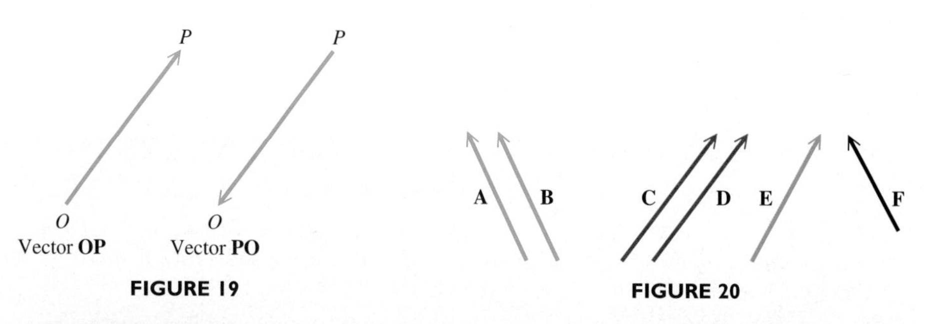


FIGURE 19 FIGURE 20

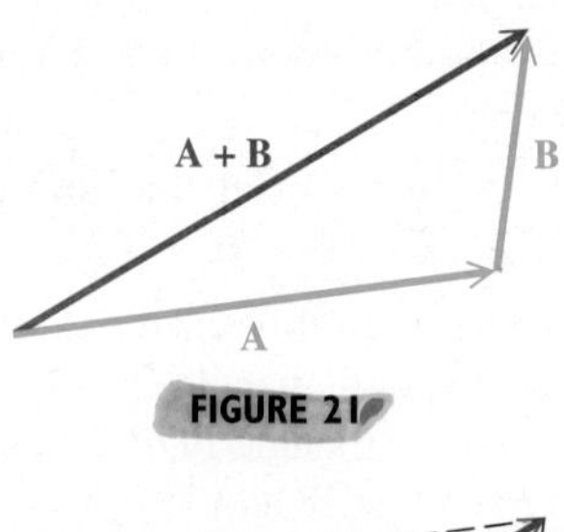


FIGURE 21

Two vectors are equal if and only if they both have the same direction and the same magnitude. In Figure 20, vectors **A** and **B** are equal, as are vectors **C** and **D**.

To find the sum of two vectors **A** and **B**, we place the initial point of vector **B** at the terminal point of vector **A**, as shown in Figure 21. The vector with the same initial point as **A** and the same terminal point as **B** is the sum **A** + **B**. The sum of two vectors is also a vector.

Another way to find the sum of two vectors is to use the **parallelogram rule.** Place vectors **A** and **B** so that their initial points coincide. Then, complete a parallelogram that has **A** and **B** as two sides. The diagonal of the parallelogram with the same initial point as **A** and **B** is the sum **A** + **B** found by the definition. Compare Figures 21 and 22. Parallelograms can be used to show that vector **B** + **A** is the same as vector **A** + **B**, or that **A** + **B** = **B** + **A**, so that vector addition is commutative. The vector sum **A** + **B** is called the **resultant** of vectors **A** and **B**.

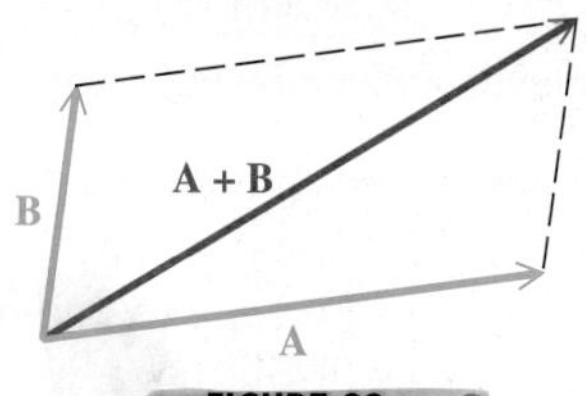


FIGURE 22

For every vector **v** there is a vector −**v** that has the same magnitude as **v** but opposite direction. Vector −**v** is called the **opposite** of **v**. (See Figure 23.) The sum of **v** and −**v** has magnitude 0 and is called the **zero vector.** As with real numbers, to subtract vector **B** from vector **A**, find the vector sum **A** + (−**B**). (See Figure 24.)

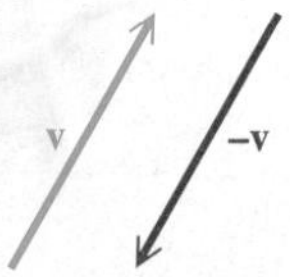

v −v

Vectors **v** and −**v** are opposites.

FIGURE 23

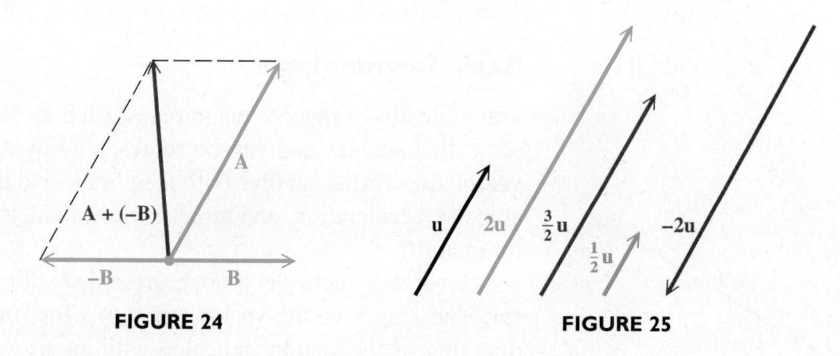


FIGURE 24 FIGURE 25

The **scalar product** of a real number (or scalar) $k$ and a vector **u** is the vector $k \cdot \mathbf{u}$, which has magnitude $|k|$ times the magnitude of **u**. As suggested by Figure 25, the vector $k \cdot \mathbf{u}$ has the same direction as **u** if $k > 0$, and opposite direction if $k < 0$.

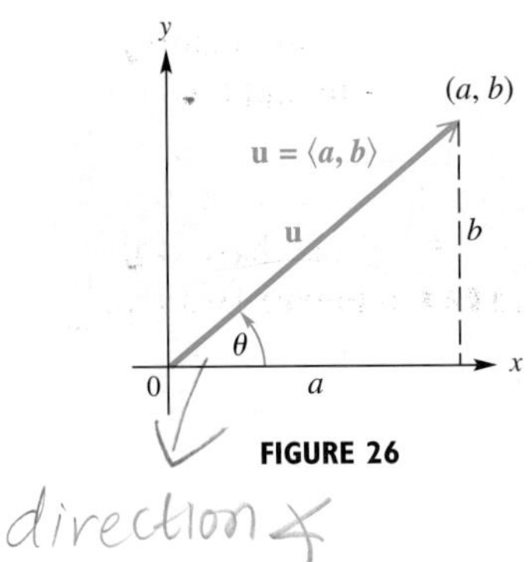

FIGURE 26

## Algebraic Interpretation of Vectors

We now consider vectors in a rectangular coordinate system. A vector with its initial point at the origin is called a **position vector.** A position vector **u** with its endpoint at the point $(a, b)$ is written $\langle a, b\rangle$, so $\mathbf{u} = \langle a, b\rangle$. This means that every vector in the real plane corresponds to an ordered pair of real numbers. Thus, geometrically a vector is a directed line segment; algebraically, it is an ordered pair. The numbers $a$ and $b$ are the **horizontal component** and **vertical component** of vector **u**. Figure 26 shows the vector $\mathbf{u} = \langle a, b\rangle$. The positive angle between the $x$-axis and a position vector is the **direction angle** for the vector. In Figure 26, $\theta$ is the direction angle for vector **u**.

From Figure 26, we can see that the magnitude and direction of a vector are related to its horizontal and vertical components.

**Looking Ahead to Calculus**

In addition to two-dimensional vectors in a plane, calculus courses introduce three-dimensional vectors in space. (See Appendix A.) The magnitude of the two-dimensional vector $\langle a, b\rangle$ is given by $\sqrt{a^2 + b^2}$. If we extend this to the three-dimensional vector $\langle a, b, c\rangle$, the expression becomes $\sqrt{a^2 + b^2 + c^2}$. Similar extensions are made for other concepts.

**Magnitude and Direction Angle of a Vector $\langle a, b\rangle$**

The magnitude (length) of vector $\mathbf{u} = \langle a, b\rangle$ is given by

$$|\mathbf{u}| = \sqrt{a^2 + b^2}.$$

The direction angle $\theta$ satisfies $\tan\theta = \frac{b}{a}$, where $a \neq 0$.

**EXAMPLE 1** *Finding Magnitude and Direction Angle*

Find the magnitude and direction angle for $\mathbf{u} = \langle 3, -2\rangle$.

**Analytic Solution**

The magnitude is $|\mathbf{u}| = \sqrt{3^2 + (-2)^2} = \sqrt{13}$. To find the direction angle $\theta$, start with

$$\tan\theta = \frac{b}{a} = \frac{-2}{3} = -\frac{2}{3}.$$

Vector **u** has a positive horizontal component and a negative vertical component, placing the position vector in quadrant IV. A calculator gives $\tan^{-1}\left(-\frac{2}{3}\right) \approx -33.7°$. Adding 360° yields the direction angle $\theta = 326.3°$. See Figure 27.

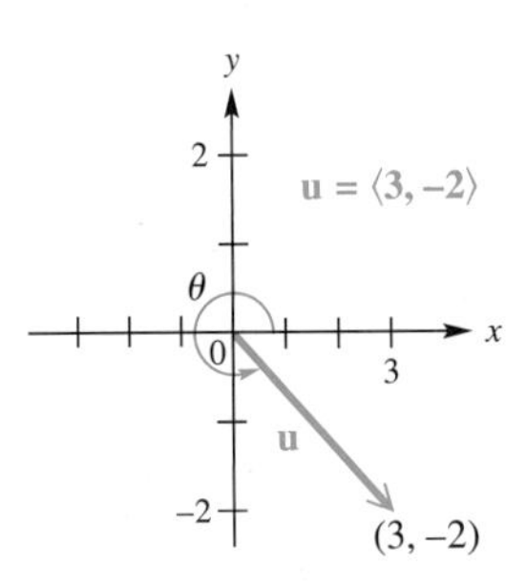

FIGURE 27

**Graphing Calculator Solution**

The calculator screen returns the magnitude and direction angle, given the horizontal and vertical components. Notice that an approximation for $\sqrt{13}$ is given, and the direction angle is returned as a measure with smallest possible absolute value. It would be necessary to add 360° to the value of $\theta$ to obtain a positive direction angle. See Figure 28.

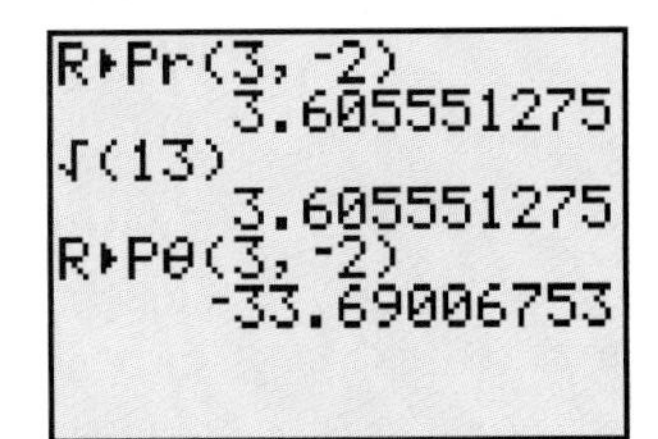

FIGURE 28

To learn more about this feature, consult your owner's manual or the graphing calculator manual that accompanies this text.

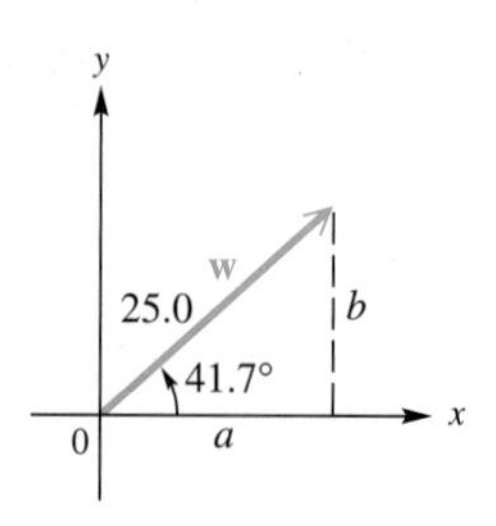

FIGURE 29

### Horizontal and Vertical Components

The horizontal and vertical components, respectively, of a vector **u** having magnitude $|\mathbf{u}|$ and direction angle $\theta$ are given by

$$a = |\mathbf{u}| \cos \theta \quad \text{and} \quad b = |\mathbf{u}| \sin \theta.$$

That is, $\mathbf{u} = \langle a, b \rangle = \langle |\mathbf{u}| \cos \theta, |\mathbf{u}| \sin \theta \rangle$.

**EXAMPLE 2** *Finding Horizontal and Vertical Components*

Vector **w** in Figure 29 has magnitude 25.0 and direction angle 41.7°. Find the horizontal and vertical components.

**Analytic Solution**

Use the two formulas in the box, with $|\mathbf{w}| = 25.0$ and $\theta = 41.7°$.

$$a = 25.0 \cos 41.7° \qquad b = 25.0 \sin 41.7°$$
$$a \approx 18.7 \qquad b \approx 16.6$$

Therefore, $\mathbf{w} = \langle 18.7, 16.6 \rangle$. The horizontal component is 18.7, and the vertical component is 16.6 (rounded to the nearest tenth).

**Graphing Calculator Solution**

See Figure 30. The results support the analytic solution.

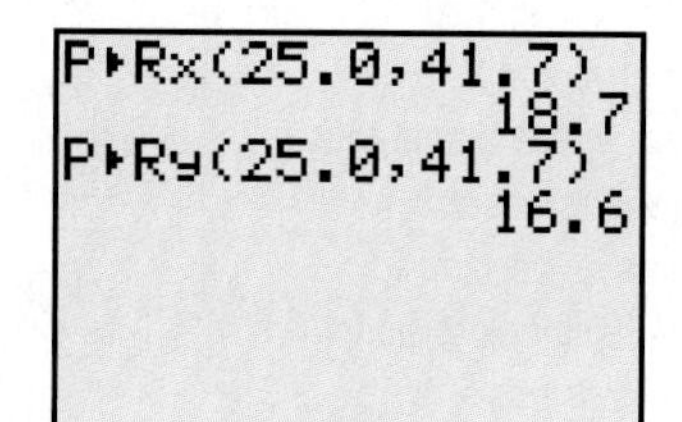

FIGURE 30

The following properties of parallelograms are helpful when studying vectors.

1. A parallelogram is a quadrilateral whose opposite sides are parallel.
2. The opposite sides and opposite angles of a parallelogram are equal, and adjacent angles of a parallelogram are supplementary.
3. The diagonals of a parallelogram bisect each other, but do not necessarily bisect the angles of the parallelogram.

**EXAMPLE 3** *Finding the Magnitude of a Resultant*

Two forces of 15 and 22 newtons act on a point in the plane. (A *newton* is a unit of force that equals .225 pound.) If the angle between the forces is 100°, find the magnitude of the resultant force.

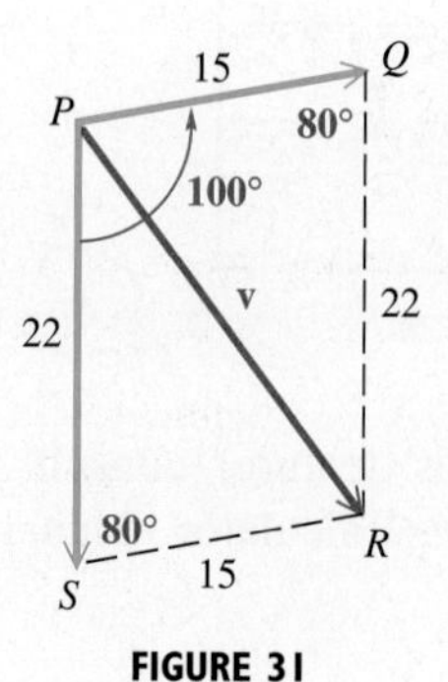

FIGURE 31

**Solution** As shown in Figure 31, a parallelogram that has the forces as adjacent sides can be formed. The angles of the parallelogram adjacent to angle $P$ measure 80°, since adjacent angles of a parallelogram are supplementary. Opposite sides of the parallelogram are equal in length. The resultant force divides the parallelogram into two triangles. Use the law of cosines with either triangle to obtain

$$|\mathbf{v}|^2 = 15^2 + 22^2 - 2(15)(22) \cos 80°$$
$$\approx 225 + 484 - 115$$
$$|\mathbf{v}|^2 \approx 594$$
$$|\mathbf{v}| \approx 24.$$

To the nearest unit, the magnitude of the resultant force is 24 newtons.

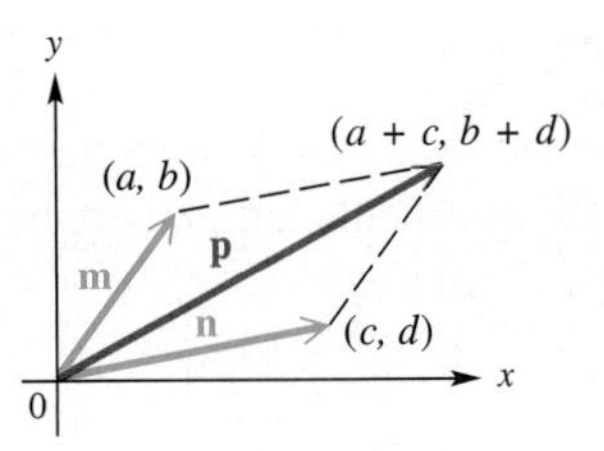

FIGURE 32

## Operations with Vectors

In Figure 32, $\mathbf{m} = \langle a, b\rangle$, $\mathbf{n} = \langle c, d\rangle$, and $\mathbf{p} = \langle a + c, b + d\rangle$. Using geometry, we can show that the endpoints of the three vectors and their origin form a parallelogram. Since a diagonal of this parallelogram gives the resultant of **m** and **n**, we have $\mathbf{p} = \mathbf{m} + \mathbf{n}$ or

$$\langle a + c, b + d\rangle = \langle a, b\rangle + \langle c, d\rangle.$$

Similarly, we could verify the following vector operations.

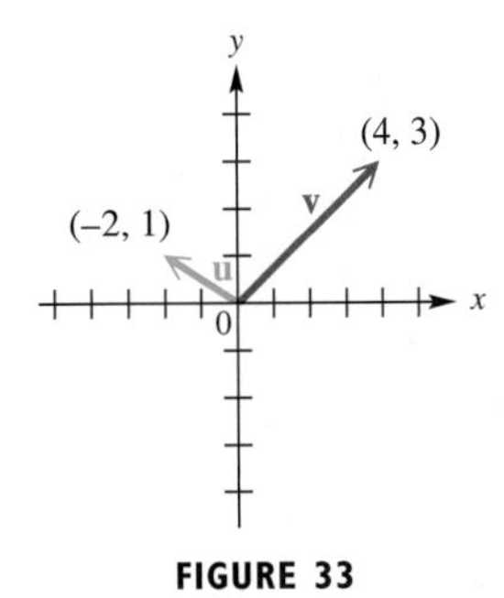

FIGURE 33

**Vector Operations**

For any real numbers $a$, $b$, $c$, $d$, and $k$,

$$\langle a, b\rangle + \langle c, d\rangle = \langle a + c, b + d\rangle$$
$$k \cdot \langle a, b\rangle = \langle ka, kb\rangle.$$
$$\text{If } \mathbf{a} = \langle a_1, a_2\rangle, \text{ then } -\mathbf{a} = \langle -a_1, -a_2\rangle.$$
$$\langle a, b\rangle - \langle c, d\rangle = \langle a, b\rangle + -\langle c, d\rangle = \langle a - c, b - d\rangle$$

**EXAMPLE 4** *Performing Vector Operations*

Let $\mathbf{u} = \langle -2, 1\rangle$ and $\mathbf{v} = \langle 4, 3\rangle$. Find the following: **(a)** $\mathbf{u} + \mathbf{v}$, **(b)** $-2\mathbf{u}$, **(c)** $4\mathbf{u} - 3\mathbf{v}$. (See Figure 33.)

**Analytic Solution**

**(a)** $\mathbf{u} + \mathbf{v} = \langle -2, 1\rangle + \langle 4, 3\rangle$
$= \langle -2 + 4, 1 + 3\rangle$
$= \langle 2, 4\rangle$

**(b)** $-2\mathbf{u} = -2 \cdot \langle -2, 1\rangle$
$= \langle -2(-2), -2(1)\rangle$
$= \langle 4, -2\rangle$

**(c)** $4\mathbf{u} - 3\mathbf{v} = 4 \cdot \langle -2, 1\rangle - 3 \cdot \langle 4, 3\rangle$
$= \langle -8, 4\rangle - \langle 12, 9\rangle$
$= \langle -8 - 12, 4 - 9\rangle$
$= \langle -20, -5\rangle$

**Graphing Calculator Solution**

Vector arithmetic can be performed with a graphing calculator, as shown in Figure 34.

```
{-2,1}+{4,3}
              {2 4}
-2{-2,1}
             {4 -2}
4{-2,1}-3{4,3}
           {-20 -5}
```

FIGURE 34

A **unit vector** is a vector that has magnitude 1. Two very useful unit vectors are defined as follows and shown in Figure 35(a).

$$\mathbf{i} = \langle 1, 0\rangle \qquad \mathbf{j} = \langle 0, 1\rangle$$

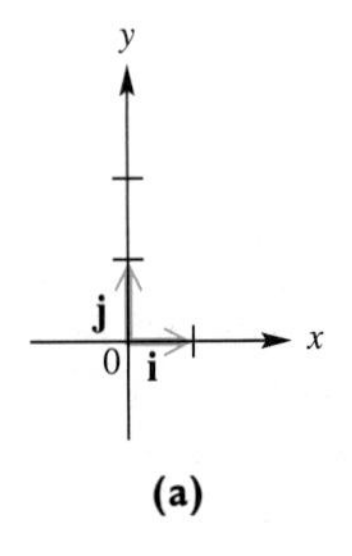

(a)

FIGURE 35

With the unit vectors **i** and **j**, we can express any other vector $\langle a, b\rangle$ in the form $a\mathbf{i} + b\mathbf{j}$, as shown in Figure 35(b) on the next page, where $\langle 3, 4\rangle = 3\mathbf{i} + 4\mathbf{j}$. The vector operations previously given can be restated, using the $a\mathbf{i} + b\mathbf{j}$ notation.

**i, j Form for Vectors**

If $\mathbf{v} = \langle a, b\rangle$, then $\mathbf{v} = a\mathbf{i} + b\mathbf{j}$.

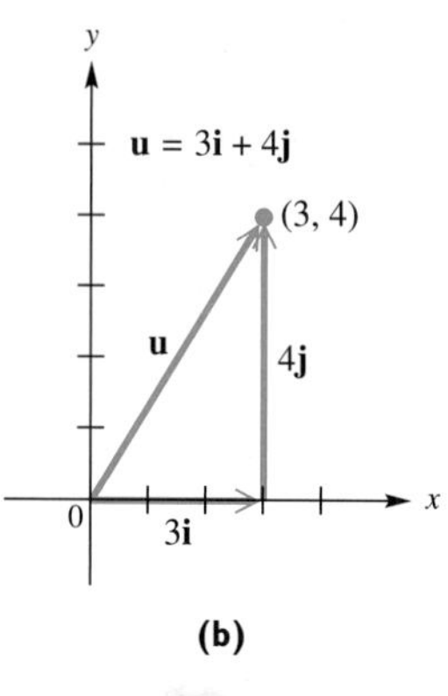


FIGURE 35

## Dot Product and the Angle between Vectors

The *dot product* of two vectors is a real number, not a vector. It is also known as the *inner product* or *scalar product.* Dot products are used to determine the angle between two vectors, derive geometric theorems, and solve physics problems.

> **Dot Product**
>
> The **dot product** of the two vectors $\mathbf{u} = \langle a, b\rangle$ and $\mathbf{v} = \langle c, d\rangle$ is denoted $\mathbf{u} \cdot \mathbf{v}$, read "**u** dot **v**," and given by
>
> $$\mathbf{u} \cdot \mathbf{v} = ac + bd.$$

**EXAMPLE 5** *Finding the Dot Product*

Find each dot product.

**(a)** $\langle 2, 3\rangle \cdot \langle 4, -1\rangle$ **(b)** $\langle 6, 4\rangle \cdot \langle -2, 3\rangle$

**Solution**

**(a)** $\langle 2, 3\rangle \cdot \langle 4, -1\rangle = 2(4) + 3(-1) = 5$

**(b)** $\langle 6, 4\rangle \cdot \langle -2, 3\rangle = 6(-2) + 4(3) = 0$

The following properties of dot products are easily verified.

> **Properties of the Dot Product**
>
> For all vectors **u**, **v**, and **w** and real numbers $k$,
>
> **(a)** $\mathbf{u} \cdot \mathbf{v} = \mathbf{v} \cdot \mathbf{u}$ **(b)** $\mathbf{u} \cdot (\mathbf{v} + \mathbf{w}) = \mathbf{u} \cdot \mathbf{v} + \mathbf{u} \cdot \mathbf{w}$
>
> **(c)** $(\mathbf{u} + \mathbf{v}) \cdot \mathbf{w} = \mathbf{u} \cdot \mathbf{w} + \mathbf{v} \cdot \mathbf{w}$ **(d)** $(k\mathbf{u}) \cdot \mathbf{v} = k(\mathbf{u} \cdot \mathbf{v}) = \mathbf{u} \cdot (k\mathbf{v})$
>
> **(e)** $\mathbf{0} \cdot \mathbf{u} = 0$ **(f)** $\mathbf{u} \cdot \mathbf{u} = |\mathbf{u}|^2.$

For example, to prove the first part of (d), we let $\mathbf{u} = \langle a, b\rangle$ and $\mathbf{v} = \langle c, d\rangle$. Then,

$$\begin{aligned}(k\mathbf{u}) \cdot \mathbf{v} &= (k\langle a, b\rangle) \cdot \langle c, d\rangle = \langle ka, kb\rangle \cdot \langle c, d\rangle \\ &= kac + kbd = k(ac + bd) \\ &= k(\langle a, b\rangle \cdot \langle c, d\rangle) = k(\mathbf{u} \cdot \mathbf{v}).\end{aligned}$$

The proofs of the remaining properties are similar.

The dot product of two vectors can be positive, 0, or negative. A geometric interpretation of the dot product explains when each of these cases occurs. This interpretation involves the angle between the two vectors. Consider the vectors $\mathbf{u} = \langle a, b\rangle$ and $\mathbf{v} = \langle c, d\rangle$, as shown in Figure 36. The **angle $\theta$ between u and v** is defined to be the angle having the two vectors as its sides for which $0° \leq \theta \leq 180°$. The following theorem relates the dot product to the angle between the vectors. Its proof is outlined in Exercise 101.

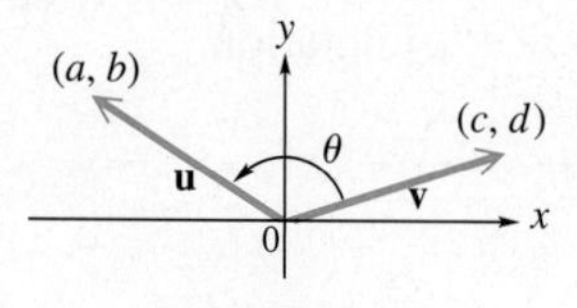


FIGURE 36

### Geometric Interpretation of Dot Product

If $\theta$ is the angle between the two nonzero vectors **u** and **v**, where $0° \le \theta \le 180°$, then

$$\mathbf{u} \cdot \mathbf{v} = |\mathbf{u}|\,|\mathbf{v}| \cos \theta \quad \text{or} \quad \cos \theta = \frac{\mathbf{u} \cdot \mathbf{v}}{|\mathbf{u}|\,|\mathbf{v}|}.$$

**EXAMPLE 6** *Finding the Angle between Two Vectors*

Find the angle $\theta$ between the two vectors $\mathbf{u} = \langle 3, 4 \rangle$ and $\mathbf{v} = \langle 2, 1 \rangle$.

**Solution** By the geometric interpretation,

$$\cos \theta = \frac{\mathbf{u} \cdot \mathbf{v}}{|\mathbf{u}|\,|\mathbf{v}|} = \frac{\langle 3,4 \rangle \cdot \langle 2,1 \rangle}{|\langle 3,4 \rangle|\,|\langle 2,1 \rangle|}$$

$$= \frac{3(2) + 4(1)}{\sqrt{9+16}\sqrt{4+1}} = \frac{10}{5\sqrt{5}} \approx .894427191.$$

Therefore, $\theta \approx \cos^{-1} .894427191 \approx 26.57°.$

FIGURE 37

For angles $\theta$ between 0° and 180°, $\cos \theta$ is positive, 0, or negative when $\theta$ is less than, equal to, or greater than 90°, respectively. Therefore, the dot product is positive, 0, or negative according to this table.

| Dot Product | Angle between Vectors |
|---|---|
| Positive | Acute |
| 0 | Right |
| Negative | Obtuse |

**FOR DISCUSSION**

Suppose that $|\mathbf{v}| = 5$ and $|\mathbf{u}| = 3$. How could you use the value of $\mathbf{v} \cdot \mathbf{u}$ to determine if **v** and **u** point in the same direction? Opposite directions?

**NOTE** If $\mathbf{a} \cdot \mathbf{b} = 0$ for two nonzero vectors **a** and **b**, then $\cos \theta = 0$ and $\theta = 90°$. Thus, **a** and **b** are perpendicular or **orthogonal vectors.** See Figure 37.

## Applications of Vectors

The law of sines and the law of cosines are often used to solve applied problems involving vectors.

**EXAMPLE 7** *Applying Vectors to a Navigation Problem*

A ship leaves port on a bearing of 28° and travels 8.2 miles. The ship then turns due east and travels 4.3 miles. How far is the ship from port? What is its bearing from port?

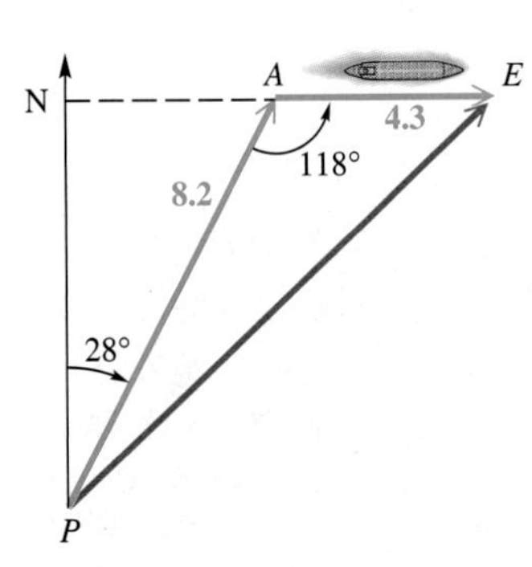

FIGURE 38

**Solution** In Figure 38, vectors **PA** and **AE** represent the ship's path. The magnitude and bearing of the resultant **PE** can be found as follows. Triangle $PNA$ is a right triangle, so angle $NAP = 90° - 28° = 62°$. Then angle $PAE = 180° - 62° = 118°$. Use the law of cosines to find $|\mathbf{PE}|$, the magnitude of vector **PE**.

$$|\mathbf{PE}|^2 = 8.2^2 + 4.3^2 - 2(8.2)(4.3)\cos 118°$$

$$|\mathbf{PE}|^2 \approx 118.84$$

Therefore, $|\mathbf{PE}| \approx 10.9$, or 11 miles. Nearest integer

To find the bearing of the ship from port, first find angle $APE$. Use the law of sines, along with the value of $|\mathbf{PE}|$, before rounding.

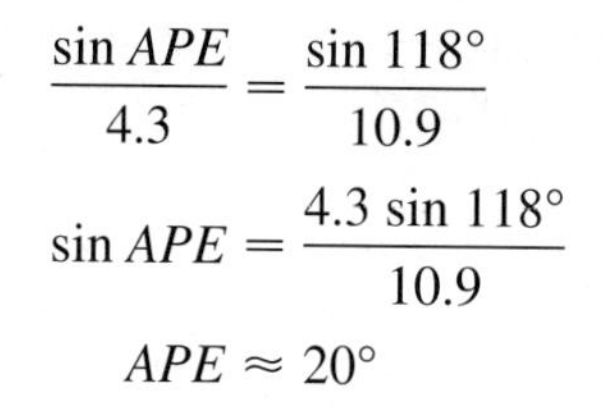

$$\frac{\sin APE}{4.3} = \frac{\sin 118°}{10.9}$$

$$\sin APE = \frac{4.3 \sin 118°}{10.9}$$

$$APE \approx 20° \qquad \text{Use the inverse sine function.}$$

The ship is 11 miles from port at a bearing of $28° + 20° = 48°$.

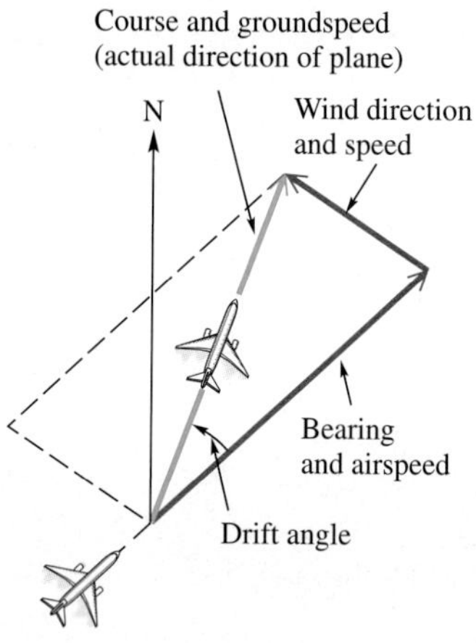


**FIGURE 39**

In air navigation, the **airspeed** of a plane is its speed relative to the air, while the **groundspeed** is its speed relative to the ground. Because of wind, these two speeds are usually different. The groundspeed of the plane is represented by the vector sum of the airspeed and windspeed vectors. See Figure 39.

## EXAMPLE 8 *Applying Vectors to a Navigation Problem*

A plane with an airspeed of 192 mph is headed on a bearing of 121°. A north wind is blowing (from north to south) at 15.9 mph. Find the groundspeed and the actual bearing of the plane.

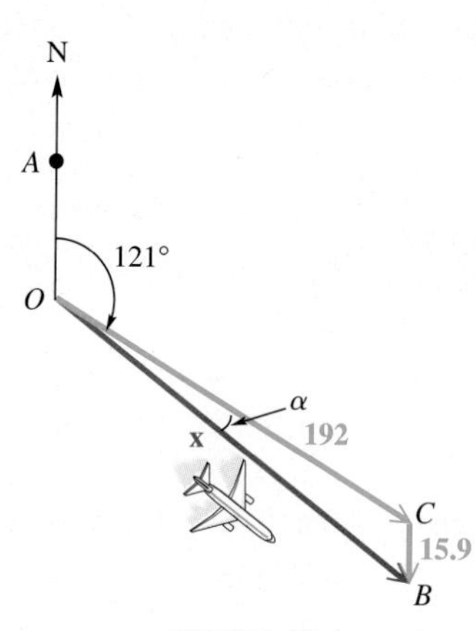


**FIGURE 40**

**Solution** In Figure 40, the groundspeed is represented by $|\mathbf{x}|$. We must find angle $\alpha$ to find the bearing, which will be $121° + \alpha$. From Figure 40, angle $BCO$ equals angle $AOC$, which equals 121°. We find $|\mathbf{x}|$ by using the law of cosines.

$$|\mathbf{x}|^2 = 192^2 + 15.9^2 - 2(192)(15.9)\cos 121°$$

$$|\mathbf{x}|^2 \approx 40{,}261$$

$$|\mathbf{x}| \approx 200.7, \quad \text{or} \quad 201 \text{ mph}$$

We find $\alpha$ by using the law of sines. As before, we use the value of $|\mathbf{x}|$ before rounding.

$$\frac{\sin \alpha}{15.9} = \frac{\sin 121°}{200.7}$$

$$\sin \alpha \approx .06792320$$

$$\alpha \approx 3.89°$$

To the nearest degree, $\alpha$ is 4°. The groundspeed is about 201 mph on a bearing of approximately $121° + 4° = 125°$.

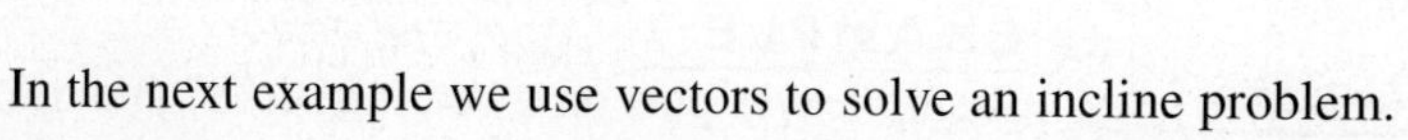

In the next example we use vectors to solve an incline problem.

## EXAMPLE 9 *Finding a Required Force*

Find the force required to pull a wagon weighing 50 pounds up a ramp inclined 20° to the horizontal. (Assume that there is no friction.)

**Solution** In Figure 41, the vertical 50-pound force **BA** represents the force of gravity. It is the sum of the vectors **BC** and **−AC**. The vector **BC** represents the force with which the wagon pushes against the ramp. The vector **BF** represents the force that would pull the wagon up the ramp. Since vectors **BF** and **AC** are equal, $|\mathbf{AC}|$ gives the magnitude of the required force.

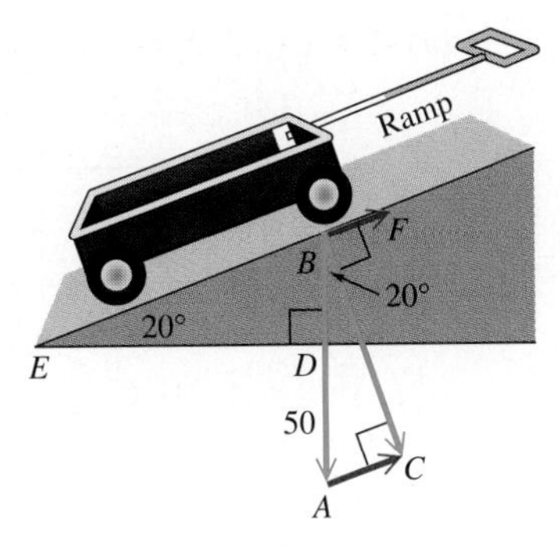


FIGURE 41

Vectors **BF** and **AC** are parallel, so angle *EBD* equals angle *A*. Since angle *BDE* and angle *C* are right angles, triangles *CBA* and *DEB* have two corresponding angles equal and so are similar triangles. Therefore, angle *ABC* equals angle *E*, which is 20°. From right triangle *ABC*,

$$\sin 20^\circ = \frac{|\mathbf{AC}|}{50}$$

$$|\mathbf{AC}| = 50 \sin 20^\circ \approx 17.$$

To the nearest pound, a 17-pound force will be required to pull the wagon up the ramp.

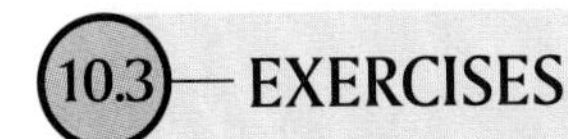

## 10.3 EXERCISES

*Concept Check* *Refer to the following vectors.*

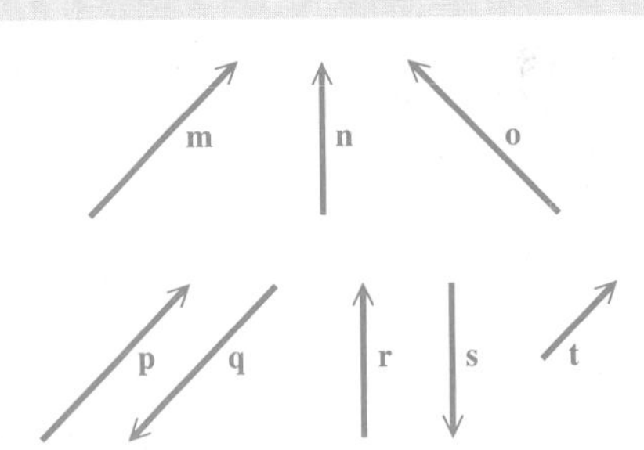


1. Name all pairs of vectors that appear to be equal.
2. Name all pairs of vectors that are opposites.
3. Name all pairs of vectors where the first is a scalar multiple of the other, with the scalar positive.
4. Name all pairs of vectors where the first is a scalar multiple of the other, with the scalar negative.

*Refer to the vectors pictured. Make a copy or a careful sketch of each, and then draw a sketch to represent each vector in Exercises 5–16.. For example, find* **a** + **e** *by placing* **a** *and* **e** *so that their initial points coincide. Then, use the parallelogram rule to find the resultant, shown in the figure to the right.*

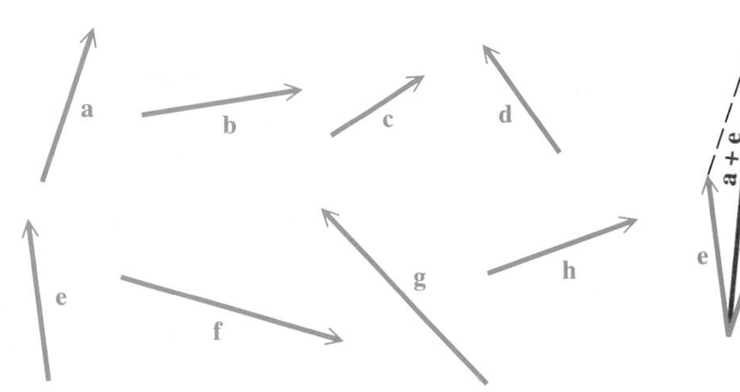


5. −**b**
6. −**g**
7. 3**a**
8. 2**h**
9. **a** + **b**
10. **h** + **g**
11. **a** − **c**
12. **d** − **e**
13. **a** + (**b** + **c**)
14. (**a** + **b**) + **c**
15. **c** + **d**
16. **d** + **c**

*Use the figure to find each vector.* **(a)** $\mathbf{a} + \mathbf{b}$ **(b)** $\mathbf{a} - \mathbf{b}$ **(c)** $-\mathbf{a}$

**17.**

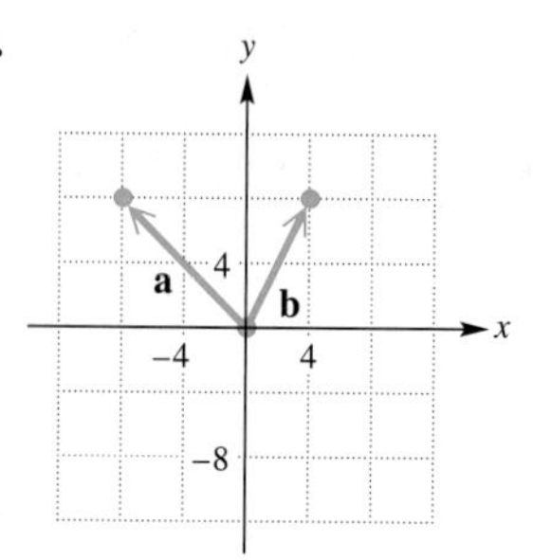


**18.**

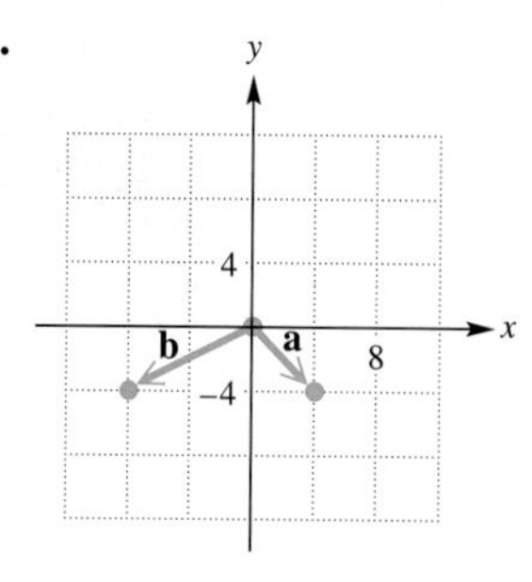


**19.**

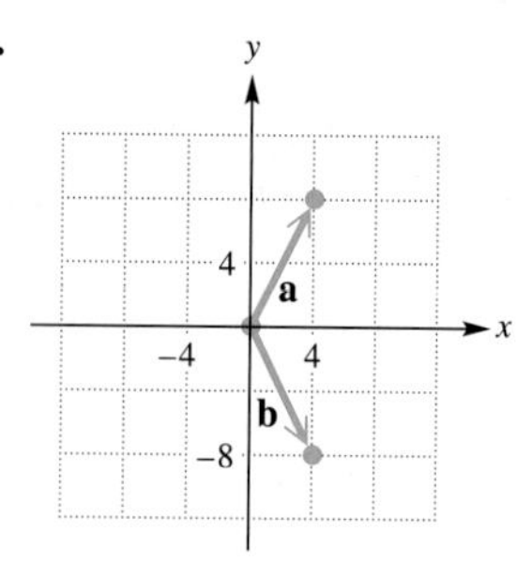


**20.**

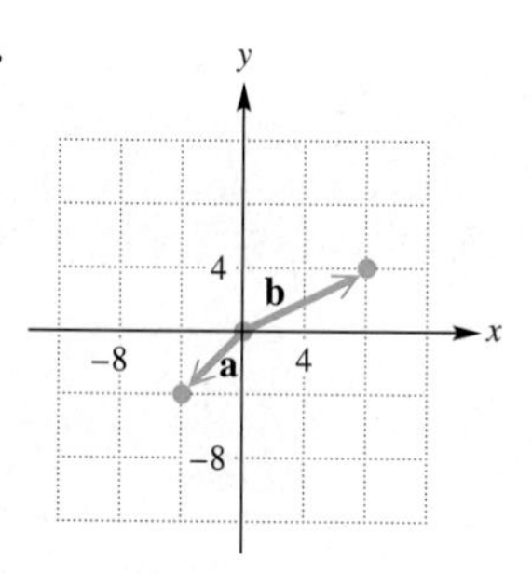


**21.**

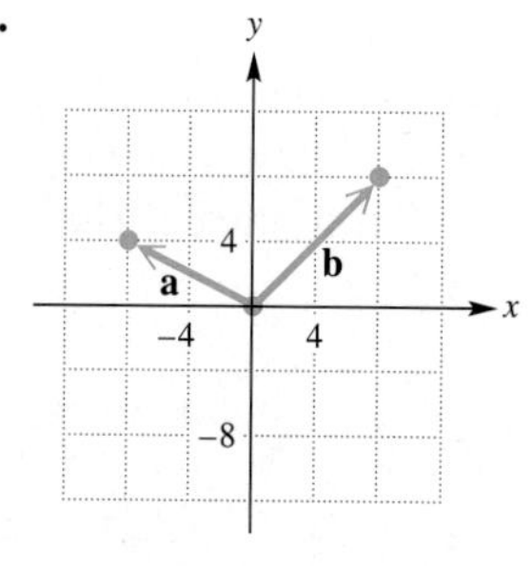


**22.**

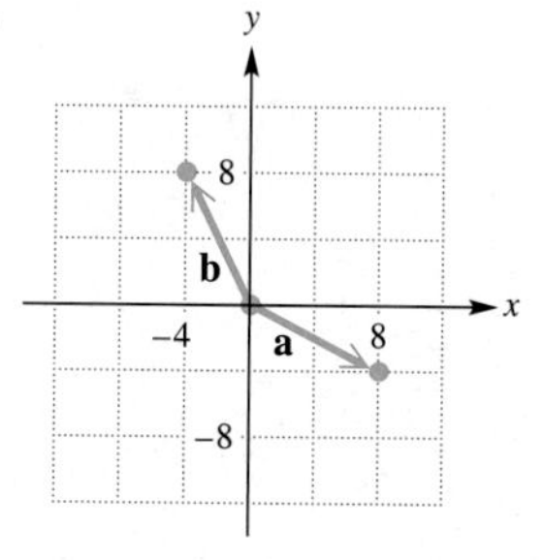


*Find each vector.* **(a)** $2\mathbf{a}$ **(b)** $2\mathbf{a} + 3\mathbf{b}$ **(c)** $\mathbf{b} - 3\mathbf{a}$

**23.** $\mathbf{a} = 2\mathbf{i}, \mathbf{b} = \mathbf{i} + \mathbf{j}$

**24.** $\mathbf{a} = -\mathbf{i} + 2\mathbf{j}, \mathbf{b} = \mathbf{i} - \mathbf{j}$

**25.** $\mathbf{a} = \langle -1, 2 \rangle, \mathbf{b} = \langle 3, 0 \rangle$

**26.** $\mathbf{a} = \langle -2, -1 \rangle, \mathbf{b} = \langle -3, 2 \rangle$

*Given* $\mathbf{u} = \langle -2, 5 \rangle$ *and* $\mathbf{v} = \langle 4, 3 \rangle$, *find each vector.*

**27.** $\mathbf{u} + \mathbf{v}$

**28.** $\mathbf{u} - \mathbf{v}$

**29.** $\mathbf{v} - \mathbf{u}$

**30.** $5\mathbf{v}$

**31.** $-5\mathbf{v}$

**32.** $3\mathbf{u} + 6\mathbf{v}$

*Approximate the horizontal and vertical component of* $\mathbf{v}$ *shown in each figure.*

**33.**

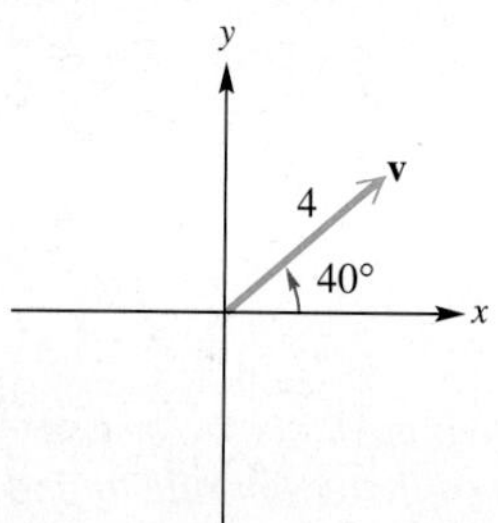


**34.**

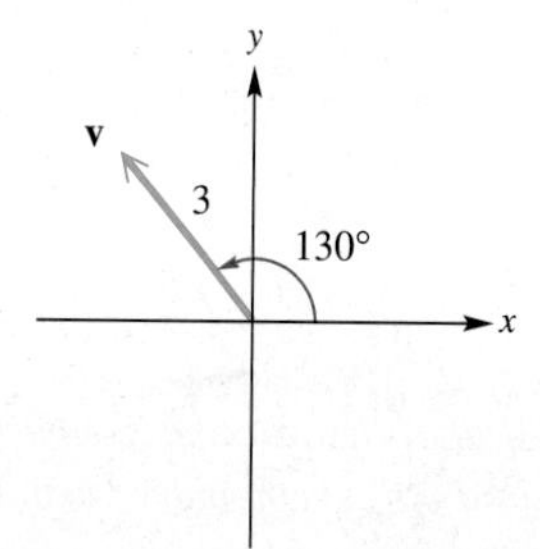


**35.**

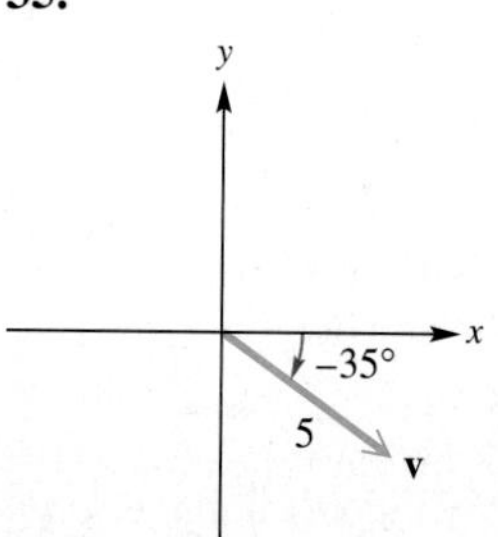


**36.**

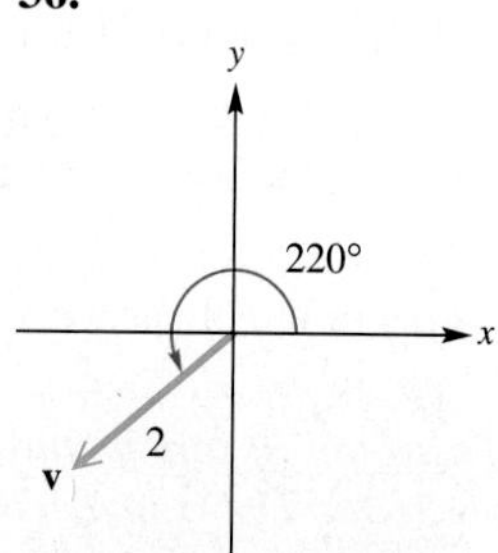


*Use the parallelogram rule to find the magnitude of the resultant force for the two forces shown in each figure.*

**37.**

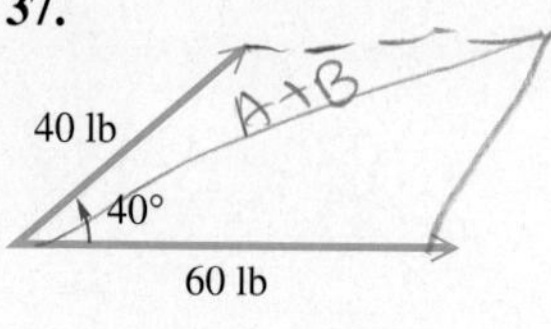


**38.**

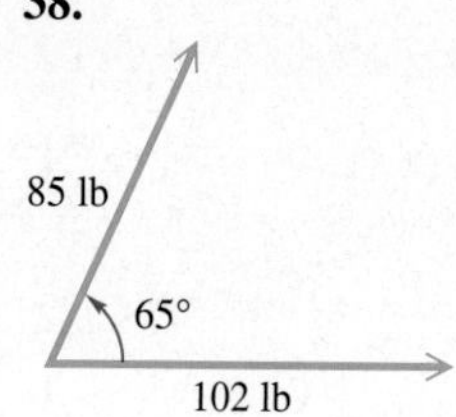


**39.**

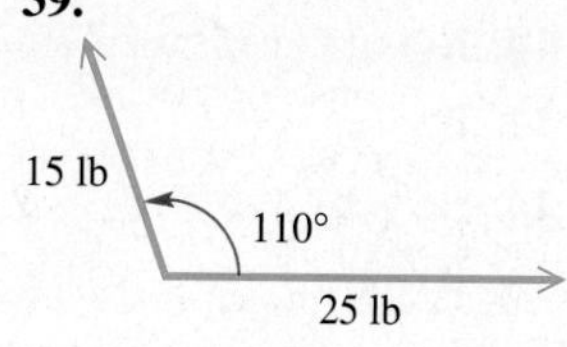


**40.**

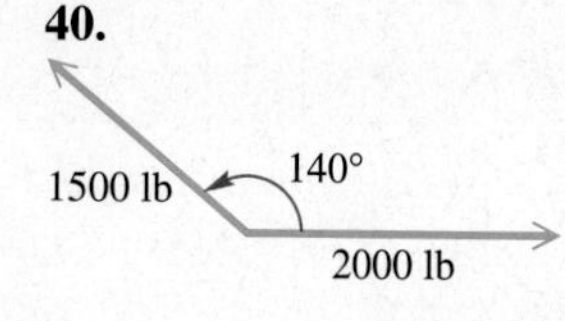

*Write each vector in the form* $a\mathbf{i} + b\mathbf{j}$. *Round a and b to three decimal places, if necessary.*

**41.** $\langle -5, 8\rangle$ **42.** $\langle 6, -3\rangle$ **43.** $\langle 2, 0\rangle$ **44.** $\langle 0, -4\rangle$

**45.** Direction angle 45°, magnitude 8 **46.** Direction angle 210°, magnitude 3

**47.** Direction angle 115°, magnitude .6 **48.** Direction angle 208°, magnitude .9

*Find the magnitude and direction angle (to the nearest tenth) for each vector. Give the measure of the direction angle as an angle in* $[0, 360°)$.

**49.** $\langle 1, 1\rangle$ **50.** $\langle -4, 4\sqrt{3}\rangle$ **51.** $\langle 8\sqrt{2}, -8\sqrt{2}\rangle$ **52.** $\langle \sqrt{3}, -1\rangle$

**53.** $\langle 15, -8\rangle$ **54.** $\langle -7, 24\rangle$ **55.** $\langle -6, 0\rangle$ **56.** $\langle 0, -12\rangle$

*Find the dot product for each pair of vectors.*

**57.** $\langle 6, -1\rangle, \langle 2, 5\rangle$ **58.** $\langle -3, 8\rangle, \langle 7, -5\rangle$ **59.** $\langle 2, -3\rangle, \langle 6, 5\rangle$

**60.** $\langle 1, 2\rangle, \langle 3, -1\rangle$ **61.** $4\mathbf{i}, 5\mathbf{i} - 9\mathbf{j}$ **62.** $2\mathbf{i} + 4\mathbf{j}, -\mathbf{j}$

*Find the angle between each pair of vectors.*

**63.** $\langle 2, 1\rangle, \langle -3, 1\rangle$ **64.** $\langle 1, 7\rangle, \langle 1, 1\rangle$ **65.** $\langle 1, 2\rangle, \langle -6, 3\rangle$

**66.** $\langle 4, 0\rangle, \langle 2, 2\rangle$ **67.** $3\mathbf{i} + 4\mathbf{j}, \mathbf{j}$ **68.** $-5\mathbf{i} + 12\mathbf{j}, 3\mathbf{i} + 2\mathbf{j}$

*Let* $\mathbf{u} = \langle -2, 1\rangle$, $\mathbf{v} = \langle 3, 4\rangle$, *and* $\mathbf{w} = \langle -5, 12\rangle$. *Evaluate each expression.*

**69.** $(3\mathbf{u}) \cdot \mathbf{v}$ **70.** $\mathbf{u} \cdot (\mathbf{v} - \mathbf{w})$ **71.** $\mathbf{u} \cdot \mathbf{v} - \mathbf{u} \cdot \mathbf{w}$ **72.** $\mathbf{u} \cdot (3\mathbf{v})$

*Determine whether each pair of vectors is orthogonal.*

**73.** $\langle 1, 2\rangle, \langle -6, 3\rangle$ **74.** $\langle 3, 4\rangle, \langle 6, 8\rangle$ **75.** $\langle 1, 0\rangle, \langle \sqrt{2}, 0\rangle$

**76.** $\langle 1, 1\rangle, \langle 1, -1\rangle$ **77.** $\sqrt{5}\mathbf{i} - 2\mathbf{j}, -5\mathbf{i} + 2\sqrt{5}\mathbf{j}$ **78.** $-4\mathbf{i} + 3\mathbf{j}, 8\mathbf{i} - 6\mathbf{j}$

*Solve each problem.*

**79.** *Heading* A plane flies 650 mph on a bearing of 175.3°. A 25 mph wind, from a direction of 266.6°, blows against the plane. Find the resulting bearing of the plane.

**80.** *Airspeed and Groundspeed* A pilot wants to fly on a course of 74.9°. By flying due east, he finds that a 42 mph wind, blowing from the south, puts him on course. Find the airspeed and the groundspeed.

**81.** *Distance Traveled* Starting at point *A*, a ship sails 18.5 kilometers on a bearing of 189°, then turns and sails 47.8 kilometers on a bearing of 317°. Find the distance of the ship from point *A*.

**82.** *Course and Groundspeed* An airline route from San Francisco to Honolulu is on a bearing of 233°. A jet flying at 450 mph on that bearing flies into a wind blowing at 39 mph from a direction of 114°. Find the resulting bearing and groundspeed of the plane.

**83.** *Course and Groundspeed* A pilot is flying at 168 mph. She wants her flight path to be on a course of 57° 40′. A wind is blowing from the south at 27.1 mph. Find the bearing the pilot should fly, and find the plane's groundspeed.

**84.** *Bearing and Airspeed* What bearing and airspeed are required for a plane to fly 400 miles due north in 2.5 hours if the wind is blowing from a direction of 328° at 11 mph?

**85.** *Groundspeed and Course* A plane is headed due south with an airspeed of 192 mph. A wind from a direction of 78° is blowing at 23 mph. Find the groundspeed and resulting bearing of the plane.

**86.** *Groundspeed and Course* An airplane is flying on a bearing of 174° at an airspeed of 240 kilometers per hour. A 30 kilometer per hour wind is blowing from a direction of 245°. Find the groundspeed and resulting bearing of the plane.

**87.** *Weight of a Box* Two people are carrying a box. One person exerts a force of 150 pounds at an angle of 62.4° with the horizontal. The other person exerts a force of 114 pounds at an angle of 54.9°. Find the weight of the box.

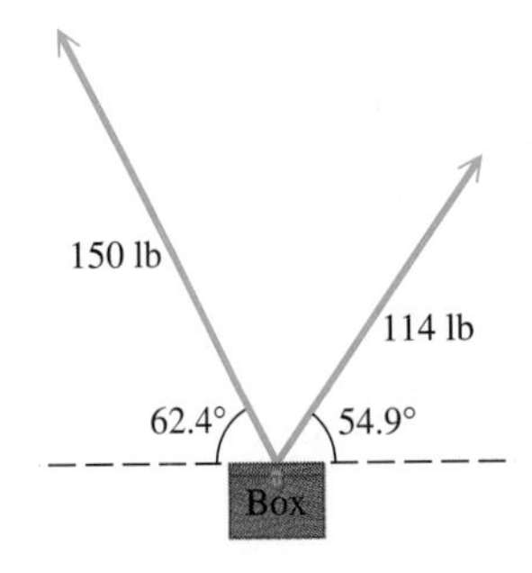

**88.** *Angle of a Hill Slope* A force of 25 pounds is required to push an 80-pound wagon up a hill. What angle does the hill make with the horizontal?

**89.** *Weight of a Boat* A force of 500 pounds is required to pull a boat up a ramp inclined at 18° with the horizontal. How much does the boat weigh?

**90.** *Force Needed to Keep a Car Parked* Find the force required to keep a 3000-pound car parked on a hill that makes an angle of 15° with the horizontal.

**91.** *Magnitudes of Forces* A force of 176 pounds makes an angle of 78° 50′ with a second force. The resultant of the two forces makes an angle of 41° 10′ with the first force. (See the figure.) Find the magnitudes of the second force and of the resultant.

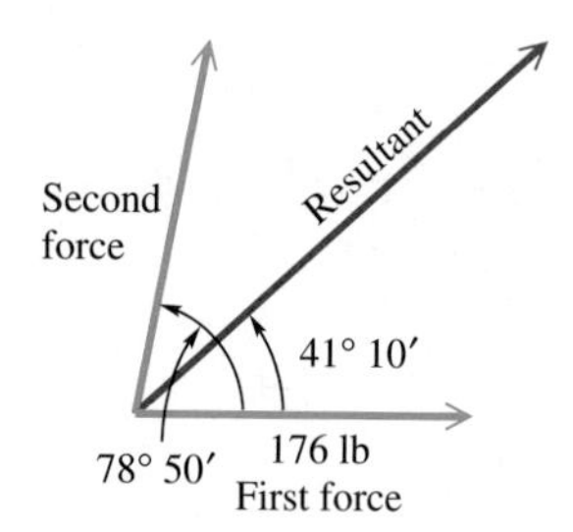


**92.** *Movement of a Motorboat* Suppose you would like to cross a 132-foot wide river in a motorboat. Assume that the motorboat can travel at 7 mph relative to the water and that the current is flowing west at the rate of 3 mph. The bearing $\theta$ is chosen so that the motorboat will land at a point exactly across from the starting point.

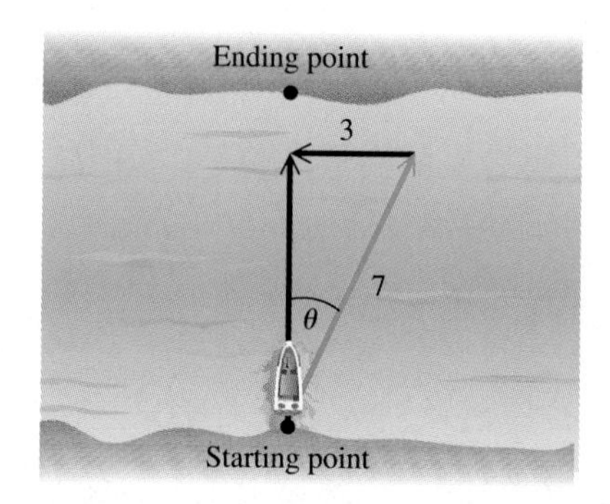


**(a)** At what speed will the motorboat be traveling relative to the banks?

**(b)** How long will it take for the motorboat to make the crossing?

**(c)** What is the measure of angle $\theta$?

## RELATING CONCEPTS

**For individual or group investigation (Exercises 93–98)**

*Consider the two vectors* **v** *and* **u** *shown. Assume all values are exact. Work Exercises 93–98 in order.*

**93.** Use trigonometry alone (without using vector notation) to find the magnitude and direction angle of **u** + **v**. Use the law of cosines and the law of sines in your work.

**94.** Find the horizontal and vertical components of **u**, using your calculator.

**95.** Find the horizontal and vertical components of **v**, using your calculator.

**96.** Find the horizontal and vertical components of **u** + **v** by adding the results you obtained in Exercises 94 and 95.

**97.** Use your calculator to find the magnitude and direction angle of the vector **u** + **v**.

**98.** Compare your answers in Exercises 93 and 97. What do you notice? Which method of solution do you prefer?

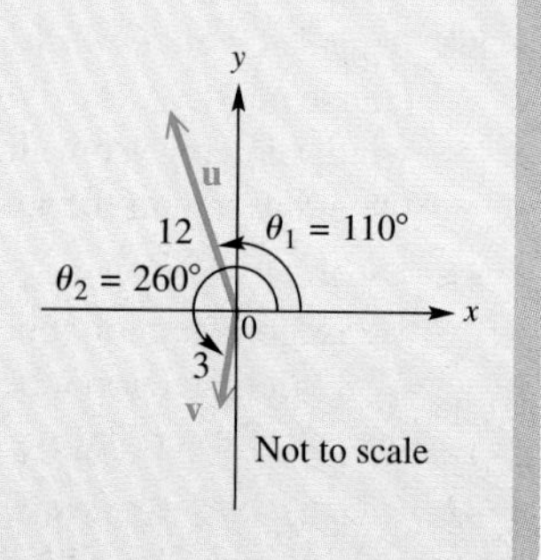


**99.** *(Modeling) Wind and Vectors* A wind can be described by $\mathbf{v} = 6\mathbf{i} + 8\mathbf{j}$, where vector **j** points north and represents a south wind of 1 mph.

**(a)** What is the speed of the wind?

**(b)** Find 3**v**. Interpret the result.

**(c)** Interpret the wind if it switches to $\mathbf{u} = -8\mathbf{i} + 8\mathbf{j}$.

**100.** *(Modeling) Measuring Rainfall* Suppose that vector **R** models the amount of rainfall in inches and the direction it falls, and vector **A** models the area in square inches and orientation of the opening of a rain gauge, as illustrated in the figure on the next page. The total volume $V$ of water collected in the rain gauge is given by $V = |\mathbf{R} \cdot \mathbf{A}|$. This formula calculates the volume of water collected even if the wind is blowing the rain in a slanted direction or the rain gauge is not exactly vertical. Let $\mathbf{R} = \mathbf{i} - 2\mathbf{j}$ and $\mathbf{A} = .5\mathbf{i} + \mathbf{j}$.

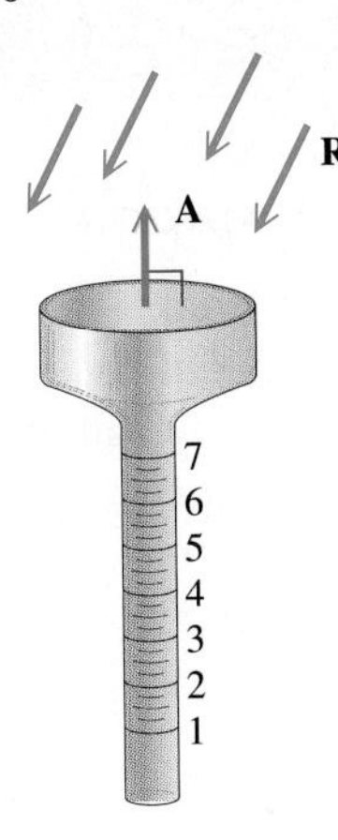


**(a)** Find $|\mathbf{R}|$ and $|\mathbf{A}|$. Interpret your results.

**(b)** Calculate $V$ and interpret this result.

**(c)** For the rain gauge to collect the maximum amount of water, what should be true about vectors **R** and **A**?

**101.** *The Dot Product* In the figure

$$\mathbf{a} = \langle a_1, a_2 \rangle, \quad \mathbf{b} = \langle b_1, b_2 \rangle$$

and $\quad \mathbf{a} - \mathbf{b} = \langle a_1 - b_1, a_2 - b_2 \rangle.$

Apply the law of cosines to the triangle and derive the equation

$$\mathbf{a} \cdot \mathbf{b} = |\mathbf{a}|\,|\mathbf{b}| \cos \theta.$$

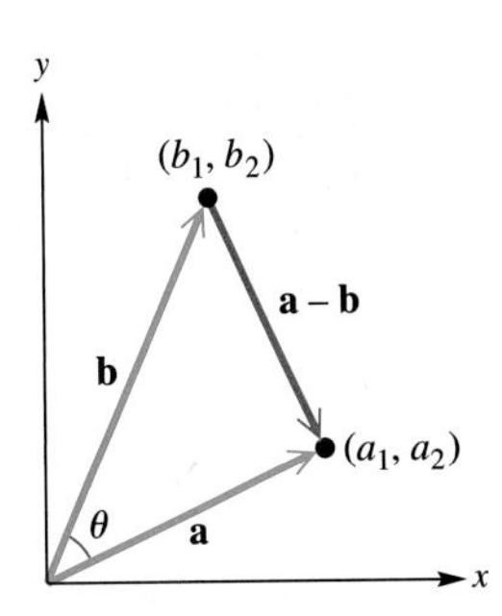


## Reviewing Basic Concepts (Sections 10.1–10.3)

*Round values to the nearest tenth in Exercises 1–5.*

**1.** Solve triangle $ABC$ if $A = 44°$, $C = 62°$, and $a = 12$.

**2.** Solve triangle $ABC$ if $A = 32°$, $a = 6$, and $b = 8$. How many solutions are there?

**3.** Solve triangle $ABC$ if $C = 41°$, $c = 12$, and $a = 7$. How many solutions are there?

**4.** Use the law of cosines to solve each triangle.

**(a)** **(b)**

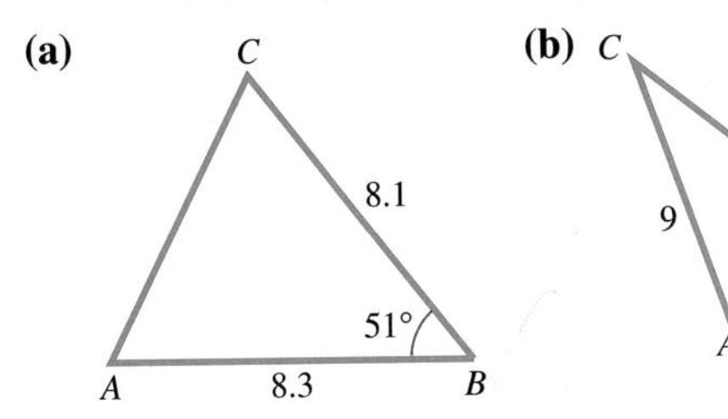


**5.** Find the area of triangle $ABC$ if $a = 4.5$, $b = 5.2$, and $C = 55°$.

**6.** Find the area of triangle $ABC$ if $a = 6$, $b = 7$, and $c = 9$.

**7.** Let $\mathbf{v} = 2\mathbf{i} - \mathbf{j}$ and $\mathbf{u} = -3\mathbf{i} + 2\mathbf{j}$. Find each expression.

**(a)** $2\mathbf{v} + \mathbf{u}$ **(b)** $2\mathbf{v}$ **(c)** $\mathbf{v} - 3\mathbf{u}$

**8.** Let $\mathbf{a} = \langle 3, -2 \rangle$ and $\mathbf{b} = \langle -1, 3 \rangle$. Find $\mathbf{a} \cdot \mathbf{b}$ and the angle between **a** and **b**, rounded to the nearest tenth of a degree.

**9.** *Resultant Force* Find the magnitude of the resultant force of the two forces shown in the figure.

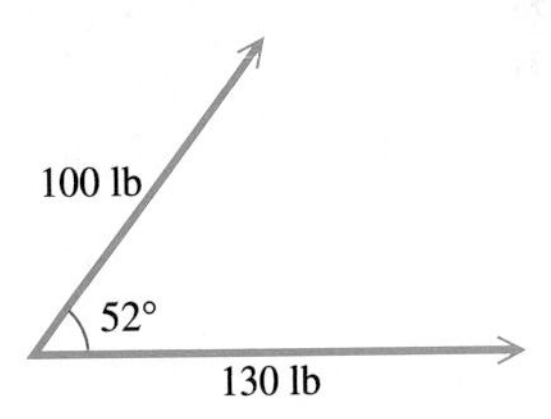


**10.** *Height of an Airplane* Two observation points $A$ and $B$ are 950 feet apart. From these points the angles of elevation of an airplane are 52° and 57°. (See the figure.) Find the height of the airplane.

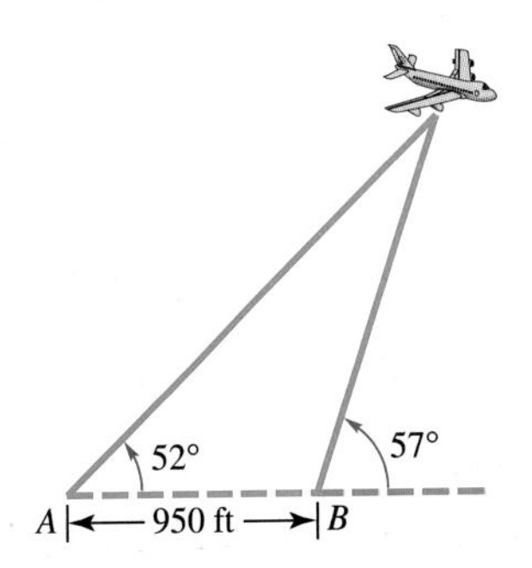

# TRIGONOMETRIC (POLAR) FORM OF COMPLEX NUMBERS

The Complex Plane and Vector Representation • Trigonometric (Polar) Form • Products of Complex Numbers in Trigonometric Form • Quotients of Complex Numbers in Trigonometric Form

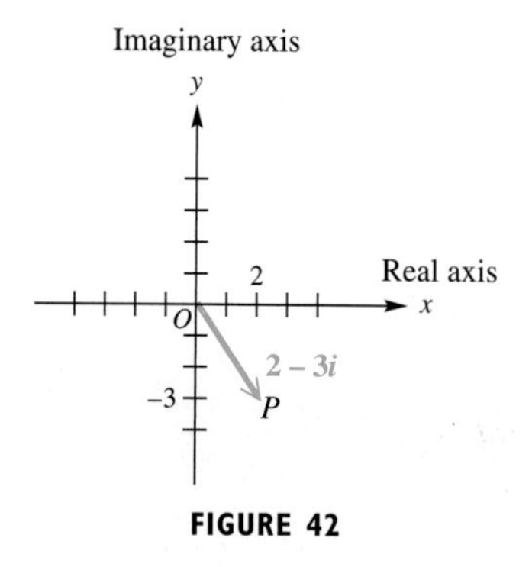


FIGURE 42

## The Complex Plane and Vector Representation

Unlike real numbers, complex numbers cannot be ordered. One way to organize and illustrate them is by using a graph. To graph a complex number such as $2 - 3i$, the familiar coordinate system must be modified. We do this by calling the horizontal axis the **real axis** and the vertical axis the **imaginary axis.** Then complex numbers can be graphed in this **complex plane,** as shown in Figure 42 for the complex number $2 - 3i$. Each complex number $a + bi$ determines a unique position vector with initial point $(0, 0)$ and terminal point $(a, b)$. This shows the close connection between vectors and complex numbers.

**NOTE** This geometric representation is the reason that $a + bi$ is called the **rectangular form** of a complex number. (*Rectangular form* is also called *standard form.*)

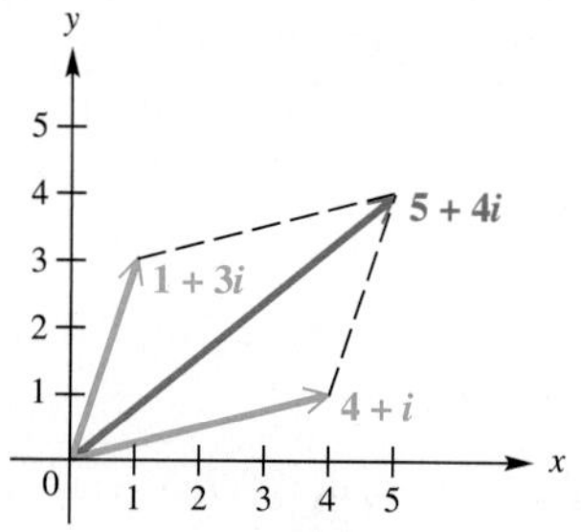

FIGURE 43

Recall from Chapter 3 that the sum of the two complex numbers $4 + i$ and $1 + 3i$ is

$$(4 + i) + (1 + 3i) = 5 + 4i.$$

Graphically, the sum of two complex numbers is represented by the vector that is the resultant of the vectors corresponding to the two numbers, as shown in Figure 43.

FIGURE 44

**EXAMPLE 1** *Expressing the Sum of Complex Numbers Graphically*

Find the sum of $6 - 2i$ and $-4 - 3i$. Graph both complex numbers and their resultant.

**Solution** The sum is found by adding the two numbers.

$$(6 - 2i) + (-4 - 3i) = 2 - 5i$$

The graphs are shown in Figure 44.

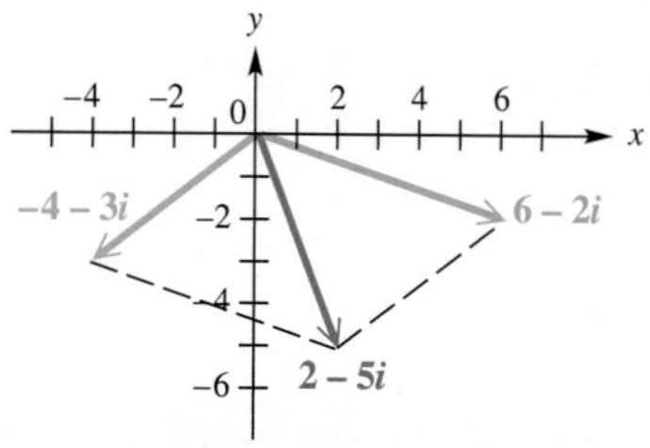

## Trigonometric (Polar) Form

Figure 45 shows the complex number $x + yi$ that corresponds to a vector **OP** with direction angle $\theta$ and magnitude $r$. The following relationships among $x$, $y$, $r$, and $\theta$ can be verified from Figure 45.

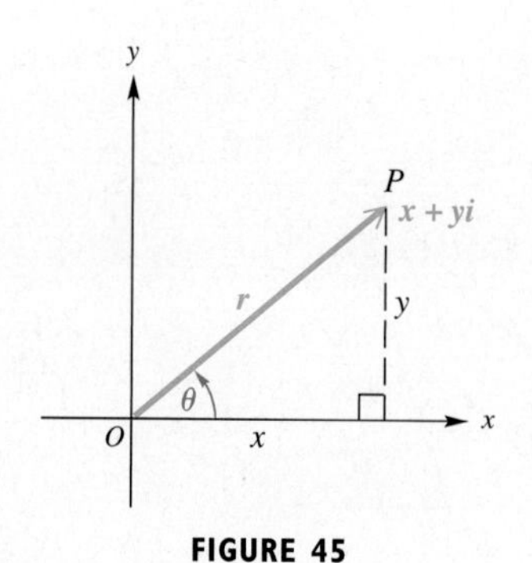


FIGURE 45

**Relationships among $x$, $y$, $r$, and $\theta$**

$$x = r\cos\theta \qquad y = r\sin\theta$$

$$r = \sqrt{x^2 + y^2} \qquad \tan\theta = \frac{y}{x}, \quad \text{if } x \neq 0$$

Substituting $x = r\cos\theta$ and $y = r\sin\theta$ from these relationships into $x + yi$ gives

$$\begin{aligned} x + yi &= r\cos\theta + (r\sin\theta)i \\ &= r(\cos\theta + i\sin\theta). \end{aligned}$$

**Trigonometric (Polar) Form of a Complex Number**

The expression

$$\mathbf{r(\cos\theta + i\sin\theta)}$$

is called the **trigonometric form** (or **polar form**) of the complex number $x + yi$. The expression $\cos\theta + i\sin\theta$ is sometimes abbreviated cis $\theta$. Using this notation,

$$\mathbf{r(\cos\theta + i\sin\theta)} \text{ is written } \mathbf{r \text{ cis } \theta}.$$

The number $r$ is called the **modulus** or **absolute value** of $x + yi$, and $\theta$ is the **argument** of $x + yi$. In this section we choose the value of $\theta$ in the interval $[0°, 360°)$. However, any angle coterminal with $\theta$ would work as the argument.

**EXAMPLE 2** *Converting from Trigonometric Form to Rectangular Form*

Express $2(\cos 300° + i\sin 300°)$ in rectangular form.

**Analytic Solution**

To evaluate cos 300° and sin 300° by hand, refer to the techniques discussed in Chapter 8. Since $\cos 300° = \frac{1}{2}$ and $\sin 300° = -\frac{\sqrt{3}}{2}$,

$$\begin{aligned} 2(\cos 300° + i\sin 300°) &= 2\left(\frac{1}{2} - i\frac{\sqrt{3}}{2}\right) \\ &= 1 - i\sqrt{3}. \end{aligned}$$

**Graphing Calculator Solution**

Figure 46 confirms the analytic solution.

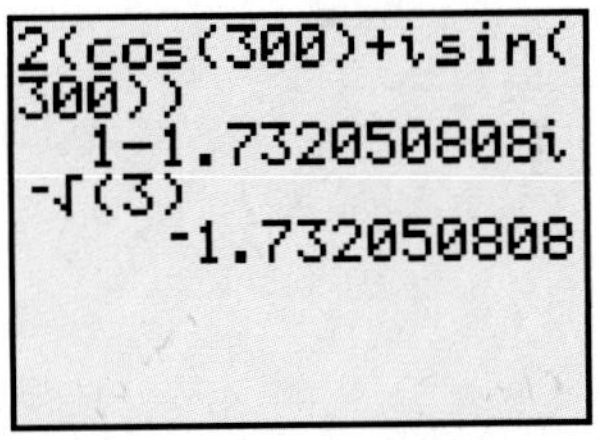

```
2(cos(300)+isin(
300))
     1-1.732050808i
-√(3)
       -1.732050808
```

The imaginary part is an approximation for $-\sqrt{3}$.

**FIGURE 46**

To convert from rectangular form to trigonometric form, we use the following procedure.

**Converting from Rectangular to Trigonometric Form**

1. Sketch a graph of the number $x + yi$ in the complex plane.
2. Find $r$ by using the equation $r = \sqrt{x^2 + y^2}$.
3. Find $\theta$ by using the equation $\tan\theta = \frac{y}{x}$, $x \neq 0$, choosing the quadrant indicated in Step 1.

**CAUTION** Errors often occur in Step 3. Be sure to choose the correct quadrant for $\theta$ by referring to the graph sketched in Step 1.

**EXAMPLE 3** *Converting from Rectangular to Trigonometric Form*

Write each complex number in trigonometric form.

**(a)** $-\sqrt{3} + i$ **(b)** $-3i$

**Solution**

**(a)** We start by sketching the graph of $-\sqrt{3} + i$ in the complex plane, as shown in Figure 47. Next, we find $r$. Since $x = -\sqrt{3}$ and $y = 1$,

$$r = \sqrt{x^2 + y^2} = \sqrt{(-\sqrt{3})^2 + 1^2} = \sqrt{3 + 1} = 2.$$

Then, we find $\theta$.

$$\tan\theta = \frac{y}{x} = \frac{1}{-\sqrt{3}} = -\frac{\sqrt{3}}{3}$$

Since $\tan\theta = -\frac{\sqrt{3}}{3}$, the reference angle for $\theta$ in radians is $\frac{\pi}{6}$. From the graph, we see that $\theta$ is in quadrant II, so $\theta = \pi - \frac{\pi}{6} = \frac{5\pi}{6}$. Therefore, in trigonometric form,

$$-\sqrt{3} + i = 2\left(\cos\frac{5\pi}{6} + i\sin\frac{5\pi}{6}\right) = 2\text{ cis }\frac{5\pi}{6}.$$

**TECHNOLOGY NOTE**

The top screen shows how to convert from rectangular $(x, y)$ form to trigonometric form. The calculator is in radian mode. The results agree with our analytic results in Example 3(a).

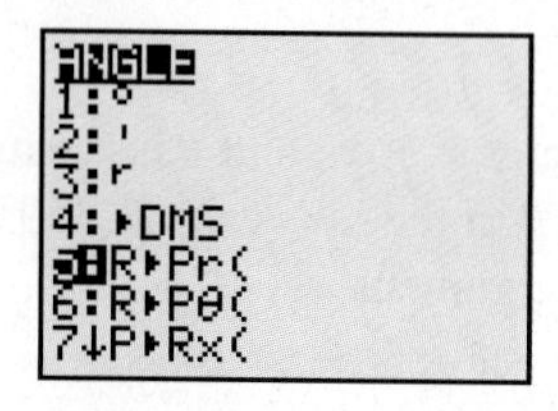


R▸Pr(-√(3),1)
2
R▸Pθ(-√(3),1)
2.617993878
5π/6
2.617993878

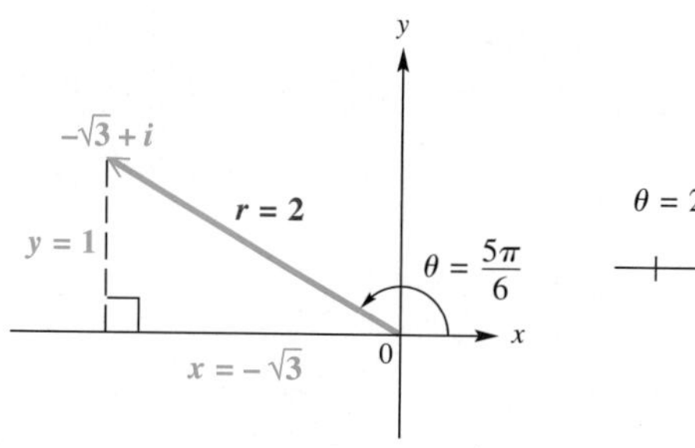


FIGURE 47

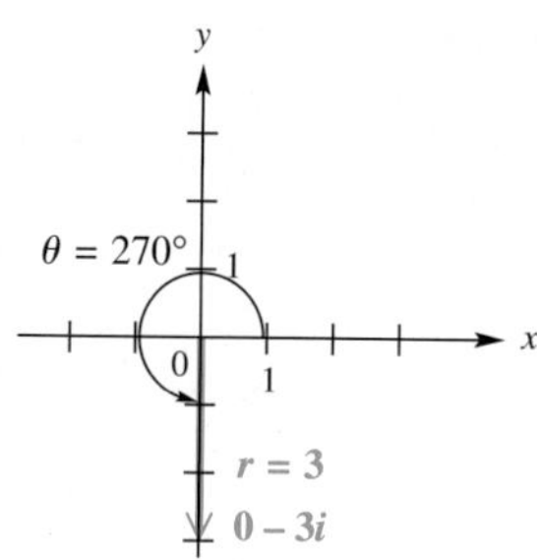


FIGURE 48

**(b)** The sketch of $-3i$ is shown in Figure 48. Since $-3i = 0 - 3i$, we have $x = 0$ and $y = -3$. We find $r$ as follows.

$$r = \sqrt{0^2 + (-3)^2} = \sqrt{0 + 9} = \sqrt{9} = 3$$

We cannot find $\theta$ by using $\tan\theta = \frac{y}{x}$, because $x = 0$. From the graph, a value for $\theta$ is 270°. In trigonometric form,

$$-3i = 3(\cos 270° + i\sin 270°) = 3\text{ cis }270°.$$

**NOTE** In Example 3, we gave answers in both forms: $r(\cos\theta + i\sin\theta)$ and $r\text{ cis }\theta$. These forms will be used interchangeably throughout the rest of this chapter.

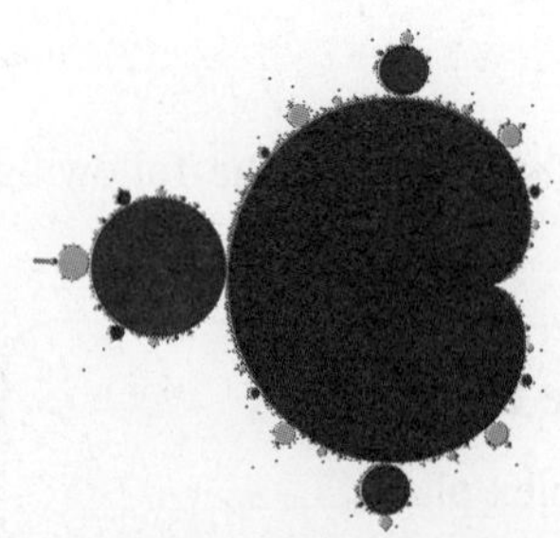

*Source:* Figure from Crownover, R., *Introduction to Fractals and Chaos.* Copyright © 1995. Boston: Jones and Bartlett Publishers. Reprinted with permission.

FIGURE 49

We can apply complex numbers to the study of *fractals,* first discussed in the chapter introduction. Another example of a fractal is the *Mandelbrot set* shown in Figure 49.

**EXAMPLE 4** *Deciding Whether a Complex Number Is in the Julia Set*

The fractal called the *Julia set* is shown in Figure 50. To determine if a complex number $z = a + bi$ is in this Julia set, perform the following sequence of calculations. Repeatedly compute the values of $z^2 - 1$, $(z^2 - 1)^2 - 1$, $[(z^2 - 1)^2 - 1]^2 - 1, \ldots.$ If the moduli of any of the resulting complex numbers exceed 2, then the complex number $z$ is not in the Julia set. Otherwise $z$ is part of this set and the point $(a, b)$ should be shaded in the graph.

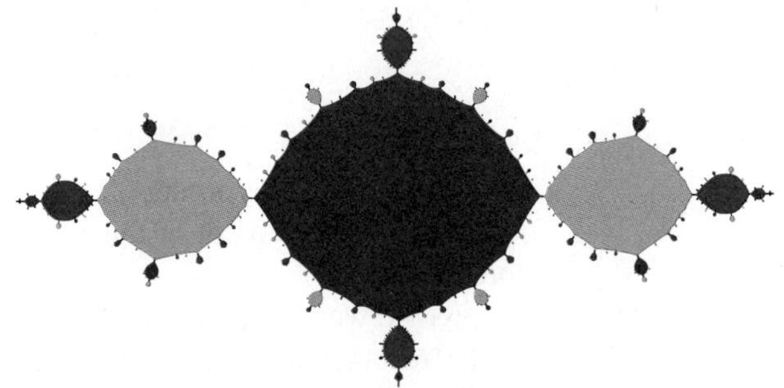

*Source:* Figure from Crownover, R., *Introduction to Fractals and Chaos.* Copyright © 1995. Boston: Jones and Bartlett Publishers. Reprinted with permission.

**FIGURE 50**

Determine whether each number belongs to the Julia set.

**(a)** $z = 0 + 0i$ **(b)** $z = 1 + 1i$

### Solution

**(a)** Here

$$\begin{aligned} z &= 0 + 0i = 0, \\ z^2 - 1 &= 0^2 - 1 = -1, \\ (z^2 - 1)^2 - 1 &= (-1)^2 - 1 = 0, \\ [(z^2 - 1)^2 - 1]^2 - 1 &= 0^2 - 1 = -1, \end{aligned}$$

and so on. We see that the calculations repeat as $0, -1, 0, -1$, and so on. The moduli are either 0 or 1, which do not exceed 2, so $0 + 0i$ is in the Julia set and the point $(0, 0)$ is part of the graph.

**(b)** We have $z^2 - 1 = (1 + i)^2 - 1 = (1 + 2i + i^2) - 1 = -1 + 2i$. The modulus is $\sqrt{(-1)^2 + 2^2} = \sqrt{5}$. Since $\sqrt{5}$ is greater than 2, $1 + 1i$ is not in the Julia set and $(1, 1)$ is not part of the graph.

TECHNOLOGY NOTE

With the calculator in complex and degree modes, the math menu can be used to find the angle and the magnitude (absolute value) of the vector that corresponds to a given complex number.

```
angle(1+√(3)i)
                60
abs(1+√(3)i)
                 2
```

## Products of Complex Numbers in Trigonometric Form

Using the FOIL method to multiply complex numbers in rectangular form, we find the product of $1 + i\sqrt{3}$ and $-2\sqrt{3} + 2i$ as follows. (See Section 3.1.)

$$\begin{aligned} (1 + i\sqrt{3})(-2\sqrt{3} + 2i) &= -2\sqrt{3} + 2i - 2i(3) + 2i^2\sqrt{3} \\ &= -2\sqrt{3} + 2i - 6i - 2\sqrt{3} \\ &= -4\sqrt{3} - 4i \end{aligned}$$

We can also find this same product by first converting the complex numbers $1 + i\sqrt{3}$ and $-2\sqrt{3} + 2i$ to trigonometric form. Using the method explained earlier in this section,

$$1 + i\sqrt{3} = 2(\cos 60° + i \sin 60°)$$

and

$$-2\sqrt{3} + 2i = 4(\cos 150° + i \sin 150°).$$

If we now multiply the trigonometric forms, and if we use the identities for the cosine and the sine of the sum of two angles, the result is

$$\begin{aligned}[2(\cos 60^\circ + i\sin 60^\circ)][4(\cos 150^\circ + i\sin 150^\circ)]\\ &= 2\cdot 4(\cos 60^\circ\cdot\cos 150^\circ + i\sin 60^\circ\cdot\cos 150^\circ\\ &\quad + i\cos 60^\circ\cdot\sin 150^\circ + i^2\sin 60^\circ\cdot\sin 150^\circ)\\ &= 8[(\cos 60^\circ\cdot\cos 150^\circ - \sin 60^\circ\cdot\sin 150^\circ)\\ &\quad + i(\sin 60^\circ\cdot\cos 150^\circ + \cos 60^\circ\cdot\sin 150^\circ)]\\ &= 8[\cos(60^\circ + 150^\circ) + i\sin(60^\circ + 150^\circ)]\\ &= 8(\cos 210^\circ + i\sin 210^\circ).\end{aligned}$$

The modulus of the product, 8, is equal to the product of the moduli of the factors, $2\cdot 4$, and the argument of the product, 210°, is the sum of the arguments of the factors, $60^\circ + 150^\circ$.

As we would expect, the product obtained when multiplying by the first method is the rectangular form of the product obtained when multiplying by the second method.

$$8(\cos 210^\circ + i\sin 210^\circ) = 8\left(-\frac{\sqrt{3}}{2} - \frac{1}{2}i\right) = -4\sqrt{3} - 4i$$

We can generalize this work in the following *product theorem.*

**Product Theorem**

If $r_1(\cos\theta_1 + i\sin\theta_1)$ and $r_2(\cos\theta_2 + i\sin\theta_2)$ are any two complex numbers, then

$$\begin{aligned}&[r_1(\cos\theta_1 + i\sin\theta_1)]\cdot[r_2(\cos\theta_2 + i\sin\theta_2)]\\ &= r_1r_2[\cos(\theta_1 + \theta_2) + i\sin(\theta_1 + \theta_2)].\end{aligned}$$

In compact form, this is written

$$(r_1 \text{ cis } \theta_1)(r_2 \text{ cis } \theta_2) = r_1r_2 \text{ cis}(\theta_1 + \theta_2).$$

**EXAMPLE 5** *Using the Product Theorem*

Find the product of

$$3(\cos 45^\circ + i\sin 45^\circ) \quad\text{and}\quad 2(\cos 135^\circ + i\sin 135^\circ).$$

**Solution**

$$\begin{aligned}&[3(\cos 45^\circ + i\sin 45^\circ)][2(\cos 135^\circ + i\sin 135^\circ)]\\ &= 3\cdot 2[\cos(45^\circ + 135^\circ) + i\sin(45^\circ + 135^\circ)] \qquad \text{Product theorem}\\ &= 6(\cos 180^\circ + i\sin 180^\circ)\end{aligned}$$

This can be expressed as $6(-1 + i\cdot 0) = 6(-1) = -6$. The two complex numbers in this example are complex factors of $-6$.

## Quotients of Complex Numbers in Trigonometric Form

The rectangular form of the quotient of the complex numbers $1 + i\sqrt{3}$ and $-2\sqrt{3} + 2i$ is

$$\begin{aligned}\frac{1 + i\sqrt{3}}{-2\sqrt{3} + 2i} &= \frac{(1 + i\sqrt{3})(-2\sqrt{3} - 2i)}{(-2\sqrt{3} + 2i)(-2\sqrt{3} - 2i)} \\ &= \frac{-2\sqrt{3} - 2i - 6i - 2i^2\sqrt{3}}{12 - 4i^2} \\ &= \frac{-8i}{16} = -\frac{1}{2}i.\end{aligned}$$

Writing $1 + i\sqrt{3}$, $-2\sqrt{3} + 2i$, and $-\frac{1}{2}i$ in trigonometric form gives

$$\begin{aligned}1 + i\sqrt{3} &= 2(\cos 60° + i \sin 60°) \\ -2\sqrt{3} + 2i &= 4(\cos 150° + i \sin 150°)\end{aligned}$$

and

$$-\frac{1}{2}i = \frac{1}{2}[\cos(-90°) + i \sin(-90°)].$$

The modulus of the quotient, $\frac{1}{2}$, is the quotient of the two moduli, $\frac{2}{4} = \frac{1}{2}$. The argument of the quotient, $-90°$, is the difference of the two arguments, $60° - 150° = -90°$. Generalizing from this example leads to the *quotient theorem.*

### Quotient Theorem

If $r_1(\cos \theta_1 + i \sin \theta_1)$ and $r_2(\cos \theta_2 + i \sin \theta_2)$ are complex numbers, where $r_2(\cos \theta_2 + i \sin \theta_2) \neq 0$, then

$$\frac{r_1(\cos \theta_1 + i \sin \theta_1)}{r_2(\cos \theta_2 + i \sin \theta_2)} = \frac{r_1}{r_2}[\cos(\theta_1 - \theta_2) + i \sin(\theta_1 - \theta_2)].$$

In compact form, this is written

$$\frac{r_1 \text{ cis } \theta_1}{r_2 \text{ cis } \theta_2} = \frac{r_1}{r_2} \text{ cis}(\theta_1 - \theta_2).$$

**FOR DISCUSSION**

In Example 6, the complex number

$$10 \text{ cis}(-60°) = 5 - 5i\sqrt{3}$$

was divided by the complex number

$$5 \text{ cis}(150°) = -\frac{5\sqrt{3}}{2} + \frac{5}{2}i.$$

Would you rather perform this division (by hand) in trigonometric form or in rectangular form? Explain your reasoning.

**EXAMPLE 6** *Using the Quotient Theorem*

Find the quotient $\dfrac{10 \text{ cis}(-60°)}{5 \text{ cis } 150°}$. Write the result in rectangular form.

**Solution**

$$\begin{aligned}\frac{10 \text{ cis}(-60°)}{5 \text{ cis } 150°} &= \frac{10}{5} \text{ cis}(-60° - 150°) && \text{Quotient theorem} \\ &= 2 \text{ cis}(-210°) && \text{Subtract.} \\ &= 2[\cos(-210°) + i \sin(-210°)] \\ &= 2\left[-\frac{\sqrt{3}}{2} + i\left(\frac{1}{2}\right)\right] && \cos(-210°) = -\tfrac{\sqrt{3}}{2};\ \sin(-210°) = \tfrac{1}{2} \\ &= -\sqrt{3} + i && \text{Rectangular form}\end{aligned}$$

## 10.4 EXERCISES

**1.** *Concept Check* The modulus of a complex number represents the ______________ of the vector representing it in the complex plane.

**2.** *Concept Check* What is the geometric interpretation of the argument of a complex number?

*Graph each complex number as a vector in the complex plane.*

**3.** $6 - 5i$ **4.** $2 - 2i\sqrt{3}$ **5.** $-4i$ **6.** $3i$ **7.** $-8$ **8.** $2$

*Give the rectangular form of the complex number represented in each graph.*

**9.**

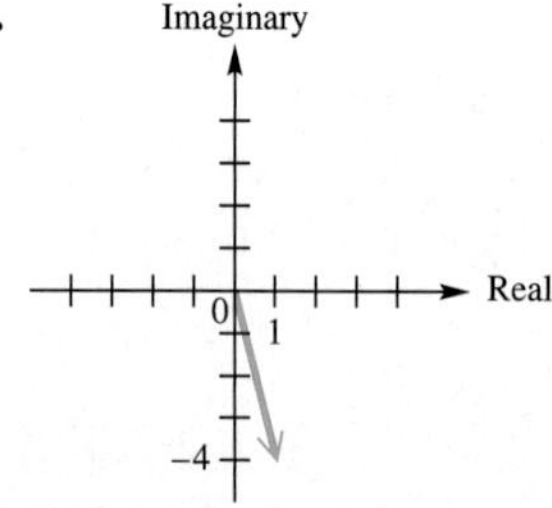


**10.**

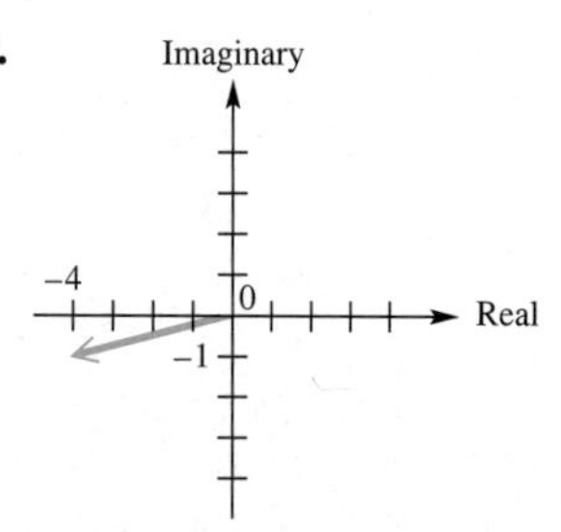


**11.** *Concept Check* What must be true for a complex number to also be a real number?

**12.** *Concept Check* If a real number is graphed in the complex plane, on what axis does the vector lie?

**13.** *Concept Check* A complex number of the form $a + bi$ will have its corresponding vector lying on the $y$-axis provided $a =$ ______________.

**14.** Are the notations $a + bi$ and $a\mathbf{i} + b\mathbf{j}$ equivalent? Explain.

*Find the resultant (sum) of each pair of complex numbers. Express in rectangular form $a + bi$.*

**15.** $4 - 3i, -1 + 2i$ **16.** $2 + 3i, -4 - i$ **17.** $-3, 3i$

**18.** $6, -2i$ **19.** $2 + 6i, -2i$ **20.** $-5 - 8i, -1$

*Find the modulus r of the number.*

**21.** $1 + i$ **22.** $3 - 4i$ **23.** $12 - 5i$ **24.** $-24 + 7i$

**25.** $-6$ **26.** $15i$ **27.** $2 - 3i$ **28.** $11 - 60i$

*Write each complex number in rectangular form. Give exact values for the real and imaginary parts.*

**29.** $2(\cos 45° + i \sin 45°)$ **30.** $4(\cos 60° + i \sin 60°)$ **31.** $10 \text{ cis } 90°$

**32.** $8 \text{ cis } 270°$ **33.** $4(\cos 240° + i \sin 240°)$ **34.** $2(\cos 330° + i \sin 330°)$

**35.** $\cos \frac{\pi}{6} + i \sin \frac{\pi}{6}$ **36.** $3\left(\cos \frac{5\pi}{6} + i \sin \frac{5\pi}{6}\right)$ **37.** $5 \text{ cis}\left(-\frac{\pi}{6}\right)$

**38.** $6 \text{ cis } \frac{3\pi}{4}$ **39.** $\sqrt{2} \text{ cis } \pi$ **40.** $\sqrt{3} \text{ cis } \frac{3\pi}{2}$

*Write each complex number in trigonometric form $r(\cos \theta + i \sin \theta)$, where r is exact, and $-180° < \theta \leq 180°$. You may wish to use your calculator to support your result.*

**41.** $3 - 3i$ **42.** $-2 + 2i\sqrt{3}$ **43.** $1 + i\sqrt{3}$

**44.** $-3 - 3i\sqrt{3}$ **45.** $-2i$ **46.** $7$

*Write each complex number in trigonometric form, where r is exact, and $-\pi < \theta \leq \pi$.*

**47.** $4\sqrt{3} + 4i$ **48.** $\sqrt{3} - i$ **49.** $-\sqrt{2} + i\sqrt{2}$

**50.** $-5 - 5i$ **51.** $-4$ **52.** $5i$

*Julia Set* *Refer to Example 4 to solve Exercises 53 and 54.*

**53.** Is $z = -.2i$ in the Julia set?

**54.** The graph of the Julia set in Figure 50 appears to be symmetric with respect to both the $x$-axis and $y$-axis. Complete the following to show that this is true.

(a) Show that complex conjugates have the same modulus.

(b) Compute $z_1^2 - 1$ and $z_2^2 - 1$, where $z_1 = a + bi$ and $z_2 = a - bi$.

(c) Discuss why if $(a, b)$ is in the Julia set then so is $(a, -b)$.

(d) Conclude that the graph of the Julia set must be symmetric with respect to the $x$-axis.

(e) Using a similar argument, show that the Julia set must also be symmetric with respect to the $y$-axis.

**55.** Use your calculator in radian mode to find the trigonometric form of $3 + 5i$. Give as many decimal places as the calculator displays.

**56.** *Concept Check* Give the smallest *positive* radian measure of $\theta$ if $r > 0$ and **(a)** $r$ cis $\theta$ has real part equal to 0, **(b)** $r$ cis $\theta$ has imaginary part equal to 0.

*Find each product in rectangular form, using exact values.*

**57.** $[3(\cos 60° + i \sin 60°)][2(\cos 90° + i \sin 90°)]$

**58.** $[4(\cos 30° + i \sin 30°)][5(\cos 120° + i \sin 120°)]$

**59.** $[2(\cos 45° + i \sin 45°)][2(\cos 225° + i \sin 225°)]$

**60.** $[8(\cos 300° + i \sin 300°)][5(\cos 120° + i \sin 120°)]$

**61.** $\left[5 \text{ cis } \frac{\pi}{2}\right]\left[3 \text{ cis } \frac{\pi}{4}\right]$

**62.** $\left[6 \text{ cis } \frac{2\pi}{3}\right]\left[5 \text{ cis}\left(-\frac{\pi}{6}\right)\right]$

**63.** $\left[\sqrt{3} \text{ cis } \frac{\pi}{4}\right]\left[\sqrt{3} \text{ cis } \frac{5\pi}{4}\right]$

**64.** $\left[\sqrt{2} \text{ cis } \frac{5\pi}{6}\right]\left[\sqrt{2} \text{ cis } \frac{3\pi}{2}\right]$

*Find each quotient in rectangular form, using exact values.*

**65.** $\dfrac{4(\cos 120° + i \sin 120°)}{2(\cos 150° + i \sin 150°)}$

**66.** $\dfrac{10\left(\cos \frac{5\pi}{4} + i \sin \frac{5\pi}{4}\right)}{5\left(\cos \frac{\pi}{4} + i \sin \frac{\pi}{4}\right)}$

**67.** $\dfrac{16(\cos 300° + i \sin 300°)}{8(\cos 60° + i \sin 60°)}$

**68.** $\dfrac{24\left(\cos \frac{5\pi}{6} + i \sin \frac{5\pi}{6}\right)}{2\left(\cos \frac{\pi}{6} + i \sin \frac{\pi}{6}\right)}$

**69.** $\dfrac{3 \text{ cis } \frac{61\pi}{36}}{9 \text{ cis } \frac{13\pi}{36}}$

**70.** $\dfrac{12 \text{ cis } 293°}{6 \text{ cis } 23°}$

## RELATING CONCEPTS

**For individual or group investigation (Exercises 71–77)**

*Consider the complex numbers*

$$w = -1 + i \quad \text{and} \quad z = -1 - i.$$

**71.** Multiply $w$ and $z$, using their rectangular forms. Leave the product in rectangular form.

**72.** Find the trigonometric forms of $w$ and $z$.

**73.** Multiply $w$ and $z$, using their trigonometric forms and the method described in this section. Leave the product in trigonometric form.

**74.** Use the result of Exercise 73 to find the rectangular form of $wz$. How does this compare to your result in Exercise 71?

**75.** Find the quotient $\frac{w}{z}$, using their rectangular forms and multiplying both the numerator and the denominator by the conjugate of the denominator. Leave the quotient in rectangular form.

**76.** Use the trigonometric forms of $w$ and $z$ found in Exercise 72 to divide $w$ by $z$, using the method described in this section.

**77.** Use the result of Exercise 76 to find the rectangular form of $\frac{w}{z}$. How does this compare to your result in Exercise 75?

**78.** *Concept Check* Without actually performing the operations, state why the products

$$[2(\cos 45° + i \sin 45°)][5(\cos 90° + i \sin 90°)]$$

and

$$[2(\cos(-315°) + i \sin(-315°))] \cdot [5(\cos(-270°) + i \sin(-270°))]$$

are the same.

**79.** Consider the equation $(r \text{ cis } \theta)^2 = (r \text{ cis } \theta)(r \text{ cis } \theta) = r^2 \text{ cis}(\theta + \theta) = r^2 \text{ cis } 2\theta$. State in your own words how we can square a complex number in trigonometric form. (In the next section, we will develop this idea further.)

**80.** *(Modeling) Alternating Current* The alternating current $I$ in amps in an electric inductor is

$$I = \frac{E}{Z},$$

where $E$ is the voltage and $Z = R + X_L i$ is the impedance. If $E = 8(\cos 20° + i \sin 20°)$, $R = 6$, and $X_L = 3$, find the current. Give the answer in rectangular form.

**81.** *(Modeling) Electric Current* The current $I$ in a circuit with voltage $E$, resistance $R$, capacitive reactance $X_c$, and inductive resistance $X_L$ is

$$I = \frac{E}{R + (X_L - X_c)i}.$$

Find $I$ if $E = 12(\cos 25° + i \sin 25°)$, $R = 3$, $X_L = 4$, and $X_c = 6$. Give the answer in rectangular form.

**82.** *Concept Check* Under what conditions is the difference between two nonreal complex numbers $a + bi$ and $c + di$ a real number?

**83.** *(Modeling) Impedance* In the parallel electric circuit shown in the figure below, the impedance $Z$ can be calculated by using the equation

$$Z = \frac{1}{\frac{1}{Z_1} + \frac{1}{Z_2}},$$

where $Z_1$ and $Z_2$ are the impedances for the branches of the circuit. If $Z_1 = 50 + 25i$ and $Z_2 = 60 + 20i$, calculate $Z$.

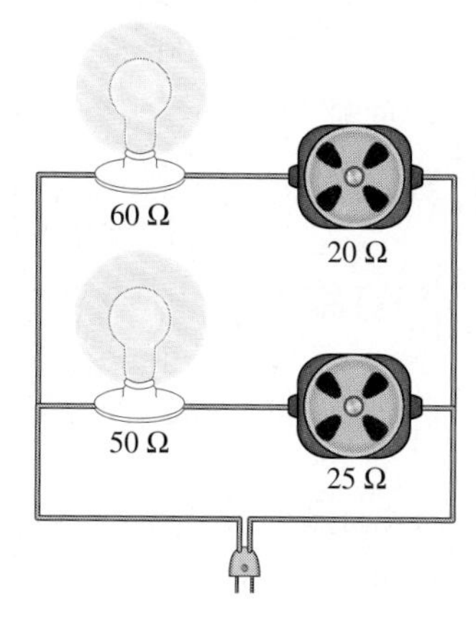


## 10.5 POWERS AND ROOTS OF COMPLEX NUMBERS

Powers of Complex Numbers (De Moivre's Theorem) • Roots of Complex Numbers

### Powers of Complex Numbers (De Moivre's Theorem)

In the previous section, we studied the product theorem for complex numbers in trigonometric form. Because raising a number to a positive integer power is a repeated application of the product rule, it would seem likely that a theorem for finding powers of complex numbers exists. This is indeed the case. For example, the square of the complex number $r(\cos \theta + i \sin \theta)$ is

$$\begin{aligned}[r(\cos \theta + i \sin \theta)]^2 &= [r(\cos \theta + i \sin \theta)][r(\cos \theta + i \sin \theta)] \\ &= r \cdot r[\cos(\theta + \theta) + i \sin(\theta + \theta)] \\ &= r^2(\cos 2\theta + i \sin 2\theta).\end{aligned}$$

In the same way,

$$[r(\cos \theta + i \sin \theta)]^3 = r^3(\cos 3\theta + i \sin 3\theta).$$

These results suggest the following theorem for positive integer values of $n$. Although the theorem is stated and can be proved for all $n$, we use it only for positive integer values of $n$ and their reciprocals.

**De Moivre's Theorem**

If $r(\cos\theta + i\sin\theta)$ is a complex number, and if $n$ is any real number, then

$$[r(\cos\theta + i\sin\theta)]^n = r^n(\cos n\theta + i\sin n\theta).$$

In compact form, this is written

$$[r\text{ cis }\theta]^n = r^n(\text{cis } n\theta).$$

This theorem is named after the French expatriate friend of Isaac Newton, Abraham De Moivre (1667–1754), although he never explicitly stated it.

**EXAMPLE 1** *Finding a Power of a Complex Number*

Find $(1 + i\sqrt{3})^8$ and express the result in rectangular form.

**Solution** First convert $1 + i\sqrt{3}$ into trigonometric form.

$$1 + i\sqrt{3} = 2(\cos 60^\circ + i\sin 60^\circ)$$

Now, apply De Moivre's theorem.

$$\begin{aligned}(1 + i\sqrt{3})^8 &= [2(\cos 60^\circ + i\sin 60^\circ)]^8 \\ &= 2^8[\cos(8\cdot 60^\circ) + i\sin(8\cdot 60^\circ)] \\ &= 256(\cos 480^\circ + i\sin 480^\circ) \\ &= 256(\cos 120^\circ + i\sin 120^\circ) && \text{480° and 120° are coterminal.} \\ &= 256\left(-\frac{1}{2} + i\frac{\sqrt{3}}{2}\right) && \cos 120^\circ = -\tfrac{1}{2};\ \sin 120^\circ = \tfrac{\sqrt{3}}{2} \\ &= -128 + 128i\sqrt{3} && \text{Rectangular form}\end{aligned}$$

## Roots of Complex Numbers

Every nonzero complex number has exactly $n$ distinct complex $n$th roots. De Moivre's theorem can be extended to find all $n$th roots of a complex number.

***n*th Root**

For a positive integer $n$, the complex number $a + bi$ is an ***n*th root** of the complex number $x + yi$ if

$$(a + bi)^n = x + yi.$$

To find the three complex cube roots of $8(\cos 135^\circ + i\sin 135^\circ)$, for example, look for a complex number, say $r(\cos\alpha + i\sin\alpha)$, that will satisfy

$$[r(\cos\alpha + i\sin\alpha)]^3 = 8(\cos 135^\circ + i\sin 135^\circ).$$

By De Moivre's theorem, this equation becomes

$$r^3(\cos 3\alpha + i \sin 3\alpha) = 8(\cos 135° + i \sin 135°).$$

To satisfy this equation, set $r^3 = 8$ and $\cos 3\alpha + i \sin 3\alpha = \cos 135° + i \sin 135°$. The first of these conditions implies that $r = 2$, and the second implies that

$$\cos 3\alpha = \cos 135° \quad \text{and} \quad \sin 3\alpha = \sin 135°.$$

For these equations to be satisfied, $3\alpha$ must represent an angle that is coterminal with 135°. Therefore, we must have

$$3\alpha = 135° + 360° \cdot k, \quad k \text{ any integer}$$

or

$$\alpha = \frac{135° + 360° \cdot k}{3}, \quad k \text{ any integer.}$$

Now, let $k$ take on the integer values 0, 1, and 2.

$$\text{If } k = 0, \text{ then } \quad \alpha = \frac{135° + 0°}{3} = 45°.$$

$$\text{If } k = 1, \text{ then } \quad \alpha = \frac{135° + 360°}{3} = \frac{495°}{3} = 165°.$$

$$\text{If } k = 2, \text{ then } \quad \alpha = \frac{135° + 720°}{3} = \frac{855°}{3} = 285°.$$

In the same way, $\alpha = 405°$ when $k = 3$. But note that $405° = 45° + 360°$ so $\sin 405° = \sin 45°$ and $\cos 405° = \cos 45°$. Similarly, if $k = 4$, $\alpha = 525°$, which has the same sine and cosine values as 165°. To continue with larger values of $k$ would just be repeating solutions already found. Therefore, all of the cube roots (three of them) can be found by letting $k = 0, 1,$ or 2.

When $k = 0$, the root is $2(\cos 45° + i \sin 45°)$.
When $k = 1$, the root is $2(\cos 165° + i \sin 165°)$.
When $k = 2$, the root is $2(\cos 285° + i \sin 285°)$.

In summary, we see that $2(\cos 45° + i \sin 45°)$, $2(\cos 165° + i \sin 165°)$, and $2(\cos 285° + i \sin 285°)$ are the three cube roots of $8(\cos 135° + i \sin 135°)$.

Generalizing our results, we state the following theorem.

**$n$th Root Theorem**

If $n$ is any positive integer, $r$ is a positive real number, and $\theta$ is in degrees, then the nonzero complex number $r(\cos \theta + i \sin \theta)$ has exactly $n$ distinct $n$th roots, given by

$$\sqrt[n]{r}(\cos \alpha + i \sin \alpha),$$

where

$$\alpha = \frac{\theta + 360° \cdot k}{n} \quad \text{or} \quad \alpha = \frac{\theta}{n} + \frac{360° \cdot k}{n}, \qquad k = 0, 1, 2, \ldots, n - 1.$$

**NOTE** In the statement of the $n$th root theorem, if $\theta$ is in radians, then

$$\alpha = \frac{\theta + 2\pi k}{n} \quad \text{or} \quad \alpha = \frac{\theta}{n} + \frac{2\pi k}{n}.$$

In the next example, we find the square roots of a complex number by writing the arguments in radian mode.

**EXAMPLE 2** *Finding Complex Roots*

Find the two square roots of $4i$. Write the roots in rectangular form, and check your results directly with a calculator.

**Solution** First write $4i$ in trigonometric form as

$$4i = 4\left(\cos\frac{\pi}{2} + i\sin\frac{\pi}{2}\right).$$

Here $r = 4$ and $\theta = \frac{\pi}{2}$. The square roots have modulus $\sqrt{4} = 2$ and arguments as follows.

$$\alpha = \frac{\frac{\pi}{2}}{2} + \frac{2\pi k}{2} = \frac{\pi}{4} + \pi k$$

Since there are two square roots, let $k = 0$ or $1$.

$$\text{If } k = 0, \text{ then } \quad \alpha = \frac{\pi}{4} + \pi \cdot 0 = \frac{\pi}{4}.$$

$$\text{If } k = 1, \text{ then } \quad \alpha = \frac{\pi}{4} + \pi \cdot 1 = \frac{5\pi}{4}.$$

Using these values for $\alpha$, the square roots are $2 \text{ cis } \frac{\pi}{4}$ and $2 \text{ cis } \frac{5\pi}{4}$, which can be written in rectangular form as

$$\sqrt{2} + i\sqrt{2} \quad \text{and} \quad -\sqrt{2} - i\sqrt{2}.$$

Check the results by squaring each answer, as shown in Figure 51.

(√(2)+i√(2))²
4i
(-√(2)-i√(2))²
4i

FIGURE 51

**EXAMPLE 3** *Finding Complex Roots*

Find all fourth roots of $-8 + 8i\sqrt{3}$. Write the roots in rectangular form.

**Solution** First write $-8 + 8i\sqrt{3}$ in trigonometric form as

$$-8 + 8i\sqrt{3} = 16 \text{ cis } 120°.$$

Here $r = 16$ and $\theta = 120°$. The fourth roots of this number have modulus $\sqrt[4]{16} = 2$ and arguments as follows.

$$\alpha = \frac{120°}{4} + \frac{360° \cdot k}{4} = 30° + 90° \cdot k$$

Since there are four fourth roots, let $k = 0, 1, 2,$ or $3$.

$$\text{If } k = 0, \text{ then } \quad \alpha = 30° + 90° \cdot 0 = 30°.$$

$$\text{If } k = 1, \text{ then } \quad \alpha = 30° + 90° \cdot 1 = 120°.$$

$$\text{If } k = 2, \text{ then } \quad \alpha = 30° + 90° \cdot 2 = 210°.$$

$$\text{If } k = 3, \text{ then } \quad \alpha = 30° + 90° \cdot 3 = 300°.$$

Using these angles, the fourth roots are

$$2 \text{ cis } 30°, \quad 2 \text{ cis } 120°, \quad 2 \text{ cis } 210°, \quad \text{and} \quad 2 \text{ cis } 300°.$$

**TECHNOLOGY NOTE**

In Example 3, $-8 + 8i\sqrt{3}$ is converted into the form $16(\cos 120° + i \sin 120°)$. In the screen below, a graphing calculator performs this conversion.

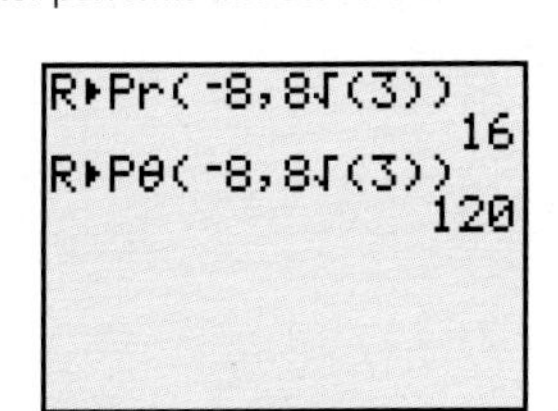


Degree Mode

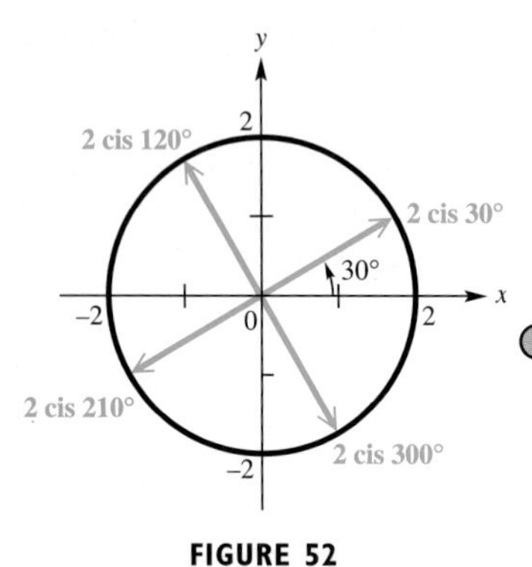


FIGURE 52

These four roots can be written in rectangular form as $\sqrt{3} + i$, $-1 + i\sqrt{3}$, $-\sqrt{3} - i$, and $1 - i\sqrt{3}$. The graphs of these roots are all on a circle that has center at the origin and radius 2, as shown in Figure 52. Notice that the roots are equally spaced about the circle, 90° apart.

**EXAMPLE 4** *Solving an Equation by Finding Complex Roots*

Find all complex number solutions of $x^5 - 1 = 0$. Graph them as vectors in the complex plane.

**Solution** Write the equation as

$$x^5 - 1 = 0$$
$$x^5 = 1.$$

While there is only one real number solution, 1, there are five complex number solutions. To find these solutions, first write 1 in trigonometric form as

$$1 = 1 + 0i = 1(\cos 0° + i \sin 0°).$$

The modulus of the fifth roots is $\sqrt[5]{1} = 1$, and the arguments are given by

$$0° + 72° \cdot k, \quad k = 0, 1, 2, 3, \text{ or } 4.$$

By using these arguments, the fifth roots are

$$1(\cos 0° + i \sin 0°), \quad k = 0$$
$$1(\cos 72° + i \sin 72°), \quad k = 1$$
$$1(\cos 144° + i \sin 144°), \quad k = 2$$
$$1(\cos 216° + i \sin 216°), \quad k = 3$$

and

$$1(\cos 288° + i \sin 288°). \quad k = 4$$

The solution set of the equation can be written as {cis 0°, cis 72°, cis 144°, cis 216°, cis 288°}. The first of these roots equals 1; the others cannot easily be expressed in rectangular form but can be approximated with a calculator. The five fifth roots all lie on a unit circle and are equally spaced around it every 72°, as shown in Figure 53.

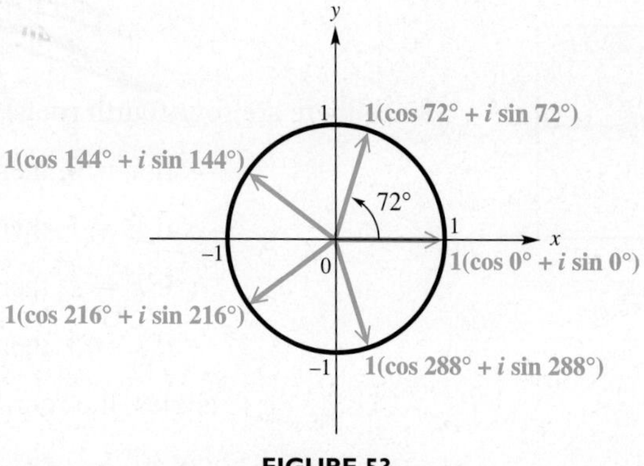


FIGURE 53

## 10.5 EXERCISES

*Find each power. Write the answer in rectangular form.*

1. $[3(\cos 30° + i \sin 30°)]^3$
2. $[2(\cos 135° + i \sin 135°)]^4$
3. $\left(\cos \frac{\pi}{4} + i \sin \frac{\pi}{4}\right)^8$
4. $\left[2\left(\cos \frac{2\pi}{3} + i \sin \frac{2\pi}{3}\right)\right]^3$
5. $[3 \text{ cis } 100°]^3$
6. $[3 \text{ cis } 40°]^3$
7. $(\sqrt{3} + i)^5$
8. $(2\sqrt{2} - 2i\sqrt{2})^6$
9. $(2 - 2i\sqrt{3})^4$
10. $\left(\frac{\sqrt{2}}{2} - \frac{\sqrt{2}}{2}i\right)^8$
11. $(-2 - 2i)^5$
12. $(-1 + i)^7$

*Find the cube roots of each complex number. Leave the answer in trigonometric form. Then, graph each cube root as a vector in the complex plane.*

13. $1$
14. $i$
15. $8 \text{ cis } 60°$
16. $27(\cos 300° + i \sin 300°)$
17. $-8i$
18. $27i$
19. $-64$
20. $27$
21. $1 + i\sqrt{3}$
22. $2 - 2i\sqrt{3}$
23. $-2\sqrt{3} + 2i$
24. $\sqrt{3} - i$

*Find each root and express it in rectangular form. Check your results with a calculator.*

25. The square roots of $4(\cos 120° + i \sin 120°)$
26. The cube roots of $27(\cos 180° + i \sin 180°)$
27. The cube roots of $\cos 180° + i \sin 180°$
28. The fourth roots of $16(\cos 240° + i \sin 240°)$
29. The square roots of $i$
30. The square roots of $-4i$
31. The cube roots of $64i$
32. The fourth roots of $-1$
33. The fourth roots of 81
34. The square roots of $-1 + i\sqrt{3}$
35. For the real number 1, find and graph all indicated roots. Give answers in rectangular form.
    **(a)** Fourth **(b)** Sixth
36. For the complex number $i$, find and graph all indicated roots. Give answers in trigonometric form.
    **(a)** Square **(b)** Fourth

37. Explain why a positive real number must have a positive real $n$th root.
38. *Concept Check* True or false: **(a)** Every real number must have two real square roots. **(b)** Some real numbers have three real cube roots.

### RELATING CONCEPTS

**For individual or group investigation (Exercises 39–44)**

*We examine how the three complex cube roots of $-8$ can be found in two different ways. Work Exercises 39–44 in order.*

39. All complex roots of the equation $x^3 + 8 = 0$ are cube roots of $-8$. Factor $x^3 + 8$ as the sum of two cubes.
40. One of the factors found in Exercise 39 is linear. Set it equal to 0, solve, and determine the real cube root of $-8$.
41. One of the factors found in Exercise 39 is quadratic. Set it equal to 0, solve, and determine the rectangular forms of the other two cube roots of $-8$.
42. Use the method described in this section to find the three complex cube roots of $-8$. Give them in trigonometric form.
43. Convert the trigonometric forms found in Exercise 42 to rectangular form.
44. Compare your results in Exercises 40 and 41, and Exercise 43. What do you notice?

*Find all complex solutions for each equation. Leave your answers in trigonometric form.*

45. $x^4 + 1 = 0$
46. $x^4 + 16 = 0$
47. $x^5 - i = 0$
48. $x^4 - i = 0$
49. $x^3 + 1 = 0$
50. $x^3 + i = 0$
51. $x^3 - 8 = 0$
52. $x^5 + 1 = 0$

**53.** *(Modeling) Basins of Attraction* The fractal shown in the figure is the solution to Cayley's problem of determining the basins of attraction for the cube roots of unity (1). The three cube roots of unity are

$$w_1 = 1, \quad w_2 = -\frac{1}{2} + \frac{\sqrt{3}}{2}i, \quad \text{and} \quad w_3 = -\frac{1}{2} - \frac{\sqrt{3}}{2}i.$$

This fractal can be generated by repeatedly evaluating the function defined by

$$f(z) = \frac{2z^3 + 1}{3z^2},$$

where $z$ is a complex number. We begin by picking $z_1 = a + bi$ and then successively computing $z_2 = f(z_1)$, $z_3 = f(z_2)$, $z_4 = f(z_3)$, .... If the resulting values of $f(z)$ approach $w_1$, color the pixel at $(a, b)$ red. If it approaches $w_2$, color it blue, and if it approaches $w_3$, color it yellow. If this process continues for a large number of different $z_1$, the fractal in the figure will appear. Determine the appropriate color of the pixel for each value of $z_1$. (*Source:* Crownover, R., *Introduction to Fractals and Chaos,* Jones and Bartlett, 1995.)

**(a)** $z_1 = i$ **(b)** $z_1 = 2 + i$ **(c)** $z_1 = -1 - i$

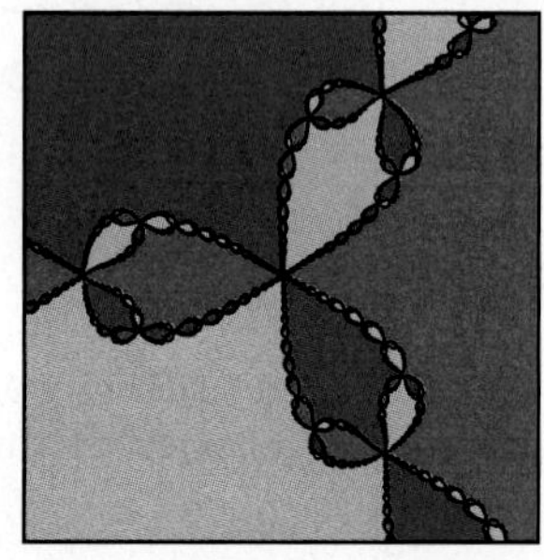

*Source:* Kincaid, D. R. and Cheney, E.W., *Numerical Analysis: Mathematics of Scientific Computing,* 1st edition, © 1991. Reprinted with permission of Brooks/Cole, an imprint of the Wadsworth Group, a division of Thomson Learning.

**54.** *(Modeling)* The screens illustrate how a pentagon can be graphed using parametric equations. Note that a pentagon has five sides, and the T-step is $\frac{360}{5} = 72$. The display at the bottom of the graph screen indicates that one fifth root of 1 is $1 + 0i = 1$. Use this technique to find all fifth roots of 1, and express the real and imaginary parts in decimal form.

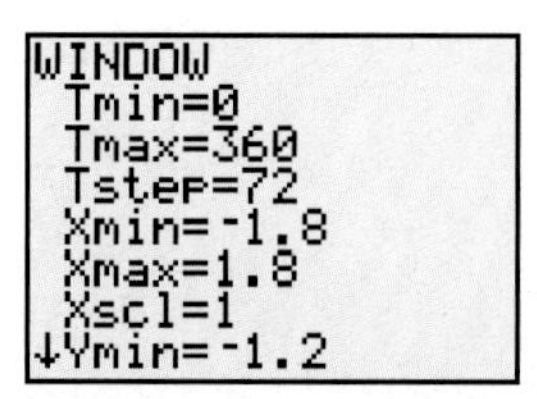


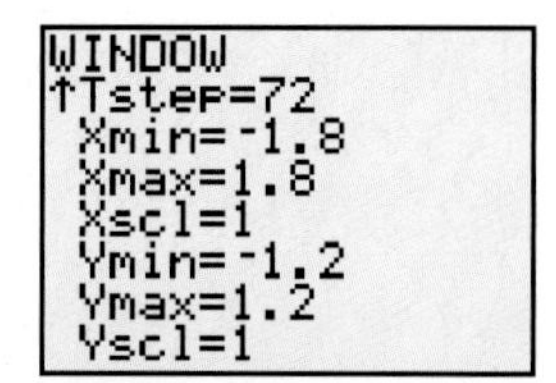


This is a continuation of the previous screen.

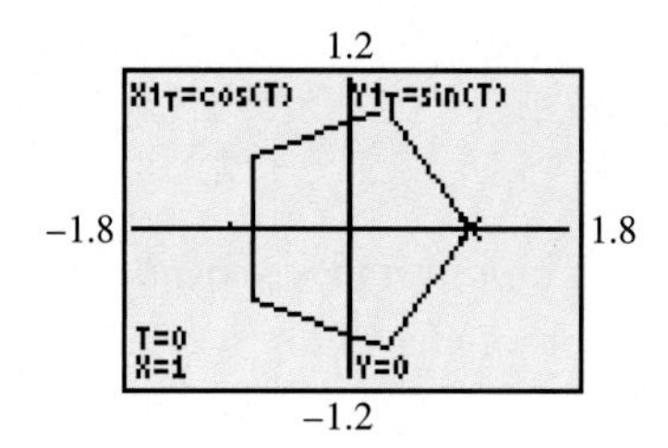


The calculator is in parametric, degree, and connected graph modes.

**55.** Use the method of Exercise 54 to find the first three of the ten 10th roots of 1.

## Reviewing Basic Concepts (Sections 10.4 and 10.5)

**1.** Write the complex number $2(\cos 60° + i \sin 60°)$ in rectangular form.

**2.** Find the modulus of the complex number $3 - 4i$.

**3.** Write $-\sqrt{2} + i\sqrt{2}$ in trigonometric form, where $0° \le \theta < 360°$.

**4.** Calculate the product $z_1 z_2$ and quotient $\frac{z_1}{z_2}$, where

$$z_1 = 4(\cos 135° + i \cos 135°)$$

and $$z_2 = 2(\cos 45° + i \sin 45°).$$

Write your answers in trigonometric and rectangular form.

**5.** Evaluate the power $[4 \text{ cis } 17°]^3$, and write your answer in trigonometric form.

**6.** Find the cube roots of $-64$, and express them in rectangular form.

**7.** Find the square roots of $2i$, and express them in rectangular form.

**8.** Find all complex solutions for the equation $x^3 = -1$. Leave your answers in trigonometric form.

# 10.6 POLAR EQUATIONS AND GRAPHS

Polar Coordinate System • Graphs of Polar Equations • Classifying Polar Equations • Converting Equations

## Polar Coordinate System

Pole

Polar axis

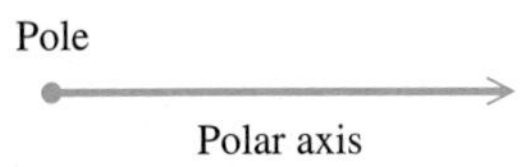

FIGURE 54

We have been using the rectangular coordinate system to graph equations. Another useful coordinate system is the **polar coordinate system.** This system is based on a point, called the **pole,** and a ray, called the **polar axis.** The polar axis is usually drawn in the direction of the positive $x$-axis, as shown in Figure 54.

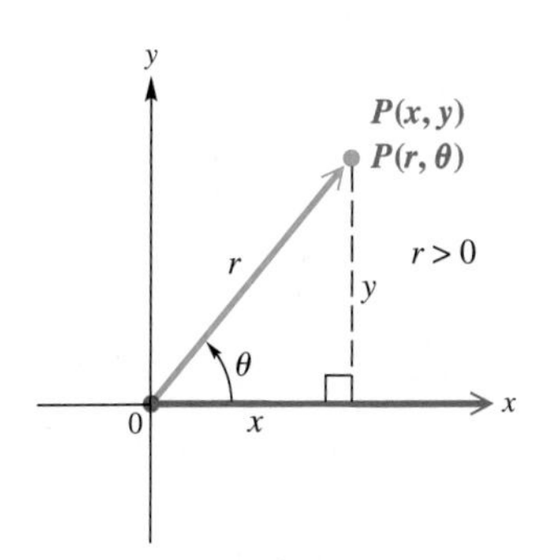


FIGURE 55

In Figure 55, the pole has been placed at the origin of a rectangular coordinate system, so that the polar axis coincides with the positive $x$-axis. Point $P$ has rectangular coordinates $(x, y)$. Point $P$ can also be located by giving the directed angle $\theta$ from the positive $x$-axis to ray $OP$ and the directed distance $r$ from the pole to point $P$. The ordered pair $(r, \theta)$ gives the **polar coordinates** of point $P$. If $r > 0$, point $P$ lies on the terminal side of $\theta$ and if $r < 0$, point $P$ lies on the ray pointing in the opposite direction of the terminal side of $\theta$, a distance $|r|$ from the origin. See Figure 56.

Using Figure 55 and trigonometry, we can establish the following relationships between rectangular and polar coordinates.

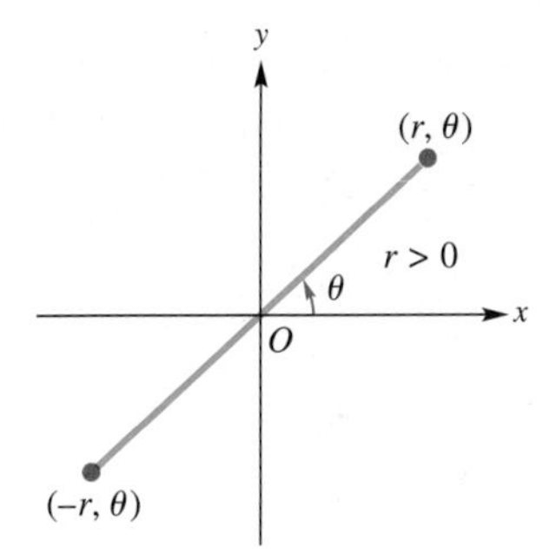


FIGURE 56

**Rectangular and Polar Coordinates**

If a point has rectangular coordinates $(x, y)$ and polar coordinates $(r, \theta)$, then these coordinates are related as follows.

$$\mathbf{x = r\cos\theta \qquad y = r\sin\theta}$$

$$\mathbf{r^2 = x^2 + y^2 \qquad \tan\theta = \frac{y}{x}}, \qquad \text{if } x \neq 0$$

**EXAMPLE 1** *Plotting Points with Polar Coordinates*

Plot each point by hand in the polar coordinate system. Then, determine the rectangular coordinates of each point.

**(a)** $P(2, 30°)$ **(b)** $Q\left(-4, \frac{2\pi}{3}\right)$ **(c)** $R\left(5, -\frac{\pi}{4}\right)$

**Solution**

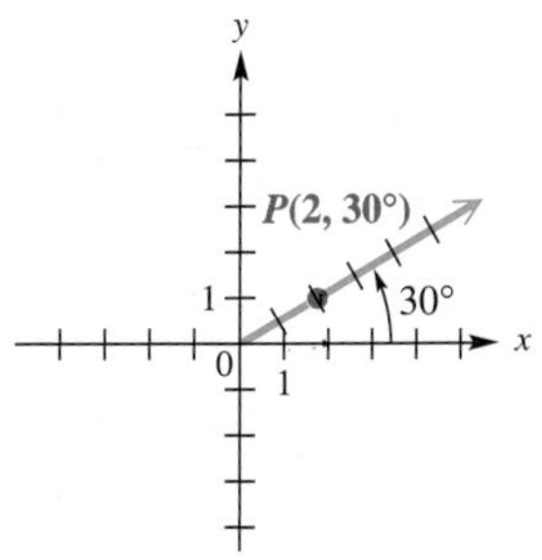

**(a)** In this case, $r = 2$ and $\theta = 30°$, so the point $P$ is located 2 units from the origin in the positive direction on a ray making a 30° angle with the polar axis, as shown in Figure 57.

Using the conversion equations, we find the rectangular coordinates as follows.

$$\begin{aligned} x &= r\cos\theta & y &= r\sin\theta \\ &= 2\cos 30° & &= 2\sin 30° \\ &= 2\left(\frac{\sqrt{3}}{2}\right) = \sqrt{3} & &= 2\left(\frac{1}{2}\right) = 1 \end{aligned}$$

The rectangular coordinates are $(\sqrt{3}, 1)$.

y

P(2, 30°)

30°

x

FIGURE 57

**(b)** Since $r$ is negative, $Q$ is 4 units in the opposite direction from the pole on an extension of the $\frac{2\pi}{3}$ ray. See Figure 58. The rectangular coordinates are

$$x = -4\cos\frac{2\pi}{3} = -4\left(-\frac{1}{2}\right) = 2 \quad \text{and} \quad y = -4\sin\frac{2\pi}{3} = -4\left(\frac{\sqrt{3}}{2}\right) = -2\sqrt{3}.$$

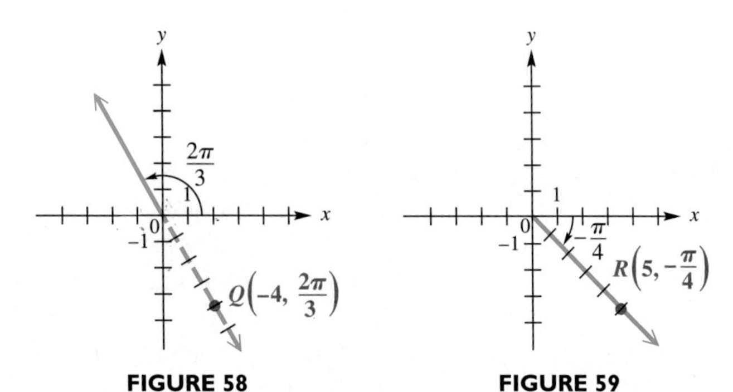


FIGURE 58 FIGURE 59

**Looking Ahead to Calculus**

Techniques studied in calculus associated with derivatives and integrals provide methods of finding slopes of tangent lines to polar curves, areas bounded by such curves, and lengths of their arcs. The equations of the conic sections (parabola, circle, ellipse, and hyperbola) can be represented in polar form using eccentricity.

**(c)** Point $R$ is shown in Figure 59. Since $\theta$ is negative, the angle is measured in the clockwise direction. Furthermore, we have

$$x = 5\cos\left(-\frac{\pi}{4}\right) = \frac{5\sqrt{2}}{2} \quad \text{and} \quad y = 5\sin\left(-\frac{\pi}{4}\right) = -\frac{5\sqrt{2}}{2}.$$

One important difference between rectangular coordinates and polar coordinates is that while a given point in the plane can have only one pair of rectangular coordinates, this same point can have an infinite number of pairs of polar coordinates. For example, $(2, 30°)$ locates the same point as $(2, 390°)$ or $(2, -330°)$ or $(-2, 210°)$.

**EXAMPLE 2** *Giving Alternative Forms for Coordinates of a Point*

**(a)** Give three other pairs of polar coordinates for the point $P(3, 140°)$.

**(b)** Determine two pairs of polar coordinates for the point with rectangular coordinates $(-1, 1)$.

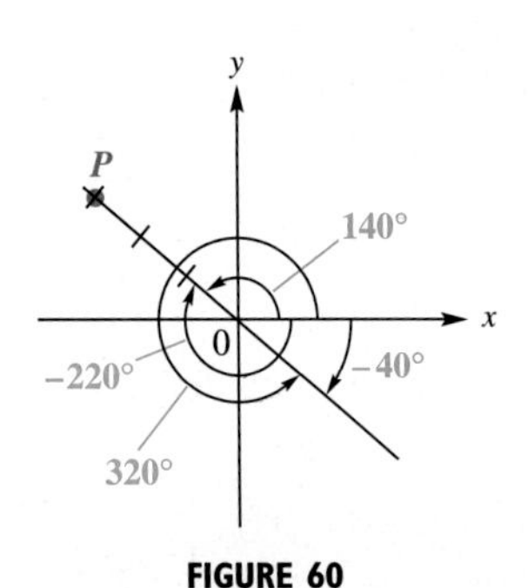


FIGURE 60

**Solution**

**(a)** Three pairs that could be used for the point are $(3, -220°)$, $(-3, 320°)$, and $(-3, -40°)$. See Figure 60.

**(b)** As shown in Figure 61, the point $(-1, 1)$ lies in the second quadrant. Since $\tan\theta = \frac{1}{-1} = -1$, one possible value for $\theta$ is $135°$. Also,

$$r = \sqrt{x^2 + y^2} = \sqrt{(-1)^2 + 1^2} = \sqrt{2}.$$

Therefore, two pairs of polar coordinates are $(\sqrt{2}, 135°)$ and $(\sqrt{2}, -225°)$. (Any angle coterminal with $135°$ could have been used for the second angle.)

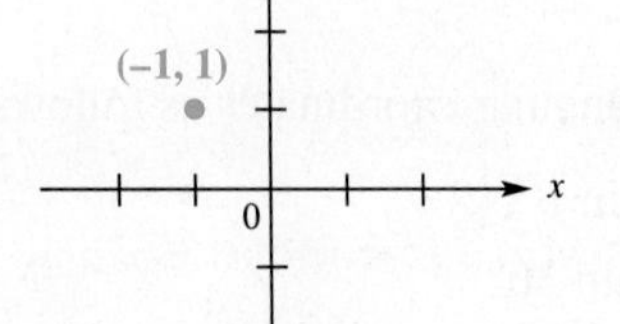


FIGURE 61

## Graphs of Polar Equations

The graphs that we have studied so far have used rectangular coordinates. We now examine how equations in polar coordinates are graphed. An equation such as

$$r = 3\sin\theta, \quad r = 2 + \cos\theta, \quad \text{or} \quad r = \theta,$$

where $r$ and $\theta$ are the variables, is a **polar equation.** The simplest equation for many types of curves turns out to be a polar equation.

**TECHNOLOGY NOTE**

When graphing in polar coordinates, be sure that your calculator is set in polar mode. Also, decide whether your calculator should be in degree or radian mode.

The traditional method of graphing polar equations is much the same as the traditional method of graphing rectangular equations (that is, equations in $x$ and $y$). We evaluate $r$ for various values of $\theta$ until a pattern appears, and then we join the points with a smooth curve. Graphing calculators can perform the task in a matter of seconds.

### EXAMPLE 3 *Graphing a Polar Equation (Cardioid)*

Graph $r = 1 + \cos \theta$.

**Analytic Solution**

To graph this equation, find some ordered pairs (as in the table) and then connect the points in order—from $(2, 0°)$ to $(1.9, 30°)$ to $(1.7, 45°)$ and so on. The graph is shown in Figure 62. This curve is called a **cardioid** because of its heart shape.

| $\theta$ | $\cos\theta$ | $r = 1 + \cos\theta$ | $\theta$ | $\cos\theta$ | $r = 1 + \cos\theta$ |
|---|---|---|---|---|---|
| 0° | 1 | 2 | 135° | −.7 | .3 |
| 30° | .9 | 1.9 | 150° | −.9 | .1 |
| 45° | .7 | 1.7 | 180° | −1 | 0 |
| 60° | .5 | 1.5 | 270° | 0 | 1 |
| 90° | 0 | 1 | 315° | .7 | 1.7 |
| 120° | −.5 | .5 | | | |

Once the pattern of values of $r$ becomes clear, it is not necessary to find more ordered pairs. That is why we stopped with the ordered pair $(1.7, 315°)$ in the table. Notice that the curve has been graphed on a **polar grid.**

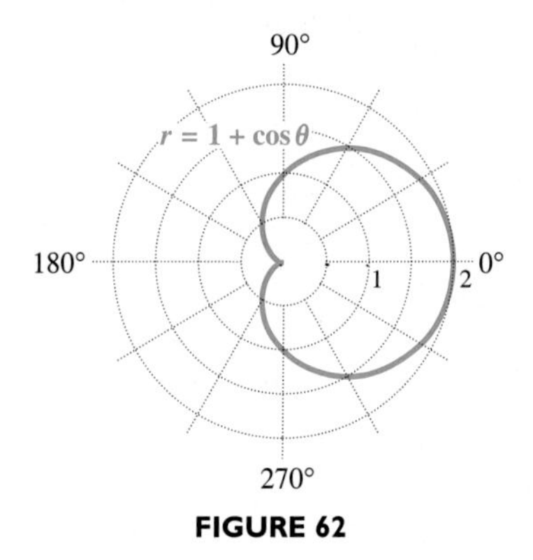


**FIGURE 62**

**Graphing Calculator Solution**

For this equation, we choose degree mode and graph it for values of $\theta$ in the interval $[0°, 360°]$. The screens in Figures 63(a) and (b) show the choices needed to generate the graph shown in Figure 63(c).

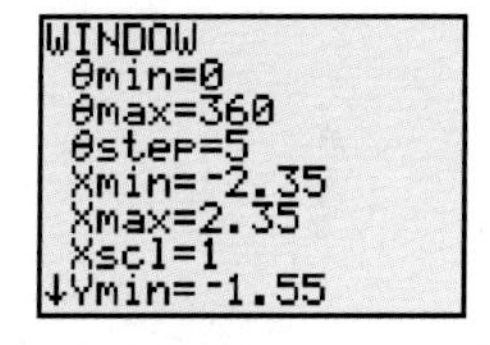


(a)

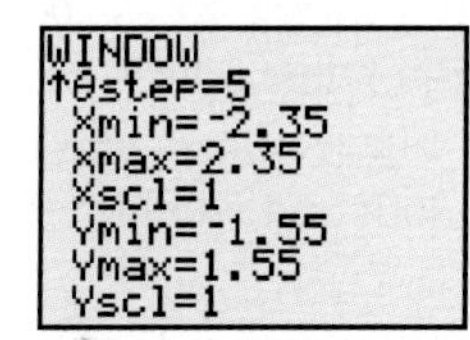


This is a continuation of the previous screen.

(b)

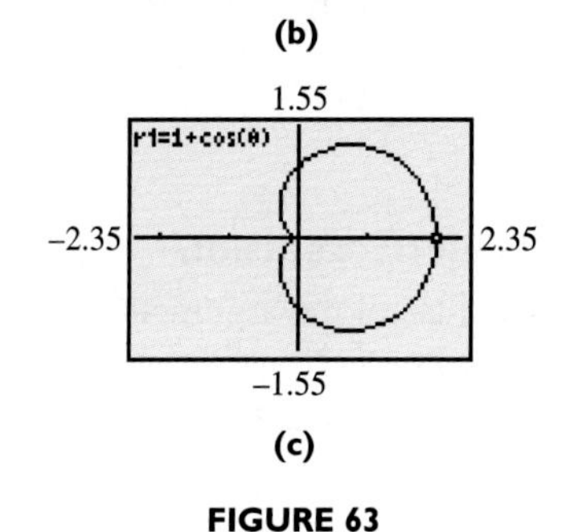


(c)

**FIGURE 63**

### EXAMPLE 4 *Graphing a Polar Equation (Lemniscate)*

Graph $r^2 = \cos 2\theta$.

**Analytic Solution**

First complete a table of ordered pairs as shown on the next page, and then sketch the graph, as in Figure 64. The point $(-1, 0°)$, with $r$ negative, may be plotted as $(1, 180°)$. Also, $(-.7, 30°)$ may be plotted as $(.7, 210°)$, and so on. This curve is called a **lemniscate.**

**Graphing Calculator Solution**

To graph $r^2 = \cos 2\theta$ with a graphing calculator, let

*(continued)*

| $\theta$ | 0° | 30° | 45° | 135° | 150° | 180° |
|---|---|---|---|---|---|---|
| $2\theta$ | 0° | 60° | 90° | 270° | 300° | 360° |
| $\cos 2\theta$ | 1 | .5 | 0 | 0 | .5 | 1 |
| $r = \pm\sqrt{\cos 2\theta}$ | ±1 | ±.7 | 0 | 0 | ±.7 | ±1 |

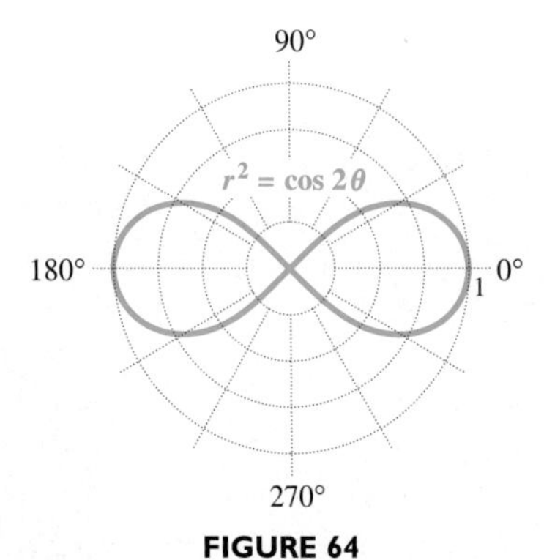


FIGURE 64

Values of $\theta$ for $45° < \theta < 135°$ are not included in the table because the corresponding values of $\cos 2\theta$ are negative (quadrants II and III) and so do not have real square roots. Values of $\theta$ larger than 180° give $2\theta$ larger than 360° and would repeat the points already found.

$$r_1 = \sqrt{\cos 2\theta}$$

and $$r_2 = -\sqrt{\cos 2\theta}.$$

See Figures 65(a) and (b).

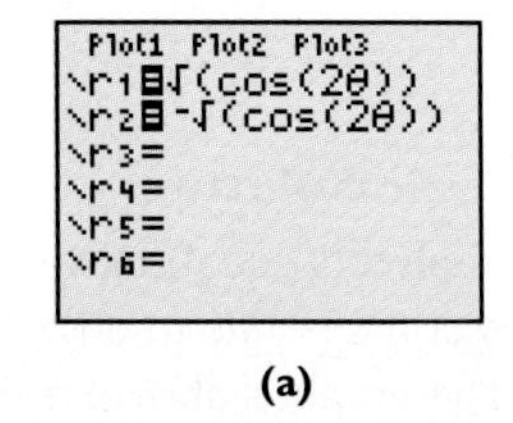


(a)

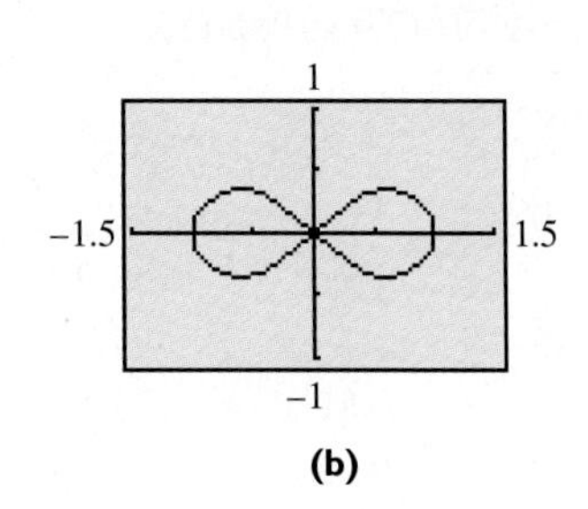


(b)

FIGURE 65

## EXAMPLE 5 *Graphing a Polar Equation (Rose)*

Graph $r = 3 \cos 2\theta$.

### Analytic Solution

Because of the argument $2\theta$, the graph requires a larger number of points. A few ordered pairs are given in the table. You should complete the table similarly through the first 180° so that $2\theta$ has values up to 360°.

| $\theta$ | 0° | 15° | 30° | 45° | 60° | 75° | 90° |
|---|---|---|---|---|---|---|---|
| $2\theta$ | 0° | 30° | 60° | 90° | 120° | 150° | 180° |
| $\cos 2\theta$ | 1 | .9 | .5 | 0 | −.5 | −.9 | −1 |
| $r = 3 \cos 2\theta$ | 3 | 2.6 | 1.5 | 0 | −1.5 | −2.6 | −3 |

Plotting these points in order gives the graph, called a **four-leaved rose.** Notice in Figure 66 how the graph is developed with a continuous curve, beginning with the upper half of the right horizontal leaf and ending with

### Graphing Calculator Solution

The screen in Figure 67 on the next page shows the graph of

$$r = 3 \cos 2\theta.$$

You can duplicate this screen and watch how the graph takes shape, comparing it to the description in Figure 66. Trace the rose curve and watch the cursor to verify that the numbering that appears in Figure 66 is indeed correct.

*(continued)*

the lower half of that leaf. As the graph is traced, the curve goes through the pole four times.

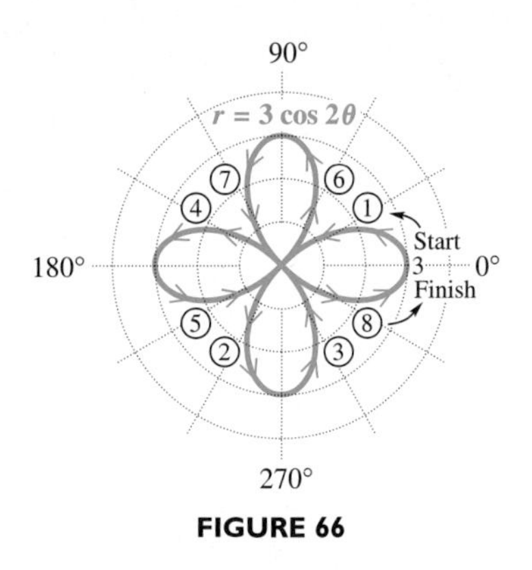

FIGURE 66

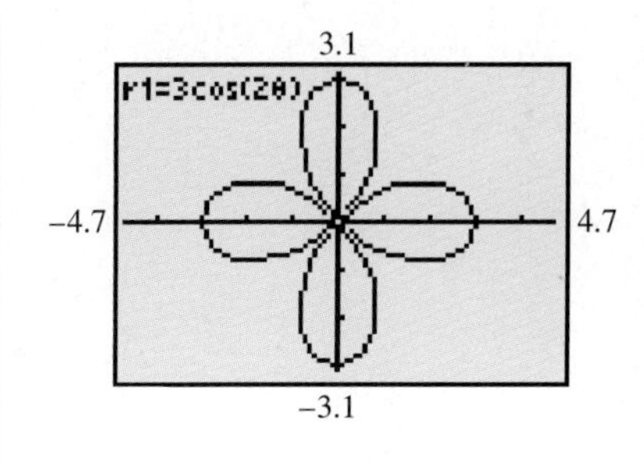

FIGURE 67

The equation in Example 5 has a graph that belongs to a family of curves called **roses.** The graphs of $r = a \sin n\theta$ and $r = a \cos n\theta$ are roses, with $n$ petals if $n$ is odd, and $2n$ petals if $n$ is even. The value of $a$ determines the length of the petals.

**EXAMPLE 6** *Graphing a Polar Equation (Spiral of Archimedes)*

Graph $r = 2\theta$ ($\theta$ measured in radians).

### Analytic Solution

Some ordered pairs are shown in the table. Since $r = 2\theta$, rather than a trigonometric function of $\theta$, we must also consider negative values of $\theta$. The radian measures have been rounded.

| $\theta$ (radians) | $r = 2\theta$ | $\theta$ (radians) | $r = 2\theta$ |
|---|---|---|---|
| $-\pi$ | $-6.2$ | $\frac{\pi}{3}$ | 2 |
| $-\frac{\pi}{2}$ | $-3.2$ | $\frac{\pi}{2}$ | 3.2 |
| $-\frac{\pi}{4}$ | $-1.6$ | $\pi$ | 6.2 |
| 0 | 0 | $\frac{3\pi}{2}$ | 9.4 |
| $\frac{\pi}{6}$ | 1 | $2\pi$ | 12.6 |

Figure 68 on the next page shows this graph, called a **spiral of Archimedes.**

### Graphing Calculator Solution

Figure 69 on the next page shows more of the spiral than is seen in the traditional graph of

$$r = 2\theta.$$

Notice how rapidly the graph of this spiral moves away from the pole (origin).

Experiment with various minimum and maximum values of $\theta$ for $r = 2\theta$. What behaviors do you observe? (As you experiment with various maximums and minimums for $\theta$, you will also need to enlarge the window to accommodate the increased size of the graph.)

*(continued)*

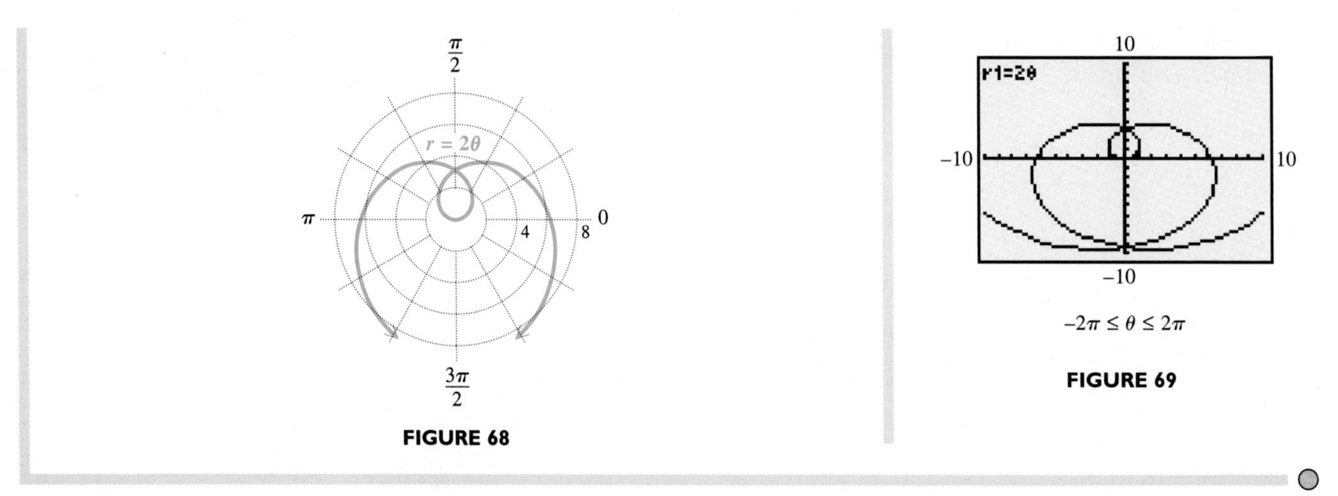


FIGURE 68

$-2\pi \le \theta \le 2\pi$

FIGURE 69

## Classifying Polar Equations

The following table summarizes some common polar graphs and forms of their equations. (In addition to circles, lemniscates, and roses, which were just presented, we include **limaçons.** Cardioids are a special case of limaçons, where $\left|\frac{a}{b}\right| \geq 1$.)

**Circles and Lemniscates**

Circles Lemniscates

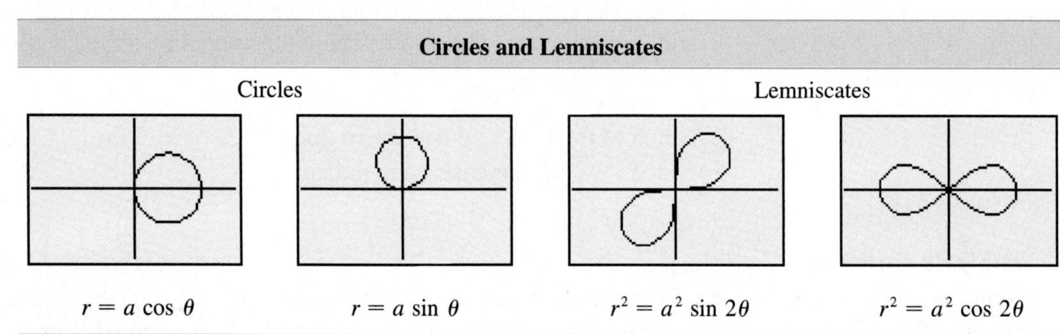

$r = a \cos \theta$ $r = a \sin \theta$ $r^2 = a^2 \sin 2\theta$ $r^2 = a^2 \cos 2\theta$

**Limaçons**

$r = a \pm b \sin \theta$ or $r = a \pm b \cos \theta$

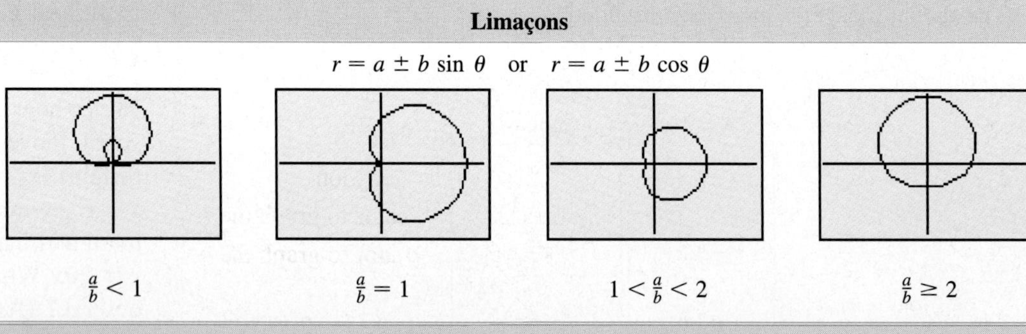

$\frac{a}{b} < 1$ $\frac{a}{b} = 1$ $1 < \frac{a}{b} < 2$ $\frac{a}{b} \geq 2$

**Rose Curves**

$2n$ petals if $n$ is even, $n \geq 2$ $n$ petals if $n$ is odd

$n = 2$ $n = 4$ $n = 3$ $n = 5$

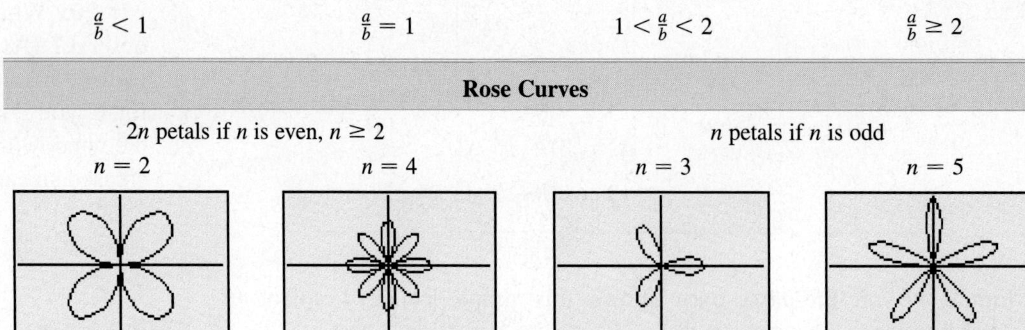

$r = a \sin n\theta$ $r = a \cos n\theta$ $r = a \cos n\theta$ $r = a \sin n\theta$

## Converting Equations

We can transform polar equations to rectangular equations and vice versa.

**EXAMPLE 7** *Converting a Polar Equation to a Rectangular Equation*

Consider the polar equation $r = \dfrac{4}{1 + \sin\theta}$.

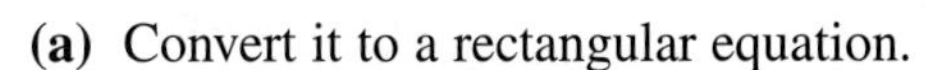

**(a)** Convert it to a rectangular equation.

**(b)** Use a graphing calculator to graph the polar equation for $0 \le \theta \le 2\pi$.

**(c)** Use a graphing calculator to graph the rectangular equation.

**Solution**

**(a)** Multiply each side of the equation by the denominator on the right, to clear the fraction.

$$r = \frac{4}{1 + \sin\theta} \qquad \text{Polar equation}$$

$$r + r\sin\theta = 4 \qquad \text{Multiply by } 1 + \sin\theta.$$

$$\sqrt{x^2 + y^2} + y = 4 \qquad \text{Let } r = \sqrt{x^2 + y^2} \text{ and } y = r\sin\theta.$$

$$\sqrt{x^2 + y^2} = 4 - y \qquad \text{Subtract } y.$$

$$x^2 + y^2 = (4 - y)^2 \qquad \text{Square each side.}$$

$$x^2 + y^2 = 16 - 8y + y^2 \qquad \text{Expand the right side.}$$

$$x^2 = -8y + 16 \qquad \text{Subtract } y^2.$$

$$x^2 = -8(y - 2) \qquad \text{Rectangular equation}$$

**(b)** Figure 70(a) shows a graph with polar coordinates.

**(c)** Solving $x^2 = -8(y - 2)$ for $y$, we obtain $y = 2 - \frac{1}{8}x^2$. Its graph is shown in Figure 70(b). Notice that the two graphs are the same parabola.

$r = \dfrac{4}{1 + \sin\theta}$

(a)

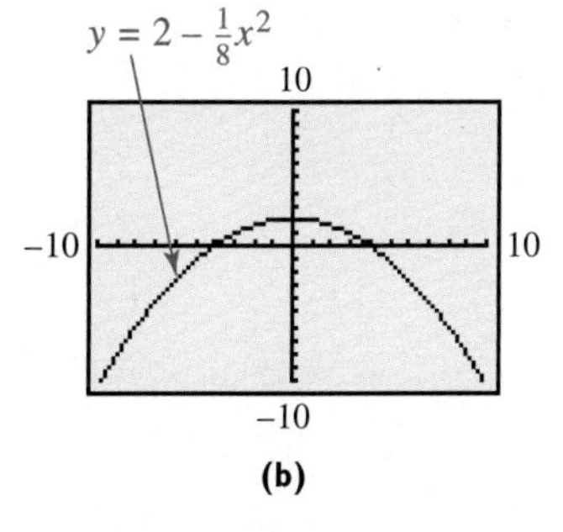


(b)

FIGURE 70

**EXAMPLE 8** *Converting a Rectangular Equation to a Polar Equation*

Consider the rectangular equation $3x + 2y = 4$.

**(a)** Convert it to a polar equation.

**(b)** Use a graphing calculator to graph the rectangular equation.

**(c)** Use a graphing calculator to graph the polar equation for $0° \le \theta \le 360°$.

**Solution**

**(a)**

$$3x + 2y = 4 \qquad \text{Rectangular equation}$$

$$3r\cos\theta + 2r\sin\theta = 4 \qquad \text{Let } x = r\cos\theta \text{ and } y = r\sin\theta.$$

$$r(3\cos\theta + 2\sin\theta) = 4 \qquad \text{Factor out } r.$$

$$r = \frac{4}{3\cos\theta + 2\sin\theta} \qquad \text{Solve for } r\text{; polar equation}$$

**(b)** We solve the given rectangular equation for $y$ to obtain $y = -\frac{3}{2}x + 2$. Its graph is shown in Figure 71(a).

**(c)** The polar equation found in part (a) is graphed in Figure 71(b).

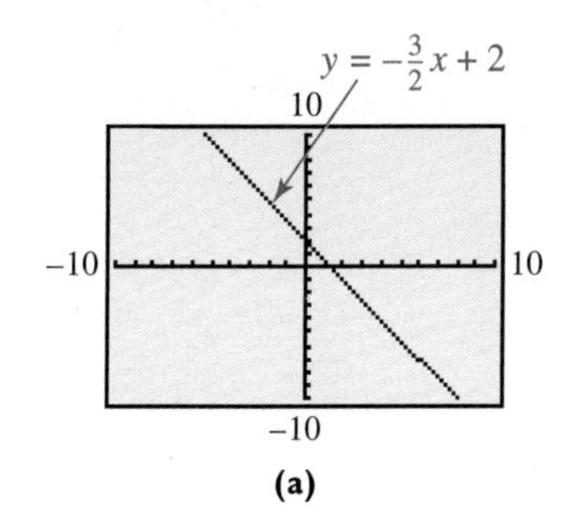


(a)

$r = \dfrac{4}{3\cos\theta + 2\sin\theta}$

(b)

FIGURE 71

## 10.6 EXERCISES

**1.** *Concept Check* For each point given in polar coordinates, state the quadrant in which the point lies if it is graphed in a rectangular coordinate system.

**(a)** $(5, 135°)$ **(b)** $(2, 60°)$

**(c)** $(6, -30°)$ **(d)** $(4.6, 213°)$

**2.** *Concept Check* For each point given in polar coordinates, state the axis on which the point lies if it is graphed in a rectangular coordinate system. Also, state whether it is on the positive portion or the negative portion of the axis. (For example, $(5, 0°)$ lies on the positive $x$-axis.)

**(a)** $(7, 360°)$ **(b)** $(4, 180°)$

**(c)** $(2, -90°)$ **(d)** $(8, 450°)$

*Plot each point, given its polar coordinates. Give two other pairs of polar coordinates for each point.*

**3.** $(1, 45°)$ **4.** $(3, 120°)$ **5.** $(-2, 135°)$ **6.** $(-4, 27°)$ **7.** $(5, -60°)$

**8.** $(2, -45°)$ **9.** $(-3, -210°)$ **10.** $(-1, -120°)$ **11.** $(3, 300°)$ **12.** $(4, 270°)$

*Plot the point whose rectangular coordinates are given. Then determine two pairs of polar coordinates for the point with $0° \le \theta < 360°$.*

**13.** $(-1, 1)$ **14.** $(1, 1)$ **15.** $(0, 3)$ **16.** $(0, -3)$ **17.** $(\sqrt{2}, \sqrt{2})$

**18.** $(-\sqrt{2}, \sqrt{2})$ **19.** $\left(\frac{\sqrt{3}}{2}, \frac{3}{2}\right)$ **20.** $\left(-\frac{\sqrt{3}}{2}, -\frac{1}{2}\right)$ **21.** $(3, 0)$

**22.** *Concept Check* Match each polar graph to its corresponding equation from choices A–D.

**(a)**

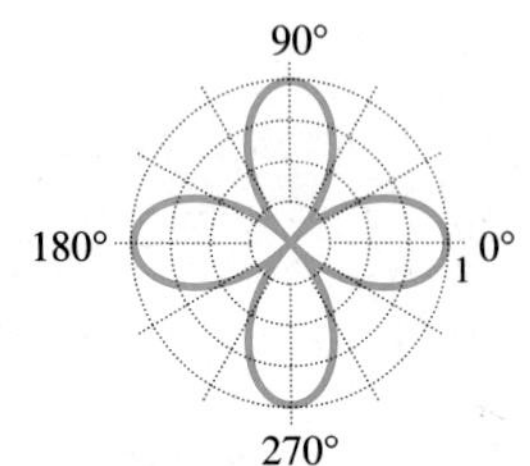


**(b)**

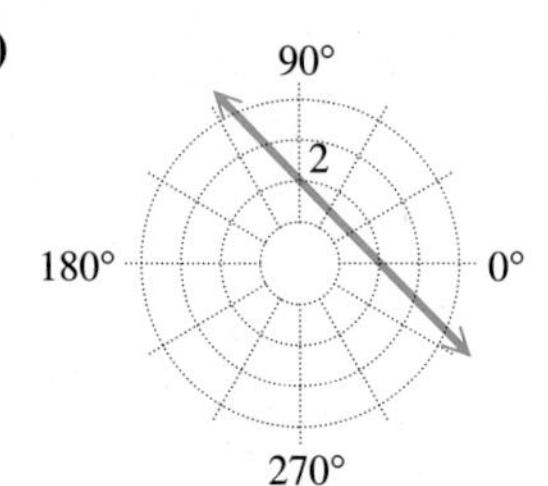


**(c)**

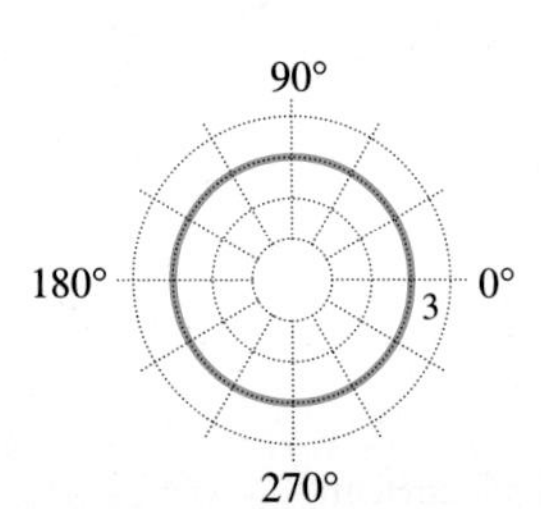


**(d)**

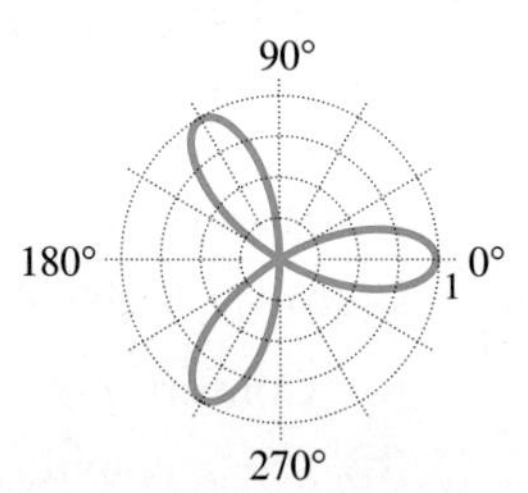


**A.** $r = 3$ **B.** $r = \cos 3\theta$

**C.** $r = \cos 2\theta$ **D.** $r = \dfrac{2}{\cos \theta + \sin \theta}$

*Graph each polar equation for $\theta$ in $[0°, 360°)$. In Exercises 23–32 identify the type of polar graph.*

**23.** $r = 2 + 2 \cos \theta$ **24.** $r = 2(4 + 3 \cos \theta)$ **25.** $r = 1 + 2 \sin \theta$

**26.** $r = 1 - 2 \cos \theta$ **27.** $r = 4 \cos 2\theta$ **28.** $r = 3 \cos 5\theta$

**29.** $r^2 = 4 \cos 2\theta$ **30.** $r^2 = 4 \sin 2\theta$ **31.** $r = 4(1 - \cos \theta)$

**32.** $r = 3(2 - \cos \theta)$ **33.** $r = 2 \sin \theta \tan \theta$ (This is a *cissoid*.) **34.** $r = \dfrac{\cos 2\theta}{\cos \theta}$ (This is a *cissoid* with a loop.)

## RELATING CONCEPTS

**For individual or group investigation (Exercises 35–42)**

*In Chapter 2, we discussed symmetry with respect to the x-axis, y-axis, and origin. Visualize an xy-plane superimposed on the polar coordinate system, with the pole at the origin and the polar axis on the positive x-axis. Then a polar graph may be symmetric with respect to the x-axis (the polar axis), the y-axis (the line $\theta = \frac{\pi}{2}$), or the origin (the pole). Work Exercises 35–42 in order.*

**35.** Complete the missing ordered pairs in each graph.

**(a)**

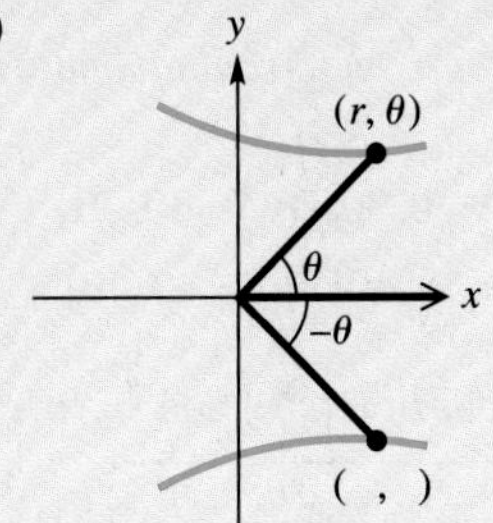


**(b)**

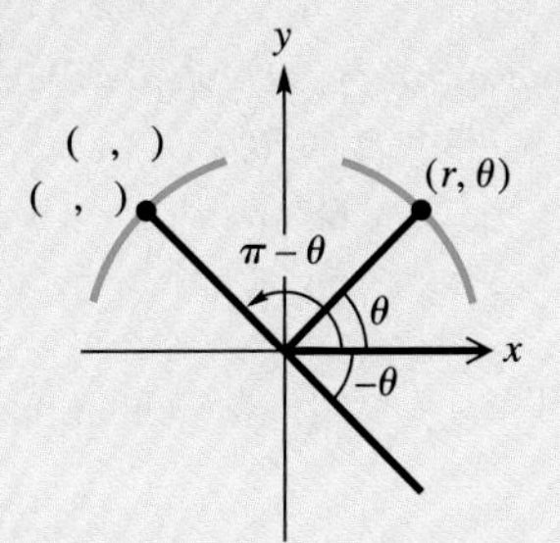


**(c)**

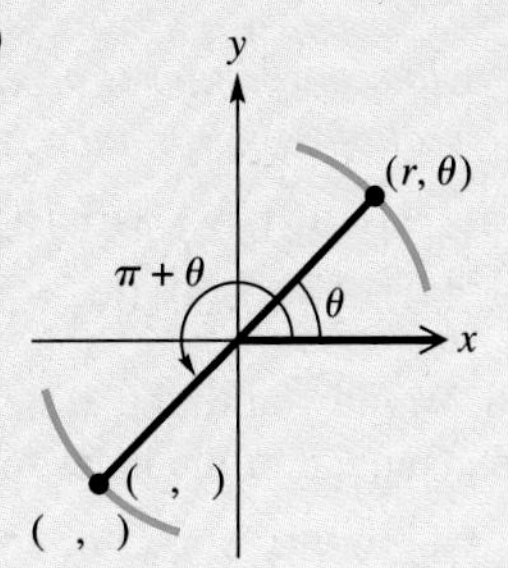


*Use your answers for Exercise 35 to complete the sentences in Exercises 36–42.*

**36.** The graph of $r = f(\theta)$ is symmetric with respect to the polar axis if substitution of ________ for $\theta$ leads to an equivalent equation.

**37.** The graph of $r = f(\theta)$ is symmetric with respect to the vertical line $\theta = \frac{\pi}{2}$ if substitution of ________ for $\theta$ leads to an equivalent equation.

**38.** Alternatively, the graph of $r = f(\theta)$ is symmetric with respect to the vertical line $\theta = \frac{\pi}{2}$ if substitution of ________ for $r$ and ________ for $\theta$ leads to an equivalent equation.

**39.** The graph of $r = f(\theta)$ is symmetric with respect to the pole if substitution of ________ for $r$ leads to an equivalent equation.

**40.** Alternatively, the graph of $r = f(\theta)$ is symmetric with respect to the pole if substitution of ________ for $\theta$ leads to an equivalent equation.

**41.** In general, the completed statements in Exercises 35–40 mean that the graphs of polar equations of the form $r = a \pm b \cos \theta$ (where $a$ may be 0) are symmetric with respect to ________.

**42.** In general, the completed statements in Exercises 35–40 mean that the graphs of polar equations of the form $r = a \pm b \sin \theta$ (where $a$ may be 0) are symmetric with respect to ________.

**43.** Explain the method you would use to graph $(r, \theta)$ by hand if $r < 0$.

**44.** Explain why, if $r > 0$, the points $(r, \theta)$ and $(-r, \theta + 180°)$ have the same graph.

*Concept Check* *Use the concepts of this section to answer each question.*

**45.** If a point lies on an axis in the rectangular plane, then what kind of angle must $\theta$ be if $(r, \theta)$ represents the point in polar coordinates?

**46.** What will the graph of $r = k$ be, for $k > 0$?

**47.** How would the graph of Figure 62 change if the equation were $r = 1 - \cos \theta$?

**48.** How would the graph of Figure 62 change if the equation were $r = 1 + \sin \theta$?

**49.** The graphs of rose curves have equations of the form $r = a \sin n\theta$ or $r = a \cos n\theta$. What does the value of $a$ determine? What does the value of $n$ determine?

**50.** In Exercise 49, if $n = 1$, what will the graph be? What will the value of $a$ determine?

*For each equation, find an equivalent equation in rectangular coordinates. Then graph the equation.*

**51.** $r = 2 \sin \theta$

**52.** $r = 2 \cos \theta$

**53.** $r = \dfrac{2}{1 - \cos \theta}$

**54.** $r = \dfrac{3}{1 - \sin \theta}$

**55.** $r + 2\cos\theta = -2\sin\theta$

**56.** $r = \dfrac{3}{4\cos\theta - \sin\theta}$

**57.** $r = 2\sec\theta$

**58.** $r = -5\csc\theta$

**59.** $r(\cos\theta + \sin\theta) = 2$

**60.** $r(2\cos\theta + \sin\theta) = 2$

*For each equation, find an equivalent equation in polar coordinates.*

**61.** $x + y = 4$

**62.** $2x - y = 5$

**63.** $x^2 + y^2 = 16$

**64.** $x^2 + y^2 = 9$

**65.** $y = 2$

**66.** $x = 4$

*The graph of $r = a\theta$ in polar coordinates is an example of the spiral of Archimedes. With your calculator set to radian mode, use the given value of a and interval of $\theta$ to graph the spiral in the window specified.*

**67.** $a = 1, 0 \le \theta \le 4\pi, [-15, 15]$ by $[-15, 15]$

**68.** $a = 2, -4\pi \le \theta \le 4\pi, [-30, 30]$ by $[-30, 30]$

**69.** $a = 1.5, -4\pi \le \theta \le 4\pi, [-20, 20]$ by $[-20, 20]$

**70.** $a = -1, 0 \le \theta \le 12\pi, [-40, 40]$ by $[-40, 40]$

**71.** Refer to Example 8. Would you find it easier to graph the equation with the rectangular or the polar form? Why?

**72.** Show that the distance between $(r_1, \theta_1)$ and $(r_2, \theta_2)$ is $\sqrt{r_1^2 + r_2^2 - 2r_1r_2\cos(\theta_1 - \theta_2)}$.

*Find the polar coordinates of the points of intersection of the given curves for the specified interval of $\theta$.*

**73.** $r = 4\sin\theta, r = 1 + 2\sin\theta; 0 \le \theta < 2\pi$

**74.** $r = 3, r = 2 + 2\cos\theta; 0° \le \theta < 360°$

**75.** $r = 2 + \sin\theta, r = 2 + \cos\theta; 0 \le \theta < 2\pi$

**76.** $r = \sin 2\theta, r = \sqrt{2}\cos\theta; 0 \le \theta < \pi$

*Solve each problem.*

**77.** ***(Modeling) Planet Orbits*** The polar equation

$$r = \frac{a(1 - e^2)}{1 + e\cos\theta}$$

can be used to graph the orbits of the planets, where $a$ is the average distance in astronomical units from the sun and $e$ is a constant called the eccentricity. (See Section 6.3.) The sun will be located at the pole. The table lists $a$ and $e$ for the planets.

| Planet | *a* | *e* |
|---|---|---|
| Mercury | .39 | .206 |
| Venus | .78 | .007 |
| Earth | 1.00 | .017 |
| Mars | 1.52 | .093 |
| Jupiter | 5.20 | .048 |
| Saturn | 9.54 | .056 |
| Uranus | 19.20 | .047 |
| Neptune | 30.10 | .009 |
| Pluto | 39.40 | .249 |

*Source:* Karttunen, H., P. Kröger, H. Oja, M. Putannen, and K. Donners (Editors), *Fundamental Astronomy*, Springer-Verlag, 1994; Zeilik, M., S. Gregory, and E. Smith, *Introductory Astronomy and Astrophysics*, Saunders College Publishers, 1992.

**(a)** Graph the orbits of the four closest planets on the same polar grid. Choose a viewing window that results in a graph with nearly circular orbits.

**(b)** Plot the orbits of Earth, Jupiter, Uranus, and Pluto on the same polar grid. How does Earth's distance from the sun compare to these planets?

**(c)** Use graphing to determine whether or not Pluto is always the farthest planet from the sun.

***(Modeling) Radio Towers and Broadcasting Patterns*** *Many times radio stations do not broadcast in all directions with the same intensity. To avoid interference with an existing station to the north, a new station may be licensed to broadcast only east and west. To create an east–west signal, two radio towers are sometimes used, as illustrated in the figure. Locations where the radio signal is received correspond to the interior of the curve defined by $r^2 = 40{,}000\cos 2\theta$, where the polar axis (or positive x-axis) points east.*

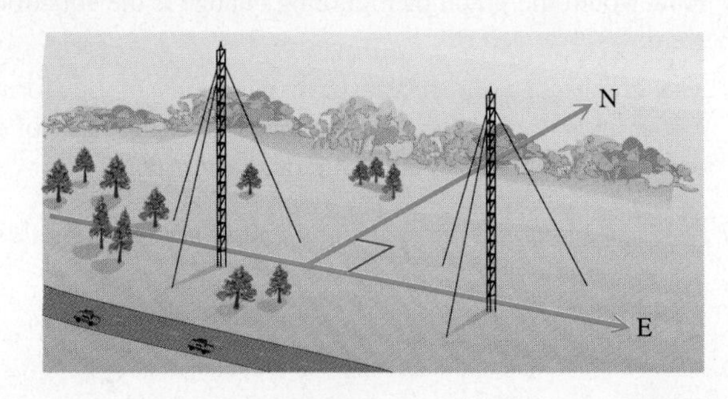

**78.** Graph $r^2 = 40{,}000 \cos 2\theta$ for $0° \leq \theta \leq 360°$, where units are in miles. Assuming the radio towers are located near the pole, use the graph to describe the regions where the signal can be received and where the signal cannot be received.

**79.** Refer to Exercise 78. Suppose a radio signal pattern is given by $r^2 = 22{,}500 \sin 2\theta$. Graph this pattern and interpret the results.

## 10.7 MORE PARAMETRIC EQUATIONS

Parametric Equations with Trigonometric Functions • The Cycloid • Applications of Parametric Equations

### Parametric Equations with Trigonometric Functions

In Section 6.4 parametric equations were introduced. However, these equations did not include trigonometric functions. If we use trigonometric functions in parametric equations, many interesting curves can be drawn. Some examples are shown in Figure 72.

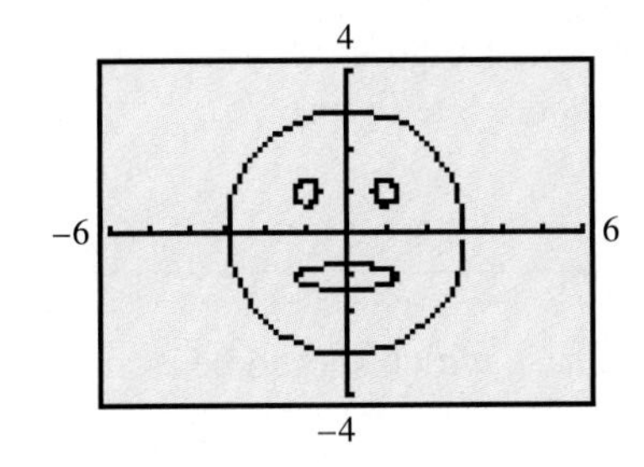

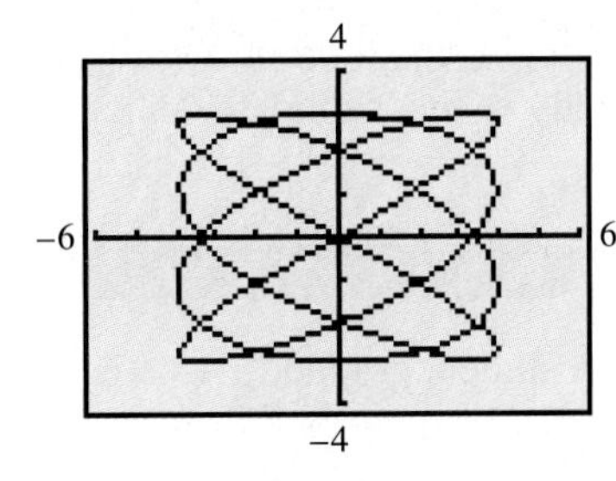

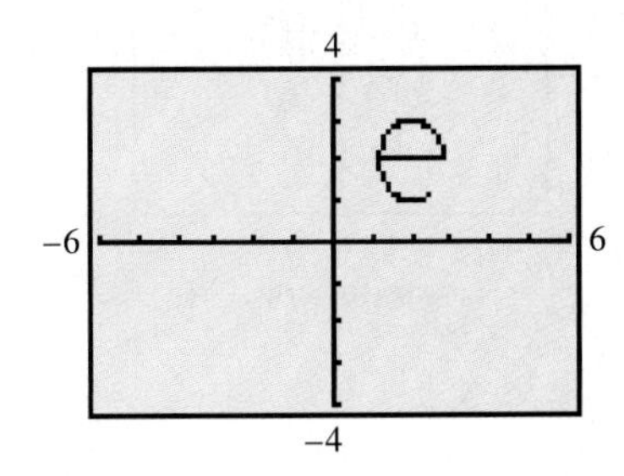

**FIGURE 72**

**EXAMPLE 1** *Graphing a Circle with Parametric Equations*

Graph $x = 2 \cos t$ and $y = 2 \sin t$, for $0 \leq \theta \leq 2\pi$. Find an equivalent equation by using rectangular coordinates.

**TECHNOLOGY NOTE**

When graphing parametric equations, be sure that your calculator is set in parametric mode. A square window is necessary for the curve in Example 1 to appear circular rather than elliptical.

**Solution** Let $X_{1T} = 2\cos(T)$ and $Y_{1T} = 2\sin(T)$, and graph these parametric equations as shown in Figure 73.

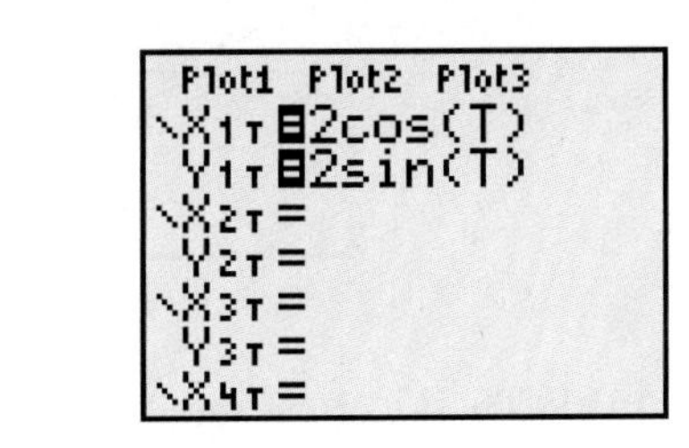


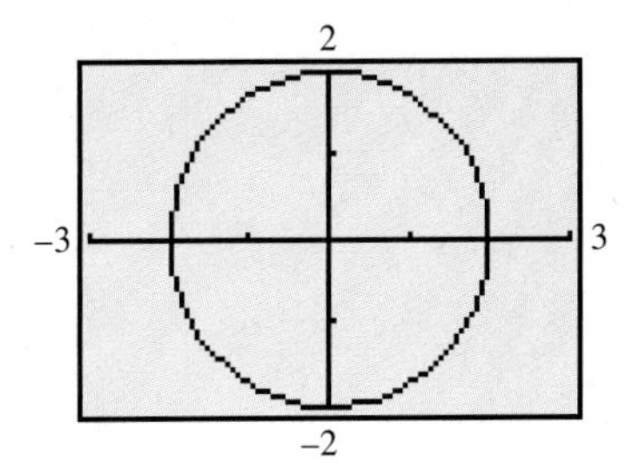

**FIGURE 73**

To verify that this graph is a circle, consider the following.

$$\begin{aligned} x^2 + y^2 &= (2 \cos t)^2 + (2 \sin t)^2 && x = 2\cos t,\ y = 2 \sin t \\ &= 4 \cos^2 t + 4 \sin^2 t && \text{Properties of exponents} \\ &= 4(\cos^2 t + \sin^2 t) && \text{Distributive property} \\ &= 4 && \cos^2 t + \sin^2 t = 1 \end{aligned}$$

The parametric equations are equivalent to $x^2 + y^2 = 4$, which is the equation of a circle with center $(0, 0)$ and radius 2.

$x = 2 \sin t$, $y = 3 \cos t$ for $t$ in $[0, 2\pi]$

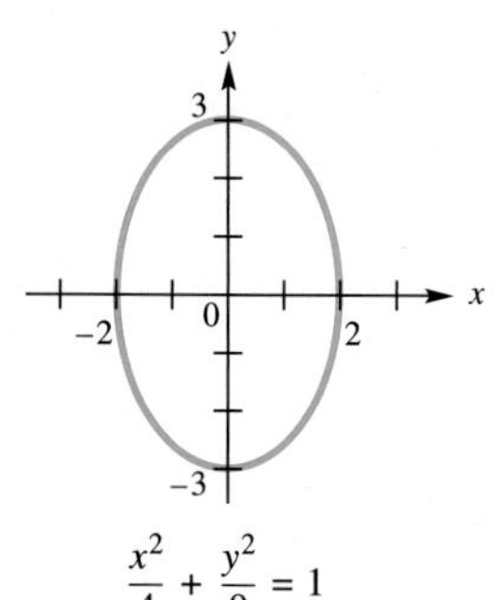


$\frac{x^2}{4} + \frac{y^2}{9} = 1$

**FIGURE 74**

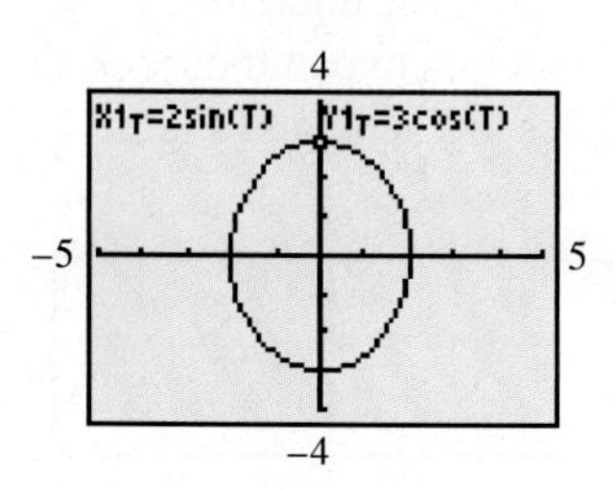


**FIGURE 75**

**EXAMPLE 2** *Graphing an Ellipse with Parametric Equations*

Graph the plane curve defined by $x = 2 \sin t$, $y = 3 \cos t$, for $t$ in $[0, 2\pi]$.

**Solution** To convert to a rectangular equation, square both sides of each equation; solve one for $\sin^2 t$, the other for $\cos^2 t$.

$$x = 2 \sin t \qquad y = 3 \cos t$$

$$x^2 = 4 \sin^2 t \qquad y^2 = 9 \cos^2 t$$

$$\frac{x^2}{4} = \sin^2 t \qquad \frac{y^2}{9} = \cos^2 t$$

Now add corresponding sides of the two equations.

$$\frac{x^2}{4} + \frac{y^2}{9} = \sin^2 t + \cos^2 t$$

$$\frac{x^2}{4} + \frac{y^2}{9} = 1$$

This is the equation of an ellipse, as shown in Figure 74. The ellipse can be graphed directly with a calculator in parametric mode. See Figure 75.

## What Went WRONG?

A student graphed $x = \cos \frac{1}{2}t$, $y = \sin \frac{1}{2}t$, which should be a circle with radius 1 because

$$x^2 + y^2 = \cos^2 \frac{1}{2}t + \sin^2 \frac{1}{2}t = 1.$$

However, the following screen shows the graph obtained by the student.

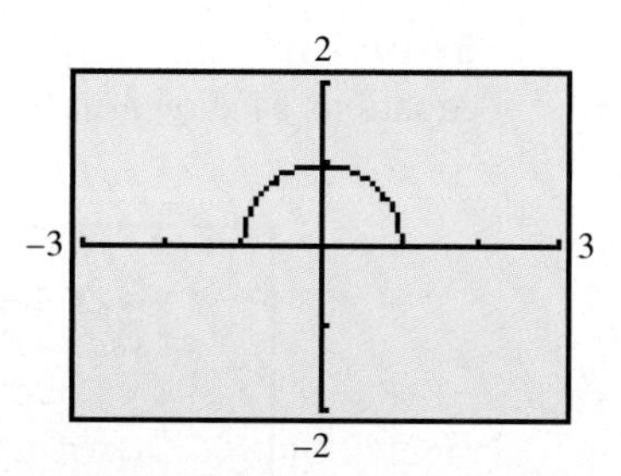


What Went WRONG? What should the student do to change the graph from a semicircle to a circle?

---

*Answer to What Went Wrong?*

The periods for $y = \cos \frac{1}{2}t$ and $y = \sin \frac{1}{2}t$ are both $4\pi$. The interval for $t$ should be $0 \le t \le 4\pi$, rather than $0 \le t \le 2\pi$, which is the interval the student used.

**Looking Ahead to Calculus**

The cycloid is a special case of a curve traced out by a point at a given distance from the center of a circle as the circle rolls along a straight line. Such a curve is called a *trochoid.* It is just one of several parametrically defined curves studied in calculus.

## The Cycloid

The path traced by a fixed point on the circumference of a circle rolling along a line is called a *cycloid.* A **cycloid** is defined by

$$x = at - a\sin t, \quad y = a - a\cos t, \quad \text{for } t \text{ in } (-\infty, \infty).$$

**EXAMPLE 3** *Graphing a Cycloid*

Graph the cycloid with $a = 1$ for $t$ in $[0, 2\pi]$.

**Analytic Solution**

There is no simple way to find a rectangular equation for the cycloid from its parametric equations. Instead, begin with a table of values.

| $t$ | $0$ | $\frac{\pi}{4}$ | $\frac{\pi}{2}$ | $\pi$ | $\frac{3\pi}{2}$ | $2\pi$ |
|---|---|---|---|---|---|---|
| $x$ | 0 | .08 | .6 | $\pi$ | 5.7 | $2\pi$ |
| $y$ | 0 | .3 | 1 | 2 | 1 | 0 |

Plotting the ordered pairs $(x, y)$ from the table of values leads to the portion of the graph in Figure 76 from 0 to $2\pi$.

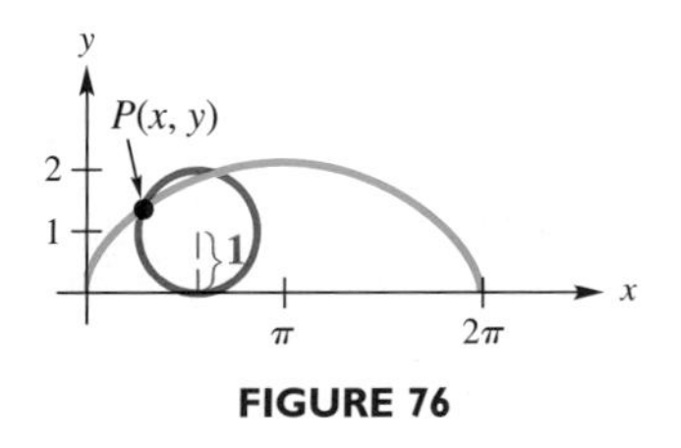

FIGURE 76

**Graphing Calculator Solution**

It is easier to graph a cycloid with a graphing calculator in parametric mode than with traditional methods. See Figure 77.

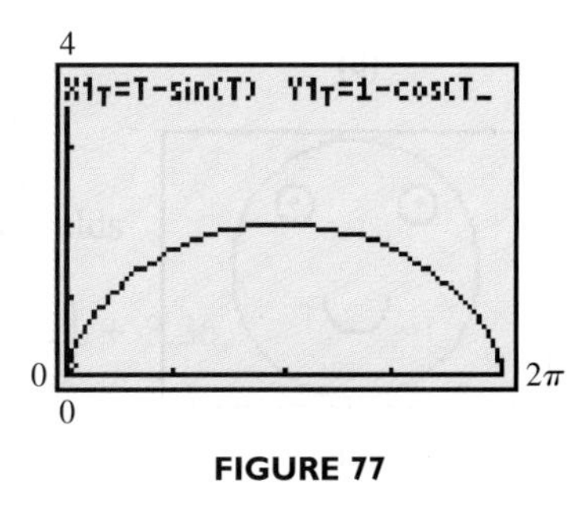

FIGURE 77

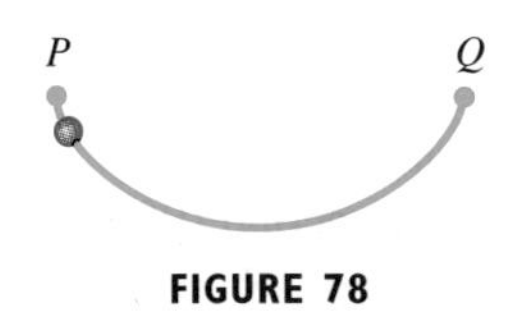

FIGURE 78

The cycloid has an interesting physical property. If a flexible cord or wire goes through points $P$ and $Q$ as in Figure 78, and a bead is allowed to slide due to the force of gravity without friction along this path from $P$ to $Q$, the path that requires the shortest time takes the shape of the graph of an inverted cycloid.

## Applications of Parametric Equations

Parametric equations are used frequently in computer graphics to design a variety of figures and letters. Computer fonts are sometimes designed using parametric equations. In the next example, we use parametric equations to design a "smiley" face consisting of a head, two eyes, and a mouth.

**EXAMPLE 4** *Creating a Drawing with Parametric Equations*

Graph a "smiley" face using parametric equations.

**Solution** *Head* We can use a circle centered at the origin for the head. If the radius is 2, then we let $x = 2\cos t$ and $y = 2\sin t$ for $0 \le t \le 2\pi$. These equations are graphed in Figure 79(a) on the next page.

*Eyes* For the eyes we can use two small circles. The eye in the first quadrant can be modeled by $x = 1 + .3\cos t$ and $y = 1 + .3\sin t$. This represents a circle centered at

## 10.7 EXERCISES

*Find a rectangular equation for each curve and describe the curve.*

**1.** $x = 3 \sin t$, $y = 3 \cos t$; for $t$ in $[-\pi, \pi]$

**2.** $x = 2 \sin t$, $y = 2 \cos t$; for $t$ in $[0, 2\pi]$

**3.** $x = 2 \cos^2 t$, $y = 2 \sin^2 t$; for $t$ in $\left[0, \frac{\pi}{2}\right]$

**4.** $x = \sqrt{5} \sin t$, $y = \sqrt{3} \cos t$; for $t$ in $[0, 2\pi]$

**5.** $x = 3 \tan t$, $y = 2 \sec t$; for $t$ in $\left(-\frac{\pi}{2}, \frac{\pi}{2}\right)$

**6.** $x = \cot t$, $y = \csc t$; for $t$ in $(0, \pi)$

*Graph each pair of parametric equations for $0 \le t \le 2\pi$. Describe any differences in the two graphs.*

**7.** **(a)** $x = 3 \cos t$, $y = 3 \sin t$
**(b)** $x = 3 \cos 2t$, $y = 3 \sin 2t$

**8.** **(a)** $x = 2 \cos t$, $y = 2 \sin t$
**(b)** $x = 2 \cos t$, $y = -2 \sin t$

**9.** **(a)** $x = 3 \cos t$, $y = 3 \sin t$
**(b)** $x = 3 \sin t$, $y = 3 \cos t$

**10.** **(a)** $x = -1 + \cos t$, $y = 2 + \sin t$
**(b)** $x = 1 + \cos t$, $y = 2 + \sin t$

*Find a rectangular equation for each curve and graph the curve.*

**11.** $x = \sin t$, $y = \csc t$; for $t$ in $(0, \pi)$

**12.** $x = \tan t$, $y = \cot t$; for $t$ in $\left(0, \frac{\pi}{2}\right)$

**13.** $x = 2 + \sin t$, $y = 1 + \cos t$; for $t$ in $[0, 2\pi]$

**14.** $x = 1 + 2 \sin t$, $y = 2 + 3 \cos t$; for $t$ in $[0, 2\pi]$

*Graph each pair of parametric equations.*

**15.** $x = 2 + \cos t$, $y = \sin t - 1$; $0 \le t \le 2\pi$

**16.** $x = -2 + \cos t$, $y = \sin t + 1$; $0 \le t \le 2\pi$

**17.** $x = \cos^3 t$, $y = \sin^3 t$; $0 \le t \le 2\pi$

**18.** $x = \cos^5 t$, $y = \sin^5 t$; $0 \le t \le 2\pi$

**19.** $x = |3 \sin t|$, $y = |3 \cos t|$; $0 \le t \le \pi$

**20.** $x = 3 \sin 2t$, $y = 3 \cos t$; $0 \le t \le 2\pi$

*Graph each cycloid for t in the specified interval.*

**21.** $x = t - \sin t$, $y = 1 - \cos t$; for $t$ in $[0, 4\pi]$

**22.** $x = 2t - 2 \sin t$, $y = 2 - 2 \cos t$; for $t$ in $[0, 8\pi]$

*Graph each set of parametric equations for $0 \le t \le 2\pi$ in the window $[0, 6]$ by $[0, 4]$. Identify the letter of the alphabet that is being graphed.*

**23.** $x_1 = 1$, $y_1 = 1 + t/\pi$
$x_2 = 1 + t/(3\pi)$, $y_2 = 2$
$x_3 = 1 + t/(2\pi)$, $y_3 = 3$

**24.** $x_1 = 1$, $y_1 = 1 + t/\pi$
$x_2 = 1 + t/(3\pi)$, $y_2 = 2$
$x_3 = 1 + t/(2\pi)$, $y_3 = 3$
$x_4 = 1 + t/(2\pi)$, $y_4 = 1$

**25.** $x_1 = 1$, $y_1 = 1 + t/\pi$
$x_2 = 1 + 1.3 \sin(.5t)$, $y_2 = 2 + \cos(.5t)$

**26.** $x_1 = 2 + .8 \cos(.85t)$, $y_1 = 2 + \sin(.85t)$
$x_2 = 1.2 + t/(1.3\pi)$, $y_2 = 2$

***(Modeling) Designing Letters*** *Find a set of parametric equations that results in a letter similar to the one shown in each figure. Use the window $[-4.7, 4.7]$ by $[-3.1, 3.1]$, and turn off the coordinate axes. Answers may vary.*

**27.**

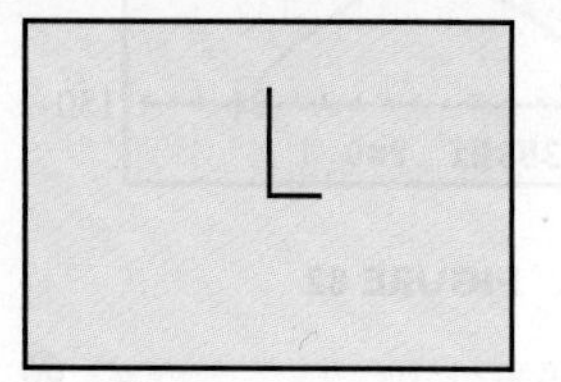

**28.**

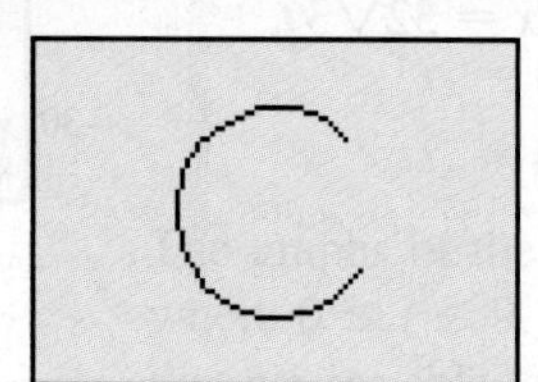

**29.**

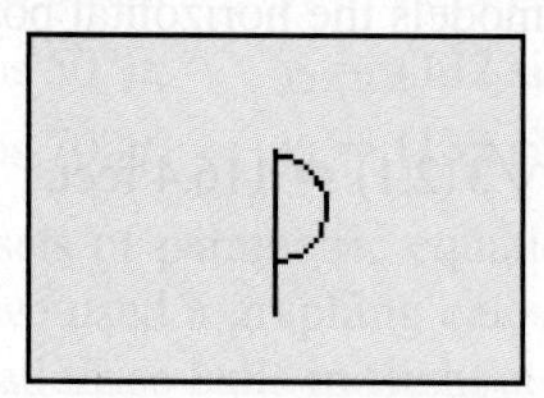

**30.**

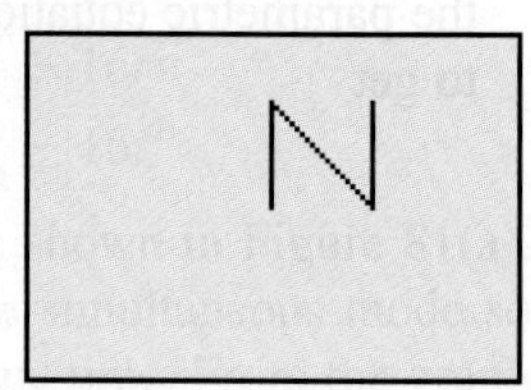

## The Cycloid

**Looking Ahead to Calculus**

The cycloid is a special case of a curve traced out by a point at a given distance from the center of a circle as the circle rolls along a straight line. Such a curve is called a *trochoid.* It is just one of several parametrically defined curves studied in calculus.

The path traced by a fixed point on the circumference of a circle rolling along a line is called a *cycloid.* A **cycloid** is defined by

$$x = at - a \sin t, \quad y = a - a \cos t, \quad \text{for } t \text{ in } (-\infty, \infty).$$

**EXAMPLE 3** *Graphing a Cycloid*

Graph the cycloid with $a = 1$ for $t$ in $[0, 2\pi]$.

**Analytic Solution**

There is no simple way to find a rectangular equation for the cycloid from its parametric equations. Instead, begin with a table of values.

| $t$ | 0 | $\frac{\pi}{4}$ | $\frac{\pi}{2}$ | $\pi$ | $\frac{3\pi}{2}$ | $2\pi$ |
|---|---|---|---|---|---|---|
| $x$ | 0 | .08 | .6 | $\pi$ | 5.7 | $2\pi$ |
| $y$ | 0 | .3 | 1 | 2 | 1 | 0 |

Plotting the ordered pairs $(x, y)$ from the table of values leads to the portion of the graph in Figure 76 from 0 to $2\pi$.

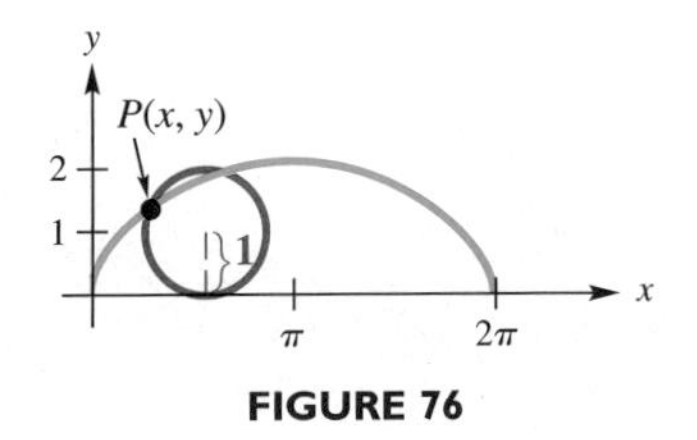


FIGURE 76

**Graphing Calculator Solution**

It is easier to graph a cycloid with a graphing calculator in parametric mode than with traditional methods. See Figure 77.

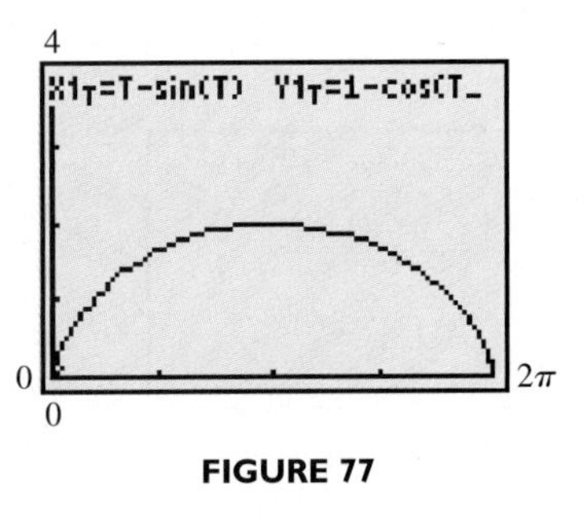


FIGURE 77

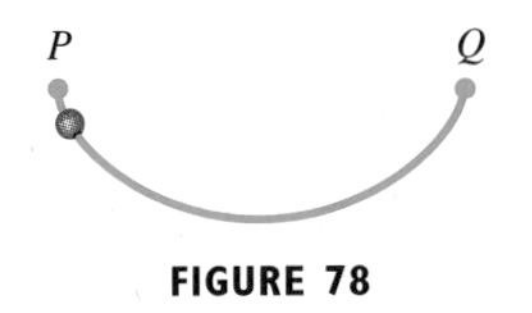


FIGURE 78

The cycloid has an interesting physical property. If a flexible cord or wire goes through points $P$ and $Q$ as in Figure 78, and a bead is allowed to slide due to the force of gravity without friction along this path from $P$ to $Q$, the path that requires the shortest time takes the shape of the graph of an inverted cycloid.

## Applications of Parametric Equations

Parametric equations are used frequently in computer graphics to design a variety of figures and letters. Computer fonts are sometimes designed using parametric equations. In the next example, we use parametric equations to design a "smiley" face consisting of a head, two eyes, and a mouth.

**EXAMPLE 4** *Creating a Drawing with Parametric Equations*

Graph a "smiley" face using parametric equations.

**Solution** *Head* We can use a circle centered at the origin for the head. If the radius is 2, then we let $x = 2 \cos t$ and $y = 2 \sin t$ for $0 \le t \le 2\pi$. These equations are graphed in Figure 79(a) on the next page.

*Eyes* For the eyes we can use two small circles. The eye in the first quadrant can be modeled by $x = 1 + .3 \cos t$ and $y = 1 + .3 \sin t$. This represents a circle centered at

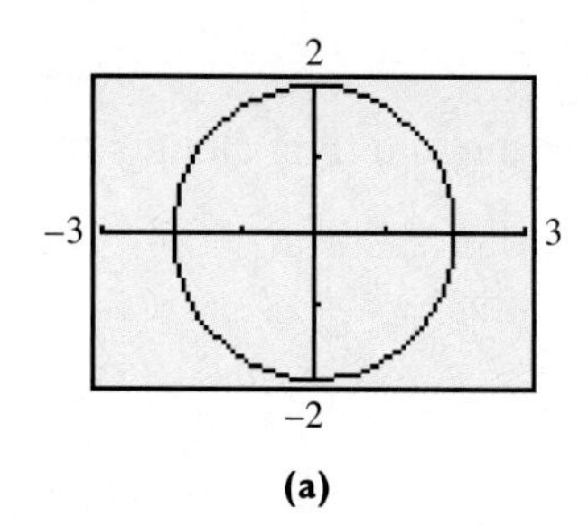

(a)

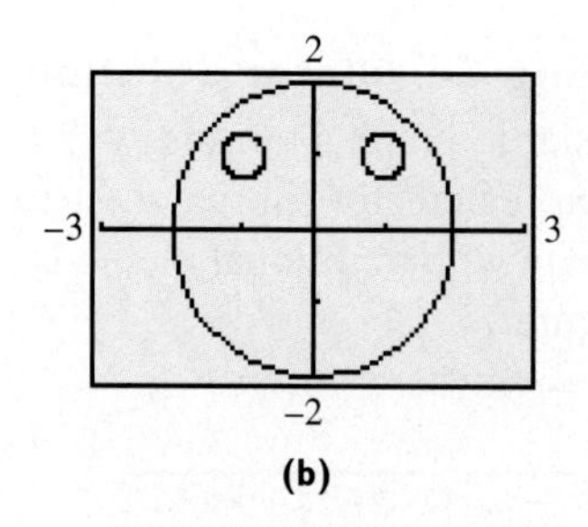

(b)

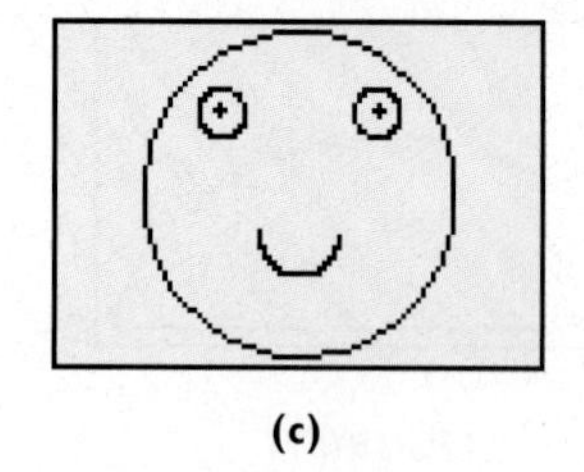
(c)

FIGURE 79

(1, 1) with radius .3. The eye in the second quadrant can be modeled by $x = -1 + .3 \cos t$ and $y = 1 + .3 \sin t$ for $0 \leq t \leq 2\pi$, which is a circle centered at $(-1, 1)$ with radius .3. These graphs are shown in Figure 79(b).

*Mouth* For the smile, we can use the lower half of a circle. Using trial and error, we might arrive at $x = .5 \cos \frac{1}{2}t$ and $y = -.5 - .5 \sin \frac{1}{2}t$. This is a semicircle centered at $(0, -.5)$ with radius .5. Since we are letting $0 \leq t \leq 2\pi$, the term $\frac{1}{2}t$ ensures that only half a circle (semicircle) is drawn. The minus sign before $.5 \sin \frac{1}{2}t$ in the $y$-equation results in the lower half of the semicircle being drawn rather than the upper half. The final result is shown in Figure 79(c). We can add the pupils by plotting the points (1, 1) and $(-1, 1)$. Turning off the coordinate axes rids the face of lines.

### FOR DISCUSSION

Modify the face in Example 4 so that it is frowning. Try to find a way to make the right eye shut rather than open.

Parametric equations are used to simulate motion. If a ball is thrown with a velocity of $v$ feet per second at an angle $\theta$ with the horizontal, its flight can be modeled by the parametric equations

$$x = (v \cos \theta)t \quad \text{and} \quad y = (v \sin \theta)t - 16t^2 + h,$$

where $t$ is in seconds and $h$ is the ball's initial height in feet above the ground. The term $-16t^2$ occurs because gravity is pulling downward. See Figure 80. These equations ignore air resistance.

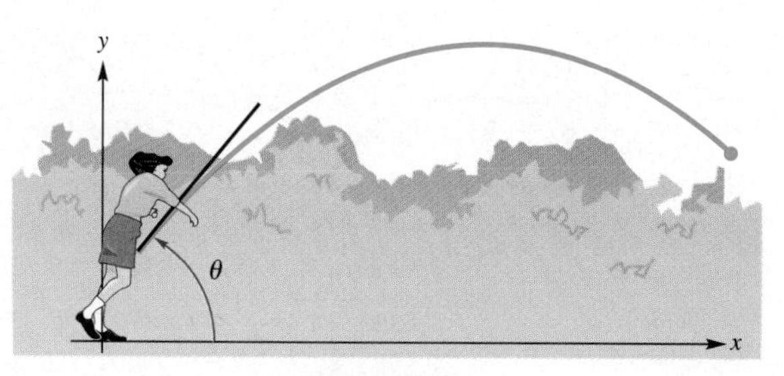

FIGURE 80

**EXAMPLE 5** *Simulating Motion with Parametric Equations*

Three golf balls are hit simultaneously into the air at 132 feet per second (90 mph) at angles of 30°, 50°, and 70° with the horizontal.

**(a)** Assuming the ground is level, determine graphically which ball travels the farthest. Estimate this distance.

**(b)** Which ball reaches the greatest height? Estimate this height.

**Solution**

**(a)** The three sets of parametric equations determined by the three golf balls are as follows since $h = 0$.

$$x_1 = (132 \cos 30°)t, \qquad y_1 = (132 \sin 30°)t - 16t^2$$
$$x_2 = (132 \cos 50°)t, \qquad y_2 = (132 \sin 50°)t - 16t^2$$
$$x_3 = (132 \cos 70°)t, \qquad y_3 = (132 \sin 70°)t - 16t^2$$

The graphs of the three sets of parametric equations are shown in Figure 81(a), where $0 \leq t \leq 9$. We have used a graphing calculator in simultaneous mode so that we are able to view all three balls in flight at the same time. From the graph in Figure 81(b), we can see that the ball hit at 50° travels the farthest distance. Using the trace feature, we estimate this distance to be about 540 feet.

**(b)** The ball hit at 70° reaches the greatest height, about 240 feet.

400

0 600

0

**(a)**

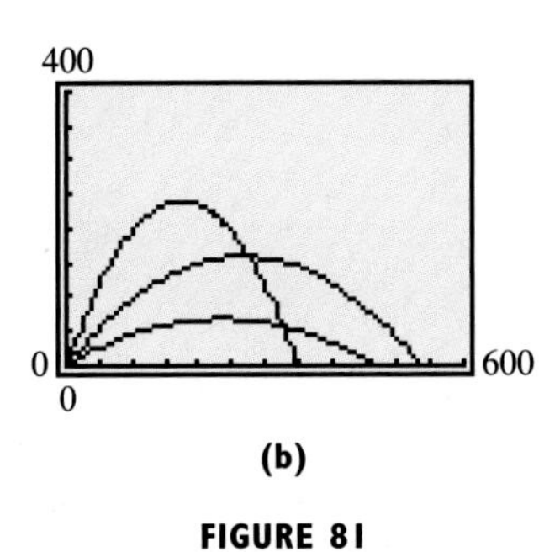


**(b)**

**FIGURE 81**

**FOR DISCUSSION**

If a golf ball is hit at 88 feet per second (60 mph), use trial and error to find the angle $\theta$ that results in a maximum distance for the ball.

**EXAMPLE 6** *Examining Parametric Equations of Flight*

A small rocket is launched from a table that is 3.36 feet above the ground. Its initial velocity is 64 feet per second, and it is launched at an angle of 30° with respect to the ground. Find the rectangular equation that models this path. What type of path does the rocket follow?

**Solution** Its path is defined by the parametric equations

$$x = (64 \cos 30°)t \quad \text{and} \quad y = (64 \sin 30°)t - 16t^2 + 3.36$$

or, equivalently,

$$x = 32\sqrt{3}\,t \quad \text{and} \quad y = -16t^2 + 32t + 3.36.$$

From $x = 32\sqrt{3}\,t$, we obtain

$$t = \frac{x}{32\sqrt{3}}.$$

Substituting into the other parametric equation for $t$ yields

$$y = -16\left(\frac{x}{32\sqrt{3}}\right)^2 + 32\left(\frac{x}{32\sqrt{3}}\right) + 3.36.$$

Simplifying, we find that the rectangular equation is

$$y = -\frac{1}{192}x^2 + \frac{\sqrt{3}}{3}x + 3.36.$$

Because this equation defines a parabola, the rocket follows a parabolic path.

**EXAMPLE 7** *Analyzing the Path of a Projectile*

Determine the total flight time and the horizontal distance traveled by the rocket in Example 6.

**Analytic Solution**

The equation $y = -16t^2 + 32t + 3.36$ tells the vertical position of the rocket at time $t$. We need to determine those values of $t$ for which $y = 0$ since these values correspond to the rocket at ground level. This yields

$$0 = -16t^2 + 32t + 3.36.$$

Using the quadratic formula, the solutions are $t = -.1$ or $t = 2.1$. Since $t$ represents time, $t = -.1$ is an unacceptable answer. Therefore, the flight time is 2.1 seconds.

The rocket was in the air for 2.1 seconds, so we can use $t = 2.1$ and the parametric equation that models the horizontal position, $x = 32\sqrt{3}\,t$, to get

$$x = 32\sqrt{3}(2.1) \approx 116.4 \text{ feet}.$$

**Graphing Calculator Solution**

Figure 82 shows that when $t = 2.1$, the horizontal distance covered is approximately 116.4 feet, which agrees with the analytic solution.

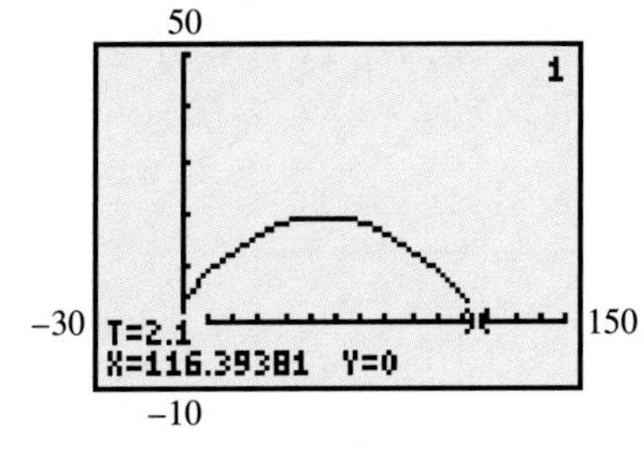


**FIGURE 82**

## 10.7 EXERCISES

*Find a rectangular equation for each curve and describe the curve.*

**1.** $x = 3 \sin t$, $y = 3 \cos t$; for $t$ in $[-\pi, \pi]$

**2.** $x = 2 \sin t$, $y = 2 \cos t$; for $t$ in $[0, 2\pi]$

**3.** $x = 2 \cos^2 t$, $y = 2 \sin^2 t$; for $t$ in $\left[0, \dfrac{\pi}{2}\right]$

**4.** $x = \sqrt{5} \sin t$, $y = \sqrt{3} \cos t$; for $t$ in $[0, 2\pi]$

**5.** $x = 3 \tan t$, $y = 2 \sec t$; for $t$ in $\left(-\dfrac{\pi}{2}, \dfrac{\pi}{2}\right)$

**6.** $x = \cot t$, $y = \csc t$; for $t$ in $(0, \pi)$

*Graph each pair of parametric equations for $0 \le t \le 2\pi$. Describe any differences in the two graphs.*

**7.** **(a)** $x = 3 \cos t$, $y = 3 \sin t$
**(b)** $x = 3 \cos 2t$, $y = 3 \sin 2t$

**8.** **(a)** $x = 2 \cos t$, $y = 2 \sin t$
**(b)** $x = 2 \cos t$, $y = -2 \sin t$

**9.** **(a)** $x = 3 \cos t$, $y = 3 \sin t$
**(b)** $x = 3 \sin t$, $y = 3 \cos t$

**10.** **(a)** $x = -1 + \cos t$, $y = 2 + \sin t$
**(b)** $x = 1 + \cos t$, $y = 2 + \sin t$

*Find a rectangular equation for each curve and graph the curve.*

**11.** $x = \sin t$, $y = \csc t$; for $t$ in $(0, \pi)$

**12.** $x = \tan t$, $y = \cot t$; for $t$ in $\left(0, \dfrac{\pi}{2}\right)$

**13.** $x = 2 + \sin t$, $y = 1 + \cos t$; for $t$ in $[0, 2\pi]$

**14.** $x = 1 + 2 \sin t$, $y = 2 + 3 \cos t$; for $t$ in $[0, 2\pi]$

*Graph each pair of parametric equations.*

**15.** $x = 2 + \cos t$, $y = \sin t - 1$; $0 \le t \le 2\pi$

**16.** $x = -2 + \cos t$, $y = \sin t + 1$; $0 \le t \le 2\pi$

**17.** $x = \cos^3 t$, $y = \sin^3 t$; $0 \le t \le 2\pi$

**18.** $x = \cos^5 t$, $y = \sin^5 t$; $0 \le t \le 2\pi$

**19.** $x = |3 \sin t|$, $y = |3 \cos t|$; $0 \le t \le \pi$

**20.** $x = 3 \sin 2t$, $y = 3 \cos t$; $0 \le t \le 2\pi$

*Graph each cycloid for t in the specified interval.*

**21.** $x = t - \sin t$, $y = 1 - \cos t$; for $t$ in $[0, 4\pi]$

**22.** $x = 2t - 2 \sin t$, $y = 2 - 2 \cos t$; for $t$ in $[0, 8\pi]$

*Graph each set of parametric equations for $0 \le t \le 2\pi$ in the window $[0, 6]$ by $[0, 4]$. Identify the letter of the alphabet that is being graphed.*

**23.** $x_1 = 1$, $y_1 = 1 + t/\pi$
$x_2 = 1 + t/(3\pi)$, $y_2 = 2$
$x_3 = 1 + t/(2\pi)$, $y_3 = 3$

**24.** $x_1 = 1$, $y_1 = 1 + t/\pi$
$x_2 = 1 + t/(3\pi)$, $y_2 = 2$
$x_3 = 1 + t/(2\pi)$, $y_3 = 3$
$x_4 = 1 + t/(2\pi)$, $y_4 = 1$

**25.** $x_1 = 1$, $y_1 = 1 + t/\pi$
$x_2 = 1 + 1.3 \sin(.5t)$, $y_2 = 2 + \cos(.5t)$

**26.** $x_1 = 2 + .8 \cos(.85t)$, $y_1 = 2 + \sin(.85t)$
$x_2 = 1.2 + t/(1.3\pi)$, $y_2 = 2$

*(Modeling) Designing Letters* *Find a set of parametric equations that results in a letter similar to the one shown in each figure. Use the window $[-4.7, 4.7]$ by $[-3.1, 3.1]$, and turn off the coordinate axes. Answers may vary.*

**27.**

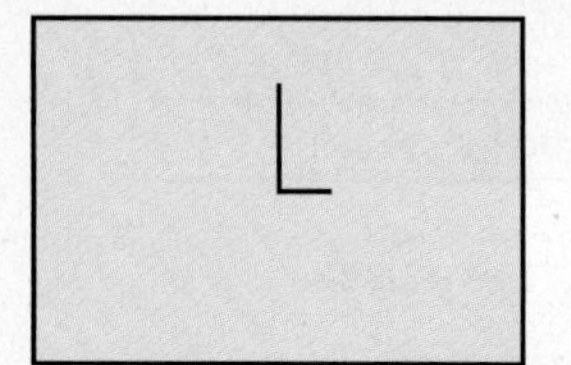

**28.**

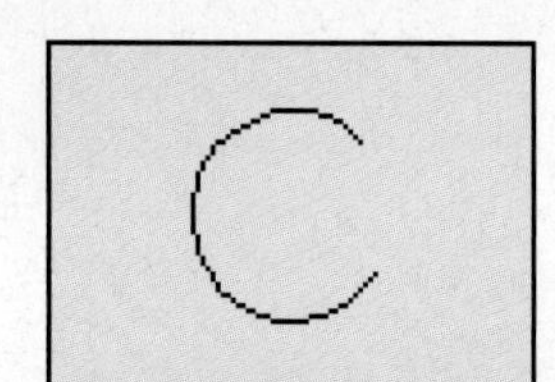

**29.**

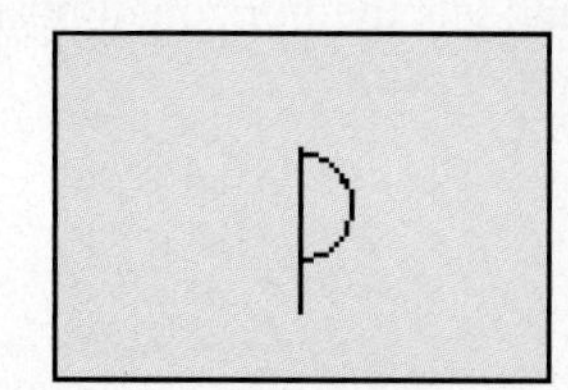

**30.**

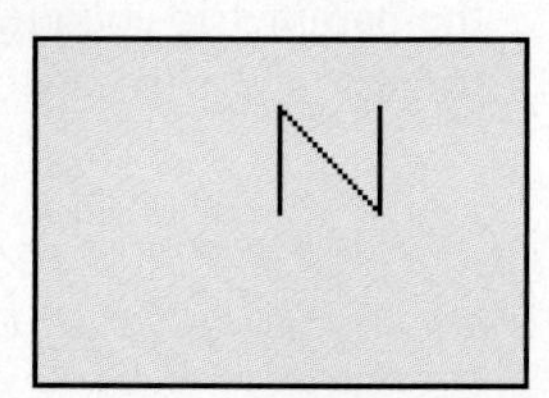

**31.** *(Modeling) Designing a Face* Refer to Example 4. Use parametric equations to create your own "smiley" face. This face should have a head, a mouth, and eyes.

**32.** *(Modeling) Designing a Face* Add a nose to the face that you designed in Exercise 31.

*Lissajous Figures The middle screen in Figure 72 is an example of a Lissajous figure. Lissajous figures occur in electronics and may be used to find the frequency of an unknown voltage. Graph each Lissajous figure for $0 \le t \le 6.5$ in the window $[-6, 6]$ by $[-4, 4]$.*

**33.** $x = 2 \cos t,\ y = 3 \sin 2t$

**34.** $x = 3 \cos 2t,\ y = 3 \sin 3t$

**35.** $x = 3 \sin 4t,\ y = 3 \cos 3t$

**36.** $x = 4 \sin 4t,\ y = 3 \sin 5t$

*(Modeling) In Exercises 37–40, do the following.*

**(a)** *Determine the parametric equations that model the path of the projectile.*

**(b)** *Determine the rectangular equation that models the path of the projectile.*

**(c)** *Determine how long the projectile is in flight and the horizontal distance covered.*

**37.** *Flight of a Model Rocket* A model rocket is launched from the ground with a velocity of 48 feet per second at an angle of 60° with respect to the ground.

**38.** *Flight of a Golf Ball* Tiger is playing golf. He hit a golf ball from the ground at an angle of 60° with respect to the ground at a velocity of 150 feet per second.

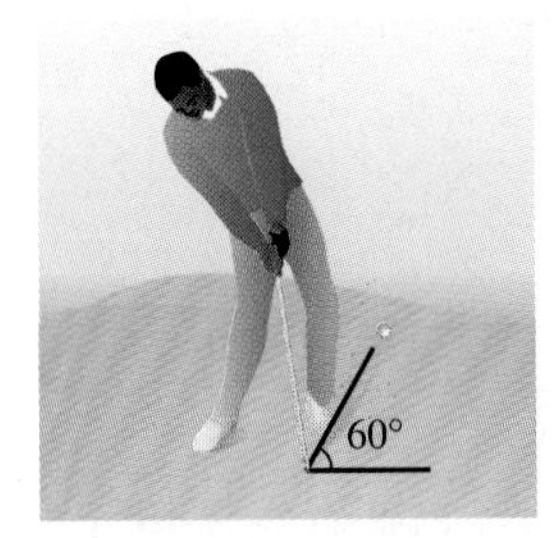


**39.** *Flight of a Softball* Sally hits a softball when it is 2 feet above the ground. The ball leaves her bat at an angle of 20° with respect to the ground at a velocity of 88 feet per second.

**40.** *Flight of a Baseball* Luis hits a baseball when it is 2.5 feet above the ground. The ball leaves his bat at an angle of 29° from the horizontal with a velocity of 136 feet per second.

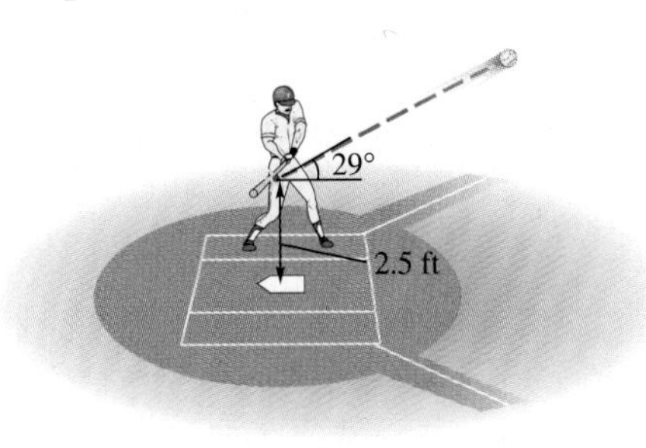


**41.** *(Modeling) Simulating Gravity on the Moon* If an object is thrown on the moon, then the parametric equations of flight are

$$x = (v \cos \theta)t \quad \text{and} \quad y = (v \sin \theta)t - 2.66t^2 + h.$$

Estimate the distance that a golf ball hit at 88 feet per second (60 mph) at an angle of 45° with the horizontal travels on the moon if the moon's surface is level.

**42.** *(Modeling) Flight of a Baseball* A baseball is hit from a height of 3 feet at a 60° angle above the horizontal. Its initial velocity is 64 feet per second.

**(a)** Write parametric equations that model the flight of the baseball.

**(b)** Determine the horizontal distance traveled by the ball in the air. Assume that the ground is level.

**(c)** What is the maximum height of the baseball? At that time, how far has the ball traveled horizontally?

**(d)** Would the ball clear a 5-foot high fence that is 100 feet from the batter?

*(Modeling) Path of a Projectile In Exercises 43 and 44, a projectile has been launched from the ground with an initial velocity of 88 feet per second. You are given parametric equations that model the path of the projectile.*

**(a)** *Graph the parametric equations.*

**(b)** *Approximate $\theta$, the angle the projectile makes with the horizontal at launch, to the nearest tenth of a degree.*

**(c)** *Based on your answer to part (b), write parametric equations for the projectile using the cosine and sine functions.*

**43.** $x = 82.69265063t,\ y = -16t^2 + 30.09777261t$

**44.** $x = 56.56530965t,\ y = -16t^2 + 67.41191099t$

**45.** The spiral of Archimedes has polar equation $r = a\theta$, where $r^2 = x^2 + y^2$. Show that a parametric representation of the spiral of Archimedes is

$$x = a\theta \cos \theta, \quad y = a\theta \sin \theta; \quad \text{for } \theta \text{ in } (-\infty, \infty).$$

**46.** Show that the hyperbolic spiral $r\theta = a$, where $r^2 = x^2 + y^2$, is given parametrically by

$$x = \frac{a \cos \theta}{\theta}, \quad y = \frac{a \sin \theta}{\theta}; \quad \text{for } \theta \text{ in } (-\infty, 0) \cup (0, \infty).$$

## Reviewing Basic Concepts (Sections 10.6 and 10.7)

1. For the point with polar coordinates $(-2, 130°)$, state the quadrant in which the point lies if it is graphed in a rectangular coordinate system.
2. Let point $P$ have rectangular coordinates $(-2, 2)$. Determine two pairs of equivalent polar coordinates for point $P$.
3. Graph the polar equation $r = 2 - 2\cos\theta$. Identify the type of polar graph.
4. Find an equivalent equation in rectangular coordinates for the polar equation $r = 2\cos\theta$.
5. Find an equivalent equation in polar coordinates for the equation $x + y = 6$.
6. Find a rectangular equation for the curve described by $x = 2\cos t$, $y = 4\sin t$, $0 \le t < 2\pi$.
7. Graph the parametric equations
$$x = 2 - \sin t, \quad y = \cos t - 1,$$
for $0 \le t < 2\pi$.
8. *(Modeling) Flight of a Golf Ball* A golf ball is hit with a 45° angle of elevation, from the top of a ridge that is 50 feet above an area of level ground. The initial velocity of the ball is 88 feet per second or 60 mph. Find the horizontal distance traveled by the ball in the air.

## CHAPTER 10 SUMMARY

**KEY TERMS & SYMBOLS** | **KEY CONCEPTS**

### 10.1 The Law of Sines

side-angle-side (SAS)
angle-side-angle (ASA)
side-side-side (SSS)
side-angle-angle (SAA)
side-side-angle (SSA)
oblique triangle
ambiguous case

**LAW OF SINES**

In any triangle $ABC$, with sides $a$, $b$, and $c$,

$$\frac{a}{\sin A} = \frac{b}{\sin B}, \quad \frac{a}{\sin A} = \frac{c}{\sin C}, \quad \text{and} \quad \frac{b}{\sin B} = \frac{c}{\sin C}.$$

This can be written in compact form as $\dfrac{a}{\sin A} = \dfrac{b}{\sin B} = \dfrac{c}{\sin C}$.

### 10.2 The Law of Cosines and Area Formulas

semiperimeter

**LAW OF COSINES**

In any triangle $ABC$, with sides $a$, $b$, and $c$,

$$a^2 = b^2 + c^2 - 2bc\cos A, \quad b^2 = a^2 + c^2 - 2ac\cos B, \quad \text{and} \quad c^2 = a^2 + b^2 - 2ab\cos C.$$

**HERON'S AREA FORMULA (SSS)**

If a triangle has sides of lengths $a$, $b$, and $c$, and if the semiperimeter is $s = \frac{1}{2}(a + b + c)$, then the area $\mathcal{A}$ of the triangle is $\mathcal{A} = \sqrt{s(s-a)(s-b)(s-c)}$.

**AREA OF A TRIANGLE (SAS)**

In any triangle $ABC$, the area $\mathcal{A}$ is given by any of the following formulas:

$$\mathcal{A} = \frac{1}{2}bc\sin A, \quad \mathcal{A} = \frac{1}{2}ab\sin C, \quad \text{and} \quad \mathcal{A} = \frac{1}{2}ac\sin B.$$

That is, the area is half the product of the lengths of two sides and the sine of the angle included between them.

| KEY TERMS & SYMBOLS | KEY CONCEPTS |
|---|---|

## 10.3 Vectors and Their Applications

scalars
vector quantities
vector, **v**, **OP**, or $\overrightarrow{OP}$
initial point
terminal point
magnitude, $|\mathbf{OP}|$
parallelogram rule
resultant vector
opposite of **v**
zero vector
scalar product
position vector, $\langle a, b\rangle$
horizontal component
vertical component
direction angle
unit vectors, **i**, **j**
dot product
angle between two vectors
orthogonal vectors
airspeed
groundspeed

**RESULTANT VECTOR**
The resultant (or sum) **A** + **B** of vectors **A** and **B** is shown in the figure.

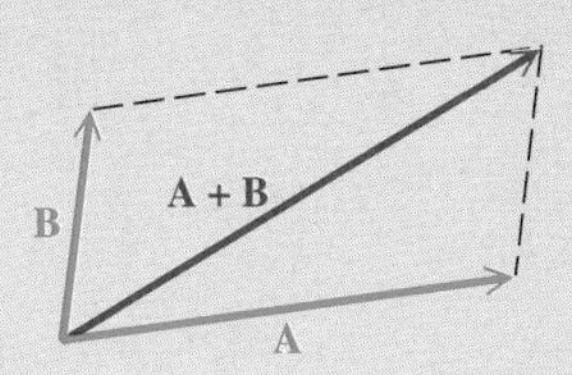


**MAGNITUDE AND DIRECTION ANGLE OF A VECTOR**
The magnitude (length) of vector $\mathbf{u} = \langle a, b\rangle$ is given by $|\mathbf{u}| = \sqrt{a^2 + b^2}$. The direction angle $\theta$ satisfies $\tan\theta = \frac{b}{a}$, where $a \neq 0$.

**VECTOR OPERATIONS**
For any real numbers $a$, $b$, $c$, $d$, and $k$,

$$\langle a, b\rangle + \langle c, d\rangle = \langle a + c, b + d\rangle$$
$$k \cdot \langle a, b\rangle = \langle ka, kb\rangle$$
$$\text{If } \mathbf{a} = \langle a_1, a_2\rangle, \text{ then } -\mathbf{a} = \langle -a_1, -a_2\rangle.$$
$$\langle a, b\rangle - \langle c, d\rangle = \langle a, b\rangle + -\langle c, d\rangle = \langle a - c, b - d\rangle$$

If $\mathbf{u} = \langle x, y\rangle$ has direction angle $\theta$, then $\mathbf{u} = \langle |\mathbf{u}|\cos\theta, |\mathbf{u}|\sin\theta\rangle$.

**i, j FORM FOR VECTORS**
If $\mathbf{v} = \langle a, b\rangle$, then $\mathbf{v} = a\mathbf{i} + b\mathbf{j}$, where $\mathbf{i} = \langle 1, 0\rangle$ and $\mathbf{j} = \langle 0, 1\rangle$.

**DOT PRODUCT**
The dot product of the two vectors $\mathbf{u} = \langle a, b\rangle$ and $\mathbf{v} = \langle c, d\rangle$, denoted $\mathbf{u} \cdot \mathbf{v}$, is given by

$$\mathbf{u} \cdot \mathbf{v} = ac + bd.$$

If $\theta$ is the angle between **u** and **v**, where $0° \leq \theta \leq 180°$, then

$$\mathbf{u} \cdot \mathbf{v} = |\mathbf{u}||\mathbf{v}|\cos\theta.$$

**PROPERTIES OF THE DOT PRODUCT**
For all vectors **u**, **v**, and **w** and real numbers $k$,

(a) $\mathbf{u} \cdot \mathbf{v} = \mathbf{v} \cdot \mathbf{u}$
(b) $\mathbf{u} \cdot (\mathbf{v} + \mathbf{w}) = \mathbf{u} \cdot \mathbf{v} + \mathbf{u} \cdot \mathbf{w}$
(c) $(\mathbf{u} + \mathbf{v}) \cdot \mathbf{w} = \mathbf{u} \cdot \mathbf{w} + \mathbf{v} \cdot \mathbf{w}$
(d) $(k\mathbf{u}) \cdot \mathbf{v} = k(\mathbf{u} \cdot \mathbf{v}) = \mathbf{u} \cdot (k\mathbf{v})$
(e) $\mathbf{0} \cdot \mathbf{u} = 0$
(f) $\mathbf{u} \cdot \mathbf{u} = |\mathbf{u}|^2$.

## 10.4 Trigonometric (Polar) Form of Complex Numbers

real axis

**TRIGONOMETRIC (POLAR) FORM OF A COMPLEX NUMBER**
The expression $r(\cos\theta + i\sin\theta)$ is called the trigonometric form (or polar form) of the complex number $x + yi$. The expression $\cos\theta + i\sin\theta$ is sometimes abbreviated cis $\theta$. Using this notation, $r(\cos\theta + i\sin\theta)$ is written $r$ **cis** $\theta$.

| KEY TERMS & SYMBOLS | KEY CONCEPTS |
|---|---|
| imaginary axis<br>complex plane<br>rectangular form<br>trigonometric (polar) form<br>modulus (absolute value)<br>argument | **PRODUCT THEOREM**<br>If $r_1(\cos\theta_1 + i\sin\theta_1)$ and $r_2(\cos\theta_2 + i\sin\theta_2)$ are any two complex numbers, then<br>$\mathbf{[r_1(\cos\theta_1 + i\sin\theta_1)] \cdot [r_2(\cos\theta_2 + i\sin\theta_2)] = r_1r_2[\cos(\theta_1+\theta_2) + i\sin(\theta_1+\theta_2)]}$,<br>or in compact form,<br>$\mathbf{(r_1 \text{ cis } \theta_1)(r_2 \text{ cis } \theta_2) = r_1r_2 \text{ cis}(\theta_1+\theta_2)}$.<br><br>**QUOTIENT THEOREM**<br>If $r_1(\cos\theta_1 + i\sin\theta_1)$ and $r_2(\cos\theta_2 + i\sin\theta_2)$ are any two complex numbers, where $r_2(\cos\theta_2 + i\sin\theta_2) \neq 0$, then<br>$\mathbf{\dfrac{r_1(\cos\theta_1 + i\sin\theta_1)}{r_2(\cos\theta_2 + i\sin\theta_2)} = \dfrac{r_1}{r_2}[\cos(\theta_1-\theta_2) + i\sin(\theta_1-\theta_2)]}$,<br>or in compact form, $\mathbf{\dfrac{r_1 \text{ cis } \theta_1}{r_2 \text{ cis } \theta_2} = \dfrac{r_1}{r_2}\text{cis}(\theta_1-\theta_2)}$. |
| **10.5** Powers and Roots of Complex Numbers<br>$n$th root of a complex number | **DE MOIVRE'S THEOREM**<br>If $r(\cos\theta + i\sin\theta)$ is a complex number, and if $n$ is any real number, then<br>$\mathbf{[r(\cos\theta + i\sin\theta)]^n = r^n(\cos n\theta + i\sin n\theta)}$.<br>In compact form, $\mathbf{[r \text{ cis } \theta]^n = r^n(\text{cis } n\theta)}$.<br><br>***n*TH ROOT**<br>For a positive integer $n$, the complex number $a + bi$ is an $n$th root of the complex number $x + yi$ if $\mathbf{(a + bi)^n = x + yi}$.<br><br>***n*TH ROOT THEOREM**<br>If $n$ is any positive integer, $r$ is a positive real number, and $\theta$ is in degrees, then the nonzero complex number $r(\cos\theta + i\sin\theta)$ has exactly $n$ distinct $n$th roots, given by<br>$\mathbf{\sqrt[n]{r}(\cos\alpha + i\sin\alpha)}$,<br>where<br>$\mathbf{\alpha = \dfrac{\theta + 360° \cdot k}{n}}$ or $\mathbf{\alpha = \dfrac{\theta}{n} + \dfrac{360° \cdot k}{n}}$, $k = 0, 1, 2, \ldots, n-1$. |
| **10.6** Polar Equations and Graphs<br>polar coordinate system<br>pole<br>polar axis<br>polar coordinates | **RECTANGULAR AND POLAR COORDINATES**<br>The following relationships hold between the point $(x, y)$ in the rectangular coordinate plane and the point $(r, \theta)$ in the polar coordinate plane.<br>$\mathbf{x = r\cos\theta}$, $\mathbf{y = r\sin\theta}$, $\mathbf{r^2 = x^2 + y^2}$, $\mathbf{\tan\theta = \dfrac{y}{x}}$, if $x \neq 0$ |

| KEY TERMS & SYMBOLS | KEY CONCEPTS |
|---|---|
| polar equation<br>cardioid<br>polar grid<br>lemniscate<br>rose curve (four-leaved rose)<br>spiral of Archimedes<br>limaçons | Polar coordinates determine a point by locating it $\theta$ degrees from the polar axis (the positive $x$-axis) and $r$ units from the pole (or origin). |
| **10.7** More Parametric Equations<br>cycloid | **PARAMETRIC EQUATIONS**<br>Parametric equations with trigonometric functions can be used to describe many different types of curves. Examples include circles, ellipses, and cycloids.<br>**HEIGHT OF AN OBJECT**<br>If an object has an initial velocity $v$, initial height $h$, and travels so that its initial angle of elevation is $\theta$, then its flight after $t$ seconds is modeled by the parametric equations<br>$\mathbf{x = (v \cos \theta)t}$ and $\mathbf{y = (v \sin \theta)t - 16t^2 + h}$. |

## CHAPTER 10 REVIEW EXERCISES

*Use the law of sines or the law of cosines to find the indicated part of each triangle ABC.*

**1.** $C = 74° 10'$, $c = 96.3$ meters, $B = 39° 30'$; find $b$

**2.** $a = 165$ meters, $A = 100.2°$, $B = 25.0°$; find $b$

**3.** $a = 86.14$ inches, $b = 253.2$ inches, $c = 241.9$ inches; find $A$

**4.** $a = 14.8$ feet, $b = 19.7$ feet, $c = 31.8$ feet; find $B$

**5.** $A = 129° 40'$, $a = 127$ feet, $b = 69.8$ feet; find $B$

**6.** $B = 39° 50'$, $b = 268$ centimeters, $a = 340$ centimeters; find $A$

**7.** $B = 120.7°$, $a = 127$ feet, $c = 69.8$ feet; find $b$

**8.** $A = 46.2°$, $b = 184$ centimeters, $c = 192$ centimeters; find $a$

*Find the area of each triangle ABC with the given information.*

**9.** $b = 840.6$ meters, $c = 715.9$ meters, $A = 149.3°$

**10.** $a = 6.90$ feet, $b = 10.2$ feet, $C = 35° 10'$

**11.** $a = .913$ kilometer, $b = .816$ kilometer, $c = .582$ kilometer

**12.** $a = 43$ meters, $b = 32$ meters, $c = 51$ meters

*Solve each problem.*

**13.** *Height of a Balloon* The angles of elevation of a balloon from two points $A$ and $B$ on level ground are $24° 50'$ and $47° 20'$, respectively. As shown in the figure, points $A$ and $B$ are in the same vertical plane and are 8.4 miles apart. Approximate the height of the balloon above the ground to the nearest tenth of a mile.

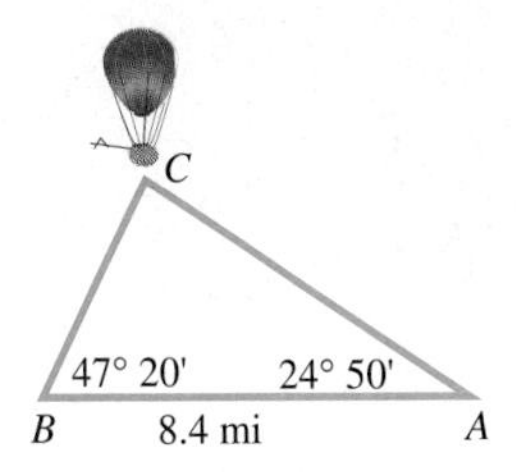


**14.** *Distance between Points on a Softball Field* The pitcher's mound on a regulation softball field is 46 feet from home plate. The distance between the bases is 60 feet, as shown in the figure. How far is the pitcher's mound at point $M$ from third base (point $T$)? Give your answer to the nearest foot.

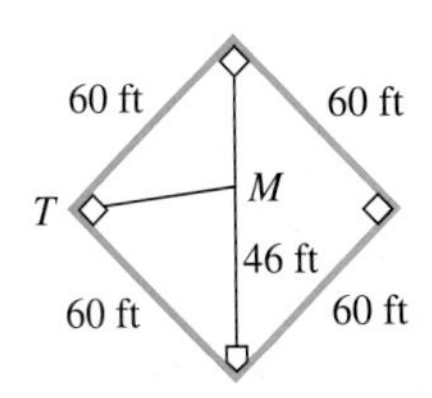

**15.** *Distance in a Boat Race* The course for a boat race starts at point $X$ and goes in the direction S 48° W to point $Y$. It then turns and goes S 36° E to point $Z$, and finally returns back to point $X$. If point $Z$ lies 10 kilometers directly south of point $X$, find the distance from $Y$ to $Z$ to the nearest kilometer.

**16.** *Radio Direction Finders* Radio direction finders are placed at points $A$ and $B$, which are 3.46 miles apart on an east–west line, with $A$ west of $B$. From $A$, the bearing of a certain illegal pirate radio transmitter is 48°, and from $B$ the bearing is 302°. Find the distance between the transmitter and $A$.

**17.** *Distance across a Canyon* To measure the distance $AB$ across a canyon for a power line, a surveyor measures angles $B$ and $C$ and the distance $BC$. (See the figure.) What is the distance from $A$ to $B$?

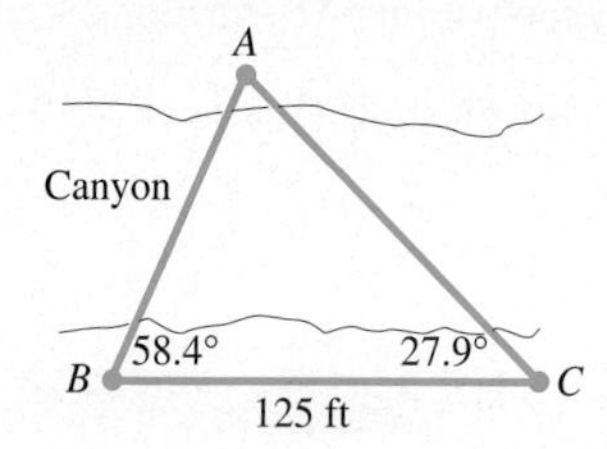


**18.** *Length of a Brace* A banner on an 8.0-foot pole is to be mounted on a building at an angle of 115°, as shown in the figure. Find the length of the brace.

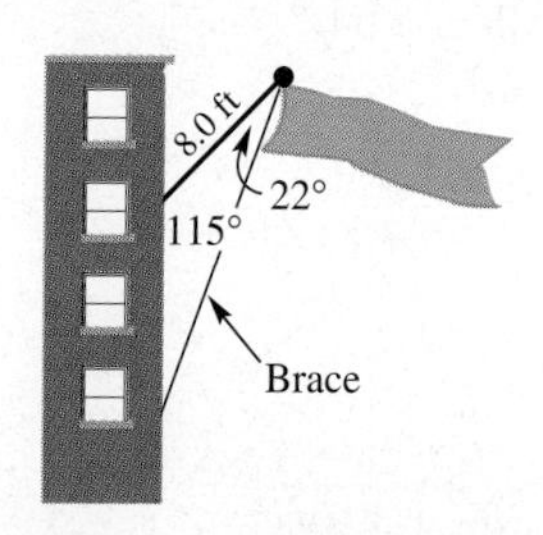


**19.** *Hanging Sculpture* A hanging sculpture in an art gallery is to be hung with two wires of lengths 15.0 feet and 12.2 feet so that the angle between them is 70.3°. How far apart should the ends of the wire be placed on the ceiling?

**20.** *Pipeline Position* A pipeline is to run between points $A$ and $B$, which are separated by a protected wetlands area. To avoid the wetlands, the pipe will run from point $A$ to $C$ and then to $B$. The distances involved are $AB = 150$ kilometers, $AC = 102$ kilometers, and $BC = 135$ kilometers. What angle should be used at point $C$?

**21.** If we are given $a$, $A$, and $C$ in a triangle $ABC$, does the possibility of the ambiguous case exist? If not, explain why.

**22.** Can a triangle $ABC$ exist if $a = 4.7$, $b = 2.3$, and $c = 7.0$? If not, explain why. Answer this question without using trigonometry.

**23.** *Concept Check* Given $a = 10$ and $B = 30°$, determine the values of $b$ for which $A$ has **(a)** exactly one value, **(b)** two values, **(c)** no value.

**24.** *Concept Check* If angle $C$ of a triangle $ABC$ measures 90°, what does the law of cosines become?

**25.** Use the vectors shown here to sketch $\mathbf{a} + 3\mathbf{c}$.

*Find the magnitude and direction angle for* $\mathbf{u}$, *rounded to the nearest tenth of a degree.*

**26.** $\mathbf{u} = \langle 21, -20 \rangle$

**27.** $\mathbf{u} = \langle -9, 12 \rangle$

*Vector* $\mathbf{v}$ *has the given magnitude and direction angle. Find the horizontal and vertical components of* $\mathbf{v}$.

**28.** $|\mathbf{v}| = 50$, $\theta = 45°$ (give exact values)

**29.** $|\mathbf{v}| = 69.2$, $\theta = 75°$

**30.** $|\mathbf{v}| = 964$, $\theta = 154°\ 20'$

*Find* **(a)** *the dot product and* **(b)** *the angle between the pairs of vectors.*

**31.** $\mathbf{u} = \langle 6, 2 \rangle$, $\mathbf{v} = \langle 3, -2 \rangle$

**32.** $\mathbf{u} = \langle 2\sqrt{3}, 2 \rangle$, $\mathbf{v} = \langle 5, 5\sqrt{3} \rangle$

**33.** Are the vectors $\mathbf{u} = \langle 5, -1 \rangle$ and $\mathbf{v} = \langle -2, -10 \rangle$ orthogonal? Explain.

*Solve each problem.*

**34.** *Weight of a Sled and Child* Paula and Steve are pulling their daughter Jessie on a sled. Steve pulls with a force of 18 pounds at an angle of 10°. Paula pulls with a force of 12 pounds at an angle of 15°. Find the magnitude of the resultant force.

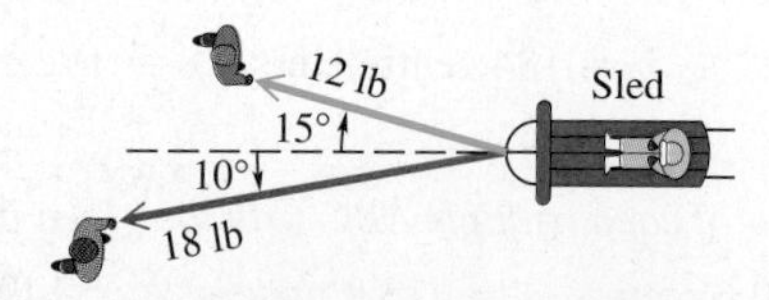


**35.** *Barge Movement* One rope pulls a barge directly east with a force of 1000 newtons. Another rope pulls the barge to the northeast with a force of 2000 newtons. Find the resultant force acting on the barge, and the angle between the resultant and the first rope.

**36.** *Direction and Airspeed* A plane has an airspeed of 520 mph. The pilot wishes to fly on a course of 310°. A wind of 37 mph is blowing from a bearing of 212°. On what bearing should the pilot fly, and what will be her actual speed?

**37.** *Bearing and Speed* A long-distance swimmer starts out swimming a steady 3.2 mph due north. A 5.1 mph current is flowing on a bearing of 12°. What is the swimmer's resulting bearing and speed?

*Graph each complex number as a vector in the complex plane.*

**38.** $5i$ **39.** $-4 + 2i$

*Find the resultant of each pair of complex numbers.*

**40.** $7 + 3i$ and $-2 + i$ **41.** $2 - 4i$ and $5 + i$

*Complete the table in Exercises 42–45.*

| | Rectangular Form | Trigonometric Form |
|---|---|---|
| 42. | $-2 + 2i$ | ________ |
| 43. | ________ | $3(\cos 90° + i \sin 90°)$ |
| 44. | ________ | $2 \text{ cis } 225°$ |
| 45. | $-4 + 4i\sqrt{3}$ | ________ |

*Perform each indicated operation. Give the answer in rectangular form.*

**46.** $[5(\cos 90° + i \sin 90°)][6(\cos 180° + i \sin 180°)]$

**47.** $[3 \text{ cis } 135°][2 \text{ cis } 105°]$

**48.** $\dfrac{2(\cos 60° + i \sin 60°)}{8(\cos 300° + i \sin 300°)}$ **49.** $\dfrac{4 \text{ cis } 270°}{2 \text{ cis } 90°}$

**50.** $(\sqrt{3} + i)^3$ **51.** $(2 - 2i)^5$

**52.** $(\cos 100° + i \sin 100°)^6$ **53.** $(\text{cis } 20°)^3$

*Find the indicated roots and graph as vectors in the complex plane. Leave your answers in polar form.*

**54.** The cube roots of $-27i$

**55.** The fourth roots of $16i$

**56.** The fifth roots of 32

**57.** Solve the equation $x^4 + i = 0$. Leave solutions in polar form.

*Convert to rectangular coordinates. Give exact values.*

**58.** $(12, 225°)$ **59.** $\left(-8, -\dfrac{\pi}{3}\right)$

*Convert to polar coordinates, with $-180° < \theta \leq 180°$. Give exact values.*

**60.** $(-6, 6)$ **61.** $(0, -5)$

*Use a graphing calculator to graph each polar equation for $0° \leq \theta \leq 360°$. Use a square window.*

**62.** $r = 4 \cos \theta$ (circle)

**63.** $r = 1 - 2 \sin \theta$ (limaçon with a loop)

**64.** $r = 2 \sin 4\theta$ (eight-leaved rose)

*Find an equivalent equation in rectangular coordinates.*

**65.** $r = \dfrac{3}{1 + \cos \theta}$ **66.** $r = \dfrac{4}{2 \sin \theta - \cos \theta}$

**67.** $r = \sin \theta + \cos \theta$ **68.** $r = 2$

*Find an equivalent equation in polar coordinates.*

**69.** $x = -3$ **70.** $y = x$

**71.** $y = x^2$ **72.** $x = y^2$

**73.** Without actually performing the operations, state why the quotients

$$[2(\cos 45° + i \sin 45°)] \div [5(\cos 90° + i \sin 90°)]$$

and

$$[2(\cos(-315°) + i \sin(-315°))] \div [5(\cos(-270°) + i \sin(-270°))]$$

are the same.

**74.** *Concept Check* Under what conditions is the difference between two imaginary numbers $a + bi$ and $c + di$ an imaginary number?

*Find a rectangular equation for each plane curve with the given parametric equations.*

**75.** $x = \cos 2t$, $y = \sin t$, for $t$ in $(-\pi, \pi)$

**76.** $x = 5 \tan t$, $y = 3 \sec t$, for $t$ in $\left(-\dfrac{\pi}{2}, \dfrac{\pi}{2}\right)$

**77.** Graph the curve defined by the parametric equations

$$x = t + \cos t, \ y = \sin t, \text{ for } t \text{ in } [0, 2\pi].$$

**78.** *(Modeling) Flight of a Baseball* A baseball is hit when it is 3.2 feet above the ground. It leaves the bat with a velocity of 118 feet per second at an angle of 27° with respect to the ground. Find the horizontal distance that the baseball travels in the air.

## CHAPTER 10 TEST

**1.** Find the indicated part of each triangle.

**(a)** $A = 25.2°$, $a = 6.92$ yards, $b = 4.82$ yards; find $C$

**(b)** $C = 118°$, $b = 132$ kilometers, $a = 75.1$ kilometers; find $c$

**(c)** $a = 17.3$ feet, $b = 22.6$ feet, $c = 29.8$ feet; find $B$

**2.** Given $a = 10$ and $B = 150°$ in triangle $ABC$, determine the values of $b$ for which $A$ has

**(a)** exactly one value **(b)** two values **(c)** no values.

**3.** What conditions determine whether or not three positive numbers can represent the sides of a triangle?

**4.** Solve each applied problem.

**(a)** *Distance between Boats* Two boats leave a dock together. Each travels in a straight line. The angle between their courses measures 54.2°. One boat travels 36.2 kilometers per hour, and the other travels 45.6 kilometers per hour. How far apart will they be after 3 hours?

**(b)** *Distance on a Baseball Field* A baseball diamond is a square, 90.0 feet on a side, with home plate and the three bases at the vertices. The pitcher's rubber is located 60.5 feet from home plate. Find the distance from the pitcher's rubber to second base.

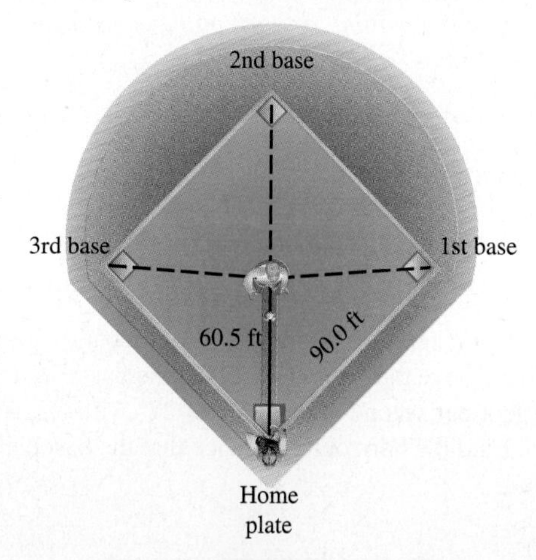


**(c)** *Horizontal and Vertical Components* Find the horizontal and vertical components of the vector with magnitude 569 that is inclined 127.5° from the horizontal.

**(d)** *Magnitude of a Resultant* Find the magnitude of the resultant of forces of 475 pounds and 586 pounds that form an angle of 78.2°.

**5.** For the vectors $\mathbf{u} = \langle -1, 3 \rangle$ and $\mathbf{v} = \langle 2, -6 \rangle$ find the following.

**(a)** $\mathbf{u} + \mathbf{v}$ **(b)** $-3\mathbf{v}$ **(c)** $\mathbf{u} \cdot \mathbf{v}$

**(d)** The angle between $\mathbf{u}$ and $\mathbf{v}$

**6.** Find the following for the complex numbers

$$4 \text{ cis } 240° \quad \text{and} \quad -4 + 4i\sqrt{3}.$$

**(a)** The rectangular form of $4 \text{ cis } 240°$

**(b)** The trigonometric form of $-4 + 4i\sqrt{3}$

**(c)** Their resultant in the form $a + bi$

**7.** Perform each indicated operation. Give the answer in rectangular form.

**(a)** $3(\cos 30° + i \sin 30°) \cdot 5(\cos 90° + i \sin 90°)$

**(b)** $\dfrac{2 \text{ cis } 315°}{4 \text{ cis } 45°}$ **(c)** $\left(1 - i\sqrt{3}\right)^5$

**(d)** Find the fourth roots of $\sqrt{3} + i$. Leave the answers in trigonometric form.

**8.** For the polar equation $r = 4 \cos \theta$, do the following.

**(a)** Graph the equation for $0° \leq \theta \leq 360°$.

**(b)** Find an equivalent equation in rectangular coordinates.

**(c)** Is the graph from part (a) what you would expect for the graph of the equation from part (b)? Explain.

**9.** Find an equivalent equation in polar coordinates for $-x + 2y = 4$ in the form $r = f(\theta)$.

**10.** Graph the parametric equations $x = 2 \cos 2t$, $y = 2 \sin 2t$, for $t$ in $[0, 2\pi]$.

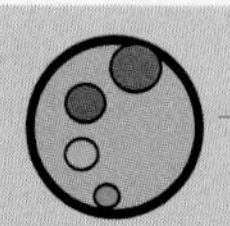

## Chapter 10 Project

### When Is a Circle Really a Polygon?

When we draw a line or circle with pencil and paper, the pencil point is drawing an infinite number of points on a continuous curve. A graphing calculator plots a series of points and connects the points with line segments. This means the polar circle $r = 8$, graphed in degree mode with the default setting for $\theta$-step (7.5°), is not really a circle, but a polygon with vertices plotted every 7.5° around the origin and connected by line segments. Since $\frac{360}{7.5} = 48$, the "circle" is really a polygon with 48 vertices (a 48-gon).

Thus, to draw a 48-gon (with 48 sides) set your calculator in degree and polar modes and set the format menu for polar graphing. Set $\theta\text{min} = 0$, $\theta\text{max} = 360$, $\theta\text{-step} = 7.5$, with $x$ in $[-15, 15]$ and $y$ in $[-10, 10]$. Enter $r_1 = 8$ in the Y-menu. The resulting 48-gon is shown in Figure A.

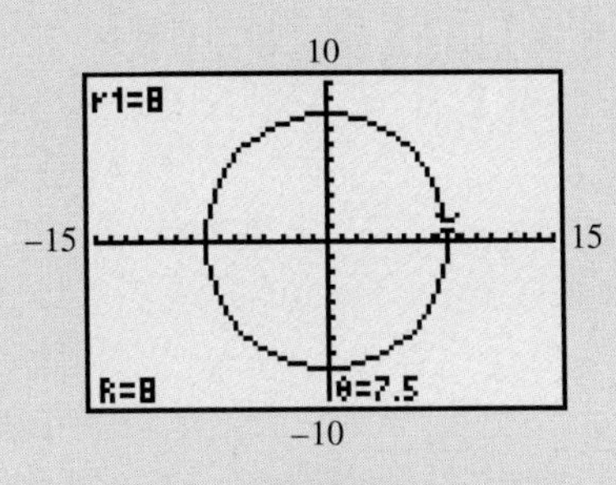


FIGURE A

## Activities

1. A smaller $\theta$-step makes a polygon with more sides, producing a more accurate-looking circle, but taking longer to draw. By changing $\theta$-step appropriately, draw a 360-gon.
2. To construct any regular polygon, only the $\theta$-step needs to be changed. For a triangle, a three-sided polygon, using $\theta\text{-step} = 120$ produces the graph in Figure B.

   Changing only the $\theta$-step, construct a square, regular pentagon, hexagon, septagon, octagon, nonagon, and decagon (figures with 4, 5, 6, 7, 8, 9, and 10 sides, respectively). We show the hexagon in Figure C. Which $\theta$-step works for each polygon? (*Hint:* How did we choose $\theta\text{-step} = 120$ for the triangle? It involves $\theta\text{max}$ as well as the number of sides.)

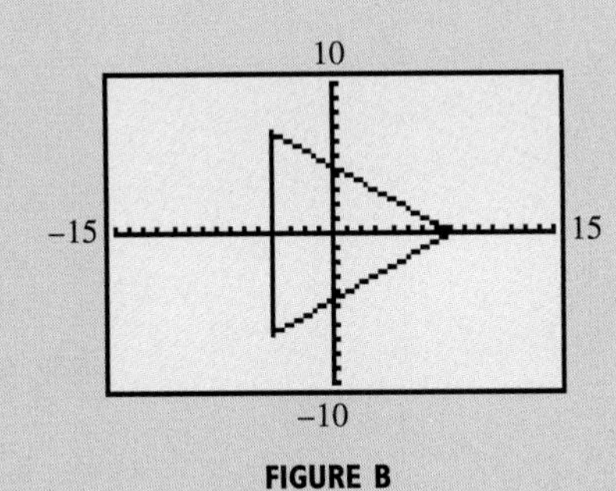


FIGURE B

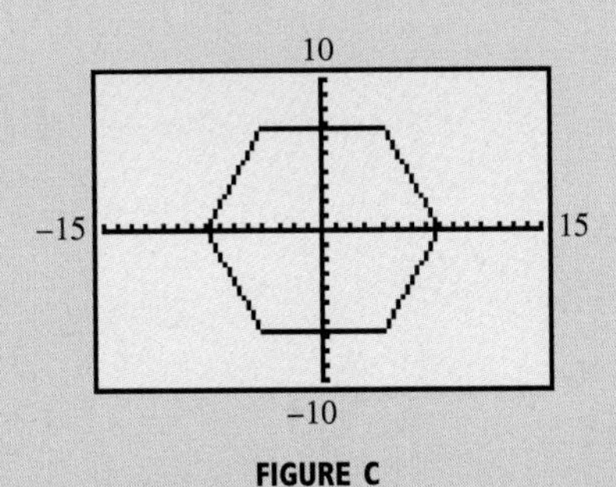


FIGURE C

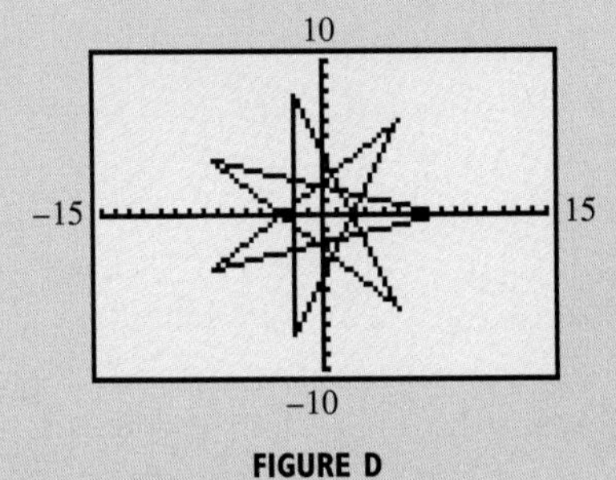


FIGURE D

3. Star polygons can be constructed similarly, changing only $\theta\text{max}$ and $\theta$-step. We show the star polygon with 7 points in Figure D, where we used $\theta\text{-step} = \left(\frac{360}{7}\right) \cdot 3 = 154.29$ and $\theta\text{max} = 360 \cdot 3 = 1080$. Construct star polygons with 5, 9, and 11 points.

   This investigation illustrates the importance of plotting enough points for an accurate representation of a curve. As demonstrated here, with "selected sampling," we can severely distort the expected results.

# 11 Further Topics in Algebra

LARGELY due to the frequent use of antibiotics, many strains of bacteria are becoming resistant to certain drugs. For example, some bacteria are resistant to both the antibiotics ampicillin and tetracycline. Genetic engineers want to predict what will happen to these bacteria after many generations. Will the bacteria remain resistant to both antibiotics indefinitely?

To answer this and other questions related to the spread of disease, further topics in algebra are needed. Bacterial growth can be modeled and simulated, using sequences. The behavior of the bacteria after a long period of time often involves an element of chance, which requires the study of probability. The solution to this genetic problem requires not only the new concepts of sequences and probability, but also previously learned concepts such as matrices. As our mathematical ability has increased during this course, so has the complexity of the applications that we can solve.

*Source:* Hoppensteadt, F. and C. Peskin, *Mathematics in Medicine and the Life Sciences,* Springer-Verlag, 1992.

CHAPTER OUTLINE

# 11.1 SEQUENCES AND SERIES

Sequences • Series and Summation Notation • Summation Properties

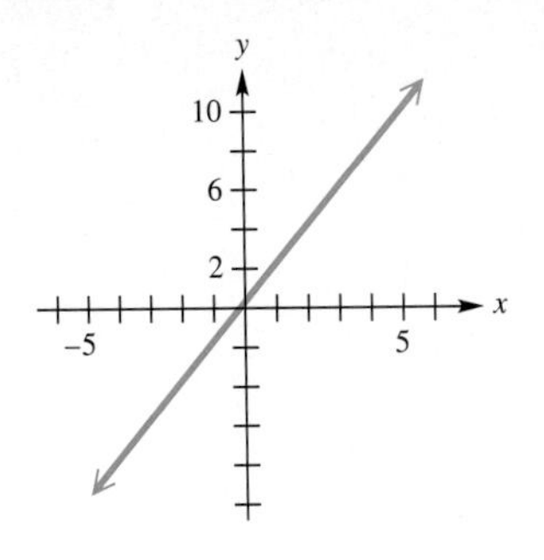


$f(x) = 2x$

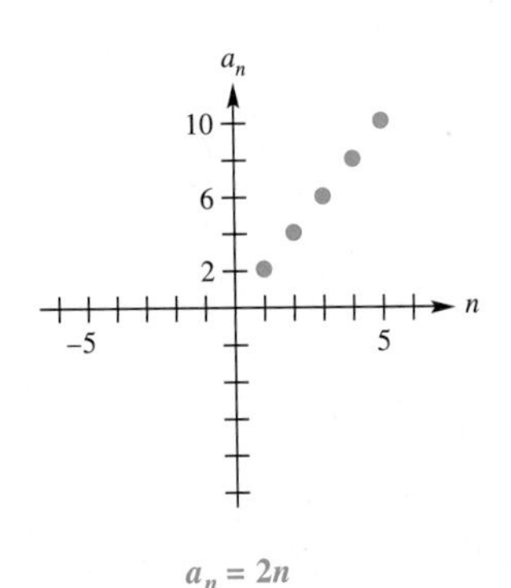


$a_n = 2n$

**FIGURE 1**

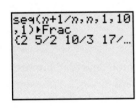


**(a)**

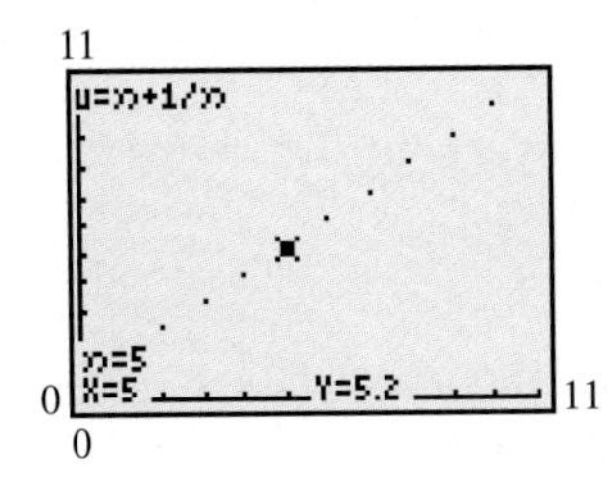


The fifth term is 5.2.

**(b)**

**FIGURE 2**

## Sequences

Sequences are a fundamental concept in mathematics and have many applications. A *sequence* is a function that computes an ordered list. For example, the average person in the United States uses 100 gallons of water each day. The function defined by $f(n) = 100n$ generates the terms of the sequence

$$100, 200, 300, 400, 500, 600, 700, \ldots,$$

when $n = 1, 2, 3, 4, 5, 6, 7, \ldots$. This function represents the gallons of water used by the average person after $n$ days.

A second example of a sequence involves investing money. If \$100 is deposited into a savings account paying 5% interest compounded annually, then the function defined by $g(n) = 100(1.05)^n$ calculates the account balance after $n$ years. The terms of the sequence are

$$g(1), g(2), g(3), g(4), g(5), g(6), g(7), \ldots.$$

and can be approximated as

$$105, 110.25, 115.76, 121.55, 127.63, 134.01, 140.71, \ldots.$$

**Sequence**

A **sequence** is a function that has a set of natural numbers as its domain.

Instead of using $f(x)$ notation to indicate a sequence, it is customary to use $a_n$, where $a_n = f(n)$. The letter $n$ is used instead of $x$ as a reminder that $n$ represents a *natural number*. The elements in the range of a sequence, called the **terms** of the sequence, are $a_1, a_2, a_3, \ldots$. The elements of both the domain and the range of a sequence are *ordered*. The first term is found by letting $n = 1$, the second term is found by letting $n = 2$, and so on. The **general term,** or ***n*th term,** of the sequence is $a_n$.

Figure 1 shows graphs of $f(x) = 2x$ and $a_n = 2n$. Notice that $f(x)$ is a continuous function, and $a_n$ is discontinuous. To graph $a_n$, we plot points of the form $(n, 2n)$ for $n = 1, 2, 3, \ldots$.

A graphing calculator can list the terms in a sequence. Using sequence mode to list the first 10 terms of the sequence with general term $a_n = n + \frac{1}{n}$ produces the result shown in Figure 2(a). Additional terms of the sequence can be seen by scrolling to the right. Sequences can also be graphed in sequence mode. Figure 2(b) shows a calculator screen with the graph of $a_n = n + \frac{1}{n}$. Notice that for $n = 5$, the term is $5 + \frac{1}{5} = 5.2$.

### EXAMPLE 1 *Finding Terms of Sequences*

Write the first five terms for each sequence.

**(a)** $a_n = \dfrac{n + 1}{n + 2}$ **(b)** $a_n = (-1)^n \cdot n$ **(c)** $a_n = \dfrac{(-1)^n}{2^n}$

TECHNOLOGY NOTE

Some graphing calculators have a designated sequence mode (similar to the modes function, polar, and parametric) to investigate and graph sequences defined in terms of *n*, where *n* is a natural number. Consult the graphing calculator manual that accompanies this text or your owner's manual.

**Solution**

**(a)** Replacing $n$, in turn, with 1, 2, 3, 4, and 5 gives

$$n = 1: \quad a_1 = \frac{1+1}{1+2} = \frac{2}{3}$$

$$n = 2: \quad a_2 = \frac{2+1}{2+2} = \frac{3}{4}$$

$$n = 3: \quad a_3 = \frac{3+1}{3+2} = \frac{4}{5}$$

$$n = 4: \quad a_4 = \frac{4+1}{4+2} = \frac{5}{6}$$

$$n = 5: \quad a_5 = \frac{5+1}{5+2} = \frac{6}{7}.$$

**(b)** Replace $n$ with 1, 2, 3, 4, and 5 to obtain

$$n = 1: \quad a_1 = (-1)^1 \cdot 1 = -1$$
$$n = 2: \quad a_2 = (-1)^2 \cdot 2 = 2$$
$$n = 3: \quad a_3 = (-1)^3 \cdot 3 = -3$$
$$n = 4: \quad a_4 = (-1)^4 \cdot 4 = 4$$
$$n = 5: \quad a_5 = (-1)^5 \cdot 5 = -5.$$

**(c)** For $a_n = \frac{(-1)^n}{2^n}$, we have $a_1 = -\frac{1}{2}$, $a_2 = \frac{1}{4}$, $a_3 = -\frac{1}{8}$, $a_4 = \frac{1}{16}$, and $a_5 = -\frac{1}{32}$.

A sequence is a **finite sequence** if the domain is the set $\{1, 2, 3, 4, \ldots, n\}$, where $n$ is a natural number. An **infinite sequence** has the set of *all* natural numbers as its domain. For example, the sequence of natural-number multiples of 2,

$$2, 4, 6, 8, 10, 12, 14, \ldots,$$

is infinite, but the sequence of days in June,

$$1, 2, 3, 4, \ldots, 29, 30,$$

is finite.

If the terms of an infinite sequence get closer and closer to some real number, the sequence is said to be **convergent** and to **converge** to that real number. For example, the sequence defined by $a_n = \frac{1}{n}$ approaches 0 as $n$ becomes large. Thus, $a_n$ is a convergent sequence that converges to 0. Both traditional and calculator graphs of this sequence for $n = 1, 2, 3, \ldots, 10$ are shown in Figure 3. Notice that the terms of $a_n$ approach the $x$-axis, or $y = 0$.

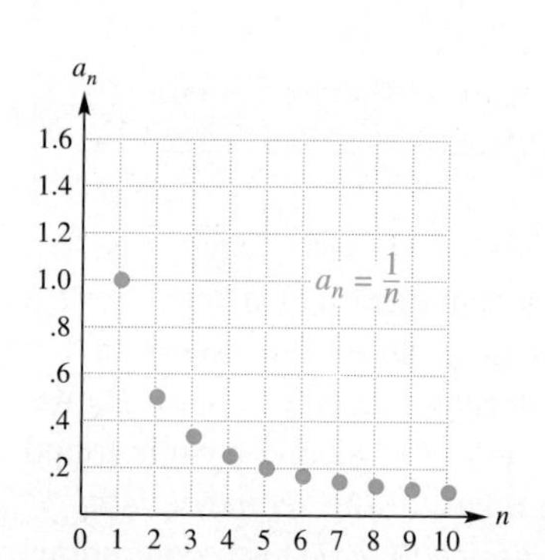


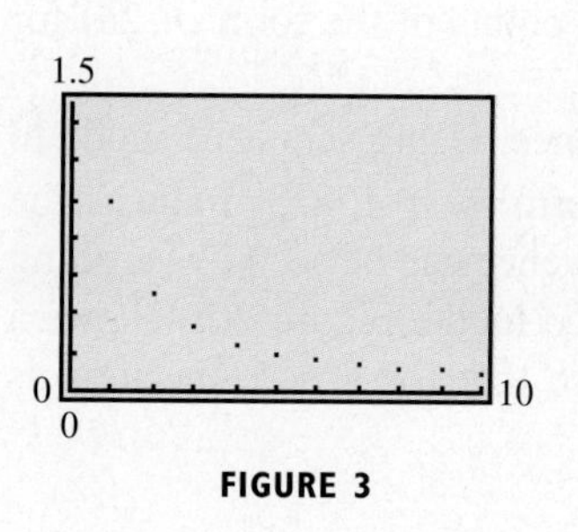


FIGURE 3

A sequence that does not converge to some number is **divergent.** The terms of the sequence $a_n = n^2$ are

$$1, 4, 9, 16, 25, 36, 49, 64, 81, \ldots.$$

This sequence is divergent because as $n$ becomes large, the values for $a_n$ do not approach a fixed number; rather, they increase without bound.

Some sequences are defined by a **recursive definition,** one in which each term is defined as an expression involving the previous term or terms. On the other hand, the sequences in Example 1 were defined *explicitly,* with a formula for $a_n$ that does not depend on a previous term.

### EXAMPLE 2 *Using a Recursion Formula*

Find the first four terms for each sequence.

**(a)** $a_1 = 4$; for $n > 1$, $a_n = 2 \cdot a_{n-1} + 1$

**(b)** $a_1 = 2$; for $n > 1$, $a_n = a_{n-1} + n - 1$

#### Solution

**(a)** This is a recursive definition. We know $a_1 = 4$. Since $a_n = 2 \cdot a_{n-1} + 1$,

$$a_2 = 2 \cdot a_1 + 1 = 2 \cdot 4 + 1 = 9$$
$$a_3 = 2 \cdot a_2 + 1 = 2 \cdot 9 + 1 = 19$$
$$a_4 = 2 \cdot a_3 + 1 = 2 \cdot 19 + 1 = 39.$$

**(b)**

$$a_1 = 2$$
$$a_2 = a_1 + 2 - 1 = 2 + 1 = 3$$
$$a_3 = a_2 + 3 - 1 = 3 + 2 = 5$$
$$a_4 = a_3 + 4 - 1 = 5 + 3 = 8$$

#### FOR DISCUSSION

One of the most famous sequences in mathematics is the **Fibonacci sequence,**

$$1, 1, 2, 3, 5, 8, 13, 21, 34, 55, \ldots,$$

named for the Italian mathematician Leonardo of Pisa (1170–1250), who was also known as Fibonacci. The Fibonacci sequence is found in numerous places in nature. For example, male honeybees hatch from eggs that have not been fertilized, so a male bee has only one parent, a female. On the other hand, female honeybees hatch from fertilized eggs, so a female has two parents, one male and one female. The number of ancestors in consecutive generations of bees follows the Fibonacci sequence. Successive terms in the sequence also appear in plants, such as the daisy head, pineapple, and pine cone.

1. Try to discover the pattern in the Fibonacci sequence.
2. Using the description given above, write a recursive definition that calculates the number of ancestors of a male bee in each generation.

### EXAMPLE 3 *Modeling Insect Population Growth*

Frequently the population of a particular insect does not continue to grow indefinitely. Instead, its population grows rapidly at first, and then levels off because of competition for limited resources. In one study, the behavior of the winter moth was modeled with a sequence similar to the following, where $a_n$ represents the population density in thousands per acre during year $n$. (*Source:* Varley, G. and G. Gradwell, "Population models for the winter moth," Symposium of the Royal Entomological Society of London 4.)

$$a_1 = 1$$
$$a_n = 2.85a_{n-1} - .19a_{n-1}^{\ 2}, \quad \text{for } n \geq 2$$

**(a)** Give a table of values for $n = 1, 2, 3, \ldots, 10$.

**(b)** Graph the sequence. Describe what happens to the population density of the winter moth.

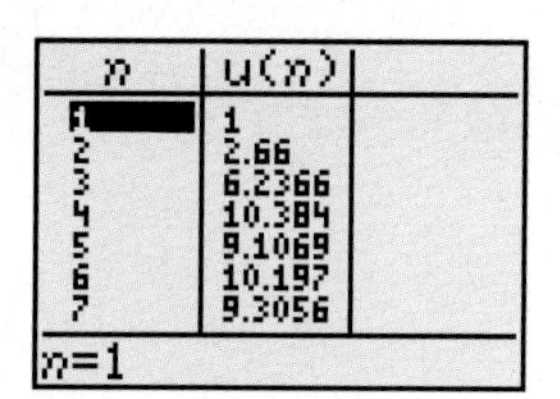


FIGURE 4

**Solution**

**(a)** Evaluate $a_1, a_2, a_3, \ldots, a_{10}$ recursively. Since $a_1 = 1$,

$$a_2 = 2.85a_1 - .19a_1^2 = 2.85(1) - .19(1)^2 = 2.66,$$

and $$a_3 = 2.85a_2 - .19a_2^2 = 2.85(2.66) - .19(2.66)^2 \approx 6.24.$$

Approximate values for other terms are shown in the table. Figure 4 shows the computation of the sequence, denoted by $u(n)$ rather than $a_n$, using a calculator.

| $n$ | 1 | 2 | 3 | 4 | 5 | 6 | 7 | 8 | 9 | 10 |
|---|---|---|---|---|---|---|---|---|---|---|
| $a_n$ | 1 | 2.66 | 6.24 | 10.4 | 9.11 | 10.2 | 9.31 | 10.1 | 9.43 | 9.98 |

**(b)** The graph of a sequence is a set of discrete points. Plot the points

$$(1, 1), (2, 2.66), (3, 6.24), \ldots, (10, 9.98),$$

as shown in Figure 5(a). At first, the insect population increases rapidly, and then oscillates about the line $y = 9.7$. (See the "For Discussion" in the margin.) The oscillations become smaller as $n$ increases, indicating that the population density may stabilize near 9.7 thousand per acre. In Figure 5(b), the first 20 terms have been plotted with a calculator.

**FOR DISCUSSION**

In Example 3, the insect population stabilizes near the value $k = 9.7$ thousand. This value of $k$ can be found by solving the quadratic equation $k = 2.85k - .19k^2$. Explain why this is true.

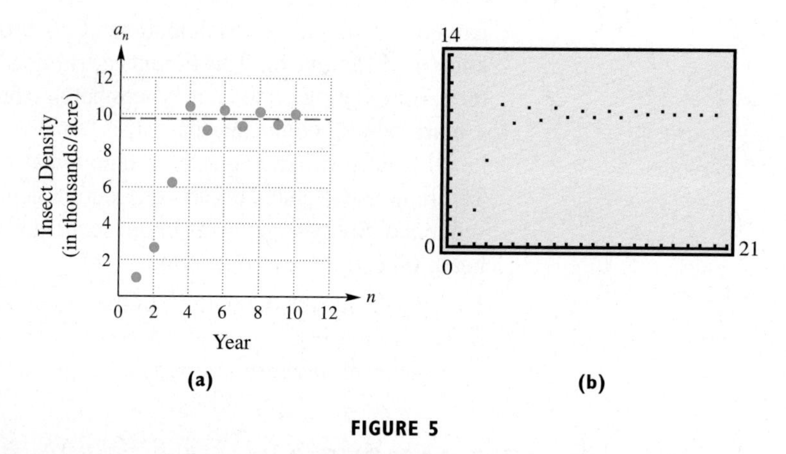


(a) (b)

FIGURE 5

## Series and Summation Notation

Suppose a person has a starting salary of \$30,000 and receives a \$2000 raise each year. Then,

$$30{,}000,\ 32{,}000,\ 34{,}000,\ 36{,}000,\ 38{,}000$$

are terms of the sequence that describe this person's salaries over a 5-year period. The total earned is given by the *finite series*

$$30{,}000 + 32{,}000 + 34{,}000 + 36{,}000 + 38{,}000,$$

whose sum is \$170,000. Any sequence can be used to define a series. For example, the infinite sequence

$$1, \frac{1}{3}, \frac{1}{9}, \frac{1}{27}, \frac{1}{81}, \frac{1}{243}, \ldots$$

**Looking Ahead to Calculus**

The use of sigma notation, $\Sigma$, is introduced in a first calculus course in conjunction with the *definite integral*, symbolized with an elongated S: $\int$. Sigma notation is used in the definition of a definite integral:

$$\int_a^b f(x)\,dx = \lim_{n \to \infty} \sum_{i=1}^{n} f(x_i)\,\Delta x_i.$$

The definite integral can be used to calculate the area between two curves.

defines the terms of the *infinite series*

$$1 + \frac{1}{3} + \frac{1}{9} + \frac{1}{27} + \frac{1}{81} + \frac{1}{243} + \cdots.$$

If a sequence has terms $a_1, a_2, a_3, \ldots,$ then $S_n$ is defined as the sum of the first $n$ terms. That is, $S_n = a_1 + a_2 + a_3 + \cdots + a_n$. The sum of the terms of a sequence is called a **series.** The symbol $\Sigma$, the Greek capital letter *sigma,* is used to indicate a sum.

**Series**

A **finite series** is an expression of the form

$$S_n = a_1 + a_2 + a_3 + \cdots + a_n = \sum_{i=1}^{n} a_i,$$

and an **infinite series** is an expression of the form

$$S_\infty = a_1 + a_2 + a_3 + \cdots + a_n + \cdots = \sum_{i=1}^{\infty} a_i.$$

The letter $i$ is called the **index of summation.**

**CAUTION** Do not confuse this use of $i$ with the use of $i$ to represent an imaginary number. Other letters may be used for the index of summation, such as $k$ and $j$.

**EXAMPLE 4** *Using Summation Notation*

Evaluate the series $\sum_{k=1}^{6} (2^k + 1)$.

**Analytic Solution**

Write each of the six terms, then evaluate the sum.

$$\begin{aligned}\sum_{k=1}^{6} (2^k + 1) &= (2^1 + 1) + (2^2 + 1) + (2^3 + 1)\\ &\quad + (2^4 + 1) + (2^5 + 1) + (2^6 + 1)\\ &= (2 + 1) + (4 + 1) + (8 + 1)\\ &\quad + (16 + 1) + (32 + 1) + (64 + 1)\\ &= 3 + 5 + 9 + 17 + 33 + 65\\ &= 132\end{aligned}$$

**Graphing Calculator Solution**

The screen in Figure 6 agrees with the analytic result. The graphing calculator stores the sequence into a list, $L_1$, and then computes the sum of the six terms in the list.

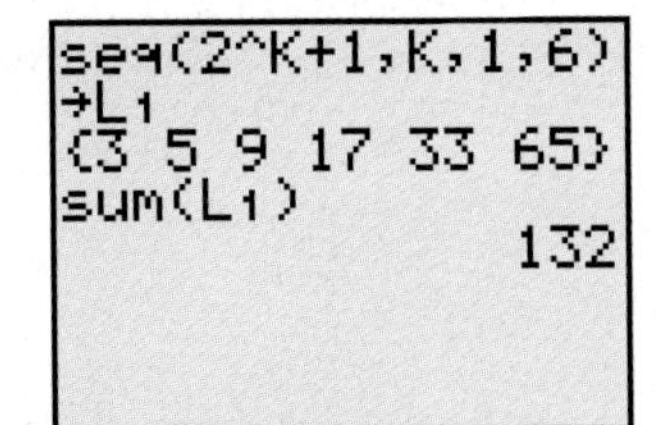


**FIGURE 6**

**EXAMPLE 5** *Using Summation Notation with Subscripts*

Write the terms for each series. Evaluate each sum, if possible.

**(a)** $\sum_{j=3}^{6} a_j$ **(b)** $\sum_{i=1}^{3} (6x_i - 2)$ if $x_1 = 2$, $x_2 = 4$, and $x_3 = 6$

**(c)** $\sum_{i=1}^{4} f(x_i)\,\Delta x$ if $f(x) = x^2$, $x_1 = 0$, $x_2 = 2$, $x_3 = 4$, $x_4 = 6$, and $\Delta x = 2$

**Looking Ahead to Calculus**

Sums like the one in Example 5(c) are frequently used in calculus.

**Solution**

**(a)** $\sum_{j=3}^{6} a_j = a_3 + a_4 + a_5 + a_6$

**(b)** Let $i = 1, 2$, and 3, respectively, to obtain

$$\sum_{i=1}^{3} (6x_i - 2) = (6x_1 - 2) + (6x_2 - 2) + (6x_3 - 2).$$

Now, substitute the given values for $x_1$, $x_2$, and $x_3$.

$$\begin{aligned}\sum_{i=1}^{3} (6x_i - 2) &= (6 \cdot 2 - 2) + (6 \cdot 4 - 2) + (6 \cdot 6 - 2)\\ &= 10 + 22 + 34\\ &= 66\end{aligned}$$

**(c)**

$$\begin{aligned}\sum_{i=1}^{4} f(x_i)\,\Delta x &= f(x_1)\,\Delta x + f(x_2)\,\Delta x + f(x_3)\,\Delta x + f(x_4)\,\Delta x\\ &= x_1^2\,\Delta x + x_2^2\,\Delta x + x_3^2\,\Delta x + x_4^2\,\Delta x\\ &= 0^2(2) + 2^2(2) + 4^2(2) + 6^2(2) \qquad f(x) = x^2;\ \Delta x = 2\\ &= 0 + 8 + 32 + 72\\ &= 112\end{aligned}$$

**EXAMPLE 6** *Estimating $\pi$ with a Series*

The infinite series given by

$$\frac{\pi^4}{90} = \frac{1}{1^4} + \frac{1}{2^4} + \frac{1}{3^4} + \frac{1}{4^4} + \frac{1}{5^4} + \cdots + \frac{1}{n^4} + \cdots$$

can be used to estimate $\pi$.

**(a)** Approximate $\pi$ by finding the sum of the first four terms.

**(b)** Use a calculator to approximate $\pi$ by summing the first 50 terms. Compare the result to the actual value of $\pi$.

**Solution**

**(a)** Summing the first four terms gives

$$\frac{\pi^4}{90} \approx \frac{1}{1^4} + \frac{1}{2^4} + \frac{1}{3^4} + \frac{1}{4^4} \approx 1.078751929.$$

This approximation can be solved for $\pi$ by multiplying by 90 and then taking the fourth root. Thus,

$$\pi \approx \sqrt[4]{90(1.078751929)} \approx 3.139.$$

```
sum(seq(1/n^4,n,
1,50))
        1.082320646
(90*Ans)^(1/4)
        3.141590776
π
        3.141592654
```

FIGURE 7

**(b)** In Figure 7, the first 50 terms of the series provide an approximation of $\pi \approx 3.141590776$. This computation matches the actual value of $\pi$ for the first five decimal places.

## Summation Properties

Properties of summation provide useful shortcuts for evaluating series.

**Summation Properties**

If $a_1, a_2, a_3, \ldots, a_n$ and $b_1, b_2, b_3, \ldots, b_n$ are two sequences, and $c$ is a constant, then for every positive integer $n$,

**(a)** $\sum_{i=1}^{n} c = nc$ **(b)** $\sum_{i=1}^{n} ca_i = c\sum_{i=1}^{n} a_i$

**(c)** $\sum_{i=1}^{n} (a_i + b_i) = \sum_{i=1}^{n} a_i + \sum_{i=1}^{n} b_i$ **(d)** $\sum_{i=1}^{n} (a_i - b_i) = \sum_{i=1}^{n} a_i - \sum_{i=1}^{n} b_i.$

To prove Property (a), expand the series to get

$$c + c + c + c + \cdots + c,$$

where there are $n$ terms of $c$, so the sum is $nc$.

Property (c) also can be proved by first expanding the series:

$$\sum_{i=1}^{n} (a_i + b_i) = (a_1 + b_1) + (a_2 + b_2) + \cdots + (a_n + b_n).$$

Now, use the commutative and associative properties to rearrange the terms.

$$\begin{aligned}\sum_{i=1}^{n} (a_i + b_i) &= (a_1 + a_2 + \cdots + a_n) + (b_1 + b_2 + \cdots + b_n)\\ &= \sum_{i=1}^{n} a_i + \sum_{i=1}^{n} b_i\end{aligned}$$

Proofs of the other two properties are similar.

The following results can be proved by mathematical induction. (See Section 5 of this chapter.)

**Summation Rules**

$$\sum_{i=1}^{n} i = 1 + 2 + \cdots + n = \frac{n(n+1)}{2}$$

$$\sum_{i=1}^{n} i^2 = 1^2 + 2^2 + \cdots + n^2 = \frac{n(n+1)(2n+1)}{6}$$

$$\sum_{i=1}^{n} i^3 = 1^3 + 2^3 + \cdots + n^3 = \frac{n^2(n+1)^2}{4}$$

**EXAMPLE 7** *Using the Summation Properties*

Use the summation properties to find each sum.

**(a)** $\sum_{i=1}^{40} 5$ **(b)** $\sum_{i=1}^{22} 2i$ **(c)** $\sum_{i=1}^{14} (2i^2 - 3)$

**Solution**

**(a)** $\sum_{i=1}^{40} 5 = 40(5) = 200$ Property (a) with $n = 40$ and $c = 5$

**(b)** $\sum_{i=1}^{22} 2i = 2\sum_{i=1}^{22} i$ Property (b) with $c = 2$ and $a_i = i$

$= 2 \cdot \frac{22(22 + 1)}{2}$ Summation rules

$= 506$ Simplify.

**(c)** $\sum_{i=1}^{14} (2i^2 - 3) = \sum_{i=1}^{14} 2i^2 - \sum_{i=1}^{14} 3$ Property (d) with $a_i = 2i^2$ and $b_i = 3$

$= 2\sum_{i=1}^{14} i^2 - \sum_{i=1}^{14} 3$ Property (b) with $c = 2$ and $a_i = i^2$

$= 2 \cdot \frac{14(14 + 1)(2 \cdot 14 + 1)}{6} - 14(3)$ Summation rules; Property (a)

$= 1988$ Simplify.

**TECHNOLOGY NOTE**

In Figures 6 and 8, the calculator was set in function mode in order to use the variables $K$ and $I$ instead of $n$, as in Figure 7.

**EXAMPLE 8** *Using the Summation Properties*

Evaluate $\sum_{i=1}^{6} (i^2 + 3i + 5)$.

**Analytic Solution**

$\sum_{i=1}^{6} (i^2 + 3i + 5) = \sum_{i=1}^{6} i^2 + \sum_{i=1}^{6} 3i + \sum_{i=1}^{6} 5$ Property (c)

$= \sum_{i=1}^{6} i^2 + 3\sum_{i=1}^{6} i + \sum_{i=1}^{6} 5$ Property (b)

$= \sum_{i=1}^{6} i^2 + 3\sum_{i=1}^{6} i + 6(5)$ Property (a)

Using the summation rules, we obtain

$$= \frac{6(6 + 1)(2 \cdot 6 + 1)}{6} + 3\left[\frac{6(6 + 1)}{2}\right] + 6(5)$$

$$= 91 + 3(21) + 6(5) = 184.$$

**Graphing Calculator Solution**

Figure 8 confirms the analytic result.

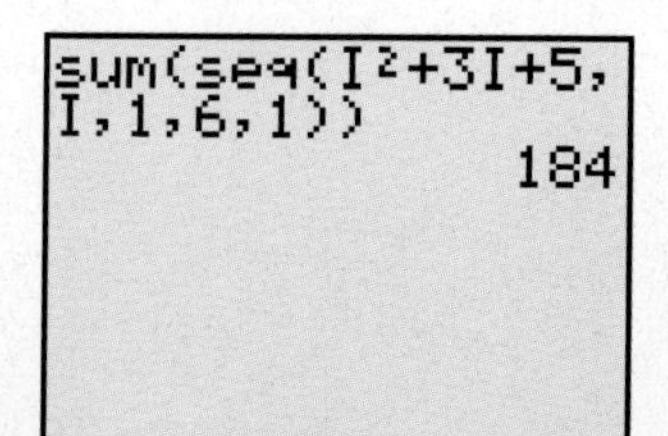


**FIGURE 8**

## 11.1 EXERCISES

*Write the first five terms of each sequence.*

**1.** $a_n = 4n + 10$

**2.** $a_n = 6n - 3$

**3.** $a_n = 2^{n-1}$

**4.** $a_n = -3^n$

**5.** $a_n = \left(\frac{1}{3}\right)^n(n - 1)$

**6.** $a_n = (-2)^n(n)$

**7.** $a_n = (-1)^n(2n)$

**8.** $a_n = (-1)^{n-1}(n + 1)$

**9.** $a_n = \frac{4n - 1}{n^2 + 2}$

**10.** $a_n = \frac{n^2 - 1}{n^2 + 1}$

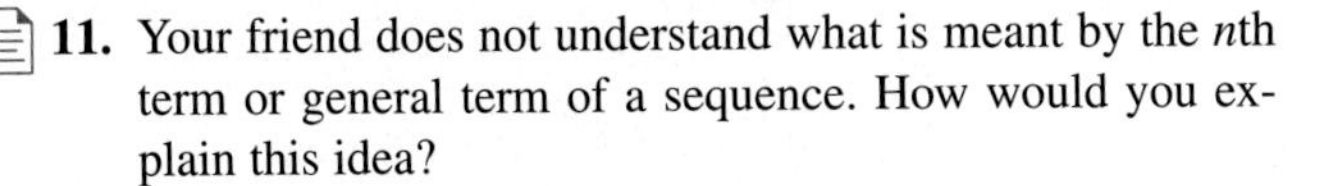

**11.** Your friend does not understand what is meant by the $n$th term or general term of a sequence. How would you explain this idea?

**12.** How are sequences related to functions?

*Decide whether each sequence is finite or infinite.*

**13.** The sequence of days of the week

**14.** The sequence of dates in the month of November

**15.** 1, 2, 3, 4

**16.** −1, −2, −3, −4

**17.** 1, 2, 3, 4, ...

**18.** −1, −2, −3, −4, ...

**19.** $a_1 = 3$; for $2 \leq n \leq 10$, $a_n = 3 \cdot a_{n-1}$

**20.** $a_1 = 1$; $a_2 = 3$; for $n \geq 3$, $a_n = a_{n-1} + a_{n-2}$

*Find the first four terms of each sequence.*

**21.** $a_1 = -2$, $a_n = a_{n-1} + 3$, for $n > 1$

**22.** $a_1 = -1$, $a_n = a_{n-1} - 4$, for $n > 1$

**23.** $a_1 = 1$, $a_2 = 1$, $a_n = a_{n-1} + a_{n-2}$, for $n \geq 3$ (the Fibonacci sequence)

**24.** $a_1 = 2$, $a_n = n \cdot a_{n-1}$, for $n > 1$

*Evaluate each series.*

**25.** $\sum_{i=1}^{5} (2i + 1)$

**26.** $\sum_{i=1}^{6} (3i - 2)$

**27.** $\sum_{j=1}^{4} \frac{1}{j}$

**28.** $\sum_{i=1}^{5} (i + 1)^{-1}$

**29.** $\sum_{i=1}^{4} i^i$

**30.** $\sum_{k=1}^{4} (k + 1)^2$

**31.** $\sum_{k=1}^{6} (-1)^k \cdot k$

**32.** $\sum_{i=1}^{7} (-1)^{i+1} \cdot i^2$

**33.** $\sum_{i=2}^{5} (6 - 3i)$

**34.** $\sum_{i=3}^{7} (5i + 2)$

**35.** $\sum_{i=-2}^{3} 2(3)^i$

**36.** $\sum_{i=-1}^{2} 5(2)^i$

**37.** $\sum_{i=-1}^{5} (i^2 - 2i)$

**38.** $\sum_{i=3}^{6} (2i^2 + 1)$

*Evaluate the terms for each sum, where $x_1 = -2$, $x_2 = -1$, $x_3 = 0$, $x_4 = 1$, and $x_5 = 2$.*

**39.** $\sum_{i=1}^{5} (2x_i + 3)$

**40.** $\sum_{i=1}^{4} x_i^2$

**41.** $\sum_{i=1}^{3} (3x_i - x_i^2)$

**42.** $\sum_{i=1}^{3} (x_i^2 + 1)$

**43.** $\sum_{i=2}^{5} \frac{x_i + 1}{x_i + 2}$

**44.** $\sum_{i=1}^{5} \frac{x_i}{x_i + 3}$

*Evaluate the terms of $\sum_{i=1}^{4} f(x_i)\,\Delta x$, with $x_1 = 0$, $x_2 = 2$, $x_3 = 4$, $x_4 = 6$, and $\Delta x = .5$, for each function.*

**45.** $f(x) = 4x - 7$

**46.** $f(x) = 6 + 2x$

**47.** $f(x) = 2x^2$

**48.** $f(x) = x^2 - 1$

**49.** $f(x) = \frac{-2}{x + 1}$

**50.** $f(x) = \frac{5}{2x - 1}$

*Use the summation properties and rules to evaluate each series.*

**51.** $\sum_{i=1}^{100} 6$ **52.** $\sum_{i=1}^{20} 5i$ **53.** $\sum_{i=1}^{15} i^2$ **54.** $\sum_{i=1}^{50} 2i^3$ **55.** $\sum_{i=1}^{5} (5i + 3)$

**56.** $\sum_{i=1}^{5} (8i - 1)$ **57.** $\sum_{i=1}^{5} (4i^2 - 2i + 6)$ **58.** $\sum_{i=1}^{6} (2 + i - i^2)$ **59.** $\sum_{i=1}^{4} (3i^3 + 2i - 4)$ **60.** $\sum_{i=1}^{6} (i^2 + 2i^3)$

*Use a graphing calculator to graph the first ten terms of each sequence. Make a conjecture as to whether the sequence converges or diverges. If you think it converges, determine the number to which it converges.*

**61.** $a_n = \dfrac{n + 4}{2n}$ **62.** $a_n = \dfrac{1 + 4n}{2n}$ **63.** $a_n = 2e^n$

**64.** $a_n = n(n + 2)$ **65.** $a_n = \left(1 + \dfrac{1}{n}\right)^n$ **66.** $a_n = 5 - \dfrac{1}{n}$

*Solve each problem involving sequences and series.*

**67.** ***Estimating*** $\pi$ Find the sum of the first six terms of the series

$$\frac{\pi^4}{90} = \frac{1}{1^4} + \frac{1}{2^4} + \frac{1}{3^4} + \frac{1}{4^4} + \frac{1}{5^4} + \cdots + \frac{1}{n^4} + \cdots$$

presented in Example 6. Use your result to estimate $\pi$. Compare your answer to the actual value of $\pi$.

**68.** ***(Modeling) Insect Population*** Suppose an insect population density in thousands per acre during year $n$ can be modeled by the recursively defined sequence

$$a_1 = 8$$
$$a_n = 2.9a_{n-1} - .2a_{n-1}^2, \quad \text{for } n > 1.$$

**(a)** Find the population for $n = 1, 2, 3$.

**(b)** Graph the sequence for $n = 1, 2, 3, \ldots, 20$. Use the window $[0, 21]$ by $[0, 14]$. Interpret the graph.

**69.** ***(Modeling) Bacteria Growth*** If certain bacteria are cultured in a medium with sufficient nutrients, they will double in size and then divide every 40 minutes. Let $N_1$ be the initial number of bacteria cells, $N_2$ the number after 40 minutes, $N_3$ the number after 80 minutes, and $N_j$ the number after $40(j - 1)$ minutes. (*Source:* Hoppensteadt, F. and C. Peskin, *Mathematics in Medicine and the Life Sciences*, Springer-Verlag, 1992.)

**(a)** Write $N_{j+1}$ in terms of $N_j$ for $j \geq 1$.

**(b)** Determine the number of bacteria after two hours if $N_1 = 230$.

**(c)** Graph the sequence $N_j$ for $j = 1, 2, 3, \ldots, 7$. Use the window $[0, 10]$ by $[0, 15{,}000]$.

**(d)** Describe the growth of these bacteria when there are unlimited nutrients.

**70.** ***(Modeling) Verhulst's Model for Bacteria Growth*** Refer to Exercise 69. If the bacteria are not cultured in a medium with sufficient nutrients, competition will ensue and the growth will slow. According to Verhulst's model, the number of bacteria $N_j$ at time $40(j - 1)$ in minutes can be determined by the sequence

$$N_{j+1} = \left[\frac{2}{1 + (N_j/K)}\right]N_j,$$

where $K$ is a constant and $j \geq 1$. (*Source:* Hoppensteadt, F. and C. Peskin, *Mathematics in Medicine and the Life Sciences*, Springer-Verlag, 1992.)

**(a)** If $N_1 = 230$ and $K = 5000$, make a table of $N_j$ for $j = 1, 2, 3, \ldots, 20$. Round values in the table to the nearest integer.

**(b)** Graph the sequence $N_j$ for $j = 1, 2, 3, \ldots, 20$. Use the window $[0, 20]$ by $[0, 6000]$.

**(c)** Describe the growth of these bacteria when there are limited nutrients.

**(d)** Make a conjecture as to why $K$ is called the *saturation constant*. Test your conjecture by changing the value of $K$ in the given formula.

*The series*

$$e^a \approx 1 + a + \frac{a^2}{2!} + \frac{a^3}{3!} + \cdots + \frac{a^n}{n!},$$

*where* $n! = 1 \cdot 2 \cdot 3 \cdot 4 \cdot \cdots \cdot n$, *can be used to estimate the value of* $e^a$ *for any real number* $a$. *Use the first eight terms of this series to approximate each expression. Compare this estimate with the actual value.*

**71.** $e$ **72.** $e^{-1}$

*The recursively defined sequence*

$$a_1 = k$$
$$a_n = \frac{1}{2}\left(a_{n-1} + \frac{k}{a_{n-1}}\right), \quad \text{for } n > 1$$

*can be used to compute $\sqrt{k}$ for any positive number k. This sequence was known to Sumerian mathematicians 4000 years ago, but it is still used today. Use this sequence to approximate the given square root by finding $a_6$. Compare your result with the actual value. (Source:* Heinz-Otto, P., *Chaos and Fractals,* Springer-Verlag, 1993.*)*

**73.** $\sqrt{2}$ **74.** $\sqrt{11}$

## 11.2 ARITHMETIC SEQUENCES AND SERIES

Arithmetic Sequences • Arithmetic Series

### Arithmetic Sequences

A sequence in which each term after the first is obtained by adding a fixed number to the previous term is an **arithmetic sequence** (or **arithmetic progression**). The fixed number that is added is the **common difference.** The sequence

$$5,\ 9,\ 13,\ 17,\ 21, \ldots$$

is an arithmetic sequence because each term after the first is obtained by adding 4 to the previous term. That is,

$$\begin{aligned} 9 &= 5 + 4 \\ 13 &= 9 + 4 \\ 17 &= 13 + 4 \\ 21 &= 17 + 4, \end{aligned}$$

and so on. The common difference is 4.

If the common difference of an arithmetic sequence is $d$, then by the definition of an arithmetic sequence,

$$d = a_{n+1} - a_n$$

for every positive integer $n$ in its domain.

**EXAMPLE 1** *Finding the Common Difference*

Find the common difference $d$ for the arithmetic sequence $-9, -7, -5, -3, -1, \ldots$.

**Solution** Since this sequence is arithmetic, $d$ can be found by choosing any two consecutive terms and subtracting the first from the second. Choosing $-7$ and $-5$ gives $d = -5 - (-7) = 2$. Choosing $-9$ and $-7$ gives $d = -7 - (-9) = 2$, the same result.

If $a_1$ and $d$ are known, then all the terms of an arithmetic sequence can be found.

**EXAMPLE 2** *Finding Terms Given $a_1$ and d*

Write the first five terms for each arithmetic sequence.

**(a)** The first term is 7, and the common difference is $-3$.

**(b)** $a_1 = -12, d = 5$

**Solution**

**(a)** Here, $a_1 = 7$ and $d = -3$, so

$$a_2 = 7 + (-3) = 4,$$
$$a_3 = 4 + (-3) = 1,$$
$$a_4 = 1 + (-3) = -2,$$
$$a_5 = -2 + (-3) = -5.$$

**(b)** Starting with $a_1$, add $d$ to each term to get the next term.

$$a_1 = -12$$
$$a_2 = -12 + 5 = -7$$
$$a_3 = -7 + 5 = -2$$
$$a_4 = -2 + 5 = 3$$
$$a_5 = 3 + 5 = 8$$

If $a_1$ is the first term of an arithmetic sequence and $d$ is the common difference, then the terms of the sequence are given by

$$a_1 = a_1$$
$$a_2 = a_1 + d$$
$$a_3 = a_2 + d = a_1 + d + d = a_1 + 2d$$
$$a_4 = a_3 + d = a_1 + 2d + d = a_1 + 3d$$
$$a_5 = a_4 + d = a_1 + 3d + d = a_1 + 4d$$
$$a_6 = a_5 + d = a_1 + 4d + d = a_1 + 5d,$$

and, in general,

$$a_n = a_1 + (n - 1)d.$$

This result can be proved by mathematical induction. (See Section 5 of this chapter.)

**$n$th Term of an Arithmetic Sequence**

In an arithmetic sequence with first term $a_1$ and common difference $d$, the $n$th term, $a_n$, is given by

$$a_n = a_1 + (n - 1)d.$$

**EXAMPLE 3** *Finding Terms of an Arithmetic Sequence*

Find $a_{13}$ and $a_n$ for the arithmetic sequence $-3, 1, 5, 9, \ldots$.

**Solution** Here, $a_1 = -3$ and $d = 1 - (-3) = 4$. To find $a_{13}$, substitute 13 for $n$ in the formula for the $n$th term.

$$a_n = a_1 + (n - 1)d$$
$$a_{13} = a_1 + (13 - 1)d$$
$$a_{13} = -3 + (12)4$$
$$a_{13} = 45$$

Find $a_n$ by substituting values for $a_1$ and $d$ in the formula for $a_n$.

$$a_n = -3 + (n - 1) \cdot 4$$
$$a_n = -3 + 4n - 4 \qquad \text{Distributive property}$$
$$a_n = 4n - 7$$

**EXAMPLE 4** *Finding Terms of an Arithmetic Sequence*

Find $a_{18}$ and $a_n$ for the arithmetic sequence having $a_2 = 9$ and $a_3 = 15$.

**Solution** Find $d$ first; $d = a_3 - a_2 = 15 - 9 = 6$.

Since
$$a_2 = a_1 + d,$$
$$9 = a_1 + 6 \qquad a_2 = 9, d = 6$$
$$a_1 = 3.$$

Then,
$$a_{18} = 3 + (18 - 1)6 \qquad \text{Formula for } a_n;\ n = 18$$
$$a_{18} = 105,$$

and
$$a_n = 3 + (n - 1)6$$
$$a_n = 3 + 6n - 6 \qquad \text{Distributive property}$$
$$a_n = 6n - 3.$$

**EXAMPLE 5** *Finding the First Term of an Arithmetic Sequence*

Suppose that an arithmetic sequence has $a_8 = -16$ and $a_{16} = -40$. Find $a_1$.

**Solution** Since $a_{16} = a_8 + 8d$, it follows that

$$8d = a_{16} - a_8 = -40 - (-16) = -24,$$

and so $d = -3$. To find $a_1$, use the equation $a_8 = a_1 + 7d$.

$$-16 = a_1 + 7d \qquad a_8 = -16$$
$$-16 = a_1 + 7(-3) \qquad d = -3$$
$$a_1 = 5$$

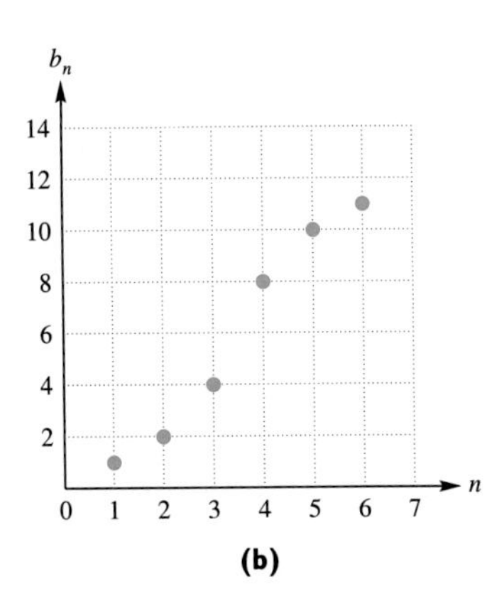


FIGURE 9

The graph of any sequence is a scatter diagram. To determine the characteristics of the graph of an arithmetic sequence, start by rewriting the formula for the $n$th term.

$$a_n = a_1 + (n - 1)d \qquad \text{Formula for } n\text{th term}$$
$$= a_1 + nd - d \qquad \text{Distributive property}$$
$$= dn + (a_1 - d) \qquad \text{Commutative and associative properties}$$
$$= dn + c \qquad c = a_1 - d$$

The points in the graph of an arithmetic sequence are determined by $f(n) = dn + c$, where $n$ is a natural number. Thus, the points in the graph of $f$ must lie on the *line* $y = dx + c$, where $d$ is the slope of the line and $c$ is its $y$-intercept. For example, the sequence $a_n$ shown in Figure 9(a) is an arithmetic sequence because the points that comprise its graph are collinear (lie on a line). The slope determined by these points is 2, so the common difference $d$ equals 2. On the other hand, the sequence $b_n$ shown in Figure 9(b) is not an arithmetic sequence because the points are not collinear.

**EXAMPLE 6** *Find the nth Term from a Graph*

Find a formula for the $n$th term of the sequence $a_n$ shown in Figure 10. What are the domain and range of this sequence?

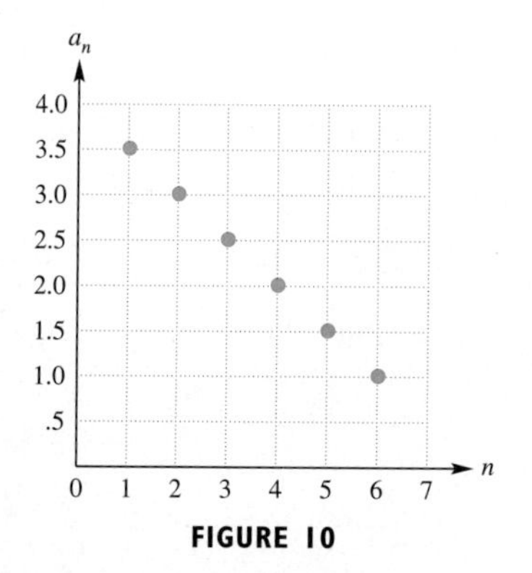

FIGURE 10

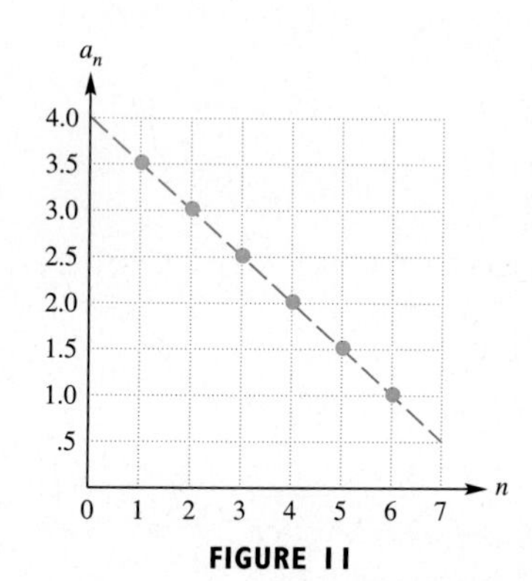

FIGURE 11

**Solution** The points in Figure 10 lie on a line, so the sequence is arithmetic. The equation of the dashed line shown in Figure 11 is $y = -.5x + 4$, so the $n$th term of this sequence is determined by $a_n = -.5n + 4$. The sequence is composed of the points

$$(1, 3.5), (2, 3), (3, 2.5), (4, 2), (5, 1.5), (6, 1).$$

Thus, the domain of the sequence is given by $\{1, 2, 3, 4, 5, 6\}$, and the range is given by $\{1, 1.5, 2, 2.5, 3, 3.5\}$.

## Arithmetic Series

The sum of the terms of an arithmetic sequence is an **arithmetic series.** To illustrate, suppose that a person borrows \$3000 and agrees to pay \$100 per month plus interest of 1% per month on the unpaid balance until the loan is paid off. The first month, \$100 is paid to reduce the loan, plus interest of $(.01)3000 = 30$ dollars. The second month, another \$100 is paid toward the loan, and $(.01)2900 = 29$ dollars is paid for interest. Since the loan is reduced by \$100 each month, interest payments decrease by $(.01)100 = 1$ dollar each month, forming the arithmetic sequence

$$30, 29, 28, \ldots, 3, 2, 1.$$

The total amount of interest paid is given by the sum of the terms of this sequence. Now we develop a formula to find this sum without adding all 30 numbers directly. Since the sequence is arithmetic, we can write the sum of the first $n$ terms as

$$S_n = a_1 + [a_1 + d] + [a_1 + 2d] + \cdots + [a_1 + (n - 1)d].$$

We used the formula for the general term in the last expression. Now we write the same sum in reverse order, beginning with $a_n$ and *subtracting d.*

$$S_n = a_n + [a_n - d] + [a_n - 2d] + \cdots + [a_n - (n - 1)d]$$

Adding respective sides of these two equations term by term, we obtain

$$S_n + S_n = (a_1 + a_n) + (a_1 + a_n) + \cdots + (a_1 + a_n)$$

or

$$2S_n = n(a_1 + a_n),$$

since there are $n$ terms of $a_1 + a_n$ on the right. Now, we solve for $S_n$ to get

$$S_n = \frac{n}{2}(a_1 + a_n).$$

Using the formula $a_n = a_1 + (n - 1)d$, we can also write this result for $S_n$ as

$$S_n = \frac{n}{2}[a_1 + a_1 + (n - 1)d]$$

or

$$S_n = \frac{n}{2}[2a_1 + (n - 1)d],$$

which is an alternative formula for the sum of the first $n$ terms of an arithmetic sequence.

**FOR DISCUSSION**

Explain why there is no formula for the sum of the terms of an infinite arithmetic sequence.

**Sum of the First $n$ Terms of an Arithmetic Sequence**

If an arithmetic sequence has first term $a_1$ and common difference $d$, then the sum of the first $n$ terms is given by

$$S_n = \frac{n}{2}(a_1 + a_n) \quad \text{or} \quad S_n = \frac{n}{2}[2a_1 + (n - 1)d].$$

The first formula is used when the first and last terms are known; otherwise, the second formula is used.

In the sequence of interest payments, $n = 30$, $a_1 = 30$, and $a_n = 1$. Choosing the first formula,

$$S_n = \frac{n}{2}(a_1 + a_n),$$

gives

$$S_{30} = \frac{30}{2}(30 + 1) = 15(31) = 465,$$

so a total of \$465 interest will be paid over the 30 months.

**EXAMPLE 7** *Using the Sum Formulas*

**(a)** Evaluate $S_{12}$ for the arithmetic sequence $-9, -5, -1, 3, 7, \ldots$.

**(b)** Use a formula for $S_n$ to evaluate the sum of the first 60 positive integers.

**Solution**

**(a)** We want the sum of the first 12 terms. Using $a_1 = -9$, $n = 12$, and $d = 4$ in the second formula,

$$S_n = \frac{n}{2}[2a_1 + (n - 1)d],$$

gives

$$S_{12} = \frac{12}{2}[2(-9) + 11(4)] = 156.$$

**(b)** In this example, $n = 60$, $a_1 = 1$, and $a_{60} = 60$, so use the first formula.

$$S_n = \frac{n}{2}(a_1 + a_n)$$

$$S_{60} = \frac{60}{2}(1 + 60) = 1830$$

### EXAMPLE 8 *Using the Sum Formulas*

The sum of the first 17 terms of an arithmetic sequence is 187. If $a_{17} = -13$, find $a_1$ and $d$.

**Solution** Use the first formula for $S_n$, with $n = 17$, to find $a_1$.

$$S_{17} = \frac{17}{2}(a_1 + a_{17}) \qquad n = 17$$

$$187 = \frac{17}{2}(a_1 - 13) \qquad S_{17} = 187,\ a_{17} = -13$$

$$22 = a_1 - 13 \qquad \text{Multiply by } \tfrac{2}{17}.$$

$$a_1 = 35$$

Since $a_{17} = a_1 + (17 - 1)d$,

$$-13 = 35 + 16d \qquad a_{17} = -13,\ a_1 = 35$$

$$-48 = 16d$$

$$d = -3.$$

Any sum of the form

$$\sum_{i=1}^{n} (di + p),$$

where $d$ and $p$ are real numbers, represents the sum of the terms of an arithmetic sequence having first term $a_1 = d(1) + p = d + p$ and common difference $d$. These sums can be evaluated using the formulas in this section.

### EXAMPLE 9 *Using Summation Notation*

Evaluate each sum.

**(a)** $\sum_{i=1}^{10} (4i + 8)$ **(b)** $\sum_{k=3}^{9} (4 - 3k)$

**Analytic Solution**

**(a)** This sum contains the first 10 terms of the arithmetic sequence having

$$a_1 = 4 \cdot 1 + 8 = 12, \qquad \text{First term}$$

and $$a_{10} = 4 \cdot 10 + 8 = 48. \qquad \text{Last term}$$

Thus, $$\sum_{i=1}^{10} (4i + 8) = S_{10} = \frac{10}{2}(12 + 48) = 300.$$

**Graphing Calculator Solution**

The screen in Figure 12 shows the sums for the series in parts (a) and (b). These results agree with the analytic solutions. Remember that a series is the sum of the terms of a sequence.

*(continued)*

**(b)** The terms of the sum are

$$[4 - 3(3)] + [4 - 3(4)] + [4 - 3(5)] + \cdots + [4 - 3(9)]$$
$$= -5 + (-8) + (-11) + \cdots + (-23).$$

If the sequence started with $k = 1$, there would be nine terms. Since it starts at 3, two of those terms are missing, so there are seven terms with $a_1 = -5$ and $a_7 = -23$.

$$\sum_{k=3}^{9} (4 - 3k) = \frac{7}{2}[(-5) + (-23)] = -98$$

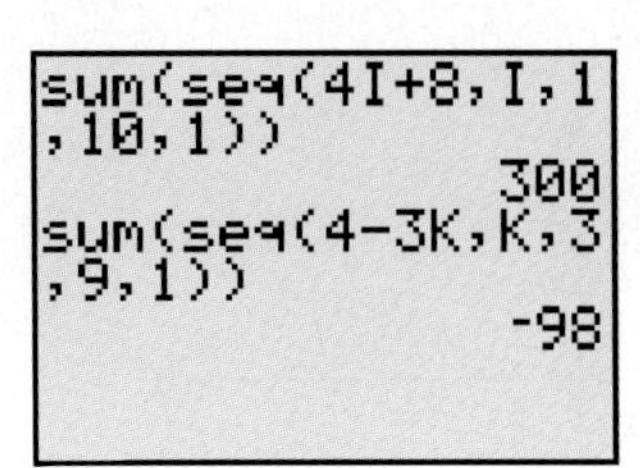


**FIGURE 12**

## 11.2 EXERCISES

*Find the common difference d for each arithmetic sequence.*

**1.** $2, 5, 8, 11, \ldots$

**2.** $4, 10, 16, 22, \ldots$

**3.** $3, -2, -7, -12, \ldots$

**4.** $-8, -12, -16, -20, \ldots$

**5.** $x + 3y, 2x + 5y, 3x + 7y, \ldots$

**6.** $t^2 + q, -4t^2 + 2q, -9t^2 + 3q, \ldots$

*Write the first five terms for each arithmetic sequence.*

**7.** The first term is 8, and the common difference is 6.

**8.** The first term is $-2$, and the common difference is 12.

**9.** $a_1 = 5, d = -2$

**10.** $a_1 = 4, d = 3$

**11.** $a_3 = 10, d = -2$

**12.** $a_1 = 3 - \sqrt{2},\ a_2 = 3$

*Find $a_8$ and $a_n$ for each arithmetic sequence.*

**13.** $a_1 = 5, d = 2$

**14.** $a_1 = -3, d = -4$

**15.** $a_3 = 2, d = 1$

**16.** $a_4 = 5, d = -2$

**17.** $a_1 = 8, a_2 = 6$

**18.** $a_1 = 6, a_2 = 3$

**19.** $a_{10} = 6, a_{12} = 15$

**20.** $a_{15} = 8, a_{17} = 2$

**21.** $a_1 = x, a_2 = x + 3$

**22.** $a_2 = y + 1, d = -3$

*Find $a_1$ for each arithmetic sequence.*

**23.** $a_5 = 27, a_{15} = 87$

**24.** $a_{12} = 60, a_{20} = 84$

**25.** $S_{16} = -160, a_{16} = -25$

**26.** $S_{28} = 2926, a_{28} = 199$

*Find the sum of the first 10 terms for each arithmetic sequence.*

**27.** $a_1 = 8, d = 3$

**28.** $a_1 = -9, d = 4$

**29.** $a_3 = 5, a_4 = 8$

**30.** $a_2 = 9, a_4 = 13$

**31.** $5, 9, 13, \ldots$

**32.** $8, 6, 4, \ldots$

**33.** $a_1 = 10, a_{10} = 5.5$

**34.** $a_1 = -8, a_{10} = -1.25$

*Find $a_1$ and d for each arithmetic sequence.*

**35.** $S_{20} = 1090, a_{20} = 102$

**36.** $S_{31} = 5580, a_{31} = 360$

**37.** $S_{12} = -108, a_{12} = -19$

**38.** $S_{25} = 650, a_{25} = 62$

*Find a formula for the nth term of the arithmetic sequence shown in each graph. Then state the domain and range of the sequence.*

**39.**

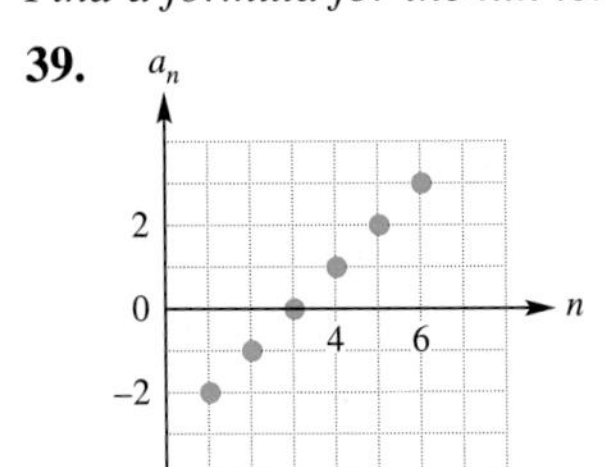


**40.**

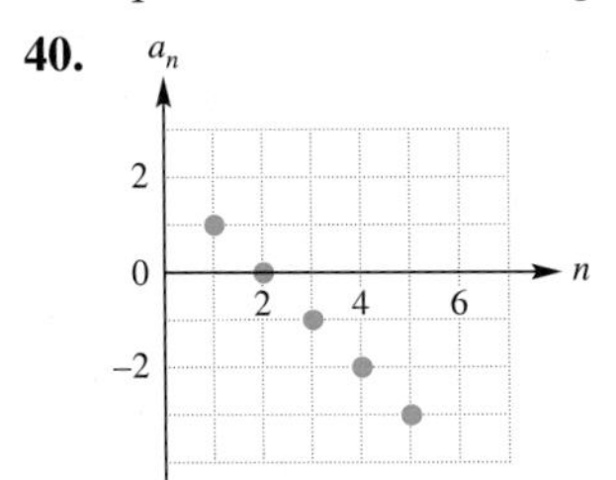


**41.**

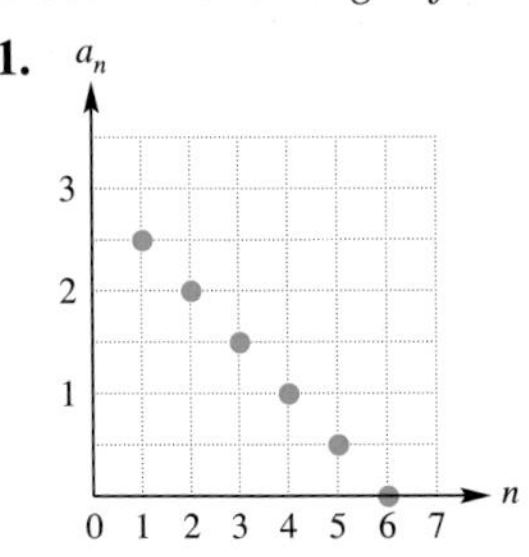

**42.**

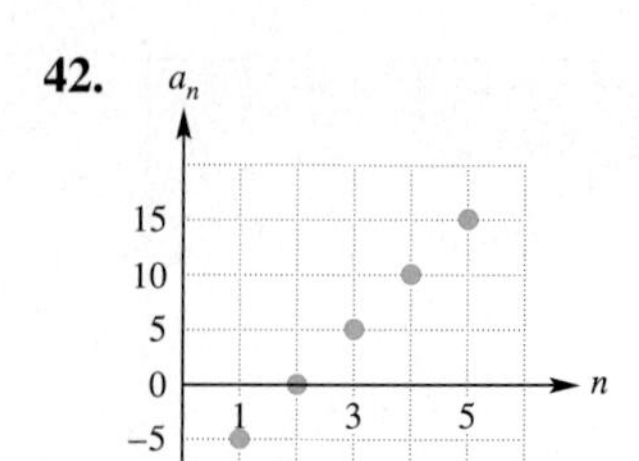


**43.**

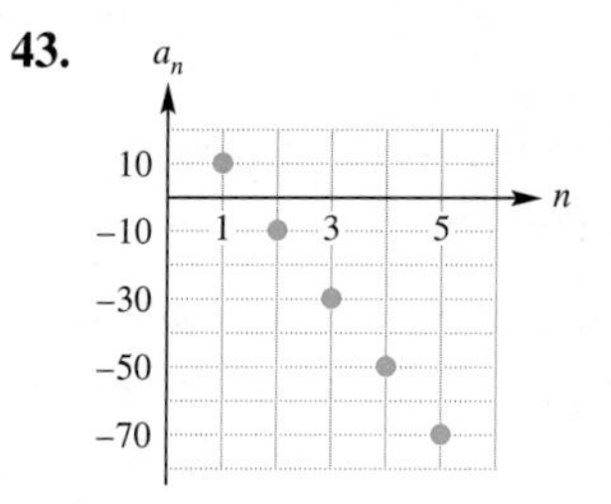


**44.**

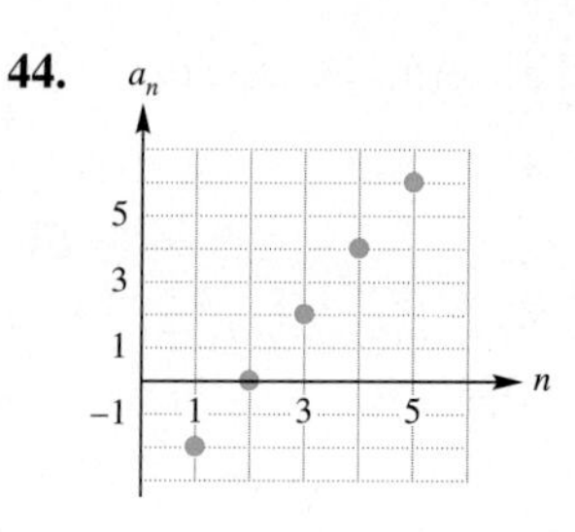


*Use a formula to find the sum of each arithmetic series.*

**45.** $3 + 5 + 7 + 9 + 11 + 13 + 15 + 17$

**46.** $7.5 + 6 + 4.5 + 3 + 1.5 + 0 + (-1.5)$

**47.** $1 + 2 + 3 + 4 + \cdots + 50$

**48.** $1 + 3 + 5 + 7 + \cdots + 97$

**49.** $-7 + (-4) + (-1) + 2 + 5 + \cdots + 98 + 101$

**50.** $89 + 84 + 79 + 74 + \cdots + 9 + 4$

**51.** The first 40 terms of the series given by $a_n = 5n$

**52.** The first 50 terms of the series given by $a_n = 1 - 3n$

*Evaluate each sum.*

**53.** $\sum_{i=1}^{3} (i + 4)$

**54.** $\sum_{i=1}^{5} (i - 8)$

**55.** $\sum_{j=1}^{10} (2j + 3)$

**56.** $\sum_{j=1}^{15} (5j - 9)$

**57.** $\sum_{i=1}^{12} (-5 - 8i)$

**58.** $\sum_{k=1}^{19} (-3 - 4k)$

**59.** $\sum_{i=1}^{1000} i$

**60.** $\sum_{k=1}^{2000} k$

## RELATING CONCEPTS

**For individual or group investigation (Exercises 61–64)**

*Let $f(x) = mx + b$. Work Exercises 61–64 in order.*

**61.** Find $f(1)$, $f(2)$, and $f(3)$.

**62.** Consider the sequence $f(1), f(2), f(3), \ldots$. Is it an arithmetic sequence?

**63.** If the sequence is arithmetic, what is the common difference?

**64.** What is $a_n$ for the sequence described in Exercise 62?

*Use the sequence feature of a graphing calculator to evaluate the sum of the first 10 terms of the arithmetic sequence. In Exercises 67 and 68, round to the nearest thousandth.*

**65.** $a_n = 4.2n + 9.73$

**66.** $a_n = 8.42n + 36.18$

**67.** $a_n = \sqrt{8}n + \sqrt{3}$

**68.** $a_n = -\sqrt[3]{4}n + \sqrt{7}$

*Solve each problem involving arithmetic sequences.*

**69.** Find the sum of all the integers from 51 to 71.

**70.** Find the sum of all the integers from $-8$ to 30.

**71.** *Clock Chimes* If a clock strikes the proper number of chimes each hour on the hour, how many times will it chime in a month of 30 days?

**72.** *Telephone Pole Stack* A stack of telephone poles has 30 in the bottom row, 29 in the next, and so on, with one pole in the top row. How many poles are in the stack?

**73.** *City Population Growth* Five years ago, the population of a city was 49,000. Each year the zoning commission permits an increase of 580 in the population. What will the maximum population be 5 years from now?

**74.** *Supports on a Slide* A super slide of uniform slope is to be built on a level piece of land. There are to be 20 equally spaced supports, with the longest support 15 meters long and the shortest 2 meters long. Find the total length of all the supports.

**75.** *Rungs of a Ladder* How much material would be needed for the rungs of a ladder of 31 rungs, if the rungs taper uniformly from 18 inches to 28 inches?

**76.** *(Modeling) Growth Pattern for Children* The normal growth pattern for children aged 3–11 follows that of an arithmetic sequence. An increase in height of about 6 centimeters per year is expected. Thus, 6 would be the common difference of the sequence. A child who measures 96 centimeters at age 3 would have his expected height in subsequent years represented by the sequence 102, 108, 114, 120, 126, 132, 138, 144. Each term differs from the adjacent terms by the common difference, 6.

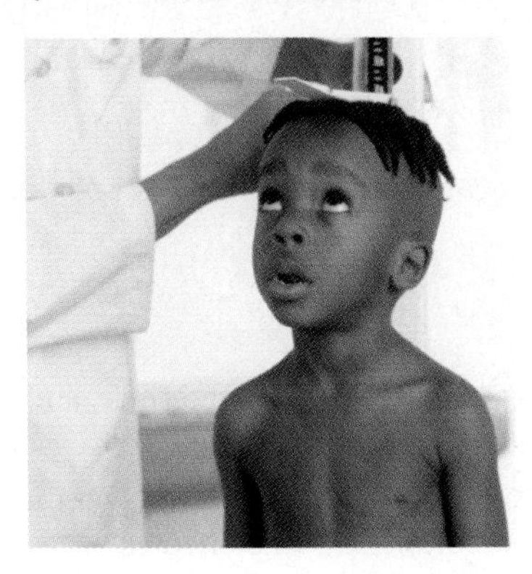

**(a)** If a child measures 98.2 centimeters at age 3 and 109.8 centimeters at age 5, what would be the common difference of the arithmetic sequence describing her yearly height?

**(b)** What would we expect her height to be at age 8?

**77.** *Concept Check* Find all arithmetic sequences $a_1, a_2, a_3, \ldots$ such that $a_1^2, a_2^2, a_3^2, \ldots$ is also an arithmetic sequence.

**78.** Suppose that $a_1, a_2, a_3, \ldots$ and $b_1, b_2, b_3, \ldots$ are both arithmetic sequences. Let $d_n = a_n + c \cdot b_n$, for any real number $c$ and every positive integer $n$. Show that $d_1, d_2, d_3, \ldots$ is an arithmetic sequence.

**79.** *Concept Check* Suppose that $a_1, a_2, a_3, a_4, a_5, \ldots$ is an arithmetic sequence. Is $a_1, a_3, a_5, \ldots$ an arithmetic sequence?

**80.** Explain why the sequence $\log 2, \log 4, \log 8, \log 16, \ldots$ is arithmetic.

# 11.3 GEOMETRIC SEQUENCES AND SERIES

Geometric Sequences • Geometric Series • Infinite Geometric Series • Annuities

## Geometric Sequences

Suppose you agreed to work for 1¢ the first day, 2¢ the second day, 4¢ the third day, 8¢ the fourth day, and so on, with your wages doubling each day. How much will you earn on day 20? How much will you have earned altogether in 20 days? These questions will be answered in this section.

A **geometric sequence** (or **geometric progression**) is a sequence in which each term after the first is obtained by multiplying the preceding term by a constant nonzero real number, called the **common ratio.** The sequence of wages,

$$1, 2, 4, 8, 16, \ldots,$$

is an example of a geometric sequence in which the first term is 1 and the common ratio is 2. Notice that if we divide any term (except the first term) by the preceding term, we obtain the common ratio $r = 2$.

$$\frac{a_2}{a_1} = \frac{2}{1} = 2$$

$$\frac{a_3}{a_2} = \frac{4}{2} = 2$$

$$\frac{a_4}{a_3} = \frac{8}{4} = 2$$

$$\frac{a_5}{a_4} = \frac{16}{8} = 2$$

If the common ratio of a geometric sequence is $r$, then by the definition of a geometric sequence,

$$r = \frac{a_{n+1}}{a_n}$$

for every positive integer $n$. Therefore, the common ratio can be found by choosing any term except the first and dividing it by the preceding term.

In the geometric sequence

$$2, 8, 32, 128, \ldots,$$

$r = 4$. Notice that

$$\begin{aligned} 8 &= 2 \cdot 4 \\ 32 &= 8 \cdot 4 = (2 \cdot 4) \cdot 4 = 2 \cdot 4^2 \\ 128 &= 32 \cdot 4 = (2 \cdot 4^2) \cdot 4 = 2 \cdot 4^3. \end{aligned}$$

To generalize this, assume that a geometric sequence has first term $a_1$ and common ratio $r$. The second term can be written as $a_2 = a_1 r$, the third as $a_3 = a_2 r = (a_1 r)r = a_1 r^2$, and so on. Following this pattern, the $n$th term is $a_n = a_1 r^{n-1}$. Again, this result can be proved by mathematical induction. (See Section 5 of this chapter.)

**$n$th Term of a Geometric Sequence**

In the geometric sequence with first term $a_1$ and common ratio $r$, the $n$th term is

$$a_n = a_1 r^{n-1}.$$

**EXAMPLE 1** *Finding the nth Term of a Geometric Sequence of Wages*

The formula for the $n$th term of a geometric sequence can be used to answer the first question posed at the beginning of this section. How much will be earned on day 20 if daily wages follow the sequence $1, 2, 4, 8, 16, \ldots$?

**Solution** To answer the question, let $a_1 = 1$ and $r = 2$, and find $a_{20}$.

$$a_{20} = a_1 r^{19} = 1(2)^{19} = 524{,}288 \text{ cents, or } \$5242.88$$

**EXAMPLE 2** *Using the Formula for the nth Term*

Find $a_5$ and $a_n$ for the geometric sequence $4, -12, 36, -108, \ldots$.

**Solution** The first term, $a_1$, is 4. Find $r$ by choosing any term except the first and dividing it by the preceding term. For example,

$$r = \frac{36}{-12} = -3.$$

Since $a_4 = -108$, $a_5 = -3(-108) = 324$. The fifth term also could be found by using the formula $a_n = a_1 r^{n-1}$, and replacing $n$ with 5, $r$ with $-3$, and $a_1$ with 4.

$$a_5 = 4 \cdot (-3)^{5-1} = 4 \cdot (-3)^4 = 324$$

By the formula, $a_n = 4 \cdot (-3)^{n-1}$.

**EXAMPLE 3** *Using the Formula for the nth Term*

Find $a_1$ and $r$ for the geometric sequence with third term 20 and sixth term 160.

**Solution** Use the formula for the $n$th term of a geometric sequence.

$$\text{For } n = 3, \quad a_3 = a_1 r^2 = 20;$$
$$\text{for } n = 6, \quad a_6 = a_1 r^5 = 160.$$

Because $a_1 r^2 = 20$, $a_1 = \frac{20}{r^2}$. Substituting in the second equation gives

$$a_1 r^5 = 160$$
$$\left(\frac{20}{r^2}\right) r^5 = 160$$
$$20r^3 = 160$$
$$r^3 = 8$$
$$r = 2.$$

Since $a_1 r^2 = 20$ and $r = 2$, it follows that $a_1 = 5$.

**EXAMPLE 4** *Modeling a Population of Fruit Flies*

A population of fruit flies is growing in such a way that each generation is 1.5 times as large as the last generation. Suppose there were 100 insects in the first generation. How many would there be in the fourth generation?

**Solution** The population can be written as a geometric sequence, with $a_1$ as the first-generation population, $a_2$ the second-generation population, and so on. Then the fourth-generation population is $a_4$. Using the formula for $a_n$, with $n = 4$, $r = 1.5$, and $a_1 = 100$, gives

$$a_4 = a_1 r^3 = 100(1.5)^3 = 100(3.375) = 337.5.$$

In the fourth generation, the population numbers about 338 insects.

## Geometric Series

A **geometric series** is the sum of the terms of a geometric sequence. In applications, it may be necessary to find the sum of the terms of such a sequence. For example, a scientist might want to know the total number of insects in four generations of the population discussed in Example 4. This population would equal $a_1 + a_2 + a_3 + a_4$, or

$$100 + 100(1.5) + 100(1.5)^2 + 100(1.5)^3 = 812.5 \approx 813 \text{ insects.}$$

To find a formula for the sum of the first $n$ terms of a geometric sequence, $S_n$, first write the sum as

$$S_n = a_1 + a_2 + a_3 + \cdots + a_n$$

or

$$S_n = a_1 + a_1 r + a_1 r^2 + \cdots + a_1 r^{n-1}. \quad (1)$$

If $r = 1$, $S_n = na_1$, which is a correct formula for this case. If $r \neq 1$, multiply each side of equation (1) by $r$ to obtain

$$rS_n = a_1 r + a_1 r^2 + a_1 r^3 + \cdots + a_1 r^n. \quad (2)$$

If equation (2) is subtracted from equation (1), we obtain

$$\begin{aligned} S_n &= a_1 + a_1r + a_1r^2 + \cdots + a_1r^{n-1} \\ rS_n &= \qquad a_1r + a_1r^2 + \cdots + a_1r^{n-1} + a_1r^n \\ \hline S_n - rS_n &= a_1 \qquad\qquad\qquad\qquad\qquad - a_1r^n \end{aligned}$$ Subtract.

or $S_n(1 - r) = a_1(1 - r^n)$, Factor.

which finally gives

$$S_n = \frac{a_1(1 - r^n)}{1 - r}, \quad \text{where } r \neq 1.$$ Divide by $1 - r$.

This discussion is summarized below.

**Sum of the First $n$ Terms of a Geometric Sequence**

If a geometric sequence has first term $a_1$ and common ratio $r$, then the sum of the first $n$ terms is given by

$$S_n = \frac{a_1(1 - r^n)}{1 - r}, \quad \text{where } r \neq 1.$$

We can use this formula to find the total fruit fly population in Example 4 over the four-generation period. With $n = 4$, $a_1 = 100$, and $r = 1.5$,

$$S_4 = \frac{100(1 - 1.5^4)}{1 - 1.5} = \frac{100(1 - 5.0625)}{-.5} = 812.5 \approx 813 \text{ insects},$$

which agrees with our previous result.

**EXAMPLE 5** *Applying the Sum of the First n Terms*

At the beginning of this section, we posed the following question: How much will you have earned altogether after 20 days?

**Analytic Solution**

To answer the second question posed at the beginning of this section, we must find the total amount earned in 20 days with daily wages of

$$1, 2, 4, 8, \ldots$$

cents. Since $a_1 = 1$ and $r = 2$,

$$S_{20} = \frac{1(1 - 2^{20})}{1 - 2} = 1{,}048{,}575 \text{ cents},$$

or \$10,485.75. Not bad for 20 days of work!

**Graphing Calculator Solution**

See Figure 13, where the sum agrees with the analytic solution.

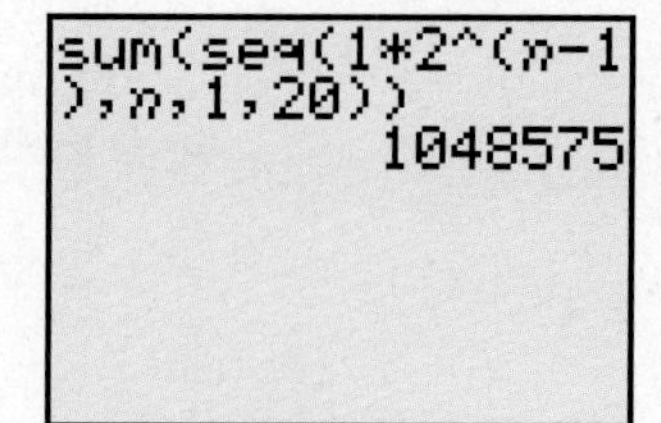


FIGURE 13

**EXAMPLE 6** *Finding the Sum of the First n Terms*

Find $\sum_{i=1}^{6} 2 \cdot 3^i$.

**Analytic Solution**

This series is the sum of the first six terms of a geometric sequence having $a_1 = 2 \cdot 3^1 = 6$ and $r = 3$. From the formula for $S_n$,

$$S_6 = \frac{6(1 - 3^6)}{1 - 3} = \frac{6(1 - 729)}{2}$$
$$= \frac{6(-728)}{-2} = 2184.$$

**Graphing Calculator Solution**

See Figure 14, where the sum agrees with the analytic solution.

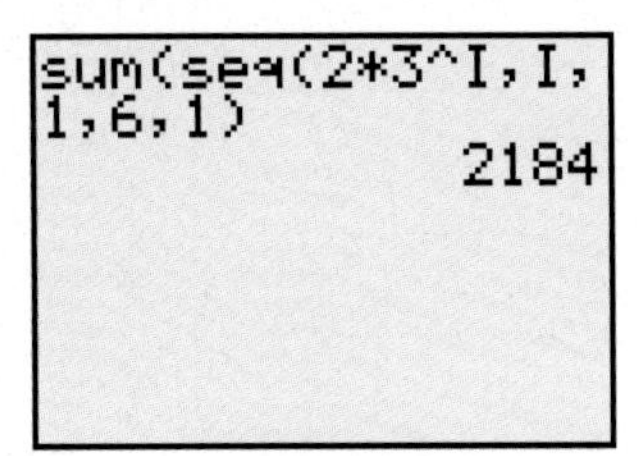


FIGURE 14

## Infinite Geometric Series

We can extend our discussion of sums of sequences to include infinite geometric sequences such as

$$2, 1, \frac{1}{2}, \frac{1}{4}, \frac{1}{8}, \frac{1}{16}, \ldots,$$

with first term 2 and common ratio $\frac{1}{2}$. Using the earlier formula gives the sequence of sums

$$S_1 = 2, \quad S_2 = 3, \quad S_3 = 3.5, \quad S_4 = 3.75, \quad S_5 = 3.875, \quad S_6 = 3.9375.$$

With a calculator in function mode, we define $Y_1$ as $S_n$.

$$Y_1 = \frac{2\left(1 - \left(\frac{1}{2}\right)^X\right)}{1 - \frac{1}{2}}$$

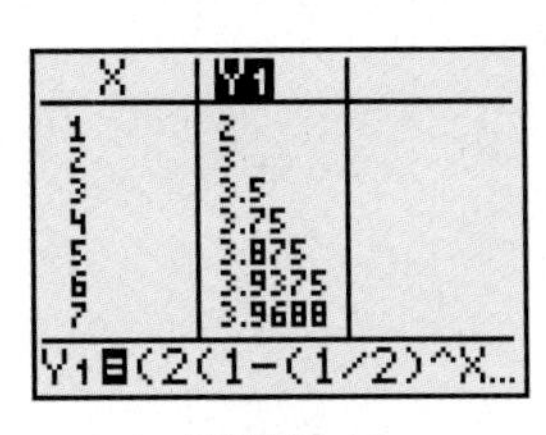


FIGURE 15

We can observe the first few terms as X takes on the values $1, 2, 3, \ldots$ using the table in Figure 15. These terms seem to be getting closer and closer to the number 4.

For no value of $n$ is $S_n = 4$. However, if $n$ is large enough, then $S_n$ is as close to 4 as desired.* As mentioned earlier, we say the sequence converges to 4. This is expressed as

$$\lim_{n \to \infty} S_n = 4.$$

(Read: "The limit of $S_n$ as $n$ increases without bound is 4.") Since

$$\lim_{n \to \infty} S_n = 4,$$

the number 4 is called the *sum of the terms* of the infinite geometric sequence

$$2, 1, \frac{1}{2}, \frac{1}{4}, \ldots,$$

and

$$2 + 1 + \frac{1}{2} + \frac{1}{4} + \cdots = 4.$$

**Looking Ahead to Calculus**

In calculus, different types of infinite series are studied. One important question to answer for each series is whether $\lim_{n \to \infty} S_n$ converges to a real number.

*The phrases "large enough" and "as close as desired" are not nearly precise enough for mathematicians; much of a standard calculus course is devoted to making them more precise.

**What Went WRONG?**

The preceding discussion justified the statement that the sum of the terms 2, 1, $\frac{1}{2}, \frac{1}{4}, \frac{1}{8}, \frac{1}{16}, \ldots$ gets closer and closer to 4 by adding more and more terms of the sequence. However, this sum will always differ from 4 by some small amount. The following figure is an extension of the table in Figure 15. According to the calculator, the sum *is* 4 when $X \geq 17$.

| X | Y1 |
|---|---|
| 14 | 3.9998 |
| 15 | 3.9999 |
| 16 | 3.9999 |
| 17 | 4 |
| 18 | 4 |
| 19 | 4 |
| 20 | 4 |

Y1■(2(1-(1/2)^X...

What Went Wrong? Discuss the limitations of technology, basing your comments on this example.

## EXAMPLE 7 *Summing the Terms of an Infinite Geometric Sequence*

Find $1 + \frac{1}{3} + \frac{1}{9} + \frac{1}{27} + \cdots$.

**Solution** Use the formula for the sum of the first $n$ terms of a geometric sequence to obtain

$$S_1 = 1, \quad S_2 = \frac{4}{3}, \quad S_3 = \frac{13}{9}, \quad S_4 = \frac{40}{27},$$

and in general

$$S_n = \frac{1\left[1 - \left(\frac{1}{3}\right)^n\right]}{1 - \frac{1}{3}}. \qquad a_1 = 1, r = \frac{1}{3}$$

The table shows the value of $\left(\frac{1}{3}\right)^n$ for larger and larger values of $n$.

| $n$ | 1 | 10 | 100 | 200 |
|---|---|---|---|---|
| $\left(\frac{1}{3}\right)^n$ | $\frac{1}{3}$ | $1.69 \times 10^{-5}$ | $1.94 \times 10^{-48}$ | $3.76 \times 10^{-96}$ |

*Answer to What Went Wrong?*

The calculator has room to display numbers to only four decimal places in the table. When the value of $Y_1$ reaches 3.99995 or higher, the calculator rounds up to 4. Try positioning the cursor on the 4 next to the 17 in the table. At the bottom of the screen a more accurate value of 3.99996948242 will appear for $Y_1$.

As $n$ gets larger and larger, $\left(\frac{1}{3}\right)^n$ gets closer and closer to 0. That is,

$$\lim_{n \to \infty} \left(\frac{1}{3}\right)^n = 0,$$

making it reasonable that

$$\lim_{n \to \infty} S_n = \lim_{n \to \infty} \frac{1\left[1 - \left(\frac{1}{3}\right)^n\right]}{1 - \frac{1}{3}} = \frac{1(1 - 0)}{1 - \frac{1}{3}} = \frac{3}{2}.$$

Hence,

$$1 + \frac{1}{3} + \frac{1}{9} + \frac{1}{27} + \cdots = \frac{3}{2}.$$

If a geometric sequence has first term $a_1$ and common ratio $r$, then

$$S_n = \frac{a_1(1 - r^n)}{1 - r} \quad (r \neq 1)$$

for every positive integer $n$. If $-1 < r < 1$, then $\lim_{n \to \infty} r^n = 0$, and

$$\lim_{n \to \infty} S_n = \frac{a_1(1 - 0)}{1 - r} = \frac{a_1}{1 - r}.$$

This quotient, $\frac{a_1}{1 - r}$, is called the **sum of the terms of an infinite geometric sequence.** The limit $\lim_{n \to \infty} S_n$ is often expressed as $S_\infty$ or $\sum_{i=1}^{\infty} a_i$.

**Looking Ahead to Calculus**

In calculus you will find sums of the terms of infinite sequences that are not geometric.

**Sum of the Terms of an Infinite Geometric Sequence**

The sum of the terms of an infinite geometric sequence with first term $a_1$ and common ratio $r$, where $-1 < r < 1$, is given by

$$S_\infty = \frac{a_1}{1 - r}.$$

If $|r| > 1$, the terms get larger and larger in absolute value, so there is no limit as $n \to \infty$. Hence, the sequence will not have a sum.

**EXAMPLE 8** *Finding Sums of the Terms of Infinite Geometric Sequences*

Find each sum.

**(a)** $\sum_{i=1}^{\infty}\left(-\frac{3}{4}\right)\left(-\frac{1}{2}\right)^{i-1}$ **(b)** $\sum_{i=1}^{\infty}\left(\frac{3}{5}\right)^{i}$

**Solution**

**(a)** Here, $a_1 = -\frac{3}{4}$ and $r = -\frac{1}{2}$. Since $-1 < r < 1$, the preceding formula applies, and

$$S_\infty = \frac{a_1}{1-r} = \frac{-\frac{3}{4}}{1-\left(-\frac{1}{2}\right)} = -\frac{1}{2}.$$

**(b)** $$\sum_{i=1}^{\infty}\left(\frac{3}{5}\right)^{i} = \sum_{i=1}^{\infty}\left(\frac{3}{5}\right)\left(\frac{3}{5}\right)^{i-1} = \frac{\frac{3}{5}}{1-\frac{3}{5}} = \frac{3}{2}$$

## Annuities

Geometric sequences and series are important in the mathematics of finance. An example is a sequence of equal payments made at equal periods of time, such as car payments or house payments, called an **annuity.** If the payments are accumulated in an account (with no withdrawals), the sum of the payments and interest on the payments is called the **future value** of the annuity.

**EXAMPLE 9** *Finding the Future Value of an Annuity*

To save money for a trip, Paula Story deposited \$1000 at the *end* of each year for 4 years in an account paying 6% interest, compounded annually. Find the future value of this annuity.

**Analytic Solution**

To find the future value, we use the formula for compound interest, $A = P(1 + r)^t$. The first payment earns interest for 3 years, the second payment for 2 years, and the third payment for 1 year. The last payment earns no interest. The total amount is

$$1000(1.06)^3 + 1000(1.06)^2 + 1000(1.06) + 1000.$$

This is the sum of the terms of a geometric sequence with first term (starting at the end of the sum as written above) $a_1 = 1000$ and common ratio $r = 1.06$. Using the formula for $S_4$, the sum of four terms, gives

$$S_4 = \frac{1000[1-(1.06)^4]}{1-1.06} \approx 4374.62.$$

The future value of the annuity is \$4374.62.

**Graphing Calculator Solution**

See Figure 16, where the sum agrees with the analytic solution.

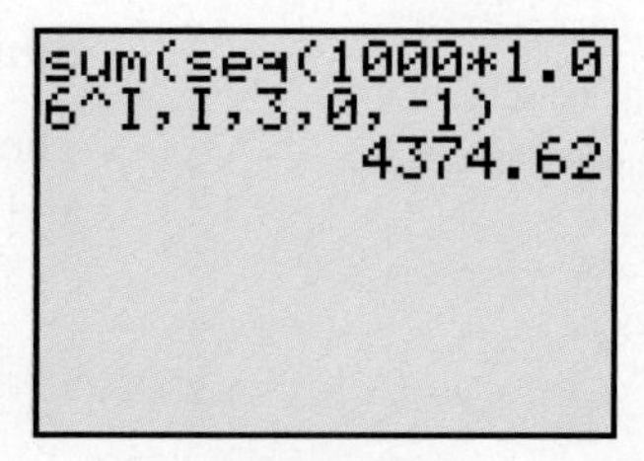


FIGURE 16

The variable T could have also been used in Figure 16.

## 11.3 EXERCISES

*Write out the terms of the geometric sequence that satisfies the given conditions.*

**1.** $a_1 = \frac{5}{3}, r = 3, n = 4$ **2.** $a_1 = -\frac{3}{4}, r = \frac{2}{3}, n = 4$ **3.** $a_4 = 5, a_5 = 10, n = 5$ **4.** $a_3 = 16, a_4 = 8, n = 5$

*Find $a_5$ and $a_n$ for each geometric sequence.*

**5.** $a_1 = 5, r = -2$ **6.** $a_1 = 8, r = -5$ **7.** $a_2 = -4, r = 3$

**8.** $a_3 = -2, r = 4$ **9.** $a_4 = 243, r = -3$ **10.** $a_4 = 18, r = 2$

**11.** $-4, -12, -36, -108, \ldots$ **12.** $-2, 6, -18, 54, \ldots$ **13.** $\frac{4}{5}, 2, 5, \frac{25}{2}, \ldots$

**14.** $\frac{1}{2}, \frac{2}{3}, \frac{8}{9}, \frac{32}{27}, \ldots$ **15.** $10, -5, \frac{5}{2}, -\frac{5}{4}, \ldots$ **16.** $3, -\frac{9}{4}, \frac{27}{16}, -\frac{81}{64}, \ldots$

### RELATING CONCEPTS

**For individual or group investigation (Exercises 17–20)**

*Using the definition of* difference *for an arithmetic sequence and* ratio *for a geometric sequence, we can find the appropriate middle term in a group of three terms so that the resulting sequence is the type desired. For example, consider the three terms*

$$5, x, .6,$$

*where x is to be determined. Work Exercises 17–20 in order.*

**17.** For these terms to form an arithmetic sequence, the difference $x - 5$ must be the same as the difference $.6 - x$. Write an equation for this statement.

**18.** Solve the equation of Exercise 17, and write the three terms of the arithmetic sequence.

**19.** For these terms to form a geometric sequence, the ratio $\frac{x}{5}$ must be the same as the ratio $\frac{.6}{x}$. Write an equation for this statement.

**20.** Solve the equation of Exercise 19 for its positive solution, and write the three terms of the geometric sequence.

*Find $a_1$ and $r$ for each geometric sequence.*

**21.** $a_3 = 5, a_8 = \frac{1}{625}$ **22.** $a_2 = -6, a_7 = -192$ **23.** $a_4 = -\frac{1}{4}, a_9 = -\frac{1}{128}$ **24.** $a_3 = 50, a_7 = .005$

*Use the formula for $S_n$ to find the sum of the first five terms for each geometric sequence. Round the answers for Exercises 29 and 30 to the nearest hundredth.*

**25.** $2, 8, 32, 128, \ldots$ **26.** $4, 16, 64, 256, \ldots$ **27.** $18, -9, \frac{9}{2}, -\frac{9}{4}, \ldots$

**28.** $12, -4, \frac{4}{3}, -\frac{4}{9}, \ldots$ **29.** $a_1 = 8.423, r = 2.859$ **30.** $a_1 = -3.772, r = -1.553$

*Find each sum.*

**31.** $\sum_{i=1}^{5} 3^i$ **32.** $\sum_{i=1}^{4} (-2)^i$ **33.** $\sum_{j=1}^{6} 48\left(\frac{1}{2}\right)^j$

**34.** $\sum_{j=1}^{5} 243\left(\frac{2}{3}\right)^j$ **35.** $\sum_{k=4}^{10} 2^k$ **36.** $\sum_{k=3}^{9} (-3)^k$

**37.** *Concept Check* Under what conditions does the sum of the terms of an infinite geometric sequence exist?

**38.** *Concept Check* The number .999... can be written as the sum of the terms of an infinite geometric sequence: $.9 + .09 + .009 + \cdots$. Here we have $a_1 = .9$ and $r = .1$. Use the formula for $S_\infty$ to find this sum. Does your intuition indicate that your answer is correct?

*Write the sum of each geometric series as a rational number. (See Exercise 38.)*

**39.** $.8 + .08 + .008 + .0008 + \cdots$

**40.** $.7 + .07 + .007 + .0007 + \cdots$

**41.** $.45 + .0045 + .000045 + \cdots$

**42.** $.36 + .0036 + .000036 + \cdots$

*Find r for each infinite geometric sequence. Identify any whose sum would not converge.*

**43.** $12, 24, 48, 96, \ldots$

**44.** $625, 125, 25, 5, \ldots$

**45.** $-48, -24, -12, -6, \ldots$

**46.** $2, -10, 50, -250, \ldots$

*Find each sum that converges.*

**47.** $16 + 2 + \frac{1}{4} + \frac{1}{32} + \cdots$

**48.** $18 + 6 + 2 + \frac{2}{3} + \cdots$

**49.** $100 + 10 + 1 + \cdots$

**50.** $128 + 64 + 32 + \cdots$

**51.** $\frac{4}{3} + \frac{2}{3} + \frac{1}{3} + \cdots$

**52.** $\frac{1}{4} - \frac{1}{6} + \frac{1}{9} - \frac{2}{27} + \cdots$

**53.** $\sum_{i=1}^{\infty} 3\left(\frac{1}{4}\right)^{i-1}$

**54.** $\sum_{i=1}^{\infty} 5\left(-\frac{1}{4}\right)^{i-1}$

**55.** $\sum_{k=1}^{\infty} (.3)^k$

**56.** $\sum_{k=1}^{\infty} 10^{-k}$

## RELATING CONCEPTS

**For individual or group investigation (Exercises 57–60)**

*Let $g(x) = ab^x$. Work Exercises 57–60 in order.*

**57.** Find $g(1)$, $g(2)$, and $g(3)$.

**58.** Consider the sequence $g(1), g(2), g(3), \ldots$. Is it a geometric sequence? If so, what is the common ratio?

**59.** What is the general term of the sequence in Exercise 58?

**60.** Explain how geometric sequences are related to exponential functions.

*Use a graphing calculator to evaluate each sum. Round to the nearest thousandth.*

**61.** $\sum_{i=1}^{10} (1.4)^i$

**62.** $\sum_{j=1}^{6} -(3.6)^j$

**63.** $\sum_{j=3}^{8} 2(.4)^j$

**64.** $\sum_{i=4}^{9} 3(.25)^i$

*Annuity Values* *Find the future value of each annuity.*

**65.** Payments of \$1000 at the end of each year for 9 years at 8% interest, compounded annually

**66.** Payments of \$800 at the end of each year for 12 years at 7% interest, compounded annually

**67.** Payments of \$2430 at the end of each year for 10 years at 6% interest, compounded annually

**68.** Payments of \$1500 at the end of each year for 6 years at 5% interest, compounded annually

*Solve each problem involving geometric sequences.*

**69.** *(Modeling) Investment for Retirement* According to T. Rowe Price Associates, a person who has a moderate investment strategy and $n$ years to retirement should have accumulated savings of $a_n$ percent of his or her annual salary. The geometric sequence with

$$a_n = 1276(.916)^n$$

gives the appropriate percent for each year $n$.

**(a)** Find $a_1$ and $r$.

**(b)** Find and interpret the terms $a_{10}$ and $a_{20}$.

**70.** *(Modeling) Investment for Retirement* Refer to Exercise 69. For someone who has a conservative investment strategy with $n$ years to retirement, the geometric sequence is

$$a_n = 1278(.935)^n.$$

(*Source:* T. Rowe Price Associates.)

**(a)** Repeat part (a) of Exercise 69.

**(b)** Repeat part (b) of Exercise 69.

**(c)** Why are the answers in parts (a) and (b) larger than in Exercise 69.

**71.** *(Modeling) Bacteria Growth* The strain of bacteria described in Exercise 69 in the first section of this chapter will double in size and then divide every 40 minutes. Let $a_1$ be the initial number of bacteria cells, $a_2$ the number after 40 minutes, and $a_n$ the number after $40(n - 1)$ minutes. (*Source:* Hoppensteadt, F. and C. Peskin, *Mathematics in Medicine and the Life Sciences*, Springer-Verlag, 1992.)

**(a)** Write a formula for the $n$th term $a_n$ of the geometric sequence $a_1, a_2, a_3, \ldots, a_n, \ldots$.

**(b)** Determine the first value for $n$ where $a_n > 1{,}000{,}000$, if $a_1 = 100$.

**(c)** How long does it take for the number of bacteria to exceed one million?

**72.** *Photo Processing* The final step in processing a black-and-white photographic print is to immerse the print in a chemical fixer. The print is then washed in running water. Under certain conditions, 98% of the fixer in a print will be removed with 15 minutes of washing. How much of the original fixer would be left after 1 hour of washing?

**73.** *Chemical Mixture* A scientist has a vat containing 100 liters of a pure chemical. Twenty liters are drained and replaced with water. After complete mixing, 20 liters of the mixture are again drained and replaced with water. What will be the strength of the mixture after nine such drainings?

**74.** *Half-Life of a Radioactive Substance* The half-life of a radioactive substance is the time it takes for half the substance to decay. Suppose the half-life of a substance is 3 years, and $10^{15}$ molecules of the substance are present initially. How many molecules will be present after 15 years?

**75.** *Depreciation in Value* Each year a machine loses 20% of the value it had at the beginning of the year. Find the value of the machine at the end of 6 years if it cost \$100,000 new.

**76.** *Sugar Processing* A sugar factory receives an order for 1000 units of sugar. The production manager thus orders production of 1000 units of sugar. He forgets, however, that the production of sugar requires some sugar (to prime the machines, for example), and so he ends up with only 900 units of sugar. He then orders an additional 100 units, and receives only 90 units. A further order for 10 units produces 9 units. Finally seeing he is wrong, the manager decides to try mathematics. He views the production process as an infinite geometric progression with $a_1 = 1000$ and $r = .1$. Using this, find the number of units of sugar that he should have ordered originally.

**77.** *Swing of a Pendulum* A pendulum bob swings through an arc 40 centimeters long on its first swing. Each swing thereafter, it swings only 80% as far as on the previous swing. How far will it swing altogether before coming to a complete stop?

**78.** *Height of a Dropped Ball* Mitzi drops a ball from a height of 10 meters and notices that on each bounce the ball returns to about $\frac{3}{4}$ of its previous height. About how far will the ball travel before it comes to rest? (*Hint:* Consider the sum of two sequences.)

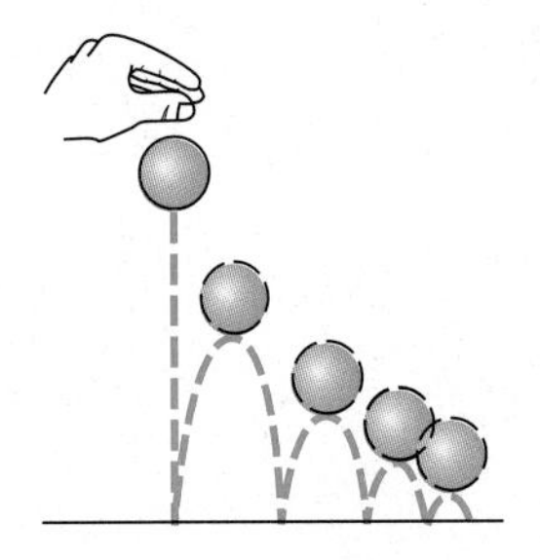

**79.** *Number of Ancestors* Each person has two parents, four grandparents, eight great-grandparents, and so on. What is the total number of ancestors a person has, going back five generations? ten generations?

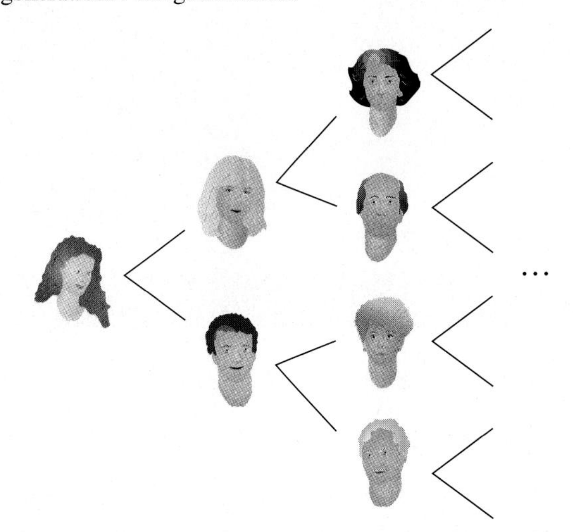

**80.** *(Modeling) Drug Dosage* Certain medical conditions are treated with a fixed dose of a drug administered at regular intervals. Suppose a person is given 2 milligrams of a drug each day, and that during each 24-hour period the body utilizes 40% of the amount of drug that was present at the beginning of the period.

**(a)** Show that the amount of the drug present in the body at the end of $n$ days is $\sum_{i=1}^{n} 2(.6)^i$.

**(b)** What will be the approximate quantity of the drug in the body at the end of each day after the treatment has been administered for a long period of time?

**81.** *Side Length of a Triangle* A sequence of equilateral triangles is constructed. The first triangle has sides 2 meters in length. To get the second triangle, midpoints of the sides of the original triangle are connected. What is the length of the side of the eighth such triangle? See the figure.

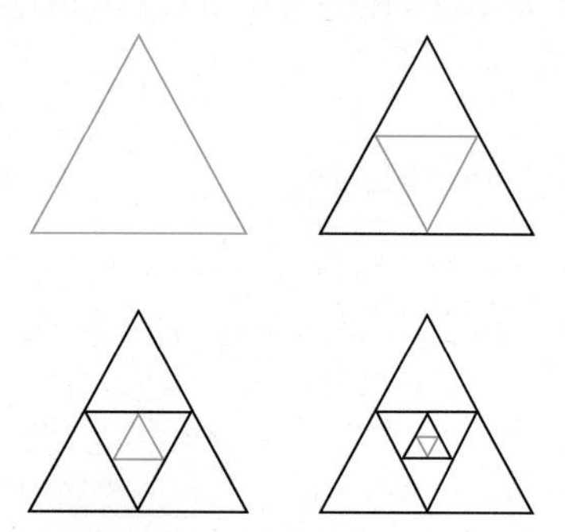

**82.** *Perimeter and Area of Triangles*

**(a)** In Exercise 81, if the process could be continued indefinitely, what would be the total perimeter of all the triangles?

**(b)** What would be the total area of all the triangles, disregarding the overlapping?

**83.** *Salaries* You are offered a 6-week summer job and are asked to select one of the following salary options.

*Option 1:* \$5000 for the first day with a \$10,000 raise each day for the remaining 29 days (that is, \$15,000 for day 2, \$25,000 for day 3, and so on)

*Option 2:* 1¢ for the first day with the pay doubled each day (that is, 2¢ for day 2, 4¢ for day 3, and so on)

Which option would you choose?

**84.** *Number of Ancestors* Suppose a genealogical Web site allows you to identify all your ancestors that lived during the last 300 years. Assuming that each generation spans about 25 years, guess the number of ancestors that would be found during the 12 generations. Then use the formula for a geometric series to find the actual value.

**85.** *Concept Check* Let $a_1, a_2, a_3, \ldots$ and $b_1, b_2, b_3, \ldots$ be geometric sequences. Let $d_n = c \cdot a_n \cdot b_n$ for any real number $c$ and every positive integer $n$. Show that $d_1, d_2, d_3, \ldots$ is a geometric sequence.

**86.** Explain why the sequence

$$\log 6, \log 36, \log 1296, \log 1{,}679{,}616, \ldots$$

is geometric.

## Reviewing Basic Concepts (Sections 11.1–11.3)

**1.** Find the first five terms of the sequence $a_n = (-1)^{n-1}(4n)$.

**2.** Evaluate the series $\sum_{i=1}^{5} (3i + 1)$.

**3.** Decide whether the sequence $a_n = 1 - \frac{2}{n}$ converges or diverges. If it converges, determine the number to which it converges.

**4.** Write the first five terms of the arithmetic sequence with $a_1 = 8$ and $d = -2$.

**5.** Find $a_1$ for the arithmetic sequence with $a_5 = 5$ and $a_8 = 17$.

**6.** Find the sum of the first 10 terms of the arithmetic sequence with $a_1 = 2$ and $d = 5$.

**7.** Find $a_3$ and $a_n$ for the geometric sequence with $a_1 = -2$ and $r = -3$.

**8.** Is the series $5 + 3 + \frac{9}{5} + \cdots + \frac{243}{625}$ arithmetic or geometric? Find the sum of the terms of this series.

**9.** Find the sum of the infinite geometric series $\sum_{k=1}^{\infty} 3\left(\frac{2}{3}\right)^k$.

**10.** *Annuity Value* Find the future value of an annuity if payments of \$500 are made at the end of each year for 13 years at 9% interest compounded annually.

# THE BINOMIAL THEOREM

A Binomial Expansion Pattern • Pascal's Triangle • $n$-Factorial • Binomial Coefficients • The Binomial Theorem • $r$th Term of a Binomial Expansion

## A Binomial Expansion Pattern

In this section, we introduce a method for writing the terms of expressions of the form $(x + y)^n$, where $n$ is a natural number. The formula for writing the terms of $(x + y)^n$ is called the *binomial theorem.* Some expansions of $(x + y)^n$, for various nonnegative integer values of $n$, are given below.

**Looking Ahead to Calculus**

Students taking calculus study the binomial series, which follows from Isaac Newton's extension to the case where the exponent is no longer a positive integer. His result led to a series for $(1 + x)^k$, where $k$ is a real number and $|x| < 1$.

$$
\begin{aligned}
(x + y)^0 &= 1 \\
(x + y)^1 &= x + y \\
(x + y)^2 &= x^2 + 2xy + y^2 \\
(x + y)^3 &= x^3 + 3x^2y + 3xy^2 + y^3 \\
(x + y)^4 &= x^4 + 4x^3y + 6x^2y^2 + 4xy^3 + y^4 \\
(x + y)^5 &= x^5 + 5x^4y + 10x^3y^2 + 10x^2y^3 + 5xy^4 + y^5
\end{aligned}
$$

Notice that after the special case $(x + y)^0 = 1$, each expression begins with $x$ raised to the same power as the binomial itself. That is, the expansion of $(x + y)^1$ has a first term of $x^1$, $(x + y)^2$ has a first term of $x^2$, $(x + y)^3$ has a first term of $x^3$, and so on. Also, the last term in each expansion is $y$ to the same power as the binomial. Thus, the expansion of $(x + y)^n$ should begin with the term $x^n$ and end with the term $y^n$.

Also, the exponent on $x$ decreases by 1 in each term after the first, while the exponent on $y$, beginning with $y$ in the second term, increases by 1 in each succeeding term. That is, the *variables* in the terms of the expansion of $(x + y)^n$ have the following pattern.

$$x^n,\ x^{n-1}y,\ x^{n-2}y^2,\ x^{n-3}y^3,\ \ldots,\ xy^{n-1},\ y^n$$

This pattern suggests that the sum of the exponents on $x$ and $y$ in each term is $n$. For example, the third term in the list above is $x^{n-2}y^2$, and the sum of the exponents is $n - 2 + 2 = n$.

## Pascal's Triangle

Now, examine the *coefficients* in the terms of the expansion of $(x + y)^n$. Writing the coefficients alone gives the following pattern.

**Pascal's Triangle**

| | Row |
|---|---|
| 1 | 0 |
| 1 1 | 1 |
| 1 2 1 | 2 |
| 1 3 3 1 | 3 |
| 1 4 6 4 1 | 4 |
| 1 5 10 10 5 1 | 5 |

With the coefficients arranged in this way, each number in the triangle is the sum of the two numbers directly above it (one to the right and one to the left). For example, in row four, 1 is the sum of 1 (the only number above it), 4 is the sum of 1 and 3, 6 is the sum of 3 and 3, and so on. This triangular array of numbers is called **Pascal's triangle,** in honor of the seventeenth-century mathematician Blaise Pascal (1623–1662), one of the first to use it extensively.

To find the coefficients for $(x + y)^6$, we need to include row six in Pascal's triangle. Adding adjacent numbers, we find that row six is

$$1 \quad 6 \quad 15 \quad 20 \quad 15 \quad 6 \quad 1.$$

Using these coefficients, we obtain the expansion of $(x + y)^6$:

$$(x + y)^6 = x^6 + 6x^5y + 15x^4y^2 + 20x^3y^3 + 15x^2y^4 + 6xy^5 + y^6.$$

## *n*-Factorial

Although it is possible to use Pascal's triangle to find the coefficients of $(x + y)^n$ for any positive integer $n$, this calculation becomes impractical for large values of $n$ because of the need to write all the preceding rows. A more efficient way of finding these coefficients uses **factorial notation.** The number $n!$ (read "$n$-factorial") is defined as follows.

> **$n$-Factorial**
>
> For any positive integer $n$,
>
> $$n! = n(n-1)(n-2) \cdots (3)(2)(1),$$
>
> and $$0! = 1.$$

### EXAMPLE 1 *Evaluating Factorials*

Evaluate each factorial.

**(a)** $5!$ **(b)** $7!$ **(c)** $2!$ **(d)** $1!$ **(e)** $0!$

**Analytic Solution**

**(a)** $5! = 5 \cdot 4 \cdot 3 \cdot 2 \cdot 1$
$= 120$

**(b)** $7! = 7 \cdot 6 \cdot 5 \cdot 4 \cdot 3 \cdot 2 \cdot 1$
$= 5040$

**(c)** $2! = 2 \cdot 1 = 2$

**(d)** $1! = 1$

**(e)** $0! = 1$

**Graphing Calculator Solution**

See Figure 17, where the results agree with the analytic solution.

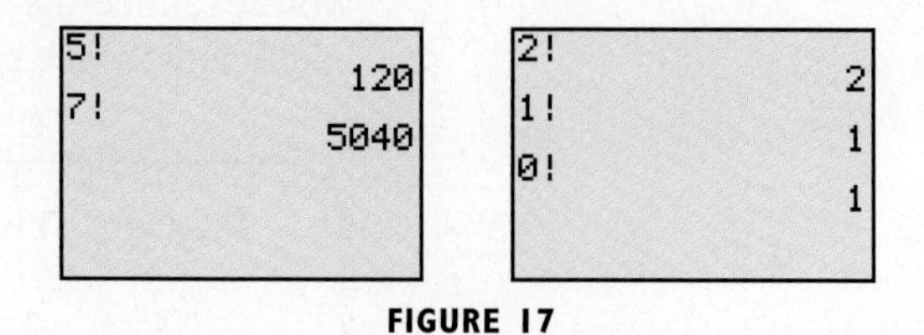


FIGURE 17

## Binomial Coefficients

Now, look at the coefficients of the expression

$$(x + y)^5 = x^5 + 5x^4y + 10x^3y^2 + 10x^2y^3 + 5xy^4 + y^5.$$

**TECHNOLOGY NOTE**

A calculator with a 10-digit display will give exact values of $n!$ for $n \le 13$ and approximate values of $n!$ for $14 \le n \le 69$. Consult the graphing calculator manual that accompanies this text or your owner's manual to learn how to calculate $n!$ on your calculator.

The coefficient of the second term, $5x^4y$, is 5, and the exponents on the variables are 4 and 1. Note that

$$5 = \frac{5!}{4!\,1!}.$$

The coefficient of the third term is 10, with exponents of 3 and 2, and

$$10 = \frac{5!}{3!\,2!}.$$

The last term (the sixth term) can be written as $y^5 = 1x^0y^5$, with coefficient 1, and exponents of 0 and 5. Since $0! = 1$,

$$1 = \frac{5!}{0!\,5!}.$$

Generalizing from these examples, the coefficient for the term of the expansion of $(x + y)^n$ in which the variable part is $x^ry^{n-r}$ (where $r \le n$) is

$$\frac{n!}{r!\,(n - r)!}.$$

This number, called a **binomial coefficient,** is often written as $\binom{n}{r}$ (read "$n$ choose $r$") or ${}_nC_r$.

**Binomial Coefficient**

For nonnegative integers $n$ and $r$, with $r \le n$,

$$_nC_r = \binom{n}{r} = \frac{n!}{r!\,(n - r)!}.$$

The binomial coefficients are numbers from Pascal's triangle. For example, $\binom{3}{0}$ is the first number in row three, and $\binom{7}{4}$ is the fifth number in row seven.

**EXAMPLE 2** *Evaluating Binomial Coefficients*

Evaluate each binomial coefficient.

**(a)** $\binom{6}{2}$ **(b)** $\binom{8}{0}$ **(c)** $\binom{10}{10}$

**Analytic Solution**

**(a)** $\binom{6}{2} = \frac{6!}{2!\,(6 - 2)!} = \frac{6!}{2!\,4!} = \frac{6 \cdot 5 \cdot 4 \cdot 3 \cdot 2 \cdot 1}{2 \cdot 1 \cdot 4 \cdot 3 \cdot 2 \cdot 1} = 15$

**(b)** $\binom{8}{0} = \frac{8!}{0!\,(8 - 0)!} = \frac{8!}{0!\,8!} = \frac{8!}{1 \cdot 8!} = 1$

**Graphing Calculator Solution**

Graphing calculators can compute binomial coefficients with the *combinations* function, denoted nCr. See Figure 18 on the next page.

*(continued)*

(c) $\binom{10}{10} = \dfrac{10!}{10!\,(10-10)!} = \dfrac{10!}{10!\,0!} = 1$

Notice in parts (b) and (c) that $0! = 1$.

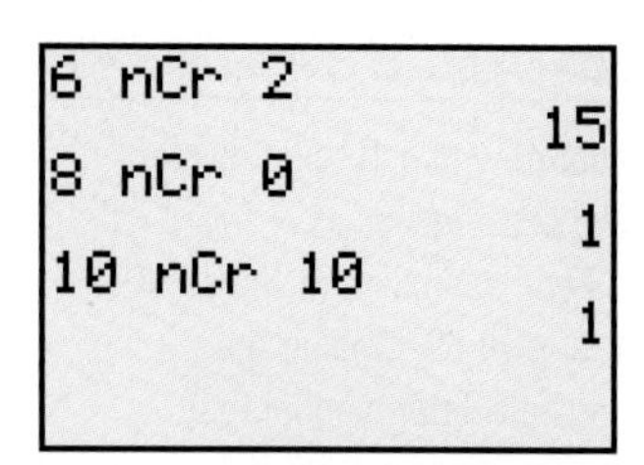


**FIGURE 18**

### FOR DISCUSSION

1. Compute pairs of binomial coefficients of the form ${}_nC_r$ and ${}_nC_{n-r}$; for example, ${}_{10}C_3$ and ${}_{10}C_7$.
2. Compare your results with Pascal's triangle.
3. Discuss your results, and express them as a generalization.

Refer again to Pascal's triangle. Notice the symmetry in each row. This suggests that the binomial coefficients should satisfy

$$\binom{n}{r} = \binom{n}{n-r}.$$

This is true, since

$$\binom{n}{r} = \frac{n!}{r!\,(n-r)!} \quad \text{and} \quad \binom{n}{n-r} = \frac{n!}{(n-r)!\,r!}.$$

## The Binomial Theorem

Our observations about the expansion of $(x + y)^n$ are summarized as follows.

1. There are $n + 1$ terms in the expansion.
2. The first term is $x^n$, and the last term is $y^n$.
3. In each succeeding term, the exponent on $x$ decreases by 1 and the exponent on $y$ increases by 1.
4. The sum of the exponents on $x$ and $y$ in any term is $n$.
5. The coefficient of the term with $x^r y^{n-r}$ or $x^{n-r} y^r$ is $\binom{n}{r}$.

These observations about the expansion of $(x + y)^n$ for any positive integer value of $n$ suggest the **binomial theorem.**

TECHNOLOGY NOTE

The two screens illustrate how the sequence and table capabilities of a graphing calculator can generate rows of Pascal's triangle.

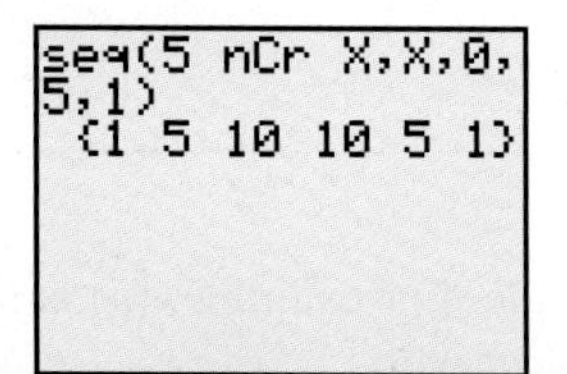


| X | Y1 |
|---|---|
| 0 | 1 |
| 1 | 6 |
| 2 | 15 |
| 3 | 20 |
| 4 | 15 |
| 5 | 6 |
| 6 | 1 |

Y1=6 nCr X

**Binomial Theorem**

For any positive integer $n$,

$$(x+y)^n = x^n + \binom{n}{1}x^{n-1}y + \binom{n}{2}x^{n-2}y^2 + \binom{n}{3}x^{n-3}y^3 + \cdots + \binom{n}{r}x^{n-r}y^r + \cdots + \binom{n}{n-1}xy^{n-1} + y^n.$$

A proof of the binomial theorem that uses mathematical induction is given in the next section.

**Looking Ahead to Calculus**

The binomial theorem is used to show that the derivative of $f(x) = x^n$ is given by the term $nx^{n-1}$. This fact is used extensively throughout the study of calculus.

**NOTE** The binomial theorem looks much more manageable written in summation notation. The theorem can be summarized as follows:

$$(x + y)^n = \sum_{r=0}^{n} \binom{n}{r} x^{n-r} y^r.$$

## EXAMPLE 3 *Applying the Binomial Theorem*

Write the binomial expansion of $(x + y)^9$.

**Solution** Use the binomial theorem.

$$(x + y)^9 = x^9 + \binom{9}{1}x^8y + \binom{9}{2}x^7y^2 + \binom{9}{3}x^6y^3 + \binom{9}{4}x^5y^4 + \binom{9}{5}x^4y^5 + \binom{9}{6}x^3y^6 + \binom{9}{7}x^2y^7 + \binom{9}{8}xy^8 + y^9$$

Now, evaluate each of the binomial coefficients.

$$\begin{aligned}(x + y)^9 &= x^9 + \frac{9!}{1!8!}x^8y + \frac{9!}{2!7!}x^7y^2 + \frac{9!}{3!6!}x^6y^3 + \frac{9!}{4!5!}x^5y^4 \\ &\quad + \frac{9!}{5!4!}x^4y^5 + \frac{9!}{6!3!}x^3y^6 + \frac{9!}{7!2!}x^2y^7 + \frac{9!}{8!1!}xy^8 + y^9 \\ &= x^9 + 9x^8y + 36x^7y^2 + 84x^6y^3 + 126x^5y^4 + 126x^4y^5 \\ &\quad + 84x^3y^6 + 36x^2y^7 + 9xy^8 + y^9\end{aligned}$$

## EXAMPLE 4 *Applying the Binomial Theorem*

Expand $\left(a - \frac{b}{2}\right)^5$.

**Solution** Write the binomial as follows.

$$\left(a - \frac{b}{2}\right)^5 = \left(a + \left(-\frac{b}{2}\right)\right)^5$$

Now, use the binomial theorem with $x = a$, $y = -\frac{b}{2}$, and $n = 5$ to get the expansion

$$\begin{aligned}\left(a - \frac{b}{2}\right)^5 &= a^5 + \binom{5}{1}a^4\left(-\frac{b}{2}\right) + \binom{5}{2}a^3\left(-\frac{b}{2}\right)^2 + \binom{5}{3}a^2\left(-\frac{b}{2}\right)^3 + \binom{5}{4}a\left(-\frac{b}{2}\right)^4 + \left(-\frac{b}{2}\right)^5 \\ &= a^5 + 5a^4\left(-\frac{b}{2}\right) + 10a^3\left(-\frac{b}{2}\right)^2 + 10a^2\left(-\frac{b}{2}\right)^3 + 5a\left(-\frac{b}{2}\right)^4 + \left(-\frac{b}{2}\right)^5 \\ &= a^5 - \frac{5}{2}a^4b + \frac{5}{2}a^3b^2 - \frac{5}{4}a^2b^3 + \frac{5}{16}ab^4 - \frac{1}{32}b^5.\end{aligned}$$

**NOTE** As Example 4 illustrates, any binomial expansion of the *difference* of two terms has alternating signs.

## EXAMPLE 5 *Applying the Binomial Theorem*

Expand $\left(\frac{3}{m^2} - 2\sqrt{m}\right)^4$. (Assume $m > 0$.)

**Solution** By the binomial theorem,

$$\left(\frac{3}{m^2} - 2\sqrt{m}\right)^4 = \left(\frac{3}{m^2}\right)^4 + \binom{4}{1}\left(\frac{3}{m^2}\right)^3(-2\sqrt{m}) + \binom{4}{2}\left(\frac{3}{m^2}\right)^2(-2\sqrt{m})^2$$
$$+ \binom{4}{3}\left(\frac{3}{m^2}\right)(-2\sqrt{m})^3 + (-2\sqrt{m})^4$$
$$= \frac{81}{m^8} + 4\left(\frac{27}{m^6}\right)(-2m^{1/2}) + 6\left(\frac{9}{m^4}\right)(4m)$$
$$+ 4\left(\frac{3}{m^2}\right)(-8m^{3/2}) + 16m^2.$$

Here, we used the fact that $\sqrt{m} = m^{1/2}$. Finally,

$$\left(\frac{3}{m^2} - 2\sqrt{m}\right)^4 = \frac{81}{m^8} - \frac{216}{m^{11/2}} + \frac{216}{m^3} - \frac{96}{m^{1/2}} + 16m^2.$$

## $r$th Term of a Binomial Expansion

Any term in a binomial expansion can be determined without writing out the entire expansion. For example, the seventh term of $(x + y)^9$ has $y$ raised to the sixth power (since $y$ has the power 1 in the second term, the power 2 in the third term, and so on). The exponents on $x$ and $y$ in each term must have a sum of 9, so the exponent on $x$ in the seventh term is $9 - 6 = 3$. Thus, writing the coefficient as given in the binomial theorem, the seventh term should be

$$\frac{9!}{6!\,3!}x^3y^6 \quad \text{or} \quad 84x^3y^6.$$

This is in fact the seventh term of $(x + y)^9$ found in Example 3. This discussion suggests the next theorem.

**$r$th Term of the Binomial Expansion**

The $r$th term of the binomial expansion of $(x + y)^n$, where $n \geq r - 1$, is

$$\binom{n}{r-1}x^{n-(r-1)}y^{r-1}.$$

**EXAMPLE 6** *Finding a Specific Term of a Binomial Expansion*

Find the fourth term of $(a + 2b)^{10}$.

**Solution** In the fourth term, $2b$ has exponent 3, and $a$ has exponent $10 - 3 = 7$. Using $n = 10$, $r = 4$, $x = a$, and $y = 2b$ in the preceding formula, we find that the fourth term is

$$\binom{10}{3}a^7(2b)^3 = 120a^7(8b^3) = 960a^7b^3.$$

## 11.4 EXERCISES

*Evaluate the following.*

1. $\frac{6!}{3!3!}$
2. $\frac{5!}{2!3!}$
3. $\frac{7!}{3!4!}$
4. $\frac{8!}{5!3!}$
5. $\binom{8}{3}$
6. $\binom{7}{4}$
7. $\binom{10}{8}$
8. $\binom{9}{6}$
9. $\binom{13}{13}$
10. $\binom{12}{12}$
11. $\binom{n}{n-1}$
12. $\binom{n}{n-2}$
13. ${}_8C_3$
14. ${}_9C_7$
15. ${}_{100}C_2$
16. ${}_{20}C_{15}$

17. *Concept Check* How many terms are there in the expansion of $(x + y)^8$?

18. *Concept Check* How many terms are there in the expansion of $(x + y)^{10}$?

19. *Concept Check* What are the first and last terms in the expansion of $(2x + 3y)^4$?

20. Describe in your own words how you would determine the binomial coefficient for the fifth term in the expansion of $(x + y)^8$.

*Write the binomial expansion for each expression.*

21. $(x + y)^6$
22. $(m + n)^4$
23. $(p - q)^5$
24. $(a - b)^7$
25. $(r^2 + s)^5$
26. $(m + n^2)^4$
27. $(p + 2q)^4$
28. $(3r - s)^6$
29. $(7p + 2q)^4$
30. $(4a - 5b)^5$
31. $(3x - 2y)^6$
32. $(7k - 9j)^4$
33. $\left(\frac{m}{2} - 1\right)^6$
34. $\left(3 + \frac{y}{3}\right)^5$
35. $\left(\sqrt{2}r + \frac{1}{m}\right)^4$
36. $\left(\frac{1}{k} - \sqrt{3}p\right)^3$

*Write the indicated term of each binomial expansion.*

37. Sixth term of $(4h - j)^8$
38. Eighth term of $(2c - 3d)^{14}$
39. Fifteenth term of $(a^2 + b)^{22}$
40. Twelfth term of $(2x + y^2)^{16}$
41. Fifteenth term of $(x - y^3)^{20}$
42. Tenth term of $(a^3 + 3b)^{11}$

*Concept Check Use the concepts of this section to work Exercises 43–46.*

43. Find the middle term of $(3x^7 + 2y^3)^8$.

44. Find the two middle terms of $(-2m^{-1} + 3n^{-2})^{11}$.

45. Find the value of $n$ for which the coefficients of the fifth and eighth terms in the expansion of $(x + y)^n$ are the same.

46. Find the term in the expansion of $(3 + \sqrt{x})^{11}$ that contains $x^4$.

## RELATING CONCEPTS

**For individual or group investigation (Exercises 47–50)**

*In this section, we saw how the factorial of a positive integer n can be computed as a product: $n! = 1 \cdot 2 \cdot 3 \cdot \cdots \cdot n$. Calculators and computers can evaluate factorials very quickly. Before the days of technology, mathematicians developed a formula, called* ***Stirling's formula,*** *for approximating large factorials. Interestingly enough, the formula involves the irrational numbers $\pi$ and $e$.*

$$n! \approx \sqrt{2\pi n} \cdot n^n \cdot e^{-n}$$

*As an example, the exact value of 5! is 120, and Stirling's formula gives the approximation as 118.019168 with a graphing calculator. This is "off" by less than 2, an error of only 1.65%.*

*Work Exercises 47–50 in order.*

47. Use a calculator to find the exact value of 10! and its approximation, using Stirling's formula.

48. Subtract the smaller value from the larger value in Exercise 47. Divide it by 10! and convert to a percent. What is the percent error?

49. Repeat Exercises 47 and 48 for $n = 12$.

50. Repeat Exercises 47 and 48 for $n = 13$. What seems to happen as $n$ gets larger?

*In later courses, it is shown that*

$$(1 + x)^n = 1 + nx + \frac{n(n-1)}{2!}x^2 + \frac{n(n-1)(n-2)}{3!}x^3 + \cdots$$

*for any real number n (not just positive integer values) and any real number x, where* $|x| < 1$. *This result, a generalized binomial theorem, may be used to find approximate values of powers and roots. For example,*

$$\begin{aligned}(1.008)^{1/4} &= (1 + .008)^{1/4}\\ &= 1 + \frac{1}{4}(.008) + \frac{\frac{1}{4}\left(-\frac{3}{4}\right)}{2!}(.008)^2 + \frac{\frac{1}{4}\left(-\frac{3}{4}\right)\left(-\frac{7}{4}\right)}{3!}(.008)^3 + \cdots\\ &\approx 1.002.\end{aligned}$$

*Use this result to approximate each quantity in Exercises 51–54 to the nearest thousandth.*

**51.** $(1.02)^{-3}$ **52.** $\frac{1}{1.04^5}$ **53.** $(1.01)^{3/2}$ **54.** $(1.03)^2$

# 11.5 MATHEMATICAL INDUCTION

Proof by Mathematical Induction • Proving Statements • Generalized Principle of Mathematical Induction • Proof of the Binomial Theorem

## Proof by Mathematical Induction

Many results in mathematics are claimed true for every positive integer. Any of these results could be checked for $n = 1$, $n = 2$, $n = 3$, and so on, but since the set of positive integers is infinite, it would be impossible to check every possible case. For example, let $S_n$ represent the statement that the sum of the first $n$ positive integers is $\frac{n(n+1)}{2}$, that is,

$$S_n\text{: } 1 + 2 + 3 + \cdots + n = \frac{n(n+1)}{2}.$$

The truth of this statement can be checked quickly for the first few values of $n$.

If $n = 1$, $S_1$ is $1 = \frac{1(1+1)}{2}$, a true statement, since $1 = 1$.

If $n = 2$, $S_2$ is $1 + 2 = \frac{2(2+1)}{2}$, a true statement, since $3 = 3$.

If $n = 3$, $S_3$ is $1 + 2 + 3 = \frac{3(3+1)}{2}$, a true statement, since $6 = 6$.

If $n = 4$, $S_4$ is $1 + 2 + 3 + 4 = \frac{4(4+1)}{2}$, a true statement, since $10 = 10$.

Since the statement is true for $n = 1, 2, 3$, and 4, and so on, can we conclude that the statement is true for all positive integers by checking a finite number of examples? The answer is no. To prove that such a statement is true for every positive integer, we use the following principle.

**Principle of Mathematical Induction**

Let $S_n$ be a statement concerning the positive integer $n$. Suppose that

1. $S_1$ is true;
2. for any positive integer $k$, $k \leq n$, if $S_k$ is true, then $S_{k+1}$ is also true.

Then, $S_n$ is true for every positive integer $n$.

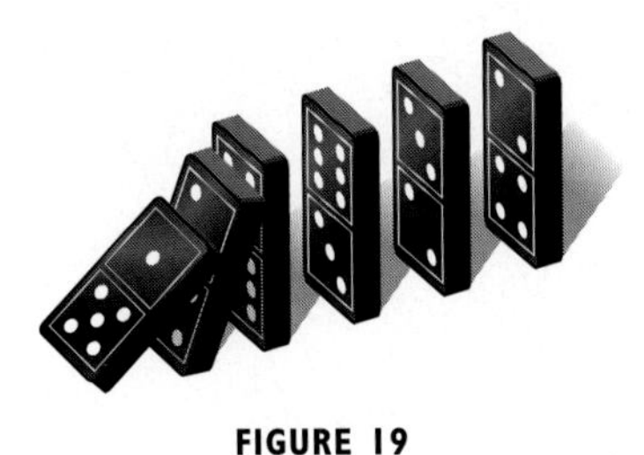

FIGURE 19

A proof by mathematical induction can be explained as follows. By assumption (1), the statement is true when $n = 1$. By assumption (2), the fact that the statement is true for $n = 1$ implies that it is true for $n = 1 + 1 = 2$. Using (2) again, the statement is thus true for $2 + 1 = 3$, for $3 + 1 = 4$, for $4 + 1 = 5$, and so on. Continuing in this way shows that the statement must be true for *every* positive integer.

The situation is similar to that of an infinite number of dominoes lined up, as suggested in Figure 19. If the first domino is pushed over, it pushes the next, which pushes the next, and so on, indefinitely.

Another example of the principle of mathematical induction might be an infinite ladder. Suppose the rungs are spaced so that, whenever you are on a rung, you know you can move to the next rung. Then *if* you can get to the first rung, you can go as high up the ladder as you wish.

Two separate steps are required for a proof by mathematical induction.

**Proof by Mathematical Induction**

**Step 1** Prove that the statement is true for $n = 1$.

**Step 2** Show that for any positive integer $k$, if $S_k$ is true, then $S_{k+1}$ is also true.

## Proving Statements

Mathematical induction is used in the next example to prove the statement $S_n$ discussed earlier.

**EXAMPLE 1** *Proving an Equality Statement*

Let $S_n$ represent the statement

$$1 + 2 + 3 + \cdots + n = \frac{n(n + 1)}{2}.$$

Prove that $S_n$ is true for every positive integer $n$.

**Solution** The proof by mathematical induction is as follows.

**Step 1** Show that the statement is true when $n = 1$. If $n = 1$, $S_1$ becomes

$$1 = \frac{1(1 + 1)}{2},$$

which is true.

**Step 2** Show that if $S_k$ is true, then $S_{k+1}$ is also true, where $S_k$ is the statement

$$1 + 2 + 3 + \cdots + k = \frac{k(k + 1)}{2},$$

and $S_{k+1}$ is the statement

$$1 + 2 + 3 + \cdots + k + (k + 1) = \frac{(k + 1)[(k + 1) + 1]}{2}.$$

Start with $S_k$ and assume it is a true statement.

$$1 + 2 + 3 + \cdots + k = \frac{k(k + 1)}{2}$$

Add $k + 1$ to each side of this equation to obtain $S_{k+1}$.

$$1 + 2 + 3 + \cdots + k + (k + 1) = \frac{k(k + 1)}{2} + (k + 1)$$

Now, factor out the common factor $k + 1$ on the right to get

$$= (k + 1)\left(\frac{k}{2} + 1\right)$$

$$= (k + 1)\left(\frac{k + 2}{2}\right)$$

$$1 + 2 + 3 + \cdots + k + (k + 1) = \frac{(k + 1)[(k + 1) + 1]}{2}.$$

This final result is the statement for $n = k + 1$; it has been shown that if $S_k$ is true, then $S_{k+1}$ is also true. The two steps required for a proof by mathematical induction have been completed, so the statement $S_n$ is true for every positive integer $n$.

### EXAMPLE 2 *Proving an Equality Statement*

Let $S_n$ represent the statement

$$2^1 + 2^2 + 2^3 + 2^4 + \cdots + 2^n = 2^{n+1} - 2.$$

Prove that $S_n$ is true for every positive integer $n$.

**Solution**

**Step 1** Show that the statement $S_1$ is true, where $S_1$ is

$$2^1 = 2^{1+1} - 2.$$

Since $2 = 4 - 2$, $S_1$ is a true statement.

**Step 2** Show that if $S_k$ is true, then $S_{k+1}$ is also true, where $S_k$ is

$$2^1 + 2^2 + 2^3 + \cdots + 2^k = 2^{k+1} - 2$$

and $S_{k+1}$ is

$$2^1 + 2^2 + 2^3 + \cdots + 2^k + 2^{k+1} = 2^{(k+1)+1} - 2.$$

Start with $S_k$ and add $2^{k+1}$ to each side of the equation. Then algebraically change the right side to look like the right side of $S_{k+1}$.

$$2^1 + 2^2 + 2^3 + \cdots + 2^k + 2^{k+1} = 2^{k+1} - 2 + 2^{k+1}$$

$$= 2 \cdot 2^{k+1} - 2$$

$$= 2^{(k+1)+1} - 2$$

The final result is the statement $S_{k+1}$. Therefore, if $S_k$ is true, then $S_{k+1}$ is also true. The two steps required for a proof by mathematical induction have been completed, so the statement $S_n$ is true for every positive integer $n$.

**EXAMPLE 3** *Proving an Inequality Statement*

Prove that if $x$ is a real number between 0 and 1, then for every positive integer $n$,

$$0 < x^n < 1.$$

**Solution**

**Step 1** Here $S_1$ is the statement

$$\text{if } 0 < x < 1, \text{ then } 0 < x^1 < 1,$$

which is true.

**Step 2** $S_k$ is the statement

$$\text{if } 0 < x < 1, \text{ then } 0 < x^k < 1.$$

To show that this implies that $S_{k+1}$ is true, multiply all three parts of $0 < x^k < 1$ by $x$ to get

$$x \cdot 0 < x \cdot x^k < x \cdot 1.$$

(Here the fact that $0 < x$ is used.) Simplify to obtain

$$0 < x^{k+1} < x.$$

Since $x < 1$,

$$0 < x^{k+1} < 1.$$

By this work, if $S_k$ is true, then $S_{k+1}$ is true. Since both steps for a proof by mathematical induction have been completed, the given statement is true for every positive integer $n$.

## Generalized Principle of Mathematical Induction

Some statements $S_n$ are not true for the first few values of $n$, but are true for all values of $n$ that are greater than or equal to some fixed integer $j$. The following slightly generalized form of the principle of mathematical induction takes care of these cases.

**Generalized Principle of Mathematical Induction**

Let $S_n$ be a statement concerning the positive integer $n$. Let $j$ be a fixed positive integer. Suppose that

**Step 1** $S_j$ is true;

**Step 2** for any positive integer $k$, $k \geq j$, $S_k$ implies $S_{k+1}$.

Then $S_n$ is true for all positive integers $n$, where $n \geq j$.

**EXAMPLE 4** *Using the Generalized Principle*

Let $S_n$ represent the statement $2^n > 2n + 1$. Show that $S_n$ is true for all values of $n$ such that $n \geq 3$.

**Solution** (Check that $S_n$ is false for $n = 1$ and $n = 2$.)

**Step 1** Show that $S_n$ is true for $n = 3$. If $n = 3$, $S_n$ is

$$2^3 > 2 \cdot 3 + 1,$$

or

$$8 > 7.$$

Thus, $S_3$ is true.

**Step 2** Now show that $S_k$ implies $S_{k+1}$, for $k \geq 3$, where

$$S_k \text{ is } 2^k > 2k + 1$$

and

$$S_{k+1} \text{ is } 2^{k+1} > 2(k + 1) + 1.$$

Multiply each side of $2^k > 2k + 1$ by 2, obtaining

$$2 \cdot 2^k > 2(2k + 1),$$

or

$$2^{k+1} > 4k + 2.$$

Rewrite $4k + 2$ as $2(k + 1) + 2k$, giving

$$2^{k+1} > 2(k + 1) + 2k.$$

Since $k$ is a positive integer greater than 3,

$$2k > 1.$$

It follows that

$$2^{k+1} > 2(k + 1) + 2k > 2(k + 1) + 1,$$

or

$$2^{k+1} > 2(k + 1) + 1,$$

as required. Thus, $S_k$ implies $S_{k+1}$, and this, together with the fact that $S_3$ is true, shows that $S_n$ is true for every positive integer $n$ greater than or equal to 3.

## Proof of the Binomial Theorem

The binomial theorem, introduced in the previous section, can be proved by mathematical induction. That is, for any positive integer $n$ and any numbers $x$ and $y$,

$$(x + y)^n = x^n + \binom{n}{1}x^{n-1}y + \binom{n}{2}x^{n-2}y^2 + \binom{n}{3}x^{n-3}y^3 + \cdots + \binom{n}{r}x^{n-r}y^r + \cdots + \binom{n}{n-1}xy^{n-1} + y^n. \tag{1}$$

***Proof*** Let $S_n$ be statement (1). Begin by verifying $S_n$ for $n = 1$,

$$S_1\text{:} \quad (x + y)^1 = x^1 + y^1,$$

which is true.

Now, assume that $S_n$ is true for the positive integer $k$. Statement $S_k$ becomes (using the definition of the binomial coefficient)

$$S_k\text{:} \quad (x + y)^k = x^k + \frac{k!}{1!\,(k-1)!}x^{k-1}y + \frac{k!}{2!\,(k-2)!}x^{k-2}y^2 + \cdots + \frac{k!}{(k-1)!\,1!}xy^{k-1} + y^k. \tag{2}$$

Multiply each side of equation (2) by $x + y$.

$$(x + y)^k \cdot (x + y)$$
$$= x(x + y)^k + y(x + y)^k$$
$$= \left[x \cdot x^k + \frac{k!}{1!(k-1)!}x^k y + \frac{k!}{2!(k-2)!}x^{k-1}y^2 + \cdots + \frac{k!}{(k-1)!1!}x^2y^{k-1} + xy^k\right]$$
$$+ \left[x^k \cdot y + \frac{k!}{1!(k-1)!}x^{k-1}y^2 + \cdots + \frac{k!}{(k-1)!1!}xy^k + y \cdot y^k\right]$$

Rearrange terms to get

$$(x + y)^{k+1} = x^{k+1} + \left[\frac{k!}{1!(k-1)!} + 1\right]x^k y + \left[\frac{k!}{2!(k-2)!} + \frac{k!}{1!(k-1)!}\right]x^{k-1}y^2$$
$$+ \cdots + \left[1 + \frac{k!}{(k-1)!1!}\right]xy^k + y^{k+1}. \qquad (3)$$

The first expression in brackets in equation (3) simplifies to $\binom{k+1}{1}$. To see this, note that

$$\binom{k+1}{1} = \frac{(k+1)(k)(k-1)(k-2)\cdots 1}{1 \cdot (k)(k-1)(k-2)\cdots 1} = k + 1.$$

Also, $$\frac{k!}{1!(k-1)!} + 1 = \frac{k(k-1)!}{1(k-1)!} + 1 = k + 1.$$

The second expression becomes $\binom{k+1}{2}$, the last $\binom{k+1}{k}$, and so on. The result of equation (3) is just equation (2) with every $k$ replaced by $k + 1$. Thus, the truth of $S_n$ when $n = k$ implies the truth of $S_n$ for $n = k + 1$, which completes the proof of the theorem by mathematical induction.

## 11.5 EXERCISES

**1.** *Concept Check* A proof by mathematical induction allows us to prove that a statement is true for all ________.

**2.** Suppose that Step 2 in a proof by mathematical induction can be satisfied, but Step 1 cannot. May we conclude that the proof is complete? Explain.

**3.** *Concept Check* For which natural numbers is the statement $2^n > 2n$ *not true*?

**4.** Write out in full and verify the statements $S_1$, $S_2$, $S_3$, $S_4$, and $S_5$ for the following. Then, use mathematical induction to prove that the statement is true for every positive integer $n$.

$$2 + 4 + 6 + \cdots + 2n = n(n + 1)$$

*Use mathematical induction to prove each statement. Assume that n is a positive integer.*

**5.** $3 + 6 + 9 + \cdots + 3n = \dfrac{3n(n+1)}{2}$

**6.** $1 + 3 + 5 + \cdots + (2n - 1) = n^2$

**7.** $5 + 10 + 15 + \cdots + 5n = \dfrac{5n(n+1)}{2}$

**8.** $4 + 7 + 10 + \cdots + (3n + 1) = \dfrac{n(3n+5)}{2}$

**9.** $3 + 3^2 + 3^3 + \cdots + 3^n = \dfrac{3(3^n - 1)}{2}$

**10.** $1^2 + 2^2 + 3^2 + \cdots + n^2 = \dfrac{n(n+1)(2n+1)}{6}$

**11.** $1^3 + 2^3 + 3^3 + \cdots + n^3 = \frac{n^2(n+1)^2}{4}$

**12.** $5 \cdot 6 + 5 \cdot 6^2 + 5 \cdot 6^3 + \cdots + 5 \cdot 6^n = 6(6^n - 1)$

**13.** $\frac{1}{1 \cdot 2} + \frac{1}{2 \cdot 3} + \frac{1}{3 \cdot 4} + \cdots + \frac{1}{n(n+1)} = \frac{n}{n+1}$

**14.** $7 \cdot 8 + 7 \cdot 8^2 + 7 \cdot 8^3 + \cdots + 7 \cdot 8^n = 8(8^n - 1)$

**15.** $\frac{4}{5} + \frac{4}{5^2} + \frac{4}{5^3} + \cdots + \frac{4}{5^n} = 1 - \frac{1}{5^n}$

**16.** $\frac{1}{2} + \frac{1}{2^2} + \frac{1}{2^3} + \cdots + \frac{1}{2^n} = 1 - \frac{1}{2^n}$

**17.** $\frac{1}{1 \cdot 4} + \frac{1}{4 \cdot 7} + \frac{1}{7 \cdot 10} + \cdots + \frac{1}{(3n-2)(3n+1)} = \frac{n}{3n+1}$

**18.** $x^{2n} + x^{2n-1}y + \cdots + xy^{2n-1} + y^{2n} = \frac{x^{2n+1} - y^{2n+1}}{x - y}$

*Find all positive integers n for which the given statement is* not *true.*

**19.** $3^n > 6n$

**20.** $3^n > 2n + 1$

**21.** $2^n > n^2$

**22.** $n! > 2n$

*Prove each statement by mathematical induction.*

**23.** $(a^m)^n = a^{mn}$ (Assume $a$ and $m$ are constant.)

**24.** $(ab)^n = a^n b^n$ (Assume $a$ and $b$ are constant.)

**25.** $2^n > 2n$, if $n \geq 3$

**26.** $3^n > 2n + 1$, if $n \geq 2$

**27.** If $a > 1$, then $a^n > 1$.

**28.** If $a > 1$, then $a^n > a^{n-1}$.

**29.** If $0 < a < 1$, then $a^n < a^{n-1}$.

**30.** $2^n > n^2$, for $n > 4$

**31.** If $n \geq 4$, then $n! > 2^n$, where $n! = n(n-1)(n-2) \cdots (3)(2)(1)$.

**32.** $4^n > n^4$, for $n \geq 5$

*Solve each problem.*

**33.** *Number of Handshakes* Suppose that each of the $n$ ($n \geq 2$) people in a room shakes hands with everyone else, but not with himself. Show that the number of handshakes is $\frac{n^2 - n}{2}$.

**34.** *Sides of a Polygon* The series of sketches starts with an equilateral triangle having sides of length 1. In the following steps, equilateral triangles are constructed on each side of the preceding figure. The length of the sides of each new triangle is $\frac{1}{3}$ the length of the sides of the preceding triangles. Develop a formula for the number of sides of the $n$th figure. Use mathematical induction to prove your answer.

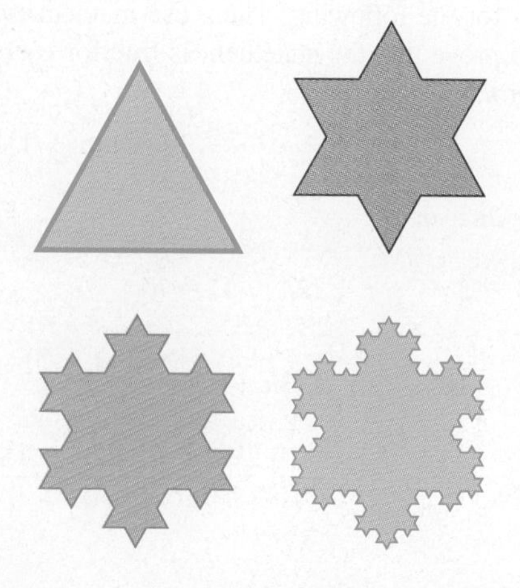

**35.** *Perimeter* Find the perimeter of the $n$th figure in Exercise 34.

**36.** *Area* Show that the area of the $n$th figure in Exercise 34 is

$$\sqrt{3}\left[\frac{2}{5} - \frac{3}{20}\left(\frac{4}{9}\right)^{n-1}\right].$$

**37.** *Tower of Hanoi* A pile of $n$ rings, each ring smaller than the one below it, is on a peg. Two other pegs are attached to a board with this peg. In the game called the *Tower of Hanoi* puzzle, all the rings must be moved to a different peg, with only one ring moved at a time, and with no ring ever placed on top of a smaller ring. Find the least number of moves that would be required. Prove your result with mathematical induction.

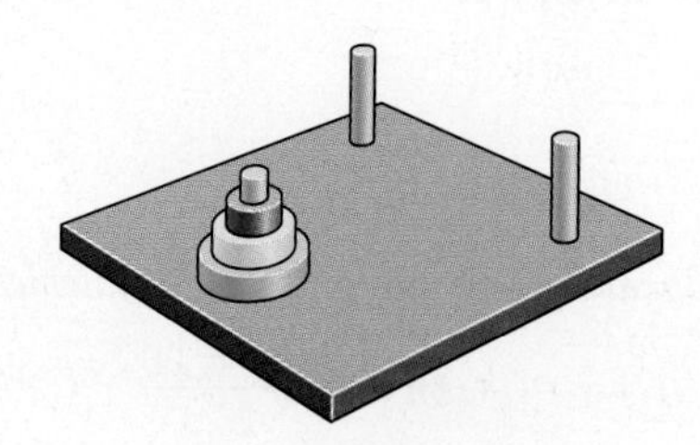

## Reviewing Basic Concepts (Sections 11.4 and 11.5)

1. Calculate ${}_5C_3$.
2. How many terms are there in the expansion of $(x + y)^n$?
3. Write the binomial expansion of $(a + 2b)^4$.
4. Write the binomial expansion of $\left(\frac{1}{m} - n^2\right)^3$.
5. Find the third term of $(x - 2y)^6$.
6. Use mathematical induction to prove that

$$4 + 8 + 12 + 16 + \cdots + 4n = 2n(n + 1).$$

7. Use mathematical induction to prove that $n^2 \leq 2^n$ for $n \geq 4$.

# COUNTING THEORY

Fundamental Principle of Counting • Permutations • Combinations • Distinguishing between Permutations and Combinations

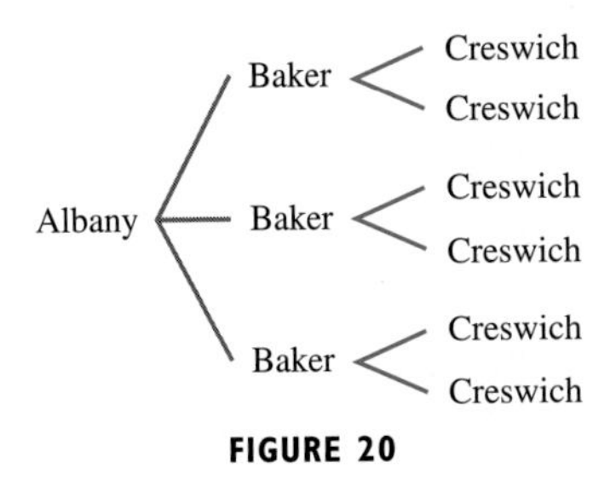


FIGURE 20

## Fundamental Principle of Counting

If there are 3 roads from Albany to Baker and 2 roads from Baker to Creswich, in how many ways can one travel from Albany to Creswich by way of Baker? For each of the 3 roads from Albany to Baker, there are 2 different roads from Baker to Creswich. Hence, there are $3 \cdot 2 = 6$ different ways to make the trip, as shown in the **tree diagram** in Figure 20.

Two events are **independent events** if neither influences the outcome of the other. The opening example illustrates the fundamental principle of counting with independent events.

**Fundamental Principle of Counting**

If $n$ independent events occur, with

$m_1$ ways for event 1 to occur,
$m_2$ ways for event 2 to occur,
⋮
and $m_n$ ways for event $n$ to occur,

then there are

$$m_1 \cdot m_2 \cdot \cdots \cdot m_n$$

different ways for all $n$ events to occur.

**EXAMPLE 1** *Using the Fundamental Principle of Counting*

A restaurant offers a choice of 3 salads, 5 main dishes, and 2 desserts. Use the fundamental principle of counting to find the number of different 3-course meals that can be selected.

**Solution** Three events are involved: selecting a salad, selecting a main dish, and selecting a dessert. The first event can occur in 3 ways, the second event can occur in 5 ways, and the third event can occur in 2 ways; thus, there are

$$3 \cdot 5 \cdot 2 = 30 \text{ possible meals.}$$

**EXAMPLE 2** *Using the Fundamental Principle of Counting*

A teacher has 5 different books that she wishes to arrange in a row. How many different arrangements are possible?

**Solution** Five events are involved: selecting a book for the first spot, selecting a book for the second spot, and so on. For the first spot, the teacher has 5 choices. After a choice has been made, the teacher has 4 choices for the second spot. Continuing in this manner, there are 3 choices for the third spot, 2 for the fourth spot, and 1 for the fifth spot. By the fundamental principle of counting, there are

$$5 \cdot 4 \cdot 3 \cdot 2 \cdot 1 = 120 \text{ arrangements.}$$

**EXAMPLE 3** *Arranging $r$ of $n$ Items ($r < n$)*

Suppose the teacher in Example 2 wishes to place only 3 of the 5 books in a row. How many arrangements of 3 books are possible?

**Solution** The teacher still has 5 ways to fill the first spot, 4 ways to fill the second spot, and 3 ways to fill the third. Since only 3 books will be used, there are only 3 spots to be filled (3 events) instead of 5, with

$$5 \cdot 4 \cdot 3 = 60 \text{ arrangements.}$$

## Permutations

Refer to Example 3. Since each ordering of 3 books is considered a different *arrangement,* the number 60 in that example is called the number of *permutations* of 5 things taken 3 at a time, written $P(5, 3) = 60$. The number of ways of arranging 5 elements from a set of 5 elements, written $P(5, 5) = 120$, was found in Example 2.

A **permutation** of $n$ elements taken $r$ at a time is one of the *arrangements* of $r$ elements from a set of $n$ elements. Generalizing from the preceding examples, the number of permutations of $n$ elements, taken $r$ at a time, denoted by $P(n, r)$, is

$$\begin{aligned} P(n, r) &= n(n-1)(n-2)\cdots(n-r+1) \\ &= \frac{n(n-1)(n-2)\cdots(n-r+1)(n-r)(n-r-1)\cdots(2)(1)}{(n-r)(n-r-1)\cdots(2)(1)} \\ &= \frac{n!}{(n-r)!}. \end{aligned}$$

In summary, we have the following result.

**Permutations of $n$ Elements Taken $r$ at a Time**

If $P(n, r)$ denotes the number of permutations of $n$ elements taken $r$ at a time, with $r \leq n$, then

$$P(n, r) = \frac{n!}{(n-r)!}.$$

An alternative notation for $P(n, r)$ is ${}_nP_r$.

**EXAMPLE 4** *Using the Permutations Formula*

Find the following.

**(a)** The number of permutations of the letters L, M, and N

**(b)** The number of permutations of 2 of the 3 letters L, M, and N

**Analytic Solution**

**(a)** By the formula for $P(n, r)$, with $n = 3$ and $r = 3$,

$$P(3,3) = \frac{3!}{(3-3)!} = \frac{3!}{0!} = \frac{3!}{1} = 3 \cdot 2 \cdot 1 = 6.$$

As shown in the tree diagram in Figure 21, the 6 permutations are

LMN, LNM, MLN, MNL, NLM, NML.

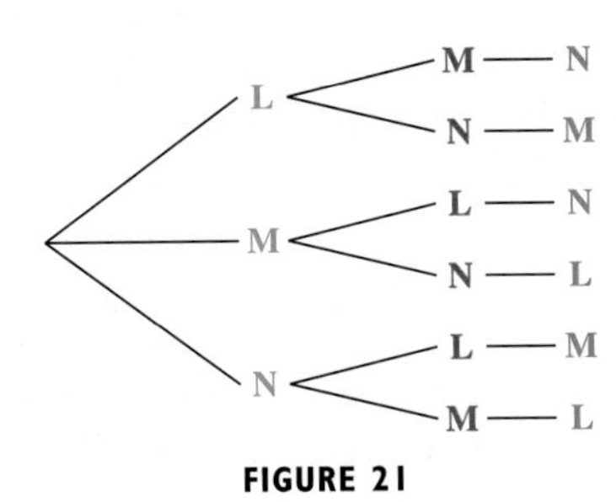


FIGURE 21

**(b)** Find $P(3, 2)$.

$$P(3,2) = \frac{3!}{(3-2)!} = \frac{3!}{1!} = \frac{6}{1} = 6$$

This result is the same as the answer in part (a) because after the first two choices are made, the third choice is already determined since only one letter is left.

**Graphing Calculator Solution**

Graphing calculators use the notation nPr for evaluating permutations. This function, like the combinations function, is often found in the math menu. The screen in Figure 22 confirms the analytic results in parts (a) and (b).

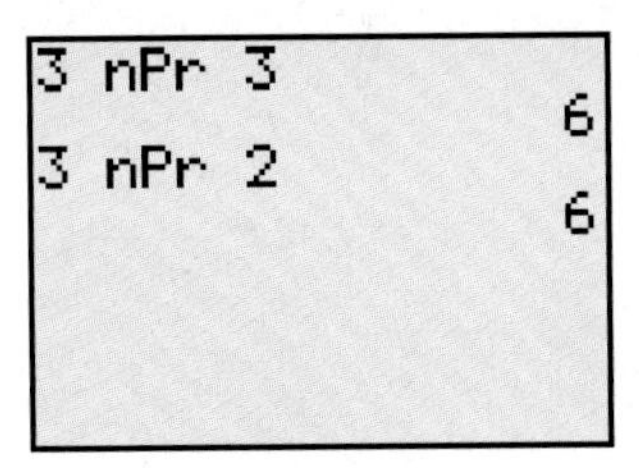


FIGURE 22

You can consult the graphing calculator manual that accompanies this text or your owner's manual to learn how to calculate permutations.

**EXAMPLE 5** *Using the Permutations Formula*

Suppose 8 people enter an event in a swim meet. In how many ways could the gold, silver, and bronze prizes be awarded?

**Analytic Solution**

Using the fundamental principle of counting, there are 3 choices to be made, resulting in $8 \cdot 7 \cdot 6 = 336$ ways. The formula for $P(n, r)$ also can be used to get the same result.

$$P(8,3) = \frac{8!}{5!} = \frac{8 \cdot 7 \cdot 6 \cdot 5 \cdot 4 \cdot 3 \cdot 2 \cdot 1}{5 \cdot 4 \cdot 3 \cdot 2 \cdot 1}$$
$$= 336 \text{ ways}$$

**Graphing Calculator Solution**

See Figure 23 for calculator support of this computation.

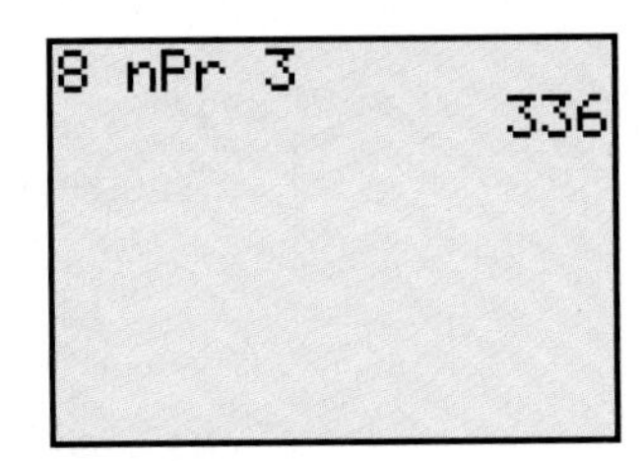


FIGURE 23

**EXAMPLE 6** *Using the Permutations Formula*

In how many ways can 5 students be seated in a row of 5 desks?

**Solution** Use $P(n, n)$ with $n = 5$ to get

$$P(5,5) = 5! = 5 \cdot 4 \cdot 3 \cdot 2 \cdot 1 = 120.$$

## Combinations

Earlier, we saw that there are 60 ways that a teacher can arrange 3 of 5 different books in a row. That is, there are 60 permutations of 5 things taken 3 at a time. Suppose that the teacher does not wish to arrange the books in a row, but rather wishes to choose, without regard to order, any 3 of the 5 books to donate to a book sale. In how many ways can this be done?

At first glance, we might say 60 again, but this is incorrect. The number 60 counts all possible *arrangements* of 3 books chosen from 5. The following 6 arrangements, however, would all lead to the same set of 3 books being given to the book sale.

| | |
|---|---|
| mystery-biography-textbook | biography-textbook-mystery |
| mystery-textbook-biography | textbook-biography-mystery |
| biography-mystery-textbook | textbook-mystery-biography |

The list shows 6 different *arrangements* of 3 books but only one *set* of 3 books. A subset of items selected *without regard to order* is called a **combination.** The number of combinations of 5 things taken 3 at a time is written $\binom{5}{3}$, or $C(5, 3)$.

**NOTE** This combinations notation also represents the binomial coefficient. That is, binomial coefficients are the combinations of $n$ elements chosen $r$ at a time.

To evaluate $\binom{5}{3}$ or $C(5, 3)$, start with the $5 \cdot 4 \cdot 3$ *permutations* of 5 things taken 3 at a time. Since order doesn't matter, and each subset of 3 items from the set of 5 items can have its elements rearranged in $3 \cdot 2 \cdot 1 = 3!$ ways, $\binom{5}{3}$ can be found by dividing the number of permutations by 3!, or

$$\binom{5}{3} = \frac{5 \cdot 4 \cdot 3}{3!} = \frac{5 \cdot 4 \cdot 3}{3 \cdot 2 \cdot 1} = 10.$$

There are 10 ways that the teacher can choose 3 books for the book sale.

Generalizing this discussion gives the formula for the number of combinations of $n$ elements taken $r$ at a time:

$$C(n, r) = \binom{n}{r} = \frac{P(n, r)}{r!}.$$

Another version of this formula is found as follows.

$$C(n, r) = \binom{n}{r} = \frac{P(n, r)}{r!} = \frac{n!}{(n-r)!} \cdot \frac{1}{r!} = \frac{n!}{(n-r)!\,r!}$$

This last version is the most useful for calculation and is the one we used earlier to calculate binomial coefficients.

**Combinations of $n$ Elements Taken $r$ at a Time**

If $C(n, r)$ or $\binom{n}{r}$ or represents the number of combinations of $n$ things taken $r$ at a time, with $r \leq n$, then

$$C(n, r) = \binom{n}{r} = \frac{n!}{(n-r)!r!}.$$

NOTE $C(n, r) = {}_nC_r = \binom{n}{r}$

**EXAMPLE 7** *Using the Combinations Formula*

How many different committees of 3 people can be chosen from a group of 8 people?

**Analytic Solution**

Since a committee is an unordered set, use combinations to get

$$\begin{aligned} C(8, 3) &= \binom{8}{3} \\ &= \frac{8!}{5!3!} \\ &= \frac{8 \cdot 7 \cdot 6 \cdot 5 \cdot 4 \cdot 3 \cdot 2 \cdot 1}{5 \cdot 4 \cdot 3 \cdot 2 \cdot 1 \cdot 3 \cdot 2 \cdot 1} \\ &= 56. \end{aligned}$$

**Graphing Calculator Solution**

The notation nCr is used by graphing calculators to find combinations. The screen in Figure 24 agrees with the analytic result.

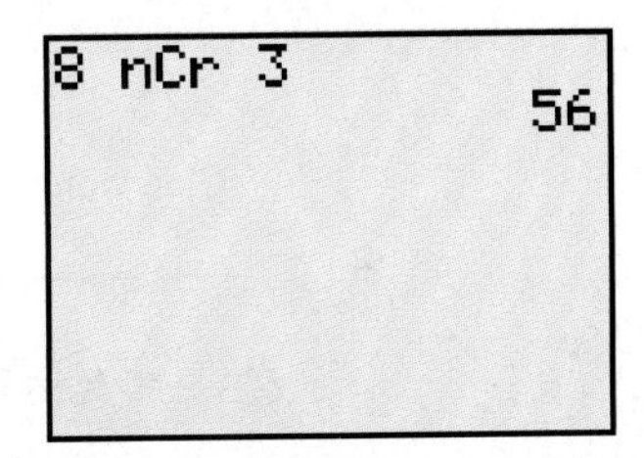


FIGURE 24

**EXAMPLE 8** *Using the Combinations Formula*

Three lawyers are to be selected from a group of 30 to work on a special project.

**(a)** In how many different ways can the lawyers be selected?

**(b)** In how many ways can the group of 3 be selected if a certain lawyer must work on the project?

**Analytic Solution**

**(a)** Here we wish to know the number of 3-element combinations that can be formed from a set of 30 elements. (We want combinations, not permutations, since order within the group does not matter.)

$$C(30, 3) = \binom{30}{3} = \frac{30!}{27!3!} = 4060$$

There are 4060 ways to select the project group.

**Graphing Calculator Solution**

The screen in Figure 25 on the next page confirms the analytic results in parts (a) and (b). You can consult the graphing calculator manual that accompanies this text or your owner's manual to learn how to calculate combinations.

*(continued)*

**(b)** Since 1 lawyer has already been selected for the project, the problem is reduced to selecting 2 more from the remaining 29 lawyers.

$$C(29, 2) = \binom{29}{2} = \frac{29!}{27!\,2!} = 406$$

In this case, the project group can be selected in 406 ways.

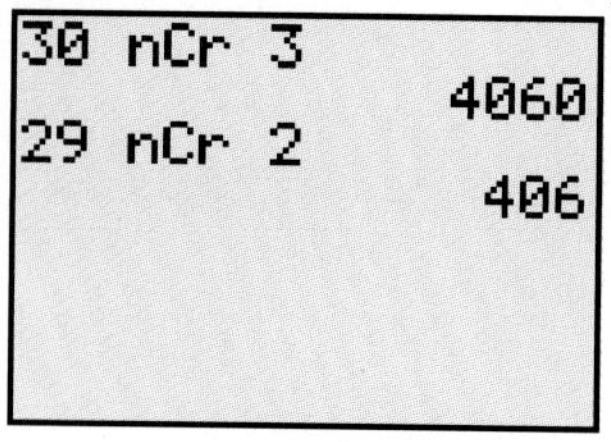

```
30 nCr 3
            4060
29 nCr 2
             406
```

FIGURE 25

## Distinguishing between Permutations and Combinations

Permutations and combinations will be used to solve probability problems in the next section. Both permutations and combinations give the number of ways to choose $r$ objects from a set of $n$ objects. However, the differences between permutations and combinations are outlined in the following chart.

| Permutations | Combinations |
|---|---|
| Different orderings or arrangements of the $r$ objects are different permutations. | Each choice or subset of $r$ objects gives one combination. Order within the group of $r$ objects does *not* matter. |
| $P(n, r) = \frac{n!}{(n-r)!}$ | $C(n, r) = \binom{n}{r} = \frac{n!}{(n-r)!\,r!}$ |
| Clue words: arrangement, schedule, order | Clue words: group, committee, sample, selection |

**EXAMPLE 9** *Distinguishing between Permutations and Combinations*

A sales representative has 10 accounts in a certain city.

**(a)** In how many ways can a group of 3 accounts be selected?

**(b)** In how many ways can calls be scheduled for 3 of the 10 accounts?

**Solution**

**(a)** Within a selection of 3 accounts, the arrangement of the accounts is not important, so use combinations.

$$C(10, 3) = \binom{10}{3} = \frac{10!}{7!\,3!} = 120$$

There are 120 ways to select a group of 3 accounts.

**(b)** To schedule calls, the sales representative must *order* each selection of 3 accounts. Use permutations here, since order is important.

$$P(10, 3) = \frac{10!}{(10-3)!} = \frac{10!}{7!} = 720$$

There are 720 orderings in which the 3 accounts can be called on.

To illustrate the differences between permutations and combinations in another way, suppose 2 cans of soup are to be selected from 4 cans on a shelf: noodle (N), bean

(B), mushroom (M), and tomato (T). As shown in Figure 26(a), there are 12 ways to select 2 cans from the 4 cans if the order matters (if noodle first and bean second is considered different from bean, then noodle, for example). On the other hand, if order is unimportant, then there are 6 ways to choose 2 cans of soup from the 4, as illustrated in Figure 26(b).

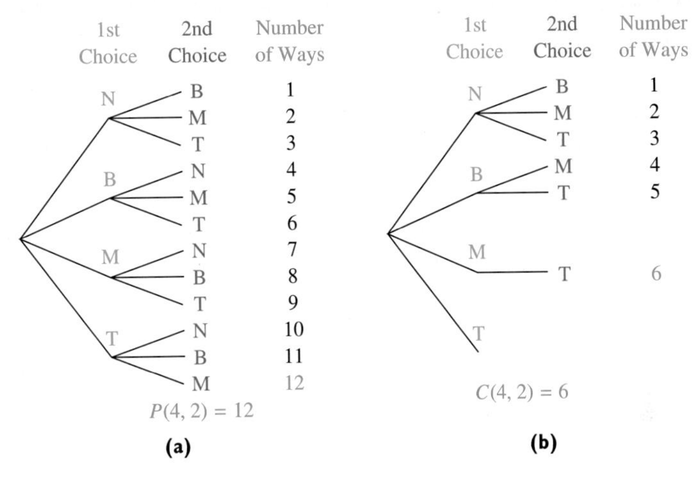


**FIGURE 26**

**CAUTION** Not all counting problems lend themselves to either permutations or combinations. Whenever a tree diagram or the multiplication principle can be used directly, do so.

## 11.6 EXERCISES

*Evaluate each expression.*

**1.** $P(12, 8)$ **2.** $P(5, 5)$ **3.** $P(9, 2)$ **4.** $P(10, 9)$ **5.** $P(5, 1)$ **6.** $P(6, 0)$

**7.** $C(4, 2)$ **8.** $C(9, 3)$ **9.** $C(6, 0)$ **10.** $C(8, 1)$ **11.** $\binom{12}{4}$ **12.** $\binom{16}{3}$

*Use a calculator to evaluate each expression.*

**13.** ${}_{20}P_5$ **14.** ${}_{100}P_5$ **15.** ${}_{15}P_8$ **16.** ${}_{32}P_4$

**17.** ${}_{20}C_5$ **18.** ${}_{100}C_5$ **19.** $\binom{15}{8}$ **20.** $\binom{32}{4}$

**21.** *Concept Check* Decide whether the situation described involves a permutation or a combination of objects.

**(a)** A telephone number

**(b)** A Social Security number

**(c)** A hand of cards in poker

**(d)** A committee of politicians

**(e)** The "combination" on a combination lock

**(f)** A lottery choice of six numbers where the order does not matter

**(g)** An automobile license plate

**22.** Explain the difference between a permutation and a combination. What should you look for in a problem to decide which of these is an appropriate method of solution?

*Use the fundamental principle of counting to solve each problem.*

**23.** *Home Plan Choices* How many different types of homes are available if a builder offers a choice of 5 basic plans, 3 roof styles, and 2 exterior finishes?

**24.** *Auto Varieties* An auto manufacturer produces 7 models, each available in 6 different colors, with 4 different upholstery fabrics, and 5 interior colors. How many varieties of the auto are available?

**25.** *Radio Station Call Letters* How many different 4-letter radio-station call letters can be made

**(a)** if the first letter must be K or W and no letter may be repeated?

**(b)** if repeats are allowed (but the first letter is K or W)?

**(c)** How many of the 4-letter call letters (starting with K or W) with no repeats end in R?

**26.** *Meal Choices* A menu offers a choice of 3 salads, 8 main dishes, and 5 desserts. How many different 3-course meals (salad, main dish, dessert) are possible?

**27.** *Names for a Baby* A couple has narrowed down the choice of a name for their new baby to 3 first names and 5 middle names. How many different first- and middle-name combinations are possible?

**28.** *Concert Program Arrangement* A concert to raise money for an economics prize is to consist of 5 works: 2 overtures, 2 sonatas, and a piano concerto. In how many ways can a program with these 5 works be arranged?

**29.** *License Plates* For many years, the state of California used 3 letters followed by 3 digits on its automobile license plates.

**(a)** How many different license plates are possible with this arrangement?

**(b)** When the state ran out of new plates, the order was reversed to 3 digits followed by 3 letters. How many additional plates were then possible?

**(c)** Several years ago, the plates described in part (b) were also used up. The state then issued plates with 1 letter followed by 3 digits and then 3 letters. How many plates does this scheme provide?

**30.** *Telephone Numbers* How many 7-digit telephone numbers are possible if the first digit cannot be 0 and

**(a)** only odd digits may be used?

**(b)** the telephone number must be a multiple of 10 (that is, it must end in 0)?

**(c)** the telephone number must be a multiple of 100?

**(d)** the first 3 digits are 481?

**(e)** no repetitions are allowed?

*Solve each problem involving permutations.*

**31.** *Seating People in a Row* In an experiment on social interaction, 6 people will sit in 6 seats in a row. In how many ways can this be done?

**32.** *Genetics Experiment* In how many ways can 7 of 10 monkeys be arranged in a row for a genetics experiment?

**33.** *Course Schedule Arrangement* A business school offers courses in keyboarding, spreadsheets, transcription, business English, technical writing, and accounting. In how many ways can a student arrange a schedule if 3 courses are taken?

**34.** *Course Schedule Arrangement* If your college offers 400 courses, 20 of which are in mathematics, and your counselor arranges your schedule of 4 courses by random selection, how many schedules are possible that do not include a math course?

**35.** *Club Officer Choices* In a club with 15 members, how many ways can a slate of 3 officers consisting of president, vice-president, and secretary/treasurer be chosen?

**36.** *Batting Orders* A baseball team has 20 players. How many 9-player batting orders are possible?

**37.** *Basketball Positions* In how many ways can 5 players be assigned to the 5 positions on a basketball team, assuming that any player can play any position? In how many ways can 10 players be assigned to the 5 positions?

**38.** *Letter Arrangement* How many ways can all the letters of the word TOUGH be arranged?

*Solve each problem involving combinations.*

**39.** *Seminar Presenters* A banker's association has 30 members. If 4 members are selected at random to present a seminar, how many different groups of 4 are possible?

**40.** *Apple Samples* How many different samples of 3 apples can be drawn from a crate of 25 apples?

**41.** *Hamburger Choices* Howard's Hamburger Heaven sells hamburgers with cheese, relish, lettuce, tomato, mustard, or ketchup. How many different hamburgers can be made that use any 3 of the extras?

**42.** *Financial Planners* Three financial planners are to be selected from a group of 12 to participate in a special program. In how many ways can this be done? In how many ways can the group that will not participate be selected?

**43.** *Card Combinations* Five cards are marked with the numbers 1, 2, 3, 4, or 5, shuffled, and 2 cards are then drawn. How many different 2-card hands are possible?

**44.** *Marble Samples* If a bag contains 15 marbles, how many samples of 2 marbles can be drawn from it? how many samples of 4 marbles?

**45.** *Marble Samples* In Exercise 44, if the bag contains 3 yellow, 4 white, and 8 blue marbles, how many samples of 2 can be drawn in which both marbles are blue?

**46.** *Apple Samples* In Exercise 40, if it is known that there are 5 rotten apples in the crate,

**(a)** how many samples of 3 could be drawn in which all 3 are rotten?

**(b)** how many samples of 3 could be drawn in which there are 1 rotten apple and 2 good apples?

**47.** *Convention Delegation Choices* A city council is composed of 5 liberals and 4 conservatives. Three members are to be selected randomly as delegates to a convention.

**(a)** How many delegations are possible?

**(b)** How many delegations could have all liberals?

**(c)** How many delegations could have 2 liberals and 1 conservative?

**(d)** If 1 member of the council serves as mayor, how many delegations are possible that include the mayor?

**48.** *Delegation Choices* Seven workers decide to send a delegation of 2 to their supervisor to discuss their grievances.

**(a)** How many different delegations are possible?

**(b)** If it is decided that a certain employee must be in the delegation, how many different delegations are possible?

**(c)** If there are 2 women and 5 men in the group, how many delegations would include at least 1 woman?

*Use any or all of the methods described in this section to solve each problem.*

**49.** *Course Schedule* If Dwight Johnston has 8 courses to choose from, how many ways can he arrange his schedule if he must pick 4 of them?

**50.** *Pineapple Samples* How many samples of 3 pineapples can be drawn from a crate of 12?

**51.** *Soup Ingredients* Velma specializes in making different vegetable soups with carrots, celery, beans, peas, mushrooms, and potatoes. How many different soups can she make with any 4 ingredients?

**52.** *Secretary/Manager Assignments* From a pool of 7 secretaries, 3 are selected to be assigned to 3 managers, 1 secretary to each manager. In how many ways can this be done?

**53.** *Musical Chairs Seatings* In a game of musical chairs, 12 children will sit in 11 chairs (1 will be left out). How many seatings are possible?

**54.** *Plant Samples* In an experiment on plant hardiness, a researcher gathers 6 wheat plants, 3 barley plants, and 2 rye plants. She wishes to select 4 plants at random.

**(a)** In how many ways can this be done?

**(b)** In how many ways can this be done if exactly 2 wheat plants must be included?

**55.** *Committee Choices* In a club with 8 men and 11 women members, how many 5-member committees can be chosen that have the following?

**(a)** All men **(b)** All women

**(c)** 3 men and 2 woman **(d)** No more than 3 women

**56.** *Committee Choices* From 10 names on a ballot, 4 will be elected to a political party committee. In how many ways can the committee of 4 be formed if each person will have a different responsibility?

**57.** *Combination Lock* A briefcase has 2 locks. The combination to each lock consists of a 3-digit number, where digits may be repeated. How many combinations are possible? (*Hint:* The word *combination* is a misnomer. Lock combinations are permutations where the arrangement of the numbers is important.)

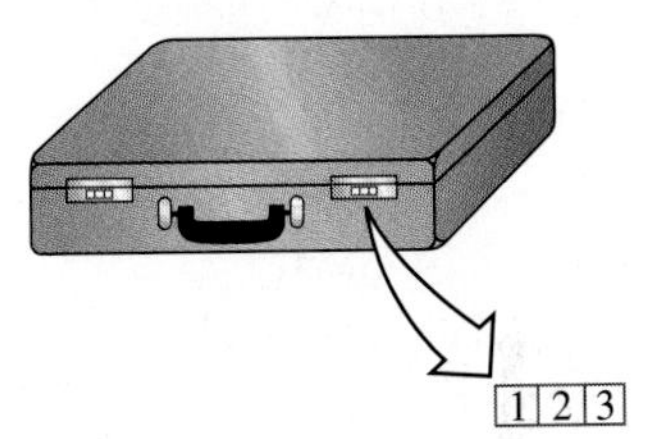


**58.** *Combination Lock* A typical combination for a padlock consists of 3 numbers from 0 to 39. Count the number of combinations that are possible with this type of lock, if a number may be repeated.

**59.** *Garage Door Openers* The code for some garage door openers consists of 12 electrical switches that can be set to either 0 or 1 by the owner. With this type of opener, how many codes are possible? (*Source:* Promax.)

**60.** *Lottery* To win the jackpot in a lottery game, a person must pick 3 numbers from 0 to 9 in the correct order. If a number can be repeated, how many ways are there to play the game?

**61.** *Keys* How many distinguishable ways can 4 keys be put on a circular key ring?

**62.** *Sitting at a Round Table* How many ways can 7 people sit at a round table? Assume that a different way means that at least 1 person is sitting next to someone different.

*Prove each statement for positive integers n and r, with $r \leq n$. (Hint: Use the definitions of permutations and combinations.)*

**63.** $P(n, n-1) = P(n, n)$

**64.** $P(n, 1) = n$

**65.** $P(n, 0) = 1$

**66.** $\binom{n}{n} = 1$

**67.** $\binom{n}{0} = 1$

**68.** $\binom{n}{n-1} = n$

**69.** $\binom{n}{n-r} = \binom{n}{r}$

**70.** Explain why the restriction $r \leq n$ is needed in the formula for $P(n, r)$.

## RELATING CONCEPTS

**For individual or group investigation (Exercises 71 and 72)**

*Series are often used in mathematics and science to make approximations. Large values of factorials often occur in counting theory. The value of n! can quickly become too large for most calculators to evaluate. To estimate n! for large values of n, we can use the property of logarithms that*

$$\begin{aligned}\log(n!) &= \log(1 \times 2 \times 3 \times \cdots \times n) \\ &= \log 1 + \log 2 + \log 3 + \cdots + \log n.\end{aligned}$$

*Using a sum and sequence utility on a calculator, we can then determine r such that $n! \approx 10^r$ since $r = \log n!$. For example, the screen shown here illustrates that a calculator will give the same approximation of 30! using the factorial function and the formula just discussed.*

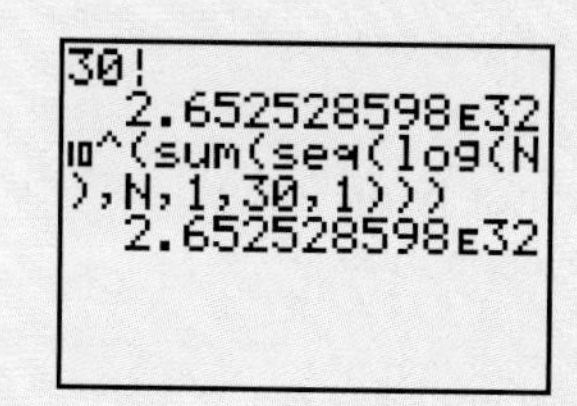


*Use this technique to approximate each quantity in Exercises 71 and 72. Then, try to compute each value directly on your calculator.*

**71. (a)** 50! **(b)** 60! **(c)** 65!

**72. (a)** $P(47, 13)$ **(b)** $P(50, 4)$ **(c)** $P(29, 21)$

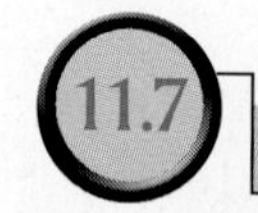

# 11.7 PROBABILITY

Basic Concepts • Complements and Venn Diagrams • Odds • Union of Two Events • Binomial Probability

## Basic Concepts

Consider an experiment that has one or more **outcomes,** each of which is equally likely to occur. For example, the experiment of tossing a fair coin has two equally likely outcomes: heads ($H$) or tails ($T$). Also, the experiment of rolling a fair die has 6 equally likely outcomes: landing so the face that is up shows a 1, 2, 3, 4, 5, or 6.

The set $S$ of all possible outcomes of a given experiment is called the **sample space** for the experiment. (In this text, all sample spaces are finite.) The sample space for the experiment of tossing a coin once consists of the outcomes $H$ and $T$. This sample space can be written in set notation as

$$S = \{H, T\}.$$

Similarly, a sample space for the experiment of rolling a single die once is

$$S = \{1, 2, 3, 4, 5, 6\}.$$

Any subset of the sample space is called an **event.** In the experiment with the die, for example, "the number showing is a 3" is an event, say $E_1$, such that $E_1 = \{3\}$. "The number showing is greater than 3" is also an event, say $E_2$, such that $E_2 = \{4, 5, 6\}$. To represent the number of outcomes that belong to event $E$, the notation $n(E)$ is used. Then, $n(E_1) = 1$ and $n(E_2) = 3$.

The notation $P(E)$ is used for the *probability* of an event $E$. If the outcomes in the sample space for an experiment are equally likely, then the probability of event $E$ occurring is found as follows.

**Probability of Event $E$**

In a sample space with equally likely outcomes, the **probability** of an event $E$, written $P(E)$, is the ratio of the number of outcomes in sample space $S$ that belong to event $E$, $n(E)$, to the total number of outcomes in sample space $S$, $n(S)$. That is,

$$P(E) = \frac{n(E)}{n(S)}.$$

To use this definition to find the probability of the event $E_1$ in the die experiment, start with the sample space $S = \{1, 2, 3, 4, 5, 6\}$, and the desired event, $E_1 = \{3\}$. Since $n(E_1) = 1$ and $n(S) = 6$,

$$P(E_1) = \frac{n(E_1)}{n(S)} = \frac{1}{6}.$$

**EXAMPLE 1** *Finding Probabilities of Events*

A single die is rolled. Write each event in set notation, and give the probability of the event.

**(a)** $E_3$: the number showing is even

**(b)** $E_4$: the number showing is greater than 4

**(c)** $E_5$: the number showing is less than 7

**(d)** $E_6$: the number showing is 7

**Solution**

**(a)** Since $E_3 = \{2, 4, 6\}$, $n(E_3) = 3$. As given earlier, $n(S) = 6$, so

$$P(E_3) = \frac{3}{6} = \frac{1}{2}.$$

**(b)** Event $E_4 = \{5, 6\}$, with $n(E_4) = 2$, so

$$P(E_4) = \frac{2}{6} = \frac{1}{3}.$$

**(c)** $E_5 = \{1, 2, 3, 4, 5, 6\}$ and $P(E_5) = \frac{6}{6} = 1.$

**(d)** $E_6 = \emptyset$ and $P(E_6) = \frac{0}{6} = 0.$

**NOTE** In Example 1(c), $E_5 = S$. Therefore, the event $E_5$ is certain to occur every time the experiment is performed. An event that is certain to occur, such as $E_5$, always has probability 1. On the other hand, $E_6 = \emptyset$ and $P(E_6) = 0$. The probability of an impossible event, such as $E_6$, is always 0. For any event $E$, $P(E)$ is between 0 and 1 inclusive.

**EXAMPLE 2** *Estimating Probability of Organ Transplants*

In 1997 there were 51,277 people waiting for an organ transplant. The table lists the number of patients waiting for the most common types of transplants. Assuming none of these people need two or more transplants, approximate the probability that a transplant patient chosen at random will need

**(a)** a kidney or a heart; **(b)** neither a kidney nor a heart.

| Organ Transplant | Patients Waiting |
|---|---|
| Heart | 3774 |
| Kidney | 35,025 |
| Liver | 7920 |
| Lung | 2340 |

*Source:* Coalition on Organ and Tissue Donation.

**Solution**

**(a)** Let each patient represent an outcome in a sample space $S$. The event $E$ of a transplant patient needing either a kidney or a heart contains $35{,}025 + 3774 = 38{,}779$ outcomes. The desired probability is

$$P(E) = \frac{n(E)}{n(S)} = \frac{38{,}779}{51{,}277} \approx .76.$$

**(b)** Let $F$ be the event of a patient waiting for an organ other than a kidney or a heart. Then,

$$n(F) = n(S) - n(E) = 51{,}277 - 38{,}779 = 12{,}498.$$

The probability of $F$ is

$$P(F) = \frac{n(F)}{n(S)} = \frac{12{,}498}{51{,}277} \approx .24.$$

## Complements and Venn Diagrams

The set of all outcomes in the sample space that do *not* belong to event $E$ is called the **complement** of $E$, written $E'$. Events $E$ and $F$ in Example 2 are said to be *complements* because an organ transplant patient must be in either $E$ or $F$, but cannot be in both. In an experiment of drawing a single card from a standard deck of 52 cards, let $E$ be the event "the card is an ace." Then, $E'$ is the event "the card is not an ace." From the definition of $E'$, for an event $E$,

$$E \cup E' = S \quad \text{and} \quad E \cap E' = \emptyset.$$

**NOTE** The **union** of two sets $A$ and $B$ is the set $A \cup B$ made up of all the elements from either $A$ or $B$, or both. The **intersection** of sets $A$ and $B$, written $A \cap B$, is made up of all the elements that belong to both sets.

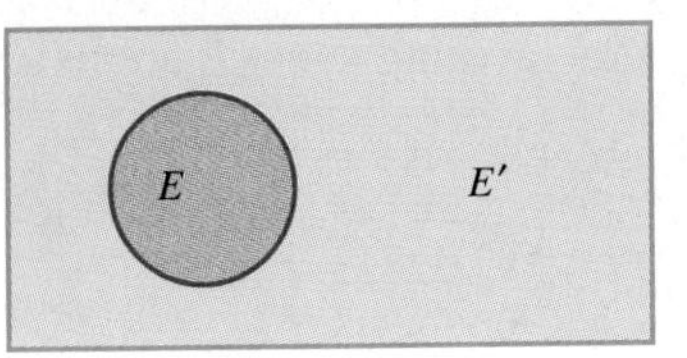


FIGURE 27

Probability concepts can be illustrated with **Venn diagrams,** as shown in Figure 27. The rectangle in Figure 27 represents the sample space in an experiment. The area inside the circle represents event $E$; and the area inside the rectangle, but outside the circle, represents event $E'$.

**NOTE** A standard deck of 52 cards has four suits: hearts ♥, clubs ♣, diamonds ♦, and spades ♠, with thirteen cards of each suit. Each suit has an ace, king, queen, jack, and cards numbered from 2 to 10. The hearts and diamonds are red and the spades and clubs are black. We will refer to this standard deck of cards in this section.

**EXAMPLE 3** *Using the Complement*

In the experiment of drawing a card from a well-shuffled deck, find the probability of event $E$, the card is an ace, and event $E'$.

**Solution** Since there are 4 aces in the deck of 52 cards, $n(E) = 4$ and $n(S) = 52$. Therefore,

$$P(E) = \frac{n(E)}{n(S)} = \frac{4}{52} = \frac{1}{13}.$$

Of the 52 cards, 48 are not aces, so

$$P(E') = \frac{n(E')}{n(S)} = \frac{48}{52} = \frac{12}{13}.$$

In Example 3, $P(E) + P(E') = \frac{1}{13} + \frac{12}{13} = 1$. This is always true for any event $E$ and its complement $E'$. That is,

$$P(E) + P(E') = 1.$$

This can be restated as

$$P(E) = 1 - P(E') \quad \text{or} \quad P(E') = 1 - P(E).$$

These two equations suggest an alternative way to compute the probability of an event. For example, if $P(E) = \frac{1}{13}$, then

$$P(E') = 1 - \frac{1}{13} = \frac{12}{13}.$$

## Odds

Sometimes probability statements are expressed in terms of odds, a comparison of $P(E)$ with $P(E')$. The **odds** in favor of an event $E$ are expressed as the ratio of $P(E)$ to $P(E')$ or as the fraction $\frac{P(E)}{P(E')}$. For example, if the probability of rain can be established as $\frac{1}{3}$, the odds that it will rain are

$$P(\text{rain}) \text{ to } P(\text{no rain}) = \frac{1}{3} \text{ to } \frac{2}{3} = \frac{\frac{1}{3}}{\frac{2}{3}} = \frac{1}{2} \quad \text{or} \quad 1 \text{ to } 2.$$

On the other hand, the odds that it will not rain are 2 to 1 $\left(\text{or } \frac{2}{3} \text{ to } \frac{1}{3}\right)$. If the odds in favor of an event are, say, 3 to 5, then the probability of the event is $\frac{3}{8}$, and the probability of the complement of the event is $\frac{5}{8}$. If the odds favoring event $E$ are $m$ to $n$, then

$$P(E) = \frac{m}{m+n} \quad \text{and} \quad P(E') = \frac{n}{m+n}.$$

**EXAMPLE 4** *Finding Odds in Favor of an Event*

A shirt is selected at random from a dark closet containing 6 blue shirts and 4 shirts that are not blue. Find the odds in favor of a blue shirt being selected.

**Solution** Let $E$ represent "a blue shirt is selected." Then,

$$P(E) = \frac{6}{10} \text{ or } \frac{3}{5} \quad \text{and} \quad P(E') = 1 - \frac{3}{5} = \frac{2}{5}.$$

Therefore, the odds in favor of a blue shirt being selected are

$$P(E) \text{ to } P(E') = \frac{3}{5} \text{ to } \frac{2}{5} = \frac{\frac{3}{5}}{\frac{2}{5}} = \frac{3}{2} \quad \text{or} \quad 3 \text{ to } 2.$$

## Union of Two Events

Since events are sets, we can use set operations to find the union of two events. Suppose a fair die is tossed. Let $H$ be the event "the result is a 3," and $K$ the event "the result is an even number." From the results earlier in this section,

$$H = \{3\} \qquad K = \{2,4,6\} \qquad H \cup K = \{2,3,4,6\}$$

$$P(H) = \frac{1}{6} \qquad P(K) = \frac{3}{6} = \frac{1}{2} \qquad P(H \cup K) = \frac{4}{6} = \frac{2}{3}.$$

Notice that $P(H) + P(K) = P(H \cup K)$.

Before assuming that this relationship is true in general, consider another event $G$ for this experiment, "the result is a 2."

$$G = \{2\} \qquad K = \{2,4,6\} \qquad G \cup K = \{2,4,6\}$$

$$P(G) = \frac{1}{6} \qquad P(K) = \frac{3}{6} = \frac{1}{2} \qquad P(G \cup K) = \frac{3}{6} = \frac{1}{2}$$

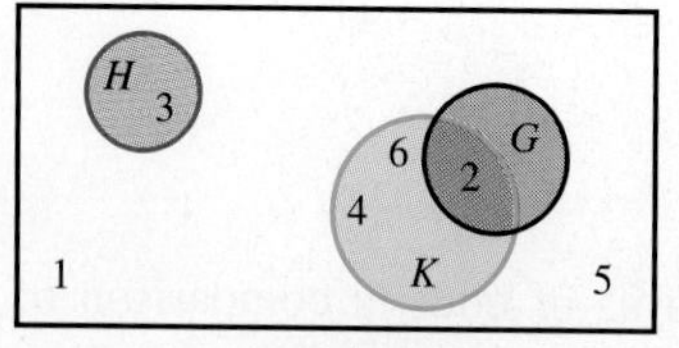


FIGURE 28

In this case, $P(G) + P(K) \neq P(G \cup K)$. As Figure 28 suggests, the difference in the two examples above comes from the fact that events $H$ and $K$ cannot occur simultaneously. Such events are called **mutually exclusive events.** In fact, $H \cap K = \emptyset$, which is always true for mutually exclusive events. Events $G$ and $K$, however, can occur simultaneously. Both are satisfied if the result of the roll is a 2, the element in their intersection ($G \cap K = \{2\}$). This example suggests the following property.

**Probability of the Union of Two Events**

For any events $E$ and $F$,

$$P(E \text{ or } F) = P(E \cup F) = P(E) + P(F) - P(E \cap F).$$

## EXAMPLE 5 *Finding Probabilities of Unions*

One card is drawn from a well-shuffled deck of 52 cards. What is the probability of each event?

**(a)** The card is an ace or a spade. **(b)** The card is a 3 or a king.

### Solution

**(a)** The events "drawing an ace" and "drawing a spade" are not mutually exclusive since it is possible to draw the ace of spades, an outcome satisfying both events. The probability is

$$P(\text{ace or spade}) = P(\text{ace}) + P(\text{spade}) - P(\text{ace and spade})$$
$$= \frac{4}{52} + \frac{13}{52} - \frac{1}{52} = \frac{16}{52} = \frac{4}{13}.$$

**(b)** "Drawing a 3" and "drawing a king" are mutually exclusive events because it is impossible to draw one card that is both a 3 and a king. Using the preceding rule,

$$P(3 \text{ or } K) = P(3) + P(K) - P(3 \text{ and } K)$$
$$= \frac{4}{52} + \frac{4}{52} - 0 = \frac{8}{52} = \frac{2}{13}.$$

## EXAMPLE 6 *Finding Probabilities of Unions*

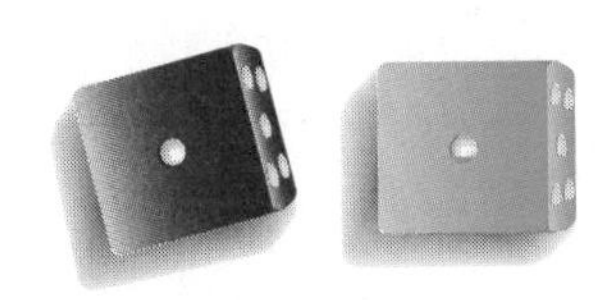

Suppose two fair dice are rolled, one at a time. Find each probability.

**(a)** The first die shows a 2, or the sum of the two dice is 6 or 7.

**(b)** The sum of the two dice is at most 4.

### Solution

**(a)** Think of the two dice as being distinguishable, the first one red and the second one green, for example. (Actually, the sample space is the same even if they are not distinguishable.) A sample space with equally likely outcomes is shown in Figure 29, where (1, 1) represents the event "the first (red) die shows a 1 and the second (green) die shows a 1," (1, 2) represents "the first die shows a 1 and the second die shows a 2," and so on.

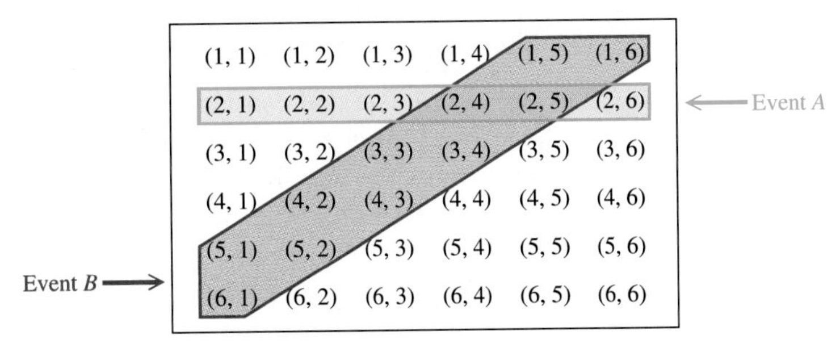


**FIGURE 29**

Let $A$ represent the event "the first die shows a 2," and $B$ represent the event "the sum of the results is 6 or 7." These events are indicated in Figure 29. From the diagram, event $A$ has 6 elements, $B$ has 11 elements, and the sample space has

36 elements. Thus,

$$P(A) = \frac{6}{36}, \quad P(B) = \frac{11}{36}, \quad \text{and} \quad P(A \cap B) = \frac{2}{36}.$$

By the union rule,

$$\begin{aligned} P(A \cup B) &= P(A) + P(B) - P(A \cap B) \\ &= \frac{6}{36} + \frac{11}{36} - \frac{2}{36} = \frac{15}{36} = \frac{5}{12}. \end{aligned}$$

**(b)** "At most 4" can be written as "2 or 3 or 4." (A sum of 1 is not possible.) Then,

$$P(\text{at most } 4) = P(2 \text{ or } 3 \text{ or } 4) = P(2) + P(3) + P(4), \quad (*)$$

since the events represented by "2," "3," and "4" are mutually exclusive.

The sample space for this experiment includes the 36 possible pairs of numbers shown in Figure 29. The pair $(1, 1)$ is the only one with a sum of 2, so $P(2) = \frac{1}{36}$. Also, $P(3) = \frac{2}{36}$ since both $(1, 2)$ and $(2, 1)$ give a sum of 3. The pairs $(1, 3)$, $(2, 2)$, and $(3, 1)$ have a sum of 4, so $P(4) = \frac{3}{36}$. Substituting into equation (*) gives

$$P(\text{at most } 4) = \frac{1}{36} + \frac{2}{36} + \frac{3}{36} = \frac{1}{6}.$$

The properties of probability are summarized as follows.

**Properties of Probability**

For any events $E$ and $F$,

**1.** $0 \le P(E) \le 1$; **2.** $P(\text{a certain event}) = 1$;
**3.** $P(\text{an impossible event}) = 0$; **4.** $P(E') = 1 - P(E)$;
**5.** $P(E \text{ or } F) = P(E \cup F) = P(E) + P(F) - P(E \cap F)$.

**CAUTION** When finding the probability of a union, remember to subtract the probability of the intersection from the sum of the probabilities of the individual events.

## Binomial Probability

An experiment that consists of repeated independent trials with only two outcomes in each trial, success or failure, is called a **binomial experiment.** Let the probability of success in one trial be $p$. Then the probability of failure is $1 - p$, and the probability of $r$ successes in $n$ trials is given by

$$\binom{n}{r} p^r (1 - p)^{n-r}.$$

**EXAMPLE 7** *Finding Probabilities, Using a Binomial Experiment*

An experiment consists of rolling a die 10 times. Find each probability.

**(a)** Exactly 4 of the tosses result in a 3.

**(b)** In 9 of the 10 tosses, the result is not a 3.

**Analytic Solution**

**(a)** The probability of a 3 on one roll is $p = \frac{1}{6}$. Here $n = 10$ and $r = 4$, so the required probability is

$$\binom{10}{4}\left(\frac{1}{6}\right)^4\left(1 - \frac{1}{6}\right)^{10-4} = 210\left(\frac{1}{6}\right)^4\left(\frac{5}{6}\right)^6$$

$$\approx .054.$$

**(b)** Using $n = 10$, $r = 9$, and $p = 1 - \frac{1}{6} = \frac{5}{6}$, we find that this probability is

$$\binom{10}{9}\left(\frac{5}{6}\right)^9\left(\frac{1}{6}\right)^1 \approx .323.$$

**Graphing Calculator Solution**

Graphing calculators that have statistical distribution functions give binomial probabilities. In Figure 30, the numbers in parentheses separated by commas represent $n$, $p$, and $r$, respectively.

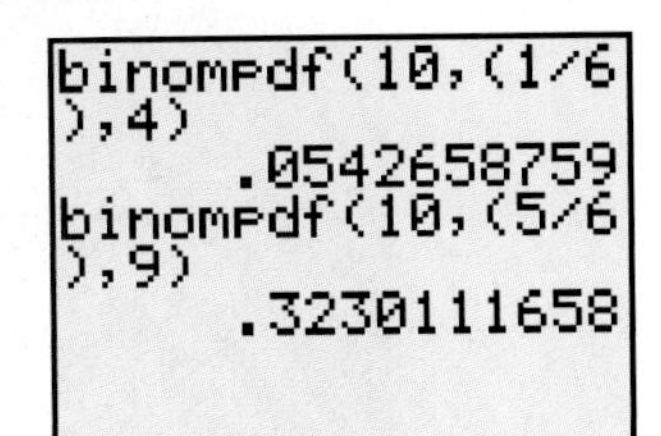


**FIGURE 30**

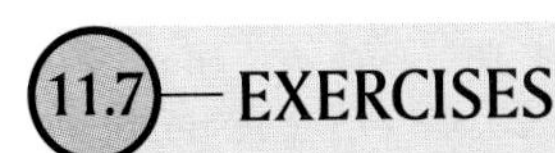

## 11.7 EXERCISES

*Write a sample space with equally likely outcomes for each experiment.*

**1.** A two-headed coin is tossed once.

**2.** Two ordinary coins are tossed.

**3.** Three ordinary coins are tossed.

**4.** Slips of paper marked with the numbers 1, 2, 3, 4, or 5 are placed in a box. After mixing well, two slips are drawn, where the order is not important.

**5.** The spinner shown here is spun twice.

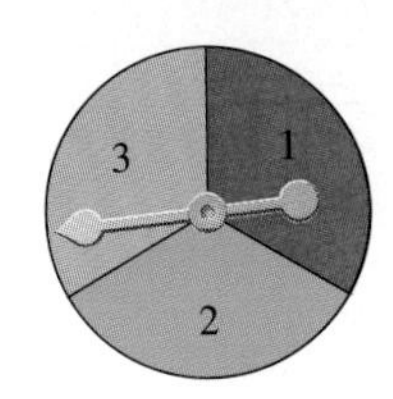


**6.** A die is rolled and then a coin is tossed.

*Write each event in set notation, and give the probability of the event.*

**7.** In Exercise 1,

**(a)** the result of the toss is heads.

**(b)** the result of the toss is tails.

**8.** In Exercise 2,

**(a)** both coins show the same face.

**(b)** at least one coin turns up heads.

**9.** In Exercise 5,

**(a)** the result is a repeated number.

**(b)** the second number is 1 or 3.

**(c)** the first number is even and the second number is odd.

**10.** In Exercise 4,

**(a)** both slips are marked with even numbers.

**(b)** both slips are marked with odd numbers.

**(c)** both slips are marked with the same number.

**(d)** one slip is marked with an odd number, the other with an even number.

**11.** A student gives the number to a probability problem as $\frac{6}{5}$. Explain why this answer must be incorrect.

**12.** *Concept Check* If the probability of an event is .857, what is the probability that the event will not occur?

*Work each problem.*

**13.** *Drawing a Marble* A marble is drawn at random from a box containing 3 yellow, 4 white, and 8 blue marbles. Find the probabilities in parts (a)–(c).

**(a)** A yellow marble is drawn.

**(b)** A black marble is drawn.

**(c)** The marble is yellow or white.

**(d)** What are the odds in favor of drawing a yellow marble?

**(e)** What are the odds against drawing a blue marble?

**14.** *Batting Average* A baseball player with a batting average of .300 comes to bat. What are the odds in favor of his getting a hit?

**15.** *Drawing Slips of Paper* In Exercise 4, what are the odds that the sum of the numbers on the two slips of paper is 5?

**16.** *Probability of Rain* If the odds that it will rain are 4 to 5, what is the probability of rain?

**17.** *Probability of a Candidate Losing* If the odds that a candidate will win an election are 3 to 2, what is the probability that the candidate will lose?

**18.** *Drawing a Card* A card is drawn from a well-shuffled deck of 52 cards. Find the probability that the card is

**(a)** a 9. **(b)** black. **(c)** a black 9.

**(d)** a heart.

**(e)** a face card (K, Q, J of any suit).

**(f)** red or a 3.

**(g)** less than a 4 (consider aces as 1s).

**19.** *Guest Arrival at a Party* Mrs. Schmulen invites 10 relatives to a party: her mother, 2 uncles, 3 brothers, and 4 cousins. If the chances of any one guest arriving first are equally likely, find each probability.

**(a)** The first guest is an uncle or brother.

**(b)** The first guest is a brother or cousin.

**(c)** The first guest is a brother or her mother.

**20.** *Dice Rolls* Two dice are rolled. Find the probability of each event.

**(a)** The sum is at least 10.

**(b)** The sum is either 7 or at least 10.

**(c)** The sum is 2 or the dice both show the same number.

**21.** *Concept Check* Match each probability in parts (a)–(g) with one of the statements in A–F.

**(a)** $P(E) = -.1$ **(b)** $P(E) = .01$

**(c)** $P(E) = 1$ **(d)** $P(E) = 2$

**(e)** $P(E) = .99$ **(f)** $P(E) = 0$

**(g)** $P(E) = .5$

**A.** The event is certain to occur.

**B.** The event cannot occur.

**C.** The event is very likely to occur.

**D.** The event is very unlikely to occur.

**E.** The event is just as likely to occur as not occur.

**F.** The event is impossible.

**22.** *Small-Business Loan* The probability that a bank with assets greater than or equal to $30 billion will make a loan to a small business is .002. What are the odds against such a bank making a small-business loan? (*Source: The Wall Street Journal* analysis of *CA1 Reports*.)

**23.** *Languages Spoken in Hispanic Households* In a 1998 survey of Hispanic households, 20.4% of the respondents said that only Spanish was spoken at home, 11.9% said that only English was spoken, and the remainder said that both Spanish and English were spoken. What are the odds that English is spoken in a randomly selected Hispanic household? (*Source: American Demographics*, May 1999.)

**24.** *U.S. Population Origins* Projected Hispanic and non-Hispanic U.S. populations (in thousands) for 2025 are given in the table. (Other populations are all non-Hispanic.)

| Type | Number |
|---|---|
| Hispanic | 58,930 |
| White | 209,117 |
| Black | 43,511 |
| Native American | 2,744 |
| Asian | 20,748 |

*Source:* U.S. Bureau of the Census.

Assume these projections are accurate. Estimate the probability that a U.S. resident selected at random in 2025 is

**(a)** Hispanic. **(b)** not White.

**(c)** Native American or Black.

**(d)** What are the odds that a randomly selected U.S. resident is Asian?

**25.** *Consumer Purchases* The table on the next page shows the probability that a customer at a department store will make a purchase in the indicated price range.

| Cost | Probability |
|---|---|
| Below $5 | .25 |
| $5–$19.99 | .37 |
| $20–$39.99 | .11 |
| $40–$69.99 | .09 |
| $70–$99.99 | .07 |
| $100–$149.99 | .08 |
| $150 or more | .03 |

Find the probability that a customer makes a purchase that is

**(a)** less than $20. **(b)** $40 or more.

**(c)** more than $99.99. **(d)** less than $100.

**26.** *State Lottery* One game in a state lottery requires you to pick 1 heart, 1 club, 1 diamond, and 1 spade, in that order, from the 13 cards in each suit. What is the probability of getting all four picks correct and winning $5000?

**27.** *State Lottery* If three of the four selections in Exercise 26 are correct, the player wins $200. Find the probability of this outcome.

**28.** *Partner Selection* The law firm of Alam, Bartolini, Chinn, Dickinson, and Ellsberg has two senior partners, Alam and Bartolini. Two of the attorneys are to be selected to attend a conference. Assuming that all are equally likely to be selected, find each probability.

**(a)** Chinn is selected.

**(b)** Alam and Dickinson are selected.

**(c)** At least one senior partner is selected.

**29.** *Opinion Survey* The management of a firm surveys its workers, classified as follows for the purpose of an interview: 30% have worked for the company more than 5 years; 28% are female; 65% contribute to a voluntary retirement plan; and $\frac{1}{2}$ of the female workers contribute to the retirement plan. Find each probability.

**(a)** A male worker is selected.

**(b)** A worker is selected who has worked for the company less than 5 years.

**(c)** A worker is selected who contributes to the retirement plan or is female.

**30.** Explain why the probability of an event must be a number between and inclusive of 0 and 1.

*Gender Makeup of a Family* *Suppose a family has 5 children. Also, suppose that the probability of having a girl is $\frac{1}{2}$. Find the probability that the family has the following children.*

**31.** Exactly 2 girls and 3 boys

**32.** Exactly 3 girls and 2 boys

**33.** No girls

**34.** No boys

**35.** At least 3 boys

**36.** No more than 4 girls

*A die is rolled 12 times. Find the probability of rolling the following.*

**37.** Exactly 12 ones

**38.** Exactly 6 ones

**39.** No more than 3 ones

**40.** No more than 1 one

*College Student Smokers* *The table gives the results of a survey of 14,000 college students who were cigarette smokers in 1997.*

| Number of Cigarettes Per Day | Percent (as a decimal) |
|---|---|
| Less than 1 | .45 |
| 1 to 9 | .24 |
| 10 to 19 | .20 |
| A pack of 20 or more | .11 |

*Using the percents as probabilities, find the probability that, out of 10 of these student smokers selected at random, the following were true.*

**41.** Four smoked less than 10 cigarettes per day.

**42.** Five smoked a pack or more per day.

**43.** Fewer than 2 smoked between 1 and 19 cigarettes per day.

**44.** No more than 3 smoked less than 1 cigarette per day.

**45.** *Color-Blind Males* The probability that a male will be color-blind is .042. Find the probabilities that in a group of 53 men, the following are true.

**(a)** Exactly 5 are color-blind.

**(b)** No more than 5 are color-blind.

**(c)** At least 1 is color-blind.

**46.** The screens shown on the next page illustrate how the table feature of a graphing calculator can be used to find the probabilities of having 0, 1, 2, 3, or 4 girls in a family of 4 children. (Note that 0 appears for X = 5 and X = 6. Why is this so?) Use this approach to determine the following.

| X | Y1 |
|---|---|
| 0 | .0625 |
| 1 | .25 |
| 2 | .375 |
| 3 | .25 |
| 4 | .0625 |
| 5 | 0 |
| 6 | 0 |

Y1■(4 nCr X)*(....

**(a)** Find the probabilities of having 0, 1, 2, or 3 boys in a family of 3 children.

**(b)** Find the probabilities of having 0, 1, 2, 3, 4, 5, or 6 girls in a family of 6 children.

**47.** *(Modeling) Disease Infection* What will happen when an infectious disease is introduced into a family? Suppose a family has $I$ infected members and $S$ members who are not infected but are susceptible to contracting the disease. The probability $P$ of $k$ people not contracting the disease during a 1-week period can be calculated by the formula

$$P = \binom{S}{k} q^k (1 - q)^{S-k},$$

where $q = (1 - p)^I$, and $p$ is the probability that a susceptible person contracts the disease from an infectious person. For example, if $p = .5$, then there is a 50% chance that a susceptible person exposed to one infectious person for 1 week will contract the disease. (*Source:* Hoppensteadt, F. and C. Peskin, *Mathematics in Medicine and the Life Sciences,* Springer-Verlag, 1992.)

**(a)** Compute the probability $P$ of 3 family members not becoming infected within 1 week if there are currently 2 infected and 4 susceptible members. Assume that $p = .1$.

**(b)** A highly infectious disease can have $p = .5$. Repeat part (a) with this value of $p$.

**(c)** Determine the probability that everyone would become sick in a large family if initially $I = 1$, $S = 9$, and $p = .5$. Discuss the results.

**48.** *(Modeling) Disease Infection* Refer to Exercise 47. Suppose that in a family $I = 2$ and $S = 4$. If the probability $P$ is .25 of there being $k = 2$ uninfected members after 1 week, estimate graphically the possible values of $p$. (*Hint:* Write $P$ as a function of $p$.)

## RELATING CONCEPTS

For individual or group investigation (Exercises 49 and 50)

*(Modeling) Drug Resistance Mathematics can be used to analyze the resistance of some types of bacteria to antibiotics. Probability, sequences, and matrices are all concepts necessary to understand drug resistance. Work the following exercises to learn more. (Refer to the chapter introduction.)*

**49.** Some bacteria, called *haploid organisms,* contain genetic material called *plasmids*. Plasmids can cause bacteria to become resistant to certain types of antibiotic drugs. Suppose the bacteria carry two plasmids that cause drug resistance. Plasmid $R_1$ is resistant to the antibiotic ampicillin, and plasmid $R_2$ is resistant to the antibiotic tetracycline. If a bacterium has both plasmids, it will be resistant to both antibiotics. $R_1$ and $R_2$ are passed in cell division to daughter cells. The type of plasmids inherited by a daughter cell is random. It can have 0, 1, or 2 plasmids of type $R_1$. The probability $P_{i,j}$ that a mother cell with $i$ plasmids of type $R_1$ produces a daughter cell with $j$ plasmids of type $R_1$ can be calculated by the formula

$$P_{i,j} = \frac{\binom{2i}{j}\binom{4-2i}{2-j}}{\binom{4}{2}}.$$

(*Source:* Hoppensteadt, F. and C. Peskin, *Mathematics in Medicine and the Life Sciences,* Springer-Verlag, 1992.)

**(a)** Compute the nine values of $P_{i,j}$ for $0 \le i \le 2$ and $0 \le j \le 2$. Assume that $\binom{0}{0} = 1$ and $\binom{i}{j} = 0$ whenever $i < j$.

**(b)** Write your results from part (a) in the matrix

$$P = \begin{bmatrix} P_{0,0} & P_{0,1} & P_{0,2} \\ P_{1,0} & P_{1,1} & P_{1,2} \\ P_{2,0} & P_{2,1} & P_{2,2} \end{bmatrix}.$$

**(c)** Describe the matrix. Where are the probabilities the greatest?

**50.** The genetic makeup of future generations of the haploid bacteria can be modeled with matrices. Let $A = [a_1 \ a_2 \ a_3]$ be a $1 \times 3$ matrix containing three probabilities. $a_1$ is the probability that a cell has two $R_1$ plasmids, $a_2$ is the probability that it has one $R_1$ plasmid and one $R_2$ plasmid, and $a_3$ is the probability that a cell has two $R_2$ plasmids. If the current generation of bacteria only has one plasmid of each type, then $A_1 = [0 \ 1 \ 0]$. The probabilities $A_{k+1}$ for plasmids $R_1$ and $R_2$ in each future generation can be calculated from the equation $A_{k+1} = A_k P$, where $k \ge 1$ and the $3 \times 3$ matrix $P$ was determined in the previous exercise. The phenomenon that results from this sequence of calculations was not well understood until fairly recently. It is now used in the genetic engineering of plasmids.

**(a)** If all bacteria initially have both plasmids, make a conjecture as to the types of plasmids future generations of bacteria will have.

**(b)** Test your conjecture by repeatedly computing with your calculator the matrix product $A_{k+1} = A_k P$, with $A_1 = [0 \ 1 \ 0]$ for $k = 1, 2, 3, \ldots, 12$. Interpret the result. It may surprise you.

## Reviewing Basic Concepts (Sections 11.6 and 11.7)

1. *Book Arrangement* A student has 4 different books that she wishes to arrange in a row. How many arrangements are there?
2. Calculate $P(7, 3)$.
3. *Basketball Player Choices* How many ways can 2 guards, 2 forwards, and 1 center be selected from a basketball team composed of 6 guards, 5 forwards, and 3 centers?
4. Calculate $\binom{10}{4}$.
5. *Home Plan Choices* How many different homes are available if a builder offers a choice of 9 basic plans, 4 roof styles, and 2 exterior finishes?
6. Write the sample space for an experiment consisting of tossing a coin twice.
7. *Dice Roll* Find the probability of rolling a sum of 11 with two dice.
8. *Drawing Cards* Find the probability of drawing 4 aces and 1 queen from a standard deck of 52 cards.
9. *Probability of Rain* If the odds that it will rain are 3 to 7, what is the probability of rain?
10. *High School Graduates* In 1998 there were 2.81 million high school graduates, of which 1.45 million were male. (*Source:* U.S. National Center for Education Statistics.) If a 1998 high school graduate is selected at random, estimate the probability that this graduate is female.

## CHAPTER 11 SUMMARY

**KEY TERMS & SYMBOLS**

### 11.1 Sequences and Series

sequence
terms of a sequence
general or $n$th term, $a_n$
finite sequence
infinite sequence
convergent
divergent
recursive definition
Fibonacci sequence
series
finite series
infinite series
summation (sigma) notation, $\sum_{n=1}^{\infty}$
index of summation

**KEY CONCEPTS**

**SEQUENCE**
A sequence is a function that has a set of natural numbers as its domain.

**SERIES**
A finite series is an expression of the form

$$S_n = a_1 + a_2 + a_3 + \cdots + a_n = \sum_{i=1}^{n} a_i,$$

and an infinite series is an expression of the form

$$S_\infty = a_1 + a_2 + a_3 + \cdots + a_n + \cdots = \sum_{i=1}^{\infty} a_i.$$

**SUMMATION PROPERTIES**
If $a_1, a_2, a_3, \ldots, a_n$ and $b_1, b_2, b_3, \ldots, b_n$ are two sequences, and $c$ is a constant, then for every positive integer $n$,

(a) $\sum_{i=1}^{n} c = nc$

(b) $\sum_{i=1}^{n} ca_i = c\sum_{i=1}^{n} a_i$

(c) $\sum_{i=1}^{n} (a_i + b_i) = \sum_{i=1}^{n} a_i + \sum_{i=1}^{n} b_i$

(d) $\sum_{i=1}^{n} (a_i - b_i) = \sum_{i=1}^{n} a_i - \sum_{i=1}^{n} b_i.$

| KEY TERMS & SYMBOLS | KEY CONCEPTS |
|---|---|
| **11.2 Arithmetic Sequences and Series**<br>arithmetic sequence (or progression)<br>common difference, $d$<br>arithmetic series | **ARITHMETIC SEQUENCE**<br>A sequence in which each term after the first is obtained by adding a fixed number to the previous term is an arithmetic sequence. The fixed number that is added is the common difference.<br><br>***n*TH TERM OF AN ARITHMETIC SEQUENCE**<br>In an arithmetic sequence with first term $a_1$ and common difference $d$, the $n$th term, $a_n$, is given by<br>$$a_n = a_1 + (n - 1)d.$$<br>**SUM OF THE FIRST *n* TERMS OF AN ARITHMETIC SEQUENCE**<br>If an arithmetic sequence has first term $a_1$ and common difference $d$, then the sum of the first $n$ terms is given by<br>$$S_n = \frac{n}{2}(a_1 + a_n) \quad \text{or} \quad S_n = \frac{n}{2}[2a_1 + (n - 1)d].$$ |
| **11.3 Geometric Sequences and Series**<br>geometric sequence (or progression)<br>common ratio, $r$<br>geometric series<br>annuity<br>future value of an annuity | **GEOMETRIC SEQUENCE**<br>A geometric sequence is a sequence in which each term after the first is obtained by multiplying the preceding term by a constant nonzero real number, called the common ratio.<br><br>***n*TH TERM OF A GEOMETRIC SEQUENCE**<br>In the geometric sequence with first term $a_1$ and common ratio $r$, the $n$th term is given by<br>$$a_n = a_1r^{n-1}.$$<br>**SUM OF THE FIRST *n* TERMS OF A GEOMETRIC SEQUENCE**<br>If a geometric sequence has first term $a_1$ and common ratio $r$, then the sum of the first $n$ terms is given by<br>$$S_n = \frac{a_1(1 - r^n)}{1 - r}, \quad \text{where } r \neq 1.$$<br>**SUM OF THE TERMS OF AN INFINITE GEOMETRIC SEQUENCE**<br>The sum of the terms an infinite geometric sequence with first term $a_1$ and common ratio $r$, where $-1 < r < 1$, is given by<br>$$S_\infty = \frac{a_1}{1 - r}.$$<br>If $r \leq -1$ or $r \geq 1$, the sum does not exist. |
| **11.4 The Binomial Theorem**<br>binomial theorem<br>factorial notation<br>$n$-factorial, $n!$<br>binomial coefficient | **PASCAL'S TRIANGLE**<br>1 (Row 0)<br>1 1 (Row 1)<br>1 2 1 (Row 2)<br>1 3 3 1 (Row 3)<br>1 4 6 4 1 (Row 4)<br>1 5 10 10 5 1 (Row 5) |

| KEY TERMS & SYMBOLS | KEY CONCEPTS |
|---|---|

**$n$-FACTORIAL**
For any positive integer $n$,

$$n! = n(n-1)(n-2)\cdots(3)(2)(1) \quad \text{and} \quad 0! = 1.$$

**BINOMIAL COEFFICIENT**
For nonnegative integers $n$ and $r$, with $r \le n$,

$$ {}_nC_r = \binom{n}{r} = \frac{n!}{r!\,(n-r)!}.$$

**BINOMIAL THEOREM**
For any positive integer $n$,

$$(x+y)^n = x^n + \binom{n}{1}x^{n-1}y + \binom{n}{2}x^{n-2}y^2 + \binom{n}{3}x^{n-3}y^3 + \cdots + \binom{n}{r}x^{n-r}y^r + \cdots + \binom{n}{n-1}xy^{n-1} + y^n.$$

**$r$TH TERM OF THE BINOMIAL EXPANSION**
The $r$th term of the binomial expansion of $(x+y)^n$, where $n \ge r-1$, is

$$\binom{n}{r-1}x^{n-(r-1)}y^{r-1}.$$

## 11.5 Mathematical Induction

**PROOF BY MATHEMATICAL INDUCTION**

**Step 1** Prove that the statement is true for $n = 1$.

**Step 2** Show that for any positive integer $k$, if $S_k$ is true, then $S_{k+1}$ is also true.

**GENERALIZED PRINCIPLE OF MATHEMATICAL INDUCTION**
Let $S_n$ be a statement concerning the positive integer $n$. Let $j$ be a fixed positive integer. Suppose that

**Step 1** $S_j$ is true;

**Step 2** for any positive integer $k$, $k \ge j$, $S_k$ implies $S_{k+1}$.

Then $S_n$ is true for all positive integers $n$, where $n \ge j$.

## 11.6 Counting Theory

tree diagram
independent events
permutation, $P(n, r)$
combination, $C(n, r)$ or $\binom{n}{r}$

**FUNDAMENTAL PRINCIPLE OF COUNTING**
If $n$ independent events occur, with

$m_1$ ways for event 1 to occur,
$m_2$ ways for event 2 to occur,
⋮
and $m_n$ ways for event $n$ to occur,

then there are

$$m_1 \cdot m_2 \cdot \cdots \cdot m_n$$

different ways for all $n$ events to occur.

| KEY TERMS & SYMBOLS | KEY CONCEPTS |
|---|---|
| | **PERMUTATIONS OF $n$ ELEMENTS TAKEN $r$ AT A TIME**<br>If $P(n, r)$ denotes the number of permutations of $n$ elements taken $r$ at a time, with $r \leq n$, then<br>$$P(n, r) = \frac{n!}{(n-r)!}.$$<br>**COMBINATIONS OF $n$ ELEMENTS TAKEN $r$ AT A TIME**<br>If $\binom{n}{r}$ or $C(n, r)$ represents the number of combinations of $n$ things taken $r$ at a time, with $r \leq n$, then<br>$$C(n, r) = \binom{n}{r} = \frac{n!}{(n-r)!r!}.$$ |
| **11.7 Probability**<br>outcomes<br>sample space<br>event<br>probability<br>complement, $E'$<br>Venn diagram<br>odds<br>mutually exclusive events<br>binomial experiment | **SAMPLE SPACE AND EVENT**<br>The set $S$ of all possible outcomes of a given experiment is called the sample space for the experiment. Any subset of $S$ is called an event.<br>**PROBABILITY OF EVENT $E$**<br>In a sample space with equally likely outcomes, the probability of an event $E$, written $P(E)$, is the ratio of the number of outcomes in sample space $S$ that belong to event $E$, $n(E)$, to the total number of outcomes in sample space $S$, $n(S)$. That is,<br>$$P(E) = \frac{n(E)}{n(S)}.$$<br>**COMPLEMENT**<br>The complement of an event $E$, written $E'$, is the set of all outcomes not in $E$. Thus,<br>$$P(E) + P(E') = 1.$$<br>**ODDS**<br>The odds in favor of an event $E$ are expressed as the ratio of $P(E)$ to $P(E')$ or $\frac{P(E)}{P(E')}$.<br>**PROBABILITY OF THE UNION OF TWO EVENTS**<br>For any events $E$ and $F$,<br>$$P(E \text{ or } F) = P(E \cup F) = P(E) + P(F) - P(E \cap F).$$<br>**PROPERTIES OF PROBABILITY**<br>For any events $E$ and $F$,<br>**1.** $0 \leq P(E) \leq 1$; **2.** $P(\text{a certain event}) = 1$;<br>**3.** $P(\text{an impossible event}) = 0$; **4.** $P(E') = 1 - P(E)$;<br>**5.** $P(E \text{ or } F) = P(E \cup F) = P(E) + P(F) - P(E \cap F)$.<br>**BINOMIAL PROBABILITY**<br>An experiment that consists of repeated independent trials with only two outcomes in each trial, success or failure, is called a binomial experiment. Let the probability of success in one trial be $p$. Then the probability of failure is $1 - p$, and the probability of $r$ successes in $n$ trials is given by<br>$$\binom{n}{r} p^r (1-p)^{n-r}.$$ |

## CHAPTER 11 REVIEW EXERCISES

*Write the first five terms for each sequence. State whether the sequence is* arithmetic, geometric, *or* neither.

**1.** $a_n = \dfrac{n}{n+1}$

**2.** $a_n = (-2)^n$

**3.** $a_n = 2(n+3)$

**4.** $a_n = n(n+1)$

**5.** $a_1 = 5;\ a_n = a_{n-1} - 3$, for $n \geq 2$

*In Exercises 6–9, write the first five terms of the sequence described.*

**6.** Arithmetic, $a_2 = 10$, $d = -2$

**7.** Arithmetic, $a_3 = \pi$, $a_4 = 1$

**8.** Geometric, $a_1 = 6$, $r = 2$

**9.** Geometric, $a_1 = -5$, $a_2 = -1$

**10.** An arithmetic sequence has $a_5 = -3$ and $a_{15} = 17$. Find $a_1$ and $a_n$.

**11.** A geometric sequence has $a_1 = -8$ and $a_7 = -\frac{1}{8}$. Find $a_4$ and $a_n$.

*Find $a_8$ for each arithmetic sequence.*

**12.** $a_1 = 6, d = 2$

**13.** $a_1 = 6x - 9, a_2 = 5x + 1$

*Find $S_{12}$ for each arithmetic sequence.*

**14.** $a_1 = 2, d = 3$

**15.** $a_2 = 6, d = 10$

*Find $a_5$ for each geometric sequence.*

**16.** $a_1 = -2, r = 3$

**17.** $a_3 = 4, r = \dfrac{1}{5}$

*Find $S_4$ for each geometric sequence.*

**18.** $a_1 = 3, r = 2$

**19.** $a_1 = -1, r = 3$

**20.** $\dfrac{3}{4}, -\dfrac{1}{2}, \dfrac{1}{3}, \ldots$

*Evaluate each sum that exists.*

**21.** $\sum_{i=1}^{7} (-1)^{i-1}$

**22.** $\sum_{i=1}^{5} (i^2 + i)$

**23.** $\sum_{i=1}^{4} \dfrac{i+1}{i}$

**24.** $\sum_{j=1}^{10} (3j - 4)$

**25.** $\sum_{j=1}^{2500} j$

**26.** $\sum_{i=1}^{5} 4 \cdot 2^i$

**27.** $\sum_{i=1}^{\infty} \left(\dfrac{4}{7}\right)^i$

**28.** $\sum_{i=1}^{\infty} -2\left(\dfrac{6}{5}\right)^i$

*Evaluate each sum that converges. If the series diverges, say so.*

**29.** $24 + 8 + \dfrac{8}{3} + \dfrac{8}{9} + \cdots$

**30.** $-\dfrac{3}{4} + \dfrac{1}{2} - \dfrac{1}{3} + \dfrac{2}{9} - \cdots$

**31.** $\dfrac{1}{12} + \dfrac{1}{6} + \dfrac{1}{3} + \dfrac{2}{3} + \cdots$

**32.** $.9 + .09 + .009 + .0009 + \cdots$

*Evaluate each sum, where $x_1 = 0$, $x_2 = 1$, $x_3 = 2$, $x_4 = 3$, $x_5 = 4$, and $x_6 = 5$.*

**33.** $\sum_{i=1}^{4} (x_i^2 - 6)$

**34.** $\sum_{i=1}^{6} f(x_i)\,\Delta x;\ f(x) = (x-2)^3,\ \Delta x = .1$

*Write each sum, using summation notation.*

**35.** $4 - 1 - 6 - \cdots - 66$

**36.** $10 + 14 + 18 + \cdots + 86$

**37.** $4 + 12 + 36 + \cdots + 972$

**38.** $\dfrac{5}{6} + \dfrac{6}{7} + \dfrac{7}{8} + \cdots + \dfrac{12}{13}$

*Use the binomial theorem to expand each expression.*

**39.** $(x + 2y)^4$

**40.** $(3z - 5w)^3$

**41.** $\left(3\sqrt{x} - \frac{1}{\sqrt{x}}\right)^5$

**42.** $(m^3 - m^{-2})^4$

*Find the indicated term or terms for each expansion.*

**43.** Sixth term of $(4x - y)^8$

**44.** Seventh term of $(m - 3n)^{14}$

**45.** First four terms of $(x + 2)^{12}$

**46.** Last three terms of $(2a + 5b)^{16}$

**47.** What kinds of statements are proved by mathematical induction? Give examples.

**48.** Describe a proof by mathematical induction.

*Use mathematical induction to prove that each statement is true for every positive integer n.*

**49.** $1 + 3 + 5 + 7 + \cdots + (2n - 1) = n^2$

**50.** $2 + 6 + 10 + 14 + \cdots + (4n - 2) = 2n^2$

**51.** $2 + 2^2 + 2^3 + \cdots + 2^n = 2(2^n - 1)$

**52.** $1^3 + 3^3 + 5^3 + \cdots + (2n - 1)^3 = n^2(2n^2 - 1)$

**53.** *Concept Check* Is a student identification number an example of a permutation or a combination?

*Find the value of each expression.*

**54.** $P(9, 2)$

**55.** $P(6, 0)$

**56.** $\binom{8}{3}$

**57.** $9!$

**58.** $C(10, 5)$

*Solve each problem.*

**59.** *Wedding Plans* Two people are planning their wedding. They can select from 2 different chapels, 4 soloists, 3 organists, and 2 ministers. How many different wedding arrangements are possible?

**60.** *Couch Styles* Bob Schiffer, who is furnishing his apartment, wants to buy a new couch. He can select from 5 different styles, each available in 3 different fabrics, with 6 color choices. How many different couches are available?

**61.** *Summer Job Assignments* Four students are to be assigned to 4 different summer jobs. Each student is qualified for all 4 jobs. In how many ways can the jobs be assigned?

**62.** *Conference Delegations* A student body council consists of a president, vice-president, secretary/treasurer, and 3 representatives at large. Three members are to be selected to attend a conference.

**(a)** How many different such delegations are possible?

**(b)** How many are possible if the president must attend?

**63.** *Tournament Outcomes* Nine football teams are competing for first-, second-, and third-place titles in a statewide tournament. In how many ways can the winners be determined?

**64.** *License Plates* How many different license plates can be formed with a letter followed by 3 digits and then 3 letters? How many such license plates have no repeats?

**65.** *Drawing a Marble* A marble is drawn at random from a box containing 4 green, 5 black, and 6 white marbles. Find each probability.

**(a)** A green marble is drawn.

**(b)** A marble that is not black is drawn.

**(c)** A blue marble is drawn.

**66.** *Drawing a Marble* Refer to Exercise 65, and answer each question.

**(a)** What are the odds in favor of drawing a green marble?

**(b)** What are the odds against drawing a white marble?

**(c)** What are the odds in favor of drawing a marble that is not white?

*Drawing a Card A card is drawn from a standard deck of 52 cards. Find the probability that the following is drawn.*

**67.** A black king

**68.** A face card or an ace

**69.** An ace or a diamond

**70.** A card that is not a diamond

*Swimming Pool Filter Samples A sample shipment of 5 swimming pool filters is chosen. The probability of exactly 0, 1, 2, 3, 4, or 5 filters being defective is given in the table.*

| Number Defective | 0 | 1 | 2 | 3 | 4 | 5 |
|---|---|---|---|---|---|---|
| Probability | .31 | .25 | .18 | .12 | .08 | .06 |

*Find the probability that the given number of filters are defective.*

**71.** No more than 3

**72.** At least 2

**73.** More than 5

**74.** *Die Rolls* A die is rolled 12 times. Find the probability that exactly 2 of the rolls result in a 5.

75. *Coin Tosses* A coin is tossed 10 times. Find the probability that exactly 4 of the tosses result in a tail.

76. *Political Orientation* The table describes the political orientations of college freshmen in the class of 2002.

| Political Orientation | Number of Freshmen (in thousands) |
|---|---|
| Far left | 44.28 |
| Liberal | 341.12 |
| Middle of the road | 926.6 |
| Conservative | 303.4 |
| Far right | 24.6 |
| Total | 1640 |

*Source: The American Freshman: National Norms for Fall 1998;* American Council on Education, UCLA.

(a) What is the probability that a randomly selected student from the class is in the conservative group?

(b) What is the probability that a randomly selected student from the class is on the far left or the far right politically?

(c) What are the odds against a randomly selected student from the class being politically middle of the road?

## CHAPTER 11 TEST

1. Write the first five terms for each sequence. State whether the sequence is arithmetic, geometric, or neither.

(a) $a_n = (-1)^n(n + 2)$ (b) $a_n = -3 \cdot \left(\frac{1}{2}\right)^n$

(c) $a_1 = 2, a_2 = 3; a_n = a_{n-1} + 2a_{n-2}$, for $n \geq 3$

2. In each sequence described, find $a_5$.

(a) An arithmetic sequence with $a_1 = 1$ and $a_3 = 25$

(b) A geometric sequence with $a_1 = 81$ and $r = -\frac{2}{3}$

3. Find the sum of the first ten terms of each sequence described.

(a) Arithmetic, with $a_1 = -43$ and $d = 12$

(b) Geometric, with $a_1 = 5$ and $r = -2$

4. Evaluate each sum that exists.

(a) $\sum_{i=1}^{30} (5i + 2)$ (b) $\sum_{i=1}^{5} (-3 \cdot 2^i)$

(c) $\sum_{i=1}^{\infty} (2^i) \cdot 4$ (d) $\sum_{i=1}^{\infty} 54\left(\frac{2}{9}\right)^i$

5. (a) Use the binomial theorem to expand $(2x - 3y)^4$.

(b) Find the third term in the expansion of $(w - 2y)^6$.

6. Evaluate the following.

(a) 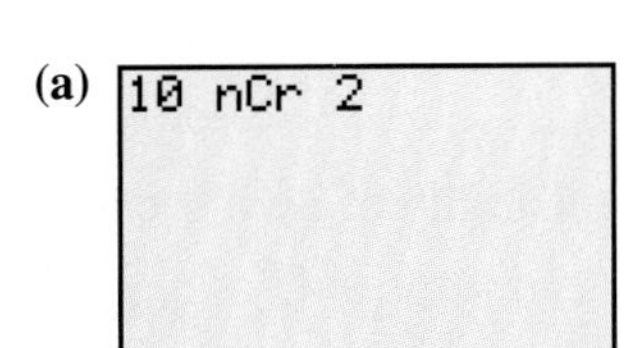


(b) $\binom{7}{3}$

(c) 7! (d) $P(11, 3)$

7. Use mathematical induction to prove that for all natural numbers $n$,

$$8 + 14 + 20 + 26 + \cdots + (6n + 2) = 3n^2 + 5n.$$

*Solve each problem involving counting theory.*

8. *Athletic Shoe Choices* A sports-shoe manufacturer makes athletic shoes in 4 different styles. Each style comes in 3 different colors, and each color comes in 2 different shades. How many different types of shoes can be made?

9. *Club Officer Choices* A club with 20 members plans to elect a president, a secretary, and a treasurer from its membership. If a member can hold at most one office, in how many ways can the three offices be filled?

10. *Club Officer Choices* Refer to Exercise 9. If there are 8 men and 12 women in the club, in how many ways can 2 men and 3 women be chosen to attend a conference?

11. *Drawing a Card* A single card is drawn from a standard deck of 52 cards.

(a) Find the probability of drawing a red 3.

(b) Find the probability of drawing a card that is not a face card.

(c) Find the probability of drawing a king or a spade.

(d) What are the odds in favor of drawing a face card?

12. *Rolling a Die* An experiment consists of rolling a die eight times. Find the probability of each event.

(a) Exactly three rolls result in a 4.

(b) All eight rolls result in a 6.

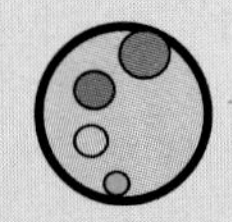

# Chapter 11 Project

## Using Experimental Probabilities to Simulate Family Makeup

In the last section of this chapter, we calculated many theoretical probabilities. Even though the probability of obtaining a head on one toss of a coin is $\frac{1}{2}$, it does not mean that a head will appear on exactly one-half of the tosses. In the experiments that follow, we simulate the makeup of a family, and experimentally determine the probability that a family of 4 children will consist of 2 boys and 2 girls. (Do you expect this probability to be $\frac{1}{2}$?)

Probabilities can be calculated in two ways. One way is *theoretically,* as described in the section on probability. The other is *experimentally*, by actually conducting the experiment and recording the successes compared with the total number of trials. The **law of large numbers** says that the larger the number of trials in an experiment, the closer the experimental probability will be to the theoretical probability.

## Activities

1. Work with a partner. One member of the group should flip two coins simultaneously 20 times, and the other should tally the results in a table similar to the one shown here. The possible events are 2 heads and 0 tails, 1 head and 1 tail, and 0 heads and 2 tails.

| Event | Tally | Relative Frequency | Experimental Probability (as a decimal) |
|---|---|---|---|
| 2 heads, 0 tails | | $\frac{?}{10}$ | |
| 1 head, 1 tail | | $\frac{?}{10}$ | |
| 0 heads, 2 tails | | $\frac{?}{10}$ | |

(a) List the sample space and calculate the *theoretical* probability of each event.

(b) Explain why the three outcomes are not equally likely.

(c) Your experimental probabilities probably do not match the theoretical probabilities. Why not? How can you obtain better results?

2. With your partner, repeat the experiment but this time flip four coins simultaneously 20 times and tally the results. The possible events are given in the table.

| Event | Tally | Relative Frequency | Experimental Probability (as a decimal) |
|---|---|---|---|
| 4 heads, 0 tails | | $\frac{?}{20}$ | |
| 3 heads, 1 tail | | $\frac{?}{20}$ | |
| 2 heads, 2 tails | | $\frac{?}{20}$ | |
| 1 head, 3 tails | | $\frac{?}{20}$ | |
| 0 heads, 4 tails | | $\frac{?}{20}$ | |

(a) List the sample space and calculate the *theoretical* probability of each event.

(b) How frequently did you get 2 heads and 2 tails? Are you surprised?

**(c)** Your experimental probabilities probably do not match the theoretical probabilities. Why not? How can you obtain better results?

After working Activity 2, you probably agree that actual coin tossing is quite time-consuming. To simulate coin tosses, we can use a random integer generator on a graphing calculator, such as the TI-83 Plus. The first line in Figure A shows the syntax for generating a list of 4 integers on the closed interval $[0, 1]$; a toss of four coins can be simulated this way. Let us agree that 0 represents tails and 1 represents heads. See the six sample lines in Figure A.

By finding the sum of the entries in the list, we can determine the outcome. For example, a sum of 0 means that there are 4 tails, a sum of 1 means that there are 3 tails and 1 head, a sum of 2 means that there are 2 tails and 2 heads, and so on. The calculator can keep a tally of the sums, as shown in the short program on the home screen in Figure B.

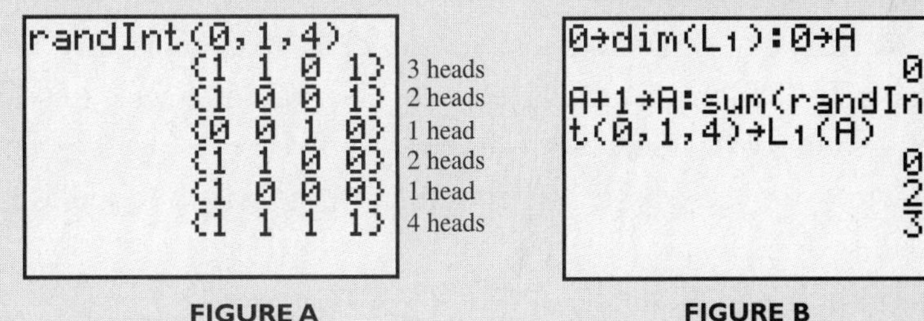


FIGURE A FIGURE B

After running the program a large number of times, we can plot the results in a histogram, as seen in Figure C. The first display of Figure D shows how many times the experiment was run (150), the second display shows the experimental probability for a sum of 2 (meaning 2 tails and 2 heads), and the third display shows the theoretical probability for 2 heads and 2 tails.

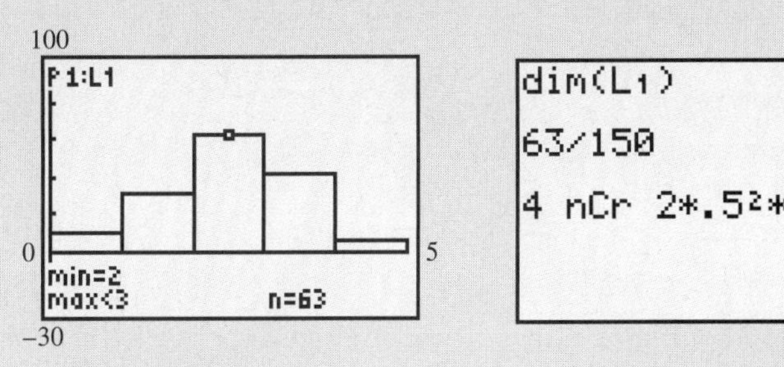


The sum 2 appears 63 times.

FIGURE C FIGURE D

**3.** With your partner, duplicate the procedure just described to determine an experimental probability of having 2 girls and 2 boys in a family of 4 children. Let 0 represent the outcome of having a girl and 1 represent the outcome of having a boy. Use the calculator to obtain 200 outcomes. How close is your result to the theoretical probability? It should be much closer than the one you found in Activity 2. Why is this so?

# Reference: Basic Algebraic Concepts and Geometry Formulas

IN THIS REFERENCE, we present a review of topics that are usually studied in beginning and intermediate algebra courses. You may wish to refer to the various sections of this reference from time to time if you need to refresh your memory on these basic concepts. A list of geometry formulas is included in Section R.6.

CHAPTER OUTLINE

# R.1 REVIEW OF EXPONENTS AND POLYNOMIALS

Rules for Exponents • Terminology for Polynomials • Adding and Subtracting Polynomials • Multiplying Polynomials

## Rules for Exponents

Work with exponents can be simplified by using the rules for exponents. By definition, the notation $a^m$ (where $m$ is a positive integer and $a$ is a real number) means that $a$ appears as a factor $m$ times. In the same way, $a^n$ (where $n$ is a positive integer) means that $a$ appears as a factor $n$ times. In the product $a^m \cdot a^n$, the number $a$ would appear $m + n$ times.

**Product Rule**

For all positive integers $m$ and $n$ and every real number $a$,

$$a^m \cdot a^n = a^{m+n}.$$

**EXAMPLE 1** *Using the Product Rule*

Find each product.

**(a)** $y^4 \cdot y^7$ **(b)** $(6z^5)(9z^3)(2z^2)$

**Solution**

**(a)** $y^4 \cdot y^7 = y^{4+7} = y^{11}$

**(b)** $(6z^5)(9z^3)(2z^2) = (6 \cdot 9 \cdot 2)(z^5z^3z^2)$ Commutative and associative properties

$= 108z^{5+3+2}$ Product rule

$= 108z^{10}$

A zero exponent is defined as follows. (Justification is found in Section R.4.)

**Zero Exponent**

For any nonzero real number $a$,

$$a^0 = 1.$$

**NOTE** The expression $0^0$ is not defined.

**EXAMPLE 2** *Using the Definition of $a^0$*

Evaluate each power.

**(a)** $3^0$ **(b)** $(-4)^0$ **(c)** $-4^0$ **(d)** $-(-4)^0$ **(e)** $(7r)^0$, if $r \neq 0$

**Solution**

**(a)** $3^0 = 1$ **(b)** $(-4)^0 = 1$ Let $a = -4$ in the definition.

**(c)** $-4^0 = -1$, since $-4^0 = -(4^0) = -1$. **(d)** $-(-4)^0 = -(1) = -1$

**(e)** $(7r)^0 = 1$, if $r \neq 0$

The expression $(2^5)^3$ can be written as

$$(2^5)^3 = 2^5 \cdot 2^5 \cdot 2^5.$$

By a generalization of the product rule for exponents, this product is

$$(2^5)^3 = 2^{5+5+5} = 2^{15}.$$

The same exponent could have been obtained by multiplying 3 and 5. This example suggests the first of the power rules given here. The others are found in a similar way.

**Power Rules**

For all positive integers $m$ and $n$ and all real numbers $a$ and $b$,

$$(a^m)^n = a^{mn}, \quad (ab)^m = a^m b^m, \quad \left(\frac{a}{b}\right)^m = \frac{a^m}{b^m} \quad (b \neq 0).$$

**EXAMPLE 3** *Using the Power Rules*

Simplify each expression.

**(a)** $(5^3)^2$ **(b)** $(3^4x^2)^3$ **(c)** $\left(\frac{2^5}{b^4}\right)^3$, if $b \neq 0$

**Solution**

**(a)** $(5^3)^2 = 5^{3(2)} = 5^6$ **(b)** $(3^4x^2)^3 = (3^4)^3(x^2)^3 = 3^{4(3)}x^{2(3)} = 3^{12}x^6$

**(c)** $\left(\frac{2^5}{b^4}\right)^3 = \frac{(2^5)^3}{(b^4)^3} = \frac{2^{15}}{b^{12}}$, if $b \neq 0$

**CAUTION** Do not confuse expressions like $mn^2$ and $(mn)^2$, which are *not* equal. The second power rule given in the box can be used only with the second expression: $(mn)^2 = m^2n^2$.

## Terminology for Polynomials

An **algebraic expression** is the result of adding, subtracting, multiplying, dividing (except by 0), or finding roots or powers of any combination of variables and constants. Examples of algebraic expressions include

$$2x^2 - 3x, \quad \frac{5y}{2y - 3}, \quad \sqrt{m^3 - 8}, \quad \text{and} \quad (3a + b)^4.$$

The simplest algebraic expressions, *polynomials,* are discussed here.

The product of a real number and one or more variables raised to powers is called a **term.** The real number is called the **numerical coefficient,** or just the **coefficient.** The coefficient in $-3m^4$ is $-3$, while the coefficient in $-p^2$ is $-1$. **Like terms** are terms with the same variables raised to the same powers. For example, $-13x^3$, $4x^3$, and $-x^3$ are like terms, while $6y$ and $6y^2$ are not.

A **polynomial** is defined as a term or a finite sum of terms, with only nonnegative integer exponents permitted on the variables. Examples of polynomials include

$$5x^3 - 8x^2 + 7x - 4, \quad 9p^5 - 3, \quad 8r^2, \quad \text{and} \quad 6.$$

The expression $9x^2 - 4x + \frac{6}{x}$ is not a polynomial because of the term $\frac{6}{x}$. The terms of a polynomial cannot have variables in a denominator.

The greatest exponent on a variable in a polynomial in one variable is the **degree** of the polynomial. A nonzero constant is said to have degree 0. (The polynomial 0 has no degree.) For example, $3x^6 - 5x^2 + 2x + 3$ is a polynomial of degree 6.

A polynomial can have more than one variable. A term containing more than one variable has degree equal to the sum of all the exponents appearing on the variables in the term. For example, $-3x^4y^3z^5$ is of degree $4 + 3 + 5 = 12$. The degree of a polynomial in more than one variable is equal to the greatest degree of any term appearing in the polynomial. By this definition, the polynomial $2x^4y^3 - 3x^5y + x^6y^2$ is of degree 8 because of the $x^6y^2$ term.

A polynomial containing exactly three terms is called a **trinomial;** one containing exactly two terms is a **binomial;** and a single-term polynomial is called a **monomial.**

## Adding and Subtracting Polynomials

Since the variables used in polynomials represent real numbers, a polynomial represents a real number. This means that all properties of real numbers hold for polynomials. In particular, the distributive property holds, so

$$3m^5 - 7m^5 = (3 - 7)m^5 = -4m^5.$$

Thus, polynomials are added by adding coefficients of like terms; polynomials are subtracted by subtracting coefficients of like terms.

### EXAMPLE 4 *Adding and Subtracting Polynomials*

Add or subtract, as indicated.

**(a)** $(2y^4 - 3y^2 + y) + (4y^4 + 7y^2 + 6y)$

**(b)** $(-3m^3 - 8m^2 + 4) - (m^3 + 7m^2 - 3)$

**(c)** $8m^4p^5 - 9m^3p^5 + (11m^4p^5 + 15m^3p^5)$

**(d)** $4(x^2 - 3x + 7) - 5(2x^2 - 8x - 4)$

#### Solution

**(a)** $(2y^4 - 3y^2 + y) + (4y^4 + 7y^2 + 6y)$

$= (2 + 4)y^4 + (-3 + 7)y^2 + (1 + 6)y$

$= 6y^4 + 4y^2 + 7y$

**(b)** $(-3m^3 - 8m^2 + 4) - (m^3 + 7m^2 - 3)$

$= (-3 - 1)m^3 + (-8 - 7)m^2 + [4 - (-3)]$

$= -4m^3 - 15m^2 + 7$

**(c)** $8m^4p^5 - 9m^3p^5 + (11m^4p^5 + 15m^3p^5) = 19m^4p^5 + 6m^3p^5$

**(d)** $4(x^2 - 3x + 7) - 5(2x^2 - 8x - 4)$

$= 4x^2 - 4(3x) + 4(7) - 5(2x^2) - 5(-8x) - 5(-4)$    Distributive property

$= 4x^2 - 12x + 28 - 10x^2 + 40x + 20$    Associative property

$= -6x^2 + 28x + 48$    Add like terms.

As shown in parts (a), (b), and (d) of Example 4, polynomials in one variable are often written with their terms in *descending order* by degree, so the term of highest degree is first, the one with the next highest degree is next, and so on.

## Multiplying Polynomials

The associative and distributive properties, together with the properties of exponents, can also be used to find the product of two polynomials. For example, to find the product of $3x - 4$ and $2x^2 - 3x + 5$, treat $3x - 4$ as a single expression and use the distributive property as follows.

$$(3x - 4)(2x^2 - 3x + 5) = (3x - 4)(2x^2) + (3x - 4)(-3x) + (3x - 4)(5)$$

Now, use the distributive property three separate times.

$$\begin{aligned} &= 3x(2x^2) - 4(2x^2) + 3x(-3x) - 4(-3x) + 3x(5) - 4(5) \\ &= 6x^3 - 8x^2 - 9x^2 + 12x + 15x - 20 \\ &= 6x^3 - 17x^2 + 27x - 20 \end{aligned}$$

It is sometimes more convenient to write such a product vertically.

$$\begin{array}{rrrr} & 2x^2 & -\ 3x & +\ 5 \\ & & 3x & -\ 4 \\ \hline & -8x^2 & +\ 12x & -\ 20 \\ 6x^3 & -\ 9x^2 & +\ 15x & \\ \hline 6x^3 & -\ 17x^2 & +\ 27x & -\ 20 \end{array}$$

Add in columns.

**EXAMPLE 5** *Multiplying Polynomials*

Multiply $(3p^2 - 4p + 1)(p^3 + 2p - 8)$.

**Solution** Multiply each term of the second polynomial by each term of the first, and add these products. It is most efficient to work vertically with polynomials of more than two terms so that like terms can be placed in columns.

$$\begin{array}{rrrrrr} & & & 3p^2 & -\ 4p & +\ 1 \\ & & & p^3 & +\ 2p & -\ 8 \\ \hline & & & -24p^2 & +\ 32p & -\ 8 \\ & & 6p^3 & -\ 8p^2 & +\ 2p & \\ 3p^5 & -\ 4p^4 & +\ p^3 & & & \\ \hline 3p^5 & -\ 4p^4 & +\ 7p^3 & -\ 32p^2 & +\ 34p & -\ 8 \end{array}$$

Multiply $3p^2 - 4p + 1$ by $-8$.
Multiply $3p^2 - 4p + 1$ by $2p$.
Multiply $3p^2 - 4p + 1$ by $p^3$.
Add in columns.

The FOIL method is a convenient way to find the product of two binomials. The memory aid FOIL (for First, Outside, Inside, Last) gives the pairs of terms to be multiplied to obtain the product.

**EXAMPLE 6** *Using FOIL to Multiply Two Binomials*

Find each product.

**(a)** $(6m + 1)(4m - 3)$ **(b)** $(2x + 7)(2x - 7)$ **(c)** $r^2(3r + 2)(3r - 2)$

**Solution**

$$\text{(a)}\ (6m + 1)(4m - 3) = \overset{F}{6m(4m)} + \overset{O}{6m(-3)} + \overset{I}{1(4m)} + \overset{L}{1(-3)}$$
$$= 24m^2 - 14m - 3$$

$$\text{(b)}\ (2x + 7)(2x - 7) = 4x^2 - 14x + 14x - 49$$
$$= 4x^2 - 49$$

$$\text{(c)}\ r^2(3r + 2)(3r - 2) = r^2(9r^2 - 4)$$
$$= 9r^4 - 4r^2$$

In part (a) of Example 6, the product of two binomials was a trinomial, while in parts (b) and (c), the product of two binomials was a binomial. The product of two binomials of the forms $x + y$ and $x - y$ is always a binomial. The squares of binomials of the forms $(x + y)^2$ and $(x - y)^2$ are also special products.

**Special Products**

| | |
|---|---|
| Product of the Sum and Difference of Two Terms | $(x + y)(x - y) = x^2 - y^2$ |
| Square of a Binomial | $(x + y)^2 = x^2 + 2xy + y^2$ |
| | $(x - y)^2 = x^2 - 2xy + y^2$ |

**EXAMPLE 7** *Using the Special Products*

Find each product.

**(a)** $(3p + 11)(3p - 11)$ **(b)** $(5m^3 - 3)(5m^3 + 3)$

**(c)** $(9k - 11r^3)(9k + 11r^3)$ **(d)** $(2m + 5)^2$ **(e)** $(3x - 7y^4)^2$

**Solution**

**(a)** Use the pattern in the box for the product of the sum and difference of two terms. Replace $x$ with $3p$ and $y$ with 11.

$$(3p + 11)(3p - 11) = (3p)^2 - 11^2$$
$$= 9p^2 - 121$$

$$\text{(b)}\ (5m^3 - 3)(5m^3 + 3) = (5m^3)^2 - 3^2$$
$$= 25m^6 - 9$$

$$\text{(c)}\ (9k - 11r^3)(9k + 11r^3) = (9k)^2 - (11r^3)^2$$
$$= 81k^2 - 121r^6$$

$$\text{(d)}\ (2m + 5)^2 = (2m)^2 + 2(2m)(5) + 5^2$$
$$= 4m^2 + 20m + 25$$

$$\text{(e)}\ (3x - 7y^4)^2 = (3x)^2 - 2(3x)(7y^4) + (7y^4)^2$$
$$= 9x^2 - 42xy^4 + 49y^8$$

**CAUTION** As shown in Examples 7(d) and (e), the square of a binomial has three terms. Students often mistakenly give $a^2 + b^2$ as the result of expanding $(a + b)^2$. Be careful to avoid this error.

## R.1 EXERCISES

*Use the properties of exponents to simplify each expression. Leave answers with exponents.*

**1.** $(-4)^3 \cdot (-4)^2$ **2.** $(-5)^2 \cdot (-5)^6$ **3.** $2^0$ **4.** $-2^0$

**5.** $(5m)^0$, if $m \neq 0$ **6.** $(-4z)^0$, if $z \neq 0$ **7.** $(2^2)^5$ **8.** $(6^4)^3$

**9.** $(2x^5y^4)^3$ **10.** $(-4m^3n^9)^2$ **11.** $-\left(\frac{p^4}{q}\right)^2$ **12.** $\left(\frac{r^8}{s^2}\right)^3$

*Concept Check* *Identify each expression as a* polynomial *or* not a polynomial. *For each polynomial, give the degree and identify it as a* monomial, binomial, trinomial, *or* none of these.

**13.** $-5x^{11}$ **14.** $9y^{12} + y^2$ **15.** $18p^5q + 6pq$

**16.** $2a^6 + 5a^2 + 4a$ **17.** $\sqrt{2}x^2 + \sqrt{3}x^6$ **18.** $-\sqrt{7}m^5n^2 + 2\sqrt{3}m^3n^2$

**19.** $\frac{1}{3}r^2s^2 - \frac{3}{5}r^4s^2 + rs^3$ **20.** $\frac{5}{p} + \frac{2}{p^2} + \frac{5}{p^3}$ **21.** $-5\sqrt{z} + 2\sqrt{z^3} - 5\sqrt{z^5}$

*Find each sum or difference.*

**22.** $(3x^2 - 4x + 5) + (-2x^2 + 3x - 2)$ **23.** $(4m^3 - 3m^2 + 5) + (-3m^3 - m^2 + 5)$

**24.** $(12y^2 - 8y + 6) - (3y^2 - 4y + 2)$ **25.** $(8p^2 - 5p) - (3p^2 - 2p + 4)$

**26.** $(6m^4 - 3m^2 + m) - (2m^3 + 5m^2 + 4m) + (m^2 - m)$ **27.** $-(8x^3 + x - 3) + (2x^3 + x^2) - (4x^2 + 3x - 1)$

*Find each product.*

**28.** $(4r - 1)(7r + 2)$ **29.** $(5m - 6)(3m + 4)$ **30.** $\left(3x - \frac{2}{3}\right)\left(5x + \frac{1}{3}\right)$

**31.** $\left(2m - \frac{1}{4}\right)\left(3m + \frac{1}{2}\right)$ **32.** $4x^2(3x^3 + 2x^2 - 5x + 1)$ **33.** $2b^3(b^2 - 4b + 3)$

**34.** $(2z - 1)(-z^2 + 3z - 4)$ **35.** $(m - n + k)(m + 2n - 3k)$ **36.** $(r - 3s + t)(2r - s + t)$

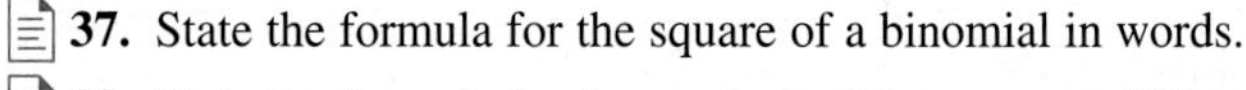

**37.** State the formula for the square of a binomial in words.

**38.** State the formula for the product of the sum and difference of two terms in words.

*Use a special products formula to find each product.*

**39.** $(2m + 3)(2m - 3)$ **40.** $(8s - 3t)(8s + 3t)$ **41.** $(4m + 2n)^2$

**42.** $(a - 6b)^2$ **43.** $(5r + 3t^2)^2$ **44.** $(2z^4 - 3y)^2$

**45.** $[(2p - 3) + q]^2$ **46.** $[(4y - 1) + z]^2$ **47.** $[(3q + 5) - p][(3q + 5) + p]$

**48.** $[(9r - s) + 2][(9r - s) - 2]$ **49.** $[(3a + b) - 1]^2$ **50.** $[(2m + 7) - n]^2$

*Perform the indicated operations.*

**51.** $(6p + 5q)(3p - 7q)$ **52.** $(2p - 1)(3p^2 - 4p + 5)$

**53.** $(p^3 - 4p^2 + p) - (3p^2 + 2p + 7)$ **54.** $(6k - 3)^2$

**55.** $y(4x + 3y)(4x - 3y)$ **56.** $(r^5 - r^3 + r) + (3r^5 - 4r^4 + r^3 + 2r)$

**57.** $(2z + y)(3z - 4y)$ **58.** $(7m + 2n)(7m - 2n)$

**59.** $(3p + 5)^2$ **60.** $2(3r^2 + 4r + 2) - 3(-r^2 + 4r - 5)$

**61.** $p(4p - 6) + 2(3p - 8)$ **62.** $m(5m - 2) + 9(5 - m)$

**63.** $-y(y^2 - 4) + 6y^2(2y - 3)$ **64.** $-z^3(9 - z) + 4z(2 + 3z)$

# R.2 REVIEW OF FACTORING

Factoring Out the Greatest Common Factor • Factoring by Grouping • Factoring Trinomials • Factoring Special Products • Factoring by Substitution

The process of finding polynomials whose product equals a given polynomial is called **factoring.** For example, since $4x + 12 = 4(x + 3)$, both 4 and $x + 3$ are called **factors** of $4x + 12$. Also, $4(x + 3)$ is called the **factored form** of $4x + 12$. A polynomial that cannot be written as a product of two polynomials with integer coefficients is a **prime** or **irreducible polynomial.** A polynomial is **factored completely** when it is written as a product of prime polynomials with integer coefficients.

## Factoring Out the Greatest Common Factor

Polynomials are factored by using the distributive property. For example, to factor $6x^2y^3 + 9xy^4 + 18y^5$, we look for the monomial that is the greatest common factor (GCF) of all the terms. The terms of this polynomial have $3y^3$ as the greatest common factor. By the distributive property,

$$\begin{aligned} 6x^2y^3 + 9xy^4 + 18y^5 &= 3y^3(2x^2) + 3y^3(3xy) + 3y^3(6y^2) \\ &= 3y^3(2x^2 + 3xy + 6y^2). \end{aligned}$$

**EXAMPLE 1** *Factoring Out the Greatest Common Factor*

Factor out the greatest common factor from each polynomial.

**(a)** $9y^5 + y^2$ **(b)** $6x^2t + 8xt + 12t$

**(c)** $14m^4(m + 1) - 28m^3(m + 1) - 7m^2(m + 1)$

**Solution**

**(a)** $9y^5 + y^2 = y^2(9y^3) + y^2(1)$ GCF $= y^2$

$= y^2(9y^3 + 1)$ Remember to include the 1.

**(b)** $6x^2t + 8xt + 12t = 2t(3x^2 + 4x + 6)$ GCF $= 2t$

**(c)** The GCF is $7m^2(m + 1)$. Use the distributive property as follows.

$$\begin{aligned} 14m^4(m + 1) - 28m^3(m + 1) - 7m^2(m + 1) &= 7m^2(m + 1)[(2m^2 - 4m - 1)] \\ &= 7m^2(m + 1)(2m^2 - 4m - 1) \end{aligned}$$

**NOTE** Factoring can always be checked by multiplying.

## Factoring by Grouping

When a polynomial has more than three terms, it can sometimes be factored using a method called **factoring by grouping.** For example, to factor

$$ax + ay + 6x + 6y,$$

collect the terms into two groups so that each group has a common factor.

$$ax + ay + 6x + 6y = (ax + ay) + (6x + 6y)$$

Factor each group.

$$ax + ay + 6x + 6y = a(x + y) + 6(x + y)$$

The quantity $(x + y)$ is now a common factor, which can be factored out, producing

$$ax + ay + 6x + 6y = (x + y)(a + 6).$$

It is not always obvious which terms should be grouped. Experience and repeated trials are the most reliable tools when factoring by grouping.

**EXAMPLE 2** *Factoring by Grouping*

Factor each polynomial by grouping.

**(a)** $mp^2 + 7m + 3p^2 + 21$ **(b)** $2y^2 - 2z - ay^2 + az$

**(c)** $4x^3 + 2x^2 - 2x - 1$

**Solution**

**(a)** Group the terms as follows.

$$mp^2 + 7m + 3p^2 + 21 = (mp^2 + 7m) + (3p^2 + 21)$$

Factor out the greatest common factor from each group.

$$\begin{aligned}(mp^2 + 7m) + (3p^2 + 21) &= m(p^2 + 7) + 3(p^2 + 7)\\ &= (p^2 + 7)(m + 3)\end{aligned}$$

$p^2 + 7$ is a common factor.

**(b)**

$$\begin{aligned}2y^2 - 2z - ay^2 + az &= (2y^2 - 2z) + (-ay^2 + az) &&\text{Group terms.}\\ &= 2(y^2 - z) + a(-y^2 + z) &&\text{Factor each group.}\end{aligned}$$

The expression $-y^2 + z$ is the negative of $y^2 - z$, so group the terms as follows.

$$\begin{aligned}2y^2 - 2z - ay^2 + az &= (2y^2 - 2z) - (ay^2 - az)\\ &= 2(y^2 - z) - a(y^2 - z) &&\text{Factor each group.}\\ &= (y^2 - z)(2 - a) &&\text{Factor out } y^2 - z.\end{aligned}$$

**(c)**

$$\begin{aligned}4x^3 + 2x^2 - 2x - 1 &= 2x^2(2x + 1) - 1(2x + 1)\\ &= (2x + 1)(2x^2 - 1)\end{aligned}$$

Later, in Example 5(d), we show another way to factor by grouping three of the four terms.

## Factoring Trinomials

Factoring is the opposite of multiplication. Since the product of two binomials is usually a trinomial, we can expect factorable trinomials (that have terms with no common factor) to have two binomial factors. Thus, factoring trinomials requires using FOIL backward.

**EXAMPLE 3** *Factoring Trinomials*

Factor each trinomial.

**(a)** $4y^2 - 11y + 6$ **(b)** $6p^2 - 7p - 5$ **(c)** $16y^3 + 24y^2 - 16y$

**Solution**

**(a)** To factor this polynomial, we must find integers $a$, $b$, $c$, and $d$ such that

$$4y^2 - 11y + 6 = (ay + b)(cy + d).$$

By using FOIL, we see that $ac = 4$ and $bd = 6$. The positive factors of 4 are 4 and 1 or 2 and 2. Since the middle term is negative, we consider only negative factors of 6. The possibilities are $-2$ and $-3$ or $-1$ and $-6$. Now we try various arrangements of these factors until we find one that gives the correct coefficient of $y$.

$$(2y-1)(2y-6) = 4y^2 - 14y + 6 \quad \text{Incorrect}$$
$$(2y-2)(2y-3) = 4y^2 - 10y + 6 \quad \text{Incorrect}$$
$$(y-2)(4y-3) = 4y^2 - 11y + 6 \quad \text{Correct}$$

The last trial gives the correct factorization.

**(b)** Again, we try various possibilities. The positive factors of 6 could be 2 and 3 or 1 and 6. As factors of $-5$ we have only $-1$ and 5 or $-5$ and 1. We try different combinations of these factors until the correct one is found.

$$(2p-5)(3p+1) = 6p^2 - 13p - 5 \quad \text{Incorrect}$$
$$(3p-5)(2p+1) = 6p^2 - 7p - 5 \quad \text{Correct}$$

Thus, $6p^2 - 7p - 5$ factors as $(3p-5)(2p+1)$.

**(c)** First factor out the greatest common factor, $8y$.

$$16y^3 + 24y^2 - 16y = 8y(2y^2 + 3y - 2) \quad \text{GCF} = 8y$$
$$= 8y(2y-1)(y+2) \quad \text{Factor the trinomial.}$$

## Factoring Special Products

Each of the special patterns for multiplication given in Section R.1 can be used in reverse to get a pattern for factoring. Perfect square trinomials can be factored as follows.

**Perfect Square Trinomials**

$$x^2 + 2xy + y^2 = (x+y)^2$$
$$x^2 - 2xy + y^2 = (x-y)^2$$

**EXAMPLE 4** *Factoring Perfect Square Trinomials*

Factor each polynomial.

**(a)** $16p^2 - 40pq + 25q^2$ **(b)** $169x^2 + 104xy^2 + 16y^4$

**Solution**

**(a)** Since $16p^2 = (4p)^2$ and $25q^2 = (5q)^2$, use the second pattern shown in the box, with $4p$ replacing $x$ and $5q$ replacing $y$, to get

$$16p^2 - 40pq + 25q^2 = (4p)^2 - 2(4p)(5q) + (5q)^2$$
$$= (4p - 5q)^2.$$

Make sure that the middle term of the trinomial being factored, $-40pq$ here, is twice the product of the two terms in the binomial $4p - 5q$.

$$-40pq = 2(4p)(-5q)$$

**(b)** $169x^2 + 104xy^2 + 16y^4 = (13x + 4y^2)^2$ since $2(13x)(4y^2) = 104xy^2$.

The pattern for the product of the sum and difference of two terms gives the following factorization.

**Difference of Squares**

$$x^2 - y^2 = (x + y)(x - y)$$

**EXAMPLE 5** *Factoring Differences of Squares*

Factor each polynomial.

**(a)** $4m^2 - 9$ **(b)** $256k^4 - 625m^4$

**(c)** $(a + 2b)^2 - 4c^2$ **(d)** $x^2 - 6x + 9 - y^4$

**Solution**

**(a)** First, we recognize that $4m^2 - 9$ is the difference of squares, since $4m^2 = (2m)^2$ and $9 = 3^2$. We can use the pattern for the difference of squares, with $2m$ replacing $x$ and 3 replacing $y$.

$$\begin{aligned} 4m^2 - 9 &= (2m)^2 - 3^2 \\ &= (2m + 3)(2m - 3) \end{aligned}$$

**(b)** We use the difference of squares pattern twice.

$$\begin{aligned} 256k^4 - 625m^4 &= (16k^2)^2 - (25m^2)^2 \\ &= (16k^2 + 25m^2)(16k^2 - 25m^2) \\ &= (16k^2 + 25m^2)(4k + 5m)(4k - 5m) \end{aligned}$$

**(c)**
$$\begin{aligned} (a + 2b)^2 - 4c^2 &= (a + 2b)^2 - (2c)^2 \\ &= [(a + 2b) + 2c][(a + 2b) - 2c] \\ &= (a + 2b + 2c)(a + 2b - 2c) \end{aligned}$$

**(d)** We group the first three terms to get a perfect square trinomial. Then, we use the difference of squares pattern.

$$\begin{aligned} x^2 - 6x + 9 - y^4 &= (x^2 - 6x + 9) - y^4 \\ &= (x - 3)^2 - (y^2)^2 \\ &= [(x - 3) + y^2][(x - 3) - y^2] \\ &= (x - 3 + y^2)(x - 3 - y^2) \end{aligned}$$

Two other special factoring patterns are listed here. Each can be verified by multiplying on the right side of the equation.

**Difference and Sum of Cubes**

| | |
|---|---|
| Difference of cubes | $x^3 - y^3 = (x - y)(x^2 + xy + y^2)$ |
| Sum of cubes | $x^3 + y^3 = (x + y)(x^2 - xy + y^2)$ |

**EXAMPLE 6** *Factoring Sums and Differences of Cubes*

Factor each polynomial.

**(a)** $x^3 + 27$ **(b)** $m^3 - 64n^3$ **(c)** $8q^6 + 125p^9$

**Solution**

**(a)** Notice that $27 = 3^3$, so the expression is a sum of cubes. Use the second pattern given in the box.

$$\begin{aligned} x^3 + 27 &= x^3 + 3^3 \\ &= (x + 3)(x^2 - 3x + 3^2) \\ &= (x + 3)(x^2 - 3x + 9) \end{aligned}$$

**(b)** Since $64n^3 = (4n)^3$, the given polynomial is a difference of cubes. To factor, use the first pattern in the box, replacing $x$ with $m$ and $y$ with $4n$.

$$\begin{aligned} m^3 - 64n^3 &= m^3 - (4n)^3 \\ &= (m - 4n)[m^2 + m(4n) + (4n)^2] \\ &= (m - 4n)(m^2 + 4mn + 16n^2) \end{aligned}$$

**(c)** First write $8q^6$ as $(2q^2)^3$ and $125p^9$ as $(5p^3)^3$, so the given polynomial is a sum of cubes. Then factor.

$$\begin{aligned} 8q^6 + 125p^9 &= (2q^2)^3 + (5p^3)^3 \\ &= (2q^2 + 5p^3)[(2q^2)^2 - (2q^2)(5p^3) + (5p^3)^2] \\ &= (2q^2 + 5p^3)(4q^4 - 10q^2p^3 + 25p^6) \end{aligned}$$

## Factoring by Substitution

Sometimes a polynomial can be factored by substituting one expression for another.

**EXAMPLE 7** *Factoring by Substitution*

Factor each polynomial.

**(a)** $6z^4 - 13z^2 - 5$ **(b)** $10(2a - 1)^2 - 19(2a - 1) - 15$
**(c)** $(2a - 1)^3 + 8$

**Solution**

**(a)** Replace $z^2$ with $y$, so $y^2 = (z^2)^2 = z^4$. This replacement gives

$$\begin{aligned} 6z^4 - 13z^2 - 5 &= 6y^2 - 13y - 5 \\ &= (2y - 5)(3y + 1). \quad \text{Factor.} \end{aligned}$$

Replacing $y$ with $z^2$ gives

$$6z^4 - 13z^2 - 5 = (2z^2 - 5)(3z^2 + 1).$$

(Some students prefer to factor this type of trinomial directly by using trial and error with FOIL.)

**(b)** Replacing $2a - 1$ with $m$ and factoring gives

$$10m^2 - 19m - 15 = (5m + 3)(2m - 5).$$

Now, replace $m$ with $2a - 1$ in the factored form and simplify.

$$\begin{aligned} &10(2a-1)^2 - 19(2a-1) - 15 \\ &= [5(2a-1)+3][2(2a-1)-5] && \text{Let } m = 2a - 1. \\ &= (10a - 5 + 3)(4a - 2 - 5) && \text{Multiply.} \\ &= (10a-2)(4a-7) && \text{Add.} \\ &= 2(5a-1)(4a-7) && \text{Factor out the common factor.} \end{aligned}$$

**(c)** Let $2a - 1 = x$ to get

$$\begin{aligned} (2a-1)^3 + 8 &= x^3 + 8 \\ &= x^3 + 2^3 \\ &= (x+2)(x^2 - 2x + 2^2). \end{aligned}$$

Replacing $x$ with $2a - 1$ gives

$$\begin{aligned} (2a-1)^3 + 8 &= (2a - 1 + 2)[(2a-1)^2 - 2(2a-1) + 4] && \text{Let } x = 2a - 1. \\ &= (2a+1)(4a^2 - 4a + 1 - 4a + 2 + 4) \\ &= (2a+1)(4a^2 - 8a + 7). && \text{Combine terms.} \end{aligned}$$

## R.2 EXERCISES

**1.** *Concept Check* Match each polynomial in Column I with its factored form in Column II.

| I | II |
|---|---|
| **(a)** $x^2 + 10xy + 25y^2$ | **A.** $(x + 5y)(x - 5y)$ |
| **(b)** $x^2 - 10xy + 25y^2$ | **B.** $(x + 5y)^2$ |
| **(c)** $x^2 - 25y^2$ | **C.** $(x - 5y)^2$ |
| **(d)** $25y^2 - x^2$ | **D.** $(5y + x)(5y - x)$ |

**2.** *Concept Check* Match each polynomial in Column I with its factored form in Column II.

| I | II |
|---|---|
| **(a)** $8x^3 - 27$ | **A.** $(3 - 2x)(9 + 6x + 4x^2)$ |
| **(b)** $8x^3 + 27$ | **B.** $(2x - 3)(4x^2 + 6x + 9)$ |
| **(c)** $27 - 8x^3$ | **C.** $(2x + 3)(4x^2 - 6x + 9)$ |

*Factor the greatest common factor from each polynomial.*

**3.** $4k^2m^3 + 8k^4m^3 - 12k^2m^4$

**4.** $28r^4s^2 + 7r^3s - 35r^4s^3$

**5.** $2(a + b) + 4m(a + b)$

**6.** $4(y - 2)^2 + 3(y - 2)$

**7.** $(2y - 3)(y + 2) + (y + 5)(y + 2)$

**8.** $(6a - 1)(a + 2) + (6a - 1)(3a - 1)$

**9.** $(5r - 6)(r + 3) - (2r - 1)(r + 3)$

**10.** $(3z + 2)(z + 4) - (z + 6)(z + 4)$

**11.** $2(m - 1) - 3(m - 1)^2 + 2(m - 1)^3$

**12.** $5(a + 3)^3 - 2(a + 3) + (a + 3)^2$

*Factor each polynomial by grouping.*

**13.** $6st + 9t - 10s - 15$

**14.** $10ab - 6b + 35a - 21$

**15.** $10x^2 - 12y + 15x - 8xy$

**16.** $2m^4 + 6 - am^4 - 3a$

**17.** $t^3 + 2t^2 - 3t - 6$

**18.** $x^3 + 3x^2 - 5x - 15$

**19.** *Concept Check* Layla factored $16a^2 - 40a - 6a + 15$ by grouping and obtained $(8a - 3)(2a - 5)$. Jamal factored the same polynomial and gave an answer of $(3 - 8a)(5 - 2a)$. Are both of these answers correct? If not, why not?

*Factor each trinomial.*

**20.** $6a^2 - 48a - 120$
**21.** $8h^2 - 24h - 320$
**22.** $3m^3 + 12m^2 + 9m$
**23.** $9y^4 - 54y^3 + 45y^2$
**24.** $6k^2 + 5kp - 6p^2$
**25.** $14m^2 + 11mr - 15r^2$
**26.** $5a^2 - 7ab - 6b^2$
**27.** $12s^2 + 11st - 5t^2$
**28.** $9x^2 - 6x^3 + x^4$
**29.** $30a^2 + am - m^2$
**30.** $24a^4 + 10a^3b - 4a^2b^2$
**31.** $18x^5 + 15x^4z - 75x^3z^2$

*Factor each perfect square trinomial.*

**32.** $9m^2 - 12m + 4$
**33.** $16p^2 - 40p + 25$
**34.** $32a^2 - 48ab + 18b^2$
**35.** $20p^2 - 100pq + 125q^2$
**36.** $4x^2y^2 + 28xy + 49$
**37.** $9m^2n^2 - 12mn + 4$
**38.** $(a - 3b)^2 - 6(a - 3b) + 9$
**39.** $(2p + q)^2 - 10(2p + q) + 25$
**40.** $(5r + 2s)^2 + 6(5r + 2s) + 9$

*Factor each difference of squares.*

**41.** $9a^2 - 16$
**42.** $16q^2 - 25$
**43.** $25s^4 - 9t^2$
**44.** $36z^2 - 81y^4$
**45.** $(a + b)^2 - 16$
**46.** $(p - 2q)^2 - 100$
**47.** $p^4 - 625$
**48.** $m^4 - 81$

**49.** *Concept Check* Which of the following is the correct complete factorization of $x^4 - 1$?

**A.** $(x^2 - 1)(x^2 + 1)$ **B.** $(x^2 + 1)(x + 1)(x - 1)$ **C.** $(x^2 - 1)^2$ **D.** $(x - 1)^2(x + 1)^2$

**50.** *Concept Check* Which of the following is the correct complete factorization of $x^3 + 8$?

**A.** $(x + 2)^3$ **B.** $(x + 2)(x^2 + 2x + 4)$ **C.** $(x + 2)(x^2 - 2x + 4)$ **D.** $(x + 2)(x^2 - 4x + 4)$

*Factor each sum or difference of cubes.*

**51.** $8 - a^3$
**52.** $r^3 + 27$
**53.** $125x^3 - 27$
**54.** $8m^3 - 27n^3$
**55.** $27y^9 + 125z^6$
**56.** $27z^3 + 729y^3$
**57.** $(r + 6)^3 - 216$
**58.** $(b + 3)^3 - 27$
**59.** $27 - (m + 2n)^3$
**60.** $125 - (4a - b)^3$

**61.** Is the following factorization of $3a^4 + 14a^2 - 5$ correct? Explain. If it is incorrect, give the correct factors.

$$3a^4 + 14a^2 - 5 = 3u^2 + 14u - 5 \quad \text{Let } u = a^2.$$
$$= (3u - 1)(u + 5)$$

*Factor each polynomial by substitution.*

**62.** $m^4 - 3m^2 - 10$
**63.** $a^4 - 2a^2 - 48$
**64.** $7(3k - 1)^2 + 26(3k - 1) - 8$
**65.** $6(4z - 3)^2 + 7(4z - 3) - 3$
**66.** $9(a - 4)^2 + 30(a - 4) + 25$
**67.** $20(4 - p)^2 - 3(4 - p) - 2$

*Factor by any method.*

**68.** $a^3(r + s) + b^2(r + s)$
**69.** $4b^2 + 4bc + c^2 - 16$
**70.** $(2y - 1)^2 - 4(2y - 1) + 4$
**71.** $x^2 + xy - 5x - 5y$
**72.** $8r^2 - 3rs + 10s^2$
**73.** $p^4(m - 2n) + q(m - 2n)$
**74.** $36a^2 + 60a + 25$
**75.** $4z^2 + 28z + 49$
**76.** $6p^4 + 7p^2 - 3$
**77.** $1000x^3 + 343y^3$
**78.** $b^2 + 8b + 16 - a^2$
**79.** $125m^6 - 216$
**80.** $q^2 + 6q + 9 - p^2$
**81.** $12m^2 + 16mn - 35n^2$
**82.** $216p^3 + 125q^3$
**83.** $4p^2 + 3p - 1$
**84.** $100r^2 - 169s^2$
**85.** $144z^2 + 121$
**86.** $(3a + 5)^2 - 18(3a + 5) + 81$
**87.** $(x + y)^2 - (x - y)^2$
**88.** $4z^4 - 7z^2 - 15$

# REVIEW OF RATIONAL EXPRESSIONS

Domain of a Rational Expression • Lowest Terms of a Rational Expression • Multiplying and Dividing Rational Expressions • Adding and Subtracting Rational Expressions • Complex Fractions

An expression that is the quotient of two polynomials is a **rational expression.** It is written in the form $\frac{P}{Q}$, where $P$ and $Q$ are polynomials, and $Q \neq 0$. Some examples of rational expressions are

$$\frac{x+6}{x+2}, \quad \frac{(x+6)(x+4)}{(x+2)(x+4)}, \quad \text{and} \quad \frac{2p^2+7p-4}{5p^2+20p}.$$

## Domain of a Rational Expression

The **domain** of a rational expression is the set of real numbers for which the expression is defined. Because of the restriction that the denominator cannot be 0, the domain consists of all real numbers except those that make the denominator 0. These numbers are found by setting the denominator equal to 0 and solving the resulting equation. For example, in the rational expression

$$\frac{x+6}{x+2},$$

the solution to the equation $x + 2 = 0$ is excluded from the domain. Since this solution is $-2$, the domain is the set of all real numbers $x$ such that $x \neq -2$, written $\{x|x \neq -2\}$.

If the denominator of a rational expression contains a product, we find the domain using the zero-product property (Section 3.3), which states that $ab = 0$ if and only if $a = 0$ or $b = 0$. For example, to find the domain of

$$\frac{(x+6)(x+4)}{(x+2)(x+4)},$$

we solve as follows.

$$\begin{aligned} (x+2)(x+4) &= 0 \\ x+2 = 0 \quad &\text{or} \quad x+4=0 \\ x = -2 \quad &\text{or} \quad x = -4 \end{aligned}$$

The domain consists of the set of real numbers $x$ such that $x \neq -2, -4$, written $\{x|x \neq -2, -4\}$.

## Lowest Terms of a Rational Expression

A rational expression is in lowest terms when the greatest common factor of its numerator and denominator is 1. Just as the fraction $\frac{6}{8}$ is written in lowest terms as $\frac{3}{4}$, rational expressions can also be written in lowest terms. This is done with the fundamental principle of fractions.

**Fundamental Principle of Fractions**

$$\frac{ac}{bc} = \frac{a}{b} \quad (b \neq 0, c \neq 0)$$

### EXAMPLE 1 *Writing Rational Expressions in Lowest Terms*

Write each rational expression in lowest terms.

**(a)** $\dfrac{2p^2 + 7p - 4}{5p^2 + 20p}$ **(b)** $\dfrac{6 - 3k}{k^2 - 4}$

**Solution**

**(a)** Factor the numerator and denominator and apply the fundamental principle.

$$\frac{2p^2 + 7p - 4}{5p^2 + 20p} = \frac{(2p - 1)(p + 4)}{5p(p + 4)} = \frac{2p - 1}{5p}$$

In the original expression, $p$ cannot be 0 or $-4$, because $5p^2 + 20p \neq 0$, so this result is valid only for values of $p$ other than 0 and $-4$. From now on, we assume such restrictions when writing rational expressions in lowest terms.

**(b)** Factor to get $\dfrac{6 - 3k}{k^2 - 4} = \dfrac{3(2 - k)}{(k + 2)(k - 2)}$. The factors $2 - k$ and $k - 2$ have opposite signs. To find a common factor, multiply the numerator and denominator by $-1$.

$$\frac{6 - 3k}{k^2 - 4} = \frac{3(2 - k)(-1)}{(k + 2)(k - 2)(-1)}$$

Since $(k - 2)(-1) = -k + 2$, or $2 - k$,

$$\frac{6 - 3k}{k^2 - 4} = \frac{3(2 - k)(-1)}{(k + 2)(2 - k)} = \frac{-3}{k + 2}.$$

Working in an alternative way would lead to the equivalent result

$$\frac{3}{-k - 2}.$$

**CAUTION** Remember that the fundamental principle requires a pair of common *factors,* one in the numerator and one in the denominator. For example,

$$\frac{2x + 4}{6} = \frac{2(x + 2)}{6} = \frac{2(x + 2)}{2 \cdot 3} = \frac{x + 2}{3}.$$

It would be *incorrect* to write $\dfrac{2x + 4}{6}$ as $\dfrac{x + 4}{3}$ (crossed out). *Factor first.*

## Multiplying and Dividing Rational Expressions

Rational expressions are multiplied and divided by using definitions from earlier work with fractions.

**Multiplying and Dividing Fractions**

For fractions $\frac{a}{b}$ and $\frac{c}{d}$ $(b \neq 0, d \neq 0)$,

$$\frac{a}{b} \cdot \frac{c}{d} = \frac{ac}{bd} \quad \text{and} \quad \frac{a}{b} \div \frac{c}{d} = \frac{a}{b} \cdot \frac{d}{c}, \quad \text{if } \frac{c}{d} \neq 0.$$

**EXAMPLE 2** *Multiplying and Dividing Rational Expressions*

Multiply or divide, as indicated.

**(a)** $\frac{2y^2}{9} \cdot \frac{27}{8y^5}$ **(b)** $\frac{3m^2 - 2m - 8}{3m^2 + 14m + 8} \cdot \frac{3m + 2}{3m + 4}$

**(c)** $\frac{5}{8m + 16} \div \frac{7}{12m + 24}$ **(d)** $\frac{3p^2 + 11p - 4}{24p^3 - 8p^2} \div \frac{9p + 36}{24p^4 - 36p^3}$

**Solution**

**(a)**

$$\frac{2y^2}{9} \cdot \frac{27}{8y^5} = \frac{2y^2 \cdot 27}{9 \cdot 8y^5}$$

$$= \frac{2 \cdot 9 \cdot 3 \cdot y^2}{9 \cdot 2 \cdot 4 \cdot y^2 \cdot y^3} \qquad \text{Factor.}$$

$$= \frac{3}{4y^3} \qquad \text{Fundamental principle}$$

The product was written in lowest terms in the last step.

**(b)**

$$\frac{3m^2 - 2m - 8}{3m^2 + 14m + 8} \cdot \frac{3m + 2}{3m + 4} = \frac{(m - 2)(3m + 4)}{(m + 4)(3m + 2)} \cdot \frac{3m + 2}{3m + 4} \qquad \text{Factor.}$$

$$= \frac{(m - 2)(3m + 4)(3m + 2)}{(m + 4)(3m + 2)(3m + 4)} \qquad \text{Multiply fractions.}$$

$$= \frac{m - 2}{m + 4} \qquad \text{Fundamental principle}$$

**(c)**

$$\frac{5}{8m + 16} \div \frac{7}{12m + 24} = \frac{5}{8(m + 2)} \div \frac{7}{12(m + 2)} \qquad \text{Factor.}$$

$$= \frac{5}{8(m + 2)} \cdot \frac{12(m + 2)}{7} \qquad \text{Definition of division}$$

$$= \frac{5 \cdot 12(m + 2)}{8 \cdot 7(m + 2)} \qquad \text{Multiply.}$$

$$= \frac{15}{14} \qquad \text{Fundamental principle}$$

**(d)**

$$\frac{3p^2 + 11p - 4}{24p^3 - 8p^2} \div \frac{9p + 36}{24p^4 - 36p^3} = \frac{(p + 4)(3p - 1)}{8p^2(3p - 1)} \div \frac{9(p + 4)}{12p^3(2p - 3)}$$

$$= \frac{(p + 4)(3p - 1)(12p^3)(2p - 3)}{8p^2(3p - 1)(9)(p + 4)}$$

$$= \frac{12p^3(2p - 3)}{9 \cdot 8p^2}$$

$$= \frac{p(2p - 3)}{6}$$

## Adding and Subtracting Rational Expressions

Adding and subtracting rational expressions also depends on definitions from earlier work with fractions.

**Adding and Subtracting Fractions**

For fractions $\frac{a}{b}$ and $\frac{c}{d}$ $(b \neq 0, d \neq 0)$,

$$\frac{a}{b} + \frac{c}{d} = \frac{ad + bc}{bd} \quad \text{and} \quad \frac{a}{b} - \frac{c}{d} = \frac{ad - bc}{bd}.$$

In practice, rational expressions are normally added or subtracted after rewriting the rational expressions with a common denominator, preferably the least common denominator.

**Finding the Least Common Denominator (LCD)**

1. Write each denominator as a product of prime factors.
2. Form a product of all the different prime factors. Each factor should have as exponent the *greatest* exponent that appears on that factor.

## EXAMPLE 3 *Adding and Subtracting Rational Expressions*

Add or subtract, as indicated.

**(a)** $\dfrac{5}{9x^2} + \dfrac{1}{6x}$ **(b)** $\dfrac{y + 2}{y^2 - y} - \dfrac{3y}{2y^2 - 4y + 2}$

**(c)** $\dfrac{3}{(x - 1)(x + 2)} - \dfrac{1}{(x + 3)(x - 4)}$

### Solution

**(a)** Write each denominator as a product of prime factors.

$$9x^2 = 3^2 \cdot x^2$$
$$6x = 2^1 \cdot 3^1 \cdot x^1$$

For the least common denominator, form the product of all the prime factors, with each factor having the greatest exponent that appears on it. Here the greatest exponent on 2 is 1, while both 3 and $x$ have greatest exponent 2. The LCD is

$$2^1 \cdot 3^2 \cdot x^2 = 18x^2.$$

Now use the fundamental principle to write the given expressions with this denominator, then add.

$$\frac{5}{9x^2} + \frac{1}{6x} = \frac{5 \cdot 2}{9x^2 \cdot 2} + \frac{1 \cdot 3x}{6x \cdot 3x}$$
$$= \frac{10}{18x^2} + \frac{3x}{18x^2}$$
$$= \frac{10 + 3x}{18x^2}$$

Always check at this point to see that the answer is in lowest terms.

**(b)** Factor each denominator.

$$\frac{y+2}{y^2-y} - \frac{3y}{2y^2-4y+2} = \frac{y+2}{y(y-1)} - \frac{3y}{2(y-1)^2}$$

The LCD is $2y(y-1)^2$. Write each rational expression with this denominator, then subtract.

$$\frac{y+2}{y(y-1)} - \frac{3y}{2(y-1)^2} = \frac{(y+2)\cdot 2(y-1)}{y(y-1)\cdot 2(y-1)} - \frac{3y\cdot y}{2(y-1)^2\cdot y}$$

$$= \frac{2(y^2+y-2)}{2y(y-1)^2} - \frac{3y^2}{2y(y-1)^2} \qquad \text{Multiply.}$$

$$= \frac{2y^2+2y-4-3y^2}{2y(y-1)^2} \qquad \text{Multiply; subtract.}$$

$$= \frac{-y^2+2y-4}{2y(y-1)^2} \qquad \text{Combine terms.}$$

**(c)** The LCD is $(x-1)(x+2)(x+3)(x-4)$. Write each fraction with this denominator, then subtract.

$$\frac{3}{(x-1)(x+2)} - \frac{1}{(x+3)(x-4)}$$

$$= \frac{3(x+3)(x-4)}{(x-1)(x+2)(x+3)(x-4)} - \frac{(x-1)(x+2)}{(x+3)(x-4)(x-1)(x+2)}$$

$$= \frac{3(x^2-x-12)-(x^2+x-2)}{(x-1)(x+2)(x+3)(x-4)} \qquad \text{Multiply; subtract.}$$

$$= \frac{3x^2-3x-36-x^2-x+2}{(x-1)(x+2)(x+3)(x-4)} \qquad \text{Be careful with signs.}$$

$$= \frac{2x^2-4x-34}{(x-1)(x+2)(x+3)(x-4)}$$

## Complex Fractions

Any quotient of two rational expressions is called a **complex fraction.** Complex fractions can usually be simplified.

**EXAMPLE 4** *Simplifying Complex Fractions*

Simplify each complex fraction.

**(a)** $\dfrac{6-\dfrac{5}{k}}{1+\dfrac{5}{k}}$ **(b)** $\dfrac{\dfrac{a}{a+1}+\dfrac{1}{a}}{\dfrac{1}{a}+\dfrac{1}{a+1}}$

**Solution**

**(a)** Multiply both numerator and denominator by the LCD of all the fractions, $k$.

$$\frac{k\left(6-\dfrac{5}{k}\right)}{k\left(1+\dfrac{5}{k}\right)} = \frac{6k-k\left(\dfrac{5}{k}\right)}{k+k\left(\dfrac{5}{k}\right)} = \frac{6k-5}{k+5}$$

**(b)** Multiply both numerator and denominator by the LCD of all the fractions, $a(a+1)$.

$$\frac{\frac{a}{a+1}+\frac{1}{a}}{\frac{1}{a}+\frac{1}{a+1}}=\frac{\left(\frac{a}{a+1}+\frac{1}{a}\right)a(a+1)}{\left(\frac{1}{a}+\frac{1}{a+1}\right)a(a+1)}$$

$$=\frac{\frac{a}{a+1}(a)(a+1)+\frac{1}{a}(a)(a+1)}{\frac{1}{a}(a)(a+1)+\frac{1}{a+1}(a)(a+1)} \quad \text{Distributive property}$$

$$=\frac{a^2+(a+1)}{(a+1)+a}$$

$$=\frac{a^2+a+1}{2a+1}$$

Alternatively, first add the terms in the numerator and denominator, and then divide.

$$\frac{\frac{a}{a+1}+\frac{1}{a}}{\frac{1}{a}+\frac{1}{a+1}}=\frac{\frac{a^2+1(a+1)}{a(a+1)}}{\frac{1(a+1)+1(a)}{a(a+1)}} \quad \text{Find the LCD; add terms in the numerator and denominator.}$$

$$=\frac{\frac{a^2+a+1}{a(a+1)}}{\frac{2a+1}{a(a+1)}} \quad \text{Combine terms in the numerator and denominator.}$$

$$=\frac{a^2+a+1}{a(a+1)}\cdot\frac{a(a+1)}{2a+1} \quad \text{Definition of division}$$

$$=\frac{a^2+a+1}{2a+1} \quad \text{Multiply fractions; write in lowest terms.}$$

## R.3 EXERCISES

*Find the domain of each rational expression.*

1. $\frac{x-2}{x+6}$
2. $\frac{x+5}{x-3}$
3. $\frac{2x}{5x-3}$
4. $\frac{6x}{2x-1}$
5. $\frac{-8}{x^2+1}$
6. $\frac{3x}{3x^2+7}$
7. $\frac{3x+7}{(4x+2)(x-1)}$
8. $\frac{9x+12}{(2x+3)(x-5)}$

*Write each rational expression in lowest terms.*

9. $\frac{25p^3}{10p^2}$
10. $\frac{14z^3}{6z^2}$
11. $\frac{8k+16}{9k+18}$
12. $\frac{20r+10}{30r+15}$
13. $\frac{3(t+5)}{(t+5)(t-3)}$
14. $\frac{-8(y-4)}{(y+2)(y-4)}$
15. $\frac{8x^2+16x}{4x^2}$
16. $\frac{36y^2+72y}{9y}$
17. $\frac{m^2-4m+4}{m^2+m-6}$
18. $\frac{r^2-r-6}{r^2+r-12}$
19. $\frac{8m^2+6m-9}{16m^2-9}$
20. $\frac{6y^2+11y+4}{3y^2+7y+4}$

*Find each product or quotient.*

21. $\dfrac{15p^3}{9p^2} \div \dfrac{6p}{10p^2}$

22. $\dfrac{3r^2}{9r^3} \div \dfrac{8r^3}{6r}$

23. $\dfrac{2k + 8}{6} \div \dfrac{3k + 12}{2}$

24. $\dfrac{5m + 25}{10} \cdot \dfrac{12}{6m + 30}$

25. $\dfrac{x^2 + x}{5} \cdot \dfrac{25}{xy + y}$

26. $\dfrac{3m - 15}{4m - 20} \cdot \dfrac{m^2 - 10m + 25}{12m - 60}$

27. $\dfrac{4a + 12}{2a - 10} \div \dfrac{a^2 - 9}{a^2 - a - 20}$

28. $\dfrac{6r - 18}{9r^2 + 6r - 24} \cdot \dfrac{12r - 16}{4r - 12}$

29. $\dfrac{p^2 - p - 12}{p^2 - 2p - 15} \cdot \dfrac{p^2 - 9p + 20}{p^2 - 8p + 16}$

30. $\dfrac{x^2 + 2x - 15}{x^2 + 11x + 30} \cdot \dfrac{x^2 + 2x - 24}{x^2 - 8x + 15}$

31. $\dfrac{m^2 + 3m + 2}{m^2 + 5m + 4} \div \dfrac{m^2 + 5m + 6}{m^2 + 10m + 24}$

32. $\dfrac{y^2 + y - 2}{y^2 + 3y - 4} \div \dfrac{y^2 + 3y + 2}{y^2 + 4y + 3}$

33. $\dfrac{2m^2 - 5m - 12}{m^2 - 10m + 24} \div \dfrac{4m^2 - 9}{m^2 - 9m + 18}$

34. $\dfrac{6n^2 - 5n - 6}{6n^2 + 5n - 6} \cdot \dfrac{12n^2 - 17n + 6}{12n^2 - n - 6}$

35. $\dfrac{x^3 + y^3}{x^2 - y^2} \cdot \dfrac{x + y}{x^2 - xy + y^2}$

36. $\dfrac{8y^3 - 125}{4y^2 - 20y + 25} \cdot \dfrac{2y - 5}{y}$

37. $\dfrac{x^3 + y^3}{x^3 - y^3} \cdot \dfrac{x^2 - y^2}{x^2 + 2xy + y^2}$

38. $\dfrac{x^2 - y^2}{(x - y)^2} \cdot \dfrac{x^2 - xy + y^2}{x^2 - 2xy + y^2} \div \dfrac{x^3 + y^3}{(x - y)^4}$

39. *Concept Check* Which of these rational expressions equal $-1$? (Assume denominators are nonzero.)

A. $\dfrac{x - 4}{x + 4}$ B. $\dfrac{-x - 4}{x + 4}$ C. $\dfrac{x - 4}{4 - x}$ D. $\dfrac{x - 4}{-x - 4}$

40. In your own words, explain how to find the least common denominator for two fractions.

*Find each sum or difference.*

41. $\dfrac{3}{2k} + \dfrac{5}{3k}$

42. $\dfrac{8}{5p} + \dfrac{3}{4p}$

43. $\dfrac{a + 1}{2} - \dfrac{a - 1}{2}$

44. $\dfrac{y + 6}{5} - \dfrac{y - 6}{5}$

45. $\dfrac{3}{p} + \dfrac{1}{2}$

46. $\dfrac{9}{r} - \dfrac{2}{3}$

47. $\dfrac{1}{6m} + \dfrac{2}{5m} + \dfrac{4}{m}$

48. $\dfrac{8}{3p} + \dfrac{5}{4p} + \dfrac{9}{2p}$

49. $\dfrac{1}{a + 1} - \dfrac{1}{a - 1}$

50. $\dfrac{1}{x + z} + \dfrac{1}{x - z}$

51. $\dfrac{m + 1}{m - 1} + \dfrac{m - 1}{m + 1}$

52. $\dfrac{2}{x - 1} + \dfrac{1}{1 - x}$

53. $\dfrac{3}{a - 2} - \dfrac{1}{2 - a}$

54. $\dfrac{q}{p - q} - \dfrac{q}{q - p}$

55. $\dfrac{x + y}{2x - y} - \dfrac{2x}{y - 2x}$

56. $\dfrac{m - 4}{3m - 4} + \dfrac{3m + 2}{4 - 3m}$

57. $\dfrac{1}{a^2 - 5a + 6} - \dfrac{1}{a^2 - 4}$

58. $\dfrac{-3}{m^2 - m - 2} - \dfrac{1}{m^2 + 3m + 2}$

59. $\dfrac{1}{x^2 + x - 12} - \dfrac{1}{x^2 - 7x + 12} + \dfrac{1}{x^2 - 16}$

60. $\dfrac{2}{2p^2 - 9p - 5} + \dfrac{p}{3p^2 - 17p + 10} - \dfrac{2p}{6p^2 - p - 2}$

61. $\dfrac{3a}{a^2 + 5a - 6} - \dfrac{2a}{a^2 + 7a + 6}$

62. $\dfrac{2k}{k^2 + 4k + 3} + \dfrac{3k}{k^2 + 5k + 6}$

*Simplify each complex fraction.*

63. $\dfrac{1 + \dfrac{1}{x}}{1 - \dfrac{1}{x}}$

64. $\dfrac{2 - \dfrac{2}{y}}{2 + \dfrac{2}{y}}$

65. $\dfrac{\dfrac{1}{x + 1} - \dfrac{1}{x}}{\dfrac{1}{x}}$

66. $\dfrac{\dfrac{1}{y + 3} - \dfrac{1}{y}}{\dfrac{1}{y}}$

67. $\dfrac{1 + \dfrac{1}{1 - b}}{1 - \dfrac{1}{1 + b}}$

68. $m - \dfrac{m}{m + \dfrac{1}{2}}$

69. $\dfrac{m - \dfrac{1}{m^2 - 4}}{\dfrac{1}{m + 2}}$

70. $\dfrac{\dfrac{3}{p^2 - 16} + p}{\dfrac{1}{p - 4}}$

# R.4 REVIEW OF NEGATIVE AND RATIONAL EXPONENTS

Negative Exponents and the Quotient Rule • Rational Exponents

## Negative Exponents and the Quotient Rule

In Section R.1 we introduced the product rule for exponents. In the product rule $a^m \cdot a^n = a^{m+n}$, the exponents are *added*. Now consider an expression such as $\frac{a^3}{a^7}$. If $a \neq 0$, then

$$\frac{a^3}{a^7} = \frac{a \cdot a \cdot a}{a \cdot a \cdot a \cdot a \cdot a \cdot a \cdot a} = \frac{1}{a \cdot a \cdot a \cdot a} = \frac{1}{a^4}.$$

This suggests that we should *subtract* exponents when dividing. Subtracting exponents gives

$$\frac{a^3}{a^7} = a^{3-7} = a^{-4}.$$

The only way to keep these results consistent is to define $a^{-4}$ as $\frac{1}{a^4}$. This example suggests the following definition.

**Negative Exponent**

If $a$ is a nonzero real number and $n$ is any integer, then

$$a^{-n} = \frac{1}{a^n}.$$

**EXAMPLE 1** *Using the Definition of a Negative Exponent*

Evaluate each expression in parts (a)–(c). In parts (d) and (e), write the expression without negative exponents.

**(a)** $4^{-2}$ **(b)** $\left(\frac{2}{5}\right)^{-3}$ **(c)** $-4^{-2}$ **(d)** $x^{-5}$ **(e)** $xy^{-3}$

**Solution**

**(a)** $4^{-2} = \frac{1}{4^2} = \frac{1}{16}$

**(b)** $\left(\frac{2}{5}\right)^{-3} = \frac{1}{\left(\frac{2}{5}\right)^3} = \frac{1}{\frac{8}{125}} = \frac{125}{8}$

**(c)** $-4^{-2} = -\frac{1}{4^2} = -\frac{1}{16}$

**(d)** $x^{-5} = \frac{1}{x^5} \quad (x \neq 0)$

**(e)** $xy^{-3} = x \cdot \frac{1}{y^3} = \frac{x}{y^3} \quad (y \neq 0)$

From part (b) of Example 1,

$$\left(\frac{2}{5}\right)^{-3} = \frac{125}{8} = \left(\frac{5}{2}\right)^3.$$

This result can be generalized. If $a \neq 0$ and $b \neq 0$, then for any integer $n$,

$$\left(\frac{a}{b}\right)^{-n} = \left(\frac{b}{a}\right)^{n}.$$

The quotient rule for exponents follows from the definition of negative exponents.

**Quotient Rule**

For all integers $m$ and $n$ and all nonzero real numbers $a$,

$$\frac{a^m}{a^n} = a^{m-n}.$$

By the quotient rule, if $a \neq 0$,

$$\frac{a^m}{a^m} = a^{m-m} = a^0.$$

On the other hand, any nonzero quantity divided by itself equals 1. This is why we defined $a^0 = 1$ in Section R.1.

**EXAMPLE 2** *Using the Quotient Rule*

Simplify each expression. Assume that all variables represent nonzero real numbers.

**(a)** $\frac{12^5}{12^2}$ **(b)** $\frac{a^5}{a^{-8}}$ **(c)** $\frac{16m^{-9}}{12m^{11}}$ **(d)** $\frac{25r^7z^5}{10r^9z}$

**Solution**

**(a)** $\frac{12^5}{12^2} = 12^{5-2} = 12^3$ **(b)** $\frac{a^5}{a^{-8}} = a^{5-(-8)} = a^{13}$

**(c)** $\frac{16m^{-9}}{12m^{11}} = \frac{16}{12} \cdot m^{-9-11} = \frac{4}{3}m^{-20} = \frac{4}{3} \cdot \frac{1}{m^{20}} = \frac{4}{3m^{20}}$

**(d)** $\frac{25r^7z^5}{10r^9z} = \frac{25}{10} \cdot \frac{r^7}{r^9} \cdot \frac{z^5}{z^1} = \frac{5}{2}r^{-2}z^4 = \frac{5z^4}{2r^2}$

The rules for exponents stated in Section R.1 also apply to negative exponents.

**EXAMPLE 3** *Using the Rules for Exponents*

Simplify each expression. Write answers without negative exponents. Assume that all variables represent nonzero real numbers.

**(a)** $3x^{-2}(4^{-1}x^{-5})^2$ **(b)** $\frac{5m^{-3}}{10m^{-5}}$ **(c)** $\frac{12p^3q^{-1}}{8p^{-2}q}$ **(d)** $\frac{(3x^2)^{-1}(3x^5)^{-2}}{(3^{-1}x^{-2})^2}$

**Solution**

**(a)** $3x^{-2}(4^{-1}x^{-5})^2 = 3x^{-2}(4^{-2}x^{-10})$ Power rule

$= 3 \cdot 4^{-2} \cdot x^{-2+(-10)}$ Rearrange factors; product rule

$= 3 \cdot 4^{-2} \cdot x^{-12}$

$= \frac{3}{16x^{12}}$ Write with positive exponents.

**(b)** $\dfrac{5m^{-3}}{10m^{-5}} = \dfrac{5}{10}m^{-3-(-5)}$ Quotient rule

$= \dfrac{1}{2}m^2$ or $\dfrac{m^2}{2}$

**(c)** $\dfrac{12p^3q^{-1}}{8p^{-2}q} = \dfrac{12}{8} \cdot \dfrac{p^3}{p^{-2}} \cdot \dfrac{q^{-1}}{q^1}$

$= \dfrac{3}{2} \cdot p^{3-(-2)}q^{-1-1}$ Quotient rule

$= \dfrac{3}{2}p^5q^{-2}$

$= \dfrac{3p^5}{2q^2}$ Write with positive exponents.

**(d)** $\dfrac{(3x^2)^{-1}(3x^5)^{-2}}{(3^{-1}x^{-2})^2} = \dfrac{3^{-1}x^{-2}3^{-2}x^{-10}}{3^{-2}x^{-4}}$ Power rule

$= \dfrac{3^{-1+(-2)}x^{-2+(-10)}}{3^{-2}x^{-4}} = \dfrac{3^{-3}x^{-12}}{3^{-2}x^{-4}}$ Product rule

$= 3^{-3-(-2)}x^{-12-(-4)} = 3^{-1}x^{-8}$ Quotient rule

$= \dfrac{1}{3x^8}$ Write with positive exponents.

**CAUTION** Notice the use of the power rule $(ab)^n = a^nb^n$ in Example 3(d): $(3x^2)^{-1} = 3^{-1}(x^2)^{-1} = 3^{-1}x^{-2}$. Remember to apply the exponent to a numerical coefficient.

## Rational Exponents

The definition of $a^n$ can be extended to rational values of $n$ by defining $a^{1/n}$ to be the $n$th root of $a$. By one of the power rules of exponents (extended to a rational exponent),

$$(a^{1/n})^n = a^{(1/n)n} = a^1 = a,$$

which suggests that $a^{1/n}$ is a number whose $n$th power is $a$.

**$a^{1/n}$**

**$n$ even** If $n$ is an *even* positive integer, and if $a > 0$, then $a^{1/n}$ is the positive real number whose $n$th power is $a$. That is, $(a^{1/n})^n = a$. (In this case, $a^{1/n}$ is the principal $n$th root of $a$. See Section R.5.)

**$n$ odd** If $n$ is an *odd* positive integer, and *$a$ is any real number,* then $a^{1/n}$ is the positive or negative real number whose $n$th power is $a$. That is, $(a^{1/n})^n = a$.

**EXAMPLE 4** *Using the Definition of $a^{1/n}$*

Evaluate each expression.

**(a)** $36^{1/2}$ **(b)** $-100^{1/2}$ **(c)** $-(225)^{1/2}$ **(d)** $625^{1/4}$ **(e)** $(-1296)^{1/4}$

**(f)** $-1296^{1/4}$ **(g)** $(-27)^{1/3}$ **(h)** $-32^{1/5}$

**Solution**

**(a)** $36^{1/2} = 6$ because $6^2 = 36$.

**(b)** $-100^{1/2} = -10$

**(c)** $-(225)^{1/2} = -15$

**(d)** $625^{1/4} = 5$

**(e)** $(-1296)^{1/4}$ is not a real number. (Why?)

**(f)** $-1296^{1/4} = -6$

**(g)** $(-27)^{1/3} = -3$

**(h)** $-32^{1/5} = -2$

In general, the notation $a^{m/n}$ must be defined so that all the rules for exponents hold. For the power rule to hold, $(a^{1/n})^m$ must equal $a^{m/n}$. Therefore, $a^{m/n}$ is defined as follows.

**Rational Exponent**

For all integers $m$, all positive integers $n$, and all real numbers $a$ for which $a^{1/n}$ is a real number,

$$a^{m/n} = (a^{1/n})^m.$$

**EXAMPLE 5** *Using the Definition of $a^{m/n}$*

Evaluate each expression.

**(a)** $125^{2/3}$ **(b)** $32^{7/5}$ **(c)** $-81^{3/2}$

**(d)** $(-27)^{2/3}$ **(e)** $16^{-3/4}$ **(f)** $(-4)^{5/2}$

**Solution**

**(a)** $125^{2/3} = (125^{1/3})^2 = 5^2 = 25$

**(b)** $32^{7/5} = (32^{1/5})^7 = 2^7 = 128$

**(c)** $-81^{3/2} = -(81^{1/2})^3 = -(9)^3 = -729$

**(d)** $(-27)^{2/3} = [(-27)^{1/3}]^2 = (-3)^2 = 9$

**(e)** $16^{-3/4} = \dfrac{1}{16^{3/4}} = \dfrac{1}{(16^{1/4})^3} = \dfrac{1}{2^3} = \dfrac{1}{8}$

**(f)** $(-4)^{5/2}$ is not a real number because $(-4)^{1/2}$ is not a real number.

**NOTE** For all real numbers $a$, integers $m$, and positive integers $n$ for which $a^{1/n}$ is a real number,

$$a^{m/n} = (a^{1/n})^m \quad \text{or} \quad a^{m/n} = (a^m)^{1/n}.$$

So $a^{m/n}$ can be evaluated either as $(a^{1/n})^m$ or as $(a^m)^{1/n}$. It is usually easier to find $(a^{1/n})^m$. For example,

$$27^{4/3} = (27^{1/3})^4 = 3^4 = 81$$

or

$$27^{4/3} = (27^4)^{1/3} = 531{,}441^{1/3} = 81.$$

The form $(27^{1/3})^4$ is easier to evaluate.

The earlier results concerning integer exponents also apply to rational exponents. These definitions and rules are summarized here.

**Definitions and Rules for Exponents**

Let $r$ and $s$ be rational numbers. The following results are valid for all positive numbers $a$ and $b$.

$$a^r \cdot a^s = a^{r+s} \qquad (ab)^r = a^r \cdot b^r \qquad (a^r)^s = a^{rs}$$

$$\frac{a^r}{a^s} = a^{r-s} \qquad \left(\frac{a}{b}\right)^r = \frac{a^r}{b^r} \qquad a^{-r} = \frac{1}{a^r}$$

**EXAMPLE 6** *Using the Definitions and Rules for Exponents*

Simplify each expression. Assume that all variables represent positive real numbers.

**(a)** $\dfrac{27^{1/3} \cdot 27^{5/3}}{27^3}$ **(b)** $81^{5/4} \cdot 4^{-3/2}$ **(c)** $6y^{2/3} \cdot 2y^{1/2}$

**(d)** $\left(\dfrac{3m^{5/6}}{y^{3/4}}\right)^2 \cdot \left(\dfrac{8y^3}{m^6}\right)^{2/3}$ **(e)** $m^{2/3}(m^{7/3} + 2m^{1/3})$

**Solution**

**(a)** $\dfrac{27^{1/3} \cdot 27^{5/3}}{27^3} = \dfrac{27^{1/3+5/3}}{27^3}$ Product rule

$= \dfrac{27^2}{27^3} = 27^{2-3}$ Quotient rule

$= 27^{-1} = \dfrac{1}{27}$

**(b)** $81^{5/4} \cdot 4^{-3/2} = (81^{1/4})^5(4^{1/2})^{-3} = 3^5 \cdot 2^{-3} = \dfrac{3^5}{2^3}$ or $\dfrac{243}{8}$

**(c)** $6y^{2/3} \cdot 2y^{1/2} = 12y^{2/3+1/2} = 12y^{7/6}$

**(d)** $\left(\dfrac{3m^{5/6}}{y^{3/4}}\right)^2 \cdot \left(\dfrac{8y^3}{m^6}\right)^{2/3} = \dfrac{9m^{5/3}}{y^{3/2}} \cdot \dfrac{4y^2}{m^4} = 36m^{5/3-4}y^{2-3/2} = \dfrac{36y^{1/2}}{m^{7/3}}$

**(e)** $m^{2/3}(m^{7/3} + 2m^{1/3}) = m^{2/3+7/3} + 2m^{2/3+1/3} = m^3 + 2m$

**EXAMPLE 7** *Factoring Expressions with Negative or Rational Exponents*

Factor out the smallest power of the variable. Assume that all variables represent positive real numbers.

**(a)** $12x^{-2} - 8x^{-3}$ **(b)** $4m^{1/2} + 3m^{3/2}$ **(c)** $y^{-1/3} + y^{2/3}$

**Solution**

**(a)** The smallest exponent on $x$ here is $-3$. Since 4 is a common numerical factor, factor out $4x^{-3}$.

$$12x^{-2} - 8x^{-3} = 4x^{-3}(3x - 2)$$

Check by multiplying on the right. The factored form can be written without negative exponents as $\dfrac{4(3x - 2)}{x^3}$.

**(b)** $4m^{1/2} + 3m^{3/2} = m^{1/2}(4 + 3m)$. To check, multiply $m^{1/2}$ by $4 + 3m$.

**(c)** $y^{-1/3} + y^{2/3} = y^{-1/3}(1 + y)$. The factored form can be written with only positive exponents as $\frac{1 + y}{y^{1/3}}$.

## R.4 EXERCISES

*Concept Check* *Match each expression from Group I with the correct choice from Group II. Choices may be used once, more than once, or not at all.*

| I | | II | |
|---|---|---|---|
| **1.** $\left(\frac{4}{9}\right)^{3/2}$ | **2.** $\left(\frac{4}{9}\right)^{-3/2}$ | **A.** $\frac{9}{4}$ | **B.** $-\frac{9}{4}$ |
| **3.** $-\left(\frac{9}{4}\right)^{3/2}$ | **4.** $-\left(\frac{4}{9}\right)^{-3/2}$ | **C.** $-\frac{4}{9}$ | **D.** $\frac{4}{9}$ |
| **5.** $\left(\frac{8}{27}\right)^{2/3}$ | **6.** $\left(\frac{8}{27}\right)^{-2/3}$ | **E.** $\frac{8}{27}$ | **F.** $-\frac{27}{8}$ |
| **7.** $-\left(\frac{27}{8}\right)^{2/3}$ | **8.** $-\left(\frac{27}{8}\right)^{-2/3}$ | **G.** $\frac{27}{8}$ | **H.** $-\frac{8}{27}$ |

*Simplify each expression. Assume that all variables represent positive real numbers.*

**9.** $(-4)^{-3}$ **10.** $(-5)^{-2}$ **11.** $\left(\frac{1}{2}\right)^{-3}$ **12.** $\left(\frac{2}{3}\right)^{-2}$

**13.** $-4^{1/2}$ **14.** $25^{1/2}$ **15.** $8^{2/3}$ **16.** $-81^{3/4}$

**17.** $27^{-2/3}$ **18.** $(-32)^{-4/5}$ **19.** $\left(\frac{27}{64}\right)^{-4/3}$ **20.** $\left(\frac{121}{100}\right)^{-3/2}$

**21.** $(16p^4)^{1/2}$ **22.** $(36r^6)^{1/2}$ **23.** $(27x^6)^{2/3}$ **24.** $(64a^{12})^{5/6}$

*Perform the indicated operations. Write your answers with only positive exponents. Assume that all variables represent positive real numbers.*

**25.** $2^{-3} \cdot 2^{-4}$ **26.** $5^{-2} \cdot 5^{-6}$ **27.** $27^{-2} \cdot 27^{-1}$ **28.** $9^{-4} \cdot 9^{-1}$

**29.** $\frac{4^{-2} \cdot 4^{-1}}{4^{-3}}$ **30.** $\frac{3^{-1} \cdot 3^{-4}}{3^2 \cdot 3^{-2}}$ **31.** $(m^{2/3})(m^{5/3})$ **32.** $(x^{4/5})(x^{2/5})$

**33.** $(1 + n)^{1/2}(1 + n)^{3/4}$ **34.** $(m + 7)^{-1/6}(m + 7)^{-2/3}$ **35.** $(2y^{3/4}z)(3y^{-2}z^{-1/3})$ **36.** $(4a^{-1}b^{2/3})(a^{3/2}b^{-3})$

**37.** $(4a^{-2}b^7)^{1/2} \cdot (2a^{1/4}b^3)^5$ **38.** $(x^{-2}y^{1/3})^5 \cdot (8x^2y^{-2})^{-1/3}$ **39.** $\left(\frac{r^{-2}}{s^{-5}}\right)^{-3}$ **40.** $\left(\frac{p^{-1}}{q^{-5}}\right)^{-2}$

**41.** $\left(\frac{-a}{b^{-3}}\right)^{-1}$ **42.** $\frac{7^{-1/3}7r^{-3}}{7^{2/3}r^{-2}}$ **43.** $\frac{12^{5/4}y^{-2}}{12^{-1}y^{-3}}$ **44.** $\frac{6k^{-4}(3k^{-1})^{-2}}{2^3k^{1/2}}$

**45.** $\frac{8p^{-3}(4p^2)^{-2}}{p^{-5}}$ **46.** $\frac{k^{-3/5}h^{-1/3}t^{2/5}}{k^{-1/5}h^{-2/3}t^{1/5}}$ **47.** $\frac{m^{7/3}n^{-2/5}p^{3/8}}{m^{-2/3}n^{3/5}p^{-5/8}}$ **48.** $\frac{m^{2/5}m^{3/5}m^{-4/5}}{m^{1/5}m^{-6/5}}$

**49.** $\frac{-4a^{-1}a^{2/3}}{a^{-2}}$ **50.** $\frac{8y^{2/3}y^{-1}}{2^{-1}y^{3/4}y^{-1/6}}$ **51.** $\frac{(k + 5)^{1/2}(k + 5)^{-1/4}}{(k + 5)^{3/4}}$ **52.** $\frac{(x + y)^{-5/8}(x + y)^{3/8}}{(x + y)^{1/8}(x + y)^{-1/8}}$

*Find each product. Assume that all variables represent positive real numbers.*

**53.** $y^{5/8}(y^{3/8} - 10y^{11/8})$
**54.** $p^{11/5}(3p^{4/5} + 9p^{19/5})$
**55.** $-4k(k^{7/3} - 6k^{1/3})$
**56.** $-5y(3y^{9/10} + 4y^{3/10})$
**57.** $(x + x^{1/2})(x - x^{1/2})$
**58.** $(2z^{1/2} + z)(z^{1/2} - z)$
**59.** $(r^{1/2} - r^{-1/2})^2$
**60.** $(p^{1/2} - p^{-1/2})(p^{1/2} + p^{-1/2})$

*Factor, using the given common factor. Assume all variables represent positive real numbers.*

**61.** $4k^{-1} + k^{-2}$; $k^{-2}$
**62.** $y^{-5} - 3y^{-3}$; $y^{-5}$
**63.** $9z^{-1/2} + 2z^{1/2}$; $z^{-1/2}$
**64.** $3m^{2/3} - 4m^{-1/3}$; $m^{-1/3}$
**65.** $p^{-3/4} - 2p^{-7/4}$; $p^{-7/4}$
**66.** $6r^{-2/3} - 5r^{-5/3}$; $r^{-5/3}$
**67.** $(p + 4)^{-3/2} + (p + 4)^{-1/2} + (p + 4)^{1/2}$; $(p + 4)^{-3/2}$
**68.** $(3r + 1)^{-2/3} + (3r + 1)^{1/3} + (3r + 1)^{4/3}$; $(3r + 1)^{-2/3}$

# R.5 REVIEW OF RADICALS

Radical Notation • Rules for Radicals • Simplifying Radicals • Operations with Radicals • Rationalizing Denominators

## Radical Notation

In Section R.4 the notation $a^{1/n}$ was used for the $n$th root of $a$ for appropriate values of $a$ and $n$. An alternative (and more familiar) notation for $a^{1/n}$ uses *radical notation*.

> **Radical Notation for $a^{1/n}$**
>
> If $a$ is a real number, $n$ is a positive integer, and $a^{1/n}$ is a real number, then
>
> $$\sqrt[n]{a} = a^{1/n}.$$

The symbol $\sqrt[n]{\ }$ is a **radical sign,** the number $a$ is the **radicand,** and $n$ is the **index** of the radical $\sqrt[n]{a}$. It is customary to use $\sqrt{a}$ instead of $\sqrt[2]{a}$ for the square root.

For even values of $n$ (square roots, fourth roots, and so on) and $a > 0$, there are two $n$th roots, one positive and one negative. In such cases, the notation $\sqrt[n]{a}$ represents the positive root, the **principal $n$th root.** The negative root is written $-\sqrt[n]{a}$.

**EXAMPLE 1** *Evaluating Roots*

Evaluate each root.

**(a)** $\sqrt[4]{16}$ **(b)** $-\sqrt[4]{16}$ **(c)** $\sqrt[4]{-16}$

**(d)** $\sqrt[5]{-32}$ **(e)** $\sqrt[3]{1000}$ **(f)** $\sqrt[6]{\frac{64}{729}}$

**Solution**

**(a)** $\sqrt[4]{16} = 16^{1/4} = (2^4)^{1/4} = 2$
**(b)** $-\sqrt[4]{16} = -16^{1/4} = -(2^4)^{1/4} = -2$
**(c)** $\sqrt[4]{-16}$ is not a real number.
**(d)** $\sqrt[5]{-32} = [(-2)^5]^{1/5} = -2$
**(e)** $\sqrt[3]{1000} = 10$
**(f)** $\sqrt[6]{\frac{64}{729}} = \frac{2}{3}$

With $a^{1/n}$ written as $\sqrt[n]{a}$, $a^{m/n}$ can also be written with radicals.

**Radical Notation for $a^{m/n}$**

If $a$ is a real number, $m$ is an integer, $n$ is a positive integer, and $\sqrt[n]{a}$ is a real number, then

$$a^{m/n} = \left(\sqrt[n]{a}\right)^m = \sqrt[n]{a^m}.$$

**EXAMPLE 2** *Converting from Rational Exponents to Radicals*

Write in radical form and simplify.

**(a)** $8^{2/3}$ **(b)** $(-32)^{4/5}$ **(c)** $-16^{3/4}$ **(d)** $x^{5/6}$ $(x \geq 0)$ **(e)** $3x^{2/3}$
**(f)** $2p^{-1/2}$ $(p > 0)$ **(g)** $(3a + b)^{1/4}$ $(3a + b \geq 0)$

**Solution**

**(a)** $8^{2/3} = \left(\sqrt[3]{8}\right)^2 = 2^2 = 4$

**(b)** $(-32)^{4/5} = \left(\sqrt[5]{-32}\right)^4 = (-2)^4 = 16$

**(c)** $-16^{3/4} = -\left(\sqrt[4]{16}\right)^3 = -(2)^3 = -8$

**(d)** $x^{5/6} = \sqrt[6]{x^5}$ $(x \geq 0)$

**(e)** $3x^{2/3} = 3\sqrt[3]{x^2}$

**(f)** $2p^{-1/2} = \dfrac{2}{p^{1/2}} = \dfrac{2}{\sqrt{p}}$ $(p > 0)$

**(g)** $(3a + b)^{1/4} = \sqrt[4]{3a + b}$ $(3a + b \geq 0)$

**EXAMPLE 3** *Converting from Radicals to Rational Exponents*

Write in exponential form.

**(a)** $\sqrt[4]{x^5}$ $(x \geq 0)$ **(b)** $\sqrt{3y}$ $(y \geq 0)$ **(c)** $10\left(\sqrt[5]{z}\right)^2$ **(d)** $5\sqrt[3]{(2x^4)^7}$
**(e)** $\sqrt{p^2 + q}$ $(p^2 + q \geq 0)$

**Solution**

**(a)** $\sqrt[4]{x^5} = x^{5/4}$ $(x \geq 0)$

**(b)** $\sqrt{3y} = (3y)^{1/2} = 3^{1/2}y^{1/2}$ $(y \geq 0)$

**(c)** $10\left(\sqrt[5]{z}\right)^2 = 10z^{2/5}$

**(d)** $5\sqrt[3]{(2x^4)^7} = 5(2x^4)^{7/3} = 5 \cdot 2^{7/3}x^{28/3}$

**(e)** $\sqrt{p^2 + q} = (p^2 + q)^{1/2}$ $(p^2 + q \geq 0)$

By the definition of $\sqrt[n]{a}$, for any positive integer $n$, if $\sqrt[n]{a}$ is defined, then

$$\left(\sqrt[n]{a}\right)^n = a.$$

If $a$ is positive, or if $a$ is negative and $n$ is an odd positive integer, then

$$\sqrt[n]{a^n} = a.$$

Because of the conditions just given, we *cannot* simply write $\sqrt{x^2} = x$. For example, if $x = -5$, then

$$\sqrt{x^2} = \sqrt{(-5)^2} = \sqrt{25} = 5 \neq x.$$

Since a negative value of $x$ can produce a positive result, we use absolute value. For any real number $a$,

$$\sqrt{a^2} = |a|.$$

For example, $\sqrt{(-9)^2} = |-9| = 9$ and $\sqrt{13^2} = |13| = 13$.

This result can be generalized to any even $n$th root.

**Evaluating $\sqrt[n]{a^n}$**

If $n$ is an even positive integer, then $\sqrt[n]{a^n} = |a|$.
If $n$ is an odd positive integer, then $\sqrt[n]{a^n} = a$.

**EXAMPLE 4** *Using Absolute Value to Simplify Roots*

Simplify each expression.

**(a)** $\sqrt{p^4}$ **(b)** $\sqrt[4]{p^4}$ **(c)** $\sqrt{16m^8r^6}$ **(d)** $\sqrt[6]{(-2)^6}$ **(e)** $\sqrt[5]{m^5}$
**(f)** $\sqrt{(2k+3)^2}$ **(g)** $\sqrt{x^2 - 4x + 4}$

**Solution**

**(a)** $\sqrt{p^4} = \sqrt{(p^2)^2} = |p^2| = p^2$

**(b)** $\sqrt[4]{p^4} = |p|$

**(c)** $\sqrt{16m^8r^6} = |4m^4r^3| = 4m^4|r^3|$

**(d)** $\sqrt[6]{(-2)^6} = |-2| = 2$

**(e)** $\sqrt[5]{m^5} = m$

**(f)** $\sqrt{(2k+3)^2} = |2k+3|$

**(g)** $\sqrt{x^2 - 4x + 4} = \sqrt{(x-2)^2} = |x-2|$

**NOTE** When working with variable radicands, we usually assume that all expressions in radicands represent only nonnegative real numbers.

## Rules for Radicals

Three rules for working with radicals are given next. These rules are just the power rules for exponents written in radical notation.

**Rules for Radicals**

For all real numbers $a$ and $b$, and positive integers $m$ and $n$ for which the indicated roots are real numbers,

$$\sqrt[n]{a} \cdot \sqrt[n]{b} = \sqrt[n]{ab} \qquad \sqrt[n]{\frac{a}{b}} = \frac{\sqrt[n]{a}}{\sqrt[n]{b}} \quad (b \neq 0) \qquad \sqrt[m]{\sqrt[n]{a}} = \sqrt[mn]{a}.$$

**EXAMPLE 5** *Using the Rules for Radicals to Simplify Radical Expressions*

Simplify each expression.

**(a)** $\sqrt{6} \cdot \sqrt{54}$ **(b)** $\sqrt[3]{m} \cdot \sqrt[3]{m^2}$ **(c)** $\sqrt{\frac{7}{64}}$

**(d)** $\sqrt[4]{\frac{a}{b^4}}$ **(e)** $\sqrt[7]{\sqrt[3]{2}}$ **(f)** $\sqrt[4]{\sqrt{3}}$

**Solution**

**(a)** $\sqrt{6} \cdot \sqrt{54} = \sqrt{6 \cdot 54} = \sqrt{324} = 18$

**(b)** $\sqrt[3]{m} \cdot \sqrt[3]{m^2} = \sqrt[3]{m^3} = m$

**(c)** $\sqrt{\frac{7}{64}} = \frac{\sqrt{7}}{\sqrt{64}} = \frac{\sqrt{7}}{8}$

**(d)** $\sqrt[4]{\frac{a}{b^4}} = \frac{\sqrt[4]{a}}{\sqrt[4]{b^4}} = \frac{\sqrt[4]{a}}{b} \quad (a \geq 0, b > 0)$

**(e)** $\sqrt[7]{\sqrt[3]{2}} = \sqrt[21]{2}$ Use the third rule.

**(f)** $\sqrt[4]{\sqrt{3}} = \sqrt[8]{3}$

## Simplifying Radicals

In work with fractions, it is customary to write a fraction in its simplest form. For example, $\frac{10}{2}$ is written as 5, $-\frac{9}{6}$ is written as $-\frac{3}{2}$, and $\frac{4}{16}$ is written as $\frac{1}{4}$. Similarly, expressions with radicals are often written in their simplest forms.

> **Simplified Radicals**
>
> An expression with radicals is simplified when the following conditions are satisfied.
>
> 1. The radicand has no factor raised to a power greater than or equal to the index.
> 2. The radicand has no fractions.
> 3. No denominator contains a radical.
> 4. Exponents in the radicand and the index of the radical have no common factor.
> 5. All indicated operations have been performed (if possible).

**EXAMPLE 6** *Simplifying Radicals*

Simplify each radical.

**(a)** $\sqrt{175}$ **(b)** $-3\sqrt[5]{32}$ **(c)** $\sqrt{288m^5}$ **(d)** $\sqrt[3]{81x^5y^7z^6}$

**Solution**

**(a)** $\sqrt{175} = \sqrt{25 \cdot 7} = \sqrt{25} \cdot \sqrt{7} = 5\sqrt{7}$

**(b)** $-3\sqrt[5]{32} = -3\sqrt[5]{2^5} = -3 \cdot 2 = -6$

**(c)** $\sqrt{288m^5} = \sqrt{144m^4 \cdot 2m} = 12m^2\sqrt{2m}$

**(d)** $\sqrt[3]{81x^5y^7z^6} = \sqrt[3]{27 \cdot 3 \cdot x^3 \cdot x^2 \cdot y^6 \cdot y \cdot z^6}$   Factor.

$= \sqrt[3]{27x^3y^6z^6(3x^2y)}$   Group all perfect cubes.

$= 3xy^2z^2\sqrt[3]{3x^2y}$   Remove all perfect cubes from under the radical.

If the index of the radical and an exponent in the radicand have a common factor, the radical can be simplified by writing it in exponential form, simplifying the rational exponent, and then writing the result as a radical again.

**EXAMPLE 7** *Simplifying Radicals by Writing Them with Rational Exponents*

Simplify each radical.

**(a)** $\sqrt[6]{3^2}$ **(b)** $\sqrt[6]{x^{12}y^3}$ $(y \geq 0)$ **(c)** $\sqrt[9]{\sqrt{6^3}}$

**Solution**

**(a)** $\sqrt[6]{3^2} = 3^{2/6} = 3^{1/3} = \sqrt[3]{3}$

**(b)** $\sqrt[6]{x^{12}y^3} = (x^{12}y^3)^{1/6} = x^2y^{3/6} = x^2y^{1/2} = x^2\sqrt{y}$ $(y \geq 0)$

**(c)** $\sqrt[9]{\sqrt{6^3}} = \sqrt[9]{6^{3/2}} = (6^{3/2})^{1/9} = 6^{1/6} = \sqrt[6]{6}$

In Example 7(a), we simplified $\sqrt[6]{3^2}$ as $\sqrt[3]{3}$. However, to simplify $(\sqrt[6]{x})^2$, the variable $x$ must be nonnegative. For example, suppose we write

$$(-8)^{2/6} = [(-8)^{1/6}]^2.$$

This result is not a real number, because $(-8)^{1/6}$ is not a real number. On the other hand,

$$(-8)^{1/3} = -2.$$

Here, even though $\frac{2}{6} = \frac{1}{3}$,

$$\left(\sqrt[6]{x}\right)^2 \neq \sqrt[3]{x}.$$

If $a$ is nonnegative, then it is always true that $a^{m/n} = a^{mp/(np)}$ (for $p \neq 0$). Simplifying rational exponents on negative bases should be considered case by case.

## Operations with Radicals

Radicals with the same radicand and the same index, such as $3\sqrt[4]{11pq}$ and $-7\sqrt[4]{11pq}$, are called **like radicals.** Like radicals are added or subtracted by using the distributive property. Only like radicals can be combined. It is sometimes necessary to simplify radicals before adding or subtracting.

### EXAMPLE 8 *Adding and Subtracting Like Radicals*

Add or subtract, as indicated. Assume all variables represent positive real numbers.

**(a)** $3\sqrt[4]{11pq} + \left(-7\sqrt[4]{11pq}\right)$ **(b)** $7\sqrt{2} - 8\sqrt{18} + 4\sqrt{72}$

**(c)** $\sqrt{98x^3y} + 3x\sqrt{32xy}$

**Solution**

**(a)** $3\sqrt[4]{11pq} + \left(-7\sqrt[4]{11pq}\right) = -4\sqrt[4]{11pq}$

**(b)** First, remove all perfect square factors from under the radical. Then, use the distributive property.

$$\begin{aligned} 7\sqrt{2} - 8\sqrt{18} + 4\sqrt{72} &= 7\sqrt{2} - 8\sqrt{9 \cdot 2} + 4\sqrt{36 \cdot 2} \\ &= 7\sqrt{2} - 8 \cdot 3\sqrt{2} + 4 \cdot 6\sqrt{2} \\ &= 7\sqrt{2} - 24\sqrt{2} + 24\sqrt{2} \\ &= 7\sqrt{2} \qquad \text{Distributive property} \end{aligned}$$

**(c)**

$$\begin{aligned} \sqrt{98x^3y} + 3x\sqrt{32xy} &= \sqrt{49 \cdot 2 \cdot x^2 \cdot x \cdot y} + 3x\sqrt{16 \cdot 2 \cdot x \cdot y} \\ &= 7x\sqrt{2xy} + 3x(4)\sqrt{2xy} \\ &= 7x\sqrt{2xy} + 12x\sqrt{2xy} \\ &= 19x\sqrt{2xy} \qquad \text{Distributive property} \end{aligned}$$

Multiplying radical expressions is much like multiplying polynomials.

### EXAMPLE 9 *Multiplying Radical Expressions*

Find each product.

**(a)** $\left(\sqrt{2} + 3\right)\left(\sqrt{8} - 5\right)$ **(b)** $\left(\sqrt{7} - \sqrt{10}\right)\left(\sqrt{7} + \sqrt{10}\right)$

**Solution**

**(a)**

$$\begin{aligned} \left(\sqrt{2} + 3\right)\left(\sqrt{8} - 5\right) &= \sqrt{2}\left(\sqrt{8}\right) - \sqrt{2}(5) + 3\sqrt{8} - 3(5) \qquad \text{FOIL} \\ &= \sqrt{16} - 5\sqrt{2} + 3\left(2\sqrt{2}\right) - 15 \qquad \text{Multiply.} \\ &= 4 - 5\sqrt{2} + 6\sqrt{2} - 15 \\ &= -11 + \sqrt{2} \qquad \text{Combine terms.} \end{aligned}$$

**(b)** $\left(\sqrt{7}-\sqrt{10}\right)\left(\sqrt{7}+\sqrt{10}\right)=\left(\sqrt{7}\right)^2-\left(\sqrt{10}\right)^2$ Product of the sum and difference of two terms

$= 7 - 10$

$= -3$

## Rationalizing Denominators

Condition 3 of the rules for simplifying radicals described earlier requires that no denominator contain a radical. The process of achieving this is called **rationalizing the denominator.** To rationalize a denominator, we multiply by a form of 1.

**EXAMPLE 10** *Rationalizing Denominators*

Rationalize each denominator.

**(a)** $\frac{4}{\sqrt{3}}$ **(b)** $\frac{2}{\sqrt[3]{x}}$ $(x \neq 0)$

**Solution**

**(a)** $\frac{4}{\sqrt{3}}=\frac{4}{\sqrt{3}}\cdot\frac{\sqrt{3}}{\sqrt{3}}=\frac{4\sqrt{3}}{3}$ Multiply by $\frac{\sqrt{3}}{\sqrt{3}}=1$.

**(b)** $\frac{2}{\sqrt[3]{x}}=\frac{2}{\sqrt[3]{x}}\cdot\frac{\sqrt[3]{x^2}}{\sqrt[3]{x^2}}$ $\frac{\sqrt[3]{x^2}}{\sqrt[3]{x^2}}=1$

$=\frac{2\sqrt[3]{x^2}}{\sqrt[3]{x^3}}=\frac{2\sqrt[3]{x^2}}{x}$ $\sqrt[3]{x}\cdot\sqrt[3]{x^2}=\sqrt[3]{x^3}=x$

In Example 9(b), we saw that

$$\left(\sqrt{7}-\sqrt{10}\right)\left(\sqrt{7}+\sqrt{10}\right)=-3,$$

a rational number. This suggests a way to rationalize a denominator that is a binomial in which one or both terms is a radical. The expressions $a\sqrt{m}+b\sqrt{n}$ and $a\sqrt{m}-b\sqrt{n}$ are called **conjugates.**

**EXAMPLE 11** *Rationalizing a Binomial Denominator*

Rationalize the denominator of $\frac{1}{1-\sqrt{2}}$.

**Solution** The best approach is to multiply both numerator and denominator by the conjugate of the denominator—in this case, $1+\sqrt{2}$.

$$\frac{1}{1-\sqrt{2}}=\frac{1\left(1+\sqrt{2}\right)}{\left(1-\sqrt{2}\right)\left(1+\sqrt{2}\right)}=\frac{1+\sqrt{2}}{1-2}=-1-\sqrt{2}$$

## R.5 EXERCISES

*Concept Check Match the rational exponent expression in Exercises 1–8 with the equivalent radical expression in A–H. Assume that $x \neq 0$.*

**1.** $(-3x)^{1/3}$ **2.** $-3x^{1/3}$ **3.** $(-3x)^{-1/3}$ **4.** $-3x^{-1/3}$ **5.** $(3x)^{1/3}$ **6.** $3x^{-1/3}$ **7.** $(3x)^{-1/3}$ **8.** $3x^{1/3}$

**A.** $\frac{3}{\sqrt[3]{x}}$ **B.** $-3\sqrt[3]{x}$ **C.** $\frac{1}{\sqrt[3]{3x}}$ **D.** $\frac{-3}{\sqrt[3]{x}}$ **E.** $3\sqrt[3]{x}$ **F.** $\sqrt[3]{-3x}$ **G.** $\sqrt[3]{3x}$ **H.** $\frac{1}{\sqrt[3]{-3x}}$

*Write each expression in radical form. Assume all variables represent positive real numbers.*

**9.** $(-m)^{2/3}$ **10.** $p^{5/4}$ **11.** $(2m + p)^{2/3}$ **12.** $(5r + 3t)^{4/7}$

*Write each expression in exponential form. Assume all variables represent nonnegative real numbers.*

**13.** $\sqrt[5]{k^2}$ **14.** $-\sqrt[4]{z^5}$ **15.** $-3\sqrt{5p^3}$ **16.** $m\sqrt{2y^5}$

*Concept Check* *Use the ideas of this section to answer each question.*

**17.** For which of the following cases is $\sqrt{ab} = \sqrt{a} \cdot \sqrt{b}$ a true statement?

**A.** $a$ and $b$ both positive **B.** $a$ and $b$ both negative

**18.** For what positive integers $n$ greater than or equal to 2 is $\sqrt[n]{a^n} = a$ always a true statement?

**19.** For what values of $x$ is $\sqrt{9ax^2} = 3x\sqrt{a}$ a true statement? Assume $a \geq 0$.

**20.** Which of the following expressions is *not* simplified? Give the simplified form.

**A.** $\sqrt[3]{2y}$ **B.** $\frac{\sqrt{5}}{2}$ **C.** $\sqrt[4]{m^3}$ **D.** $\sqrt{\frac{3}{4}}$

*Simplify each radical expression. Assume that all variables represent positive real numbers.*

**21.** $\sqrt[3]{125}$ **22.** $\sqrt[4]{81}$ **23.** $\sqrt[5]{-3125}$ **24.** $\sqrt[3]{343}$ **25.** $\sqrt{50}$

**26.** $\sqrt{45}$ **27.** $\sqrt[3]{81}$ **28.** $\sqrt[3]{250}$ **29.** $-\sqrt[4]{32}$ **30.** $-\sqrt[4]{243}$

**31.** $-\sqrt{\frac{9}{5}}$ **32.** $-\sqrt[3]{\frac{3}{2}}$ **33.** $-\sqrt[3]{\frac{4}{5}}$ **34.** $\sqrt[4]{\frac{3}{2}}$ **35.** $\sqrt[3]{16(-2)^4(2)^8}$

**36.** $\sqrt[3]{25(3)^4(5)^3}$ **37.** $\sqrt{8x^5z^8}$ **38.** $\sqrt{24m^6n^5}$ **39.** $\sqrt[3]{16z^5x^8y^4}$ **40.** $-\sqrt[6]{64a^{12}b^8}$

**41.** $\sqrt[4]{m^2n^7p^8}$ **42.** $\sqrt[4]{x^8y^7z^9}$ **43.** $\sqrt[4]{x^4 + y^4}$ **44.** $\sqrt[3]{27 + a^3}$ **45.** $\sqrt{\frac{2}{3x}}$

**46.** $\sqrt{\frac{5}{3p}}$ **47.** $\sqrt{\frac{x^5y^3}{z^2}}$ **48.** $\sqrt{\frac{g^3h^5}{r^3}}$ **49.** $\sqrt[3]{\frac{8}{x^2}}$ **50.** $\sqrt[3]{\frac{9}{16p^4}}$

**51.** $\sqrt[4]{\frac{g^3h^5}{9r^6}}$ **52.** $\sqrt[4]{\frac{32x^5}{y^5}}$ **53.** $\frac{\sqrt[3]{mn} \cdot \sqrt[3]{m^2}}{\sqrt[3]{n^2}}$ **54.** $\frac{\sqrt[3]{8m^2n^3} \cdot \sqrt[3]{2m^2}}{\sqrt[3]{32m^4n^3}}$ **55.** $\frac{\sqrt[4]{32x^5y} \cdot \sqrt[4]{2xy^4}}{\sqrt[4]{4x^3y^2}}$

**56.** $\frac{\sqrt[4]{rs^2t^3} \cdot \sqrt[4]{r^3s^2t}}{\sqrt[4]{r^2t^3}}$ **57.** $\sqrt[3]{\sqrt{4}}$ **58.** $\sqrt[4]{\sqrt[3]{2}}$ **59.** $\sqrt[6]{\sqrt[3]{x}}$ **60.** $\sqrt[8]{\sqrt[4]{y}}$

*Simplify each expression, assuming that all variables represent nonnegative numbers.*

**61.** $4\sqrt{3} - 5\sqrt{12} + 3\sqrt{75}$ **62.** $2\sqrt{5} - 3\sqrt{20} + 2\sqrt{45}$ **63.** $3\sqrt{28p} - 4\sqrt{63p} + \sqrt{112p}$

**64.** $9\sqrt{8k} + 3\sqrt{18k} - \sqrt{32k}$ **65.** $2\sqrt[3]{3} + 4\sqrt[3]{24} - \sqrt[3]{81}$ **66.** $\sqrt[3]{32} - 5\sqrt[3]{4} + 2\sqrt[3]{108}$

**67.** $\frac{1}{\sqrt{3}} - \frac{2}{\sqrt{12}} + 2\sqrt{3}$ **68.** $\frac{1}{\sqrt{2}} + \frac{3}{\sqrt{8}} + \frac{1}{\sqrt{32}}$ **69.** $\frac{5}{\sqrt[3]{2}} - \frac{2}{\sqrt[3]{16}} + \frac{1}{\sqrt[3]{54}}$

**70.** $\frac{-4}{\sqrt[3]{3}} + \frac{1}{\sqrt[3]{24}} - \frac{2}{\sqrt[3]{81}}$ **71.** $(\sqrt{2} + 3)(\sqrt{2} - 3)$ **72.** $(\sqrt{5} + \sqrt{2})(\sqrt{5} - \sqrt{2})$

**73.** $(\sqrt[3]{11} - 1)(\sqrt[3]{11^2} + \sqrt[3]{11} + 1)$ **74.** $(\sqrt[3]{7} + 3)(\sqrt[3]{7^2} - 3\sqrt[3]{7} + 9)$ **75.** $(\sqrt{3} + \sqrt{8})^2$

**76.** $(\sqrt{2} - 1)^2$ **77.** $(3\sqrt{2} + \sqrt{3})(2\sqrt{3} - \sqrt{2})$ **78.** $(4\sqrt{5} - 1)(3\sqrt{5} + 2)$

*Rationalize the denominator of each radical expression. Assume that all variables represent nonnegative numbers and that no denominators are 0.*

**79.** $\frac{\sqrt{3}}{\sqrt{5} + \sqrt{3}}$ **80.** $\frac{\sqrt{7}}{\sqrt{3} - \sqrt{7}}$ **81.** $\frac{1 + \sqrt{3}}{3\sqrt{5} + 2\sqrt{3}}$ **82.** $\frac{\sqrt{7} - 1}{2\sqrt{7} + 4\sqrt{2}}$

**83.** $\dfrac{p}{\sqrt{p} + 2}$ **84.** $\dfrac{\sqrt{r}}{3 - \sqrt{r}}$ **85.** $\dfrac{a}{\sqrt{a + b} - 1}$ **86.** $\dfrac{3m}{2 + \sqrt{m + n}}$

## GEOMETRY FORMULAS

The following table provides a summary of some important formulas from geometry.

| Figure | Formula | Example |
|---|---|---|
| **Square** | Perimeter: $P = 4s$<br>Area: $A = s^2$ | |
| **Rectangle** | Perimeter: $P = 2L + 2W$<br>Area: $A = LW$ | |
| **Triangle** | Perimeter: $P = a + b + c$<br>Area: $A = \frac{1}{2}bh$ | |
| **Pythagorean Theorem (for Right Triangles)** | $c^2 = a^2 + b^2$ | |
| **Sum of the Angles of a Triangle** | $A + B + C = 180°$ | |
| **Circle** | Diameter: $d = 2r$<br>Circumference:<br>$C = 2\pi r = \pi d$<br>Area: $A = \pi r^2$ | |
| **Parallelogram** | Area: $A = bh$<br>Perimeter: $P = 2a + 2b$ | |

| Figure | Formula | Example |
|---|---|---|
| **Trapezoid** | Area: $A = \frac{1}{2}h(b_1 + b_2)$<br>Perimeter:<br>$P = a + b_1 + c + b_2$ | $b_1$, $a$, $h$, $c$, $b_2$ |
| **Sphere** | Volume: $V = \frac{4}{3}\pi r^3$<br>Surface area: $S = 4\pi r^2$ | 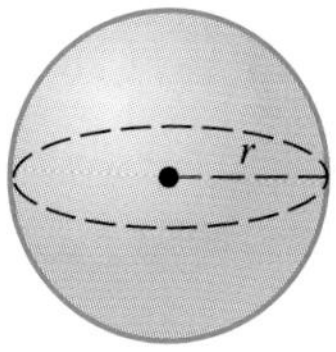 $r$ |
| **Cone** | Volume: $V = \frac{1}{3}\pi r^2 h$<br>Surface area:<br>$S = \pi r\sqrt{r^2 + h^2}$<br>(excludes the base) | 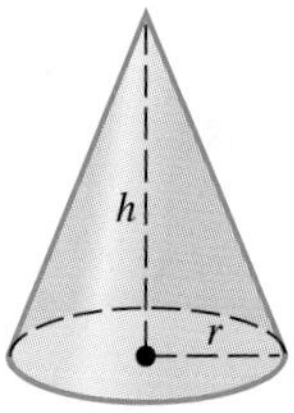 $h$, $r$ |
| **Cube** | Volume: $V = e^3$<br>Surface area: $S = 6e^2$ | 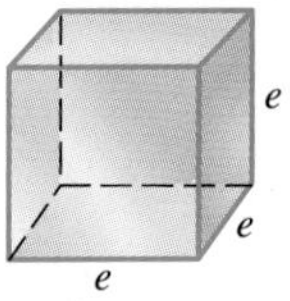 $e$, $e$, $e$ |
| **Rectangular Solid** | Volume: $V = LWH$<br>Surface area:<br>$S = 2HW + 2LW + 2LH$ | 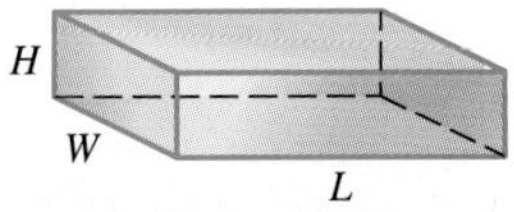 $H$, $W$, $L$ |
| **Right Circular Cylinder** | Volume: $V = \pi r^2 h$<br>Surface area:<br>$S = 2\pi rh + 2\pi r^2$<br>(includes top and bottom) | 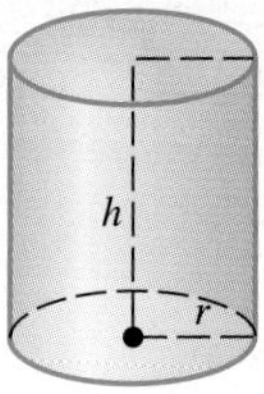 $h$, $r$ |
| **Right Pyramid** | Volume: $V = \frac{1}{3}Bh$<br>$B$ = area of the base | 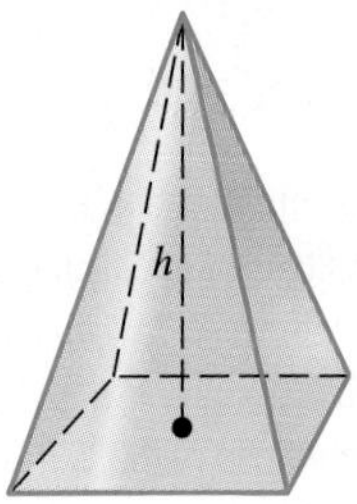 $h$ |

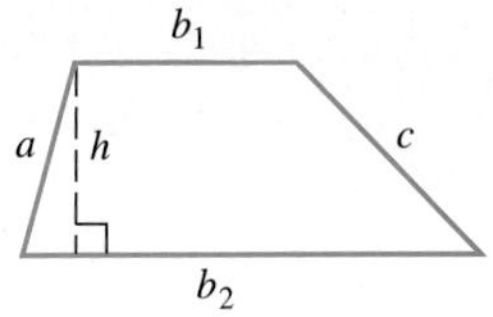

# Appendix A

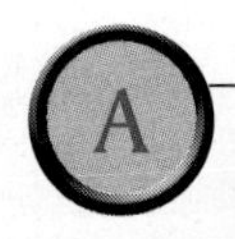

## VECTORS IN SPACE

Rectangular Coordinates in Space • Vectors in Space • Vector Definitions and Operations • Direction Angles in Space • An Application of the Dot Product

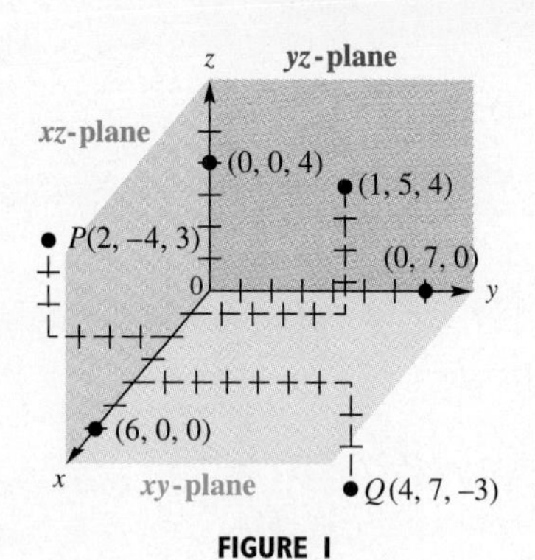


**FIGURE 1**

### Rectangular Coordinates in Space

Since vector quantities occur in space as well as in a plane, the properties developed earlier for vectors in a two-dimensional plane can be extended to three-dimensional space. The plane determined by the $x$- and $y$-axes is called the ***xy*-plane.** A third axis is needed to locate points in space; the $z$-axis goes through the origin in the $xy$-plane and is perpendicular to both the $x$-axis and the $y$-axis. Figure 1 shows one way to orient the three axes, with the $yz$-plane as the plane of the page and the $x$-axis perpendicular to it.

We can associate each point in space with an **ordered triple** $(x, y, z)$. Several ordered triples are shown in Figure 1. The region of three-dimensional space where all coordinates are positive is called the **first octant.**

The distance formula given in Chapter 1 is a special case of the distance formula for points in space, which can be proved similarly.

> **Distance Formula (Three-Dimensional Space)**
>
> If $P_1 = (x_1, y_1, z_1)$ and $P_2 = (x_2, y_2, z_2)$ are two points in a three-dimensional coordinate system, then the distance between $P_1$ and $P_2$ is given by
>
> $$d(P_1, P_2) = \sqrt{(x_2 - x_1)^2 + (y_2 - y_1)^2 + (z_2 - z_1)^2}.$$

**EXAMPLE 1** *Finding the Distance between Two Points in Space*

Find the distance between the points $P(2, -4, 3)$ and $Q(4, 7, -3)$ shown in Figure 1.

**Solution** By the distance formula,

$$d(P, Q) = \sqrt{(4 - 2)^2 + [7 - (-4)]^2 + (-3 - 3)^2} = \sqrt{161}.$$

**Looking Ahead to Calculus**

The concepts introduced in this section are extended in calculus to define a *vector-valued function,* a function whose domain is a set of real numbers and whose range is a set of vectors. For example,

$$\mathbf{R}(t) = f_1(t)\mathbf{i} + f_2(t)\mathbf{j} + f_3(t)\mathbf{k}$$

is a vector-valued function.

### Vectors in Space

Extending the notation used to represent vectors in the plane, we denote a vector $\mathbf{v}$ in space with initial point at the origin as

$$\mathbf{v} = \langle a, b, c \rangle. \qquad \text{Component form}$$

With the unit vectors $\mathbf{i} = \langle 1,0,0\rangle$, $\mathbf{j} = \langle 0,1,0\rangle$, and $\mathbf{k} = \langle 0,0,1\rangle$ in the directions of the positive $x$-axis, $y$-axis, and $z$-axis, respectively, we can also represent $\mathbf{v}$ as

$$\mathbf{v} = a\mathbf{i} + b\mathbf{j} + c\mathbf{k}, \qquad \text{i, j, k form}$$

where $a$, $b$, and $c$ are scalars. The scalars $a$, $b$, and $c$ are the components of vector $\mathbf{v}$.

**EXAMPLE 2** *Writing Vectors in* **i, j, k** *Form*

Write the vectors **OP** and **OQ** using the components for points $P$ and $Q$ defined in Example 1.

**Solution** Since $P = (2, -4, 3)$, the components of **OP** are 2, −4, and 3. Similarly, the components of **OQ** are 4, 7, and −3. Then

$$\mathbf{OP} = 2\mathbf{i} - 4\mathbf{j} + 3\mathbf{k} \quad \text{and} \quad \mathbf{OQ} = 4\mathbf{i} + 7\mathbf{j} - 3\mathbf{k}.$$

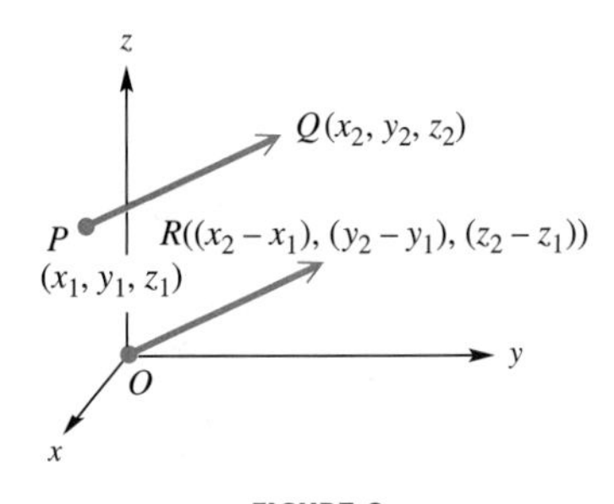


FIGURE 2

Vectors **OP** and **OQ** are position vectors because each has initial point at the origin. For $P = (x_1, y_1, z_1)$ and $Q = (x_2, y_2, z_2)$, the component form of vector **PQ** (which is not a position vector) is represented as

$$\mathbf{PQ} = \langle x_2 - x_1, y_2 - y_1, z_2 - z_1\rangle.$$

As Figure 2 suggests, **PQ** is equal to the position vector

$$\mathbf{OR} = (x_2 - x_1)\mathbf{i} + (y_2 - y_1)\mathbf{j} + (z_2 - z_1)\mathbf{k}.$$

**EXAMPLE 3** *Finding a Position Vector that Corresponds to a Given Vector*

Find the position vector corresponding to **PQ** with $P = (2, -4, 3)$ and $Q = (4, 7, -3)$.

**Solution** Using the definition above, the position vector is

$$\begin{aligned}\mathbf{v} &= (4 - 2)\mathbf{i} + [7 - (-4)]\mathbf{j} + (-3 - 3)\mathbf{k}\\ &= 2\mathbf{i} + 11\mathbf{j} - 6\mathbf{k}.\end{aligned}$$

## Vector Definitions and Operations

The vector concepts discussed in Section 10.3 can be extended to vectors in space.

**Vector Definitions and Operations**

If $\mathbf{v} = a\mathbf{i} + b\mathbf{j} + c\mathbf{k}$ and $\mathbf{w} = d\mathbf{i} + e\mathbf{j} + f\mathbf{k}$ are vectors and $g$ is a scalar, then

$$\begin{aligned}\mathbf{v} &= \mathbf{w} \text{ if and only if } a = d, b = e, \text{ and } c = f,\\ \mathbf{v} + \mathbf{w} &= (a + d)\mathbf{i} + (b + e)\mathbf{j} + (c + f)\mathbf{k},\\ \mathbf{v} - \mathbf{w} &= (a - d)\mathbf{i} + (b - e)\mathbf{j} + (c - f)\mathbf{k},\\ g\mathbf{v} &= ga\mathbf{i} + gb\mathbf{j} + gc\mathbf{k},\\ |\mathbf{v}| &= \sqrt{a^2 + b^2 + c^2},\\ \text{and} \quad \mathbf{v} \cdot \mathbf{w} &= ad + be + cf.\end{aligned}$$

**EXAMPLE 4** *Performing Vector Operations*

Find the following if $\mathbf{v} = 2\mathbf{i} + 6\mathbf{j} - 4\mathbf{k}$ and $\mathbf{w} = -\mathbf{i} + 5\mathbf{k}$.

(a) $\mathbf{v} + \mathbf{w}$ (b) $\mathbf{w} - \mathbf{v}$ (c) $-10\mathbf{w}$

(d) $|\mathbf{v}|$ (e) $\mathbf{v} \cdot \mathbf{w}$ (f) $\mathbf{w} \cdot \mathbf{v}$

**Solution**

(a) $\mathbf{v} + \mathbf{w} = (2 - 1)\mathbf{i} + (6 + 0)\mathbf{j} + (-4 + 5)\mathbf{k} = \mathbf{i} + 6\mathbf{j} + \mathbf{k}$

(b) $\mathbf{w} - \mathbf{v} = (-1 - 2)\mathbf{i} + (0 - 6)\mathbf{j} + [5 - (-4)]\mathbf{k} = -3\mathbf{i} - 6\mathbf{j} + 9\mathbf{k}$

(c) $-10\mathbf{w} = -10(-1)\mathbf{i} + (-10)5\mathbf{k} = 10\mathbf{i} - 50\mathbf{k}$

(d) $|\mathbf{v}| = \sqrt{2^2 + 6^2 + (-4)^2} = \sqrt{56} = 2\sqrt{14}$

(e) $\mathbf{v} \cdot \mathbf{w} = 2(-1) + 6(0) + (-4)5 = -22$

(f) $\mathbf{w} \cdot \mathbf{v} = -1(2) + 0(6) + 5(-4) = -22$

Properties of the dot product for vectors in two dimensions can be extended to vectors in three dimensions, as can the geometric interpretation of the dot product.

If $\theta$ is the angle between two nonzero vectors $\mathbf{v}$ and $\mathbf{w}$ in space, where $0 \le \theta \le \pi$, then $\mathbf{v} \cdot \mathbf{w} = |\mathbf{v}||\mathbf{w}| \cos \theta$.

Dividing each side of the equation by $|\mathbf{v}||\mathbf{w}|$ gives the following result.

**Angle between Two Vectors**

If $\theta$ is the angle between two nonzero vectors $\mathbf{v}$ and $\mathbf{w}$, where $0 \le \theta \le \pi$, then

$$\cos \theta = \frac{\mathbf{v} \cdot \mathbf{w}}{|\mathbf{v}||\mathbf{w}|}.$$

**EXAMPLE 5** *Finding the Angle between Two Vectors*

Find the angle between $\mathbf{v} = 7\mathbf{i} - 3\mathbf{j} - 5\mathbf{k}$ and $\mathbf{w} = 4\mathbf{i} + 6\mathbf{j} - 8\mathbf{k}$.

**Solution** First we must find $|\mathbf{v}|$ and $|\mathbf{w}|$.

$$|\mathbf{v}| = \sqrt{7^2 + (-3)^2 + (-5)^2} = \sqrt{83}$$
$$|\mathbf{w}| = \sqrt{4^2 + 6^2 + (-8)^2} = \sqrt{116} = 2\sqrt{29}$$

Now, we use the formula for $\cos \theta$.

$$\cos \theta = \frac{\mathbf{v} \cdot \mathbf{w}}{|\mathbf{v}||\mathbf{w}|} = \frac{7(4) + (-3)6 + (-5)(-8)}{\sqrt{83} \cdot 2\sqrt{29}} \approx .5096$$

A calculator gives $\theta \approx 1.036$ radians or 59.4°.

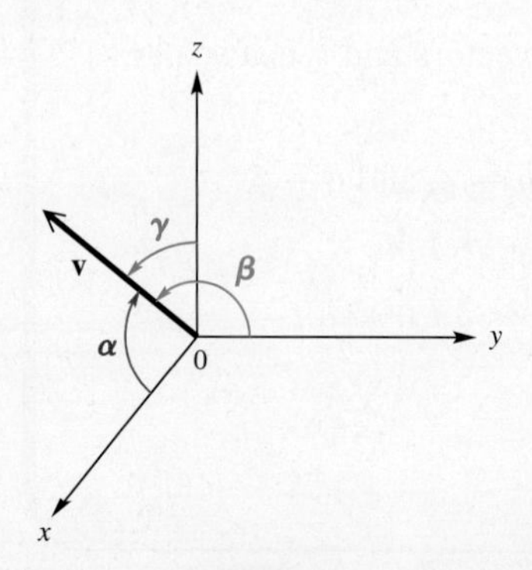

FIGURE 3

## Direction Angles in Space

A vector in two dimensions is determined by its magnitude and direction angle. In three dimensions, a vector is determined by its magnitude and three direction angles, that is, the angles between the vector and each positive axis. As shown in Figure 3,

$\alpha$ is the direction angle between $\mathbf{v}$ and the positive $x$-axis,
$\beta$ is the direction angle between $\mathbf{v}$ and the positive $y$-axis,
and $\gamma$ is the direction angle between $\mathbf{v}$ and the positive $z$-axis.

We can evaluate these angles using the expression for the cosine of the angle between two vectors. Recall, $\mathbf{i} = \langle 1, 0, 0 \rangle$, $\mathbf{j} = \langle 0, 1, 0 \rangle$, $\mathbf{k} = \langle 0, 0, 1 \rangle$, and each has magnitude 1. For $\mathbf{v} = a\mathbf{i} + b\mathbf{j} + c\mathbf{k}$, the three direction angles are determined as follows.

**Direction Angles**

$$\cos\alpha = \frac{\mathbf{v}\cdot\mathbf{i}}{|\mathbf{v}||\mathbf{i}|} = \frac{a}{|\mathbf{v}|}, \quad \cos\beta = \frac{\mathbf{v}\cdot\mathbf{j}}{|\mathbf{v}||\mathbf{j}|} = \frac{b}{|\mathbf{v}|}, \quad \cos\gamma = \frac{\mathbf{v}\cdot\mathbf{k}}{|\mathbf{v}||\mathbf{k}|} = \frac{c}{|\mathbf{v}|}.$$

The quantities $\cos\alpha$, $\cos\beta$, and $\cos\gamma$ are called **direction cosines.** In Exercise 49 you are asked to prove that

$$\cos^2\alpha + \cos^2\beta + \cos^2\gamma = 1.$$

**EXAMPLE 6** *Finding Direction Angles of a Vector*

Find the direction angles of $\mathbf{w} = 2\mathbf{i} - 3\mathbf{j} + 6\mathbf{k}$. Give answers in degrees.

**Solution** We have $|\mathbf{w}| = \sqrt{4 + 9 + 36} = 7$. Then

$$\cos\alpha = \frac{a}{|\mathbf{w}|} = \frac{2}{7}, \quad \cos\beta = \frac{b}{|\mathbf{w}|} = -\frac{3}{7}, \quad \text{and} \quad \cos\gamma = \frac{c}{|\mathbf{w}|} = \frac{6}{7},$$

so $\alpha \approx 73.4°$, $\beta \approx 115.4°$, and $\gamma \approx 31.0°$.

**EXAMPLE 7** *Verifying the Sum of the Squares of the Direction Angle Cosines*

Verify that $\cos^2\alpha + \cos^2\beta + \cos^2\gamma = 1$ for the vector in Example 6.

**Solution**

$$\left(\frac{2}{7}\right)^2 + \left(-\frac{3}{7}\right)^2 + \left(\frac{6}{7}\right)^2 = \frac{4}{49} + \frac{9}{49} + \frac{36}{49} = 1$$

**EXAMPLE 8** *Finding a Direction Angle*

Figure 4 shows that the angle between a vector $\mathbf{u}$ and the positive $z$-axis is 120°. The angle between $\mathbf{u}$ and the positive $x$-axis is 90°. What angle does $\mathbf{u}$ make with the positive $y$-axis?

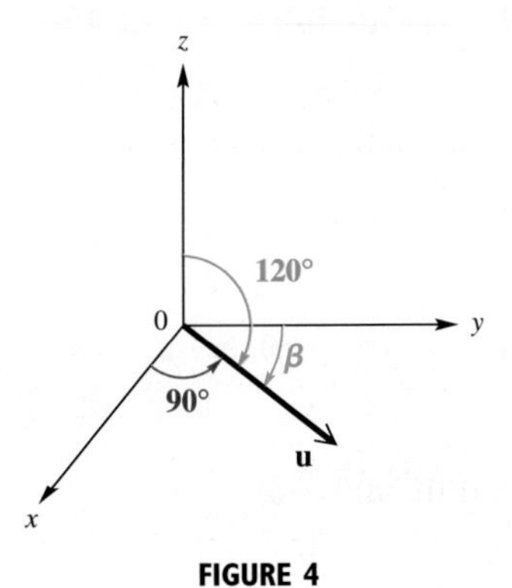

FIGURE 4

**Solution** Here, $\cos\gamma = \cos 120° = -\frac{1}{2}$ and $\cos\alpha = \cos 90° = 0$. We know that $\cos^2\alpha + \cos^2\beta + \cos^2\gamma = 1$, so

$$\cos^2\beta = 1 - \cos^2\alpha - \cos^2\gamma$$

$$\cos^2\beta = 1 - 0 - \frac{1}{4} \qquad \text{Substitute; } \left(-\frac{1}{2}\right)^2 = \frac{1}{4}.$$

$$\cos^2\beta = \frac{3}{4}$$

$$\cos\beta = \frac{\sqrt{3}}{2} \qquad 0° < \beta < 90°$$

$$\beta = 30°.$$

## An Application of the Dot Product

The **work** done by a constant force **F** as it moves a particle from point $P$ to point $Q$ is defined as $\mathbf{F} \cdot \mathbf{PQ}$.

**EXAMPLE 9** *Applying the Dot Product to Work*

Find the work (in units) done by a force $\mathbf{F} = \langle 3, 4, 2 \rangle$ that moves a particle from $P(1, 0, 5)$ to $Q(3, 3, 8)$.

**Solution** We must find **PQ**. Since $P$ is the initial point and $Q$ is the terminal point,

$$\mathbf{PQ} = \langle 3 - 1, 3 - 0, 8 - 5 \rangle = \langle 2, 3, 3 \rangle.$$

Then the work done is

$$\begin{aligned} \mathbf{F} \cdot \mathbf{PQ} &= \langle 3, 4, 2 \rangle \cdot \langle 2, 3, 3 \rangle \\ &= 6 + 12 + 6 = 24 \text{ (work units).} \end{aligned}$$

## A APPENDIX A EXERCISES

*Concept Check Fill in the blanks to complete each statement.*

1. The plane determined by the $x$-axis and the $z$-axis is called the ________.
2. A position vector has its ________ point at the origin.
3. The component form of the position vector with terminal point $(5, 3, -2)$ is ________.
4. The **i**, **j**, **k** form of the position vector with terminal point $(6, -1, -3)$ is ________.

*Find the distance between the points P and Q.*

5. $P = (0, 0, 0)$; $Q = (2, -2, 5)$
6. $P = (0, 0, 0)$; $Q = (7, 4, -1)$
7. $P = (10, 15, 9)$; $Q = (8, 3, -4)$
8. $P = (5, 4, -4)$; $Q = (3, 7, 2)$
9. $P = (20, 25, 16)$; $Q = (5, 5, 6)$
10. $P = (14, 10, 18)$; $Q = (-2, 4, 9)$

*For each pair of points, find the position vector that corresponds to* **PQ** *in component form and in* **i**, **j**, **k** *form.*

11. $P = (0, 0, 0)$; $Q = (2, -2, 5)$
12. $P = (0, 0, 0)$; $Q = (7, 4, -1)$
13. $P = (10, 15, 0)$; $Q = (8, 3, -4)$
14. $P = (0, 4, -4)$; $Q = (3, 7, 2)$
15. $P = (20, 25, 6)$; $Q = (5, 5, 16)$
16. $P = (14, 10, 18)$; $Q = (-2, 4, 9)$
17. Using the points in Exercise 11, write **QP** in component form. How does it compare to **PQ**?
18. Compare the distance between points $P$ and $Q$ in three-dimensional space and the magnitude of vector **PQ**.

*Given* $\mathbf{u} = 2\mathbf{i} + 4\mathbf{j} + 7\mathbf{k}$, $\mathbf{v} = -3\mathbf{i} + 5\mathbf{j} + 2\mathbf{k}$, *and* $\mathbf{w} = 4\mathbf{i} - 3\mathbf{j} - 6\mathbf{k}$, *find the following.*

19. $\mathbf{u} - \mathbf{w}$
20. $\mathbf{v} + \mathbf{w}$
21. $4\mathbf{u} + 5\mathbf{v}$
22. $-\mathbf{v} + 3\mathbf{u}$
23. $|\mathbf{u}|$
24. $|\mathbf{w}|$
25. $|\mathbf{w} + \mathbf{u}|$
26. $|2\mathbf{v}|$
27. $\mathbf{v} \cdot \mathbf{w}$
28. $\mathbf{u} \cdot \mathbf{w}$
29. $\mathbf{v} \cdot \mathbf{v}$
30. $\mathbf{u} \cdot \mathbf{u}$

*Find the angle between each pair of vectors.*

31. $\langle 2, -2, 0 \rangle, \langle 5, -2, -1 \rangle$
32. $\langle 4, 0, 0 \rangle, \langle 5, 3, -2 \rangle$
33. $\langle 6, 0, 0 \rangle, \langle 8, 3, -4 \rangle$
34. $\langle -1, 2, -3 \rangle, \langle 0, -2, 1 \rangle$
35. $\langle 1, 0, 0 \rangle, \langle 0, 1, 0 \rangle$
36. $\langle 0, 0, 1 \rangle, \langle 0, 1, 0 \rangle$

*Find the direction angles of each vector. Give answers in degrees. Check your work by showing that the sum of the squares of the cosines equals* 1.

37. $\mathbf{u} = 2\mathbf{i} + 4\mathbf{j} + 7\mathbf{k}$
38. $\mathbf{v} = -3\mathbf{i} + 5\mathbf{j} + 2\mathbf{k}$
39. $\mathbf{w} = 4\mathbf{i} - 3\mathbf{j} - 6\mathbf{k}$
40. $\mathbf{y} = 2\mathbf{i} - 3\mathbf{j} + 4\mathbf{k}$

*Work each problem.*

**41.** The angle between vector **u** and the positive $x$-axis is 45°. The angle between **u** and the positive $y$-axis is 120°. Find the angle between **u** and the positive $z$-axis.

**42.** The direction angle between vector **v** and the $y$-axis is 135°. The direction angle between **v** and the $z$-axis is 90°. What is the direction angle between **v** and the $x$-axis?

**43.** What must be true for two vectors in space to be parallel?

**44.** Decide whether the vectors $\langle 3, 5, -1 \rangle$ and $\langle -12, -20, 4 \rangle$ are parallel. Explain your answer.

Work *Find the work (in units) done by a force* **F** *in moving a particle from P to Q.*

**45.** $\mathbf{F} = \langle 2, 0, 5 \rangle$; $P = (0, 0, 0)$; $Q = (1, 3, 2)$

**46.** $\mathbf{F} = \langle 3, 2, 0 \rangle$; $P = (0, 0, 0)$; $Q = (4, -2, 5)$

**47.** $\mathbf{F} = \mathbf{i} + 2\mathbf{j} - \mathbf{k}$; $P = (2, -1, 2)$; $Q = (5, 7, 8)$

**48.** $\mathbf{F} = 3\mathbf{i} + 2\mathbf{j}$; $P = (4, 7, 6)$; $Q = (10, 15, 12)$

**49.** Prove that for direction cosines $\cos \alpha$, $\cos \beta$, and $\cos \gamma$,

$$\cos^2 \alpha + \cos^2 \beta + \cos^2 \gamma = 1.$$

# Appendix B

## POLAR FORM OF CONIC SECTIONS

Up to this point, we have worked with equations of conic sections in rectangular form. If the focus of a conic section is at the pole, the polar form of its equation is

$$r = \frac{ep}{1 \pm e \cdot f(\theta)},$$

where $f$ is either the sine or cosine function.

**Polar Forms of Conic Sections**

A polar equation of the form

$$r = \frac{ep}{1 \pm e \cos \theta} \quad \text{or} \quad r = \frac{ep}{1 \pm e \sin \theta}$$

has a conic section as its graph. The eccentricity is $e$ (where $e > 0$), and $|p|$ is the distance between the pole (focus) and the directrix.

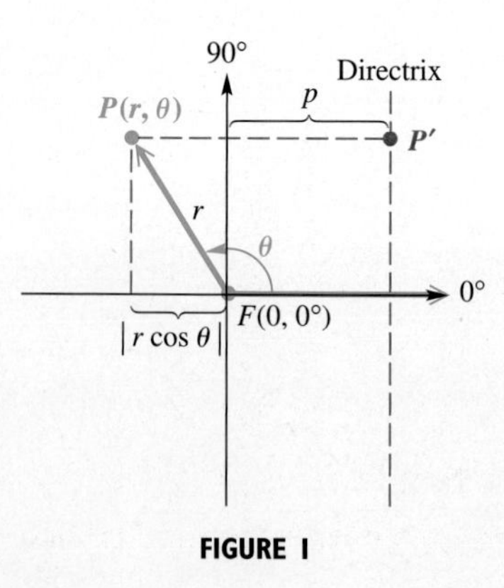


FIGURE 1

We can verify that $r = \frac{ep}{1 + e \cos \theta}$ does indeed satisfy the definition of a conic section. Consider Figure 1, where the directrix is vertical and $p > 0$ units to the right of the focus $F(0, 0°)$.

Let $P(r, \theta)$ be a point on the graph. Then the distance between $P$ and the directrix is

$$\begin{aligned}
PP' &= |p - x| \\
&= |p - r \cos \theta| && x = r \cos \theta \\
&= \left| p - \left( \frac{ep}{1 + e \cos \theta} \right) \cos \theta \right| && \text{Use the equation for } r. \\
&= \left| \frac{p(1 + e \cos \theta) - ep \cos \theta}{1 + e \cos \theta} \right| && \text{Use a common denominator.} \\
&= \left| \frac{p + ep \cos \theta - ep \cos \theta}{1 + e \cos \theta} \right| && \text{Distributive property} \\
PP' &= \left| \frac{p}{1 + e \cos \theta} \right|.
\end{aligned}$$

Since

$$r = \frac{ep}{1 + e\cos\theta},$$

we can multiply each side by $\frac{1}{e}$ to obtain

$$\frac{p}{1 + e\cos\theta} = \frac{r}{e}.$$

We substitute $\frac{r}{e}$ for the expression in the absolute value bars above.

$$PP' = \left|\frac{r}{e}\right| = \frac{|r|}{|e|} = \frac{|r|}{e}$$

The distance between the pole and $P$ is $PF = |r|$, so the ratio of $PF$ to $PP'$ is

$$\frac{PF}{PP'} = \frac{|r|}{\frac{|r|}{e}} = e.$$

Thus, by the definition, the graph has eccentricity $e$ and must be a conic.

In the preceding discussion, we assumed a vertical directrix to the right of the pole. There are three other possible situations, and all four are summarized in the table.

| If the equation is: | then the directrix is: |
|---|---|
| $r = \frac{ep}{1 + e\cos\theta}$ | *vertical*, $p$ units to the *right* of the pole. |
| $r = \frac{ep}{1 - e\cos\theta}$ | *vertical*, $p$ units to the *left* of the pole. |
| $r = \frac{ep}{1 + e\sin\theta}$ | *horizontal*, $p$ units *above* the pole. |
| $r = \frac{ep}{1 - e\sin\theta}$ | *horizontal*, $p$ units *below* the pole. |

### EXAMPLE 1 *Graphing a Conic Section with Equation in Polar Form*

Graph $r = \frac{8}{4 + 4\sin\theta}$.

**Analytic Solution**

Divide both numerator and denominator by 4 to get

$$r = \frac{2}{1 + \sin\theta}.$$

Based on the preceding table, this is the equation of a conic with $ep = 2$ and $e = 1$. Thus $p = 2$. Since $e = 1$, the graph is a parabola. The focus is at the pole, and the directrix is horizontal, 2 units *above* the pole. The vertex must have polar coordinates $(1, 90°)$. Letting $\theta = 0°$ and $\theta = 180°$ gives the additional points $(2, 0°)$ and $(2, 180°)$. See Figure 2 on the next page.

**Graphing Calculator Solution**

Enter

$$r_1 = \frac{8}{4 + 4\sin\theta},$$

with the calculator in polar and degree modes. The first two screens in Figure 3 on the next page show the window settings, and the third screen shows the graph. Notice that the point $(1, 90°)$ is indicated at the bottom of the third screen.

*(continued)*

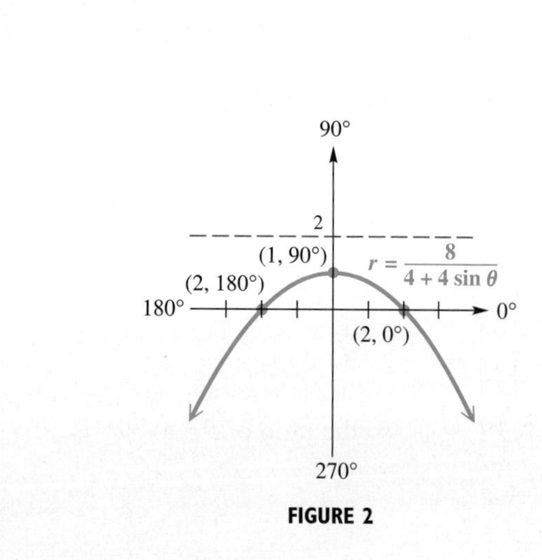


FIGURE 2

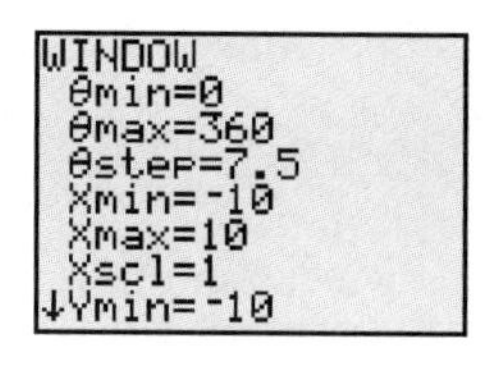


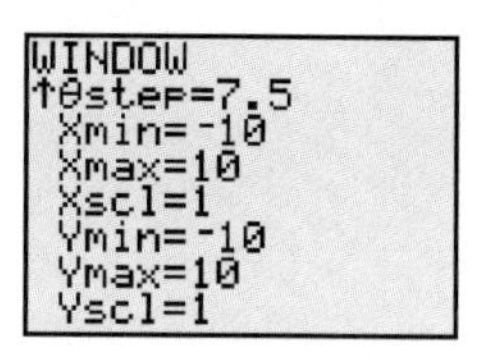


This is a continuation of the screen to the left.

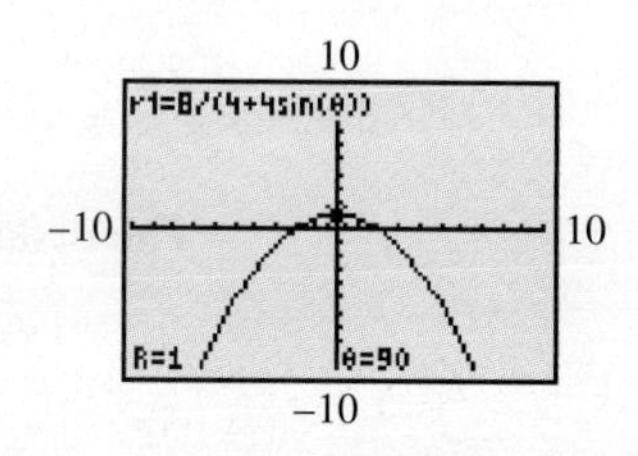


Degree Mode

FIGURE 3

## EXAMPLE 2 *Finding a Polar Equation*

Find the polar equation of a parabola with focus at the pole and vertical directrix 3 units to the left of the pole.

**Solution** The eccentricity $e$ must be 1, $p$ must equal 3, and the equation must be of the form

$$r = \frac{ep}{1 - e\cos\theta}.$$

Thus, we have

$$r = \frac{1 \cdot 3}{1 - 1\cos\theta} = \frac{3}{1 - \cos\theta}.$$

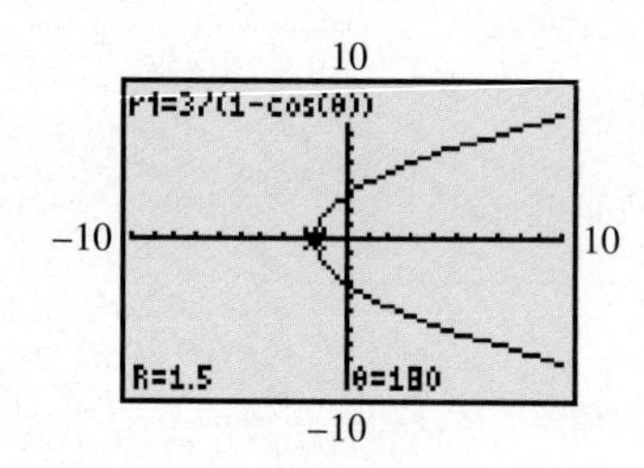


Degree Mode

FIGURE 4

The calculator graph in Figure 4 supports our result. When $\theta = 180°$, $r = 1.5$. The distance from $F(0, 0°)$ to the directrix is $2r = 2(1.5) = 3$ units, as required.

## EXAMPLE 3 *Identifying and Converting from Polar to Rectangular Form*

Identify the type of conic represented by

$$r = \frac{8}{2 - \cos\theta}.$$

Then convert the equation to rectangular form.

**Solution** To identify the type of conic, we divide both the numerator and the denominator on the right side by 2 to obtain

$$r = \frac{4}{1 - \frac{1}{2}\cos\theta}.$$

From the table, we see that this is a conic that has a vertical directrix with $e = \frac{1}{2}$; thus it is an ellipse.

To convert to rectangular form, we start with the given equation.

$$
\begin{aligned}
r &= \frac{8}{2 - \cos\theta} \\
r(2 - \cos\theta) &= 8 && \text{Multiply by } 2 - \cos\theta. \\
2r - r\cos\theta &= 8 && \text{Distributive property} \\
2r &= r\cos\theta + 8 && \text{Add } r\cos\theta \text{ to each side.} \\
(2r)^2 &= (r\cos\theta + 8)^2 && \text{Square each side.} \\
(2r)^2 &= (x + 8)^2 && r\cos\theta = x \\
4r^2 &= x^2 + 16x + 64 \\
4(x^2 + y^2) &= x^2 + 16x + 64 && r^2 = x^2 + y^2 \\
4x^2 + 4y^2 &= x^2 + 16x + 64 && \text{Distributive property} \\
3x^2 + 4y^2 - 16x - 64 &= 0 && \text{Standard form}
\end{aligned}
$$

The coefficients of $x^2$ and $y^2$ are both positive and are not equal, further supporting our assertion that the graph is an ellipse.

## B APPENDIX B EXERCISES

*Graph each conic whose equation is given in polar form. Use a traditional or a calculator graph, as directed by your instructor.*

**1.** $r = \dfrac{6}{3 + 3\sin\theta}$ **2.** $r = \dfrac{10}{5 + 5\sin\theta}$ **3.** $r = \dfrac{-4}{6 + 2\cos\theta}$ **4.** $r = \dfrac{-8}{4 + 2\cos\theta}$

**5.** $r = \dfrac{2}{2 - 4\sin\theta}$ **6.** $r = \dfrac{6}{2 - 4\sin\theta}$ **7.** $r = \dfrac{4}{2 - 4\cos\theta}$ **8.** $r = \dfrac{6}{2 - 4\cos\theta}$

**9.** $r = \dfrac{-1}{1 + 2\sin\theta}$ **10.** $r = \dfrac{-1}{1 - 2\sin\theta}$ **11.** $r = \dfrac{-1}{2 + \cos\theta}$ **12.** $r = \dfrac{-1}{2 - \cos\theta}$

*Find a polar equation of the parabola with focus at the pole, satisfying the given conditions.*

**13.** Vertical directrix 3 units to the right of the pole

**14.** Vertical directrix 4 units to the left of the pole

**15.** Horizontal directrix 5 units below the pole

**16.** Horizontal directrix 6 units above the pole

*Find a polar equation for the conic with focus at the pole, satisfying the given conditions. Also identify the type of conic represented.*

**17.** $e = \frac{4}{5}$; vertical directrix 5 units to the right of the pole

**18.** $e = \frac{2}{3}$; vertical directrix 6 units to the left of the pole

**19.** $e = \frac{5}{4}$; horizontal directrix 8 units below the pole

**20.** $e = \frac{3}{2}$; horizontal directrix 4 units above the pole

*Identify the type of conic represented, and convert the equation to rectangular form.*

**21.** $r = \dfrac{6}{3 - \cos\theta}$ **22.** $r = \dfrac{8}{4 - \cos\theta}$ **23.** $r = \dfrac{-2}{1 + 2\cos\theta}$ **24.** $r = \dfrac{-3}{1 + 3\cos\theta}$

**25.** $r = \dfrac{-6}{4 + 2\sin\theta}$ **26.** $r = \dfrac{-12}{6 + 3\sin\theta}$ **27.** $r = \dfrac{10}{2 - 2\sin\theta}$ **28.** $r = \dfrac{12}{4 - 4\sin\theta}$

# Appendix C

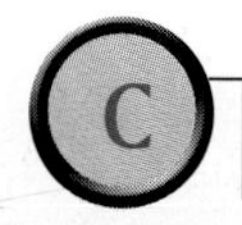


## ROTATION OF AXES

Derivation of Rotation Equations • Applying a Rotation Equation • Summary of Conics with an $xy$-Term

**Looking Ahead to Calculus**

Rotation of axes is a topic traditionally covered in calculus texts, in conjunction with parametric equations and polar coordinates. The coverage in calculus is typically the same as that seen in this section.

### Derivation of Rotation Equations

If we begin with an $xy$-coordinate system having origin $O$ and rotate the axes about $O$ through an angle $\theta$, the new coordinate system is called a **rotation** of the $xy$-system. Trigonometric identities can be used to obtain equations for converting the coordinates of a point from the $xy$-system to the rotated $x'y'$-system. Let $P$ be any point other than the origin, with coordinates $(x, y)$ in the $xy$-system and $(x', y')$ in the $x'y'$-system. See Figure 1. Let $OP = r$, and let $\alpha$ represent the angle made by $OP$ and the $x'$-axis. As shown in Figure 1,

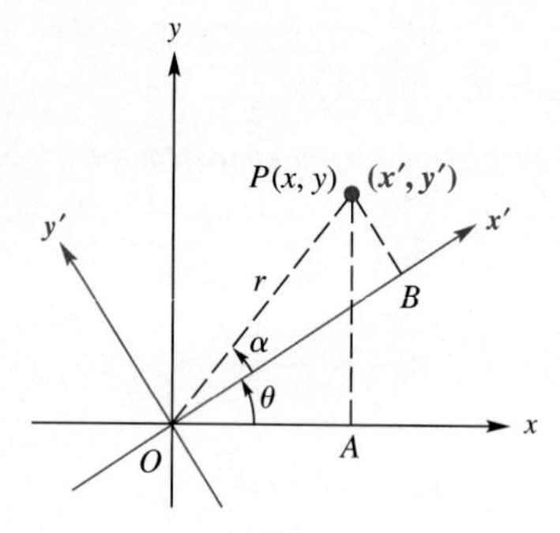


FIGURE 1

$$\cos(\theta + \alpha) = \frac{OA}{r} = \frac{x}{r}, \quad \sin(\theta + \alpha) = \frac{AP}{r} = \frac{y}{r},$$

$$\cos \alpha = \frac{OB}{r} = \frac{x'}{r}, \quad \sin \alpha = \frac{PB}{r} = \frac{y'}{r}.$$

These four statements can be rewritten as

$$x = r\cos(\theta + \alpha), \quad y = r\sin(\theta + \alpha), \quad x' = r\cos\alpha, \quad y' = r\sin\alpha.$$

Using the trigonometric identity for the cosine of the sum of two angles gives

$$\begin{aligned} x &= r\cos(\theta + \alpha) \\ &= r(\cos\theta\cos\alpha - \sin\theta\sin\alpha) \\ &= (r\cos\alpha)\cos\theta - (r\sin\alpha)\sin\theta \\ &= x'\cos\theta - y'\sin\theta. \end{aligned}$$

In the same way, by using the identity for the sine of the sum of two angles, $y = x'\sin\theta + y'\cos\theta$. This proves the following result.

**Rotation Equations**

If the rectangular coordinate axes are rotated about the origin through an angle $\theta$, and if the coordinates of a point $P$ are $(x, y)$ and $(x', y')$ with respect to the $xy$-system and the $x'y'$-system, respectively, then the **rotation equations** are

$$\mathbf{x = x'\cos\theta - y'\sin\theta} \quad \text{and} \quad \mathbf{y = x'\sin\theta + y'\cos\theta.}$$

## Applying a Rotation Equation

**EXAMPLE 1** *Finding an Equation after a Rotation*

The equation of a curve is $x^2 + y^2 + 2xy + 2\sqrt{2}x - 2\sqrt{2}y = 0$. Find the resulting equation if the axes are rotated 45°. Graph the equation.

**Solution** If $\theta = 45°$, then $\sin\theta = \frac{\sqrt{2}}{2}$ and $\cos\theta = \frac{\sqrt{2}}{2}$, and the rotation equations become

$$x = \frac{\sqrt{2}}{2}x' - \frac{\sqrt{2}}{2}y' \quad \text{and} \quad y = \frac{\sqrt{2}}{2}x' + \frac{\sqrt{2}}{2}y'.$$

Substituting these values into the given equation yields

$$x^2 + y^2 + 2xy + 2\sqrt{2}x - 2\sqrt{2}y = 0$$

$$\left[\frac{\sqrt{2}}{2}x' - \frac{\sqrt{2}}{2}y'\right]^2 + \left[\frac{\sqrt{2}}{2}x' + \frac{\sqrt{2}}{2}y'\right]^2 + 2\left[\frac{\sqrt{2}}{2}x' - \frac{\sqrt{2}}{2}y'\right]\left[\frac{\sqrt{2}}{2}x' + \frac{\sqrt{2}}{2}y'\right] + 2\sqrt{2}\left[\frac{\sqrt{2}}{2}x' - \frac{\sqrt{2}}{2}y'\right] - 2\sqrt{2}\left[\frac{\sqrt{2}}{2}x' + \frac{\sqrt{2}}{2}y'\right] = 0.$$

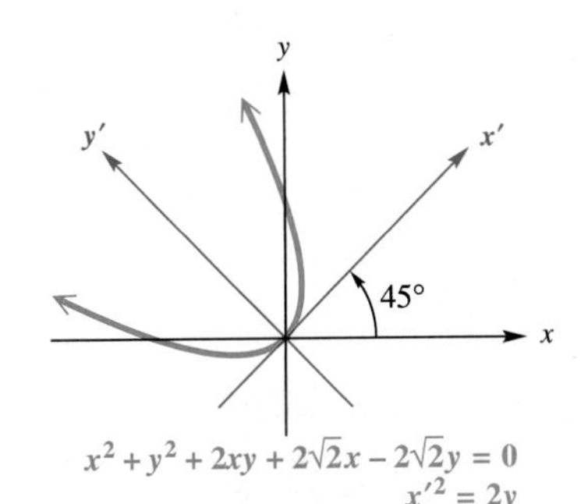


FIGURE 2

Expanding these terms yields

$$\frac{1}{2}x'^2 - x'y' + \frac{1}{2}y'^2 + \frac{1}{2}x'^2 + x'y' + \frac{1}{2}y'^2 + x'^2 - y'^2 + 2x' - 2y' - 2x' - 2y' = 0.$$

Collecting terms gives

$$2x'^2 - 4y' = 0$$

$$x'^2 - 2y' = 0 \qquad \text{Divide by 2.}$$

or, finally

$$x'^2 = 2y',$$

the equation of a parabola. The graph is shown in Figure 2.

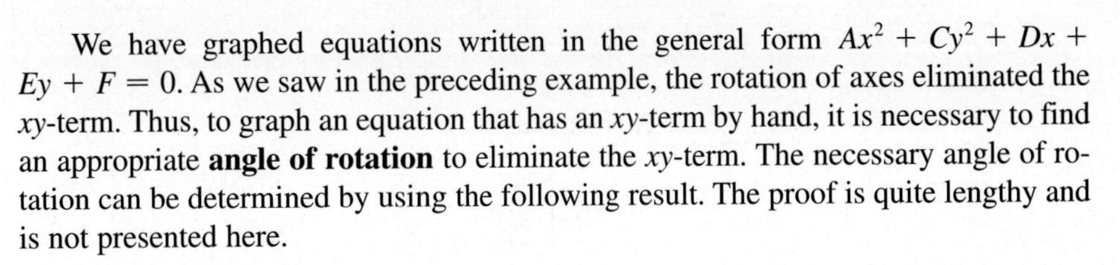

We have graphed equations written in the general form $Ax^2 + Cy^2 + Dx + Ey + F = 0$. As we saw in the preceding example, the rotation of axes eliminated the $xy$-term. Thus, to graph an equation that has an $xy$-term by hand, it is necessary to find an appropriate **angle of rotation** to eliminate the $xy$-term. The necessary angle of rotation can be determined by using the following result. The proof is quite lengthy and is not presented here.

### Angle of Rotation

The $xy$-term is removed from the general equation

$$Ax^2 + Bxy + Cy^2 + Dx + Ey + F = 0$$

by a rotation of the axes through an angle $\theta$, $0° < \theta < 90°$, where

$$\cot 2\theta = \frac{A - C}{B}.$$

This result can be used to find the appropriate angle of rotation, $\theta$. To find the rotation equations, first find $\sin\theta$ and $\cos\theta$. The following example illustrates a way to obtain $\sin\theta$ and $\cos\theta$ from $\cot 2\theta$ without first identifying the angle $\theta$.

**EXAMPLE 2** *Rotating and Graphing*

Rotate the axes and graph $52x^2 - 72xy + 73y^2 = 200$.

**Solution** Here $A = 52$, $B = -72$, and $C = 73$. By substitution,

$$\cot 2\theta = \frac{52 - 73}{-72} = \frac{-21}{-72} = \frac{7}{24}.$$

To find $\sin\theta$ and $\cos\theta$, use the trigonometric identities

$$\sin\theta = \sqrt{\frac{1 - \cos 2\theta}{2}} \quad \text{and} \quad \cos\theta = \sqrt{\frac{1 + \cos 2\theta}{2}}.$$

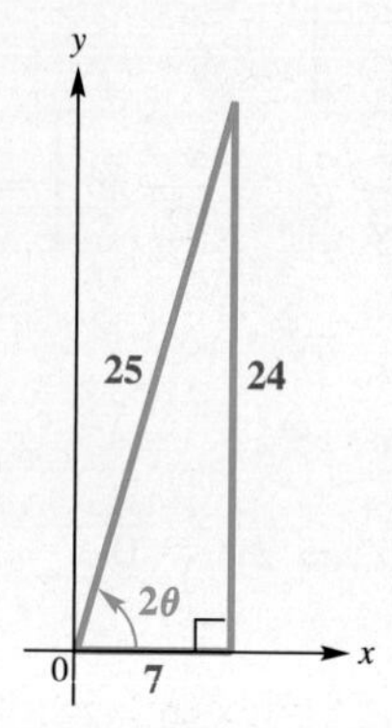


**FIGURE 3**

Sketch a right triangle and label it as in Figure 3, to see that $\cos 2\theta = \frac{7}{25}$. (Recall that in the two quadrants for which we are concerned, cosine and cotangent have the same sign.) Then

$$\sin\theta = \sqrt{\frac{1 - \frac{7}{25}}{2}} = \sqrt{\frac{9}{25}} = \frac{3}{5} \quad \text{and} \quad \cos\theta = \sqrt{\frac{1 + \frac{7}{25}}{2}} = \sqrt{\frac{16}{25}} = \frac{4}{5}.$$

Use these values for $\sin\theta$ and $\cos\theta$ to obtain

$$x = \frac{4}{5}x' - \frac{3}{5}y' \quad \text{and} \quad y = \frac{3}{5}x' + \frac{4}{5}y'.$$

Substituting these expressions for $x$ and $y$ into the original equation yields

$$52\left[\frac{4}{5}x' - \frac{3}{5}y'\right]^2 - 72\left[\frac{4}{5}x' - \frac{3}{5}y'\right]\left[\frac{3}{5}x' + \frac{4}{5}y'\right] + 73\left[\frac{3}{5}x' + \frac{4}{5}y'\right]^2 = 200.$$

This becomes

$$52\left[\frac{16}{25}x'^2 - \frac{24}{25}x'y' + \frac{9}{25}y'^2\right] - 72\left[\frac{12}{25}x'^2 + \frac{7}{25}x'y' - \frac{12}{25}y'^2\right] + 73\left[\frac{9}{25}x'^2 + \frac{24}{25}x'y' + \frac{16}{25}y'^2\right] = 200.$$

Combining terms gives

$$25x'^2 + 100y'^2 = 200.$$

Divide each side by 200 to get

$$\frac{x'^2}{8} + \frac{y'^2}{2} = 1,$$

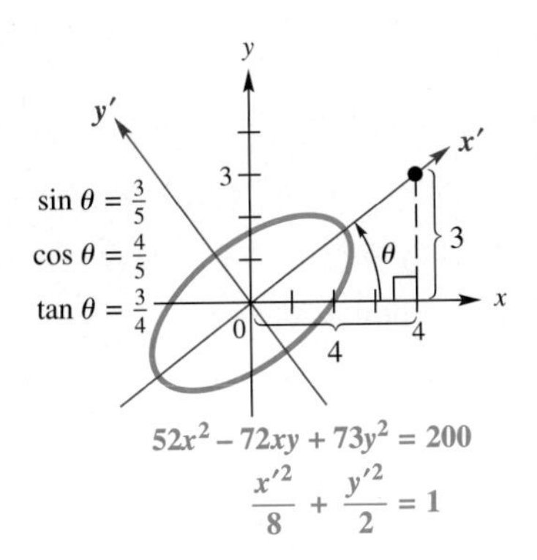


FIGURE 4

an equation of an ellipse having $x'$-intercepts $\pm 2\sqrt{2}$ and $y'$-intercepts $\pm\sqrt{2}$. The graph is shown in Figure 4. To find $\theta$, use the fact that

$$\frac{\sin\theta}{\cos\theta} = \frac{\frac{3}{5}}{\frac{4}{5}} = \frac{3}{4} = \tan\theta,$$

from which $\theta \approx 37°$. 

## Summary of Conics with an xy-Term

The following summary enables us to use the general equation to decide on the type of graph to expect.

**Equations of Conics with *xy*-Term**

If the general second-degree equation

$$Ax^2 + Bxy + Cy^2 + Dx + Ey + F = 0$$

has a graph, it will be one of the following:

**(a)** a circle or an ellipse (or a point) if $B^2 - 4AC < 0$;
**(b)** a parabola (or one line or two parallel lines) if $B^2 - 4AC = 0$;
**(c)** a hyperbola (or two intersecting lines) if $B^2 - 4AC > 0$;
**(d)** a straight line if $A = B = C = 0$, and $D \neq 0$ or $E \neq 0$.

## C APPENDIX C EXERCISES

*Use the summary in this section to predict the graph of each second-degree equation.*

**1.** $4x^2 + 3y^2 + 2xy - 5x = 8$
**2.** $x^2 + 2xy - 3y^2 + 2y = 12$
**3.** $2x^2 + 3xy - 4y^2 = 0$
**4.** $x^2 - 2xy + y^2 + 4x - 8y = 0$
**5.** $4x^2 + 4xy + y^2 + 15 = 0$
**6.** $-x^2 + 2xy - y^2 + 16 = 0$

*Find the angle of rotation $\theta$ that will remove the xy-term in each equation.*

**7.** $2x^2 + \sqrt{3}xy + y^2 + x = 5$
**8.** $4\sqrt{3}x^2 + xy + 3\sqrt{3}y^2 = 10$
**9.** $3x^2 + \sqrt{3}xy + 4y^2 + 2x - 3y = 12$
**10.** $4x^2 + 2xy + 2y^2 + x - 7 = 0$
**11.** $x^2 - 4xy + 5y^2 = 18$
**12.** $3\sqrt{3}x^2 - 2xy + \sqrt{3}y^2 = 25$

*Use the given angle of rotation to remove the xy-term and graph each equation.*

**13.** $x^2 - xy + y^2 = 6;\ \theta = 45°$
**14.** $2x^2 - xy + 2y^2 = 25;\ \theta = 45°$
**15.** $8x^2 - 4xy + 5y^2 = 36;\ \sin\theta = \frac{2}{\sqrt{5}}$
**16.** $5y^2 + 12xy = 10;\ \sin\theta = \frac{3}{\sqrt{13}}$

*Remove the xy-term from each equation by performing a suitable rotation. Graph each equation.*

**17.** $3x^2 - 2xy + 3y^2 = 8$
**18.** $x^2 + xy + y^2 = 3$
**19.** $x^2 - 4xy + y^2 = -5$
**20.** $x^2 + 2xy + y^2 + 4\sqrt{2}x - 4\sqrt{2}y = 0$

21. $7x^2 + 6\sqrt{3}xy + 13y^2 = 64$

22. $7x^2 + 2\sqrt{3}xy + 5y^2 = 24$

23. $3x^2 - 2\sqrt{3}xy + y^2 - 2x - 2\sqrt{3}y = 0$

24. $2x^2 + 2\sqrt{3}xy + 4y^2 = 5$

*In each equation, remove the xy-term by rotation. Then translate the axes and sketch the graph.*

25. $x^2 + 3xy + y^2 - 5\sqrt{2}y = 15$

26. $x^2 - \sqrt{3}xy + 2\sqrt{3}x - 3y - 3 = 0$

27. $4x^2 + 4xy + y^2 - 24x + 38y - 19 = 0$

28. $12x^2 + 24xy + 19y^2 - 12x - 40y + 31 = 0$

29. $16x^2 + 24xy + 9y^2 - 130x + 90y = 0$

30. $9x^2 - 6xy + y^2 - 12\sqrt{10}x - 36\sqrt{10}y = 0$

31. Look at the box titled "Angle of Rotation." Explain why no rotation is applicable if the value of $B$ is 0.

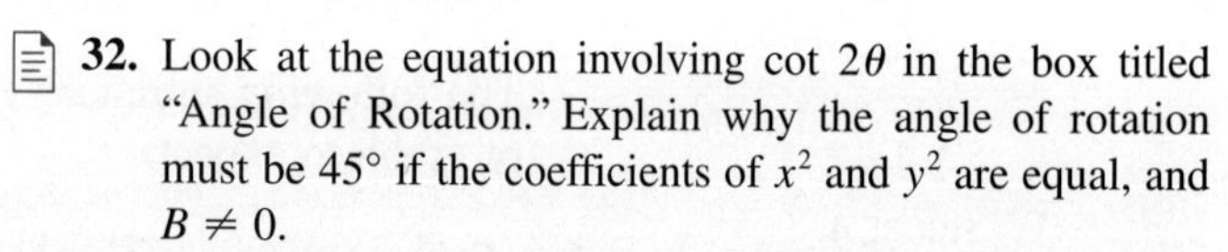

32. Look at the equation involving cot $2\theta$ in the box titled "Angle of Rotation." Explain why the angle of rotation must be 45° if the coefficients of $x^2$ and $y^2$ are equal, and $B \neq 0$.

# Answers to Selected Exercises

## To The Student

If you need further help with algebra, you may want to obtain a copy of the *Student's Solution Manual* that goes with this book. It contains solutions to all the odd-numbered section and Chapter Review exercises and all the Relating Concepts, Reviewing Basic Concepts, and Chapter Test exercises. Your college bookstore either has the *Manual* or can order it for you.

In this section we provide the answers that we think most students will obtain when they work the exercises using the methods explained in the text. If your answer does not look exactly like the one given here, it is not necessarily wrong. In many cases there are equivalent forms of the answer. For example, if the answer section shows $\frac{3}{4}$ and your answer is .75, you have obtained the correct answer but written it in a different (yet equivalent) form. Unless the directions specify otherwise, .75 is just as valid an answer as $\frac{3}{4}$. In general, if your answer does not agree with the one given in the text, see whether it can be transformed into the other form. If it can, then it is the correct answer. If you still have doubts, talk with your instructor.

## CHAPTER 1 LINEAR FUNCTIONS, EQUATIONS, AND INEQUALITIES

### 1.1 Exercises *(page 8)*

**1. (a)** 10 **(b)** 0, 10 **(c)** $-6, -\frac{12}{4}$ (or $-3$), 0, 10 **(d)** $-6, -\frac{12}{4}$ (or $-3$), $-\frac{5}{8}$, 0, .31, $.\overline{3}$, 10 **(e)** $-\sqrt{3}, 2\pi, \sqrt{17}$ **(f)** All are real numbers. **3. (a)** None are natural numbers. **(b)** None are whole numbers. **(c)** $-\sqrt{100}$ (or $-10$), $-1$ **(d)** $-\sqrt{100}$ (or $-10$), $-\frac{13}{6}$, $-1$, 5.23, $9.\overline{14}$, 3.14, $\frac{22}{7}$ **(e)** None are irrational numbers. **(f)** All are real numbers. **5.** −4 −3 −2 −1 0 1

**7.** −.5 0 .75 $\frac{5}{3}$ 3.5

**9.** A rational number can be written as a fraction $\frac{p}{q}$, $q \neq 0$, where $p$ and $q$ are integers, whereas an irrational number cannot.

**11.** I **13.** III **15.** none **17.** II **19.** none **21.** I or III **23.** II or IV **25.** $y$-axis **27.** Answers will vary. **29.** $[-5, 5]$ by $[-25, 25]$ **31.** $[-60, 60]$ by $[-100, 100]$ **33.** $[-500, 300]$ by $[-300, 500]$

Graph for Exercises 11–19:

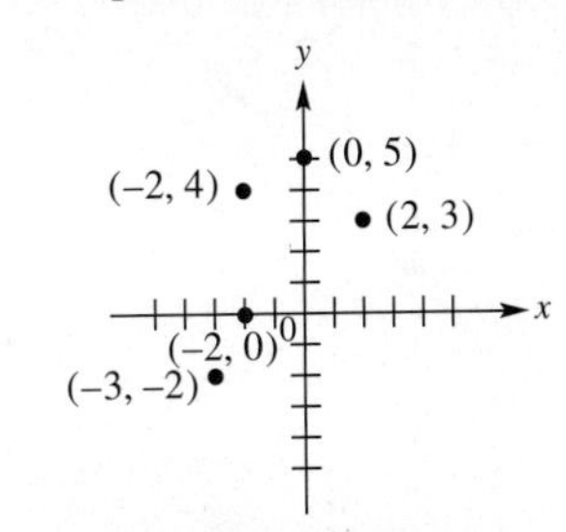


**35.** 10, −10, 10, −10

**37.** 10, −5, 10, −5

**39.**

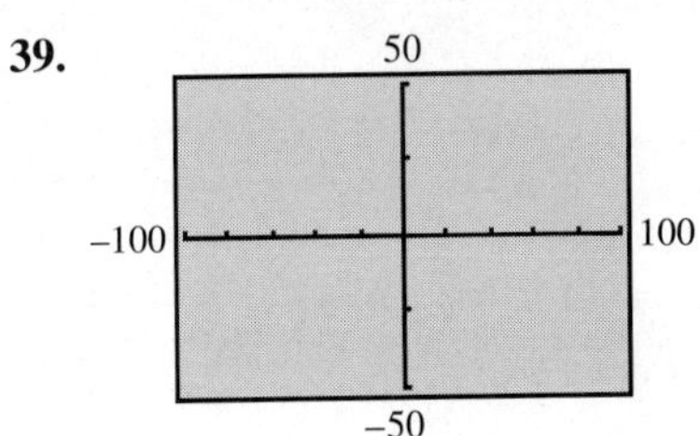

**41.** There are no tick marks. To set a screen with no tick marks on the axes, use Xscl = 0 and Yscl = 0. **43.** 7.616 **45.** 3.208 **47.** 3.045 **49.** 4.359 **51.** .25 **53.** 5.66 **55.** 98.63 **57.** 8.25 **59.** .72 **61.** $-2.82$ **63.** $c = 17$ **65.** $b = 84$ **67.** $c = \sqrt{89}$ **69.** $a = 4$ **71. (a)** $\sqrt{40} = 2\sqrt{10}$ **(b)** $(-1, 4)$ **73. (a)** $\sqrt{128} = 8\sqrt{2}$ **(b)** $(9, 3)$ **75. (a)** $\sqrt{34}$ **(b)** $\left(-\frac{11}{2}, -\frac{7}{2}\right)$ **77.** $12,698.50 **79.** 1965: $3495; 1985: $10,886.50 **81.** yes

**83. (a)**

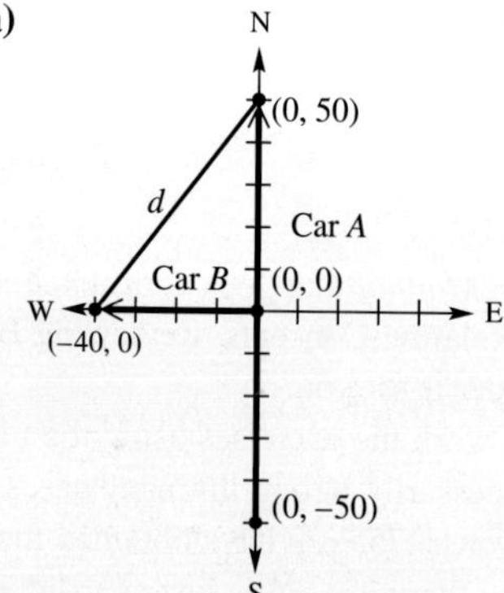


**(b)** $d = \sqrt{4100} \approx 64.0$ mi **85. (a)** $a + b$; $a + b$; $a^2 + 2ab + b^2$ **(b)** $\frac{1}{2}ab$; $2ab$; $c^2$ **(c)** $2ab + c^2$ **(d)** $a^2 + 2ab + b^2$; $2ab + c^2$ **(e)** $a^2 + b^2$; $c^2$

## 1.2 Exercises *(page 20)*

**1.** $(-1, 4)$ **3.** $(-\infty, 0)$ **5.** $[1, 2)$

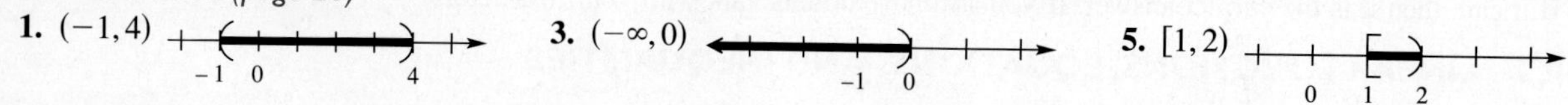


**7.** $\{x \mid -4 < x < 3\}$ **9.** $\{x \mid x \le -1\}$ **11.** $\{x \mid -2 \le x < 6\}$ **13.** $\{x \mid x \le -4\}$ **15.** Use a parenthesis if the symbol is $<$ or $>$, and use a square bracket if the symbol is $\le$ or $\ge$. **17.** domain: $\{5, 3, 4, 7\}$; range: $\{1, 2, 9, 6\}$; function **19.** domain: $\{4, 3, -2\}$; range: $\{1, -5, 3, 7\}$; not a function **21.** domain: $\{1, 2, 3, 4\}$; range: $\{-5\}$; function **23.** domain: $(-\infty, \infty)$; range: $(-\infty, \infty)$; function **25.** domain: $[-4, 4]$; range: $[-3, 3]$; not a function **27.** domain: $[2, \infty)$; range: $[0, \infty)$; function **29.** domain: $[-9, \infty)$; range: $(-\infty, \infty)$; not a function **31.** domain: $\{-5, -2, -1, -.5, 0, 1.75, 3.5\}$; range: $\{-1, 2, 3, 3.5, 4, 5.75, 7.5\}$; function **33.** domain: $\{2, 3, 5, 11, 17\}$; range: $\{1, 7, 20\}$; function **35.** 3 **37.** 7 **39.** $-6$ **41.** 5 **43.** 36 **45.** 10 **47.** $(-2, 3)$ **49.** 7; 8 **51. (a)** 0 **(b)** 4 **(c)** 2 **(d)** 4 **53. (a)** $-3$ **(b)** $-2$ **(c)** 0 **(d)** 2 **55. (a)–(d)** Answers will vary; see the definitions.

**57.**

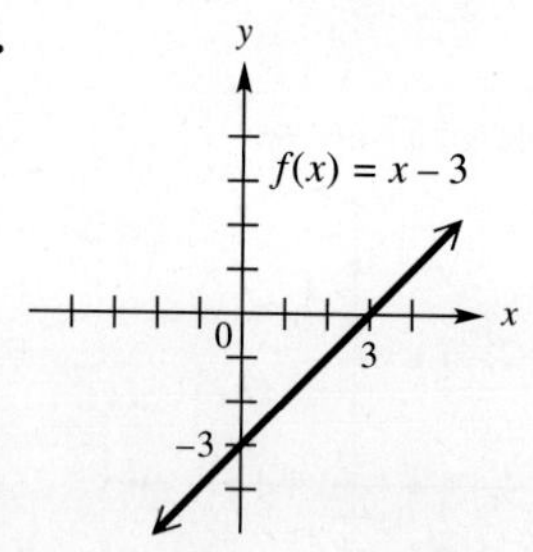


**59.**

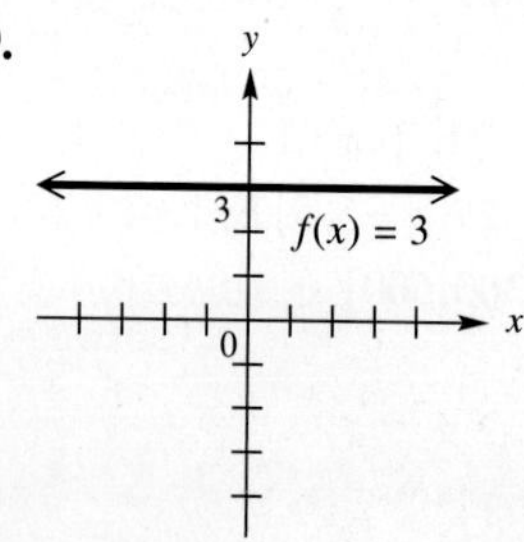


**61.**

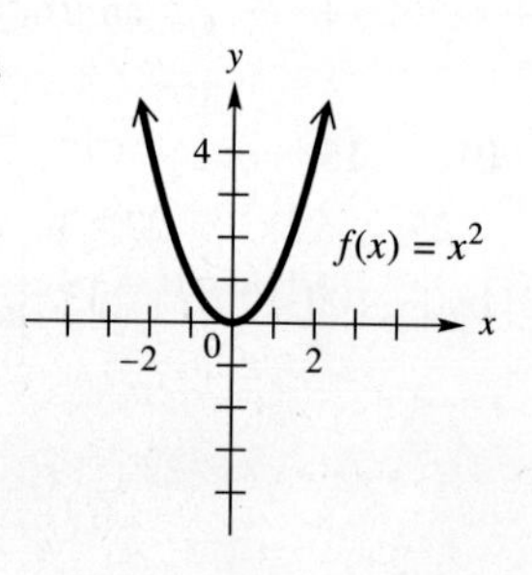


**63. (a)** $f(15) = 3$; if the delay is 15 sec, the lightning is 3 mi away. **(b)**

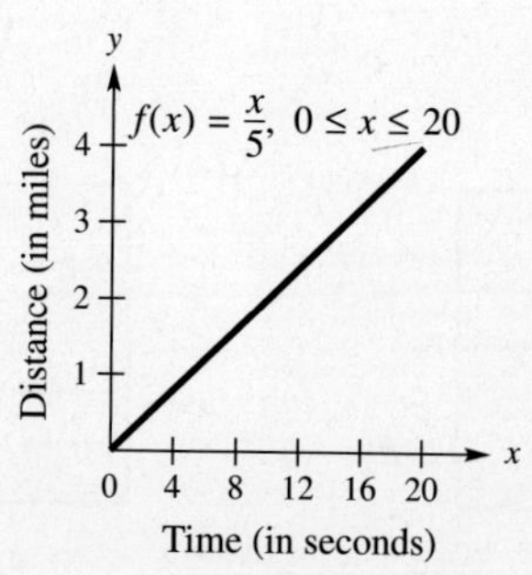

**65.** $f(x) = .075x; f(86) = \$6.45$ **67.** $f(x) = 92x + 75$; \$1087

### Reviewing Basic Concepts *(page 23)*

**1.**

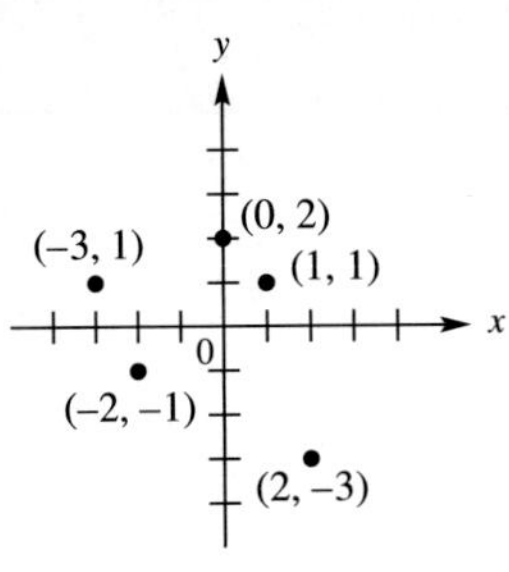


**2.** $d(P, Q) = \sqrt{149}; M = \left(1, \frac{3}{2}\right)$ **3.** 1.168 **4.** 60 in. **5.** $(-2, 5]; [4, \infty)$

**6.** not a function; domain: $[-2, 2]$; range: $[-3, 3]$ **7.** 23 **8.** $f(2) = 3; f(-1) = -3$ **9.**

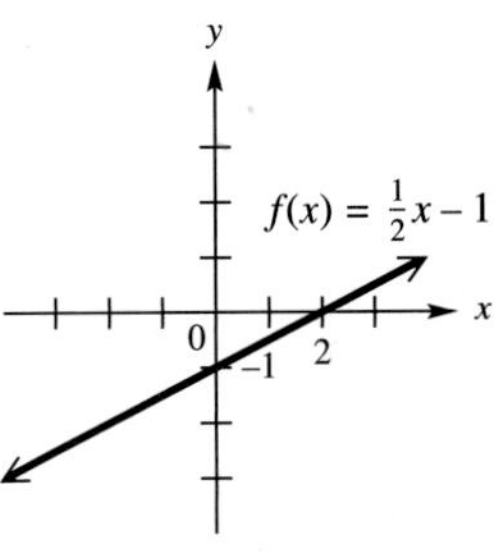


### 1.3 Exercises *(page 33)*

**1. (a)** $f(-2) = 0; f(4) = 6$ **(b)** $-2$
**(c)** The $x$-intercept is $-2$ and corresponds to the zero of $f$.

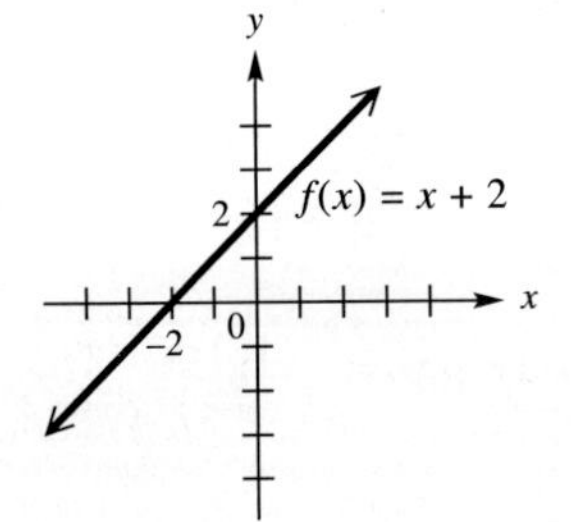


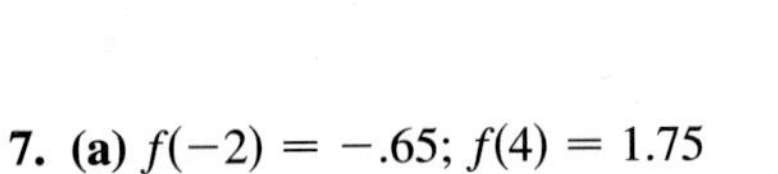

**3. (a)** $f(-2) = 8; f(4) = -10$ **(b)** $\frac{2}{3}$
**(c)** The $x$-intercept is $\frac{2}{3}$ and corresponds to the zero of $f$.

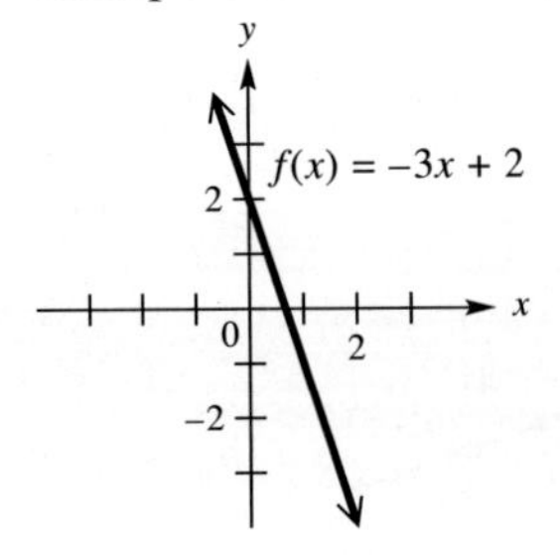


**5. (a)** $f(-2) = -\frac{2}{3}; f(4) = \frac{4}{3}$ **(b)** 0
**(c)** The $x$-intercept is 0 and corresponds to the zero of $f$.

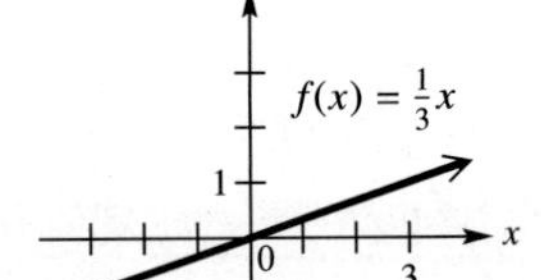


**7. (a)** $f(-2) = -.65; f(4) = 1.75$
**(b)** $-.375$ **(c)** The $x$-intercept is $-.375$ and corresponds to the zero of $f$.

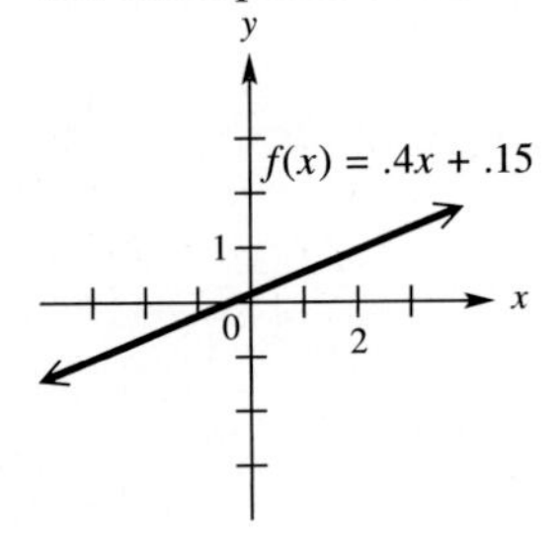


**9. (a)** $f(-2) = 1; f(4) = -\frac{1}{2}$ **(b)** 2
**(c)** The $x$-intercept is 2 and corresponds to the zero of $f$.

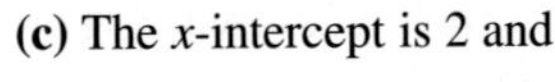

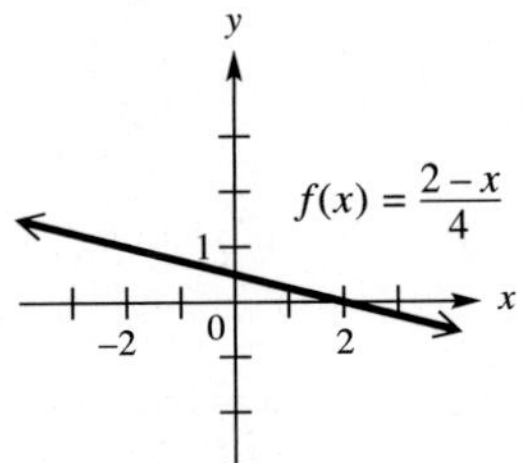


**11. (a)** 4
**(b)** $-4$
**(c)** $(-\infty, \infty)$
**(d)** $(-\infty, \infty)$
**(e)** 1

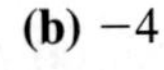

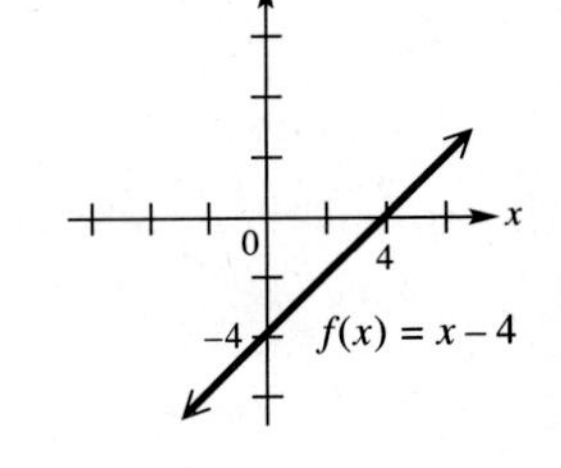

**13.** **(a)** 2 **(b)** $-6$ **(c)** $(-\infty, \infty)$ **(d)** $(-\infty, \infty)$ **(e)** 3

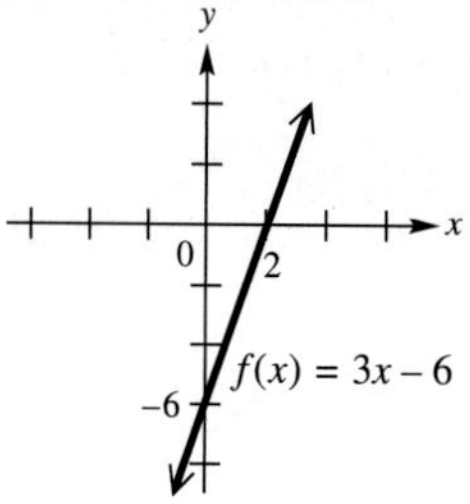

$f(x) = 3x - 6$

**15.** **(a)** 5 **(b)** 2 **(c)** $(-\infty, \infty)$ **(d)** $(-\infty, \infty)$ **(e)** $-\frac{2}{5}$

$f(x) = -\frac{2}{5}x + 2$

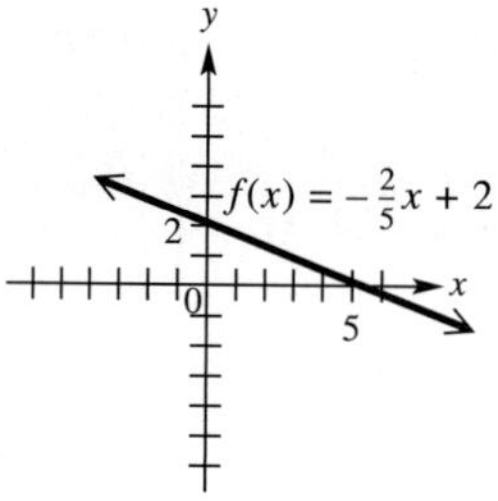

**17.** **(a)** 0 **(b)** 0 **(c)** $(-\infty, \infty)$ **(d)** $(-\infty, \infty)$ **(e)** 3

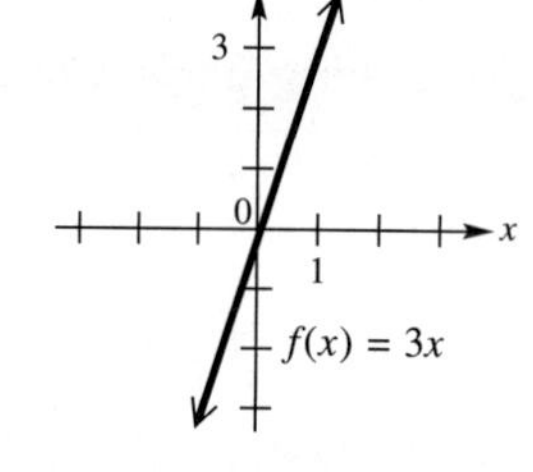


**19.** The point $(0, 0)$ must lie on the line.

**21.** **(a)** none **(b)** $-3$ **(c)** $(-\infty, \infty)$ **(d)** $\{-3\}$ **(e)** 0

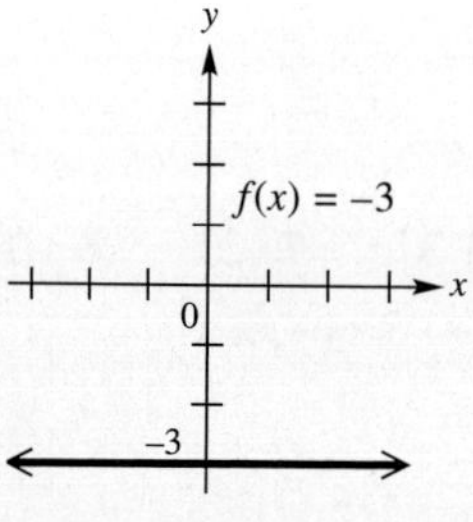

$f(x) = -3$

**23.** **(a)** $-1.5$ **(b)** none **(c)** $\{-1.5\}$ **(d)** $(-\infty, \infty)$ **(e)** undefined

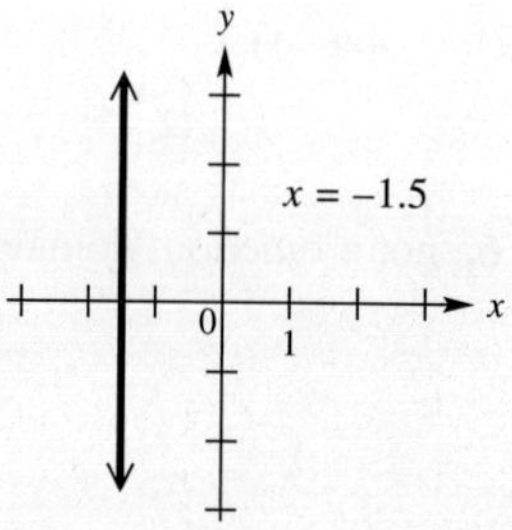

$x = -1.5$

**25.** **(a)** 2 **(b)** none **(c)** $\{2\}$ **(d)** $(-\infty, \infty)$ **(e)** undefined

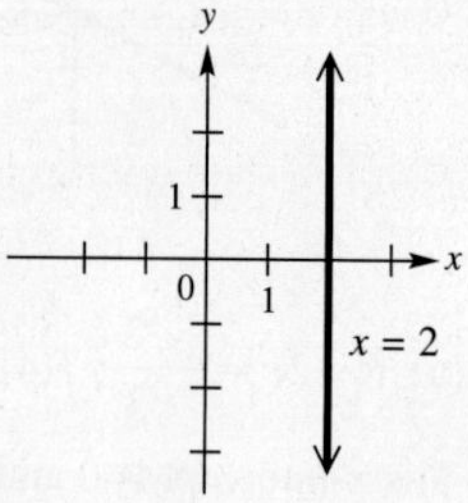

$x = 2$

**27.** constant function **29.** $y = 0$ **31.** window B **33.** window B

**35.** $\frac{1}{5}$ **37.** $\frac{7}{9}$ **39.** 0 **41.** 2 **43.** $-\frac{2}{3}$ **45.** 0 **47.** $m = 4$; $b = 2$; $Y_1 = 4X + 2$ **49.** **(a)** A **(b)** C **(c)** D **(d)** B

**51.**

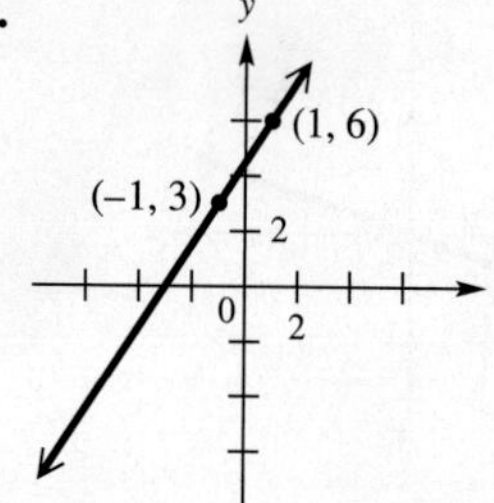


**53.**

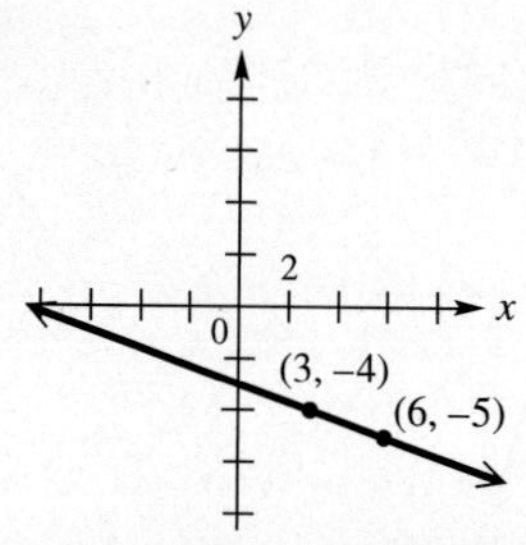


**55.**

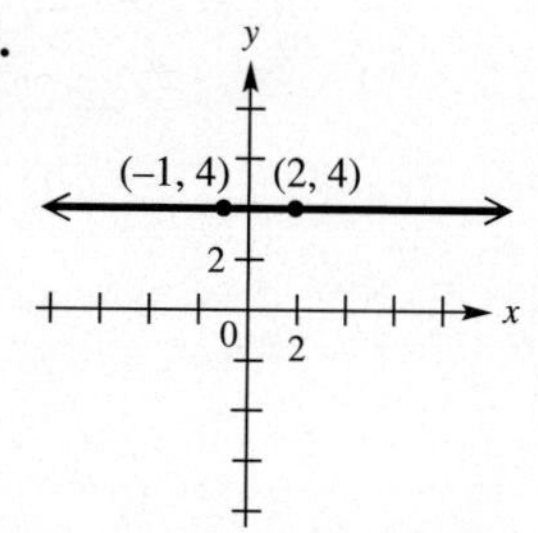


**57.**

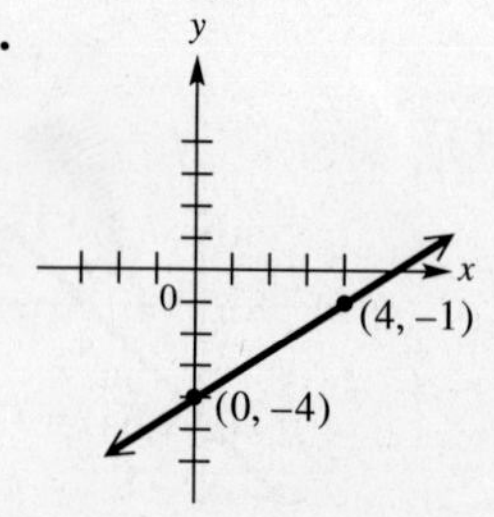


**59.** $y = \frac{3}{4}x - 4$ **61.** $f(x) = \frac{1}{4}x + 3$; 3.625 in.

**63.** **(a)** $f(x) = 500x + 2000$ **(b)** Water is entering the pool at 500 gal per hr; the pool has 2000 gal in it initially. **(c)** 5500 gal

## 1.4 Exercises *(page 46)*

**1.** $y = -2x + 5$ **3.** $y = -1.5x - 3.5$ **5.** $y = -.5x - 3$ **7.** $y = .8x - 2.8$ **9.** $y = -1.5x + 6.5$ **11.** $y = 3x + 8$ **13.** $y = 8x + 12$

*Note:* The graphs of the lines in Exercises 15–19 are the same as the graphs in Exercises 11–15 in Section 1.3. In the answers here, we give the $x$- and $y$-intercepts. Refer to the graphs given previously.

**15.** $x$: 4; $y$: $-4$ **17.** $x$: 2; $y$: $-6$ **19.** $x$: 5; $y$: 2 **21.**

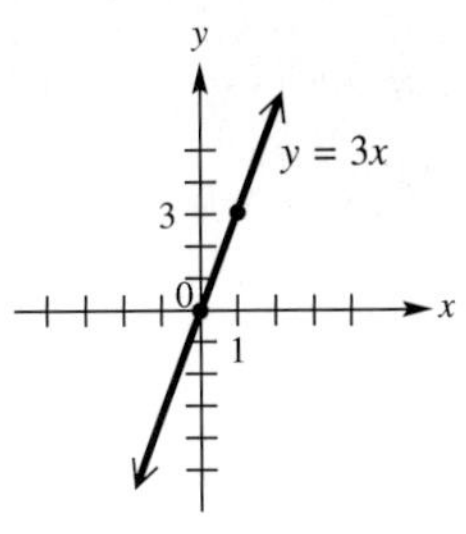


We used $(0, 0)$ and $(1, 3)$.

**23.**

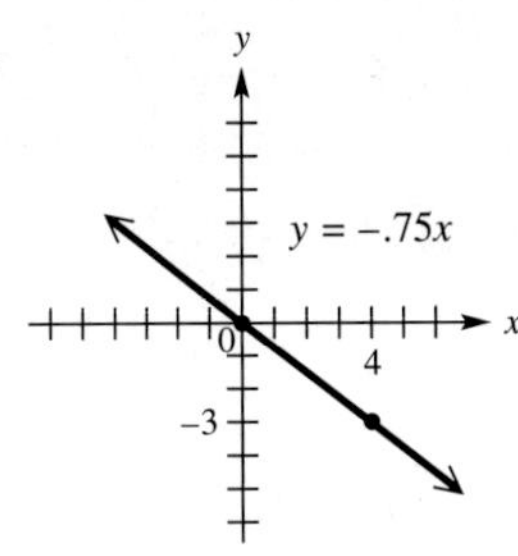


We used $(0, 0)$ and $(4, -3)$.

**25.** $y = -\frac{5}{3}x + 5$ **27.** $y = \frac{2}{7}x + \frac{4}{7}$ **29.** $y = -\frac{3}{4}x + \frac{25}{8}$ **31.** $y = -\frac{1}{3}x + \frac{11}{3}$ **33.** $y = \frac{5}{3}x + \frac{13}{3}$

**35.** $x = -5$ **37.** $y = -.2x + 7$ **39.** $y = \frac{1}{2}x$ **41.** They are *not* parallel. Line $y_1$ has slope 2.3, and line $y_2$ has slope 2.3001. Since $2.3 \neq 2.3001$, and nonvertical parallel lines must have *equal* slopes, they are not parallel. They only appear to be parallel in the figure because the slopes differ by .0001, a very small number. **43. (a)** the Pythagorean theorem and its converse **(b)** $\sqrt{x_1^2 + m_1^2x_1^2}$ **(c)** $\sqrt{x_2^2 + m_2^2x_2^2}$ **(d)** $\sqrt{(x_2 - x_1)^2 + (m_2x_2 - m_1x_1)^2}$ **(f)** $-2x_1x_2(m_1m_2 + 1) = 0$ **(g)** Since $x_1 \neq 0$ and $x_2 \neq 0$, this implies $m_2m_1 + 1 = 0$, so $m_1m_2 = -1$. **(h)** The product of the slopes of perpendicular lines, neither of which is parallel to an axis, is $-1$. **45.** 3 **46.** 3 **47.** equal **48.** $\sqrt{10}$ **49.** $2\sqrt{10}$ **50.** $3\sqrt{10}$ **51.** The sum is $3\sqrt{10}$, which is equal to the answer in Exercise 50. **52.** *B*; *C*; *A*; *C* **53.** The midpoint is $(3, 3)$, which is the same as the middle entry in the table. **54.** 7.5 **55. (a)** $y = 11x + 117$ **(b)** 11 mph **(c)** 117 mi **(d)** 130.75 mi

**57. (a)** $y = 280x + 1480$ **(b)** 4000

**59. (a)** a linear relation

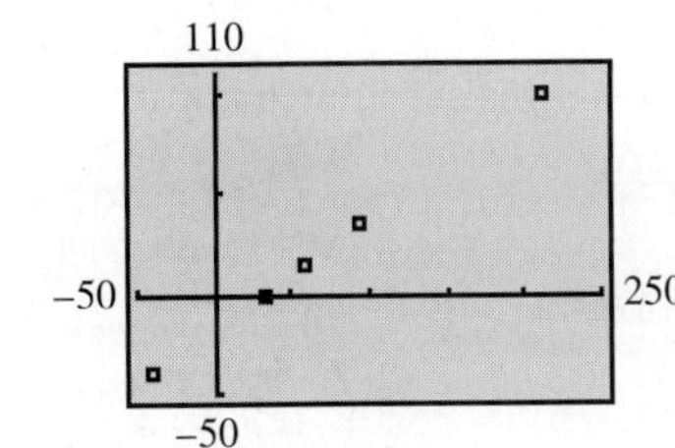


**(b)** $C(x) = \frac{5}{9}(x - 32)$; a slope of $\frac{5}{9}$ means that the Celsius temperature changes 5° for every 9° change in the Fahrenheit temperature.

**(c)** $28\frac{1}{3}$°C

**61. (a)** $y - 37.7 = \frac{223}{390}(x - 1960)$ **(b)** 1970: 43.4%; 1980: 49.1%; 1990: 54.9%; the value for 1970 nearly agrees with the table, but the other two values are slightly low. **63. (a)** $y \approx 635.042x - 1{,}254{,}358.2$ **(b)**

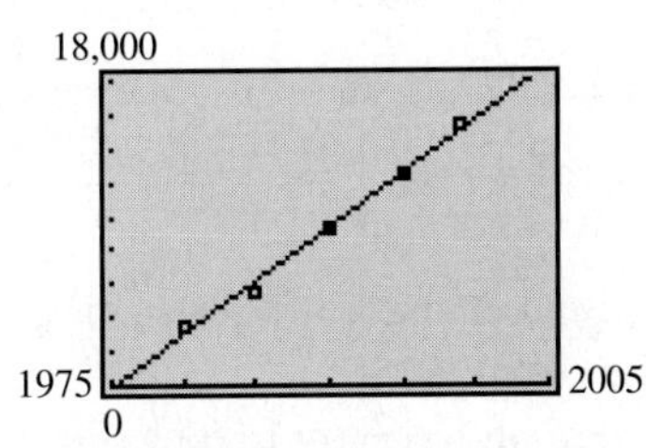


**(c)** It is about $10,645, which is quite close to the actual value of $10,498.

**65. (a)** $y \approx 14.68x + 277.82$ **(b)** about 2500 light-years **67.** $y \approx .101x + 11.6$; $r \approx .909$; strong, positive correlation since $r$ is close to 1

## Reviewing Basic Concepts *(page 51)*

**1.** $f(x) = 1.4x - 3.1; f(1.3) = -1.28$ **2.** $x$-intercept: $\frac{1}{2}$; $y$-intercept: 1; domain: $(-\infty, \infty)$; range: $(-\infty, \infty)$; slope: $-2$

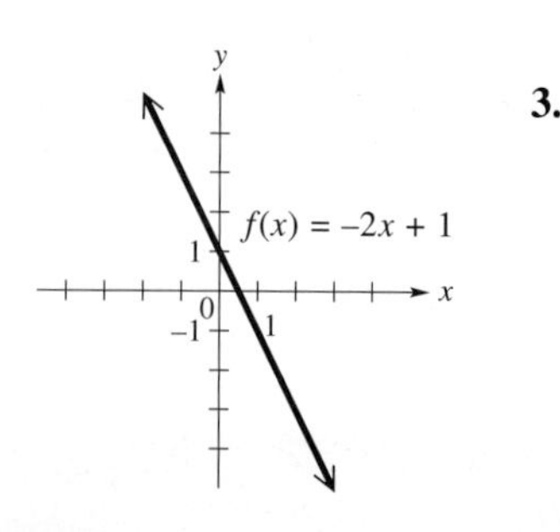


**3.** $\frac{2}{7}$

**4.** $x = -2; y = 10$ **5.**

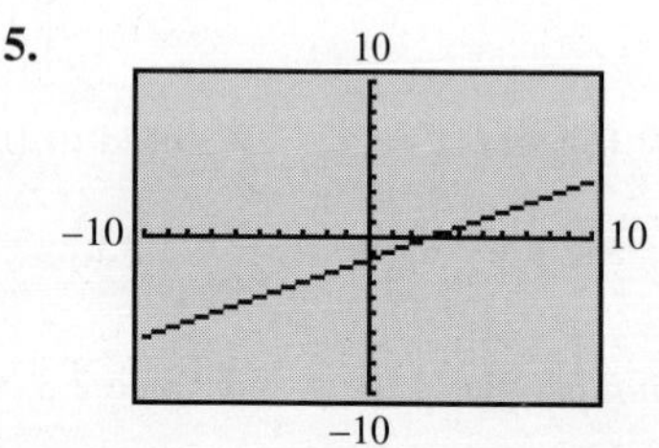


| X | Y1 |
|---|---|
| -3 | -2.9 |
| -2 | -2.4 |
| -1 | -1.9 |
| 0 | -1.4 |
| 1 | -.9 |
| 2 | -.4 |
| 3 | .1 |

Y1=.5X-1.4

**6.** $y = 2x - 3$ **7.** $y = -\frac{2}{7}x + \frac{24}{7}$ **8.** $y = -\frac{2}{3}x + \frac{7}{3}$

**9. (a)**

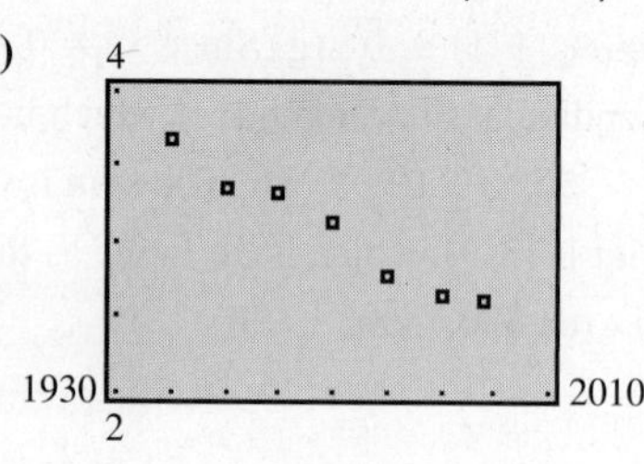


**(b)** negative **(c)** $y \approx -.01904x + 40.577$; $r \approx -.9792$ **(d)** 3.05; the estimate is close to the actual value of 2.94.

## 1.5 Exercises *(page 63)*

**1.** $-4$ **3.** 0 **5.** $-\frac{23}{19}$ **7.** $\frac{8}{3}$ **9.** $-\frac{29}{6}$ **11.** $\{10\}$ **13.** $\{-.8\}$ **15.** When $x = 10$ is substituted into $y_1$ or $y_2$, the result is $y = 20$. **17.** $\emptyset$; contradiction **19.** $\{12\}$ **21.** $\{3\}$ **23.** $\{-2\}$ **25.** $\{7\}$ **27.** $\{75\}$ **29.** $\{-1\}$ **31.** $\{0\}$ **33.** $\left\{\frac{5}{4}\right\}$ **35.** $\{4\}$ **37.** $\{1.5\}$ **39.** $\{16.07\}$ **41.** $\{-1.46\}$ **43.** $\{-3.92\}$ **45.** identity **47.** identity **49.** contradiction **51.** $\{3\}$ **53.** $(3, \infty)$ **55.** $(-\infty, 3]$ **57. (a)** $(20, \infty)$ **(b)** $(-\infty, 20)$ **(c)** $[20, \infty)$ **(d)** $(-\infty, 20]$ **59. (a)** $\{4\}$ **(b)** $(4, \infty)$ **(c)** $(-\infty, 4)$ **61. (a)** $(8, \infty)$ **(b)** $(-\infty, 8]$ **63. (a)** $(5, \infty)$ **(b)** $(-\infty, 5]$ **65. (a)** $(-\infty, -3]$ **(b)** $(-3, \infty)$ **67. (a)** $\left(-\frac{3}{4}, \infty\right)$ **(b)** $\left(-\infty, -\frac{3}{4}\right]$ **69.** $(-\infty, 15]$ **71.** $(-6, \infty)$ **73.** $[-8, \infty)$ **75.** $(25, \infty)$ **77. (a)** away, since distance is increasing **(b)** after 1 or 3 hr **(c)** $[1, 3]$ **(d)** $(1, 6]$ **79.** $[1, 4]$ **81.** $(-6, -4)$ **83.** $(-16, 19]$ **85.** $\left[\frac{3\sqrt{2} - 1}{2}, \frac{3\sqrt{5} - 1}{2}\right]$ **87.** $\{3\}$; one; one **88.** $(-\infty, 3)$; $(3, \infty)$; the value of $x = 3$ represents the boundary between the sets of real numbers given by $(-\infty, 3)$ and $(3, \infty)$. **89.** $\{1.5\}$; $(1.5, \infty)$; $(-\infty, 1.5)$ **90. (a)** $\left\{-\frac{b}{a}\right\}$ **(b)** $\left(-\infty, -\frac{b}{a}\right)$; $\left(-\frac{b}{a}, \infty\right)$ **(c)** $\left(-\frac{b}{a}, \infty\right)$; $\left(-\infty, -\frac{b}{a}\right)$ **91. (a)** below .65 mi or $[0, .65)$ **(b)** $\left[0, \frac{15}{23.2}\right)$ **93.** $3.98\pi \le C \le 4.02\pi$ **95. (a)** $f(x) = 3x - 1.5$ **(b)** $(1.25, \infty)$

## 1.6 Exercises *(page 73)*

**1.** $P = \frac{I}{RT}$ **3.** $W = \frac{P - 2L}{2}$ or $W = \frac{P}{2} - L$ **5.** $h = \frac{2A}{b_1 + b_2}$ **7.** $h = \frac{S - 2\pi r^2}{2\pi r}$ or $h = \frac{S}{2\pi r} - r$

**9.** $C = \frac{5}{9}(F - 32)$ **11.** 10 cm by 17 cm **13.** $\frac{13{,}079.6}{36} \approx 363.3$ yd **15.** $266\frac{2}{3}$ gal **17.** 2 L **19.** 4 L

**21.** $30\frac{1}{3}$ lb per in.$^2$ **23.** $51\frac{3}{7}$ ft **25.** 12,500 **27.** $k = 2.5$; $y = 20$ when $x = 8$ **29.** $k = .06$; $x = \$85$ when $y = \$5.10$ **31.** \$1048; $k = 65.5$ **33.** $14\frac{1}{6}$ in.

**35. (a)** $C(x) = .02x + 200$ **(b)** $R(x) = .04x$ **(c)** 10,000 **(d)** for $x < 10{,}000$, a loss; for $x > 10{,}000$, a profit

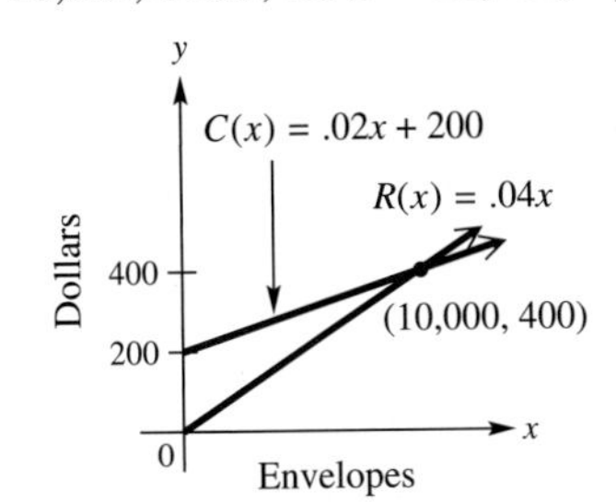


**37. (a)** $C(x) = 3.00x + 2300$ **(b)** $R(x) = 5.50x$ **(c)** 920 **(d)** for $x < 920$, a loss; for $x > 920$, a profit

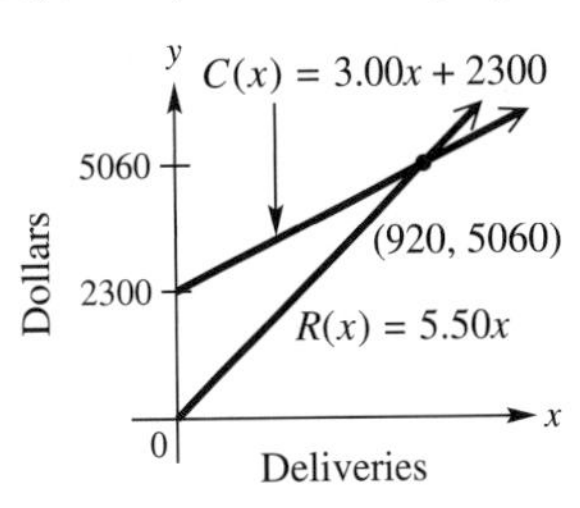


**39.** 96 **41.** \$160 **43.** 65 mph **45. (a)** $y = 640x + 1100$ **(b)** \$17,100 **(c)** Locate the point (25, 17,100) on the graph of $y = 640x + 1100$. **47. (a)** $y = 2600x + 120{,}000$ **(b)** \$156,400 **(c)** The value of the house increased, on average, by \$2600 per year. **49. (a)** $V(x) = 900x$ **(b)** $\frac{3}{50}x$ **(c)** $A = 2.4$ ach **(d)** $3.\overline{3}$ times **51. (a)** about .000021 for each individual **(b)** $C(x) = .000021x$ **(c)** about 2.1 cases **(d)** about 446,000 deaths per year **53. (a)** 500 cm$^3$ **(b)** 90°C **(c)** −273°C

## Reviewing Basic Concepts *(page 77)*

**1.** {2} **2.** {.757} **3.** $\left\{-\frac{1}{4}\right\}$ **4. (a)** contradiction **(b)** identity **(c)** conditional equation **5.** $\left(-\infty, -\frac{5}{7}\right)$

**6.** $\left(-\frac{5}{2}, 3\right]$ **7.** $\{2\}$; $[2, \infty)$ **8.** 40.5 ft **9. (a)** $R(x) = 5.5x$ **(b)** $C(x) = 1.5x + 2500$ **(c)** 625 discs **10.** $h = \frac{V}{\pi r^2}$

## Chapter 1 Review Exercises *(page 82)*

**1.** $\sqrt{612} = 6\sqrt{17}$ **3.** −4 **5.** $-\frac{3}{4}$ **7.** 36 **9.** 11 **11. (a)** −3 **(b)** $y = -3x + 2$ **(c)** (.5, .5) **(d)** $\sqrt{90} = 3\sqrt{10}$ **13.** C **15.** A **17.** E **19.** domain: $[-6, 6]$; range: $[-6, 6]$ **21.** I **23.** B **25.** I **27.** O **29.** $\left\{\frac{46}{7}\right\}$ **31.** $\left\{-\frac{5}{7}\right\}$ **33.** $\left\{\frac{7}{5}\right\}$ **35.** $(-\infty, 3)$ **37.** $\left(-\frac{10}{3}, \frac{46}{3}\right]$ **39.** $C(x) = 30x + 150$ **41.** 20

**43.** $f = \frac{AB(p + 1)}{24}$ **45. (a)** 41°F **(b)** about 21,000 ft **(c)** Graph $y = -3.52x + 58.6$. Then find the coordinates of the point where $x = 5$ to support the answer in (a). Finally, find the coordinates of the point where $y = -15$ to support the answer in (b).

**47. (a)** Answers will vary; linear regression gives $y \approx .08191x - 159.05$. **(b)** From the figure, we see that the line gives a reasonable approximation of the data. **(c)** Answers will vary; the regression line gives about \$4.52, which is too small.

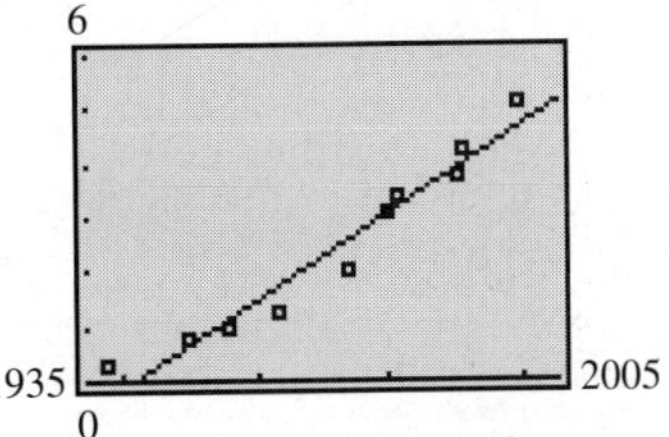


**49.** $y = 4x + 120$; 4 ft **51.** 1 hr, 8 min, 41.3 sec; this time is much faster. **53.** 7500 kg per m$^2$ **55. (a)** Using the first

the table gives $f(x) = \frac{5}{3}(x - 1994) + 504$; regression gives $y \approx 1.707x - 2899$. **(b)** 509; answers will

**59.** $[300, \infty)$; 300 videotapes or more

*(page 86)*

**(ii)** $[2, \infty)$ **(iii)** no $x$-intercepts **(iv)** 3 **(b) (i)** $(-\infty, \infty)$ **(ii)** $(-\infty, 0]$ **(iii)** 3 **(iv)** $-3$ **(c) (i)** $[-4, \infty)$ **(ii)** $[0, \infty)$ **(iii)** $-4$ **(iv)** 2 **2. (a)** $\{-4\}$ **(b)** $(-\infty, -4)$ **(c)** $[-4, \infty)$ **(d)** $\{-4\}$ **3. (a)** $\{5.5\}$ **(b)** $(-\infty, 5.5)$ **(c)** $(5.5, \infty)$ **(d)** $(-\infty, 5.5]$

**4. (a)** $\{-1\}$; the check leads to $-3 = -3$. **(b)**

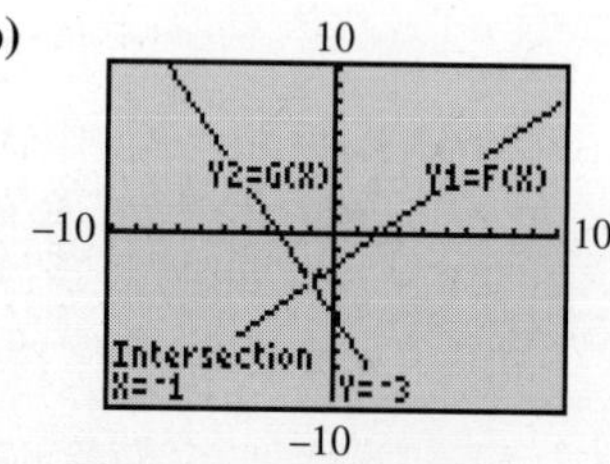


$(-1, \infty)$; the graph of $y_1 = f(x)$ is *above* the graph of $y_2 = g(x)$ for domain values greater than $-1$. **(c)** $(-\infty, -1)$; the graph of $y_1 = f(x)$ is *below* the graph of $y_2 = g(x)$ for domain values less than $-1$.

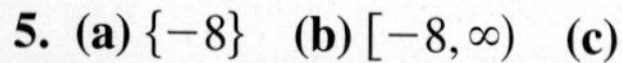

**5. (a)** $\{-8\}$ **(b)** $[-8, \infty)$ **(c)**

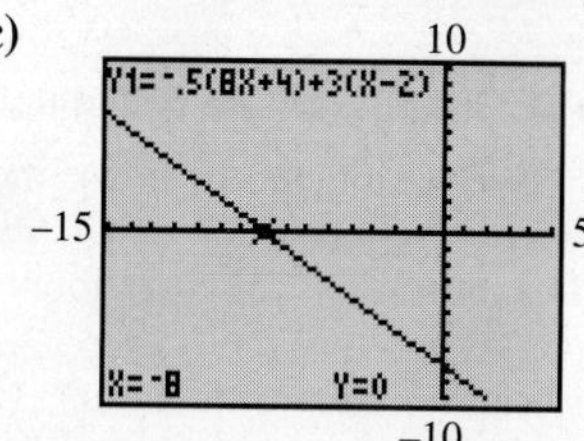


The $x$-intercept is $-8$, supporting the result of part (a). The graph of the linear function lies below or on the $x$-axis for domain values greater than or equal to $-8$, supporting the result of part (b).

**6. (a)** \$19.11 **(b)** $m = 1.1255$; This means that during the years 1980 to 2000, the monthly rate increased, on the average, by about \$1.13 per year. **7. (a)** $y = -2x - 1$ **(b)** $y = -\frac{1}{2}x + \frac{7}{2}$ **8.** $x$-intercept: 2; $y$-intercept: $-\frac{3}{2}$; slope: $\frac{3}{4}$ **9.** horizontal: $y = 7$; vertical: $x = -3$ **10. (a)** $y \approx -.65x + 32.7$; $r \approx -.98$ **(b)** $y = -.65(40) + 32.7 = 6.7$°F, which is 3.3°F lower than the actual value. **11. (a)** $57{,}600\pi \approx 180{,}956$ in.$^3$ **(b)** $g(x) = \frac{180{,}956}{231}x$ **(c)** 1958 gal **(d)** no; 2 downspouts

## CHAPTER 2 ANALYSIS OF GRAPHS OF FUNCTIONS

### 2.1 Exercises *(page 100)*

**1.** $(-\infty, \infty)$ **3.** $(0, 0)$ **5.** increases **7.** $x$-axis **9.** odd **11.** $(-\infty, \infty)$ **13.** $[0, \infty)$ **15.** $(-\infty, -3)$; $(-3, \infty)$ **17. (a)** $[3, \infty)$ **(b)** $(-\infty, 3]$ **(c)** none **(d)** $(-\infty, \infty)$ **(e)** $[0, \infty)$ **19. (a)** $(-\infty, 1]$ **(b)** $[4, \infty)$ **(c)** $[1, 4]$ **(d)** $(-\infty, \infty)$ **(e)** $(-\infty, 3]$ **21. (a)** none **(b)** $(-\infty, -2]$; $[3, \infty)$ **(c)** $(-2, 3)$ **(d)** $(-\infty, \infty)$ **(e)** $(-\infty, 1.5] \cup [2, \infty)$ **23.** increasing **25.** decreasing **27.** increasing **29. (a)** no **(b)** yes **(c)** no **31. (a)** yes **(b)** no **(c)** no **33. (a)** yes **(b)** yes **(c)** yes **35. (a)** no **(b)** no **(c)** yes **37.** $(-1.625, 2.0352051)$ **39.** $(.5, -.84089642)$ **41.** $(-5.092687, .9285541)$ **43.** $-12$; $-25$; 21 **45. (a)** **(b)**

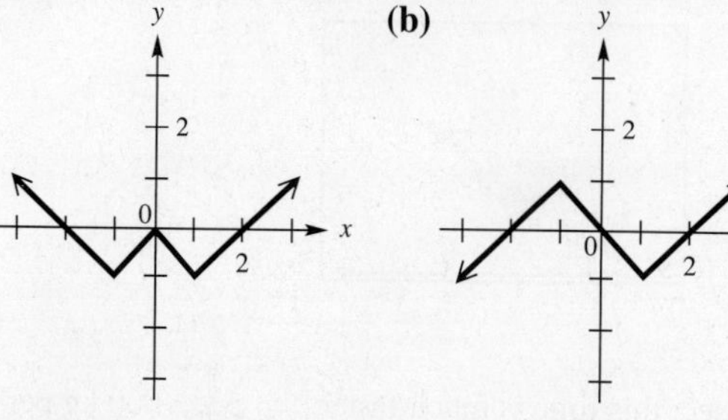


**47.** even

**53.** symmetric with respect to the origin **55.** symmetric with respect to the $y$-axis **57.** neither **59. (a)** Exercises 55 and 56 **(b)** Exercises 53 and 54 **(c)** Exercises 57 and 58

In Exercise 61 at the right, we give a calculator graph of what the sketch should look like.

**61.**

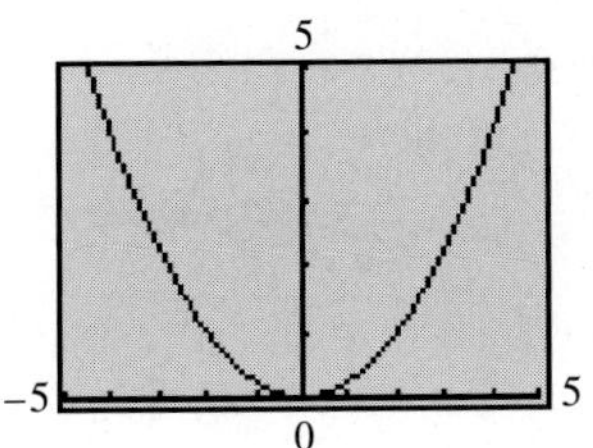


**63.** negative **64.** positive **65.** zero **66.** The slope of a tangent line is positive when the function is increasing, while the slope is negative when it is decreasing. The slope is zero when it changes from one to the other.

**67.** When $y = \dfrac{x^2 - 9}{x + 3}$ is graphed in the standard window, it may appear that there is a point for which $x = -3$. However, an error message will appear (or no $y$-value will be displayed) since the rational expression is not defined when $x = -3$.

## 2.2 Exercises *(page 110)*

**1.** $y = x^2 + 3$ **3.** $y = \sqrt{x} - 4$ **5.** $y = |x - 4|$ **7.** $y = (x + 7)^3$ **9.** To obtain the graph of $g$, shift the graph of $f$ 4 units upward. **11.** B **13.** A **15.** B **17.** C **19.** B **21. (a)** $(-\infty, \infty)$ **(b)** $[-3, \infty)$ **23. (a)** $(-\infty, \infty)$ **(b)** $[-3, \infty)$ **25.** 4 **27.** $-3$ **29.** B **31.** A **33.**

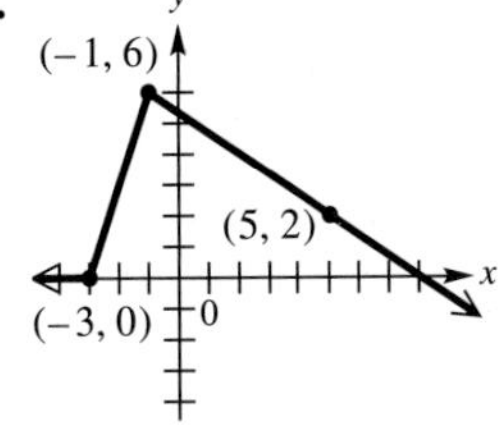


**35.**

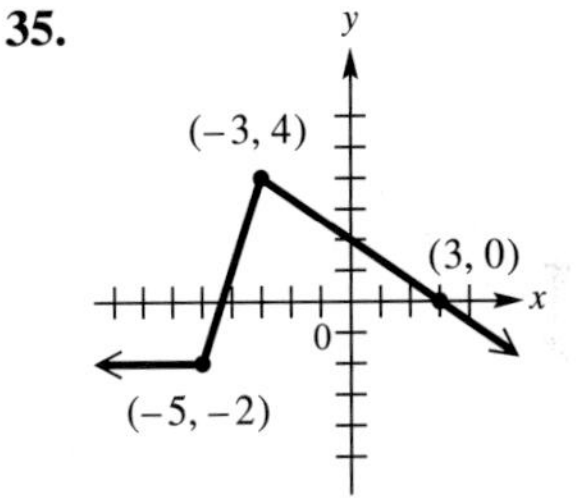


**37.** $y = (x + 4)^2 + 3$; **(a)** $[-4, \infty)$ **(b)** $(-\infty, -4]$ **39.** $y = x^3 - 5$; **(a)** $(-\infty, \infty)$ **(b)** none **41. (a)** $\{3, 4\}$ **(b)** $(-\infty, 3) \cup (4, \infty)$ **(c)** $(3, 4)$ **42. (a)** $\{\sqrt{2}\}$ **(b)** $(\sqrt{2}, \infty)$ **(c)** $(-\infty, \sqrt{2})$ **43. (a)** $\{-4, 5\}$ **(b)** $(-\infty, -4] \cup [5, \infty)$ **(c)** $[-4, 5]$ **44. (a)** $\emptyset$ **(b)** $[1, \infty)$ **(c)** $\emptyset$ **45.** $h = -3$; $k = 1$ **47.** $y = 661.4(x - 1995) + 5459.6$ **49. (a)** $y = 153.4x + 1101.6$ **(b)** $y = 153.4(x - 1984) + 1101.6$

**51.**

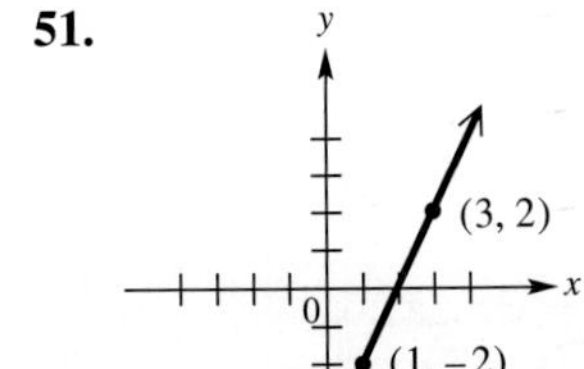
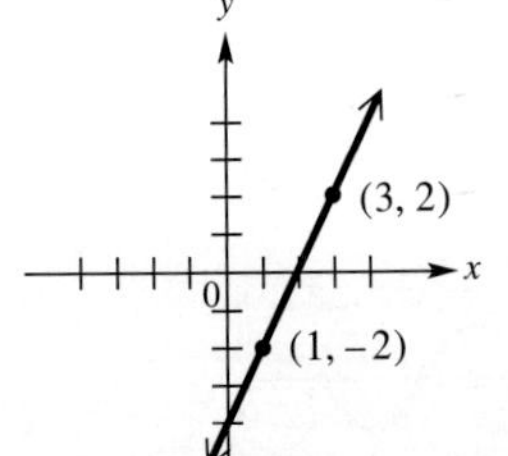


**52.** 2 **53.** $y_1 = 2x - 4$ **54.** $(1, 4)$ and $(3, 8)$ **55.** 2 **56.** $y_2 = 2x + 2$

**57.**

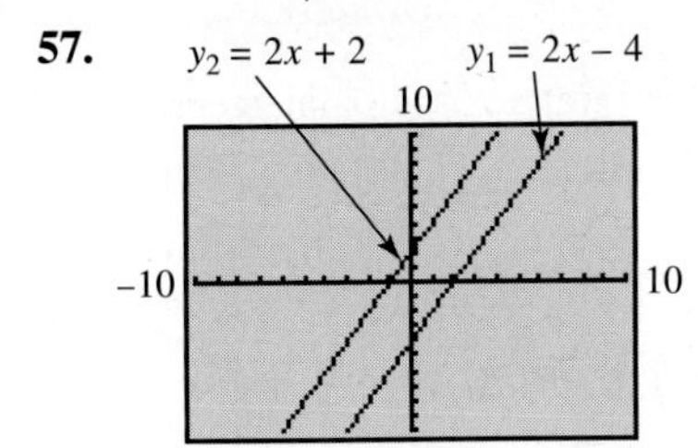


The graph of $y_2$ can be obtained by shifting the graph of $y_1$ upward 6 units. The constant, 6, comes from the 6 we added to each $y$-value in Exercise 54.

**58.** $c$; $c$; the same as; $c$; upward (or positive vertical)

## 2.3 Exercises *(page 120)*

**1.** $y = 2x^2$ **3.** $y = \sqrt{-x}$ **5.** $y = -3|x|$ **7.** $y = .25(-x)^3$ or $y = -.25x^3$ **9.**

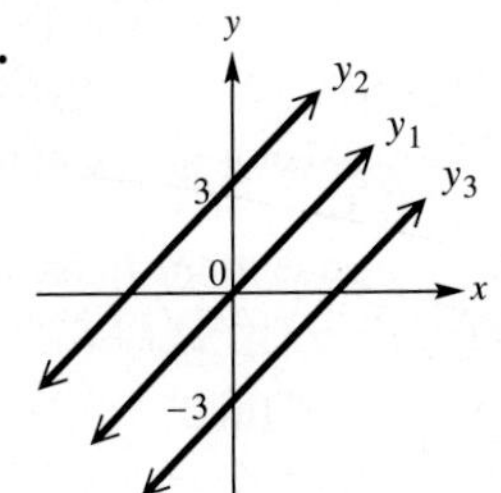

**11.**

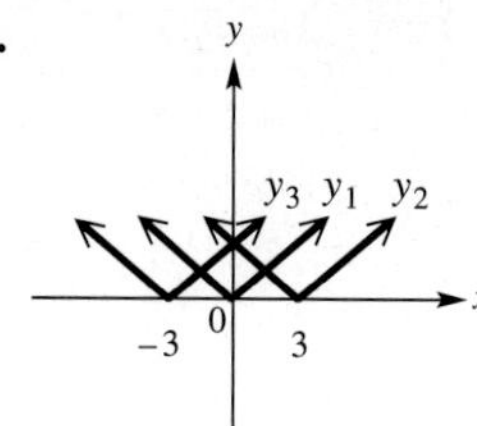


**13.**

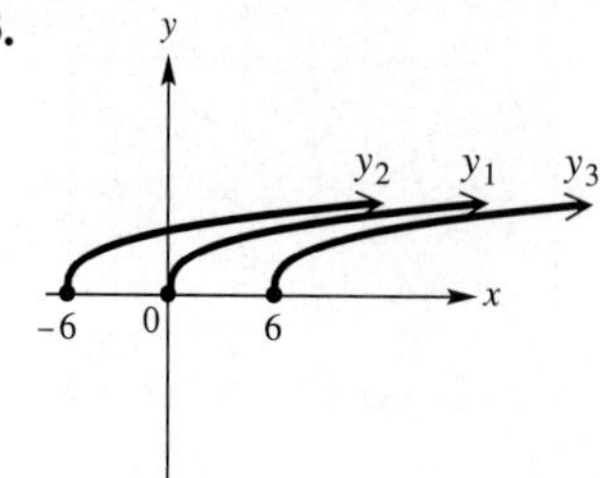


**15.**

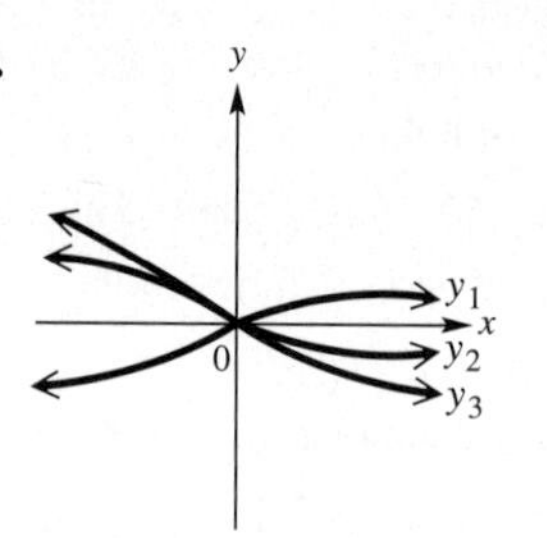


**17.**

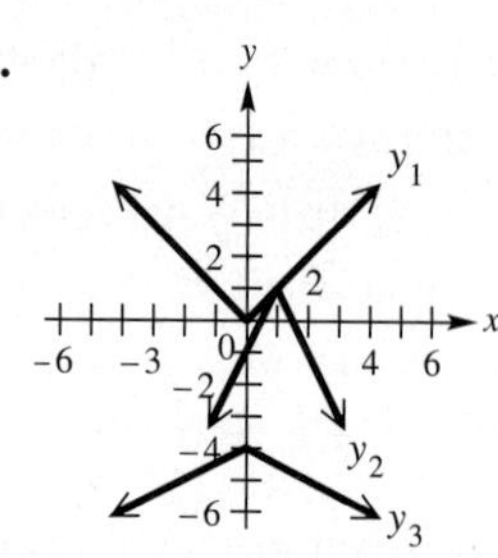


**19.** 4; $x$ **21.** 2; left; $\frac{1}{4}$; $x$; 3; downward (or negative) **23.** 3; right; 6 **25.** $y = \frac{1}{2}x^2 - 7$ **27.** $y = 4.5\sqrt{x-3} - 6$

**29.** F **31.** D **33.** B **35.** $g(x) = -(x-5)^2 - 2$

**37. (a)**

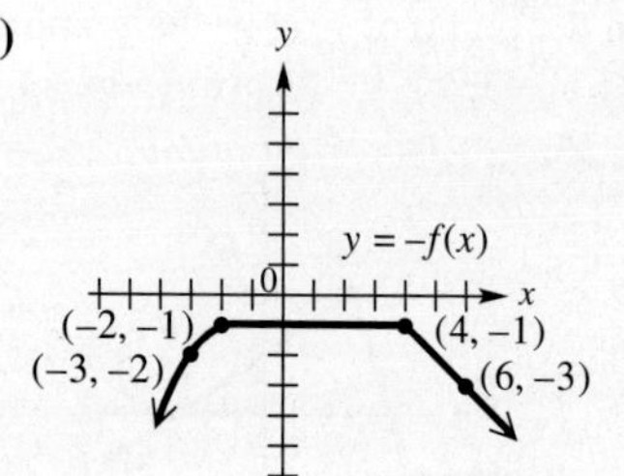


**(b)**

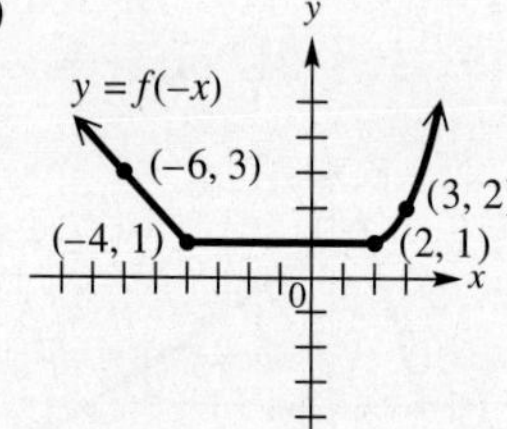


**(c)**

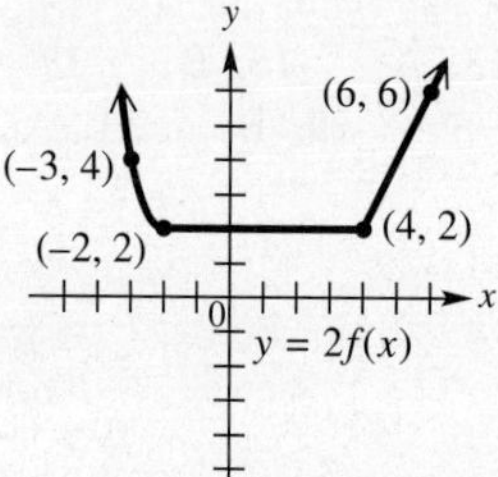


**(d)** $f(0) = 1$

**39. (a)**

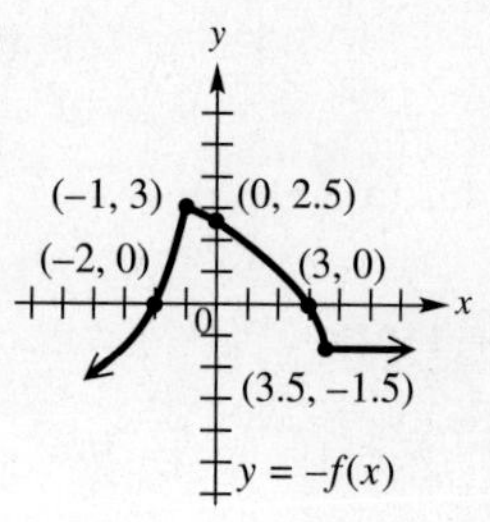


**(b)**

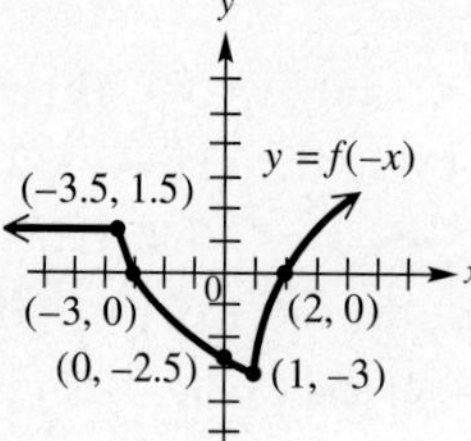


**(c)**

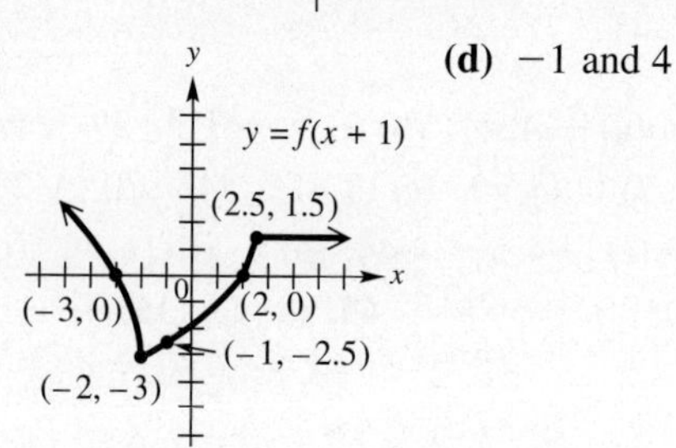


**(d)** −1 and 4

**41. (a)**

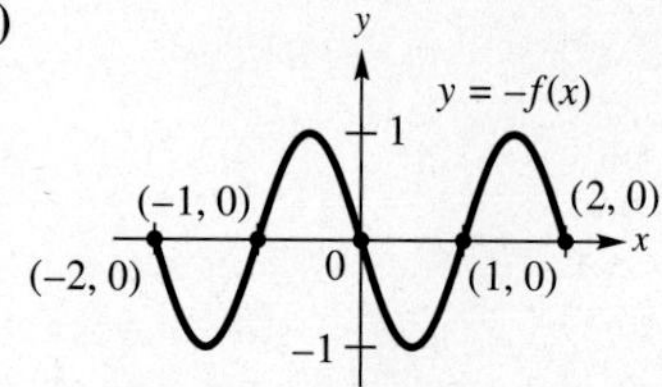


**(b)**

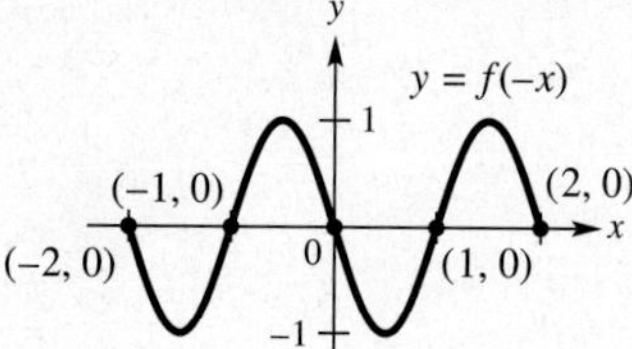


**(c)**

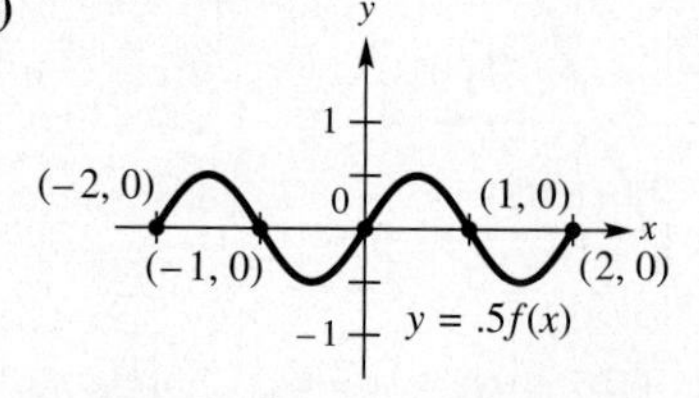


**(d)** symmetry with respect to the origin **43. (a)** $r$ is an $x$-intercept. **(b)** $-r$ is an $x$-intercept. **(c)** $-r$ is an $x$-intercept.

**45.** decreases **47.** increases **49. (a)** $[-1, 2]$ **(b)** $(-\infty, -1]$ **(c)** $[2, \infty)$ **51. (a)** $[1, \infty)$ **(b)** $[-2, 1]$ **(c)** $(-\infty, -2]$

**53.** domain: $[20, \infty)$; range: $[5, \infty)$ **55.** domain: $[-10, \infty)$; range: $(-\infty, 5]$ **57.** $(8, 10)$

## Reviewing Basic Concepts *(page 124)*

**1. (a)** $f(-3) = -6$ **(b)** $f(-3) = 6$ **(c)** $f(-3) = -6$ **(d)** $f(-3) = 6$ **2. (a)** B **(b)** D **(c)** E **(d)** A **(e)** C

**3. (a)** B **(b)** A **(c)** G **(d)** C **(e)** F **(f)** D **(g)** H **(h)** E **4. (a)** B **(b)** E **(c)** C **(d)** D **(e)** A

**5. (a)** It is the graph of $f(x) = |x|$ translated 1 unit to the left, reflected across the $x$-axis, and translated 3 units upward. The equation is $y = -|x+1| + 3$. **(b)** It is the graph of $g(x) = \sqrt{x}$ translated 4 units to the left, reflected across the $x$-axis, and translated 2 units upward. The equation is $y = -\sqrt{x+4} + 2$. **(c)** It is the graph of $g(x) = \sqrt{x}$ translated 4 units to the left, stretched vertically by a factor of 2, and translated 4 units downward. The equation is $y = 2\sqrt{x+4} - 4$. **(d)** It is the graph of $f(x) = |x|$ translated 2 units to the right, shrunken vertically by a factor of $\frac{1}{2}$, and translated 1 unit downward. The equation is $y = \frac{1}{2}|x-2| - 1$. **6. (a)** 2 **(b)** 4 **7.** The graph of $y = F(x+h)$ is a horizontal translation of the graph of $y = F(x)$.

The graph of $y = F(x) + h$ is not the same as the graph of $y = F(x + h)$ because the graph of $y = F(x) + h$ is a *vertical* translation of the graph of $y = F(x)$. **8.** The effect is either a stretch or a shrink, and perhaps a reflection across the $x$-axis. If $c > 0$, there is a stretch or shrink by a factor of $c$. If $c < 0$, there is a stretch or shrink by a factor of $|c|$, and a reflection across the $x$-axis. If $|c| > 1$, a stretch occurs; when $|c| < 1$, a shrink occurs.

**9. (a)**

| $x$ | $f(x)$ |
|---|---|
| $-3$ | 4 |
| $-2$ | $-6$ |
| $-1$ | 5 |
| 1 | 5 |
| 2 | $-6$ |
| 3 | 4 |

**(b)**

| $x$ | $f(x)$ |
|---|---|
| $-3$ | 4 |
| $-2$ | $-6$ |
| $-1$ | 5 |
| 1 | $-5$ |
| 2 | 6 |
| 3 | $-4$ |

**10.** $g(x) = f(x - 1982) = -.012053(x - 1980.06658)^2 + 9.07994$

## 2.4 Exercises *(page 133)*

**1.** 5 **3.** $[0, \infty)$ **5.** $[2, \infty)$ **7.**

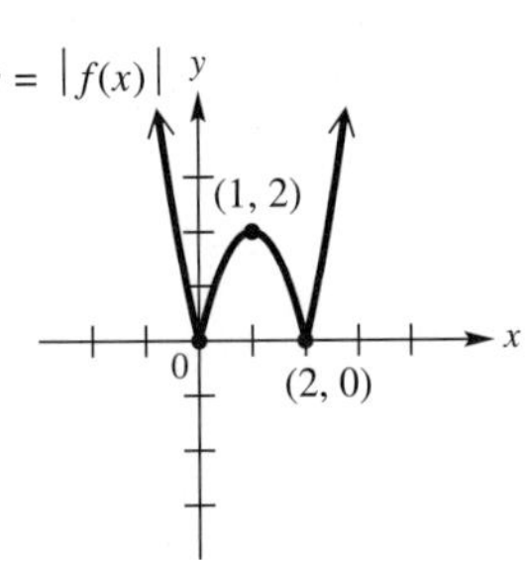


**9.**

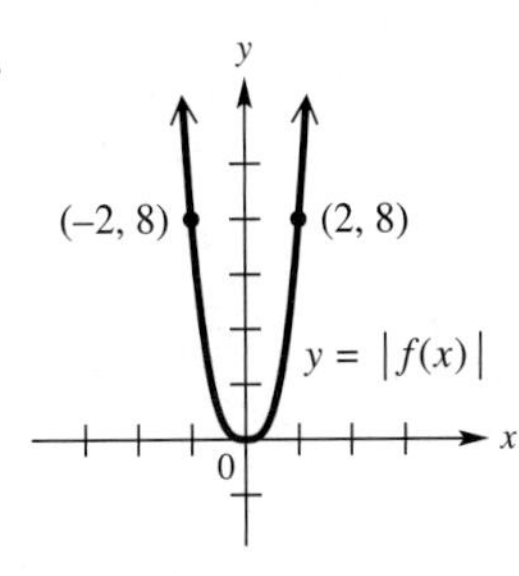


**11.** $y = |f(x)|$ has the same graph as $y = f(x)$. **13.**

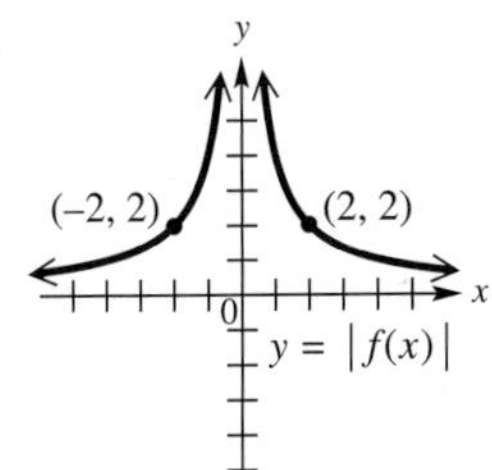


**15.**

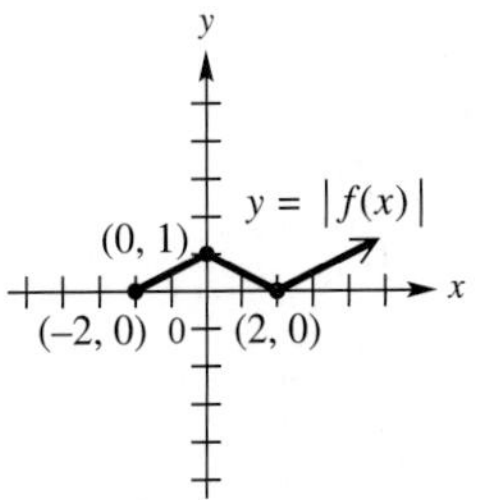


**17.** D **19.** C **21. (a)**

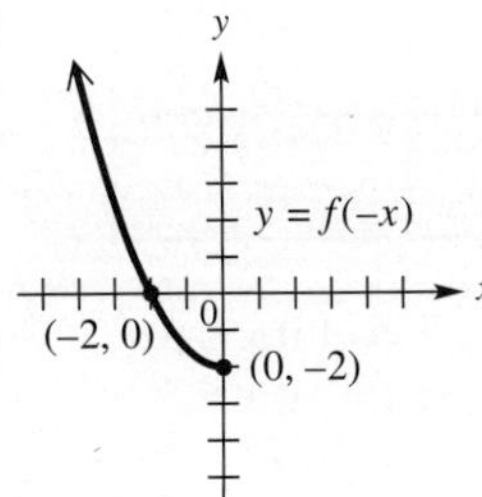


**(b)**

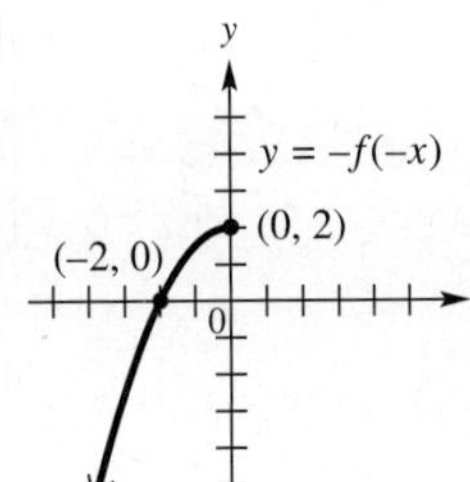


**(c)**

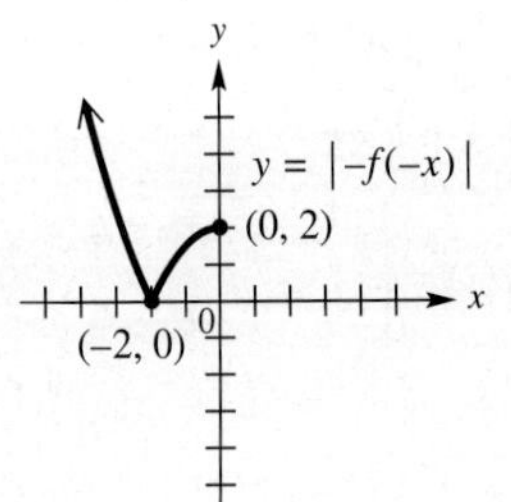


**23.** Figure A shows the graph of $y = f(x)$, while Figure B shows the graph of $y = |f(x)|$. **25. (a)** $\{-1, 6\}$ **(b)** $(-1, 6)$ **(c)** $(-\infty, -1) \cup (6, \infty)$ **27. (a)** $\{4\}$ **(b)** $\emptyset$ **(c)** $(-\infty, 4) \cup (4, \infty)$ **29.** The V-shaped graph is that of $f(x) = |.5x + 6|$, since this shape is typical of the graphs of absolute value functions of the form $f(x) = |ax + b|$. **30.** The straight line graph is that of $g(x) = 3x - 14$, which is a linear function. **31.** $\{8\}$ **32.** $(-\infty, 8)$ **33.** $(8, \infty)$ **34.** $\{8\}$ **35. (a)** $\{-13, 5\}$ **(b)** $(-\infty, -13) \cup (5, \infty)$ **(c)** $(-13, 5)$ **37. (a)** $\{2, 5\}$ **(b)** $(-\infty, 2] \cup [5, \infty)$ **(c)** $[2, 5]$ **39. (a)** $\left\{-\frac{3}{2}, \frac{1}{2}\right\}$ **(b)** $\left[-\frac{3}{2}, \frac{1}{2}\right]$ **(c)** $\left(-\infty, -\frac{3}{2}\right] \cup \left[\frac{1}{2}, \infty\right)$ **41. (a)** $\left\{\frac{5}{7}\right\}$ **(b)** $(-\infty, \infty)$ **(c)** $\left\{\frac{5}{7}\right\}$ **43. (a)** $\emptyset$ **(b)** $\emptyset$ **(c)** $(-\infty, \infty)$ **45. (a)** $\left\{-8, \frac{6}{5}\right\}$ **(b)** $(-\infty, -8) \cup \left(\frac{6}{5}, \infty\right)$ **(c)** $\left(-8, \frac{6}{5}\right)$ **47. (a)** $\left\{\frac{2}{3}, 8\right\}$ **(b)** $\left(-\infty, \frac{2}{3}\right) \cup (8, \infty)$ **(c)** $\left(\frac{2}{3}, 8\right)$

**49. (a)** $\left\{-3, \frac{5}{3}\right\}$ **(b)** $(-\infty, -3) \cup \left(\frac{5}{3}, \infty\right)$ **(c)** $\left(-3, \frac{5}{3}\right)$ **51. (a)** $\left\{-\frac{7}{8}\right\}$ **(b)** $\left(-\infty, -\frac{7}{8}\right)$ **(c)** $\left(-\frac{7}{8}, \infty\right)$

**53. (a)** $\{2, 8\}$ **(b)** $(2, 8)$ **(c)** $(-\infty, 2) \cup (8, \infty)$ **55.** $\{-3, 8\}$ **57.** $\{-2, 6\}$ **59. (a)** $19 \leq T \leq 67$ **(b)** The average monthly temperatures in Marquette vary between a low of 19°F and a high of 67°F. The monthly averages are always within 24° of 43°F. **61. (a)** $28 \leq T \leq 72$ **(b)** The average monthly temperatures in Boston vary between a low of 28°F and a high of 72°F. The monthly averages are always within 22° of 50°F. **63. (a)** $49 \leq T \leq 74$ **(b)** The average monthly temperatures in Buenos Aires vary between a low of 49°F (possibly in July) and a high of 74°F (possibly in January). The monthly averages are always within 12.5° of 61.5°F. **65. (a)** 9 **(b)** 113 or 147 **67.** $[6.5, 9.5]$ **69.** 42°F **71.** 36°F **73.** $(-.05, .05)$ **75.** $(2.74975, 2.75025)$ **77.** $\{2\}$ **79.** $(-\infty, \infty)$ **81.** $\emptyset$

## 2.5 Exercises *(page 144)*

**1. (a)** highest: 55 mph; lowest: 30 mph **(b)** 12 mi **(c)** $f(4) = 40$; $f(12) = 30$; $f(18) = 55$ **(d)** $x = 4, 6, 8, 12$, and 16. The speed limit changes at each discontinuity. **3. (a)** 50,000 gal; 30,000 gal **(b)** during the first and fourth days **(c)** $f(2) = 45$ thousand; $f(4) = 40$ thousand **(d)** 5000 gal per day **5. (a)** $-10$ **(b)** $-2$ **(c)** $-1$ **(d)** 2 **7. (a)** $-3$ **(b)** 1 **(c)** 0 **(d)** 9 **9.**

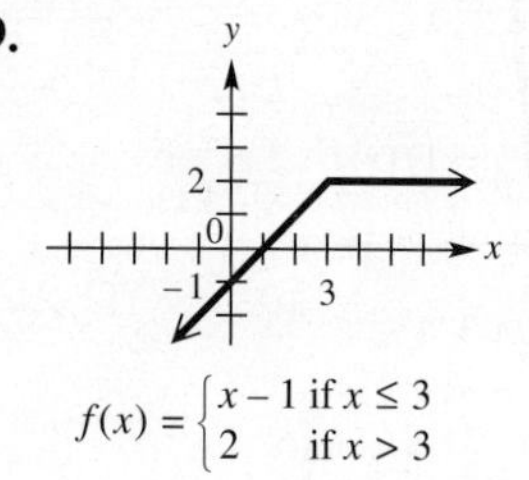


**11.**

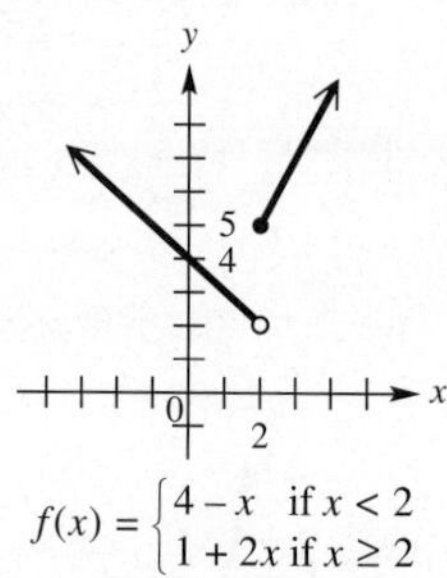


**13.**

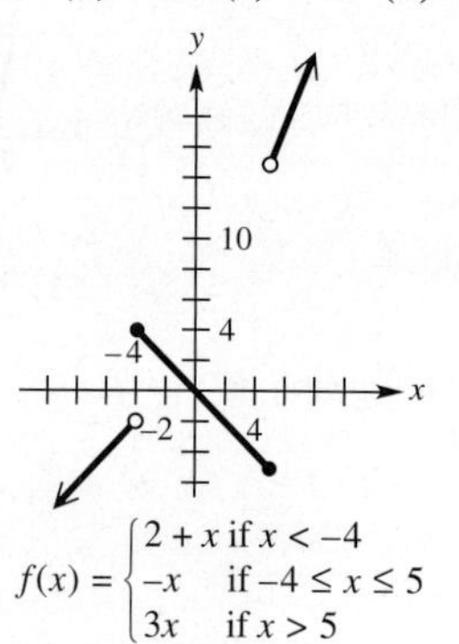


**15.**

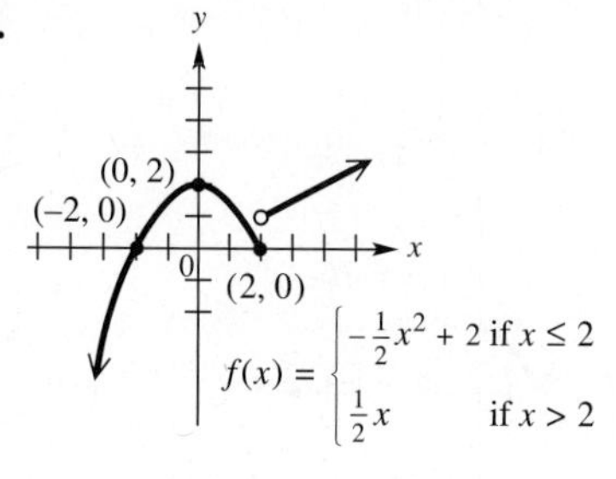


**17.** B **19.** D **21.**

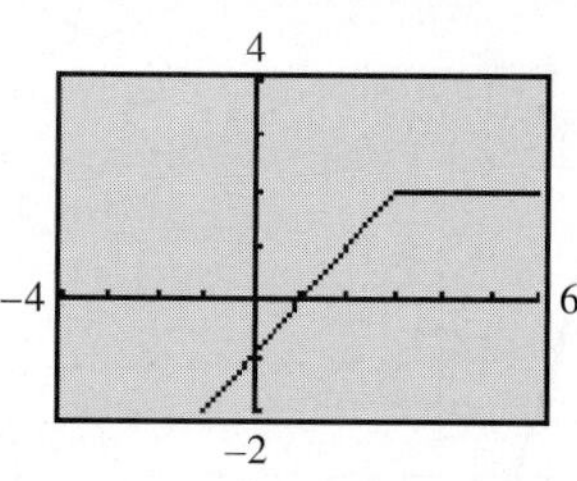


**23.**

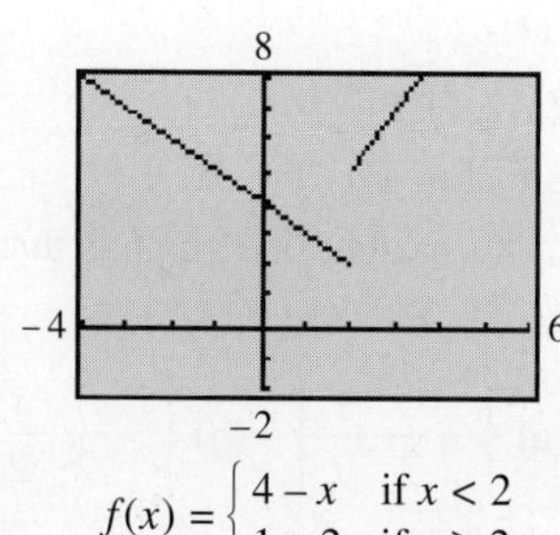


**25.**

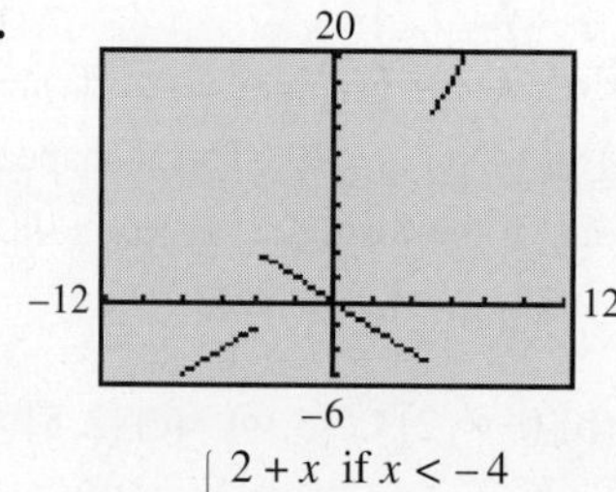


**27.**

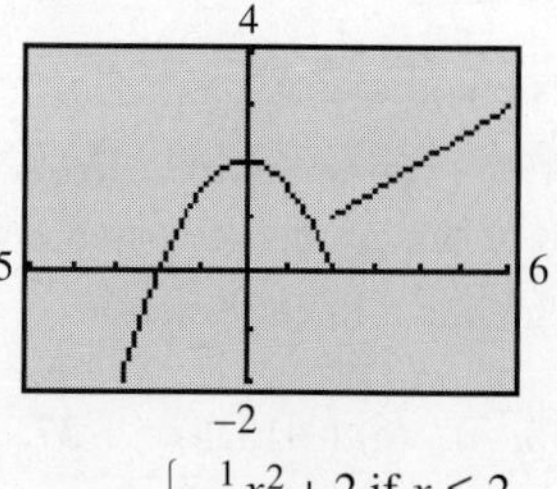


**29.** $f(x) = \begin{cases} 2 & \text{if } x \leq 0 \\ -1 & \text{if } x > 1 \end{cases}$; domain: $(-\infty, 0] \cup (1, \infty)$; range: $\{-1, 2\}$ **31.** $f(x) = \begin{cases} x & \text{if } x \leq 0 \\ 2 & \text{if } x > 0 \end{cases}$; domain: $(-\infty, \infty)$; range: $(-\infty, 0] \cup \{2\}$ **33.** $f(x) = \begin{cases} \sqrt[3]{x} & \text{if } x < 1 \\ x + 1 & \text{if } x \geq 1 \end{cases}$; domain: $(-\infty, \infty)$; range: $(-\infty, 1) \cup [2, \infty)$ **35.** There is an *overlap* of intervals since the number 4 satisfies both conditions. To be a function, every $x$-value is used only once. **37.** The graph of $y = [\![x]\!]$ is shifted 1.5 units downward. **39.** The graph of $y = [\![x]\!]$ is reflected across the $x$-axis.

**41.**

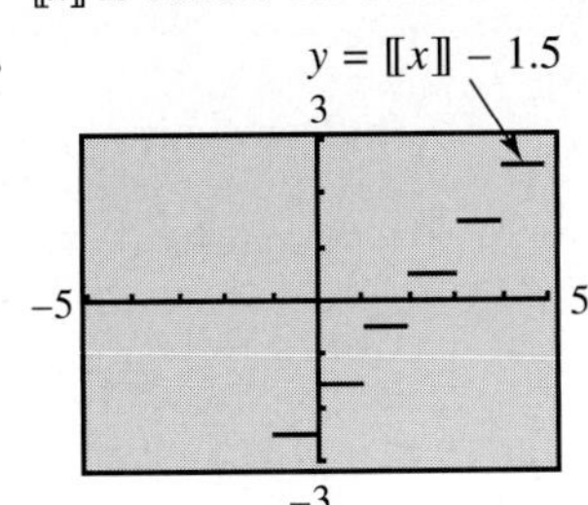


**43.**

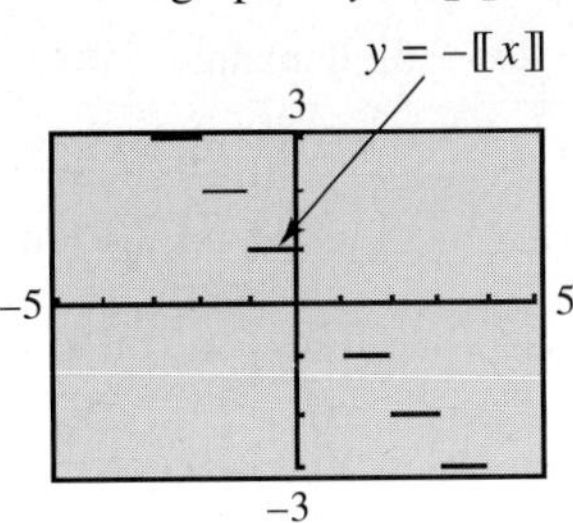


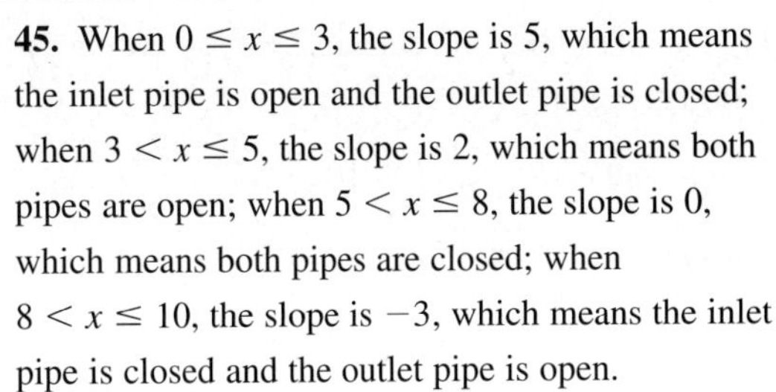

**45.** When $0 \le x \le 3$, the slope is 5, which means the inlet pipe is open and the outlet pipe is closed; when $3 < x \le 5$, the slope is 2, which means both pipes are open; when $5 < x \le 8$, the slope is 0, which means both pipes are closed; when $8 < x \le 10$, the slope is $-3$, which means the inlet pipe is closed and the outlet pipe is open.

**47.** **(a)** 140 **(b)** 220 **(c)** 220 **(d)** 220 **(e)** 220 **(f)** 60 **(g)** 60 **(h)**

**49.** **(a)** In each year, the number of cases increases. From 1988 to 1990, there is an increase of 100 cases per year. From 1990 to 1992, there is an increase of 1700 cases per year. **(b)** $f(x) = \begin{cases} 100x + 4800 & \text{if } 0 \le x \le 2 \\ 1700x + 1600 & \text{if } 2 < x \le 4 \end{cases}$

**51.** **(a)**

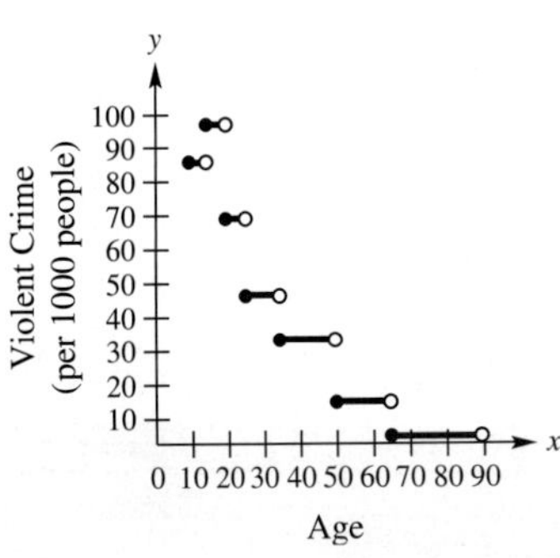


**(b)** The likelihood of being a victim of crime peaks from age 16 up to age 19, and then it decreases.

**53.** **(a)** \$1.25

**(b)** $f(x) = \begin{cases} .50 & \text{if } 0 < x \le 1 \\ .50 + .25[\![x]\!] & \text{if } 1 < x \le 5 \end{cases}$

**55.**

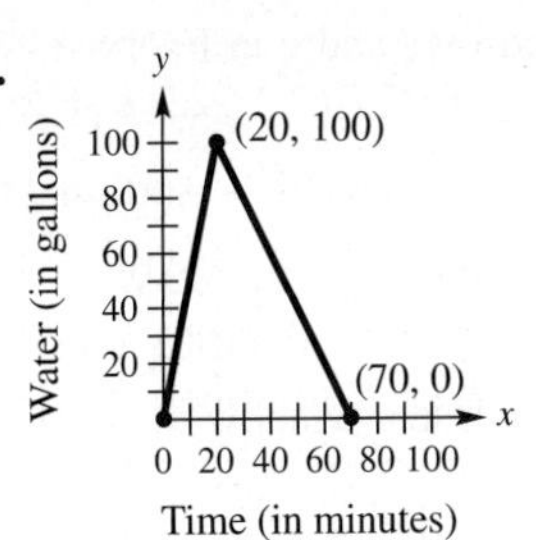


### 2.6 Exercises *(page 156)*

**1.** E **3.** F **5.** A **7.** 2450 **9.** $256x^2 + 48x + 2$ **11.** 55 **13.** 1848 **15.** $-\frac{6}{7}$ **17.** 1122 **19.** 97

**21.** **(a)** $(f + g)(x) = 10x + 2$; $(f - g)(x) = -2x - 4$; $(fg)(x) = 24x^2 + 6x - 3$ **(b)** Domain is $(-\infty, \infty)$ in all cases.
**(c)** $\left(\frac{f}{g}\right)(x) = \frac{4x - 1}{6x + 3}$; domain: $\left(-\infty, -\frac{1}{2}\right) \cup \left(-\frac{1}{2}, \infty\right)$ **(d)** $(f \circ g)(x) = 24x + 11$; domain: $(-\infty, \infty)$
**(e)** $(g \circ f)(x) = 24x - 3$; domain: $(-\infty, \infty)$ **23.** **(a)** $(f + g)(x) = |x + 3| + 2x$; $(f - g)(x) = |x + 3| - 2x$;

$(fg)(x) = |x + 3|(2x)$ **(b)** Domain is $(-\infty, \infty)$ in all cases. **(c)** $\left(\frac{f}{g}\right)(x) = \frac{|x+3|}{2x}$; domain: $(-\infty, 0) \cup (0, \infty)$
**(d)** $(f \circ g)(x) = |2x + 3|$; domain: $(-\infty, \infty)$ **(e)** $(g \circ f)(x) = 2|x + 3|$; domain: $(-\infty, \infty)$
**25.** **(a)** $(f + g)(x) = \sqrt[3]{x+4} + x^3 + 5$; $(f - g)(x) = \sqrt[3]{x+4} - x^3 - 5$; $(fg)(x) = \left(\sqrt[3]{x+4}\right)(x^3 + 5)$ **(b)** Domain is $(-\infty, \infty)$ in all cases. **(c)** $\left(\frac{f}{g}\right)(x) = \frac{\sqrt[3]{x+4}}{x^3+5}$; domain: $\left(-\infty, \sqrt[3]{-5}\right) \cup \left(\sqrt[3]{-5}, \infty\right)$ **(d)** $(f \circ g)(x) = \sqrt[3]{x^3 + 9}$; domain: $(-\infty, \infty)$
**(e)** $(g \circ f)(x) = x + 9$; domain: $(-\infty, \infty)$ **27.** **(a)** $(f + g)(x) = \sqrt{x^2+3} + x + 1$; $(f - g)(x) = \sqrt{x^2+3} - x - 1$; $(fg)(x) = \left(\sqrt{x^2+3}\right)(x + 1)$ **(b)** Domain is $(-\infty, \infty)$ in all cases. **(c)** $\left(\frac{f}{g}\right)(x) = \frac{\sqrt{x^2+3}}{x+1}$; domain: $(-\infty, -1) \cup (-1, \infty)$
**(d)** $(f \circ g)(x) = \sqrt{x^2 + 2x + 4}$; domain: $(-\infty, \infty)$ (*Note:* To see that this is the domain, graph $y = x^2 + 2x + 4$ and note that $y > 0$ for all $x$.) **(e)** $(g \circ f)(x) = \sqrt{x^2+3} + 1$; domain: $(-\infty, \infty)$ **29.** **(a)** 2 **(b)** 4 **(c)** 0 **(d)** $-\frac{1}{3}$ **31.** **(a)** 3 **(b)** $-5$ **(c)** 2 **(d)** undefined **33.** **(a)** 5 **(b)** 5 **(c)** 0 **(d)** undefined

**35.**

| $x$ | $(f+g)(x)$ | $(f-g)(x)$ | $(fg)(x)$ | $\left(\frac{f}{g}\right)(x)$ |
|---|---|---|---|---|
| −2 | 6 | −6 | 0 | 0 |
| 0 | 5 | 5 | 0 | undefined |
| 2 | 5 | 9 | −14 | −3.5 |
| 4 | 15 | 5 | 50 | 2 |

**37.** **(a)** $-4$ **(b)** 2 **(c)** $-4$

**39.** **(a)** $-3$ **(b)** $-2$ **(c)** 0 **41.** **(a)** 5 **(b)** undefined **(c)** 4 **43.** 4; 2 **45.** $-3$ **47.** 93
**53.** The graph of $y_2$ can be obtained by *reflecting* the graph of $y_1$ across the line $y_3 = x$.

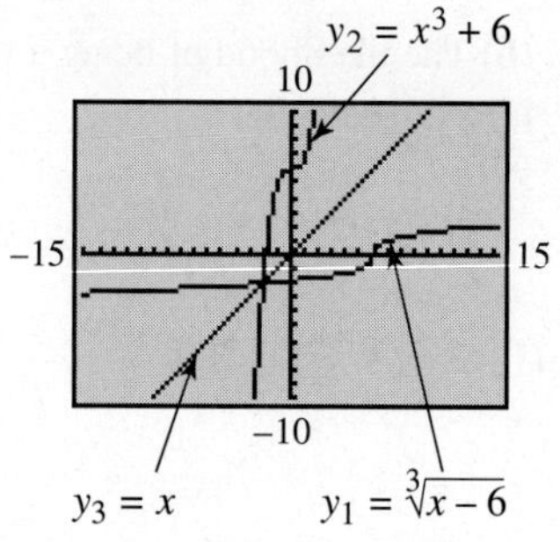


**55.** 4 **57.** $-12x - 1 - 6h$ **59.** $3x^2 + 3xh + h^2$
We give only one of many possible correct pairs of functions $f$ and $g$ in Exercises 61–65.
**61.** $f(x) = x^2$, $g(x) = 6x - 2$ **63.** $f(x) = \sqrt{x}$, $g(x) = x^2 - 1$ **65.** $f(x) = \sqrt{x} + 12$, $g(x) = 6x$
**67.** **(a)** $C(x) = 10x + 500$ **(b)** $R(x) = 35x$ **(c)** $P(x) = 35x - (10x + 500)$ or $P(x) = 25x - 500$ **(d)** 21 items **(e)** The smallest whole number for which $P(x) > 0$ is 21. Use a window of $[0, 30]$ by $[-1000, 500]$, for example.
**69.** **(a)** $C(x) = 100x + 2700$ **(b)** $R(x) = 280x$ **(c)** $P(x) = 280x - (100x + 2700)$ or $P(x) = 180x - 2700$ **(d)** 16 items **(e)** The smallest whole number for which $P(x) > 0$ is 16. Use a window of $[0, 30]$ by $[-3000, 500]$, for example.
**71.** **(a)** $V(r) = \frac{4}{3}\pi(r + 3)^3 - \frac{4}{3}\pi r^3$ **(b)**

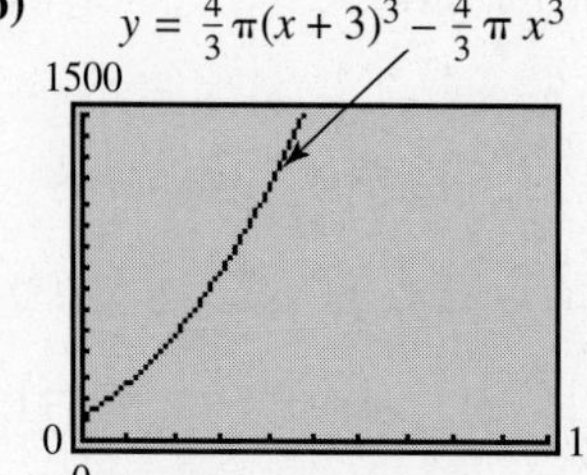


This appears to be a portion of a parabola, formed by translating the squaring function. **(c)** 1168.67 in.$^3$
**(d)** $V(4) = \frac{4}{3}\pi(7)^3 - \frac{4}{3}\pi(4)^3 \approx 1168.67$

**73.** **(a)** $P = 6x$; $P(x) = 6x$; This is a linear function. **(b)** 4 represents the width of the rectangle and 24 represents the perimeter.

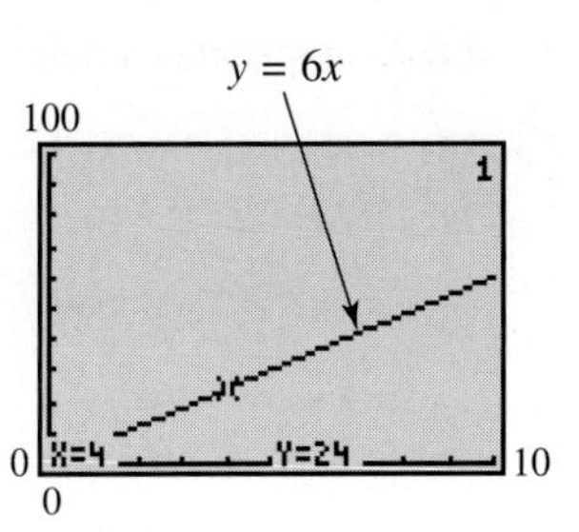


**(c)** (See graph for part (b).) perimeter $= 24$; This is the $y$-value shown on the screen for the integer $x$-value 4.

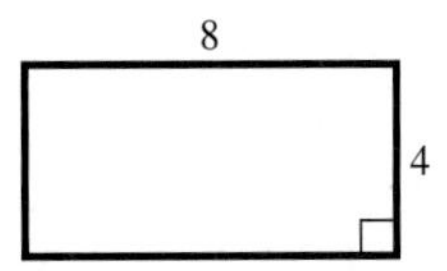


**(d)** (Answers may vary.) If the perimeter $y$ of a rectangle satisfying the given conditions is 36, then the width $x$ is 6.

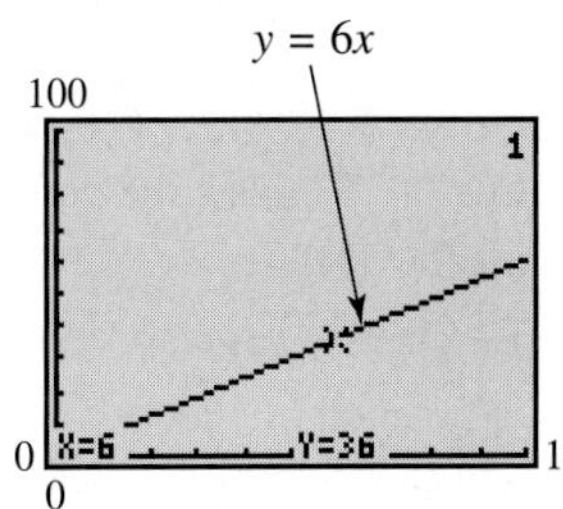


**75.** **(a)** $A(2x) = \sqrt{3}x^2$ **(b)** $A(16) = 64\sqrt{3}$ square units **(c)** On the graph of $y = \frac{\sqrt{3}}{4}x^2$, locate the point where $x = 16$ to find $y \approx 110.85$, an approximation for $64\sqrt{3}$.

**77.** **(a)**

| $x$ | 1991 | 1992 | 1993 | 1994 | 1995 | 1996 |
|---|---|---|---|---|---|---|
| $h(x)$ | 38 | 42 | 44 | 50 | 54 | 61 |

**(b)** $h(x) = f(x) + g(x)$

**79.** **(a)** 50 **(b)** $(f + g)(x)$ computes the total $SO_2$ emissions from burning coal and oil during year $x$.

**(c)**

| $x$ | 1860 | 1900 | 1940 | 1970 | 2000 |
|---|---|---|---|---|---|
| $(f + g)(x)$ | 2.4 | 12.8 | 26.5 | 50.0 | 78.0 |

**81.** **(a)** $h(x) = g(x) - f(x)$ **(b)** $h(1995) = 200$; $h(2000) = 250$ **(c)** $h(x) = 10(x - 1995) + 200$ or $h(x) = 10x - 19{,}750$

## Reviewing Basic Concepts *(page 161)*

**1.** **(a)** $\{-12, 4\}$ **(b)** $(-\infty, -12) \cup (4, \infty)$ **(c)** $[-12, 4]$

**2.**

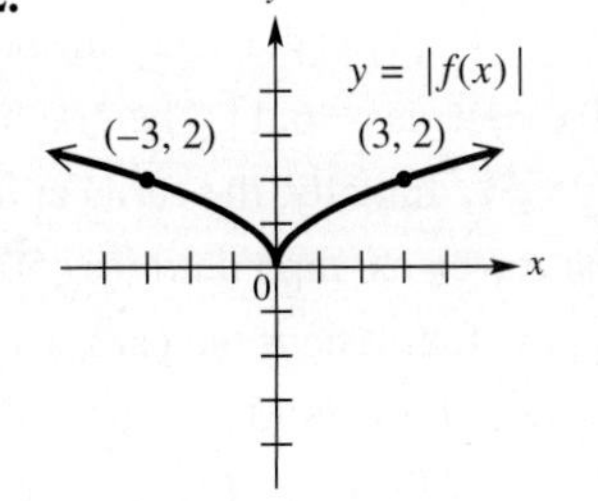


**3.** **(a)** $25.33 \le R_L \le 28.17$; $36.58 \le R_E \le 40.92$ **(b)** $5699.25 \le T_L \le 6338.25$; $8230.5 \le T_E \le 9207$

**4.** **(a)** $-3$ **(b)** 4 **(c)** 8

**5.** **(a)**

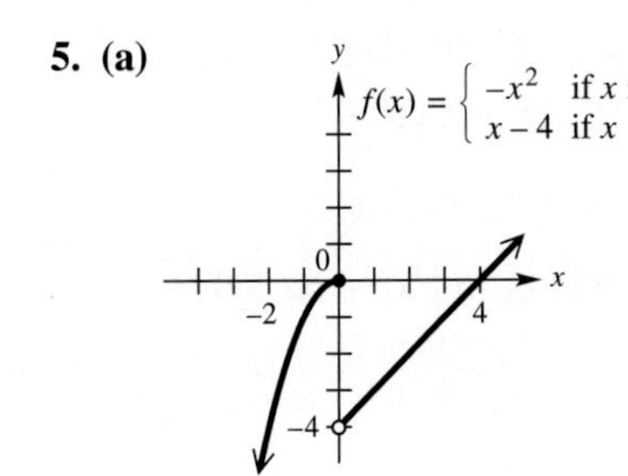


**(b)**

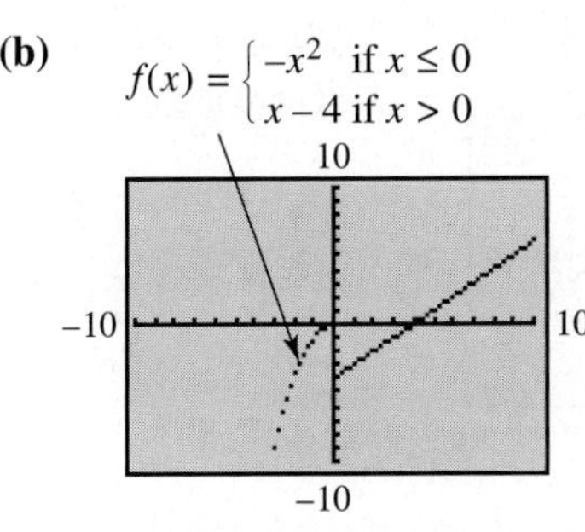


**6.** **(a)** $-6$ **(b)** $-22$ **(c)** 8 **(d)** $\frac{5}{9}$ **(e)** $-3x^2 - 4$ **(f)** $9x^2 + 24x + 16$

**7.** $f(x) = x^4$, $g(x) = x + 2$ (There are other possible choices for $f$ and $g$.)

**8.** $-4x - 2h + 3$

**9.** **(a)** $y_1 = .04x$ **(b)** $y_2 = .025x + 12.5$

**(c)** It represents the total interest earned in both accounts for 1 year. **(d)** $y_1 + y_2 = .04x + .025x + 12.5$ $28.75 is earned.

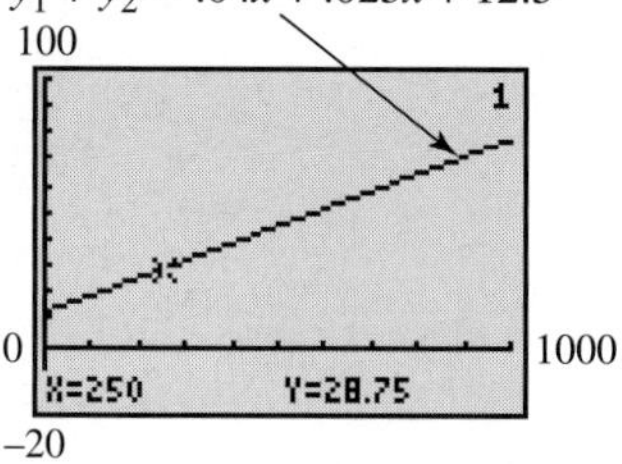


**(e)** Evaluate $y_1 + y_2$ at $x = 250$ to get 28.75 (dollars). **10.** $S = \pi r^2\sqrt{5}$

## Chapter 2 Review Exercises (*page 165*)

**1.** true **3.** false; the domain of $f(x) = \sqrt{x}$ is $[0, \infty)$, while the domain of $f(x) = \sqrt[3]{x}$ is $(-\infty, \infty)$. **5.** true **7.** true **9.** true **11.** $[0, \infty)$ **13.** $(-\infty, \infty)$ **15.** $(-\infty, \infty)$ **17.** $[0, \infty)$ **19. (a)** $(-\infty, -2), [-2, 1], (1, \infty)$ **(b)** $[-2, 1]$ **(c)** $(-\infty, -2)$ **(d)** $(1, \infty)$ **(e)** $(-\infty, \infty)$ **(f)** $\{-2\} \cup [-1, 1] \cup (2, \infty)$ **21.** $x$-axis symmetry, $y$-axis symmetry, origin symmetry; not a function **23.** $y$-axis symmetry; even function **25.** $x$-axis symmetry; not a function **27.** true **29.** true **31.** false; For example, $f(x) = x^3$ is odd, and $(2, 8)$ is on the graph, but $(-2, 8)$ is not. **33.** Start with the graph of $y = x^2$. Shift it 4 units to the left, stretch vertically by a factor of 3, reflect across the $x$-axis, and shift 8 units downward.

**35.**

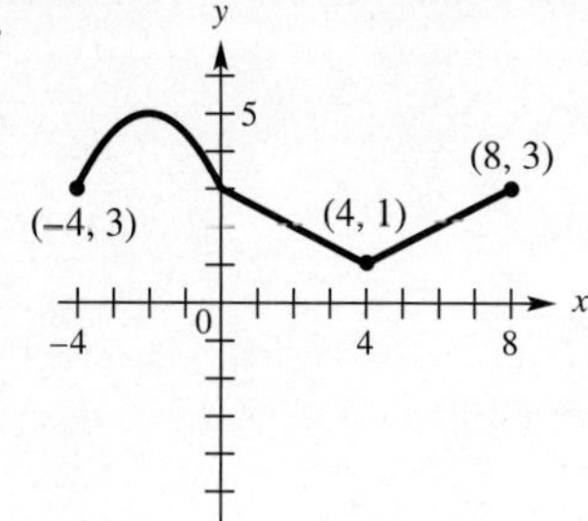


**37.**

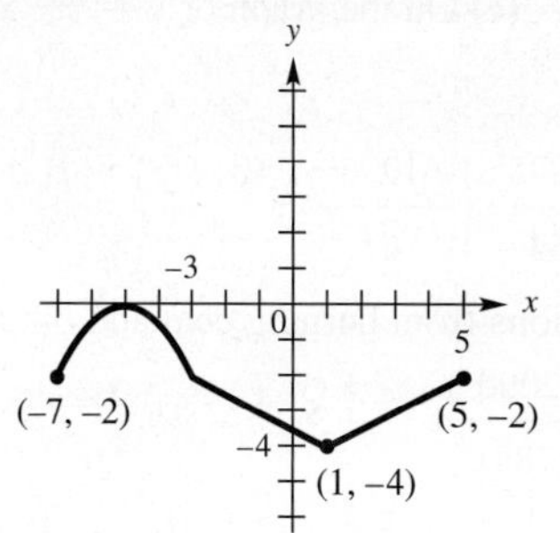


**39.**

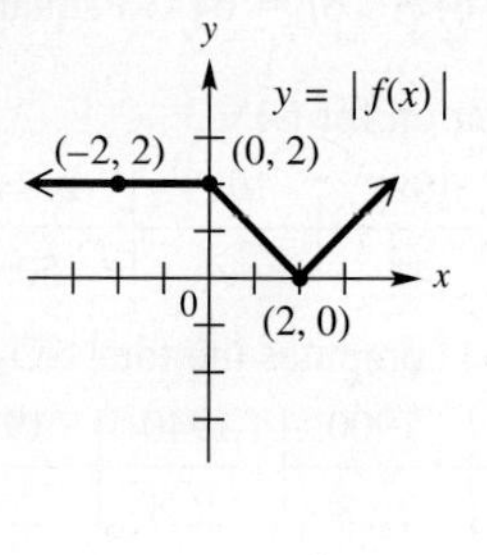


**41.** The graph of $y = |f(x)|$ is the same as that of $y = f(x)$. **43.** $\left\{-\frac{15}{4}, \frac{9}{4}\right\}$ **45.** $\left\{-\frac{7}{3}, 2\right\}$ **47.** $[-6, 1]$ **49.** The $x$-coordinates of the points of intersection of the graphs are $-6$ and 1. Thus, $\{-6, 1\}$ is the solution set of $y_1 = y_2$. The graph of $y_1$ lies on or below the graph of $y_2$ between $-6$ and 1, so the solution set of $y_1 \leq y_2$ is $[-6, 1]$. The graph of $y_1$ lies above the graph of $y_2$ everywhere else, so the solution set of $y_1 \geq y_2$ is $(-\infty, -6] \cup [1, \infty)$. **51.** Initially, the car is at home. After traveling 30 mph for 1 hr, the car is 30 mi away from home. During the second hour the car travels 20 mph until it is 50 mi away. During the third hour the car travels toward home at 30 mph until it is 20 mi away. During the fourth hour the car travels away from home at 40 mph until it is 60 mi away from home. During the last hour, the car travels 60 mi at 60 mph until it arrives home.

**53.**

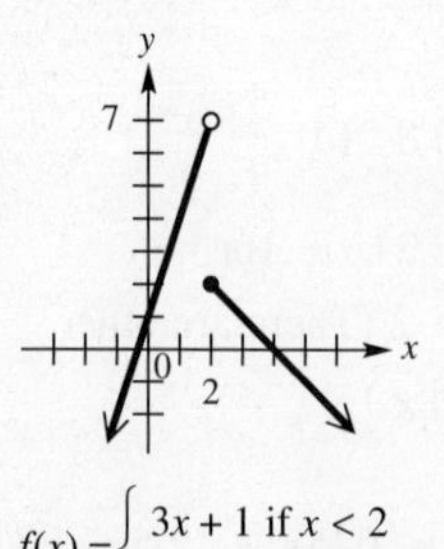


**55.**

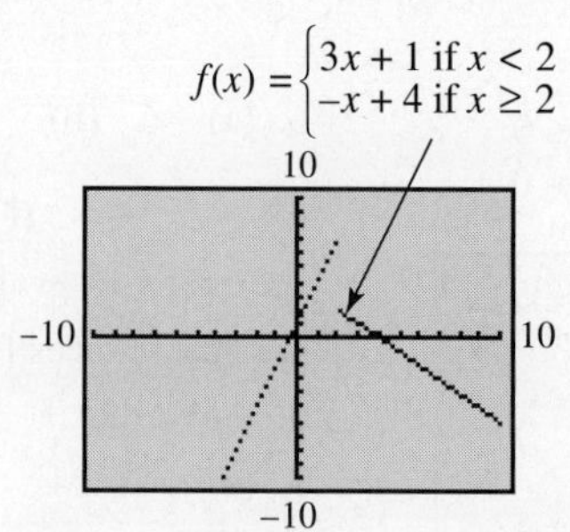


Dot mode

**57.** 5 **59.** 0 **61.** 3 **63.** 2 **65.** 8 **67.** $-6$ **69.** 2 **71.** 2 **73.** $f(x) = x^2$ and $g(x) = x^3 - 3x$ (There are other possible choices for $f$ and $g$.)

**75.** $V(r) = \frac{4}{3}\pi(r + 4)^3 - \frac{4}{3}\pi r^3$

**77.** $f(x) = 36x$; $g(x) = 1760x$; $(g \circ f)(x) = 63{,}360x$

## Chapter 2 Test *(page 168)*

**1.** **(a)** D **(b)** D **(c)** C **(d)** B **(e)** C **(f)** C **(g)** C **(h)** D **(i)** D **(j)** C

**2.** **(a)**

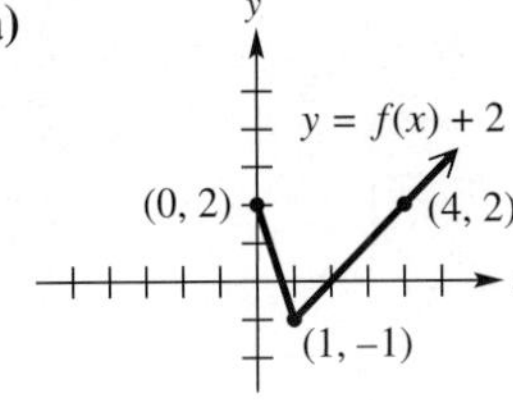


**(b)**

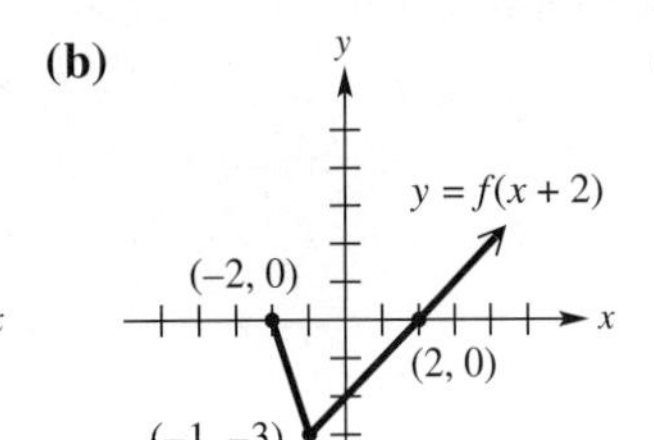


**(c)**

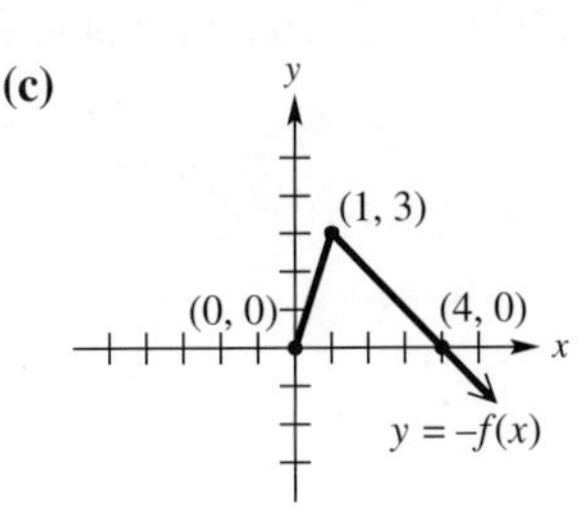


**(d)**

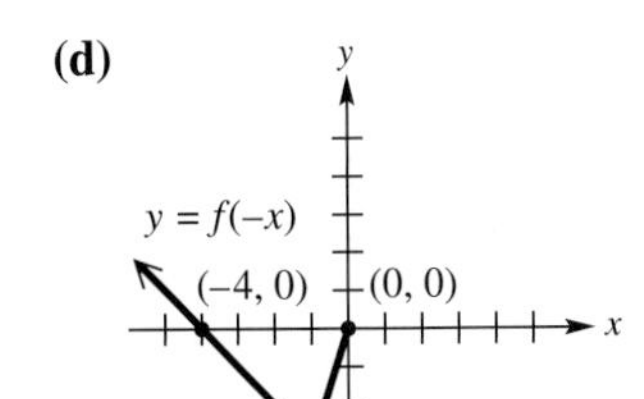


**(e)**

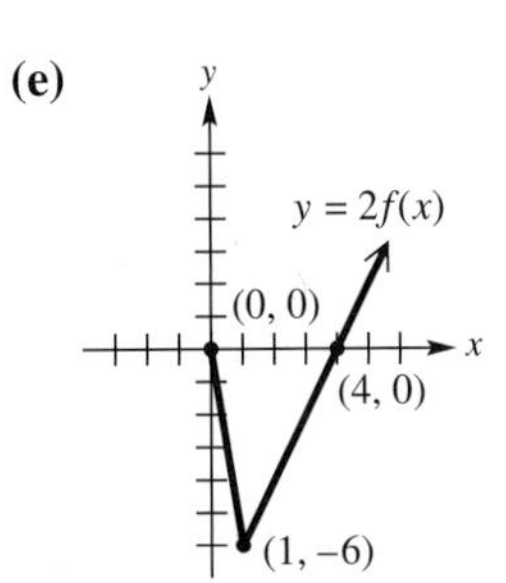


**(f)**

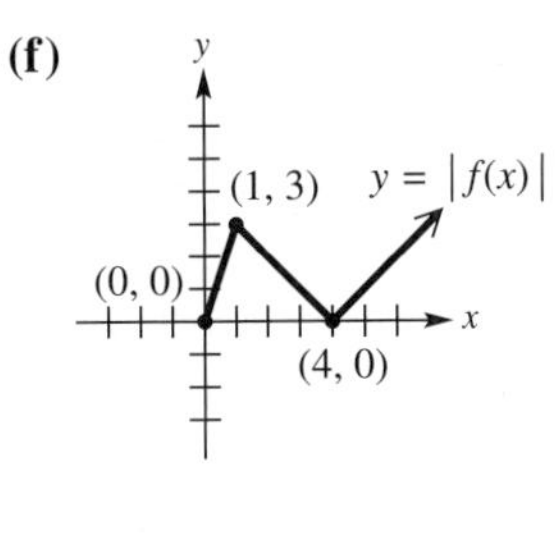


**3.** **(a)** $(-3, 6)$ **(b)** $(-3, -6)$ **(c)**

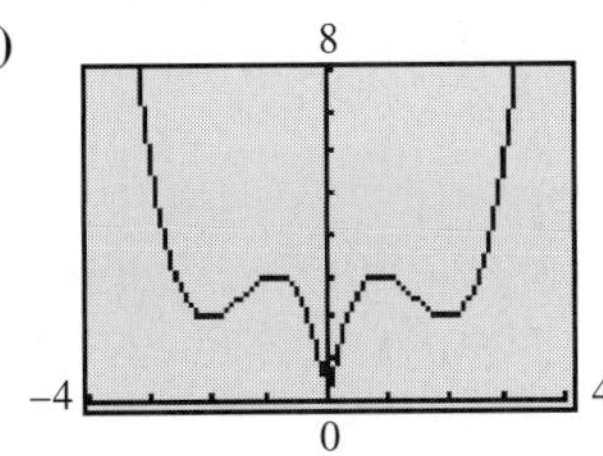


(We give an actual screen here. The drawing should resemble it.)

**4.** **(a)** Shift the graph of $y = \sqrt[3]{x}$ to the left 2 units, stretch by a factor of 4, and shift 5 units downward.

**(b)**

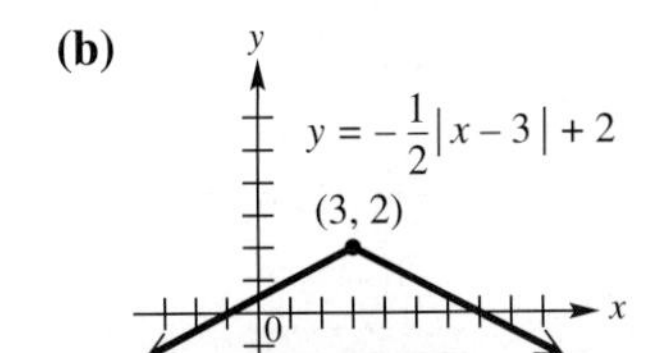


domain: $(-\infty, \infty)$; range: $(-\infty, 2]$

**5.** **(a)** $(-\infty, -3)$ **(b)** $(4, \infty)$ **(c)** $[-3, 4]$ **(d)** $(-\infty, -3)$, $[-3, 4]$, $(4, \infty)$ **(e)** $(-\infty, \infty)$ **(f)** $(-\infty, 2)$

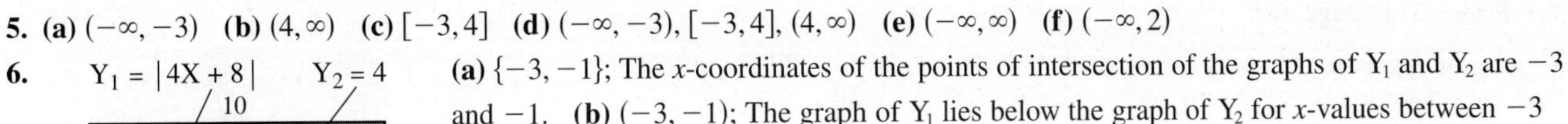

**6.**

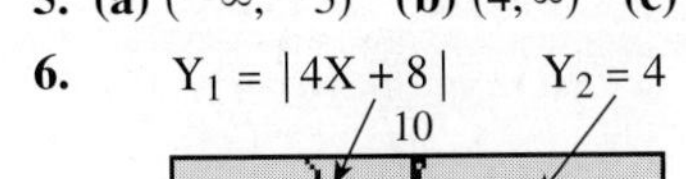


**(a)** $\{-3, -1\}$; The $x$-coordinates of the points of intersection of the graphs of $Y_1$ and $Y_2$ are $-3$ and $-1$. **(b)** $(-3, -1)$; The graph of $Y_1$ lies below the graph of $Y_2$ for $x$-values between $-3$ and $-1$. **(c)** $(-\infty, -3) \cup (-1, \infty)$; The graph of $Y_1$ lies above the graph of $Y_2$ for $x$-values less than $-3$ or for $x$-values greater than $-1$.

**7.** **(a)** $2x^2 - x + 1$ **(b)** $\dfrac{2x^2 - 3x + 2}{-2x + 1}$ **(c)** $\left(-\infty, \dfrac{1}{2}\right) \cup \left(\dfrac{1}{2}, \infty\right)$ **(d)** $8x^2 - 2x + 1$ **(e)** $4x + 2h - 3$

**8. (a)**

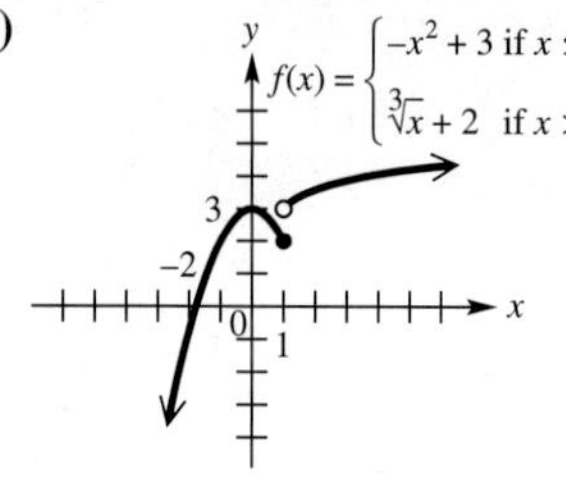


**(b)**

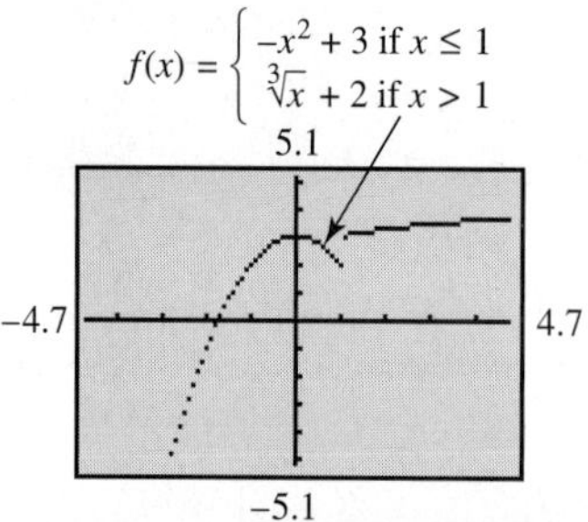


**9. (a)**

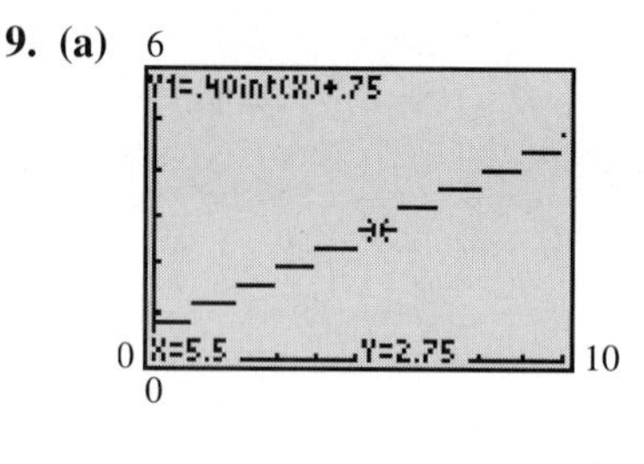


**(b)** \$2.75 is the cost for a 5.5-min call. See the display at the bottom of the screen. **10. (a)** $C(x) = 3300 + 4.50x$ **(b)** $R(x) = 10.50x$ **(c)** $P(x) = R(x) - C(x) = 6.00x - 3300$ **(d)** 551 **(e)**

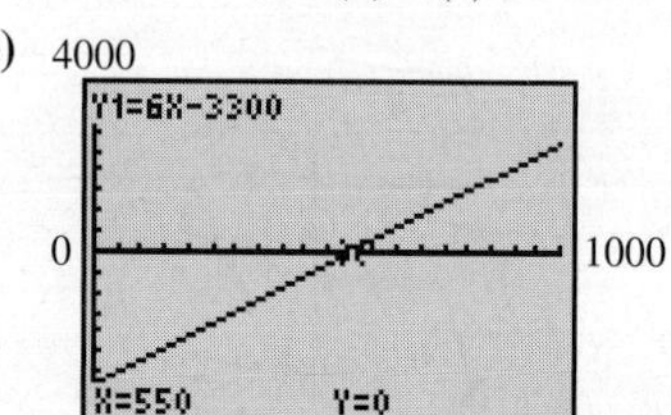


The first integer $x$-value for which $P(x) > 0$ is 551.

## CHAPTER 3 POLYNOMIAL FUNCTIONS

### 3.1 Exercises *(page 178)*

**1. (a)** 0 **(b)** −9 **(c)** imaginary **3. (a)** $\pi$ **(b)** 0 **(c)** real **5. (a)** 0 **(b)** $\sqrt{6}$ **(c)** imaginary **7. (a)** 2 **(b)** 5 **(c)** imaginary **9.** $10i$ **11.** $-20i$ **13.** $-i\sqrt{39}$ **15.** $5 + 2i$ **17.** $9 - 5i\sqrt{2}$ **19.** −5 **21.** 2 **23.** A real number $a$ may be written in the form $a + 0i$, a complex number; however, $a + bi$, $b \neq 0$, cannot be written without the imaginary term, so it is not a real number. **25.** $7 - i$ **27.** 2 **29.** $1 - 10i$ **31.** $-14 + 2i$ **33.** $5 - 12i$ **35.** 13 **37.** 7 **39.** $25i$ **41.** $2 + 12i$ **43.** $i$ **45.** $-i$ **47.** 1 **49.** −1 **51.** $-i$ **53.** $\left(\frac{\sqrt{2}}{2} + \frac{\sqrt{2}}{2}i\right)^2 = i$ **55.** $5 + 3i$ **57.** $18i$ **59.** $-5 - i$ **61.** $-1 - 2i$ **63.** $2i$ **65.** $\frac{7}{25} - \frac{24}{25}i$ **67.** $\frac{7}{3}i$ **69.** We are multiplying by 1, the multiplicative identity. **71.** $(-3)^3 - (-3)^1 - 7(-3) + 15 = 0$ **72.** $(2 - i)^3 - (2 - i)^2 - 7(2 - i) + 15 = 0$ **73.** $(2 + i)^3 - (2 + i)^2 - 7(2 + i) + 15 = 0$ **74.** They are conjugates.

### 3.2 Exercises *(page 188)*

**1.** B **3.** D **5. (a)** $P(x) = 2\left(x - \frac{1}{2}\right)^2 - \frac{49}{2}$ **(b)** $\left(\frac{1}{2}, -\frac{49}{2}\right)$ or $(.5, -24.5)$ **7. (a)** $y = (x - 1)^2 - 16$ **(b)** $(1, -16)$ **9. (a)** $f(x) = -2\left(x - \frac{3}{2}\right)^2 + \frac{9}{2}$ **(b)** $\left(\frac{3}{2}, \frac{9}{2}\right)$ or $(1.5, 4.5)$ **11. (a)** $P(x) = 4\left(x - \frac{11}{4}\right)^2 - \frac{169}{4}$ **(b)** $\left(\frac{11}{4}, -\frac{169}{4}\right)$ or $(2.75, -42.25)$ **13. (a)** D **(b)** B **(c)** C **(d)** A **15. (a)** $(2, 0)$ **(b)** domain: $(-\infty, \infty)$; range: $[0, \infty)$ **(c)** $x = 2$ **(d)** $[2, \infty)$ **(e)** $(-\infty, 2]$ **(f)** minimum: 0 **17. (a)** $(-3, -4)$ **(b)** domain: $(-\infty, \infty)$; range: $[-4, \infty)$ **(c)** $x = -3$ **(d)** $[-3, \infty)$ **(e)** $(-\infty, -3]$ **(f)** minimum: −4 **19. (a)** $(-3, 2)$ **(b)** domain: $(-\infty, \infty)$; range: $(-\infty, 2]$ **(c)** $x = -3$ **(d)** $(-\infty, -3]$ **(e)** $[-3, \infty)$ **(f)** maximum: 2 **21. (a)** $(5, -4)$ **(b)**

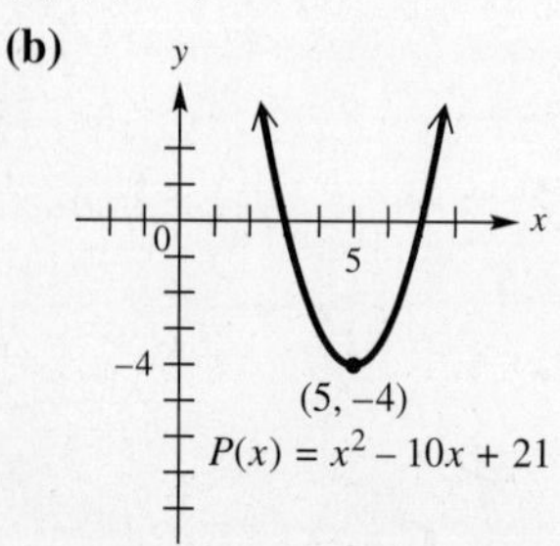


**23. (a)** $(2, 2)$ **(b)**

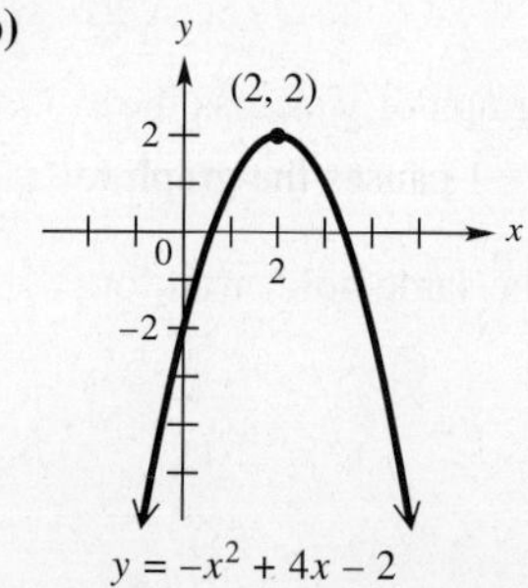


**25. (a)** $(1, 3)$ **(b)**

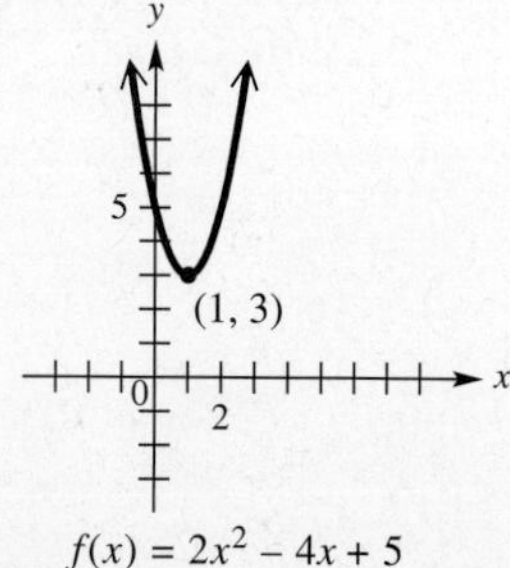

**27.** $(2.71, 5.20)$; $-1.33, 6.74$ **29.** $(1.12, .56)$; none **31.** $-3$ **33.** **(a)** $(4, -12)$ **(b)** minimum **(c)** $-12$ **(d)** $[-12, \infty)$ **35.** **(a)** $(1.5, 2)$ **(b)** maximum **(c)** 2 **(d)** $(-\infty, 2]$ **37.** quadratic; $a < 0$ **39.** quadratic; $a > 0$ **41.** linear; positive **43.** **(a)** quadratic; positive **(b)** $(4, 74)$ **(c)** In 1994 company bankruptcy filings reached the minimum number of 74. The number 74 is a little high since the actual value from the table is 70. **45.** **(a)** maximum **(b)** 1994; 41 million **47.** **(a)** The value of $t$ cannot be negative since $t$ represents time elapsed from the throw. **(b)** Since the rock was projected from ground level, $s_0$, the initial height of the rock is 0. **(c)** $s(t) = -16t^2 + 90t$ **(d)** 99 ft **(e)** After 2.8125 sec, the maximum height, 126.5625 ft, is attained. Locate the vertex $(2.8125, 126.5625)$. **(f)** 5.625 sec **49.** **(a)** The ball will not reach 355 ft because the graph of $y_1 = -16x^2 + 50x$ does not intersect the graph of $y_2 = 355$. **(b)** The ball reaches a height of 355 ft at $t \approx 1.4$ sec and $t \approx 14.2$ sec. **51.** $P(x) = 3x^2 + 6x - 1$ **53.** $P(x) = \frac{1}{2}x^2 - 8x + 35$ **55.** $P(x) = -\frac{2}{3}x^2 - \frac{16}{3}x - \frac{38}{3}$

**57.**

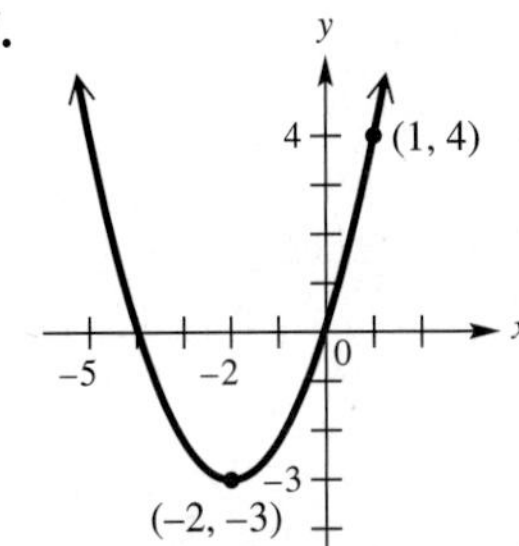


**59.**

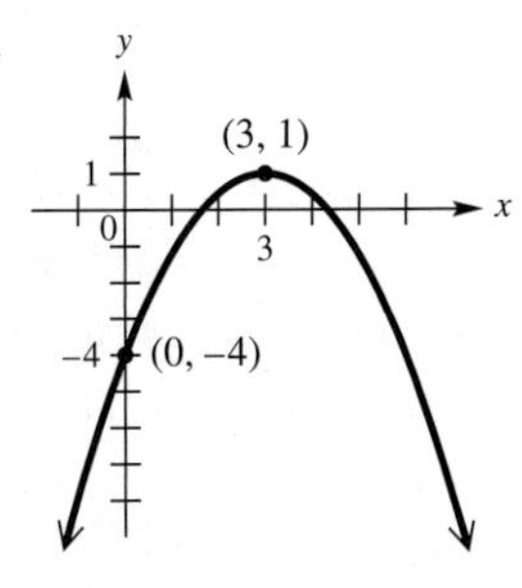


### 3.3 Exercises *(page 205)*

**1.** D; $\left\{-\frac{1}{3}, 7\right\}$ **3.** C; $\{-4, 3\}$ **5.** $\{\pm 4\}$ **7.** $\{\pm 3\}$ **9.** $\{\pm 4i\}$ **11.** $\{\pm 3i\sqrt{2}\}$ **13.** $\left\{\frac{1 \pm 2\sqrt{3}}{3}\right\}$ **15.** $\left\{\frac{3}{5} \pm \frac{\sqrt{3}}{5}i\right\}$ **17.** $\{1 \pm \sqrt{5}\}$ **19.** $\left\{-\frac{1}{2} \pm \frac{1}{2}i\right\}$ **21.** $\left\{\frac{1 \pm \sqrt{5}}{2}\right\}$ **23.** $\{3 \pm \sqrt{2}\}$ **25.** $\left\{\frac{3}{2} \pm \frac{\sqrt{2}}{2}i\right\}$ **27.** $\left\{\frac{-3 \pm 3\sqrt{65}}{8}\right\}$ **29.** 0; one real solution; rational **31.** 84; two real solutions; irrational **33.** $-23$; no real solutions **35.** $a = 1$; $b = -9$; $c = 20$ **37.** $a = 1$; $b = -2$; $c = -1$

In Exercises 39–43, answers may vary. We give one example for each exercise.

**39.**

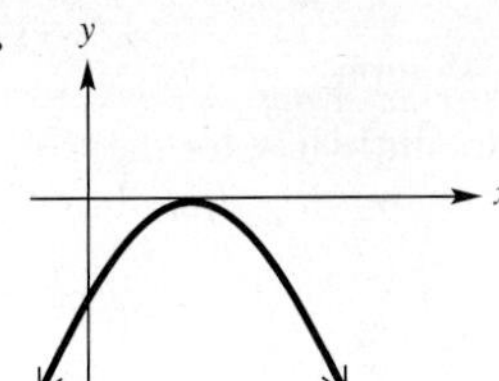


**41.**

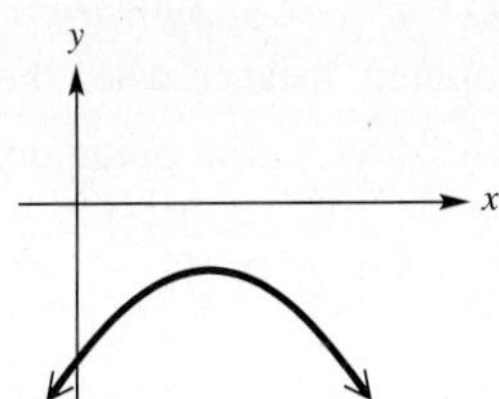


**43.**

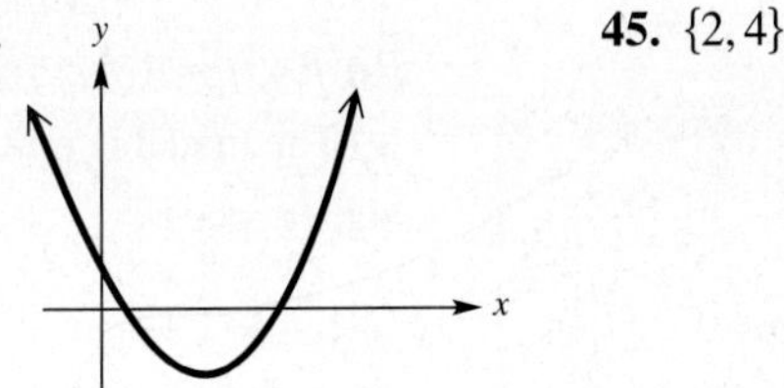


**45.** $\{2, 4\}$

**47.** $(-\infty, 2) \cup (4, \infty)$ **49.** $(-\infty, 3) \cup (3, \infty)$ **51.** $(-\infty, \infty)$ **53.** no real solutions; two imaginary complex solutions **55.** 3 **57.** yes; negative **59.** **(a)** $(-\infty, -3] \cup [-1, \infty)$ **(b)** $(-3, -1)$ **61.** **(a)** $\left(-\infty, \frac{1}{2}\right) \cup (4, \infty)$ **(b)** $\left[\frac{1}{2}, 4\right]$ **63.** **(a)** $\left[1 - \sqrt{2}, 1 + \sqrt{2}\right]$ **(b)** $\left(-\infty, 1 - \sqrt{2}\right) \cup \left(1 + \sqrt{2}, \infty\right)$ **65.** $-4$ and 2 **66.** the interval $(-4, 2)$ **67.** The graph of $y_2$ is obtained by reflecting the graph of $y_1$ across the $x$-axis. **68.** the interval $(-4, 2)$ **69.** They are the same. **70.** Because multiplying a function $f$ by $-1$ causes the graph to reflect across the $x$-axis, the original solution set of $f(x) < 0$ is the same as the solution set of $-f(x) > 0$. The same holds true for the solution sets of $f(x) > 0$ and $-f(x) < 0$. **71.** $t = \frac{\pm\sqrt{2sg}}{g}$ **73.** $v = \frac{\pm\sqrt[4]{Frk^3M^3}}{kM}$ **75.** $R = \frac{E^2 - 2Pr \pm E\sqrt{E^2 - 4Pr}}{2P}$ **77.** $x = \frac{y \pm \sqrt{8 - 11y^2}}{4}$; $y = \frac{x \pm \sqrt{6 - 11x^2}}{3}$ **79.** $x \approx 3.3$; This means that nobody lives beyond age 98. (There are, of course, exceptions.)

**Reviewing Basic Concepts** *(page 207)*

**1.** $3 + 7i$ **2.** $26i$ **3.** $-\frac{50}{13} + \frac{10}{13}i$ **4.**

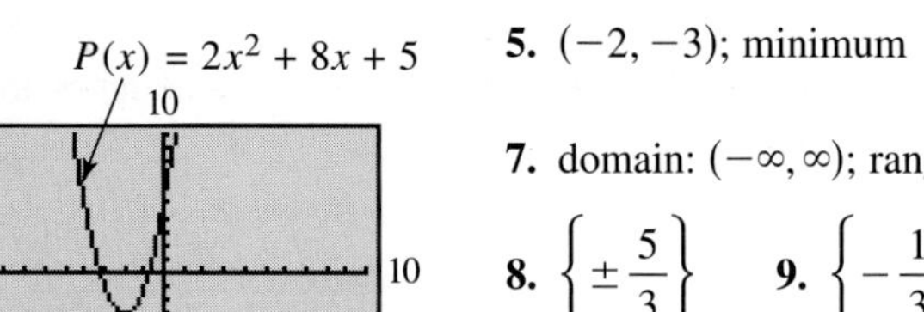


**5.** $(-2, -3)$; minimum **6.** $x = -2$

**7.** domain: $(-\infty, \infty)$; range: $[-3, \infty)$

**8.** $\left\{\pm\frac{5}{3}\right\}$ **9.** $\left\{-\frac{1}{3}, 2\right\}$

**10.** $\left\{\frac{1 \pm \sqrt{13}}{2}\right\}$ **11.** $\left[-\frac{1}{3}, 2\right]$

**12.** $\left(-\infty, \frac{1 - \sqrt{13}}{2}\right) \cup \left(\frac{1 + \sqrt{13}}{2}, \infty\right)$

**3.4 Exercises** *(page 212)*

**1.** A **3. (a)** $30 - x$ **(b)** $0 < x < 30$ **(c)** $P(x) = -x^2 + 30x$ **(d)** 15 and 15; The maximum product is 225.
**5. (a)** $640 - 2x$ **(b)** $0 < x < 320$ **(c)** $A(x) = -2x^2 + 640x$ **(d)** between 57.04 ft and 85.17 ft or 234.83 ft and 262.96 ft **(e)** 160 ft by 320 ft; The maximum area is 51,200 ft$^2$. **7. (a)** $2x$ **(b)** length: $2x - 4$; width: $x - 4$; $x > 4$ **(c)** $V(x) = 4x^2 - 24x + 32$ **(d)** 8 in. by 20 in. **(e)** 13.0 to 4.2 in. **9.** 7.875 in. **11.** 20 ft **13.** 5 ft
**15.** a 17-ft ladder **17. (a)** $80 - x$ **(b)** $400 + 20x$ **(c)** $R(x) = -20x^2 + 1200x + 32{,}000$ **(d)** 5 or 55 **(e)** \$1000
**19. (a)** 23.32 ft per sec **(b)** yes **(c)** 12.88 ft **21. (a)** 3.5 ft **(b)** about .2 ft and 2.3 ft **(c)** 1.25 ft **(d)** about 3.78 ft
**23.** about 104.5 ft per sec; 71.25 mph **25. (a)** 19.2 hr **(b)** 84.3 ppm
**27. (a)** 120.2

3 15
39.8

**(b)** $f(x) = .6(x - 4)^2 + 50$ **(c)** There is a good fit.

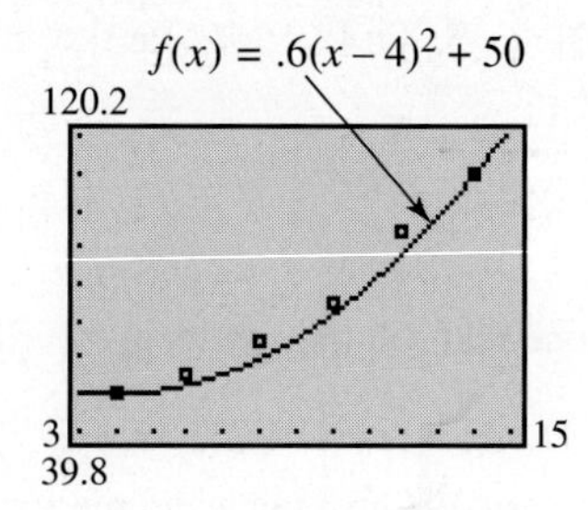


**(d)** $g(x) = .402x^2 - 1.175x + 48.343$ **(e)** $f(16) = 136.4$ thousand; $g(16) \approx 132.5$ thousand

**29. (a)**

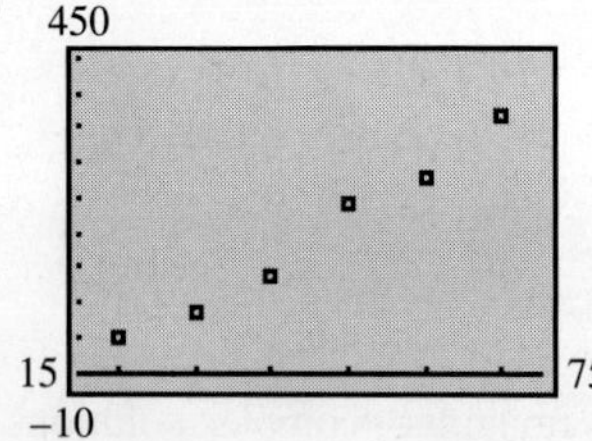


**(b)** $f(45) \approx 161.5$ ft is the stopping distance when the speed is 45 mph.
**(c)** The model is quite good, although the stopping distances are a little low for the higher speeds.

**3.5 Exercises** *(page 227)*

**1.** As the odd exponent $n$ gets larger, the graph *flattens out* in the window $[-1, 1]$ by $[-1, 1]$.

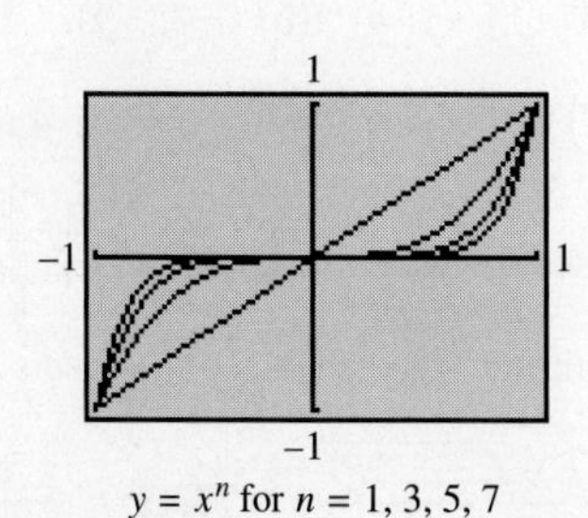


$y = x^n$ for $n = 1, 3, 5, 7$

The graph of $y = x^7$ will be between $y = x^5$ and the $x$-axis in this window.

**3.** The graph of $f(x) = x^n$ for $n \in$ {positive odd integers} will take the shape of the graph of $f(x) = x^3$, but gets steeper as $n$ gets larger.

**5.** Shift the graph of $y = x^4$ three units to the left, stretch vertically by a factor of 2, and shift downward 7 units.

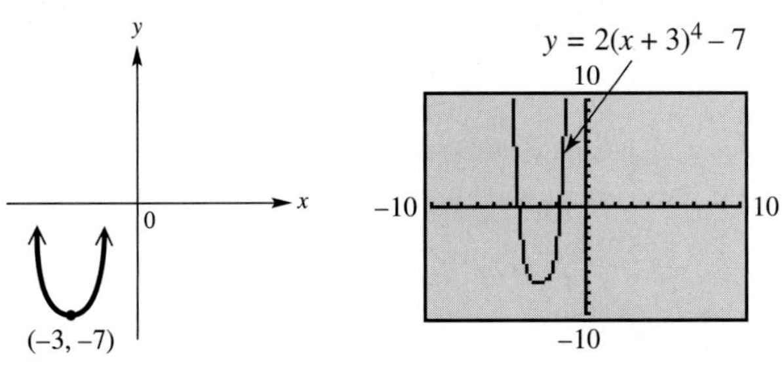


**6.** Shift the graph of $y = x^4$ one unit to the left, stretch vertically by a factor of 3, reflect across the $x$-axis, and shift upward 12 units.

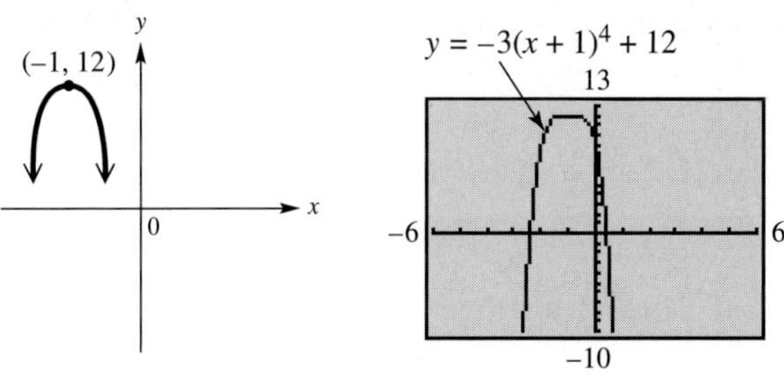


**7.** Shift the graph of $y = x^3$ one unit to the right, stretch vertically by a factor of 3, reflect across the $x$-axis, and shift upward 12 units.

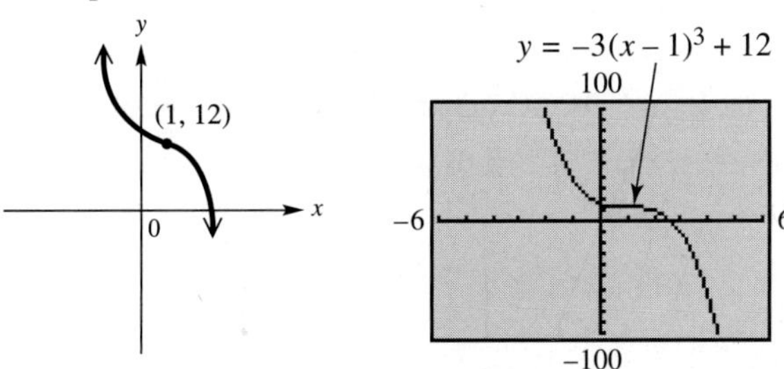


**8.** Shift the graph of $y = x^5$ one unit to the right, shrink vertically by a factor of .5, and shift 13 units upward.

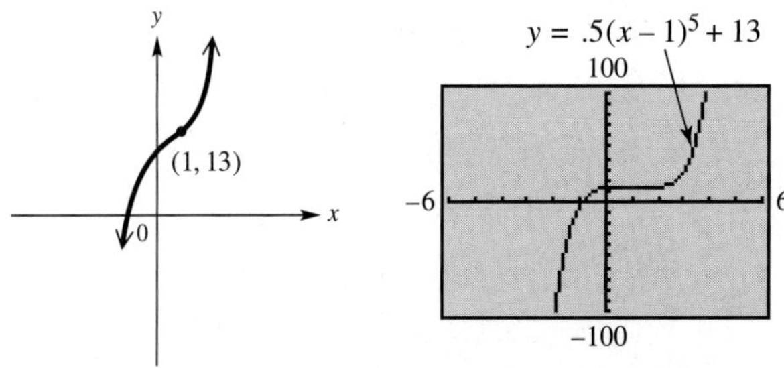


**9.** at least 4 **11.** $(a, b)$ and $(c, d)$, local maxima; $(e, t)$, a local minimum **13.** $(a, b)$, absolute maximum **15.** local maximum values of $b$ and $d$; local minimum of $t$; absolute maximum of $b$ **17.** local maximum: $(2, 3.67)$; local minimum: $(3, 3.5)$ **19.** local maximum: $(-3.33, -1.85)$; local minimum: $(-4, -2)$ **21.** two: 2.10 and 2.15 **23.** none

**25.** **27.** **29.** **31.** **33.** D **35.** B **37.** always concave up **38.** concave down for $x < 0$ and concave up for $x > 0$ **39.** concave down for $x < 0$ and concave up for $x > 0$ **40.** The end behavior must be , indicating downward concavity as $|x| \to \infty$. **41.** false **43.** true **45.** true **47.** false **49.** A **51.** one **53.** B and D **55.** one **57.** B **59.**

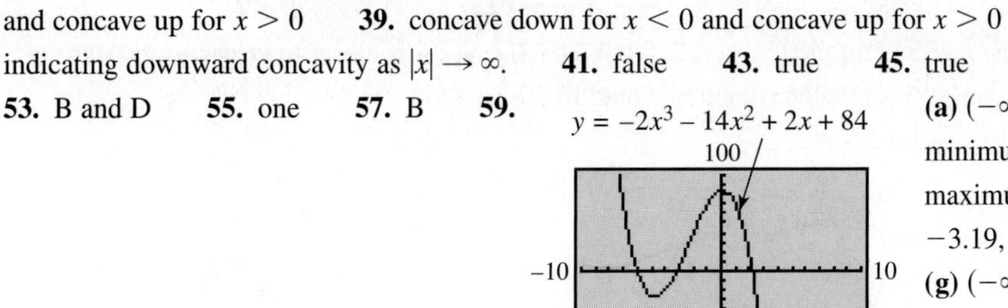


**(a)** $(-\infty, \infty)$ **(b)** $(-4.74, -27.03)$; not an absolute minimum point **(c)** $(.07, 84.07)$; not an absolute maximum point **(d)** $(-\infty, \infty)$ **(e)** $x$-intercepts: $-6$, $-3.19$, $2.19$; $y$-intercept: 84 **(f)** $[-4.74, .07]$ **(g)** $(-\infty, -4.74]$; $[.07, \infty)$

**61.** $y = x^5 + 4x^4 - 3x^3 - 17x^2 + 6x + 9$

20

–4 4

–20

**(a)** $(-\infty, \infty)$ **(b)** $(-1.73, -16.39)$, $(1.35, -3.49)$; Neither is an absolute minimum point. **(c)** $(-3, 0)$, $(.17, 9.52)$; Neither is an absolute maximum point. **(d)** $(-\infty, \infty)$ **(e)** $x$-intercepts: $-3$, $-.62$, $1$, $1.62$; $y$-intercept: 9 **(f)** $(-\infty, -3]$; $[-1.73, .17]$; $[1.35, \infty)$ **(g)** $[-3, -1.73]$; $[.17, 1.35]$

**63.**

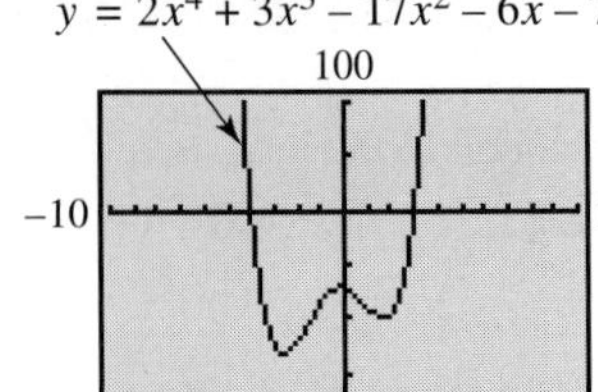


**(a)** $(-\infty, \infty)$ **(b)** $(-2.63, -132.69)$ is an absolute minimum point; $(1.68, -99.90)$ **(c)** $(-.17, -71.48)$; not an absolute maximum point **(d)** $[-132.69, \infty)$ **(e)** $x$-intercepts: $-4$, 3; $y$-intercept: $-72$ **(f)** $[-2.63, -.17]$; $[1.68, \infty)$ **(g)** $(-\infty, -2.63]$; $[-.17, 1.68]$

**65.**

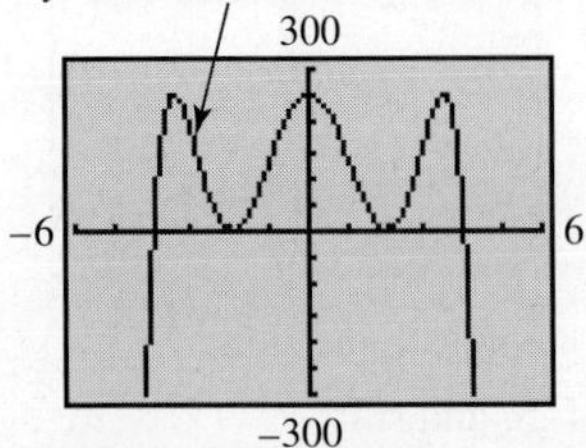


**(a)** $(-\infty, \infty)$ **(b)** $(-2, 0)$; $(2, 0)$; Neither is an absolute minimum point. **(c)** $(-3.46, 256)$, $(0, 256)$, and $(3.46, 256)$; All are absolute maximum points. **(d)** $(-\infty, 256]$ **(e)** $x$-intercepts: $-4, -2, 2, 4$; $y$-intercept: 256 **(f)** $(-\infty, -3.46]$; $[-2, 0]$; $[2, 3.46]$ **(g)** $[-3.46, -2]$; $[0, 2]$; $[3.46, \infty)$

There are many possible valid windows in Exercises 67–71. We give only one in each case.

**67.** $[-10, 10]$ by $[-40, 10]$ **69.** $[-10, 20]$ by $[-1500, 500]$ **71.** $[-10, 10]$ by $[-20, 500]$

**73. (a)**

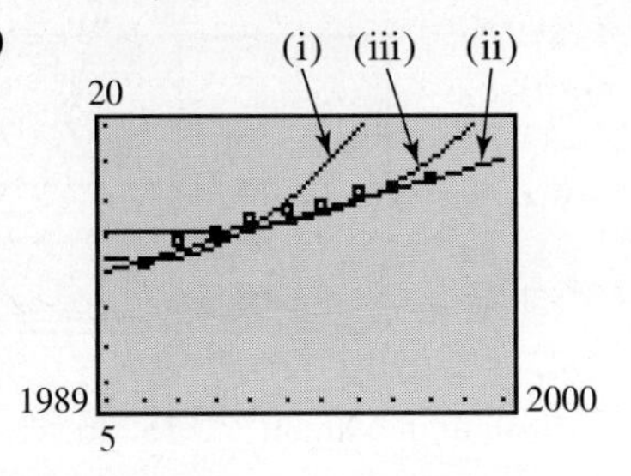


**(b)** All three approximate the data up to 1994, but only function (ii), the linear function, approximates the data for 1995–2000.

## Reviewing Basic Concepts *(page 231)*

**1. (a)** The total length of the fence must equal $2L + 2x = 300$. From this equation, $L = 150 - x$. **(b)** $A(x) = x(150 - x)$ **(c)** $0 < x < 150$ **(d)** 50 m by 100 m

**2. (a)**

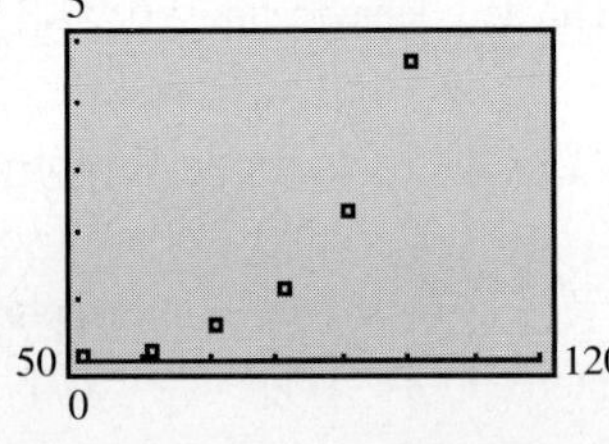


**(b)** $f(x) = .0018(x - 51)^2 + .1$ **(c)** $g(x) = .0026x^2 - .3139x + 9.426$ **(d)** The regression function fits slightly better because it is closer to or passes through more data points. Neither function would fit the data for $x < 51$.

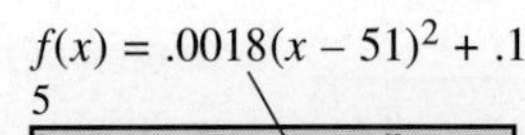


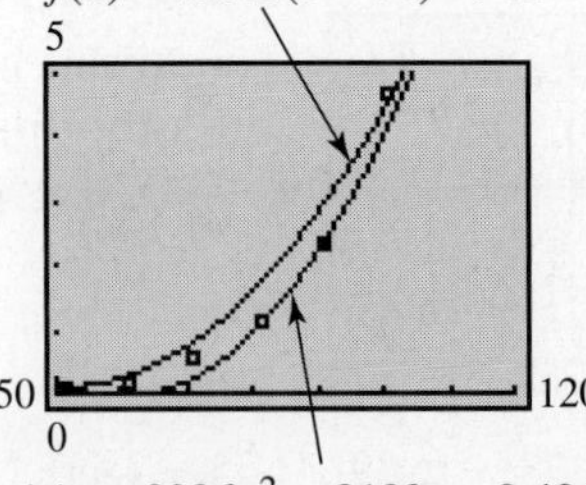

**3.** ↗ **4.** two; three **5.** ↖↗ **6.**

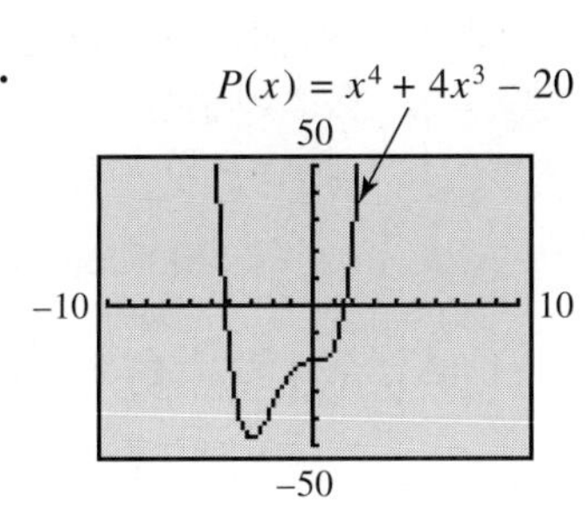


**7.** The only extreme point is $(-3, -47)$, an absolute minimum. **8.** $x$-intercepts: $-4.26$, $1.53$; $y$-intercept: $-20$

### 3.6 Exercises *(page 240)*

**1.** $P(1) = -5$ and $P(2) = 2$ differ in sign. The zero is approximately 1.79. **3.** $P(2) = 2$ and $P(2.5) = -.25$ differ in sign. The zero is approximately 2.39. **5.** $P(2) = 16$ and $P(1.5) = -.375$ differ in sign. The zero is approximately 1.52. **7.** There is at least one zero between 2 and 2.5. **9.** $x^2 - 3x - 2$ **11.** $3x^2 + 4x + \dfrac{3}{x-5}$ **13.** $x^3 - x^2 - 6x$ **15.** 2 **17.** $-25$ **19.** $-5$ **21.** yes **23.** no **25.** yes **27.** $x + 3, x - 1, x - 4$ **28.** $-3, 1, 4$ **29.** $-3, 1, 4$ **30.** $-10$; $-10$ **31.** $(-3, 1) \cup (4, \infty)$ **32.** $(-\infty, -3) \cup (1, 4)$ **33.** $\dfrac{-1-\sqrt{5}}{2}, \dfrac{-1+\sqrt{5}}{2}$ **35.** $\dfrac{1-\sqrt{13}}{6}, \dfrac{1+\sqrt{13}}{6}$ **37.** $P(x) = (x-2)(2x-5)(x+3)$ **39.** $P(x) = (x+4)(3x-1)(2x+1)$ **41.** possible: 0 or 2 positive real zeros, 1 negative real zero; actual: 0 positive, 1 negative **43.** possible: 1 positive real zero, 1 negative real zero; actual: 1 positive, 1 negative **45.** possible: 0 or 2 positive real zeros, 1 or 3 negative real zeros; actual: 0 positive, 1 negative

### 3.7 Exercises *(page 249)*

Answers may vary in Exercises 1–5.

**1.** $P(x) = x^3 - 8x^2 + 21x - 20$ **3.** $P(x) = x^3 - 5x^2 + x - 5$ **5.** $P(x) = x^3 - 6x^2 + 10x$ **7.** $P(x) = -\dfrac{1}{6}x^3 + \dfrac{13}{6}x + 2$ **9.** $P(x) = -\dfrac{1}{2}x^3 - \dfrac{1}{2}x^2 + x$ **11.** $P(x) = -x^3 + 6x^2 - 10x + 8$ **13.** $-1 + i, -1 - i$ **15.** $-1 + \sqrt{2}, -1 - \sqrt{2}$ **17.** $-3i, \dfrac{1}{2} + \dfrac{\sqrt{3}}{2}i, \dfrac{1}{2} - \dfrac{\sqrt{3}}{2}i$

Answers may vary in Exercises 19–29.

**19.** $P(x) = x^2 - x - 20$ **21.** $P(x) = x^4 + x^3 - 5x^2 + x - 6$ **23.** $P(x) = x^3 - 3x^2 + 2$ **25.** $P(x) = x^4 - 7x^3 + 17x^2 - x - 26$ **27.** $P(x) = x^3 - 11x^2 + 33x + 45$ **29.** $P(x) = x^4 + 2x^3 - 10x^2 - 6x + 45$ **31.** Use synthetic division twice, with $k = -2$. Zeros are $-2, 3, -1$; $P(x) = (x+2)^2(x-3)(x+1)$. **33.** 1, 3, or 5 **35.** The polynomial must also have a zero of $1 - i$, which makes it impossible for it to have degree 3.

**37.**

$P(x) = 2x^3 - 5x^2 - x + 6$
$= (x+1)(2x-3)(x-2)$

**39.**

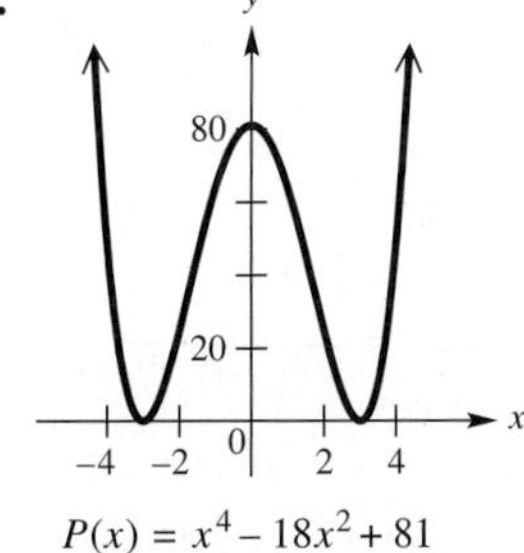


$P(x) = x^4 - 18x^2 + 81$
$= (x-3)^2(x+3)^2$

**41.**

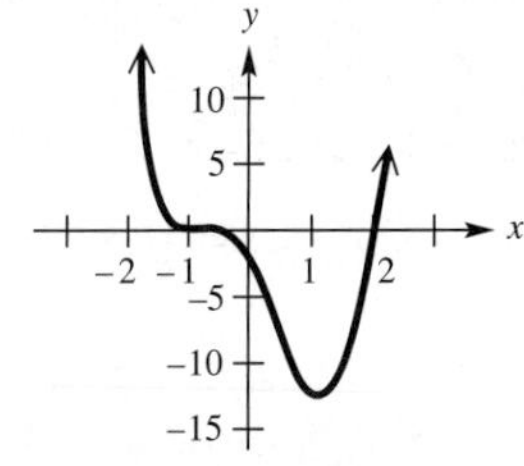


$P(x) = 2x^4 + x^3 - 6x^2 - 7x - 2$
$= (2x+1)(x-2)(x+1)^2$

**43.** **(a)** $\pm 1, \pm 2, \pm 5, \pm 10$ **(b)** Eliminate values less than $-2$ or greater than 5. **(c)** $-2, -1, 5$ **(d)** $P(x) = (x + 2)(x + 1)(x - 5)$ **45.** **(a)** $\pm 1, \pm 2, \pm 3, \pm 5, \pm 6, \pm 10, \pm 15, \pm 30$ **(b)** Eliminate values less than $-5$ or greater than 2. **(c)** $-5, -3, 2$ **(d)** $P(x) = (x + 5)(x + 3)(x - 2)$

**47.** **(a)** $\pm 1, \pm 2, \pm 3, \pm 4, \pm 6, \pm 12, \pm \frac{1}{2}, \pm \frac{3}{2}, \pm \frac{1}{3}, \pm \frac{2}{3}, \pm \frac{4}{3}, \pm \frac{1}{6}$ **(b)** Eliminate values less than $-4$ or greater than $\frac{3}{2}$. **(c)** $-4, -\frac{1}{3}, \frac{3}{2}$ **(d)** $P(x) = (x + 4)(3x + 1)(2x - 3)$ **49.** **(a)** $\pm 1, \pm 2, \pm 3, \pm 6, \pm \frac{1}{2}, \pm \frac{3}{2}, \pm \frac{1}{3}, \pm \frac{2}{3}, \pm \frac{1}{6}, \pm \frac{1}{12}, \pm \frac{1}{4}, \pm \frac{3}{4}$ **(b)** Eliminate values less than $-\frac{3}{2}$ or greater than $\frac{1}{2}$. **(c)** $-\frac{3}{2}, -\frac{2}{3}, \frac{1}{2}$ **(d)** $P(x) = (3x + 2)(2x + 3)(2x - 1)$

**51.** $-2, -1, \frac{5}{2}$ **53.** $\frac{3}{2}, 4$ **57.** Not all coefficients of $P(x)$ are real numbers.

### 3.8 Exercises *(page 258)*

**1.** $\left\{0, \pm \frac{\sqrt{7}}{7}i\right\}$ **3.** $\left\{-\frac{2}{3}, \pm 1\right\}$ **5.** $\{\pm 1, \pm\sqrt{10}\}$ **7.** $\{-2, -1.5, 1.5, 2\}$ **9.** $\{-4, 4, -i, i\}$ **11.** $\{-8, 1, 8\}$

**13.** $\{-1.5, 0, 1\}$ **15.** $\{-1 \pm \sqrt{7} (\approx -1 \pm 2.65)\}$ **17.** $\left\{0, \frac{-1 \pm \sqrt{73}}{6}\right\}$ or $\{0, \approx 1.26, \approx -1.59\}$

**19.** $\left\{0, -\frac{1}{2} - \frac{\sqrt{3}}{2}i, -\frac{1}{2} + \frac{\sqrt{3}}{2}i\right\}$ **21.** $\{-4i, -i, i, 4i\}$ **23.** $\left\{-3, 2, -1 - i\sqrt{3}, -1 + i\sqrt{3}, \frac{3}{2} + \frac{3\sqrt{3}}{2}i, \frac{3}{2} - \frac{3\sqrt{3}}{2}i\right\}$

**25.** It is not a comprehensive graph because the positive $x$-intercepts and end behavior as $x \to \infty$ are not shown. **26.** symmetry with respect to the $y$-axis **27.** $\{-5, -\sqrt{3}, \sqrt{3}, 5\}$ **28.** There are no imaginary solutions. Because the polynomial is fourth degree, there are at most four solutions, all of which are real.

**29.**

$P(x) = x^3 - 3x^2 - 6x + 8$
$= (x - 4)(x - 1)(x + 2)$

**(a)** $\{-2, 1, 4\}$
**(b)** $(-\infty, -2) \cup (1, 4)$
**(c)** $(-2, 1) \cup (4, \infty)$

**31.**

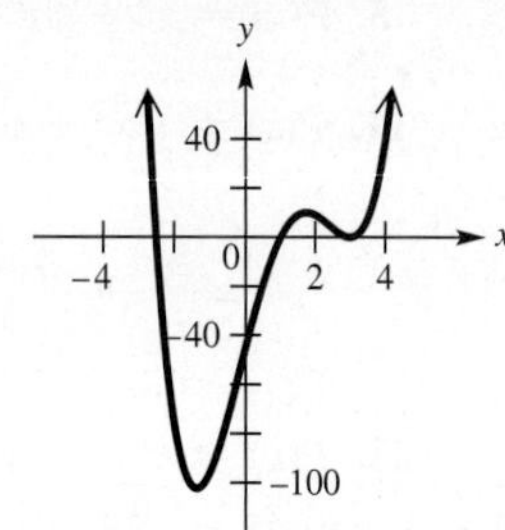


$P(x) = 2x^4 - 9x^3 - 5x^2 + 57x - 45$
$= (x - 3)^2(2x + 5)(x - 1)$

**(a)** $\{-2.5, 1, 3 \text{ (multiplicity 2)}\}$
**(b)** $(-2.5, 1)$
**(c)** $(-\infty, -2.5) \cup (1, 3) \cup (3, \infty)$

**33.**

$P(x) = -x^4 - 4x^3 + 3x^2 + 18x$
$= x(2 - x)(x + 3)^2$

**(a)** $\{-3 \text{ (multiplicity 2)}, 0, 2\}$
**(b)** $\{-3\} \cup [0, 2]$
**(c)** $(-\infty, 0] \cup [2, \infty)$

**35.** $\{-.88, 2.12, 4.86\}$ **37.** $\{1.52\}$ **39.** $\{-.40, 2.02\}$ **41.** $\{-i, i\}$ **43.** $\left\{-1, \frac{1}{2} + \frac{\sqrt{3}}{2}i, \frac{1}{2} - \frac{\sqrt{3}}{2}i\right\}$

**45.** $\left\{3, -\frac{3}{2} - \frac{3\sqrt{3}}{2}i, -\frac{3}{2} + \frac{3\sqrt{3}}{2}i\right\}$ **47.** $\{-2, 2, -2i, 2i\}$ **49.** **(a)** about 7.13 cm; The ball floats partly above the surface. **(b)** The sphere is more dense than water and sinks below the surface. **(c)** 10 cm; The balloon floats even with the surface. **51.** about 11.34 cm **53.** **(a)** $0 < x < 6$ **(b)** $V(x) = 4x^3 - 60x^2 + 216x$ **(c)** about 2.35; about 228.16 in.$^3$ **(d)** $.42 < x < 5$ **55.** about 2.61 in. **57.** **(a)** $x - 1$ **(b)** $\sqrt{x^2 - (x - 1)^2}$ or $\sqrt{2x - 1}$ **(c)** $2x^3 - 5x^2 + 4x - 28{,}225 = 0$ **(d)** 7 in., 24 in., and 25 in. **59.** **(a)** about 66.15 in.$^3$ **(b)** $.54 \text{ in.} < x < 2.92 \text{ in.}$

**61. (a)**

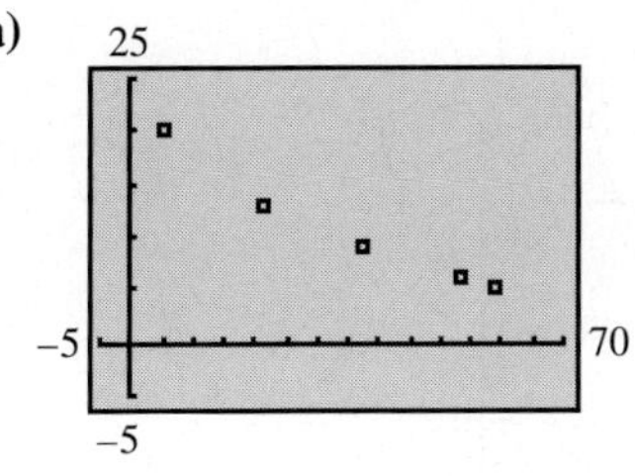


**(b)** $C(x) = .0035x^2 - .49x + 22$ **(c)** $C(x) = -.000068x^3 + .00987x^2 - .653x + 23$

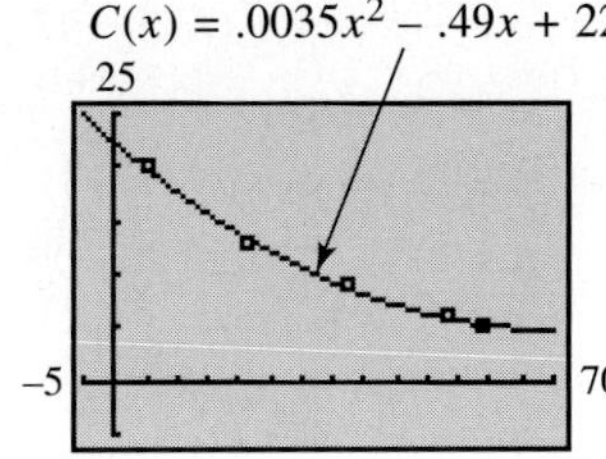


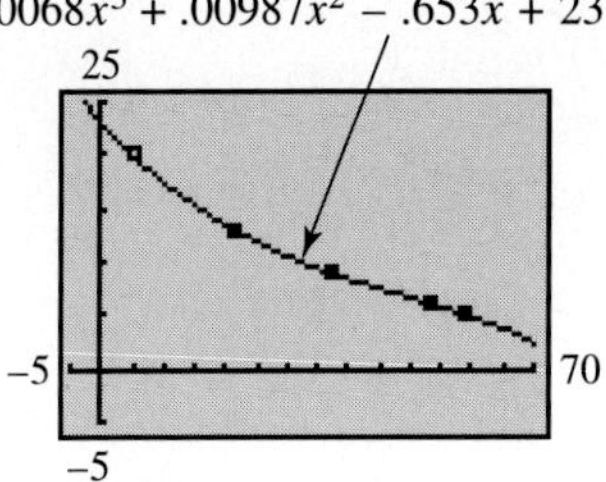


**(d)** The cubic function is a slightly better fit. **(e)** $0 \le x < 31.92$ **63.** at approximately 3.4 sec

## Reviewing Basic Concepts *(page 261)*

**1.** 30 **2.** no **3.** $P(x) = (x + 2)(2x - 3)(x - 2 - i)(x - 2 + i)$ **4.** $P(x) = x^3 - \frac{3}{2}x^2 + x - \frac{3}{2}$

**5.** $P(x) = x^4 + 6x^3 + 5x^2 + 8x + 80$ **6.** $-2, -1, \frac{5}{2}$ **7.** $\left\{\pm\sqrt{\frac{6 + \sqrt{33}}{3}}, \pm\sqrt{\frac{6 - \sqrt{33}}{3}}\right\}$

**8.** $P(x) = -.31x^3 + 5.8x^2 - 15x + 9$; $P(9) \approx 118$, which means that in 1999, about 118 million debit cards were issued.

## Chapter 3 Review Exercises *(page 265)*

**1.** $18 - 4i$ **3.** $14 - 52i$ **5.** $\frac{1}{10} + \frac{3}{10}i$ **7.** $(-\infty, \infty)$ **9.** ↖↗ **11.** $-8$ **13.** $\left[\frac{3}{2}, \infty\right); \left(-\infty, \frac{3}{2}\right]$

**15.** The graph intersects the $x$-axis at $-1$ and 4, supporting the answer in (a). It lies above the $x$-axis when $x < -1$ or $x > 4$, supporting the answer in (b). It lies on or below the $x$-axis when $x$ is between $-1$ and 4 inclusive, supporting the answer in (c). **17.** Since the discriminant is greater than 0, there are two $x$-intercepts. **19.** $(1.04, 6.37)$ **21. (a)** 4 **(b)** 1 **(c)** two **(d)** none **23.** 25 sec **25.** 6.3 sec and 43.7 sec **27. (a)** $V(x) = 12x^2 - 128x + 256$ **(b)** 20 in. by 60 in. **(c)** One way is to graph $y_1 = V(x)$ and $y_2 = 2496$ and show that the graphs intersect at $x = 20$.

**29.** $Q(x) = x^2 + 4x + 1$; $R = -7$ **31.** $-1$ **33.** 28 **35.** $7 - 2i$

In Exercises 37 and 39, other answers are possible.

**37.** $P(x) = x^3 - 13x^2 + 46x - 48$ **39.** $P(x) = x^4 + 5x^3 + x^2 - 9x + 2$ **41.** yes

**43.** Any polynomial that can be factored as $P(x) = a(x - b)^2(x - c)^2$ works. One example is $P(x) = 2(x - 1)^2(x - 3)^2$.

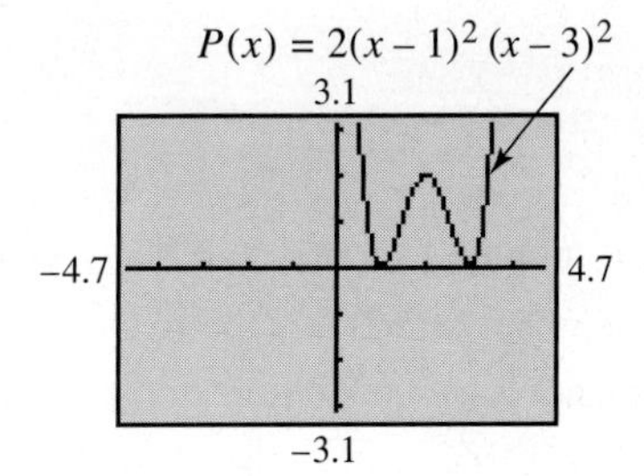


**45.** $-3, 4, 1 - i, 1 + i$ **47.** $-2, \frac{1}{3}, 4$ **49.** $(x - 3)(x^2 + x - 1)$ **53.** $\{0, -1 + 2i, -1 - 2i\}$; 0 is the only $x$-intercept.

**55.** even **57.** positive **59.** $(-\infty, a) \cup (b, c)$ **61.** $\{d, h\}$ **63.** Since $f(x)$ has three real zeros and a polynomial of degree 3 can have at most three zeros, there can be no other zeros, real or imaginary. **65.** Because $Y_1 = -X^3 - 5X + 4$ is a polynomial function, the intermediate value theorem indicates that there is a real zero between .7 and .8, as there is a sign change: $Y_1 > 0$ for $X = .7$, $Y_1 < 0$ for $X = .8$. **67.** true **69.** true **71.** false **73.** $(2, 0)$ **75.** $(-\infty, \infty)$

**77.** $\left\{-\sqrt{7}, -\frac{2}{3}, \sqrt{7}\right\}$ **79.** 4 in. by 4 in. by 4 in. **81.** 2.9 million

## Chapter 3 Test (page 269)

**1.** **(a)** $20 - 9i$ **(b)** $4 - i$ **(c)** $i$ **(d)** $12 + 16i$ **2.** **(a)** $(-1, 8)$ **(b)**

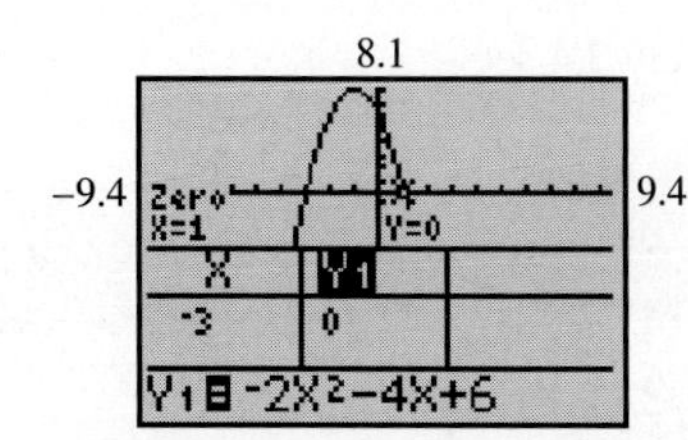


**(c)** $-3, 1$

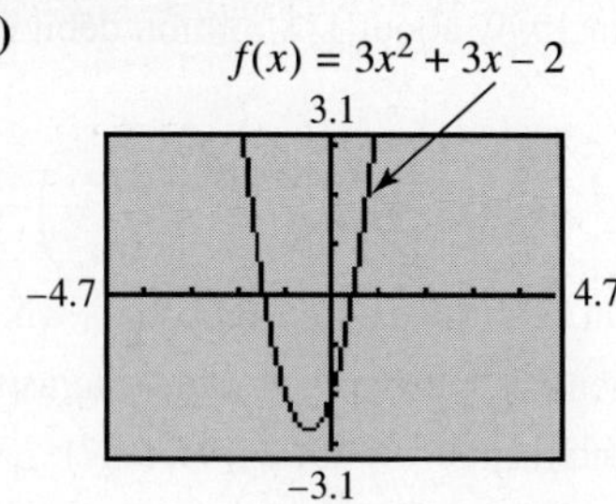


**(d)** 6 **(e)** domain: $(-\infty, \infty)$; range: $(-\infty, 8]$ **(f)** increasing: $(-\infty, -1]$; decreasing: $[-1, \infty)$

**3.** **(a)** $\left\{\frac{-3 \pm \sqrt{33}}{6}\right\}$ **(b)**

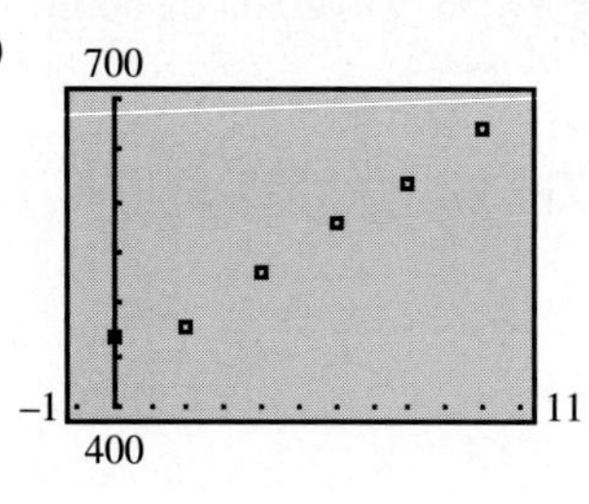


**(i)** $\left(\frac{-3 - \sqrt{33}}{6}, \frac{-3 + \sqrt{33}}{6}\right)$

**(ii)** $\left(-\infty, \frac{-3 - \sqrt{33}}{6}\right] \cup \left[\frac{-3 + \sqrt{33}}{6}, \infty\right)$

**4.** 15 in. by 12 in. by 4 in.

**5.** **(a)**

700
−1
11
400

**(b)** $f(x) = 2x^2 + 470$ **(c)** $g(x) = .737x^2 + 13.8x + 461$; function $g$

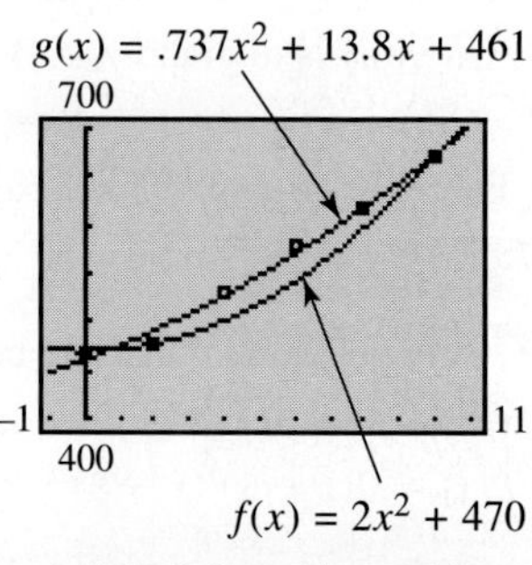


**6.** **(a)** The turning points are approximately (1.6, 3.6), (3, 1.2), (4.4, 3.6). **(b)** After 1.6 min the runner is 360 ft from the starting line. The runner turns and runs toward the starting line. After 3 min the runner is 120 ft from the starting line, turns, and runs away from the starting line. After 4.4 min the runner is again 360 ft from the starting line. The runner turns and runs back to the starting line.

**7.** **(a)** $-1 + i\sqrt{3}, -1 - i\sqrt{3}$

**(b)** ↰ ↱

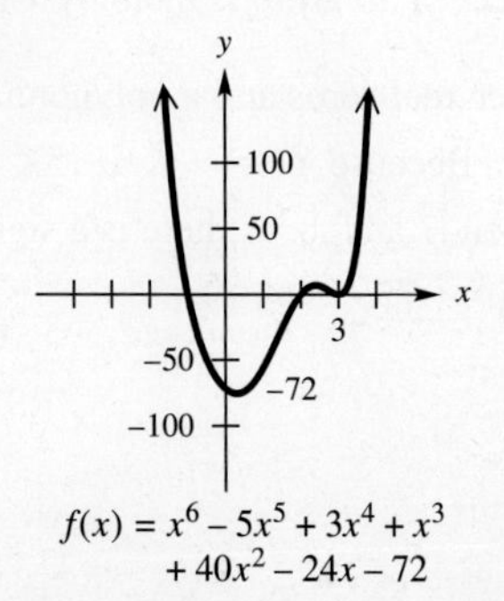


**8.** **(a)** $\left\{-\frac{5}{2}, \frac{5}{2}, -i, i\right\}$ **(b)**

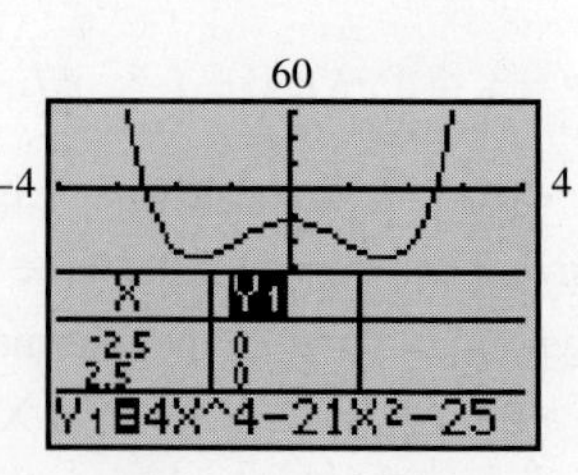


**(c)** It is symmetric with respect to the $y$-axis.

**(d)** **(i)** $\left(-\infty, -\frac{5}{2}\right] \cup \left[\frac{5}{2}, \infty\right)$ **(ii)** $\left(-\frac{5}{2}, \frac{5}{2}\right)$

**9.** **(a)** approximately .189, 1, approximately 3.633 **(b)** two

**10.** **(a)** $f(x) = -.249x^3 + 4.47x^2 + .212x + 467$ **(b)** $g(x) = .0977x^4 - 2.20x^3 + 16.5x^2 - 22.1x + 470$

**(c)**

**(d)** cubic: 679 million; quartic: 726 million; The quartic function is a better estimate because it continues to increase, while the cubic function turns downward.

## CHAPTER 4 RATIONAL, POWER, AND ROOT FUNCTIONS

### 4.1 Exercises *(page 278)*

**1.** $(-\infty, 0) \cup (0, \infty)$; $(-\infty, 0) \cup (0, \infty)$ **3.** none; $(-\infty, 0) \cup (0, \infty)$; none **5.** $x = 3$; $y = 2$ **7.** even; symmetry with respect to the $y$-axis **9.** A, B, C **11.** A **13.** A **15.** C

In Exercises 17–27, we give the domain and then the range below the traditional graph.

**17.** To obtain the graph of $f$, stretch the graph of $y = \frac{1}{x}$ vertically by a factor of 2.

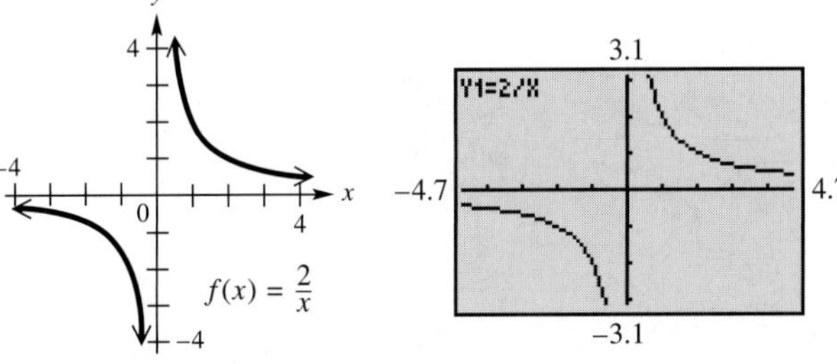

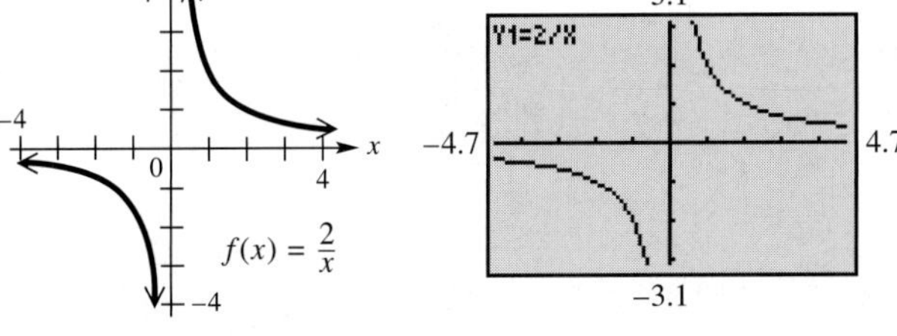

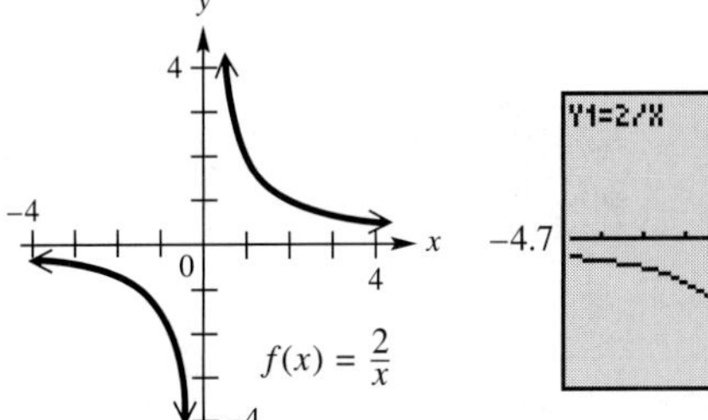


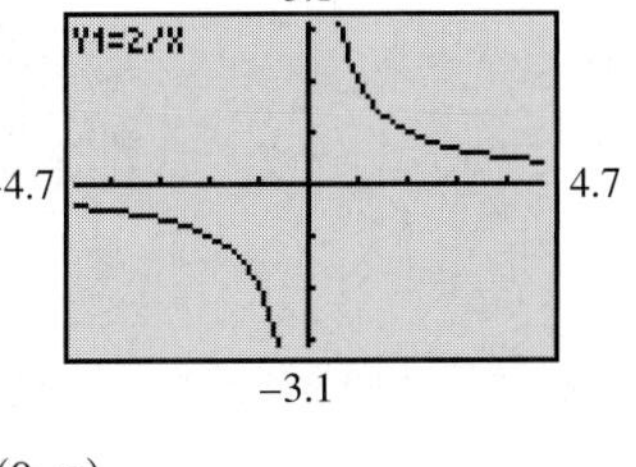


$(-\infty, 0) \cup (0, \infty)$; $(-\infty, 0) \cup (0, \infty)$

**19.** To obtain the graph of $f$, shift the graph of $y = \frac{1}{x}$ to the left 2 units.

$(-\infty, -2) \cup (-2, \infty)$; $(-\infty, 0) \cup (0, \infty)$

**21.** To obtain the graph of $f$, shift the graph of $y = \frac{1}{x}$ upward 1 unit.

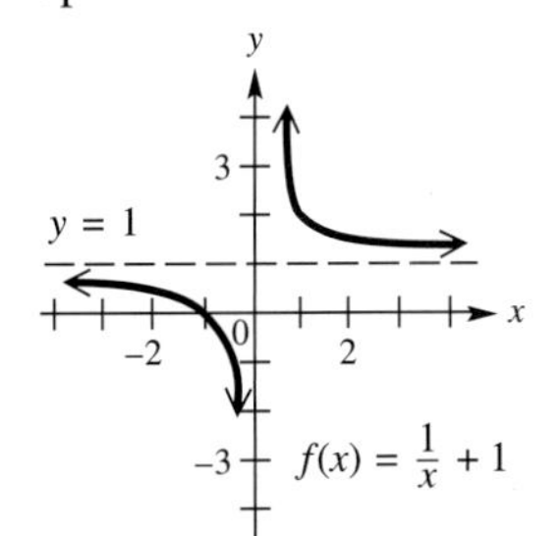


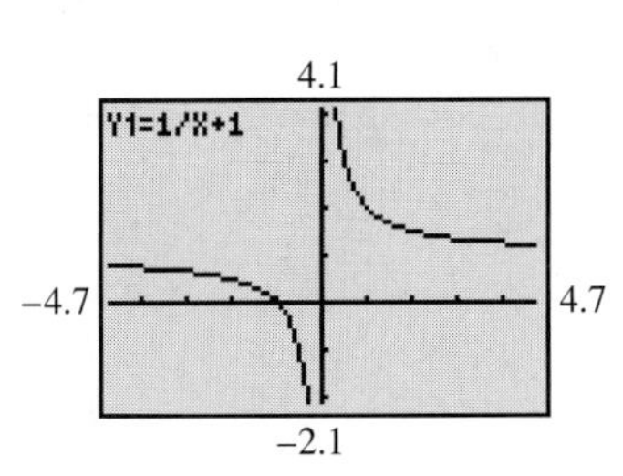


$(-\infty, 0) \cup (0, \infty)$; $(-\infty, 1) \cup (1, \infty)$

**23.** To obtain the graph of $f$, stretch the graph of $y = \frac{1}{x^2}$ vertically by a factor of 2, and reflect across the $x$-axis.

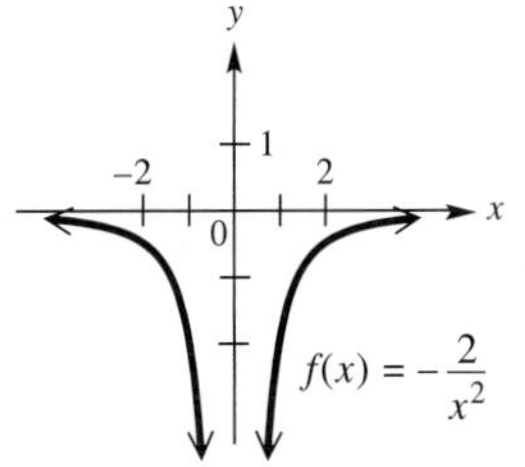


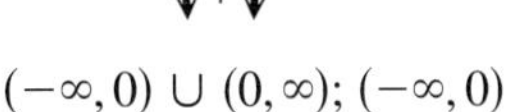


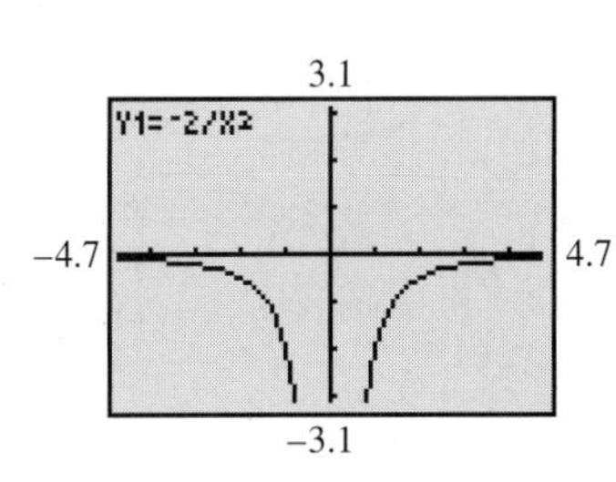


$(-\infty, 0) \cup (0, \infty)$; $(-\infty, 0)$

**25.** To obtain the graph of $f$, shift the graph of $y = \frac{1}{x^2}$ to the right 3 units.

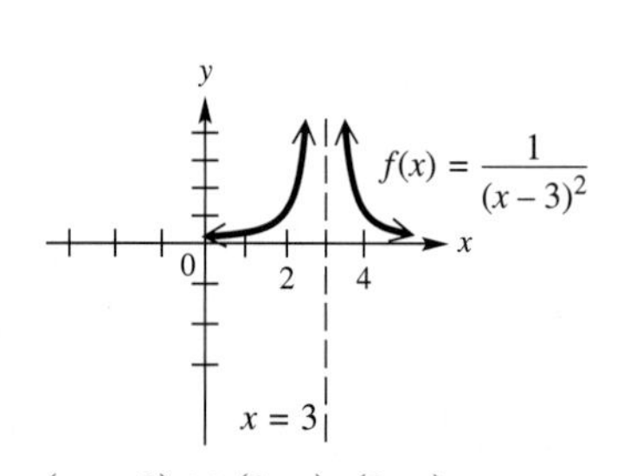


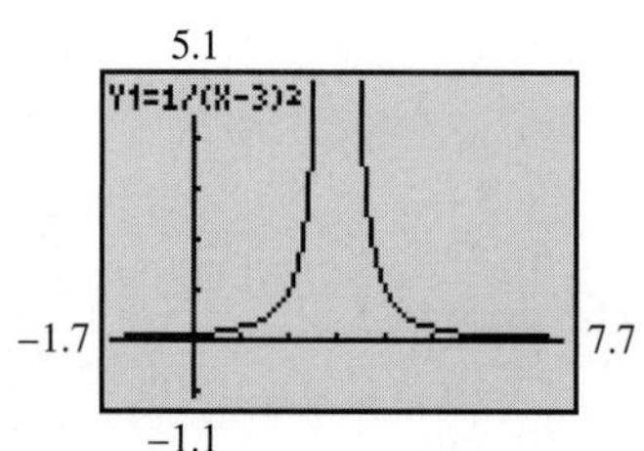


$(-\infty, 3) \cup (3, \infty)$; $(0, \infty)$

**27.** To obtain the graph of $f$, shift the graph of $y = \frac{1}{x^2}$ to the left 2 units, reflect across the $x$-axis, and shift 3 units downward.

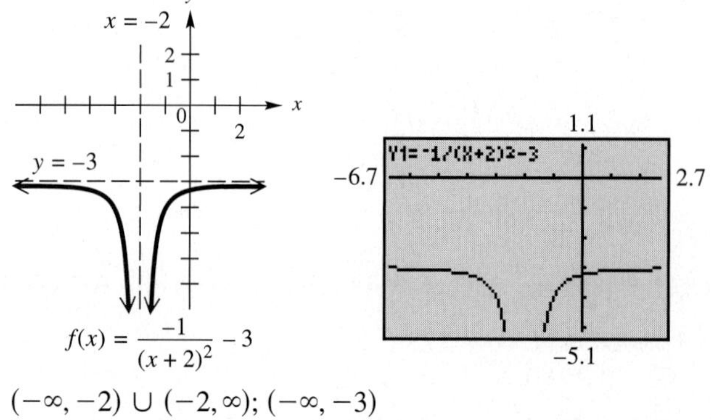


$(-\infty, -2) \cup (-2, \infty)$; $(-\infty, -3)$

**29.** C **31.** B

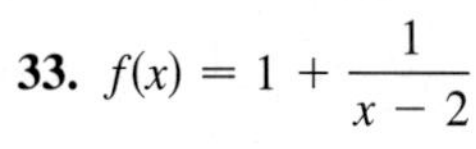
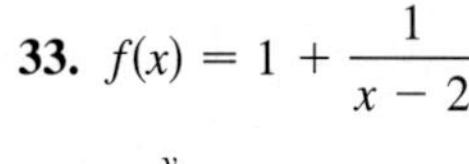

**33.** $f(x) = 1 + \frac{1}{x - 2}$

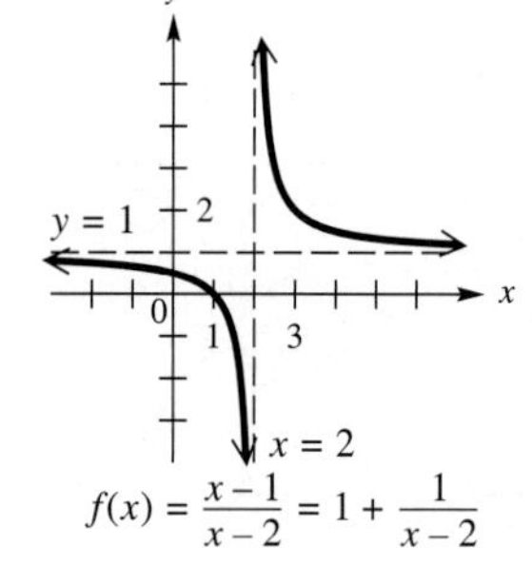


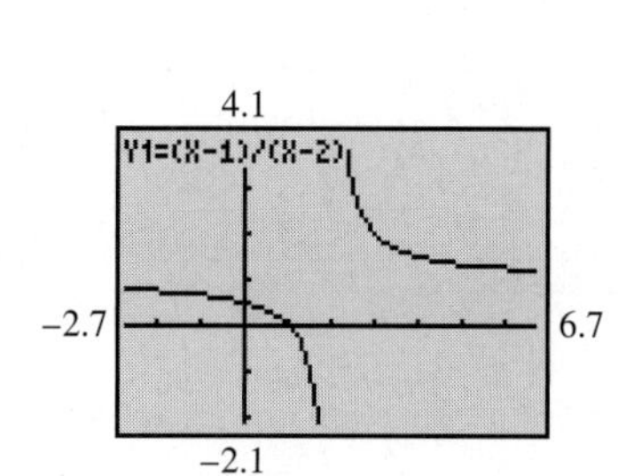


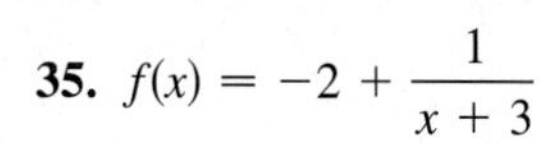

**35.** $f(x) = -2 + \frac{1}{x + 3}$

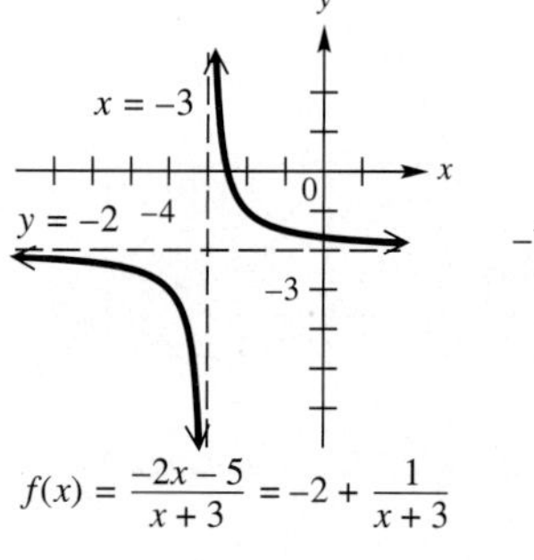


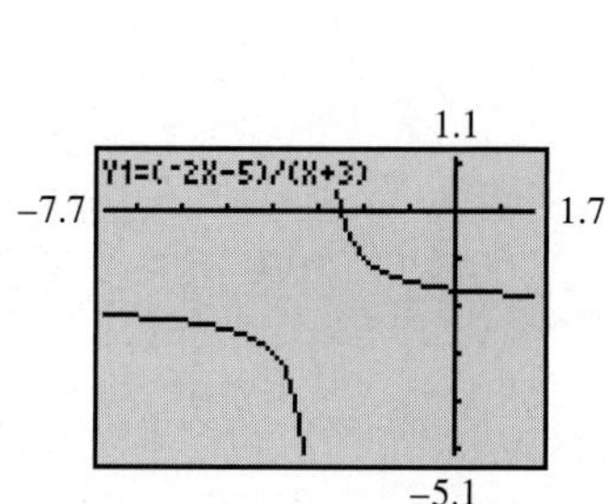


**37.** $f(x) = 2 + \frac{1}{x - 3}$

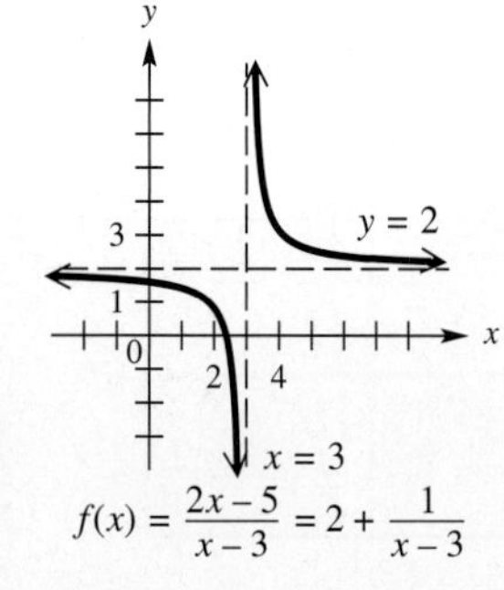


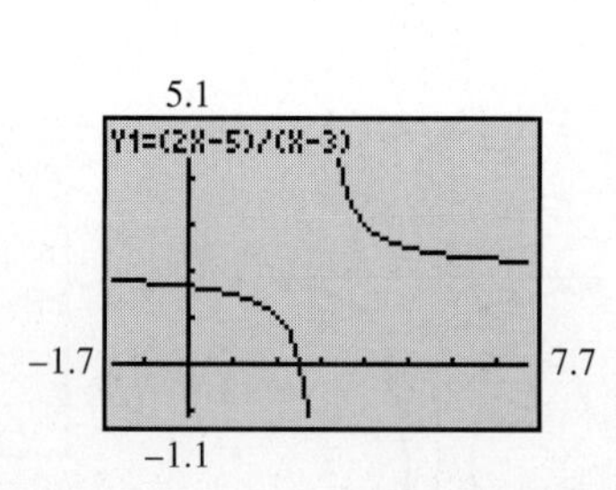


## 4.2 Exercises *(page 290)*

**1.** D **3.** G **5.** E **7.** F

In Exercises 9–15, V.A. represents vertical asymptote, H.A. represents horizontal asymptote, and O.A. represents oblique asymptote.

**9.** V.A.: $x = 5$; H.A.: $y = 0$ **11.** V.A.: $x = -\frac{1}{2}$; H.A.: $y = -\frac{3}{2}$ **13.** V.A.: $x = -3$; O.A.: $y = x - 3$

**15.** V.A.: $x = -2$, $x = \frac{5}{2}$; H.A.: $y = \frac{1}{2}$ **17.** A

**19.**

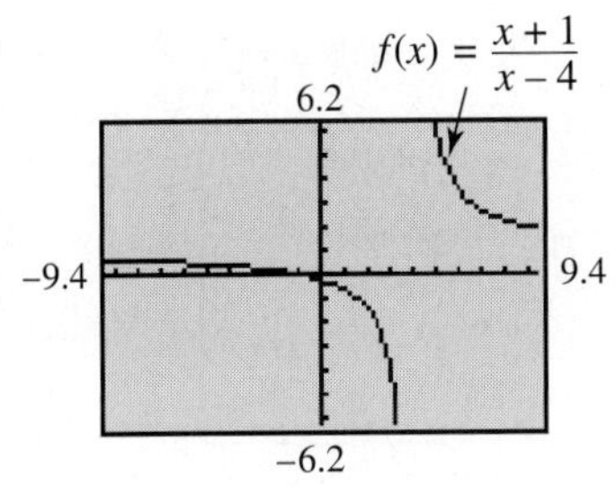


**21.**

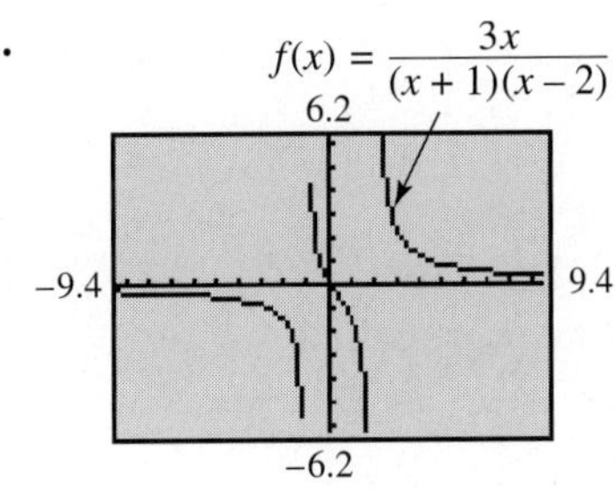


**23.**

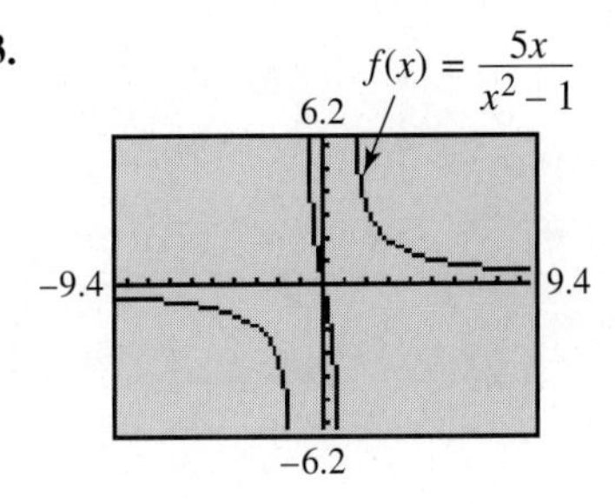


**25.**

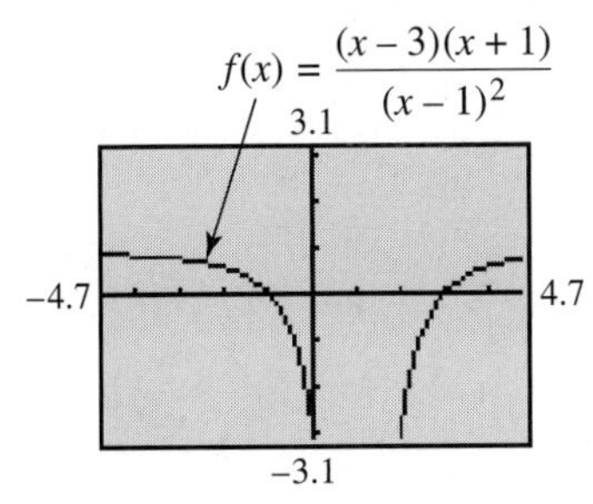


**27.**

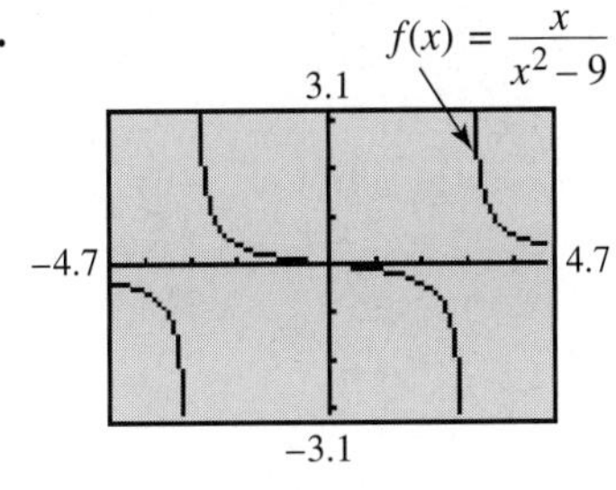


**29.**

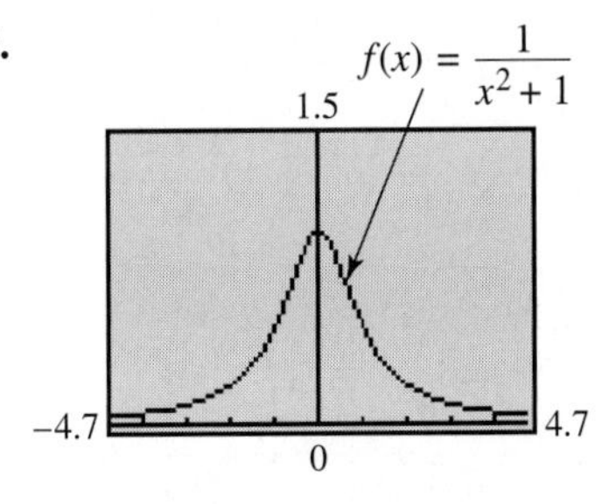


**31.**

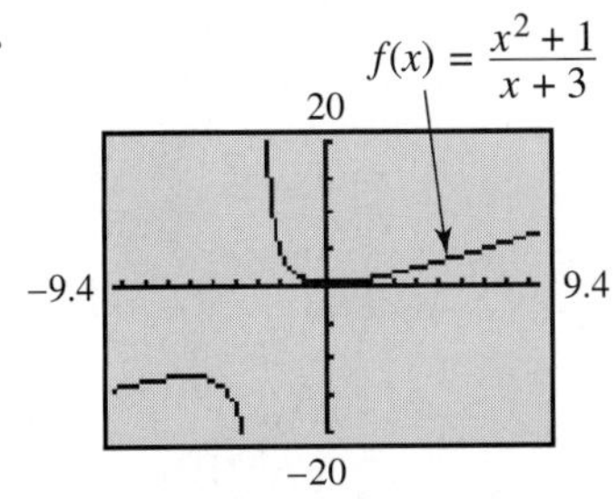


**33.**

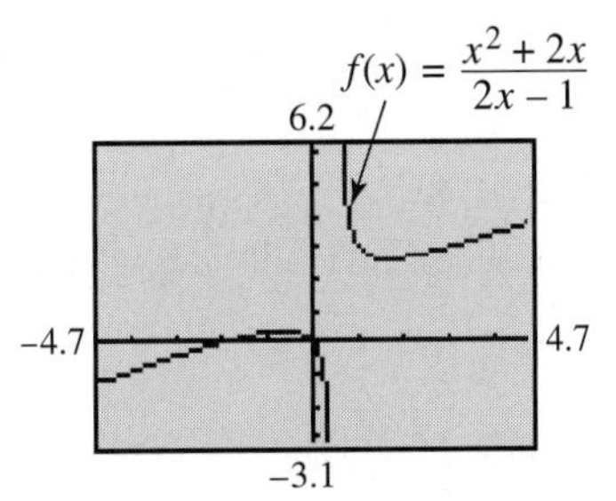


**35.**

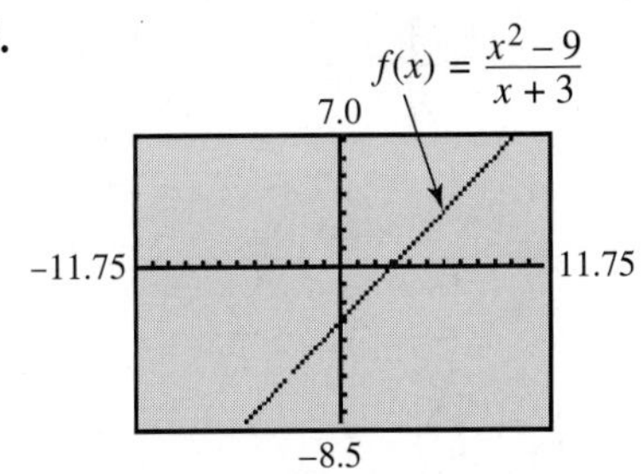


**37.** $p = 4; q = 2$ **39.** $p = -2; q = -1$

Answers may vary in Exercises 41 and 43.

**41.** $f(x) = \dfrac{(x-3)(x+2)}{(x-2)(x+2)} = \dfrac{x^2 - x - 6}{x^2 - 4}$ **43.** $f(x) = \dfrac{x-2}{x(x-4)} = \dfrac{x-2}{x^2 - 4x}$ **45.** The complex solutions are $-\dfrac{1}{2} \pm \dfrac{\sqrt{15}}{2}i$.

There are no real solutions and no vertical asymptotes.

**47. (a)**

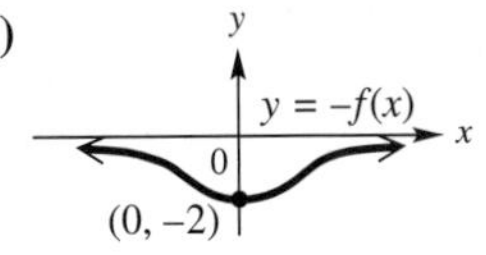


**(b)**

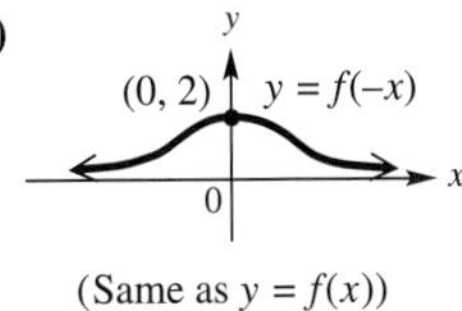


(Same as $y = f(x)$)

**48. (a)**

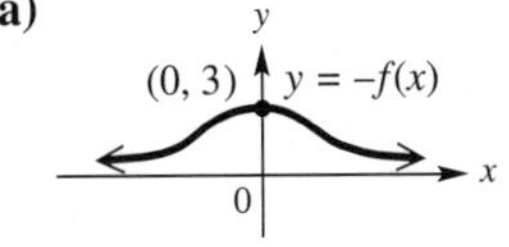


**(b)**

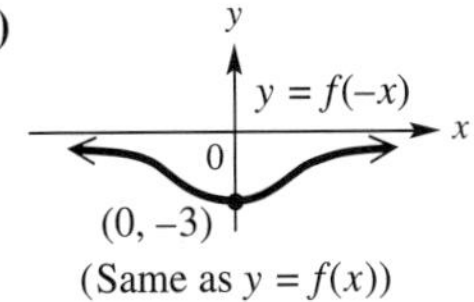


(Same as $y = f(x)$)

**49. (a)**

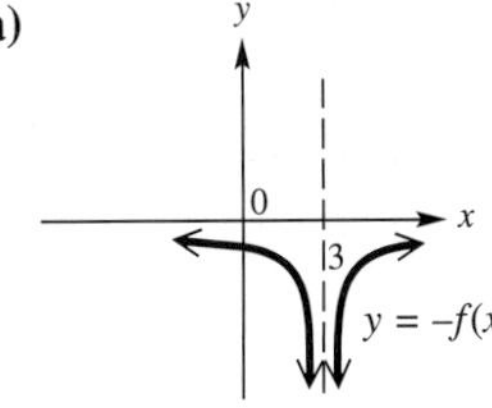


**(b)**

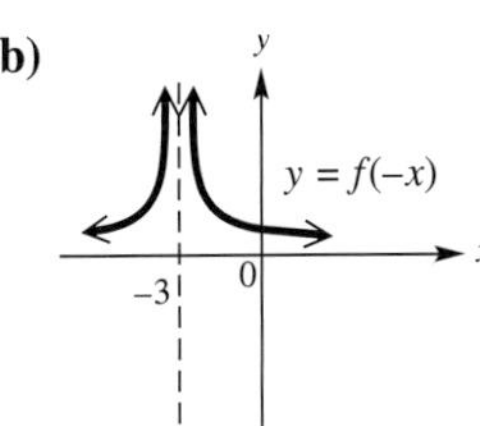

**50. (a)**

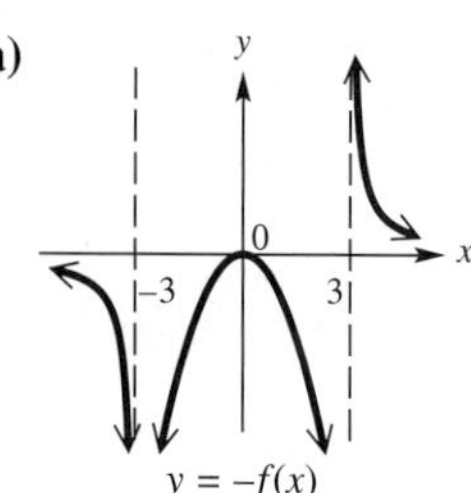


**(b)**

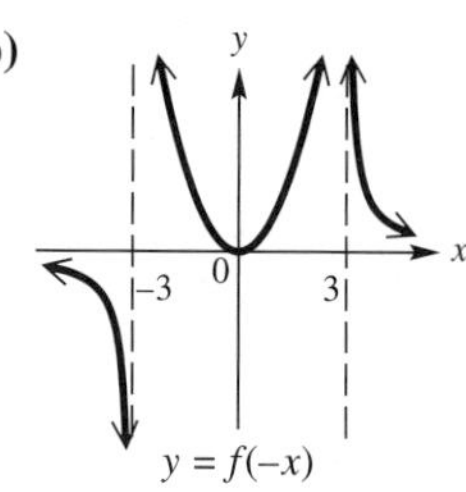


**51.** $y = -2x - 8$

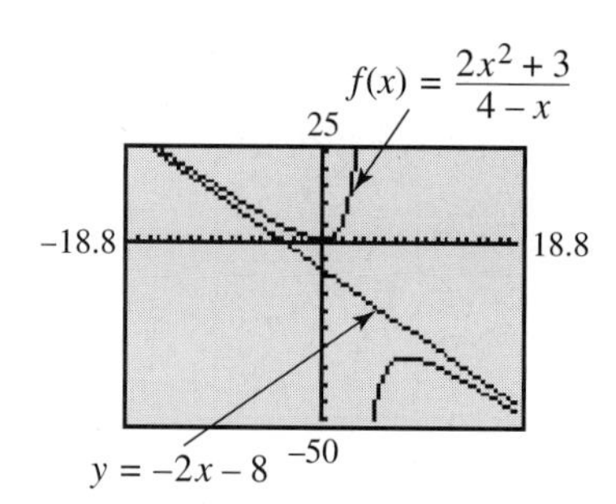


**53.** $y = -x + 3$

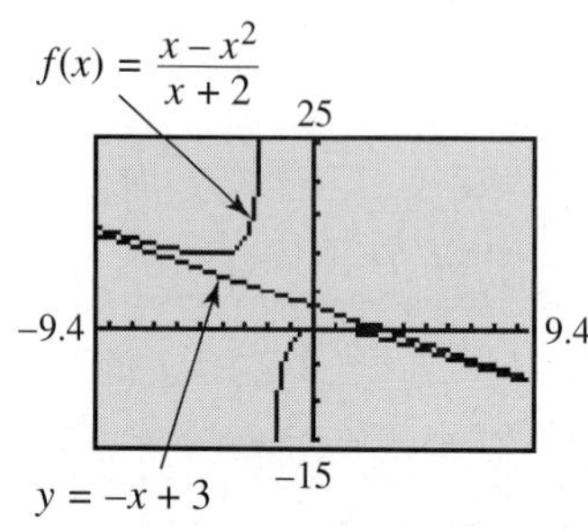


**55. (a)** $y = x + 1$ **(b)** at $x = 0$ and $x = 1$ **(c)** above

**57.**

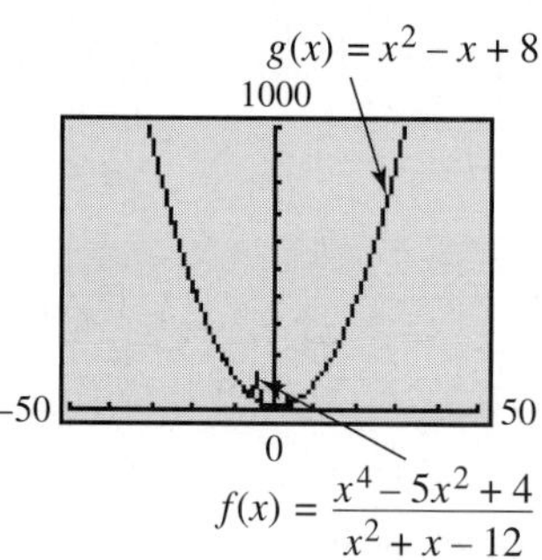


In this window, the two graphs seem to overlap (coincide), suggesting that as $|x| \to \infty$, the graph of $f$ approaches the graph of $g$, giving an asymptotic effect.

**59. (a)** $g(x) = x - 3$ **(b)** $f$ is undefined at X $= -3$, as indicated by the error message. **(c)** The "hole" is at $(-3, -6)$.

| X | Y1 | Y2 |
|---|---|---|
| -6 | -9 | -9 |
| -5 | -8 | -8 |
| -4 | -7 | -7 |
| -3 | ERROR | -6 |
| -2 | -5 | -5 |
| -1 | -4 | -4 |
| 0 | -3 | -3 |

Y1=ERROR

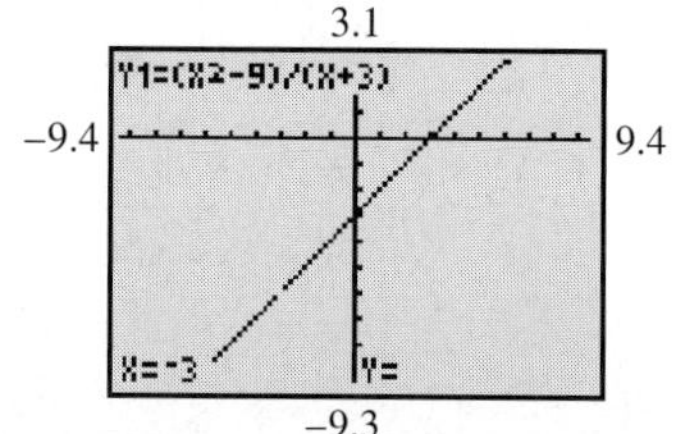


**61. (a)** The horizontal asymptote has equation $y = 0$ (the $x$-axis). **(b)** The horizontal asymptote has equation $y = \frac{a}{b}$, where $a$ is the leading coefficient of $p(x)$ and $b$ is the leading coefficient of $q(x)$. **(c)** The oblique asymptote has equation $y = ax + b$, where $ax + b$ is the quotient (with remainder disregarded) found by dividing $p(x)$ by $q(x)$.

## 4.3 Exercises *(page 301)*

**1. (a)** $\emptyset$ **(b)** $(-\infty, -2)$ **(c)** $(-2, \infty)$ **3. (a)** $\{-1\}$ **(b)** $(-1, 0)$ **(c)** $(-\infty, -1) \cup (0, \infty)$ **5. (a)** $\{0\}$ **(b)** $(-2, 0) \cup (2, \infty)$ **(c)** $(-\infty, -2) \cup (0, 2)$ **7. (a)** $\emptyset$ **(b)** $(-\infty, 0) \cup (0, \infty)$ **(c)** $\emptyset$ **9. (a)** $\{.75\}$ **(b)** $(-\infty, .75) \cup (2, \infty)$ **(c)** $(.75, 2)$ **11. (a)** $\{3\}$ **(b)** $(-5, 3]$ **(c)** $(-\infty, -5) \cup [3, \infty)$ **13. (a)** $\emptyset$ **(b)** $(-\infty, -2)$ **(c)** $(-2, \infty)$ **15. (a)** $\left\{\frac{9}{5}\right\}$

**(b)** $(-\infty, 1) \cup \left(\frac{9}{5}, \infty\right)$ **(c)** $\left(1, \frac{9}{5}\right)$ **17. (a)** $\{-2\}$ **(b)** $(-\infty, -2] \cup (1, 2)$ **(c)** $[-2, 1) \cup (2, \infty)$ **19. (a)** $\emptyset$ **(b)** $\emptyset$ **(c)** $(-\infty, 2) \cup (2, \infty)$ **21. (a)** $\emptyset$ **(b)** $(-\infty, -1)$ **(c)** $(-1, \infty)$ **23.** $(-\infty, \infty)$ **25.** $\emptyset$ **27.** $\emptyset$ **29.** $(-\infty, \infty)$

**31.** $\{1\}$ **33.** $\{-3\}$ **35.** $\{-4\}$ **37.** $\{4, 9\}$ **39.** $\left\{\frac{-3 \pm \sqrt{29}}{2}\right\}$ **41.** $\left\{\frac{3 \pm \sqrt{3}}{3}\right\}$ **43.** $\left\{\pm\frac{1}{2}, \pm i\right\}$ **45.** $\emptyset$

**47.** $\{-10\}$ **49.** $\left\{\frac{27}{56}\right\}$

**51.**

$$\frac{x-2}{x+3} \leq 2 \quad (x \neq -3)$$
$$(x+3)^2\left(\frac{x-2}{x+3}\right) \leq 2(x+3)^2$$
$$(x+3)(x-2) \leq 2(x^2 + 6x + 9)$$
$$x^2 + x - 6 \leq 2x^2 + 12x + 18$$
$$0 \leq x^2 + 11x + 24$$
$$x^2 + 11x + 24 \geq 0$$

**52.** $(3, \infty)$ **53.** $(-1, 1]$ **54.** $\left(-5, \frac{1}{2}\right]$ **55.** $\emptyset$ **56.** $\emptyset$

**57. (a)** $\{-3.54\}$ **(b)** $(-\infty, -3.54) \cup (1.20, \infty)$ **(c)** $(-3.54, 1.20)$ **59.** C **61.** D

**63. (a)** $y = 10$



**(b)** When $x = 0$, there are 1,000,000 insects. **(c)** It starts to level off at 10,000,000. **(d)** The horizontal asymptote $y = 10$ represents the limiting population after a long time.

**65. (a)** about 12.4 cars per min **(b)** 3 **67.** There are two possible solutions: width = 7 in., length = 14 in., height = 2 in.; width ≈ 2.266 in., length ≈ 4.532 in., height ≈ 19.086 in.

**69. (a)** $f(400) = \frac{2540}{400} = 6.35$ in.; a curve designed for 60 mph with a radius of 400 ft should have the outer rail elevated 6.35 in. **(b)**



As the radius $x$ of the curve increases, the elevation of the outer rail decreases. **(c)** The horizontal asymptote is $y = 0$. As the radius of the curve increases without bound ($x \to \infty$), the tracks become straight and no elevation or banking ($y \to 0$) is necessary. **(d)** 200 ft

**71. (a)** $P(9) = \frac{9-1}{9} \approx .89 = 89\%$ **(b)** $P(1.9) = \frac{1.9-1}{1.9} \approx .47 = 47\%$ **73. (a)** $D(-.1) \approx 417$; the braking distance for a car traveling at 50 mph on a 10% downhill grade is about 417 ft. **(b)** As the downhill grade $x$ increases, the braking distance increases, which agrees with driving experience. **(c)** As the downhill grade $x$ gets very close to a 30% grade, the braking distance increases without a maximum, which means stopping in time becomes impossible. **(d)** $x \approx -.062$ or 6.2% downhill **75.** $\frac{32}{15}$

**77.** $\frac{18}{125}$ **79.** increases; decreases **81.** It becomes half as much. **83.** It becomes 27 times as much. **85.** 21

**87.** $\frac{2}{9}$ ohm **89.** 17.8 lb **91.** 799.5 cm$^3$ **93.** 5.1 **95.** $-1.4$

## Reviewing Basic Concepts *(page 308)*

**1.** 

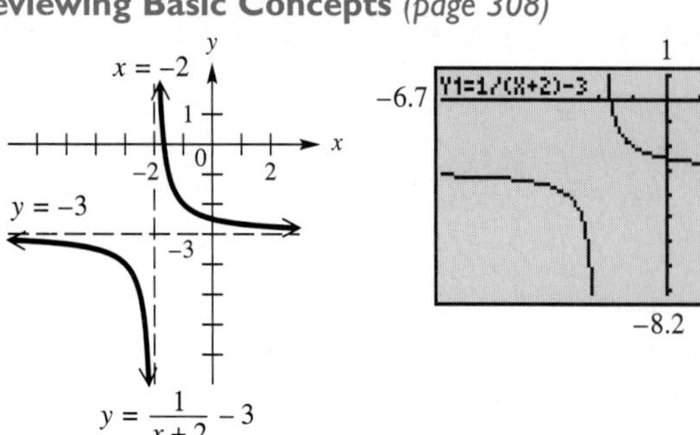


**2.** $(-\infty, -1) \cup (-1, 1) \cup (1, \infty)$ **3.** $x = 6$ **4.** $y = 1$
**5.** $y = x - 2$

**6.**

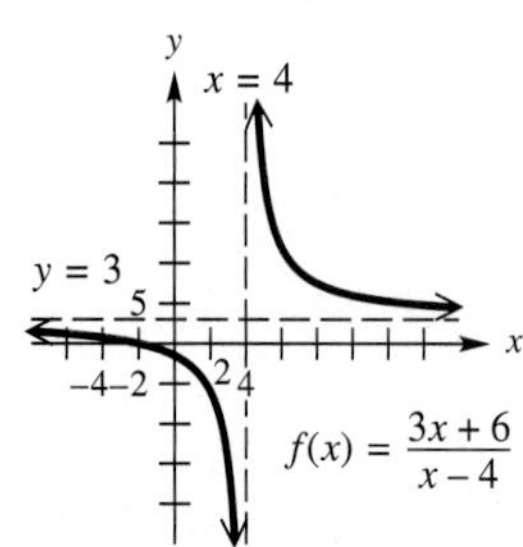


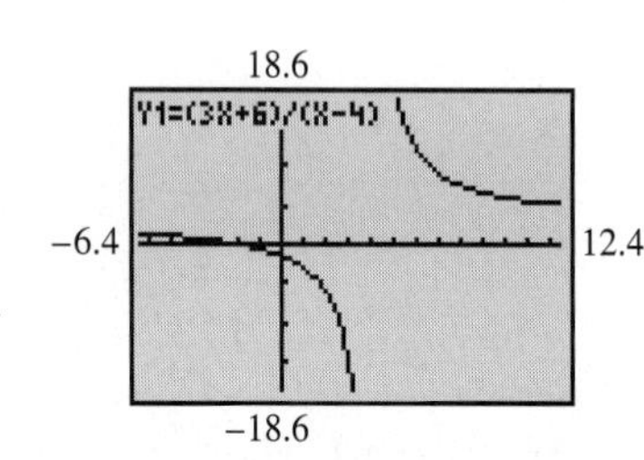


**7.** **(a)** $\{-4\}$ **(b)** $(-\infty, -4) \cup (2, \infty)$ **(c)** $(-4, 2)$
**8.** $\left(-\frac{1}{3}, \frac{3}{2}\right)$ **9.** inversely; height; 24
**10.** 4.3 vibrations per sec

## 4.4 Exercises *(page 317)*

**1.** 13 **3.** −2 **5.** 729 **7.** $\frac{1}{25}$ **9.** 100 **11.** rational approximation; −1.587401052 **13.** exact value; −5
**15.** rational approximation; 1.44224957 **17.** exact value; 2.65 **19.** rational approximation; −2.571281591
**21.** rational approximation; 1.464591888 **23.** rational approximation; .4252903703
**25.** exact value; 2 **27.** rational approximation; 1.174618943 **29.** rational approximation; 1.267463962
**31.** rational approximation; .0322515344 **33.** rational approximation; 1.181352075 **35.** **(a)** 6.2449979984
**(b)** 11.9916637711 **(c)** 95.2417975471 **37.** **(a)** .125; $\frac{1}{8}$ **(b)** $\left(\sqrt[4]{16}\right)^{-3}$; $\sqrt[4]{16^{-3}}$ (There are other expressions.) Each is equal to .125. **(c)** Show that $.125 = \frac{1}{8}$.

**39.**

**40.** 

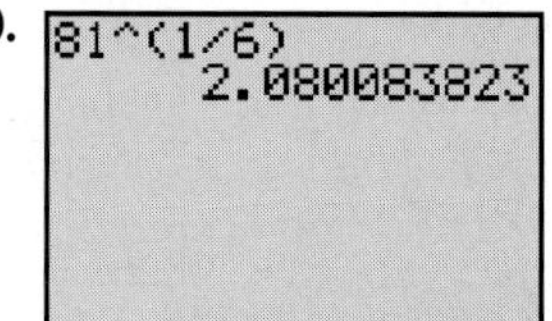


**41.**

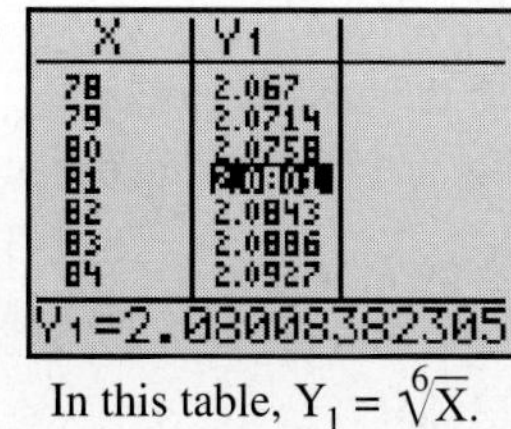


In this table, $Y_1 = \sqrt[6]{X}$.

**42.**

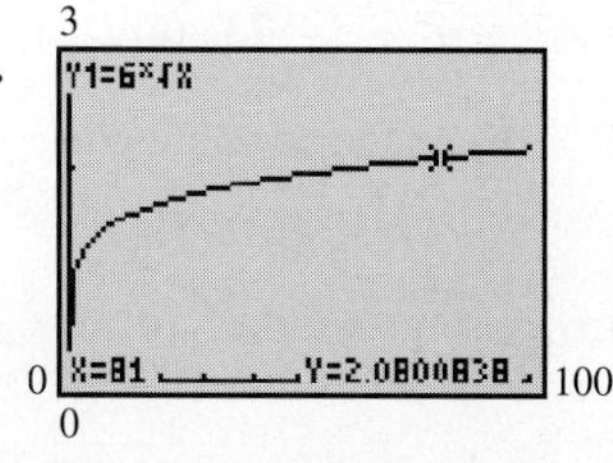


**43.** 3.2 ft² **45.** approximately 58.1 yr
**47.** **(a)** $a = 1960$ **(b)** $b \approx -1.2$
**(c)** $f(4) = 1960(4)^{-1.2} \approx 371$; if the zinc ion concentration reaches 371 mg per L, a rainbow trout will survive, on average, 4 min.

**49.** 1.06 g **51.** $a \approx 874.54$, $b \approx -.49789$ **53.** $\left[-\frac{5}{4}, \infty\right)$ **55.** $(-\infty, 6]$ **57.** $(-\infty, \infty)$ **59.** $[-7, 7]$
**61.** $[-1, 0] \cup [1, \infty)$

In Exercises 63–69, we give only the answers to parts (a)–(d), not the graphs.

**63.** **(a)** $[0, \infty)$ **(b)** $\left[-\frac{5}{4}, \infty\right)$ **(c)** none **(d)** $\{-1.25\}$ **65.** **(a)** $(-\infty, 0]$ **(b)** $(-\infty, 6]$ **(c)** none **(d)** $\{6\}$ **67.** **(a)** $(-\infty, \infty)$ **(b)** $(-\infty, \infty)$ **(c)** none **(d)** $\{3\}$ **69.** **(a)** $[0, 7]$ **(b)** $[-7, 0]$ **(c)** $[0, 7]$ **(d)** $\{-7, 7\}$ **71.** Rewrite as $y = 3\sqrt{x + 3}$. Shift the graph of $y = \sqrt{x}$ to the left 3 units and stretch vertically by a factor of 3. **73.** Rewrite as $y = 2\sqrt{x + 4} + 4$. Shift the graph of $y = \sqrt{x}$ to the left 4 units, stretch vertically by a factor of 2, and shift upward 4 units. **75.** Rewrite as $y = 3\sqrt[3]{x + 2} - 5$. Shift the graph of $y = \sqrt[3]{x}$ to the left 2 units, stretch vertically by a factor of 3, and shift downward 5 units.

**77.** The graph is a circle.
$y_1 = \sqrt{100 - x^2}$; $y_2 = -\sqrt{100 - x^2}$

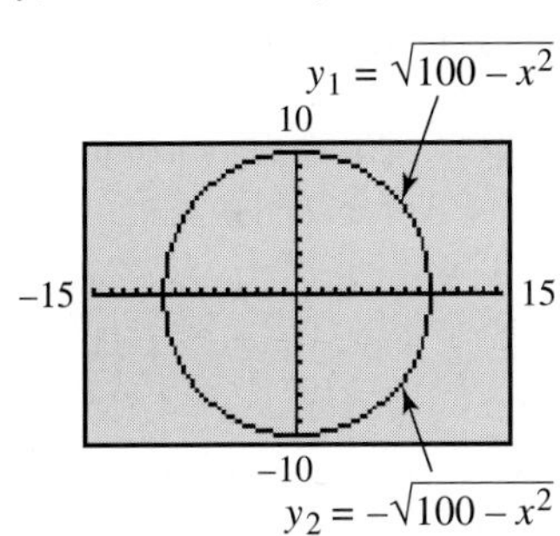


**79.** The graph is a (shifted) circle.
$y_1 = \sqrt{9 - (x - 2)^2}$; $y_2 = -\sqrt{9 - (x - 2)^2}$

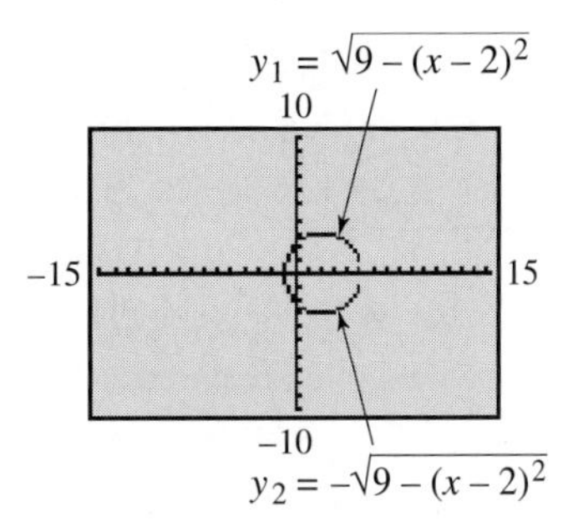


**81.** The graph is a horizontal parabola.
$y_1 = -3 + \sqrt{x}$; $y_2 = -3 - \sqrt{x}$

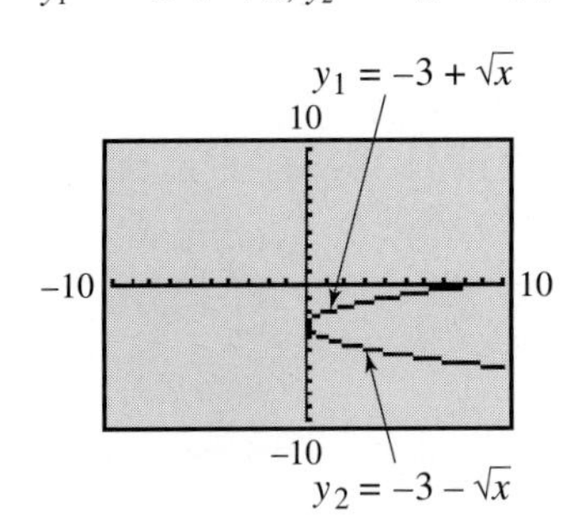


**83.** The graph is a horizontal parabola.
$y_1 = -2 + \sqrt{.5x + 3.5}$; $y_2 = -2 - \sqrt{.5x + 3.5}$

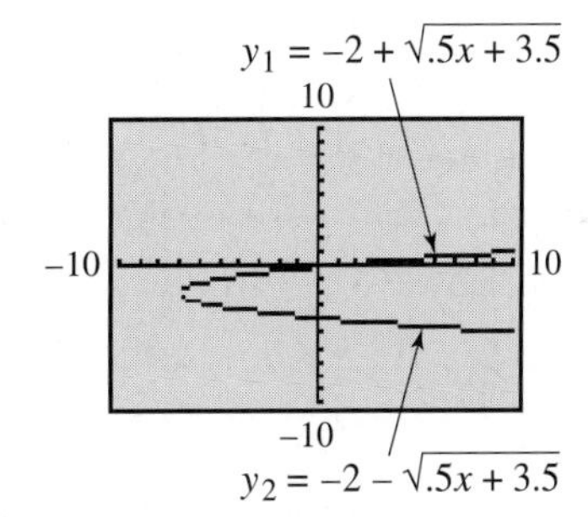


## 4.5 Exercises *(page 326)*

**1.** **(a)** $\{4\}$ **(b)** $[4, \infty)$ **(c)** $[-5, 4]$ **3.** **(a)** $\{-3\}$ **(b)** $(-\infty, -3)$ **(c)** $(-3, \infty)$

**5.**

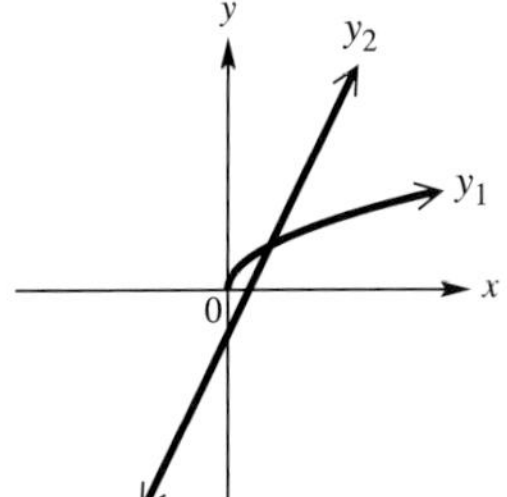


one real solution; $\{1\}$; $\frac{1}{4}$ is extraneous.

**7.**

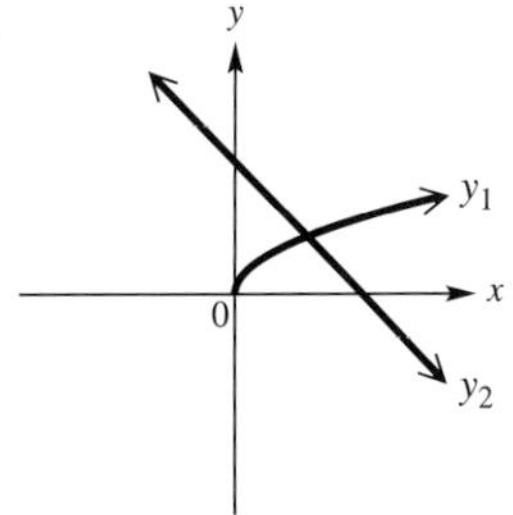


one real solution; $\left\{\frac{7 - \sqrt{13}}{2}\right\}$; $\frac{7 + \sqrt{13}}{2}$ is extraneous.

**9.**

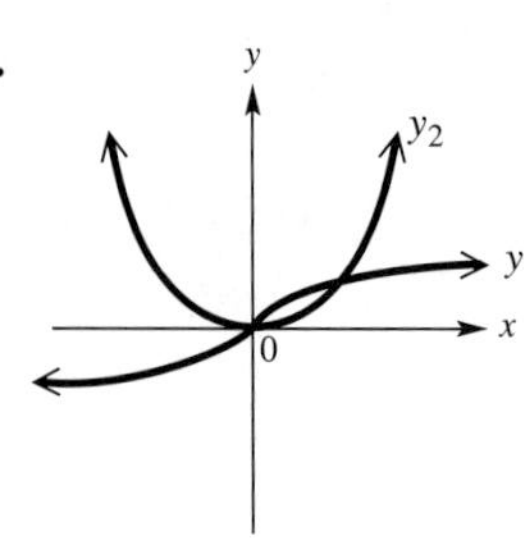


two real solutions; $\{0, 1\}$; no extraneous values

**11. (a)** $\{-1\}$ **(b)** $(-1, \infty)$ **(c)** $\left[-\frac{7}{3}, -1\right)$ **13. (a)** $\{3\}$ **(b)** $\left[-\frac{13}{4}, 3\right)$ **(c)** $(3, \infty)$ **15. (a)** $\{3\}$ **(b)** $\left[-\frac{1}{5}, 3\right)$ **(c)** $(3, \infty)$ **17. (a)** $\{2, 14\}$ **(b)** $(2, 14)$ **(c)** $(14, \infty)$ **19. (a)** $\{0, 3\}$ **(b)** $(-\infty, 0) \cup (3, \infty)$ **(c)** $(0, 3)$ **21. (a)** $\{0\}$ **(b)** $(0, \infty)$ **(c)** $\left[-\frac{1}{3}, 0\right)$ **23. (a)** $\{27\}$ **(b)** $[27, \infty)$ **(c)** $\left[\frac{5}{2}, 27\right]$ **25. (a)** $\{-8, 2\}$ **(b)** $(-\infty, -8) \cup (2, \infty)$ **(c)** $(-8, -6] \cup [0, 2)$ **27. (a)** $\left\{\frac{1}{4}, 1\right\}$ **(b)** $\left(-\infty, \frac{1}{4}\right) \cup (1, \infty)$ **(c)** $\left(\frac{1}{4}, 1\right)$

**29. (a)** $-60.9°$F **(b)** $-64.4°$F **30. (a)** $-63.4°$F **(b)** $-46.7°$F **31. (a)** $-63°$F **(b)** $-47°$F **32.** the formula in Exercise 30 **33.** 4.5 km per sec **35.** 2.5 sec **37.** 91 mph **39. (a)** $20 - x$ **(b)** $x$ must be between 0 and 20. **(c)** $AP = \sqrt{x^2 + 12^2}$; $BP = \sqrt{(20 - x)^2 + 16^2}$ **(d)** $f(x) = \sqrt{x^2 + 12^2} + \sqrt{(20 - x)^2 + 16^2}$, $0 < x < 20$

**(e)**

$f(x) = \sqrt{x^2 + 12^2} + \sqrt{(20 - x)^2 + 16^2}$
50
0
X=4 Y=35.276528
20
0

$f(4) \approx 35.28$; when the stake is 4 ft from the base of the 12-ft pole, approximately 35.28 ft of wire will be required. **(f)** When $x \approx 8.57$ ft, $f(x)$ is a minimum (approximately 34.41 ft). **(g)** This problem has examined how the total amount of wire used can be expressed in terms of the distance from the stake at $P$ to the base of the 12-ft pole. We find that the amount of wire used can be minimized when the stake is approximately 8.57 ft from the 12-ft pole.

**41.** Since $x \approx 1.31$, the hunter must travel $8 - x \approx 8 - 1.31 = 6.69$ mi along the river. **43.** After 1.38 hr (1:23 P.M.), the ships are 33.28 mi apart. **45.** $(4x - 4)^{1/3} = (x + 1)^{1/2}$ **46.** 6 **47.** $(4x - 4)^2 = (x + 1)^3$ **48.** $16x^2 - 32x + 16 = x^3 + 3x^2 + 3x + 1$, and thus, $x^3 - 13x^2 + 35x - 15 = 0$

**49.** three real roots

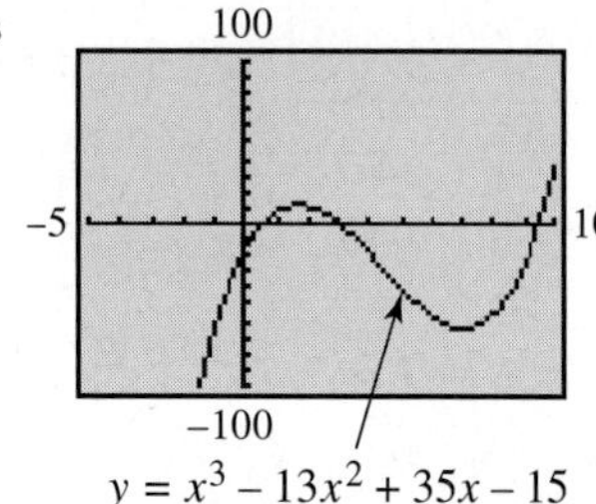


**50.**

$$\begin{array}{r|rrrr} 3 & 1 & -13 & 35 & -15 \\ & & 3 & -30 & 15 \\ \hline & 1 & -10 & 5 & 0 \end{array}$$

$P(3) = 0$

**51.** $P(x) = (x - 3)(x^2 - 10x + 5)$ **52.** $\{5 \pm 2\sqrt{5}\}$ **53.** 3; $5 + 2\sqrt{5}$; $5 - 2\sqrt{5}$

**54.** two real solutions

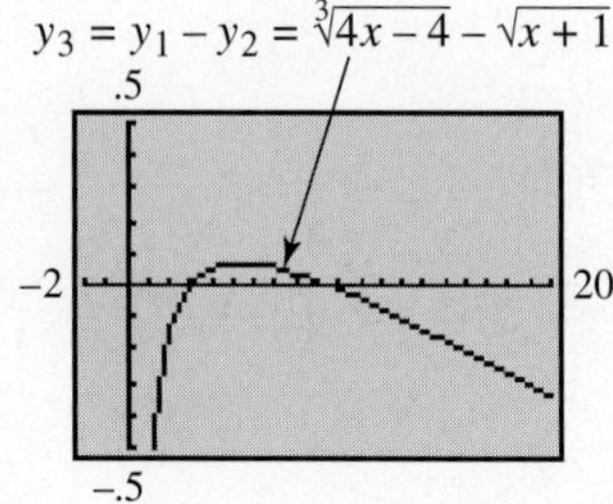


**55.** $\{3, 5 + 2\sqrt{5}\}$; for the calculator solution, $5 + 2\sqrt{5} \approx 9.47$. **56.** The solution set of the original equation is a subset of the solution set of the equation in Exercise 48. The extraneous solution $5 - 2\sqrt{5}$ was obtained when each side of the original equation was raised to the sixth power. **57.** $\{0, 1\}$ **59.** $\left\{-\frac{2}{9}, 2\right\}$ **61.** $\{2, 18\}$

### Reviewing Basic Concepts *(page 330)*

**1.**

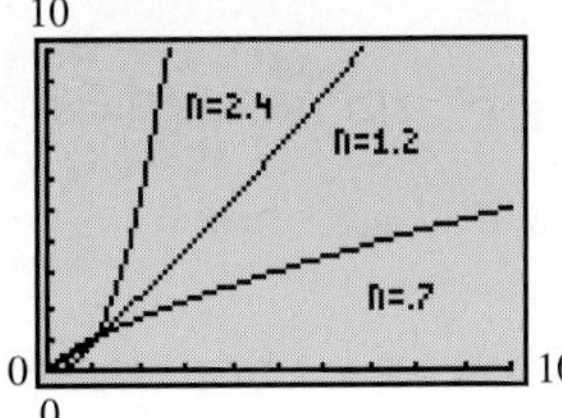


As the exponent increases in value, the curve rises more rapidly. **2.** .24 $m^2$

**3.** $y_1 = \sqrt{16 - x^2}$; $y_2 = -\sqrt{16 - x^2}$

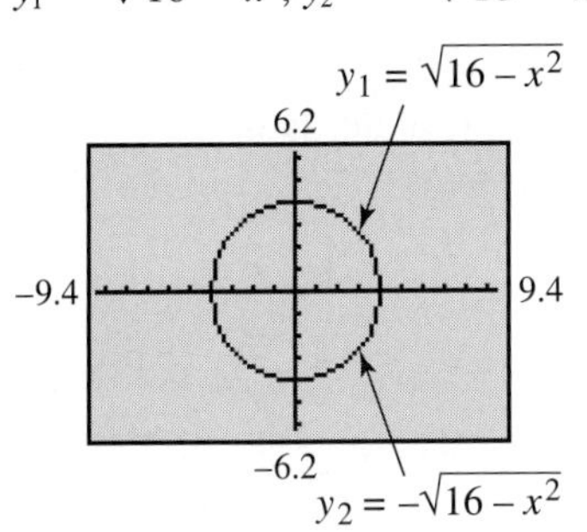


**4.** $y_1 = -2 + \sqrt{x - 2}$; $y_2 = -2 - \sqrt{x - 2}$

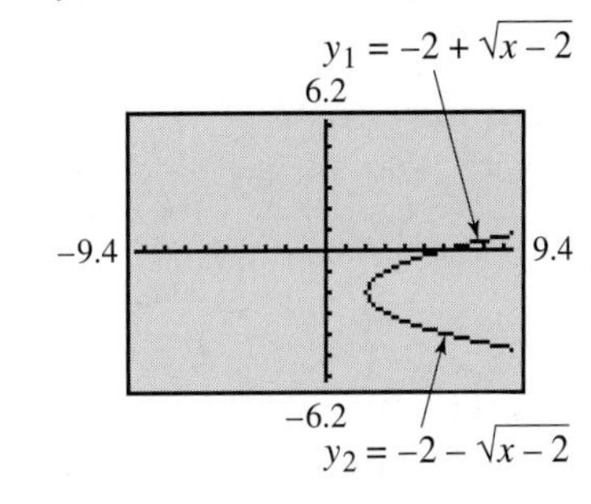


**5.** $\{4\}$ **6.** $(4, \infty)$ **7.** $\left[-\frac{4}{3}, 4\right)$ **8.** $\{-1\}$ **9.** Any value less than 15 causes the radicand to be negative, and the calculator will not return square roots of negative numbers in the table. **10.** **(a)** dog: 148; person: 69 **(b)** 6.4 in.

### Chapter 4 Review Exercises *(page 333)*

**1.** **(a)** Reflect the graph of $y = \frac{1}{x}$ across the $x$-axis, and shift upward 6 units.

**(b)**

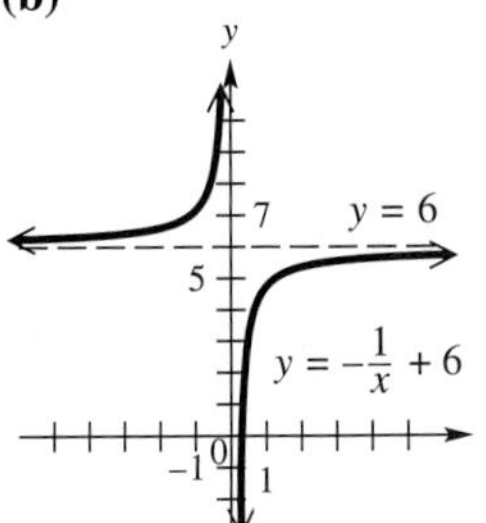


**(c)**

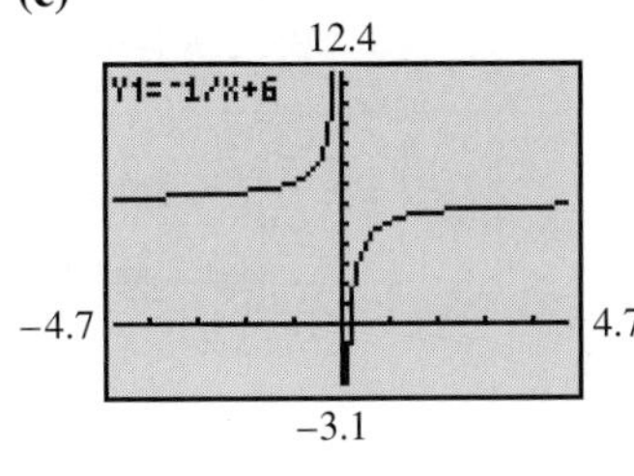


**3.** **(a)** Shift the graph of $y = \frac{1}{x^2}$ to the right 2 units, and reflect across the $x$-axis.

**(b)**

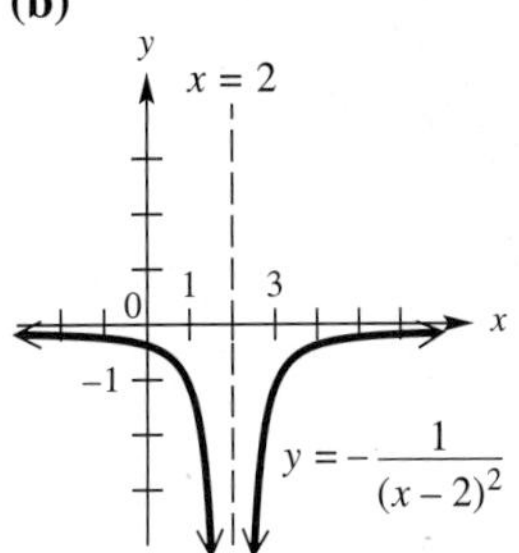


**(c)**

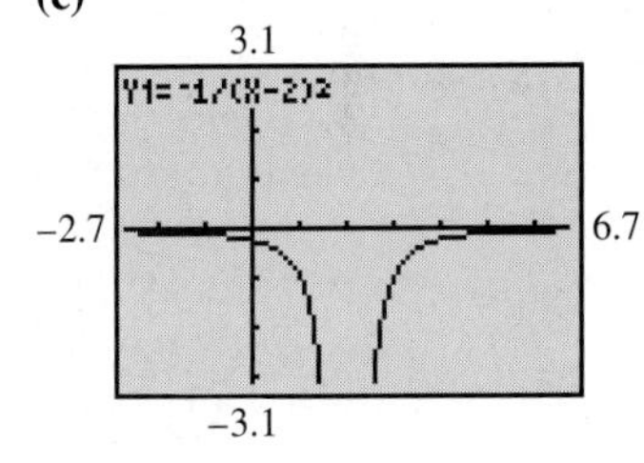


**5.** The degree of the numerator will be exactly 1 greater than the degree of the denominator.

**7.**

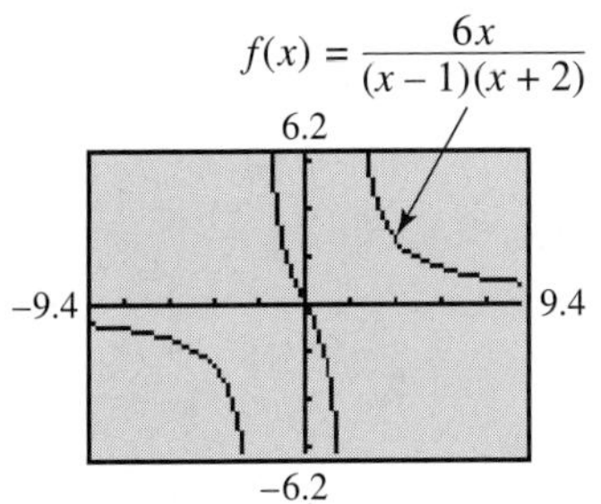


H.A.: $y = 0$;
V.A.: $x = 1$, $x = -2$

**9.**

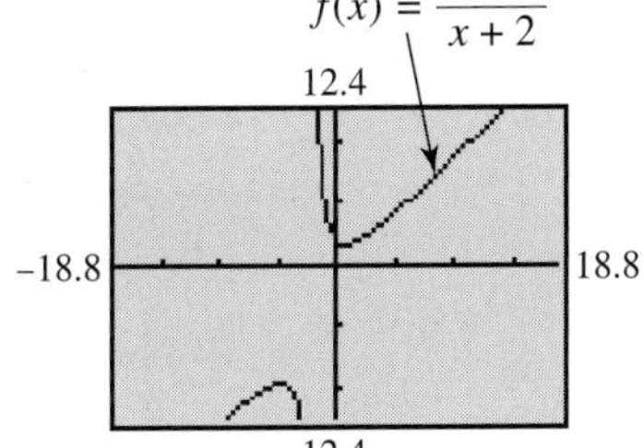


V.A.: $x = -2$;
O.A.: $y = x - 2$

**11.**

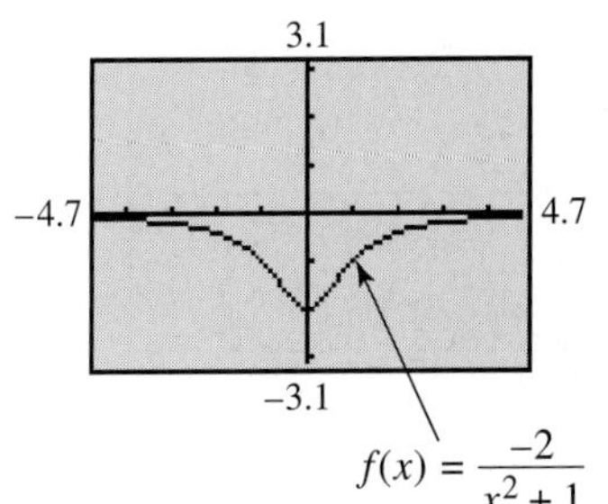


H.A.: $y = 0$

**13.** $f(x) = \dfrac{-3x + 6}{x - 1}$ **15.** **(a)** $\left\{\dfrac{2}{3}\right\}$ **(b)** $\left(-1, \dfrac{2}{3}\right)$ **(c)** $(-\infty, -1) \cup \left(\dfrac{2}{3}, \infty\right)$ **17.** **(a)** $\{0\}$ **(b)** $(-\infty, -1) \cup [0, 2)$ **(c)** $(-1, 0] \cup (2, \infty)$ **19.** $-1.30$, 1, 2.30

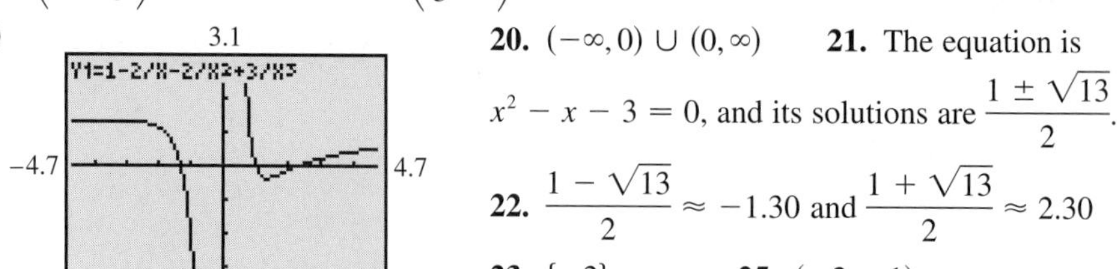


**20.** $(-\infty, 0) \cup (0, \infty)$ **21.** The equation is $x^2 - x - 3 = 0$, and its solutions are $\dfrac{1 \pm \sqrt{13}}{2}$.
**22.** $\dfrac{1 - \sqrt{13}}{2} \approx -1.30$ and $\dfrac{1 + \sqrt{13}}{2} \approx 2.30$
**23.** $\{-2\}$ **25.** $(-2, -1)$

**27.** **(a)**

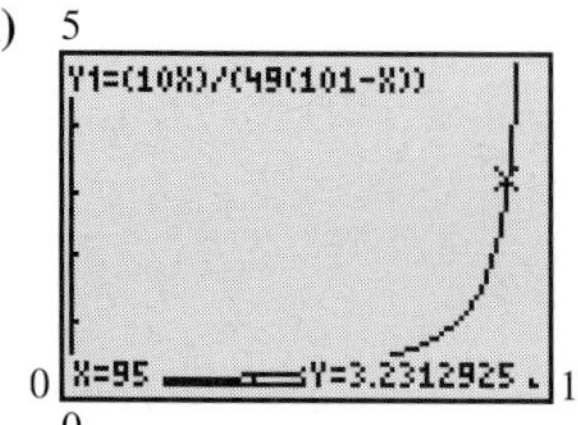


**(b)** approximately \$3231 (See the graph in part (a).) **29.** **(a)** $[0, 36]$ **(b)** The average line length is less than or equal to 8 cars when the average arrival rate is 36 cars per hr or less. **31.** 2 **33.** 847 **35.** 12.15 candela **37.** \$1375
**39.** $\dfrac{8}{9}$ metric ton **41.** $71\dfrac{1}{9}$ kg

**43.**

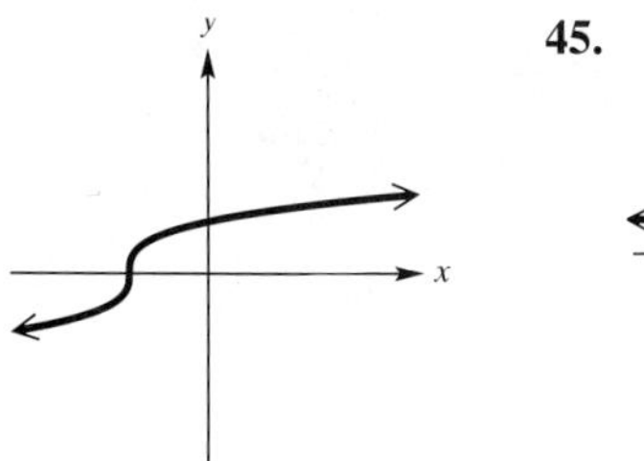


**45.**

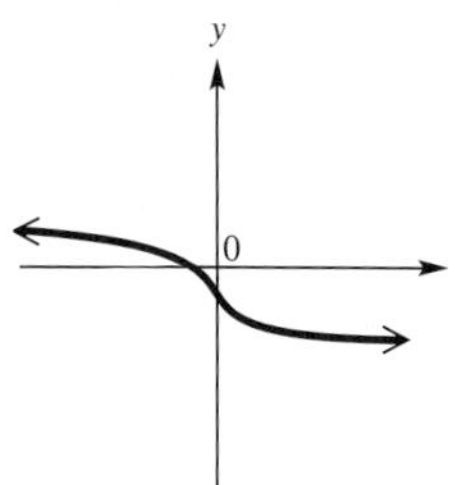

**47.** 2 **49.** $-100$ **51.** 8 **53.** .5; exact value
**55.** 2.28942848511; rational approximation **57.** $(-\infty, 0]$
**59.** $[2, \infty)$ **61.** **(a)** $\{2\}$ **(b)** $[-2.5, 2)$ **(c)** $(2, \infty)$
**63.** **(a)** $\{-1\}$ **(b)** $[-1, \infty)$ **(c)** $(-\infty, -1]$

**65.** no solutions

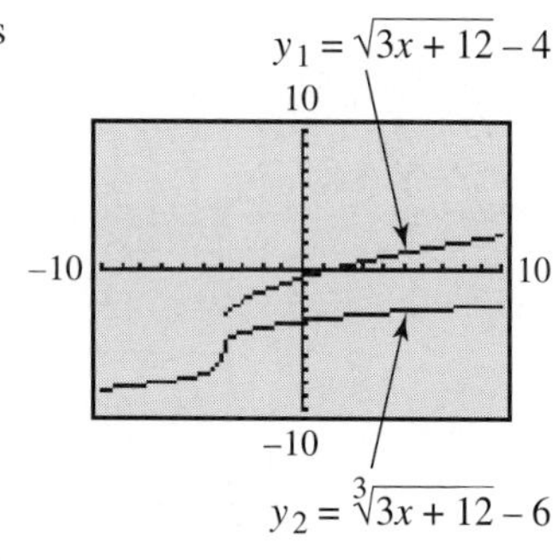


**66.** no $x$-intercepts

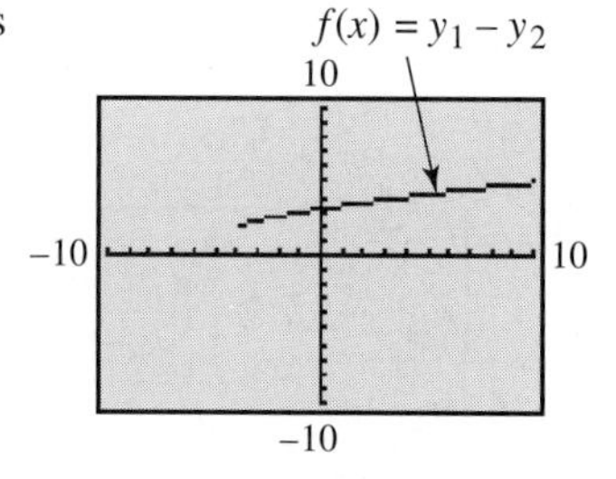


**67.** The graph of $-f(x) = y_2 - y_1$ is the reflection of the graph of $f(x) = y_1 - y_2$ across the $x$-axis.

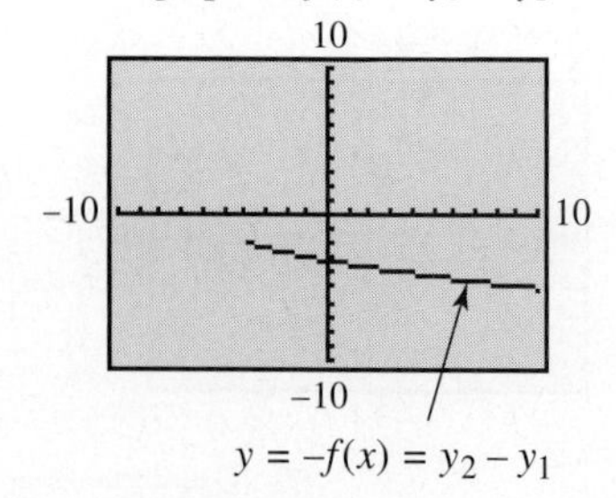


**68.** The graph of $y = f(-x)$ is the reflection of the graph of $y = f(x)$ across the $y$-axis.

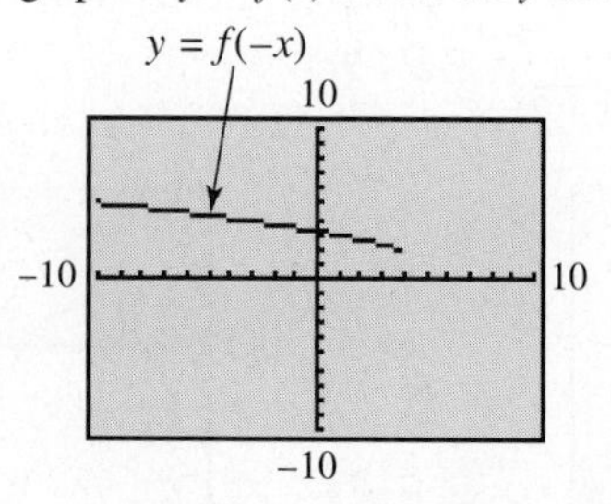


**69.** **(a)** If the length $L$ of the pendulum increases, so does the period of oscillation $T$. **(b)** There are a number of ways. One way is to realize that $k = \dfrac{L}{T^n}$ for some integer $n$. The ratio should be the constant $k$ for each data point when the correct $n$ is found. Another way is to use regression. **(c)** $k \approx .81$; $n = 2$ **(d)** 2.48 sec **(e)** $T$ increases by a factor of $\sqrt{2} \approx 1.414$.

## Chapter 4 Test *(page 337)*

**1. (a)**

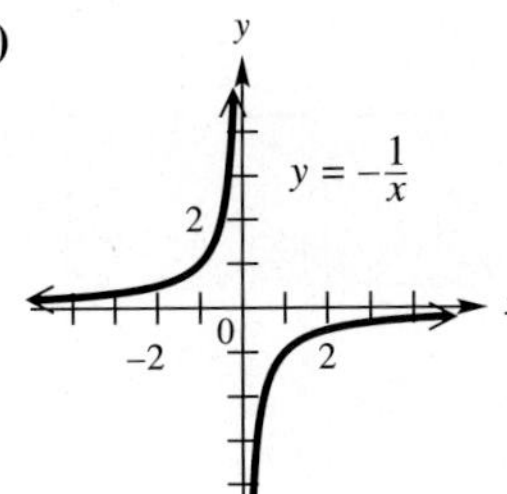


**(b)** The graph of $y = \frac{1}{x}$ is reflected across the $x$-axis or the $y$-axis.

**(c)**

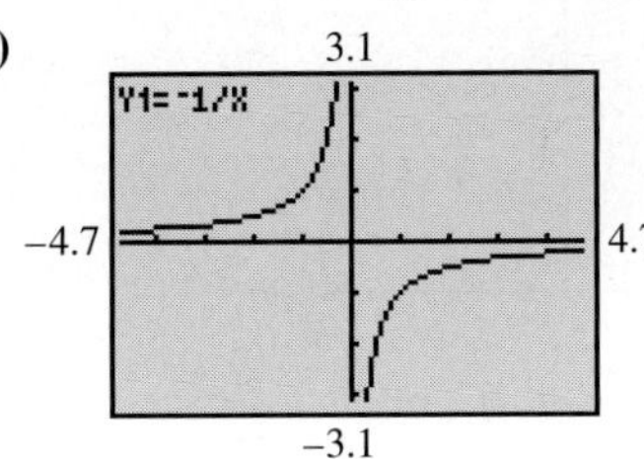


**2. (a)**

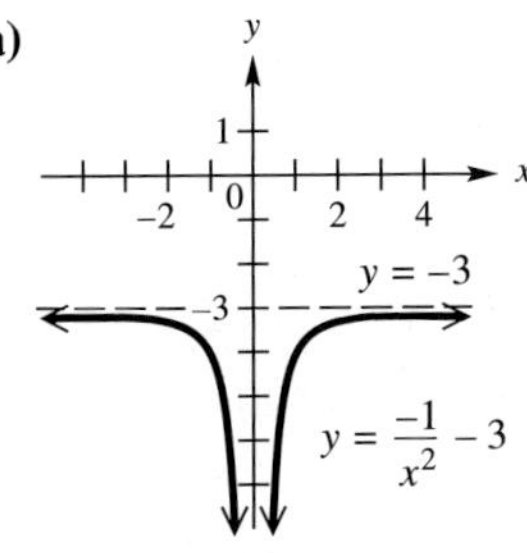


**(b)** The graph of $y = \frac{1}{x^2}$ is reflected across the $x$-axis and shifted 3 units downward.

**(c)**

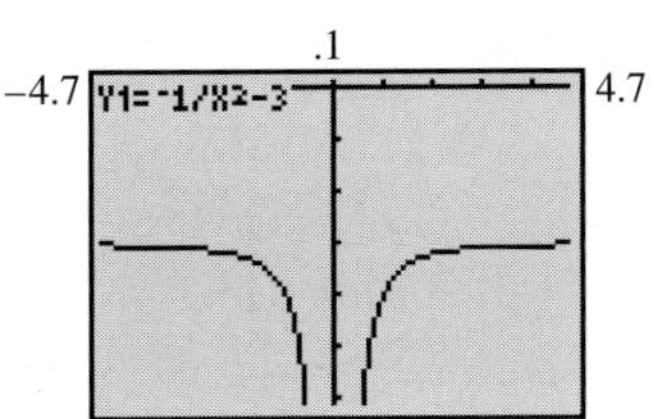


**3. (a)** $x = -1, x = 4$ **(b)** $y = 1$ **(c)** 1.5 **(d)** $-3, 2$ **(e)** $(.5, 1)$ **(f)**

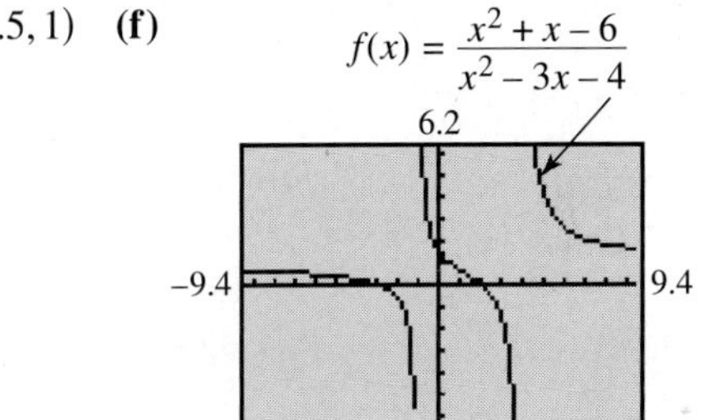


**4.** $y = 2x + 5$

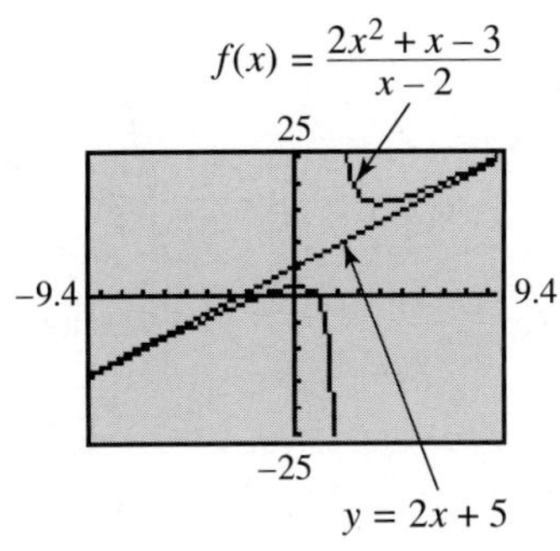


**5. (a)** $-4$ **(b)**

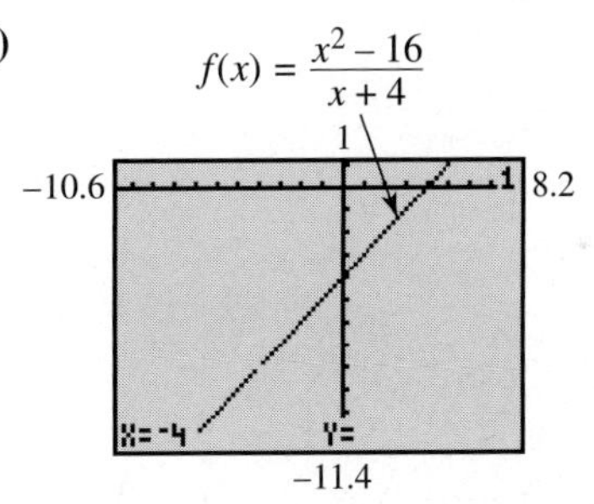


**6. (a)** $\{5\}$ **(b)** $(-\infty, -2) \cup (2, 5]$ **7. (a)** $W(30) = \frac{1}{10}$; $W(39) = 1$; $W(39.9) = 10$; when the rate is 30 vehicles per min, the average wait time is $\frac{1}{10}$ min (6 sec). The other results are interpreted similarly.

**(b)**

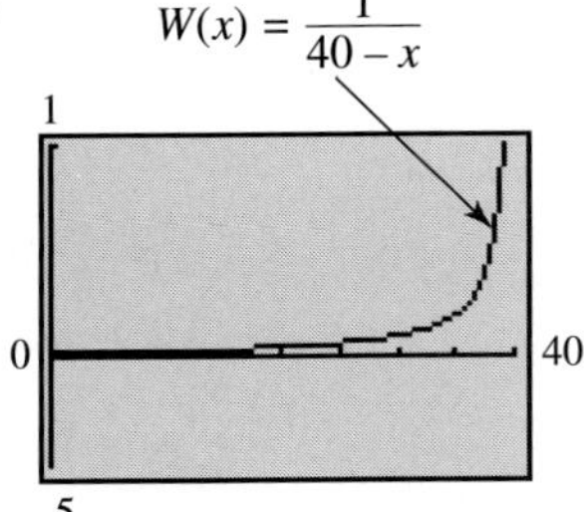


The vertical asymptote has equation $x = 40$. As $x$ approaches 40, $W$ gets larger and larger. **(c)** 39.8 **8.** 92; undernourished

**9.** radius: approximately 8.6 cm; amount: approximately 1394.9 cm$^2$

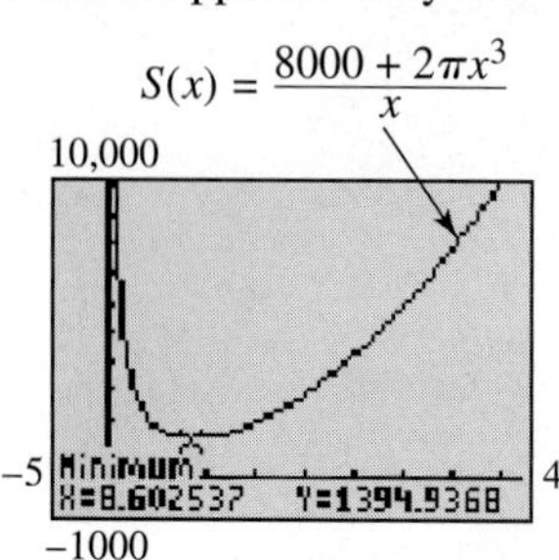


**10.**

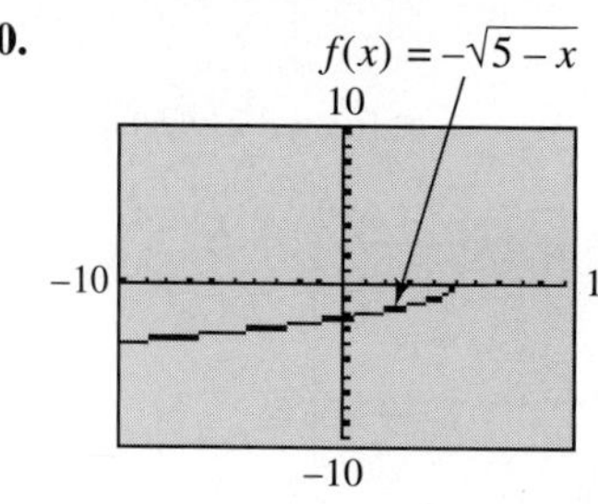


**(a)** $(-\infty, 5]$ **(b)** $(-\infty, 0]$ **(c)** increases

**11.** **(a)** $\{0\}$

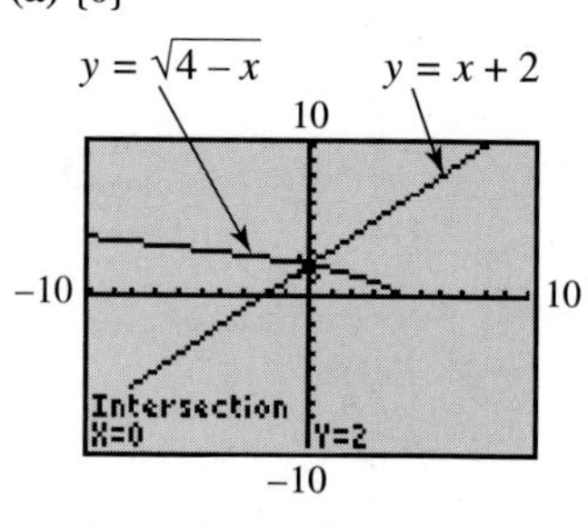


**(b)** $(-\infty, 0)$ **(c)** $[0, 4]$

**12.** The cable should be laid underwater from $P$ to a point $S$, which is on the bank 400 yd away from $Q$ in the direction of $R$.

## CHAPTER 5 INVERSE, EXPONENTIAL, AND LOGARITHMIC FUNCTIONS

### 5.1 Exercises *(page 349)*

**1.** inverse functions **3.** not inverse functions **5.** one-to-one **7.** $x$; $(g \circ f)(x)$ **9.** $(b, a)$ **11.** $y = x$ **13.** does not; it is not one-to-one **15.** one-to-one **17.** not one-to-one **19.** one-to-one **21.** not one-to-one **23.** one-to-one **25.** not one-to-one **27.** not one-to-one **29.** one-to-one **31.** one-to-one **33.** The graph fails the horizontal line test because the end behavior is either ↖ ↗ or ↙ ↘. **35.** untying your shoelaces **37.** leaving a room **39.** unwrapping a package **41.** 8 **43.** −8 **45.** 0 **47.** yes **49.** yes **51.** yes **53.** yes **55.** yes **57.** You should see the graph of the inverse of the function with which you started. **59.** **61.**

**63.**

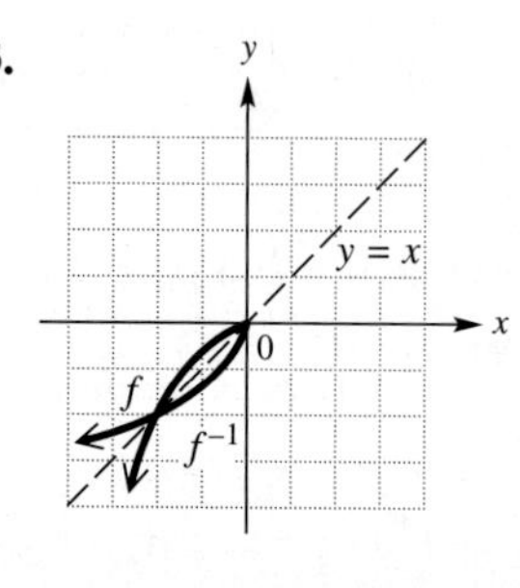


**65.** $f^{-1}(x) = \dfrac{x + 5}{4}$

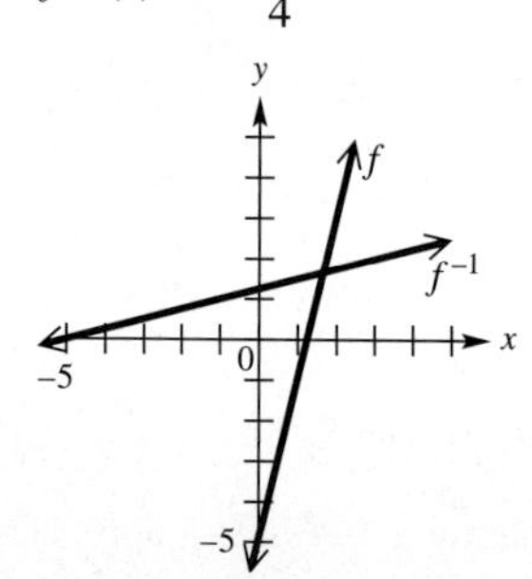


Domains and ranges of both $f$ and $f^{-1}$ are $(-\infty, \infty)$.

**67.** $f^{-1}(x) = \sqrt[3]{-x - 2}$

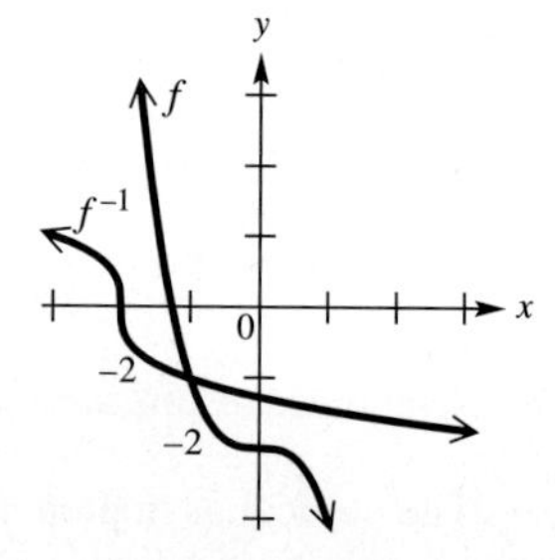


Domains and ranges of both $f$ and $f^{-1}$ are $(-\infty, \infty)$.

**69.** not one-to-one

**71.** $f^{-1}(x) = \frac{4}{x}$

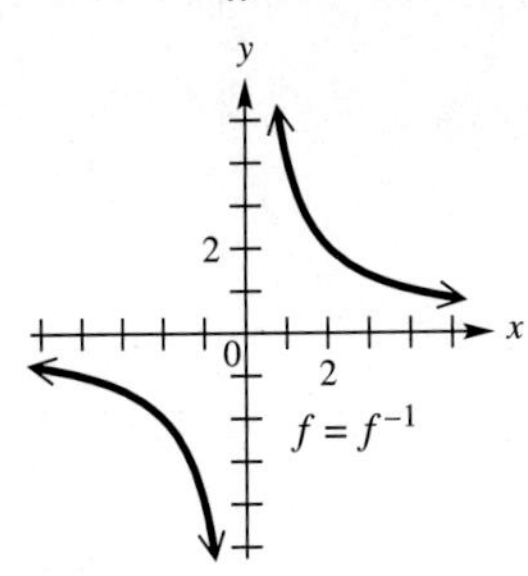


Domains and ranges of both $f$ and $f^{-1}$ are $(-\infty, 0) \cup (0, \infty)$.

**73.** $f^{-1}(x) = \sqrt{x^2 + 16}, x \leq 0$

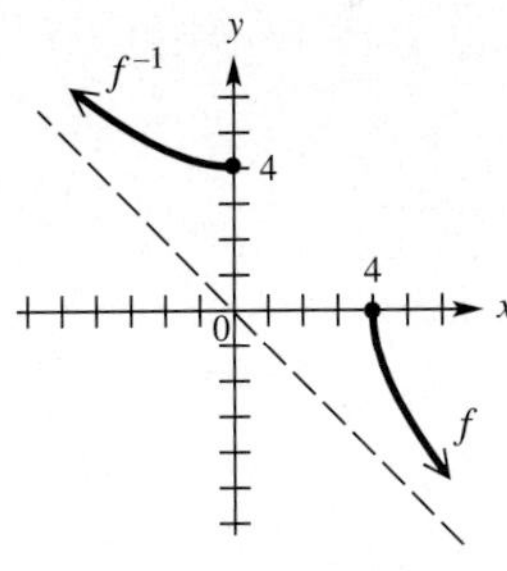


domain of $f$ = range of $f^{-1} = [4, \infty)$; domain of $f^{-1}$ = range of $f = (-\infty, 0]$

**75.** the radius of a sphere with volume 5 in.$^3$

**77.** 2 **79.** not one-to-one **81.** one-to-one; $f^{-1}(x) = \frac{4x}{x+1}$ **83.** $f^{-1}(x) = \frac{1}{2}x + 4$

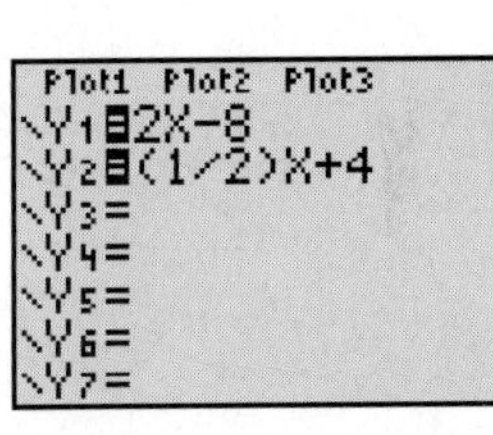


**84.** the identity function

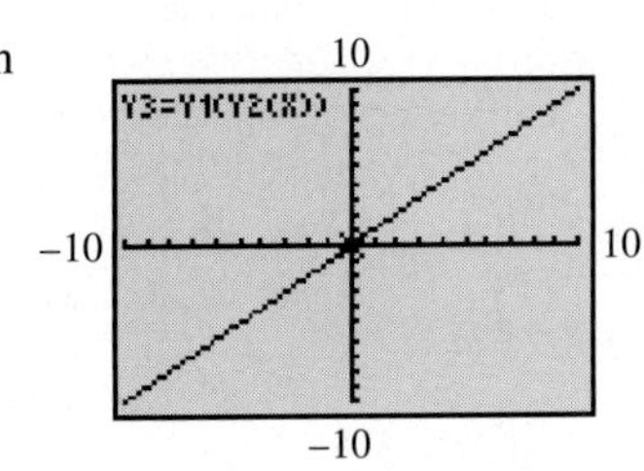


**85.** For each input $x$, the output is also $x$.

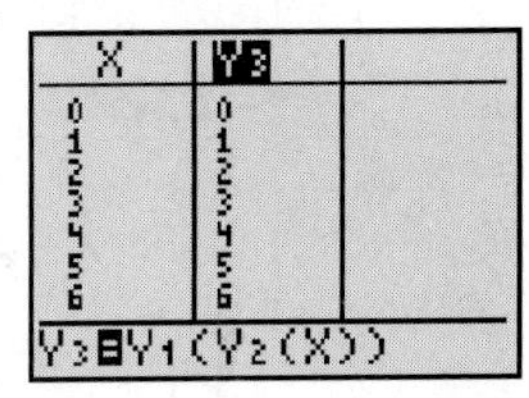


In Exercises 87–91, we give only one of several possible choices for domain restriction.

**87.** $[1, \infty)$ **89.** $[0, \infty)$ **91.** $[4, \infty)$ **93.** $f^{-1}(x) = \sqrt{x} + 1$ **95.** $f^{-1}(x) = \sqrt[4]{x}$ **97.** $f^{-1}(x) = \frac{x+9}{2}$; BIG GIRLS DONT CRY **99.** 8000 8 1000 2197 4096 6859 27 216 13824 8 6859 216; $f^{-1}(x) = \sqrt[3]{x} - 1$

## 5.2 Exercises *(page 362)*

**1.** 8.952419619 **3.** .3752142272 **5.** .0868214883 **7.** 13.1207791 **9.** The point $(\sqrt{10}, 8.9524196)$ lies on the graph of $y$. **11.** The point $(\sqrt{2}, .37521423)$ lies on the graph of $y$. **13.** 2.3 **15.** .75 **17.** .31 **19.** yes; an inverse function **20.**

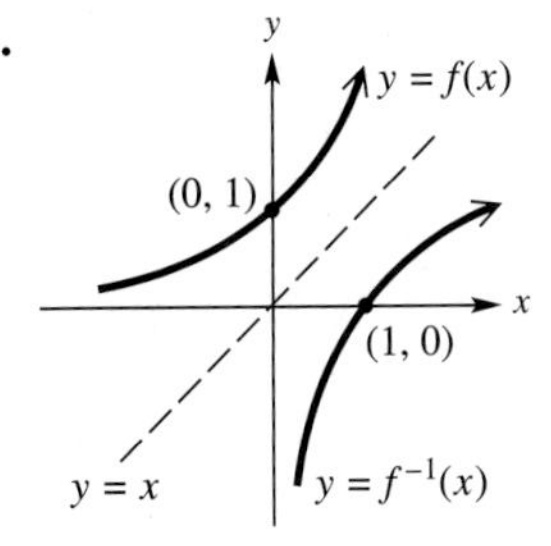


**21.** $x = a^y$ **22.** $x = 10^y$ **23.** $x = e^y$ **24.** $(q, p)$

In Exercises 25–33, we give domain, range, asymptote, and a traditional graph.

**25.** $(-\infty, \infty)$; $(0, \infty)$; $y = 0$

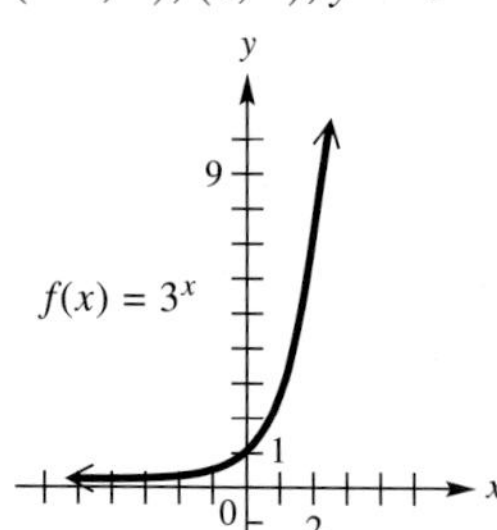


**27.** $(-\infty, \infty)$; $(0, \infty)$; $y = 0$

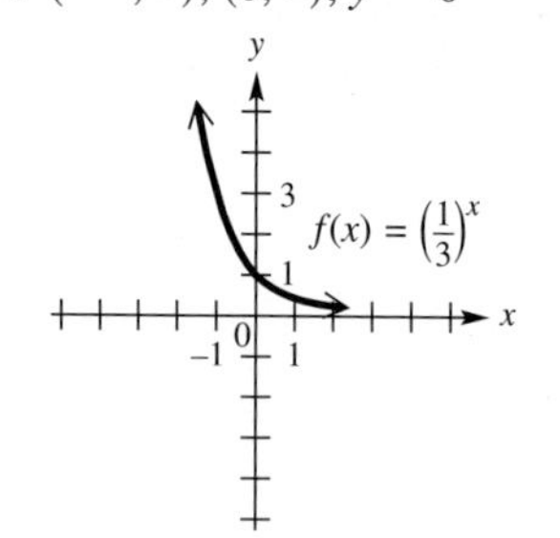


**29.** $(-\infty, \infty)$; $(0, \infty)$; $y = 0$

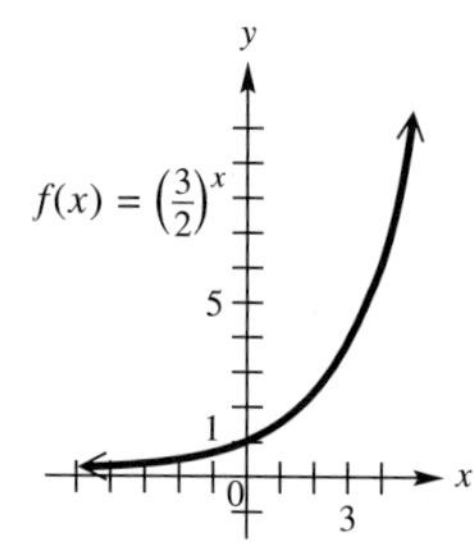


**31.** $(-\infty, \infty)$; $(0, \infty)$; $y = 0$

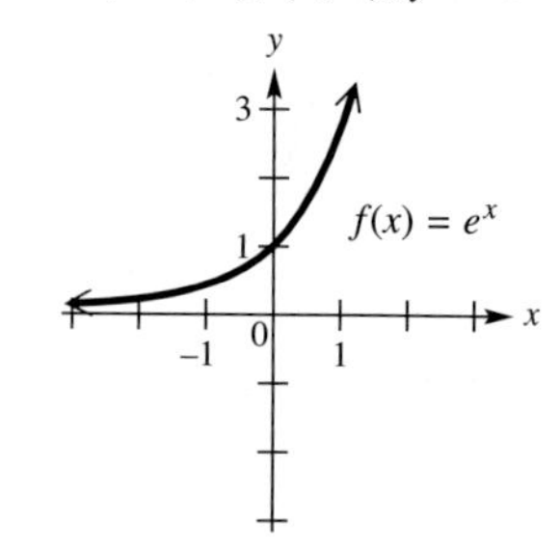


**33.** $(-\infty, \infty)$; $(0, \infty)$; $y = 0$

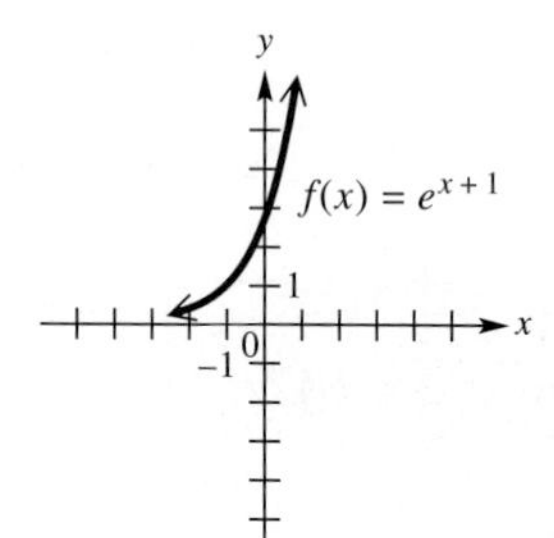


**35. (a)** $a > 1$ **(b)** $(-\infty, \infty)$; $(0, \infty)$; $y = 0$ **(c)**

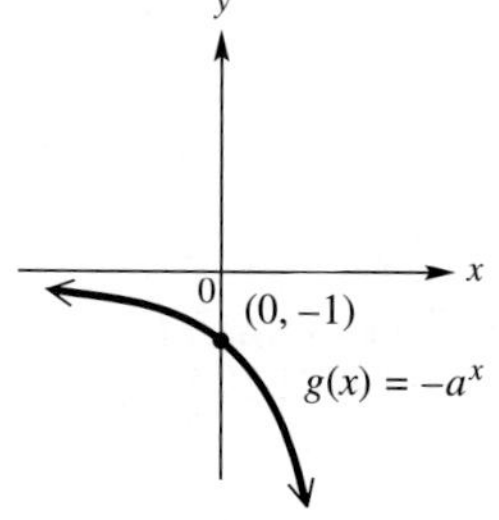


**(d)** $(-\infty, \infty)$; $(-\infty, 0)$; $y = 0$

**(e)**

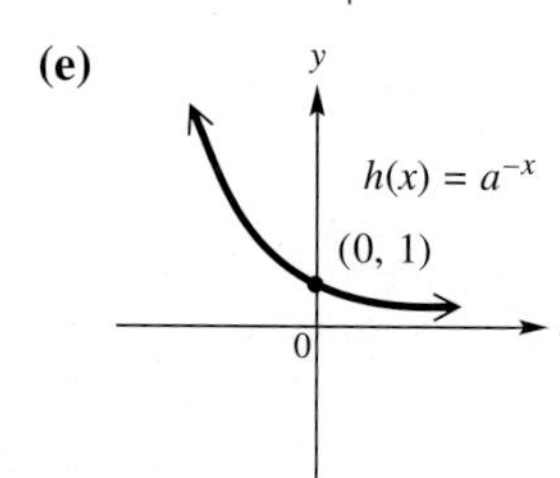


**(f)** $(-\infty, \infty)$; $(0, \infty)$; $y = 0$ **37.** Shift the graph of $y = 2^x$ to the left 5 units and 3 units downward. **39.** Reflect the graph of $y = 2^x$ across the $y$-axis, and shift 1 unit upward. **41.** Stretch the graph of $y = 2^x$ by a factor of 3, and reflect across the $x$-axis. **43.** $\left\{\frac{1}{2}\right\}$

**45.** $\{-2\}$ **47.** $\{0\}$ **49.** $\{3\}$ **51. (a)** $\left\{\frac{1}{5}\right\}$ **(b)** $\left(\frac{1}{5}, \infty\right)$ **(c)** $\left(-\infty, \frac{1}{5}\right)$

**53. (a)** $\left\{-\frac{2}{3}\right\}$ **(b)** $\left[-\frac{2}{3}, \infty\right)$ **(c)** $\left(-\infty, -\frac{2}{3}\right]$ **55. (a)** \$22,510.18 **(b)** \$22,529.85 **57. (a)** \$33,504.35 **(b)** \$33,504.71 **59.** Plan A is better by \$102.65. **61.** \$1.16; \$2.44; \$7.08; \$18.11; \$59.34; \$145.80; \$318.43 **63.** $P(1500) \approx 828$ millibars; $P(11{,}000) \approx 232$ millibars; The calculated values are close to the actual values. **65. (a)** $f(2) = 1 - e^{-.5(2)} = 1 - e^{-1} \approx .63$; There is a 63% chance that at least one car will enter the intersection during a 2-min period. **(b)** As time progresses, the probability increases and begins to approach 1. That is, it is almost certain that at least one car will enter the intersection during a 60-min period.

## 5.3 Exercises *(page 374)*

**1.** $\log_3 81 = 4$ **3.** $\log_{1/2} 16 = -4$ **5.** $\log_{10} .001 = -3$ or $\log .001 = -3$ **7.** $\ln 1 = 0$ **9.** $6^2 = 36$

**11.** $(\sqrt{3})^8 = 81$ **13.** $10^{-3} = .001$ **15.** $10^{.5} = \sqrt{10}$ **17.** $\log_a x$ represents the power to which $a$ must be raised to get $x$.

**19.** $\{3\}$ **21.** $\{\sqrt{3}\}$ or $\{3^{1/2}\}$ **23.** $\left\{\frac{1}{216}\right\}$ **25.** $\{8\}$ **27.** $\{10\}$ **29.** $\left\{-\frac{1}{8}\right\}$ **31. (a)** 7 **(b)** 9 **(c)** 4 **(d)** $k$

**33.** **(a)** 0 **(b)** 0 **(c)** 0 **(d)** 0 **35.** 4.3 **37.** $\sqrt{3}$ **39.** .5 **41.** $\sqrt{6}$ **43.** 1.633468456 **45.** −.1062382379 **47.** 4.341474094 **49.** 3.761200116 **51.** −.244622583 **53.** 9.996613531 **55.** **(a)** .3741982579; 1.3741982579; 2.3741982579; 3.3741982579 **(b)** $2.367 \times 10^0$; $2.367 \times 10^1$; $2.367 \times 10^2$; $2.367 \times 10^3$ **(c)** The decimal digits in (a) are the same. The whole number part corresponds to the exponent on 10 in scientific notation. **57.** 1.8 **59.** 13.5 **61.** $4 \times 10^{-4}$ **63.** $3.2 \times 10^{-7}$ **65.** 7.9 yr **67.** 11.6 yr **69.** $\log_4 6 - \log_4 7$ **71.** $\log_3 4 + \log_3 p - \log_3 q$

**73.** $1 + \frac{1}{2}\log_2 3 - \log_2 5 - \log_2 p$ **75.** cannot be rewritten **77.** $5 \log_z x + 3 \log_z y - \log_z 3$

**79.** $\frac{1}{3}(5 \log_p m - \log_p k - 2 \log_p t)$ **81.** $\log_b \frac{k}{ma}$ **83.** $\log_y p^{-7/6}$ or $-\log_y p^{7/6}$ **85.** $\log_b \frac{2y + 5}{\sqrt{y + 3}}$ **87.** $\log_3 \frac{1}{32p^5}$

**89.** 1.130929754 **91.** −1.584962501 **93.** .9747973963 **95.** 1.445851777 **97.** 1.59 **99.** Reflect the graph of $y = 3^x$ across the $x$-axis and shift 7 units upward. **100.**

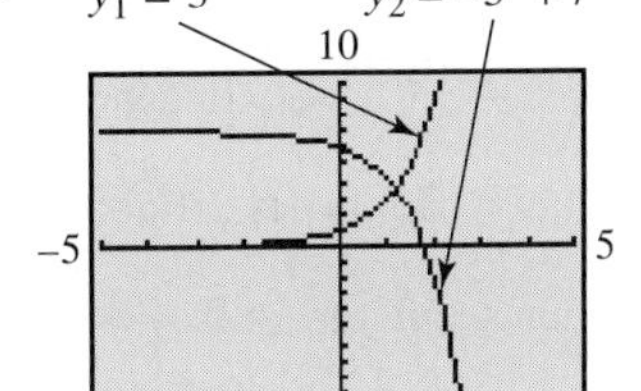


**101.** 1.7712437 **102.** $\{\log_3 7\}$

**103.** $\frac{\log 7}{\log 3} = \frac{\ln 7}{\ln 3} \approx 1.771243749$

**104.** The approximations are close enough to support the conclusion that the $x$-intercept is equal to $\log_3 7$.

## Reviewing Basic Concepts *(page 376)*

**1.** No, because the $x$-values −2 and 2 both correspond to the $y$-value 4. In a one-to-one function, each $y$-value must correspond to exactly one $x$-value (and each $x$-value to exactly one $y$-value). **2.** **(a)**

| $x$ | 12 | 21 | 32 | 45 |
|---|---|---|---|---|
| $y$ | 7 | 8 | 9 | 10 |

**(b)** $y = 4x - 5$

**3.**

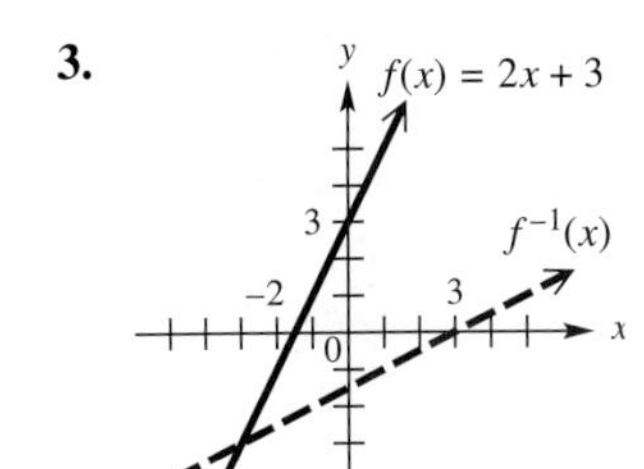


**4.**

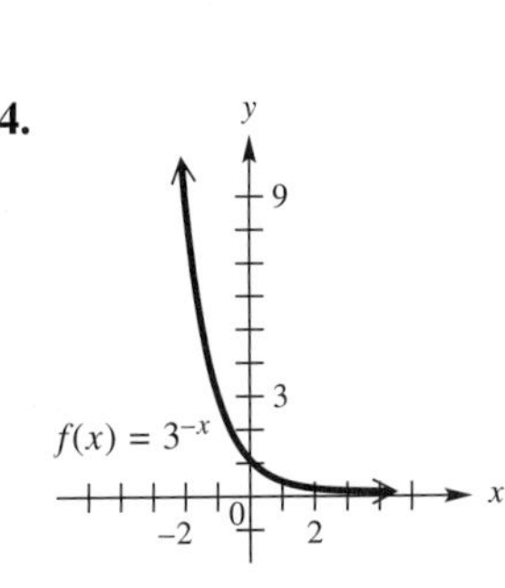


**5.** $\left\{\frac{3}{4}\right\}$ **6.** $76.10 **7.** **(a)** $-\frac{1}{2}$ **(b)** 3 **(c)** 2

**8.** $\log 3 + 2 \log x - \log 5 - \log y$ **9.** $\ln \frac{x}{2}$ **10.** 2.1 yr

## 5.4 Exercises *(page 383)*

**1.** $(0, \infty)$; $(-\infty, \infty)$; increases; $x = 0$

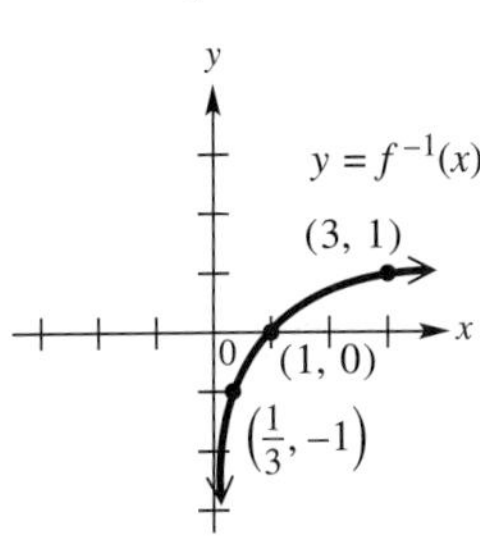


**3.** $(0, \infty)$; $(-\infty, \infty)$; decreases; $x = 0$

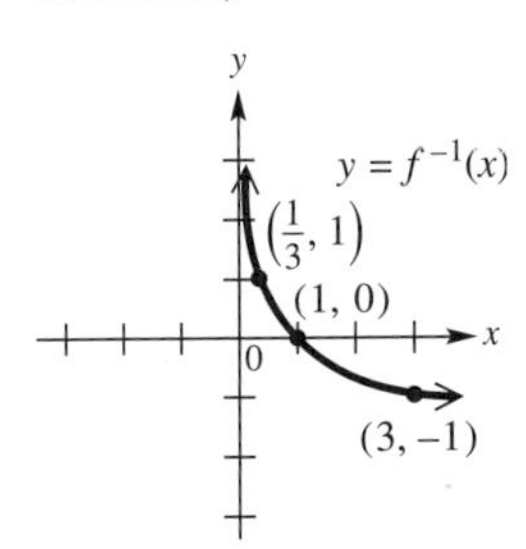


**5.** $(1, \infty)$; $(-\infty, \infty)$; increases; $x = 1$

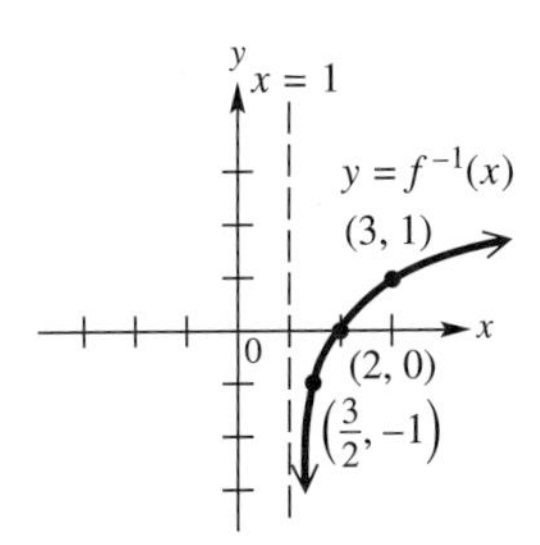


**7.** logarithmic **9.** $(0, \infty)$ **11.** $(-\infty, \infty)$ **13.** $(-\infty, -3) \cup (7, \infty)$ **15.** $(-1, 0) \cup (1, \infty)$

**17.**

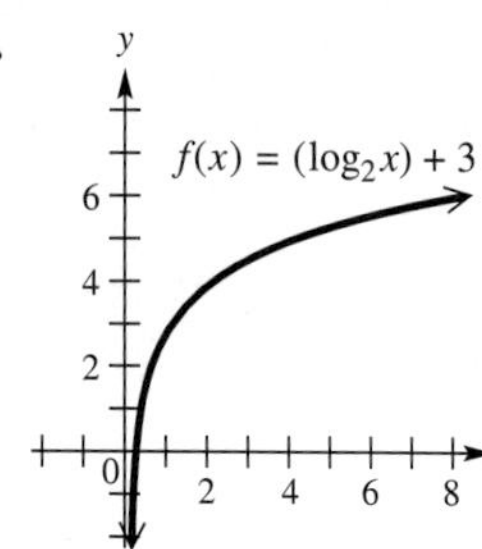


**19.**

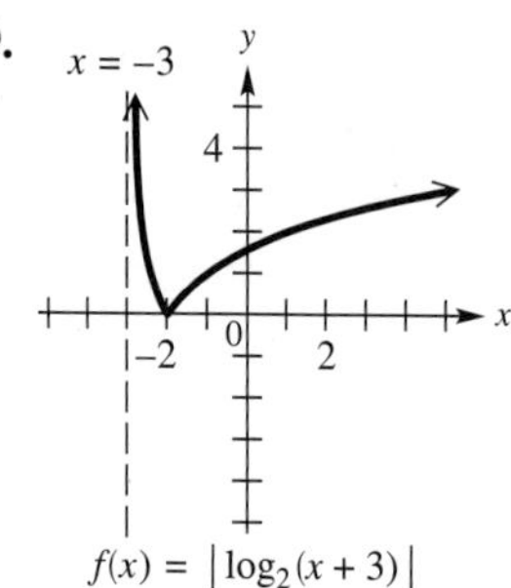


**21.**

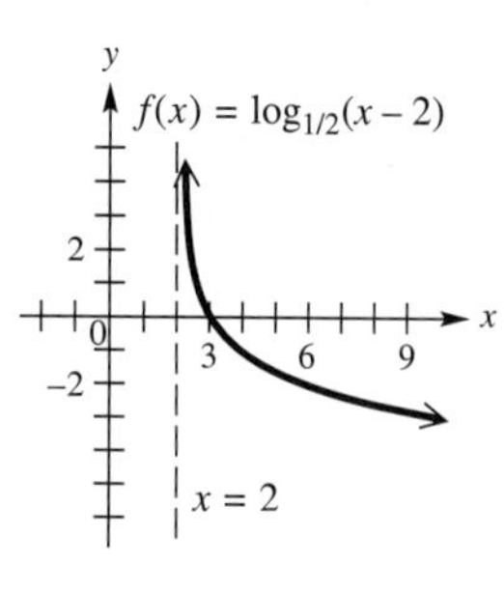


**23.** B **25.** D **27.** A **29.** C

**31.** The graph is shifted 4 units to the left. **33.** The graph is stretched vertically by a factor of 3 and shifted 1 unit upward. **35.** The graph is reflected across the $y$-axis and shifted 1 unit upward. **37.** $(-\infty, 0) \cup \left(\frac{1}{2}, \infty\right)$

**38.** $y = f(x) = \dfrac{\log(2x^2 - x)}{\log 4}$ **39.** $x = 0, x = \frac{1}{2}$ **40.** In general, to find the $y$-intercept for the graph of $y = f(x)$, we evaluate $f(0)$. Because 0 is not in the domain of $f$, $f(0)$ is not defined, and there is no $y$-intercept.

**41.**

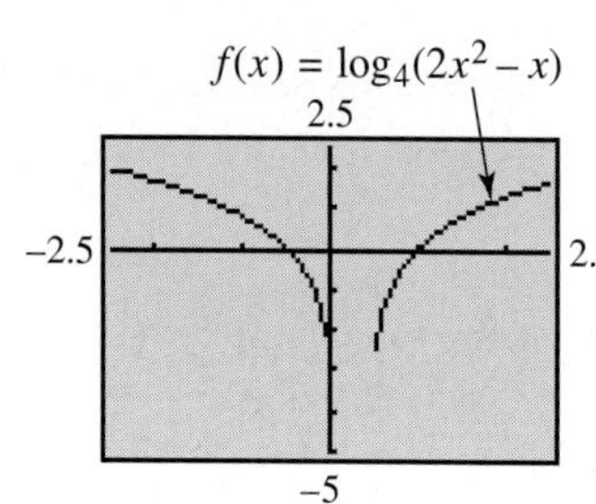


solution set of $f(x) = 0$: $\left\{-\frac{1}{2}, 1\right\}$

solution set of $f(x) < 0$: $\left(-\frac{1}{2}, 0\right) \cup \left(\frac{1}{2}, 1\right)$

solution set of $f(x) > 0$: $\left(-\infty, -\frac{1}{2}\right) \cup (1, \infty)$

**42.** $(-\infty, \infty)$ **43.** The graphs are not the same because the domain of $y = \log x^2$ is $(-\infty, 0) \cup (0, \infty)$, while the domain of $y = 2 \log x$ is $(0, \infty)$. The power rule does not apply if the argument is nonpositive. **45.** $\frac{3}{2}$ **47.** $-\frac{3}{4}$

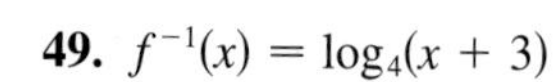

**49.** $f^{-1}(x) = \log_4(x + 3)$

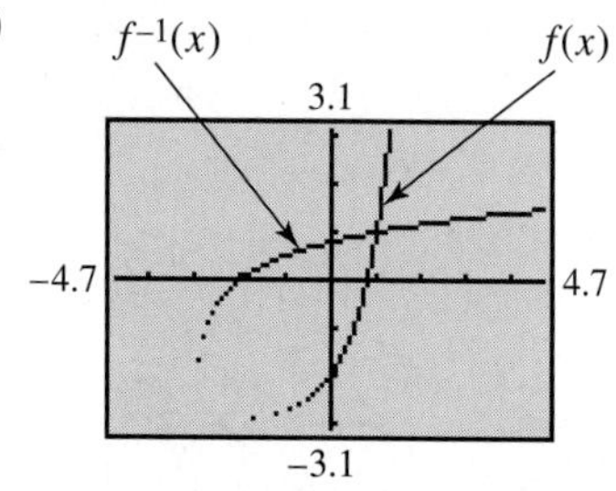


Dot Mode

**51.** $f^{-1}(x) = \log(4 - x)$

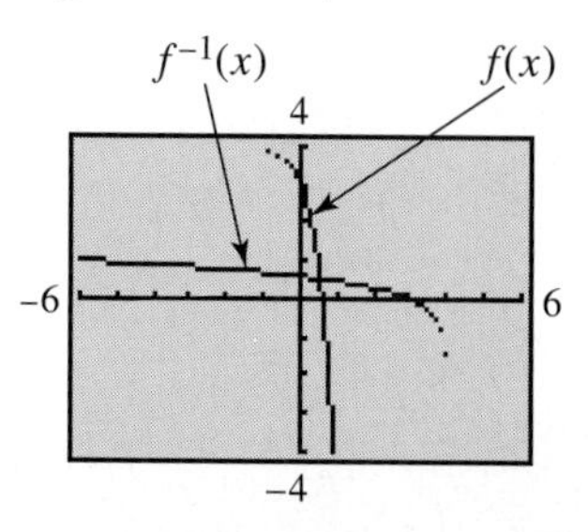


Dot Mode

**53.** **(a)** The left side is a reflection of the right side with respect to the axis of the tower. The graph of $f(-x)$ is the reflection of $f(x)$ with respect to the $y$-axis. **(b)** 984 ft **(c)** 39 ft **55.** **(a)** 28.105 in. **(b)** It tells us that at 99 mi from the eye of a typical hurricane, the barometric pressure is 29.21 in.

## 5.5 Exercises *(page 393)*

**1.** **(a)** $\{1.4036775\}$ **(b)** $(1.4036775, \infty)$ **3.** **(a)** $\{-1\}$ **(b)** $(-1, \infty)$ **5.** $\left\{\frac{1}{\log 3}\right\}$ or $\left\{\frac{\log 10}{\log 3}\right\}$ **7.** $\left\{\frac{\log 8}{\log \frac{5}{2}}\right\}$ or $\left\{\frac{\log \frac{1}{8}}{\log \frac{2}{5}}\right\}$ **9.** $\{1 + \ln 4\}$ **11.** $\left\{\frac{\ln 4 - 2}{5}\right\}$ **13.** $\emptyset$ **15.** $\{2\}$ **17.** $\{-41\}$ **19.** $\{3.9\}$ **21.** $\{25\}$ **23.** $\{2.5\}$ **25.** $\{3\}$ **27.** $\{8\}$ **29.** $\{4\}$ **31.** **(a)** $(0, 4)$ **(b)** $(4, \infty)$ **33.** The statement is incorrect. We must reject any solution that is not in the domain of any logarithmic function in the equation. **35.** The graph of $y = \ln x - \ln(x + 1) - \ln 5$ does not intersect the $x$-axis. **37.** $\{17.106\}$

The answers in Exercises 39–47 may have other equivalent forms.

**39.** $t = e^{(p-r)/k}$ **41.** $t = -\frac{1}{k}\log\left(\frac{T - T_0}{T_1 - T_0}\right)$ **43.** $k = \frac{\ln\left(\frac{A - T_0}{C}\right)}{-t}$ **45.** $x = \frac{\ln\left(\frac{A + B - y}{B}\right)}{-C}$ **47.** $A = \frac{B}{x^C}$

**49.** It is true because $(a^m)^n = a^{mn}$, and so $(e^x)^2 = e^{x \cdot 2} = e^{2x}$. **50.** $(e^x - 3)(e^x - 1) = 0$ **51.** $\{0, \ln 3\}$

**52.**

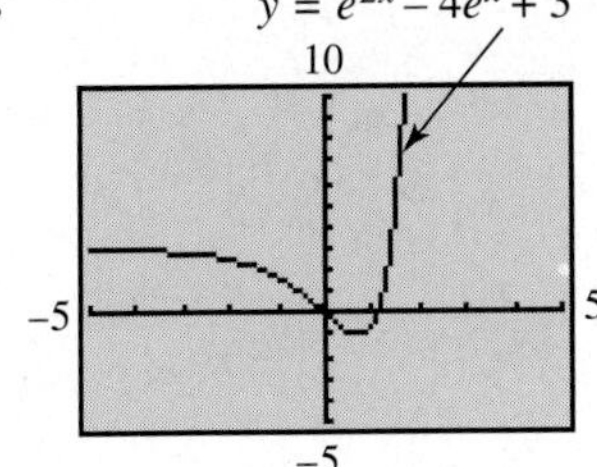


The graph intersects the $x$-axis at 0 and $\ln 3 \approx 1.099$. **53.** $(-\infty, 0) \cup (\ln 3, \infty)$ **54.** $(0, \ln 3)$ **55.** $\{-.767, 2, 4\}$ **57.** $\{2.454, 5.659\}$ **59.** $\{-.443\}$ **61.** $\{-2, 2\}$ **63.** $\{-3\}$ **65.** 85.5% **67.** 22 m

## Reviewing Basic Concepts *(page 395)*

**1.** inverse; symmetric; $y = x$; range **2.**

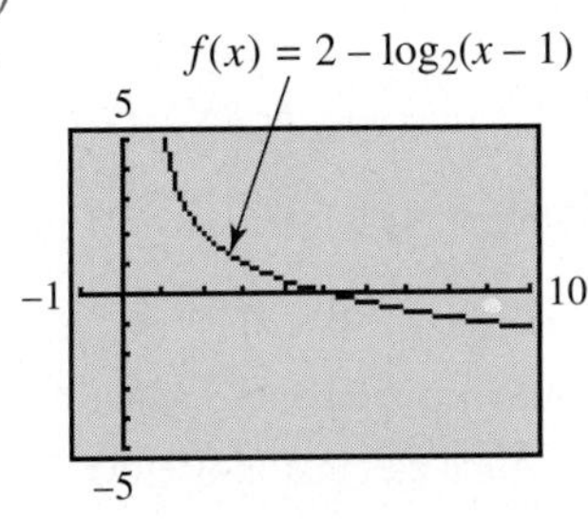


**3.** $x = 1$; $x$-intercept: 5, no $y$-intercept **4.** The graph of $f(x)$ is the same as the graph of $g(x)$ reflected across the $x$-axis and shifted 1 unit to the right and 2 units upward. **5.** $f^{-1}(x) = 1 + 2^{2-x}$

**6.** $x = \left\{\frac{\log 3}{2\log 3 - \log 4}\right\}$ or $\left\{\frac{\log 3}{\log \frac{9}{4}}\right\}$ **7.** $\{3\}$ **8.** $\{2\}$ **9.** $N = -\frac{1}{k}\ln\left(1 - \frac{H}{1000}\right)$ **10.** about 2.8 acres

## 5.6 Exercises *(page 404)*

**1.** 9000 yr ago **3.** 16,000 yr old **5. (a)** $A = A_0e^{-.032t}$ **(b)** 6.97 yr **(c)** 363 g **7. (a)** less **(b)** $A = 2e^{-.005t}$ **(c)** about 140 days **9. (a)** $4{,}000{,}000I_0$ **(b)** $3{,}200{,}000I_0$ **(c)** The quake measuring 6.6 was 1.25 times as strong. **11.** about 1.126 billion yr old **13.** Magnitude 1 is approximately 6.3 times as great as magnitude 3. **15. (a)** $f(t) = 20 + 80e^{-.693t}$ **(b)** 76.6°C **(c)** 1.415 hr or about 1 hr and 25 min **17. (a)** $C \approx .72$, $a \approx 1.041$ **(b)** 1.21 ppb **19. (a)** $(T \circ R)(x) = .206e^{.0124x}$ **(b)** $(T \circ R)(100) = .7119$; In 1900 radiative forcing caused an approximately .7°F increase in the average global temperature. **21. (a)** 46.2 yr **(b)** 46.0 yr **23.** 1.8 yr **25.** The better investment is 7% compounded quarterly; it will earn \$800.31 more. **27.** 6.14% **29.** \$5583.95 **31.** 12% **33. (a)** \$205.52 **(b)** \$1364.96 **35. (a)** \$852.72 **(b)** \$181,979.20 **37. (a)** 19 yr, 39 days **(b)** 11 yr, 166 days **39. (a)** 2,700,000 **(b)** 3,000,000 **(c)** 9,500,000 **41.** 23 days **43.** 59 mg

**45. (a)**

| $x$ | 0 | 15 | 30 | 45 | 60 | 75 | 90 | 125 |
|---|---|---|---|---|---|---|---|---|
| $g(x)$ | 7 | 21 | 57 | 111 | 136 | 158 | 164 | 178 |

**(b)** $g(x) = 261 - f(x)$ **(c)** $y_1$ models $g(x)$ better. **(d)** $f(x) = 261 - \frac{171}{1 + 18.6e^{-.0747x}}$ **47. (a)** $f(x) = 10.98(1.14)^x$ **(b)** **(c)** about \$60 billion

**49. (a)** 44.9 billion **(b)** 1992 **51. (a)** $C = 131$; $a = \left(\frac{34}{131}\right)^{1/6} \approx .799$ (Answers may vary.) **53. (a)** .065; .82; Among people age 25, 6.5% have some CHD, while among people age 65, 82% have some CHD. **(b)** 48 yr

## Chapter 5 Review Exercises *(page 413)*

**1.** not one-to-one **3.** not one-to-one **5.** $(-\infty, \infty)$ **7.** $f^{-1}(x) = \frac{x^3 + 7}{2}$ **11.** A **13.** B **15.** $(-\infty, \infty)$ **17.** 1 **19.** $f^{-1}(x) = \log_a x$ **21. (a)** $\left\{\frac{3}{7}\right\}$ **(b)** $\left(\frac{3}{7}, \infty\right)$ **23. (a)** $\emptyset$ **(b)** $(-\infty, \infty)$ **25.** 1.7657 **27.** 4.0656 **29.** 0 **31.** 12 **33.** 5 **35.** E **37.** B **39.** F **41.** 3 **43.** $\log_3 m + \log_3 n - \log_3 5 - \log_3 r$ **45.** $2 \log_5 x + 4 \log_5 y + \frac{3}{5} \log_5 m + \frac{1}{5} \log_5 p$ **47. (a)** $\{2\}$ **(b)** $(2, \infty)$ **49. (a)** $\left\{\frac{7}{5}\right\}$ or $\{1.4\}$ **(b)** $(1, 1.4)$ **51.** $\left\{\frac{3}{2}\right\}$ **53.** $\{1.303\}$ **55.** $\{2\}$ **57.** $\left\{\frac{1}{2}\right\}$ **59.** $c = de^{(N-a)/b}$ **61.** $\{.01, 2.376\}$ **63.** $\{.001, 1.805\}$ **65.** 4.0 yr **67. (a)** \$2322.37 **(b)** \$2323.67 **(c)** 36.6 yr **69.** 1.67 acres **71. (a)** 207 **(b)** 249

## Chapter 5 Test *(page 416)*

**1. (a)** B **(b)** A **(c)** C **(d)** D **(e)** the equations of the functions in (a) and (d) and those in (b) and (c)

**2. (a)** $f(x) = -2^{x-1} + 8$

**(b)** domain: $(-\infty, \infty)$; range: $(-\infty, 8)$ **(c)** Yes, it has a horizontal asymptote with equation $y = 8$. **(d)** $x$-intercept: 4; $y$-intercept: 7.5 **(e)** $f^{-1}(x) = 1 + \frac{\log(8 - x)}{\log 2}$ (Any base logarithm can be used.) **3.** $\{.5\}$ **4. (a)** \$12,442.11 **(b)** \$12,460.77 **5.** The expression $\log_5 27$ is the exponent to which 5 must be raised in order to obtain 27. To find an approximation with a calculator, use the change-of-base rule. **6. (a)** 1.659 **(b)** 6.153 **(c)** 6.049 **7.** $3 \log m + \log n - \frac{1}{2} \log y$

**8. (a)** $\{2\}$; The extraneous solution is $-4$. **(b)** $y_1 = \frac{\log x}{\log 2} + \frac{\log(x + 2)}{\log 2} - 3$

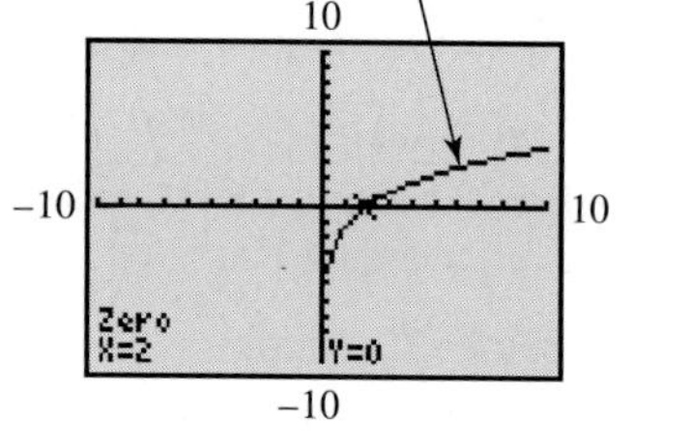


**(c)** $(2, \infty)$

**9. (a)** $\left\{\frac{\log 18}{\log 48}\right\}$ **(b)** $\{.747\}$ **10. (a)** B **(b)** D **(c)** C **(d)** A

**11. (a)** For $t = x$, $A(x) = x^2 - x + 350$ **(b)** For $t = x$, $A(x) = 350 \log(x + 1)$ **(c)** For $t = x$, $A(x) = 350(.75)^x$

**(d)** For $t = x$, $A(x) = 100(.95)^x$ Function (c) best describes $A(t)$. **12. (a)** $A(t) = 2e^{-.000433t}$ **(b)** .03 g **(c)** about 3200 yr

## CHAPTER 6 ANALYTIC GEOMETRY

### 6.1 Exercises *(page 430)*

**1.** The graph is the point (3, 3). **3.** D **5.** E **7.** H **9.** B **11.** $(x-1)^2+(y-4)^2=9$ **13.** $x^2+y^2=1$
**15.** $\left(x-\frac{2}{3}\right)^2+\left(y+\frac{4}{5}\right)^2=\frac{9}{49}$ **17.** $(x+1)^2+(y-2)^2=25$ **19.** $(x+3)^2+(y+2)^2=4$ **21.** $(2,-3)$
**22.** $3\sqrt{5}$ **23.** $(x-2)^2+(y+3)^2=45$ **24.** $(x+2)^2+(y+1)^2=41$ **25.** domain: $[-6,6]$; range: $[-6,6]$

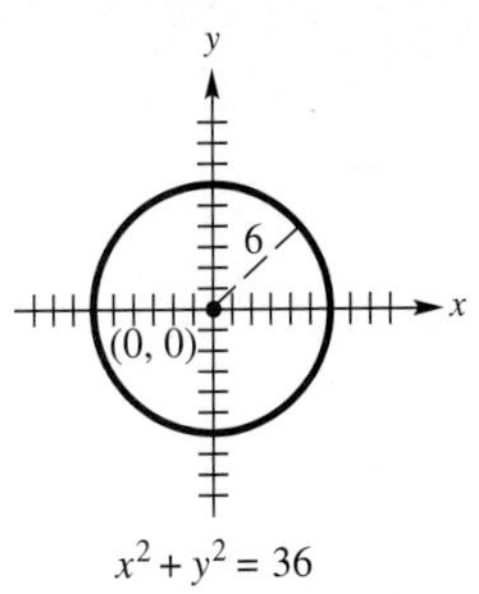


**27.** domain: $[-6,2]$; range: $[1,9]$ **29.** domain: $[-1,9]$; range: $[-2,8]$ **31.** domain: $[-9,9]$; range: $[-9,9]$

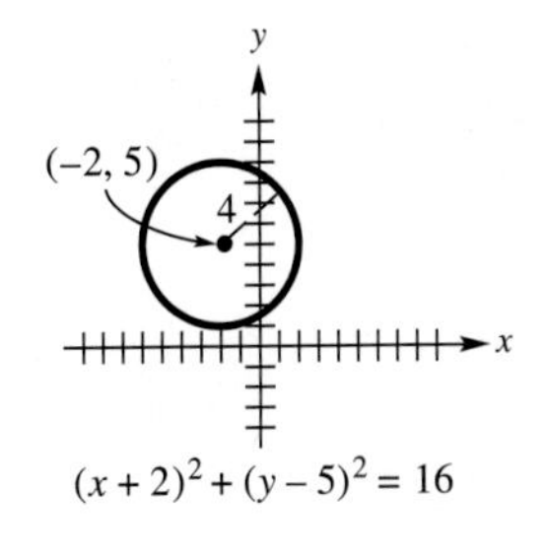


$(x-4)^2+(y-3)^2=25$

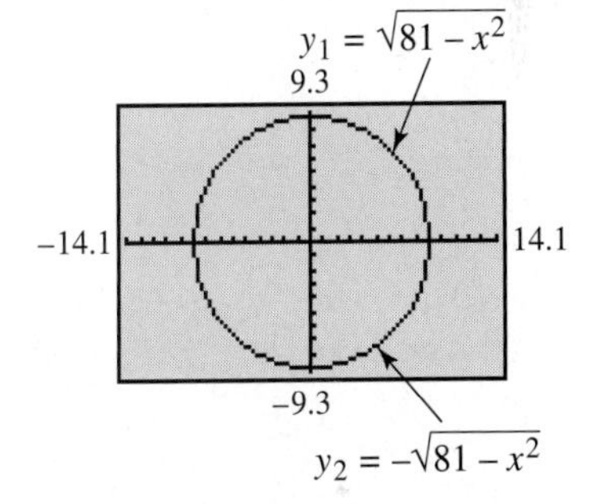


**33.** domain: $[-2,8]$; range: $[-3,7]$

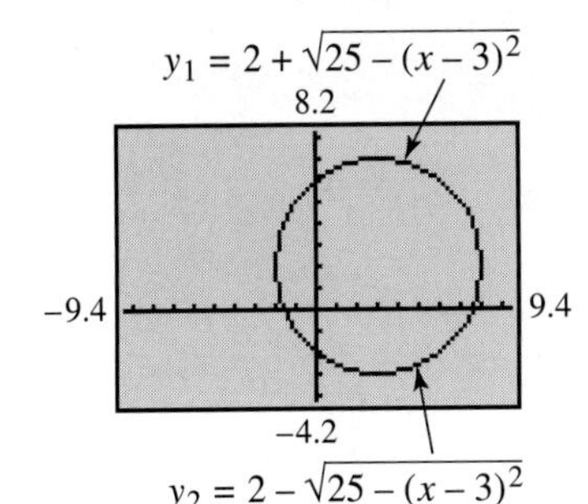


**35.** yes; center: $(-3,-4)$; radius: 4 **37.** yes; center: $(2,-6)$; radius: 6
**39.** yes; center: $\left(-\frac{1}{2},2\right)$; radius: 3 **41.** no **43.** no **45.** $(-2,-2)$
**47.** B **49.** A **51.** H **53.** G **55.** **(a)** III **(b)** II **(c)** IV **(d)** I
**57.** $(0,1)$; $y=-1$; $y$-axis **59.** $\left(0,\frac{1}{36}\right)$; $y=-\frac{1}{36}$; $y$-axis
**61.** $\left(-\frac{1}{128},0\right)$; $x=\frac{1}{128}$; $x$-axis **63.** $(-1,0)$; $x=1$; $x$-axis **65.** $y^2=20x$
**67.** $x^2=y$ **69.** $x^2=y$ **71.** $y^2=-3x$ **73.** $y^2=\frac{4}{3}x$

**75.** vertex: $(5,-4)$; axis: $x=5$; domain: $(-\infty,\infty)$; range: $[-4,\infty)$

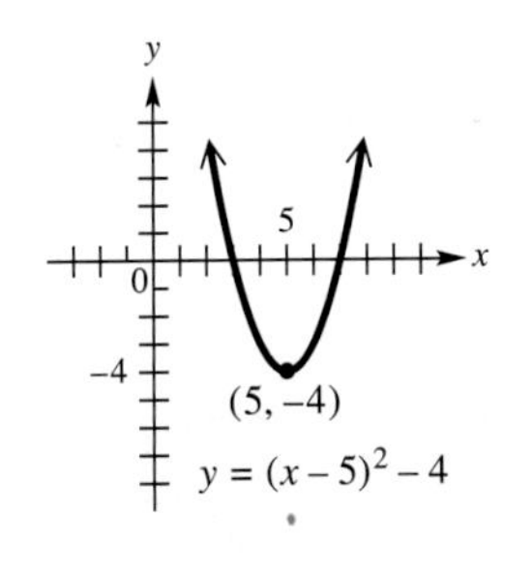


**77.** vertex: $(2,-1)$; axis: $x=2$; domain: $(-\infty,\infty)$; range: $[-1,\infty)$

$y=\frac{2}{3}(x-2)^2-1$

**79.** vertex: $(-3,-4)$; axis: $x=-3$; domain: $(-\infty,\infty)$; range: $[-4,\infty)$

$y=x^2+6x+5$

**81.** vertex: $(4, 2)$; axis: $x = 4$; domain: $(-\infty, \infty)$; range: $(-\infty, 2]$

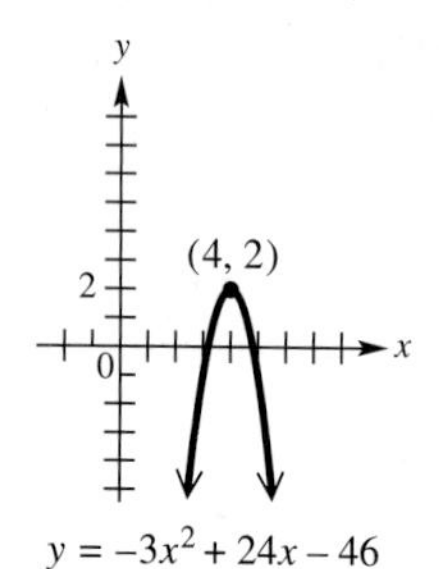


**83.** vertex: $(0, -1)$; axis: $y = -1$; domain: $[0, \infty)$; range: $(-\infty, \infty)$

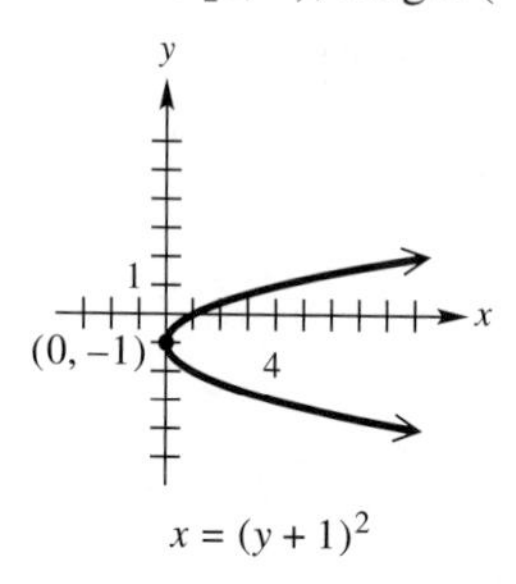


**85.** vertex: $(-1, -2)$; axis: $y = -2$; domain: $[-1, \infty)$; range: $(-\infty, \infty)$

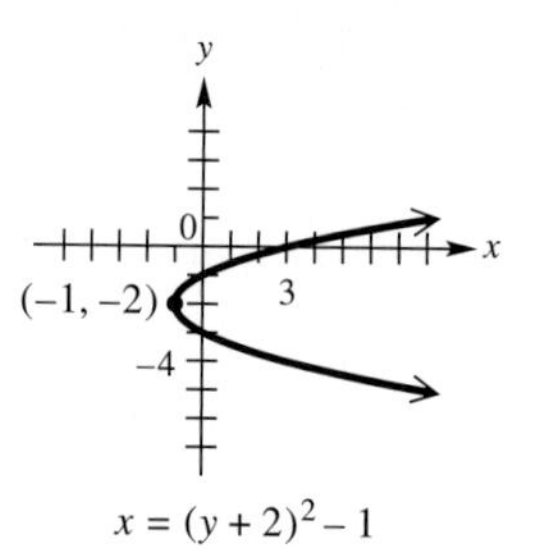


**87.** vertex: $(0, -3)$; axis: $y = -3$; domain: $(-\infty, 0]$; range: $(-\infty, \infty)$

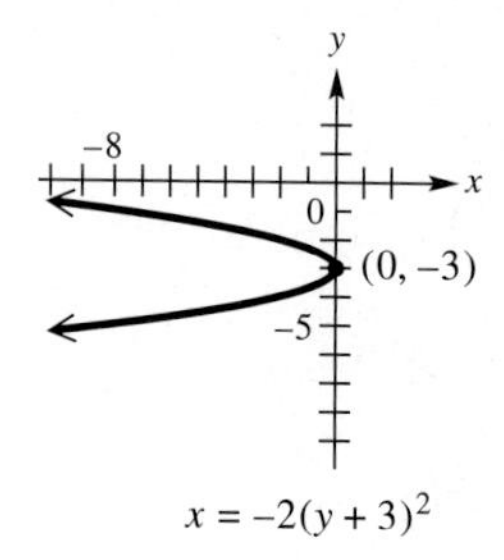


**89.** vertex: $(-9, -1)$; axis: $y = -1$; domain: $[-9, \infty)$; range: $(-\infty, \infty)$

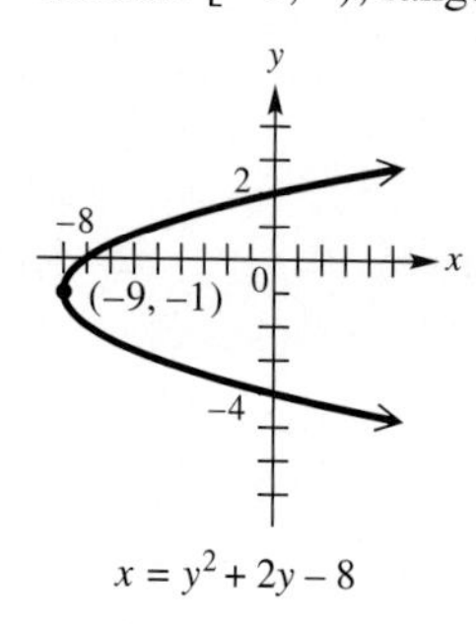


**91.** vertex: $(4, 1)$; axis: $y = 1$; domain: $[4, \infty)$; range: $(-\infty, \infty)$

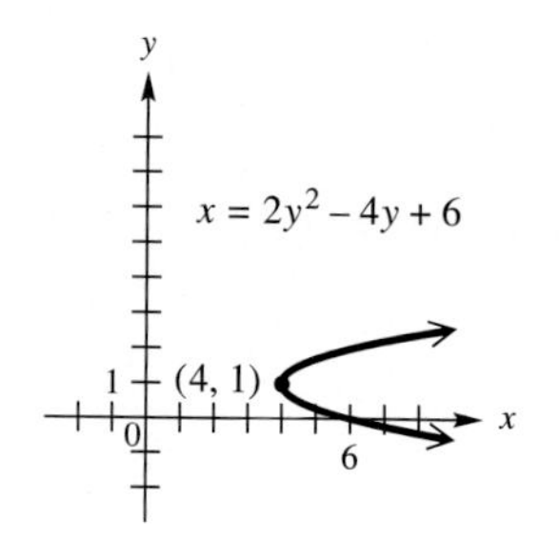


**93.** vertex: $(1, 2)$; axis: $y = 2$; domain: $[1, \infty)$; range: $(-\infty, \infty)$

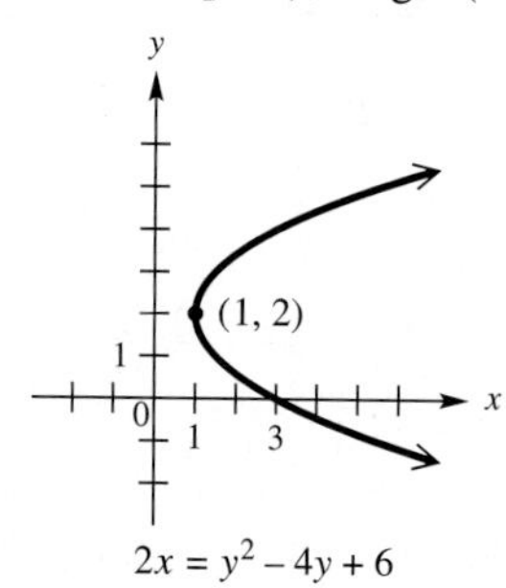


**95.** vertex: $(2, 1)$; axis: $y = 1$; domain: $(-\infty, 2]$; range: $(-\infty, \infty)$

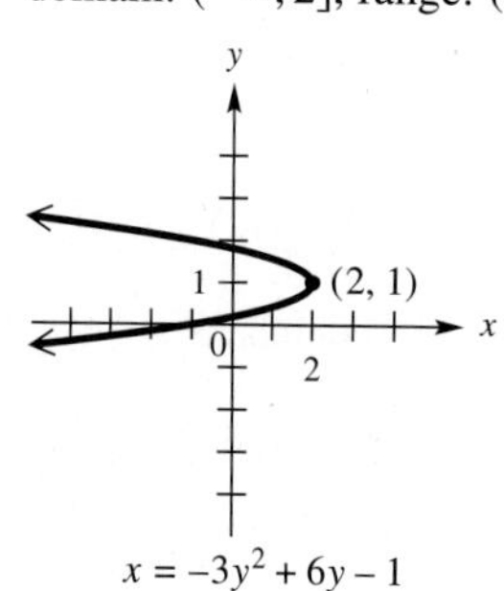


**97.** **(a)** Earth

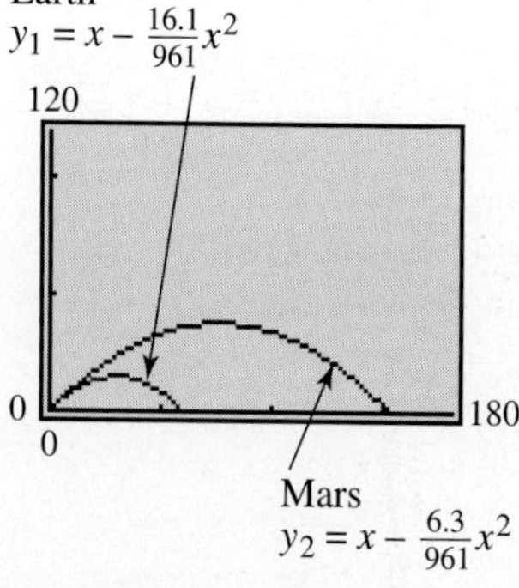


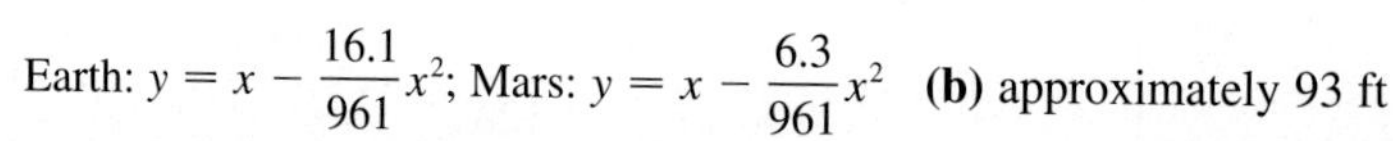

Earth: $y = x - \frac{16.1}{961}x^2$; Mars: $y = x - \frac{6.3}{961}x^2$ **(b)** approximately 93 ft

**99.** approximately 43.8 ft **101.** 60 ft

## 6.2 Exercises *(page 442)*

**1.** C **3.** F **5.** A **7.** G **9.** A circle can be interpreted as an ellipse whose two foci have the same coordinates; the "coinciding foci" give the center of the circle.

**11.** foci: $(-\sqrt{5}, 0), (\sqrt{5}, 0)$;

domain: $[-3, 3]$; range: $[-2, 2]$

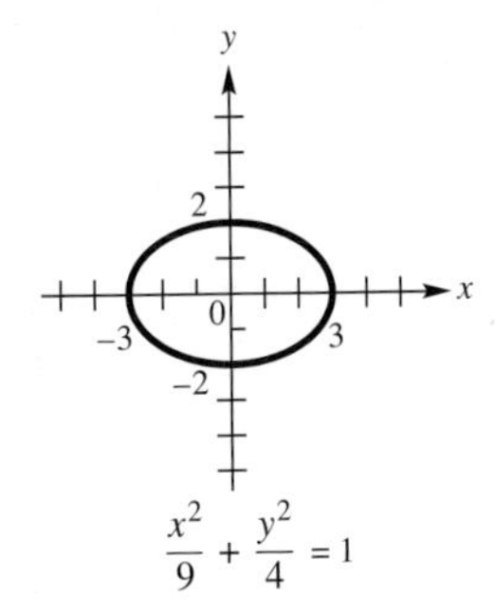


**13.** foci: $(0, -\sqrt{3}), (0, \sqrt{3})$;

domain: $[-\sqrt{6}, \sqrt{6}]$; range: $[-3, 3]$

$9x^2 + 6y^2 = 54$

**15.** domain: $\left[-\frac{3}{8}, \frac{3}{8}\right]$;

range: $\left[-\frac{6}{5}, \frac{6}{5}\right]$

$\frac{25y^2}{36} + \frac{64x^2}{9} = 1$

**17.** domain: $[-2, 4]$; range: $[-8, 2]$

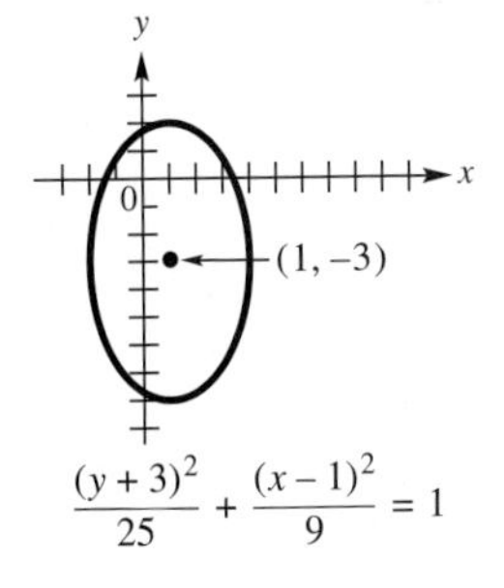


**19.** domain: $[-2, 6]$; range: $[-2, 4]$

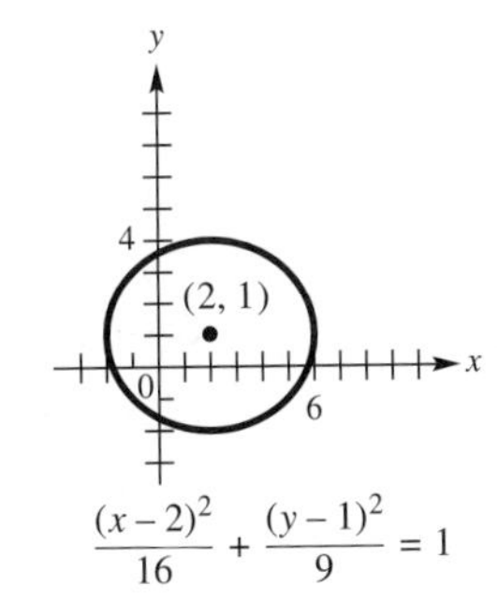


**21.** domain: $[-9, 7]$; range: $[-5, 9]$

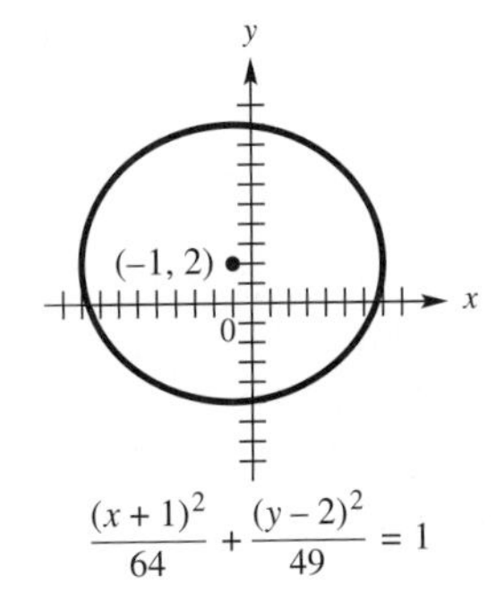


**23.** horizontal: $y = -1$; vertical: $x = 2$

**24.** $16 - \frac{16(x-2)^2}{9} \geq 0$

**25.** a parabola

**26.**

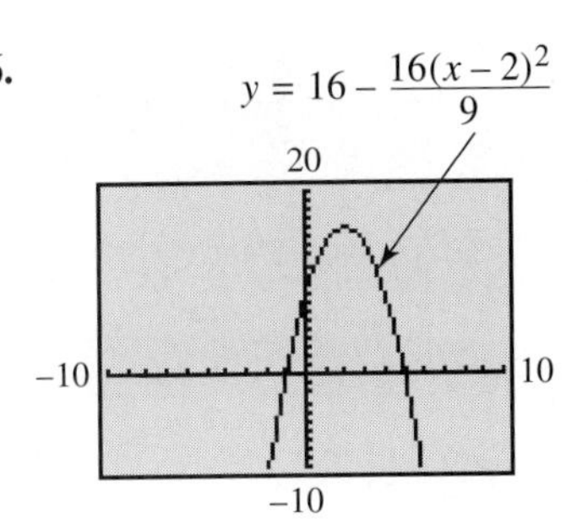


**27.** The graph of $y = 16 - \frac{16(x-2)^2}{9}$ lies above or on the $x$-axis in the interval $[-1, 5]$.

**28.** In Figure 28, we see that the domain is $[-1, 5]$. This corresponds to the solution set found graphically in Exercise 27.

**29.** $[-1, 5]$

**31.** domain: $(-\infty, -4] \cup [4, \infty)$;
range: $(-\infty, \infty)$

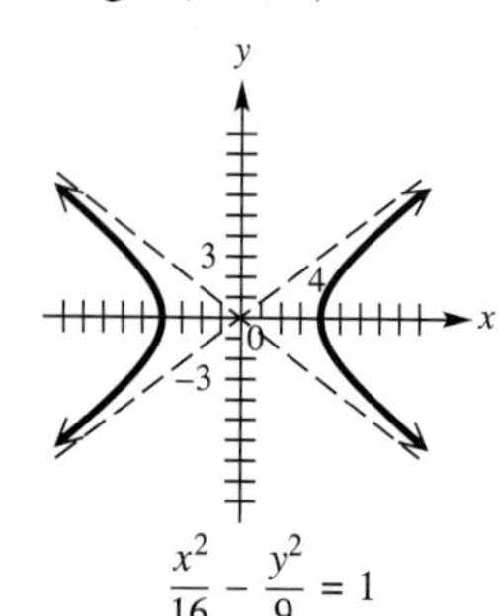


$\frac{x^2}{16} - \frac{y^2}{9} = 1$

**33.** domain: $(-\infty, \infty)$;
range: $(-\infty, -6] \cup [6, \infty)$

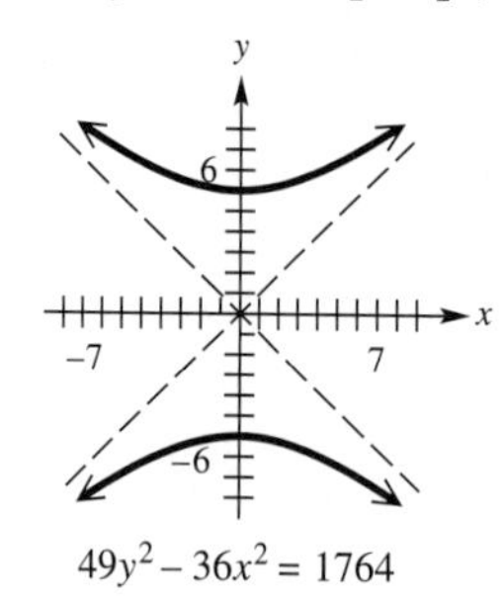


$49y^2 - 36x^2 = 1764$

**35.** domain: $\left(-\infty, -\frac{3}{2}\right] \cup \left[\frac{3}{2}, \infty\right)$;
range: $(-\infty, \infty)$

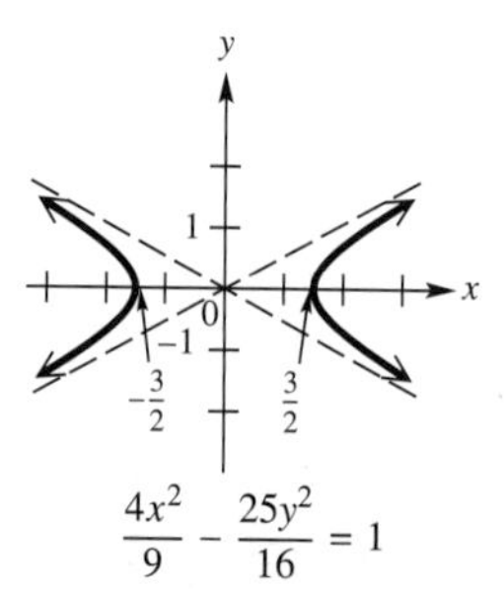


$\frac{4x^2}{9} - \frac{25y^2}{16} = 1$

**37.** center: $(1, -3)$;
domain: $(-\infty, -2] \cup [4, \infty)$;
range: $(-\infty, \infty)$

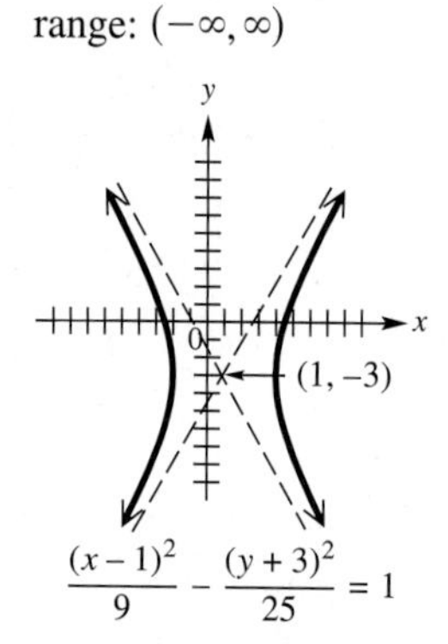


$\frac{(x-1)^2}{9} - \frac{(y+3)^2}{25} = 1$

**39.** center: $(-1, 5)$;
domain: $(-\infty, \infty)$;
range: $(-\infty, 3] \cup [7, \infty)$

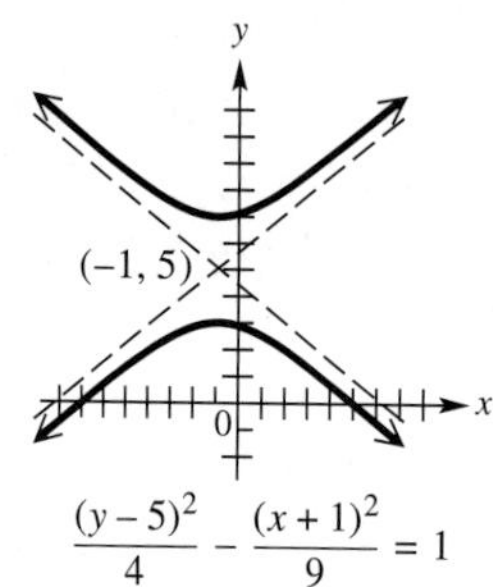


$\frac{(y-5)^2}{4} - \frac{(x+1)^2}{9} = 1$

**41.** $\frac{x^2}{16} + \frac{y^2}{12} = 1$ **43.** $\frac{x^2}{36} + \frac{y^2}{20} = 1$

**45.** $\frac{(y+2)^2}{25} + \frac{(x-3)^2}{16} = 1$ **47.** $\frac{x^2}{9} - \frac{y^2}{7} = 1$

**49.** $\frac{y^2}{9} - \frac{x^2}{25} = 1$ **51.** $(-2, 0), (2, 0)$

**52.**

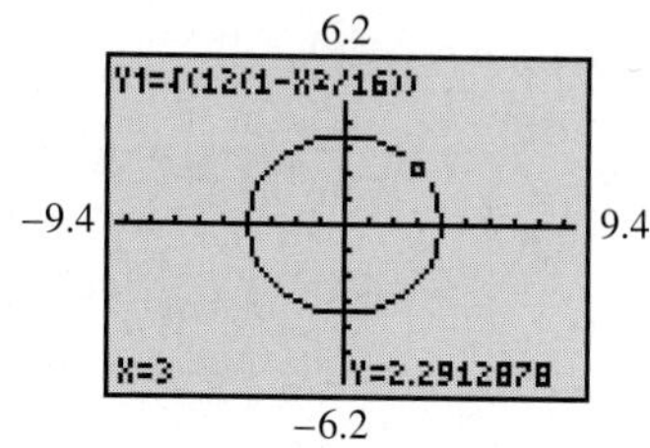


In addition to $(3, 2.2912878)$ shown on the screen, other points are $(0, 3.4641016)$ and $(-3, -2.291288)$.

**53.** The points satisfy the equation.

**54.**

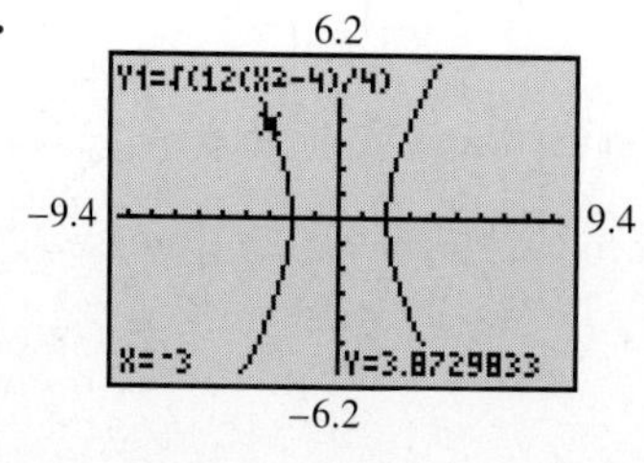


$(-4, 0), (4, 0)$; In addition to the point shown on the screen, other points are $(-2, 0)$, $(2, 0)$, and $(4, 6)$. **55.** The points satisfy the equation. **56.** Exercise 53 demonstrates that the points on the graph satisfy the definition of the ellipse for that particular ellipse. Exercise 55 demonstrates similarly that the definition of the hyperbola is satisfied for that hyperbola.

**57.** $\dfrac{x^2}{100} + \dfrac{y^2}{64} = 1$ **59.** 348.2 ft **61.** just under 12 ft

**63.** **(a)**

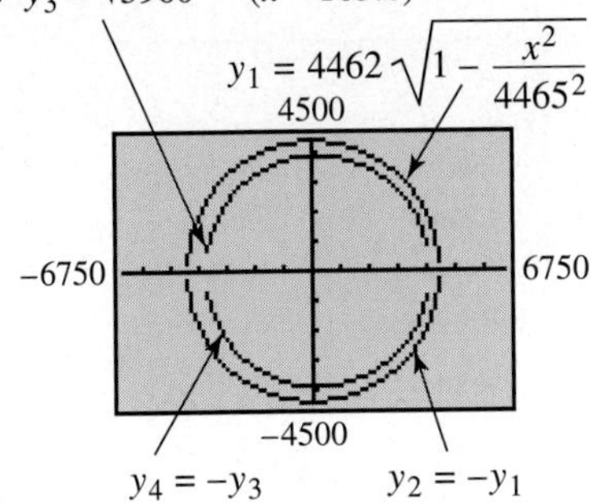


**(b)** minimum: approximately 341 mi; maximum: approximately 669 mi

**65.** **(a)** $x = \sqrt{y^2 + 2.5 \times 10^{-27}}$ **(b)** $1.2 \times 10^{-13}$ m **69.** $x^2 - \dfrac{y^2}{3} = 1$

## Reviewing Basic Concepts *(page 446)*

**1.** **(a)** B **(b)** D **(c)** A **(d)** C **2.**

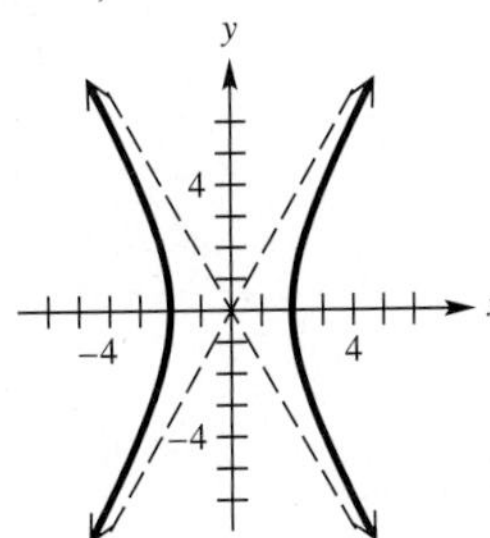


**3.**

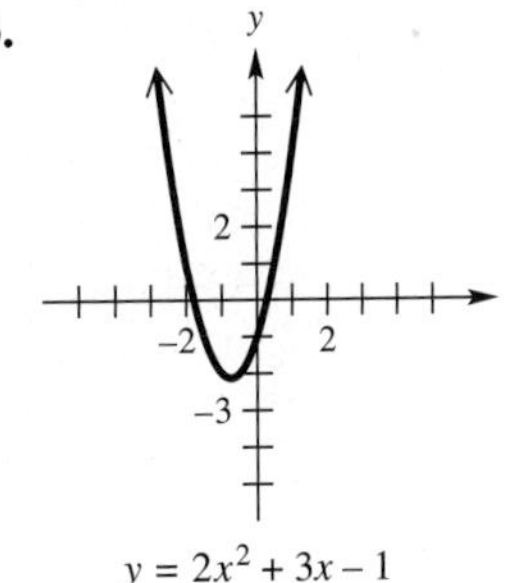


**4.**

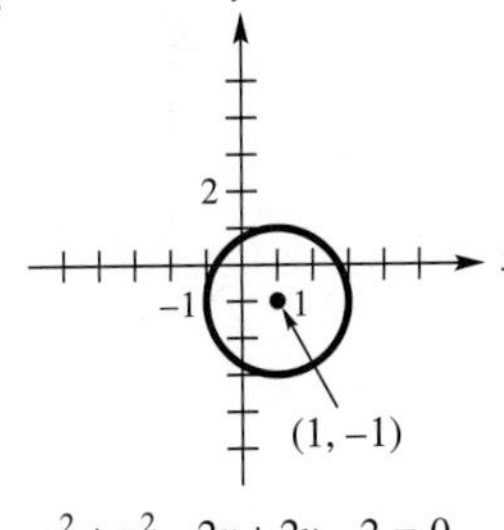

**5.**

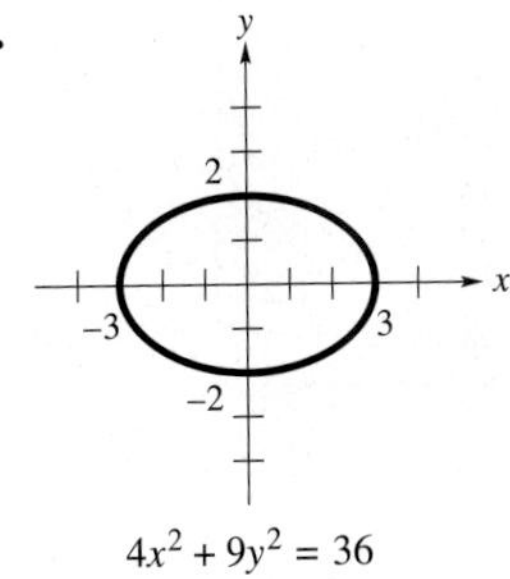


**6.** If $c < a$, it is an ellipse, and if $c > a$, it is a hyperbola. **7.** $(x - 2)^2 + (y + 1)^2 = 9$

**8.** $\dfrac{x^2}{36} + \dfrac{y^2}{20} = 1$ **9.** $\dfrac{y^2}{4} - \dfrac{x^2}{12} = 1$ **10.** $x^2 = 2y$

## 6.3 Exercises *(page 452)*

**1.** circle **3.** parabola **5.** parabola **7.** ellipse **9.** hyperbola **11.** hyperbola **13.** ellipse **15.** no graph **17.** circle **19.** parabola **21.** point **23.** no graph **25.** circle **27.** parabola **29.** hyperbola **31.** ellipse **33.** no graph **35.** line **37.** ellipse **39.** hyperbola **41.** $\dfrac{1}{2}$ **43.** $\sqrt{2}$ **45.** $\dfrac{\sqrt{21}}{7}$ **47.** $\dfrac{\sqrt{10}}{3}$ **49.** $x^2 = 32y$

**51.** $\dfrac{x^2}{36} + \dfrac{y^2}{27} = 1$ **53.** $\dfrac{x^2}{36} - \dfrac{y^2}{108} = 1$ **55.** $x^2 = -4y$ **57.** $\dfrac{y^2}{9} + \dfrac{25x^2}{81} = 1$ **59.** C, A, B, D

**61.** **(a)** Neptune: $\dfrac{(x-.2709)^2}{30.1^2}+\dfrac{y^2}{30.1^2}=1$; Pluto: $\dfrac{(x-9.8106)^2}{39.4^2}+\dfrac{y^2}{38.16^2}=1$ **(b)**

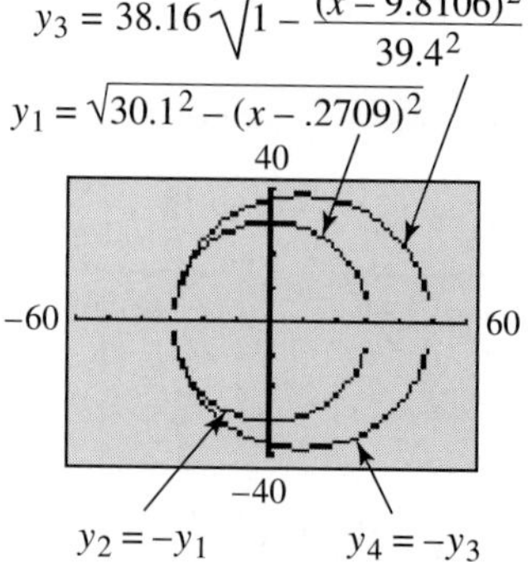


**63.** approximately 55 million mi

## 6.4 Exercises *(page 457)*

**1.** $x=2t,\ y=t+1,\ t$ in $[-2,3]$

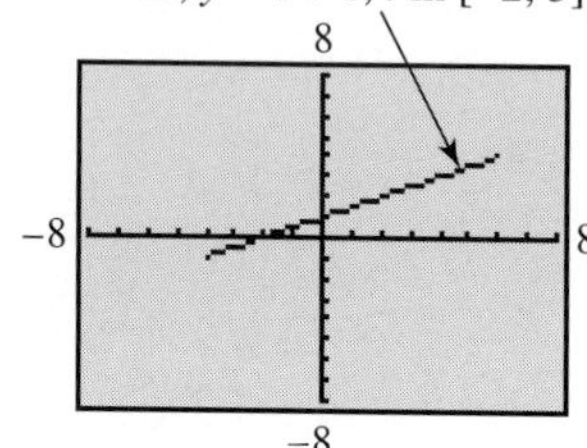


$y=\dfrac{1}{2}x+1,\ x$ in $[-4,6]$

**3.** $x=\sqrt{t},\ y=3t-4,\ t$ in $[0,4]$

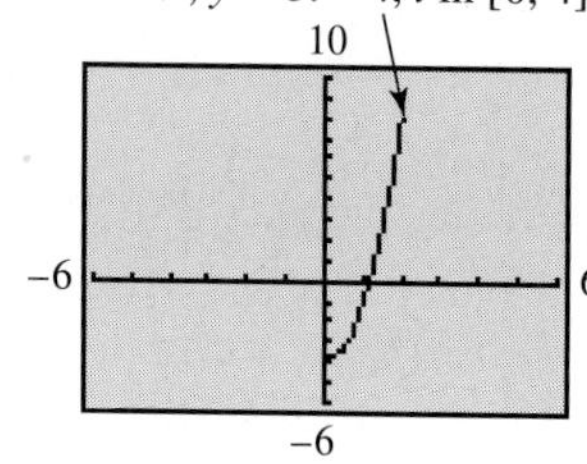


$y=3x^2-4,\ x$ in $[0,2]$

**5.** $x=t^3+1,\ y=t^3-1,\ t$ in $[-3,3]$

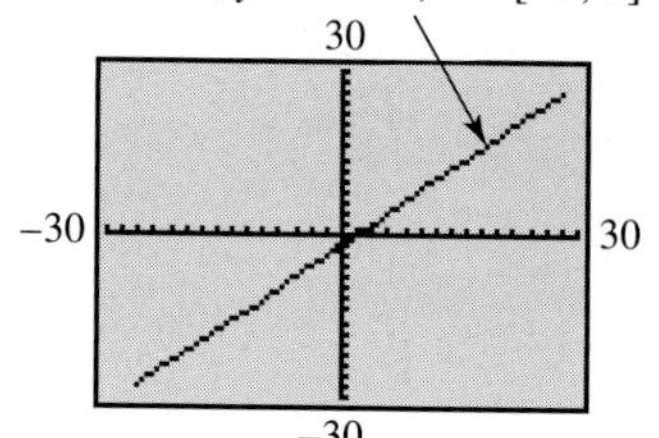


$y=x-2,\ x$ in $[-26,28]$

**7.** $x=2^t,\ y=\sqrt{3t-1},\ t$ in $\left[\frac{1}{3},4\right]$

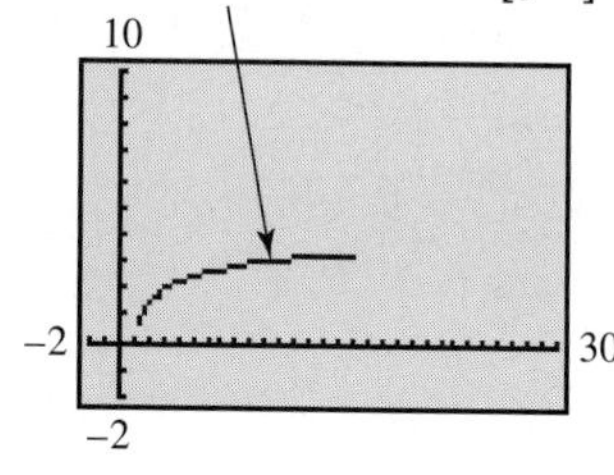


$x=2^{(y^2+1)/3},\ y$ in $\left[0,\sqrt{11}\right]$

**9.** $x=t+2,\ y=-\frac{1}{2}\sqrt{9-t^2},$ $t$ in $[-3,3]$

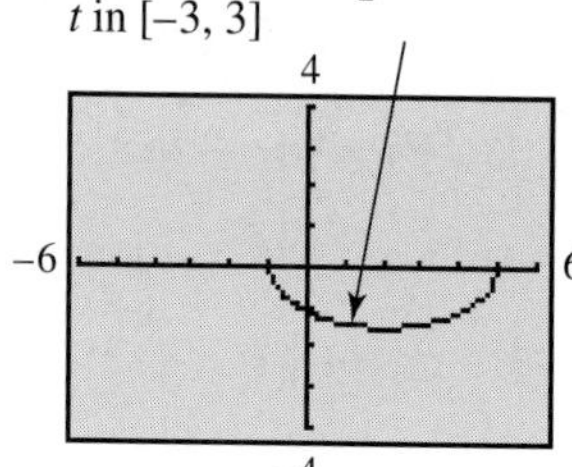


$y=-\dfrac{1}{2}\sqrt{9-(x-2)^2},\ x$ in $[-1,5]$

**11.** $x=t,\ y=\frac{1}{t},\ t$ in $(-\infty,0)\cup(0,\infty)$

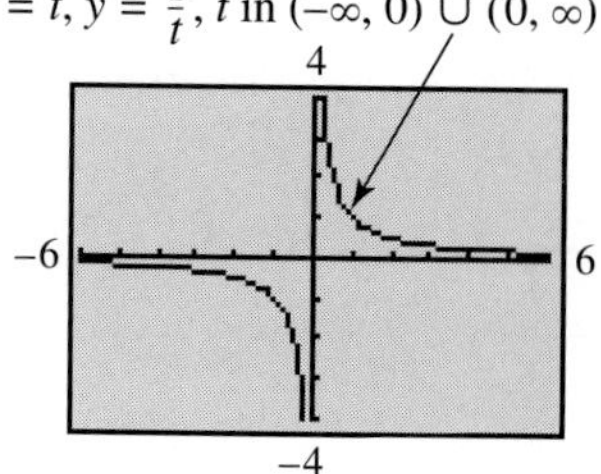


$y=\dfrac{1}{x},\ x$ in $(-\infty,0)\cup(0,\infty)$

**13.** $x=3\left(\dfrac{y}{4}\right)^{2/3},\ y$ in $(-\infty,\infty)$ **15.** $y=\sqrt{x^2+2},\ x$ in $(-\infty,\infty)$ **17.** $y=\dfrac{1}{x},\ x$ in $(0,\infty)$ **19.** $y=1-2x^2,\ x$ in $(0,\infty)$

**21.** $y=\dfrac{1}{x},\ x\neq 0$ **23.** $y=\ln x,\ x$ in $(0,\infty)$

Other answers are possible for Exercises 25 and 27.

**25.** $x=\dfrac{1}{2}t,\ y=t+3;\ x=\dfrac{t+3}{2},\ y=t+6$ **27.** $x=\dfrac{1}{3}t,\ y=\sqrt{t+2},\ t$ in $[-2,\infty)$; $x=\dfrac{t-2}{3},\ y=\sqrt{t},\ t$ in $[0,\infty)$

**29.** **(a)** 17.7 sec **(b)** 5000 ft **(c)** 1250 ft

**31.**

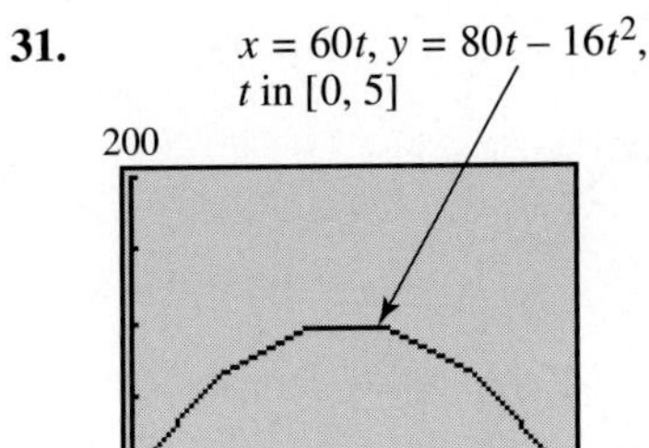


$y = \frac{4}{3}x - \frac{1}{225}x^2$

**35.** Many answers are possible, two of which are $x = t$, $y - y_1 = m(t - x_1)$ and $t = x - x_1$, $y = mt + y_1$.

## Reviewing Basic Concepts *(page 459)*

**1.** $\frac{(x-1)^2}{4} + \frac{(y+3)^2}{12} = 1$; ellipse **2.** $\frac{(y+4)^2}{12} - \frac{(x+2)^2}{6} = 1$; hyperbola **3.** $(y+2)^2 = -\frac{5}{3}(x-3)$; parabola

**4.** $\frac{2\sqrt{6}}{5}$ **5.** $\sqrt{3}$ **6.** $y^2 = -8x$ **7.** $\frac{x^2}{25} + \frac{y^2}{16} = 1$ **8.** $\frac{y^2}{16} - \frac{x^2}{9} = 1$ **9.** $\frac{9\sqrt{21}}{5} \approx 8.25$ ft

**10.** **(a)**

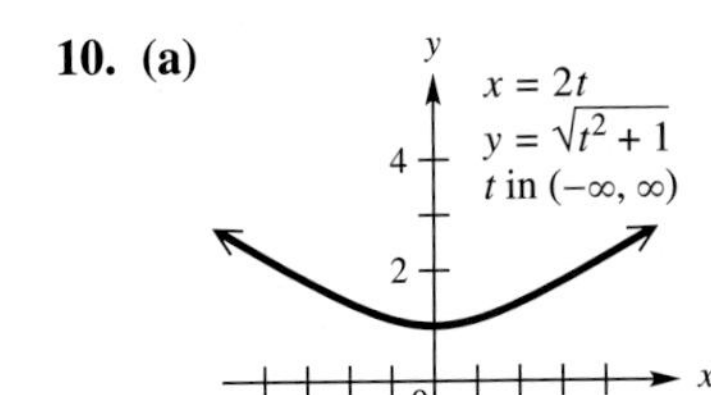


**(b)** $y = \sqrt{\frac{x^2}{4} + 1}$

## Chapter 6 Review Exercises *(page 461)*

**1.** $(x+2)^2 + (y-3)^2 = 25$;
domain: $[-7, 3]$;
range: $[-2, 8]$

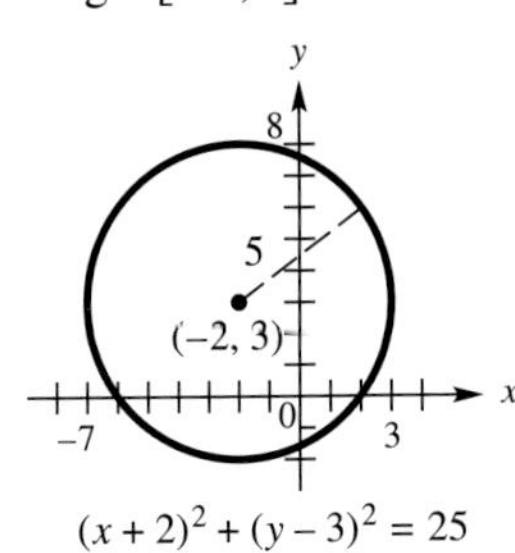


$(x + 2)^2 + (y - 3)^2 = 25$

**3.** $(x+8)^2 + (y-1)^2 = 289$;
domain: $[-25, 9]$;
range: $[-16, 18]$

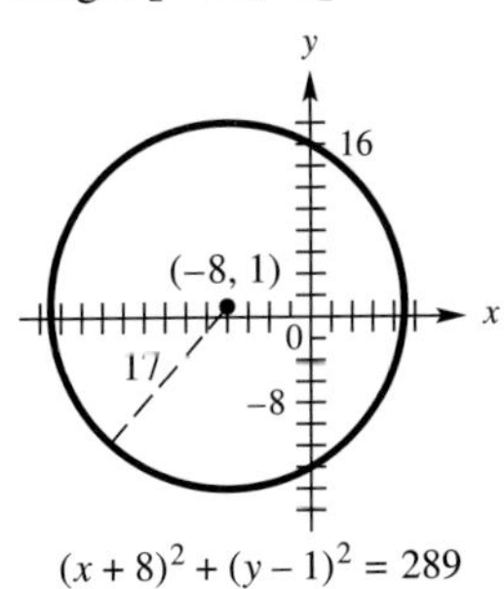


$(x + 8)^2 + (y - 1)^2 = 289$

**5.** $(2, -3)$; 1 **7.** $\left(-\frac{7}{2}, -\frac{3}{2}\right)$; $\frac{3\sqrt{6}}{2}$

**9.** The graph consists of the single point $(4, 5)$.

**11.**

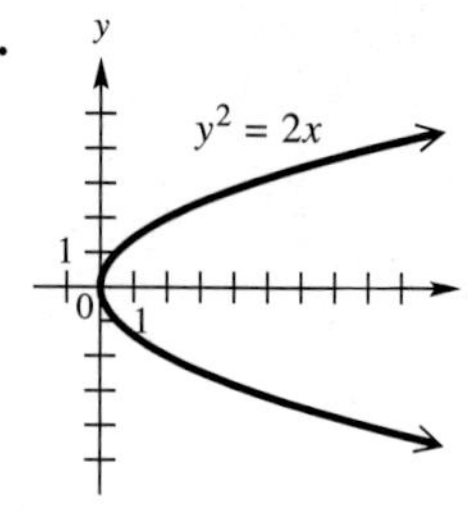


$\left(\frac{1}{2}, 0\right)$; $x = -\frac{1}{2}$; $x$-axis;
domain: $[0, \infty)$;
range: $(-\infty, \infty)$

**13.**

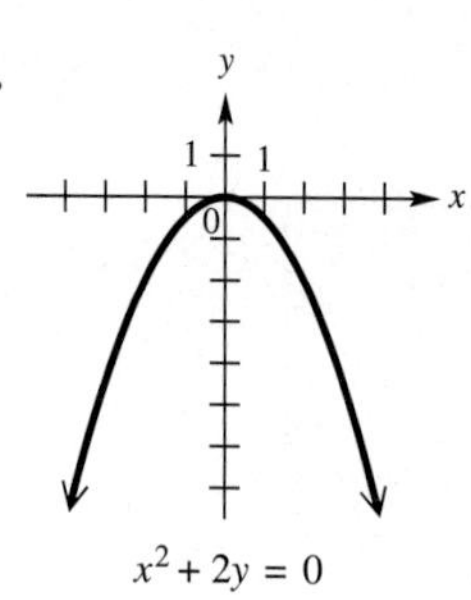


$\left(0, -\frac{1}{2}\right)$; $y = \frac{1}{2}$; $y$-axis;
domain: $(-\infty, \infty)$;
range: $(-\infty, 0]$

**15.** $y^2 = \frac{25}{2}x$ **17.** $(y - 6)^2 = 28(x + 5)$

**19.**

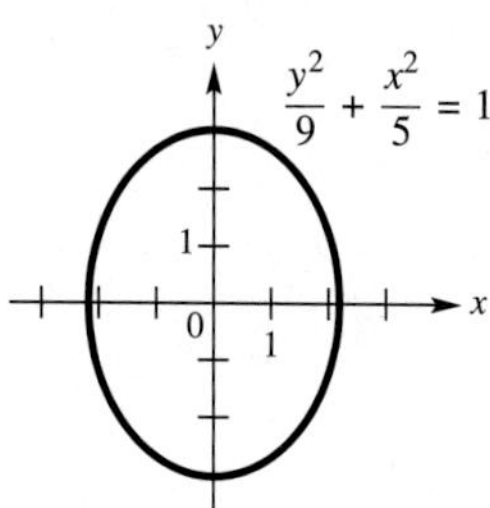


$[-\sqrt{5}, \sqrt{5}]$; $[-3, 3]$; $(0, -3)$, $(0, 3)$

**21.**

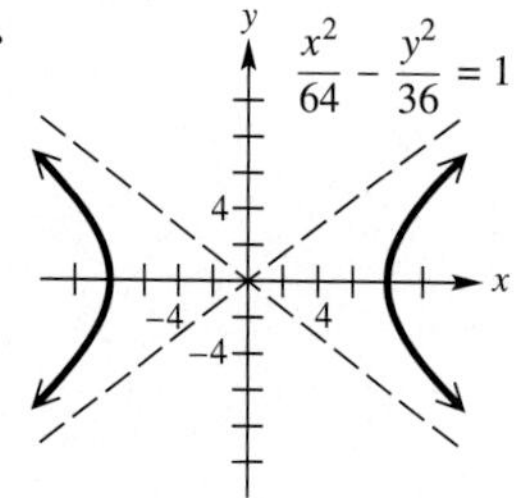


$(-\infty, -8] \cup [8, \infty)$; $(-\infty, \infty)$; $(-8, 0)$, $(8, 0)$

**23.**

y, 5, x, 0, (3, –1), –2, $\frac{(x-3)^2}{4} + (y+1)^2 = 1$

$[1, 5]$; $[-2, 0]$; $(1, -1)$, $(5, -1)$

**25.**

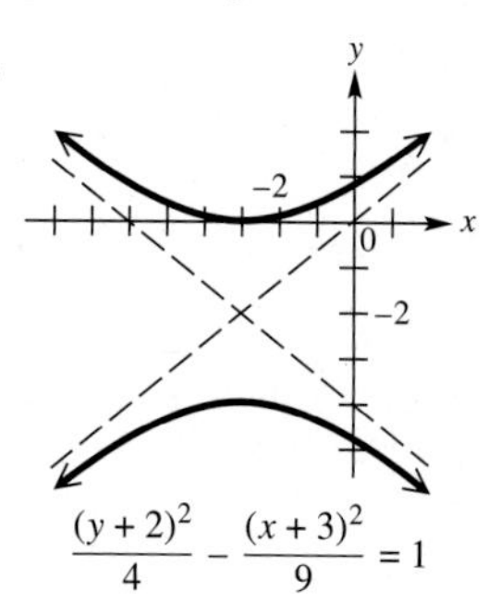


$(-\infty, \infty)$; $(-\infty, -4] \cup [0, \infty)$; $(-3, 0)$, $(-3, -4)$

**27.** $\frac{y^2}{16} + \frac{x^2}{12} = 1$ **29.** $\frac{y^2}{16} - \frac{x^2}{9} = 1$ **31.** $\frac{4y^2}{81} + \frac{4x^2}{45} = 1$

**33.** **(a)** $(-1, -3)$ **(b)** 5 **(c)** $y_1 = -3 + \sqrt{25 - (x + 1)^2}$; $y_2 = -3 - \sqrt{25 - (x + 1)^2}$ **35.** C **37.** E **39.** D **41.** $\frac{\sqrt{5}}{3}$

**43.** $x = \frac{1}{3}(y - 2)^2 - 3$ **45.** $\frac{(x - 2)^2}{16} + \frac{y^2}{12} = 1$

**47.**

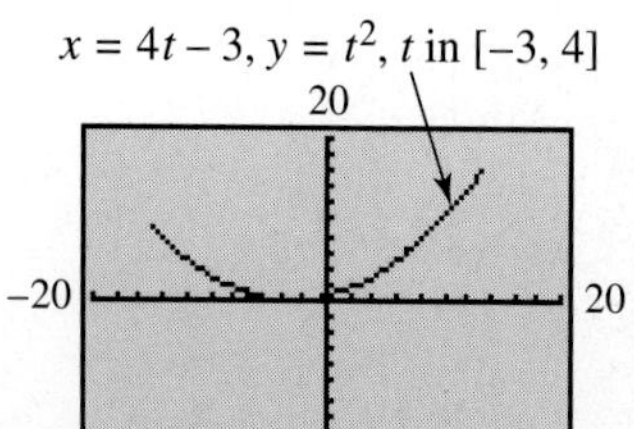


**49.**

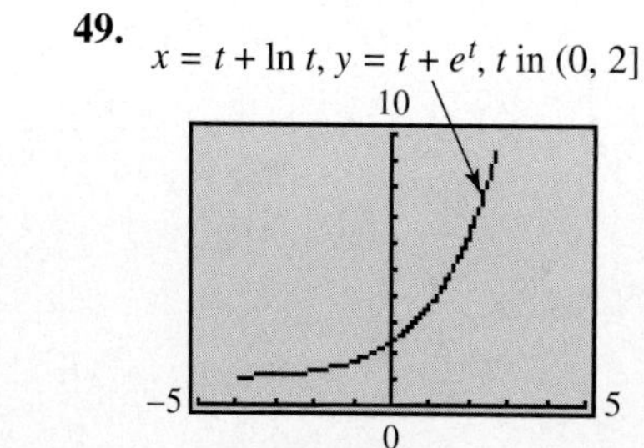


**51.** $y^2 - x^2 = 1$, $x$ in $[0, \infty)$

**53.** 66.8 and 67.7 million mi **55.** elliptic

**57.** The required increase in velocity is less when $D$ is larger.

## Chapter 6 Test (*page 463*)

**1.** **(a)** B **(b)** E **(c)** F **(d)** A **(e)** C **(f)** D **2.** $\left(\frac{1}{32}, 0\right)$; $x = -\frac{1}{32}$

**3.**

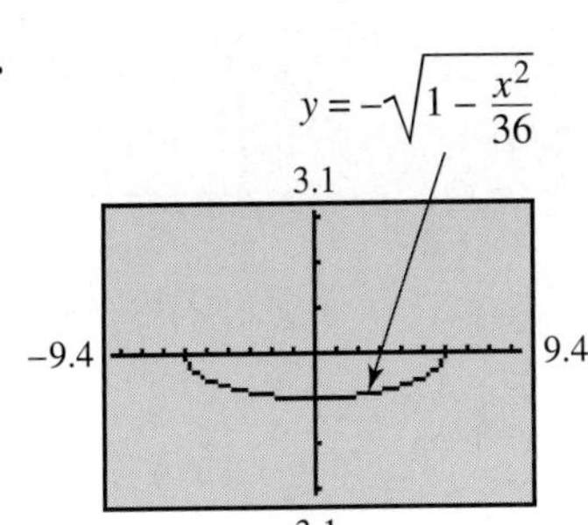


yes; domain: $[-6, 6]$; range: $[-1, 0]$ **4.** $y_1 = 7\sqrt{\frac{x^2}{25} - 1}$; $y_2 = -7\sqrt{\frac{x^2}{25} - 1}$

**5.**

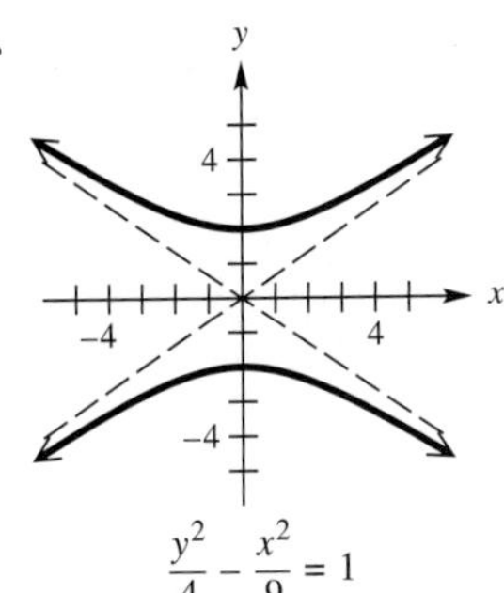

$\frac{y^2}{4} - \frac{x^2}{9} = 1$

hyperbola; center: $(0, 0)$; vertices: $(0, -2)$, $(0, 2)$; foci: $(0, -\sqrt{13})$, $(0, \sqrt{13})$

**6.**

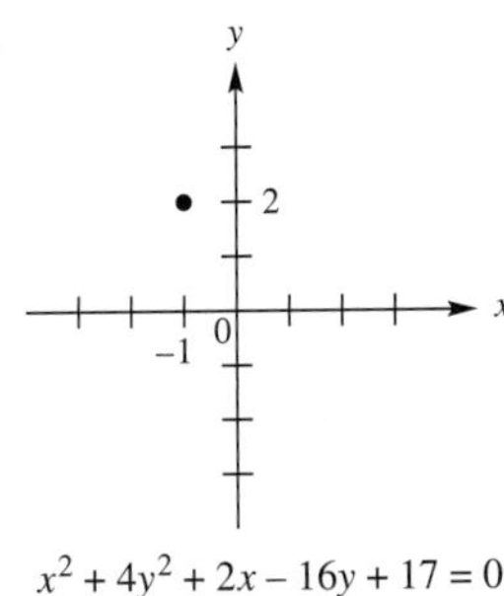

$x^2 + 4y^2 + 2x - 16y + 17 = 0$

the point $(-1, 2)$

**7.**

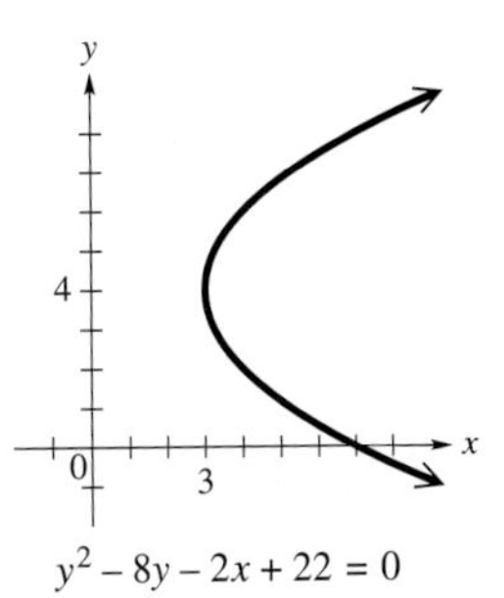

$y^2 - 8y - 2x + 22 = 0$

parabola; vertex: $(3, 4)$; focus: $(3.5, 4)$

**8.**

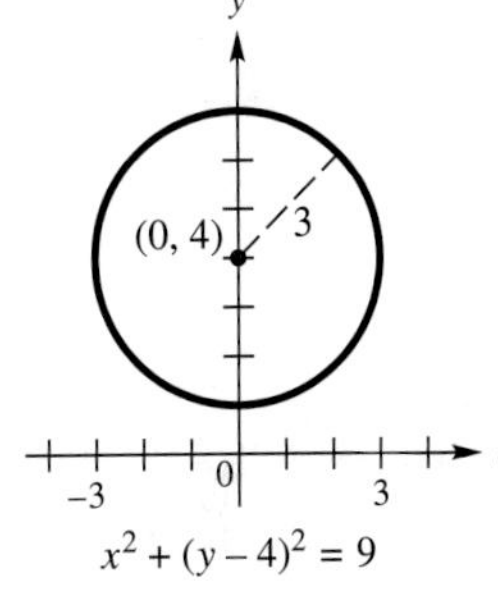

$x^2 + (y - 4)^2 = 9$

circle; center: $(0, 4)$; radius: 3

**9.**

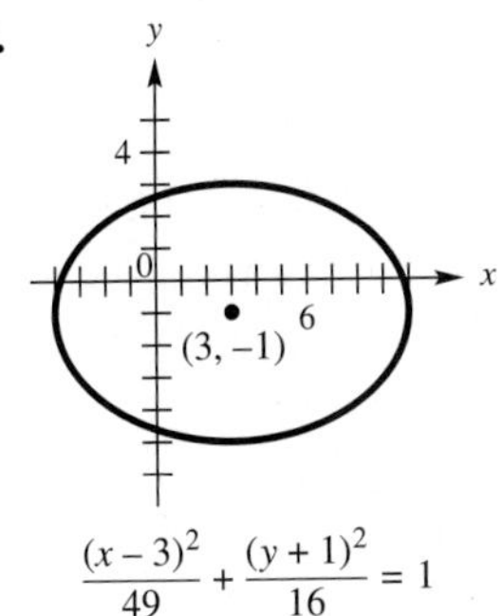

$\frac{(x-3)^2}{49} + \frac{(y+1)^2}{16} = 1$

ellipse; center: $(3, -1)$; vertices: $(-4, -1)$, $(10, -1)$; foci: $(3 + \sqrt{33}, -1)$, $(3 - \sqrt{33}, -1)$

**10.**

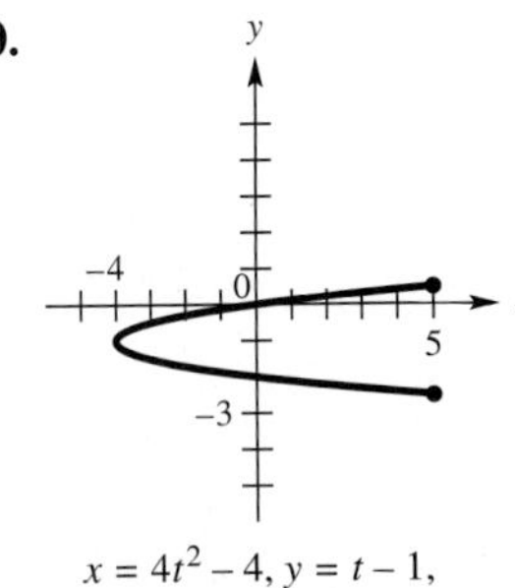

$x = 4t^2 - 4$, $y = t - 1$, $t$ in $[-1.5, 1.5]$

parabola; vertex: $(-4, -1)$; focus: $\left(\frac{-63}{16}, -1\right)$

**11. (a)** $y = -\frac{1}{8}x^2$ **(b)** $\frac{x^2}{\frac{11}{4}} + \frac{y^2}{9} = 1$ or $\frac{4x^2}{11} + \frac{y^2}{9} = 1$

**12.** $\frac{x^2}{400} + \frac{y^2}{144} = 1$; approximately 10.39 ft

## CHAPTER 7 MATRICES AND SYSTEMS OF EQUATIONS AND INEQUALITIES

### 7.1 Exercises *(page 475)*

**1.** approximately 2002 **3.** (2002, 3.1 million) **5.** the year; the number of migrants **7.** $4(1) - 1 = 3$ and $-2(1) + 3(1) = 1$ are both true. **9.** $5 - 3 = 2$ and $3^2 + 5^2 = 34$ are both true. **11.** $\{(1, 1)\}$ **13.** $\{(3, -2)\}$ **15.** $\{(6, 15)\}$ **17.** $\{(2, -3)\}$ **19.** $\{(1, 3)\}$ **21.** $\{(-2, 3)\}$ **23.** $\{(5, 0)\}$ **25.** A **27.** An inconsistent system will conclude with no variables and a false statement, such as $0 = 1$. A system with dependent equations will conclude with no variables and a true statement, such as $0 = 0$. **29.** $\{(0, 4)\}$ **31.** $\{(-1, 1)\}$ **33.** $\{(2, 3)\}$ **35.** $\left\{\left(\frac{y+9}{4}, y\right)\right\}$ **37.** $\emptyset$ **39.** $\emptyset$ **41.** $\{(12, 6)\}$ **43.** $\{(5, 2)\}$ **45.** $\{(.138, -4.762)\}$ **47.** $\{(-8.708, 15.668)\}$

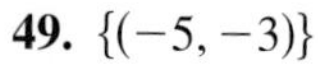

**49.** $\{(-5, -3)\}$

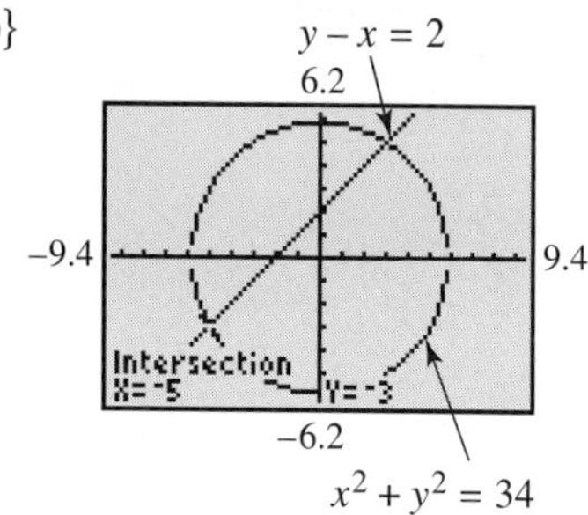


**51.**

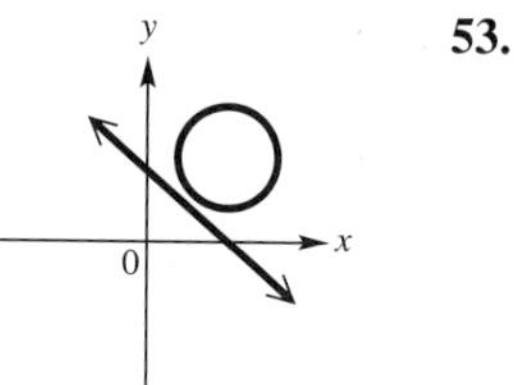


**53.**

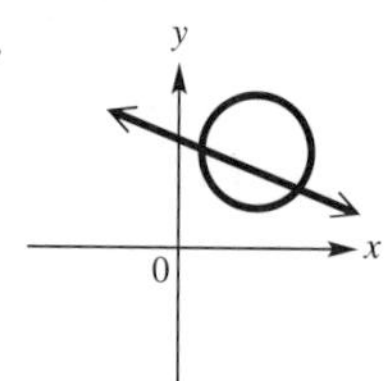


**55.**

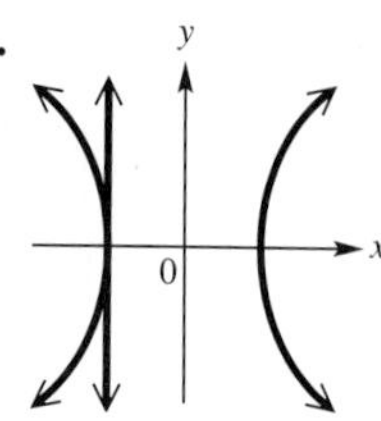


**57.**

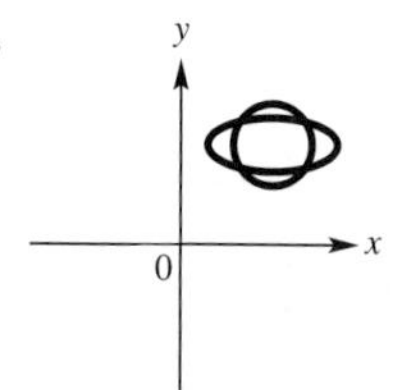


**59.**

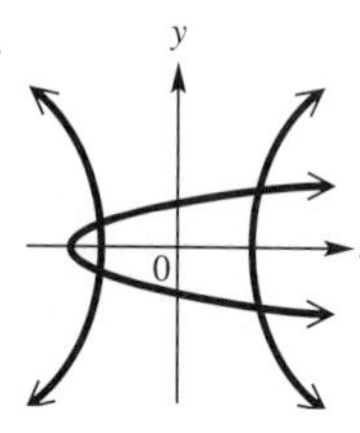


**61.** $\{(-2, -2), (1, 1)\}$ **63.** $\left\{\left(-\frac{3}{5}, \frac{7}{5}\right), (-1, 1)\right\}$ **65.** $\{(3, 1), (3, -1), (-3, 1), (-3, -1)\}$ **67.** $\{(3, 0), (-3, 0)\}$ **69.** $\left\{\left(x, \pm\sqrt{10 - x^2}\right)\right\}$

**71.** $\left\{\left(\frac{1 + \sqrt{13}}{2}, \frac{-1 + \sqrt{13}}{2}\right), \left(\frac{-1 - \sqrt{21}}{2}, \frac{3 + \sqrt{21}}{2}\right)\right\}$

**73.** $\{(-.79, .62), (.88, .77)\}$ **75.** $\{(.06, 2.88)\}$ **77.**

$$\begin{array}{r|rrrr} 5 & 1 & 0 & -85 & 300 \\ & & 5 & 25 & -300 \\ \hline & 1 & 5 & -60 & 0 \end{array} \leftarrow 5 \text{ is a solution.}$$

**78.** $x^3 - 85x + 300 = (x - 5)(x^2 + 5x - 60)$ **79.** Use the quadratic formula with $a = 1$, $b = 5$, and $c = -60$.

**80.**

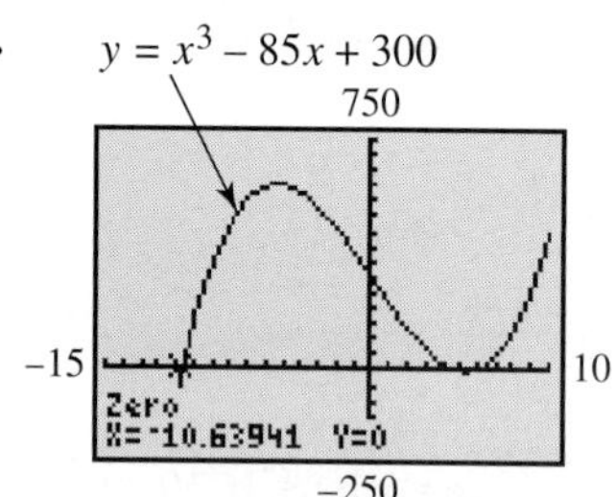


**81.** $\frac{-5 - \sqrt{265}}{2}$ **82.** Because $x$ represents a length, it must be a positive number.

**83.** **(a)** Let $x$ represent the number of bank robberies in 1994 and $y$ the number in 1995.

$x + y = 13{,}787$

$x - y = 271$

**(b)** $\{(7029, 6758)\}$

**(c)** In 1994 there were 7029 bank robberies, and in 1995 there were 6758 bank robberies.

**85.** 1.2 million foreign tourists; 5.0 million American tourists **87.** $x \approx 177.1$, $y \approx 174.9$; if an athlete's maximum heart rate is 180 beats per minute (bpm), then it will be about 177 bpm after 5 sec and 175 bpm after 10 sec. **89. (a)** $\{(5.8, 857.8)\}$ **(b)** During 1995, 857.8 million lb of canned tuna and fresh shrimp were available. **(c)**

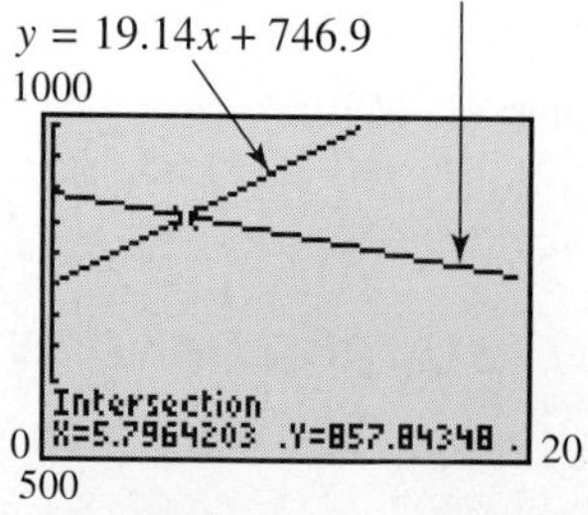


**91.** $r \approx 1.538$ in., $h \approx 6.724$ in. **93.** $W_1 = \frac{300}{1 + \sqrt{3}} \approx 109.8$ lb; $W_2 = \frac{300\sqrt{3}}{\sqrt{6} + \sqrt{2}} \approx 134.5$ lb; $\{(109.8, 134.5)\}$

**95.** $x = 300$; $y = 350$ **97.** $x = 65$; $R = C = \$385$ **99.** price: \$315; supply/demand: 420

**101.** $5t + 15u = 16$

$5t + 4u = 5$

**102.** $t = \frac{1}{5}$; $u = 1$ **103.** $x = 5$; $y = 1$ **104.** $y = \frac{-15x}{5 - 16x}$ **105.** $y = \frac{-4x}{5 - 5x}$

**106.** $y = \dfrac{-15x}{5 - 16x}$ $\quad y = \dfrac{-4x}{5 - 5x}$

2

Intersection X=5 Y=1

0 10

0

Dot Mode

**107.** $\{(2, 2)\}$

## 7.2 Exercises *(page 485)*

**1. (a)** Multiply equation (2) by 2 and add to get $7x - z = 0$. **(b)** Multiply equation (1) by $-2$ and add to get $-3y + 2z = 2$. **(c)** Multiply equation (2) by 4 and add to get $14x - 3y = 2$. **3.** The first equation yields $-6 = -6$, the second $-\frac{1}{12} = -\frac{1}{12}$, and the third $1 = 1$. **5.** $\{(1, 2, -1)\}$ **7.** $\{(2, 0, 3)\}$ **9.** $\emptyset$ **11.** $\{(1, 2, 3)\}$ **13.** $\{(-1, 2, 1)\}$ **15.** $\{(4, 1, 2)\}$ **17.** $\left\{\left(\frac{1}{2}, \frac{2}{3}, -1\right)\right\}$ **19.** $\{(2, 2, 2)\}$ **21.** $\{(1, 0, 6)\}$ **23.** $\left\{\left(\frac{1}{2}, \frac{1}{3}, \frac{1}{4}\right)\right\}$ **25.** $\{(2, 4, 2)\}$ **27.** $\left\{\left(-1, 1, \frac{1}{3}\right)\right\}$ **29.** (There are other possibilities as well.) **(a)** $\begin{aligned} x + 2y + z &= 5 \\ 2x - y + 3z &= 4 \end{aligned}$ **(b)** $\begin{aligned} x + y + z &= 5 \\ 2x - y + 3z &= 4 \end{aligned}$ **(c)** $\begin{aligned} 2x + 2y + 2z &= 8 \\ 2x - y + 3z &= 4 \end{aligned}$

**31.** For example, two perpendicular walls meeting in a corner with the floor intersect in a point. **33.** $\left\{\left(\frac{4 - z}{3}, \frac{4z - 7}{3}, z\right)\right\}$ **35.** $\left\{\left(\frac{69 - z}{9}, \frac{z + 66}{9}, z\right)\right\}$ **37.** $\left\{\left(\frac{1 - 2z}{5}, \frac{3z + 31}{5}, z\right)\right\}$ **39.** \$9.00 water: 120 gal; \$3.00 water: 60 gal; \$4.50 water: 120 gal **41.** "up close": \$24; "in the middle": \$18; "farther back": \$15 **43.** 75°, 65°, 40° **45.** at 4%: \$3000; at 4.5%: \$6000; at 2.5%: \$1000 **47.** There are three possibilities: 12 EZ, 16 compact, 0 commercial; 10 EZ, 8 compact, 3 commercial; 8 EZ, 0 compact, 6 commercial. **49.** $y = x^2 + 2x + 1$

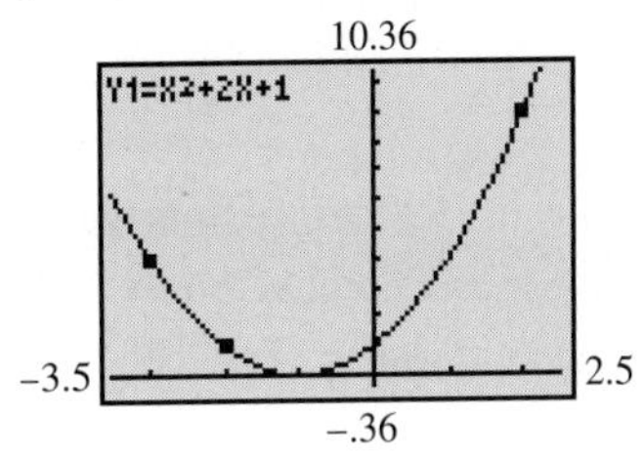


**51.** $y = 2x^2 + 3x$

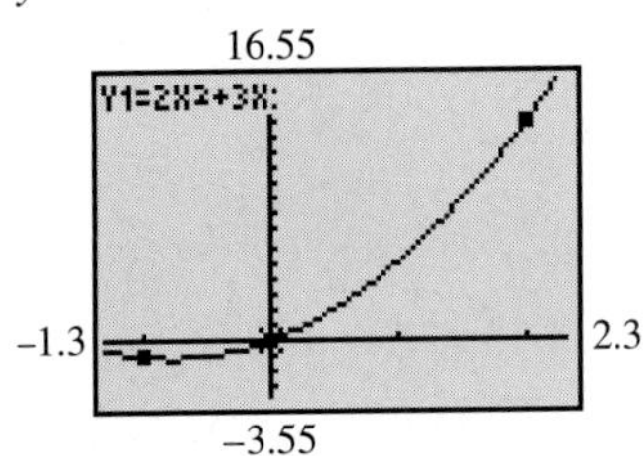


**53.** $x^2 + y^2 + x - 7y = 0$ **55.** $s(t) = -2t^2 + 20t + 5$; $s(8) = 37$

## 7.3 Exercises *(page 497)*

**1.** $\begin{bmatrix} -4 & -8 \\ 4 & 7 \end{bmatrix}$ **3.** $\begin{bmatrix} 1 & 5 & 6 \\ -1 & 8 & 5 \\ 4 & 7 & 0 \end{bmatrix}$ **5.** $\begin{bmatrix} -3 & 1 & -4 \\ 2 & 1 & 3 \\ -17 & 0 & -13 \end{bmatrix}$ **7.** $\left[\begin{array}{cc|c} 2 & 3 & 11 \\ 1 & 2 & 8 \end{array}\right]$ **9.** $\left[\begin{array}{cc|c} 1 & 5 & 6 \\ 1 & 0 & 3 \end{array}\right]$

**11.** $\left[\begin{array}{ccc|c} 2 & 1 & 1 & 3 \\ 3 & -4 & 2 & -7 \\ 1 & 1 & 1 & 2 \end{array}\right]$ **13.** $\left[\begin{array}{ccc|c} 1 & 1 & 0 & 2 \\ 0 & 2 & 1 & -4 \\ 0 & 0 & 1 & 2 \end{array}\right]$ **15.** $2x + y = 1$, $3x - 2y = -9$ **17.** $x = 2$, $y = 3$, $z = -2$ **19.** $3x + 2y + z = 1$, $2y + 4z = 22$, $-x - 2y + 3z = 15$

**21.** $\{(2, 3)\}$ **23.** $\{(-3, 0)\}$ **25.** $\left\{\left(\frac{7}{2}, -1\right)\right\}$ **27.** $\emptyset$ **29.** $\left\{\left(x, \frac{6x-1}{3}\right)\right\}$ or $\left\{\left(\frac{3y+1}{6}, y\right)\right\}$ **31.** $\{(-2, 1, 3)\}$

**33.** $\{(-1, 23, 16)\}$ **35.** $\{(3, 2, -4)\}$ **37.** $\{(2, 1, -1)\}$ **39.** $\{(5.211, 3.739, -4.655)\}$ **41.** $\{(-.250, 1.284, -.059)\}$

**43.** In both cases, we simply write the coefficients and do not write the variables. This is possible because we agree on the order in which the variables appear (descending degree). **45.** $\left\{\left(\frac{5z+14}{5}, \frac{5z-12}{5}, z\right)\right\}$ **47.** $\left\{\left(\frac{12-z}{7}, \frac{4z-6}{7}, z\right)\right\}$ **49.** $\emptyset$

**51.** $\{(0, 2, -2, 1)\}$ **53. (a)** $F = .5714N + .4571R - 2014$ **(b)** \$5700 **55.** model *A*: 5 bicycles; model *B*: 8 bicycles

**57.** at 8%: \$10,000; at 10%: \$7000; at 9%: \$8000

**59. (a)** $1990^2a + 1990b + c = 11$
$2010^2a + 2010b + c = 10$
$2030^2a + 2030b + c = 6$
**(b)** $f(x) = -.00375x^2 + 14.95x - 14,889.125$ **(c)**

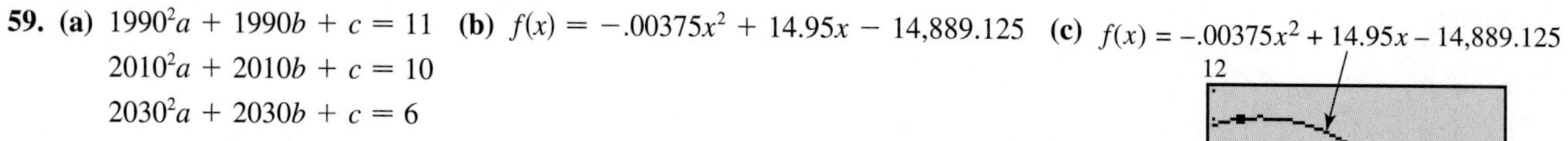

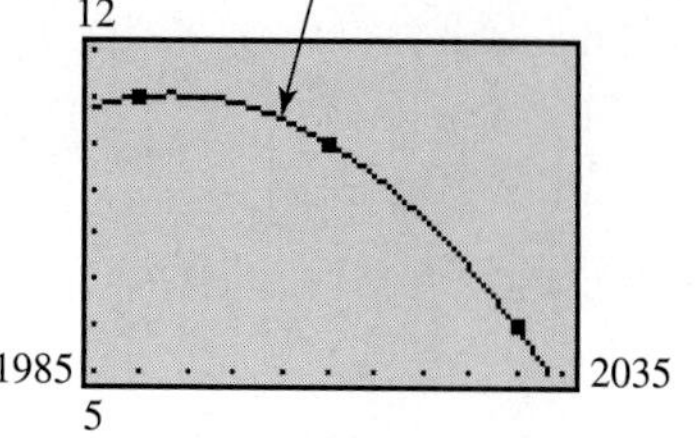


**(d)** Answers will vary. For example, in 2015 the predicted ratio is $f(2015) \approx 9.3$.

**61. (a)** $3^2a + 3b + c = 2.2$
$18^2a + 18b + c = 10.4$
$26^2a + 26b + c = 12.8$
**(b)** $f(x) = -.010725x^2 + .77188x - .019130$ **(c)**

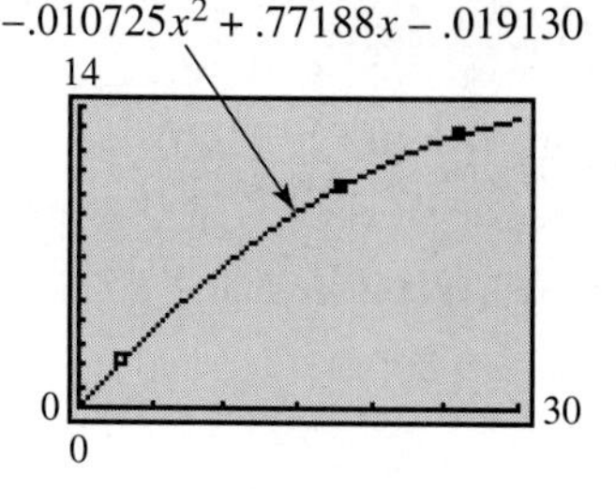


**(d)** Answers will vary. For example, in 1990 this percentage was $f(20) \approx 11.1$.

**63. (a)** At intersection A, incoming traffic is equal to $x + 5$. The outgoing traffic is given by $y + 7$. Therefore, $x + 5 = y + 7$. The other equations can be justified in a similar way.

**(b)** The three equations can be written as

$x - y = 2$
$x - z = 3$
$y - z = 1.$

The solution set can be written as $\{(z + 3, z + 1, z) \mid z \geq 0\}$.

**(c)** There are infinitely many solutions since some cars could be driving around the block continually.

**65. (a)** $a + 871b + 11.5c + 3d = 239$
$a + 847b + 12.2c + 2d = 234$
$a + 685b + 10.6c + 5d = 192$
$a + 969b + 14.2c + d = 343$
**(b)** $\left[\begin{array}{cccc|c} 1 & 871 & 11.5 & 3 & 239 \\ 1 & 847 & 12.2 & 2 & 234 \\ 1 & 685 & 10.6 & 5 & 192 \\ 1 & 969 & 14.2 & 1 & 343 \end{array}\right]$ The values are $a \approx -715.457$, $b \approx .34756$, $c \approx 48.6585$, and $d \approx 30.71951$.

**(c)** $F = -715.457 + .34756A + 48.6585P + 30.71951W$ **(d)** approximately 323, which is very close to 320

## Reviewing Basic Concepts *(page 501)*

**1.** $\{(3, -4)\}$ **2.** $\{(-2, 0)\}$ $2x + y = -4$ $-x + 2y = 2$

10

−10 10

Intersection X=-2 Y=0

−10

**3.** $\left\{\left(x, \frac{2-x}{2}\right)\right\}$ or $\{(2 - 2y, y)\}$ **4.** $\emptyset$

**5.** $\{(2, -1), (-6.5, 24.5)\}$ $6x + 2y = 10$ $2x^2 - 3y = 11$

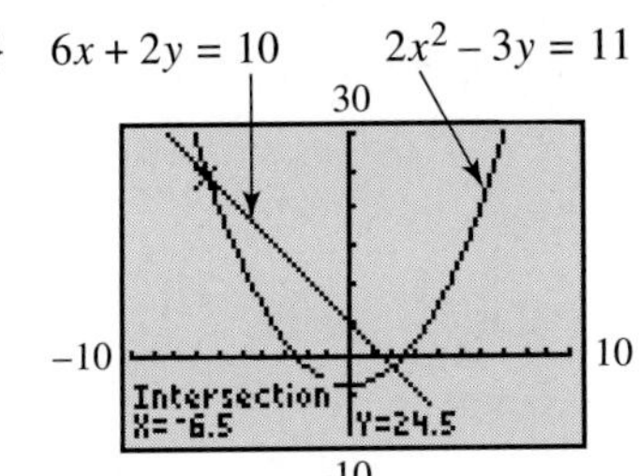


The other point of intersection is (2, −1).

**6.** $\{(-2, 1, 2)\}$ **7.** $\{(2, 1, -1)\}$ **8.** $\{(3, 2, 1)\}$ **9.** about 11.03 million with stereo sound; about 21 million without stereo sound **10.** at 8%: \$1000; at 11%: \$1500; at 14%: \$2500

## 7.4 Exercises *(page 512)*

**1.** $2 \times 2$; square **3.** $3 \times 4$ **5.** $2 \times 1$; column **7.** $1 \times 1$; square, row, column **9.** $w = 3; x = 2; y = -1; z = 4$ **11.** $w = 2; x = 6; y = -2; z = 8$ **13.** $z = 18; r = 3; s = 3; p = 3; a = \frac{3}{4}$ **15.** The two matrices must have the same dimension. To find the sum, add the corresponding entries. The sum will be a matrix with the same dimension.

**17.** $\begin{bmatrix} -2 & -7 & 7 \\ 10 & -2 & 7 \end{bmatrix}$ **19.** $\begin{bmatrix} -6 & 8 \\ 4 & 2 \end{bmatrix}$ **21.** $\begin{bmatrix} 5 & 5 \\ 12 & 0 \end{bmatrix}$ **23.** cannot be added **25.** $\begin{bmatrix} 13 & 3 & 0 & -2 \\ 9 & -12 & 4 & 8 \\ 12 & -11 & -1 & 9 \end{bmatrix}$

**27.** $\begin{bmatrix} 8 & -43 & -18 \\ 26 & 29 & 6 \\ -2 & 10 & 43 \end{bmatrix}$ **29.** $\begin{bmatrix} -4 & 8 \\ 0 & 6 \end{bmatrix}$ **31.** $\begin{bmatrix} 2 & 6 \\ -4 & 6 \end{bmatrix}$ **33.** $\begin{bmatrix} -13 & 21 \\ 2 & 15 \end{bmatrix}$ **35.** $\begin{bmatrix} 2 & 6 & 5 \\ -4 & -7 & 9 \end{bmatrix}$ **37.** $AB$: $4 \times 4$; $BA$: $2 \times 2$ **39.** $AB$: $3 \times 2$; $BA$: not defined **41.** Neither $AB$ nor $BA$ is defined. **43.** columns; rows

**45.** $\begin{bmatrix} pa + qb & pc + qd \\ ra + sb & rc + sd \end{bmatrix}$ **47.** $\begin{bmatrix} -17 \\ -1 \end{bmatrix}$ **49.** $\begin{bmatrix} 17 & -10 \\ 1 & 2 \end{bmatrix}$ **51.** $\begin{bmatrix} -2 & 10 \\ 0 & 8 \end{bmatrix}$ **53.** $\begin{bmatrix} -15 & -16 & 3 \\ -1 & 0 & 9 \\ 7 & 6 & 12 \end{bmatrix}$

**55.** $[2 \quad 7 \quad -4]$ **57.** $\begin{bmatrix} 23 & -9 \\ -6 & -2 \\ 33 & 1 \end{bmatrix}$ **59.** $\begin{bmatrix} -25 & 23 & 11 \\ 0 & -6 & -12 \\ -15 & 33 & 45 \end{bmatrix}$ **61.** cannot be multiplied **63.** $\begin{bmatrix} 10 & -10 \\ 15 & -5 \end{bmatrix}$ **65.** no

**67.** **(a)** $\begin{bmatrix} 11{,}375 & 316 & 83{,}000 \\ 6970 & 115 & 73{,}000 \\ 5446 & 159 & 35{,}700 \\ 4534 & 141 & 36{,}700 \\ 4059 & 9 & 27{,}364 \end{bmatrix}$; $\begin{bmatrix} 15{,}307 & 511 & 90{,}000 \\ 13{,}363 & 436 & 85{,}000 \\ 11{,}778 & 372 & 77{,}000 \\ 10{,}395 & 321 & 68{,}000 \\ 9235 & 282 & 61{,}900 \end{bmatrix}$ **(b)** $\begin{bmatrix} 26{,}682 & 827 & 173{,}000 \\ 20{,}333 & 551 & 158{,}000 \\ 17{,}224 & 531 & 112{,}700 \\ 14{,}929 & 462 & 104{,}700 \\ 13{,}294 & 291 & 89{,}264 \end{bmatrix}$; the total number of employees for the two companies in 1997

**(c)** $\begin{bmatrix} 3932 & 195 & 7000 \\ 6393 & 321 & 12{,}000 \\ 6332 & 213 & 41{,}300 \\ 5861 & 180 & 31{,}300 \\ 5176 & 273 & 34{,}536 \end{bmatrix}$; the amount, in 1998, that Walgreen's net income exceeded Rite Aid's net income (in millions of dollars)

**69.** **(a)** $d_{n+1} = -.05m_n + 1.05d_n$; 1.05 **(b)** after 1 yr: 3020 mountain lions, 515,000 deer; after 2 yr: 3600 mountain lions, 525,700 deer

### 7.5 Exercises (*page 524*)

**1.** −31 **3.** 7 **5.** 0 **7.** −26 **9.** 2, −6, 4 **11.** −6, 0, −6 **13.** 186 **15.** 17 **17.** 166 **19.** 0 **21.** 0 **23.** −5.5 **25.** $\left\{-\frac{4}{3}\right\}$ **27.** $\{-1, 4\}$ **29.** $\{-4\}$ **31.** $\{13\}$ **33.** 1 **35.** 9.5 **37.** 3.5 **39.** 298 **41.** −88 **43.** 0 **45.** 0 **47.** 16 **49.** $-x - 3y + 11 = 0$ or $x + 3y - 11 = 0$ **50.** $-x - 3y = -11$ or $x + 3y = 11$; equivalent **51.** $y - y_1 = \frac{y_2 - y_1}{x_2 - x_1}(x - x_1)$ **52.** $y - y_1 = \frac{y_2 - y_1}{x_2 - x_1}(x - x_1)$; equivalent **53. (a)** D **(b)** A **(c)** C **(d)** B **55.** $\{(2, 2)\}$ **57.** $\{(2, -5)\}$ **59.** $\left\{\left(\frac{4 - 2y}{3}, y\right)\right\}$ **61.** $\{(2, 3)\}$ **63.** $\{(-1, 2, 1)\}$ **65.** $\{(-3, 4, 2)\}$ **67.** $\{(0, 4, 2)\}$ **69.** $\emptyset$ **71.** $\{(-1, 2, 5, 1)\}$ **73.** If $D = 0$, Cramer's rule cannot be applied because there is no unique solution. There are either no solutions or infinitely many solutions.

### 7.6 Exercises (*page 533*)

**1.** yes **3.** no **5.** no **7.** yes **9.** It will not exist if its determinant is equal to 0. **11.** $\begin{bmatrix} 5 & -7 \\ -2 & 3 \end{bmatrix}$

**13.** $\begin{bmatrix} 2 & 1 \\ -1.5 & -.5 \end{bmatrix}$ **15.** does not exist **17.** $\begin{bmatrix} -2.5 & 5 \\ 12.5 & -15 \end{bmatrix}$ **19.** $\begin{bmatrix} 1 & 0 & 0 \\ 0 & -1 & 0 \\ -1 & 0 & 1 \end{bmatrix}$ **21.** $\begin{bmatrix} 15 & 4 & -5 \\ -12 & -3 & 4 \\ -4 & -1 & 1 \end{bmatrix}$

**23.** $\begin{bmatrix} -\frac{10}{3} & \frac{5}{9} & -\frac{10}{9} \\ \frac{20}{3} & \frac{5}{9} & \frac{80}{9} \\ -5 & \frac{5}{6} & -\frac{20}{3} \end{bmatrix}$ **25.** does not exist **27.** $\begin{bmatrix} .0543058761 & -.0543058761 \\ 1.846399787 & .153600213 \end{bmatrix}$

**29.** $\begin{bmatrix} .9987635516 & -.252092087 & -.3330564627 \\ -.5037783375 & 1.007556675 & -.2518891688 \\ -.2481013617 & -.2556769758 & 1.003768868 \end{bmatrix}$ **31.** $\begin{bmatrix} 2x + 4z & 2y + 4w \\ x - z & y - w \end{bmatrix}$

**32.** $2x + 4z = 1, 2y + 4w = 0, x - z = 0, y - w = 1$ **33.** $x = \frac{1}{6}, y = \frac{2}{3}, z = \frac{1}{6}, w = -\frac{1}{3}$ **34.** $A^{-1} = \begin{bmatrix} \frac{1}{6} & \frac{2}{3} \\ \frac{1}{6} & -\frac{1}{3} \end{bmatrix}$

**35.** the determinant of $A$ **36.** $\begin{bmatrix} \frac{d}{\det A} & \frac{-b}{\det A} \\ \frac{-c}{\det A} & \frac{a}{\det A} \end{bmatrix}$ **37.** $A^{-1} = \frac{1}{\det A}\begin{bmatrix} d & -b \\ -c & a \end{bmatrix}$ **38.** Interchange the entries in row 1, column 1 and row 2, column 2. Then change the entries in row 1, column 2 and row 2, column 1 to their negatives. Then multiply the resulting matrix by the scalar $\frac{1}{\det A}$. **39.** $A^{-1} = \begin{bmatrix} -\frac{3}{2} & 1 \\ \frac{7}{2} & -2 \end{bmatrix}$ **40.** 0 **41.** $\{(-2, 4)\}$ **43.** $\{(4, -6)\}$

**45.** $\{(2.5, -1)\}$ **47.** $\{(3, 0, 2)\}$ **49.** $\left\{\left(12, -\frac{15}{11}, -\frac{65}{11}\right)\right\}$ **51.** $\{(0, 2, -2, 1)\}$ **53.** $\{(-3.542308934, -4.343268299)\}$

**55.** $\{(-.9704156959, 1.391914631, .1874077432)\}$ **57.** $P(x) = -2x^3 + 5x^2 - 4x + 3$

**59.** $P(x) = x^4 + 2x^3 + 3x^2 - x - 1$ **61. (a)** soft drink: \$1.50; popcorn: \$2.00 **(b)** No, $A^{-1}$ does not exist.

**63. (a)**
$$\begin{aligned} 113a + 308b + c &= 10{,}170 \\ 133a + 622b + c &= 15{,}305 \\ 155a + 1937b + c &= 21{,}289 \end{aligned}$$

**(b)** $a \approx 251, b \approx .346, c \approx -18{,}300$; $T = 251A + .346I - 18{,}300$

**(c)** 11,426,000

**65. (a)**
$$\begin{aligned} a + 1500b + 8c &= 122 \\ a + 2000b + 5c &= 130 \\ a + 2200b + 10c &= 158 \end{aligned}$$
or
$$\begin{bmatrix} 1 & 1500 & 8 \\ 1 & 2000 & 5 \\ 1 & 2200 & 10 \end{bmatrix}\begin{bmatrix} a \\ b \\ c \end{bmatrix} = \begin{bmatrix} 122 \\ 130 \\ 158 \end{bmatrix}$$

**(b)** \$130,000

**67.** $\begin{bmatrix} 2 & 9 \\ 1 & 5 \end{bmatrix}$ **69.** $\begin{bmatrix} 1 & 0 & 1 \\ -1 & 0 & 2 \\ -2 & 1 & 3 \end{bmatrix}$ **71.** $\begin{bmatrix} \frac{1}{a} & 0 & 0 \\ 0 & \frac{1}{b} & 0 \\ 0 & 0 & \frac{1}{c} \end{bmatrix}$

## Reviewing Basic Concepts (*page 537*)

**1.** $\begin{bmatrix} -5 & 6 \\ -1 & 3 \end{bmatrix}$ **2.** $\begin{bmatrix} 0 & 6 \\ -9 & 12 \end{bmatrix}$ **3.** $\begin{bmatrix} 33 & -24 \\ -12 & 9 \end{bmatrix}$ **4.** $\begin{bmatrix} 1 & 3 & -3 \\ 0 & -6 & 0 \\ 4 & 2 & 2 \end{bmatrix}$ **5.** $-3$ **6.** $-14$ **7.** $\begin{bmatrix} \frac{1}{3} & \frac{4}{3} \\ \frac{2}{3} & \frac{5}{3} \end{bmatrix}$

**8.** $\begin{bmatrix} -\frac{2}{7} & -\frac{11}{14} & \frac{1}{14} \\ -\frac{4}{7} & -\frac{4}{7} & \frac{1}{7} \\ -\frac{1}{7} & -\frac{1}{7} & \frac{2}{7} \end{bmatrix}$ **9.** $W_1 = W_2 = \frac{100\sqrt{3}}{3} \approx 57.7$ lb **10.** $\{(3, 2, 1)\}$

## 7.7 Exercises (*page 545*)

**1.**
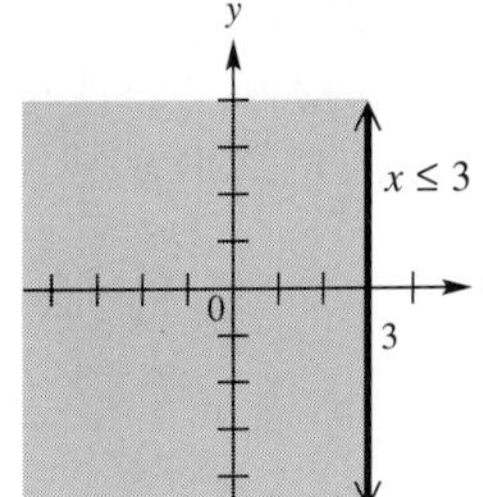


**3.**
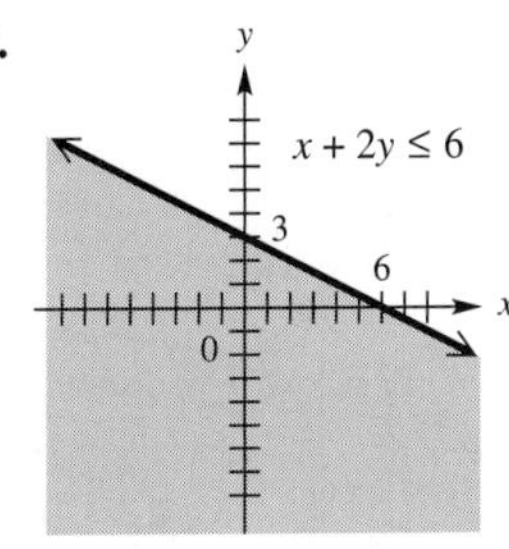


**5.**
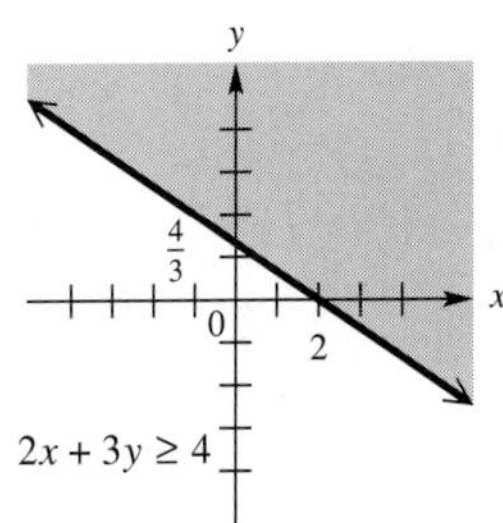


**7.**
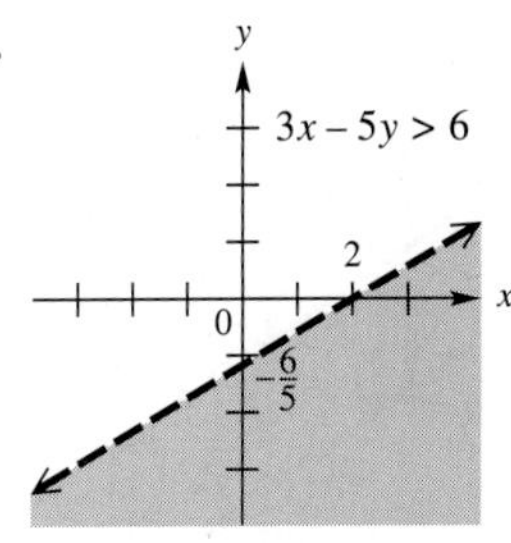


**9.**
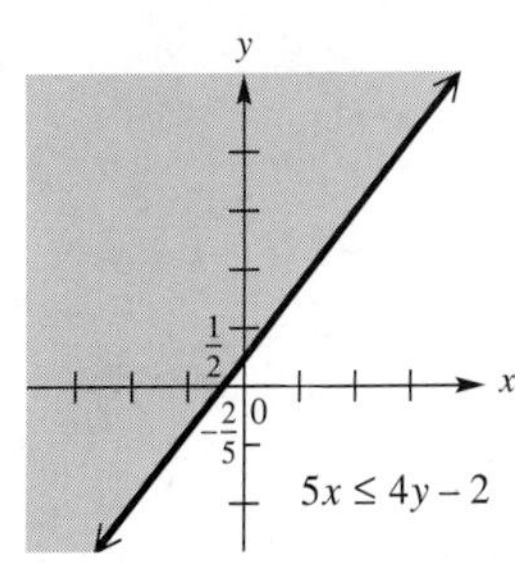


**11.**
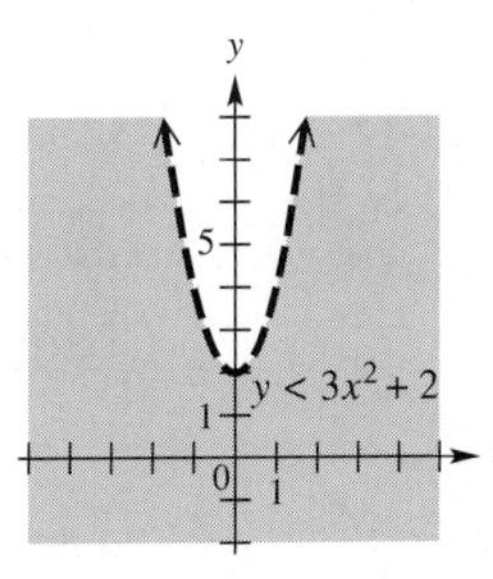


**13.**
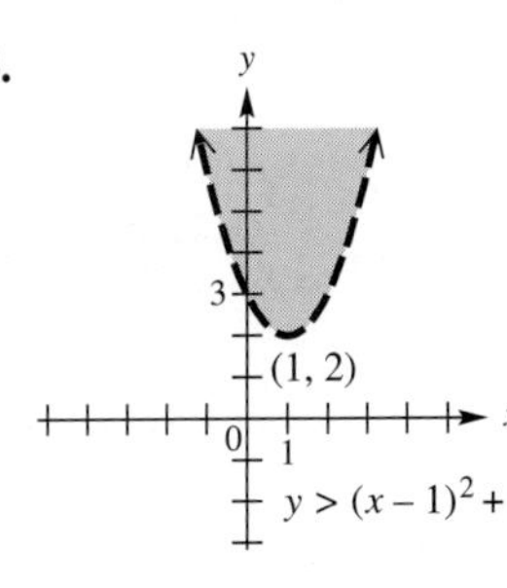


**15.**
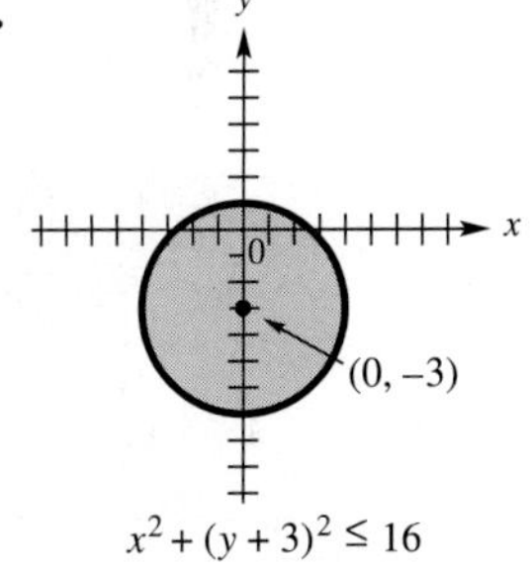


**17.** The boundary is solid if the symbol is $\geq$ or $\leq$ and dashed if the symbol is $>$ or $<$. 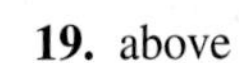 **19.** above **21.** B **23.** C
**25.** A

**27.**
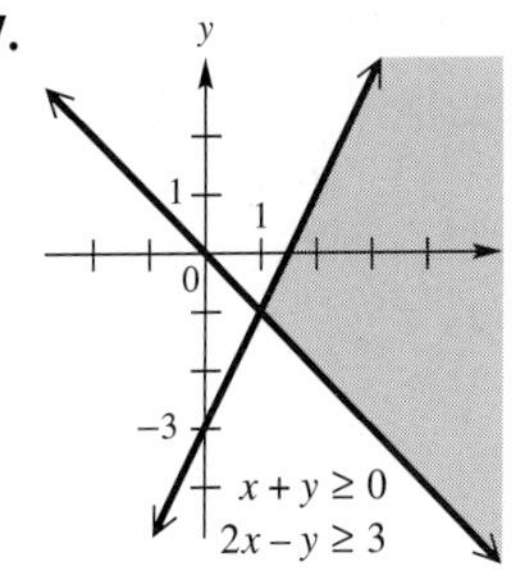


**29.**
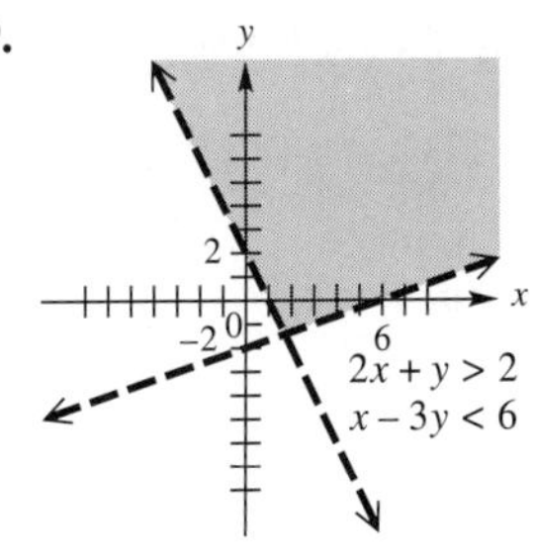


**31.**
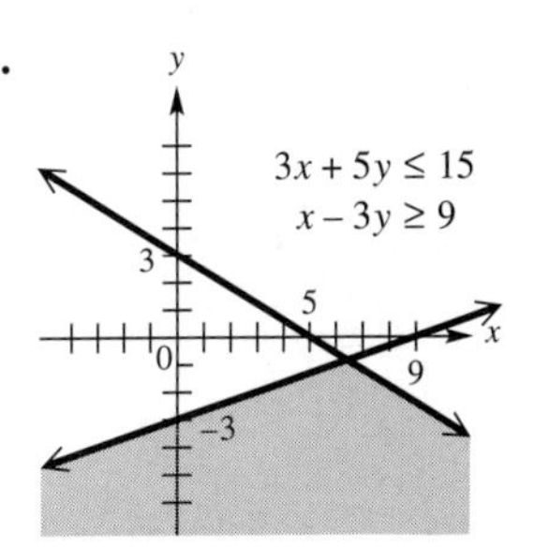


**33.**
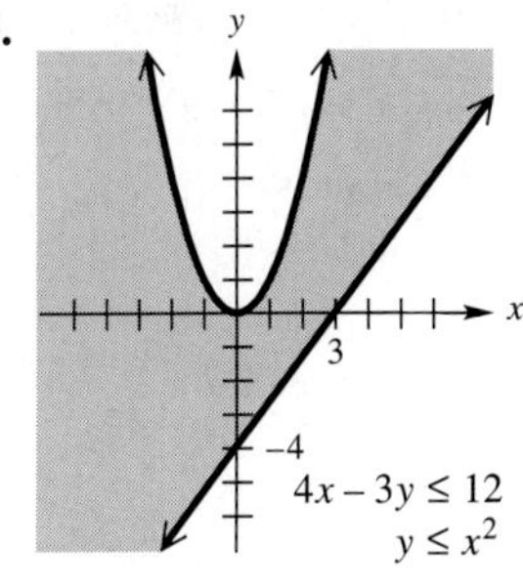


**35.**
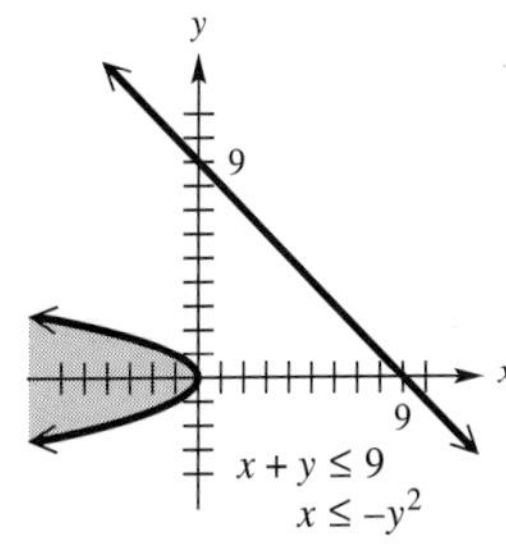


**37.**
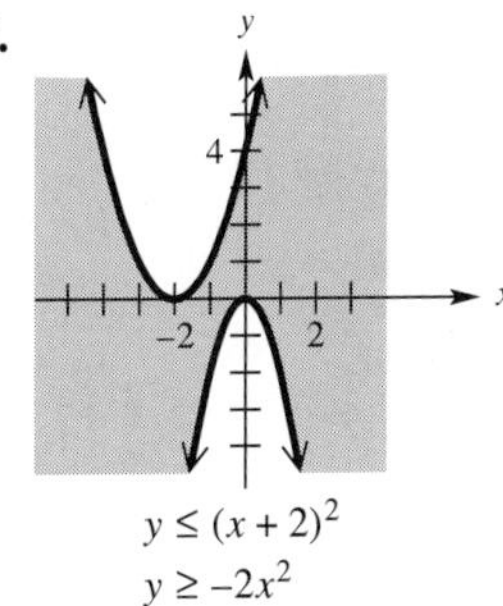


**39.**
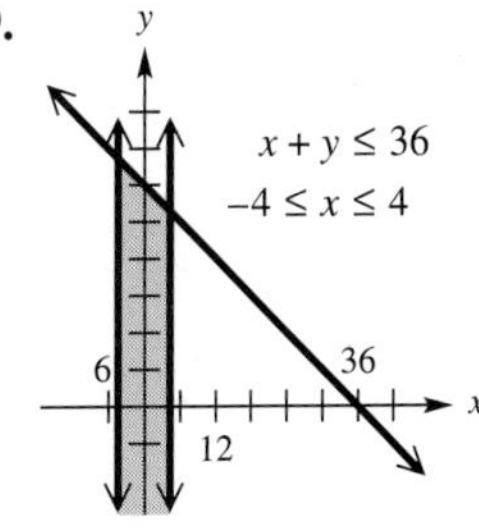


**41.**
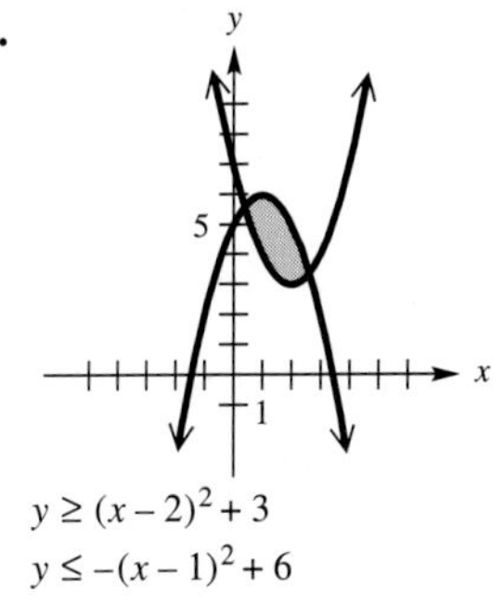

**43.**

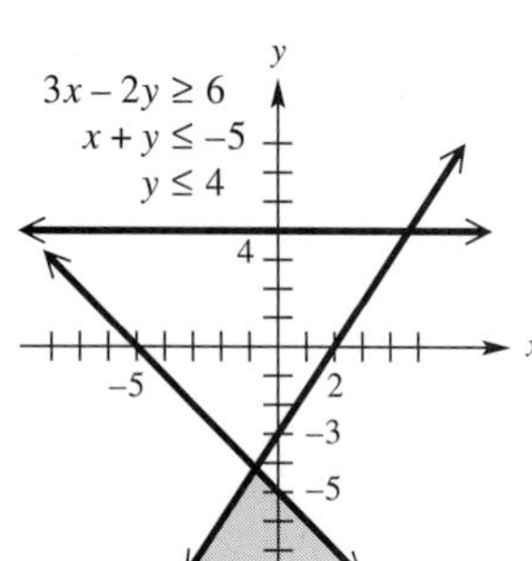


**45.**

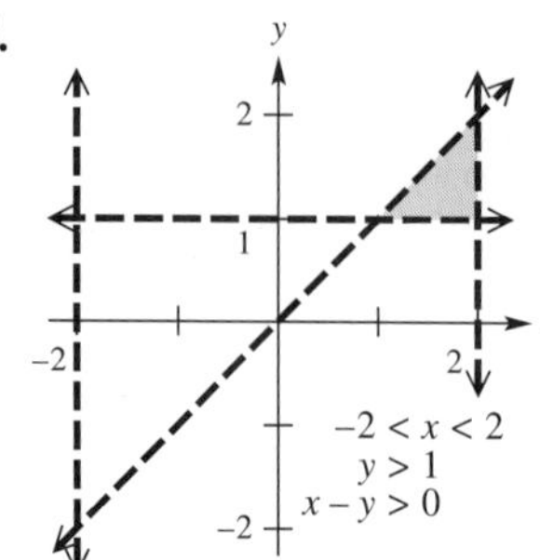


**47.**

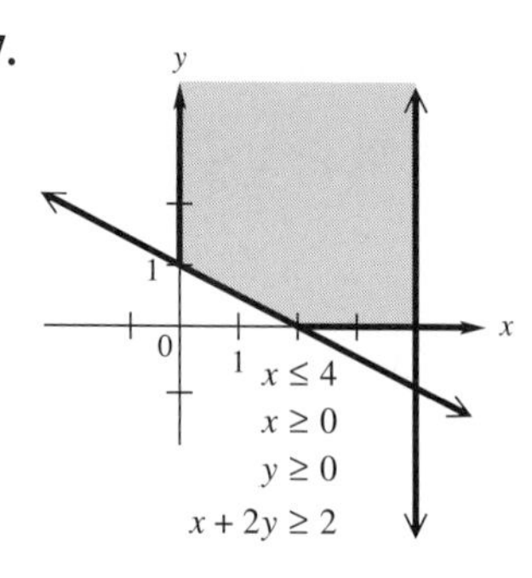


**49.**

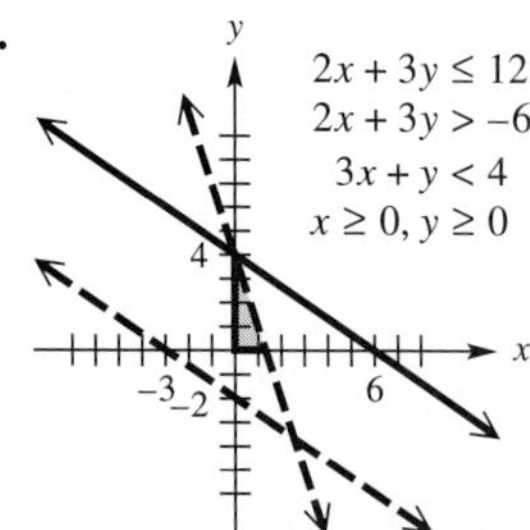


**51.**

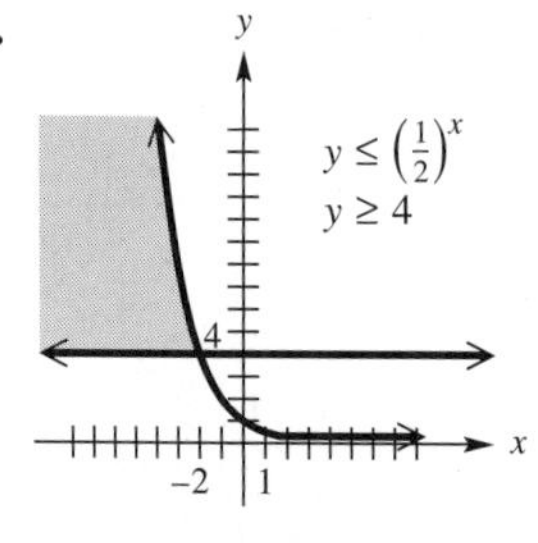


**53.**

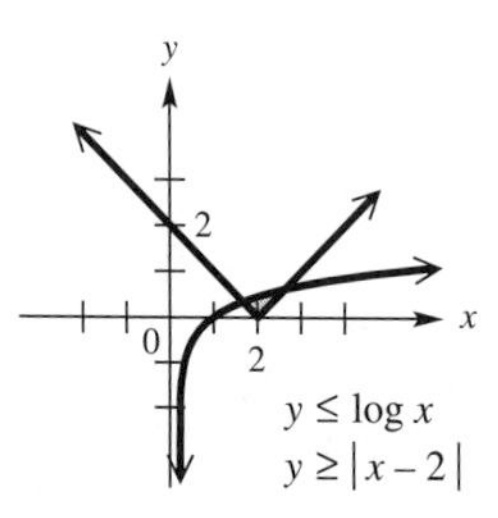


**55.** D **57.** A **59.** B

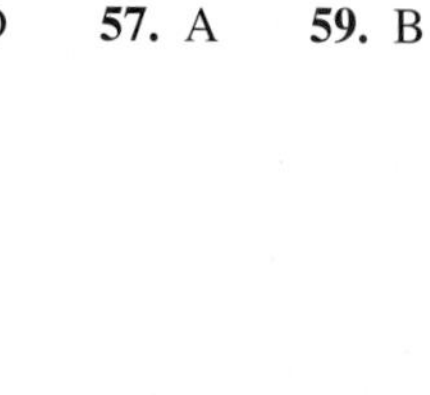

**61.**

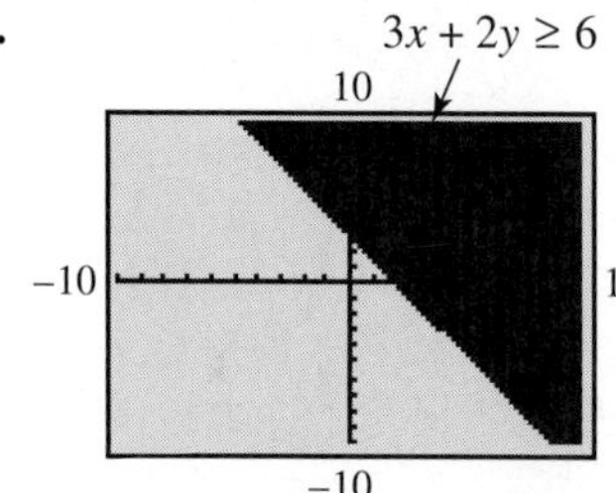


**63.**

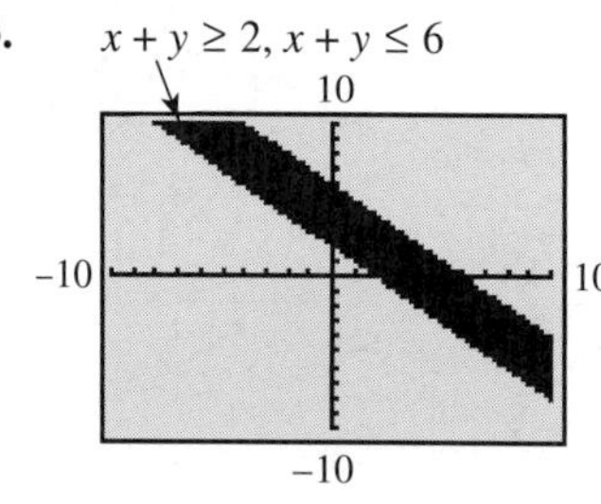


**65.**

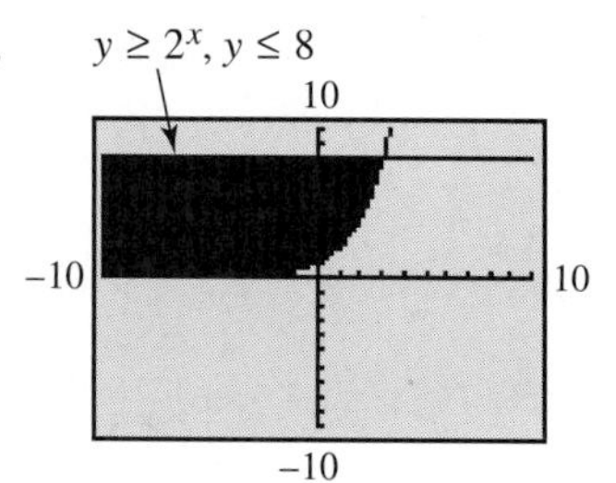


**67.** $x + 2y - 8 \geq 0,\ x + 2y \leq 12,\ x \geq 0,\ y \geq 0$ **69.** maximum of 65 at $(5, 10)$; minimum of 8 at $(1, 1)$ **71.** maximum of 66 at $(7, 9)$; minimum of 3 at $(1, 0)$ **73.** maximum of 100 at $(1, 10)$; minimum of 0 at $(1, 0)$ **75.** hat units: 5; whistle units: 0; maximum inquiries: 15 **77.** to A: 20; to B: 80; minimum cost: \$1040 **79.** gasoline: 6,400,000 gal; fuel oil: 3,200,000 gal; maximum revenue: \$16,960,000 **81.** medical kits: 0; containers of water: 4000; people aided: 400

## 7.8 Exercises (*page 554*)

**1.** $\frac{5}{3x} + \frac{-10}{3(2x+1)}$ **3.** $\frac{6}{5(x+2)} + \frac{8}{5(2x-1)}$ **5.** $\frac{5}{6(x+5)} + \frac{1}{6(x-1)}$ **7.** $\frac{-2}{x+1} + \frac{2}{x+2} + \frac{4}{(x+2)^2}$

**9.** $\frac{4}{x} + \frac{4}{1-x}$ **11.** $\frac{15}{x} + \frac{-5}{x+1} + \frac{-6}{x-1}$ **13.** $1 + \frac{-2}{x+1} + \frac{1}{(x+1)^2}$ **15.** $x^3 - x^2 + \frac{-1}{3(2x+1)} + \frac{2}{3(x+2)}$

**17.** $\frac{1}{9} + \frac{-1}{x} + \frac{25}{18(3x+2)} + \frac{29}{18(3x-2)}$ **19.** $\frac{-3}{5x^2} + \frac{3}{5(x^2+5)}$ **21.** $\frac{-2}{7(x+4)} + \frac{6x-3}{7(3x^2+1)}$

**23.** $\frac{1}{4x} + \frac{-8}{19(2x+1)} + \frac{-9x-24}{76(3x^2+4)}$ **25.** $\frac{-1}{x} + \frac{2x}{2x^2+1} + \frac{2x+3}{(2x^2+1)^2}$ **27.** $\frac{-1}{x+2} + \frac{3}{(x^2+4)^2}$

**29.** $5x^2 + \frac{3}{x} + \frac{-1}{x+3} + \frac{2}{x-1}$ **31.** coincide; correct **33.** do not coincide; not correct

**Reviewing Basic Concepts** (*page 555*)

**1.**

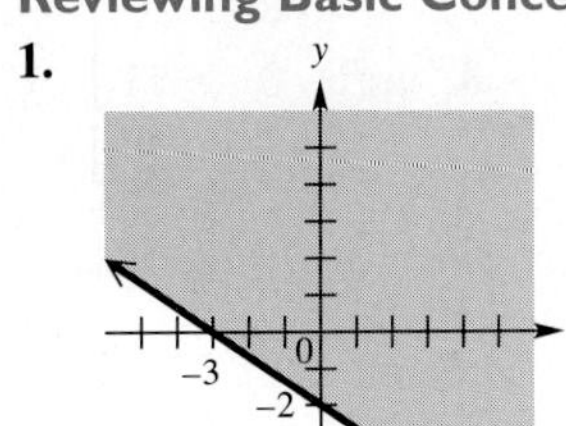


**2.**

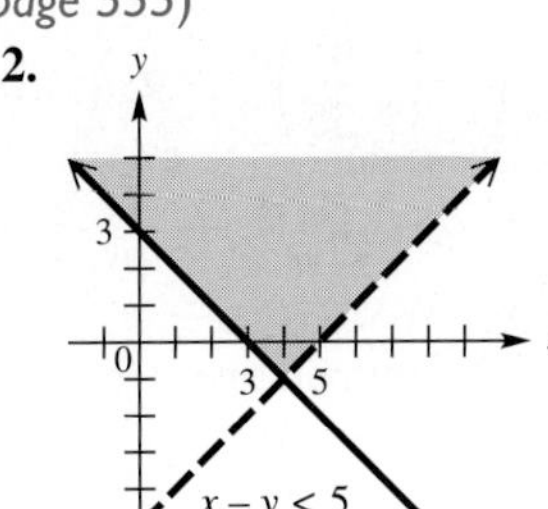


**3.**

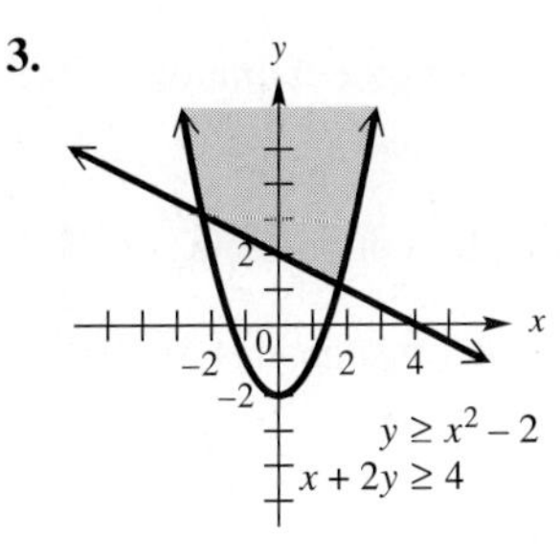


**4.**

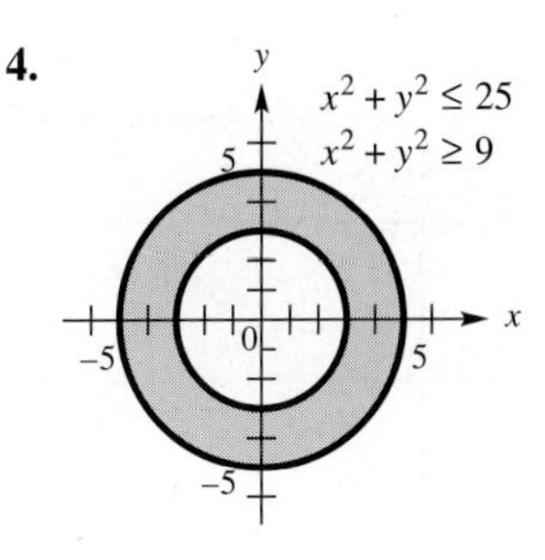


**5.** A **6.** Minimum value is 8 at $(4, 0)$. **7.** maximum: 65; minimum: 8 **8.** substance X: 130 lb; substance Y: 450 lb; minimum cost: \$1610 **9.** $\frac{7}{x-5}+\frac{3}{x+4}$ **10.** $\frac{-1}{x+2}+\frac{6}{x+4}+\frac{-3}{x-1}$

**Chapter 7 Review Exercises** (*page 558*)

**1.** $\{(5,7)\}$ **3.** $\{(5,-3)\}$ **5.** $\left\{\left(\frac{7}{5},-\frac{1}{5}\right),(1,1)\right\}$ **7.** $\{(-3\sqrt{3},\sqrt{2}),(-3\sqrt{3},-\sqrt{2}),(3\sqrt{3},\sqrt{2}),(3\sqrt{3},-\sqrt{2})\}$

**9.** $\{(6,2)\}$ **11.** **(a)** $y_1=\sqrt{2-x^2}$, $y_2=-\sqrt{2-x^2}$ **(b)** $y_3=-3x+4$ **(c)** The standard viewing window shows the points of intersection. Other windows are possible. **13.** No, a system consisting of two equations in three variables is represented by two planes in space. There will be no solutions or infinitely many solutions. **15.** $\{(6,-2,1)\}$ **17.** $\emptyset$; inconsistent system

**19.** $\{(-3,2)\}$ **21.** $\{(0,1,0)\}$ **23.** $\begin{bmatrix}-4\\6\\1\end{bmatrix}$ **25.** $\begin{bmatrix}-4&-4\\-7&16\end{bmatrix}$ **27.** $\begin{bmatrix}14x+28y&42x+22y\\18x-46y&70x+18y\end{bmatrix}$ **29.** $\begin{bmatrix}18&20\\29&-1\end{bmatrix}$

**31.** $\begin{bmatrix}-3\\10\end{bmatrix}$ **33.** $\begin{bmatrix}-2&22&31\\25&12&9\end{bmatrix}$ **35.** yes **37.** no **39.** does not exist **41.** $\begin{bmatrix}\frac{1}{2}&0\\\frac{1}{10}&\frac{1}{5}\end{bmatrix}$ **43.** $\begin{bmatrix}\frac{2}{3}&0&-\frac{1}{3}\\\frac{1}{3}&0&-\frac{2}{3}\\-\frac{2}{3}&1&\frac{1}{3}\end{bmatrix}$

**45.** $\{(2,2)\}$ **47.** $\{(2,1)\}$ **49.** $\{(1,-1,2)\}$ **51.** $\{(-1,0,2)\}$ **53.** $-25$ **55.** $-44$ **57.** $\left\{-\frac{7}{3}\right\}$

**59.** {all real numbers} **61.** **(a)** $D=5$ **(b)** $D_x=30$ **(c)** $D_y=-50$ **(d)** $x=6$; $y=-10$; solution set: $\{(6,-10)\}$
**63.** If $D=0$, there would be division by 0, which is undefined. The system will have no solutions or infinitely many solutions.

**65.** $\{(-4,2)\}$ **67.** $\{(-4,6,2)\}$ **69.** $\left\{\left(\frac{172}{67},-\frac{14}{67},-\frac{87}{67}\right)\right\}$ **71.** CDs: 80; 3.5-in. diskettes: 20 **73.** 5%: 11 mL; 15%: 3 mL; 10%: 6 mL **75.** $P(x)=\frac{1}{2}x^2+2x-2$ **77.**

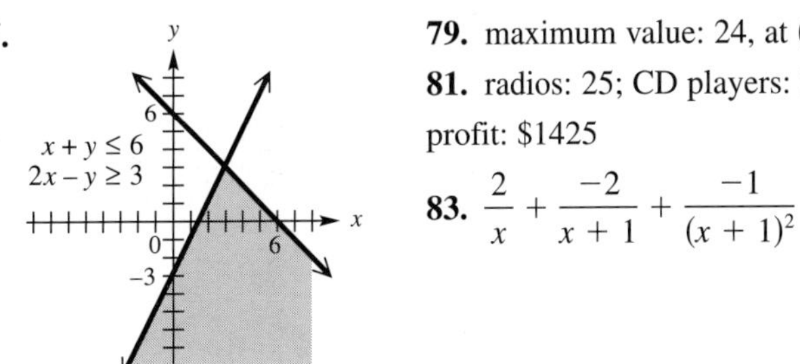


**79.** maximum value: 24, at $(0, 6)$
**81.** radios: 25; CD players: 30; maximum profit: \$1425
**83.** $\frac{2}{x}+\frac{-2}{x+1}+\frac{-1}{(x+1)^2}$

**Chapter 7 Test** (*page 562*)

**1.** **(a)** first equation: a hyperbola; second equation: a line **(b)** 0, 1, or 2 **(c)** $\left\{(1,-2),\left(-\frac{11}{35},\frac{68}{35}\right)\right\}$

**(d)**

$y_3 = 1 - 3x$ $\quad y_1 = \sqrt{\dfrac{-15x - x^2}{-4}}$

10

−10 10

Intersection X=1 Y=-2

−10

$y_2 = -y_1$

The other point of intersection is $\left(-\frac{11}{35}, \frac{68}{35}\right)$. **2.** $\{(2, 0, -1)\}$ **3. (a)** $\begin{bmatrix} 8 & 3 \\ 0 & -11 \\ 15 & 19 \end{bmatrix}$ **(b)** not possible **(c)** $\begin{bmatrix} -5 & 16 \\ 19 & 2 \end{bmatrix}$ **4. (a)** yes **(b)** yes **(c)** In general $AB \neq BA$ because matrix multiplication is not commutative. **(d)** $AC$ cannot be found, but $CA$ can. **5. (a)** 1 **(b)** $-844$ **6.** $\{(-6, 7)\}$

**7. (a)** $A = \begin{bmatrix} 1 & 1 & -1 \\ 2 & -3 & -1 \\ 1 & 2 & 2 \end{bmatrix}, X = \begin{bmatrix} x \\ y \\ z \end{bmatrix}, B = \begin{bmatrix} -4 \\ 5 \\ 3 \end{bmatrix}$ **(b)** $A^{-1} = \begin{bmatrix} \frac{1}{4} & \frac{1}{4} & \frac{1}{4} \\ \frac{5}{16} & -\frac{3}{16} & \frac{1}{16} \\ -\frac{7}{16} & \frac{1}{16} & \frac{5}{16} \end{bmatrix}$ **(c)** $\{(1, -2, 3)\}$ **(d)** $A = \begin{bmatrix} .5 & 1 & 1 \\ 2 & -3 & -1 \\ 1 & 2 & 2 \end{bmatrix}$ and $\det A = 0$, so $A^{-1}$ does not exist. **8.** $f(x) \approx -.010764x^2 + 1.2903x + 24.133$

$f(x) \approx -.010764x^2 + 1.2903x + 24.133$

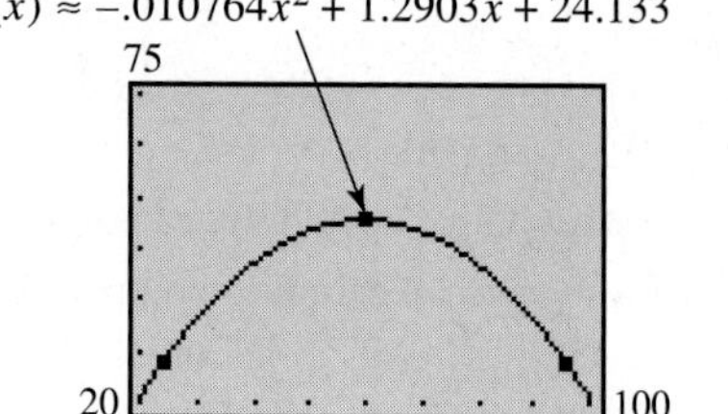


**9.** B **10.** cabinet X: 8; cabinet Y: 3 **11.** $\dfrac{4}{x-3} + \dfrac{3}{x+2}$ **12.** $\dfrac{-1}{x-2} + \dfrac{2}{x+2} + \dfrac{-3}{(x-2)^2}$

## CHAPTER 8 TRIGONOMETRIC FUNCTIONS AND APPLICATIONS

### 8.1 Exercises *(page 577)*

**1. (a)** 60° **(b)** 150° **3. (a)** 45° **(b)** 135° **5. (a)** $\dfrac{\pi}{4}$ **(b)** $\dfrac{3\pi}{4}$ **7. (a)** $(90 - x)°$ **(b)** $(180 - x)°$ **9.** 150° **11.** 70°; 110° **13.** 80°; 100° **15.** 83° 59′ **17.** 23° 49′ **19.** 17° 1′ 49″ **21.** 20.9° **23.** 91.598° **25.** 31° 25′ 47″ **27.** 89° 54′ 1″

Angles other than those given are possible in Exercises 29–33.

**29.**

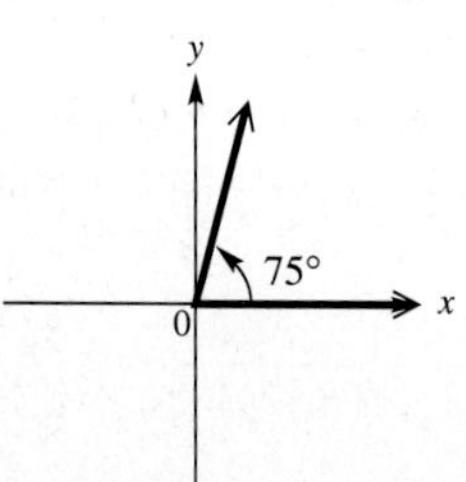


435°; −285°; quadrant I

**31.**

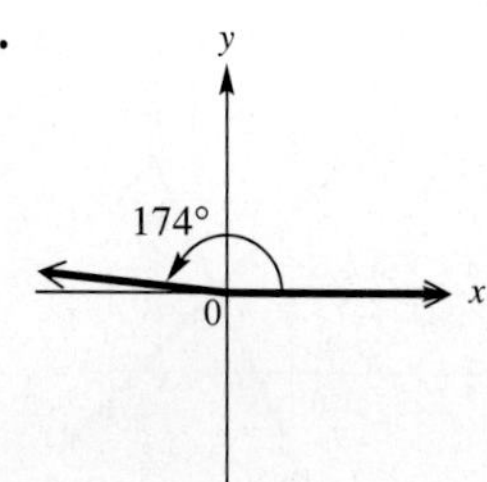


534°; −186°; quadrant II

**33.**

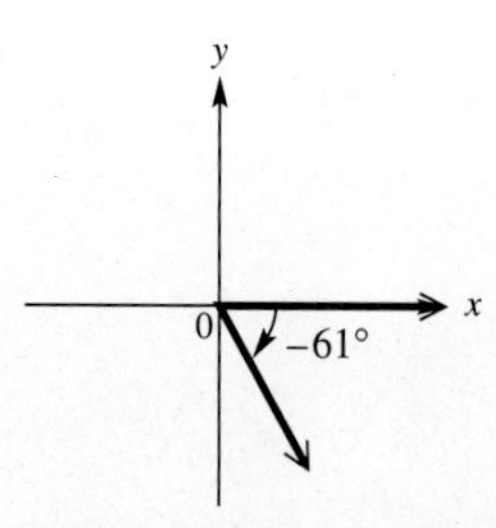


299°; −421°; quadrant IV

**35.** 320° **37.** 90° **39.** $\dfrac{7\pi}{4}$ **41.** $\dfrac{\pi}{2}$ **43.** $30° + n \cdot 360°$ **45.** $-90° + n \cdot 360°$ **47.** $\dfrac{\pi}{4} + 2n\pi$ **49.** $-\dfrac{3\pi}{4} + 2n\pi$ **51.** $\dfrac{\pi}{3}$ **53.** $\dfrac{5\pi}{6}$ **55.** 60° **57.** 315° **59.** 330° **61.** .68 **63.** 2.43 **65.** 1.12 **67.** 114° 35′ **69.** 99° 42′ **71.** We begin the answers with the blank next to 30°, and then proceed counterclockwise from there: $\dfrac{\pi}{6}$; 45; $\dfrac{\pi}{3}$; 120; 135; $\dfrac{5\pi}{6}$; $\pi$; $\dfrac{7\pi}{6}$; $\dfrac{5\pi}{4}$; 240; 300; $\dfrac{7\pi}{4}$; $\dfrac{11\pi}{6}$. **73.** $2\pi$ **75.** 8 **77.** 1 **79.** 25.8 cm

**81.** 5.05 m **83.** 1200 km **85.** 5900 km **87.** **(a)** 11.6 in. **(b)** 37° 5′ **89.** 38.5° **91.** 146 in. **93.** 1120 m$^2$
**95.** 114 cm$^2$ **97.** $6\pi$ **99.** approximately 3700 mi$^2$ **101.** 75.4 in.$^2$ **103.** **(a)** 13.85° **(b)** approximately 76 m$^2$
**105.** The area $A$ of a circle of radius $r$ is given by $A = \pi r^2$. **107.** $16\frac{1}{4}$ ft per sec, or about 11.1 mph **109.** .24 radian per sec **111.** $\frac{500\pi}{3}$ radians per sec; $\frac{2500\pi}{3}$ in. per sec **113.** **(a)** $\frac{2\pi}{365}$ radian **(b)** $\frac{\pi}{4380}$ radian per hr **(c)** approximately 66,700 mph **115.** radius: 3947 mi; circumference: 24,800 mi

## 8.2 Exercises *(page 592)*

**1.**

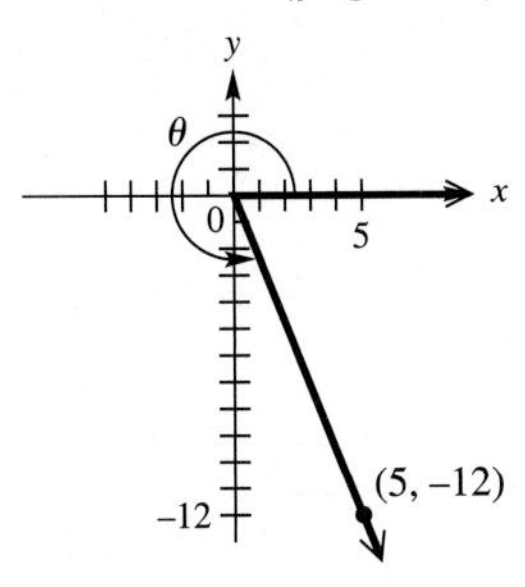


In Exercises 3–7 and 15–17, we give, in order, sine, cosine, tangent, cotangent, secant, and cosecant.

**3.** $\frac{4}{5}; -\frac{3}{5}; -\frac{4}{3}; -\frac{3}{4}; -\frac{5}{3}; \frac{5}{4}$ **5.** 1; 0; undefined; 0; undefined; 1 **7.** $\frac{\sqrt{3}}{2}; \frac{1}{2}; \sqrt{3}; \frac{\sqrt{3}}{3}; 2; \frac{2\sqrt{3}}{3}$ **9.** The sine and cosecant functions are reciprocals, and reciprocals always have the same sign, since their product is 1, a positive number.
**11.** negative **13.** negative

**15.**

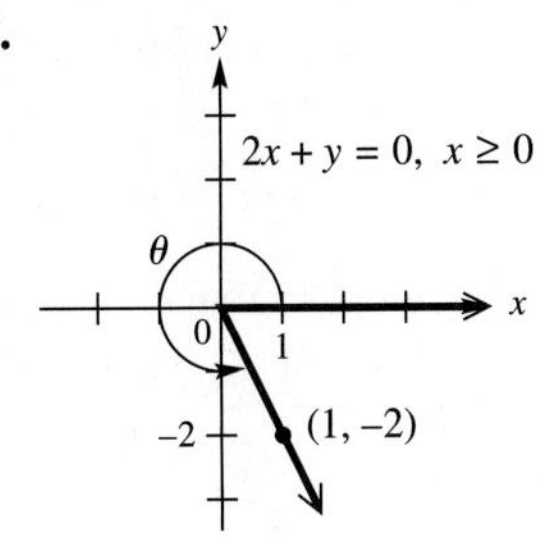


$-\frac{2\sqrt{5}}{5}; \frac{\sqrt{5}}{5}; -2; -\frac{1}{2}; \sqrt{5}; -\frac{\sqrt{5}}{2}$

**17.**

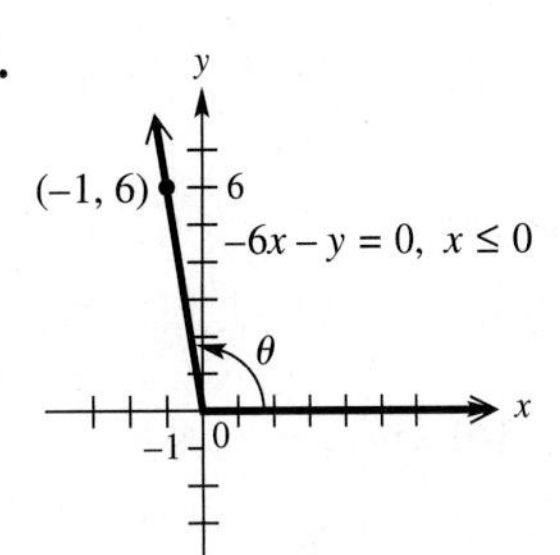


$\frac{6\sqrt{37}}{37}; -\frac{\sqrt{37}}{37}; -6; -\frac{1}{6}; -\sqrt{37}; \frac{\sqrt{37}}{6}$

**19.** sin 540° = 0; cos 540° = −1; tan 540° = 0; csc 540° is undefined; sec 540° = −1; cot 540° is undefined. **21.** −7
**23.** 3 **25.** 1 **27.** 0 **29.** 0 **31.** They are equal. **33.** They are equal. **35.** 40° **37.** 45° **39.** decrease; decrease **41.** −1; $\theta = 180°$ **43.** −5 **45.** $-\frac{3\sqrt{5}}{5}$ **47.** .10199657 **49.** The cosine of any angle cannot be greater than 1 (or less than −1). **51.** $\sqrt{3}$ **53.** 2° **55.** II **57.** I or III **59.** +; −; − **61.** −; +; −
**63.** −; +; − **65.** tan 30° **67.** sec 33° **69.** impossible **71.** possible **73.** possible **75.** impossible
**77.** $\frac{\sqrt{15}}{4}$ **79.** $-\frac{4}{3}$ **81.** $-\frac{\sqrt{3}}{2}$ **83.** −.56616682 **85.** 5

In Exercises 87–91, we give, in order, sine, cosine, tangent, cotangent, secant, and cosecant.

**87.** $\frac{15}{17}; -\frac{8}{17}; -\frac{15}{8}; -\frac{8}{15}; -\frac{17}{8}; \frac{17}{15}$ **89.** $-\frac{\sqrt{3}}{2}; -\frac{1}{2}; \sqrt{3}; \frac{\sqrt{3}}{3}; -2; -\frac{2\sqrt{3}}{3}$ **91.** −.555762; .831342; −.668512; −1.49586; 1.20287; −1.79933 **95.** false; For example, sin 30° + cos 30° ≈ .5 + .8660 = 1.3660 ≠ 1.

## Reviewing Basic Concepts (*page 594*)

**1.** **(a)** complement: 55°; supplement: 145° **(b)** complement: $\frac{\pi}{4}$; supplement: $\frac{3\pi}{4}$ **2.** 35° 15′ 0″ **3.** $59.591\overline{6}°$

**4.** **(a)** 200° **(b)** $\frac{4\pi}{3}$ **5.** **(a)** $\frac{4\pi}{3}$ **(b)** 135° **6.** **(a)** $2\pi$ cm **(b)** $3\pi$ cm² **7.** $\sin\theta = \frac{5\sqrt{29}}{29}$; $\cos\theta = -\frac{2\sqrt{29}}{29}$; $\tan\theta = -\frac{5}{2}$; $\cot\theta = -\frac{2}{5}$; $\sec\theta = -\frac{\sqrt{29}}{2}$; $\csc\theta = \frac{\sqrt{29}}{5}$ **8.** $\sin 270° = -1$; $\cos 270° = 0$; $\tan 270°$ is undefined; $\cot 270° = 0$; $\sec 270°$ is undefined; $\csc 270° = -1$ **9.** **(a)** impossible **(b)** possible **(c)** possible **10.** $\cos\theta = -\frac{\sqrt{5}}{3}$; $\tan\theta = \frac{2\sqrt{5}}{5}$; $\cot\theta = \frac{\sqrt{5}}{2}$; $\sec\theta = -\frac{3\sqrt{5}}{5}$; $\csc\theta = -\frac{3}{2}$

## 8.3 Exercises (*page 603*)

**1.** $\sin A = \frac{21}{29}$; $\cos A = \frac{20}{29}$; $\tan A = \frac{21}{20}$; $\cot A = \frac{20}{21}$; $\sec A = \frac{29}{20}$; $\csc A = \frac{29}{21}$ **3.** $\sin A = \frac{n}{p}$; $\cos A = \frac{m}{p}$; $\tan A = \frac{n}{m}$; $\cot A = \frac{m}{n}$; $\sec A = \frac{p}{m}$; $\csc A = \frac{p}{n}$

The number of digits in part (b) of Exercises 5–17 may vary.

**5.** **(a)** $\frac{\sqrt{3}}{3}$ **(b)** .5773502692 **7.** **(a)** $\frac{1}{2}$ **9.** **(a)** $\frac{2\sqrt{3}}{3}$ **(b)** 1.154700538 **11.** **(a)** $\sqrt{2}$ **(b)** 1.414213562 **13.** **(a)** $\frac{\sqrt{2}}{2}$ **(b)** .7071067812 **15.** **(a)** $\frac{\sqrt{3}}{2}$ **(b)** .8660254038 **17.** **(a)** $\sqrt{3}$ **(b)** 1.732050808 **19.** **(a)** 2 **21.** The value he gave was an *approximation,* not the exact value of sin 45°, which is $\frac{\sqrt{2}}{2}$. **23.** tan 17° **25.** cos 52° **27.** cot 64° 17′ **29.** $\sin\frac{3\pi}{10}$ **31.** $\cot\left(\frac{\pi}{2} - .5\right)$ **33.** 82° **35.** 50° **37.** 45° **39.** 30° **41.** $\frac{\pi}{3}$ **43.** $\frac{\pi}{3}$ **45.** It is easy to find one-half of 2, which is 1. This is, then, the measure of the side opposite the 30° angle, and the ratios are easily found. Yes, any positive number could have been used. **47.** $\frac{\sqrt{3}}{3}$; $\sqrt{3}$ **49.** $\frac{\sqrt{3}}{2}$; $\frac{\sqrt{3}}{3}$; $\frac{2\sqrt{3}}{3}$ **51.** $-1$; $-1$ **53.** $-\frac{\sqrt{3}}{2}$; $-\frac{2\sqrt{3}}{3}$

In Exercises 55–61, we give, in order, sine, cosine, tangent, cotangent, secant, and cosecant.

**55.** $-\frac{\sqrt{3}}{2}$; $\frac{1}{2}$; $-\sqrt{3}$; $-\frac{\sqrt{3}}{3}$; 2; $-\frac{2\sqrt{3}}{3}$ **57.** $\frac{\sqrt{2}}{2}$; $\frac{\sqrt{2}}{2}$; 1; 1; $\sqrt{2}$; $\sqrt{2}$ **59.** $-\frac{1}{2}$; $\frac{\sqrt{3}}{2}$; $-\frac{\sqrt{3}}{3}$; $-\sqrt{3}$; $\frac{2\sqrt{3}}{3}$; $-2$ **61.** $\frac{\sqrt{2}}{2}$; $\frac{\sqrt{2}}{2}$; 1; 1; $\sqrt{2}$; $\sqrt{2}$ **63.** .5543090515 **65.** 1.134277349 **67.** $-1.002203376$ **69.** $-5.729741647$ **71.** .5984721441 **73.** $-3.380515006$ **75.** **(a)** $-\sin\frac{\pi}{6}$ **(b)** $-\frac{1}{2}$ **(c)** $\sin\frac{7\pi}{6} = -\sin\frac{\pi}{6} = -.5$ **77.** **(a)** $-\tan\frac{\pi}{4}$ **(b)** $-1$ **(c)** $\tan\frac{3\pi}{4} = -\tan\frac{\pi}{4} = -1$ **79.** **(a)** $-\cos\frac{\pi}{6}$ **(b)** $-\frac{\sqrt{3}}{2}$ **(c)** $\cos\frac{7\pi}{6} = -\cos\frac{\pi}{6} \approx -.8660254038$ **81.** 30°; 150° **83.** 120°; 300° **85.** 120°; 300° **87.** 46.59388121°; 313.4061188° **89.** 24.39257624°; 155.6074238° **91.** 41.24818261°; 221.2481826° **93.** .2095206607; 3.351113314 **95.** 1.27979966; 4.421392314 **97.** It represents the distance from the point $(x_1, y_1)$ to the origin. **98.** 60° **99.** 60° **100.** 60° **101.** It is a measure of the angle formed by the positive $x$-axis and the ray $y = \sqrt{3}x$, $x \geq 0$. **102.** $\left(\frac{y_1}{x_1}\right)^2 = 3$; exact value: $\sqrt{3}$ **103.** slope; tangent **104.** 1; It illustrates the identity $\cos^2\theta + \sin^2\theta = 1$. **105.** They agree, as they are both approximately 1.154700538.

**106.** $x$-coordinate: .5; $y$-coordinate: .8660254 (approximately) **107.** $\cos 60° = .5$; This is the $x$-coordinate of the point found in Exercise 106. Because $r = 1$, here $\cos 60° = \frac{x}{1} = x = .5$. **108.** $\sin 60° \approx .86602540$; This is the $y$-coordinate of the point found in Exercise 106. Because $r = 1$, here $\sin 60° = \frac{y}{1} = y \approx .86602540$. **109.** $2 \times 10^8$ m per sec **111.** 19° **113.** 48.7° **115.** **(a)** approximately 155 ft **(b)** approximately 194 ft **(c)** As the grade decreases from uphill to downhill, the braking distance increases, which corresponds to driving experience.

## 8.4 Exercises *(page 614)*

**1.** $B = 53° 40'$; $a = 571$ m; $b = 777$ m **3.** $M = 38.8°$; $n = 154$ m; $p = 198$ m **5.** $A = 47.9108°$; $c = 84.816$ cm; $a = 62.942$ cm **7.** $B = 62.00°$; $a = 8.17$ ft; $b = 15.4$ ft **9.** $A = 17.00°$; $a = 39.1$ in.; $c = 134$ in. **11.** $c = 85.9$ yd; $A = 62° 50'$; $B = 27° 10'$ **13.** The other acute angle requires the least work to find. **15.** 38.6598° **17.** Because $AD$ and $BC$ are parallel, angle $DAB$ is congruent to angle $ABC$, as they are alternate interior angles of the transversal $AB$. (A theorem of elementary geometry assures us of this.) **19.** It is measured clockwise from the north. **21.** 9.35 m **23.** 13.3 ft **25.** 128 ft **27.** 26.3° or 26° 20′ **29.** $A = 35.987°$ or $35° 59' 10''$; $B = 54.013°$ or $54° 00' 50''$ **31.** 114 ft **33.** 5.18 m **35.** 84.7 m **37.** 148 mi **39.** 1.48 mi **41.** 150 km **43.** 583 ft **45.** 33.4 m **47.** **(a)** 23.4 ft **(b)** 48.3 ft **(c)** The faster the speed, the more land needs to be cleared inside the curve. **49.** **(a)** $\tan \theta = \frac{y}{x}$ **(b)** $x = \frac{y}{\tan \theta}$ **51.** $a = 12$; $b = 12\sqrt{3}$; $d = 12\sqrt{3}$; $c = 12\sqrt{6}$ **53.** $m = \frac{7\sqrt{3}}{3}$; $a = \frac{14\sqrt{3}}{3}$; $n = \frac{14\sqrt{3}}{3}$; $q = \frac{14\sqrt{6}}{3}$ **55.** $A = \frac{s^2\sqrt{3}}{4}$

## Reviewing Basic Concepts *(page 619)*

**1.** $\sin A = \frac{8}{17}$; $\cos A = \frac{15}{17}$; $\tan A = \frac{8}{15}$; $\cot A = \frac{15}{8}$; $\sec A = \frac{17}{15}$; $\csc A = \frac{17}{8}$ **2.** Row 1: $\frac{1}{2}, \frac{\sqrt{3}}{2}, \frac{\sqrt{3}}{3}, \sqrt{3}, \frac{2\sqrt{3}}{3}, 2$; Row 2: $\frac{\sqrt{2}}{2}, \frac{\sqrt{2}}{2}, 1, 1, \sqrt{2}, \sqrt{2}$; Row 3: $\frac{\sqrt{3}}{2}, \frac{1}{2}, \sqrt{3}, \frac{\sqrt{3}}{3}, 2, \frac{2\sqrt{3}}{3}$ **3.** **(a)** cos 63° **(b)** $\cot \frac{3\pi}{10}$ **4.** **(a)** 80° **(b)** 5° **(c)** $\frac{\pi}{3}$ **5.** $\sin 315° = -\frac{\sqrt{2}}{2}$; $\cos 315° = \frac{\sqrt{2}}{2}$; $\tan 315° = -1$; $\cot 315° = -1$; $\sec 315° = \sqrt{2}$; $\csc 315° = -\sqrt{2}$ **6.** **(a)** .725374371 **(b)** 5.67128182 **(c)** −1.321348709 **7.** 150°; 330° **8.** .75; 2.391592654 **9.** 34 ft **10.** 19,600 ft

## 8.5 Exercises *(page 625)*

**1.** $-\frac{1}{2}$ **3.** −1 **5.** −2 **7.** $-\sqrt{3}$ **9.** $-\frac{1}{2}$ **11.** $\frac{2\sqrt{3}}{3}$ **13.** .5736 **15.** .4068 **17.** 1.2065 **19.** 14.3338 **21.** −1.0460 **23.** −3.8665 **25.** $\sin \theta = \frac{\sqrt{2}}{2}$; $\cos \theta = \frac{\sqrt{2}}{2}$; $\tan \theta = 1$; $\cot \theta = 1$; $\sec \theta = \sqrt{2}$; $\csc \theta = \sqrt{2}$ **27.** $\sin \theta = -\frac{12}{13}$; $\cos \theta = \frac{5}{13}$; $\tan \theta = -\frac{12}{5}$; $\cot \theta = -\frac{5}{12}$; $\sec \theta = \frac{13}{5}$; $\csc \theta = -\frac{13}{12}$ **29.** .7 **31.** 4 or 2.3 **33.** .2095 **35.** 1.4430 **37.** 1.4747 **39.** .9846 **41.** (−.8011, .5985) **43.** (.4385, −.8987) **45.** I **47.** II **49.** $f(7) \approx 14$; On July 1, there are about 14 hr of daylight. **51.** **(a)** 30° **(b)** 60° **(c)** 75° **(d)** 86° **(e)** 86° **(f)** 60° **53.** full moon: $t = \pm\pi, \pm 3\pi, \pm 5\pi, \ldots$; new moon: $t = 0, \pm 2\pi, \pm 4\pi, \ldots$

## 8.6 Exercises *(page 640)*

**1.** G **3.** E **5.** B **7.** F **9.** D **11.** H **13.** B **15.** F **17.** B **19.** C **21.** $y = 4 \sin \frac{1}{2}x$ (There are other correct answers.) **23.** **(a)** 1; 1; 3; −2 **(b)** −1; −1; −3; −8 **24.** [−8, −2] **25.** The trig viewing window as defined in the text has Ymin $= -4$ and Ymax $= 4$. Because the range of $f$ includes values less than $-4$, the local minimum points will not appear in the trig viewing window. **26.** $-2$; $-8$ **27.** The period of $f$ is $\frac{2\pi}{2} = \pi$, and the distance between $-2\pi$ and $2\pi$ is $4\pi$ units. Since $\frac{4\pi}{\pi} = 4$, the interval $[-2\pi, 2\pi]$ will show 4 periods of the graph.

**28.** $f(x) = -5 + 3\sin\left[2\left(x - \frac{\pi}{2}\right)\right]$

**29.** Ymin $= -8$; Ymax $= -2$ **30.** $f(-2) \approx -7.270407$; $f(-2 + \pi) \approx -7.270407$; Because the period of $f$ is $\pi$, $f(p) = f(p + \pi)$ for any value of $p$. In this case, $p = -2$.

**31.** **(a)** 2 **(b)** $2\pi$ **(c)** $\pi$ **(d)** none **(e)** $[-2, 2]$

**33.** **(a)** 4 **(b)** $4\pi$ **(c)** $-\pi$ **(d)** none **(e)** $[-4, 4]$ **35.** **(a)** 1 **(b)** $\frac{2\pi}{3}$ **(c)** $\frac{\pi}{15}$ **(d)** upward 2 units **(e)** $[1, 3]$

**37.** $y = \sin\left(x - \frac{\pi}{4}\right)$

**39.** $y = 2\cos\left(x - \frac{\pi}{3}\right)$

**41.** $y = -4\sin(2x - \pi)$

**43.** $y = \frac{1}{2}\cos\left(\frac{1}{2}x - \frac{\pi}{4}\right)$

**45.** $y = 1 - \frac{2}{3}\sin\frac{3}{4}x$

**47.** $y = 1 - 2\cos\frac{1}{2}x$

**49.** $y = -3 + 2\sin\left(x + \frac{\pi}{2}\right)$

**51.** $y = \frac{1}{2} + \sin\left[2\left(x + \frac{\pi}{4}\right)\right]$

**53.** 24 hr **55.** approximately 6:00 P.M.; approximately .2 ft **57.** approximately 2:00 A.M.; approximately 2.6 ft

**59.** **(a)** 80°; 50° **(b)** 15° **(c)** about 35,000 yr **(d)** downward **61.** **(a)** about 2 hr **(b)** 1 yr

**63.** **(a)** 3.8; $\frac{1}{20}$ **(b)** 20 **(c)** $-3.074$; 1.174; $-3.074$; $-3.074$; 1.174 **(d)**

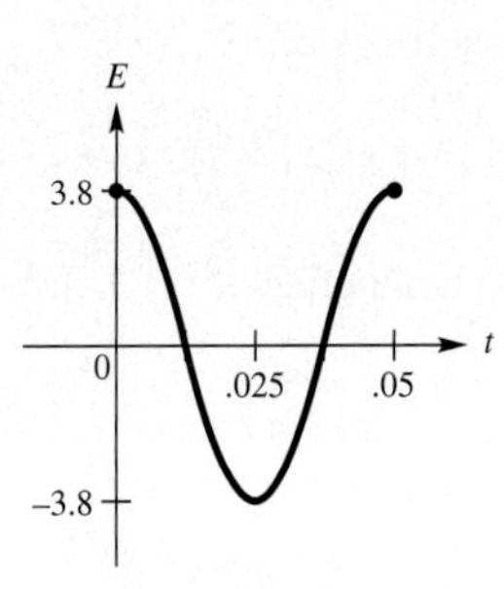


$E = 3.8 \cos 40\pi t$

**65.** **(a)** $C(x) = .04x^2 + .6x + 330 + 7.5 \sin(2\pi x)$

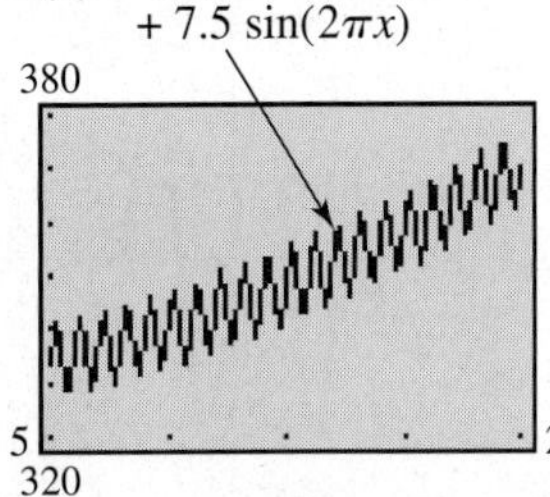


**(b)** Answers will vary. The seasons are more dramatic in Alaska.
**(c)** $C(x) = .04(x - 1970)^2 + .6(x - 1970) + 330 + 7.5 \sin[2\pi(x - 1970)]$

**67.** **(a)** 70.4° **(b)**

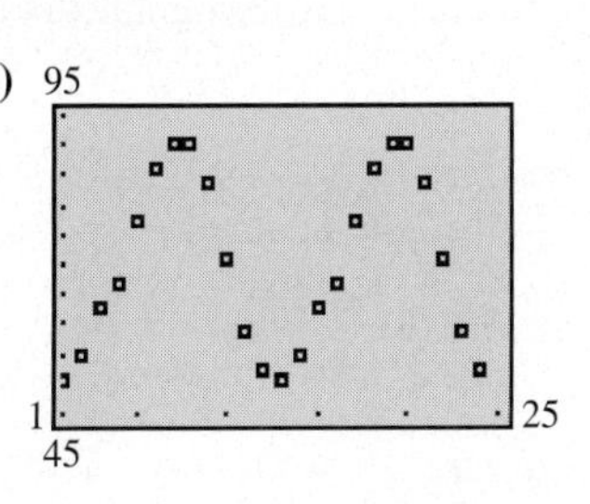


**(c)** $f(x) = 19.5 \cos\left[\frac{\pi}{6}(x - 7.2)\right] + 70.5$

**(d)** The function gives an excellent model for the data.

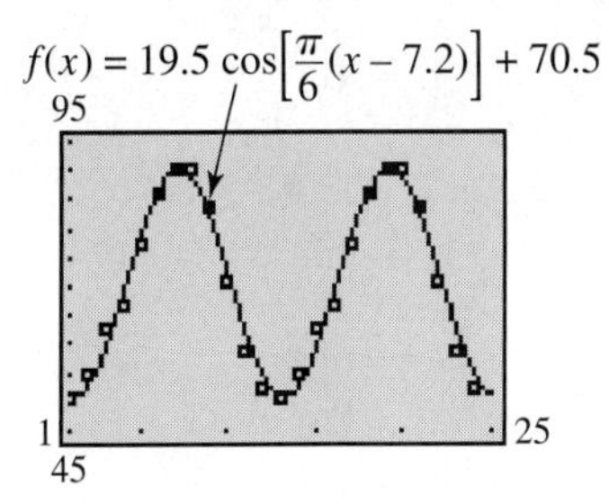


**(e)**

SinReg
y=a*sin(bx+c)+d
a=19.72
b=.52
c=-2.17
d=70.47

TI-83 fixed to the nearest hundredth

**69.**

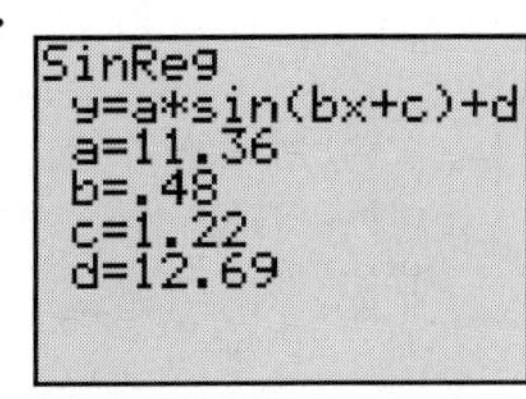
SinReg
y=a*sin(bx+c)+d
a=11.36
b=.48
c=1.22
d=12.69

TI-83 fixed to the nearest hundredth

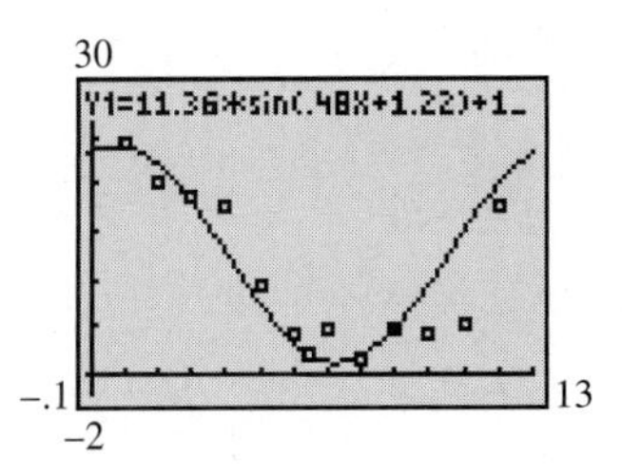

## Reviewing Basic Concepts (*page 645*)

**1. (a)** $(1, 0)$ **(b)** $\left(-\frac{\sqrt{2}}{2}, -\frac{\sqrt{2}}{2}\right)$ **(c)** $(0, 1)$ **2.** $\cos\left(-\frac{5\pi}{2}\right) = 0$; $\sin\left(-\frac{5\pi}{2}\right) = -1$; $\cot\left(-\frac{5\pi}{2}\right) = 0$; $\csc\left(-\frac{5\pi}{2}\right) = -1$; $\tan\left(-\frac{5\pi}{2}\right)$ and $\sec\left(-\frac{5\pi}{2}\right)$ are undefined. **3. (a)** $\cos\frac{7\pi}{6} = -\frac{\sqrt{3}}{2}$; $\sin\frac{7\pi}{6} = -\frac{1}{2}$; $\tan\frac{7\pi}{6} = \frac{\sqrt{3}}{3}$; $\cot\frac{7\pi}{6} = \sqrt{3}$; $\sec\frac{7\pi}{6} = -\frac{2\sqrt{3}}{3}$; $\csc\frac{7\pi}{6} = -2$ **(b)** $\cos\left(-\frac{2\pi}{3}\right) = -\frac{1}{2}$; $\sin\left(-\frac{2\pi}{3}\right) = -\frac{\sqrt{3}}{2}$; $\tan\left(-\frac{2\pi}{3}\right) = \sqrt{3}$; $\cot\left(-\frac{2\pi}{3}\right) = \frac{\sqrt{3}}{3}$; $\sec\left(-\frac{2\pi}{3}\right) = -2$; $\csc\left(-\frac{2\pi}{3}\right) = -\frac{2\sqrt{3}}{3}$ **4.** $\cos 2.25 \approx -.6281736227$; $\sin 2.25 \approx .7780731969$; $\tan 2.25 \approx -1.238627616$; $\cot 2.25 \approx -.8073451511$; $\sec 2.25 \approx -1.591916572$; $\csc 2.25 \approx 1.285226125$ **5.** quadrant IV

**6.** period: $2\pi$; amplitude: 1

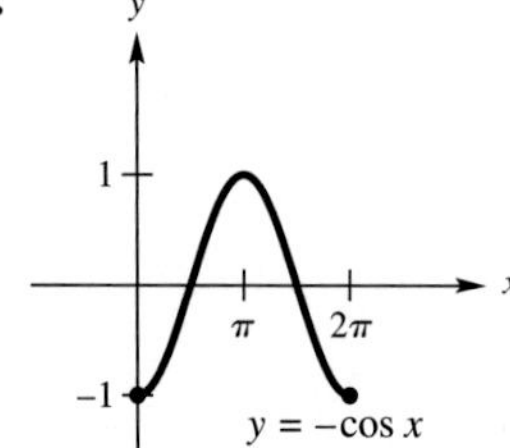


**7.** amplitude: 3; period: 2; phase shift: $-1$

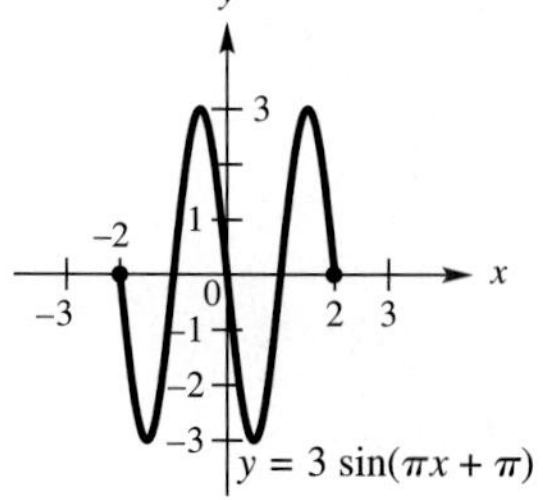


**8. (a)** maximum: 18.9 hr; minimum: 5.9 hr **(b)** The amplitude represents half the difference in the daylight hours between the longest and shortest days. The period represents 12 months or one year.

## 8.7 Exercises (*page 654*)

**1.** B **3.** E **5.** D **7.** true **9.** true **11.** false; $\tan(-x) = -\tan x$ for all $x$ in the domain. **13. (a)** $4\pi$ **(b)** none **(c)** $(-\infty, -2] \cup [2, \infty)$ **15. (a)** $2\pi$ **(b)** $-\frac{\pi}{2}$ **(c)** $(-\infty, -2] \cup [2, \infty)$ **17. (a)** $3\pi$ **(b)** $\frac{\pi}{2}$ **(c)** $(-\infty, \infty)$ **19. (a)** $\pi$ **(b)** $-\frac{\pi}{2}$ **(c)** $\left(-\infty, -\frac{1}{2}\right] \cup \left[\frac{1}{2}, \infty\right)$ **21. (a)** $\pi$ **(b)** $-\frac{\pi}{4}$ **(c)** $(-\infty, \infty)$

**23.**

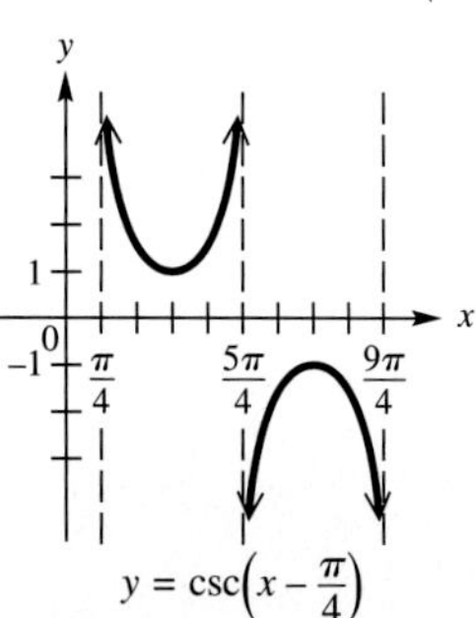


**25.**

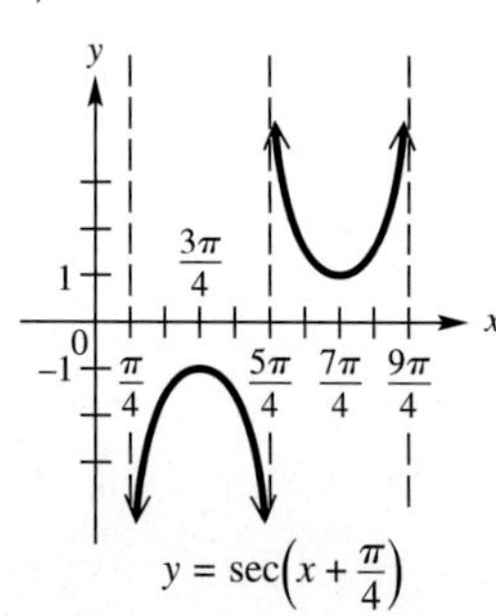


**27.**

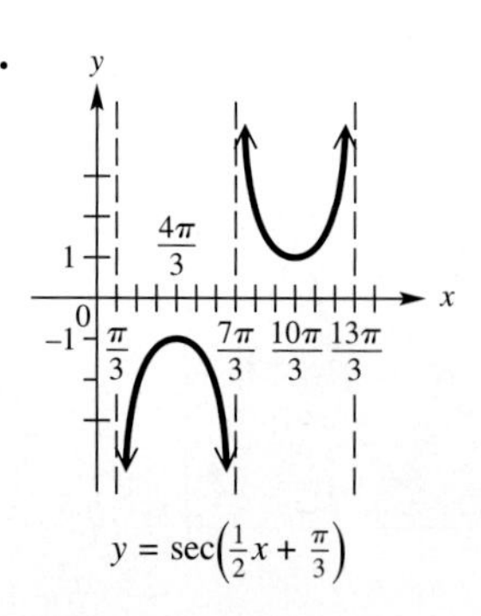


**29.**

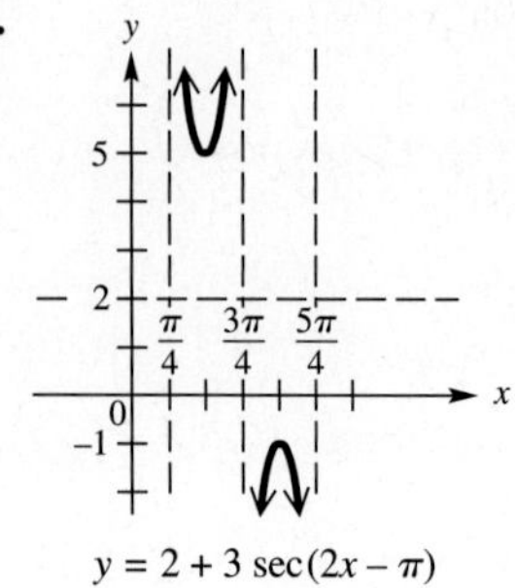


**31.**

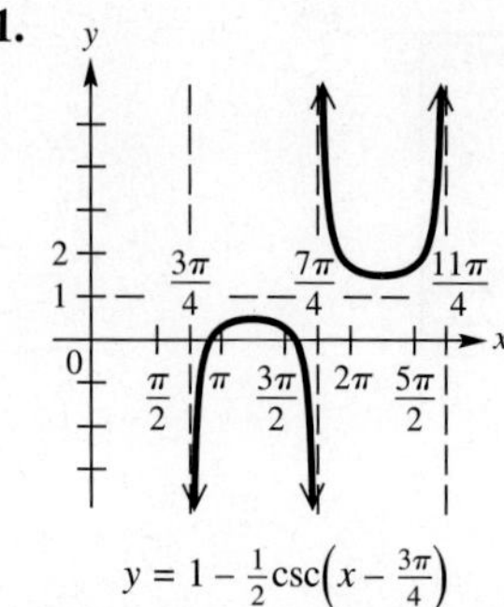


**33.**

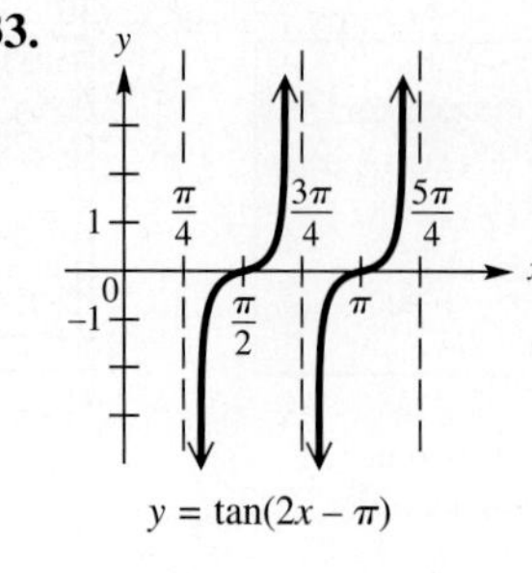

**35.**

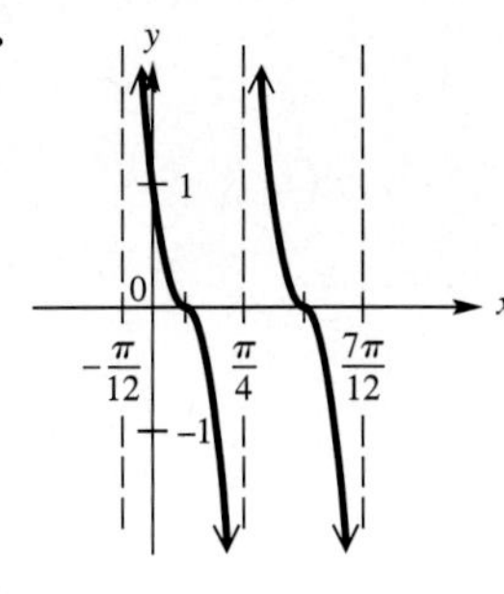


$y = \cot\left(3x + \frac{\pi}{4}\right)$

**37.**

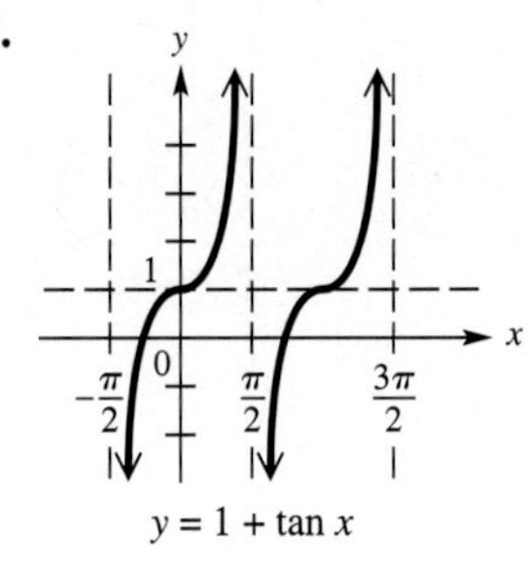


$y = 1 + \tan x$

**39.**

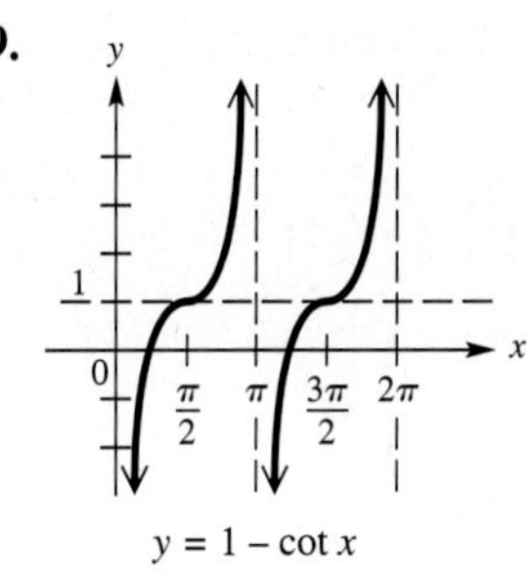


$y = 1 - \cot x$

**41.**

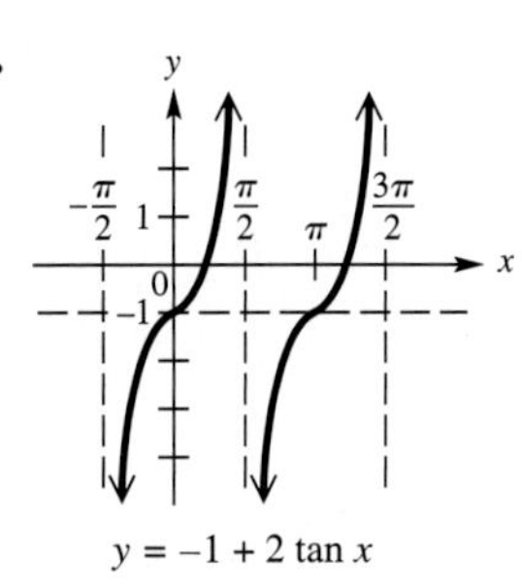


$y = -1 + 2 \tan x$

**43.**

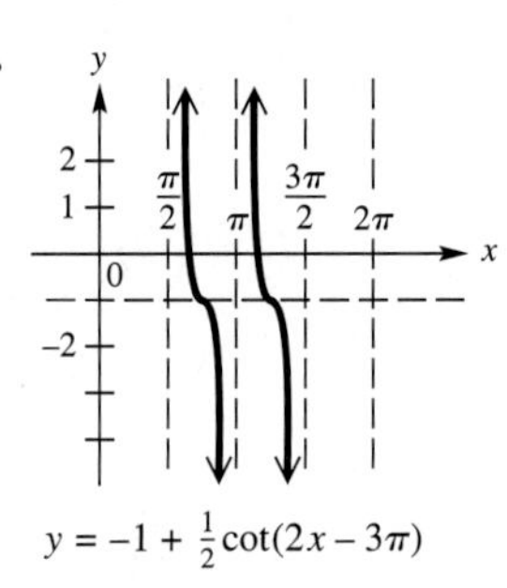


$y = -1 + \frac{1}{2}\cot(2x - 3\pi)$

**45.** For values in the interval $-1 \leq x \leq 1$, $x$ and $\tan x$ are approximately equal. As $x \to 0$, $\tan x$ approaches the value $x$.

**47.** **(a)** 0 m **(b)** −2.9 m **(c)** −12.3 m **(d)** 12.3 m **(e)** It leads to $\tan \frac{\pi}{2}$, which is undefined. **49.** In both cases, the result is 1.366025404. **51.** In both cases, the result is 1.83195123.

## 8.8 Exercises *(page 658)*

**1.** **(a)** $s(t) = 2\cos(4\pi t)$ **(b)** $s(1) = 2$; The weight is neither moving upward nor downward. At $t = 1$ the motion of the weight is changing from up to down. **3.** **(a)** $s(t) = -3\cos(2.5\pi t)$ **(b)** $s(1) = 0$; upward

**5.** $s(t) = .21\cos(55\pi t)$

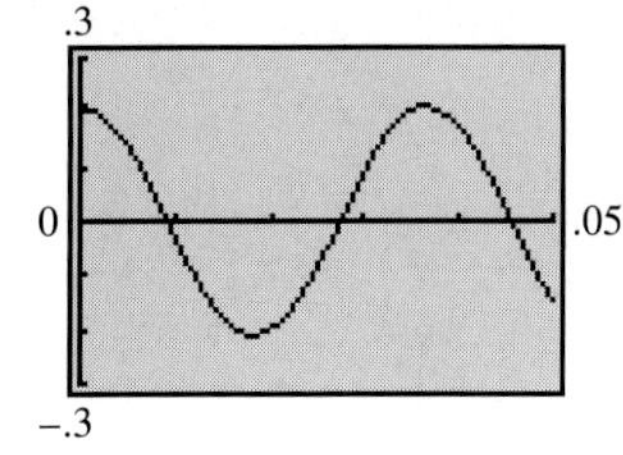


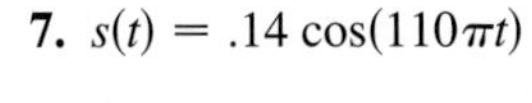

**7.** $s(t) = .14\cos(110\pi t)$

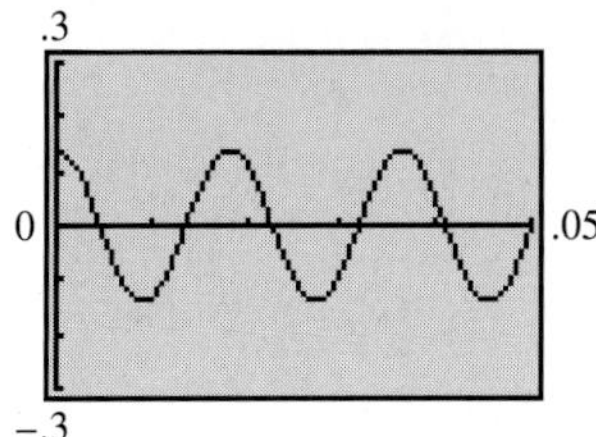


**9.** **(a)** $s(t) = 2\sin 2t$; amplitude: 2; period: $\pi$; frequency: $\frac{1}{\pi}$ **(b)** $s(t) = 2\sin 4t$; amplitude: 2; period: $\frac{\pi}{2}$; frequency: $\frac{2}{\pi}$

**11.** $\frac{8}{\pi^2}$ **13.** **(a)** amplitude: $\frac{1}{2}$; period: $\sqrt{2}\pi$; frequency: $\frac{\sqrt{2}}{2\pi}$ **(b)** $s(t) = \frac{1}{2}\sin\sqrt{2}t$ **15.** **(a)** 4 in. **(b)** $\frac{5}{\pi}$ cycles per sec; $\frac{\pi}{5}$ sec **(c)** after $\frac{\pi}{10}$ sec **(d)** approximately 2; After 1.466 sec, the weight is about 2 in. above the equilibrium position.

**17.** **(a)** $s(t) = -2\cos 6\pi t$ **(b)** 3 cycles per sec

## Reviewing Basic Concepts *(page 659)*

**1.**

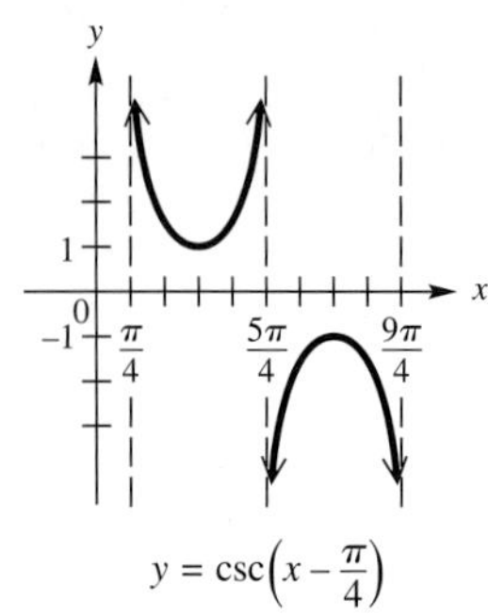


$y = \csc\left(x - \frac{\pi}{4}\right)$

period: $2\pi$; phase shift: $\frac{\pi}{4}$;

domain: $\left\{x \,\middle|\, x \neq \frac{\pi}{4} + n\pi, \text{ where } n \text{ is an integer}\right\}$;

range: $(-\infty, -1] \cup [1, \infty)$

**2.**

y
x
1
0
−1
$\frac{3\pi}{4}$
$\frac{\pi}{4}$
$\frac{5\pi}{4}$
$\frac{7\pi}{4}$
$\frac{9\pi}{4}$

$y = \sec\left(x + \frac{\pi}{4}\right)$

period: $2\pi$; phase shift: $-\frac{\pi}{4}$;

domain: $\left\{x \,\middle|\, x \neq \frac{\pi}{4} + n\pi, \text{ where } n \text{ is an integer}\right\}$;

range: $(-\infty, -1] \cup [1, \infty)$

**3.**

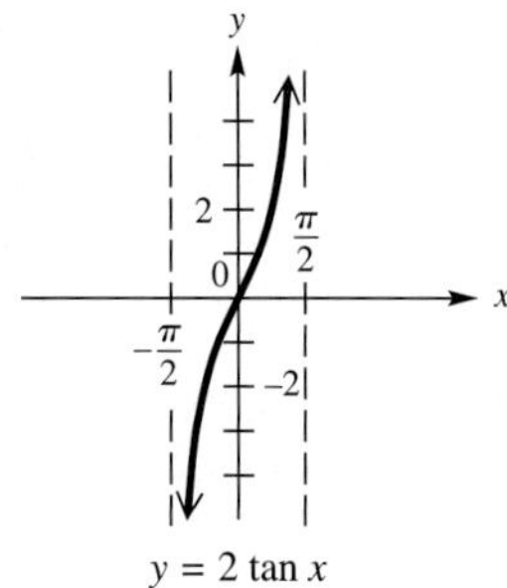


$y = 2 \tan x$

period: $\pi$; domain: $\left\{x \,\middle|\, x \neq \frac{\pi}{2} + n\pi, \text{ where } n \text{ is an integer}\right\}$;

range: $(-\infty, \infty)$

**4.**

y
x
2
0
−2
$\frac{\pi}{2}$
$\pi$

$y = 2 \cot x$

period: $\pi$; domain: $\{x \mid x \neq n\pi, \text{ where } n \text{ is an integer}\}$;

range: $(-\infty, \infty)$

**5.** **(a)** 4 in. **(b)** after $\frac{1}{8}$ sec **(c)** 4 cycles per sec; $\frac{1}{4}$ sec

## Chapter 8 Review Exercises *(page 664)*

**1.** 186° **3.** 1280° **5.** 1 radian $\approx 57.3° > 1°$ **7.** $\frac{2\pi}{3}$ **9.** 225° **11.** $\frac{4\pi}{3}$ in. **13.** 35.8 cm **15.** 41 yd

In Exercises 17–21, we give, in order, sine, cosine, tangent, cotangent, secant, and cosecant.

**17.** $-\frac{\sqrt{2}}{2}; -\frac{\sqrt{2}}{2}; 1; 1; -\sqrt{2}; -\sqrt{2}$ **19.** $0; -1; 0;$ undefined; $-1;$ undefined

**21.** $-\frac{2\sqrt{85}}{85}; \frac{9\sqrt{85}}{85}; -\frac{2}{9}; -\frac{9}{2}; \frac{\sqrt{85}}{9}; -\frac{\sqrt{85}}{2}$ **23.** tangent and secant **25.** possible

In Exercise 27, we give, in order, sine, cosine, tangent, cotangent, secant, and cosecant.

**27.** $-\frac{\sqrt{39}}{8}; -\frac{5}{8}; \frac{\sqrt{39}}{5}; \frac{5\sqrt{39}}{39}; -\frac{8}{5}; -\frac{8\sqrt{39}}{39}$ **29.** Evaluate $\frac{1}{1.6778490} \approx .5960011896$

In Exercises 31–35, we give, in order, sine, cosine, tangent, cotangent, secant, and cosecant.

**31.** $\frac{20}{29}; \frac{21}{29}; \frac{20}{21}; \frac{21}{20}; \frac{29}{21}; \frac{29}{20}$ **33.** $-\frac{\sqrt{3}}{2}; \frac{1}{2}; -\sqrt{3}; -\frac{\sqrt{3}}{3}; 2; -\frac{2\sqrt{3}}{3}$ **35.** $-\frac{1}{2}; \frac{\sqrt{3}}{2}; -\frac{\sqrt{3}}{3}; -\sqrt{3}; \frac{2\sqrt{3}}{3}; -2$

**37.** $-1.3563417$ **39.** $.20834446$ **41.** $55.7°$ **43.** No, it will give an angle whose tangent equals 25.

**45.** $B = 31° 30'$; $a = 638$; $b = 391$ **47.** 73.7 ft **49.** 1200 m **51.** 140 mi **53.** 419 **55.** $-\frac{1}{2}$ **57.** 2

**59.** $-11.4266$ **61.** **(b)** $\frac{\pi}{6}$ **(c)** less ultraviolet light when $\theta = \frac{\pi}{3}$ **63.** .5528 **65.** 2; $2\pi$; none; none **67.** $\frac{1}{2}$; $\frac{2\pi}{3}$; none; none **69.** 2; $8\pi$; 1 up; none **71.** 3; $2\pi$; none; $-\frac{\pi}{2}$ **73.** not applicable; $\pi$; none; $\frac{\pi}{8}$

**75.**

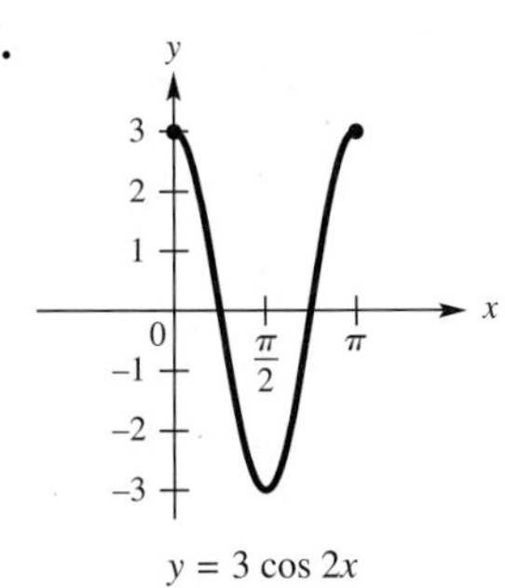


**77.**

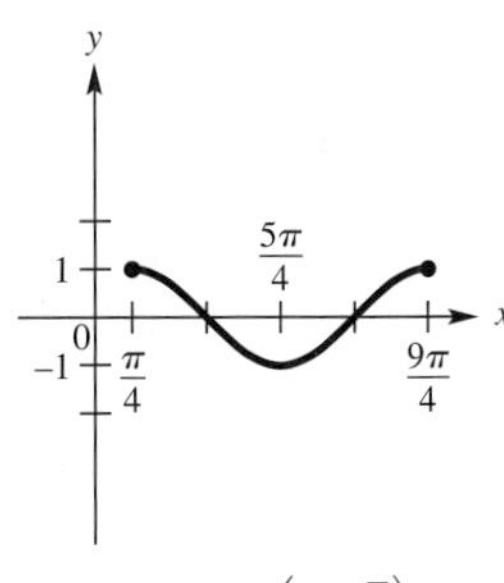


**79.**

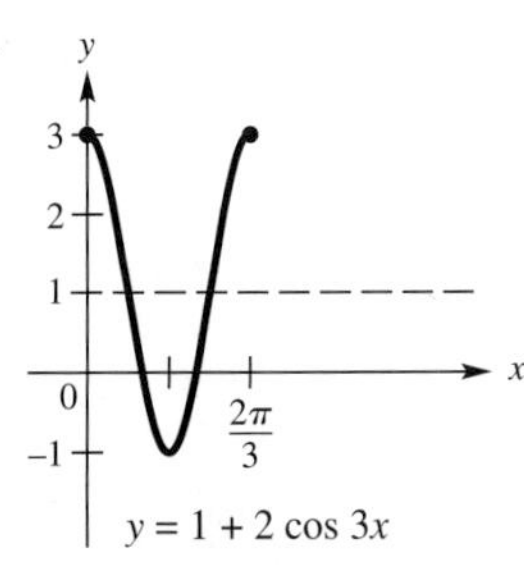


**81.** tangent **83.** cosine **85.** cotangent **87.** $y = 3 \sin\left[2\left(x - \frac{\pi}{4}\right)\right]$

**89.** $y = \frac{1}{3} \sin \frac{\pi}{2} x$ **91.** **(a)**

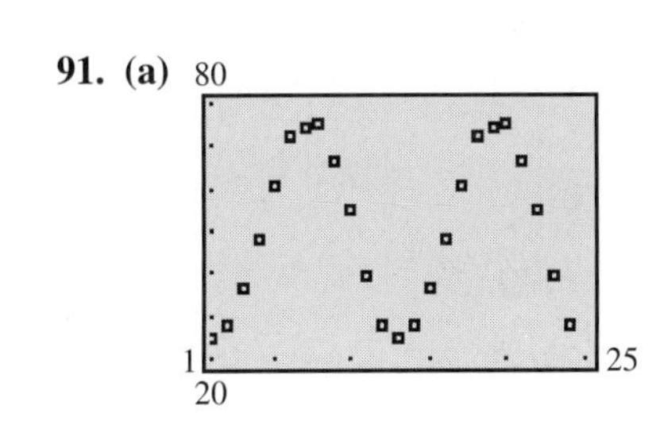


**(b)** $f(x) = 25 \sin\left[\frac{\pi}{6}(x - 4.2)\right] + 50$

**(c)** $a$ represents the amplitude, $b$ the period, $c$ the vertical shift, and $d$ the phase shift.

**(d)** The function gives an excellent model for the data.

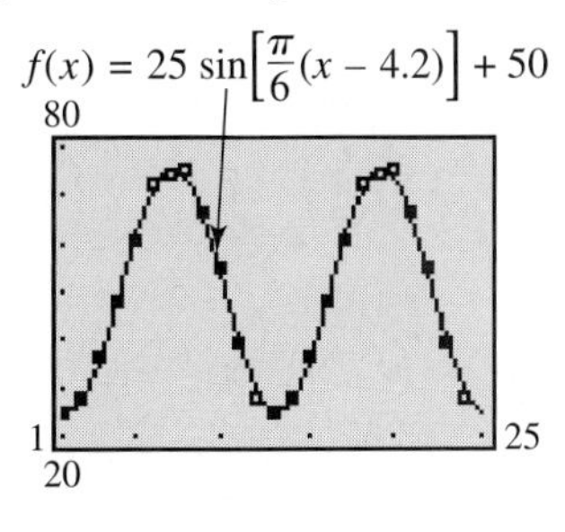


**(e)**

SinReg
y=a*sin(bx+c)+d
a=25.77
b=.52
c=-2.19
d=50.57

**93.** **(a)** 100 **(b)** 258 **(c)** 122 **(d)** 296 **95.** amplitude: 4; period: 2; frequency: $\frac{1}{2}$ **97.** The frequency is the number of cycles in one second; 4 in. below; at the initial point; $2\sqrt{2}$ in. below

## Chapter 8 Test *(page 669)*

**1.** 203° **2.** 2700° **3. (a)** $\frac{2\pi}{3}$ **(b)** 162° **4. (a)** $\frac{4}{3}$ **(b)** 15,000 $\text{cm}^2$ **5.** $\frac{12\pi}{5} \approx 7.54$ in. **6. (a)** 210° **(b)** $x = -\sqrt{3}$; $y = -1$ **(c)** $\sin(-150°) = -\frac{1}{2}$; $\cos(-150°) = -\frac{\sqrt{3}}{2}$; $\tan(-150°) = \frac{\sqrt{3}}{3}$; $\cot(-150°) = \sqrt{3}$; $\sec(-150°) = -\frac{2\sqrt{3}}{3}$; $\csc(-150°) = -2$ **(d)** $-\frac{5\pi}{6}$ **7.** .97169234 **8.** III **9.** $\sin\theta = -\frac{5\sqrt{29}}{29}$; $\cos\theta = \frac{2\sqrt{29}}{29}$; $\tan\theta = -\frac{5}{2}$ **10.** $\sin\theta = -\frac{3}{5}$; $\tan\theta = -\frac{3}{4}$; $\cot\theta = -\frac{4}{3}$; $\sec\theta = \frac{5}{4}$; $\csc\theta = -\frac{5}{3}$ **11.** $x = 4$; $y = 4\sqrt{3}$; $z = 4\sqrt{2}$; $w = 8$ **12.** $-\sqrt{3}$ **13. (a)** .97939940 **(b)** .20834446 **(c)** 1.9362132 **14.** $B = 31°\ 30'$; $a = 638$; $b = 391$
**15.** 15.5 ft **16.** 110 km

**17.**

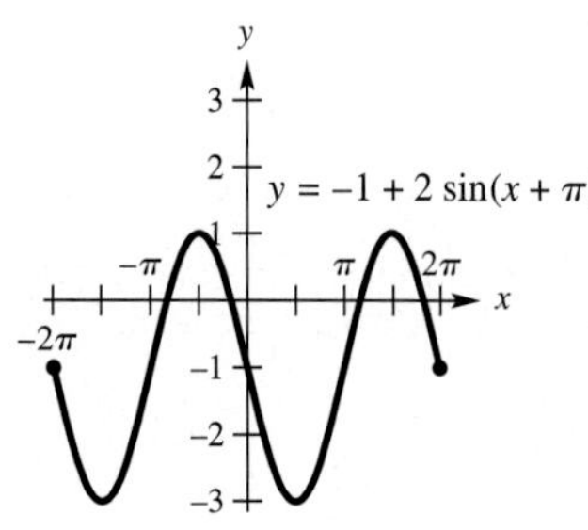


**18.**

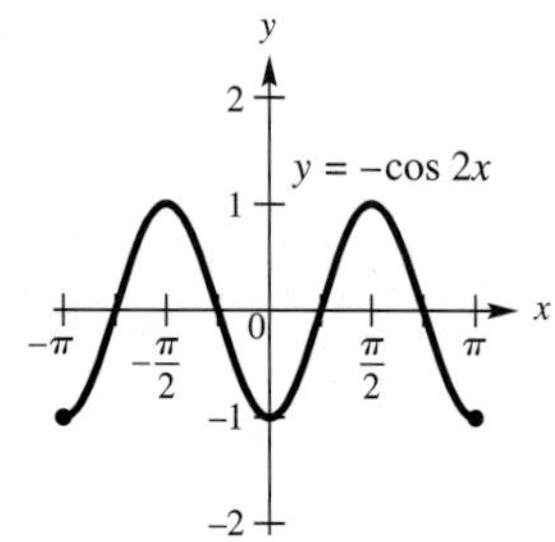


**19.**

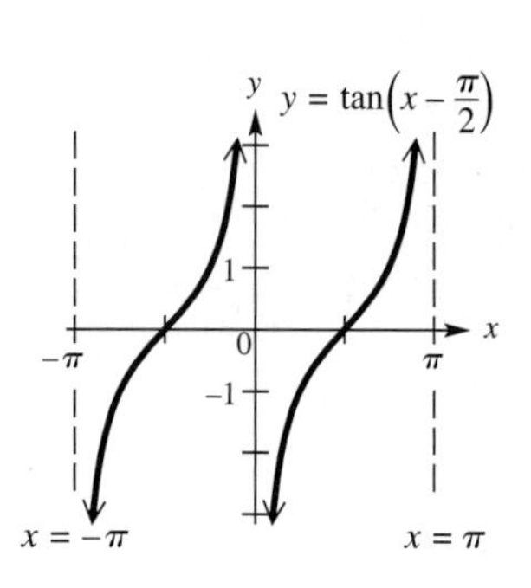


**20. (a)** $f(x) = 17.5 \sin\left[\frac{\pi}{6}(x-4)\right] + 67.5$

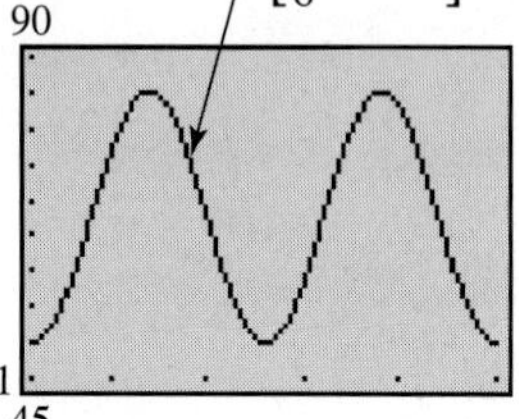


**(b)** 17.5; 12; 4; 67.5 upward
**(c)** approximately 52°F
**(d)** 50°F in January; 85°F in July
**(e)** approximately 67.5°; This is the vertical translation.

**21.** 3 in. **22.** after $\frac{1}{2}$ sec

# CHAPTER 9 TRIGONOMETRIC IDENTITIES AND EQUATIONS

## 9.1 Exercises *(page 681)*

**1.** odd **3.** odd **5.** even **7.** B **9.** E **11.** A **13.** A **15.** D **17.** cos 4.38 **19.** −sin .5 **21.** $-\tan\frac{\pi}{7}$ **23.** The student has neglected to write the argument of the functions (for example, $\theta$ or $t$).
**25.** $\frac{\pm\sqrt{1+\cot^2\theta}}{1+\cot^2\theta}$; $\frac{\pm\sqrt{\sec^2\theta - 1}}{\sec\theta}$ **27.** $\frac{\pm\sin\theta\sqrt{1-\sin^2\theta}}{1-\sin^2\theta}$; $\frac{\pm\sqrt{1-\cos^2\theta}}{\cos\theta}$; $\pm\sqrt{\sec^2\theta - 1}$; $\frac{\pm\sqrt{\csc^2\theta - 1}}{\csc^2\theta - 1}$
**29.** $\frac{\pm\sqrt{1-\sin^2\theta}}{1-\sin^2\theta}$; $\pm\sqrt{\tan^2\theta + 1}$; $\frac{\pm\sqrt{1+\cot^2\theta}}{\cot\theta}$; $\frac{\pm\csc\theta\sqrt{\csc^2\theta - 1}}{\csc^2\theta - 1}$ **31.** $\sin\theta$ **33.** $\tan^2\beta$ **35.** $\tan^2 x$

**37.** $\sec^2 x$ **39.**

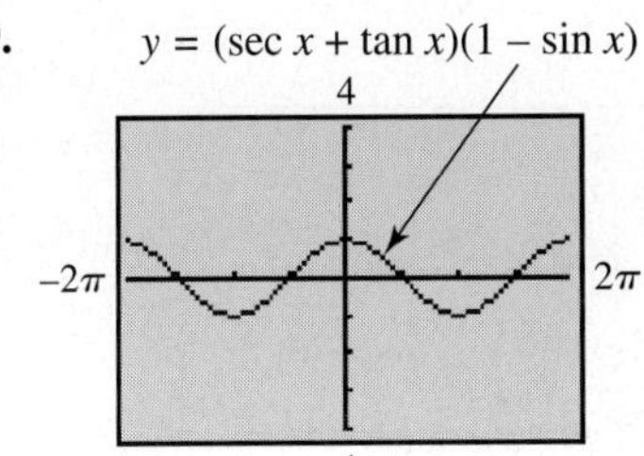


**40.** $y = \cos x$ **41.**

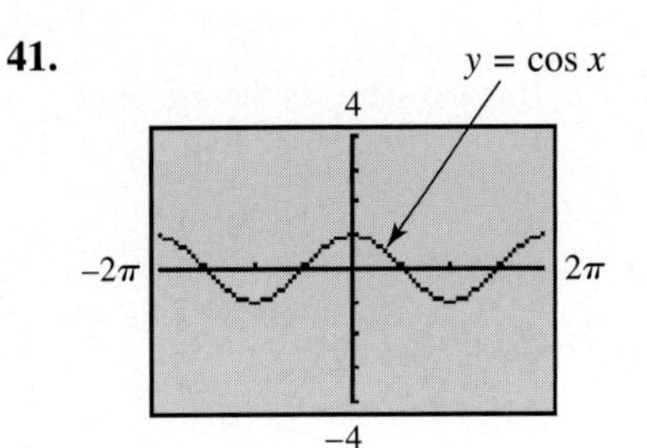


**42.** yes; $(\sec x + \tan x)(1 - \sin x) = \cos x$ **44.** $\dfrac{\cos x + 1}{\sin x + \tan x} = \cot x$ **63.** While the equation is true for the *particular* value $\theta = \dfrac{\pi}{2}$, it is not true in *general.* To be an identity, the equation must be true in *all* cases for which the functions involved are defined. **65.** when $\sin x \geq 0$ **67.** **(a)** $I = k(1 - \sin^2 \theta)$ **(b)** For $\theta = 2\pi n$ for all integers $n$, $\cos^2 \theta = 1$, its maximum value, and $I$ attains a maximum value of $k$.

**69.** **(a)** The total mechanical energy $E$ is always 2. The spring has maximum potential energy when it is fully stretched but not moving. The spring has maximum kinetic energy when it is not stretched but is moving fastest.

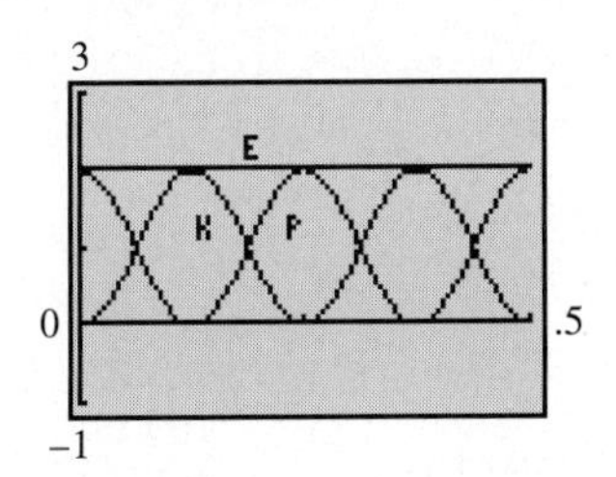


**(b)** Let $Y_1 = P(t)$, $Y_2 = K(t)$, and $Y_3 = E(t)$. $Y_3 = 2$ for all inputs. The spring is stretched the most (has greatest potential energy) when $t = .25, .5, .75, \ldots$. At these times kinetic energy is 0.

| X | Y2 | Y3 |
|---|---|---|
| 0 | 0 | 2 |
| .05 | .69098 | 2 |
| .1 | 1.809 | 2 |
| .15 | 1.809 | 2 |
| .2 | .69098 | 2 |
| .25 | 0 | 2 |
| .3 | .69098 | 2 |

Y3=Y1+Y2

**(c)** $E(t) = k$ or $E(t) = 2$

## 9.2 Exercises *(page 691)*

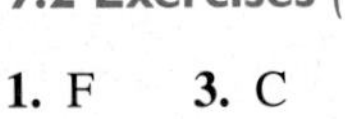

**1.** F **3.** C **5.** $\dfrac{\sqrt{6} - \sqrt{2}}{4}$ **7.** $\dfrac{-\sqrt{2} - \sqrt{6}}{4}$ **9.** $\dfrac{-\sqrt{6} + \sqrt{2}}{4}$ **11.** $\dfrac{\sqrt{6} - \sqrt{2}}{4}$ **13.** $-\sqrt{3} - 2$ **15.** $\dfrac{\sqrt{2} + \sqrt{6}}{4}$ **17.** $-1$ **19.** $\dfrac{\sqrt{2}}{2}$ **21.** $-1$ **23.** $\sin x$ **25.** $\sin x$ **27.** $-\sin x$ **29.** $\dfrac{\sqrt{2}(\sin x - \cos x)}{2}$ **31.** $\dfrac{\sqrt{2}(\cos x + \sin x)}{2}$ **33.** $-\tan x$ **35.** Find each cofunction of $A + B$, and then use the reciprocal identity.

**36.**

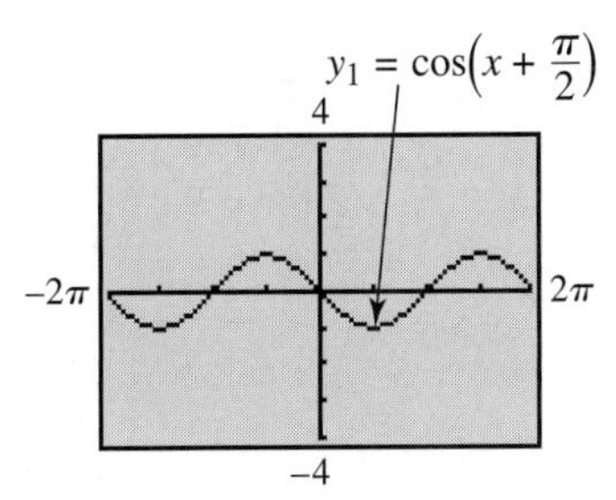


**37.** The graph of $y_1$ is obtained by shifting the graph of $y = \cos x$ to the left $\dfrac{\pi}{2}$ units.

**38.** $-.841471$ **40.** It is the same as the graph of $y_1 = \cos\left(x + \dfrac{\pi}{2}\right)$.

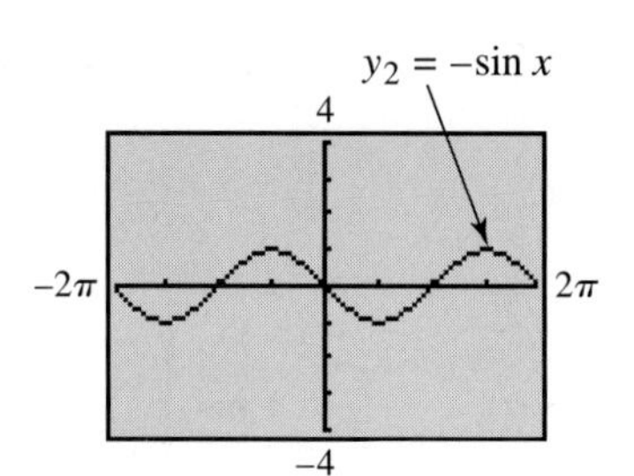


**41.** $-.841471$; It is the same as the approximation in Exercise 38. **43.** **(a)** $-\dfrac{63}{65}$ **(b)** $\dfrac{33}{65}$ **(c)** $-\dfrac{63}{16}$ **(d)** $-\dfrac{33}{56}$ **(e)** IV **(f)** II

**45.** **(a)** $-\dfrac{36}{85}$ **(b)** $\dfrac{84}{85}$ **(c)** $\dfrac{36}{77}$ **(d)** $-\dfrac{84}{13}$ **(e)** III **(f)** II **53.** 3 **55.** **(a)** 425 lb **(c)** 0°

**57.** **(a)** The pressure $P$ is oscillating. **(b)** The pressure oscillates and amplitude decreases as $r$ increases. **(c)** $P = \dfrac{a}{n\lambda}\cos(ct)$

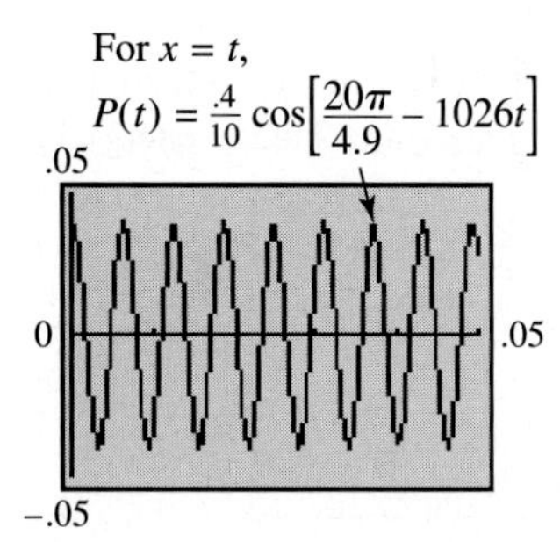


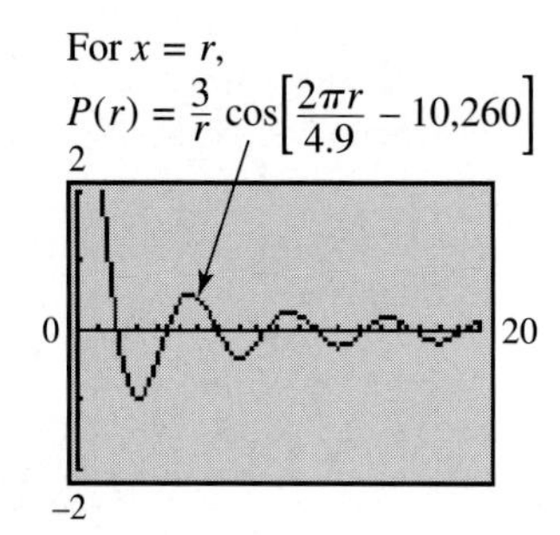


## Reviewing Basic Concepts *(page 693)*

**1.** $\dfrac{\sin^2 x}{\cos x}$ **2.** $\dfrac{\sqrt{3} - 3}{3 + \sqrt{3}}$ or $\sqrt{3} - 2$ **3.** 0 **4.** $\dfrac{\sqrt{2}}{2}(\sin x - \cos x)$ **5.** $\dfrac{-2 + \sqrt{15}}{6}$; $\dfrac{\sqrt{5} - 2\sqrt{3}}{6}$; $\dfrac{-2 - \sqrt{15}}{\sqrt{5} - 2\sqrt{3}}$ or $\dfrac{8\sqrt{5} + 9\sqrt{3}}{7}$ **10.** $-20\cos\dfrac{\pi t}{4}$

## 9.3 Exercises *(page 703)*

**1.** $\sin 2\theta = -\dfrac{4\sqrt{21}}{25}$; $\cos 2\theta = \dfrac{17}{25}$ **3.** $\sin 2x = \dfrac{4}{5}$; $\cos 2x = \dfrac{3}{5}$ **5.** $\sin 2\alpha = -\dfrac{4\sqrt{55}}{49}$; $\cos 2\alpha = \dfrac{39}{49}$ **7.** $\dfrac{\sqrt{3}}{2}$

**9.** $\dfrac{\sqrt{3}}{2}$ **11.** $-\dfrac{\sqrt{2}}{2}$ **13.** $\dfrac{1}{2}\tan 102°$ **15.** $\dfrac{1}{4}\cos 94.2°$ **17.** $\cos^4 x - \sin^4 x = \cos 2x$

Forms of answers may vary in Exercises 19 and 21.

**19.** $\cos 3x = 4\cos^3 x - 3\cos x$ **21.** $\tan 4x = \dfrac{4\tan x - 4\tan^3 x}{1 - 6\tan^2 x + \tan^4 x}$ **23.** $\dfrac{\sqrt{2 - \sqrt{3}}}{2}$ **25.** $1 - \sqrt{2}$ $\left(\text{or } -\sqrt{3 - 2\sqrt{2}}\right)$

**27.** $\dfrac{\sqrt{2 + \sqrt{2}}}{2}$ **29.** $\dfrac{\sqrt{10}}{4}$ **31.** 3 **33.** $-\sqrt{7}$ **39.** sin 20° **41.** tan 73.5° **43.** tan 29.87°

**55.** sin 160° − sin 44° **57.** sin 225° + sin 55° **59.** $-2\sin 3x \sin x$ **61.** −2 sin 11.5° cos 36.5°

**63.** $2\cos 6x \cos 2x$ **65.** **(a)** $\cos\dfrac{\theta}{2} = \dfrac{R - b}{R}$ **(b)** $\tan\dfrac{\theta}{4} = \dfrac{b}{50}$ **(c)** 54°

**67.** **(a)** For $x = t$, $W = VI =$ $(163 \sin 120\pi t)(1.23 \sin 120\pi t)$

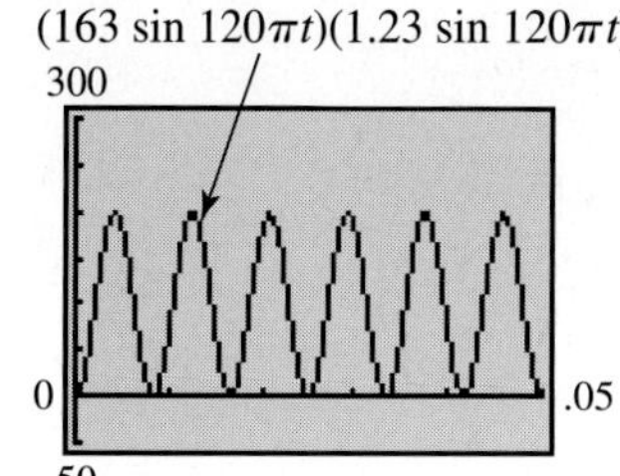


**(b)** maximum: 200.49 watts; minimum: 0 watts
**(c)** $a = -100.245$, $\omega = 240\pi$, $c = 100.245$
**(e)** 100.245 watts

## 9.4 Exercises *(page 714)*

**1.** **(a)** $[-1, 1]$ **(b)** $\left[-\frac{\pi}{2}, \frac{\pi}{2}\right]$ **(c)** increasing **(d)** $-2$ is not in the domain. **3.** **(a)** $(-\infty, \infty)$ **(b)** $\left(-\frac{\pi}{2}, \frac{\pi}{2}\right)$ **(c)** increasing **(d)** no **5.** $\cos^{-1}\frac{1}{a}$ **7.** $\frac{\pi}{4}$ **9.** $\pi$ **11.** $-\frac{\pi}{2}$ **13.** 0 **15.** $\frac{\pi}{2}$ **17.** $\frac{\pi}{4}$ **19.** $\frac{5\pi}{6}$ **21.** $\frac{3\pi}{4}$ **23.** $-\frac{\pi}{6}$ **25.** $\frac{\pi}{6}$ **27.** $-45°$ **29.** $-60°$ **31.** $120°$ **33.** $-30°$ **35.** $-7.6713835°$ **37.** $113.500970°$ **39.** $30.987961°$ **41.** .83798122 **43.** 2.3154725 **45.** 1.1900238

**47.**

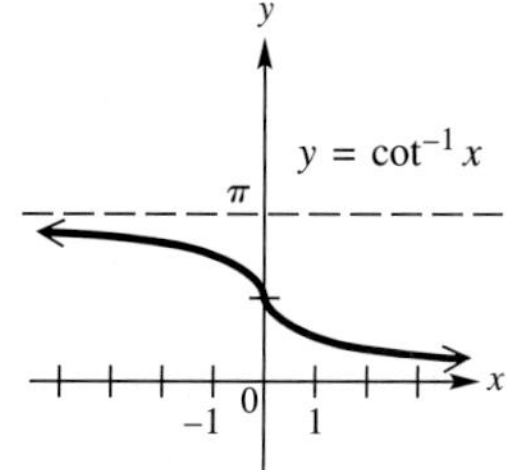


**49.**

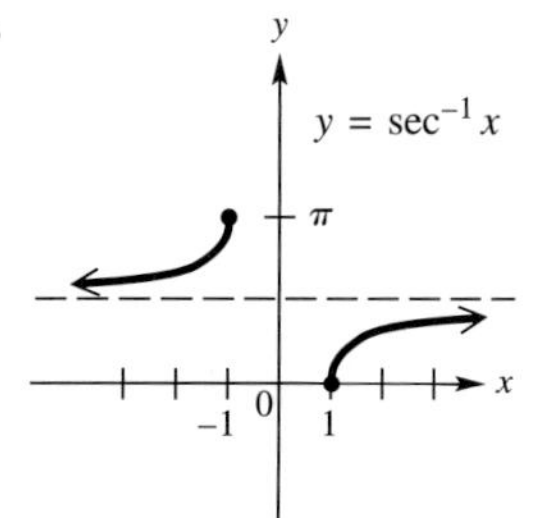


**51.**

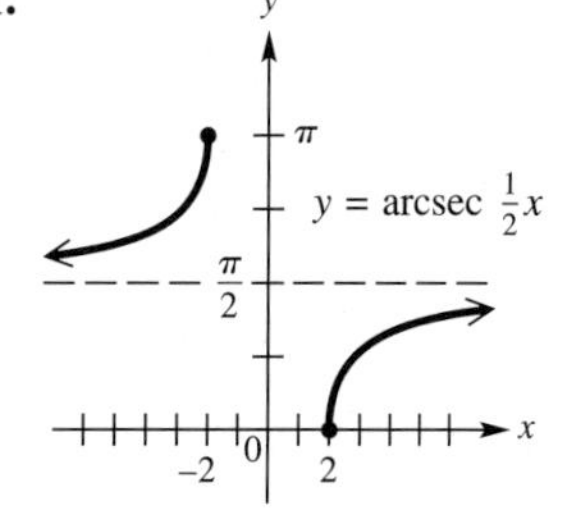


**53.** The domain of $y = \tan^{-1} x$ is $(-\infty, \infty)$. **54.** In both cases, the result is $x$. In each case, the graph is a straight line bisecting quadrants I and III (i.e., the line $y = x$).

**55.** It is the graph of $y = x$.

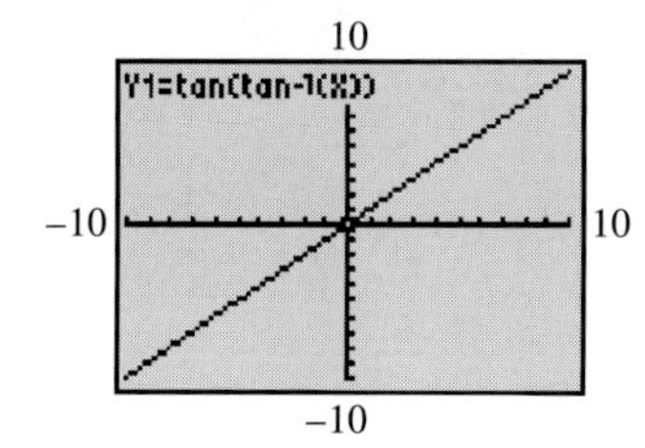


**56.** It does not agree because the range of the inverse tangent function is $\left(-\frac{\pi}{2}, \frac{\pi}{2}\right)$, not $(-\infty, \infty)$, as was the case in Exercise 55.

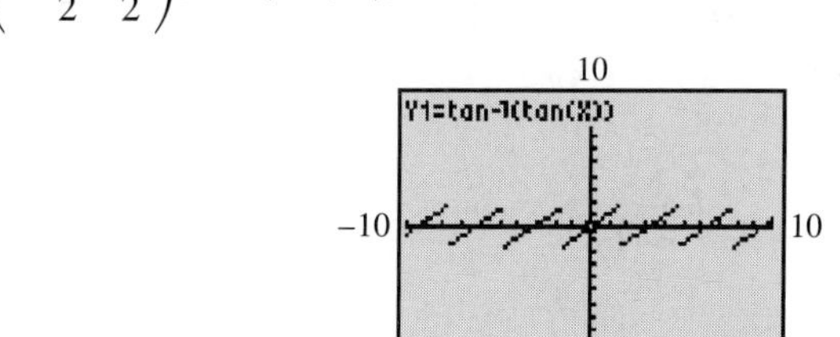


**57.** $\frac{\sqrt{7}}{3}$ **59.** $\frac{\sqrt{5}}{5}$ **61.** $\frac{120}{169}$ **63.** $-\frac{7}{25}$ **65.** $\frac{4\sqrt{6}}{25}$ **67.** 2 **69.** $\frac{63}{65}$ **71.** $\frac{\sqrt{10} - 3\sqrt{30}}{20}$ **73.** .894427191 **75.** .1234399811 **77.** $\sqrt{1 - u^2}$ **79.** $\frac{\sqrt{1 - u^2}}{u}$ **81.** $\frac{\sqrt{u^2 - 4}}{u}$ **83.** $\frac{u\sqrt{2}}{2}$ **85.** $\frac{2\sqrt{4 - u^2}}{4 - u^2}$ **87.** **(a)** $45°$

**(b)** $y = x$ **89. (a)** 35° **(b)** 24° **(c)** 18° **(e)** about 2 ft

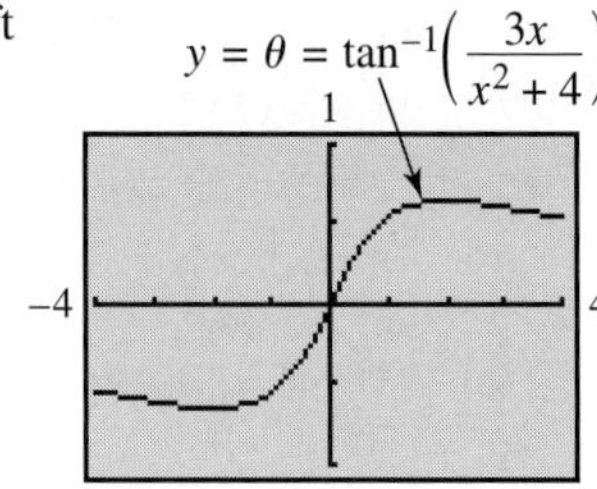


**91.** about 44.7%

### Reviewing Basic Concepts *(page 717)*

**1.** $-\dfrac{\sqrt{119}}{7}$ **2.** $\sin 2\theta = \dfrac{4\sqrt{2}}{9}$; $\cos 2\theta = \dfrac{7}{9}$; $\tan 2\theta = \dfrac{4\sqrt{2}}{7}$ **3.** $\dfrac{\sqrt{2} + \sqrt{6}}{4}$ **4.** $\sin 175° - \sin 125°$

**6. (a)** $\dfrac{\pi}{6}$ **(b)** $-\dfrac{\pi}{4}$ **7. (a)** 60° **(b)** 135° **8.**

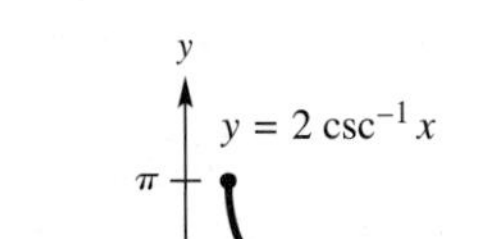


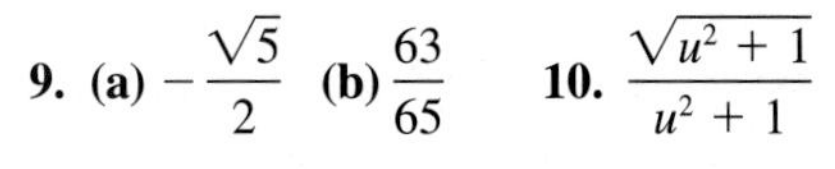

**9. (a)** $-\dfrac{\sqrt{5}}{2}$ **(b)** $\dfrac{63}{65}$ **10.** $\dfrac{\sqrt{u^2 + 1}}{u^2 + 1}$

### 9.5 Exercises *(page 722)*

**1.** $-1 < b < 1$; $b = \pm 1$; $b > 1$ or $b < -1$ **3.** $\left\{\dfrac{3\pi}{4}, \dfrac{7\pi}{4}\right\}$ **5.** $\left\{\dfrac{\pi}{6}, \dfrac{5\pi}{6}\right\}$ **7.** $\left\{\dfrac{\pi}{4}, \dfrac{2\pi}{3}, \dfrac{5\pi}{4}, \dfrac{5\pi}{3}\right\}$ **9.** $\{\pi\}$

**11.** $\left\{\dfrac{\pi}{6}, \dfrac{7\pi}{6}, \dfrac{4\pi}{3}, \dfrac{5\pi}{3}\right\}$ **13.** $\left\{\dfrac{\pi}{4}, \dfrac{3\pi}{4}, \dfrac{5\pi}{4}, \dfrac{7\pi}{4}\right\}$ **15.** $\left\{\dfrac{\pi}{4}, \dfrac{3\pi}{4}, \dfrac{5\pi}{4}, \dfrac{7\pi}{4}\right\}$ **17.** $\left\{\dfrac{\pi}{4}, \dfrac{5\pi}{4}\right\}$ **19. (a)** $\left\{\dfrac{7\pi}{6}, \dfrac{11\pi}{6}\right\}$

**(b)** $\left[0, \dfrac{7\pi}{6}\right) \cup \left(\dfrac{11\pi}{6}, 2\pi\right)$ **(c)** $\left(\dfrac{7\pi}{6}, \dfrac{11\pi}{6}\right)$ **21. (a)** $\{0, \pi\}$ **(b)** $\left(0, \dfrac{\pi}{2}\right) \cup \left(\dfrac{\pi}{2}, \pi\right) \cup \left(\pi, \dfrac{3\pi}{2}\right) \cup \left(\dfrac{3\pi}{2}, 2\pi\right)$ **(c)** $\emptyset$

**23. (a)** $\left\{\dfrac{7\pi}{6}, \dfrac{3\pi}{2}, \dfrac{11\pi}{6}\right\}$ **(b)** $\left[0, \dfrac{7\pi}{6}\right) \cup \left(\dfrac{11\pi}{6}, 2\pi\right)$ **(c)** $\left(\dfrac{7\pi}{6}, \dfrac{3\pi}{2}\right) \cup \left(\dfrac{3\pi}{2}, \dfrac{11\pi}{6}\right)$ **25. (a)** $\left\{0, \dfrac{\pi}{2}, \pi, \dfrac{3\pi}{2}\right\}$

**(b)** $\left(0, \dfrac{\pi}{2}\right) \cup \left(\dfrac{\pi}{2}, \pi\right) \cup \left(\pi, \dfrac{3\pi}{2}\right) \cup \left(\dfrac{3\pi}{2}, 2\pi\right)$ **(c)** $\emptyset$ **27.** $\tan^3 x - 3\tan x = 0$ **28.** $\left\{0, \dfrac{\pi}{3}, \dfrac{2\pi}{3}, \pi, \dfrac{4\pi}{3}, \dfrac{5\pi}{3}\right\}$

**29.** They all satisfy the equation. **30.** $\left\{\dfrac{\pi}{3}, \dfrac{2\pi}{3}, \dfrac{4\pi}{3}, \dfrac{5\pi}{3}\right\}$ **31.** No. The solutions 0 and $\pi$ were lost when dividing by $\tan x$, because both $\tan 0$ and $\tan \pi$ equal 0, and division by 0 is undefined. **33.** $\{.94, 2.21, 3.28, 6.14\}$ **35.** $\{.67, 1.83, 3.81, 4.97\}$ **37.** $\emptyset$ **39.** $\{57.7°, 159.2°\}$ **41.** $\{1.01, 2.78\}$ **43.** $\{.75, 2.39\}$ **45.** $\{1.38\}$ **47.** In the second line, dividing by $\sin x$ causes two solutions to be lost. Instead, factor in line 1, set each factor equal to 0, and then solve to get $x = 0$ or $x = \dfrac{\pi}{2}$.

**49.** about 83.6° **51.** 14°

### 9.6 Exercises *(page 728)*

**1.** $\dfrac{\pi}{3}, \pi, \dfrac{4\pi}{3}$ **3. (a)** $\left\{\dfrac{\pi}{12}, \dfrac{11\pi}{12}, \dfrac{13\pi}{12}, \dfrac{23\pi}{12}\right\}$ **(b)** $\left[0, \dfrac{\pi}{12}\right) \cup \left(\dfrac{11\pi}{12}, \dfrac{13\pi}{12}\right) \cup \left(\dfrac{23\pi}{12}, 2\pi\right)$ **5. (a)** $\left\{\dfrac{\pi}{2}, \dfrac{7\pi}{6}, \dfrac{11\pi}{6}\right\}$ **(b)** $\emptyset$

**7. (a)** $\left\{\dfrac{3\pi}{8}, \dfrac{5\pi}{8}, \dfrac{11\pi}{8}, \dfrac{13\pi}{8}\right\}$ **(b)** $\left[\dfrac{3\pi}{8}, \dfrac{5\pi}{8}\right] \cup \left[\dfrac{11\pi}{8}, \dfrac{13\pi}{8}\right]$ **9. (a)** $\left\{\dfrac{\pi}{2}, \dfrac{3\pi}{2}\right\}$ **(b)** $\left[0, \dfrac{\pi}{2}\right) \cup \left(\dfrac{3\pi}{2}, 2\pi\right)$ **11.** $\left\{\dfrac{\pi}{2}\right\}$

**13.** $\{\pi\}$ **15.** $\left\{\frac{\pi}{4}, \frac{\pi}{2}, \frac{5\pi}{4}, \frac{3\pi}{2}\right\}$ **17.** $\left\{\frac{\pi}{3}, \frac{\pi}{2}, \frac{3\pi}{2}, \frac{5\pi}{3}\right\}$ **19.** $\{15°, 45°, 135°, 165°, 255°, 285°\}$ **21.** $\{0°\}$
**23.** $\{120°, 240°\}$ **25.** $\{30°, 150°, 270°\}$ **27.** $\{11.8°, 78.2°, 191.8°, 258.2°\}$ **29.** $\{30°, 90°, 150°, 210°, 270°, 330°\}$
**31.** $\{.262, 1.309, 1.571, 3.403, 4.451, 4.712\}$ **33.** $\{1.047, 3.142, 5.236\}$ **35.** $\{.259, 1.372, 3.142, 4.911, 6.024\}$
**37.** $\frac{\tan 2\theta}{2} \neq \tan \theta$, because the 2 in $2\theta$ is not a factor of the entire numerator. **39.** **(a)** 2 sec **(b)** $\frac{10}{3}$ sec
**41.** **(a)** For $x = t$,
$P(t) = .004 \sin\left(2\pi(261.63)t + \frac{\pi}{7}\right)$

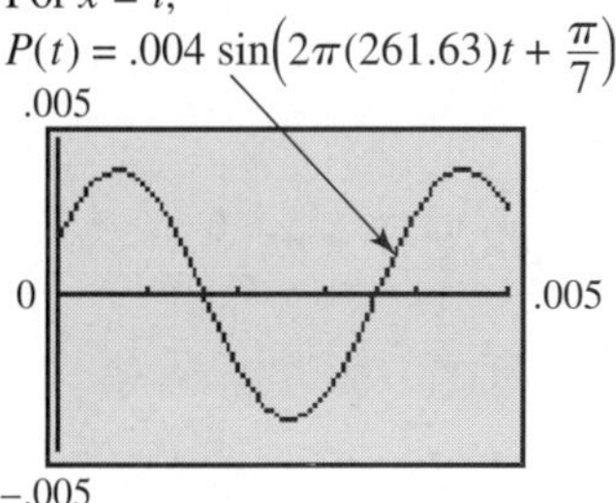


**(b)** .00164 and .00355 **(c)** $.00164 \leq t \leq .00355$ **(d)** outward

**43.** **(a)** 3 beats per sec

For $x = t$,
$P(t) = .005 \sin 440\pi t + .005 \sin 446\pi t$

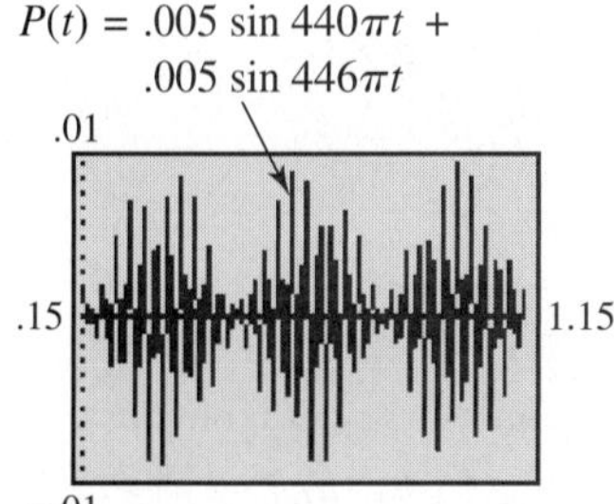


**(b)** 4 beats per sec

For $x = t$,
$P(t) = .005 \sin 440\pi t + .005 \sin 432\pi t$

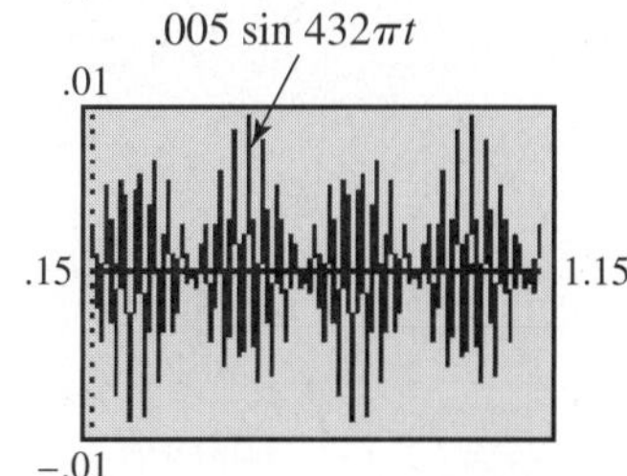


**(c)** The number of beats is equal to the absolute value of the difference in the frequencies of the two tones.

## Reviewing Basic Concepts *(page 730)*

**1.** $\left\{\frac{\pi}{12}, \frac{11\pi}{12}, \frac{13\pi}{12}, \frac{23\pi}{12}\right\}$ **2.** $\left\{\frac{5\pi}{6}, \frac{11\pi}{6}\right\}$ **3.** $\left\{0, \frac{\pi}{4}, \frac{5\pi}{4}\right\}$ **4.** $\left\{\frac{\pi}{6}, \frac{\pi}{2}, \frac{3\pi}{2}, \frac{11\pi}{6}\right\}$ **5.** $\{38.4°, 218.4°, 104.8°, 284.8°\}$
**6.** $\{78.0°, 282.0°\}$ **7.** $\{90°, 210°, 330°\}$ **8.** $\{90°, 270°\}$ **9.** $\{.68058878, 1.4158828\}$ **10.** $\{0, .37600772\}$

## Chapter 9 Review Exercises *(page 733)*

**1.** sine, tangent, cotangent, cosecant **3.** cos 3 **5.** $-\tan 3$ **7.** $-\csc 3$ **9.** B **11.** F **13.** D **15.** $\frac{\cos^2 \theta}{\sin \theta}$
**17.** $\frac{1 + \cos \theta}{\sin \theta}$ **19.** $\sec x = -\frac{\sqrt{41}}{4}$; $\cos x = -\frac{4\sqrt{41}}{41}$; $\cot x = -\frac{4}{5}$; $\sin x = \frac{5\sqrt{41}}{41}$; $\csc x = \frac{\sqrt{41}}{5}$ **21.** B **23.** A
**25.** C **27.** D **41.** $\frac{\pi}{4}$ **43.** $\frac{\pi}{6}$ **45.** $\frac{\pi}{3}$ **47.** 60° **49.** 0° **51.** 12.51631252° **53.** Cos $x$ is defined for every real number, but arccos $x$ is defined only on the interval $[-1, 1]$. Arccos(cos $x$) $= x$ only for $x$ in the interval $[0, \pi]$.
**55.** $\frac{\sqrt{7}}{4}$ **57.** $\frac{9}{7}$ **59.** $\frac{294 + 125\sqrt{6}}{92}$ **61.** $\sqrt{1 - u^2}$ **63.** $\left\{\frac{\pi}{2}, \frac{3\pi}{2}\right\}$ **65.** $\left\{.7297276562, \frac{\pi}{2}, 2.411864997\right\}$
**67.** $\left\{1.19028995, \frac{3\pi}{4}, 4.331882603, \frac{7\pi}{4}\right\}$ **69.** $\left\{\frac{\pi}{4}, \frac{\pi}{2}, \frac{5\pi}{4}, \frac{3\pi}{2}\right\}$ **71.** $\left\{0, \frac{\pi}{3}, \frac{\pi}{2}, \frac{2\pi}{3}, \pi, \frac{4\pi}{3}, \frac{3\pi}{2}, \frac{5\pi}{3}\right\}$
**73.** $\left\{\frac{\pi}{6}, \frac{2\pi}{3}, \frac{7\pi}{6}, \frac{5\pi}{3}\right\}$ **75.** **(a)** $\left(\frac{\pi}{4}, \frac{\pi}{2}\right) \cup \left(\frac{5\pi}{4}, \frac{3\pi}{2}\right)$ **(b)** $\left[0, \frac{\pi}{4}\right) \cup \left(\frac{\pi}{2}, \frac{5\pi}{4}\right) \cup \left(\frac{3\pi}{2}, 2\pi\right)$
**77.** **(a)** $t = \frac{3 \sin^{-1} 3y}{4\pi}$ **(b)** .27 sec **79.** .0007 sec **81.** If $\theta_1 > 48.8°$, then $\theta_2 > 90°$ and the light beam will stay completely

under water. **82.** $\tan\theta = \frac{b}{a}$; $m = \frac{b}{a}$; $m = \tan\theta$ **83.** $\beta = \theta + \alpha$ **84.** $\theta = \beta - \alpha$ **85.** $\tan\theta = \tan(\beta - \alpha)$; $\tan\theta = \frac{\tan\beta - \tan\alpha}{1 + \tan\beta\tan\alpha}$ **86.** $\tan\theta = \frac{m_2 - m_1}{1 + m_1 m_2}$

### Chapter 9 Test *(page 736)*

**1. (a)** $\frac{2 - 2\sqrt{30}}{15}$ **(b)** $\frac{\sqrt{5} - 4\sqrt{6}}{15}$ **(c)** $\frac{-3 - \sqrt{5}}{2}$ **(d)** $-\frac{23}{25}$ **2.** $-1$ **3.** $\sec x - \sin x\tan x = \cos x$
**5. (a)** $-\sin\theta$ **(b)** $-\sin\theta$

**6.** **(a)** domain: $[0, 2]$; range: $\left[-\frac{\pi}{2}, \frac{\pi}{2}\right]$
**(b)** $x$-intercept: 1; $y$-intercept: $-\frac{\pi}{2}$
**(c)** No; The domain of $y = \sin^{-1} x$ is $[-1, 1]$. 2 is not in this interval.

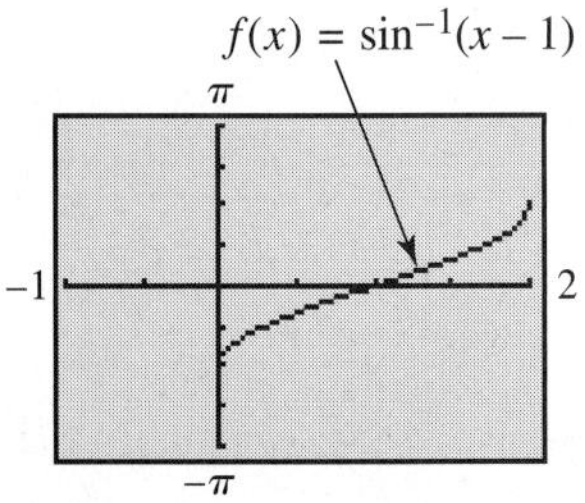

**7. (a)** $\frac{2\pi}{3}$ **(b)** 0 **(c)** $\frac{\pi}{3}$ **(d)** $\frac{\sqrt{5}}{3}$ **(e)** $\frac{4\sqrt{2}}{9}$ **8.** $\frac{u\sqrt{1 - u^2}}{1 - u^2}$ **9. (a)** $\left\{\frac{\pi}{2}, \frac{3\pi}{2}\right\}$ **(b)** $\{18.4°, 135°, 198.4°, 315°\}$
**(c)** $\left\{0, \frac{2\pi}{3}, \frac{4\pi}{3}\right\}$ **(d)** $\{120°, 240°\}$

**10. (a)** domain: $[0, 11]$; range: $[0, 100]$ (in hundreds) **(b)** 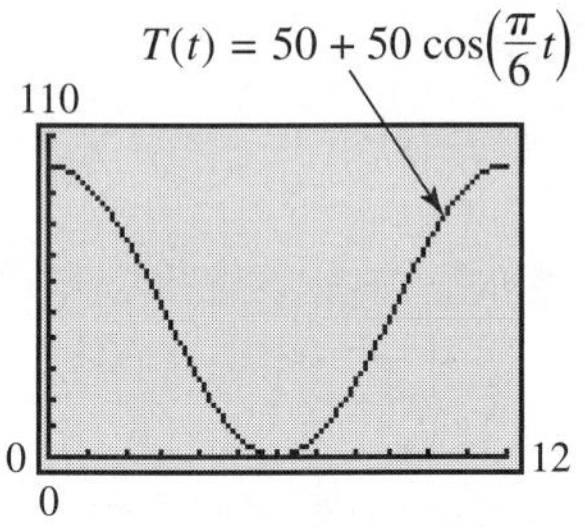
**(c)** The maximum of 10,000 animals occurs at 0 months (July). The minimum of 0 animals occurs at 6 months (January).
**(d)** There are 5000 animals at 3 months (October) and at 9 months (April).
**(e)** at 2 months (September) and 10 months (May)
**(f)** $t = \frac{6}{\pi}\arccos\left(\frac{T - 50}{50}\right)$

## CHAPTER 10 APPLICATIONS OF TRIGONOMETRY; VECTORS

### 10.1 Exercises *(page 748)*

**1.** C **3.** no **5.** no **7.** yes **9.** yes **11.** $C = 95°$, $b \approx 13$ m, $a \approx 11$ m **13.** $B = 37.3°$, $a \approx 38.5$ ft, $b \approx 51.0$ ft **15.** $B = 48°$, $a \approx 11$ m, $b \approx 13$ m **17.** $A = 36.54°$, $a \approx 28.10$ m, $b \approx 44.17$ m **19.** $A = 49°\ 40'$, $b \approx 16.1$ cm, $c \approx 25.8$ cm **21.** $C = 91.9°$, $a \approx 490$ ft, $c \approx 847$ ft **23.** $B_1 \approx 49.1°$, $C_1 \approx 101.2°$, $B_2 \approx 130.9°$, $C_2 \approx 19.4°$ **25.** no such triangle **27.** $B \approx 27.19°$, $C \approx 10.68°$ **29.** $B \approx 20.6°$, $C \approx 116.9°$, $c \approx 20.6$ ft **31.** no such triangle **33.** $B_1 \approx 49°\ 20'$, $C_1 \approx 92°\ 00'$, $c_1 \approx 15.5$ km; $B_2 \approx 130°\ 40'$, $C_2 \approx 10°\ 40'$, $c_2 \approx 2.88$ km **35.** $A_1 \approx 52°\ 10'$, $C_1 \approx 95°\ 00'$, $c_1 \approx 9520$ cm; $A_2 \approx 127°\ 50'$, $C_2 \approx 19°\ 20'$, $c_2 \approx 3160$ cm **37.** The Pythagorean theorem only applies to right triangles. **39.** If we are given only three sides, then any equation from the law of sines will contain two unknowns. At least one angle measure must be given. **41.** 118 m **43.** 1.93 mi **45.** 10.4 in. **47.** 111° **49.** 5.1 mi, 7.2 mi **51.** 26.5 km **53.** 2.18 km **55.** 38.3 cm **57. (a)** $4 < a < 5$ **(b)** $a = 4$ or $a \geq 5$ **(c)** $a < 4$ **59.** $\sin C = 1$; $C = 90°$; right triangle **61.** The longest side must be opposite the largest angle. Thus, $B$ must be larger than $A$, which is impossible because $A$ and $B$ cannot both be obtuse. **63.** The property could not exist. **65.** increasing **66.** If $B < A$, then $\sin B < \sin A$ because $y = \sin x$ is increasing on $\left[0, \frac{\pi}{2}\right]$. **67.** $b = \frac{a\sin B}{\sin A}$ **68.** $b = \frac{a\sin B}{\sin A} = a \cdot \frac{\sin B}{\sin A}$. Since $\frac{\sin B}{\sin A} < 1$, $b = a \cdot \frac{\sin B}{\sin A} < a \cdot 1 = a$, so $b < a$. **69.** If $B < A$ then $b < a$, but $b > a$ in triangle $ABC$. **71.** about 10,285 ft

## 10.2 Exercises *(page 759)*

**1.** **(a)** SAS **(b)** law of cosines **3.** **(a)** SSA **(b)** law of sines **5.** **(a)** ASA **(b)** law of sines **7.** **(a)** ASA **(b)** law of sines **9.** $a \approx 5.4, B \approx 40.7°, C \approx 78.3°$ **11.** $A \approx 22.3°, B \approx 108.2°, C \approx 49.5°$ **13.** $A \approx 33.6°, B \approx 50.7°, C \approx 95.7°$ **15.** $c \approx 2.83$ in., $A \approx 44.9°, B \approx 106.8°$ **17.** $c \approx 6.46$ m, $A \approx 53.1°, B \approx 81.3°$ **19.** $A \approx 82°, B \approx 37°, C \approx 61°$ **21.** $C \approx 102° 10', B \approx 35° 50', A \approx 42° 00'$ **23.** $C \approx 84° 30', B \approx 44° 40', A \approx 50° 50'$ **25.** $a \approx 156$ cm, $B \approx 64° 50', C \approx 34° 30'$ **27.** $b \approx 9.529$ in., $A \approx 64.59°, C \approx 40.61°$ **29.** $a \approx 15.7$ m, $B \approx 21.6°, C \approx 45.6°$ **31.** $A \approx 30°, B \approx 56°, C \approx 94°$ **33.** The value of $\cos \theta$ will be greater than 1; your calculator will give you an error message (or a complex number) when using the inverse cosine function. **35.** $AB \approx 257$ m **37.** 281 km **39.** 10.9 mi **41.** 115 km **43.** 18 ft **45.** about 5500 m **47.** 438.14 ft **49.** 350° **51.** 2000 km **53.** 163.5° **55.** 22 ft **57.** Since $A$ is obtuse, $90° < A < 180°$. The cosine of a quadrant II angle is negative. **58.** In $a^2 = b^2 + c^2 - 2bc \cos A$, $\cos A$ is negative, so $a^2 = b^2 + c^2 +$ (a positive quantity). Thus, $a^2 > b^2 + c^2$. **59.** $b^2 + c^2 > b^2$ and $b^2 + c^2 > c^2$. If $a^2 > b^2 + c^2$, then $a^2 > b^2$ and $a^2 > c^2$ from which $a > b$ and $a > c$ because $a$, $b$, and $c$ are nonnegative. **60.** Because $A$ is obtuse, it is the largest angle, so the longest side should be $a$, not $c$. **61.** 46.4 $m^2$ **63.** 78 $m^2$ **65.** 3650 $ft^2$ **67.** 228 $yd^2$ **69.** 100 $m^2$ **71.** 33 cans **73.** The area and perimeter are both 36.

**75.** y, P(2, 5), Q(−1, 3), R(4, 0), 0, x

**76.** $PQ$: $\sqrt{13}$; $QR$: $\sqrt{34}$; $PR$: $\sqrt{29}$ **77.** $\dfrac{\sqrt{13} + \sqrt{34} + \sqrt{29}}{2}$ **78.** 9.5 ($units^2$) **79.** 9.5 ($units^2$); yes **80.** 64.65° (approximately) **81.** 9.5 ($units^2$); yes **82.** Answers will vary. One answer might be that computing the area of a triangle by using different methods must lead to the same answer, because the area is unique.

## 10.3 Exercises *(page 771)*

**1.** **m** and **p**; **n** and **r** **3.** **m** and **p** equal 2**t**, or **t** is one half **m** or **p**; also **m** = 1**p** and **n** = 1**r** **5.**

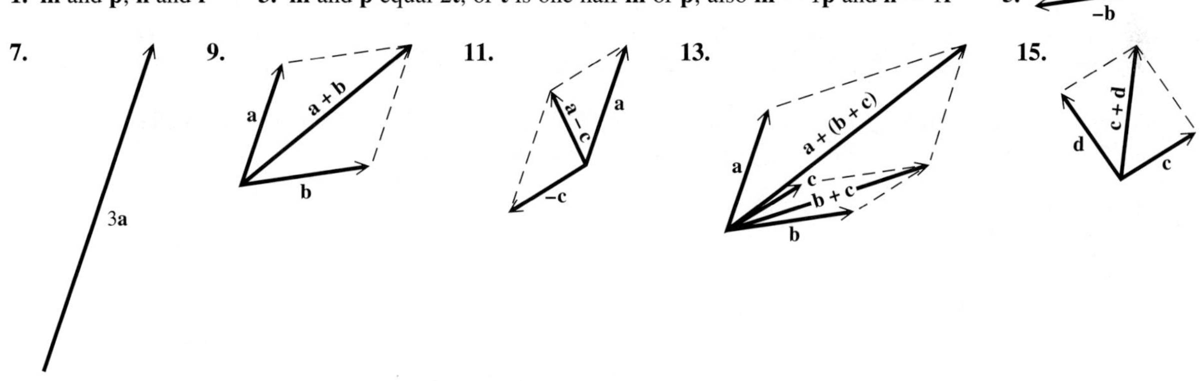


**17.** **(a)** $\langle -4, 16\rangle$ **(b)** $\langle -12, 0\rangle$ **(c)** $\langle 8, -8\rangle$ **19.** **(a)** $\langle 8, 0\rangle$ **(b)** $\langle 0, 16\rangle$ **(c)** $\langle -4, -8\rangle$ **21.** **(a)** $\langle 0, 12\rangle$ **(b)** $\langle -16, -4\rangle$ **(c)** $\langle 8, -4\rangle$ **23.** **(a)** 4**i** **(b)** 7**i** + 3**j** **(c)** −5**i** + **j** **25.** **(a)** $\langle -2, 4\rangle$ **(b)** $\langle 7, 4\rangle$ **(c)** $\langle 6, -6\rangle$ **27.** $\langle 2, 8\rangle$ **29.** $\langle 6, -2\rangle$ **31.** $\langle -20, -15\rangle$ **33.** horizontal: $4 \cos 40° \approx 3.06$; vertical: $4 \sin 40° \approx 2.57$ **35.** horizontal: $5 \cos(-35°) \approx 4.10$; vertical: $5 \sin(-35°) \approx -2.87$ **37.** 94.2 lb **39.** 24.4 lb **41.** −5**i** + 8**j** **43.** 2**i** **45.** $4\sqrt{2}$**i** + $4\sqrt{2}$**j** **47.** −.254**i** + .544**j** **49.** $\sqrt{2}$; 45° **51.** 16; 315° **53.** 17; 331.9° **55.** 6; 180° **57.** 7 **59.** −3 **61.** 20 **63.** 135° **65.** 90° **67.** $\cos^{-1} \dfrac{4}{5} \approx 36.87°$ **69.** −6 **71.** −24 **73.** orthogonal **75.** not orthogonal **77.** not orthogonal **79.** 173.1° **81.** 39.2 km **83.** bearing: 65° 30′; groundspeed: 181 mph **85.** bearing: 187°; groundspeed: 198 mph **87.** 226 lb **89.** 1600 lb **91.** 190 lb and 283 lb, respectively **93.** magnitude: 9.52082827; direction angle: 119.0646784° **94.** $\langle -4.10424172, 11.27631145\rangle$ **95.** $\langle -.520944533, -2.954423259\rangle$

**96.** $\langle -4.625186253, 8.321888191\rangle$ **97.** magnitude: 9.52082827; direction angle: 119.0646784° **98.** They are the same. Preference of method is an individual choice. **99.** **(a)** 10 mph **(b)** $3\mathbf{v} = 18\mathbf{i} + 24\mathbf{j}$; this represents a 30 mph wind in the direction of **v**. **(c)** **u** represents a southeast wind of $\sqrt{128} \approx 11.3$ mph.

## Reviewing Basic Concepts *(page 775)*

**1.** $B = 74°, b \approx 16.6, c \approx 15.3$ **2.** two solutions: $B_1 \approx 45.0°, C_1 \approx 103°, c_1 \approx 11.0$ or $B_2 \approx 135.0°, C_2 \approx 13°, c_2 \approx 2.5$ **3.** one solution: $A \approx 22.5°, B \approx 116.5°, b \approx 16.4$ **4.** **(a)** $b \approx 7.1, A \approx 63.0°, C \approx 66.0°$ **(b)** $A \approx 110.7°, B \approx 37.0°, C \approx 32.3°$ **5.** 9.6 **6.** 21 **7.** **(a)** **i** **(b)** $4\mathbf{i} - 2\mathbf{j}$ **(c)** $11\mathbf{i} - 7\mathbf{j}$ **8.** $-9$; 142.1° **9.** 207 lb **10.** 7200 ft

## 10.4 Exercises *(page 782)*

**1.** magnitude (length) **3.**

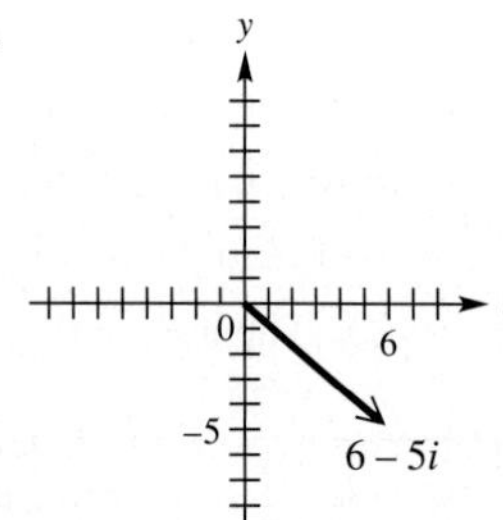


**5.** **7.**

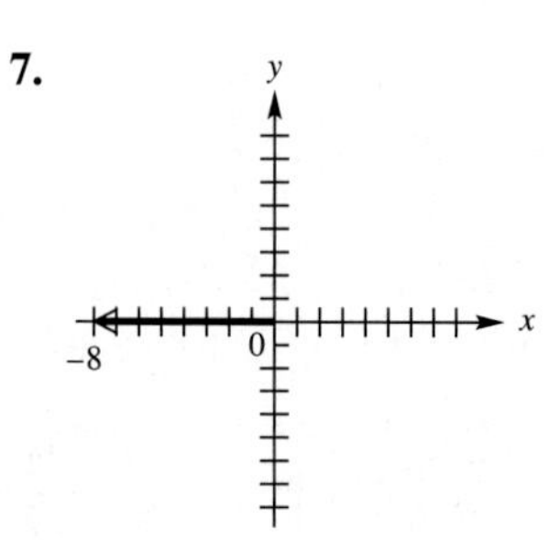


**9.** $1 - 4i$ **11.** The imaginary part must be 0. **13.** 0 **15.** $3 - i$ **17.** $-3 + 3i$ **19.** $2 + 4i$ **21.** $\sqrt{2}$ **23.** 13 **25.** 6 **27.** $\sqrt{13}$ **29.** $\sqrt{2} + i\sqrt{2}$ **31.** $10i$ **33.** $-2 - 2i\sqrt{3}$ **35.** $\frac{\sqrt{3}}{2} + \frac{1}{2}i$ **37.** $\frac{5\sqrt{3}}{2} - \frac{5}{2}i$ **39.** $-\sqrt{2}$ **41.** $3\sqrt{2}[\cos(-45°) + i\sin(-45°)]$ **43.** $2(\cos 60° + i\sin 60°)$ **45.** $2[\cos(-90°) + i\sin(-90°)]$ **47.** $8\left(\cos\frac{\pi}{6} + i\sin\frac{\pi}{6}\right)$ **49.** $2\left(\cos\frac{3\pi}{4} + i\sin\frac{3\pi}{4}\right)$ **51.** $4(\cos\pi + i\sin\pi)$ **53.** yes **55.** 5.830951895 cis 1.030376827 **57.** $-3\sqrt{3} + 3i$ **59.** $-4i$ **61.** $-\frac{15\sqrt{2}}{2} + \frac{15\sqrt{2}}{2}i$ **63.** $-3i$ **65.** $\sqrt{3} - i$ **67.** $-1 - i\sqrt{3}$ **69.** $-\frac{1}{6} - \frac{\sqrt{3}}{6}i$ **71.** 2 **72.** $w = \sqrt{2}$ cis 135°; $z = \sqrt{2}$ cis 225° **73.** 2 cis 0° **74.** 2; It is the same. **75.** $-i$ **76.** cis$(-90°)$ **77.** $-i$; It is the same. **79.** To square a complex number in trigonometric form, square the modulus $r$ and double the argument $\theta$. **81.** $1.7 + 2.8i$ **83.** $27.43 + 11.5i$

## 10.5 Exercises *(page 789)*

**1.** $27i$ **3.** 1 **5.** $\frac{27}{2} - \frac{27\sqrt{3}}{2}i$ **7.** $-16\sqrt{3} + 16i$ **9.** $-128 + 128i\sqrt{3}$ **11.** $128 + 128i$

**13.** $\cos 0° + i\sin 0°$, $\cos 120° + i\sin 120°$, $\cos 240° + i\sin 240°$

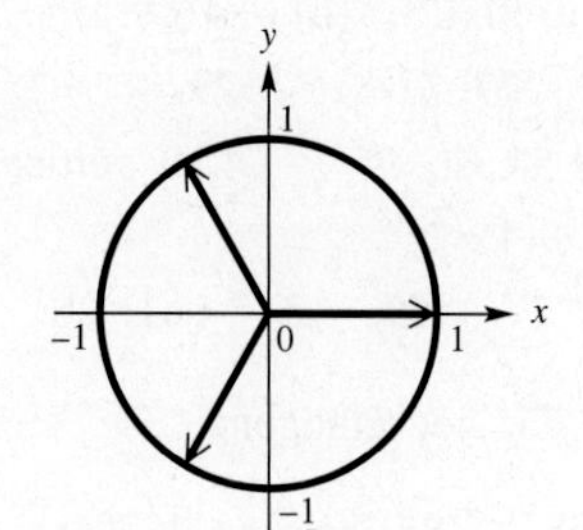


**15.** 2 cis 20°, 2 cis 140°, 2 cis 260°

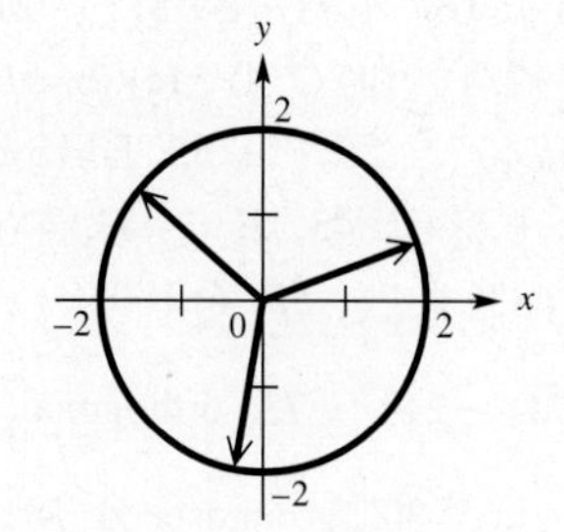


**17.** $2(\cos 90° + i\sin 90°)$, $2(\cos 210° + i\sin 210°)$, $2(\cos 330° + i\sin 330°)$

**19.** $4(\cos 60° + i\sin 60°)$, $4(\cos 180° + i\sin 180°)$, $4(\cos 300° + i\sin 300°)$

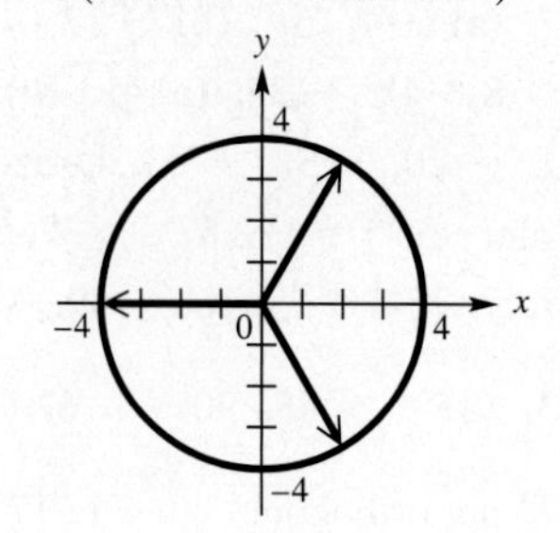

**21.** $\sqrt[3]{2}(\cos 20° + i \sin 20°)$, $\sqrt[3]{2}(\cos 140° + i \sin 140°)$, $\sqrt[3]{2}(\cos 260° + i \sin 260°)$

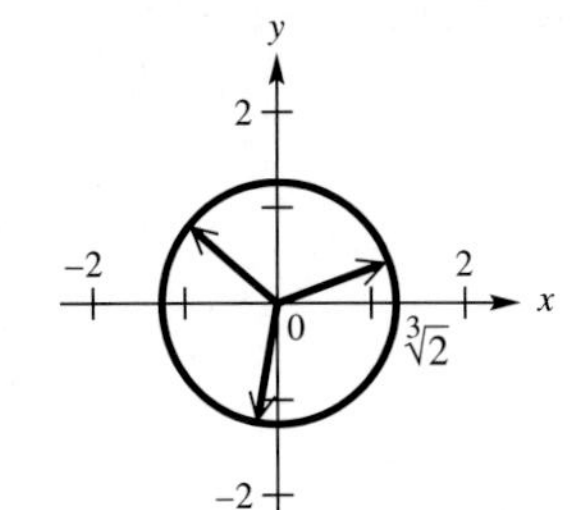


**23.** $\sqrt[3]{4}(\cos 50° + i \sin 50°)$, $\sqrt[3]{4}(\cos 170° + i \sin 170°)$, $\sqrt[3]{4}(\cos 290° + i \sin 290°)$

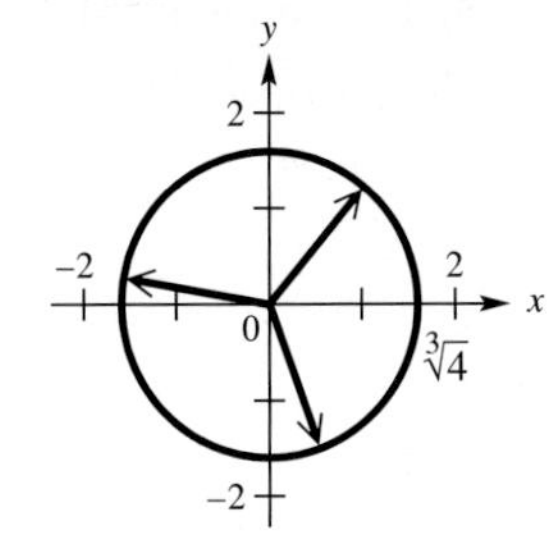


**25.** $1 + i\sqrt{3}, -1 - i\sqrt{3}$ **27.** $-1, \frac{1}{2} + \frac{\sqrt{3}}{2}i, \frac{1}{2} - \frac{\sqrt{3}}{2}i$

**29.** $\frac{\sqrt{2}}{2} + \frac{\sqrt{2}}{2}i, -\frac{\sqrt{2}}{2} - \frac{\sqrt{2}}{2}i$

**31.** $-4i, 2\sqrt{3} + 2i, -2\sqrt{3} + 2i$ **33.** $\pm 3, \pm 3i$

**35. (a)**

$\cos 0° + i \sin 0°$,
$\cos 90° + i \sin 90°$,
$\cos 180° + i \sin 180°$,
$\cos 270° + i \sin 270°$

**(b)**

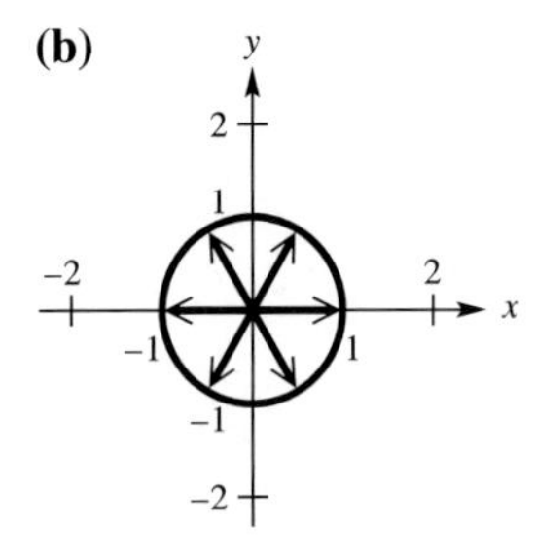


$\cos 0° + i \sin 0°$,
$\cos 60° + i \sin 60°$,
$\cos 120° + i \sin 120°$,
$\cos 180° + i \sin 180°$,
$\cos 240° + i \sin 240°$,
$\cos 300° + i \sin 300°$

**37.** The argument for a positive real number is $\theta = 0°$. By the $n$th root theorem with $k = 0$, $\alpha = 0°$. Thus, one $n$th root has an argument of 0° and it must be real. **39.** $(x + 2)(x^2 - 2x + 4)$ **40.** $x + 2 = 0$ implies $x = -2$. **41.** $x^2 - 2x + 4 = 0$ implies $x = 1 + i\sqrt{3}$ or $x = 1 - i\sqrt{3}$. **42.** 2 cis 60°, 2 cis 180°, 2 cis 300° **43.** $1 + i\sqrt{3}, -2, 1 - i\sqrt{3}$ **44.** They are the same.

**45.** {$\cos 45° + i \sin 45°$, $\cos 135° + i \sin 135°$, $\cos 225° + i \sin 225°$, $\cos 315° + i \sin 315°$} **47.** {$\cos 18° + i \sin 18°$, $\cos 90° + i \sin 90°$, $\cos 162° + i \sin 162°$, $\cos 234° + i \sin 234°$, $\cos 306° + i \sin 306°$} **49.** {$\cos 60° + i \sin 60°$, $\cos 180° + i \sin 180°$, $\cos 300° + i \sin 300°$} **51.** {$2(\cos 0° + i \sin 0°)$, $2(\cos 120° + i \sin 120°)$, $2(\cos 240° + i \sin 240°)$}

**53. (a)** blue **(b)** red **(c)** yellow **55.** $1, .80901699 + .58778525i, .30901699 + .95105652i$

## Reviewing Basic Concepts *(page 790)*

**1.** $1 + i\sqrt{3}$ **2.** 5 **3.** $2(\cos 135° + i \sin 135°)$ **4.** $z_1z_2 = 8(\cos 180° + i \sin 180°) = -8$; $\frac{z_1}{z_2} = 2(\cos 90° + i \sin 90°) = 2i$ **5.** 64 cis 51° **6.** $-4, 2 + 2i\sqrt{3}, 2 - 2i\sqrt{3}$ **7.** $1 + i, -1 - i$

**8.** {cis 60°, cis 180°, cis 300°}

## 10.6 Exercises *(page 798)*

**1.** **(a)** II **(b)** I **(c)** IV **(d)** III

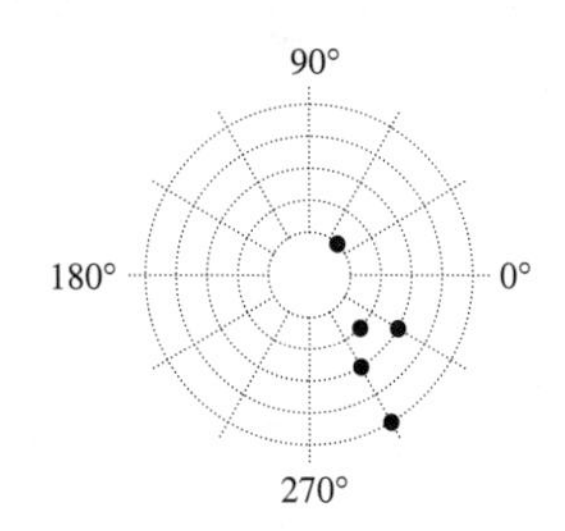


Answers may vary in Exercises 3–11.

**3.** $(1, 405°), (-1, 225°)$ **5.** $(-2, 495°), (2, 315°)$ **7.** $(5, 300°), (-5, 120°)$ **9.** $(-3, 150°), (3, -30°)$ **11.** $(3, 660°), (-3, 120°)$

Graphs for Exercises 3, 5, 7, 9, 11

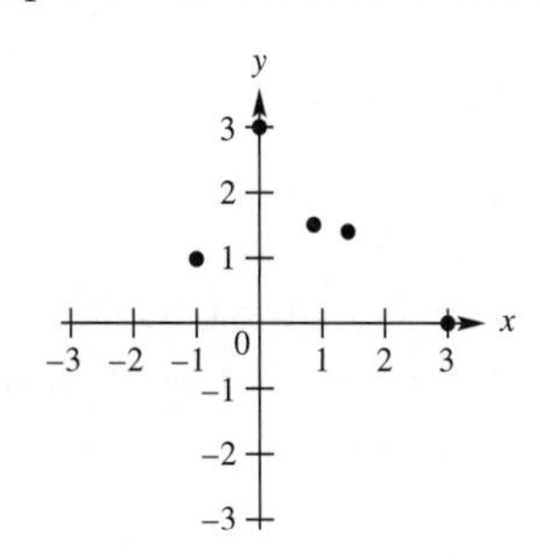


Answers may vary in Exercises 13–21.

**13.** $\left(\sqrt{2}, 135°\right), \left(-\sqrt{2}, 315°\right)$ **15.** $(3, 90°), (-3, 270°)$ **17.** $(2, 45°), (-2, 225°)$ **19.** $\left(\sqrt{3}, 60°\right), \left(-\sqrt{3}, 240°\right)$ **21.** $(3, 0°), (-3, 180°)$

Graphs for Exercises 13, 15, 17, 19, 21

**23.** cardioid

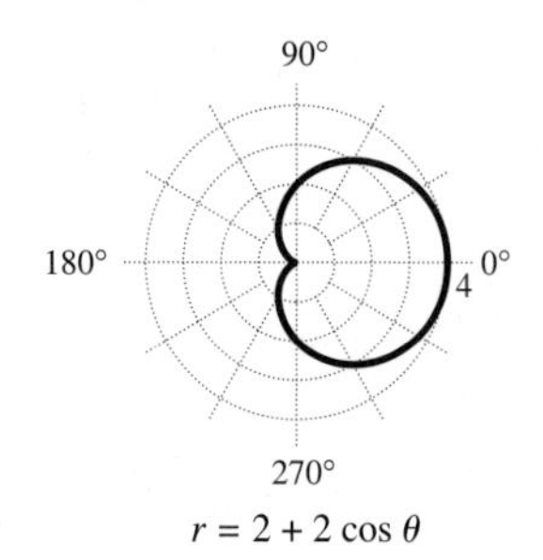


$r = 2 + 2\cos\theta$

**25.** limaçon with a loop

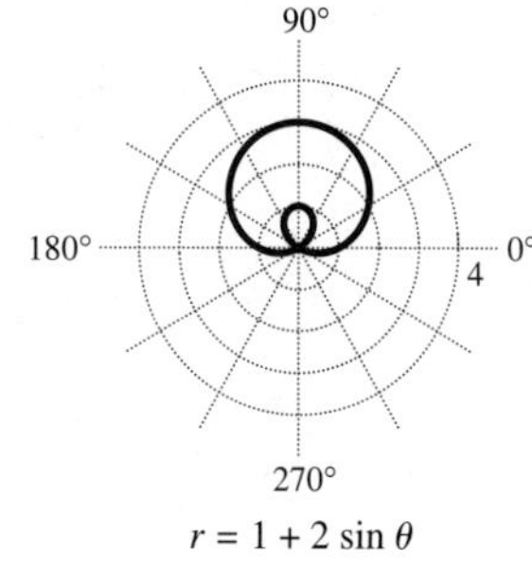


$r = 1 + 2\sin\theta$

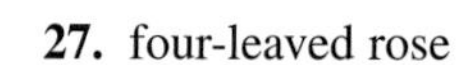

**27.** four-leaved rose

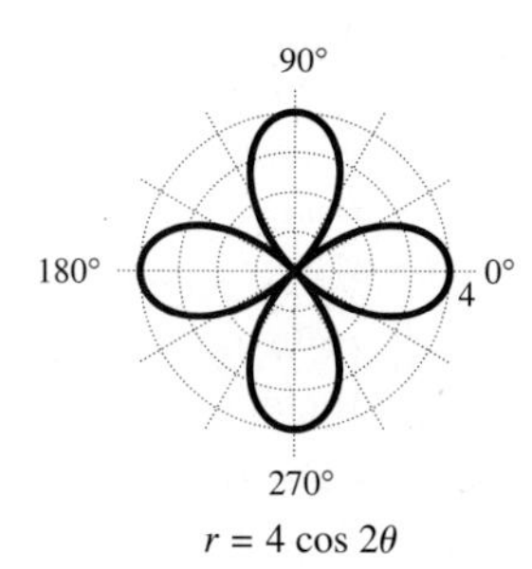


$r = 4\cos 2\theta$

**29.** lemniscate

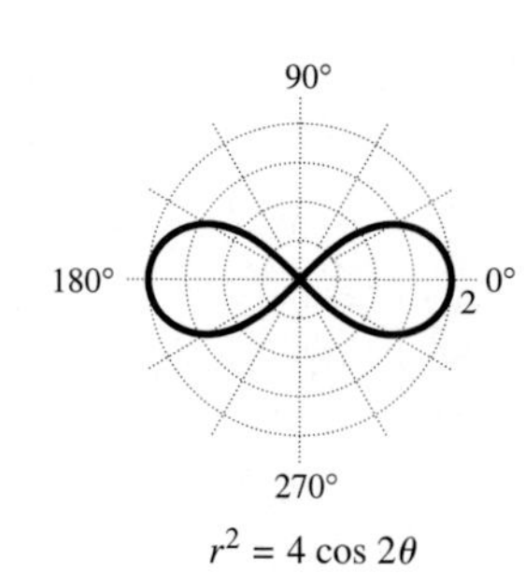


$r^2 = 4\cos 2\theta$

**31.** cardioid

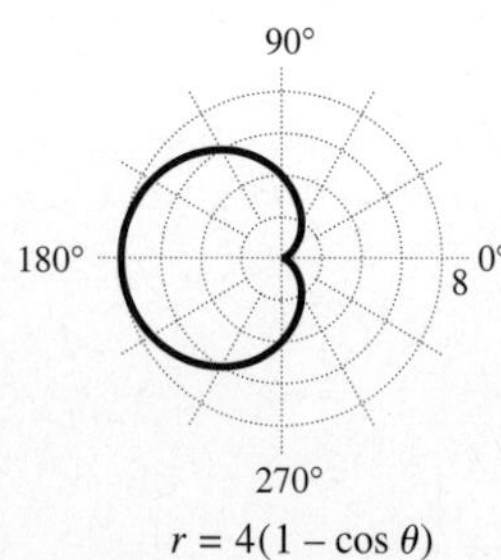


$r = 4(1 - \cos\theta)$

**33.**

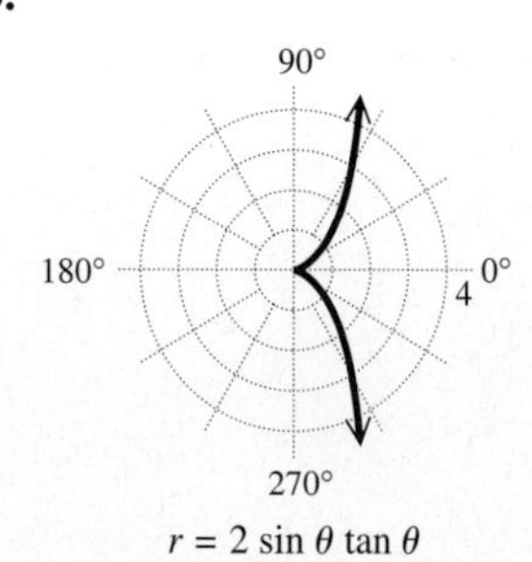


$r = 2\sin\theta\tan\theta$

**35.** **(a)** $(r, -\theta)$ **(b)** $(r, \pi - \theta)$ or $(-r, -\theta)$ **(c)** $(r, \pi + \theta)$ or $(-r, \theta)$ **36.** $-\theta$ **37.** $\pi - \theta$ **38.** $-r; -\theta$ **39.** $-r$ **40.** $\pi + \theta$ **41.** the polar axis

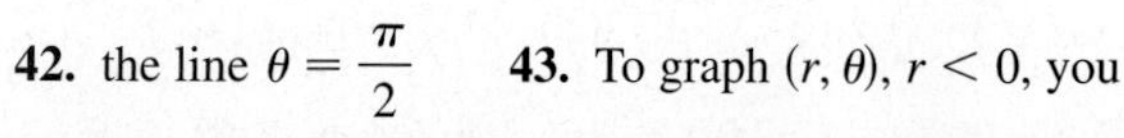

**42.** the line $\theta = \frac{\pi}{2}$ **43.** To graph $(r, \theta)$, $r < 0$, you could locate $\theta$, add 180° to it, and move $|r|$ units along the terminal ray of $\theta + 180°$ in standard position.

**45.** $\theta$ must be coterminal with 0°, 90°, 180°, or 270°; quadrantal **47.** reflected across the line $\theta = \frac{\pi}{2}$ ($y$-axis)

**49.** The value of $a$ determines the length of the petals. The value of $n$ determines the number of petals—$n$ petals if $n$ is odd and $2n$ petals if $n$ is even.

**51.** $x^2 + (y - 1)^2 = 1$

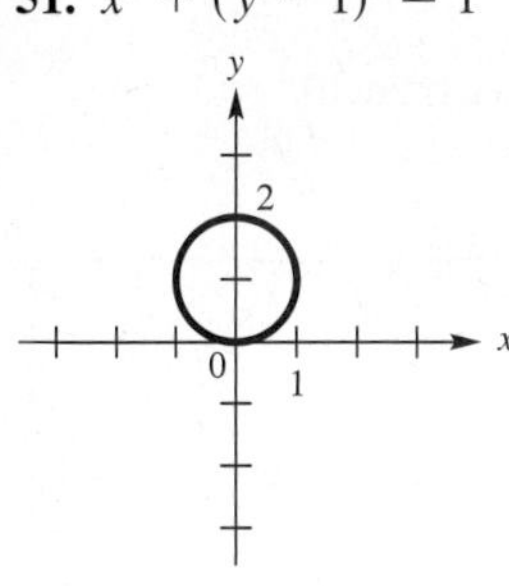


$r = 2 \sin \theta$
$x^2 + (y - 1)^2 = 1$

**53.** $y^2 = 4(x + 1)$

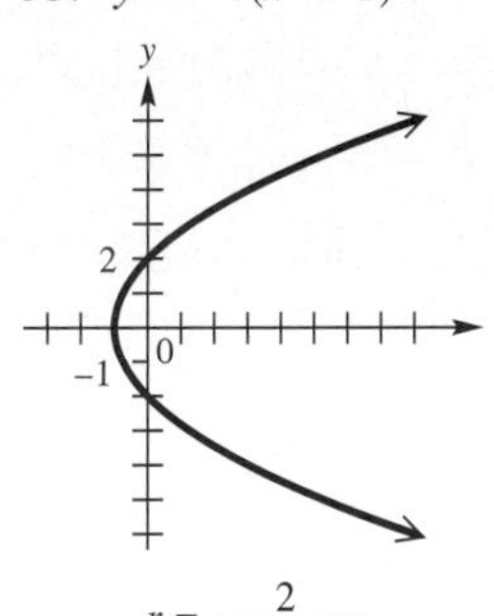


$r = \dfrac{2}{1 - \cos \theta}$
$y^2 = 4(x + 1)$

**55.** $(x + 1)^2 + (y + 1)^2 = 2$

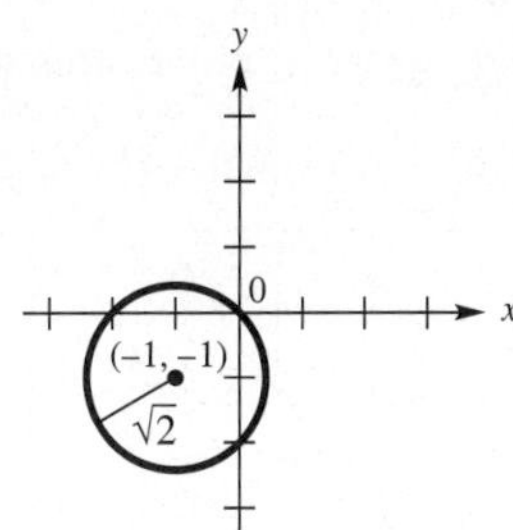


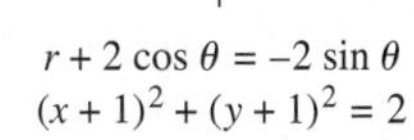

$r + 2 \cos \theta = -2 \sin \theta$
$(x + 1)^2 + (y + 1)^2 = 2$

**57.** $x = 2$

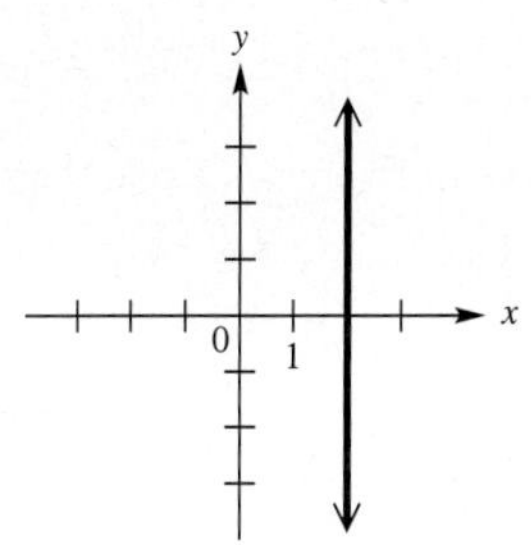


$r = 2 \sec \theta$
$x = 2$

**59.** $x + y = 2$

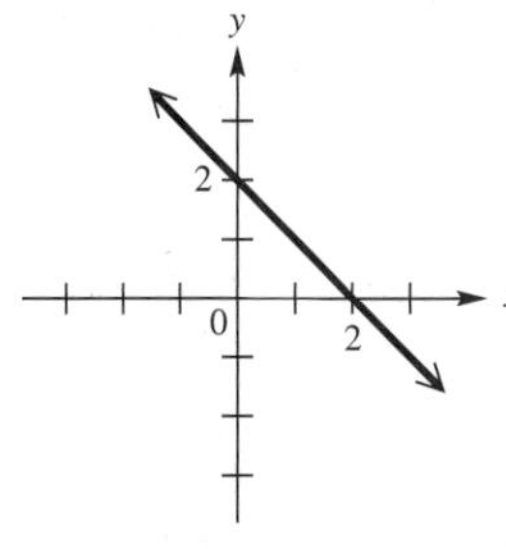


$r(\cos \theta + \sin \theta) = 2$
$x + y = 2$

**61.** $r = \dfrac{4}{\cos \theta + \sin \theta}$ **63.** $r = 4$ **65.** $r = 2 \csc \theta$ or $r = \dfrac{2}{\sin \theta}$

**67.**

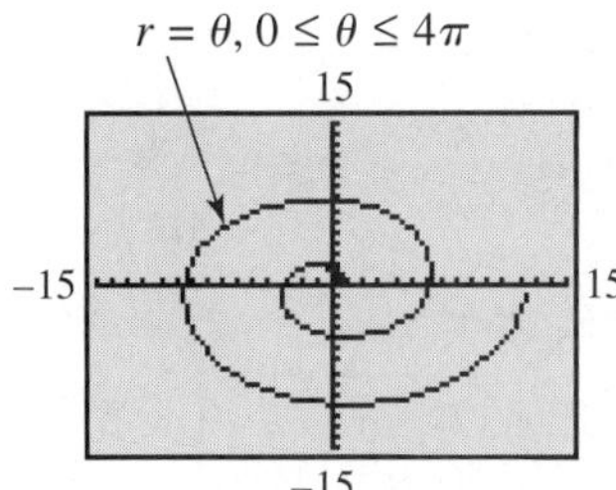


**69.**

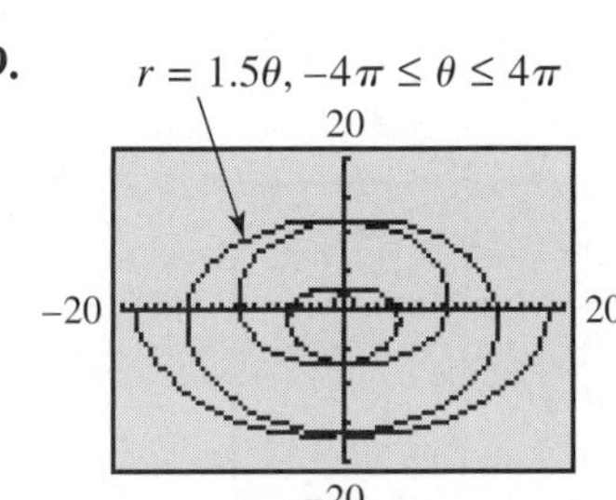


**71.** Answers will vary. **73.** $\left(2, \dfrac{\pi}{6}\right), \left(2, \dfrac{5\pi}{6}\right)$

**75.** $\left(\dfrac{4 + \sqrt{2}}{2}, \dfrac{\pi}{4}\right), \left(\dfrac{4 - \sqrt{2}}{2}, \dfrac{5\pi}{4}\right)$

**77.** **(a)**

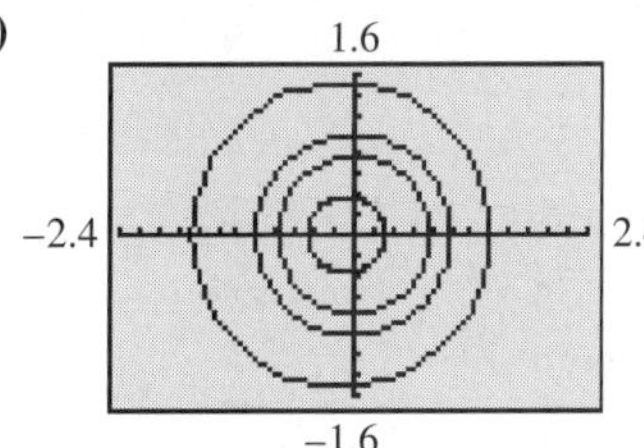


**(b)**

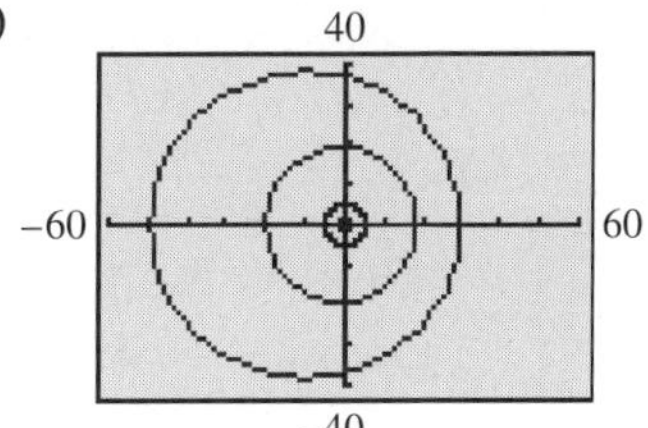


Earth is closest to the sun. **(c)** no

**79.** The radio signal is received inside the "figure eight." This region is generally in a southwest–northeast direction from the two towers with maximum distance 150 mi.

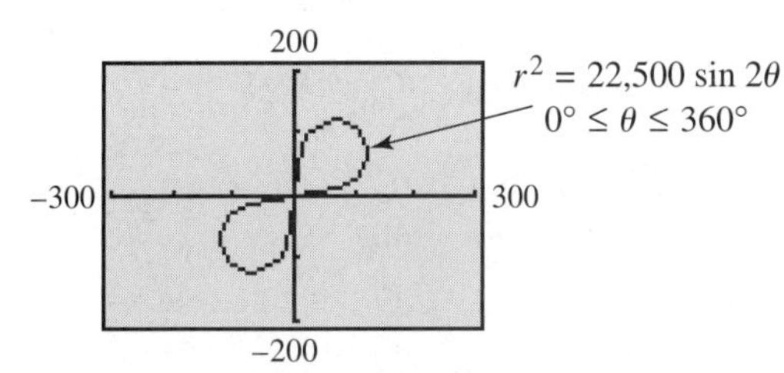

## 10.7 Exercises *(page 806)*

**1.** $x^2 + y^2 = 9$; circle with radius 3 **3.** $y = 2 - x$; line segment **5.** $\frac{y^2}{4} - \frac{x^2}{9} = 1$; hyperbola (upper branch)

Windows in Exercises 7 and 9 are $[-4.7, 4.7]$ by $[-3.1, 3.1]$.

**7.** **(a)** traces a circle of radius 3 once

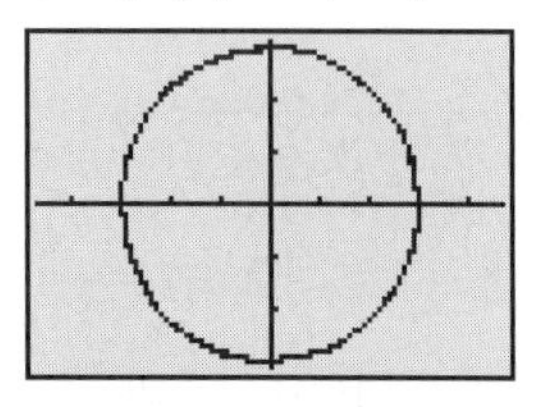

**(b)** traces a circle of radius 3 twice

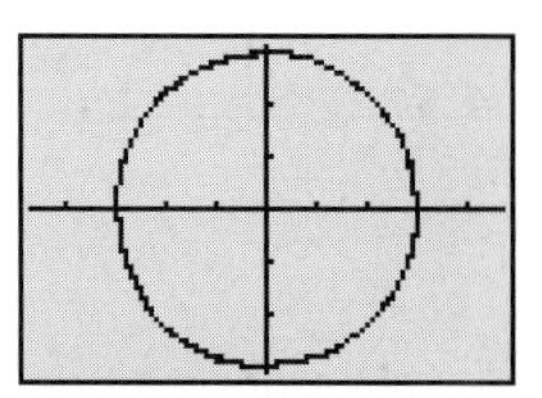

**9.** **(a)** traces a circle of radius 3 once counterclockwise, starting at $(3, 0)$

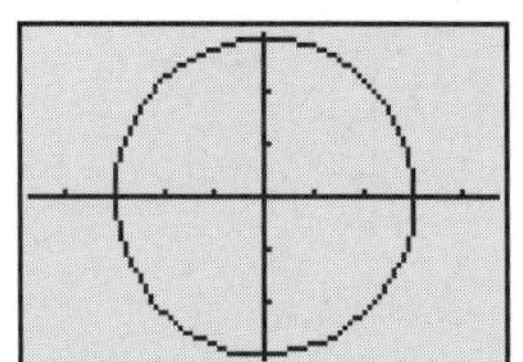

**(b)** traces a circle of radius 3 once clockwise, starting at $(0, 3)$

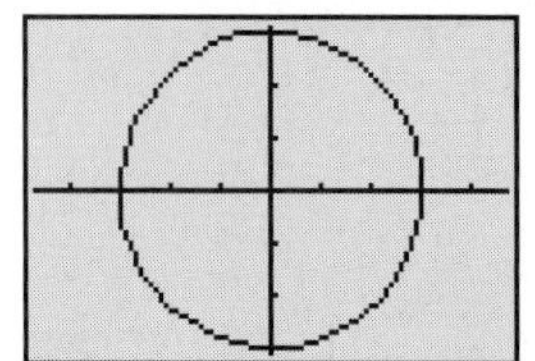

**11.**

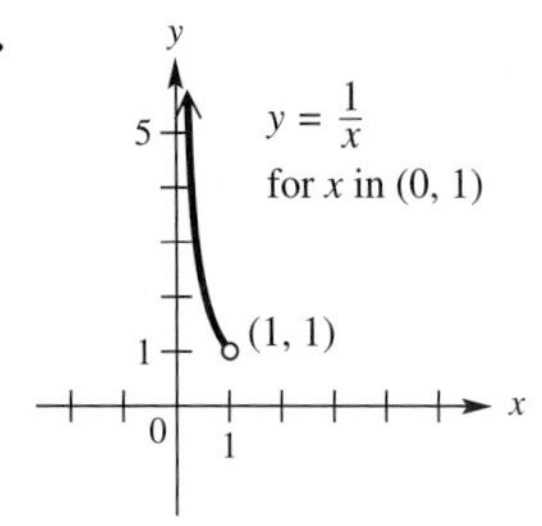


**13.**

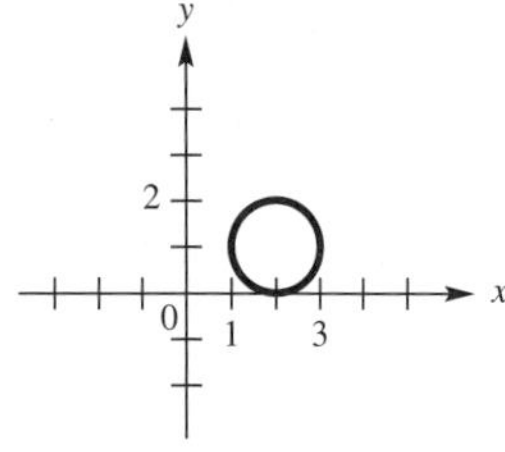


**15.**

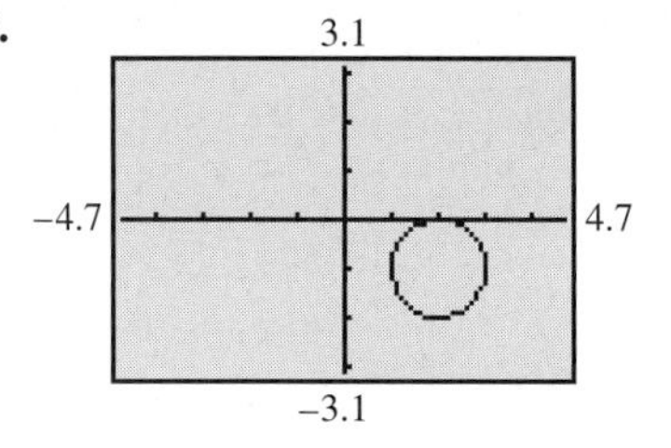


**17.**

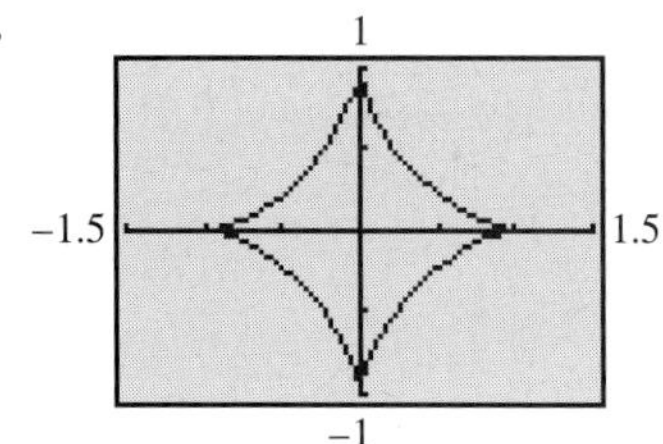


**19.**

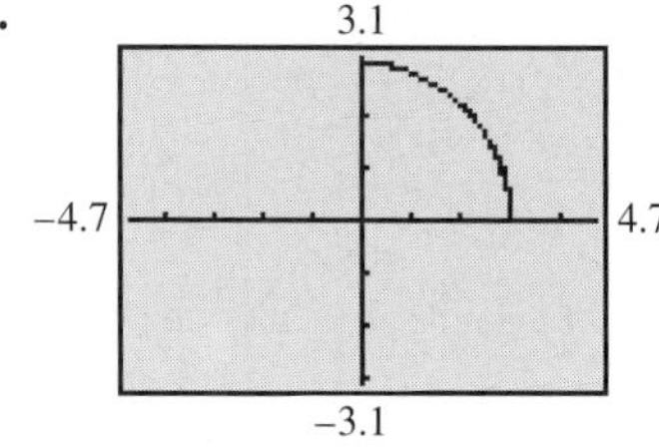


**21.**

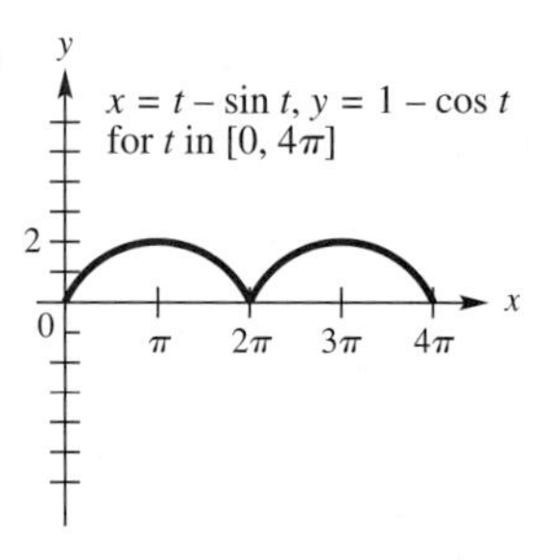


The windows in Exercises 23 and 25 are $[0, 6]$ by $[0, 4]$.

**23.** F

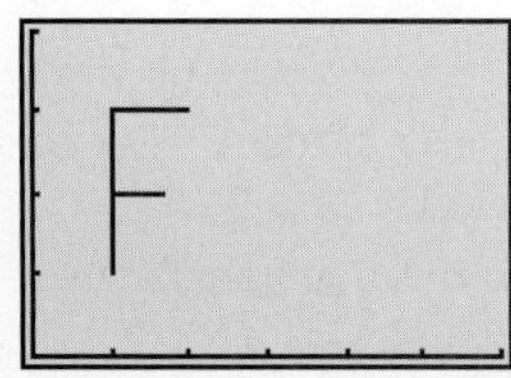

**25.** D

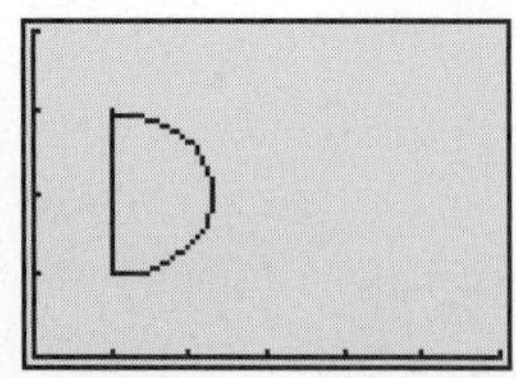

**27.** $x_1 = 0$, $y_1 = 2t$; $x_2 = t$, $y_2 = 0$; $0 \le t \le 1$; answers may vary. **29.** $x_1 = \sin t$, $y_1 = \cos t$; $x_2 = 0$, $y_2 = t - 2$; $0 \le t \le \pi$; answers may vary. **31.** Answers may vary.

The windows for Exercises 33 and 35 are $[-6, 6]$ by $[-4, 4]$.

**33.**

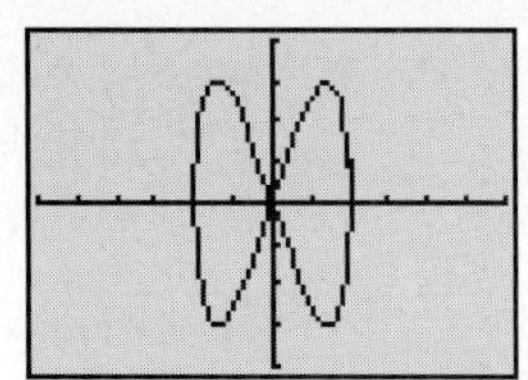

**35.**

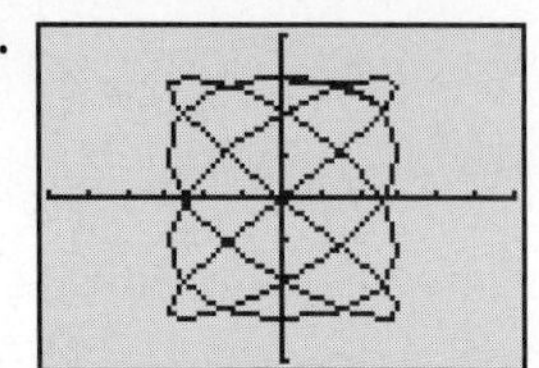

**37.** **(a)** $x = 24t$, $y = -16t^2 + 24\sqrt{3}t$
**(b)** $y = -\frac{1}{36}x^2 + \sqrt{3}x$ **(c)** 2.6 sec; 62 ft

**39.** **(a)** $x = (88 \cos 20°)t$, $y = 2 - 16t^2 + (88 \sin 20°)t$ **(b)** $y = 2 - \dfrac{x^2}{484 \cos^2 20°} + (\tan 20°)x$ **(c)** 1.9 sec; 161 ft

**41.** about 1456 ft **43.** **(a)**

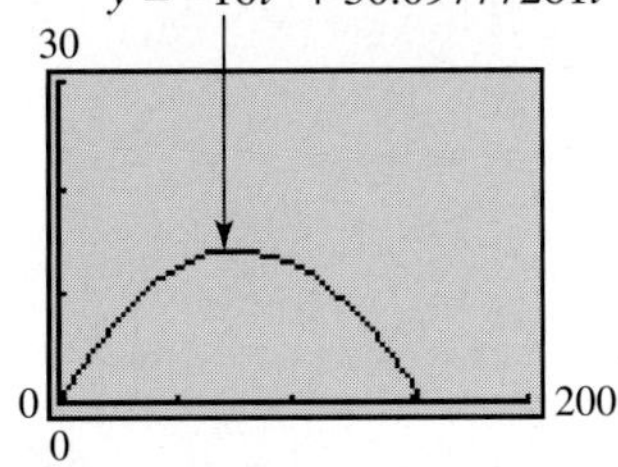


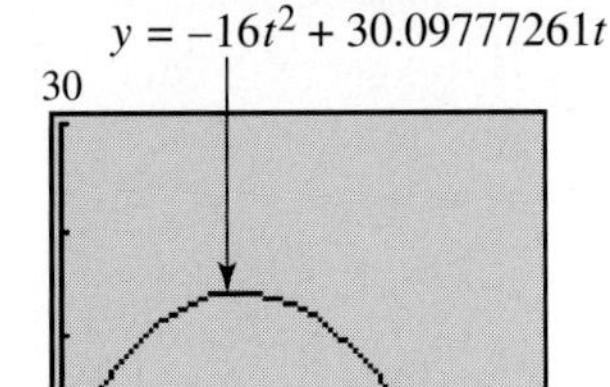

**(b)** 20.0° **(c)** $x = (88 \cos 20.0°)t$, $y = -16t^2 + (88 \sin 20.0°)t$

## Reviewing Basic Concepts *(page 808)*

**1.** IV **2.** $(2\sqrt{2}, 135°)$, $(-2\sqrt{2}, -45°)$; answers may vary.

**3.** cardioid

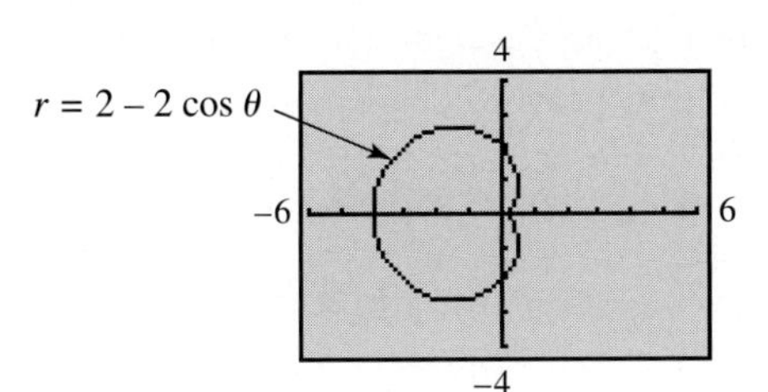


**4.** $(x - 1)^2 + y^2 = 1$ **5.** $r = \dfrac{6}{\cos \theta + \sin \theta}$

**6.** $\dfrac{x^2}{4} + \dfrac{y^2}{16} = 1$

**7.**

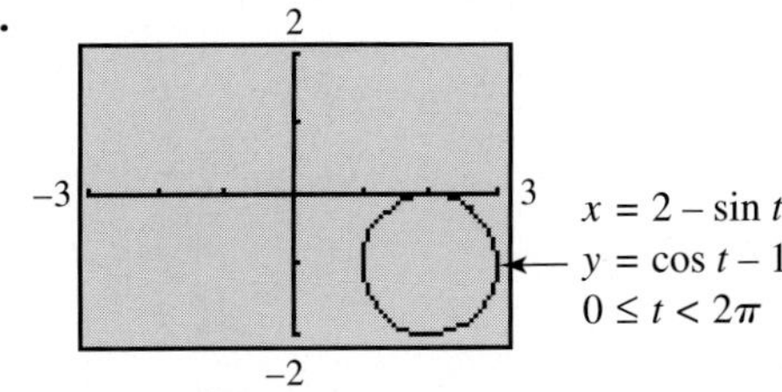


**8.** 285 ft

## Chapter 10 Review Exercises *(page 811)*

**1.** 63.7 m **3.** 19.87° or 19° 52′ **5.** 25° 00′ **7.** 173 ft **9.** 153,600 m² **11.** .234 km² **13.** 2.7 mi **15.** 7 km **17.** 58.6 ft **19.** 15.8 ft **21.** No, we are not given SSA. **23.** **(a)** $b = 5$, $b \geq 10$ **(b)** $5 < b < 10$ **(c)** $b < 5$

**25.**

**27.** 15; 126.9°

**29.** horizontal: $69.2 \cos 75° \approx 17.9$; vertical: $69.2 \sin 75° \approx 66.8$

**31.** **(a)** 14 **(b)** 52.13°

**33.** Yes, because $\mathbf{u} \cdot \mathbf{v} = 0$.

**35.** 2800 newtons; 30.4°

**37.** bearing: 7° 20′; speed: 8.3 mph

**39.**

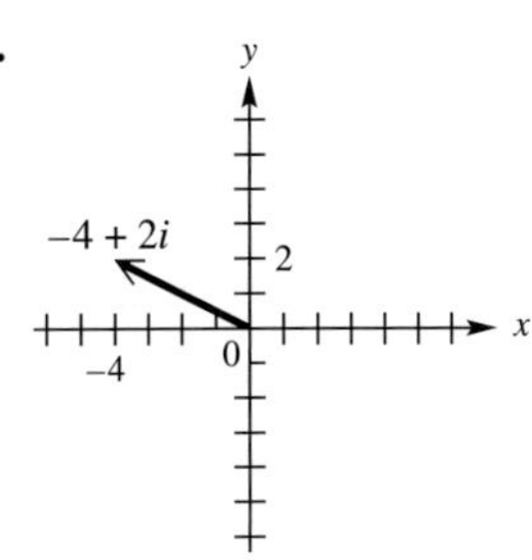


**41.** $7 - 3i$ **43.** $3i$ **45.** 8 cis 120° **47.** $-3 - 3i\sqrt{3}$ **49.** $-2$

**51.** $-128 + 128i$ **53.** $\dfrac{1}{2} + \dfrac{\sqrt{3}}{2}i$

**55.** 2 cis 22.5°, 2 cis 112.5°, 2 cis 202.5°, 2 cis 292.5°

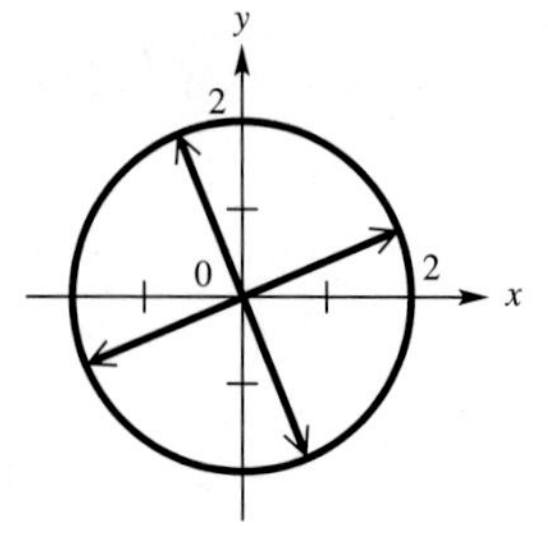


**57.** {cis 67.5°, cis 157.5°, cis 247.5°, cis 337.5°} **59.** $\left(-4, 4\sqrt{3}\right)$ **61.** (5, −90°)

**63.**

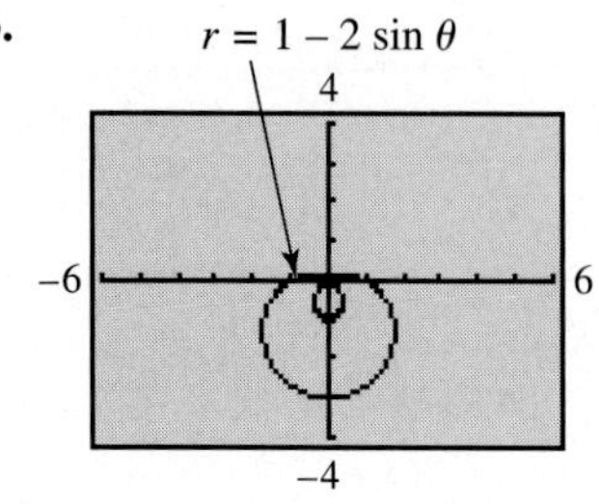


**65.** $y^2 + 6x - 9 = 0$ **67.** $x^2 - x + y^2 - y = 0$ **69.** $r = -\dfrac{3}{\cos\theta}$

**71.** $r = \dfrac{\tan\theta}{\cos\theta}$ **73.** They are the same because the two factors in each product are the same. They only appear to be different. (Since the argument 45° is coterminal with −315°, and 90° is coterminal with −270°, the quotients will have coterminal arguments.)

**75.** $y^2 = -\dfrac{1}{2}(x - 1)$ or $2y^2 + x - 1 = 0$, for $x$ in $[-1, 1]$

**77.**

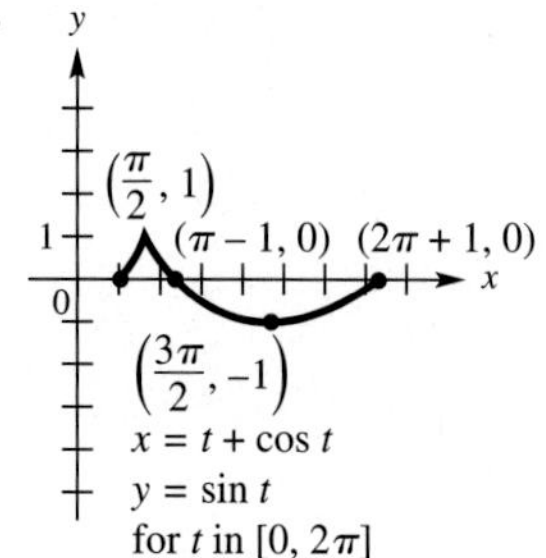


## Chapter 10 Test *(page 814)*

**1. (a)** 137.5° **(b)** 180 km **(c)** 49.0° **2. (a)** $b > 10$ **(b)** not possible **(c)** $b \leq 10$ **3.** The sum of the two smaller numbers must be greater than the largest number. **4. (a)** 115 km **(b)** 66.8 ft **(c)** −346; 451 **(d)** 826 lb **5. (a)** $\langle 1, -3\rangle$ **(b)** $\langle -6, 18\rangle$ **(c)** −20 **(d)** 180° **6. (a)** $-2 - 2i\sqrt{3}$ **(b)** 8 cis 120° **(c)** $-6 + 2i\sqrt{3}$ **7. (a)** $-\dfrac{15}{2} + \dfrac{15\sqrt{3}}{2}i$ **(b)** $-\dfrac{1}{2}i$ **(c)** $16 + 16i\sqrt{3}$ **(d)** $\sqrt[4]{2}$ cis 7.5°, $\sqrt[4]{2}$ cis 97.5°, $\sqrt[4]{2}$ cis 187.5°, $\sqrt[4]{2}$ cis 277.5°

**8. (a)**

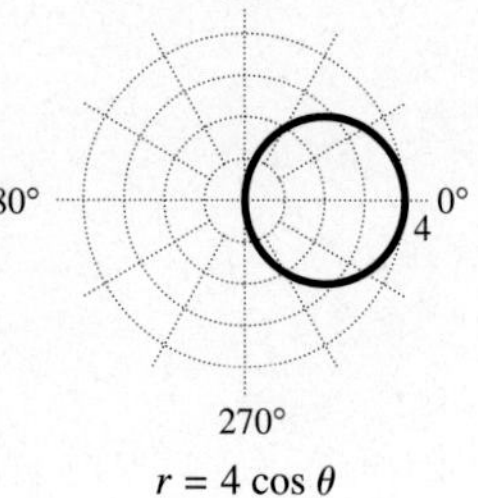


**(b)** $x^2 - 4x + y^2 = 0$

**(c)** Yes, the graph of $x^2 - 4x + y^2 = 0$ is a circle with center (2, 0) and radius 2.

**9.** $r = \dfrac{4}{2 \sin \theta - \cos \theta}$ **10.**

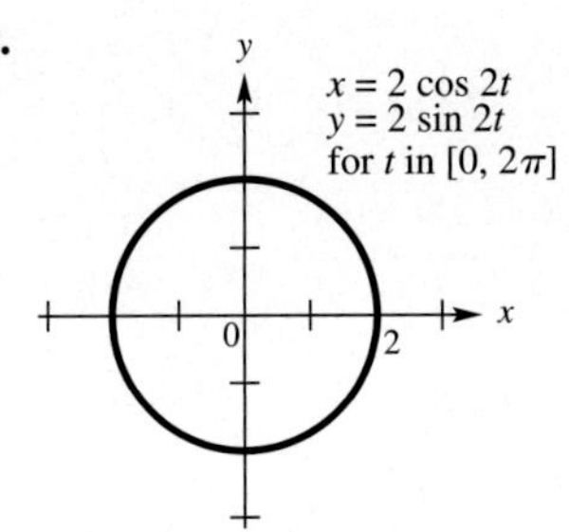


## CHAPTER 11 FURTHER TOPICS IN ALGEBRA

### 11.1 Exercises *(page 825)*

**1.** 14, 18, 22, 26, 30 **3.** 1, 2, 4, 8, 16 **5.** $0, \frac{1}{9}, \frac{2}{27}, \frac{1}{27}, \frac{4}{243}$ **7.** $-2, 4, -6, 8, -10$ **9.** $1, \frac{7}{6}, 1, \frac{5}{6}, \frac{19}{27}$ **11.** The $n$th term is the term that is in position $n$. For example, in the sequence 2, 4, 6, 8, . . . , the *third* term is 6, which is in position 3. The $n$th term here is given by $2n$. **13.** finite **15.** finite **17.** infinite **19.** finite **21.** $-2, 1, 4, 7$ **23.** 1, 1, 2, 3 **25.** 35 **27.** $\frac{25}{12}$ **29.** 288 **31.** 3 **33.** $-18$ **35.** $\frac{728}{9}$ **37.** 28 **39.** $-1 + 1 + 3 + 5 + 7$ **41.** $-10 - 4 + 0$ **43.** $0 + \frac{1}{2} + \frac{2}{3} + \frac{3}{4}$ **45.** $-3.5 + .5 + 4.5 + 8.5$ **47.** $0 + 4 + 16 + 36$ **49.** $-1 - \frac{1}{3} - \frac{1}{5} - \frac{1}{7}$ **51.** 600 **53.** 1240 **55.** 90 **57.** 220 **59.** 304 **61.** converges to $\frac{1}{2}$ **63.** diverges **65.** converges to $e \approx 2.71828$ **67.** 1.081123534; $\pi \approx 3.1407$; accurate to three decimal places rounded **69. (a)** $N_{j+1} = 2N_j$ for $j \geq 1$ **(b)** 1840 **(c)**

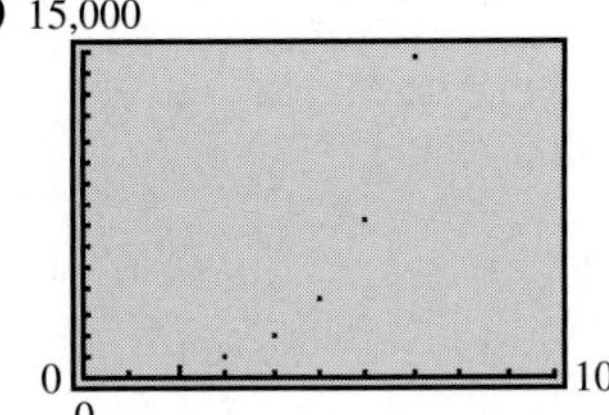


**(d)** The growth is very rapid. Since there is a doubling of the bacteria at equal intervals, their growth is exponential.

**71.** 2.718254, $e \approx 2.718282$ **73.** $a_6 \approx 1.414213562$; $\sqrt{2} \approx 1.414213562$

### 11.2 Exercises *(page 833)*

**1.** 3 **3.** $-5$ **5.** $x + 2y$ **7.** 8, 14, 20, 26, 32 **9.** $5, 3, 1, -1, -3$ **11.** 14, 12, 10, 8, 6 **13.** $a_8 = 19$; $a_n = 2n + 3$ **15.** $a_8 = 7$; $a_n = n - 1$ **17.** $a_8 = -6$; $a_n = -2n + 10$ **19.** $a_8 = -3$; $a_n = 4.5n - 39$ **21.** $a_8 = x + 21$; $a_n = x + 3n - 3$ **23.** 3 **25.** 5 **27.** 215 **29.** 125 **31.** 230 **33.** 77.5 **35.** $a_1 = 7$; $d = 5$ **37.** $a_1 = 1$; $d = -\frac{20}{11}$

In Exercises 39–43, $D$ is the domain and $R$ is the range.

**39.** $a_n = n - 3$; $D$: $\{1, 2, 3, 4, 5, 6\}$; $R$: $\{-2, -1, 0, 1, 2, 3\}$ **41.** $a_n = 3 - \frac{1}{2}n$; $D$: $\{1, 2, 3, 4, 5, 6\}$; $R$: $\{0, .5, 1, 1.5, 2, 2.5\}$ **43.** $a_n = 30 - 20n$; $D$: $\{1, 2, 3, 4, 5\}$; $R$: $\{-70, -50, -30, -10, 10\}$ **45.** 80 **47.** 1275 **49.** 1739 **51.** 4100 **53.** 18 **55.** 140 **57.** $-684$ **59.** 500,500 **61.** $f(1) = m + b$; $f(2) = 2m + b$; $f(3) = 3m + b$ **62.** yes **63.** $m$ **64.** $a_n = mn + b$ **65.** 328.3 **67.** 172.884 **69.** 1281 **71.** 4680 **73.** 54,800 **75.** 713 in. **77.** All terms are the same constant. **79.** yes

## 11.3 Exercises (*page 843*)

**1.** $\frac{5}{3}$, 5, 15, 45 **3.** $\frac{5}{8}, \frac{5}{4}, \frac{5}{2}$, 5, 10 **5.** $a_5 = 80; a_n = 5(-2)^{n-1}$ **7.** $a_5 = -108; a_n = -\frac{4}{3}(3)^{n-1}$

**9.** $a_5 = -729; a_n = -9(-3)^{n-1}$ **11.** $a_5 = -324; a_n = -4(3)^{n-1}$ **13.** $a_5 = \frac{125}{4}; a_n = \frac{4}{5}\left(\frac{5}{2}\right)^{n-1}$

**15.** $a_5 = \frac{5}{8}; a_n = 10\left(-\frac{1}{2}\right)^{n-1}$ **17.** $x - 5 = .6 - x$ **18.** solution: 2.8; terms: 5, 2.8, .6 **19.** $\frac{x}{5} = \frac{.6}{x}$

**20.** solution: $\sqrt{3}$; terms: 5, $\sqrt{3}$, .6 **21.** $a_1 = 125; r = \frac{1}{5}$ **23.** $a_1 = -2; r = \frac{1}{2}$ **25.** 682 **27.** $\frac{99}{8}$ **29.** 860.95

**31.** 363 **33.** $\frac{189}{4}$ **35.** 2032 **37.** The sum exists if $|r| < 1$. **39.** $\frac{8}{9}$ **41.** $\frac{5}{11}$ **43.** 2; does not converge

**45.** $\frac{1}{2}$ **47.** $\frac{128}{7}$ **49.** $\frac{1000}{9}$ **51.** $\frac{8}{3}$ **53.** 4 **55.** $\frac{3}{7}$ **57.** $g(1) = ab; g(2) = ab^2; g(3) = ab^3$ **58.** yes; the common ratio is $b$. **59.** $a_n = ab^n$ **60.** The independent variable is the exponent. **61.** 97.739 **63.** .212 **65.** \$12,487.56 **67.** \$32,029.33 **69. (a)** $a_1 = 1169; r = .916$ **(b)** $a_{10} = 531; a_{20} = 221$; This means that a person who is 10 yr from retirement should have savings of 531% of his or her annual salary; a person 20 yr from retirement should have savings of 221% of his or her annual salary. **71. (a)** $a_n = a_1 \cdot 2^{n-1}$ **(b)** 15 (rounded from 14.28 since $n$ must be a natural number)

**(c)** 560 min, or 9 hr and 20 min **73.** 13.4% **75.** \$26,214.40 **77.** 200 cm **79.** 62; 2046 **81.** $\frac{1}{64}$ m

**83.** Option 2 pays better.

## Reviewing Basic Concepts (*page 846*)

**1.** 4, −8, 12, −16, 20 **2.** 50 **3.** converges to 1 **4.** 8, 6, 4, 2, 0 **5.** −11 **6.** 245 **7.** $a_3 = -18$; $a_n = -2(-3)^{n-1}$ **8.** geometric; $\frac{7448}{625}$ **9.** 6 **10.** \$11,476.69

## 11.4 Exercises (*page 853*)

**1.** 20 **3.** 35 **5.** 56 **7.** 45 **9.** 1 **11.** $n$ **13.** 56 **15.** 4950 **17.** 9 **19.** $16x^4$; $81y^4$

**21.** $x^6 + 6x^5y + 15x^4y^2 + 20x^3y^3 + 15x^2y^4 + 6xy^5 + y^6$ **23.** $p^5 - 5p^4q + 10p^3q^2 - 10p^2q^3 + 5pq^4 - q^5$

**25.** $r^{10} + 5r^8s + 10r^6s^2 + 10r^4s^3 + 5r^2s^4 + s^5$ **27.** $p^4 + 8p^3q + 24p^2q^2 + 32pq^3 + 16q^4$

**29.** $2401p^4 + 2744p^3q + 1176p^2q^2 + 224pq^3 + 16q^4$

**31.** $729x^6 - 2916x^5y + 4860x^4y^2 - 4320x^3y^3 + 2160x^2y^4 - 576xy^5 + 64y^6$ **33.** $\frac{m^6}{64} - \frac{3m^5}{16} + \frac{15m^4}{16} - \frac{5m^3}{2} + \frac{15m^2}{4} - 3m + 1$

**35.** $4r^4 + \frac{8\sqrt{2}r^3}{m} + \frac{12r^2}{m^2} + \frac{4\sqrt{2}r}{m^3} + \frac{1}{m^4}$ **37.** $-3584h^3j^5$ **39.** $319{,}770a^{16}b^{14}$ **41.** $38{,}760x^6y^{42}$ **43.** $90{,}720x^{28}y^{12}$

**45.** 11 **47.** exact: 3,628,800; approximate: 3,598,695.619 **48.** about .830% **49.** exact: 479,001,600; approximate: 475,687,486.5; about .692% **50.** exact: 6,227,020,800; approximate: 6,187,239,475; about .639%; As $n$ gets larger, the percent error decreases. **51.** .942 **53.** 1.015

## 11.5 Exercises (*page 859*)

**1.** positive integers **3.** $n = 1$ and $n = 2$

Although we do not usually give proofs, the answers to Exercises 5 and 13 are shown here.

**5.** *Step 1:* $3(1) = 3$ and $\frac{3(1)(1+1)}{2} = \frac{6}{2} = 3$, so $S$ is true for $n = 1$. *Step 2:* $S_k$: $3 + 6 + 9 + \cdots + 3k = \frac{3(k)(k+1)}{2}$

$S_{k+1}$: $3 + 6 + 9 + \cdots + 3(k+1) = \frac{3(k+1)((k+1)+1)}{2}$; Add $3(k+1)$ to each side of $S_k$ and simplify until you obtain $S_{k+1}$.

Since $S$ is true for $n = 1$ and $S$ is true for $n = k + 1$ when it is true for $n = k$, $S$ is true for every positive integer $n$.

**13.** *Step 1:* $\frac{1}{1 \cdot 2} = \frac{1}{2}$ and $\frac{1}{1+1} = \frac{1}{2}$, so $S$ is true for $n = 1$. *Step 2:* $S_k$: $\frac{1}{1 \cdot 2} + \frac{1}{2 \cdot 3} + \frac{1}{3 \cdot 4} + \cdots + \frac{1}{k(k+1)} = \frac{k}{k+1}$

$S_{k+1}$: $\frac{1}{1 \cdot 2} + \frac{1}{2 \cdot 3} + \cdots + \frac{1}{(k+1)[(k+1)+1]} = \frac{k+1}{(k+1)+1}$; Add $\frac{1}{(k+1)[(k+1)+1]}$ to each side of $S_k$ and simplify until you obtain $S_{k+1}$. Since $S$ is true for $n = 1$ and $S$ is true for $n = k + 1$ when it is true for $n = k$, $S$ is true for every positive integer $n$. **19.** $n = 1$ or $2$ **21.** $n = 2, 3,$ or $4$ **35.** $\frac{4^{n-1}}{3^{n-2}}$ or $3\left(\frac{4}{3}\right)^{n-1}$ **37.** $2^n - 1$

## Reviewing Basic Concepts (*page 861*)

**1.** 10 **2.** $n + 1$ **3.** $a^4 + 8a^3b + 24a^2b^2 + 32ab^3 + 16b^4$ **4.** $\frac{1}{m^3} - \frac{3n^2}{m^2} + \frac{3n^4}{m} - n^6$ **5.** $60x^4y^2$

## 11.6 Exercises (*page 867*)

**1.** 19,958,400 **3.** 72 **5.** 5 **7.** 6 **9.** 1 **11.** 495 **13.** 1,860,480 **15.** 259,459,200 **17.** 15,504 **19.** 6435 **21.** **(a)** permutation **(b)** permutation **(c)** combination **(d)** combination **(e)** permutation **(f)** combination **(g)** permutation **23.** 30 **25.** **(a)** 27,600 **(b)** 35,152 **(c)** 1104 **27.** 15 **29.** **(a)** 17,576,000 **(b)** 17,576,000 **(c)** 456,976,000 **31.** 720 **33.** 120 **35.** 2730 **37.** 120; 30,240 **39.** 27,405 **41.** 20 **43.** 10 **45.** 28 **47.** **(a)** 84 **(b)** 10 **(c)** 40 **(d)** 28 **49.** 1680 **51.** 15 **53.** 479,001,600 **55.** **(a)** 56 **(b)** 462 **(c)** 3080 **(d)** 8526 **57.** 1,000,000 **59.** 4096 **61.** 6 **71.** **(a)** $3.04140932 \times 10^{64}$ **(b)** $8.320987113 \times 10^{81}$ **(c)** $8.247650592 \times 10^{90}$ **72.** **(a)** $8.759976613 \times 10^{20}$ **(b)** 5,527,200 **(c)** $2.19289732 \times 10^{26}$

## 11.7 Exercises (*page 877*)

**1.** $S = \{H\}$ **3.** $S = \{(H,H,H), (H,H,T), (H,T,T), (H,T,H), (T,T,T), (T,T,H), (T,H,T), (T,H,H)\}$
**5.** $S = \{(1,1), (1,2), (1,3), (2,1), (2,2), (2,3), (3,1), (3,2), (3,3)\}$

**7.** **(a)** $\{H\}$; 1 **(b)** $\emptyset$; 0 **9.** **(a)** $\{(1,1), (2,2), (3,3)\}$; $\frac{1}{3}$ **(b)** $\{(1,1), (1,3), (2,1), (2,3), (3,1), (3,3)\}$; $\frac{2}{3}$ **(c)** $\{(2,1), (2,3)\}$; $\frac{2}{9}$

**11.** $P = \frac{6}{5} > 1$ is impossible. **13.** **(a)** $\frac{1}{5}$ **(b)** 0 **(c)** $\frac{7}{15}$ **(d)** 1 to 4 **(e)** 7 to 8 **15.** 1 to 4 **17.** $\frac{2}{5}$ **19.** **(a)** $\frac{1}{2}$

**(b)** $\frac{7}{10}$ **(c)** $\frac{2}{5}$ **21.** **(a)** F **(b)** D **(c)** A **(d)** F **(e)** C **(f)** B **(g)** E **23.** 199 to 51 **25.** **(a)** .62 **(b)** .27 **(c)** .11

**(d)** .89 **27.** $\frac{48}{28{,}561} \approx .001681$ **29.** **(a)** .72 **(b)** .70 **(c)** .79 **31.** $\frac{5}{16} = .3125$ **33.** $\frac{1}{32} = .03125$ **35.** $\frac{1}{2}$

**37.** approximately $4.6 \times 10^{-10}$ **39.** approximately .875 **41.** .042246 **43.** .026864 **45.** **(a)** .047822 **(b)** .976710 **(c)** .897110 **47.** **(a)** approximately 40.4% **(b)** approximately 4.7% **(c)** approximately .2%; this means that in a large family or group of people, it is unlikely that everyone will become sick even though the disease is highly infectious.

**49.** **(a)** $P_{00} = 1$; $P_{01} = 0$; $P_{02} = 0$; $P_{10} = \frac{1}{6}$; $P_{11} = \frac{2}{3}$; $P_{12} = \frac{1}{6}$; $P_{20} = 0$; $P_{21} = 0$; $P_{22} = 1$

**(b)** $P = \begin{bmatrix} P_{00} & P_{01} & P_{02} \\ P_{10} & P_{11} & P_{12} \\ P_{20} & P_{21} & P_{22} \end{bmatrix} = \begin{bmatrix} 1 & 0 & 0 \\ \frac{1}{6} & \frac{2}{3} & \frac{1}{6} \\ 0 & 0 & 1 \end{bmatrix}$ **(c)** The matrix exhibits symmetry. The sum of the probabilities in each row is equal to 1. The greatest probabilities lie along the diagonal. This means that a mother cell is most likely to produce a daughter cell like itself. **50.** **(a)** Answers will vary. **(b)** $A_{k+1}$ approaches $[.5 \quad 0 \quad .5]$. Half of the bacteria are resistant to one antibiotic, half are resistant to the other antibiotic, and none are resistant to both antibiotics.

### Reviewing Basic Concepts (page 881)

**1.** 24 **2.** 210 **3.** 450 **4.** 210 **5.** 72 **6.** $S = \{HH, HT, TH, TT\}$ **7.** $\frac{2}{36} = \frac{1}{18}$ **8.** $\frac{4}{2{,}598{,}960} \approx .0000015$ **9.** $\frac{3}{10}$ **10.** approximately .484

### Chapter 11 Review Exercises (page 885)

**1.** $\frac{1}{2}, \frac{2}{3}, \frac{3}{4}, \frac{4}{5}, \frac{5}{6}$; neither **3.** 8, 10, 12, 14, 16; arithmetic **5.** 5, 2, $-1$, $-4$, $-7$; arithmetic **7.** $3\pi - 2, 2\pi - 1, \pi, 1, -\pi + 2$ **9.** $-5, -1, -\frac{1}{5}, -\frac{1}{25}, -\frac{1}{125}$ **11.** $-1$; $-8\left(\frac{1}{2}\right)^{n-1} = -\left(\frac{1}{2}\right)^{n-4}$ or 1; $-8\left(-\frac{1}{2}\right)^{n-1} = \left(-\frac{1}{2}\right)^{n-4}$ **13.** $-x + 61$ **15.** 612 **17.** $\frac{4}{25}$ **19.** $-40$ **21.** 1 **23.** $\frac{73}{12}$ **25.** 3,126,250 **27.** $\frac{4}{3}$ **29.** 36 **31.** diverges **33.** $-10$

In Exercises 35 and 37, other answers are possible.

**35.** $\sum_{i=1}^{15}(-5i + 9)$ **37.** $\sum_{i=1}^{6} 4(3)^{i-1}$ **39.** $x^4 + 8x^3y + 24x^2y^2 + 32xy^3 + 16y^4$ **41.** $243x^{5/2} - 405x^{3/2} + 270x^{1/2} - 90x^{-1/2} + 15x^{-3/2} - x^{-5/2}$ **43.** $-3584x^3y^5$ **45.** $x^{12} + 24x^{11} + 264x^{10} + 1760x^9$ **47.** Statements containing the set of natural numbers as their domains are proved by mathematical induction. **53.** permutation **55.** 1 **57.** 362,880 **59.** 48 **61.** 24 **63.** 504 **65.** **(a)** $\frac{4}{15}$ **(b)** $\frac{2}{3}$ **(c)** 0 **67.** $\frac{1}{26}$ **69.** $\frac{4}{13}$ **71.** .86 **73.** 0 **75.** approximately .205

### Chapter 11 Test (page 887)

**1.** **(a)** $-3, 4, -5, 6, -7$; neither **(b)** $-\frac{3}{2}, -\frac{3}{4}, -\frac{3}{8}, -\frac{3}{16}, -\frac{3}{32}$; geometric **(c)** 2, 3, 7, 13, 27; neither **2.** **(a)** $a_5 = 49$ **(b)** $a_5 = 16$ **3.** **(a)** 110 **(b)** $-1705$ **4.** **(a)** 2385 **(b)** $-186$ **(c)** does not exist **(d)** $\frac{108}{7}$ **5.** **(a)** $16x^4 - 96x^3y + 216x^2y^2 - 216xy^3 + 81y^4$ **(b)** $60w^4y^2$ **6.** **(a)** 45 **(b)** 35 **(c)** 5040 **(d)** 990 **8.** 24 **9.** 6840 **10.** 6160 **11.** **(a)** $\frac{1}{26}$ **(b)** $\frac{10}{13}$ **(c)** $\frac{4}{13}$ **(d)** 3 to 10 **12.** **(a)** $\binom{8}{3}\left(\frac{1}{6}\right)^3\left(1 - \frac{1}{6}\right)^5 \approx .104$ **(b)** $\binom{8}{8}\left(\frac{1}{6}\right)^8\left(1 - \frac{1}{6}\right)^0 \approx .000000595$

## CHAPTER R REFERENCE: BASIC ALGEBRAIC CONCEPTS AND GEOMETRY FORMULAS

### R.1 Exercises (page 896)

**1.** $(-4)^5$ **3.** 1 **5.** 1 **7.** $2^{10}$ **9.** $2^3x^{15}y^{12}$ **11.** $-\frac{p^8}{q^2}$ **13.** polynomial; degree 11; monomial **15.** polynomial; degree 6; binomial **17.** polynomial; degree 6; binomial **19.** polynomial; degree 6; trinomial **21.** not a polynomial **23.** $m^3 - 4m^2 + 10$ **25.** $5p^2 - 3p - 4$ **27.** $-6x^3 - 3x^2 - 4x + 4$ **29.** $15m^2 + 2m - 24$ **31.** $6m^2 + \frac{1}{4}m - \frac{1}{8}$ **33.** $2b^5 - 8b^4 + 6b^3$ **35.** $m^2 + mn - 2km - 2n^2 + 5kn - 3k^2$ **37.** Find the sum of the square of the first term, twice the product of the two terms, and the square of the last term. **39.** $4m^2 - 9$ **41.** $16m^2 + 16mn + 4n^2$ **43.** $25r^2 + 30rt^2 + 9t^4$ **45.** $4p^2 - 12p + 9 + 4pq - 6q + q^2$ **47.** $9q^2 + 30q + 25 - p^2$ **49.** $9a^2 + 6ab + b^2 - 6a - 2b + 1$ **51.** $18p^2 - 27pq - 35q^2$ **53.** $p^3 - 7p^2 - p - 7$ **55.** $16x^2y - 9y^3$ **57.** $6z^2 - 5yz - 4y^2$ **59.** $9p^2 + 30p + 25$ **61.** $4p^2 - 16$ **63.** $11y^3 - 18y^2 + 4y$

**R.2 Exercises** *(page 902)*

**1.** **(a)** B **(b)** C **(c)** A **(d)** D **3.** $4k^2m^3(1 + 2k^2 - 3m)$ **5.** $2(a + b)(1 + 2m)$ **7.** $(y + 2)(3y + 2)$ **9.** $(r + 3)(3r - 5)$ **11.** $(m - 1)(2m^2 - 7m + 7)$ **13.** $(2s + 3)(3t - 5)$ **15.** $(2x + 3)(5x - 4y)$ **17.** $(t + 2)(t^2 - 3)$ **19.** Both are correct. **21.** $8(h - 8)(h + 5)$ **23.** $9y^2(y - 1)(y - 5)$ **25.** $(7m - 5r)(2m + 3r)$ **27.** $(3s - t)(4s + 5t)$ **29.** $(5a + m)(6a - m)$ **31.** $3x^3(2x + 5z)(3x - 5z)$ **33.** $(4p - 5)^2$ **35.** $5(2p - 5q)^2$ **37.** $(3mn - 2)^2$ **39.** $(2p + q - 5)^2$ **41.** $(3a + 4)(3a - 4)$ **43.** $(5s^2 + 3t)(5s^2 - 3t)$ **45.** $(a + b + 4)(a + b - 4)$ **47.** $(p^2 + 25)(p + 5)(p - 5)$ **49.** B **51.** $(2 - a)(4 + 2a + a^2)$ **53.** $(5x - 3)(25x^2 + 15x + 9)$ **55.** $(3y^3 + 5z^2)(9y^6 - 15y^3z^2 + 25z^4)$ **57.** $r(r^2 + 18r + 108)$ **59.** $(3 - m - 2n)(9 + 3m + 6n + m^2 + 4mn + 4n^2)$ **61.** It is incomplete because $a^2$ has not been substituted back for $u$. **63.** $(a^2 - 8)(a^2 + 6)$ **65.** $2(6z - 5)(8z - 3)$ **67.** $(18 - 5p)(17 - 4p)$ **69.** $(2b + c + 4)(2b + c - 4)$ **71.** $(x + y)(x - 5)$ **73.** $(m - 2n)(p^4 + q)$ **75.** $(2z + 7)^2$ **77.** $(10x + 7y)(100x^2 - 70xy + 49y^2)$ **79.** $(5m^2 - 6)(25m^4 + 30m^2 + 36)$ **81.** $(6m - 7n)(2m + 5n)$ **83.** $(4p - 1)(p + 1)$ **85.** prime **87.** $4xy$

**R.3 Exercises** *(page 909)*

**1.** $\{x \mid x \neq -6\}$ **3.** $\left\{x \mid x \neq \frac{3}{5}\right\}$ **5.** $(-\infty, \infty)$ **7.** $\left\{x \mid x \neq -\frac{1}{2}, 1\right\}$ **9.** $\frac{5p}{2}$ **11.** $\frac{8}{9}$ **13.** $\frac{3}{t - 3}$ **15.** $\frac{2x + 4}{x}$ **17.** $\frac{m - 2}{m + 3}$ **19.** $\frac{2m + 3}{4m + 3}$ **21.** $\frac{25p^2}{9}$ **23.** $\frac{2}{9}$ **25.** $\frac{5x}{y}$ **27.** $\frac{2(a + 4)}{a - 3}$ **29.** 1 **31.** $\frac{m + 6}{m + 3}$ **33.** $\frac{m - 3}{2m - 3}$ **35.** $\frac{x + y}{x - y}$ **37.** $\frac{x^2 - xy + y^2}{x^2 + xy + y^2}$ **39.** B, C **41.** $\frac{19}{6k}$ **43.** 1 **45.** $\frac{6 + p}{2p}$ **47.** $\frac{137}{30m}$ **49.** $\frac{-2}{(a + 1)(a - 1)}$ **51.** $\frac{2m^2 + 2}{(m - 1)(m + 1)}$ **53.** $\frac{4}{a - 2}$ or $\frac{-4}{2 - a}$ **55.** $\frac{3x + y}{2x - y}$ or $\frac{-3x - y}{y - 2x}$ **57.** $\frac{5}{(a - 2)(a - 3)(a + 2)}$ **59.** $\frac{x - 11}{(x + 4)(x - 3)(x - 4)}$ **61.** $\frac{a^2 + 5a}{(a + 6)(a - 1)(a + 1)}$ **63.** $\frac{x + 1}{x - 1}$ **65.** $\frac{-1}{x + 1}$ **67.** $\frac{(2 - b)(1 + b)}{b(1 - b)}$ **69.** $\frac{m^3 - 4m - 1}{m - 2}$

**R.4 Exercises** *(page 916)*

**1.** E **3.** F **5.** D **7.** B **9.** $-\frac{1}{64}$ **11.** 8 **13.** $-2$ **15.** 4 **17.** $\frac{1}{9}$ **19.** $\frac{256}{81}$ **21.** $4p^2$ **23.** $9x^4$ **25.** $\frac{1}{2^7}$ **27.** $\frac{1}{27^3}$ **29.** 1 **31.** $m^{7/3}$ **33.** $(1 + n)^{5/4}$ **35.** $\frac{6z^{2/3}}{y^{5/4}}$ **37.** $2^6a^{1/4}b^{37/2}$ **39.** $\frac{r^6}{s^{15}}$ **41.** $-\frac{1}{ab^3}$ **43.** $12^{9/4}y$ **45.** $\frac{1}{2p^2}$ **47.** $\frac{m^3p}{n}$ **49.** $-4a^{5/3}$ **51.** $\frac{1}{(k + 5)^{1/2}}$ **53.** $y - 10y^2$ **55.** $-4k^{10/3} + 24k^{4/3}$ **57.** $x^2 - x$ **59.** $r - 2 + r^{-1}$ or $r - 2 + \frac{1}{r}$ **61.** $k^{-2}(4k + 1)$ or $\frac{4k + 1}{k^2}$ **63.** $z^{-1/2}(9 + 2z)$ or $\frac{9 + 2z}{z^{1/2}}$ **65.** $p^{-7/4}(p - 2)$ or $\frac{p - 2}{p^{7/4}}$ **67.** $(p + 4)^{-3/2}(p^2 + 9p + 21)$ or $\frac{p^2 + 9p + 21}{(p + 4)^{3/2}}$

**R.5 Exercises** *(page 922)*

**1.** F **3.** H **5.** G **7.** C **9.** $\sqrt[3]{(-m)^2}$ or $(\sqrt[3]{-m})^2$ **11.** $\sqrt[3]{(2m + p)^2}$ or $(\sqrt[3]{2m + p})^2$ **13.** $k^{2/5}$ **15.** $-3 \cdot 5^{1/2}p^{3/2}$ **17.** A **19.** $x \geq 0$ **21.** 5 **23.** $-5$ **25.** $5\sqrt{2}$ **27.** $3\sqrt[3]{3}$ **29.** $-2\sqrt[4]{2}$ **31.** $-\frac{3\sqrt{5}}{5}$ **33.** $-\frac{\sqrt[3]{100}}{5}$ **35.** $32\sqrt[3]{2}$ **37.** $2x^2z^4\sqrt{2x}$ **39.** $2zx^2y\sqrt[3]{2z^2x^2y}$ **41.** $np^2\sqrt[4]{m^2n^3}$ **43.** cannot be simplified further **45.** $\frac{\sqrt{6x}}{3x}$ **47.** $\frac{x^2y\sqrt{xy}}{z}$ **49.** $\frac{2\sqrt[3]{x}}{x}$ **51.** $\frac{h\sqrt[4]{9g^3hr^2}}{3r^2}$ **53.** $\frac{m\sqrt[3]{n^2}}{n}$ **55.** $2\sqrt[4]{x^3y^3}$ **57.** $\sqrt[6]{2}$ **59.** $\sqrt[18]{x}$ **61.** $9\sqrt{3}$

**63.** $-2\sqrt{7p}$ **65.** $7\sqrt[3]{3}$ **67.** $2\sqrt{3}$ **69.** $\dfrac{13\sqrt[3]{4}}{6}$ **71.** $-7$ **73.** 10 **75.** $11 + 4\sqrt{6}$ **77.** $5\sqrt{6}$

**79.** $\dfrac{\sqrt{15} - 3}{2}$ **81.** $\dfrac{3\sqrt{5} - 2\sqrt{3} + 3\sqrt{15} - 6}{33}$ **83.** $\dfrac{p(\sqrt{p} - 2)}{p - 4}$ **85.** $\dfrac{a(\sqrt{a + b} + 1)}{a + b - 1}$

## Appendix A Exercises *(page 930)*

**1.** $xz$-plane **3.** $\langle 5, 3, -2\rangle$ **5.** $\sqrt{33}$ **7.** $\sqrt{317}$ **9.** $5\sqrt{29}$ **11.** $\langle 2, -2, 5\rangle$, $2\mathbf{i} - 2\mathbf{j} + 5\mathbf{k}$
**13.** $\langle -2, -12, -4\rangle$, $-2\mathbf{i} - 12\mathbf{j} - 4\mathbf{k}$ **15.** $\langle -15, -20, 10\rangle$, $-15\mathbf{i} - 20\mathbf{j} + 10\mathbf{k}$ **17.** $\langle -2, 2, -5\rangle$; $\mathbf{QP} = -\mathbf{PQ}$
**19.** $-2\mathbf{i} + 7\mathbf{j} + 13\mathbf{k}$ **21.** $-7\mathbf{i} + 41\mathbf{j} + 38\mathbf{k}$ **23.** $\sqrt{69}$ **25.** $\sqrt{38}$ **27.** $-39$ **29.** 38 **31.** 25.4°
**33.** 32.0° **35.** 90° **37.** $\alpha = 76.1°$; $\beta = 61.2°$; $\gamma = 32.6°$ **39.** $\alpha = 59.2°$; $\beta = 112.6°$; $\gamma = 140.2°$ **41.** 60°
**43.** They have the same position vector. **45.** 12 work units **47.** 13 work units

## Appendix B Exercises *(page 935)*

In Exercises 1–11, we give only calculator graphs.

**1.**

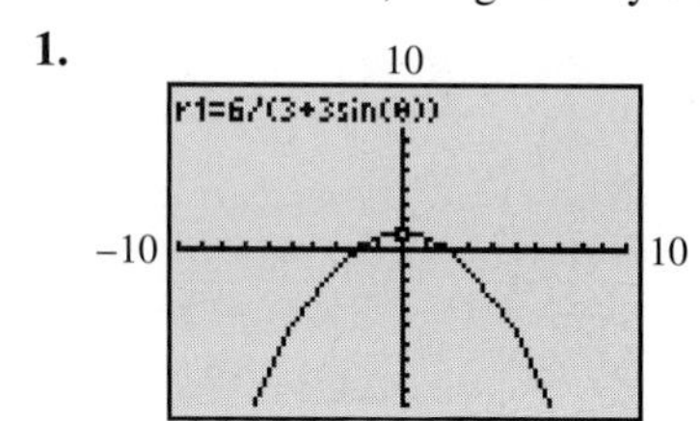


**3.**

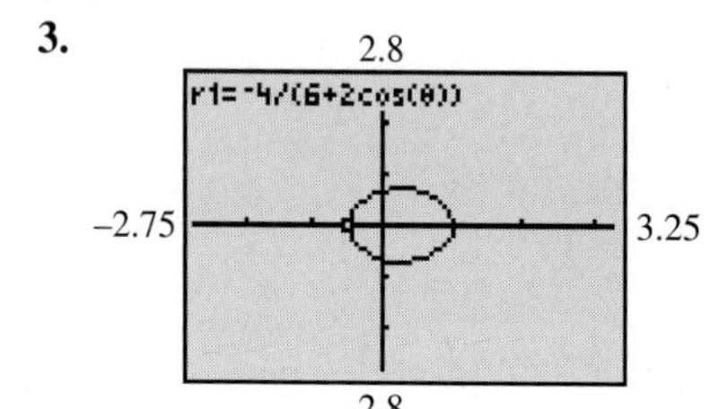


**5.**

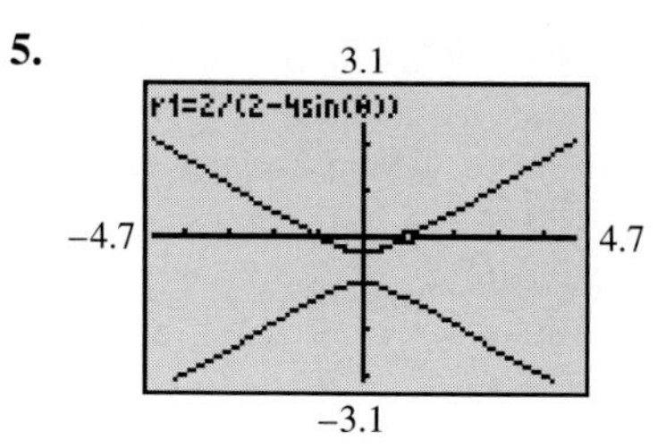


**7.**

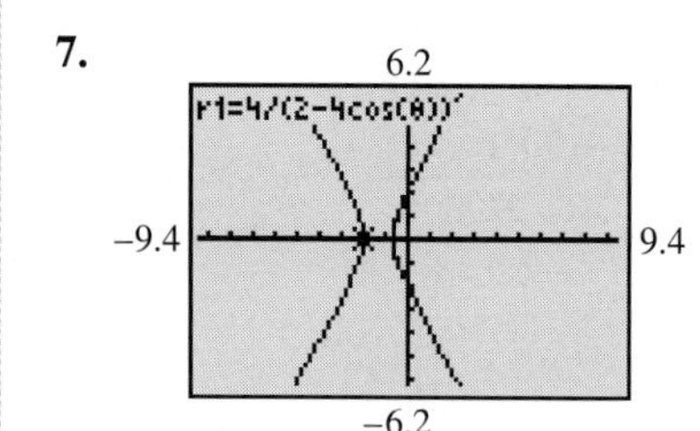


**9.**

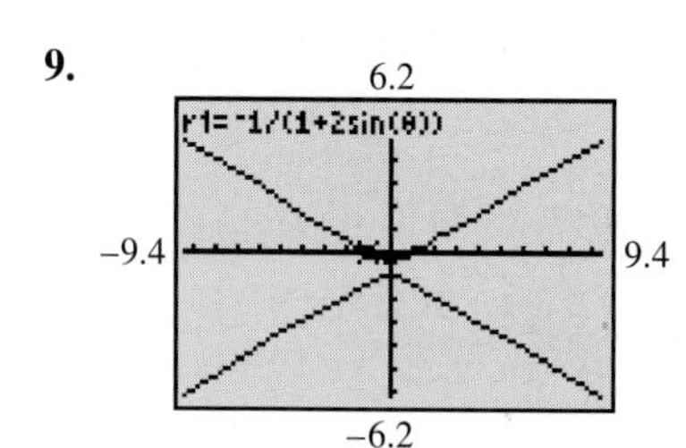


**11.**

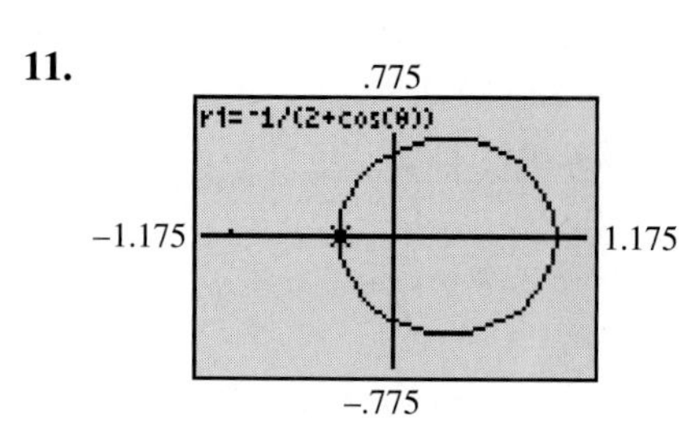


**13.** $r = \dfrac{3}{1 + \cos\theta}$ **15.** $r = \dfrac{5}{1 - \sin\theta}$ **17.** $r = \dfrac{20}{5 + 4\cos\theta}$; ellipse **19.** $r = \dfrac{40}{4 - 5\sin\theta}$; hyperbola
**21.** ellipse; $8x^2 + 9y^2 - 12x - 36 = 0$ **23.** hyperbola; $3x^2 - y^2 + 8x + 4 = 0$ **25.** ellipse; $4x^2 + 3y^2 - 6y - 9 = 0$
**27.** parabola; $x^2 - 10y - 25 = 0$

## Appendix C Exercises *(page 939)*

**1.** circle or ellipse or a point **3.** hyperbola or two intersecting lines **5.** parabola or one line or two parallel lines
**7.** 30° **9.** 60° **11.** 22.5°

**13.**

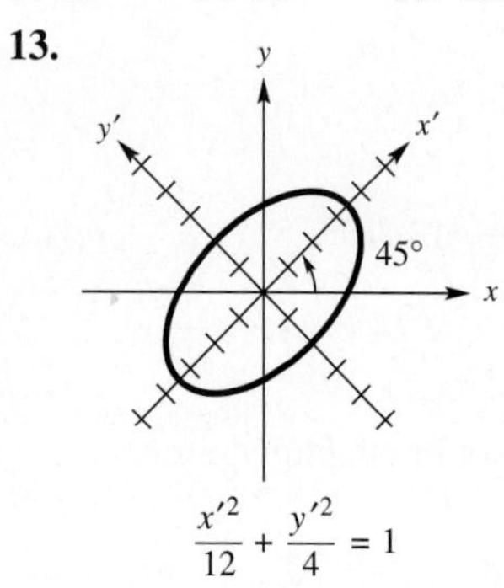


**15.**

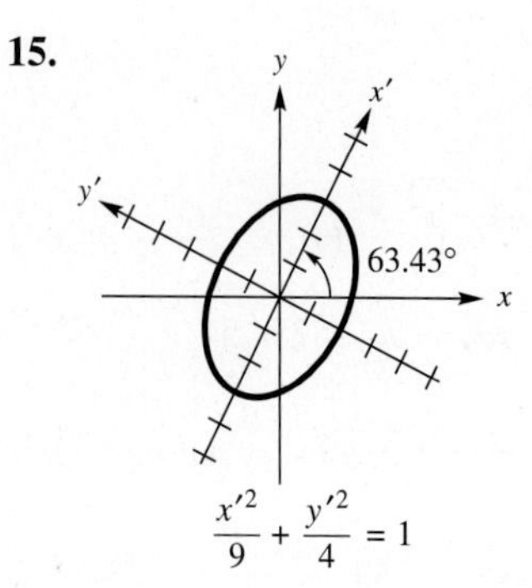


**17.**

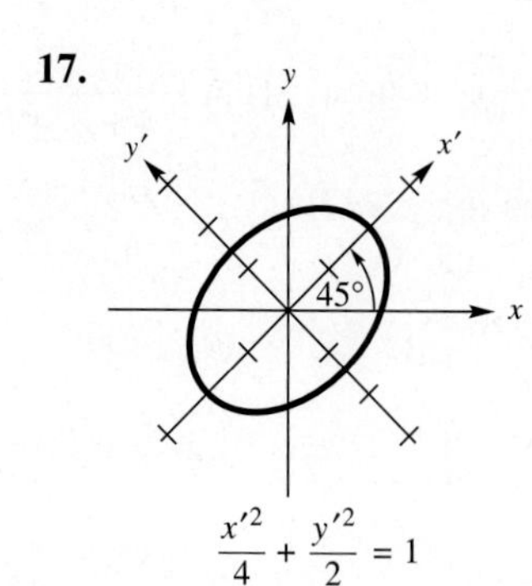

**19.**

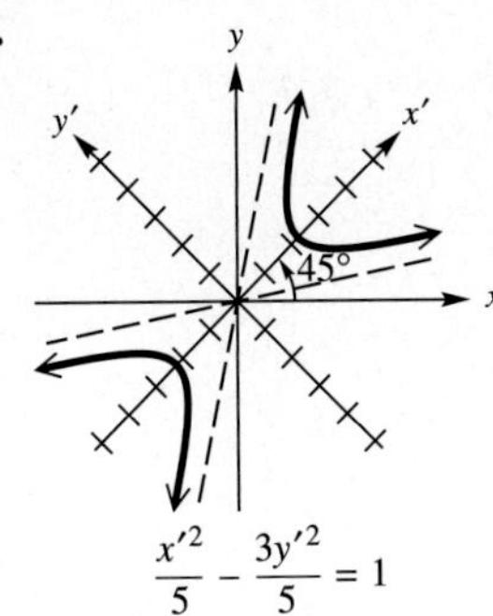


**21.**

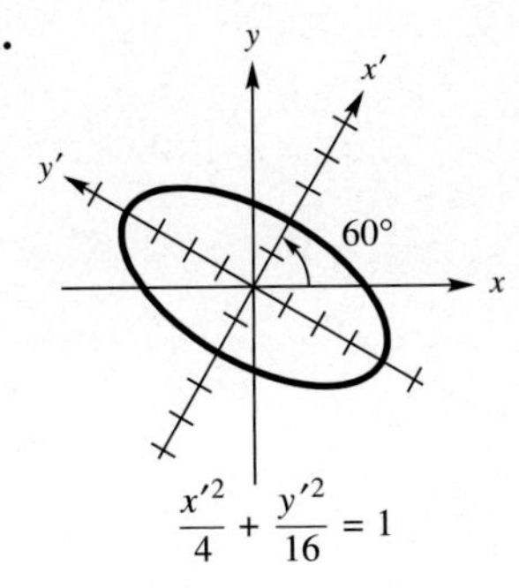


**23.**

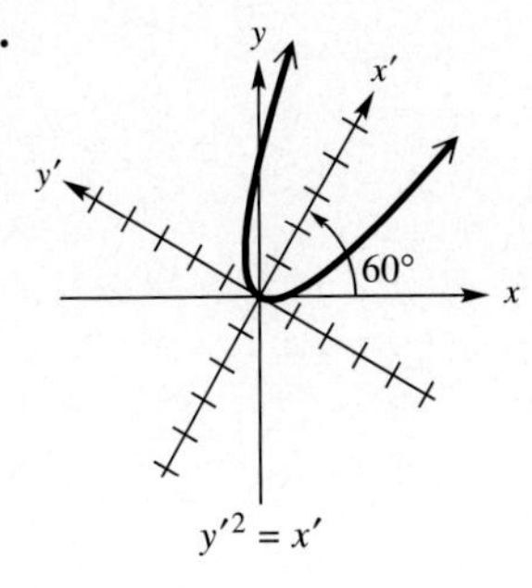


**25.**

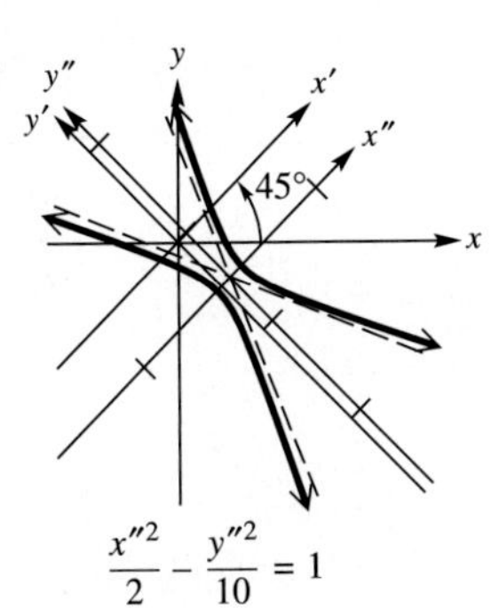


**27.**

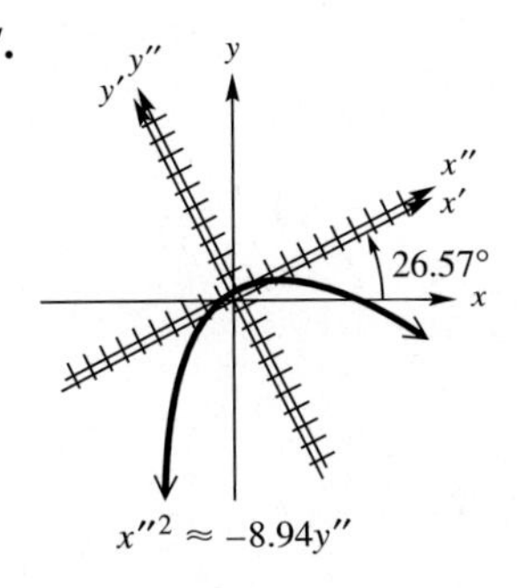


**29.**

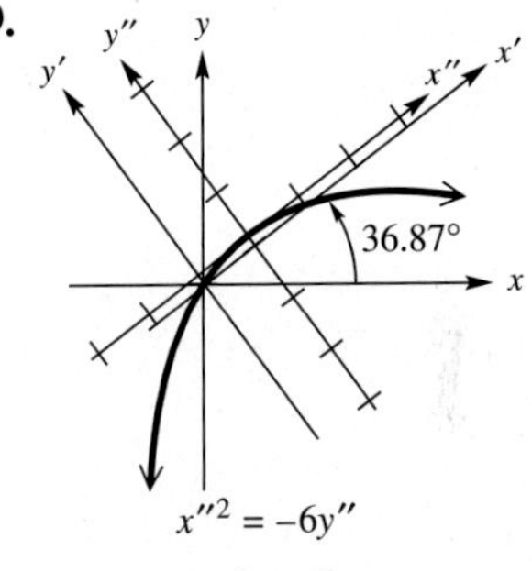


**31.** If $B = 0$, $\cot 2\theta$ is undefined. The graph may be translated but is not rotated.

# Index of Applications

## Astronomy/Aerospace



## Automotive



## Biology/Life Sciences



## Business

## Chemistry



## Construction



## Consumer



## Economics



## Education



## Energy



## Engineering



## Environment



## Finance

## General Interest



## Geology



## Geometry



## Government



## Labor

## Medical



## Physics



## Probability



## Sports/Entertainment



## Statistics/Demographics



## Technology



## Transportation

## Weather

# Index

## D



## E

## F

## G



## H

## M



## N

## Q



## R

## S

## T

Z

*Photo Credits:* **p. 1 top** © 2002 PhotoDisc; **p. 1** Malcolm McKenna/The New York Times; **p. 8** © 2002 PhotoDisc; **p. 23** © 2002 PhotoDisc; **p. 33** © Phil Schermeister/CORBIS; **p. 45** © corbisstockmarket.com; **p. 77** © corbisstockmarket.com; **p. 88** AP/Wide World Photos; **p. 89 top** PictureQuest; **p. 109** © Richard Hutchings/PhotoEdit; **p. 112** © Brett Patterson/CORBIS; **p. 136** © corbisstockmarket.com; **p. 148** © 2002 PhotoDisc; **p. 154** © Jeff Zaruba/CorbisStockMarket; **p. 169** Courtesy of John Hornsby; **p. 172 top** PictureQuest; **p. 172** © 2001 NBA Entertainment, photo by Walter Iooss, Jr.; **p. 186** © corbisstockmarket.com; **p. 191** Iowa State University; **p. 204** © 2002 PhotoDisc; **p. 216** © corbisstockmarket.com; **p. 257** © CORBIS; **p. 268** © CORBIS; **p. 272** © CorbisStockMarket; **p. 297** © 2002 PhotoDisc; **p. 301** © 2002 PhotoDisc; **p. 307** Courtesy of NASA; **p. 312** © 2002 PhotoDisc; **p. 319** © Frank Lane Picture Agency/CORBIS; **p. 328** © Tony Freeman/PhotoEdit; **p. 334** © Joseph Sohm; Chromo Sohm Inc./CORBIS; **p. 342** © CorbisStockMarket; **p. 362** © 2002 PhotoDisc; **p. 385** © Temp Sport/CORBIS; **p. 392** © Lightscapes/corbisstockmarket.com; **p. 395** © Lightscapes/corbisstockmarket.com; **p. 404** © Archivo Iconografico, S.A./CORBIS; **p. 415** © Russell Munson/The Stock Market; **p. 416** © corbisstockmarket.com; **p. 417** © Alan Schein/corbisstockmarket.com; **p. 419** Magellan/NASA; **p. 433** © 2002 PhotoDisc; **p. 444** Courtesy of NASA; **p. 457** Courtesy of NASA; **p. 458** Courtesy of NASA; **p. 463** Courtesy of NASA; **p. 466** ABC Computer/Ames Laboratory, Department of Energy, ISU, Iowa; **p. 473** © corbisstockmarket.com; **p. 501** © David Young-Wolff/PhotoEdit; **p. 510** © Alan Schein/Corbis Stock Market; **p. 566** © 2002 PhotoDisc; **p. 575** © 2002 PhotoDisc; **p. 582** © Duomo/CORBIS; **p. 624** © 2002 PhotoDisc; **p. 627** © Tim Thompson/CORBIS; **p. 628** © MacNeal Hospital/Stone; **p. 673** © 2002 PhotoDisc; **p. 693** © Rob Gage/FPG; **p. 698** © 2002 PhotoDisc; **p. 716** © Eyewire; **p. 724** © 2002 PhotoDisc; **p. 730** © Eyewire; **p. 739** © Stephen Johnson/Stone; **p. 761** © Dave G. Houser/CORBIS; **p. 800** Courtesy of NASA; **p. 803** © B. Busco/The Image Bank; **p. 813** AP/Wide World Photos; **p. 816** © CorbisStockMarket; **p. 819** © Karen Tweedy-Holmes/CORBIS; **p. 835** LWA-Dann Tardif/CorbisStockMarket; **p. 878** AP/Wide World Photos

# FUNCTION CAPSULES

**IDENTITY FUNCTION** $f(x) = x$

Domain: $(-\infty, \infty)$ Range: $(-\infty, \infty)$

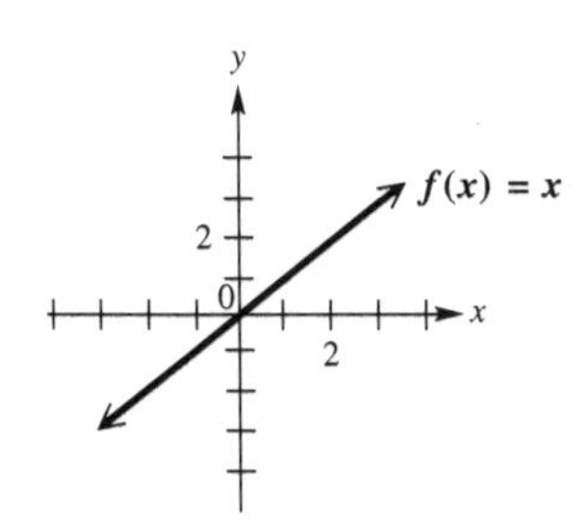


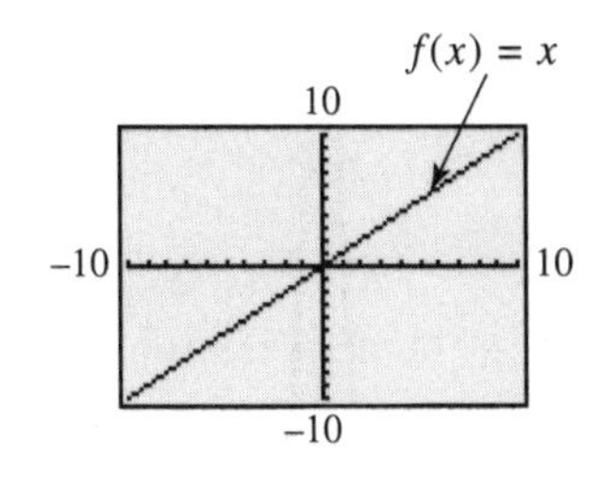


**SQUARING FUNCTION** $f(x) = x^2$

Domain: $(-\infty, \infty)$ Range: $[0, \infty)$

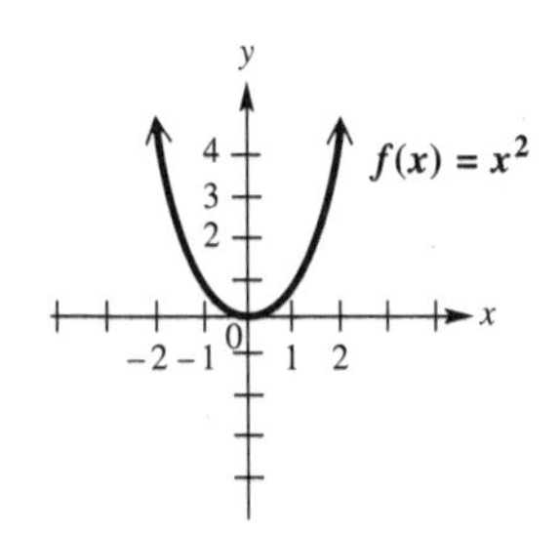


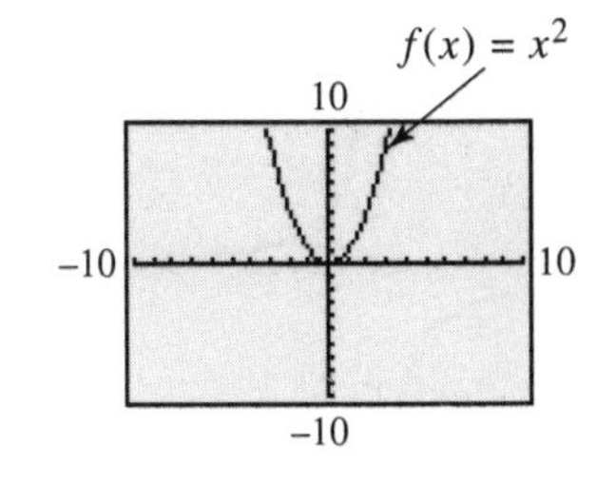


**CUBING FUNCTION** $f(x) = x^3$

Domain: $(-\infty, \infty)$ Range: $(-\infty, \infty)$

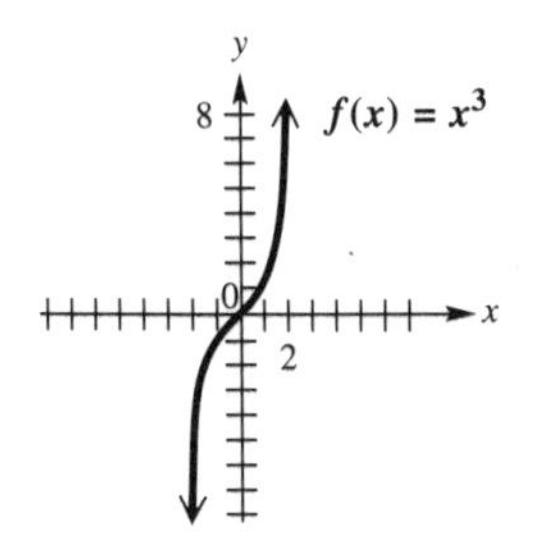


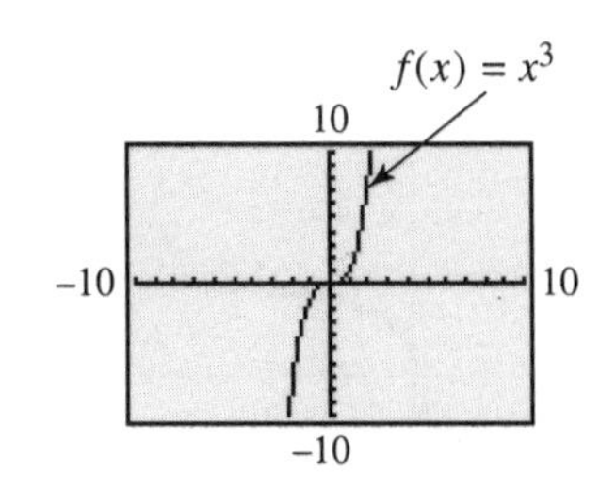


**SQUARE ROOT FUNCTION** $f(x) = \sqrt{x}$

Domain: $[0, \infty)$ Range: $[0, \infty)$

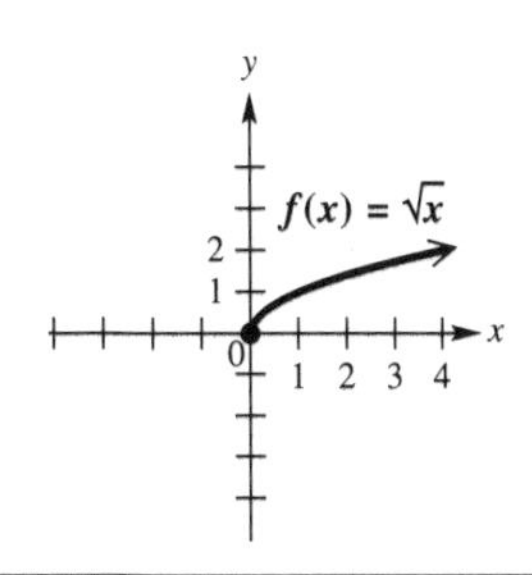


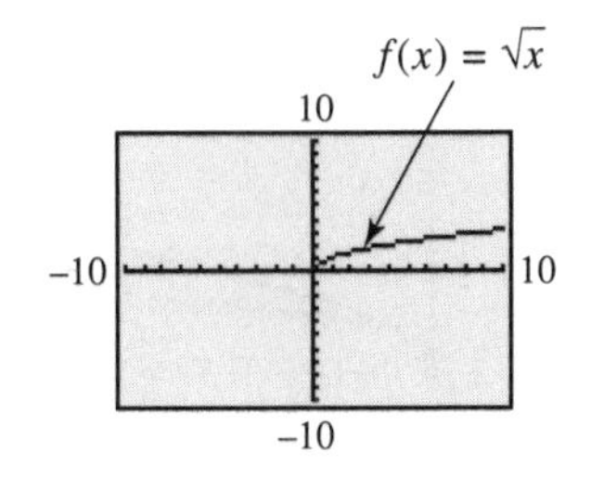


**CUBE ROOT FUNCTION** $f(x) = \sqrt[3]{x}$

Domain: $(-\infty, \infty)$ Range: $(-\infty, \infty)$

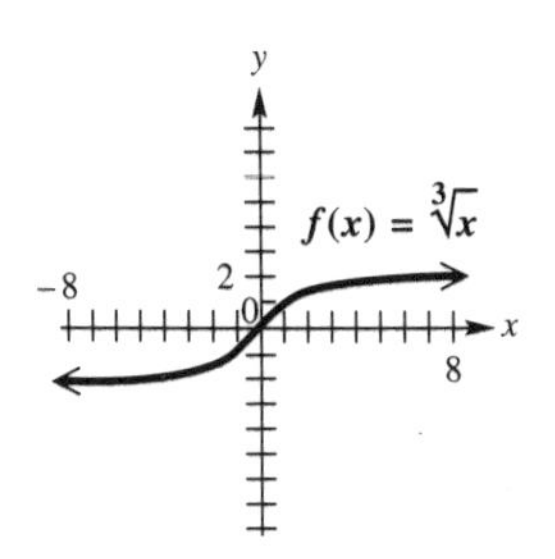


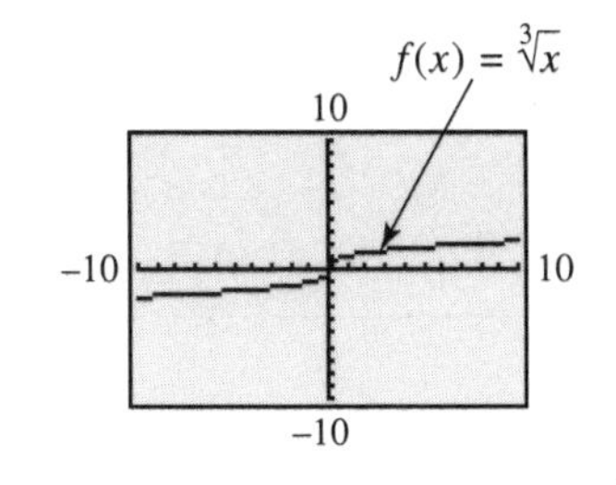


**ABSOLUTE VALUE FUNCTION** $f(x) = |x|$

Domain: $(-\infty, \infty)$ Range: $[0, \infty)$

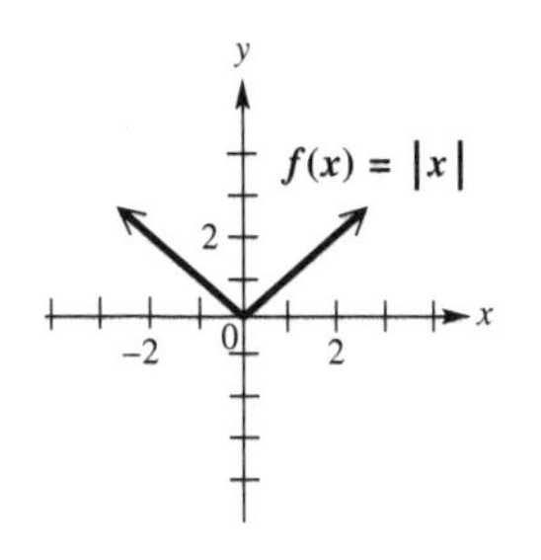


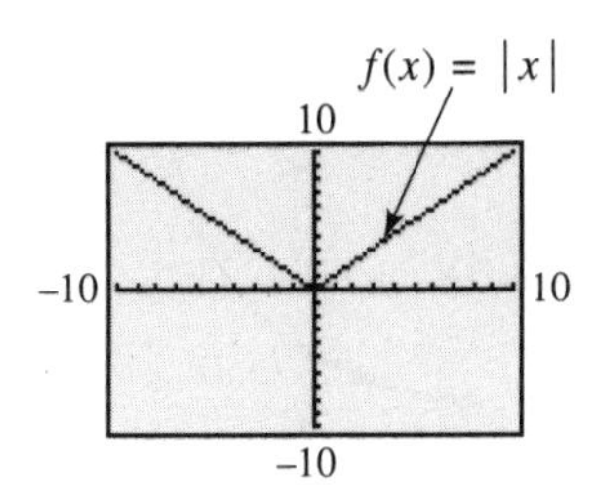


**GREATEST INTEGER FUNCTION** $f(x) = [\![x]\!]$

Domain: $(-\infty, \infty)$

Range: $\{x \mid x \text{ is an integer}\} = \{\ldots, -3, -2, -1, 0, 1, 2, 3, \ldots\}$

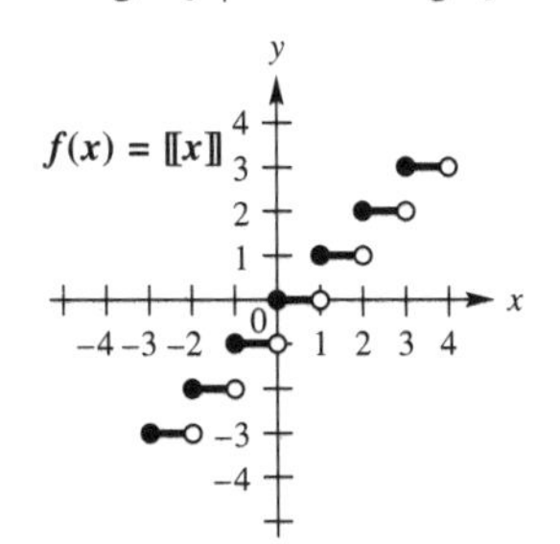


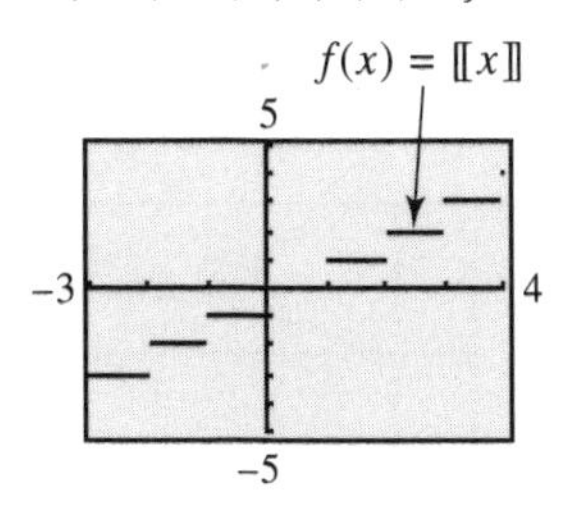


Dot Mode

# FUNCTION CAPSULES

**RECIPROCAL FUNCTION** $f(x) = \frac{1}{x}$

Domain: $(-\infty, 0) \cup (0, \infty)$ Range: $(-\infty, 0) \cup (0, \infty)$

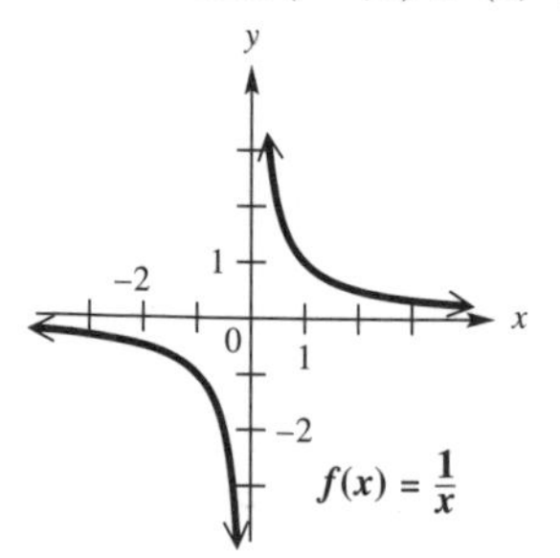
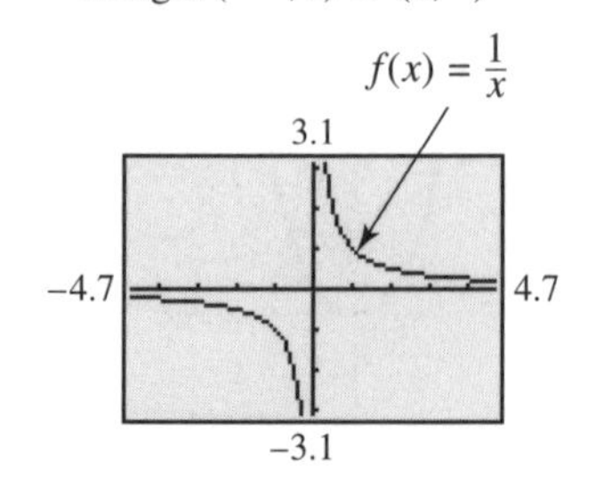

**RATIONAL FUNCTION** $f(x) = \frac{1}{x^2}$

Domain: $(-\infty, 0) \cup (0, \infty)$ Range: $(0, \infty)$

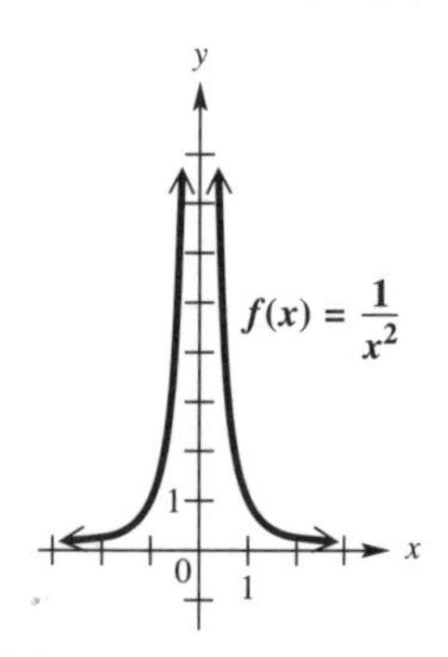
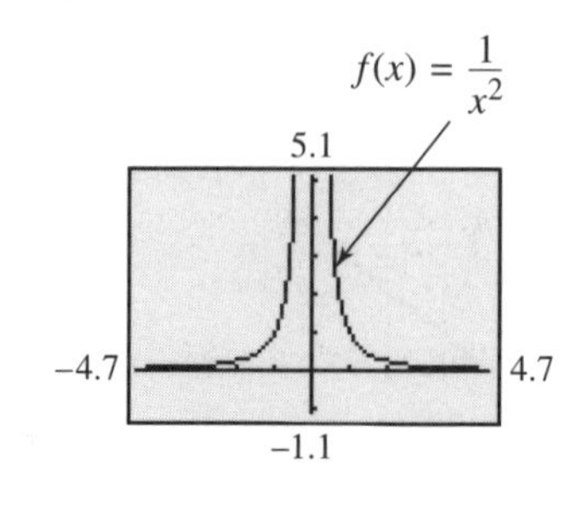

**ROOT FUNCTION, $n$ EVEN** $f(x) = \sqrt[n]{x}$

Domain: $[0, \infty)$ Range: $[0, \infty)$

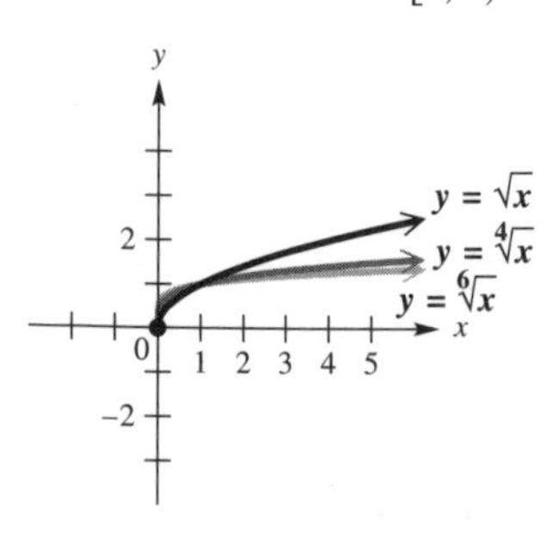
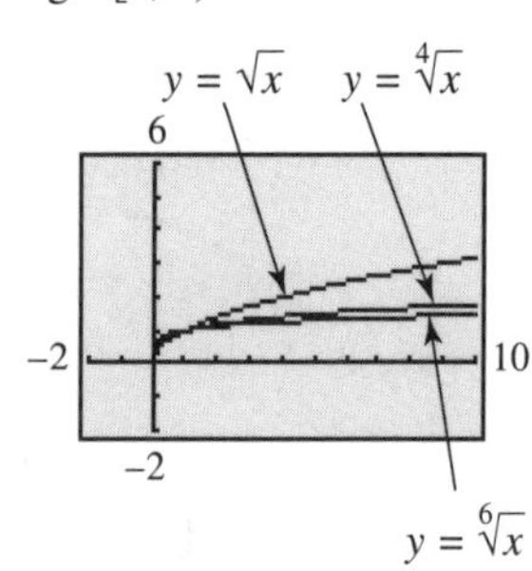

**ROOT FUNCTION, $n$ ODD** $f(x) = \sqrt[n]{x}$

Domain: $(-\infty, \infty)$ Range: $(-\infty, \infty)$

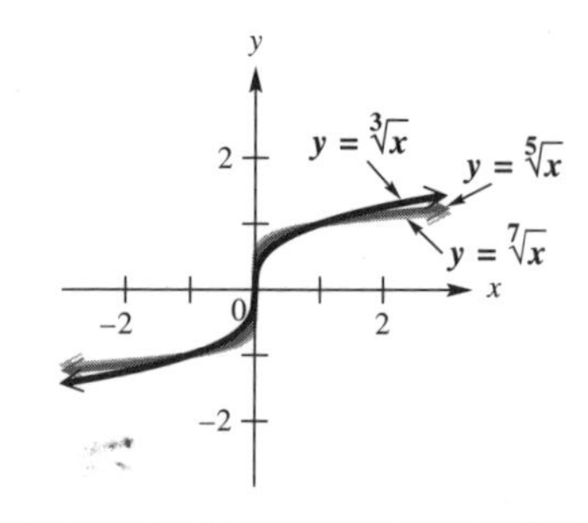
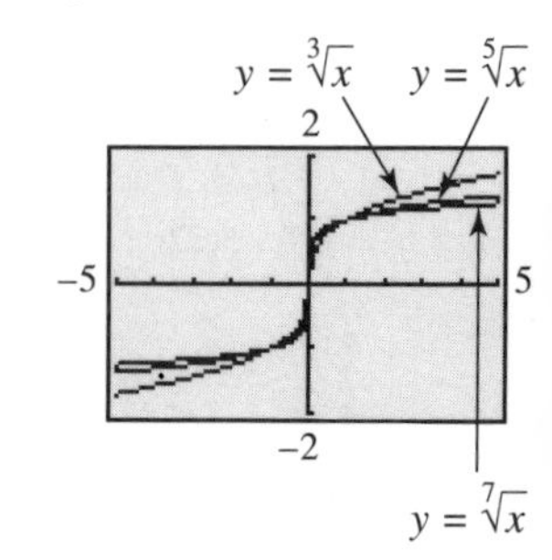
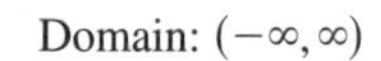
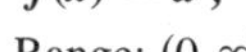
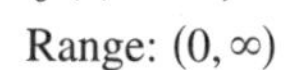

**EXPONENTIAL FUNCTION** $f(x) = a^x, \quad a > 1$

Domain: $(-\infty, \infty)$ Range: $(0, \infty)$

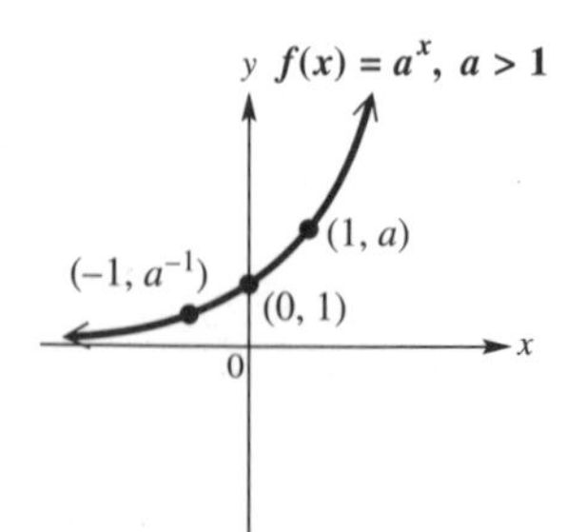
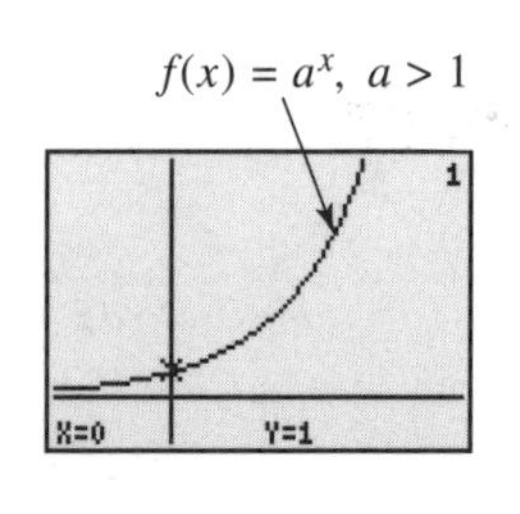

**EXPONENTIAL FUNCTION** $f(x) = a^x, \quad 0 < a < 1$

Domain: $(-\infty, \infty)$ Range: $(0, \infty)$

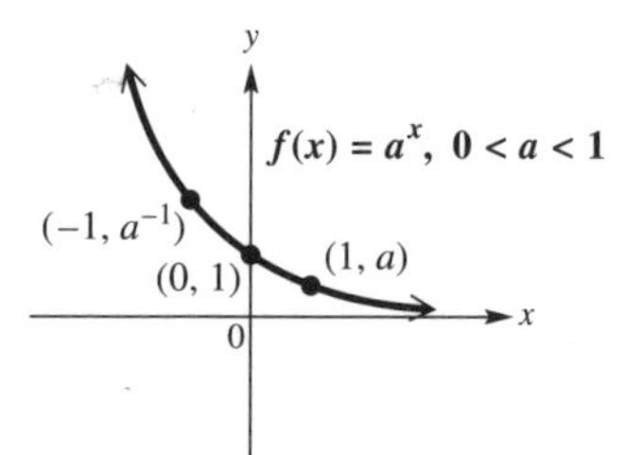
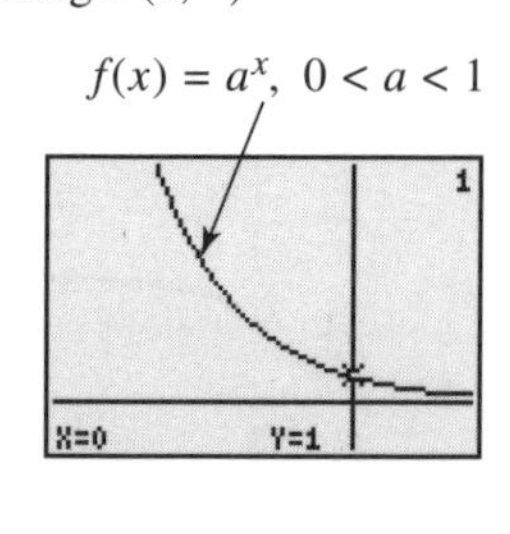

**LOGARITHMIC FUNCTION** $f(x) = \log_a x, \quad a > 1$

Domain: $(0, \infty)$ Range: $(-\infty, \infty)$

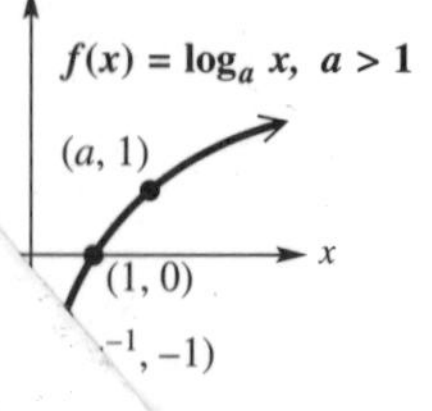
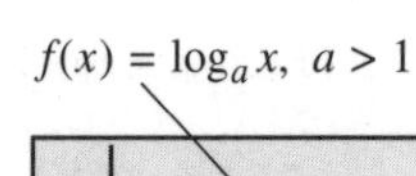
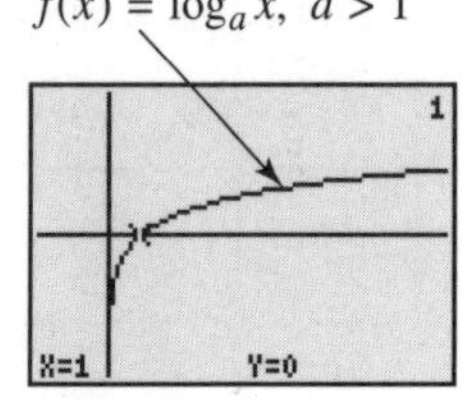

**LOGARITHMIC FUNCTION** $f(x) = \log_a x, \quad 0 < a < 1$

Domain: $(0, \infty)$ Range: $(-\infty, \infty)$

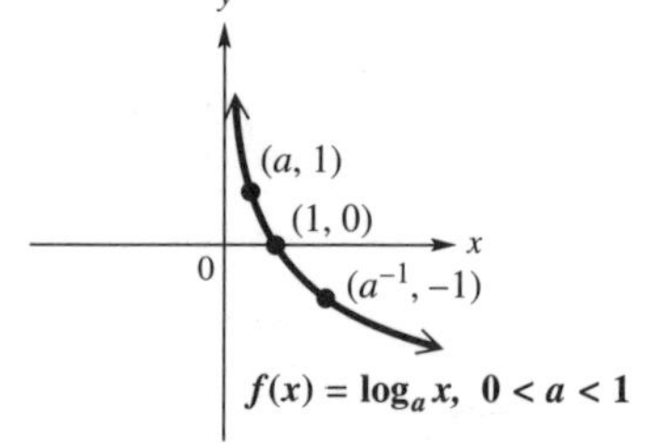
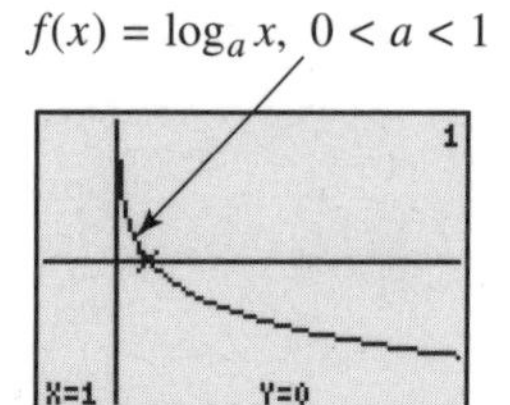